Biological Chemistry

SECOND EDITION

HENRY R. MAHLER
EUGENE H. CORDES
Department of Chemistry
Indiana University

Harper & Row
Publishers
New York / Evanston / San Francisco / London

The photograph appearing on the cover is a precession diagram of myoglobin. [Reprinted from F. H. Crick and J. C. Kendrew, Advan. Protein Chem., **12**. *133 (1957). Courtesy of the authors and Academic Press, New York.*]

BIOLOGICAL CHEMISTRY
Second Edition

Standard Book Number: 06-044172-0

Library of Congress Catalog Card Number: 76-141169

Contents

Some Symbols and Abbreviations Used in This Book

A	absorbance
c	concentration in grams per cm^3
D	translational diffusion coefficient
D	dielectric constant
DFP	diisopropylfluorophosphate
E	extinction
E	voltage
f	frictional coefficient
f/f_0	frictional ratio
F	Faraday constant
Fd	ferredoxin
F_{hkl}	Fourier components
fp	flavoprotein
G	Gibbs free energy
g	force of gravity
g	spectroscopic splitting constant
H	enthalpy
$H_{(\lambda)}$	hypochromicity
$h_{(\lambda)}$	hyperchromicity
h,k,l	Miller indices
hr	hour
i_θ	scattered light intensity
I_θ	incident light intensity
K	equilibrium constant
K_a	acid dissociation constant
K_s'	salting out constant
K_s	Michaelis constant
$\bar{K}_s$	Enzyme-substrate dissociation constant
k	Boltzmann's constant
m	particle mass
M	molar
M	molecular weight
M_{min}	minimum molecular weight
$\bar{M}_n$	number average molecular weight
$\bar{M}_w$	weight average molecular weight
$\bar{M}_z$	z-average molecular weight
n	index of refraction
pK	negative logarithm of K
pI	isoelectric point
PMS	phenazine methosulfate
PRPP	5-phospho-1-pyrophosphoryl ribose
R	gas constant
R_G	radius of gyration
$[R]_\lambda$	reduced mean residue rotation
r	mol ligand bound/mol macromolecule

s	sedimentation constant (coefficient)
S	entropy
S	Svedberg unit
t_m	melting point in degrees C
T_m	melting point degrees K
Z	mean net charge
α_λ	rotation at wavelength λ
$[\alpha]_\lambda$	specific rotation
γ	activity coefficient
ε	molar extinction coefficient
$\Gamma/2$	ionic strength
η	viscosity
$[\eta]$	intrinsic viscosity
λ	wavelength
μ	ionic strength
μ	magnetic susceptibility
$\bar{v}$	partial specific volume
ρ	solvent density
ρ_B	buoyant density
σ	transition width
$^\circ$	degrees Centigrade
Π	osmotic pressure

Preface

One of the many attractions of biochemistry is that it is a far from static subject. New and exciting discoveries—at times bewildering and mystifying, at times providing a true flash of insight—strike the interested observer with ever increasing frequency. This makes for constant awareness and stimulation in research but renders difficult the task of the student and the teacher of the subject, who sometimes wish that it would hold still long enough to be seized, perceived, and understood.

This book presents our view of what biological chemistry is like at the time of writing, within the limitations set out in the preface to the first edition. The response to its predecessor by students and colleagues alike was to us most gratifying, in particular in demonstrating that the need for the kind of treatment we had adopted was real and in pointing out to us weaknesses and areas of possible improvement. This new edition has been written as much in response to these suggestions as to the new developments in the field. These combined pressures have convinced us that what we had to do was not to restrict ourselves to a mere updating and revising but instead to write what amounts to a new book. The reader will observe that we have changed the organization by now starting with the chapter dealing with first principles (Chapter 2, Equilibria and Thermodynamics), by including a chapter on the chemistry of carbohydrates (Chapter 10), and by dealing with the central area of contemporary molecular biology in three separate chapters to end the book (Chapters 19, 20, 21). Quite in general, all the topics dealing with the structure, metabolism, and function of macromolecules of biological importance—nucleic acids, proteins, polysaccharides—have been written in a new and expanded form; in order to keep the volume within an acceptable size we have tried while revising to condense somewhat a few of the other topics, leaving them, however, essentially intact.

A word about the philosophy of the treatment and its application to teaching: First, we have considered it essential to go into sufficient experimental and factual detail in many instances to bring home to the student just how and why biochemists draw the inferences and the conclusions that they do. Second, we have tried to be this comprehensive in more than one area, because we hope that this volume will serve as the backbone and support for teachers in a wide variety of courses differing in approach, content, and level of difficulty. By choosing which topics, chapters, and pages to use in their totality, which ones to omit or condense, upon which ones to expand, the teacher of biochemistry makes his vital and crucial imprint on the presentation of the subject. We hope that with this book we can open the door to show the many paths toward an understanding of how chemistry and biology serve one another. Which of these paths the student will follow depends on who points and leads the way.

As before, this book could not have been written without the constant and devoted help of many people—our secretaries Mrs. Anne Noll and Mrs. Marcie Lindvall, our students who helped us with the proofreading, and the editorial and production staff at Harper & Row. We especially appreciate the help of our many colleagues who provided us with comments and suggestions

—a practice we fervently hope they will continue—and who, with their publishers, permitted us to reproduce tabular and pictorial material. We owe a special debt of gratitude to Professors R. Abeles, L. Anderson, T. P. Bennett, M. Coon, R. A. Cooper, R. Estabrook, F. R. N. Gurd, J. Hurwitz, W. P. Jencks, M. E. Jones, J. Kirsch, H. L. Kornberg, D. Koshland, W. Lands, H. Lardy, W. Lennarz, J. H. Quastel, A. San Pietro, G. Schatz, P. Slonimski, and C. Thomas, who read the various chapters in detail and without whom, as was true of their predecessors, this book could not have been written.

Henry R. Mahler
Eugene H. Cordes

Preface to the First Edition

This book attempts to provide a clear, thorough, and up-to-date treatment of the areas of knowledge fundamental to biochemistry. To this end it presents detailed studies of the physical chemistry of biomacromolecules; of the thermodynamics, kinetics, and mechanisms of enzymatic reactions; of subcellular organization; of the intermediary metabolism of the major classes of chemical substances; and of selected aspects of molecular biology.

The authors have had to use their judgment about the content, organization, and relative depth of coverage that would make the book most useful to the student. Some topics normally discussed in biochemistry texts have been omitted. A specific discussion of carbohydrate chemistry, for instance, is absent because an increasing amount of information on this subject is being presented in basic chemistry courses and excellent discussions can be found in standard textbooks of organic chemistry and biochemistry. Other peripheral topics have been excluded, regretfully, in order to keep the size of the book within reasonable boundaries. We have not been able to deal with such important aspects of physiological chemistry as, for instance, vision, muscle contraction, nerve conduction, and hormonal action. We feel confident, however, that the student who has mastered the fundamentals presented here will be equipped to study these related topics intelligently from other sources.

We feel it desirable to provide the student with an understanding of the chemical and physical properties of biomacromolecules and of the characteristics of the chemical reactions in which they are involved prior to a specific discussion of intermediary metabolism. For this reason the material is organized in the sequence outlined above. Those topics that draw heavily on physical and organic chemistry precede those that are more closely related to the biological disciplines. Treatment of topics in this order also permits the convenient organization of a two- or three-semester course in biochemistry.

Certain concepts introduced early in the book, namely, the physical properties of proteins, enzymatic mechanisms, and the control of enzymatic activity, are discussed at greater length in later chapters. For example, the treatment of enzyme mechanisms in Chapter 7 is rather general and, for the most part, the individual mechanisms are introduced subsequently. Thus, a plausible mechanism for nearly every enzyme encountered in a discussion of carbohydrate metabolism is presented in Chapter 10.

Chapter 10 also includes our first discussion of metabolic sequences and their control. The integration of metabolism and mechanism here has produced a chapter that may appear abnormally long compared with, for example, the chapter concerned with lipid metabolism. It is important to recognize that this disparity in length is largely a result of the integration of topics and does not reflect an unevenness in the depth of coverage of metabolism. On the other hand Chapter 18 appears to be long in proportion to the others because it contains the bulk of our discussion of molecular biology.

This book is intended essentially for seniors or first-year graduate students who have completed the basic courses in organic and physical chemistry but who have had no previous exposure to biochemistry. For students less sophisticated in physical chemistry, certain portions of some of the early chapters will have to be deleted.

On the whole, the text mirrors rather closely the biochemistry courses we have taught at Indiana University over the past few years. In a two-semester course in biochemistry, one would ordinarily cover the material in Chapters 1 through 8 during the first semester and the remainder of the book during the second. For a one-semester course in intermediary metabolism, this book may be used by introducing portions of the chapters on enzymes and coenzymes, followed by Chapter 9 and the subsequent material. Similarly, appropriate selections can be made for an introductory course in Macromolecules of Biological Importance.

One particular problem in writing this book was the question of enzyme nomenclature—whether to use the tried and true trivial names or to adhere rigorously to the new systematic forms proposed by the Commission on Enzymes of the International Union of Biochemistry. After considerable contemplation and with the advice of several of our colleagues, we elected to retain the trivial nomenclature, since it is the one the student would meet with in most of his reading. We could not, for instance, bring ourselves to abandon "aconitase" entirely, in favor of "citrate (isocitrate) hydrolyase," with "aconitate hydratase" as the only trivial alternative allowed. However, the systematic nomenclature is introduced and discussed, and systematic names are provided for most of the enzymes encountered.

In presenting the material, the authors have attempted, almost on a page by page basis, to treat each topic with the appropriate depth, steering between the Scylla of superficiality and the Charybdis of unwieldiness. In order to make the book useful as a reference work as well as a text, we have provided substantial collections of data relevant to important topics. Pertinent references to the research literature are made throughout, and a bibliography of books, monographs, and review articles has been placed at the end of each chapter. Here also difficult choices had to be made, and in selecting the articles to be cited, we have tried to choose those that were written most recently or are of particular note, those that summarize a series of related developments, and those that would be available to most students. We extend our sincere apologies to all authors whose important contributions may not have been explicitly recognized in this context.

In closing, we should like to acknowledge the invaluable assistance of many of our colleagues. In particular, we are indebted to Professors R. Baldwin, J. M. Buchanan, R. Clayton, W. W. Cleland, C. Frieden, F. R. N. Gurd, H. O. Halvorson, F. Haurowitz, F. M. Huennekens, W. P. Jencks, E. P. Kennedy, H. L. Kornberg, J. Marmur, A. Meister, W. J. Moore, A. Novikoff, W. Rutter, and A. San Pietro for suggesting major improvements in organization and content and for correcting many errors of fact. If this book has any merit, it is due largely to their contributions. Shortcomings still remaining are, of course, completely the responsibility of the authors.

Henry R. Mahler
Eugene H. Cordes

1
Introduction

THE NATURE OF BIOCHEMISTRY

The science of biological chemistry deals directly with the most complex and fascinating phenomenon within the purview of man: life. Therein lies the heart of the difficulty of defining just what biological chemistry, or biochemistry, is and of saying just what it is that a biochemist does. Life itself has never received a satisfactory definition, although a good many have been offered at one time or another. At one level of sophistication, attempts to define life in terms of its external manifestations have failed. Thus, most of us recognize things like growth, metabolism, differentiation, reproduction, and the like as associated with the living state. But it is trivial to show that definitions couched in these terms fall short of the mark; exceptions abound. Thus, it is difficult to deny that the jackass is a living thing; yet the jackass is sterile and cannot reproduce. At another level of sophistication, it has been argued that all living things are involved in the intake of energy in one form and its utilization in another. Thus, energy conversion may be regarded as the crucial matter which distinguishes living from nonliving. This argument is probably a much better one than those based on more visible phenomena and a great many kinds of energy conversion are to be found in things recognizable as living: The conversion of the radiant energy of the sun into chemical bond energy by plants is one example. However, once again objections may be raised such as the interesting one to be found in "The Andromeda Strain" in the following passage:

> "He then challenged the team to prove that they (three inanimate objects) were not living. He placed the black cloth in the sunlight; it became warm. This, he announced, was an example of energy-conversion—radiant energy to heat.
>
> "It was objected that this was merely passive energy absorption, not conversion. It was also objected that the conversion, if it could be called that, was not purposeful. It served no function.
>
> " 'How do you know it is not purposeful?' Leavitt had demanded."[1]

Happily, our inability to properly define life and, hence, to be able to define biochemistry causes us no great problems. Like a species of bird, most biochemists can be recognized in the field even if one does not have all of the distinguishing characteristics at his fingertips. At the outset, it may prove helpful to point out that biochemists are generally more concerned with subtler aspects of life, such as energy conversion, than they are with movement, growth, reproduction, and so forth. If one wanted to erect fences, it

[1] Michael Crichton, "The Andromeda Strain," Alfred A. Knopf, Inc. New York. Copyright © 1969 by Centesis Corporation.

would probably be justifiable to place the latter properties on that side of the fence occupied by physiologists, developmental biologists, and the like. However, erecting these fences is distinctly contrary to the contemporary mood of science and, at any event, no one really has a clear idea of what a physiologist or developmental biologist is either.

Basically, biological chemistry is an interface between two ways of looking at nature: that of the chemist and that of the biologist. The central distinguishing feature of chemistry is its emphasis on interactions between molecules. Chemistry is concerned with the relationship between the molecular structure of a substance and its functions and reactions. Biology, on the other hand, has centered on the general properties of the many features of living nature without being overly concerned about their molecular basis. It is the general task of the biochemist to search for explanations for the phenomena of life in terms of interactions between the molecules characteristic of living organisms. This search can take a great many forms and biochemists are to be found in some unlikely places within the scientific sphere.

A RATIONAL APPROACH TO BIOLOGICAL CHEMISTRY

The concept of biochemistry as an interface between chemistry and biology defines our approach to the subject matter which is from the point of view of the chemist, probing in detail the structures and interactions of molecules typical of living systems and, in so doing, providing the foundation for a better understanding of their functions as individuals and as members of more complex structures.

To understand the contributions that the methodology and philosophy of chemistry have to offer the biological scientist, let us briefly consider the nature of the systems of interest. *Living systems are invariably highly organized structures.* This organization takes a number of forms. In the first place, all living systems contain at least two types of biomacromolecules: proteins and nucleic acids. Choosing the former as an example, we may note (1) that proteins are very large molecules, having molecular weights between about 10,000 and a few million, and (2) that a great many kinds of proteins are to be found in even very simple living forms. Proteins are polymers and their most basic structure consists of a chain of monomers, the amino acids of which there are twenty types, possessing perhaps 100–200 units. But the functional state of these molecules is one in which this chain (or several of them) adopts a single unique three-dimensional conformation. Thus, the molecular organization is reflected not only in a specific covalent structure but in higher-order structural forms as well. The consequence is that proteins have properties which are not readily discernible in the amino acids from which they are formed. These comments apply with almost equal force to the nucleic acids.

In the second place, biomacromolecules are themselves organized into much larger and more complex structures. A variety of these appear crucial for many of the metabolic processes characteristic of living systems. Efficient coupling of oxidative metabolic reactions with generation of metabolic energy (in the form of ATP) is associated with membrane structures in all cells. These structures contain a great many kinds of proteins as well as other constituents and all of these components are arranged in a highly organized way. Consequently, the assembly exhibits many properties not found in its individual constituents and, indeed, even the properties of individual constituents are no longer the same as a result of incorporation into the more complex structure.

At a higher level of organization, cells consist of a variety of types of such subcellular structures as well as of free protein molecules and other constituents. Again, all these constituents of cells must be structurally organized in a highly specific way to generate the special properties of cells. The further extensions from cells to tissues and from tissues to whole organisms are evident. But we must recognize that living systems also exhibit levels of organization complementary to and extending beyond that which is directly specified in terms of the structures of biomacromolecules and the more complicated assemblies derived from them. This is *functional* organization. The structural organization itself generates specific and interacting functions. But the life process also requires that these functions be integrated into specific patterns which may, moreover, be time-dependent. That is, built into the structures of the macromolecules or arising from their interactions one must have the possibility of *control* of function and this control must be able to be extended sequentially. These controls generate and regulate orderly sequences of metabolic reactions required for the organism to develop, differentiate, respond, recover, and reproduce; in short, to live. Once again, organization, this time at higher levels, has resulted in the appearance of new properties of the system. The phenomenon of life can be defined in terms of the functional integration of simple structures that are themselves devoid of all of the characteristics of living systems.

Nevertheless, the physicochemical properties of these inanimate molecules must, ultimately, account for all the features of the life process; the alternative view that living things possess a "life-force" not understandable in terms of the laws of physics and chemistry has been abandoned for a good many years as is discussed in the next section. It follows that understanding the life process in detail involves pinpointing how *function arises from structure*.

If we return to proteins as an example, we note that proteins serve several functions in living organisms including that of *catalyst* for substantially all the chemical reactions making up the metabolic processes. That is, some proteins, enzymes, have the capacity to dramatically increase the rates of these reactions without themselves undergoing chemical change. Moreover, the activity of many of these proteins is itself subject to control mechanisms. The chemist adopts the point of view that all facets of enzyme function derive from and may be understood on the basis of the three-dimensional structure of the proteins involved. The same point of view maintains that the three-dimensional structure can be understood in terms of the nature and arrangement of the amino acids constituting the molecular backbone.

In fact, the philosophy and methodology of the chemist have already yielded substantial insight into precisely these relationships. The English chemist F. Sanger and subsequent workers have developed methods for the determination of the amino acid sequences of proteins, supplying a crucial piece of structural information. Several brilliant studies employing the methodology of X-ray diffraction, following the lead of J. C. Kendrew and M. F. Perutz, have established the three-dimensional structures of several proteins. Efforts are already underway, though none has been successful as yet, to calculate such three-dimensional structures from the amino acid sequence. Such calculations are based only on the properties of the amino acids themselves and on known types of chemical interaction between them. At the same time that structural information concerning the proteins was accumulating, detailed studies of their function as catalysts were underway in other laboratories. Many answers have been provided concerning the pathway of the reactions, the involvement of certain groups on the enzyme surface, and the like. The result of both lines of effort is that enzyme function is coming into focus in terms of protein structure. We stand at the threshold of a fundamental understanding of how these crucial molecules make their contribution to the life process and it is upon this and similar foundations that insight into function of the more complex structures will develop.

Despite these triumphs, it is evident that our understanding of protein structure, for example, is still not adequate to fully understand the structure–function relationship. The static structure of protein molecules is in hand but the dynamic, time-dependent, ones are not. Further development along these lines requires that one understand something about flexibility of various parts of the molecule, how structure changes on binding of small molecules and the like. Once again, the increasing sophistication of chemical methodology appears capable of providing the necessary information. On page 172 of this book, there is presented a ^{13}C nuclear magnetic resonance spectrum of a protein containing only the natural abundance of this isotope. This spectrum represents a stunning achievement in nuclear magnetic resonance spectroscopy and promises much for the future of protein chemistry. While the details are not important at this point, it is sufficient to say that this technique is capable of providing insight precisely into the dynamic aspects of protein structure which have been lacking until now. This will inexorably yield additional understanding of structure–function relationships.

On the basis of this one example, it is clear that chemistry has a great deal to contribute to understanding of the biological sciences. In passing, it is worth noting that elucidation of the structure of another kind of biomacromolecule, a nucleic acid, has actually generated a whole new field in the biological sciences, *molecular biology*.

We hope that the above examples suffice to illustrate the contributions of chemistry to the biological sciences and to justify treating biological chemistry from this point of view whenever possible.

THE ORIGINS OF BIOCHEMISTRY

Pointing out where biochemistry came from and defining the nature of the subject matter until the decade between 1940 and 1950 is more straightforward than defining the scope of the subject matter currently.

Research along the lines of modern biochemistry clearly goes back 200 years. During the latter half of the eighteenth century and throughout the nineteenth, a substantial effort was made to understand both the structural and the metabolic aspects of the life process, efforts that met with some success. Because of the inherent complexity of the problems under attack, a sophisticated appreciation of the chemicals and chemical reactions associated with living material had to wait for the development of other chemical disciplines, particularly of analytical and organic chemistry.

Early structural biochemistry is indissolubly linked with the development of organic chemistry. In fact the pioneering researches of the great Swedish chemist Karl Scheele (1742–1786) concerning the chemical composition of plant and animal tissues proved, eventually, to be the impetus required for this development of biochemistry. Scheele isolated a number of natural substances, including tartaric, lactic, uric, oxalic, citric, and malic acids as well as glycerol, several esters, and casein. With the development of techniques for quantitative elemental analysis in the laboratories of Jöns Berzelius and Justus Liebig during the early stages of the nineteenth century, the substances isolated by Scheele were found to contain carbon. Efforts to synthesize carbon-containing, that is, organic, compounds followed. This was a development of more significance than might appear at first glance, for a belief termed *vitalism,* which maintained that organic compounds could be synthesized only through the agency of a vital force, supposed to be present only in living tissues, was widespread during this period. Although not many investigators realized it at the time, vitalism was rendered untenable by the synthesis of urea by Friedrich Wöhler (1800–1882) in 1828. This achievement was the unexpected result of attempts to prepare ammonium cyanates through the treatment of metal cyanates with ammonium salts. Wöhler's work was followed by the synthesis of acetic acid by Adolf Kolbe in 1844 and the synthesis of several organic compounds by Marcellin Berthelot in the 1850s. Vitalism was quietly laid to rest. Organic synthesis remains very much alive.

The structural chemistry of lipids received considerable attention in the nineteenth century through the work of Michel Chevreul (1786–1889). He demonstrated through studies on saponification, that fats were composed of fatty acids, of which he isolated several, and glycerol. Several others made modest contributions to structural biochemistry during this period. The most elegant and significant work in this field, however, came from the laboratory of the German chemist Emil Fischer (1852–1919). In the course of his studies, this remarkable man completely revolutionized research concerning the structures of carbohydrates, amino acids, and fats.

The nineteenth century also saw the initiation of work on the isolation and characterization of the high-molecular-weight materials so prominent in living matter. Of particular note is the work of Mulder, Liebig, Schützenberger, and others on the isolation of amino acids from protein hydrolyzates and again that of Fischer, who deduced how these amino acids are linked in the intact protein. In 1868, F. Miescher discovered nucleic acid in the nuclei of pus cells obtained from discarded surgical bandages and initiated a series of investigations concerning the distribution and properties of the nucleic acids.

Certain facets of the metabolic aspect of biochemistry were also pursued prior to the twentieth century, usually as a result of attention focused on agricultural or medical problems. Of particular note are the basic studies of Antoine Lavoisier on respiration in the period 1779–1784. As a result of the calorimetric measurement of the heat evolved in combustion and in respiration of living cells, Lavoisier concluded that respiration can be equated with combustion, slower but not essentially different from that of charcoal. These findings initiated extensive research into energy metabolism, and, as a result, the caloric values per gram of carbohydrate, fat, and protein were established early in the nineteenth century. During the same period the process of fermentation was recognized by Theodor Schwann as being of biological origin. Schwann clearly recognized yeast as a plant capable of converting sugar to alcohol and carbon dioxide. Many of the leading chemists of the day, including Berzelius, Wöhler, and Liebig, considered yeast to be nonliving and fermentation to be caused solely by oxygen.

Research on fermentation was subsequently pursued by several important chemists, the greatest of whom was Louis Pasteur. He identified several organisms that carried out various fermentations, including that leading to butyric acid, a type performed by organisms that function without oxygen. Thus Pasteur introduced the concept of aerobic and anaerobic organisms and their associated fermentations. These conclusions again aroused the ire of Liebig, who had failed to modulate his earlier feelings concerning these processes. These studies culminated in the demonstration by Eduard Büchner in 1897 of fermentation of sugars by a yeast extract. This observation permitted the subsequent identification of the individual reactions comprising the overall process and that of the enzymes that catalyzed each of them. It also expunged, quietly, yet once and for all, any traces of vitalism still remaining.

Other important developments of the nineteenth century included investigations of photosynthesis and carbon dioxide fixation in plants by Horace de Saussure, studies on the process of digestion by Lazaro Spallanzani, René de Réaumur, William Beaumont, and Claude Bernard, and the development of surgical techniques for the study of animal physiology and biochemistry. One of the most important conclusions derived from work in this era was that of the basic unity of biochemistry throughout nature. Although each species, of course, exhibits biochemical individuality, there proved to be striking similarities in the fashions in which widely differing life forms accomplished closely related tasks. This fact enormously simplifies the overall problem of the understanding of life processes.

With the advent of the twentieth century, biochemistry burst into full bloom. Important developments occurred rapidly on many fronts. From a medical standpoint, an understanding of human nutritional requirements is of great importance. The significance of unknown food factors was clearly recognized by Frederick Hopkins at Cambridge University and his associates, who developed the concept of deficiency diseases. Extensive series of feeding experiments utilizing synthetic diets followed in this country, principally under the direction of Babcock, McCollum, Osborne, Sherman, and Mendel. Pellagra, scurvy, rickets, and beriberi became widely recognized as such deficiency diseases, and the curative agents, vitamins, a term introduced by the Polish biochemist Funk, were isolated and subsequently characterized. The work of McCollum, Szent-Györgyi, Steenbock, and Elvehjem is of particular note.

The work of Büchner on cell-free systems capable of carrying out fermentations was energetically followed up in many laboratories, including those of Harden and Young, Embden, and Meyerhof, with the result that the complete biochemical pathway was eventually elucidated. The work of Warburg during this period resulted in the elucidation of the structure of certain cofactors required for the fermentation process. The beautiful work of Krebs on the oxidative metabolism of carbohydrates followed, as did important developments in other areas of intermediary metabolism, particularly in the laboratories of Green, Lynen, Leloir, Bloch, Kennedy, Krebs, Davis, and Shemin. The results of these investigations occupy a considerable portion of this book. Most of the remainder of these pages is devoted to results of studies concerning the properties of macromolecules, particularly proteins and nucleic acids.

Interest in the structure and biochemical properties of proteins was dramatically stimulated by the classic contribution of Sumner—the discovery in 1926 that biocatalysts, enzymes, are proteins. The phenomenon of catalysis was clearly recognized by Berzelius in 1835. At that time, he pointed out potato diastase, an enzyme that catalyzes the hydrolysis of starch, as an example of a biocatalyst and reasoned that all materials of living tissues are formed under the influence of catalytic action, a conclusion thoroughly vindicated by subsequent work. Following, and in some cases preceding, the recognition of biocatalysts, many such entities were isolated, purified to some extent, and the associated reactions investigated kinetically. This work, together with the development of the dynamic aspects of biochemistry indicated above, focused great attention on enzymes. Nevertheless, the chemical composition of these substances was completely unknown prior to Sumner's work. Early workers had suggested that enzymes were protein in nature, but the prevailing opinion in the early twentieth century was that enzymes belonged to none of the recognized classes of organic compounds. The revelation of Sumner was greeted with skepticism and derision, particularly by Willstätter and his students. Sumner's contention rested on his crystallization of the enzyme urease, and its chemical constitution and degradability by proteolytic enzymes, with resultant loss of enzymatic activity, of the crystalline material. Sumner was a persistent and tenacious man armed with a body of convincing evidence and did not capitulate to authority. Subsequent work on the purification of enzymes by Northrup and Kunitz corroborated the protein nature of enzymes and established Sumner as the father of modern enzymology. Invaluable work in this field has been accomplished by du Vigneaud, Sanger, Stein, Moore, Perutz, Kendrew, and Phillips. Concurrently, the work of Chargaff, Watson, Crick, and Wilkins led to the formulation of the structure of deoxyribonucleic acid, and this marked the beginning of the field of molecular biology.

LITERATURE OF BIOCHEMISTRY

The rapidity with which new and important developments in biochemistry occur makes a thorough understanding of the biochemical literature mandatory. A table of the most important review and research publications, and their customary abbreviations, follows.

Review Publications

Publication	Abbreviation
Accounts of Chemical Research	Accounts Chem. Res.
Advances in Carbohydrate Chemistry and Biochemistry	Advan. Carbohyd. Chem.
Advances in Enzyme Regulation	Advan. Enzyme Reg.
Advances in Enzymology	Advan. Enzymol.
Advances in Lipid Research	Advan. Lipid Res.
Advances in Protein Chemistry	Advan. Protein Chem.
Annals of the New York Academy of Sciences	Ann. N.Y. Acad. Sci.
Annual Reports of the Chemical Society	Ann. Reps.
Annual Review of Biochemistry	Ann. Rev. Biochem.
Annual Review of Microbiology	Ann. Rev. Microbiol.
Annual Review of Physiology	Ann. Rev. Physiol.
Annual Review of Plant Physiology	Ann. Rev. Plant Physiol.
Bacteriological Reviews	Bacteriol. Rev.
Biochemical Journal	Biochem. J.
Biochemical Society Symposia	Biochem. Soc. Symp.
Biological Reviews	Biol. Rev.
British Medical Bulletin	Brit. Med. Bull.
Chemical Reviews	Chem. Rev.
Cold Spring Harbor Symposia on Quantitative Biology	Cold Spring Harbor Symp. Quant. Biol.
Essays in Biochemistry	Essays in Biochem.
Harvey Lectures	Harvey Lectures
Physiological Reviews	Physiol. Rev.
Progress in Biophysics and Molecular Biology	Progr. Biophys. Mol. Biol.
Progress in Medicinal Chemistry	Progr. Med. Chem.
Progress in Molecular and Subcellular Biology	Progr. Mol. Subcell. Biol.
Progress in Nucleic Acid Research and Molecular Biology	Progr. Nucl. Acid Res. Mol. Biol.
Quarterly Review of Biology	Quart. Rev. Biol.
Quarterly Reviews of the Chemical Society	Quart. Rev. (London)
Vitamins and Hormones	Vitamins and Hormones

Research Publications

Publication	Abbreviation
Acta Chemica Scandinavica	Acta Chem. Scand.
American Journal of Physiology	Am. J. Physiol.
Analytical Biochemistry	Anal. Biochem.
Archives of Biochemistry and Biophysics (until 1951, Archives of Biochemistry)	Arch. Biochem. Biophys. (until 1951, Arch. Biochem.)

Journal	Abbreviation
Berichte der Deutschen Chemischen Gesellschaft	Chem. Ber.
Biochemische Zeitschrift	Biochem. Z.
Biochimica et Biophysica Acta	Biochim. Biophys. Acta
Biochemical and Biophysical Research Communications	Biochem. Biophys. Res. Commun.
Biochemistry	Biochemistry
Biochemistry (Eng. trans. of Biokhimiya)	Biochemistry (U.S.S.R.) (English trans.)
Biophysical Journal	Biophys. J.
Biopolymers	Biopolymers
Bulletin de la Société Chimie Biologique	Bull. Soc. Chim. Biol.
Canadian Journal of Biochemistry	Can. J. Biochem.
Carbohydrate Research	Carbohyd. Res.
Chemistry and Physics of Lipids	Chem. Phys. Lipids
Comparative Biochemistry and Physiology	Comp. Biochem. Physiol.
Comptes Rendus	Compt. Rend.
Enzymologia	Enzymologia
European Journal of Biochemistry	Europ. J. Biochem.
F.E.B.S. Letters	FEBS Lett.
Federation Proceedings	Federation Proc.
Helvetica Chimica Acta	Helv. Chim. Acta
(Hoppe-Seyler's) Zeitschrift für Physiologische Chemie	Z. Physiol. Chem.
Journal of the American Chemical Society	J. Amer. Chem. Soc.
Journal of Bacteriology	J. Bacteriol.
Journal of Bioenergetics	J. Bioenerg.
Journal of Biological Chemistry	J. Biol. Chem.
Journal of Cell Biology	J. Cell. Biol.
Journal of Cellular Physiology	J. Cellular Physiol.
Journal of Chromatography	J. Chromatog.
Journal of Evolutionary Biochemistry and Physiology	J. Evol. Biochem. Physiol.
Journal of Lipid Research	J. Lipid Res.
Journal of Membrane Biology	J. Memb. Biol.
Journal of Molecular Biology	J. Mol. Biol.
Journal of Biochemistry, Tokyo	J. Biochem. (Tokyo)
Journal of the Chemical Society	J. Chem. Soc.
Journal of Experimental Medicine	J. Exp. Med.
Journal of General Microbiology	J. Gen. Microbiol.
Journal of General Physiology	J. Gen. Physiol.
Journal of Immunology	J. Immunol.
Journal of Neurochemistry	J. Neurochem.
Journal of Nutrition	J. Nutr.
Journal of Organic Chemistry	J. Org. Chem.
Journal of Pharmacology and Experimental Therapeutics	J. Pharmacol. Exp. Therap.
Journal of Theoretical Biology	J. Theoret. Biol.
Journal of Physical Chemistry	J. Phys. Chem.
Journal of Physiology	J. Physiol.
(Liebig's) Annalen der Chemie	Ann. Chem.
Life Sciences	Life Sci.
Nature	Nature
Naturwissenschaften	Naturwissenschaften
Plant Physiology	Plant Physiol.
Proceedings of the National Academy of Sciences	Proc. Natl. Acad. Sci. U.S.
Proceedings of the Royal Society	Proc. Roy. Soc. (London)
Proceedings of the Society for Experimental Biology and Medicine	Proc. Soc. Exp. Biol. Med.
Science	Science
Scientific American	Sci. Am.
Zeitschrift für Naturforschung	Z. Naturforsch.

ABBREVIATIONS OF BIOCHEMISTRY

Perhaps to a greater extent than in any other discipline, the language of biochemistry is expressed by abbreviations. The tables of accepted abbreviations indicated below are hardly comprehensive but will serve as an introduction to those which are most frequently encountered in the literature. Additional information may be found in: *J. Biol. Chem.*, **241:** 527, 2491 (1966).

TABLE 1-1
Symbols for monomeric units in macromolecules (or in phosphorylated compounds)

Symbol	*Monomeric unit in macromolecule*	*Symbol*	*Monomeric unit in macromolecule*
A, Ado	Adenosine	Hyl	Hydroxylysine
Ala	Alanine	Hyp	Hydroxyproline
Arg	Arginine	I, Ino	Inosine
Asp	Aspartic Acid	Ile	Isoleucine
$Asp(NH_2)$, Asn	Asparagine	Leu	Leucine
C, Cyd	Cytidine	Lys	Lysine
Cys[a]	Cystine (half)	Man	Mannose
Cys	Cysteine	Met	Methionine
de, d	(indicates "deoxy" in carbohydrates and nucleotides)	Nir	Ribosylnicotinamide
		Orn	Ornithine
		P, p	Phosphate
f	(Suffix) furanose	*p*	(Suffix) pyranose
Fru	Fructose	Phe	Phenylalanine
Gal	Galactose	Pro	Proline
G, Glc[b]	Glucose	Rib	Ribose
G, Guo[b]	Guanosine	Ser	Serine
GlcA	Gluconic acid	Thr	Threonine
GlcN	Glucosamine	Trp	Tryptophan
GlcNAc	*N*-acetylglucosamine	T, Thd	Ribosylthymine
GlcUA	Glucuronic acid	Tyr	Tyrosine
Glu	Glutamic acid	U, Urd	Uridine
$Glu(NH_2)$, Gln	Glutamine	Val	Valine
Gly	Glycine	Xao	Xanthosine
His	Histidine		

[a] With vertical bond above or below "s."

[b] The one-letter symbol G must *not* be used if confusion between its two meanings can arise.

TABLE 1-2
Abbreviations for semisystematic or trivial names

ACTH	Adrenocorticotropin, adrenocorticotropic hormone, or corticotropin
ADP	Adenosine 5′-diphosphate (pyro)
AMP	Adenosine 5′-phosphate
ATP	Adenosine 5′-triphosphate (pyro)
BAL	2,3-Dimercaptopropanol
CDP	Cytidine 5′-diphosphate (pyro)
CM-cellulose	*O*-(Carboxymethyl)cellulose
CMP	Cytidine 5′-phosphate
CoA (or CoASH)	Coenzyme A
CoASAc	Acetyl coenzyme A
CTP	Cytidine 5′-triphosphate (pyro)
DEAE-cellulose	*O*-(Diethylaminoethyl) cellulose
DDT	1,1,1-Trichloro-2, 2-bis(*p*-chlorophenyl)ethane
DFP	Diisopropyl phosphorofluoridate
DNA	Deoxyribonucleic acid
DNP-	2,4-Dinitrophenyl-
DOPA	3,4-Dihydroxyphenylalanine
DPN[a]	Diphosphopyridine nucleotide
DPT	Diphosphothiamine (thiamine pyrophosphate, cocarboxylase)
EDTA	Ethylenediaminetetraacetate
FAD	Flavin-adenine dinucleotide
FDNB	1-Fluoro-2,4-dinitrobenzene
FMN	Riboflavin 5′-phosphate
GDP	Guanosine 5′-diphosphate (pyro)
GMP	Guanosine 5′-phosphate
GSH	Glutathione
GSSG	Oxidized glutathione

TABLE 1-2 (Continued)

GTP	Guanosine 5′-triphosphate (pyro)
Hb, HbCO, HbO_2	Hemoglobin, carbon monoxide hemoglobin, oxyhemoglobin
IDP	Inosine 5′-diphosphate (pyro)
IMP	Inosine 5′-phosphate
ITP	Inosine 5′-triphosphate (pyro)
Mb, MbCO, MbO_2	Myoglobin, carbon monoxide myoglobin, oxymyoglobin
MetHb, MetMb	Methemoglobin, metmyoglobin
MSH	Melanocyte-stimulating hormone
NAD	Nicotinamide-adenine dinucleotide (cozymase, Coenzyme I, diphosphopyridine nucleotide)
NADP	Nicotinamide-adenine dinucleotide phosphate (Coenzyme II, triphosphopyridine nucleotide)
NMN	Nicotinamide mononucleotide
P_i	Inorganic orthophosphate
PP_i	Inorganic pyrophosphate
RNA	Ribonucleic acid
TEAE-cellulose	*O*-(Triethylaminoethyl)cellulose
TPN[b]	Triphosphopyridine nucleotide
Tris	Tris(hydroxymethyl)aminomethane (2-amino-2-hydroxymethylpropane-1,3-diol)
UDP	Uridine diphosphate (pyro)
UDPG	Uridine diphosphate glucose
UMP	Uridine monophosphate
UTP	Uridine triphosphate (pyro)

[a] Replaced by NAD.
[b] Replaced by NADP.

References

Chittenden, R. H.: "The Development of Physiological Chemistry in the United States," Chemical Catalogue Company, New York, 1930.

Handler, P. (ed.): "Biology and the Future of Man," Harper & Row, New York, 1970.

Ihde, A. J.: "The Development of Modern Chemistry," Harper & Row, New York, 1964.

Partington, J. R.: "A Short History of Chemistry," Harper & Row, New York, 1960.

2

Equilibria and Thermodynamics in Biochemical Transformations

An understanding of equilibria and thermodynamic behavior for reactions or systems of reactions is frequently a prerequisite to the full appreciation of biochemical processes. In this chapter, treatment of protonic and metal-ion equilibria and methods for calculation of changes in the basic thermodynamic functions—the Gibbs free energy G (or F), the enthalpy H, and the entropy S—are outlined. Several examples of biochemically important calculations are presented.

PROTONIC EQUILIBRIA

Many important biochemical phenomena are sensitive functions of the state of ionization of the molecular species concerned. These include rates of enzyme-catalyzed reactions, protein conformations and transconformations, ion-binding properties of proteins, electrophoretic and chromatographic behavior of small and large molecules, hemoglobin-mediated transport of oxygen, and so forth. At this point, techniques of dealing with typical protonic equilibria problems are discussed.

*p*H Scale

It is frequently more convenient to express the "acidity" of aqueous solutions in terms of the negative logarithm of the hydrogen-ion concentration, pH, rather than in terms of this quantity itself. If we ignore the difference between concentrations and activities (this usually involves an error less than 0.1 pH unit), we may write $p\text{H} \equiv -\log (\text{H}^+)$. Thus a solution 0.1 M in a strong acid is characterized by a value of pH of 1, pure water by a value of 7, and a solution 0.1 M in a strong base by a value of 13. Values of pH are ordinarily measured with the aid of a commercial pH meter equipped with a glass electrode. Such measurements are accurate to ± 0.01 pH unit if carried out carefully with an instrument of good quality. More accurate measurements usually require the use of electromotive-force cells, which do not involve a liquid junction.

Calculation of *p*H in Systems of Known Acid or Base Strength

In the case of strong (i.e., completely dissociated) acids or bases in aqueous solution, values of pH for solutions containing known concentrations of acid or base may be arrived at immediately. Clearly, for the case of strong monobasic acids, $p\text{H} = -\log C$, and, for the case of strong bases, $p\text{H} = 14 + \log C$, in which C is the molar concentration of acid or base and 14 is the negative logarithm of the dissociation constant for water, $K_w = (\text{H}^+)\,(\text{OH}^-) = 10^{-14}$.

The calculation is slightly more complicated for the case of a weakly dissociating acid (or base) in aqueous solution. Let K_a be the dissociation constant for a weak acid HA and C be the total concentration of A-containing species; that is,

$$C = (\mathrm{H}A) + (A-)$$

$$K_a = \frac{(\mathrm{H}+)(A-)}{(\mathrm{H}A)} \tag{2-1}$$

The condition that the solution be electrically neutral overall requires the equality

$$(A-) + (\mathrm{OH}-) = (\mathrm{H}+) \tag{2-2}$$

Since the concentration of hydroxide ion in mildly acidic solutions is very small indeed, Eq. (2-2) can ordinarily be simplified to

$$(A-) = (\mathrm{H}+) \tag{2-3}$$

Inserting Eq. (2-3) into Eq. (2-1) yields

$$K_a = \frac{(\mathrm{H}+)^2}{[C - (\mathrm{H}+)]} \tag{2-4}$$

This quadratic equation may be solved by the usual methods to give an explicit relationship between the hydrogen-ion concentration, K_a, and C:

$$(\mathrm{H}+) = \tfrac{1}{2}[-K_a \pm (K_a^2 + 4K_aC)^{1/2}] \tag{2-5}$$

Under most conditions, a somewhat simpler equation may be employed to yield the same information. This equation may be derived by returning to Eq. (2-4) and realizing that usually $C \gg (\mathrm{H}+)$. For example, a 0.1 M solution of acetic acid has $C = 0.1$ and $(\mathrm{H}+) \simeq 0.002$. Employing the above assumption, solution of Eq. (2-4) yields

$$(\mathrm{H}+)^2 = CK_a \tag{2-6}$$

Taking logarithms, rearranging, and defining $-\log K \equiv pK_a$, we obtain

$$p\mathrm{H} = \frac{pK_a - \log C}{2} \tag{2-7}$$

Returning to our example of a solution of 0.1 M acetic acid, Eq. (2-5) yields a calculated pH of 2.878, and Eq. (2-7) yields a calculated pH of 2.875 (K_a for acetic acid is 1.76×10^{-5}, $pK_a = 4.75$). This calculation clearly indicates that, provided C is not too small or the acid too strong, Eq. (2-7) is quite adequate for calculation of pH values for most purposes. Clearly, similar equations may be derived for the case of solutions of weakly dissociating bases, The analog of Eq. (2-7) for this case is

$$p\mathrm{H} = \frac{14 + pK_a + \log C}{2} \tag{2-8}$$

in which pK_a refers to the conjugate acid of the added base.

Titration of a Weak Acid with a Strong Base

Henderson–Hasselbalch Equation. Values of pH for a solution of a weak acid to which a known amount of strong base, sodium or potassium hydroxide, for example, has been added may be arrived at by considerations similar to those employed above. The principal distinction lies in the fact that the equation of electrical neutrality must be rewritten so as to include the cation of the strong base. Thus, if we define C as before and X as the amount of strong base added, the equation of electrical neutrality is

$$(M+) + (\mathrm{H}+) = (A-) + (\mathrm{OH}-) \simeq (A-) \tag{2-9}$$

Realizing that $(M^+) = X$, the expression for K_a becomes

$$K_a = \frac{(\mathrm{H}+)(A-)}{(\mathrm{H}A)} = \frac{(\mathrm{H}+)(\mathrm{H}+ + X)}{C - X - (\mathrm{H}+)} \tag{2-10}$$

Although this equation may be solved directly, it is convenient to realize that ordinarily $X, C \gg (\mathrm{H}^+)$, in which case Eq. (2-10) simplifies to

$$K_a = \frac{(\mathrm{H}+)(X)}{C - X} \tag{2-11}$$

Taking logarithms, we obtain

$$p\mathrm{H} = pK_a + \log \frac{X}{C - X} = pK_a + \log \frac{(\mathrm{base})}{(\mathrm{acid})} \tag{2-12}$$

the well-known and very useful *Henderson–Hasselbalch equation.* In terms of this equation, it is easy to see that the pK_a is just that pH at which HA is present half associated and half dissociated; that is, (base) = (acid). Clearly, values of pK_a yield quantitative information concerning acid strength, strong acids being characterized by small values of pK_a and vice versa. Values of pK_a for the common amino acids are presented in Table 3-1; other values of biochemical interest, in Table 2-1.

A plot of pH against equivalents of added hydroxide ion for the titration of acetic acid is shown in Fig. 2-1. Values of pK_a are frequently evaluated from the midpoint of titration curves. At values of pH removed from the vicinity of the pK_a of acetic acid (4.75) the pH changes very rapidly with addition of small amounts of base. However, near the pK_a, addition of base results in only modest alterations in pH; that is, solutions of acetic acid are *buffered* at values of pH near 4.75. Buffering is very important biochemically both in the

TABLE 2-1
Values of pK_a for the dissociation of some important acids

Acid	*Conjugate base*	pK_a
Acetic acid	Acetate	4.75
Chloroacetic acid	Chloroacetate	2.86
Trichloroacetic acid	Trichloroacetate	0.65
Propionic acid	Propionate	4.88
Lactic acid	Lactate	3.86
Pyruvic acid	Pyruvate	2.50
Acetoacetic acid	Acetoacetate	3.58
Malic acid	Malate monoanion	3.40
Malate monoanion	Malate dianion	5.20
Fumaric acid	Fumarate monoanion	3.03
Fumarate monoanion	Fumarate dianion	4.54
Citric acid	Citrate monoanion	3.09
Citrate monoanion	Citrate dianion	4.75
Citrate dianion	Citrate trianion	5.45
Ammonium ion	Ammonia	9.24
Methylammonium ion	Methylamine	10.62
Imidazolium ion	Imidazole	7.12
Tris (hydroxymethyl) methyl ammonium ion	Tris (hydroxymethyl) amino methane	8.10
H_3PO_4	$H_2PO_4^-$	2.12
$H_2PO_4^-$	HPO_4^{2-}	7.21
HPO_4^{2-}	PO_4^{3-}	12.32

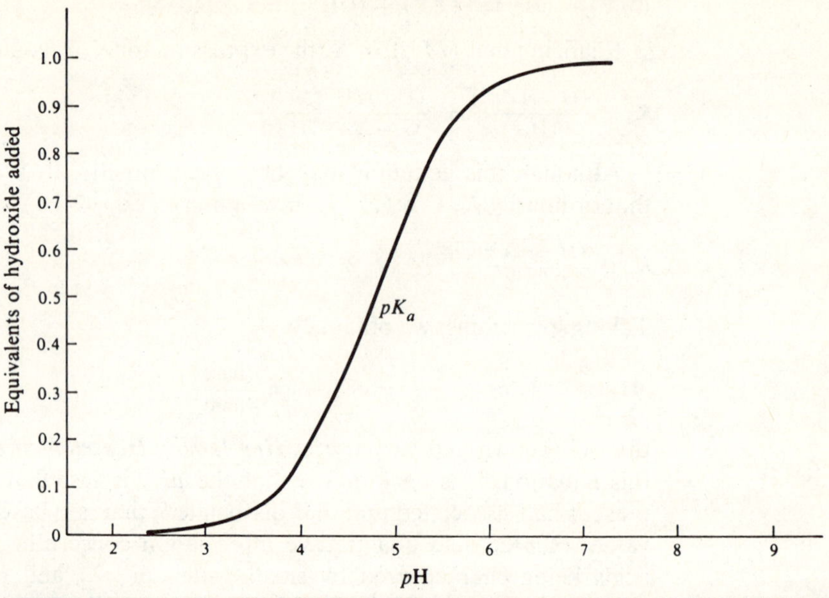

FIGURE 2-1
Titration of acetic acid.

sense that organisms must maintain reasonably constant values of pH in the intra- and intercellular milieu and in the sense that proper handling of biochemical materials in the laboratory requires the maintenance of constant pH. It is easy to demonstrate that buffering capacity is a maximum at a value of pH equal to the pK_a of the acid component of the buffering system. Defining $dX/dp\text{H}$ as *buffering value* and α as (base)/[(acid) + (base)], it is clear

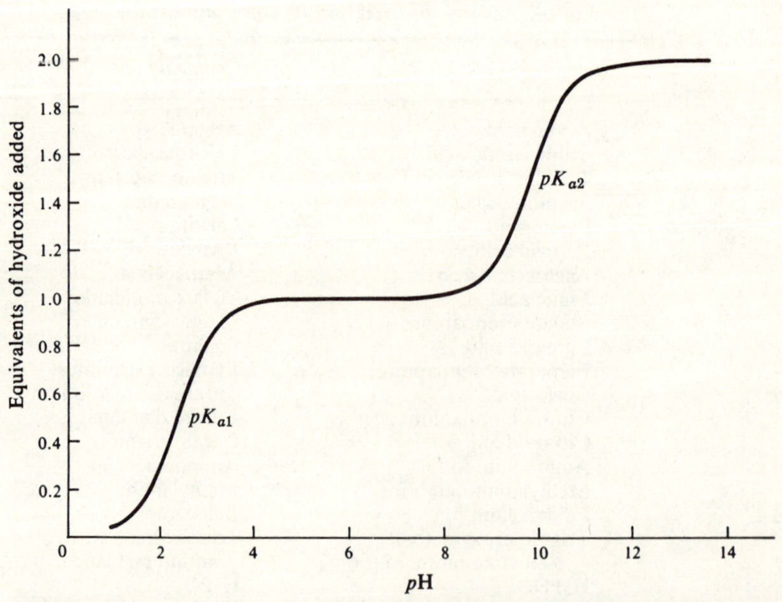

FIGURE 2-2
Titration of glycine.

that $dX/dp\text{H}$ = constant × $d\alpha/dp\text{H}$. Writing the Henderson–Hasselbalch equation in terms of α, and differentiating we obtain

$$dp\text{H} = dpK_a + d\log\frac{\alpha}{1-\alpha} = \frac{1}{2.3}\,d\ln\frac{\alpha}{1-\alpha} = \frac{d\alpha}{2.3\alpha(1-\alpha)} \tag{2-13}$$

from which we conclude that

$$\frac{d\alpha}{dp\text{H}} = 2.3\alpha(1-\alpha) \quad \text{or} \quad \frac{dX}{dp\text{H}} = \text{constant}\cdot 2.3\alpha(1-\alpha) \tag{2-14}$$

Because the product $\alpha(1 - \alpha)$ is a maximum for $\alpha = 0.5$, which corresponds to a value of pH equal to the pK_a, it follows directly that buffering capacity is a maximum at the pK_a. In practice, a given acid–base pair may usually be employed as a buffer in a pH range that extends from 1 pH unit below the associated pK_a to 1 pH unit above this value.

Dibasic Acids: Protonic Equilibria of Amino Acids. In Fig. 2-2, the titration curve for glycine is shown. This curve is characterized by two dissociations, having pK_a values of 2.35 and 9.78. Reference to Table 2-1 would suggest that $pK_{a1}(2.35)$ corresponds to the ionization of the glycine carboxyl group and that $pK_{a2}(9.78)$ corresponds to the ionization of the glycine ammonium ion. Although such a description actually yields a very nearly correct picture of glycine ionization, it is oversimplified and may introduce appreciable errors in related cases. The ionization of the glycine molecule is correct in Eq. (2-15):

$$\begin{array}{ccccc} & \overset{k_a}{\nearrow} & H_3\overset{+}{N}{-}CH_2{-}CO_2^- \; (HA) & \overset{k_c}{\searrow} & \\ H_3\overset{+}{N}{-}CH_2{-}CO_2H \; (HAH) & & & & NH_2{-}CH_2{-}CO_2^- \; (A) \\ & \underset{k_b}{\searrow} & NH_2{-}CH_2{-}CO_2H \; (AH) & \underset{k_d}{\nearrow} & \end{array} \tag{2-15}$$

The complete ionization pattern is described by four microscopic dissociation constants, k_a, k_b, k_c, and k_d, one for each of the four microscopic dissociations of the system. These are related by $k_ak_c = k_bk_d$. Thus, if three of the microscopic constants are known, the fourth is uniquely defined. The observed macroscopic dissociation constants, K_1 and K_2, are clearly related to the microscopic constants, as indicated below:

$$K_1 = \frac{(\text{H}+)[(\text{H}A) + (A\text{H})]}{(\text{H}A\text{H})} = k_a + k_b \tag{2-16}$$

$$K_2 = \frac{(\text{H}+)(A)}{[(\text{H}A) + (A\text{H})]} = \frac{k_ck_d}{k_c + k_d} \tag{2-17}$$

Thus the macroscopic constant K_1 includes contributions from loss of a proton from both the carboxyl and ammonium functions, and K_2 contains similar contributions. Because we have three independent microscopic constants and only two measurable macroscopic constants, the latter are insufficient to determine the former. However, the magnitude of K_b may be estimated from the dissociation constant of, for example, glycine ethyl ester, $K_a = 2 \times 10^{-8}$. Assuming that $k_b = K_a$ and employing Eqs. (2-16) and 2-17), we find that

$$\begin{aligned} k_a &= K_1 - k_b = K_1 - K_a = 4.45 \times 10^{-3} - 2 \times 10^{-8} \simeq 4.45 \times 10^{-3} \\ k_b &= K_a = 2 \times 10^{-8} \\ k_c &= K_2\,\frac{k_a + k_b}{k_a} \simeq K_2 \simeq 1.66 \times 10^{-10} \\ k_d &= \frac{k_ak_c}{k_b} \simeq 3.70 \times 10^{-5} \end{aligned} \tag{2-18}$$

Thus, for the case of glycine, K_1 is very nearly equal to k_a, and K_2 is very nearly equal to k_c, in accordance with the assignments of the macroscopic constants to the individual ionizations indicated above. Only in cases in which k_a and k_b or k_c and k_d are of comparable magnitudes are appreciable differences found between microscopic and macroscopic constants.

The ratio of k_a to k_b yields the pH-independent ratio of the concentrations of the zwitterionic and uncharged forms of the glycine molecule:

$$\frac{k_a}{k_b} = \frac{(\mathrm{H}A)}{(A\mathrm{H})} = \frac{(\mathrm{Gly}^{\pm})}{(\mathrm{Gly}^{0})} = 223{,}000 \tag{2-19}$$

The magnitude of this ratio clearly indicates that, at all values of pH, the zwitterionic form of glycine greatly predominates over the uncharged form (page 52).

METAL ION–LIGAND EQUILIBRIA

The association of protonic acids (Brønsted acids) with anionic or neutral bases may be regarded as a special case of the association of acids in general (Lewis and Brønsted acids) with bases in general. A very important additional special case is the association of metal ions with anionic and neutral ligands. These two cases are closely related, as may be appreciated from the fact that the hydrated proton and metal ions frequently compete for available binding sites in suitable bases such as ammonia and imidazole.

Metal ions play a prominent role in biochemical phenomena, resulting, in part, from their association with both large and small molecules. In the former class are a great many metal ion–protein complexes, which fall into two general categories, and the metal ion–nucleic acid complexes, which will be discussed later (page 233). In some cases, mental ions are an integral part of protein structure and cannot be removed without the concomitant destruction of this structure. Such proteins are termed *metalloproteins.* In addition, there are a large number of reversible metal ion–protein interactions known. Formation of such complexes usually results in the stabilization of a particular conformation of the protein moiety. The association of adenosine triphosphate (ATP) with metal ions, which is absolutely required for the integrity of most ATP-dependent enzymatic reactions, is a prominent example of the importance of interactions between metal ions and small organic molecules. A detailed discussion of the role of metal ions in biochemical systems has been prepared by Gurd and Wilcox and by Gurd (see references at end of chapter).

The principal difference in the handling of problems concerning metal–ion equilibria and protonic equilibria arises from the fact that metal ions generally interact with more than one basic site. Thus, metal ion–ligand systems may be described by assuming the stepwise formation of complexes, MA, MA_2, . . ., MA_n, in which M denotes the metal ion, A the ligand, and n is the maximum number of ligands that may associate with M. We may then write a series of equilibrium constants, K_1, K_2, . . ., K_n for the successive addition of ligands to the metal ion (see page 18):

$$K_1 = \frac{(MA)}{(M)(A)};\ K_2 = \frac{(MA_2)}{(MA)(A)};\ldots;\ K_n = \frac{(MA_n)}{(MA_{n-1})(A)} \tag{2-20}$$

Alternatively these systems may be described in terms of overall association constants, β_1, β_2, . . ., β_n, in which $\beta_n = (MA_n)/(M)(A)^n$. These constants are related to the successive constants by $\beta_1 = K_1$, $\beta_2 = K_1K_2$, $\beta_n = K_1K_2 \ldots K_n$. Such constants may be measured conductometrically, spectrophotometrically, and by a variety of additional techniques.

Thus, the standard free energy change is a measure of the useful chemical work that may be obtained from the conversion of 1 mol of reactant to product under conditions in which the $(P)/(R)$ ratio is unity.

The calculation of standard free energy changes from an equilibrium constant and Eq. (2-30) is deceptively nontrivial. In cases involving the loss or uptake of a proton, which are exceedingly common biochemically, there are three ways, all correct, of making such a calculation. The first equilibrium constant measured, that for the hydrolysis of ethyl acetate, was determined by Berthelot and St. Giles in 1863 (1). This reaction provides a suitable case for the discussion of the three conventions employed to obtain standard free energy changes (2).

Convention I. The hydrolysis of ethyl acetate may be described by the equation

$$CH_3\overset{\overset{O}{\|}}{C}{-}OCH_2CH_3 + H_2O \rightleftharpoons CH_3CO_2H + CH_3CH_2OH \tag{2-32}$$

with the associated equilibrium constant

$$K = \frac{(CH_3CO_2H)(CH_3CH_2OH)}{(CH_3CO_2CH_2CH_3)(H_2O)} = 0.33$$

the activity of water being taken as 55 M. It follows directly that at 25°

$$\Delta G^0 = -RT \ln 0.33 = (-RT \ln 10) \log 0.33$$

$$= -2.3 \times 1.99 \times 298 \log 0.33 = 650 \text{ cal/mol}$$

Rather than include the concentration of water in the equilibrium expression for reactions occurring in aqueous solution, one generally takes, by convention, the activity of water as unity. If we reformulate the equilibrium expression for the hydrolysis of ethyl acetate employing unit activity for water instead of 55 M, the equilibrium constant becomes $0.33 \times 55 = 18$, and the standard free energy change at 25° becomes -1710 cal/mol:

$$\Delta G^0 = -RT \ln 18 = (-RT \ln 10) \log 18$$

$$= -1710 \text{ cal/mol}$$

Thus, by taking the activity of water as unity, we automatically make the standard free energy of hydrolysis more negative by $-RT \ln 55 = -2360$ cal/mol. We shall continue to employ unit activity for water in discussions of the other conventions below.

Convention II. Under physiological conditions (*p*H 7), the product of ethyl acetate hydrolysis is not acetic acid but acetate ion. Hence, an alternative formulation of ethyl acetate hydrolysis is given by

$$CH_3\overset{\overset{O}{\|}}{C}{-}OCH_2CH_3 + H_2O \rightleftharpoons CH_3CO_2^- + H^+ + CH_3CH_2OH \tag{2-33}$$

with the associated equilibrium constant

$$K' = \frac{(CH_3CO_2^-)(H^+)(CH_3CH_2OH)}{(CH_3CO_2CH_2CH_3)} = 18 \times 1.76 \times 10^{-5} = 3.20 \times 10^{-4}$$

The value of K' is obtained from the value of K indicated above and the dissociation constant for acetic acid, $K_{diss} = (CH_3COO^-)(H^+)/(CH_3COOH) = 1.76 \times 10^{-5}$. Employing this formulation, the standard free energy change for ethyl acetate hydrolysis is

$$\Delta G^0 = -2.3 \times 1.99 \times 298 \times \log 3.20 \times 10^{-4}$$

$$= -4900 \text{ cal/mol}$$

Convention III. The two conventions above are both based on *p*H-independent equilibrium constants and yield *p*H-independent values for the standard free energy change for this reaction. The calculated values may be employed, together with Eq. (2-29), to calculate the available free energy at any concentration of reactants and any *p*H by substituting into Eq. (2-29) the actual concentration of the indicated species present at that *p*H. Alternatively, the hydrolysis reaction may be formulated for a particular *p*H, usually *p*H 7, and in terms of the total concentration of each reactant:

$$CH_3\overset{\overset{\displaystyle O}{\|}}{C}OCH_2CH_3 + H_2O \rightleftharpoons \begin{pmatrix} CH_3CO_2^- \\ + \\ CH_3CO_2H \end{pmatrix} + CH_3CH_2OH + (H^+;\ pH\ 7) \quad (2\text{-}34)$$

For this equation, the appropriate equilibrium constant is

$$K'' = \frac{[(CH_3CO_2H) + (CH_3CO_2^-)](CH_3CH_2OH)}{(CH_3CO_2CH_2CH_3)} = K' \times 10^7 = 3.20 \times 10^3$$

and the standard free energy change, denoted by $\Delta G^{0\prime}$,

$$\Delta G^{0\prime} = -2.3 \times 1.99 \times 298 \times \log 3.20 \times 10^3$$
$$= -4780 \text{ cal/mol}$$

To recapitulate, each of these conventions is correct and useful. The first two yield *p*H-independent standard free energy changes and the last a *p*H-dependent standard free energy change; that is, the last is directly useful only at the *p*H for which it is defined. In employing the first two conventions, the actual concentrations of species present in acid–base equilibrium must be employed to calculate free energy changes according to Eq. (2-29). With the third convention, total concentrations are employed. Convention III is probably the most generally useful procedure for biochemical problems. In the first place, biochemical reactions generally occur at or near *p*H 7. In the second place, the calculation of free energy changes according to conventions I or II frequently requires the knowledge of the dissociation constants of acids participating in the reaction. These are often unknown for compounds of biological interest. Finally, convention I is mildly displeasing on esthetic grounds, because it frequently does not accurately represent the actual reaction taking place in solution.

Prior to taking leave of the hydrolysis of ethyl acetate, two additional points are of importance. First, the available free energy from this reaction is *p*H dependent. Because, above *p*H 5, this reaction liberates one proton, the free energy change for this reaction becomes more negative by $\Delta G = -2.3 \times 1.98 \times 298 \log 10 = -1364$ cal/mol per *p*H unit. We return to this point below. Second, the free energy change for this reaction depends, not only on the relative concentrations of reactants and products, but on their absolute concentrations as well. As indicated above, $\Delta G^{0\prime}$ for all components present in their standard state at 25° is −4780 cal/mol. In contrast, if all components are present at 0.01 M (more accurately at activities of 0.01), $\Delta G'$ becomes −7560 cal/mol.

Another example is as follows. The standard free energy of hydrolysis of adenosine triphosphate (ATP) to adenosine diphosphate (ADP) and inorganic phosphate,

$$H_2O + ATP \rightleftharpoons ADP + \text{phosphate} \quad (2\text{-}35)$$

an important biochemical reaction (Chapter 8), has been evaluated indirectly from the measured equilibrium constant, K_{eq}, for the reaction catalyzed by glutamine synthetase (3):

$$\text{Glutamate} + NH_4{}^+ + \text{ATP} \rightleftharpoons \text{glutamine} + \text{ADP} + P_i + H_2O \tag{2-36}$$

Expressing K_{eq} in terms of convention III, we obtain

$$K_{eq} = \frac{(\text{glutamine})(\Sigma\ \text{ADP})(\Sigma\ P_i)}{(\text{glutamate})\ (NH_4{}^+)(\Sigma\ \text{ATP})} = 1200 \qquad p\text{H 7 and } 37° \tag{2-37}$$

In the presence of magnesium ion, a requirement for biochemical ATP-dependent reactions, Eq. (2-37) must be reformulated as

$$K_{eq}{}' = \frac{(\text{glutamine})(\Sigma\ \text{Mg ADP})(\Sigma\ P_i)}{(\text{glutamate})(NH_4{}^+)(\Sigma\ \text{MgATP})} = 442 \tag{2-38}$$

to account for complex formation between this metal ion and the polyphosphates. The overall glutamine synthetase reaction may be expressed as the sum of two reactions:

$$\Sigma\ \text{Mg ATP} \rightleftharpoons \Sigma\ \text{Mg ADP} + \Sigma\ P_i \qquad K_1{}', \Delta G_1{}^{0\prime} \tag{2-39}$$

$$\text{Glutamate} + NH_4{}^+ \rightleftharpoons \text{glutamine} \qquad K_2{}', \Delta G_2{}^{0\prime} \tag{2-40}$$

It follows directly that

$$K_{eq}{}' = K_1{}' \cdot K_2{}' \qquad \Delta G_{eq}{}^{0\prime} = \Delta G_1{}^{0\prime} + \Delta G_2{}^{0\prime} \tag{2-41}$$

$\Delta G_1{}^{0\prime}$ can now be calculated by substituting the experimentally determined values for $K_{eq}{}'$ [Eq. (2-38)] and $K_2{}'$ [Eq. (2-40)] into Eq. (2-41). $K_2{}'$ has been measured at 37° and pH 7 as equal to $2.5 \times 10^{-3}\ \text{M}^{-1}$. This value, plus the one given earlier for $K_{eq}{}'$, allows the following calculation for $\Delta G_1{}^{0\prime}$, the free energy of hydrolysis of the magnesium chelate of ATP at 37° and pH 7 (4–7):

$$\Delta G_2{}^{0\prime} = -1.98 \times 2.3 \times 310 \times \log 2.5 \times 10^{-3} = 3750\ \text{cal/mol}$$

$$\Delta G_{eq}{}^{0\prime} = -1.98 \times 2.3 \times 310 \times \log 442 = -3670\ \text{cal/mol}$$

$$\Delta G_1{}^{0\prime} = \Delta G_{eq}{}^{0\prime} - \Delta G_2{}^{0\prime} = -3670 - 3750 = -7420\ \text{cal/mol}$$

Consequently, the standard free energy change for the hydrolysis of the magnesium chelate of ATP at 37° and pH 7 is slightly more negative than 7 kcal/mol. This value is in good agreement with estimates obtained by other means.

The actual available free energy from the hydrolysis of 1 mole of ATP under the intracellular conditions is considerably more negative than the standard free energy change and is probably in the range −10 to −12 kcal/mol.

The standard free energy changes for the hydrolysis of a number of substances of biochemical interest are summarized in Table 2-4.

Suppose that we wish to convert the free energy change for hydrolysis of ethyl acetate or ATP calculated according to convention III to one based on a pH-independent convention: How do we proceed? For any calculation of this type, one can proceed readily bearing in mind the following two rules (8):

1. The equilibrium expression for the pH-independent equilibrium constant and the free energy of hydrolysis may include any desired ionic species of the reactants but must be based on a balanced equation for the reaction, including any hydrogen ions that are given off or taken up.

2. The actual concentrations of the particular ionic species given in this expression and which are present at a given value of pH must be substituted into the expression for the pH-independent equilibrium constant for free energy change.

TABLE 2-4
Standard free energy of hydrolysis at pH 7 and 25° for several compounds of biochemical interest[a]

Substrate	$-\Delta G^{0\prime}$ (*p*H 7)
Ethyl acetate	4,720
Glycine ethyl ester (39°)	8,400
Acetylcholine	6,000
Acetylcarnitine (35°)	7,210
Valyl-tRNA (30°)	8,400
p-Nitrophenyl acetate	13,010
S-Acetylmercaptoacetate	7,200
Acetyl Coenzyme A	7,520
Acetoacetyl Coenzyme A	10,500
Acetyl phosphate	10,300
Phosphoenolpyruvate	14,800
Carbamyl phosphate (*p*H 9.5)	12,300
Creatine phosphate (37°)	10,300
Phosphoarginine (*p*H 8)	7,700
Adenosine triphosphate (AMP, PP_i, excess Mg)	7,700
Adenosine triphosphate (ADP, P_i, excess Mg)	7,300
Pyrophosphate	8,000
Glucose-1-phosphate	5,000
Glucose-6-phosphate	3,300
Uridine diphosphate glucose (*p*H 7.6)	7,300
Acetyl adenylate	13,300
Peptides	ca. 500
Glutamine	3,400
Acetylimidazole	12,970
Maltose	4,000
Sucrose (dihemiacetal)	7,000
Glycogen	4,000

[a] Standard free energy of hydrolysis at *p*H 7.0 based on a standard state of 1 M total stoichiometric concentration of reactants and products, except for hydrogen ion, and on an activity of water of 1.0 (convention III).

SOURCE: Modified from the extensive collection of W. P. Jencks, *Handbook of Biochemistry*, H. Sober (ed.), Chemical Rubber Co., Cleveland, 1968, pp. J144 ff.

Variation in $\Delta G^{0\prime}$ with *p*H

As noted above, the use of $\Delta G^{0\prime}$ values, which are based on total analytical concentrations at the *p*H at which they are measured and are, therefore, *p*H-dependent, is frequently to be preferred to the use of *p*H-independent ΔG^0 values. The problem thus arises how to evaluate the variation of $\Delta G^{0\prime}$ values with *p*H. If H^+ is a reactant in an equation and the stoichiometry does not change in the *p*H range of interest (i.e., no reactants ionize in this *p*H range, or, if they do, they are on opposite sides of the equation with identical pK_a values), we can represent the reaction as

$$A + B = P + Q + nH^+ \tag{2-42}$$

For such a reaction at equilibrium

$$\Delta G^0 = -RT \ln \frac{(P)(Q)(H^+)^n}{(A)(B)} \tag{2-43}$$

or if

$$\Delta G^{0\prime} = -RT \ln \frac{(P)(Q)}{(A)(B)}$$

we get

$$\Delta G^0 = \Delta G^{0\prime} - RT \ln (\mathrm{H}^+)^n \tag{2-44}$$

which can be rearranged to

$$\Delta G^{0\prime} = \Delta G^0 - n(RT \ln 10)p\mathrm{H} \tag{2-45}$$

To correct $\Delta G^{0\prime}$ from one pH to another,

$$\Delta G_2{}^{0\prime} = \Delta G_1{}^{0\prime} - n(\mathrm{RT} \ln 10)\ \Delta p\mathrm{H} \tag{2-46}$$

where $\Delta p\mathrm{H} = p\mathrm{H}_2 - p\mathrm{H}_1$. If the H^+ in Eq. (2-42) were on the left side of the equation, the signs of the last terms would be positive in Eqs. (2-44) to (2-46). We thus see that for reactions with a single H^+ involved, $\Delta G^{0\prime}$ changes RT ln 10 calories per pH unit.

Although many reactions can be handled by the equations above, there are many common biochemicals that ionize in the neutral pH range, so that we must consider the energy involved in the ionization process. To evaluate these changes in $\Delta G^{0\prime}$ with pH, we need to know only the difference in energy between the ionization states of the reactants and products at the two pH values. This is expressed by the equation

$$\Delta G_2{}^{0\prime} = \Delta G_1{}^{0\prime} + \sum_{\text{products}} (G_{\text{ion }2} - G_{\text{ion }1}) - \sum_{\text{reactants}} (G_{\text{ion }2} - G_{\text{ion }1}) \tag{2-47}$$

in which numerical subscripts refer to the two pH values, and ΔG_{ion} represents the ΔG of ionization for a given molecule at that pH, that is, the energy involved in losing protons from the molecule to give the equilibrium mixture of ionized forms as it occurs at the given pH. Note that the difference in ΔG_{ion} at the two pH values is determined for each molecule in the reaction, and then these energy differences are summed, with appropriate signs, to get the overall energy difference.

The expression for ΔG_{ion} for any compound may be derived as follows. We consider the change:

$$A\mathrm{H}_n = \text{equilibrium mixture of } A\mathrm{H}_n, A\mathrm{H}_{n-1}, \ldots, A\mathrm{H}, \text{ and } A \tag{2-48}$$

after ionization, in which $A\mathrm{H}_n$ is the unionized form and A is the fully ionized form. (We will not indicate charges on the various forms of A, because these depend on whether the ionizing group is COOH, phosphate, $\mathrm{RNH_3}^+$, or of some other type.) This can be written

$$A\mathrm{H}_n = \sum_i A_i \tag{2-49}$$

The energy involved in this process is the difference in chemical potentials of the entities on the two sides of this equation after equilibrium has been reached. Because this potential is given by the expression $-RT \ln C$, where C is the concentration or activity of the entity involved, we can write the following expression for ΔG_{ion}:

$$\Delta G_{\text{ion}} = -RT \ln \sum_i A_i + RT \ln A\mathrm{H}_n \tag{2-50}$$

in which the first term is the chemical potential of the sum of all forms of A as they exist in the ionized state, and the second term is the potential of the unionized $A\mathrm{H}_n$ present. This can be written as shown in Eq. (2–51).

$$\Delta G_{\text{ion}} = -RT \ln \frac{\sum_i A_i}{AH_n} \tag{2-51}$$

If K_1, K_2, K_3, etc., are the successive dissociation constants of protons from AH_n, we can write equations of the type

$$AH_{(n-1)} = \left(\frac{K_1}{\text{H+}}\right) AH_n \qquad AH_{(n-2)} = \left(\frac{K_2}{\text{H+}}\right) AH_{(n-1)} = \left(\frac{K_1}{\text{H+}}\right)\left(\frac{K_2}{\text{H+}}\right) AH_n$$

and thus

$$AH_{(n-3)} = \left(\frac{K_1}{\text{H+}}\right)\left(\frac{K_2}{\text{H+}}\right)\left(\frac{K_3}{\text{H+}}\right) AH_n, \text{ etc.}$$

Then

$$\sum A_i = AH_n \left[1 + \frac{K_1}{\text{H+}} + \left(\frac{K_1}{\text{H+}}\right)\left(\frac{K_2}{\text{H+}}\right) + \left(\frac{K_1}{\text{H+}}\right)\left(\frac{K_2}{\text{H+}}\right)\left(\frac{K_3}{\text{H+}}\right) + \ldots + \left(\frac{K_1}{\text{H+}}\right)\left(\frac{K_2}{\text{H+}}\right)\left(\frac{K_3}{\text{H+}}\right)(\cdots)\left(\frac{K_n}{\text{H+}}\right)\right] \tag{2-52}$$

Note that AH_n itself is one of the forms present after ionization and thus must be included. Equation (2-51) can now be written

$$\begin{aligned}\Delta G_{\text{ion}} &= -RT \ln \left[\frac{AH_n}{AH_n}(1 + \ldots +)\right] \\ &= -RT \ln \left[1 + \frac{K_1}{\text{H+}} + \left(\frac{K_1}{\text{H+}}\right)\left(\frac{K_2}{\text{H+}}\right) + \ldots + \left(\frac{K_1}{\text{H+}}\right)\left(\frac{K_2}{\text{H+}}\right) \times (\cdots)\left(\frac{K_n}{\text{H+}}\right)\right]\end{aligned} \tag{2-53}$$

This is the general equation for ΔG_{ion} at any pH.

As an example of the calculation of the variation of $\Delta G^{0\prime}$ with pH, we consider the case of ATP hydrolysis. Both ATP and ADP have strongly acidic ionizations which we shall ignore. To the extent that the pK_a values for these ionizations are the same, they cancel anyway on both sides of the expression in Eq. (2-47). The dissociation constants of interest are those for the conversion of the ATP trianion to the ATP tetraanion, $K^{\text{ATP}} = 1.12 \times 10^{-7}$; for the conversion of the ADP dianion to the ADP trianion, $K^{\text{ADP}} = 2.09 \times 10^{-7}$; for the conversion of the phosphate monoanion to the phosphate dianion, $K^{\text{P}_1} = 1.51 \times 10^{-7}$; and for conversion of the phosphate dianion to the phosphate trianion, $K^{\text{P}_2} = 3.16 \times 10^{-13}$. Substituting these values into Eq. (2-53) and then into Eq. (2-47), we obtain

$$\frac{\Delta G_{p\text{H}}{}^{0\prime} - \Delta G_7{}^{0\prime}}{RT} = \ln \left[\frac{1 + K^{\text{ADP}}/10^{-7}}{1 + K^{\text{ADP}}/(\text{H+})}\right] + \ln \left[\frac{1 + K^{\text{P}_1}/10^{-7} + K^{\text{P}_1}K^{\text{P}_2}/10^{-14}}{1 + K^{\text{P}_1}/(\text{H+}) + K^{\text{P}_1}K^{\text{P}_2}/(\text{H+})^2}\right] - \ln \left[\frac{1 + K^{\text{ATP}}/10^{-7}}{1 + K^{\text{ATP}}/(\text{H+})}\right] \tag{2-54}$$

in which $(\Delta G_{p\text{H}}{}^{0\prime} - \Delta G_7{}^{0\prime})$ is the difference in $\Delta G^{0\prime}$ values at the pH in question and at pH 7. A plot of the quantity on the left side of Eq. (2-54) against pH is shown in Fig. 2-3. The $\Delta G^{0\prime}$ for ATP hydrolysis is seen to be nearly constant from pH 4 to pH 6, to decrease linearly from pH 7 to pH 11, and to decrease more rapidly under more alkaline conditions.

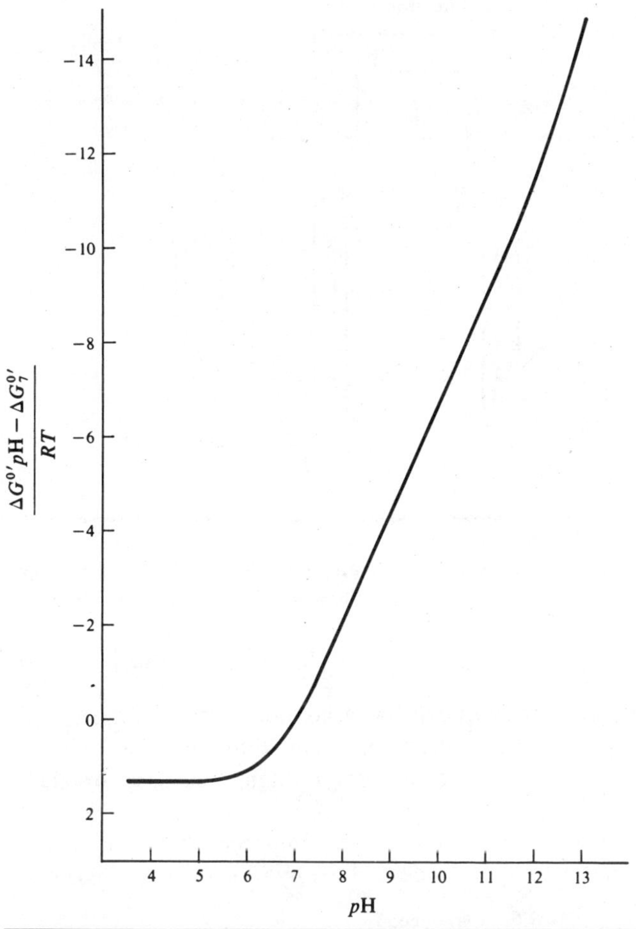

FIGURE 2-3
Variation of $\Delta G^{0\prime}$ for ATP hydrolysis as a function of pH.

FREE ENERGY CHANGES: OXIDATION–REDUCTION REACTIONS

The calculation of free energy changes for oxidation–reduction reactions requires a quantitative measure of the tendency of substances to donate or accept electrons. Such a quantitative measure is termed an *oxidation–reduction potential.*

The establishment of a scale of oxidation–reduction potentials requires the choice of a common standard, whose potential is arbitrarily set equal to zero, to which other potentials may be referred. This common standard is the hydrogen electrode, $\frac{1}{2}H_2 \rightarrow H^+ + e^-$, whose potential is taken as zero for a solution containing hydrogen ions at unit activity in equilibrium with hydrogen gas at 1 atm pressure. The oxidation–reduction potential of any electrode may then be conveniently measured relative to the hydrogen electrode by use of an apparatus depicted schematically in Fig. 2-4. Inert metallic electrodes are placed in both half-cells and connected through a *potentiometer,* a device capable of measuring the potential difference between the two half-cells, ordinarily expressed in volts. The two half-cells are then connected by means of a salt bridge that permits the migration of ions but prevents direct chemical reaction between the components of the individual half-cells. Electrons now tend to flow either into or out of the hydrogen electrode, depending on whether

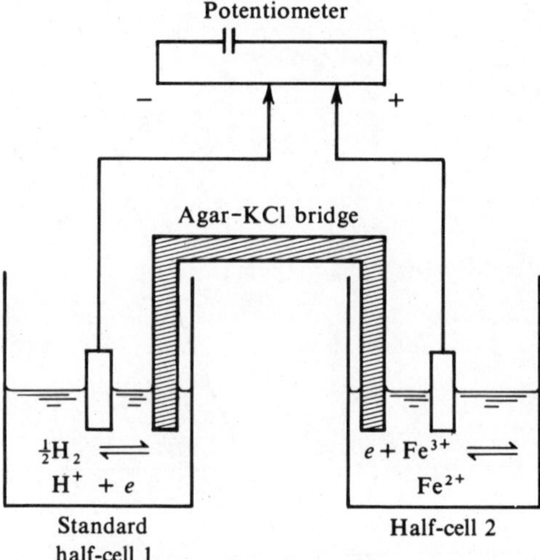

FIGURE 2-4
Schematic representation of an apparatus suitable for the determination of oxidation–reduction potentials. J. Fruton and S. Simmonds, *General Biochemistry,* Wiley, New York, 1958, p. 297.

the alternate half-cell has a greater or lesser tendency than the reference electrode to donate electrons. The difference in oxidation–reduction potential between the two electrodes is then measured by determining, with the potentiometer, the voltage just necessary to prevent such an electron flow.

In quantitative terms, the difference between the oxidation–reduction potential of any electrode and the hydrogen electrode is given by

$$E = E_0 - \frac{RT}{nF} \ln \frac{(\text{reduced})}{(\text{oxidized})} \tag{2-55}$$

in which R is the gas constant; T is the absolute temperature; F is the faraday, equal to 23,000 cal/absolute volt equivalent; and n is the number of electrons per gram equivalent transferred in the reaction. E is the observed potential difference measured in volts, and E_0 is the standard oxidation–reduction potential for the electrode in question. It is clear that, when the activity of the oxidized species is equal to that of the reduced species, the measured potential difference is equal to the standard oxidation–reduction potential. The variation of E with the percentage of oxidation is shown schematically in Fig. 2-5. Equation (2-55) is more conveniently written as

$$E = E_0 - \frac{2.3RT}{nF} \log \frac{(\text{red})}{(\text{ox})} \tag{2-56}$$

At 30°, $2.3RT/nF$ has the value 0.06 for $n = 1$ and 0.03 for $n = 2$.

Two conventions exist for the determination of the sign of oxidation–reduction potentials: one that assigns more negative potentials to systems that have an increasing tendency to donate electrons (reduction potentials) and one that assigns more negative potentials to systems that have an increasing tendency to accept electrons (oxidation potentials). In accordance with the recommendation of the International Union of Pure and Applied Chemistry, the former convention is adopted in this book.

Just as it is frequently convenient to obtain standard free energy changes for hydrolytic reactions for pH 7, it is often necessary to know standard oxidation–

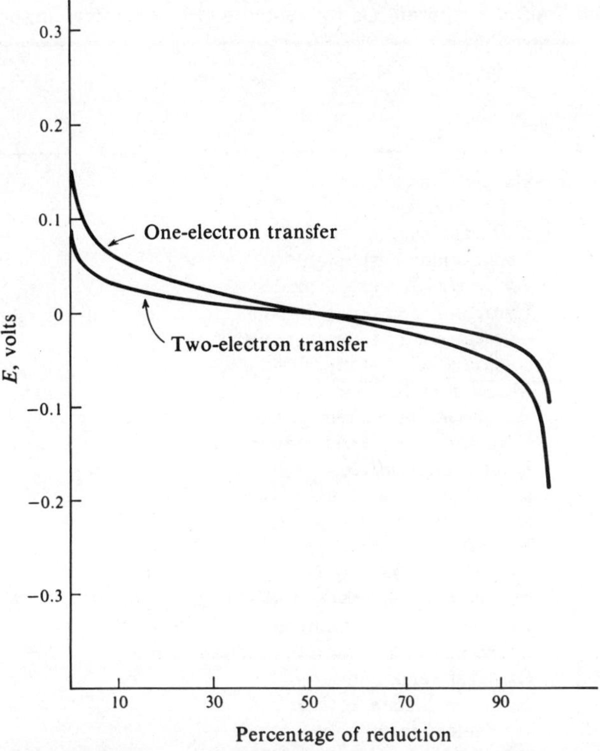

FIGURE 2-5
Representation of typical potentiometric titration for reactions involving the transfer of one and two electrons. E_0 is arbitrarily taken as zero volts.

reduction potentials at this *p*H rather than at *p*H 0. Such electrode potentials are indicated by the symbol E_0'. Several electrode potentials of biochemical interest are collected in Table 2-5.

Oxidation–reduction reactions that involve the uptake or liberation of hydrogen ions are characterized by electrode potentials that are a function of *p*H. The simplest such case is the hydrogen electrode itself. For this instance, because the standard potential for this electrode is zero,

$$E = \frac{RT}{F} \ln \frac{(\mathrm{H}+)}{\mathrm{P_{H_2}}^{1/2}} \tag{2-57}$$

which reduces to

$$E = \frac{2.3RT}{F} \log (\mathrm{H}+) = 0.06 \log (\mathrm{H}+) = -0.06\, p\mathrm{H} \tag{2-58}$$

because the standard hydrogen electrode is defined with respect to hydrogen gas at a pressure of unity. Thus, at *p*H 7, the electrode potential for the hydrogen electrode is −0.420 V.

A somewhat more complicated example is provided by the *p*H dependence of the potential for half-oxidation of anthraquinone-2,7-disulfonate (10):

(2-59)

TABLE 2-5
Standard reduction potentials for systems of biochemical importance

System	E_0' (pH 7), *volts*
Oxygen/water	0.816
Ferric/ferrous	0.77
Nitrate/nitrite	0.42
Ferricyanide/ferrocyanide	0.36
Oxygen/hydrogen peroxide	0.30
Cytochrome *a;* ferric/ferrous	0.29
Cytochrome *c;* ferric/ferrous	0.25
Crotonyl-SCoA/butyryl-SCoA	0.19
Methemoglobin/hemoglobin	0.17
Adrenodoxin; ox/red	0.15
Cytochrome b_2; ferric/ferrous	0.12
Ubiquinone; ox/red	0.10
Dehydroascorbic acid/ascorbic acid	0.06
Metmyoglobin/myoglobin	0.046
Fumarate/succinate	0.03
Methylene blue, ox/red	0.01
Yellow enzyme, FMN/$FMNH_2$	−0.122
Pyruvate + ammonium/alanine	−0.13
α-Oxoglutarate + ammonium/glutamate	−0.14
Oxalacetate/malate	−0.17
Pyruvate/lactate	−0.19
Acetaldehyde/ethanol	−0.20
Riboflavin, ox/red	−0.21
Glutathione, ox/red	−0.23
Acetoacetate/β-hydroxybutyrate	−0.27
Lipoic acid, ox/red	−0.29
NAD^+/NADH	−0.32
Pyruvate/malate	−0.33
Cystine/cysteine	−0.34
Uric acid/xanthine	−0.36
Ferredoxin; ox/red	−0.41
Carbon dioxide/formate	−0.42
H^+/H_2	−0.42
Acetate/acetaldehyde	−0.60
Succinate/α-oxoglutarate	−0.67
Acetate + carbon dioxide/pyruvate	−0.70

Defining the two dissociation constants for the reduced species

$$K_1 = \frac{(H+)(Hred-)}{(H_2red)} \qquad K_2 = \frac{(H+)(red^{2-})}{(Hred-)}$$

and realizing that (total red) = H_2red + $Hred^-$ + red^{2-}, we can express (total red) in terms of H_2red and these equilibrium constants:

$$(\text{total red}) = (H_2red)\left[1 + \frac{K_1}{(H+)} + \frac{K_1K_2}{(H+)^2}\right] \tag{2-60}$$

Solving Eq. (2-60) for (H_2red) and substituting into the equation describing the electrode potential for this system,

$$E = E_0 + \frac{RT}{nF}\ln\frac{(ox)(H+)^2}{(H_2red)} \tag{2-61}$$

we obtain

$$E = E_0 + \frac{RT}{2F} \ln \frac{(\text{ox})}{(\text{total red})} + \frac{RT}{2F} \ln [(\text{H}+)^2 + K_1(\text{H}+) + K_1K_2] \tag{2-62}$$

At low values of pH, $(\text{H}^+)^2$ is greater than $K_1(\text{H}^+)$ and K_1K_2, and Eq. (2-62) reduces to

$$E = E_0 + \frac{RT}{nF} \ln \frac{(\text{ox})}{(\text{red})_T} + \frac{RT}{2F} \ln (\text{H}+)^2 = E_0 + 0.03 \log \frac{(\text{ox})}{(\text{red})_T} - 0.06\, p\text{H} \tag{2-63}$$

Provided the ratio of concentrations of oxidized and reduced species is constant, the electrode potential for this reaction varies in a linear manner with pH, characterized by a slope of -0.06. Similar considerations indicate that under less acidic conditions in which $K_1(\text{H}^+) \gg (\text{H}^+)^2$ and K_1K_2, a linear dependence of E on pH is again observed, characterized by a slope of -0.03, and, finally, under basic conditions in which $K_1K_2 \gg (\text{H}^+)^2$ and $K_1(\text{H}^+)$, E becomes independent of pH. This behavior is indicated for 9,10-anthraquinone-2,7-disulfonate in Fig. 2-6.

In general, for an oxidation reaction involving the release of a protons and the transfer of n electrons, the expression for the electrode potential

$$\text{H}_a\text{red} \rightleftharpoons \text{ox} + a\text{H}+ \qquad n \text{ electron transfer}$$

has the form

$$E = E_0 + \frac{RT}{nF} \ln \frac{(\text{ox})}{(\text{red})_T} + \frac{RT}{nF} \ln (\text{H}+)^a \tag{2-64}$$

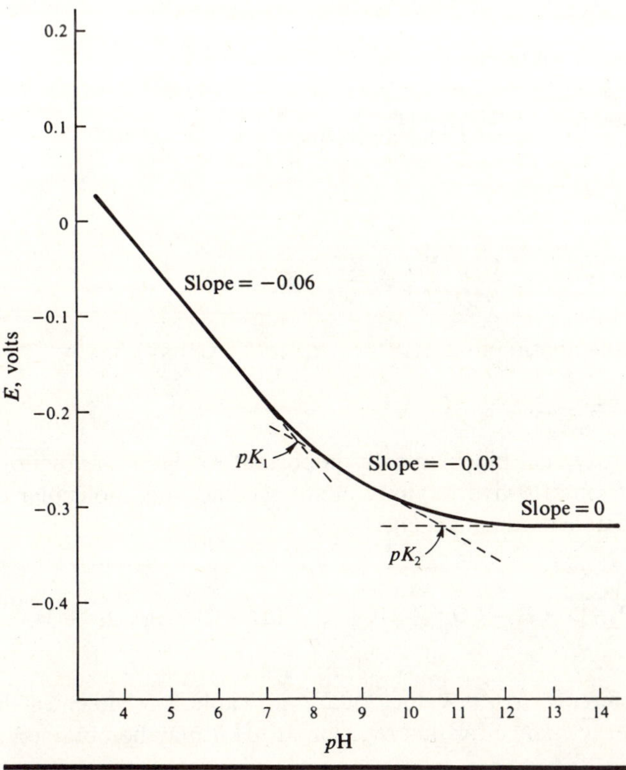

FIGURE 2-6
Dependence of the electrode potential for the half-oxidation of anthraquinone-2,7-disulfonate upon pH.

Hence, the variation of electrode potential with pH is described by

$$\frac{\Delta E}{\Delta p\text{H}} = -0.06\,\frac{a}{n} \tag{2-65}$$

With the above considerations in mind, we are now in a position to turn to the actual evaluation of free energy changes for oxidation–reduction reactions. Because, once the standard free energy change is known, Eq. (2-29) suffices to calculate the actual free energy change under any given set of conditions, the problem is to evaluate ΔG^0 from the standard oxidation–reduction potentials. The method for accomplishing this evaluation is developed in terms of the following example.

Consider the reversible oxidation–reduction reaction

$$A_{\text{red}} + B_{\text{ox}} \rightleftharpoons B_{\text{red}} + A_{\text{ox}} \tag{2-66}$$

composed of the two individual electrode reactions $A_{\text{red}} \rightleftharpoons A_{\text{ox}}$ and $B_{\text{red}} \rightleftharpoons B_{\text{ox}}$, which are characterized by standard oxidation–reduction potentials $E_0{}^A$ and $E_0{}^B$, respectively. At equilibrium, the electrode potential is clearly given by either of the following two expressions:

$$E = E_0{}^A - \frac{RT}{nF}\ln\frac{(A_r)}{(A_o)} \tag{2-67}$$

or

$$E = E_0{}^B - \frac{RT}{nF}\ln\frac{(B_r)}{(B_o)} \tag{2-68}$$

Hence, we may write

$$E_0{}^B - E_0{}^A = \frac{RT}{nF}\left[\ln\frac{(B_r)}{(B_o)} - \ln\frac{(A_r)}{(A_o)}\right] \tag{2-69}$$

and

$$\Delta E^0 = \frac{RT}{nF}\ln\frac{(A_o)(B_r)}{(A_r)(B_o)} = \frac{RT}{nF}\ln K \tag{2-70}$$

Remembering that $\Delta G^0 = -RT\ln K$, we have finally

$$\Delta G^0 = -nF\,\Delta E_0 \tag{2-71}$$

A typical biochemically important oxidation–reduction reaction is the conversion of β-hydroxybutyrate to acetoacetate, molecular oxygen serving as the ultimate oxidizing agent:

$$CH_3\text{—}\underset{\substack{|\\H}}{\overset{\substack{OH\\|}}{C}}\text{—}CH_2\text{—}CO_2^- + \tfrac{1}{2}O_2 \rightleftharpoons CH_3\text{—}\overset{\substack{O\\\|}}{C}\text{—}CH_2CO_2^- + H_2O \tag{2-72}$$

Reference to the values of E_0' in Table 2-5 indicates that the standard free energy change for this reaction at pH 7 may be obtained as follows:

$$\Delta G^{0\prime} = -2 \times 23{,}000 \times [0.815 - (-0.293)]$$

$$= -46{,}000 \times (1.11) = -51{,}000 \text{ cal/mol}$$

STANDARD FREE ENERGY CHANGES FROM STANDARD FREE ENERGIES OF FORMATION

In cases in which equilibrium-constant data are unavailable, standard free energy changes may frequently be approximately evaluated from values for the standard free energy of formation $\Delta G^0 f$ of the various reactants from the elements in their standard states. Values of $\Delta G^0 f$ are evaluated from enthalpies of formation $\Delta H^0 f$, entropies of formation $\Delta S^0 f$, and the relationship $\Delta G^0 f = \Delta H^0 f - T\,\Delta S^0 f$. The determination of the latter quantities has been considered in detail in several standard textbooks and is not repeated here. Suffice it to say that the standard enthalpies of formation are derived from heat-of-combustion data and the standard entropies of formation from variation in heat capacities with temperature. The standard free energies of formation of several compounds of biochemical interest are included in Table 2-6. It should be noted that the values of $\Delta G^0 f$ refer to the standard free energies of formation of the pure substance. In general, the corresponding values for aqueous solutions at unit activity are a good deal more useful. Such values are denoted by the symbol $\Delta \bar{G}^0$ and may be derived simply from the $\Delta G^0 f$ values.

Standard free energy changes for reactions may then be evaluated from data of the type included in Table 2-6 and the relationship

$$\Delta G^0 = \Sigma\, \Delta \bar{G}^0_{\text{products}} - \Sigma\, \Delta \bar{G}^0_{\text{reactants}} \tag{2-73}$$

For example, consider the reversible hydration of fumarate to malate:

$$\begin{matrix} CO_2^- & & & & CO_2^- \\ | & & & & | \\ CH & + & H_2O & \rightleftharpoons & H{-}C{-}OH \\ \| & & & & | \\ CH & & & & CH_2 \\ | & & & & | \\ CO_2^- & & & & CO_2^- \end{matrix} \tag{2-74}$$

The standard free energy change for this reaction is given by

$$\Delta G^0 = 201.98 - (144.41 + 56.69) = -0.88 \text{ kcal/mol}$$

This value is in good agreement with that obtained from the measured equilibrium constant for this reaction. Note that, because the value of $\Delta G^0 f$ for pure water is employed in this calculation, the ΔG^0 and K_{eq} obtained are in terms of water at unit activity rather than 55 M. In a similar fashion the standard free energy change for the complete oxidation of glucose to carbon dioxide and water may be obtained:

$$\text{Glucose} + 6O_2(g) \rightleftharpoons 6CO_2(g) + 6H_2O$$

$$\Delta G^0 = 6(56.69 + 94.26) - 219.22 = -686.48 \text{ kcal/mol} \tag{2-75}$$

It should be recognized that calculation of standard free energy changes from standard free energies of formation frequently involves obtaining a small difference between two large numbers. Very small percentage errors in the free energies of formation, therefore, often lead to large percentage errors in the calculated free energy changes. Consequently, results of such calculations must be interpreted with reasonable caution.

STANDARD ENTHALPY AND ENTROPY CHANGES

The magnitude of the change in enthalpy or entropy for biochemical reactions is sometimes of considerable value for the understanding of the reaction. The standard enthalpy changes associated with reactions are usually evaluated from measurement of the equilibrium constant as a function of temperature and the *van't Hoff equation:*

$$\frac{d \ln K}{dT} = \frac{\Delta H^0}{RT^2} \tag{2-76}$$

TABLE 2-6
Standard free energies of formation of several biochemically important substances in the pure state ($\Delta G^0 f$) and in aqueous solution at unit activity ($\Delta \tilde{G}^0$)

Substance	$-\Delta G^0 f$ (*kcal/mol, 25°*)	$-\Delta \tilde{G}^0$ (*kcal/mol, 25°*)
Acetic acid (liquid)	93.75	33.38
$Acetate^-$	.....	95.48
$Acetoacetate^-$	.....	118.00
L-Alanine	88.40	88.75
Ammonia (gas)	3.98	6.37
Ammonium ion	.....	19.00
L-Asparagine	183.50 (monohydrate)	125.86
L-Aspartic acid	174.76	172.31
$L\text{-}Aspartate^{+2-}$	.....	166.99
n-Butyric acid (liquid)	90.65	90.86
n-$Butyrate^-$	.....	84.28
Carbon dioxide (gas)	94.26	92.31
Bicarbonate	.....	140.31
$Citrate^{3-}$	.....	279.24
$Isocitrate^{3-}$	.....	277.65
Creatine	63.19	63.17 (monohydrate)
Creatinine	6.90	6.91
Cysteine	82.08	81.21
Ethanol (liquid)	41.77	43.39
Formaldehyde (gas)	26.30	31.2
Fumaric acid	156.49	154.67
$Fumarate^{2-}$	.....	144.41
α-D-Glucose	217.56	219.22
L-Glutamic acid	173.81	171.76
$L\text{-}Glutamate^{+2-}$	.....	165.87
Glycine	88.61	89.14
Glycogen (per glucose unit)	.....	158.3
Hydroxide ion	.....	37.60
Hydrogen ion	.....	0.00
β-Hydroxybutyric acid	.....	127.00
β-$Hydroxybutyrate^-$	.....	121.00
α-$Oxoglutarate^{2-}$	.....	190.62
L-Leucine	82.63	81.68
DL-Leucylglycine	.....	110.90
$Malate^{2-}$	.....	201.98
Methanol (liquid)	39.73	41.88
Nitrate	.....	26.41
Nitrite	.....	8.25
$Oxoglutarate^{2-}$	.....	190.53
$Pyruvate^-$	.....	133.44
Succinic acid	178.68	178.39
$Succinate^{2-}$	.....	164.97
L-Threonine	.....	123.0
L-Tyrosine	96.10	92.55
Urea	47.12	48.72
Water (liquid)	56.69	

SOURCE: J. T. Edsall and J. Wyman, *Biophysical Chemistry*, Academic Press, New York, 1958, p. 236.

Assuming that ΔH^0 is independent of temperature over the range of interest, this equation may be integrated to yield

$$\ln K = -\frac{\Delta H^0}{RT} + \text{constant} \tag{2-77}$$

TABLE 2-7
Thermodynamics of protein denaturation

	ΔG, *kcal/mol*	ΔH, *kcal/mol*	ΔS, *cal/deg/mol*
Ribonuclease thermal transition, *p*H 2.5, 30°	0.9	57	185
Chymotrysinogen, thermal transition, *p*H 3, 0.01 M Cl^-, 25°	7.3	39	105
Myoglobin, thermal transition, *p*H 9, 25°	13.6	42	95

A plot of log K against the reciprocal of the absolute temperature, provided the assumption of the temperature independence of ΔH^0 is correct, yields a straight line of slope $-\Delta H^0/2.3R$, from which ΔH^0 may be easily evaluated.

Once ΔG^0 and ΔH^0 for a reaction at a given temperature have been established, the standard entropy change ΔS^0 may be obtained immediately from the relationship $\Delta S^0 = (\Delta H^0 - \Delta G^0)/T$.

The use of enthalpy and entropy changes as aids in the understanding of biochemical reactions is illustrated by a consideration of studies on the temperature dependence of the denaturation of proteins. In Table 2-7, the derived thermodynamic parameters for denaturation of several proteins are collected. The entropy changes are exceedingly large, the ΔS^0 values for most chemical reactions falling in the range $+10$ to -30 eu/mol, and emphasizes, in a dramatic fashion, the degree of structural organization of these proteins in the native state and the loss of this organization in the thermal denaturation process (see also the more complete discussion on page 177).

Values of ΔH^0 and ΔS^0 are also helpful in identifying the nature of acidic or basic residues required for enzymatic activity (Chapter 7) and as an indication of the nature of the forces binding substrates or inhibitors to enzymes.

HIGH-ENERGY COMPOUNDS

The understanding of the essential role of molecules such as ATP as a driving force for a very large number of endergonic reactions has led to a classification of biochemical substances as *high-energy* or *low-energy* compounds. The alternative terms *energy-rich* and *energy-poor* are similarly employed. Molecules such as ATP, which are characterized by a large negative free energy of hydrolysis under physiological conditions are considered energy rich, and molecules such as ordinary phosphate esters are considered energy poor. This somewhat nebulous classification has introduced a fair amount of confusion into the biochemical literature because these terms do not mean the same to all workers. In this book, we adopt the definition suggested by Jencks: "An energy-rich compound may be defined as one whose reaction with a substance commonly present in the environment is accompanied by a large negative free energy change at physiological *p*H" (11). Quantitatively, how does one ascertain if a particular compound falls into the energy-rich class?

Because biochemically prominent compounds exhibit a wide spectrum of reactivities, there is no clear line of demarcation between the energy-rich and energy-poor classes. As an approximate guide, compounds exhibiting a free energy of hydrolysis at *p*H 7 more negative than 7 kcal/mol and compounds exhibiting an oxidation–reduction potential at *p*H 7 more negative than that of the oxygen electrode by 0.5 V may be considered energy rich. Because, as emphasized in the definition above, we are interested in free energy changes at physiological *p*H, it is clear that, for hydrolysis and related reactions, only free energy changes calculated according to convention III can be considered as a measure of the energy-rich nature of a compound. Similarly, electrode potentials for oxidation–reduction reactions should be those referred to a standard state of *p*H 7.

Energy-rich compounds fall into several distinct groups, the most prominent of which are those derived from phosphate. The latter compounds fall into four categories: phosphoric acid anhydrides, phosphoric-carboxylic anhydrides, phosphoguanidines, and enol phosphates. In addition, thiol esters, amino acid esters, and dihydropyridines, among others, are all regarded as energy rich. Each of these classes of compounds is discussed below.

Phosphoric Acid Anhydrides. As indicated above, the prime example of an energy-rich compound is the phosphoric acid anhydride ATP. Reference to Table 2-4 indicates that the hydrolysis of 1 mol of this substance to ADP and inorganic phosphate or to AMP and pyrophosphate is accompanied, in both cases, by the release of more than 7 kcal of free energy. Other phosphoric acid anhydrides exhibit similar standard free energies of hydrolysis at *p*H 7. The energy-rich nature of compounds of this type is most easily understood through consideration of the chemically more familiar, and directly related, case of acetic anhydride. The large negative free energy change that accompanies the hydrolysis of acetic anhydride at *p*H 7 is principally the result of two factors. First, the electrophilic carbonyl carbon atom is stabilized by electron donation and destabilized by electron withdrawal. Because the acetyl group is a reasonably strong electron-withdrawing substituent, it tends to destabilize acetic anhydride with respect to the products of the hydrolysis reaction. The destabilizing effect of electron-withdrawing substituents is easily grasped by considering the reactivities of acetyl chloride, acetic anhydride, acetyl phosphate, ethyl acetate, and acetamide. Although other factors are certainly involved, the order of reactivity of these derivatives of acetic acid decreases with the decreasing ability of the substituent on the carbonyl carbon to withdraw electrons. Second, the resonance energy of acetic anhydride, 29 kcal, is considerably less than 36 kcal, the resonance energy of two acetate anions, the hydrolysis products at *p*H 7. Consequently, acetic anhydride is unstable with respect to its hydrolysis products. The loss of resonance energy upon the formation of the anhydride results from the fact that the pi electrons of the oxygen atom linking the two carbonyl carbons cannot satisfy the electronic demand of both carbonyl functions simultaneously:

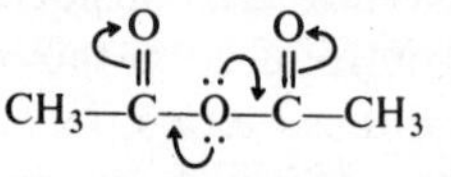

$$CH_3-\overset{\overset{O}{\|}}{C}-\ddot{O}-\overset{\overset{O}{\|}}{C}-CH_3$$

This situation has been termed *competing resonance,* because both carbonyl functions are competing for the electrons on a single oxygen. This competition is clearly relieved in the hydrolysis products. These considerations apply with equal validity to the hydrolysis of phosphoric anhydrides such as ATP, although the gain in resonance energy on hydrolysis is probably less important. In the latter case, an additional factor is also probably important. Because ATP and related substances exist as either the trianion or tetraanion at neutral *p*H, electrostatic repulsions, which are at least partially relieved in the products, may tend to destabilize polyphosphates with respect to the hydrolytic products.

One driving force for the hydrolysis of ATP and other polyphosphates whose importance is particularly difficult to assess is the differential solvation of the reactants and products. Clearly, if the products are solvated better than

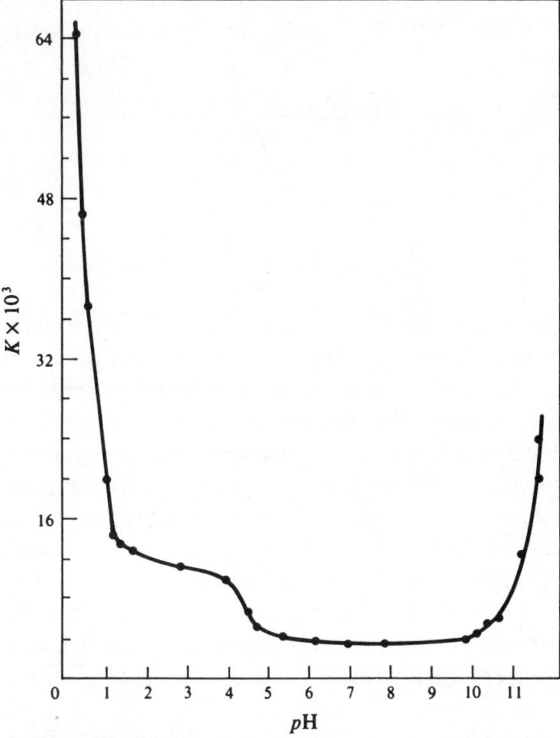

FIGURE 2-7
The *p*H-rate profile for the hydrolysis of acetyl phosphate. D. E. Koshland, Jr., *J. Am. Chem. Soc.*, **74:** 2286 (1952); by permission of the copyright owners, the American Chemical Society.

the reactants, impetus will be provided for the hydrolytic reaction. Although all the pertinent data are not available, it does appear entirely possible that differential solvation will prove to be the single most important driving force for these reactions.

Metal-ion catalysis is of great importance for the reactions of ATP and related substances. As indicated throughout the chapters of this book dealing with intermediary metabolism, the transfer of the terminal phosphate from the polyphosphate chain of ATP to a variety of acceptors in enzyme-catalyzed reactions is metal ion–dependent. In a similar fashion, the nonenzymatic transfer of this phosphate to acceptors such as water, phosphate, and acetate is also dependent on the presence of metal ions (12,13).

Carboxylic-Phosphoric Anhydrides. The biochemically most prominent acyl phosphates (carboxylic phosphoric anhydrides) are acetyl phosphate and amino acyl adenylates, which are characterized by standard free energies of hydrolysis at *p*H 7 of 10.3 and 13.3 kcal, respectively (Table 2-4). The factors that account for these free energies of hydrolysis are similar to those discussed above for acetic anhydride: the destabilizing influence of the electron-withdrawing phosphate group on the carbonyl function and the decreased resonance energy of the starting material relative to the hydrolysis products.

The nonenzymatic hydrolysis of acetyl phosphate exhibits a most interesting *p*H-rate profile (Fig. 2-7) (14). Under very acidic conditions, an acid-catalyzed hydrolysis of the undissociated substrate occurs. Under less acidic conditions, reactions that depend kinetically on the concentrations of the acetyl phosphate monoanion and dianion occur, the former reaction being considerably more rapid. These reactions almost certainly occur via unimolecular reaction pathways with the expulsion of monomeric metaphosphate (15):

$$CH_3C(=O)\text{—}O\text{—}P(=O)(OH)\text{—}O^- \longrightarrow CH_3CO_2H + PO_3^- \tag{2-78}$$

$$CH_3\text{—}C(=O)\text{—}O\text{—}P(=O)(O^-)\text{—}O^- \longrightarrow CH_3CO_2^- + PO_3^- \tag{2-79}$$

The greater reactivity of the monoanion is presumably due to two factors. First, the monoanion probably decomposes with proton transfer to the leaving group [Eq. (2-78)]. This is a case of internal general acid catalysis (Chapter 7). Second, because the monoanion is more electron-withdrawing than the dianion, the former species experiences greater destabilization than the latter. The hydrolysis of acetyl phosphate is subject to catalysis by metal ions (16). In contrast to the reactivities for the uncatalyzed reaction, it is the dianion, which is a much better chelating agent than the monoanion, whose hydrolysis is most strikingly subject to such catalysis.

Phosphoguanidines. Creatine phosphate, which serves as an energy storehouse in muscle, and arginine phosphate are important examples of naturally occurring phosphoguanidines:

$$H_2N^+\!=\!C(\text{—}NH\text{—}PO_3^{2-})\text{—}N(CH_3)\text{—}CH_2\text{—}CO_2^-$$

Creatine phosphate

$$H_2N^+\!=\!C(\text{—}NH\text{—}PO_3^{2-})\text{—}NH\text{—}(CH_2)_3\text{—}CH(NH_3^+)\text{—}CO_2^-$$

Arginine phosphate

The standard free energy of hydrolysis of such compounds at physiological pH is approximately -7 to -10 kcal (Table 2-4). The energy-rich nature of such compounds is largely due to the inhibition of the normal resonance of the guanidinium ion. The guanidinium ion, or alkyl guanidinium ion, is a nearly symmetrical species with roughly equal contributions to its stability from each of the valence-bond forms that may be written for it. The large resonance energy associated with this symmetrical species accounts for the very high basicity of the guanidines. The conversion of guanidines to guanidine phosphates destroys the symmetry of this molecule. In particular, the resonance form with the positive charge adjacent to the positively charged phosphorus atom is quite unfavorable:

$$R\text{—}NH\text{—}C(=N^+H_2)\text{—}NH_2 \leftrightarrow R\text{—}N^+H\!=\!C(NH_2)\text{—}NH_2 \leftrightarrow R\text{—}NH\text{—}C(NH_2)\!=\!N^+H_2$$

$$R\text{—}NH\text{—}C(=N^+H_2)\text{—}NH\text{—}P^+(O^-)_3 \leftrightarrow R\text{—}N^+H\!=\!C(NH_2)\text{—}NH\text{—}P^+(O^-)_3 \leftrightarrow \left[R\text{—}NH\text{—}C(NH_2)\!=\!N^+H\text{—}P^+(O^-)_3\right]$$

Enol Phosphates. Phosphoenol pyruvate, an important intermediate in the metabolism of carbohydrates, is characterized by such a large negative free energy of hydrolysis (15 kcal) that its reaction with ADP to form ATP and pyruvic acid proceeds very nearly to completion:

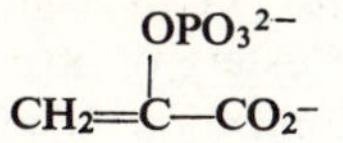

$$CH_2{=}\overset{\overset{\displaystyle OPO_3^{2-}}{|}}{C}{-}CO_2^-$$

Phosphoenol pyruvate

Compounds of this general nature are excellent phosphorylating agents and have recently found wide application in the synthesis of organic phosphates. The energy-rich nature of phosphoenol pyruvate is the result of two factors. First, the enol of pyruvic acid, in which form the phosphate is constrained to exist, is less stable than the keto form, which is adopted by the hydrolysis product, by approximately 10 to 12 kcal/mol. Thus, the conversion of the enol to the keto form provides a strong driving force for the hydrolysis reaction. Second, the hybridization of the O—P bond in phosphoenol pyruvate is sp_2. Such bonds are inherently less strong than sp_3 bonds, and, consequently, enol phosphates are less stable than ordinary phosphate esters. An additional example of behavior of this type is provided by the acyl cyanides, in which the bond to be cleaved is *sp*. Acyl cyanides are excellent acyl group donors.

Thiol Esters. Thiol esters, particularly those involving coenzyme A (Chapter 8), play a very significant role in metabolism, particularly in acyl-group transfer reactions. As indicated in Table 2-4, the standard free energy of hydrolysis under physiological conditions is approximately −8 kcal/mol. Thus, thiol esters clearly fall into the category of high-energy compounds. The diminished resonance interaction between the pi electrons of the sulfur atom and the carbonyl group, compared with such interaction in ordinary oxygen esters or in carboxylate anions, provides a driving force for the hydrolysis reaction and accounts for the increased standard free energy of hydrolysis compared with oxygen esters (page 392). Indeed, some studies have actually suggested that the sulfur atom not only does not act as an effective electron donor to the carbonyl group, but may actually act as an electron acceptor (17):

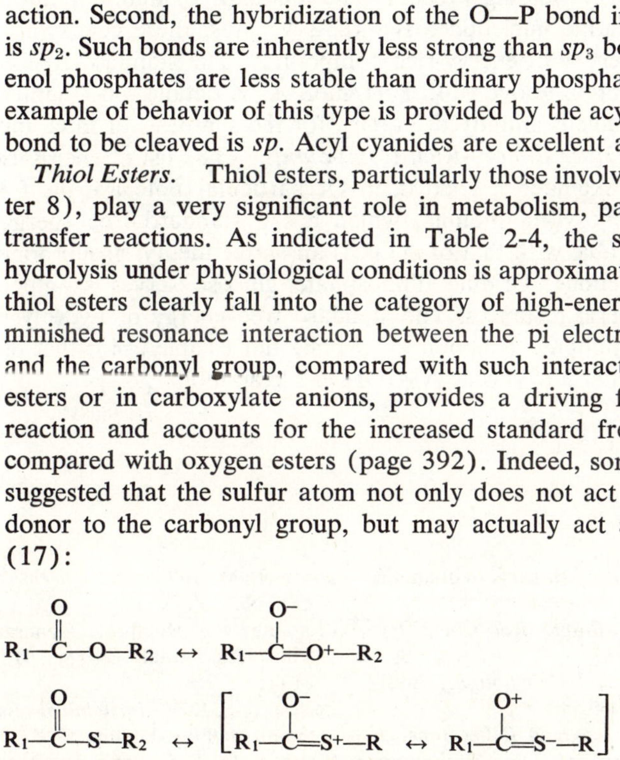

$$R_1{-}\overset{\overset{\displaystyle O}{\|}}{C}{-}O{-}R_2 \leftrightarrow R_1{-}\overset{\overset{\displaystyle O^-}{|}}{C}{=}O^+{-}R_2$$

$$R_1{-}\overset{\overset{\displaystyle O}{\|}}{C}{-}S{-}R_2 \leftrightarrow \left[R_1{-}\overset{\overset{\displaystyle O^-}{|}}{C}{=}S^+{-}R \leftrightarrow R_1{-}\overset{\overset{\displaystyle O^+}{|}}{C}{=}S^-{-}R\right]$$

Relatively poor

In any event, the inhibition of resonance in the thiol esters provides a straightforward explanation for the energy-rich nature of these compounds. The increased free energy released on the hydrolysis of acetoacetyl thiol esters compared with ordinary thiol esters results from destabilization of the thiol esters by the strongly electron-withdrawing acetyl substituent in the acyl portion of the molecule.

Amino Acid Esters. Amino acid esters, which are important intermediates in the biosynthesis of proteins, exhibit a standard free energy of hydrolysis comparable with that of ATP. It should be noted that the free energy of hydrolysis of glycine ethyl ester, for example, to the free acid is nearly the same as that for ethyl acetate. However, the high acidity of glycine, $pK_a = 2.3$, compared with that of acetic acid, $pK_a = 4.76$, results in the standard free energy of hydrolysis of the amino acid ester at pH 7 being greater than that of ethyl acetate by $298 \times 1.98 \times 2.3 \times \Delta pK_a = -3300$ cal/mol at 25°. Although the division of ester hydrolysis at pH 7 into a hydrolysis reaction to the free acid and an ionization of the acid is convenient for the purposes of calculating free energy changes, it is improper to regard the large free energy changes accompanying the hydrolysis of esters of strongly acidic acids as resulting from a large ΔG term for the ionization. The free energy is a state function and does not depend on the course arbitrarily chosen for the reaction. The thermodynamic instability of amino acid esters is properly considered to result from destabilization of the ester by the strongly electron-withdrawing ammonium group. In this respect, amino acid esters are similar to esters of acetoacetic acid, mentioned above.

Reduced Pyridine Nucleotides. The transfer of a pair of electrons from the pyridine nucleotide coenzymes, nicotinamide adenine dinucleotide and nicotinamide adenine dinucleotide phosphate (Chapter 8), to oxygen is an energy-yielding process. The standard oxidation–reduction potential for the pyridine nucleotides is −0.32 V; thus, these coenzymes clearly fall into the category of energy-rich compounds. The standard free energy change for the reaction above is 40 kcal/mol. As is detailed in Chapter 15, this free energy change is employed, in part, for the synthesis of three molecules of ATP.

Other Energy-Rich Compounds. The list of energy-rich compounds might be extended a good deal. Of particular note are the following examples. *S*-adenosyl-methionine, which has a standard free energy of hydrolysis comparable with that of ATP, is an active methyl group donor in many metabolic reactions. Uridine diphosphate glucose serves as glucosyl group donor in several instances. The standard free energy of hydrolysis of this and related compounds is about −7.3 kcal/mol (Table 2-4); that of the important cyclic diester 3′:5′-cyclic AMP is −12 kcal/mol.

REFERENCES

Books and Review Articles

Clark, W. M.: *Topics in Physical Chemistry,* Williams & Wilkins, Baltimore, 1948.

Edsall, J. T., and J. Wyman: *Biophysical Chemistry,* Academic Press, New York, 1958.

Finlayson, J. S.: *Basic Biochemical Calculations,* Addison-Wesley, Reading, Mass., 1969.

Gurd, F. R. N.: Binding of Protons and Other Ions, in S. J. Leach (ed.), *Physical Principles and Techniques of Protein Chemistry,* part B, Academic Press, New York, 1970.

Gurd, F. R. N., and P. E. Wilcox: Complex Formation Between Metallic Cations and Proteins, Peptides, and Amino Acids, *Advan. Protein Chem.,* **11:** 311 (1956).

Hill, T. L., *Thermodynamics for Chemists and Biologists,* Addison-Wesley, Reading, Mass., 1968.

Klotz, I. M.: *Chemical Thermodynamics,* W. A. Benjamin, New York, 1964.

Lipmann, F.: Metabolic Generation and Utilization of Phosphate Bond Energy, *Advan. Enzymol.,* **1:** 99 (1941).

Patton, A. R.: *Biochemical Energetics and Kinetics,* Saunders, Philadelphia, 1965.

Rossotti, F. J. C., and H. Rossotti: *Determination of Stability Constants,* McGraw-Hill, New York, 1961.

Segel, I. H., *Biochemical Calculations,* Wiley, New York, 1968.

Sober, H. A. (ed.), *Handbook of Biochemistry,* Chemical Rubber Co., Cleveland, 1968, pp. J150ff.

Specific References

1. Berthelot, A., and P. de St.-Giles: *Ann. Chem. Phys.,* **68:** 225 (1863).
2. Jencks, W. P., S. Cordes, and J. Carriuolo: *J. Biol. Chem.,* **235:** 3608 (1960).
3. Levintow, L., and A. Meister: *J. Biol. Chem.,* **209:** 265 (1954).
4. Ratner, S.: *The Enzymes,* vol. 6, Academic Press, New York, 1962, p. 495.

5. Benzinger, T., C. Kitzinger, R. Hems, and K. Burton: *Biochem. J.,* **71:** 400 (1959).
6. Robbins, E. A., and P. D. Boyer: *J. Biol. Chem.,* **224:** 121 (1957).
7. Atkinson, M. R., E. Johnson, and R. K. Morton: *Nature,* **184:** 1925 (1959).
8. Jencks, W. P.: in H. A. Sober (ed.), *Handbook of Biochemistry,* Chemical Rubber Co., Cleveland, 1968, pp. J144ff.
9. Carpenter, F. H.: *J. Am. Chem. Soc.,* **82:** 1111 (1960).
10. Conant, J. B., H. M. Kahn, L. F. Fieser, and S. S. Kurtz, Jr.: *J. Am. Chem. Soc.,* **44:** 1382 (1922).
11. Jencks, W. P.: ref. 4, vol. 6, p. 339.
12. Lowenstein, J. M.: *Biochem. J.,* **70:** 222 (1958); **75:** 269 (1960).
13. Lowenstein, J. M., and M. N. Schatz: *J. Biol. Chem.,* **236:** 305 (1961).
14. Koshland, D. E., Jr.: *J. Am. Chem. Soc.,* **73:** 4103 (1951).
15. Di Sabato, G., and W. P. Jencks: *J. Am. Chem. Soc.,* **83:** 4400 (1961).
16. Koshland, D. E., Jr.: *J. Am. Chem. Soc.,* **74:** 2286 (1952).
17. Baker, A. W., and G. H. Harris: *J. Am. Chem. Soc.,* **82:** 1923 (1960).

3

Proteins: Structure and Properties

Proteins are essential components of all living systems from the most elementary to the most complex. At one extreme, the simplest viruses possess none of the usual attributes associated with the "living" state save the ability to reproduce under the proper set of conditions. These viruses are composed of just two types of molecules: proteins and nucleic acids. The protein portion serves as an enclosure which is occupied by the nucleic acid, the infective agent. At the other extreme, complex organisms such as man make a great many uses of proteins. They serve as *enzymes,* which have catalytic properties; *antibodies,* which serve as a weapon in the defense arsenal of the organism; *structural elements,* which serve to define and maintain the architectural constructs and to govern changes in these (as in contractile processes); *transport devices,* which are charged with the responsibility of carrying crucial materials such as oxygen, lipids, and many ions to the site of their utilization; and *metabolic regulators,* including certain *hormones,* which serve to coordinate and integrate the chemistry of the organism in ways necessary to support life. Thus, proteins are concerned with all physiological processes.

Some organisms employ proteins in quite exotic ways. For example, some fishes are able to live in exceptionally cold (supercooled) water by virtue of a protein that has a remarkable capacity to lower the freezing point of the fishes blood! That so many diverse functions are served by one class of molecules is striking. We can anticipate the chemistry of the proteins to be both complex and fascinating.

As one may judge from these comments, the chemistry of proteins and the chemistry of life are intimately related. The relationship is close enough to warrant the assertion that one requirement for the specification that a given object is "living" is that it contains protein.

Proteins are macromolecules and, like all biological macromolecules, are *polymers.* The distinguishing feature of any polymer is that it is constructed by the repetition of one or more structural elements, termed the monomers. In the case of the proteins, the monomers are a group of about 20 amino acids. These are linked together to form chains of varying length. Although no clear line exists, proteins are generally considered to have minimal chain lengths of about 50 amino acids corresponding to molecular weights near 5000. The most complicated proteins contain several thousand amino acids and have molecular weights of several million. Thus, a range of structural complexity underlies the functional diversity noted above.

We begin our consideration of the chemistry of proteins with a brief discussion of the chemistry of the monomeric units, the α-amino acids, and the nature of the bond that unites them in proteins. Later, some of the physical methods of interest for the characterization of proteins are considered together with properties of these molecules. Finally, some aspects of the isolation and purification of proteins from natural sources are considered. In Chapter 4, a more detailed examination of the structure of proteins is presented.

AMINO ACIDS: STRUCTURAL UNITS OF PROTEINS

Structures

The complete hydrolysis of proteins yields a mixture of about 20 *amino acids*. With two exceptions, these substances have a primary amino function and a carboxyl function joined to the same carbon atom; hence they are termed α-amino acids and are derivatives of the general structure NH_2—CHR—COOH. The two exceptions, proline and hydroxyproline, are α-imino acids. The structures of the amino acids that are commonly found in the hydrolyzates of proteins are shown in Table 3-1. These amino acids are classified into seven groups: those which have (1) aliphatic side chains, (2) hydroxylic side chains, (3) carboxylic side chains and their amides, (4) basic side chains, (5) aromatic side chains, (6) sulfur-containing side chains, and (7) the imino acids. In addition to those listed in Table 3-1, several other amino acids have been reported to be present in the hydrolyzates of individual proteins. As these are distinctly isolated cases and are not apparently intimately concerned with either the structure or function of these proteins, they will not be considered further.

Both the portion of the amino acids that is common to all and the portion that varies from one to the other are of interest to us. As is detailed below, the proteins are constructed of linear chains of amino acids: The amino and carboxyl groups are the moieties through which one amino acid is linked to another. Hence, these two groups plus the α-carbon atom between them form the backbone of the molecule. The side chains, upon which our classification in Table 3-1 is based, are appended to this backbone and are responsible for the individual characteristics of protein molecules. Clearly, the nature and arrangement of these side chains serve to differentiate one protein from another at the most basic level of structural organization. It will prove of much interest later to understand the types of interactions in which each of the amino acids participates. Let us, therefore, consider the properties of the amino acids in terms of the nature of their side chains.

We may distinguish three classes of amino acids: nonpolar, uncharged polar, and charged polar. Into the nonpolar class we may place Gly, Ala, Val, Leu, Ile, Phe, Met, and Pro. In addition, Tyr and Trp have substantial nonpolar characteristics but also have the capacity to interact with water in a favorable way. These side chains are related to compounds which themselves have little or no solubility in water. Placing such groups in an aqueous environment has the effect of inducing the water molecules in the immediate vicinity to adopt an ordered structure. An extreme example is provided by the water clathrates of hydrocarbons. Methane, for example, forms a crystalline hydrate with water, in which the water forms a lattice into which methane molecules may be inserted. The water is frequently referred to as the "host" and the methane as the "guest." This hydrate has remarkable stability: It remains crystalline above the melting point of ice despite the extreme volatility of methane, bp = −161°.

TABLE 3-1
Amino acids that commonly occur in proteins

Name	*Structure*	*Values of pK_a*	*pI*	*Remarks*
I. Aliphatic amino acids				
1. Glycine (Gly)	$H_2N—CH_2—CO_2H$	2.34; 9.6	5.97	
2. Alanine (Ala)	$H_2N—CH(CH_3)—CO_2H$	2.35; 9.69	6.02	
3. Valine (Val)	$H_2N—CH(CH(CH_3)_2)—CO_2H$	2.32; 9.62	5.97	
4. Leucine (Leu)	$H_2N—CH(CH_2—CH(CH_3)_2)—CO_2H$	2.36; 9.60	5.98	
5. Isoleucine (Ile)	$H_2N—CH(CH(CH_3)—CH_2—CH_3)—CO_2H$	2.36; 9.68	6.02	Has two asymmetric carbon atoms
II. Hydroxyamino acids				
6. Serine (Ser)	$H_2N—CH(CH_2OH)—CO_2H$	2.21; 9.15	5.68	
7. Threonine (Thr)	$NH_2—CH(CH(OH)—CH_3)—CO_2H$	2.63; 10.43	6.53	Has two asymmetric carbon atoms

III. Dicarboxylic amino acids and their amides				
8. Aspartic acid (Asp)	$HO_2C-CH_2-CH(NH_2)-CO_2H$	2.09 (α-carboxyl); 3.86 (β-carboxyl); 9.82	2.97	
9. Asparagine ($AspNH_2$) (or Asn)	$H_2NOC-CH_2-CH(NH_2)-CO_2H$	2.02; 8.8	5.41	
10. Glutamic acid (Glu)	$HO_2C-CH_2-CH_2-CH(NH_2)-CO_2H$	2.19 (α-carboxyl); 4.25 (γ-carboxyl); 9.67	3.22	
11. Glutamine ($GluNH_2$) (or Gln)	$H_2NOC-CH_2-CH_2-CH(NH_2)-CO_2H$	2.17; 9.13	5.65	
IV. Amino acids having basic functions				
12. Lysine (Lys)	$H_2N-(CH_2)_4-CH(NH_2)-CO_2H$	2.18; 8.95 (α-amino); 10.53 (ϵ-amino)	9.74	
13. Hydroxylysine (Hyl)	$H_2N-CH_2-CH(OH)-(CH_2)_2-CH(NH_2)-CO_2H$	2.13; 8.62 (α-amino); 9.67 (ϵ-amino)	9.15	Has two asymmetric carbon atoms; occurs only in collagen and gelatin

TABLE 3-1 (Continued)

Name	*Structure*	*Values of pK_a*	*pI*	*Remarks*
14. Histidine (His)	imidazole ring (HC–N, C–N, H) – CH_2; NH_2—CH—CO_2H	1.82; 6.0 (imidazole); 9.17	7.58	
15. Arginine (Arg)	$(CH_2)_3$—N(H)—C(=NH)—NH_2; NH_2—CH—CO_2H	2.17; 9.04 (α-amino); 12.48 (guanidino)	10.76	
V. Aromatic amino acids (histidine included in category IV)				
16. Phenylalanine (Phe)	C_6H_5—CH_2; NH_2—CH—CO_2H	1.83; 9.13	5.98	λ_{max} = 259 nm; ε_{259} = 2×10^2
17. Tyrosine (Tyr)	OH—C_6H_4—CH_2; NH_2—CH—CO_2H	2.20; 9.11 (α-amino); 10.07 (phenolic hydroxyl)	5.65	λ_{max} = 278 nm; ε_{278} = 1.1×10^3

18. Tryptophan (Trp)	(indole ring)–CH_2; NH_2—CH—CO_2H	2.38; 9.39	5.88	$\lambda_{max} = 279$ nm $\varepsilon_{279} = 5.2 \times 10^3$
19. Thyroxine	NH_2—CH(CO_2H)—CH_2—(ring, I, I)—O—(ring, I, I)—OH			Occurs only in the protein thyroglobulin (elaborated by the thyroid gland)
VI. Sulfur-containing amino acids				
20. Cysteine (Cys)	CH_2—SH; NH_2—CH—CO_2H	1.71; 8.33 (sulfhydryl); 10.78 (α-amino)	5.02	
21. Cystine Cys–Cys	(CH_2—S—; NH_2—CH—CO_2H)$_2$	1.65; 2.26 (carboxyls) 7.85; 9.85 (aminos)	5.06	
22. Methionine (Met)	$(CH_2)_2$—SCH_3; NH_2—CH—CO_2H	2.28; 9.21	5.75	
VII. Imino acids				
23. Proline (Pro)	(pyrrolidine ring, N–H)—CO_2H	1.99; 10.60	6.10	
24. Hydroxyproline (Hyp)	HO—(pyrrolidine ring, N–H)—CO_2H	1.92; . 9.73	5.83	Has two asymmetric carbon atoms; occurs only in collagen and gelatin

The ordering of water that results from placing the side chain of a nonpolar amino acid in an aqueous environment causes the entropy of the system to decrease and is, therefore, energetically unfavorable. Thus, such nonpolar groups will tend to cluster together in a way which involves minimal interaction with the aqueous environment and maximal interaction with each other.

The individual amino acids differ of course, in their degree of hydrophobicity: the larger the nonpolar side chain, the more hydrophobic. In Fig. 3-1, space-filling models of the pertinent side chains are shown. Clearly, Phe is the bulkiest of the group and Gly the smallest. Two of the amino acids in this category have special properties. Phenylalanine can interact favorably with other aromatic systems by virtue of overlap between the two pi-electron systems. Methionine can interact with metal ions because the sulfur atom possesses unshared electron pairs. Nature appears to have chosen the nonpolar amino

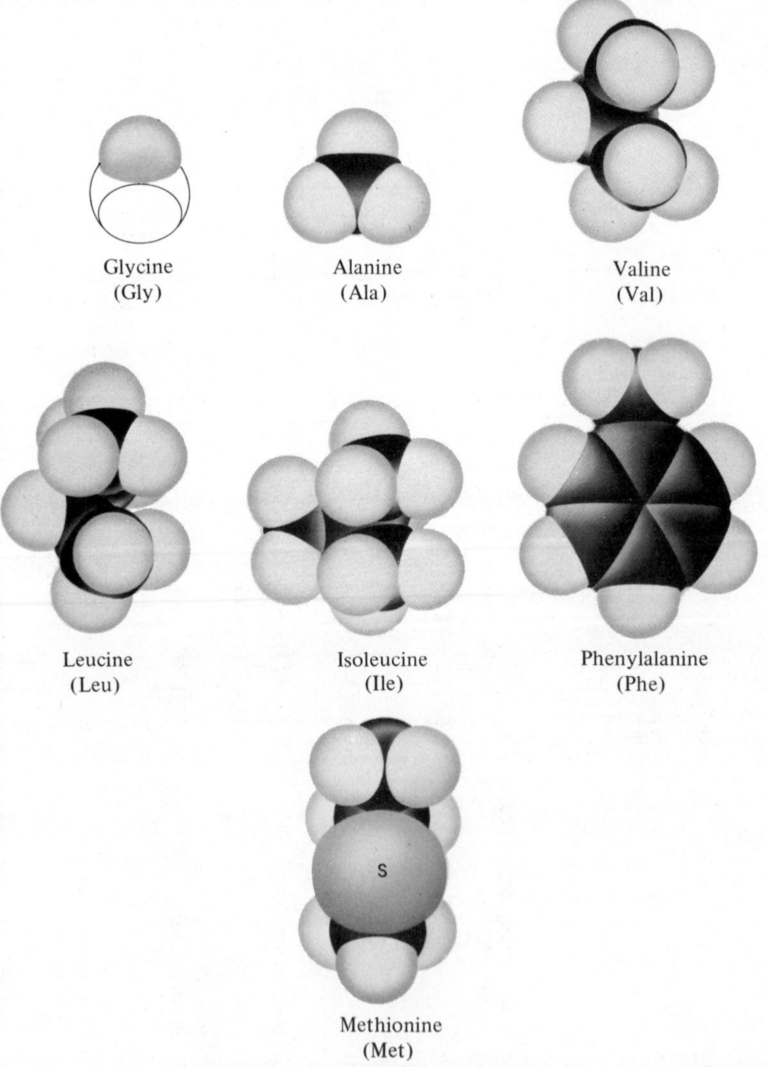

FIGURE 3-1

Space-filling drawings of the side chain of the nonpolar amino acids. The drawings give an accurate representation of the relative sizes of the atoms. From *The Structure and Action of Proteins* by Richard E. Dickerson and Irving Geis. Copyright © 1969 by Richard E. Dickerson and Irving Geis. Reprinted by permission of the authors and Harper & Row, Publishers, Inc.

acids both on the basis of the variety of hydrophobic bulk and molecular shape and on the basis of ability to participate in specialized interactions. Note, specifically, that the amino acid having a hydrocarbon structure corresponding to the side chain of methionine does not occur in proteins; methionine is capable of providing the necessary bulk and yields the metal-binding capacity as a bonus.

The uncharged polar group of amino acids contains a variety of structures, being comprised by Ser, Thr, Asn, Gln, Trp, and Tyr. Each of these side chains is capable of interacting with water molecules through formation of hydrogen bonds. Similar interactions with other polar molecules are possible as well and, in fact, occur regularly in proteins, as we shall see in Chapter 4. Here again, a variety of sizes and shapes is present in the side chains and both Tyr and Trp can interact with other aromatic systems, as noted above for the case of Phe. Moreover, the hydroxylic amino acids can act as nucleophilic reagents and all members of the groups can interact with metal ions.

Finally, several amino acids bear a full charge at neutral *p*H and fall into the charged polar category: Asp and Glu are negatively charged and His, Lys, and Arg bear a positive charge (refer to the discussion of acid–base reactions of amino acids in Chapter 2). These side chains can be involved in electrostatic interactions as well as hydrogen-bond formation. They will strongly prefer an aqueous environment and, insofar as is possible, will occupy positions in which they are fully solvated by the available water molecules. One will not expect to find a charged polar side chain surrounded on all sides by nonpolar ones. Later we shall see how these general considerations are translated into specific protein structures.

At this early stage, we can already begin to see why it is that proteins are capable of performing such a variety of functions. Since there are about 20 different amino acids, we can form 400 separate structures which contain just two amino acids, 8000 structures with three, 160,000 structures with four, and so on. A small protein might have a chain composed of 100 amino acids: 20^{100} possible structures could be formed. Since the individual amino acids possess such distinct properties, one can expect enormous variability in characteristics of such a collection of structures. This variability is reflected in the diversity of function observed.

Many amino acids and their derivatives that are not constituents of proteins are, nevertheless, biochemically important. A brief compilation of some of the most prominent such substances, together with a few comments concerning their biochemical importance, is presented in Table 3-2.

Optical Activity of Amino Acids

With the single exception of glycine, all amino acids obtained from proteins are optically active since the α-carbon of each is asymmetric. The absolute configuration of the amino acids has been related to that of L-glyceraldehyde. Without exception, amino acids obtained from protein digested in such a way that racemization does not occur are of the L configuration. The stereochemical relationship of L-glyceraldehyde to L-alanine is depicted below:

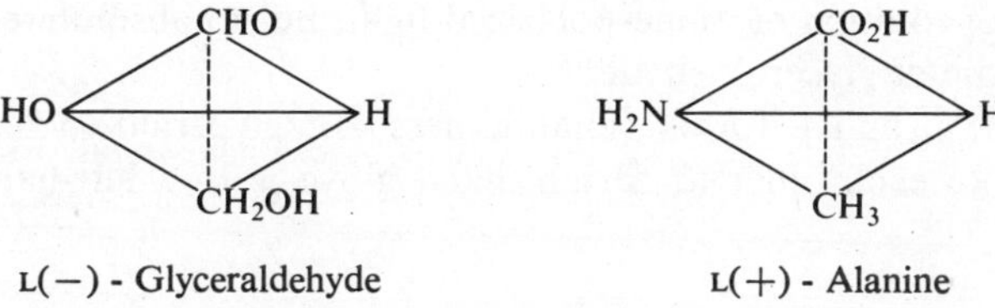

L(−) - Glyceraldehyde L(+) - Alanine

TABLE 3-2
Amino acids and derivatives of biological importance

Name	*Structure*	*Occurrence, Function, etc.*
1. β-Alanine	$NH_2{-}CH_2CH_2{-}COOH$	A component of the vitamin pantothenic acid and the derived coenzyme, coenzyme A; also occurs in the naturally occurring peptides carnosine and anserine
2. γ-Amino butyric acid	$NH_2{-}(CH_2)_3{-}COOH$	A constituent of plant tissue and of brain tissue in mammals and some amphibia and birds
3. Sarcosine	*N*-Methyl glycine	An intermediate in the metabolism of one-carbon compounds; a constituent of the actinomycins, a group of potent antimetabolites
4. Betaine	*N,N,N*-Trimethyl glycine	A constituent of plant and animal tissues; an intermediate in the metabolism of lipids
5. *O*-diazoacetylserine (azaserine)	$CH_2O{-}COCHN_2$ $NH_2{-}CH{-}COOH$	Antimetabolite isolated from certain molds; inhibits the synthesis of nucleic acids
6. Homoserine	CH_2CH_2OH $NH_2{-}CH{-}COOH$	An important intermediate in metabolism; occurs in plant and animal tissues
7. Ornithine	$(CH_2)_3{-}NH_2$ $NH_2{-}CH{-}COOH$	An important intermediate in the biosynthesis of urea
8. Citrulline	$(CH_2)_3{-}NH{-}C(=O){-}NH_2$ $NH_2{-}CH{-}COOH$	An important intermediate in urea biosynthesis; the immediate precursor of arginine
9. Histamine	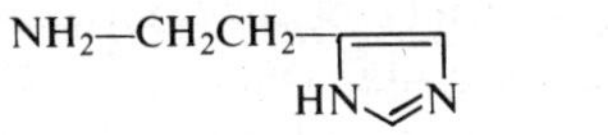 $NH_2{-}CH_2CH_2{-}$ HN N	The decarboxylation product of histidine; a powerful vasodepressor
10. 5-Hydroxytryptamine (serotonin)	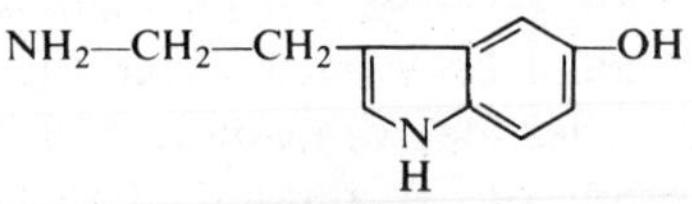 $NH_2{-}CH_2{-}CH_2{-}$ N H —OH	A substance important in the transmission of nerve impulses
11. Epinephrine (adrenalin)	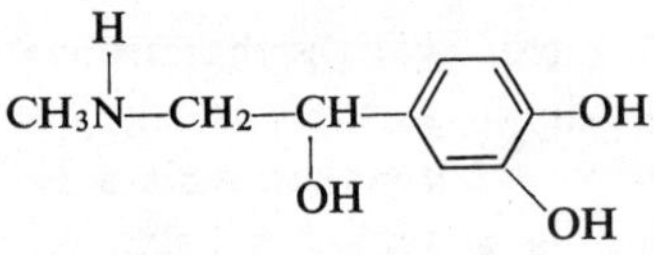 H $CH_3N{-}CH_2{-}CH{-}$ —OH OH OH	A derivative of tyrosine; an important hormone involved in many regulatory processes
12. Taurine	$NH_2{-}CH_2CH_2{-}SO_3H$	An oxidation product of cysteine metabolism; a component (in conjugation with a steroid) of bile salts
13. Penicillamine	$(CH_3)_2C{-}SH$ $NH_2{-}CH{-}COOH$	A component of the antibiotic penicillin

It is worth emphasizing that the designations L and D refer, not to the direction of rotation of plane-polarized light, but to absolute configurations relative to those of glyceraldehyde.

The D and L forms (enantiomers) of glyceraldehyde and alanine are related to each other as a right-hand glove is to a left-hand one. This point is

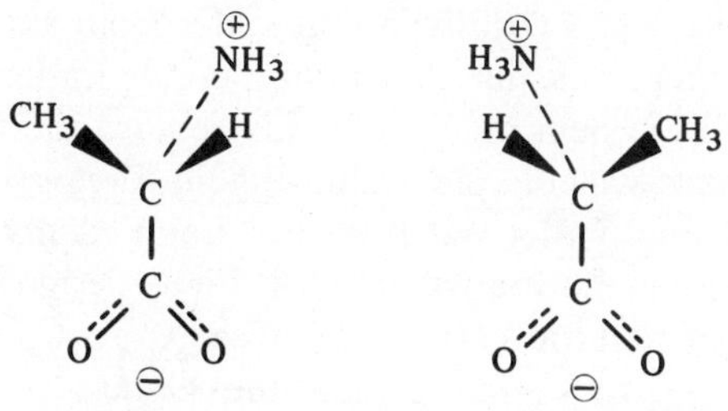

FIGURE 3-2
L-Alanine and its enantiomorph (mirror image), D-alanine.

made graphically in Fig. 3-2, in which space-filling models of the two enantiomorphs of alanine are provided.

Just why all the amino acids that occur in proteins are exclusively L is not clear. A great many hypotheses have been advocated at one time or another but none has proved persuasive. It may well be that the choice of the L forms was dictated by chance rather than by a distinct preference based on one or more physical forces. This may continue to prove a difficult matter to resolve.

The sign and magnitude of the optical rotation of the amino acids is, of course, a sensitive function of the nature of the side chain. In addition, as both the amino and carboxyl functions undergo protolytic reactions, the optical rotation is also a function of the *p*H of the solution in which it is measured.

In addition to the asymmetric center at the α-carbon atom, threonine, isoleucine, hydroxyproline, hydroxylysine, and cystine have a second asymmetric center. Thus, there exist, in the laboratory if not in nature, four *diastereomers* of these amino acids. These are designated L, D, L-*allo*, and D-*allo*. The configurations of threonine are indicated below:

COOH	COOH	COOH	COOH
H_2N—C—H	H—C—NH_2	H_2N—C—H	H—C—NH_2
H—C—OH	HO—C—H	HO—C—H	H—C—OH
CH_3	CH_3	CH_3	CH_3
L-Threonine	D-Threonine	L-Allothreonine	D-Allothreonine

Cystine is clearly a special case, because the two asymmetric centers are identical. In cases of this type, the configuration of one center may be the mirror image of the other, and the molecule is "internally compensated" and optically inactive. Such forms are termed *meso* (*m*) forms:

COOH COOH	COOH COOH
H_2N—C—H H_2N—C—H	H_2N—C—H H—C—NH_2
CH_2—S—S—CH_2	CH_2—S—S—CH_2
L-Cystine	*m*-Cystine

Although, as noted above, D-amino acids do not occur naturally in proteins, they are found widely distributed in living tissue. D-Amino acids are particularly abundant as constituents of bacterial cell walls. In addition, they are found incorporated into the structures of many antibiotics, including bacitracin, gramidicin, and the actinomycins (see page 907).

In some cases, the D and L designations suffer from ambiguities. To remedy these, a more systematic means of denoting configurations has been devised; it is possible that this system may eventually replace the D and L designations completely. The new system, the Cahn–Ingold–Prelog convention, employs the designations *R* and *S* [derived from the Latin *rectus* (right) and *sinister* (left)] and is based on the following rules. First, designate the four different groups about an asymmetric carbon atom as A, B, C, and D and assign a priority to each of them so that A has the highest priority, B the second highest, and so forth. Priorities are determined on the basis of the atomic number of the substituents. Thus, an oxygen substituent has priority over a nitrogen one, which has priority over a carbon one. Should two substituents have the same atomic number, precedence is given to the substituent with a second atom of higher atomic number. Thus, a carboxyl group has priority over a methyl group. A sequence of priorities of common substituents is I, Br, SO_3H, F, OH, NH_2, CO_2H, CHO, CH_2OH, CH_3. Second, examine the tetrahedral arrangement of the four groups about the asymmetric carbon from the side remote to the substituent of lowest priority, D. Third, if the sequence of substituents A → B → C traces a clockwise direction, the asymmetric carbon is designated *R;* if the sequence traces a counterclockwise direction, it is designated *S*. Thus, D-glyceraldehyde may also be described as *R*-glyceraldehyde and L-alanine as *S*-alanine:

CHO | C---H (HO, CH_2OH) ≡ CHO | H—C—OH | CH_2OH

R-Glyceraldehyde D-Glyceraldehyde

CO_2H | C---H (CH_3, NH_2) ≡ CO_2H | H_2N—C—H | CH_3

S-Alanine L-Alanine

Note that if a compound contains several asymmetric carbon atoms, the configuration about each of them is specified. Thus, L-threonine may also be described as 2(*S*), 3(*R*)-threonine:

CO_2H | H_2N—C—H | H—C—OH | CH_3 ≡ CO_2H | H_2N C H | C (H, OH) | CH_3 ≡ H | HO_2C C NH_2 | C (HO, CH_3) | H

L-Threonine 2(*S*), 3(*R*)-Threonine

Acid–Base Behavior of Amino Acids

A discussion of the protonic and metal-ion equilibria in which amino acids are involved was presented in Chapter 2. The following brief comments will suffice at this point.

All amino acids have at least two groups that are involved in protolytic reactions in aqueous solution. The α-carboxyl group of amino acids is characterized by a value of pK_a (that pH at which the group under consideration is half associated and half dissociated) in the range 2 to 3. The α-amino group, in contrast, is characterized by a pK_a (for the conjugate acid) near 10. Thus at pH values between about 4 and 9, the amino acid exists in a *dipolar ion* or *zwitterionic* form, the carboxyl group being dissociated and the amino group associated.

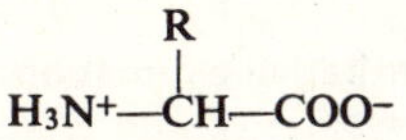

$$H_3N^+—\underset{}{\overset{R}{\overset{|}{C}}}H—COO^-$$

Over this entire pH range, the amino acids bear little net charge, and at one pH, the isoelectric point (see page 93), pI, bear no net charge. A little reflection will indicate that, for amino acids having no additional acidic or basic groups, the pI is midway between the two pK_a values; $pI = (pK_a^{NH_2} + pK_a^{CO_2H})/2$. The pK_a values and isoelectric points for protein amino acids are included in Table 3-1.

The side chains of several of the amino acids, including those of aspartic acid, glutamic acid, lysine, histidine, arginine, cysteine, and tyrosine, have appreciable acidic or basic properties. The appropriate values of pK_a are also included in Table 3-1. The isoelectric structures for these amino acids vary according to the basicity or acidity of the side-chain group. The isoelectric structures of aspartic acid and lysine, determined from the pK_a data, are predominantly

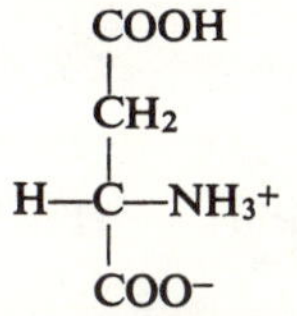

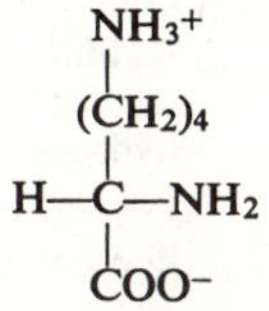

$$\begin{array}{c} COOH \\ | \\ CH_2 \\ | \\ H—C—NH_3^+ \\ | \\ COO^- \end{array} \qquad\qquad \begin{array}{c} NH_3^+ \\ | \\ (CH_2)_4 \\ | \\ H—C—NH_2 \\ | \\ COO^- \end{array}$$

The isoelectric point for aspartic acid must be very nearly midway between the pK_a values for the α- and β-carboxyls. At this pH, 2.97, the amino function exists almost completely as the ammonium ion and the α-carboxyl exists largely as the carboxylate, the β-carboxyl being largely undissociated. Hence, aspartic acid at pH 2.97 has one positively charged group and one negatively charged group and is isoelectric. Similar considerations indicate that the isoelectric point for lysine must be midway between the pK_a values for the α-amino and ε-amino functions.

The dipolar-ion structures for amino acids suffice to account for many of their properties. These include (1) a relatively high solubility in water, (2) a relatively low solubility in organic solvents, (3) large dipole moments, and (4) high melting points. The last point may be attributed to substantial electrostatic attractions between the oppositely charged groups in the crystal lattice.

Chemical Reactions of Amino Acids

All the amino acids have a series of reactions in common—those which occur at the amino or carboxyl functions, or both. In addition, if the amino acid contains a reactive side chain, it exhibits a series of specific reactions. Such specific reactions are of interest in two particular respects. First, reactions of those groups that result in the production of color are frequently useful for the detection and semiquantitative estimation of proteins and of the individual amino acids. These reactions will not be discussed in detail here, but a compilation of them is presented in Table 3-3. Second, reactions that occur at the side chains of amino acids are frequently of use in efforts to modify proteins chemically. Such reactions are discussed in Chapter 4.

Many of the reactions at the amino or carboxyl group are quite familiar (alkylation, acylation, ester and amide formation) and need not be reiterated here. Instead we consider a few reactions that are of general utility in amino acid and protein chemistry.

Ninhydrin. The reaction of ninhydrin (triketohydrindene hydrate) with amino acids is of particular importance for the detection and quantitative

TABLE 3-3
Reactions useful for detection and semiquantitative estimation of amino acids and proteins

Name	*Reactants*	*Amino acid detected*	*Color*
Millon reaction	$HgNO_3$ in nitric acid with a trace of nitrous acid	Tyrosine	Red
Xanthoproteic reaction	Boiling concentrated nitric acid	Tyr, try, phe	Yellow
Glyoxylic acid reaction (Hopkins-Cole reaction)	Glyoxylic acid in conc H_2SO_4	Tryptophan	Purple
Ehrlich reaction	*p*-Dimethylaminobenz-aldehyde in conc HCl	Tryptophan	Blue
Sakaguchi reaction	α-Naphthol and sodium hypochlorite	Arginine	Red
Nitroprusside reaction	Sodium nitroprusside in dil NH_3	Cysteine	Red
Sullivan reaction	Sodium 1,2 naphthoquinone-4-sulfonate and sodium hydrosulfite	Cysteine	Red
Pauly reaction	Diazotized sulfanilic acid in alkaline solution	Histidine, tyrosine	Red
Folin-Ciocalteu reaction	Phosphomolybdotungstic acid	Tyrosine	Blue

estimation of the latter. Ninhydrin is a powerful oxidizing agent and elicits the oxidative deamination of the α-amino group, liberating ammonia, carbon dioxide, the corresponding aldehyde, and a reduced form of the ninhydrin:

$$\text{Ninhydrin} + NH_2\text{—CHR—}CO_2H \rightarrow \text{Hydrindantin} + RCHO + CO_2 + NH_3 \qquad (3\text{-}1)$$

$$\text{Ninhydrin} + \text{Hydrindantin} + NH_3 \rightarrow \text{Purple} \qquad (3\text{-}2)$$

The ammonia then reacts with an additional mole of ninhydrin and the reduced ninhydrin to yield a purple substance that absorbs maximally near 570 nm. Because this absorption is nearly a linear function of the amount of amino groups originally present, this reaction provides a convenient and quantitative colorimetric assay for amines. Although the ninhydrin reaction is quite general for primary amines, the evolution of carbon dioxide, which may be followed manometrically, is reasonably diagnostic for α-amino acids. In the cases of the imino acids, an alternative product is formed that is bright yellow (absorption maximum near 440 nm). Hence, the ninhydrin reaction serves to distinguish these substances from the bulk of the protein amino acids.

1-Fluoro-2,4-dinitrobenzene. The arylation of amino acids with 1-fluoro-2,4-dinitrobenzene (FDNB) is of great importance in structural protein chemistry (see page 112):

$$O_2N\text{-}C_6H_3(NO_2)\text{-}F + NH_2\text{—}R \rightarrow O_2N\text{-}C_6H_3(NO_2)\text{-}\overset{H}{N}\text{—}R + HF \qquad (3\text{-}3)$$

The value of dinitrophenylation lies (1) in the quantitative production of yellow crystalline derivatives which are readily separated by chromatographic techniques and which may be quantified spectrophotometrically and (2) in the resistance of the DNP-amino acid linkage to acid hydrolysis. The importance of the latter point for structural studies is discussed in Chapter 4.

Phenylisothiocyanate. Phenylisothiocyanate reacts readily with amino acids to yield the corresponding phenylthiohydantoic acids:

$$\phi\text{—N=C=S} + NH_2\text{—}\overset{R}{CH}\text{—}CO_2H \rightarrow \phi\text{—}\overset{H}{N}\text{—}\overset{S}{\overset{\|}{C}}\text{—}\overset{H}{N}\text{—}\overset{R}{CH}\text{—}CO_2H \qquad (3\text{-}4)$$

Phenylthiohydantoic acid

These products, on treatment with acid in nonhydroxylic solvents, cyclize, yielding the phenylthiohydantoins:

$$\phi\text{—}\overset{H}{N}\text{—}\overset{S}{\overset{\|}{C}}\text{—}\overset{H}{N}\text{—}\overset{R}{CH}\text{—}CO_2H \xrightarrow[\text{Nitromethane}]{H^+} \phi\text{—N}\langle C(=S)\text{—NH—CH(R)—C(=O)}\rangle + H_2O \qquad (3\text{-}5)$$

Phenylthiohydantoin

These amino acid derivatives are easily characterized and quantitated. This reaction, like that of FDNB, is quite useful in structural studies.

Cyanate. Cyanate reacts with amino groups in a manner similar to that of phenylisothiocyanate:

$$NCO^- + H_2N\text{—}\overset{R}{CH}\text{—}CO_2H \xrightarrow{+H^+} H_2N\text{—}\overset{O}{\overset{\|}{C}}\text{—}\overset{H}{N}\text{—}\overset{R}{CH}\text{—}CO_2H \qquad (3\text{-}6)$$

This reaction is useful in studies elucidating the sequence of amino acids in proteins (page 113).

Phosgene. The reaction of phosgene with α-amino acids yields N-carboxy anhydrides (Leuchs anhydrides):

$$R\text{—}\overset{H}{\underset{CO_2H}{C}}\text{—}NH_2 + Cl\text{—}\overset{O}{\overset{\|}{C}}\text{—}Cl \rightarrow R\text{—}\overset{H}{C}\langle \overset{H}{N}\text{—C(=O)—O—C(=O)}\rangle + 2HCl \qquad (3\text{-}7)$$

These componds react readily with nucleophilic reagents and are important intermediates in the synthesis of proteinlike polymers. A similar reaction occurs with carbon disulfide:

$$R\text{—}\overset{H}{\underset{CO_2H}{C}}\text{—}NH_2 + CS_2 \rightarrow R\text{—}\overset{H}{C}\langle \overset{H}{N}\text{—C(=S)—S—C(=O)}\rangle + H_2O \qquad (3\text{-}8)$$

CLASSIFICATION OF PROTEINS

The currently employed classification for proteins is essentially that suggested by national committees organized for the purpose in 1907. Ideally, such a classification should be based on structural aspects of the proteins, like those which are common for the simpler organic chemicals. However, too little was known concerning the details of protein structure in 1907 to permit this, and, indeed, the present understanding of protein structure remains inadequate to form a basis for such a classification. The scheme of 1907 is based principally on solubility properties. Unfortunately, distinct boundaries between the various solubility classes do not exist, and only limited use is made of this classification system. A general understanding of this scheme is of some importance, however, because (1) the terms employed are still widely used in the biochemical literature and (2) the prospect of a more nearly satisfactory classification system in the immediate future appears remote.

The proteins are divided into two general classes: (1) the simple proteins, defined as those yielding only α-amino acids and their derivatives on complete hydrolysis, and (2) the conjugated proteins, defined as those which contain a protein molecule bound to an organic, nonprotein *prosthetic group*. The various subclasses of the simple proteins are described below.

Albumins. The albumins are characterized by solubility in water and in dilute aqueous salt solutions. Proteins in this subclass are found widely distributed in living material. Many have been obtained in crystalline form.

Globulins. Like the albumins, the globulins are soluble in dilute aqueous salt solutions but, in contrast, are insoluble or sparingly soluble in distilled water. Those proteins which are insoluble in water, the euglobulins, are frequently distinguished from those which are sparingly soluble, the pseudoglobulins. It should be clear that, on the basis of solubility properties, no clear distinction can be drawn between the pseudoglobulin and albumin designations. Alternatively, globulins have been defined as proteins salted out of aqueous solution by one-half saturation with ammonium sulfate. The globulins are an important class of proteins and are widely distributed throughout the plant and animal kingdoms. Many have been prepared in crystalline form.

Prolamins. The prolamins are insoluble in water but dissolve in 50 to 90 percent aqueous ethanol solutions and in aqueous solutions of other low-molecular-weight alcohols. The prolamins have been detected only in plant sources and have a limited distribution therein.

Glutelins. The glutelins are insoluble in all the solvents above but are soluble in dilute solutions of acids or bases. These proteins also occur only in plant materials and are poorly understood.

Protamines. The distinctive features of the protamine group of proteins are not their solubility properties but their size and amino acid content. They are among the smallest proteins, having molecular weights in the neighborhood of 5000. The protamines are exceedingly rich in the basic amino acid arginine, which may account for 70 to 80 percent of the total amino acid content. The protamines do not contain sulfur. Their high basicity (*pI* in the range 10 to 12) accounts for their ready formation of salts and their tendency to associate with acidic proteins and the nucleic acids (also acidic molecules). Protamines are found in mature spermatozoa and have been isolated from the sperm of salmon (salmine), herring (clupeine), sardines (sardinine), and other fish.

Histones. Histones, like the protamines, are characterized by a high content of basic amino acids but contain a greater variety of amino acids than the latter group does. The histones appear to occur in combination with the nucleic acid in the nuclei of the somatic cells of many organisms. Their role is incompletely established, but it appears likely that they fulfill some function in the regulation of gene expression (Chapters 5 and 21).

Scleroproteins. Returning to classification based on solubility properties, the scleroproteins are characterized by their insolubility in most ordinary solvents. The scleroproteins are employed principally for architectural purposes and are localized in connective tissue, bone, hair, skin, wool, silk, and the like. The two principal classes are the collagens and the keratins.

The *collagens,* major constituents of cartilage and other connective tissue, are characterized by high contents of glycine, proline, and hydroxyproline and are distinctly lacking in sulfur-containing amino acids. Collagens are the only proteins known to contain significant amounts of hydroxyproline. Collagens may be solubilized in the presence of dilute acidic buffers, *p*H 3 to 4, and reprecipitated at neutral *p*H. Prolonged boiling in water converts collagens to soluble proteins termed *gelatins.*

The *keratins,* major constituents of skin, hair, and other epidermal structures, contain variable amounts of the sulfur-containing amino acids and are frequently rich in the basic amino acids. Keratins may be solubilized by treatment with a variety of agents, such as calcium bisulfite and thioglycolic acid.

The conjugated proteins are classified according to the nature of the prosthetic group to which they are linked. The principal categories are indicated below.

Nucleoproteins. As the name suggests, the prosthetic group of the nucleoproteins is nucleic acid (Chapter 5). These acidic substances are frequently found in combination, through salt linkages, with the basic proteins of the protamine and histone classes.

Lipoproteins. The lipid moiety of the lipoproteins is quite variable both qualitatively and quantitatively. For example, the α_1-lipoprotein of serum contains glyceride, phosphatide, and cholesterol, which together constitute about 30 to 40 percent of the total complex. The β_1-lipoprotein of serum contains some glyceride, but the phosphatide and cholesterol account for the greater part of the lipid and represent nearly 75 percent of the total material. The nature of the linkages between the lipid and protein is uncertain.

Mucoproteins. The prosthetic group of the mucoproteins is carbohydrate in nature. Aside from the fact that all mucoproteins contain hexosamine, the nature and quantity of the carbohydrate component is quite variable. The mucoproteins are widely distributed. The carbohydrate–protein linkages are, in some cases, covalent and, in others, are of the salt type.

Chromoproteins. The prosthetic group of the chromoproteins is undefined except for the stipulation that it be a pigment. Many protein-bound pigments are known, and several such examples will be encountered throughout the course of this book.

THE PEPTIDE BOND

A very important structural feature of any polymer is the nature of the linkages that unite the monomeric units. In 1902, Hofmeister and Fischer independently suggested that proteins were assembled through the formation of secondary amide linkages between the α-carboxyl and α-amino functions of adjacent amino acids. Such bonds are commonly termed *peptide bonds.* The structures resulting from the formation of peptide bonds are termed *peptides.* The individual amino acids of peptides are termed *residues.* Thus, a peptide containing two amino acid residues is a dipeptide, one containing three such residues is a tripeptide, and so forth. The structure of a typical tetrapeptide is indicated below:

```
                                   CH3
                                  /
                                 S
                                 |
     CH3  O  H          O  H  (CH2)2O  H  CH2OH
     |    ‖  |          ‖  |     |   ‖  |   |
NH2—C————C—N—CH2—C—N—CH—C—N—CH—CO2H
     |
     H
```

Peptides are usually written with the terminal amino group on the left and the terminal carboxyl group on the right. They are named as acyl derivatives of the C-terminal amino acid. Thus the tetrapeptide indicated above is alanylglycylmethionylserine.

Evidence favoring the peptide bond as the basic structural feature of proteins is compelling. We summarize the principal lines of evidence requiring this conclusion below:

1. Proteins have relatively few titratable amino and carboxyl functions. Those that are found are accounted for in terms of the terminal residues and the side chains of aspartic acid, glutamic acid, arginine, and lysine. In the course of the hydrolysis of proteins, the titratable amino and carboxylic groups rapidly increase in number and do so in concert; that is, the rate of appearance of carboxyl groups is equal to the rate of appearance of amino groups. Thus, one carboxyl and one amino group appear to be involved per peptide bond in the intact protein.

2. Analysis of incomplete protein hydrolyzates reveals the presence of di- and tripeptides and other oligopeptides. The structures of these materials may be unambiguously established by comparison with synthetically prepared products.

3. Enzymes (proteins with catalytic activities) that degrade proteins specifically hydrolyze amide or closely related bonds, as revealed from studies on model compounds of known structure. A particularly favorable case is provided by the enzyme pepsin. Pepsin is a proteolytic enzyme that, as a result of numerous studies, appears to be almost completely specific for the peptide bond.

4. Biuret, whose structural formula is

```
       H
NH2CONCONH2
```

and related materials containing peptide bonds give a purple color when treated with copper sulfate in alkaline solution (the biuret reaction); proteins and peptides give a particularly strong biuret reaction, suggesting that a great many such bonds are present. The crystal structures for some biuret complexes have been worked out; one example is provided by the dipotassium bis-biuretocuprate (II) tetrahydrate molecule:

```
[                                   O      ] (2+)
                                   //
   O                     NH—C
    \\                  /      \NH
      C—HN—Cu—NH—C/
     /          /           \\
HN—C—HN                      O
    //
   O
]
```

Note that each amide group has lost one proton in the process of coordinating with the copper (II) ion. This complex is violet-pink. Related complexes are formed by peptides.

5. Proteins exhibit both infrared spectra and far-ultraviolet spectra which are consistent with and give support for the presence of peptide bonds.

6. In the case of two proteins, insulin and ribonuclease, and several smaller polypeptides, the structures have been unambiguously established by total synthesis. In all cases, the amino acids were linked by peptide bonds.

7. Finally, the elegant X-ray diffraction studies of Perutz and Kendrew (Chapter 4) established the peptide bond as the inter–amino acid linkage for the proteins myoglobin and hemoglobin. Their conclusions have been amply confirmed for other proteins in several other laboratories.

Thus, the peptide-bond hypothesis must be regarded as firmly established. This conclusion is not intended to rule out the occasional presence of covalent bonds of other types. Clearly, closely related bonds might be formed involving the side-chain carboxyl and amino functions of aspartic acid and lysine, for example. Or interresidue ester bonds might be formed between glutamic acid and serine or threonine. Indeed, some examples of bonds other than amide linkages involving α-carboxyl and α-amino functions are known to occur in proteins. However, the common occurrence of such bonds is clearly ruled out by the considerations above.

What are the important properties of the peptide bond from the standpoint of protein structure? In this regard, two properties are dominant. First, the six atoms of the peptide group C_α—CO—NH—C_α (in which C_α refers to the α-carbon atoms of adjacent residues in the peptide chain) are coplanar. This coplanarity results from the substantial resonance energy stabilization derived from the maximum interaction of the lone electron pair on nitrogen with the pi electrons of the C=O. The C—N bond of the peptide group is about 1.32 Å long, corresponding to about 50 percent double-bond character. Second, the α-carbon atoms of the peptide group lie *trans* to the C—N. The configuration and dimensions of an extended polypeptide chain are shown in Fig. 3-3. As will be shown later, the coplanarity of the peptide group and the *trans* configuration of α-carbon atoms with respect to the C—N play a most important structural role in proteins.

PEPTIDE SYNTHESIS

The chemical synthesis of specific peptides is of great interest. Such syntheses provide an excellent opportunity to probe the relationship between chemical structure and biological activity. This most often finds its expression in the synthesis of peptides structurally related to naturally occurring ones. For example, substitutions of one amino acid for another at distinct sites, inversion in the sequence of two amino acids along the chain, and the like may reveal which amino acids are crucial to biological activity and which are not. Beyond probing structure–function relationships, peptide synthesis is useful for obtaining specifically labeled (with a radioisotope usually) peptides for studies of their distribution within an organism or route of degradation. Their synthesis also provides the final proof of the structure of naturally occurring peptides.

Quite aside from peptides identical or related to natural ones, the synthesis of peptides can lead to important new insights. For example, the construction of synthetic antigens has proved an important step in the understanding of the basis for antigenicity, that is, the capacity to elicit the formation of antibodies (1). For all these reasons, much attention has been devoted to improv-

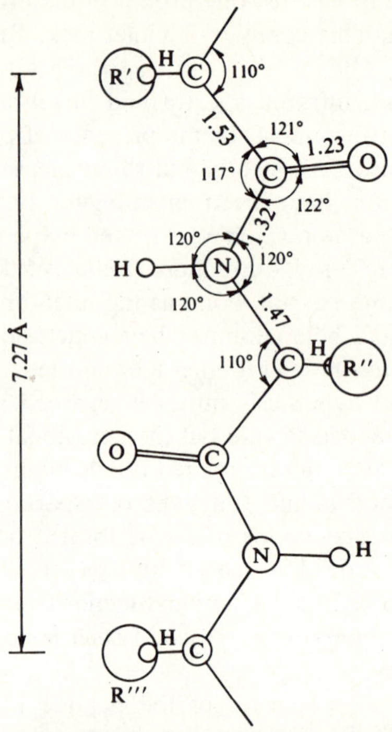

FIGURE 3-3
Dimensions and configuration of a fully extended polypeptide chain. L. Pauling, R. B. Corey, and H. R. Branson, *Proc. Natl. Acad. Sci. U.S.*, **37:** 205 (1951).

ing the tactics and strategy for the synthesis of peptides. Since the syntheses are often lengthy and several syntheses may be required to establish a point, particular emphasis has been given to devising methods that are simple, rapid, and efficient.

Basically, peptide synthesis consists of three steps. First, those amino and carboxyl groups which are not to participate in the reaction must be chemically blocked. Second, the free amino and carboxyl functions must be linked. Third, the blocking groups must be removed. In addition, if amino acids, such as cysteine, lysine, and aspartic acid, which contain side chains that may react during the condensation reaction, are involved in the synthesis, their side chains must be blocked as well. The qualifications of potential blocking agents that render them suitable for use in peptide synthesis include: (1) their introduction should proceed quantitatively and without racemization of the amino acid; (2) the blocked amino acid must be stable to the conditions employed in the condensation reaction; (3) the group must be readily removed under conditions in which the peptide bond is stable and in which racemization of the peptide does not occur. The most generally useful blocking groups for peptide synthesis are considered below.

Carboxyl groups are generally blocked through formation of the corresponding methyl, ethyl, or benzyl esters. The esterifications are accomplished by classical means. Once the peptide bond is formed, the methyl and ethyl groups are hydrolyzed by base. The benzyl group may be removed by hydrolysis or, preferably, by catalytic hydrogenation over palladium.

A greater range of reagents is employed for the blocking of the amino function. The most prominent of them include benzylchlorocarbonate, triphenylmethyl chloride, *t*-butylchlorocarbonate, and trifluoroacetic anhydride.

Reaction of amino acids with benzylchlorocarbonate yields the corresponding benzyloxycarbonyl (or carbobenzoxy, symbolized by CBZ or Z) derivatives:

$$\phi\text{—CH}_2\text{—O—}\overset{\overset{\text{O}}{\|}}{\text{C}}\text{—Cl} + \text{NH}_2\text{—R} \rightarrow \phi\text{—CH}_2\text{—O—}\overset{\overset{\text{O}}{\|}}{\text{C}}\text{—}\overset{\overset{\text{H}}{|}}{\text{N}}\text{—R} + \text{HCl} \tag{3-9}$$

The benzyloxycarbonyl function is readily removed by treatment with HBr in glacial acetic acid, catalytic hydrogenation over palladium, or with sodium in liquid ammonia. Triphenylmethyl (trityl) derivatives of amino acids, synthesized employing triphenylmethyl chloride:

$$\phi\text{—}\underset{\underset{\phi}{|}}{\overset{\overset{\phi}{|}}{\text{C}}}\text{—Cl} + \text{H}_2\text{N—R} \rightarrow \phi\text{—}\underset{\underset{\phi}{|}}{\overset{\overset{\phi}{|}}{\text{C}}}\text{—}\overset{\overset{\text{H}}{|}}{\text{N}}\text{—R} + \text{HCl} \tag{3-10}$$

have also proved of value for peptide synthesis. This group is readily removed by catalytic hydrogenation over palladium. Related reactions are available for the introduction and removal of other amino-group blocking agents.

Once the appropriate amino and carboxyl functions are blocked, formation of the peptide bond is the next problem. This may be accomplished through activation of the carboxyl group in any of several fashions. The principal pitfall to be avoided is racemization of the amino acids during the reaction. Racemization yields several products that are frequently extremely difficult to separate. Suitable activated forms of the carboxyl group include the acyl chlorides (racemization is sometimes a problem here), *p*-nitrophenyl esters, cyanomethyl esters, and thiolesters. In each case, these derivatives are readily susceptible to nucleophilic attack and react smoothly with the amino functions.

In addition to these methods, the acid azides are very useful in peptide synthesis. These are formed from acyl hydrazides by treatment with nitrous acid:

$$\text{R}\overset{\overset{\text{O}}{\|}}{\text{C}}\text{—OEt} \xrightarrow{\text{NH}_2\text{NH}_2} \text{R—}\overset{\overset{\text{O}}{\|}}{\text{C}}\text{—}\overset{\overset{\text{H}}{|}}{\text{N}}\text{—NH}_2 \xrightarrow{\text{HNO}_2} \text{R—}\overset{\overset{\text{O}}{\|}}{\text{C}}\text{—N=N}^+\text{=N}^- \tag{3-11}$$

Peptide synthesis from acid azides occurs without detectable racemization and is, in this respect, the preferred synthetic route.

Several methods of peptide synthesis are available that avoid the synthesis and isolation of carboxyl-activated intermediates. The most widely used of these methods involves carbodiimides, such as dicyclohexylcarbodiimide, as dehydrating agents. The probable pathway for such reactions is

$$\text{C}_6\text{H}_{11}\text{—N=C=N—C}_6\text{H}_{11} + \text{RCOOH} \rightarrow \text{C}_6\text{H}_{11}\text{—}\overset{\overset{\text{H}}{|}}{\text{N}}\text{—}\underset{\underset{\text{O—}\underset{\underset{\text{O}}{\|}}{\text{C}}\text{—R}}{|}}{\text{C}}\text{=N—C}_6\text{H}_{11}$$

$$\text{C}_6\text{H}_{11}\text{—}\overset{\overset{\text{H}}{|}}{\text{N}}\text{—}\underset{\underset{\text{O—}\underset{\underset{\text{O}}{\|}}{\text{C}}\text{—R}}{|}}{\text{C}}\text{=N—C}_6\text{H}_{11} + \text{R'NH}_2 \rightarrow \tag{3-12}$$

$$\text{C}_6\text{H}_{11}\text{—}\overset{\text{H}}{\text{N}}\text{—}\overset{\overset{\text{O}}{\|}}{\text{C}}\text{—}\overset{\text{H}}{\text{N}}\text{—C}_6\text{H}_{11} + \text{R—}\overset{\overset{\text{O}}{\|}}{\text{C}}\text{—}\overset{\text{H}}{\text{N}}\text{—R'}$$

Finally, use of *N*-carboxyanhydrides [Eq. (3-7)] both blocks the amino function and activates the carboxyl one. Thus, the addition of 1 mol of *N*-carboxyanhydride to 1 mol of an amino acid yields a dipeptide:

$$\underset{\text{(N-carboxyanhydride: } R_1\text{—}\overset{|}{\text{CH}}\text{—C(=O)—O—C(=O)—NH—, ring)}}{} + NH_2\text{—}\overset{R_2}{\overset{|}{C}}H\text{—}CO_2H \longrightarrow NH_2\text{—}\overset{R_1}{\overset{|}{C}}H\text{—}\overset{O}{\overset{\|}{C}}\text{—}\overset{H}{\overset{|}{N}}\text{—}\overset{R_2}{\overset{|}{C}}H\text{—}CO_2H + CO_2 \qquad (3\text{-}13)$$

This procedure may be repeated several times by the subsequent additions of new *N*-carboxyanhydrides. It is a most convenient method of peptide synthesis, because the repeated removal of blocking groups is effectively avoided. Care is required with this process, because polymer formation can be a problem (see below).

With the necessary tools for blocking, condensation, and deblocking steps in hand, the next matter is that of the strategy of peptide synthesis. It is in this area that particularly striking progress has been made recently. Classically, one carried out the necessary steps in homogeneous solution. This approach requires, usually, the isolation and purification of each product along the reaction pathway. This is generally a difficult matter requiring both skill and patience on the part of the experimentalist. The classic work of Vincent du Vigneaud in the 1940s on the synthesis of some nonapeptide hormones (see page 66) was rewarded with the Nobel Prize. Subsequent workers have, with the aid of improved methodology, succeeded in synthesizing peptides a good deal larger; Herculean efforts were frequently required. One example of a classic peptide synthesis, that of glutathione, is diagrammed in Fig. 3-4.

Because glycine is not optically active, it is clearly advantageous to initiate this synthesis from the C-terminal end to minimize the possibilities of racemization of the optically active residues. Thus, the condensation of the amino

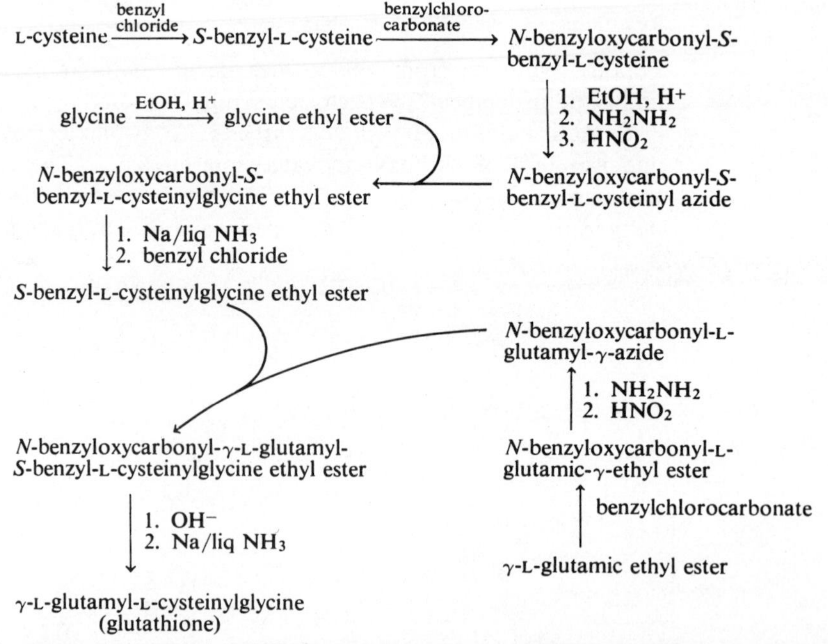

FIGURE 3-4
Synthetic route to glutathione.

group of glycine with the carboxyl group of cysteine must be accomplished prior to the formation of the other peptide bond. To this end, glycine is converted to the ethyl ester, and cysteine is treated first with benzyl chloride, blocking the sulfhydryl group, and then with benzylchlorocarbonate, blocking the amino group. The *N*-benzyloxycarbonyl-*S*-benzyl-L-cysteine is converted to the acid azide [Eq. (3-11)] and condensed with glycine ethyl ester, yielding the fully blocked dipeptide. Treatment with sodium in liquid ammonia removes both the benzyl and benzyloxycarbonyl groups; the former is replaced by subsequent reaction with benzyl chloride. The product, *S*-benzyl-L-cysteinylglycine ethyl ester, is now suitably blocked for reaction with *N*-benzyloxycarbonyl-L-glutamic acid-γ-azide (synthesized as shown in Fig. 3-4). The fully blocked tripeptide is converted to glutathione through saponification, removing the C-terminal ester function, and treatment with sodium in liquid ammonia, removing the benzyl and benzyloxycarbonyl functions.

This classical strategy of peptide synthesis has now been augmented by a *solid-phase* technique, largely due to the efforts of Merrifield and his coworkers. The basic idea here is to synthesize the peptide in a stepwise way while the growing chain is attached to a solid support. The principal simplification introduced is that the purification of each product is avoided. For example, if the peptide is attached to a synthetic polymer support in the form of a bead, purification consists simply of filtering off the beads following each step. Resuspension of the beads readies the system for the addition of the next amino acid. The total saving in time and energy introduced by use of the solid-phase strategy can be exceptional for synthesis of a large peptide.

The overall procedure consists of the following steps. First, one attaches a suitable functional group to a solid particle. Chloromethyl groups are often employed, as these react with the free carboxyl group of amino acids linking the two together. Second, a suitably blocked amino acid is added to the functional group of the particle. In principle, one can attach either end of the amino acid, but the carboxyl group has been generally employed. Third, the protecting agent is removed. Finally, a second blocked amino acid is condensed with the first, usually through use of dicyclohexylcarbodiimide as the dehydrating reagent. Repetition of the latter three steps generates longer peptides. This procedure is shown schematically in Fig. 3-5. At the conclusion

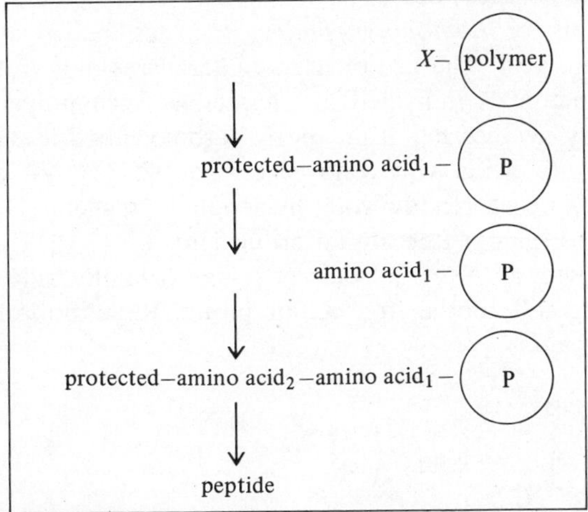

FIGURE 3-5
Schematic representation of solid-phase peptide synthesis.

of the synthetic process, the completed peptide is cleaved from the particle and recovered from the solution.

Two points are worth noting explicitly. First, the solid-phase method relies strongly on classical developments in terms of blocking agents and condensing agents. Second, the homogeneous and solid-phase methods are not mutually exclusive processes and, in fact, it will frequently prove useful to combine the two methods. Thus, short peptides generated by the solid-phase technique may be joined together in a homogeneous system, or vice versa. Each peptide synthesis is likely to pose its individual problems, which will best be dealt with by adapting the classical and solid-phase methods in a specialized way.

Finally, we wish to note one added advantage of the solid-phase method: It is susceptible to automation. Since the manual purification of the peptide intermediates is avoided, the other steps may be mechanized. In fact, an instrument capable of about six complete cycles of synthesis per day is commercially available.

One nice example of the use of the solid-phase peptide synthesis has dealt with structure-reactivity correlations of bradykinin and its derivatives (2–4). This nonapeptide (page 68) is a smooth muscle hypotensive agent. In Table 3-4, the structures of some of the modifications of bradykinin together with indications of the biological activity are collected.

Of all peptide syntheses thus far achieved, the most impressive may be the complete synthesis of ribonuclease. This protein is composed of a single chain of 124 amino acids (see Fig. 4-5, and pages 122–123). It has, in fact, been synthesized twice: once by Merrifield and his associates at The Rockefeller University employing the solid-state methods just described, and once by Denkewalter and Hirschmann at Merck and Company employing a modification of the classical methods. The latter synthesis relied on the construction of fragments of ribonuclease employing *N*-carboxyanhydride synthesis [Eq. (3-13)] followed by coupling of the fragments to form the complete molecule. The precise fragments and their arrangement in ribonuclease are shown in Fig. 3-6. The fragment coupling strategy of peptide synthesis is established as an important and potent one, although it does not appear capable of automation.

Methods indicated above permit the ordered synthesis of many peptides, including some of considerable size and complexity. In addition, unordered synthesis of polypeptides is also of great value. Of particular importance is the synthesis of *homopolypeptides,* such as polyglycine, polyserine, and polyglutamic acid. The use of these materials as aids in the understanding of protein structure is indicated in Chapter 4. Such polyamino acids are most conveniently synthesized from the *N*-carboxyanhydrides formed in the reaction of amino acids with phosgene [Eq. (3-7)]. *N*-Carboxyanhydrides of amino acids react readily with nucleophilic reagents, liberating carbon dioxide and forming a free amino group [Eq. (3-13)]. This amino group may then react with another molecule of *N*-carboxyanhydride, extending the chain and forming still another free amino group. Repetition of this process clearly yields polymers:

$$n\ \text{(N-carboxyanhydride: ring with H–C(R), N–H, C=O, O, C=O)} \xrightarrow{\text{initiator}} \left(\begin{array}{c} \text{H}\quad \text{R}\quad \text{O} \\ |\quad\ \ |\quad\ \ \| \\ \text{—N—CH—C—} \end{array} \right)_n + nCO_2 \qquad (3\text{-}14)$$

The degree of polymerization obtained in these reactions is a function of the nature and concentration of the reaction initiator (that molecule which per-

TABLE 3-4
Immunologic and biologic activity of alanine-substituted bradykinin variants

Peptide	*Amino acid sequence*									*50 % Inhibitory conc., μg*	*Binding activity*	*Biological activity (rat uterus)*
	1	2	3	4	5	6	7	8	9			
Bradykinin	H-Arg	Pro	Pro	Gly	Phe	Ser	Pro	Phe	Arg-OH	0.070	1	1
Ala^9-bradykinin									Ala	0.045	1.6	1/500
Ala^6-bradykinin						Ala				0.100	1/1.4	1/10
Ala^8-bradykinin								Ala		0.160	1/2	1/1500
Ala^7-bradykinin							Ala			0.250	1/4	1/80
Ala^2-bradykinin		Ala								1.00	1/14	1/300
Ala^5-bradykinin					Ala					3.10	1/44	1/1000
Ala^4-bradykinin				Ala						4.00	1/57	1/1500
Ala^3-bradykinin			Ala							19.00	1/270	1

SOURCE: J. E. Spragg, J. M. Schroeder, J. M. Stewart, K. F. Austen, and E. Haber: *Biochemistry,* **6:** 3933 (1967); by permission of the copyright owners, the American Chemical Society.

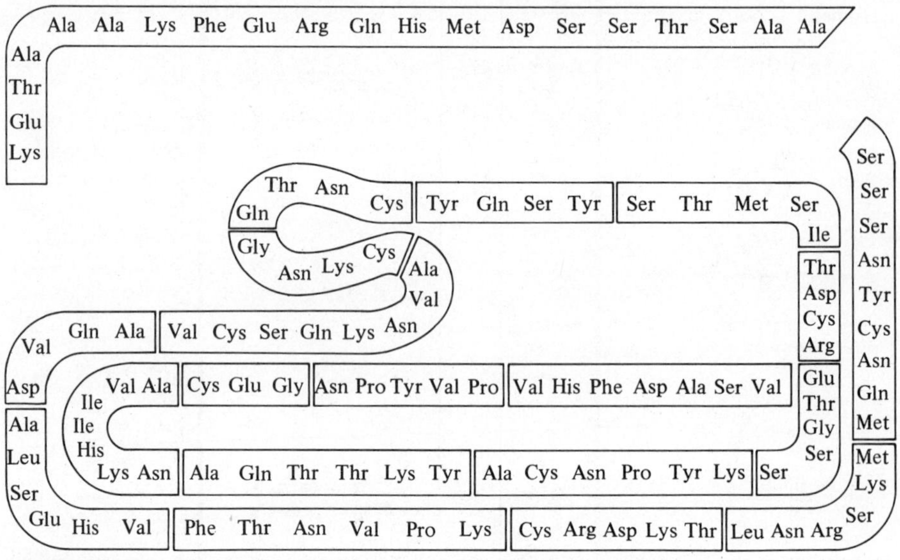

FIGURE 3-6

Fragment-coupling method as applied to the synthesis of ribonuclease. R. G. Denkewalter and R. Hirschmann, *Am. Scientist,* **4:** 389 (1969).

forms the initial attack on the *N*-carboxyanhydride), solvent, and temperature. Thus, the desired degree of polymerization may frequently be obtained within reasonably narrow limits by proper choice of reaction conditions and initiator.

NATURALLY OCCURRING PEPTIDES

Nearly all living tissues appear to contain appreciable quantities of low-molecular-weight peptides. For example, glutathione (see above) is a constituent of most plant and animal tissues, carnosine (β-alanyl-L-histidine) and anserine (β-alanyl-1-methyl-L-histidine) are present in vertebrate muscle tissue:

$$H_2N{-}CH_2{-}CH_2{-}\overset{\overset{O}{\|}}{C}{-}\overset{\overset{H}{|}}{N}{-}\underset{\underset{\text{imidazol-4-yl (NH)}}{\underset{|}{CH_2}}}{CH}{-}CO_2H$$

Carnosine

$$H_2N{-}CH_2{-}CH_2{-}\overset{\overset{O}{\|}}{C}{-}\overset{\overset{H}{|}}{N}{-}\underset{\underset{\text{imidazol-4-yl (N}{-}CH_3)}{\underset{|}{CH_2}}}{CH}{-}CO_2H$$

Anserine

More interesting classes of peptides are those which exhibit well-known physiological or antibacterial activity.

The mammalian neurohypophysis elaborates several peptide hormones, including the nonapeptides oxytocin and vasopressin:

```
NH2
|
Cys——Tyr——Ile
|          |
S          |
|          |
S          |
|          |
Cys—AspNH2—GluNH2
|
|
Pro—Leu—Gly—CONH2
   Beef oxytocin
```

```
NH2
|
Cys——Tyr——Phe
|          |
S          |
|          |
S          |
|          |
Cys—AspNH2—GluNH2
|
|
Pro—Arg—Gly—CONH2
  Beef vasopressin
```

Human	Ala·Glu·Lys·Lys·Asp·Glu·Gly·Pro·Tyr·Arg·Met·Glu·His·Phe·Arg·Trp·Gly·Ser·Pro·Pro·Lys·Asp
Monkey	Asp·Glu·Gly·Pro·Tyr·Arg·Met·Glu·His·Phe·Arg·Trp·Gly·Ser·Pro·Pro·Lys·Asp
Horse	Asp·Glu·Gly·Pro·Tyr·Lys·Met·Glu·His·Phe·Arg·Trp·Gly·Ser·Pro·Arg·Lys·Asp
Beef	Asp·Ser·Gly·Pro·Tyr·Lys·Met·Glu·His·Phe·Arg·Trp·Gly·Ser·Pro·Pro·Lys·Asp
Pig	Asp·Glu·Gly·Pro·Tyr·Lys·Met·Glu·His·Phe·Arg·Trp·Gly·Ser·Pro·Pro·Lys·Asp

FIGURE 3-7

Amino acid sequences of β-melanocyte–stimulating hormone from five species. Amino acid substitutions among these species, as compared with the human, are indicated by the underlined residues. Abraham White et al., *Principles of Biochemistry,* McGraw-Hill, New York, 3rd ed., 1964, p. 902.

Oxytocin elicits the contraction of smooth muscle and the ejection of milk in lactating females, and vasopressin elicits pressor and antidiuretic effects. The structure and chemical synthesis of these hormones were elegantly established in a classic series of investigations by du Vigneaud and his collaborators. Oxytocin and vasopressin are structurally related and differ only in the nature of the residues at positions 3 and 8. *Vasopressin* refers to a class of compounds in which the hormones performing the physiological role of vasopressin in a variety of mammals differ in the nature of the amino acid at position 8 and, occasionally, at position 3.

In addition to oxytocin and vasopressin, the neurohypophysis also liberates two polypeptide hormones that stimulate the pigment-forming activity of the melanocytes. Accordingly, these are termed *α- and β-melanocyte-stimulating hormones.* The β-melanocyte-stimulating hormone of several species has been characterized structurally and, like vasopressin, exhibits some species specificity. The structures of the hormones isolated from five species are shown in Fig. 3-7.

The adenohypophysis secretes, among others, the *adrenocorticotropic hormone (ACTH)* that stimulates the growth and metabolic activity of the adrenal cortex. ACTH is a polypeptide consisting of 39 amino acid residues, the sequence of the first 24 of which being common to the hormone of all species investigated. The last 15 show several variations from species to species. The structure of human ACTH is indicated below:

Ser·Tyr·Ser·Met·Glu·His·Phe·Arg·Trp·Gly·Lys·Pro·Val·Gly·Lys·Lys·Arg·
Arg·Pro·Val·Lys·Val·Tyr·Pro·Asp·Ala·Gly·Glu·Asp·$GluNH_2$·Ser·Ala·Glu·
Ala·Phe·Pro·Leu·Glu·Phe

Human ACTH

One of the most interesting groups of biologically active peptides is derived from proteins, normally present in the plasma, that are themselves biologically inert. These include the angiotensins, bradykinin, and kallidin.

The *angiotensins* are a group of octapeptides that have pressor activity. The individual angiotensins are distinguished by the nature of the residue in position 5, which varies from species to species. These substances are derived from a plasma protein termed *angiotensinogen* by successive cleavage of particular peptide bonds by the proteolytic enzymes trypsin, renin, and converting enzyme. The reaction pathway leading to the formation of active angiotensins in the horse is summarized in Fig. 3-8.

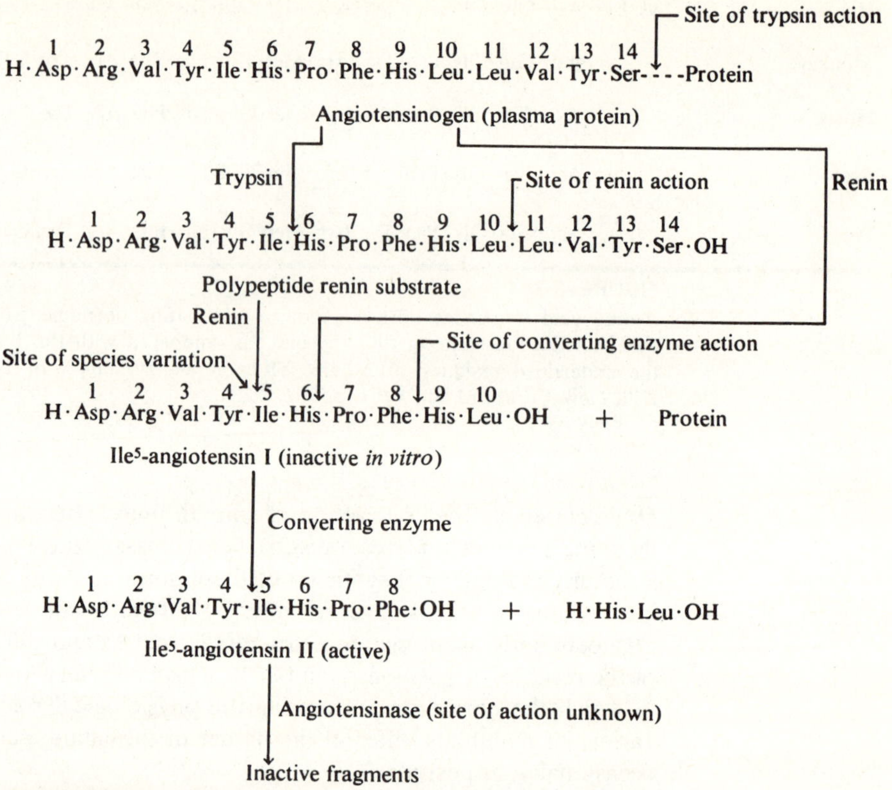

Site of trypsin action

1 2 3 4 5 6 7 8 9 10 11 12 13 14
H·Asp·Arg·Val·Tyr·Ile·His·Pro·Phe·His·Leu·Leu·Val·Tyr·Ser-↓--Protein

Angiotensinogen (plasma protein)

Trypsin | Site of renin action | Renin

1 2 3 4 5 ↓6 7 8 9 10 ↓ 11 12 13 14
H·Asp·Arg·Val·Tyr·Ile·His·Pro·Phe·His·Leu·Leu·Val·Tyr·Ser·OH

Polypeptide renin substrate

Renin

Site of converting enzyme action

Site of species variation

1 2 3 4 5 6↓ 7 8 ↓ 9 10
H·Asp·Arg·Val·Tyr·Ile·His·Pro·Phe·His·Leu·OH + Protein

Ile[5]-angiotensin I (inactive *in vitro*)

Converting enzyme

1 2 3 4 ↓5 6 7 8
H·Asp·Arg·Val·Tyr·Ile·His·Pro·Phe·OH + H·His·Leu·OH

Ile[5]-angiotensin II (active)

Angiotensinase (site of action unknown)

Inactive fragments

FIGURE 3-8

Enzymic steps in conversion of horse angiotensinogen into "pressor" peptides. K. Hofmann and P. G. Katsoyannis, in H. Neurath (ed.), *The Proteins,* vol. 1, Academic Press, New York, 1963, p. 142.

Bradykinin, a nonapeptide liberated upon treatment of plasma with trypsin or snake venom, is a potent smooth-muscle hypotensive agent. The structure of bradykinin is indicated below:

Arg·Pro·Pro·Gly·Phe·Ser·Pro·Phe·Arg

Bradykinin

Kallidin, a decapeptide with physiological properties similar to those of bradykinin, is liberated from a plasma protein termed *kallidinogen:*

Lys·Arg·Pro·Pro·Gly·Phe·Ser·Pro·Phe·Arg

Kallidin

The structure of this peptide differs from that of bradykinin only by the addition of one lysine residue to the N-terminal end of the peptide chain.

Several cyclic peptides are potent antibacterial agents. These include *tyrocidin* and *gramicidin,* isolated from the bacterium *Bacillus brevis,* and *bacitracin,* which is isolated from *B. licheniformis.*

(-L-Val·L-Orn·L-Leu·D-Phe·L-Pro)$_2$

Gramicidin S

L-Val·L-Orn·L-Leu·D-Phe·L-Pro
L-Tyr·L-GluNH$_2$·L-AspNH$_2$·D-Phe·L-Phe

Tyrocidin

```
CH3
 |
CH2
 |
CH3—CH    S—CH2   O                                  D-Asp
     |    /    \   ||                                   ·
NH2—CH·C═N—CH—C·L-Leu·D-Glu·L-Ile·L-Leu·L-Cys·L-Asp·L-His
                                                     ·         ·
                                                  D-Orn·L-Ile·D-Phe
```

Bacitracin

Each of these peptides is unusual in that it contains ornithine, an amino acid not found in natural proteins, and one or more D-amino acids. Bacitracin is further distinguished by the presence of a thiazoline ring, involving the sulfhydryl group of cysteine and the carbonyl carbon of the adjacent isoleucine, and by the occurrence of an amide linkage involving the β-carboxyl function of aspartic acid.

DETECTION AND ESTIMATION OF PROTEINS

The quantitative or semiquantitative estimation of protein is a frequently encountered problem, one which, for example, arises continually in devising methods for the purification of a particular protein (see below). In the course of discussion regarding chemical reactions of the amino acids, a number of color tests were described that are due to reaction of one or more of the amino acid side chains with an appropriate reagent (Table 3-3). Unhydrolyzed proteins give these reactions as well, indicating that, in the intact protein, these side chains are not in covalent linkage. Each of these reactions has some utility for the detection of proteins. However, they are not generally useful for the quantitative estimation of proteins, because the frequency with which each of the side chains appears, and hence the strength of the reaction observed, varies from protein to protein. For example, the protamine salmine, which is composed of 85 percent arginine by weight, gives a great deal more red color per gram of protein in the Sakaguchi reaction (Table 3-3) than does the milk protein β-lactoglobulin, which contains less than 3 percent arginine by weight. It is clear that a colorimetric determination of protein employing this reaction would be subject to considerable uncertainty, a reservation that applies with equal force to the remainder of the color reactions in Table 3-3. These reactions are of some use in assaying very heterogeneous mixtures of proteins, because, in this case, it is unlikely that the mixture would be strikingly rich or poor in particular amino acids.

A commonly employed method of protein estimation, measurement of the optical density of protein solutions, also suffers from this disadvantage. Most proteins exhibit ultraviolet-light absorption spectra with absorption maxima near 278 nm. Reference to Table 3-1 indicates that tryptophan, tyrosine, and phenylalanine absorb light in this region of the spectrum, the absorption of the former two being stronger than that of the last. The complete ultraviolet absorption spectra of these amino acids are presented in Fig. 3-9.

In addition to these chromophores, the disulfide bond of cystine absorbs light in the ultraviolet region of the spectrum. The extent to which such absorption will contribute to that of a given protein is difficult to predict; both the wavelength of maximum absorption and the intensity of the absorption are sensitive functions of the dihedral angle of the disulfide bond, β.

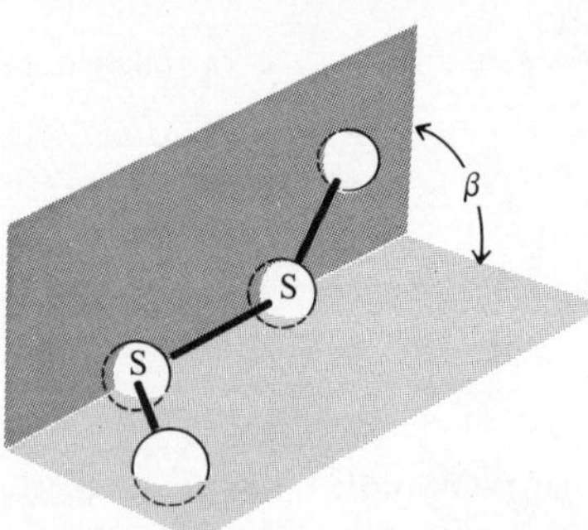

The wavelength of maximum absorption can vary all the way from about 230 nm to 370 nm as a function of β. In Table 3-5, some typical data for light absorption by the disulfide chromophore are collected. In favorable cases it may prove possible to establish the dihedral angle of disulfide groups in proteins from their ultraviolet spectra.

Absorption of monochromatic light by a dissolved substance may be conveniently described by the Beer–Lambert law:

$$\log\left(\frac{I_0}{I}\right) = \varepsilon cl \tag{3-15}$$

in which I_0 and I refer to the incident and transmitted light intensities, respectively, ε is a proportionality constant called the *molar absorptivity* or *extinction coefficient*, c is concentration in moles per liter, and l is the length of the optical path in centimeters. The term log (I_0/I) is readily measured by means of a spectrophotometer and is usually called A, the *absorbance*, or E,

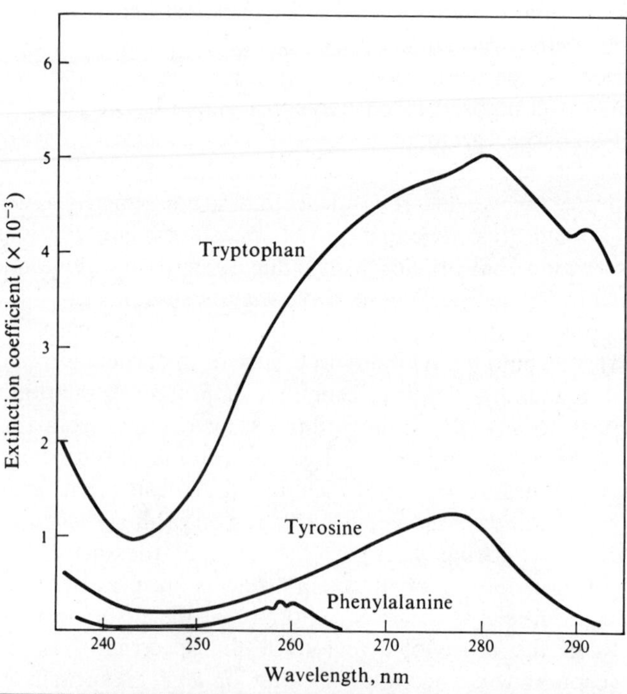

FIGURE 3-9
Ultraviolet absorption spectra of tryptophan, tyrosine, and phenylalanine.

TABLE 3-5
Maximum wavelength, λ, and dihedral angle, β, for light absorption by selected disulfides

Compound	λ, *nm*	β
1. Dimethyl disulfide $(CH_3)_2S_2$	253	84.7°
2. 1,2-Dithiane $CH_2CH_2CH_2CH_2SS$ (ring)	286	
3. 1,2-Dithiane-3,6-dicarboxylic acid $SSCH(CO_2H)CH_2CH_2CHCO_2H$ (ring)		60°
4. α-Lipoic acid $CH_2CH_2SSCH(CH_2)_4CO_2H$ (ring)	320	35°
5. 1,2-Dithiolane-4-carboxylic acid $CH_2SSCH_2CHCO_2H$ (ring)	330	27°
6. 1,2-Dithiolane $CH_2CH_2CH_2SS$ (ring)	330	
7. Acetylaranotin	345	15.2° 18.2°

By permission of M. Carmack and L. Neubert, Indiana University.

the *extinction*. Measurements of A are usually made in optical cells of path length 1 cm; under these conditions, the Beer–Lambert law may be rewritten as

$$A \equiv E = \varepsilon c \tag{3-16}$$

where ε is then the absorbance of a 1 M solution of the material of interest at the chosen wavelength and has the dimensions liter/mol/cm $=$ M^{-1}/cm. Clearly if ε is known, measurements of A yield concentrations directly. Because many values of ε are known or may be easily measured and because determination of A is simple, this method is exceptionally widely employed for concentration determination.

The difficulty of this method for protein determination is, of course, the variability of the abundance of tyrosine and tryptophan, upon which values of ε for protein preparations depend, from protein to protein. This method is reasonably reliable if a heterogeneous mixture of proteins is to be analyzed, for reasons indicated above, and may be used to good advantage in dealing with relatively pure protein preparations whose extinction coefficients may be accurately measured (or calculated from the amino acid composition, if known).

A more generally accurate, and correspondingly more tedious, method of protein determination is based on the quantitative estimation of protein nitrogen. Nearly all proteins contain approximately 16 percent nitrogen by weight. The analysis is accomplished by conversion of the protein nitrogen to ammonia through digestion with concentrated sulfuric acid in the presence of copper sulfate, mercuric sulfate, or other suitable catalyst. The ammonia formed may be quantitated by removal from the digestion mixture through steam distillation followed by titration or addition of Nessler's reagent (mercuric potassium iodide in aqueous sodium hydroxide), which permits ammonia to be determined colorimetrically.

One of the most widely employed colorimetric tests for proteins is the *Lowry method,* in which the Folin phenol reagent is combined with the biuret test. Specifically, slightly basic cupric sulfate is added to the protein solution followed a few minutes later by a concentrated phenol solution. The resulting color is much more intense than either of the two reagents would have elicited alone. This fact gives the Lowry method great sensitivity; as little as 10 μg of protein is detectable. The method suffers somewhat from nonlinearity between protein concentration and color intensity.

The biuret reaction (page 58) may also be used to quantitate protein concentration, because this reaction detects peptide bonds. These occur with approximately the same frequency per gram of material for most proteins.

Other methods employed for the estimation of protein include (1) measurement of refractive index, (2) measurement of turbidity following precipitation with trichloroacetic acid, and (3) determination of dry weight—the basic method employed to standardize the more frequently employed techniques.

AMINO ACID ANALYSIS OF PROTEINS

The accurate determination of the amino acid composition of proteins is of great importance in their characterization and for an understanding of their structure. Such an analysis consists of two steps: hydrolysis of the protein to the constituent amino acids, followed by quantitative estimation of the hydrolysis products.

Protein Hydrolysis

The complete hydrolysis of proteins may be accomplished by the use of strong acid, strong base, or proteolytic enzymes as catalysts. The use of strong acid is ordinarily the method of choice.

Although proteins may be hydrolyzed in the presence of a variety of strong acids, constant-boiling hydrochloric acid, about 6 N, is most frequently employed. The reaction is usually carried out in evacuated sealed tubes at about 110° for periods ranging from 12 to 96 hr. Under these conditions, the peptide bonds are quantitatively hydrolyzed (relatively long periods of hydrolysis are required for the complete liberation of valine, leucine, and isoleucine), yielding the amino acid hydrochlorides. Among the amino acids that usually occur in proteins, tryptophan is completely destroyed in this process, and small losses of serine and threonine, which may be corrected for, are encountered. The amino acids are not racemized by acid hydrolysis under these conditions.

The complete hydrolysis of proteins may also be achieved by incubating them with 2 to 4 N sodium hydroxide at 100° for 4 to 8 hr. Such hydrolysis is of limited utility for routine analysis, because (1) cysteine, cystine, serine, threonine, and arginine are decomposed in the process, (2) other amino acids may be partially destroyed by deamination; and (3) racemization of the amino acids occurs. The principal use of alkaline hydrolysis is for the estimation of tryptophan, the amino acid destroyed by acid hydrolysis.

The complete enzymatic hydrolysis of proteins is difficult, because most enzymes attack only specific types of linkages rapidly (see Chapter 17). Often a combination of enzymes is employed, and extended time periods are required. Certain bacterial proteases cause very extensive hydrolysis by themselves. This method is further complicated, owing to possible contamination resulting from proteolysis of the enzymes themselves. Although of limited use for complete hydrolysis, enzymatic methods have been employed with great success for the partial hydrolysis of proteins, an important step for amino acid sequence determination (see Chapter 4).

Amino Acid Analysis

Following the complete hydrolysis of a protein, the analysis is completed by obtaining a quantitative analysis for each of the amino acids. Such analysis is ordinarily accomplished chromatographically either on a suitable column or on paper. Any of several monographs may be consulted for details concerning chromatographic procedures.

The separation of a mixture of amino acids is most readily accomplished by employing *ion-exchange chromatography*. The column material of choice is a sulfonated polystyrene resin, a cation exchanger. The amino acid mixture is placed on the column at *p*H 3, conditions under which the individual amino acids are positively charged. The cationic amino acids displace some of the sodium ions from the $—SO_3^-Na^+$ form of the column and thus are bound to the column material by electrostatic forces. Clearly, those amino acids which are most basic are bound most tightly and those which are least basic are bound least tightly. The column is then developed by gradually increasing the *p*H and ionic strength of the buffers with which the column is washed. These conditions gradually cause neutralization of the positive charges on the amino acids and, in addition, weaken salt linkages. Thus, as the column is developed, the acidic amino acids, such as glutamic and aspartic acids, are removed readily from the resin, followed by the neutral amino acids, followed, finally, by the basic amino acids. This method serves to separate all the amino acids commonly found in proteins, because it depends on adsorption of the amino acids to the resin material as well as on ion-exchange properties. The column material is sufficiently selective in its adsorption properties so that all the neutral amino acids, which cannot be separated on the basis of ion exchange, are eluted separately from the column. As the amino acids are eluted from the column, they are collected individually in an automatic fraction collector. Once separated, the individual amino acids are quantitated by reaction with ninhydrin and measurement of the resultant color intensity. Several automatic amino acid analyzers are commercially available. These instruments automatically separate the amino acids on an ion-exchange column, collect them, add ninhydrin, heat to develop the color, record the color intensity, and plot the intensity on a graph. A typical automatically recorded chromatographic analysis of an amino acid mixture is shown in Fig. 3-10. The identity of each amino acid is established on the basis of its position on the chromatogram, and it is estimated quantitatively on the basis of the area under each curve. This method is frequently accurate to within 2 percent. Such analyses may be routinely accomplished in a few hours. Needless to say, other cation-exchange resins and anion-exchange resins may also be employed for the separation of amino acids.

Amino acid mixtures may also be separated by *paper chromatographic techniques,* a method of great historical significance. Paper chromatographic separations principally reflect differences in partition coefficients of the various amino acids between an organic and an aqueous phase. Thus, the mixture to be separated is applied to a suitable paper as support that is then irrigated with a developing liquid, usually composed of aqueous and organic components. The aqueous phase serves as stationary phase, the organic phase as mobile phase. Hence, amino acids that are relatively soluble in the organic phase and insoluble in the aqueous phase move well to the front of the paper, which is

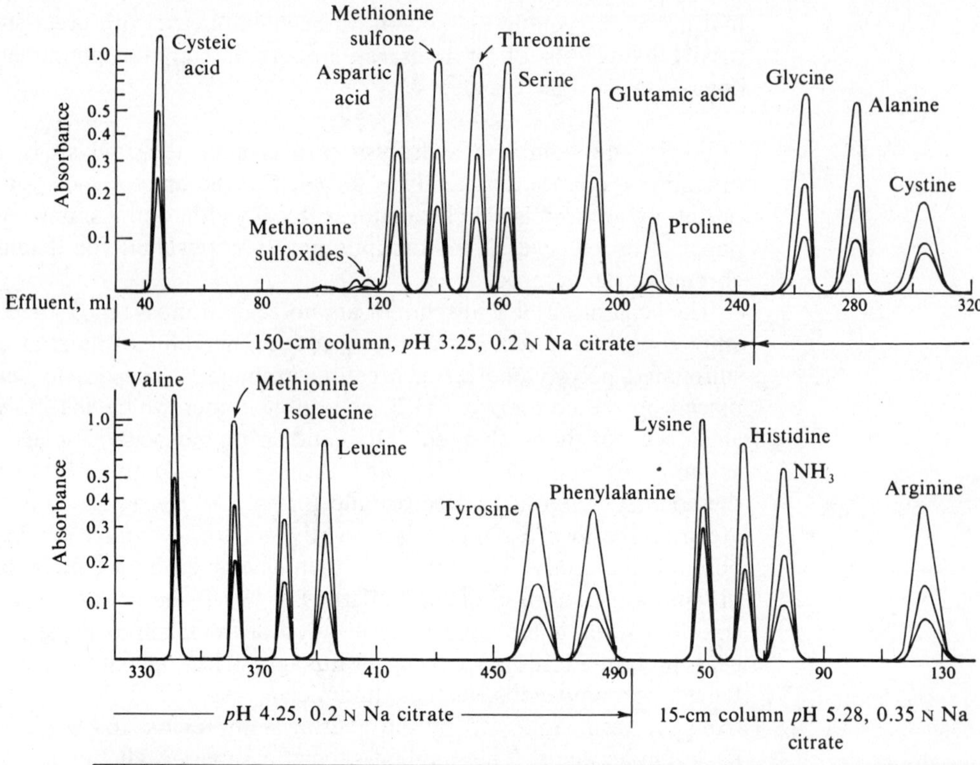

FIGURE 3-10
Automatically recorded chromatographic analysis of a synthetic mixture of amino acids on a sulfonated polystyrene resin. D. H. Spackman, W. H. Stein, and S. Moore, *Anal. Chem.*, **30:** 1190 (1958); by permission of the copyright owners, the American Chemical Society.

maintained in a vertical position, and those which are more soluble in the aqueous phase remain near the origin. In practice, development of a chromatogram with a single solvent system is not adequate for the complete separation of all the amino acids. Consequently one resorts to *two-dimensional chromatography.* In this technique, the chromatogram is developed in one direction with a first solvent system, rotated 90°, and developed a second time with a second solvent system. A typical two-dimensional chromatogram illustrating separation of amino acids from a protein hydrolyzate is presented in Fig. 3-11. The individual amino acids are located by treatment of the chromatogram with ninhydrin, the amino acids appearing as purple spots. These spots may be cut out and eluted. Measurement of the optical density of the eluate affords a quantitative evaluation of the quantity of amino acid present. This method is clearly somewhat more tedious, and less accurate, than automatic analysis employing column chromatography but is often more sensitive.

A variant of this method is the chromatography of the 2,4-dinitrophenyl derivatives of the amino acids. These may be formed directly in the protein hydrolyzate by treatment with 1-fluoro-2,4-dinitrobenzene at *p*H 8 and 40° for 1 hr. The derivatives may be extracted from the hydrolyzate with ether and subjected to two-dimensional paper chromatography. Because the dinitrophenyl derivatives are yellow, they may be detected visually on the chromatogram, cut out and eluted, and assayed spectrophotometrically.

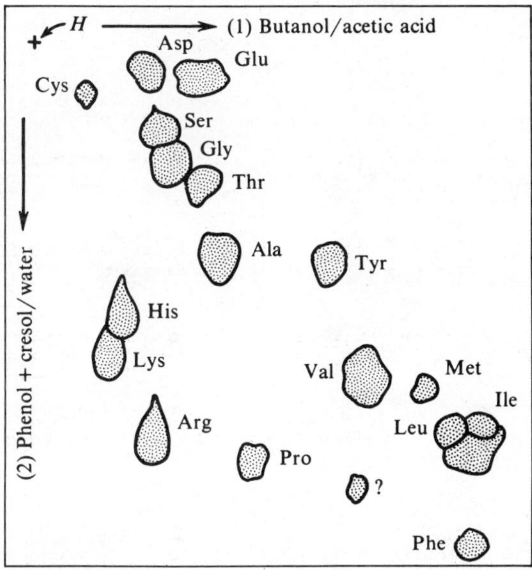

FIGURE 3-11
Two-dimensional chromatogram of amino acids. F. Haurowitz, *The Chemistry and Function of Proteins,* Academic Press, New York, 2nd ed., 1963, p. 31.

In addition to the chromatographic methods indicated above, amino acid analysis by *electrophoresis* and *isotope dilution* techniques are of importance.

Electrophoretic migration in an electric field, like ion-exchange methods, takes advantage of charge differences between the amino acids at appropriate values of *p*H. For example, if an amino acid mixture is placed on a suitable support, frequently paper saturated with an appropriate buffer in aqueous solution, and subjected to an electric field, glutamic and aspartic acid, at neutral *p*H, migrate toward the anode, and histidine, lysine, and arginine migrate toward the cathode, thus effecting separation. Amino acids may be detected and quantitated as in the case of paper chromatography. Chromatography and electrophoresis on paper are frequently combined—chromatography in one direction and electrophoresis in the other.

The amino acid analyses of several proteins, derived by one or more of the methods above, are collected in Table 3-6.

MOLECULAR-WEIGHT DETERMINATION FOR MACROMOLECULES

A basic item of information concerning a protein is its molecular weight. The measurement of molecular weights for proteins and other macromolecules is not trivial. Methods ordinarily employed for molecular-weight determinations on small molecules, notably boiling-point elevation, freezing-point depression, and vapor-pressure measurements, are of little or no use for macromolecules, owing to their size and instability. For example, to obtain a freezing-point depression in water of 0.1° with a protein of molecular weight 100,000 would require a solution containing nearly 5½ kg of protein per liter. One is, of course, considerably worse off employing measurements of alteration in vapor pressure or boiling point.

In the discussion that follows, several of the most important methods for molecular-weight determinations for macromolecules are outlined. Some methods are also developed in Chapter 5; these are particularly pertinent to

TABLE 3-6
Amino acid composition of proteins[a]

Protein Source Molecular weight Total N	*Insulin* *Bovine* *5,733* *nd*	*Cytochrome c* *Horse heart* *12,100* *15.98*	*Serum albumin* *Human* *68,000* *nd*		*Glucagon* *Bovine* *3,647* *17.45*		*Hemoglobin α-chain* *Human* *15,120* *nd*	*Hemoglobin β-chain* *Human* *15,860* *nd*	*Histone* *Calf thymus* *15,500* *17.4*		*Collagen* *Elastoidin (sharkfin)* *nd* *18.2*	
	(2)	(2)	(1)	(2)	(1)	(2)	(2)	(2)	(1)	(2)	(1)	(2)
Glycine	70.5	90.1	1.60	21.3	2.08	27.7	46.1	82.4	5.07	67.6	25.53	330.4
Alanine	44.8	43.0	nd	nd	2.68	30.1	139.0	95.0	9.68	108.8	11.39	128.0
Serine	53.0	0	3.70	35.2	11.60	110.5	71.4	29.6	4.92	46.9	4.43	42.2
Threonine	17.3	72.7	5.0	42.0	9.93	83.4	56.8	42.8	5.47	46.0	2.87	24.1
Proline	16.7	24.0	5.1	44.3	0	0	47.5	46.0	4.62	40.2	14.11	122.7
Hydroxyproline	0	0	0	0	0	0	0	0	0	0	9.68	73.9
Valine	71.0	17.3	7.7	65.8	3.22	27.5	86.7	112.5	5.79	49.5	2.02	17.3
Isoleucine	10.0	42.1	1.7	13.0	0	0	<0.06	<0.06	4.59	35.0	2.46	18.8
Leucine	101.0	43.0	11.9	90.8	7.42	56.6	121.0	115.0	8.18	62.4	2.50	19.1
Phenylalanine	49.8	26.4	7.8	47.3	9.42	57.1	46.4	50.2	2.79	16.9	2.45	14.8
Tyrosine	66.0	31.4	4.66	25.7	10.60	58.6	18.7	17.8	3.69	20.4	1.72	9.5
Tryptophan	0	8.3	0.19	0.9	5.71	28.0	5.9	11.1	0	0	0	0
Cystine/2	52.0	nd	5.58	46.5	0	0	6.1	12.3	0	0	0.22	1.8
Cysteine	0	18.2	0.7	5.8	0	0	nd	nd	0	0	0	0
Methionine	0	nd	1.28	8.6	3.77	25.3	12.6	6.11	1.14	7.6	2.05	13.8
Aspartic acid	53.0	58.7	10.4	78.2	15.40	115.8	78.0	81.9	5.12	38.5	6.13	46.1
Glutamic acid	115.8	89.3	17.4	11.8	12.40	84.4	33.8	76.3	9.71	66.0	12.07	82.1
Amide-N	nd	70.2	0.88	62.9	1.60	114.3	33.3	62.6	0.71	50.7	0.56	40.8
Arginine	16.3	14.9	6.15	35.3	9.52	54.7	35.6	18.6	11.61	66.7	9.02	51.8
Histidine	34.5	21.5	3.50	22.6	3.81	24.5	65.8	56.0	2.48	16.0	0.70	4.5
Lysine	14.5	147.1	12.30	84.2	4.06	27.8	72.6	69.6	17.35	118.8	3.59	24.6
Hydroxylysine	0	0	0	0	0	0	0	0	0	0	1.17	7.1
Totals	786.2	748.0		780.0		812.1	944.0	978.2		807.3		1,042.7

[a] nd, not determined; (1), grams of amino acid in 100 g of protein; (2), moles of amino acid in 100,000 g of protein.
SOURCE: G. R. Tristram and R. H. Smith, in H. Neurath (ed.), *The Proteins*, vol. 1, Academic Press, New York, 1963, pp. 46–50.

the nucleic acids. At the outset, it is worth bearing in mind that several of the physical techniques employed depend upon the shape, as well as the size, of the protein or other macromolecule. When this is the case, one has the necessity of sorting out the shape contribution from the size contribution; at the same time, one has the possibility of obtaining some useful information concerning shape from the measurements. Finally, not only are molecular weights of interest in themselves but changes in molecular weights, even if absolute numbers are not available, can be very useful. The physical interaction of proteins with themselves or other protein molecules can be approached by studying the changes in molecular weight that accompany the association or dissociation processes.

Analytical Data

The metal-ion (or prosthetic group, in general) content, for cases in which there is any, or the amino acid composition of a protein may be employed to calculate *minimum molecular weights.* Such calculations are made on the assumption that each protein molecule contains only a single metal atom, amino acid residue or other constituents of the chosen type. If more than one such constituent is present, the calculated molecular weight will be too small by a factor equal to the number actually present. Quantitatively, we may write

$$M_{\text{min}} = \frac{\text{constituent weight} \times 100}{\text{percentage of constituent}} \qquad (3\text{-}17)$$

Thus, the knowledge that hemoglobin, an oxygen-transporting protein, contains 0.335 percent iron by weight indicates a minimum molecular weight $(55.85/0.335) \times 100 = 16{,}700$ for this protein. Physical measurements of the type described below demonstrate that the actual molecular weight is near 66,000. Hence, each molecule of hemoglobin must contain four atoms of iron. When attempting similar calculations employing amino acid composition data, one chooses, naturally, that amino acid which appears least frequently in the protein. This method is principally employed as a check on molecular-weight determinations obtained by physical methods.

Electron Microscopy

The electron microscope is an instrument capable of obtaining, for properly prepared samples and under optimal conditions, an optical resolution of about 10 Å. Because most proteins are larger than 20 Å in their shortest dimension, they may be detected directly in the electron microscope and photographed. If one knows (1) the number of particles in a given field under the electron microscope, (2) the volume of solution from which it was originally derived, and (3) the dry weight of protein per milliliter of the original solution, one can evidently calculate a value for the molecular weight. Conditions (1) and (3) are readily satisfied, the former by counting the number of particles in an electron micrograph and the latter by analytical means discussed earlier in this chapter. The volume from which a particular field is derived may be obtained by the prior addition of a known number of polystyrene latex particles to the original solution. For example, suppose a solution containing 1 mg/ml of protein and 10^{13} latex particles per milliliter is examined under the electron microscope and a field is observed to contain 10 latex particles and 10 protein particles. A simple calculation reveals that each protein particle must weigh about 10^{-18} g, from which a molecular weight of about 6×10^5 is obtained. Of course, several such fields must be examined to avoid serious uncertainties, owing to statistical fluctuations in the number of particles that appear.

Electron microscopy, for sufficiently large molecules, also clearly yields information concerning particle shape. It must be borne in mind, however, that electron microscopy requires desiccated preparations, precluding the examination of proteins in their normal hydrated state. Thus, uncertainties are introduced regarding the relationship of protein shape and dimensions as seen by the electron microscope and those which exist in solution. Nevertheless, the electron microscope is a potent tool in revealing particle shapes, and its utility should not be underestimated. It has been of great value, for example, in the understanding of the morphology of viruses. An electron micrograph of individual T2 bacteriophage particles is presented in Fig. 3-12.

Osmotic Pressure

Among the colligative properties of matter, only osmotic pressure is of utility for the determination of macromolecular weights. The theory underlying measurement of osmotic pressure is straightforward. If, for example, a protein solution is separated from a protein-free buffer solution by a semipermeable membrane through which buffer components and solvent may pass but which is impermeable to protein, a nonequilibrium situation will exist. The chemical activity of water in the protein solution will be lower than that in the solution on the other side of the membrane; consequently, water will tend to flow into the protein solution. The strength of this tendency is termed the *osmotic pressure* and may be conveniently measured by the amount of hydrostatic pressure just required to prevent the net transport of water across the membrane. Such measurements are accomplished with the aid of an osmometer, such as the one illustrated diagrammatically in Fig. 3-13.

For a two-component system, the osmotic pressure π is related to concentration m (moles per liter) by the van't Hoff law:

$$\lim_{m \to 0} \frac{\pi}{m} = RT \tag{3-18}$$

FIGURE 3-12

Electron micrograph of T2 bacteriophage. Courtesy of The Virus Laboratory, University of California, Berkeley, California.

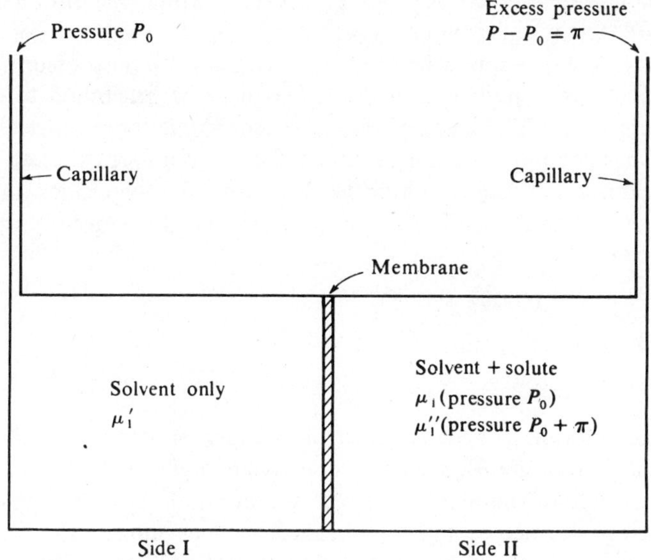

FIGURE 3-13
Schematic representation of an osmotic-pressure apparatus. C. Tanford, *Physical Chemistry of Macromolecules,* Wiley, New York, 1961.

This equation is more conveniently written, for our purpose, in terms of the molecular weight M and concentrations in grams per liter c, yielding

$$\lim_{c \to 0} \frac{\pi}{RTc} = \frac{1}{M} \tag{3-19}$$

This equation provides the basis for the determination of molecular weights by osmotic-pressure measurements, because c, T, and π are all susceptible to experimental determination. Because a limiting value of π/RTc is required, osmotic-pressure measurements must be carried out at a variety of protein concentrations and the values carefully extrapolated to zero concentration. This is a fairly tedious procedure and constitutes one of the major drawbacks to this method. Alternatively, the osmotic pressure may be expressed in terms of a power series in c:

$$\frac{\pi}{RTc} = \frac{1}{M} + Bc + Cc^2 + \ldots \tag{3-20}$$

and the coefficients determined from measurements of π as a function of c. Values of M may then be calculated without extrapolation to zero concentration.

In addition to the large amount of time required for an accurate molecular-weight determination, the principal drawbacks to this method include (1) the possibility of protein aggregation or precipitation during the time period, often extended, required to attain equilibrium; (2) the fact that the measurement is best made under conditions, frequently unknown, in which the protein carries no net charge (see below); and (3) the method yields no information regarding purity of the sample. In light of these disadvantages and the availability of more suitable techniques, osmotic-pressure measurements are now seldom used to determine the molecular weights of proteins. Nevertheless, the method is of historical importance and still finds use in some circumstances. What is perhaps more interesting is the situation at the membrane if the measurements are not carried out at a pH at which the protein has no net charge. We turn to this point now.

If the protein carries a net charge, osmotic-pressure measurements yield values of π higher than those that may be attributed to the presence of protein alone. The basis of this phenomenon is termed the *Gibbs–Donnan effect.* Consider the situation in which the protein carries a net negative charge in a solution of sodium chloride. Denoting the two sides of the semipermeable membrane by the subscripts a and b and protein concentration by (P^-), we write

$$(Na^+)_b = (Cl^-)_b + (P^-)_b \tag{3-21}$$

for solution b, and

$$(Na^+)_a = (Cl^-)_a \tag{3-22}$$

for solution a, because electroneutrality must be maintained on both sides of the membrane. Because the negative charges of the protein are present on only one side of the membrane, the sodium and chloride ions distribute themselves unequally across the membrane. Such asymmetric distribution increases the observed value of the osmotic pressure. For the reaction

$$(Na^+)_a + (Cl^-)_a \rightleftharpoons (Na^+)_b + (Cl^-)_b \tag{3-23}$$

at equilibrium (page 20)

$$\Delta G = 0 = -RT \ln \frac{(Na^+)_b(Cl^-)_b}{(Na^+)_a(Cl^-)_a} \tag{3-24}$$

Rewriting and solving, we find that

$$(Na^+)_a(Cl^-)_a = (Na^+)_b(Cl^-)_b \tag{3-25}$$

Now, employing Eqs. (3-21), (3-22), and (3-23), it is easy to show that

$$(Cl^-)_a = (Cl^-)_b \left[1 + \frac{(P^-)}{(Cl^-)_b} \right]^{1/2} \tag{3-26}$$

Thus, the extent of asymmetric distribution of a diffusible ion across the membrane is accentuated by a high concentration of protein salt and is minimized by a high ionic strength. Consequently, particularly if the pH at which the protein bears no net charge is not known, osmotic-pressure measurements with proteins or other charged macromolecules should be conducted in the presence of high salt concentrations. For many purposes a salt concentration of about 0.2 M is sufficient.

The comments above may, of course, be readily generalized for any protein charge and for electrolytes that are other than uni-univalent.

Sedimentation Techniques

Unquestionably, those techniques which focus on the behavior of proteins under the influence of intense gravitational fields provide some of the most sensitive and useful methods for the determination of molecular weights. Moreover, the importance of sedimentation techniques transcends molecular-weight measurements. The shape of the molecules and the interactions between them are reflected in their behavior in gravitational fields. Hence, ideally, we may obtain information concerning size, shape, and protein–protein interactions from appropriate measurements. These are carried out with the aid of ultracentrifuges. The first ultracentrifuge was constructed by Svedberg and his associates in 1925. The *ultracentrifuge* consists basically of a rotor, in which cells containing the material under investigation are placed, a refrigerated and evacuable chamber in which the rotor turns, a high-speed electrical motor, and an optical system suitable for measuring the concentration of protein, or other material, at each point in the cell throughout the duration of a run. A cross section through an electrically driven ultracentrifuge is shown in Fig. 3-14. Modern ultracentrifuges operate at speeds up to 70,000 rpm, generating gravitational fields up to 500,000 times the force of gravity. Such intense forces rapidly sediment macromolecules and are sufficient to cause asymmetric distribution of low-molecular-weight materials, such as sucrose and cesium chloride, within the ultracentrifuge cell. The use of the ultracentrifuge lends itself to several independent methods for the determination of molecular weights. We discuss the most important of these below.

All the various methods utilizing the ultracentrifuge depend on how the system under study is redistributed under the influence of a gravitational field. The design of both the cell and optical systems reflects this point; that is, the designs permit one to determine the concentration of protein at any point in the centrifuge cell at any point in time. A very large number of optical

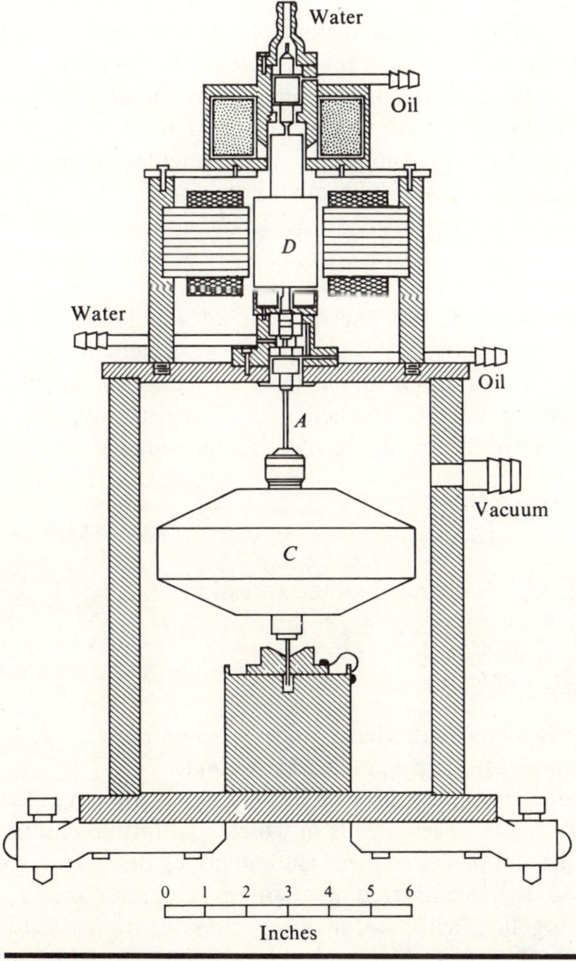

FIGURE 3-14
Cross section through an electrically driven ultracentrifuge; the rotor (C) is suspended from a motor armature (D) by means of the flexible shaft (A). C. Skarstrom and J. W. Beams, *Rev. Sci. Instr.*, **11:** 398 (1940).

systems and cell designs, each suited to a particular application, are currently available. Consideration of these is beyond the scope of the present treatment and the interested student is referred to the excellent reviews of the subject which are cited in the references.

There are three basic types of studies which we may carry out with the ultracentrifuge: *equilibrium methods, transport methods,* and *transient-state methods.*

Equilibrium Methods

In a two-component system, containing just one solute in a solvent, the equilibrium distribution of solute as a function of the distance from the center of rotation depends on balancing the tendency of the protein to sediment to the bottom of the cell due to the influence of the gravitational field and the tendency of the protein to distribute itself uniformly throughout the cell due to diffusion down its concentration gradient. Theoretical consideration of this situation generates the following equation, which underlies all further considerations:

$$\frac{1}{r}\frac{dc}{dr} = \frac{M(1-\bar{v}\rho)\omega^2}{RT(1 + d\ln\gamma/d\ln c)} \tag{3-27}$$

In this equation, r is the distance from the center of rotation to the point of observation in the centrifuge cell, dc/dr is the concentration gradient at that point, c is the protein concentration, M is the molecular weight, $\bar{v}$ is the partial specific volume of the protein, ρ is the density of the solvent, γ is the activity coefficient of the protein, and ω is the angular velocity of the centrifuge head in radians per second. The quantities directly determinable from ultracentrifuge measurements include c, r, dc/dr, and ω. Values of $\bar{v}$ are determined by density measurements in a pycnometer or may be calculated, approximately, from amino acid composition data. Typical values of $\bar{v}$ for proteins are in the range 0.70 to 0.75 cm^3/g. The solvent density is usually near unity but may be measured quite simply and accurately if necessary. Thus, the only quantity not readily available that is needed for calculation of the molecular weight is the activity coefficient γ. Consequently, Eq. (3-27) is frequently written in terms of an *apparent* molecular weight:

$$\frac{1}{r}\frac{dc}{dr} = \frac{M_{app}(1-\bar{v}\rho)\omega^2}{RT} \tag{3-28}$$

in which M_{app} is defined by the expression

$$M_{app} = \frac{M}{1 + BMc} \tag{3-29}$$

where B is a constant. Hence, measurement of M_{app} as a function of c suffices to determine M, the true molecular weight.

One of the simplest uses of Eq. (3-27) is direct application at the midpoint of short (1 mm or less) cells in which equilibrium can be reached in relatively short time periods. From measurements of indices of refraction, the concentration and concentration gradient at the midpoint are determined. These values, together with the ancillary ones mentioned earlier, and Eq. (3-29) suffice to determine the molecular weight. A large number of alternative methods are also used; all are derived from the same basic considerations.

How is the situation changed if we have a multicomponent system in which several solutes are present? Clearly, we will expect to obtain some average molecular weight for all the species present. In fact, there are several different kinds of averages and different techniques yield different averages. The three most common ones are the number average, M_n, weight average, M_w, and z average, M_z, defined in the following ways:

$$M_n = \frac{\Sigma n_i M_i}{n_i} \tag{3-30}$$

$$M_w = \frac{\Sigma n_i M_i^2}{\Sigma n_i M_i} \tag{3-31}$$

$$M_z = \frac{\Sigma n_i M_i^3}{\Sigma n_i M_i^2} \tag{3-32}$$

Values of n refer to the mol fractions of the various solute species. These averages differ to the extent that they give varying importance to heavier components. Thus, the number average gives the least and the z average the most importance to the heaviest components of the system.

Depending on how the data collected are treated, it is usually possible to obtain the weight-average or z-average values from single measurements at equilibrium. The number-average molecular weight cannot be obtained in this way but is available, for example, from osmotic-pressure measurements.

A special use of average molecular weights obtained for multicomponent systems is for the case of chemically reacting species. A simple example is provided by the self-association of a protein to form dimers, trimers, tetramers, or larger species:

$$n\text{P} \overset{K_{\text{eq}}}{\rightleftharpoons} (\text{P})_n \tag{3-33}$$

Thus, the multicomponent nature of the system derives from the existence of one and the same chemical entity in more than one state of aggregation. The distribution of each of the components within the centrifuge cell is independent of the fact that an equilibrium exists. That is, each component has the distribution that it would have if it existed alone. This observation leads to a particularly simple conclusion: the usual methods for finding weight-average molecular weights for mixtures will suffice for this case as well.

The above considerations suggest a straightforward means for determining the equilibrium constant for such association processes [Eq. (3-33)]. Since we should have more of the associated species as the concentration of the protein is increased, the weight-average molecular weight is expected to increase with increasing protein concentration. In fact, it is possible to derive quantitative relationships between the protein concentration and the weight-average molecular weight for a number of situations. For example, for a monomer–dimer equilibrium, we have the expression

$$\frac{2M_1}{M_w} - 1 = \left(\frac{1}{1 + 4K_{\text{eq}}c}\right)^{1/2} \tag{3-34}$$

In this expression, M_1 is the molecular weight of the monomeric unit. Clearly, measurements of M_w as a function of c will yield a value of K_{eq}. It is also possible to distinguish among a number of the possible cases that can arise: monomer–dimer, monomer–dimer–trimer, and so forth. Thus equilibrium sedimentation measurements provide a convenient and powerful tool for the study of protein–protein interactions in aqueous solution.

Isodensity Equilibrium Sedimentation

In the above discussion, we have dealt with the distribution of a protein throughout a homogeneous medium in accordance with the demands of gravitational sedimentation and diffusion. One obtains a continuous gradient of protein from the top of the centrifuge cell to the bottom. It is possible to modify the experimental setup for equilibrium sedimentation experiments in an interesting and useful way which involves the replacement of the homogeneous medium with one that varies in density throughout the length of the centrifuge cell. This matter is discussed in some detail in Chapter 5 (pp. 210–215). At the moment, we can note that substances such as cesium chloride form continuous density gradients under substantial gravitational forces so that the density at the top of the cell is less and the density at the bottom of the cell is greater than that for most proteins. Under these conditions, the protein sample will move until it reaches the point where the density of the protein and solvent are equal, that is, the point where $(1 - \bar{v}\rho)$ is zero. At this density the protein will form a Gaussian band. From the shape of this band, it is possible to derive the molecular weight of the solvated species. More about these matters follows in Chapter 5.

Transport Methods

A fundamentally different approach to some of the same problems discussed above involves study of the *rate of sedimentation* of proteins, and other macromolecules, through a medium under the influence of an applied gravitational force. In this case, we focus our attention on a nonequilibrium situation within the centrifuge cell rather than on the equilibrium one. The transport properties of a protein depend on quantities other than the molecular weight; density and shape are important as well. However, a combination of measurements of sedimentation rates and diffusion properties does serve to sort out the molecular weight from the other factors. The basic equation employed to obtain the molecular weight from such studies is

$$M = \frac{RT}{D(1 - \bar{v}\rho)}\left[1 + \frac{d(\ln \gamma)}{dc}\right]\frac{dr/dt}{\omega^2 r} \tag{3-35}$$

in which M is the molecular weight. Equation (3-35) is termed the *Svedberg equation.* The quantity in the equation that is determined with the aid of the ultracentrifuge is $(dr/dt)/\omega^2 r$, which is ordinarily termed the *sedimentation constant* and designated s. Equation (3-35) then assumes the form

$$M = \frac{RTs}{D(1 - \bar{v}\rho)}\left[1 + \frac{d(\ln \gamma)}{dc}\right] \tag{3-36}$$

The units of s are reciprocal seconds; the basic unit, for convenience, is taken as 10^{-13} sec, which is termed *one Svedberg* (S). Employing these units, the sedimentation constants for most proteins fall in the range 1 to 50 S.

The velocity dr/dt with which the protein sediments in the gravitational field is obtained by observing, photographically, the position of the maximum concentration of protein throughout the course of the run. A typical sedimentation pattern is shown in Fig. 3-15. These photographs were taken by employing the *Schlieren optical system,* which measures the refractive index gradient at each point in the cell. As the protein moves down the cell, the peak height diminishes and the peak width increases. This is a normal consequence of the diffusion of the protein molecules. Because ω is a known quantity and r may be measured directly from the photographs as a function of time, the determination of dr/dt suffices to determine s. Sedimentation constants are ideally obtained at a variety of protein concentrations and the values of s extrapolated

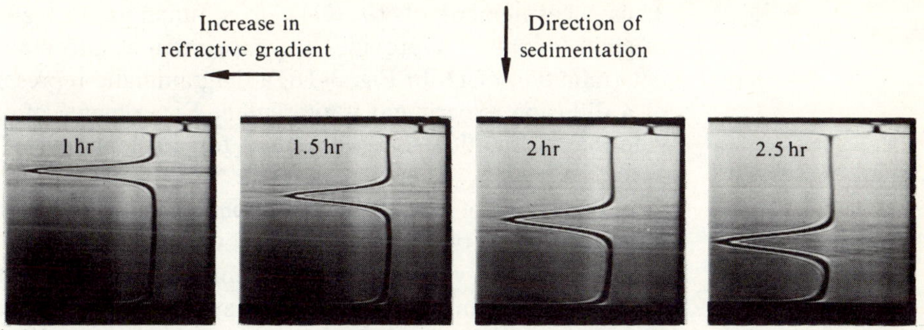

FIGURE 3-15
Sedimentation of triose phosphate dehydrogenase. The set of four photographs was recorded by the refractive-index method.

to zero concentration, conditions in which the activity coefficient term becomes unity. In addition, it is customary to correct sedimentation constants to a standard state with reference to solvent viscosity taken as that of water at 20° (page 208).

Sedimentation-velocity studies frequently provide information regarding homogeneity of the protein preparation. Preparations that contain more than one component may exhibit several peaks in the ultracentrifuge or asymmetry in a single peak. The criterion for homogeneity in the ultracentrifuge, then, is a single sharp, symmetrical peak whose width is not greater than can be accounted for in terms of diffusion of the protein.

Clearly, use of Eq. (3-36) for the determination of a molecular weight requires values for $\bar{v}$ and the diffusion constant, D, as well as the sedimentation constant. The diffusion constant may sometimes be measured directly in the centrifuge cell from the rate of spreading of the boundary or it may be determined separately. In either case, the essential considerations are the following:

The basic equation that describes the process of diffusion is due to Fick and is usually referred to as *Fick's first law of diffusion*:

$$\frac{ds}{dt} = -DA\left(\frac{\partial c}{\partial x}\right)_t \qquad (3\text{-}37)$$

This equation states that the amount ds of a substance crossing a given area A in time dt is proportional to the concentration gradient $\partial c/\partial x$ across that area. The diffusion constant, D, which is related to molecular size and shape, is the proportionality constant. Beginning with Eq. (3-37), it is possible to derive *Fick's second law of diffusion,* which is a more useful statement for the purpose of determining D:

$$\left(\frac{\partial c}{\partial t}\right)_x = D\left(\frac{\partial^2 c}{\partial x^2}\right)_t \qquad (3\text{-}38)$$

For particular geometrical cases, Eq. (3-38) may be solved directly. For example, for diffusion from an infinitely long column of solution at concentration C_0 into an infinitely long column of pure solvent (in a container of uniform cross section), solution of Eq. (3-38) yields

$$\left(\frac{\partial c}{\partial t}\right)_x = \frac{C_0 x}{4\pi^{1/2}D^{1/2}t^{3/2}}\, e^{-x^2/4Dt} \qquad (3\text{-}39)$$

Thus measurement of $(\partial c/\partial t)_x$ as a function of t and distance from the original boundary x provides the experimental information necessary for the determination of D. In Fig. 3-16, a diagrammatic representation of the results of a diffusion experiment is presented. Experiments of this kind clearly provide the information that we require for a calculation of M according to Eq. 3-36.

A collection of partial molal volumes, diffusion constants, and molecular weights for some typical proteins is provided in Table 3-7.

Sedimentation-velocity measurements are also important for the study of chemically reacting systems. Those systems which react at rates slow with respect to the length of time it requires to make the experimental observations are almost trivial; the number of distinct sedimenting boundaries is equal to the number of physically distinct species. Areas under individual peaks may be corrected to yield estimates of the relative concentration of each species. What is more challenging is the case in which the rate of the chemical reaction is rapid in comparison with the time necessary to make the experimental observations. These systems may be identified in several ways, depending on the characteristics of the individual reaction.

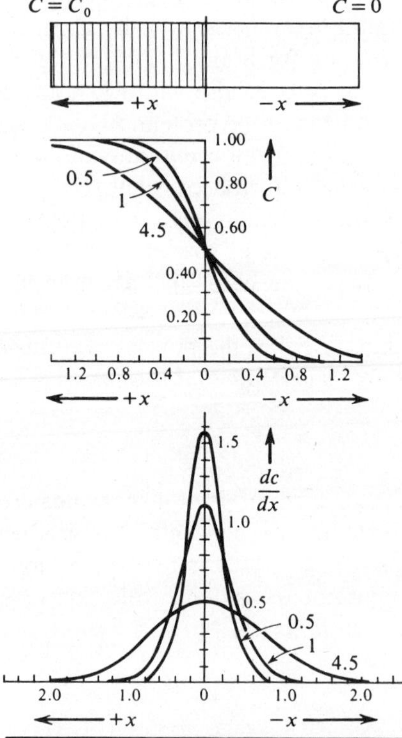

FIGURE 3-16

Relationship between concentration and distance of migration in a diffusion column. (*Top*) Graphical illustration of the diffusion column; (*center*) relationship between concentration and distance of migration; (*bottom*) relationship between concentration gradient and distance of migration. The curves as drawn refer to diffusion times of 0.5, 1, and 4.5 hr., respectively. The whole diagram should be rotated counterclockwise about an angle of 90° in order to conform to the proper space directions. H. Neurath, *Chem. Rev.*, **30:** 357 (1942); by permission of the copyright owners, the American Chemical Society.

TABLE 3-7
Molecular weights, diffusion constants, and partial specific volumes for several proteins

Protein	*M*	$\bar{v}$	$D^0_{20,w} \times 10^7$
Ribonuclease	13,683	0.728	11.9
Lysozyme	14,100	0.688	10.4
Chymotrypsinogen	23,200	0.721	9.5
β-Lactoglobulin	35,000	0.751	7.82
Ovalbumin	45,000	0.748	7.76
Serum albumin	65,000	0.734	5.94
Hemoglobin	68,000	0.749	6.9
Catalase	250,000	0.73	4.1
Urease	480,000	0.73	3.46
Tropomyosin	93,000	0.71	2.24
Fibrinogen	330,000	0.710	2.02
Collagen	345,000	0.695	0.69
Myosin	493,000	0.728	1.16

SOURCE: C. Tanford, *Physical Chemistry of Macromolecules,* Wiley, New York, 1961.

One example is provided by studies with two species of β-lactoglobulin, β-A and β-B (4). Under the conditions of these experiments, the former species is involved in a monomer–tetramer equilibrium but the latter is not. These facts show up very clearly in the dependence of the sedimentation constant upon protein concentration. The results are shown graphically in Fig. 3-17. Note that for β-B, the values of s decline slowly with increasing protein concentration, as usual, but that those for β-A increase quite rapidly with increasing protein concentration, reflecting a transition from predomi-

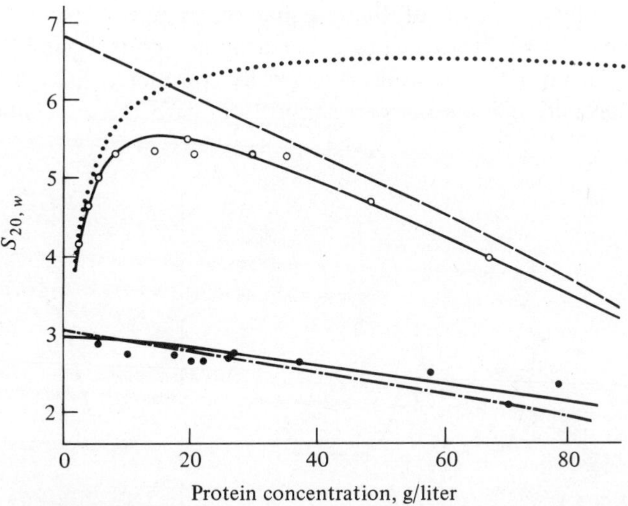

FIGURE 3-17
Dependence of the sedimentation constants for two forms of β-lactoglobulin, β-A and β-B, on protein concentration at pH 4.65, 2°. Open circles are the experimental points for β-A, and the closed ones are for β-B. S. N. Timasheff and R. Townend, *J. Am. Chem. Soc.,* **83:** 464 (1963); by permission of the copyright owners, the American Chemical Society.

nantly monomer to predominantly tetramer in solution, and then decrease with further concentration increases.

Just as use of gradients proves useful in the case of equilibrium sedimentation measurements, so it does for the case of sedimentation-velocity studies. In *boundary sedimentation,* we have the disadvantage that, except for the slowest component, each component in a mixture sediments in the presence of the other ones. Use of density gradients permits one to do *zone sedimentation,* in which case each component migrates as a band, rather than a single boundary, through the medium. The difference is shown schematically in Fig. 3-18.

Zone sedimentation proves particularly useful in work with relatively impure preparations inasmuch as it provides an opportunity to examine the properties of a component that might not even be detected using boundary sedimentation.

Transient-State Methods

An example of a transient-state sedimentation method is provided by the widely used *Archibald method.* The basic point here is simply that a condition of equilibrium always exists in the centrifuge cell at two places: the meniscus and the bottom. Clearly, no material transfer takes place across these boundaries, so the equilibrium relationship always holds even though the conditions in the immediate vicinity may be far from equilibrium. Thus, for the meniscus and bottom of the cell, we may write

$$M_m = \frac{RT\left(1 + \dfrac{d(\ln \gamma)}{dc}\right)}{(1 - \bar{v}\rho)\omega^2} \; \frac{(dc/dr)_m}{c_m r_m} \tag{3-40}$$

and

$$M_b = \frac{RT\left(1 + \dfrac{d(\ln \gamma)}{dc}\right)}{(1 - \bar{v}\rho)\omega^2} \; \frac{(dc/dr)_b}{c_b r_b} \tag{3-41}$$

Note that the use of these equations requires that both the concentration gradient and the absolute concentration at the boundary be known. Use of data from the cell bottom is generally considered risky, because small amounts of protein aggregation can seriously disturb the concentration gradient there.

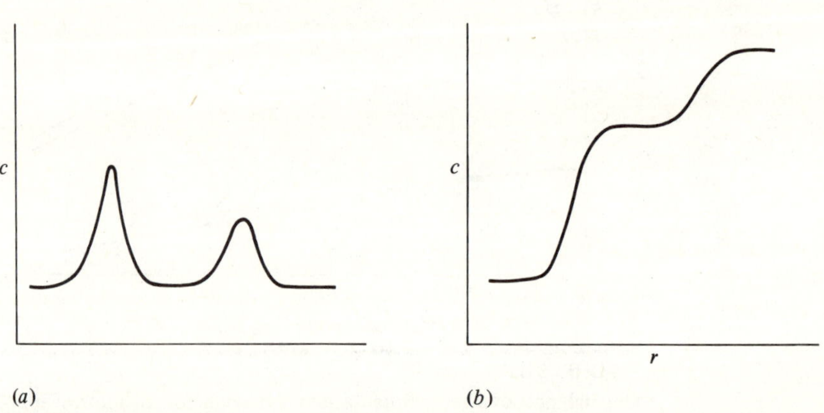

FIGURE 3-18

Concentration profiles in sedimentation velocity experiments: (*a*) zone sedimentation velocity (two solutes); (*b*) boundary sedimentation velocity (two solutes).

Gel-Filtration Techniques

A distinct means of determining the molecular weight of macromolecules is based on the "molecular-sieve" effect. The molecular sieves themselves are three-dimensional networks of polymer chains which are prepared by cross-linking linear polymers. The most common and useful commercial preparations are marketed under the trade names Sephadex and Bio-Gel. Sephadex consists of cross-linked polysaccharide chains; Bio-Gel of cross-linked polyacrylamide chains. When these molecular sieves are placed in water, they swell and form a gel. The particles forming this gel have pores, the size of which is determined by the degree of cross-linking of the individual polymer chains: a great deal of cross-linking leads to very small pores, whereas little cross-linking yields large pores. It is the variation in pore size that makes these molecular sieves so useful for molecular-weight determinations and other purposes.

To understand what happens to a mixture of macromolecules when applied to the top of a column of molecular sieves, refer to Fig. 3-19. Note that the molecules which are small enough to penetrate the pores of the gel particles will enter these particles and their rate of elution from the column will be retarded. Large molecules, in contrast, cannot pass through the pores to enter the particles and are rapidly eluted from the column. This provides, of course, a very nice separation technique for mixtures of macromolecules, a subject to which we shall return later. It also provides a means for molecular-weight determination. It is worth noting that commercially available preparations come in a variety of useful pore sizes. Sephadex G-50, for example, has rather small pores and excludes molecules with molecular weights greater than 10,000 and is useful for fractionating molecules in the molecular-weight range 500 to 10,000. Sephadex G-200, in contrast, excludes only those molecules having weights greater than 200,000 and is useful for fractionation purposes over the molecular-weight range 5,000 to 200,000.

Since molecular sieves have the capacity to sort out molecules according to size, some relationship should exist between the *elution volume,* that volume of solvent passing through the column required to elute the macromolecule, and the molecular weight. Such relationships do exist, although they are different for different classes of macromolecules. In Fig. 3-20, a plot of elution volume for a series of globular proteins against the logarithm

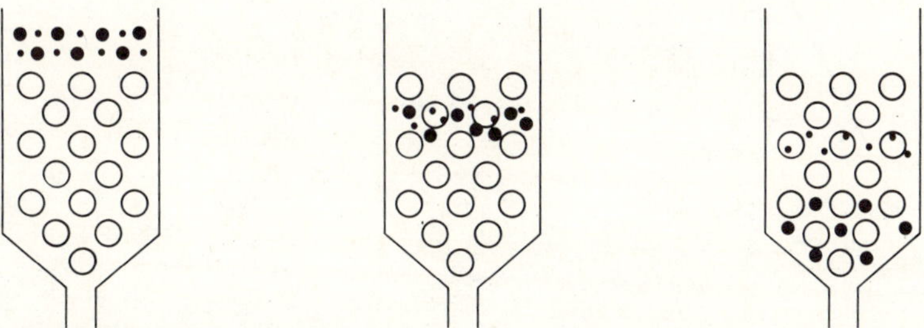

FIGURE 3-19
Schematic view of chromatography on molecular sieves. Molecules larger than the largest pores of the gel cannot penetrate the gel particles and therefore pass through the bed in the liquid phase outside the particles. Smaller molecules, however, penetrate the gel particles to a varying extent depending on their shape and size. Molecules are therefore eluted from the column in the order of decreasing molecular size. The large dots represent the large molecules; small dots, the small molecules; and circles, gel particles.

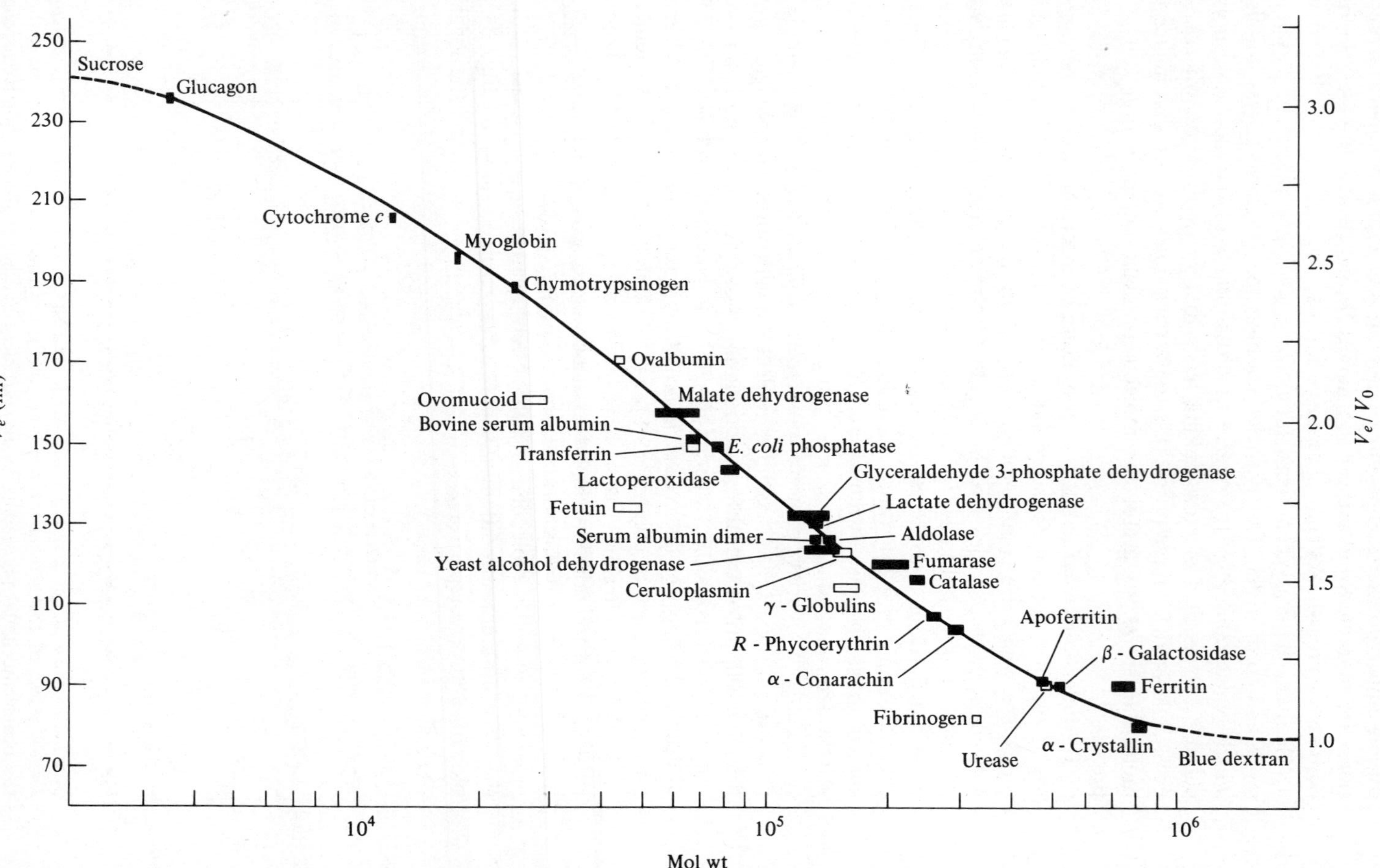

FIGURE 3-20

Relationship of elution volume V_e to molecular weight for proteins on a column of cross-linked dextran (Sephadex G-200) (2.5 × 50 cm) at *p*H 7.5. Open bars represent glycoproteins. V_0 represents the void volume. After P. Andrews, *Biochem. J.*, **96:** 595 (1965).

of the molecular weight is provided. The relationship is approximately linear over a substantial molecular-weight range. This method has two primary advantages: It is simple and rapid and it may be applied to quite impure samples. It does suffer from some uncertainties, however, in that some proteins are preferentially adsorbed onto the gel material so that factors other than molecular size can be important.

Finally, chromatography on molecular sieves is useful for the study of chemically interacting systems (5). Clearly, if a monomer–n-mer equilibrium accompanies the gel filtration, we shall not expect the protein to elute in a symmetrical peak. This follows, because the local concentration at any point in the eluting material will affect both the degree of association and the retention time on the column. Thus, at the forward edge, we expect to find high protein concentrations associated with a high degree of association and at the trailing one, lower concentrations together with a lower degree of dissociation. Hence, the forward edge ought to be markedly sharper than the trailing one. A typical example is provided in Fig. 3-21 for the molecular-sieve chromatography of carboxyhemoglobin, which undergoes a reversible monomer–tetramer equilibration.

From the elution volume of the forward boundary, the minimum in gradient of total solute eluted, and the shape of the trailing edge it is possible to calculate the stoichiometry of the association process and the equilibrium constant for it (5).

PROTEINS AS ELECTROLYTES

Many of the interesting and important properties of proteins derive from the fact that, under most conditions, they are multivalent electrolytes. The groups in proteins that are involved in acid–base equilibria are principally the side chains of glutamic and aspartic acid, lysine, histidine, arginine, cysteine, and tyrosine (see Table 3-1). The α-amino and α-carboxyl groups of the constituent amino acids are, for the most part, bound in peptide linkage and, hence, contribute little to the electrolyte character of proteins.

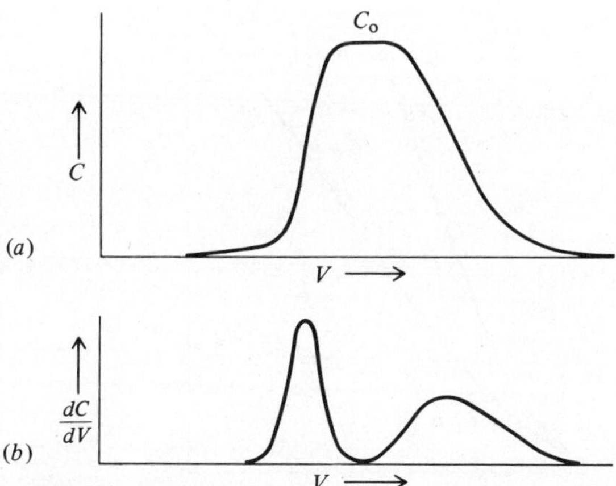

FIGURE 3-21

Characteristic elution patterns for carboxyhemoglobin undergoing simultaneous molecular-sieve transport and reversible dissociation. Diagram (*a*) is a tracing from the pattern obtained with $C_0 = 5.0 \times 10^{-6}$ mol/liter and (*b*) is the concentration gradient across the zone. G. K. Ackers and T. E. Thompson, *Proc. Natl. Acad. Sci. U.S.*, **53:** 342 (1965).

Titration of Proteins

Titration curves for proteins are complex and difficult to interpret theoretically. This complexity is the result of two factors. First, there is ordinarily a large number of titratable groups present in each protein molecule; 50 to 60 such groups per 100,000 molecular weight is a typical figure. Second, the value of pK_a for each of the titratable groups of proteins may differ by one or more pH units from that for the simple amino acid (Table 3-1). Thus, for example, if a protein contains 10 aspartic acid residues, each β-carboxyl is, in general, characterized by a distinct value of pK_a, and these might range from 3 to 6. Such variation in pK_a for groups present in proteins, or other charged macromolecules, may be ascribed to three factors: (1) electrostatic effects resulting from the ionization of other groups on the protein, (2) medium effects due to the proximity of hydrophobic residues, and (3) hydrogen bonding.

Electrostatic effects on the hydrogen-ion equilibria of groups present in macromolecules may be treated theoretically to a certain approximation. For example, assuming the protein is a uniformally charged sphere of radius b, Eq. (3-42) may be deduced:

$$pH = pK_{int} + \log \frac{\alpha}{1-\alpha} - 0.868\omega\bar{Z} \tag{3-42}$$

In this equation pK_{int} is the intrinsic pK_a(pK_a in the absence of charges) of a particular group, α is the fraction of acid groups that have ionized, $\bar{Z}$ is the mean net charge on the molecule, and ω is a complicated function of b, the dielectric constant and other variables. The use, and success, of this equation in accounting for titration curves of macromolecules is illustrated by the simple case of the ionization of poly-L-tyrosine. In Fig. 3-22, α, as determined spec-

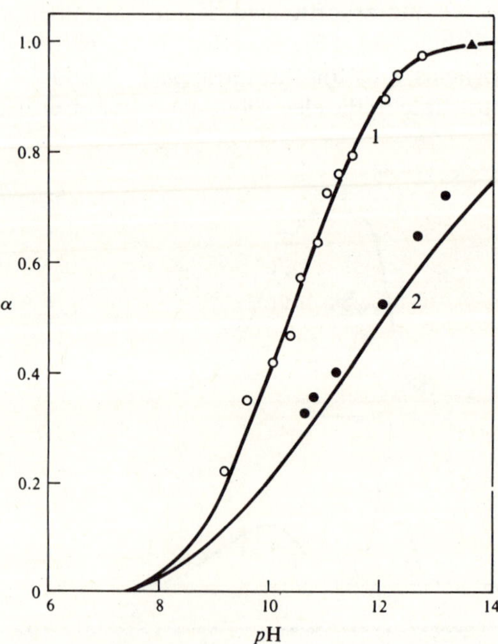

FIGURE 3-22

Spectrophotometric titrations of poly-L-tyrosine: pH versus fraction of phenolic groups ionized. Open circles are experimental points with 0.20 M NaCl; closed circles are experimental points with no added salt, extrapolated to zero ionic strength. E. Katchalski and M. Sela, *J. Am. Chem. Soc.*, **75:** 5284 (1953); by permission of the copyright owners, the American Chemical Society.

trophotometrically, is plotted against pH for titration of a sample containing, on the average, 30 L-tyrosine residues per molecule. Taking b as 17.5 Å, in accord with the mean molecular weight, $\bar{Z}$ as 30α, pK_{int} as 9.5, and ω as 0.066, the solid line in the figure was calculated and found to agree perfectly with the experimental data. Needless to say, similar calculations for proteins, for which neither the assumption of spherical shape nor that of uniform charge distribution is usually satisfactory, seldom have such a pleasing conclusion.

Quantitative calculations concerning medium effects on ionization equilibria for proteins are very difficult, because the dielectric constant in the immediate vicinity of ionizable groups is seldom known. Qualitatively, ionizations of the carboxylic acid type, in which an uncharged molecule dissociates to yield two ions, are known to be strongly repressed by a decreasing medium polarity, because the ionic products are destabilized relative to the uncharged reactant:

$$R{-}CO_2H \rightleftharpoons RCO_2^- + H^+ \tag{3-43}$$

On the other hand, ionizations of the ammonium acid type are rather insensitive to alterations in the nature of the medium, because ionic species are present on both sides of the reaction. Hence, medium effects on their stability tend to cancel:

$$R_3NH^+ \rightleftharpoons R_3N + H^+ \tag{3-44}$$

Clearly, hydrogen-bond formation may, by stabilizing either the acid or the base form of an ionizable group, markedly alter its tendency to add or lose a proton. Because the sites of intramolecular hydrogen bonding in proteins are just beginning to be known, such effects have not yet been included in a general quantitative theory.

The titration curve for ribonuclease, a small, nearly spherical protein of known amino acid composition, is shown in Fig. 3-23. A comparison of the number of titratable groups in each pH region with the amino acid composition is presented in Table 3-8. Excellent agreement is found. Of particular interest is the behavior of the phenolic groups of tyrosine. Of the six present in ribonuclease, three dissociate normally with an intrinsic pK_a of 9.95. The other three dissociate only above pH 12 and are, hence, almost certainly buried in the interior of the protein in its native state and, therefore, not readily accessible to solvent. In accordance with this concept, titrations of these groups lead to irreversible effects on the structure of the protein, as revealed by the fact that the back titration does not follow the same course as the forward one.

In general, one does not expect such good agreement between the amino acid composition and the number of titratable groups in each pH range as that illustrated in Table 3-8, particularly if extremes in acid and base concentrations are avoided. However, because most of the ionizable groups of proteins appear to be near the protein surface, fair agreement is usually found. Back titration of proteins is always essential to assure that irreversible structural alteration, revealing formerly hidden groups, has not occurred or to detect it if it has.

Isoelectric Point. In the course of titration of a protein from the completely acidic to the completely basic form, at some pH the mean charge on the protein must be zero. This pH is termed the *isoelectric point.* Experimentally the isoelectric point is defined as that pH at which the protein does not migrate in an electric field (see the discussion below on electrophoresis). The isoelectric point may be approximately calculated from amino acid composition data. Because the condition of electroneutrality of proteins is affected by the presence of salts, as a result of the capacity of proteins to bind ions and the alteration in

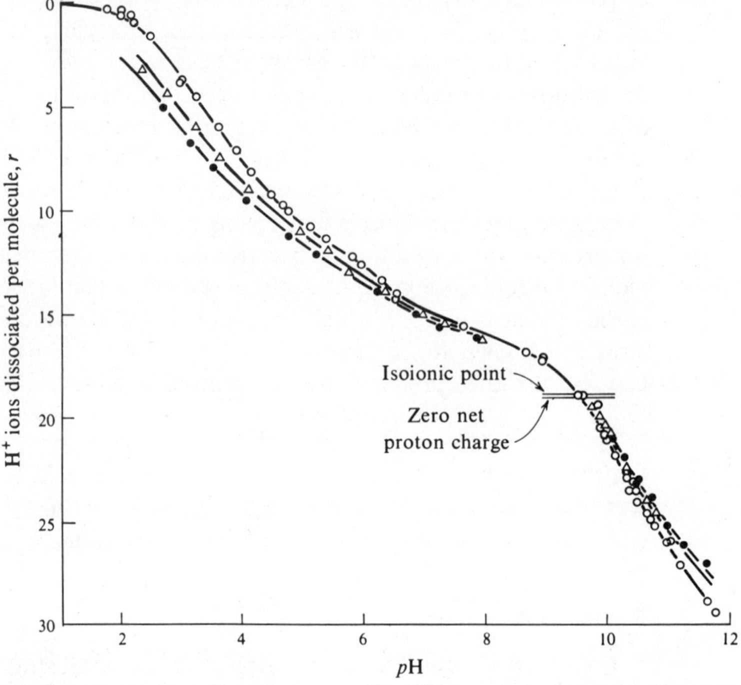

FIGURE 3-23

Titration data for ribonuclease at 25°, ionic strengths 0.01 (●), 0.03 (△), and 0.15 (○). The figure shows the point of zero net charge (corresponding to $r = 19$) and also the isoionic pH at the particular concentration that was used. C. Tanford and J. D. Hauenstein, *J. Am. Chem. Soc.,* **78:** 5287 (1956); by permission of the copyright owners, the American Chemical Society.

pK_a due to ionic-strength effects, the isoelectric point is a function of the nature and concentration of the buffer, and, in general, of any other solutes present, in which the isoelectric point is measured. On the other hand, isoelectric points are not a function of protein concentration. The isoelectric points of most proteins are close to neutrality, reflecting approximately equal numbers of acidic and basic residues as constituents of the protein. In some cases, however, distinctly different behavior is noted. For example, pepsin, an enzyme found in the stomach, where conditions are normally very acid, has an isoelectric point near 1, and protamines typically have isoelectric points near 12. The isoelectric points for several proteins are collected in Table 3-9.

Isoionic Point. A protein solution that contains no ion other than those intrinsic to the protein itself and those derived from the dissociation of water is called an isoionic solution. Such solutions may be prepared by passage of the protein solution through a column containing both anion- and cation-exchange resins (a mixed-bed resin) so that all ions other than hydroxyl and hydrogen are removed. Isoionic solutions may also be prepared by dialysis against distilled water (see below) for extended periods of time. This method is ordinarily less satisfactory than the use of mixed-bed resins. The pH of an isoionic protein solution is termed the *isoionic point* of that protein. Because electroneutrality requires that

$$(\mathrm{H}+) + (\mathrm{P})\bar{Z} = (\mathrm{OH}-) \tag{3-45}$$

TABLE 3-8
Titratable groups of ribonuclease

Group	pK_{int} expected from data on small molecules	Number of sites from amino acid analysis	Data obtained from titration curve		
			(group total)	Number of sites	pK_{int}
α-COOH	3.75	1	11 {	(1)	
Side chain COOH	4.6	10		10	4.7
Imidazole	7.0	4	5 {	4	6.5
α-NH_2	7.8	1		(1)	7.8
Phenolic	9.6	6	16 {	3	9.95
				3	>12
Side chain NH_2	10.2	10		10	10.2
Guanidyl	>12	4		4	>12

SOURCE: C. Tanford and J. D. Hauenstein, *J. Am. Chem. Soc.*, **78:** 5287 (1956), by permission of the copyright owners, the American Chemical Society.

an isoionic solution is not isoelectric unless the isoelectric point happens to be exactly *p*H 7. Equation (3-45) also indicates that the isoionic point is a function of protein concentration. Clearly, for sufficiently dilute solutions all proteins have an isoionic point at *p*H 7. However, at moderate protein concentrations, the isoionic point is an insensitive function of protein concentration and closely approximates the isoelectric point unless the latter is measured in the presence of strongly bound ions.

Interaction of Proteins with Small Ions

The binding of protons by proteins, just discussed, is but one example of the general ability of proteins to interact with charged species. Such species include other small cations and anions as well as complex polyelectrolytes including other proteins. In particular, the interaction of proteins with small ions is of great importance and has been thoroughly studied. The importance of these processes stems from several considerations: the requirements of many enzymes for charged species to ensure the integrity of the catalytic process, the necessity of actively mediating the transport of ions through the agency of specific protein carriers, and the like.

As we can judge from our discussion above, the chemistry of ion binding to proteins is complex. In the first place, ion binding occurs at discrete loci (binding sites) on the protein, and each of these may be unique. We are just beginning to understand some details of the structure of individual loci. Second, each binding site may contain several functional groups. For example, transition-metal ions ordinarily make chemical associations with several groups simultaneously (page 17). The protein might donate as few as one of the interacting groups or as many as six. Note that it is not required that the binding sites on the protein be themselves charged; the imidazole side chain of histidine is quite active in binding to metal ions under conditions in which it is not charged. Other examples can readily be cited. Third, the binding site for an ion does not necessarily preexist as such in the protein structure. Thus, a transition metal ion might interact with several groups in a sequential fashion leading to the formation of a structure in which it is surrounded by a

TABLE 3-9
Isoelectric points for several proteins

Protein	*pI*[a]	$\Gamma/2$[b]
Protamines:		
Salmine	12.1	
Clupeine	12.1	
Sturine	11.71	
Histones:		
Thymohistone	10.8	
Globin (human)	7.5	0.10
Albumins:		
Ovalbumin	4.59	0.10
	4.71	0.01
Conalbumin	6.8	0.10
	7.1	0.01
Serum albumin	4.7.4.9	0.02
Myoalbumin	3.5	
Myogen A	6.3	
Globulins:		
β-Lactoglobulin	5.1.5.3	0.10
	4.7.5.1	0.01
Livetin	4.8.5.0	
γ_1-Globulin (human)	5.8	0.10
	6.6	0.01
γ_2-Globulin (human)	7.3	0.10
	8.2	0.01
Myosin A	5.2.5.5	
Tropomyosin	5.1	
Thyroglobulin	4.58	
Siderophilin	5.9	
Fetuin	3.4.3.5	
Fibrinogen	5.5.5.8	
α-Crystallin	4.8	
β-Crystallin	6.0	
Arachin	5.1	
Conarachin	3.9	
Scleroproteins:		
Keratins	3.7.5.0	
Keratein	4.6.4.7	
Collagen	6.6.6.8	
Ichthyocol	4.8.5.2	
Gelatin	4.7.5.0	

Protein	*pI*[a]	$\Gamma/2$[b]
Phosphoproteins:		
α-Casein	4.0.4.1	0.10
β-Casein	4.5	0.10
γ-Casein	5.8.6.0	0.05
Mucoproteins:		
α-Ovomucoid	3.83.4.41	
α_1-Mucoprotein	1.8.2.7	
Vitellomucoid	5.5	
Urinary gonadotropin	3.2.3.3	
Lysozyme	11.0.11.2	0.01
Chromoproteins:		
Myoglobin	6.99	
Hemoglobin (human)	7.07	
Hemoglobin (hen)	7.23	
Hemoglobin (horse)	6.92	
Hemocyanins	4.6.6.4	
Hemerythrin	5.60	
Chlorocruorin	4.3.4.5	
Erythrocruorins	4.6.6.2	
Cytochrome *c*	9.8.10.1	
Rhodopsin	4.47.4.57	
Lipoproteins:		
Thromboplastin	5.2	
α_1-Lipoprotein	5.5	
β_1-Lipoprotein	5.4	
β-Lipovitellin	5.9	
Nucleoproteins:		
Thymonucleohistone	ca. .4	
Bushy stunt virus	4.11	0.02
Turnip yellow virus	3.75	
Vaccinia virus	5.3	
Unclassified:		
Somatotropin	6.85	
Prolactin	5.73	
Insulin	5.35	
Pepsin	ca. 1.0	

[a] *pI*, isoelectric point or range.
[b] $\Gamma/2$, ionic strength of solvent (buffer).
SOURCE: E. Gordon Young, in M. Florkin and E. H. Stotz (eds.), *Comprehensive Biochemistry*, Elsevier, New York, 1963, vol. 7, p. 25.

number of functional groups on the protein that were previously present in quite a different geometry. Finally, the structure and charge of each of the binding sites may be sensitive functions of the *p*H, ionic strength, temperature, and other environmental factors.

The experimental methods for study of the binding of ions to proteins are many and varied. For example, the activity of the unbound ion may be measured; this is typically done in the case of the binding of protons since the activity measurement is a simple one requiring only a *p*H meter equipped with a glass electrode. Recently, a number of other ion-specific electrodes have become available so that the activities of sodium, potassium, copper, and other ions in aqueous solution are readily measurable. Alternatively, material-balance studies may be employed. One typical technique is *equilibrium dialysis,* in which the protein is restrained within a semipermeable membrane to which the ions of interest are permeable. At equilibrium, the concentrations of the ion within and outside the protein-containing compartment yield quantitative information concerning the number of ions bound. Gel filtration (molecular-sieve chromatography) provides a similar means of partitioning proteins from small ions and may be used for the same kind of study. Finally, one may directly observe the bound ions. These may change color, magnetic resonance properties, and the like, or they may change some property of the protein which can be observed. Most of the methods of spectroscopy have been applied to one case or another within this area of study.

The effect of ion concentration on binding by proteins is subject to quantitative treatment. Consider a protein P that has n binding sites for an ion A. We may then write n equilibrium constants:

$$\begin{aligned} K_1 &= \frac{(PA)}{(P)(A)} \\ K_2 &= \frac{(PA_2)}{(PA)(A)} \\ &\vdots \\ K_n &= \frac{(PA_n)}{(PA_{n-1})(A)} \end{aligned} \tag{3-46}$$

Making the further assumption that each binding site is uninfluenced by the state of occupancy of the other sites (i.e., interactions between the sites do not occur) and that the intrinsic affinity of each site for A is identical, it is then easy to show that

$$K_1 = nK \tag{3-47}$$

in which K is the intrinsic equilibrium constant for each site. Equation (3-47) may also be arrived at intuitively since, with all sites unoccupied, the equilibrium constant for binding the first A must equal the intrinsic constant times the number of available sites. Equation (3-47) may be generalized to yield all the equilibrium constants of Eq. (3-46):

$$K_i = \frac{n - i + 1}{i} K \tag{3-48}$$

Employing Eq. (3-48), a relationship between (A) and r, the ratio of moles of A bound to total moles of protein, may be derived:

$$r = \frac{\text{moles } A \text{ bound}}{\text{moles total protein}} \tag{3-49}$$

We obtain

$$\frac{r}{(A)} = Kn - Kr \quad \text{or} \quad K = \frac{r}{(n-r)(A)} \tag{3-50}$$

Thus, a plot of $r/(A)$ against r yields a straight line whose intercept on the r axis is a measure of n and whose intercept on the $r/(A)$ axis is a measure of Kn. Such a plot clearly permits the evaluation of both K and n. A typical plot of this type is shown in Fig. 3-24.

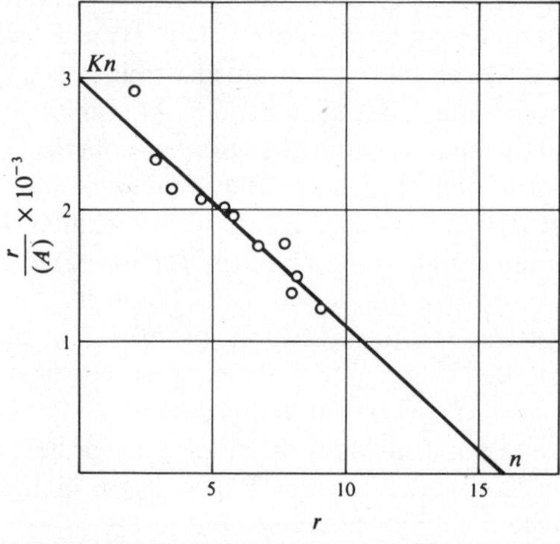

FIGURE 3-24
Type of extrapolation to determine the maximum number of small molecules (or ions) that may be bound by a protein molecule. I. Klotz, in H. Neurath and K. Bailey (eds.), *The Proteins,* vol. 1, Academic Press, New York, 1953, p. 773.

The structures of a number of metal ion–protein complexes are encountered in later chapters. These include the iron atom in myoglobin (page 150), the several atoms of iron in ferredoxin (page 657) and rubredoxin (page 657), and the zinc atom in carboxypeptidase (page 158). An analysis of the protolytic behavior of the imidazole groups of ribonuclease is also considered later (page 164). At this point let us consider the single example of the binding of anions to the enzyme acetoacetic decarboxylase.

The acetoacetic decarboxylase of *Clostridium butylicum* is subject to inhibition by anions generally (6). This inhibition involves a mobile equilibrium between the anions free in solution and bound to a single site on the protein because (1) the inhibition occurs immediately on mixing the anion and the

TABLE 3-10
Concentration of several anions required to produce 50 percent inhibition of acetoacetate decarboxylase at *p*H 5.2 and 30°

Anion	*Concentration for 50% inhibition,* M
HSO_3^-	7×10^{-5}
SCN^-	11×10^{-5}
ClO_4^-	73×10^{-5}
I^-	100×10^{-5}
NO_3^-	126×10^{-5}
ClO_3^-	400×10^{-5}
Br^-	$1{,}000 \times 10^{-5}$
Cl^-	$5{,}000 \times 10^{-5}$
BrO_3^-	$7{,}600 \times 10^{-5}$
F^-	$10{,}000 \times 10^{-5}$
IO_3^-	$10{,}000 \times 10^{-5}$
Trichloroacetate	$33{,}000 \times 10^{-5}$

SOURCE: I. Fridovich, *J. Biol. Chem.*, **238**, 592 (1963).

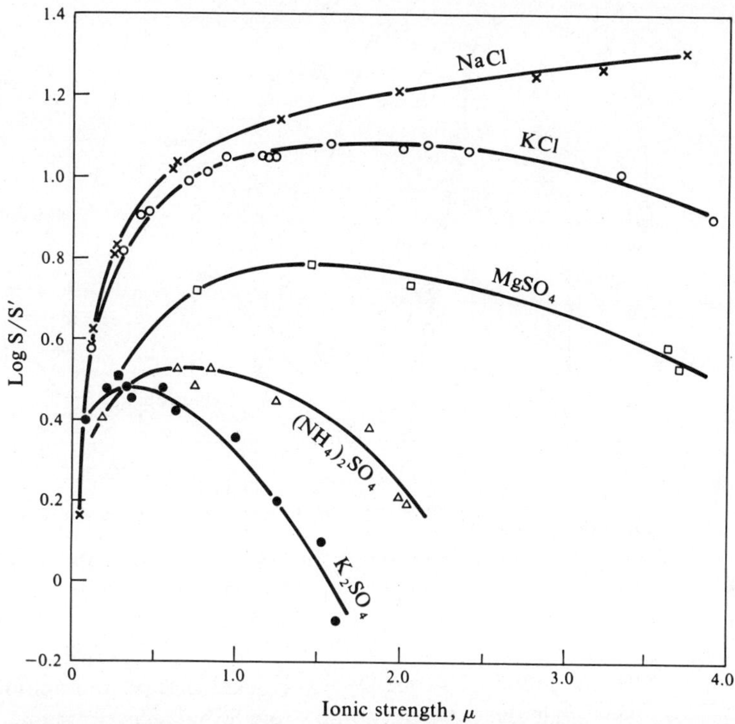

FIGURE 3-25

Solubility of isoelectric carboxyhemoglobin as a function of ionic strength and ion type. A. A. Green, *J. Biol. Chem.*, **95:** 47 (1932).

protein, (2) the inhibition is reversible upon dilution or dialysis, and (3) the anions compete with each other as inhibitors. Binding of univalent anions to the enzyme depends upon having a group with the pK_a of 5.8 in the protonated form. Thus, it seems likely that a protonated histidine is part of the binding site. In Table 3-10, the concentrations of the anions required to produce 50 percent inhibition of the enzyme are collected. If one assumes that the enzyme complexed with the protein is not active, then these concentrations are directly related to equilibrium constants for anion binding. Note that the affinity of the monovalent anions for the enzyme parallels that usually found for anion-exchange resins, micellar surfaces, and the like, as well as the tendency to ion-pair with large cations in aqueous solution.

Solubility Properties of Proteins

The solubility of most proteins is a sensitive function of ionic strength, *p*H, and the concentration of organic solvents. We consider each of these variables in turn.

Ionic Strength. In Fig. 3-25, the solubility of isoelectric carboxyhemoglobin is shown as a function of the ionic strength, μ, for several salts. Ionic strength is defined as

$$\mu = \tfrac{1}{2} \sum c_i Z_i^2 \qquad (3\text{-}51)$$

in which c_i is the concentration of the ith component and Z_i is its charge. From the data in this figure, it is clear that at low ionic strengths the protein is *salted in* and at high ionic strengths the protein is *salted out*. This behavior is typical

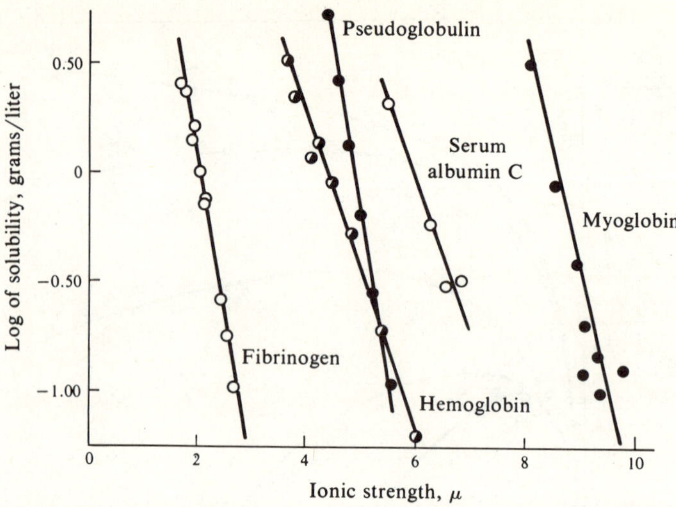

FIGURE 3-26

Solubility of proteins in ammonium sulfate solutions. E. Cohn and J. T. Edsall, *Proteins, Amino Acids, and Peptides,* Academic Press, New York, 1943.

of most proteins. From Debye–Hückel theory, the activity coefficient γ for a given ion is related to the ionic strength by the expression

$$-\log \gamma = \frac{1.81 \times 10^6}{D^{3/2}T^{3/2}} Z_1 Z_2 \sqrt{\mu} \tag{3-52}$$

in which D is the dielectric constant of the medium and Z_1 and Z_2 are the charges on the ion and its counterion. Proteins, as multivalent electrolytes, behave in a fashion qualitatively similar to that predicted for simple ions by Eq. (3-52) at low salt concentrations; that is, increasing concentrations of salt stabilize (decrease the activity coefficient of) the charged groups on the protein, thereby increasing its solubility. The phenomenon of salting out, observed at higher ionic strengths, is probably the result of competition between the protein and the salt ions for the available water molecules for purposes of solvation. Thus, at sufficiently high concentrations of salt, insufficient molecules of water are available for full solvation of the protein, and protein–protein interactions become more important than protein–water interactions; that is, precipitation occurs. In this region of salt concentration, protein solubility is often logarithmically related to ionic strength as indicated in Eq. (3-53):

$$\log S = \beta' - K_s'\mu \tag{3-53}$$

In this equation, S is solubility in grams per liter, β' is the logarithm of the hypothetical solubility (not the actual solubility) at zero ionic strength, and K_s' is termed the *salting-out constant.*

The logarithm of the solubility of several proteins that obey this relation in ammonium sulfate solution is plotted against ionic strength in Fig. 3-26. Small changes in ionic strength are seen to cause large changes in protein solubility. This fact is frequently taken advantage of for the purification of proteins (see below). K_s' is seen to depend only slightly on the nature of the protein but varies markedly with the nature of the salt (Fig. 3-25). In contrast, β' is very dependent on the nature of the protein.

*p*H. The solubilities of most proteins, when plotted against *p*H at constant ionic strength, exhibit U-shaped curves, with the minimum near the isoelectric point. Such behavior is illustrated for β-lactoglobulin in Fig. 3-27. Although the solubilities are markedly dependent on the ionic strength, the minimum occurs in each case near *p*H 5.2 to 5.4, the isoelectric point. This behavior may be rationalized on the basis that, at the isoelectric point, intermolecular electrostatic repulsions between solute molecules are at a minimum and crystal-lattice forces in the solid are at a maximum. For many proteins, the effect of *p*H on solubility is more complicated (one case even exhibits a double minimum), indicating that the naive explanation presented above is not completely adequate.

Organic Solvents. Most organic solvents are good protein precipitants. Because increasing concentrations of acetone, for example, decrease the capacity of aqueous solvents to solvate the charged groups of proteins, this phenomenon receives a ready explanation. Some organic solvents, which are characterized by relatively high dielectric constants, are, in contrast, rather good protein solvents. These include dimethyl sulfoxide, formamide, and dichloroacetic acid.

Electrophoresis

The movement of charged species under the influence of an external electric field is termed *electrophoresis*. The electrophoresis of proteins was introduced by Tiselius in 1937 and has been widely employed, for a variety of purposes, ever since. Because one of the more distinctive features of proteins is their

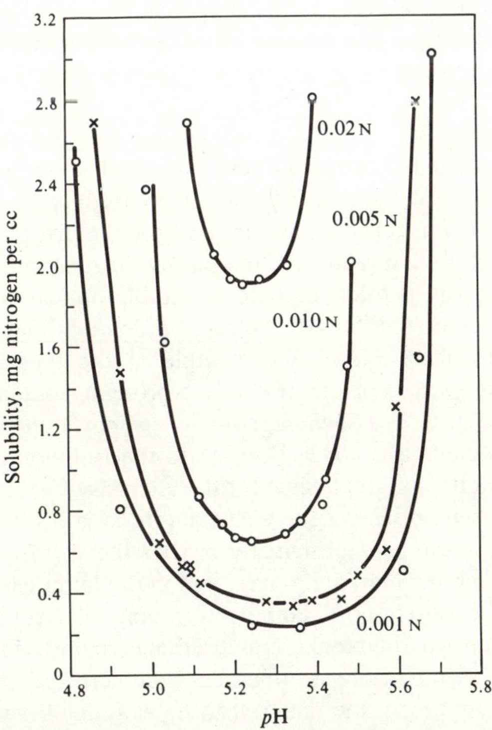

FIGURE 3-27
Solubility of β-lactoglobulin as a function of *p*H at four different concentrations of sodium chloride. S. Fox and J. S. Foster, *Introduction to Protein Chemistry*, Wiley, New York, 1957.

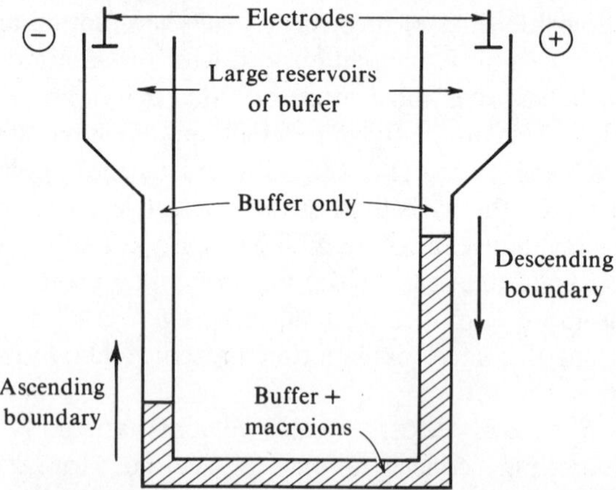

FIGURE 3-28
Schematic diagram of the Tiselius cell for measuring electrophoretic mobility by the moving-boundary method. The macroions are assumed to have a positive charge; for negatively charged ions, the electrodes would be reversed. C. Tanford, *Physical Chemistry of Macromolecules,* Wiley, New York, 1961.

difference in charge (compare the isoelectric points in Table 3-9) and because electrophoretic mobilities are principally a function of charge, electrophoresis is one of the best methods for the study of the composition of complex mixtures of proteins and for their separation. Two electrophoretic techniques are commonly employed in the study of proteins: the moving-boundary type and the zone type. We consider each of these in turn.

Moving-Boundary Electrophoresis. A schematic representation of an apparatus for moving-boundary electrophoresis experiments is shown in Fig. 3-28. The apparatus is first partially filled with the protein solution of interest, over which is carefully layered a protein-free buffer solution, forming a sharp boundary between the two. Electrodes are then immersed in the buffer solution, and the entire apparatus is placed in a carefully thermostated bath for the maintenance of constant temperature and for the dissipation of heat generated by the apparatus. The run is initiated by turning on the current, and the progress of the run is followed with a suitable optical system. If several components are present and if their electrophoretic mobilities differ, several moving boundaries are observed. As an example of the type of data obtained, the electrophoretic pattern of the mixture of proteins contained in human plasma is shown in Fig. 3-29. Because complete separation of the proteins is not achieved, this method is not well suited to the isolation of the individual components but is often of great value for the analysis of complex mixtures.

Zone Electrophoresis. Zone electrophoresis is distinguished from moving-boundary electrophoresis principally in that the former involves the use of a support on which the material moves, and in the latter case the ions move freely in solution. Typical supports employed in zone electrophoresis include paper, cellulose, starch gels or blocks, polyurethane foam, and polyacrylamide gels. A typical zone electrophoresis apparatus is depicted schematically in Fig. 3-30. In zone electrophoresis, the material to be examined is initially applied to the support in a narrow band approximately midway between the buffer reservoirs in which the electrodes are immersed. As electrophoresis proceeds, each component migrates toward the cathode or anode, depending on its charge, at its characteristic rate. Thus, at the termination of a run, each component is,

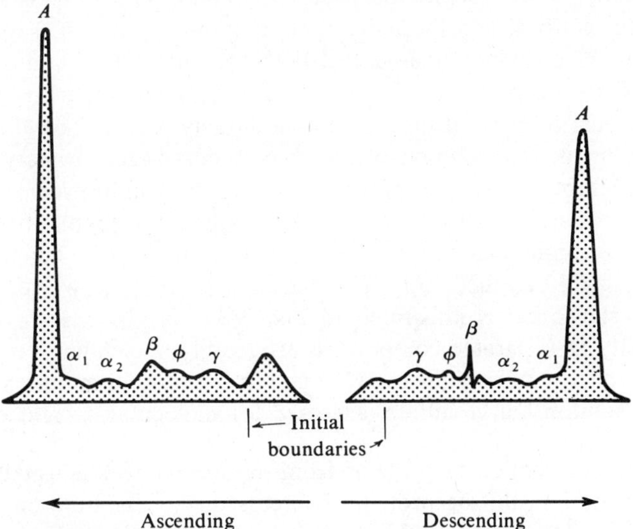

FIGURE 3-29
Electrophoretic refractive index gradient pattern for normal human blood plasma, measured at pH 8.6, where all the proteins are negatively charged; the peak labeled A represents serum albumin; α_1, α_2, β, and γ are mixtures of serum globulins, and ϕ represents fibrinogen. The two unlabeled peaks are the stationary peaks near the positions of the initial boundary. R. A. Alberty, *J. Chem. Educ.*, **25:** 619 (1948); by permission of the copyright owners, the American Chemical Society.

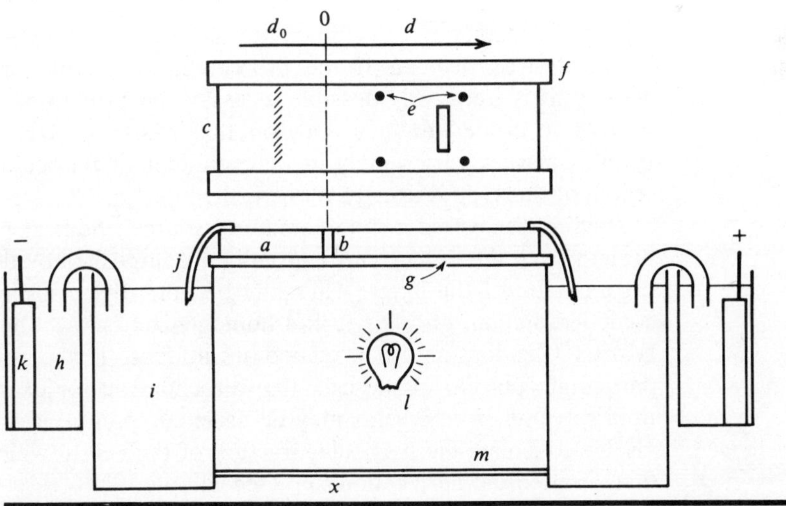

FIGURE 3-30
Block electrophoresis by moving-zone method. (a) Side view of starch block at time t_0; (b) initial zone; (c) top view of starch block after time t; (d) distance from origin to migrated zone; (d_0) distance from origin to substance with zero mobility (moved by solvent flow); (e) reference marker substances along each border; (f) strip used in molding block; (g) glass support; (h), (i) two-chamber electrode vessel; (j) connecting bridge to block; (k) electrode; (l) lamp; (m) connecting tube to equalize levels in electrode vessels. R. Trautman, in M. Florkin and E. H. Stotz (eds.), *Comprehensive Biochemistry*, vol. 7, Elsevier, New York, 1963, p. 134.

ideally, clearly separated from the others. The support may then be divided into sections, and the individual components may be separately eluted. Thus, zone electrophoresis is useful both for analysis of a multicomponent mixture and for preparation of purified proteins from such a mixture. The choice of support in zone electrophoresis is largely dictated by the purpose of the experiment. If resolution of the components is the primary objective, paper or, preferably, starch or polyacrylamide gel is employed as support. If, on the other hand, the experiment is principally for preparative purposes, a starch block might well be employed, because rather large quantities of material may be readily applied to it. The extreme resolving power of zone electrophoresis on starch gel is illustrated in Fig. 3-31 for the case of the plasma proteins. Fully 18 separated components are readily detectable. Polyacrylamide gels also have extremely high resolving power for protein mixtures and, in the presence of soduim lauryl sulfate, are used for molecular-weight determinations.

PURIFICATION OF PROTEINS

The purification of proteins from natural sources is usually a formidable task. To begin with, the protein of interest may constitute only 0.1 percent or less of the dry weight of the starting material, and much of the remainder of the material may be composed of other proteins some of whose properties probably closely resemble those of the desired product. In addition, many of the usual methods of purification of organic materials, such as distillation and solvent extraction, are not suitable for the purification of proteins, owing to their size and instability. It is difficult to imagine an organic chemist attempting to isolate a desired product from a reaction that went in only 0.1 percent yield and produced several hundred side products. This is, nevertheless, the problem frequently faced by the biochemist who desires to study the physical or biological properties of a particular protein.

The initial problem in protein purification is the choice, if indeed one has a choice, of a natural material from which isolation is to be attempted. The principal consideration in this matter is to find a source that is relatively rich in the desired protein. If the protein has a definite measurable biological activity, it is frequently possible to assay the protein in a variety of potential sources in crude mixtures and, on this basis, to select a favorable source. Because tissues vary widely in their content of particular proteins, an initial search of this type is usually very worthwhile.

Having discovered a suitable natural source, the next problem is to free the protein from the tissue. This is usually accomplished by destroying the cellular organization of the latter. This may be done by exposing the tissue to ultrasonic irradiation, grinding it in a homogenizer, subjecting it to the action of a Waring Blendor, or other related techniques. If the protein is bound to a particulate portion of the cell, the mitochondria for example, treatment with lipid solvents or detergents may be necessary to free it. In such cases it is advisable to isolate the particular fraction of the cell in which the protein occurs first, as an initial purification step (see Chapter 9).

Several precautions must be observed during the procedures necessary for release of the protein from the tissue and during the subsequent fractionation steps. These include careful control of temperature, *p*H, and protein concentration. Proteins are fragile molecules. In general, exposure to even moderate temperatures, for example 37°, elicits their slow denaturation. Consequently, most protein-purification procedures should be carried out near the freezing point of the solvent in use. On the other hand, if the protein of interest happens to be unusually stable to heat, a brief heat treatment may serve to precipitate

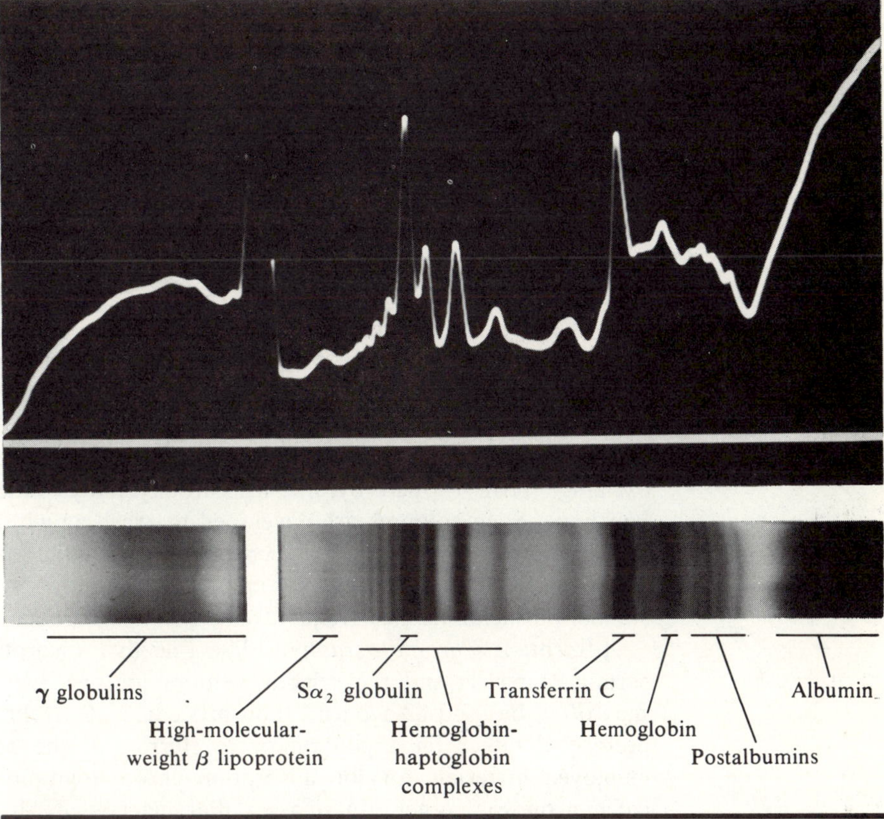

FIGURE 3-31

Possibility of quantitative serum protein analyses by starch-gel electrophoresis is illustrated in this figure. (*Bottom*) A photograph taken by transmitted light of a stained, unsliced 3-mm starch gel made transparent in glycerol; (*top*) a photograph of the corresponding trace appearing on the cathode-ray tube of an electronic analyzer. O. Smithies, *Advan. Protein Chem.*, **14:** 65 (1959).

the bulk of the contaminating material, thus effecting a substantial purification. In addition to heat, proteins are sensitive to extremes of acid and base concentration. Hence, protein purification should ordinarily be carried out in the presence of buffers that maintain the *p*H in the neighborhood of 7. Finally, protein concentration should always be maintained as high as reasonably possible, because many proteins tend to denature in dilute solution. Proteins, if they are enzymes, may often be protected by the presence of their substrates, which tend to confer additional stability on them.

Many fractionation methods are available for the purification of proteins. We only discuss a few of these.

Salt Precipitation. As indicated in Fig. 3-26, proteins vary markedly in their solubilities in concentrated salt solutions. Consequently, purification may be achieved by addition, for example, of sufficient salt to precipitate some of the contaminating proteins, which may then be collected by centrifugation and discarded, a further addition precipitating the protein of interest but not all the remaining contaminating proteins. The protein of interest, together with other proteins that may have precipitated over the same concentration range, may then be collected. The salts most frequently employed for this purpose are ammonium sulfate and sodium sulfate; the former is usually preferable.

Isoelectric Precipitation. Because, as noted above, most proteins have a minimum solubility near their isoelectric point, suitable adjustment of the *p*H may result in the precipitation of the protein of interest, most of the contaminating proteins remaining in solution.

Precipitation with Organic Solvents. A variant of the methods above involves the use of organic solvents, particularly acetone, methanol, and ethanol, in attempts to precipitate certain proteins selectively from a mixture. These solvents often easily denature proteins, and the use of very low temperatures is advisable.

Ion-Exchange Chromatography. Proteins, like amino acids, may frequently be purified by column chromatography on ion-exchange resins. Resins with hydrophilic backbones, usually cellulose, are the most satisfactory for this purpose. Two commonly used column materials for protein purification include DEAE-cellulose, an anion-exchange resin formed by linking diethylaminoethyl functions to a cellulose backbone, and CM-cellulose, a cation-exchange resin formed by linking carboxymethyl functions to a cellulose backbone. Such columns are developed by washing with buffers of increasing ionic strength or increasing or decreasing values of *p*H, as appropriate. Properly employed, the resolving power of these columns is great, and very substantial purifications may frequently be obtained through their use.

Adsorption Chromatography. The selective adsorption of proteins onto certain materials and the selective elution of proteins following adsorption may often be used as a purification procedure. This chromatography may be performed either on a column or in slurries of the adsorbant. Frequently employed materials for the adsorption chromatography of proteins include calcium phosphate gel, alumina gel, diatomaceous earth (Celite), starch, and hydroxylapatite.

Electrophoresis. As mentioned above, electrophoresis of protein mixtures by the zone method employing a high-capacity support, usually starch blocks, may effect excellent separations and permits near quantitative recovery of the protein of interest. Thus, zone electrophoresis is a good purification procedure.

Molecular Sieves. The use of molecular sieves, which sort out molecules according to size, may be adapted to protein purification. Recently, diethylaminoethyl and carboxymethyl functions have been linked to Sephadex, so that a single run through a column may fractionate both on the basis of size and charge.

Crystallization. Provided that the use of other procedures has yielded a reasonably pure protein preparation, crystallization may sometimes be achieved. Repeated recrystallizations usually make significant contributions to the overall purity of the product. Proteins exhibit wide variations in their ease of crystallizability. Most albumins and globulins can be crystallized; in contrast, no glutelin has ever been crystallized. Conditions for the crystallization of a given protein must be arrived at empirically. Careful control of *p*H, ionic strength, protein concentration, and temperature is essential.

The purification of proteins is frequently more an art than a science. The selection of methods for a particular case must be arrived at by trial and error. The evaluation of a particular procedure for inclusion in an overall purification scheme rests on two criteria: the degree of purification and the yield. Clearly a procedure that results in an overall purification of 100-fold but yields only 1 percent of the product is of little use. The degree of purification is calculated by comparing the specific activities of the protein under study before and after each step. The specific activity is defined as the ratio of a convenient measure of the protein, enzymatic activity for example, to the total

TABLE 3-11
Summary of a hypothetical protein-purification procedure

Preparation	*Volume, ml*	*Protein, mg/ml*	*Activity, arbitrary units/ml*	*Specific activity, units/ mg protein*	*Yield, percent (overall)*	*Fold purification (overall)*
Crude extract	200	2.5	100	40	100	1
$(NH_4)_2SO_4$ precipitation	80	1.25	200	160	80	4
DEAE-cellulose chromatography	20	1.0	500	500	50	12.5
Preparative electrophoresis	5	1.0	1,500	1,500	37.5	37.5
Isoelectric precipitation	3	0.3	2,000	6,700	30	167
Crystallization	1	0.5	5,000	10,000	25	250

amount of protein present. In Table 3-11, a hypothetical purification procedure is presented together with the specific activities, yields, and degrees of purification calculated for each step. A careful study of this table should result in a thorough understanding of the fashion in which the various quantities are derived.

The purification of proteins employs several auxiliary procedures in addition to the fractionation methods discussed above. Two of these, dialysis and lyophilization, are of particular importance. The removal of salt from a protein solution by dialysis, the equilibration of a protein solution with a buffer solution across a protein-impermeable membrane, has been employed for many years. This procedure is of obvious importance, following, for example, a step involving protein precipitation with ammonium sulfate. Alternatively, desalting may be achieved by passing the protein solution through a suitable molecular sieve.

Lyophilization, freeze drying, is an important method of concentrating protein solutions or reducing them to dryness. In this procedure, the protein solution is frozen in an acetone–Dry Ice bath and connected to a vacuum line. The rate of cooling due to evaporation of solvent from the frozen solution is sufficient to maintain the solution in the frozen state throughout the course of the concentration. Most proteins are not denatured by this procedure and may be stored in the dry state, in the cold, for long periods of time without deterioration of the sample. Frozen solutions are, in general, less stable. Owing to the low surface tension of protein solutions, they may not be concentrated by evaporation under vacuum in the liquid state. Pumping on a protein solution results in extensive foaming, so that most of the protein ends up in the pump oil; furthermore, proteins are denatured on surfaces, and hence foaming must be avoided.

CRITERIA OF PURITY

One of the most difficult questions that arises in the course of purification of a protein is when to stop. That is, how does one know when a protein preparation is homogeneous? Several criteria are employed to answer this question. These are briefly discussed below.

Crystallinity. Small organic compounds that have been crystallized and recrystallized a few times may ordinarily be considered to be pure. Although this generalization was originally thought to apply to proteins as well, numerous examples of heterogeneous crystalline protein preparations have been identified. Consequently, although the crystallization of a protein is certainly a happy event, it cannot be considered an adequate criterion of purity of the preparation.

Solubility Curves. One of the more satisfactory criteria for protein purity is the nature of its solubility curves. Solubility curves are constructed by carefully measuring the amount of protein in solution as a function of the amount of solid protein added to the solution. The ionic strength, temperature, and volume of the solution must, of course, be kept constant, and complete equilibration between the solid and solute phases must be reached. For a pure preparation, a plot of amount of protein in solution against amount of solid protein added yields, below the saturation point, a straight line of finite slope, a sharp break at the saturation point, and, above the saturation point, a line of slope zero. Solubility curves for a pure and an impure preparation of the proteolytic enzyme pepsin are shown in Fig. 3-32. The presence of impurities may alter the shape of the solubility curves from the ideal in a number of ways, depending on whether the impurity is more or less soluble than the protein and on its relative abundance. Even if this test is carefully applied, a 5 percent impurity can be overlooked.

Ultracentrifuge Studies. The shape, width, and rate of spreading of the protein peak as it sediments in the ultracentrifuge can provide evidence concerning protein purity. Impurities are detected as distinct peaks, shoulders on the main peak, asymmetry of the main peak, or a peak width greater than that which may be accounted for in terms of the diffusion constant of the protein. Such studies are best carried out at a variety of values of *p*H and ionic strength.

Electrophoretic Studies. Homogeneity on electrophoresis, particularly employing starch or polyacrylamide gels as the supporting medium, at several values of *p*H is a good criterion of purity. As above, one should observe only a single symmetrical peak whose width is consistent with the diffusion constant.

Comparison of Molecular-Weight Determinations. Independent determinations of the molecular weight by methods that yield number-average molecular weights for heterogeneous systems, such as osmotic pressure, and those which yield weight-average molecular weights for the same systems, such as light scattering, may be employed as a criterion of purity. Clearly, if the preparation is heterogeneous, distinct molecular weights are likely to be obtained for the two determinations.

Analytical Data. Amino acid composition data may be employed to detect impurities in protein preparations. For example, suppose such an

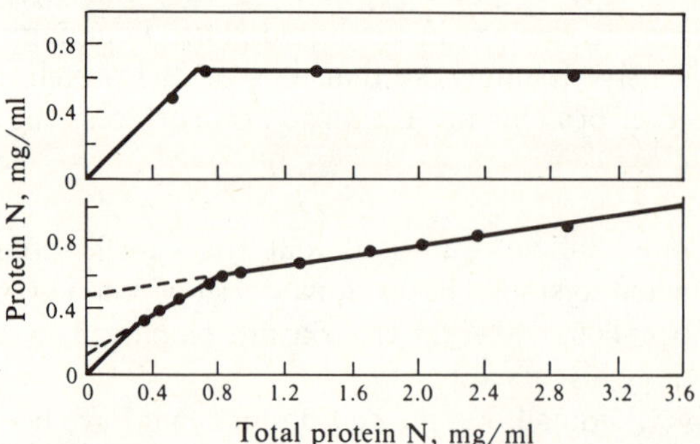

FIGURE 3-32
Solubility diagrams of pure A pepsin (*upper*) and of crystalline pepsin prepared from Cudahy pepsin (*lower*) in half-saturated magnesium sulfate in 0.05 M acetate buffer, *p*H 4.6, 22°. Ordinates: protein N in solution; abscissas: total protein N in the mixture. J. H. Northrop, M. Kunitz, and R. M. Herriott, *Crystalline Enzymes,* Columbia University Press, New York, 2nd ed., 1948.

analysis revealed that a particular protein contains 0.3 mol of histidine per mole of protein. One is entitled to conclude that the protein probably contains no histidine but that an impurity in the preparation does.

Detection of Impurities by Functional Tests. Minute amounts of impurities may frequently be detected if they have specific biological properties, such as enzymatic or hormonal activity. If a sample of protein is found to contain several such activities and if they are all due to a single protein, the ratio of their activities should remain unchanged throughout the course of the purification. If such ratios change as the enzyme is fractionated, the individual functional activities must be considered to belong to distinct proteins, and the protein of interest cannot be considered pure until the extraneous activities have been eliminated.

Homogeneity on Chromatography in Several Solvent Systems and with a Variety of Supports. None of the criteria of purity above are, in themselves, adequate to establish the homogeneity of protein preparations. Careful work requires that several independent criteria be employed and that, even if each provides a satisfactory result, carefully qualified conclusions be drawn.

REFERENCES

Books and Review Articles

Ackers, G. K.: Analytical Gel Chromatography of Proteins, *Advan. Protein Chem.*, **24:** 343 (1970).

Creeth, J. M., and R. H. Pain: The Determination of Molecular Weights of Biological Macromolecules by Ultracentrifuge Methods, *Progr. Biophys. Mol. Biol.*, **17:** 217 (1967).

Dickerson, R. E., and I. Geis: *The Structure and Action of Proteins,* Harper & Row, New York, 1969.

Edsall, J. T., and J. Wyman: *Biophysical Chemistry,* vol. 1, Academic Press, New York, 1958.

Florkin, M., and E. H. Stotz (eds.): *Comprehensive Biochemistry,* vols. 2, 3, 7, and 8, Elsevier, New York, 1963.

Fowden, L., D. Lewis, and H. Tristram: Toxic Amino Acids: Their Action as Antimetabolites, *Advan. Enzymol.*, **29:** 89 (1967).

Freeman, H. C.: Crystal Structures and Metal–Peptide Complexes, *Advan. Protein Chem.*, **22:** 257 (1967).

Gurd, F. R. N.: Binding of Protons and Other Ions, in S. J. Leach (ed.), *Physical Principles and Techniques of Protein Chemistry,* part B, Academic Press, New York, 1970.

Hill, R. L.: Hydrolysis of Proteins, *Advan. Protein Chem.*, **20:** 37 (1965).

Merrifield, R. B.: Solid-Phase Peptide Synthesis, *Advan. Enzymol.*, **32:** 221 (1969).

Meister, A.: *Biochemistry of the Amino Acids,* Academic Press, New York, 1965.

Schachman, H. K.: The Ultracentrifuge: Problems and Prospects, *Biochemistry,* **2:** 887 (1963).

Tanford, C.: *Physical Chemistry of Macromolecules,* Wiley, New York, 1961.

Tristram, G. R., and R. H. Smith: Amino Acid Composition of Proteins, *Advan. Protein Chem.*, **18:** 227 (1963).

Waley, S. G.: Naturally Occurring Peptides, *Advan. Protein Chem.*, **21:** 2 (1966).

Specific References

1. Richards, F. F., R. W. Sloane, Jr., and E. Haber: *Biochemistry,* **6:** 476 (1967).
2. Stewart, J. M., and D. W. Woolley: *Nature,* **206:** 619 (1965); **207:** 1160 (1965).
3. Spragg, J., E. Schröder, J. M. Stewart, K. F. Austen, and E. Haber: *Biochemistry,* **6:** 3933 (1967).
4. Timasheff, S. N., and R. Townend: *J. Am. Chem. Soc.*, **83:** 464 (1961); T. F. Kumosinski and S. N. Timasheff: *J. Am. Chem. Soc.*, **88:** 5635 (1966).
5. Ackers, G. K., and T. E. Thompson: *Proc. Natl. Acad. Sci. U.S.*, **53:** 342 (1965).
6. Fridovich, I.: *J. Biol. Chem.*, **238:** 592 (1963).

4
Structural Organization of Proteins

In Chapter 3 we outlined certain basic features of protein structure. These include structures for the constituent amino acids and the nature of the covalent linkage, the peptide bond, through which they are linked in linear arrays. We now want to consider some of the structural characteristics of proteins in more detail, together with a description of the physical and chemical methods employed for their elucidation.

ORGANIZATIONAL LEVELS OF PROTEIN STRUCTURE

The complexity of the detailed structure of proteins is inherent in their size. Discussions of this topic are rendered a good deal more manageable, and perhaps more meaningful, by considering several levels of structural organization. These are termed *primary, secondary, tertiary,* and *quaternary structure.* The distinctions among these levels of organization may be made in terms of the nature of the interactions necessary for their maintenance. The primary structure of a protein is defined by its covalent bonding, that is, the number and sequence of amino acid residues linked by peptide bonds. Thus the primary structure corresponds to the usual structural formula written for an organic compound. For small organic compounds, which have a relatively small number of conformations available to them, such structural formulas tell most of the story, because the spatial relationships between atoms are approximately defined by such formulas. Clearly, the situation is hardly so simple for molecules composed of one or more long flexible chains. The number of possible conformations is staggering, and the specification of higher orders of structural organization is made obligatory.

Ordering of the peptide chains resulting from the formation of hydrogen bonds between the carbonyl oxygen and amide nitrogen atoms of the polypeptide backbone

$$\rangle C{=}O \cdots H{-}N\langle$$

generates secondary structure. Formation of such hydrogen bonds may impose a variety of conformations on the polypeptide. These fall into two broad classes: *helical structures* and *sheet structures.* These are discussed below.

It is clear that, for example, a long helical segment of a polypeptide might itself assume a variety of conformations, owing to interactions between the side chains of the various amino acid residues. Such interactions might include hydrogen-bond formation, van der Waals interactions, charge-transfer forces, salt linkages, and so forth. Protein structure that results from interactions between the residue side chains is termed tertiary structure.

Finally, intermolecular interactions between protein molecules—themselves folded or superfolded—can yield aggregates of individual molecules, and define quaternary structure. Aggregate formation depends on surface interactions between residue side chains as well as between exposed portions of the peptide backbone of separate monomeric polypeptides. These organizational levels of protein structure will now be considered sequentially.

PRIMARY STRUCTURE OF PROTEINS

In 1955, the elegant work of F. Sanger and his associates at Cambridge culminated in the determination of the complete primary structure of the polypeptide hormone insulin, molecular weight 5700. This signal achievement in the history of protein chemistry, which certainly stands as one of the most important of all structure elucidations, was rewarded with the Nobel Prize in chemistry in 1957. In the course of his studies, Sanger developed several important concepts and techniques for dealing with the problems involved in primary structure determinations for macromolecules. These developments were furthered by the work of Hirs, Stein, and Moore at The Rockefeller Institute (now University) and Anfinsen at the National Institutes of Health, who determined the amino acid sequence of ribonuclease, molecular weight 13,700, in 1960 (1,2). These two seminal studies have served to open the floodgates for success in this area; an encyclopedic collection now houses the known data concerning amino acid sequences of proteins and the collection grows rapidly. As we shall see, this work has not only had the most important consequences for protein chemistry itself but also has yielded insight into a variety of related matters in the fields of biochemistry and biology and the interface between them.

Elucidation of primary structure is a multistep process, consisting basically of (1) determination of the number of independent polypeptide chains of the protein, (2) isolation of the individual polypeptide chains, (3) specific cleavage of each of the polypeptide chains of the native molecule into smaller chains of manageable size, (4) sequence determination for each of the cleavage products, (5) fitting the individual segments into a unique sequence for each of the native chains, and (6) identifying the sites of linkage holding individual peptide chains together. Thus the molecule is degraded in an orderly fashion, the structure of the degradation products determined, and, employing this information, the structure of the native molecule deduced. Preliminary to performing this sequence of operations, the molecular weight and amino acid composition for the protein under consideration are usually determined by the methods outlined in Chapter 3.

Determination of N- and C-Terminal Amino Acids

Each peptide chain of a protein molecule, provided the chain is not circular or blocked at either of the ends, has an N-terminal amino acid, bearing a free α-amino group, and a C-terminal amino acid, bearing a free α-carboxyl group. Thus, the number of polypeptide chains may be determined from analysis of the number of such residues. If a protein is found to contain one alanine, one lysine, and one serine in N-terminal positions per mole of protein, the conclusion that three peptide chains are present is warranted.

The N-terminal amino acids are usually determined by one of four methods. First, the polypeptide may be arylated by treatment with 1-fluoro-2,4-

dinitrobenzene [Eq. 3-3] yielding the 2,4-dinitrophenyl (DNP) derivative of the N-terminal residue. Complete hydrolysis of the protein, extraction of the DNP-amino acid into organic solvents, and chromatographic or other determination of the amino acid involved reveals the nature of the N terminus. Of course, other groups of the protein, for example the ε-amino group of lysine, also react with this reagent, but only one α-amino group, that of the N-terminal residue, is so modified. This method was developed by Sanger, and, in consequence, 1-fluoro-2,4-dinitrobenzene is frequently referred to as the *Sanger reagent*.

The second important reagent employed for determination of N-terminal residues is phenylisothiocyanate (the *Edman reagent*). This reagent reacts with the free α-amino groups in dilute alkali to yield the phenylthiocarbamyl peptide [Eq. 3-4]. Treatment of this product with acid yields the phenylthiohydantoin (PTH) of the N-terminal amino acid and forms a new N-terminal amino acid (see Fig. 4-1 for details):

$$\phi\text{—NH—C(=S)—NH—CHR}_1\text{—C(=O)—NH—R}_2 \xrightarrow{H^+} \underset{\text{(PTH: ring } \phi\text{—N—C(=S)—NH—CH(R}_1\text{)—C(=O)—)}}{\text{phenylthiohydantoin}} + H_2N\text{—}R_2 \qquad (4\text{-}1)$$

Phenylisothiocyanate provides a versatile and powerful tool for determination of amino acid sequences of proteins. Two advantages derive from the fact that the peptide minus the original N terminus may be recovered intact. First, recovery of the product peptide permits identification of the N terminus by a "subtractive" method. Thus, determination of the amino acid composition of the product peptide and comparison with the amino acid composition of the original will reveal the nature of the missing amino acid—the N terminus. Alternatively, the PTH-amino acid may be identified directly, most readily by gas–liquid chromatography (3,4). Both methods are in active current use.

Second, recovery of the peptide minus the original N-terminal amino acid permits the reaction cycle to be repeated, yielding the identity of the new N terminus and a product peptide that may again be subjected to analysis, and so forth. Thus, careful work will permit the experimenter to work his way down the polypeptide chain sequentially, identifying the amino acids as he goes (Fig. 4-1). In most instances, sequences of not more than 10 amino acids have been identified in this way. Very careful work has permitted identification of sequences 20 amino acids long in some cases (5–7); the most impressive example of the manual use of this technique is the complete elucidation of the structure of a peptide hormone consisting of a single chain of 32 amino acids (8).

The importance of phenylisothiocyanate for protein chemistry evoked a series of studies, principally by Edman, concerning the basic chemistry of the reaction steps. Methods resulted for improving the efficiency with which each cycle of reactions can be repeated. This work has culminated in the development of an automated Edman degradation procedure (9,10). In fact, an instrument capable of carrying out repeated Edman degradations without any manual manipulations at all is currently available commercially. In his own laboratory, Edman has succeeded in using an automated system for a remarkable 60 cycles of degradation on the myoglobin from the humpback whale (9). Although problems still exist with the use of the automated system, it

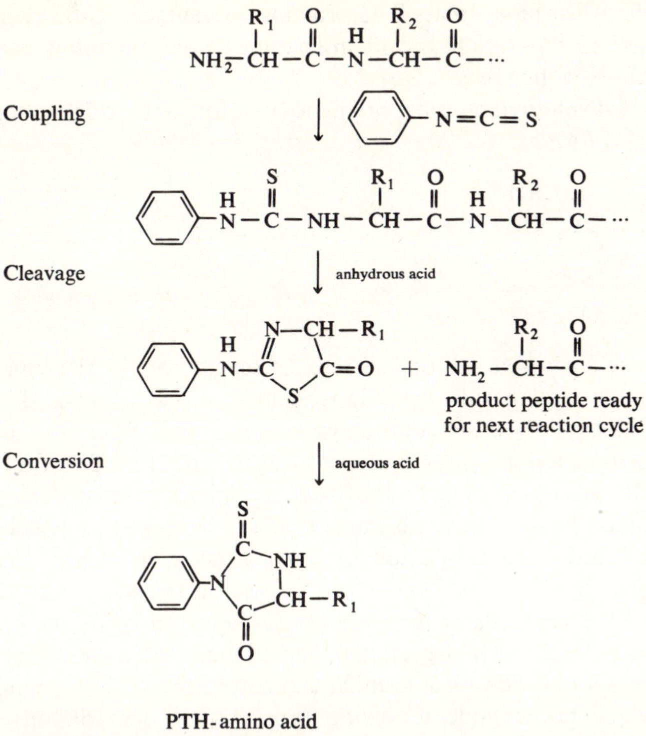

$$NH_2-CH(R_1)-C(=O)-NH-CH(R_2)-C(=O)-\cdots$$

Coupling $\downarrow$ $C_6H_5-N{=}C{=}S$

$$C_6H_5-NH-C(=S)-NH-CH(R_1)-C(=O)-NH-CH(R_2)-C(=O)-\cdots$$

Cleavage $\downarrow$ anhydrous acid

$$\text{anilinothiazolinone (}C_6H_5-NH-C\text{ ring: N=C, N—CH—}R_1\text{, C=O, S)} \;+\; NH_2-CH(R_2)-C(=O)-\cdots$$

product peptide ready for next reaction cycle

Conversion $\downarrow$ aqueous acid

$$\text{PTH ring: } C_6H_5-N<\;C(=S)-NH-CH(R_1)-C(=O)\;>$$

PTH-amino acid

FIGURE 4-1

Three-stage Edman degradation. The phenylthiocarbamyl derivative of the initial peptide is converted into the PTH-amino acid plus a new peptide in a two-step sequence involving the anilinothiazolinone as intermediate.

seems certain that the next few years will see widespread use of these techniques coupled with on-line detection systems. Direct printout of the sequence information obtained can then follow. Mass-spectrometric techniques mentioned below should complement the automated Edman degradation techniques.

A third chemical reagent well suited to determination of N-terminal residues is potassium cyanate (11). Treatment of peptides with this reagent in slightly alkaline solution results in carbamylation of amino functions:

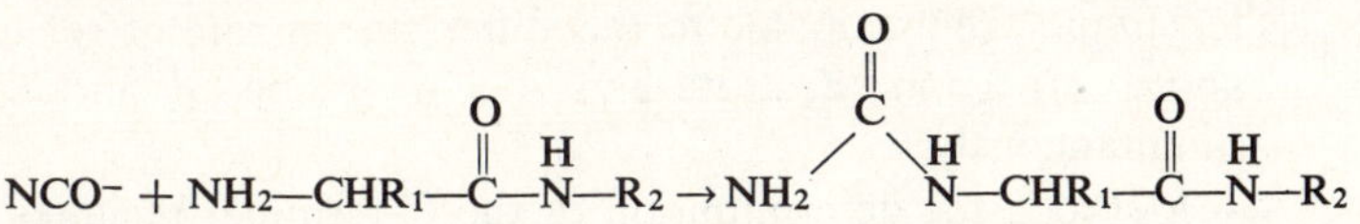

$$NCO^- + NH_2-CHR_1-\overset{\overset{O}{\|}}{C}-\overset{H}{N}-R_2 \rightarrow NH_2-\overset{\overset{O}{\|}}{C}-\overset{H}{N}-CHR_1-\overset{\overset{O}{\|}}{C}-\overset{H}{N}-R_2 \tag{4-2}$$

Exposure of the carbamylated peptides to dilute acid at elevated temperatures results in cleavage of the N-terminal residue from the chain with its cyclization to the hydantoin:

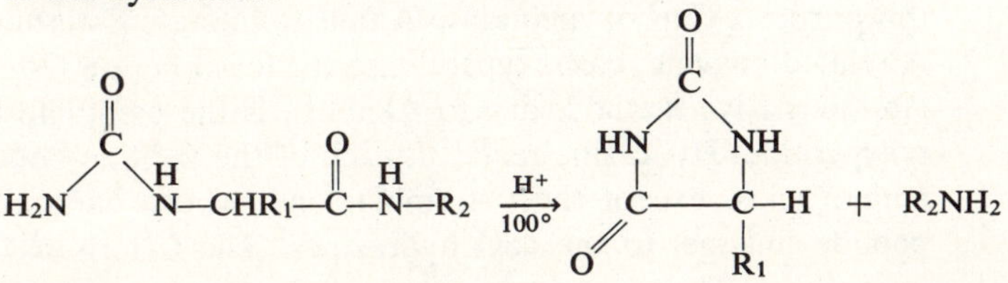

$$H_2N-\overset{\overset{O}{\|}}{C}-\overset{H}{N}-CHR_1-\overset{\overset{O}{\|}}{C}-\overset{H}{N}-R_2 \xrightarrow[100^\circ]{H^+} \text{hydantoin (HN—C(=O)—NH, C(=O)—C(H)(}R_1\text{))} + R_2NH_2 \tag{4-3}$$

These reactions are clearly closely related to those involved in the use of phenylisothiocyanate. The principal advantage of the cyanate procedure is the ease of separation and identification of the hydantoin by chromatography on ion-exchange resins.

1-Dimethylaminonaphthalene-5-sulfonyl chloride, commonly known as *dansyl chloride,*

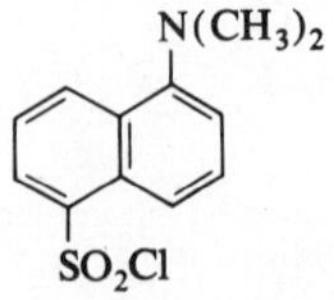

$N(CH_3)_2$

SO_2Cl

provides a fourth potent chemical means for the determination of N-terminal residues. This sulfonylating reagent reacts with exposed amino groups rapidly under mildly basic conditions, yielding derivatives that are markedly resistant to destruction under the usual conditions of acid hydrolysis of polypeptides. The principal advantage of the use of this reagent derives from the fact that the derivatives have an intense yellow fluorescence which permits their detection in exceedingly small amounts. Indeed, the dansyl chloride method is the most sensitive one currently available and it finds much use as a result.

One particularly nice use of the dansyl chloride method is its combination with the Edman degradation technique. One application involves the use of dansyl chloride to determine the nature of the N terminus of the newly generated polypeptide following one cycle of degradation: the great sensitivity of the method permits the use of a very small aliquot of this product. Alternatively, the phenylthiohydantoin of the liberated amino acid can be converted to the dansyl derivative for confirmation of the structural assignment.

Finally, use of the *exopeptidase,* leucine aminopeptidase, is applicable to N-terminal amino acid determinations. This enzyme requires the presence of a free α-amino function for activity and hence splits only the N terminus from the polypeptide chain. As with the Edman reagent, cleavage of the N-terminal amino acid from the chain results in the unmasking of a new such residue. By following the kinetics of the release of amino acids from a polypeptide, information as to their sequence may be obtained. For example, for a peptide of structure $H_2N—A \cdot B \cdot C \cdot D \cdots$, *A* would be liberated more rapidly than *B,* which is liberated more rapidly than *C,* and so forth. Leucine aminopeptidase, as the name implies, cleaves N-terminal leucine residues most rapidly but splits off all amino acids (with the exception of proline), although many only slowly. Most of the difficulties and ambiguities experienced in the use of this technique are ascribable to this difference in rate of release; for example, if *B* above is a good substrate and *A* a poor one, *A* and *B* are released almost simultaneously.

Methods for determination of the C-terminal residues are less satisfactory. In a fashion quite analogous to that described for leucine aminopeptidase, the enzyme carboxypeptidase A (actually a class of closely related enzymes), which requires a free α-carboxyl function for activity, may be employed for this purpose. Carboxypeptidase A fails to liberate C-terminal Arg, Lys, or Pro. A related enzyme, carboxypeptidase B, does liberate C-terminal Arg and Lys. An alternative method, due to Akabori, is the complete hydrazinolysis of the polypeptide. Hydrazine reacts at each of the peptide bonds, converting all the amino acids, except the C-terminal one, whose carboxyl is not involved in peptide linkage, to the acyl hydrazines. The C-terminal amino acid may be separated and identified chromatographically. A less widely employed method involves reduction of the C-terminal carboxyl to the corresponding alcohol through the use of appropriate hydride ion donors. Complete hydrolysis then yields the amino alcohol of the C-terminal residue, which may be isolated and identified.

Cleavage of Disulfide Bridges

Independent polypeptide chains comprising a protein molecule are frequently held together by the disulfide bridges of cystine residues. For proteins composed of a single chain, these bridges may bring together two quite distant points in terms of linear amino acid sequence. Hydrolysis of chains containing such disulfide linkages may lead to complex mixtures of peptides. In addition, the disulfide bridges are unstable to most hydrolytic conditions and, furthermore, readily undergo a number of rearrangement reactions. Consequently, the disulfide bridges are usually destroyed prior to the initiation of sequence studies.

Two general methods are employed: oxidation and reduction. Oxidation of disulfides with performic acid, a technique employed in studies on both insulin and ribonuclease, yields the corresponding cysteic acid residues:

$$\begin{array}{c} | \\ S \\ | \\ S \\ | \end{array} \xrightarrow{H-\overset{O}{\overset{\|}{C}}-O_2H} \begin{array}{c} | \\ SO_3^- \\ \\ SO_3^- \\ | \end{array} \tag{4-4}$$

These residues are quite stable to hydrolytic and other procedures, and their polar character may be employed to advantage for separation of peptides. Performic acid also converts methionine to the corresponding sulfone, a derivative happily less susceptible to air oxidation than methionine itself. The principal disadvantages in the use of performic acid are the destruction of tryptophan and the less than quantitative oxidation of the disulfide bridges.

Treatment of 1 mol of disulfide with sulfite yields 1 mol of sulfhydryl group and 1 mol of *S*-sulfonate:

$$\begin{array}{c} | \\ S \\ | \\ S \\ | \end{array} \xrightarrow{SO_3^{2-}} \begin{array}{c} | \\ SH \\ \\ SSO_3^- \\ | \end{array} \tag{4-5}$$

This reagent does not destroy (nor in the case of methionine, protect) any other amino acids. Its principal disadvantages are the necessity for blocking the reactive sulfhydryl group and the instability of the S—SO_3 function under acidic conditions.

Disulfides may be reduced to the sulfhydryl form by a variety of reagents, including mercaptoethanol, dithiothreitol, and sodium borohydride:

$$\begin{array}{c} | \\ S \\ | \\ S \\ | \end{array} + 2HSCH_2CH_2OH \rightarrow \begin{array}{c} | \\ SH \\ \\ SH \\ | \end{array} + \begin{array}{l} S-CH_2CH_2OH \\ | \\ S-CH_2CH_2OH \end{array} \tag{4-6}$$

Mercaptoethanol and dithiothreitol are the most suitable reagents for this purpose. Because the liberated sulfhydryl groups are quite reactive and susceptible to reoxidation to the disulfide, they are generally blocked before one proceeds further in the degradation of the peptide chain. Early studies employed iodoacetamide or iodoacetate, which alkylate the sulfhydryl groups. More recent experience indicates that acrylonitrile is a better reagent for this purpose. Cyanoethylation of protein sulfhydryl groups with acrylonitrile is quantitative and proceeds without detectable side reactions:

$$R—SH + CH_2{=}CH—CN \rightarrow R—S—CH_2CH_2CN \quad (4\text{-}7)$$

The number of chains and disulfide bridges and nature of the N terminus for several proteins are collected in Table 4-1.

Recapitulating up to this point: Methods for ascertaining the amino acid composition, molecular weight, number of peptide chains, and nature of the terminal residues for each chain of a protein have been described. In addition, methods for the destruction of intra- and interchain disulfide bridges and their conversion to stable derivatives have been indicated. Having performed these operations, one is in a position to proceed with degradation of the peptide chains. If more than one chain is involved, these, of course, must be separated chromatographically and each attacked separately.

Specific Cleavage of Polypeptides

Amino acid sequence determination of a long polypeptide requires that the polypeptide be specifically cleaved at a limited number of sites, yielding products containing several amino acids. Specificity of cleavage is important, in fact, crucial, to ensure that a limited number of peptides is obtained in reproducible fashion, each in sufficient quantity to permit its further investigation. For the specific cleavage of peptide chains, the proteolytic enzymes trypsin, chymotrypsin, and pepsin are frequently employed.

TABLE 4-1
Amino-terminal groups, disulfides, and number of chains of some proteins

Protein	*Source*	*Mol. wt.*	*Amino-terminal groups*	*No. of S—S bonds*	*No. of chains*
Insulin[a]	Bovine, porcine, etc., pancreas	5,700	1 Phe, 1 Gly	3	2
Ribonuclease T1[a]	*Aspergillus oryzae*	11,100	1 Ala	2	1
Cytochrome *c*[a]	Vertebrates, heart	12,400	Acetyl-Gly	0	1
Ribonuclease[a]	Bovine pancreas	13,700	1 Lys	4	1
Lysozyme[a]	Chicken egg white	14,400	1 Lys	4	1
Myoglobin[a]	Human muscle	17,000	1 Gly	0	1
Myoglobin[a]	Whale muscle	17,000	1 Val	0	1
Papain[a]	Papaya	21,000	1 Ile	3	1
Trypsinogen[a]	Bovine pancreas	24,400	1 Val	6	1
Trypsin[a]	Bovine pancreas	24,000	1 Ile	6	1
Chymotrypsinogen[a]	Bovine pancreas	24,700	1 Half-Cys	5	1
α-Chymotrypsin[a]	Bovine pancreas	24,300	1 Half-Cys, 1 Ala, 1 Ile	5	3
Subtilisin[a]	*Bacillus subtilis*	27,500	1 Ala	0	1
Serum albumin	Human serum	66,500	1 Asp	17	1
Enolase	Rabbit muscle	67,000	2 Acetyl-Ala	0	2
Hemoglobin[a]	Human erythrocytes	65,000	4 Val	0	4
Glyceraldehyde-3-phosphate dehydrogenase	Rabbit muscle	140,000	4 Val	0	4
Lactic acid dehydrogenase	Bovine muscle	140,000	?	0	4
Alcohol dehydrogenase	Yeast	140,000	?	0	4
Alcohol dehydrogenase[a]	Equine liver	75,000		0	2
γ-Globulin	Human serum	160,000		17	4

[a] Amino acid sequences of these proteins have been determined.

SOURCE: A. White, P. Handler, and E. L. Smith, *Principles of Biochemistry,* McGraw-Hill, New York, 4th ed., 1968.

Each of these enzymes exhibits specificity for the type of peptide bonds it attacks. Trypsin catalyzes the hydrolysis of peptide bonds whose carbonyl function is donated by a basic amino acid, usually arginine or lysine. Pepsin and chymotrypsin prefer to catalyze the hydrolysis of peptide linkages involving the aromatic amino acids, particularly tryptophan, tyrosine, and phenylalanine. Some aspects of the specificity of these and other proteolytic enzymes are collected in Table 4-2.

Among the proteolytic enzymes, trypsin is the most specific and hence the enzyme of choice for such studies. It is clear, however, that the use of a single perfectly specific enzyme does not suffice for the determination of the complete sequence. If, for example, trypsin cleavage of a polypeptide resulted in the production of five peptides, which comprise the entire length of the native chain, and if the sequence of amino acids within each of the five segments were established, one would still need to know the order in which the segments appeared in the native peptide. For this purpose other peptides must be obtained that overlap those initially obtained.

The principal advantage of the use of enzymatic hydrolysis, the specificity of the reaction in terms of the nature of the peptide bonds cleaved, also imposes a stringent limitation on the versatility of this method. Ideally one might well desire cleavage at a point not ordinarily susceptible to the action of trypsin, for example. Alternatively, one might wish to prevent cleavage at a susceptible site. Recent developments have partially fulfilled these objectives. For example, reaction of the ε-amino groups of lysine with ethyltrifluorothioacetate in mildly basic solution yields a blocked amino acid residue whose adjacent peptide bond is no longer susceptible to trypsin-catalyzed hydrolysis (12):

$$R{-}NH_2 + CF_3{-}\overset{\overset{\displaystyle O}{\|}}{C}{-}S{-}CH_2CH_3 \rightarrow R{-}\overset{H}{N}{-}\overset{\overset{\displaystyle O}{\|}}{C}{-}CF_3 + C_2H_5SH \qquad (4\text{-}8)$$

TABLE 4-2
Specificity of some proteolytic enzymes

Proteolytic enzyme	*Source*	*Major sites of action*[a]	*Other sites of action*
Trypsin	Pancreas	Arg, Lys	
Chymotrypsin	Pancreas	Trp, Phe, Tyr	Leu, Met, $AspNH_2$, His
Pepsin	Gastric mucosa	Trp, Phe, Tyr, Met, Leu	Various; acidic, etc.
Carboxypeptidase A	Pancreas	C-terminal bond of Tyr, Trp, Phe, etc.	Does not act at Arg, Lys, Pro
Carboxypeptidase B	Pancreas	Arg, Lys	None.
Leucine aminopeptidase	Kidney, intestinal mucosa, etc.	N-terminal bond of various residues	No action at X-Pro bond[b]
Papain	Papaya	Arg, Lys, Gly, etc.	Wide specificity; does not act at acidic residues
Subtilisin	*Bacillus subtilis*	Aromatic and aliphatic residues	Various
Elastase	Pancreas	Neutral residues	
Thermolysin	*Bacillus thermoproteolyticus*	Hydrophobic residues, Leu, Ile, Phe, etc.	Various

[a] Except for the carboxypeptidases, the sites of action refer to the residues bearing the carbonyl group of the peptide bond, for example trypsin-catalyzed hydrolysis of arginyl and lysyl bonds.
[b] X, any other amino acid residue.

SOURCE: A. White, P. Handler, and E. L. Smith, *Principles of Biochemistry*, McGraw-Hill, New York, 4th ed., 1968.

Thus, treatment of such a peptide with trypsin would lead to cleavage only at the arginyl residues. Following removal of the blocking group under mildly alkaline conditions, trypsin catalyzes a subsequent cleavage at lysine residues. In a similar fashion, alkylation of cysteine residues with β-haloethylamines yields peptide linkages that are susceptible to trypsin-catalyzed hydrolysis (13):

$$R{-}SH + Br{-}CH_2CH_2{-}NH_2 \rightarrow R{-}S{-}CH_2CH_2{-}NH_2 + HBr \quad (4\text{-}9)$$

Hence the versatility and power of trypsin as a specific reagent for polypeptide cleavage is markedly increased, and overlapping peptides may be obtained by successive treatments of suitably modified substrates with the enzyme.

Much attention has been paid to the development of chemical reagents that would affect the cleavage of polypeptide chains at specific residues. Of particular importance in this regard is the cleavage of polypeptide chains at methionine residues by treatment with cyanogen bromide. This remarkably specific reaction has been employed to good advantage in structure determinations for ribonuclease, myoglobin, trypsinogen, and other proteins.

Sequence Determination of Short Peptide Chains

Following specific cleavage of polypeptide chains and separation of the hydrolysis products chromatographically, the next task is the sequence determination for each of the degradation products. In general this is accomplished by (1) determination of the amino acid composition of each peptide, (2) determination of the N- and C-terminal residues, and (3) further degradations, if necessary. Clearly, for the case of tripeptides, knowledge of the composition and nature of the terminal residues suffices to establish the amino acid sequence uniquely. For longer peptides, use of the Edman reagent is of great value. Sequential analysis from the N-terminal end, if used in conjunction with knowledge of the amino acid composition, may suffice to establish the sequence of peptides containing several amino acids. Similar experiments may be attempted, generally less successfully, with the use of leucine aminopeptidase and carboxypeptidase. In the early experiments of Sanger on insulin, partial acid hydrolysis was employed for the formation of very short peptides that were sequenced and, ultimately, yielded sufficient information for establishment of the sequence of the longer peptide. Such methods have been largely superseded by the refinement of use of the Edman reagent (page 113).

When all the sequences of the degradation products have been determined, together with those for a suitable number of overlapping peptides, the complete sequence of the native chain is established. This point is illustrated below. The final step is the localization of the disulfide bridges. This may be accomplished by the cleavage of the protein with the disulfide groups intact. A limited amount of additional sequence work usually suffices to localize the groups involved in disulfide linkage.

As detailed above, careful use of the Edman degradation technique can overcome much of the experimental drudgery inherent in amino acid sequence determinations by the classical means just described. Although little used in past work, mass spectrometry promises to provide an alternative tool for sequence studies. This method has proved of value for structure determinations of rather short peptides. In favorable cases, the complete amino acid sequence of a peptide can be determined from a single mass spectrum (14). Bombardment of the peptide with electrons kicks one or more electrons out of the peptide, generating a positively charged species whose molecular weight is accurately measured by the spectrometer. Moreover, this positively charged species fragments at the various peptide bonds so that positively charged species corresponding to the complete peptide minus a terminal amino acid, minus the two terminal amino acids, and so forth, are formed. From the masses of the resulting fragments, it is possible to calculate the molecular weight of the amino acid lost and hence its identity. This method proves particularly useful with peptides available in small amounts because only microgram quantities may be required for a complete analysis.

This technique in the form just described will not suffice to determine the sequences of large peptides, because these are insufficiently volatile to exist in the gas phase in quantities suitable for analysis. It is true that if one were to specifically cleave large fragments into smaller ones, and separate these as described above, mass spectrometry provides a means of sequencing them. This serves no great purpose, however, because (1) the separations are generally more time consuming than the sequence determinations on the separated peptides and (2) the Edman degradation technique already provides a potent tool for performing this task. What is attractive is the possible use of mass spectrometry to deduce a sequence from direct analysis of an *unfractionated* mixture of peptides derived from a longer one by nonspecific cleavage. Since each fragment will have its own molecular weight, it is possible to know the compositions of all the fragments in the mixture. From molecular weights of a sufficient number of fragments it will be possible to deduce a unique structure for the original peptide. Such efforts utilize on-line computational facilities to record, store, and decipher the information collected. The technique is sufficiently complicated so that analyses of this type will be restricted to a relatively few laboratories, at least initially, but some of these may be regional or national facilities available to many researchers. This method should reach fruition in the early 1970s.

Structure of Insulin

Insulin is a polypeptide hormone concerned with the regulation of the rate of carbohydrate metabolism (Chap. 11). The bovine hormone contains 51 amino acids arrayed in two chains, as evidenced by the presence of N-terminal glycine and phenylalanine. The chain terminating in glycine is termed the A chain and contains 21 amino acids. The chain terminating in phenylalanine is termed the B chain and contains 30 amino acids. Sanger and his group oxidized insulin with performic acid and separated the two chains chromatographically. Each chain was degraded both enzymatically and by acid hydrolysis. In Figs. 4-2 and 4-3, the major peptides obtained from the hydrolysis of each of the chains are shown together with the unique structures that were deduced. The sites of cleavage by trypsin, chymotrypsin, and pepsin, as indicated in these figures, conform in large part to expectations based on the known specificity of these enzymes toward synthetic materials (Table 4-2). Cleavage at some additional sites was also found, particularly in the case of the pepsin-catalyzed hydrolyses. Note how the overlapping peptides obtained with the use of different hydrolytic techniques complement each other and permit an unambiguous assignment of the overall sequence. Each of the major peptides shown in Figs. 4-2 and 4-3 was sequenced by nonspecific hydrolysis with acid, determination of the sequences of the resultant di-, tri-, and tetrapeptides, and correlating this information into a unique sequence. As indicated above, this task is currently markedly simplified by use of the Edman reagent. The complete sequences of the A and B chains, together with the localization of the disulfide bridges and amide groups of aspartic and glutamic acids (these are destroyed by acidic hydrolysis), yielded the complete structure for insulin, as shown in Fig. 4-4.

```
Sequences deduced from lower peptides   Gly·Ile·Val·Glu·Glu·CySO3H·CySO3H·Ala                Ser·Leu·Tyr·Glu·Leu·Glu·Asp·Tyr·CySO3H
                                                                              Ser·Val·CySO3H                                 CySO3H·Asp

Peptides from peptic hydrolysate        Gly·Ile·Val·Glu·Glu·CySO3H·CySO3H·Ala·Ser·Val·CySO3H·Ser·Leu         Leu·Glu·Asp·Tyr·CySO3H·Asp
                                        Gly·Ile·Val·Glu·Glu·CySO3H·CySO3H·Ala·Ser·Val                Tyr·Glu      Glu·Asp·Tyr·CySO3H·Asp
                                        Gly·Ile·Val·Glu                                              Tyr·Glu·Leu     Asp·Tyr·CySO3H·Asp
                                                                                                     Tyr·Glu·Leu·Glu
                                                                                                             Leu·Glu

Peptides from chymotryptic hydrolysate  Gly·Ile·Val·Glu·Glu·CySO3H·CySO3H·Ala·Ser·Val·CySO3H·Ser·Leu·Tyr                     CySO3H·Asp
                                        Gly·Ile·Val·Glu·Glu·CySO3H·CySO3H·Ala·Ser·Val·CySO3H             Glu·Leu·Glu·Asp·Tyr·CySO3H·Asp
                                                                                             Ser·Leu·Tyr
                                                                                                         Glu·Leu·Glu·Asp·Tyr

Structure of fraction A                 Gly·Ile·Val·Glu·Glu·CySO3H·CySO3H·Ala·Ser·Val·CySO3H·Ser·Leu·Tyr·Glu·Leu·Glu·Asp·Tyr·CySO3H·Asp
                                        1   2   3   4   5   6      7      8   9   10  11     12  13  14  15  16  17  18  19  20     21
Bonds split by pepsin                                  ⇡                             ⇡              ↑   ⇡   ⇡   ↑   ↑
Bonds split by chymotrypsin                                                                 ⇡           ↑                          ⇡
---------------------------------------------------------------------------------------------------------------------------------------
                                        ↑ major sites of action of enzymes;  ⇡ other bonds split by enzymes.
```

FIGURE 4-2
Structure of the A chain of oxidized insulin. J. I. Harris and V. M. Ingram in P. Alexander and R. J. Block (eds.), *The Composition, Structure, and Reactivity of Proteins*, vol. 2, Pergamon Press, New York, 1960.

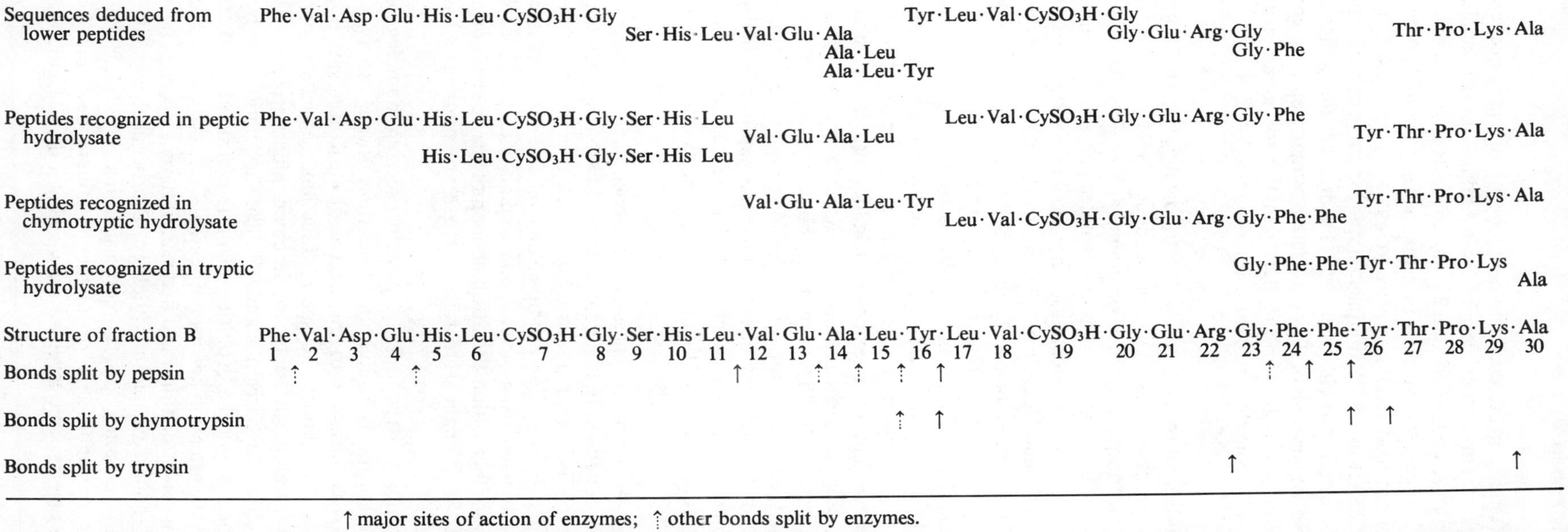

Sequences deduced from lower peptides
Phe·Val·Asp·Glu·His·Leu·CySO$_3$H·Gly
Ser·His·Leu·Val·Glu·Ala
Ala·Leu
Ala·Leu·Tyr
Tyr·Leu·Val·CySO$_3$H·Gly
Gly·Glu·Arg·Gly
Gly·Phe
Thr·Pro·Lys·Ala

Peptides recognized in peptic hydrolysate
Phe·Val·Asp·Glu·His·Leu·CySO$_3$H·Gly·Ser·His Leu
His·Leu·CySO$_3$H·Gly·Ser·His Leu
Val·Glu·Ala·Leu
Leu·Val·CySO$_3$H·Gly·Glu·Arg·Gly·Phe
Tyr·Thr·Pro·Lys·Ala

Peptides recognized in chymotryptic hydrolysate
Val·Glu·Ala·Leu·Tyr
Leu·Val·CySO$_3$H·Gly·Glu·Arg·Gly·Phe·Phe
Tyr·Thr·Pro·Lys·Ala

Peptides recognized in tryptic hydrolysate
Gly·Phe·Phe·Tyr·Thr·Pro·Lys
Ala

Structure of fraction B
Phe·Val·Asp·Glu·His·Leu·CySO$_3$H·Gly·Ser·His·Leu·Val·Glu·Ala·Leu·Tyr·Leu·Val·CySO$_3$H·Gly·Glu·Arg·Gly·Phe·Phe·Tyr·Thr·Pro·Lys·Ala
1 2 3 4 5 6 7 8 9 10 11 12 13 14 15 16 17 18 19 20 21 22 23 24 25 26 27 28 29 30

Bonds split by pepsin: ⇡ (1–2), ⇡ (4–5), ↑ (11–12), ⇡ (13–14), ⇡ (14–15), ⇡ (15–16), ↑ (16–17), ⇡ (23–24), ↑ (24–25), ↑ (25–26)

Bonds split by chymotrypsin: ⇡ (15–16), ↑ (16–17), ↑ (25–26), ↑ (26–27)

Bonds split by trypsin: ↑ (22–23), ↑ (29–30)

↑ major sites of action of enzymes; ⇡ other bonds split by enzymes.

FIGURE 4-3
Structure of the B chain of oxidized insulin. J. I. Harris and V. M. Ingram, in P. Alexander and R. J. Block (eds.), *The Composition, Structure, and Reactivity of Proteins,* vol. 2, Pergamon Press, New York, 1960.

Gly·Ile·Val·Glu·Glu(NH_2)·Cy·Cy·Ala·Ser·Val·Cy·Ser·Leu·Tyr·Glu(NH_2)·Leu·Glu·Asp(NH_2)·Tyr·Cy·Asp(NH_2)
(S—S between Cy 6 and Cy 11 of the upper chain; S—S from Cy 7 of the upper chain to Cy 7 of the lower chain; S—S from Cy 20 of the upper chain to Cy 19 of the lower chain)
Phe·Val·Asp(NH_2)·Glu(NH_2)·His·Leu·Cy·Gly·Ser·His·Leu·Val·Glu·Ala·Leu·Tyr·Leu·Val·Cy·Gly·Glu·Arg·Gly·Phe·Phe·Tyr·Thr·Pro·Lys·Ala

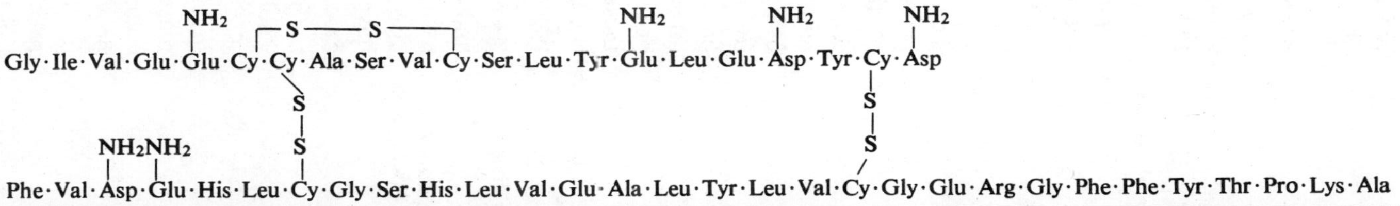

FIGURE 4-4
Covalent structure of bovine insulin. J. I. Harris and V. M. Ingram, in P. Alexander and R. J. Block (eds.), *The Composition, Structure, and Reactivity of Proteins,* vol. 2, Pergamon Press, New York, 1960.

Structure of Ribonuclease

The second major achievement in the establishment of amino acid sequences for proteins was the work of Hirs, Stein, Moore, and Anfinsen on ribonuclease. This single-chain enzyme, composed of 124 amino acids and containing four intrachain disulfide bridges, was studied in a manner similar to that employed for insulin; performic acid oxidation was followed by enzymatic hydrolysis. The principal advances in techniques employed to establish the primary structure of this larger protein were (1) the use of ion-exchange resins for the quantitative separation of low-molecular-weight peptides and (2) the use of quantitative methods for the determination of amino acid composition of peptides. The complete structure for ribonuclease is shown in Fig. 4-5.

Generalizations

Amino acid sequence studies allow us to draw the following generalizations. First, proteins have unique structures. That is, ribonuclease from a given species, for example, has a unique sequence of amino acids along the polypeptide chain; it is not a collection of individual molecular species which differ in primary structure. This statement does not preclude differences in primary structure for functionally identical proteins from different species. This topic is dealt with immediately below. Second, there do not seem to be preferences for particular amino acid sequences resulting from the properties of the amino acids themselves. Thus, discernible patterns of amino acid sequences do not recur throughout a broad spectrum of proteins. Third, proteins are heterogeneous with respect to the number and type of disulfide bridges, although the number and arrangement of disulfide bridges is unique for a particular protein. Certain proteins have no disulfide bridges, others have either inter- or intrachain disulfides, or have both types. Fourth, sequence studies have largely confirmed the specificity for proteolytic enzymes originally deduced from studies on low-molecular-weight peptides. Finally, these studies have confirmed our earlier conclusions that only amino acids of the L configuration occur in proteins and that covalent linkages other than the peptide bond are rare.

Individuality of Proteins: Species Specificity and Mutational Alterations

As emphasized in Chapter 3, certain polypeptide hormones (i.e., vasopressin and β-MSH) that perform functionally identical tasks in different species may exhibit subtle differences in their primary structure. Such species specificity also exists for the case of proteins. For example, Sanger and his associates have sequenced insulins obtained from several mammalian sources. In each case, amino acid variations either within the disulfide loop of the A chain or at the carboxyl terminus of the B chain were found. These are summarized in Table 4-3.

The comparison of primary structures of proteins serving homologous functions in different organisms is useful and important for a variety of reasons. First, it will provide insight into the degree of structural variation compatible with the same (or almost the same) function. Second, it may provide clues to those areas along the polypeptide chain which are essential for function. Finally, such a comparison should indicate a relationship between the evolution of species and the evolution of protein structure. It is to this last point that most of the attention has been directed.

Classically, distinctions and relationships between different species have been made on the basis of macroscopic morphology. Biochemical capacities have proved useful for the classification of microorganisms. It is clear that the underlying basis for morphological or metabolic differences must be molecular. That is, at the macroscopic level we are viewing a manifestation of

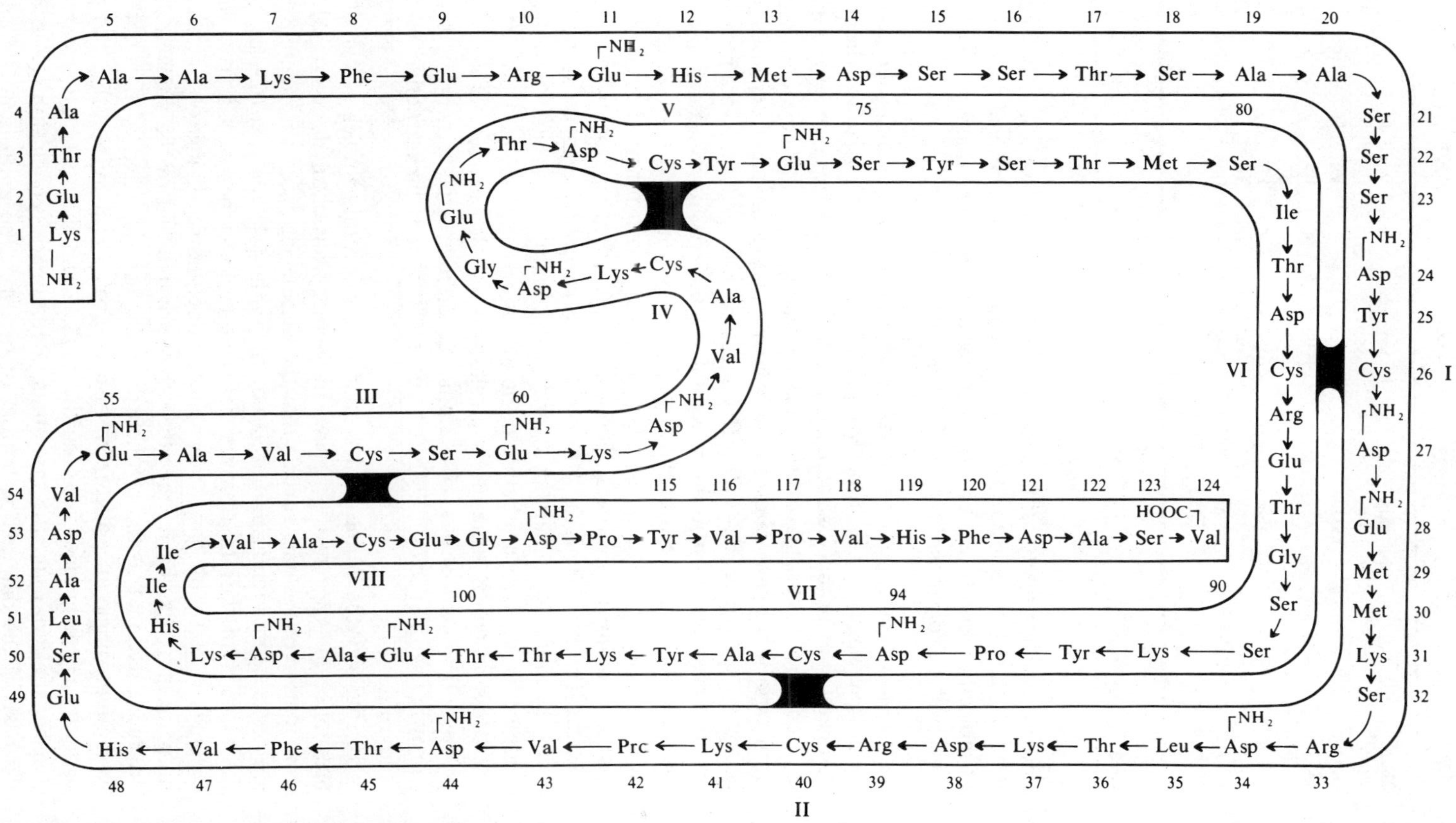

FIGURE 4-5
Covalent structure of ribonuclease. D. G. Smyth, W. H. Stein, and S. Moore, *J. Biol. Chem.*, **238:** 227 (1963).

TABLE 4-3
Species variation in primary structure of insulin

Source	*A8**	*A9**	*A10**	*B30**
Beef	Ala	Ser	Val	Ala
Pig	Thr	Ser	Ile	Ala
Sheep	Ala	Gly	Val	Ala
Horse	Thr	Gly	Ile	Ala
Sperm whale	Thr	Ser	Ile	Ala
Sei whale	Ala	Ser	Thr	Ala
Man	Thr	Ser	Ile	Thr
Dog	Thr	Ser	Ile	Ala
Rabbit	Thr	Ser	Ile	Ser

[a] Positions refer to the residue number from the amino terminus of the A or B chain (see Fig. 4-4).

one, or perhaps many, differences at the molecular level. Ideally, one would wish to classify species and construct evolutionary relationships on the basis of molecular structures. The species variation in the primary structure of proteins provides one possibility of doing just this.

Cytochrome c

There have been a number of important studies in this area, but perhaps the most revealing are those of Margoliash and his associates, who have investigated the primary structures of *cytochrome c* from an amazing variety of species. Cytochrome *c* functions as an electron carrier along a pathway from oxidizable substrates to molecular oxygen or other terminal electron acceptor (see pages 680–689). It has associated with it a nonprotein heme group containing an iron atom capable of undergoing redox reactions. Thus, the iron atom is reduced to the ferrous state when it accepts electrons coming from the substrate but is reoxidized to the ferric form when it passes the electrons along to the next member of the chain on the way to the final electron acceptor. This function is a very general one in living species and many of them do contain cytochrome *c,* making it a suitable protein for a study of the relationship between the evolution of species and proteins. Moreover, this protein is structurally a relatively simple one. It consists of a single polypeptide chain about 104 amino acids long (the exact number varies somewhat with species). For this reason, too, it is an attractive protein for study.

In Fig. 4-6, the amino acid sequences for several samples of cytochrome *c* are collected. Clearly differences in primary structure do exist. Moreover, examination of the data reveals that the number of differences increases as the species become increasing unrelated. In Table 4-4, comparisons are made of the number of variant residues between various species. Clearly, the cytochromes *c* of rabbit, pig, cow, sheep, dog, horse, kangaroo, monkey, and man are more like each other than they are like that of the chicken and they are more like that of the chicken than they are like that of the moth. Thus, assuming a single ancestral form of cytochrome *c*, it follows that the point of evolutionary divergence between the chicken and the mammals occurred after that for the moth and the mammals, a reasonable conclusion. Note that, on the basis of the data at the bottom of Fig. 4-7, the moth is not more closely related to yeast than man is. Thus, the time of evolutionary divergence from yeast to the moth and man is comparable. On the basis of considerations of this type, one can produce an evolutionary tree based solely on information

Horse	Acetyl-Gly-Asp -Val-Glu-Lys -Gly-Lys-Lys-Ile-Phe-Val-GluNH$_2$-Lys
Pig or Cow	Acetyl-Gly-Asp -Val-Glu-Lys -Gly-Lys-Lys-Ile-Phe-Val-GluNH$_2$-Lys
Kangaroo	Acetyl-Gly-Asp -Val-Glu-Lys -Gly-Lys-Lys-Ile-Phe-Val-GluNH$_2$-Lys
Human	Acetyl-Gly-Asp -Val-Glu-Lys -Gly-Lys-Lys-Ile-Phe- Ile -Met -Lys
Chicken	Acetyl-Gly-Asp - Ile-Glu-Lys -Gly-Lys-Lys-Ile-Phe-Val-GluNH$_2$-Lys
Tuna	Acetyl-Gly-Asp -Val-Ala-Lys -Gly-Lys-Lys-Thr-Phe-Val-GluNH$_2$-Lys
Moth	Gly-Val-Pro-Ala-Gly-AspNH$_2$-Ala-Glu-AspNH$_2$-Gly-Lys-Lys-Ile-Phe-Val-GluNH$_2$-Arg
	−4
Yeast	Thr-Glu-Phe-Lys-Ala-Gly-Ser -Ala-Lys-Lys -Gly-Ala-Thr-Leu-Phe-Lys-Thr -Arg
	−5 1 5 10

Horse	CyS-Ala-GluNH$_2$-CyS-His-Thr-Val-Glu-Lys -Gly-Gly-Lys-His-Lys-Thr-Gly-Pro-AspNH$_2$-Leu
	└—HEME—┘
Pig or Cow	CyS-Ala-GluNH$_2$-CyS-His-Thr-Val-Glu-Lys -Gly-Gly-Lys-His-Lys-Thr-Gly-Pro-AspNH$_2$-Leu
	└—HEME—┘
Kangaroo	CyS-Ala-GluNH$_2$-CyS-His-Thr-Val-Glu-Lys -Gly-Gly-Lys-His-Lys-Thr-Gly-Pro-AspNH$_2$-Leu
	└—HEME—┘
Human	CyS-Ser-GluNH$_2$-CyS-His-Thr-Val-Glu-Lys -Gly-Gly-Lys-His-Lys-Thr-Gly-Pro-AspNH$_2$-Leu
	└—HEME—┘
Chicken	CyS-Ser-GluNH$_2$-CyS-His-Thr-Val-Glu-Lys -Gly-Gly-Lys-His-Lys-Thr-Gly-Pro-AspNH$_2$-Leu
	└—HEME—┘
Tuna	CyS-Ala-GluNH$_2$-CyS-His-Thr-Val-Glu-AspNH$_2$-Gly-Gly-Lys-His-Lys-Val-Gly-Pro-AspNH$_2$-Leu
	└—HEME—┘
Moth	CyS-Ala-GluNH$_2$-CyS-His-Thr-Val-Glu-Ala -Gly-Gly-Lys-His-Lys-Val-Gly-Pro-AspNH$_2$-Leu
	└—HEME—┘
Yeast	CyS-Glu-Leu -CyS-His-Thr-Val-Glu-Lys -Gly-Gly-Pro-His-Lys-Val-Gly-Pro-AspNH$_2$-Leu
	└—HEME—┘
	15 20 25 30

Horse	His -Gly-Leu-Phe-Gly-Arg-Lys-Thr-Gly-GluNH$_2$-Ala-Pro -Gly-Phe-Thr-Tyr-Thr-Asp
Pig or Cow	His -Gly-Leu-Phe-Gly-Arg-Lys-Thr-Gly-GluNH$_2$-Ala-Pro -Gly-Phe-Ser -Tyr-Thr-Asp
Kangaroo	AspNH$_2$-Gly- Ile -Phe-Gly-Arg-Lys-Thr-Gly-GluNH$_2$-Ala-Pro -Gly-Phe-Thr-Tyr-Thr-Asp
Human	His -Gly-Leu-Phe-Gly-Arg-Lys-Thr-Gly-GluNH$_2$-Ala-Pro -Gly-Tyr-Ser -Tyr-Thr-Ala
Chicken	His -Gly-Leu-Phe-Gly-Arg-Lys-Thr-Gly-GluNH$_2$-Ala-Glu -Gly-Phe-Ser -Tyr-Thr-Asp
Tuna	Trp -Gly-Leu-Phe-Gly-Arg-Lys-Thr-Gly-GluNH$_2$-Ala-Glu -Gly-Tyr-Ser -Tyr-Thr-Asp
Moth	His -Gly-Phe-Tyr-Gly-Arg-Lys-Thr-Gly-GluNH$_2$-Ala-Pro -Gly-Phe-Ser -Tyr-Ser-AspNH$_2$
Yeast	His -Gly- Ile -Phe-Cly-Arg-His-Ser -Gly-GluNH$_2$-Ala-GluNH$_2$ -Gly-Tyr-Ser -Tyr-Thr-Asp
	35 40 45 50

Horse	Ala-AspNH$_2$-Lys-AspNH$_2$-Lys-Gly - Ile -Thr-Trp-Lys -Glu -Glu -Thr -Leu-Met
Pig or Cow	Ala-AspNH$_2$-Lys-AspNH$_2$-Lys-Gly - Ile -Thr-Trp-Gly -Glu -Glu -Thr -Leu-Met
Kangaroo	Ala-AspNH$_2$-Lys-AspNH$_2$-Lys-Gly - Ile - Ile -Trp-Gly -Glu -Asp -Thr -Leu-Met
Human	Ala-AspNH$_2$-Lys-AspNH$_2$-Lys-Gly - Ile - Ile -Trp-Gly -Glu -Asp -Thr -Leu-Met
Chicken	Ala-AspNH$_2$-Lys-AspNH$_2$-Lys-Gly - Ile -Thr-Trp-Gly -Glu -Asp -Thr -Leu-Met
Tuna	Ala-AspNH$_2$-Lys-Ser -Lys-Gly - Ile -Val-Trp-AspNH$_2$-AspNH$_2$-Asp -Thr -Leu-Met
Moth	Ala-AspNH$_2$-Lys-Ala -Lys-Gly - Ile -Thr-Trp-Gly -Asp -Asp -Thr -Leu-Phe
Yeast	Ala-AspNH$_2$- Ile -Lys -Lys-AspNH$_2$-Val-Leu-Trp-Asp -Glu -AspNH$_2$-AspNH$_2$-Met-Ser
	55 60 65

Horse	Glu-Tyr-Leu-Glu-AspNH$_2$-Pro-Lys-Lys-Tyr- Ile -Pro-Gly-Thr-Lys-Met- Ile -Phe-Ala-Gly- Ile
Pig or Cow	Glu-Tyr-Leu-Glu-AspNH$_2$-Pro-Lys-Lys-Tyr- Ile -Pro-Gly-Thr-Lys-Met- Ile -Phe-Ala-Gly- Ile
Kangaroo	Glu-Tyr-Leu-Glu-AspNH$_2$-Pro-Lys-Lys-Tyr- Ile -Pro-Gly-Thr-Lys-Met- Ile -Phe-Ala-Gly- Ile
Human	Glu-Tyr-Leu-Glu-AspNH$_2$-Pro-Lys-Lys-Tyr- Ile -Pro-Gly-Thr-Lys-Met- Ile -Phe-Val-Gly- Ile
Chicken	Glu-Tyr-Leu-Glu-AspNH$_2$-Pro-Lys-Lys-Tyr- Ile -Pro-Gly-Thr-Lys-Met- Ile -Phe-Ala-Gly- Ile
Tuna	Glu-Tyr-Leu-Glu-AspNH$_2$-Pro-Lys-Lys-Tyr- Ile -Pro-Gly-Thr-Lys-Met- Ile -Phe-Ala-Gly- Ile
Moth	Glu-Tyr-Leu-Glu-AspNH$_2$-Pro-Lys-Lys-Tyr- Ile -Pro-Gly-Thr-Lys-Met-Val -Phe-Ala-Gly-Leu
Yeast	Glu-Tyr-Leu-Thr-AspNH$_2$-Pro-Lys-Lys-Tyr- Ile -Pro-Gly-Thr-Lys-Met-Ala -Phe-Gly-Gly-Leu
	70 75 80 85

Horse	Lys-Lys-Lys-Thr -Glu-Arg-Glu -Asp-Leu- Ile -Ala-Tyr-Leu-Lys-Lys-Ala-Thr -AspNH$_2$-GluCOOH
Pig or Cow	Lys-Lys-Lys-Gly -Glu-Arg-Glu -Asp-Leu- Ile -Ala-Tyr-Leu-Lys-Lys-Ala-Thr -AspNH$_2$-GluCOOH
Kangaroo	Lys-Lys-Lys-Gly -Glu-Arg-Ala -Asp-Leu- Ile -Ala-Tyr-Leu-Lys-Lys-Ala-Thr -AspNH$_2$-GluCOOH
Human	Lys-Lys-Lys-Glu -Glu-Arg-Ala -Asp-Leu- Ile -Ala-Tyr-Leu-Lys-Lys-Ala-Thr -AspNH$_2$-GluCOOH
Chicken	Lys-Lys-Lys-Ser -Glu-Arg-Val -Asp-Leu- Ile -Ala-Tyr-Leu-Lys-Asp-Ala-Thr -Ser -LysCOOH
Tuna	Lys-Lys-Lys-Gly -Glu-Arg-GluNH$_2$ -Asp-Leu-Val -Ala -Tyr-Leu-Lys-Ser -Ala-Thr -Ser - D
Moth	Lys-Lys-Ala-AspNH$_2$-Glu-Arg-Ala -Asp-Leu- Ile -Ala-Tyr-Leu-Lys-Glu-Ser- Thr - D -LysCOOH
Yeast	Lys-Lys-Glu-Lys -Asp-Arg-AspNH$_2$ -Asp-Leu- Ile -Thr-Tyr-Leu-Lys-Lys-Ala- D -CySH -GluCOOH
	90 95 100 104

FIGURE 4-6

Amino acid sequences of cytochromes *c* from horse, pig, cow, human, kangaroo, chicken, and tuna hearts, and from the moth (*Samia cynthia*) and baker's yeast. The "D" at residues 102, 103, and 104 indicates that the amino acids occupying these positions are absent. The residues underlined are those which differ from the amino acids in the corresponding positions of horse heart cytochome *c*. E. Margoliash and A. Schejter, *Advan. Protein Chem.*, **21:** 114 (1966).

TABLE 4-4
Evolution of cytochromes c

Species comparison	Variant residues: Number	Span	Average	Divergence of lines in millions of years
Rabbit—chicken	8			
Pig, cow, sheep—chicken	9			
Dog—chicken	10			
Horse—chicken	11	8–12	10.7	280 (assumed)
Kangaroo—chicken	12			
Monkey—chicken	12			
Man—chicken	13			
Rabbit—tuna	17			
Pig, cow, sheep—tuna	17			
Dog—tuna	18			
Horse—tuna	19			
Kangaroo—tuna	20	17–21	18.7	490 (calculated)
Man—tuna	21			
Monkey—tuna	21			
Chicken—tuna	17			
Rabbit—moth	26			
Dog—moth	26			
Pig, cow, sheep—moth	27			
Kangaroo—moth	29			
Horse—moth	29	26–33	28.7	750 (calculated)
Monkey—moth	30			
Man—moth	31			
Chicken—moth	27			
Tuna—moth	33			
Rabbit—yeast	44			
Pig, cow, sheep—yeast	44			
Dog—yeast	44			
Monkey—yeast	44			
Man—yeast	44	44–48	45.2	1180 (calculated)
Horse—yeast	45			
Kangaroo—yeast	45			
Chicken—yeast	46			
Tuna—yeast	48			
Moth—yeast	48			

SOURCE: E. Margoliash and A. Schejter, *Advan. Protein Chem.*, **21:** 113 (1966).

contained in molecular structures within the organisms. One such example is provided in Fig. 4-7.

It is important to return now to the data in Fig. 4-6 and examine the kind of amino acid substitutions which occur between the various samples of cytochrome *c*. About half of the positions show no variability at all from species to species, and it is possible that the exact nature of the amino acids at these positions is crucial for function. The changes that do occur are generally of a "conservative" type. That is, a hydrophobic amino acid is replaced by another hydrophobic one or a hydrophilic amino acid is replaced by another hydrophilic one. Thus interchange of leucine for isoleucine or valine is a conservative change. Interchange of valine for lysine is not. Nonconservative changes might reasonably be expected to introduce substantial structure alterations at the three-dimensional level inconsistent with maintenance of the same function. More about this point follows later in the chapter.

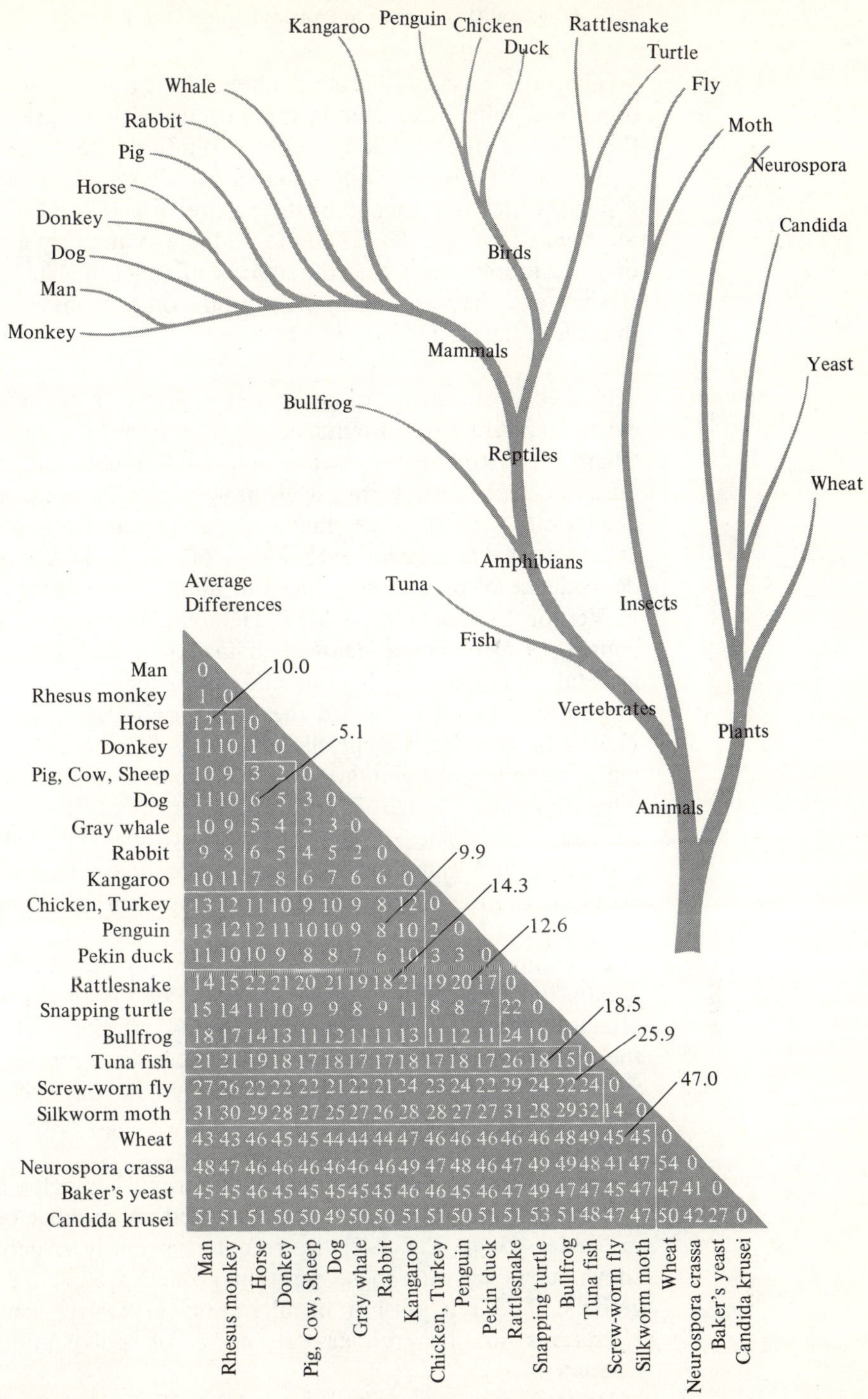

	Man	Rhesus monkey	Horse	Donkey	Pig, Cow, Sheep	Dog	Gray whale	Rabbit	Kangaroo	Chicken, Turkey	Penguin	Pekin duck	Rattlesnake	Snapping turtle	Bullfrog	Tuna fish	Screw-worm fly	Silkworm moth	Wheat	Neurospora crassa	Baker's yeast	Candida krusei
Man	0																					
Rhesus monkey	1	0																				
Horse	12	11	0																			
Donkey	11	10	1	0																		
Pig, Cow, Sheep	10	9	3	2	0																	
Dog	11	10	6	5	3	0																
Gray whale	10	9	5	4	2	3	0															
Rabbit	9	8	6	5	4	5	2	0														
Kangaroo	10	11	7	8	6	7	6	6	0													
Chicken, Turkey	13	12	11	10	9	10	9	8	12	0												
Penguin	13	12	12	11	10	10	9	8	10	2	0											
Pekin duck	11	10	10	9	8	8	7	6	10	3	3	0										
Rattlesnake	14	15	22	21	20	21	19	18	21	19	20	17	0									
Snapping turtle	15	14	11	10	9	9	8	9	11	8	8	7	22	0								
Bullfrog	18	17	14	13	11	12	11	11	13	11	12	11	24	10	0							
Tuna fish	21	21	19	18	17	18	17	17	18	17	18	17	26	18	15	0						
Screw-worm fly	27	26	22	22	22	21	22	21	24	23	24	22	29	24	22	24	0					
Silkworm moth	31	30	29	28	27	25	27	26	28	28	27	27	31	28	29	32	14	0				
Wheat	43	43	46	45	45	44	44	44	47	46	46	46	46	46	48	49	45	45	0			
Neurospora crassa	48	47	46	46	46	46	46	46	49	47	48	46	47	49	49	48	41	47	54	0		
Baker's yeast	45	45	46	45	45	45	45	45	46	46	45	46	47	49	47	47	45	47	47	41	0	
Candida krusei	51	51	51	50	50	49	50	50	51	51	50	51	51	53	51	48	47	47	50	42	27	0

FIGURE 4-7

Family tree of the cytochromes *c*. The species differences shown in the table lead to a tree of family relatedness. Note that there is no ascending hierarchy. From the viewpoint of a yeast (if it had one, and therein lies a real if anthropocentric distinction), a moth, a man, and a bullfrog are equally far away. Note also how provincial is the view that we usually take of the living kingdom. The differences between fungi are greater than those between insects and vertebrates. From *The Structure and Action of Proteins* by Richard E. Dickerson and Irving Geis. Copyright © 1969 by Richard E. Dickerson and Irving Geis. Reprinted by permission of the authors and Harper & Row, Publishers, Inc.

The only unsatisfying point about the evolution of cytochrome *c* is that functional differences among the various proteins are difficult to detect. On the basis of most thinking about evolution, one would expect a change in structure to be reflected in a functional alteration of the molecule generating a pressure toward change in the nature of the species. Thus, changes at the molecular level are the driving force for evolutionary change, not the product of it. It follows then that differences in functional behavior ought to occur. If they do, they are sufficiently subtle ones to have escaped our attention thus far in the case of the cytochromes *c*.

Abnormal Hemoglobins

The species specificity of protein structure is a manifestation of mutational events (essentially instantaneous alterations in an organismal property, Chapter 19) that occurred long ago. Such mutational events also show up in terms of differences within a given species. Since mutations frequently have deleterious consequences, much attention has been paid to the immediate effects at the molecular level. Much of our early knowledge concerning the dependence of protein structure on mutation was based on the elegant work of Vernon Ingram, now at MIT. Hemoglobin, as noted below in more detail, consists of two pairs of identical chains: the α and β chains (or α and γ chains in fetal hemoglobin). Ingram and his associates were able to demonstrate that normal hemoglobin and the hemoglobin of victims of sickle-cell anemia (HbS), a genetically controlled disease, differ only in the substitution of a valine residue for a glutamic acid residue at a unique position in the β chain. Note that this change, in contrast to most of those which distinguish the various cytochromes *c*, is a nonconservative one; valine has hydrophobic properties and glutamic acid hydrophilic ones. This single nonconservative change has dramatic consequences inasmuch as it markedly reduces the solubility of the hemoglobin. In the deoxy state this protein actually precipitates within the red blood cells, causing a distortion in their shape and, finally, greatly increasing the viscosity of the blood. With the viscosity increase, the blood flows more slowly through the veins, arteries, and capillaries, generating the anemic state. Thus, the single nonconservative amino acid substitution distinguishing normal hemoglobin and hemoglobin S has much greater functional consequences than do 40 or more conservative substitutions in cytochrome *c*.

Subsequently, a large number of abnormal hemoglobins have been studied, and many amino acid replacements noted. A number of examples are given in Table 4.5. In Fig. 4-8, the complete primary structure of hemoglobin is shown together with those positions which are altered in abnormal hemoglobins. The relationship of alteration in DNA structure (mutation) to alterations in this primary structure of polypeptides is considered in Chapter 19.

Related, and very important, work has been carried out by Yanofsky on the tryptophan synthetase of *Escherichia coli*. This enzyme, which consists of two chains, has been the subject of extensive structural and genetic investigations. For example, a set of eight distinct α chains has been obtained from mutationally altered bacteria that differ in the nature of the amino acid at one position but are otherwise identical (Fig. 19-9A). Another protein studied extensively in this context is the coat protein of tobacco mosaic virus, investigated by Wildmann and by Fraenkel-Conrat and their collaborators. Again simple mutational events lead to the replacement of individual amino acids.

TABLE 4-5
Amino acid substitutions in human hemoglobins

α-Chain abnormalities				*β-Chain abnormalities*			
Residue number	*Abnormal Hb*	*Residue in Hb-A*	*Residue in abnormal Hb*	*Residue number*	*Abnormal Hb*	*Residue in Hb-A*	*Residue in abnormal Hb*
16	I	Lys	Asp	6	C	Glu	Lys
30	G_{Honolulu} =	Glu	Gln	6	S	Glu	Val
	$G_{\text{Singapore}}$ =			7	$G_{\text{San Jose}}$	Glu	Gly
	$G_{\text{Hong Kong}}$ =			26	E	Glu	Lys
57	Norfolk	Gly	Asp	63	$M_{\text{Saskatoon}}$ =	His	Tyr
58	M_{Boston}	His	Tyr		M_{Emory}		
68	$G_{\text{Phil.}}$ =	Asn	Lys	63	Zürich	His	Arg
	G_{Bristol} =			67	$M_{\text{Milwaukee-1}}$	Val	Glu
	$D_{\alpha\ \text{St. Louis}}$			121	D_{Punjab}	Glu	Gln
116	$O_{\text{Indonesia}}$	Glu	Lys				

Subtle differences in the primary structure of closely related proteins may often be readily detected by the technique of *fingerprinting* (15). Obtaining the fingerprint of a protein involves partial digestion by one or more proteolytic enzymes followed by determination of the pattern of the product peptides on electrophoresis or chromatography. Fingerprints for normal and sickle-cell hemoglobin are shown in Fig. 4-9. Close examination of these fingerprints indicates that, with the exception of one peptide, the patterns are identical. Hence, that structural unit which is genetically altered is pinpointed immediately, and the nature of the structural alteration may be defined without the necessity of sequencing the entire molecule. Indeed, in favorable cases, strong evidence for the nature of the amino acid substitution may be gained simply from comparison of the amino acid compositions of the corresponding peptides isolated from the two proteins. Definitive evidence for the substitution depends, of course, on obtaining the sequence of the peptides concerned.

Immunoglobulins

Considerations concerning the primary structure of one important group of proteins, the *immunoglobulins,* stand in marked contrast to those just described. The immunoglobulins include the antibodies that circulate in the blood serum of animals and certain structurally related species. *Antibodies* are proteins elicited by the lymphoid cell system in response to invasion by a foreign species, the *antigen,* and have the capacity to react specifically with the antigen and, in so doing, to provide for its removal from the organism. The synthesis of antibodies underlies all our immunization procedures against disease, and antigen–antibody reactions are probably involved in all allergic responses. Consequently, much attention has been directed to understanding the structure, function, and synthesis of the immunoglobulins.

The immunoglobulins are a structurally heterogeneous group of proteins. On electrophoresis, they migrate generally with the γ-globulins but are sufficiently diverse so that they range into the β-globulin fraction as well. It is precisely this heterogeneity that has proved a major stumbling block to understanding the primary structure of these proteins.

The general organization of the immunoglobulins is shown in Fig. 4-10. The complete molecule, of molecular weight 160,000, is composed of a pair of light (L) chains, molecular weight 23,000, and a pair of heavy (H) chains, molecular weight 50,000, bound together by interchain disulfide bonds as shown in Fig. 4-10. There are three types of heavy chain and two types of

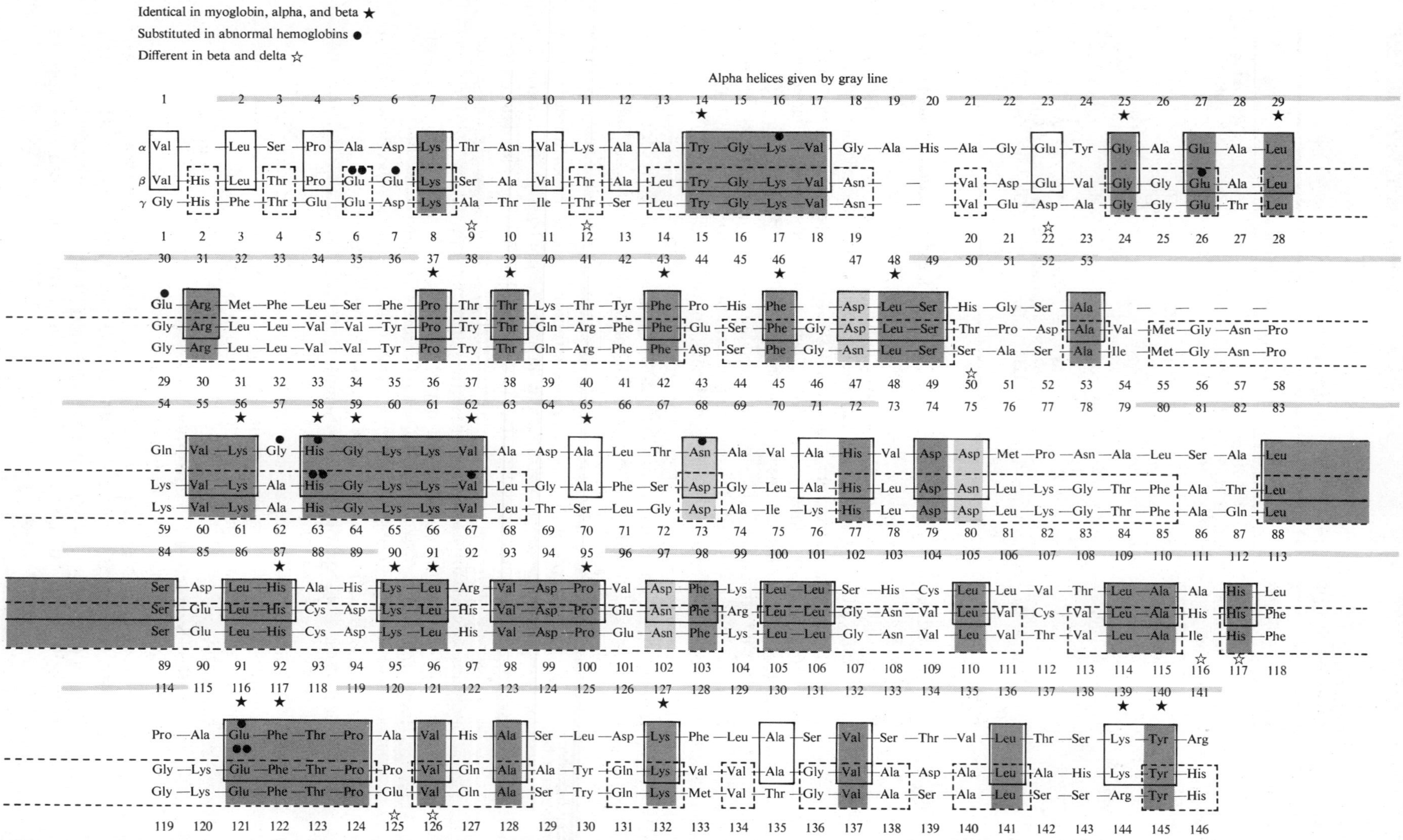

Identical in myoglobin, alpha, and beta ★
Substituted in abnormal hemoglobins ●
Different in beta and delta ☆

Alpha helices given by gray line

1 2 3 4 5 6 7 8 9 10 11 12 13 14 15 16 17 18 19 20 21 22 23 24 25 26 27 28 29
α Val — Leu — Ser — Pro — Ala — Asp — Lys — Thr — Asn — Val — Lys — Ala — Ala — Try — Gly — Lys — Val — Gly — Ala — His — Ala — Gly — Glu — Tyr — Gly — Ala — Glu — Ala — Leu
β Val — His — Leu — Thr — Pro — Glu — Glu — Lys — Ser — Ala — Val — Thr — Ala — Leu — Try — Gly — Lys — Val — Asn — — Val — Asp — Glu — Val — Gly — Gly — Glu — Ala — Leu
γ Gly — His — Phe — Thr — Glu — Glu — Asp — Lys — Ala — Thr — Ile — Thr — Ser — Leu — Try — Gly — Lys — Val — Asn — — Val — Glu — Asp — Ala — Gly — Gly — Glu — Thr — Leu
1 2 3 4 5 6 7 8 9 10 11 12 13 14 15 16 17 18 19 20 21 22 23 24 25 26 27 28

30 31 32 33 34 35 36 37 38 39 40 41 42 43 44 45 46 47 48 49 50 51 52 53
Glu — Arg — Met — Phe — Leu — Ser — Phe — Pro — Thr — Thr — Lys — Thr — Tyr — Phe — Pro — His — Phe — — Asp — Leu — Ser — His — Gly — Ser — Ala — — — — —
Gly — Arg — Leu — Leu — Val — Val — Tyr — Pro — Try — Thr — Gln — Arg — Phe — Phe — Glu — Ser — Phe — Gly — Asp — Leu — Ser — Thr — Pro — Asp — Ala — Val — Met — Gly — Asn — Pro
Gly — Arg — Leu — Leu — Val — Val — Tyr — Pro — Try — Thr — Gln — Arg — Phe — Phe — Asp — Ser — Phe — Gly — Asn — Leu — Ser — Ser — Ala — Ser — Ala — Ile — Met — Gly — Asn — Pro
29 30 31 32 33 34 35 36 37 38 39 40 41 42 43 44 45 46 47 48 49 50 51 52 53 54 55 56 57 58

54 55 56 57 58 59 60 61 62 63 64 65 66 67 68 69 70 71 72 73 74 75 76 77 78 79 80 81 82 83
Gln — Val — Lys — Gly — His — Gly — Lys — Lys — Val — Ala — Asp — Ala — Leu — Thr — Asn — Ala — Val — Ala — His — Val — Asp — Asp — Met — Pro — Asn — Ala — Leu — Ser — Ala — Leu
Lys — Val — Lys — Ala — His — Gly — Lys — Lys — Val — Leu — Gly — Ala — Phe — Ser — Asp — Gly — Leu — Ala — His — Leu — Asp — Asn — Leu — Lys — Gly — Thr — Phe — Ala — Thr — Leu
Lys — Val — Lys — Ala — His — Gly — Lys — Lys — Val — Leu — Thr — Ser — Leu — Gly — Asp — Ala — Ile — Lys — His — Leu — Asp — Asp — Leu — Lys — Gly — Thr — Phe — Ala — Gln — Leu
59 60 61 62 63 64 65 66 67 68 69 70 71 72 73 74 75 76 77 78 79 80 81 82 83 84 85 86 87 88

84 85 86 87 88 89 90 91 92 93 94 95 96 97 98 99 100 101 102 103 104 105 106 107 108 109 110 111 112 113
Ser — Asp — Leu — His — Ala — His — Lys — Leu — Arg — Val — Asp — Pro — Val — Asp — Phe — Lys — Leu — Leu — Ser — His — Cys — Leu — Leu — Val — Thr — Leu — Ala — Ala — His — Leu
Ser — Glu — Leu — His — Cys — Asp — Lys — Leu — His — Val — Asp — Pro — Glu — Asn — Phe — Arg — Leu — Leu — Gly — Asn — Val — Leu — Val — Cys — Val — Leu — Ala — His — His — Phe
Ser — Glu — Leu — His — Cys — Asp — Lys — Leu — His — Val — Asp — Pro — Glu — Asn — Phe — Lys — Leu — Leu — Gly — Asn — Val — Leu — Val — Thr — Val — Leu — Ala — Ile — His — Phe
89 90 91 92 93 94 95 96 97 98 99 100 101 102 103 104 105 106 107 108 109 110 111 112 113 114 115 116 117 118

114 115 116 117 118 119 120 121 122 123 124 125 126 127 128 129 130 131 132 133 134 135 136 137 138 139 140 141
Pro — Ala — Glu — Phe — Thr — Pro — Ala — Val — His — Ala — Ser — Leu — Asp — Lys — Phe — Leu — Ala — Ser — Val — Ser — Thr — Val — Leu — Thr — Ser — Lys — Tyr — Arg
Gly — Lys — Glu — Phe — Thr — Pro — Pro — Val — Gln — Ala — Ala — Tyr — Gln — Lys — Val — Val — Ala — Gly — Val — Ala — Asp — Ala — Leu — Ala — His — Lys — Tyr — His
Gly — Lys — Glu — Phe — Thr — Pro — Glu — Val — Gln — Ala — Ser — Try — Gln — Lys — Met — Val — Thr — Gly — Val — Ala — Ser — Ala — Leu — Ser — Ser — Arg — Tyr — His
119 120 121 122 123 124 125 126 127 128 129 130 131 132 133 134 135 136 137 138 139 140 141 142 143 144 145 146

FIGURE 4-8

Alpha, beta, and gamma chains of normal human hemoglobins. Copyright © 1963, by Frank W. Putnam. Prepared by Frank W. Putnam, Division of Biology. Indiana University.

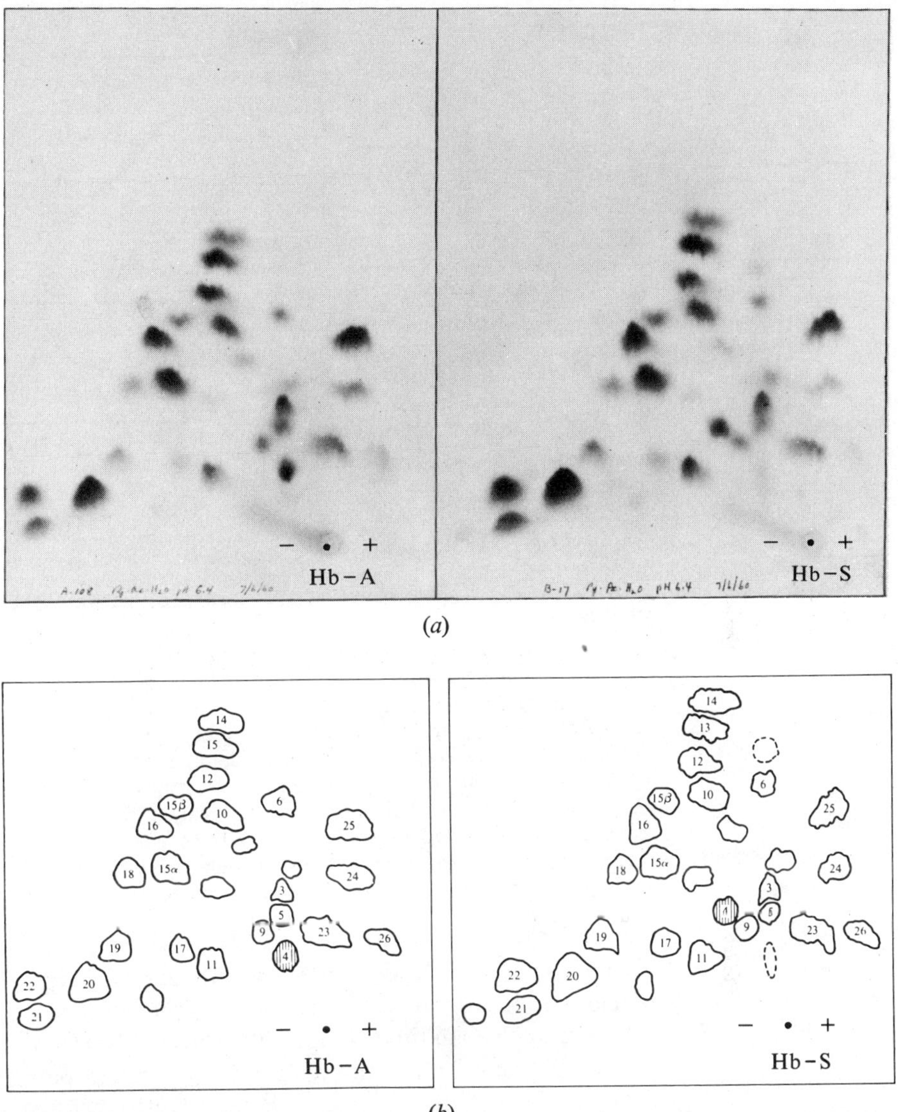

FIGURE 4-9
Fingerprints of trypsin digests of hemoglobins A and S. (*a*) Photograph of a pair of fingerprints. (*b*) Tracings of a pair of fingerprints showing the system for numbering peptide pairs. Dashed lines indicate peptides that only become visible after heating the chromatogram. C. Baglioni, *Biochim. Biophys. Acta,* **48:** 392 (1961).

light chain normally present in circulating immunoglobulins; these are distinguished on the basis of their antigenic behavior. The nature of the heavy chain provides a classification into three main groups; these have been termed γG, γA, and γM or, alternatively, IgG, IgA, and IgM. Each of these classes is subdivided into two parts, depending on the nature of the light chain. The classification of the six subgroups of the immunoglobulins is depicted in Fig. 4-11.

The heterogeneity of the immunoglobulins far transcends that based on the above properties. That is, the λ light chains are not homogeneous as to amino acid sequence but are themselves a group of polypeptides of variable

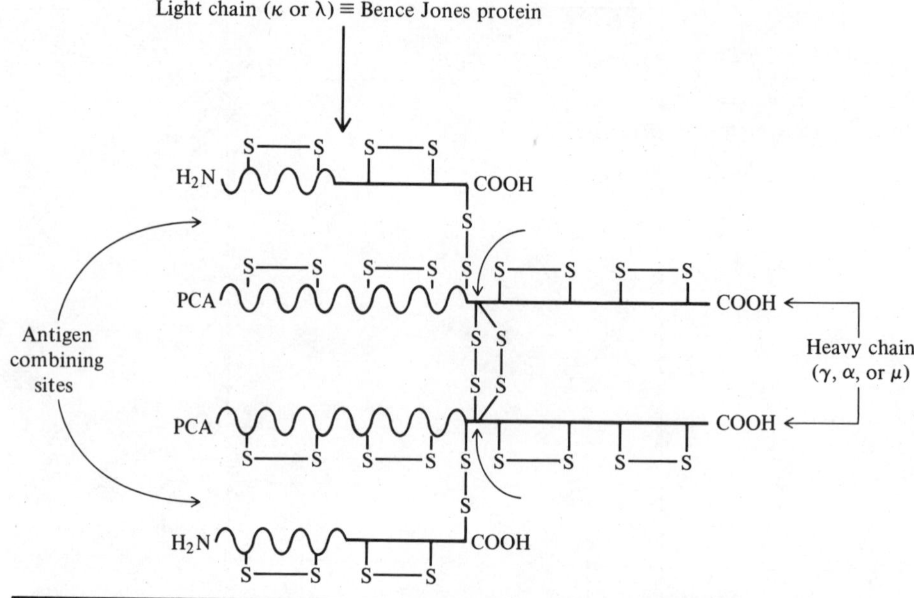

FIGURE 4-10
Diagram drawn to scale of the linear polypeptide chain structure of γG-globulin; γA- and γM-globulins probably have an analogous structure. Two identical half-molecules are symmetrically arranged; each contains a light chain and a heavy chain. The position and number of the intrachain disulfide bonds are correctly shown in the light chains as deduced from sequence analysis of human and mouse Bence-Jones proteins but are hypothetical in the heavy chains. Human γG has two, but rabbit γG has only one heavy–heavy interchain bond at the location shown. The variable portion of the sequence is symbolized by wavy lines and the constant portion by straight lines. The location and relative length of these regions are correctly shown in light chains but are hypothetical in heavy chains. F. W. Putnam, *Science,* **163:** 633 (1969).

primary structure. It is known that at least 40 types of light chains exist in man (16). How is one to determine primary sequences for immunoglobulins when such a multitude of individual species is present?

Such determinations are possible due to certain pathological conditions. In contrast to healthy humans, those afflicted with multiple myeloma or macroglobulinemia secrete large amounts of homogeneous immunoglobulins into the serum. Incomplete immunoglobulins, termed *Bence-Jones proteins,* may be excreted in the urine in multiple myeloma. These abnormal proteins provide a suitably homogeneous source of immunoglobulins to make sequence work feasible. It is worth noting that each patient excretes a unique Bence-Jones protein.

It has proved possible to obtain complete amino acid sequences for both heavy (17,18) and light (19–21) immunoglobulin chains. Both have the remarkable property that the C-terminal part of the molecule has an invariant structure common to all, whereas the N-terminal part has a structure that differs from individual to individual and from molecule to molecule within a given individual. Focusing attention on the light chains, the primary structure of the human λ-type Bence-Jones protein Sh is shown in Fig. 4-12 together with comparisons for two closely related proteins (22). The relative dispositions of constant and variable parts of the light and heavy chains is included in Fig. 4-10; note that the antigen combining sites are made up entirely of parts of the variable regions.

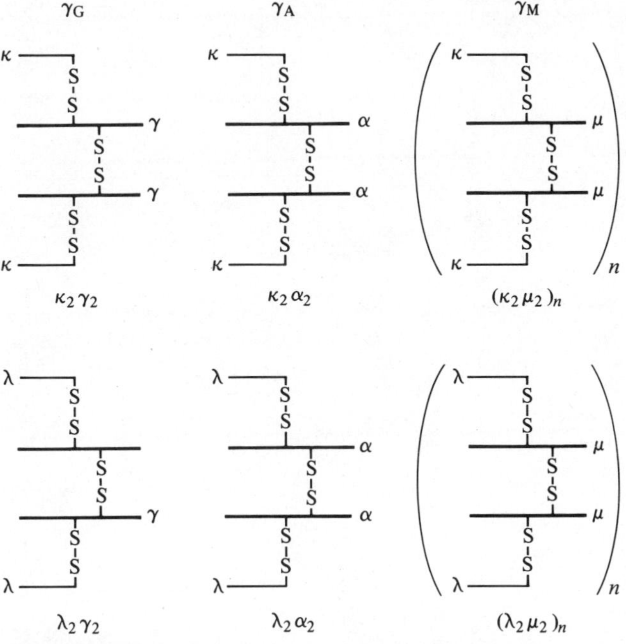

FIGURE 4-11
Polypeptide chain structure of the three major classes of the immunoglobulins (γG, γA, and γM). The light chains are denoted κ and λ and the heavy chains, γ, α, or μ. The chain formula is given under each subgroup. Although only one disulfide bond between the heavy chains is depicted in this schematic figure, certain human γG- and γM-globulins have more than one such bond per subunit. F. W. Putnam, *Nobel Symp.*, **3:** 45 (1967).

These findings constitute a unique problem in protein synthesis. As detailed in Chapter 19, one believes that protein structures are uniquely determined by the structure of genes (nucleic acids) which are established without respect to contact with antigens. If this is true, how can we reconcile this with the formation of specific immunoglobulins, differing in primary structure and capable of recognizing particular antigens, after exposure to these agents.

SECONDARY STRUCTURE OF PROTEINS

Both the carbonyl oxygen and amide hydrogen of the peptide bond have the capacity to form hydrogen bonds. In aqueous solution, these atoms are usually free to form hydrogen bonds with water molecules. However, in both the cases of fibrous and globular proteins, a substantial portion of the polypeptide chain is ordinarily isolated from direct contact with the solvent through being buried in the interior of the total structure. Under these circumstances, and frequently even when the peptide bonds are exposed to solvent, the hydrogen-bonding capacities of these groups are satisfied through the linkage of one peptide group to another:

$$\begin{array}{l} | \qquad\qquad\quad | \\ \mathrm{C{=}O \cdots H{-}N} \\ | \qquad\qquad\quad | \\ \mathrm{N{-}H \cdots O{=}C} \\ | \qquad\qquad\quad | \end{array}$$

Structures resulting from such interactions constitute secondary structure for proteins. Hydrogen bonds between different peptide groups can result in

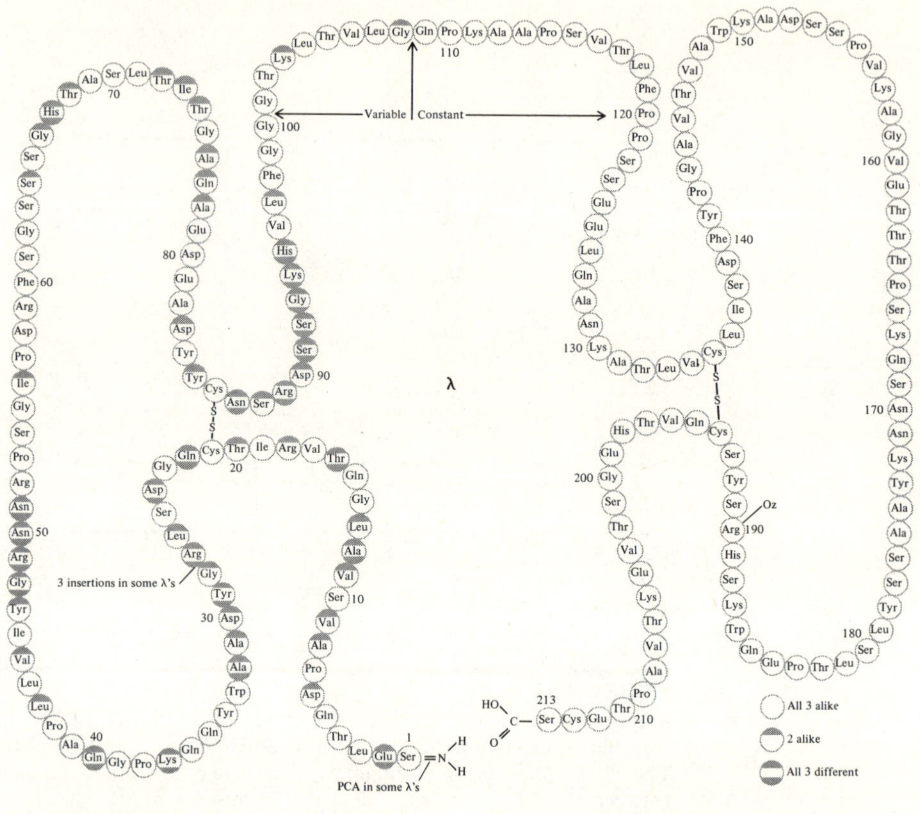

FIGURE 4-12
Amino acid sequence of the human λ-type Bence-Jones protein Sh. Positions given in white circles are identical in the human λ-type Bence-Jones proteins Ha and Bo. Where the circle is black at the top, two of the three λ chains have the same amino acid but differ from the third. All three proteins differ in positions where the circles are black at the top and bottom. F. W. Putnam, *Nobel Symp.,* **3:** 45 (1967).

the formation of two general structural types: helices and sheets. Let us consider the two cases separately.

The classic work in the field of helical structures in proteins is that of L. Pauling and R. B. Corey. From an extensive series of X-ray diffraction studies on crystalline low-molecular-weight amides, Pauling, Corey, and Branson established a set of criteria for formation of the most stable secondary structures (23). These requirements include (1) that the peptide group be planar and have bond lengths and angles identical to those found in crystals of simple secondary amides (see Fig. 3-1), (2) that every carbonyl oxygen and amide nitrogen be involved in hydrogen-bond formation, (3) that the hydrogen-bonded hydrogens lie close to a line joining the oxygen and nitrogen atoms involved in formation of the bond, and (4) that the operation, in going from one residue to the next (in terms, for example, of translation along or rotation around a central axis), be the same for every residue. Although recent studies reveal that some helices are formed that do not fully meet these requirements, Pauling and Corey identified the one structure that does, and this one has proved to be much the most important helical structure that occurs in proteins. It is known as the α helix.

Since the peptide bond itself is planar, the polypeptide chain has just two

FIGURE 4-13
Definition of the rotation angles ϕ and ψ for a pair of peptide units; $\phi = \psi = 0$ corresponds to the conformation shown in the diagram, which is that of a fully extended chain. J. T. Edsall, P. J. Flory, J. C. Kendrew, A. M. Liquori, G. Némethy, G. N. Ramachandran, and H. A. Scheraga, *J. Mol. Biol.,* **15:** 339 (1966).

degrees of freedom: rotation about the Cα—N bond, ϕ, and rotation about the Cα—C bond, ψ. This situation is shown diagrammatically in Fig. 4-13. Specification of both of these angles for each of the residues along a polypeptide chain would completely define the three-dimensional structure of that chain. Any repeating structure such as a helix can be specified by a value for each of these angles; each residue must have the same angles for the structure to be a repeating one. Specifically, the α helix is generated by the specification that ϕ be 132° and ψ be 123°. These angles can be translated into more usual parameters of helical structure. These include the number of units per turn of the helix, n, the distance traveled parallel to the helix axis per residue, d, and the pitch of the helix, $p = n \cdot d$. For the case of the α helix, there are 3.6 residues per turn and a translation per residue near 1.5 Å. Thus, the pitch of the helix is 5.4 Å. Drawings of the α-helical structures for polypeptides are shown in Fig. 4-14. Examination of the figure reveals that the hydrogen bonds

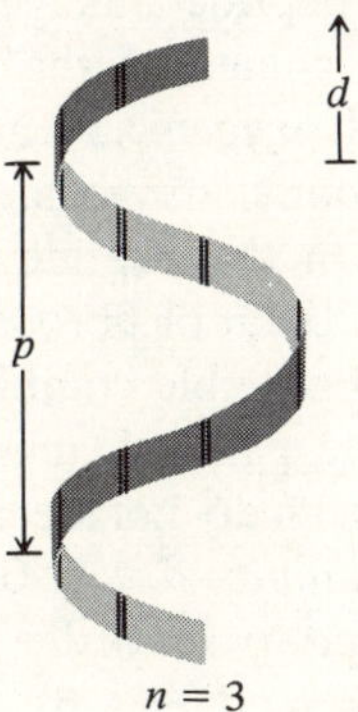

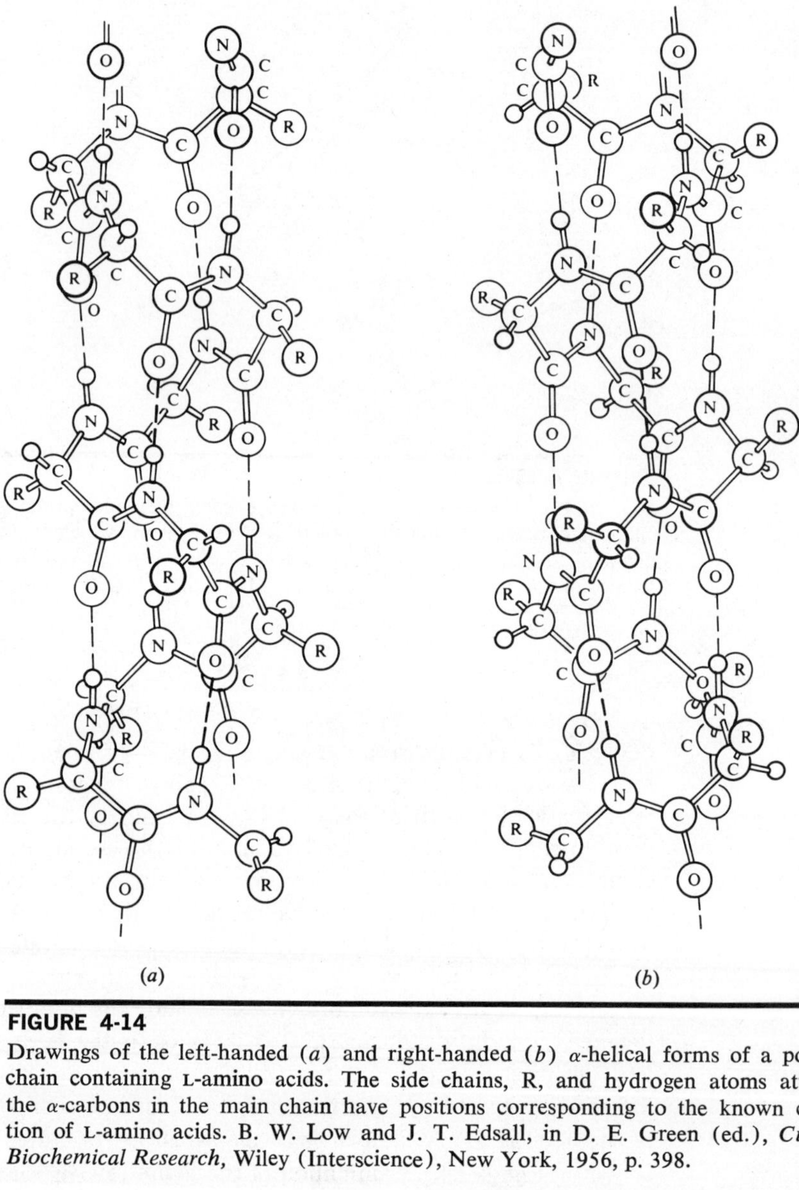

(a) (b)

FIGURE 4-14
Drawings of the left-handed (*a*) and right-handed (*b*) α-helical forms of a polypeptide chain containing L-amino acids. The side chains, R, and hydrogen atoms attached to the α-carbons in the main chain have positions corresponding to the known configuration of L-amino acids. B. W. Low and J. T. Edsall, in D. E. Green (ed.), *Currents in Biochemical Research,* Wiley (Interscience), New York, 1956, p. 398.

formed between the components of the peptide linkages are nearly parallel to the long axis of the helix. The side chains of the amino acid residues bristle out from the helix. Each carbonyl oxygen and amide hydrogen is involved in hydrogen-bond formation: carbonyl oxygens are hydrogen-bonded to amide hydrogens three residues back in the peptide chain. Proteins containing α-helical structure may be either globular or fibrous.

It is important to recognize that not all possible combinations of the angles ϕ and ψ yield reasonable structures. Some pairs of angles do provide stable structures, such as the α helix, while others do not because they bring nonbonded atoms too close together. For example, if ϕ is 0° and ψ is 180°, the two adjacent carbonyl oxygen atoms overlap so badly that the structure is an impossible one. Similarly, if ϕ is 180° and ψ is 0°, the amide hydrogen atoms overlap too much and the structure is also impossible. This entire topic

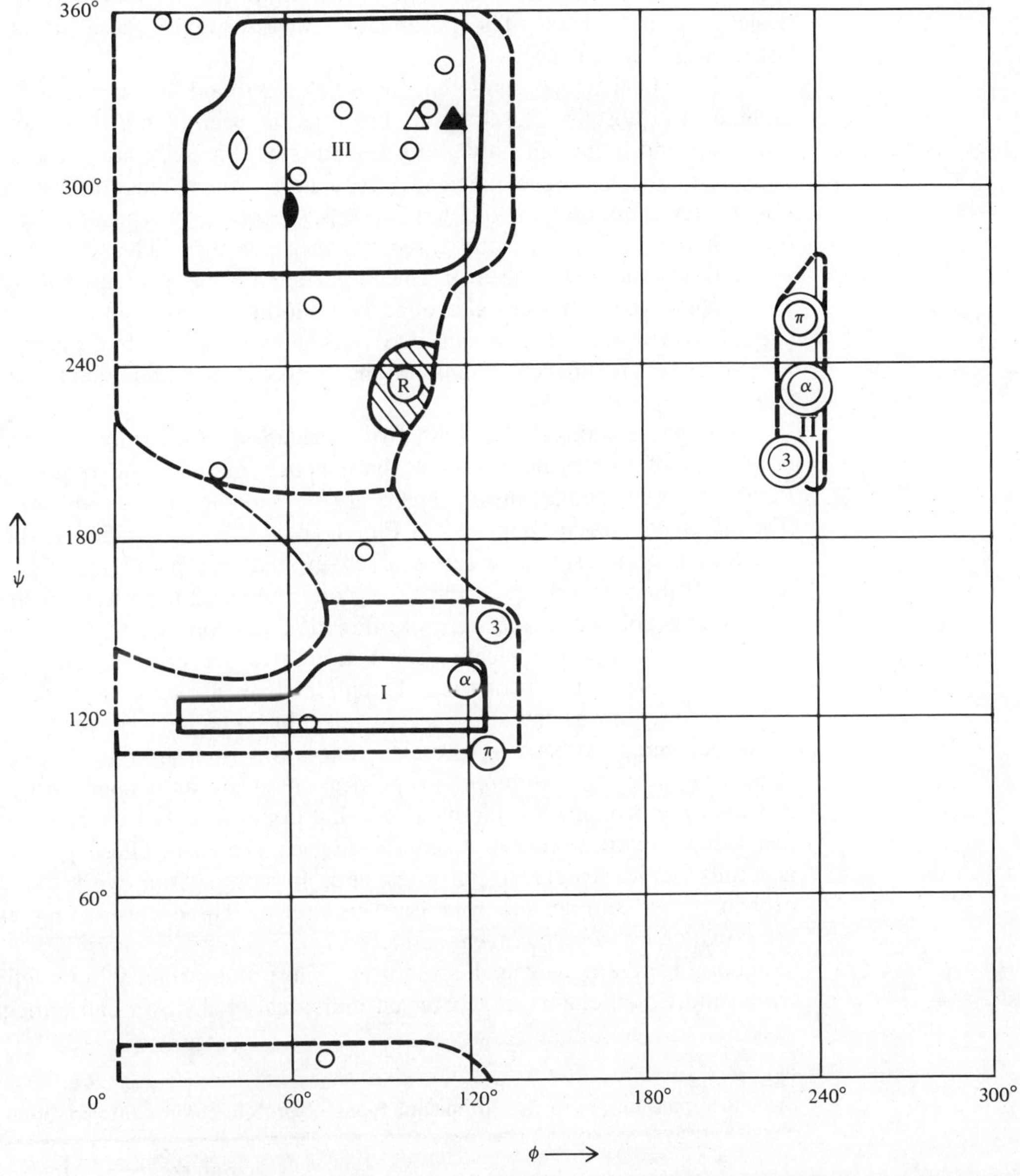

FIGURE 4-15
Ramachandran plot for a pair of peptide units with β carbon atoms attached the α carbons in the L configuration. The N—C$^\alpha$—C′ bond angle has been taken as 110°, near the ideal tetrahedral value. Fully allowed conformations are outlined by a solid line and partially allowed ones by a dotted line. Conformations plotted are ⓐ = α_P (right-handed), ◎ = α_M (left-handed), ⓟ = π_P, ◎ π_M, ③ = $3_{10,P}$, ◎ = $3_{10,M}$, ◆ = β-structure, ◇ = silk, Ⓡ = Ribbon structure, 2.2_7-helix, ▲ = collagen, △ = poly-Gly, -Pro, -Hypro, o = observed in simple peptides. From G. N. Ramachandran and V. Sasisekharan, *Advan. Prot. Chem.*, **23**, 283 (1968).

has been thoroughly investigated by G. N. Ramachandran and his research group in Madras. These workers have constructed plots of ϕ against ψ showing which areas are allowed and which are not allowed on the basis of forbidden nonbonded interactions. A Ramachandran map is shown in Fig. 4-15 together with locations on the map of those structures, both helical and sheet, that have been identified in proteins or artificial polypeptides. Note that nearly all the observed structures do fall into the regions that have been designated as allowed. The parameters for several of these structures are collected in Table 4-6.

Of the helical structures noted, the α helix is found in a variety of proteins, globular and fibrous. The 3_{10} helix has thus far been identified in hemoglobin and lysozyme, although only short segments of this helix are present. The 2_7 helix (ribbon) has never been identified in a protein but does occur in one known tetrapeptide. In the artificial polypeptide polyproline, a helix is encountered that has exactly three residues per turn. This is a particularly interesting structure because in the important fibrous protein collagen, three polyproline-type helices twist around one another to generate a three-stranded superhelix. In this case about two out of every three hydrogen bonds is formed between different chains of the triplex. More about collagen follows later.

The sheet structures that result from formation of intermolecular hydrogen bonds are of two principal types, both consistent with the requirements of Pauling, Corey, and Branson. These are termed the *parallel* and *antiparallel pleated sheets* and are depicted in Fig. 4-16. As the names imply, the former structure is composed of a series of chains that are parallel in the sense of having all their N termini at the same end, and the latter structure has every other chain pointed in the opposite direction. In contrast to the α helix, the hydrogen bonds of the pleated sheet structures are oriented nearly perpendicularly to the long axis of the polypeptide chain. Proteins that contain sheet structures are fibrous and are generally insoluble in aqueous solvents.

As one might anticipate, proteins that form different types of secondary structures have quite different properties; these are associated with the uses to which the proteins are directed in living organisms. Let us briefly consider just three examples chosen from the fibrous proteins. These proteins quite generally serve structural purposes and, in comparison with the globular proteins, have simple and repetitive structures. Three contrasting examples are provided by *wool, collagen,* and *silk.*

Wool fibers are α-helical structures. Since the primary interactions are within individual chains, not between individual chains, we can anticipate the

TABLE 4-6
Structural parameters of five important types of protein-chain conformations

	α helix	3_{10} *helix*	2_7 *ribbon*	*Polyproline helix*	*Antiparallel β pleated sheet*
ϕ	132° (113°)[a]	131° (106°)	105°	103°	40°
ψ	123° (136°)	154° (176°)	250°	326°	315°
n	3.61	3.00	2.00	−3.00	2.00
d(Å)	1.50	2.00	2.80	3.12	3.47
p(Å)	5.41	6.00	5.60	9.36	6.95

[a] Values in parentheses are alternate values obtained in independent investigations.
SOURCE: R. E. Dickerson and I. Geis, *The Structure and Action of Proteins,* Harper & Row, New York, 1969.

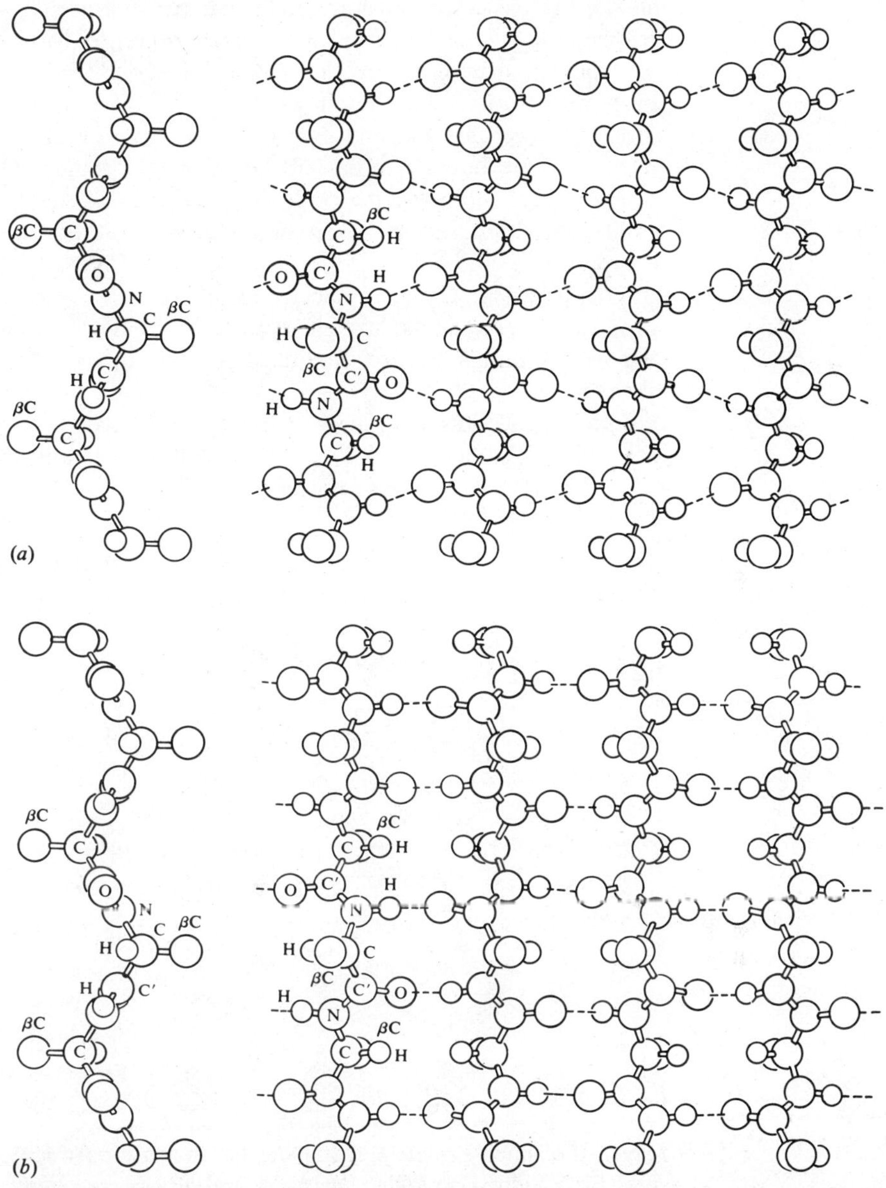

FIGURE 4-16

Representation of the parallel (*a*) and antiparallel (*b*) pleated sheet structures for polypeptides. L. Pauling and R. B. Corey, *Proc. Natl. Acad. Sci. U.S.*, **37:** 729 (1951).

properties of wool. The fibers are flexible and elastic, reflecting the deformability of the helical structures and lack of strong interchain interactions. The latter point is responsible for the fact that wool fibers are not particularly strong.

Collagen is the most important insoluble fiber of connective tissues. It comprises about 6 percent of the body weight of man and is the single most abundant protein. Collagen has a very unusual amino acid composition; about one third of the residues are glycine and another third are either proline or hydroxyproline. Long stretches of the polypeptide chain contain repeating sequences of the type Gly-X-Pro or Gly-X-Hyp. As mentioned above, the basic structure of collagen is triple-stranded, three helically coiled polypeptide chains winding about each other to form a superhelix. These superhelical structures have a diameter of about 15 Å and are about 2800 Å long, with a molecular weight near 30,000. This basic unit is frequently referred to as *tropocollagen*. Two of every three carbonyl oxygens and amide hydrogens are bonded to one of the other two chains. This interchain hydrogen bonding generates a molecule that is strong and rigid.

The structural organization of collagen as it occurs in tendons is quite complex. The individual stages are shown in Fig. 4-17. Basically, individual tropocollagen molecules associate in an overlapping way to form a collagen fibril; the characteristic striated pattern of the fibrils is believed to result from this overlapping structure. The fibrils, in turn, are organized into macroscopic structures.

Finally, silk has a β-sheet structure. Silk, like collagen, has a repeating amino acid sequence. In this case, the repeat unit contains six residues:

(Gly · Ser · Gly · Ala · Gly · Ala)

The individual sheets are packed together so that glycine faces glycine and alanine (or serine) faces alanine:

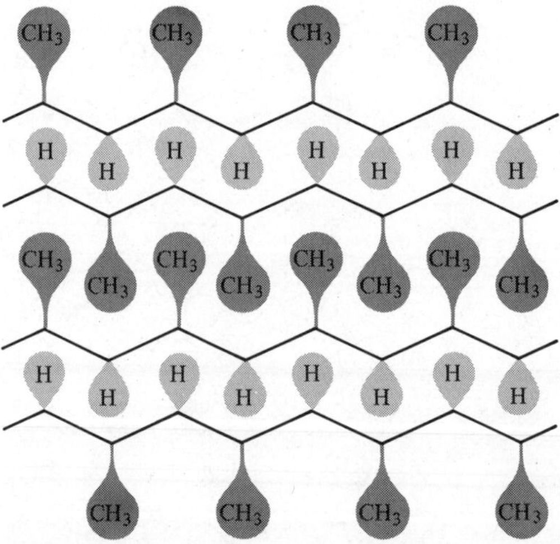

Thus, the spacings between the sheets alternate between 3.5 and 5.7 Å. Since the silk structure involves strong interchain interactions, it has great strength and little extensibility or elasticity. Since the individual sheets are deformable, silk has great flexibility.

We now direct attention to the matter of the experimental differentiation among the various types of secondary structures.

Differentiation Between Secondary Structures

Infrared Spectroscopy and Infrared Dichroism. Polypeptides exhibit characteristic infrared spectra. The positions and number of bands observed is a function of the secondary structure of the polypeptide. To begin with, let us consider some of the features of the infrared spectra of randomly coiled molecules of this type. The absorption bands of particular interest are found near 3290, 1650, and 1535 cm^{-1}. These absorptions are due, respectively, to the N—H stretching mode, the stretching mode of the

$$\rangle C{=}O$$

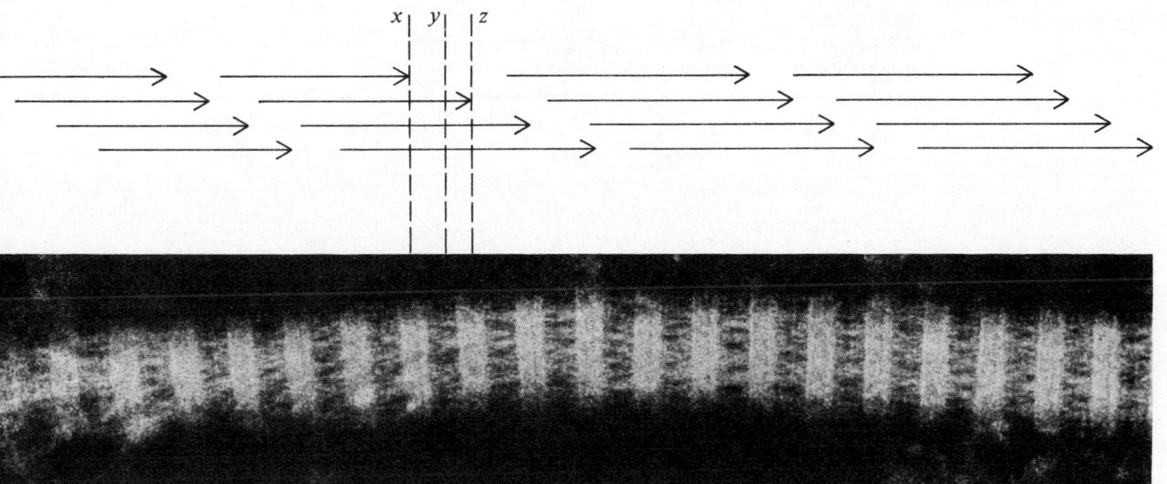

FIGURE 4-17
Electron micrograph of native collagen fibrils, negatively stained with phosphotungstic acid and a schematic representation of the alignment of the tropocollagen molecules in the native fibril. Dashed lines indicate hole zone *xy* (approximately 400Å) and overlap zone *yz* (approximately 250Å). (Courtesy of Dr. F. B. Merk.)

bond, and the deformation mode (admixed with the C—N stretching mode) of the N—H bond. The latter two bands are termed the *amide I* and *amide II* bands. The transition from a randomly coiled molecule to an α helix, parallel-chain pleated sheet, or antiparallel-chain pleated sheet is accompanied by characteristic alterations in the nature of the spectra. In Table 4-7, the observed infrared frequencies for the amide I and II bands of polypeptides in various conformations are recorded (24). As a result of a vibrational interaction among peptide groups in the chains and other factors, each of these bands now exhibits two maxima in the ordered structures. Inspection of the data in Table 4-7 reveals that preliminary structural assignments may be made on the basis of the positions of the amide I and II bands.

More definitive conclusions may be drawn from infrared dichroism studies on ordered solids employing plane-polarized infrared light. Infrared dichroism is the dependence of infrared absorption on the orientation of the solid sample. Infrared dichroism arises as a result of the fact that maximum absorption of light is observed when the transition moment of the concerned vibration is parallel to the electric vector of the light. To the extent that these are not parallel, diminished light absorption is found, and, for the perfectly perpendicular case, light absorption is abolished. How do these considerations apply to differentiating between helical and pleated sheet conformations for polypeptides?

As noted above, the α-helical structure involves the formation of hydrogen bonds that are approximately parallel to the long axis of the molecules. Hence the transition moments for the

$$\gt N{-}H \quad \text{and} \quad \gt C{=}O$$

TABLE 4-7
Observed frequencies of amide I and II bands of polypeptides in various conformations (cm^{-1})

Conformation	*Amide I frequencies*	*Amide II frequencies*
Random coil	1,656 (s)[a]	1,535 (s)
Helix	1,650 (s) 1,652 (m)	1,516 (w) 1,546 (s)
Parallel-chain pleated sheet	1,645 (w) 1,630 (s)	1,530 (s) 1,550 (w)
Antiparallel-chain pleated sheet	1,685 (w) 1,632 (s)	1,530 (s)

[a] The letter symbols in parentheses indicate observed intensities: s = strong, m = medium, and w = weak.

SOURCE: T. Miyazawa and E. R. Blout, *J. Am. Chem. Soc.,* **83:** 712 (1961); by permission of the copyright owners, the American Chemical Society.

stretching modes are parallel to the long axis of the molecule and that for the N—H deformation mode is perpendicular to the axis. Thus, if the molecules of a sample are ordered in a two-dimensional array with all the long axes parallel, absorption due to the stretching modes should be most intense when the plane of the polarized light is parallel to the long axis (parallel dichroism) and absorption due to the deformation mode should be most intense when the plane of the polarized light is perpendicular to the long axis (perpendicular dichroism). These considerations are illustrated in Fig. 4-18 for the case of poly-γ-benzyl-L-glutamate (24). This sample, which is known to be α-helical, exhibits parallel dichroism for the amide I band and perpendicular dichroism for the amide II band as predicted. The band at 1731 cm^{-1} is the C=O stretching mode for the benzyl ester. It exhibits little dichroism, indicating that this carbonyl function is randomly oriented with respect to the long axis of the molecule. Precisely the opposite behavior is expected, and found, for molecules in sheet structures. As shown in Fig. 4-16, the intermolecular hydrogen bonds of sheet structures are oriented approximately perpendicular to the long axis of the molecule. (See also (25).)

Optical Rotatory Dispersion and Circular Dichroism

Two useful methods for distinguishing between the various types of secondary structure of proteins derive from the way that plane-polarized light behaves on passage through a dissymmetric medium. One can resolve plane-polarized light into two circularly polarized beams that trace out helical paths of opposite chiralities: one beam is polarized in a left-handed sense (L) and the other in a right-handed one (R). When plane-polarized light passes through a dissymmetric medium, the left-handed and right-handed circularly polarized beams have different indices of refraction ($n_R \neq n_L$) and different molar extinction coefficients ($\varepsilon_R \neq \varepsilon_L$). The former fact is responsible for the phenomenon of circular birefringence and the latter for that of *circular dichroism* (CD). The dependence of the circular birefringence on wavelength of the light is *optical rotatory dispersion* (ORD).

Optical rotatory dispersion spectra are recorded through measurement of the optical rotation, $[\alpha]$, as a function of wavelength, λ. For proteins, the most useful quantity is the reduced mean residue rotation, $[m']$:

$$[m'] = \frac{3}{n^2 + 2} \frac{\text{MRW}}{100} [\alpha]_\lambda \tag{4-10}$$

in which n is the index of refraction and MRW is the mean residue weight.

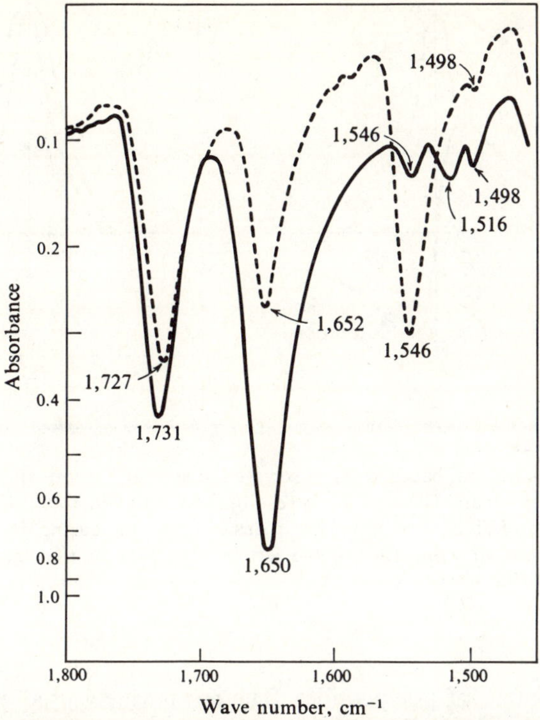

FIGURE 4-18

Polarized infrared spectra of poly-γ-L-glutamate (film unidirectionally oriented from chloroform solution). Solid line, electric vibration direction parallel to orientation direction; dashed line, electric vibration direction perpendicular to orientation direction. T. Miyazawa and E. R. Blout, *J. Am. Chem. Soc.,* **83:** 712 (1961); by permission of the copyright owners, the American Chemical Society.

Circular dichroism spectra are recorded as an ellipticity of the transmitted light; the useful quantity is the mean residue ellipticity, $[\theta]$. Both ORD and CD spectra are related to asymmetric or dissymmetric electronic transitions. The relationships are shown diagrammatically in Fig. 4-19. Note that the CD band has very nearly the same shape as the absorption band and is, therefore, rather sharp, and that the ORD band is, in contrast, infinitely broad.

ORD and CD spectra contain basically the same information. In fact, the values of $[m']$ and $[\theta]$ can be directly related to one another mathematically. Thus, it is possible, but not always easy, to construct a CD spectrum from an experimental ORD spectrum, and vice versa. In practice, both types of measurement are widely employed for sorting out various types of structures and for observing changes in them.

Since both optical rotation and ellipticity depend upon the environment of the absorbing chromophore, it follows that these quantities should be dependent on the structure of the molecule. One example is provided in Fig. 4-20, in which ORD spectra for poly-L-lysine are provided as a function of *p*H. Clearly, the spectra for the three conformations are quite distinct and provide a basis for making a distinction between them. The CD spectra exhibit substantial differences as well. Through collection of data of this type, it has been possible to characterize the ORD and CD spectra for several chain conformations; a collection of such data is provided in Table 4-8.

Both ORD and CD data have been employed in efforts to estimate the

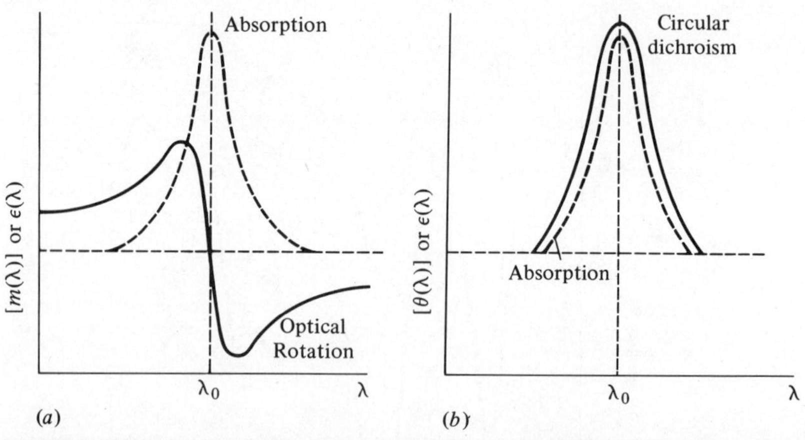

FIGURE 4-19

Isolated absorption bands with associated negative Cotton effect in the optical rotary dispersion spectrum (*a*) and positive ellipticity band in the circular dichroism spectrum (*b*). The symbol λ is a wavelength designation, increasing in value toward the right and decreasing in value toward the left in each half of the figure. S. Beychok, *Science,* **154:** 1288 (1966).

helical content of polypeptides. The most widely used criterion has been the b_0 value from the Moffitt–Yang equation:

$$[m'] = a_0 \frac{\lambda_0^2}{\lambda^2 - \lambda_0^2} + b_0 \frac{\lambda_0^4}{(\lambda^2 - \lambda_0^2)^2} \tag{4-11}$$

TABLE 4-8
Optical rotatory dispersion and circular dichroism of various conformations[a]

λ, nm	*α helix*[b]		*β forms*[c]		*Coil*[d]	
ORD	λ, nm	[m']	λ, nm	[m']	λ, nm	[m']
Trough	232.5	−15,000	229–30	(5000)	239	−2,000
Peak	198.5	68,000	205	(24,000)	231	−1,800
Trough	182–184	—	~190	(−17,000)	205	−19,000
Peak					189	16,000
CD	λ, nm	[θ]	λ, nm	[θ]	λ, nm	[θ]
Minimum	222.5	−41,000	218	(−20,000)	238	−300
Minimum	210	−38,000				
Maximum	191.5	90,000	195–197	(48,000)	218	4,900
Minimum					197.5	−38,000

[a] The magnitude of the extrema listed here is still tentative. Dimensions: reduced mean residue rotations, [m'], in deg-cm²/decimol and mean residue ellipticity, [θ], in deg-cm²/decimol (uncorrected for refractive index).

[b] Poly-L-glutamic acid in water at *p*H 4.75.

[c] Silk-fibroin in 50% methanol (extrapolated to 100% β form). The values in the parentheses are only rough estimates and vary with the solvent composition.

[d] Poly-L-glutamic acid in water at *p*H 7.39.

SOURCE: J. T. Yang, in P. Alexander and H. P. Lundgren (eds.), *A Laboratory Manual of Analytical Methods of Protein Chemistry,* vol. 5, Pergamon Press, New York, 1969, p. 25.

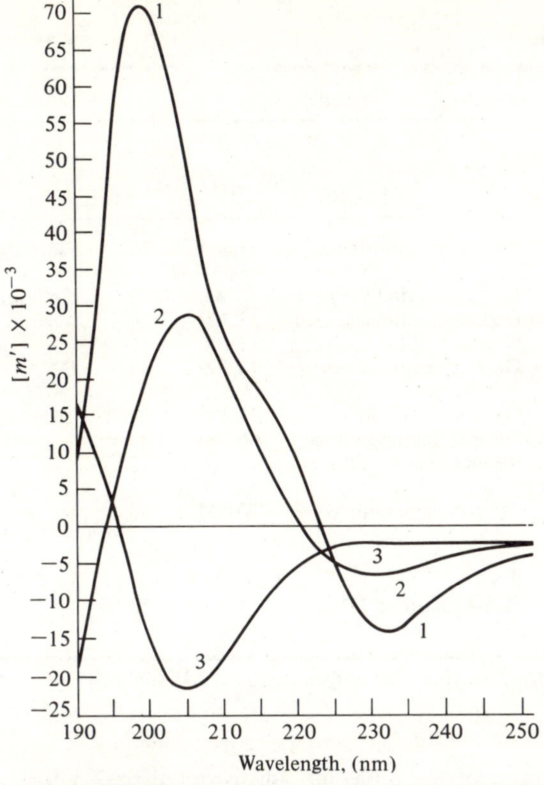

FIGURE 4-20
ORD curves of poly-L-lysine in the far-ultraviolet region, in the α-helical, β, and random coil conformations. Curve 1, α helix; 2, β structure; 3, random chain.

In this equation, a_0, b_0, and λ_0 (= 212 nm) are constants. Values of b_0, which may be obtained graphically, are known to vary from about −630 for perfectly α-helical polymers to zero for randomly coiled ones. Estimation of helical content by interpolation between these two values works reasonably well for synthetic homo- and copolymers of the amino acids, but is likely to fail for native proteins, for which a number of complications may arise. Helical content has also been estimated on the basis of the optical rotation values at 233 nm and on the basis of the ellipticity at 222 nm. As above, these values may yield approximate answers but cannot be firmly relied upon. The only unambiguous method for the determination of helical content of a protein is wide-angle X-ray diffraction, to which we turn shortly.

TERTIARY STRUCTURE OF PROTEINS

It has been recognized for quite some time that most proteins, in their native state, exist as rather tight compact structures. This can be appreciated directly from a consideration of the hydrodynamic behavior of proteins. Consider, for example, the *intrinsic viscosities* of a number of macromolecules. The intrinsic viscosity is defined as

$$[\eta] = \frac{V}{m} \Lambda \qquad (4\text{-}12)$$

in which V is the volume of the particle, m is its mass, and Λ is a function of

TABLE 4-9
Intrinsic viscosities for macromolecules

Compact globular particles			*Randomly coiled chains*			*Rodlike particles*		
	Molecular weight	$[\eta]$, cm^3/g		*Molecular weight*	$[\eta]$, cm^3/g		*Molecular weight*	$[\eta]$, cm^3/g
Polystyrene latex particles	10^9	2.4	Polystyrene in toluene	45,000	28	Fibrinogen	330,000	27
			Polystyrene in toluene	70,000	37	Collagen	345,000	1,150
Ribonuclease	13,700	3.3	Reduced ribonuclease	13,700	14.4	Myosin	620,000	230
Lysozyme	14,400	3.0	Oxidized ribonuclease	14,100	11.6			
Myoglobin	17,000	3.1	Oxidized ribonuclease in urea	14,100	13.9	DNA	5×10^6	5,000
β-Lactoglobulin	35,000	3.4						
Ovalbumin	44,000	4.0	Ovalbumin in urea	44,000	34	TMV	4×10^7	29
Serum albumin	65,000	3.7	Serum albumin in urea	66,000	22			
Hemoglobin	67,000	3.6	Reduced serum albumin in urea	66,000	53			
Liver alcohol dehydrogenase	83,000	4.0	Myosin in guanidine-hydrochloride	200,000	93			
Hemerythrin	107,000	3.6						
Aldolase	142,000	3.8						
			RNA	1.5×10^6	100			
Ribosomes (yeast)	3.5×10^6	5.0	Heat-denatured DNA	5×10^6	150			
Bushy stunt virus	8.9×10^6	4.0						

SOURCE: H. K. Schachman, *Cold Spring Harbor Symp. Quant. Biol.*, **28**: 409 (1963).

the axial ratio of the molecule and is equal to 2.5 for rigid spheres (a result due to Einstein) and increases with increasing axial ratio. In Table 4-9, values of the function $[\eta]$ are collected for several macromolecules that are grouped into three classes: compact globular particles, randomly coiled chains, and rodlike structures. As predicted by Eq. (4-12), the values for most of the native globular proteins are independent of molecular weight. Intrinsic viscosities are dependent on molecular weight for molecules that are not spherical, but the variations in $[\eta]$ in Table 4-9 are principally the result of shape differences. Note particularly the marked increase in $[\eta]$ that occurs on unfolding the polypeptide chain by cleavage of disulfide bridges (ribonuclease, serum albumin) or by treatment with concentrated solutions of urea (ovalbumin, serum albumin). Finally, note the very large values of $[\eta]$ for molecules, such as collagen and DNA (discussed in Chapter 5), which have a helical secondary structure but which lack superfolding of the helix. Data of the type shown in Table 4-9 lead to the conclusion that most proteins exhibit such superfolding (tertiary structure).

In addition, the folded structures for proteins must be highly specific; that is, the polypeptide chain or chains do not merely collapse into a roughly spherical structure but undergo a precise series of folding operations yielding a unique, or nearly unique, structure. This conclusion follows directly from the extreme sensitivity of the biological activity (i.e., enzymatic activity) of proteins to alterations in their tertiary structure (see below). In terms of the protein as a highly and specifically folded structure, we wish to ascertain, first, detailed information concerning the intimate details of such structures and, second, to account for such structures in terms of the nature of the forces responsible for their maintenance. A sophisticated approach to the understanding of the tertiary structure of proteins is afforded by X-ray diffraction.

X-Ray Diffraction and Tertiary Structure of Proteins

X-ray diffraction is one of the most potent general methods for the determination of molecular structures. Its successes in unraveling complicated chemical structures are numerous (vitamin B_{12}, for example). It should be recognized that the classical uses of X-ray diffraction involved molecules much less complicated than even the simplest and smallest proteins. The brilliant work of M. F. Perutz, J. C. Kendrew, and others, described below, has convincingly established that this method may be successfully applied to obtain detailed information about the latter class of molecules as well.

The basic equation describing the diffraction of a monochromatic beam of X rays by scattering centers (atoms) is due to Bragg:

$$n\lambda = 2d \sin \theta \tag{4-13}$$

This equation relates the crystal spacing d to the angle θ of the incident and reflected X ray; λ is the wavelength of the incident beam, and n, the order of the reflection, takes on integral values. Under conditions in which the Bragg equation is satisfied, maxima in the X-ray diffraction pattern are obtained.

There are several types of preparations and several ways of recording diffraction maxima that are useful in X-ray diffraction studies. For the specific case of proteins, detailed structural information is obtained only through the use of single crystals at least 300 μm in all dimensions. These are large crystals by protein-chemistry standards and obtaining them is frequently a primary experimental difficulty. They are mounted in a quartz capillary tube with some liquid of crystallization on a stage in line with the beam of X rays, and either cameras or counting devices are employed to record the positions and intensities of the diffraction maxima. The stage upon which the crystal is mounted is constructed in such a way that it may be rotated about three perpendicular axes:

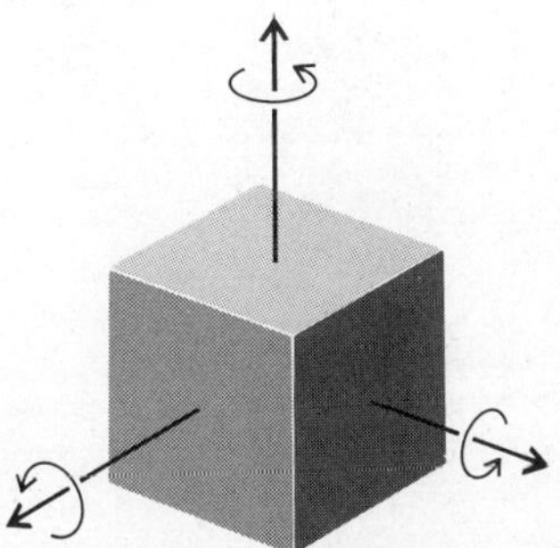

Thus, all possible orientations of the crystal with respect to the incident X-ray beam may be achieved.

The modern means of obtaining the diffraction pattern of a crystal is highly automated. The functioning of the diffractometer is controlled by a computer programmed in a way so that the various diffraction maxima are ferreted out through controlled rotations of the crystals about the three axes as noted above. Once a diffraction maximum is located, the crystal is held in a proper position while the intensity of the maximum is recorded through counting devices; the intensity of the maximum together with its location are then recorded. Once again the crystal is rotated with respect to the X-ray beam until another maximum is located and so forth. It is possible to record the positions and intensities of as many as 1000 maxima per day in this way without the physical intervention of the experimenter. A detailed analysis of X-ray diffraction methods as they apply to protein structure determinations has been prepared by Holmes and Blow (see references), to which the reader may refer for details.

In Fig. 4-21, an X-ray precession diagram of myoglobin is illustrated.

From the general features of the X-ray diffraction patterns, the dimensions of the unit cell, the number of molecules per unit cell, and an estimate of the molecular weight of the protein may be obtained (provided the density is known). We now wish to indicate briefly how the diffraction patterns may be employed to establish the positions of the individual atoms within the unit cell.

Consider a continuous function $\rho(x,y,z)$ whose value at every point in the crystal is a measure of the electron density at that point. Because electron-dense regions correspond to the position of atoms, a detailed knowledge of $\rho(x,y,z)$ suffices for the localization of all the atoms in the unit cell. A theorem due to Fourier states that irregular continuous functions, such as $\rho(x,y,z)$,

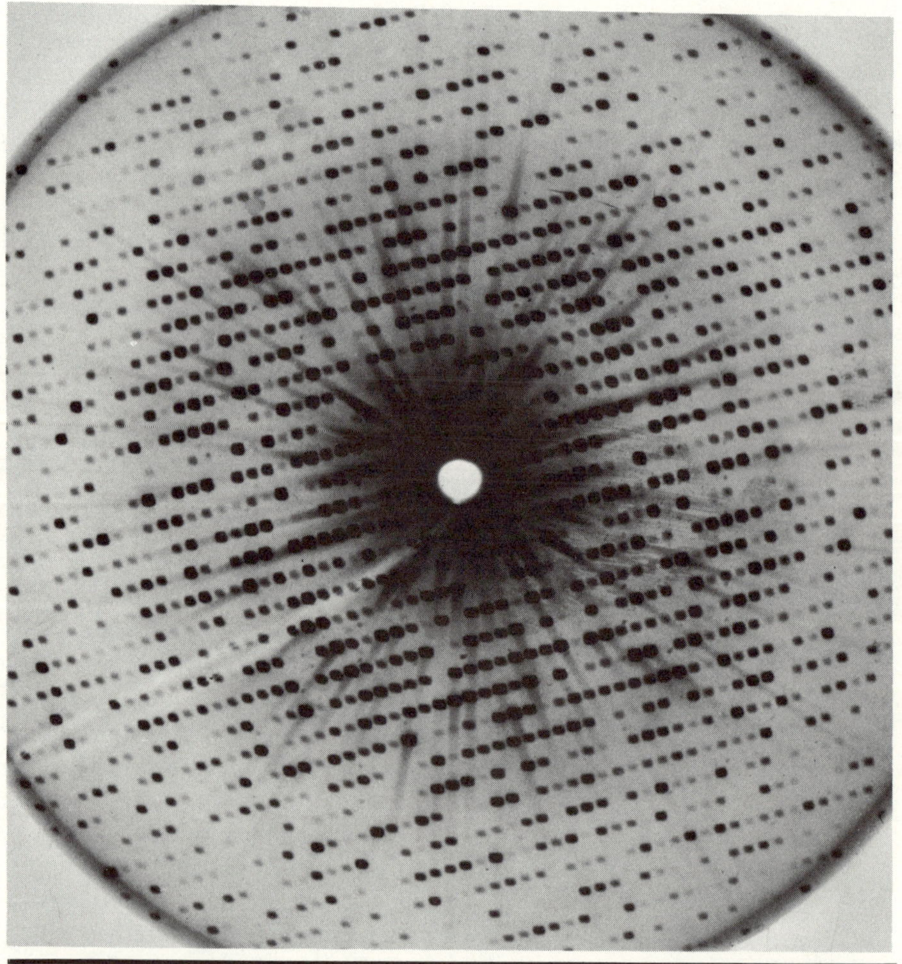

FIGURE 4-21
Precession diagram, myoglobin. F. H. Crick and J. C. Kendrew, *Advan. Protein Chem.*, **12:** 133 (1957).

may be expressed in terms of a summation of harmonic functions. In our case, we may write

$$\rho(x,y,z) = \frac{1}{V} \sum_{h,k,l=-\infty}^{h,k,l=\infty}\sum\sum F_{hkl} \exp\,[i\alpha(hkl)] \exp\,[-2\pi i(hx + ky + lz)] \tag{4-14}$$

in which the F_{hkl} are the observable structure amplitudes of the crystal reflection hkl, $\alpha(hkl)$ are the associated phases, and V is the volume of the unit cell. The summation is taken over the observable reflections. The indices h, k, and l (Miller indices) define a particular plane in the unit cell and may be associated with each individual diffraction maximum. Clearly, if all the F_{hkl} and $\alpha(hkl)$ were known, one could calculate $\rho(x,y,z)$ and hence obtain the complete structure. The amplitudes of each F_{hkl} may be obtained directly from the intensities of each of the diffraction maxima. The relationship between the intensity of a maximum and its F_{hkl} is expressed thus:

$$I = KF_{hkl}F^{*}_{hkl} \tag{4-15}$$

A consideration of this equation reveals that the amplitudes of F_{hkl} may be directly calculated but that the phase of the F_{hkl} may not be obtained from intensity measurements, because $Fe^{i\alpha} \cdot F^{*}e^{-i\alpha} = FF^{*}$. Thus, the Fourier components obtained from diffraction patterns are indeterminate as to phase.

The resolving power of X-ray diffraction techniques is slightly less than 1 Å. Because the interatomic distance for most covalent bonds is longer than this, one should be able to identify each of the atoms in the crystal. Diffraction patterns yield a reciprocal view of the crystal; that is, the diffraction maxima near the center of the pattern correspond to long spacings and those near the periphery correspond to short spacings in this crystal. Hence, as implied by Eq. (4-14), fine resolution of the structure requires the measurement of intensities, and the calculation of Fourier components, of maxima corresponding to small spacings. Thus, the summation indicated in Eq. (4-14) can become quite complicated if one wishes to obtain a highly resolved structure. For example, the work of Kendrew on the structure of myoglobin involved the calculation of more than 20,000 Fourier components. A resolution of less than 2 Å was obtained. The development of very high speed electronic computers markedly simplifies the treatment of data of this complexity.

Method of Isomorphous Replacements. A workable solution to the phase problem indicated above has been provided by Kendrew and Perutz. The solution is termed the method of *isomorphous replacements.* This method involves the specific introduction of (preferably) one heavy atom per unit cell into the crystal structure of the protein. Comparison of the diffraction patterns for the crystalline protein and crystalline heavy-atom derivative permits the localization of the heavy atom within the unit cell. Once the heavy atom has been localized, many of the phase angles associated with the Fourier components of the three-dimensional electron density function may be unambiguously determined. Determination of all such phase angles requires the use of more than one heavy-atom derivative. Perutz employed six for the hemoglobin studies, and Kendrew required four for the myoglobin studies. Heavy atoms are employed because they act as intense scattering centers and, consequently, are relatively easy to locate in the unit cell. Scattering intensity is roughly a function of the square of the atomic number.

It is to be emphasized that formation of the heavy-atom derivatives must yield isomorphous replacements; that is, introduction of the heavy atom must not alter the size or symmetry of the unit cell nor alter the coordinates of the atoms within the unit cell. Formation of suitable heavy-atom derivatives of proteins is frequently the greatest stumbling block to the determination of the structure of macromolecules by means of X-ray diffraction. Preparation of such derivatives may be attempted in several ways: (1) introduction of the heavy atom by covalent linkage to the protein followed by crystallization of the derivative in a form isomorphous with the unmodified protein, (2) crystallization of the protein from solutions containing the heavy-metal ion, or (3) diffusion of the ions into the protein crystal by prolonged contact of the crystals with solutions containing the heavy-atom ions.

Ideally, one wishes to introduce the heavy atoms by taking advantage of the existence of reactive groups within the protein molecule. For example, it proved possible in the case of hemoglobin and certain other proteins to introduce a mercury atom through reaction of *p*-chloromercuribenzoate with sulfhydryl groups of the protein. A related example has found use in studies on the proteolytic enzyme papain; *p*-iodobenzeneisothiocyanate reacts readily with the α-amino group, resulting in the incorporation of one atom of iodine into the protein.

In studies with metalloenzymes, the resident metal ion can frequently be replaced by others generating a series of heavy metal derivatives. This technique has proved useful in the study of carboxypeptidase A and carbonic anhydrase, both of which occur normally as zinc proteins.

In spite of these successes, the number of derivatives prepared in this way is not large. Probably, the hit-or-miss method of diffusing many heavy metal species into the protein crystals in the hope that some of them will form suitable derivatives will remain the most important technique in the near future.

Note, finally, that the preparation of heavy-atom derivatives of proteins is not a theoretically necessary procedure for resolving the phase-angle problem. Other possible approaches are being studied and may ultimately prove superior to current practice. But for the moment, the preparation of isomorphous derivatives remains the best solution to this problem.

To recapitulate: The diffraction patterns of proteins are sufficiently detailed to permit localization of most of the atoms in the unit cell. Such localization may be accomplished by the measurement of a sufficient number of diffraction maxima for the crystalline protein and several isomorphous heavy-atom derivatives. The accompanying calculations require the use of a battery of high-speed computers. We now consider the studies that have revealed the structures of myoglobin, hemoglobin, and other proteins.

Structure of Myoglobin

Myoglobin is the oxygen-storage protein of muscle. It consists of a single chain of 153 amino acid residues and one iron-containing heme group and has a molecular weight of 17,500. It is a somewhat atypical protein in two respects: possession of a relatively high content of α helix (77 percent) and a complete lack of disulfide bridges or free sulfhydryl groups.

The X-ray diffraction studies of Kendrew and his associates on sperm-whale myoglobin have revealed the structural aspects of this molecule in intimate detail. The reader is referred to the original publications and to reviews of this outstanding achievement (26–28). Early work on this molecule involved the determination of the amplitudes and phases of 400 Fourier components and yielded a structure with a 6 Å resolution. At this resolution, the polypeptide chain was recognizable as a region of high electron density that was folded in a highly intricate fashion, yielding a rather compact structure. Determination of the amplitude and phase of 9600 Fourier components yielded a structure at a 2 Å resolution, and subsequent work has led to a refined structure with a resolution of 1.5 Å. At this resolution, most of the atoms in the molecule may be unambiguously located. Such resolution, of course, permits most of the amino acid sequence to be determined directly from the X-ray diffraction data. Comparison of the sequence derived from X-ray data with that derived by classical chemical means yields concordant results. In addition, knowledge of the structure at this resolution permits reliable calculations of interatomic distances, from which a great deal of information concerning the interactions of residue side chains may be derived. We shall return to this point later.

An electron-density map of myoglobin is illustrated in Fig. 4-22. This three-dimensional map, constructed by tracing out two-dimensional electron maps on Lucite sheets for a number of planes in the crystal and superimposing the plates, is a graphic representation of the function $\rho(x,y,z)$ for this molecule. Close examination of the electron-density map reveals the position of the heme group and some of the helical segments.

The function of myoglobin is the storage of oxygen. The oxygen molecule is attached directly to the ferrous iron atom at the center of the heme structure (pages 419–422):

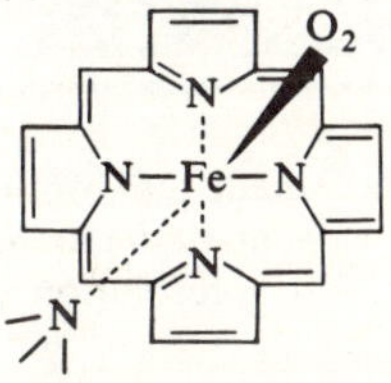

The myoglobin molecule is arrayed in space in a way so as to provide a compact enclosure for the heme group. Specifically, eight helical segments of varying length interspersed with segments of random coil surround the heme group, as shown in Fig. 4-23. Note that the helical segments are designated A, B, C, and so on, beginning with the N terminus, and the interhelical regions are designated AB (the region between helices A and B), BC, and so on. Note that the histidine at position F8 forms the fifth ligand to the iron atom and that the one at position E7 is nearly in contact with the oxygen atom on the other side of the ring. A complete model of myoglobin is provided in Fig. 4-24.

In Table 4-10, data concerning the length of the helical regions and the axial rotation and translation per residue are presented. Note that these values are in essential agreement with those of Pauling and Corey for the α helix (28).

Kendrew has summarized the principal features of the myoglobin molecule, and we reiterate them at this point. (1) The molecule is compact and contains, at most, four molecules of water in its interior. (2) Nearly all the polar groups (Lys, Arg, Glu, Asp, His, Ser, Thr, Tyr, Trp) are on the surface of the molecule and hence exposed to solvent. (3) Bound water molecules are attached to all polar groups at the surface. (4) Apart from the water molecules bound to the polar groups, the liquid regions surrounding the crystal show no sign of ordering. (5) The interior of the molecule is composed of nonpolar residues. (6) The fifth coordination site of the iron atom in the

FIGURE 4-22
Electron-density map of the myoglobin crystal, made up of Lucite sheets, on each of which are traced the contours of electron density at that depth. The dark band in the middle is the heme group. This map was made at a resolution of 2 Å. George Rodger, Magnum Photos.

tetrapyrrole prosthetic group is occupied by a histidine residue. The sixth such coordination site, on the other side of the tetrapyrrole structure, is probably occupied by a water molecule that is presumably displaced by O_2 when the myoglobin is converted to its oxygenated form.

The distribution of myoglobin strongly reflects its function. Thus the best sources are from the diving mammals, such as whale, seal, and dolphin. High concentrations of myoglobin in these mammals permit them to store sufficient oxygen to remain underwater for long periods. In man, myoglobin is found in high concentration only in the heart muscle, which, by virtue of its function, must have the capacity to continue functioning under conditions of substantial oxygen debt.

Following the structure determination for sperm-whale myoglobin, a number of interesting related studies have been performed. A complete structure determination at a 2 Å resolution has been carried out for the *p*-chloromercuribenzene sulfonate derivative of myoglobin (29). This study has established that the small molecule is bound in a region where two myoglobin molecules come together, taking advantage of favorable electrostatic interactions, at least. This provides an explanation for the fact that this ion forms a good complex with crystalline myoglobin but not with myoglobin in solution.

Both xenon and cyclopropane form specific complexes with myoglobin. It has been established that, in both cases, the small molecule occupies an otherwise empty region within the protein bounded in part by one of the

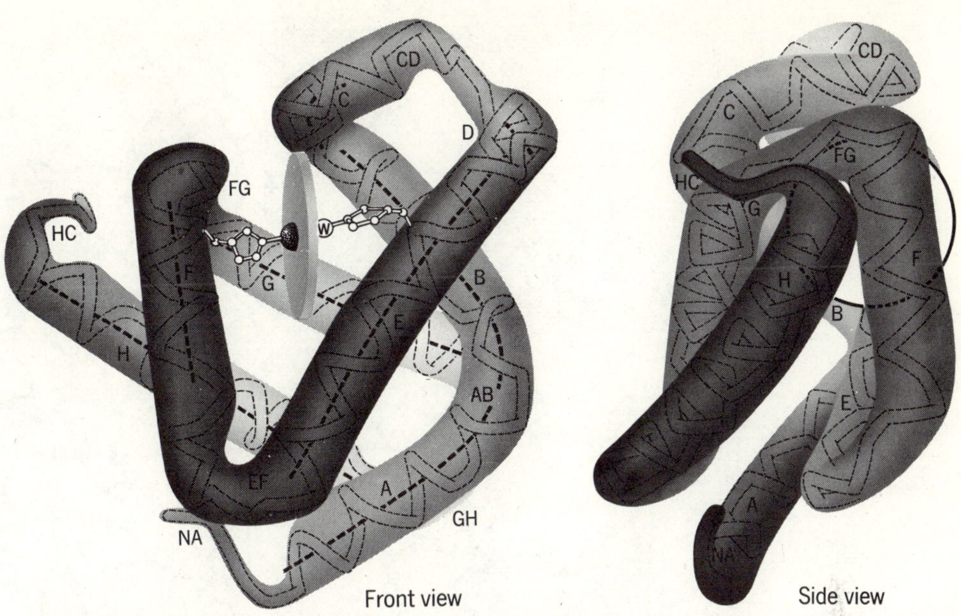

FIGURE 4-23
The myoglobin molecule is built up from eight stretches of α helix that form a box for the heme group. Histidines interact with the heme to the left and right, and the oxygen molecule sits at W. Helixes E and F form the walls of a box for the heme; B, G, and H are the floor; and the CD corner closes the open end. The course of the main chain is outlined inside the envelopes. From *The Structure and Action of Proteins* by Richard E. Dickerson and Irving Geis. Copyright © 1969 by Richard E. Dickerson and Irving Geis. Reprinted by permission of the authors and Harper & Row, Publishers, Inc.

pyrrole rings of the heme group and by the imidazole ring attached to the heme iron atom (histidine F8) (30,31). The xenon atom appears not to disturb the normal myoglobin structure in a detectable way, but the slightly larger cyclopropane molecule does cause small changes in the position of several side chains.

Of great significance is a study by Perutz, Kendrew, and Watson concerning the nature of the interactions responsible for the molecular integrity of the globin chain (32). The primary structures for some 18 different globin chains (including those of both hemoglobin and myoglobin) were examined; on the basis of structure determinations thus far carried out, all are thought to have the same basic three-dimensional conformation. Of the 150 (more or less) residues, only 9 were found to be invariant in all globin chains. Of these, 4 are directly connected with the heme function and 4 more are directly involved in chain conformation, usually at corners of helices. A summary of the secondary structure of globin is provided in Fig. 4-25.

Beyond these crucial positions, an interesting pattern develops for the type of change in amino acid that is tolerable in terms of localization of the residue with respect to the molecular surface. For those residues which lie on the molecular surface of myoglobin, marked (nonconservative) changes occur with regularity. Thus acidic groups are sometimes interchanged for basic ones, polar for nonpolar, and the like. But in the molecular interior, a consistent pattern does emerge: All substitutions involve the replacement of nonpolar

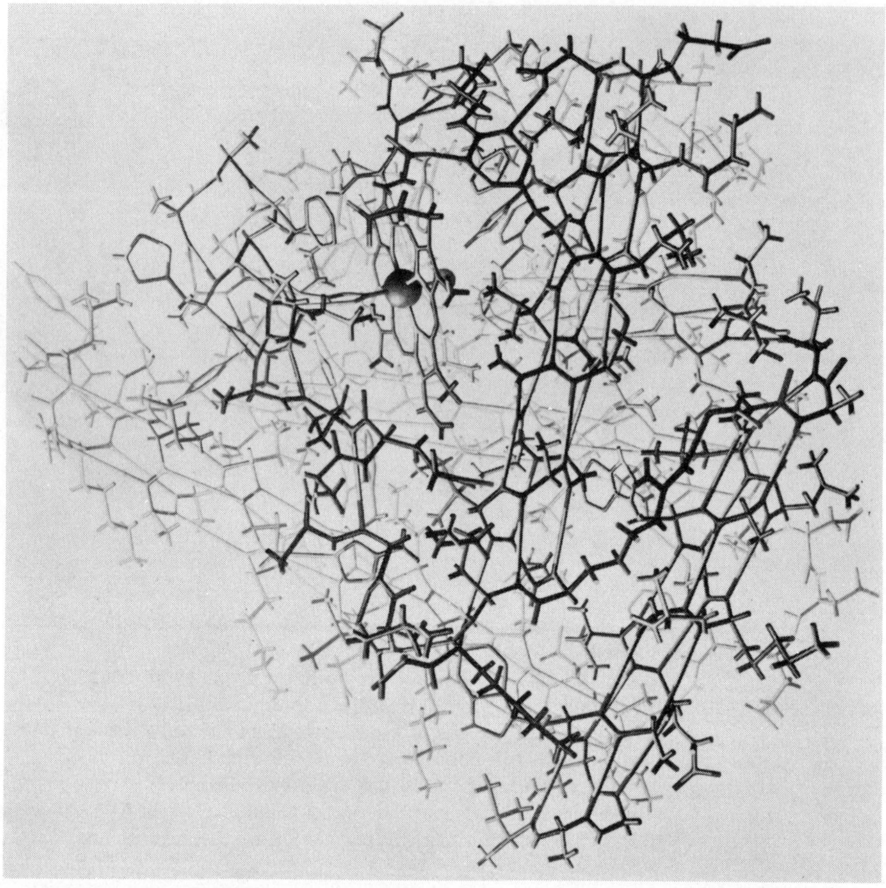

FIGURE 4-24
Model of the complete myoglobin molecule. Photo from painting by Irving Geis. J. C. Kendrew, *Sci. Am.,* **205:** 98 (1961).

amino acids by other nonpolar ones. These findings are consistent with our previous conclusions drawn from the data on cytochromes *c* (page 125) but add an important new insight: Nonconservative changes appear acceptable on the molecular surface but not in the molecular interior.

TABLE 4-10
Parameters characterizing the helical segments of the myoglobin molecule

Helix	*Number of residues*	*φ, radians*[a]	*h, Å/residue*[b]
A	16	1.73	1.50
B	16	1.69	1.47
D	7	1.73	1.45
E_1	10	1.74	1.52
E_2	10	1.71	1.49
F	9	1.70	1.46
G	19	1.75	1.53
H	24	1.73	1.49
α-helix	..	1.74	1.50

[a] φ, axial rotation per residue.
[b] *h,* axial translation per residue.
SOURCE: J. C. Kendrew, *Brookhaven Symp. Biol.,* **15:** 216 (1963).

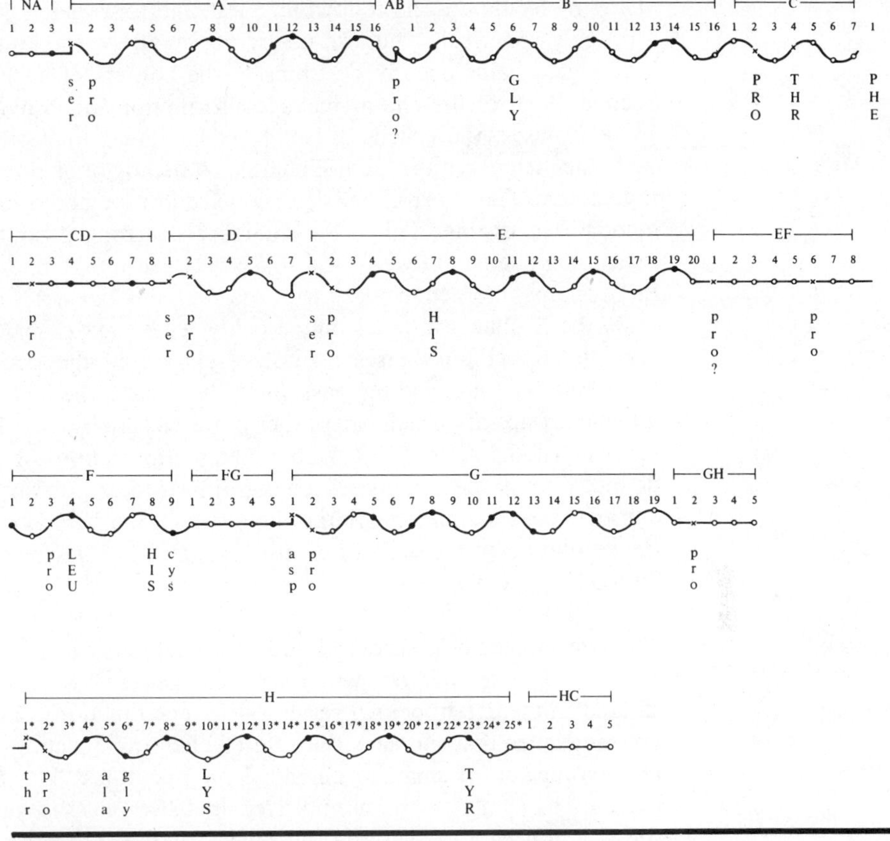

FIGURE 4-25

Secondary structure of globin. α-Helical segments are represented as sine waves, nonhelical segments as straight lines. Black circles mark sites where only nonpolar residues occur; crosses, where prolines or combinations of prolines with serine, threonine, aspartic acid, or asparagine occur. All other sites are represented by open circles. Residues in capital letters are invariant, those in lowercase letters mark other residues of special interest, such as the reactive cysteine of the β chain, which must be on the surface. Note that this diagram shows the total incidence of proline in all species analyzed so far and that no single species contains proline at all the points indicated here. The sine waves are drawn so that the top of each wave points to the inside of the globin chain. M. F. Perutz, J. C. Kendrew, and H. C. Watson, *J. Mol. Biol.*, **13:** 676 (1965).

Structure of Hemoglobin

Hemoglobin, the oxygen-carrying pigment of mammalian serum, is a more complicated molecule than myoglobin, although the two do have structural similarities. The structure of hemoglobin at a 2.8-Å resolution was deduced by the X-ray diffraction work of Perutz and his associates (33–36). Owing to the complexity of hemoglobin, its structure was completed after that of myoglobin despite the fact that work on hemoglobin was initiated several years earlier. Kendrew, as a student of Perutz, did some of the earlier work on hemoglobin prior to turning his attention to the myoglobin problem. In recognition of their stunning achievements in the field of protein chemistry, Perutz and Kendrew shared the Nobel Prize in chemistry for 1962.

Hemoglobin is composed of four separate polypeptides. These occur as two α chains and two β chains. The function of hemoglobin is discussed in some detail on pages 419–420 and 666–671. The principal structural features of the molecule are the following. First, the four chains are arranged to form a roughly spherical structure, 65 by 55 by 50 Å, as shown in Fig. 4-26. The individual chains occupy positions at the corners of an irregular tetrahedron. Second, each of the chains has a conformation much like that of myoglobin. In both cases, the crucial E7 and F8 histidines make the same interactions with the heme group. The α chains of hemoglobin do lack the D helix of myoglobin. The short C helix, an α helix in myoglobin, is a 3_{10} helix in both hemoglobin chains. Third, the chains are arranged so that interactions between like chains are weak and those between unlike ones are strong. That is, the α chains interact strongly with the β chains but not with each other. Likewise, the β chains interact only weakly with each other. Finally, deoxygenation of hemoglobin causes a marked change in the structure, which results from one $\alpha\beta$ half sliding past the other, with the net result that the two α heme groups approach each other more closely by 1 Å, but the two β hemes separate by 6.5 Å (see Fig. 4-26). The conformations of the individual chains do not change detectably. A change in structure of this type has profound consequences for understanding the oxygenation behavior of hemoglobin, and the regulation of the activity of other proteins, as described in detail on pages 299–311.

Structure of Lysozyme

The two molecules considered above, myoglobin and hemoglobin, function as carriers of molecular oxygen. Lysozyme and other enzymatic proteins, considered immediately below, serve a different function: catalysis. That is, they are molecules that increase the rate of chemical reactions without themselves undergoing any permanent chemical modifications. The principal features of proteins as catalysts are summarized in Chapter 7. In the case of lysozyme, the reaction catalyzed is the hydrolysis of linkages between sugars involved in cell wall structures (page 480). The best substrate for lysozyme is an alternating copolymer of *N*-acetyl-D-glucosamine and *N*-acetyl muramic acid:

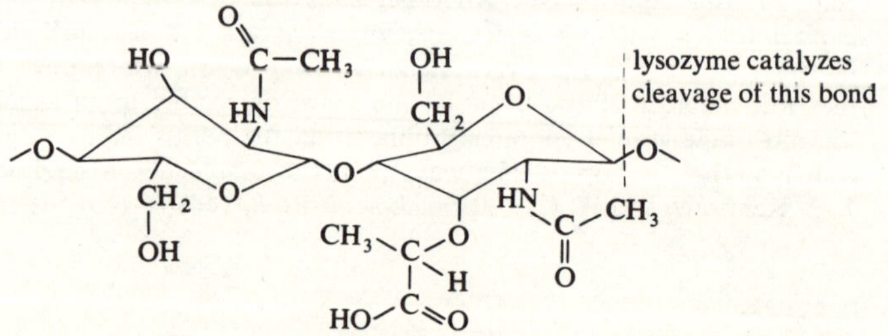

lysozyme catalyzes cleavage of this bond

Just how lysozyme speeds up the cleavage of the bond linking these two entities is described on page 368.

The structure of lysozyme was established through the work of Phillips and his coworkers (37–39). The amino acid–sequence studies, which were crucial to the interpretation of the X-ray diffraction data, came from the studies of Canfield (40) and Jolles (41). A pair of stereo photographs of a model of the lysozyme molecule is provided in Fig. 4-27. The positioning of the substrate on the protein molecule is shown. Investigation of these figures reveals the central aspects of the lysozyme structure, which, we may note at the outset, is quite different from that of myoglobin. The molecule is ellipsoidal, with dimensions of 45 by 30 by 30 Å. It is a single polypeptide chain of 129

FIGURE 4-26
Completed model of a single molecule of reduced human hemoglobin (*left*), compared with the horse oxyhemoglobin model (*right*). H. Muirhead and M. F. Perutz, *Cold Spring Harbor Symp. Quant. Biol.,* **28:** 451 (1963).

amino acid residues. Only about 25 percent of these occur in α-helical structures. In further contrast to myoglobin, lysozyme contains a region of β structure: residues 41 through 48 run in a straight line and then residues 49 through 54 double back to form a parallel chain hydrogen-bonding with the first one. Thus, we form an antiparallel pleated sheet structure. A cleft can be found on the surface of the lysozyme molecule, and this runs for most of the length of one surface. Finally, we can note that, as in myoglobin, most of the interior of the molecule is composed of hydrophobic residues and most of the surface by hydrophilic ones. There are one or two exceptions to

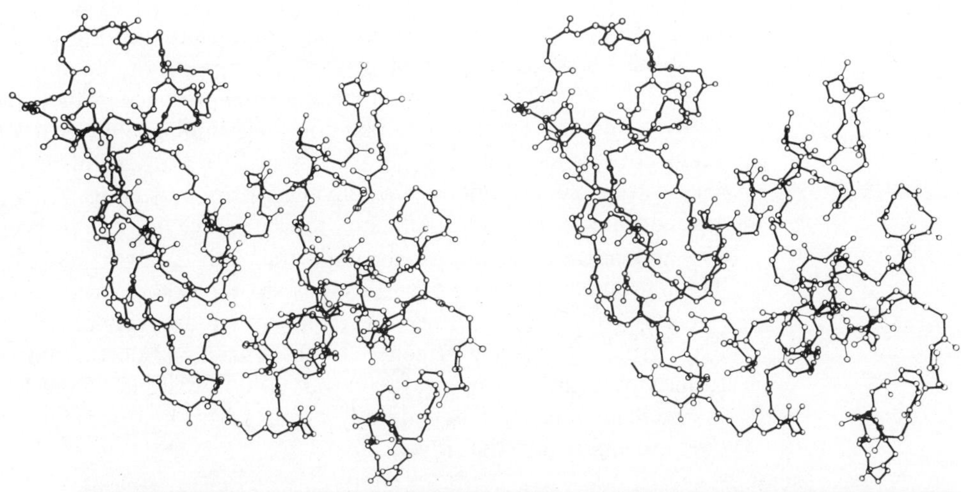

FIGURE 4-27
A computer-generated stereo pair showing the main chain conformation of the lysozyme molecule. Courtesy of Professor D. C. Phillips.

this statement, and these residues turn out to be important for the catalytic process, as detailed in Chapter 7.

One of the attractive features of the structural study of proteins by X-ray diffraction is that once the detailed structure of the native protein is established, it requires relatively little additional work to determine slightly modified structures. This fact has been put to good use in the case of lysozyme and other enzymes. Specifically, inhibitors of lysozyme which are structurally related to the normal carbohydrate substrates can be introduced into the crystals, either through cocrystallization or by diffusion into the preformed crystals, and their localization within the crystal determined. Because of the structural similarity between substrate and inhibitor, it is likely that the inhibitor occupies the same site (or part of it) that is occupied by the normal substrates. Studies of this type with trisaccharides have established a number of interactions between the inhibitor and enzyme. It has been found that the groove on the molecular surface is sufficiently large to accommodate a chain of six sugar units; trisaccharide inhibitors occupy the first three of these. Model building based on the localization of the inhibitor permits one to tentatively assess the interactions between a hexasaccharide substrate and the complete binding site of the protein. These interactions are shown in Fig. 4-28. A detailed discussion of the reaction mechanism associated with lysozyme-promoted reactions is deferred until Chapter 7, but it is worth noting the proximity of residues 35 and 52, both acidic amino acids, and the junction between the fourth and fifth rings of the substrate. These residues are believed to play an important role in cleaving the glycosidic bond at this point.

Structure of Chymotrypsin

As noted in Chapter 3, chymotrypsin is a proteolytic enzyme, having the capacity to catalyze the hydrolytic cleavage of peptide bonds. The mechanism of the reaction catalyzed by chymotrypsin is discussed on pages 369–370 and the activation of this enzyme is considered on pages 764–767.

The structure of chymotrypsin has been established by Blow, Sigler, et al. (42,43). The main-chain conformations are shown in Fig. 4-29. The molecule consists of three polypeptide chains held together by disulfide bonds. It is derived from a precursor, chymotrypsinogen, which is a single chain of 246 amino acids, by the selective cleavage of certain peptide bonds (page 765).

There are several points of interest concerning the chymotrypsin structure. First, the overall conformation of chymotrypsinogen, which has no catalytic activity, and chymotrypsin, which does, are similar. The small changes that do occur are specified on page 765. Second, there are several regions of a twisted sheet structure present in the chymotrypsin molecule. Third, a substantial amount of evidence has implicated the serine residue at position 195, histidine 57, and aspartic acid 102 as important for catalytic activity (page 369). These groups do occupy positions close to each other in the three-dimensional structure, as shown in Fig. 4-29, and are, moreover, close to an inhibitory tosyl group covalently attached to the enzyme. Fourth, the only α helix is formed by eight residues at the carboxyl-terminal end of the C chain. Finally, the molecular interior is generally occupied by hydrophobic groups; charged ones are exposed to the solvent.

Structure of Carboxypeptidase

Lipscomb and coworkers have established the three-dimensional structure of carboxypeptidase (44–48). Like chymotrypsin, this enzyme catalyzes the hydrolysis of peptide bonds but, unlike chymotrypsin, does so beginning at

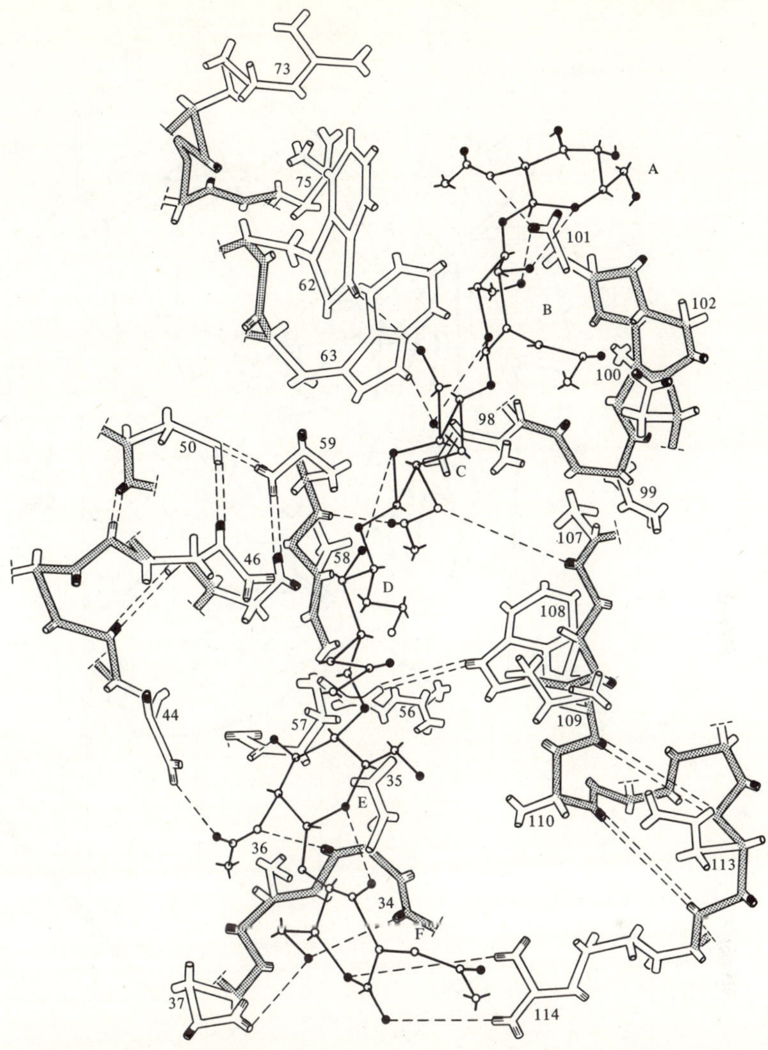

FIGURE 4-28
Atomic arrangement in the lysozyme molecule in the neighborhood of the cleft with a molecule of hexa-*N*-acetylchitohexanose shown bound to the enzyme. The polypeptide chain is shown speckled and NH and O atoms are indicated by line and full-shading, respectively. Sugar residues A, B, and C are observed in the binding of tri-*N*-acetylchitotriose (and β-*N*-acetylglucosamine for residue C). Residues D, E, and F occupy positions that have been inferred from model building. It is suggested that the linkage hydrolyzed by the action of the enzyme is between residues D and E. D. C. Phillips, *Proc. Natl. Acad. Sci. U.S.*, **57:** 493 (1967).

the C terminus and working down the chain sequentially (page 372). The protein is a single polypeptide 307 amino acid residues long, molecular weight 34,600. It is ellipsoidal with dimensions of 52 by 44 by 40 Å. The main-chain conformation of carboxypeptidase is shown in Fig. 4-30. Note that the molecule contains one atom of zinc, located in a depression in the surface which is adjacent to a pocket extending into the molecular interior. It is into this pocket that a portion of the substrate molecule fits.

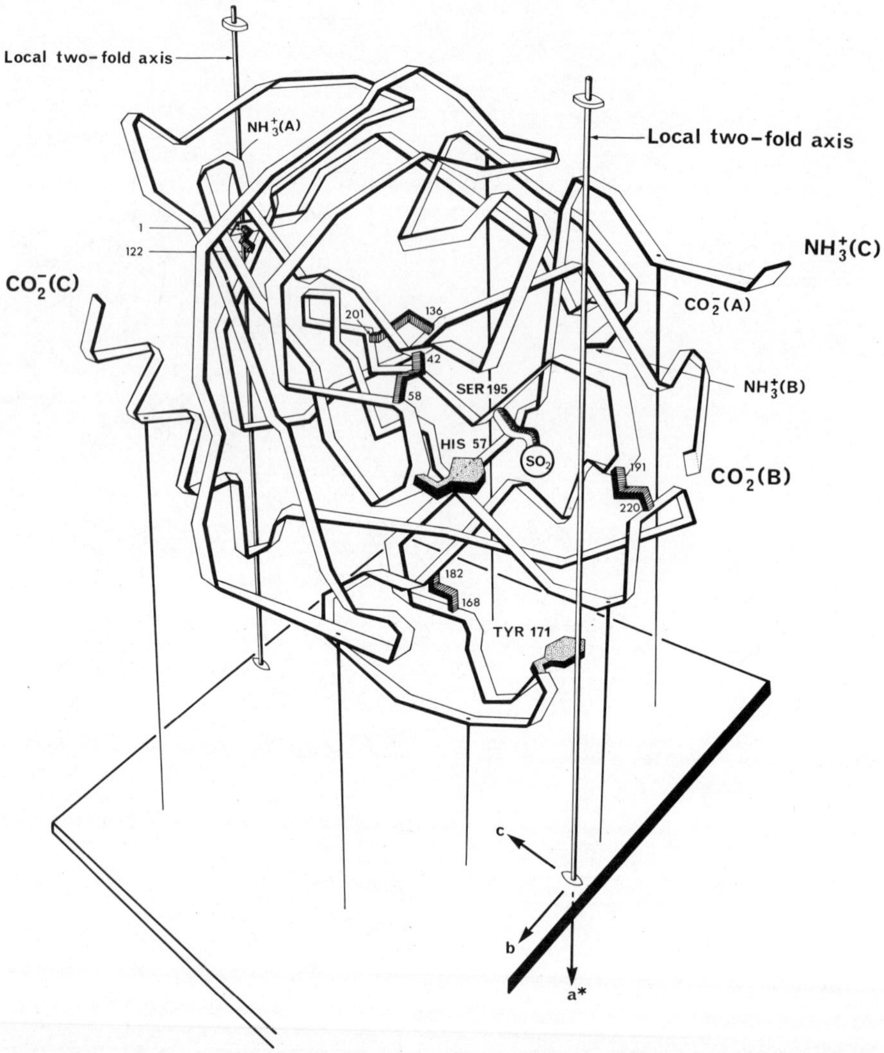

FIGURE 4-29
Schematic drawing representing the conformation of the polypeptide chain in α-chymotrypsin. Positions of the disulfide bridges have been deleted for the sake of clarity. The side chains of His 57 and the sulfonated serine 195 are shown. Reproduced from: P. Sigler, B. W. Matthews, R. Henderson, and D. M. Blow: *J. Mol. Biol.,* **35:** 143 (1968). Courtesy of Dr. D. M. Blow. Investigation supported by the Medical Research Council of England.

About 30 percent of the amino acid residues are involved in α-helical structures. These are organized into four helical segments, which are 15, 16, 17, and 18 residues long. All of the helical structure lies on the outside of the molecule. In addition, about 55 residues are involved in formation of a twisted pleated sheet conformation. For the most part, this structure occupies the interior of the molecule.

As in the case of lysozyme, the structure of complexes between carboxypeptidase and inhibitors has been established. The nature of the interactions between enzyme and inhibitor are considered more fully in Chapter 7. But at

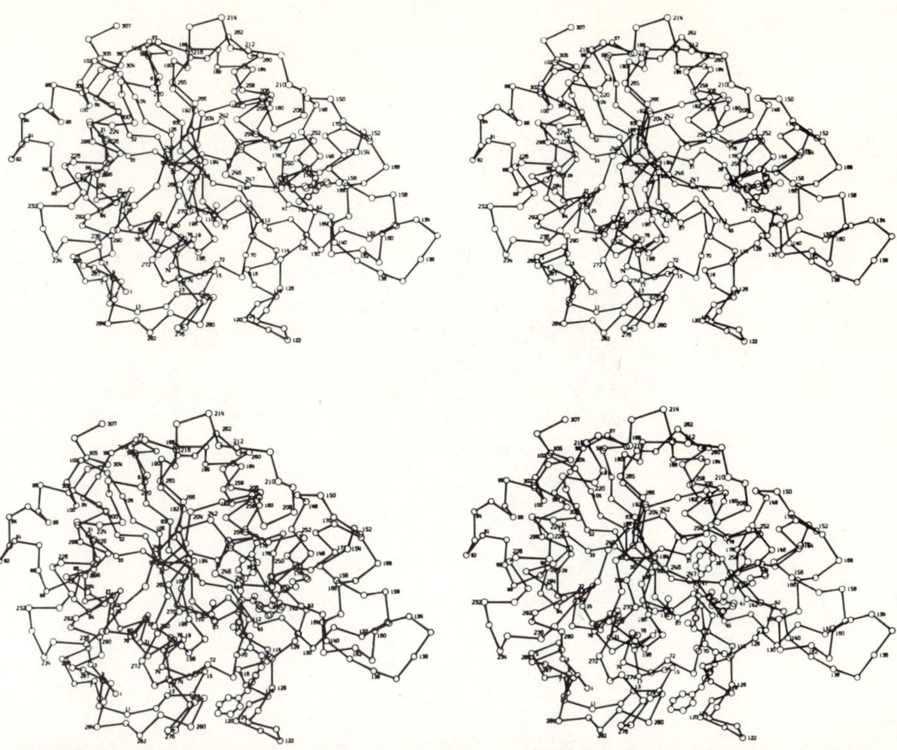

FIGURE 4-30
Computer-generated stereo pairs of the main chain conformation of carboxypeptidase A. In the upper pair, the conformation that the protein assumes in the absence of interaction with substrate is shown. Note particularly the positions of Arg 145, Tyr 248, and Glu 270. In the lower pair, the conformation adopted by the protein following interaction with glycyl-L-tyrosine is shown. In this pair, the substrate CBZ-Ala-Ala-Tyr has been positioned by model building, starting with conclusions drawn from the Gly-Tyr-carboxypeptidase map. Courtesy of Professor W. N. Lipscomb.

this point it is worth pointing out a striking structural change in carboxypeptidase that accompanies the binding of small molecules such as glycyl-L-tyrosine. Although the overall structure remains about constant, large changes in the positions of some residue side chains do occur in the immediate vicinity of the binding site. In Fig. 4-31, stereo pairs of the protein structure in the vicinity of the binding site of carboxypeptidase are provided for the native protein and for the protein complexed with an inhibitor. Note specifically the striking alteration of the side chain of Tyr 248 caused by substrate binding. This binding induces the phenolic hydroxyl group of this side chain to swing into a position near the peptide bond to be split where it is considered to play an active role in the catalytic process.

Structure of Ribonuclease

Ribonuclease is an enzyme that is capable of catalyzing the hydrolysis of the phosphodiester bonds in ribonucleic acid in a specific way (page 197). The amino acid sequence of this small protein has been discussed previously (page 123).

One of the striking aspects of the chemistry of ribonuclease is the observation of F. M. Richards that the molecule may be cleaved, through treatment

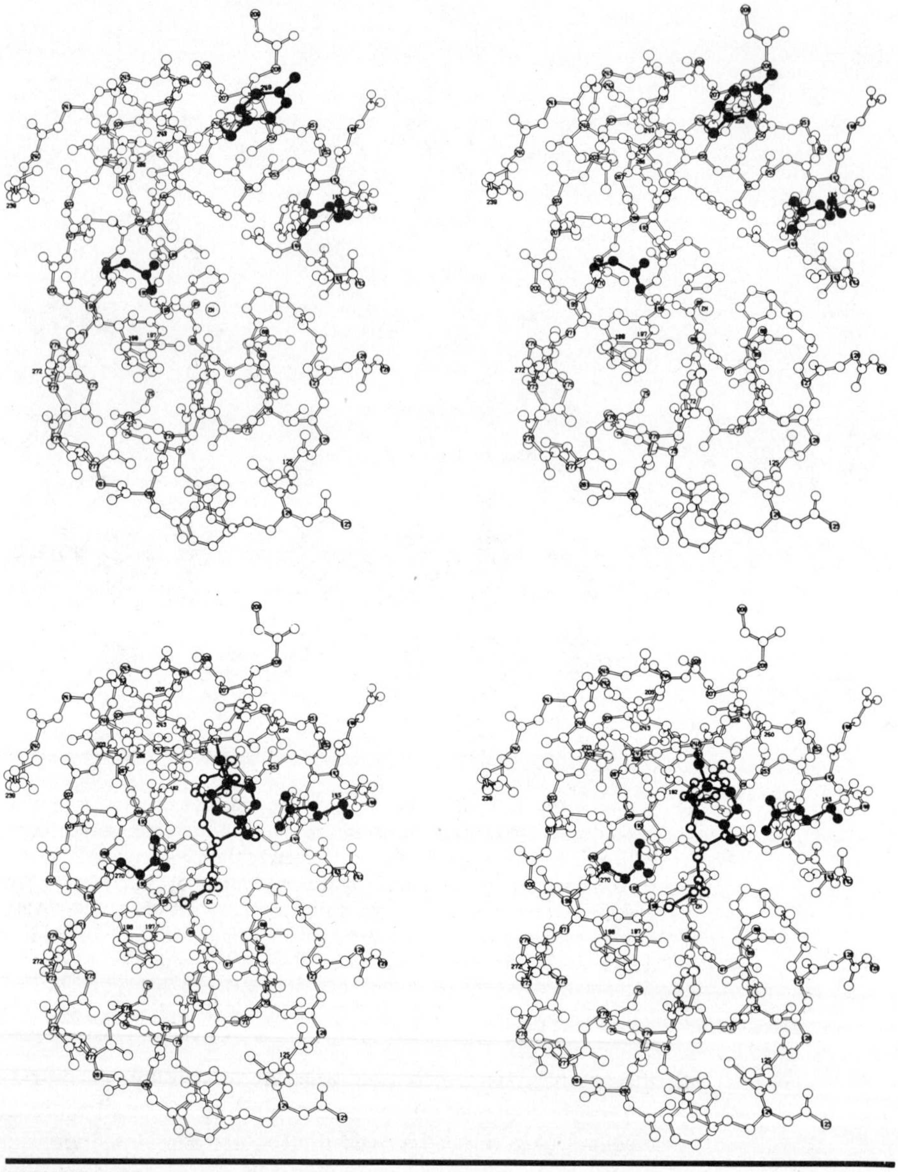

FIGURE 4-31

In the upper part, a model of the active site region of carboxypeptidase A in the negative state is shown. The side chains of Arg 145, Tyr 248, and Glu 270 are emphasized. In the bottom part, the active site of the molecule is again shown this time with a molecule of Gly-Tyr added and with the above residues in the positions they assume following complexion with substrate. Compare this figure with Figure 4-30. Courtesy of Professor W. N. Lipscomb.

with subtilisin, at one bond to yield two fragments, the S-protein and the S-peptide:

$$\underset{(124\ \text{residues})}{\text{Ribonuclease A}} \xrightarrow{\text{subtilisin}} \underset{(104\ \text{residues})}{\text{ribonuclease S}} + \underset{(\text{N terminal 20 residues})}{\text{S-peptide}} \qquad (4\text{-}16)$$

Simply mixing stoichiometric amounts of ribonuclease S and the S-peptide generates a fully active enzyme molecule, even though the covalent bond has not been reformed and despite the fact that neither of the fragments has detectable catalytic activity. This is certainly a dramatic example of the importance of weak interactions in maintaining the three-dimensional structures of proteins.

The complete structures of both ribonuclease A and S have been determined, the former by Harker (49) and the latter by Richards (50,51). A skeletal model for ribonuclease S is provided in Fig. 4-32. For those residues that are common to the two proteins, the conformations are quite similar.

Ribonuclease has three regions of α-helical structure, including residues 2 through 13, 26 through 33, and 50 through 58. These include about 15 percent of the total molecule. In addition, there is a substantial region of antiparallel β-pleated sheet structure which helps to form a crevice in the molecule into which the substrate fits, just as in the case of lysozyme. The interior of the molecule is very strongly hydrophobic and contains 18 CH_3—, 8—CH_2—, 2 C_6H_5—CH_2—, 2 CH_3—S—CH_2—CH_2—, 1 proline, and 1 disulfide bridge.

Crystallization of ribonuclease in the presence of an inhibitor has helped to locate the site of substrate binding. Two residues implicated in the catalytic

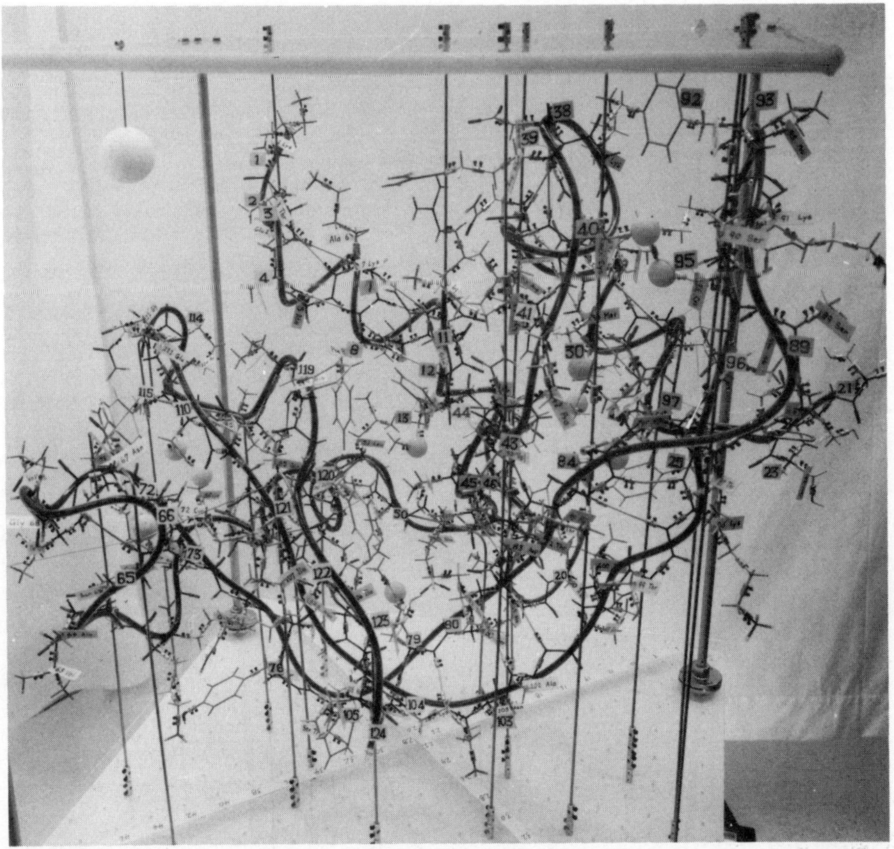

FIGURE 4-32
Photograph of a skeletal model of ribonuclease S. The break between the S-peptide (residues 1 through 20) and S-protein (residues 21 through 124) is at the lower left-hand corner of the photograph. The small balls represent sulfur atoms (of Cys and Met residues) and the large ball hanging above the model shows the van der Waals diameter of a paraffinic hydrogen atom. Reproduced by courtesy of H. C. Wyckoff and F. M. Richards.

process by earlier studies, histidine 12 and histidine 119, are, in fact, located at the binding site.

Several additional protein structures have been established, and work in this area is proceeding apace. It is certain that many additional insights into protein chemistry will be gained from these efforts. However, the discussion to this point has sufficed to indicate the principal features of protein structure.

There remains, however, one nagging uncertainty about the structures of proteins determined by X-ray diffraction that must be considered. The technique of X-ray diffraction requires the use of protein crystals; yet what we really wish to know is the structure of the molecule in solution. What is the relationship of the protein crystal structure to that in solution?

There are two lines of evidence that, happily, suffice to indicate that the three-dimensional structures of proteins change little in going from solution to the crystal. The first of these relies on a comparison of the chemical reactivity of groups on the protein in the crystal and in solution and the second on comparison of the functional characteristics of the protein under these circumstances.

The most careful and conclusive of the former type of study has been carried out by Gurd and coworkers, who have studied the alkylation of sperm-whale myoglobin by bromoacetate (52). This reagent reacts most readily with the imidazole side chains of the histidine residues of the protein. Provided that the reaction is not forced, it is possible to selectively alkylate those histidines which are at or near the protein surface and leave unreacted those which are in the interior of the molecule and hence less accessible to the alkylating reagent. Provided that the conformation of myoglobin is the same in the crystalline and solution states, the same histidines ought to react in the two cases. With one exception, this is exactly what is observed. It is true that the C1 histidine is alkylated in the crystalline preparation but not in solution, suggesting that the two structures are not identical. But the most important point is that they cannot be very different.

The second line of attack has been pursued principally by Richards and Quiocho, who have examined the catalytic properties of several enzymes in the crystalline and solution states (53–56). The crystalline preparations were extensively cross-linked by reaction with glutaraldehyde, which has the effect of linking up the various molecules within the crystal so that dissolution cannot occur upon suspension in water. It has been observed that ribonuclease, chymotrypsin, and carboxypeptidase, among other enzymes, are catalytically active in the crystalline state and, moreover, that many of the features of the catalytic process are similar to those observed for the solution reactions. Since the catalytic properties of enzymes are sensitive functions of the exact three-dimensional structure, it follows that these must be nearly the same for the proteins in solution and in the crystals. Consequently, one may have considerable confidence that the structures deduced through the methodology of X-ray diffraction are closely related to those characteristic of the same proteins in solution.

FORCES INVOLVED IN MAINTENANCE OF SECONDARY AND TERTIARY STRUCTURES

Four types of interactions are considered principally responsible for the maintenance of the secondary and tertiary structures of proteins. These include interpeptide hydrogen bonds, side-chain hydrogen bonds, ionic bonds, and apolar or hydrophobic bonds (Fig. 4-33). It is certainly possible that interactions of other types may contribute to the stability of the folded form of certain polypeptides, but little evidence concerning them is available at this time.

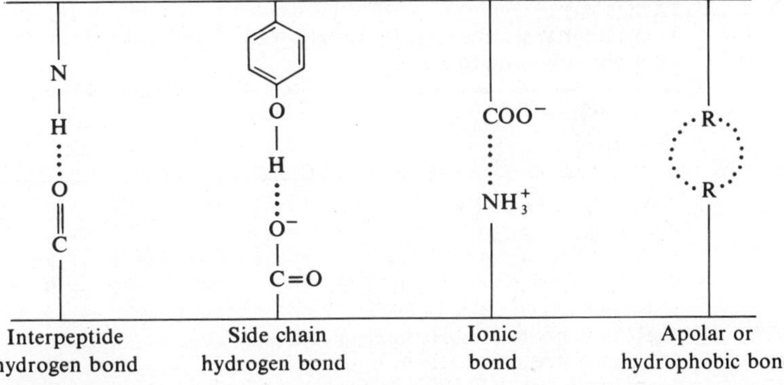

FIGURE 4-33
Intramolecular noncovalent bonds responsible for secondary and tertiary structures. H. K. Schachman, *Cold Spring Harbor Symp. Quant. Biol.,* **28:** 409 (1963).

Apolar Bonds

The tendency of hydrocarbons or other nonpolar molecules to associate in aqueous solution is termed *apolar* (or hydrophobic) bond formation. There is a good deal of evidence that formation of apolar bonds is one of the most important, if not the most important, factors influencing the higher-order structures of proteins. Our most detailed information concerning the relative importance of apolar and polar interactions comes from the studies on myoglobin. Employing computer methods, Kendrew calculated the interatomic distances, pairwise, for all the atoms of myoglobin. Those atoms which were more than 2 and less than 3.1 Å apart were considered to be involved in polar interactions, and those more than 3.1 and less than 4.1 Å apart to be involved in apolar interactions. The number of interactions in the latter category exceeds those in the former by an order of magnitude. Thus, for myoglobin, at least, formation of apolar bonds appears to be of primary significance for formation of the compact structure of this molecule.

This conclusion has been amply strengthened by the structures of other proteins. Note that the thermodynamics of protein unfolding are consistent with exposure of large numbers of hydrophobic groups to the solvent (see below).

Two lines of evidence establish that factors other than apolar bond formation are certainly important for maintenance of the three-dimensional structure of proteins. First, because the ΔH values for the formation of apolar bonds are probably slightly positive (see discussion below), the strength of apolar bonds increases in the region 0 to 60°. Over this temperature range most proteins tend to lose their secondary and tertiary structure (denature). Second, the effectiveness of chemical reagents in unfolding proteins does not parallel their hydrophobic characteristics. If only apolar bond formation were important for stabilization of folded structures, one might well expect that molecules that could best stabilize hydrophobic residues exposed on unfolding would be most effective at promoting such unfolding.

The driving force for formation of apolar bonds is largely entropic, as may be judged from the data in Table 4-11, in which the thermodynamic changes accompanying the transfer of hydrocarbons from nonpolar solvents to water are collected. The enthalpy changes for such transfers are actually slightly negative but are more than counterbalanced by large negative entropy changes. To the extent that transfer of hydrocarbons from an aqueous to a nonpolar environment (the reverse reaction) is a good model for the transfer

TABLE 4-11
Thermodynamic changes in transfer of hydrocarbons from nonpolar solvents to water

Process	*Temp, °K*	ΔS_u, *eu*	ΔH, *cal/mole*	ΔG_u, *cal/mole*
CH_4 in benzene → CH_4 in H_2O	298	−18	−2800	+2600
CH_4 in ether → CH_4 in H_2O	298	−19	−2400	+3300
CH_4 in CCl_4 → CH_4 in H_2O	298	−18	−2500	+2900
C_2H_6 in benzene → C_2H_6 in H_2O	298	−20	−2200	+3800
C_2H_6 in CCl_4 → C_2H_6 in H_2O	298	−18	−1700	+3700
C_2H_4 in benzene → C_2H_4 in H_2O	298	−15	−1610	+2920
C_2H_2 in benzene → C_2H_2 in H_2O	298	−7	−190	+1870
Liquid propane → C_3H_8 in H_2O	298	−23	−1800	+5050
Liquid *n*-butane → C_4H_{10} in H_2O	298	−23	−1000	+5850
Liquid benzene → C_6H_6 in H_2O	291	−14	0	+4070
Liquid toluene → C_7H_8 in H_2O	291	−16	0	+4650
Liquid ethyl benzene → C_8H_{10} in H_2O	291	−19	0	+5500
Liquid *m*- or *p*-xylene → C_8H_{10} in H_2O	291	−20	0	+5800

SOURCE: W. Kauzmann, *Advan. Protein Chem.*, **14:** 1 (1959).

of the hydrophobic side chains from an aqueous environment encountered in the unfolded state to a nonpolar environment encountered in the folded state, these results suggest that entropic changes are principally responsible for apolar bond formation in proteins. In support of this conclusion, the driving force for the association of small hydrophobic molecules with proteins and for the formation of micelles from detergents in aqueous solution is largely entropic.

What is the source of the favorable entropy change for the formation of apolar bonds? It is known that exposed hydrophobic residues in the extended, less highly structured, form of the protein have an organizing effect on the surrounding water molecules. This organization restricts the rotational and translational freedom of the water molecules and hence is accompanied by a negative entropy change. Association of the hydrocarbon-like moieties in the interior of the native molecule thus tends to free this organized solvent shell. To recapitulate, the driving force for formation of apolar bonds is largely the positive entropy change resulting from the liberation of solvent molecules organized about the apolar residues that accompanies the transfer of these residues from an aqueous to a nonaqueous environment.

Polar Interactions

As in the case of the apolar, our most detailed information concerning polar interactions in proteins is provided by X-ray diffraction studies on myoglobin. These are summarized in Table 4-12. The most striking aspect of these data is the fact that relatively few of the total possible strong polar interactions are realized. In most cases, the polar groups are bound to solvent. The preliminary conclusion appears to be warranted that, at least in the case of myoglobin, the protein molecule assumes a conformation in which a maximum number of the polar groups are exposed on the surface and that strong interresidue polar interactions occur only when allowed by restrictions imposed by a structure otherwise largely dictated by apolar interactions. It remains to be established whether this generalization is true for most other proteins as well.

The formation of interpeptide hydrogen bonds (α helix) deserves further

TABLE 4-12
Polar interactions in myoglobin

				Bonding to other than solvent		
Residue	*Total number*	*Number in solvent or surface*	*Number "buried"*	*None, intermolecular or weak intramolecular*	*Strong intramolecular*	*Partners*
Lys	19	19	0	16	3	Glu
Arg	4	4	0	2	1	Asp, propionyl
						Asp, Asp
Glu or Gln	19	19	0	12–13	2–3	Lys
					1	Lys, Trp
					1–3	Chain, own NH
Asp or Asn	8	8	0	3–4	3	Arg
					1	His
Ser	6	5	1	2	3	Chain CO, 4 back
					1	Chain NH
Thr	5	3	1–2	1–2	1	Chain CO, 4 back
					1	Chain CO, 3 back
					1	Chain NH
His	11	7–8	3	7–8	2	Chain CO, $Fe^{2+}(H_2O)$
					1	Asp
Trp	2	2	0	1	1	Glu
Tyr	3	3	0	2	1	Chain CO

SOURCE: F. M. Richards, *Ann. Rev. Biochem.*, **32:** 269 (1963).

comment. What factors influence the stability of the α helix? There appears to be little inherent stability imparted to helical structures from the formation of interamide hydrogen bonds in aqueous solution. This is true because (1) the components of the amide linkages are involved in hydrogen bonding in aqueous solution whether a helical structure is formed or not and (2) thermodynamic evidence indicates that interamide hydrogen bonds are no stronger than those between amides and water. In a very nice study, Klotz and Franzen observed that the self-association of *N*-methyl acetamide, a model for the peptide bond, in water occurs only in very concentrated solutions (57). The overall enthalpy change for intermolecular hydrogen bond formation between two *N*-methyl acetamide molecules is essentially zero. A dramatic change in the self-association behavior of *N*-methyl acetamide was found when solvents of poorer hydrogen-bonding capacity were employed. Thus, in carbon tetrachloride, intermolecular hydrogen-bond formation was essentially complete at *N*-methyl acetamide concentrations an order of magnitude lower than those for detectable association in water. There is clearly a moral to this story. Although we have concluded that interamide hydrogen-bond formation itself contributes little to the overall stability of the molecule, it is important to recognize that if, as a result of apolar forces, for example, the amide linkages are constrained to exist in the interior of a folded molecule isolated from solvent, there exists a strong driving force for formation of helical structures (just as in the case of *N*-methyl acetamide in carbon tetrachloride solution). Thus, a protein assumes a structure in which not only the sum of the polar and apolar side-chain interactions but also the formation of hydrogen bonds to the peptide linkages, whether with solvent or amides of the chain, are maximized simultaneously. Put another way, the tendency of the peptide linkages to form hydrogen bonds may markedly perturb a hypothetical structure derived from a consideration of interactions between the residue side chains only.

The effects of the hydrogen-bonding capacity of the chemical environment on the formation of helical structures may also be appreciated from a consideration of the solvent effects on the secondary structure of synthetic polypeptides. For example, we find that poly-γ-benzyl-L-glutamic acid adopts a completely helical structure in dioxane, chloroform, and other poor hydrogen-bonding solvents. Thus, if the hydrogen-bonding capacities of the polypeptide are not satisfied by the solvent, they are satisfied by formation of the helices. Conversely, poly-γ-benzyl-L-glutamic acid exists as a random coil in hydrazine and dichloroacetic acid, solvents of high hydrogen-bonding capacity. In such solvents the driving force for interamide hydrogen bond formation has been lost.

In addition to solvent effects, electrostatic considerations play an important role in helix stability. For example, at neutral *p*H poly-L-glutamic acid exists as a random coil. This may be readily rationalized, because under these conditions the γ-carboxyls of the glutamic residues are ionized. Electrostatic repulsions are minimized by permitting these anionic centers to be as far apart as possible, a condition met by a random coil rather than a helical conformation. Under acidic conditions, in which the carboxyl functions are uncharged, poly-L-glutamic acid assumes a helical conformation.

RELATIONSHIP BETWEEN PRIMARY AND HIGHER-ORDER STRUCTURES

As indicated above, most, if not all, proteins have a well-defined structure, in terms of both the sequence of amino acids and the folding of the polypeptide chain. The important question immediately arises as to the origin of the higher-order structures of proteins. The simplest possibility is that the three-dimensional structure of proteins is completely determined by the primary structure. This is equivalent to the statement that a protein assumes that conformation which is thermodynamically most stable under the existing conditions. Alternatively, it is possible that, at the time of its synthesis, a specific three-dimensional structure is imparted to the protein that persists after the release of the protein from the synthetic apparatus. This conformation need not be the thermodynamically most stable one; it is only required that the energy barrier for transconformation reactions be sufficiently high so that the occurrence of such processes is rare and their speed slow. Several lines of evidence, summarized below, strongly suggest that the former alternative is correct.

Perhaps the most convincing experiments suggesting that primary structure determines secondary and tertiary structures are those which demonstrate restoration of proteins to their native states following an initial loss of the native structure. For example, if the ribonuclease molecule is completely unfolded by reduction of the four disulfide bridges with mercaptoethanol in 8 M urea, reoxidation of the molecule under properly controlled conditions results in the recovery of 95 to 100 percent of the molecules in the native conformation, as judged by recovery of enzymatic activity and by comparison of physical properties (58). Such an experiment is illustrated diagrammatically in Fig. 4-34. Statistical calculations indicate that, if reforming of

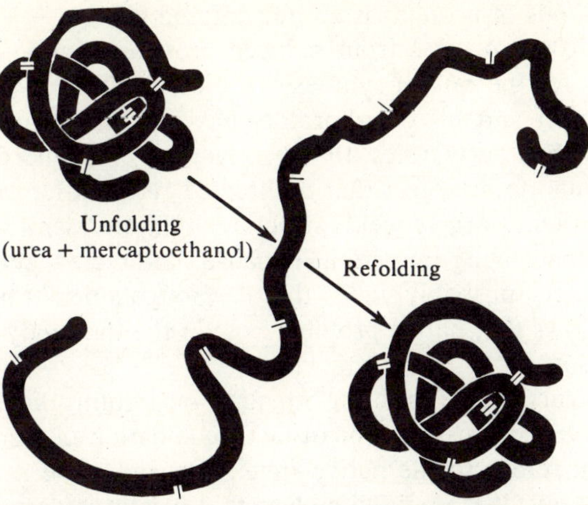

FIGURE 4-34
Model system for studying the conversion of a linear polypeptide chain to a folded protein. A native protein, with intramolecular disulfide bonds, is reduced and unfolded with β-mercaptoethanol and 8 *M* urea. After removal of these reagents, the reduced protein is allowed to undergo spontaneous refolding and reoxidation. C. J. Epstein, R. F. Goldberger, and C. B. Anfinsen, *Cold Spring Harbor Symp. Quant. Biol.,* **28:** 439 (1963).

the disulfide bridges were completely random, only about 1 percent of the molecules would have assumed their native conformations. A collection of data of this type is presented in Table 4-13. In each case, with the exception of insulin, reformation of native structures greatly exceeds that expected on the basis of random recovery. These results do not imply that formation of

TABLE 4-13
Refolding of disulfide-bond-containing proteins (following reduction and unfolding)

Protein	*Disulfide bonds*	*Recovery of activity, percent*	*Random recovery, percent*
Ribonuclease (bovine pancreatic)	4	95–100	ca. 1
Lysozyme (egg white)	4	50–80	ca. 1
Taka-amylase A (*Aspergillus oryzae*)	4	48	ca. 0.3
Trypsin (bovine):	6		ca. 0.01
CM-cellulose trypsin (insoluble)		4	
Poly-DL-alanyl trypsin (soluble)		8	
Insulin (bovine)	3	5–10	ca. 6.7
Alkaline phosphatase (*E. coli*)	2	80	ca. 33
Pepsinogen (swine)	3	50	ca. 6.7
Serum albumin (human)	17	50	?

SOURCE: C. J. Epstein, R. F. Goldberger, and C. B. Anfinsen, *Cold Spring Harbor Symp. Quant. Biol.,* **28:** 439 (1963).

disulfide bonds of proteins is an uncatalyzed process *in vivo*. Indeed, reformation of native structures from reduced proteins may be a good deal too slow to account for the rate of synthesis of biologically active proteins in the living cell. Anfinsen and his collaborators have identified an enzymatic system in beef liver that accelerates the reactivation of reduced ribonuclease (58). These results do establish that undirected reformation of disulfide bridges of reduced proteins largely yields structures that are identical to those of proteins isolated from living material. Because undirected reformation of tertiary structures very probably yields the thermodynamically most stable conformation, it follows that native proteins exist in the thermodynamically most stable conformation.

It is instructive to focus our attention on insulin, the apparent exception to the generalization just drawn. In fact, reoxidation of insulin yields only about a statistical return to the native state. How does one rationalize this finding? As it turns out, it is now quite clear that insulin is derived from a proinsulin, which consists of a single polypeptide chain. The route for conversion of proinsulin to insulin is shown in Fig. 4-35 (59). Thus, the disulfide bonds of insulin are formed biogenetically prior to formation of the double-stranded structure. If one reduces proinsulin and then permits the disulfide bridges to reform spontaneously, a quantitative yield of native proinsulin is regenerated (59). Hence, the native proinsulin structure is the thermodynamically most stable one, in accord with our previous conclusions about other proteins. Insulin itself, then, is not an exception to the conclusion that primary structure determines higher-order structures but a consequence of a chemical change subsequent to locking in certain geometrical restrictions deriving from disulfide-bond formation.

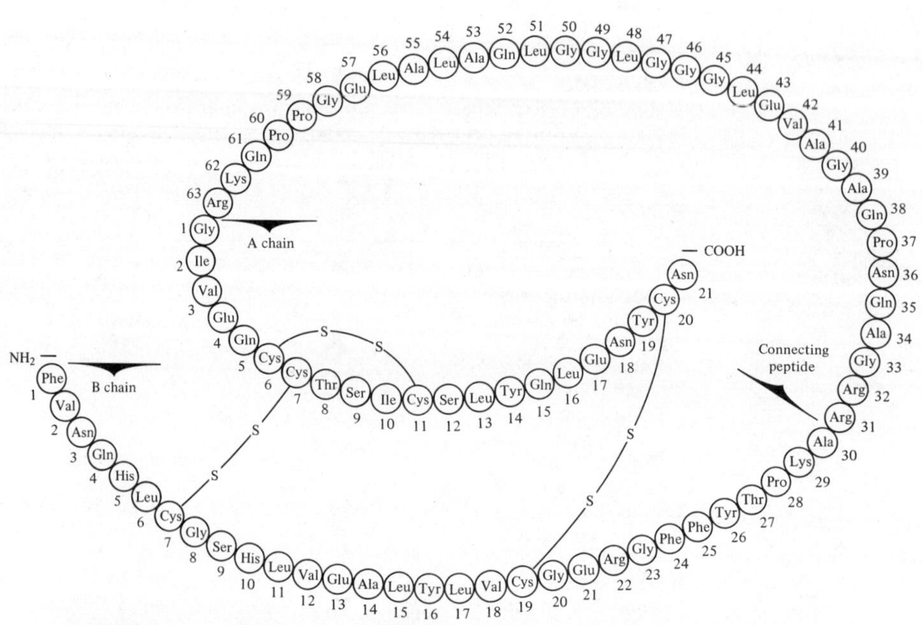

FIGURE 4-35

Proposed primary structure of porcine proinsulin. R. E. Chance, R. M. Ellis, and W. W. Bromer, *Science,* **161:** 165 (1968).

PROTON MAGNETIC RESONANCE SPECTRA OF PROTEINS

As can be appreciated from the close inspection of any three-dimensional model of a protein, several chemically identical groups may be expected to behave in different ways as a consequence of their individual local environments. We have already noted one specific example: the various histadines of myoglobin react with bromoacetate at different rates, reflecting the extent to which they are isolated from the aqueous environment. A number of related examples are known and are cited in Chapter 7. In addition to accessibility to the aqueous solution, the behavior of groups can be affected by the proximity of charged or hydrophobic groups, by hydrogen-bonding to a neighboring function, and so forth. One of the principal interests in protein chemistry is to examine the chemical behavior of individual groups within the complete protein molecule. As is detailed in Chapter 7 and implied earlier in this chapter, enzymatic activity involves the cooperation of a few groups near or in contact with the substrate. Consequently, it becomes of great importance to understand what the chemical behavior of these groups is and how it depends on the immediate surroundings. Other aspects of protein function may also depend on the detailed chemical behavior of specific groups.

Information of this type is not easy to come by. Classically, protein chemistry has focused on properties of the whole molecule and not on those of individual groups within it. Most physical techniques are insufficiently sensitive or have insufficient resolving power to sort out one group from others of the same kind in the same molecule. Recent advances in proton magnetic resonance spectroscopy have proved equal to this task in a limited sense at least; further developments should permit still greater understanding and insight.

Obtaining meaningful proton magnetic resonance spectra of proteins is complicated by several factors. First, the protein molecule contains a great many different kinds of protons. Therefore, an instrument of great resolving power is required to make distinctions between them. Second, since there will be only one or a few protons in exactly the same chemical environment, signals are necessarily weak, and marked sensitivity is required to detect them. Finally, the slow reorientation of biopolymers in solution, a result of their size, tends to make line widths larger, increasing both the difficulty of resolution and that of detection. Until the last few years, these difficulties proved so formidable that proton magnetic resonance spectra of proteins consisted of a few broad bands uninterpretable in a detailed way.

This situation has been altered by the advent of 220-MHz nmr spectrometers. These high-frequency instruments have much greater resolving power and sensitivity than those previously available and are capable of yielding useful information about the behavior of individual groups within a protein molecule. Moreover, proton resonance spectroscopy is in principle capable of yielding direct information about the secondary and tertiary structure of macromolecules, because the chemical shifts depend upon conformation. Determined efforts have been made to deduce the conformation of gramicidin S, a cyclic polypeptide with 10 amino acid residues, by this means (60–63).

Let us focus attention on one specific example: the proton magnetic resonance spectrum of bovine pancreatic ribonuclease. We have previously indicated the amino acid sequence (page 123) and three-dimensional structure (page 163) for this molecule. Of all the protons that occur in proteins, the easiest to sort out are the C-2 protons of the imidazole moiety of the histidine residues (see page 172).

This proton absorbs at very low fields almost isolated from other peaks in protein spectra. In ribonuclease, there are four histidines, at positions 12, 48, 105, and 119. It ought, therefore, to be possible to detect four peaks, one for each of the C-2 protons, assuming that the immediate environment of each of the four is different. Not only has it been possible to detect all four peaks but to assign them as well, in order of increasing field, to residues 105, 12, 119, and 48 (64). The nmr spectrum is shown in Fig. 4-36a.

The assignment of these peaks to individual residues is a central achievement in protein chemistry. It provides a means for examining the effect of external perturbations on each of them through examining the chemical shift and/or line width as a function of a given variable. For example, protonation of the imidazole residues causes the resonances of the C-2

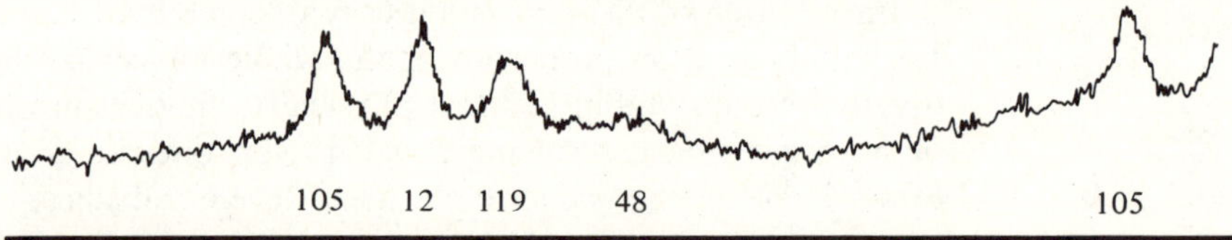

FIGURE 4-36a

100 MHz proton magnetic resonance spectrum of the histidine absorption region of 0.0065 M ribonuclease. Peak 105 in the right-hand part of the spectrum is the imidazole C_4H peak of His 105. Other peaks labeled 105, 12, 119, and 48 are imidazole C_2H resonances. Solutions are 0.2 *M* NaCl (D_2O) at pH 5.5 and 32°. D. H. Meadows and O. Jardetzky, *Proc. Natl. Acad. Sci. U.S.*, **61:** 406 (1968).

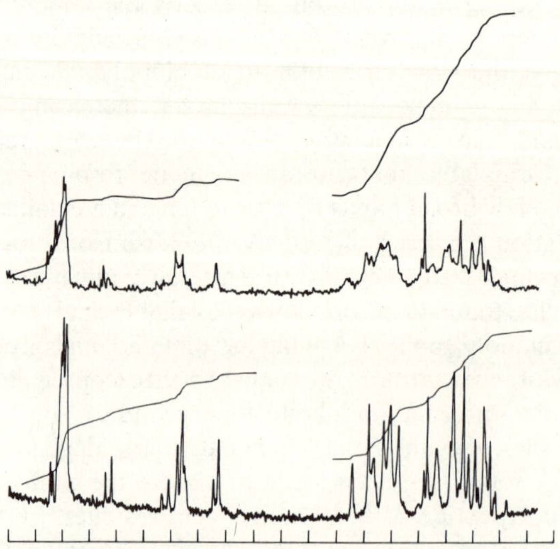

FIGURE 4-36b

Proton decoupled natural abundance carbon-13 Fourier transform nmr spectra of ribonuclease A at 45° and 15.08 MHz. The scale runs from 0 to 200 part per million upfield from CS_2. (Top): Normal spectrum of 0.017 M native protein at pH 4.12, resulting from 32,768 scans in 6 hours. (Bottom): Normal spectrum of 0.015 M denatured protein at pH 1.46, from 31,284 scans in 12 hours. Courtesy of A. Allerhand, D. Doddrell, V. Glushko, D. W. Cochran, E. Wenkert, P. J. Lawson, and F. R. N. Gurd.

protons to shift to lower field. By studying the dependence of the chemical shifts on pH, one can obtain titration curves for each imidazole group and, from them, values for each of the pK_a's (65,66). The data are collected in Table 4-14. Clearly, the local environment of the imidazole groups does affect the acid–base behavior of the side chains.

Similar methods suffice to localize those histidines which occupy sites close to that to which the substrate binds. Thus, the chemical shifts of histidines 12 and 119 are much more affected than are those for histidines 48 and 105 by the addition of inhibitors to the enzyme (67–69). This observation is in agreement with results of previous work, which indicates that histidines 12 and 119 play an active role in the catalytic process.

In addition to proton nmr, magnetic resonance properties of other nuclei will certainly find important applications in protein chemistry. Among other nuclei, ^{13}C seems particularly promising. One example, ribonuclease, is included in Fig. 4-36 along with the proton spectrum. This stunning achievement in nmr, employing natural abundances of ^{13}C, indicates the potency of the method.

QUATERNARY STRUCTURE OF PROTEINS

Of all the proteins that we have considered thus far, hemoglobin is unique in the sense that it contains not one but four independent polypeptide chains, that is, chains which are not held together by covalent bonds. Each of these chains is characterized by its own secondary and tertiary structure. The interactions between the exposed groups of the folded polypeptide chains are responsible for holding together the complete hemoglobin molecule. Proteins such as hemoglobin, which consist of more than one independent polypeptide chain, are said to possess *quaternary structure.*

The concept of protein subunits was initially championed by Sørensen and Svedberg in the 1930s but later fell into obscurity, only to be resurrected again 30 years later. Quaternary structure for proteins is now recognized to be widespread; a compilation of pertinent data is presented in Table 4-15 (70). Additional examples continue to be discovered with regularity. It now appears that the great majority of proteins with molecular weights above 50,000 are composed of subunits rather than a single chain.

The subunit structure of proteins is of interest in two respects. First, it is a novel facet of protein chemistry, and questions about the organization of the subunits naturally arise. Second, the subunit structure appears to underlie the regulation of the activity of enzymes by small molecules, typically the substrates or products of the reactions which they catalyze. This topic is explored in some detail in Chapter 6 (pages 299ff). For the moment, let us consider briefly the structural organization of these proteins.

Other than hemoglobin, wide-angle X-ray diffraction of several subunit-containing proteins are in various stages of completion. These include lactic

TABLE 4-14
Values of pK_a for the conjugate acids of the four histidine side chains of bovine pancreatic ribonuclease

Residue Number	pK_a
12	6.2
48	6.4
105	6.7
119	5.8

dehydrogenase, glyceraldehyde-3-phosphate dehydrogenase, malic dehydrogenase, aspartic transcarbamylase, and others. The first three of the enzymes catalyze oxidation–reduction reactions and form part of a class of enzymes known as nicotinamide nucleotide–containing dehydrogenases (page 662). Lactic dehydrogenase provides a representative example. This protein consists of four subunits having a molecular weight near 35,000 (Table 4-15). The subunits are somewhat asymmetrical themselves but form a rather cubelike aggregate by occupying positions at the corners of a square. An electron micrograph and schematic views of the organization of the complete molecule are shown in Fig. 4-37 (71).

Structural Basis for Isozymes

Quaternary structure of proteins may provide a structural basis for at least some cases of *isozymes*. Isozymes or isoenzymes are enzymes performing the same function, but present in more than one molecular form within the same tissue or species. Studies of N. O. Kaplan and C. Markert indicate that lactic dehydrogenase provides a good example of such an enzyme. In the chicken, lactic dehydrogenase occurs in two principal forms: that characteristic of breast muscle and that characteristic of heart muscle (page 513). These two forms are distinguishable on the basis of amino acid composition and distinct physical, immunologic, and catalytic properties. The number of different lactic dehydrogenases in the chicken and in several other species is five. Three of these isozymes exhibit properties that are intermediate between those of muscle and heart types. For example, electrophoresis yields five distinct and equally spaced bands with the muscle and heart types at the extremes of the pattern. Furthermore, the amino acid compositions of the five isozymes show a regular variation from that characteristic of the muscle to that of the heart type. Likewise, the enzymatic and immunologic properties exhibit a similar gradation from one of the extreme types to the other.

The structure of lactic dehydrogenase provides a ready explanation for these results. Since lactic dehydrogenase is a tetramer, the assumption of two gene products (two types of subunit) accounts for the observed findings. Thus, designating one type of subunit as H (heart or α) and the other as M (muscle or β), five structurally distinct lactic dehydrogenases may be formed: HHHH (α_4), and HHHM ($\alpha_3\beta$), HHMM ($\alpha_2\beta_2$), HMMM ($\alpha\beta_3$), MMMM (β_4). This formulation clearly assumes that all the subunit sites in the tetramer are equivalent; otherwise one might have, for example, four molecules corresponding to the structure HHHM. Consideration of the lactic dehydrogenase isozymes as composed of varying amounts of each of two subunit types obviously accounts for the gradual variation in the properties of the isozymes between two extremes (the HHHH and MMMM molecules).

Multienzyme Complexes

Formation of multienzyme complexes is closely related to formation of quaternary structure for proteins. Several multienzyme complexes are discussed in modest detail in later sections of this book. A particularly good example is the pyruvate dehydrogenase of *E. coli* (see page 519 and Fig. 4-38). This particle, of molecular weight 4,000,000, is composed of three types of enzymes, one of which, at least, is in turn composed of subunits. The complete particle contains a minimum of 88 separate polypeptides. This complex system is remarkably self-organizing and can be spontaneously formed by mixing the component parts. This reassociation proceeds in high yields.

TABLE 4-15
Subunit constitution of proteins

Protein	Mol. wt.	Subunits No.	Subunits Mol. wt.
insulin	11,466	2	5,733
thrombin	31,000	(3)	(10,000)
β-lactoglobulin	35,000	2	17,500
rhodanese	37,000	2	18,500
bovine growth hormone	48,000	2	25,000
Neurospora malate dehydrogenase	54,000	4	13,500
hemoglobin	64,500	4	16,000
thiogalactoside transacetylase	65,300	2	29,700
rat liver malate dehydrogenase	66,300	2	37,500
O-acetylserine sulfhydrylase A	68,000	2	34,000
tropomyosin B	68,000	2	33,500
avidin	68,300	4	18,000
concanavalin A	71,000	4	17,500
glycerol-1-phosphate dehydrogenase	78,000	2	40,000
uridine diphosphogalactose-4-epimerase	79,000	2	39,000
alkaline phosphatase	80,000	2	40,000
creatine kinase	80,000	2	40,000
liver alcohol dehydrogenase	80,000	4	20,000
yeast aldolase	80,000	2	40,000
enolase	82,000	2	41,000
haptoglobin 1–1	85,000	2	40,000
procarboxypeptidase	87,000	1	34,500
		2	25,000
firefly luciferase	92,000	2	52,000
methionine-transfer RNA synthetase	96,000	2	48,000
α-amylase	97,600	2	48,200
aspartate aminotransferase	100,000	2	50,000
hexokinase	102,000	4	27,500
hemerythrin	108,000	8	13,500
spinach leaf aldolase	120,000	4	30,000
tyrosinase	128,000	4	32,000
C-reactive protein	129,000	6	21,500
fructose diphosphatase	130,000	2	29,000
		2	37,000
mammary glucose-6-phosphate dehydrogenase	130,000	2	63,000
ornithine amino transferase	132,000	4	33,000
L-amino acid oxidase	135,000	2	70,000
glyceraldehyde-3-phosphate	140,000	2	72,000
dehydrogenase	72,000	2	37,000
mouse nerve growth factor protein	140,000	4–6	30,000
tartaric acid dehydrase	145,000	4	39,000
lactic dehydrogenase	150,000	4	35,000
	35,000	2	18,000
pyridoxamine pyruvate transaminase	150,000	4	38,000
yeast alcohol dehydrogenase	150,000	4	37,000
ceruloplasmin	151,000	8	18,000
tryptophan synthetase	159,000	2	49,500
		2	29,500
muscle aldolase	160,000	4	40,000
cystathionine γ-synthetase	160,000	4	40,000
threonine deaminase	160,000	4	40,000
carboxylesterase	167,000	2	85,500

TABLE 4-15 (Continued)

Protein	Mol. wt.	Subunits No.	Subunits Mol. wt.
thetin homocysteine methyl-pherase	180,000	3–4	50,000
histidine decarboxylase	190,000	10	19,000
fumarase	194,000	4	48,500
Salmonella threonine deaminase	194,000	4	48,500
phosphoenolpyruvate carboxylase	198,000	4	49,200
plasma high-density lipoprotein	210,000	4	58,000
phosphoribosyl adenosine triphosphate: pyrophosphate phosphoribosyl transferase	215,000	6	36,000
tryptophanase	220,000	2	110,000
	110,000	2	55,000
paramyosin	220,000	2	110,000
formyltetrahydrofolate synthetase	230,000	4	58,000
catalase	232,000	4	57,500
pyruvate kinase	237,000	4	57,200
anthranilate synthetase complex	240,000	6	40,000
glucose-6-phosphate dehydrogenase	240,000	6	43,000
phytochrome	252,000	6	42,000
phycocyanin	266,000	2	134,000
	134,000	4	28,000
glycolate oxidase	270,000	2	140,000
mitochondrial adenosine triphosphatase	284,000	10	26,000
cysteine synthetase	309,000	1	160,000
		2	68,000
aspartyl transcarbamylase	310,000	2	100,000
		2	50,000
	100,000	3	33,000
	50,000	3	17,000
acetoacetate decarboxylase	340,000	6	62,000
	62,000	2	29,000
arachin	345,000	2	180,000
	180,000	6	30,000
phosphorylase A	370,000	4	92,500
lipovitellin	400,000	2	200,000
phosphoenolpyruvate carboxytrans-phosphorylase	430,000	(3–4)	(120,000)
fatty acid synthetase	450,000	2	230,000
apoferritin	480,000	20	24,000
urease	483,000	6	83,000
fraction 1 protein, carboxydismutase	515,000	24	22,000
myosin	468,000	2	212,000
		2–3	20,000
β-galactosidase	520,000	4	130,000
	130,000	3–4	(40,000)
glutamine synthetase	592,000	12	48,500
pyruvate carboxylase	660,000	4	165,000
	165,000	4	45,000
thyroglobulin	669,000	2	335,000
propionyl carboxylase	700,000	4	175,000
α-crystallin	810,000	(30)	26,000
arginine decarboxylase	850,000	5	165,000
	165,000	2	85,000
lipoic reductase-transacetylase	1,600,000	60	27,000
glutamic dehydrogenase	2,000,000	8	250,000
	250,000	5	50,000

TABLE 4-15 (Continued)

Protein	*Mol. wt.*	*Subunits*	
		No.	*Mol. wt.*
hemocyanin	300,000– 9,000,000		385,000 70,000 35,000
chlorocruorin	2,750,000	12	250,000
bromegrass mosaic virus	4,600,000	180	20,000
turnip-yellow mosaic virus	5,000,000	150	21,000
poliomyelitis virus	5,500,000	130	27,000
cucumber mosaic virus	6,000,000	185	21,500
alfalfa mosaic virus	7,400,000	160	35,000
liver acetyl coenzyme A carboxylase	8,300,000	2	4,100,000
	4,100,000	10	409,000
bushy stunt virus	9,000,000	120	60,000
potato virus X	35,000,000	650	52,000
tobacco mosaic virus	40,000,000	2,130	17,500

SOURCE: I. M. Klotz and D. W. Darnall, *Science,* **169:** 127 (1969).

PROTEIN DENATURATION

The folded conformations of proteins are generally fragile structures. Treatment of proteins with extremes of acid or base concentration, high concentrations of salt or organic solvents, heat, radiation, and the like cause the higher-order protein structures to be lost to a variable degree. This process is termed denaturation and has been defined by Tanford as a "major change from the original native structure without alteration of the amino acid sequence, i.e., without severance of any of the primary chemical bonds which join one amino acid to another."

The denaturation of proteins may be studied in any number of ways, because changes in the shape of a protein are reflected in most aspects of protein behavior. Hence, study of sedimentation behavior, viscosity, electrophoresis, optical rotatory dispersion, circular dichroism, light scattering, thermodynamic properties, reactivity of individual groups, immunochemical properties, and biological properties may all yield useful measures of protein denaturation. These various methods vary in their sensitivity. Thus, a change in a biological activity (catalytic power for example) may be elicited by a structural change too small to be picked up in terms of a change in sedimentation constant or other physical measurement. Thus, results of denaturation studies may reflect the type of measurement that is employed to study the phenomenon.

Let us consider (1) the chemical basis for denaturing activity, (2) the nature of the denaturated state, (3) the denaturation process, and (4) the thermodynamics of denaturation. Denaturation may be either a reversible or irreversible process. Where only unfolding and refolding of the molecule occurs, denaturation is generally reversible. If changes in the arrangement of disulfide bonds occur, the process is usually irreversible. We shall be concerned only with reversible processes.

Studies of the relative effectiveness of various chemicals as protein denaturants measure (provided the denaturation is of the reversible type) the effect of the various agents on the equilibrium constant for interconversion of the native and denatured states. Thus, a given denaturant may function by destabilizing the native state, stabilizing the denatured state, or both. Employing the myoglobin molecule as a model, one may tentatively conclude that

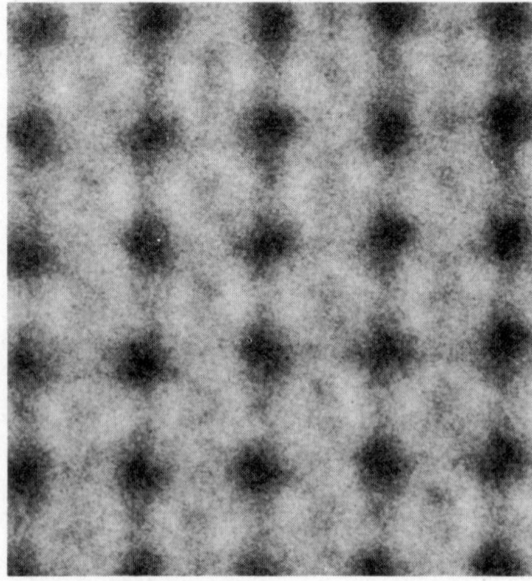

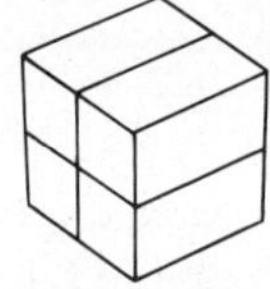

One molecule

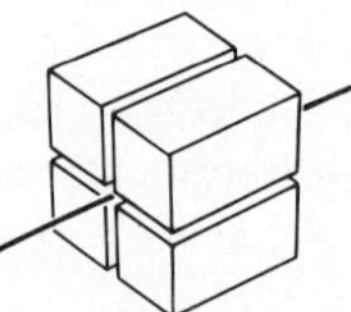

Exploded view of four subunits

FIGURE 4-37

(*Top*) Electron micrograph of a crystal of LDH, showing individual molecules. The view is that indicated by an arrow in the bottom drawing. (*Bottom*) Packing of tetrameric molecules of LDH in the crystal. The arrow shows the view in the electron micrograph. From *The Structure and Action of Proteins* by Richard E. Dickerson and Irving Geis. Copyright © 1969 by Richard E. Dickerson and Irving Geis. Reprinted by permission of the authors and Harper & Row, Publishers, Inc.

most of the polar groups are exposed to the solvent in both the native and denatured states but that the peptide backbone and nonpolar groups are largely isolated from solvent in the native state and exposed in the denatured state. Because the effects of the denaturant on the stability of the polar groups

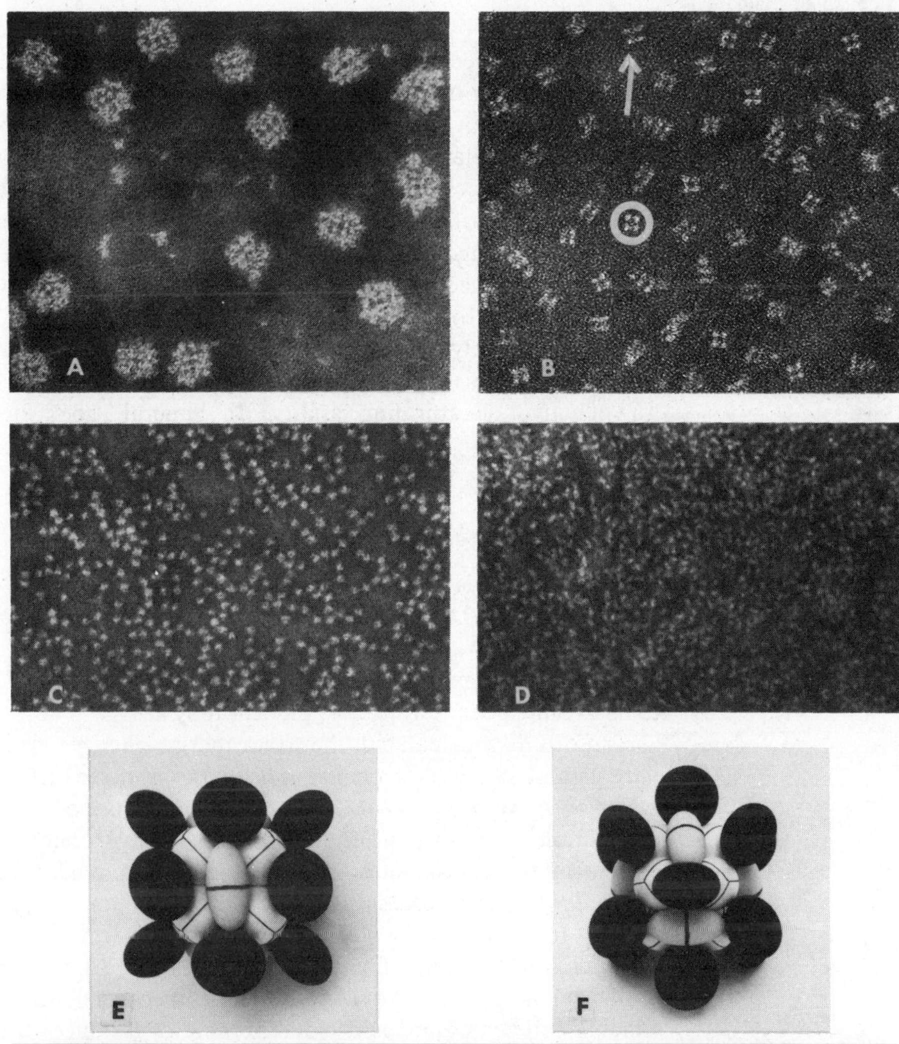

FIGURE 4-38
(*a*) Electron micrograph of the *E. coli* pyruvate dehydrogenase complex negatively stained with phosphotungstate. ×200,000. (*b*) Electron micrograph of dihydrolipoyl transacetylase component. ×200,000. Two orientations of the transacetylase particle can be seen, corresponding to a view down a fourfold axis (circled) and a twofold axis (arrow) of a cube. (*c, d*) Electron micrographs of the pyruvate dehydrogenase and dihydrolipoyl dehydrogenase components, respectively. ×200,000. (*e, f*) Tentative model of the *E. coli* pyruvate dehydrogenase complex. The 12 molocules of pyruvate dehydrogenase (Black oblate spheroids) and 6 molecules of dihydrolipoyl dehydrogenase (prolate spheroids) are aligned, respectively, on the 12 edges and on the 6 faces of the transacetylase cube. Courtesy of L. J. Reed and R. M. Oliver, University of Texas.

should be largely the same in the native and denatured states and because the peptide backbone and the nonpolar groups are largely isolated from the denaturant in the native state, it is reasonable to consider the denaturing ability of urea and related compounds as primarily reflecting stabilization of the exposed peptide backbone and nonpolar residues by the denaturant. Studies on the ability of various denaturants to solubilize the model compounds acetyl tetraglycine and CBZ-diglycinamide suggest that stabilization of both peptide linkages and nonpolar residues is important for denaturing activity (72).

The chemical basis of denaturing activity is not well understood, despite the fact that it has been the subject of thorough investigation. A particularly comprehensive study of denaturation has been carried out by Gordon and Jencks on the protein bovine serum albumin (73). Employing optical rotation at 579 nm as a quantitative measure of denaturation, these workers studied the relative efficiencies of 163 compounds as denaturants for this protein. Not surprisingly, no direct correlation was found between denaturing effectiveness and (1) monofunctional hydrogen-bonding capacity, (2) bifunctional hydrogen-bonding capacity, (3) the hydrophobic character, or (4) alterations in solvent polarity for these denaturants. A combination of these factors, which vary quantitatively from protein to protein, must be responsible for denaturing effectiveness. This study rather clearly delineated the structural requirements for effective denaturating ability. In general, good denaturants have the structure

```
  X
  |\ H
R—C—N—Y
```

in which for X, $NH_2 > S > O$; for R, $NH_2 > CH_3S \sim CH_3O > S > O \sim H \sim CH_3$; and for Y, $H > NH_2 >$ anion. On this basis, urea ($X = O$, $R = NH_2$, $Y = H$), guanidinium salts ($X = NH_2$, $R = NH_2$, $Y = H$), and *S*-methyl isothiourea ($X = NH_2$, $R = CH_3S$, $Y = H$) would be expected to be excellent denaturants, as, indeed, they are.

Tanford has calculated the free energy changes for the transfer of the various amino acid side chains and for the transfer of the peptide group from water to 8 M urea (74). In accordance with the conclusion above, these calculations indicate that urea stabilizes both hydrophobic and peptide groups.

One of the central questions relating to protein denaturation is how one can tell when the process has reached completion. That is, what are criteria which establish that the denaturated state is a random coil (a conformation in which internal rotation can occur about every single bond in the molecule)? Statistical mechanics provide us with a theoretical *radius of gyration* for a random coil containing a specified number of residues. X-ray and light scattering techniques provide a method for determining this quantity experimentally. Specifically, we have

$$\langle L^2 \rangle_0 = 6R_G{}^2 = Cn \tag{4-17}$$

in which $\langle L^2 \rangle_0^{\frac{1}{2}}$ is the unperturbed end-to-end distance, R_G the radius of gyration, n the number of residues, and C a coefficient that depends on the amino acid composition. Provided that the theoretical and experimental radii of gyration agree, one has evidence that the denaturation process has yielded a random-coil conformation.

Viscosity measurements provide another criterion for the random-coil structure. Theory again predicts viscosity behavior for this case:

$$[\eta]M_0 = K'n^x \tag{4-18}$$

in which $[\eta]$ is the specific viscosity, M_0 the mean residue weight, K' a coefficient that is almost independent of molecular weight, and x an exponent that takes on values between 0.5 and 0.8. Again, comparison of experimental and theoretical results can provide evidence for the nature of the denatured state.

As noted above, ORD curves for proteins are sensitive to the nature of the secondary structure of the molecule. Reference to Fig. 4-20 will indicate the type of ORD spectrum in the far ultraviolet which is characteristic of the random-coil conformation and how it differs from those for other conformations (75).

The use of these and other techniques has provided much insight into how individual denaturants vary in terms of their ability to convert proteins to the completely unfolded state. In studies largely from Tanford's laboratories, it has been established that 6 to 8 M guanidinium chloride is quite generally sufficient to produce the random-coil conformation. On the other hand, solutions of urea, the most widely employed denaturant, as concentrated as 10 M do not always lead to complete denaturation.

Guanidinium chloride is the most effective salt as a denaturant. In addition, however, several anions are quite effective in disrupting protein structure. The order of anion effectiveness is $SCN^- > ClO_4^- > I^- > Br^- > Cl^-$. This order parallels the abilities of the anions to disrupt water structure (the chaotropic series); this factor may well be the most important one in establishing this order.

Surfactants are unique among denaturants in that they cause marked structural disruption of most proteins at very low concentrations, 10^{-5} to 10^{-2} M. This reflects the fact that the surfactant molecules are capable of interacting strongly with hydrophobic residues of the protein. Surfactants themselves, above a crucial concentration, form *micelles,* which bear some structural resemblance to typical globular proteins in the sense of having an interior hydrophobic in character surrounded by a predominantly hydrophilic exterior (page 151).

Protein denaturation is a cooperative process (see page 299). That is, the transition from the native to denatured state is sharp, occurring over a narrow range of denaturant concentration, temperature, and so forth. One example is provided in Fig. 4-39, in which the optical rotation of lysozyme is plotted against the concentration of guanidinium chloride (76). Note that no unfold-

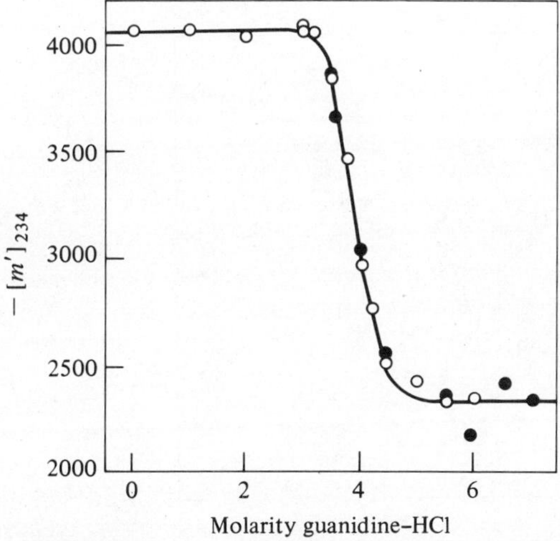

FIGURE 4-39
Denaturation of lysozyme by guanidine hydrochloride, as followed by measurement of optical rotation. Open circles represent extrapolated end points of kinetic experiments. C. Tanford, R. H. Pain, and N. S. Otchin, *J. Mol. Biol.,* **15:** 489 (1969).

ing is evident up to at least 3.5 M guanidinium chloride and that unfolding is complete at a concentration of 4.5 M.

Some denaturation reactions appear to be "two-state" processes; that is, there is no evidence for intermediates along the reaction pathway. It is possible that these reactions are all or none, but it may also be true that we have simply failed to detect intermediates that do occur. In other cases, it is quite possible to detect intermediates: one example is provided in Fig. 4-40. Thus denaturation of paramyosin as followed by optical rotation clearly exhibits at least two intermediate stages of unwinding prior to arrival at the completely unfolded form (77).

The thermodynamics of protein denaturation is complicated. In terms of a two-state formalism,

$$N \rightleftharpoons D \tag{4-19}$$

we can write an equilibrium constant, K_D, for the process. Experimental quantities that relate directly to the degree of denaturation are then measures of K_D. A plot of optical rotation of β-lactoglobulin against temperature is shown in Fig. 4-41 (78). This particular case has been well studied and it has been established that only two states, native and denaturated, are present under the conditions examined. Thus, the bell-shaped plots obtained can only reflect the fact that ΔH for the reaction must be negative at low temperatures and positive at high temperatures. For simple chemical reactions, this is a very rare situation over such a narrow temperature range.

The variation in enthalpy changes with temperature defines the heat capacity for the reaction:

$$\left(\frac{\partial \Delta H}{\partial T}\right)_p = \Delta C_{p,D} \tag{4-20}$$

Our qualitative conclusions above require that values of $\Delta C_{p,D}$ be large; for β-lactoglobulin a value of -2100 cal/deg/mol is obtained. This is a typical value for protein denaturation. Although a precise understanding of why

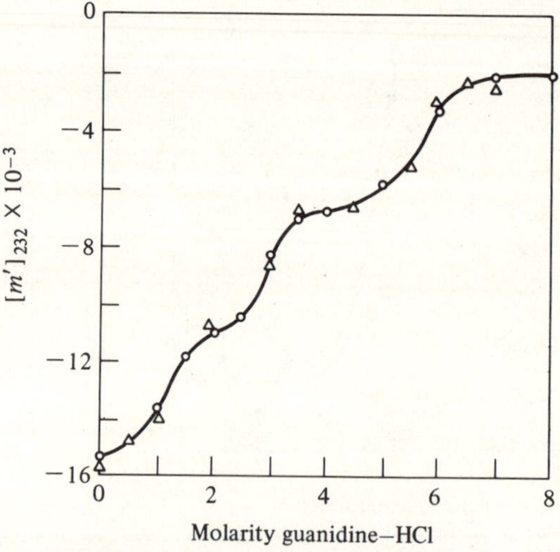

FIGURE 4-40
The denaturation of paramyosin by GuHCl, at room temperature and pH 7.3, as followed by optical rotation at 232 nm. L. M. Riddiford, *J. Biol. Chem.*, **241:** 2792 (1966).

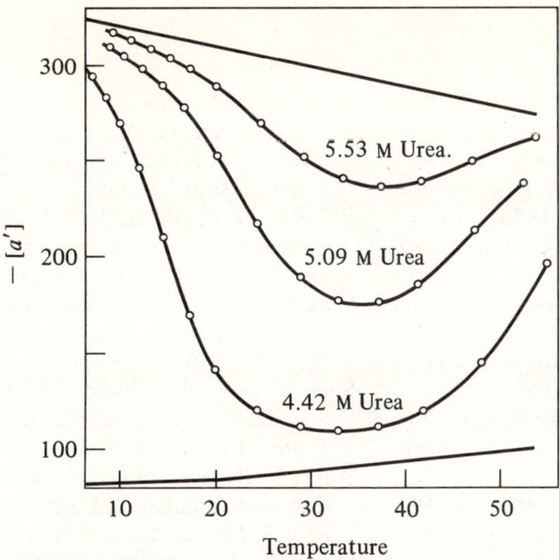

FIGURE 4-41
Effect of temperature on the optical rotation (365 nm) of β-lactoglobulin, at constant urea concentration, *p*H 3. The lines at the top and bottom of the figure represent the rotations for completely denatured and completely native protein, respectively. N. C. Pace and C. Tanford, *Biochemistry,* **7:** 198 (1968); by permission of the copyright owners, the American Chemical Society.

values of $\Delta C_{p,D}$ are so large is not available, it is known that this is typical for processes in which aliphatic groups are exposed to an aqueous environment.

The variation in ΔH with temperature is accompanied by a large change in ΔS with temperature. Some typical data are collected in Table 4-16. Note that with the temperature increase the values may change from markedly negative to markedly positive. This behavior presumably reflects the way in which two opposing factors add up as a function of temperature changes. The unfolding of the protein itself is an order → disorder transition accompanied by an entropy increase, but, as hydrophobic side chains are exposed to solvent, the water undergoes a disorder → order transition with a negative entropy change.

TABLE 4-16
Effect of temperature on the entropy of denaturation

	Value of ΔS_D, *cal/deg/mol*			
	0°	25°	50°	75°
Ribonuclease, thermal transition, *p*H 2.5	31	155	340	—
Chymotrypsinogen, thermal transition, *p*H 2.5	−80	105	360	680
β-Lactoglobulin, 5 M urea, *p*H 3	−260	−72	100	260

SOURCE: C. Tanford, *Advan. Protein Chem.*, **24:** 1 (1970).

REFERENCES

Books and Review Articles

Ackers, G. K.: Analytical Gel Chromatography of Proteins, *Advan. Protein Chem.*, **24:** 343 (1970).

Allerhand, A., and E. A. Trull: Nuclear Magnetic Resonance, *Ann. Rev. Phys. Chem.*, **21:** 317 (1970).

Beychok, S.: Circular Dichroism of Biological Macromolecules, *Science*, **154:** 1288 (1966).

Brookhaven Symposia in Biology, Structure, Function and Evolution in Proteins, **21:** (1968).

Dickerson, R. E., and I. Geis: *The Structure and Action of Proteins,* Harper & Row, New York, 1969.

Florkin, M., and E. H. Stotz (eds.): *Comprehensive Biochemistry,* vols. 2, 3, 7, and 8, Elsevier, New York, 1963.

Holmes, K. C., and D. M. Blow *The Use of X-ray Diffraction in the Study of Protein and Nucleic Acid Structure,* Wiley (Interscience), New York, 1966.

Lipscomb, W. N.: Structure and Mechanism in the Enzymatic Activity of Carboxypeptidase A and Relations to Chemical Sequence, *Accounts Chem. Res.*, **3:** 81 (1970).

Margoliash, E., and A. Schejter: Cytochrome *c, Advan. Protein Chem.,* **21:** 113 (1966).

Marsh, R. E., and J. Donohue: Crystal Structure Studies of Amino Acids and Peptides, *Advan. Protein Chem.*, **22:** 235 (1967).

North, A. C. T., and D. C. Phillips: X-Ray Studies of Crystalline Proteins, *Progr. Biophys. Mol. Biol.*, **19:** 5 (1969).

Putnam, F. W.: Immunoglobulin Structure: Variability and Homology, *Science,* **163:** 633 (1969).

Ramachandran, G. N., and V. Sasisekharan: Conformation of Polypeptides and Proteins, *Advan. Protein Chem.*, **23:** 283 (1968).

Roberts, G. C. K., and O. Jardetzky: Nuclear Magnetic Resonance Spectroscopy of Amino Acids, Peptides, and Proteins, *Advan. Protein Chem.*, **24:** 447 (1970).

Schachman, H.: Considerations on the Tertiary Structure of Proteins, *Cold Spring Harbor Symp. Quant. Biol.*, **28:** 409 (1963).

Spande, T. F., B. Witkop, Y. Degani, and A. Patchornik: Specific Cleavage and Modification of Proteins, *Advan. Protein Chem.*, **24:** 97 (1970).

Stark, G. R.: Recent Developments in Chemical Modification and Sequential Degradation of Proteins, *Advan. Protein Chem.*, **24:** 261 (1970).

Stryer, L.: Implications of X-Ray Crystallographic Studies of Protein Structure, *Ann. Rev. Biochem.*, **37:** 25 (1968).

Tanford, C.: *Physical Chemistry of Macromolecules,* Wiley, New York, 1961.

Tanford, C.: Protein Denaturation, *Advan. Protein Chem.*, **23:** 121 (1968); **24:** 1 (1969).

Timasheff, S. N., and M. J. Gorbunoff: Conformation of Proteins, *Ann. Rev. Biochem.*, **36:** 13 (1967).

Venkatachalam, C. M., and G. N. Ramachandran: Conformation of Polypeptide Chains, *Ann. Rev. Biochem.*, **38:** 45 (1969).

Witkop, B.: Chemical Cleavage of Proteins, *Science,* **162:** 318 (1968).

Specific References

1. Smyth, D. G., W. H. Stein, and S. Moore: *J. Biol. Chem.*, **238:** 227 (1963).
2. Hirs, C. H. W., S. Moore, and W. H. Stein: *J. Biol. Chem.*, **235:** 633 (1960).
3. Harman, R. E., J. L. Patterson, and W. J. A. VandenHeuvel: *Anal. Biochem.*, **25:** 452 (1968).
4. Pisano, J. J., and T. J. Bronzert: *J. Biol. Chem.*, **244:** 5597 (1969).
5. Blombäck, B., and R. F. Doolittle: *Acta, Chem. Scand.*, **17:** 1819 (1963).
6. Doolittle, R. F.: *Biochem. J.*, **94:** 742 (1965).
7. Jentsch, J.: *Z. Naturforsch,* **23B:** 1613 (1968).
8. Niall, H. D., H. T. Keutmann, D. H. Copp, and J. T. Potts, Jr.: *Proc. Natl. Acad. Sci. U.S.*, **64:** 771 (1969).
9. Edman, P., and G. Begg: *European J. Biochem.*, **1:** 80 (1967).
10. Niall, H. D., H. Penhasi, P. Gilbert, R. C. Myers, F. G. Williams, and J. T. Potts, Jr.: *Federation Proc.*, **28:** 661 (1969).
11. Stark, G. R., and D. G. Smyth: *J. Biol. Chem.*, **238:** 214 (1963).
12. Goldberger, R. F., and C. B. Anfinsen: *Biochemistry,* **1:** 401 (1962).
13. Lindley, H.: *Nature,* **178:** 647 (1956).
14. Barber, M., P. Jolles, E. Vilkas, and E. Lederer: *Biochem. Biophys. Res. Commun.*, **18:** 469 (1965).
15. Ingram, V. M.: *Biochim. Biophys. Acta,* **28:** 539 (1958).
16. Cohen, S. and R. R. Porter: *Biochem. J.*, **90:** 278 (1964).
17. Cunningham, B. A., M. N. Pflumm, U. Rutishauser, and G. M. Edelman: *Proc. Natl. Acad. Sci. U.S.*, **64:** 997 (1969).
18. Edelman, G. M., B. A. Cunningham, W. E. Gall, P. D. Gottlieb, U. Rutishauser, and M. J. Waxdal: *Proc. Natl. Acad. Sci. U.S.*, **63:** 78 (1969).
19. Wikler, M., K. Titani, T. Shinoda, and F. W. Putnam: *J. Biol. Chem.*, **242:** 1668 (1967).
20. Putnam, F. W., T. Shinoda, K. Titani, and M. Wikler: *Science,* **157:** 1050 (1967).
21. Dreyer, W. J., W. R. Gray, and L. Hood: *Cold Spring Harbor Symp. Quant. Biol.*, **32:** 353 (1967).
22. Putnam, F. W.: *Nobel Symp.*, **3:** 45 (1967).
23. Pauling, L., R. B. Corey, and H. R. Branson: *Proc. Natl. Acad. Sci. U.S.*, **37:** 205 (1951).
24. Miyazawa, T., and E. R. Blout: *J. Am. Chem. Soc.*, **83:** 712 (1961).
25. Ambrose, E. J., A. Elliott and R. B. Temple: *Proc. Roy. Soc. (London)*, **A206:** 192 (1951).

26. Kendrew, J. C., R. E. Dickerson, B. E. Strandberg, R. G. Hart, D. R. Davies, D. C. Phillips, and V. C. Shore: *Nature,* **185:** 422 (1960).
27. Kendrew, J. C., H. C. Watson, B. E. Strandberg, R. E. Dickerson, D. C. Phillips, and V. C. Shore: *Nature,* **190:** 666 (1961).
28. Kendrew, J. C.: *Brookhaven Symp. Biol.,* **15:** 216 (1962).
29. Watson, H. C., J. C. Kendrew, and L. Stryer: *J. Mol. Biol.* **8:** 166 (1964).
30. Schoenborn, B. P., H. C. Watson, and J. C. Kendrew: *Nature,* **207:** 28 (1965).
31. Schoenborn, B. P., *Nature,* **214:** 1120 (1967).
32. Perutz, M. F., J. C. Kendrew, and H. C. Watson: *J. Mol. Biol.,* **13:** 669 (1965).
33. Perutz, M. F., M. G. Rossmann, A. F. Cullis, H. Muirhead, G. Will, and A. C. T. North: *Nature,* **185:** 416 (1960).
34. Cullis, A. F., H. Muirhead, M. F. Perutz, M. G. Rossmann, and A. C. T. North: *Proc. Roy. Soc. (London),* **A265:** 15 (1961).
35. Cullis, A. F., H. Muirhead, M. F. Perutz, M. G. Rossmann, and A. C. T. North: *Proc. Roy. Soc. (London),* **A265:** 161 (1961).
36. Perutz, M. F.: *J. Mol. Biol.,* **13:** 646 (1965).
37. Phillips, D. C.: *Proc. Natl. Acad. Sci. U.S.,* **57:** 484 (1967).
38. Blake, C. C. F., G. A. Mair, A. C. T. North, D. C. Phillips, and V. R. Sarma: *Proc. Roy. Soc. (London),* **B167:** 365 (1967).
39. Blake, C. C. F., L. N. Johnson, G. A. Mair, A. C. T. North, D. C. Phillips, and V. R. Sarma: *Proc. Roy. Soc. (London),* **B167:** 378 (1967).
40. Canfield, R. E., and A. K. Liu: *J. Biol. Chem.,* **240:** 1997 (1965).
41. Jolles, J., J. Jauregui-Adell, I. Bernier, and P. Jolles: *Biochim. Biophys. Acta,* **78:** 668 (1963).
42. Sigler, P. B., D. M. Blow, B. W. Matthews, and R. Henderson: *J. Mol. Biol.,* **35:** 143 (1968).
43. Matthews, B. W., P. B. Sigler, R. Henderson, and D. M. Blow: *Nature,* **214:** 652 (1967).
44. Lipscomb, W. N., J. C. Coppola, J. A. Hartsuck, M. L. Ludwig, H. Muirhead, J. Searl, and T. A. Steitz: *J. Mol. Biol.,* **19:** 423 (1966).
45. Ludwig, M. L., J. A. Hartsuck, T. A. Steitz, H. Muirhead, J. C. Coppola, G. N. Reeke, and W. N. Lipscomb: *Proc. Natl. Acad. Sci. U.S.,* **57:** 511 (1967).
46. Lipscomb, W. N., J. A. Hartsuck, G. N. Reeke, Jr., F. A. Quiocho, P. H. Bethge, M. L. Ludwig, T. A. Steitz, H. Muirhead, and J. C. Coppola: *Brookhaven Symp. Biol.,* **21:** 24 (1968).
47. Lipscomb, W. N., J. A. Hartsuck, F. A. Quiocho, and G. N. Reeke, Jr.: *Proc. Natl. Acad. Sci. U.S.,* **64:** 28 (1969).
48. Lipscomb, W. N., G. N. Reeke, Jr., J. A. Hartsuck, F. A. Quiocho, and P. H. Bethge: *Phil. Trans. Roy. Soc. (London),* **B257:** 177 (1970).
49. Kartha, G., J. Bello, and D. Harker: *Nature,* **213:** 862 (1967).
50. Wyckoff, H. W., K. D. Hardman, N. M. Allewell, T. Inagami, D. Tsernoglou, L. N. Johnson, and F. M. Richards: *J. Biol. Chem.,* **242:** 3749 (1967).
51. Wyckoff, H. W., K. D. Hardman, N. M. Allewell, T. Inagami, L. N. Johnson, and F. M. Richards: *J. Biol. Chem.,* **242:** 3984 (1967).
52. Hugli, T. E., and F. R. N. Gurd: *J. Biol. Chem.,* **245:** 1939 (1970).
53. Quiocho, F. A., and F. M. Richards: *Biochemistry,* **5:** 4062 (1966).
54. Quiocho, F. A., and F. M. Richards: *Proc. Natl. Acad. Sci. U.S.,* **52:** 833 (1964).
55. Doscher, M. S., and F. M. Richards: *J. Biol. Chem.,* **238:** 2399 (1963).
56. Quiocho, F. A., W. H. Bishop, and F. M. Richards: *Proc. Natl. Acad. Sci. U.S.,* **57:** 525 (1967).
57. Klotz, I. M., and J. S. Franzen: *J. Amer. Chem. Soc.,* **84:** 3461 (1962).
58. Epstein, C. J., R. F. Goldberger, and C. B. Anfinsen: *Cold Spring Harbor Symp. Quant. Biol.,* **28:** 439 (1963).
59. Chance, R. E., R. M. Ellis, and W. W. Bromer: *Science,* **161:** 165 (1968).
60. Liquori, A. M., and F. Conti: *Nature,* **217:** 635 (1968).
61. Schwyzer, R., and U. Ludescher: *Biochemistry,* **7:** 2519 (1968).
62. Stern, A., W. A. Gibbons, and L. C. Craig: *Proc. Natl. Acad. Sci. U.S.,* **61:** 734 (1968).
63. Conti, F.: *Nature,* **221:** 777 (1969).
64. Meadows, D. H., O. Jardetzky, R. M. Epand, H. H. Rüterjans, and H. A. Scheraga: *Proc. Natl. Acad. Sci. U.S.,* **60:** 766 (1968).
65. Roberts, G. C. K., D. H. Meadows, and O. Jardetzky: *Biochemistry,* **8:** 2053 (1969).
66. Rüterjans, H., and H. Witzel: *European J. Biochem.,* **9:** 118 (1969).
67. Meadows, D. H., and O. Jardetzky: *Proc. Natl. Acad. Sci. U.S.,* **61:** 406 (1968).
68. Meadows, D. H., G. C. K. Roberts, and O. Jardetzky: *J. Mol. Biol.,* **45:** 491 (1969).
69. Roberts, G. C. K., J. Hannah, and O. Jardetzky: *Science,* **165:** 504 (1969).
70. Klotz, I. M., and D. W. Darnall: *Science,* **166:** 127 (1969).
71. Rossmann, M. G., B. A. Jeffery, P. Main, and S. Warren: *Proc. Natl. Acad. Sci. U.S.,* **57:** 515 (1967).
72. Robinson, D. R., and W. P. Jencks: *J. Biol. Chem.,* **238:** PC1558 (1963).
73. Gordon, J. A., and W. P. Jencks: *Biochemistry,* **2:** 47 (1963).
74. Nozaki, Y., and C. Tanford: *J. Biol. Chem.,* **238:** 4074 (1963).
75. Greenfield, N., B. Davidson, and G. D. Fasman: *Biochemistry,* **6:** 1630 (1967).
76. Tanford, C., R. H. Pain, and N. S. Otchin: *J. Mol. Biol.,* **15:** 489 (1966).
77. Riddiford, L. M.: *J. Biol. Chem.,* **241:** 2792 (1966).
78. Pace, N. C., and C. Tanford: *Biochemistry,* **7:** 198 (1968).

5 Nucleic Acids

The nucleic acids provide for some of the most essential functions in the economy of a living organism: the storage and transmission of its genetic constitution and the way in which this information is expressed in the synthesis of all cellular proteins, and hence of all cellular constituents. We shall concern ourselves with these functions of nucleic acids in Chapters 19 and 21 and with their modes of biosynthesis in Chapter 20; here we shall consider some aspects of their chemistry.

GENERAL CONSIDERATIONS

Nucleic acids, like proteins, are macromolecules. As a matter of fact, among them are the largest molecules known: there is now good reason to believe that all the deoxyribonucleic acid (DNA) of bacteria and viruses is comprised of a single molecule with a molecular weight ranging from 10^6 in small viruses to 10^8 to 10^9 or more in the larger bacteria (Table 5-1). As the name indicates, nucleic acid molecules are strongly acidic, and at physiological *p*H carry a high density of negative charge. For this reason they are found associated in the cell with various species of cations: frequently with basic proteins, such as histones or histone-like entities; sometimes with oligoamines[1] such as cadaverine, $[H_2N(CH_2)_5NH_2]$; putrescine, $[H_2N(CH_2)_4NH_2]$; spermidine, $[H_2N(CH_2)_4NH(CH_2)_3NH_2]$; or spermine, $[H_2N(CH_2)_3NH(CH_2)_4NH(CH_2)_3NH_2]$; and very commonly with the alkaline earth cations, especially Mg^{2+}. In order to break the strong multiple electrostatic bonds that link the cationic protein to the anionic nucleic acid, a variety of methods have been employed. One of the most generally useful is to treat a buffered solution of the complex at neutrality with aqueous phenol, frequently in the presence of an added protein denaturant such as sodium dodecyl (lauryl) sulfate or sodium salicylate, sometimes at elevated temperatures.[2] The denatured protein is extracted into the phenol phase, and the nucleic acid remains behind in the aqueous layer, from which it can be precipitated in the cold by the addition of 2 to 3 volumes of ethanol. Another method takes direct advantage of this insolubility of nucleic acids in aqueous ethanol; the complex is dis-

[1] At *p*H values around neutrality each of the nitrogenous groups in these compounds carries a unit positive charge.

[2] Higher temperatures have to be employed in the case of preparations from animal sources to bring the DNA into solution from the interphase.

TABLE 5-1
Sizes of DNA genomes

Organism	*Molecular weight, daltons*	*Nucleotide pairs*	*Experimental contour length, μm*
Viruses			
ϕX174	16×10^5	4.6×10^3[a]	1.77[b], 1.64[a,b]
polyoma (animal virus)	30×10^5	4.5×10^3	1.56[b]
λ coliphage	320×10^5	49×10^3	17.2
T5 coliphage	850×10^5	130×10^3	39
T2 coliphage	1300×10^5	200×10^3	56
Bacteria			
Mycoplasma gallisepticum	2×10^8	3×10^5	
Haemophilus influenzae	7×10^8	10×10^5	
Bacillus subtilis	13×10^8	20×10^5	
Escherichia coli	26×10^8	40×10^5	1400[b]
Fungi			
Saccharomyces cerevisiae	130×10^8	200×10^5	
S. cerevisiae mitochondria	600×10^5	75×10^3	25
Animals			
Drosophila melanogaster[c]	1.2×10^{11}	2×10^8	
Sea urchin sperm	6.0×10^{11}	8×10^8	
Sea urchin mitochondria	100×10^5	15×10^3	4.5[b]
Chicken[c]	13×10^{11}	21×10^8	
Mouse[c]	30×10^{11}	47×10^8	
Mouse mitochondria	100×10^5	15×10^3	4.74–5.24[b]
Human[c]	30×10^{11}	56×10^8	
Human mitochondria	100×10^5	15×10^3	4.81[b]

[a] Double-stranded, replicative form.
[b] Circular molecule.
[c] Diploid genome; all others haploid.
SOURCE: Most of the data are from B. J. McCarthy, *Bacteriol. Rev.*, **31:** 215 (1967); P. Borst and A. M. Kroon: *Int. Rev. Cytol.*, **26:** 107 (1969); P. Borst, *Symp. Soc. Gen. Microbiol.*, **24:** in press; C. A. Thomas, Jr., and L. A. MacHattie: *Ann. Rev. Biochem.*, **36:** 485 (1967).

sociated with 1 M NaCl or $NaClO_4$ and the nucleic acid fractionally precipitated with alcohol. Alternatively, the proteins may be removed first by precipitation after repeated denaturation by shaking with a solution of octanol and chloroform (the Sevag method). Pure nucleic acids contain approximately 15 percent nitrogen and 10 percent phosphorus. They contain heterocyclic bases that cause a strong absorption in the ultraviolet, with a peak around 260 nm (see below). Per gram-atom of phosphorus they contain one molecule of a sugar: either D-*ribose* in *ribo*nucleic acid (RNA) or *2-deoxy*-D-*ribose* in *deoxy*ribonucleic acid (DNA). These sugars can be identified and assayed quantitatively by characteristic color reactions: ribose by reaction with orcinol and $FeCl_3$ in HCl; deoxyribose by reaction with diphenylamine or indole in HCl. Histochemically DNA is estimated by the Feulgen reaction, which is based on the ability of its partial hydrolysis products to restore a magenta color to basic fuchsin, previously rendered colorless with sulfurous acid (Schiff's reagent). Alternative procedures make use of basic dyes (methyl green, pyronine, acridine orange) that react strongly with nucleic acids. Under carefully controlled conditions the two types of nucleic acid can be stained selectively. They can also be distinguished readily by their susceptibility to digestion by specific nucleases (DNases versus RNases).

MONOMERIC UNITS

When nucleic acids are subjected to complete chemical hydrolysis, there are obtained, ideally, mixtures containing 1 mol of phosphate, 1 mol of sugar, and 1 mol of a mixture of heterocyclic bases.[3] If the chemical hydrolysis is performed under milder conditions, or by enzymatic means, we may obtain either an equimolar mixture of inorganic phosphate plus a group of nucleosides, which are the *N*-ribosyl (or deoxyribosyl) derivatives of the bases, or the corresponding set of nucleotides, which are sugar-phosphate esters of the latter. Nucleotides are the true monomeric units of the nucleic acids (Fig. 5-1).

Bases

The bases found in nucleic acids are either pyrimidines or purines—the latter containing a fused pyrimidine–imidazole ring system (Fig. 5-2). In DNA the common bases are the pyrimidines thymine (T) and cytosine (C), and the purines adenine (A) and guanine (G). Some 5-methylcytosine occurs occasionally, especially in the DNA from higher plants (e.g., wheat germ) and certain bacteria; 6-methylaminopurine is a minor constituent of the DNA of bacteria and bacterial viruses (bacteriophages); and uracil and 5-hydroxymethyluracil have been reported as occurring in certain bacteriophage DNA's (1). A particular class of bacteriophages, the T-even phages of *Escherichia coli,* contain 5-hydroxymethylcytosine (sometimes linked to glucose residues in glycoside linkage) in place of cytosine. The structure of these bases is shown in Fig. 5-2.

Most RNA's also contain only four bases: the purines adenine and guanine, and the pyrimidines uracil (not thymine, as in DNA) and cytosine. In some RNA's, especially the low-molecular-weight ("soluble") amino acid transfer RNA (tRNA) of the cytoplasm (page 247), hypoxanthine and various methylated bases such as thymine, 5-methylcytosine, 6-methylaminopurine (and its 6,6-dimethyl derivative), 1-methylguanine, and 2-methylamino-6-hydroxypurine (and its 2,2-dimethyl derivative) as well as some more complex derivatives replace some of the normal constituents. These and the rare methylated bases in DNA are probably formed as a result of methylation of intact polymeric nucleic acids rather than of the component monomeric units.[4] Certain bases (and their nucleosides and nucleotides) not usually occurring in nature have recently aroused interest as chemotherapeutic and mutagenic agents, because they prove to be readily incorporated into nucleic acids instead of the "natural" bases. Among them are the uracil analogs 2-thio- and 5-fluorouracil, the thymine analog 5-bromouracil, and the guanine analogs 8-azaguanine and 2,6-diaminopurine.

Other purine and pyrimidine bases also occur in living organisms without, however, being components of nucleic acids: among them are the pyrimidine orotic acid, an intermediate in pyrimidine biosynthesis (Chapter 18), and hypoxanthine,[5] xanthine, and uric acid, which are catabolites of the purine bases. Their nucleotides inosinic and xanthylic acid, on the other hand, are key intermediates in purine biosynthesis (Chapter 18). The substituted oxygenated purines theophylline, theobromine, and caffeine are important constituents of plants.

Three sets of properties of the bases are of concern to us here:

1. They are all capable of lactam-lactim and/or enamine-ketimine tautom-

[3] Except for certain DNA's that contain glucose residues linked to one of the bases.
[4] Separate, specific enzymes are responsible for these methylations.
[5] As already stated, hypoxanthine is also found in transfer RNA.

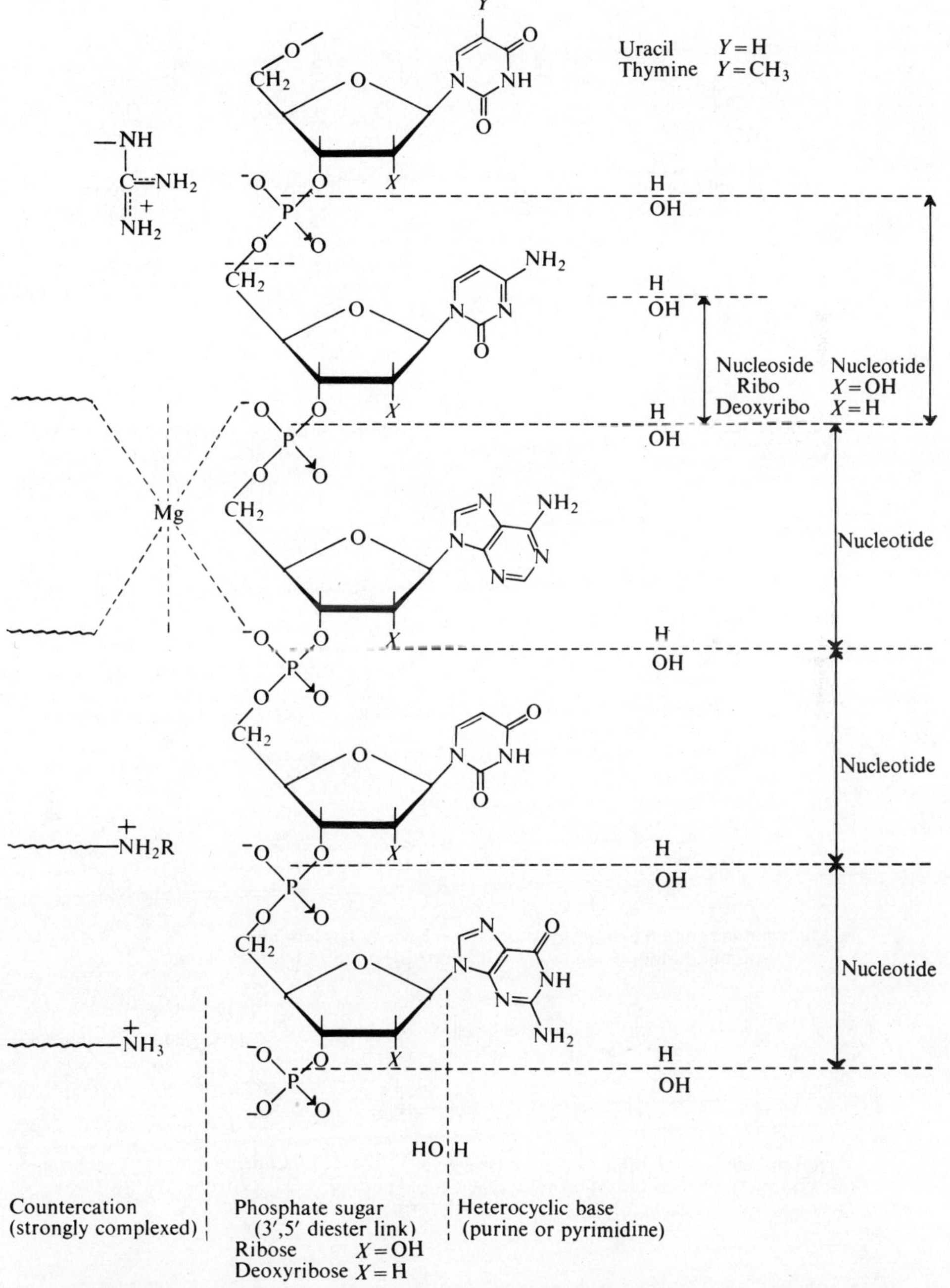

FIGURE 5-1
Random segment of a hypothetical nucleoprotein chain and its constituent parts.

Pyrimidine (Chemical Abstracts numbering system; an older convention used a system analogous to that indicated for the pyrimidine portion of purine)

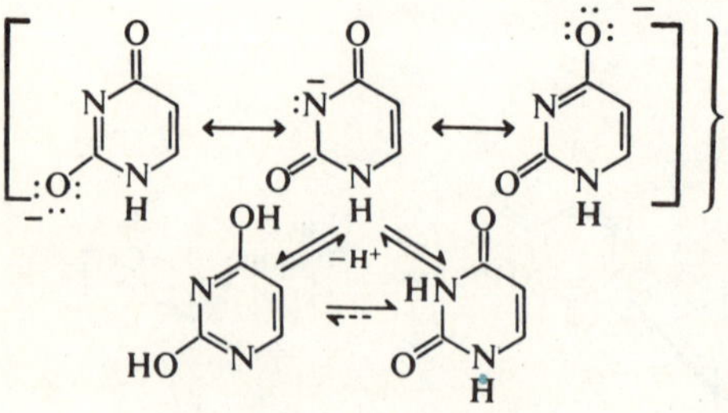

Alternative possible forms of the enolate ion

Uracil (2,4-dioxypyrimidine)
Lactim Lactam

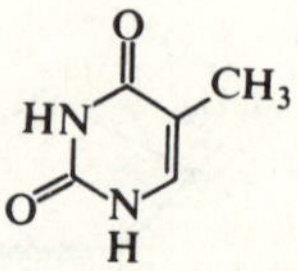

Thymine

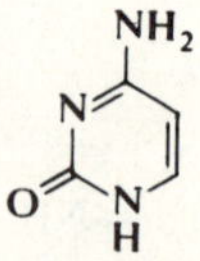

Cytosine
(2-oxy-4-aminopyrimidine)

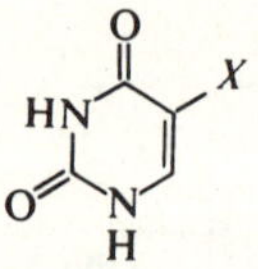

Halogenated uracil
X = F, Br, I

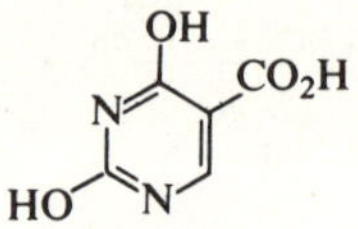

Orotic acid
(5-carboxyuracil)
lactim form

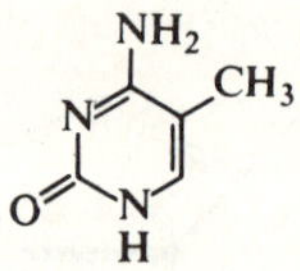

5-Methylcytosine

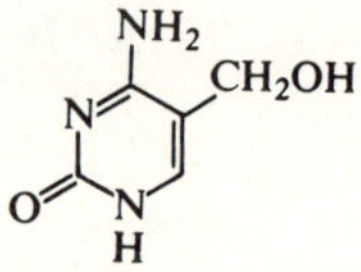

5-Hydroxymethylcytosine

(*a*)

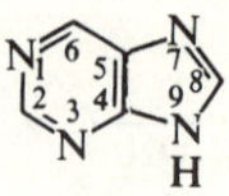

Purine

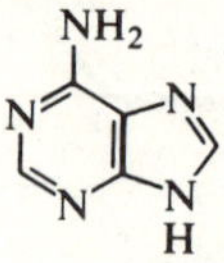

Adenine

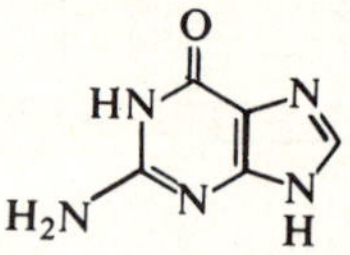

Guanine

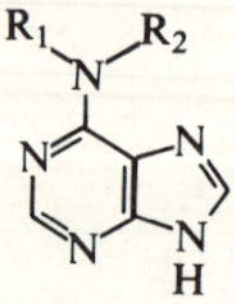

6-Methylamino- and 6,6-dimethyl aminopurine

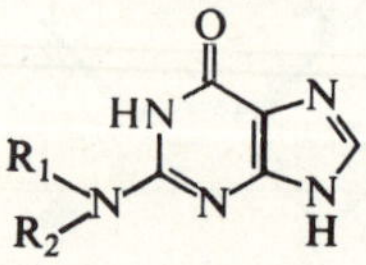

2-Methylamino- and 2,2-dimethylamino-6-hydroxypurine

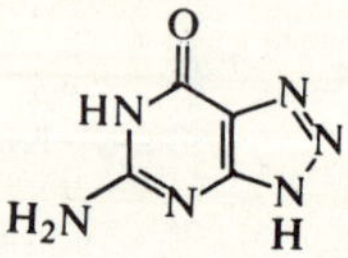

8-Azaguanine

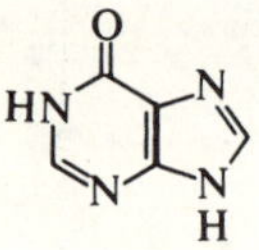

Hypoxanthine
(6-oxypurine)
lactam form

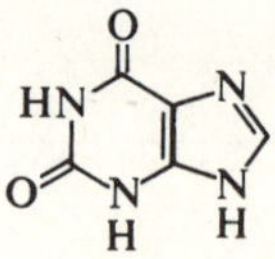

Xanthine
(2,6-dioxypurine)
lactam form

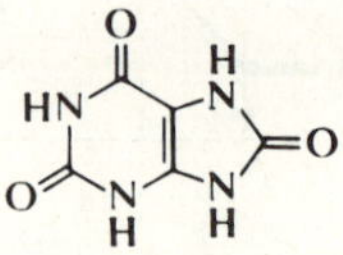

Uric acid
(2,6,8-trioxypurine)
lactam form

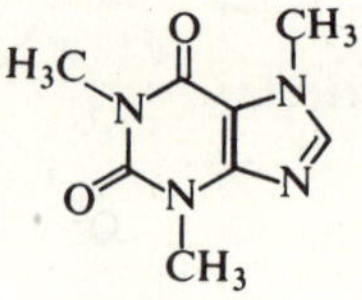

Caffeine
(1,3,7-trimethyl xanthine)

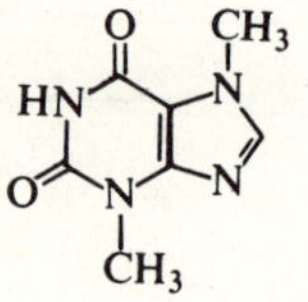

Theobromine
(3,7-dimethylxanthine)

(*b*)

FIGURE 5-2

Structures of (*a*) common pyrimidines and (*b*) common purines (the preponderant tautomer found in aqueous solution at *p*H ~ 7 is indicated).

erism. The case of uracil is shown as an example. The predominant form in aqueous solution at pH ~7 is indicated in all cases (it predominates, not only in the bases, but in their more complex derivatives as well)[6]; yet the rarer tautomers may be of great importance in biology, as potential causes of mutation (cf. page 231 and Chapter 19).

2. The purines and pyrimidines are weak bases: typical pK_a values are for uracil N^1, 9.5; for thymine N^1, 9.9; and for guanine (N^1—C^6—OH), 9.5; in addition we find for cytosine N^3, 4.5; for adenine N^1, 4.2; and for guanine N^7, 3.2.

3. They all absorb light in the ultraviolet, which accounts for that property of the nucleic acids (cf. page 187). Typical absorption spectra for the five most common bases are shown in Fig. 5-3 (2). This intense absorption is useful, not only for identifying the bases and their derivatives directly in solution, but also for their spotting on paper or cellulose acetate chromatograms or electropherograms. Since it leads to complete "quenching" of the fluorescence of the paper irradiated with ultraviolet light, the bases appear as dark spots against a bright background.

Nucleosides A nucleoside is an *N*-glycosyl derivative of a heterocyclic base. As indicated above, the two sugars of interest are D-ribose and D-2-deoxyribose; the glycosyl linkage is β- from N^1 of pyrimidines and N^9 of purines, and the sugar is in the furanose configuration. Thus adenosine is 9-β-D-ribofuranosyladenine, deoxyadenosine is 9-β-D-2-deoxyribofuranosyladenosine, guanosine is 9-β-D-ribofuranosylguanine, deoxyguanosine is 9-β-D-2-deoxyribofuranosylguanine, uridine is 1-β-D-ribofuranosyluracil, thymidine is 1-β-D-2-deoxyribofuranosylthymine, cytidine is 1-β-D-ribofuranosylcytosine, and deoxycytidine is 1-β-D-2-deoxyribofuranosylcytosine:

NH₂ N N N N HOCH₂ O 1′ OH OH ≡ OH OH 1′ N N O CH₂OH N N NH₂

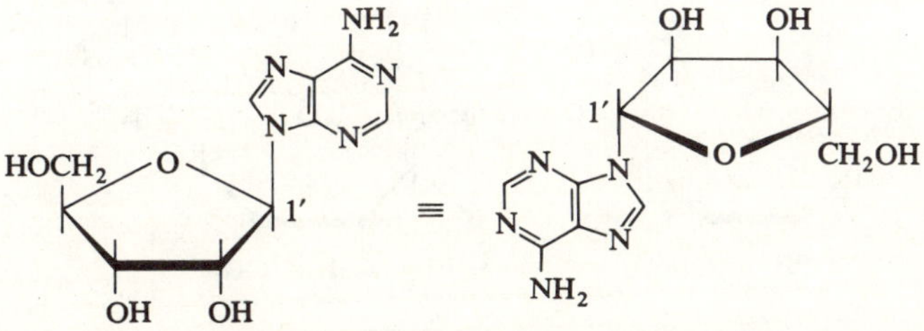

Adenosine (A), 9-β-D-ribofuranosyladenine

O CH₃ NH N O HOCH₂ O 1′ OH OH ≡ OH OH O CH₂OH O N HN CH₃ O

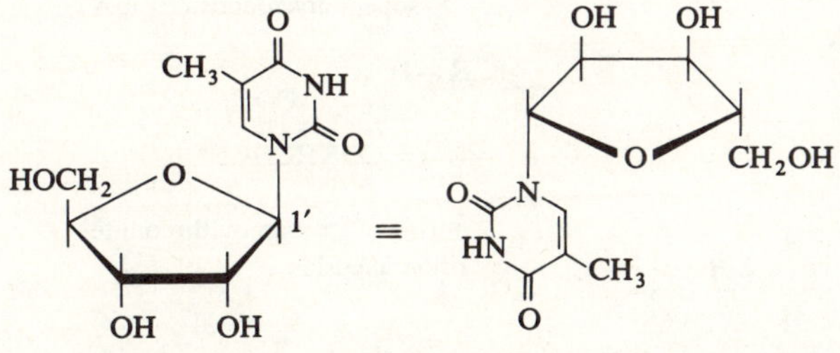

Thymidine (T), 1-β-D-2-deoxyribofuranosylthymine

[6] Inosine, the nucleoside of hypoxanthine, appears to be an exception; according to Wolfenden (3) the enol (lactim) form is present in significant amounts at neutral pH. This fact may of significance in accounting for its prevalence and base-pairing proclivities in the anticodon position of tRNA's (page 251 and Chapter 21).

Except for deoxyguanosine, the conformation (3,4) of the glycosyl bond in isolated nucleosides is predominantly *anti* (as shown), not *syn* (as below for pseudouridine). The conformation of the furanose ring is coplanar for four of the atoms, with the fifth (either C-2′ or C-3′) displaced from the plane by ~0.5 Å. If this displacement is on the same side as the protruding C-5′, the resultant conformation is designated as *endo;* if in the opposite direction, the conformation is *exo*. For all nucleosides except deoxyadenosine the conformation is *endo*. The same rules (i.e., *endo* and *anti*) appear to be obeyed by all nucleotides. In addition to a large number of methylated bases, certain RNA's, especially tRNA, also contain, in significant amounts, an unusual nucleoside of uracil called pseudouridine (ψrd or ψ) with a *C*-glycosyl bond:

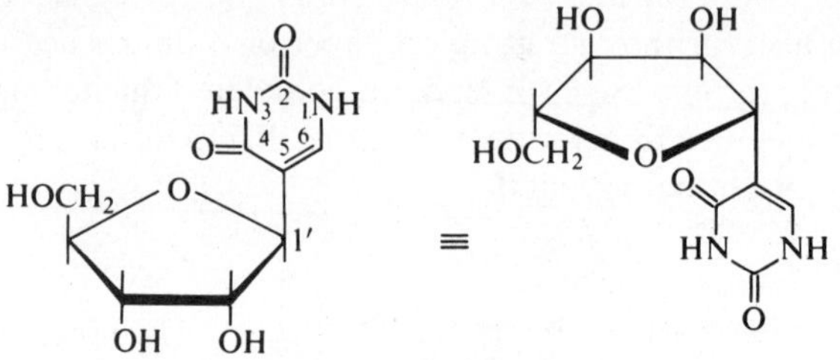

(ψ) Pseudouridine,
5-β-D-ribofuranosyluridine

Other nucleosides found in tRNA include 4-thiouridine (abbreviated as s^4U or ^4S) and certain of its derivatives, N^6-isopentenyladenosine, (ipA or iA), the corresponding carbamoylthreonine derivative; 2′-*O*-methyl (Nm) derivatives; and others.

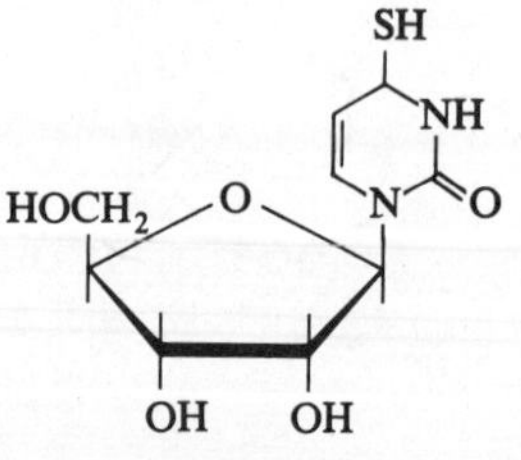

4-thiouridine (^4S)

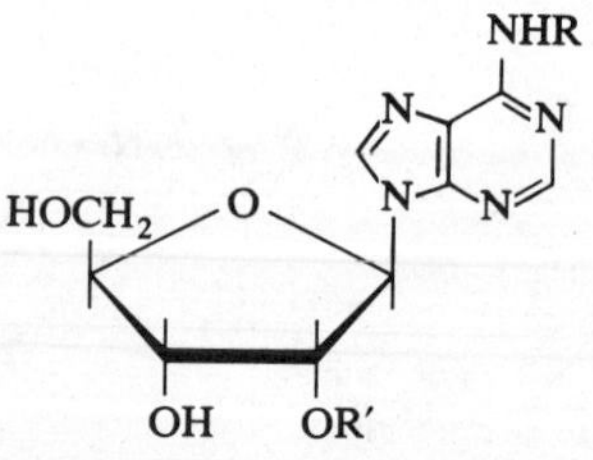

R′=H

$R{=}{-}CH_2{-}CH{=}C(CH_3)_2$

N^6-isopentenyladenosine (ipA)

R′=H

$R{=}{-}CONHCH(CO_2H)CHOHCH_3$

purine (6-carbamoylthreonine) ribonucleoside

$R'{=}CH_3$

R=H

2′-*O*-methyladenosine (Am)

Other nucleosides of natural origin (but not derived from nucleic acids) include the antibiotics cordycepin, nebularine, puromycin, and tubercidin, all elaborated by molds and fungi, and a group of compounds isolated from sponges, among which spongothymidine, spongouridine, the corresponding cytosine derivative ("cytosine arabinoside"), and spongosine are examples:

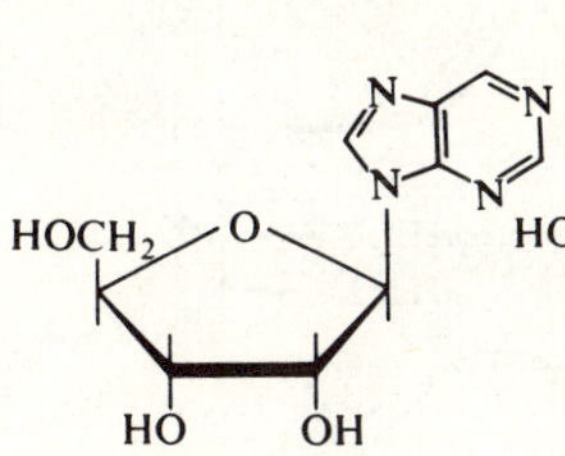

Nebularine
9-β-D-ribofuranosyl*purine*

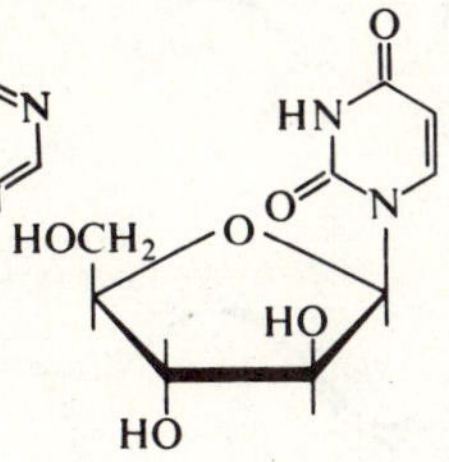

Spongouridine
1-β-D-*arabino*furanosyluracil

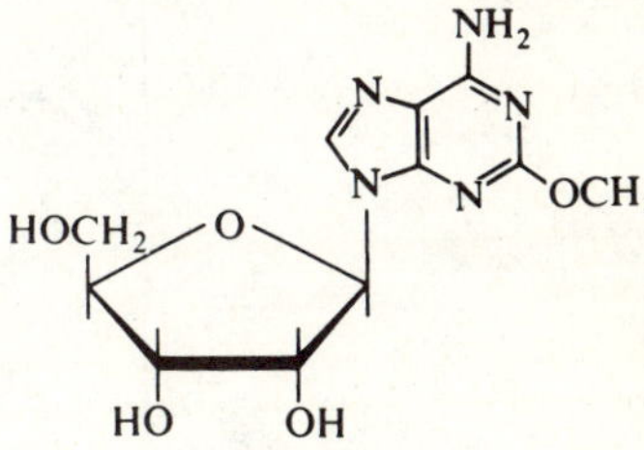

Spongosine
9-β-D-ribofuranosyl-
2-*methoxyadenine*

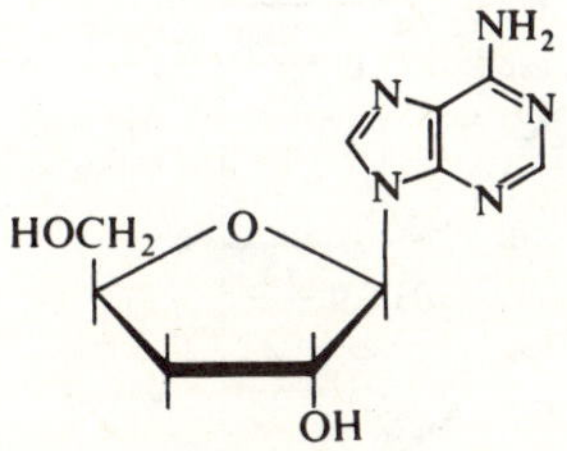

Cordycepin
9-β-(3′-*deoxy*-D-ribofuranosyl)adenine

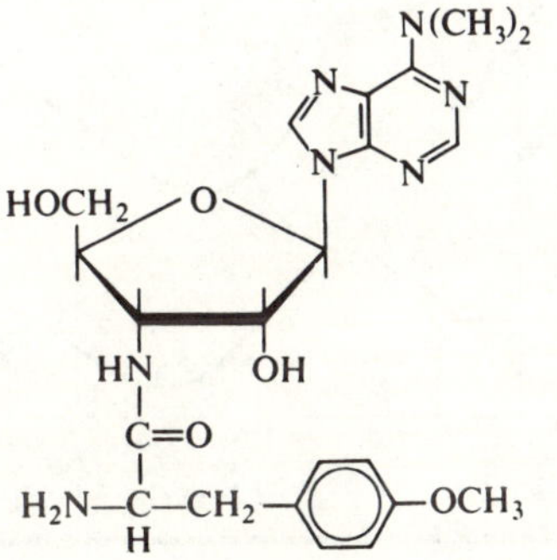

Puromycin
6-dimethylamino-9-(3′-deoxy-3′-*p*-methoxy-L-phenylalanylamino-β-D-ribofuranosyl)-purine

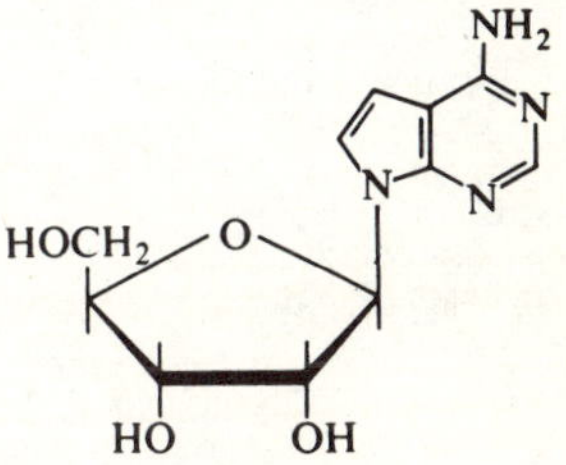

Tubercidin
7-*deaza*adenosine

Nucleotides Nucleotides are sugar-phosphate esters of the nucleosides and can be obtained by mild chemical or enzymatic hydrolysis of the nucleic acids. In the case of the ribonucleotides we might expect to find 2′- and 3′- as well as 5′-mononucleotides (or ribonucleoside-2′-, 3′-, or 5′-phosphates; the prime indicates esterification on the sugar moiety), but only the 3′ and 5′ isomers can exist in the deoxy series. These compounds are indeed found in nature, not only as products or substrates of nucleic acid metabolism, but also, at least some of them, as constituents of nucleotide coenzymes (Chapter 8). In addition, the cyclical-2′:3′-phosphates of all the ribonucleotides are formed as hydrolytic products (see page 197), the cyclical-3′:5′-phosphate of adenosine is an important metabolic regulator (Chapters 8 and 11), and adenosine 2′-5′-diphosphate and adenosine 3′-5′-diphosphate occur as parts of the coenzymes NADP and coenzyme A, respectively (Chapter 8). The latter two nucleotides are also obtained from the terminal residues of polynucleotide chains by appropriate hydrolytic procedures (see the next section). Nucleotides can also be regarded as monoesterified derivatives of phosphoric acids. They are, therefore, strong dibasic acids with pK_a values at ~1.0 and 6.0. Note that these ionizations are in addition to those ascribable to acid functions in their

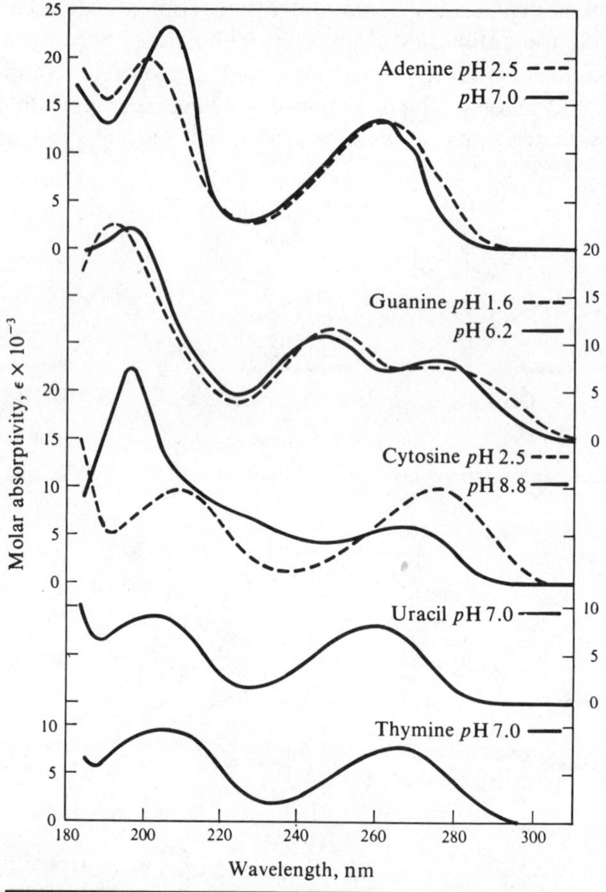

FIGURE 5-3

Absorption spectra of the common bases. D. Voet, W. B. Gratzer, R. A. Cox, and P. Doty, *Biopolymers,* **1:** 193 (1963).

heterocyclic rings. The structure of some of these compounds is shown in Fig. 5-4, and in Table 5-2 we summarize the nomenclature and abbreviations for the monomeric units commonly employed in the field of nucleic acid chemistry.

POLYNUCLEOTIDES

Primary Structure—The Phosphodiester Backbone

An overwhelming body of evidence, based on careful enzymatic and chemical degradations, suggests that the naturally occurring polynucleotides, and especially the two types of nucleic acids proper, all have the same basic structure (Fig. 5-1): a linear polymer in which successive monomeric nucleotide residues are linked to one another by phosphate groups leading to 3′,5′-phosphodiester bridges between the pentoses of adjacent nucleoside units. For DNA this is the only plausible structure on the basis of the stoichiometry of its composition (bases : deoxyribose : phosphate = 1 : 1 : 1), because the absence of −OH on the 2′ position does not permit isomeric linear or branched polymeric structures. In the case of RNA, such alternatives do exist on paper but do not appear to occur in the structure of any natural polymer examined so far. Three shorthand notations have been employed to represent the structure of such linear polymers; the conventions implicit are self-evident.

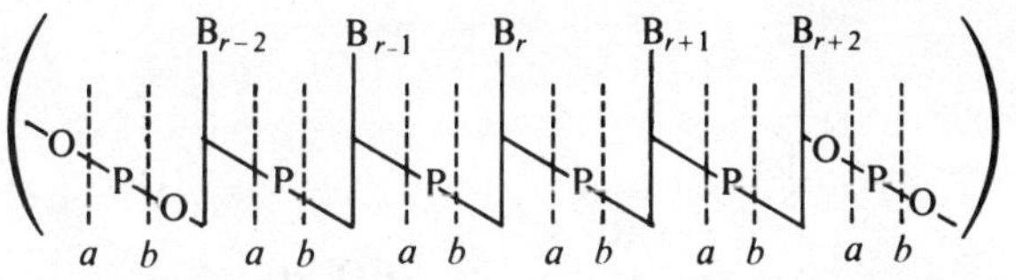

$$\equiv -N_{r-2}-N_{r-1}-N_r-N_{r+1}-N_{r+2}- \equiv -p\,N_{r-2}\,p\,N_{r-1}\,p\,N_r\,p\,N_{r+1}\,p\,N_{r+2}\,p-$$

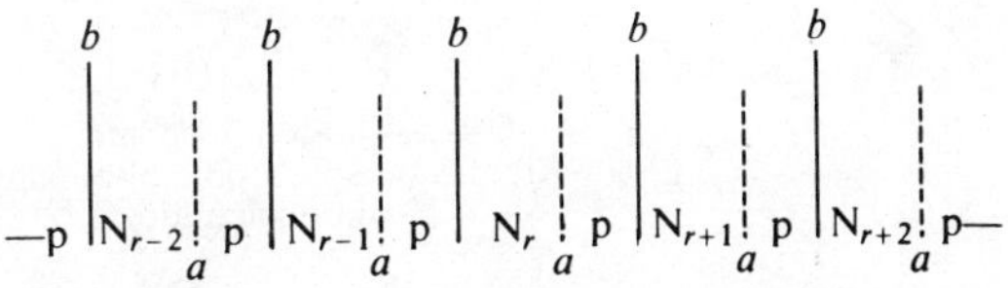

where B = A, C, G, U, or T, respectively; N are the corresponding nucleosides; and a and b indicate points of attachment of P to O, pN ≡ N-5′-*P* and Np ≡ N-3′-*P*.

As with any such polymer there are two singular regions—the two end groups. In the case of DNA the following four possibilities exist:

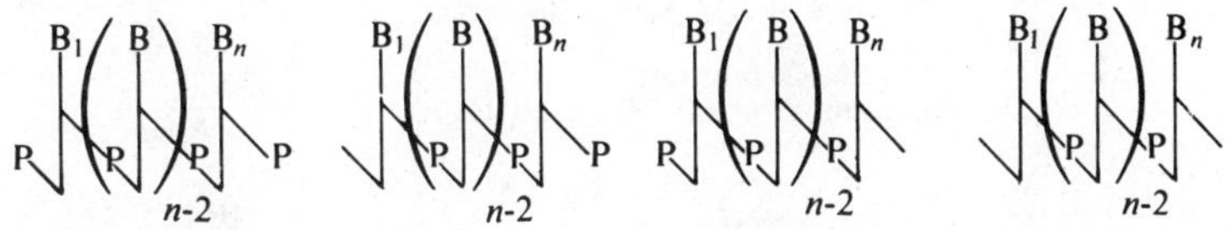

	$pN_1(pN)_{n-2}pN_np$	$N_1(pN)_{n-2}pN_np$	$pN_1(pN)_{n-2}pN_n$	$N_1(pN)_{n-2}pN_n$
Head	5′-*P* or pN_1	5′-OH, N_1	5′-*P*, pN_1	5′-OH, N_1
Tail	3′-*P* or N_np	3′-*P*, N_np	3′-OH, N_n	3′-OH, N_n
	(*a*)	(*b*)	(*c*)	(*d*)

Observe that only structures (*b*) and (*c*) show strict stoichiometry between phosphate and nucleoside residues. Structure (*a*) has one excess phosphate; it has probably been formed by phosphorylation of a polynucleotide. On the other hand, structure (*d*) is deficient in one phosphate group; therefore, it was probably formed from one of the other structures as a result of hydrolytic action, catalyzed by a phosphomonoesterase (phosphatase). Since in polyribonucleotides the 2′ position is also available, we observe cyclic 2′:3′-phosphates [symbol $(p)N_1(pN)_{n-2}N_n > p$].

If we are dealing with a homogeneous polynucleotide and have available methods of quantitative analysis with respect to its component monomers, a determination of the ratio intrachain mononucleotides/end group(s) is sufficient to determine n, the chain length, and hence the molecular weight. If, in addition, we are able to determine the molecular weight M by an independent technique, the ratio of M (individual strand) : M (parent molecule), or simply the number of unique end groups per M units of parent molecule affords a measure of the number of polynucleotide chains in the parent molecule. End-group determinations can be performed either after complete hydrolysis of the polynucleotide, or on the macromolecular species itself, for instance by measuring the number of phosphate residues liberated by specific 5′- or 3′-phosphomonoesterases completely free of any (diesterase) activity capable of degrading the polymer (5). A particularly elegant method for the determination of end groups and M consists of labeling a free 5′-OH [(*b*) or (*d*) in the diagram]—if necessary after removing any residual 5′-*P* [structure (*a*) or (*c*)] by phosphatase action—by means of radioactive phosphate of

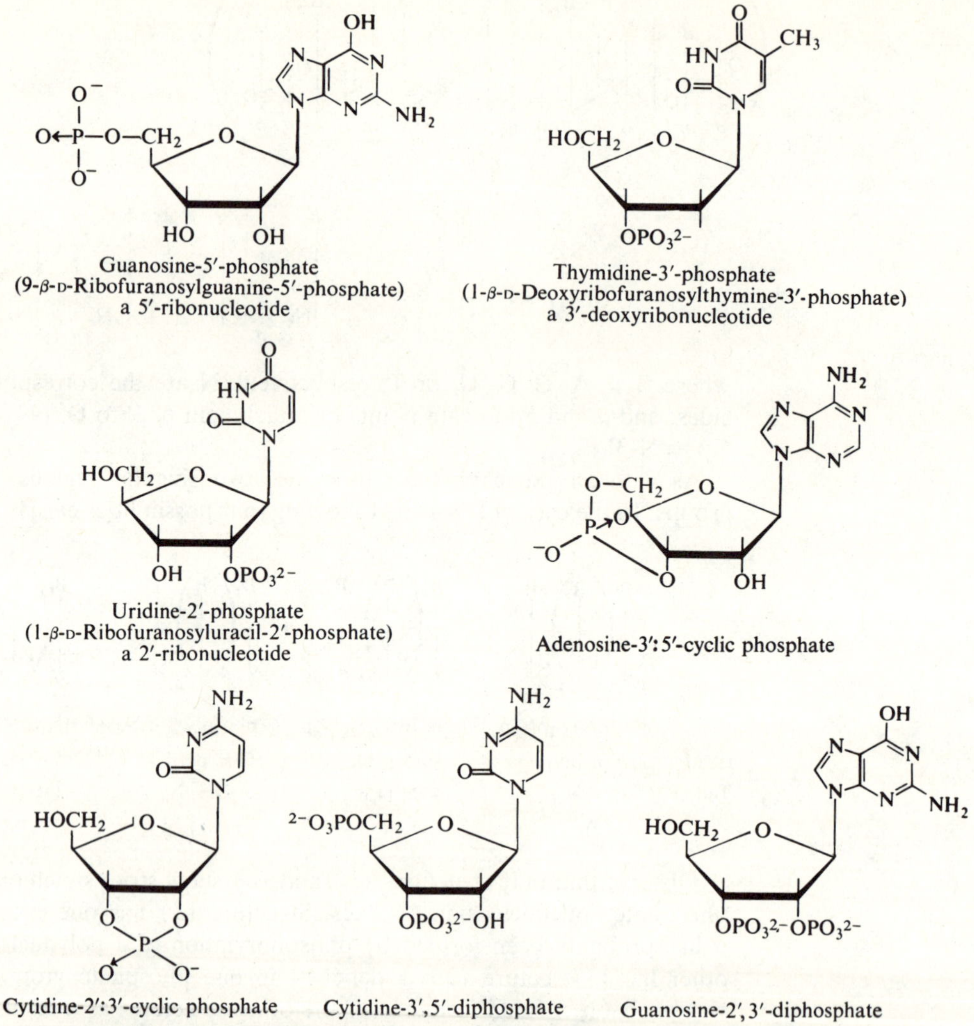

FIGURE 5-4

Structures of some common nucleotides (different bases are used as examples).

known specific radioactivity; the latter is transferred from labeled ATP in the presence of polynucleotide kinase (6) (page 862).

$$pN_1(N)_{n-1} \xrightarrow[\text{phosphatase}]{-P_i} N_1(N)_{n-1} \xrightarrow[\text{kinase}]{\text{ADP}{\sim}\text{P}^* \quad \text{ADP}} p^*N_1(N)_{n-1}$$

Because two of the three ionizable hydroxyl groups in phosphoric acid are esterified when the residue is incorporated into a polynucleotide structure, the internal nucleotide units $(Np)_{n-2}$ exhibit only the primary phosphate ionization with $pK_a' \simeq 1$. Hence, as pointed out earlier, nucleic acids occur commonly as salts.

Hydrolytic Procedures

Chemical Hydrolysis. Polyribonucleotides are degraded by alkali (0.1 N KOH for 20 min at 100°; 0.3 N KOH for 16 hr at 37°) to yield $n - 2$ mixed 2′- and 3′-ribonucleotides (Np) from the interior of the chain. A summary of the products from the head and tail of a number of different chains with different degrees of phosphorylation is given in the diagram on page 200. The cleavage of the phosphodiester linkages, therefore, occurs more or less at random, in the interior, in position (*b*). The cyclic 2′:3′-phosphates are reaction intermediates, and attack on them by OH^- brings about the cleavage of either of the C–O–P bonds with equal facility and hence produces a mixture of the 2′- and 3′-phosphates. The reaction therefore proceeds as follows:

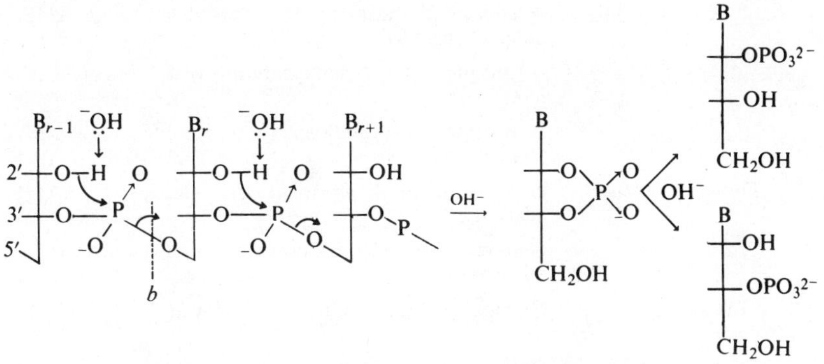

This participation by the neighboring *cis*-2′-OH group is impossible in the deoxy series. These facts account for the alkali lability of polyribo- contrasted with the well-established stability of polydeoxyribonucleotides. In effect, degradation of this type (*b* cleavage) leads to the transfer of a phosphate residue from a 5′-nucleotide to its *nearest neighbor* on the left. Consider the polynucleotide to be made up of 5′-nucleotide residues as during biosynthesis, when polynucleotides are formed by the condensation of nucleotide-5′-triphosphates (Chapter 20). Now assume that the phosphate residue linking nucleosides N_{r-1} and N_r had originally been attached to base B_r and carried a radioactive phosphorus label. After cleavage at point *b* this phosphate has been transferred to the ribose linked to base B_{r-1}. This procedure forms the basis for the *nearest-neighbor analysis* of polynucleotides originally developed by Josse, Kaiser, and Kornberg (7) for DNA[7] and since applied successfully to a variety of polynucleotides synthesized enzymatically *in vitro* (see Chapter 20).

Enzymatic Hydrolysis (8–10). The enzyme *ribonuclease* (RNase) from bovine pancreas (page 123) exhibits a similar mode of action and cleavage. It is an *endonuclease* attacking intranucleotide bonds, provided the phosphate of the susceptible P–O bond is attached to the 3′-OH of a pyrimidine nucleoside. The first step, the formation of cyclic phosphates from polynucleotides, appears to be reversible. The same type of cleavage (*b*) is also catalyzed by *phosphodiesterases* isolated from various plant sources (peas, tobacco leaves, and rye grass) all of which are, however, relatively nonspecific with respect to the nature of the bases in the susceptible nucleotides. An enzyme from *Bacillus subtilis,* on the other hand, appears to have a specificity complementary to that of the pancreatic one; it preferentially catalyzes hydrolysis adjacent to purine-3′-phosphate residues. Even more specific enzymes have been isolated from the crude enzyme *Takadiastase; ribonuclease T1* catalyzes the analogous cleavage of internucleotide linkages adjacent to a 3′-guanylate residue exclusively; *ribonuclease T2* effects only the cleavage adjacent to 3′-adenylate groups.

[7] Which uses an enzymatic method, since, as just mentioned, deoxypolymers are alkali stable.

TABLE 5-2
Common nucleotides

Base	*Nucleosides*	*Nucleotides (as salts)*	*Abbreviations*[a]
Adenine	Adenosine	Adenosine-5′-phosphate or 5′-adenylate[b]	5′-AMP,[b] Ado-5′-*P*, pA
		Adenosine-3′-phosphate or 3′-adenylate (adenylic acid b)	3′-AMP, Ado-3′-*P*, Ap
		Adenosine-2′-phosphate or 2′-adenylate (adenylic acid a)	2′-AMP, Ado-2′-*P*, (Ap)
		Adenosine-2′:3′-*cyc*-phosphate or *cyc* 2′:3′-adenylate	*cyc* 2′:3′-AMP, Ado-2′:3′-*P*, A>(p)
		Adenosine-3′:5′-*cyc*-phosphate or *cyc* 3′:5′-adenylate	*cyc* 3′:5′-AMP, Ado-3′:5′-*P*
	Deoxyadenosine	Deoxyadenosine-5′-phosphate or 5′-deoxy-adenylate	5′-dAMP,[b] dAdo-5′-*P*, pdA
		Deoxyadenosine-3′-phosphate or 3′-deoxy-adenylate	3′-dAMP, dAdo-3′-*P*, dAp
Guanine	Guanosine	Guanosine-5′-phosphate or 5′-guanylate,[b] etc.	5′-GMP,[b] Guo-5′-*P*, pG
	Deoxyguanosine	Deoxyguanosine-5′-phosphate or 5′-deoxyguanylate etc.	5′-dGMP,[b] dGuo-5′-*P*, pdG
Cytosine	Cytidine	Cytidine-5′-phosphate or 5′-cytidylate,[b] etc.	5′-CMP,[b] Cyd-5′-*P*, pC
	Deoxycytidine	Deoxycytidine-5′-phosphate or 5′-deoxy-cytidylate,[b] etc.	5′-dCMP,[b] dCyd-5′-*P*, pdC
Uracil	Uridine	Uridine-5′-phosphate or 5′-uridylate,[b] etc.	5′-UMP,[b] Urd-5′-*P*, pU
	[Deoxyuridine][c]	Deoxyuridine-5′-phosphate or 5′-deoxy-uridylate,[b] etc.	5′-dUMP,[b] dUrd-5′-*P*, pdU
Thymine	[Ribosylthymine]	Ribosylthymine-5′-phosphate or 5′-ribosyl-thymidylate,[b] etc.	5′-(r)TMP,[d] Thd-5′-*P*, pT
	Thymidine	Thymidine-5′-phosphate or 5′-thymidylate,[b] etc.	5′-(d)TMP,[d] Thd-5′-*P*, pdT
Hypoxanthine	Inosine	Inosine-5′-phosphate or 5′-inosinate,[b] etc.	5′-IMP,[b] Ino-5′-*P*, pI
Xanthine	Xanthosine	Xanthosine-5′-phosphate or 5′-xanthylate[b]	5′-XMP,[b] Xao-5′-*P*, pX

[a] Official abbreviations as suggested by the tentative rules of IUPAC and IUB.
[b] The omission of the 5′ designation (e.g., adenylate or AMP) is quite common in this series.
[c] Occurs only infrequently.
[d] 5′-Thymidylate (i.e., the deoxyderivative) is usually abbreviated TMP; if this designation is used, the ribonucleotide is rTMP.

All the enzymes discussed so far appear to be specific for the ribo series and to require free −OH groups in the 2′ position. A less demanding enzyme has been isolated from bovine spleen: this diesterase appears to be nonspecific with respect not only to the nature of the base but also to the sugar moiety and attacks both polyribonucleotides and oligodeoxyribonucleotides. It is an *exonuclease* of type *b,* liberates nucleoside-3′-phosphates stepwise, starting preferentially at the head end that carries a *free* 5′-OH group [structure (*b*), page 195].

An equally useful enzyme, but of opposite specificity, has been isolated in highly purified form from the venom of poisonous snakes (Russell's viper, rattlesnake, etc.). It, too, is an exonuclease relatively nonspecific with respect to bases and catalyzes the hydrolysis of both polyribo- and oligodeoxyribonucleotides. As mentioned, its specificity is of type *a,* and it therefore liberates nucleoside-5′-phosphates in a stepwise manner, starting at the tail end, where it requires a free 3′-OH group [structure (*c*) above]. The procedures useful for the elucidation of polyribonucleotide structure are summarized in Fig. 5-5, using a hypothetical decanucleotide. They have been successfully employed for the elucidation of the complete nucleotide sequences of specific transfer RNA's (page 250).

Polydeoxyribonucleotides are stable to the action of alkali but can be hydrolytically degraded by a variety of enzymatic procedures. A common type of *deoxyribonuclease* (*DNase I*) is isolated from bovine pancreas; it is an endonuclease requiring bivalent metal ions (usually Mg^{2+}), effective at neutral *p*H, and catalyzing random *a* cleavage of intranucleotide bonds preferentially between adjacent purines and pyrimidines. The products are oligonucleotides of the general structure (*c*) (page 195). Subsequent or simultaneous attack by purified snake venom phosphodiesterase, freed of contaminating enzymes, then leads to almost quantitative liberation of the constituent deoxyribonucleoside-5′-phosphates.

When the complementary specificity (*b* cleavage) is desired, as it is, for instance, in the determination of nearest neighbors, one makes use of DNases with an acidic *p*H optimum and no metal requirement, frequently called *DNases II*. Such enzymes have been isolated from mammalian (spleen, thymus) and bacterial sources (*Staphylococcus aureus,* formerly called *Micrococcus pyogenes* var. aureus). The products are oligonucleotides of structure (*b*) (page 195), which can then be further degraded by the phosphodiesterase from spleen to yield the desired deoxyribonucleoside-3′-phosphates. Particularly useful for a variety of studies are DNases, or phosphodiesterases in general, which can distinguish between polynucleotides in single-stranded and double-stranded conformation (8–10) (cf. page 242). Pancreatic RNase is of this type: It rapidly hydrolyzes single- but not double-stranded regions in RNA, or RNA-DNA hybrid duplexes, and can therefore be used as a diagnostic reagent for the identification of such structures. Most ordinary DNases are not similarly conformation-specific. However, enzymes specific for single-stranded regions have been isolated both from microbial (*Escherichia coli* exonuclease I) and mammalian sources, and under carefully controlled conditions the micrococcal DNase II also appears to behave in this manner. Conversely, the exonuclease III from *E. coli* attacks double-stranded DNA from the 3′ termini (removing any P present by virtue of intrinsic phosphatase action), producing 5′-nucleotides progressing toward the center of the molecule.

Other Degradations. Simultaneous exposure of polydeoxyribonuleotides to moderate heat and low *p*H ($\geqq 60°$; *p*H $\leqq 3$) or to concentrated HCO_2H at 37° leads to an interesting reaction (Fig. 5-5). Most features of the primary structure are retained (that is, only relatively few scissions of the phosphodiester backbone occur), but the sugar–purine bonds are cleaved specifically, leading to the formation of *apurinic acids.* Similarly, treatment of DNA with hydrazine produces *apyrimidinic acids,* which are structures similar to that shown on the bottom left but with all purines (dG, dA, dA; dG; dA) retained and the pyrimidines eliminated. Treatment of RNA's with 10 M hydroxylamine at *p*H 10 removes uracil residues quantitatively to yield *deuridylic* derivatives.

Circular Structures

Many naturally occurring polynucleotides, particularly in the deoxy series, contain neither a free 3′ nor a free 5′ terminus (—OH, or P); they are *covalently circular*. The requisite intrastrand linkage

$$pN_1N_{n-2}N_n \xrightarrow{\sim P \rightarrow P_i} \underbrace{N_1(N)_{n-2}N_{\bar{n}}}_{p}$$

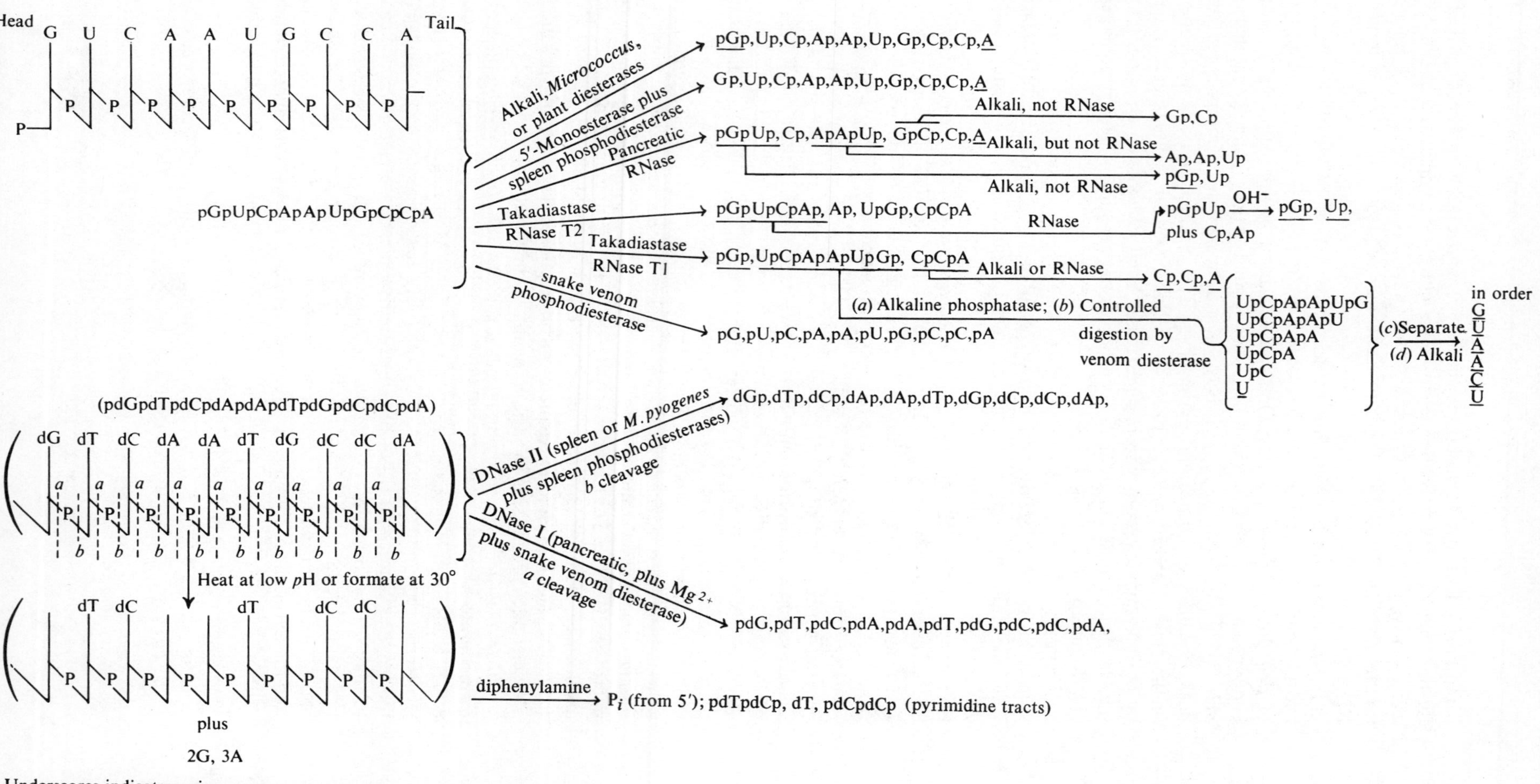

FIGURE 5-5

Hydrolytic procedures for polynucleotides. A decaribonucleotide and an analogous segment of a polydeoxyribonucleotide are shown. Underscores indicate unique sequences.

is formed enzymatically. The enzymes responsible are referred to as polynucleotide-joining enzymes or ligases (page 894), and the reaction requires the simultaneous hydrolysis of a pyrophosphate linkage in a coreactant (ATP or NAD^+). Among such circular structures found in nature are the DNA's from a number of bacterial and animal viruses, from bacteria, and from the mitochondria of multicellular organisms (Table 5-1). Sometimes the DNA can also exist as a *catenane,* that is, an oligomer containing two, three, or even four interlocked monomeric rings (11).

Higher-Order Structures

Bases in polynucleotides are capable of mutual interactions to impart a number of ordered structures varying greatly in stability. Depending on whether they involve intra- or interstrand links, they have been referred to as possible conformations or configurations, respectively. Three principal limiting structures—characteristic of whole or parts of molecules—can be distinguished: disordered (random coil), single-stranded (non-hydrogen-bonded) helical, and multistranded (hydrogen-bonded) helical; among the latter, double-stranded (duplex) helices are the most prevalent and important. In addition, we must conceive of linear and circular structures, actually identified so far only in the cases of duplexes, where the ends are joined by virtue of *non*covalent interactions; these belong to a class of structures referred to as *(con)catemers.*[8] Furthermore, it has been shown that tertiary structure, in the sense of a twisting of the molecule as a whole, can be introduced into a helical, circularly covalent duplex molecule (called a *superhelix* or *supercoil*). Since such a molecule (Fig. 5-6)—unlike its relaxed linear, or noncovalently circular, counterpart—is topologically constrained from responding to changes in its environment by expanding or contracting its component helix, it can only adjust by the removal or addition of twists to the molecule as a whole (12,13). The introduction of such twists may, in turn, lead to folding of the twisted circle into condensed, highly compact conformations that may be of importance in packing long, linear polynucleotide molecules into highly restricted spaces such as viral shells (capsids) or chromosomal structures (14). We shall introduce and discuss these various secondary and higher-order structures within the context of actual cases.

POLYDEOXYRIBONUCLEOTIDES

Localization and Function

In eucaryotic organisms most of the cellular DNA is concentrated in the nucleus (Chapter 9). Extranuclear DNA has been reported to occur in chloroplasts, mitochondria, and other large organelles, which are at least potentially self-reproducing (15–17). In bacteria the DNA appears to be present in a small number (one to three) of aggregates (nucleoids) in the interior of the cell but connected to its membrane, whereas in DNA-containing viruses it occupies the central region of their spherical or polyhedral head structures. Nuclei of animal cells contain on the order of 2 mg of DNA per gram of fresh tissue. This is equivalent to about 4 to 8 $\times 10^{-12}$ g (picogram) of DNA per nucleus (Table 5-1). These values (for any one cell type) are quite invariant with the physiological state of the aminal but are directly dependent on the multiplicity of chromosome units (ploidy) in the cell. Bacterial cells contain about one thousandth of this amount (of the order of 4×10^{-3} pg), and DNA-containing viruses (animal and bacterial) range all the way from 2×10^{-4} pg down to 2×10^{-6} pg. For all these organisms there is now

[8] Concatemers are oligomers of complete genomes held together by *either* covalent or noncovalent bonds.

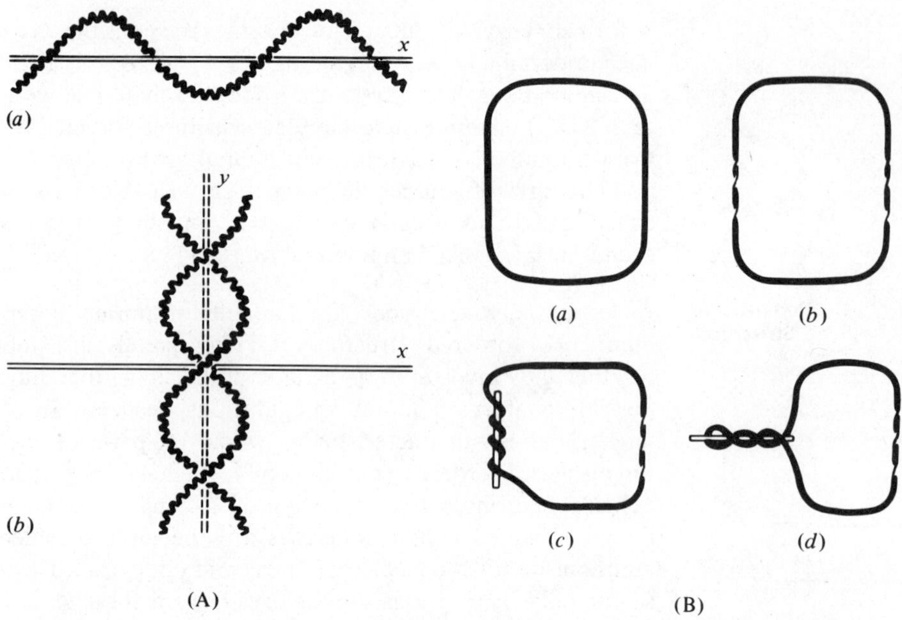

FIGURE 5-6
(A) Sections of the two possible first-order superhelical models for closed circular DNA. Either of these models, or a combination of them, formally represents the closed circular molecule which contains negative superhelices, $\tau < 0$. (*a*) A section of the toroidal model shown about its superhelical axis, *x*. The winding is left-handed about this axis. (*b*) A section of the interwound model shown twisted about its superhelical axis, *y*. The winding is right-handed about this axis, but remains left-handed if viewed about the axis *x*. Both models contain the correct number of duplex turns per superhelical turn in native SV40 DNA. W. Bauer and J. Vinograd, *J. Mol. Biol.,* **33:** 141 (1968).
(B) An illustration of the formation of superhelical turns in a closed duplex consisting of two noninterwound strands. The topological winding number, α, in this figure has a value of zero. (*a*) Two noninterwound strands. The duplex winding number, β, and the superhelical winding number, τ, have values of zero. (*b*) Three right-handed duplex turns and three left-handed duplex turns have been wound into the duplex in (*a*). The right- and left-handed duplexes cancel each other; $\alpha = \beta = \tau = 0$. (*c*) Three right-handed duplex turns and three left-handed superhelical turns have been introduced into the duplex in (*a*). The separate strands in the superhelical region rotate about each other. In this arrangement $\alpha = 0$, $\beta = +3$, and $\tau = -3$. The superhelical turns may be regarded as forming a section of a toroidal superhelix. (*d*) Three right-handed duplex turns and three right-handed interwound superhelical turns have been introduced into the duplex in (*a*). The paired strands in the interwound superhelix do not rotate around each other. In this arrangement $\alpha = 0$, $\beta = +3$, and $\tau = -3$. J. Vinograd, J. Lebowitz, and R. Watson, *J. Mol. Biol.,* **33:** 173 (1968).

absolutely compelling evidence that DNA is the genetic material, that is, that the genetic information is both stored in and transmitted by these macromolecules. Further discussion of this point is deferred to Chapters 19 through 21.

Base Composition

The base composition of a large variety of DNA's isolated from all types of organisms has been investigated intensively and thoroughly. The pioneering investigations of this kind, which belong to the fundamental chemical contributions to the contemporary flowering of *molecular biology,* were carried out by E. Chargaff and his collaborators in the period 1950–1953. Some of the generalizations which emerged from these investigations and which have been amply confirmed subsequently (see Table 5-3) may be stated as follows.

1. The base composition of its DNA is a characteristic of the organism (species or strain).

2. Different cells or tissues of the same organism have identical or at least closely similar base compositions; the latter is unaffected by age, development, nutritional state, and all other physiological or environmental factors.

3. The actual base composition exhibits a wide variation from organism to organism; this variation is much greater in the case of bacteria (where it ranges from about 25 to 75 mole percent G + C) than in higher organisms. It is most clearly expressed by the *dissymmetry ratio* (A + T)/(G + C).

ABLE 5-3
ase composition of various DNA's

pecies	*Organ*	*A*	*G*	*C*[a]	*T*	*A/T*	*G/C*	*Pu/Py*	*(A + T)/(G + C*[b]*)*
		Animals and plants							
Ian	Thymus	30.9	19.9	19.8	29.4	1.05	1.00	1.04	1.52
ig	Thymus	30.0	20.4	20.7	28.9				1.43
alf	Thymus	28.0	22.0	21.0 (1.3)	28.0				1.35
	Spleen	28.2	21.2	21.0 (1.3)	28.2				1.30
	Pancreas	27.8	21.9	21.9	27.5				1.29
	Kidney	28.3	22.6	20.9	28.2				1.30
	Sperm	28.7	27.2	20.7 (1.3)	27.2				1.26
Iouse	All tissues	29.0	21.5	20.7	29.0	1.00	1.04	1.01	1.37
	All tissues (satellite component)[c]	32.2	16.4	17.8	33.3	0.97	0.92	0.95	1.92
	Heavy strand	20.7	12.5	22.1	44.7	0.46	0.56	0.50	1.87
	Light strand	44.2	19.9	14.3	21.6	2.04	1.39	1.79	1.92
Ien	Erythrocytes	28.8	20.5	21.5	29.2	1.02	0.95	0.97	1.38
urtle	Erythrocytes	29.7	22.0	21.3	27.9	1.05	1.03	1.00	1.31
Ierring	Sperm	27.8	22.5	20.7 (1.9)	27.5				1.23
almon	Sperm	29.7	20.8	20.4	29.1				1.43
ocusta migratoria (locust)	Whole	29.3	20.5	20.7 (0.2)	29.3	1.00	1.00	1.00	1.41
rbacia punctulata (sea urchin)	Sperm	28.4	19.5	19.3	32.8				1.58
aracentrotus lividus (sea urchin)	Sperm	32.8	17.7	17.3 (1.1)	32.1				1.85
Iarine crabs	All tissues (minor component)[d]	47.3	2.7	2.7	47.3	1.00	1.00	1.00	17.5
Vheat germ		27.3	22.7	22.8	27.1				1.19
spergillus nidulans (mold)		25.0	25.1	25.0	24.9	1.00	1.00	1.00	1.00
accharomyces cerevisiae (baker's yeast)	Whole cell	31.3	18.7	17.1	32.9	0.95	1.10	1.00	1.79
	Mitochondria[d] (wild type)	41.3	11.3	10.0	37.5	1.23	1.13	1.10	4.75
	Mitochondria[d] (cytoplasmic mutant)	48.5	1.5	1.5	48.5	1.0	1.0	1.0	32.2
uglena gracilis	Whole cell	22.6	27.7	25.8[e]	24.4	0.93	1.07	1.01	0.88
	Chloroplast	38.2	12.3	11.3	38.1	1.00	1.09	1.02	3.23
	Mitochondria	~18	33	32	18	1.0	1.0	1.0	0.55

TABLE 5-3 (Continued)

Species	A	G	C[a]	T	A/T	G/C	Pu/Py	(A + T)/ (G + C[b])
			Bacteria and viruses					
Sarcina lutea	13.4	37.1	37.1	12.4	1.08	1.00	1.02	0.35
Alcaligenes faecalis	16.5	33.9	32.8	16.8	0.98	1.03	1.02	0.50
Brucella abortus	21.0	29.0	28.9	21.1	1.00	1.00	1.00	0.73
Escherichia coli W	24.7	26.0	25.7	23.6	1.04	1.01	1.02	0.93
Vibrio cholerae	28.8	20.0	23.3	27.9	1.03	0.86	0.95	1.31
Staphylococcus aureus SA-B	30.8	21.0	19.0	29.2	1.05	1.11	1.07	1.50
Bacillus cereus alesti	33.5	17.3	16.0	33.2	1.01	1.08	1.03	2.00
Micrococcus pyogenes Oxf.	34.1	15.0	15.7	32.2	1.06	0.96	1.02	2.26
Clostridium perfringens (Fred)	36.9	14.0	12.8	36.3	1.02	1.09	1.04	2.70
Coliphage T2	32.5[f]	18.2	16.8[g]	32.5	1.00	1.09	1.04	1.86
Coliphage T7	26.0	24.0	24.0	26.0	0.98	1.01	0.99	1.11
Coliphage λ	21.3	28.6	22.9	22.9	1.01	0.98	0.99	1.00
Coliphage φX174 (isolated virus)	24.6	24.1	18.5	32.7	0.75	1.30	0.95	1.35
Coliphage φX174 (replicative form)[h]	26.3	22.3	22.3	26.4	1.00	1.00	1.00	1.18
B. megaterium phage α (virus)	28.4	22.1	22.3[i]	27.4	1.04	0.98	1.01	1.25
Chain A (light)	30.1	24.1	21.3	24.5	1.23	1.13	1.29	1.20
Chain B (heavy)	24.0	19.9	24.2	32.1	0.75	0.82	0.78	1.27

[a] Values for 5-methylcytosine are shown in parentheses.
[b] Includes various substituted cytosines.
[c] W. G. Flamm, P. M. B. Walker, and M. McCallum: *J. Mol. Biol.,* **40:** 423 (1969).
[d] Approximately 0.2 of the total DNA is of this type.
[e] Contains 10 mol of 5-methylcytosine/100 mol of cytosine.
[f] Contains 0.45 mol of 6-methylaminopurine/100 mol of adenine.
[g] Hydroxymethylcytosine completely replaces cytosine.
[h] Isolated from phage-infected cells.
[i] Contains 5-hydroxymethyluracil.

4. Closely related organisms exhibit similar base composition; this has been made part of a chemical taxonomy of bacteria (18).

5. All "normal" DNA's exhibit certain chemical regularities that are defined by *Chargaff's rules*[9]; A = T and G = C; as a corollary Σ purines (A + G) = Σ pyrimidines (C + T); also (A + C) = (G + T)—this used to be 6-amino = 6-keto in the old numbering system for pyrimidines.

6. Because A + G + C + T = 1, a corollary to generalization 5 states that any nucleic acid not exhibiting the regularities just described must contain certain unusual structural features (cf., e.g., the case of the DNA from bacteriophage φX174, which has been shown to be single stranded).

The alkali metal salts of DNA can be spun into fibers (i.e., unidimensional crystallites) that give sharp and regular X-ray diffraction patterns. Such patterns were originally obtained in 1938 by Astbury and Bell, who observed strong reflections, corresponding to a regular spacing of 3.34 Å along the fiber axis, interpreted by them as being due to flat nucleotides stacked on top of one another and arranged perpendicular to the fiber axis. Another set of reflections suggested that the structure pattern of DNA was composed of units that had a repeating distance of approximately 27 Å. Structure parameters for ribonu-

[9] If the DNA contains unusual bases, such as substituted cytosines, then C in the identities stands for the sum of C plus all substituted cytosines, and T stands for uracil plus all substituted uracils (including thymine).

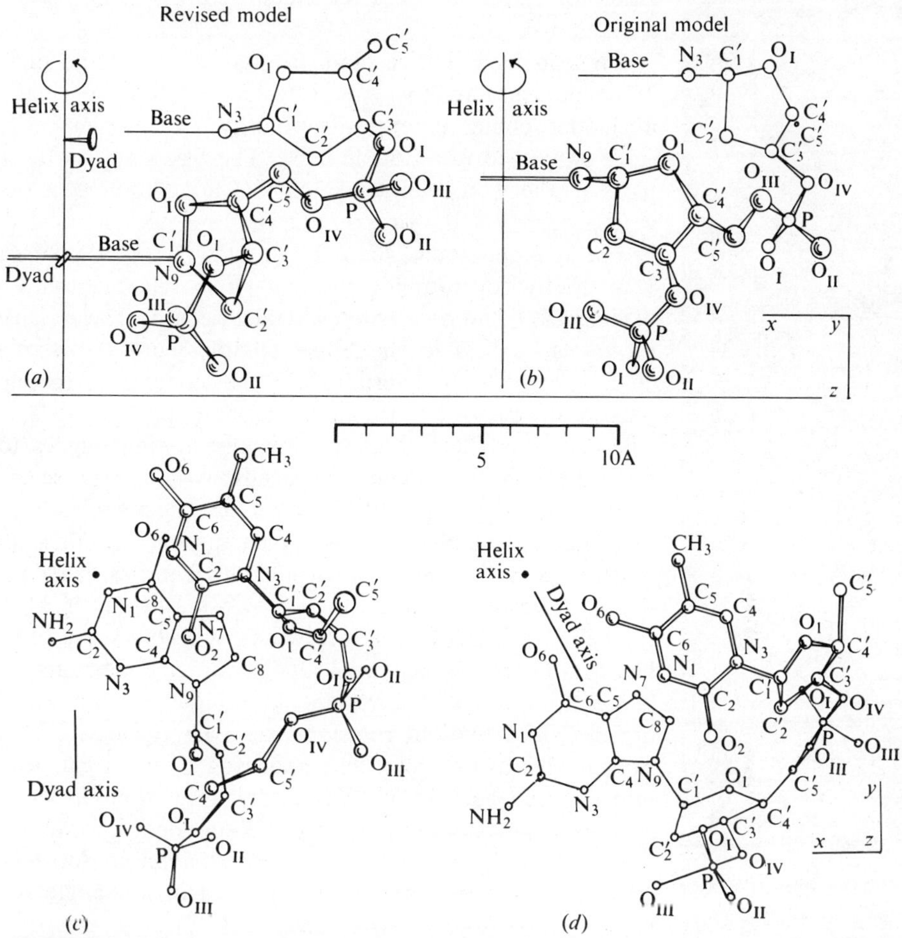

FIGURE 5-7
Structural features of the DNA model. The revised model of DNA of Langridge et al. (*a*) shows a view of two nucleotides looking along one of the dyad axes of symmetry toward the helix axis; (*c*) shows the same two nucleotides looking down the helix axis; the bases are thymine and guanine. Coordinates for these drawings from R. Langridge et al., *J. Mol. Biol.*, **2:** 19, 38 (1960). The original model of Watson and Crick (*b*) shows a view along an arbitrary direction (*y*) perpendicular to the helix axis; (*d*) shows the two nucleotides looking down the helix axis (*z*). Coordinates from F. H. C. Crick and J. D. Watson, *Proc. Roy. Soc.* (*London*), **A223:** 80 (1954). The dyad axes in (*d*) refer to the upper one of the nucleotides, that is, the thymidylic acid residues.

cleosides based on X-ray data were obtained by Furberg during the period 1949–1952. On the basis of these items of information, plus new X-ray diffraction measurements obtained by Wilkins on the lithium salt of DNA, which showed the repeat distance to be 34 Å and strong maximal reflections at 3.4 Å, plus the chemical data discussed in the last paragraph, Crick and Watson (19) proposed their *new model for DNA* in 1953. It provided, for the first time, a means of expressing basic biological phenomena in precise structural chemical terms and has proved to be the starting point for the revolution in biological thinking during the past decade. For these contributions, Watson, Crick, and Wilkins received the Nobel Prize in medicine in 1962. This model has since been revised on the basis of more refined X-ray data by Wilkins and his col-

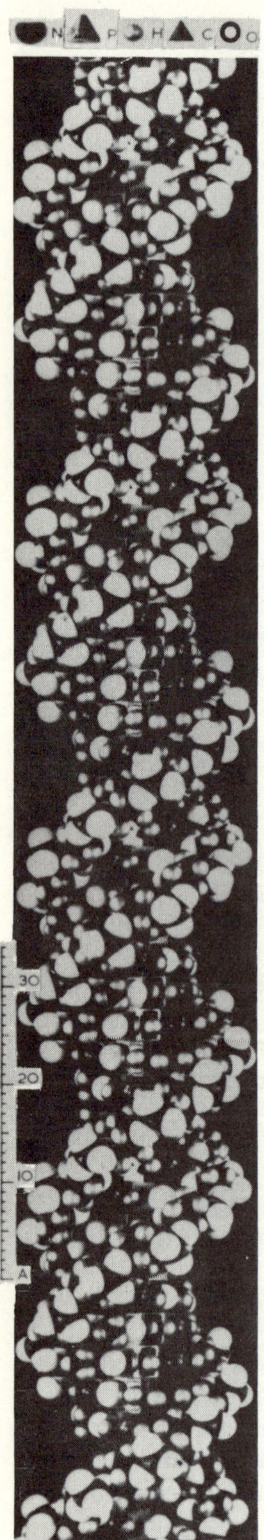

laborators, and its characteristic structural features are summarized in Figs. 5-7 and 5-8.

DNA in its B lattice configuration, which is the one stable at high (i.e., > 66 percent) relative humidity, consists of two right-handed helical polynucleotide chains of opposite polarity,[10] plectonemically coiled[11] around the same axis to form a double helix. The bases are on the inside of the helix in pairs arranged in such a fashion that a pyrimidine of one chain always pairs with a purine of the opposite strand, and vice versa. Only certain base pairs can be accommodated spatially; these allowed base pairs are A with T, and G with C (hydrogen-bonded, as shown in Fig. 5-9, with substituted bases replacing their parents,[9] wherever they occur). These pairing rules require that the bases in the two chains are complementary so that the sequence of one chain is completely determined by the sequence of its partner. The nucleoside pairs in the interior of the helix measure approximately 11 Å from C_1' to C_1'. The planes of their aromatic rings are at right angles to the helix axis, and their apolar bulk is stacked vertically, with a distance of 3.4 Å between base pairs. The highly negatively charged phosphodiester backbone conversely faces outward, and its strongly polar groups can thus interact with the components of the aqueous environment. There are 10 bases per turn of the double helix, and because each turn has a height of 34 Å, these features give the helix an exact 10-fold screw axis with the sense of the screw being right-handed. There are also additional symmetry elements that have the effect of leaving the structure invariant (looking the same when it is turned upside down), since twofold rotation axes pass through each of the base pairs at right angles to the helix axis and each of them can bring one of the chains into congruence with its partner by a rotation of 180°. The helix has one shallow (approximately 12 Å) and one deep groove (approximately 22 Å) across. In the fiber the double helices tend to form hexagonal arrays; the same type of structure is also observed in low-angle X-ray patterns of concentrated DNA solutions or gels (20). Electron micrographs (Fig. 5-10), as well as other physical evidence summarized below, of many viral and certain bacterial DNA's indicate that these molecules are covalently circular.

Direct experimental evidence for the Watson–Crick structure is of the following type:

1. Infrared spectroscopy (21–23) under a variety of conditions, but perhaps most conclusively on thin sheets of Li or NaDNA, equilibrated with D_2O to unmask regions ordinarily obscured by hydration, shows characteristic bands at 1685, 1660 (pyrimidine), and 1644 cm^{-1} [weaker bands at 1620 (purine) and 1575] that have been assigned to C=O, C=N, and C=C stretching and NH_2 deformation vibrations of the bases in the helical array. Most characteristic is the carbonyl band at 1685 (sometimes reported at 1680) cm^{-1}. This band is considerably reduced (and those at 1660 and 1620 are enhanced) when the molecular structure is disrupted by heat, alkali, formamide, or DNase.

2. The structure defines a precise relationship between molecular weight

[10] In chemical terms this means the internucleotide linkage in both strands is 3′ → 5′ *but* starting at opposite ends.

[11] This means that the double helix cannot be separated into its constituent coils without unwinding.

FIGURE 5-8

Structure of DNA in the B configuration. (*Left*) Space-filling model with van der Waals' radii indicated; (*right*) projection of the model. Courtesy of M. H. F. Wilkins.

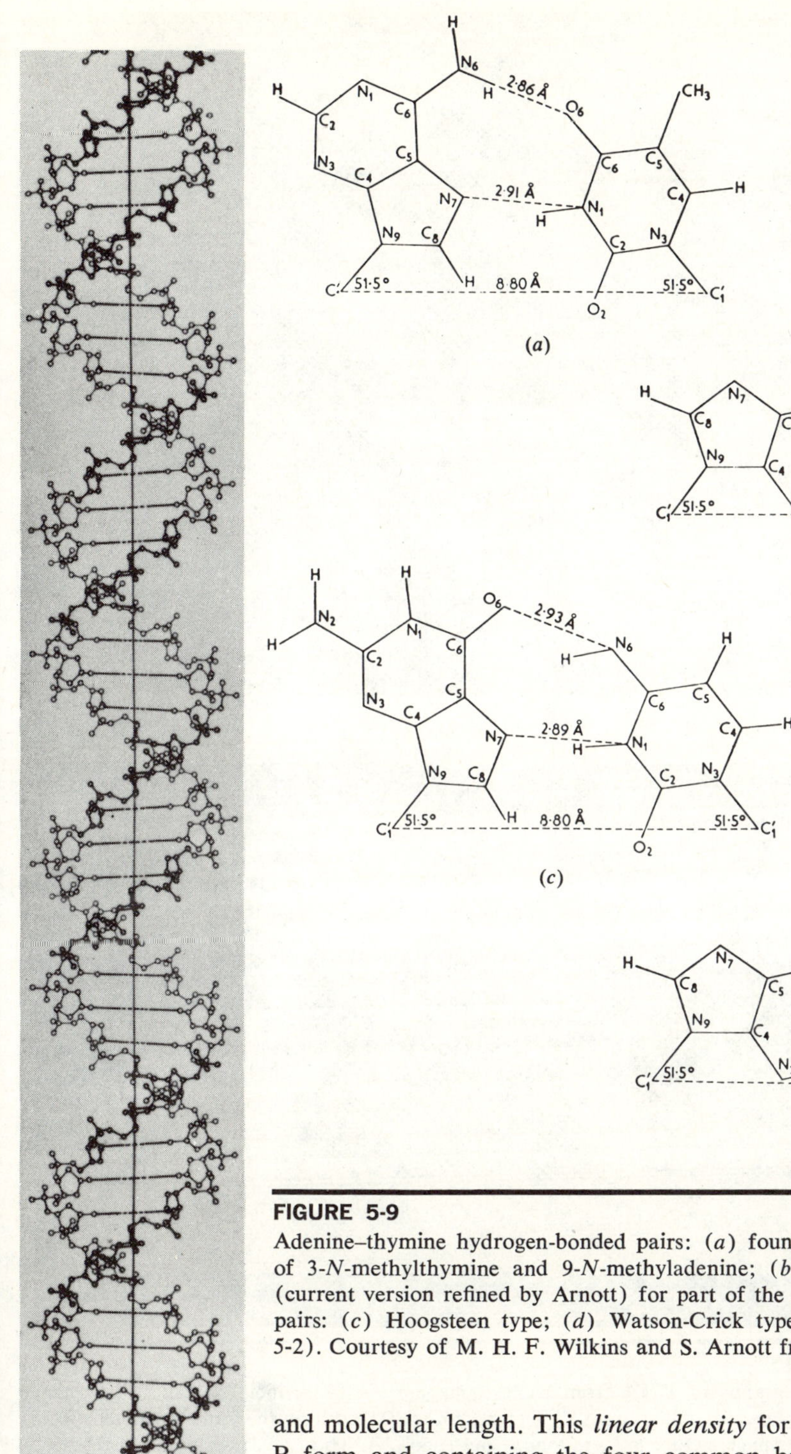

FIGURE 5-9

Adenine–thymine hydrogen-bonded pairs: (*a*) found by Hoogsteen in a 1 : 1 complex of 3-*N*-methylthymine and 9-*N*-methyladenine; (*b*) postulated by Watson and Crick (current version refined by Arnott) for part of the structure of DNA. Guanine–cytosine pairs: (*c*) Hoogsteen type; (*d*) Watson-Crick type. Note pyrimidine numbering (Fig. 5-2). Courtesy of M. H. F. Wilkins and S. Arnott from *J. Mol. Biol.*, **11:** 391 (1965).

and molecular length. This *linear density* for the Na^+ salt of a DNA, in the B form and containing the four common bases, is 192 daltons/Å or 0.52 μm/10^6 daltons. Actual length measurements, usually by electron microscopy, occasionally by radioautography, agree remarkably well with this prediction; for instance, the DNA of the replicative form of coliphage ϕX174 ($M = 3.2 \times 10^6$) has a length of 1.64 μm, that of coliphage λ ($M = 32 \times 10^6$) one of 17.2 μm, that of coliphage T2 ($M = 130 \times 10^6$) one of 56 μm, and finally that of *E. coli* itself ($M = 2.6 \times 10^9$) one of 1400 μm (1.4 mm!).

FIGURE 5-10
Electronmicrographs of DNA from bacteriophage. (*a*) T2 artificial circle. (*b*) $\lambda_c b_2 b_5$ and λ_c circles, naked preparations ("Beer technique"); both at the same magnification; the bar in the upper picture is 2 μm long. Courtesy C. A. Thomas, Jr., and L. MacHattie.

3. Evidence for complementariness plus the opposing polarity of the two strands comes from a study of nearest-neighbor frequencies; opposite polarity generates a pattern different from that produced by strands of identical polarity (e.g, TpG=CpA and GpA=TpC for antiparallel, but TpG=ApC and GpA=

CpT for parallel chains). The experiments of Josse and Kornberg clearly indicate that the first alternative is correct.

4. The structure predicts that the two strands, although complementary, need not have identical base compositions[12] and that it should, therefore, be possible to separate the two chains either chemically or physically.[13] This has been accomplished first in the case of bacteriophage α (Table 5-3) and more recently for a number of other DNA's.

5. The chemical and physical properties of DNA in solution (cf. below) accord well with those predicted from the model.

6. Some evidence has been obtained for correct base pairing in certain crystals and even in solution (21–25). Among the crystals that form Watson–Crick structures with the correct hydrogen-bond distances predicted from Fig. 5-9 are 9-ethylguanine · 1-methyl cytosine and 9-ethylguanine · 5-bromo-1-methylcytosine. However, adenine · uracil, 9-ethyladenine · 1-methyluracil, 9-methyladenine · 1-methylthymine, and adenosine · 5-bromouridine (B), form hydrogen-bonded complexes, different from those in DNA, with hydrogen bonds between adenine-NH_2 and O^4 of uracil and between N^3 of uracil and the imidazole N^7 of adenine (the Hoogsteen pairs of Fig. 5-9).

Base pairing in organic solvents has been observed by means of ir and nmr spectroscopy. It occurs in organic solvents ($HCCl_3$, CCl_4) that effectively eliminate the nonspecific "stacking" interactions and therefore reveals primarily the associations resulting from the hydrogen bonding. Magnitudes of association constants are G with C ($\sim 10^5$ M^{-1}) >> A with U ($\sim 10^2$ M^{-1}) >> U with U (~ 6.7 M^{-1}) > A with A (~ 3.3 M^{-1}). No interaction is observed between the "false" pairs G and U (or T), C and U (or T), or A and G.

Noncomplementary and Other Unusual Polydeoxyribonucleotides

The first DNA that did not obey Chargaff's rules was isolated from the small *E. coli* bacteriophage ϕX174. It has a molecular weight of 1.6×10^6 daltons and the base composition shown in Table 5-3. Its properties indicated that its conformation was quite different from that of most "normal" DNA's and resembled much more closely those characteristic of some RNA's. That it was, in fact, circular became evident soon thereafter, first on the basis of its resistance to attack by exonucleases, and later on direct examination in the electron microscope. It is of interest that, like some RNA viruses, it undergoes a transformation during its replication *in vivo* to a so-called *replicative form,* which contains the complementary strand (Table 5-3) and appears to be a supercoiled circular duplex helix.

When *E. coli* DNA polymerase is allowed to react with either of the nucleoside triphosphate pairs (dATP + dTTP) or (dGTP + dCTP), two unusual duplex polymers result. From the first pair is obtained the alternating self-complementary copolymer poly d(A-T) [also abbreviated as $d(A\text{-}T)_n$], from the second the homopolymer duplex[14] poly dG · poly dC [also abbreviated as $(dG)_n \cdot (dC)_n$]. Naturally occurring DNA's resembling the alternating dAdT polymer, but containing a small proportion of G and C residues

[12] The base pairs in the two orientations $A_1 \cdot T_2 \neq T_1 \cdot A_2$ and/or $G_1 \cdot C_2 \neq C_1 \cdot G_2$, where subscripts refer to the two chains.

[13] A strand with a high (A + G) content would, for instance, be lighter, exhibit a lower density in concentrated CsCl, and might respond differentially to certain adsorbants.

[14] Deoxy-5-bromouridine (dBPT) triphosphate can substitute for the thymidine, deoxyinosine-5′-triphosphate for the guanosine, and deoxy-5-bromocytidine-5′-triphosphate for the cytidine derivatives in all these reactions (i.e., in the polymerization and duplex formations), and thus duplexes with these "unusual" bases can be readily obtained.

as well, have been identified in the nuclei of marine crabs (26,27) and in the mitochondria of certain respiration-deficient, cytoplasmically inherited mutant strains of baker's yeast (28,29).

Many other "biosynthetic" polydeoxyribonucleotides have been prepared by the action of *E. coli* DNA polymerase (30) and of the terminal deoxynucleotide transferase from calf thymus (31). The former, with appropriate oligodeoxynucleotide duplexes, functioning as both templates and chain initiators or primers (Chapter 20), produces the alternating, non-self-complementary, duplexes poly d(G-T) · poly d(C-A) and poly d(G-A) · poly d(C-T).[15] The latter, with short oligonucleotides as initiators, can produce all desired single-stranded homopolymers from the appropriate substrates. Some of the structures, as established for both the deoxy and ribo series by X-ray crystallography on isolated fibers, are summarized in Table 5-4; the geometrical features can best be analyzed in terms of the parameters of Fig. 5-11.

Physical Chemistry of High-Molecular-Weight DNA in Solution (32)

Sedimentation in Homogeneous Media (33). One of the most useful techniques for the study of substances of very high molecular weight consists in observing their behavior in a gravitational field, that is, in the ultracentrifuge. Runs must be made at a high dilution of the polymer (to minimize aggregation and other concentration effects), under conditions in which it assumes its most stable conformation, that is, in the presence of a counterion (usually a monovalent inorganic cation such as Na^+) at a relatively high ionic strength (> 0.1 M) and at moderate speeds of the ultracentrifuge rotor ($< 15{,}000$ rpm). The low concentration (~ 10 μg/ml) necessitates the use of the ultraviolet optics of the ultracentrifuge, and the rate of migration of the concentration boundary with time is then measured and substituted in Eq. (5-1):

$$s_{\text{obs}} = \frac{1}{\omega^2 r}\frac{dr}{dt} = \frac{\ln (r_2/r_1)}{\omega^2(t_2 - t_1)} \tag{5-1}$$

where s, which has the dimensions of time, equals the sedimentation coefficient, the r_i's are the distances in centimeters migrated by the molecule from the axis of rotation at the times t_2 and t_1 (expressed in seconds), and ω is the angular velocity in radians per second or 0.105 rpm. Usually the observed s is corrected to $s_{20,\,w}$, the value it would have if the density and viscosity of the solvent were that of water at 20°, and is expressed in Svedberg units (S) equal to 10^{-13} sec:

$$s_{20,w} = \left[s_{\text{obs}} \left(\frac{\eta_t}{\eta_{20}} \frac{\eta}{\eta_0} \right) \right] \frac{1 - \bar{v}\rho_{20,w}}{1 - \bar{v}\rho} \simeq [\;]\, \frac{\rho_B - \rho_{20,w}}{\rho_B - \rho} \tag{5-2}$$

where η_t and η_{20} are viscosities of water at t, the temperature of the run and at 20°, η/η_0 is the viscosity of the solvent relative to that of water, $\rho_{20,\,w}$ and ρ are the densities of water at 20° and the solvent at t°, respectively, and $\bar{v}$[16] is the partial specific volume of the solute. The latter is equal to $1/\rho_B$, where ρ_B is the buoyant density of the macromolecule in the system under investigation and is a function of composition (see below). On the average it is close to 0.55 cm^3/g for native double-stranded DNA and slightly lower for single-stranded polynucleotides; for RNA, values around 0.51 ± 0.02 cm^3/g have been used.

[15] The former pair appears to resemble DNA more closely in its stability; the latter is less stable.

[16] This expression assumes that $\bar{v}$ is itself temperature independent, since more accurately it should be $\bar{v}$ at 20° in the numerator, $\bar{v}$ at the temperature of the run in the denominator; for DNA $\bar{v} = 0.553$ at 5° and 0.556 at 21° (43).

TABLE 5-4
Dimensions of helical polynucleotide fibers

Polynucleotide	*Pitch* $Z \times N$, Å	*Residues per turn (N)*	*Translation per residue (Z)*	*Rotation per residue (α)*	*Tilt angle*[a] (γ)	*Structure*
Na DNA, A form (humidity = 75%)	28.2	11	2.55	32.7°	20°	Antiparallel double helix
Na DNA, B form (humidity = 92%)	34.6	10	3.46	36°		Antiparallel double helix
Li DNA, B form (humidity = 66%)	33.7	10	3.37	36°	2°	Antiparallel double helix
Li DNA, C form	31.0	9.3	3.32	39°	6°	Antiparallel double helix
poly d(A-T) and poly d(A-B)	33.4	10	3.34	36°		Antiparallel double helix
Na RNA, A form[b]	30.9	11	2.73	32.7°	~20°	Antiparallel double helix
Na RNA, A′ form[c]	36.0	12	3.0	30°	~20°	Antiparallel double helix
Na RNA, A″ form[d]		11.3			~20°	Antiparallel double helix
poly rA · poly rU; poly rI · poly rC	two forms similar to A and A′ forms of RNA					
"Acid" poly rA						
B form	36.2	10.0	3.6	36°		Antiparallel double helix
A form	31.8	8.30	3.8	43°		Parallel double helix
"Acid" poly rC	37.3	12	3.11	30°		Double helix
poly rA · (poly rU)$_2$	31.1	10	3.1	36°		Triple helix
Hybrid DNA · RNA[e]	28.5	11	2.6		20°	Double helix

[a] Angle between perpendicular to helix axis and individual bases; see also Fig. 5-11.
[b] Found in double-standed RNA's, mainly of viral origin (reovirus, wound tumor virus, rice dwarf virus, and replicative intermediate of MS2) as well as in poly rA · poly rU and poly rI · poly rC at low salt concentration; Z and N may equal 3.0 and 10, respectively.
[c] Found in poly rA · poly rU at high salt concentration.
[d] May be a mixture of several forms.
[e] Formed by using the single-tranded DNA from coliphage f1 as a template for DNA-dependent RNA polymerase.
SOURCE: D. R. Davies, *Ann. Rev. Biochem.* **36:** 348 (1967); S. Arnott, W. Fuller, A. Hodgson, and I. Prutton, *Nature,* **220:** 561 (1968); G. Milman, R. Langridge, and M. J. Chamberlin, *Proc. Natl. Acad. Sci. U.S.,* **57:** 1804 (1967).

(For proteins, the values are close to 0.725.) The sedimentation coefficients so determined are dependent on the concentration of the DNA added and frequently also on the speed of the ultracentrifuge rotor. The concentration dependence, which is usually of the form $1/s = 1/s^0 + Kc$ can be established either by performing measurements at different concentrations and extrapolating to $c = 0$ or by using an empirical equation due to Eigner:

$$s^0 = \begin{cases} s(1 + 0.90[\eta]c) & \text{for native DNA} \\ s(1 + 1.70[\eta]c) & \text{for single-stranded, randomly coiled polymers} \end{cases} \qquad (5\text{-}3)$$

where s^0 is the desired sedimentation coefficient at zero concentration and $[\eta]$ is the intrinsic viscosity (page 215).

In general the s^0 values for a polymer of any given molecular weight increase the more extended and rigid (i.e., less flexible) its conformation.

Centrifugation in Density Gradients. Convective disturbances, interactions between solute molecules, and other bothersome experimental difficulties are minimized or eliminated if centrifugation is carried out in the presence of a continuous gradient of a dense material. Two types of experiment can be performed here also: sedimentation-velocity experiments and equilibrium experiments.

For sedimentation runs the preferred medium is a buffered ($\sim$0.2 M Na^+, pH $\sim$ 7) linear gradient of sucrose, often between 5 and 20 percent, that can be generated in the centrifuge tube before the start of the run by means of simple mixing devices. An alternative is the use of self-forming gradients of CsCl (see below). The material to be investigated is layered on top of a centrifuge tube or cell which is then placed in the analytical or preparative swinging bucket rotor, and the centrifugal field is applied at a rate and for a length of time sufficient to force significant migration of the macromolecule into the gradient:[17] from 3 to 4 hr at 20,000 rpm for materials with s values in the range 40 to 55 S up to 10 to 20 hr at 38,000 rpm for smaller molecules with s values of 2 to 10 S. The material sedimenting in the gradient can be monitored continuously by means of the ultraviolet optical system of the analytical ultracentrifuge. Alternatively, the preparative tube may simply be pierced at the bottom, or the fluid column extruded continuously through the top, and fractions of fluid collected, or monitored continuously, and the concentration of material measured by light absorption (uv at 260 nm), chemical reactivity (deoxyribose content), biological activity, content of a selective radioactive label (^{32}P), and so on. Clearly if a selective property is used for monitoring, the material need not even be pure. The sedimentation coefficient may then be calculated by application of Eqs. (5-1) and (5-2) or, provided the conditions used are exactly those described by Burgi and Hershey (34), by use of the simple relation

$$s_{20,w} = \frac{6.45 \times 10^{10} \times (\text{distance from meniscus in cm})}{\text{rpm}^2 \times \text{hr}} \tag{5-4}$$

A most convenient method is a comparison of the unknown with a standard of similar composition and known s value under identical conditions, preferably in the same tube. Such standards are now readily available, some even commercially; thus a number of enzymes can serve as protein standards (lysozyme, M = 17,200, $s_{20,w}$ = 2.15 S; alcohol dehydrogenase from yeast, M = 150,000, $s_{20,w}$ = 7.6 S; catalase, M = 250,000, $s_{20,w}$ = 11.35 S). The tRNA's from liver, yeast, or *E. coli* have an s value of 4.2 S; in 0.01 M Mg^{2+} the three ribosomal RNA's of *E. coli* show s values of 5.0, 16.0, and 23.0 S, respectively, and can be used as standards for RNA; the DNA from bacteriophages T4, λ, or T7 can be obtained relatively easily in an undegraded form with s values of 62.0, 36.2, and 32 S, respectively, and are used as standards for "normal" DNA. Then the following simple relationship holds for any one pair of standard and unknown of identical $\bar{v}$:

$$s_{20,w}\,(\text{unknown}) = s_{20,w}\,(\text{standard}) \times R \tag{5-5}$$

where

$$R = \left(\frac{\text{distance traveled from meniscus by unknown}}{\text{distance traveled from meniscus by standard}}\right)_{\bar{v}=\text{constant}}$$

[17] The distance traveled is very nearly directly proportional to the time of centrifugation and the square of the rotor speed.

A useful variation of the technique suitable especially for analytical runs has been introduced by Vinograd (35). A thin lamella of a solution of the macromolecule is layered onto a denser miscible liquid (50 percent D_2O, or 0.5 to 2.0 M alkali metal chlorides) in a specially constructed ultracentrifuge cell. Sedimentation through the liquid occurs as before in the form of a band of narrow concentration distribution that is determined photographically as a function of time by means of the ultraviolet optics. The concentration gradient required is generated continuously by diffusion of small molecules between the lamella and the bulk solution and occasionally by sedimentation of small solute molecules in the bulk. Extremely small amounts of nucleic acids ($\sim$1 μg) can be used (35, 36) and, as shown in Fig. 5-12, the method, as do band sedimentations in general, yields, on monitoring by absorption, the derivative of the concentration distribution (band) rather than the integrated distribution itself (boundary), obtained in ordinary sedimentation runs.

Equilibrium sedimentations of DNA in CsCl gradients were first studied by Meselson, Stahl, and Vinograd (37). Here a concentrated solution of a salt of high density (such as 7.7 M CsCl, concentrated Cs_2SO_4, or concentrated potassium tartrate) is spun in the ultracentrifuge, which leads, after establishment of sedimentation diffusion equilibrium, to a continuous, stable, and linear gradient of the solute molecules with the most concentrated, densest region at the bottom of the tube. For CsCl solutions the density may be estimated from the refractive index:

$$\rho^{25.0^\circ} = 10.8601\, n_D{}^{25.0^\circ} - 13.4974 \tag{5-6a}$$

or from their composition by weight:

$$\text{wt}\,\% = 137.48 - 138.11\,\frac{1}{\rho^{25.0^\circ}} \tag{5-6b}$$

If a dilute solution of a macromolecule such as DNA is placed in such a system, it forms, at equilibrium, a relatively narrow band centered at r_0, the density position corresponding to ρ_B, its own buoyant density under these conditions. The density at r_0, the midpoint of a band, can be determined from Eq. (5-6a). If ρ_{std}, the density of a standard, is known (migration to r_{std}), then

$$\rho_B = \rho_{std} + 4.2\omega^2(r_0{}^2 - r^2{}_{std}) \times 10^{-10}\ \text{g/cm}^3 \tag{5-7a}$$

or, at 44,700 rpm and 25°,

$$\rho_B = \rho_{std} + 0.00922(r_0{}^2 - r^2{}_{std})\ \text{g/cm}^3 \tag{5-7b}$$

The DNA of *E. coli* (50 percent G + C) is a commonly used standard,[18] with $\rho = 1.710$ g/cm^3.

The density of double-standard DNA solutions in concentrated CsCl is a linear function of the sum of the mol fractions X of G plus C. Quantitatively the relation (38) is given by Eq. (5-8).

[18] The correct value probably equals 1.7035 g/cm^3 at 25°, but many authors still use the earlier value as a matter of convenience; for example, the correlations (5-8a) and (5-8b) are based on it. For certain applications, especially for the determination of the extent of glucosylation of viral DNA, for sedimentation of RNA, and polyribonucleotides in general, Cs_2SO_4 appears to offer distinct advantages: The correlation between glucose content and ρ_B is linear, and the ρ_B of DNA lies between 1.420 and 1.435 and that of RNA around 1.6 in this solvent.

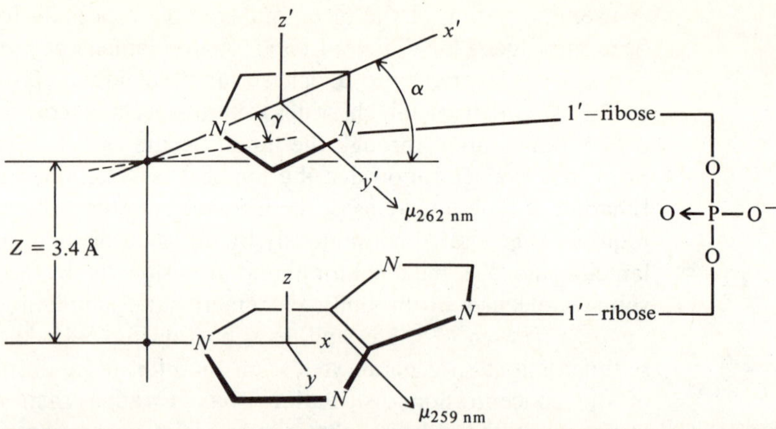

FIGURE 5-11
Incipient helix in a dinucleotide XpY, where X = Py and Y = Pu with a "helix" repeat Z = 3.4 Å similar to that observed in many polynucleotides (Table 5-4). The coordinate system for the first base (X) (x, y, z) is drawn with the z axis parallel to the helix axis and y and x axes at right angles thereto. The direction of the dipole moment of the principal electronic transition responsible for the strong uv absorption at 259 nm is also indicated. The second base (Y) is described by the coordinate system x', y', z'. The latter is rotated with respect to x, y, z by α, the helix increment angle, and in certain structures (e.g., the A forms of DNA and RNA) is tilted with respect to the perpendicular to the helix axis by the tilt angle γ. The direction of the electric dipole is again shown. C. A. Bush and J. Brahms, *J. Chem. Phys.,* **46:** 79 (1967); C. A. Bush and I. J. Tinoco, Jr., *J. Mol. Biol.,* **23:** 601 (1967).

$$\rho_B = 1.660 + 0.098(X_G + X_C) \tag{5-8a}$$

or

$$X_G + X_C = \frac{\rho_B - 1.660}{0.098} \tag{5-8b}$$

Twisted superhelices exhibit buoyant densities (at pH 12.5) that are ~0.018 g/cm³ greater than their relaxed counterparts. Since the heavy Cs^+ ions (and their counterions) are bound more tightly to single-stranded than to double-stranded DNA,[19] the former exhibits a buoyant density that is 0.016 g/cm³ higher than the latter in CsCl and 0.024 g/cm³ higher in Cs_2SO_4, both at neutrality. This shift is due predominantly to the presence of pyrimidines in the polynucleotide. When all the protonated groups have been titrated with alkali, the two strands of the duplex separate and the single strands exhibit a further density increment of the order of 0.060 g/cm³ at this high pH, owing now principally to their T + G content. Thus, strands that differ sufficiently in their composition can be separated in neutral or, more easily, in alkaline CsCl (or Cs_2SO_4) gradients (39). Another method for separating DNA's or DNA fragments (this time containing the same number of strands) depends on their differential ability to bind heavy metal ions reversibly; thus, regions that are rich in (G + C) bind Ag^+ and those that are rich in (A + T) bind Hg^{2+} preferentially. These metal complexes may be separated in Cs_2SO_4 and the metals removed by ion exchange or dialysis against EDTA. Finally, density differences may be accentuated by allowing dC-rich, dG-rich, or dA-rich, single-stranded polynucleotides to form duplexes with poly rG (or

[19] As shown in Fig. 5-13, for normal DNA of *B. subtilis* in CsCl, ρ_B = 1.703; for denatured DNA, ρ_B = 1.717; for the same DNA labeled completely with ^{15}N and ^{3}H in the native form ρ_B = 1.743; in its denatured form ρ_B = 1.760.

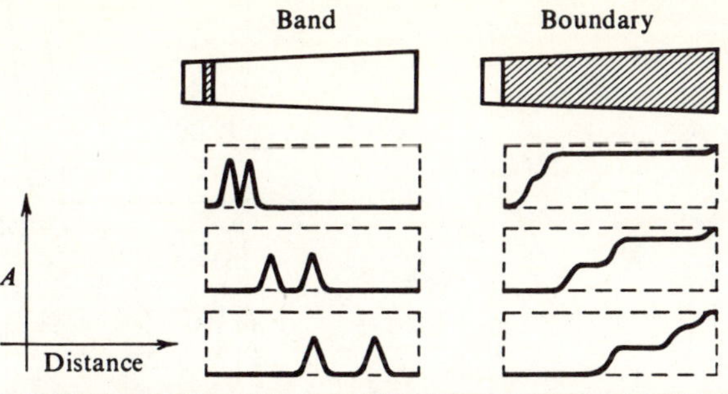

FIGURE 5-12
Comparison of band and boundary centrifugation. J. Vinograd, R. Bruner, R. Kent, and J. Weigle, *Proc. Natl. Acad. Sci. U.S.*, **49:** 962 (1963).

rI), poly rC, and poly rU, respectively, which can then be easily separated from the noncomplexed strand in density gradients.

If the macromolecules in the density gradient at equilibrium are homogeneous, the distribution of their concentration about r_0 is Gaussian. At r_0 itself there are no centrifugal forces operating on the molecules, and their spreading away from r_0 is due to diffusion alone. This spreading is opposed by both centrifugal and centripetal forces, and these opposing forces produce the Gaussian distribution of concentrations. Because large macromolecules diffuse less rapidly than small ones, they produce narrower bands. The variance, σ^2, of the Gaussian distribution is given by

$$\sigma^2 = \frac{RT\rho_B}{M_s(d\rho/dr)\omega^2 r_0} \tag{5-9}$$

where the only quantity not defined previously is $d\rho/dr$, the density gradient around r_0, and the subscript s in M, the molecular weight, refers to the solvated forms actually present under the conditions of the experiment. Mixtures of macromolecules of identical buoyant density but differing in molecular weight form symmetrical but non-Gaussian bands. Therefore, molecular weights can be calculated from band shapes in density equilibrium experiments.

Viscosity. Solutions of native DNA, because of the great size and asymmetry of the molecule, show a very high viscosity. This property, then, provides a means for a relatively easy, inexpensive, and convenient method for the characterization of the polymer from the point of view of its size, shape, and conformation.

The parameter to be determined is $[\eta]$, the *limiting viscosity number,* defined by

$$[\eta] = \lim_{c\to 0}\left[\frac{(\eta/\eta_0) - 1}{c}\right] = \lim_{c\to 0}\left[\frac{\ln(\eta/\eta_0)}{c}\right] \tag{5-10a}$$

where c is the concentration of the polymer in grams per milliliter, η the viscosity of the solution, and η_0 the viscosity of the pure solvent. The fraction η/η_0 is frequently called the *viscosity ratio* or the *relative viscosity,* η_{rel}, and the quantity $[(\eta/\eta_0) - 1]$ is referred to as the *specific viscosity,* η_{sp}. When the concentration c is expressed in grams per deciliter, the parameter defined by Eq. (5-10a) is usually called the *intrinsic viscosity,* which is therefore given in units of deciliters per gram.

In practice, determinations of viscosity entail comparisons, either of the time

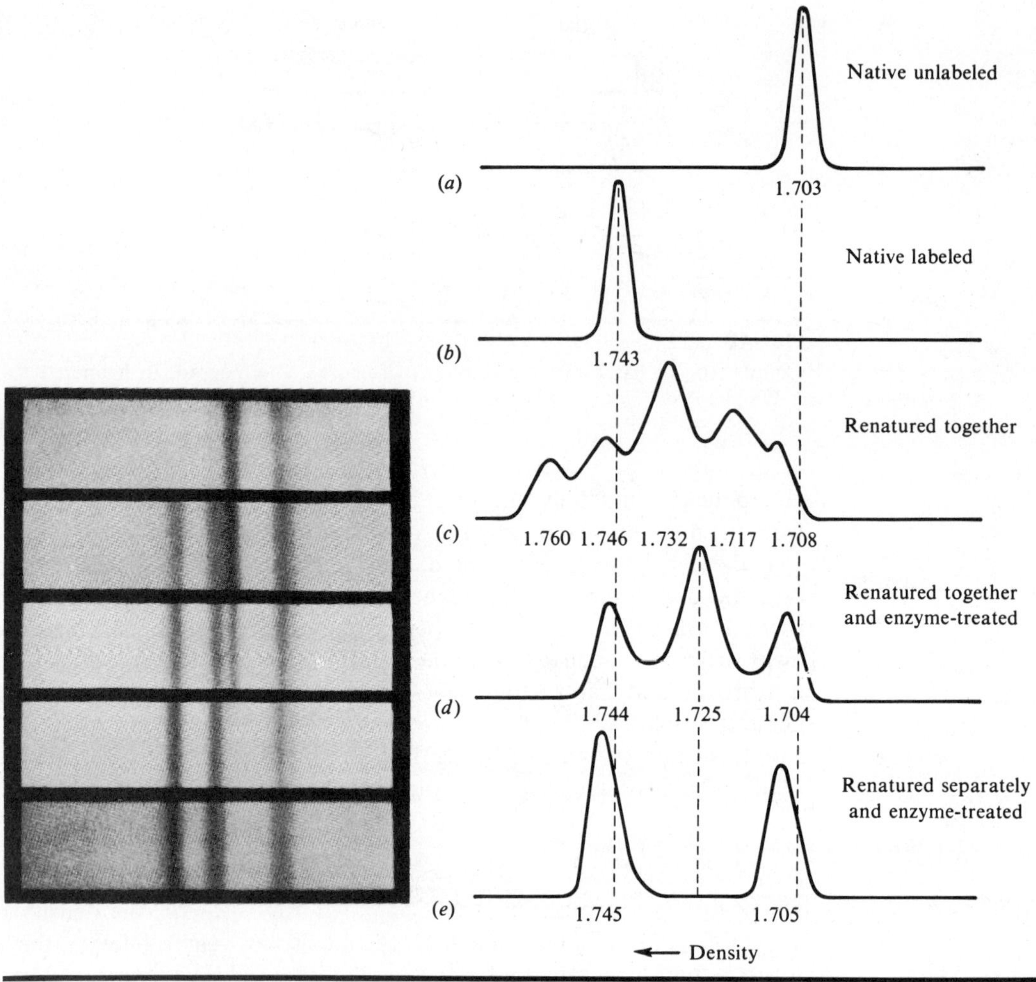

FIGURE 5-13

Density configurations in CsCl. (*Left*) Ultra-violet absorption photographs of ^{14}N–^{15}N-labeled DNA, showing different stages of the thermally induced separation into subunits. The band at the far right has been used as a standard and is DNA isolated from *D. pneumoniae*. The other band in the top photograph is the biologically formed hybrid DNA. The second photograph shows the stability of the hybrid to a 20-min exposure at 93.8°. At 100° the number of molecules separating increases rapidly with time of exposure, as shown in the next three photographs. The samples were heated in SSC at 30 μg/ml for 30 sec, 1 min, and 10 min, respectively. (*Right*) The effect of *E. coli* phosphodiesterase on a heated and annealed mixture of heavy labeled and normal *B. subtilis* DNA. (*a*) and (*b*) Native samples; (*c*) mixed so that the final concentration of each was 5 μg/ml and then heated and annealed; (*d*) the sample shown in (*c*) was dialyzed against 0.067 M glycine buffer and incubated with the *E. coli* phosphodiesterase; (*e*) heated and annealed separately, treated with the phosphodiesterase and then mixed. C. L. Schildkraut, J. Marmur, and P. Doty, *J. Mol. Biol.*, **3:** 595 (1961).

required for a floating body such as a sphere or a cylinder to complete a revolution in the solution to be examined with the time required in the pure solvent, or of the respective times of flow through a capillary. Viscometers of the first type (Couette) can now be constructed relatively easily (40), and the ratio of the revolution times t/t_0, gives η/η_0 directly. A great advantage of this rotating-cylinder viscometer is that it can be operated under conditions approaching virtually zero shear rate. This is important for two reasons:

1. Easy degradation by hydrodynamic shear is one of the most characteristic properties of DNA of high molecular weight; as a matter of fact, the large bacteriophage DNA's can be degraded with double-strand scission near the center of the molecule by simple pipetting of dilute solutions to produce half, quarter, and eighth molecules.

2. Care must be taken to extrapolate first to zero shear before performing the extrapolation to zero concentration. This operation is performed by plotting the quantity in brackets in Eq. (5-10a) (i.e., η_{sp}/c) against c, because usually the expression is of the form

$$\left[\frac{(\eta/\eta_0) - 1}{c}\right] = [\eta] + k' [\eta]^2 c \qquad (5\text{-}10b)$$

Here k', the Huggins constant, is close to ~0.6 for all DNA's in 0.2 M Na^+. In general, the larger the volume of the macromolecule in solution, the higher its intrinsic viscosity.

Determination of Molecular Weight and Shape from Hydrodynamic Measurements. On page 83 we saw how, in polymer chemistry, different measurements can yield different types of average molecular weight if the solution in question is made up, as is most frequently the case, of a population of molecules differing somewhat in their properties. The three most common averages used are the weight-average molecule weight $\bar{M}_w$, the number-average molecular weight $\bar{M}_n$, and the z-average molecular weight $\bar{M}_z$, here defined again by

$$\bar{M}_n = \frac{n_1M_1 + n_2M_2 + n_3M_3 + \cdots}{n_1 + n_2 + n_3 + \cdots} = \frac{\Sigma\, n_iM_i}{\Sigma\, n_i} \qquad (5\text{-}11)$$

$$\bar{M}_w = \frac{c_1M_1 + c_2M_2 + c_3M_3}{c_1 + c_2 + c_3} = \frac{\Sigma\, c_iM_i}{\Sigma\, c_i} = \frac{\Sigma\, n_iM_i^2}{\Sigma\, n_iM_i} \qquad (5\text{-}12)$$

$$\bar{M}_z = \frac{\Sigma\, n_iM_i^3}{\Sigma\, n_iM_i^2} \qquad (5\text{-}13)$$

In all these, n_i is the number of moles of species i (with molecular weight M_i) per unit volume. The weight concentration in grams per unit volume is $c_i = n_iM_i$. Let us also remember that $\bar{M}_n$ is the most "democratic" measure: Each molecule is counted once, regardless of weight; $\bar{M}_w$ and $\bar{M}_z$ give increased representation to the heaviest molecules—heaviest weight to the heavyweights, most of all in the z-average. $\bar{M}_n$ is obtained from a measure of colligative properties such as osmotic pressure. $\bar{M}_w$ is the molecular-weight average obtained from light scattering, dielectric dispersion, fluorescence depolarization, or sedimentation measurements; viscosity measurements yield an average close to, but usually not identical with, $\bar{M}_w$.

For monodisperse systems (i.e., those containing one and only one species), all three molecular weights are identical. For others the ratio $\bar{M}_w/\bar{M}_n$ is a measure of their polydispersity. Fortunately, in the field of nucleic acid chemistry viruses provide a source of remarkably monodisperse macromolecules of high M (Tables 5-1 and 5-10).

As we have seen, in protein chemistry and other branches of polymer chemistry, M can be determined from s plus D, a measure of the shape of the molecule. Experimentally this is commonly accomplished by determining D, the diffusion coefficient, which depends on molecular size and shape: $D =$

RT/Nf, where R is the gas constant, T the absolute temperature, N Avogadro's number, and f the frictional coefficient, itself determined by molecular shape and size.[20] Because $s^0 = M(1 - \bar{v}\rho)/Nf$, we can eliminate f by substitution and obtain the Svedberg equation[21] (page 84).

$$M = \frac{s^0/RT}{D(1 - \bar{v}\rho)} \tag{5-14}$$

The determination of D is tedious at best and very difficult in the case of polymers of very high M. An alternative method is to measure $[\eta]$ and to utilize an equation by Scheraga, Mandelkern, and Flory:[21]

$$M = \left[\frac{s^0[\eta]^{1/3}\eta_0 N 10^{-13}}{\beta(1 - \bar{v}\rho)}\right]^{3/2} \tag{5-15}$$

All terms have the usual meaning: s^0 is the sedimentation coefficient corrected to water at 20° and extrapolated to zero concentration, $[\eta]$ is the intrinsic viscosity of the polymer in dl/g, η_0 is the viscosity of water at 20°, N is Avogadro's number, $\bar{v}$ is the partial specific volume of the polymer, and ρ is the density of water at 20°; β is a constant for any given polymer but is itself a function of the shape (specifically the axial ratio) of the molecule. It equals 2.12×10^6 for a sphere; 2.6×10^6 for a random, free-draining coil; and 3.6×10^6 for a rigid rod of axial ratio approaching infinity. The reason for this apparent insensitivity of β to shapes is the following: Eq. (5-15) was originally derived by combining two semiempirical equations relating s and $[\eta]$ to M, each containing a parameter that is itself extremely sensitive to shape. But their resultant combination into the single parameter β serves to render the latter much less responsive.

For most nucleic acids, and specifically for both single- and double-stranded DNA, experimentally and theoretically,[22] a value of $\beta = 2.4 \times 10^6$ appears adequate (43). Substituting this and the other appropriate values in Eq. (5-15) we obtain Eq. (5-16), in which s^0 must be expressed in Svedberg units:

$$M = 13{,}200\,(s^0_{20,w}\,[\eta]^{1/3})^{3/2} \tag{5-16}$$

It is not always convenient to obtain both s and $[\eta]$ independently in order to determine M, or alternatively one may wish to extract some information concerning molecular shape from these two hydrodynamic measurements. This requires one to establish the dependence of M on s and $[\eta]$ for nucleic acids in different configurations and conformations. This dependence is of the form

$$s^0_{20,w} \quad \text{or} \quad [\eta] = KM^a \tag{5-17}$$

For double-stranded, linear DNA of high molecular weight, at a pH ~7, in 0.2 M Na^+ at 22°: $K_s = 0.034$, $a_s = 0.405$, $K_\eta = 6.9 \times 10^4$, and $a_\eta = 0.70$. In sucrose gradients under analogous conditions $a_s \approx 0.35$, and hence Eq. (5-5) can be recast to read

$$R = \frac{s_{\text{unknown}}}{s_{\text{standard}}} = \left(\frac{M_{\text{unknown}}}{M_{\text{standard}}}\right)^{0.35} \tag{5-18}$$

[20] f is equal to $6\pi\eta_0 a/F$, where η_0 is the coefficient of viscosity of the solvent and a is the radius of a sphere having the same volume V_e as a hydrodynamically equivalent ellipsoid of revolution of shape factor F that has been calculated in terms of only the axial ratio a/b. For spheres $1/F = 1.00$; for ellipsoids it is > 1.

[21] The value for M determined from Eq. (5-14) lies between $\bar{M}_n$ and $\bar{M}_w$; from Eq. (5-15) it is close to, but not identical with, $\bar{M}_w$.

[22] Using a model of a "wormlike" coil (41, 42).

Another set of useful equations (44) for such DNA is

$$0.445 \log M = 1.819 + \log (s^0_{20,w} - 2.7) \tag{5-19a}$$

$$0.665 \log M = 2.863 + \log ([\eta] + 5) \tag{5-19b}$$

For relaxed (hydrogen-bonded, nicked) circular DNA, s and $[\eta]$ are larger than those characteristic of the linear form (40, 45); for any one value of M, increases of 1.18- and 1.6-fold for s and $[\eta]$, respectively, would be expected on theoretical grounds and have indeed been observed. An additional increment is to be anticipated by the introduction of supertwists into covalently linked circles: values of 1.25- to 2.0-fold in s^0 have been reported for various superhelices, depending on the number of turns introduced, a parameter that is itself dependent on ionic strength (46–48): 3.8 turns per 10^6 daltons at 0.06 M Na^+ (close to maximum) versus 3.8 per 10^7 daltons at 1 M. On exposure to alkali convalently circular DNA first unwinds partially to a relaxed form equal in sedimentation coefficient to 1.16 the linear form, and after complete titration with alkali is converted to a collapsed form with a very large s value, ~3- to 4-fold that of a single strand with half the molecular weight. For linear, or circular, single-stranded DNA at neutrality and 0.2 M Na^+, $K_s = 0.022$, $a_s = 0.48$, $K_\eta = 4.9 \times 10^{-4}$, and $a_\eta = 0.55$. These values are strongly dependent on ionic strength and/or pH. At a higher ionic strength the molecule contracts further and becomes more spherically symmetrical,[23] as attested by values of $K_s = 0.0105$ and $a_s = 0.55$ in 1.0 M Na^+ at neutrality. On the other hand, by raising the pH to 13 or lowering the $[Na^+]$ to 0.01 M, the molecule expands and becomes more asymmetrical, with $K_s \simeq 0.055$, $a_s = 0.40$, $K_\eta = 3.1 \times 10^{-5}$, and $a_\eta = 0.91$. This set of constants also appears to be applicable to RNA of analogous conformation, that is, when all interbase interactions have been minimized, as for instance in the presence of formaldehyde. Some of the hydrodynamic parameters for various configurations and conformations actually observed with DNA are summarized in Fig. 5-14.

Light Scattering (49). When a beam of light is allowed to impinge on solute molecules in dilute solution, it is scattered or diffracted in all directions, owing to secondary emission by oscillating dipoles induced in the solute molecules by the electric vector of the radiation. If the solute is an asymmetric molecule, for example a nucleic acid, its dimension in at least one direction is greater than 0.05 times the wavelength of the incident light (usually a mercury arc that provides monochromatic radiation at 4358 or 5461 Å). Under these circumstances the solute particles no longer act as point dipoles, but instead it is necessary to consider them as containing several different scattering centers. The consequence is that the amount of light scattered to and measured at angles different from that of the incident light is not equal, and the greatest scattering is in the forward direction (with $\theta = 0$), and the back scattering ($\theta = 180°$) is least, owing to destructive interference.[24] Advantage can be taken of these facts to determine, not only the $\bar{M}_w$, but also to gain some information concerning the molecular shape of macromolecules from one and the same type of measurement. Unfortunately, the methodological difficulties (measurements at small angles, 10° or less are required) be-

[23] For a spherically symmetrical molecule in a *theta solvent*, $a_s = a_\eta = 0.50$.

[24] The angular dependence of scattering by such a molecule can be described in terms of a function P_θ, [see Eq. (5-23)] the particle-scattering factor, closely related to the expression used to describe the scattering of X-rays or electrons by gas molecules, $P_\theta = 1$ at $\theta = 0°$.

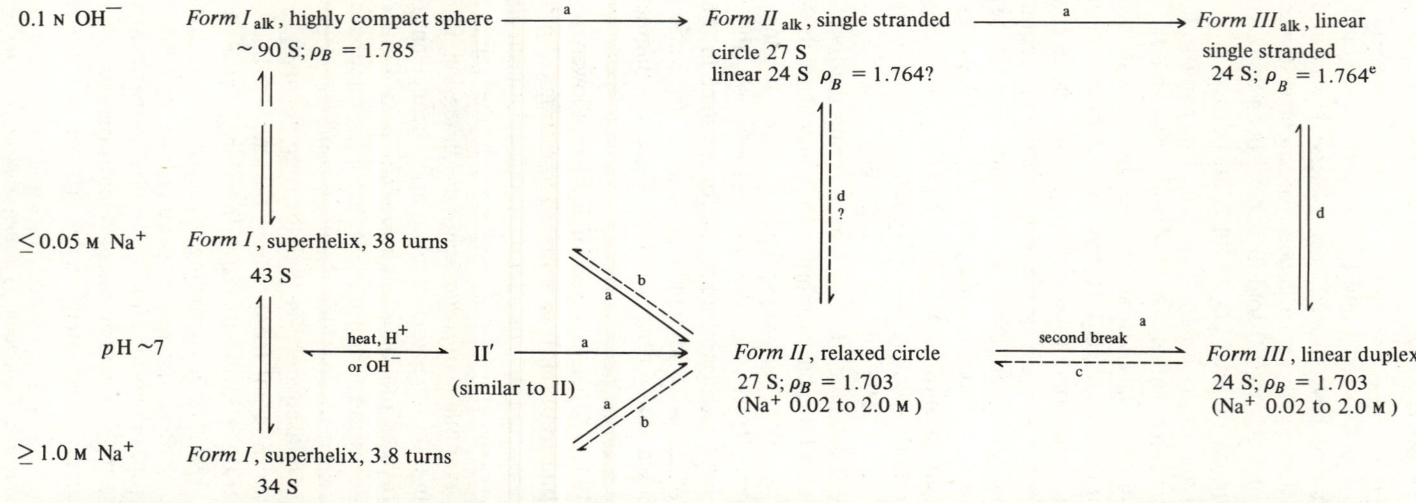

FIGURE 5-14

Hydrodynamic parameters of various configurations of DNA. A hypothetical DNA of $M = 10 \times 10^6$ daltons ($\sim 5\ \mu m$) and a base composition of 25 mol % of each of the four common bases is taken as an example. Composite data from Refs. All sedimentation coefficients are $s^0_{20,w}$ for Na^+ salt, except for I_{alk}.

[a] Introduced with endonuclease (very low concentration); by hydroquinone or dithiothreitol.

[b] By means of polynucleotide ligase plus ATP (NAD^+).

[c] Annealing (65° in 0.2 M Na^+) of complementary ends if present.

[d] Reneutralization plus annealing; without the latter, especially at low ionic strengths or in presence of formaldehyde, strands remain separated; $s = 25$ S, $\rho_B = 1.719$ g/cm^3.

[e] May separate into two components: $\rho_B = 1.758$ and 1.771, respectively.

I. R. David and D. R. Wolstenholme, *J. Mol. Biol.,* **28:** 233 (1967); J. Vinograd, J. Lebowitz, R. Radloff, R. Watson, and P. Laipis, *Proc. Natl. Acad. Sci. U.S.,* **53:** 1104 (1965); V. C. Bode and L. A. MacHattie, *J. Mol. Biol.,* **32:** 673 (1968); L. V. Crawford and M. J. Waring, *ibid.,* **25:** 23 (1967).

come most severe precisely in the region of M of greatest interest in DNA chemistry, that is, for M of $\geqslant 3 \times 10^6$, unless special instruments are used, but for smaller DNA molecules and for most species of RNA as well, the technique is of great power and precision. In any event the technique is valid only for macromolecules for which 0.05 × long dimension $< \lambda <$ 0.5 × long dimension.

In practice one determines the scattered or diffracted intensity i_θ obtained by irradiating the solution with a beam of incident intensity I_0 and wavelength λ at different angles θ with a phototube placed at a distance r from the solution. One thus obtains the Rayleigh ratio $R = i_\theta r^2/I_0$ or, more commonly, the parameter R_θ defined by Eq. (5-20), which equals Rayleigh's ratio at $\theta = 90°$:

$$R_\theta = \frac{r^2 i_\theta}{I_0(1 + \cos^2 \theta)} \tag{5-20}$$

In the limit when $\theta \to 0$ Eq. (5-21) describes the dependence of M on R_θ:

$$\left(\lim_{\theta \to 0} \frac{Kc}{R_\theta}\right)_{c=\text{const}} = \frac{1}{M} + 2Bc + 3Cc^2 + \cdots \tag{5-21}$$

where B and C are the second and third virial coefficients, respectively, which lead to nonlinearity in the dependence of M^{-1} on solute concentration, and K is an optical constant for the system.[25] On the other hand, it can be shown that, as $c \to 0$, Eq. (5-22) describes the dependence of Kc/R_θ on shape and thus on P_θ:

$$\left(\lim_{c \to 0} \frac{Kc}{R_\theta}\right)_{\theta=\text{const}} = \frac{1}{MP_\theta} = \frac{1}{M}\left(1 + \frac{h^2 R_G^2}{3} + \cdots\right) \tag{5-22}$$

$$h = \frac{4\pi}{\lambda} \sin \frac{\theta}{2}$$

$$P_\theta^{-1} = 1 + \frac{h^2 R_G^2}{3} + \cdots \tag{5-23}$$

Thus one performs a graphical double extrapolation by means of a *Zimm plot,* which involves plotting Kc/R_θ against $\sin^2 (\theta/2) + kc$, where k is an arbitrary constant chosen to provide a convenient spread of the data (Fig. 5-15). Extrapolating to $\theta = 0$, at constant c [curve (a)] one obtains a plot of Kc/R_θ versus kc, which affords M^{-1} as intercept and $2B/k$ as limiting slope. On the other hand, extrapolation to $c = 0$ but holding θ constant [curve (b)] provides a plot of Kc/R_θ versus $\sin^2 (\theta/2)$ with an intercept again of M^{-1} but now with a limiting slope equal to $(16\pi^2/3\lambda^2)R_G^2M^{-1}$. We thus find that the *radius of gyration* R_G is given by

$$R_G^2 = (\text{limiting slope/intercept})_{c=0} \times \frac{3\lambda^2}{16\pi^2} \tag{5-24}$$

where λ is the wavelength in the medium actually employed and equals $\lambda_0 \times n_0$, the refractive index of the medium.

[25] Given by $\dfrac{2\pi^2 n_0^2 (n - n_0)^2}{\lambda_0^4 \bar{N}}$, where n_0 is the refractive index of the solvent, n the refractive index of the solution, λ_0 the wavelength in a vacuum, and $\bar{N}$ the number of scattering centers per cubic centimeter.

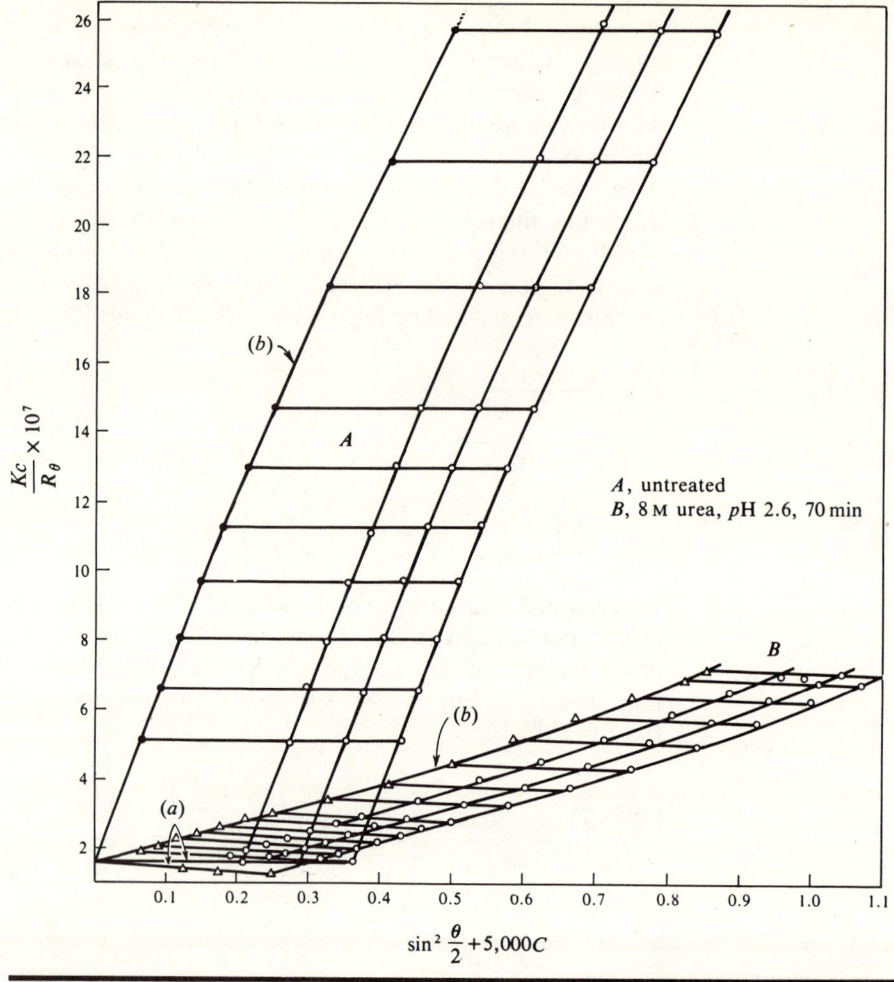

FIGURE 5-15
Light-scattering (Zimm) plots for two conformations of DNA. Lines (*a*): results of extrapolation at constant *c* to $\theta = 0$ [Eq. (5-21)]; lines (*b*): results of extrapolation at constant θ to $c = 0$ [Eq. (5-22)]. S. A. Rice and P. Doty, *J. Am. Chem. Soc.*, **79:** 3937 (1957); by permission of the copyright owners, the American Chemical Society.

$R_G{}^2$ is related to the dimensions of various hydrodynamic shapes in a simple manner:

For spheres,

$$R_G{}^2{}_{(S)} = \frac{3r^2}{5} \tag{5-25a}$$

For rods,

$$R_G{}^2{}_{(R)} = \frac{l^2}{12} \tag{5-25b}$$

For random coils,

$$R_G{}^2{}_{(C)} = \frac{\langle r^2 \rangle}{6} \tag{5-25c}$$

where r is the radius of the sphere, l the length of the rod, and $<r^2>^{1/2}$ the root-mean-square end-to-end distance of the random coil. Quite generally the more flexible a polymer, the greater its R_G. Thus, if the shape (or the dimensions) of the molecule is known or can be hypothesized on the basis of additional information (such as hydrodynamic data or electron microscopy), light-scattering data provide a test for the proposed model.[26] Quite analogous conclusions regarding the shape and dimensions of the DNA molecule in solution have been reached by Luzzati and his collaborators on the basis of the closely related method of X-ray scattering at very small angles (20).

Direct Physical Methods. A great deal of information concerning the dimensions and conformation of DNA molecules *in vitro* has been obtained recently by direct examination of the material by a variety of techniques. Perhaps the oldest and most straightforward method is examination by measurement in the electron microscope. This technique led to somewhat ambiguous answers as long as the experimenter was restricted to the use of isolated purified DNA shadowed with heavy metals (Pt, Pd, U) and spread by one means or another on the electron-microscope grid. However, more recently Kleinschmidt discovered that nucleic acid either liberated *in situ* by osmotic shock or after prior isolation could be spread and stabilized on water as a monomolecular film by means of complexation with a basic protein such as cytochrome *c* (50). The electron micrographs shown in Fig. 5-10 were produced by this technique; both the circular nature of the molecules and their dimensions are clearly visualized[27]: The length and width of the chains are those demanded by the Watson–Crick–Wilkins B helix. With the DNA from many coliphages the change from linear to circular form is readily reversible and hence must be due to noncovalent interactions. In the case of phage λ the molecule is known to possess "sticky ends" in the form of complementary single-stranded regions at opposite ends and these can join to form hydrogen-bonded structures: intermolecularly linked linear oligomeric chains (*concatemers*) by thermal treatment at relatively high DNA concentration, or circles within the same molecule by heating to 75° followed by slow cooling in 0.6 M NaCl. In the case of phage T2, such ends must first be exposed by limited exonuclease III treatment.

Another extremely useful set of techniques is that of *autoradiography* (53) (sometimes also called radioautography). Two types of experiments have been performed. The first is similar to that frequently employed for the solution of other biological problems and uses ^3H, a soft-beta emitter, usually in the form of a tritiated thymidine, incorporated into the DNA to generate an image in a suitably sensitive photographic emulsion. In the hands of Cairns this technique has provided a magnificent insight into the molecular dimensions of DNA's of viral and even bacterial sources (54) and has given results that are neatly supplementary to those obtained with electron micrography and completely consistent with the model.

[26] For instance, for a random coil the radius of gyration can be calculated from $[\eta]$ and M by the Flory–Fox equation:

$$R_G = \frac{M^{1/3}[\eta]^{1/3}}{6^{1/2}\phi^{1/3}} = \frac{5.5 \times 10^{-3}\,M}{s^0_{20,w}}$$

where $\phi = 1.32$ and $[\eta]$ is expressed in deciliters per gram.

[27] The technique is so powerful as to permit visualization not only of local imperfections due to partial denaturation or other treatments but even of nonhomologous regions due to genetic lesions (deletions or substitutions) in incompletely matched, annealed, heteroduplex molecules (51, 52).

The second autoradiographic technique, developed by Levinthal and Thomas, has been particularly useful for the development of primary molecular weight standards for very large DNA's. In this technique ^{32}P at a high concentration is incorporated into the DNA of a bacteriophage. The virus particle or its DNA is then embedded in a sensitive emulsion, and after radioactive decay for a suitable length of time and photographic development, the number of β-ray tracks emanating from single-point sources (corresponding to the viral DNA) are measured. These patterns are essentially star-shaped, and the method has frequently been referred to as *star gazing.* Because every track represents the decay of a single ^{32}P atom, the total phosphorus content of each point source (i.e., each viral DNA molecule) can be estimated from the number of tracks emanating from it, the exposure time, the known half-life of ^{32}P, and the experimentally determined specific radioactivity of the virus used. The relevant equation is

$$M \simeq \frac{10^4 N^*}{A_0} \tag{5-26}$$

where N^* equals the average number of tracks per star and A_0 is the specific activity of the radioactive phosphorus (in curies per gram) in the growth medium.

A related technique requires only biological measurements; it depends on the inactivation of infectiveness or some other biological activity of DNA as a consequence of radioactive decay of some atom (^3H or ^{32}P) incorporated into the macromolecule. The resultant *suicide rate* is given by

$$\ln (s/s_0) = 34 \times 10^{-6} \alpha A_0 n(1 - e^{-\lambda t}) \tag{5-27}$$

Here s_0 and s represent the titer of biological activity to times t_0 and t, respectively, A_0 is specific radioactivity at t_0 (in curies per gram), α is a constant (0.10 for ^{32}P, 0.14 for ^3H incorporated into double-stranded DNA), n represents the total number of atoms (P or H) in the virus molecule or other particle potentially capable of being labeled, and λ is the fractional rate (in days^{-1}) of radioactive decay (0.0485 day^{-1} for ^{32}P).

Optical Properties of Polynucleotides in Solution

Each monomeric unit in a polynucleotide represents a chromophore with strong absorption in the ultraviolet region of the spectrum (Fig. 5-3). The electronic transitions responsible are mainly π-π^* with, probably, some contribution by n-π^* to the long-wavelength band around 290 nm. In qualitative terms the spectra of all polynucleotides reflect this fact. Those containing all four bases (including all DNA's and RNA's) show a broad absorption maximum in the region between 256 and 265 nm, a trough around 230 nm, and a second peak in the far ultraviolet around 195 nm. However, when each and every chromophore in such a structure is surrounded in a sterically defined fashion by other chromophores, as is the case for all polynucleotides in aqueous solutions at low temperature and an ionic strength corresponding to physiologic conditions, it can no longer be treated as an independent oscillator. Instead, coupling between transition dipoles will take place, with nearest vertical neighbors to a first approximation, with more distant neighbors for a more complete description. This coupling produces two important effects that are currently under intensive investigation because of their great utility in providing insight into nucleic acid structure: the phenomena of *hypochromism* and *optical activity* (dissymmetry). Some of the underlying principles can be appreciated by a consideration of Fig. 5-11, which summarizes the structural parameters of a simple dinucleotide, the minimal structure that might be expected to be subject to analogous constraints and hence to exhibit the relevant phenomena—an expectation that is, in fact, fulfilled. Coupling of the electric transition moments leads to a decrease in absorbance relative to those exhibited by the sum of X and Y (hypochromism); for greatest effectiveness it requires maximum overlap between orbitals of adjacent bases accomplished by decreasing Z and maintaining base planes parallel ($\gamma \simeq 0$, $\alpha \neq 90°$). On the other hand, interaction of the electric with the magnetic transition moments will generate optical activity not inherent in the sum of X and Y *provided* the orientation of the two bases are neither parallel nor antiparallel, that is, α is neither 0 nor 180°. The functional dependence of optical activity on wavelength, as measured by circular dichroism (CD) or optical rotatory dispersion (ORD), is exquisitely dependent on the precise geometry of the repeating unit, that is, the parameters Z, α, and γ shown in the figure; conversely, these CD and ORD measurements are capable, at least in theory, of providing precise information concerning the geometry of polynucleotides in solution.

Hypochromism (55–61). Since all that is required for measurements of hypochromism is an ordinary UV spectrophotometer, usually with a thermostated cell compartment and a constant-temperature bath to permit measurements over a range of temperatures as well as of wavelength, this parameter has been investigated most extensively over a number of years for a wide variety of poly- and oligonucleotides. It provides the most convenient tool for the identification and, under appropriate conditions, also the quantitative estimation of structural alterations important in polynucleotide chemistry such as the destruction and reformation of oriented structures (denaturation and renaturation), the formation of double- and multistranded homo- and heteropolymers from their components, as well as of other association–dissociation processes. Hypochromism refers to the decrease in f, the oscillator strength taken equal to $4.32 \times 10^{-9} \int_{\lambda_1}^{\lambda_2} \frac{\varepsilon_\lambda}{\lambda^2}\, d\lambda$ [i.e., the areas under the absorption band (between λ_1 and λ_2, the cutoff wavelengths[28]) with the ordinate expressed in centimeters] between some reference and the experimental curve (55). Data are, however, usually expressed in one of two ways (59–61): as *hypochromicities* (H_λ) at a stated wavelength (i.e., the fractional decrease in going from a disordered, reference state to the ordered, experimental one) or as *hyperchromicities* (h_λ), which is the inverse (i.e., the fractional increase in going from the latter to the former):

$$H_\lambda \equiv \varepsilon^O{}_\lambda/\varepsilon^D{}_\lambda = A^O{}_\lambda/A^D{}_\lambda \tag{5-28a}$$

$$h_\lambda \equiv [H_\lambda{}^{-1} - 1] = \varepsilon^D{}_\lambda/\varepsilon^O{}_\lambda - 1 = A^D{}_\lambda/A^O{}_\lambda - 1 \tag{5-28b}$$

Here $A^D{}_\lambda$ and $A^O{}_\lambda$ are the absorbances and $\varepsilon^D{}_\lambda$ and $\varepsilon^O{}_\lambda$ the extinction coefficients, at wavelength λ, of the disordered and ordered forms, respectively. The reference-disordered states are taken either as the isolated component single strands (h), preferably in a conformation that minimizes interactions between chromophores,[29] or the library values of the ε of the individual nucleotides in their correct stoichiometry (h'). Data for a number of naturally occurring and biosynthetic polynucleotides are presented in Table 5-5.

[28] More precisely, the integral is between zero and infinity.

[29] *Not* a random coil, since in this relatively condensed conformation electronic interaction between bases is still possible, but a completely extended one at very low ionic strength.

TABLE 5-5
Hyperchromicities of representative polynucleotides[a]

Polynucleotide	*Percentage G + C*	λ, *nm*	h[b]	h'[b]	ε(*P*)[b]
dG · dC in DNA (calc)	100	260	0.18 (0.41)[c]	0.42 (0.79)[c]	7,000
Sarcina lutea DNA	73	260	0.29	0.52	6,690
Escherichia coli DNA	50	260	0.39	0.60	6,740
Calf thymus DNA	41	260	0.41	0.66	6,600
Bacillus megaterium DNA	38	260	0.42	0.65	6,860
dA · dT in DNA (calc)	0	260	0.56	0.78	6,650
φX DNA	cf. Table 5-3	260	0.31	0.54	7,500
$(dG)_n \cdot (dC)_n$	100	253	0.21 (0.60)[c]	0.42 (0.75)[c]	7,400
$d(A\text{-}T)_n$	0	260	0.42	0.84	6,650
$(dA)_n \cdot (dT)_n$	0	260	0.43	0.94	5,900
$d(T\text{-}G)_n \cdot d(A\text{-}C)_n$	50	260	0.24		6,970
$d(T\text{-}C)_n \cdot d(A\text{-}G)_n$	50	260	0.26		
$(dA)_n$		257	0.16	0.39	8,600
$(dT)_n$		260			8,700
$(dG)_n$		260			8,990
$(dC)_n$		270			7,200
$(rG)_n \cdot (rC)_n$	100	252	0.60		7,500
$(rA)_n \cdot (rU)_n$	0	260	0.58	0.98	6,250
$(rA)_n$		257	0.19 (0.25)[d]	0.47 (0.76)[d]	10,700 (8,600)[d]
$(rU)_n$		260	~0.10	0	9,200
$(rG)_n$		252		0.39	9,500
$(rC)_n$		270	0.24		6,500 (7,400)[d]

[a] All values at a [Na+] of 0.1 to 0.2 M at *p*H 7 to 8; at [Na+] ≤ 10^{-3} M, especially in the presence of denaturants (e.g., methanol, ethylene glycol), h approaches h'.

[b] ε(*P*), h, and h' are the extinction coefficients (monomer basis), and the hyperchromicities at the indicated wavelengths relative to the high-temperature form (≥ 95°) or library spectra, respectively.

[c] At 280 nm.

[d] In acid solution (*p*H 4).

SOURCE: Most data from A. M. Michelson, *The Chemistry of Nucleosides and Nucleotides,* Academic Press, New York, 1963; Mahler, Kline, and Mehrotra (61); Inman and Baldwin (62); Inman (63); Chamberlin (64); Wells, Ohtsuka, and Khorana (65); Bollum (31); Ts'o, Rapaport, and Bollum (66); and Green and Mahler (67).

The extinction coefficients are presented as the quantity ε(*P*) introduced by Chargaff and Zamenhof, which is expressed on a mononucleotide (phosphorus) basis and therefore allows a direct comparison of various polymers, oligomers, and monomers with one another. The following generalizations can be derived from data of this sort:

1. All native DNA's, regardless of base composition and *M,* give ε(*P*)[30] values of about the same magnitude.

2. Hyperchromism is dependent on base composition; conversely, base compositions can be determined from hyperchromic spectra (59–61).

3. Hyperchromism is dependent on configuration and conformation: Single-stranded, ordered polynucleotides are, in general, less hyperchromic than are double-stranded ones; among single-stranded homopolymers hyper-

[30] An ε(*P*) in the range of 6100 to 6900 is generally considered normal for ordinary, double-stranded DNA in a buffer containing Na+ ion at 0.2 to 0.5 M. This corresponds to an absorption of 18.0 to 19.0 per mg of DNA/ml.

chromism is related to the extent of self-association ($G > C \geqq A \gg U$). The presence or absence of the 2′-OH group in a polymer appears relatively unimportant (62–66); this is in sharp contrast to its influence on rotational spectra (see below).

4. Even polymers that behave hydrodynamically as random coils, such as single-stranded polynucleotides (most types of RNA, denatured DNA after strand separation, ϕX DNA, or single-stranded homopolymeric nucleotides) exhibit considerable residual hypochromism relative to the mononucleotides (66,67). This observation is closely related to the fact that even in oligonucleotides, down to the dinucleotides, the opportunity for electronic coupling remains and such structures exhibit relevant optical phenomena (see Table 5-6 and Fig. 5-11).

ORD and CD (68–71). Among the most characteristic phenomena that set oligomeric and polymeric nucleotides apart from their monomers are their interaction with polarized light of wavelengths in the region of their absorption maxima. The phenomenology involved and the experimental techniques have been discussed in Chapter 4 and are described in (68–73). ORD and CD are equivalent phenomena and the two types of spectra are interconvertible by the *Kronig–Kramers transformation.* Poly- and oligonucleotides generate spectra that are of the general type shown in Fig. 5-16*a*. Curve 1 here is referred to as a "conservative," curve 2 as a "nonconservative" pattern (75). Some actual examples are shown in Fig. 5-16*b*. Three facts stand out from a number of investigations: (1) native DNA generates a conservative, RNA, even when double stranded, generates a nonconservative spectrum; (2) the magnitude of the relevant effects are greater in the latter than in the former case; and (3) oligonucleotides exhibit considerable optical activity (76) which may be of the conservative or nonconservative type (55, 56, 73–75). Again interactions appear stronger in the ribo than in the deoxyribo series.

ORD spectra are usually expressed as mean residue rotations $[m]_\lambda$ or reduced mean residue rotations $[m']_\lambda$:

$$[m']_\lambda = [m]_\lambda \frac{3}{n^2+2} = \frac{\text{mean residue weight}}{100} [\alpha]_\lambda \frac{3}{n^2+2} \tag{5-29}$$

where $[\alpha]_\lambda$ is the specific rotation at wavelength $\lambda = \dfrac{(\alpha_{\text{obs}} \times 100)}{(d \times c)}$; α_{obs} = observed rotation in degrees, d = path length of cell in decimeters, c = concentration in g/100 ml, and n is the refractive index of the solvent. The dimensions of $[m]$ or $[m']$ are then deg-cm²/decimol. CD spectra are frequently expressed in terms of mean residue ellipticities $[\theta]_\lambda$ or reduced mean residue ellipticities:

$$[\theta]_\lambda = 3300(\varepsilon_L - \varepsilon_R) = \frac{3300(A_L - A_R)}{md} = \text{mean residue weight}\left(330\,\frac{A_L - A_R}{Cd}\right) \tag{5-30}$$

Here the experimentally determined quantity is ($A_L - A_R$), the differential absorbance for left and right polarized light; ($\varepsilon_L - \varepsilon_R$) is the corresponding differential molar absorptivity or extinction coefficient, m and C are concentrations in moles/liter or g/ml, respectively, and d is the path length in centimeters. An alternative method for expressing CD data, particularly for nonconservative spectra, is to calculate the rotational strength R_K of the optically active electronic transition (CD band):

$$R_K = 22.95 \times 10^{-40} \int_0^\infty (\varepsilon_L - \varepsilon_R)\,d\nu/\nu = 0.696 \times 10^{-42} \int_{\lambda_1}^{\lambda_2} \frac{[\theta]_\lambda}{\lambda}\,d\lambda \tag{5-31}$$

TABLE 5-6
Optical parameters of oligocytidylic acids[a]

	$pH \geq 7.5$					pH 4.0		
	$\varepsilon_{max} \times 10^{-3}$		$[m'] \times 10^{-3}$		$R_{max} \times 10^{40}$,	$\varepsilon_{max} \times 10^{-3}$,	$R_{max} \times 10^{40}$,	$R_{min} \times 10^{40}$,
n	rC_n[b]	dC_n[b]	rC_n[b]	dC_n[b]	rC_n[c]	rC_n[d]	rC_n[d]	rC_n[d]
1	9.26	9.23	16.0	12.4	12	9	27	0
2	8.26 (7.92)[c]	8.66	26.3	13.9	25.9	8.65	27	0
3	8.14	8.49			31			
4	7.96	8.37		20.4	34		27	0
6	7.80	8.22		19.0		7.60	27	0
8	7.43	8.08	32.9	20.5	34.5		27.5	− 8.7
10	7.37 (6.2)[c]	8.04		20.1	35	6.99	27.2	− 7.8
12	7.27							
14	7.20							
Polymer	6.32 (6.09)[c]	7.47	61.0	26.6	57.2	6.5	39.2	−13.7

[a] ε_{max} is the molar extinction coefficient per residue at the absorption maximum in $cm^{-1} \times M^{-1}$; $[m']$ is the reduced molar peak-to-trough rotation per residue for the ORD spectrum in deg-mc²/decimol; R_{max} and R_{min} are the rotational strength per residue at the maxima and minima of the CD spectrum in cgs units.

[b] In 0.02 M Tris–0.08 M NaCl, pH 8.5 at 10°; A. Adler, L. Grossman, and G. D. Fasman, *Proc. Natl. Acad. Sci. U.S.*, **57:** 423 (1967).

[c] In 0.01 M Tris–4.7 M KF, pH 7.5 to 8.0 except for polymer (in 0.1 M KF); J. Brahms, J. C. Maurizot, and A. M. Michelson, *J. Mol. Biol.*, **25:** 465 (1967).

[d] In 0.1 M NaCl–0.05 M Na acetate, pH 4.0.

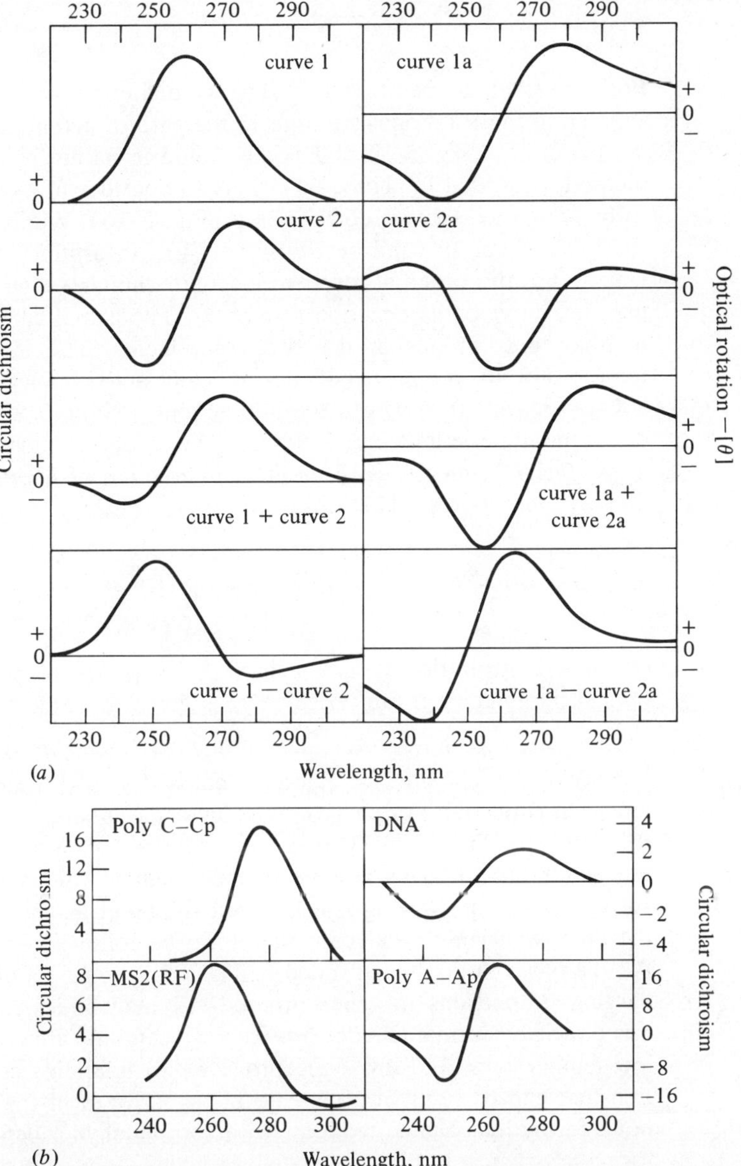

FIGURE 5-16

(*a*) Shape of circular dichroism curves (*left*) and optical rotatory dispersion curves (*right*) for a polymer absorbing at 260 nm. The top two curves are the characteristic curves which combine to give all others. Each curve can be positive or negative. (*b*) Experimental circular dichroism for various polynucleotides. The upper left curve is single-stranded polycytidylic acid minus cytidylic acid [C. A. Bush and J. Brahms, *J. Chem. Phys.*, **46:** 73 (1967)]; upper right curve is double-stranded DNA [J. Brahms and W. F. H. M. Mommaerts, *J. Mol. Biol.*, **10:** 79 (1964)]; lower left curve is the double-stranded replicative form of MS 2 ribonucleic acid (J. Thiéry and S. R. Jaskunas, personal communication); lower right curve is single-strand polyadenylic acid minus adenylic acid (75). All measurements were at neutral *p*H. I. Tinoco, Jr., *J. Chem. Phys.*, **65:** 91 (1968).

For a modified Gaussian band with band center at $\nu_{\max}$ (wavelength $\lambda_{\max}$), $[\theta] \simeq k\,\exp-(\nu - \nu_{\max})^2/\Delta^2$, where Δ is the band width (i.e., that span of frequencies for which $[\theta] \geqq 0.368[\theta]_{\max}$), Eq. (5-31) can be approximated by (with λ and Δ in cm)

$$R_K \simeq \frac{0.696 \times 10^{-42}}{\lambda_{max}} \sqrt{\pi[\theta]_{max}\Delta} \tag{5-31a}$$

For conservative spectra R_K is zero by definition, and alternative means for estimating the rotational strength in the case of actual cases corresponding to the patterns of Fig. 5-16 and others found in nature have to be devised. One method, proposed by Tinoco (70), is to consider any experimental spectrum to be composed of two components (Fig. 5-16), one a nonconservative one similar in shape to (and derived from) the absorption curve of the monomer (curve 1), the other a conservative one, similar to (and generated by) its first derivative or its negative (curve 2). The actual curve will depend on the relative contributions of the two components. The interactions responsible for the first are properties of any optically active molecule and are therefore always present: it measures the interactions among *different* absorption bands of the monomeric units in a polymer. The second is characteristic of oligomers or polymers; it measures the mutual interaction of *identical* (or closely similar) chromophores. The observed curve can be defined by a rotational *oscillator strength* R_K', equal to $22.95 \times 10^{-40} \int_0^\infty (\varepsilon_L - \varepsilon_R)d\nu$. The conservative contribution then equals $R_K' - \nu_{max}R_K$. A comparison of various optical parameters in the series $(C)_n$ is given in Table 5-6.

Chemical Reactions of DNA in Solution

Reagents Capable of Reacting with Free Amino Groups. A reagent frequently employed in nucleic acid chemistry is formaldehyde, which can react with the free amino groups of cytosine, guanine, and adenine to form the corresponding hydroxymethyl derivatives (or perhaps, after dehydration, the Schiff bases). This change is accompanied by an alteration in the absorption spectrum, giving rise to a red shift and an increase in absorption at the peak wavelength and can, therefore, be followed spectrophotometrically. All the bases in double-stranded helical DNA in its native conformation, and varying proportions in other ordered polynucleotides, are protected against this attack by formaldehyde, provided precautions are observed (neutral *p*H, low temperature, and low concentration) to minimize conformational alteration that can be brought about by this agent even in the absence of covalent bond formation. Hence reaction with formaldehyde, before and after a conformation change has been induced, provides a measure of the degree of order in the native polymer. Formaldehyde has also been used as a "chaperone" to prevent reassociation of strands subsequent to their dissociation.

Another very useful reagent, especially in studies of mutagenesis *in vitro,* is nitrous acid: 0.05 to 1 M $NaNO_2$ in acetate buffers at *p*H 4.0 to 5.0. Under these conditions adenine is transformed to hypoxanthine, guanine to xanthine, and cytosine to uracil. The base-pairing properties (Fig. 5-17) are such as to make likely a *transition* to the other pair in the course of the replication process.[31] The deamination rate has been estimated (77) to be 4.2×10^{-5} mol total base deaminated $\times$ min^{-1} $\times$ mol DNA nucleotide^{-1} at 25°, *p*H 4.15, in 0.25 M NO_2^-. Deamination of cytosine can also be brought about (78) simply by heating at *p*H 3.5. Another common process, occurring at about one fourth the rate of deamination, is one that leads (probably via the

[31] $A \cdot \xrightarrow{HNO_2} I \cdot T \underbrace{\rightarrow I \cdot C \rightarrow}_{\text{replication}} G \cdot C$ and $G \cdot C \xrightarrow{HNO_2} G \cdot U \underbrace{\rightarrow A \cdot U \rightarrow}_{\text{replication}} A \cdot T$.

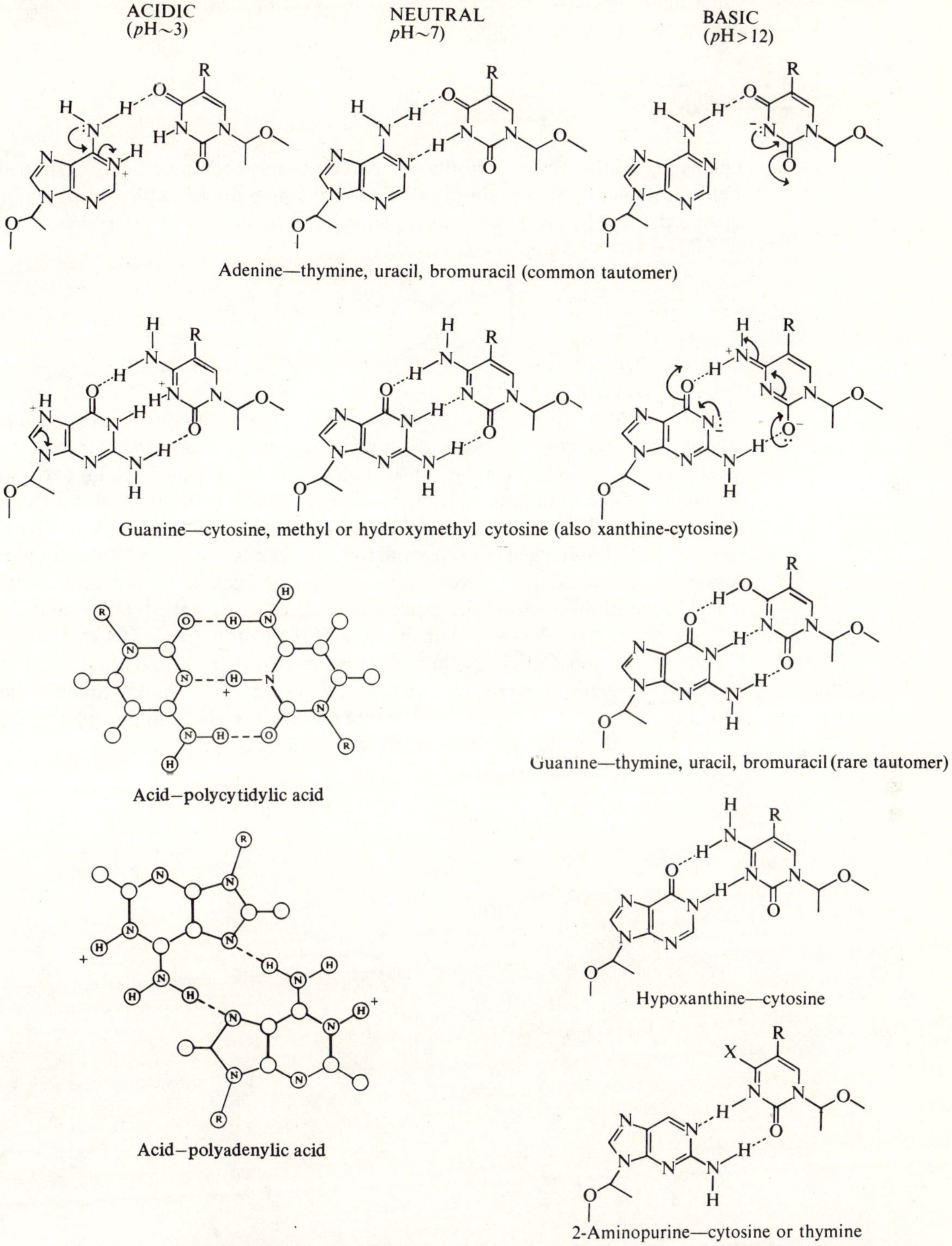

FIGURE 5-17
Base-pairing properties under various conditions.

formation of a common intermediate, the diazonium salt of the amino bases, followed by N_2 elimination and simultaneous alkylation at susceptible sites) to the cross-linking of the two chains by covalent bonds.

Similar cross links can be obtained by treating DNA with acid, with ultraviolet light at relatively high intensities, or with bifunctional alkylating agents such as the nitrogen mustards (79),

$$ClCH_2CH_2\underset{\displaystyle R}{\underset{|}{N}}CH_2CH_2Cl$$

The nature of the products in the last two cases has been tentatively identified; with ultraviolet light it is surmised to be a thymine dimer, with the alkylating agent a di(guanin-7-yl) derivative, formed on elimination of the ribose:

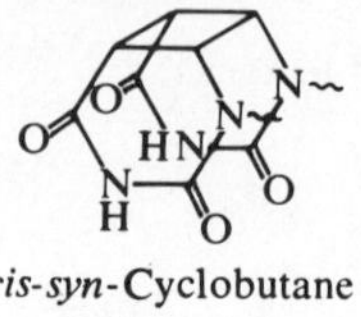

cis-syn-Cyclobutane dimer

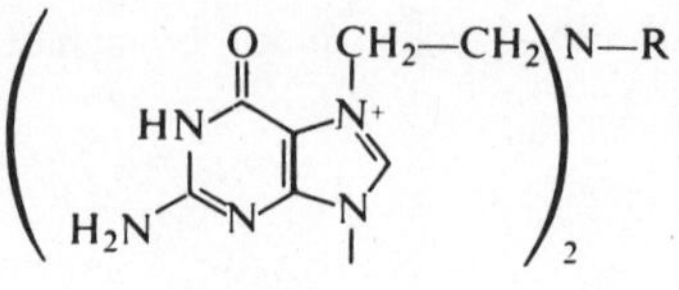

Thus, cross-linking takes place between appropriately placed *thymine* residues on the *same* strand and *guanine* residues on *opposite* strands, respectively. Cross-link formation also appears to be responsible, in part, for the ability of the antibiotic Mitomycin C and related agents (all of which are reduced intracellularly to active derivatives) in inhibiting DNA synthesis (page 902). Topologically, cross-linking between strands and formation of covalently joined duplex circles are equivalent; complete unwinding and strand separation cannot take place. Therefore, cross-linked DNA and covalently circular DNA are similar in many of their properties (79).

One of the most effective mutagens is hydroxylamine, NH_2OH, and its alkylated derivatives NH_2OR, such as methoxyamine. Unlike many other agents their action can be rendered highly specific: at pH 6, in pyrophosphate, 1 M NH_2OH reacts almost exclusively with cytosine and its derivatives.[32] The sequence of events is probably

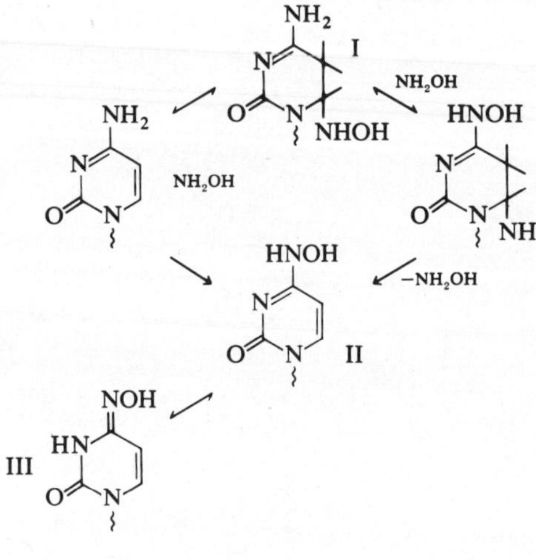

I

II

III

It is not yet certain which of the three products I, II, and III is responsible for the mutagenic action by exhibiting base-pairing properties similar to U, or

[32] At lower concentrations ($< 10^{-2}$ M) hydroxylamine exerts a strong inactivating effect. The cause of this phenomenon is due to the introduction of single-strand breaks in the DNA. A free $\gt$N—OH group in the reagent appears to be necessary for the inactivating reaction; hence H_2NOH is exclusively mutagenic.

whether the latter is, in fact, produced. The mutagenic effect is due to the *transition*

$$G \cdot C \xrightarrow{HONH_2} \begin{matrix} G \cdot C^* \\ G \cdot U(?) \end{matrix} \rightarrow A \cdot U \rightarrow A \cdot T$$

Protonation and Deprotonation Reactions. The protonation of DNA in solution occurs at pH values that differ considerably from those of either the free bases or the nucleotides in the same solvent.[33] In general the pK_a values for amino groups in DNA are found to be approximately 0.4 pH unit lower at high salt concentration and ~1.0 pH unit higher at low salt concentration than those of the corresponding mononucleotides. If the solution is neutral to start with, adenine and cytosine are protonated first (with pK_a values between 5 and 4), followed by guanine (with a $pK_a' \simeq 3$). A consideration of Fig. 5-17 indicates that the first set of protonations leaves base-pairing properties and helix dimensions virtually unchanged; on the contrary, the monoprotonated [A^+(H) · T] and especially [G · C$(H)^+$] pairs may actually afford greater stability. But on protonation of guanine we are faced with a powerful charge repulsion within one and the same base pair, which can only lead to profound, possibly irreversible, change in structure occurring over a very narrow pH range. As might be expected, single-stranded DNA's are titrated at somewhat higher pH values. Conversely the removal of protons from guanine and thymine bases in DNA can be observed by titration with alkali. With double-stranded DNA (46) this occurs over an exceedingly narrow pH range (<0.2 unit) around pH 11.8, with an attendant increase in ρ_B of ~0.060 g/cm^3. Superhelices are titrated over a somewhat wider range at a $p\text{H} > 12$, while single-stranded DNA's require a lower and even wider pH range between 10 and 11. These titrations are accompanied by rises in ρ_B of 0.080 and ~0.047, respectively (12) (Fig. 15-14).

Interactions with Small Molecules and Ions. Polynucleotides, being polyanions, must be accompanied by an appropriate number of cations in their vicinity in order to maintain electrical neutrality. The extent of binding (or, what is equivalent, their average distance from the phosphodiester backbone) of these small ions can vary greatly, however. On the one hand, we meet inorganic countercations exemplified by the alkali metals, such as Na^+ and Li^+, which serve only to neutralize the negative charges on the phosphate residues electrostatically; this allows the nucleic acid to assume its most compact conformation, as shown by increased s^0, h, and $[\theta]$ and a decrease in R_G. The changes in these parameters bear a direct functional dependence on ionic strength and are much more pronounced for the more flexible, single-stranded than for the intrinsically more rigid, double-stranded polymers. Divalent cations such as the alkaline earths, certain transition metals (Co^{2+}, Mn^{2+}, Ni^{2+}, and Zn^{2+}), or organic diamines such at putrescine and cadaverine, which frequently accompany DNA *in vivo,* show a different behavior; they are bound stoichiometrically, probably to the phosphate residues. Thus, within this stoichiometric range they are far more effective than the monovalent ions (at the equivalence point 1 Mg^{2+} is equivalent to 10^2 to 10^3 Na^+) in affecting

[33] pK_a values are different because of the high negative charge density around the phosphate; this fact also accounts for the variation in base pK_a values in DNA with ionic strength of added counterion.

the conformation of the macromolecules.[34] In addition, the aliphatic oligoamines and especially other relatively large organic oligo- and polycations may exert other effects on DNA structure not duplicated by the inorganic ions (80–85).

Other metal cations, such as Ag^+, Hg^{2+}, Cu^{2+}, Fe^{2+}, and perhaps Pb^+, react, not only with the phosphate residues, but with some of the bases as well and profoundly and frequently irreversibly affect the structure of nucleic acids (86).[35] Of great interest is the interaction of nucleic acids with certain polynuclear, aromatic planar cationic (basic) dyes of which the acridines (proflavine, acridine orange) are examples. Dyes of this type have long been used as biological stains capable of binding selectively to and thus rendering visible DNA and RNA (or rather double-stranded versus less highly organized polyanions) by virtue of their absorption or fluorescence. More recently they have come into prominence as highly selective mutagens, capable apparently of producing additions or deletions of single nucleotides in the course of DNA replication (87) as well as of specific inhibition of this process and or RNA synthesis (88). A number of interactions are possible between such dyes and highly organized polyanions such as DNA: at low dye/monomer ratios ($< 1 : 3$) we may find complexes in which (1) a single positive charge on the dye is lined up opposite to one on the phosphate, with the dye essentially parallel to the bases (i.e., at right angles to the helix axis), increasing the diameter of the helix; (2) a dye molecule still in parallel has slipped in between two adjacent base residues and in this manner increases the length of the helix, a process called *intercalation* by Lerman, its discoverer (89); (3) a dipositively charged dye molecule lines up along the helix axis and cross links two phosphate residues on either the same or on opposite strands, thus enhancing the stability of the helix. As the dye/monomer ratio increases, (4) dimeric and (5) polymeric dye aggregates are stacked on the outside of the helix opposite to adjacent phosphate residues and again at right angles to the helix axis.

Among the various interactions mentioned, intercalation would appear to be the most significant and has received a great deal of attention recently (89–92). In a rigid polymer the consequences of introducing a flat aromatic molecule between two adjacent bases would produce a linear expansion along the helix axis (Figs. 5-8 and 5-18) and increase $[\eta]$, decrease s^0, alter the large-angle X-ray pattern—but with retention of the 3.4-Å meridional spacing—decrease the mass per unit length ratio (electron microscopy), and increase R_G. It may also produce, perhaps, a greater stability toward further disruption of secondary structure and lead to strong electronic coupling between intercalator and bases, with clear-cut effects on light absorption, emission, and rotation (in two wavelength regions—that characteristic of the ligand and that of the bases). All these effects have indeed been observed with the three important classes of intercalating molecules shown in Fig. 5-18.

In the case of actinomycin D (C_1) and related molecules there is some question whether the observed strong interaction ($K_D \approx 5 \times 10^{-7}$ in > 0.2

[34] Their stoichiometry differs. Alkaline earths and Co^{2+} show a stoichiometry of 1 ion^{2+} per 2 phosphates; the diamines are bound in the ratio 1 : 3.

[35] Hg and Ag binding, which are strictly reversible, has been found to be extremely useful in distinguishing AT-rich from GC-rich segments and, in conjunction with Cs_2SO_4 centrifugation, is of great value in separating DNA species differing in overall composition (page 212).

M Na^+) is due entirely or predominantly to intercalation because alternative models for the structure of the complex can also be entertained (see Fig. 5-18, top left). In any case, effective interaction requires (1) the presence of a 2-NH_2 purine in the DNA [ordinarily guanine, but a DNA with adenine replaced by 2,6-diaminopurine is effective (93)]; (2) a carboxamido substituent on the phenoxazone ring; and (3) the cyclic peptide. The latter is necessary for biological effects perhaps by assuming a conformation bridging the two backbone strands and blocking the minor groove, thus impeding the progress of the DNA-dependent RNA polymerase; the specific mode of action of actinomycin is blocking the synthesis of RNA on a DNA template.

Acridines (proflavin is the example shown) and phenanthridines (such as ethidium bromide) are proven intercalating agents; the former is effective only at low ionic strength and is a potent frame-shift mutagen (page 907); the latter is effective even at very high ionic strength and hence must be capable of very strong interactions. These result in three important applications, particularly with covalently circular helical DNA, which because of its topological constraints is capable of tolerating only a limited amount of intercalation, a process that requires local untwisting of the duplex.

1. With saturating concentrations of the dye, a smaller amount of this ligand is bound to superhelical than to relaxed or linear DNA. This results in a difference in buoyant density of ~0.040 g/cm^3 between the two forms, which is ample to separate them cleanly in analytical or preparative gradients (94).

2. In the course of titrating superhelical DNA with the dye, as a result of the topological constraints mentioned, the twists first become unwound at low dye concentrations to a point of no twists (s^0 of superhelix = s^0 of relaxed helix)—where superhelical DNA binds dye more tenaciously than any of the other forms. As more dye is bound, new twists are introduced, but now in the opposite direction (47, 95). From the amount of dye bound at the equivalence point the number of superhelical twists can be calculated, providing the angle required to unwind the double helix per intercalative event (~12°) is known. The type of result obtained (94, 95) is shown in Fig. 5-18B for the DNA from the animal virus SV40, a molecule with $M = 3.2 \times 10^6$, and by this technique known to contain 15 left-handed twists at high ionic strength.

3. At low concentrations the agent interacts specifically with the mitochondrial DNA of eucaryotic cells.

Denaturation

General Considerations. When solutions of DNA, originally at *p*H values around neutrality and in the presence of counterions of moderate ionic strength, are exposed to any of a number of agents—such as increase in temperature, addition of hydrogen or hydroxyl ions, removal of the counterion, or the addition of a large variety of organic reagents (alcohols or glycols, amides, aliphatic and heterocyclic amines, substituted ureas, phenols, sulfoxides), or certain anions ($CCl_3CO_2^-$, CNS^-, ClO_4^-)—they undergo a number of structural (conformational) changes, varying in the extent of their reversibility. These changes have frequently been lumped together and, by analogy with the somewhat similar processes in protein chemistry discussed in Chapter 4, have been termed *denaturation,* and the agents responsible have been called *denaturants* or *denaturing agents.* Any one of the hydrodynamic (s^0, $[\eta]$, R_G, etc.) or optical properties (h, $[\theta]$, etc.) of the macromolecule may be used to follow the process qualitatively and if only two states, a *native* and a *de-*

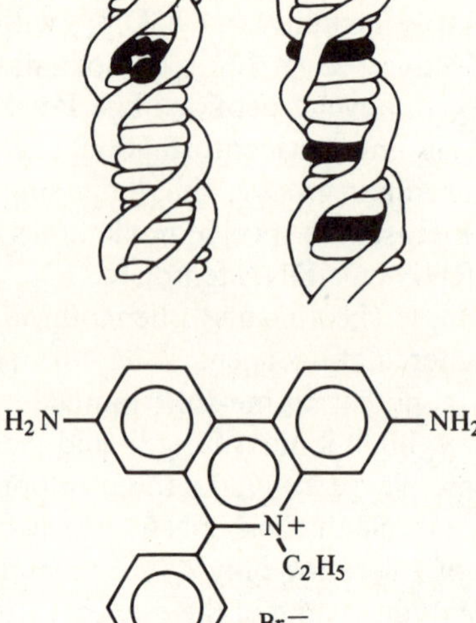

H_2N — NH_2 — N^+ — C_2H_5 — Br^-

Ethidium

(2,7-diamino-9-phenyl-10-ethyl phenanthridinium bromide)

A phenanthridine

H_2N — N^+ — H — NH_2 — Cl^-

Proflavine

(2,8-diaminoacridine hydrochloride)

An acridine

Sarcosine — L-Pro — L-N-Me Val — D-Val — O — L-Thr — C=O (×2) — N — O — NH_2 — O — CH_3 — CH_3

Actinomycin C_1 or D

A substituted aminophenoxazone

(*A*)

(*a*) (*b*) (*c*)

(*d*) (*e*) (*f*)

(*B*)

natured one, need to be distinguished, quantitatively as well.[36] Agents of various classes are frequently potentiating and mutually complementary—for example, addition of various denaturing agents lowers the temperature at which denaturation takes place, as does a decrease in ionic strength; at an elevated temperature the concentration of denaturant required to give a desired degree of denaturation is lower than at room temperature, and so on. Some of these factors are brought out in the simplified and schematic Fig. 5-19, as are some of the following additional points:

1. The transitions observed can frequently be represented by simple sigmoids defined by two parameters: the position of their midpoint and the steepness of the slope at this point (or, by an equivalent parameter, the transition width). The steepness and the shape of these transitions are analogous to those observed in highly cooperative order–disorder phase transitions ("melting") of crystallites or other polymers.

2. If the system under study involves just two components [i.e., if intermolecular heterogeneity is absent, and the only two species involved are the native and those denatured form(s) in equilibrium which are to be measured by the changes observed] and if the transition is reversible over a considerable region beyond the midpoint, one can regard the latter as that point at which statistically 50 percent of the molecules have been converted from the native to the denatured forms (i.e., some combination between the two extremes of all the molecules having been denatured to the extent of 50

[36] In order to utilize the quantitative relationships obtained for the calculation of the thermodynamic parameters (ΔG, ΔS, and ΔH) of the reaction, it is, of course, necessary to demonstrate that the observed changes are strictly reversible and to have available an exact measure of the relevant parameter for the two limiting forms (96). Other parameters besides the ones mentioned have also been employed (e.g., ir or nmr spectroscopy), and, in the case of the DNA's of the T-even bacteriophages, which because of their glycosyl residues are antigenic in the denatured form, this biological activity can also be used. The destruction of other biological activities of certain DNA's (i.e., their transforming ability or infectivity) have similarly been utilized as a quantitative measure of denaturation.

FIGURE 5-18

(A) Intercalation in a DNA double helix. (*Top*) Schematic representation of the consequence of intercalating (right) planar aromatic molecules (shaded structures shown at the bottom) into duplex DNA in the B conformation (left). An alternative binding scheme for actinomycin is also shown on the left. After Refs. 89–92.

(B) Representation of the effects of the binding of ethidium bromide to SV40 DNA I and II. The upper part of the diagram presents three stages in the reversible binding of dye to SV40 DNA I. (*a*) Represents the dye-free molecule with 14 superhelical turns, $\tau_0 = -14$. The addition of 420 molecules of ethidium bromide completely unwinds the superhelical turns to form the relaxed molecule, (*b*). The addition of a further 720 dye molecules, which occurs at a free dye concentration of 100 μg/ml, leads to the formation of a positive superhelical molecule, $\tau = 24$, shown in (*c*). The lower part of the diagram shows three stages in the reversible binding of dye to SV40 DNA II, which remains relaxed throughout. (*e*) Represents the relaxed, nicked molecule with the same number of dye molecules bound as to the relaxed, intact molecule in (*b*). The arrows joining (*b*) and (*e*) indicate that the nick may be introduced or repaired without change of the 420 molecules of ethidium bromide bound. The nicked molecule is nearly saturated, as shown in (*f*), with 1860 molecules of ethidium bromide bound. Introduction of a single-strand scission into I in the presence of a high dye concentration results in an irreversible unwinding of the superhelix accompanied by an approximately 65 percent increase of the amount of dye bound. In this diagram each block represents 30 molecules of dye. All the dye molecules are not shown in (*c*). The number of dye molecules in (*b*), (*e*), and (*f*) represent experimental values determined in this study. After W. Bauer and J. Vinograd, *J. Mol. Biol.*, **33:** 141 (1968).

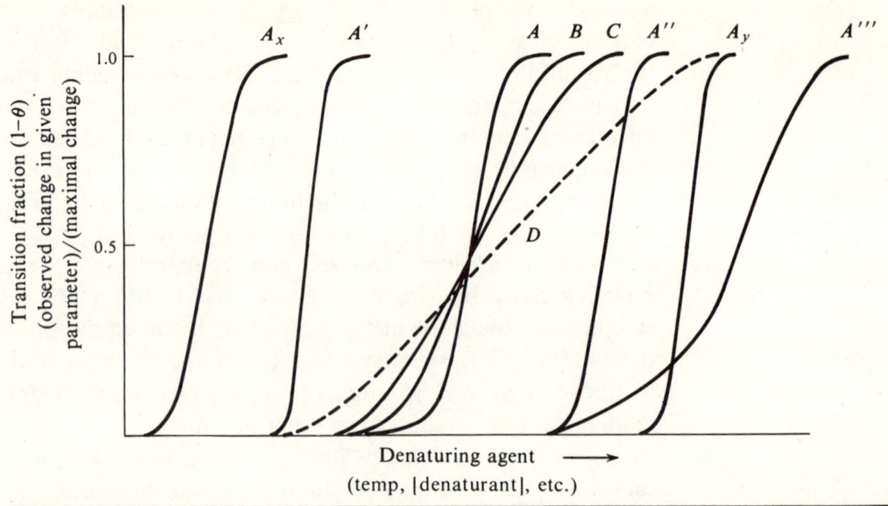

FIGURE 5-19
Schematic transition profiles for denaturation. Curve A: DNA (a percent G + C) from a bacterial virus; B: DNA (a percent G + C) from a bacterium; C: DNA (a percent G + C) from animal or plant; A': DNA (b percent G + C, $b \ll a$) from a bacterial virus, $M \simeq$ that of DNA in A; A'': DNA (c percent G + C, $c > a$) from a bacterial virus, $M \simeq$ that of DNA in A or A'; A''': the DNA of A'' is in the form of a superhelix; A_x: same DNA as in A but in the presence of additional denaturing conditions or agents; A_y: same DNA as in A but in the presence of a stabilizing agent; D: single-stranded DNA.

percent or half of the molecules denatured 100 percent). That means for the transition DNA (native) $\rightleftharpoons$ DNA (denatured), with an equilibrium constant

$$K_{\text{tr}} = \frac{[\text{denatured}]}{[\text{native}]} \equiv \frac{\text{fraction of base pairs in disordered regions}}{\text{fraction in helical regions}} \tag{5-32}$$

this is the point at which $\Delta G_{\text{tr}} = 0$ and $\Delta H_{\text{tr}} = T_{1/2}\Delta S_{\text{tr}}$. Since ΔH_{tr} can be determined from the dependence of K_{tr} on temperature by means of the van't Hoff equation (Chapter 2), all the thermodynamic parameters can be calculated. If temperature is the independent variable in bringing about the transition, the latter becomes strictly analogous to the melting of a one-dimensional crystallite, and the temperature at the transition midpoint $t_{1/2}$ (on the absolute scale $T_{1/2}$) becomes equivalent to a melting point (t_m or T_m). ΔH_{tr} has also been determined directly by calorimetry and in 0.1 M Na^+ equals ~8 kcal/mol base pair for all double-helical polynucleotides examined; the corresponding entropy changes are of the order of 15 to 25 cal/deg/mol base pair.

3. The magnitude of T_m shows a direct and, in most solvents, linear dependence (97) on (G + C) content. In neutral solution with Na^+ [37] as the counterion the relation obeyed for DNA's with mol fraction (G + C) in the range 30 to 70 percent is

$$(\text{G} + \text{C}) = 2.44(t_m - 81.5 - 16.6 \log [\text{Na}^+]) \tag{5-33}$$

In standard saline citrate (0.20 M Na^+ containing 0.015 M citrate pH 7.5) this relation becomes

[37] Di- and polyvalent cations are much more effective because, as already discussed, they are capable of *stoichiometric* rather than ion atmosphere charge neutralization.

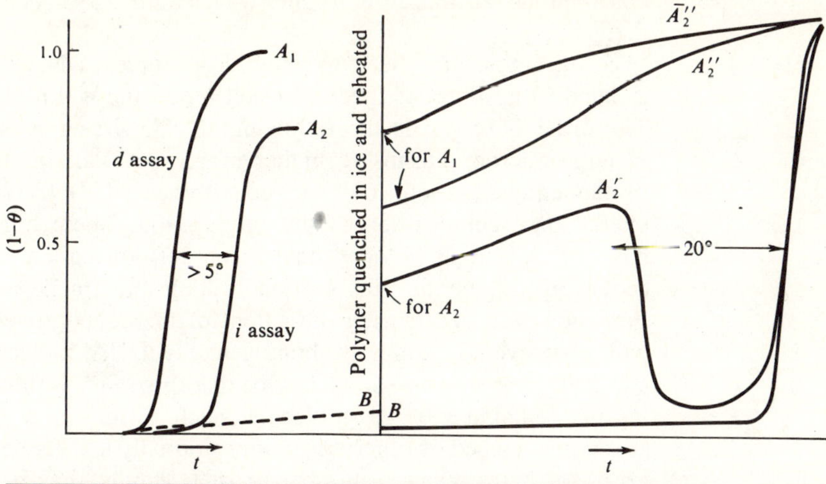

FIGURE 5-20
Reversible and irreversible denaturation (schematic). Note that ionic strength at the right is greater than at the left. Symbols similar to those in Fig. 5-19.

$$(\mathrm{G}+\mathrm{C}) = 2.44(t_m - 69.3) \tag{5-33a}$$

If it is desired to study denaturation of high-melting DNA's without, however, lowering the ionic strength, 7.2 M $NaClO_4$ is a useful solvent (12, 98) inducing the transition (midpoint $= t_m'$) at a considerably lower temperature than does SSC (midpoint $= t_m$):

$$(\mathrm{G}+\mathrm{C}) = 1.78(t_m' - 19.0) \tag{5-33b}$$

t_m and t_m' are related by

$$t_m = 53.9 + 0.732t_m' \tag{5-33c}$$

t_m is relatively insensitive to the molecular weight provided the polynucleotide used is sufficiently long so that internal residues greatly outweigh those at the ends—which may contain frayed, single-stranded regions. Cross-linked or covalently circular DNA is protected against denaturation and shows a biphasic response (96) (curve A‴): first the twists are released at a low temperature, followed by a relatively noncooperative melting at higher temperatures.

4. The transition width, usually abbreviated as σ, provides a measure of cooperativeness in the transition and depends principally on the following parameters[38] (Fig. 5-19):

a. Configuration and conformation. Single-stranded polynucleotides show much less cooperativity (large σ) than do double-helical polynucleotides. In general, cooperativity requires hydrogen bonding. The effect of covalent cross-linking has already been mentioned.

b. Intermolecular heterogeneity. It is apparent from 2 that if two different homo- or heteropolymers, or even two DNA's varying sufficiently in base composition, become more and more similar, the resultant sigmoids will overlap to an increasing extent until visual resolution into two separate curves is no longer possible. [This resolution may however, be improved by making

[38] Various spans have been employed for σ, depending on the assumed integral distribution function; the two most commonly used are $\sigma_{2/3}$, the width corresponding to 0.67 of the total transition, and $\sigma_{1/2}$, the width for 0.50 of the total, both centered about the transition midpoint.

observations at more than one wavelength (59) or by plotting the first derivative.]

c. Intramolecular heterogeneity. For duplex DNA's of the same M but isolated from different species or cell types, the width of the transition varies. The order observed is the following: DNA viruses < small microorganisms < large microorganisms < higher organisms. Part of this is indubitably due to the equivalent of (b), the coexistence of DNA fragments of varying overall base composition in the preparation. These different fragments may have preexisted *in vivo* or have been introduced during the isolation procedure by fragmenting the extremely long and fragile native DNA molecule. Part, however, is intrinsic; the extent of the cooperativeness of the transition will vary even in different homogeneous DNA molecules, and certain intrinsically less stable regions may be disrupted first without immediately affecting the remainder of the molecule. Such regions may be "frayed" ends, or small mismatched or otherwise compositionally less stable (i.e., rich in A + T) arrays of bases. Once a number of such "faults" has been introduced on increasing the amount of the disruptive force, the remainder of the molecular structure may collapse in the characteristic highly cooperative, "all-or-none" manner over an exceedingly narrow range (99). This latter may be as little as $\leqslant 1°$ for highly homogeneous polymers of large M.

d. Other factors affecting conformation or configuration, such as covalent joining, an increase in ionic strength, and the presence of divalent cations capable of nonspecific or specific and stoichiometric interactions, can affect the transition width. In general these effects are more pronounced the more flexible the polymer, and the most compact and rigid conformations always exhibit the most cooperativeness in the transition and hence the smallest transition width.

5. Although a number of experimentally observable parameters can be used to follow the transition and these various parameters frequently give qualitatively the same result (i.e., a relatively narrow sigmoid), their exact shape and midpoints need not, and usually do not, coincide. It is clear why this should be so if we compare hydrodynamic and optical measurements, for changes in dimensions and shape certainly need not be directly reflected by, nor reflect, changes in symmetry $[\theta]$, base–base interactions (h), proton vibrations (ir), or proton precessions (nmr). But even changes in the last four parameters do not necessarily coincide; dissymmetry may be lost before most of the stacking is affected, and there may be essentially complete stacking even in the absence of extensive hydrogen bonding as in single-stranded polymers such as $(A)_n$ and $(C)_n$ or single-stranded DNA (100).

6. Transition profiles have usually been determined in one of two ways (Fig. 5-20): at the ambient conditions (temperature, pH, etc.) or after rapid return to some standard condition of temperature, pH, etc. (usually a temperature around 20°, a pH around 6 to 7, an ionic strength of 0.1 to 0.2, etc.). The former, curve A_1, called *d assay* by Geiduschek (101), measures the extent of the transition actually observed (i.e., the position of the equilibrium) at the moment of measurement; the latter, curve A_2, termed *i assay,* the extent of rapid thermodynamic reversibility of the process attainable under standard conditions. If the new value is identical with the one originally obtained under standard conditions, then the change has been prima facie reversible and the phenomenon is a true denaturation–renaturation; if not, then some irreversible process must have supervened. The nature and the potential reversibility of these processes is discussed below.

Thermodynamics of Transitions in DNA and Other Complementary Duplexes with Random Monomers. Although a variety of studies of this problem have been performed, no completely satisfactory treatment has yet been proposed that would serve to analyze the contribution of the relevant, frequently potentiating, and overlapping contributions to helix stability and maintenance. A number of these and various approaches thereto have been suggested: charge distribution (monopole), or London dispersion dipole–dipole and dipole–induced dipole interactions between vertical and horizontal (hydrogen-bonded) neighbors; electrostatic interactions between phosphate residues and the surroundings; contributions by the solvent, probably largely entropic, leading to the formation of solvophobic interactions within the polymer; conformational alterations in the nucleosides, and restriction of their degrees of freedom in going from the more random (coil) to the constrained helical conformation. Three sets of conclusions appear to be relatively firm. One concerns the importance of base stacking interactions as contributors to helix stability. Single-stranded homopoly- and oligomeric molecules, including ones completely incapable of hydrogen bonding, do form stable, stacked, probably helical conformations in solution. But these do not undergo cooperative transitions. Hence hydrogen bonding is required for recognition and maintenance of proper base pairs [those predicted by Chargaff's rules and the Watson–Crick pairing scheme in neutral solution; in acid solution $(A)_n$ and $(C)_n$ form strong homopolymer duplexes by virtue of an additional proton per chain or a shared proton, respectively, (Fig. 5-17)], as well as for cooperativeness.[39] There is evidence that ribopolymers are more stable than are analogous ones in the deoxy series, while the 2′OH group as such is probably not a contributory factor. Second, the helix-coil transitions in a large, duplex DNA proceeds in two steps: one is predominantly enthalpic, corresponds to the reversible (d) phase described earlier and leads to virtually complete disruption of order (see also Fig. 5-21). The second, occurring at a higher temperature over an exceedingly narrow temperature range, is slow, predominantly entropic, and corresponds to the irreversible (i) phase; it involves the untwisting of the two strands and constitutes "melting" proper. The experimentally determined transition width (σ) is thus a reflection of the homogeneity of different portions of the molecule or the preparation with respect to the first, enthalpic part (102). The microscopic stability of these regions may vary greatly and depends on the actual constellation of base pairs present. Third, interactions with the solvent and conformational factors, also probably largely entropic, make important, but quantitatively hard to assess, contributions to helix stability.

Native and Denatured Conformations. A number of different conformations have been identified in denaturation studies. On the one extreme we have, of course, the completely ordered perfectly double-helical and complementary conformation II A in Fig. 5-21. At the other extreme we find an essentially completely disordered conformation: two separated, completely extended random coils with no extensive structure or order remaining (conformation I D). This corresponds to the completely melted (76), "irreversibly" de-

[39] Quantum mechanical calculations (Pullman and Pullman) indicate that *in vacuo* the vertical (stacking) and horizontal (H-bond) interactions make almost equal contributions to the total interaction (stabilization) energy and that the latter varies for sets of nearest neighbors in the order G—C next to a homologous pair (~ −30 kcal/mol base pair) > G—C next to A—T (~ −20 kcal/mol base pair) > A—T next to a homologous pair (~ −kcal/mol base pair).

(A·T)
(G·C)
(A·T)
(1) (2) (3) (4) (5) (6) (7)

(*A*)

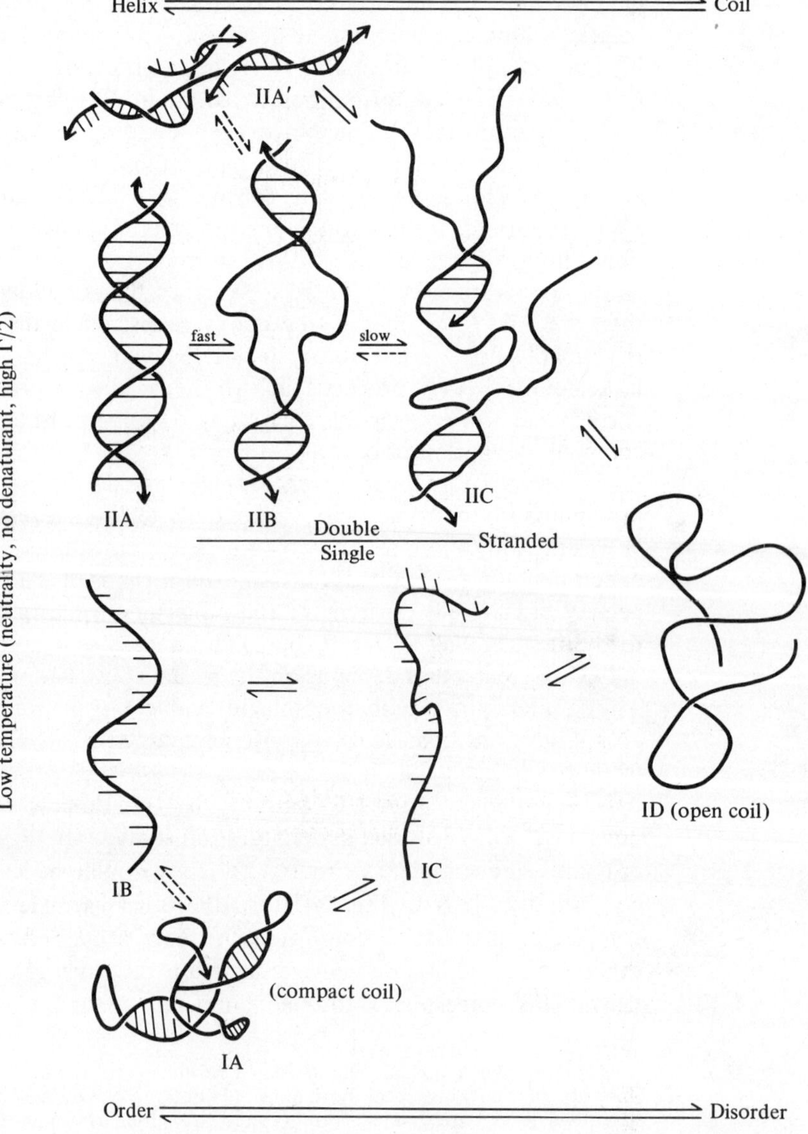

High temperature (extremes of *p*H, denaturant present, low Γ/2)

(*B*)

natured form obtained at high temperatures or extremes of pH, and so on, where not only the d assay, but also the i assay, has indicated the completion of the transition (curve A_2 in Fig. 5-20). On removal of the denaturing agent and return to standard condition, a certain amount of order is restored. Possible conformations are indicated as I A, I B, or II A′ in Fig. 5-21. The first two correspond to the extent of order restored to single-stranded polynucleotides (either homogeneous DNA or RNA species—that is, consider only strand A present). The denaturation profiles of such polymers, which are identical to those observed when an "irreversibly" denatured, melted, double-stranded polynucleotide is subject to a second cycle of denaturation, is given by curve $\bar{A}_2''$ or A_2'' of Fig. 5-20 ($\bar{A}_2''$ for minimal and A_2'' for relatively extensive complementariness inherent in the single strand). Conformations similar to II B, that is, extensively disoriented regions still held together by very short complementary ones in register and identical to those present originally in the native form (G · C-rich "nuclei" or "sticky regions"), probably occur as intermediates in the course of most denaturation processes. Their existence explains the differences observed between "reversible" and "irreversible" processes (that is, denaturation versus melting as measured by the d versus i assay described above). For covalently cross-linked or circular DNA, some regions always remain in register and denaturation is reversible as long as the links remain intact (curve B in Fig. 5-20). This process is very rapid even at low temperatures and is referred to as "snapback." Finally, forms intermediate between II A and II B and containing both extensively ordered as well as disordered regions capable of instantaneous reversion to the completely ordered starting conformation have been identified in denaturation brought about by heat, high pH, or certain organic solvents as well as by H/T exchange with the solvent, even at low temperatures.

Irreversible Denaturation (*Melting Out*). The separation and recombination of complementary strands in double-stranded polynucleotides have been studied extensively in recent years. In the case of DNA it was first observed by Marmur, Doty, and Schildkraut and has since been under active investigation in a large number of laboratories (32, 57). Appropriate conditions bring about the complete separation of the two complementary strands; this process of strand disentanglement may be quite slow (of the order of minutes) and require relatively extreme conditions of temperature, pH, and so on, well beyond the midpoint of the reversible transition (Fig. 5-20). In order to bring

FIGURE 5-21

(A) Details of the rapid phase in thermal transitions of duplex helices [II A and II B of part (b)]. Schematic drawing showing the proposed mechanism of DNA denaturation. State (1) is identical to II A of part (b), an intact double helix containing regions locally rich in A · T and G · C pairs, as indicated. Nucleation of melting occurs in the A · T-rich regions, producing at equilibrium a distribution over states like (2) and (3). A further temperature perturbation is made when the molecule is in state (3). The instantaneous response ($<$10 ms) is melting of a few base pairs at the end of helical sections, with the twist taken up in adjacent coil regions as in (4). Unwinding of the molecule follows, (4) $\rightleftharpoons$ (5), with melting of further base pairs at the ends of the helical sections. This reaction is proposed as a fast effect, limited by viscous resistance to unwinding which varies with the 2.3 power of M. At the final temperature some helical regions will be near their critical transition temperature and will undergo a slow transition from state (5) through the intermediate (6) to (7). After H.-Ch. Spatz and D. M. Crothers, *J. Mol. Biol.*, **42:** 191 (1969).
(B) Helix-coil transitions in polynucleotides.

about the re-formation of an ordered duplex from its separated strands [*re-zippering* (79)] several conditions must obtain:

1. The reaction must be thermodynamically favored; that is, it must be allowed to take place in a region (15 to 30°) below t_m at a relatively high ionic strength.

2. It must be favored kinetically as well. Strand recombination is second order and, therefore, is favored by high concentrations of the nucleic acid. The strands must be able to approach closely; this again is favored by high ionic strength and/or the presence of specific counterions such as Mg^{2+} as well as an absence of charges in the bases (neutrality).

3. For complete re-formation of order all base pairs must be perfectly matched. This is easily accomplished in the case of homopolymers [i.e., the formation of $(dG)_n \cdot (dC)_n$, $(rU)_n \cdot (rA)_n$ or duplex $d(A\text{-}T)_n$]. But with a natural polynucleotide the probability is much greater that extensive mismatching will occur and only short, imperfectly matched regions will be present originally (structures similar to II C in Fig. 5-21). For extensive reordering to take place, these imperfect regions must be ruptured; this will be reflected in a high free energy of activation, restrict the permissible temperatures to those just below the t_m, and require a relatively long period of time. The process just described has aptly been termed *annealing*. Its occurrence can be followed by a number of techniques, including the determination of ρ_B by centrifugation in CsCl (Fig. 5-13), chromatographic techniques, or absorbancy–temperature profiles (curve A_2' in Fig. 5-20).

4. Annealing and restoration of order are, therefore, favored by conditions that permit homologous matching segments to be sorted out and combined. This means that for chains in free solution, molecular homogeneity is the prime consideration, and renaturation is most easily accomplished with DNA from small viruses, somewhat less easily with DNA from large viruses, still less easily with that from bacteria, and least easily, if at all, with that from higher organisms. It also means that the process is favored if one of the strands can be immobilized, say in an agar gel, on hydroxylapatite columns, or on nitrocellulose membranes and the other is broken down to segments of relatively short length ($M \simeq 10^4$ to 10^5).

5. The kinetics of this reassociation process has been studied quantitatively (103–107). The second-order rate constants in 1.0 M [Na^+] at ($t_m - 25°$), the optimum, is given by (103)

$$k_2 = 3 \times 10^5 L^{0.5}/\bar{N} \text{ M}^{-1} \text{ sec}^{-1} \tag{5-34}$$

Here $\bar{N}$ is the *complexity* of the DNA, equal to the total number of base pairs in nonrepeating sequences within a given DNA *genome* (virus, bacterium, or eucaryotic nucleus). In the case of viruses or bacteria (animal mitochondria?) repeating sequences are virtually absent and $\bar{N}$ equals the total number of base pairs; in the case of a biosynthetic duplex such as $(dA)_n \cdot (dT)_n$ or $d(A\text{-}T)_n$, $\bar{N} = 1$. L equals the average number of bases per single strand of the denatured DNA used (i.e., it is proportional to molecular weight). In practice strands are sheared (i.e., cut at random) prior to the start of the experiment; then $\bar{N} \gg L$ and k_2^{obs}, the observed rate, ≈ 0.67 of the rate that would be observed with perfectly matched strands.

The energy and entropy of activation for renaturation are ~6 to 7 kcal and –40 eu/mol base pair, respectively. Its mechanism involves the initial formation of small nuclei (consisting of G + C-rich complements between the strands) followed by a rapid first-order zippering up. The rate constant k_2 in Eq. (5-34) is inversely proportional to solvent viscosity. From our earlier statements we recognize that the forward and reverse processes involve the identical path.

6. Britten and Kohne (106) have investigated equivalent kinetic parameters governing renaturation in genomes varying in size from 4000 (MS 2) to 3.2×10^9 (calf thymus or bull sperm) nucleotide pairs. In the latter case, as in higher organisms in general, the genome appears to consist of three fractions: ~70 percent of the sequences appear virtually unique; ~20 percent of the sequences are reiterated, at least over a part of the total, some 1000- to 10,000-fold (e.g., those for ribosomal RNA); and ~10 percent of the sequences correspond to a small region (~300 nucleotides, identified first as the so-called satellite DNA of mouse) that is reiterated with considerable fidelity some 10^6 times!

7. Association to form complementary duplexes is not restricted to DNA from the same organism; by performing analogous *cross-hybridization* experiments the extent of relatedness of different genomes can be established. It is not even restricted to both strands being contributed by DNA (105). Since all cellular RNA appears to be synthesized on DNA templates, one can establish the relatedness between any given RNA and the corresponding DNA regions of appropriate genomes. Identification and isolation of such hybrid molecules are greatly aided by their characteristic properties, which include insensitivity to pancreatic and T1 RNase, a banding position in CsCl intermediate between that of DNA and RNA, and retention by cellulose nitrate filters under conditions where RNA is passed through.

Similar experiments can also be performed to study the relative stabilities of polydeoxyribonucleotides versus polyribonucleotides and their complexes. In general the order observed is ribo · ribo > deoxy · deoxy ≧ deoxy · ribo with the possible hybrid isomers exhibiting different stabilities.

The "minimal stable length" required for duplex formation (105), that is, the number of adjacent base pairs that must be formed with perfect fidelity, not permitting even a single mismatch, and which may also correspond to the recognition length within the nuclei for renaturation, may be as low as 12 for tracts rich in G + C; for all except those virtually pure A + T it is ≦ 20.

Finally, the formation of hybrid DNA's, for example as described in Fig. 5-13, where a hybrid duplex is formed from heavy ^{15}ND and light ^{14}NH duplexes, provides conclusive evidence of the whole sequence of denaturation, melting, and zippering up described in these paragraphs.

POLYRIBONUCLEOTIDES

Localization and Function

Ribonucleic acid (RNA) is found in all major cell fractions. Quantitatively the most important contribution [amounting to 80 percent of the total in rat liver (107)] is made up of the RNA (rRNA) of ribonucleoprotein particles called *ribosomes*. Whatever their source these contain ~60 percent RNA by weight and consist of a large subunit (~50 S; with a RNA chain of $M \geqq 1 \times 10^6$ daltons, 23 to 30 S, plus one containing 120 nucleotides, 5 S)[40] and a small subunit (~30 S; with RNA of $M \geqq 5 \times 10^5$ daltons, 16 to 18 S). Ribosomes occur free in the cytosol, particularly in microorganisms, as well as attached to membranous components of the cell such as the plasma membrane (in bacteria) or the endoplasmic reticulum (of eucaryotic cells) (see

[40] In the ribosomes derived from the cytosol of animal cells an additional RNA species of sedimentation coefficient ~7 S appears to be associated with the large subunit.

Chapter 9). Usually ribosomes are found in close association with one another in the form of linear or rosette-shaped aggregates called polyribosomes or *polysomes*. Ribonucleoprotein particles are also frequently found in intimate association with the chromosomal or nuclear apparatus. In cells containing nucleoli such particles are found in that organelle as well; they are now known to constitute the precursors of the cytoplasmic ribosomes. DNA-associated RNA is at least in part messenger RNA (mRNA), in part rRNA (in bacteria), and in part (in eucaryotes) a species that is synthesized and degraded rapidly but never leaves the nucleus (108). In rat liver these various forms of nuclear RNA account for 4.3 percent of the total. RNA is also found in the cytosol (i.e., that fraction that cannot be sedimented after 4 hr $\times$ 105,000 *g* in 0.25 M sucrose, and cannot be precipitated by 1 M NaCl). This fraction accounts for 15 percent of the total and consists mainly of amino acid transfer RNA (tRNA). Finally, there are small amounts of RNA associated with the mitochondria of animal and plant tissues and the chloroplasts of plants. Functionally all these RNA's appear to be involved in one phase or another of protein synthesis, which is discussed in Chapter 21. The ribosomes provide the machinery by means of which the amino acids are polymerized to polypeptide chains. The template for this process is provided by messenger RNA synthesized on and complementary to one of the two strands of the appropriate DNA segment, eventually transferred and attached to ribosomes. The amino acids, in turn, are brought to and arranged on this template by means of adapters provided by different aminoacyl tRNA's.

Another class of RNA is provided by a group of viruses (page 253). Four different types may be distinguished: small, spherical plant and bacterial viruses (such as the *E. coli* viruses f2 or MS2, and turnip yellow virus) yielding RNA molecules of $M = 1$ to 2×10^6; small rodlike viruses such as tobacco mosaic virus with an RNA of approximately the same size; and large animal viruses such as Newcastle disease containing larger RNA molecules. These three groups contain single-stranded RNA. The fourth contains double-stranded RNA; reoviruses of animals and wound tumor viruses of plants are examples.

Base Composition

The base composition of a variety of different RNA's falling into the various classes just described is summarized in Table 5-7. In general it will be seen that the striking complementariness found in DNA and described by Chargaff's rules is absent even in the viral RNA's, the only regularity remaining being the fact that $G + U \simeq A + C$. Messenger RNA's, especially if relatively large segments of the DNA have been transcribed, are frequently more DNA-like in their overall base composition. Yet, extensive second-order structure and helicity in the molecule prevents its proper function. As pointed out by Sueoka, most ribosomal RNA's from microorganisms contain $\sim$52 mol percent G + C and most soluble RNA's $\sim$58 mol percent G + C, regardless of the base composition of the DNA from the same cell. Methylated nucleosides (both *N*-Me and 2′-O-Me derivatives) as well as ψ are found in the 16 S and 23 S rRNA (but not in 5 S RNA) and in tRNA. The latter also contain a profusion of other unusual nucleosides such as hU (5,6-dihydrouridine), T, ψ, I, iA, ^{4}S, etc. (page 192); some 35 have been identified so far.

Structure

Phosphodiester Backbone. The backbone structure of all the polyribonucleotides so far examined is analogous to that of DNA; that is, it is one in which the monomeric nucleotide units are arranged in a linear fashion linked to one another by a 3′,5′-phosphodiester bond. Branching, although possible in principle, is not observed. From their mode of biosynthesis (Chapter 20) RNA's would be expected to exhibit the generalized structure $pppN_1 \cdots N_n$ that is, a triphosphate residue at the 5′ terminus and a free hydroxyl group at the 3′ terminus. This pattern is actually observed in the case of RNA synthesized *in vitro* and of viral RNA synthesized *in vivo*. In most other isolated RNA's the 5′-terminal triphosphate has been hydrolyzed to a 5′-p residue.

Base Sequence. The first sequence of a nucleic acid to be analyzed was that of an alanine transfer RNA ($tRNA^{Ala}$) from yeast, a brilliant tour de force, performed by R. Holley (109), which required a number of years for its completion, was announced in 1965, and recognized by the award of the Nobel Prize three years later. By then the structure of at least eight additional tRNA's had already been determined, and this number had risen to a score (110) by 1969. Some of these structures are shown in Fig. 5-22 and Table 5-8. The task of sequencing an RNA has been greatly simplified by several

TABLE 5-7
Base composition of various RNA's

Species [DNA composition in brackets]	*Type*	*A*	*G*	*C*	*U*
	Eucaryotes				
Rat liver [42% G + C] mammal	Nuclear (total)	20.2	25.7	29.5	24.6
	Nuclear "messenger"[a]	26.0	24.3	23.8	25.9
	Nucleolar	16.3	33.0	29.9	20.7
	Total ribosomal	17.3	33.8	29.3	20.5
	18 S Ribosomal	20.1	31.2	28.1	22.5
	28 S Ribosomal	15.5	36.3	30.3	17.9
	5 S Ribosomal[b]	18.2	33.9	25.7	22.2
	Cytoplasmic "messenger"[a]	32.2	23.0	22.4	22.4
	Cytoplasmic "messenger"[c]	29.6	24.6	20.0	25.6
	Hemoglobin messenger[d]	21.2	20.7	34.5	23.7
	Mitochondrial	17.8	31.8	28.4	20.9
	Bulk tRNA[e]	20.9	30.9	28.5	24.1[f]
Duck red cells, immature[c] [43.5% G + C] bird	Nuclear "messenger"	25.3	23.1	21.8	29.8
	Nuclear rRNA precursor	19.9	29.9	27.7	22.5
	18 S Ribosomal	22.7	28.8	26.5	22.6
	28 S Ribosomal	17.7	33.2	31.2	17.9
	Cytoplasmic "messenger"	25.6	26.3	27.4	20.7
Xenopus laevis [64% G + C] (clawed toad), amphibian	Nuclear rRNA precursor[g,h]	21.3	36.6	27.2	15.0
	18 S Ribosomal (av)	24.1	29.0	24.1	22.9
	28 S Ribosomal (av)	19.7	35.0	27.9	17.4
Drosophila [39% G + C] (fruitfly), insect	16 S Ribosomal	28.8	23.5	20.3	27.4
	28 S Ribosomal	30.8	22.5	19.6	27.1
Arbacia [40% G + C] (sea urchin), echinoderm	17 S Ribosomal	23.1	29.9	23.8	23.2
	28 S Ribosomal	19.9	34.9	26.0	19.3
Neurospora crassa [54% G + C] (bread mold) fungus [mito ~40% G + C]	17 S Ribosomal (cytoplasmic)	25.2	28.9	19.9	26.0
	26 S Ribosomal (cytoplasmic)	26.5	28.9	20.6	24.0
	16 S Ribosomal (mitochondria)	31.8	20.4	16.0	31.7
	23 S Ribosomal (mitochondria)	33.9	19.1	15.0	31.9
Saccharomyces cerevisiae[i] [~40% G + C] (baker's yeast) [mito ~20% G + C]	17 S Ribosomal (cytoplasmic)	26.6	26.1	19.1	28.1
	26 S Ribosomal (cytoplasmic)	26.4	28.4	19.2	26.0
	15 S Ribosomal (mitochondria)	38.4	16.1	11.0	34.5
	22 S Ribosomal (mitochondria)	40.3	14.0	11.0	34.6

TABLE 5-7 (Continued)

Species [DNA composition in brackets]	*Type*	*A*	*G*	*C*	*U*
	Bacteria, Rickettsiae, and Viruses[j]				
Pseudomonas aeruginosa	16 S Ribosomal	25.1	32.8	21.6	20.5
[67% G + C]	23 S Ribosomal	26.3	31.2	21.2	21.3
Eschericha coli [50% G + C]	16 S Ribosomal	24.8	31.0	22.7	21.5
	23 S Ribosomal	25.4	33.5	21.5	19.6
	Cellular "messenger"	24.1	27.7	24.7	23.5
[T2 or T4 DNA = 40% G + C]	Bacteriophage "messenger"	~30	~20	~20	~30
Bacillus subtilis	16 S Ribosomal	26.5	29.6	22.3	21.6
	23 S Ribosomal	26.5	32.0	22.5	19.3
Psittacosis organism	16 S Ribosomal	26.4	32.2	17.1	24.3
(*Meningopneumonitis* organism)	23 S Ribosomal	28.3	30.6	16.7	24.4
Reovirus (animal virus, double-stranded RNA)	Total viral ($M \simeq 14 \times 10^6$ daltons)	37.7	17.5	16.9	27.9
	Components L + M + S[k] ($M = 2.3, 1.3, 0.8 \times 10^6$ daltons)	27.5	21.9	21.9	28.7
	Component A (low M)[k]	81.3	2.8	3.2	12.7
Wound tumor (plant virus, double-stranded RNA)	Viral ($M = 10 \times 10^6$ daltons)	31.1	18.6	19.1	31.3
Turnip yellow mosaic (spherical plant virus)	Viral ($M = 1.7 \times 10^6$ daltons)	22.6	17.2	38.1	22.1
Tobacco necrosis (spherical plant virus)	Viral ($M = 1.5 \times 10^6$ daltons)	27.9	24.4	22.0	25.0
Tobacco necrosis satellite (growth dependent on tobacco necrosis)	Defective viral ($M = 4 \times 10^5$ daltons)	28.0	24.0	22.1	24.9
Tobacco mosaic (rod-shaped plant virus)	Viral ($M = 2.2 \times 10^6$ daltons)	29.8	25.3	18.5	26.3
Qβ	Viral ($M = 0.9 \times 10^6$ daltons)	22.3	23.7	22.3	29.4
f2	Viral ($M = 1.1 \times 10^6$ daltons)	22.1	26.8	25.9	25.1
MS2, R17 (spherical bacterial viruses)	Viral ($M = 1.1 \times 10^6$ daltons)	23.0	26.7	24.9	25.5
Poliomyelitis	Viral ($M \simeq 2 \times 10^6$ daltons)	29.0	24.0	27.0	25.0
Encephalomyocarditis	Viral ($M \simeq 2 \times 10^6$ daltons)	27.0	24.0	23.0	26.0
Influenza (animal viruses, containing single-stranded RNA)	Viral ($M = 2 \times 10^6$ daltons)	23.1	20.1	24.0	32.8

[a] A. Hadjivassiliou, and G. Brawerman, *Biochemistry,* **6:** 1934 (1967).

[b] For HeLa cells; E. Knight, Jr., and J. E. Darnell, *J. Mol. Biol.,* **28:** 491 (1967).

[c] G. Attardi, H. Parnas, M.-I. H. Hwang, and B. Attardi, *J. Mol. Biol.,* **20:** 145 (1966).

[d] From rabbit reticulocytes (DNA base comp. identical to rat); D. G. Laycock, and J. A. Hunt, *Nature,* **221:** 1118 (1969).

[e] Besides ψ, also contains 5-Me-C; 6-Me-, 1-Me-G; 2-Me-, and 2,2-Me_2-6-hydroxypurine.

[f] Includes 3.8%ψ.

[g] From the properties of a distinct DNA species, located in the nucleolar organizer region of the chromosome and isolated as a separate satellite component in CsCl gradients.

[h] M. Birnstiel, J. Speirs, I. Purdom, K. Jones, and U. E. Loening, *Nature,* **219:** 454 (1968).

[i] M. Fauman, M. Rabinowitz, and G. S. Getz, *Biochim. Biophys. Acta,* **182:** 355 (1969).

[j] R. A. Cox, *Quart. Rev.,* **22:** 499 (1968); F. Amaldi, *Nature,* **221:** 95 (1969); A. R. Bellamy, L. Shapiro, J. T. August, and W. K. Joklik, *J. Mol. Biol.,* **29:** 1 (1967).

[k] The high-molecular-weight components contribute ~80%, component A, ~20% of the total RNA of the virion.

related advances (111–114): (1) the development of reliable chromatographic [e.g., benzoylated DEAE-cellulose, DEAE-Sephadex] and electrophoretic (acrylamide/agarose gel) techniques for the separation of polyribonucleotides of any desired size class; (2) the development of a two-dimensional technique (using electrophoresis on cellulose acetate at *p*H 3.5 in one direction and combined electrophoresis plus paper or thin-layer chromatography on DEAE in the other) for the separation and identification of nuclease digests of ^{32}P-labeled RNA, together with a "library" of R_F values defining the composition of the various spots obtained; and (3) the realization that hydrolysis by pancreatic and T1 ribonuclease, when performed first to a limited extent, and then to completion, could yield all the information, including the requisite overlaps for the structure determination. This methodology, suitable modified, has now been extended to ever-longer RNA molecules. First the sequence of 5 S ribosomal RNA was completely elucidated; today so much progress has already been reported for other rRNA's and viral RNA's that by the time this volume is published complete success may have been achieved also in these endeavors (115).

Higher-Order Structure and Conformation. Except for RNA viruses, which exhibit a double-stranded helical structure (Table 5-10) either in the isolated particle (*virion*) or as an obligatory intermediate in the course of replication (page 908), all intracellular RNA's appear to be single stranded. They nevertheless exhibit a high degree of secondary structure, which, especially at low concentration, must be predominantly due to interactions within one and the same chain. These are believed to be of two kinds:

1. Hydrogen bonds between appropriate bases arranged in a "hairpin loop." This conformation will lead to cooperative transitions analogous to those seen in double-helical structures, such as DNA with, however, several provisos: (a) the bases are tilted relative to the helix axis, similar to the A rather than the B conformation in DNA (Table 5-4); (b) since in general the loops and particularly the hydrogen-bonded regions will be rather short, end effects will be of much greater importance and the structure will be characterized by much less conformational permanence than in DNA; particularly since (c) occasional mismatched pairs probably occur within a segment; and (d) base pairing other than that observed in DNA (e.g., the Hoogsteen pairs, Fig. 5-19, and particularly G · U pairing) is possible in the absence of the severe steric constraints operating in DNA fibers.

2. Single-stranded helices with bases stacked on top of one another; these will show strong electronic coupling and considerable stability but without cooperativity, so the whole array can essentially be regarded as a sum of successive dinucleotide pairs. The extent to which these two ordered plus the completely disordered conformation contribute to the structure of any isolated RNA in solution has not yet been unambiguously determined, not even in those instances, such as tRNA and 5 S RNA, where the primary structure is known with certainty. That these molecules assume definite conformations that are important, and perhaps determinant, in their biological function is attested to by the fact that they can undergo changes from inactive forms when annealed in the presence of divalent metal ions in low concentration (116) and by the occurrence of *temperature-sensitive mutations* (114) (page 859) in tRNA. These mutations lead to tRNA molecules that differ from the parental type in a single base and as a consequence are active at a low (*permissive*), but inactive as a higher (*restrictive*), temperature, which, however, is perfectly acceptable in the the case of the parental (i.e., unchanged) component.

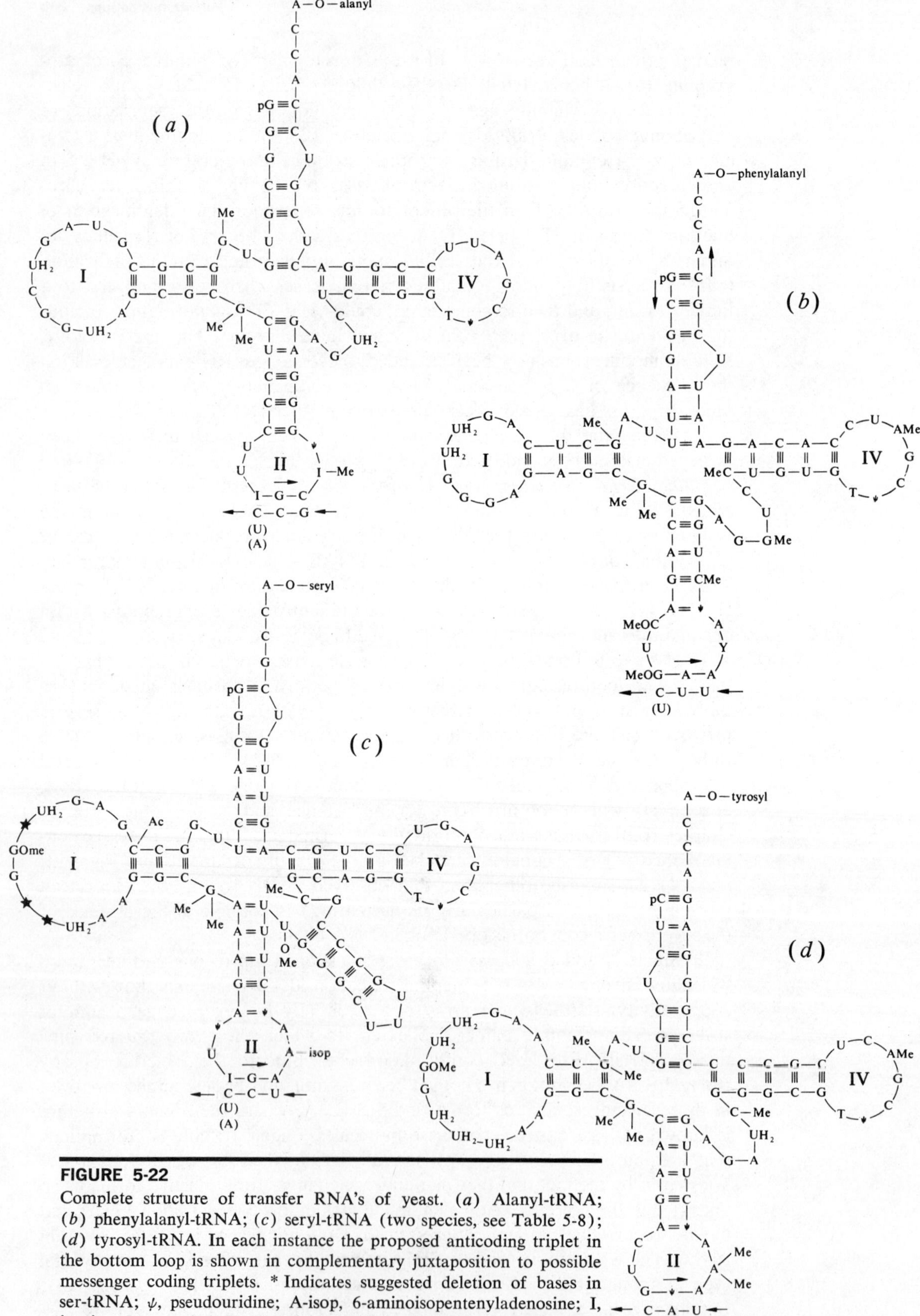

FIGURE 5-22

Complete structure of transfer RNA's of yeast. (*a*) Alanyl-tRNA; (*b*) phenylalanyl-tRNA; (*c*) seryl-tRNA (two species, see Table 5-8); (*d*) tyrosyl-tRNA. In each instance the proposed anticoding triplet in the bottom loop is shown in complementary juxtaposition to possible messenger coding triplets. * Indicates suggested deletion of bases in ser-tRNA; ψ, pseudouridine; A-isop, 6-aminoisopentenyladenosine; I, inosine; Me, methyl. T. H. Jukes, *Biochem. Biophys. Res. Commun.* **24:** 744 (1966); J. T. Madison, *Ann. Rev. Biochem.*, **37:** 131 (1968).

TABLE 5-8
Homologous alignment of base sequences in yeast alanine, serine, tyrosine, and phenylalanine transfer RNA's[a]

	Ala	*Ser*	*Tyr*	*Phe*
	pG	pG	pC	pG
	G	G	U	C
	G	C	C	G
	C	A	U	G
H_1	G	A	C	A(G)
	U	C	G	U(G)
	G	U	G	U(A)
	*U	U	U	U
a				
	GMe	G	A	A
10	*G	G	GMe	GMe
	*C	C	C	C
H_2				
	G	CAc	C	U
	C	G	A	C
	G	—	—	—
	U	—	—	—
	*A	A	A	A
	*G	G	G	G
I	*UH_2	UH_2	UH_2	UH_2
	C	—	UH_2	UH_2
20	*G	GOMe	GOMe	G
	*G	G	G	G
	—	—	UH_2	—
	—	—	UH_2	—
	UH_2	UH_2	UH_2	G
	*A	A	A	A
	G	A	A	G
	C	G	G	A
H_2'	*G	G	G	G
	*C	C	C	C
b *30*	*GMe_2	GMe_2	GMe_2	GMe_2
	C	A	C	C
	U	A	A	C
H_3	C	A	A	A
	C	G	G	G
	C	A	A	A
	U	ψ	C	COMe
	*U	U	U	U
	I	I	G	GOMe
II	G	G	ψ	A
40	C	A	A	A
	IMe	Aisop	AMe_2	Y(Y′)
	ψ	A	A	A
	G	ψ	ψ	ψ
	G	C	C	CMe(C)
H_3'	G	U	U	U
	A	U	U	G
	G	U	G	G(A)
	A	UOMe	A	A
	*G	G	G	G
50	—	G	A	GMe
	—	G	—	—
	—	C	—	—
	—	U	—	—
III	—	U(C)	—	—
	—	U	—	—
	—	G	—	—
	—	C	—	—
	—	C	—	—
	—	C	—	—
60	UH_2	G	UH_2	U(UH_2)
	*C	CMe	CMe	C
	U	G	G	CMe(G)
	C	C	G	U(C)
H_4	C	A	G	G
	G	G	C	U
	*G	G	G	G
	*T	T	T	T
	*ψ	ψ	ψ	ψ
	*C	C	C	C
IV *70*	*G	G(A)	G	G
	*A	A	AMe	AMe
	U	G(A)	C	U
	U	U	U	C
	*C	C	C	C
	C	C	G	A
H_4'	G	U	C	C
	G	G	C	A(G)
	A	C	C	G(C)
	C	A	C	A(U)
80	U	G	C	A(C)
	C	U	G	U(A)
H_1'	G	U	G	U(C)
	U	G	G	C
	C	U	A	G
	C	C	G	C
d *86*	A	G	A	A
	C			
	C			
	A-OH			

[a] H_1, H_1', first helical region; H_2, H_2', second helical region; etc. a, b, connecting regions; I, II, III, IV, loops: I, H_2U loop; II: anticodon loop; IV: TψCG loop (see also Fig. 21-2). * Denotes common residues. Base substitutions in the corresponding tRNA from wheat germ are in parentheses.

SOURCES: T. H. Jukes, with permission; J. T. Madison, *Ann. Rev. Biochem.*, **37:** 131 (1968); B. S. Dudock, G. Katz, E. K. Taylor, and R. W. Holley, *Proc. Natl. Acad. Sci. U.S.* **62:** 941 (1969).

In their paper (109) on the structure of $tRNA^{Ala}$, Holley et al. also suggested several possible secondary structures, all of them maximizing base pairing, that this molecule might assume. Of these, one in particular, the cloverleaf structure shown in Fig. 5-22, has found favor by workers in the field (110). Although many of the properties of the molecules in solution are consistent with the models, it is by no means certain that alternative ones have been rigorously excluded, and assignment of a unique structure must await a great deal of additional work, including X-ray analysis of crystalline forms now underway in a number of laboratories (115). One of the more surprising recent findings is that in spite of the individuality and differences in primary structure among the tRNA's from a single organism (e.g., yeast), their three-dimensional structures must be very similar, because many (if not the majority) can be accommodated within the *same* crystal lattice (117). Tertiary structure (i.e., hyperfolding of the hydrogen-bonded molecule) has also been demonstrated, for instance by showing that residues 8 (sU) and 13 (C) in a valine tRNA of *E. coli* can be photochemically linked and that this can be used to construct a three-dimensional model that corresponds quite closely to the experimental small-angle X-ray scattering curve (118).

Reactivity. The greater accessibility of the bases in single-stranded RNA makes them more accessible to a number of reagents; conversely, absence of reactivity of a particular base provides good evidence for conformational screening, although not *necessarily* by hydrogen bonding. This is particularly true of the reaction of the amino bases with formaldehyde and HNO_2. Other reagents that have been frequently employed are HONO, NH_2OH at *p*H 6, and particularly $MeONH_2$ (methoxylamine) for cytosine, NH_2OH at *p*H 10 for uracil, and kethoxal (β-ethoxy-α-oxobutyraldehyde) for the ^1N–^2C–NH_2 grouping in guanine (119).

Hydrodynamic Properties. Single-stranded polyribonucleotides appear to exhibit a greater tendency for self-association, owing to base stacking, than do the deoxy analogs. Hence, hydrodynamically these polymers do not collapse to random coils at temperature $\leq 20°$ in neutral solution of moderate ionic strength. This is evident from the values of K_s, a_s, and K_η, a_η for Eq. (5-17) that have been reported. Thus, for poly rU these are 3.29×10^{-2}, 0.42, and 8.98×10^{-5}, 0.75; for poly rA 2.1×10^{-2}, 0.45, and 6.2×10^{-4}, 0.65, respectively. Only in 1.1 M HCHO–0.2 M phosphate do the molecules assume a conformation that is reminiscent of the behavior of single-stranded DNA with $K_s = 0.05$ and $a_s = 0.40$. Incidentally, all RNA's appear to obey the modified Scheraga–Mandelkern equation [Eq. (5-16)] but with a value of 14,300 for the constant term. In sucrose gradients Eq. (5-18) is obeyed with a value of 0.46 for the coefficient.

Ribosomal RNA. Ribosomal RNA's appear to exhibit variations in their sedimentation coefficient and molecular size depending on their origin. In addition, close attention has to be paid to the methods of isolation, the ionic strength, and particularly the presence of divalent metal ions such as Mg^{2+}, since sedimentation coefficients are sensitive to this parameter and the metal at high concentration can lead to aggregation artifacts. The commonly accepted $s^0{}_{20,w}$ values, under roughly equivalent conditions, are for the two large ribosomal RNA's: from bacteria, 16 and 22.5 S ($M = 5.5 \times 10^5$ daltons and 1.07×10^6 daltons); from protozoa (*Paramecium*), 18 and 25.5 S (6.9 and 1.25×10^6 daltons); from cytosol of higher animals,[41] 18 and 30 S (7.0 ×

[41] These ribosomes contain in addition to the common 5 S, an additional small RNA component with $s^0 = 7$ S. A component of similar size has also been reported in nuclei (123).

10^5 and 1.8×10^6); from unicellular-plant (yeast, *Neurospora,* etc.) cytosol, 17 and 26 S (7.2×10^5 and 1.5×10^6 daltons); from their mitochondria 16 and 23 S (6.6×10^5 and 1.3×10^6 daltons); from higher-plant cytosol, 17 and 26 S (7.0×10^5 and 1.4×10^6 daltons); and from higher-plant chloroplasts, 16 and 22 S (5.5×10^5 and 1×10^6 daltons).

Viral RNA's (115, 124–127). The small bacteriophage RNA's appear to present a particularly favorable target for experiments designed to establish complete nucleotide sequences and correlate them with the information encoded therein as expressed in the amino acid sequence of the (three) proteins specified (see Chapter 19 and 21). For the viruses of the MS2 to R17 class, a possible partial structure consistent with the available information is shown in Fig. 5-23. The ribosomal binding and initiation sites [the three loops containing AUG, the initiator codon (Chapters 19 and 21)] for the three proteins have been isolated and sequenced by Steitz, while large tracts corresponding to the center and to the C-terminal end of the coat (*capsid*) protein, as well as the region between the latter and the one specifying the next protein have been determined by other workers in Sanger's laboratory. Analogous investigations in the unrelated but equally small virus Qβ has led to the assignment of the complete primary sequence for some 150 nucleotides from the 5′ terminus (starting with $\text{pppG(G)}_{2\text{ or }3}\text{GAC}_4\text{CU}_3\text{AG}_4\text{UC–}$), for some 20 nucleotides ending in the 3′-OH terminus ($\text{–GC}_3\text{UC}_3\text{UCUCUC}_3\text{A–OH}$), and the region around the initiator for the coat protein ($\text{–PyA}_2\text{U}_3\text{GAUC}$*AUG*-GCAAAAAUUGAGAC–).

Synthetic Polyribonucleotides

Just as in the case of polydeoxyribonucleotides, a great deal of information concerning structure and function of RNA's has been obtained by studying polymers of defined sequence, obtained by the action of various enzymes on appropriate substrates *in vitro.* As a matter of fact, chronologically such studies on homopolymeric ribonucleotides preceded corresponding studies in the deoxy series, because the enzyme polynucleotide phosphorylase first isolated and studied by Grunberg-Manago, Ortiz, and Ochoa in 1956 is capable of synthesizing *any* homoribopolymer (128) under appropriate conditions (see page 909) by the reaction

$$n\,\text{rNDP} \xrightarrow{\text{Me}^{2+}} (\text{rN})_n + \text{r}\,\text{P}_i \qquad \text{N} = \text{A, U, G, C, T, }\psi\text{, I, B, etc.} \tag{5-34}$$

Homo- and heteropolymers of defined sequence can also be synthesized by action of the DNA-dependent RNA polymerase using single-stranded polydeoxy or, occasionally, polyribonucleotides of known sequence (cf. the previous section) (115, 129–131):

$$\begin{aligned}(\text{d or rN, N}', \text{N}'', \text{N}''')_n + n(\text{r}\overline{\text{N}}\text{TP} + \text{r}\overline{\text{N}'}\text{TP} + \text{r}\overline{\text{N}''}\text{TP} + \text{r}\overline{\text{N}'''}\text{TP}) &\xrightarrow{\text{Me}^{2+}} \\ (\text{d or rN, N}', \text{N}'', \text{N}''')_n \cdot (\text{r}\overline{\text{N}}, \overline{\text{N}'}, \overline{\text{N}''}, \overline{\text{N}'''})_n + 4n\,\text{PP}_i\end{aligned} \tag{5-35}$$

Here the r$\overline{\text{N}}$TP's represent the ribonucleoside triphosphates complementary in the Watson–Crick sense (i.e., A · T, U, ψ, etc.; C · G, I, MeG; G · C, MeC, HO-MeC) to the N's in the template. The hybrid duplexes formed can be dissociated by heat or alkali and separated in CsCl gradients or chromatographic columns.

Homopolymers. Poly rA, poly rC, and perhaps poly rU form single-stranded helical structures of somewhat greater stability than the deoxy analogs at neutral *p*H in solutions of moderate ionic strength. The order of helix stability is rA > rC >> rU; poly rU exhibits evidence for structure only at low temperature. Base–base interactions generate considerable hypo-

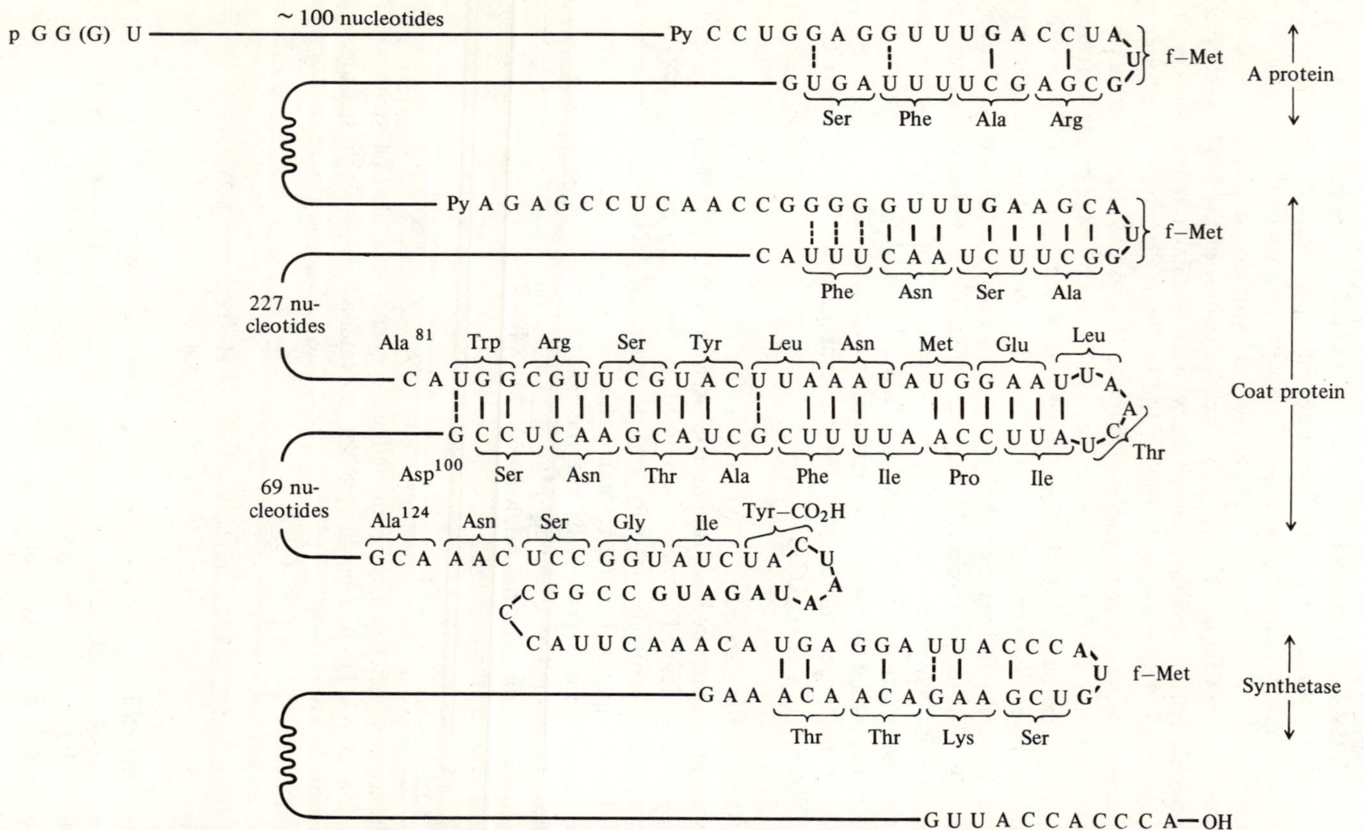

FIGURE 5-23

Known primary sequences in R17 RNA. Terminator and initiator codons (Chapters 19 and 21) are shown in heavy type. Correspondence to the three nascent N-terminal sequences as well as to others [for the coat protein in which the complete sequence has been determined—K. Weber, *Biochemistry*, **6:** 3144 (1967); K. Weber and W. Konigsberg, *J. Biol. Chem.*, **242:** 3563 (1967)] are also shown. Secondary structures suggested are stabilized by Watson–Crick and GU base pairs.

chromism and rotational strength (Table 5-5) but are noncooperative, melt out over a wide range of temperatures, and are reasonably independent of ionic strength. The free energy of base stacking has been estimated as 1 kcal/mol bases for $(rA)_n$. Hydrodynamically, the structure is close to that of a random coil and the molecules are to be regarded as ones in which fluctuating regions of stacked, helically arranged, probably tilted, bases (average number in a stack ~8 at 5°) alternate with others that are randomly oriented.

When the pH is lowered, poly rA and poly rC assume rigid double-helical conformations with interbase hydrogen bonds as indicated in Fig. 5-17. Crystalline fibers can be obtained from these molecules and X-ray analysis discloses the structure parameters summarized in Table 5-4, that is, parallel strands with base planes tilted relative to the helix axes. Poly rG and poly rI exhibit strong interstrand interactions even at neutral pH, but only in the latter instance has it been possible to obtain evidence for a defined structure, containing in all likelihood three strands within the helix.

Homopolymer Duplexes. A great deal of effort has been devoted to studies of a wide variety of such helical duplexes consisting of every conceivable combination of naturally occurring bases and their analogs in both the ribo and deoxyribo series. Structural parameters for some of the better characterized ones are given in Table 5-4. The formation of such duplexes or complexes containing more than two strands can be monitored and their stoichiometry determined most readily by application of the method of continuous variations. Here we plot the absolute magnitude of some parameter characteristic of either component of a binary system (a variety of hydrodynamic, optical, or other measurements have actually been employed) against its mol fraction X_A or X_B (Fig. 5-24). If there is no interaction, then strict additivity is observed (e.g., Beer's law is followed in the case of absorbancy measurements), and a straight line connecting the two points, which correspond to the magnitude of the parameter in question for the pure components $X_A = 1.0$ and $X_B = 1 - X_A = 1.0$, is obtained. If there is a stoichiometric interaction between them, however, then the plot consists of two or more straight-line segments that intersect at values of X_A corresponding to its mol fraction in the complex. Similar methods can also be used in the study of complex formation between polynucleotides and oligonucleotides of varying size.

The alternating heteropolymer duplexes $(\text{N-}\bar{\text{N}})_n \cdot (\bar{\text{N}}\text{-N})_n$ are generally less stable than the isomeric homopolymer pairs $\text{N}_n \cdot \bar{\text{N}}_n$ in the deoxy series, of equal or greater stability in the ribo series; T-containing polymers are always more stable than ones containing U.

Disproportionation Reactions. The formation of an rA · rU or a dA · dT base pair in an oligo- or polymer liberates ~ 10 kcal/mol at room temperature and low ionic strength where the duplex is the only stable species. As the temperature and/or the ionic strength is raised the triple-stranded species $[(rA)_n \cdot 2(rU)_n]$ becomes more stable relative to $(rA)_n$ plus the duplex. Hence we may observe the disproportionation equilibrium

$$2\,\text{N}_n \cdot \bar{\text{N}}_n \rightleftharpoons \text{N}_n \cdot (\bar{\text{N}}_n)_2 + \text{N}_n \tag{5-36}$$

Although the duplexes in the ribo series are always more stable than the corresponding ones in the deoxy series, the stability of mixed (hybrid) duplexes depends on the base. Thus, we find the order of stability $rA_n \cdot dT_n <$ $dA_n \cdot dT_n < rA_n \cdot rT_n < rC_n \cdot dI_n < dC_n \cdot dI_n < dC_n \cdot rI_n < rC_n \cdot rI_n$ and

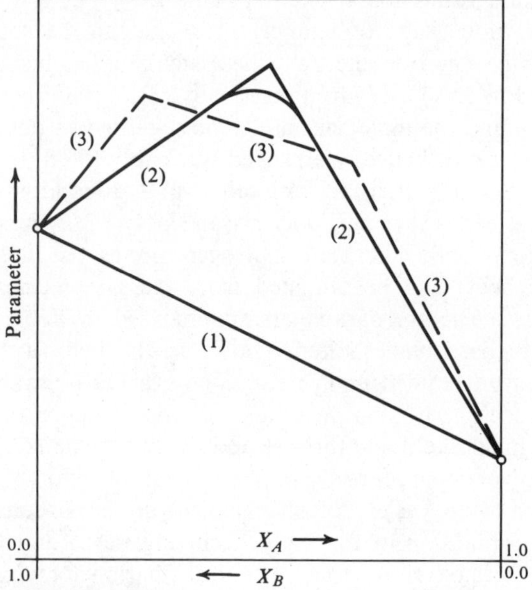

FIGURE 5-24
Method of continuous variations applied to the interaction of polynucleotides. Curve (1): hypothetical curve observed in complete absence of interactions between *A* and *B*. Curve (2): solid line—hypothetical curve observed for strong interaction forming a stoichiometric complex *A*:*B*, dashed line—hypothetical curve for formation of complex *A*:*B* with measurable dissociation at the equivalence point. Curve (3): hypothetical curve obtained for strong interactions forming *two* stoichiometric complexes *A*:*B*:*A* (or A_2:*B*) and *B*:*A*:*B* (or *A*:B_2).

probably $dG_n \cdot dC_n < rG_n \cdot dC_n \leqq dG_n \cdot rC_n < rG_n \cdot rC_n$. Therefore there exists the possibility of another type of dismutation reaction,

$$d\bar{N}_n \cdot rN_n + rN_n \cdot d\bar{N}_n \rightleftharpoons rN_n \cdot r\bar{N}_n + dN_n \cdot d\bar{N}_n \tag{5-37}$$

NUCLEOPROTEINS

Proteins Associated with Chromosomal DNA (Chromatin) (132, 133)

Histones. These constitute a family of heterogeneous basic (positively charged at neutral *p*H) proteins rich in lysine and/or arginine, of relatively low molecular weight. They appear to be strongly, but noncovalently, and stoichiometrically complexed to the DNA of the somatic cells of all multicellular organisms, animals as well as plants.[42] They were first discovered by Miescher and later studied by Kossel during the second half of the last century. The histones may be dissociated from DNA by means of acid, plus alcohol or high ionic strength, and are then fractionated by chromatography on a cation exchanger (either on CM-cellulose or Amberlite CG50 in the presence of guanidinium hydrochloride). In this manner four main fractions (I through IV in Murray's nomenclature, see Table 5-9) are obtained which differ in their relative content of lysine and arginine, as shown (133–135). The lysine-rich fractions I and II may then be further subdivided, and in this manner it could be shown (134) that the most lysine-rich fraction of calf thymus is comprised of four homogeneous polypeptides, each with $M = 21{,}000$, which appear to exhibit considerable overlap, although not an identity, of primary structure.

Of special interest is histone IV (also called histone F2A-1), which has

[42] There are recurrent reports of the occurrence of histone-like proteins in association with the DNA of microorganisms as well.

TABLE 5-9
Amino acid composition of chromatin associated proteins of calf thymus

	Histones[a]							*Acidic proteins*			
Fraction	*Ia*	*Ib*	*IIa*	*IIb_1*	*IIb_2*	*III*	*IV*	*Total*[b]	*R-PNP*[c]	*$(NH_4)_2SO_4$ ppt*[c]	*Mitotic protein*[d]
Ala	24.0	24.2	11.2	12.1	10.4	11.7	12.2	7.65	7.67	8.18	7.5
Arg	3.0	2.6	11.6	8.8	6.7	12.8	12.7	5.60	5.76	4.58	5.1
Asp	2.5	2.5	4.8	6.1	4.9	4.4	4.5	9.48	8.63	9.76	9.6
Cys	0	0	0	0	0	0	0	1.29	1.51	0.85	1.5
Glu	4.5	4.3	9.6	9.2	8.2	9.8	10.5	12.39	12.4	15.2	14.1
Gly	6.7	7.3	9.8	10.0	6.0	8.7	7.8	7.54	7.15	8.25	7.9
His	0.4	0.2	1.9	2.8	2.4	1.6	1.6	1.51	2.31	2.11	2.8
Ile	1.3	1.2	5.4	4.6	5.0	5.4	5.4	4.31	5.25	4.73	4.5
Leu	5.3	5.0	8.7	2.3	4.4	8.6	8.9	9.91	9.89	7.70	7.9
Lys	25.3	26.2	9.6	11.5	15.9	10.0[e]	9.7[e]	6.25	7.38	7.70	6.1
Met	0	0	1.2	0.2	1.6	1.2	1.2	1.94	2.18	1.85	2.5
Phe	0.6	0.6	2.3	1.2	1.6	2.5	2.7	3.88	4.10	4.23	3.5
Pro	8.6	9.1	3.7	4.5	4.5	3.8	4.2	4.96	4.86	4.82	5.2
Ser	6.4	6.5	4.6	5.6	10.0	4.1	4.1	7.54	5.78	5.22	5.3
Thr	5.8	5.4	6.8	4.5	6.2	7.3	7.3	5.50	5.02	5.25	6.2
Trp	0	0	0	0		0	0	2.05			0.8
Tyr	0.7	0.7	2.8	3.3	4.4	2.4	2.3	2.59	3.20	2.52	3.0
Val	4.9	4.0	6.5	6.4	7.2	5.8	5.6	5.60	6.92	7.08	6.3

[a] K. Murray, Histone Nomenclature, in *The Nucleohistones,* J. Bonner and P. Ts'o (eds.), Holden-Day, San Francisco, 1964.

[b] K. Marushige, D. Brutlag, and J. Bonner, *Biochemistry,* **7:** 3149 (1968).

[c] T. Y. Wang, *J. Biol. Chem.,* **242:** 1220 (1967).

[d] Highly purified protein (two subunits of $M = 57{,}000$) isolated from central core of sperm tail of sea urchin, known to be very similar in composition and properties to protein(s) of the mitotic apparatus (and of the protein of other microtubules); R. C. Weisenberg, G. G. Borisy, and E. W. Taylor, *Biochemistry,* **7:** 4466 (1968).

[e] Contains ~1 mol % ε-*N*-methyllysine.

been obtained in homogeneous form from calf thymus and pea buds (136–137). Both molecules are proteins of $M = 11{,}280$ containing 102 residues with only two amino acid replacements, Arg ↔ Lys and Ile ↔ Val. This conservation of sequence, which is unique in the by-now extensive catalog of primary structure homologies (and differences) between species (here in two different kingdoms!), suggests an overriding requirement for some essential and precisely defined function of the molecule. It, together with the relative paucity of molecular species among the histones of any one cell, also implies that this function probably cannot be an exquisitely specific one in the regulation of the expression of *individual* genes or gene clusters, as had at one time been suggested. The molecule has other interesting features as well. Most of the basic residues appear to be concentrated in the N-terminal half, most of the hydrophobic and especially the aromatic ones in the C-terminal half. Hence for nucleohistone formation protein–DNA interactions appear concentrated in the former half, protein–protein interactions in the latter half. The sequence -Lys^{16}-Arg-His-Arg-Lys^{20}- constitutes the largest single cluster of basic residues. Both of the Lys residues are susceptible to modification: Lys^{16} carries an ε-*N*-methyl group in all chains whether derived from calf or pea; Lys^{20} carries an ε-*N*-acetyl group in half the chains from calf only. This acetylation is of metabolic, and possible regulatory, significance because rapid incorporation of acetyl groups into the histones of a number of cell types is observed *in vivo* and histone IV appears to be the principal target (138).

Acidic Proteins. Not all the proteins associated with the chromosomal apparatus are basic and thus subsumed under the histone classification. Approximately one third of the proteinaceous material requires even higher ionic strength than do histones for dissociation from DNA and is constituted by a class of moderately acidic proteins, which in turn are susceptible to further fractionation (Table 5-9). As might be expected, these proteins interact strongly not only with DNA, but also with histones, free or complexed to DNA. Unlike histones, their association with DNA does not lead to a gross inhibition of the latter in functioning as a template in RNA synthesis (Chapter 20) and thus they might be capable to affect this process by more subtle and specific means, in a manner analogous to the bacterial repressor proteins, which are of similar composition and fulfill precisely this function (Chapter 21). In the table we also present the composition of another class of proteins, one member of which is known to fulfill an essential function in cell division—the protein component of the mitotic spindle; proteins of this group (the *microtubule proteins*) are related to the muscle protein actin; are capable of aggregating into long, linear, fibrous polymers; and are capable of binding GTP as well as the mitotic poison colchicine.

Protamines. In the mature sperm cells of some families of fishes, the DNA is associated not with histones, but with other basic proteins called *protamines.* These protamines are very basic proteins, containing a high proportion of arginine and in this respect bear some resemblance to the arginine-rich histones. They are, however, proteins of smaller size, less heterogeneous in size and composition, and appear to be species specific. The replacement of histones by protamines in the course of spermatogenesis represents an interesting chapter in the chemical aspects of subcellular differentiation.

Viruses

Viruses constitute an exceedingly interesting example of homogeneous nucleoprotein complexes with a defined biological function. They are capable of vegetative existence and replication only inside the cells of their hosts (which, depending on virus species, may be bacteria, plants, or animals), fashioning and controlling the host's metabolism to their own purposes. In their ability to reproduce their own kind, to multiply, to be capable of mutation, and to exchange genetic material provided by more than one parent, they exhibit the most salient characteristics of living organisms. Yet they can be isolated in pure, frequently crystalline forms, stored, examined, analyzed, and manipulated and, in short, treated as distinct chemical entities. Thus, although themselves forming part of living nature, they have also provided investigators with more easily comprehensible models or analogs for life processes characteristic of more complex forms. Ever since W. M. Stanley's isolation of tobacco mosaic virus (TMV) in 1935 and the demonstration of its ribonucleoprotein nature in the very next year by Bowden and Pirie, they have been the favorite subject of investigation in molecular biology, and some of the most illustrious names in this young branch of science have been identified with the study of viruses. Among them we should mention Delbrück,* Luria,* Hershey,* Watson,* Fraenkel-Conrat, Benzer, Stent, Sinsheimer, Spiegelman, Levinthal, Thomas, and Zinder in the United States; Lwoff,* Jacob,* and Monod* in France; Schramm and Gierer in Germany; Crick* and Brenner in England; and many others. (* Denotes Nobel laureates.)

Some of the characteristics of isolated viruses are summarized in Table

5-10. Their genetic material is provided by either DNA or RNA. The fact that the protein, which is invariably found in all viruses [and may contribute anywhere from 33 to 99 percent of the total mass of the intact particle (virion)] only provides protection for the nucleic acid that it surrounds—and sometimes assists in the infection process—but is itself nonessential for purposes of viral replication, was clearly demonstrated by two classical experiments. In 1952, Hershey and Chase demonstrated by double labeling (^{35}S for protein, ^{32}P for DNA) that the vast majority of the protein of bacteriophage T2 never entered the bacterial cell and could be removed from the infected cell by shearing in a Waring Blendor, and an equally impressive amount of the nucleic acid had to penetrate into the interior in order to initiate the process of cellular infection and viral self-duplication. In 1956–1957, Schramm, and independently Fraenkel-Conrat and their collaborators, demonstrated that TMV RNA alone, in the complete absence of protein, was necessary and sufficient to produce infection, serve as the initiator of viral replication, and as genetic determinant. Since then similar experiments have been performed with the nucleic acids (DNA as well as RNA) of numerous viruses, including several of those shown in Table 5-10. The ability of some of these isolated nucleic acids to carry out the exchange of selected items of genetic information, known as *transfection,* will be discussed on page 853.

Let us briefly consider the different classes of viruses and their hosts. Bacterial viruses (bacteriophages) are frequently fairly large. Their nucleic acid, usually regular double-stranded DNA, is contained in a crystalline head of polyhedral symmetry that is made up of many small ($M \sim 2 \times 10^4$) protein subunits, most of them of one kind. Frequently they also have a more or less flexible tail that serves as an attachment and injection organ. Because this tail assembly may itself be made up of many and diverse functional and structural components, the number of different proteins present in any one virus particle may, in certain species (T-even coliphages) be quite large ($\geqq 20$) (Fig. 5-25). These larger phages contain double-stranded DNA which may exhibit a number of peculiarities. Although in most instances the chains are continuous, those of phage T5 appear to be interrupted (nicked) at specific, nonoverlapping points ══════ ═══ ═══ . Most of the DNA's examined appear to be terminantly repetitious (e.g., ABC . . . XYZABC, where ABC . . . are nucleotide sequences corresponding to specific genetic regions). The latter may, in turn, be either exposed as in phage λ,

```
ABC————————XYZ ---, or
   ————————————ABC
```

latent as in other phage DNA molecules,

```
ABC————————ABC
---        ---
ABC————————ABC
```

(where the bars represent complementary base sequences). In some phages, such as T2 or T4, the terminally repetitious region differs for different molecules; the collection of molecules in the population is circularly permuted, that is, we find individuals that carry ABC, others that carry BCD, still others with CDE, DEF, and so on all the way to XYZ, YZA, and ZAB:

```
ABCDEF————————————————————XYZABC
     DEFGHI————————————————————ABCDEF
          GHIJKL———————————————————— DEFGHI
```

The plant viruses all contain RNA and are of three main types. The first type contains double-stranded RNA (wound tumor viruses; see below); the

TABLE 5-10
Some properties of viruses[a]

Name of virus (indicating host)	*Nature of nucleic acid (number of chains)*	*Nucleic acid content*		*Particle weight × 10⁶*		*Approximate dimensions, Å*
		Percentage	*10^{-17}g/ particle*	*Of intact virus*	*Of nucleic acid*	
E. coli bacteriophage T2, T4, T6	DNA (2)	61	22	220	130	Head 65 × 95 Tail 20 × 95
E. coli bacteriophage λ	DNA (2)	50	5.4	50	33	Head 54 × 54 Tail 10 × 140
E. coli bacteriophage T7	DNA (2)	41	4.4	38	25	Head 47 × 47 Tail 10 × 15
E. coli bacteriophage fd	DNA (1)[c]	12	0.20	10.8	1.3	Rod 50 × 7000
E. coli bacteriophage ϕX174, S13	DNA (1)[c]	26	0.28	6.2	1.70	RSP,[b] $d \simeq 200$
E. coli bacteriophage MS2, R17	RNA (1)	32	0.17	3.6	1.05	RSP,[b] $d = 175$
Wound tumor (plants)	RNA (2)	14.3	1.15	70	10	RSP,[b] $d \sim 700$
Tobacco mosaic	RNA (1)	5–6	0.36	40	2.2	Rod 150 × 3000
Tomato bushy stunt	RNA (1)	15	0.27	10.6	1.65	RSP,[b] $d = 280$
Tomato ring spot	RNA (1)	20	0.112	1.5	0.66	RSP[b]
Tomato necrosis satellite	RNA (1)	20	0.066	1.97	0.40	RSP,[b] $d \sim 210$
Fowl pox	DNA (2)	10	33	2000	200	Brick 2300 × 2500
Herpes simplex (man, etc.)	DNA (2)		10		60	RSP,[b] $d \simeq 1000$
Adenovirus (man, etc.)	DNA (2)	5.0	1.7	200	10	RSP,[b] $d \simeq 700$
Polyoma (vertebrates)	DNA (2)[c]	13.4	0.57	21.1	3.1	RSP,[b] $d = 450$
SV40 (monkey kidney)	DNA (2)[c]		0.50		3.2	RSP,[b] $d \sim 400$
Reovirus (vertebrates)	RNA (2)		>1.6		>10	RSP,[b] $d = 750$
Newcastle disease (vertebrates)	RNA (1)	4	5.3		32	
Poliomyelitis (vertebrates)	RNA (1)	22–30	~0.8	6.7	~2.0	RSP,[b] $d = 300$
Equine encephalitis	RNA (1)	4.4	0.36	50	2.2	RSP,[b] $d \simeq 500$
Influenza, type A (man, etc.)	RNA (1)	~0.8	~0.4	280	~2.2	RSP,[b] $d \simeq 750$

[a] Klug and Caspar have shown that viruses of the RSP[b] type contain (10 T + 2) capsomeres (C) made up in toto of 60 T structure units (SU) arranged as 10(T − 1) hexamers + 12 pentamers. Thus, for ϕX, T = 1, and the capsid is made up of 60 SU in 12 pentameric clusters. For turnip yellow T = 3, and the capsid contains 180 SU in 20 hexameric plus 12 pentameric clusters; for polyoma T = 4, and the arrangement is 30 hexamers plus 12 pentamers. For the viruses in the wound tumor (plant)–reovirus (animal) group T = 9, and the capsid consists of 450 SU arranged in 80 hexameric plus 12 pentameric clusters. For the big animal DNA viruses in the Herpes–chickenpox group T = 16, and the capsids contain 150 hexamers plus 12 pentameric clusters. In the case of Herpes the composition of the total virus has been determined: protein, phospholipid, carbohydrate, and DNA are present in the proportions 70 : 22 : 16 : 6.5. For the DNA-containing animal viruses exemplified by the so-called adenoviruses and infectious canine hepatitis T = 25, and the total number of SU reaches 1500, arranged in a total of 252 capsomeres; for the largest virus discovered so far, the RNA-containing virus of insects called *Tipula iridescens* T = 81, the number of capsomeres equals 812, and the number of SU reaches the staggering total of 4860.

[b] RSP, roughly spherical, polyhedral particle with diameter indicated.

[c] Covalently circular molecule.

SOURCE: A. M. Rauth, *Biophys. J.*, **5:** 257 (1965); C. A. Thomas, Jr., and L. A. MacHattie, *Ann. Rev. Biochem.*, **36:** 485 (1967).

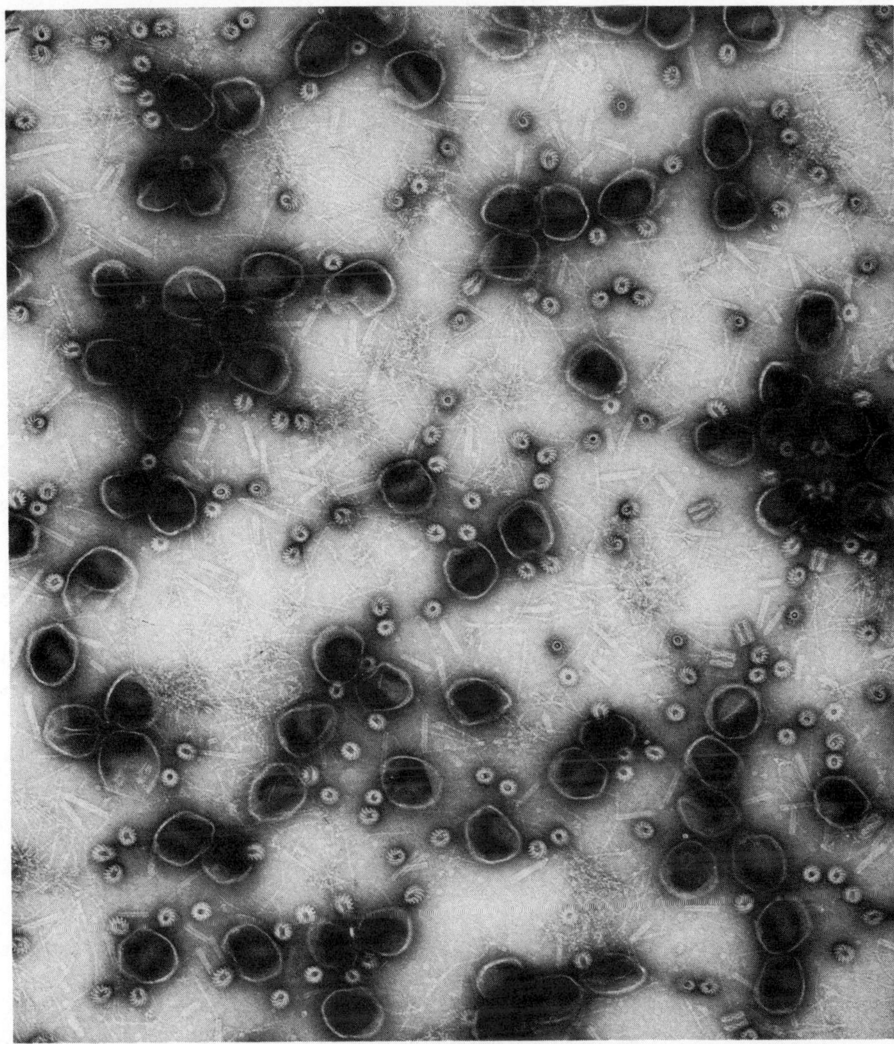

FIGURE 5-25

Structural components of bacteriophage T2. Ghosted T2 coliphage disrupted into its separate morphological parts by a 2-sec exposure to 0.1 N HCl followed by addition of 2 N Tris at *p*H 7.6. Head membranes, necks, cores, sheaths, and end plates with attached tail fibers can be seen as separate structures. This work was done by M. B. Baylor, V. B. Peters, and P. F. Roslansky. Courtesy of M. B. Baylor.

second type (exemplified by TMV) has a linear structure in which the capsid is made up of many (2100) identical structure units of $M = 18{,}000$ arranged in helical fashion around a central core, 40 Å in diameter, containing a fully extended RNA molecule. Thus, the RNA is encased in a sheath or tubing of protein producing a rodlike virus about 3000 Å long and 150 Å across. Incidentally, such rods, identical in all respects to the parent virus, have been reconstituted *in vitro* by mixing together isolated highly purified TMV RNA and homogeneous TMV protein (of which the complete primary structure is now known). The helical type of structural arrangement, as pointed out by Caspar and Klug, is, therefore, capable of formation by straightforward aggregation of the components, produces the maximal amount of surface area exposed and of nucleic acid–protein interaction, and probably necessitates complete disassembly prior to or in the course of the infection process.

The third type is also a stable and symmetrical arrangement. It is found in the icosahedral viruses and is exemplified by bushy stunt virus. The shell of this virus is made up of about 60 capsomeres containing ~200 structure units arranged in an isometric capsid, like one of the geodesic domes designed by Buckminster Fuller, but only 200 Å in diameter. This arrangement must result from a more complex process than simple aggregation, exposes the minimum of surface area to the environment, and probably requires that not all the nucleic acid can be in contact with the protein shell, which, however, need not be disassembled in order for the genetic material to be expelled.

Animal viruses may contain either RNA or DNA; in the former case they frequently reproduce in the cytoplasm, in the latter in the nuclei of the host cell. Of the small RNA viruses, those producing poliomyelitis and the Cocksackie (after the place of its discovery) viruses are simple polyhedral nucleoproteins, similar to the plant viruses described. Others, such as those responsible for various forms of encephalitis and for influenza, contain RNA of about the same small molecular weight but are themselves much larger though still polyhedral, thanks in part to the inclusion of other components, such as lipids or polysaccharides, respectively. One of the most intriguing groups of animal viruses is that of the reoviruses, which contain double-stranded helical, complementary RNA of high molecular weight, which, therefore, resembles ordinary DNA in many of its chemical, structural, and functional aspects (see also Table 5-10). At present the reoviruses are infectious agents in search of a disease, because so far no definite association between virus and any disease of man, monkeys, cattle, or dogs has been proved; yet these are all species from which similar viruses have been isolated and which form antibodies to the agents. As mentioned, in plants a group of viruses with very similar properties with regard to morphology, size (capsids, about 700 Å in diameter, containing 92 capsomeres), RNA content, and structure (Tables 5-10 and 5-7) are known that cause disease in a large number of species and produce tumors in some—the so-called wound tumor viruses. The virus is transmitted from plant to plant by an insect, the leaf hopper.

In addition to the RNA viruses, animal viruses are also known that contain DNA. Among them are those causing herpes, chickenpox, mumps, measles, fowl pox, smallpox (vaccinia), psittacosis, rabies, and polyoma—an exceedingly malignant tumor in rodents. The DNA of polyoma has a molecular weight of 3.0×10^6 and was the first double-helical DNA with a superhelical structure, as shown by chemical and physical techniques.

The smallest viral RNA isolated so far belongs to the tobacco necrosis satellite virus. The virus was given this name because it invariably occurs together with various strains of ordinary tobacco necrosis virus and is incapable of multiplying in plants in their absence. It constitutes, however, a separate and immunologically distinct entity. Its RNA has a $M = 4.0 \times 10^5$ (1200 nucleotide residues). The molecular weight of its capsid protein subunit equals 39,000 (372 residues). Thus, the RNA can code, at the most, for that one protein (page 870).

REFERENCES

Books

Bonner, J., and P. Ts'o (eds.): *The Nucleohistones,* Holden-Day, San Francisco, 1964.

Busch, H.: *Histones and Other Nuclear Proteins,* Academic Press, New York, 1965.

Cantoni, G. L., and D. R. Davies (eds.): *Procedures in Nucleic Acid Research,* vol. I, Harper & Row, New York, 1966.

Chargaff, E.: *Essays on Nucleic Acids,* Elsevier, New York, 1963.

Chargaff, E., and J. N. Davidson (eds.): *The Nucleic Acids,* vols. 1–3, Academic Press, New York, 1955, 1960.

Cold Spring Harbor Symposium on Quantitative Biology: *Replication of DNA in Micororganisms,* vol. 33, 1968.

Cold Spring Harbor Symposium on Quantitative Biology: *The Genetic Code.* vol. 31, 1966.

Cold Spring Harbor Symposium on Quantitative Biology: *The Mechanism of Protein Synthesis,* vol. 34, 1969.

Cold Spring Harbor Symposium on Quantitative Biology: *Synthesis and Structure of Macromolecules,* vol. 28, 1963.

Davidson, J. N.: *Biochemistry of the Nucleic Acids,* Wiley, New York, 5th ed., 1965.

Florkin, M., and E. H. Stotz (eds.): *Comprehensive Biochemistry,* vol. 8, part B (see Nucleic Acids), Elsevier, New York, 1963.

Fraenkel-Conrat, H.: *The Chemistry and Biology of Viruses,* Academic Press, New York, 1969.

Grossman, L., and K. Moldave (eds.): *Methods in Enzymology, Nucleic Acids,* vol. XIIA, 1967; vol. XIIB, 1968, Academic Press, New York.

Jordan, D. O.: *The Chemistry of Nucleic Acids,* Butterworths, London, 1960.

Martin, R. B.: *Introduction to Biophysical Chemistry,* McGraw-Hill, New York, 1964, chaps. 6, 8, 10, 11, 13, and 17–20.

Michelson, A. M.: *The Chemistry of Nucleosides and Nucleotides,* Academic Press, New York, 1963.

Perutz, M. F.: *Proteins and Nucleic Acids,* Elsevier, New York, 1962.

Shugar, D. (ed.): *Genetic Elements: Properties and Function,* Academic Press, New York, 1967.

Spirin, A. S.: *Macromolecular Structure of Ribonucleic Acids,* Van Nostrand Reinhold, New York, 1964.

Steiner, R. F., and R. F. Beers, Jr.: *Polynucleotides,* Elsevier, New York, 1961.

Tanford, C.: *Physical Chemistry of Macromolecules,* Wiley, New York, 1961, chaps. 5 and 6.

Taylor, J. H. (ed.): *Molecular Genetics,* part I, 1963; part II, 1967, Academic Press, New York.

Vogel, H. J., V. Bryson, and J. O. Lampen (eds.): *Informational Macromolecules,* Academic Press, New York, 1963.

Review Articles

Attardi, G., and F. Amaldi: Structure and Synthesis of Ribosomal RNA, *Ann. Rev. Biochem.,* **39:** 183 (1970).

Baldwin, R. L.: Molecular Aspects of the Gene: Replication Mechanisms, in I. C. Gunsalus and R. Y. Stanier (eds.), *The Bacteria,* vol. 5, Academic Press, New York, 1964, p. 327.

Borst, P., and A. M. Kroon: Mitochondrial DNA: Physicochemical Properties, Replication and Genetic Function, *Intern. Rev. Cytol.,* **26:** 107 (1969).

Burton, K.: Sequence Determination in Nucleic Acids, in P. N. Campbell and G. D. Greville (eds.), *Essays in Biochemistry,* vol. 1, Academic Press, New York, 1965, p. 57.

Chamberlin, M. J.: Comparative Properties of DNA, RNA and Hybrid Homopolymer Pairs, *Federation Proc.,* **24:** 1446 (1965).

Cox, R. A.: Macromolecular Structure and Properties of Ribonucleic Acid, *Quart. Rev. Chem. Soc.,* **22:** 499 (1968).

Davies, D. R.: X-Ray Diffraction Studies of Macromolecules, *Ann. Rev. Biochem.,* **36:** 321 (1967).

Felsenfeld, G., and H. T. Miles: The Physical and Chemical Properties of Nucleic Acids, *Ann. Rev. Biochem.,* **36:** 407 (1967).

Gilham, P. T.: RNA Sequence Analysis, *Ann. Rev. Biochem.,* **39:** 227 (1970).

Holley, R. W.: Experimental Approaches to the Determination of the Nucleotide Sequences of Large Oligonucleotides and Small Nucleic Acids, *Progr. Nucl. Acid Res. Mol. Biol.,* **8:** 37 (1969).

Josse, J., and J. Eigner.: Physical Properties of Deoxyribonucleic Acid, *Ann. Rev. Biochem.,* **35:** 789 (1966).

Khorana, H. G.: Synthesis in the Study of Nucleic Acids, *Biochem. J.,* **109:** 709 (1968).

Kochetkov, N. K., and E. I. Budowsky: The Chemical Modification of Nucleic Acids, *Progr. Nucl. Acid Res. Mol. Biol.,* **9:** 403 (1969).

Lehman, I. R., S. Linn, and C. C. Richardson: Nucleases as Reagents in the Study of Nucleic Acid Structure, *Federation Proc.,* **24:** 1466 (1965).

Madison, J. T.: Primary Structure of RNA, *Ann. Rev. Biochem.,* **37:** 131 (1968).

McCarthy, B. J., and R. B. Church: The Specificity of Molecular Hybridization Reactions, *Ann. Rev. Biochem.,* **39:** 131 (1970).

Michelson, A. M., J. Massoulié, and W. Guschlbauer: Synthetic Polynucleotides, *Progr. Nucl. Acid Res. Mol. Biol.,* **6:** 83 (1967).

RajBhandary, U. L., and A. Stuart: Nucleic Acids—Sequence Analysis, *Ann. Rev. Biochem.,* **35:** 759 (1966).

Sueoka, N.: Compositional Variation and Heterogeneity of Nucleic Acids and Proteins, in I. C. Gunsalus and R. Y. Stanier (eds.), *The Bacteria,* vol. 5, Academic Press, New York, 1964, p. 419.

Szybalski, W.: Effects of Elevated Temperatures on DNA and on Some Polynucleotides: Denaturation, Renaturation and Cleavage of Glycosidic and Phosphate Ester Bonds, in A. H. Rose (ed.), *Thermobiology,* Academic Press, New York, 1967, p. 73.

Thomas, C. A., Jr., and L. A. MacHattie: The Anatomy of Viral DNA Molecules, *Ann. Rev. Biochem.,* **36:** 485 (1967).

Specific References

1. Takahashi, I.: *J. Gen. Microbiol.,* **31:** 211 (1963); Takahashi, I., and J. Marmur: *Biochem. Biophys. Res. Commun.,* **10:** 289 (1963); Langridge, R., and J. Marmur: *Science,* **143:** 1450 (1964); Marmur, J., and S. Cordes: in H. J. Vogel, V. Bryson, and J. O. Lampen (eds.), *Informational Macromolecules,* Academic Press, New York, 1963, p. 79.
2. Borek, E., and P. R. Srinivasan: *Ann. Rev. Biochem.,* **35:** 275 (1966); Starr, J. L., and B. H. Sells: *ibid.,* **39:** in press.
3. Wolfenden, R. V.: *J. Mol. Biol.,* **40:** 307 (1969); cf. Davies, D. R.: *loc. cit.*
4. Haschemeyer, A. E. V., and A. Rich: *J. Mol. Biol.,* **27:** 369 (1967).
5. Richards, O. C., and P. D. Boyer: *J. Mol. Biol.,* **11:** 327 (1965).
6. Richardson, C. C.: *J. Mol. Biol.,* **15:** 49 (1966); Richardson, C. C., and B. Weiss: *J. Gen. Physiol.,* **49:** 81 (1966).
7. Josse, J., A. D. Kaiser, and A. Kornberg: *J. Biol. Chem.,* **236:** 864 (1961).
8. Lehman, I. R., and A. L. Nussbaum: *J. Biol. Chem.,* **239:** 2628 (1964); Lehman, I. R.: *Progr. Nucl. Acid Res. Mol. Biol.,* **2:** 84 (1963).
9. Laskowski, M., Sr.: *Advan. Enzymol.,* **29:** 165 (1967).
10. Sober, H. A. (ed.): *Handbook of Biochemistry,* Chemical Rubber Co., Cleveland, 1968, H-20 to H-26.
11. Hudson, B., and J. Vinograd: *Nature,* **216:** 647 (1967); **221:** 332 (1969).
12. Vinograd, J., J. Lebowitz, and R. Watson: *J. Mol. Biol.,* **33:** 173 (1968).
13. Glaubiger, D., and J. E. Hearst: *Biopolymers,* **5:** 691 (1967).
14. Fong, P.: *J. Theoret. Biol.,* **15:** 230 (1967).
15. Gibor, A., and S. Granick: *Science,* **145:** 890 (1964); Granick, S., and A. Gibor: *Progr. Nucl. Acid Res. Mol. Biol.,* **6:** 43 (1967).
16. Borst, P., A. M. Kroon, and G. J. C. M. Ruttenberg: in D. Shugar (ed.), *Genetic Elements, Properties and Function,* Academic Press, New York, 1967, p. 81.
17. Borst, P., and A. M. Kroon: *Intern. Rev. Cytol.,* **26:** 107 (1969).
18. Marmur, J., S. Falkow, and M. Mandel: *Ann. Rev. Microbiol.,* **17:** 329 (1963).
19. Watson, J. D., and F. H. C. Crick: *Nature,* **171:** 737, 964 (1953).
20. Luzzati, V.: *Progr. Nucl. Acid Res. Mol. Biol.,* **1:** 347 (1963); Luzzati, V., J. Witz, and A. Mathis: in D. Shugar (ed.), *Genetic Elements, Properties and Function,* Academic Press, New York, 1967, p. 41.
21. Davies, D. R.: *loc. cit.*
22. Felsenfeld, G., and H. T. Miles: *loc. cit.*
23. Miles, H. T.: *Methods Enzymol.,* **12B:** 256 (1968).
24. Kyogoku, Y., R. C. Lord, and A. Rich: *J. Am. Chem. Soc.,* **89:** 496 (1967); Küchler, E., and J. Derkosch: *Z. Naturforsch.,* **21b:** 209 (1966).
25. Funck, T.: quoted in M. Eigen, *Nobel Symp.,* **5:** 333 (1967).
26. Sueoka, N.: *J. Mol. Biol.,* **3:** 31 (1961); Sueoka, N., and T-Y. Cheng: *ibid.,* **4:** 161 (1962); *Proc. Natl. Acad. Sci. U.S.,* **48:** 1851 (1963).
27. Smith, M.: *Biochem. Biophys. Res. Commun.,* **10:** 67 (1963); *J. Mol. Biol.,* **9:** 17 (1964).
28. Mehrotra, B. D., and H. R. Mahler: *Arch. Biochem. Biophys.,* **128:** 685 (1968).
29. Bernardi, G., F. Carnevali, A. Nicolaieff, G. Piperno, and G. Tecce: *J. Mol. Biol.,* **37:** 493 (1968).
30. Khorana, H. G.: *Biochem. J.,* **109:** 709 (1968).
31. Bollum, F. J.: *Methods Enzymol.,* **12B:** 591 (1968).
32. Eigner, J.: *Methods Enzymol.,* **12B:** 386 (1968); Josse, J., and J. Eigner: *loc. cit.*
33. Schachman, H. K.: *Methods Enzymol.,* **4:** 32 (1957); *Ultracentrifugation in Biochemistry,* Academic Press, New York, 1959.
34. Burgi, E., and A. D. Hershey: *Biophys. J.,* **3:** 309 (1963).
35. Vinograd, J., R. Bruner, R. Kent, and J. Weigle: *Proc. Natl. Acad. Sci. U.S.,* **49:** 902 (1963).
36. Studier, F. W.: *J. Mol. Biol.,* **11:** 373 (1965).
37. Meselson, M., F. W. Stahl, and J. Vinograd: *Proc. Natl. Acad. Sci. U.S.,* **43:** 581 (1957).
38. Schildkraut, C. L., J. Marmur, and P. Doty: *J. Mol. Biol.,* **4:** 430 (1962); Mandel, M., C. L. Schildkraut, and J. Marmur: *Methods Enzymol.,* **12B:** 184 (1968).
39. Szybalski, W.: *Methods Enzymol.,* **12B:** 330 (1968).
40. Zimm, B. H., and D. M. Crothers: *Proc. Natl. Acad. Sci. U.S.,* **48:** 905 (1962).
41. Peterlin, A.: *Nature,* **171:** 259 (1953); *J. Polymer Sci.,* **10:** 425 (1953).
42. Hearst, J. E., and W. Stockmayer: *J. Chem. Phys.,* **37:** 1425 (1962).
43. Gray, H. B., Jr., V. A. Bloomfield, and J. E. Hearst: *J. Chem. Phys.,* **46:** 1493 (1967); Gray, H. B., Jr., and J. E. Hearst: *J. Mol. Biol.,* **35:** 111 (1968).
44. Crothers, D. M., and B. H. Zimm: *J. Mol. Biol.,* **12:** 525 (1965).
45. Bloomfield, V. A., and B. H. Zimm: *J. Chem. Phys.,* **44:** 315 (1966).
46. Vinograd, J., J. Morris, N. Davidson, and W. F. Dove, Jr.: *Proc. Natl. Acad. Sci. U.S.,* **49:** 12 (1963).
47. Bode, V. C., and L. A. MacHattie: *J. Mol. Biol.,* **32:** 673 (1968).
48. Crawford, L. V., and M. J. Waring: *J. Mol. Biol.,* **25:** 23 (1967).
49. Hearst, J. E., and J. Vinograd: *Proc. Natl. Acad.*

Sci. U.S., **47:** 825, 1005 (1961); *Arch. Biochem. Biophys.*, **92:** 206 (1961).
50. Kleinschmidt, A. K.: *Methods Enzymol.*, **12B:** 361 (1968).
51. Thomas, C. A., Jr.: *J. Cell. Phys.*, **70:** 13 (1967).
52. Westmoreland, B. C., W. Szybalski, and H. Ris: *Science*, **163:** 1343 (1969).
53. Lark, C.: *Methods Enzymol.*, **12B:** 616 (1968).
54. Cairns, J.: *J. Mol. Biol.*, **6:** 208 (1963).
55. DeVoe, H., and I. Tinoco, Jr.: *J. Mol. Biol.*, **4:** 500 (1962); Tinoco, I., Jr.: in B. Pullman and M. Weissbluth (eds.), *Molecular Biophysics*, Academic Press, New York, 1965, p. 269; Cantor, C. R., and I. Tinoco, Jr.: *J. Mol. Biol.*, **13:** 65 (1965); *Biopolymers*, **5:** 821 (1967).
56. Michelson, A. M.: *loc. cit.;* Michelson, A. M., J. Massoulié, and W. Guschlbauer: *Progr. Nucl. Acid Res. Mol. Biol.*, **6:** 83 (1967).
57. Marmur, J., R. Rownd, and C. L. Schildkraut: *Progr. Nucl. Acid Res. Mol. Biol.*, **1:** 231 (1963).
58. Hirschman, S. Z., and G. Felsenfeld: *J. Mol. Biol.*, **16:** 347 (1966); Felsenfeld, G., and S. Z. Hirschman: *J. Mol. Biol.*, **13:** 407 (1965).
59. Felsenfeld, G.: *Methods Enzymol.*, **12B:** 247 (1968).
60. Fresco, J. R., L. C. Klotz, and E. G. Richards: *Cold Spring Harbor Symp. Quant. Biol.*, **28:** 83 (1963).
61. Mahler, H. R., B. E. Kline, and B. D. Mehrotra: *J. Mol. Biol.*, **9:** 801 (1964).
62. Inman, R. B., and R. L. Baldwin: *J. Mol. Biol.*, **8:** 452 (1964).
63. Inman, R. B.: *J. Mol. Biol.*, **9:** 624 (1964).
64. Chamberlin, M. J.: *Federation Proc.*, **24:** 1446 (1965).
65. Wells, R. D., E. Ohtsuka, and H. G. Khorana: *J. Mol. Biol.*, **14:** 221 (1965).
66. Ts'o, P. O. P., S. A. Rapaport, and F. J. Bollum: *Biochemistry*, **5:** 4153 (1966).
67. Green, G., and H. R. Mahler: *Biochemistry*, **9:** 368 (1970).
68. Beychok, S.: *Science*, **154:** 1288 (1966); *Ann. Rev. Biochem.*, **37:** 437 (1968).
69. Eyring, H., H. Liu, and D. Caldwell: *Chem. Rev.*, **68:** 525 (1968).
70. Tinoco, I., Jr.: *J. Chim. Phys.*, **65:** 91 (1968).
71. Brahms, J.: *J. Chim. Phys.*, **65:** 105 (1968).
72. Adler, A. J., and G. D. Fasman: *Methods Enzymol.*, **12B:** 268 (1968).
73. Mommaerts, W. F. H. M.: *Methods Enzymol.*, **12B:** 302 (1968).
74. Yang, J. T., and T. Samejima: *Progr. Nucl. Acid Res. Mol. Biol.*, **9:** 223 (1969).
75. Bush, C. A., and J. Brahms: *J. Chem. Phys.*, **46:** 79 (1967); Van Holde, K. E., J. Brahms, and A. M. Michelson: *J. Mol. Biol.*, **12:** 726 (1965); Brahms, C. A.: *J. Chim. Phys.*, **65:** 105 (1968).
76. Sober, H. A. (ed.): *Handbook of Biochemistry*, Chemical Rubber Co., Cleveland, 1968, G-93 to G-103.
77. Becker, E. F., Jr., B. K. Zimmerman, and E. P. Geiduschek: *J. Mol. Biol.*, **8:** 377 (1964).
78. Shapiro, R., and R. S. Klein: *Biochemistry*, **5:** 2358 (1966).
79. Kohn, K. W., C. L. Spears, and P. Doty: *J. Mol. Biol.*, **19:** 266 (1966).
80. Mahler, H. R., G. Green, R. Goutarel, and Q. Khuong-Huu: *Biochemistry*, **7:** 1568 (1968).
81. Gabbay, E. J.: *J. Am. Chem. Soc.*, **90:** 5257 (1968); **91:** 5136 (1969).
82. Liquori, A. M., L. Costantino, V. Crescenzi, V. Elia, E. Giglio, R. Puliti, M. De Santis Savino, and V. Vitagliano: *J. Mol. Biol.*, **24:** 113 (1967).
83. Olins, D. E., A. L. Olins, and P. H. von Hippel: *J. Mol. Biol.*, **24:** 157 (1967); **33:** 265 (1968).
84. Shapiro, J. T., M. Leng, and G. Felsenfeld: *Biochemistry*, **8:** 3219 (1969).
85. Latt, S. A., and H. A. Sober: *Biochemistry*, **6:** 3293, 3307 (1967).
86. Weser, U.: *Struct. Bonding (Berlin)*, **5:** 41 (1968).
87. Brenner, S., L. Barnett, F. H. C. Crick, and A. Orgel: *J. Mol., Biol.*, **3:** 121 (1961); Crick, F. H. C., L. Barnett, S. Brenner, and R. J. Watts-Tobin: *Nature*, **193:** 1227 (1961)
88. Hurwitz, J., A. Evans, C. Babinet, and A. Skalka: *Cold Spring Harbor Symp. Quant. Biol.*, **28:** 59 (1963).
89. Lerman, L. S.: *J. Mol. Biol.*, **3:** 18 (1961); *Proc. Natl. Acad. Sci. U.S.*, **49:** 94 (1963).
90. Waring, M. J.: *Nature*, **219:** 1320 (1968); *Proc. Soc. Gen. Microbiol.*, **16:** 235 (1966).
91. Reich, E.: *Proc. Soc. Gen. Microbiol.*, **16:** 266 (1966); Hamilton, L. D., W. Fuller, and E. Reich: *Nature*, **198:** 538 (1963).
92. Müller, W., and D. M. Crothers: *J. Mol. Biol.*, **35:** 251 (1968).
93. Cerami, A., E. Reich, D. C. Ward, and I. H. Goldberg: *Proc. Natl. Acad. Sci. U.S.*, **57:** 1036 (1967).
94. Radloff, R., W. Bauer, and J. Vinograd: *Proc. Natl. Acad. Sci. U.S.*, **57:** 1514 (1967).
95. Bauer, W., and J. Vinograd: *J. Mol. Biol.*, **33:** 141 (1968).
96. Richards, E. G.: *European J. Biochem.*, **6:** 88 (1968).
97. Mandel, M., and J. Marmur: *Methods Enzymol.*, **12B:** 195 (1968).
98. Hamaguchi, K., and E. P. Geiduschek: *J. Am. Chem. Soc.*, **84:** 1329 (1962).
99. Beers, W., A. Cerami, and E. Reich: *Proc. Natl. Acad. Sci. U.S.*, **58:** 1624 (1967).
100. Studier, F. W.: *J. Mol. Biol.*, **41:** 189, 199 (1969).
101. Geiduschek, E. P.: *J. Mol. Biol.*, **4:** 467 (1962).
102. Crothers, D. M., N. R. Kallenbach, and B. H. Zimm: *J. Mol. Biol.*, **11:** 802 (1965).
103. Wetmur, J. G., and N. Davidson: *J. Mol. Biol.*, **31:** 349 (1968).
104. Thrower, K. J., and A. R. Peacocke: *Biochem. J.*, **109:** 543 (1968).
105. Walker, P. M. B.: *Progr. Nucl. Acid Res. Mol. Biol.*, **9:** 301 (1969).
106. Britten, R. J., and D. E. Kohne: *Science*, **161:** 529 (1968).
107. Blobel, G., and V. R. Potter: *J. Mol. Biol.*, **26:** 279 (1967).
108. Shearer, R. W., and B. J. McCarthy: *Biochemistry*, **6:** 283 (1967).
109. Holley, R. W., J. Apgar, G. A. Everett, J. T. Madi-

son, M. Marquisee, S. H. Merrill, J. R. Penswick, and A. Zamir: *Science,* **147:** 1462 (1965); Holley, R. W., J. T. Madison, and A. Zamir: *Biochem. Biophys. Res. Commun.,* **17:** 389 (1964).

110. Philipps, G. R.: *Nature,* **223:** 374 (1969).
111. Gillam, I., S. Millward, D. Blew, M. von Tigerstrom, E. Wimmer, and G. M. Tener: *Biochemistry,* **6:** 3043 (1967); Nishimura, S., F. Harada, U. Narushima, and T. Seno: *Biochim. Biophys. Acta,* **142:** 133 (1967).
112. Sanger, F., and G. G. Brownlee: *Methods Enzymol.,* **12A:** 361 (1967); Sanger, F., G. G. Brownlee, and B. G. Barrell: *J. Mol. Biol.,* **13:** 373 (1965); Brownlee, G. G., and F. Sanger: *ibid.,* **23:** 337 (1967); Brownlee, G. G., F. Sanger, and B. G. Barrell: *ibid.,* **34:** 379 (1968); Fellner, P., and F. Sanger: *Nature,* **219:** 236 (1968); Dube, S. K., K. A. Marker, B. F. C. Clark, and S. Cory: *ibid.,* **218:** 232 (1968); **220:** 1039 (1968).
113. Loening, U. E.: *Biochem. J.,* **102:** 251 (1967); Peacock, A. C., and C. W. Dingman: *Biochemistry,* **6:** 1818 (1967).
114. Gefter, M. L., and R. L. Russell: *J. Mol. Biol.,* **39:** 145 (1969).
115. See numerous references in *Cold Spring Harbor Symp. Quant. Biol.,* **34** (1969) and *Nature,* **223** (1969); also Fellner, P., C. Ehresmann, and J. P. Ebel: *ibid.,* **225:** 26 (1970).
116. Fresco, J. R., A. Adams, R. Ascione, D. Henley, and T. Lindahl: *Cold Spring Harbor Symp. Quant. Biol.,* **31:** 527 (1966); Sueoka, N., T. Kano-Sueoka, and W. J. Gartland: *ibid.,* **31:** 571 (1966); Aubert, M., J. F. Scott, M. Reynier, and R. Monier: *Proc. Natl. Acad. Sci. U.S.,* **61:** 292 (1968).
117. Blake, R. D., J. R. Fresco, and R. Langridge: *Nature,* **225:** 32 (1970).
118. Yaniv, M., A. Favre, and B. G. Barrell: *Nature,* **223:** 1331 (1969); Ninio, J., A. Favre, and M. Yaniv: *ibid.,* **223:** 1333 (1969).
119. Litt, M., and V. Hancock: *Biochemistry,* **6:** 1848 (1967).
120. Stanley, W. M., Jr., and R. M. Bock: *Biochemistry,* **4:** 1302 (1965).
121. Stutz, E., and H. Noll: *Proc. Natl. Acad. Sci. U.S.,* **57:** 774 (1967); *Methods Enzymol.,* **12B:** 129 (1968); Küntzel, H., and H. Noll: *Nature,* **215:** 1340 (1967).
122. Loening, U. E., and J. Ingle: *Nature,* **215:** 363 (1967).
123. Hodnett, J. L., and H. Busch: *J. Biol. Chem.,* **243:** 6334 (1968).
124. Dahlberg, J. E.: *Nature,* **220:** 548 (1968); Billeter, M. A., J. E. Dahlberg, H. M. Goodman, J. Hindley, and C. Weissman: *ibid.,* **224:** 1083 (1969).
125. De Wachter, R., and W. Fiers: *J. Mol. Biol.,* **30:** 507 (1967); *Nature,* **221:** 233 (1969).
126. Steitz, J. A.: *Nature,* **224:** 957 (1969); Roblin, R.: *J. Mol. Biol.,* **31:** 51 (1968).
127. Adams, J. M., P. G. N. Jeppesen, F. Sanger, and B. G. Barrell: *Nature,* **223:** 1009 (1969); Nichols, J. L.: *ibid.,* **225:** 147 (1970).
128. Thang, M. N., and M. Grunberg-Manago: *Methods Enzymol.,* **12B:** 522 (1968); Kimhi, Y., and U. Z. Littauer: *ibid.,* **12B:** 513 (1968).
129. Weiss, S. B.: *Methods Enzymol.,* **12B:** 555 (1968).
130. Richardson, J.: *Progr. Nucl. Acid Res. Mol. Biol.,* **9:** 75 (1969).
131. Krakow, J. S., and W. J. Hortley: *Progr. Nucl. Acid Res. Mol. Biol.,* **8:** 566 (1968).
132. Bonner, J. et al.: *Methods Enzymol.,* **12B:** 3 (1968).
133. Busch, H.: *Methods Enzymol.,* **12B:** 65 (1968).
134. Kinkade, J. M., Jr., and R. D. Cole: *J. Biol. Chem.,* **241:** 5790, 5798 (1967).
135. Phillips, D. M. P., and E. W. Johns: *Biochem. J.,* **94:** 127 (1965).
136. DeLange, R. J., D. M. Fambrough, E. L. Smith, and J. Bonner: *J. Biol. Chem.,* **244:** 319 (1969).
137. Quagliarotti, G., Y. Ogawa, C. W. Taylor, P. Sautiere, J. Jordan, W. C. Starbuck, and H. Busch: *J. Biol. Chem.,* **244:** 1796 (1969).
138. Allfrey, V. G., R. Faulkner, and A. E. Mirsky: *Proc. Natl. Acad. Sci. U.S.,* **51:** 786 (1964); Pogo, B. G. T., A. O. Pogo, V. G. Allfrey, and A. E. Mirsky: *ibid.,* **59:** 1337 (1968).
139. Rauth, A. M.: *Biophys. J.,* **5:** 257 (1965).
140. Thomas, C. A., Jr., and L. A. MacHattie: *loc. cit.*

6 Enzyme Kinetics

Living cells are successful in carrying out thermodynamically favorable reactions under conditions of concentrations, temperature, *p*H, pressure, and environment that differ greatly from those employed by the preparative organic chemist; cells are equally successful in integrating and regulating these reactions. The key to these phenomena is provided by the fact that virtually all intracellular processes are mediated by enzymes and thus it now becomes necessary to inquire into the *rate behavior* or *kinetics* of such enzyme-catalyzed reactions and into the means available for affecting their kinetic control.

CHEMICAL KINETICS

Kinetics of Uncatalyzed Reactions

Before discussing the kinetics of enzyme-catalyzed reactions, we shall briefly review the salient features of ordinary (that is, uncatalyzed) reactions. Let us consider the simplest chemical reaction,

$$A \xrightarrow{k} P \tag{6-1}$$

one in which A is converted spontaneously and essentially irreversibly to P in, say, an isomerization reaction, then its rate or velocity v at any time t equals $-da/dt$, the rate of disappearance of substrate, or dp/dt, the rate of appearance of product. Now, ideally, if this is a true monomolecular process, by the law of mass action this rate must be proportional to a, the concentration of A at any time t:

$$v = -\frac{da}{dt} = ka \tag{6-2}$$

This statement and its algebraic formulation constitute the *velocity, rate,* or *kinetic law* (or better *equation*) characteristic of the reaction in question. We observe that on the right side we have a function containing a to the first power, and the reaction is, therefore, said to be *first order in A*. The proportionality constant k is called the *rate constant* (or *coefficient*) or *specific reaction rate;* clearly it has the dimensions of time^{-1}. Now let us consider the slightly more complex and much more general case of a bimolecular metathetical reaction in which one molecule of A must interact in the rate-determining step with a molecule of B to yield the following products.

$$A + B \xrightarrow{k} P + (Q) + \cdots \tag{6-3}$$

The rate equation in this instance is given by

$$v = -\frac{da}{dt} = -\frac{db}{dt} = \frac{dp}{dt} = kab \tag{6-4a}$$

This reaction is *second-order overall,* for its rate is proportional to the product of two concentration terms (a and b, respectively) and, therefore, is first order in a as well as in b. The rate constant k in this instance has the dimensions of concentration^{-1} time^{-1}. For the special case of two molecules of A interacting to yield products, the rate equation is

$$v = -\frac{da}{dt} = ka^2 \tag{6-4b}$$

This form of the equation is also obeyed if we start with identical concentrations of A and B.

In the general case

$$qA + rB + \cdots \xrightarrow{k} \text{products} \tag{6-5}$$

$$v = -\frac{da}{dt} = ka^qb^r \cdots \qquad n = q + r + \cdots \tag{6-6}$$

where n molecules must interact with each other in the rate-limiting step of a reaction, the rate equation for this nth-order reaction contains n concentration terms, and the rate constant has the dimensions of concentration^{1-n} time^{-1}. Practically this general case need not concern us particularly, because even termolecular reactions, that is, those in which three different substrate molecules must participate simultaneously in a rate-limiting step, are statistically highly improbable, and reactions involving three or more participants usually proceed by a series of bimolecular and/or monomolecular steps. Reactions of this type are kinetically of first, second, third, or mixed order overall—depending on which steps are rate-limiting—and even fractional orders are not uncommon. From the foregoing it is evident that the determination of the order of a reaction is the most important part in describing its kinetics. The simplest and most straightforward procedure (the *differential method*) is to vary the concentration of each reactant and to observe the resulting effect on the reaction rate [Eqs. (6-6) and (6-7)]. We do this by plotting ln v versus ln a and obtain q, the order in A, as the slope of the resulting straight line [Eq. (6-7)]. It is to be noted that the intercept equals ln k, if A is the only reactant and if $n = q$ [Eq. (6-8)] or, alternatively, if the stoichiometry is such that one molecule of B (and C) must react per molecule of A, and the concentration of all such reactants are equal initially:

$$\ln v = \ln -\frac{da}{dt} = \ln k + q \ln a + \cdots \tag{6-7}$$

$$\ln v = \ln k + n \ln a \tag{6-8}$$

Most conveniently we measure v_0, the initial rate during the very early stages of the reaction (the tangent at $t = 0$ of the *progress curve* obtained by plotting the dependence of a or p on t, as shown in Fig. 6-1, as a function of a_0, the concentration of substrate added initially). The order so obtained (by plotting log v_0 versus log a_0) has frequently been called the *order with respect to concentration* or the *true order*. In principle it is also possible to use a single

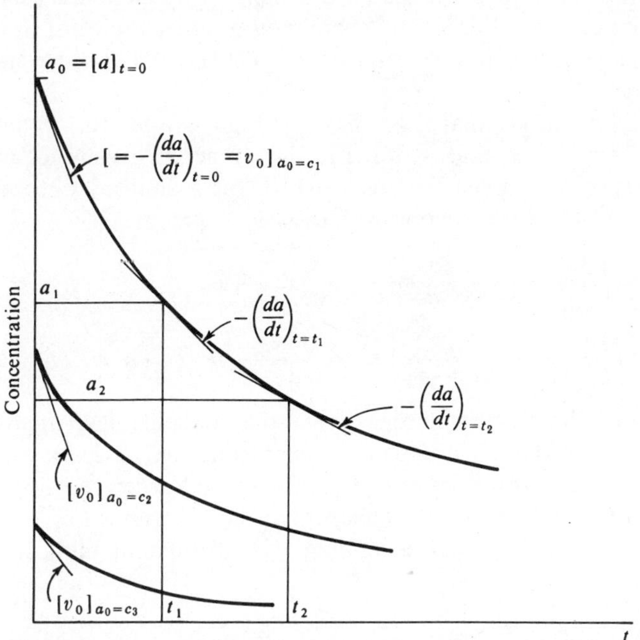

FIGURE 6-1

Progress curves showing the change of a, the concentration of A as a function of time. The tangents at $t = 0$ are used to determine y_0 and those at t_1 and t_2 determine v at those times.

concentration of substrate, because in the course of the reaction it varies continuously from a_0, at the start, to a_∞ at the end of the reaction. The latter is either virtually zero—in the case of irreversible reactions (or when care is taken to remove the product as soon as it is formed)—or, more commonly, especially in biochemistry, where the $-\Delta G^0$ for the reactions in question are frequently not very large—a finite and significant number a_e, the concentration of a when equilibrium has been established. The order determined in this fashion is then called the *order with respect to time*. In the case of competition by products or of equilibrium reactions it is not nearly so accurate or useful as that determined with respect to concentration.

Integration of Eqs. (6-2), (6-4a), (6-4b), or (6-6) provides us with alternatives for the description of simple reactions and with two additional methods for the determination of the reaction orders. The integrated forms, with the limits $t = 0$ and $t = t$, with concentrations of A given by a_0 and a, respectively, are shown as Eq. (6-9), corresponding to (6-2); (6-10a), corresponding to (6-4a); or (6-10b), corresponding to (6-4b). Here p is the concentration of product at time t formed stoichiometrically from A; hence $p + a = a_0$:

$$kt = \ln \frac{a_0}{a} = \ln \frac{a_0}{a_0 - p} \tag{6-9}$$

$$kt = \frac{1}{a_0 - b_0} \ln \frac{b_0(a_0 - p)}{a_0(b_0 - p)} = \frac{1}{i} \ln \frac{a(a_0 - i)}{a_0(a - i)} \qquad a_0 = b_0 + i, \quad i > 0 \tag{6-10a}$$

$$kt = \frac{p}{a_0(a_0 - p)} = \frac{a_0 - a}{a_0 a} = \frac{1}{a} - \frac{1}{a_0} \qquad a_0 = b_0 \tag{6-10b}$$

These equations may be used, whenever a reaction is suspected to be of either first or second order, by examining whether a plot of ln or log a [Eq. (6-9)], $1/a$ [Eq. 6-10b)], or log $[a/(a-i)]$ [Eq. (6-10a)], versus t yields a straight line.

We observe that τ, the time interval required for obtaining any given extent of reaction x, that is, for a to have reached a value equal to $a_0\ (1-x)$, are respectively given by Eqs. (6-11), for a first-order reaction, and (6-12), for a second-order reaction with $a_0 = b_0$:

$$\tau = \frac{1}{k} \ln \frac{a_0}{a_0 - a_0 x} = \frac{1}{k} \ln \frac{1}{(1-x)} = \frac{2.3}{k} \log \frac{1}{(1-x)} \tag{6-11}$$

$$\tau = \frac{1}{k}\left[\frac{1}{a_0(1-x)} - \frac{1}{a_0}\right] = \frac{1}{k}\frac{x}{a_0(1-x)} \tag{6-12}$$

Thus the fractional time (e.g., the so-called half-time or half-period when $1 - x = 0.5$) is a constant, independent of a_0 for a first-order reaction and inversely proportional to a_0 for a second-order reaction. In general τ is proportional to $1/a_0^{n-1}$. Thus, n can be determined if we know τ_1 and τ_2, the fractional times corresponding to two different initial concentrations $a_0 = c_1$ and $a_0 = c_2$:

$$\frac{\tau_1}{\tau_2} = \left(\frac{c_2}{c_1}\right)^{n-1} \quad \text{or} \quad \log \frac{\tau_1}{\tau_2} = (n-1) \log \frac{c_2}{c_1} \tag{6-13}$$

$$n = 1 + \frac{\log \tau_1 - \log \tau_2}{\log c_2 - \log c_1} \tag{6-14}$$

One more fact remains to be considered at this point. If, throughout the course of a rate-determining reaction, the concentration of any one reactant is much larger than that of all the others, its concentration remains virtually constant. This means that, as long as this condition holds, we can recast the rate equations so as to combine this concentration term with the rate constant and in effect reduce the kinetic order of the reaction by one; for example, for reaction (6-3) with $b_0 \gg a_0$, Eq. (6-4) reduces to Eq. (6-15):

$$v \simeq kab_0 = k'a \tag{6-15}$$

Thus the reaction, although still bimolecular, has become kinetically *apparently* or *pseudo* first order overall. This is an important consideration, not only in chemical kinetics, but especially in enzyme kinetics, where we frequently deal only with initial velocities such that essentially $a = a_0$.

Let us now consider a somewhat more complicated reaction sequence[1]:

$$A + B \underset{k_2}{\overset{k_1}{\rightleftharpoons}} P \tag{6-16}$$

in which A plus B combine *reversibly* to yield P. The net rate of formation of product is then given by Eq. (6-17), which contains two terms corresponding to the rate of generation and disappearance of P, respectively:

$$v = \frac{dp}{dt} = -\frac{da}{dt} = -\frac{db}{dt} = k_1ab - k_2p \tag{6-17}$$

[1] In this book we use, for convenience's sake, the convention that $k_1, k_3, k_5, \ldots, k_{2n-1}$ designate the rate constants for the individual steps in the forward direction; $k_2, k_4, k_6, \ldots, k_{2n}$ those for the corresponding ones in the reverse direction, where n = the step in question. The Enzyme Commission of the International Union of Biochemistry has recommended that $k_{+1}, k_{+2}, k_{+3}, \ldots, k_{+n}$, etc., and $k_{-1}, k_{-2}, k_{-3}, \ldots, k_{-n}$ be used for this purpose. We shall have occasional recourse to this system for didactic purposes.

At equilibrium $dp/dt = 0$, and the concentrations of the participants are a_e, b_e, and p_e:

$$k_1 a_e b_e = k_2 p_e \tag{6-18}$$

This allows us to express k_2 in terms of k_1 and the various equilibrium concentrations. Substituting back into Eq. (6-17) yields

$$v = k_1 \left(ab - \frac{p}{p_e} a_e b_e \right) \tag{6-19}$$

Because $a = a_0 - p$ and $b = b_0 - p$, we can rewrite Eq. (6-19):

$$v = \frac{dp}{dt} = \frac{k_1}{p_e} [p_e(a_0 - p)(b_0 - p) - p(a_0 - p_e)(b_0 - p_e)]$$

$$= \frac{k_1}{p_e} [(p_e - p)(a_0 b_0 - p p_e)] \tag{6-20}$$

This equation can be integrated; it provides us with an explicit description of the extent of reaction at any time.[2] For the case $a_0 = b_0$ we find

$$k_1 t = \frac{p_e}{a_0^2 - p_e^2} \ln \frac{a_0^2 p_e - p_e^2 p}{a_0^2 p_e - a_0^2 p} \tag{6-21}$$

But if we go to the situation summarized by Eq. (6-22), the reaction of A plus B to yield an *intermediate* X, which then breaks down to yield product(s) P:

$$A + B \underset{k_2}{\overset{k_1}{\rightleftharpoons}} X \overset{k_3}{\longrightarrow} P \tag{6-22}$$

we find that we require more than one kinetic (differential) equation to describe the reaction system:

$$-\frac{da}{dt} = -\frac{db}{dt} = k_1 ab - k_2 x \tag{6-23a}$$

$$\frac{dx}{dt} = k_1 ab - (k_2 + k_3)x \tag{6-23b}$$

$$\frac{dp}{dt} = k_3 x \tag{6-23c}$$

These three differential equations plus the conservation equation

$$a_0 - a = b_0 - b \tag{6-23d}$$

provide us with four equations in four unknowns. But in the most general case, where there exists no restriction with respect to the relative magnitude of the rate constants k_1, k_2, and k_3, the relative concentration of a_0 and b_0, and the particular time t for which we wish to establish the rate law, their explicit solution is not possible, and they can only be integrated numerically, preferably with the help of a computer. A solution provided by W. W. Cleland is given in Fig. 6-2.

There are, however, several limiting cases for which it is possible to convert the set of differential equations to forms that are capable of a simple interpretation.

[2] Equation (6-17) can also be integrated, at least in principle, in all cases in terms of p and k_1 above. If K_{eq} is known as well, this then provides a means for calculating the rate constants in both directions.

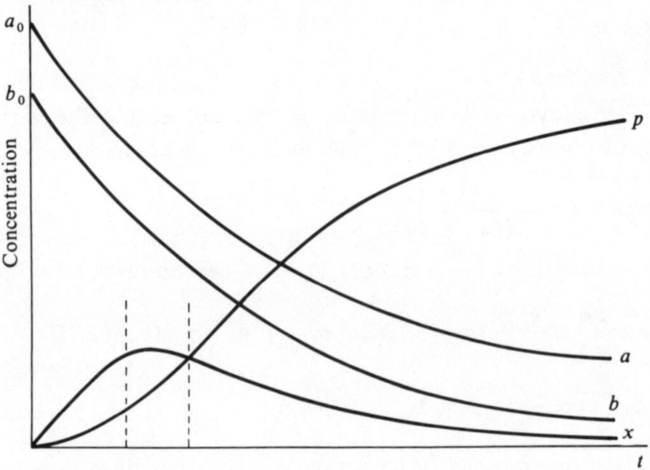

FIGURE 6-2

Progress curve for the reaction sequence $A + B \underset{k_2}{\overset{k_1}{\rightleftharpoons}} X \overset{k_3}{\longrightarrow} P$ with $k_1 \simeq k_2 \simeq k_3$; $a_0 \simeq b_0$. Adapted from W. W. Cleland.

1. $k_3 \ll k_2$. Under this condition the rate of product formation is governed by the monomolecular breakdown of X, which in turn is formed rapidly and reversibly from A plus B; kinetically the reaction is first order in either reactant.

The rate equation is

$$-\frac{da}{dt} = \frac{dp}{dt} = k_3 K_{\text{eq}} ab \tag{6-24a}$$

where

$$K_{\text{eq}} = \frac{x}{ab} = \frac{k_1}{k_2}$$

2. $k_3 \gg k_1 \simeq k_2$. X is converted to P as rapidly as it is formed; the rate equation (6-24b) has the same form as Eq. (6-24a), and the kinetics are, therefore, similar to those in the case above, despite the differences in rate-limiting steps:

$$v = \frac{dp}{dt} = k_1 ab \tag{6-24b}$$

that is, identical to Eq. (6-4), with $k = k_1$. We recall that its integrated form is given by Eq. (6-10a) or (6-10b).

3. $dx/dt = 0$. There is a short time interval, indicated approximately by dashed lines in Fig. 6-2, during which the slope of the plot of x versus $t \simeq 0$, or $x \simeq$ constant. While this condition obtains, during the so-called *steady* or *stationary state,* Eq. (6-23b) now reads (because $dx/dt \simeq 0$)

$$k_1 ab \simeq (k_2 + k_3) x \tag{6-23b'}$$

which permits us to express x in terms of a and b. This converts this differential into an algebraic equation and allows us to find explicit solutions for Eqs. (6-23a–d).

From Eq. (6-23b′),

$$x = \frac{k_1ab}{k_2 + k_3} \tag{6-23e}$$

Substitution into Eq. (6-23c):

$$\frac{dp}{dt} = \frac{k_1k_3ab}{k_2 + k_3} = \frac{k_1(a_0 - p)(b_0 - p)}{1 + k_2/k_3} \tag{6-25}$$

which may be integrated to provide an explicit value of p at any time t.

We see that Eqs. (6-24a) and (6-24b) are special cases derivable from Eq. (6-25) by setting $k_2 \gg k_3$ in one case, $k_3 \gg k_2$ in the other. Thus, for these two limiting cases, the steady-state assumption gives valid results at all times.

Kinetics of Catalyzed Reactions

Most ordinary homogeneous chemical reactions can be accelerated in the presence of an additional reaction component, a *catalyst,* that remains unaltered in concentration during the progress of the reaction. The rate law for a typical reaction involving reactant A and catalyst C at instantaneous concentration a and c, respectively, is given by

$$v = kac = k'a \tag{6-26}$$

where $k' = kc$. The reaction is bimolecular but exhibits pseudo-first-order kinetics with the observed first-order rate constant k' equal to the true second-order constant multiplied by the catalyst concentration c.

A somewhat different situation obtains if the catalyst actually combines with a reactant at one stage and is regenerated at a subsequent stage of the reaction:

$$A + C \underset{k_2}{\overset{k_1}{\rightleftharpoons}} X \overset{k_3}{\longrightarrow} P + C \tag{6-27}$$

Note that this differs from Eq. (6-22) solely by virtue of the regeneration of C in the second half of the reaction. The species X is to be regarded as an intermediate compound or complex, retaining all the atoms originally present in both the reactant A and the catalyst C. By analogy with Eq. (6-23) the kinetic equations are

$$-\frac{da}{dt} = k_1ac - k_2x \tag{6-28a}$$

$$\frac{dx}{dt} = k_1ac - (k_2 + k_3)x \tag{6-28b}$$

$$\frac{dc}{dt} = (k_2 + k_3)x - k_1ac \tag{6-28c}$$

$$\frac{dp}{dt} = k_3x \tag{6-28d}$$

plus the conservation equation

$$c_0 = c + x \tag{6-28e}$$

Once again, no explicit solution is possible, and numerical integration requires a computer (Fig. 6-3). But we observe that now the slopes of the progress curves for both X and C versus time are virtually equal to zero over a considerable portion of the plot. Thus the steady-state approximation is valid for all times t after the dashed line, that is, throughout the whole period ex-

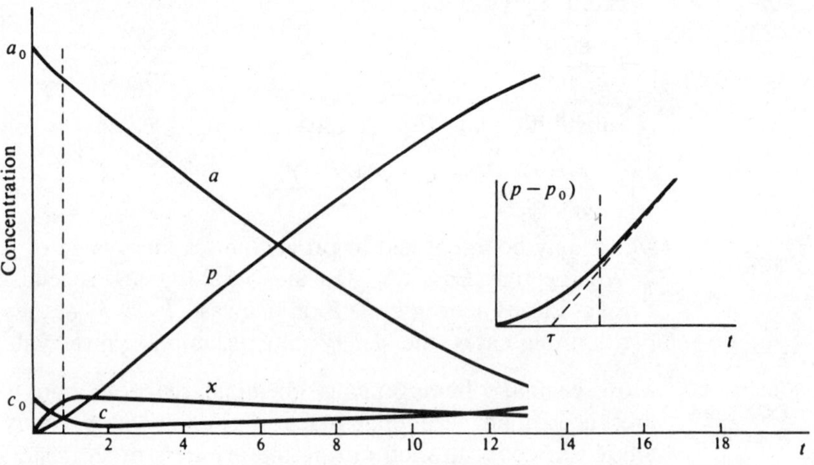

FIGURE 6-3

Progress curve for $A + C \underset{k_2}{\overset{k_1}{\rightleftharpoons}} X \xrightarrow{k_3} P + C$ with $k_1 \simeq k_2 \simeq k_3$. Adapted from W. W. Cleland.

cept for a very brief period prior to the establishment of the steady state, where $dx/dt \gg 0$ and $dc/dt \ll 0$.

The larger the ratio a_0/c_0, the smaller the initial period during which the steady-state approximation does not hold. Thus, we have the equation

$$\frac{dx}{dt} \simeq 0 \simeq k_1 ac - (k_2 + k_3)x \qquad \text{or} \qquad k_1 ac \simeq (k_2 + k_3)x \tag{6-28b'}$$

Substituting in Eq. (6-28e), we obtain

$$x = \frac{k_1 a c_0}{k_1 a + k_2 + k_3} \tag{6-29}$$

and substituting this in Eq. (6-28d),

$$v = -\frac{da}{dt} = \frac{dp}{dt} = k_3 x = \frac{k_1 k_3 a c_0}{k_1 a + k_2 + k_3} = \frac{k_3 c_0 a}{(k_2 + k_3)/k_1 + a} \tag{6-30}$$

This reduces to the fundamental equation

$$v = \frac{Va}{K_a + a} \tag{6-31}$$

where $V = k_3 c_0$ and $K_a = (k_2 + k_3)/k_1$. In its integrated form this becomes

$$Vt = K_a \ln \frac{a_0}{a} + (a_0 - a) \tag{6-32}$$

which is the equation for curve a of Fig. 6-3. This rate law for the reaction of one reactant with a recycling catalyst is obviously of mixed first and zero order. Two limiting cases may be distinguished.

4. $a \gg K_a$. The reaction is zero order with

$$v = V \tag{6-31a}$$

where V is a limiting, *maximal velocity*. If $a = 100K_a$, the deviation from zero order is < 1 percent; even for $a = 10K_a$, the deviation is only 9 percent.

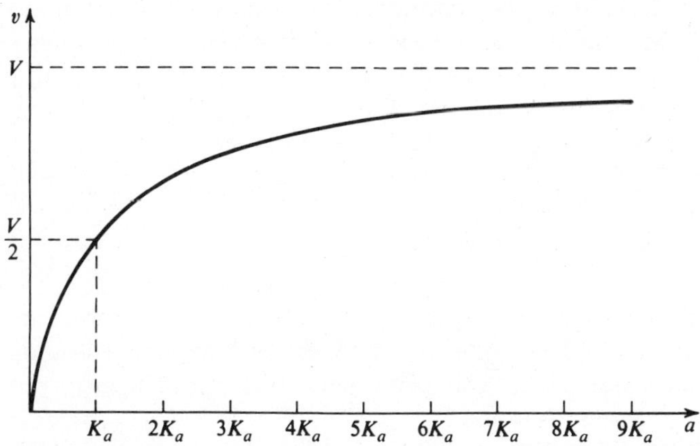

FIGURE 6-4
Graph for the determination of the order with respect to concentration for $v = Va/(K_a + a)$.

5. $a \ll K_a$. The reaction is first order with

$$v = k'a \tag{6-31b}$$

where $k' = V/K_a$. Again if $a = K_a/100$, the deviation from first order is < 1 percent; for $a = K_a/10$, the deviation is < 10 percent.

For all other values of a, and quite in general, the rate law is uniquely determined and the shape of the plot of v versus a, a rectangular hyperbola (Fig. 6-4), completely defined by the two kinetic constants, or parameters, V and K_a. The former, as we have already seen, is the limiting velocity as $a \to \infty$ and thus has the dimensions of conc/time; the latter equals that concentration of A, (a_K) for which $v = V/2$. Substituting in Eq. (6-31), $-V/2 = Va_K/(K_a + a_K)$ or $K_a + a_K = 2a_K$; K_a has the dimensions of concentrations, or M.

KINETICS OF ENZYME-CATALYZED REACTIONS

General Observations

Provided the reaction is performed under carefully controlled conditions, the variation of rate with substrate concentration for a large number of enzyme-catalyzed reactions is described by Eq. (6-31), the *Michaelis–Menten equation,* or, alternatively, by Fig. 6-4. To distinguish it from the more general one we shall now use S, and its concentration s (for substrate) instead of A and a, as well as E and e (for enzyme) instead of C.[3] The two kinetic parameters V and K_s are thus both necessary and sufficient to define the rate law for these reactions explicitly, provided that, as we shall show in subsequent sections:

1. They involve only a single substrate, or, in the case of multisubstrate reactions, the concentrations of all other substrates are held constant.

2. True initial velocities v_0 are measured using different starting concentrations s_0 of S. v_0 is defined as the extrapolated rate $(-ds/dt)_{t=0} = (dp/dt)_{t=0}$.

[3] In this edition we use S and s for A and a to conform more closely to common usage in the enzymological literature, which, however, frequently employs S indiscriminately for both "substrate" and the concentration of the latter. The expressions used previously are easily converted by substituting "s" for "a" wherever it occurs in this (one substrate) section.

3. $s_0 \gg e_0$, the concentration of enzyme added initially, for all values of s_0 used, and the same constant concentration e_0 is used in all measurements. However, for all cases in which Eq. (6-31) is obeyed, and provided the restrictions being discussed are observed, Eq. (6-31) can always be rewritten for $c_0 \equiv e_0$ and *any* initial substrate concentration a_0:

$$(v_0)_{s=s_0} = \frac{k_3 e_0 s_0}{K_s + s_0} = k e_0 \tag{6-33}$$

where $k = k_3 s_0/(K_s + s_0)$ and thus $(dv_0/de)_{s=\text{const}} = k$. Therefore, v_0 is proportional to enzyme concentration, and this forms the basis of all enzyme assays and permits variation of enzyme concentration during initial rate measurements at the convenience of the investigator, provided the validity of Eq. (6-33) is tested first.

4. All other possible variables, that is, the temperature and the buffer used (with respect to pH, composition, and ionic strength), are defined and maintained constant in all measurements.

It thus becomes of the greatest importance to determine V and K_s as accurately as possible. The hyperbolic plot (Fig. 6-4) is not particularly useful for this purpose, because it involves the determination of an asymptote as $s \to \infty$, a notoriously difficult if not impossible task. Two methods for recasting Eq. (6-31) in *linear* forms, which transform this equation of a hyperbola into one of a straight line, are possible, however. They are due to Lineweaver and Burk, and Eadie and Hofstee and are given below as Eqs. (6-34) and (6-35) and Fig. 6-5, respectively; they are frequently referred to by the names of these investigators.

Double reciprocal form of Lineweaver and Burk—plot $1/v$ versus $1/s$:

$$\frac{1}{v} = \frac{K_s}{V}\frac{1}{s} + \frac{1}{V} \tag{6-34}$$

Single reciprocal form of Eadie and Hofstee—plot v/s versus v or s/v versus s:

$$\frac{v}{s} = -\frac{v}{K_s} + \frac{V}{K_s} \qquad \text{or} \qquad v = V - \frac{K_s v}{s} \tag{6-35}$$

also

$$\frac{s}{v} = \frac{K_s}{V} + \frac{s}{V} \tag{6-36}$$

The way in which these equations may be employed for the determination of V and K_s is also indicated in the figure. Plotting of data by Eq. (6-35) appears to have some advantages, but a really rigorous determination of actual values for V and K_s requires statistical treatment, preferably by means of a computer program.[4]

Although V thus has a well-defined meaning, both operationally and physically, as the extrapolated maximum velocity under the specified conditions, the same cannot be said for K_s. Operationally it is defined in terms of V as stated above:

$$K_s = (s)_{v=V/2} = s_{0.5} \tag{6-37}$$

[4] Data can be fitted to Eqs. (6-34) and (6-36) by the least-squares method, using v^4 weighting factors.

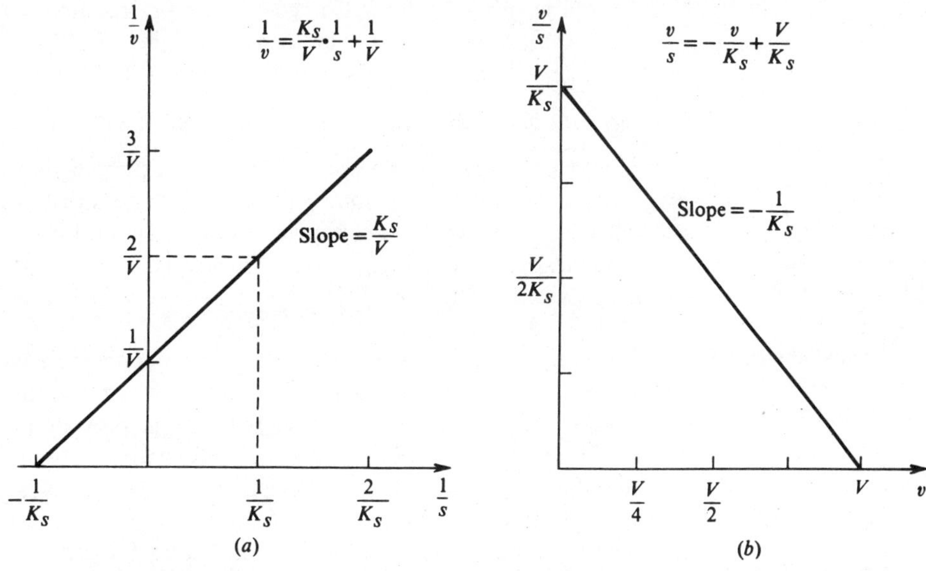

FIGURE 6-5
Reciprocal forms of the rate equation for an enzyme-catalyzed reaction: (*a*) Lineweaver and Burk; (*b*) Eadie and Hofstee.

but its physical significance depends on the mechanism, which first has to be established for each reaction by other means. In the succeeding sections we shall discuss a number of plausible mechanisms all of which give rise to the identical rate law [Eqs. (6-31) to (6-33)] and this does not exhaust the number of possibilities. Experimentally, values of K_s vary generally between 10^{-2} M and 10^{-5} M for most single-substrate reactions. In bisubstrate cases, values as low as 10^{-8} M or 10^{-9} M are occasionally observed.

Equation (6-33) gives us an opportunity to define a number of terms frequently found in the literature of enzyme chemistry.

An *enzyme unit* is now usually defined as the amount of substrate converted to product per unit time under certain specified conditions of *p*H, substrate concentration, and so on, because, as stated above, the concentration of enzyme is proportional to its catalytic activity; for example,

enzyme unit = mol product formed/min

specific enzymatic activity = units/mg enzyme protein
= mols product/min/mg protein

total enzymatic activity = specific activity × total amount of enzymatic protein
= total number of enzyme units

purification (enrichment) $= \dfrac{\text{specific activity (step } n)}{\text{specific activity (reference)}}$

yield (recovery) $= \dfrac{\text{total activity (step } n)}{\text{total activity (reference)}}$

catalytic constant = V/e_0 product formed/min/mol pure enzyme

This follows directly from Eq. (6-33), but it requires (1) a pure enzyme and (2) a knowledge of its molecular weight. If (2) is unavailable, the catalytic constant is frequently expressed per 100,000 g of enzyme;

turnover number = catalytic constant/ number of active centers

It is equal to the catalytic constant provided the enzyme is monomeric and each enzyme molecule can interact with only one substrate molecule at any one time; many enzymes are, however, polymeric and contain more than one independent, catalytically active site. Turnover numbers for enzymes can vary all the way from 10^2 to 10^6.

One-Substrate Reactions

In discussing mechanisms of this type, we are concerned with those in which either a single substrate participates exclusively, or the much more common case where the second participating reactant is present in such a large and constant concentration as to be kinetically without significance. The latter includes the large class of reactions catalyzed by hydrolases, of many transferases in the presence of a large amount of acceptor, of many isomerases and lyases (i.e., the enzymes of classes 2, 3, 4, and 5 of the classification scheme on page 326), a very large number of instances indeed.

Michaelis–Menten Mechanism. The simplest mechanism that can be written is the one already described—Eq. (6-27) with $C \equiv E$ for enzyme, $A \equiv S$ for substrate—that is, the reversible formation of a complex or compound between S and E, followed by the virtually irreversible breakdown of this complex to yield product and regenerate the free enzyme. In the special (and infrequent) case that $k_3 \ll k_2$, Eq. (6-30) reduces to

$$v_0 \simeq \frac{k_3 e_0 s_0}{k_2/k_1 + s_0} = \frac{V s_0}{\bar{K}_s + s_0} \tag{6-38}$$

where $\bar{K}_s = k_2/k_1$, the dissociation constant for the reaction

$$S + E \underset{k_2}{\overset{k_1}{\rightleftharpoons}} X$$

In this instance, originally discussed by Michaelis and Menten in 1913, K_s, the kinetically defined *Michaelis constant,* equals $\bar{K}_s$ the dissociation constant of the enzyme–substrate complex.[5]

Reversible Formation of the Enzyme–Substrate Complex. Let us now remove the restriction that the complex X breaks down irreversibly and instead admit the possibility that it can be formed from the product side. Thus we have

$$S + E \underset{k_2}{\overset{k_1}{\rightleftharpoons}} X \underset{k_4}{\overset{k_3}{\rightleftharpoons}} E + P \tag{6-39a}$$

$$-\frac{ds}{dt} = k_1 se - k_2 x \tag{6-39b}$$

$$\frac{dx}{dt} = (k_1 s + k_4 p)e - (k_2 + k_3)x \tag{6-39c}$$

[5] The Enzyme Commission has recommended that K_M be used for the kinetically determined Michaelis constant and that K_S be used for the enzyme–substrate dissociation constant. This leads to cumbersome and/or ambiguous expressions in cases where more than one substrate participates in an enzymatic reaction. In the notation suggested by Cleland, dissociation constants are designated as K_{ia}, etc. (Table 6-2).

$$\frac{de}{dt} = (k_2 + k_3)x - (k_1 s + k_4 p)e \tag{6-39d}$$

$$\frac{dp}{dt} = k_3 x - k_4 pe \tag{6-39e}$$

$$e_0 = e + x \tag{6-39f}$$

In the steady state, which obtains when $e_0 \ll s$ and/or p, $dx/dt \simeq de/dt \simeq 0$ and

$$(k_1 s + k_4 p)e = (k_2 + k_3)x \tag{6-39c$'$}$$

which with Eq. (6-39f) gives us two equations in e and x:

$$x = \frac{(k_1 s + k_4 p)e_0}{D} \qquad e = \frac{(k_2 + k_3)e_0}{D} \tag{6-40}$$

where $D = k_1 s + k_4 p + k_2 + k_3$. Substituting these values into either Eq. (6-39b) or (6-39e), we obtain

$$v = \frac{-ds}{dt} = \frac{dp}{dt} = \frac{(k_1 k_3 s - k_2 k_4 p)e_0}{(k_2 + k_3) + k_1 s + k_4 p} \tag{6-41}$$

Equation (6-41) reduces to the Michaelis–Menten equation in terms of the initial velocity v_1 for the forward reaction, that is Eq. (6-31), if we set $p = 0$, and to that for the reverse reaction with initial velocity v_2 if we set $s = 0$:

$$(v_1)_{p=0} = \frac{k_1 k_3 s e_0}{(k_2 + k_3) + k_1 s} = \frac{V_1 s}{K_s + s} = \frac{V_1}{(K_s/s) + 1} \tag{6-42a}$$

$$(v_2)_{s=0} = \frac{k_2 k_4 p e_0}{(k_2 + k_3) + k_4 p} = \frac{k_2 e_0 p}{(k_2 + k_3)/k_4 + p} = \frac{V_2 p}{K_p + p} = \frac{V_2}{(K_p/p) + 1} \tag{6-42b}$$

We can also, in this instance, calculate the values of the individual rate constants from the experimentally determined parameters K_s, K_p, V_1, and V_2 provided e_0 is known and expressed as mols of catalytic sites per liter.

$$k_1 = \frac{V_1 + V_2}{K_s e_0} \qquad k_2 = \frac{V_2}{e_0} \qquad k_3 = \frac{V_1}{e_0} \qquad k_4 = \frac{V_1 + V_2}{K_p e_0} \tag{6-43}$$

By definition $p_{eq}/s_{eq} \equiv K_{eq}$, the equilibrium constant for the overall reaction $S \rightleftharpoons P$, and at equilibrium $v = 0$; thus

$$K_{eq} = \frac{k_1 k_3}{k_2 k_4} \tag{6-44}$$

the equilibrium constant for step 1 divided by the equilibrium constant for step 2.

We can cast the rate equation for this mechanism [Eq. (6-41)] into another useful form. In general we can write

$$v = \frac{(N_1 s - N_2 p)e_0}{C_0 + C_1 s + C_2 p} \tag{6-45}$$

where in this instance $V_1 = N_1/C_1$, $V_2 = N_2/C_2$, $K_s = C_0/C_1$, $K_p = C_0/C_2$, and $K_{eq} = N_1/N_2$. Now multiply by N_2/C_1C_2 and obtain

$$v = \frac{V_1 V_2 s - V_1 V_2 p/K_{eq}}{K_s V_2 + s V_2 + V_1 p/K_{eq}} \tag{6-46}$$

This is the general form of the rate equation, and equations similar to it are obtained, not only for more complicated one-substrate or pseudo-one-substrate mechanisms, but for mechanisms involving two or three reactants as well. Unlike Eqs. (6-31) and (6-33)—or (6-42)—it is quite general and describes the effect of reactant concentration on rate, not only at the start of the reaction in the complete absence of P (in the forward direction) or of S (in the reverse direction), but at any time during the approach to equilibrium.

It must of course reduce to Eq. (6-42a) and (6-42b) in the special cases discussed earlier. The first case, $p = 0$, is self-evident. All we have to do is set $p = 0$ (provided that $V_2 \neq 0$) and then cancel out V_2. On the other hand, for the case $s = 0$ we obtain, after multiplying both numerator and denominator by K_{eq}/V_1,

$$v = \frac{V_2 p}{K_{eq} K_s V_2 / V_1 + p} \tag{6-47}$$

which at first glance is not equal to Eq. (6-42b). But from the definitions given above [Eq. (6-45)] we see that

$$\frac{K_{eq} K_s V_2}{V_1} = \frac{C_0}{C_2} = K_p \tag{6-48}$$

or

$$K_{eq} = \frac{V_1 K_p}{V_2 K_s} \tag{6-49}$$

the *Haldane relationship.* Now substituting K_p for the constant denominator term in Eq. (6-47), we obtain Eq. (6-42b).

The Haldane relationship is an exceedingly interesting one, because it states that the kinetic parameters of a reversible enzyme-catalyzed reaction are not independent of one another but are limited by the thermodynamic equilibrium constant for the overall reaction; this is, of course, equivalent to saying that the rate constants for specific steps must also exhibit a definite relationship to the overall equilibrium constant [Eq. (6-44)]. It has several interesting implications. The principle of microscopic reversibility states that any enzyme capable of catalyzing a reaction in the forward direction must of necessity also catalyze the reverse reaction. Yet on numerous occasions it would be of advantage to an organism to ensure that the reaction proceeds in one direction only, and that product not back up by reconversion to substrate. The objective is frequently reached by introducing steps that have a relatively large $-\Delta G^0$, that is, a large K_{eq}, together with a provision that $K_s \leq K_p$. Table 6-1 shows how these changes affect the ratio of maximal velocities V_1/V_2.

TABLE 6-1
Constraints on V_1/V_2 imposed by the Haldane relation: $p_{eq}/s_{eq} = K_{eq} = V_1 K_p / V_2 K_s$

	V_1/V_2 for K_p/K_a equal to				
K_{eq}	*10^4*	*10^2*	*1*	*10^{-2}*	*10^{-4}*
10^4	1	10^2	10^4	10^6	10^8
10^2	10^{-2}	1	10^2	10^4	10^6
1	10^{-4}	10^{-2}	1	10^2	10^4
10^{-2}	10^{-6}	10^{-4}	10^{-2}	1	10^2
10^{-4}	10^{-8}	10^{-6}	10^{-4}	10^{-2}	1

These two conditions cooperate to ensure a minimum of back reaction in spite of possible accumulation of product. As the concentration of product increases, most enzymatic reactions slow down; this is due to the phenomenon of product inhibition. To see this most clearly, let us rewrite the rate equation once again in a slightly different form:

$$v = \frac{K_p V_1 s - K_s V_2 p}{K_s K_p + K_p s + K_s p} = v_1 - v_2 \tag{6-50a}$$

where v_1 is the velocity in the forward direction, v_2 the velocity in the reverse direction:

$$v_1 = \frac{V_1 s}{K_s + s + pK_s/K_p}$$

$$\frac{1}{v_1} = \frac{K_s}{V_1}\frac{1}{s}\left(1 + \frac{p}{K_p}\right) + \frac{1}{V_1} \tag{6-50b}$$

$$v_2 = \frac{-V_2 p}{K_p + p + sK_p/K_s} \quad \text{(the minus sign indicates direction only)}$$

$$\frac{1}{v_2} = \frac{K_p}{V_2}\frac{1}{p}\left(1 + \frac{s}{K_s}\right) + \frac{1}{V_2} \tag{6-50c}$$

Once again we find a completely symmetrical set of equations, which now state that an enzymatic reaction slows down as equilibrium is approached, not only by virtue of the thermodynamic back reaction—the negative numerator term in Eq. (6-50a)—but also because, as p increases, an increasing proportion of the enzyme is immobilized as an *EP* complex. This kinetic effect of product inhibition is thus an intrinsic property of any realistic, that is, reversible, mechanism of enzymatic catalysis. Its effect on a double reciprocal plot is shown in Fig. 6-6. We observe that, compared with Fig. 6-5*a* and Eq. (6-34), the inter-

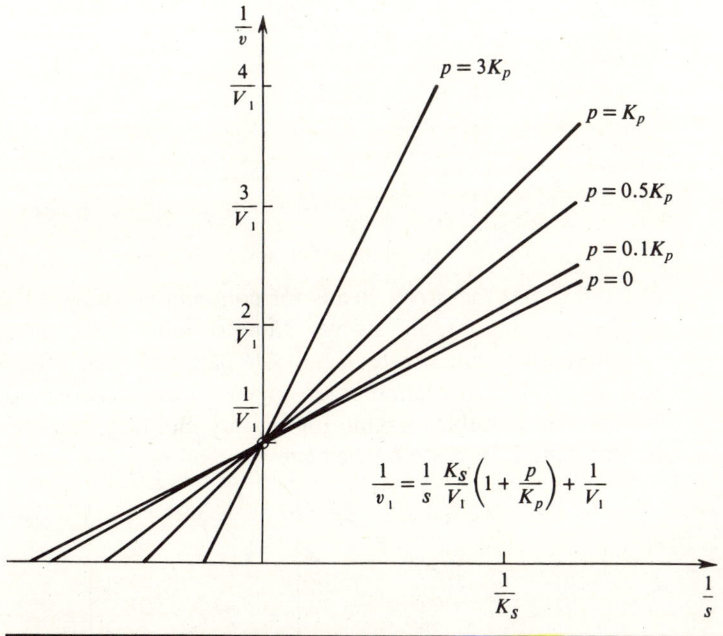

FIGURE 6-6
Rate equation in the presence of product.

cept $(1/V)$ is unaffected but the slope is increased by the constant factor $(1 + p/K_p)$. Behavior of this type is called *competitive inhibition.* It results whenever an inhibitor I combines only with the same form of the enzyme as does the substrate under investigation, and as far as this site is concerned, the enzyme can exist in just two forms, ES and EI. In this simple instance both P and S combine with E, but neither combines with X.

Reversible Formation of Two Complexes. It is not very realistic to expect that an enzyme capable of forming a complex or compound with S and capable of converting it reversibly to P is unable to form a reversible complex with P as well. A mechanism that takes this objection into account is given by

$$S + E \underset{k_2}{\overset{k_1}{\rightleftharpoons}} X \underset{k_4}{\overset{k_3}{\rightleftharpoons}} Y \underset{k_6}{\overset{k_5}{\rightleftharpoons}} E + P \tag{6-51}$$

The pertinent steady-state and conservation equations are

$$v = \frac{dp}{dt} = k_5 y - k_6 pe \tag{6-52a}$$

$$\frac{dx}{dt} = k_1 se - (k_2 + k_3)x + k_4 y \tag{6-52b}$$

$$\frac{dy}{dt} = k_6 pe + k_3 x - (k_4 + k_5)y \tag{6-52c}$$

$$e_0 = e + x + y \tag{6-52d}$$

In the steady state $dx/dt = dy/dt = de/dt = 0$. This yields three linear equations to be solved for three unknowns and substituted into the velocity equation (6-52a). This can be done by solving a third-order determinant (a determinant of order n for n forms of the enzyme in more general mechanisms) or more simply by a direct graphic method, due to King and Altman.

First, we write a simple geometric figure describing the mechanism; in this instance

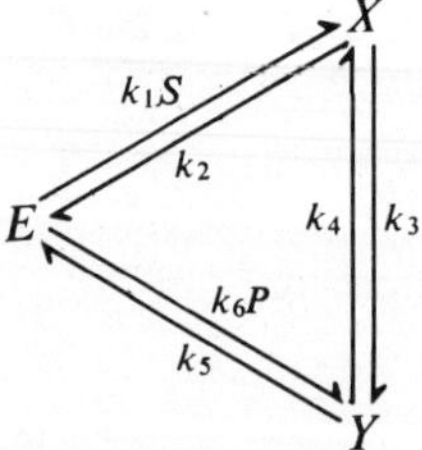

observing that the arrow heads for consecutive steps, 1→3→5 (forward) and 6→4→2 (reverse), are always made to point in the same direction.

Second, we write down all possible patterns (consisting of a number of lines one fewer than the number of enzyme forms: two in our case) that interconvert the possible enzyme forms; any closed loops are forbidden and must be eliminated. In our case we have

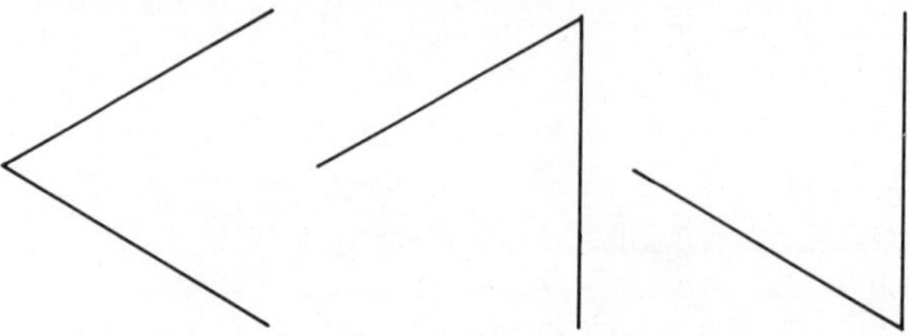

In general there will be $m!/(n-1)!(m-n+1)!$ patterns,[6] where m is the number of interconversion steps (lines in the basic figure) and n is the number of enzyme forms (corners in the basic figure).

Third, we write down, for each enzyme form, an expression of the following type: form/e_0 = [sum, for all patterns, of the product of the rate constant (multiplied by concentration terms, if any) on the path along the pattern leading to that form]/Σ, where Σ represents the sum of all numerator terms. In our case we have

$$\frac{e}{e_0} = \frac{k_2k_5 + k_2k_4 + k_3k_5}{\Sigma} \tag{6-53a}$$

$$\frac{x}{e_0} = \frac{k_1k_5s + k_1k_4s + k_4k_6p}{\Sigma} \tag{6-53b}$$

$$\frac{y}{e_0} = \frac{k_2k_6p + k_1k_3s + k_3k_6p}{\Sigma} \tag{6-53c}$$

$\sum$ = sum of all numerator terms

This, then, provides us with a set of partition equations that define the proportion of enzyme in any one form in terms of the individual rate constants. This is useful, not only for the present purpose, but for the determinations of inhibition patterns, of isotope exchange equilibria and rates, and so on.

We now substitute these partition equations into the rate equation (6-52a) and obtain

$$v = \frac{(k_1k_3k_5s - k_2k_4k_6p)e_0}{k_2k_5 + k_2k_4 + k_3k_5 + k_1(k_3 + k_4 + k_5)s + k_6(k_2 + k_3 + k_4)p} \tag{6-54}$$

In coefficient form this can be rewritten as

$$v = \frac{(N_1s - N_2p)e_0}{C_0 + C_1s + C_2p} \tag{6-54a}$$

which is identical to Eq. (6-45), derived for the simpler mechanism. Thus, no distinction between the two mechanisms can be made on the basis of steady-state rate measurements; indeed, it can be demonstrated quite rigorously that the introduction of any number of unimolecular steps between different enzyme forms, all containing the same number of substrate (or product) molecules, does not alter the form of the rate equation, whether expressed in the form of kinetic parameters (V_i's and K_i's) or coefficients. What does change is the composition of these parameters or coefficients in terms of individual rate constants.

There is an important difference inherent in the two mechanisms, described by Eqs. (6-39) and (6-51), corresponding to the rate laws of Eqs. (6-41) and (6-54), respectively. The former contains just four rate constants; the latter is expressed in terms of six of these entities. Now experimentally we can obtain only four kinetic parameters: V_1, K_s, V_2, and K_p. It is thus evident that only the simpler mechanism permits an explicit solution in terms of all its constituent rate constants [Eq. (6-43)]; for the more complex case we find ourselves faced with more unknowns than we have experimental parameters or equations. In general only certain restricted mechanisms permit the complete evaluation of all individual rate constants in terms of the kinetic parameters, that is, without simplifying assumptions or additional information.

[6] Not all these appear in the final calculation, which eliminates all patterns containing loops.

Reactions with One Substrate and Two Products (Covalent Intermediate). As discussed in Chapter 7, there is now compelling evidence for the formation of acyl enzymes, that is, covalent enzyme–substrate compounds, in the case of several hydrolytic enzymes of different specificity such as chymotrypsin, trypsin, thrombin, cholinesterase, and a number of phosphatases (1). The complete mechanism for such a reaction is

$$E + S \underset{k_2}{\overset{k_1}{\rightleftharpoons}} (ES) \underset{k_4}{\overset{k_3}{\rightleftharpoons}} EQ' + P$$
$$EQ' \underset{k_6}{\overset{k_5}{\rightleftharpoons}} EQ \underset{k_8}{\overset{k_7}{\rightleftharpoons}} E + Q \quad (6\text{-}55)$$

Provided K_{eq} is large, that is, the reaction favors product formation, the complete rate equation is (2)

$$v = \frac{Vs}{K_s + s + K_s p/\bar{K}_p + K_s q/\bar{K}_q + sp/\bar{K}_p + pqK_s/\bar{K}_p\bar{K}_q} \quad (6\text{-}56)$$

In the absence of products the expression for v_0 is therefore of the form of Eq. (6-31),

$$v = \frac{Vs_0}{K_s + s_0}$$

If Q, the second product released, is present, but P, the first one, is not, the equation becomes

$$v = \frac{Vs}{K_s(1 + q/\bar{K}_q) + s} \quad (6\text{-}56a)$$

If P is present and Q is not, it becomes

$$v = \frac{Vs}{K_s(1 + K_s p/K_s\bar{K}_p) + s(1 + p/\bar{K}_p)} \quad (6\text{-}56b)$$

Thus, the second product released gives competitive, the first *noncompetitive* inhibition, in which both the slope and the intercept of a v^{-1} versus s^{-1} plot vary with inhibitor concentration (Fig. 6-6). Rate measurements in the presence of product thus permit determination of the order of release of reactants as well as of the kinetic constants V, K_s, $\bar{K}_s$, $\bar{K}_p$, and $\bar{K}_q$.

In this mechanism

$$K_s = \frac{(k_2 + k_3)k_5k_7}{k_1k_3(k_5 + k_6 + k_7)} \qquad \bar{K}_s = \frac{k_2}{k_1}\ , \quad \bar{K}_p = \frac{k_4}{k_3}\ , \quad \bar{K}_q = \frac{k_8}{k_7}$$

provided that $k_7 \gg k_6$, that is, the intermediate EQ is not covalently linked to the enzyme and breaks down rapidly. This restriction appears to be fulfilled in the majority, if not all, the cases so far investigated and we can then formulate the following simplified mechanism:

$$E + S \underset{k_2}{\overset{k_1}{\rightleftharpoons}} ES \overset{k_3}{\rightleftharpoons} EQ' \underset{}{\overset{k_5(H_2O)}{\rightleftharpoons}} E + Q \quad (6\text{-}57)$$
$$\qquad\qquad\qquad + P$$

with the rate law

$$v = \frac{Vs}{K_s + s} = \frac{(k_3k_5/k_3 + k_5)e_0s}{\bar{K}_s(k_5/k_3 + k_5) + s} \tag{6-58}$$

Here V is the experimentally determined maximal velocity, sometimes called k_{cat}/e_0, and K_s is the experimentally determined Michaelis constant (or $s_{0.5}$) and equals $\bar{K}_s$, the dissociation constant of the enzyme–substrate complex, multiplied by the ratio of rate constants shown.

If $e_0 \ll s_0$, p, the amount of P produced at any time t, is given by (3)

$$p = \frac{Ve_0s_0}{s_0 + K_s} t + e_0\left(\frac{k_3/(k_3 + k_5)}{1 + K_s/s_0}\right)^2 \left\{1 - \exp - \left[\frac{(k_3 + k_5)s_0 + k_5\bar{K}_s}{\bar{K}_s + s_0}\right]t\right\} \tag{6-59}$$

For large values of t the exponential term approaches zero and the formation of p becomes a linear function of t:

$$p = \tilde{S}t + \pi \tag{6-59a}$$

where the slope $\tilde{S} = \dfrac{Ve_0s_0}{s_0 + K_s}$

and the intercept $\pi = e_0\left(\dfrac{k_3/(k_3 + k_5)}{1 + K_s/s_0}\right)^2$

Thus, if a plot of the concentration of the first product released against time is linear, the data conform to Eq. (6-59a); this provides experimental support for the existence of a covalent intermediate and, as long as $k_2 \gg k_3$, a determination of the enzyme concentration e_0, since

$$1/\sqrt{\pi} = 1/\sqrt{e_0} + K_s/(\sqrt{e_0})s_0 \tag{6-60}$$

Hence, a plot of $1/\sqrt{\pi}$, the reciprocal intercept of Eq. (6-59a), versus $1/s_0$ is again a straight line; its intercept with the ordinate affords $1\sqrt{e_0}$ and that with the abscissa $-K_s$.

Bisubstrate Reactions

General Considerations. We mentioned earlier that most enzymatic reactions involve more than one substrate, and any accurate description must take account of this fact. Consider the general equation for a reversible reaction of this type:

$$A + B \rightleftharpoons P + Q \tag{6-61}$$

The development of a comprehensive kinetic theory for this bisubstrate reaction has been due mainly to the work of Alberty and his collaborators, Dalziel, Cleland, and Hanes and Wong. The experimental facts on which much of this theory is based are quite analogous to those for one-substrate reactions: the variation of the initial velocity as a function of the variation of the concentration of one of the two substrates (for example, a, then called the *variable* substrate) is given by a Michaelis–Menten equation [(6-31), (6-34), and (6-35)] provided that (1) the same restrictions are observed as have been discussed for that case and (2) that the second substrate (here b, the *fixed* substrate) is held constant at a concentration of $b_1 \gg e_0$. If a similar set of measurements is then carried out at a different concentration $b_2 \gg e_0$, again the variation of the initial velocity as a function of a follows the Michaelis–Menten equation, but now it usually yields a different value for the two kinetic parameters V and K_a; they are, therefore, functions of the concentration of b as well. In general the pattern of the reciprocal plots is that of two families of straight lines [Eqs. (6-62) and (6-63) and Fig. 6-7] since, of course, the symmetry of

Eq. (6-61) demands that the same consideration hold for an inversion of fixed and variable substrate and for the reverse reaction as well. In general we shall use, in what is to follow, A, B; a, b; K_a, $\bar{K}_a$; k_{2n-1}; V_1, K_1, for substrates; their concentrations; their Michaelis (K_a) and dissociation constants ($\bar{K}_a$); individual rate constants; the maximal velocity and the equilibrium constant for the overall reaction, all in the forward direction, and P, Q; p, q; K_p, $\bar{K}_p$; k_{2n}; V_2, K_2 for the corresponding parameters in the reverse direction:

$$v_1 = \frac{V_1 ab}{\bar{K}_a K_b + K_b a + K_a b + ab} \tag{6-62a}$$

$$v_2 = \frac{V_2 pq}{\bar{K}_q K_p + K_q p + K_p q + pq} \tag{6-62b}$$

$$\frac{V_1}{v_1} = 1 + \frac{K_a}{a} + \frac{K_b}{b} + \frac{\bar{K}_a K_b}{ab} \tag{6-63a}$$

$$\frac{V_2}{v_2} = 1 + \frac{K_p}{p} + \frac{K_q}{q} + \frac{K_p \bar{K}_q}{pq} \tag{6-63b}$$

We see that for the two-substrate case the kinetic equation requires a total of four kinetic parameters: V_1, K_a, K_b, and $\bar{K}_a$ or K_{ia}.[7] This set is both necessary and sufficient for a unique description of the kinetics. Compare this with the one-substrate case, where just two parameters are required.[8]

Ordinarily one determines $n \times m$ values of the initial velocity v_1, corresponding to n different concentrations of a, and m of b (with $n \simeq m$), and then plots them in the manner indicated in Fig. 6-7a(1) and (2) as functions of $1/a$ and $1/b$ (see Dalziel equation, Table 6-2). The transformations of Eq. (6-63a) for these *primary* plots are then replotted as shown in Fig. 6-7b to give four *secondary* plots: two—6-7b(1) and (2)—the *intercept* replots, with identical intercepts $= 1/V_1$; and the other two [6-7b(3) and (4)] the two *slope* replots, with identical slopes $= \bar{K}_a K_b / V_1$. In toto the four replots

[7] Other conventions have also been used. Their relationship in terms of Eqs. (6-62) and (6-63) is shown in Table 6-2.

[8] Note that $\bar{K}_a$ or $\bar{K}_q$, the dissociation constant for the appropriate complex, can always be calculated, either from the slope of the plot of $1/v$ versus $1/b$ or directly from the point of intersection of the primary $1/v$ versus $1/a$ plots.

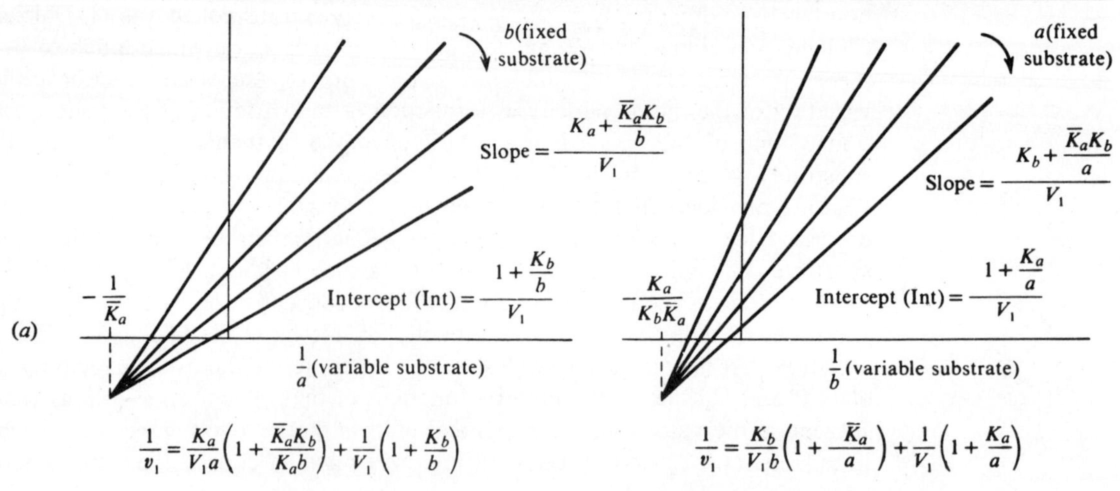

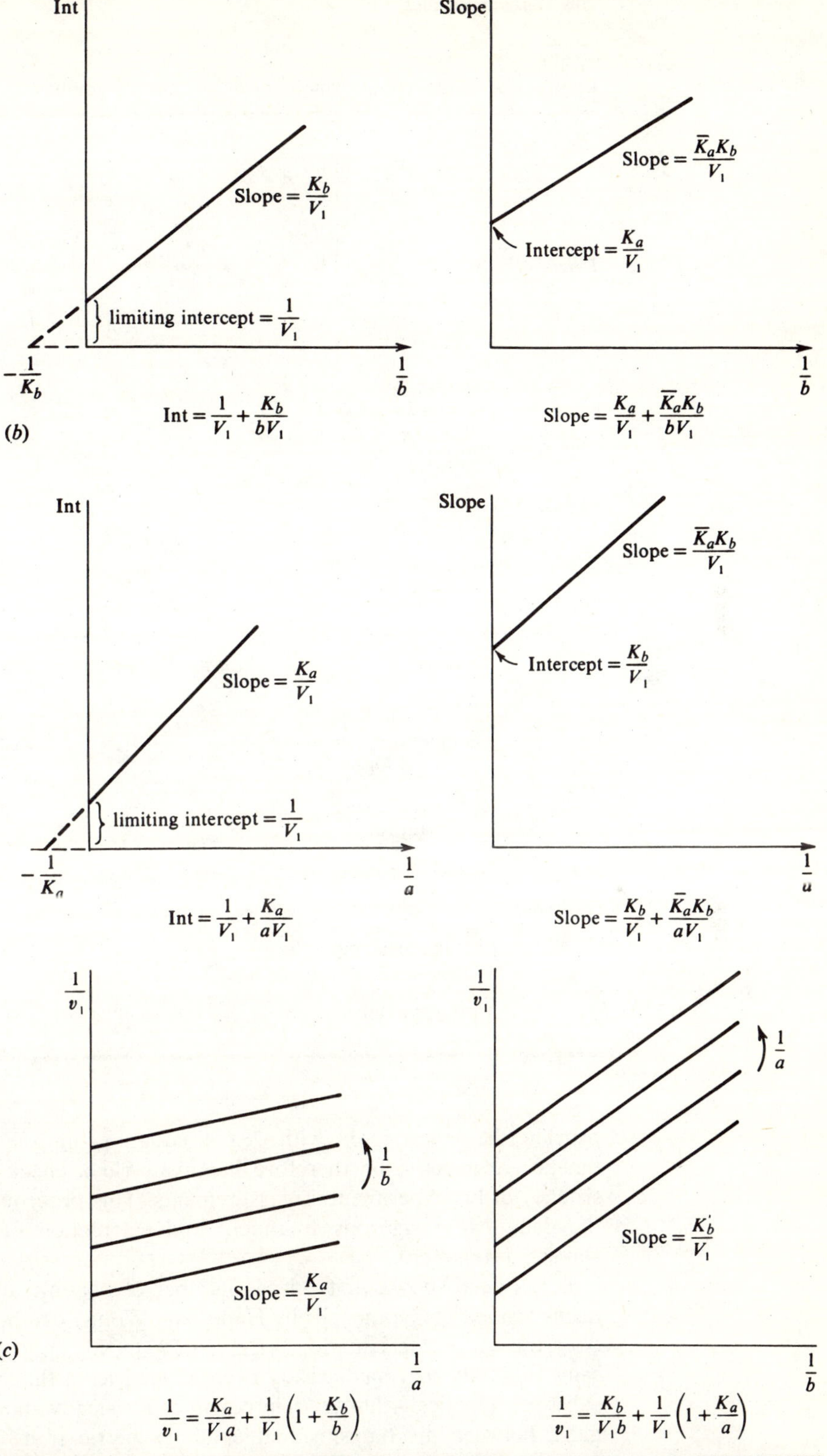

FIGURE 6-7

Reciprocal plots for bisubstrate reactions [Eq. (6-63)]. (*a*) Primary plots (reciprocal initial velocity versus 1/variable substrate); (*b*) secondary plots (intercepts and slopes of primary plots versus 1/fixed substrate); (*c*) primary plots for $\bar{K}_a K_b = 0$ (ping-pong mechanism), that is, $V_1/v_1 = 1 + K_a/a + K_b/b$.

TABLE 6-2
Comparison of different symbolisms of bisubstrate kinetics

Alberty (1956)

$$\frac{V_f}{v_f} = 1 + \frac{K_A}{A} + \frac{K_B}{B} + \frac{K_{AB}}{AB} \qquad K_A = \frac{\phi_1}{\phi_0} \quad K_{AB} = \frac{\phi_{12}}{\phi_0} \quad K_B = \frac{\phi_2}{\phi_0} \quad V_f = \frac{e}{\phi_0}$$

Dalziel (1957)

$$\frac{e}{v_f} = \phi_0 + \frac{\phi_1}{S_1} + \frac{\phi_2}{S_2} + \frac{\phi_{12}}{S_1S_2}$$

Bloomfield et al. (1963)

$$\frac{V_{AB}}{v_f} = 1 + \frac{K_{AB}}{K_B}\frac{1}{A} + \frac{K_{AB}}{K_A}\frac{1}{B} + \frac{K_{AB}}{AB} \qquad K_A = \frac{\phi_2}{\phi_{12}} \quad K_B = \frac{\phi_1}{\phi_{12}} \quad K_{AB} = \frac{\phi_{12}}{\phi_0} \quad V_f = \frac{e}{\phi_0}$$

Cleland (1963)

$$\frac{V_1}{v_1} = 1 + \frac{K_a}{A} + \frac{K_b}{B} + \frac{K_{ia}K_b}{AB} \qquad K_a = \frac{\phi_1}{\phi_0} \quad K_{ia} = \frac{\phi_{12}}{\phi_2} \quad K_b = \frac{\phi_2}{\phi_0} \quad V_1 = \frac{e}{\phi_0}$$

Our designation	*Meaning*	*Dalziel*	*Alberty*	*Bloomfield et al.*	*Cleland*	*Enzyme Commission*
K_a	Limiting Michaelis constant for A	$\frac{\phi_1}{\phi_0}$	K_A	$\frac{K_{AB}}{K_B}$	K_a	K_m^A
K_b	Limiting Michaelis constant for B	$\frac{\phi_2}{\phi_0}$	K_B	$\frac{K_{AB}}{K_A}$	K_b	K_m^B
$\bar{K}_a$	Dissociation constant for A	$\frac{\phi_{12}}{\phi_2}$	$\frac{K_{AB}}{K_B}$	K_A	K_{ia}	K_s^A
V_1	Limiting maximum velocity	$\frac{e}{\phi_0}$	V_f	V_{AB}	V_1	V
$\frac{V_1}{e_0}$	Turnover number	$\frac{1}{\phi_0}$	$\frac{V_1}{E_0}$	$\frac{V_{AB}}{E_0}$	$\frac{V_1}{E_t}$	None

provide the investigator with 2×4 equations in the four unknowns, the kinetic parameters; he therefore has available a check on the internal consistency of his experimental measurements. This procedure is used to visualize the data. More rigorous treatment and calculation of parameters requires statistical treatment.

It has been shown that a large number of possible and reasonable mechanisms (called homeomorphs by Hanes and Wong) can be devised, all of which obey the same rate equations [Eq. (6-63)]. Not only are we faced with the same difficulty that has already been alluded to in the simpler case of single-substrate kinetics—that it is impossible, by steady-state methods, to distinguish between mechanisms that entail isometric transformations of enzyme forms containing the same number of substrate (or transformed substrate ≡ product) molecules—but now there are also intrinsic differences in overall mechanism to contend with. For this reason considerable ingenuity and a great deal of discussion has gone into devising diagnostic tests that permit a selection of the most reasonable mechanism. Fortunately, the majority of the enzymatic reactions that have been studied in detail appear to fall into just a few categories[9]:

1. *Ping-pong* mechanisms of the type

$$\begin{array}{c} A \downarrow \quad P \uparrow \quad B \downarrow \quad Q \uparrow \\ \hline E \quad (EA\ E'P) \quad E' \quad (E'B\ EQ) \quad E \end{array}$$

$$A + E \underset{k_2}{\overset{k_1}{\rightleftharpoons}} (EA \rightleftharpoons E'P) \underset{k_4}{\overset{k_3}{\rightleftharpoons}} P + E'$$

or

$$B + E' \underset{k_6}{\overset{k_5}{\rightleftharpoons}} (E'B \rightleftharpoons EQ) \underset{k_8}{\overset{k_7}{\rightleftharpoons}} Q + E$$

or

$$E \underset{k_2}{\overset{k_1 A}{\rightleftharpoons}} (EA \rightleftharpoons E'P) \underset{k_4 P}{\overset{k_3}{\rightleftharpoons}} E' \underset{k_6}{\overset{k_5 B}{\rightleftharpoons}} (EQ \rightleftharpoons E'B) \underset{k_8 Q}{\overset{k_7}{\rightleftharpoons}} E \tag{6-64}$$

Here products are formed prior to reaction with the second substrate at the expense of conversion of the enzyme (or more usually of a coenzyme or prosthetic group tightly linked to the enzyme) from a form E to a form E'. Note that P is now the designation of the product form of A, and Q of the product form of B.

Among reactions that apparently obey this mechanism are those catalyzed by various pyridoxal phosphate (abbreviated below as pyr-CHO) requiring transaminases (4) (aminotransferases, Chapter 17) and those catalyzed by certain flavoproteins (5) (abbreviated as fp, Chapter 15).

$$\underset{A}{\underset{+NH_3}{R_1CHCO_2^-}} + \underset{E}{\text{Enz-pyrCHO}} \rightleftharpoons EA \rightleftharpoons E'P \rightleftharpoons \underset{E'}{\text{Enz-pyrCH}_2\text{NH}_3^+} + \underset{P}{R_1\overset{O}{\overset{\|}{C}}CO_2^-}$$

$$\underset{B}{R_2\underset{O}{\underset{\|}{C}}CO_2^-} + \underset{E'}{\text{Enz-pyrCH}_2\text{NH}_3^+} \rightleftharpoons E'B \rightleftharpoons EQ \rightleftharpoons \underset{E}{\text{Enz-pyrCHO}} + \underset{Q}{\underset{+NH_3}{R_2CHCO_2^-}} \tag{6-64a}$$

and

$$\underset{A}{DH_2} + \underset{E}{\text{fp}} \rightleftharpoons EA \rightleftharpoons E'P \rightleftharpoons \underset{E'}{\text{fpH}_2} + \underset{P}{D}$$

$$\underset{B}{Ox} + \underset{E'}{\text{fpH}_2} \rightleftharpoons E'B \rightleftharpoons EQ \rightleftharpoons \underset{E}{\text{fp}} + \underset{Q}{Ox\text{H}_2} \tag{6-64b}$$

where DH_2 is an oxidizable substrate and Ox an electron acceptor or oxidant.[10]

Kinetic analysis shows that for a mechanism of this kind the last term in Eq. (6-63a) or (6-63b) (i.e., the one in $a \cdot b$) is missing and rate equations are of the type

$$\frac{V_1}{v_1} = 1 + \frac{K_a}{a} + \frac{K_b}{b} \tag{6-63c}$$

This means that the double reciprocal plot exhibits an unusual feature: all plots of $1/v$ versus either $1/a$ or $1/b$ as variable substrates are parallel lines

[9] We show three methods of representing the reaction mechanism, including the graphic shorthand notation due to Cleland.

[10] Frequently the flavin is only reduced by one electron equivalent to the half-reduced semiquinone state. This requires the presence on the enzyme of a second reducible group such as a metal ion, a second flavin, or a disulfide (Chapter 15).

of identical slope, regardless of the concentration of the fixed substrate [Fig. 6-7c(1) and (2)]. Consideration of the complete rate equation for the reversible overall reaction shows another characteristic of this mechanism: Q (and not the product P) is a competitive inhibitor (affects slope of double reciprocal plot only, leaving identical intercept) for A, and P (not Q) for B. Several different Haldane relations are obtained:

$$K_{eq} = \frac{\bar{K}_p \bar{K}_q}{\bar{K}_a \bar{K}_b} = \frac{V_1}{V_2}\frac{\bar{K}_p K_q}{\bar{K}_a K_b} = \frac{V_1}{V_2}\frac{K_p \bar{K}_q}{K_a \bar{K}_b} = \frac{V_1^2}{V_2^2}\frac{K_p K_q}{K_a K_b} \tag{6-64c}$$

2. *Ordered* or sequential mechanisms of the type

$$\begin{array}{ccccccccc} & A\downarrow & & B\downarrow & & P\uparrow & & Q\uparrow & \\ \hline E & & AE & & (AEB\ \ QEP) & & QE & & E \end{array}$$

$$A + E \underset{k_2}{\overset{k_1}{\rightleftharpoons}} AE$$

$$\text{or}\quad AE + B \underset{k_4}{\overset{k_3}{\rightleftharpoons}} (AEB \rightleftharpoons QEP) \underset{k_6}{\overset{k_5}{\rightleftharpoons}} QE + P \qquad \text{or} \qquad \begin{array}{c} E \underset{k_2}{\overset{k_1 A}{\rightleftharpoons}} AE \underset{k_4}{\overset{k_3 B}{\rightleftharpoons}} AEB \\ E \underset{k_7}{\overset{k_8 Q}{\rightleftharpoons}} QE \underset{k_5}{\overset{k_6 P}{\rightleftharpoons}} QEP \\ AEB \rightleftharpoons QEP \end{array} \tag{6-65a}$$

$$QE \underset{k_8}{\overset{k_7}{\rightleftharpoons}} E + Q$$

Here A and its product Q are called the *leading* or *obligatory* reactants and are frequently the coenzyme in the reaction.

Among the most important reactions obeying this mechanism are those catalyzed by a variety of NAD- and NADP-requiring dehydrogenases (6–10). Now A = NAD(P)$^+$ and Q = NAD(P)H (Chapter 8). Ordered mechanisms generate rate laws of Eq. (6-62) or (6-63), but because of their symmetry it is not immediately obvious which substrate is to be identified as A, the "leading" substrate, which must interact with the enzyme before the reaction can proceed. This identification can, however, usually be accomplished by means of equilibrium binding studies (page 97). Furthermore, only A, and not B, is competitively inhibited by its own product (Q).

The main Haldane relation for this mechanism is

$$K_{eq} = \frac{V_1}{V_2}\frac{K_p \bar{K}_q}{K_b \bar{K}_a} \tag{6-65b}$$

If in the mechanism of reaction (6-65) the central complexes are rapidly interconvertible it will be described by only eight rate constants, and we can, therefore, express rate constants and kinetic parameters in terms of one another. The expressions obtained are

$$\begin{array}{llll} k_1 = \dfrac{V_1}{e_0 K_a} & k_2 = \dfrac{V_1 \bar{K}_a}{e_0 K_a} & k_3 = \dfrac{V_1}{e_0 K_b}\left(1 + \dfrac{k_4}{k_5}\right) & k_4 = \dfrac{V_2 k_2}{k_2 e_0 - V_2} \\ k_5 = \dfrac{V_1 k_7}{k_7 e_0 - V_1} & k_6 = \dfrac{V_2}{e_0 K_p}\left(1 + \dfrac{k_5}{k_4}\right) & k_7 = \dfrac{\bar{K}_q V_2}{K_q e_0} & k_8 = \dfrac{V_2}{e_0 K_q} \end{array} \tag{6-65c}$$

Random Mechanisms. For a mechanism in which all possible binary enzyme substrate complexes are formed rapidly and reversibly, the magnitude of any dissociation constant for a reactant is unaffected by the prior attachment of any other reactant to the enzyme, and the only slow step is the interconversion of the two ternary complexes:

$$\begin{array}{ccccc} A + E \underset{}{\overset{K_1}{\rightleftharpoons}} AE & \searrow^{K_3} & & {}^{K_4}\nearrow & PE \overset{K_2}{\rightleftharpoons} E + P \\ & & AEB \underset{k_2}{\overset{k_1}{\rightleftharpoons}} PEQ & & \\ B + E \overset{K_5}{\rightleftharpoons} EB & \nearrow_{K_7} & & {}_{K_8}\searrow & EQ \overset{K_6}{\rightleftharpoons} E + Q \end{array} \tag{6-66a}$$

The rate law will again be given by Eq. (6-63), but now

$$\begin{array}{lllll} V_1 = k_1 e_0 & \bar{K}_a = K_1 & \bar{K}_b = K_5 & K_a = K_7 & K_b = K_3 \\ V_2 = k_2 e_0 & \bar{K}_p = K_2 & \bar{K}_q = K_6 & K_p = K_8 & K_q = K_4 \end{array} \tag{6-66b}$$

Depending on the magnitude of the ratio $\bar{K}_a/K_a$, double reciprocal plots for reactions obeying this mechanism may show intersection points above, below, or on the abscissa. If on this axis, the plots become identical to those for the sequential mechanism, because the dissociation constant for any one reactant is unaffected by the presence of the second. The usual case is for the crossover point to be above the abscissa, which indicates that addition of one substrate facilitates addition of the second (dissociation constant decreased). Although thus indistinguishable from the sequential mechanism in certain instances on the basis of initial rate studies alone, this mechanism exhibits features that permit its identification by means of tracer experiments. In isotope-exchange studies with the tracer added at equilibrium (11,12), the random mechanism requires that the rate of the appearance of tracer in A when Q is labeled, and in B when P is labeled (and vice versa), be faster than that of A into Q (and the reverse). Furthermore, for dehydrogenases, those obeying random mechanisms exhibit larger H/D or H/T isotope effects on initial rates than do ones with ordered mechanisms (13). The reaction catalyzed by many kinases (phosphotransferases), enzymes that catalyze the reaction A—H + RO—PO_3^- → APO_3^{2-} + ROH, appear to behave in this fashion. A good example is creatine kinase.

Special Cases. In addition to the three important mechanisms just described, the rate law of Eqs. (6-62) and (6-63) is also generated by other more restrictive models. For instance, if we assume that in mechanism (6-65a) $k_5 \gg k_7$, $k_4 \gg k_2$, $(k_5 + k_7) \gg k_6$, and $(k_2 + k_4) \gg k_3$, then we obtain as a special case the *Theorell–Chance mechanism.* In this mechanism the steady-state concentrations of the ternary complexes are very low, and the first product P appears to be formed directly from the substrate B by interaction with an E—A complex. Experimentally the distinguishing feature for this mechanism is that the reciprocal plots show the pattern of Fig. 6-7*a* and *b, but* the product inhibition pattern is that yielded by ping-pong mechanisms, that is, competitive inhibition of A by Q (not P) and of B by P (not Q). Under some conditions the reaction catalyzed by the lactic dehydrogenase of mammalian heart and the alcohol dehydrogenase of mammalian liver may obey this mechanism (6, 9):

$$A + E \rightleftharpoons AE \underset{-B+P}{\overset{+B-P}{\rightleftharpoons}} QE \rightleftharpoons E + Q \tag{6-67}$$

Another mechanism that gives rise to double reciprocal plots identical to those just described may be applicable to a number of general transferase reactions, that is, reactions that represent an extension of those described by

Eqs. (6-55) and (6-56). The enzymes, for example various phosphatases and other hydrolases, also catalyze metathetical reactions of the general type

$$A{-}X + Y{-}H \rightleftharpoons A{-}H + Y{-}X$$

where Y—H may be either HO—H or some other acceptor RO—H. Many of these reactions have also been shown to involve enzyme-bound covalent intermediates, and Arion and Nordlie (14) have suggested that in that case the mechanism may be

$$\begin{array}{ccccc} & k_1AX \nearrow & E{-}AX & \searrow k_3 & \\ & \swarrow k_2 & & k_4A \nwarrow & \\ E & & \xleftarrow{\;kH{-}OH\;} & & E{-}X \\ & \searrow k_8YX & & k_6 \nearrow & \\ & k_7 \nwarrow & E{-}YX & \swarrow k_5HY & \end{array} \tag{6-68}$$

Here the upper branch describes hydrolysis of AX in the absence of transfer; the lower branch is applicable to the analogous hydrolysis of XY. The rate law for either of these reactions is described by Eqs. (6-58). Transfer, without hydrolysis, a reaction that is not likely with hydrolytic enzymes, would then be represented by the reaction in the outer rim. It is clear that this path is equivalent to a ping-pong mechanism [Eq. 6-64)] and its rate equation is therefore given by Eq. (6-63c), and graphically represented by Fig. 6-7*c*.

If we measure the progress of the overall reaction by following the formation of XY, then the rate law is given by Eqs. (6-62) and (6-63) (Fig. 6-7) with (ax), (yx), etc., substituted for a, b, etc. If, as occasionally observed, the double reciprocal plots of $1/v$ versus $1/ax$ or versus $1/yx$ all intersect on the x axis, this means that $K_{ax}K_{yx} = \bar{K}_a\bar{K}_b \equiv \bar{K}_{ax}\bar{K}_{yx}$ or, in other words, that the kinetically determined Michaelis constants for the various substrates are identical with their respective dissociation constants.

Enzyme Activation

By an activator we mean a small molecule, frequently an inorganic ion, which is required for, or which at least stimulates, the activity of an enzyme but which, unlike a coenzyme, is not itself an explicit participant in the reaction. Although many kinetic schemes involving the interactions of such activators with various components of the system may be written, only a limited number appear to be of practical importance. We shall restrict our attention to the two most common ones: obligatory (or exclusive) interaction with the free enzyme and obligatory (or exclusive) interaction with the free substrate.

Interaction with the Free Enzyme. Many enzymes contain dissociable metal ions as part of their active sites. Among them we may mention Mg^{2+} (or Mn^{2+}) in enolase, β-methylaspartase, isocitrate dehydrogenase, malic enzyme, and so on. In addition, enzymes may also require the presence of a monovalent cation, usually Na^+, K^+, or NH_4^+, for the stabilization of the particular conformation responsible for maximal catalytic activity.

One possible reaction path for activation is

$$\begin{gathered} E + M \underset{k_2}{\overset{k_1}{\rightleftharpoons}} EM \\ S + EM \underset{k_4}{\overset{k_3}{\rightleftharpoons}} SEM \underset{k_6}{\overset{k_5}{\rightleftharpoons}} PEM \underset{k_8}{\overset{k_7}{\rightleftharpoons}} EM + P \end{gathered} \tag{6-69}$$

The analogy between this mechanism and that of Eq. (6-65) is only apparent; they become kinetically equivalent only if the *EM* complexes with which *E* and *P* combine are *not* identical. The rate law for this mechanism is given by

$$v_1 = \frac{V_1 sm}{\bar{K}_m K_s + K_s m + sm} \qquad v_2 = \frac{V_2 pm}{\bar{K}_m K_p + K_p m + pm} \tag{6-70}$$

If the same mechanism is extended to include formation of an activator-free *ES* complex and its breakdown to yield product by a slower reaction with specific rate constant k', solution by steady-state methods is still possible but becomes quite complicated. If we assume all reactions to be rapid in comparison with the breakdown of the two complexes *SEM* and *PEM* (the *equilibrium assumption*), we obtain for the scheme

$$\begin{array}{ccc} E + S & \overset{\bar{K}_s}{\rightleftharpoons} & ES \xrightarrow{k} \\ + & & + \\ M & & M \\ \bar{K}_m \Big\Updownarrow & & \Big\Updownarrow \bar{K}_{sm} \\ ME + S & \overset{\bar{K}_{ms}}{\rightleftharpoons} & MES \xrightarrow{k'} \end{array} \tag{6-71a}$$

the rate law

$$\begin{aligned} \text{rate} = v + v' &= \frac{V_1(s/\bar{K}_s) + V'_1 ms/\bar{K}_m\bar{K}_{ms}}{1 + s/\bar{K}_s + m/\bar{K}_m + ms/\bar{K}_m\bar{K}_{ms}} \\ &= \frac{V_1 + V'_1 m\bar{K}_s/\bar{K}_m\bar{K}_{ms}}{1 + \bar{K}_s/s[1 + m/\bar{K}_m(1 + s/\bar{K}_{ms})]} \end{aligned} \tag{6-71b}$$

where $V_1 - ke_0$ and $V_1' = k'e_0$. An enzyme that appears to obey this mechanism, but with $k' = 0$, is methylaspartase (15).

The dissociation constant for the enzyme–metal complex can, therefore, be determined by Eq. (6-70) or (6-71b) from kinetic measurements in the steady state, provided the activation is essentially compulsory. Direct measurements can also frequently be applied either as alternatives to or in confirmation of these kinetic results. Among them[11] are (1) ultraviolet difference spectrophotometry (16), applicable if there is a distinction, however slight, between the spectra of *E, EM,* and *EMS;* (2) electron spin resonance, provided the ion to be used is paramagnetic (e.g., Mn^{2+}); (3) proton relaxation rate, a nuclear magnetic resonance technique; and (4) equilibrium binding studies (e.g., equilibrium dialysis, sedimentation, or gel exclusion) using sensitive analytical techniques (e.g., radioisotope measurements) for the determination of the metal in either the free or the bound state (17). The last two methods named are applicable, at least in principle, to all cases of interest.

Interaction with the Substrate. A large number, and perhaps all of the important reactions involving nucleoside di- and triphosphates, require the simultaneous participation of a divalent metal ion (usually Mg^{2+}) in stoichiometric amounts. Investigation has shown that the probable mechanism for this activation is the obligate combination of the cation with the substrate prior to its

[11] These methods will, of course, measure any combination of metal with enzyme, including sites other than the enzymatically active one.

interaction with the enzyme. Thus this substrate–metal complex rather than the free substrate itself is the true reactant. The reaction scheme is

$$S + M \underset{k_2}{\overset{k_1}{\rightleftharpoons}} SM$$

$$SM + E \underset{k_4}{\overset{k_3}{\rightleftharpoons}} (ESM \rightleftharpoons EPM) \underset{k_6}{\overset{k_5}{\rightleftharpoons}} E + PM \tag{6-72}$$

$$PM \underset{k_8}{\overset{k_7}{\rightleftharpoons}} P + M$$

Unlike all other mechanisms discussed so far, this implies that S exists in more than one kinetically important form (as S and SM) and that s, the concentration of free substrate, can no longer be regarded as being essentially constant and equal to s_0, the stoichiometric concentration of substrate added.

Therefore, in these instances one has to use the concentration of SM rather than of free S in all relevant rate equations (18). If the equilibrium between substrate and metal is established rapidly (dissociation constant $\bar{K}_m = k_2/k_1$) and the initial rate as a function of substrate concentration is examined in the presence of a large excess of metal ($m_0 > s$) we observe the rate law given by Eq. (6-72a):

$$\frac{V_1}{v_1} = 1 + \frac{K_s}{s}\left(1 + \frac{\bar{K}_m}{m_0}\right) \tag{6-72a}$$

Conversely, the initial rate as a function of metal concentration in the presence of substrate in large excess ($s_0 > m$) is given by Eq. (6-72b):

$$\frac{V_1}{v_1} = 1 + \frac{K_s}{m}\left(1 + \frac{\bar{K}_m}{s_0}\right) \tag{6-72b}$$

The complete symmetry of Eqs. (6-72a) and (6-72b) indicates that (1) plots of v_1 (or $1/v_1$) versus s (or $1/s$) with M at a given concentration c should be completely superimposable on plots of v (or $1/v$) versus m (or $1/m$) with S at the identical concentration c, (2) the maximal velocity V_1 can be obtained and is equal in magnitude from either plot, and (3) a replot of the slope from either plot can only yield the complex constant $K_s\bar{K}_m$ and neither the dissociation constant $\bar{K}_m$ nor the Michaelis constant K_s.

The stability constants for ATP complexed with various metal ions have been measured with care (19) and shown to be in the order Cu > Zn $\simeq$ Ni > Co > Mn $\approx$ Mg $\geqq$ Ca for both the reaction at an alkaline and an acid pH, that is, $[\mathrm{Me \cdot ATP^{2-}}]/[\mathrm{Me^{2+}}][\mathrm{ATP^{4-}}]$ and $[\mathrm{MeHATP^{-}}]/[\mathrm{Me^{2+}}][\mathrm{HATP^{3-}}]$ (stability constants for Mg^{2+} 4.88 and 2.70, respectively). On the other hand, their *rate* of formation has also been measured by means of temperature-jump kinetics (20) (page 321). The second-order association rates (in $\mathrm{M^{-1}\ sec^{-1}}$) are for Ni^{2+} – 4.1×10^6; for Mg^{2+} – 1.2×10^7; for Co^{2+} – 9.2×10^7; for Mn^{2+} $> 10^9$; and for Ca^{2+} $> 10^9$; the first-order dissociation rates (in $\mathrm{sec^{-1}}$) are for the metals in the same order: 0.38×10^2, 1.2×10^3, 2.0×10^3, $>10^4$, and $>10^5$, respectively. The implication is that, although the dissociation constant for Mg^{2+} ATP^{-4} is of the same order of magnitude ($\sim 10^{-4}$) as that for Ca^{2+} (see above), the kinetics of their formation (and dissociation) are greatly different, the rate constants for the Ca^{2+} complex being some two orders of magnitude faster in both directions. These dissociation rates indicate that Ni^{2+} would not be an effective activator for those enzymes in which the dissociation of an enzyme–substrate–metal complex is the rate-limiting elementary step. On the other hand, if the role of the metal is primarily concerned with a bond-breaking step, its efficiency may be related to its acidity, that is, its ability to dissociate a proton from the innermost hydration shell; in this case the order should be $Ca^{2+} < Mg^{2+} \simeq Mn^{2+} < Co^{2+}$.

Enzyme Inhibition

General Principles. A substance which decreases the rate of a biochemical reaction is called an inhibitor. Classically, the study of inhibitory effects on isolated enzymic reactions and on metabolic sequences in general has been of the greatest importance in establishing the nature of the free reactants, the nature of their binding site on the enzyme, and the specificity and mechanism of the reaction. In the cell, inhibition of key reactions by substances that may be products of the reaction itself or of the same (or even of a different) metabolic sequence provide a ready and delicately poised control mechanism for the maintenance of a relatively constant intracellular environment and for its response to alterations in the external milieu. Finally, such selective inhibitions by natural or synthetic compounds (*antimetabolites*) must form the base of at least one broad approach toward a rational pharmacology and chemotherapy, especially if we take into account the following facts: (1) advantage can be taken of the differences in specificities of different enzymes such that the "unnatural" products of a proximal may become the inhibitors of a distal reaction, a process that has been called "lethal synthesis" by Peters (21); (2) enzymes are themselves also products of metabolism, and their synthesis may be similarly manipulated.

The kinetics of most enzyme reactions in the presence of varying concentrations of inhibitors are as follows. The most characteristic features are linear double reciprocal plots of $1/v$ versus $1/s$, but (1) the slope, (2) the intercept, or (3) both are altered by factors $(1 + i/K_{ij})$, where i is the concentration of inhibitor and K_{ij} is a characteristic constant. The first case is called *competitive,* the second *uncompetitive,* and the third *noncompetitive* (Fig. 6-8).

Replots of slopes or intercepts are usually linear but may be parabolic, hyperbolic, or of even greater complexity. *Dead-end inhibitors* (see below) normally produce linear plots, unless the inhibitor combines twice with the same enzyme form or unless this inhibitor is a product that combines "downstream" and backs the reactants up to its point of combination with the enzyme.

Reactions with Single Substrates. 1. The inhibitor may resemble the substrate sufficiently so as to be bound in its stead at the active site. It forms an *EI* instead of an *ES* complex and thus competes with substrate for the free enzyme. This *EI* complex is either incapable of being converted to any product, in which case we speak of a *dead-end inhibitor,* or the rate may be so slow as to be kinetically insignificant. This type of inhibition is referred to as *competitive.* The kinetic diagnostic is that in the double reciprocal plot only the slope is affected; the intercept $(1/V)$ remains unaffected (Fig. 6-8*a*). That products can act as competitive inhibitors has been discussed on page 281.

2. The kinetic diagnostic for *uncompetitive inhibition* is that in the double reciprocal plot the slopes remain constant (i.e., the lines are parallel), but the intercepts vary with i (Fig. 6-8*b*). This form of inhibition is rare with single substrates where one possible but not highly likely mechanism entails combi-

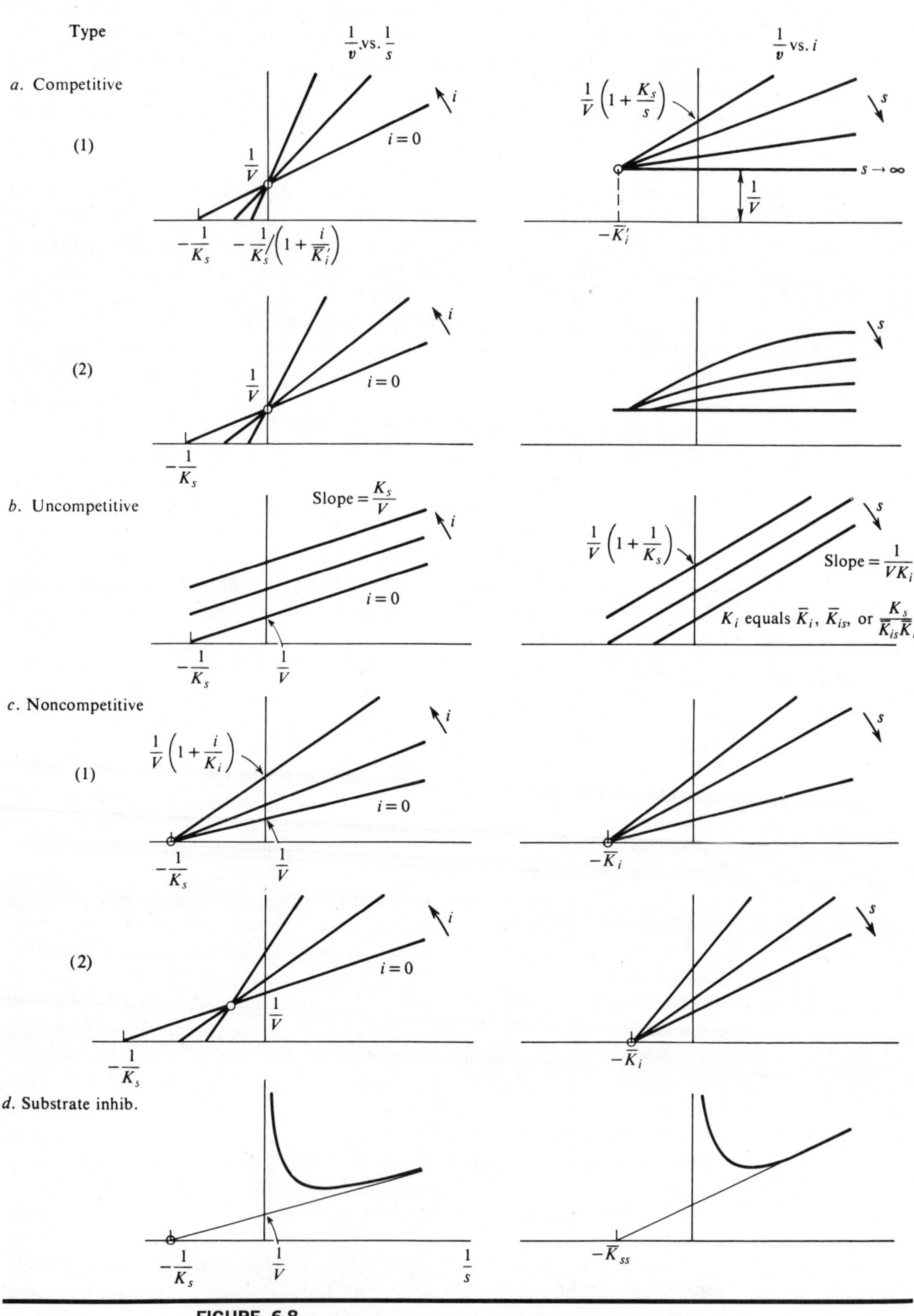

FIGURE 6-8
Kinetic patterns for various inhibition types (see Table 6-3).

nation of I exclusively with the ES complex. It becomes quite common in more complex cases.

3. The inhibitor can combine with the enzyme at a second site, not identical wholly or in part to the substrate-binding site. We obtain, therefore, the formation of IE as well as IES complexes, with the result that the rate of conversion $S \rightarrow P$ is slowed but not stopped. The effect may be on the affinity of the enzyme for S, in which case K_s is increased, or on the catalytic rate, in which case V is decreased. Two different alternatives can then be distinguished.

The IES complex is converted to products at the same rate as the ES complex, but the presence of I on the enzyme affects its affinity for S; that is, $\bar{K}_{is} > \bar{K}_s$. Kinetically this mechanism is indistinguishable from case 1 above so far as the $1/v$ versus $1/s$ plot is concerned, but it leads to a different $1/v$ versus i and slope versus i plot (hyperbolic instead of linear). We may call this case *apparently* or *hyperbolic competitive* [Fig. 6-8a(2)] and it may be regarded as an instance of allosteric inhibition (see the next section). An example is given in Ref. 22.

The presence of I on the enzyme affects its affinity for S as before, but also the IES complex is converted to products at a rate that is less than that of the ES complex. This has frequently been cited as the explanation for *noncompetitive* inhibition, that is, the pattern in which we observe effects on the slope as well as on the intercept of the double reciprocal plot. The straight lines always intersect at one point to the left of the origin, which may be above, below, or on the $1/s$ axis (Fig. 6-8c). Replots of slopes or intercepts versus i are hyperbolic. In the special case where this point is on the axis, we have *pure noncompetitive* inhibition, and this has been interpreted in the literature in terms of a complete absence of any influence of I on the affinity of E for S, that is, that $\bar{K}_{is} = \bar{K}_s$ [Fig. 6-8c(1)], a most unlikely event. Actually noncompetitive inhibition in the more realistic multisubstrate cases may perfectly well be caused by interaction with the substrate binding site.

Many instances of inhibition by excess substrate due to its binding at a secondary site, with a somewhat smaller affinity, are known (Fig. 6-8d). If the ternary complex is completely inactive the rate law is given by

$$v \simeq \frac{V'}{1 + K_s/s + s/\bar{K}'_{ss}} \tag{6-73}$$

Plots of v (not $1/v$) versus log s give symmetrical bell-shaped curves with inflection points $\simeq K_s$ and $\bar{K}'_{ss}$. The concentration s_{opt} at which v reaches its maximal value is readily obtained by differentiating Eq. (6-73) with respect to s and setting the resulting equation equal to zero and solving for s: $s_{\text{opt}} = (K_s\bar{K}'_{ss})^{1/2}$.

Multisubstrate Cases. When more than one substrate participates in an enzymatic reaction, the kinetic effects exerted by inhibitors can become quite complex. Among other things, slopes and/or intercepts of the double reciprocal plot may no longer be linear but may become higher- (usually second-) order functions of i. In addition, different mechanisms generate quite different patterns for product inhibition and substrate inhibition as well as for the effects observed on addition of competitive dead-end inhibitors. To obtain a complete description, one must, of course, derive the complete rate equation, which, in turn, must take into account the appropriate complexes between all enzyme forms and inhibitor, and then recast the equation to determine the effect on rate produced by varying any particular component. If the rate equation, in the absence of inhibitor, is already known, then the procedure is relatively simple, particularly in the case of dead-end inhibitors. One simply multiplies all those terms in the denominator of the rate equation which correspond to the enzyme forms combining with the inhibitor by $(1 + i/\bar{K}_i)$, where $\bar{K}_i$ is the dissociation constant of the complex between that particular enzyme form and inhibitor. For example, in the ordered mechanism of reaction (6-65), the rate law of which is given by Eq. (6-62), the first and third terms in the denominator represent the enzyme in its free form (E), the second term represents the EA complex, and the fourth term corresponds to the sum of the ternary AEB and PEQ complexes plus the EQ complex.

If all that is required is a qualitative picture of the inhibition type to be expected (i.e., competitive, uncompetitive, or noncompetitive), this can be accomplished by inspection of the proposed mechanism with the use of the following simple rules due to Cleland:

1. An inhibitor affects the slope of a reciprocal plot ($1/v$ versus $1/s$) when it and the variable substrate S either compete directly for the same form of the enzyme or react with forms that are separated from each other in the reaction path only by reversible steps, so that either component can affect the concentration of enzyme available to the other by simple displacement of the various equilibria intervening. Thus, an effect of I may be overcome by raising S.

2. An inhibitor affects the intercept of such a reciprocal plot when it combines with a form of the enzyme that differs from that responsible for combination with free S and thus lowers the amount of total enzyme available for distribution among the usual enzyme forms in a manner that cannot be overcome by saturation with S.

3. These effects can occur separately, in which case they produce competitive or uncompetitive patterns or, jointly, the noncompetitive pattern. If a substrate or inhibitor combines with more than one form of an enzyme, the resulting pattern is due to the sum of the different inhibitions produced singly. Any patterns can be linear or of higher order (e.g., parabolic or hyperbolic).

As a useful example we shall now consider the case of adding an inhibitor (including the product Q) competing for the free enzyme with A, the first substrate bound in the case of the ordered bireactant mechanism [Eq. (6-65)]. If we look at the mechanism as a whole, that is,

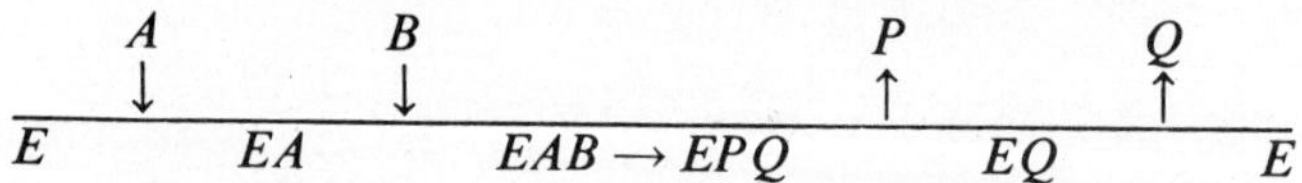

and consider three different types of inhibitors, a dead-end inhibitor combining with either EA or EQ, the product P, and a dead-end inhibitor combining with free enzyme, application of Cleland's rules predicts the following patterns:

Inhibitor form	*A as variable substrate*	*B as variable substrate*
EAI	Uncompetitive	Competitive
EQI	Uncompetitive	Uncompetitive
EQP	Noncompetitive	Noncompetitive
EI	Competitive	Noncompetitive

Observe that the mechanisms involving EAI, EQI, or EI complexes can easily be distinguished. Incidentally a linear noncompetitive pattern in which the various lines cross on the x axis is a purely fortuitous circumstance. Nonlinear (e.g., parabolic) inhibitions are observed in certain instances when I can combine with more than one enzyme form. With a dead-end inhibitor this must mean that both an EI and an EI_2 complex is formed; with products the situation is not quite so straightforward. Product inhibition patterns are also exceedingly useful as criteria for the selection of permissible mechanisms; as already mentioned, Q added to the forward reaction gives inhibition competitive to A, noncompetitive to B; P added is noncompetitive to both A and B for the ordered mechanism, but for the random equilibrium mechanisms both P and Q give competitive patterns with either A or B.

Important Inhibitors. Table 6-3 gives a summary of some important inhibitors of interest to the biochemist.

Allosteric Regulation (23–27)

General Features. One of the most remarkable recent developments in enzymology and molecular biology has been the discovery that a large number of key enzymes in metabolic sequences (Chapter 11) appear to exhibit anomalous kinetic behavior. First documented for enzymes responsible for the regulation of amino acid (28) and pyrimidine (29) biosynthesis in bacteria the relevant phenomenology has since been demonstrated to be of quite general and pervasive applicability. Its most explicit statement, by Monod, Wyman, and Changeux (24) and Blangy, Buc, and Monod (25) postulates that the catalytic function of enzymes may be affected and controlled by interaction with small molecules, not only directly, at or close to the active site, but also indirectly, at distant, secondary, *allosteric* sites. This type of interaction may be recognized and distinguished from the more common direct action of such modifiers (activators or inhibitors) at the catalytic site by some or all of the following characteristics:

1. Anomalous kinetic order with respect to substrate (plots of v versus s yield sigmoid rather than the usual, hyperbolic curves; Fig. 6-9, heavy lines) suggesting multiple interacting binding sites. Allosteric effects resulting from interactions between identical ligands (i.e., a substrate binding to the enzyme at more than one site in the absence of other modifiers) have been termed *homotropic* effects. These effects are always cooperative ($\bar{n}$ in Hill plots > 1; Fig. 6-11).

2. Certain agents called *allosteric effectors, modifiers,* or *determinants,* which are usually completely unrelated structurally to substrate and which frequently represent the end (or key) products of an important metabolic sequence, are bound to the enzyme and profoundly influence its catalytic activity. Allosteric effects resulting from interactions between different ligands are termed *heterotropic* effects. These effects may be either cooperative or antagonistic; that is, they may result in either activation or inhibition (Fig. 6-9). The allosteric effectors may influence the affinity of the active site for substrate (K systems), the rate of conversion of substrate to product following formation of the enzyme–substrate complex (V systems), or both. Kinetically, heterotropic inhibitory effects are "partially competitive"; that is, they can be only partially overcome by high concentrations of substrate. "Normal," hyperbolic kinetic behavior is observed at high concentrations of effector.

3. Other agents, which bear at least some resemblance to the substrate structurally, elicit a biphasic response; they activate at low and inhibit at high concentrations, thus they appear to act as both allosteric effectors in the former and substrate-competitive inhibitors in the latter range.

4. The interaction of these various molecules with the allosteric site on the enzyme is frequently correlated with alterations in the quaternary structure of the protein. Reversible association–dissociation equilibria of the type polymer $\rightleftharpoons$ monomer (*protomer*) $\rightleftharpoons$ subunits (in which protomer represents the mini-

mal structure still having an active site, i.e., capable of effecting the catalytic conversion of substrate to product) may be profoundly affected by such interactions.

5. The effects exerted on the catalytic site, although not necessarily the

TABLE 6-3
Important inhibitors[a]

Compound		*Postulated mode of action*	*Reaction or enzyme affected*
Organophosphorus compounds	$(R_1O)(R_2O)P(\rightarrow O)X$	Covalent bond to serine-OH: protein—O—P(→O)(—OR_1)—OR_2	Trypsin, cholinesterase, phosphoglucomutase, etc.
Organic mercurials	XHgAr	Covalent mercaptide bond to cysteine-SH = protein-*S*-HgAr, reversed by addition of R′SH in excess	Large number of enzymes
Trivalent arsenicals	$R_1As{=}O$	Formation of As-mercaptide with *vic*-disulfides, e.g., sterically favorable cysteines or lipoate (cyclic S—As(R_1)—S)	Large number of enzymes, e.g., pyruvate and α-ketoglutarate dehydrogenase complexes, dihydrolipoate dehydrogenase
Cd^{+2}		Formation of Cd-disulfide with similar proteins as above	See above
Derivatives of iodoacetate	ICH_2COR	Formation of *S*-alkyl derivatives of cysteine = protein-S-CH_2COR	Large number of enzymes, e.g., glyceraldehyde-*P*-dehydrogenase
Heavy metals	Ag^+, Hg^+, etc.	Formation of covalent metal salts; protein-SH, -CO_2^-, -histidine	Large number of enzymes, e.g., β-fructofuranosidase (Ag^+-histidine)
Metal-complexing agents	CN^-, S^{2-}, N_3^- (low concentrations)	Formation of complex with metallo enzymes, especially metalloporphyrins and copper enzymes	Cytochrome oxidase, catalase, urate oxidase, etc.
	CO	Same as above	Cytochrome oxidase, polyphenol oxidase
	CN^-, S^{2-}, N_3^-, F^- (higher concentrations)	Formation of complex with metallo enzymes as above, or removal of metal (Cu, Fe, Mo, Zn, Mg, etc.) from enzyme	Large number of metallo enzymes
Metal-chelating agents	Ethylenediaminetetraacetate;	Same as above;	Large number of metallo enzymes
	o-phenanthroline; α,α'-dipyrridyl;	same as above, especially for Fe, Zn;	
	diethyldithiocarbamate;	same as above, especially for Cu;	
	8-hydroxyquinoline	same as above, especially for Mg, Zn	
Malonate	$CH_2(CO_2^-)_2$	Competitive inhibitor for succinate	Succinic dehydrogenase, citric acid cycle
Pyrophosphate	$^-HO_2P(\rightarrow O)—O—P(\rightarrow O)O_2H^-$	Same as above	Same as above

TABLE 6-3 (Continued)

Compound		*Postulated mode of action*	*Reaction or enzyme affected*
Fluorocitrate	$HO{-}C(CHF\,CO_2^-)(CH_2CO_2^-){-}CO_2^-$	Competitive inhibitor for citrate	Aconitase, citric acid cycle
Oxamate	$H_2NCOCO_2^-$	Competitive inhibitor for L-lactate	L-Lactate dehydrogenase
Long-chain fatty acid amides	$RCONH_2$	Competitive inhibitor for alcohols, especially ethanol	Alcohol dehydrogenase
D-threose-2,4-diP	CHO–CH(OP)–CH(OH)–CH_2OP	Noncompetitive inhibitor for D-glyceraldehyde-3-*P*; forms covalent enzyme complex	D-glyceraldehyde-*P*-dehydrogenase, glycolysis
Monoguanidines	$H_2NC(=NH)NH(CH_2)_nCH_3$	Competitive inhibitor for monoamines	Monoamine oxidase, brain and nerve function
Diguanidines	$H_2NC(=NH)NH(CH_2)_nNHC(=NH)NH_2$	Competitive inhibitor for diamines	Diamine oxidase
Substituted tetra-alkyl-ammonium compounds	$CH_3N^+(R_1)(R_2){-}R_3$	Competitive inhibitors for acetylcholine; if $R_1 = CH_3NCO{-}$ (with R′ on N), alkylation of enzyme also	(Acetyl) choline esterase, brain and nerve function
Sulfonamides	$RNH{-}SO_2{-}C_6H_4{-}NH_2$	Competitive inhibitor of *p*-aminobenzoate	Formation of folic acid and hence of all folic acid coenzyme-dependent reactions in susceptible organisms
Folic acid analogs	Aminopterin, amethopterin	Competitive inhibitor of folate	Reduction of folate and dihydrofolate and hence of formation of methionine, thymine, purines
Pyridine-3-sulfonate	pyridine ring with SO_3^- at position 3	Competitive inhibitor of nicotinic acid	Biosynthesis of nicotinamide coenzymes
Riboflavin monosulfate	$Rb{-}SO_3^-$ (page 416 for structure of Rb)	Competitive inhibitor of riboflavin monophosphate (RMP)	Biosynthesis of flavin adenine dinucleotide and RMP-dependent reactions
Deoxypyridoxine phosphate	pyridine ring with HO, CH_3, $CH_2OPO_3^{2-}$, H_3C substituents	Competitive inhibitor of pyridoxal-*P*	Most reactions involving amino acids

TABLE 6-3 (Continued)

Compound		*Postulated mode of action*	*Reaction or enzyme affected*
Adenosine-5′-phosphorothioate[b]	$^{-}O-P(=S)(O^{-})-OCH_2$–(ribose)–Ade	Analog of 5′-AMP	Activates isocitrate dehydrogenase (NAD) fructose, 1, 6-diphosphatase, phosphorylase *b*; inhibits 5′-nucleotidase action
Pyrithiamine	H_3C, N, N, NH_2; CH_2CH_2OH, N^+, CH_3	Competitive inhibitor of thiamine	Formation of thiamine pyrophosphate and hence of the latter's function as coenzyme
Avidin	Protein from egg white; $M = 4.5 \times 10^4$	Binds to biotin	Inhibits all biotin enzymes
5-Methyltryptophan	CH_3, N, H	Competitive inhibitor of tryptophan	Blocks formation of tryptophanyl-tRNA and hence protein synthesis
Azaserine	$N_2{=}CH{-}C(=O){-}OCH_2CH(NH_3^+)CO_2^-$	Competitive inhibitor of glutamine	Competes with glutamine in $-NH_2$ transfer reactions of purine synthesis
6-Azauridine-5′-*P*	O, HN, N, N, O, R	Competitive inhibitor of orotidylate	Competes in orotidylate decarboxylase and blocks pyrimidine and hence nucleic acid biosynthesis
5-Fluoro-deoxyuridine-5′-*P*	O, F, HN, N, O, R	Competitive inhibitor of uridylate	Competes in thymidylate synthetase and hence blocks DNA synthesis
Cytosine arabinoside	$(P)OCH_2$–O–Cyt	Competes with cytidylate	Competes in reduction of cytidylate to 2′-deoxycytidylate and DNA synthesis
Puromycin	See page 955	Competitive with aminoacyl tRNA's	Prevents completion of nascent polypeptide chains in protein synthesis

[a] Most of these inhibitors and their action are described in more detail in the appropriate chapter, as are additional ones not covered here; the latter is especially true for the inhibitors of electron transport and oxidative phosphorylation (Chapter 15), of carbohydrate metabolism (Chapters 11 and 14), of photosynthesis (Chapter 12), and of nucleic acid and protein synthesis (Chapters 20 and 21).
[b] A. W. Murray and M. R. Atkinson, *Biochemistry,* **7:** 4023 (1968).

binding itself of the various effectors at the allosteric site, can be diminished or eliminated entirely by treatments that lead to alterations of protein conformation (*desensitization*): reaction with mercurials and other protein side-chain reagents, high or low *p*H, denaturation by heat, urea, and so on. Treatments that alter heterotropic interactions alter homotropic interactions simultaneously (all or nearly all systems that exhibit heterotropic effects exhibit homotropic effects as well).

6. Most or all allosteric proteins are polymers (or oligomers) involving several protomeric units. The proteins themselves give evidence of particular conformational flexibility and a resultant ease of inactivation; hence they are difficult to purify and obtain in homogeneous form. They are also frequently cold labile; dissociation into protomers is favored by low temperatures, indicating the importance of hydrophobic interactions in their association (page 173).

Subunit Model. Monod, Wyman, and Changeux have analyzed the kinetics of allosteric interaction in terms of their model (*concerted transition, all or none*):

They assume a polymer consisting of n protomers (and containing n homologous sites for binding an effector M) in two conformational states with degrees of polymerization $R_0, R_1, R_2, \ldots, R_n$ and $T_0, T_1, T_2, \ldots, T_n$, respectively, differing in their affinity for the ligand M, and all related by the following equilibria:

$$R_0 \overset{L}{\rightleftharpoons} T_0$$

$$R_0 + M \overset{\bar{K}_r}{\rightleftharpoons} R_1 \qquad T_0 + M \overset{\bar{K}_t}{\rightleftharpoons} T_1$$

$$R_1 + M \overset{\bar{K}_r}{\rightleftharpoons} R_2 \qquad T_1 + M \overset{\bar{K}_t}{\rightleftharpoons} T_2$$

$$\vdots \qquad\qquad\qquad \vdots$$

$$R_{n-1} + M \overset{\bar{K}_r}{\rightleftharpoons} R_n \qquad T_{n-1} + M \overset{\bar{K}_t}{\rightleftharpoons} T_n$$

Transitions between the two states tend to be fully concerted; that is, intermediate or hybrid states such as R_iT_j $(i + j = n)$ do not exist or are present in such low concentrations as to be insignificant, either in the absence or in the presence of M. It follows from these assumptions that the presence of the modifier M will affect the relative amounts of the two conformers at equilibrium; hence both homotropic and heterotropic effects are mediated through alterations of this equilibrium.

Let us define the following parameters:

n = number of protomers and, hence, of equivalent binding sites

$\alpha = m/\bar{K}^R$, *normalized concentration* of a *single* allosteric modifier M, at concentration m

$\bar{K}^R$, dissociation constant of ligand with protein in the R conformation

$\bar{K}^T$, dissociation constant of ligand with protein in the T conformation

$c = \bar{K}^R/\bar{K}^T < 1$, the *nonexclusive binding coefficient;* that is, R is the conformation with higher affinity (more *affine*) for M

$L = R/T$, the ratio of protein in the two states in the *absence* of *any* modifier

$\beta = i/\bar{K}_i{}^T$ and $\beta' = i/\bar{K}_i{}^R$, normalized concentrations of an added allosteric inhibitor added at concentration i

$\gamma = a/\bar{K}_a{}^R$ and $\gamma' = a/\bar{K}_a{}^T$, normalized concentrations of an added allosteric activator, added at concentration a

$\bar{K}_i{}^T$, $\bar{K}_i{}^R$; $\bar{K}_a{}^T$, $\bar{K}_a{}^R$, the relevant dissociation constants for complexes between I or A and the two conformations. We can now derive the relevant equations:

$\bar{R}$, the *state function,* that is, the fraction of protein in the R conformation:

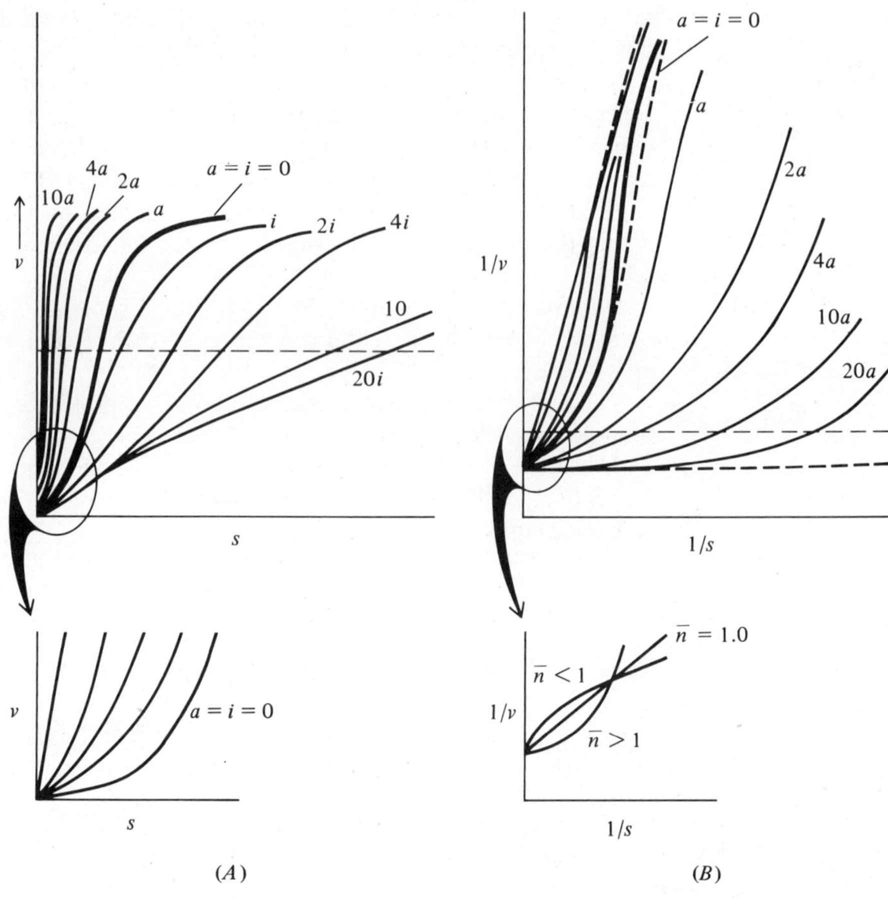

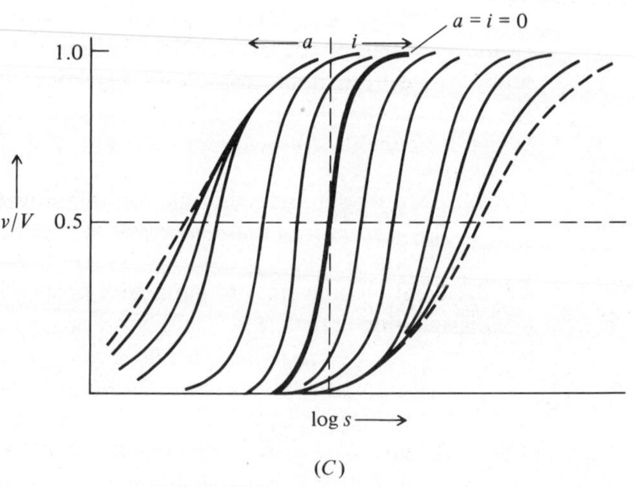

FIGURE 6-9

Three different representations of kinetic effects of allosteric inhibitors (added at concentrations *i*, 2*i*, etc.) or activators (added at concentrations *a*, 2*a*, etc.) on plots relating initial velocity to substrate concentration. The intercepts of all curves with the horizontal dashed lines indicate (on the *x* axis) the values of $s_{0.5}$—the concentration of the substrate required for half maximal velocity. The lines drawn correspond to cases actually

$$\bar{R} = \frac{R_0 + R_1 + R_2 + \cdots + R_n}{(R_0 + R_1 + \cdots + R_n) + (T_0 + T_1 + \cdots + T_n)} = \frac{(1+\alpha)^n}{L'(1+c\alpha)^n + (1+\alpha)^n} \tag{6-74}$$

$\bar{Q}$, the *quotient function,* the ratio of R states to T states:

$$\bar{Q} = \frac{\bar{R}}{\bar{T}} = \frac{\bar{R}}{1-\bar{R}} = \frac{(1+\alpha)^n}{L'(1+c\alpha)^n} \tag{6-75}$$

$\bar{Y}$, the *saturation function,* the fraction of the protein occupied by M:

$$\bar{Y} = \frac{\alpha}{1+\alpha}\bar{R} + \frac{c\alpha}{1+c\alpha}(1-\bar{R}) \tag{6-76}$$

If all equilibria including the ones between R and T are established rapidly, the rate of an enzymatic reaction is proportional to the concentration of the *EM* complex (i.e., if the system under study is a K system, so that it exhibits allosteric effects only on the apparent K_s, usually now called $s_{0.5}$).

$$\bar{Y} = \frac{v}{V} \quad \text{or} \quad \frac{v}{nV'} \qquad \text{where } V' \text{ is the maximal velocity } \textit{per site} \tag{6-76a}$$

L' in the above equations is the equilibrium constant between R and T states and will depend on the concentration of the added effectors I and A as follows:

$$L' = L\left[\frac{(1+\beta)(1+\bar{\gamma})}{(1+\gamma)(1+\bar{\beta})}\right]^n \tag{6-77}$$

Homotropic Effects. The model predicts that cooperative *homotropic* effects of the modifier M, in the simplest case equal to the substrate S, depends on the magnitude of L', in this case equal to L, and c as shown in Fig. 6-10. The *cooperativity* expresses itself in the curvature of the lower parts of the curves; also compare the predictions of the model with the situation actually observed (Fig. 6-9 versus Fig. 6-10). If the binding is *non*exclusive, that is, $c \neq 0$, there will be three cases where a "normal" hyperbolic v versus s or linear v^{-1} versus s^{-1} plot is obtained: (1) when $L \to 0$, that is, when there exists only a single, kinetically significant conformer, the T state; (2) when $L \to \infty$, the converse holds and all the protein has been converted to the R state; and (3) when $c = 1$ and hence $\bar{K}_s^R = \bar{K}_s^T$. In these cases $\bar{n}$, the Hill coefficient [Fig. 6-11 and Eq. (6-79)], equals unity. For intermediate but larger values of c the curves will be sigmoidal and $\bar{n} > 1$. If the binding is exclusive (i.e., $c = 0$), then strictly speaking $\bar{n}$ can become unity and the kinetic behavior normal only for $L \to 0$, although, in practice deviations will be small for values of $L < 1$. In this case Eq. (6-76) becomes

$$\bar{Y} = \frac{\alpha}{1+\alpha}\bar{R} = \frac{\alpha(1+\alpha)^{n-1}}{L + (1+\alpha)^n} \tag{6-76b}$$

---

reported in the literature (e.g., Ref. 25). The heavy dashed line on the double reciprocal plot corresponds to a simpler case susceptible to straightforward theoretical analysis (Fig. 6-10). In the insert to this plot is shown a comparison of the patterns obtained for homotropic effects: absent (straight line), or occurring with positive (concave upward) or negative (convex upward) cooperativity. The first corresponds to Hill coefficients (Fig. 6-11) equal to one, the second > 1, the last < 1.

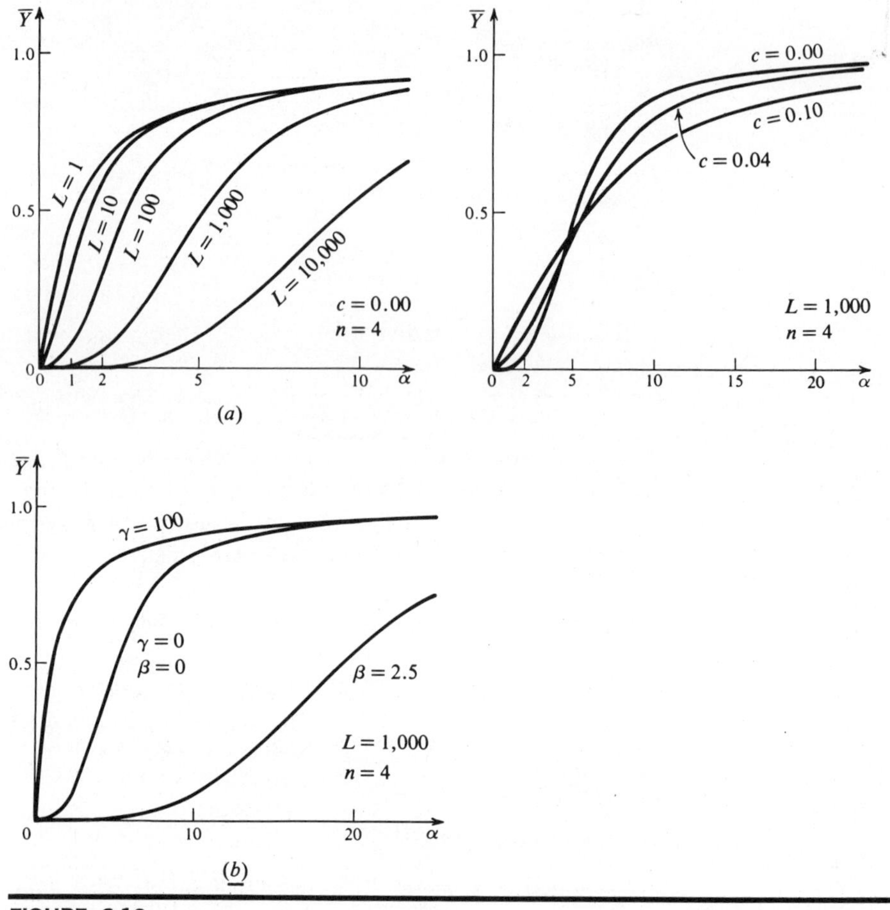

FIGURE 6-10

(a) Theoretical curves of the saturation function $\bar{Y}$ (equal to v/V) for a perfect K system drawn to various values of L and c. (b) Theoretical curves showing the heterotropic effects of an allosteric activator (γ) or inhibitor (β) upon the shape of the saturation function for substrate (α). J. Monod, J. Wyman, and J.-P. Changeux, *J. Mol. Biol.*, **12:** 88 (1965).

Heterotropic Effects. Such effects will be observed when an additional effector, either an activator or inhibitor, can bind to one or the other of the two states, or to both independently, with the appropriate ratio of binding constants given above. Thus, it will affect the equilibrium between R and T states (and hence L), *without, however, affecting the value of any of the other constants* (Fig. 6-10b). Allosteric *activation* will be observed whenever a molecule exhibits the same preference between the two states as does the substrate (i.e., for the R state); the effect here will be to *lower* $s_{0.5}$, the concentration of substrate required to yield a rate exactly one half that of V (i.e., the apparent Michaelis constant). Conversely, allosteric *inhibition* will be observed whenever the molecule in question exhibits a preference opposite to that of the substrate (i.e., for the T state); the effect will be to *raise* $s_{0.5}$. These effects on $s_{0.5}$ for the three methods of representation given in Fig. 6–9 are shown by the s values corresponding to the intersections of the dashed horizontal lines with the experimental curves. Incidentally, the various effects are completely symmetrical; thus the activator or the inhibitor could be regarded as the primary ligand M, and then the kinetic effects of adding substrate as an additional effector studied, and so on.

When $c = 0$, addition of an activator reduces homotropic effects so that in the limit all cooperativity is lost, the plots become normal, and $\bar{n} = 1$. Conversely, the addition of an inhibitor increases homotropic effects, the curves become maximally cooperative (i.e., sigmoidal), and $\bar{n}$ reaches its maximum value. For nonexclusive binding (i.e., $c \neq 0$) cooperativity is lost and normal behavior with $\bar{n} = 1$ is observed at high concentrations of *either* activator or inhibitor, because the enzyme will be entirely in the R or T state, respectively. The predictions of the model with $c = 5 \times 10^{-4}$ M and $n = 4$ have been verified in the case of phosphofructokinase from *E. coli* by showing excellent agreement between experimental and calculated curves of $\bar{Y}$ versus α, for different (experimentally determined) values for $\alpha_{0.5}$.

Determination of Binding Constants for a Homotropic Effector (Substrate). When both allosteric activation and inhibition can be observed, then, as was just stated, the limiting curves for the saturation function $\bar{Y}$ (dashed curves in Fig. 6-9c) provide a measure of this function for the R (left) and T state, respectively. Thus $\bar{K}^R$ and $\bar{K}^T$ can be read off directly from the curves simply by taking the antilogs of the appropriate value of $\log s_{0.5}$. By definition, c is then the ratio of $\bar{K}^R/\bar{K}^T$. Alternatively, since in most instances $\sqrt{c}$ can be shown to be equal to $s_{0.5}$ in the absence of an additional modifier (30), determination of either $\bar{K}^R$ or $\bar{K}^T$ suffices for the calculation of the other binding constant. Incidentally, for a tetramer the individual microscopic binding and equilibrium constants are defined in terms of L, C, and $\bar{K}^R$ (or $\bar{K}^T$) as follows (the subscript indicates the number of substrate molecules bound, $n = 4$):

$$K_1{}^R = \frac{r_0 s}{r_1} = \frac{\bar{K}^R}{4} \quad K_2{}^R = \frac{r_1 s}{r_2} = \frac{2\bar{K}^R}{3} \quad K_3{}^R = \frac{r_2 s}{r_3} = \frac{3\bar{K}^R}{2} \quad K_4{}^R = \frac{r_3 s}{r_4} = 4\bar{K}^R \tag{6-78a}$$

$$K_1{}^T = \frac{t_0 s}{t_1} = \frac{\bar{K}^T}{4} \quad K_2{}^T = \frac{t_1 s}{t_2} = \frac{2\bar{K}^T}{3} \quad K_3{}^T = \frac{t_2 s}{t_3} = \frac{3\bar{K}^T}{2} \quad K_4{}^T = \frac{t_3 s}{t_4} = 4\bar{K}^T \tag{6-78b}$$

$$\frac{t_0}{r_0} = L \quad \frac{t_1}{r_1} = Lc \quad \frac{t_2}{r_2} = Lc^2 \quad \frac{t_3}{r_3} = Lc^3 \quad \frac{t_4}{r_4} = Lc^4 \tag{6-78c}$$

Relation to Hill Coefficient (30, 31). Until the recent spate of interest and speculation regarding regulatory enzymes and allosteric and other conformational models, the only well-authenticated instance of cooperativity in a biochemical reaction was that of the oxygen (and CO) saturation curves for hemoglobin in contrast to myoglobin (page 670). These curves were interpreted by Brown and Hill in terms of Eq. (6-79):

$$y = \frac{Kx^n}{1 + Kx^n} \tag{6-79}$$

or

$$\log \frac{y}{1-y} = \log (-Kx^n) \tag{6-79a}$$

where y is the fractional saturation, x the partial pressure of O_2, n the number of oxygen molecules bound per hemoglobin molecule, and K a constant. The sigmoidal shape of the curve demands that the binding be cooperative, that is, that the binding of each oxygen molecule increase the affinity of the protein for the next. By analogy, then, cooperative phenomena in enzyme kinetics have frequently been expressed in terms of an equivalent *Hill equation:*

$$\log \frac{v}{V - v} = n \log s - \log K \tag{6-79b}$$

Equation (6-79) presupposes an ideal case of infinite cooperativity. In that case the appropriate plots (Fig. 6-11) are straight lines throughout and $\bar{n}$ the slope of the straight line corresponds to n in the equation and equals the number of binding sites. In actual cases the curves frequently deviate from linearity, and even in their central portion, where $\bar{n}_{0.5}$ equals the slope corresponding to substrate concentration $s_{0.5}$, this parameter is a function not only of n but also the strength of interaction between sites. Furthermore, $\bar{n}_{\max}$ is not necessarily equal to $\bar{n}_{0.5}$. For the allosteric model of Monod, Wyman, and Changeux with $c = 0$, $\bar{n}$ becomes equal to n at large values of L', that is, in the presence of a large concentration of an allosteric inhibitor. For $c \neq 0$, $\bar{n}_{\max}$ is bell-shaped as a function of L' and reaches a maximum value of $L' = c^{-n/2}$. At this point cooperativity is maximal, the greatest homotropic effects are achieved, and $\bar{n}_{\max} \to n$.

Determination of Binding Constants for Allosteric Activators and Inhibitors. Provided c, the nonexclusive binding coefficient for the substrate, is

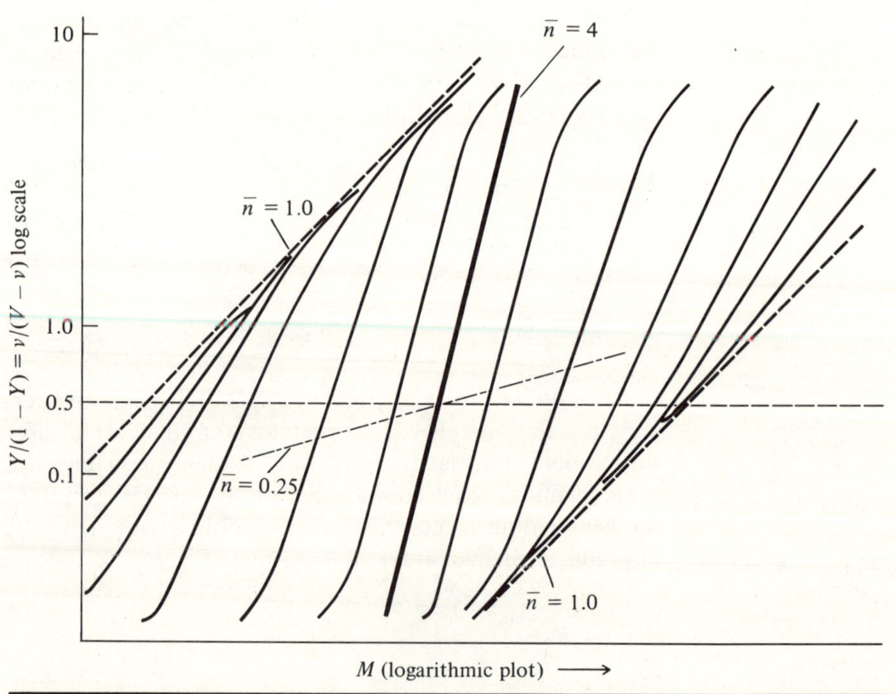

FIGURE 6-11

Possible shapes of Hill plots for enzyme-catalyzed reactions showing either homotropic effects (s increasing from left to right across the whole range), or heterotropic effects (with s held constant at any one value). Addition of A in increasing concentrations will generate the family of curves to the right, addition of I the one to the left; $\bar{n}$ is the slope and $\bar{n}_{0.5}$ is the slope of each line at the "midpoint," that is, for $v/(V - v) = 0.5$. Shown also is a curve for an enzyme exhibiting "negative cooperativity" (Ref. 41), that is, a value of n and $\bar{n} < 1$.

small, the second term in Eq. (6-74) can be neglected and Eq. (6-76b) holds. We can rewrite this Equation as

$$\bar{R} = \frac{v}{V}\frac{1+\alpha}{\alpha} = \frac{v}{V'} \tag{6-80}$$

where V' is the maximal velocity that can be attained for *any one* concentration of the primary ligand (e.g., the substrate) by varying the concentration of a second one with an affinity for this same enzyme form (i.e., an activator). We now rewrite Eqs. (6-75) and (6-77) for the ratio between the two conformations:

$$\bar{Q} = \frac{v}{V' - v} = \frac{(1+\alpha)^n(1+\gamma)^n}{L(1+\gamma')^n} \tag{6-81}$$

At constant α this becomes

$$\bar{Q} = \frac{1}{L_\alpha}\frac{(1+\gamma)^n}{(1+\gamma')^n} \tag{6-81a}$$

Thus, once a value of n has been determined or can be assumed from physical data, a plot of $\sqrt[n]{v/(V'-v)}$ against the activator concentration will yield a family of straight lines with x intercept equal to $-\bar{K}_m{}^R$ (or $-\bar{K}_a{}^R$), the negative of the dissociation constant for the activator–enzyme (in the R conformation) complex. Analogous treatment for an effector binding to the enzyme in the opposite (T) conformation (i.e., an inhibitor) yields

$$\bar{Q} = \frac{1}{L_\alpha}\frac{(1+\beta)^n}{(1+\beta')^n} \tag{6-81b}$$

and its dissociation constant $-\bar{K}_m{}^T$ (or $\bar{K}_i{}^T$) is obtained in the same manner. An interesting prediction of, and thus a test for, the model follows from the definition of $\bar{Q}$, which with the proper substitution for L' in Eq. (6-77) becomes

$$\bar{Q} = \frac{(1+\alpha)^n}{L}\frac{(1+\gamma)^n}{(1+\beta)^n} \tag{6-81c}$$

Thus, provided the *ratio* of concentrations of inhibitor and activator is held constant at $\bar{K}_i{}^T/\bar{K}_a{}^R$, the initial velocity, for any one concentration of the substrate, will be constant for all actual concentrations of activator and inhibitor used.

Other Models (27–33, 36)

The fully concerted (all-or-none) model of Monod, Wyman, and Changeux which we have just described interprets the kinetic behavior of regulatory enzymes—for the case of homotropic interactions—in terms of three parameters, $\bar{K}^R$, $\bar{K}^T$, and L. Now it is in the nature of things that three parameters (and their number is of course increased further in the case of heterotropic interactions) are both necessary *and* sufficient to fit the experimental data. Therefore, in the absence of subsidiary information a number of alternative models of equal a priori plausibility can be devised and shown to fit appropriate sets of data. As always, the formulation of models—and the rate laws predicted by them which can then be compared with those experimentally observed—are useful in excluding variants, but are never unique.

Sequential (Induced-Fit) Model. In 1966 Koshland, Némethy, and Filmer (33) proposed an alternative to the concerted model that postulates that conformational changes occur *only* as a consequence of the binding of effector, each molecule bound affecting the affinity for the next to be bound. Retaining the earlier designations we then find the following series of exclusive equilibria for the homotropic case with $n = 4$:

$$T_4 + S \underset{K_1}{\overset{}{\rightleftharpoons}} T_3RS \underset{K_2}{\overset{+S}{\rightleftharpoons}} T_2R_2S_2 \underset{K_3}{\overset{+S}{\rightleftharpoons}} TR_3S_3 \underset{K_4}{\overset{+S}{\rightleftharpoons}} R_4S_4$$

Formally, in this particular instance this model becomes equivalent to one proposed earlier by Adair (37) for the case of hemoglobin and generalized recently by Weber and Anderson (38) to protein–ligand binding in general. Although Koshland and his colleagues based their expressions on interaction constants (R with R, T with T, and R with T), these can be shown to be interconvertible with the microscopic equilibrium constants for the four steps and hence, *formally*, to the constants of the concerted model. The sequential model is similar to one proposed by Pauling (40), which postulates equivalence of binding sites, each with intrinsic dissociation constant $\bar{K}$ for a ligand, which are altered on binding by the first substrate molecule to become $\bar{\alpha}\bar{K}$, by the second molecule to become $\bar{\beta}\bar{K}$, by the third to become $\bar{\gamma}\bar{K}$, and so on. The following relations then hold (again for $n = 4$):

$$\frac{v}{V} = \frac{s/\bar{K}[1 + 3\bar{\alpha}s/\bar{K} + 3\bar{\alpha}^2\bar{\beta}(s/\bar{K})^2 + \bar{\alpha}^3\bar{\beta}^2\bar{\gamma}(s/\bar{K})^3]}{1 + 4s/\bar{K} + 6\bar{\alpha}(s/\bar{K})^2 + 4\bar{\alpha}^2\bar{\beta}(s/\bar{K})^3 + \bar{\alpha}^3\bar{\beta}^2\bar{\gamma}(s/\bar{K})^4} \tag{6-82}$$

and

$$\frac{\bar{K}}{\bar{K}^R} = \frac{1+L}{1+Lc}; \quad \bar{\alpha} = 1 - \frac{(1-c)^2/2c}{1 + (1/2Lc) + (Lc/2)}$$

$$\bar{\beta} = 1 - \frac{(1-c)^2/2c}{1 + (1/2Lc^2) + (Lc^2/2)} \tag{6-83}$$

$$\bar{\gamma} = 1 - \frac{(1-c)^2/2c}{1 + (1/2Lc^3) + (Lc^3/2)}$$

Thus, in this limiting case, the two models become formally equivalent three-parameter models. Actually, it has been shown by Eigen that these represent simply two special cases of a much more general model, which is depicted in Fig. 6-12.

However, in more complicated cases, and with heterotropic effects added, the models become distinct (34), although the number of adjustable parameters now becomes of necessity larger and the interpretation murkier. Other models, which do not depend on conformational transitions but postulate alternative interaction sites (35) or branched reaction paths (36), have also been proposed.

Negative Cooperativity. Just as there exist positive homotropic effects for certain proteins (i.e., a facilitation of binding of subsequent effector molecules by those that precede them), so do there appear the inverse effects for certain others. This concept of negative homotropism or cooperativity appears to hold for the binding of NAD^+ to the D-glyceraldehyde-3-*P* dehydrogenase from rabbit muscle which has been studied intensively by Conway and Koshland (39), who have interpreted their binding and —SH reactivity data in terms of a sequential change of the four subunits. The experimentally determined macroscopic dissociation constants for the four coenzyme molecules added sequentially are $K_1 \lesssim 10^{-11}$ M, $K_2 \lesssim 10^{-9}$ M, $K_3 = 3 \times 10^{-7}$ M, and $K_4 = 2.6 \times 10^{-5}$ M. Other enzymes for which nega-

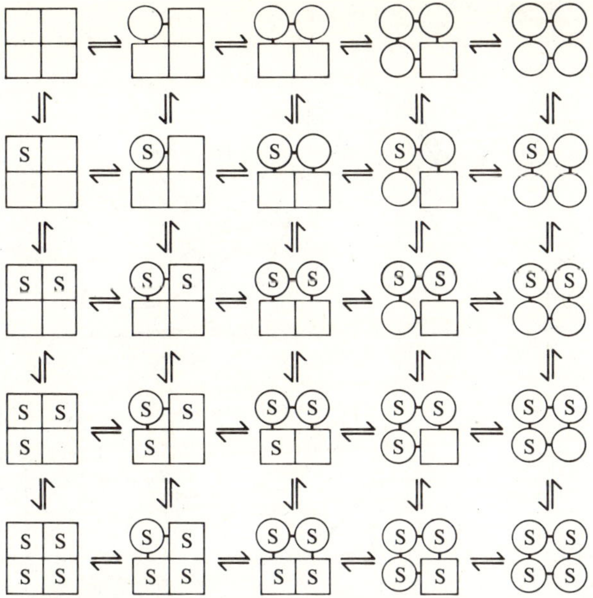

FIGURE 6-12

General reaction scheme of allosteric binding to a four-subunit protein. Some of the intermediates in which *S* can bind to either the square or the circle are just represented by one state. There would be all together 44 nonequivalent forms. M. Eigen, *Nobel Symp.,* **5:** 341 (1967).

tive cooperativity has been observed are phosphoenolpyruvate carboxylase and CTP synthetase (41). Negative cooperativity is easily distinguished from cases exhibiting no or positive cooperativity in double reciprocal (Fig. 6-9) and Hill (Fig. 6-11) plots—the value of $\bar{n}$, the Hill coefficient being less than unity (41).

Effects of *p*H

General Considerations. Almost all enzymes are extremely sensitive to *p*H; their activity is diminished at either side of a relatively narrow range. These effects are due to a combination of three factors: (1) effects of extremes of *p*H on protein structure, including alterations on the strength and mode of binding of prosthetic groups; (2) effects on the ionization of the substrate; and (3) effects on its binding to the enzyme and on reactivity in catalysis. It is only with the third class that we shall be concerned here; the first two can usually be determined independently of the reaction under kinetic study and corrections made for their effects.

The initial velocity of the enzymic reaction proper frequently exhibits three distinct phases as a function of *p*H (Fig. 6-13a): a region of *p*H (at low values) where there is an increase, a region (at high values) where there is a decrease, and an intermediate range (usually around neutrality) where the activity is maximal and approximately constant—the *p*H optimum—leading to a characteristic bell-shaped curve *in toto.*[12] In a number of instances these curves have been interpreted in terms of certain proton ionization constants

[12] In other instances one or the other branch of the curve may be missing. In these cases the curve consists of only two segments; the branch (either ascending or descending) in question is joined to a horizontal portion.

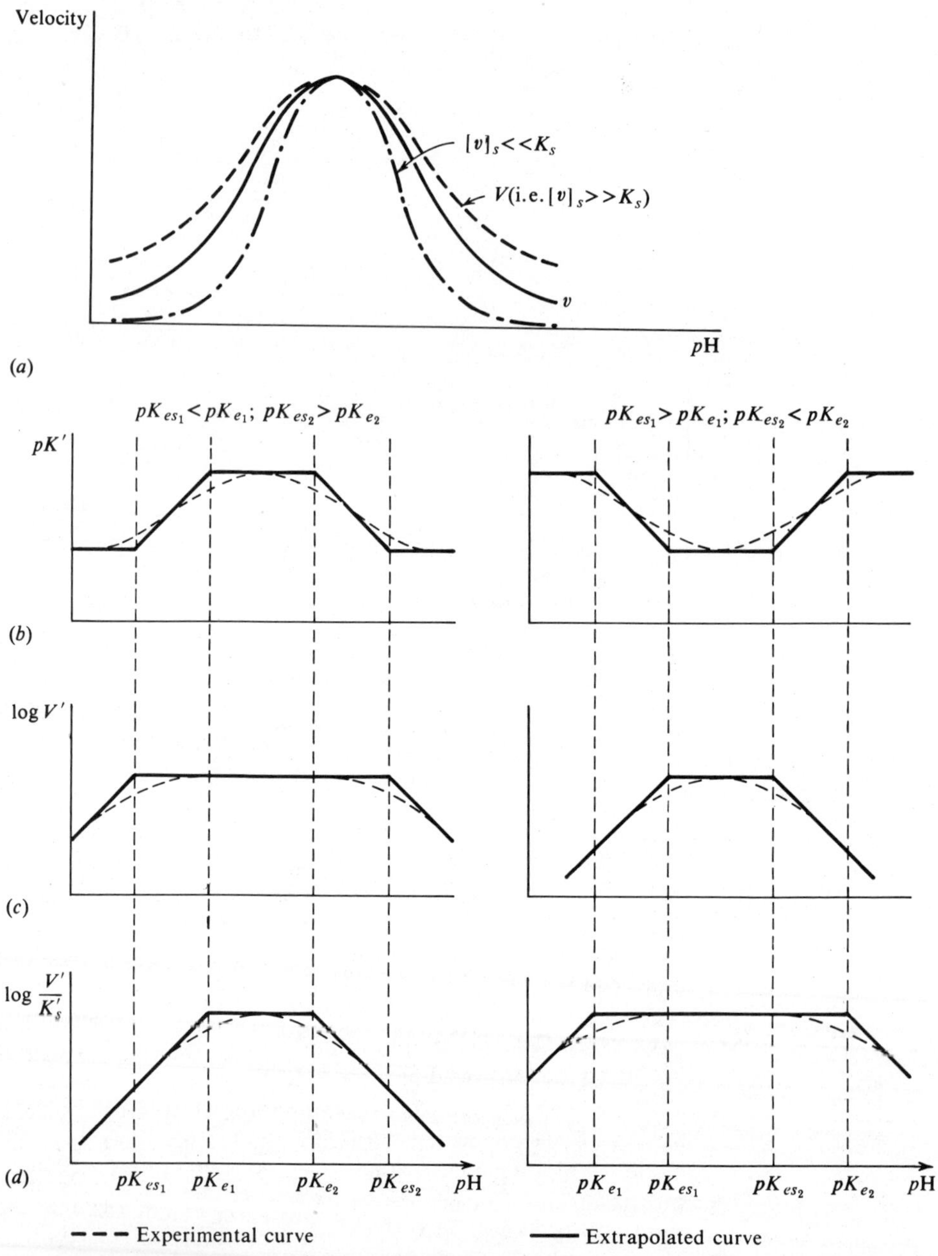

FIGURE 6-13
Effect of pH on a simple enzymatic reaction. Dashed line denotes experimental curves; solid line denotes theoretical curves.

either on the free enzyme or on the enzyme–substrate complex. In general these interpretations are not really correct; as we shall see, the dependence of the rate on pH can be so interpreted only in its pure first-order (i.e., $s \ll K_s$; $v = sV/K_s$) [Eq. (6-31b)] and zero-order regions (i.e., $s \gg K_s$; $v \simeq V$) [Eq. (6-31a)]. These two cases are indicated by the appropriate curves in Fig. (6-13a). At these extremes the inflection points, if calculated properly, may indeed correspond to such ionizations—on the enzyme in the former instance, on the enzyme–substrate complex in the latter. But the initial velocity, as ordinarily determined, lies between these two extremes, and its pH dependence, similarly, is intermediate and cannot be explained in such simple terms.

pH Dependence of a Simple One-Substrate Reaction. The simplest mechanism may be expanded in the following manner to include the effects of pH:

$$\begin{array}{ccccc} E & & ES & & \\ +H^+ \Big\Downarrow\Big\Uparrow \bar{K}_{e2} & & +H^+ \Big\Downarrow\Big\Uparrow \bar{K}_{es2} & & \\ S + E\mathrm{H} & \underset{k_2}{\overset{k_1}{\rightleftharpoons}} & E\mathrm{H}S & \xrightarrow{k_3} & E\mathrm{H} + \text{products} \\ +H^+ \Big\Downarrow\Big\Uparrow \bar{K}_{e1} & & +H^+ \Big\Downarrow\Big\Uparrow \bar{K}_{es1} & & \\ E\mathrm{H}_2 & & E\mathrm{H}_2S & & \end{array} \tag{6-84}$$

The assumptions are that only one state of ionization of substrate S and of enzyme EH is capable of reaction; that the enzyme ionizes as a dibasic acid, in which the completely deprotonated E and diprotonated $E\mathrm{H}_2$ forms are completely inactive with regard to substrate binding and that similar considerations apply with regard to the catalytic function of the active enzyme–substrate complexes ($E\mathrm{H}S$); and finally that the appropriate ionization constants are, in order of the number of protons concerned, for the free enzyme $\bar{K}_{e1}$ and $\bar{K}_{e2}$ and correspondingly for the complex $\bar{K}_{es1}$ and $\bar{K}_{es2}$. The steady-state rate equation for the mechanism is the familiar

$$v = \frac{V's}{K' + s} \tag{6-85a}$$

where V' and K', the observed parameters, are now functions of pH as follows ($\mathrm{H}^+ \equiv$ hydrogen-ion concentration):

$$V' = \frac{V}{1 + \mathrm{H}^+/\bar{K}_{es1} + \bar{K}_{es2}/\mathrm{H}^+} \tag{6-85b}$$

where V is the pH-independent maximal velocity $= k_3 e_0$;

$$K' = K_s \frac{1 + \mathrm{H}^+/\bar{K}_{e1} + \bar{K}_{es2}/\mathrm{H}^+}{1 + \mathrm{H}^+/\bar{K}_{es1} + \bar{K}_{es2}/\mathrm{H}^+} \tag{6-85c}$$

where K_s is the pH-independent Michaelis constant $= (k_2 + k_3)/k_1$;

$$\frac{V'}{K'} = \frac{V}{K_s} \frac{1}{1 + \mathrm{H}^+/\bar{K}_{e1} + \bar{K}_{e2}/\mathrm{H}^+} \tag{6-85d}$$

where V/K_s is the pH-independent velocity at low s [Eq. (6-31b)].

Plots of V' (or some function thereof) versus H^+ (or some function) can thus provide information only about $\bar{K}_{es1}$ and $\bar{K}_{es2}$; plots of K' versus H^+ depend on both sets, $\bar{K}_{e1}$, $\bar{K}_{e2}$ and $\bar{K}_{es1}$, $\bar{K}_{es2}$; plots of the ratio of the two parameters, or of v at $s \ll K_s$, are required for a determination of $\bar{K}_{e1}$ and $\bar{K}_{e2}$.

The evaluation of these constants can best be accomplished by the use of double logarithmic plots of $\log_{10} V'$ versus pH; $\log_{10} (K')^{-1} \equiv pK'$ versus pH; and $\log_{10} [V'/K']$ (or v at $s \ll K_s$) versus pH. Idealized plots are shown in Fig. 6-13*b* to *d*. Most of them follow either the left or the right pattern with, occasionally, a combination of the left branch of the left pattern with the right branch of the right one, and vice versa. If only $\bar{K}_{e1}$ (or $\bar{K}_{e2}$) are operative, this leads, of course, to an elimination of the opposite branch. The use of the curves for the evaluation of the four ionization constants is reasonably self-explanatory. The same treatment used for K'_s can also be applied to K_i', the pH-dependent inhibition constant. Some simple rules for the interpretation of such pK versus pH curves, due to Dixon, follow[13]:

1. The graph consists of straight-line sections joined by short curved parts.
2. The straight-line portions have integral slopes, that is, zero or one-unit or two-unit, positive or negative.
3. Each bend indicates the pK of an ionizing group in one of the components, and the straight portions intersect at a pH corresponding to the pK.
4. Each pK produces a change of one unit in the slope.
5. Each pK of a group situated in the ES complex produces an increase in (positive) slope; each pK in the free enzyme produces a decrease in slope.
6. The curvature at the bends is such that the (actual) graph misses the (theoretical) intersection point of the neighboring straight parts by a vertical distance of 0.3 unit (= log 2); if two pK values occur together, the distance is equal to log 3.
7. The slope of any straight section is numerically equal to the change of charge occurring in that pH range where the complex dissociates into free enzyme and substrate.

The ionization constants obtained in this manner are not true group ionization constants but are instead *molecular ionization constants;* the two types are related as follows:

$$\begin{array}{ccccc} & \overset{\bar{K}_3}{\rightleftharpoons} & E & \overset{\bar{K}_4}{\rightleftharpoons} & \\ EH & & & & HE \\ & \underset{\bar{K}_1}{\rightleftharpoons} & HEH & \underset{\bar{K}_2}{\rightleftharpoons} & \end{array} \qquad \begin{aligned} \bar{K}_{e1} &= \bar{K}_1 + \bar{K}_2 \\ \bar{K}_{e2} &= \frac{\bar{K}_3\bar{K}_4}{\bar{K}_3 + \bar{K}_4} \end{aligned} \tag{6-86}$$

pH Dependence of Reversible One-Substrate Reactions. For the reversible formation of ES and EP, we can formulate the corresponding pH-dependent mechanism:

$$\begin{array}{ccccccc} E & & ES & & EP & & \\ {+H^+}\Big\updownarrow \bar{K}_{e2} & & {+H^+}\Big\updownarrow \bar{K}_{es2} & & {+H^+}\Big\updownarrow \bar{K}_{ep2} & & \\ S + EH & \underset{k_2}{\overset{k_1}{\rightleftharpoons}} & EHS & \underset{k_4}{\overset{k_3}{\rightleftharpoons}} & EHP & \underset{k_6}{\overset{k_5}{\rightleftharpoons}} & EH + P \\ {+H^+}\Big\updownarrow \bar{K}_{e1} & & {+H^+}\Big\updownarrow \bar{K}_{es1} & & {+H^+}\Big\updownarrow \bar{K}_{ep1} & & \\ EH_2 & & EH_2S & & EH_2P & & \end{array} \tag{6-87}$$

The variation of the observed parameters V' and K' with s is still given by Eq. (6-84), but now their pH dependence is

$$v = \frac{V_1'V_2's - V_1'V_2'p}{K_s'V_2' + sK_p' + K_s'p} \qquad v_1 = \frac{V_1's}{K_s' + s} \qquad v_2 = \frac{V_2'p}{K_p' + p} \tag{6-88a}$$

$$V_1' = \frac{V_1}{1 + K'_{es1}/\mathrm{H}^+ + \mathrm{H}^+/K'_{es2}} \qquad K_s' = \frac{K_s(1 + \bar{K}_{e1}/\mathrm{H}^+ + \mathrm{H}^+/\bar{K}_{e2})}{1 + K'_{es1}/\mathrm{H}^+ + \mathrm{H}^+/K'_{es2}} \tag{6-88b}$$

$$V_2' = \frac{V_2}{1 + K'_{ep1}/\mathrm{H}^+ + \mathrm{H}^+/K'_{ep2}} \qquad K_p' = \frac{K_p(1 + \bar{K}_{e1}/\mathrm{H}^+ + \mathrm{H}^+/\bar{K}_{e2})}{1 + K'_{ep1}/\mathrm{H}^+ + \mathrm{H}^+/K'_{ep2}} \tag{6-88c}$$

[13] In more general cases log (V/K) versus pH depends on true dissociation constants of the variable substrate or of the enzyme forms with which it reacts. If we are interested in the latter, we must make certain that the former do not interfere.

We observe that the numerator terms are analogous to those in the previous case (and involve constants for the free enzyme), but the denominators now contain new constants which are no longer the simple acid dissociation constants of the enzyme–substrate complex and which also differ in the forward and in the reverse directions. They are defined as follows in terms of the true acid dissociation constants and of certain rate constants for their interconversion:

$$K'_{es1} = \frac{k_3 + k_4 + k_5}{(k_4 + k_5)\bar{K}_{es1} + k_3/\bar{K}_{ep1}} \qquad K'_{es2} = \frac{(k_4 + k_5)\bar{K}_{es2} + k_3\bar{K}_{ep2}}{k_3 + k_4 + k_5}$$

$$K'_{ep1} = \frac{k_2 + k_3 + k_4}{(k_2 + k_3)\bar{K}_{ep1} + k_4/\bar{K}_{es1}} \qquad K'_{ep2} = \frac{(k_2 + k_3)\bar{K}_{ep1} + k_4\bar{K}_{es2}}{k_2 + k_3 + k_4} \tag{6-88d}$$

It is, therefore, no longer possible to obtain the acid dissociation constants of the enzyme–substrate and enzyme–product complexes from the pH-rate profiles, but plots of either $(v_1)_{s \ll K'_s}$ or V'_1/K'_s versus pH can still be used for the determination of K_{e1} and K_{e2}, and identical values for these constants must result from plots of $(v_2)_{p \ll K'_p}$ or V_2'/K'_p versus pH. Values of pK for ionizable groups on enzymes that may form the basis of assignments for the $\bar{K}_{e1}$ and $\bar{K}_{e2}$ observed in kinetic studies are given in Table 6-4.

Effects of Temperature

General Principles. In the older literature there frequently appears a description and definition of a *temperature optimum* for an enzymatic reaction analogous to the pH optimum just discussed. But although the latter can be explained in terms of thermodynamics alone—of two or more reversibly interconvertible forms of the enzyme (and its complexes)—the former is generally due to a combination of two phenomenologically different events: one related to enzyme stability and due to the interconversion of two forms of the enzyme (at least in principle reversible also) between a "native,"

TABLE 6-4
Acid pK values of side-chain groups in proteins and their ΔH^0 values

	pK (25°)				
		Protein			
Group	*Small molecule*[a]	*Bovine Serum Albumin*[b]	*Likely Range*[c]	*Likely Range*[d]	$\Delta H°$ *(cal/mol) Protein*[c]
α-Carboxyl	3.6	3.75	3.0–3.2	3.1–3.6[f]	0 ± 1500
β- or γ-carboxyl	4.6	4.0	3.0–4.7	4.6	0 ± 1500
Imidazole	6.0	6.9	5.6–7.0	. . .	6900–7500
α-Amino	7.9	7.75	7.6–8.4	7.8	10,000–13,000
α-Amino in cystine	. . .	. . . .	6.5–8.5	. . .	. . .
Sulfhydryl	9.2	. . . .	8.3–8.6[e]	. . .	. . .
Phenolic OH	9.8	10.35	9.8–10.4	9.6	6000
ε-Amino	<10.5	9.8	9.4–10.6	10.2	10,000–12,000
Guanidino	<14.0	>12	11.6–12.6	>12	12,000–13,000

[a] Compiled by F. R. N. Gurd and P. E. Wilcox, *Advan. Protein Chem.*, **11:** 346 (1956).
[b] C. Tanford, S. A. Swanson, and W. S. Shore, *J. Am. Chem. Soc.*, **77:** 6414 (1955).
[c] E. J. Cohn and J. T. Edsall, *Proteins, Amino Acids and Peptides,* Van Nostrand Reinhold, New York, 1943.
[d] J. Steinhardt and S. Beychok, in H. Neurath (ed.), *The Proteins,* vol. 2, Academic Press, New York, 2nd ed., 1964, p. 171.
[e] R. E. Benesch and R. Benesch, *J. Am. Chem. Soc.*, **77:** 5877 (1955).
[f] Actual values are 4.08 (β) and 4.50 (γ) at $\mu = 25°$; Y. Nozaki and C. Tanford, *J. Biol. Chem.*, **242:** 4731 (1967).

enzymatically active, and a "denatured" one, devoid of or vastly decreased in enzymatic activity; the other related to enzyme activity, that is, a kinetic effect of temperature on reaction rates. Because enzyme denaturation generally leads to a decrease in activity while the kinetic effect produces an increase in activity with increasing temperature, it is evident that for any one time of exposure to a given temperature the rate of an enzymic reaction might appear to increase at first, go through a maximum, and eventually decline.

Effect of Temperature on Ordinary Chemical Reactions. Let us briefly return to a simple, reversible chemical reaction, the equilibrium described by Eq. (6-16):

$$S + B \underset{k_{-1}}{\overset{k_{+1}}{\rightleftharpoons}} P$$

with an equilibrium constant

$$K_{\text{eq}} = \frac{p_{\text{eq}}}{s_{\text{eq}} \times b_{\text{eq}}} = \frac{k_{+1}}{k_{-1}}$$

The effect of raising the temperature is to affect the position of the equilibrium and the magnitude of K_{eq}. Quantitatively this behavior is described by the van't Hoff equation (Chapter 2):

$$\Delta H_f{}^0 = 2.303RT^2 \frac{d \log K_{\text{eq}}}{dT} = -2.303 \times 1.98 \frac{d \log K_{\text{eq}}}{dT^{-1}} \text{ cal/mol} \tag{6-89a}$$

where $\Delta H_f{}^0$ is the standard enthalpy for the reaction as written and $-\Delta H_f{}^0 \equiv \Delta H_r{}^0$ is the corresponding quantity for the reverse reaction (i.e., that described by $K_{\text{eq}}' = k_{-1}/k_{+1}$). Equation (6-89) is made use of in the following way. If the decadic logarithms of K_{eq} are plotted against the reciprocal of absolute temperature (Fig. 6-14d), the slope multiplied by $-(2.303 \times 1.98)$ equals $\Delta H_f{}^0$. (If we use $\log K_{\text{eq}}'$ instead, then the slope of its variation with T^{-1} multiplied by the same factor gives $\Delta H_r{}^0$.) For any two values, i.e., K_1 and K_2 of K_{eq} at two temperatures T_1 and T_2, $\Delta H_f{}^0$ is given by

$$\Delta H_f{}^0 = \frac{4.560(\log K_2 - \log K_1)}{1/T_1 - 1/T_2} \text{ cal/mol}$$

$$= \frac{4.560T_1T_2(\log K_2 - \log K_1)}{T_2 - T_1} \text{ cal/mol} \tag{6-89b}$$

Thus an increase in extent of reaction with increasing temperature (i.e., an endothermic reaction) is characterized by positive value of $\Delta H_f{}^0$, and, conversely, a decrease with an increasing temperature (i.e., an exothermic reaction) corresponds to a negative value for $\Delta H_f{}^0$.

Once ΔH^0 and K_{eq} are known for any reaction, the other thermodynamic functions may be calculated, if we remember that

$$\Delta G^0 = -2.303RT \log K \tag{6-90}$$

and

$$\Delta S^0 = \frac{\Delta H^0 - \Delta G^0}{T} \tag{6-91}$$

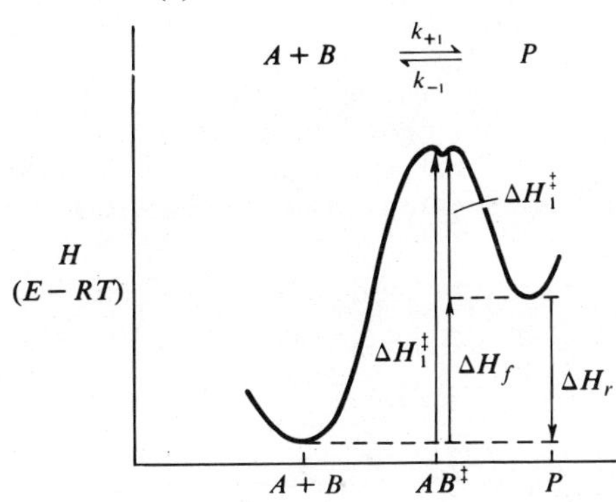

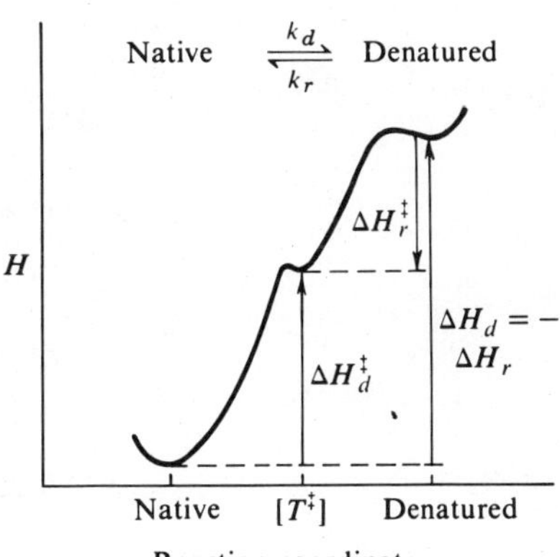

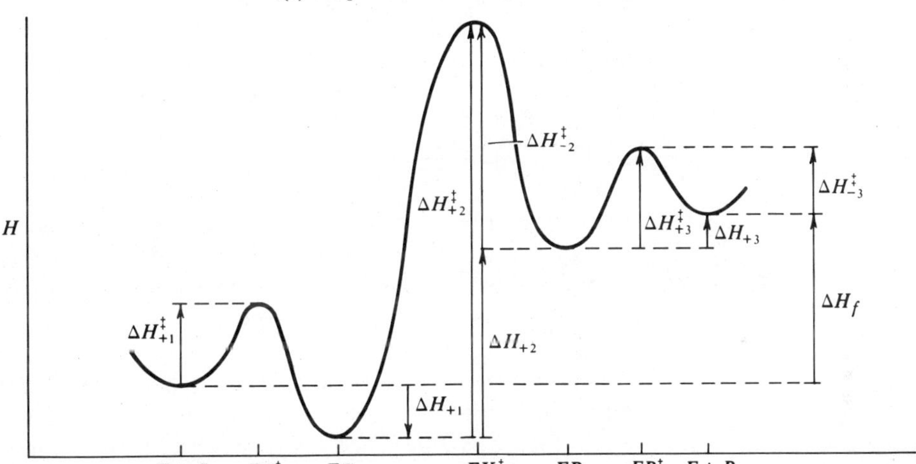

$$E + S \underset{k_{-1}}{\overset{k_{+1}}{\rightleftharpoons}} ES \underset{k_{-2}}{\overset{k_{+2}}{\rightleftharpoons}} EP \underset{k_{-3}}{\overset{k_{+3}}{\rightleftharpoons}} E + P$$

$$K_f = \frac{p_{eq}\, k_{+1}\, k_{+2}\, k_{+3}}{s_{eq}\, k_{-1}\, k_{-2}\, k_{-3}} \qquad K_1 = \frac{(es)_{eq}}{e_{eq}\, s_{eq}} = \frac{k_{+1}}{k_{-1}} \qquad K_2 = \frac{(ep)_{eq}}{(es)_{eq}} = \frac{k_{+2}}{k_{-2}} \qquad K_3 = \frac{e_{eq}\, p_{eq}}{(ep)_{eq}} = \frac{k_{+3}}{k_{-3}}$$

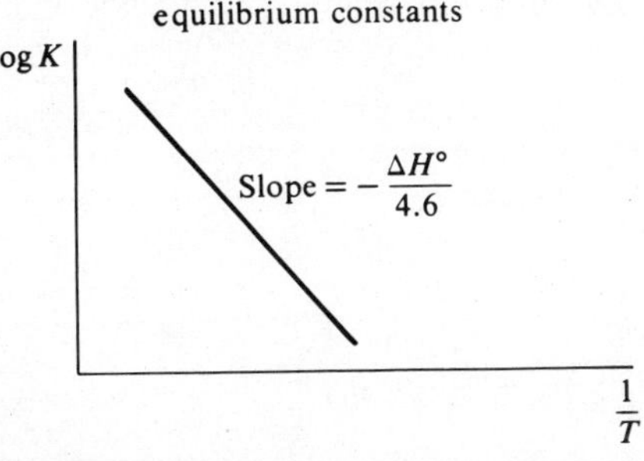

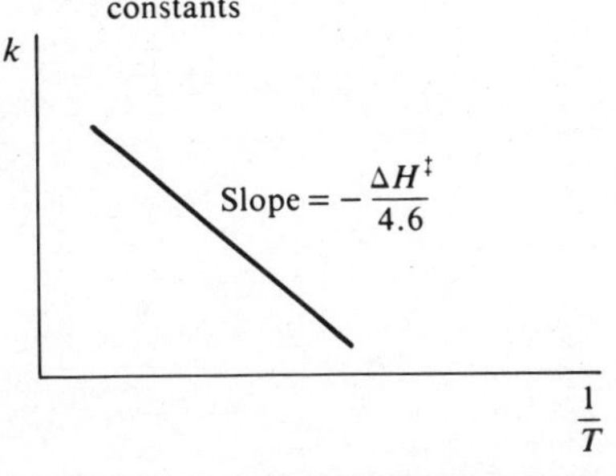

FIGURE 6-14
Effect of temperature on various reaction parameters.

What is the effect of temperatures on k_{+1} or k_{-1}, the rate constants of the forward or reverse reaction, respectively? This is given by the Arrhenius equation which states that

$$E_s = 2.303RT^2 \frac{d \log k}{dT} = -2.303R \frac{d \log k}{dT^{-1}} \tag{6-92a}$$

E_s, the Arrhenius activation energy, is obtained from a plot of the log of the rate constant against $1/T$ (Fig. 6-14*c*) and multiplying the slope by $-(2.303 \times 1.98)$. From rates k'' and k' at any two temperatures (T_2 and T_1), E_s—frequently called Q in the older biochemical literature—may be calculated by

$$E_s = \frac{4.560T_1T_2(\log k'' - \log k')}{T_2 - T_1} \text{ cal/mol} \tag{6-92b}$$

There is a striking similarity between Eqs. (6-89a) and (6-89b) on the one hand and (6-92a) and (6-92b) on the other. The first set, however, deals with a thermodynamic quantity ΔH^0 and is, therefore, a precise description of a property of the system in question; as a matter of fact, the van't Hoff equation is merely an alternative formulation of the Gibbs–Helmholtz equation:

$$\left(\frac{\partial \Delta G/T}{\partial T}\right)_{p=\text{const}} = -\frac{\Delta H}{T^2} \tag{6-92c}$$

The second set [Eqs. (6-92a) and (6-92b)] is purely empirical and provides a statement of an experimental observation. In order to interpret this observation, additional assumptions are required. One such set of assumptions, which has proved extremely fruitful, is the *theory of absolute reaction rates,* which postulates that the rate of any chemical reaction is proportional to the concentration of an *activated complex* or *transition state* that is in equilibrium with the reactants and thus governed by the usual laws of thermodynamics. Figure 6-14*a* is a typical energy diagram: $AB^{\ddagger}$ is the transition state; in general $(\quad)^{\ddagger}$ denotes a property of that state. Therefore,

$$k = \frac{\kappa \bar{k} T}{h} K^{\ddagger} \tag{6-93}$$

where κ is a *transmission coefficient,* usually equal to unity, $\bar{k}$ = Boltzmann's constant = 1.38×10^{-16} erg/deg, h = Planck's constant = 6.62×10^{27} erg·sec, T = absolute temperature, and $K^{\ddagger}$ is the equilibrium constant for the reaction (reactants) $\rightleftharpoons$ (activated complex).

Because the laws of thermodynamics apply, we can write, assuming that $\kappa = 1$,

$$k = \frac{\bar{k}T}{h} \exp\left(\frac{-\Delta G^{\ddagger}}{RT}\right) = \frac{\bar{k}T}{h} \exp\left(\frac{-\Delta H^{\ddagger}}{RT} + \frac{\Delta S^{\ddagger}}{R}\right) \tag{6-94a}$$

or

$$\Delta G^{\ddagger} = RT\left(\ln \frac{\bar{k}T}{h} - \ln k\right) = 2.303RT \log \frac{\bar{k}T}{hk} \tag{6-94b}$$

If $\Delta S^{\ddagger}$ does not vary with temperature, differentiation of Eq. (6-93), with $(\Delta H^{\ddagger} - T\Delta S^{\ddagger})$ substituted for $\Delta G^{\ddagger}$, yields

$$\frac{d \ln k}{dT} = \frac{1}{T} + \frac{\Delta H^\ddagger}{RT^2} = \frac{\Delta H^\ddagger + RT}{RT^2} \tag{6-95a}$$

or

$$(\Delta H^\ddagger + RT) = 2.303RT^2 \frac{d \log k}{dT} \tag{6-95b}$$

Equation (6-95b) is identical with the empirical equation (6-92a), with the experimental quantity E_s approximately 600 cal larger than $\Delta H^\ddagger$, the enthalpy difference between the transition state and the reactant:

$$E_s = \Delta H^\ddagger + RT \tag{6-96}$$

The relationships between the kinetic and thermodynamic functions ΔH_f, ΔH_r, $\Delta H^\ddagger{}_{+1}$, and $\Delta H^\ddagger{}_{-1}$ are all also shown in Fig. 6-14*a*.

Protein Denaturation. The thermodynamics of this process—described in Chapter 4—has already been discussed in Chapters 2 and 4. Its most characteristic features are (1) the high positive values for $\Delta S_d{}^0$ (above ambient temperatures), which means that, although the process is strongly endothermic, its ΔG^0 and hence its equilibrium position may be such as to favor denaturation at a relatively low temperature and over quite a narrow temperature range, and (2) the fact that $\Delta H^\ddagger{}_d < \Delta H_d$, and hence $\Delta H^\ddagger{}_r \ll 0$. Therefore, provided the temperature is sufficiently high to favor denaturation thermodynamically, its rate will not be enhanced by a further increase in temperature. In these and other respects, denaturation is qualitatively different from an ordinary exothermic chemical reaction; these differences are brought out clearly by a comparison of Fig. 6-14*a* and *b*.

Enzymatic Reactions Proper. The principles discussed in the preceding paragraphs are directly applicable to enzymatic reactions, although in practice the determination of the various parameters may not always be simple. In Fig. 6-14*c* the energy diagram for the simple, reversible one-substrate mechanism of Eq. (6-51) is presented. For each of the functions ΔH, ΔG, and ΔS (only the presentation for ΔH is shown) a total of six parameters (e.g., ΔH_f, ΔH_{+1}, ΔH_{+2}, $\Delta H^\ddagger{}_{+1}$, $\Delta H^\ddagger{}_{+2}$, $\Delta H^\ddagger{}_{+3}$) need to be determined for a complete description. Once this has been accomplished, all others can be calculated as simple sums or differences. The values required can be obtained by the application of Eqs. (6-92) and (6-95) to the temperature dependence of the six rate constants (and their appropriate ratios to yield four equilibrium constants). The intrinsic difficulty here is the fact, already discussed, that steady-state measurements provide only four parameters and hence four equations, but we have six unknown rate constants. Therefore, additional determinations are essential. These may be provided by independent measurements of either the equilibrium or rate constant for the formation (or dissociation) of two of the complexes.

In the figure, the highest activation barrier is given by $\Delta H^\ddagger{}_{+2}$, that corresponding to k_{+2}, the rate constant for the conversion of the enzyme–substrate to the enzyme–product complex. This fact makes this the smallest k_+ and this step the rate-limiting reaction. This need not always be so. There are known instances when either k_{+1} or k_{+3}, one the rate constant for the formation of the enzyme–substrate and the other the rate constant for dissociation of the enzyme–product complex, is smaller than the other two, and hence the rate-limiting step in question. This is reflected by $\Delta H^\ddagger{}_{+1}$ or $\Delta H^\ddagger{}_{+3}$ being the highest barrier in the two instances. In all these cases, when one of the constants is clearly smaller than all the rest, and the energy barrier for the corresponding reaction clearly the largest, this is the reaction influencing V_f (V_1 in the earlier discussion) to the greatest extent: to a first approximation $V_f \simeq ke$, where k is the rate constant in question, and $\Delta H_f^{\ddagger} \simeq -2.3R\, d \log V_f/dT^{-1} - RT$.

Measurements of Individual Rate Constants

We have pointed out earlier that the calculation of individual rate constants from steady-state kinetic measurements is possible only for a restricted number of mechanisms and may fail even in the simplest possible case, that involving a single substrate, if the reversible formation of an enzyme–product complex is included in the mechanism. In any event it is desirable to have independent measurements of these constants available, as a check on any postulated steady-state mechanism, and because of their intrinsic interest and importance in formulating theories of enzyme action.

By far the largest number of measurements have been concerned with determinations of k_1 (or k_{+1}), the bimolecular rate constant for the "on" reaction in the formation of the enzyme–substrate complex. If the equilibrium constant for the dissociation can also be determined, this then permits the computation of k_2 (or k_{-1}), the monomolecular "off" velocity for dissociation. Two different types of techniques have been employed: rapid-flow techniques in conjunction with pre-steady-state measurements and relaxation techniques after steady state or equilibrium has been attained.

Pre-Steady-State Kinetics. Let us return to Eq. (6-27) and Fig. 6-3. It will be recalled that $v = dp/dt = k_3x$; therefore, by differentiating,

$$\frac{d^2p}{dt^2} = k_3\frac{dx}{dt}$$

From this equation and Eqs. (6-28b) and (6-28e) we obtain

$$\frac{d^2p}{dt^2} + \frac{dp}{dt}(k_1a + k_2 + k_3) - k_1k_3ac_0 = 0 \tag{6-97}$$

For an enzymatic reaction, $c \equiv e$ and $a \simeq s_0$, the solution of this differential equation yields

$$p = p_0 + \frac{k_3s_0e_0t}{s_0 + K_s} + \frac{k_1k_3s_0e_0}{(k_1s_0 + k_2 + k_3)^2} \times \{\exp[-(k_1s_0 + k_2 + k_3)t] - 1\} \quad \text{(full reaction)} \tag{6-98a}$$

(if $p_0 = 0$, this, of course, eliminates the first term). This equation describes the whole initial part of the progress curve (p on Fig. 6-3) both during the initial lag period (to the left of the dashed line) and its almost linear increase subsequently, once the steady state has been reached. In the smaller graph in Fig. 6-3 we have magnified the scales to focus on the initial *pre-steady-state* period.

During this period (i.e., for very small values of t) we can use a series expansion of the exponential term, and, by neglecting all higher-order terms of this expansion, we obtain the approximate solution

$$p - p_0 = \frac{k_1k_3e_0s_0t^2}{2} \quad \text{(pre-steady state)} \tag{6-98b}$$

Because we know from steady-state measurement that $V_1 = k_3e_0$ and $K_s = (k_2 + k_3)/k_1$, measurement of p during this early phase (of the order of

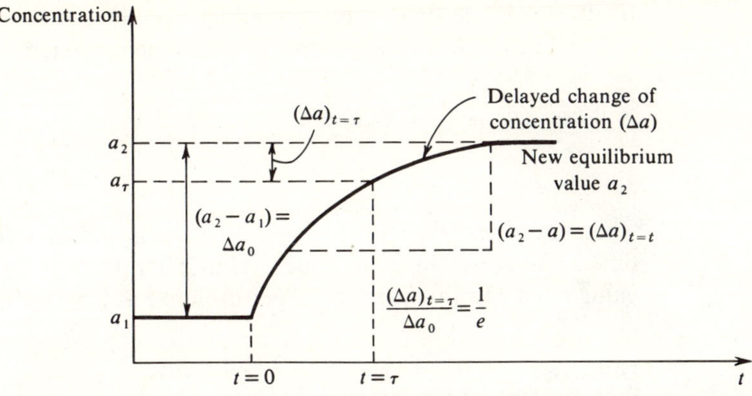

FIGURE 6-15
Kinetics of relaxation. Apply sudden change of external parameter, such as *T, P,* etc., to a system originally at equilibrium with $a = a_1$.

10^{-2} sec) by rapid-flow or stopped-flow techniques, usually by means of absorption or fluorescence spectrophotometry, permits the calculation of the desired constants.

In addition, we may make use of the following relations: The exponential term in Eq. (6-98a) describes the pre-steady-state situation; in the steady state it can be neglected. We thus see that now

$$p - p_0 = \frac{k_3 s_0 e_0 t}{s_0 + K_s} - \frac{k_3 s_0 e_0}{k_1(s_0 + K_s)^2}$$

$$\simeq k_3 e_0 t - \frac{k_3}{k_1}\frac{e_0}{s_0} \qquad \text{for } s_0 > K_s \qquad \text{(steady state)} \tag{6-98c}$$

which is the equation for the tangential dashed line in the smaller graph in Fig. 6-3. Setting $(p - p_0) = 0$, to obtain the x intercept τ, yields k_1 directly in terms of s_0 and this parameter:

$$\tau = \frac{1}{k_1 s_0} \tag{6-99}$$

Relaxation Kinetics. The principle of these methods is the following: An external perturbation of some physical parameter (i.e., of temperature, pressure, electric field density, etc.) that affects a system at chemical equilibrium is rapidly applied to such a system; hence, we speak of temperature, or *T*; pressure, or *P*; etc., jumps. We then measure, by suitably rapid techniques (optical measurements such as spectrophotometry, polarimetry, or fluorometry in the case of *T* and *P* jumps), the rate of delayed response (relaxation) of the system to the altered conditions. In Fig. 6-15 we depict the time course of such a reaction schematically when the external parameter has been altered very rapidly as compared with the response time of the chemical system, which is relatively slow. Temperature, for example, can be raised 10° in $<10^{-6}$ sec by passing a high-voltage pulse (10^4 to 10^5 V) through the solution; thus the time range for *T*-jump experiments is 1 sec down to 1 μsec.

Because equilibria are shifted only a small fraction, a few percent or less, in such experiments the kinetic equations governing the response are always relatively simple first-order differential equations. For example, in the simple case described in Fig. 6-2, with a single elementary step,

$$A + B \underset{k_2}{\overset{k_1}{\rightleftharpoons}} AB \quad \text{or} \quad S + E \underset{k_2}{\overset{k_1}{\rightleftharpoons}} SE \tag{6-16}$$

$$\Delta a = \Delta a_0 e^{-t/\tau} \tag{6-100}$$

where $\Delta a = a_2 - a$ and $\Delta a_0 = a_2 - a_1$; a is the concentration of A at any time t, a_1 is its old equilibrium value prior to, and a_2 the new equilibrium value reached subsequent to, a perturbation at $t = 0$. τ is the so-called *relaxa-*

TABLE 6-5
Rate constants of enzyme–substrate reactions[a]

Protein or enzyme	*Substrate*	k_f, M^{-1} *sec* $^{-1}$	k_d, *sec* $^{-1}$	*Method*
Fumarase	Fumarate	$>10^9$	$>4.5 \times 10^4$	SS
Fumarase	L-Malate	$>10^9$	$>4 \times 10^4$	SS
Acetylcholinesterase	Acetylcholine	$>10^9$	. . .	SS
Urease	Urea	$>5 \times 10^6$	. . .	SS
Myosin ATPase	ATP	$>8 \times 10^6$	. . .	SS
Hexokinase	Glucose	3.7×10^6	1.5×10^3	SS
Hexokinase	MgATP	$>4 \times 10^6$	$>6 \times 10^2$	SS
β-Amylase	Amylose	$>5.8 \times 10^7$	. . .	SS
Liver alcohol dehydrogenase	NAD	5.3×10^5	74	SS
Liver alcohol dehydrogenase	NADH	1.1×10^7	3.1	SS
Liver alcohol dehydrogenase-NAD	C_2H_5OH	$>1.2 \times 10^4$	>74	SS
Liver alcohol dehydrogenase-NADH	CH_3CHO	$>3.7 \times 10^5$	>3.1	SS
Malate dehydrogenase	NADH	6.8×10^8	2.4×10^2	TJ
Old yellow enzyme	FMN	1.5×10^6	$\sim 10^{-4}$	F
Catalase	H_2O_2	5×10^6	. . .	F
Catalase-H_2O_2	H_2O_2	1.5×10^7	. . .	F
Peroxidase	H_2O_2	9×10^6	<1.4	F
Peroxidase-H_2O_2(II)	Hydroquinone	2.5×10^6	. . .	F
Peroxidase	Cytochrome *c*	1.2×10^8	. . .	F
Glutamic-aspartic transaminase (aldehydic)	Glutamate	3.3×10^7	2.8×10^3	TJ
Glutamic-aspartic transaminase (aldehydic)	Aspartate	$>10^7$	$>5 \times 10^3$	TJ
Glutamic-aspartic transaminase (aldehydic)	NH_2OH	3.7×10^6	38	TJ
Glutamic-aspartic transaminase (aldehydic)	Ketoglutarate, oxalacetate	$>5 \times 10^8$	$>5 \times 10^4$	TJ
Glutamic-aspartic transaminase (aminic)	Oxalacetate	7×10^7	1.4×10^2	TJ
Glutamic-aspartic transaminase (aminic)	Ketoglutarate	2.1×10^7	70	TJ
BSA	NR′	2×10^6	35	TJ
BSA	NSR′	3.5×10^5	2.5	TJ
Anti-R-antibodies[b]	NR′	2.2×10^7	50	TJ
Anti-DNP-antibodies[b]	DNP-lysine	8.1×10^7	1.1	F
$Hb(O_2)_3$	O_2	2×10^7	36	F
Globin	Carboxyheme	7×10^7	. . .	F

[a] ATP, adenosine triphosphate; NAD, nicotinamide adenine dinucleotide; FMN, flavin mononucleotide; BSA, bovine serum albumin; NR′, 1-naphthol-4-(4′-azobenzene azo-phenyl) arsonic acid; NSR′, (naphthol-3-sulfonic acid 4-(4′-azobenzene azo)) phenyl arsonic acid; DNP, 2,4-dinitrophenyl; Hb, hemoglobin; SS, steady state; F, flow method; TJ, temperature jump method, k_f, rate constant for complex formation; k_d, rate constant for complex dissociation.

[b] Rabbit γ-globulin.

SOURCE: Eigen and Hammes (42), except for malate dehydrogenase, which is from G. H. Czerlinski and G. Schreck (43).

tion time which is a function of the rate constants and the equilibrium concentrations of the reactants defined as follows (b_2 is the new equilibrium concentration of B or E):

$$\frac{1}{\tau} = k_2 + k_1(a_2 + b_2) \tag{6-101}$$

Therefore, determination of the dependence of τ on the equilibrium concentrations of the reactants permits a ready calculation of k_1 and k_2. In Table 6-5, taken largely from the review by Eigen and Hammes (42), are summarized values of k_1 and k_2 for a number of different enzymatic and enzyme-like reactions. Most of the values for k_1 lie in the range 10^6 to 10^8 M^{-1}/sec. The latter value[14] approaches 10^9 M^{-1}/sec, the estimated range for a purely diffusion-controlled process involving relatively large, slowly diffusing species. The fact that k_1 is smaller than this value, especially in the case of some poorer substrates, has been interpreted in terms of a rate-determining fit of substrate and enzyme, involving orientation of the substrate on the enzyme surface, conformational changes of the latter, or both. The relatively small values of k_2 reflect the strength of the binding between substrate and protein; because k_2 frequently also exhibits large enthalpy and entropy increments, this provides evidence for the importance of structural effects in either the protein part of the active site or the solvent shell surrounding it, or both.

[14] Not all the values are *p*H independent; the actual experimental values for k_1 in these cases are less than the *p*H independent ones. The latter have to be compared with the diffusion-controlled value.

REFERENCES

Books, Reviews and Key References

Alberty, R. A.: Enzyme Kinetics, *Advan. Enzymol.*, **17:** 1 (1956); The Interpretation of Steady State Kinetic Data on Enzymatic Reactions, *Brookhaven Symp. Biol.*, **15:** 18 (1962).

Alberty, R. A.: The Rate Equation for an Enzymic Reaction, in P. D. Boyer, H. Lardy, and K. Myrbäck (eds.), *The Enzymes*, vol. 1, Academic Press, New York, 2nd ed., 1959, p. 143.

Bloomfield, V., L. Peller, and R. A. Alberty: Multiple Intermediates in Steady-State Enzyme Kinetics. II. Systems Involving Two Reactants and Two Products, *J. Am. Chem. Soc.*, **84:** 4367 (1962); III. Analysis of the Kinetics of Some Reactions Catalyzed by Dehydrogenases, *J. Am. Chem. Soc.*, **84:** 4375 (1962).

Cleland, W. W.: The Kinetics of Enzyme-Catalyzed Reactions with Two or More Substrates or Products. I. Nomenclature and Rate Equations, *Biochim. Biophys. Acta*, **67:** 104 (1963); II. Inhibition: Nomenclature and Theory, *Biochim. Biophys. Acta*, **67:** 173 (1963); III. Prediction of Initial Velocity and Inhibition Patterns by Inspection, *Biochim. Biophys. Acta*, **67:** 188 (1963). Enzyme Kinetics, in *Ann. Rev. Biochem.*, **36:** 77 (1967).

Dawes, E. A.: Enzyme Kinetics, in M. Florkin and E. H. Stotz (eds.), *Comprehensive Biochemistry*, vol. 12, Elsevier, New York, 1964, p. 89.

Dixon, M., and E. C. Webb: *Enzymes*, Academic Press, New York, 3rd ed., 1967, especially chaps. 2, 4, 8, and 9.

Eigen, M.: Kinetics of Reaction Control and Information Transfer in Enzymes and Nucleic Acids, in S. Claesson (ed.), *Fast Reactions and Primary Processes in Chemical Kinetics* (Nobel Symposium, vol. 5), Wiley (Interscience), New York, 1967, p. 333.

Gutfreund, H.: *Enzyme Kinetics*, Benjamin, New York, 1968.

Hearon, J. Z., S. A. Bernhard, S. L. Friess, D. J. Botts, and M. F. Morales: Enzyme Kinetics, in P. D. Boyer, H. Lardy, and K. Myrbäck (eds.), *The Enzymes*, vol. 1, Academic Press, New York, 2nd ed., 1959, p. 49.

Laidler, K. J.: *The Chemical Kinetics of Enzyme Action*, Clarendon Press, Oxford, 1958.

Peller, L., and R. A. Alberty: Multiple Intermediates in Steady State Enzyme Kinetics. I. The Mechanism Involving a Single Substrate and Product, *J. Am. Chem. Soc.*, **81:** 5907 (1959).

Reiner, J. M.: Quantitative Aspects of Enzymes and Enzyme Systems, in M. Florkin and E. H. Stotz (eds.), *Comprehensive Biochemistry*, vol. 12, Elsevier, New York, 1964, p. 126.

Segal, H. L.: The Development of Enzyme Kinetics, in P. D. Boyer, H. Lardy, and K. Myrbäck (eds.), *The Enzymes*, vol. 1, Academic Press, New York, 2nd ed., 1959, p. 1.

Walter, C.: *Enzyme Kinetics: Open and Closed Systems*, Ronald Press, New York, 1966.

Webb., J. L.: *Enzyme and Metabolic Inhibitors*, vols. I–III, Academic Press, New York, 1963–1966.

Specific References

1. Bender, M. L., and F. J. Kézdy: *Ann. Rev. Biochem.*, **34:** 49 (1965).
2. Cleland, W. W.: Private communication.
3. Kézdy, F. J., and M. L. Bender: *Biochemistry,* **1:** 1097 (1962): Kézdy, F. J., L. Lorand, and K. D. Miller: *Biochemistry,* **4:** 2302 (1965).
4. Henson, C. P., and W. W. Cleland: *Biochemistry,* **3:** 338 (1964).
5. Hinkson, J. W., and H. R. Mahler: *Biochemistry,* **2:** 209 (1963); Nygaard, A. P., in P. D. Boyer, H. Lardy, and K. Myrbäck (eds.), *The Enzymes,* vol. 7, Academic Press, New York, 2nd ed., 1963, p. 557.
6. Dalziel, K.: *J. Biol. Chem.*, **238:** 2850 (1963); *Biochem. J.*, **84:** 244 (1962); Dalziel, K., and F. M. Dickinson: *Biochem. J.*, **100:** 34 (1966).
7. Wratten, C. C., and W. W. Cleland: *Biochemistry,* **2:** 935 (1963).
8. Raval, D. N., and R. G. Wolfe: *Biochemistry,* **1:** 1112 (1962); *Biochemistry,* **2:** 220 (1963).
9. Zewe, V., and H. J. Fromm: *J. Biol. Chem.,* **237:** 1668 (1962); *Biochemistry,* **4:** 782 (1965).
10. Schwert, G. W., B. R. Miller, and R. J. Peanasky: *J. Biol. Chem.*, **242:** 3245 (1967).
11. Boyer, P. D., and E. Silverstein: *Acta Chem. Scand.*, **17:** 5195 (1963); Silverstein, E., and P. D. Boyer: *J. Biol. Chem.*, **239:** 3901, 3908 (1964).
12. Morrison, J. F., and W. W. Cleland: *J. Biol. Chem.*, **241:** 673 (1966).
13. Simon, H., and D. Palm: *Angew. Chem. (Intern. Engl. Ed.)*, **5:** 920 (1966); Palm, D.: *Z. Naturforsch.*, **21B:** 540, 547 (1966); Palm, D., T. Fiedler, and D. Ruhrseitz: *ibid.*, **23B:** 623 (1968).
14. Arion, W. J., and R. C. Nordlie: *J. Biol. Chem.*, **239:** 2752 (1964).
15. Bright, H. J.: *J. Biol. Chem.*, **240:** 1198 (1965); *Biochemistry,* **6:** 1191 (1967).
16. Suelter, C. H., and W. Melander: *J. Biol. Chem.*, **238:** PC4108 (1963).
17. Mildvan, A. S., and M. Cohn: *J. Biol. Chem.*, **240:** 238 (1965); **241:** 1178 (1966); Mildvan, A. S., J. S. Leigh, and M. Cohn: *Biochemistry,* **6:** 1805 (1967).
18. Alben, J. O.: *J. Biol. Chem.*, **242:** 1827 (1967).
19. Perrin, D. D., and V. S. Sharma: *Biochim. Biophys. Acta,* **127:** 35 (1966).
20. Hammes, G. G., and S. A. Levison: *Biochemistry,* **3:** 1504 (1964).
21. Peters, R. A.: *Advan. Enzymol.*, **18:** 113 (1957).
22. Worcel, A., D. S. Goldman, and W. W. Cleland: *Federation Proc.*, **24:** 357 (1965); *J. Biol. Chem.*, **240:** 3399 (1965).
23. Monod, J., and F. Jacob: *Cold Spring Harbor Symp. Quant. Biol.*, **26:** 389 (1961).
24. Monod, J., J. Wyman, and J-P. Changeux: *J. Mol. Biol.*, **12:** 88 (1965).
25. Blangy, D., H. Buc, and J. Monod: *J. Mol. Biol.*, **31:** 13 (1968).
26. Stadtman, E. R.: *Advan. Enzymol.*, **28:** 41 (1966).
27. Frieden, C.: *J. Biol. Chem.*, **242:** 4045 (1967).
28. Umbarger, H. E.: *Science,* **123:** 848 (1956); Umbarger, H. E., and B. Brown: *J. Biol. Chem.*, **233:** 415 (1958).
29. Yates, R. A., and A. B. Pardee: *J. Biol. Chem.*, **221:** 757 (1956).
30. Rubin, M. M., and J-P. Changeux: *J. Mol. Biol.*, **21:** 265 (1966).
31. Atkinson, D. E.: *Ann. Rev. Biochem.*, **35:** 85 (1966).
32. Brown, W. E. L., and A. V. Hill: *Proc. Roy. Soc. (London)*, **B94:** 297 (1922–23).
33. Koshland, D. E., G. Némethy, and D. Filmer: *Biochemistry,* **5:** 365 (1966).
34. Kirtley, M. E., and D. E. Koshland, Jr.: *J. Biol. Chem.*, **242:** 4192 (1967).
35. Sweeny, J. R., and J. R. Fisher: *Biochemistry,* **7:** 561 (1968).
36. Griffin, C. C., and L. Brand: *Arch. Biochem. Biophys.*, **126:** 856 (1968).
37. Adair, G. S.: *J. Biol. Chem.*, **63:** 529 (1925).
38. Weber, G., and S. R. Anderson: *Biochemistry,* **4:** 1942 (1965).
39. Conway, A., and D. E. Koshland, Jr.: *Biochemistry,* **7:** 4011 (1968).
40. Pauling, L.: *Proc. Natl. Acad. Sci. U.S.*, **21:** 186 (1935).
41. Levitzki, A., and D. E. Koshland, Jr.: *Proc. Natl. Acad. Sci. U.S.*, **62:** 1121 (1969).
42. Eigen, M., and G. G. Hammes: *Advan. Enzymol.*, **25:** 1 (1963).
43. Czerlinski, G. H., and G. Schreck: *Biochemistry,* **3:** 89 (1964).

7
Mechanism of Enzyme Action

Protein molecules serve several functions in living systems. Perhaps their most striking biochemical role is their ability to affect, in a specific and efficient manner, the rates of the wide spectrum of reactions that constitute the dynamic aspect of the life process and to be themselves susceptible to a variety of controls. Proteins having such catalytic activity (Chapter 6) are called *enzymes*.

That enzymes are, in fact, protein in nature was initially established in 1926 by Sumner, who succeeded in obtaining the enzyme urease in crystalline form from jack bean meal and subsequently identifying it as a protein. This enzyme catalyzes the conversion of urea to carbon dioxide and ammonia. Although this classic contribution to the understanding of biochemistry was initially met with criticism and even derision, the crystallization and characterization of several additional enzymes by Northrup and Kunitz established the protein nature of enzymes beyond doubt. Since then more than 100 enzymes have been obtained in crystalline form. All are proteins. The protein nature of enzymes is so well established that, as indicated above, the identification as protein is inherent in the definition of the word "enzyme."

Enzymes are exceptional catalysts in four respects. First, they are exceedingly efficient. An appreciation of the catalytic efficiency of enzymes may be gained by realizing that, under optimal conditions, most enzymatic reactions proceed 10^8 to 10^{11} times more rapidly than the corresponding nonenzymatic reactions. The turnover number (number of substrate molecules metabolized per enzyme molecule per minute) for most enzymes is approximately 1000 and, in extreme cases, may be greater than 1,000,000. The rates of certain steps in enzymatic reactions are diffusion controlled, or nearly so. Thus, many reactions that ordinarily occur only under extreme conditions of temperature or acid or base concentration proceed rapidly and quantitatively at values of *p*H near neutrality and at room temperature in the presence of the appropriate enzymes. Second, most enzymatic reactions are specific in terms of the nature of the reaction catalyzed and the structure of the substrate utilized. Third, the spectrum of reactions catalyzed by enzymes is broad. Enzymes are responsible for catalyzing hydrolytic reactions, polymerizations, oxidation–reductions, dehydrations, aldol condensations, acyl transfer reactions, and free-radical reactions, for example. Proteins, consequently, are exceptionally versatile catalysts. Finally, the enzymes themselves are subject to a variety of cellular controls. Their rate of synthesis, as well as their final concentration, is under genetic control and is influenced by small molecules such as the very substrates and products of the reaction in which they themselves participate. Furthermore, they may be present in both inactive and active forms, and the rate and extent of that interconversion is influenced by the composition of the environment at any particular instant. It is worth noting at this point that nearly all biological reactions are enzyme catalyzed and hence subject to the controls indicated above. The biosynthesis of the enzymes themselves is enzyme catalyzed.

A recent monograph by Jencks provides an elegant description of the many considerations relevant to enzymatic reactions (see the references).

CLASSIFICATION AND NOMENCLATURE FOR ENZYME-CATALYZED REACTIONS

A systematic classification and nomenclature for enzyme-catalyzed reactions has been established by the Commission on Enzymes of the International Union of Biochemistry. According to this classification, enzymes are divided into six general groups:

1. Oxidoreductases—catalyzing oxidation–reduction reactions
2. Transferases—catalyzing group-transfer reactions
3. Hydrolases—catalyzing hydrolytic reactions
4. Lyases—catalyzing the addition of groups to double bonds or vice versa
5. Isomerases—catalyzing isomerizations
6. Ligases (synthetases)—catalyzing the condensation of two molecules coupled with the cleavage of a pyrophosphate bond of ATP or similar triphosphate.

Each broad class of enzymes is further divided into subclasses and sub-subclasses on the basis of the nature of the individual transformations involved. Such subclassifications describe the nature of the coenzyme involved, if any, the type of isomerization, the type of bond hydrolyzed, and so on. This classification scheme provides a method of identifying individual enzymes by a series of four numbers. The first number indicates the main class into which the enzyme falls, the second and third indicate the subclass and sub-subclass, respectively, and the fourth is the serial number of the enzyme in its particular sub-subclass. This numbering and classification scheme is illustrated in Table 7-1, which is an abbreviated version of the comprehensive compilation but contains most of the important enzymes with which we shall deal in this book.

The naming of individual enzymes is a more formidable task; first, because it is difficult to establish a systematic nomenclature that accurately describes the nature of the reaction involved and, second, because an extensive and very widely used trivial nomenclature has already evolved. A systematic nomenclature has been developed (together with the systematic classification), and one must be familiar with both the systematic and trivial systems.

A careful examination of the systematic nomenclature presented in Table 7-1 provides familiarity with the rules for arriving at a systematic name for a given enzyme. Most trivial names attempt to give some information concerning the nature of the substrates involved and the nature of the catalyzed reaction, and most such names end with the suffix "ase." A few of the commonly employed enzyme categories are indicated below:

1. Dehydrogenases—enzymes that catalyze dehydrogenation of their substrates with a molecule other than molecular oxygen as hydrogen acceptor. If hydrogen transfer from the donor molecule is not readily demonstrated, the term *reductase* may be employed to describe such enzymes.

TABLE 7-1
Classification and nomenclature for enzyme-catalyzed reactions

Number	*Systematic Name*	*Trivial Name*	*Reaction*
1. Oxidoreductases			
1.1. Acting on CH—OH *group of donors*			
1.1.1. With NAD or NADP as acceptor			
1.1.1.1	Alcohol : NAD oxidoreductase	Alcohol dehydrogenase	Alcohol + NAD^+ = aldehyde or ketone + NADH
1.1.1.8	L-Glycerol-3-phosphate : NAD oxidoreductase	Glycerolphosphate dehydrogenase	L-Glycerol-3-phosphate + NAD^+ = dihydroxyacetone phosphate + NADH
1.1.1.23	L-Histidinol : NAD oxidoreductase	Histidinol dehydrogenase	L-Histidinol + $2NAD^+$ = L-histidine + 2NADH
1.1.1.25	Shikimate : NADP oxidoreductase	Shikimate dehydrogenase	Shikimate + $NADP^+$ = 5-dehydroshikimate + NADPH
1.1.1.27	L-Lactate : NAD oxidoreductase	Lactate dehydrogenase	L-Lactate + NAD^+ = pyruvate + NADH
1.1.1.33	Mevalonate : NADP oxidoreductase	Mevaldate reductase	Mevalonate + $NADP^+$ = mevaldate + NADPH
1.1.1.35	L-3-Hydroxyacyl-CoA : NAD oxidoreductase	3-Hydroxyacyl-CoA dehydrogenase	L-3-Hydroxyacyl-CoA + NAD^+ = 3-oxo-acyl-CoA + NADH
1.1.1.37	L-Malate : NAD oxidoreductase	Malate dehydrogenase	L-Malate + NAD^+ = oxalacetate + NADH
1.1.1.42	L_s-Isocitrate : NADP oxidoreductase (decarboxylating)	Isocitrate dehydrogenase	L_s-Isocitrate + $NADP^+$ = 2-oxoglutarate + CO_2 + NADPH
1.1.1.49	D-Glucose-6-phosphate : NADP oxidoreductase	Glucose-6-phosphate dehydrogenase	D-Glucose-6-phosphate + $NADP^+$ = D-glucono-δ-lactone-6-phosphate + NADPH
1.1.3. With O_2 as acceptor			
1.1.3.4	β-D-Glucose : oxygen oxidoreductase	Glucose oxidase	β-D-Glucose + O_2 = D-glucono-δ-lactone + H_2O_2
1.2. Acting on the aldehyde- or keto-group of donors			
1.2.1. With NAD or NADP as acceptor			
1.2.1.2	Formate : NAD oxidoreductase	Formate dehydrogenase	Formate + NAD^+ = CO_2 + NADH
1.2.1.12	D-Glyceraldehyde-3-phosphate : NAD oxidoreductase (phosphorylating)	Glyceraldehydephosphate dehydrogenase, triosephosphate dehydrogenase	D-Glyceraldehyde-3-phosphate + orthophosphate + NAD^+ = 1,3-diphospho-D-glyceric acid + NADH
1.2.3. With O_2 as acceptor			
1.2.3.2	Xanthine : oxygen oxidoreductase	Xanthine oxidase	Xanthine + H_2O + O_2 = urate + H_2O_2
1.2.4. With lipoate as acceptor			
1.2.4.1	Pyruvate : lipoate oxidoreductase (acceptor-acetylating)	Pyruvate dehydrogenase	Pyruvate + oxidized lipoate = 6-*S*-acetyldihydrolipoate + CO_2
1.3. Acting on the CH—CH *group of donors*			
1.3.1. With NAD or NADP as acceptor			
1.3.1.1	4,5-Dihydro-uracil : NAD oxidoreductase	Dihydro-uracil dehydrogenase	4,5-Dihydro-uracil + NAD^+ = uracil + NADH
1.4. Acting on the $CH—NH_2$ *group of donors*			
1.4.1. With NAD or NADP as acceptor			
1.4.1.2	L-Glutamate : NAD oxidoreductase (deaminating)	Glutamate dehydrogenase	L-Glutamate + H_2O + NAD^+ = 2-oxoglutarate + NH_3 + NADH
1.4.3. With O_2 as acceptor			
1.4.3.2	L-Aminoacid : oxygen oxidoreductase (deaminating)	L-Aminoacid oxidase	An L-amino acid + H_2O + O_2 = a 2-oxo-acid + NH_3 + H_2O_2
1.4.3.3	D-Aminoacid : oxygen oxidoreductase (deaminating)	D-Aminoacid oxidase	A D-amino acid + H_2O + O_2 = a 2-oxo-acid + NH_3 + H_2O_2

TABLE 7-1 (Continued)

Number	*Systematic Name*	*Trivial Name*	*Reaction*
	1.5. Acting on the C—NH group of donors		
	1.5.1. With NAD or NADP as acceptor		
1.5.1.5	5,10-Methylenetetrahydrofolate : NADP oxidoreductase	Methylenetetrahydrofolate dehydrogenase	5,10-Methylenetetrahydrofolate + H^+ + $NADP^+$ = 5,10-methenyltetrahydrofolate + NADPH
	1.6. Acting on NADH *or* NADPH *as donor*		
	1.6.2. With a cytochrome as an acceptor		
1.6.2.1	NADH : cytochrome *c* oxidoreductase	NADH cytochrome *c* reductase	NADH + 2 oxidized cytochrome *c* = NAD^+ + 2 reduced cytochrome *c*
	1.6.4. With a disulfide compound as acceptor		
1.6.4.3	NADH : lipoamide oxidoreductase	Lipoamide dehydrogenase	NADH + oxidized lipoamide = NAD^+ + dihydro-lipoamide
	1.6.6. With a nitrogenous group *as acceptor*		
1.6.6.2	NAD(P)H : nitrate oxidoreductase	Nitrate reductase	NAD(P)H + nitrate = $NAD(P)^+$ + nitrite + H_2O
1.6.6.4	NAD(P)H : nitrite oxidoreductase	Nitrite reductase	3 NAD(P)H + nitrite = 3 $NAD(P)^+$ + NH_4OH + H_2O
	1.9. Acting on heme groups of donors		
	1.9.3. With O_2 as acceptor		
1.9.3.1	Cytochrome *c* : oxygen oxidoreductase	Cytochrome oxidase	4 Reduced cytochrome *c* + O_2 = 4 oxidized cytochrome *c* + $2H_2O$
	1.11. *Acting on* H_2O_2 as *acceptor*		
1.11.1.6	Hydrogen-peroxide : hydrogen-peroxide oxidoreductase	Catalase	$H_2O_2 + H_2O_2 = O_2 + 2H_2O$
	1.99. Other enzymes using O_2 *as oxidant*		
	(Reaction not yet fully understood; no systematic names recommended)		
	1.99.1. Hydroxylases		
1.99.1.2	...	Phenylalanine 4-hydroxylase	Hydroxylates L-phenylalanine to L-tyrosine; a reduced pteridine is one of the reactants
1.99.1.5	...	Kynurenine 3-hydroxylase	Hydroxylates L-kynurenine to 3-hydroxykynurenine; NADPH is one of the reactants
	2. Transferases		
	2.1. Transferring one-carbon groups		
	2.1.1. Methyltransferases		
	(For the trivial name, "methyltransferase" may be replaced by "transmethylase")		
2.1.1.2	*S*-Adenosylmethionine : guanidinoacetate *N*-methyltransferase	Guanidinoacetate methyltransferase	*S*-Adenosylmethionine + guanidinoacetate = *S*-adenosylhomocysteine + creatine
2.1.1.5	Betaine : L-homocysteine *S*-methyltransferase	Betaine homocysteine methyltransferase	Betaine + L-homocysteine = dimethylglycine + L-methionine
	2.1.2. Hydroxymethyl-, formyl-, and related transferases		
	(For the trivial name, "hydroxymethyltransferase," "formyltransferase," and "formiminotransferase" may be replaced by "transhydroxymethylase," "transformylase," and "transformiminase," respectively)		
2.1.2.1	L-Serine : tetrahydrofolate 5,10-hydroxymethyltransferase	Serine hydroxymethyltransferase	L-Serine + tetrahydrofolate = glycine + 5,10-methylenetetrahydrofolate
2.1.2.4	*N*-Formiminoglycine : tetrahydrofolate 5-formiminotransferase	Glycine formiminotransferase	*N*-Formiminoglycine + tetrahydrofolate = glycine + 5-formiminotetrahydrofolate
	2.1.3. Carboxyl- and carbamoyltransferases		
	(For the trivial name, "carboxyltransferase" and "carbamoyltransferase" may be replaced by "transcarboxylase" and "transcarbamoylase," respectively)		

TABLE 7-1 (Continued)

Number	*Systematic Name*	*Trivial Name*	*Reaction*
2.1.3.2	Carbamoylphosphate : L-aspartate carbamoyltransferase	Aspartate carbamoyltransferase	Carbamoyl phosphate + L-aspartate = orthophosphate + *N*-carbamoyl-L-aspartate
2.2. Transferring aldehydic or ketonic residues			
2.2.1.1	D-Sedoheptulose-7-phosphate : D-glyceraldehyde-3-phosphate glycolaldehydetransferase	Transketolase, glycolaldehyde-transferase	D-Sedoheptulose-7-phosphate + D-glyceraldehyde-3-phosphate = D-ribose-5-phosphate + D-xylulose-5-phosphate
2.2.1.2	D-Sedoheptulose-7-phosphate : D-glyceraldehyde-3-phosphate dihydroxyacetonetransferase	Transaldolase, dihydroxyacetone-transferase	D-Sedoheptulose-7-phosphate + D-glyceraldehyde-3-phosphate = D-erythrose-4-phosphate + D-fructose-6-phosphate
2.3. Acyltransferases			
2.3.1. Acyltransferases			
(For the trivial name, "acetyltransferase," etc., may be replaced by "transacetylase," etc.)			
2.3.1.7	Acetyl-CoA : carnitine *O*-acetyltransferase	Carnitine acetyltransferase	Acetyl-CoA + carnitine = CoA + *O*-acetylcarnitine
2.3.1.8	Acetyl-CoA : orthophosphate acetyltransferase	Phosphate acetyltransferase	Acetyl-CoA + orthophosphate = CoA + acetylphosphate
2.3.1.12	Acetyl-CoA : dihydrolipoate *S*-acetyltransferase	Lipoate acetyltransferase	Acetyl-CoA + dihydrolipoate = CoA + *S*-6-acetylhydrolipoate
2.4. Glycosyltransferases			
(For the trivial name, "glucosyltransferase," "fructosyltransferase," etc., may be replaced by "transglucosylase," "transfructosylase," etc.)			
2.4.1. Hexosyltransferases			
2.4.1.8	Maltose : orthophosphate glucosyltransferase	Maltose phosphorylase	Maltose + orthophosphate = β-D-glucose-1-phosphate + D-glucose
2.4.2. Pentosyltransferases			
2.4.2.1	Purine nucleoside : orthophosphate ribosyltransferase	Purine nucleoside phosphorylase	Purine nucleoside + orthophosphate = α-D-ribose-1-phosphate + purine
2.4.2.10	Orotidine-5′-phosphate : pyrophosphate phosphoribosyltransferase	Orotate phosphoribosyltransferase, orotidine-5′-phosphate pyrophosphorylase	Orotidine 5′-phosphate + pyrophosphate = orotate-5-phospho-α-D-ribosylpyrophosphate
2.4.2.13	ATP : L-methionine *S*-adenosyltransferase	Methionine adenosyltransferase	ATP + L-methionine + H_2O = *S*-adenosylmethionine + trimetaphosphate
2.6. Transferring nitrogenous groups			
2.6.1. Aminotransferases			
(For the trivial name, "aminotransferase" may be replaced by "transaminase")			
2.6.1.1	L-Aspartate : 2-oxoglutarate aminotransferase	Aspartate aminotransferase	L-Aspartate + 2-oxoglutarate = oxalacetate + L-glutamate
2.6.1.2	L-Alanine : 2-oxoglutarate aminotransferase	Alanine aminotransferase	L-Alanine + 2-oxoglutarate = pyruvate + L-glutamate
2.7. Transferring phosphorus-containing groups			
2.7.1. Phosphotransferases with an alcohol group as acceptor			
2.7.1.2	ATP : glucose-6-phosphotransferase	Glucokinase	ATP + D-glucose = ADP + D-glucose-6-phosphate
2.7.1.15	ATP : D-ribose-5-phosphotransferase	Ribokinase	ATP + D-ribose = ADP + D-ribose-5-phosphate
2.7.1.28	ATP : D-glyceraldehyde-3-phosphotransferase	Triokinase	ATP + D-glyceraldehyde = ADP + D-glyceraldehyde-3-phosphate
2.7.1.40	ATP : pyruvate phosphotransferase	Pyruvate kinase	ATP + pyruvate = ADP + phosphoenolpyruvate
2.7.2. Phosphotransferases with a carboxyl group as acceptor			

TABLE 7-1 (Continued)

Number	*Systematic Name*	*Trivial Name*	*Reaction*
2.7.2.1	ATP : acctate phosphotransferase	Acetate kinase	ATP + acetate = ADP + acetyl phosphate
	2.7.4. Phosphotransferases with a phospho- group as acceptor		
2.7.4.3	ATP : AMP phosphotransferase	Adenylate kinase	ATP + AMP = ADP + ADP
	2.7.7. Nucleotidyl transferases		
2.7.7.9	UTP : α-glucose-1-phosphate uridylyltransferase	Glucose-1-phosphate uridylyltransferase, UDPG pyrophosphorylase	UTP + α-D-glucose-1-phosphate = pyrophosphate + UDPglucose
2.7.7.13	GTP : α-D-mannose-1-phosphate guanylyltransferase	Mannose-1-phosphate guanylyltransferase	GTP + α-D-mannose-1-phosphate = pyrophosphate + GDPmannose
2.7.7.14	CTP : ethanolaminephosphate cytidylyltransferase	Ethanolaminephosphate cytidylyltransferase	CTP + ethanolamine phosphate = pyrophosphate + CDPethanolamine
	2.8. Transferring sulfur-containing groups		
	2.8.3. CoA-transferases		
2.8.3.1	Acetyl-CoA : propionate CoA-transferase	Propionate CoA-transferase	Acetyl-CoA + propionate = acetate + propionyl-CoA
2.8.3.5	Succinyl-CoA : 3-oxo-acid CoA-transferase	3-Ketoacid CoA-transferase	Succinyl-CoA + a 3-oxo-acid = succinate + a 3-oxo-acyl-CoA
	3. Hydrolases		
	3.1. Acting on ester bonds		
	3.1.1. Carboxylic ester hydrolases		
3.1.1.7	Acetylcholine acetyl-hydrolase	Acetylcholinesterase	Acetylcholine + H_2O = choline + acetic acid
3.1.1.17	D-Glucono-δ-lactone hydrolase	Gluconolactonase	D-Glucono-δ-lactone + H_2O = D-gluconic acid
	3.1.2. Thiolester hydrolases		
3.1.2.1	Acetyl-CoA hydrolase	Acetyl-CoA hydrolase	Acetyl-CoA + H_2O = CoA + acetic acid
	3.1.3. Phosphoric monoester hydrolases		
3.1.3.1	Orthophosphoric monoester phosphohydrolase	Alkaline phosphatase	An orthophosphoric monoester + H_2O = an alcohol + H_3PO_4
3.1.3.2	Orthophosphoric monoester phosphohydrolase	Acid phosphatase	An orthophosphoric monoester + H_2O = an alcohol + H_3PO_4
3.1.3.9	D-Glucose-6-phosphate phosphohydrolase	Glucose-6-phosphatase	D-Glucose-6-phosphate + H_2O = D-glucose + H_3PO_4
	3.1.4. Phosphoric diester hydrolases		
3.1.4.1	Orthophosphoric diester phosphohydrolase	Phosphodiesterase	A phosphoric diester + H_2O = a phosphoric monoester + an alcohol
	3.2. Acting on glycosyl compounds		
	3.2.1. Glycoside hydrolases		
3.2.1.1	α-1,4-Glucan-4-glucanohydrolase	α-Amylase	Hydrolyzes α-1,4-glucan links in polysaccharides containing three or more α-1,4-linked D-glucose units
	3.4. Acting on peptide bonds (peptide hydrolases)		
	3.4.1. α-Aminopeptide aminoacidohydrolases		
3.4.1.1	. . .	Leucine aminopeptidase	Hydrolyses L-peptides, splitting off N-terminal residues with a free α-amino group, especially when the N-terminal residue is leucine or a related amino acid

TABLE 7-1 (Continued)

Number	Systematic Name	Trivial Name	Reaction
	3.4.2. α-Carboxypeptide aminoacidohydrolases		
3.4.2.1	...	Carboxypeptidase A	Hydrolyses peptides, splitting off the C-terminal L-amino acid residue unless this is a basic residue or proline
	3.4.4. Peptide peptidohydrolases		
3.4.4.1	...	Pepsin	Hydrolyses peptides, including those with bonds adjacent to aromatic or dicarboxylic L-amino acid residues
3.4.4.4	...	Trypsin	Hydrolyses peptides, amides, esters etc., at bonds involving the carboxy groups of L-arginine or L-lysine
3.4.4.5	...	Chymotrypsin	Hydrolyses peptides, amides, esters, etc., especially at bonds involving the carboxyl groups of aromatic L-amino acids
3.4.4.10	...	Papain	Hydrolyses peptides, amides, and esters, especially at bonds involving basic amino acids, or leucine or glycine
3.4.4.13	...	Thrombin	Hydrolyses peptides, amides, and esters of L-arginine; converts fibrinogen into fibrin
	3.5. Acting on C—N bonds other than peptide bonds		
	3.5.1. In linear amides		
3.5.1.5	Urea amidohydrolase	Urease	Urea + H_2O = CO_2 + $2NH_3$
	3.6. Acting on acid anhydride bonds		
	3.6.1. In phosphoryl-containing anhydrides		
3.6.1.1	Pyrophosphate phosphohydrolase	Inorganic pyrophosphatase	Pyrophosphate + H_2O = 2 orthophosphate
3.6.1.3	ATP phosphohydrolase	ATPase	ATP + H_2O = ADP + orthophosphate
	4. Lyases		
	4.1. Carbon-carbon lyases		
	4.1.1. Carboxy-lyases		
4.1.1.1	2-Oxo-acid carboxy-lyase	Pyruvate decarboxylase	A 2-oxo-acid = an aldehyde + CO_2
4.1.1.4	Acetoacetate carboxy-lyase	Acetoacetate decarboxylase	Acetoacetate = acetone + CO_2
4.1.1.12	L-Aspartate 4-carboxy-lyase	Aspartate 4-decarboxylase	L-Aspartate = L-alanine + CO_2
4.1.1.15	L-Glutamate 1-carboxy-lyase	Glutamate decarboxylase	L-Glutamate = 4-aminobutyrate + CO_2
4.1.1.17	L-Ornithine carboxy-lyase	Ornithine decarboxylase	L-Ornithine = putrescine + CO_2
4.1.1.22	L-Histidine carboxy-lyase	Histidine decarboxylase	L-Histidine = histamine + CO_2
4.1.1.23	Orotidine-5′-phosphate carboxy-lyase	Orotidine-5′-phosphate decarboxylase	Orotidine-5′-phosphate = UMP + CO_2
	4.1.2. Aldehyde-lyases		
4.1.2.5	L-Threonine acetaldehyde-lyase	Threonine aldolase	L-Threonine = glycine + acetaldehyde
4.1.2.7	Ketose-1-phosphate aldehyde-lyase	Aldolase	A ketose-1-phosphate = dihydroxyacetone phosphate + an aldehyde
	4.1.3. Ketoacid-lyases		
4.1.3.1	L_s-Isocitrate glyoxylate-lyase	Isocitrate lyase	L_s-Isocitrate = succinate + glyoxylate
4.1.3.7	Citrate oxalacetate-lyase (CoA-acetylating)	Citrate synthase	Citrate + CoA = acetyl-CoA + H_2O + oxalacetate

TABLE 7-1 (Continued)

Number	Systematic Name	Trivial Name	Reaction
4.2. Carbon-oxygen lyases			
4.2.1. Hydro-lyases			
4.2.1.4	Citrate hydro-lyase	Citrate dehydratase	Citrate = *cis*-aconitate + H_2O
4.2.1.10	5-Dehydroquinate hydro-lyase	5-Dehydroquinate dehydratase	5-Dehydroquinate = 5-dehydroshikimate + H_2O
4.2.1.13	L-Serine hydro-lyase (deaminating)	L-Serine dehydratase	L-Serine + H_2O = pyruvate + NH_3 + H_2O
4.3. Carbon-nitrogen lyases			
4.3.1. Ammonia-lyases			
4.3.1.1	L-Aspartate ammonia-lyase	Aspartate ammonia-lyase	L-Aspartate = fumarate + NH_3
4.3.1.3	L-Histidine ammonia-lyase	Histidine ammonia-lyase	L-Histidine = urocanate + NH_3
4.3.2. Amidine-lyases			
4.3.2.1	L-Argininosuccinate arginine-lyase	Argininosuccinate lyase	L-Argininosuccinate = fumarate + L-arginine
4.3.2.2	Adenylosuccinate AMP-lyase	Adenylosuccinate lyase	Adenylosuccinate = fumarate + AMP
5. Isomerases			
5.1. Racemases and epimerases			
5.1.1. Acting on amino acids and derivatives			
5.1.1.1	Alanine racemase	Alanine racemase	L-Alanine = D-alanine
5.1.1.5	Lysine racemase	Lysine racemase	L-Lysine = D-lysine
5.1.2. Acting on hydroxyacids and derivatives			
5.1.2.1	Lactate racemase	Lactate racemase	L-Lactate = D-lactate
5.1.3. Acting on carbohydrates and derivatives			
5.1.3.1	D-Ribulose-5-phosphate 3-epimerase	Ribulose phosphate epimerase	D-Ribulose-5-phosphate = D-xylulose-5-phosphate
5.1.3.2	UDPglucose 4-epimerase	UDPglucose epimerase	UDPglucose = UDPgalactose
6. Ligases			
6.1. Forming C—O bonds			
6.1.1. Amino acid-RNA ligases			
6.1.1.1	L-Tyrosine : tRNA ligase (AMP)	Tyrosyl-tRNA synthetase	ATP + L-tyrosine + tRNA = AMP + pyrophosphate + L-tyrosyl-tRNA
6.1.1.4	L-Leucine : tRNA ligase (AMP)	Leucyl-tRNA synthetase	ATP + L-leucine + tRNA = AMP + pyrophosphate + L-leucyl-tRNA
6.1.1.7	L-Alanine : tRNA ligase (AMP)	Alanyl-tRNA synthetase	ATP + L-alanine + tRNA = AMP + pyrophosphate + L-alanyl-tRNA
6.2. Forming C—S bonds			
6.2.1. Acid-thiol ligases			
6.2.1.1	Acetate : CoA ligase (AMP)	Acetyl-CoA synthetase	ATP + acetate + CoA = AMP + pyrophosphate + acetyl-CoA
6.2.1.4.	Succinate : CoA ligase (GDP)	Succinyl-CoA synthetase	GTP + succinate + CoA = GDP + orthophosphate + succinyl-CoA
6.3. Forming C—N bonds			
6.3.4. Other C—N ligases			
6.3.4.1.	Xanthosine-5′-phosphate : ammonia ligase (AMP)	GMP synthetase	ATP + xanthosine 5′-phosphate + NH_3 = AMP + pyrophosphate + GMP
6.3.4.2.	UTP : ammonia ligase (ADP)	CTP synthetase	ATP + UTP + NH_3 = ADP + orthophosphate + CTP
6.3.4.3.	Formate : tetrahydrofolate ligase (ADP)	Formyltetrahydrofolate synthetase	ATP + formate + tetrahydrofolate = ADP + orthophosphate + 10-formyltetrahydrofolate

TABLE 7-1 (Continued)

Number	*Systematic Name*	*Trivial Name*	*Reaction*
6.4. Forming C—C bonds			
6.4.1.1.	Pyruvate : carbon-dioxide ligase (ADP)	Pyruvate carboxylase	ATP + pyruvate + CO_2 + H_2O = ADP + orthophosphate + oxalacetate
6.4.1.2.	Acetyl-CoA : carbon-dioxide ligase (ADP)	Acetyl-CoA carboxylase	ATP + acetyl-CoA + CO_2 + H_2O = ADP + orthophosphate + malonyl-CoA
6.4.1.3.	Propionyl-CoA : carbon-dioxide ligase (ADP)	Propionyl-CoA carboxylase	ATP + propionyl-CoA + CO_2 + H_2O = ADP + orthophosphate + methylmalonyl-CoA

2. Oxidases—enzymes that catalyze oxidation of their substrates, with molecular oxygen serving as electron acceptor.

3. Hydroxylases—enzymes that catalyze the introduction of hydroxyl functions into their substrates, with molecular oxygen serving as oxygen donor.

4. Oxygenases—enzymes that catalyze the incorporation of an entire oxygen molecule into their substrates in the course of the oxidative cleavage of a carbon–carbon bond.

5. Kinases—enzymes that catalyze the transfer of phosphate from ATP or, much more infrequently, from another nucleoside triphosphate to their substrates.

6. Thiokinases—enzymes that catalyze the formation of thiol esters (of coenzyme A, page 387) from carboxylic acid substrates in reactions involving the cleavage of ATP.

7. Phosphatases—enzymes that catalyze the hydrolytic cleavage of phosphate esters.

8. Phosphorylases—enzymes that catalyze the addition of the elements of phosphoric acid across glycosyl or related linkages.

9. Transferases—enzymes that catalyze the transfer of a particular group between two substrates. Subcategories include the transacetylases (acetyl group transfer), transcarbamylases (carbamyl group transfer), and transcarboxylases (carboxyl group transfer).

10. Mutases—enzymes that catalyze the apparent migration of a phosphate group from one hydroxyl group to another within the same molecule.

11. Synthetases—enzymes that catalyze the condensation of two molecules coupled with the cleavage of ATP or other nucleoside triphosphate.

In several cases, enzymes are named by simply adding the suffix "-ase" to the name of the substrate (e.g., ATPase, ribonuclease, aconitase). In addition, many trivial names bear no relationship to either the substrates or type of reaction catalyzed. Examples include many proteolytic enzymes such as chymotrypsin, trypsin, pepsin, papain, rennin, and ficin. Finally, one must recognize that several names may be employed for one and the same enzyme. The enzyme that catalyzes the formation of citrate from oxalacetate and a two-carbon fragment related to acetate (page 607) has been called condensing enzyme, citrate condensing enzyme, citrogenase, and citrate synthetase, in addition to the systematic names. Most other names in the trivial nomenclature are reasonably self-explanatory. Some, but by no means all, of them have been approved by the Enzyme Commission as acceptable alternatives (Table 7-1).

ACTIVE SITE OF ENZYMES

The central questions in the study of reaction mechanisms concern the structure of the transition state and the nature of the intermediates for the reaction under consideration. In the case of enzyme-catalyzed reactions, a knowledge of the structure of the transition state involves not only an understanding of the geometry of the substrate molecule, but also of the three-dimensional conformation of the enzyme. The high catalytic efficiency and marked specificity exhibited by enzymatic reactions imply the participation of several distinct functional groups of the enzyme in the catalytic process. This implication has been fully borne out by, for example, the X-ray diffraction studies of lysozyme, chymotrypsin, and carboxypeptidase, which establish the involvement of several groups, usually distant from each other along the polypeptide backbone but near one another in space, in the associated catalytic processes (pages 368–372). This fact, together with the structural complexities inherent in the chemistry of molecules of the size of proteins, suffices to indicate that a precise definition of the transition-state structure for enzymatic reactions is a formidable task. Nevertheless, the mutually supportive investigations concerning protein structure and the various facets of enzyme activity have in several cases, including the three enzymes mentioned above, brought our goals within reach. Later, we shall consider these three enzymes in some detail, developing what is known concerning their catalytic activities. For the great preponderance of proteins for which the complete three-dimensional structure is not known, one must be content to work toward less sophisticated goals.

The immediate goals in the study of the mechanism of enzyme-catalyzed reactions may be formulated as follows: (1) the elucidation of the pathway for enzyme-catalyzed reactions; (2) the identification of amino acids involved in substrate binding and in bond-making and bond-breaking reactions; (3) determination of their approximate arrangement in space; and (4) development of a rationale, through assigning specific catalytic roles to the groups involved, that explains the velocity of enzyme-catalyzed reactions to within an order of magnitude. It is within the context of this approach that the succeeding discussion is couched.

In many instances, the integrity of the enzymatic process requires the participation of one or more coenzymes. A discussion of mechanisms for enzymatic reactions involving coenzymes is presented in Chapter 8.

Most enzymatic reactions involve substrates that are small compared with the size of the catalyst molecule. Consequently, only a small portion of the amino acid side chains and peptide bonds are near or in direct contact with the substrate molecule in the enzyme–substrate complex. This realization gives rise to the concept of an *active site* for enzymatic reactions. The active site of an enzyme is taken to include those side chains and peptide bonds which are in direct physical contact with the substrate (perhaps through intervening water molecules) and other side chains or peptide bonds that, although not in direct contact with the substrate, perform a direct function in the catalytic process. A schematic representation of an active site is pictured in Fig. 7-1. Part or all of the remainder of the protein serves the less direct function of providing a structural backbone suitable for maintaining the components of the active site in the three-dimensional conformation required for efficient, specific catalysis. The requirement for such a structural backbone is empha-

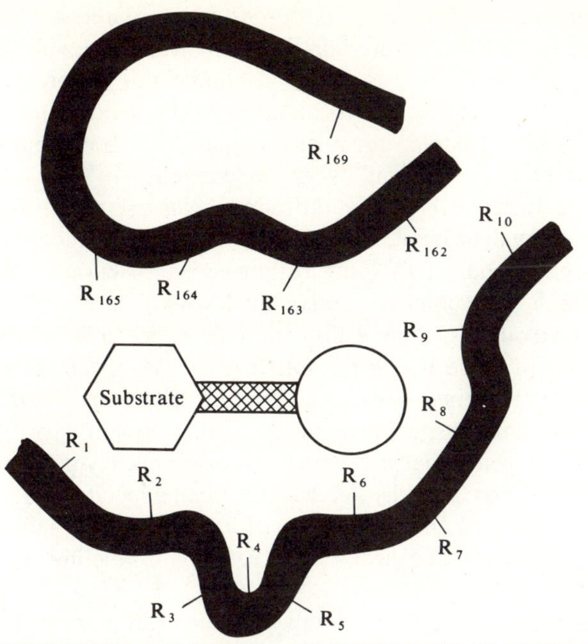

FIGURE 7-1
Schematic active site. The crosshatched area indicates a bond to be broken in the enzyme action. The R's represent some side chains and the heavy lines the backbone of two segments of the protein chain. D. E. Koshland, Jr., *Advan. Enzymol.,* **22:** 45 (1960).

sized by the dramatic alterations in enzymatic activity that frequently result when the conformation of the protein is altered through one of the denaturing processes (heat, urea, organic solvents, etc.). Part of the enzyme molecule may, of course, play no role in the catalytic process.

Identification of Amino Acids at the Active Site of Enzymes

The identification of amino acids present at the active site of enzymes is clearly of major importance for the understanding of the basic facets of the mechanism of enzyme-catalyzed reactions. Several methods, of varying utility, have been employed in attempts to identify at least some of the residues that constitute these active sites.

The most elegant means for identification of those groups at the active sites of enzymes is the use of wide-angle X-ray diffraction studies on enzyme–substrate or enzyme–inhibitor complexes. Provided that the substrate or inhibitor is associated with the crystalline enzyme in the same way as it is during the catalytic process in aqueous solution, we can "see" directly those groups which comprise the active site and their relationship to the substrate. Thus, we have already observed that Glu^{35} and Asp^{52} appear to be part of the active site of lysozyme (page 159) and that Arg^{145}, Tyr^{248}, and Glu^{270} are part of the active site of carboxypeptidase (page 161). The qualifying statement made above ought not to be taken lightly. It is entirely possible that stable crystalline complexes of enzymes and substrates or inhibitors may be formed in which the small molecule occupies a position not in contact with the active site. Thus, attention may be directed completely toward the wrong part of the molecule. A more likely danger, but one with less serious consequences, is that a substrate or inhibitor may form a complex at the active site of an enzyme but in a catalytically nonproductive way. There is substantial evidence for formation of *nonproductive complexes* in enzymatic reactions, that is, enzyme–substrate complexes in which the substrate is bound in such a way that catalysis is impossible. Should this occur in a crystalline preparation under structural study, attention would be directed to the correct part of the enzyme molecule, but misassignment of those groups actually involved in the catalytic process would very likely result.

Aside from such X-ray diffraction studies, the most direct method is to attach a covalent label, which itself is stable to procedures required to degrade polypeptide chains, to some residue present in the active site. For this purpose, it is convenient to distinguish two classes of enzymes: those which involve covalent enzyme–substrate intermediates in the course of the catalytic process and those which do not. In the case of the former class, the problem of specifically labeling the active site, rather than introducing a label into the protein in a nonspecific fashion, may frequently be solved by employing labeling reagents that are either the normal enzymatic substrates or structurally related compounds. As a result of the specific binding of these materials to the active site, the probability of introducing the label at alternative sites is greatly reduced. The problem of active-site labeling of enzymes of the latter class is more difficult.

The existence of covalent enzyme–substrate intermediates has been demonstrated for reactions catalyzed by an increasingly large number of enzymes. This list now includes chymotrypsin (1), trypsin (2), triose phosphate dehydrogenase (D-glyceraldehyde-3-phosphate : NAD oxidoreductase) (3), phosphoglucomutase (α-D-glucose-1,6-diphosphate : α-D-glucose-1-phosphate phosphotransferase) (4), phosphoglyceric acid mutase (5), several phosphatases (6), several aldolases (7), and acetoacetic decarboxylase (acetoacetic carboxy-lyase) (8), among others. In a few cases, the enzyme–substrate intermediate is sufficiently stable thermodynamically to exist in amounts sufficient for isolation and study. Phosphorylated phosphoglucomutase, an intermediate in the interconversion of glucose-1-phosphate and glucose-6-phosphate, may be prepared through incubation of the dephosphoenzyme with its phosphorylated glucose substrates (4):

$$\text{Phosphoenzyme} + \text{glucose-1-phosphate} \rightleftharpoons \text{dephosphoenzyme} + \text{glucose-1,6-diphosphate} \quad (7\text{-}1)$$

$$\text{Dephosphoenzyme} + \text{glucose-1,6-diphosphate} \rightleftharpoons \text{phosphoenzyme} + \text{glucose-6-phosphate} \quad (7\text{-}2)$$

Chemical and enzymatic degradation of the phosphorylated enzyme yields serine phosphate, which is flanked on its C-terminal side by a histidine residue. With the exception of this histidine residue, the amino acid composition in the neighborhood of the reactive serine residue resembles that for proteases and esterases (see below). Phosphorylated phosphoglyceric acid mutase, which is quite similar to phosphoglucomutase in its overall reaction pattern, may be similarly prepared and studied (5,9).

In the majority of cases, enzyme–substrate intermediates are not sufficiently stable with respect to starting materials or products to permit their isolation in appreciable quantities. In some cases, this difficulty may be overcome by stabilizing the intermediate by chemical modification as it is being formed. For example, incubation of aldolase (fructose-1,6-diphosphate: D-glyceraldehyde-3-phosphate-lyase) with radioactive dihydroxyacetone phosphate, or transaldolase (D-sedoheptulose-7-phosphate: D-glyceraldehyde-3-phosphate dihydroxyacetone transferase) with radioactive fructose-6-phosphate in the presence of sodium borohydride, a reducing agent, results in the irreversible incorporation of radioactivity into the protein (10,11). In both cases, complete acid hydrolysis of the protein has led to the identification of N-ε-glyceryllysine among the hydrolysis products. Thus, these reactions must involve the formation of a Schiff base between the keto function of the substrate molecule and the ε-amino group of a lysine residue, which is then reduced by sodium borohydride, yielding a stable secondary amine:

$$\text{Enzyme—NH}_2 + \text{O}{=}\!\!< \;\rightleftharpoons\; \text{enzyme—N}{=}\!\!< \tag{7-3}$$

$$\text{Enzyme—N}{=}\!\!< + \text{NaBH}_4 \rightarrow \text{enzyme—}\overset{\text{H}}{\underset{\text{H}}{\text{N}}}\text{—}\!\!< \xrightarrow{\text{acid}} \text{lysine—}\overset{\text{H}}{\underset{\text{H}}{\text{N}}}\text{—}\!\!< \tag{7-4}$$

Similar results have been obtained in the case of acetoacetic decarboxylase (8). This enzyme also appears to catalyze the formation of a Schiff base between the keto function of the substrate and the ε-amino group of lysine, which can be stabilized either by reduction with sodium borohydride or by the addition of cyanide to the imine double bond.

Of more general utility, although less elegant, are methods that involve labeling of the active site of enzymes by *quasi-substrates.* A quasi-substrate is a substance that is sufficiently similar to the normal substrate molecule to form the covalent enzyme–substrate complex but sufficiently different so that this intermediate decomposes slowly if at all. This method has been of particular importance in the labeling of the active site of esterases and proteases with diisopropylphosphofluoridate. Several enzymes in this class react stoichiometrically with this reagent to yield a diisopropylphosphoryl enzyme that is catalytically inert (Table 7-2). With this reagent, which is not closely related structurally to the normal substrates for these enzymes, the danger of introducing a label at some position other than the active site is considerably greater than in the cases described above. Yet the rapidity, stoichiometry, and specificity of this reaction strongly suggest that the active site is in fact being labeled. For example, DFP phosphorylates one of the two serine residues of chymotrypsin specifically. In corroboration of the conclusion above, chymotrypsin, for example, may also be labeled by many additional reagents including p-nitrophenyl acetate; in each case the same serine hydroxyl function, and no other group, is acylated. In the case of many, but not all, of the esterases and proteases studied, the group phosphorylated by DFP is a serine hydroxyl group flanked on the N-terminal side by either alanine or glycine. These results are summarized in Table 7-2.

Attaching a covalent label specifically to the active site of those enzymes which do not catalyze the intermediate formation of a covalent enzyme–substrate complex is, as mentioned above, a difficult task. The basic difficulty may be illustrated by the following example. Many enzymes are known to react with p-hydroxymercuribenzoate or p-chloromercuribenzenesulfonate through mercuration of the exposed sulfhydryl groups, yielding a catalytically inactive product. However, in most such cases it is unclear whether an essential sulfhydryl group in the active site has been blocked or whether mercuration of a sulfhydryl group present in the structural backbone has caused conformation changes, which may be quite small, that have led to an inactive enzyme. The lack of specificity of such amino acid reagents for residues located in the active site may, in theory at least, be overcome by the technique of *affinity labeling* (12). The idea underlying affinity labeling is

TABLE 7-2
Amino acid sequence around the reactive serine residue of some esterases and proteases

Sequence	*Enzyme*
Gly-Val-Ser-Ser-Cys-Met-Gly-Asp-*Ser*-Gly-Gly-Pro-Leu-Val-Cys-Lys	Chymotrypsin
NH_2 | Asp-Ser-Cys-Glu-Gly-Gly-Asp-*Ser*-Gly-Pro-Val-Cys-Ser-Gly-Lys	Trypsin
Asp-*Ser*-Gly	Thrombin
Asp-*Ser*-Gly	Elastase
Phe-Gly-Glu-*Ser*-Ala-Gly(Ala,Ala,Ser)	Butyryl cholinesterase
Glu-*Ser*-Ala	(Eel)acetylcholinesterase
Gly-Glu-*Ser*-Ala-Gly-Gly	Liver aliesterase (horse)
NH_2 | Asp-Gly-Thr-*Ser*-Met-Ala-Ser-Pro-His	Subtilisin (*B. subtilis*)
Thr-*Ser*-Met-Ala	Mold protease (*Aspergillus oryzae*)
Thr-Ala-*Ser*-His-Asp	Phosphoglucomutase
NH_2 | Lys-Glu-Ile-*Ser*-Val-Arg	Phosphorylase
Thr-Gly-Lys-Pro-Asp-Tyr-Val-Thr-Asp-*Ser*-Ala-Ala-Ser-Ala	Alkaline phosphatase (*E. coli*)

SOURCE: R. A. Oosterbaan and J. A. Cohen, in T. W. Goodwin, J. B. Harris, and B. S. Hartley (eds.), *Structure and Activity of Enzymes,* Academic Press, New York, 1964.

simple. Since the problem it to introduce the label specifically at the active site, the enzyme is reacted with some reagent, Sub—*X,* in which Sub is structurally related to the normal substrate for the enzyme and *X* is some reactive function, covalently attached to Sub, which can react with one or more amino acid side chains. Thus, as a result of the structural similarity of the reagent to the enzyme substrate, Sub—*X* should be specifically bound at the active site, and the reaction of *X* with available residues should accomplish the desired labeling.

The use of affinity labeling is, of course, not limited to those enzymes which do not form covalent enzyme–substrate intermediates. In fact, some of the most elegant work employing affinity labeling has been done with proteolytic enzymes that do form such intermediates. The point is that affinity-labeling techniques provide an opportunity to identify additional residues in the active site that do not directly form covalent bonds with the substrate. For example, histidines have been identified as components of the active sites of several proteolytic enzymes through use of alkylating reagents which bear structural similarity to the normal substrates. Some of the pertinent data are collected in Table 7-3. Note the correspondence between the structure of a typical substrate and that of the labeling reagent.

An alternative method useful for active-site labeling with reagents not related structurally to the normal enzymatic substrate is the following. In the presence of substrate or competitive inhibitor, the enzyme is treated with an

TABLE 7-3
Localization of histidine residues at the active sites of several proteolytic enzymes through the use of affinity labeling

Enzyme(s)	*Typical substrate*	*Affinity label*	*Ref.*
α-Chymotrypsin	$R{-}NH{-}CH(CH_2C_6H_5){-}C(=O){-}NH{-}R$	$CH_3{-}C_6H_4{-}SO_2{-}NH{-}CH(CH_2C_6H_5){-}C(=O){-}CH_2Cl$	13
Trypsin and cocoonase	$R{-}NH{-}CH(CH_2CH_2CH_2CH_2{-}NH_3^+){-}C(=O){-}NH{-}R$	$CH_3{-}C_6H_4{-}SO_2{-}NH{-}CH(CH_2CH_2CH_2CH_2{-}NH_3^+){-}C(=O){-}CH_2Cl$	14, 15
Chymotrypsin C and *Sorangium* protease C	$R{-}NH{-}CH(CH_2CH(CH_3)_2){-}C(=O){-}NH{-}R$	$CH_3{-}C_6H_4{-}SO_2{-}NH{-}CH(CH_2CH(CH_3)_2){-}C(=O){-}CH_2Cl$	16, 17

excess of reagent. Unreacted reagent and substrate or competitive inhibitor are then removed, and the enzyme is reacted with the same reagent, this time isotopically labeled. The substrate or competitive inhibitor is assumed to protect the active site of the enzyme from reaction with the reagent in the initial reaction. This reaction then leads to modification of all residues sensitive to the particular reagent chosen that are not present at the active site. Following removal of the protective agent, the sensitive residues in the active site are exposed and may react with the labeled reagent in the second reaction. Thus, all residues sensitive to the chosen reagent are reacted, but those, and only those, at the active site bear an isotopic label, and these residues may, therefore, be easily identified.

In the case of coenzyme-dependent reactions, a covalent label is sometimes built into the active site as a result of the covalent binding of coenzyme (prosthetic group) to enzyme. Particular examples include cytochrome *c*, in which the metalloporphyrin is covalently linked to two cysteine residues through thioether bonds by the vinyl side chains of the tetrapyrrole (page 678) and pyridoxal phosphate–dependent enzymes (page 393), in which the coenzyme may be attached covalently to amino groups of lysine residues in the form of a Schiff base. These cases are considered more fully in Chapter 8.

Indirect methods, which do not involve covalent labeling of the active site, may be employed for the tentative identification of amino acids involved in the catalytic process. The great advantage of covalent labeling of the active site, the direct identification and location in the peptide chain of at least one residue at the active site (frequently a residue intimately involved in the catalytic process), is lost in the use of indirect methods. Such methods can never locate the residue in the peptide chain, and it is difficult to know if the residue under study is at the active site or in the structural backbone.

The most widely used indirect method of identifying residues required for the catalytic process is the variation of some parameter of the reaction, such as the maximum velocity or the Michaelis constant, with *p*H. As emphasized in Chapter 6, such parameters frequently vary with *p*H in a manner suggesting the titration of a single ionizable group (page 310). From such titration curves, the pK_a for the group involved may be calculated, and its identification attempted by comparing this value with the known pK_a values for the amino acid side chains. This process involves several difficulties. First, as a result of interactions with neighboring groups in the protein or with substrate or buffer, the pK_a for an ionizable group in a protein may differ significantly from the corresponding value for the group free in solution. Furthermore, the pK_a values for several titratable groups present in proteins overlap considerably. For instance, a group with a pK_a of 10 might be assigned to an amino group, to a phenolic hydroxyl group, or possibly to a sulfhydryl group. The assignment of a pK_a to a particular group is sometimes aided by determination of ΔH for the ionization. However, a unique assignment of pK_a to a functional group is frequently not possible. Second, the *p*H behavior may reflect the titration of several residues rather than of an individual group. Finally, breaks in *p*H-rate profiles may result from other factors, such as a change in the rate-determining step, in addition to titrations. Fortunately, all these difficulties need not always occur. Frequently an assignment made on the basis of *p*H effects may be corroborated by other techniques. For example, several, but not all, of the enzymes listed in Table 7-2 appear to involve a histidine residue in the catalytic process on the basis of the effect of *p*H on the maximum velocities. In the case of chymotrypsin, this conclusion is supported by the finding that a chloroketone analog of a chymotrypsin substrate reacts specifically with one histidine residue with resultant inactivation (Table 7-3).

An additional method that has been widely employed in attempts to identify residues required for enzymatic activity is the loss of catalytic power with amino acid modification. The reaction of *p*-hydroxymercuribenzoate with sulfhydryl groups is an example of such a modification. This method is hampered a great deal by the lack of specificity of most amino acid reagents for a single type of residue and by the variability of reactivities of functional groups in proteins. The lack of specificity of most reagents used to modify amino acid residue side chains in proteins may be appreciated from a consideration of the collection of data in Table 7-4. Here many of the commonly used reagents are classified in terms of those residue side chains with which they may react. These difficulties may perhaps be overcome in many cases by the use of a residue modification–enzyme activity technique developed by Koshland and Ray. This technique involves the treatment of the enzyme with an excess of amino acid reagent. The kinetics of the modification of one or several amino acids may be followed simultaneously with the kinetics of enzyme activity loss by withdrawing samples periodically and assaying for enzymatic activity and amino acid composition. These experiments may yield pseudo-first-order rate constants both for the modification of the affected residues and for the loss of enzymatic activity. The results of a model experiment in which several amino acids are modified are shown in Fig. 7-2. It is clear that amino acids that are lost more rapidly than enzymatic activity cannot be required for catalysis, but those which are lost less rapidly may be required. Residue X_4 thus cannot be involved in the catalytic process, since it is destroyed at a rate greater than that for the loss of activity, but X_1 and X_2 may be involved. The finding that the sum of the rate constants for the loss of X_1 and X_2 is equal to the rate constant for the loss of enzymatic activity strongly suggests that these residues are actually involved. Amino acid X_5 is not modified, and, consequently, no conclusions concerning its role in the catalytic process may be drawn. This technique has, for example, been applied to phosphoglucomutase (18). In this case, a methionine residue has been implicated as necessary for the catalytic process, and a histidine residue is required for full activity, although its loss does not completely inactivate the enzyme. It is tempting to consider that this histidine is the one located adjacent to the reactive serine in this enzyme.

Note that plots of the type shown in Fig. 7-2 are not always first order. If several amino acids of a given type have different degrees of exposure to reactant, non-first-order kinetics for their loss will be observed (see page 164). If one of these is involved in catalysis, a correlation between loss of residue and loss of activity may be difficult to find. These considerations place limitations on the general use of the Koshland–Ray technique.

The above is a brief summary of some of the techniques available for the determination of the composition of the active site of enzymes together with a few of the results that have been obtained.

Conformation of the Active Site

The three-dimensional constellation of amino acids that constitutes the active site of an enzyme is not known in the great majority of cases. Indeed, it is not even certain that, in the absence of substrate, the three-dimensional conformation of the amino acids at the active site is unique, and if it is, that it corresponds to the catalytically active conformation. In this regard, two theories have been proposed regarding active-site conformation. Late in the nineteenth century, Emil Fischer, in order to explain the specificity of enzyme-catalyzed reactions, proposed that the enzyme is disposed in a reasonably rigid template to which the substrate could be bound. This *lock–key hypothesis* accounts for enzyme specificity, since template rigidity would cause steric repulsions to molecules differing structurally from the normal substrate. In contrast, Koshland has advocated an *induced-fit theory* to account for the specificity of enzymic reactions (19). This theory involves three postulates: "(1) that a considerable alteration in the geometry of the protein is caused by the substrate as it fits into the active site; (2) that a delicate orientation of catalytic groups is required for enzyme action; (3) that the substrate induces this proper orientation by the change it causes in geometry."

That is, a good substrate will interact strongly with the enzyme and the enzyme–substrate binding forces are employed to cause the enzyme to adopt an energetically unfavorable but catalytically active conformation. This point is shown schematically in Fig. 7-3. Although a poor substrate might

TABLE 7-4
Specificity of reagents commonly used to chemically modify proteins

	Peptide–Gln, Asn	*α-Amino, Lys*	*α-Carboxyl, Glu, Asp*	*Cystine*	*Arg*	*Ser, Thr*	*His*	*Trp*	*Tyr*	*Cysteine*	*Met*
Acetic anhydride		+				+			+	+	
Acetylimidazole		±							+	+	
Acrylonitrile		+								+	
Aldehydes, formaldehyde	+	+			+		+	±		+	
Aryl sulfonyl chlorides		+				±	+		+	+	
N-Bromosuccinimide							+	+	+	+	
Carbodiimides			+								
Cyanate		+				+				+	
Diazonium salts		+	+				+		+	+	
Dimethyl sulfate		+	+						+	+	+
Ethanethiol trifluoroacetate		+									
Ethyl acetimidate, Imido esters		+									
N-Ethyl maleimide		±					±			+	
Ferricyanide, iodosobenzoate, porphyrindin										+	

Fluorodinitro-benzene	+					+		+	+	
Haloacetate, amide	+	+				+			+	+
Heavy metals						±			+	
Iodine-KI, iodine monochloride						±	±	+	+	
Ketene	+				+		+	+	+	
Malonaldehyde	+			+		+		+		+
Mercuribenzoate	±					±			+	
Methyl iodide	+	+						+	+	
O-Methylisourea	+									
Nitrophenyl acetate	+				±			+	+	
Nitrous acid	+	+					+	+	+	
Phenylisothio-cyanate	+							+	+	
Phosphorus oxychloride	+				+			±		
Photooxidation			+			+	+	+	+	±
Propylene, ethylene oxide	+	+						+	+	
Tetranitromethane								+	+	±

SOURCE: B. L. Vallee and J. F. Riordan, *Ann. Rev. Biochem.*, 38: 733 (1969).

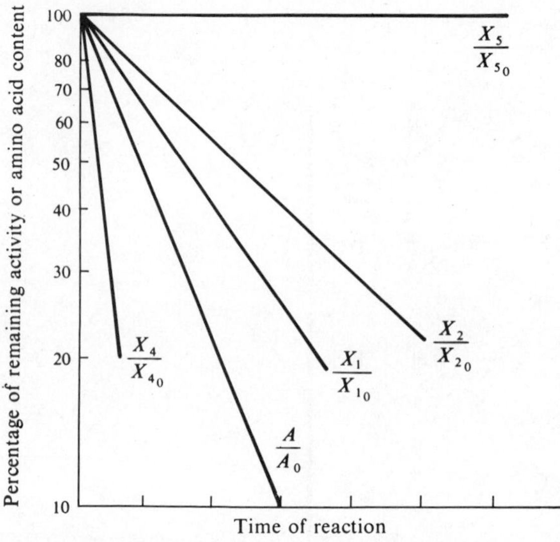

FIGURE 7-2

Loss of activity in a hypothetical enzyme on treatment with a reagent that destroys some amino acids and does not affect others. A/A_0 represents the fraction of enzyme activity and X/X_0 the fraction of a given amino acid that remains at a given time. D. E. Koshland, Jr., *Advan. Enzymol.,* **22:** 45 (1960).

interact with the enzyme, the binding forces would be too small to force the enzyme into the catalytically active form. Alternatively, the enzyme–substrate binding forces may be employed to induce strain in the substrate and labilize it for reaction. These points are developed more fully below.

The induced-fit theory was designed in an attempt to account for several findings not easily explained by the lock–key hypothesis. Consider for example, the reaction catalyzed by sucrose phosphorylase:

$$\text{Glucose-1-phosphate} + \text{fructose} \rightleftharpoons \text{glucosyl-fructose (sucrose)} + P_i \tag{7-5}$$

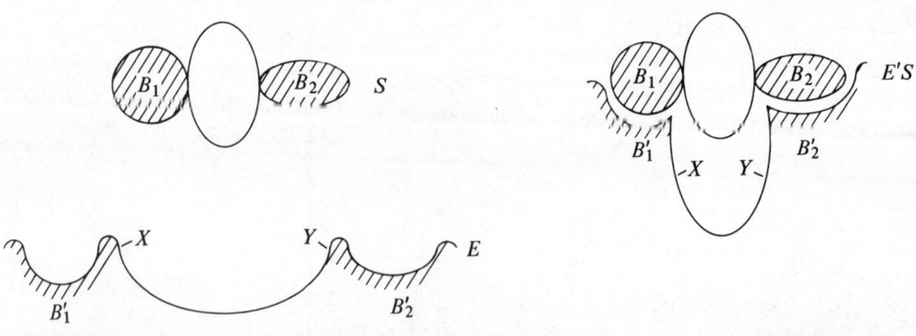

FIGURE 7-3

Induced-fit mechanism, in which binding of substrate through groups B_1 and B_2 causes a change in the conformation of the enzyme so that the groups X and Y in the active site, which are required for catalysis, are properly positioned relative to the substrate. W. P. Jencks, *Catalysis in Chemistry and Enzymology,* McGraw-Hill, New York, 1969.

A lock–key hypothesis would predict, that, in the absence of fructose as a glucosyl group acceptor, a measurable reaction of glucose-1-phosphate with water should occur. Glucose-1-phosphate is bound to the enzyme in the absence of fructose, and water is almost certainly present, owing to its high concentration, at the active site, and, consequently, should function as a glucosyl group acceptor. However, there is only a very slow reaction of glucose-1-phosphate with water in the presence of sucrose phosphorylase. Perhaps the most straightforward explanation of this finding is that fructose, by binding to the active site, causes an alteration in active-site geometry to the catalytically active conformation. Water is too small a molecule to possess the groups necessary to provide the required binding energy.

Several additional findings related to the above also suggest the inadequacy of the classical theory. The induced-fit theory receives support from chemical findings suggesting a modification of enzyme conformation upon substrate binding (20).

X-ray diffraction studies of enzyme-inhibitor complexes have provided direct evidence supporting the induced-fit hypothesis for some cases at least. The most dramatic example is that of carboxypeptidase (21). The binding of glycyl-L-phenylalanine to this enzyme causes Tyr^{248} to move about 12 Å, to a position in which it is in physical contact with bound substrate and in which it can play a direct catalytic role. Stereo pairs showing the structure of the active-site region in the presence and absence of this dipeptide were provided in Fig. 4-31. Smaller substrate- or inhibitor-induced conformation changes have been observed for other enzymes. In contrast, other enzymes, for instance chymotrypsin, seem to undergo very little structural alteration on substrate binding.

It seems probable that, for some enzymes, the lock–key hypothesis will prove capable of describing the specificity attributes of enzymatic reactions, and, for others, the induced-fit hypothesis will more accurately describe the situation.

The lock–key hypothesis may not prove as simple a concept as implied in the foregoing discussion. Although a good substrate may associate with the enzyme active site on the enzyme surface in only one way, it is possible that other related molecules, which lack one of the binding sites, might fit in a number of ways. Figure 7-4 provides a diagrammatic example of such behavior. Enzyme specificity for poor substrates would then be expressed in the maximal reaction velocity in terms of the fraction of time that the substrate forms a productive complex with the enzyme. However, the lock–key hypothesis modified to include formation of nonproductive complexes still seems inadequate to account for the failure of water to act as a glucosyl acceptor in the sucrose phosphorylase [Eq. (7-5)] and related reactions. Water has too few orientational possibilities available to it to account for its unreactivity in terms of formation of nonproductive complexes with the enzyme.

PATHWAY OF ENZYME-CATALYZED REACTIONS

An obvious prerequisite for a basic understanding of a chemical reaction, enzymatic or not, is a knowledge of the bonds broken and formed and the nature of the intermediates, if any, produced in the course of the reaction. Such information constitutes a knowledge of the *pathway* for the reaction. One of the most important methods of obtaining information relevant to reaction pathways is from analysis of initial velocity and inhibition patterns from

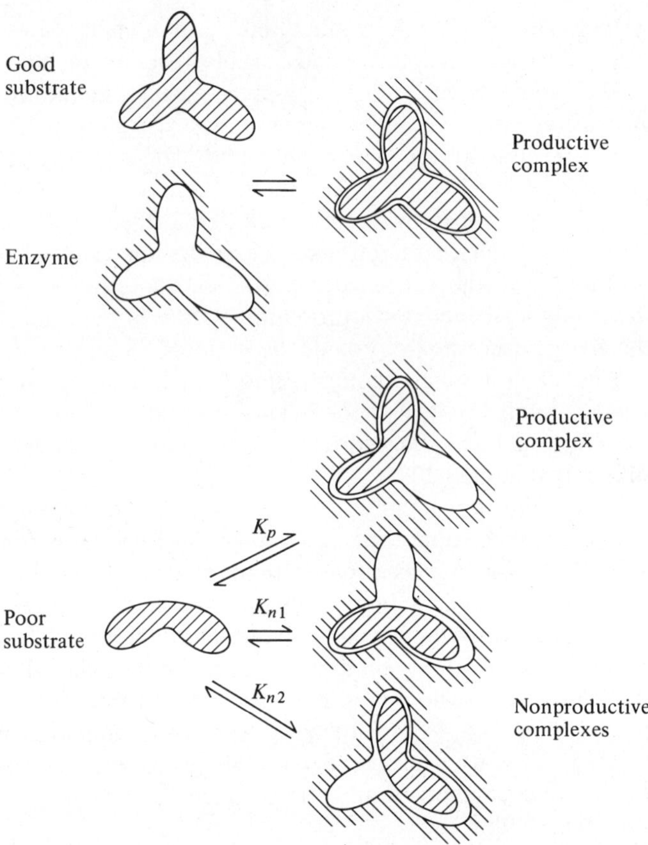

FIGURE 7-4
Productive and nonproductive binding of good and poor substrates. W. P. Jencks, *Catalysis in Chemistry and Enzymology,* McGraw-Hill, New York, 1969.

steady-state kinetic data. This topic has been fully explored in Chapter 6. In addition, valuable information pertinent to these points can frequently be obtained by means of isotopic labeling studies. Many isotopes, such as ^{14}C, ^{3}H, ^{32}P, and ^{35}S, are radioactive and may be distinguished from the normal isotopes on this basis. Others, such as ^{2}H, ^{15}N, and ^{18}O, differ from the ordinary isotopes only in mass and may, therefore, be assayed by recording mass spectra. Infrared and magnetic resonance spectra are also useful tools for distinguishing among isotopes.

Two types of experiments with isotopes are frequently performed: *Isotope incorporation studies,* which primarily yield information concerning the position of bond breakage and formation but which may be used in attempts to identify intermediates as well, and *exchange studies,* which involve replacement of a resident group with a similar labeled group and which primarily yield information concerning the existence of reaction intermediates. As an example of the former type of study, consider the cleavage of glucose-1-phosphate as catalyzed by alkaline phosphatase (orthophosphoric monoester phosphohydrolase). This reaction could conceivably occur with cleavage of either the C—O bond or the P—O bond. If this reaction is carried out in the presence of $H_2{}^{18}O$-enriched water, the former reaction path yields glucose containing one atom of ^{18}O (Eq. 7-6).

$$\text{Glucose-1-phosphate (O—PO}_3\text{H}^- \text{, cleavage at C—O)} + H_2{}^{18}O \longrightarrow \text{glucose (}{}^{18}OH\text{)} + H_2PO_4^- \qquad (7\text{-}6)$$

The latter is characterized by phosphate containing one atom of ^{18}O:

$$\text{Glucose-1-phosphate (O}\,\vdots\,\text{PO}_3\text{H}^-\text{, cleavage at P—O)} + H_2{}^{18}O \longrightarrow \text{glucose (OH)} + H_2P^{18}O_4^- \qquad (7\text{-}7)$$

The appearance of one atom of ^{18}O in inorganic phosphate identifies the P—O bond as the site of cleavage of this substrate. Nearly all phosphatase reactions appear to proceed in this fashion. On the other hand, the reaction of fructose with glucose-1-phosphate—sucrose phosphorylase reaction, Eq. (7-5)—in ^{18}O-enriched water yields sucrose and inorganic phosphate, which does not contain excess ^{18}O. Hence this reaction must occur with C—O bond cleavage, a general characteristic of phosphorylase reactions. In summary, these experiments indicate that phosphatases are phosphoryl group-transfer enzymes and phosphorylases are glycosyl group-transfer enzymes.

Sucrose phosphorylase (disaccharide glucosyltransferase) also provides an example of the use of isotope-exchange reactions. This enzyme catalyzes the following exchanges, each carried out in the absence of the second substrate:

$$\text{Glucose-1-}P + {}^{32}P_i \rightleftharpoons \text{glucose-1-}{}^{32}P + P_i \qquad (7\text{-}8)$$

$$\text{Glucose-fructose} + \text{fructose-}{}^{14}C \rightleftharpoons \text{glucose-fructose-}{}^{14}C + \text{fructose} \qquad (7\text{-}9)$$

These findings are consistent with the possibility of a covalently linked glucosyl-enzyme intermediate; that is, the reaction may proceed via a two-step reaction path:

$$\text{Glucose-1-}P + \text{enzyme} \rightleftharpoons \text{glucosyl-enzyme} + P_i \qquad (7\text{-}10)$$

$$\text{Glucosyl-enzyme} + \text{fructose} \rightleftharpoons \text{glucosyl-fructose} + \text{enzyme} \qquad (7\text{-}11)$$

It is important to realize that such results are suggestive but certainly not conclusive evidence for the existence of such an intermediate, since these exchange reactions may be equally well accounted for on the basis of direct displacement reactions. Furthermore, the absence of exchange reactions does not rule out the existence of a covalent enzyme–substrate intermediate, since both substrates may have to be present for catalytic activity. For example, the reaction catalyzed by the citrate-condensing enzyme (page 607) almost certainly proceeds with attack of the enol or carbanion of acetyl-SCoA on the carbonyl function of oxalacetate. Yet incubation of acetyl-SCoA alone with this enzyme does not result in the exchange of the methyl hydrogens of acetyl-SCoA with the hydrogens of solvent. Thus formation of the enol or carbanion of this species requires the presence of the second substrate, oxalacetate. Thus, although suggestive findings may be obtained in exchange experiments, they must be interpreted with caution.

An example of an isotope-exchange reaction that permits more definite conclusions to be drawn is provided by the aldolase reaction.

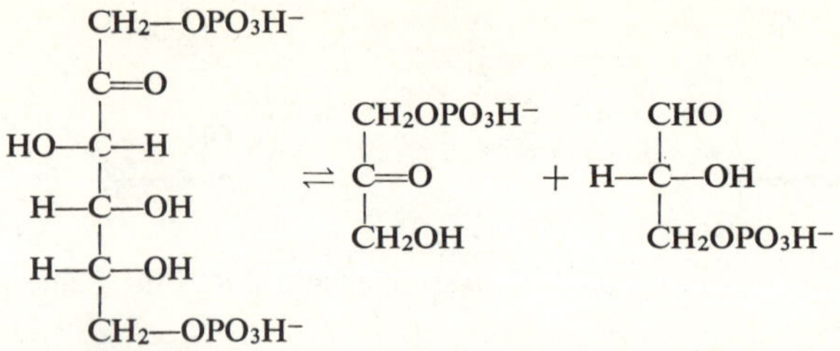

$$
\begin{array}{l}
CH_2\text{—}OPO_3H^- \\
\;| \\
C\text{=}O \\
\;| \\
HO\text{—}C\text{—}H \\
\;\;\;\;| \\
H\text{—}C\text{—}OH \\
\;\;\;\;| \\
H\text{—}C\text{—}OH \\
\;| \\
CH_2\text{—}OPO_3H^-
\end{array}
\rightleftharpoons
\begin{array}{l}
CH_2OPO_3H^- \\
| \\
C\text{=}O \\
| \\
CH_2OH
\end{array}
+
\begin{array}{l}
\quad\;\; CHO \\
\quad\;\;\; | \\
H\text{—}C\text{—}OH \\
\quad\;\;\; | \\
\quad\;\; CH_2OPO_3H^-
\end{array}
\qquad (7\text{-}12)
$$

Several workers have observed that the incubation of dihydroxyacetone phosphate with aldolase in either deuterium oxide or in water containing tritium oxide results in the incorporation of one atom of deuterium or tritium into this molecule. Similar incorporation is not observed if glyceraldehyde-3-phosphate is employed as substrate. These findings indicate that the dihydroxyacetone phosphate, and not the glyceraldehyde-3-phosphate, is activated by aldolase and strongly suggest that the carbanion, enol, or enolate of the former substrate is a (enzyme-bound) reaction intermediate. The pathway of the aldolase reaction may then be considered as the sum of two reactions (see page 507):

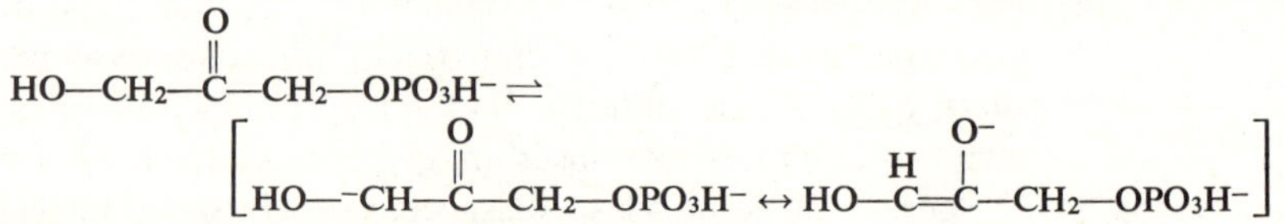

$$
HO\text{—}CH_2\text{—}\overset{\overset{\displaystyle O}{\|}}{C}\text{—}CH_2\text{—}OPO_3H^- \rightleftharpoons
\left[HO\text{—}{}^{-}CH\text{—}\overset{\overset{\displaystyle O}{\|}}{C}\text{—}CH_2\text{—}OPO_3H^- \leftrightarrow HO\text{—}\overset{H}{C}\text{=}\overset{\overset{\displaystyle O^-}{|}}{C}\text{—}CH_2\text{—}OPO_3H^- \right]
\qquad (7\text{-}13)
$$

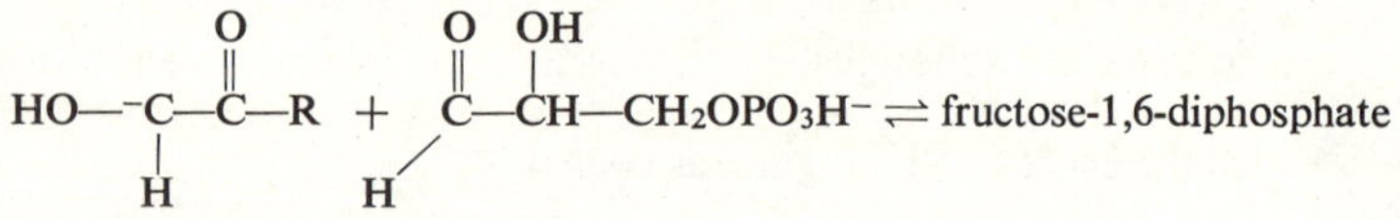

$$
HO\text{—}{}^{-}\underset{\underset{\displaystyle H}{|}}{C}\text{—}\overset{\overset{\displaystyle O}{\|}}{C}\text{—}R + \underset{H\,/}{\overset{\overset{\displaystyle O}{\|}}{C}}\text{—}\overset{\overset{\displaystyle OH}{|}}{CH}\text{—}CH_2OPO_3H^- \rightleftharpoons \text{fructose-1,6-diphosphate}
\qquad (7\text{-}14)
$$

The few examples cited above indicate the type of information concerning the pathway of enzyme-catalyzed reactions that may be obtained from studies employing isotopes. Additional examples are contained in the chapters of this volume dealing with intermediary metabolism.

MECHANISMS RELATED TO ENZYMATIC CATALYSIS

Experimental techniques for identifying at least some of the constituents at the active sites of enzymes as well as methods for the elucidation of the principal features of the pathway of enzyme-catalyzed reactions have been considered above. It is now necessary to organize this information into plausible reaction mechanisms; that is, the structure of the transition state must be approximated through assigning roles to the functional groups involved in the catalytic process. The catalytic mechanisms involved in enzymatic reactions are largely unknown. However, insight into processes involved in enzymatic catalysis may be obtained from studies concerning catalysis of model reactions. These studies fall into two categories. The first, more general, is concerned with an overall understanding of the susceptibility to and mechanisms of catalysis of reactions occurring in aqueous solution. The second category concentrates on obtaining detailed information about catalysis and mechanisms of reactions that are either identical to or closely related to actual enzymatic reactions. A survey of results obtained from studies of the former type and their application to enzyme-catalyzed reactions is presented below. Several studies of the latter type are discussed in Chapter 8.

Proximity and Orientation Effects

The initial steps in enzyme-catalyzed reactions involve the binding of one or more reactants to the enzyme surface. Such a binding process raises the possibility that one function of the enzyme may be to approximate and orient reactive sites in such a fashion as to permit the reaction to occur more readily. Such catalysis may involve *proximity effects,* that is, rate increases due solely to bringing the reactants more closely together, as well as *orientation effects,* which involve orienting groups into a configuration favoring reaction.

Are rate increases expected to result from proximity and orientation effects sufficient to account for the increased rates of enzymatic, as compared with the corresponding nonenzymatic, reactions? We can gain substantial insight into this question from quite basic considerations. Looking first at proximity effects, it is clear that the largest rate acceleration would arise from completely surrounding one reactant with another. The largest possible local concentration of one reactant in the presence of another in aqueous solution is 55 M, the concentration of water in water. We shall make use of this value below. Turning to orientation effects, considerable experience indicates that for ordinary bimolecular reactions about 1 collision out of 10 to 100 between molecules with energy sufficient for reaction actually leads to reaction. Thus, orientation requirements for reaction are ordinarily not too severe and limit the amount of rate acceleration that one can expect from them. In summary, maximal catalysis to be expected from proximity and orientation effects is perhaps 100 times greater than the effect of increasing reactant concentrations to 55 M.

Taking another approach to the same question, Koshland has calculated rate increases expected from proximity effects alone by making the following assumptions: (1) the rates of adsorption and desorption of substrates and products from the enzyme surface are infinite, (2) concentrations of reactants on the enzyme surface are the same as in hypothetical solids, and (3) the concentration of active sites is the same as the enzyme concentration (22). Results of these calculations, presented in Table 7-5, clearly indicate that rather small rate increases, at best, result from proximity effects. This result is principally the consequence of the very low concentration of active sites in dilute enzyme solutions. These results indicate decisively that proximity effects alone cannot account for the rate of enzyme-catalyzed reactions. Similar calculations performed after the introduction of reasonable orientation parameters (see above) do not alter this conclusion. It is true that, in the case of reactions involving a large number of reactants and catalysts (i.e., five), very large rate increases are predicted for a carefully oriented reaction. However, the rates of such reactions in solution are so slow that they have never been detected. In any event, the inclusion of several catalyst molecules emphasizes the role of catalytic processes other than proximity and orientation effects.

Although we must invoke additional factors, it does appear reasonable that proximity and orientation effects do contribute something to the rates for enzyme-catalyzed reactions. This point is emphasized by a comparison of rates for reactions occurring via intra- or intermolecular reaction paths. For the sake of quantitative comparisons, pseudo-first-order rate constants for intermolecular reactions calculated by assuming a concentration of 1 M for the attacking reagent may be compared with the first-order rate constants for the intramolecular reaction. For example, the first-order rate constant for loss of phenolate from phenyl-γ-(4-imidazolyl)-butyrate is 20-fold larger than the pseudo-first-order rate constant for loss of phenolate from *p*-nitrophenyl acetate in the presence of 1 M imidazole:

$k = 0.043\ \text{sec}^{-1}$ (7-15)

$k = 0.0018\ \text{sec}^{-1}$ (1M) (7-16)

That is, the intramolecular reaction behaves as if the imidazole function were present at an effective concentration of 20 M. This value seems reasonable in terms of our limitations discussed above. More striking examples include the relative rates of intermolecular and intramolecular attack of the amide function on esters:

$k_2 = 1.2 \times 10^6\ \text{M}^{-1}\cdot\text{min}^{-1}$, OH^- (7-17)

$k_2 < 20\ \text{M}^{-1}\cdot\text{min}^{-1}$, OH (1M) (7-18)

and the relative rates of anhydride formation from the mono-*p*-methylphenyl esters of glutarate and *exo*-3,6-*endoxo*-Δ^4-tetrahydrophthalate:

k_{19} (7-19)

$$\frac{k_{19}}{k_{20}} = 53{,}000$$

k_{20} (7-20)

In these cases, the rates of the intramolecular reactions would be obtained in the intermolecular (or improperly oriented intramolecular) cases only in the presence of impossibly high concentrations of the attacking reagent. Since these values fall outside the limitations that appear to be pertinent for reactions in aqueous solution, factors other than proximity and orientation effects must be important in determining the rates of these intramolecular reactions. Thus, rates of intramolecular reactions must be employed with caution as

TABLE 7-5
Comparison of observed enzymic rates with those calculated by collision frequency theory[a]

Enzyme	*Substrate and concentration*	$[E_T]$	*Observed* V_0	*Observed* V_E	V_E/V_0 *calculated from theory*	V_E/V_0 *observed* / V_E/V_0 *theory*
		equiv/l	*(moles/l)/min*			
Hexokinase	0.0003 M Glucose; 0.002 M ATP	10^{-7}	$<1 \times 10^{-13}$	1.3×10^{-3}	1.8	$>7 \times 10^{9}$
Phosphorylase	0.016 M Glucose-1-phosphate; 10^{-5} M Glycogen	6×10^{-8}	$<5 \times 10^{-15}$	1.6×10^{-3}	2.5	$>1.4 \times 10^{11}$
Alcohol dehydrogenase	0.0004 M NAD; 0.04 M Ethanol	4×10^{-7}	$<6 \times 10^{-12}$	2.7×10^{-3}	0.5	$>9 \times 10^{8}$
Creatine kinase	0.024 M Creatine; 0.004 M ATP	3×10^{-9}	$<3 \times 10^{-9}$	4×10^{-5}	0.5×10^{-3}	$>2.7 \times 10^{7}$

[a] $[E_T]$, total concentration of active sites; V_0, nonenzymatic velocity; V_E, enzymatic velocity.
SOURCE: D. E. Koshland, Jr., *J. Cellular Comp. Physiol.*, **47:** Suppl. 1, 217 (1956).

indicators of rate increases available from proximity and orientation effects. The steric factors which are built into the intramolecular models may decrease solvation, overcome van der Waals' repulsions, and change electronic structures relative to the same reactive groups existing independently in solution. All these factors may contribute substantially to the rate differences observed.

Although intramolecular reactions may reasonably be cited, within limitations, as models for enzyme-catalyzed reactions, it must be remembered that the approximation and orientation of the reactants have been built into the reaction by chemical synthesis. The enzyme, on the other hand, must perform this function itself through the use of the available binding forces between enzyme and substrate. Thus, a more satisfactory model for enzyme-catalyzed reactions would be a reaction in which weak forces served to bind substrate to catalyst or other reactant prior to bond-changing reactions. Such pre-equilibrium binding should facilitate bond-changing reactions and, hopefully, confer some specificity on the reaction. Rather few reactions in aqueous solution are known that exhibit these characteristics.

Let us consider three examples: catalysis of hydrolysis of leucine esters by benzaldehyde (23), catalysis of ester aminolysis by surfactants (24), and catalysis of ester hydrolysis by cyclodextrins (25).

Catalysis of hydrolysis of the *p*-nitrophenyl ester of leucine by benzaldehyde is striking. The catalysis is observable in solutions as dilute as 10^{-4} M benzaldehyde; moreover, benzaldehyde is a very poor nucleophilic reagent. It seems likely that the reaction pathway involves formation of an amino alcohol through addition of the amino group of leucine to the carbonyl group of benzaldehyde.

$$C_6H_5\text{-}CHO + NH_2\text{-}CHR\text{-}\overset{O}{\overset{\|}{C}}\text{-}OR' \rightleftharpoons \text{[cyclic adduct: HN–CHR–C(=O)(OR')–O}^-\text{–CH(}C_6H_5\text{)]} \xrightarrow{-OR'} \text{[oxazolidinone: HN–CHR–C(=O)–O–CH(}C_6H_5\text{)]}$$

$$\longrightarrow H^+N(=CH\text{-}C_6H_5)\text{-}CHR\text{-}CO_2^- \xrightarrow{H_2O} C_6H_5\text{-}CHO + NH_2\text{-}CHR\text{-}\overset{O}{\overset{\|}{C}}OH \qquad (7\text{-}21)$$

Formation of the amino alcohol makes possible a rapid intramolecular reaction involving the former carbonyl oxygen of benzaldehyde as nucleophile [Eq. (7-21)]. Subsequent reactions regenerate the catalyst and yield leucine and alcohol as products. This reaction is particularly noteworthy in that association of catalyst with substrate not only converts an intermolecular reaction into an intramolecular one but also generates the active form of the catalyst at the same time.

Distinct but equally interesting is the reaction of myristoyl histidine with *p*-nitrophenyl esters in the presence of cationic surfactants (24):

$$CH_3(CH_2)_{12}\text{-}\overset{O}{\overset{\|}{C}}\text{-}\overset{H}{N}\text{-}CH(CH_2\text{-imidazolyl})\text{-}CO_2^- + R\text{-}\overset{O}{\overset{\|}{C}}\text{-}O\text{-}C_6H_4\text{-}NO_2 \xrightarrow[-\,^-O\text{-}C_6H_4\text{-}NO_2]{CH_3(CH_2)_{12}\text{-}\overset{+}{N}(CH_3)_3}$$

$$R'\text{-}\overset{O}{\overset{\|}{C}}\text{-}\overset{H}{N}\text{-}CH(CH_2\text{-imidazolyl-}N\text{-}C(=O)R)\text{-}CO_2^- \xrightarrow{H_2O} R\text{-}CO_2^- + R'\text{-}\overset{O}{\overset{\|}{C}}\text{-}\overset{H}{N}\text{-}CH(CH_2\text{-imidazolyl})\text{-}CO_2^- \qquad (7\text{-}22)$$

The reaction proceeds through the attack of the imidazole moiety of the histidine on the ester with intermediate formation of an acylimidazole, which then undergoes hydrolysis to complete the reaction [Eq. (7-22)]. One function of the surfactant is to form *micelles,* which are the site of reaction. Ability to form micelles is a general property of molecules having both hydrophilic and hydrophobic properties (page 713). The structure of a simple micelle is shown diagrammatically in Fig. 7-5. Like proteins, the surfactants adopt a structure in which the hydrophobic portions occupy the surface. These micelles can interact with both the histidine and the ester, by virtue of the hydrophobic properties of these species, and in so doing bring

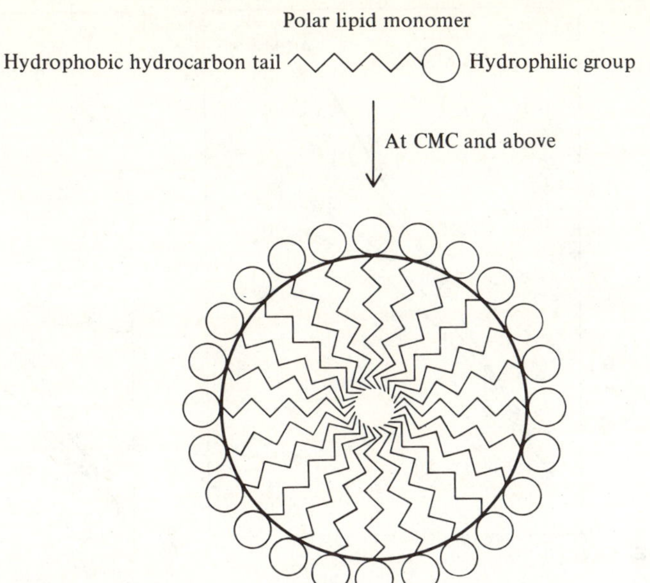

FIGURE 7-5
Schematic view of micelle formation.

them together, increasing their rate of reaction. Typical results are shown in Fig. 7-6. In addition to the large rate increases observed, the system is of interest in two respects. First, note the specificity of the reaction: the more hydrophobic the ester, the greater the rate enhancement. Second, the rate is linear in histidine concentration only when the ratio of histidine to cationic surfactant is relatively high. When too much surfactant is present, there is the possibility of formation of nonproductive complexes, that is, association of the ester with the micelle in a region where no myristoyl histidine is located.

Cyclodextrins are chains of sugar molecules which are closed to form roughly circular structures with a cavity in the center. An example is shown in Fig. 7-7. These molecules have the capacity to include properly sized organic molecules in the cavity while present in aqueous solution. As a result of such inclusion, the reactivity of the guest molecule is generally altered, frequently by reaction with the cyclodextrin itself. These reactions may show marked substrate specificity. For example, the acylation of cyclohexaamylose by *m-t*-butylphenyl acetate is 180 times more rapid than the same reaction with the *p-t*-butylphenyl ester (25). Note that the reaction is one of acylation, not overall hydrolysis; reactions in cyclodextrins are more notable for their specificity than for overall rate enhancement. The association constants for inclusion of the two esters in the cavity are similar; the *meta* derivative must form a complex in which the hydroxyl groups of the cyclodextrin are close to the acyl function of the ester while the *para* derivative must not. As in the case of enzymatic reactions, saturation kinetics with respect to substrate concentration are observed. Reactions of this sort may serve as models not only for enzymatic reactions but also for transport processes across membranes.

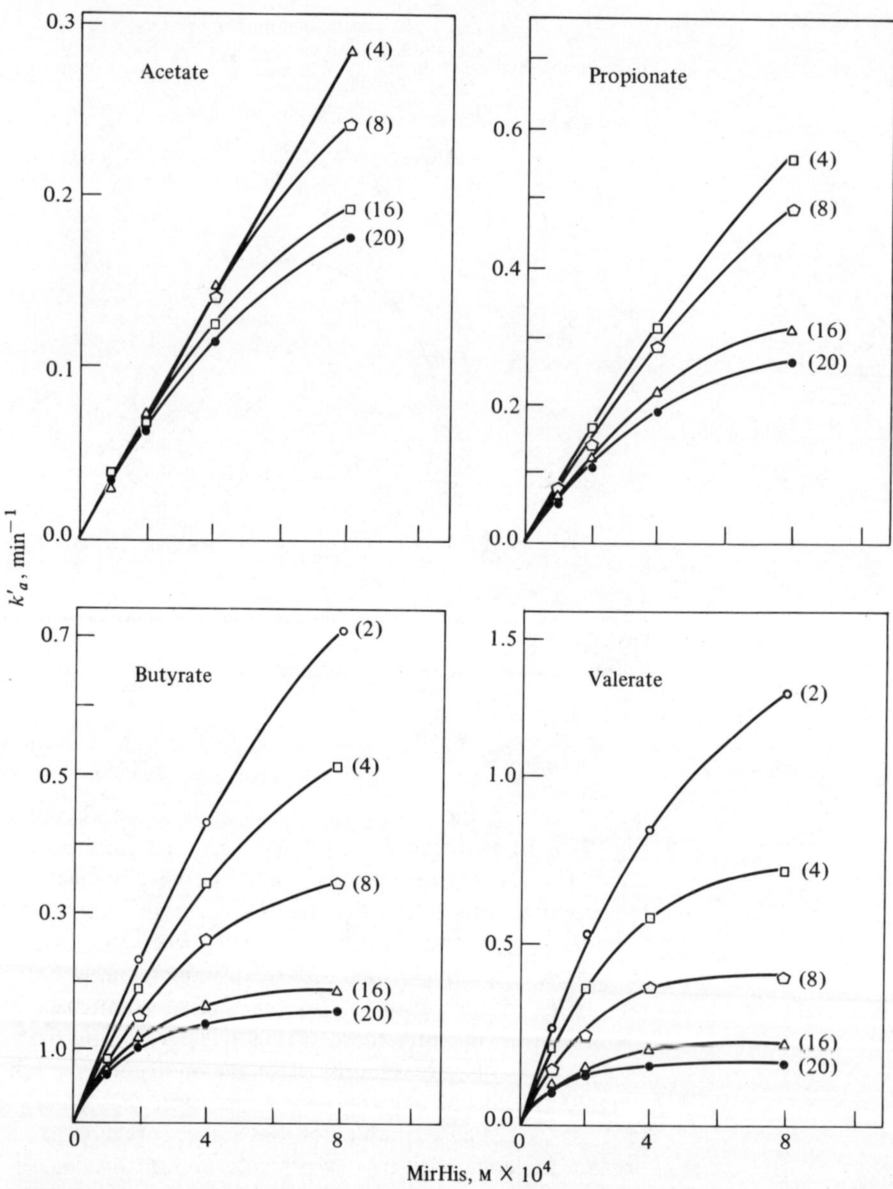

FIGURE 7-6

Pseudo-first-order rate constants for the liberation of *p*-nitrophenol in the reaction of mixed micelles of *N*-myristoylhistidine and cetyltrimethylammonium bromide (at the ratio indicated in parentheses) with *p*-nitrophenyl acetate, propionate, butyrate, and valerate. C. Gitler and A. Ochoa-Solano, *J. Am. Chem. Soc.,* **90:** 5004 (1968); by permission of the copyright owners, the American Chemical Society.

General Acid–Base Catalysis

Catalysis of reactions by acids and bases is a general phenomenon in organic chemistry. Such catalysis may be classified into two categories based on empirical observation rather than on necessary mechanistic distinction. Reactions whose rates vary in a linear fashion with hydronium or hydroxyl ion concentration, but which do not vary with the concentration of other acidic

FIGURE 7-7
Molecular model of α-cyclodextrin (cyclohexaamylose). Courtesy of F. Cramer.

or basic species in solution, are subject to *specific acid* or *specific base catalysis*. Reactions whose rates are a function of the concentration of all acids or bases in solution, rather than just that of hydronium ion or hydroxyl ion, are subject to *general acid* or *general base catalysis*. The concept of general acid–base catalysis was introduced into classical kinetic theory by J. N. Brønsted in the 1920s.

The mutarotation of glucose provides an excellent example of a reaction subject to general acid–base catalysis:

CH_2OH, O, OH, HO, OH, OH $\rightleftharpoons$ CH_2OH, O, OH, OH, HO, OH (7-23)

In Table 7-6, first-order rate constants for the conversion of α-D-glucose to β-D-glucose are presented as a function of the concentration of a 1 : 1 acetic acid–sodium acetate buffer. The rate constants for this reaction clearly increase in a linear fashion with increasing buffer concentration. Since the *p*H of the solution is fixed by the buffer ratio, the observed catalysis cannot be ascribed to hydronium ions or hydroxyl ions and must, consequently, be due to acetic acid, acetate ion, or both. Further studies of the buffer catalysis at a variety of buffer ratios have revealed that the observed catalysis is due to both species. Thus the rate law for glucose mutarotation in acetic acid–acetate buffers has the form

$$k_{obs} = k_0 + k_H(H^+) + k_{OH}(OH^-) + k_{HAc}(HAc) + k_{Ac}(Ac^-) \quad (7\text{-}24)$$

in which the "uncatalyzed" term k_0 is probably the consequence of catalysis by water molecules. In general, the rate law for any reaction subject to both general acid and general base catalysis may be expressed in the form

$$k_{obs} = \sum_i k_i^A(A_i) + \sum_j k_j^B(B_j) \tag{7-25}$$

in which the summations are taken over all the basic and acidic species present in solution.

General acid–base catalysis appears to perform an important function in enzymatic catalysis. The side chains of several of the amino acids, including particularly glutamic acid, aspartic acid, histidine, tyrosine, and cysteine, have appreciable acidic or basic character and might function as general acid–base catalysts. It is certainly not unreasonable to envision the preferential employment of these amino acid side chains as catalysts rather than hydronium or hydroxyl ions, which are present in vanishingly small concentrations under conditions in which most enzymes are maximally active. Furthermore, the *p*H-rate profiles exhibited by many enzymatic reactions indicate dependence of the catalytic process on the ionization state of one or more acidic or basic groups. Finally, solvent deuterium isotope effects on enzymatic reactions are, in several cases, consistent with proton transfer in the rate-determining step, a feature of general acid–base catalysis (see below).

Four questions concerning general acid–base catalysis are pertinent to an assessment of the role of such catalysis in enzymatic processes. First, what reaction types are susceptible to such catalysis? Second, what factors determine the efficiency of a particular molecule as a general acid or base catalyst? Third, what is the mechanism of general acid–base catalysis? And, finally, how does the susceptibility of a reaction to general acid–base catalysis vary as the reactivity of the reactants is varied? These questions are discussed individually and sequentially below.

To begin, many types of reactions of considerable biochemical interest are susceptible to general acid–base catalysis. These include many carbonyl addition reactions, the hydrolysis of some carboxylic and phosphoric esters, the aminolysis of esters, and tautomerizations. Several such reactions are collected in Table 7-7. In addition, just as the concept of acids may be extended to include nonprotonic compounds capable of coordinating with an unshared

TABLE 7-6
Effect of buffer concentration on first-order rate constants for mutarotation of glucose

Acetic acid (M)	*Acetate* (M)	k_{obs} (min^{-1})
0.00	0.00	8.8×10^{-5}
0.01	0.01	9.3×10^{-5}
0.02	0.02	9.8×10^{-5}
0.05	0.05	11.2×10^{-5}
0.10	0.10	13.6×10^{-5}
0.20	0.20	18.4×10^{-5}
0.50	0.50	32.8×10^{-5}
1.0	1.0	56.8×10^{-5}

electron pair (*Lewis acids*), so may the concept of general acid catalysis be extended to include catalysis by such acids, particularly metal ions. Metal-ion catalysis (*superacid catalysis*) extends the list of reactions susceptible to general acid–base catalysis to many additional cases, including decarboxylations and many reactions of phosphates and polyphosphates.

The efficiency of a particular species as a general acid or general base catalyst is primarily a function of its acid or base strength. This statement is expressed quantitatively in the *Brønsted catalysis laws:*

$$k_A = G_A(K)^\alpha \tag{7-26}$$

$$k_B = G_B\left(\frac{1}{K}\right)^\beta \tag{7-27}$$

in which k_A and k_B are catalytic constants for acid and base catalysis, K is the dissociation constant of the acid catalyst or of the conjugate acid of the base catalyst, and G_A, G_B, α, and β are constants that depend on the nature of the reaction, the solvent, and the temperature. The Brønsted catalysis laws are the earliest examples of linear free energy relationships. Particular interest is focused on the magnitude of the exponents, α and β, in the Brønsted catalysis laws, since, as indicated below, these parameters measure, indirectly, the susceptibility of the reaction in question to general acid–base catalysis. Rewriting Eqs. (7-26) and (7-27), we obtain

$$\log k_A = \log G_A - \alpha pK_a \tag{7-28}$$

$$\log k_B = \log G_B + \beta pK_a \tag{7-29}$$

from which it is clear that plots of catalytic constants against the corresponding values of pK_a yield straight lines whose slopes define the magnitude of α or β. These exponents are, by definition, positive and take on values between zero and unity (in unusual cases, values greater than unity have been observed). The Brønsted exponents for several general-acid- and general-base-catalyzed reactions are included in Table 7-7.

The magnitude of the Brønsted exponent measures the sensitivity of a reaction rate to changes in the strength of the catalyzing acid or base, large values indicating a high sensitivity and small values indicating a relative insensitivity. For example, if an acid-catalyzed reaction is characterized by an α value near unity, the reaction is highly sensitive to the strength of the acid catalyst, and hydronium ion is the only effective catalyst detectable. Therefore, only a specific-acid-catalyzed reaction is observed experimentally. In contrast, an α value near zero indicates that all acids are of approximately equal efficiency as catalysts. Since water is present in very high concentrations in aqueous solution, it serves as the only effective catalyst, and the reaction appears to be uncatalyzed. The relative contributions to the observed rate from catalysis by hydronium ion, acetic acid, and water are compared for a variety of values of α in Table 7-8. From these data and the discussion above, it is clear that general acid catalysis is detected only for reactions characterized by intermediate values of α. These comments apply, of course, with equal validity to general-base-catalyzed reactions. There is, in general, no simple kinetic method for distinguishing between preequilibrium transfer of a proton between catalyst and substrate and general acid–base catalysis by hydronium ion or

TABLE 7-7
Brief compilation of reactions exhibiting general acid–base catalysis

Reaction	*Catalysis*	α	β
$CH_3CHO + H_2O$	General acid, general base	0.54	
$\phi\text{—}CHO + NH_2NHCONH_2$	General acid	0.25	
$Cl\text{—}CH_2\text{—}C(=O)\text{—}CH_2Cl + H_2O$	General acid, general base	0.27 . . .	 0.50
Glucose mutarotation	General acid, general base	0.27 . . .	 0.34
$R_2\overset{+}{N}{=}C\phi_2 + H_2O$	General base	. . .	0.27
Methyl orthobenzoate hydrolysis	General acid	0.74	
$CH_3\text{—}C(=O)\text{—}O\text{—}C(=O)\text{—}CH_3 + H_2O$	General base		
$Cl_2HC\text{—}CO_2Et + H_2O$	General base	. . .	0.47
$CH_3\text{—}C(=O)\text{—}O\text{—}\phi + NH_3$	General base		
$CH_3\text{—}C(=O)\text{—}O\text{—}\phi + NH_2OCH_3$	General acid, general base	~0 . . .	 0.4
$CH_3\text{—}C(=O)\text{—}CH_3 \rightarrow CH_3\text{—}C(OH){=}CH_2$	General acid, general base		
cyclic $\text{—}O\text{—}P(=O)(OCH_3)\text{—}O\text{—}$ (ring) $+ H_2O$	General base		
$^{-}O_2C\text{—}C(=O)\text{—}CH_2\text{—}CO_2^{-} \rightarrow {}^{-}O_2C\text{—}C(=O)\text{—}CH_3 + CO_2$	Metal ion		
$(RO)_2P(=O)\text{—}O\text{—}P(=O)(OR)_2$ solvolysis	Metal ion, general base		

TABLE 7-8
Relative magnitude of catalysis due to various species in 0.1 N acetic acid–0.1 N sodium acetate buffer as function of α

	Proportion of catalysis due to		
	H_3O^+	H_2O	CH_3COOH
0.1	0.002%	98%	2%
0.5	3.6%	0.01%	96.4%
1.0	99.8%	5×10^{-12}%	0.2%

SOURCE: R. P. Bell, *Acid–Base Catalysis,* Oxford University Press, New York, 1941, p. 94.

hydroxyl ion for reactions with large α or β values. Thus, reactions for which no general acid–base catalysis may be experimentally demonstrated may, nevertheless, occur with such catalysis.

The Brønsted catalysis laws are generally obeyed accurately provided that a series of structurally similar catalysts is employed. Variation in the charge type of the catalyst or in the nature of the atom involved in proton transfer may cause a significant deviation from the Brønsted relation. Such deviations are, however, usually smaller than those found for the attack of nucleophilic reagents (see below). Bell and Higginson have conducted a rather extensive study of general acid catalysis for the hydration of acetaldehyde (26). The catalytic constants obtained in this study for 63 acids fall reasonably close to the line drawn through the points for carboxylic acids, but several significant deviations do occur. The magnitude and even the direction of the deviation for a particular catalyst may change from reaction to reaction. Consequently, the factors that determine catalytic efficiency are, in part, a function of the reaction type and cannot be wholly contained in the structures of the catalysts themselves. Factors that do appear to be of importance in determining catalytic efficiency include (1) the equilibrium constant for the formation of the hydrogen-bonded complex between catalyst and substrate, (2) the degree of bond polarization induced in the substrate as a result of such complex formation, and possibly (3) the rate of proton transfer between catalyst and substrate (27).

As suggested above, the distinguishing feature of general acid–base catalysis is the involvement of proton transfer in the transition state. Rates of reactions occurring with preequilibrium proton transfer depend only on the pH of the reaction medium and, consequently, exhibit only specific acid–base catalysis. The understanding of the mechanism of general acid–base catalysis is complicated by the fact that it is usually possible to write two or more kinetically indistinguishable mechanisms for this catalysis. Consider, for example, the general acid catalysis of the attack of an amine on a carbonyl compound. Two kinetically indistinguishable mechanisms for such a reaction are indicated in I and II:

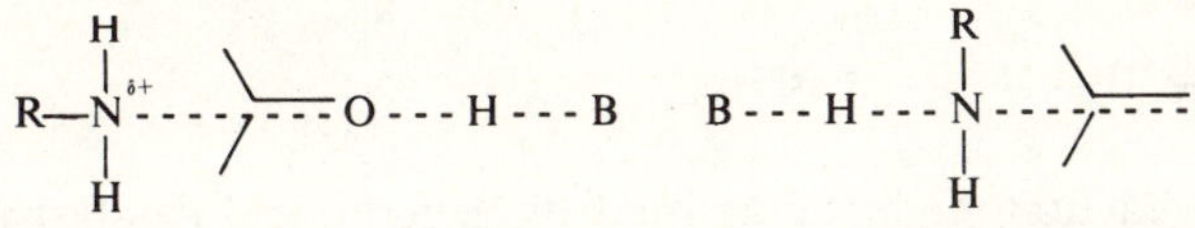

```
    H                                        R
    |δ+                                      |
R—N- - - - C- - - O - - - H - - - B   B - - - H - - - N - - - C - - - - OHδ+
    |                                        |
    H                                        H

          I                                       II
```

Transition state I exhibits true general acid catalysis involving a partial donation of a proton to the carbonyl oxygen atom concerted with attack of the nucleophilic reagent. Transition state II exhibits specific acid–general base catalysis involving preequilibrium protonation of the carbonyl oxygen atom and partial proton removal from the nucleophilic reagent as it attacks the carbonyl carbon. The rate law for mechanism I is given by $k_{obs} = k'(RNH_2)(HB)$, and that for mechanism II by $k_{obs} = k''(RNH_2)(H^+)(B^-)$. Since $(HB) = K(H^+)(B^-)$, these rate laws are kinetically identical. In a few cases, distinction between these mechanistic alternatives has been possible, but the detailed results do not concern us here. In any event, whatever the mechanism is in one direction, the principle of microscopic reversibility requires that it is the opposite in the reverse direction. Thus, if a reaction proceeds via true general acid catalysis in one direction, it must proceed via a specific acid–general base pathway in the other.

Although the quantitative relationships between susceptibility to general acid–base catalysis and substrate reactivity are not completely clear, there does exist a fair amount of evidence that suggests that such catalysis is more important for difficult reactions than for facile ones. For example, the attack of water on carbonyl functions is almost completely dependent on general acid catalysis, but the attack of the stronger nucleophilic reagent semicarbazide exhibits not only a marked acid-catalyzed reaction, but also an uncatalyzed reaction. The attack of the still better nucleophilic reagent hydroxylamine is only slightly subject to acid catalysis, and that of strongly basic amines exhibits no acid catalysis at all (28). The suggestion that general acid–base catalysis is important for difficult reactions has clear implications for enzyme-catalyzed reactions, since most biological substrates are nonactivated.

Just as the partial donation of a proton to a carbonyl oxygen atom increases the electrophilicity of the carbonyl carbon atom, the coordination of a metal ion with an unshared electron pair accomplishes the same purpose. Thus, no clear distinction exists between metal-ion catalysis and classical general acid catalysis. Metal-ion-catalyzed reactions, of course, show a high degree of specificity in terms of both the nature of the metal ion and the structural features of the substrate.

Concerted Acid–Base Catalysis

The ambiguity in mechanism for most general acid–base-catalyzed reactions discussed above emphasizes that, for most reactions, a mechanistic role may be assigned to both acidic and basic catalysts. The concept of *concerted* general acid–general base catalysis thus arises in a natural fashion. The simultaneous action of several catalytic activities in a single reaction is appealing from the standpoint of enzyme-catalyzed processes, since it appears unlikely that any single catalytic activity is sufficient to account for the high efficiency of such reactions.

As an example of a reaction which might profit from concerted acid–base catalysis, let us consider the enolization of a ketone:

$$-\overset{\overset{\displaystyle O}{\|}}{C}-CH_2-H \rightleftharpoons -\overset{\overset{\displaystyle OH}{|}}{C}=CH_2 \qquad (7\text{-}30)$$

This reaction might be assisted by general acid catalysis through donation of a proton to the carbonyl oxygen and by general base catalysis through removal of a proton from carbon. If both types of catalysis occur at the same time, we can visualize the transition state as shown:

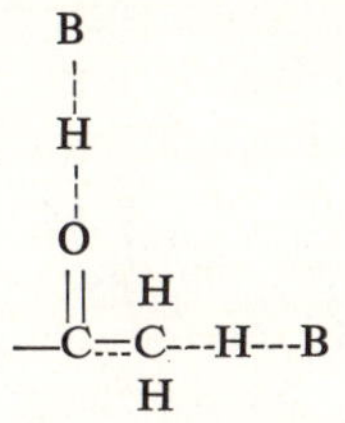

```
 B
 ¦
 H
 ¦
 O
 ||  H
—C==C---H---B
     H
```

Such catalysis would be experimentally manifested by the appearance of the terms $(HB)(B)$ in the rate law for the reaction. Such terms for reactions in aqueous solution have been identified quite rarely and those that have been, including enolizations (29), generally make only a small contribution to the total rate. There is, then, rather little direct evidence establishing the importance of concerted general acid–base catalysis for reactions in water.

The most striking examples of the consequences of bifunctional acid–base catalysis come not from effects upon rates but upon courses of reactions. Schmir and his associates have provided several examples of pronounced effects of buffers having both acidic and basic properties, such as the phosphate dianion and the bicarbonate ion, on the mode of partitioning of the tetrahedral intermediates that are formed in the course of imidate and thioimidate hydrolysis (30). Since formation of this intermediate is the rate-determining step, the buffer influence appears only in the product distribution. The buffers are considered to act in the following manner:

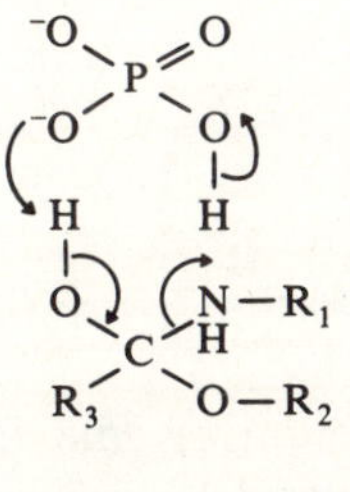

```
 ⁻O     O
   \P//
 ⁻O/  \O
  H     H
  O    N—R1
   \C/ H
 R3/  \O—R2
```

Thus, the buffer functions by facilitating the departure of the amine from the intermediate with formation of the ester product. The other alternative, expulsion of alcohol with formation of amide as product, is not subject to such facilitation.

It should be clearly recognized that the failure to observe terms in the rate law for a given reaction that are proportional to the product of acid and base concentrations does not suffice to indicate that the reaction does not proceed via a concerted pathway. One or more molecules of water may serve the function of either general acid or general base catalyst and would, of course, not appear in the rate law. However, the importance of concerted catalysis for reactions in aqueous solution is yet to be established. The role of concerted catalysis in enzymatic reactions, therefore, remains a speculation, although an appealing one.

Covalent Catalysis

In contrast to general acid–base catalysis, covalent catalysis, as the name implies, is characterized by the existence of a covalently bound substrate–catalyst intermediate. Since, as noted above, an increasingly large number of enzyme-catalyzed reactions are known to involve covalent enzyme–substrate intermediates and hence occur with covalent catalysis, considerable attention has been directed to the understanding of catalysis of this type.

The demonstration of covalent catalysis for a reaction requires the fulfillment of two criteria: (1) the demonstration of the existence of a covalent substrate–catalyst intermediate and (2) proof that the rate of formation and decomposition of this intermediate accounts for the overall rate of the catalyzed reaction. Reactions exhibiting these characteristics fall into two categories: those occurring with nucleophilic and those occurring with electrophilic attack of catalyst on substrate. Since most of those enzymatic reactions which involve covalent catalysis occur via a nucleophilic route, attention is directed to this subclass of covalent catalyses. Two questions are of particular interest: what types of reaction are subject to nucleophilic catalysis, and what factors determine the efficiency of a particular molecule as a nucleophilic catalyst?

Several examples of nucleophilic catalysis involving substrates at the aldehyde level of oxidation are known. For example, semicarbazone formation from *p*-chlorobenzaldehyde is markedly catalyzed by aniline. This reaction occurs with the initial formation of a Schiff base between the aldehyde and aniline followed by a rapid attack of semicarbazide on the Schiff base. The Schiff base is particularly reactive toward nucleophilic attack, since it is easily protonated, yielding a highly reactive cationic species:

$$\text{Cl}-C_6H_4-\overset{O}{\overset{\|}{C}}-H + NH_2-C_6H_5 \overset{H^+}{\rightleftharpoons} \text{Cl}-C_6H_4-\overset{H}{C}=\overset{H}{\overset{+}{N}}-C_6H_5$$

$$\text{Cl}-C_6H_4-\overset{H}{C}=\overset{H}{\overset{+}{N}}-C_6H_5 + NH_2NH\overset{O}{\overset{\|}{C}}NH_2 \rightarrow \text{Cl}-C_6H_4-\overset{H}{C}=N\overset{H}{N}\overset{O}{\overset{\|}{C}}NH_2 + \overset{+}{N}H_3-C_6H_5 \qquad (7\text{-}31)$$

More directly relevant to enzymatic reactions is the finding that semicarbazone formation from pyridoxal and pyridoxal phosphate is subject to nucleophilic catalysis by both primary and secondary amines (31). Pyridoxal phosphate–dependent enzymes contain this coenzyme bound to an ε-amino group of a lysine residue in the form of a Schiff base (Chapter 8). Since pyridoxal phosphate reactions involve the formation of a Schiff base with amino acids, the finding of nucleophilic catalysis for such a reaction clearly indicates that it is advantageous to the enzyme to have this coenzyme bound in a Schiff base form. Reactions of this type are discussed in more detail in Chapter 8 (page 393).

In addition to nucleophilic catalysis aiding carbonyl addition reactions, such catalysis is also important for reactions requiring activation of the carbon atom α to the carbonyl group. Two examples recognized for many years include the decarboxylation of acetoacetic acid and the dealdolization of diacetone alcohol.

$$CH_3-\overset{O}{\overset{\|}{C}}-CH_2-CO_2H + RNH_2 \rightleftharpoons CH_3-\overset{H\overset{+}{N}R}{\overset{\|}{C}}-CH_2-CO_2^-$$

$$CH_3-\overset{H\overset{+}{N}R}{\overset{\|}{C}}-CH_2-CO_2^- \longrightarrow CH_3-\overset{HNR}{\overset{|}{C}}=CH_2 + CO_2 \qquad (7\text{-}32)$$

$$CH_3-\overset{HNR}{\overset{|}{C}}=CH_2 \rightleftharpoons CH_3-\overset{NR}{\overset{\|}{C}}-CH_3 \xrightarrow{H_2O} CH_3-\overset{O}{\overset{\|}{C}}-CH_3 + RNH_2$$

$$(CH_3)_2-\overset{OH}{\overset{|}{C}}-CH_2-\overset{O}{\overset{\|}{C}}-CH_3 + RNH_2 \rightleftharpoons (CH_3)_2-\overset{O^-}{\overset{|}{C}}-CH_2-\overset{H\overset{+}{N}R}{\overset{\|}{C}}-CH_3$$

$$(CH_3)_2-\overset{O^-}{\overset{|}{C}}-CH_2-\overset{H\overset{+}{N}R}{\overset{\|}{C}}-CH_3 \longrightarrow (CH_3)_2-C=O + CH_2=\overset{HNR}{\overset{|}{C}}-CH_3 \qquad (7\text{-}33)$$

$$CH_2=\overset{HNR}{\overset{|}{C}}-CH_3 \rightleftharpoons CH_3-\overset{NR}{\overset{\|}{C}}-CH_3 \xrightarrow{H_2O} CH_3-\overset{O}{\overset{\|}{C}}-CH_3 + RNH_2$$

In both cases, the observed catalysis results from the action of the protonated Schiff base serving as an electron sink. This type of catalysis for these reactions appears to be directly related to the corresponding enzymatic catalysis. As noted earlier in this chapter, acetoacetic decarboxylase and some enzymes catalyzing aldol condensations (aldolases) involve the formation of Schiff bases between enzyme and substrate. In addition, such catalysis also serves as a model for pyridoxal phosphate-dependent and thiamine pyrophosphate-dependent enzymatic reactions (see Chapter 8).

A very large number of reactions involving substrates at the acyl level of oxidation are subject to nucleophilic catalysis. Examples are provided by the imidazole-catalyzed acyl transfer and hydrolysis reactions. Such reactions occur via the intermediate formation of acyl imidazoles:

$$R-\overset{O}{\overset{\|}{C}}-X + N\underset{\smile}{\frown}NH \rightleftharpoons R-\overset{O}{\overset{\|}{C}}-\overset{+}{N}\underset{\smile}{\frown}NH + X^- \qquad (7\text{-}34)$$

$$R-\overset{O}{\overset{\|}{C}}-\overset{+}{N}\underset{\smile}{\frown}NH + Y^- \rightleftharpoons R-\overset{O}{\overset{\|}{C}}-Y + N\underset{\smile}{\frown}NH \qquad (7\text{-}35)$$

Catalysis of reactions of this type is interesting from a biochemical standpoint, since the number of related enzymatic reactions is large and several enzymes catalyzing such reactions involve the intermediate formation of enzyme–substrate compounds.

Since nucleophilic catalysis requires that the catalyst have a higher reactivity toward the substrate than the final acceptor and, in addition, yield an intermediate that is more reactive than the substrate toward the final acceptor, nucleophilic catalysts must have unusual properties. In general, factors that increase kinetic instability of a reactant tend to increase the kinetic stability of the product. For example, amide ion, a very powerful nucleophilic reagent toward esters yields a very stable product, and chloride ion, which would yield a very reactive intermediate, is an exceptionally poor nucleophilic reagent toward esters. Thus, neither of these species would function as a nucleophilic catalyst for acyl transfer reactions. In contrast, imidazole is very much more nucleophilic toward esters than predicted on the basis of basicity alone (see below) and yields products, acyl imidazoles, that are highly susceptible to nucleophilic attack.

Several detailed studies of the factors that influence nucleophilicity have been carried out, particularly with substrates at the alkyl or acyl level of oxidation. Since substrates of the latter type are of particular biochemical interest, attention is directed to these results. Jencks and Carriuolo have studied the reactivities of a large number of nucleophilic reagents toward *p*-nitrophenyl acetate (32). These results are presented in the form of a Brønsted plot in Fig. 7-8. As expected, the nucleophilic reactivities are roughly in accord with the basicities of the attacking reagent, but many large deviations are observed. Hence, factors other than basicity are clearly of importance in determining nucleophilicity. Compounds with abnormally high polarizabilities, such as the peroxide anion and the sulfhydryl anion, exhibit abnormally high nucleophilicities. Other factors that influence nucleophilicity include steric effects, solvation effects, internal general acid catalysis, and unusually strong bonding to carbon relative to hydrogen. Many compounds with an unshared electron pair adjacent to the attacking atom exhibit unusually high nucleophilicities (*α effect*). This has been attributed to the donation of electrons to the attacking atom in the transition state, although other explanations are also possible.

From data of the type exhibited in Fig. 7-8 and from studies on model systems, the sulfhydryl group of cysteine and the imidazole group of histidine appear to be the prime candidates as nucleophilic reagents in enzymatic covalent catalysis. Cysteine has been implicated as the site of transient acylation in reactions catalyzed by papain, triose phosphate dehydrogenase (this enzyme functions as an esterase as well as a dehydrogenase), and, probably, ficin. Intermediate formation of phosphohistidine enzymes has been implicated in the reaction catalyzed by succinic thiokinase (33) (page 613) and in the enzymatic transfer of a phosphate group from phosphoenolpyruvate to hexoses (34). These rather few cases in which the sulfhydryl and imidazole functions of enzymes are acylated or phosphorylated may reflect the thermodynamic barrier to the formation of the intermediates. Acyl and phosphoryl imidazoles and thiol esters fall into the energy-rich category (page 35), and hence their formation from typical nonactivated biochemical substrates is endergonic.

For several enzymatic reactions that occur with covalent catalysis, the hydroxyl group of a serine residue appears to be the primary site of acylation. The serine anion is, in comparison with hydroxide or ethoxide ion, abnormally reactive in base-catalyzed reactions. This high reactivity is princi-

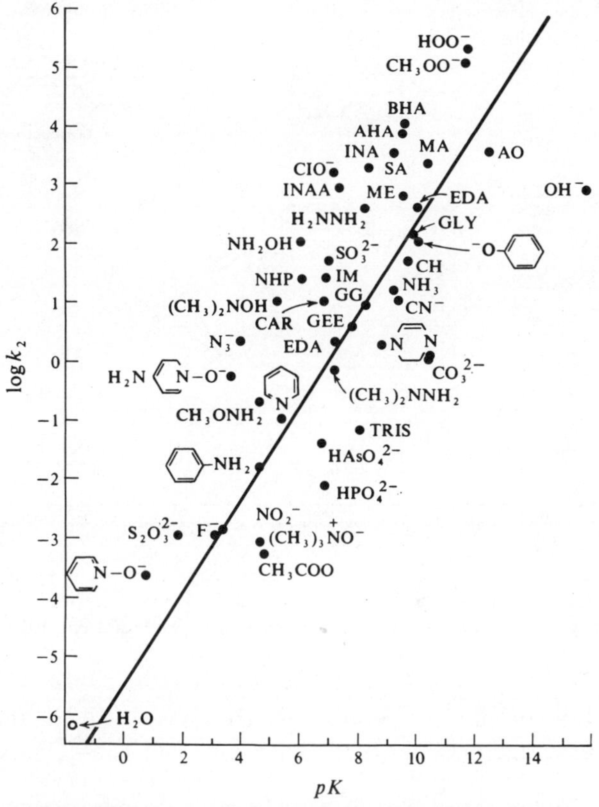

FIGURE 7-8
Rates of nucleophilic reactions with *p*-nitrophenyl acetate in aqueous solution at 25° plotted against the basicity of the attacking reagent. Abbreviations: GEE, glycine ethyl ester; GLY, glycine; GG, glycylglycine; IM, imidazole; AHA, acetohydroxamic acid; BHA, *n*-butyrylhydroxamic acid; CH, chloral hydrate anion at 30°. W. P. Jencks and J. Carriuolo, *J. Am. Chem. Soc.*, **82:** 1778 (1960): by permission of the copyright owners, the American Chemical Society.

pally the result of the relatively high acidity of this alcohol and the unexpectedly low reactivities of strongly basic anions.

Catalysis by Distortion

There have been repeated suggestions that part of the efficiency of enzymes as catalysts may be due to utilization of binding forces between substrate and enzyme to induce strain in the substrate. One visualization of catalysis by distortion is provided in Fig. 7-9. Here we view the reaction as simply a pulling apart of the substrate into its two halves. The most stable form of the enzyme is one that does not properly fit the substrate but corresponds to one in which the substrate is partially pulled apart. Association of enzyme with substrate causes the former to adopt a less stable conformation. Its tendency to reassume its stable state will tend to distort the substrate in a way which makes it approach the transition state geometry. Thus, binding forces between enzyme and substrate are employed to destabilize the substrate with respect to the transition state for its decomposition and, hence, catalyze that decomposition.

Such catalysis by distortion is reasonable and many examples from organic chemistry might be cited to demonstrate its importance. Thus, the alkaline

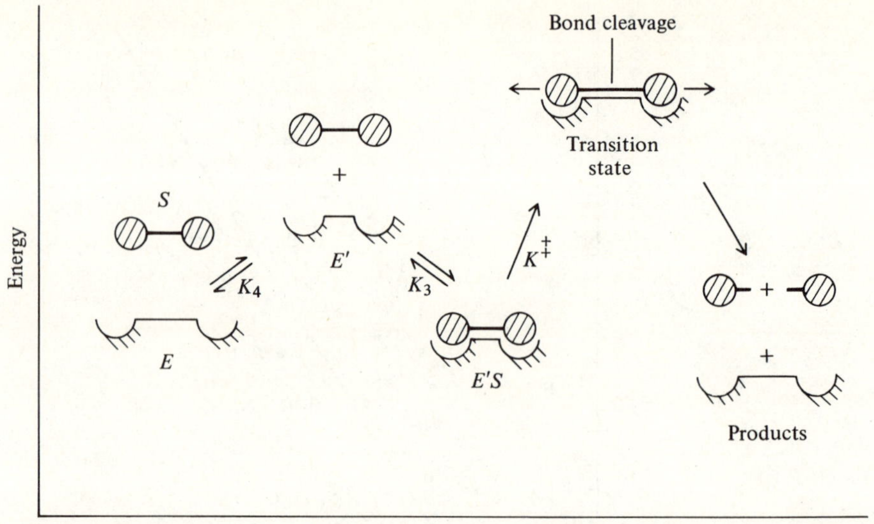

FIGURE 7-9

Scheme to show how a substrate-induced conformation change in an enzyme may be utilized to facilitate reaction of the substrate by simply helping to pull it apart. W. P. Jencks, *Catalysis in Chemistry and Enzymology,* McGraw-Hill, New York, 1969.

hydrolysis of ethylene phosphate (III) is at least 10^8 times more rapid than the same reaction of dimethyl phosphate (IV) (35–37).

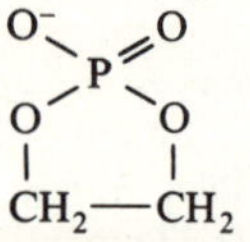

III: cyclic O^-–P(=O)–O–CH_2—CH_2–O ; IV: O^-–P(=O)(OCH_3)(OCH_3)

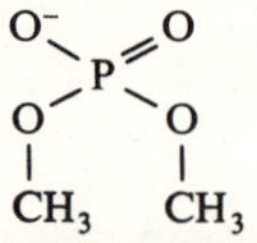

This rate difference must reflect destabilization of the cyclic structure through strain introduced in closing the ring.

In reactions of this sort, the destabilization has been built into the substrate in the course of its synthesis. In enzymatic reactions, this destabilization must occur only with or following formation of the enzyme–substrate complex. Consequently, it is difficult to demonstrate the existence of catalysis of this kind for enzymatic reactions. Nevertheless, several indirect lines of evidence do suggest that it may be important. Let us consider three of these.

In the first place, there are some enzymatic reactions which seem susceptible to no other kind of catalysis. Consider, for example, formation of trigonelline according to Eq. (7-36):

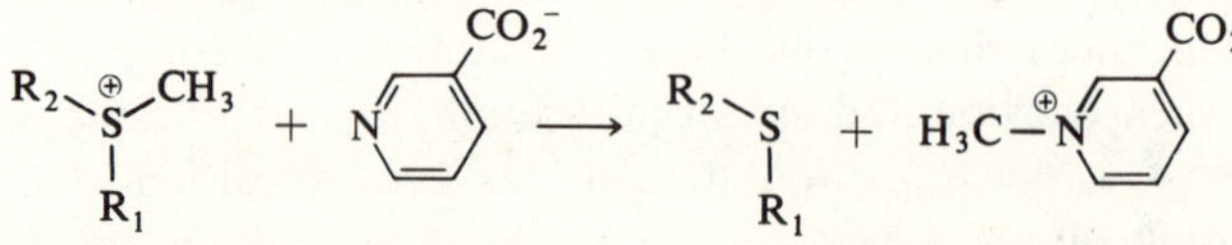

$$R_2\overset{\oplus}{S}(CH_3)R_1 + \text{nicotinate } (CO_2^-) \longrightarrow R_2SR_1 + H_3C\!-\!\overset{\oplus}{N}\text{-pyridine-}CO_2^- \qquad (7\text{-}36)$$

The simple fact is that none of the other catalytic mechanisms discussed thus far are of much use for this reaction. Yet one can imagine that the enzyme compresses together the two reactants moving them toward the transition-state structure and hence facilitating the reaction. Several other reactions fall into the same category.

In the second place, the specificity of enzymes is sometimes manifested in the maximal rates of reaction rather than in binding specificity. Thus, the forces that contribute to binding of the substrate may also contribute to lowering the activation energy for its reaction. As one example, Michaelis and catalytic constants for hydrolysis of a series of peptides by pepsin are collected in Table 7-9 (38). Note that the former constants are essentially independent of the substrate, whereas the latter vary by three orders of magnitude. One may suppose that the greater binding forces of the good substrates have been employed to induce strain in them; these forces then appear as increased rates, not as increased affinities. It is, of course, also possible to explain this set of data in terms of the induced-fit hypothesis. Utilization of binding forces between enzyme and substrate to acquire specificity or to reduce activation energy yield the same kind of experimental observations.

Finally, we may note that if the most stable form of the enzyme is that which resembles the transition state for the catalyzed reaction, then transition-state analogs ought to be particularly powerful inhibitors of the reaction. They will fit the enzyme without conformation change. As an example of such behavior consider the proline racemase reaction:

CO_2^- H N H   N sp^2 CO_2^- H   H N H CO_2^- (7-37)

The transition state must have the α-carbon in an essentially planar sp^2 state. Thus the transition state resembles furan-2-carboxylic acid (V) more closely

TABLE 7-9
Hydrolysis of peptides by pepsin at pH 4.0

Peptide[a]	K_s, mM	k_{cat}, sec^{-1}
Cbz·Gly·His·Phe·Phe·OEt	0.80	2.43
Cbz·His·Phe·Trp·OEt	0.23	0.51
Cbz·His·Phe·Phe·OEt	0.18	0.31
Cbz·His·Phe·Tyr·OEt	0.23	0.16
Cbz·His·Tyr·Phe·OMe	0.68	0.013
Cbz·His·Tyr·Tyr·OEt	0.24	0.0094
Cbz·His·Phe·Leu·OMe	0.56	0.0025

[a] Cbz, carbobenzyloxy.
SOURCE: W. P. Jencks, *Catalysis in Chemistry and Enzymology*, McGraw-Hill, New York, 1969.

than tetrahydrofuran-2-carboxylic acid (VI). Significantly, the former is much the better inhibitor of the two.

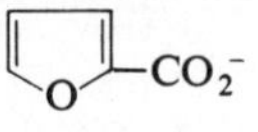

V 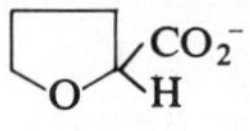 VI

MECHANISM OF ACTION OF LYSOZYME

Combinations of structural, kinetic, and related studies on enzymes with those on model systems have provided considerable insight into the catalytic processes of a few enzymes. Let us consider three of these: lysozyme, chymotrypsin, and carboxypeptidase.

Lysozyme is an enzyme that catalyzes the hydrolysis of glycosidic bonds, particularly those which are involved in the peptidoglycans of bacterial cell walls. Returning to structural studies of this enzyme (page 156), we know that there exists a cleft in the enzyme into which six hexose units will fit in such a way as to interact favorably with the enzyme. Kinetic and product-analysis studies employing artificial substrates establish that hydrolysis of the oligosaccharide chain occurs between the residues 4 and 5 (39):

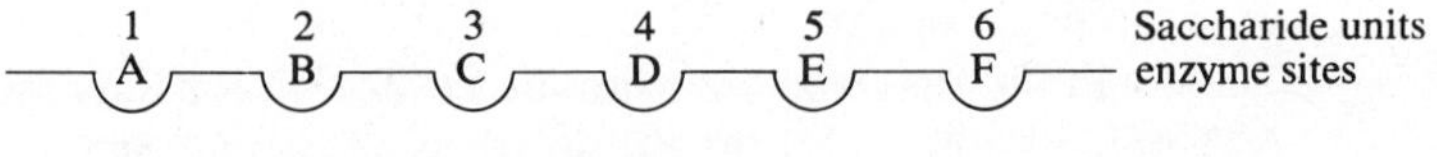

1 2 3 4 5 6 Saccharide units
A B C D E F enzyme sites

Thus, the enzyme catalyzes the transfer of C-1 of hexose residue 4 from the C-4 hydroxyl group of residue 5 to water (or other suitable acceptor). Written in its most basic form, such a reaction probably occurs in the following way (40):

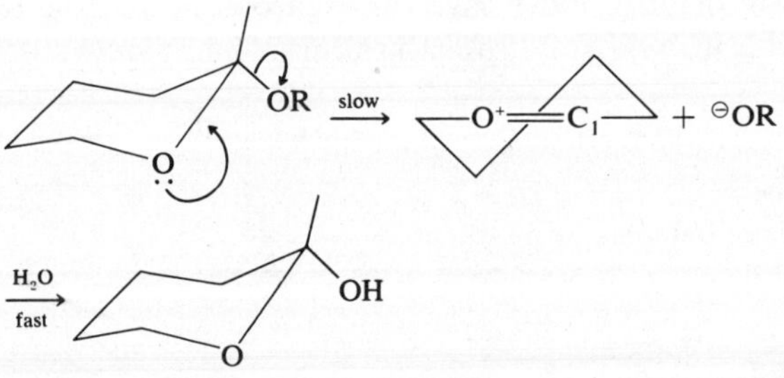

OR $\xrightarrow{\text{slow}}$ $-O^{+}{=}C_1-$ + $^{\ominus}OR$ (7-38)

$\xrightarrow[\text{fast}]{H_2O}$ OH

Expulsion of the leaving group with participation of the unshared electron pair on oxygen will cause the hexose ring to change from its normal chair form to a half-chair conformation to gain maximum overlap with the vacant orbital on C-1. Now let us think about how we might aid this reaction in terms of the general discussion above. First, we might imagine that departure of the leaving group is aided through donation of a proton to it from a general acid catalyst. Second, the developing positive charge might be stabilized by a properly located negative charge on the enzyme. Finally, the enzyme might distort the substrate itself into the half-chair conformation of the transition state. On the basis of this conjecture, we might picture the transition state for the enzymatic reaction as shown on page 369.

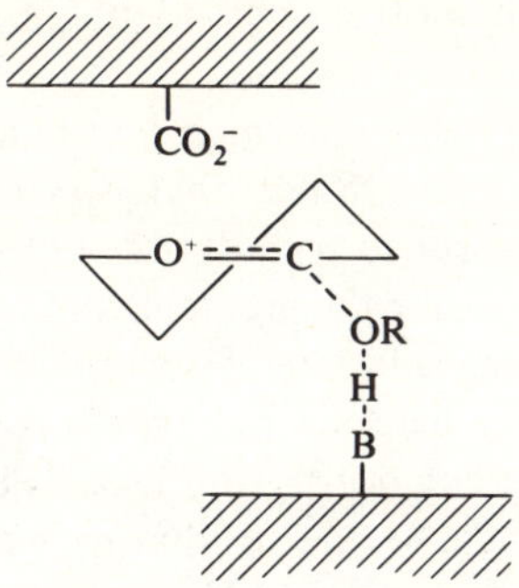

As accurately as one can tell, this is exactly what does happen (41,42). The structural studies establish that two carboxylic amino acids Glu[35] and Asp[52] lie in the vicinity of the bond to be broken in the enzyme–substrate complex. The former group exists in a region of rather hydrophobic character, which would tend to repress the ionization of the carboxyl group. This residue, then, is a prime candidate to function as a general acid catalyst, which requires that it exist as a carboxylic acid, aiding the departure of the leaving alcohol. Asp[52], in contrast, exists in a hydrophilic environment and is thus readily dissociated to form a carboxylate ion. As such, it may function to electrostatically stabilize the carboxonium ion. Finally, the X-ray data show that residue 4 of the oligosaccharide substrate can fit into its position on the enzyme surface only through distortion to a half-chair form. This distortion must be achieved at the expense of the conversion of binding energy to strain energy in the substrate. The overall effect is to diminish the energy input necessary to convert the substrate into the transition state geometry.

Although these conclusions must not be regarded as certain, they do seem likely and form a reasonable and consistent picture of the catalytic process of lysozyme.

MECHANISM OF ACTION OF CHYMOTRYPSIN

One of the most thoroughly understood enzymes, from the standpoint of the mechanism of the catalytic process, is chymotrypsin. This enzyme also provides a suitable example of the employment of the various techniques discussed above in an effort to unravel enzymatic reaction mechanisms.

Chymotrypsin is an acyl group-transfer enzyme. It transfers the acyl group from a number of donors, including esters, amides, acids, hydroxamates, etc., to a number of acceptors such as water, alcohols, and amines. As indicated below, chymotrypsin-catalyzed reactions appear to proceed with the intermediate formation of an acyl enzyme. Thus, this reaction is of the *double-displacement type*. Our present state of knowledge regarding the intimacies of this process derives principally from the following observations:

1. The active site of chymotrypsin has been covalently labeled with *quasi-substrates,* including *p*-nitrophenyl acetate, diisopropylfluorophosphate, and the nonactivated ester methyl cinnamate. These observations suggest the intermediate formation of an acyl-enzyme.

2. The maximal rates of chymotryptic hydrolysis of several derivatives of *N*-acetyl-L-tryptophan are identical. Since the inherent reactivities of these derivatives are quite different, this observation suggests that an acyl-enzyme is indeed formed from these, more or less, specific substrates and that deacylation is the rate-determining step in this catalytic process. Since the acyl-en-

zyme *N*-acetyl-L-tryptophanyl chymotrypsin is identical for all the substrates, this hypothesis accounts for the identity in rates.

3. The active site of chymotrypsin may also be covalently labeled by the specific substrate *N*-acetyl tryptophan. Since chymotrypsin has, as its normal substrate, the peptide bonds adjacent to tryptophan and tyrosine residues, this experiment provides strong support for the acyl-enzyme hypothesis.

4. Chemical and enzymatic degradation of several acyl chymotrypsins reveals that the acyl group is attached to the hydroxyl function of a serine residue (one of the two present in chymotrypsin) (see Table 7-2).

5. The conversion of this crucial serine residue to a dehydroalanyl residue renders chymotrypsin completely inactive (43), consistent with an obligatory acyl-enzyme intermediate involving this serine hydroxyl.

6. Studies concerning the effect of *p*H on the rate of chymotrypsin-catalyzed reactions have implicated the free base of a histidine residue, or its kinetic equivalent, as necessary for the catalysis. This conclusion is substantiated by the loss of enzymatic activity with histidine alkylation by the chloromethyl ketone of a chymotrypsin substrate analog (affinity labeling).

7. The hydrolysis of the model substrate *N,O*-diacetylserineamide is subject to classical general base catalysis by imidazole.

8. The effect of polar substituents on the rate of deacylation of benzoyl chymotrypsins suggests that basic catalysis is important for this reaction (44). This conclusion is strongly supported by the observation that the rate of reaction of a series of nitrogen nucleophiles with furoyl chymotrypsin is independent of the basicity of the nucleophile (45). This indicates that proton removal from nitrogen must accompany the attack reaction.

9. The rate of chymotrypsin-catalyzed reactions is decreased from two- to threefold in deuterium oxide compared with water. This observation is consistent with and offers support for proton transfer in the rate-determining step (i.e., general acid–base catalysis).

These and related observations are consistent with a mechanism of action of chymotrypsin shown in Fig. 7-10. This mechanism, which is not the only one consistent with the data, takes explicit account of the formation of tetrahedral intermediates that are normal intermediates for esterolytic reactions.

Happily, X-ray diffraction studies of chymotrypsin are consistent with the mechanism of Fig. 7-10 based on the more standard approaches noted. The acylated serine, Ser^{195}, is located very close to His^{57} in the crystalline structure (page 160). It is precisely this histidine that is alkylated by the chloromethylketone substrate analog with enzyme inactivation. It seems unlikely that the indicated mechanism is adequate to account for the catalytic efficiency of chymotrypsin. At the same time, we may have some confidence that what we have written, or a kinetically indistinguishable variant, forms an important part of the total picture.

MECHANISM OF ACTION OF CARBOXYPEPTIDASE

Finally, let us direct our attention to carboxypeptidase A and its catalytic process. The X-ray-deduced structure of this enzyme together with that for a complex with Gly-Tyr has been discussed earlier (pages 161 and 345), and we shall wish to refer to some of the details of those structures.

Carboxypeptidase A catalyzes hydrolysis of that peptide bond which is adjacent to the terminal *free* carboxyl group. It thus catalyzes protein hydrolysis in a sequential way beginning at the C terminus. The best substrates for carboxypeptidase A have terminal residues containing a bulky aromatic or branched-chain aliphatic group.

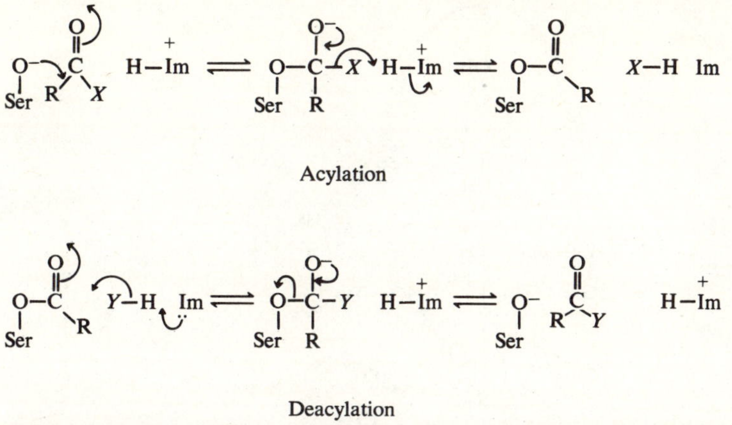

Acylation

Deacylation

FIGURE 7-10

Possible mechanism for chymotrypsin-catalyzed acyl transfer reactions. The attack of the serine hydroxyl groups on the acyl substrate is viewed as involving the conjugate base of this group. Departure of the leaving group, *X*, from the tetrahedral intermediate involves general acid catalysis by the imidazolium ion of a histidine residue. Deacylation of this enzyme, involving *Y*—H as nucleophile, is just the microscopic reverse of the acylation process.

Kinetic studies of the peptidase activity have revealed bell-shaped *p*H-rate profiles with inflection points at *p*H 6.7 and 8.5 (46,47). This behavior suggests the importance of both acidic and basic catalysis in the hydrolysis process. Moreover, kinetic studies employing ester substrates reveal that the reaction is twice as fast in water as in deuterium oxide, which suggests the involvement of proton transfer in the transition state, consistent with general acid or general base catalysis, or both.

Finally, carboxypeptidase A contains a single atom of zinc per mol of enzyme and the catalytic activity of the enzyme requires the presence of this ion (see also page 769).

If we now examine the environment of a substrate bound to carboxypeptidase A (page 162), four protein functional groups in the vicinity of the susceptible peptide bond are evident: Arg^{145}, Tyr^{248}, Glu^{270}, and Zn. How these residues may function in the catalytic process is shown in Fig. 7-11. Arg^{145} is assigned the function of binding the terminal carboxylate group of the substrate. Tyr^{248} is viewed as a general acid catalyst aiding the departure of the leaving amine. Glu^{270} may function either directly as a nucleophilic reagent toward the peptide, as shown, or as a general base catalyst aiding the nucleophilic attack of H_2O on the peptide. Clearly, the ionizations of Glu^{270} and Tyr^{248} must account for the *p*H-rate behavior noted above. The zinc ion, which is bound to two histidines and one glutamate and therefore has a net charge of +1, is considered to have the carbonyl oxygen atom of the susceptible peptide bond as a fourth ligand and, consequently, to increase the susceptibility of the substrate to nucleophilic attack at the carbonyl carbon. Finally, the protein environment at the binding site possesses a pocket into which the bulky side chain of the terminal residue may fit. This picture provides a reasonable view of the events that may occur leading to substrate hydrolysis. As noted above (page 345), one of the most striking aspects of

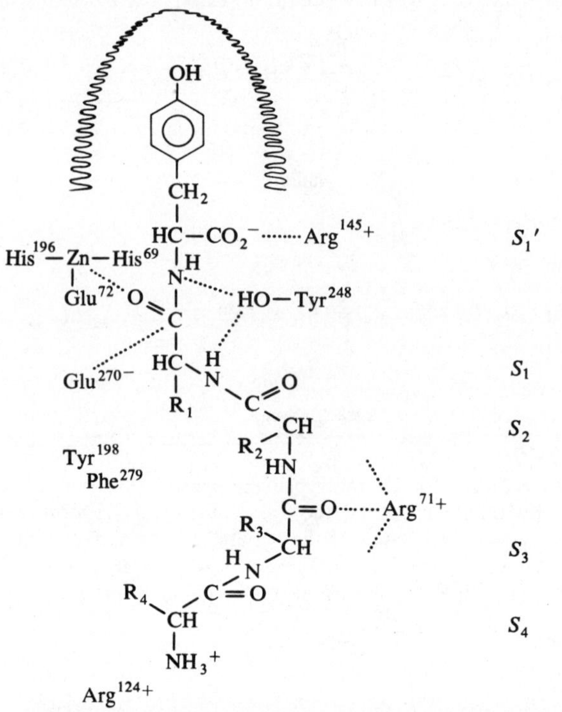

FIGURE 7-11

View of the interactions between carboxypeptidase and a substrate, illustrating principal features of the catalytic process. W. N. Lipscomb, J. A. Hartsuck, G. N. Reeke, Jr., F. A. Quiocho, P. H. Bethge, M. L. Ludwig, T. A. Steitz, H. Muirhead, and J. C. Coppola, *Brookhaven Symp. Biol.*, **21:** 24 (1968).

the behavior of carboxypeptidase A is the large conformation change induced in the enzyme as a result of substrate binding. A small displacement of Arg^{145} permits a large displacement of the side chain of Tyr^{248} to occur. The center of the phenolic ring of this residue moves some 12 Å during this process. This movement can be seen very clearly in the two sets of stereo pairs centering on the active site of carboxypeptidase A shown in Fig. 4-31.

ASPECTS OF ENZYME SPECIFICITY

One of the chemically most striking, and biochemically most important, aspects of enzyme-catalyzed reactions is their marked specificity. It is clear that, in the absence of a strict limitation on the range of substrates acted upon by individual enzymes, an orderly metabolic pattern is impossible to achieve.

A large number of studies of enzyme specificity have been performed. Ideally, such studies are carried out by making a series of limited structural alterations at each of several sites on the normal substrate for the reaction under consideration. An examination of the pertinent kinetic parameters for the reaction of the altered substrate serves as a measure of the degree to which the alteration has affected binding and susceptibility to reaction. The investigation of the specificity of an enzyme requires the use of pure enzyme, or at least an enzyme preparation not contaminated with enzymes catalyzing the same type of reaction as that of interest, to ensure that small amounts of reaction are not the result of the action of a second enzyme. In addition, such a study requires the use of carefully purified substrates to avoid the possible detection of the reaction of an impurity. Unfortunately, a large percentage of specificity studies do not meet one or both of these criteria. In addition, many studies have not been performed at the variety of substrate concentrations necessary to sort out observed kinetic behavior into effects on the Michaelis constant and on the maximum velocity.

The basic goal of specificity studies is to elucidate the structural requirements for catalysis, and on this basis to obtain a picture of the catalytically active conformation of the active site, whether a rigid template or substrate induced. The majority of such studies are either carried out insufficiently carefully, as noted above, or in insufficient detail to permit meaningful conclusions to be drawn. However, considerable progress has been made in isolated cases (48). Frequently, such studies cannot be performed at all, since many enzymes appear to be absolutely specific for a single substrate or single pair of substrates. Only in cases in which the enzyme tolerates at least modest alterations in substrate structure is a detailed investigation of the number and spatial arrangement of groups within binding sites on the enzyme surface possible.

As many examples of the specificity of individual enzymes are presented later in this book, these will not be considered at this point. Dixon and Webb have published an extensive compilation of information of this type. One aspect of enzyme specificity does, however, merit special consideration. This is the ability of enzymes to distinguish, in some cases, between chemically identical groups. Three examples of such specificity are considered below.

The first example concerns the glycerokinase reaction:

$$\begin{array}{c}CH_2OH\\|\\HO{-}C{-}H\\|\\CH_2OH\end{array} + ATP \rightarrow \begin{array}{c}CH_2OH\\|\\HO{-}C{-}H\\|\\CH_2OPO_3H^-\end{array} + \quad ADP \tag{7-39}$$

L-α-Glycerophosphate

The product of this reaction is exclusively L-α-glycerophosphate. This observation suffices to indicate that this enzyme treats the chemically identical hydroxymethyl groups of glycerol in an asymmetric fashion, since, if this were not so, a mixture of L-α- and D-α-glycerophosphate would have been formed. In corroboration of this conclusion, glycerol-1-^{14}C, formed enzymatically from glucose-3,4-^{14}C, is converted enzymatically to fructose-1,6-diphosphate-3,4-^{14}C by the pathway shown in Fig. 7-12. This finding indicates that the phosphorylation of glycerol occurs on the hydroxymethyl group not derived from the 3 and 4 positions of glucose, since if this molecule were handled symmetrically, the fructose-1,6-diphosphate would have been labeled at the 1 and 6 positions as well as the 3 and 4.

Related behavior is exhibited by the enzyme aconitase, which metabolizes citric acid (page 609). Citric acid, labeled in a primary carboxyl group, prepared from labeled oxalacetate and unlabeled acetate, as shown in Fig. 7-13, is converted enzymatically to α-oxoglutarate. The α-oxoglutarate is labeled only in the carboxyl group adjacent to the carbonyl function, indicating that aconitase distinguishes between the chemically identical carboxymethyl groups of citrate as shown.

Finally, aldolase, as noted above, catalyzes the exchange of only one hydrogen of dihydroxyacetone phosphate with solvent deuterium or tritium. If aldo-

$$\begin{array}{c}{}^{*}CH_2OH\\|\\HO{-}C{-}H\\|\\CH_2OH\end{array} + ATP \xrightarrow{ADP} \begin{array}{c}{}^{*}CH_2OH\\|\\HO{-}C{-}H\\|\\CH_2OPO_3H^-\end{array} \xrightarrow{-2[H]} \begin{array}{c}{}^{*}CH_2OH\\|\\C{=}O\\|\\CH_2OPO_3H^-\end{array} \rightleftharpoons \begin{array}{c}{}^{*}CHO\\|\\H{-}C{-}OH\\|\\CH_2OPO_3H^-\end{array}$$

$$\begin{array}{c}CH_2OPO_3H^-\\|\\C{=}O\\|\\HO{-}{}^{*}C{-}H\\|\\H{-}{}^{*}C{-}OH\\|\\H{-}C{-}OH\\|\\CH_2OPO_3H^-\end{array}$$

FIGURE 7-12
Labeling pattern of fructose-1,6-diphosphate derived enzymatically from glycerol-1-^{14}C.

lase were unable to discriminate between the two hydrogens on the pertinent carbon atom, reversal of this reaction would lead to exchange of both hydrogen atoms. This finding was confirmed and the absolute stereochemistry of the dihydroxyacetone phosphate carbanion was determined by Rose (49).

$$\begin{array}{c}CH_2OPO_3H^-\\|\\C{=}O\\|\\HO{-}C{-}H\\|\\T\end{array} \xrightarrow{HIO_4} \begin{array}{c}CO_2H\\|\\HO{-}C{-}H\\|\\T\end{array} \xrightarrow[\text{oxidase}]{\text{glycolic}} \begin{array}{c}CO_2H\\|\\O{=}C\\ \diagdown\\ T\end{array} \qquad (7\text{-}40)$$

Dephosphorylation and periodate oxidation of tritiated substrate, formed in the aldolase reaction, yielded glycolic acid, which was oxidized to glyoxylic acid enzymatically without the loss of tritium. Glycolic oxidase oxidizes L-lactate but not D-lactate; consequently, the tritiated dihydroxyacetone phosphate must have been labeled in the same relative position as that occupied by the proton in D-lactic acid.

Ogston has pointed out that the enzymatic differentiation of chemically identical groups can be rationalized on the basis of a three-point attachment of substrate to the enzyme surface (50). Such a situation is depicted schematically for the aconitase reaction in Fig. 7-14. It has been argued that the three-point-attachment hypothesis is unnecessary, since the two *a* groups on a

$$\begin{array}{l}CH_3\overset{O}{\overset{\|}{C}}{-}SCoA\\O{=}C{-}CO_2H\\\quad|\\\quad CH_2{-}{}^{*}CO_2H\end{array} \xrightarrow{-HSCoA} \begin{array}{c}CH_2{-}CO_2H\\|\\HO{-}C{-}CO_2H\\|\\CH_2{-}{}^{*}CO_2H\end{array} \xrightarrow{\text{aconitase}} \begin{array}{c}CH_2{-}CO_2H\\|\\H{-}C{-}CO_2H\\|\\HO{-}C{-}C^{*}O_2H\\|\\H\end{array}$$

$$\xrightarrow[-2H]{-CO_2} \begin{array}{c}CO_2H\\|\\CH_2\\|\\CH_2\\|\\C{=}O\\|\\{}^{*}CO_2H\end{array}$$

FIGURE 7-13
Labeling pattern of α-oxoglutarate derived enzymatically from citrate-1-^{14}C.

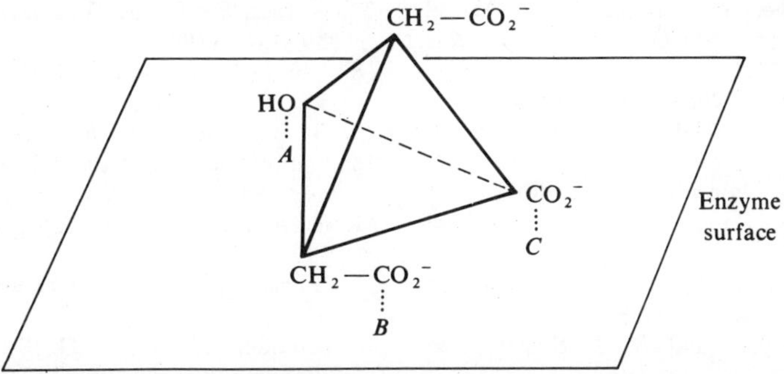

FIGURE 7-14
Schematic representation of the attachment of citrate to aconitase.

C_{aabc} atom inherently appear distinct to an unsymmetrical reagent. The point is, in any event, that a three-dimensional match of a substrate having a C_{aabc} atom with an asymmetric enzyme surface places each of the *a* groups in a unique position.

REFERENCES

Books and Review Articles

Bell, R. P.: *Acid–Base Catalysis,* The Clarendon Press, Oxford, 1941.

Bender, M. L., and R. Breslow: Mechanisms of Organic Reactions, in M. Florkin and E. H. Stotz (eds.), *Comprehensive Biochemistry,* vol. 2, Elsevier, New York, 1962, p. 1.

Bernhard, S. A.: *The Structure and Function of Enzymes,* Benjamin, New York, 1968.

Boyer, P. D., H. Lardy, and K. Myrbäck (eds.): *The Enzymes,* Academic Press, New York, 1959–1963.

Bruice, T. C., and S. J. Benkovic: *Bioorganic Mechanisms,* Benjamin, New York, 1966.

Chipman, D. M., and N. Sharon: Mechanism of Lysozyme Action, *Science,* **165:** 454 (1969).

Dixon, M., and E. C. Webb: *Enzymes,* Academic Press, New York, 2nd ed., 1964.

Eigen, M., and G. G. Hammes: Elementary Steps in Enzyme Reactions, *Advan. Enzymol.,* **25:** 1 (1963).

Florkin, M., and E. H. Stotz (eds.): *Comprehensive Biochemistry,* vol. 13, Elsevier, New York, 1964.

Goodwin, T. W., J. I. Harris, and B. S. Hartley (eds.): *Structure and Activity of Enzymes,* Academic Press, New York, 1964.

Gutfreund, H.: *An Introduction to the Study of Enzymes,* Wiley, New York, 1965.

Haldane, J. B. S.: *Enzymes,* Longmans Green, London, 1930; reprinted by M.I.T. Press, Cambridge, 1965.

Jencks, W. P.: *Catalysis in Chemistry and Enzymology,* McGraw-Hill, New York, 1969.

Jencks, W. P.: Mechanism of Enzyme Action, *Ann. Rev. Biochem.,* **32:** 639 (1963).

Koshland, D. E., Jr.: The Active Site and Enzyme Action, *Advan. Enzymol.,* **22,** 45 (1960).

Koshland, D. E., Jr., and K. E. Neet: The Catalytic and Regulatory Properties of Proteins, *Ann. Rev. Biochem.,* **37:** 359 (1968).

Kosower, E. M.: *Molecular Biochemistry,* McGraw-Hill, New York, 1962.

Northrup, J. H., M. Kunitz, and R. M. Herriott: *Crystalline Enzymes,* Columbia University Press, New York, 2nd ed., 1948.

Rose, I. A.: Mechanisms of Enzyme Action, *Ann. Rev. Biochem.,* **35:** 23 (1966).

Vallee, B. L., and J. F. Riordan: Chemical Approaches to the Properties of the Active Sites of Enzymes: *Ann. Rev. Biochem.,* **38:** 733 (1969).

Waley, S. G.: *Mechanisms of Organic and Enzymic Reactions,* Oxford University Press, New York, 1962.

Westley, J.: *Enzymic Catalysis,* Harper & Row, New York, 1969.

Williams, A.: Mechanism of Action and Specificity of Proteolytic Enzymes, *Quart. Rev.,* **23:** 1 (1969).

Wolstenholme, G. E. W., and C. M. O'Connor: *Steric Course of Microbiological Reactions,* Ciba Foundation Study Group No. 2, Little, Brown, Boston, 1959.

Specific References

1. Bender, M. L.: *J. Am. Chem. Soc.,* **84,** 2582 (1962).
2. Bender, M. L., and E. T. Kaiser: *J. Am. Chem. Soc.,* **84:** 2556 (1962).
3. Harris, I., B. P. Meriwether, and J. H. Park: *Nature,* **198:** 154 (1963).
4. Koshland, D. E., Jr., and M. J. Erwin: *J. Am. Chem. Soc.,* **79:** 2657 (1957).
5. Pizer, L. I.: *J. Am. Chem. Soc.,* **80:** 4432 (1958).
6. Schwartz, J. H., and F. Lipmann: *Proc. Natl. Acad. Sci. U.S.,* **47:** 1996 (1961).

7. Grazi, E., H. Meloche, G. Martinez, W. A. Wood, and B. L. Horecker: *Biochem. Biophys. Res. Commun.,* **10:** 4 (1963).
8. Westheimer, F. H.: *Proc. Chem. Soc.,* 253 (1963).
9. Grisolia, S., and W. W. Cleland: *Biochemistry,* **7:** 1115 (1968).
10. Horecker, B. L., S. Pontremoli, C. Ricci, and T. Cheng: *Proc. Natl. Acad. Sci. U.S.,* **47:** 1949 (1961).
11. Speck, J. C., Jr., P. T. Rowley, and B. L. Horecker: *J. Am. Chem. Soc.,* **85:** 1012 (1963).
12. Wofsy, L., H. Metzger, and S. J. Singer: *Biochemistry,* **1:** 1031 (1962).
13. Ong, E. B., E. Shaw, and G. Schoellmann: *J. Am. Chem. Soc.,* **86:** 1271 (1964).
14. Shaw, E., M. Mares-Guia, and W. Cohen: *Biochemistry,* **4:** 2219 (1965).
15. Kafatos, F. C., J. H. Law, and A. M. Tartakoff: *J. Biol. Chem.,* **242:** 1488 (1967).
16. Tobita, T., and J. E. Folk: *Biochem. Biophys. Acta,* **147:** 15 (1967).
17. Smillie, L. B., and D. R. Whitaker: *J. Am. Chem. Soc.,* **89:** 3350 (1967).
18. Ray, W. J., Jr., and D. E. Koshland, Jr.: *J. Am. Chem. Soc.,* **85:** 1977 (1963).
19. Koshland, D. E., Jr.: *The Enzymes,* vol. 1, P. D. Boyer, H. Lardy, and K. Myrbäck (eds.), Academic Press, New York, 1959, p. 305.
20. Yankeelov, J. A., Jr., and D. E. Koshland, Jr.: *J. Biol. Chem.,* **240:** 1593 (1965).
21. Lipscomb, W. N., J. A. Hartsuck, G. N. Reeke, Jr., F. A. Quiocho, P. H. Bethge, M. L. Ludwig, T. A. Steitz, H. Muirhead, and J. C. Coppola: *Brookhaven Symp. Biol.,* **21:** 24 (1968).
22. Koshland, D. E., Jr.: *J. Cell. Comp. Physiol.,* **47:** Suppl. 1, 217 (1956).
23. Capon, B. and R. Capon: *Chem. Commun.,* **1965:** 502.
24. Gitler, C., and A. Solano-Ochoa: *J. Am. Chem. Soc.,* **90:** 5004 (1968).
25. VanEtten, R. L., J. F. Sebastian, G. A. Clowes, and M. L. Bender: *J. Am. Chem. Soc.,* **89:** 3242 (1967); VanEtten, R. L., G. A. Clowes, J. F. Sebastian, and M. L. Bender; *ibid.,* 3253 (1967).
26. Bell, R. P., and W. C. E. Higginson: *Proc. Roy. Soc.* (*London*), **A197**, 141 (1949).
27. Cordes, E. H., and W. P. Jencks: *J. Am. Chem. Soc.,* **84:** 4319 (1962).
28. Cordes, E. H., and W. P. Jencks: *J. Am. Chem. Soc.,* **85:** 2843 (1963).
29. Banks, B. E. C.: *J. Chem. Soc.,* **1963:** 63.
30. Cunningham, B. A., and G. L. Schmir: *J. Am. Chem. Soc.,* **88:** 551 (1966).
31. Cordes, E. H., and W. P. Jencks: *Biochemistry,* **1:** 773 (1962).
32. Jencks, W. P., and J. Carriuolo: *J. Am. Chem. Soc.,* **82:** 1778 (1960).
33. Mitchell, R. A., L. G. Butler, and P. D. Boyer: *Biochem. Biophys. Res. Commun.,* **16:** 545 (1964).
34. Kundig, W., S. Ghosh, and S. Roseman: *Proc. Natl. Acad. Sci. U.S.,* **52:** 1067 (1964).
35. Kumamoto, J., J. R. Cox, Jr., and F. H. Westheimer: *J. Am. Chem. Soc.,* **78:** 4858 (1956).
36. Haake, P. C., and F. H. Westheimer: *J. Am. Chem. Soc.,* **83:** 1102 (1961).
37. Kaiser, E. T., M. Panar, and F. H. Westheimer: *J. Am. Chem. Soc.,* **85:** 602 (1963).
38. Inouye, K., and J. S. Fruton: *Biochemistry,* **6:** 1765 (1967).
39. Rupley, J. A., and V. Gates: *Proc. Natl. Acad. Sci. U.S.,* **57:** 496 (1967).
40. Dahlquist, F. W., T. Rand-Meir, and M. A. Raftery: *Proc. Natl. Acad. Sci. U.S.,* **61:** 1194 (1968).
41. Blake, C. C. F., L. N. Johnson, G. A. Mair, A. C. T. North, D. C. Phillips, and V. R. Sarma: *Proc. Roy. Soc.* (*London*), **B167:** 378 (1967).
42. Phillips, D. C.: *Proc. Natl. Acad. Sci. U.S.,* **57:** 484 (1967).
43. Strumeyer, D. H., W. N. White, and D. E. Koshland, Jr., *Proc. Natl. Acad. Sci. U.S.,* **50:** 931 (1963).
44. Caplow, M., and W. P. Jencks: *Biochemistry,* **1:** 883 (1962).
45. Inward, P. W., and W. P. Jencks: *J. Biol. Chem.,* **240:** 1986 (1965).
46. Lumry, R., E. L. Smith, and R. L. Glantz: *J. Am. Chem. Soc.,* **73:** 4330 (1951).
47. Carson, F. W., and E. T. Kaiser: *J. Am. Chem. Soc.,* **88:** 1212 (1966).
48. Hein, G. E., and Niemann: *J. Am. Chem. Soc.,* **84:** 4487 (1962).
49. Rose, I. A.: *J. Am. Chem. Soc.,* **80:** 5835 (1958).
50. Ogston, A. G.: *Nature,* **162:** 963 (1948).

8
Coenzymes

Numerous enzymes, catalyzing a broad spectrum of biochemical reactions, require the presence of a small nonprotein *prosthetic group,* more or less tightly linked to the protein, for efficient performance of catalytic function. Since these prosthetic groups are frequently intimately concerned with the overall reaction, they are also termed cofactors or *coenzymes.* A hazy distinction between prosthetic groups and coenzymes is sometimes drawn, the term "prosthetic group" being reserved for those molecules which are very tightly bound to the enzyme. Even the distinction between a substrate and a coenzyme for an enzymatic reaction is often rendered unclear, since the coenzyme, as well as the substrate, may be structurally altered in the course of the reaction. However, in subsequent reactions, the original structure of the coenzyme is usually regenerated, and the substrate usually undergoes further chemical modification. In those cases in which the coenzyme emerges from a reaction step in a structurally altered form, the distinction between coenzyme and substrate may be made on this basis of eventual regeneration. Also, in coenzyme-dependent reactions the binding of coenzyme to enzyme frequently precedes substrate–enzyme interaction in an obligatory manner (page 290).

In this chapter, the most prominent coenzymes are discussed individually. Emphasis is directed toward understanding the type or types of enzymatic reactions for which each coenzyme is required in terms of the basic reaction mechanisms involved. Extended and detailed discussions of the coenzymes treated below, and several others in addition, are presented in *The Enzymes,* vols. 2 and 3 (see the references). The discussion below complements the previous comments on the mechanism of enzyme action (Chapter 7) and is an introduction to intermediary metabolism as well.

ADENOSINE TRIPHOSPHATE

Biological oxidations are the principal source of energy necessary to drive a multitude of endergonic biological processes. The free energy derived from the transfer of a pair of electrons from a substrate to molecular oxygen or other terminal electron acceptor is largely converted, by reactions not completely understood (Chapter 15), into chemical energy in the form of the energy-rich intermediate adenosine triphosphate (ATP).

Adenosine triphosphate (ATP)

The free energy derived from the ultimate hydrolysis of the pyrophosphate linkages of ATP is then available, in a coupled enzymatic reaction (Chapter 2) to drive an otherwise unfavorable reaction more nearly toward completion. The fundamental role of the hydrolysis of ATP as the driving force for biochemical processes was first clearly recognized by Lipmann. Biochemical processes known to be dependent on ATP hydrolysis include muscle contraction, photosynthesis, bioluminescence, discharge of electric organs, and the biosynthesis of proteins, nucleic acids, complex carbohydrates, and lipids, among many others.

The number of individual enzymes known to require ATP as substrate or energy source is very large, and many such reactions are described in the chapters of this volume dealing with intermediary metabolism. Some of the most important are summarized in Table 8-1 and Fig. 8-1. The free energy of hydrolysis of the pyrophosphate linkages of ATP was discussed in some detail in Chapter 2, and the production of ATP coupled with biological oxidations is considered in Chapter 15. ATP-dependent reactions may be placed into two

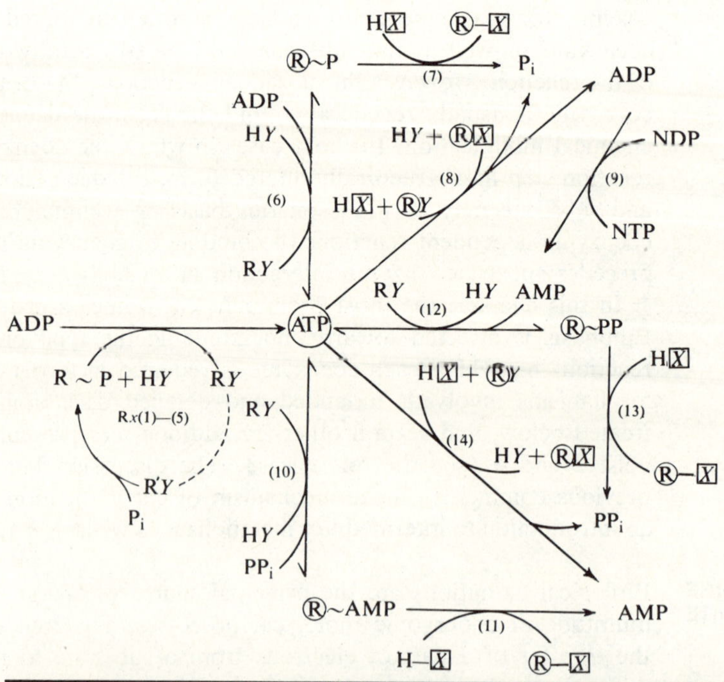

FIGURE 8-1

Schematic representation of the functions of ATP. Numbers in parentheses refer to reactions in Table 8-1. Bonds designated by ~ are characterized by large negative $\Delta G'$ of hydrolysis. N refers to a nucleoside.

broad categories: the first involves the transfer of some portion of the ATP molecule to a suitable acceptor molecule, and, in the second class, the cleavage of ATP provides the driving force for otherwise unfavorable reactions. Reactions in the former category may be further classified as to the nature of the group donated from the ATP molecule: the phosphoryl, pyrophosphoryl, adenyl, and adenosinyl moieties. The site of cleavage of the ATP molecule associated with each of these group transfers, together with representative examples, is presented in Table 8-2. Some additional examples are found in Table 8-1 and throughout the latter portion of this book. Those reactions in which ATP functions as an energy source are catalyzed by the ligases (or synthetases) and deserve further discussion at this point.

All ATP-dependent biosynthetic processes involve the formation of a covalent bond between two substrate molecules coupled with the cleavage of one of the pyrophosphate linkages of ATP. Most reactions of this type appear to be in one of three classes. The first involves the consecutive action of two enzymes, a phosphotransferase or pyrophosphotransferase in the first step and a phosphorylase or pyrophosphorylase in the second. For example, one route for the biosynthesis of inosine from ribose and hypoxanthine is via the initial formation of ribose-1-phosphate followed by formation of the *N*-glycosidic linkage:

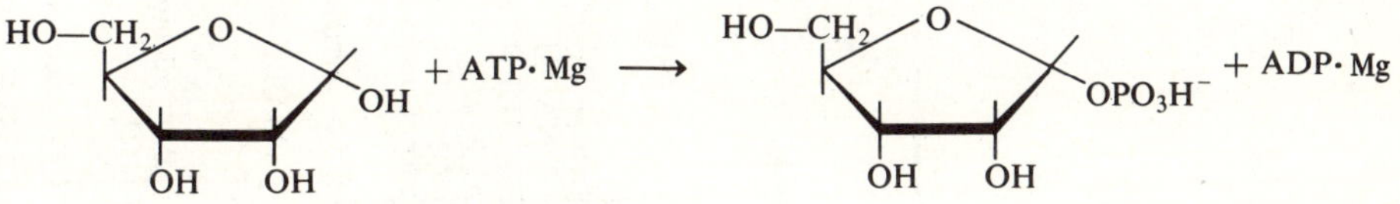

HO—CH₂ (ribose, OH, OH, OH) + ATP·Mg ⟶ HO—CH₂ (ribose, OH, OH, OPO_3H^-) + ADP·Mg (8-1)

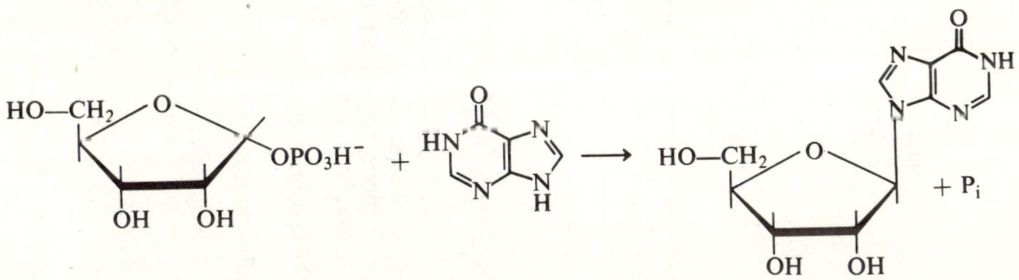

HO—CH₂ (ribose, OH, OH, OPO_3H^-) + hypoxanthine ⟶ HO—CH₂ (inosine, OH, OH) + P_i (8-2)

On the other hand, the biosynthesis of adenosine-5′-phosphate from adenine and ribose-5-phosphate involves an initial pyrophosphorylation, forming 5-phospho-α-D-ribosylpyrophosphate, followed by formation of the *N*-glycosidic bond:

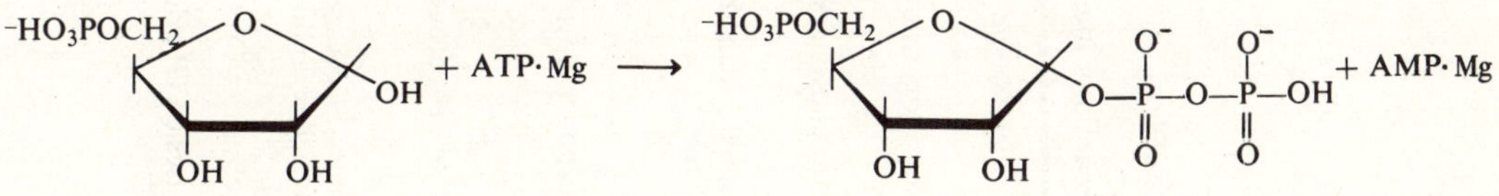

$^-HO_3POCH_2$ (ribose, OH, OH, OH) + ATP·Mg ⟶ $^-HO_3POCH_2$ (ribose, OH, OH, O—P(O⁻)(=O)—O—P(O⁻)(=O)—OH) + AMP·Mg (8-3)

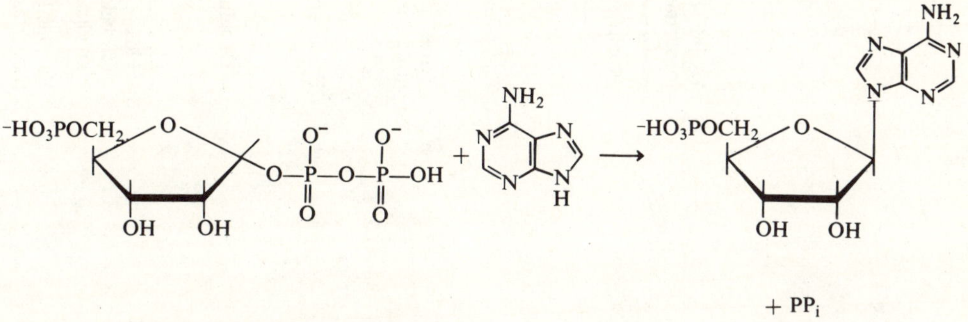

$^-HO_3POCH_2$ (ribose, OH, OH, O—P(O⁻)(=O)—O—P(O⁻)(=O)—OH) + adenine (NH_2) ⟶ $^-HO_3POCH_2$ (adenosine, NH_2, OH, OH) + PP_i (8-4)

TABLE 8-1
Important reactions involving ATP

Reaction or reaction type	*Example of stoichiometry*	*Nature of R group Fig. 8-1*
Oxidative phosphorylation	$A_{red} + B_{ox} + P_i + ADP \rightarrow A_{ox} + B_{red} + ATP$	Unknown
Photosynthetic phosphorylation	$A^1_{red} + B^1_{ox} + ADP + P_i \xrightarrow{h\nu} A^1_{ox} + B^1_{red} + ATP$	Unknown
Triose phosphate dehydrogenase plus glycerate kinase	D-Glyceraldehyde-3-phosphate + P_i + $NAD^+ \rightleftharpoons$ 1,3-diphospho-D-glycerate + NADH 1,3-Diphospho-D-glycerate + ADP $\rightleftharpoons$ 3-phospho-D-glycerate + ATP	$-\overset{O}{\overset{\Vert}{C}}-O-$
Enolase (phosphoenolpyruvate hydratase)	2-Phospho-D-glycerate $\rightleftharpoons$ phosphoenolpyruvate + H_2O Phosphoenolpyruvate + ADP $\rightleftharpoons$ pyruvate + ATP	$=\overset{H}{C}-O-$
α-Oxoglutarate dehydrogenase	α-Oxoglutarate + NAD^+ + CoASH $\rightarrow$ succinyl-SCoA + NADH	
Plus succinate : CoA ligase (GDP) Plus nucleoside diphosphate kinase (cf. below)	Succinyl-SCoA + GDP + $P_i \rightleftharpoons$ succinate + GTP + CoASH GTP + ADP $\rightleftharpoons$ GDP + ATP	$-\overset{O}{\overset{\Vert}{P}}-OH$ with O^- below P
or plus succinate : CoA ligase (ADP)	Succinyl-SCoA + ADP + $P_i \rightleftharpoons$ succinate + ATP + CoASH	
Various kinases (ATP : donor phosphotransferases)	ATP + acetate $\rightleftharpoons$ acetyl phosphate + ADP (+H_2O)	$-\overset{O}{\overset{\Vert}{C}}-O-$
	ATP + creatine $\rightleftharpoons$ creatine phosphate + ADP (+H_2O)	$-\overset{^+NH_2}{\overset{\Vert}{C}}-\overset{H}{N}-$

Acyl transferases (e.g., phosphotransacetylase)	Acetyl phosphate + HSCoA $\rightleftharpoons$ acetyl-SCoA + P_i	$-\overset{O}{\overset{\Vert}{C}}-O-$
Various synthetases (X : R ligases)	ATP + L-glutamate + L-cysteine $\rightleftharpoons$ γ-L-glutamyl-L-cysteine + ADP + P_i	$-\overset{O}{\overset{\Vert}{C}}-O-$
Nucleoside diphosphate kinases (ATP : nucleoside diphosphate phosphotransferase)	ATP + NDP $\rightleftharpoons$ ADP + NTP	P_i
Various nucleotidyl transferases	ATP + FMN $\rightleftharpoons$ FAD + PP_i	$X-\overset{O^-}{\overset{\vert}{\underset{\underset{O}{\Vert}}{P}}}-O-$
Various ligases (synthetases of groups 6.1 and 6.2, first step)	ATP + enz + RCO_2^- $\rightleftharpoons$ $\left[\text{enz}\cdot R-\overset{O}{\overset{\Vert}{C}}-O-AMP\right]$	$\left[\text{enz}\cdot R-\overset{O}{\overset{\Vert}{C}}-O-\right]$
Various ligases (second step)	$\left[\text{enz}\cdot R-\overset{O}{\overset{\Vert}{C}}-O-AMP\right]$ + HX $\rightleftharpoons$ AMP + $R-\overset{O}{\overset{\Vert}{C}}-O-X$	$\left[\text{enz}\cdot R-\overset{O}{\overset{\Vert}{C}}-O-\right]$
ATP : D-ribose-5-phosphate pyrophosphotransferase	ATP + D-ribose-5-phosphate $\rightleftharpoons$ AMP + 5-phospho-α-D-ribosylpyrophosphate (PRPP)	$-O \backslash CH-O- /$
Various NMP pyrophosphorylases, pentosyltransferases	PRPP + orotate $\rightleftharpoons$ orotidine-5′-phosphate + PP_i	$-O \backslash CH-O- /$
Various ligases (group 6.3)	ATP + L-pantoate + β-alanine $\rightleftharpoons$ L-pantothenate + AMP + PP_i	$-\overset{O}{\overset{\Vert}{C}}-O-$
	ATP + xanthosine-5′-phosphate + NH_3 $\rightleftharpoons$ GMP + AMP + PP_i	

TABLE 8-2
Donor functions of ATP

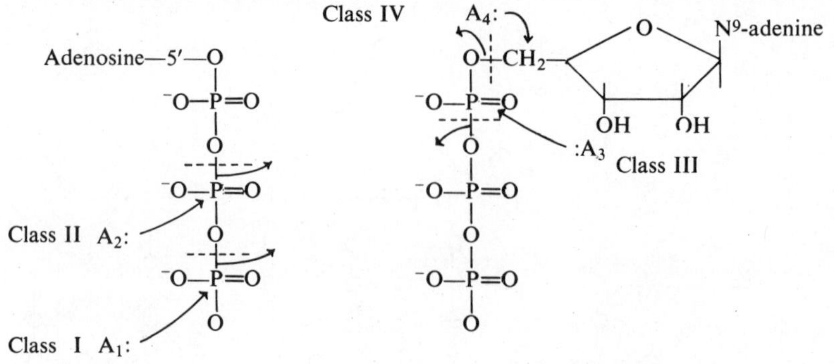

Class IV A_4:

Adenosine—5′—O—P(=O)(O^-)—O—P(=O)(O^-)—O—P(=O)(O^-)—O

O—CH_2—(ribose, OH OH)—N^9-adenine

:A_3 Class III

Class II A_2:

Class I A_1:

Class ATP acts as donor of	Formation of	Example
I Phosphate (phosphotransferases, 2.7–2.7.4)	*A.* Anhydride or equivalent ($\Delta G' \geq 0$ kcal/mole) *B.* Ester, amide, or equivalent ($\Delta G' < 0$ kcal/mole)	Acetate + ATP ⇌ acetyl phosphate + ADP D-glucose + ATP ⇌ D-glucose-6-phosphate + ADP
II Pyrophosphate (pyrophosphotransferases, 2.7.6)	*A.* Acceptor-pyrophosphate	ATP + D-ribose-5-phosphate ⇌ AMP + PRPP
III Adenosine-5′-phosphate	*A.* Enzyme-bound acyl adenylate and transfer to acceptor (6.1.1,6.2.1)	ATP + RCO_2^- + enz ⇌ [enz·RCO_2^-·AMP] + PP_i [enz·R—C(=O)—AMP] + CoASH ⇌ R—C(=O)—SCoA + AMP + enz
	B. Dinucleotide coenzymes (2.7.7)	ATP + FMN ⇌ FAD + PP_i
	C. Polyadenylate (2.7.7)	(nATP) ⇌ $(AMP)_n$ + nPP_i
IV Adenosine	*S*-adenoysl methionine (2.4.2.13)	ATP + methionine ⇌ *S*-adenosyl methionine + trimetaphosphate
V Driving force for reaction	Synthetases, ligases (Class 6)	

The second and third classes of ATP-dependent synthetase reactions involve only one enzyme, rather than two, and are usually concerned with reactions involving carboxyl group activation. They differ in the nature of the products derived from the hydrolytic cleavage of ATP. For example, the enzyme converting acetate to acetyl CoA, a thiol ester (see below), cleaves ATP into AMP and pyrophosphate, and that converting glutamate into glutamine cleaves ATP into ADP and phosphate.

Enzymes in the former class appear to catalyze a two-step reaction sequence involving the intermediate formation of enzyme-bound acyl adenylates.

$$R-\overset{O}{\overset{\|}{C}}-O-\underset{O}{\underset{\|}{\overset{O^-}{\overset{|}{P}}}}-O-CH_2-\text{(ribose with OH, OH; adenine, } NH_2\text{)}$$

Acyl adenylate

In this regard, the acetyl-SCoA synthetase reaction mentioned above has been particularly well studied and appears to proceed as follows:

$$\text{Enz} + CH_3-C^{18}O_2H + \text{ATP} \overset{Mg^{2+}}{\rightleftharpoons} \left[CH_3-\overset{O}{\overset{\|}{C}}-{}^{18}O-\text{AMP}\right]\cdot\text{enz} + PP_i \tag{8-5}$$

$$\left[CH_3-\overset{O}{\overset{\|}{C}}\vdots{}^{18}O-\text{AMP}\right]\cdot\text{enz} + \text{HSCoA} \rightleftharpoons CH_3-\overset{O}{\overset{\|}{C}}-\text{SCoA} + \text{enz} + \text{Ade}-\text{R}-5'-O-\underset{O}{\underset{\|}{\overset{O^-}{\overset{|}{P}}}}-{}^{18}OH \tag{8-6}$$

Evidence for this pathway may be summarized as follows. First, acetyl-AMP is converted enzymatically into acetyl CoA in the presence of CoA [Eq. (8-6)] and into acetate and ATP in the presence of pyrophosphate [reverse of Eq. (8-5)]. Second, incubation of the enzyme with acetate, ATP, and radioactive pyrophosphate, in the absence of CoA, results in the incorporation of radioactivity into ATP. This finding is interpreted in terms of the reversibility of Eq. (8-5), which accounts for the CoA-independent incorporation of pyrophosphate into ATP. Third, radioactive AMP is incorporated into ATP, in a similar type of experiment, but only in the presence of both acetate and CoA. This finding indicates the AMP is liberated only in the final step of the reaction. This incorporation requires the reversibility of the overall reaction sequence. Finally, experiments in which the carboxyl oxygens of acetate are labeled with ^{18}O indicate that one of the acetate oxygens is incorporated into AMP. Added synthetic acetyl-AMP reacts reversibly either with pyrophosphate to yield ATP and acetate or with CoASH to yield AMP and acetyl-SCoA. These findings are most easily interpreted in terms of an intermediate enzyme-bound acetyl adenylate that, in its reaction with CoASH, undergoes cleavage at the acyl-oxygen bond, indicated by the dotted line in Eq. (8-6).

The third class of synthetase reactions, yielding ADP and phosphate as products in reactions catalyzed by single enzymes, are less well understood. Some reactions catalyzed by enzymes in this class are known to occur with the incorporation of one oxygen of the substrate carboxyl group into the liberated phosphate, suggesting the intermediate formation of acyl phosphates:

$$R-CO_2^- + \text{ATP} \rightleftharpoons R-\overset{O}{\overset{\|}{C}}-O-\underset{OH}{\underset{|}{\overset{O}{\overset{\|}{P}}}}-O^- + \text{ADP} \tag{8-7}$$

$$R-\overset{O}{\overset{\|}{C}}-O-\underset{OH}{\underset{|}{\overset{O}{\overset{\|}{P}}}}-O^- + \text{acceptor} \rightleftharpoons R-\overset{O}{\overset{\|}{C}}-\text{acceptor} + P_i \tag{8-8}$$

One important aspect of ATP-dependent reactions is that they all require metal ions, usually magnesium or manganese, for activity. Although metal ions play a host of different roles in enzymatic reactions, their principal function in this case is to form chelates with ATP and other phosphates. A compilation of binding constants for several substrates of interest with Mg^{2+} and Mn^{2+} has been presented in Chapter 2 (page 19). There are several possible structures for metal-ion complexes of ATP. These include the following four:

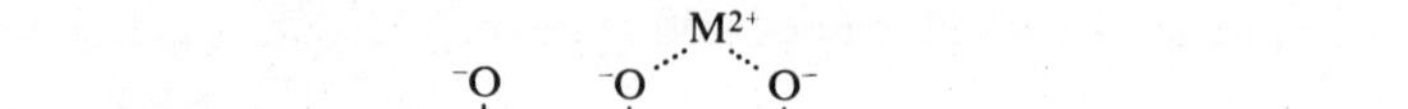

Adenine — Ribose —O—P(=O)(O⁻)—O—P(=O)(O⁻)—O—P(=O)(O⁻)—O⁻ (M^{2+} chelated between the β- and γ-phosphate O⁻)

Adenine — Ribose —O—P(=O)(O⁻)—O—P(=O)(O⁻)—O—P(=O)(O⁻)—O⁻ (M^{2+} chelated between the α- and β-phosphate O⁻)

Adenine — Ribose —O—P(=O)(O⁻)—O—P(=O)(O⁻)—O—P(=O)(O⁻)—O⁻ with ^{2+}M chelated to all three phosphates

Adenine ⋯ ^{2+}M ⋯ triphosphate chain — Ribose (metal chelated to phosphates and adenine)

Although a number of specific explanations can be introduced to account for the dependence of ATP reactions on formation of structures such as these, the central idea is clear. Metal-ion chelation increases the electrophilicity of the ATP molecule and, hence, increases its susceptibility to attack by nucleophilic reagents. Since substantially all the reactions of ATP (Tables 8-1 and 8-2) are nucleophilic in character, this is an important consideration indeed. The point is well illustrated by some corresponding nonenzymatic reactions which exhibit a related dependency on metal ions.

In addition to the reactions involving ATP itself, the adenine nucleotides serve an important function in acting as donors of glycosyl groups for polysaccharide synthesis. Related reactions involving the uridine nucleotides are considered below in more detail.

URIDINE NUCLEOTIDES

One of the most important recent advances in the understanding of metabolic processes has come from the recognition of the role of *uridine nucleotide coenzymes* in carbohydrate metabolism. This discovery derives basically from the isolation of three substances containing uridine diphosphate, amino sugars, and amino acids from penicillin-treated bacteria by Park (1) and from pioneering studies by Leloir. Numerous additional *uridine diphosphate sugars* have subsequently been isolated from a variety of natural sources and are probably of universal occurrence in nature. The structure of these compounds is basically the same, differing only in the nature of the sugar moiety, and involves a bond between the terminal phosphate of uridine diphosphate and the aldehydic carbon of the sugar:

General structure for UDP sugars

Biosynthetically, UDP-sugars are formed in enzymatic reactions between UTP and the sugar-1-phosphate (2):

$$\text{Sugar-1-phosphate} + \text{UTP} \rightleftharpoons \text{UDP-sugar} + \text{pyrophosphate} \tag{8-9}$$

Enzymes catalyzing such reactions are called *UDP-sugar pyrophosphorylases* or *uridyltransferases*. A variety of such enzymes exist that catalyze the synthesis of a spectrum of UDP-sugars.

The role of UDP-sugar coenzymes in enzymatic catalysis is twofold. First, such coenzymes are involved in a variety of glycosyl transfer reactions. For example, UDP-glucose (UDPG) is involved in glucosyl group-transfer reactions with fructose as acceptor, yielding sucrose; with anthranilic acid as acceptor, yielding *o*-aminobenzoyl glucoside; with a terminal glucose of a glycogen molecule as acceptor, lengthening the glucose chain:

$$\text{UDPG}\begin{cases} + \text{fructose} \rightarrow \text{sucrose} + \text{UDP} \\ + \text{anthranilic acid (}NH_2\text{, }CO_2H\text{)} \rightarrow \text{o-aminobenzoyl–}C(=O)\text{–O–glucose} + \text{UDP} \\ + (\text{glucose})_n \rightarrow (\text{glucose})_{n+1} + \text{UDP} \end{cases} \tag{8-10}$$

Similarly, UDP-acetylglucosamine and UDP-glucuronic acid are intermediates in the synthesis of the complex polysaccharide, hyaluronic acid (page 550) (3). In glycosyl group-transfer reactions UDP-sugars are analogous to acyl adenylates, which, as noted above, are involved in acyl group-transfer reactions also involving a variety of acceptors.

The second general type of reaction involving UDP-sugars is characterized by chemical transformation of the sugar itself. For example, UDPG-4-epimerase (UDP-glucose-4-epimerase) catalyzes the interconversion of UDPG and UDP-galactose (4):

$$\text{UDP-glucose} \underset{}{\overset{NAD^+}{\rightleftharpoons}} \text{UDP-galactose} \tag{8-11}$$

Other examples include the nicotinamide adenine dinucleotide–dependent (see below) oxidation of UDPG to UDP-glucuronic acid:

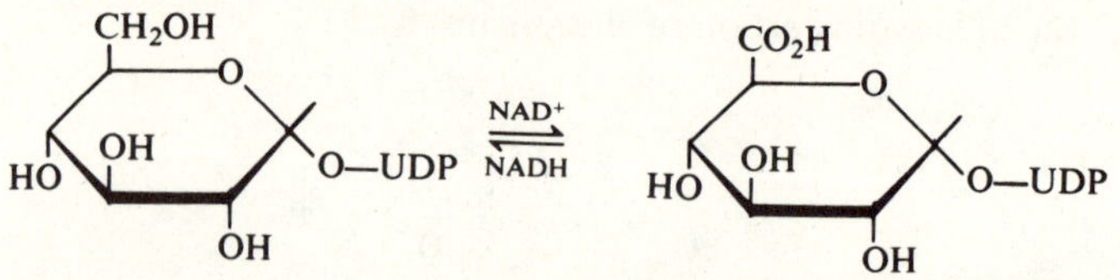

CH_2OH … O—UDP $\underset{NADH}{\overset{NAD^+}{\rightleftharpoons}}$ CO_2H … O—UDP (8-12)

and the decarboxylation of UDP-glucuronic acid, yielding UDP-xylose:

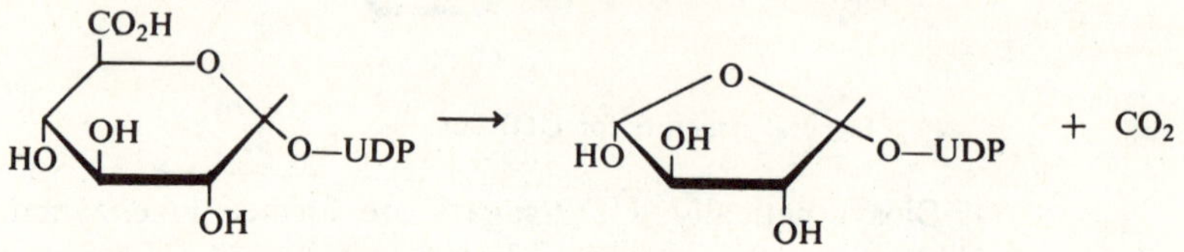

CO_2H … O—UDP ⟶ O—UDP + CO_2 (8-13)

Work in recent years has established that the types of reactions of the UDP-sugars discussed are shared by other nucleoside diphosphate sugars as well. These include particularly the derivatives of ADP, GDP, and TDP. The function of these species in the biosynthesis of polysaccharides is discussed on page 550.

CYTIDINE NUCLEOTIDES

It was noted above that adenosine nucleotides and uridine nucleotides are involved as the most prominent group-transfer agents in reactions with substrates at the acyl and aldehyde level of oxidation, respectively. At the alcohol level of oxidation, group-transfer reactions frequently involve *cytidine nucleotides.* This coenzymatic function of cytidine nucleotide derivatives was first recognized by Kennedy and Weiss, who discovered that CDP-choline and CDP-ethanolamine are essential intermediates in the biosynthesis of lecithins, an important class of lipids (5,6):

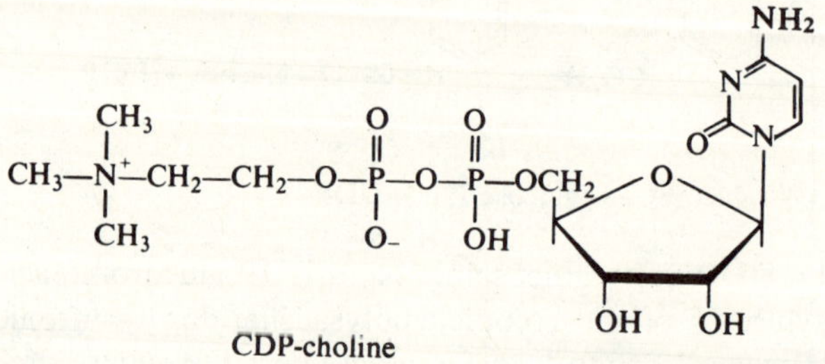

$(CH_3)_3N^+—CH_2—CH_2—O—P(O)(O^-)—O—P(O)(OH)—OCH_2$-ribose(OH)(OH)-cytosine(NH_2)

CDP-choline

$H_3N^+—CH_2—CH_2—O—P(O)(O^-)—O—P(O)(OH)—OCH_2$-ribose(OH)(OH)-cytosine(NH_2)

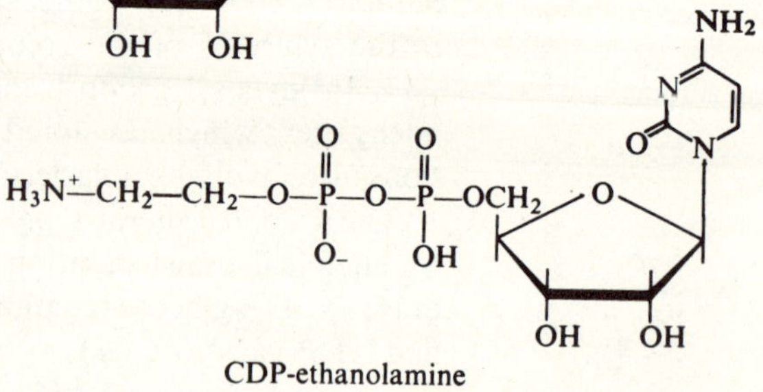

CDP-ethanolamine

CDP-alcohols are formed biosynthetically from CTP and alcohol-1-phosphate (7):

$$\text{CTP + alcohol-phosphate} \rightleftharpoons \text{CDP-alcohol + pyrophosphate} \quad (8\text{-}14)$$

These coenzymes transfer an alkyl phosphate group to a variety of acceptors. For example, CDP-choline reacts with α,β-diglycerides to yield lecithin and CMP:

$$\text{CDP-choline} + \begin{array}{l} CH_2O{-}\overset{\overset{O}{\|}}{C}{-}R \\ | \\ HC{-}O{-}\overset{\overset{O}{\|}}{C}{-}R \\ | \\ CH_2OH \end{array} \rightarrow \begin{array}{l} CH_2O{-}\overset{\overset{O}{\|}}{C}{-}R \\ | \\ HC{-}O{-}\overset{\overset{O}{\|}}{C}{-}R \\ | \\ H_2C{-}O{-}\overset{\overset{O}{\|}}{\underset{\underset{O^-}{|}}{P}}{-}OCH_2CH_2{-}N^+(CH_3)_3 \end{array} + \text{CMP} \qquad (8\text{-}15)$$

Note that the transfer reactions involving adenosine and uridine nucleotides occur with cleavage of acyl-phosphate and glucosyl-phosphate bonds, respectively, whereas those involving cytidine nucleotides occur with cleavage of a pyrophosphate linkage.

GUANOSINE AND INOSINE NUCLEOTIDES

Relatively few coenzymatic functions for guanosine or inosine nucleotides have been discovered. The following are the most important: (1) GTP reacts with mannose-1-phosphate, in a manner completely reminiscent of the reaction between UTP and glucose-1-phosphate, to yield GDP-mannose as an intermediate for the transfer of mannosyl residues; (2) GTP (or ITP) is formed in the substrate-level phosphorylation accompanying the oxidation of α-oxoglutarate by the following sequence of reactions:

$$\alpha\text{-Oxoglutarate} + \text{NAD}^+ + \text{CoASH} \rightarrow \text{succinyl-SCoA} + \text{NADH} + CO_2 \qquad (8\text{-}16)$$

$$\text{Succinyl SCoA} + \text{GDP(IDP)} + P_i \rightleftharpoons \text{succinate} + \text{CoASH} + \text{GTP(ITP)} \qquad (8\text{-}17)$$

and (3) GTP (or ITP) is consumed in the oxalacetic carboxykinase reaction that is responsible for the interconversion of oxalacetate and phosphoenolpyruvate:

$$\text{Oxalacetate} + \text{GTP} \rightleftharpoons \text{phosphoenolpyruvate} + CO_2 + \text{GDP} \qquad (8\text{-}18)$$

In addition, GTP acts as an energy donor in the formation of adenylosuccinate from IMP and L-aspartate, a key step in the biosynthesis of AMP (page 823). The important role of GTP in the biosynthesis of proteins is fully discussed in Chapter 21.

COENZYME A

Coenzyme A is the most prominent acyl group transfer coenzyme in living systems. Its universal occurrence, the large number of enzymatic reactions in which it is involved, and the variety of reaction types in which its derivatives are concerned testify to its importance.

Coenzyme A was originally identified as a heat-stable cofactor required for certain biological acetylations (8,9) and was subsequently implicated as a cofactor for the incorporation of acetate into acetoacetate and citrate (10,11). The structure of coenzyme A was elucidated in the laboratories of Lipmann, Snell, and Baddiley. This molecule is quite complex structurally and contains a multiplicity of possible functional groups.

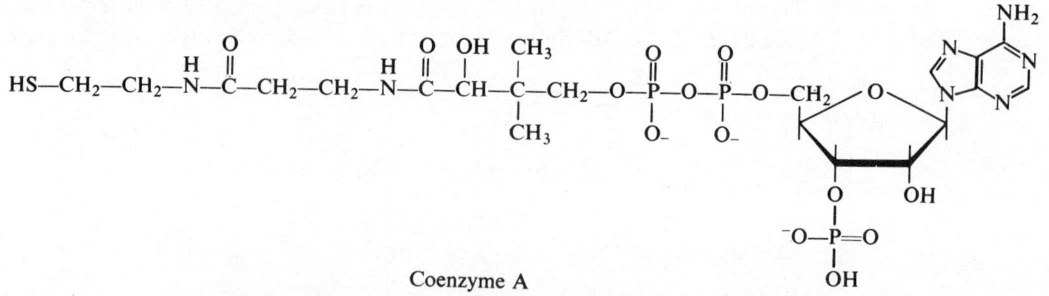

$$\text{HS—CH}_2\text{—CH}_2\text{—}\overset{\text{H}}{\text{N}}\text{—}\overset{\text{O}}{\overset{\|}{\text{C}}}\text{—CH}_2\text{—CH}_2\text{—}\overset{\text{H}}{\text{N}}\text{—}\overset{\text{O}}{\overset{\|}{\text{C}}}\text{—}\overset{\text{OH}}{\overset{|}{\text{CH}}}\text{—}\underset{\text{CH}_3}{\overset{\text{CH}_3}{\text{C}}}\text{—CH}_2\text{—O—}\underset{\text{O}^-}{\overset{\text{O}}{\overset{\|}{\text{P}}}}\text{—O—}\underset{\text{O}^-}{\overset{\text{O}}{\overset{\|}{\text{P}}}}\text{—O—CH}_2\text{—(ribose-adenine; 3'-O—P(=O)(O}^-\text{)OH)}$$

Coenzyme A

The isolation of acetyl coenzyme A from yeast by Lynen and his associates and the identification of this substance as a thiol ester demonstrate that the terminal sulfhydryl group of coenzyme A is the reactive site of this molecule in biochemical reactions (12,13). All subsequently identified acyl derivatives of coenzyme A have proved to be thiol esters as well. Hereafter we shall abbreviate coenzyme A as CoASH.

Acyl derivatives of CoASH, generally formed in ATP-dependent synthetase reactions, undergo a variety of metabolic reactions that may be classified into four groups: (1) reactions involving the attack of nucleophilic reagents on the acyl carbon atom, with transfer of the acyl function to the attacking reagent and release of CoASH; (2) reactions involving additions to those species which have an olefinic linkage in the acyl moiety; (3) reactions involving condensation at the α-carbon of the acyl-SCoA (such reactions almost certainly proceed through the formation of the carbanion at this position); and (4) reactions involving acyl group interchange. Examples of reactions in each of these classes are given in Table 8-3 and are summarized schematically in Fig. 8-2.

The choice of thiol esters rather than oxygen esters as substrates for nucleophilic reactions at the carbonyl carbon atom, for addition reactions at the β carbon atom of α,β-unsaturated substrates, and for condensation reactions at a carbon α to a carbonyl group is a natural one. In each case thiol esters are more reactive than oxygen esters. This fact may be understood in terms of the energy diagrams shown in Fig. 8-3. As shown in Fig. 8-3*a*, oxygen esters are kinetically more stable than thiol esters in reactions at the acyl carbon atom as a result of greater stabilization of the ground state by resonance interactions of the sort

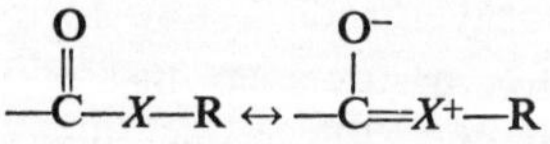

$$\text{—}\overset{\text{O}}{\overset{\|}{\text{C}}}\text{—}X\text{—R} \leftrightarrow \text{—}\overset{\text{O}^-}{\overset{|}{\text{C}}}\text{=}X^+\text{—R}$$

Such interactions are largely absent in thiol esters but important in oxygen esters; that is, ΔE_1 is principally a measure of the difference of resonance energies of oxygen and thiol esters. To the extent that the acyl carbon atom has become tetrahedral in the transition state, interactions of this sort are lost and, consequently, the energies of the transition states for reactions of oxygen and thiol esters are not greatly different; $\Delta E_1 > \Delta E_2$. The net result is that the energy of activation for reactions of oxygen esters, $\Delta E_{oe}{}^*$, is substantially greater than that for reactions of thiol esters, $\Delta E_{te}{}^*$. Simply stated, oxygen esters are less reactive than thiol esters, since (1) they are stabilized to a greater extent by resonance and (2) such resonance is largely lost in the transition state. These arguments apply with equal force to addition reactions at the β-carbon atom of these substrates, since such reactions are just vinylogous additions to the carbonyl group. Figure 8-3*b* is an energy diagram for

TABLE 8-3
Acyl-SCoA reactions

Attacking reagent	*Example*	*Stoichiometry*
Class I. Reactions involving nucleophilic attack at the acyl carbon atom		
Phosphate	Succinyl CoA synthetase (Succinate : CoA Ligase)	Succinyl-SCoA + P_i + GDP $\rightleftharpoons$ succinate + HSCoA + GTP
Amine	Glucosamine acetyl transferase	Acetyl-SCoA + D-2-glucosamine $\rightleftharpoons$ *N*-acetyl-D-glucosamine + HSCoA
Water	Acetyl-SCoA hydrolase	Acetyl-SCoA + H_2O $\rightleftharpoons$ acetate + HSCoA
Alcohol	Glycerophosphate acetyl transferase	2 fatty acyl-SCoA + L-α-glycerophosphate $\rightleftharpoons$ diglyceride phosphate + 2 HSCoA
Thiol	Lipoic transacetylase	Acetyl-SCoA + dihydrolipoate $\rightleftharpoons$ HSCoA + 6-*S*-acetyl dihydrolipoate
Hydride ion	Acyl-SCoA reductases	Acyl-SCoA + NADH + H^+ $\rightleftharpoons$ aldehyde + HSCoA + NAD^+
Carbanion	β-Ketothiolase	Acetoacetyl-SCoA + HSCoA $\rightleftharpoons$ 2 acetyl-SCoA
Class II. Reactions involving additions to unsaturated acyl derivatives of CoASH		
Water	Enoyl-SCoA hydratase	α, β-*trans*-Enoyl-SCoA + H_2O $\rightleftharpoons$ L-β-hydroxyacyl-SCoA
Class III. Reactions involving condensation at the α-carbon of the acyl-SCoA		
Acetyl-SCoA (carbanion)	Citrate synthetase (condensing enzyme)	Oxalacetate + acetyl-SCoA $\rightleftharpoons$ citrate + HSCoA
Class IV. Reactions involving acyl group interchange		
RCO_2^-	Acetoacetic-SCoA transferase	Acetoacetyl-SCoA + succinate $\rightleftharpoons$ acetoacetate + succinyl-SCoA

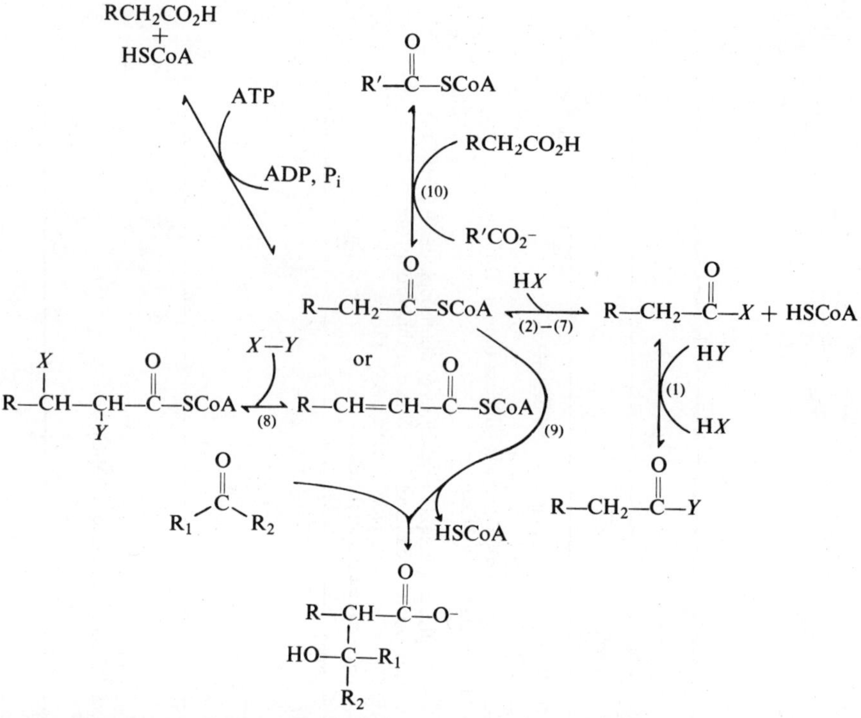

FIGURE 8-2
Schematic representation of acyl-SCoA reaction pathways. Numbers in parentheses refer to reactions in Table 8-3.

the formation of carbanions from oxygen and thiol esters. The overall equilibrium constant for formation of the thiol ester carbanion is greater than that for formation of a similar structure from an oxygen ester, since the difference in resonance energy of the reactants, ΔE_1, is substantially larger than the difference in resonance energy of the products, $\Delta E_2'$. The resonance energy of the thiol ester carbanion is nearly as large as that for the oxygen ester carbanion, since the unshared electron pair on carbon fully satisfies the electron-delocalization capacity of the adjacent carbonyl function:

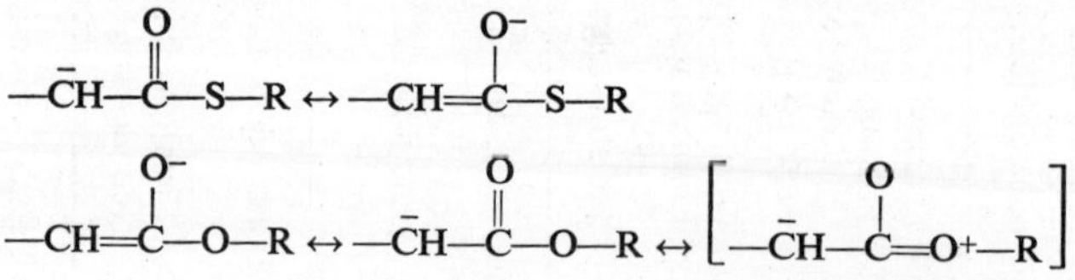

$$-\bar{C}H-C(=O)-S-R \leftrightarrow -CH=C(O^-)-S-R$$

$$-CH=C(O^-)-O-R \leftrightarrow -\bar{C}H-C(=O)-O-R \leftrightarrow \left[-\bar{C}H-C(O^-)=O^+-R\right]$$

Not an important contributing structure

That is, the structure analogous to that which is important for stabilization of oxygen esters but not thiol esters assumes much less importance in the carbanion.

LIPOIC ACID

Lipoic acid was initially isolated in crystalline form from a water-insoluble extract of beef liver (14). This isolation is one of the more spectacular achievements in the purification of natural products, since only 30 mg of crystalline lipoic acid was obtained from 10 tons of the liver residue. The structure of lipoic acid was subsequently demonstrated to be 1,2-dithiolane-3-valeric acid.

S–S $CH_2CH_2CH_2CH_2CO_2H$ H

Lipoic acid

The basic coenzymatic function of lipoic acid appears to be involved with the oxidative decarboxylation of α-keto acids such as the conversion of pyruvate to acetate and carbon dioxide. Reactions of this type are catalyzed by multienzyme units and involve many steps and several coenzymes (see Chapters 11 and 14). The role of lipoic acid in the overall process is one of oxidation and acyl group transfer. The initial decarboxylation reaction yields a thiamine pyrophosphate aldehyde intermediate (see below). In the subsequent step the acetaldehyde is transferred to lipoic acid, with the formation of 6-*S*-acetyl dihydrolipoic acid:

R–$\overset{+}{N}$ S; H–O–C–H (R) + lipoic acid (S–S, R) → [R–$\overset{+}{N}$ S; H–O–C(R)–S–(R) HS] → HS S–C(=O)–R (R) + R–$\overset{+}{N}$ S **(8-19)**

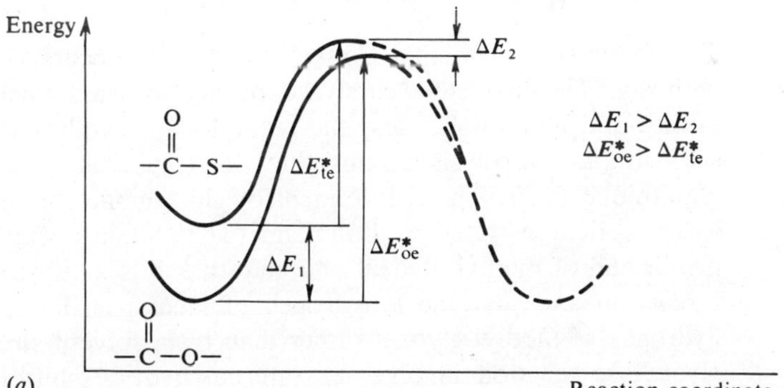

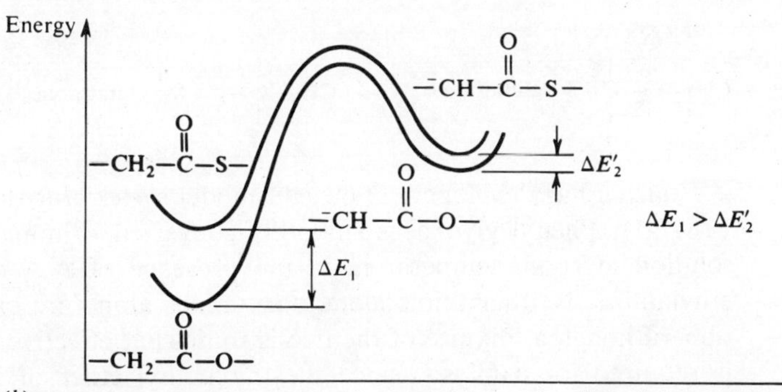

FIGURE 8-3

Energy diagrams for (*a*) attack of nucleophilic reagents on oxygen and thiol esters and (*b*) carbanion formation from oxygen and thiol esters. See text for full explanation.

and the acetyl group is then transferred to HSCoA, yielding acetyl-SCoA and dihydrolipoic acid. The oxidized form of lipoic acid is regenerated by a specific dihydrolipoic dehydrogenase. Lipoic acid is covalently bound to the dehydrogenation complex involved in the overall reaction through an amide linkage between the carboxyl function of the coenzyme and an ε-amino group of a lysine residue. The formation and hydrolysis of this amide linkage are catalyzed by a specific ATP-dependent synthetase and a hydrolase, respectively.

GLUTATHIONE

The structure of glutathione, a tripeptide of universal occurrence in living material, was established in 1929 by Hopkins as γ-L-glutamyl-L-cysteinylglycine:

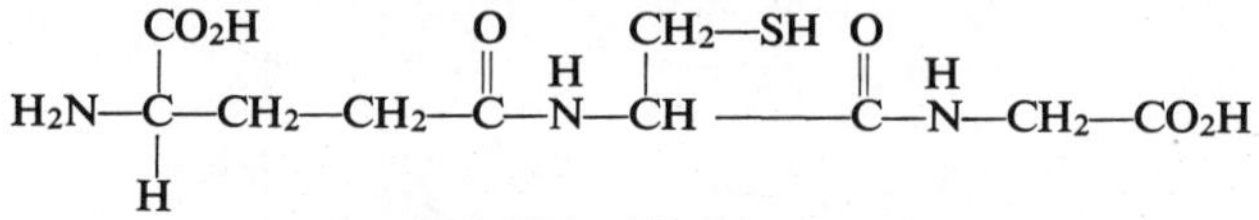

$$H_2N{-}\underset{\underset{H}{|}}{\overset{\overset{CO_2H}{|}}{C}}{-}CH_2{-}CH_2{-}\overset{\overset{O}{\|}}{C}{-}\overset{H}{N}{-}\overset{\overset{CH_2-SH}{|}}{CH}{-}\overset{\overset{O}{\|}}{C}{-}\overset{H}{N}{-}CH_2{-}CO_2H$$

Glutathione

Although a large number of biological and biochemical functions have been suggested for glutathione since its discovery, the number of instances in which glutathione is known to function as a coenzyme are rather few. Of these, the most widely studied is the glyoxylase reaction:

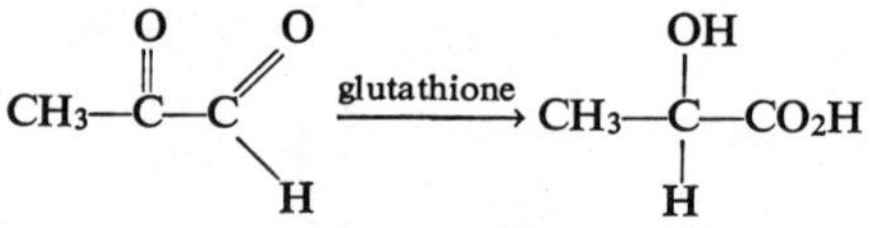

$$CH_3{-}\overset{\overset{O}{\|}}{C}{-}\overset{O}{\underset{H}{C}} \xrightarrow{\text{glutathione}} CH_3{-}\underset{\underset{H}{|}}{\overset{\overset{OH}{|}}{C}}{-}CO_2H \qquad (8\text{-}20)$$

The conversion of methyl glyoxal into lactate occurs via a two-step reaction pathway. The first step, catalyzed by glyoxylase I, yields the thiol ester of lactate and glutathione; and the second step, catalyzed by glyoxylase II, results in the hydrolysis of this thiol ester, yielding lactate and regenerating glutathione (15). The requirement of glutathione for the glyoxylase reaction was first demonstrated by Lohmann (16). Studies establishing that the thiohemiacetal of methyl glyoxal and glutathione is a substrate for the glyoxylase I reaction and that the α-hydrogen of lactate is derived from the carbonyl hydrogen of methyl glyoxal rather than from solvent strongly suggest that the glyoxylase reaction involves an internal hydride shift in the thiohemiacetal intermediate (17,18):

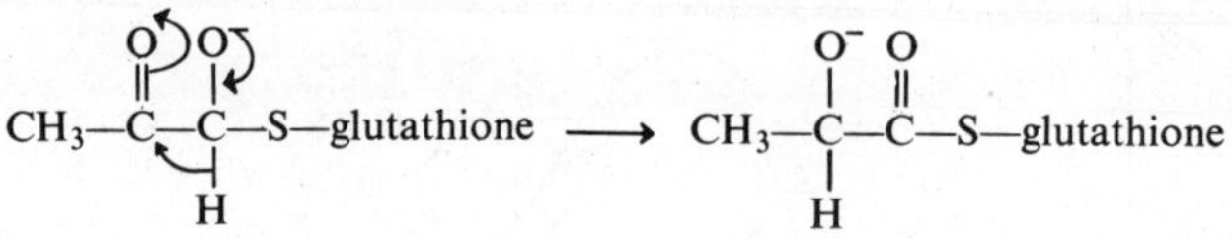

$$CH_3{-}\overset{\overset{O}{\|}}{C}{-}\underset{\underset{H}{|}}{\overset{\overset{O^-}{|}}{C}}{-}S{-}\text{glutathione} \longrightarrow CH_3{-}\underset{\underset{H}{|}}{\overset{\overset{O^-}{|}}{C}}{-}\overset{\overset{O}{\|}}{C}{-}S{-}\text{glutathione} \qquad (8\text{-}21)$$

Franzen has developed an elegant model system for the glyoxylase reaction (18–20). Phenyl glyoxal is smoothly converted to mandelic acid in aqueous solution at room temperature in the presence of *N,N*-dimethyl-β-mercaptoethylamine. Neither thiols alone nor amines alone are catalysts for this reaction, although a mixture of the two is somewhat effective. The free thiol group is required for catalysis, since the thiomethyl ether of the catalyst above is inactive. Like wise, the trimethylammonium derivative of the same catalyst is devoid of catalytic activity, indicating the necessity of a basic site on the catalyst molecule. The model reaction, as is the case with glyoxylase I, proceeds with internal hydrogen transfer. On the basis of this information, the following mechanism appears reasonable for the model system:

$$\phi\text{—C(=O)—C(=O)H} + \text{HS—CH}_2\text{—CH}_2\text{—N(CH}_3)_2 \rightleftharpoons \phi\text{—C(=O)—CH(O}^-)\text{—S—CH}_2\text{—CH}_2\text{—N}^{+}\text{H(CH}_3)_2$$

$$\rightleftharpoons \phi\text{—CH(O}^-)\text{—C(=O)—S—CH}_2\text{—CH}_2\text{—}\overset{+}{\text{N}}\text{H(CH}_3)_2 \rightleftharpoons \phi\text{—CH(OH)—CO}_2\text{H} + \text{HS—CH}_2\text{—CH}_2\text{—N(CH}_3)_2 \qquad (8\text{-}22)$$

In the presence of optically active thiolamine catalysts, a small amount of optical activity may be induced into the mandelic acid product. These findings suggest that a basic site is present at the active site of glyoxylase I.

Glutathione also functions as a coenzyme in the formaldehyde dehydrogenase reaction:

$$H_2C{=}O + NAD^+ + H_2O \rightarrow HCO_2H + NADH + H^+ \qquad (8\text{-}23)$$

and in the maleyl acetoacetate isomerase reaction:

$$\text{HO}_2\text{C—CH=CH—CO—CH}_2\text{—CO—CH}_2\text{—CO}_2\text{H}\ (cis) \rightleftharpoons \text{HO}_2\text{C—CH=CH—C(OH)=CH—C(OH)=CH—CO}_2\text{H} \qquad (8\text{-}24)$$

PYRIDOXAL PHOSPHATE

Pyridoxal phosphate is the most striking of the known coenzymes in terms of the multiplicity of different enzymatic reactions that are dependent on its presence. Nearly all these reactions are concerned with transformations of amino acids. (See, however, phosphorylase, page 501). It is not surprising, then, to learn that numerous nutritional studies in animals and microorganisms have established that a deficiency of pyridoxol (pyridoxine), a biochemical precursor of pyridoxal phosphate, results in many lesions in protein metabolism. The development of our understanding of pyridoxal phosphate–dependent reactions began in 1934 with the identification of a nutritional factor required for the cure of a specific dermatitis in young rats that developed on a deficient diet. Gyorgy called this nutritional factor vitamin B_6, a term presently used to refer to the family of substances of biochemical interest that are structurally related to pyridoxal phosphate. Vitamin B_6 was partially characterized chemically by Birch and Gyorgy in 1936 and was subsequently isolated in crystalline form in several laboratories. Its structure was established as pyridoxol, 3-hydroxy-4,5-dihydroxymethyl-2-methyl pyridine.

Following the discovery that the urine of animals that had ingested pyridoxol contained metabolites of this substance that were a great deal more effective than pyridoxol as a growth factor for the lactic acid bacteria, the metabolites were characterized by Snell in 1944 as pyridoxal and pyridoxamine (21). It has been subsequently demonstrated that the biocatalytically active form of pyridoxal (and in one case pyridoxamine) is the phosphorylated derivative.

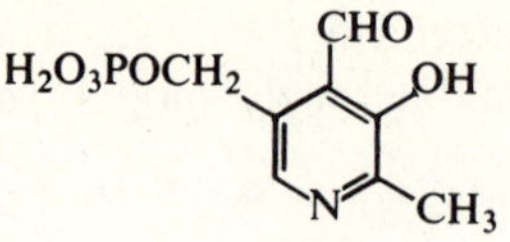

$$\text{Pyridoxal phosphate: } H_2O_3POCH_2\text{–pyridine ring with CHO, OH, }CH_3$$

Pyridoxal phosphate

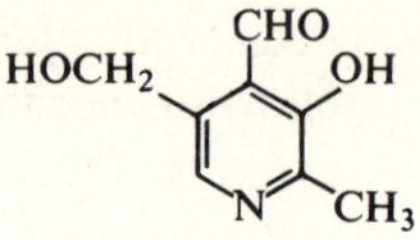

$$\text{Pyridoxal: } HOCH_2\text{–pyridine ring with CHO, OH, }CH_3$$

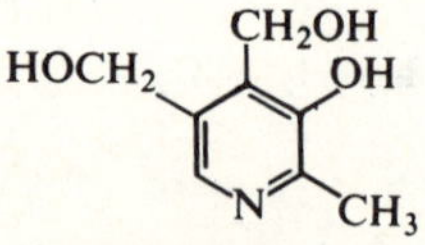

Pyridoxal

$$\text{Pyridoxol: } HOCH_2\text{–pyridine ring with }CH_2OH\text{, OH, }CH_3$$

Pyridoxol

$$\text{Pyridoxamine: } HOCH_2\text{–pyridine ring with }CH_2NH_2\text{, OH, }CH_3$$

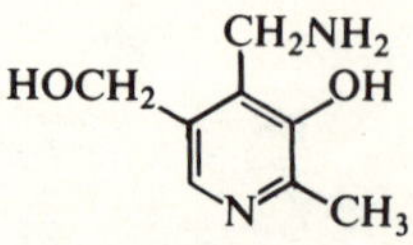

Pyridoxamine

As noted above, pyridoxal phosphate–dependent enzymes catalyze a striking variety of distinct reactions. A comprehensive compilation of these reactions has been assembled by Braunstein. They include the racemization of optically active amino acids such as alanine,

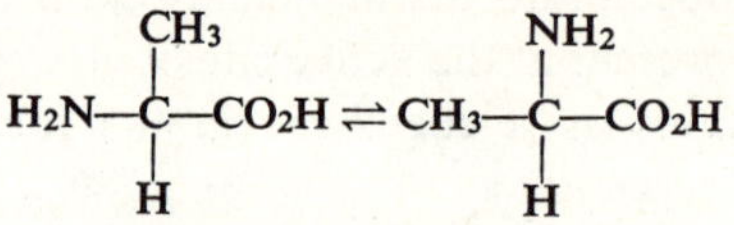

$$H_2N\text{—}C(CH_3)(H)\text{—}CO_2H \rightleftharpoons CH_3\text{—}C(NH_2)(H)\text{—}CO_2H \tag{8-25}$$

transaminations such as the interconversion of glutamate and oxalacetate with aspartate and α-oxoglutarate,

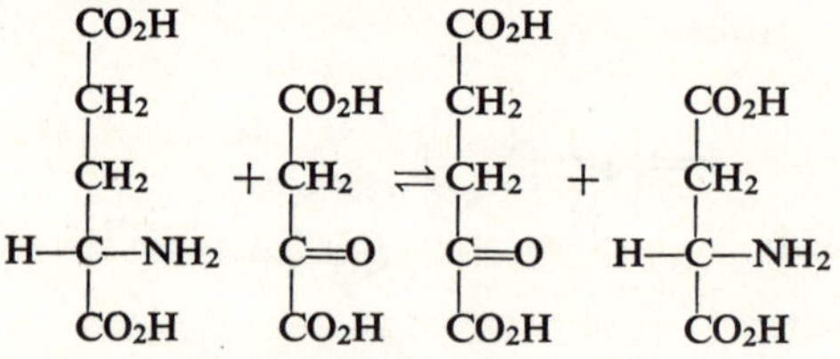

$$HO_2C\text{—}CH_2\text{—}CH_2\text{—}CH(NH_2)\text{—}CO_2H + HO_2C\text{—}CH_2\text{—}C(=O)\text{—}CO_2H \rightleftharpoons HO_2C\text{—}CH_2\text{—}CH_2\text{—}C(=O)\text{—}CO_2H + HO_2C\text{—}CH_2\text{—}CH(NH_2)\text{—}CO_2H \tag{8-26}$$

the decarboxylation of amino acids, including aspartate,

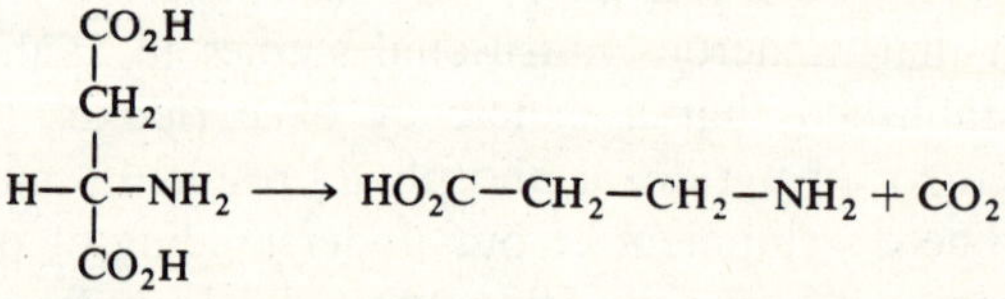

$$HO_2C\text{—}CH_2\text{—}CH(NH_2)\text{—}CO_2H \longrightarrow HO_2C\text{—}CH_2\text{—}CH_2\text{—}NH_2 + CO_2 \tag{8-27}$$

α,β-elimination reactions such as the dehydration of serine,

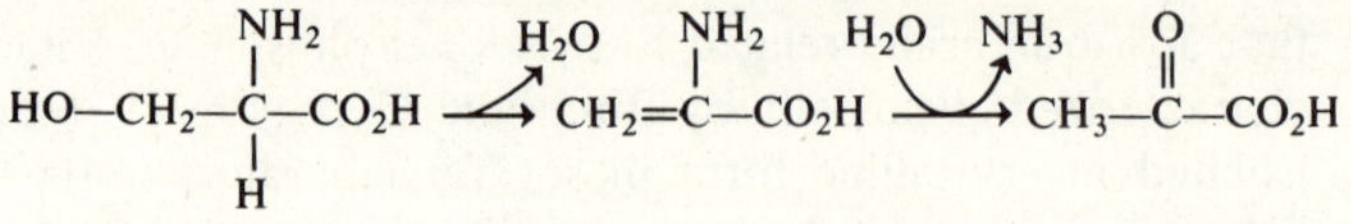

$$HO\text{—}CH_2\text{—}C(NH_2)(H)\text{—}CO_2H \xrightarrow{-H_2O} CH_2{=}C(NH_2)\text{—}CO_2H \xrightarrow{H_2O,\ -NH_3} CH_3\text{—}C(=O)\text{—}CO_2H \tag{8-28}$$

β,γ-elimination reactions such as the desulfuration of homocysteine,

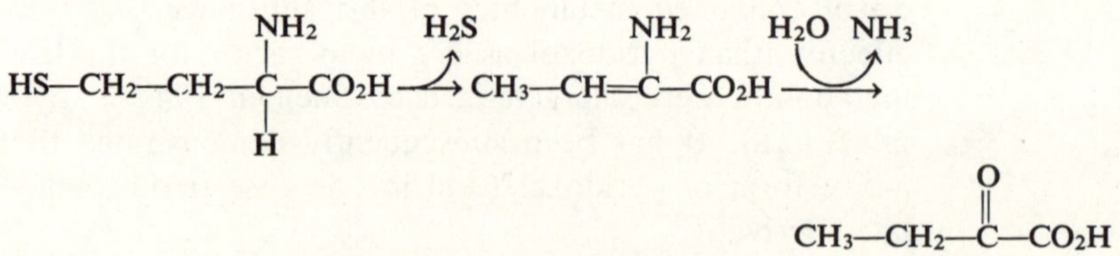

$$HS\text{—}CH_2\text{—}CH_2\text{—}C(NH_2)(H)\text{—}CO_2H \xrightarrow{-H_2S} CH_3\text{—}CH{=}C(NH_2)\text{—}CO_2H \xrightarrow{H_2O,\ -NH_3} CH_3\text{—}CH_2\text{—}C(=O)\text{—}CO_2H \tag{8-29}$$

the synthesis of tryptophan from serine and indole,

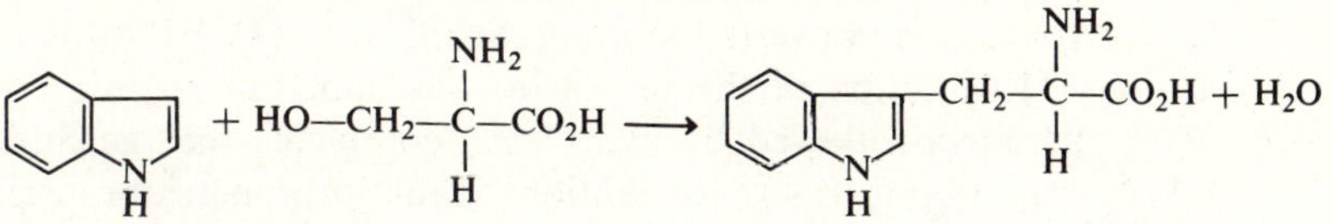

$$\text{Indole} + HO\text{—}CH_2\text{—}\underset{\displaystyle H}{\overset{\displaystyle NH_2}{C}}\text{—}CO_2H \longrightarrow \text{Indole}\text{—}CH_2\text{—}\underset{\displaystyle H}{\overset{\displaystyle NH_2}{C}}\text{—}CO_2H + H_2O \quad (8\text{-}30)$$

and the dealdolization of serine, yielding glycine and formaldehyde,

$$HOCH_2\text{—}\underset{\displaystyle H}{\overset{\displaystyle NH_2}{C}}\text{—}CO_2H + FH_4 \rightleftharpoons NH_2\text{—}CH_2\text{—}CO_2H + (HCHO)\text{—}FH_4 \quad (8\text{-}31)$$

The last reaction also requires the participation of an additional coenzyme, tetrahydrofolic acid (FH_4), which acts as an acceptor for the formaldehyde moiety (see below).

A common feature of pyridoxal phosphate–dependent amino acid transformations is the initial combination of these two components to form a Schiff base:

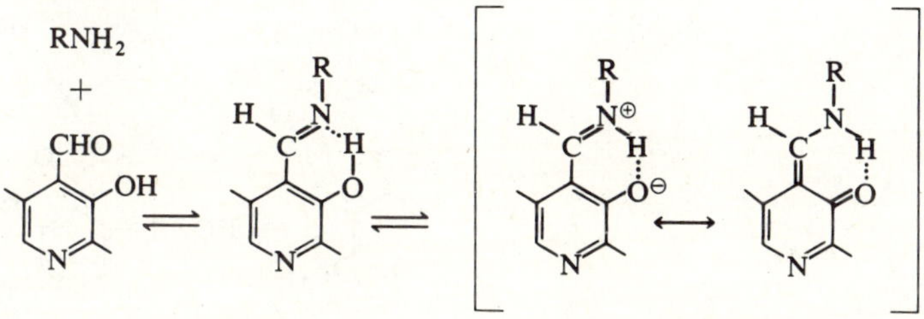

I II

Such Schiff bases exhibit characteristic spectra with absorption maxima near 330 and 410 nm. Form I of the Schiff base is responsible for the former maximum and form II for the latter, which imparts a distinct yellow color to these species (22,23).

Regardless of whether one considers form I or II of the Schiff base, the crucial point for an understanding of the mechanism of pyridoxal-dependent reactions is that the protonated structure provides a conjugated system of double bonds extending from the reaction site to the strongly electrophilic pyridinium nitrogen atom. This fact provides for facilitation of the displacement of a pair of electrons from three sites adjacent to the α-carbon atom of the amino acid:

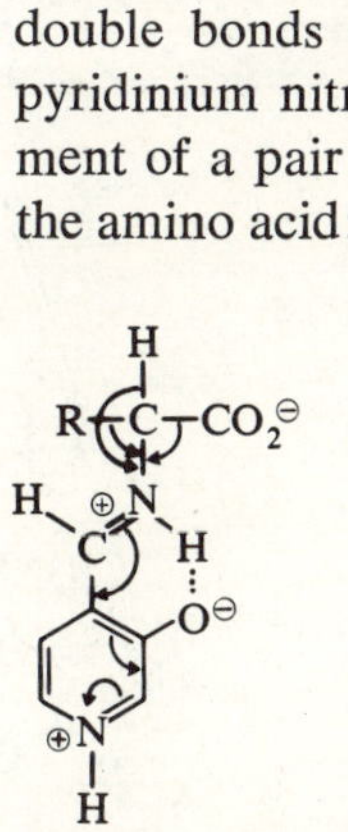

Let us view the consequences of displacing each of these three electron-pair types individually to see how they account for the observed reactions which are dependent on pyridoxal phosphate. In Fig. 8-4 it is shown how the ionization of the proton attached to the α-carbon of serine, as an example, leads to *racemization, transamination,* and *dehydration* of this amino acid. Ionization converts the initial Schiff base (IV) into its tautomer (V).

Readdition of the proton to the same site in this optically inactive intermediate, followed by hydrolysis, completes the racemization reaction (pathway 1, Fig. 8-4). Readdition of this proton to the original carbonyl carbon atom, followed by hydrolysis, yields the transamination products, pyridoxamine and the keto acid (pathway 2, Fig. 8-4). The overall transamination reaction, as indicated in Eq. (8-26), may be completed by the reversal of this process, pyridoxamine combining with the second keto acid in the initial step.

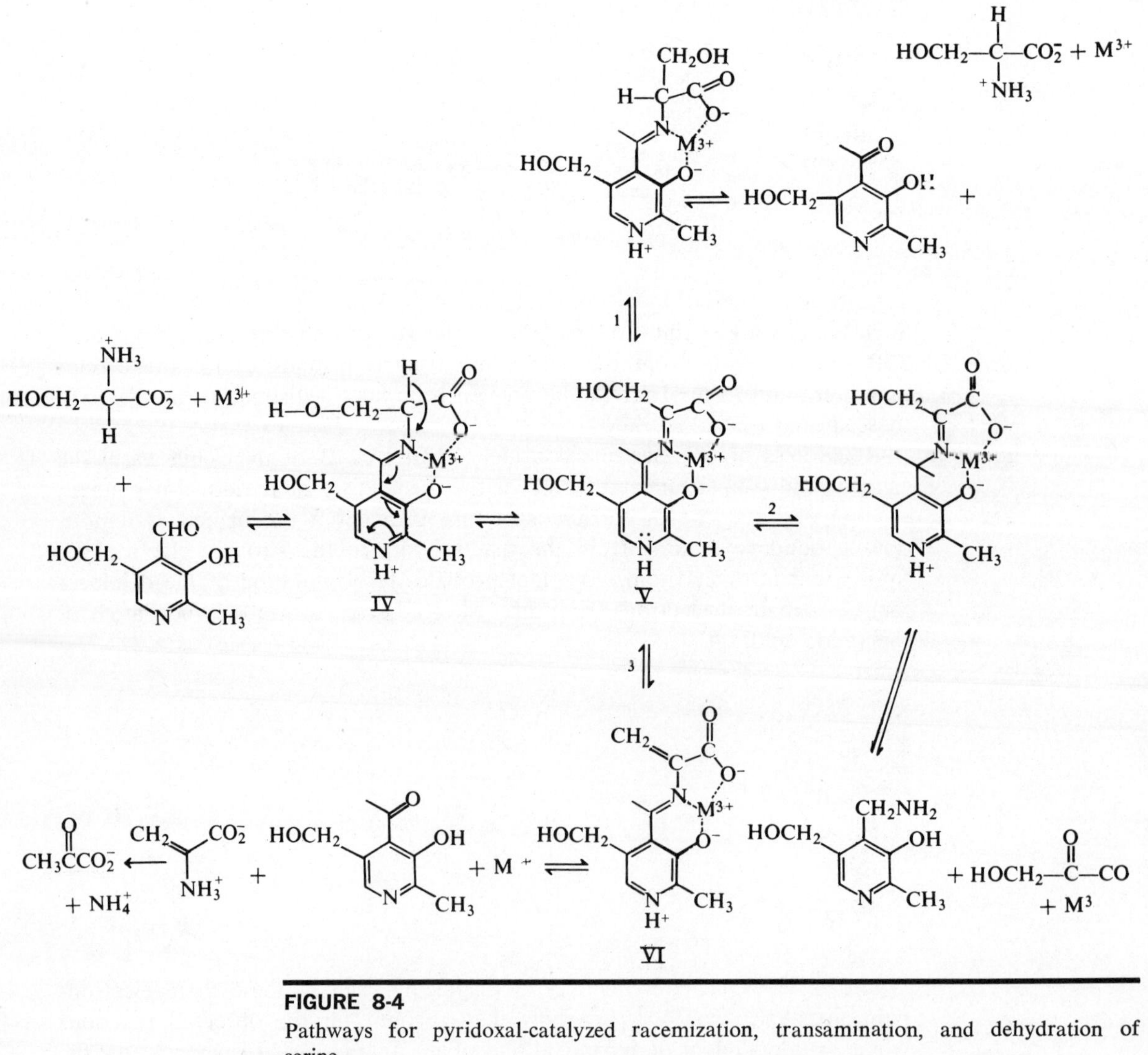

FIGURE 8-4
Pathways for pyridoxal-catalyzed racemization, transamination, and dehydration of serine.

This reaction regenerates pyridoxal and converts the keto acid into the amino acid product. Finally, the unshared pair of electrons may be employed to expel hydroxide ion from (V), yielding (VI) (pathway 3, Fig. 8-4). Hydrolysis of this Schiff base chelate yields α-amino acrylic acid, which spontaneously hydrolyzes to pyruvate.

The desulfuration of homocysteine, a *β,γ-elimination* reaction, may be formulated in the same general fashion, although the reaction is somewhat more complicated (Fig. 8-5). The first two steps, identical to those for transamination, are followed by a tautomerization, elimination, another tautomerization, and finally hydrolysis. The condensation of serine and indole to yield

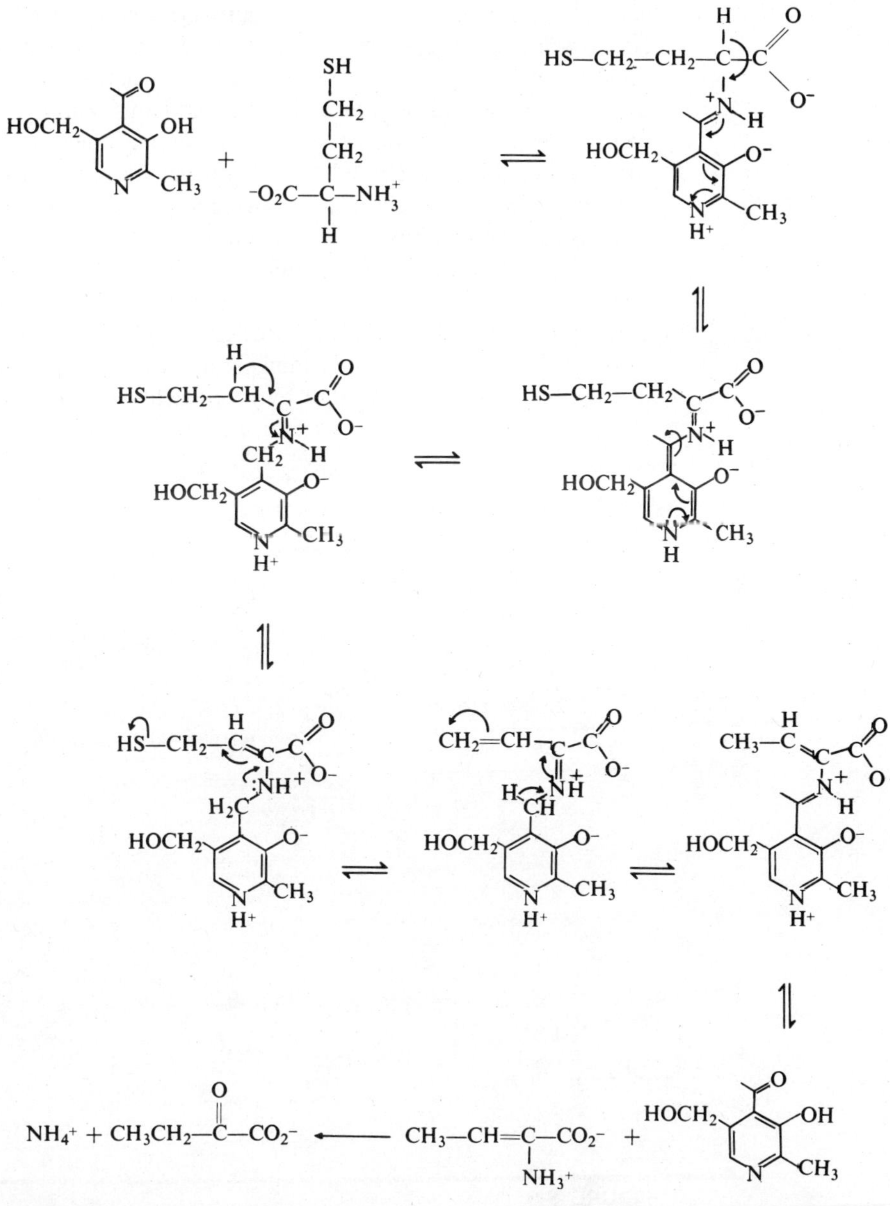

$$NH_4^+ + CH_3CH_2-\overset{O}{\overset{\|}{C}}-CO_2^- \longleftarrow CH_3-CH{=}\underset{NH_3^+}{\underset{|}{C}}-CO_2^- + \text{pyridoxal}$$

FIGURE 8-5
Pathway for pyridoxal-catalyzed desulfuration of homocysteine.

tryptophan simply involves the addition of the nucleophilic reagent indole to the unsaturated intermediate (VI) in serine dehydration followed by tautomerization and hydrolysis (Fig. 8-6).

Let us turn our attention now to the other electron pairs at the α-carbon atom of the amino acid moiety in III: those linking the carboxyl and R functions to this carbon atom. Displacement of these electron pairs can result in *decarboxylation* and *dealdolization* reactions, as shown in Fig. 8-7.

The understanding of the catalytic role of pyridoxal phosphate in these enzymatic reactions has been greatly aided by the finding that, under the proper conditions, they may nearly all be duplicated in nonenzymatic systems. In 1945, Snell observed that, at elevated temperatures, pyridoxal reacts nonenzymatically with glutamic acid to yield pyridoxamine and α-ketoglutaric acid (24). There followed an elegant series of papers by Snell and his associates demonstrating that in the presence of pyridoxal or a similar catalyst molecule and polyvalent metal cations the reactions indicated in Eqs. (8-25) to (8-31) occurred readily at 100° in aqueous solution under slightly acidic, neutral, or slightly basic conditions. Although some specificity for individual reactions could be achieved by the proper choice of *p*H and metal ion, the reactions were relatively nonspecific and tended to occur simultaneously.

These studies on model systems culminated in the formulation by Snell and coworkers (25) and by Braunstein and Shemyakin (26) of a series of similar reaction mechanisms accounting for the role of pyridoxal phosphate in these reactions. These mechanisms are essentially those outlined in Figs. 8-4 through 8-7. That these mechanisms are correct for nonenzymatic reactions at least has been substantiated by more detailed studies in other laboratories,

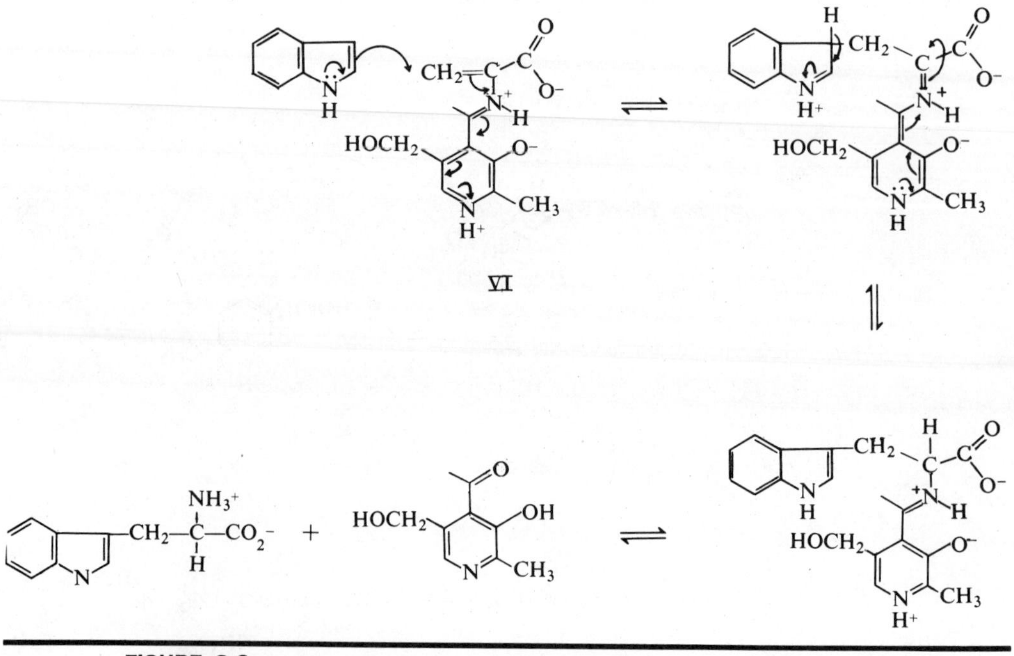

FIGURE 8-6
Tryptophan synthesis from serine and indole.

particularly those of Bruice and his associates (27–30). Their work differed in three respects from the preceding: (1) ambient temperatures rather than 100° were utilized, (2) metal ions were not employed, and (3) the reactions were followed kinetically. These studies have tended to confirm the conclusions reached earlier but have added new insight and confidence to these conclusions. Of particular interest are the investigations of the transamination between pyridoxal and α-aminophenylacetic acid in the presence of imidazole buffers. In this case, kinetic and spectrophotometric evidence was obtained for the formation of the two tautomeric Schiff bases considered to be involved in transamination (Fig. 8-4). Moreover, interconversion of these Schiff bases was demonstrated to be the rate-determining step in the overall reaction process and was demonstrated to be subject to general acid–base catalysis by imidazolium ion and imidazole. The suggested mechanism for the imidazole catalysis is

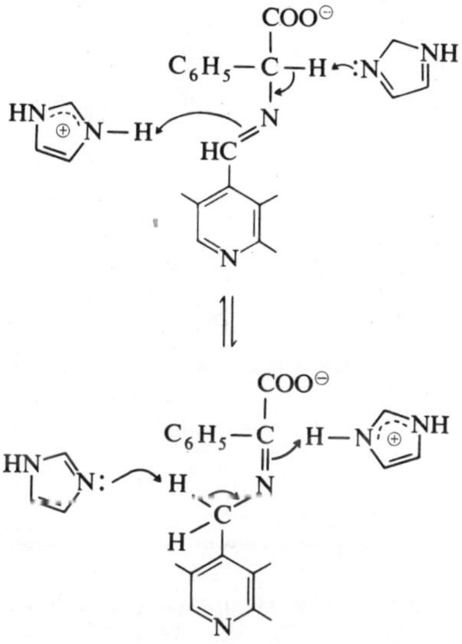

Pyridoxal-dependent enzymatic and nonenzymatic reactions differ in three important respects. First, the enzymatic reactions are specific in terms of both substrate structure and type of reaction catalyzed, but, as noted above, the nonenzymatic reactions are more nonspecific. Second, the enzymatic reactions occur more rapidly by many orders of magnitude than do their nonenzymatic counterparts. This reaction specificity and rate enhancement for the enzymatic reactions must be attributed to the protein moiety of the enzyme. Finally, the enzymatic reactions appear to involve a *transimination* reaction, rather than Schiff base formation, as the initial reaction step. Optical studies that suggested that enzyme-bound pyridoxal phosphate existed as the Schiff base rather than as the free aldehyde (31) have been confirmed by reduction of pyridoxal phosphate enzymes with sodium borohydride. This reagent converts Schiff bases into stable secondary amines:

$$\mathrm{>\!\!=N-} + \mathrm{NaBH_4} \longrightarrow \mathrm{>\!\!C(H)-N(H)-} \qquad (8\text{-}32)$$

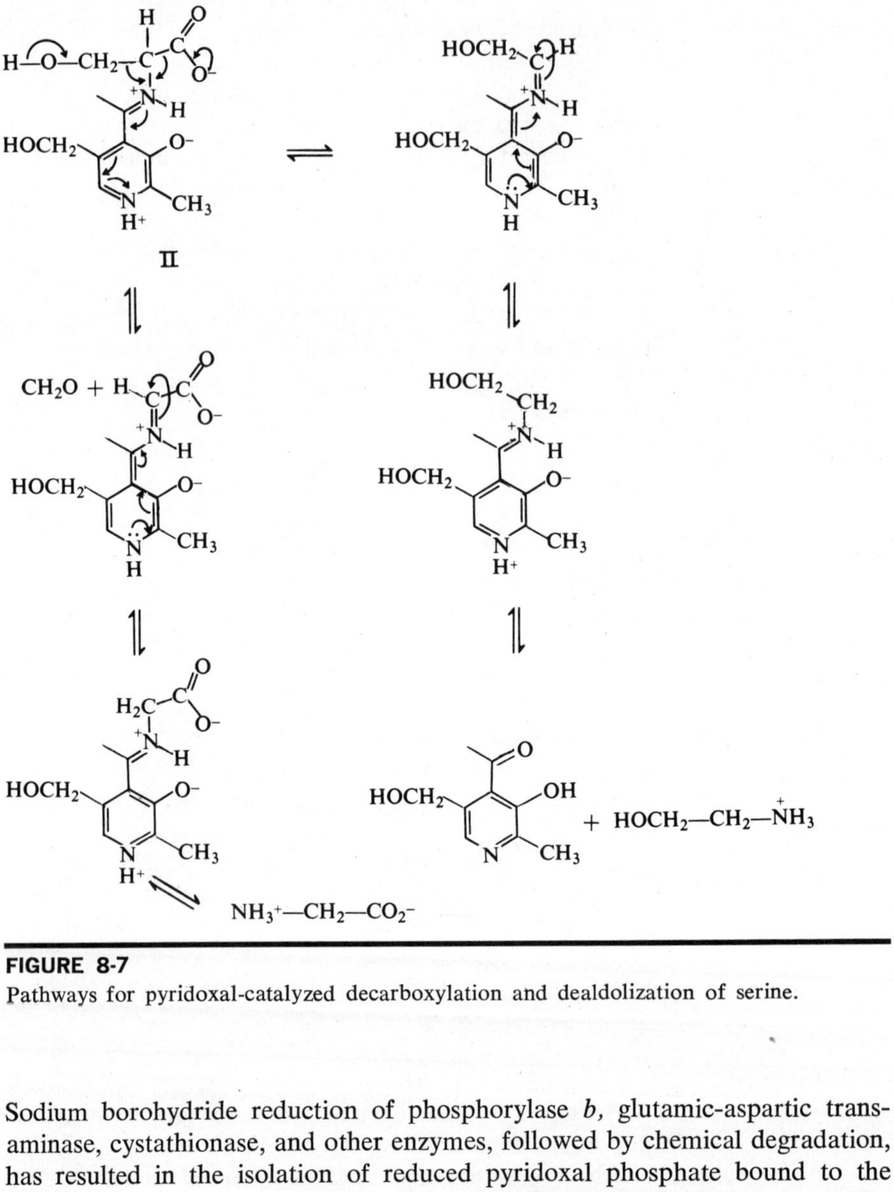

FIGURE 8-7
Pathways for pyridoxal-catalyzed decarboxylation and dealdolization of serine.

Sodium borohydride reduction of phosphorylase *b*, glutamic-aspartic transaminase, cystathionase, and other enzymes, followed by chemical degradation, has resulted in the isolation of reduced pyridoxal phosphate bound to the ε-amino group of a lysine residue. Assuming that the enzymatic reactions, like their nonenzymatic analogs, occur with the intermediate formation of a Schiff base between pyridoxal phosphate and substrate, an early step in the enzyme-catalyzed reactions may be considered to involve the conversion of a pyridoxal phosphate–enzyme Schiff base into a pyridoxal phosphate–substrate Schiff base, a trans*imination* reaction (32):

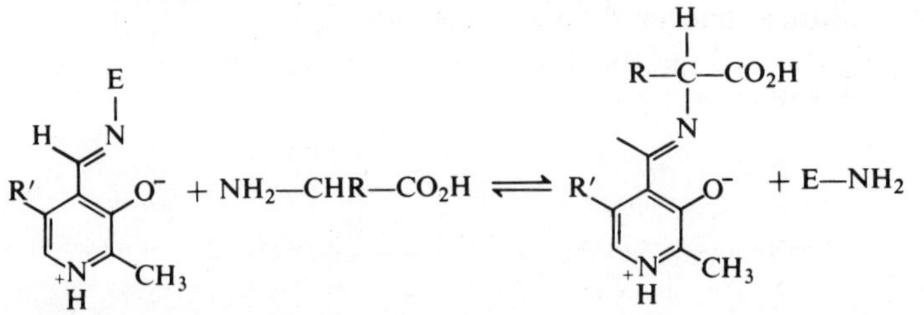

(8-33)

Such transimination reactions occur somewhat more readily than Schiff base formation involving the free carbonyl compound (33). The existence of enzyme-bound pyridoxal phosphate as the Schiff base may, therefore, be advantageous to pyridoxal phosphate–dependent enzymes.

A series of elegant studies on glutamic-aspartic transaminase has established that this enzyme catalyzes a series of reactions closely related to those deduced for the nonenzymatic models. The two tautomeric Schiff bases for this reaction, corresponding to (IV) and (V) in Fig. 8-4, have been tentatively identified as reaction intermediates, and their rate of formation, interconversion, and decomposition has been measured. The rate-determining step in enzymatic transamination, as well as in the nonenzymatic case, is the interconversion of the intermediate Schiff bases (34).

THIAMINE PYROPHOSPHATE

Thiamine pyrophosphate, a coenzyme of universal occurrence in living systems, was originally detected as a nutritional factor required for the prevention of polyneuritis in birds and beriberi in man. The crystalline vitamin (thiamine) was obtained in 1925 by Jansen and Donath, and its structure was elucidated a decade later by Williams and Cline. Thiamine consists of a substituted pyrimidine linked by a methylene group to a substituted thiazole:

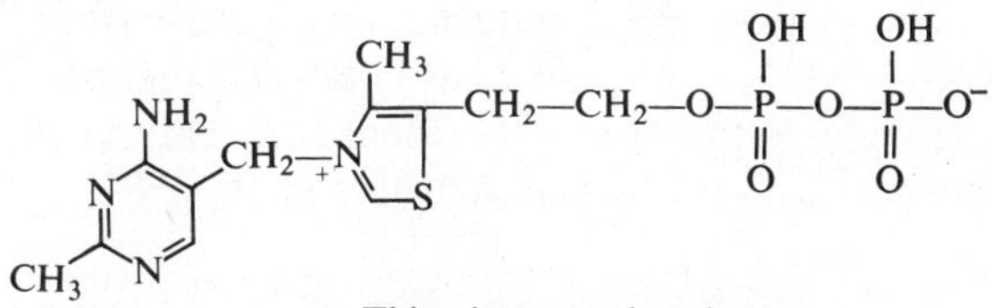

Thiamine pyrophosphate

Thiamine pyrophosphate serves as coenzyme for three types of enzymatic reactions: *nonoxidative decarboxylations of α-keto acids, oxidative decarboxylations of α-keto acids,* and *formation of α-ketols* (acyloins). The first implication of thiamine pyrophosphate in an enzymatic process was derived from the finding of Auhagen that the decarboxylation of pyruvate by yeast cells required a thermostable organic cofactor, called *cocarboxylase* by the discoverer:

$$CH_3{-}\overset{\displaystyle O}{\overset{\|}{C}}{-}CO_2H \rightarrow CH_3{-}\overset{\displaystyle O}{\overset{\|}{C}}{-}H + CO_2 \qquad (8\text{-}34)$$

This cofactor was isolated and crystallized from yeast cells by Lohmann and Schuster and demonstrated to be thiamine pyrophosphate (35). The enzyme required for this reaction, carboxylase, was subsequently purified and demonstrated to contain loosely bound thiamine pyrophosphate and magnesium, both of which are required for enzymatic activity. Thiamine-dependent carboxylases of varying specificity have been detected in a wide range of living matter and are probably of universal occurrence.

The oxidative decarboxylation by α-keto acid dehydrogenases is also thiamine pyrophosphate–dependent. These reactions, with oxygen as final acceptor, can be formulated thus:

$$R{-}\overset{\displaystyle O}{\overset{\|}{C}}{-}CO_2H + \tfrac{1}{2}O_2 \rightarrow R{-}\overset{\displaystyle O}{\overset{\|}{C}}{-}OH + CO_2 \qquad (8\text{-}35)$$

As indicated below, the role of thiamine pyrophosphate in oxidative decarboxylations is similar to that described for the nonoxidative process. The former reactions are generally quite complicated, involving several coenzymes, lipoic acid usually serving as the actual immediate oxidizing reagent. Such reactions are quite common and occur with a variety of α-keto acids (pages 519 and 612).

The enzymatic formation of α-ketols is a common reaction apparently catalyzed by many carboxylases and α-keto acid dehydrogenases. Typical examples include the reaction of pyruvate with acetaldehyde to yield acetoin:

$$CH_3\overset{O}{\overset{\|}{C}}—CO_2H + CH_3—\overset{O}{\overset{\|}{C}}—H \rightarrow CH_3—\overset{O}{\overset{\|}{C}}—\underset{H}{\underset{|}{\overset{OH}{\overset{|}{C}}}}—CH_3 + CO_2 \tag{8-36}$$

or the reaction of 2 mol of pyruvate to yield acetoin:

$$2CH_3\overset{O}{\overset{\|}{C}}—CO_2H \rightarrow CH_3—\overset{O}{\overset{\|}{C}}—\underset{H}{\underset{|}{\overset{OH}{\overset{|}{C}}}}—CH_3 + 2CO_2 \tag{8-37}$$

A somewhat different example of thiamine pyrophosphate–dependent α-ketol formation is provided by the enzyme transketolase, an important enzyme in the pentose phosphate pathway and carbohydrate metabolism in general (page 533). Transketolase catalyzes the transfer of a two-carbon fragment from a ketose onto the aldehydic carbon atom of an aldose. A typical reaction catalyzed by transketolase is

$$\begin{array}{c} CH_2OH \\ | \\ C{=}O \\ | \\ HO{—}CH \\ | \\ H{—}C{—}OH \\ | \\ CH_2OPO_3H_2 \end{array} + \begin{array}{c} CHO \\ | \\ H{—}C{—}OH \\ | \\ H{—}C{—}OH \\ | \\ H{—}C{—}OH \\ | \\ CH_2OPO_3H_2 \end{array} \rightarrow \begin{array}{c} CH_2OH \\ | \\ C{=}O \\ | \\ HO{—}CH \\ | \\ H{—}C{—}OH \\ | \\ H{—}C{—}OH \\ | \\ H{—}C{—}OH \\ | \\ CH_2OPO_3H_2 \end{array} + \begin{array}{c} CHO \\ | \\ H{—}C{—}OH \\ | \\ CH_2OPO_3H_2 \end{array} \tag{8-38}$$

Xylulose-5-phosphate   Ribose-5-phosphate   Sedoheptulose-7-phosphate   Glyceraldehyde-3-phosphate

A final distinct example is provided by the glyoxylate carboligase reaction, in which 2 mol of glyoxylate are converted to tartronic semialdehyde and carbon dioxide:

$$2CHO—CO_2H \longrightarrow \underset{\text{tartronic semialdehyde}}{CHO—CHOH—CO_2H} + CO_2 \tag{8-39}$$

In this case, the product is an α-hydroxyaldehyde rather than an α-hydroxyketone, as in the cases noted above.

All thiamine pyrophosphate–dependent reactions are basically similar. In each case a carbon–carbon bond adjacent to a keto group is cleaved, yielding a stable product such as a carboxylate group or an aldehyde plus an active aldehyde anion presumably bound to thiamine pyrophosphate. Reaction of this anion with a proton generates an aldehyde; an example is the nonoxidative decarboxylation of pyruvate [Eq. (8-34)]. Reaction with an oxidizing agent, such as oxidized lipoic acid, generates acetate; the oxidative decarboxylation of pyruvate [Eq. (8-35)] is an example. And reaction with a carbonyl group generates an acyloin, as in acetoin formation [Eqs. (8-36) and (8-37)] or in the transketolase reaction [Eq. (8-38)]. These transformations are presented diagrammatically in Fig. 8-8.

An understanding of the coenzymatic function of thiamine pyrophosphate–dependent enzymatic reactions is the result of the study of analogous reactions in nonenzymatic systems. Since the initial studies of Ugai et al. in 1943 (36), several thiamine-dependent nonenzymatic reactions have been discovered that closely resemble the enzymatic processs (37–39). The mechanism of the nonenzymatic reactions was established with the striking discovery by Breslow (40) that the aromatic hydrogen at position 2 of the thiazolium ring in thiamine exchanges readily with solvent deuterium:

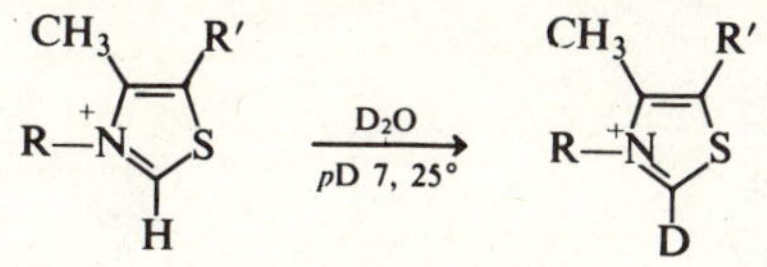

$$\xrightarrow[pD\ 7,\ 25^\circ]{D_2O} \qquad (8\text{-}40)$$

This exchange occurs with a half-time of approximately 2 min in aqueous solution at room temperature and pH 5. At pH 7 the reaction is too fast to follow by ordinary techniques (40,41). This finding indicates that the ylid (VII) is particularly stable:

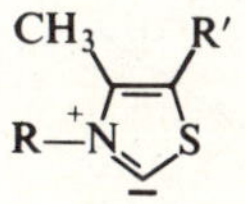

VII 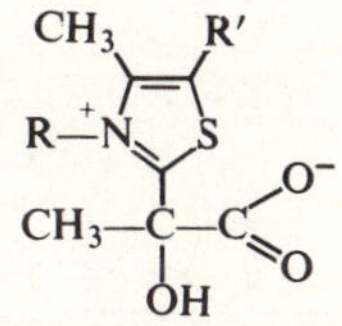 VIII

On this basis, the following reasonable mechanisms may be constructed to account for thiamine-dependent reactions (42). Consider, for example, re-

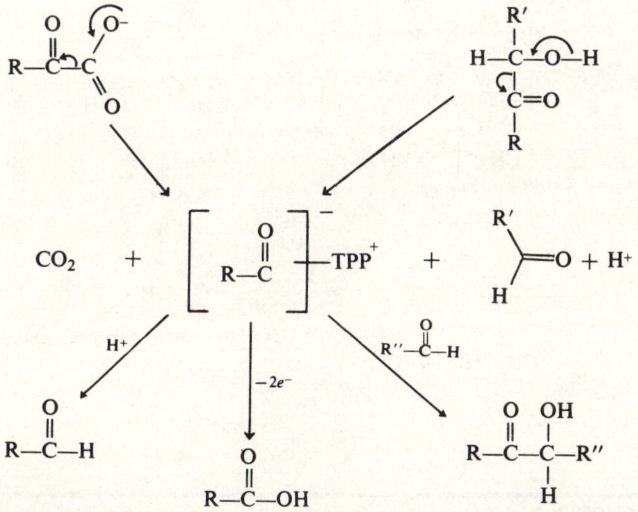

FIGURE 8-8

Basic pathway for thiamine-dependent reactions.

actions of pyruvate to yield acetaldehyde [Eq. (8-34)], acetate [Eq. (8-35)], acetoin [Eq. (8-37)], and the reaction of pyruvate with acetaldehyde to yield acetoin [Eq. (8-36)]. Suggested mechanisms for these reactions are shown in Fig. 8-9. In all cases the initial reaction involves the addition of the carbanion to the carbonyl carbon atom of pyruvate, forming compound (VIII). The strongly electrophilic nitrogen atom in (VIII) aids the decarboxylation of this species, yielding (IX). Subsequent addition of one of a variety of electrophilic reagents (see Fig. 8-8) to the former carbonyl carbon atom followed by expulsion of the original carbanion yields the required reaction products. This scheme can clearly be extended to account for the transketolase reaction as well (Fig. 8-9).

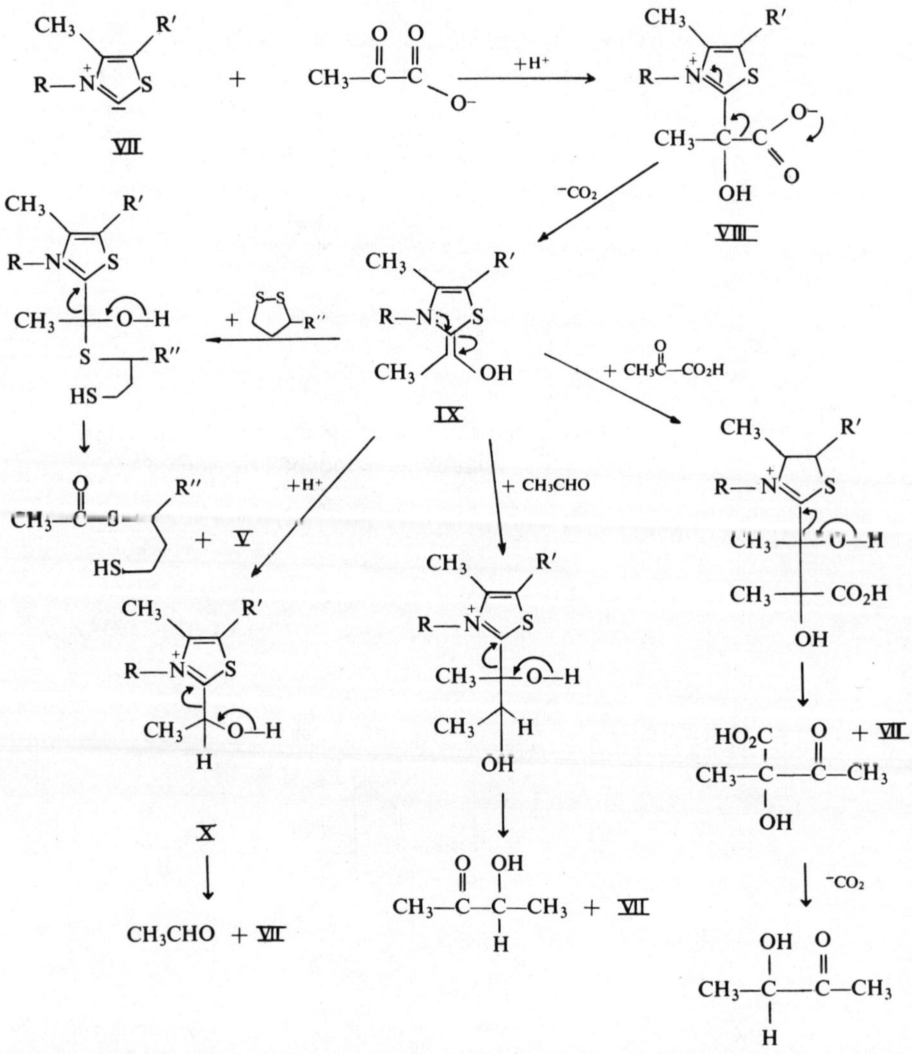

FIGURE 8-9
Mechanisms for thiamine-dependent reactions of pyruvate.

In addition to the observed H-D exchange, the internal consistency of the proposed mechanisms suggests their basic correctness. Further support for these mechanisms is derived from the findings that the intermediate hydroxyethyl thiazolium salt (X, Fig. 8-9) participates in the nonenzymatic reaction, yielding acetoin, and that otherwise effective catalysts are rendered inactive when methylated at the 2 position of the thiazole ring (43). Moreover, the mechanisms predict that the α-hydrogen atom of hydroxyethyl thiamine pyrophosphate (X) ought to exchange with solvent deuterium through its interconversion with IX:

$$\underset{\text{X}}{\overset{\oplus}{\text{N}}\text{-thiazole-}\text{H—C—OH (CH}_3)} \underset{}{\overset{D_2O}{\rightleftharpoons}} \underset{\text{IX}}{\left[\text{—N:}\ \text{C=C(CH}_3\text{)OH} \longleftrightarrow \text{—}\overset{\oplus}{\text{N}}\ \text{C}^{\ominus}\text{(CH}_3\text{)OH} \right]} \overset{D_2O}{\rightleftharpoons} \overset{\oplus}{\text{N}}\text{-thiazole-}\text{D—C—OH (CH}_3) \qquad (8\text{-}41)$$

This hydrogen–deuterium exchange has been observed experimentally, lending additional support to the suggested mechanisms (44).

Considerable evidence exists which suggests that the reaction mechanisms indicated in Fig. 8-9 are substantially correct for the corresponding enzymatic reactions as well. This evidence includes the following points. First, incubation of pyruvic decarboxylase [Eq. (8-34)] with pyruvate yields isolable quantities of hydroxyethyl thiamine pyrophosphate (X) and, possibly, the adduct of TPP with pyruvate (VIII) (45). Second, pyruvic decarboxylase utilizes hydroxyethyl thiamine pyrophosphate for the synthesis of acetaldehyde and, in the presence of acetaldehyde, for the synthesis of acetoin (46). Third, hydroxyethyl thiamine pyrophosphate is synthesized by pyruvic oxidase [Eq. (8-35)] and converted to an acetate derivative by this enzyme (47–49). Finally, transketolase [Eq. (8-38)] catalyzes the synthesis of dihydroxyethyl thiamine pyrophosphate from fructose-6-phosphate and transfers the two-carbon fragment to suitable acceptor molecules (50,51). All of these observations provide strong evidence for the mechanisms of Fig. 8-9.

In one case, the oxidative conversion of pyruvate to acetate, the suggested reaction mechanism is ambiguous. We may visualize the transfer of the two-carbon fragment bound to thiamine pyrophosphate to lipoic acid in one of two ways: (1) a direct transfer involving the electrophilic attack of oxidized lipoic acid [Fig. 8-9 and Eq. (8-42)]:

$$\text{S—S (lipoyl)} + \text{H—O—}\bar{\text{C}}\text{(R)—C(thiazolium, }\overset{\oplus}{\text{N}}) \longrightarrow \text{H—O—C(R)(S—)—C(thiazolium)} \longrightarrow \text{HS}\ \text{S—C(=O)R} + \bar{\text{C}}\text{(thiazolium, }\overset{\oplus}{\text{N}}) \qquad (8\text{-}42)$$

or (2) a preliminary oxidation of the intermediate to yield acetyl thiamine pyrophosphate followed by nucleophilic attack of reduced lipoic acid.

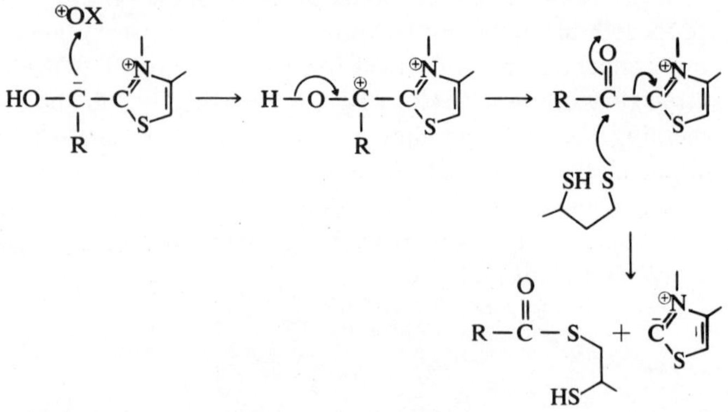

$$\text{HO}-\overset{\ominus}{\underset{\text{R}}{\text{C}}}-\text{(thiazolium)} \xrightarrow{\oplus\text{OX}} \text{H}-\text{O}-\overset{\oplus}{\underset{\text{R}}{\text{C}}}-\text{(thiazolium)} \longrightarrow \text{R}-\overset{\text{O}}{\overset{\|}{\text{C}}}-\text{(thiazolium)} \xrightarrow{\text{lipoyl (SH S)}} \text{R}-\overset{\text{O}}{\overset{\|}{\text{C}}}-\text{S}-\text{(dihydrolipoyl, HS)} + \text{(thiazolium ylide)} \qquad (8\text{-}43)$$

There is some evidence that the latter route may be the correct one. Thus, pyruvic oxidase appears capable of oxidizing hydroxyethyl thiamine pyrophosphate to acetate in the presence of artificial electron donors (45). However, the existence of acetyl thiamine pyrophosphate as a reaction intermediate remains conjectural. It is clear that this substance is a good acetyl group donor (52).

TETRAHYDROFOLIC ACID

The transfer of one-carbon fragments constitutes an important class of metabolic reactions. Such transformations generally occur via coenzymes derived from tetrahydrofolic acid (FH_4):

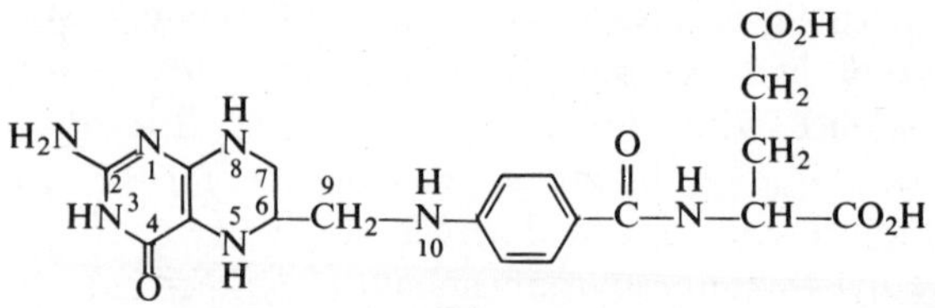

$$H_2N\text{-(tetrahydropteridine, positions 1–8)}-\overset{9}{CH_2}-\overset{H}{\underset{10}{N}}-C_6H_4-\overset{O}{\overset{\|}{C}}-\overset{H}{N}-\underset{}{CH}(CH_2CH_2CO_2H)-CO_2H$$

Tetrahydrofolic acid

FH_4 is composed of a reduced pteridine, *p*-aminobenzoic acid, and L-glutamic acid. Polyglutamic derivatives of FH_4, containing up to seven glutamate residues linked via γ-glutamyl peptide bonds, also occur in living systems and perform biochemical functions similar to those of FH_4 itself. The development of the understanding of the biochemical role of FH_4 coenzymes began with the recognition of folic acid, containing a completely aromatized pteridine ring system, as a nutritional factor for the lactic acid bacteria. The dependence of these organisms on folic acid for growth was employed as an assay system for the purification and isolation of this substance, subsequently characterized structurally. The ready oxidation of FH_4 to folic acid accounts for the isolation of the latter substance in the early studies. Folic acid, in turn, is enzymatically reduced to FH_4. There are three possible structural isomers (7,8-, 5,6-, and 5,8-) for dihydrofolate, but only the 7,8 isomer has been identified unequivocally in chemical or enzymatic reactions (53,54).

The reactions of FH_4 center about the nitrogen atoms at positions 5 and 10. In fact, all the one-carbon transfer reactions that depend on FH_4 involve covalent-bond formation between the one-carbon fragment and one or both of these nitrogen atoms. N^5 is the more basic of the two sites with a pK_a for the conjugate acid of 4.8; N^{10} has a pK_a for the conjugate acid of -1.25 and is, therefore, very weakly basic (55).

FH_4 coenzymes are concerned with the transfer of one-carbon fragments at the oxidation level of formate, formaldehyde, and methanol. At the formate level of oxidation, FH_4-formate adducts exist in four structural forms: N^{10}-formyl FH_4, N^5-formyl FH_4, N^5-formimino FH_4, and N^5,N^{10}-methenyl FH_4:

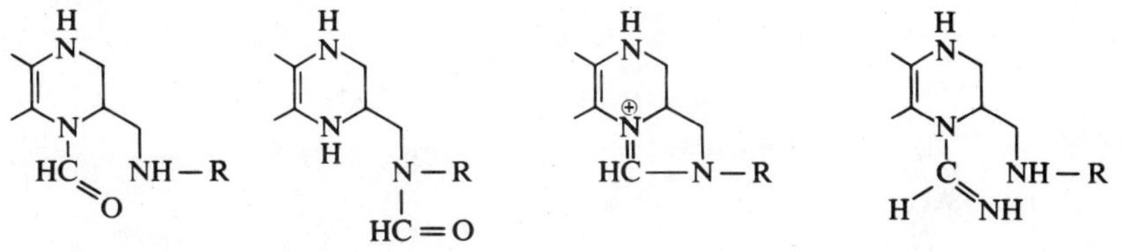

N^5-formyl FH_4 N^{10}-formyl FH_4 N^5, N^{10}-methenyl FH_4 N^5-formimino FH_4

Numerous examples of metabolic reactions involving these coenzymes are presented later in this volume. For the present we are concerned principally with the synthesis and interconversions of these coenzymes.

The synthesis of *N^{10}-formyl FH_4* is catalyzed by formyltetrahydrofolate synthetase—formate : tetrahydrofolate ligase (ADP)—in an ATP-dependent reaction:

$$HCO_2H + FH_4 + ATP \cdot Mg \xrightarrow{NH_4^+} N^{10}\text{-formyl } FH_4 + ADP \cdot Mg + P_i \quad (8\text{-}44)$$

As noted, this reaction is dependent upon ammonium ion, although potassium ions, but not sodium ions, will support the reaction (56). The function of the monovalent cations is to prevent the disassociation of the active enzyme into its catalytically inactive monomeric subunits (57,58). The hydrolysis of N^{10}-formyl FH_4 is catalyzed by a specific deacylase.

Formylglutamate formyl transferase (*N*-formimino-L-glutamate : tetrahydrofolate 5-formimino transferase) catalyzes the formation of *N^5-formyl FH_4*:

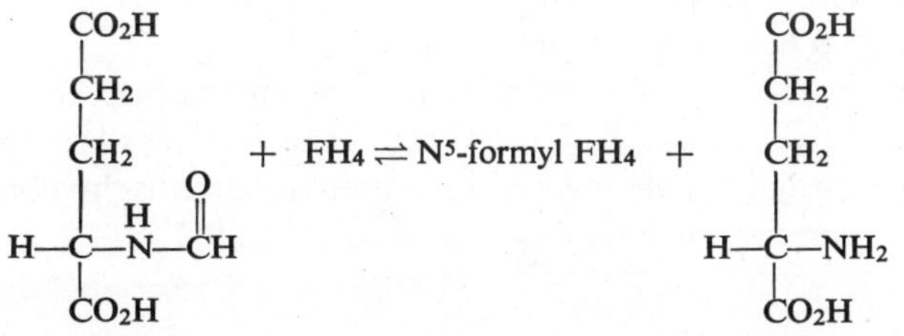

$$HO_2C-CH_2-CH_2-CH(CO_2H)-NH-CHO + FH_4 \rightleftharpoons N^5\text{-formyl } FH_4 + HO_2C-CH_2-CH_2-CH(NH_2)-CO_2H \quad (8\text{-}45)$$

The closely related coenzyme, *N^5-formimino FH_4*, may be synthesized from FH_4 and either formiminoglycine or formiminoglutamate:

$$HN{=}CH{-}NH{-}R + FH_4 \rightarrow N^5\text{-formimino } FH_4 + RNH_2 \quad (8\text{-}46)$$

The latter reaction is catalyzed by the same enzyme that promotes the reaction in Eq. (8-45).

N^5-formyl FH_4 is converted to the *N^{10} isomer* in the ATP-dependent N^5-formyltetrahydrofolate isomerase reaction:

$$N^5\text{-formyl } FH_4 + ATP \longrightarrow N^{10}\text{-formyl } FH_4 + P_i + ADP \quad (8\text{-}47)$$

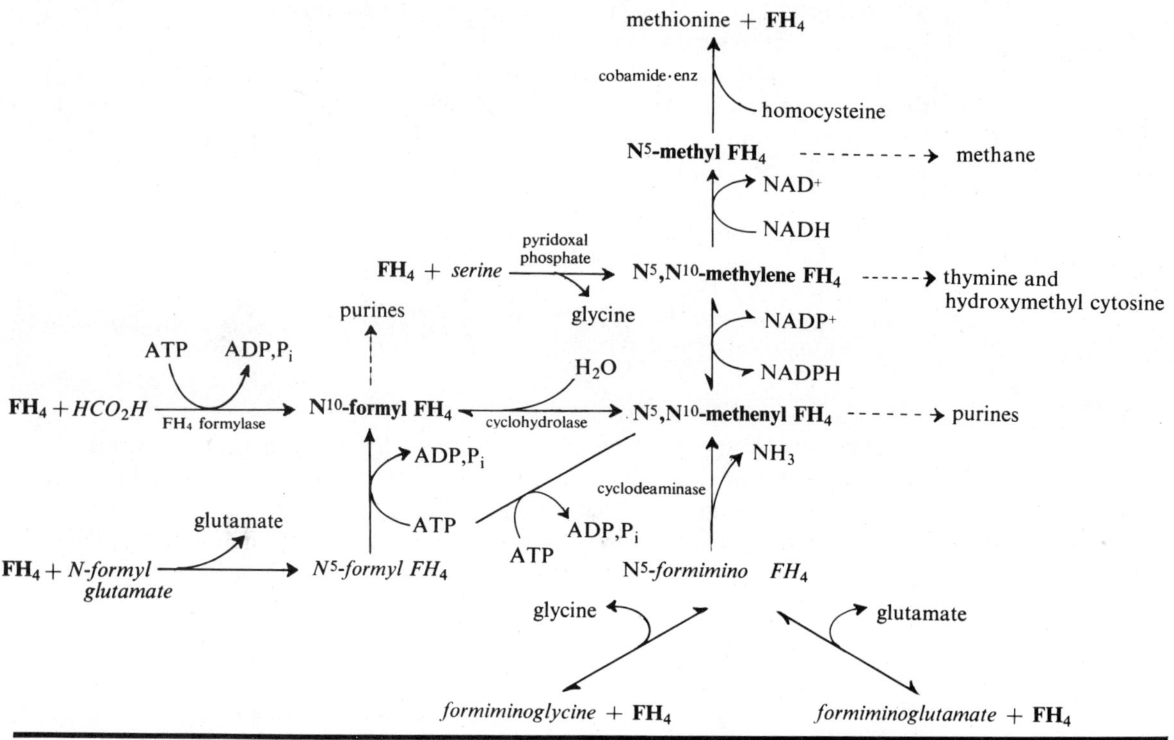

FIGURE 8-10
Metabolic interconversions of tetrahydrofolate coenzymes.

N^5,N^{10}-methenyl FH_4 may arise in several fashions: from N^{10}-formyl FH_4, N^5-formimino FH_4, and N^5-formyl FH_4. The metabolic routes for the synthesis and interconversion of the formyl derivatives of FH_4 and the other coenzymes described below are summarized in Fig. 8-10.

In contrast to transfer reactions of one-carbon fragments at the formate level of oxidation in which several FH_4 coenzymes are involved, such reactions at the formaldehyde level of oxidation apparently involve only one coenzyme, *N^5,N^{10}-methylene FH_4* (59):

5 10
—N N—R
CH$_2$

N^5,N^{10}-methylene FH_4

The principal sources of this coenzyme are the reduction of N^5,N^{10}-methenyl FH_4 by a specific dehydrogenase and the L-serine hydroxymethyltransferase reaction (L-serine : tetrahydrofolate-5,10-hydroxymethyltransferase) (Fig. 8-10). The latter enzyme requires pyridoxal phosphate, as noted in the discussion of that coenzyme (see above). This enzyme has been of considerable historical significance in the understanding of the role of FH_4 in the metabolism of one-carbon fragments.

The nonenzymatic reaction of FH_4 with formaldehyde to form N^5,N^{10}-methylene FH_4 has been subjected to close experimental scrutiny (60). Under acidic conditions, the attack of the more basic N^5 on formaldehyde, which is subject to general and specific acid catalysis, is the rate-determining step; under more basic conditions, dehydration of hydroxymethyl FH_4, also subject to general and specific acid catalysis, becomes rate determining. The overall sequence is

$$-\overset{H}{N_5}\ \overset{H}{N_{10}}- + HCHO \rightleftharpoons -\overset{OHCH_2}{N_5}\ \overset{H}{N_{10}}- \longrightarrow -\overset{\oplus}{N_5}(=CH_2)\ \overset{H}{N_{10}}- \xrightarrow{\text{fast}} -N_5(CH_2)N_{10}- + H^{\oplus} \quad (8\text{-}48)$$

These observations may shed light on the related enzymatic reaction in which the formaldehyde moiety is derived from serine.

The reduction of N^5,N^{10}-methylene FH_4 by NADPH or NADH through the intermediacy of a flavoprotein (see below) yields *N^5-methyl FH_4* (Fig. 8-10), a coenzyme for the transfer of one-carbon fragments at the oxidation level of methanol (61,62):

$$\underset{\underset{CH_3}{|}}{\overset{5}{N}} \quad \underset{H}{\overset{10}{N}}-R$$

N^5-methyl FH_4

Solvent water can transfer hydrogen to the methyl group in this reduction, probably by preequilibrium exchange with labile protons on the reduced flavin coenzyme (63). N^5-methyl FH_4 serves as the methyl group donor for the biosynthesis of methionine from homocysteine (page 781) and for the synthesis of methane (64).

N^5-methyl FH_4 is also formed, in the soil microorganism *Achromobacter* MS, by the transfer of a methyl group to FH_4 from the trimethylsulfonium ion (65):

$$CH_3-\overset{+}{\underset{\underset{CH_3}{|}}{S}}-CH_3 + FH_4 \rightleftharpoons CH_3-S-CH_3 + N^5\text{-methyl } FH_4 \quad (8\text{-}49)$$

The generality of such reactions has not been established.

NICOTINAMIDE NUCLEOTIDES

The nicotinamide nucleotides were the first coenzymes to be recognized. Harden and Young, in 1904, found the conversion of glucose to ethanol by cell-free extracts of yeast required the participation of a dialyzable cofactor, called *cozymase*. Structural studies of cozymase, particularly by Von Euler and his associates, revealed that this molecule was structurally related to adenylic acid. The further elucidation of the structure of cozymase was greatly aided by the isolation of a dialyzable cofactor from mammalian erythrocytes, required for the aerobic oxidation of glucose-6-phosphate, by Warburg and Christian in 1934 (66). These workers demonstrated that this coenzyme was composed of 1 mol of adenine, 1 mol of nicotinamide, 2 mol of pentose (D-ribose), and 3 mol of inorganic phosphate. The erythrocyte coenzyme was able to replace cozymase in a cell-free extract similar to that employed by

Harden and Young, thus establishing a close relationship between the two coenzymes. Cozymase was subsequently demonstrated to be composed of nicotinamide mononucleotide and adenylic acid:

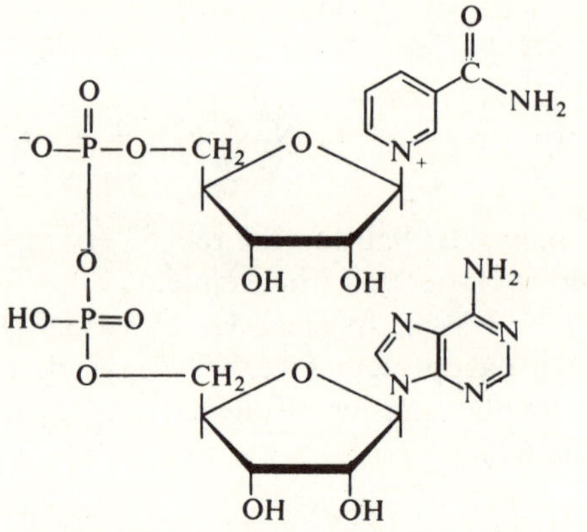

Nicotinamide adenine dinucleotide (NAD^+)

This coenzyme is now usually called nicotinamide adenine dinucleotide (NAD^+), although the older terminology of diphosphopyridine nucleotide (DPN) and coenzyme I (CoI) is still employed. The coenzyme isolated by Warburg and Christian differs from NAD^+ only in having an additional phosphate esterified at the C-2 hydroxyl group of the ribose moiety of the adenine portion of the molecule. This coenzyme is called nicotinamide adenine dinucleotide phosphate ($NADP^+$) or, less frequently, triphosphopyridine nucleotide (TPN) or coenzyme II (CoII).

NAD^+ and $NADP^+$ are derived biosynthetically, in the mammal, from nicotinamide (niacin) or nicotinic acid (page 801). A nutritional deficiency of these precursors results in pellagra (in the human), a disease characterized by dermatitis, diarrhea, and dementia.

Coenzymatically, NAD^+ and $NADP^+$ are involved in a large number of oxidation–reduction reactions catalyzed by dehydrogenases. A typical reaction is that catalyzed by alcohol dehydrogenase:

$$CH_3CH_2OH + NAD^+ \rightleftharpoons CH_3CHO + NADH + H^+ \tag{8-50}$$

Most pyridine nucleotide–dependent enzymes are highly specific for either NAD^+ or $NADP^+$, although some lack this degree of specificity for coenzyme.

The conversion of NAD^+ to the reduced form (NADH) is accompanied by a marked alteration in the spectrophotometric properties of the coenzyme. NAD^+ absorbs maximally near 260 nm, the absorption being slightly less than the summation of that due to adenine and that due to nicotinamide. On reduction, the absorbance at 260 nm decreases considerably, and a new band having a maximum at 340 nm, characteristic of a dihydronicotinamide, appears. Thus, the nicotinamide moiety of these coenzymes is the site of the oxidation–reduction reactions. The alteration in spectrophotometric properties is frequently employed in assays of nicotinamide nucleotide–dependent enzymes. Electronically, reduction might take place at either the 2, 4, or 6 position of the nicotinamide ring, yielding the corresponding dihydro derivatives (XI, XII, and XIII).

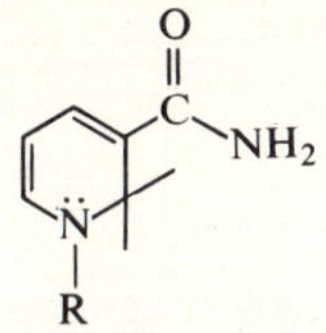

XI 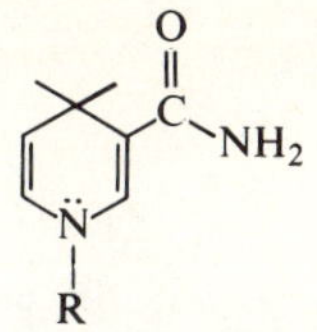 XII 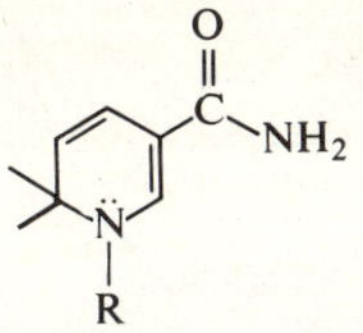 XIII

Taking advantage of the fact that the ferricyanide oxidation of N^1-methyl nicotinamide is known to yield predominantly the 2- and 6-pyridones (67). Pullman, San Pietro, and Colowick established that position 4 is the actual site of reduction as follows (68). NADD was prepared by hydrosulfite reduction of NAD^+ in deuterium oxide and converted enzymatically to deuterium-labeled NAD^+:

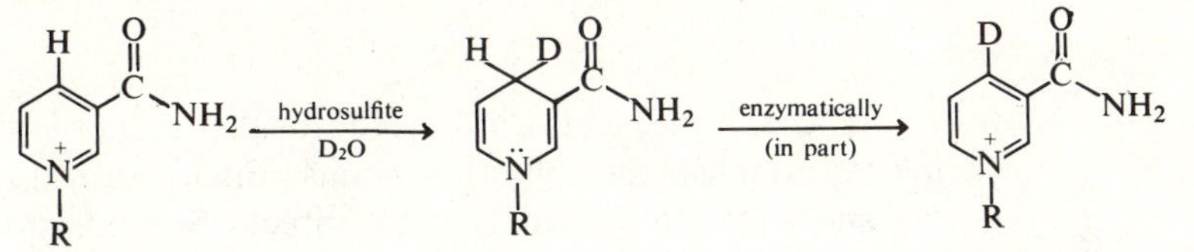

$$\xrightarrow[D_2O]{\text{hydrosulfite}} \quad \xrightarrow[\text{(in part)}]{\text{enzymatically}} \qquad (8\text{-}51)$$

This NAD^+ was cleaved enzymatically to nicotinamide, which was converted to the methiodide with methyl iodide. Oxidation of this substance with ferricyanide did not result in loss of deuterium from the molecule. Consequently, the initial reduced species must have been the 1,4-dihydro derivative:

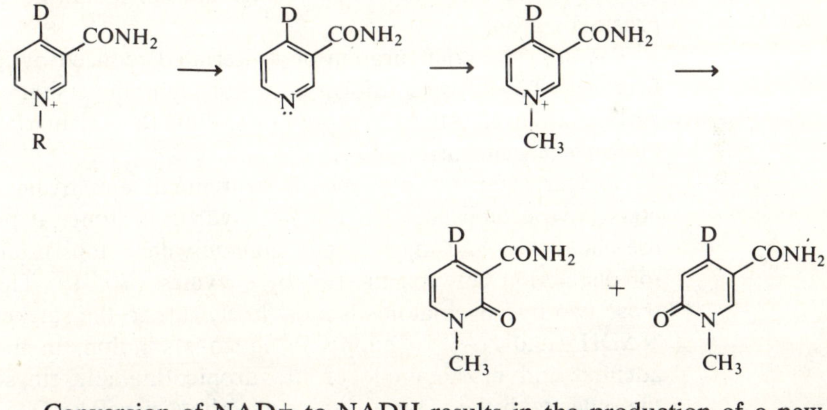

(8-52)

Conversion of NAD^+ to NADH results in the production of a new asymmetric center. Dithionite reduction of NAD^+ in deuterium oxide yields the two stereoisomers shown in Fig. 8-11, which differ in terms of the position occupied by the deuterium atom with respect to the plane of the reduced nicotinamide ring (68). NAD^+- and NADP-linked dehydrogenases exhibit stereospecificity for the A or B form of the reduced coenzyme (Table 8-4). This was convincingly demonstrated by Vennesland and Westheimer for alcohol dehydrogenase (ADH) and lactic dehydrogenase (LDH). Deuterium-labeled reduced NAD^+ was prepared enzymatically with ADH and 1,1-dideuterioethanol:

$$CH_3CD_2OH + NAD^+ \xrightarrow{ADH} CH_3CDO + NADD \qquad (8\text{-}53)$$

When the reduced coenzyme was reoxidized with either acetaldehyde and ADH or pyruvic acid and LDH, the deuterium was quantitatively removed

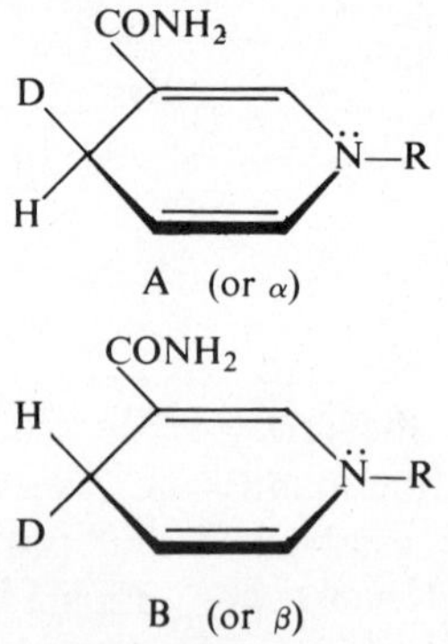

$$\text{A (or } \alpha\text{)}$$

$$\text{B (or } \beta\text{)}$$

FIGURE 8-11
Stereoisomers of monodeuterio-reduced nicotinamide adenine dinucleotide.

from the coenzyme and transferred to the reduced substrate. These experiments establish the following points. First, the transfer of hydrogen between substrate and coenzyme is direct. Second, ADH and LDH are stereospecific with respect to coenzyme. Finally, both ADH and LDH transfer hydrogen to and from the same side of the NAD^+ molecule (side A). The stereospecificity of many dehydrogenases has subsequently been established. Some enzymes utilize form A, some form B (Table 8-4). In only one known case, that of diaphorase (lipoyl dehydrogenase), is hydrogen exchanged on both sides of the nicotinamide ring, and this enzyme contains a flavin as prosthetic group.

The absolute configuration of deuteriated reduced pyridine nucleotides has been established by Cornforth et al. through conversion of the 3, 4, 5, and 6 carbon atoms of the dihydropyridine ring to a deuteriated succinic acid of known stereochemistry (69).

Nuclear magnetic resonance measurements employing 220-MHz spectrometers have established that the two hydrogen atoms at position 4 of NADH are chemically and magnetically nonequivalent, thus establishing a firm basis for distinction between the two by enzymes (70,71). The nonequivalence of these two hydrogen atoms is most likely due to the stacked conformation that NADH (and NAD^+) adopts in aqueous solution. In this conformation, the adenine and nicotinamide or dihydronicotinamide rings lie one on top of the other with the planes of the rings about parallel. Two such conformations are possible: one with the adenine ring on top of the nicotinamide ring and one with the nicotinamide ring on top of the adenine ring. Space filling models of these two structures are shown in Fig. 8-12. That these structures are at least approximately correct is established by fluorescence (72,73), nmr (70,71,74,75), and circular dichroism (76) measurements on both the oxidized and reduced coenzymes. In terms of these structures, the nonequivalence of the hydrogens at the 4 position of NADH can be attributed to their distinct dispositions with respect to the adenine ring. One ought to note that the stacked structures are in rapid equilibrium with unstacked ones and with each other in aqueous solution. At any given moment, depending upon temperature and *p*H, perhaps 50 percent of the coenzyme will exist in a stacked form.

In addition to stereospecificity with respect to coenzyme, Vennesland and Westheimer have elegantly demonstrated that the reduction of acetaldehyde

TABLE 8-4
Stereospecificity of some nicotinamide nucleotide-linked dehydrogenases[a]

Enzymes showing A stereospecificity		*Enzymes showing B stereospecificity*	
Dehydrogenase	*Nicotinamide nucleotide*	*Dehydrogenase*	*Nicotinamide nucleotide*
Formate	NAD	Hydrogen[c]	NAD
Alcohol	NAD	α-Glycerophosphate	NAD
Acetaldehyde	NAD	Triosephosphate[b]	NAD *and* NADP
Glyoxylate[b]	NAD *and* NADP	L-Glutamate[d]	NAD *or* NADP
D-Lactate	NAD	D-Glucose	NAD
L-Lactate	NAD	D-Glucose-6-phosphate	NADP
D-Glycerate	NAD	6-Phospho-D-gluconate	NADP
L-Malate	NAD	Dihydrolipoate (="diaphorase")[c]	NAD
L-Malate (decarboxylating)	NADP	Glutathione (reductase)[b,d]	NAD *and/or* NADP
D_s-Isocitrate	NADP	L-β-Hydroxybutyryl CoA	NAD
Dihydro-orotate[b,c]	NAD *and* NADP	UDPG	NAD
Dihydrofolate (reductase)	NADP	Transhydrogenase	NAD, NADP
Methylene tetrahydrofolate	NADP	β-Hydroxysteroid	NAD
Cytochrome b_5 (reductase)	NAD	3-α-Hydroxysteroid	NAD
		17-β-Hydroxysteroid[d]	NAD *or* NADP
		Cytochrome *c* (reductase)[c]	NAD

[a] In each column, enzymes are listed in approximate order of increasing size of substrate. Sources of enzymes are not given, since the results to date have been the same for a given enzyme from different sources.
[b] Different enzymes acting on NAD and NADP, respectively.
[c] Hydrogen exchange occurs between water and reduced coenzyme and shows same stereospecificity.
[d] Single enzyme acting with either NAD or NADP.
SOURCE: S. P. Colowick, J. Van Eys, and J. H. Park, in M. Florkin and E. H. Stotz (eds.), *Comprehensive Biochemistry,* vol. 14, Elsevier, New York, 1967, p. 1.

to ethanol by ADH is also stereospecific with respect to substrate. The pertinent reactions and results are presented in Fig. 8-13. Carbonyl carbon–labeled deuterioacetaldehyde was reduced enzymatically with unlabeled NADH, yielding the 1-monodeuterioethanol, which was isolated in aqueous solution by distillation. Enzymatic reoxidation of this substance yielded acetaldehyde containing 1 mol of deuterium per mol of acetaldehyde. Thus, the same hydrogen that was transferred to acetaldehyde in the reduction step was quantitatively removed in the reoxidation. Thus, ADH is able to distinguish, in a quantitative fashion, between the two hydrogens at the 1 position of ethanol. This finding was substantiated by conversion of the monodeuterioethanol to the tosylate, followed by an S_N2 hydrolysis. Since the latter reaction proceeds with stereochemical inversion, the product monodeuterioethanol must be the opposite enantiomorph of the starting material. Enzymatic reoxidation of this product yielded unlabeled acetaldehyde and NADD containing 1 mol of deuterium per mol of reduced coenzyme.

Formally, NAD^+- and NADP-dependent reactions involve the transfer of a *hydride ion,* a hydrogen nucleus with two electrons, between substrate and coenzyme. Such reactions may proceed via an actual transfer of a hydride ion or via a two-step reaction pathway involving the transfer of a hydrogen atom preceded or succeeded by the transfer of an electron. Although it is not known with certainty which of these alternatives, singly or in combination, correctly describes NAD^+-dependent enzymatic reactions, the hydride-ion transfer mechanism appears more likely. Westheimer and his collaborators have carefully investigated model reactions for NADH reductions (77,78).

One such model involves the reduction of thiobenzophenone by 1-benzyl-1,4-dihydronicotinamide:

$$H^+ + \phi C(=S)\phi + \text{1-benzyl-1,4-dihydronicotinamide } (CONH_2, N\text{—}CH_2\text{—}\phi) \longrightarrow \phi_2CH\text{—}SH + \text{1-benzylnicotinamide}^+ (CONH_2, N^+\text{—}CH_2\text{—}\phi) \qquad (8\text{-}54)$$

This reaction exhibits second-order kinetics and proceeds with direct hydrogen transfer. The reaction rate is pH-independent and is increased by increasing salt concentrations and increasing solvent polarity and by the presence of electron-withdrawing polar substituents in the thiobenzophenone moiety. The rate or course of this reaction is unaffected by the presence of free-radical traps. The 4,4-dideuterionicotinamide reacts four times less rapidly than the corresponding protio compound. These results strongly suggest, but do not prove, that the reduction of thiobenzophenones by reduced nicotinamides proceeds with the transfer of a hydride ion, rather than by a free-radical route. The overall kinetics and magnitude of the kinetic deuterium-isotope effect in some of the enzyme-catalyzed cases also are consistent with the hydride-ion mechanism.

The addition of a hydride ion to the 4 position of the nicotinamide moiety of NAD^+ is only one example of a group of reactions involving nucleophilic addition to rings of this type. In addition, cyanide, amines, carbanions, bisulfite, and other nucleophilic reagents add to NAD^+ and similar compounds. These addition reactions almost certainly proceed via polar, rather than free-radical, mechanisms.

A number of analogs of NAD^+ have been prepared and are of considerable interest to the enzymologist. Of particular note are NAD^+ analogs substituted in the pyridine moiety. These compounds are synthesized by an NADase from animal tissues that catalyzes a base-exchange reaction between NAD^+ and other pyridine bases (79):

$$\text{3-}CONH_2\text{-pyridinium-}N^+\text{—}R + \text{3-acetylpyridine } (C(=O)\text{—}CH_3) \rightleftharpoons \text{nicotinamide } (CONH_2) + \text{3-acetylpyridinium-}N^+\text{—}R \qquad (8\text{-}55)$$

Thus, coenzyme analogs containing a pyridine base substituted in the 3 position with an acetyl, carboxyl, or thiocarboxamide group, in place of the naturally occurring carboxamide group, have been prepared. Their relative reactivity in enzymatic reactions depends on the nature and source of the enzyme under consideration. Kaplan and his coworkers have employed these coenzymes to trace the evolutionary development of lactic dehydrogenases in a strikingly large number of species (80). A collection of some of these analogs is presented in Table 8-5.

FLAVIN COENZYMES

The discovery of riboflavin and the development of an understanding of its function in living systems constitutes an important chapter in the history of biochemistry. The initial detection of riboflavin or a riboflavin derivative was the isolation of an impure resinous pigment, called *lactochrome,* by Blyth in 1879. Following a dormant period of nearly half a century, several investigators isolated impure yellowish pigments from a variety of natural sources.

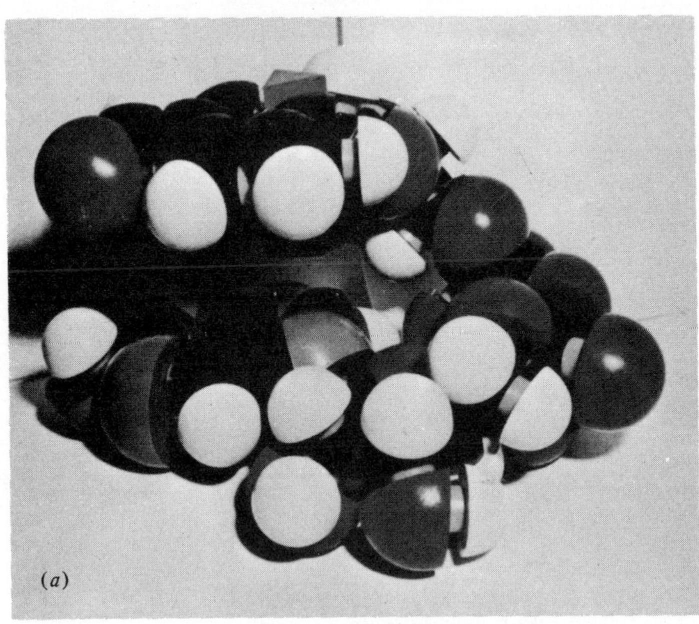

FIGURE 8-12
Space-filling models showing the two stacked conformations of NAD. (*a*) Folded conformation for β-NAD in which the pyridine ring lies above the plane of the adenine ring, arbitrarily chosen as the right-handed folding. (*b*) Folded conformation for β-NAD, in which the pyridine ring lies below the plane of the adenine ring, the left-handed folding. R. H. Sarma, V. Ross, and N. O. Kaplan, *Biochemistry,* **7:** 3052 (1968); by permission of the copyright owners, the American Chemical Society.

These included *cytoflav,* isolated by Banga and Szent-Györgyi; *lyochrom,* isolated by Ellinger and Koschara; and *lactoflavin,* isolated by Ellinger. Wagner-Jauregg and his associates succeeded in isolating a few milligrams of pure crystalline *ovoflavin* from egg white, an accomplishment that led to the dis-

$$\underset{(1.04\ \text{D/molecule})}{CH_3{-}\overset{\overset{\displaystyle D}{|}}{C}{=}O} + NADH + H^+ \longrightarrow \underset{(0.96\ \text{D/molecule})}{CH_3{-}\overset{\overset{\displaystyle D}{|}}{\underset{\underset{\displaystyle H}{|}}{C}}{-}OH}$$

$$\downarrow \text{distill}$$

$$\underset{(0.77\ \text{D/molecule})}{CH_3{-}\overset{\overset{\displaystyle D}{|}}{C}{=}O} + \underset{(0.00\ \text{D/molecule})}{NADH} + H^+ \longleftarrow CH_3{-}\overset{\overset{\displaystyle D}{|}}{\underset{\underset{\displaystyle H}{|}}{C}}{-}OH + NAD^+ \searrow CH_3{-}\overset{\overset{\displaystyle D}{|}}{\underset{\underset{\displaystyle H}{|}}{C}}{-}O\overset{\overset{\displaystyle O}{\|}}{\underset{\underset{\displaystyle O}{\|}}{S}}{-}\phi{-}CH_3$$

$$\underset{(0.00\ \text{D/molecule})}{CH_3{-}\overset{\overset{\displaystyle H}{|}}{C}{=}O} + \underset{(1\ \text{D/molecule})}{NADD} + H^+ \longleftarrow CH_3{-}\overset{\overset{\displaystyle H}{|}}{\underset{\underset{\displaystyle D}{|}}{C}}{-}OH \xleftarrow[\text{inversion}]{OH^-} \quad +NAD^+$$

FIGURE 8-13

covery that this substance, later called vitamin B_2, was a nutritional factor for mammals. Subsequently, these workers prepared relatively large amounts of pure crystalline lactoflavin from whey, setting the stage for the structure determination of this substance. The structure of lactoflavin, now known as *riboflavin,* was established by synthesis in 1935 by Kuhn, Karrer, and their coworkers. Riboflavin is composed of a substituted isoalloxazine ring linked to D-ribitol:

$$CH_2{-}\overset{\overset{\displaystyle H}{|}}{\underset{\underset{\displaystyle OH}{|}}{C}}{-}\overset{\overset{\displaystyle H}{|}}{\underset{\underset{\displaystyle OH}{|}}{C}}{-}\overset{\overset{\displaystyle H}{|}}{\underset{\underset{\displaystyle OH}{|}}{C}}{-}CH_2OH$$

(attached at N-10 of the 7,8-dimethylisoalloxazine ring: CH_3, CH_3, N, N, O, NH, O)

Riboflavin

A discovery of exceptional importance, both for the structure determination and for the understanding of the biochemical role of riboflavin, was the observation that the oxidation of glucose-6-phosphate by glucose-6-phosphate dehydrogenase (D-glucose-6-phosphate : NADP oxidoreductase) yielded NADPH, a coenzyme whose reoxidation was catalyzed by "the yellow enzyme." Methanol treatment of the yellow enzyme resolved it into a colorless protein and a protein-free pigment (81), later established by Kuhn as structurally related to riboflavin. Theorell demonstrated that the prosthetic group of the yellow enzyme was not riboflavin itself, but riboflavin-5′-phosphate. This substance is usually referred to as *riboflavin mononucleotide* (FMN), although this designation is not entirely correct, since the linkage between the base and sugar components is not a glycosidic one. Crystalline yellow enzyme has a molecular weight of approximately 100,000 and contains 2 mol of FMN per mol of enzyme.

In 1938, Warburg and Christian discovered that an enzyme catalyzing the air oxidation of D-amino acids, called D-amino acid oxidase, was a flavoprotein containing a prosthetic group distinct from FMN. The structure of

TABLE 8-5
Collection of analogs of NAD in which the structure of the nicotinamide portion of the molecule has been altered

Name	*Structure*
Pyridine-3-aldehyde analog	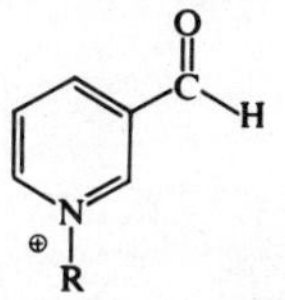
3-Acetylpyridine analog	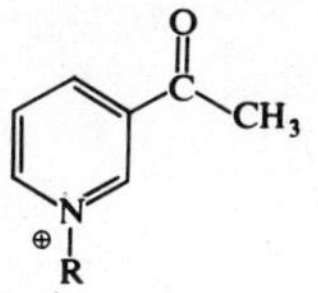
Thionicotinamide analog	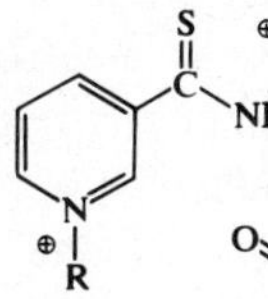
Isonicotinic acid hydrazide analog	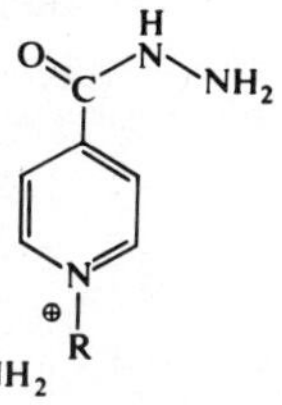
Tetrahydronicotinamide adenine dinucleotide	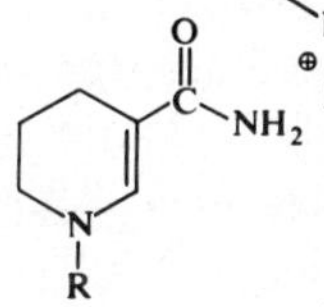
Hexahydronicotinamide adenine dinucleotide	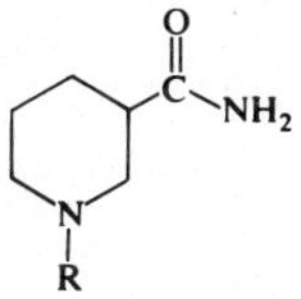
Imidazole adenine dinucleotide	

this new coenzyme, *flavin adenine dinucleotide* (FAD), was established by degradation studies and confirmed by synthesis by Todd in 1954:

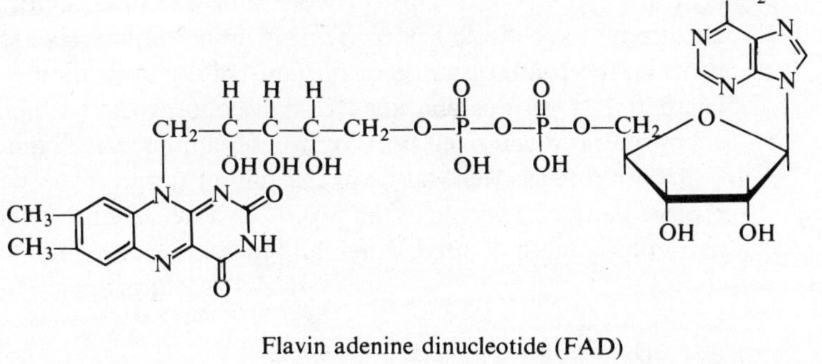

Flavin adenine dinucleotide (FAD)

Several lines of evidence established that, as in the case of the pyridine nucleotide coenzymes, the two ring systems of FAD interact (82–84). Thus, a stacked conformation in which the two rings are spatially close appears to be substantially populated at low temperatures at least.

The number of known flavoproteins utilizing either FMN or FAD as coenzyme is large, and such enzymes are universally distributed in nature.

Flavoproteins are dehydrogenation catalysts. Typical substrates dehydrogenated by flavoproteins include pyridine nucleotides, α-amino acids, α-hydroxy acids, aldehydes, mono- and dithiols, and substances containing saturated carbon–carbon bonds that are converted to olefins. In the course of such oxidations, the isoalloxazine ring system of the appropriate coenzyme, FMN or FAD, becomes reduced:

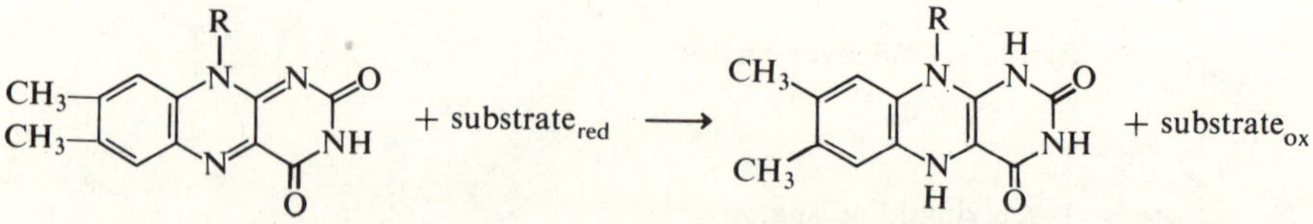

$$+ \text{substrate}_{red} \longrightarrow \; + \text{substrate}_{ox} \qquad (8\text{-}56)$$

The reduced flavoproteins, in turn, become substrates for reactions involving other electron acceptors, regenerating the oxidized form of the coenzyme. Since the natural electron acceptors for these reactions are frequently unavailable, flavoproteins are often assayed with the use of artificial electron acceptors (Table 15-6) such as methylene blue, indophenol, phenazine or tetrazolium dyes, or molecular oxygen.

Both enzymatic and nonenzymatic reductions of flavoproteins are characterized by the appearance of transient electron paramagnetic resonance signals and absorption bands in the 550- to 700-nm range. These are consistent with the reduction occurring via two consecutive one-electron transfers involving the flavin semiquinone as intermediates:

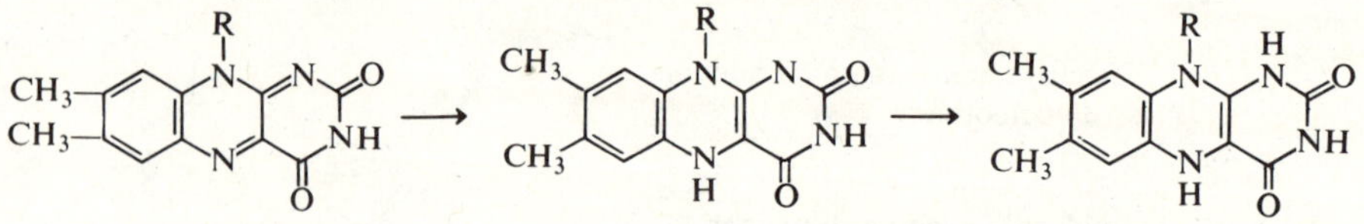

$$\longrightarrow \qquad \longrightarrow \qquad (8\text{-}57)$$

(λ_{max} = 280, ~ 370, ~ 450 nm) (λ_{max} = 280, ~ 370, ~ 450, ~ 590 nm) (λ_{max} = 280, 370 nm)

It appears likely that many enzymatic dehydrogenation reactions catalyzed by flavoproteins proceed via the flavin semiquinones.

Details of individual flavoproteins are considered in Chapter 15.

HEME COENZYMES

Coenzymes derived from a metal ion and the porphyrin nucleus occupy a central position in the life process. For example, chlorophyll, a magnesium chelate of a substituted porphyrin, is intimately involved in the photosynthetic process, the primary mode of radiant-energy utilization for living systems (see Chapter 12). In this chapter we shall concentrate on biochemical functions of coenzymes derived from porphyrins containing iron as the chelated metal.

Porphyrins are derived from the parent tetrapyrrole porphin (XIV) by the replacement of hydrogens at positions 1 to 8 with a variety of side chains, particularly methyl, ethyl, vinyl, and propionic acid functions:

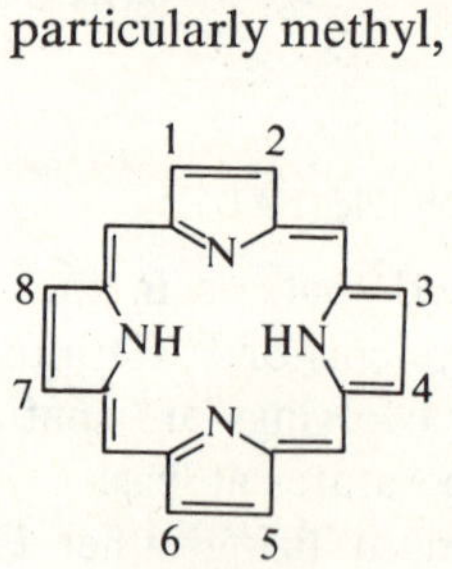

XIV

Porphyrins are classified on the basis of the nature of these side chains. Particularly important classes include the *etioporphyrins,* which have four methyl and four ethyl groups as ring substituents; the *mesoporphyrins,* which have four methyl, two ethyl, and two propionic acid functions; the *protoporphyrins,* characterized by four methyl, two vinyl, and two propionic acid functions; and the *coproporphyrins,* which contain four methyl and four propionic acid groups as side chains.

Porphyrins that contain equal numbers of two types of side chains, such as the etioporphyrins or coproporphyrins, may exist in 4 isomeric forms, depending on the manner in which the side chains are disposed about the porphin nucleus. Porphyrins such as the protoporphyrins, which contain three types of side chains, may exist in 15 isomeric forms.

Porphyrins were mentioned in the literature as early as the 1840s, and the first pure porphyrin was isolated at about the turn of the century. The correct ring structure of the porphyrin nucleus was initially proposed by Kuster in 1913, was abandoned by him in later years, and was finally established by the brilliant synthetic efforts of H. Fischer and his coworkers. A monumental monograph summarizing the developments in the chemistry and biochemistry of porphyrins and their derivatives up to 1949 has been prepared by Lemberg and Legge.

Porphyrins form metal chelates with a variety of metal ions, including those of magnesium, iron, zinc, nickel, cobalt, copper, and silver. In such chelates, the metal ion lies in the center of the porphin nucleus, four of its ligand sites being occupied by the pyrrole nitrogen atoms. Of these complexes, that consisting of iron and protoporphyrin IX (XV), generally referred to as *heme* if the iron is in the ferrous state and *hematin* if the iron is in the ferric state, are the most important biochemically:

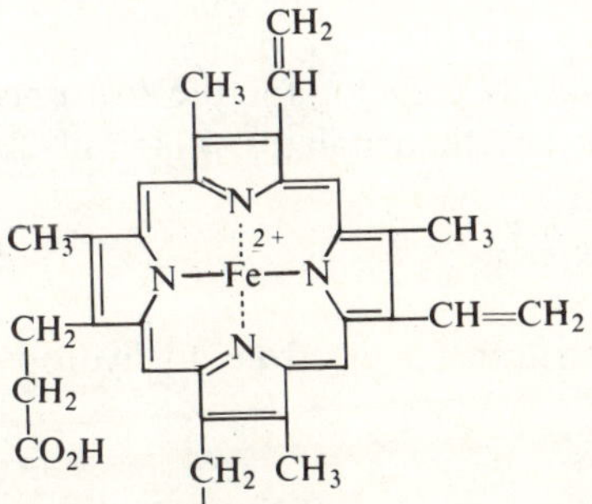

Protoporphyrin IX

Heme serves as coenzyme for proteins involved in the transfer of oxygen from one site in the organism to another, for enzymes catalyzing a variety of oxidation reactions, and for enzymes catalyzing the cleavage of peroxides (see Chapter 15).

Human hemoglobins, a class of proteins of distinct molecular structure that perform the important function of transporting oxygen from the lungs to various tissues in the body in which oxidative metabolism occurs, are, without question, the most thoroughly investigated proteins, as to both their chemistry and their biological function. The detailed structural chemistry of hemoglobin was considered in Chapter 4. It will be recalled that hemoglobin is composed of four polypeptide chains of known primary structure, which usually occur in pairs, together with four heme groups bound to the globin moiety through weak forces. The suggestion that the fifth site of the heme iron atom is occupied by an imidazole group of a histidine residue of the globin moiety has been confirmed by wide-angle X-ray diffraction studies (85). Hemoglobin has the remarkable property of combining in reversible fashion with molecular oxygen, yielding oxyhemoglobin, in which oxygen becomes the sixth ligand of the iron atom, without the simultaneous oxidation of the iron atom. Indeed, *methemoglobin,* in which the iron has been oxidized to the ferric state, is incapable of combining with oxygen. In addition to oxygen, hemoglobin also combines with several other small molecules or ions, and forms a very tight complex with carbon monoxide, accounting for the known toxicity of this substance. Heme also serves as coenzyme for other proteins performing essentially the same biochemical function, such as myoglobin, the oxygen-carrying protein of muscle, which has a molecular weight approximately one fourth that of hemoglobin and consists of a single polypeptide chain.

The *cytochromes,* a class of oxidation–reduction enzymes that are principally concerned with the transfer of electrons from flavoproteins to oxygen or other electron acceptors, also employ heme as coenzyme. These enzymes are discussed in some detail in Chapter 15. In contrast to the case of hemoglobin, in which the iron remains in the ferrous state throughout the biochemical process, the iron atom in the heme moiety of cytochromes undergoes cyclic oxidation and reduction in the course of the reactions catalyzed by these enzymes. Wang and Brinigar have constructed an elegant chemical model for the cytochrome oxidase reaction (86). They have demonstrated that an insoluble polymer formed from heme, 4,4-bipyrridyl, and polylysine is an exceptionally effective catalyst for the transfer of electrons from cytochrome *c* to molecular oxygen.

Finally, heme also serves as cofactor for the *hydroperoxidases.* These enzymes occur in two categories: the catalases, which liberate molecular oxygen from hydrogen peroxide:

$$H_2O_2 \rightarrow H_2O + \tfrac{1}{2}O_2 \tag{8-58}$$

and the peroxidases, which transfer electrons to hydroperoxides from appropriate electron donors:

$$AH_2 + H_2O_2 \rightarrow A + 2H_2O \tag{8-59}$$

In contrast to the behavior of the heme iron atom in both hemoglobins and cytochromes, the catalytic process of the hydroperoxidases appears to result in the transient oxidation of ferric iron to higher valence (ferryl) states. Many complexes of ferric iron are known to catalyze the decomposition of hydrogen peroxide and may reasonably be cited as model reactions for the catalases (87). Of particular interest is the ferric chelate of triethylenetetramine, $H_2NCH_2CH_2NHCH_2CH_2NHCH_2CH_2NH_2$, studied by Wang. In aqueous solution at 25°, this chelate catalyzes the decomposition of 11,000 mol of hydrogen peroxide per minute per mol of chelate. Although this reaction is still about 200 times slower that that of catalase, it does indicate that the turnover number for model reactions may reach enzymelike magnitudes.

Since for all these reactions, involving oxygen transport, oxidation–reduction, and the cleavage of peroxides, the same heme coenzyme is employed, the specificity of these enzymes for reaction type, as well as substrate, must reside in the protein portion of the holoenzyme. This is certainly one of the most striking examples of the specificity attribute of enzyme-catalyzed reactions. Some of the aspects controlling this specificity have been discussed by Orgel and Williams in terms of ligand field theory (88).

Details of the hemoproteins and their reactions are considered in Chapter 15; however, it is appropriate here to consider briefly how interaction of the heme group with proteins can elicit such a diversity of behavior from this group. There are at least three factors which may be important in this regard: (1) the nature of ligands at the fifth and sixth coordination sites, (2) distortion of the heme structure, and (3) involvement of neighboring groups.

Almost certainly the most important influence that a protein can have on the behavior of an associated heme group comes from direct interaction of basic groups of that protein with the heme iron atom. The heme iron atom has an octahedral coordination; four of the coordination sites are occupied by the nitrogen atoms of the tetrapyrrole structure. The other two, one directly above and the other directly below the iron atom and the plane of the tetrapyrrole ring, can be occupied by either a group on the protein or by a component of the solvent or environment. In oxygenated myoglobin, for example, the fifth site is occupied by an imidazole nitrogen of a histidine side chain and the sixth site by an oxygen molecule (page 151). The nature of these ligands is extremely important in determining the electron distribution within the heme group and hence its behavior in chemical reactions. Let us briefly consider how the ligand nature influences this electron distribution (89).

In ferric complexes there are five 3*d* electrons and in ferrous complexes six such electrons which must be allocated to molecular orbitals in the complex. There are two types of these molecular orbitals available: three nonbonding orbitals (t_{2g}) and two antibonding orbitals (e_g). Into which of these orbitals the 3*d* electrons of the iron atom will go is determined by two opposing factors. In the first place, the nonbonding orbitals are at a lower energy than the antibonding ones, and, hence, the electrons will have a natural tendency to occupy these. But, in the second place, the electrons will wish to distribute themselves as widely as possible among the available orbitals because of their mutual electrostatic repulsions. The balance between these two factors is determined by the nature of the ligand. Ligand field strength increases in the order $Cl^- < OH^- < RCO_2^- < F^- < H_2O < SCN^- < NH_3 < NO_2 < CN^-$. The stronger the ligand field, the greater the separation in energy between the nonbonding and antibonding molecular orbitals. Thus, a strong ligand field will tend to force all the electrons to occupy the nonbonding molecular orbitals. In this case, ferric complexes will have just one unpaired electron and ferrous complexes will have none. These situations are termed *low-spin states.* In contrast, with a weak ligand field, the electrons will occupy, singly, all the available molecular orbitals; ferric complexes will have five unpaired electrons and ferrous complexes will have four. These are termed *high-spin states.* Low-spin states are closely analogous to a description of the iron-ligand bond as covalent while high-spin states to one in which this bond is considered to be ionic.

The spin state of a protein may change on reaction. Thus, hemoglobin with one strong field ligand, imidazole, and one weak field one, water, is in a high-spin state. Replacement of the weak field ligand by a strong one, molecular oxygen, results in oxyhemoglobin being a low-spin state. We shall say more about these matters in Chapter 15.

The spin state of a hemoprotein (or other complex) is determined by magnetic susceptibility measurements which provide information concerning the number of unpaired electrons. Electron-spin-resonance spectroscopy is also an important tool in these studies.

In addition to determining the nature of the fifth and sixth iron ligands, proteins may influence the chemical behavior of the heme group by distorting it. Thus, in myoglobin, the iron atom is forced out of the plane of the tetrapyrrole moiety. Spectroscopic evidence suggests that in some hemoproteins, the heme group may be bent. Finally, groups of the protein may influence heme-group behavior by steric hindrance or by participating directly in the reaction as reactive functionalities.

How these various possibilities combine to determine the behavior of a given hemoprotein remains incompletely understood, although considerable insight has been gained in some cases, particularly for myoglobin and hemoglobin.

BIOTIN

Biotin, a growth factor for both yeast and humans (vitamin H), was first isolated by Kögl in 1935, 500 lb of dried egg yolk yielding 1 mg of this substance. The structure of biotin was established by du Vigneaud in 1942:

HN–C(=O)–NH fused bicyclic ring with S; side chain $CH_2CH_2CH_2CH_2CO_2H$

Biotin

Pioneering studies by Lardy and others strongly suggested that biotin functions as a coenzyme for carboxylation reactions, a suggestion now amply confirmed.

Biotin serves as coenzyme for two types of enzymatic reactions. The first class includes ATP-dependent carboxylations, with the cleavage of the ATP to ADP and inorganic phosphate. Typical examples include the acetyl-SCoA carboxylase [acetyl-CoA:carbon dioxide ligase (ADP)] reaction (90,91),

$$CH_3{-}\overset{O}{\overset{\|}{C}}{-}SCoA + CO_2 + ATP \rightarrow HO_2C{-}CH_2{-}\overset{O}{\overset{\|}{C}}{-}SCoA + ADP + P_i \tag{8-60}$$

the propionyl-SCoA carboxylase [propionyl-CoA:carbon dioxide ligase (ADP)] reaction (92),

$$CH_3{-}CH_2{-}\overset{O}{\overset{\|}{C}}{-}SCoA + CO_2 + ATP \rightarrow CH_3{-}\underset{H}{\overset{HO_2C}{C}}{-}\overset{O}{\overset{\|}{C}}{-}SCoA + ADP + P_i \tag{8-61}$$

and the β-methylcrotonyl-SCoA carboxylase [β-methylcrotonyl-CoA : carbon dioxide ligase (ADP)] reaction (93):

$$(CH_3)_2C{=}CH{-}\overset{O}{\overset{\|}{C}}{-}SCoA + CO_2 + ATP \rightarrow (HO_2C{-}CH_2)(CH_3)C{=}CH{-}\overset{O}{\overset{\|}{C}}{-}SCoA + ADP + P_i \tag{8-62}$$

In addition, biotin is involved in the carboxylation reaction leading to the formation of carbamyl phosphate (page 774), a reaction that is crucial to the synthesis of pyrimidines and in amino acid metabolism. The second type of biotin-dependent enzymatic reaction is exemplified by the methyl malonyl–oxalacetic transcarboxylase (methylmalonyl-CoA:pyruvate carboxyl-transferase) reaction (94):

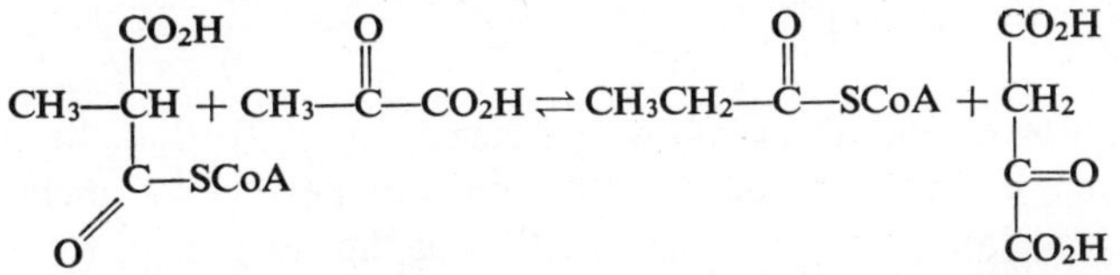

$$CH_3\text{—}\underset{\substack{|\\ C\text{—}SCoA \\ \| \\ O}}{\overset{\substack{CO_2H\\|}}{C}}H + CH_3\text{—}\overset{\substack{O\\ \|}}{C}\text{—}CO_2H \rightleftharpoons CH_3CH_2\text{—}\overset{\substack{O\\ \|}}{C}\text{—}SCoA + \underset{\substack{|\\ C=O\\ |\\ CO_2H}}{\overset{\substack{CO_2H\\|}}{C}}H_2 \qquad (8\text{-}63)$$

in which a carboxyl group is exchanged between substrates.

Studies indicating that biotin is very tightly bound to biotin-dependent enzymes, and the existence of ε-*N*-biotinyllysine (biocytin)

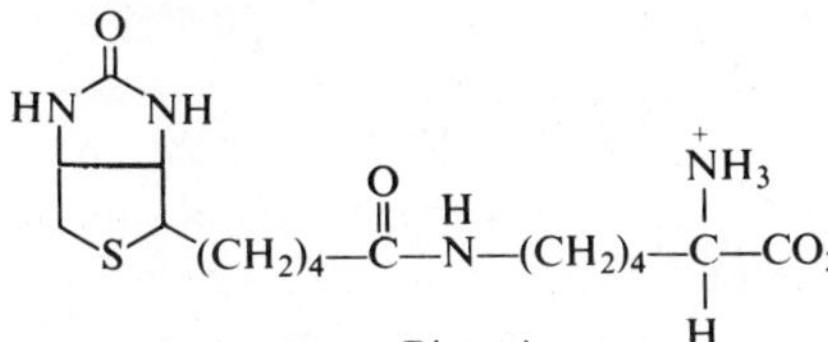

$$\text{[biotin ring: HN–C(=O)–NH, S]}\text{—}(CH_2)_4\text{—}\overset{\substack{O\\ \|}}{C}\text{—}\overset{H}{N}\text{—}(CH_2)_4\text{—}\underset{\substack{|\\ H}}{\overset{\substack{\overset{+}{N}H_3\\ |}}{C}}\text{—}CO_2^-$$

Biocytin

in natural sources suggested that biotin may be covalently bound to its enzymes through an amide linkage between the ε-amino group of a lysine residue of the enzymes and the carboxyl group of the valeric acid side chain of biotin. Lane and Lynen first definitely established that this is the case for propionyl carboxylase (95), and it is true for other biotin enzymes, as well.

The key to the understanding of the mechanism of action of biotin-dependent enzymatic reactions was provided by Lynen and his coworkers, who found that incubation of free (+)-biotin with β-methyl crotonyl carboxylase, in the absence of substrates other than CO_2, resulted in the production of a *carboxy-(+)-biotin* (96). Carboxybiotin is a labile molecule, particularly in the presence of acids, but may be stabilized by conversion to its dimethyl ester with diazomethane. The dimethyl ester of carboxy-(+)-biotin, synthesized by β-methyl crotonyl carboxylase, is identical to the product formed from the reaction of (+)-biotin methyl ester and methyl chloroformate, suggesting structure XVI for carboxyl-(+)-biotin, now known to be correct (97):

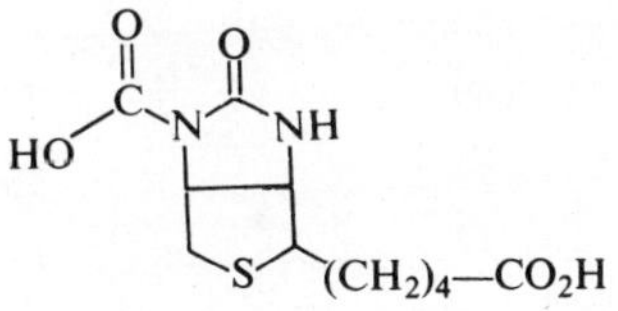

$$HO\text{—}\overset{\substack{O\\ \|}}{C}\text{—}N\text{[biotin ring: N–C(=O)–NH, S]}\text{—}(CH_2)_4\text{—}CO_2H$$

XVI

Although acetyl carboxylase and propionyl carboxylase do not catalyze the carboxylation of free biotin, they do, in the absence of substrate, fix radioactive carbon dioxide in amounts stoichiometric with the amount of bound biotin (93,97), as does β-methyl crotonyl carboxylase. Degradation of these, and related, enzymes has revealed that the structure of enzyme-bound carboxybiotin is the same as that determined for the free substance (XVI) (97).

As a result of these and related studies, biotin-dependent carboxylations may be considered to proceed by a two-step reaction path involving the intermediate formation of enzyme-bound carboxybiotin.

$$CO_2 + MgATP + \text{biotinyl-enzyme} \rightarrow MgADP + P_i + \text{carboxy-biotinyl-enzyme} \quad (8\text{-}64)$$

$$\text{Carboxy-biotinyl-enzyme} + \text{substrate} \rightarrow \text{biotinyl-enzyme} + \text{carboxy-substrate} \quad (8\text{-}65)$$

Transcarboxylations probably occur via a two-step sequence as well, enzyme-bound biotin serving as the intermediate carrier for the carboxyl group being transferred (98).

VITAMIN B_{12} COENZYMES

Following the observation by Murphy in 1926 that the liver contains a nutritional element required for the cure of pernicious anemia, more than twenty years elapsed prior to the isolation of the substance, vitamin B_{12}, in pure crystalline form. In the meantime, numerous biological manifestations of the nutritional lack of vitamin B_{12} were discovered, and, consequently, this material has received an abundance of names, including *anti-pernicious anemia factor, Factor X, L.L.D. factor,* and so forth. Early attempts by Cohn and others to purify vitamin B_{12} from liver met with only partial success in spite of many years of intensive effort. Purification procedures were severely hampered by the minute quantities of vitamin B_{12} present in the liver and by the necessity of employing pernicious anemia patients as a method of assay for this vitamin. In view of the fact that one must process factory quantities of liver in order to obtain workable amounts of purified B_{12}, it is not surprising that the preparation of the first pure crystals of this material was accomplished in industrial laboratories. In 1948, a group headed by Folkers in the United States and one headed by E. L. Smith in England announced, almost simultaneously, the isolation and crystallization of pure vitamin B_{12} from liver (99,100).

The determination of the structure of vitamin B_{12} was nearly as difficult as its isolation. Chemical studies, particularly in the laboratories of Folkers, Smith, and Todd, together with the brilliant X-ray diffraction studies of Hodgkin and her associates (101), eventually led to the formulation of B_{12} as

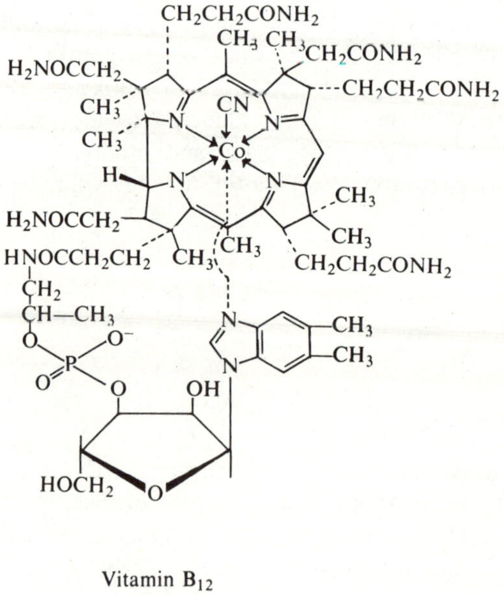

Vitamin B_{12}

Vitamin B_{12} is composed of two principal parts: the highly substituted, reduced, porphyrin-like *corrin* ring and the nucleotide, which, unlike those obtained from nucleic acids, contains an α-glycosidic linkage. The corrin ring contains tervalent cobalt chelated to the four nitrogen atoms of this ring, a nitrogen atom of the 5,6-dimethylbenzimidazole ring, and a cyanide ion, which is an artifact of the isolation procedure. The entire structure, except for the cyanide, is termed cobalamin; hence vitamin B_{12} is frequently referred to as *cyanocobalamin*. A number of additional cobalamins have been prepared, including hydroxocobalamin, aquocobalamin, and nitritocobalamin, in which the cyanide is replaced, respectively, by hydroxide ion, water, and nitrous acid (102).

In addition to these vitamin B_{12} analogs, another analog series exists in which the nitrogen base of the nucleotide moiety is varied. In addition to 5,6-dimethylbenzimidazole, adenine, 2-methyladenine, and guanine, among others, are found in naturally occurring vitamin B_{12} analogs. Many other biosynthetic analogs of the vitamin have been isolated from microorganisms grown on the desired base and factor B, a B_{12} analog containing no nucleotide.

Vitamin B_{12} itself is not active as a coenzyme for any known enzymatic reaction, but there exist two coenzymatically active derivatives. The first is a Co-5′-deoxyadenosylcorrinoid. Members of this family of coenzymes were first isolated by Barker and his coworkers in 1958 (103). Chemical and X-ray diffraction studies established that the cyanide group is displaced as a cobalt ligand by the 5-deoxyadenosyl group:

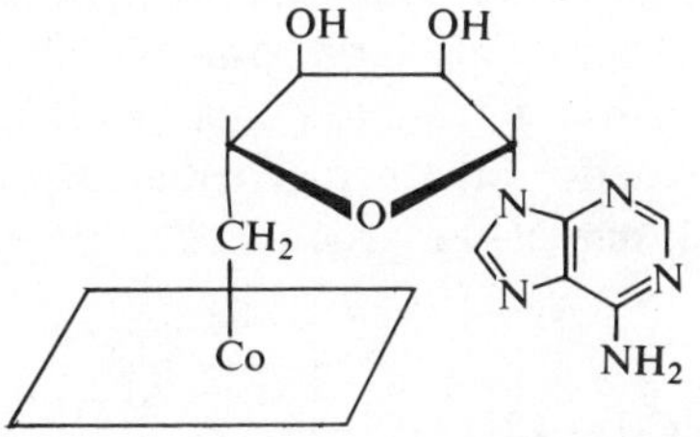

OH OH
CH$_2$ O N N N N NH$_2$
Co

There are several known enzymatic reactions which require this type of cofactor. All but one of these involves the intramolecular 1,2 shift of hydrogen coupled with a 1,2 shift of some other group. Consequently, such reactions can be represented by

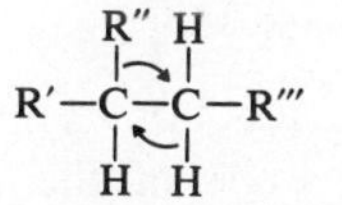

$$R'-\overset{R''}{\underset{H}{C}}-\overset{H}{\underset{H}{C}}-R''' \tag{8-66}$$

Enzymes catalyzing such reactions together with the reaction components are collected in Table 8-6. The only other enzyme requiring Co-5′-deoxyadenosylcorrinoids is ribonucleotide reductase from certain microorganisms (page 832); this reaction is similar to those listed above except that the donor of hydrogen and the acceptor of hydrogen in the reaction are different molecules.

Rearrangements of the type shown in Eq. (8-66) are not common reactions in organic chemistry and have posed a difficult mechanistic problem. Nevertheless, thanks largely to the work of Abeles and his associates, substantial insight into these reactions has now been gained. Let us focus on one reaction in particular to begin with and generalize our findings later. The diol dehydrase reaction has been particularly well studied (104–108):

$$CH_3-\underset{HO}{CH}-\underset{OH}{CH_2} \longrightarrow CH_3-CH_2-CHO \tag{8-67}$$

The crucial facts concerning this reaction are the following: First, the hydrogen atoms of substrate and product do not exchange with solvent. Second, the

TABLE 8-6
Reactions in which a Co-5′-deoxyadenosylcorrinoid participates as a coenzyme

$$\begin{array}{c} R'' \quad H \\ | \curvearrowright | \\ R'\!-\!C\!-\!C\!-\!R''' \\ | \curvearrowleft | \\ H \quad H \end{array}$$

	R′	*R″*	*R‴*
Glutamate mutase	H	$CH(NH_2)COOH$	COOH
Methylmalonyl-CoA mutase	H	COCoA	COOH
Dioldehydrase	CH_3	OH	OH
	H	NH_2	OH
Glycerol dehydrase	CH_2OH	OH	OH
Ethanolamine deaminase	H	NH_2	OH
β-Lysine isomerase	H	NH_2	$CH_2CH(NH_2)CH_2COOH$

reaction rate is about 12-fold slower when the deuterio substrate is employed instead of the protio substrate. This large primary deuterium-isotope effect suggests that the reaction is not a 1,2-hydride shift. Third, when the reaction is conducted in the simultaneous presence of tritiated propylene diol and unsubstituted ethylene glycol, one of the products is tritiated acetaldehyde:

$$\begin{array}{ccc} CH_3\text{—}CH\text{—}\overset{\overset{\large T}{|}}{C}\text{—}OH & & CH_3\text{—}CHT\text{—}CHO \\ \quad\ \ \underset{OH}{|} \quad \underset{H}{|} & & \\ + & \longrightarrow & + \\ \underset{\underset{OH}{|}}{CH_2}\text{—}\underset{\underset{OH}{|}}{CH_2} & & CH_2T\text{—}CHO \end{array} \tag{8-68}$$

This finding clearly demonstrates that the reaction is not necessarily intramolecular (although it does not preclude some of the reaction from being intramolecular). Fourth, double-labeling experiments have shown that the reaction can occur intramolecularly; this rules out concerted intermolecular processes and, together with the third point, indicates that the reaction must be stepwise. Fifth, incubation of tritiated propylene diol with the enzyme–coenzyme complex yields isolable quantities of a tritiated coenzyme. Incubation of the tritiated coenzyme with substrate results in the formation of tritiated product. Thus, the coenzyme evidently is an intermediate carrier of hydrogen between substrate and product. Sixth, chemical degradation of enzymatically prepared tritiated coenzyme reveals that the tritium is located specifically in the 5′ position. Thus, the path of hydrogen in the reaction is

$$CH_3\text{—}\underset{\underset{OH}{|}}{\overset{\overset{H}{|}}{C}}\text{—}CHTOH \longrightarrow \underset{\underset{Co}{|}}{\overset{\overset{R}{|}}{CH_2T}} \longrightarrow CH_3\text{—}CHT\text{—}CHO \tag{8-69}$$

Seventh, when L-propylene-1,2-diol labeled with ^{18}O in the 2 position is enzymatically converted to the propionaldehyde product, the ^{18}O is retained; similar experiments with the D isomer result in loss of the label. Thus, the reaction must proceed with intramolecular transfer of the hydroxyl function yielding a gem diol, which then undergoes stereospecific dehydration:

$$\text{L-}CH_3\text{—}\underset{^{18}OH}{\underset{|}{C}}HCH_2OH \longrightarrow CH_3\text{—}CH_2\text{—}\overset{OH}{\overset{|}{C}}H^{18}OH \longrightarrow CH_3\text{—}CH_2CH^{18}O + H_2O \qquad (8\text{-}70)$$

All these data are consistent with a mechanism involving abstraction of a hydride ion from the substrate by the coenzyme, followed by migration of the hydroxyl group and readdition of the hydride ion. Schematically we may write this mechanism in a form in which the carbon–cobalt bond of the coenzyme is broken with formation of a Co^{I}–carbonium intermediate: R^{+}—Co^{I}.

On the other hand, some of the reactions involving these coenzymes are more readily visualized as involving proton abstraction from the substrate together with formation of a Co^{III}—R^{-} complex. It has been suggested that the mode of cleavage of the carbon–cobalt bond in the B_{12} coenzymes may be dominated by the length of the cobalt–nitrogen bond at the fifth coordination site (109). Thus, if this bond is long, a Co^{I}—R^{+} species would be stabilized by the absence of strong electron donation from the ligand; a short bond would stabilize a Co^{III}—R^{-} species through electron donation. At any event, an open mind must be maintained concerning the details of the reaction mechanism; there is, in fact, no really firm evidence that the carbon–cobalt bond is broken at all in the reaction process.

The second class of B_{12} coenzymes are aquocorrinoids, in which the sixth coordination site is occupied by a water molecule. These coenzymes are involved in three biochemical processes: the synthesis of methionine (page 780), methane formation, and acetate synthesis. In each case, the B_{12} coenzyme appears to undergo transient methylation at the Co atom to form a structure analogous to the Co-5′-deoxyadenosylcorrinoids. In summary, these processes can be considered to occur by the following pathways:

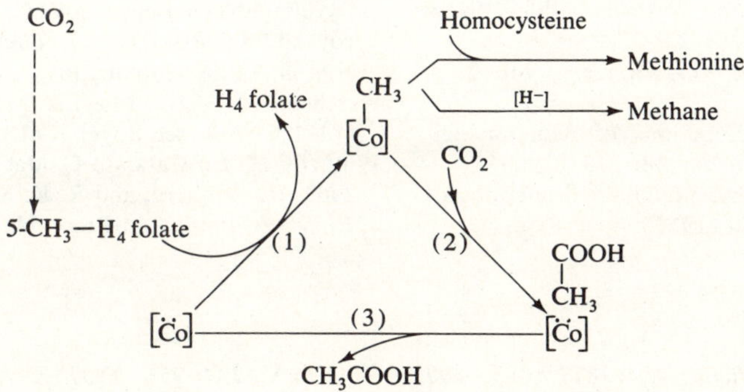

REFERENCES

Books Monographs and Review Articles

Baddiley, J.: The Structure of Coenzyme A, *Advan. Enzymol.,* **16:** 1 (1955).

Beinert, H., and R. H. Sands: Semiquinone Formation of Flavins and Flavoproteins, in M. S. Blois, Jr., et al. (eds.), *Free Radicals in Biological Systems,* Academic Press, New York, 1961, p. 17.

Bock, R. M.: Adenine Nucleotides and Properties of Pyrophosphate Compounds, in *The Enzymes,* vol. 2, p. 3 (see Boyer et al.).

Boyer, P. D., H. Lardy, and K. Myrbäck (eds.): *The Enzymes,* Academic Press, New York, 1959–1963.

Braunstein, A. E.: Pyridoxal Phosphate, in *The Enzymes,*

vol. 2, p. 113 (see Boyer et al.).

Brown, G. M.: Biosynthesis of Water-Soluble Vitamins and Derived Coenzymes, *Physiol. Rev.,* **40:** 331 (1960).

Bruice, T. C., and S. J. Benkovic: *Bioorganic Mechanisms,* vol. II, Benjamin, New York, 1966, chap 8.

Caughey, W. G.: Porphyrin Proteins and Enzymes, *Ann. Rev. Biochem.,* **36:** 611 (1967).

Chance, B., R. W. Estabrook, and T. Yonetani (eds.): *Hemes and Hemoproteins,* Academic Press, New York, 1966.

Chaykin, S.: Nicotinamide Coenzymes, *Ann. Rev. Biochem.,* **36:** 149 (1967).

Colowick, S. P., J. Van Eys, and J. H. Park: Dehydrogenation, in M. Florkin and E. H. Stotz (eds.), *Comprehensive Biochemistry,* vol. 14, Elsevier, New York, 1966, p. 1.

Dixon, M., and E. C. Webb: *Enzymes,* Academic Press, New York, 2nd ed., 1964.

Friedkin, M.: Enzymatic Aspects of Folic Acid, *Ann. Rev. Biochem.,* **32:** 185 (1963).

Hogenkamp, H. P. C.: Enzymatic Reactions Involving Corrinoids, *Ann. Rev. Biochem.,* **37:** 225 (1968).

Huennekens, F. M.: The Role of Nucleotides and Coenzymes in Enzymatic Processes, in D. E. Green (ed.), *Currents in Biochemical Research,* Wiley (Interscience), New York, 1956, p. 493.

Huennekens, F. M., and M. J. Osborn: Folic Acid Coenzymes and One Carbon Metabolism, *Advan. Enzymol.* **21:** 369 (1959).

Hutchinson, D. W.: *Nucleotides and Coenzymes,* Wiley, New York, 1964.

Jaenicke, L., and F. Lynen: Coenzyme A, in *The Enzymes,* vol. 3, p. 3 (see Boyer et al.).

Jencks, W. P.: *Catalysis in Chemistry and Enzymology,* McGraw-Hill, New York, 1969.

Kaplan, N. O.: The Pyridine Coenzymes, in *The Enzymes,* vol. 3, p. 105 (see Boyer et al.).

King, T. E., H. S. Mason, and M. Morrison (eds.): *Oxidases and Related Redox Systems,* Wiley, New York, 1965.

Knox, W. E.: Glutathione, in *The Enzymes,* vol. 2, p. 253 (see Boyer et al.).

Krampitz, L. O.: Catalytic Functions of Thiamin Diphosphate, *Ann. Rev. Biochem.,* **38:** 213 (1969).

Lardy, H. A., and R. Peanasky: Metabolic Functions of Biotin, *Physiol. Rev.,* **33:** 560 (1953).

Lemberg, R., and J. W. Legge: *Hematin Compounds and Bile Pigments,* Wiley (Interscience), New York, 1949.

Lipmann, F.: Metabolic Generation and Utilization of Phosphate Bond Energy, *Advan. Enzymol.,* **1:** 99 (1941).

Lynen, F.: The Role of Biotin-Dependent Carboxylations in Biosynthetic Reactions, *Biochem. J.,* **102:** 381 (1967).

Metzler, D. E.: Thiamine Coenzymes, in *The Enzymes,* vol. 2, p. 295 (see Boyer et al.).

Rabinowitz, J. C.: Folic Acid, in *The Enymes,* vol. 2, p. 185 (see Boyer et al.).

Reed, L. J.: Lipoic Acid, in *The Enzymes,* vol. 3, p. 195 (see Boyer et al.).

Reed, L. J.: Metabolism and Function of Lipoic Acid, *Proceedings of the International Symposium on Enzyme Chemistry,* Maruzen, Tokyo, 1958, p. 71.

Schmidt, U., P. Grafen, K. Altland, and H. W. Goedde: Biochemistry and Chemistry of Lipoic Acids, *Advan. Enzymol.,* **32:** 423 (1969).

Sherman, H., et al.: Pyridoxine and Related Compounds, in W. H. Sebrell, Jr., and R. S. Harris (eds.), *The Vitamins,* vol. III, Academic Press, New York, 1954, chap. 14.

Smith, E. L.: *Vitamin* B_{12}, Methuen, London, 1960.

Snell, E. E.: Chemical Structure in Relation to Biological Activities of Vitamin B_6, *Vitamins and Hormones,* **16:** 77 (1958).

Snell, E. E., P. M. Fasella, A. Braunstein, and A. Rossi Fanelli (eds.): *Chemical and Biological Aspects of Pyridoxal Catalysis,* Pergamon Press, New York, 1963.

Strominger, J. L.: Nucleotide Intermediates in the Biosynthesis of Polysaccharides, *Angew. Chem.* (*Intern. Engl. Ed.*), **1:** 134 (1962).

Utter, M. F.: Guanosine and Inosine Nucleotides, in *The Enzymes,* vol. 2, p. 75 (see Boyer, et al.).

Vennesland, B., and F. H. Westheimer: Hydrogen Transport and Steric Specificity in Reactions Catalyzed by Pyridine Nucleotide Dehydrogenases, in W. D. McElroy and B. Glass (eds.), *The Mechanism of Enzyme Action,* Johns Hopkins Press, Baltimore, 1954.

Westheimer, F. H.: Enzyme Models, in *The Enzymes,* vol. 1, p. 259 (see Boyer et al.).

Winfield, M. E.: Catalase Oxidation Mechanisms, in J. E. Falk, R. Lemberg, and R. K. Morton (eds.), *Haematin Enzymes,* Pergamon Press, New York, 1961, p. 245.

Specific References

1. Park, J. T.: *J. Biol. Chem.,* **194:** 877, 885, 897 (1952).
2. Munch-Petersen, A., H. M. Kalckar, E. Cutolo, and E. E. B. Smith: *Nature,* **172:** 1036 (1953).
3. Glaser, L., and D. H. Brown: *Proc. Natl. Acad. Sci. U.S.,* **41:** 253 (1955).
4. Caputto, R., L. F. Leloir, C. E. Cardini, and A. C. Paladini: *J. Biol. Chem.,* **184:** 333 (1950).
5. Kennedy, E. P., and S. B. Weiss: *J. Am. Chem. Soc.,* **77:** 250 (1955).
6. Kennedy, E. P., and S. B. Weiss: *J. Biol. Chem.,* **222:** 193 (1956).
7. Borkenhagen, L. F., and E. P. Kennedy: *J. Biol. Chem.,* **227:** 951 (1957).
8. Lipmann, F.: *J. Biol. Chem.,* **160:** 173 (1945).
9. Nachmansohn, D., and M. Berman: *J. Biol. Chem.,* **165:** 551 (1946).
10. Soodak, M., and F. Lipmann: *J. Biol. Chem.,* **175:** 999 (1948).
11. Stern, J. R., and S. Ochoa: *J. Biol. Chem.,* **179:** 491 (1949).
12. Lynen, F., E. Reichert, and L. Rueff: *Ann.,* **574:** 1 (1951).
13. Lynen, F., and E. Reichert: *Angew. Chem.,* **63:** 47 (1951).
14. Reed, L. J., I. C. Gunsalus, G. H. F. Schnakenberg,

Q. F. Soper, H. E. Boaz, S. F. Kern, and T. V. Parke: *J. Am. Chem. Soc.*, **75:** 1267 (1953).

15. Racker, E.: *J. Biol. Chem.*, **190:** 685 (1951).
16. Lohmann, K.: *Biochem. Z.*, **254:** 332 (1932).
17. Cliffe, E. E., and S. G. Waley: *Biochem. J.*, **73:** 25P (1959).
18. Franzen, V.: *Chem. Ber.*, **89:** 1020 (1956).
19. Franzen, V.: *Chem. Ber.*, **90:** 623 (1957).
20. Rose, I. A.: *Biochim. Biophys. Acta*, **25:** 214 (1957).
21. Snell, E. E.: *J. Biol. Chem.*, **154:** 313 (1944).
22. Heinert, D., and A. E. Martell: *J. Am. Chem. Soc.*, **84:** 3257 (1962); **85:** 183 (1963).
23. Dudek, G., and E. P. Dudek: *Chem. Commun.*, 464 (1965).
24. Snell, E. E.: *J. Am. Chem. Soc.*, **67:** 194 (1945).
25. Metzler, D. E., M. Ikawa, and E. E. Snell: *J. Am. Chem. Soc.*, **76:** 648 (1954).
26. Braunstein, A. E., and M. M. Shemyakin: *Biokhimiya*, **18:** 393 (1953).
27. Banks, B. E. C., A. A. Diamantis, and C. A. Vernon: *J. Chem. Soc.*, **1961:** 4235.
28. Bruice, T. C., and R. M. Topping: *J. Am. Chem. Soc.*, **85:** 1480, 1488, 1493 (1963).
29. Thanassi, J. W., A. R. Butler, and T. C. Bruice: *Biochemistry*, **4:** 1463 (1965).
30. Auld, D. S., and T. C. Bruice: *J. Am. Chem. Soc.*, **89:** 2083, 2090, 2098, 4250, 4251 (1967).
31. Jenkins, W. T., and I. W. Sizer: *J. Biol. Chem.*, **234:** 1179 (1959).
32. Snell, E. E., and W. T. Jenkins: *J. Cellular Comp. Physiol.*, **54:** suppl. 1, 161 (1959).
33. Cordes, E. H., and W. P. Jencks: *Biochemistry*, **1:** 773 (1962).
34. Hammes, G. G., and P. Fasella: *J. Am. Chem. Soc.*, **84:** 4644 (1962).
35. Lohmann, K., and P. Schuster: *Biochem. Z.*, **294:** 188 (1937).
36. Ugai, T., S. Tanaka, and S. Dokowa: *J. Pharm. Soc. Japan*, **63:** 269 (1943).
37. Mizuhara, S., and P. Handler: *J. Am. Chem. Soc.*, **76:** 571 (1954).
38. Koffler, H., and L. O. Krampitz: *Bacteriol. Proc.*, **1955:** 133.
39. Manzo, E. Y., F. Roddy, R. G. Yount, and D. E. Metzler: *J. Biol. Chem.*, **234:** 733 (1959).
40. Breslow, R.: *J. Am. Chem. Soc.*, **79:** 1762 (1957).
41. Ullrich, J., and A. Mannschreck: *Biochim. Biophys. Acta*, **115:** 46 (1966).
42. Breslow, R. E.: *J. Am. Chem. Soc.*, **80:** 3719 (1958).
43. Downes, J., and P. Sykes: *Chem. Ind.*, **1957:** 1095.
44. Mieyal, J. J., R. G. Votaw, L. O. Krampitz, and H. Z. Sable: *Biochim. Biophys. Acta*, **141:** 205 (1967).
45. Krampitz, L. O., I. Suzuki, and G. Greull: *Federation Proc.*, **20:** 971 (1961).
46. Holzer, H., and K. Beauchamp: *Angew Chem.*, **71:** 776 (1959); *Biochim. Biophys. Acta*, **46:** 225 (1961).
47. Holzer, H., H. W. Goedde, K. H. Göggel, and B. Ulrich: *Biochem. Biophys. Res. Commun.*, **3:** 599 (1960).
48. Goedde, H. W., H. Inouye, and H. Holzer: *Biochim. Biophys. Acta*, **50:** 41 (1961).
49. Scriba, P., and H. Holzer: *Biochem. Z.*, **334:** 473 (1961).
50. Holzer, H., R. Kattermann, and D. Busch: *Biochem. Biophys. Res. Commun.*, **7:** 167 (1962).
51. Datta, A. G., and E. Racker: *J. Biol. Chem.*, **236:** 624 (1961).
52. Lienhard, G. E.: *J. Am. Chem. Soc.*, **88:** 5642 (1966).
53. Hillcoat, B. L. and R. L. Blakley: *Biochem. Biophys. Res. Commun.*, **15:** 303 (1964).
54. Pastore, E. J., M. Friedkin, and O. Jardetzky: *J. Am. Chem. Soc.*, **85:** 3058 (1963).
55. Kallen, R. G., and W. P. Jencks, *J. Biol. Chem.*, **241:** 5845 (1966).
56. Whitely, H. R., and F. M. Huennekens: *J. Biol. Chem.*, **237:** 1290 (1962).
57. Welch, W. H., C. L. Irwin, and R. H. Himes: *Biochem. Biophys. Res. Commun.*, **30:** 255 (1968).
58. Scott, J. M., and J. C. Rabinowitz: *Biochem. Biophys. Res. Commun.*, **29:** 418 (1967).
59. Osborn, M. J., P. T. Talbert, and F. M. Huennekens: *J. Am. Chem. Soc.*, **82:** 4921 (1960).
60. Kallen, R. G., and W. P. Jencks: *J. Biol. Chem.*, **241:** 5851 (1966).
61. Larrabee, A. R., S. Rosenthal, R. E. Cathou, and J. M. Buchanan: *J. Am. Chem. Soc.*, **83:** 4094 (1961).
62. Larrabee, A. R., S. Rosenthal, R. E. Cathou, and J. M. Buchanan: *J. Biol. Chem.*, **238:** 1025 (1963).
63. Kisliuk, R. L., *J. Biol. Chem.*, **238:** 397 (1963).
64. Wood, J. M., and P. S. Wolfe, *Biochem. Biophys. Res. Commun.*, **22:** 119 (1966).
65. Wagner, C., S. M. Lusty, Jr., H-F. Kung, and N. L. Rogers: *J. Biol. Chem.*, **241:** 1923 (1966).
66. Warburg, O., W. Christian, and A. Griese: *Biochem. Z.*, **282:** 157 (1935).
67. Pullman, M. E., and S. P. Colowick: *J. Biol. Chem.*, **206:** 121 (1954).
68. Pullman, M. E., A. San Pietro, and S. P. Colowick: *J. Biol. Chem.*, **206:** 129 (1954).
69. Cornforth, J. W., G. Ryback, G. Popják, C. Donninger, and G. Schroepfer, Jr.: *Biochem. Biophys. Res. Commun.*, **9:** 371 (1962).
70. Sarma, R. H., and N. O. Kaplan: *J. Biol. Chem.*, **244:** 771 (1969).
71. Ratel, D. J.: *Nature*, **221:** 1239 (1969).
72. Weber, G.: *Nature*, **180:** 1409 (1957).
73. Velick, S. F.: *J. Biol. Chem.*, **233:** 1455 (1958).
74. Jardetzky, O., and N. G. Wade-Jardetzky: *J. Biol. Chem.*, **241:** 85 (1966).
75. Sarma, R. H., V. Ross, and N. O. Kaplan: *Biochemistry*, **7:** 3052 (1968).
76. Miles, D. W., and D. W. Urry: *J. Biol. Chem.*, **243:** 4181 (1968).
77. Abeles, R. H., R. F. Hutton, and F. H. Westheimer: *J. Am. Chem. Soc.*, **79:** 712 (1957).
78. Abeles, R., and F. H. Westheimer: *J. Am. Chem. Soc.*, **80:** 5459 (1958).
79. Anderson, B., C. J. Ciotti, and N. O. Kaplan: *J. Biol. Chem.*, **234:** 1219 (1959).
80. Kaplan, N. O., M. M. Ciotti, M. Hamolsky, and R. E. Bieber: *Science*, **131:** 392 (1960).
81. Warburg, O., and W. Christian: *Biochem. Z.*, **254:** 438 (1932).
82. Listowsky, I., S. Englard, J. J. Betheil, and S. Seifter: *Biochemistry*, **5:** 2548 (1966).

83. Chassy, B. M., and D. B. McCormick: *Biochemistry,* **4:** 2612 (1965).
84. Miles, D. W., and D. W. Urry: *Biochemistry,* **7:** 2791 (1968).
85. Cullis, A. F., H. Muirhead, M. F. Perutz, M. G. Rossmann, and A. C. T. North: *Proc. Roy. Soc.* (*London*), **A265:** 161 (1961–1962).
86. Wang, J. H., and W. S. Brinigar: *Proc. Natl. Acad. Sci. U.S.,* **46:** 958 (1960).
87. Wang, J. H.: *J. Am. Chem. Soc.,* **77:** 822 (1955).
88. Orgel, L. E.: in *Haematin Enzymes,* p. 17 (see Winfield); R. J. P. Williams, in *The Enzymes,* vol. 1 p. 391 (see Boyer et al.).
89. Ehrenberg, A.: *Svensk Kem. Tidskr.,* **74:** 103 (1962).
90. Wakil, S. J., E. B. Titchener, and D. M. Gibson: *Biochem. Biophys. Acta,* **29:** 225 (1958).
91. Wakil, S. J., and D. M. Gibson: *Biochem. Biophys. Acta,* **41:** 122 (1960).
92. Ochoa, S., and Y. Kaziro: *Federation Proc.,* **20:** 982 (1961).
93. Bachhawat, B. K., W. G. Robinson, and M. J. Coon: *J. Biol. Chem.,* **219:** 539 (1956).
94. Wood, H. G., S. H. G. Allen, R. Stjernholm, and B. Jacobson: *J. Biol. Chem.,* **238:** 547 (1963).
95. Lane, M. D., and F. Lynen: *Proc. Natl. Acad. Sci. U.S.,* **49:** 379 (1963).
96. Lynen, F.: *J. Cellular Comp. Physiol.,* **54:** suppl. 1, 33 (1959).
97. Knappe, J., K. Biederbick, and W. Brümmer: *Angew. Chem.,* **74:** 432 (1962).
98. Wood, H. G., and R. Stjernholm: *Proc. Natl. Acad. Sci. U.S.,* **47:** 289 (1961).
99. Rickes, E. L., N. G. Brink, F. R. Koniuszy, T. R. Wood, and K. Folkers: *Science,* **107:** 396 (1948).
100. Smith, E. L., and L. F. J. Parker: *Biochem. J.,* **43:** viii (1948).
101. Hodgkin, D. C., J. Kamper, J. Lindsey, M. MacKay, J. Pickworth, J. H. Robertson, C. B. Shoemaker, J. G. White, R. J. Prosen, and K. N. Trueblood: *Proc. Roy. Soc.* (*London*), **A242:** 228 (1957).
102. Smith, E. L., S. Ball, and D. M. Ireland: *Biochem. J.,* **52:** 395 (1952).
103. Barker, H. A., H. Weissbach, and R. D. Smyth: *Proc. Natl. Acad. Sci. U.S.,* **44:** 1093 (1958).
104. Frey, P. A., M. K. Essenberg, and R. H. Abeles: *J. Biol. Chem.,* **242:** 5369 (1967).
105. Wagner, O. W., H. A. Lee, Jr., P. A. Frey, and R. H. Abeles: *J. Biol. Chem.,* **241:** 1751 (1966).
106. Zagalak, B., P. A. Frey, G. L. Karabatsos, and R. H. Abeles: *J. Biol. Chem.,* **241:** 3028 (1966).
107. Rétey, J., A. Umani-Ronchi, J. Seibl, and D. Arigoni: *Experientia,* **22:** 502 (1966).
108. Schrauzer, G. N., and R. J. Windgassen: *J. Am. Chem. Soc.,* **89:** 143, 1999 (1967).
109. Brodie, J. D.: *Proc. Natl. Acad. Sci. U.S.,* **62:** 461 (1969).

9
Intracellular Organization and Particles

The study of cellular organization and attempts to correlate structure and function at the various hierarchical levels involved—from simple molecules to macromolecules, to aggregates such as membranes or particles, to subcellular units, to cells—comprise some of the most exciting, challenging, and rewarding fields of inquiry in contemporary biology. For the biochemist and cellular physiologist an understanding in chemical terms of the various complex structural elements that have been identified inside cells is not only of fundamental importance in its own right but must of necessity form the first step in his attempts at studying their modes of synthesis, breakdown, and interaction. These complex structures, we are beginning to suspect, hold much of the secret of how cellular processes are controlled in both space and time: the secret may consist, at least in part, of isolating and maintaining the different cellular constituents—mainly enzymes, together with their substrates, products, and modifiers (activators, inhibitors)—in different compartments; sometimes allowing, sometimes denying mutual accessibility. Implicit in this statement are two premises—amply verified experimentally in the recent past—(1) that there is a distinct and discernible *localization* of some key cellular constituents, especially of enzymes, in (or on) certain cellular structures (or "organelles," i.e., minuscule intracellular organ analogs) and (2) that it should be possible to separate the latter and, *mutatis mutandis,* the former as well, by the suitably mild techniques of *cell disruption* and *fractionation,*[1]

Such procedures then eliminate the elegant but complex, and frequently laborious, techniques of *histochemistry* and *cytochemistry* and substitute for them the methodology of the ordinary biochemical laboratory. Yet ideally they lead to a complete chemical and functional characterization of the fractions. It must be emphasized, however, that any real understanding of cellular processes must always be based on two independent lines of exploration, both converging toward the central problem: Structure as well as function must be explored, and phenomena observed with isolated components must constantly be referred to and checked against their counterparts in the intact cell. It is axiomatic that the very act of disruption and isolation of necessity

[1] This is the term commonly employed (see the references); grammatically it would probably be more correct to speak of fractionation of *subcellular components.*

does violence to the hierarchical order and may disrupt any existing levels of organization. An added, practical advantage of exploring fractionation procedures is that, once we know the intracellular distribution of an enzyme or other biologically active component, its initial purification is greatly facilitated by starting not with the total intracellular content but with the fraction of interest.

In the remainder of this chapter (1) we describe briefly what has been learned by examining the anatomy of the cell by means of the electron microscope and describe the structures observed in these investigations, (2) we discuss the methodology of cell fractionation and some of the properties of the fractions obtained by this technique, and (3) we try to correlate some of these findings in terms of biochemistry and enzymology.

STRUCTURE OF CELLS

It may surprise the reader to learn that an unambiguous identification of the cell as the fundamental unit of higher (eucaryotic, chromosomal) organisms goes back only to the year 1839, the year that the botanist Schleiden and the zoologist Schwann independently produced their *cell theory*. The next important finding in this context came in 1859, when Virchow demonstrated that all cells derived from other preexisting cells. Since then the structure of cells of a large number of animals and plants has been subject to the scrutiny of microscopic examination of ever-increasing sophistication and resolving power—first with the light microscope and more recently with the electron microscope. From these studies there has emerged a composite picture of cells as highly complicated entities, consisting of, not just a limiting membrane enclosing a blob of cytoplasm in which the chromatin-containing nucleus is suspended, but a number of variegated, highly organized, structured, convoluted, and interconnected elements, which vary from organism to organism, tissue to tissue, and cell type to cell type. Yet within all the complexity, some order can be discerned; although there is no such thing as a typical cell, almost all cells appear to have certain typical features. Some common subcellular structures can be identified that appear to bear a morphological and topological, and perhaps a functional, homology from cell to cell, regardless of origin. Let us then examine such a generalized animal cell on the basis of the electron micrograph of Fig. 9-1 and the drawing of Fig. 9-2.

Such a cell is an exceedingly small object, with an average diameter of the order of 20 μm (2×10^5 Å) and a volume of $\sim$5000 μm^3; the maximum resolution attainable with the electron microscope is 5 to 10 Å.

The Nucleus

The largest structure visible in most cells, the one first and most easily identified with the light microscope (especially after suitable staining, or in phase contrast), and the first intracellular organelle to be isolated in bulk (by Miescher in 1871) is the nucleus: a roughly spherical body, usually one per cell, approximately 5 μm in diameter and therefore with a volume of 40 μm^3. It is surrounded by and isolated from the cytoplasm by the triple-layered[2] nuclear membrane or envelope approximately 75 Å thick. This separation, however, is not absolute, since the membrane may possess numerous pores or apertures and is itself connected in many cells, such as in the liver and the pancreas, to another, cytoplasmic structure, the endoplasmic reticulum (see below). It is through these pores perhaps that certain macromolecules, syn-

[2] A dense double membrane with a less dense intramembranal space or separation in between. This *unit membrane* appears to be the fundamental feature of almost all membrane systems in nature (see the section on plasma membranes).

thesized within the nucleus but operational in the cytoplasm (RNA and perhaps even ribonucleoprotein particles), make their way from the one compartment to the other.

The interior of the nucleus (the *nucleoplasm*) appears itself organized, and usually one or more distinct, denser, spherical structures, called *nucleoli,* can be distinguished against it. These bodies are particularly rich in RNA, and the bulk of the nuclear RNA (which itself accounts for 10 to 20 percent of the total cellular RNA) appears to be localized within them. Almost all cellular DNA ($\geqq$ 95 percent) is found in the nucleus, distributed throughout the nucleoplasm as *chromatin* while the cell is in the "resting state" (i.e., engaged in maintenance and growth between divisions). In anticipation of division and during its course, the chromatin becomes highly organized into distinct linear structures called *chromosomes.* The number of chromosomes per somatic cell is constant, and this constant complement is passed on to a daughter cell as a result of mitotic division. In the course of the formation of the gametes this number is halved (meiosis). It is restored as a result of their fusion in zygote formation.

Mitochondria

The next set of structures to be identified chronologically (around 1900) and, in most cells, the next largest in size are the large granules, or *mitochondria.* When stained with such dyes as Janus green, they are close to the limit of resolution of light microscopy but are easily seen under phase contrast. The real advance of our understanding of their structure had to wait for the advent of the electron microscope and has only taken place during the past 20 years. Although the particles may vary considerably in number, size, and shape in different cells, their ultrastructure is sufficiently similar and distinct from that of other organelles so as to make their unambiguous identification easy in most cases and to point out their fundamental similarity, whether derived from protozoa, fungi, higher plants, or animals, including man. Their number varies from a dozen or so in yeast to several hundred per animal cell. Their dimensions are roughly those of an ellipsoid of revolution, with long and short axes of 1.5 and 0.5 μm, respectively, and with an average volume of some 0.8 μm^3. The most salient feature of their structure is a complex membrane system. It consists of a relatively smooth outer membrane, an intermembrane space, and a particularly highly structured inner membrane. The last forms numerous folds or invaginations called *cristae,* which penetrate deeply into the intramitochondrial *matrix* or inner compartment. Linked to the inner membrane and its cristae are numerous (several thousand) distinct small knoblike attachments. The main advantages of such a structure would appear to be the tremendous increase in surface area ($\geqq$ 10,000 μm^2 for the dimensions indicated), which permits unidirectional (vectorial) access of some reactants from one and others from the opposite side. Mitochondria have been subject to exhaustive and penetrating investigations over a good many years. These studies have resulted in a model that pictures these particles as the principal sites of energy production and transduction in all eucaryotic cells. The enzymes of the citric acid cycle (Chapter 14) as well as certain accessory enzymes such as the pyruvic dehydrogenase complex, the pyruvate carboxylase (Chapter 11), L-glutamate dehydrogenase, and certain transaminases (Chapter 17) which link carbohydrate and protein metabolism to this essential cycle, as well as the β-oxidation sequence for fatty acids (Chapter 13), all appear to be localized either in the intermembraneous (intercristael) space or inside the matrix confined by the inner

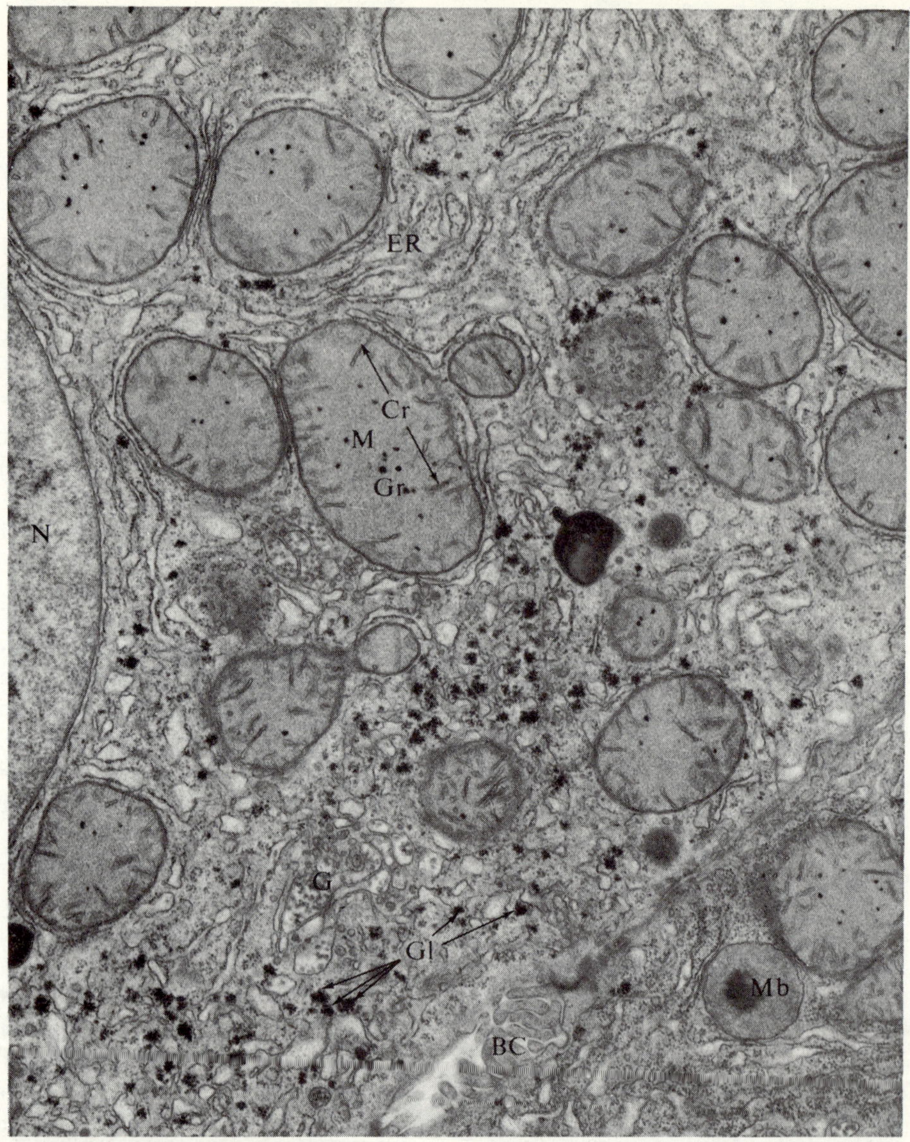

FIGURE 9-1

Cell organelles. This electron micrograph, which shows part of a rat liver parenchymal cell, is reproduced at a magnification sufficiently high (27,000 times) to show the structural details of several common cell components. The mitochondria (M), for example, are seen to be limited by two lines, which represent the two membranes constituting the wall of these tiny organelles. The inner of the two is folded into shelflike structures, the cristae (Cr), which project into the material (matrix) filling the cavity of the mitochondrion. It is probable that such membrane foldings provides expanded surfaces for the organized distribution of the many enzymes resident in these organelles. Where the energy requirements of the cell are greater, as in heart muscle, the number of cristae is greater, presumably to provide a larger surface area. Dense granules (Gr), 20 to 30 nm in diameter, also evident in the matrix, are now thought to represent accumulations of bound, divalent, metallic ions required for mitochondrial enzyme systems.

A small part of the cell nucleus (N) is included at the left of the micrograph. It is limited by a membranous envelope consisting of two membranes and an intervening space of low density. The outer of the two membranes has repeatedly been shown to be continuous with the membranes limiting the vesicular elements of the endoplasmic reticulum (ER). It is therefore reasonable to regard the cytoplasmic reticulum as an extension of the nuclear envelope and to consider both as parts of a single membranous system.

In the cytoplasm of the hepatic cell, two forms of this membranous system are regu-

membrane. That membrane itself appears to be the site of electron transport from NADH and succinate to molecular oxygen; and it together with its attachments also provides the sites for energy transduction (Chapter 15).

Endoplasmic Reticulum

Extending from the nuclear envelope outward all the way to the cell membrane and permeating large regions of the cytoplasm there exists in many cell types of a complex system of membranes (cytomembranes), the *endoplasmic reticulum.* It consists of flattened sacs (cisternae), tubules, and vesicles. Together they enclose and define a compartment called the *intracisternal space.* On electron micrographs two principal forms can be distinguished, depending on the number of attached, small, electron-dense, spherical particles: the *smooth-surfaced reticulum,* with few or no particles, and the *rough-surfaced reticulum,* which is lined with these particles (Figs. 9-3 and 9-4). The functions of the endoplasmic reticulum appear to be at least fourfold: first, to carry out a large number of important complex biosynthetic processes concerned with the elaboration of proteins and probably of complex lipids and polysaccharides as well; second, to transport these products to their sites of utilization or storage; third, to effect and control the utilization of newly synthesized and perhaps also of ingested complex molecules; and fourth (perhaps not unrelated to the first two), to function as a precursor for other membrane systems, such as the Golgi apparatus and possibly the plasma, the nuclear, and the outer mitochondrial membranes. Protein synthesis appears to be the special province of the rough-surfaced reticulum, and to be more precise, of the attached ribonucleoprotein granules or particles, called *ribosomes.* When isolated from eucaryotic cells their particle weight is $\sim 4.3 \times 10^6$ daltons, their sedimentation coefficient ~ 78 S and they contain ~ 47 percent protein and 53 percent RNA; those from bacteria have particle weights of 2.7×10^6 daltons, sedimentation coefficients of 70 S, and consist of RNA to the extent of 60 percent.

---

larly observed. In one (ER), the cisternae have ribosomes, small, dense granules (about 150 Å in diameter), on their surfaces. This is the configuration found in cells in which protein is synthesized for secretion. In the other form, the long, flat profiles of thin cisternae give way to shorter, more nearly circular profiles and are particle-free, or smooth. This part of the ER is in some places continuous with the particle-covered, or rough-surfaced units. In the hepatic cell, this smooth form is found associated with deposits of stored glycogen (Gl), and it is thought that enzyme systems important for storage and release of glycogen are associated with the limiting membranes. Ribosomes are seen throughout the field, some associated with the ER, others free on the cytoplasmic matrix.

The Golgi complex (G) also consists of vesicles limited by smooth-surfaced membranes, but in this instance they are flattened sacs or vesicles closely packed into piles or stacks. Within the Golgi vesicles, proteins and possibly other substances synthesized elsewhere in the cell are assembled into secretion granules. Rather small granules can be identified in a few sacs (near letter G), and these may represent plasma proteins known to be produced by hepatic cells. After cell homogenization, fragments of the Golgi complex, plus fragments derived from the reticulum, make up the "microsomal" fraction.

Other granular components of the liver cell, called microbodies (Mb), possess a homogeneous matrix and a dense nucleoid rich in the enzyme uricase.

In this micrograph, the structure of the bile canaliculi (BC) is clearly seen. This channel, formed by portions of the surfaces of adjacent hepatic cells, is studded with microvilli, which project from the cell surfaces into this extracellular space. Tight junctions and desmosomes seal off the canaliculus from adjoining intercellular spaces. Original micrograph courtesy Keith R. Porter. Legend adapted from that of Keith R. Porter and Mary A. Bonneville, *An Introduction to the Fine Structure of Cells and Tissues,* Lea and Febiger, Philadelphia, 1964.

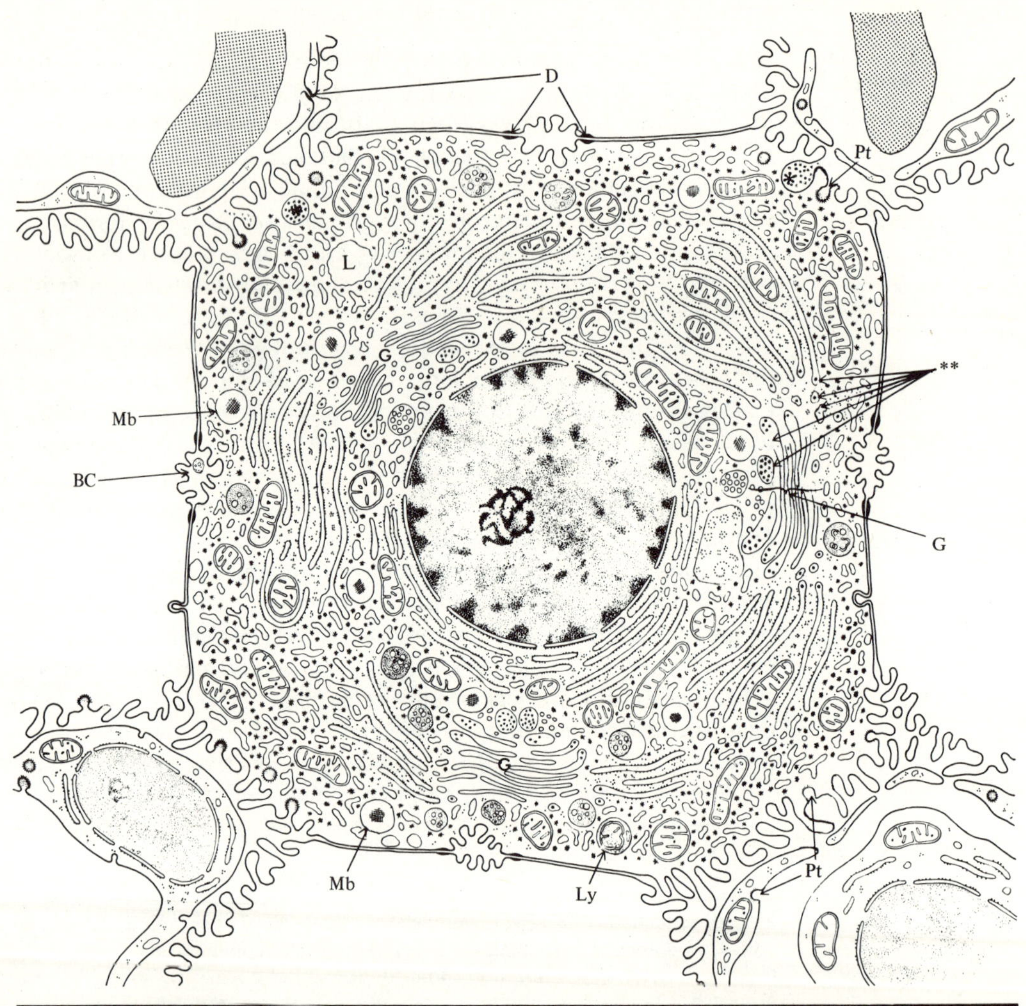

FIGURE 9-2

Diagrammatic representation of a liver cell. Arrangement of the organelles shown in Fig. 9-1 within a parenchymal cell of rat liver. We observe the nucleus, mitochondria, and the long, slender profiles (cisternae) belonging to the endoplasmic reticulum, or ER. Small particles or ribosomes are abundant on the surfaces of the latter and in the ground substance between them. The cisternae are ordinarily arranged in stacks of 6 to 12 units. One margin of such an assemblage frequently lies adjacent to the Golgi complex, and it will be noted that small granules of a particular type occupy the ends of these cisternae and the vesicles intervening between ER and Golgi, as well as the expanded ends of cisternae and spherical vesicles belonging to the Golgi proper (G). These images (**) are designed to show the mechanism of protein transport from ER to Golgi, where packaging for export takes place. The other margin of the stacks of ER cisternae borders on masses of glycogen and associated vesicles belonging to the so-called smooth endoplasmic reticulum, or SER. These two forms are often continuous, as though the smooth form may develop from the rough. Such continuities are indicated by thick arrows. It is thought by some, on incomplete evidence, that the smooth ER is involved in the transport of glucose from the liver cell during glycogenolysis. Other prominent components of the hepatic cell cytoplasm include lysosomes (Ly), containing remnants of organelles and ground substance apparently set aside for hydrolysis, and microbodies (Mb), of uncertain significance though probably rich in enzymes. A single lipid granule is indicated at L. From Keith R. Porter and Mary A. Bonneville. *An Introduction to the Fine Structure of Cells and Tissues,* Lea and Febiger, Philadelphia, 1964.

In cells designed for rapid protein synthesis and for the export of the product to other parts of the organism, such as the exocrine cells of the pancreas, most of the ribosomes are attached to the reticulum in linear arrays, and these structures are so common as to form the predominant feature of cellular architecture (Fig. 9-3). In other cell types, where protein synthesis is largely for internal consumption (e.g., in red blood cells, neurons, and embryonic and cancer cells), we find considerable smooth reticulum, some rough-surfaced reticulum, and numerous ribosomes unattached in the cytoplasm, where they are arranged in groups or clusters (called polyribosomes or polysomes). Polysomes and single ribosomes are, however, found also in cells that do contain a rough reticulum. Regardless of their origin or structural history (i.e., free or attached originally), ribosomes appear to be of remarkable uniformity throughout living nature. They consist of two unequal subunits, those of animal or plant origin[3] with particle weights of ≤ 3.1 and 1.2×10^6 daltons (1,2), and those of bacteria of 1.8 and 0.9×10^6 daltons. The properties of their RNA are discussed in Chapter 5, their biosynthesis and function in protein synthesis in Chapter 21.

Another smooth membranous organelle, functionally and perhaps structurally related to the endoplasmic reticulum, is the *Golgi apparatus* or complex. It is usually localized quite close to the nucleus in the region of the so-called *centrosomes* or *centrospheres* and consists of a stacked array of smooth sacs and of variable numbers of smooth-surfaced cisternae, vesicles, or vacuoles. Golgi complexes are most prevalent and well developed in secretory cells, such as the exocrine cells of the pancreas. There is good evidence of connections between the cisternae of the rough-surfaced reticulum and the vesicles of the Golgi complex, which in turn are linked to large vacuoles in the complex. Golgi vacuoles develop into secretory granules such as *zymogen granules,* which contain and store the proteins produced by the ribosomes of the rough-surfaced reticulum (Fig. 9-4).

Other Cytoplasmic Organelles

Other quite large cytoplasmic bodies surrounded by a membrane of the "unit" type (page 457), the *lysosomes,* are found in a large number of cell types but are especially abundant in leukocytes and in mammalian liver and kidney. They contain a sizable portion of the hydrolytic enzymes of the cell and may be the site of *intracellular digestion,* both of material ingested from the outside by the processes known as *pinocytosis* and *phagocytosis,* and—in cells undergoing rapid turnover and replacement or those subject to pathological degeneration—of cell organelles themselves. In the latter instance, they are then known as *autophagic vacuoles.* Lysosomes and bodies derived from them are also, in conjunction with the Golgi complex, concerned with the transport and secretion of newly synthesized macromolecular materials (Figs. 9-4 and 9-5).

Most specialized, highly differentiated cells contain additional spherical or vesicular bodies designed for specialized functions. Among them are the *peroxi(do)somes* (formerly called *microbodies*), concerned with metabolic reactions consuming and producing hydrogen peroxide,[4] found in a variety of animal and plant cells (9); the *synaptic vesicles* (10), storage granules for transmitter substances in the nervous system of animals; and others.

[3] Ribosomal RNA of the large subunit from plant cytosol may be somewhat smaller (3–5). Similar particles are also found in semiautonomous organelles such as chloroplasts and mitochondria—these ribosomes appear to resemble the bacterial type (4–8) in many of their properties (Chapter 21).

[4] Closely related bodies carrying key enzymes of the glyoxylate cycle (p. 527) are found in plants and protozoa; they are then known as *glyoxysomes* (Table 9-2).

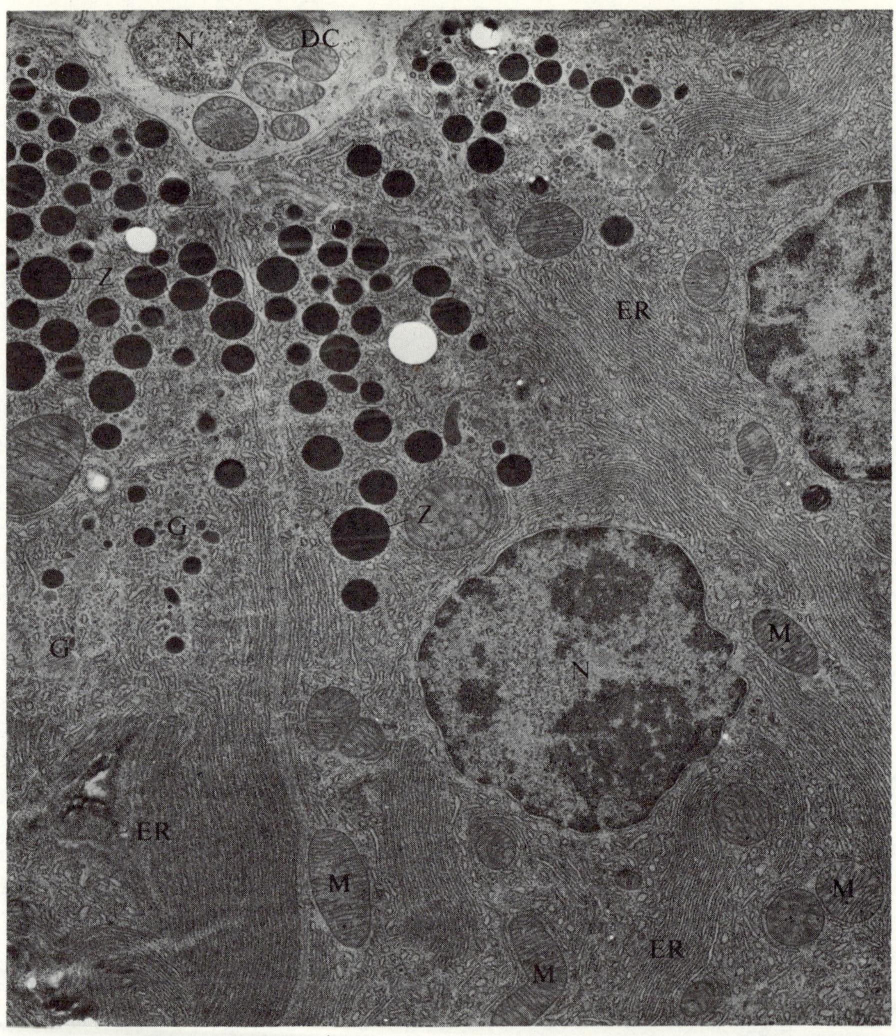

FIGURE 9-3

Pancreatic exocrine cell. Pancreas of the bat, magnification 11,500 times. The intracellular architecture of a cell designed for the manufacture and export of digestive enzymes. We see concentrated layers of endoplasmic reticulum (ER), studded with ribosomes (R in insert). G, Golgi complex; Z, zymogen granule; M, mitochondria; N, nucleus. The cells lining the ducts of the exocrine pancreas seem relatively inactive when compared to their secretory relatives. In this micrograph the cytoplasm and a small part of the nucleus (N′) of one such centroacinar duct cell (DC) lie near the tips (apices) of the glandular cells, and it is evident that this cell type contains only a few organelles and no complex membranous system. Original micrograph courtesy of Keith R. Porter.

FIGURE 9-4 (Right)

Structures and interrelationships in a pancreatic cell. (*a*) Diagram of the dispositions of organelles in a pancreatic acinar cell (see Fig. 9-3). (*b*) Schematic representation of the current views concerning the secretory pathway in the pancreatic acinar cell and the functional interrelations of its organelles. New protein synthesized at the ribosomes is segregated in the lumen of the endoplasmic reticulum and transported to the Golgi region. There it is concentrated and formed into zymogen granules that are stored in the apical cytoplasm and ultimately released at the free surface of the cell. Don W. Fawcett, Structural and Functional Variations in the Membranes of the Cytoplasm, in S. Seno and E. V. Cowdry (eds.), *Intracellular Membrane Structures,* Japan Society for Cell Biology, Okayama, Japan, 1963, p. 15.

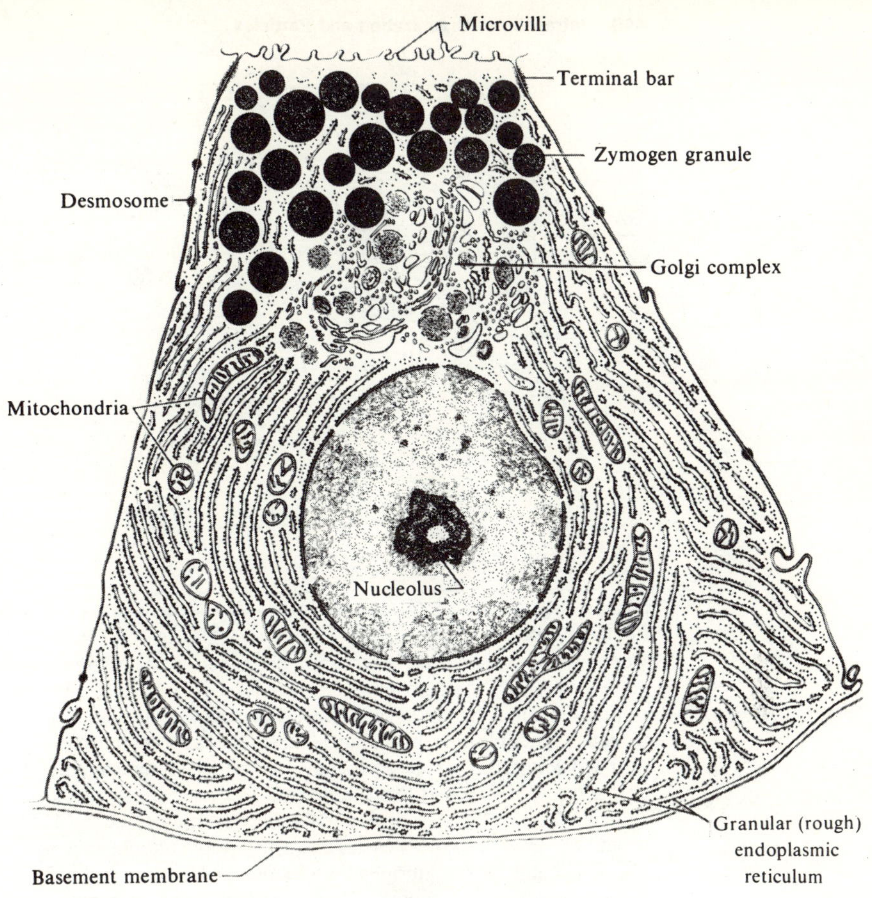
Microvilli
Terminal bar
Zymogen granule
Desmosome
Golgi complex
Mitochondria
Nucleolus
Granular (rough)
endoplasmic
reticulum
Basement membrane

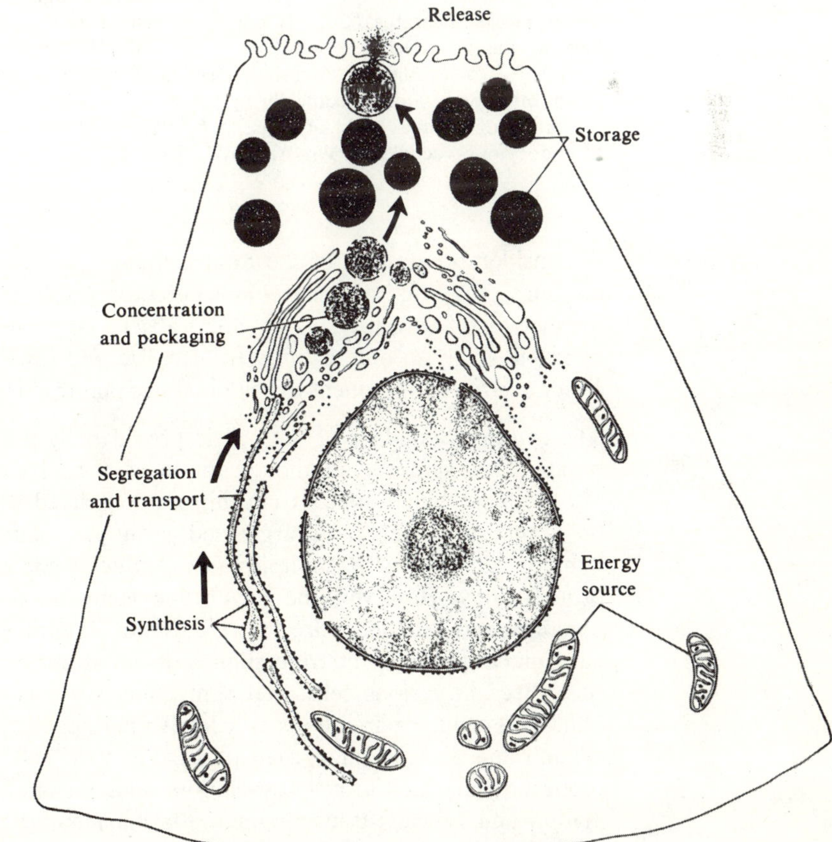
Release
Storage
Concentration
and packaging
Segregation
and transport
Energy
source
Synthesis

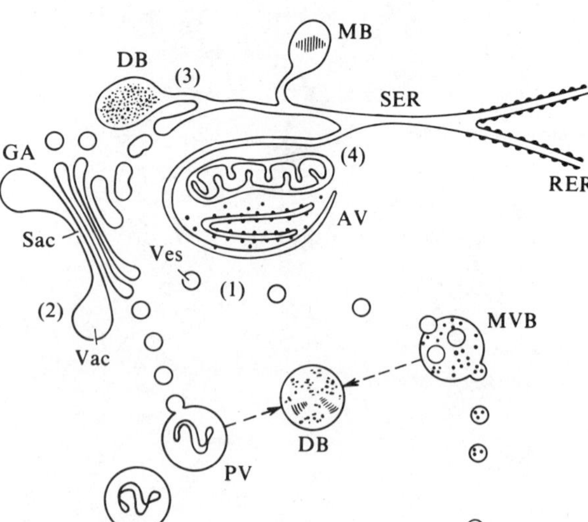

FIGURE 9-5

Current picture relating structure and function of various cytoplasmic elements. Suggested pathways for movement of materials from endoplasmic reticulum and external milieu. Proteins synthesized at the ribosomes (upper right) are considered to gain access to rough endoplasmic reticulum (RER) and move through smooth endoplasmic reticulum (SER), from which they may be routed to microbodies (MB) or lysosomes. Four pathways to lysosomes are indicated: (1) via Golgi saccules (Sac), to Golgi vesicles (Ves); (2) via Golgi saccules to Golgi vacuoles (Vac); (3) directly from endoplasmic reticulum into dense bodies (DB); (4) from endoplasmic reticulum into areas of cytoplasm isolated as an autophagic vacuole (AV). Endoplasmic reticulum, ribosomes, and a mitochondrion are shown within the forming autophagic vacuole. The polarized nature of the Golgi apparatus (GA) is suggested, with forming face to the right and mature face to the left. Golgi vesicles are shown moving into a pinocytotic or phagocytic vacuole (PV) forming from cell surface and into a multivesicular body (MVB) to which micropinocytotic vacuoles bring protein molecules. Broken arrows indicate possible formation of residual body of dense body variety (DB) from the multivesicular bodies or endocytotic vacuoles (PV). A. Novikoff et al., *Federation Proc.,* **23:** 1010 (1964).

Cytoplasmic Fluid or Cytosol

A considerable portion of the total cellular constituents, including numerous enzymes, nutrients, and macromolecular products, occurs apparently free in the cytosol or cell sap. This apparent absence of structure may be an illusion, referable solely to our inability to visualize and investigate their organization either *in situ* or subsequent to cell breakage and fractionation.

Cell or Plasma Membrane

The exterior boundary of the cell is provided by the cell (or plasma) membrane (or envelope), another example of a double "unit" membrane 80 to 100 Å thick, which appears (in thin sections fixed with $KMnO_4$ or OsO_4) to be a relatively rigid and organized structure, similar in appearance to a bimolecular layer of polar lipids (75 Å thick) coated on both sides by thin protein films. However, the membrane is by no means to be regarded as homogeneous over its entire exterior. On the contrary, it is a mosaic of different functional units, differing slightly in gross structure, highly selective and specialized in various cells, and controlling such widely divergent but highly important cellular parameters as selective permeability to, and active transport of, nutrients and ions (i.e., their passage into the cell), excitability, contractile properties, and cell-to-cell associations and recognitions (i.e., in organ formation and nervous transmission). Plasma membranes may also be the site of certain integrated enzymatic processes and, in bacteria—centered in a region called the *mesosome*—of protein synthesis, oxidative phosphorylation, and DNA synthesis (see the next section).

Other Cell Types

This rough exposition of the various characteristic cellular elements is applicable, not only to most animal cells, whether highly differentiated or not, but also to those of primitive unicellular plants such as molds and fungi (e.g., *Neurospora crassa*—the common bread mold, and yeast). Those of unicellular green plants (i.e., algae) or of green plants in general contain, in addition, another highly specialized and organized structure, the *chloroplast* (see Chapter 12).

The cells of bacteria and blue-green algae, the pro(to)caryotic or chromonemal organisms, appear grossly different. In the first place they are much smaller—their dimensions being roughly those of a cylinder 0.2 to 1 μm in diameter and 0.3 to 5 μm long. That is, they are in the same size class as animal mitochondria and have a volume some 500- to 1000-fold smaller than that of animal cells. Gone are most structural features; there is no distinct membrane-bound nucleus, no endoplasmic reticulum, and no mitochondria. The only readily discernible features inside the wall of these primitive cells are numerous ribosomes scattered throughout the cytoplasm and a cell (or protoplast) membrane with numerous attachments, which may penetrate deep into the interior of the cell. It is this set of structures that appears of utmost importance in providing what structural and functional organization there is in a bacterial cell and thus it acts as an analog and perhaps even a precursor of the more complex and highly specialized membranous elements of the more complex type of cells. In line with this hypothesis are the observations (1) that the oxidative and phosphorylative machinery of the bacterial cell is integrated into or associated with its membrane; (2) that membrane-bound ribosomes appear to be the site of the most active and effective protein synthesis; (3) that the bacterial genetic material (i.e., its DNA) appears organized with and attached to the mesosome region of the membrane; (4) that the cell membrane of marine autotrophs, particularly under conditions of severe physiological stress can become highly organized and vesicularized; (5) that the membrane in certain rapidly growing plant cells appears to be capable of pinching off mitochondria-like structures; and (6) that the composition, structure, and some of the properties of all membranous elements appear to exhibit certain incontestable similarities.

METHODS OF CELL FRACTIONATION

Just as there are no typical cells, so are there no standard methods of cell fractionation that are equally applicable to all cell types. Nevertheless, the methodology that had been originally developed for just one tissue of one animal, the liver of the rat, has proved remarkably versatile and adaptable to numerous other tissues and organisms as well. Some of the pitfalls and the reservations concerning conclusions to be reached will be discussed in the next section. Here we shall discuss first the principles and then the application in the present-day context of the centrifugal techniques worked out by such pioneers as Claude, Hogeboom, Schneider, Palade, de Duve, and Anderson.

Principles of Centrifugation

Once a tissue or population of cells has been disrupted, usually by mechanical means in a tissue grinder, a glass homogenizer with glass or Teflon pestle, a Waring Blendor or other high-speed mixer, sometimes with the aid of abrasives such as alumina or glass beads, the problem to be solved consists of the separation of the various intracellular components (or organelles) from each other. The method of choice for this purpose today is centrifugation, followed by ultracentrifugation in a supporting medium that is usually close to isotonic.[5] We remember that, in a centrifuge, particles that differ in density or in size (or shape) sediment at different rates in an applied centrifugal (i.e., gravitational) field. Their sedimentation rate (dr/dt) is directly proportional to this field G:

$$dr/dt = s_{\text{obs}}G \qquad G = \omega^2 R = \frac{4\pi^2(\text{rpm})^2}{3600}\, r \tag{9-1}$$

expressed, as relative centrifugal force in multiples of g, the gravitational constant (980 cm/sec^2). Here r is the radial distance from the axis of rotation to the particle, ω = the angular velocity (in radians per second), and the rpm are the revolutions per minute of the centrifuge rotor. A most useful nomogram, posted in all biochemical laboratories where centrifuges are in use, is due to Dole and Kotzias and relates rpm and G; it is reproduced as Fig. 9-6. For rigid spherical particles—and remember that other ellipsoids of revolution and long cylinders will behave very differently—t, the time required to sediment in a given medium from the top to the bottom of a centrifuge tube spinning in a rotor, is given by

$$t = \frac{9}{2}\,\frac{\eta}{\omega^2 r_p^{\,2}(\rho_b - \rho)} \ln \frac{r_t}{r_b} = \frac{1}{s_{\text{obs}}\omega^2} \ln \frac{r_t}{r_b} \tag{9-2}$$

where ω is again the angular velocity, η the viscosity of the medium, r_p the radius of the particle, ρ_b its density, ρ the density of the medium, and r_t and r_b radial distances from the center of rotation to the top and to the bottom of the column of fluid in the tube; the reciprocal of the constant term equals s_{obs}, the observed sedimentation coefficient.

We see then that, at a given rotor speed, the time required to centrifuge out a population of homogeneous spherical particles is inversely proportional to the square of their radii and to the first power of the differences between their density and that of the medium, and is directly proportional to the viscosity of the medium. Thus, relatively homogeneous, roughly spherical particles, of approximately equal densities, can be put into different classes solely on the basis of the time required for their sedimentation, a fact that forms the basis of the classical scheme of cell fractionation (11). First there are removed whole cells and large debris, then the fraction containing the large and dense nuclei (the density of DNA at 4° equals ~1.75, whereas that of most proteins is only 1.188), followed by one containing large particles or granules (usually largely mitochondria, but containing other structures in specialized tissues, e.g., the nerve-end particles or synaptosomes in brain, and lysosomes of a variety of tissues), followed by small mitochondria plus peroxisomes, followed in turn by the so-called "microsomal fraction," which contains fragments of both the smooth and the rough endoplasmic reticulum, as well as other vesicular elements of similar sedimentation characteristics (originating in the Golgi, the plasma and nuclear membranes, etc.), followed by free ribosomes. The media used for these separations have varied widely,

[5] An isotonic medium is one with an osmolarity equal to that of human blood serum (or sea water), that is, 0.25 M sucrose, 0.9 percent KCl, and so on.

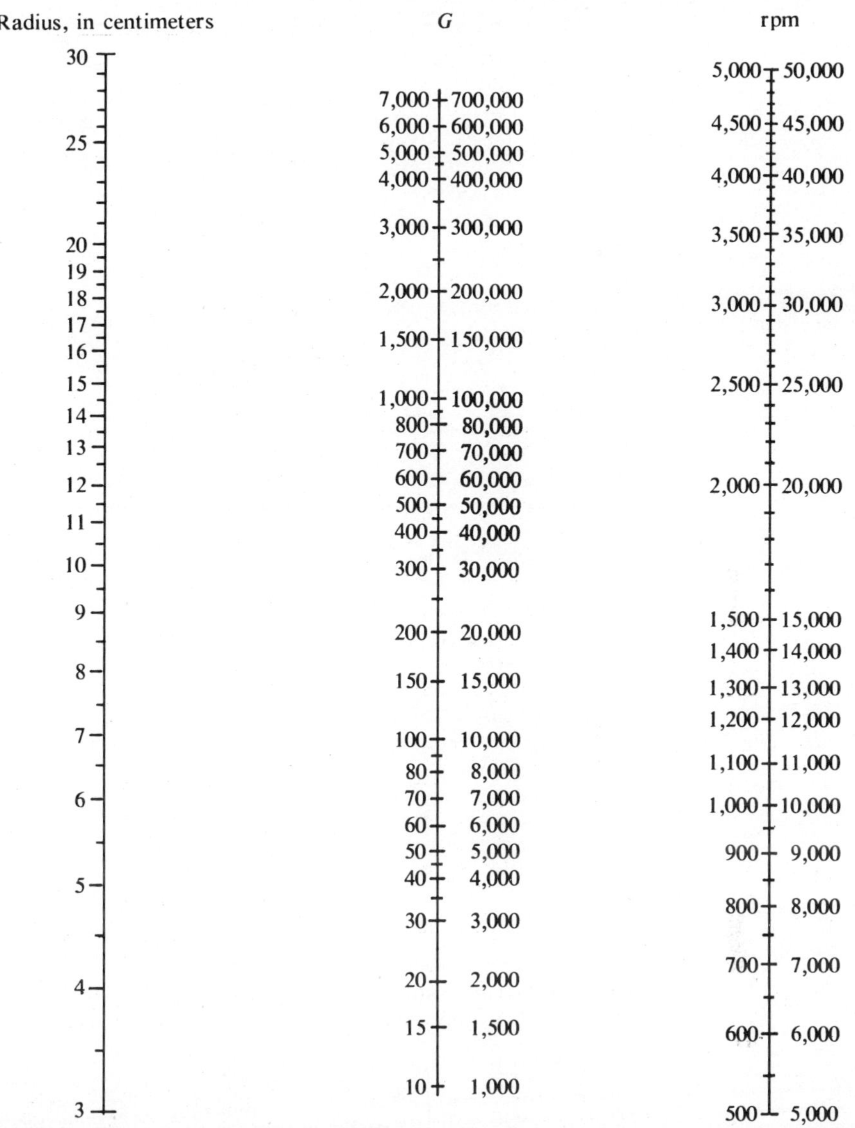

FIGURE 9-6

Nomograph relating relative centrifugal force (G) to rotor speed (revolutions per minute) and radius of the centrifuge head. A thread or straightedge stretched across the chart so as to intercept the known radius and rpm will also intercept the equivalent relative centrifugal force, G. V. P. Dole and G. C. Cotzias, *Science,* **113:** 552 (1951).

depending on a variety of considerations. Pure salt media are rarely used nowadays; the preferred one consists of sucrose in the isotonic range (from 0.15 M to about 0.33 M, with many methods employing 0.25 M or 0.44 M sugar), frequently in the presence of dilute tris(hydroxymethyl)aminomethane or phosphate buffer. Sometimes 10^{-3} M Mg^{2+} is added as well, and occasionally EDTA is included as a means of chelating heavy metal impurities. Glycerol, mannitol, or sorbitol sometime substitute for sucrose. The development of high-speed titanium rotors (e.g., No. 50 for the Spinco centrifuge)

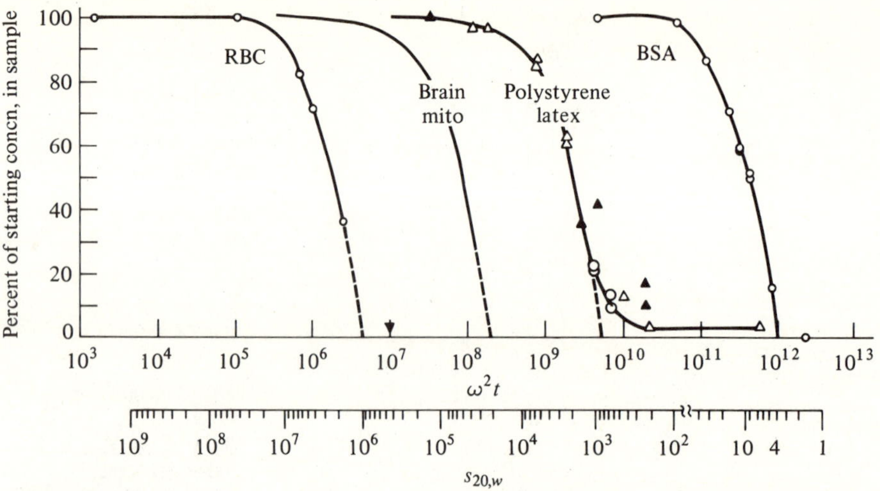

FIGURE 9-7
Sedimentation of sheep red blood cells, polystyrene latex particles, and BSA. Points indicated by squares obtained at 1 *g* using gravity; all other points obtained in Spinco No. 50 angle-head rotor. Points indicated with octagons obtained with small cotton plug in bottom of centrifuge tube. Along abscissa are plotted the integral of $\omega^2 t$ and the sedimentation coefficient in Svedberg units corrected to water at 20°. The left-hand portion of this scale is for data taken at 5°, the right-hand segment for data taken at 10°. Arrows indicate observed sedimentation coefficient for BSA and calculated sedimentation coefficient for polystyrene latex. N. G. Anderson, *Anal. Biochem.,* **23:** 72 (1968). Data for brain mitochondria, in a 30 rotor, are from (13).

and of digital integrators for the direct and accurate determination of $\omega^2 t$[6] in Eqs. (9-1) and (9-2) permit the measurement of s_{obs} and s [cf. Eq. (5-2)] directly from measurements of some characteristic properties of the particle remaining unsedimented. As shown by Anderson (12), s is then the intercept of the experimentally measured curve with the abscissa of Fig. 9-7. An ideal particle obeys the relation

$$s_{obs} = -\frac{1}{2\omega^2 t} + \ln\left[\frac{r_0^2}{r_p^2} + \frac{C_f}{C_i}\left(1 - \frac{r_0^2}{r_p^2}\right)\right] \tag{9-3}$$

where C_i is the concentration of particles (or a representative marker) at the beginning and C_f their concentration in the supernatant at the end of the run, and r_0 and r_p are the radius of the meniscus and of the plane of the surface of an arbitrarily defined pellet volume, for example, for a No. 50 Spinco rotor, 9 ml of sample and 2 ml of "pellet," $r_0 = 4.88$, $r_p = 6.8$ cm; for a No. 30 rotor, 20 ml of sample and 5 ml of "pellet," $r_0 = 7.07$ and $r_p = 8.88$ cm.

One of the most significant advances in cell fractionation has been the introduction of density-gradient techniques (11,14). Two types of gradient can be distinguished: *continuous gradients,* constructed by means of gradient building devices of varying degrees of refinement and sophistication, and *discontinuous gradients,* produced by carefully layering zones of different density by hand, one on top of the other, usually with the heaviest at the bottom of the tube. Density gradients are used for two distinct purposes, as discussed below.

[6] $\omega^2 t = \dfrac{(G \times \text{min} \times 60 \times 980)}{r_{av}}$ where r_{av} is the distance to the center of the fluid column.

Zone or Band Centrifugation

In this version, the preparation is placed in a thin layer on top of the tube and centrifuged into the gradient of a medium usually less dense than the particles to be isolated. As a result of the application of the field, the particles migrate as a distinct band through the gradient, which merely serves the purpose of preventing the spreading of the zone or band through convection. The usual equations and parameters of velocity sedimentation in the ultracentrifuge apply, and sedimentation coefficients can readily be obtained (11,14–16). This is an exceedingly useful technique of very wide applicability for both analytical and preparative purposes and has been used for the isolation and characterization of particles varying all the way from large viruses, nuclei, and mitochondria, through ribosomes, to pure proteins and nucleic acids. Usually the swinging bucket rotors of the ultracentrifuge are employed, and fractions are collected either from the bottom or the top of the tube—in the latter case after forcing a denser fluid upward from the bottom. Anderson and his collaborators have shown recently that regular angle heads can also be used (12,13), and this investigator has also pioneered in the design of a wholly novel zonal centrifuge, which allows the continuous fractionation and separation of various subcellular fractions from single large-scale runs based on either their rate or their isopycnic behavior (Fig. 9-8).

Equilibrium Density or Isopycnic Centrifugation

When particles are allowed to sediment in a density gradient so designed as to include the value of the particle density, then, after equilibrium has been established, the particles will come to rest in a band centered around their buoyant density. Since under carefully controlled conditions the bands can be made quite narrow, particles of similar size and shape but differing only slightly in density can be separated with good efficiency.

The preferred media for all these techniques for particle fractionation are buffered sucrose solutions. In Table 9-1 we show some of the properties of some of these solutions, as well as the densities characteristic of certain subcellular particles. Frequently the resolution can be improved by carrying out the separations in D_2O rather than H_2O. Glycerol, sorbitol, mannitol, and the polysaccharides Ficoll and glyogen have been substituted for sucrose in special instances.

Typical Fractionation Scheme

In Fig. 9-9 is presented, as a practical example of the above considerations, a composite scheme that has proved useful in the biochemical laboratories of Indiana University. It incorporates a number of different standard procedures and leads to reasonably pure particles of various types. Their composition, function, and content with respect to characteristic enzymes and other components will be discussed in the next section.

INTRACELLULAR LOCALIZATION OF ENZYMES AND BIOCHEMICAL ACTIVITIES

The principles and some of the limitations inherent in the approach outlined in the first section of this chapter have received a great deal of attention recently, particularly by de Duve, who has presented a cogent analysis of the conceptual approach involved. He points out that any experiment of this type consists of three steps: (1) a *destructive* one, which converts a tissue or cell suspension into a homogenate; (2) a *reordering* of the components of the homogenate into "fractions" on the basis of physical properties (i.e., sedimentation coefficient, density, etc.); and (3) an *analysis* of these fractions. Interpretation of the results then requires three stages of reconstruction in return: (1) *assignment* of morphological and biochemical *properties* to the fractions; (2) *reinterpretation of the data* in (1) in terms of the components

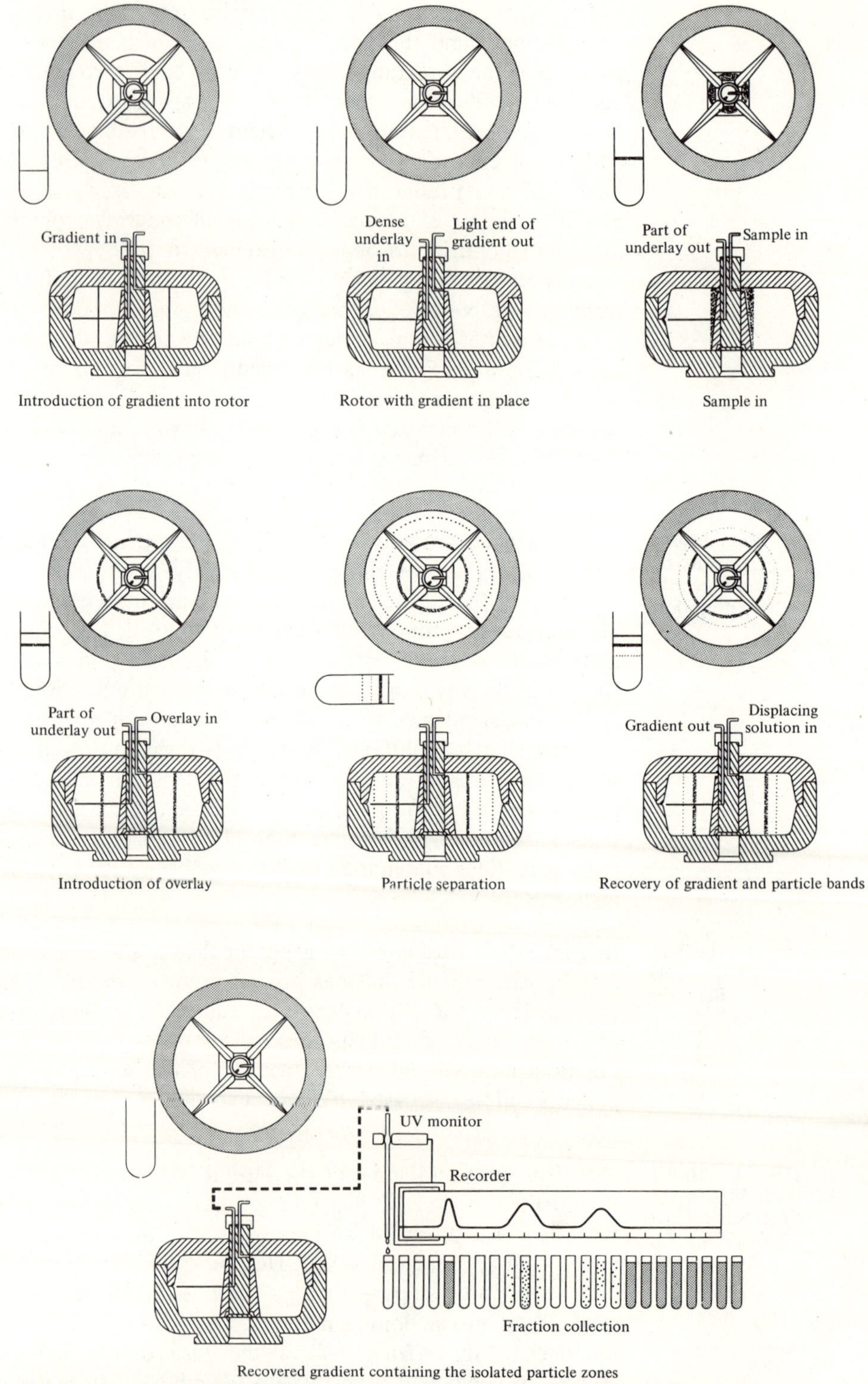

Gradient in
Introduction of gradient into rotor
Dense underlay in
Light end of gradient out
Rotor with gradient in place
Part of underlay out
Sample in
Sample in
Part of underlay out
Overlay in
Introduction of overlay
Particle separation
Gradient out
Displacing solution in
Recovery of gradient and particle bands
UV monitor
Recorder
Fraction collection
Recovered gradient containing the isolated particle zones

FIGURE 9-8

Continuous-fractionation zonal centrifuge. N. G. Anderson, *Quart. Rev. Biophys.,* **1:** 217 (1968); C. A. Price, Zonal Centrifugation, in W. W. Umbreit, R. H. Burris, and J. F. Stauffer (eds.), *Manometric Techniques,* Burgess, Minneapolis, 5th ed., 1970.

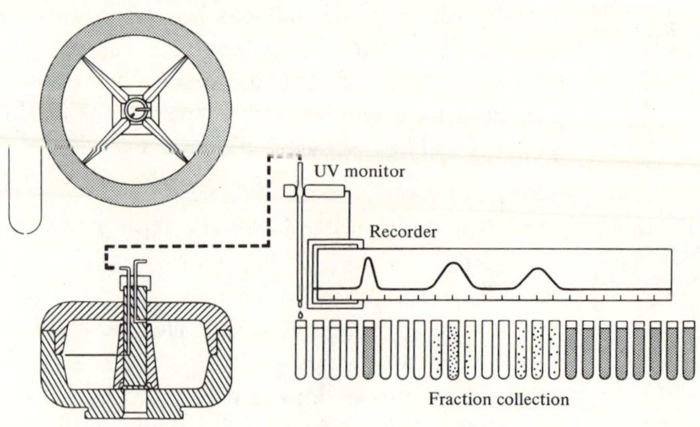

TABLE 9-1
Properties of sucrose solutions

Density		*Concentration*			*Viscosity*		
0°	*20°*	*Percent w/w*	*Molarity at 20°*	*Molality*	*0°*	*5°*	*20°*
1.010	1.008	2.49	0.073	0.075	1.90	1.60	1.06
1.020	1.017	4.92	0.146	0.151	2.04	1.72	1.13
1.0339*	1.0314*	....	0.250*	....	2.31*	....	1.2663*
1.050	1.047	12.01	0.367	0.399	2.64	2.22	1.43
1.100	1.095	23.15	0.741	0.880	4.45	3.67	2.24
1.150	1.144	33.48	1.119	1.470	8.50	6.85	3.92
1.200	1.193	43.11	1.502	2.214	19.8	15.3	7.94
1.250	1.242	52.17	1.892	3.187	60.6	44.2	19.7
1.300	1.291	60.74	2.290	4.519	275	184	66.1
1.350	1.340	68.86	2.696	6.461	2440	1410	357

[a] Isopycnic densities (g/cm^3) of particles from rat liver at 4° in sucrose are: mitochondria, 1.21; lysosomes, 1.12 (broad); peroxisomes, 1.24; ribosomes, ~1.6.
In 0.25 M sucrose the densities (g/cm^3) are mitochondria, 1.099; lysosomes, 1.101; and peroxisomes, 1.090.
[b] Data of Keller et al. (17).
SOURCE: Adapted from C. de Duve, J. Berthet, and H. Beaufay, *Progr. Biophys. Biophys. Chem.*, **9:** 325 (1959).

of the original homogenate and the fractionation techniques employed originally; and (3) an attempt to equate these *components* of a homogenate with certain *intracellular entities.*

Assumptions and Pitfalls

What is done in general is to use some scheme similar to that presented and then select certain so-called *marker* or *indicator enzymes, activities,* or *substances* that, experience suggests, may be useful for the identification of certain intracellular particles or components. The results obtained are then used to draw up a *distribution pattern* or curve and thus *define particles in terms of characteristic biochemical activities,* and conversely to assign different characteristic biochemical properties to different particle types. Extension of the method then to a whole spectrum of enzymes, etc., allows the assignment of definite functions in the cellular economy to known ultracellular components, and conversely to define and later identify new or at least hitherto unrecognized morphological components on the basis of biochemical evidence—the identification of the "acid phosphate particles" with *lysosomes* and the "urate oxidase particles" with *peroxisomes* (formerly called microbodies), both in mammalian liver, are examples of the success of this approach. Two fundamental implicit assumptions, also pointed out by de Duve, are (1) that enzymes are localized in but one intracellular site—not always easy to prove—and (2) that populations of subcellular particles are enzymatically homogeneous. Let us now examine some of the inherent difficulties and pitfalls.

Invariance. Uncritical application of any set fractionation scheme implies that analogous particles in different cells, tissues, or organisms should have the same shape, size, or density and will therefore be found in the same fraction. This is clearly untenable, and therefore fractionation schemes have to be modified and adapted to fit each individual case.

Another untenable assumption is that the intracellular localization and certain properties of particulate enzymes, such as latency (cf. below), remain

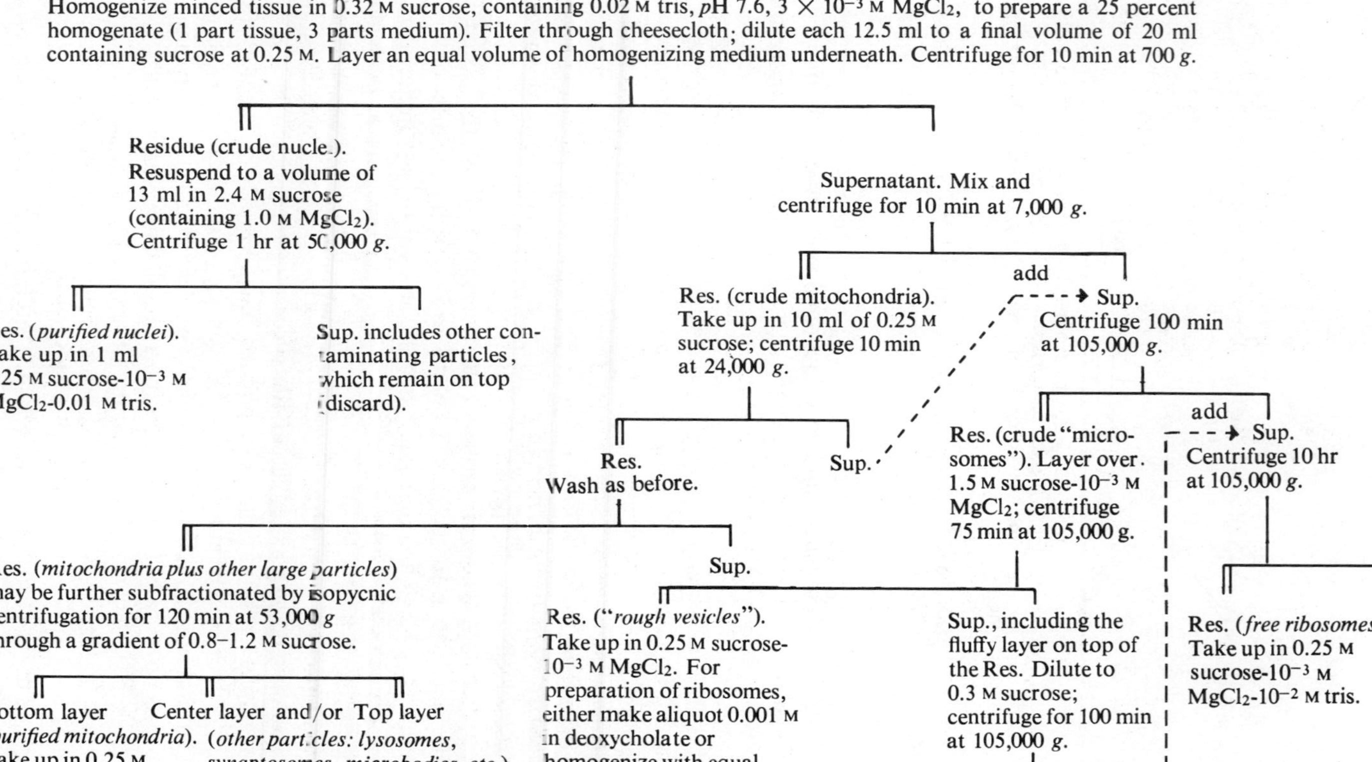

FIGURE 9-9
Typical cell-fractionation scheme. All operations at 0 to 4°.

invariant from cell to cell, etc.—all these properties, including the localization pattern, have to be reestablished for each new application.

Finally, it used to be frequently assumed that the morphological structures and features found in a purified fraction have to be identical with, or at least closely correspond to, those preexisting in the intact cell. This has turned out to be grossly erroneous, largely because of the proclivity of membranous or reticular structures to pinch off and form spherical or vesicular elements in the course of even "mild" homogenization and fractionation procedures. In this manner, "microsomes" were obtained from the endoplasmic reticulum, various "electron-transport particles" from the membrane and cristae of mitochondria, "synaptosomes" or "nerve-end particles" from the presynaptic ends of nerve cells, and a variety of particles of various stripes ("oxidosomes," "chromatophores") from the protoplast membrane of bacteria.

Latency and Crypticity. Enzymes bound to particles, especially those localized in places other than the surface of the outer limiting membrane, frequently exhibit the phenomena of *crypticity* and *latency.* These terms mean that in their native state, *in situ,* while integrated in the particulate matrix, they are occluded and inaccessible to varying degrees to substrates or other reactants and in general may well exhibit different properties with respect to kinetics, stability, conformation, etc., from those exhibited by their more accessible solubilized counterparts. Before a satisfactory and meaningful assay can be performed, membrane permeability must be altered, accessibility increased, the cryptic or latent enzyme rendered exposed or even detached by mechanical means, osmotic shock, or by treatment with detergents. Thus, almost by definition, the properties of particulate enzymes are affected by the conditions of assay employed and even more profoundly by those of solubilization—yet the latter is an obligatory first step in any purification procedure. This behavior is the bane and despair of the enzymologist, who frequently wonders just how to approach the problem of the "true" properties of a particulate enzyme and recognizes that in some instances he can create and unmask, modify as well as destroy, enzymatic activities almost at will.

Ease of Detachment and Degree of Disruption. This constitutes the obverse of what we have just discussed. Not all particulate enzymes are cryptic, and enzymes vary greatly in the ease with which they may be detached from the particle. Furthermore, their degree of integration into membrane or particle may vary from cell to cell. Therefore, the absence of a certain activity from a certain fraction (and hence, by extrapolation, its presence or absence from a particle) may not be real; the activity may have become detached or liberated in the course of the fractionation process, yet be characteristic of the particle *in situ.* Many nuclear and some mitochondrial activities are of this kind. Conversely, the presence of a component or an activity in a fraction does not guarantee its original association with a particle *in situ*—it may have originally been soluble, or detached from one particle type, only to become attached to a second type of particle. The attachment of basic proteins such as cytochrome *c* to the RNA of the ribosomes is an example of this phenomenon. One must remember that methods vigorous enough to break up cells are frequently also vigorous enough to rupture membranes and detach enzymes. What is frequently done is to perform an extrapolation to "zero disruption," that is, to check localization with ever milder methods and shorter exposures, rupturing fewer and fewer cells. Useful checks are also provided by ultracytochemical observations on electron micrographs of the intact cell (18).

With these provisions in mind, let us now consider briefly the results obtained by fractionation techniques. The student is urged to return to this section after having familiarized himself with most of the reactions and processes in the remainder of this volume.

Indicator Enzymes and Enzyme Distributions

In Table 9-2 we have collected data concerning a number of enzymes and other components that have frequently been employed as indicators or monitors for (1) certain intracellular particles and (2) for certain important biochemical processes, sequences, and so on, which in turn are outlined in Table 9-3. It must be pointed out, however, that truly integrated metabolic activities frequently require the collaboration of two or even more fractions. It is this phenomenon, perhaps, that governs many of the parameters of metabolic control. A classical example is provided by protein synthesis (Chapter 21), the *activation* of amino acids and their attachment to transfer RNA takes place within the *cytosol;* their *assembly* to polypeptides on *ribosomal clusters,* some of them forming part of the *rough endoplasmic reticulum,* is programmed by a species of RNA (*messenger RNA*), which was originally synthesized in the *nucleus* under direct control of its *DNA*. The *energy* for this process is provided by ATP formed largely in the *mitochondria,* and after the process is completed, the finished proteins in secretory cells at least are detached and "concentrated" by the *Golgi apparatus* and finally separated into *storage granules.*

The use of these enzymes and of the principles in the earlier sections is outlined in Table 9-4, which gives two examples for rat tissues: one for liver and one for brain.

Some Properties of Isolated Cell Fractions

Nuclei (19). Highly purified nuclear fractions have been obtained, by the procedures described, that contain 1 percent or less of the total homogenate activity in succinic dehydrogenase, cytochrome oxidase, urate oxidase, glucose-6-*P*-ase (1.5 percent), glucose-6-*P* dehydrogenase, and phospholipid. They contain 60 to 65 percent of the total cellular DNA, 10 percent of the total cellular protein, and exhibit ratios of protein to DNA of 5, and of RNA to DNA of 0.15 to 0.22.

Mitochondria (9,11,22–24). Mitochondria, when isolated with the greatest care from liver, are ellipsoids of revolution, which average 3.3 μm in their major axes and slightly less than 1 μm in their minor axes. In other vertebrate cells, for instance the exocrine cells of pancreas, they may assume a highly elongated configuration some 10 μm in length. Their dry weight is of the order of 1.1×10^{-13} gram per particle. There are some 800 mitochondria per rat liver cell, and they account for approximately 20 percent of the total cell nitrogen or protein. The density of most mitochondria suspended in 0.25 M sucrose equals 1.10, with their true isopycnic density as determined in sucrose gradients at $\sim$1.21 g/cm^3 and a sedimentation coefficient of the order of 10^4 S. Mitochondria of unicellular eucaryotes (yeast, *Neurospora*) are frequently both smaller and lighter. Approximately 40 percent of the total dry weight of the mitochondrion is phospholipid, of which cardiolipin is the most characteristic in its almost purely mitochondrial localization. Heart muscle mitochondria contain 1.46 μatoms of nonheme Cu, between 3.3 and 6.4 μatoms of nonheme Fe, as well as about 2.5 μatoms of heme Fe (cytochrome) per gram of protein. They also contain about 0.5 and 0.6 μmol of flavin and 4 μmol of coenzyme Q (ubiquinone) per gram of protein. Small amounts of RNA ($\sim$1 percent of protein) and DNA ($<$1 percent of protein) associated even with the most highly purified preparations are not due to impurities but fulfill a functional role, probably in the biosynthesis of certain mitochondrial proteins.

Mitochondria can assume a variety of structural and functional forms, depending on conditions of isolation. A number of agents, such as hypotonicity, higher fatty acids, Ca^{2+}, the thyroid hormone thyroxin, inorganic phosphate and arsenate, thiols such as glutathione, heavy metals, mercurials, arsenite, and CCl_4, as well as a variety of detergents, all bring about mitochondrial swelling. Of these, thyroxine, Ca^{2+}, and perhaps others act through the release of higher fatty acids such as oleic acid from the mitochondrial membranes. These fatty acids are uncouplers of oxidative phosphorylation and stimulate hydrolysis of ATP to ADP plus P_i. A spontaneous change of shape into a more spherical form is observed even in isotonic sucrose.

On the other hand, swelling can be inhibited and some contraction induced by hypertonic sucrose, albumin, Mg^{2+}, EDTA, and ATP. These agents, singly or in combination, are frequently included in media used for the isolation and storage of mitochondria. Often the reaction produced by ATP (or ADP plus P_i plus an oxidizable substrate) is a highly specific, true contraction involving the extrusion from the particle of several hundred water molecules. The states of mitochondria produced under a variety of conditions are summarized in Table 9-5 and more details are provided in Chapter 15.

The existence of the mitochondrial membrane system presents a formidable permeability barrier to the penetration of various charged species such as cations (Na^+, K^+), organic anions (di- and tricarboxylic acid substrates), NADH, ADP/ATP, cytochrome *c*, and fatty acyl CoA's. The means for overcoming this barrier in the first two instances is provided by the existence of specific, energy-driven uptake mechanisms (pumps, discussed in Chapter 15). For oxidation of the nicotinamide coenzymes the mechanism involves the participation of a substrate, itself relatively readily permeable, and of two dehydrogenases, one extra- and the other intramitochondrial in localization:

$$\left.\begin{array}{ll}\text{cytosol:} & \text{substrate (ox)} + \text{NAD(P)H} + \text{H}^+ \rightleftharpoons \text{substrate H}_2 + \text{NAD(P)}^+ \\ \text{mitochondria:} & \text{substrate H}_2 + \tfrac{1}{2}\text{O}_2 \rightleftharpoons \text{substrate} + \text{H}_2\text{O}\end{array}\right\}$$

$$\text{NAD(P)H} + \text{H}^+ + \tfrac{1}{2}\text{O}_2 \xrightarrow[\text{mitochondria}]{\text{cytosol plus}} \text{NAD(P)}^+ + \text{H}_2\text{O}$$

In the case of ADP/ATP the barrier is probably constituted by the inner membrane and its operation imposes a requirement for a vectorial transfer of ATP from its site of synthesis (i.e., the inside) to the outside and of ADP in the opposite direction; this transfer is inhibited by certain specific agents such as atractylate (22,25) (see also Chapter 15). Similarly, the uptake of fatty acyl CoA's by mitochondria prior to their oxidation is impeded by the inner membrane (26,27) and has to be circumvented by conversion to the carnitine derivatives (Chapter 13).

Endoplasmic Reticulum and Ribosomes (28). The total "microsomal" fraction, consisting of bits and pieces of the endoplasmic reticulum and associated ribosomes plus other membrane fragments of similar sedimentation characteristics, accounts for approximately 20 percent of the total cellular protein and usually better than 60 percent (sometimes as high as 80 percent) of the cellular RNA. This fraction from liver contains about 20 to 30 mg of protein, about 5 mg of RNA, and 5 to 8 mg of phospholipid per gram of tissue. The ratios of RNA to protein and phospholipid to protein are of the order of 0.20 and 0.30, respectively.

TABLE 9-2
Distribution of certain key enzymes and other components suitable as indicators or monitors[a]

Nuclei	*Mitochondria*	*Smaller particles, including lysosomes*[b]	*Microsomal fraction* — *Ribosomes*	*Microsomal fraction* — *Reticulum, Golgi, etc.*	*Soluble*
DNA	*Succinic deyhydrogenase* (for CAC and ET)	*Urate oxidase*[c] *Catalase* D-amino acid oxidase 2-hydroxy acid oxidase (P)	RNA	***Glucose-6-P-ase***[d]	NADP-linked isocitric dehydrogenase
	Cytochrome oxidase[c] (for ET)		Ribosomal proteins	***Nucleoside diphosphatase***[c,d]	Aminoacyl-tRNA synthetase
DNA-dependent RNA polymerase	NAD-linked isocitric deh[c,i] (for CAC)	Isocitrate lyase[e] Malate synthase[e]		***NADPH-linked lipid peroxidase***[d]	
Histones	Cardiolipin (for IM) Copper (ET)	"Acid" ribonuclease[c] "Acid" phosphatase[c] "Acid" deoxyribonuclease[c]		***NADPH-O_2-linked mixed-function oxidases;*** P450[f]	Acetyl CoA carboxylase (for fatty acid biosynthesis)
NMN-adenyl transferase	EPR signal at $g = 1.94$ (for ET)	β-Galactosidase[c] β-Glucuronidase[c] β-Mannosidase[c] Neuraminidase[c] Cathepsins (L)		Squalene cyclohydroxylase (for steroid biosynthesis)	Fatty acyl CoA synthetase (n malonyl CoA + acetyl CoA $\xrightarrow{\text{NADPH}}$ fatty acid + nCO_2)
	Mg^{2+} and dinitrophenol stimulated *oligomycin-sensitive ATPase*[c] (for oxidative phosphorylation and IM)	Acetyl choline: S		NADPH-linked acyl and alkoxy aryl hydroxylases	***Glucose-6-P dehydrogenase*** plus ***6-P-gluconate dehydrogenase*** (ox. pathway of carbohydrate metabolism)
	Acetyl CoA hydrolase[c]; fatty acyl CoA synthetases[g] (for fatty acid oxidation)	Trypsinogen Chymotrypsinogen Lipase Amylase Ribonuclease (Z)		Cytochrome b_5 NADPH: cytochrome ***c*** reductase	

Monoamine oxidase[g] Kynurenine hydroxylase[g]	Na$^+$-K$^+$ stimulated *ouabain-sensitive ATPase*[c,j(?)]	*Transketolase* *Transaldolase* (pentose cycle)
Adenylate kinase[h]		
Fumarase[i] (CAC) Citrate synthase[i] (CAC) Glutamate dehydrogenase[c,i]	5′-Nucleotidase[j] Nucleoside triphosphate pyrophosphatase[j]	*Phospho-fructokinase* (glycolysis)
	UDPase[k]	*Lactic dehydrogenase*
Carnitine-palmityl CoA transferase[l] α-Oxoglutaric dehydrogenase complex[l] Pyruvate dehydrogenase complex[l]	DT diaphorase[m]	
β-Hydroxybutyrate dehydrogenase[c]		
Aspartate amino transferase[c] (transaminase)		

[a] Most frequently used indicator enzymes are italicized; these are enzymes where more than 60 percent of the total homogenate activity is localized in the fraction indicated. Metabolic pathways are in parentheses. ET, electron transport; also marker for inner membrane (IM). CAC, citric acid cycle.

[b] P, *peroxisomes* of various cells; L, *lysosomes* of vertebrate cells; **S**, *synaptic vesicles* of brain tissues; Z, *zymogen granules* of glandular tissue (pancreas).

[c] Latent enzyme: particles must be disrupted for full activity.

[d] Found in rough reticulum.

[e] *Glyoxysomes* of plants; may be identical to the peroxisomes.

[f] Steroid hydroxylases plus P450 in mitochondria from adrenals.

[g] Marker for outer membrane.

[h] Marker for intermembrane space.

[i] Marker for matrix.

[j] Marker for plasma membranes.

[k] Marker for Golgi.

[l] Can be readily detached from inner membranes of mitochondria as a single, homogeneous species.

[m] Marker for smooth reticulum.

TABLE 9-3
Multienzyme systems and sequences predominantly localized in different cell fractions

Nuclei, including nucleoli (DNA-rich fractions in microorganisms)	*Mitochondria*	*Small particles (lysosomes, peroxisomes)*	*Microsomal fraction*		*"Soluble"*
			Ribosomes	*Reticulum*	
DNA replicase	Electron transfer from enzyme-bound NADH[a] → O_2 succinate[a] → O_2 choline → O_2 α-Glycerol-*P* → O_2 proline → O_2 pyruvate[a] → O_2 Citric acid cycle	Various hydrolytic processes with acid *p*H optima concerned in breakdown of nucleic acids, proteins, glycogen, and mucopolysaccharides (Lysosomes)	Assembly of amino acids into specific cytoplasmic proteins	NADH $\xrightarrow{\text{cyt } b_5}$ O_2 NADPH $\xrightarrow{\text{cyt } b_5}$ O_2	Amino acid activation and transfer to tRNA
DNA-dependent RNA-polymerase				Mucopolysaccharide side synthesis	Glutathione biosynthesis
Biosynthesis of certain basic proteins, such as histones				Glucuronide synthesis	Complete glycolytic sequence from glycogen to pyruvate and all enzymes in the reverse direction from *P*-enol pyruvate to glycogen
Biosynthesis of NAD	Oxidative phosphorylation[a] (oxidation of NADH for provision of energy)	O_2-linked dehydrogenases producing H_2O_2 (peroxisomes)		Sulfur metabolism	
	Citrulline biosynthesis	Glyoxylate cycle (peroxisomes of plants and protozoa)		Biosynthesis of cholesterol from squalene	Pentose cycle, especially the dehydrogenases (rabbit liver and kidney)
	δ-Amino levulinate biosynthesis coproporphyrinogen III → heme			Biosynthesis of triglycerides from fatty acids	
				Phospholipid biosynthesis	
	Oxidation of fatty acids			NADPH and O_2 requiring hydroxylases[b]	Biosynthesis of fatty acids by enzyme complex
	Glucuronate and ascorbate[c] biosynthesis(?)				Most NADPH-linked dehydrogenases; NADPH is generated by one set and utilized for biosynthetic reactions by a second set
	Pyruvate $\xrightarrow{+CO_2}$ oxalacetate $\xrightarrow{-CO_2}$ phosphoenolpyruvate				δ-Amino levulinate → coprophyrin III
					Tryptophan catabolism
					Catabolism of pyrimidines

[a] These multienzyme systems are associated with mitochondria in all organs of all animal species examined, as well as in higher and unicellular plants; they are associated with cell membranes and their fragments in bacteria.
[b] Steroid hydroxylases for the cleavage of the cholesterol side chain, for 18β and for 11β hydroxylation are localized in the mitochondria of the adrenal gland (20, 21) (e.g., for reactions cholesterol → pregnenolone, 17α,21-dihydroxyprogesterone → cortisol; deoxycorticosterone → corticosterone).
[c] In chicken kidney; in other mammalian tissues this may be localized in smaller particles.

After fractionation into rough- and smooth-surfaced vesicles, the analytical data are as follows, again for liver: smooth vesicles—6.4 mg of protein, 0.29 mg of RNA, and 1.71 mg of phospholipid per gram of tissue; RNA to protein, 0.05; phospholipid to protein, 0.27; rough vesicles—11.2 mg of protein, 3.56 mg of RNA, and 1.41 mg of phospholipid per gram of tissue; RNA to protein, 0.31; phospholipid to protein, 0.13.

The reticular and free ribosomes that can be obtained by further fractiona-

TABLE 9-4
Distribution of indicator enzymes: rat liver and brain (all expressed as percent of total)

Activity	*Cyto-chrome oxidase*	*Glucose-6-P-ase*	*Acid P-ase*	*β-Glucu-ronidase*	*β-N-acetyl glucu-ronidase*	*Protein*	*RNA*
Liver fraction[a]							
Nuclear (crude)	10	7	4	4	12	15	10
Mitochondrial	58	3	24	19	33	25	4
Lysosomal	17	7	41	24	39	5	2
Microsomal	4	74	20	37	12	20	70
Cytosol	0	3	13	12	6	30	20
Recovery	89	94	102	96	102	95	106
	Succinic dehydro-genase	*Lactic dehydro-genase*	*Acetyl choline*	*Na+/K+ ATPase*[b]			
Brain fraction[c]							
Crude nuclear	10	25	12				
Large particles (crude mito-chondrial)	70	20	65				
Myelin and membranes[d]	5	15	18	15			
Synaptosomes[d]	30	70	75	75			
Mitochondria[d]	60	18	5	<10			
Microsomes and cytosol	5	70	25				

[a] Data from various sources in J. K. Grant (ed.), The Structure and Function of Subcellular Components, *Biochem. Soc. Symp.*, **23:** (1963).
[b] Data of C. Cotman and W. McBride (private communication).
[c] Data from V. P. Whittaker in J. K. Grant (ed.), The Structure and Function of Subcellular Components, *Biochem. Soc. Symp.*, **23:** (1963), and V. P. Whittaker, *Progr. Biophys. Mol. Biol.*, **15:** 39 (1965).
[d] Activities expressed as percent of the total in the *crude mitochondrial fraction,* subfractions obtained by isopycnic centrifugation in 0.8 to 1.2 M sucrose.

tion of the microsomal component and prolonged centrifugation of the cytosol, respectively, appear to be identical in their properties for all practical intents and purposes.

A large number of highly important biosynthetic reactions are catalyzed by enzymes localized on the membranous part of the endoplasmic reticulum. Among them are key reactions in the biosynthesis of steroids, of other lipids, and of complex polysaccharides. One of the most characteristic sets of reactions localized in this cell fraction is the hydroxylation of a large variety of organic molecules (various aliphatic and aromatic amines, steroids, etc.) requiring the simultaneous participation of O_2 and NADPH in the reaction

$$DH_2 + O_2 + NADPH + H^+ \rightarrow DHOH + NADP^+ + H_2O$$

catalyzed by the hydroxylases or mixed-function oxidases (29) discussed on page 637.

Peroxisomes These particles with an isopycnic density in sucrose gradients ~1.25 g/cm³ and a sedimentation coefficient of ~5×10^3 are widely distributed in nature (9), having been isolated from mammalian liver and

TABLE 9-5
Metabolic states and swelling: contraction of mitochondria

	Metabolic state				
	1	*2*	*3*	*4*	*5*
Conditions	Endogenous substrate; no ADP or ATP	Endogen. + ADP	Active phase, substrate + ADP	Resting, substrate + ATP	Anaerobic, substrate + ADP (ATP) present
Initial state	Swollen	Contracted	Contracted	Swollen	Contracted
Electron-transport inhibitors (Chap. 15)	Slow contraction	None	None	Slow contraction	None
Uncoupling agents (Chap. 15)					
Low concentration	Rapid contraction	None	None	Rapid contraction	None
High concentration	Slow swelling	None	Slow swelling	Rapid swelling	Slow swelling
Effect on transition	1 → 2: rapid contraction	2 → 3: no change	3 → 4: swelling	4 → 5: contraction 4 → 3: contraction	5 → 4 or 3: swelling
Respiration rate	Slow	Slow	Fast	Slow	None
Rate limited by	ADP	Substrate	Resp. chain	ADP	O_2
% of carrier reduced					
NADH	~90	0	53	>99	100
Flavoprotein	21	0	20	40	~100
Cytochrome *b*	17	0	16	35	100
Cytochrome $(c + c_1)$	7	0	6	14	100
Cytochrome $(a + a_3)$	0	0	<4	0	100

SOURCE: Adapted from A. L. Lehninger, *The Mitochondrion,* Benjamin, New York, 1965.

kidney (where they were first identified and designated as "microbodies"), protozoa, leaves of higher plants, plant seedlings, and yeast. The bulk of the cellular activities either generating or utilizing H_2O_2 (i.e., catalase, D- and L-amino and hydroxy acid oxidases, and glycolate oxidase), when present at all, are localized in these particles. Those of mammalian cells also contain urate oxidase (Chapter 18), while the ones from plants (30,31) and protozoa (9) have been reported to be the site of glyoxylate cycle activity (Chapter 15), since they contain the key enzymes isocitrate lyase and malate synthase.

Lysosomes (9,11,32,33). Lysosomes are most prevalent in animal cells. They band isopycnically at about 1.14 g/cm^3 in sucrose but appear heterogeneous from cell to cell or even within any one cell; some show sedimentation coefficients $> 2 \times 10^4$ S and others $\leqq 5 \times 10^3$ S. These particles appear to be the intracellular scavengers since they contain a number of hydrolytic enzymes (all with acid-*p*H optima) for a wide variety of metabolites: proteins, lipids, carbohydrates, and nucleic acids (see Table 9-2).

Plasma Membranes

Composition and Structure. Most plasma membranes (also called plasmalemma) from the animal sources most intensively investigated (erythrocytes, liver cells, neoplastic cells grown in culture such as HeLa and Ehrlich ascites, synaptosomes) appear to contain (by weight) 35 to 40 percent lipid, 55 percent protein, $\leqq 5$ percent carbohydrates, and < 0.1 percent RNA. This is a composition similar to that found for most other membranes but is somewhat higher in lipid than that of mitochondrial inner membranes. It is lower in lipid than are the mitochondrial outer membranes, membranes of the smooth endoplasmic reticulum, synaptic membranes, and myelin (the degenerate plasma membrane of the Schwann cells, which is wrapped around and acts as insulator for nerve axons), all of which contain $\geqq 50$ percent lipid. Most membranes exhibit a distinct and characteristic pattern with regard to their component lipids (34–37); in general, however, polar lipids (mainly phosphatidylcholine, phosphatidylethanolamine, phosphatidylinositol plus phosphatidylserine and significant amounts of sphingomyelin) make up ~70 percent, neutral lipids (with a very high content of free cholesterol) the remainder. Among the carbohydrates are found neutral sugars, hexosamines, and characteristically sialic acid (page 547). However, most of these carbohydrates probably originate in a layer immediately adjacent to the membrane proper, the *outer coat* of the cell, a structure $\geqq 75$ Å thick that separates one cell from the next and provides for contact and recognition features. These cell coats are made up of two types of glycoproteins: one type contains ~50 percent carbohydrate arranged in long, branched oligosaccharide side chains starting with peptide linkages to asparagine residues of the backbone, and terminating in sialic acid residues; the other type contains a much larger number of shorter oligosaccharides, starting with an ester linkage at serine, threonine, and perhaps hydroxylysine and again tipped in the main by sialic acid residues. A possible arrangement is shown in Fig. 9-10. (See also Table 10-1.)

The structure of the membrane shown is a modification of the *unit-membrane model* first proposed by Danielli and Davson (38) and refined by Robertson (39). Its crucial feature is the arrangement of the membrane lipids as a bilayer (bimolecular leaflet), with their hydrophobic side chains pointing toward the interior forming a continuous hydrocarbon phase. Their hydrophilic heads point outward, where they interact with a monolayer of proteins, which in turn face the aqueous environment. The model is consistent with the electron micrographs of membranes stained with permanganate (a very harsh

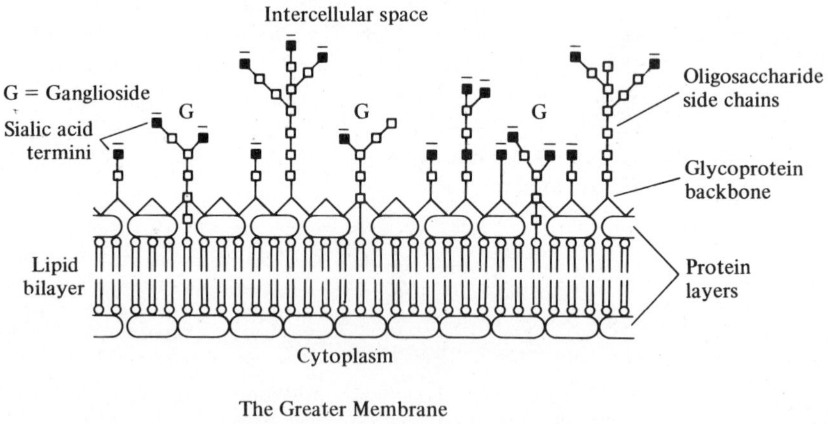

Intercellular space

G = Ganglioside

Sialic acid termini

Oligosaccharide side chains

Glycoprotein backbone

Lipid bilayer

Protein layers

Cytoplasm

The Greater Membrane

FIGURE 9-10
Model of the plasma membrane and cell coat. Adapted from A. L. Lehninger, *Proc. Natl. Acad. Sci. U.S.*, **60:** 1069 (1968).

procedure): These images consist of two electron-dense lines each ~20 Å across, separated by a transparent space ~35 Å in diameter. It is also consistent with X-ray diffraction pictures of myelin sheaths, but myelin is a quite unusual membrane. It has received strong support from model studies (40) (page 703). The modification shown proposes that the protein component does not form a continuous extended sheath or layer (i.e., wholly a β structure) but is composed instead of globular subunits. The latter may differ in their structure and function (although maintaining certain overall similarities, particularly in amino acid composition), some designed mainly for support and architectonic purposes [structural proteins (41–43), tectins (44)], others concerned with a variety of specific enzymatic and other biological functions (catalytic proteins). A variant of this model has been favored by Lucy (45): globular clusters of lipids (micelles) occasionally replaced by a globular protein within a sandwich of proteins. Quite different models have been proposed by Benson, who has postulated polypeptide chains immersed in the lipid layer (i.e., a polymeric lipoprotein structure), and by Green and Perdue, who have proposed an array of globular (lipoprotein) subunits.[7] Combinations between the various extremes (unit membrane, extended lipoprotein, and lipoprotein globules) have also been considered. It is not impossible that all membranes are anisotropic, not only vertically (i.e., "inside" versus "outside") but horizontally (along their surface) as well, and consist of a mosaic of patches varying in composition and structure—furthermore, that during function any patch or group of patches may undergo highly cooperative structural and configurational changes. Except for compositional details, particularly with respect to lipid, the main features of bacterial membranes do not appear to be fundamentally different from those of eucaryotic cells.

Function. One of the most characteristic functions, shared by all active membranes, is their ability to accumulate and maintain certain metabolites against a concentration gradient, a reaction that requires the continuous expenditure of energy (i.e., ATP) and has aptly been likened to a pump. Most

[7] A different model due to Vanderkooi and Green (45a) postulates that membranes are constituted by two-dimensional arrays of globular proteins, two molecules thick, with the interstices between them filled with lipid bilayers.

cells exhibit a nonequilibrium distribution of alkali cations (principally Na^+ and K^+) with a K^+ concentration inside greater than that outside, with the reverse for Na^+. Thus, a K^+/Na^+ pump is expected to be localized in the plasma membrane and such an activity is indeed found almost universally. One of its components is believed to be an activity measured *in situ* and partially purified from a number of sources which hydrolyzes ATP and ADP and requires Na^+ plus K^+ (the latter in a narrow concentration range) ions in addition to Mg^{2+}; this (Na^+ plus K^+)–requiring ATPase, as well as the Na^+/K^+ pump of the intact membrane, is completely inhibited by low and probably stoichiometric concentration of the cardiac glycoside ouabain (strophantidin), with the structure

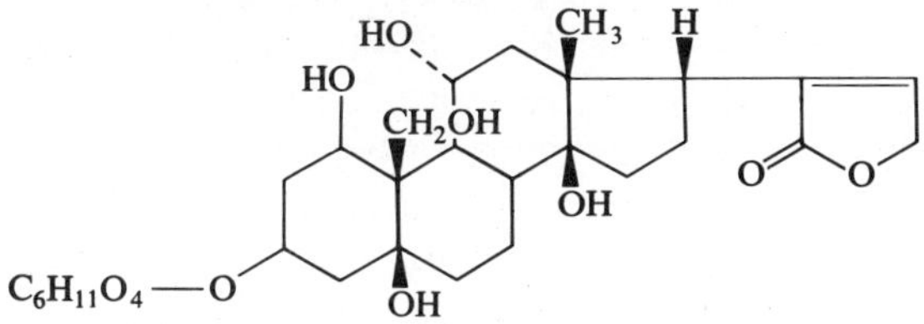

There is evidence that phosphorylation of a residue in a membrane protein (an alkali metal carrier?) is part of the reaction sequence and that the pump may operate somewhat as follows [see also Eqs. (15-57a) through Eq. (15-57d) for a more general mechanism]:

in membrane	$Na_3Prot + MgATP \rightleftharpoons Na_3Prot \sim P + MgADP$	(9-4a)
conformation change, extrusion of Na^+, K^+ uptake	Mg^{2+} $Na_3Prot \sim P + 2K^+ \text{ (out)} \rightleftharpoons K_2Prot\text{-}P + 3Na^+ \text{ (out)}$	(9-4b)
hydrolysis	$K_2Prot\text{-}P \rightleftharpoons K_2Prot + P_i$	(9-4c)
Na^+/K^+ exchange (inside)	$K_2Prot + 3Na^+ \rightleftharpoons Na_3Prot + 2K^+$	(9-4d)
Net	$3Na^+$ inside $\rightarrow$ outside; $2K^+$ outside $\rightarrow$ inside; ATP $\rightarrow$ ADP + P_i	(9-4e)

Phosphorylation, step (9-4a), requires Na^+ and is insensitive to ouabain, while dephosphorylation, steps (9-4b) and (9-4c), requires K^+ and is ouabain sensitive. Together they account for the properties of the membrane-bound ATPase. A phosphatase, measured frequently as a nitrophenylphosphatase, also appears to be a manifestation of the overall reaction, and phosphorylation of protein(s) has been demonstrated as part of the overall reaction. The maintenance of K^+ and Na^+ gradients sets up an electric potential (analogous to the one in a concentration cell) across the membrane. Some membranes (nerve, muscle, plants such as mimosa) are *excitable;* excitability is initiated by a brief controlled breakdown or partial reversal of the alkali-ion gradients due to a highly cooperative alteration of membrane structure and is propagated down the membrane unidimensionally by the same mechanism.

In addition to their ability to concentrate alkali metal ions, membranes also exhibit capabilities for the specific uptake of a wide variety of metabolites (amino acids, sugars, sulfate, etc.). The details have been studied mainly with bacterial (*E. coli*) membranes. Two different types of systems appear to be involved. One is composed of selective binding proteins in the membranes capable of interacting in a highly specific manner with the particular metabolite involved such as sulfate (46); Leu, Ile, and Val (47); or a β-D-galactoside (48) (i.e., β-thiodigalactoside or melibiose). The protein binding the latter has been isolated and purified by Fox and Kennedy, and genetic evidence suggests that it (the *M protein*) is the one specified by the *y*(*permease*) gene of the *lac* operon (Chapters 19 and 21). However, the M protein is only a part of the complete system required for galactoside uptake; the latter requires ATP hydrolysis, whereas binding of galactoside to the M protein does not, and certain galactosides, such as lactose or isopropylthiogalactoside, which are taken up are not bound. Since phosphorylation of the protein does not appear to be involved in this instance, it has been suggested that conformational changes driven by ATP form an integral part of the vectorial translocation of the complex from the outside to the inside.

The second type of uptake system (involving sugar-6-*P*) was discovered by Kundig, Ghosh, and Roseman (49) and implicated in generalized sugar transport (for lactose, sucrose, maltose, trioses, Me-glucose, etc.) by Tanaka and Lin (50). The reactions involved are

$$\text{Phosphoenolpyruvate} + \text{H-Prot} \underset{}{\overset{\text{Prot I}}{\rightleftharpoons}} \text{pyruvate} + \text{H-Prot}{\sim}P \tag{9-5a}$$

$$\text{H-Prot}{\sim}P + [\text{sugar-OH}]_{\text{out}} \xrightarrow{\text{Fraction II}} \text{H-Prot} + [\text{sugar-}P]_{\text{in}} \tag{9-5b}$$

Here *H-Prot,* a heat stable polypeptide ($M = 9400$), and *Prot I,* a soluble enzyme catalyzing the phosphorylation at the N-1 of one of its His residues at the expense of phosphoenolpyruvate (Eq. 9-5a), comprise the common system. Specificity is conferred by Fraction II, localized in membrane fragments and containing separate enzymes, each specific for phosphate transfer from H-Prot$\sim P$ to one of the nine carbohydrates so far implicated.

Membrane and Organelle Biosynthesis

In considering this problem we must distinguish two related but distinct questions: To what extent are membranes autonomous? To what extent are they capable of self-assembly? As far as autonomy is concerned, it now seems established that large organelles such as mitochondria, kinetoplasts of protozoa, and chloroplasts of green plants contain a separate genetic and protein-synthesizing system distinct from the classical one consisting of nucleus plus ribosomes (51). It is thus reasonable to asume that they are capable of at least partial autonomy in providing their substance. Their proteins, as are the proteins of all other membranes, are capable of (probably asynchronous) turnover even in nondividing, nongrowing cells. Their lipids and their carbohydrates are also continuously renewed, again asynchronously, but more rapidly. How far this autonomy extends is not at all certain. Some crucial regulatory proteins important in the formation and integration of key structural elements of the highly insoluble membranes of the organelle are almost certainly specified and synthesized by it *in situ,* as are many components of the protein-synthetic system itself, but beyond this not much is known. Other membranes appear to be assembled with the intervention of the endoplasmic reticulum (52); whether extranuclear genetic elements specify some part of their structure at this or indeed at any stage is being hotly debated.

In considering self-assembly, the crucial question is the following: To what extent is the final structure of a membrane, or of an organelle, completely specified and determined by the primary structure of the individual macromolecules (mainly protein) of which it is composed, perhaps with the additional intervention of certain others required catalytically for the assembly process itself? Conversely, do membranes (or organelles) require an additional component(s) as a template or jig?

Self-assembly processes have been studied most extensively in the case of viruses, and three different levels of complexity distinguished. The simplest is provided by a virus that may be either polyhedral or filamentous and consists of only a single monomeric polypeptide structure unit (Chapter 5). The small RNA-containing bacterial and plant viruses (Qβ and MS 2 phage, polyhedral; tobacco mosaic virus, filamentous) and small DNA-containing bacterial viruses (ϕX174, polyhedral; fd, filamentous) specify only one type of polypeptide, which can be shown to aggregate into the complete viral capsid even in the absence of nucleic acid. The next level of complexity is that observed in the capped polyhedral capsid of the T-even bacteriophages. Here assembly requires a large number of one type of polypeptide monomer (the majority species) and a smaller number of one or more additional types (minority components) (53). In their absence the monomer of the majority species assemble into abnormal, less highly organized structures (sheets or tubes) instead of the true shape. Finally, from studies of the assembly of the various tail structures and their integration with the capsid in these phages a new principle emerges: a requirement for catalytic or regulatory proteins, *not* themselves part of the final structure as integral participants in a sequentially controlled assembly process (54).

No doubt all these considerations also apply to the assembly of membranes, a process studied *in vitro* mainly with components of the inner mitochondrial membrane. Complete success has been obtained in restructuring enzyme complexes, even of high molecular weight, such as the α-oxoglutarate dehydrogenase (55) and pyruvate dehydrogenases (56) ($\bar{M} = 9.0 \times 10^6$ and 4.8×10^6, respectively; see pages 519 and 612). Limited success has also been achieved in reconstructing assemblies of proteins which appear functionally and perhaps also structurally analogous to the electron-transport complexes (Chapter 15) used as the starting material (57,58). Whether this indicates that the whole organelle is indeed capable of self-assembly in the same sense as is a complex virus, even in the absence of an organizer or template, only time will tell.

REFERENCES

Books and Symposium Volumes

Anderson, N. G. (ed.): The Development of Zonal Centrifuges and Ancillary Systems for Tissue Fractionation and Analysis, *Natl. Cancer Inst. Monograph,* **21** (1966).

Biological and Artificial Membranes, *Federation Proc.,* **27** (6), 1249–1309 (1968).

Biosynthesis and Function of Supramolecular Structures, *Federation Proc.,* **27** (5), 1160–1193 (1968).

Bloom, W., and D. W. Fawcett: *A Textbook of Histology,* Saunders, Philadelphia, 1968.

Bolis, L., and B. A. Pethica (eds.): *Membrane Models and the Formation of Biological Membranes,* North-Holland, Amsterdam, 1969.

Brachet, J.: *Biochemical Cytology,* Academic Press, New York, 1957.

Cohn, N. S.: *Elements of Cytology,* Harcourt, Brace & World, New York, 1964.

Crook, E. M. (ed.): The Structure and Function of Subcellular Components, *Biochem. Soc. Symp.,* **16** (1959).

Dalton, A. J., and F. Haguenau (eds.): *Ultrastructure in Biological Systems,* vol. 4, Academic Press, New York, 1968.

Davis, B. D., and L. Warren (eds.): *Specificity of Cell Surfaces,* Prentice-Hall, Englewood Cliffs, N.J., 1967.

Ernster, L., and Z. Drahota (eds.): *Mitochondria, Structure and Function,* Academic Press, New York, 1969.

Grant, J. K. (ed.): Methods of Separation of Subcellular Structural Components, *Biochem. Soc. Symp.,* **23:** (1963).

Green, D. E., and R. F. Goldberger: *Molecular Insights into the Living Process,* Academic Press, New York, 1967.

Hogg, J. F. (ed.): The Nature and Function of Peroxisomes (Microbodies, Glyoxysomes), *Ann. N.Y. Acad. Sci.,* **168:** (part 2) (1969).

Lehninger, A.: *The Mitochondrion,* Benjamin, New York, 1965.

Linnane, A. W., and N. K. Boardman (eds.): *Autonomy and Biogenesis of Mitochondria and Chloroplasts,* North-Holland, Amsterdam, 1970.

Loewenstein, W. R. (ed.): Biological Membranes: Recent Progress, *Ann. N.Y. Acad. Sci.,* **137** (2) (1966).

Northcote, D. H. (ed.): Structure and Function of Membranes, *Brit. Med. Bull.,* **24** (2) (1968).

Petermann, M. L.: *The Physical and Chemical Properties of Ribosomes,* Elsevier, New York, 1964.

Racker, E. (ed.): *Membranes of Mitochondria and Chloroplasts,* Van Nostrand Reinhold, New York, 1970.

Roodyn, D. B. (ed.): *Enzyme Cytology,* Academic Press, New York, 1967.

Roodyn, D. B., and D. Wilkie: *The Biogenesis of Mitochondria,* Methuen, London, 1968.

Salton, M. R. J.: *The Bacterial Cell Wall,* Elsevier, Amsterdam, 1964.

Seno, S., and E. V. Cowdry: *Intracellular Membraneous Structure,* Japan Society of Cell Biology, Okayama, Japan, 1965.

Slater, E. C., J. M. Tager, S. Papa, and E. Quagliarello (eds.): *Biochemical Aspects of the Biogenesis of Mitochondria,* Adriatica Editrice, Bari, 1968.

Slater, E. C., Z. Kaniuga, and L. Wojtczak (eds.): *Biochemistry of Mitochondria,* Academic Press, New York, 1967.

Vogel, H. J., J. O. Lampen, and V. Bryson (eds.): *Organizational Biosynthesis,* Academic Press, New York, 1967.

Review Articles

Albers, R. W.: Biochemical Aspects of Active Transport, *Ann. Rev. Biochem.,* **36:** 727 (1967).

Allfrey, V.: The Isolation of Subcellular Components, in J. Brachet and A. E. Mirsky (eds.), *The Cell,* vol. I, Academic Press, New York, 1959, p. 193.

Anderson, N. G.: Preparative Particle Separation in Density Gradients, *Quart. Rev. Biophys.,* **1:** 217 (1968).

Anderson, N. G.: Preparative Zonal Centrifugation, *Methods Biochem. Anal.,* **15:** 271 (1967).

Anderson, N. G.: Techniques for the Mass Isolation of Cellular Components, in G. Oster and A. W. Pollister (eds.), *Physical Techniques in Biological Research,* vol. III, Academic Press, New York, 1956, p. 299.

Benson, A. A.: On the Orientation of Lipids in Chloroplast and Cell Membranes, *J. Am. Oil Chem. Soc.,* **43:** 265 (1966).

Claude, A.: Studies on Cells: Morphology, Chemical Constitution and Distribution of Biochemical Functions, *Harvey Lectures,* **43:** 121 (1950).

Dalton, A. J.: Golgi Apparatus and Secretion Granules, in J. Brachet and A. E. Mirsky (eds.), *The Cell,* vol. II, Academic Press, New York, 1961, p. 603.

de Duve, C.: Principles of Tissue Fractionation, *J. Theoret. Biol.,* **6:** 33 (1964).

de Duve, C.: The Separation and Characterization of Subcellular Particles, *Harvey Lectures,* **59:** 49 (1965).

de Duve, C., and P. Baudhuin: Peroxisomes (Microbodies and Related Particles), *Physiol. Rev.,* **46:** 323 (1966).

de Duve, C., J. Berthet, and H. Beaufay: Gradient Centrifugation of Cell Particles: Theory and Applications, *Progr. Biophys. Biophys. Chem.,* **9:** 325 (1959).

de Duve, C., R. Wattiaux, and P. Baudhuin: Distribution of Enzymes Between Subcellular Fractions in Animal Tissues, *Advan. Enzymol.,* **24:** 291 (1962).

Green, D. E., and D. H. MacLennan: Structure and Function of the Mitochondrial Cristael Membrane, *BioScience,* **19:** 213 (1969).

Green, D. E., and J. F. Perdue: Membranes as Expressions of Repeating Units, *Proc. Natl. Acad. Sci. U.S.,* **55:** 1295 (1966).

Hall, D. O., and J. M. Palmer: Mitochondrial Research Today, *Nature,* **221:** 717 (1969).

Korn, E. D.: Structure of Biological Membranes, *Science,* **153:** 1491 (1966).

Lehninger, A. L.: The Neuronal Membrane, *Proc. Natl. Acad. Sci. U.S.,* **60:** 1069 (1968).

Lehninger, A. L.: Supramolecular Organization of Enzyme and Membrane Systems, *Naturwissenschaften,* **53:** 57 (1966).

Mirsky, A. E., and S. Osawa: The Interphase Nucleus, in J. Brachet and A. E. Mirsky (eds.), *The Cell,* vol. II, Academic Press, New York, 1961, p. 677.

Novikoff, A. B.: Lysosomes and Related Particles, in J. Brachet and A. E. Mirsky (eds.), *The Cell,* vol. II Academic Press, New York, 1961, p. 423.

Novikoff, A. B.: Mitochondria (Chondriosomes), in J. Brachet and A. E. Mirsky (eds.), *The Cell,* vol. II Academic Press, New York, 1961, p. 299.

Osawa, S.: Ribosome Formation and Structure, *Ann. Rev. Biochem.,* **37:** 109 (1968).

Palade, G. E.: The Organization of Living Matter, *Proc. Natl. Acad. Sci. U.S.,* **52:** 613 (1964).

Parsons, D. F.: *Ultrastructural and Molecular Aspects of Cell Membranes,* Pergamon Press, Oxford, 1967.

Porter, K. R.: The Endoplasmic Reticulum, in T. W. Goodwin and O. Lindberg (eds.), *Biological Structure and Function,* vol. I, Academic Press, New York, 1961, p. 127.

Racker, E.: The Membrane of the Mitochondrion, *Sci. Am.,* **218** (2): 32 (1968).

Racker, E.: Resolution and Reconstitution of the Inner Mitochondrial Membrane, *Federation Proc.,* **26:** 1335 (1967).

Robertson, J. D.: Unit Membranes, in M. Locke (ed.), *Cellular Membranes in Development,* Academic Press, New York, 1964, p. 1.

Rothfield, L., and A. Finkelstein: Membrane Biochemistry, *Ann. Rev. Biochem.,* **37:** 463 (1968).

Salton, M. R. J.: Structure and Function of Bacterial Cell Membranes, *Ann. Rev. Microbiol.*, **21:** 417 (1967).

Shumaker, V.: Zone Centrifugation, in J. H. Lawrence and J. W. Gofman (eds.), *Advances in Biological and Medical Physics*, Academic Press, New York, 1967, p. 245.

Whittaker, V. P.: The Application of Subcellular Fractionation Techniques to the Study of Brain Function, *Progr. Biophys. Mol. Biol.*, **15:** 39 (1965).

Descriptions of Various Fractionation Schemes, in S. P. Colowick and N. O. Kaplan (eds.), *Methods in Enzymology*, vol. I, 1955; vol. V, 1962; R. W. Estabrook (ed.), vol. X, 1967.

Specific References

1. Hamilton, M. G., and M. E. Ruth: *Biochemistry,* **8:** 851 (1969).
2. McConkey, E. H., and J. W. Hopkins: *J. Mol. Biol.,* **39:** 545 (1969).
3. Hoober, J. K., and G. Blobel: *J. Mol. Biol.,* **41:** 121 (1969).
4. Stutz, E., and H. Noll: *Proc. Natl. Acad. Sci. U.S.,* **57:** 774 (1967); Küntzel, H., and H. Noll: *Nature,* **215:** 1340 (1967).
5. Loening, U. E., and J. Ingle: *Nature,* **215:** 363 (1967).
6. Küntzel, H.: *J. Mol. Biol.,* **40:** 315 (1969).
7. Wintersberger, E.: *Z. Physiol. Chem.,* **348:** 1701 (1967).
8. Leon, S., and H. R. Mahler: *Arch. Biochem. Biophys.,* **126:** 305 (1968).
9. Müller, M., J. F. Hogg, and C. de Duve: *J. Biol. Chem.,* **243:** 5385 (1968).
10. De Robertis, E.: *Science,* **156:** 907 (1967); De Robertis, E., G. Rodriguez de Lores Arnaiz, L. Salganicoff, A. Pellegrino de Iraldi, and L. M. Zieher: *J. Neurochem.,* **10:** 225 (1963).
11. de Duve, C., B. C. Pressman, R. Gianetto, R. Wattiaux, and F. Appelmans: *Biochem. J.,* **60:** 604 (1955); Baudhuin, P., and H. Beaufay: *Arch. Intern. Physiol. Biochim.,* **71:** 119 (1963); Leighton, F., B. Poole, H. Beaufay, P. Baudhuin, J. W. Coffey, S. Fowler, and C. de Duve: *J. Cell Biol.,* **37:** 482 (1968).
12. Anderson, N. G.: *Anal. Biochem.,* **23:** 72 (1968).
13. Cotman, C., D. H. Brown, B. W. Harrell, and N. G. Anderson: *Arch. Biochem. Biophys.,* **136:** 436 (1970).
14. Martin, R. G., and B. N. Ames: *J. Biol. Chem.,* **236:** 1372 (1961).
15. Noll, H.: *Nature,* **215:** 360 (1967).
16. Schuel, H., S. R. Tipton, and N. G. Anderson: *J. Cell Biol.,* **22:** 317 (1964); Anderson, *loc. cit.*
17. Keller, P. J., E. Cohen, and R. D. Wade: *Biochemistry,* **2:** 315 (1963).
18. Goldfischer, S., E. Essner, and A. B. Novikoff: *J. Histochem. Cytochem.,* **12:** 72 (1964); Novikoff, A.: *Abstr. Sixth Intern. Congr. Biochem.,* New York, 1964.
19. Widnell, C. C., and J. R. Tata: *Biochem. J.,* **92:** 313 (1964).
20. Sweat, M. L.: *J. Am. Chem. Soc.,* **73:** 4056 (1951); Brownie, A. C., J. K. Grant, and D. W. Davidson: *Biochem. J.,* **58:** 218 (1954); Psychoyos, S., H. H. Tallan, and P. Greengard: *J. Biol. Chem.,* **241:** 2949 (1966).
21. Simpson, E. R., and R. W. Estabrook: *Arch. Biochem. Biophys.,* **129:** 384 (1969).
22. Schnaitman, C., and J. W. Greenawalt: *J. Cell Biol.,* **38:** 158 (1968).
23. Brdiczka, D., D. Pette, G. Brunner, and F. Miller: *European J. Biochem.,* **5:** 294 (1968).
24. Winkler, H. H., F. L. Bygrave, and A. L. Lehninger: *J. Biol. Chem.,* **243:** 20 (1968).
25. Chappell, J. B.: in Northcote, *loc. cit.,* p. 150.
26. Tubbs, P. K., and P. B. Garland: in Northcote, *loc. cit.,* p. 158.
27. Beattie, D. S.: *Biochem. Biophys. Res. Commun.,* **30:** 57 (1968); see also Norum, K. R., M. Farstad, and J. Bremer: *ibid.,* **24:** 797 (1966); Chappell, J. B., and A. R. Crofts: *Biochem. J.,* **95:** 707 (1965).
28. Dallner, G.: *Acta Pathol. Microbiol. Scand.,* Suppl.: 166 (1963); Dallner, G., and R. Nilsson: *J. Cell Biol.,* **31:** 181 (1966).
29. Gillette, J. R.: *Advan. Pharmacol.,* **4:** 219 (1966); Holtzman, J. L., T. E. Gram, P. L. Gigon, and J. R. Gillette: *Biochem. J.,* **110:** 407 (1968).
30. Breidenbach, R. W., and H. Beevers: *Biochem. Biophys. Res. Commun.,* **27:** 462 (1967).
31. Tolbert, N. E., A. Oeser, T. Kisaki, R. H. Hageman, and R. K. Yamazaki: *J. Biol. Chem.,* **243:** 5179 (1968).
32. Coffey, J. W., and C. de Duve: *J. Biol. Chem.,* **243:** 3255 (1968); Aronson, N. N., Jr., and C. de Duve: *ibid.,* **243:** 4564 (1968); Fowler, S., and C. de Duve: *ibid.,* **244:** 471 (1969).
33. Horvat, A., and O. Touster: *J. Biol. Chem.,* **243:** 4380 (1968); Arsenis, C., and O. Touster: *ibid.,* **243:** 5702 (1968).
34. Pfleger, R. C., N. G. Anderson, and F. Snyder: *Biochemistry,* **7:** 2826 (1968).
35. Bosmann, H. B., A. Hagopian, and E. H. Eylar: *Arch. Biochem. Biophys.,* **128:** 51 (1968).
36. Zwaal, R. F. A., and L. L. M. van Deenen: *Biochim. Biophys. Acta,* **163:** 44 (1968).
37. Lapetina, E. G., E. F. Soto, and E. de Robertis: *J. Neurochem.,* **15:** 437 (1968).
38. Danielli, J. F., and H. Davson: *J. Cellular Comp. Physiol.,* **5:** 495 (1935).
39. Robertson, J. D.: *Ann. N.Y. Acad. Sci.,* **137** (2): 421 (1966).
40. Mueller, P., and D. O. Rudin: *Nature,* **217:** 713 (1968).
41. Green, D. E., N. F. Haard, G. Lenaz, and H. I. Silman: *Proc. Natl. Acad. Sci. U.S.,* **60:** 277 (1968).
42. Mahler, H. R., and C. Cotman: in A. Lajtha (ed.), *Protein Metabolism of the Nervous System,* Plenum Press, New York, 1970.
43. Mazia, D., and A. Ruby: *Proc. Natl. Acad. Sci. U.S.,*

61: 1005 (1968); Kiehn, E. D., and J. J. Holland: *ibid.*, **61:** 1370 (1968).

44. Schnaitman, C. A.: *Proc. Natl. Acad. Sci. U.S.*, **63:** 412 (1969).
45. Lucy, J. A.: in Northcote, *loc. cit.*, p. 127.

45a. Vanderkooi, G., and D. E. Green: *Proc. Natl. Acad. Sci. U.S.*, **66:** 615 (1970).

46. Pardee, A. B.: *J. Biol. Chem.*, **241:** 5886 (1966); *Science*, **162:** 632 (1968).
47. Piperno, J. R., and D. L. Oxender: *J. Biol. Chem.*, **241:** 5732 (1966).
48. Fox, C. F., J. R. Carter, and E. P. Kennedy: *Proc. Natl. Acad. Sci. U.S.*, **57:** 698 (1967); Fox, C. F., and E. P. Kennedy: *ibid.*, **54:** 891 (1965).
49. Kundig, W., S. Ghosh, and S. Roseman: *Proc. Natl. Acad. Sci. U.S.*, **52:** 1067 (1964); Simoni, R. D., M. Levinthal, F. D. Kundig, W. Kundig, B. Anderson, P. E. Hartman, and S. Roseman: *ibid.*, **58:** 1963 (1967).
50. Tanaka, S., and E. C. C. Lin: *Proc. Natl. Acad. Sci. U.S.*, **57:** 913 (1967); Tanaka, S., D. G. Fraenkel, and E. C. C. Lin: *Biochem. Biophys. Res. Commun.*, **27:** 63 (1967).
51. Reviews in Rabinowitz, M.: *Bull. Soc. Chim. Biol.*, **50:** 311 (1968); Borst, P., A. M. Kroon, and G. J. C. M. Ruttenberg: in D. Shugar (ed.), *Genetic Elements; Properties and Functions,* Academic Press, New York, 1967; Nass, M. M. K.: *Science*, **165:** 25 (1969); Borst, P., and A. M. Kroon: *Intern. Rev. Cytol.*, **26:** 107 (1969).
52. See, e.g., Bosmann, H. B., A. Hagopian, and E. H. Eylar: *Arch. Biochem. Biophys.*, **130:** 573 (1969); Kuriyama, Y., T. Omura, P. Siekevitz, and G. E. Palade: *J. Biol. Chem.*, **244:** 2017 (1969).
53. Kellenberger, E.: Control Mechanisms of Bacteriophage Morphopoieses, in D. Wolstenholme (ed.), *Principles of Biomedical Organization,* Ciba Foundation, London, 1965.
54. Wood, W. B., R. S. Edgar, J. King, I. Lielausis, and M. Henninger: *Federation Proc.*, **27:** 1160 (1968).
55. Hayakawa, T., and M. Koike: *J. Biol. Chem.*, **242:** 1356 (1967).
56. Henney, H. R., Jr., C. R. Willms, T. Muramatsu, B. B. Mukherjee, and L. J. Reed: *J. Biol. Chem.*, **242:** 898 (1967); Willms, C. R., R. M. Oliver, H. R. Henney, Jr., B. B. Mukherjee, and L. J. Reed: *ibid.*, **242:** 889 (1967).
57. Tzagoloff, A., D. H. MacLennan, D. G. McConnell, and D. E. Green: *J. Biol. Chem.*, **242:** 2051 (1967).
58. Yamashita, S., and E. Racker: *J. Biol. Chem.*, **244:** 1220 (1969).

10 Chemistry of Carbohydrates

Carbohydrates are an important class of naturally occurring substances and are found universally distributed among plants, animals, and microorganisms. The designation "carbohydrate" derives from the fact that these organic compounds have empirical formulas $(C{\cdot}H_2O)_n$, in which n is 3 or greater, or are closely related to compounds that do. The metabolism of carbohydrates is of central importance to organisms, individually and collectively. Basically all organic foodstuffs are ultimately derived from the synthesis of carbohydrates through photosynthesis (see Chapter 12). Conversely, catabolism of carbohydrates provides the major share of the energy requirement for maintenance of life and performance of work. In addition, carbohydrates act as energy reservoirs and serve architectural functions. Prior to a consideration of the biosynthesis and metabolism of the carbohydrates, it is necessary to develop the structural chemistry of these substances in modest detail. Many excellent reference works are available and all standard textbooks of organic chemistry contain chapters devoted to it. In addition *Advances in Carbohydrate Chemistry,* a yearly publication of collected review articles, provides up-to-date and detailed accounts of particular aspects of the subject. Additional references are to be found at the end of this chapter. The student is referred to these for more detailed accounts of the structural chemistry of carbohydrates and for information regarding all but the most basic reactions; only these are included here.

Carbohydrates are divided into three basic categories: monosaccharides, oligosaccharides, and polysaccharides. The *monosaccharides* have three to nine, usually either five or six, carbon atoms and contain only one aldehyde or ketone function or are derivatives of molecules that do. The *oligosaccharides* are oligomers of monosaccharides linked by formation of glycosidic bonds. These generally contain two to eight or ten monomeric units. For carbohydrates characterized by higher degrees of polymerization, the designation *polysaccharide* is reserved. Polysaccharides are frequently molecules of great size and may have molecular weights of many million. Included among them are such important substances as glycogen, cellulose, and starch. We begin our discussion with an account of the structural chemistry of the monosaccharides and, with this foundation, proceed subsequently with the chemistry of the more complex substances.

MONOSACCHARIDES

The most common monosaccharide in living material is the simple six-carbon sugar (*hexose*) known as glucose. Several of the features of the structure and chemistry of this and related molecules may be appreciated from a consideration of the various means of representation of D-glucose. These are collected in Fig. 10-1. In structure I, D-glucose is indicated in a straight-chain form with an aldehydic function at C-1 and alcoholic functions at carbons 2 through 6. Sugars that terminate in a aldehydic function are called *aldoses.* Examination of structure I for D-glucose indicates the presence of four asymmetric carbon atoms. Thus, there exist $2^4 = 16$ optical isomers of aldohexoses, such as glucose. Assignment of the structure of each of the optical isomers of glucose then known was completed by the brilliant carbohydrate chemist Emil Fischer in 1896. In a similar fashion, we find $2^3 = 8$ optical isomers for the aldopentoses and $2^2 = 4$ for the aldotetroses. For each class of sugar, these isomers are subgrouped into D and L classes, depending on the relationship of the configuration at the penultimate carbon atom to that in D-glyceraldehyde. In Fig. 10-2, the steric relationships of the D-sugars are indicated. Similarly, the structures of the L-aldoses may be derived from L-glyceraldehyde and are, in each case, the mirror image of the corresponding D compounds.

In structure II, D-glucose is written in a cyclic hemiacetal form. This is formed by the addition of the hydroxyl function at C-5 to the aldehyde at C-1, a general property of aldoses. This raises three additional considerations. First, formation of the hemiacetal introduces a new asymmetric center at C-1 of D-glucose. There are thus two forms of this molecule and a total of 32 optical isomers for the cyclic forms of the hexoses. The former are designated α and β forms. In formulations such as II, the hydroxyl at C-1 is written to the right in α forms of D-sugars and to the left in β forms. Thus, for the sake of comparison, β-D-glucose is written as

```
      ┌────┐
HO—C—H     │
    |      │
 H—C—OH    │
    |      │
HO—C—H     │
    |      │
 H—C—OH    │
    |      │
 H—C—O─────┘
    |
   CH₂OH
```

β-D-Glucose

In the second place, we have the problem of ring size. In addition to six-membered rings of the type indicated in structures II and III of Fig. 10-1, several sugars form the related five-membered structures. These are designated *pyranose* and *furanose* forms, respectively, since they bear a formal resemblance to the six- and five-membered cyclic ethers pyran and furan:

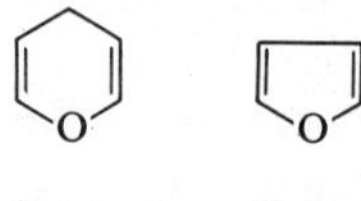

Pyran Furan

Thus, the complete name for structure II or III in Fig. 10-1 is α-D-glucopyranose. We encounter some examples of furanoses below. Finally, to what extent do sugars exist as cyclic hemiacetals as opposed to open-chain compounds in solution? Although the extent of hemiacetal formation is certainly

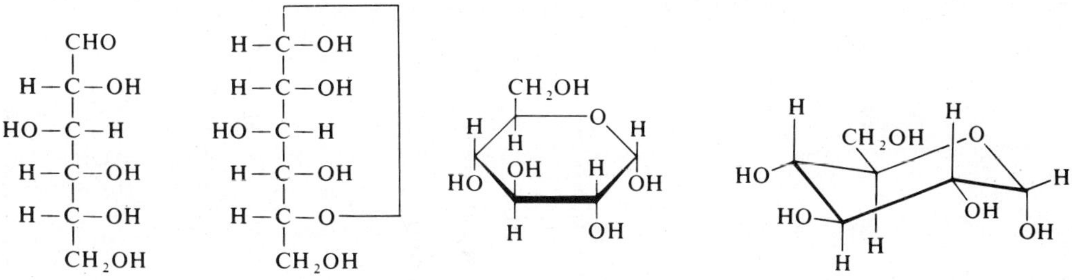

```
      CHO
       |
   H—C—OH
       |
  HO—C—H
       |
   H—C—OH
       |
   H—C—OH
       |
     CH2OH
```

(I)
D-Glucose

```
    ┌─────────┐
   H—C—OH     │
       |      │
   H—C—OH     │
       |      │
  HO—C—H      │
       |      │
   H—C—OH     │
       |      │
   H—C—O──────┘
       |
     CH2OH
```

(II)
α-D-Glucose

(III)
α-D-Glucose

(IV)
α-D-Glucose

FIGURE 10-1
Representations of the α-D-glucose molecule.

a function of the nature of the sugar, for hexoses and pentoses the portion of acyclic forms is less than 1 percent of the total.

Although structure II indicates the basic features of α-D-glucose, it provides little insight as to the actual shape of the molecule and the spatial relationships of the various functional groups to one another. A great improvement in representation of structural formulas in carbohydrates was introduced by Haworth many years ago. An example of a Haworth projection formula is provided by structure III in Fig. 10-1. In this view, the backbone carbon atoms together with the ethereal oxygen are considered to be planar, and the configuration of each of the ring substituents is indicated as being either above or below the plane of the ring. The translation from formulas of type II to Haworth projection formulas (III) is accomplished according to the following rules, which apply when the numbering in the projection proceeds clockwise: (1) substituents written to the right side of the backbone in the linear representation are written below the plane of the ring in the cyclic representation; similarly, substituents written to the left occupy positions above the ring; (2) for that single carbon atom whose hydroxyl function is involved in formation of the cyclic hemiacetal, the opposite rule applies. Thus, for D-sugars, the CH_2OH function is written in an above position, and the H atom on the same carbon is written below, despite the fact that it is to the left in the linear representation. This puzzling situation obtains because the linear representations of type II are actually misleading. The structural identity of the linear and cyclic representations is much easier to appreciate if we write the former as (this does not involve a change of configuration at carbon 5):

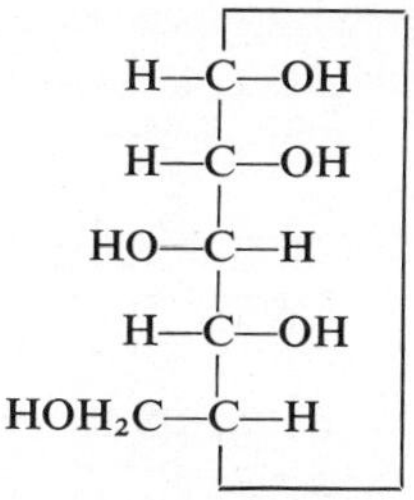

```
            ┌──────┐
       H—C—OH      │
           |       │
       H—C—OH      │
           |       │
      HO—C—H       │
           |       │
       H—C—OH      │
           |       │
 HOH2C—C—H         │
            └──────┘
```

In this formulation, the disposition of the substituents at carbon 5 follows the rule stated above for all other carbon atoms. Note that in the cyclic forms of D-sugars, the hydroxyl function at carbon 1 will be below the ring in the α form and above it in the β form.

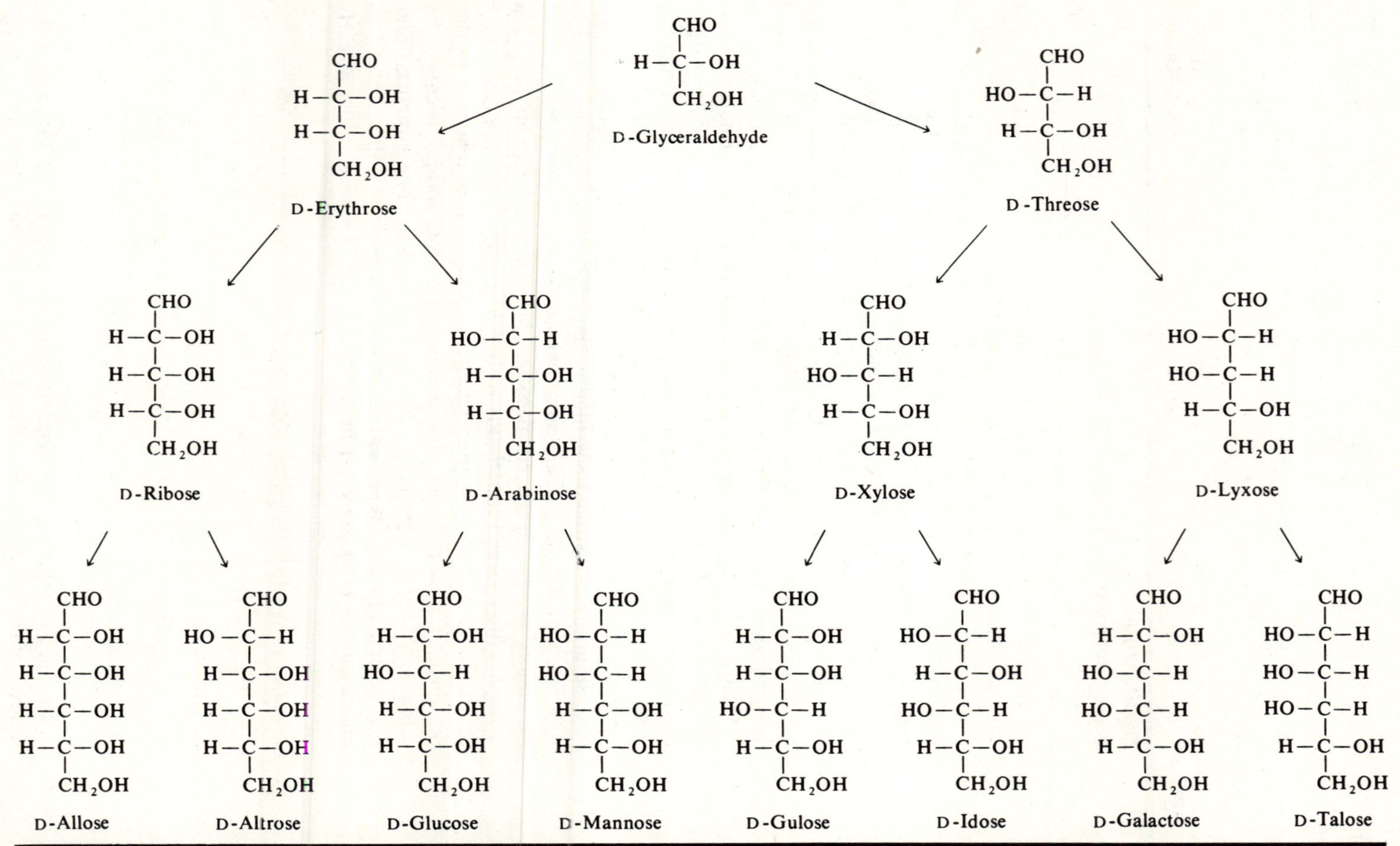

FIGURE 10-2
Configurational relationships of the D-aldoses.

The most accurate representation of the D-glucose molecule is provided in structure IV of Fig. 10-1. In this case, the conformation of the sugar is defined as a chair form, and the substituents appear in either axial or equatorial positions. This is certainly the correct structure for the crystalline species and probably obtains in solutions as well. Even though the correct conformations for many of the sugars have been established, we shall employ the Haworth projection formulas in preference to conformational structures, for reasons of convenience.

Sugars that have a ketone function, usually at position C-2, rather than an aldehyde function, are known as *ketoses*. The most prominent examples are provided by the ubiquitous hexose fructose and by the pentoses ribulose and xylulose:

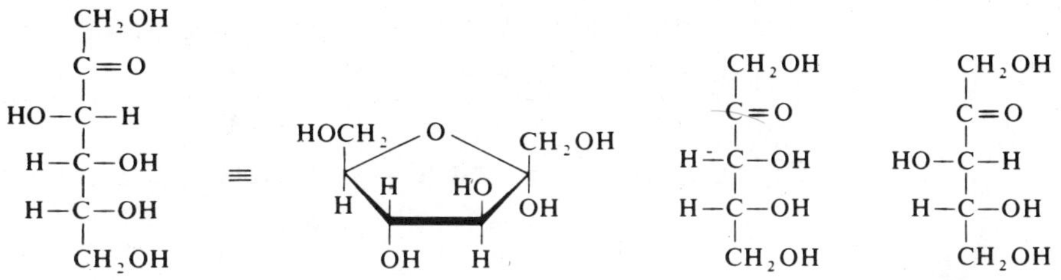

$$\begin{array}{c} CH_2OH \\ | \\ C{=}O \\ | \\ HO{-}C{-}H \\ | \\ H{-}C{-}OH \\ | \\ H{-}C{-}OH \\ | \\ CH_2OH \end{array} \equiv \text{(furanose ring: } HOCH_2,\ O,\ CH_2OH;\ H,\ H,\ HO,\ OH;\ OH,\ H) \qquad \begin{array}{c} CH_2OH \\ | \\ C{=}O \\ | \\ H{-}C{-}OH \\ | \\ H{-}C{-}OH \\ | \\ CH_2OH \end{array} \qquad \begin{array}{c} CH_2OH \\ | \\ C{=}O \\ | \\ HO{-}C{-}H \\ | \\ H{-}C{-}OH \\ | \\ CH_2OH \end{array}$$

D-Fructose D-Ribulose D-Xylulose

Fructose exists in solution at equilibrium as a mixture of furanose and pyranose forms; the exact proportions of the forms are not known. The aldopentoses generally exist as a mixture of furanose and pyranose forms, with the latter predominating. For example, xylose has negligible amounts of furanose form at equilibrium, arabinose is about 15 percent furanose, and ribose is about 25 percent furanose. The fraction of the sugars existing in the furanose form increases with increasing temperature.

The chemistry of the monosaccharides is, of course, dominated by the transformations of the carbonyl and alcohol moieties. Certain of these lead us to related classes of carbohydrates of biochemical interest.

Although, as noted above, most sugars exist predominantly in cyclic hemiacetal forms, the equilibrium between these and the aldehydic form is very mobile, and, in addition, sufficiently significant fractions of the sugars do exist in the aldehydic form at equilibrium, so that these substances undergo typical reactions with carbonyl reagents. Thus, aldoses reduce silver ion in ammonia solution (Tollens' reagent) and cupric ion complexed with citrate (Benedict's solution). Both of these reactions are of significant value in the detection and estimation of reducing sugars. In addition, sugars react with amines. A particularly important example is provided by the reaction with excess phenylhydrazine. Although the mechanism of this reaction is not quite clear, it has three noteworthy features: (1) initial formation of the phenylhydrazone of, for example, glucose; (2) an Amadori rearrangement giving rise to a carbonyl function at C-2; and (3) reductive cleavage, at some stage of the reaction, of an N—N bond in phenylhydrazine or a bound phenylhydrazino group. The overall stoichoimetry is indicated in Eq. (10-1):

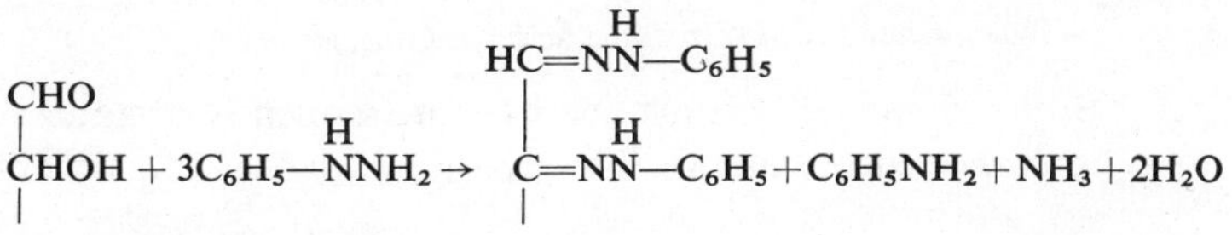

$$\begin{array}{l} CHO \\ | \\ CHOH \\ | \end{array} + 3C_6H_5{-}\overset{H}{N}NH_2 \rightarrow \begin{array}{l} HC{=}N\overset{H}{N}{-}C_6H_5 \\ | \\ C{=}N\overset{H}{N}{-}C_6H_5 \\ | \end{array} + C_6H_5NH_2 + NH_3 + 2H_2O \qquad (10\text{-}1)$$

Osazones are ordinarily crystalline compounds. This reaction was utilized with great success by Fischer in the work that established the configurations of the sugars. For example, the observation that glucose, mannose, and fructose all yield the same osazone indicates that the configuration of each of these sugars about carbons 3, 4, and 5 must be identical.

Reduction under mild conditions, with sodium borohydride, for instance, converts sugars to polyhydric alcohols. That is, the aldehyde or ketone function is reduced to the alcohol level of oxidation. These alcohols are termed *pentitols, hexitols,* and so on. Certain of them are found naturally. The most important are mannitol, glucitol (or sorbitol), and ribitol:

D-Mannitol	D-Glucitol (Sorbitol)	D-Ribitol
CH_2OH	CH_2OH	
HO—C—H	H—C—OH	CH_2OH
HO—C—H	HO—C—H	H—C—OH
H—C—OH	H—C—OH	H—C—OH
H—C—OH	H—C—OH	H—C—OH
CH_2OH	CH_2OH	CH_2OH

Closely related to these substances are the cyclic hexitols, such as the inositols. Two of these are indicated below:

myo-Inositol

scyllo-Inositol (Scyllitol)

myo-Inositol is found widely distributed in living material.

Monosaccharides undergo a variety of oxidation reactions; among the products of these are three classes of acids. Treatment of aldoses with an oxidizing agent, such as bromine, converts the aldehydic function at C-1 to a carboxylic acid function and gives rise to the *aldonic* acids. Use of stronger oxidizing agents, such as nitric acid, induces oxidation at both C-1 and the primary alcohol function and yields the *aldaric* acids. Finally, specific oxidation of the primary alcohol group yields the *uronic* acids. These three types of acids as derived from glucose are indicated below:

D-Gluconic acid	D-Glucuronic acid	D-Glucaric acid
CO_2H	CHO	CO_2H
H—C—OH	H—C—OH	H—C—OH
HO—C—H	HO—C—H	HO—C—H
H—C—OH	H—C—OH	H—C—OH
H—C—OH	H—C—OH	H—C—OH
CH_2OH	CO_2H	CO_2H

Both aldonic and uronic acids occur frequently in nature both as intermediates in carbohydrate metabolism and the latter as constituents of polysaccharides.

In addition to the polyols derived from monosaccharides by reduction and the various acids derived by oxidation, there are two additional classes of sugar derivatives that are of exceptional importance in biochemistry. These include the amino sugars and the 6-deoxyhexoses. The two most important amino sugars are D-glucosamine and D-galactosamine:

```
     CHO               CHO
     |                 |
  H—C—NH2           H—C—NH2
     |                 |
 HO—C—H            HO—C—H
     |                 |
  H—C—OH           HO—C—H
     |                 |
  H—C—OH            H—C—OH
     |                 |
     CH2OH             CH2OH
```

D-Glucosamine D-Galactosamine

Both of these occur as important constituents of many polysaccharides (usually as their *N*-acetyl derivatives; see below). The most prominent 6-deoxyhexoses include L-rhamnose and L-fucose:

```
     CHO               CHO
     |                 |
  H—C—OH           HO—C—H
     |                 |
  H—C—OH            H—C—OH
     |                 |
 HO—C—H             H—C—OH
     |                 |
 HO—C—H            HO—C—H
     |                 |
     CH3               CH3
```

L-Rhamnose L-Fucose

These are widely distributed in plants, animals, and microorganisms, frequently as constituents of polysaccharides.

The interconversion of α and β forms of sugars, *mutarotation,* occurs readily in aqueous solution and is subject to catalysis by both acids and bases. These reactions are most easily followed by observing changes in optical rotation as a function of time. For example, α-D-glucose has a specific rotation of 112°; the β anomer has a rotation of 18.7°. At equilibrium, the observed rotation is about 52.5°, corresponding to 64 percent of the β anomer and 36 percent of the α. The mutarotation of certain sugars is subject to enzymatic catalysis.

An exceptionally important reaction of the monosaccharides is *glycoside formation.* For example, if D-glucose is treated with methanol in the presence of HCl, two products are formed:

```
 CH2OH                        CH2OH                    CH2OH
      O                             O                        O   OMe
HO OH    ~OH   --MeOH/HCl-->  HO OH    OMe      +      HO OH
    OH                            OH                       OH
```
(10-2)

In each case the hydroxyl function as position 1 has been replaced by methoxyl. By analogy with the terminology developed above, these products

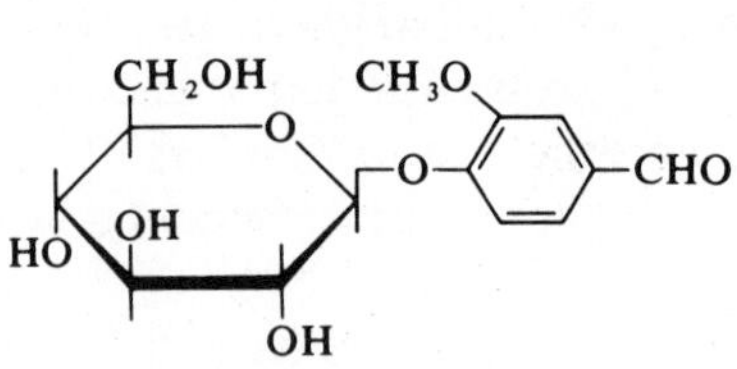

Vanillin-β-D-glucoside
(natural source of vanilla flavor)

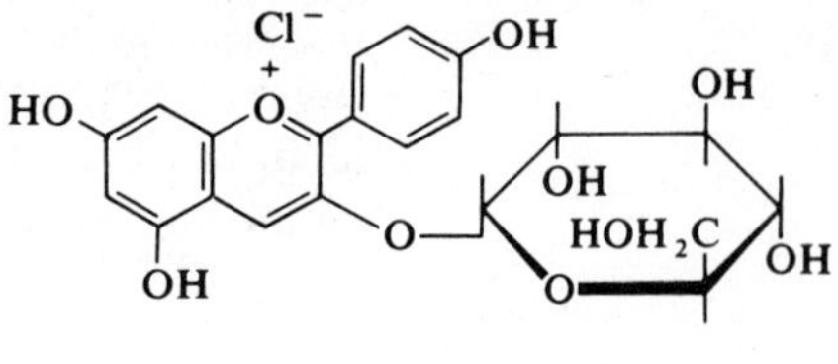

Pelargonidin-β-D-glucoside chloride
(a plant pigment)

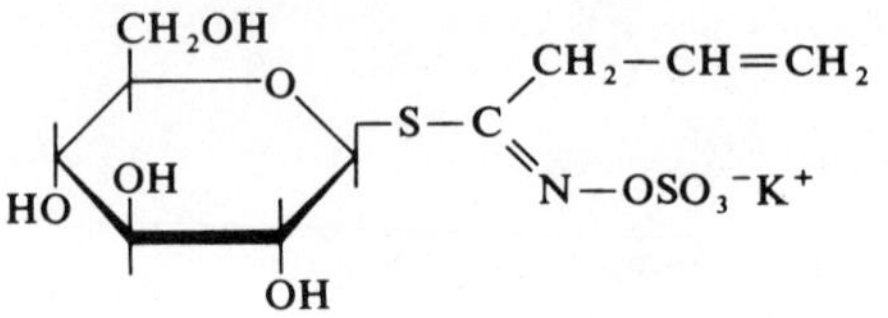

Sinigrin
(a constituent of horseradish)

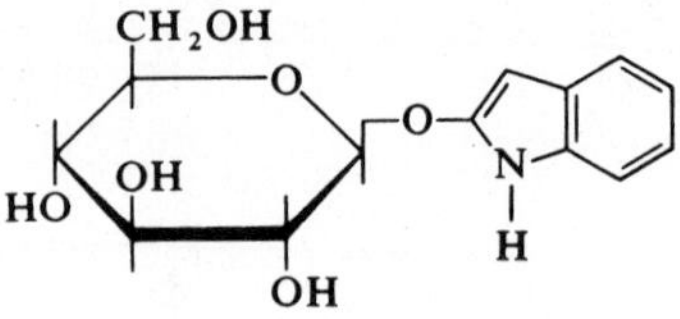

Indican
(a source of indigo dye)

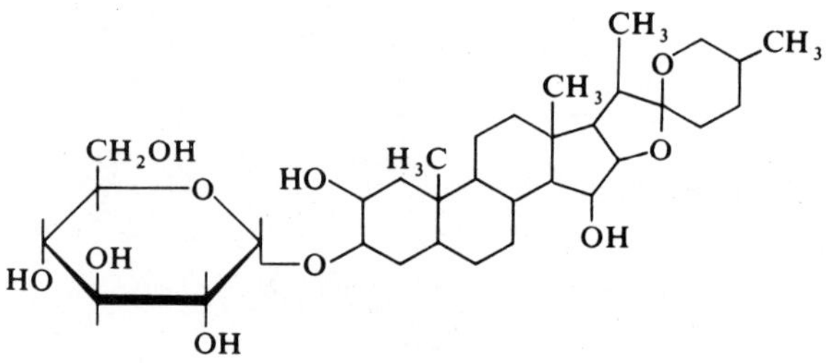

Digitogenin-α-D-glucoside
(a saponin)

FIGURE 10-3
Some naturally occurring glycosides.

are termed methyl α-D-glucoside and methyl β-D-glucoside. More complete designations would be methyl α- and β-D-glucopyranosides. The nonsugar moiety of these compounds (i.e., the group substituted for the proton of the hydroxyl group at position C-1) is termed the *aglycone.* Compounds having one or more glycosidic linkages are of great importance biochemically. This is the basic linkage through which all polysaccharides and oligosaccharides are formed. Thus, the glycosidic bond is to carbohydrate chemistry what the peptide bond is to protein chemistry and the phosphodiester linkage is to that of nucleic acids. Glycosidic bonds are also involved in the linkages of sugars to steroids, pigments, and other substances. Some examples of such conjugates are collected in Fig. 10-3. Finally, we have already encountered an exceptionally important related group of compounds, the glycosylamines, in which the linkage is through nitrogen rather than oxygen. This is the type of linkage through which D-ribose and 2-deoxy-D-ribose are bound to the nitrogen bases in such molecules as RNA, DNA, ATP, and NAD^+ (see Chapters 5 and 8).

OLIGOSACCHARIDES

The oligosaccharides are subdivided into classes termed disaccharides, trisaccharides, and so forth, depending on whether they have two, three, or more carbohydrate units. Perhaps the most abundant disaccharide is sucrose, common table sugar. Sucrose is composed of one D-glucose unit and one D-fructose unit; its structure is indicated below:

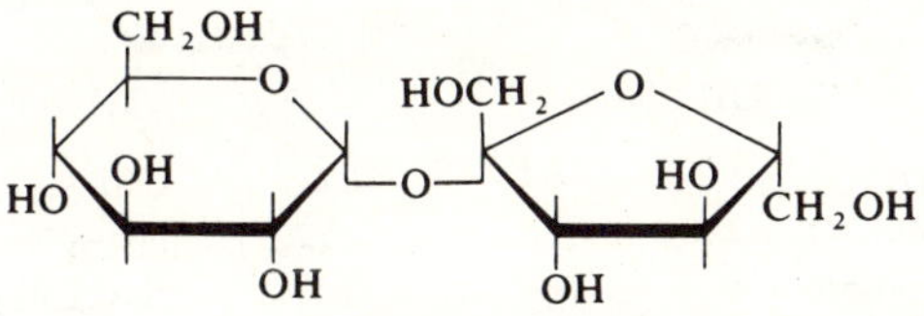

Sucrose

The complete specification of this disaccharide or of any disaccharide requires four pieces of information. First, we must specify the nature of the monosaccharides from which the disaccharide is constituted; second, for each monosaccharide we must know the ring type, furanose or pyranose, as it exists in the disaccharide; third, those carbon atoms through which the linkage is effected must be specified; and, finally, the anomeric nature of this linkage is required. Thus, a complete description of the sucrose molecule is α-D-glucopyranosyl-$(1 \rightarrow 2)$-β-D-fructofuranoside. Notice that sucrose is not a reducing sugar, since both carbon atoms at the oxidation level of the carbonyl group are involved in the glycosidic linkage. In Fig. 10-4, several other examples of prominent disaccharides are collected together with trivial and systematic names.

Few trisaccharides are of importance as naturally occurring organic compounds. Among the more prominent are raffinose [α-D-galactopyranosyl-$(1 \rightarrow 6)$-α-D-glucopyranosyl-$(1 \rightarrow 2)$-β-D-fructofuranoside] and gentianose [β-D-glucopyranosyl-$(1 \rightarrow 6)$-α-D-glucopyranosyl-$(1 \rightarrow 2)$-β-D-fructofuranoside]:

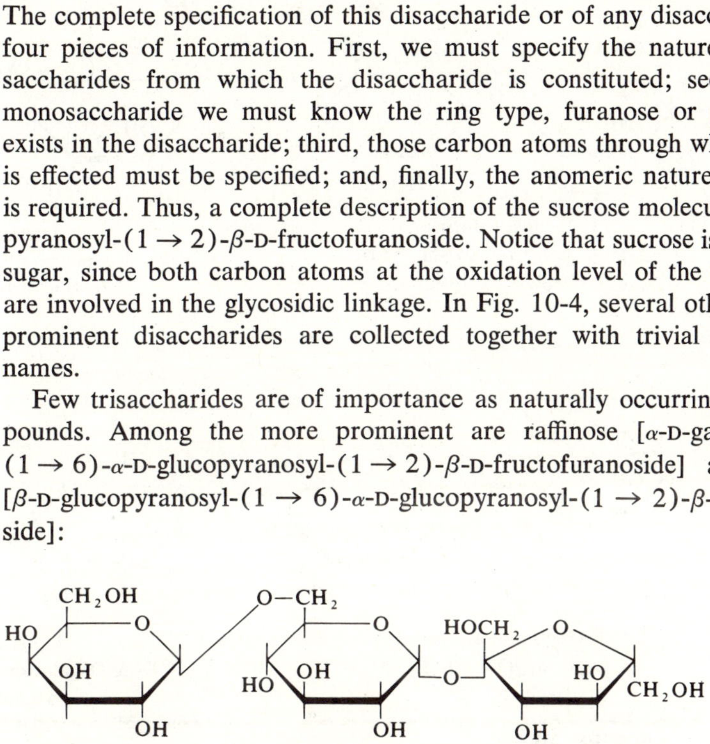

Raffinose

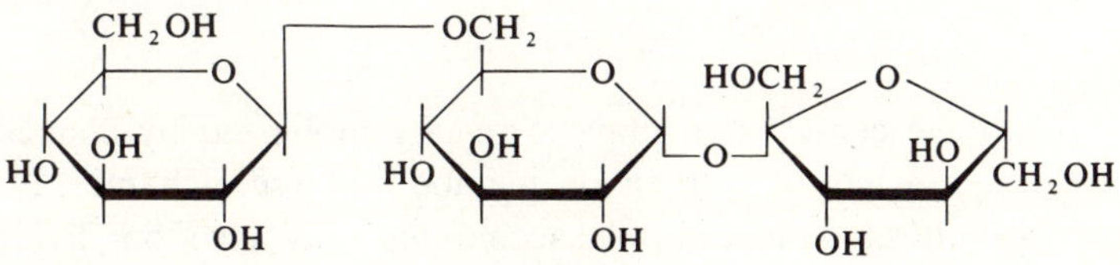

Gentianose

These sugars are constituents of plant tissues which, quite generally, have a greater variety of oligosaccharides than are found in animal tissues.

POLYSACCHARIDES

The bulk of all carbohydrate in nature exists in the form of polysaccharides. In terms of function, polysaccharides may be divided into two principal groups. The first, which includes cellulose, for example, serves principally architectural purposes; the second, which includes substances such as glycogen, functions as nutrients. That is, molecules in this group are primarily

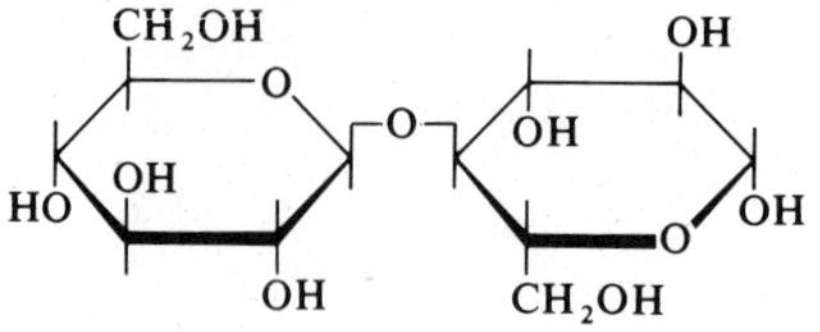

Cellobiose (β form)
O-β-D-glucopyranosyl-(1 → 4)-β-D-glucopyranose

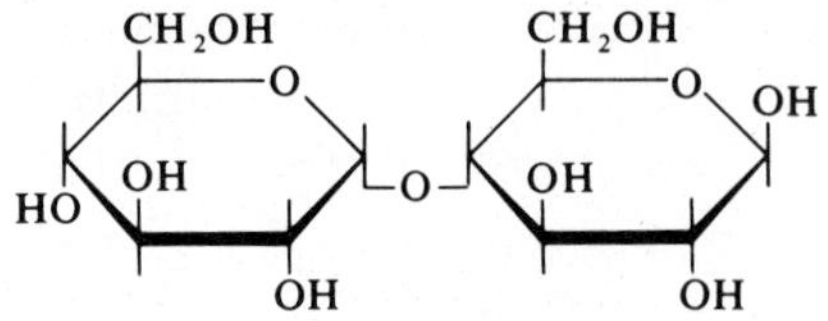

Maltose (β form)
O-α-D-glucopyranosyl-(1 → 4)-β-D-glucopyranose

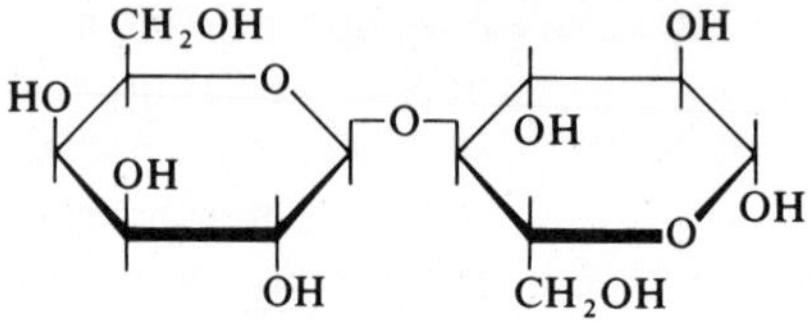

Lactose (β form)
O-β-D-galactopyranosyl-(1 → 4)-β-D-glucopyranose

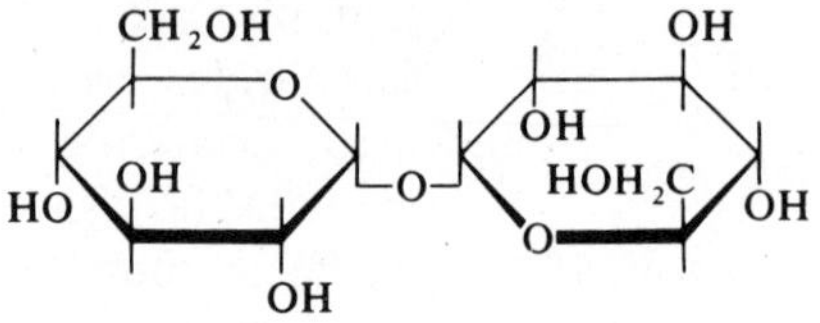

Trehalose
O-α-D-glucopyranosyl-(1 → 1)-α-D-glucopyranoside

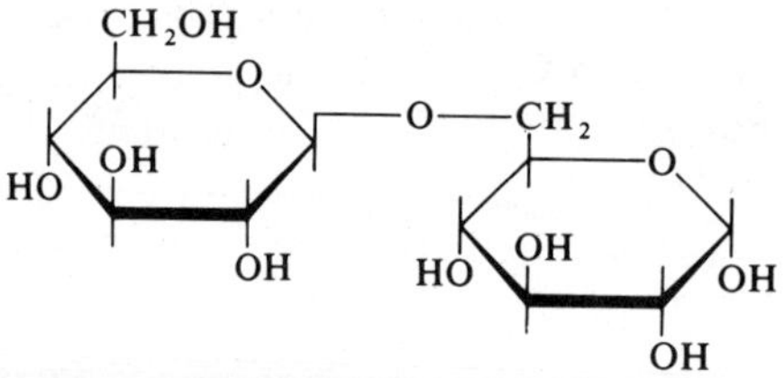

Gentiobiose (α form)
O-β-D-glucopyranosyl-(1 → 6)-α-D-glucopyranose

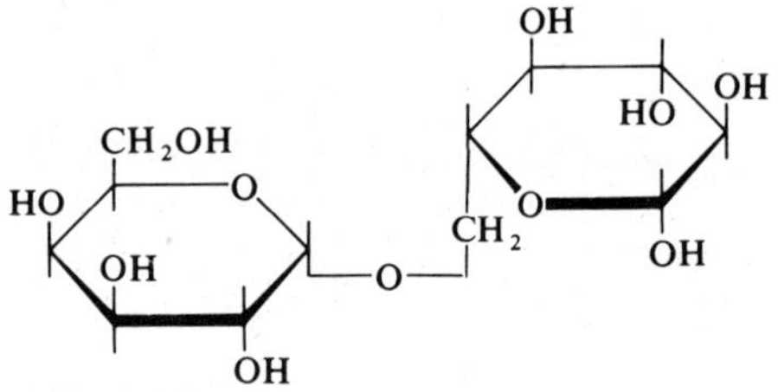

Melibiose (β form)
O-α-D-galactopyranosyl-(1 → 6)-β-D-glucopyranose

FIGURE 10-4
Structures and nomenclature for several disaccharides.

storage centers and may be readily mobilized by conversion to monosaccharides, which may then be metabolized (see Chapter 11). In terms of gross structural features, polysaccharides may also be divided into two groups: homopolysaccharides and heteropolysaccharides. The former is characterized by the occurrence of only one sort of monosaccharide within the molecule (although different sorts of intersugar linkages may be present) and the latter by the occurrence of two or more types of monomeric units. The chemistry of the former is somewhat simpler than that of the latter, in the same sense that the chemistry of polyamino acids is simpler than that of proteins, and we turn to some examples of polysaccharides in this class first.

Before actually discussing occurrence, structure, and function for particular polysaccharides, a few words about the general state of structural studies on these molecules is in order. Considerable effort is being devoted to such studies, and great progress has been made in recent years. Perhaps 10 to 20 new polysaccharides are isolated each year. The determination of the order of occurrence of monosaccharides in polysaccharides is in some ways simpler and in some ways more difficult than the sequencing of polypeptides or nucleic acids. The basic simplifying feature in polysaccharide chemistry, compared with that of proteins or nucleic acids, is that the former substances are usually composed of reasonably small repeating units. That is, each monomeric unit occurs in a regular repeating fashion throughout the molecule. Individual amino acids or nucleotides, in contrast, appear to be distributed randomly or nearly so within the corresponding polymeric products. This is also true of a number of polysaccharides. But for those which are, in fact, perfectly regular, determination of the structure of the repeating unit together with a knowledge of the molecular weight suffices to establish the complete primary structure. In most cases, however, a certain number of singularities occur, such as branch points, which complicate matters considerably. A basic complicating factor in polysaccharide chemistry is the existence of a number of types of intersugar linkages. Although all amino acid residues of proteins are connected by peptide linkages and the nucleotides of nucleic acids are always joined by 3′,5′-phosphodiester bonds, intersugar linkages may be α-$(1 \rightarrow 2)$, β-$(1 \rightarrow 3)$, α-$(1 \rightarrow 4)$, and so on. In terms of the number of types of monomeric units within a particular polysaccharide, the situation is more like that of nucleic acids than of proteins. Seldom do more than four monomer types occur within a single polysaccharide molecule. It is worth noting that, quite in general, it is more difficult to establish the sequence of a random polymer that contains few types of monomeric units than of one having many types.

We now direct our attention to the properties of individual polysaccharides.

Cellulose

The most abundant structural polysaccharide of plants, and indeed the most abundant organic substance on earth, is *cellulose*. About 10^{11} tons of cellulose are synthesized per year. It occurs almost exclusively in plants, in which it accounts for about 50 percent of the total content of carbon. The following facts suffice to establish its structure. First, total hydrolysis yields only D-glucose. Second, partial hydrolysis yields cellobiose (see page 474). Third, hydrolysis of fully methylated cellulose yields only 2,3,6-tri-*O*-methylglucose. It follows that cellulose is a linear array of D-glucopyranose units linked by β-$(1 \rightarrow 4)$-glycosidic bonds, as indicated below:

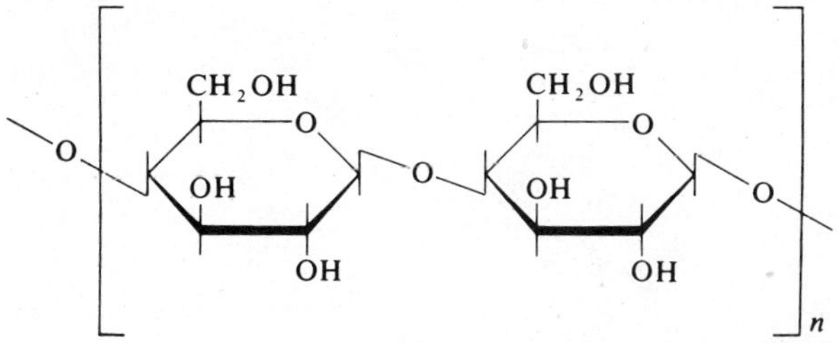

The molecular weight of isolated cellulose preparations varies from about 50,000 to more than 1 million. It is worth noting that these molecules, like the nucleic acids, may be subject to ready degradation in isolation procedures, and hence these molecular weights may be much too small. In any event, cellulose molecules exist in the cell walls of plants, not as individual molecules, but in the form of microfibrils that are several hundred angstroms long. These are formed from numerous cellulose chains arranged parallel to one another.

Chitin

Chitin, a polysaccharide closely related structurally to cellulose, performs similar functions in the lower forms of plant life, particularly the fungi, and in invertebrates, particularly the arthropods. It has wide distribution within these groups and is probably the second most abundant organic substance on earth. The copepods alone are considered to synthesize about 10^9 tons of chitin per year.

Chitin is composed of a linear array of β-linked *N*-acetyl-D-glucosamine units, as indicated below:

CH₂OH CH₂OH O O O O O OH OH NHCOCH₃ NHCOCH₃ n

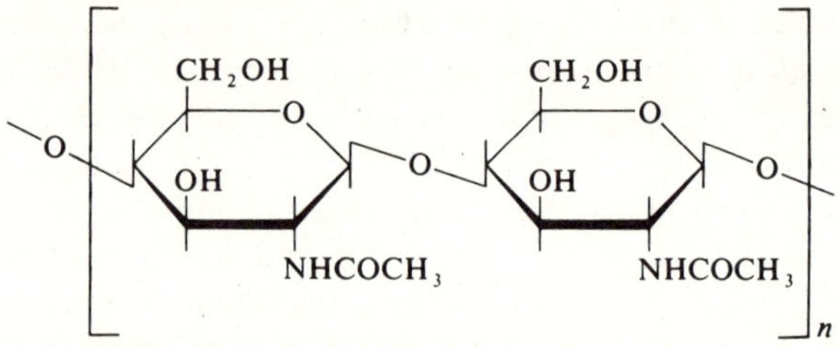

The repeating unit is identical to that for cellulose except that the hydroxyl function at C-2 has been replaced by the acetamido function. Those polysaccharides which contain amino sugars or their derivatives, such as chitin, are collectively known as *mucopolysaccharides.* It is difficult to estimate the molecular weight of chitins as they exist in nature, since they are invariably tightly linked to nonpolysaccharide material, particularly proteins and inorganic salts. Drastic methods are required to free chitin from these materials, and substantial degradation very likely occurs in these processes.

Other Structural Homopolysaccharides

In many species, neither cellulose nor chitin is employed for purposes of cell wall construction and for related architectural functions. Their place is taken by a variety of alternative polysaccharides, including *glucans, mannans, xylans,* and so forth. For example, the glucan isolated from certain species of *Agrobacter* and *Rhizobium* has a structure that is largely, but not completely, formed by β-$(1 \rightarrow 2)$-glycosidic linkages between adjacent D-glucose units, as indicated below:

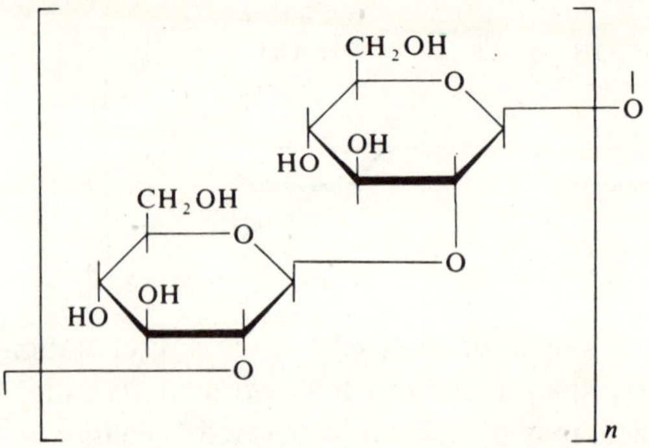

The polysaccharide is thus closely related to cellulose, differing only in the nature of the intersugar linkages. A more complicated example of a structural homopolysaccharide is provided by the mannan from baker's yeast, *Saccharomyces cerevisiae*. This polysaccharide is probably formed from the repeating units indicated below:

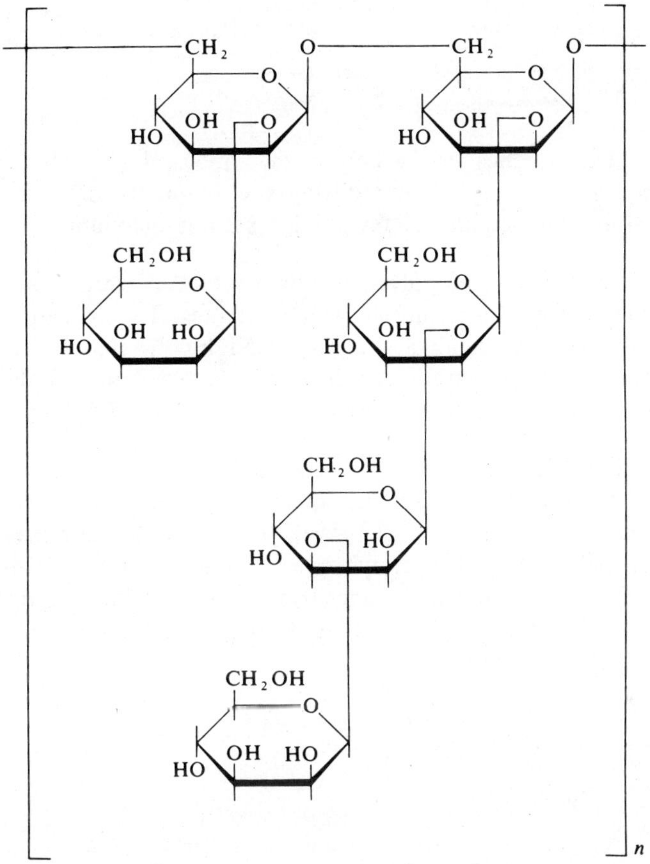

Starches Starches provide one example of the function of polysaccharides as nutritional reservoirs. They serve such a function in plants. The basic repeating unit from which the starches are synthesized involves the linkage of successive D-glucose molecules by α-$(1 \rightarrow 4)$-glycosidic bonds (contrast with cellulose):

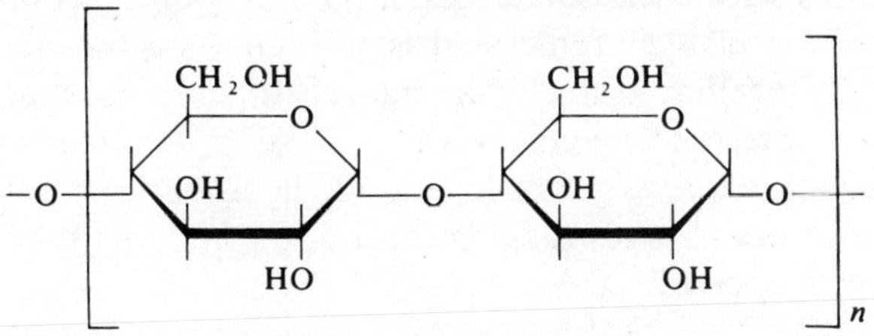

Starches fall into two distinct structural classes. Isolated samples of starch are almost always mixtures of the two. The first, *amyloses,* is constituted of long unbranched chains. In contrast, the *amylopectins* are branched-chain polysaccharides. The branch points are formed from α-$(1 \rightarrow 6)$-glycosidic linkages. A typical branch point for the amylopectins may be formulated as:

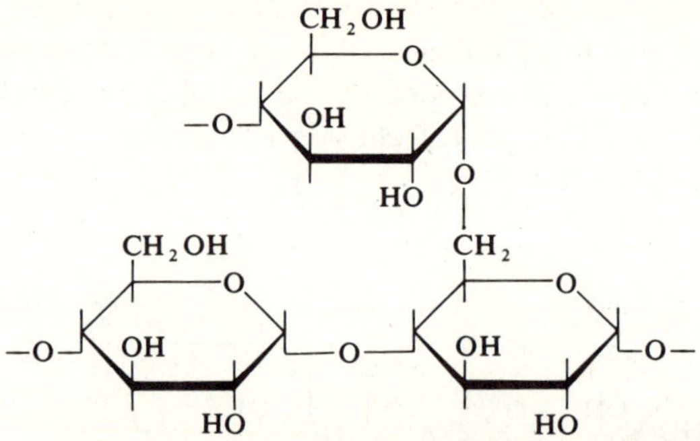

The starches vary widely in their molecular weights. This may be, in part, a result of degradation during the processes of isolation and purification. Molecular weights of a few million are not uncommon for these molecules.

Glycogen

Glycogen is also a polyglucose that serves as a nutritional reservoir but, unlike the starches, is found in animal tissues. Its structure is similar to that of amylopectin, being a branched and highly elaborated molecule. Molecular weights of preparations range from a few hundred thousand to about 100 million. Considerable polydispersity within a single preparation is common.

Hyaluronic Acid

Our first example of a heteropolysaccharide is *hyaluronic acid.* This material is composed of equimolar quantities of D-glucuronic acid and *N*-acetyl-D-glucosamine and is, therefore, a mucopolysaccharide as well. The D-glucuronic acid and *N*-acetyl-D-glucosamine occupy alternating positions in the polysaccharide. The linkages from the amino sugar to the acid are β-$(1 \rightarrow 4)$; those from the acid to the amino sugar are β-$(1 \rightarrow 3)$. Thus, hyaluronic acid has the structure

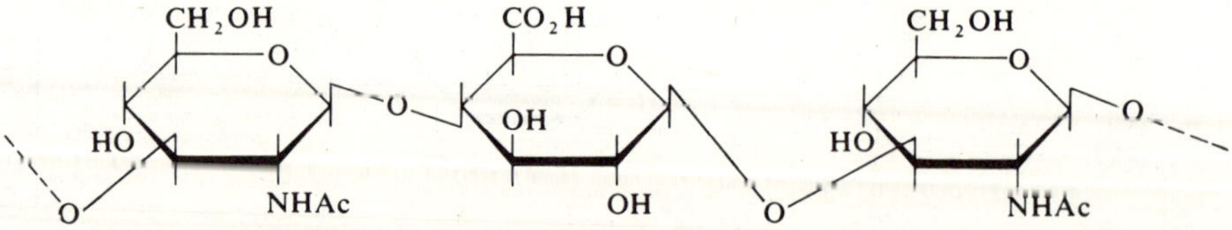

Hyaluronic acid is a widely distributed polysaccharide. It is universally present in connective tissues of animals and in their vitreous and synovial fluids as well. It is also synthesized by various strains of bacteria. Hyaluronic acid is usually found in association with protein, and hyaluronic acid–protein complexes have been isolated from natural sources. It has been suggested that the function of hyaluronic acid is to bind water in interstitial spaces and to hold cells together in a jellylike matrix. In addition, the material may provide the fluids present in joints with lubricating and shock-absorbing properties.

The molecular weight of hyaluronic acids varies considerably as a function of their source. Values ranging from a few hundred thousand to a few million have been reported.

Chondroitin Sulfates

Chondroitin, a polysaccharide identical with hyaluronic acid, except that D-galactosamine replaces D-glucosamine, is the parent substance for three polysaccharides found widely distributed in connective tissue. These are termed the *chondroitin sulfates A, B,* and *C.* Chondroitin sulfate B is frequently referred to as *dermatan sulfate.* The structures for these molecules are indicated on the following page.

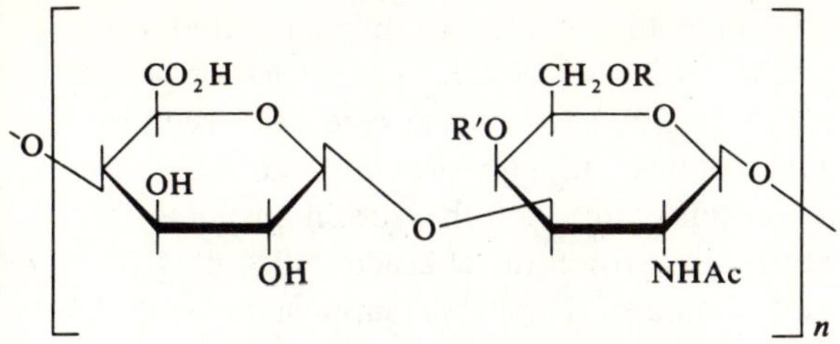

Chondroitin sulfate A: $R=H, R'=SO_3H$
Chondroitin sulfate C: $R=SO_3H, R'=H$

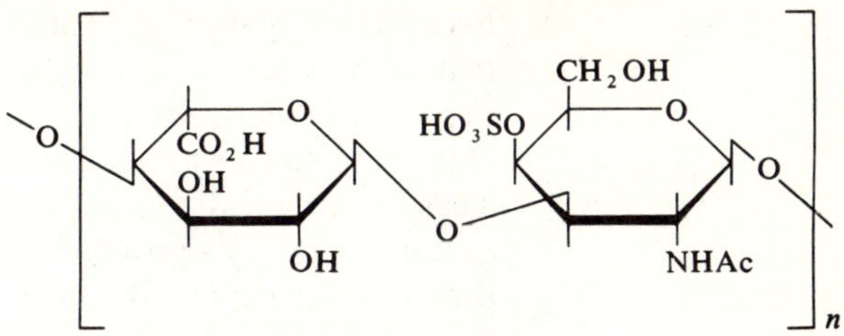

Chondroitin sulfate B
(Dermatan sulfate)

Both chondroitin sulfate A and C are composed of equimolar quantities of D-glucuronic acid, *N*-acetyl-D-galactosamine, and sulfate. The structures differ only in the position of the sulfate function. In chondroitin sulfate B, L-iduronic acid replaces the D-glucuronic acid of the A and C polysaccharides. The glycosidic bonds are alternating β-$(1 \rightarrow 3)$ and β-$(1 \rightarrow 4)$ linkages for the A and C polysaccharides and alternating α-$(1 \rightarrow 3)$ and β-$(1 \rightarrow 4)$ for chondroitin sulfate B. The chondroitin sulfates are among the smaller mucopolysaccharides, ordinarily having molecular weights in the range 50,000 to 100,000.

Keratosulfate

Keratosulfate is related to the chondroitin sulfates both structurally and in terms of distribution. It is the final principal polysaccharide component of connective tissue and occurs elsewhere as well. Keratosulfate consists of alternating D-galactose and 2-acetamido-2-deoxy-D-glucose-6-sulfate connected by alternating β-$(1 \rightarrow 4)$ and β-$(1 \rightarrow 3)$ linkages:

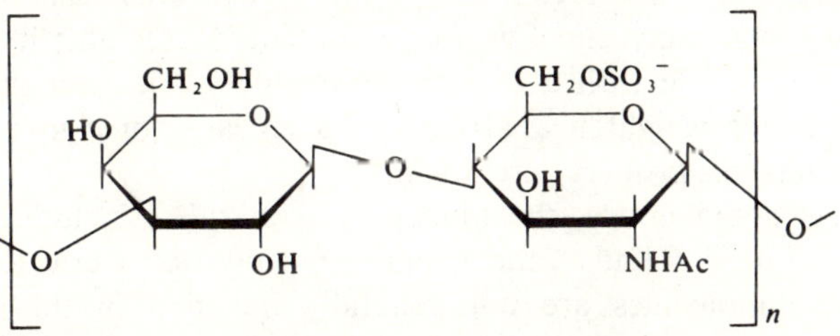

Heparin

Heparin is an interesting but not well understood mucopolysaccharide. It has a long history, having been discovered half a century ago, and has important biological properties, including that of an anticoagulant. Nevertheless, neither the structure nor biological function of this substance is completely clear. Heparin occurs principally in the mammalian circulatory system and in numerous organs where mast cells occur. These are the apparent site of synthesis and storage of heparin. It occurs firmly bound to protein. Rather drastic methods are usually required to isolate it from natural sources. The molecular weights of heparin preparations are usually in the range 10,000 to 20,000.

In addition to the polysaccharides specifically discussed above, these substances occur in the blood-group substances, in the antigenic coats of the encapsulated bacteria, as components of bacterial cell wall materials (mureins), and in numerous glycoproteins and lipopolysaccharides.

GLYCOPROTEINS

Carbohydrates not only occur free in living systems but also in combined form with proteins. Those proteins containing a covalently linked carbohydrate moiety are termed *glycoproteins.* Although such structures occur in both bacteria and plants, they are particularly prevalent and important in mammalian tissues. In Table 10-1, some of the glycoproteins isolated from mammalian sources are listed together with some of the properties of glycopeptides obtained from them through degradation of the original structures. Note that this list includes serum proteins, an egg protein, and a mucin. These are typical sources of glycoproteins; most of the serum proteins, for example, are glycoproteins. Moreover, the structural characteristics of glycoproteins that may be deduced from the data of Table 10-1 appear to be quite general. Specifically, these include the following:

1. The carbohydrate chains of glycoproteins are quite short, consisting of perhaps 8 or 10 individual saccharide residues.

2. Seven sugars comprise the preponderant portion of the carbohydrate moiety of mammalian glycoproteins. These are D-galactose, *N*-acetyl-D-galactosamine, D-glucose, *N*-acetyl-D-glucosamine, D-mannose, L-fucose, and sialic acid (*N*-acetylneuraminic acid). Of these, glucose is relatively rare. In those glycoproteins in which they are present, L-fucose and sialic acid tend to occupy terminal nonreducing positions in the carbohydrate structure.

3. Third, the linkage of carbohydrate to protein is of one of three types: (a) the amide nitrogen of asparagine linked to *N*-acetyl-D-glucosamine, (b) the hydroxyl group of serine or threonine linked to *N*-acetyl-D-galactosamine, and (c) the hydroxyl group of hydroxylysine linked to D-galactose. The first linkage type is the most common one and the last is important only for collagen.

One very interesting serum glycoprotein is that of certain Antarctic fishes who survive near-freezing water temperatures as a result of freezing-point depression of the serum due to this glycoprotein. This particular glycoprotein is a remarkably effective freezing-point depressant; it behaves in this regard as a molecule with a molecular weight of 15, even though it actually has a molecular weight near 10,000. The means through which this glycoprotein exhibits this property are unknown.

As noted above, the mucopolysaccharides, including hyaluronic acid, the chondroitin sulfates, and heparin, usually occur covalently bound to protein. These molecules are not generally included in the glycoprotein category, although the distinction is an arbitrary one. Structurally they are distinct in the sense that the mucopolysaccharide–protein structures are much richer in carbohydrates that the glycoproteins described above. At the reducing end of the polysaccharide chain, there is a special linkage region through which the polysaccharide is bound to the protein. The linkage region has the structure Gal-Gal-Xyl and the terminal xylose is attached to a serine hydroxyl of the protein.

CELL WALL STRUCTURES

Recent advances in knowledge of cell wall structures constitutes one of the most interesting chapters in modern biochemistry. Here we consider the cell wall structures that are found in bacteria and plants. As these structures perform distinct functions in these two groups of living organisms, one might expect that they would be structurally distinct as well. This proves to be true and we shall see how structure and function are related in the two cases.

Baterial Cell Walls

The cell wall of a bacterium, together with the cell membrane associated with it, performs several important functions. By permitting the diffusion of certain chemical species in and out of the bacterium, by facilitating the diffusion of others, and by impeding or preventing that of still others, these structures dominate the interaction of the interior milieu of the bacterium with the

TABLE 10-1
Properties of some glycoproteins and the associated glycopept des

Glycoprotein	*Number of residues per mol of glycopeptide*						*Amino acid associated with glycopeptide*	*Molecular weight*	*Number of glycopeptides per mol of protein*
	Man	GlcNAc	Gal	Fuc	Sial	GalNAc			
Ovalbumin, hen	5	3					Asp	1560	1
Ribonuclease B, bovine	6	2					Asp, Leu	1600	1
Soybean hemagglutinin	9	2–3					Asp	2200	3
Thyroglobulin, calf	5	1					Asp	1250	9
Transferrin, human	4	4	2		2		Asp (Ser, Gly)	3100	2
Fibrinogen, bovine	4	3	2		1		Asp	2100	2
γ-Globulin, human	2–3	4	1–2	1	1		Asp (Ser, Thr)	2300	2
Ovine submaxillary mucin[a]						1.3	Ser, Thr, Gly	660	

[a] Several hundred disaccharide groups, *N*-acetylneuraminyl-(2→6)-*N*-acetylgalactosamine, are present in this molecule; the glycopeptides isolated are free of sialic acid as a result of previous neuraminidase treatment.

SOURCE: N. Sharon, *Ann. Rev. Biochem.*, **35:** 485 (1966).

environment and so help to regulate the internal composition of the bacterium. Many of the biochemical capacities of bacteria are associated with the bacterial membrane and cell wall. Indeed, most of the capacities associated with subcellular organelles in mammaliam systems are included in this group (Chapter 9). Moreover, the bacterial cell wall protects the organism against damage due to osmotic shock or mechanical agitation and must, therefore, have substantial strength. Interactions between individual bacteria are also controlled by the nature of the bacterial cell wall or excretions from it. In addition to performing these functions, the bacterial cell wall must have the capacity to conform to the requirements of cell division (i.e., growth, rupture, and resealing).

To be consistent with all these functions, bacteria have evolved cell wall structures that are the most complex organic molecules known. The details of the cell wall structure vary, of course, from species to species. Substantial differences occur between Gram-positive and Gram-negative bacteria, for example. But the important point is that one structure appears common to all bacterial cell walls. This structure has been referred to by several names, the most common being *peptidoglycan* and *murein*. *The cell walls of bacteria are composed, in part at least, of single, bag-shaped molecules of peptidoglycan.*

Although the detailed structure of the peptidoglycan varies somewhat from bacterial species to species, the following structural features appear to be invariant. First, there exist linear polysaccharide structures composed of alternating residues of *N*-acetyl-D-glucosamine and *N*-acetylmuramic acid joined by β-(1 → 4) linkages:

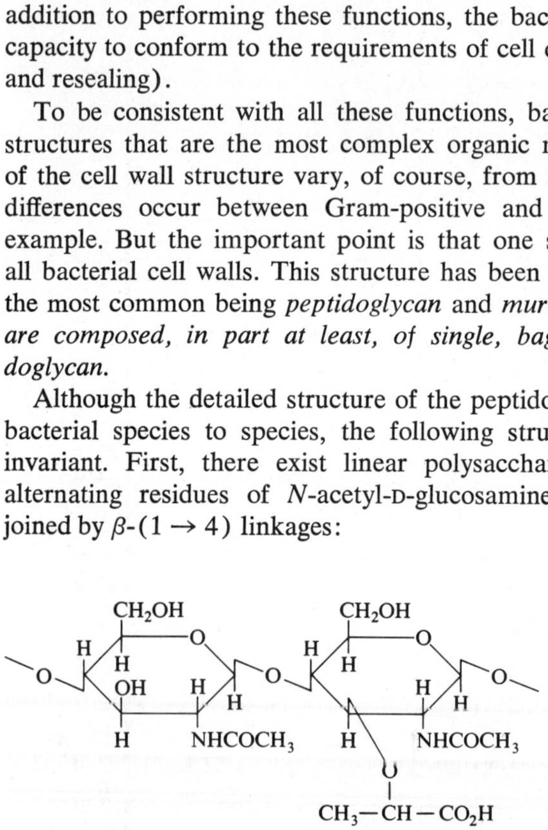

CH₂OH ... NHCOCH₃ ... $CH_3—CH—CO_2H$

Second, associated with these polysaccharide strands are tetrapeptide chains linked by amide bonds between the carboxyl groups of the muramic acid residues and the N-terminal amino function of the tetrapeptide. The sequence of the tetrapeptide is L-Ala-γ-D-Glu-L-Lys (or *meso*-diaminopimelic acid)-D-Ala. Third, these peptide subunits are cross-linked through joining the terminal carboxyl group of D-Ala to the ε-amino function of the lysine (or diaminopimelic acid) residue of another chain. This linkage may be either direct or through an intervening peptide.

Let us consider the structure of the cell wall of a particular bacterium, that of *Staphylococcus aureus,* which is particularly well known through the efforts of Strominger and coworkers (1–5). Later, we shall compare this structure with that of other bacteria in order to understand the variability that does occur. The structure of the cell wall of *S. aureus* is shown in Fig. 10-5. Note specifically the following details. First, the α-carboxyl group of the D-glutamic acid residue is substituted with an amino group, converting it into an α-glutamine residue. Second, all the muramyl groups are substituted with tetrapep-

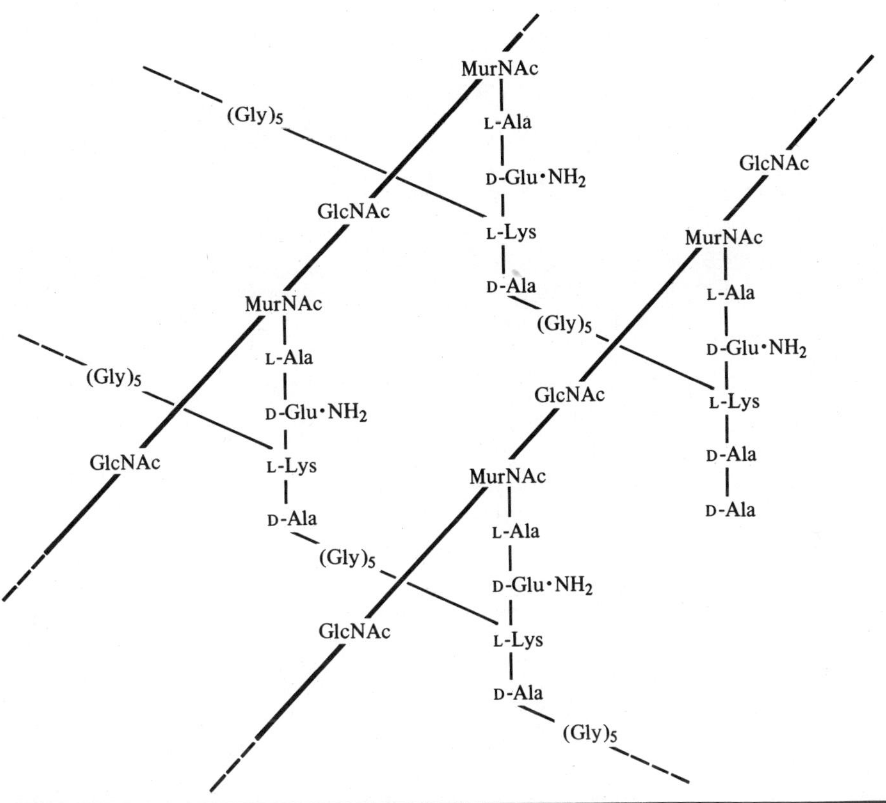

FIGURE 10-5
Peptidoglycan of *Staphylococcus aureus*.

tides. Third, the individual tetrapeptides are joined by pentaglycine bridges. Fourth, essentially all the tetrapeptides are joined together by these crosslinks. Fifth, about 50 percent of the 6-OH groups of the muramyl residues are acetylated.

The bacterial cell wall of *Micrococcus lysodeikticus* exhibits several clear differences from that of *S. aureus*. In fact, these two walls are about as different as any two which are currently known. The structure of the wall of the former bacterium is shown in Fig. 10-6 (6–11). The important differences include the following. First, the α-carboxyl group of the D-glutamic acid residue is substituted with a glycine rather than an amino group. Thus, the side chain is really a branched pentapeptide. Second, only about 20 percent of the muramyl groups are substituted with this pentapeptide. This cell wall structure is, therefore, much less tightly organized than that of *S. aureus*. Third, the individual pentapeptides are joined by that same pentapeptide, not by pentaglycine bridges. Finally, there is no acetylation of the 6-OH groups of the muramyl residues.

Other bacterial cell walls have their own individual peculiarities. For example, in the cell wall of *Escherichia coli,* the L-lysine residues are replaced by *meso*-diaminopimelic acid and the interpeptide linkages are direct peptide bonds between the amino function of diaminopimelic acid and the carboxyl

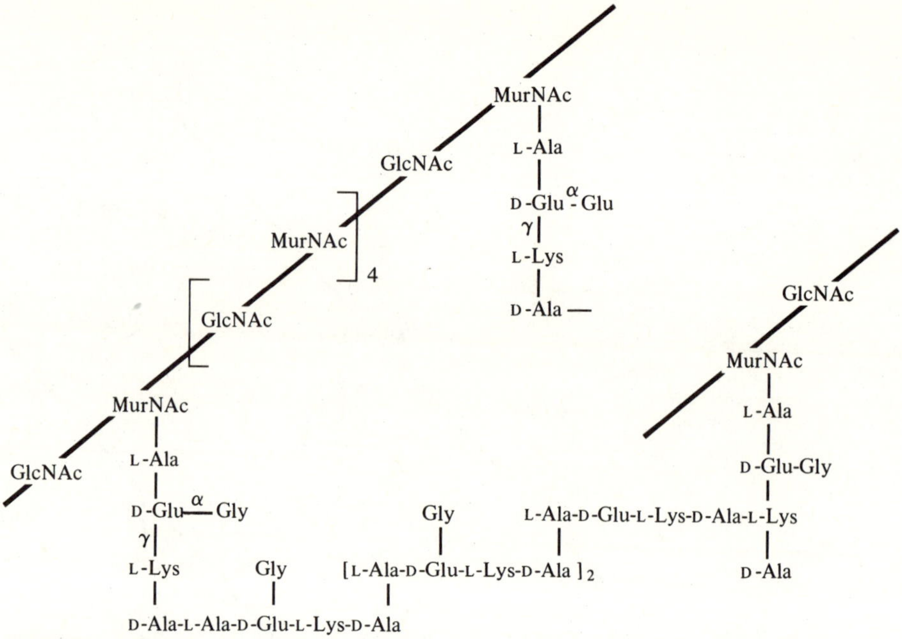

MurNAc
L-Ala
GlcNAc
D-Glu $\overset{\alpha}{-}$ Glu
MurNAc
γ
L-Lys
4
D-Ala —
GlcNAc
GlcNAc
MurNAc
MurNAc
L-Ala
L-Ala
GlcNAc
D-Glu-Gly
D-Glu $\overset{\alpha}{—}$ Gly
Gly
L-Ala-D-Glu-L-Lys-D-Ala-L-Lys
γ
L-Lys
Gly
[L-Ala-D-Glu-L-Lys-D-Ala]$_2$
D-Ala
D-Ala-L-Ala-D-Glu-L-Lys-D-Ala

FIGURE 10-6
Peptidoglycan of *Micrococcus lysodeikticus* (*M. luteus*).

group of the C-terminal D-alanine (Fig. 10-7). Nevertheless, it is important to remember that basically all of the peptidoglycan structures are alike; there are variations on the theme but the theme itself does not change.

The biosynthesis of these cell wall structures is detailed in Chapter 11.

Particularly in Gram-positive organisms, polyphosphate polymers termed *teichoic acids* are found in association with the peptidoglycans of the cell walls. Two main types may be distinguished: glycerol teichoic acids, in which a linear chain of glycerol units is linked by 1,3-phosphodiester bridges, and ribitol teichoic acids, in which ribitol units are joined together by 1,5-phosphodiester bridges. The polyphosphate backbone carries various substituents, including both amino acid and monosaccharide residues. These vary in nature and number with the bacterial strain. For example, the *Bacillus subtilis* cell wall teichoic acid is based on ribitol and has the following structure:

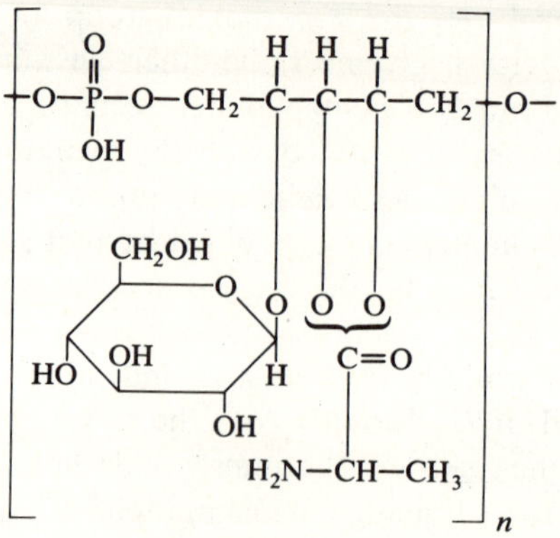

$$\left[-O-\underset{OH}{\overset{O}{\overset{\|}{P}}}-O-CH_2-\overset{H}{C}-\overset{H}{C}-\overset{H}{C}-CH_2-O- \right]_n$$

CH$_2$OH, O, OH, HO, H, OH, O, O, O, C=O, $H_2N-CH-CH_3$

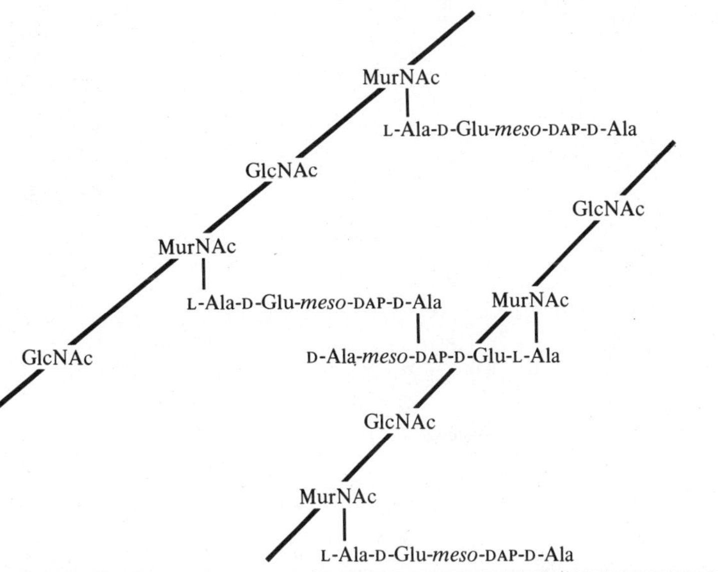

FIGURE 10-7
Peptidoglycan of *Escherichia coli.*

That of *Lactobacillus arabinosus* is based on glycerol and has the structure

$$-O-\left[CH_2-\underset{\substack{| \\ O-C(=O)-CH(CH_3)-NH_2}}{CH}-CH_2-O-\underset{OH}{\overset{O}{\overset{\|}{P}}}-O-CH_2-\underset{\substack{| \\ O-\text{glucosyl}\ (CH_2OH,\ HO,\ OH,\ OH)}}{CH}-CH_2-O\right]_n-\underset{OH}{\overset{O}{\overset{\|}{P}}}-O-$$

Gram-negative bacteria have cell walls which contain a unique lipopolysaccharide of marvelously complicated structure, not yet completely elucidated. The general structural organization of this lipopolysaccharide is shown in Fig. 10-8. The lipid portion of the molecule is termed lipid A and is composed of 3-deoxy-D-mannooctulosonic acid, glucosamine, fatty acids, and phosphate. Attached to lipid A is the "core" of the lipopolysaccharide, which is subdivided into two parts: the backbone and the outer core. The former structure contains the octulosonic acid, L-glycero-D-mannoheptose, ethanolamine, and phosphate; the latter is composed of short oligosaccharide chains constructed from glucose, galactose, and glucosamine. To the core region are attached polysaccharides which contain the determinants of the O antigen. These are responsible for the serological classification of the Enterobacteriaceae. Complete structures of these polysaccharides are known in several cases: that for *Salmonella typhimurium,* composed of abequose (3,6-dideoxy-D-*xylo*hexose), mannose, rhamnose, and galactose, is shown in Fig. 10-8. The nature of this polysaccharide, composed of repeating oligosaccharide units, appears general for all the O antigens.

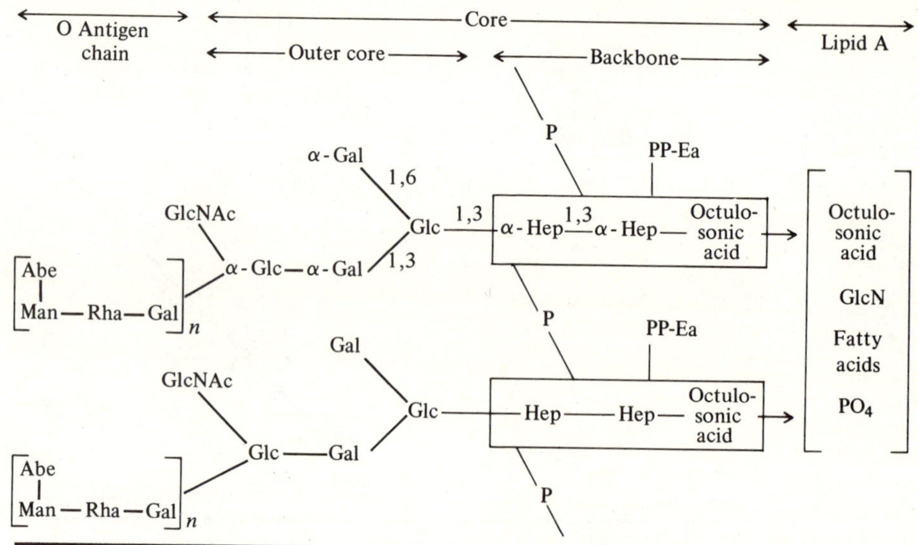

FIGURE 10-8
Structure of the lipopolysaccharide of *Salmonella typhimurium.*

The cell envelope of the Gram-negative bacteria is composed of an inner membrane, a layer of peptidoglycan, and an outer layer which appears membranous in nature and contains the lipopolysaccharide together with other components.

Plant Cell Walls

Plant cell walls share many of the functions of bacterial cell walls. Nevertheless, distinctive and important differences do exist. One important difference is in the mechanical strength of the cell wall. Plants may experience extremes of environment in terms of osmotic pressure; these are generally more severe than those experienced by bacteria and, indeed, those which bacteria can withstand. Moreover, the cell walls of plants form the skeleton of plants and must, therefore, be competent to withstand tremendous physical weights. We shall see shortly how the structure of the cell wall is related to these considerations. Finally, the cell wall is the central means of protection of plants against invasion by pathogens.

The cell walls of bacteria and animal cells grow but do not differentiate. Those of plants cells both grow and differentiate. Not only do plant cell walls differentiate, but this differentiation dominates the differentiation of the whole plant. Plant-cell-wall differentiation may be divided into several stages. Following nuclear division, cytoplasmic vesicles differentiate to form a *cell plate,* a thin disc which enlarges until it reaches the side walls of the dividing cell. The cell plate then enlarges and forms the *primary cell wall.* The basic structural element of the primary cell wall is microfibrils of cellulose (page 475). These microfibrils form the skeleton of the cell wall and are the only organized structure within it.

Cellulose, in the form of microfibrils accounts for about 20 percent of the weight of the primary cell wall. The remainder is largely composed of other polysaccharides; a little protein is also present. The protein component does contain numerous enzymes. The noncellulose polysaccharides of the primary cell wall are not structured; they form an amorphous gel within the interstices of the cellulose matrix.

Conversion of primary cell walls into secondary cell walls is accompanied by increased thickness of the wall resulting from additional deposition of cellulose and lignin. It is this secondary cell wall which accounts for the ability of plants to withstand substantial physical pressures.

REFERENCES

Books and Review Articles

Albersheim, P.: in J. Bonner and J. E. Varner (eds.), *Plant Biochemistry,* Academic Press, New York, 1965, chap. 8.

Bell, D. J.: Natural Monosaccharides and Oligosaccharides: Their Structure and Occurrence, in M. Florkin and H. S. Mason (eds.), *Comparative Biochemistry,* vol. III, part A, Academic Press, New York, 1962, pp. 287–354.

Brimacombe, J. S., and J. M. Webber: *Mucopolysaccharides,* Elsevier, New York, 1964.

Clark, F., and J. K. Grant (eds.): *The Biochemistry of Mucopolysaccharides of Connective Tissue,* Cambridge University Press, New York, 1961.

Florkin, M., and E. H. Stotz (eds.), *Comprehensive Biochemistry,* vol. 5, Elsevier, New York, 1963.

Ghuysen, J. M.: *Bacteriol. Rev.,* **32:** 425 (1968).

Ghuysen, J. M., J. L. Strominger, and D. J. Tipper: in M. Florkin and E. H. Stotz (eds.), *Comprehensive Biochemistry,* vol. 26A, Elsevier, New York, 1968, p. 53.

Ginsburg, V., and E. F. Neufeld: Complex Heterosaccharides of Animals, *Ann. Rev. Biochem.,* **38:** 371 (1969).

Gottschalk, A.: *The Chemistry and Biology of Sialic Acids,* Cambridge University Press, New York, 1960.

Hirst, E. L.: The Structure of Polysaccharides, in D. J. Bell and J. K. Grant (eds.), *The Structure and Biosynthesis of Macromolecules,* Cambridge University Press, New York, 1962, pp. 45–62.

Jeanloz, R. W.: Recent Developments in the Biochemistry of Amino Sugars, *Advan. Enzymol.,* **25:** 433 (1963).

Martin, H. H.: Biochemistry of Bacterial Cell Walls, *Ann. Rev. Biochem.,* **35:** 457 (1966).

Osborn, M. J.: Structure and Biosynthesis of the Bacterial Cell Wall, *Ann. Rev. Biochem.,* **38:** 501 (1969).

Pigman, W. W. (ed.): *The Carbohydrates: Chemistry, Biochemistry, Physiology,* Academic Press, New York, 1957.

Roseman, S.: Metabolism of Connective Tissue, *Ann. Rev. Biochem.,* **28:** 545 (1959).

Salton, M. R. J.: *The Bacterial Cell Wall,* Elsevier, New York, 1964.

Sharon, N.: Polysaccharides, *Ann. Rev. Biochem.,* **35:** 485 (1966).

Stacey, M., and S. A. Barker: *Carbohydrates of Living Tissues,* Van Nostrand Reinhold, New York, 1962.

Stacey, M., and S. A. Barker: *Polysaccharides of Microorganisms,* Clarendon Press, Oxford, 1960.

Symposium, Synthesis of Complex Polysaccharides, *Federation Proc.,* **21:** 1064 (1962).

Whistler, R. L., and C. L. Smart: *Polysaccharide Chemistry,* Academic Press, New York, 1953.

Whistler, R. L., and M. L. Wolfrom (eds.): *Methods in Carbohydrate Chemistry,* Academic Press, New York, 1962–1965.

Specific References

1. Veerkamp, J. H., R. Lambert, and Y. Saito: *Arch. Biochem. Biophys.,* **112:** 120 (1965).
2. Ghuysen, J. M., L. Dierickx, M. Leyh-Bouille, J. L. Strominger, E. Bricas, and C. Nicot: *Biochemistry,* **4:** 2237 (1965).
3. Ghuysen, J. M., D. J. Tipper, C. H. Birge, and J. L. Strominger: *Biochemistry,* **4:** 2245 (1965).
4. Jarvis, D., and J. L. Strominger: *Biochemistry,* **6:** 2591 (1967).
5. Kato, K., and J. L. Strominger: *Biochemistry,* **7:** 2754 (1968).
6. Muñoz, E., J. M. Ghuysen, M. Leyh-Bouille, J. F. Petit, and R. Tinelli: *Biochemistry,* **5:** 3091 (1966).
7. Mirelman, D., and N. Sharon: *J. Biol. Chem.,* **242:** 3414 (1967).
8. Ghuysen, J. M., E. Bricas, M. Lache, and M. Leyh-Bouille: *Biochemistry,* **7:** 1450 (1968).
9. Katz, W., and J. L. Strominger: *Biochemistry,* **6:** 930 (1967).
10. Pickering, B. T.: *Biochem. J.,* **100:** 430 (1966).
11. Campbell, J. N., M. Leyh-Bouille, and J. M. Ghuysen: *Biochemistry,* **8:** 193 (1969).

11 Metabolism of Carbohydrates

SOME COMMENTS CONCERNING METABOLISM

General Features

A cell's or an organism's *metabolism* may be defined as the totality of the chemical processes that it is capable of performing. Therefore, the metabolism of even a simple, unicellular organism is not time invariant; that is, all its aspects are not actually expressed at any given instant. This immediately raises the question of what factors control this expression, and the subject of *metabolic control* or regulation occupies much current biochemical experimentation, thought, and speculation. Regulation is performed by a delicate set of interrelated checks and balances, which have both *intrinsic* (i.e., hereditary, genetic) and *extrinsic* (environmental, physiological) components. Since all metabolism forms an interconnected whole in both time and space, these effects necessarily pervade metabolism as a whole, even though it may be convenient to focus on a particular reaction and its participants as a first approximation. It is axiomatic that all metabolism and its control must be the consequence of, and be explicable either directly or indirectly in terms of, the enzymatic makeup of the organism.

In this chapter we are concerned for the first time with *metabolic pathways* (sequences, or routes), that is, arrays of enzyme-catalyzed chemical reactions that bring about transformations of certain organic compounds vital to the organism. The compounds involved are referred to as *metabolites*. Classically, metabolic pathways have been divided into two types: *catabolic* and *anabolic*.

Metabolic Routes

Catabolic routes define essentially degradative processes in which large organic molecules, frequently supplied as food for the organism, are broken down (usually with the participation of *oxidative* reactions) to simple cellular constituents, with the attendant release of chemical free energy. This energy is then utilized by the organism for its maintenance, growth, and replication, and for transduction into other forms of energy—mechanical or osmotic work, the generation of electrical impulses, or heat. *Anabolic routes* on the other hand define *synthetic* processes; they produce complex organic cellular constituents from simpler precursors, frequently involving the participation of *reductive* steps, and require the expenditure of chemical free energy. Inquiries into the metabolism of a wide variety of organisms have shown that there is a remarkable order and simplicity in metabolic patterns in spite of an equally remarkable diversity and individuality of metabolites consumed and produced.

Of these general patterns, none is more significant than the discovery of a *central area of metabolism* joint to, and providing a direct link between, catabolic and anabolic routes.

Central Pathways

During the initial stages of catabolism, each one characteristic of a particular foodstuff, large molecules are broken down to yield—apart from CO_2 and H_2O—a quite *restricted group* of small organic molecules, liberating about one-third of the available free energy in the process. These product molecules are as follows: *for carbohydrates,* triose phosphates and/or pyruvate; *for fats,* acetyl CoA (see page 388), propionyl CoA, and glycerol; *for proteins,* acetyl CoA, oxalacetate, α-oxoglutarate, fumarate, and succinate. The same compounds are also produced even when the organism, as is the case with certain bacteria, uses much more exotic carbon sources, such as certain aromatic (e.g., benzoic or mandelic acids), aliphatic (e.g., γ-aminobutyric, or itaconic acids), or heterocyclic (e.g. purines or uric acid) compounds (see Fig. 11-1*a*).

One of the major unifying and simplifying discoveries of modern, *dynamic* biochemistry was the finding that one and the same set of reactions was involved in three crucial phases of metabolism: (1) the *interconversion* of the various products of catabolism just mentioned, rendering them equivalent, metabolically speaking; (2) their *complete oxidation* to CO_2 plus H_2O, which furnishes the organism with the remaining two thirds of its energy supply; (3) the supply of *crucial intermediates* for biosynthetic, anabolic processes.

These central [called "amphibolic" by B. Davis (1)] pathways are composed of relatively few reactions. In essence, they consist of the steps triose phosphate $\rightleftharpoons$ pyruvate, pyruvate $\rightarrow$ acetyl CoA; oxalacetate $\rightleftharpoons$ aspartate, α-oxoglutarate $\rightleftharpoons$ glutamate; and the cyclic set of reactions designed to catalyze the complete combustion of acetyl CoA (C_2 in the figure) to CO_2 plus H_2O. This device, known as the *citric acid cycle,* fulfills such a crucial role that a whole chapter (Chapter 14) is devoted to a description of its operation.

Anabolism versus Catabolism

As put rather neatly by H. L. Kornberg (2): "Whereas catabolic routes effect the transformations of nutrient carbon sources into the (interconvertible) intermediates of the central pathways, anabolic routes are sequences of enzymatic steps which enable the component building blocks of macromolecules to be constructed from the intermediates. Thus, while catabolic routes have clearly defined beginnings but no unambiguously identifiable end products, anabolic routes lead to clearly distinguishable end products from diffuse beginnings."

Another important distinction is that anabolic and catabolic routes rarely, if ever, follow the same pathways in detail. This is readily apparent when the product of catabolism is not identical with the carbon source in anabolism, as is the case with many amino acids. Thus, while aromatic amino acids are catabolized to acetyl CoA plus fumarate (or succinate), their synthesis utilizes phosphoenolpyruvate and an aldotetrose phosphate. Similarly, serine and cysteine are easily transformed catabolically to pyruvate, but their synthesis originates with 3-phosphoglycerate and aspartate, respectively.

Another example is provided by the fatty acids. Here catabolism leads to acetyl CoA as the end product, and biosynthesis commences with the same intermediate and proceeds by what appeared initially a simple reversal of the catabolic sequence. But closer attention to detail disclosed that this was far from being the case: Acetyl CoA first has to be converted to the more re-

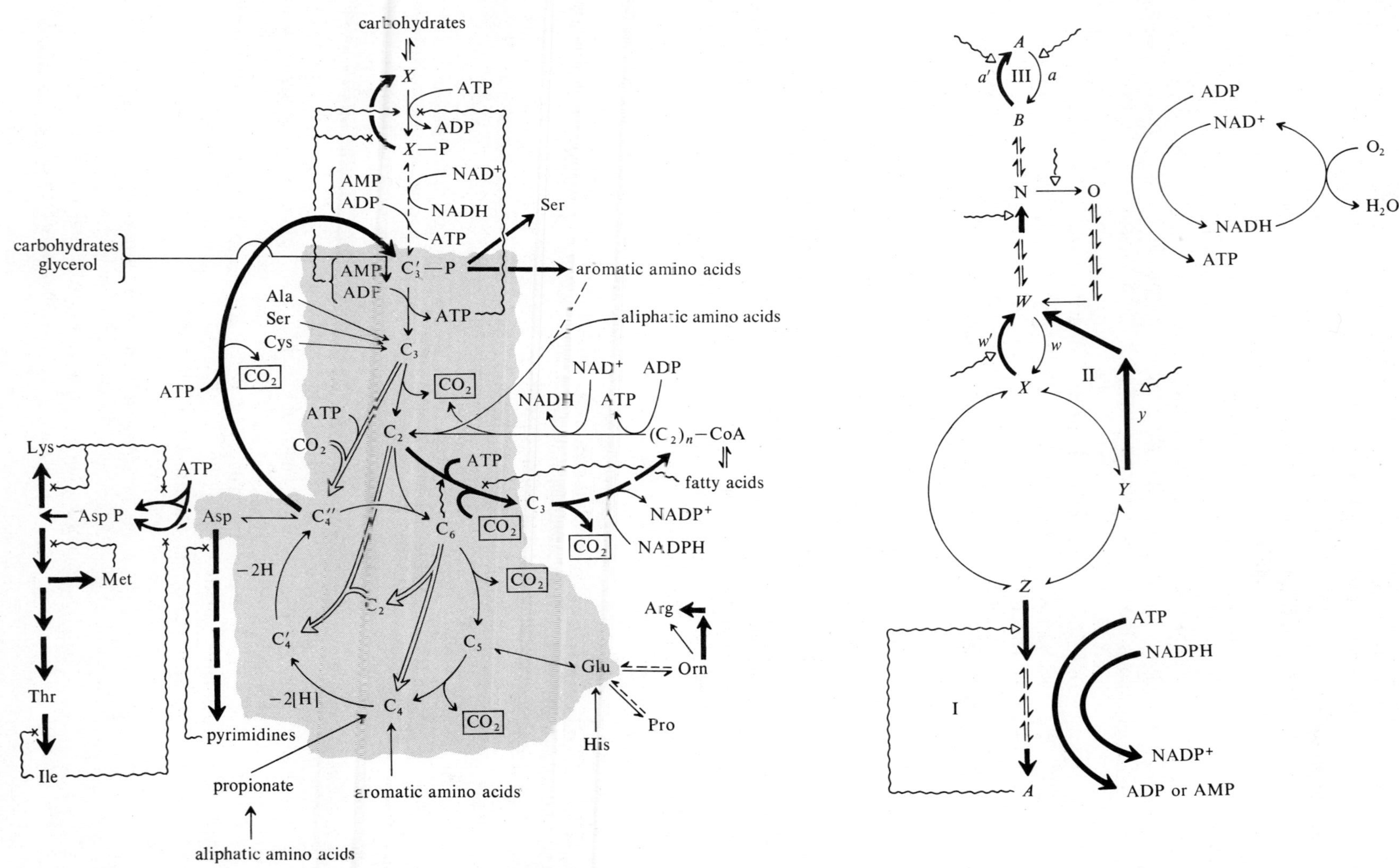

carbohydrates
X
ATP
ADP
X—P
AMP
ADP
NAD+
NADH
ATP
Ser
carbohydrates
glycerol
AMP
ADP
C_3'—P
aromatic amino acids
Ala
Ser
Cys
ATP
aliphatic amino acids
C_3
CO_2
ATP
CO_2
NAD+
ADP
NADH
ATP
ATP
CO_2
C_2
$(C_2)_n$—CoA
Lys
ATP
fatty acids
Asp P
Asp
C_4''
C_3
NADP+
C_6
CO_2
CO_2
NADPH
−2H
Met
CO_2
Arg
C_2
C_4'
C_5
Glu
Orn
Thr
−2[H]
C_4
CO_2
Pro
pyrimidines
His
Ile
propionate
aromatic amino acids
aliphatic amino acids
A
III
a′
a
B
ADP
NAD+
O_2
N
O
H_2O
NADH
ATP
W
w′
w
X
II
y
Y
Z
ATP
NADPH
I
NADP+
A
ADP or AMP

active malonyl CoA, which is not an intermediate in catabolism, and the whole sequence of enzymes responsible for the conversion of the latter to long-chain acyl CoA's (the precursors of the corresponding fatty acids—see page 714) is not identical with that devoted to catabolism and is not even located in the same cellular compartment.

Even in the case of the biosynthesis of glucose, which proceeds, in large part, by a reversal of a number of easily reversible enzymatic reactions, biosynthesis and degradation differ, as we shall see, at the two most critical points in the whole sequence, that is, at either end. Thus, for instance, glucose is converted to glucose-6-phosphate during catabolism by a phosphotransferase reaction utilizing ATP; yet it is formed in anabolism from the phosphate ester by simple hydrolysis. Pyruvate is produced catabolically from its enol phosphate by a phosphate transfer to ADP; it is utilized anabolically in most organisms by virtue of two linked reactions, which first carboxylate it to oxalacetate and only then allow its transformation to the enol phosphate. In certain organisms (bacteria, plants) in which the conversion *is* direct, the reactions still differ: In the forward direction, it is phosphoenolpyruvate + ADP → pyruvate + ATP; in the reverse, pyruvate + ATP (+ P_i) → phosphoenolpyruvate + AMP + P_i (or PP_i).

These last few examples also serve to bring home another point. The initial and final reactions in most metabolic sequences, anabolic and catabolic, are frequently rigged in such a fashion as to render them virtually irreversible thermodynamically; that is, they have $\Delta G^{0\prime}$ values (which we recall as the symbol for the standard free energy change at *p*H 7) equal to $\geqq -4$ kcal/mol. Teleologically, the reason for this is not hard to understand. It provides

FIGURE 11-1
Metabolic principles.
Left: pathways of metabolism

→ catabolic reactions and sequences
➡ anabolic reactions and sequences
⇨ anaplerotic reactions and sequences
⇝ activation by end products
⇝✕ inhibition by end products
central area of metabolism is indicated by shading

Right: generalized scheme: Note that (1) a thermodynamically or kinetically irreversible step frequently intervenes as the first reaction of a sequence or branch; not uncommonly the last reaction in a sequence is also distinguished by the same feature. (2) These irreversible reactions provide the most sensitive loci for exertion of metabolic control (*a*) of synthesis of enzyme (induction/repression and hormonal control); (*b*) of activity of enzyme [inhibition, notably feedback (end product, retro) inhibition]; (*c*) of availability of reactants and cofactors. (3) For a reversal of the sequence, one or more separate and irreversible steps must participate, giving rise to partially or completely distinct pathways for degradation (catabolic) or synthesis (anabolic). (4) Linkage between anabolism and catabolism is provided by (*a*) common sequences; (*b*) compounds (ribonucleoside triphosphates, largely ATP, and reduced nicotinamide coenzymes) which are generated during catabolism and utilized during anabolism. (5) One and the same set of reactions and enzymes may participate in more than one metabolic route.

→ specific catabolic step or sequence
➡ specific anabolic step or sequence
I separate sequence
II separate starting point (loop)
III separate initial and terminal step
⇌ reversible, joint steps
⇝ control point—control may be: effects on enzyme *synthesis;* effects on enzyme *activity; availability* of substrates and cofactors

for easy flux through the pathway and minimizes the possibility of a logjam of intermediates somewhere along the line. The enzymes responsible for these essentially irreversible and unique steps have often been referred to as *pacemaker enzymes.*

Is there then no *link* between anabolism and catabolism? Quite the contrary; the connection is very close indeed and is manifest on three levels:

1. *On that of the carbon sources.* As already mentioned, the products of catabolism, through the intervention of the central routes, become the substrates of anabolism.

2. *On that of energy supply.* Catabolism produces metabolic energy in the form of ATP or of compounds easily convertible to ATP; anabolism requires energy and consumes ATP.

3. *On that of reducing power.* Catabolism is essentially oxidative; it therefore consumes oxidizing and produces reducing power; the reverse is true for anabolism. Much of the reducing power generated during catabolism and consumed during anabolism is provided by the pyridine (nicotinamide) nucleotides NAD^+/NADH and $NADP^+$/NADPH (see page 408). But an important difference prevails—catabolism produces both NADH and NADPH; anabolism requires and consumes the latter almost exclusively.

Anaplerotic Routes

Kornberg has also directed attention to, and provided a designation for, yet a fourth type of metabolic pathway; the final stages of catabolism usually lead to complete removal of most metabolites from the common pathways, usually as CO_2 plus H_2O, plus NH_4^+ (or urea) and some other nitrogenous bases; anabolism also provides a constant drain on the pools of the intermediates of the common pathways. Thus, provision must be made for their replenishment. These ancillary sequences are called *anaplerotic.* Usually they involve the insertion of either a one-carbon (as CO_2) or a two-carbon fragment (as acetyl CoA) into the common pool.

Metabolic Control: Regulation of the Flow of Metabolites

The flux of materials through the various pathways of metabolism does not proceed in a haphazard or random fashion but rather is attuned in a most precise manner to the needs of the organism (see Fig. 11-1*b*). These are largely determined by the composition of the medium in which it finds itself and are, in higher organisms, also responsive to the signals received by any one cell population or tissue from all other cell populations or tissues either directly or indirectly by the circulatory system, the nervous system, and so on. Regulation can be exerted at many levels, but two of the most important ones are the following: One, a relatively slow and coarse control device, is to allow the composition of the environment to affect the rate and extent of synthesis of various enzymes, in other words to alter the amount or level of an enzyme in the cell. This control may be exerted either on the reactions responsible for the recognition of different genetic regions, or on their availability to, and rate of, expression. Usually subsumed by the terms *induction* and *repression,* control devices of this sort are described in the last three chapters of this book. The other, much more rapid and finely poised, is to exert *control* on the *rates* of one or more susceptible *enzyme-catalyzed reactions* (i.e., on metabolism proper). Frequently the most sensitive points for controls of this general nature are those which stand at the beginning or the end of *specific* metabolic sequences[1] (i.e., the *pacemaker enzymes* mentioned above).

[1] These may be constituted by either the first or the branch-point enzymes in the case of metabolic sequences common to more than one metabolite. Umbarger (see the references) has provided a useful classification of various logical and experimental alternatives, many of which involve multiple functional sites on one and the same enzyme or else a multiplicity of enzymes catalyzing the same step.

Frequently also, this control, which may be either *positive* (and thus constitutes *activation*) or *negative* (and thus reflects itself in *inhibition*), is exerted by one of the end products of the sequence in question. For this reason the inhibitory type, which was the first to be discovered in studies on the inhibition of one of the initial steps in a biosynthetic sequence by its end product, was called *feedback* or *retroinhibition.* Since these inhibitions, and their counterpart, the activation of early steps in catabolic sequences, involve compounds far removed metabolically and structurally from the substrates of the reactions affected, they would be expected to involve *second-site* or *allosteric* effects, that is, conformational changes of the respective enzyme proteins. Furthermore, interactions of this type are frequently mediated by ATP or its metabolites (3′:5′-cyclic AMP = Ado-3′:5′-*P*; ADP, AMP). Hence, as pointed out by Atkinson (3), there exists a delicately poised regulatory device that relates the "energy charge" of the adenylate pool [(concn of ATP + ½ ADP)/(concn of ATP + ADP + AMP)] to the concentration of modifiers capable of affecting susceptible key enzymes.

Application to Carbohydrate Metabolism

Some of the principles just enumerated may be applied to the specific problem of carbohydrate metabolism, as shown in schematic form in Fig. 11-2. Krebs and Kornberg recognize three phases in the metabolism of all three common foods. In *phase I* of catabolism, the sugars are mobilized for their subsequent transformations by conversion to phosphorylated hexoses. This entails either simple (but virtually irreversible) phosphorylation of free hexoses at the expense of ATP or the phosphorolysis of oligo- or polysaccharides to the same type of product.[2] *No useful energy* is provided by the reactions of phase I; on the contrary, energy at the expense of ATP frequently has to be provided. Both the phosphorylation and the reverse *de*phosphorylations are virtually irreversible thermodynamically and proceed at a significant rate only in one direction, and so they constitute separate reactions, catalyzed by separate enzymes. The equilibria for sugar phosphate hydrolysis as well as for sugar phosphorylation lie very far to the right—especially considering the concentration of reactants obtaining in the cell.

In *phase II* the intermediates produced in phase I are incompletely degraded, with the attendant liberation of about one third of the free energy potentially available by their complete combustion. This degradation generates three-carbon and two-carbon compounds (trioses, pyruvate, acetyl CoA), all of which are also participants in the central pathways comprised by the citric acid cycle and certain ancillary steps. The complete oxidation (to CO_2 + H_2O) of the intermediates produced in phase II or alternatively the formation of other defined end products, such as alcohol or other compounds produced during the anaerobic fermentations with microorganisms) by means of these central pathways then constitute *phase III*. For our purposes it will be convenient to subdivide phase II into two portions. One leads to the interconversion of the various six-carbon sugar phosphate esters and certain other intermediates readily derivable from, or leading to, this pool. The second consists of the degradative conversion of the hexose phosphates by several distinct pathways, specifically so designated in Fig. 11-2. These degradative routes vary in their distribution and relative importance from cell to cell and organism to organism. We shall now examine them in turn.

[2] The glycosidic bonds of *some* oligosaccharides are cleaved hydrolytically; under these circumstances an ATP-dependent phosphorylation of the product sugar must intervene before it can be metabolized further. In Enterobacteriaceae the phosphorylation of hexoses takes place at the expense of PEP (page 525).

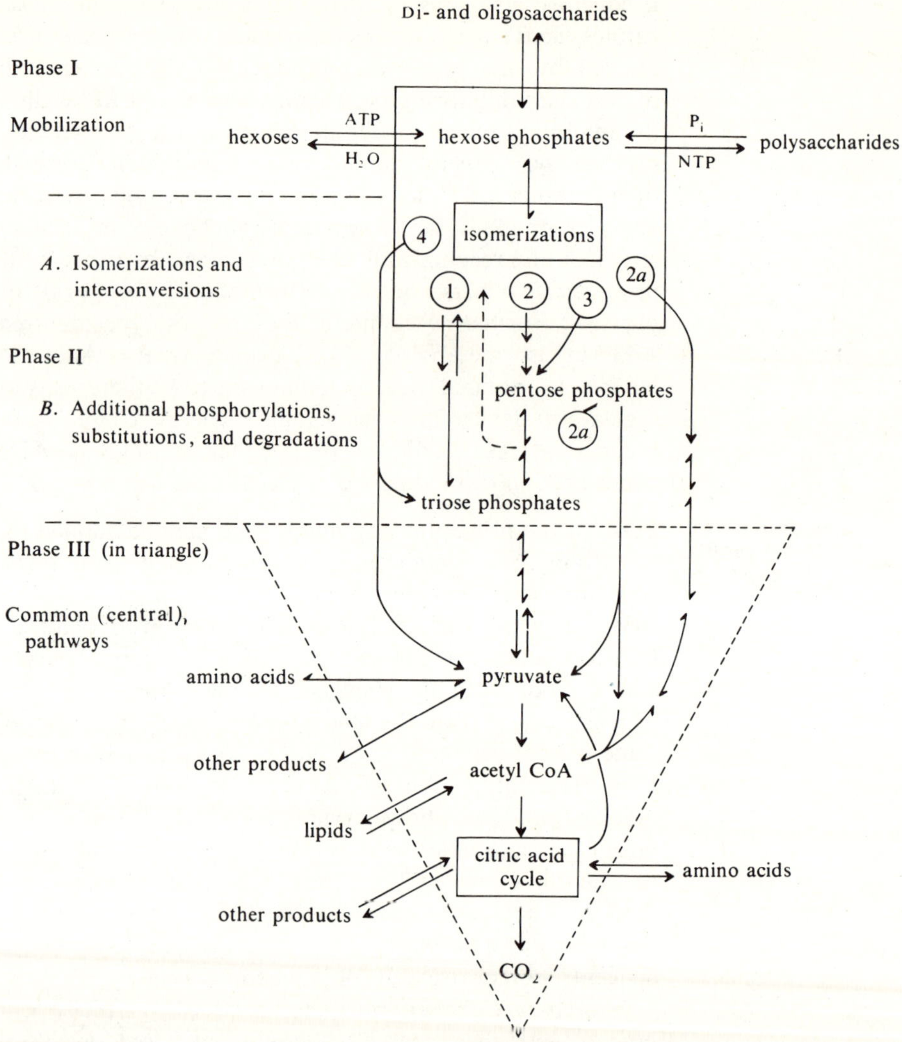

Di- and oligosaccharides

Phase I

Mobilization

hexoses ATP H_2O hexose phosphates P_i NTP polysaccharides

isomerizations

(4) (1) (2) (3) (2a)

A. Isomerizations and interconversions

Phase II

B. Additional phosphorylations, substitutions, and degradations

pentose phosphates (2a)

triose phosphates

Phase III (in triangle)

Common (central), pathways

amino acids

pyruvate

other products

acetyl CoA

lipids

citric acid cycle

amino acids

other products

CO_2

FIGURE 11-2

Schematic representation of carbohydrate metabolism.

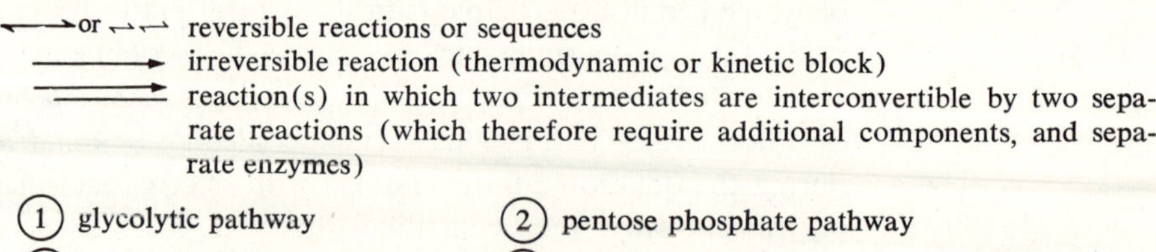

⟷ or ⇀↽ reversible reactions or sequences

⟶ irreversible reaction (thermodynamic or kinetic block)

⇄ reaction(s) in which two intermediates are interconvertible by two separate reactions (which therefore require additional components, and separate enzymes)

(1) glycolytic pathway (2) pentose phosphate pathway

(2a) phosphoketolase pathways (3) glucuronate pathway

(4) Entner–Doudoroff pathway

GLYCOLYTIC PATHWAY

History and Significance

The primordial and still the major route of glucose catabolism in most cells is encompassed by the series of reactions that convert it to two molecules of pyruvate, $CH_3COCO_2^-$. Such diverse processes as the generation of the ATP, which then furnishes the energy for the contraction of skeletal muscles in an environment low in oxygen; the generation of ATP for the continuous and stringent energy demand by heart muscle, operating in a highly aerobic environment and leading to the complete oxidation of glucose; and many of the microbial fermentations of glucose to ethanol, lactate, glycerol, glycols, and a wide variety of other products, which must provide all the intermediates and most of the energy required for the growth of these organisms, all make use of this pathway. It is therefore understandable that this was the first such sequence to be completely described on the level of the interaction of the individual pure enzymes, their substrates, and cofactors. And so modern biochemistry and enzymology can properly be said to have arisen as a consequence of these studies. They had their start in the observations of Harden and Young in the early years of this century on the alcoholic fermentation of glucose by a yeast extract, which clearly showed that (1) the reaction was absolutely dependent on inorganic phosphate; (2) in the presence of P_i, a hexose diphosphate accumulated; (3) this accumulation could be prevented and all the glucose converted to ethanol plus CO_2 if arsenate was added; and (4) the extract could be fractionated into a heat-labile, nondialyzable component ("zymase") and a dialyzable fraction, the latter containing a heat-stable component ("cozymase"). Another pioneering set of observations was provided by Meyerhof in the 1920s, who showed that muscle extracts were able to convert glycogen to lactic acid (a process that A. V. Hill had previously demonstrated to be ultimately related to muscular contraction) and that this conversion appeared to make use of intermediates and reactions similar to those of alcoholic fermentation. Meyerhof himself, Embden, Parnas, Neuberg, Warburg, and Carl and Gertie Cori are only a few of the illustrious scientists (the first, and the last three Nobel laureates) who have contributed to this field and have created, through their studies and their training of others, biochemistry as we know it today.

Overall Pathway of Glycolysis [Embden–Meyerhof–Parnas (EMP) Scheme]

For the moment we restrict our attention to the set of reactions for which this scheme was first devised. They encompass the conversion of glucose to ethanol plus CO_2 in the alcoholic fermentation by yeasts, and the conversion of glucose or of a glucose unit (provided by glycogen) to lactic acid (or rather to its salt lactate) in the muscles of animals. Neither of these processes requires oxygen, and both can take place and provide the requisite cellular energy supply under completely anaerobic conditions. The sequence of intermediates and reactions is shown in Fig. 11-3, and some of their characteristics, thermodynamic and enzymatic, are summarized in Table 11-1 and 11-2. Before considering the individual steps, let us take a closer look at the most important features of the overall scheme.

1. The two pathways make use of the identical reactions from glucose-6-*P* to pyruvate and differ only in the eventual fate of the latter and hence in the means employed for the *regeneration of NAD+ from NADH* (see item 3). These two considerations also govern all those other metabolic sequences in which a hexose unit is initially converted to two molecules of pyruvate. In muscle glycolysis, pyruvate and NADH interact directly in the presence of *lactate dehydrogenase* to yield lactate and regenerate the NAD+ (Table 11-1, reaction 13). In alcoholic fermentation the pyruvate is first *decarboxylated* to acetaldehyde (reaction 14) and the latter *reduced* by NADH to give alcohol.

2. The unique *C—C bond-scission reaction,* which characterizes the scheme, is reaction 6, catalyzed by phosphofructose (or fructose-di-*P*) aldolase, which converts a ketohexose-diphosphate into two molecules of triose-phosphate.

3. The unique *oxidation–reduction reaction* of the scheme is reaction 8,

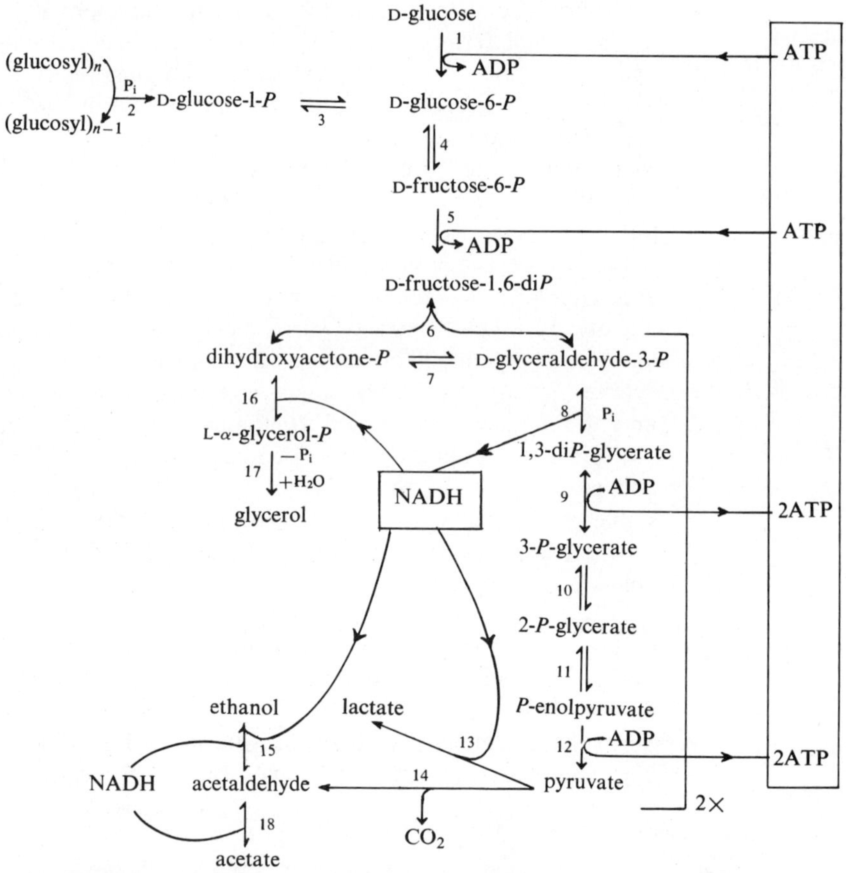

FIGURE 11-3

Embden–Meyerhof–Parnas scheme of glycolysis.

glycolysis: glucose + $2P_i$ + 2ADP → 2 lactate + 2ATP (classical, mammalian muscle or brain)

glucose + P_i → α-glycerol-*P* + pyruvate (insect flight muscle, striated muscle)

fermentations: glucose + $2P_i$ → 2 ethanol + $2CO_2$ + 2ATP + $2H_2O$ (first form)

glucose + HSO_3^- → glycerol + acetaldehyde · HSO_3 + CO_2 (second form, no net ATP)

2 glucose → 2 glycerol + ethanol + acetate + $2CO_2$ (third form, no net ATP)

glucose + (P_i) → α-glycerol-*P* + acetaldehyde + CO_2

glycerol + P_i (third form, no net ATP)

catalyzed by D-glyceraldehyde-3-*P* dehydrogenase; is requires NAD^+ and P_i and produces 1,3-diphosphoglycerate (a compound with a high $-\Delta G^{0\prime}$ of hydrolysis) and NADH. Note that by virtue of reaction 7, which interconverts the two triose phosphates, both the products of reaction 6 can be channeled through reaction 8.

4. For each molecule of free glucose traversing the pathway, two molecules of ATP are consumed (in reactions 1 and 5) and four molecules are produced (one each per triose in reaction 8 plus 9, and 11 plus 12, respectively). If we start with glycogen, one molecule of ATP is required per hexose unit (no ATP requirement in reaction 2). Thus, the former process generates two

TABLE 11-1
Reactions and thermodynamics of carbohydrate metabolism

Reaction number	*Equation (Cosubstrates or coenzymes shown above, activators below arrow)*	*Name of enzyme*	*Characteristic inhibitor*	$\Delta G^{0\prime}$,[a] *kcal/mol*
1	Glucose + ATP $\xrightarrow{Mg^{2+}}$ glucose-6-*P* + ADP	Hexokinase Glucokinase	Glucose-6-*P*(P_i)	−3.42
1*a*	Glucose-6-*P* + H_2O $\xrightarrow{Mg^{2+}}$ glucose + P_i	Glucose-6-phosphatase Nonspecific phosphatases	Glucose, orinase	−4.02
2	Glycogen + $nP_i \rightleftharpoons n$ glucose-1-*P*	(α-1,4-Glucan) phosphorylase		0.73
3	Glucose-1-*P* $\xrightleftharpoons{\text{glucose-1,6-}P}$ glucose-6-*P*	Phosphoglucomutase	F, organophosphorus inhibitors	−1.74
4	Glucose-6-*P* $\rightleftharpoons$ fructose-6-*P*	Phosphoglucose (glucose phosphate) isomerase	2-Deoxyglucose-6-*P*	0.50
5	Fructose-6-*P* + ATP $\xrightarrow[(ADP,\ AMP)K^+]{Mg^{2+}}$ fructose-1,6-di*P* + ADP + H^+	Phosphofructokinase	ATP, citrate	−3.4
5*a*	Fructose-1,6-di*P* + H_2O $\xrightarrow{Mg^{2+}}$ fructose-6-*P* + P_i	Fructose diphosphatase Nonspecific phosphatases	AMP, fructose-1,6-di*P*, F^-, Zn^{2+}, Fe^{2+}	−4.0
6	Fructose-1,6-di*P* $\rightleftharpoons$ dihydroxyacetone-*P* + glyceraldehyde-3-*P*	(Fructose-*P*) aldolase	Chelating agents[b] AMP,[c] ATP[d]	+5.73
7	Dihydroxyacetone-*P* $\rightleftharpoons$ glyceraldehyde-3-*P*	Triose phosphate isomerase		+1.83
8	2 × (Glyceraldehyde-3-*P* + P_i + NAD^+ $\rightleftharpoons$ 1,3-diphosphoglycerate + NADH + H^+)	Glyceraldehyde-*P*-dehydrogenase; triose-*P*-dehydrogenase	ICH_2COR D-threose-2,4-di*P*	2 × (+1.50)
9	2 × (1,3-Diphosphoglycerate + ADP + H^+ $\xrightarrow{Mg^{2+}}$ 3-*P*- glycerate + ATP)	Phosphoglycerate kinase		2 × (−6.78)
10	2 × (3-*P*-Glycerate $\xrightleftharpoons{\text{glycerate-2,3-di}P}$ 2-*P*-glycerate)	Phosphoglyceromutase		2 × (+1.06)
11	2 × (2-*P*-Glycerate $\xrightleftharpoons{Mg^{2+}\text{ or }Mn^{2+}}$ phosphoenolpyruvate)	Enolase (phosphopyruvate hydratase)	Ca^{2+} F^- plus P_i	2 × (+0.44)
12	2 × (phosphoenolpyruvate + ADP + H^+ $\xrightarrow[K^+(Rb^+,\ Cs^+)]{Mg^{2+}}$ pyruvate + ATP)	Pyruvate kinase	Ca^{2+} *vs* Mg^{2+} Na^+ *vs* K^+	2 × (−5.72)
13	2 × (Pyruvate + NADH + H^+ $\rightleftharpoons$ lactate + NAD^+)	Lactate dehydrogenase	Oxamate	2 × (−6.00)
14	2 × (Pyruvate + H^+ → acetaldehyde + CO_2)	Pyruvate (de)carboxylase		2 × (−4.72)
15	2 × (Acetaldehyde + NADH + H^+ $\rightleftharpoons$ ethanol + NAD^+)	Alcohol dehydrogenase	(HSO_3^-)	2 × (−5.15)

Sums

$(\text{glucose})_n + H_2O \rightarrow 2\ \text{lactate} + 2H^+ + (\text{glucose})_{n-1}$	−52.43[e]	
$(\text{glucose})_n + 3P_i + 3ADP \rightarrow 2\ \text{lactate} + 3ATP + (\text{glucose})_{n-1}$	glycolysis (muscle)	−27.32[f]
glucose → 2 lactate + $2H^+$	−47.4[e]	
glucose + $2P_i$ + 2ADP → 2 lactate + 2ATP	glycolysis or lactate fermentation	−29.74[f]
glucose → 2 ethanol + CO_2	−56.1[e]	
glucose + $2P_i$ + 2ADP → 2 ethanol + $2CO_2$ + 2ATP	alcoholic fermentation	−37.48[f]

[a] $\Delta G^{0\prime}$ values refer to *p*H 7.0 with all other reactants including H_2O at unit activity; most data from Johnson (4). The free energy of formation of glucose in aqueous solution equals 217.56 kcal/mol, its ΔG^0 of combustion to $CO_2 + H_2O$ is 686 kcal/mol, and $\Delta G^{0\prime}$ for $(\text{glucose})_n + H_2O \rightarrow (\text{glucose})_{n-1}$ equals −5.03 kcal/mol.

[b] Microbial enzymes.

[c] Liver.

[d] Muscle (activated by ADP).

[e] From $\Delta\tilde{G}^0$ (page 34).

[f] From this table.

TABLE 11-2
Molecular properties of glycolytic enzymes

Reaction number	*Enzyme*	*Source*	*M*	*Remarks*
1	Hexokinase	Yeast	96,600	
2	Phosphorylase *a*	Muscle	360,000	Tetramer
	Phosphorylase *b*	Muscle	180,000	Dimer
	Phosphorylase	Liver	200,000	Dimer (active)
3	Phosphoglucomutase	Yeast	112,000	
		Muscle	62,000	
		E. coli	60,000	
4	Phosphoglucose isomerase	Yeast	145,000	
5	Phosphofructokinase	Muscle, yeast	380,000	Dimer
6	Aldolase	Muscle	150,000	Tetramer
		Liver, brain	150,000	Differ from muscle enzyme
		Microorganisms	~80,000	Metal-activated, dimer
7	Triosephosphate isomerase	Muscle	43,000	
8	Triosephosphate dehydrogenase	Muscle	140,000	Tetramer
9	. . .			
10	Phosphoglyceromutase (phosphotransferase type)	Muscle	57,000	
		Yeast	112,000	(Dimer?)
	Phosphoglyceromutase (mutase type)	Wheat germ	30,000	
11	Enolase	Muscle	85,000	Dimer
		Yeast	60,000	Monomer
12	Pyruvate kinase	Muscle	230,000	Tetramer
13	Lactate dehydrogenase	Vertebrate tissues	140,000	Tetramers of varying structure

molecules of ATP/hexose, the latter three molecules of ATP/hexose. Of course, glycogen formation itself requires ATP (page 549).

5. As a corollary to item 4, catabolism of free glucose requires the consumption of two molecules of inorganic phosphate, that of a hexose unit three molecules of inorganic phosphate.

6. The overall stoichiometries, therefore, are those indicated. Since the presently accepted value of $\Delta G^{0\prime}$ for ATP $\rightarrow$ ADP + $H_2O \simeq -7.5$ kcal/mol,[3] we see that (a) only a small proportion of the total free energy potentially derivable from the catabolism of a hexose molecule is actually rendered available by the two anaerobic processes described. (b) This is in part because both lactate and ethanol are compounds in which carbon is still at a relatively reduced level and only in part because of inherent inefficiencies. (c) The values of $\Delta G^{0\prime}$ assumed for the individual steps are probably nearly correct, since their sum added to the appropriate multiple of $\Delta G^{0\prime}$ for ATP hydrolysis pretty nearly equals the values calculated from $\Delta G^{0\prime}$ (formation) plus $\Delta G^{0\prime}$ (ATP hydrolysis).

7. The two sets of end products (lactate or ethanol plus CO_2) accumulate under anaerobic conditions. The most efficient means for the removal of the two organic compounds is their complete oxidation to CO_2 plus H_2O. These processes require aerobic conditions, which can obtain either when one and the same cell is able to function in, and interact with, different physiological

[3] $\Delta G'$ is considerably more negative: with the concentrations present in resting muscle it is probably close to -14.5 kcal/mol; hence, here the efficiency is considerably higher and may approach $\sim 43.5/52 \simeq 0.83$ for glycogen $\rightarrow$ lactate.

environments (e.g., yeast and certain mammalian cells) or, in higher organisms, when different environments interact with different specialized cells (e.g., when the lactate produced in skeletal muscle is removed by the circulation and transported to the liver for oxidation).

Reaction (11-1)

Conversion of D-*Glucose to* D-*Glucose*-6-*P*. Since glucose-6-*P* is the "mobilized" form of glucose (i.e., that in which this sugar enters not only the glycolytic pathway, but all other pathways of carbohydrate metabolism), enzymes catalyzing reaction 1 are found extremely widely distributed throughout living nature:

$$\text{Hexose} + \text{Mg} \cdot \text{ATP} \rightarrow \text{hexose-6-}P + \text{Mg} \cdot \text{ADP} \qquad (11\text{-}1)$$

These enzymes vary greatly in their specificity: some, which may be properly referred to as *hexokinases* (ATP : D-hexose-6-phosphotransferases) are relatively nonspecific and catalyze the phosphorylation of a wide variety of hexoses of appropriate configuration, including D-fructose, D-mannose, 2-deoxy-D-glucose, and D-glucosamine. For instance, the crystalline enzyme from yeast reacts with all sugars that have the general structure

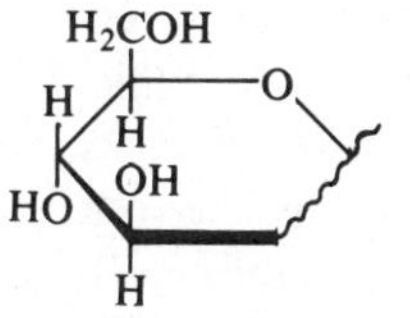

H_2COH, H, H, OH, HO, H, O

The α-pyranose form of the aldoses is acted on slightly more effectively than the β anomer.

Some typical values of K_s and V are for glucose: 1.5×10^{-4} M and 13,000 min^{-1}; glucosamine: 1×10^{-4} M, 12,000 min^{-1}; mannose: 1.0×10^{-4} M, 6000 min^{-1}; and fructose: 1.5×10^{-3} M and 26,000 min^{-1}.

Vertebrates contain in their tissues varying proportions of four different forms (isoenzymes) of this enzyme, distinct in their electrophoretic mobilities and in their catalytic properties. The two principal forms are (1) a *hexokinase* (three isozymes) ("low-K_s form"), K_s (glucose) = 0.01 to 0.04 mM, specifically inhibited by glucose-6-*P*; and (2) a *glucokinase* ("high-K_s form"), K_s (glucose) = 20 mM, unaffected by glucose-6-*P*. In liver, the latter is the predominating form consistent with a K_s in the range of glucose concentrations found in blood; in glycolyzing tissues such as brain or muscle the other forms predominate (5,6).

All kinases (ATP : substrate phosphotransferases) that form an ordinary *O*-phosphate ester utilize a nucleoside triphosphate (usually ATP) as a phosphate donor and require a divalent cation (usually Mg^{2+}) as an obligatory coreactant. Evidence based on kinetic studies and magnetic resonance spectroscopy indicates that the metal cofactor interacts first with ATP to yield the binary complex, which then functions as the true substrate and becomes attached to the enzyme. Experiments with ^{18}O-labeled substrate have shown that the P-transfer step proper proceeds as shown:

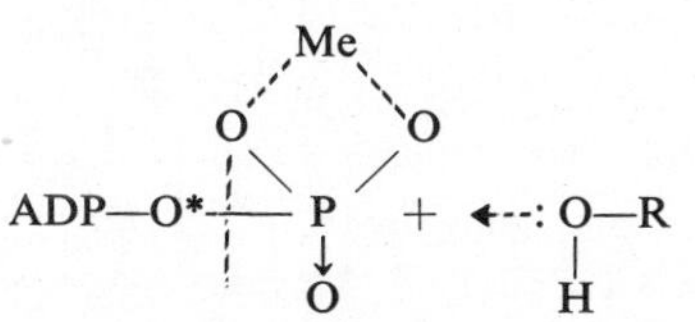

Me, O, O, ADP—O*—P + ←: O—R, O, H

perhaps through chelates such as the following:

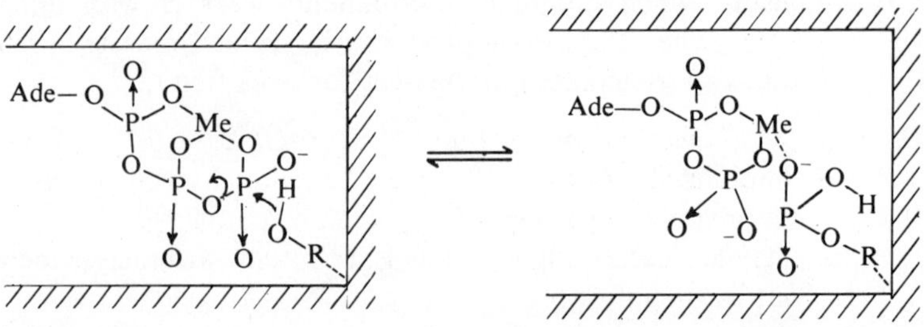

Ade—O P O⁻ Me O O O P O P O⁻ H O R ⇌ Ade—O P O Me O P O O⁻ O H P O R

Reaction (11-1a)

Hydrolysis of D-*Glucose-6-P.* The hydrolysis of glucose-6-*P* (reaction 1*a* of Table 11-1) is a key reaction in gluconeogenesis (see page 529, the formation of glucose during anabolism) and in the conversion of liver glycogen to blood glucose in mammals. In liver, as well as in kidney and intestine—but not in muscle—there occurs an enzyme, D-*glucose-6-phosphatase* (D-glucose-6-*P*-phosphohydrolase) [localized in the endoplasmic reticulum and thus concentrated in the microsomal fraction of tissue homogenates (Chapter 9)], capable of catalyzing this conversion. The enzyme requires the presence of phospholipids (7) and is relatively specific with respect to the carbohydrate moiety; the only other substrate susceptible to rapid hydrolysis is D-glucosamine-6-*P*. The same enzyme (8) is, however, also capable of hydrolyzing inorganic pyrophosphate and of catalyzing the reaction glucose + $PP_i \rightarrow$ glucose-6-*P* + P_i. It is activated by Mg^{2+}, inhibited by D-glucose, PP_i, ammonium molybdate, and by the oral antidiabetic drug Orinase [1-butyl-3-(*p*-tolyl sulfonyl)-urea]. In addition, the nonspecific "acid" and "alkaline" phosphatases (so called because of the position of their *p*H optima) that are found in almost all cells (but again not in muscle) are capable of catalyzing the rapid hydrolysis of this ester.

Reaction (11-2) Control of Glycogenolysis

Conversion of Glycogen and Starch to D-*Glucose-1-P.* The conversion of starch and other starchlike glucose-containing polysaccharides, such as glycogen, is catalyzed by *phosphorylases,* more correctly referred to as *α-1,4-glucan phosphorylases,* widely distributed in a variety of organisms. The reaction catalyzed by the enzyme may be represented as

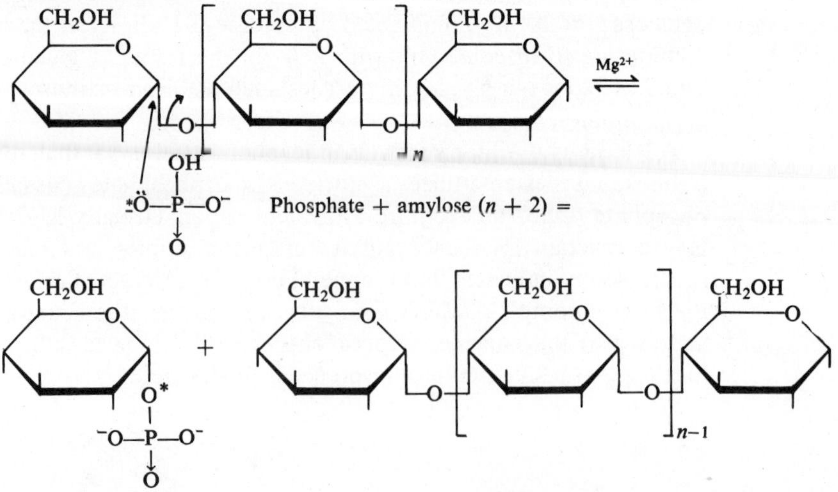

CH₂OH O [CH₂OH O O]ₙ CH₂OH O O ⇌ (Mg^{2+})

OH *Ö—P—O⁻ O Phosphate + amylose (*n* + 2) =

CH₂OH O O* ⁻O—P—O⁻ O + CH₂OH O O [CH₂OH O O]ₙ₋₁ CH₂OH O (11-2)

α-D-Glucopyranosyl-1-*P* + (glucose-1-*P*) Amylose (*n* + 1) etc.

As written, this is a phosphorolysis commencing at the free, nonreducing, end of an *amylose chain* (see page 477), removing one glucose unit at a time, and yielding a total of $(n + 1)$ glucose-1-*P* molecules until the reducing end is reached (which results in the formation of free glucose). If the substrate is amylopectin (see page 477), phosphorolysis continues only until the branch points are reached and the product is a *limit dextrin.* This is then acted on by an *oligo-1,4 → 1,4-glucantransferase,* and then the 1 → 6 bond at the branch is cleaved by an *amylo-1,6-glucosidase,*

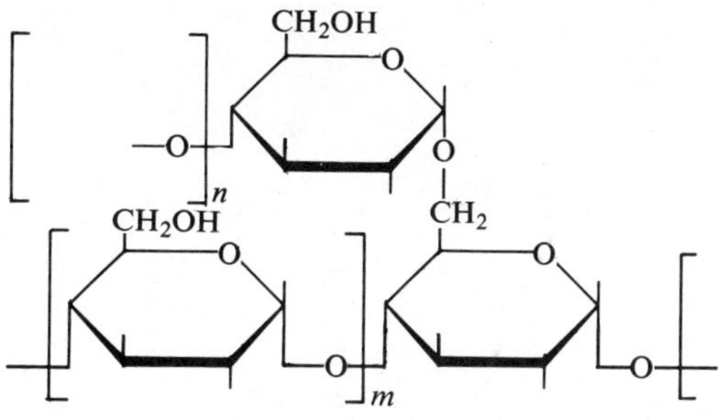

yielding free glucose, whereupon phosphorylase is able to act again until the next branch is reached, and so on. Phosphorylases are key enzymes in the mobilization of carbohydrates, converting them from their storage into their metabolically active forms.

The phosphorylases from mammalian liver and skeletal muscle have been studied extensively. The enzyme from muscle has been crystallized: it is a tetramer ($M = 360{,}000$) composed of four identical subunits. Each of these contains one serine residue esterified to orthophosphate[4] and essential for enzymatic activity, as well as one molecule of pyridoxal phosphate. The latter, as the Schiff base, is linked to a lysine residue and is also essential for enzymatic function.

As the enzyme responsible for the mobilization of storage carbohydrates in vertebrates, phosphorylase would be expected to provide for fine tuning by metabolic control devices and this expectation is amply borne out by recent experiments (9-12). In muscle, regulation is affected at five levels (Scheme 11-1).

1. The enzyme proper can exist in two forms, an active phosphorylated one (as shown in footnote 4), called *phosphorylase a,* and a relatively inactive, dephosphorylated form, called *phosphorylase b.*

2. This interconversion is mediated in the direction $a \rightarrow b$ by a specific *phosphorylase a phosphatase,* in the reverse direction by the enzyme *phosphorylase b kinase,* catalyzing the transfer of the terminal phosphate residues from four ATP molecules to *phosphorylase b.*

3. The kinase is itself a phosphoenzyme and it in turn can exist in two forms: an active phosphorylated form and an inactive dephosphorylated form. Again the conversion active → inactive is mediated by a *phosphatase,* the reverse by a specific *phosphorylase kinase kinase.* This enzyme is itself subject to a profound activation by Ado-3′:5′-*P* (cyclic AMP). The mediator is derived from ATP in a reaction catalyzed by *adenyl(ate) cyclase,* and can be

[4] The sequence around this residue is -Lys-Gln-Ile-Ser-Val-Arg-
|
P

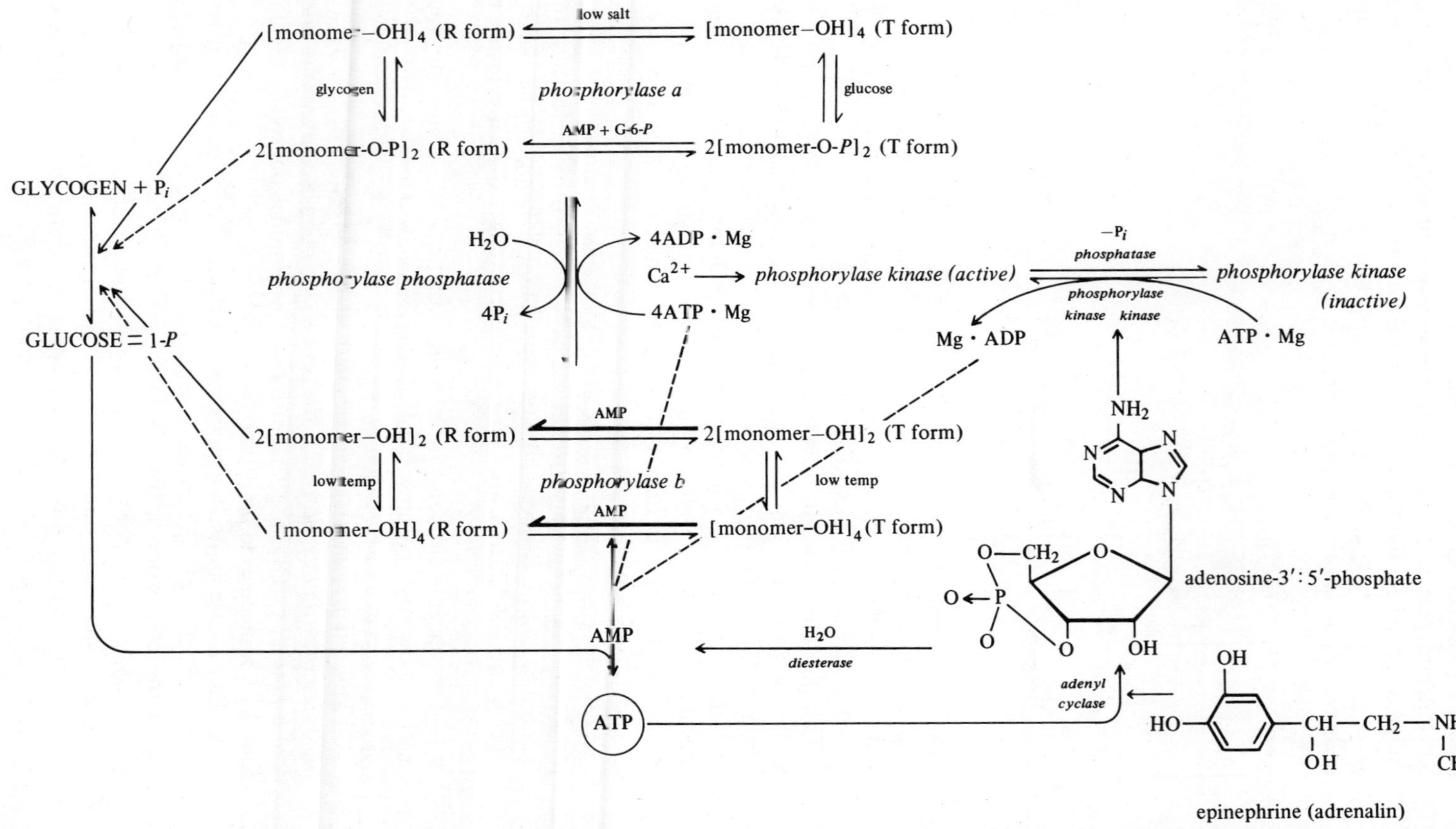

Scheme 11-1
Interconversions and regulation of muscle phosphorylase.

broken down hydrolytically by a *phosphodiesterase*. And at the apex of this inverted pyramid of activation reactions influenced by effectors stands the activation of the cyclase by the catecholamine hormone *epinephrine* (adrenalin).

4. Both phosphorylase *a* and *b* can exist as dimeric and tetrameric forms, and although the position of the equilibrium is affected by the intracellular concentration of the enzymes, the temperature and, most profoundly, the composition of the medium, it is probably safe to state that, in warm-blooded animals at least, it is probably the *tetra*meric phosphorylase *a* and the *di*meric phosphorylase *b* that are the preponderant forms.

5. The position of the equilibrium and, more importantly, the activity of the two enzymes is affected by certain mediators in a manner describable by the Monod–Changeux–Wyman cooperativity model: homotropically by the three substrates, P_i, glycogen, and glucose-1-*P*, particularly the latter; heterotropically by the activator adenosine-5′-phosphate. Interaction with the latter and allosterism in general is more pronounced in the case of phosphorylase *b* —so that this enzyme form is completely inactive in the absence of effectors— it, unlike phosphorylase *a,* is also inhibited by glucose as an allosteric inhibitor. All these interactions are designed to effect an instantaneous and massive response to changes in the energy charge of the adenylate pool (i.e., the supply and demand for ATP and its dephosphorylated derivatives). Although other cells and tissues may differ in the details of their response, particularly to the *"second messenger"* Ado-3′:5′-*P,* the overall features of these regulatory events appear to be retained. Thus, in liver a similar, but *not* identical, set of reactions can be demonstrated; but the phosphorylases of the two tissues are genetically and immunochemically distinct. Furthermore, liver phosphorylase is probably a *dimer* with a molecular weight approximately equal to that of *muscle phosphorylase b*. It is converted to an inactive form by the loss of two molecules of P_i per molecule without any attendant change in molecular weight; this inactive form cannot be activated by AMP. It can, however, be rephosphorylated by a specific kinase, which requires cyclic AMP as a cofactor. This enzymatic activation is stimulated not only by epinephrine but also by the polypeptide hormone *glucagon,* a hyperglycemic factor (elaborated in the pancreas) that is inactive in muscle either *in vitro* or *in vivo*.

Even though the reaction catalyzed by phosphorylase is readily reversible ($K_{eq} = [C_6H_{10}O_5]_{n+1}[P_i]/[C_6H_{10}O_5]_n$ [glucose-1-*P*] $\simeq 3$ at *p*H 7 and 25°), it is only concerned with glycogen breakdown and is not responsible for *glycogenesis,* the process of glycogen resynthesis. The latter still takes place even when the ratio P_i/glucose-1-*P* is as high as 300, and in certain hereditary disorders (McArdle's and Hers' syndrome) in which, respectively, muscle or liver phosphorylase is completely lacking. Furthermore, in the presence of epinephrine, which, as has just been described, stimulates phosphorylase activity, glycogen is broken down rather than synthesized. Finally and conclusively, the pathway of glycogen synthesis, thanks to the investigation of Leloir and others, is now known to involve the polymerization not of glucose-1-*P* itself, but instead of UDPG, the mixed anhydride of this compound with UMP, and a specific enzyme *glycogen synthetase* (UDPG-glycogen glycosyltransferase) is involved (see page 549).

Reaction (11-3) *Interconversion of* D-*Glucose-1-P and* D-*Glucose-6-P*. This reaction is important not only in the context of glycolysis but also for that of the metabolism of oligo- and polysaccharides in general. It is catalyzed by the enzyme *phosphoglucomutase* (α-D-glucose-1,6-di*P* : α-D-glucose-1-*P* phosphotransferase), which has been obtained in pure form from a variety of sources, such as muscle, yeast, and bacteria. The enzyme is specific for the α-D-pyranose-*P* and the K_{eq} at *p*H 7 and 25° for [glucose-6-*P*]/[glucose-1-*P*] = 19, and thus favors the formation of the slightly more stable ester over the hemiacetal phosphate. An enzyme from rabbit skeletal muscle contains one serine-OH group essential for enzymatic activity.[5] Glucose-1,6-di*P* (or glycerate-1,3-di*P*) is required for the reaction, which in addition requires Mg^{2+} and is inhibited by F^-. In its course the serine at the active site of the enzyme becomes phosphorylated (13,14) and the resultant phosphoenzyme is the reactive and kinetically significant intermediate, without, however, implicating *free* glucose-1,6-di*P* in an analogous role (15,16):

$$\text{(glucose-1-}O\text{—P}^{\dagger}O_3^{2-}\text{; CH}_2\text{OH)} + \text{E—O—P}^{*}O_3^{2-} \underset{}{\overset{Mg^{2+}}{\rightleftharpoons}} \text{E—O—P}^{\dagger}O_3^{2-} + \text{(glucose-6-CH}_2\text{OP}^{*}O_3^{2-}\text{; 1-OH)}$$

$$\text{E—OH} \cdot \text{glucose}\begin{cases}\text{P}^{*}(6)\\ \text{P}^{\dagger}(1)\end{cases}$$

$$\rightleftharpoons \text{(glucose with CH}_2\text{OPO}^{*}O_3^{2-}\text{ and O—P}^{\dagger}O_3^{2-}\text{)} + \text{E—OH}$$

At *p*H 7.4 the ΔG^0 values for the hydrolysis of glucose-1-*P*, the 1-*P* from glucose-1-6-di*P* and the *P* of the phosphoserine of the enzyme are −4.2, −5.1, and −1.1 kcal/mol, respectively.

The labeling pattern starting with the phosphoenzyme and proceeding from left to right is indicated with an asterisk (*) and in the reverse direction with a dagger (†). The required coenzyme is formed in reaction (11-3a) catalyzed by *phosphoglucokinase* (ATP-D-glucose-1-*P*:6-phosphotransferase):

$$\text{Glucose-1-}P + \text{ATP} \cdot \text{Mg} \longrightarrow \text{glucose-1,6-di}P + \text{ADP} \cdot \text{Mg} \qquad (11\text{-}3a)$$

Phosphoglucomutase is absolutely specific for the α anomer of glucose-1-*P*, but it also catalyzes analogous reactions with phosphate esters other than those of glucose [e.g., reaction (11-3b)], but at rates 100 to 10,000 times slower (i.e., mannose > ribose > fructose > galactose). The mannose-di*P* so formed can then function in the interconversion of mannose-1-*P* and mannose-6-*P*:

$$\text{Glucose-1,6-di}P + \text{mannose-1-}P \rightleftharpoons \text{mannose-1,6-di}P + \text{glucose-6-}P \qquad (11\text{-}3b)$$

Reaction (11-4) *Interconversion of* D-*Glucose-6-P and* D-*Fructose-6-P*. This reaction is catalyzed by *phosphoglucoisomerase* (glucose-phosphate isomerase, D-glu-

[5] The sequence around this residue is (see page 338) -Thr-Ala-Ser-His-Asp- in enzymes from rabbit muscle or *E. coli* (17). The former enzyme appears to be particularly susceptible to conformational alterations. It, as well as the enzyme from yeast, exists in the form of several, equally active species.

cose-6-*P* ketol-isomerase), an enzyme that has been obtained in highly purified form from rabbit muscle. The enzyme is quite specific, since the analogous anomerization mannose-6-*P* $\rightleftharpoons$ fructose-6-*P* so readily brought about by dilute alkali (the Lobry de Bruyn–Alberta van Eckenstein equilibrium) requires a separate enzyme, *phosphomannoisomerase* (mannose phosphate isomerase, D-mannose-6-*P* ketol-isomerase), also found in muscle. The two hydrogens attached to the C-1 of the fructose-*P* must be activated selectively by the two enzymes, since glucose-6-*P*-1-*d* does not exchange its deuterium with H_2O in the presence of the glucoisomerase but does so when both enzymes are present. Studies on the mechanism of isomerase reactions of this type favor a concerted mechanism with a *cis* ene-diol as the intermediate (18):

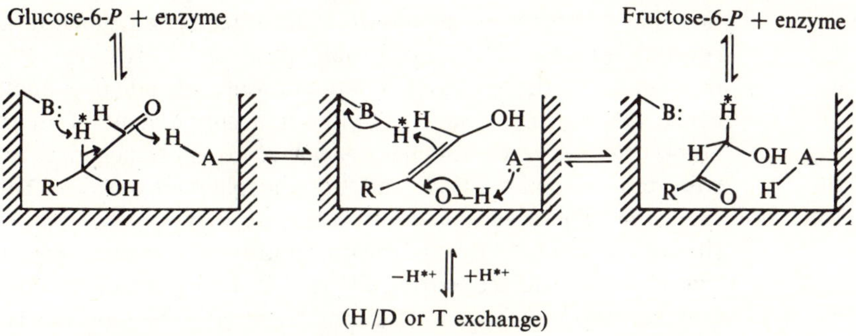

Glucose-6-*P* + enzyme ⇌ [B: H* H C=O ... H–A] ⇌ [B H* ... OH ... A ... O–H] ⇌ [B: H* ... H–C–OH ... C=O ... H–A] ⇌ Fructose-6-*P* + enzyme (11-4)

$-H^{*+}$ ⇅ $+H^{*+}$
(H/D or T exchange)

Note that this provides the required means for distinguishing the stereochemistry of the C-1 of fructose-1-*P*; for glucose-*P* isomerase it is

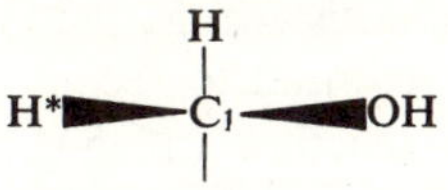

H
H*–C_1–OH

versus

H*
H–C_1–OH.

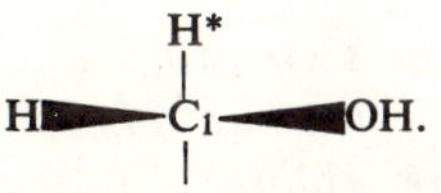

for mannose-*P* isomerase. The resulting configurations around the C-2 of the aldohexose are then

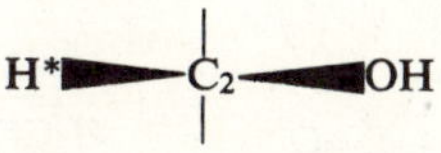

H*–C_2–OH and HO–C_2–H*

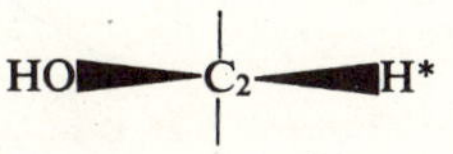

respectively. With the yeast enzyme, the order of activity is open-chain form $>$ α-glucopyranose $\gg$ β-glucopyranose. In addition, this enzyme appears to catalyze the equilibrium between the α anomer and the free aldose.

The enzyme is specifically and competitively inhibited by 2-deoxy-D-glucose-6-*P*, which is readily formed from the corresponding free deoxysugar in the presence of hexokinases.

Reaction (11-5) *Formation of* D-*Fructose-1,6-diP from Fructose-6-P. Control of Glycosis.* The reaction is catalyzed by a highly specific enzyme called *phosphofructokinase* (ATP : D-fructose-6-*P* 1-phosphotransferase) and shares the other properties of kinase reactions, that is, requirements for ATP (which in this instance can be replaced by UTP or ITP) and Mg^{2+} and a negative $\Delta G^{0\prime}$.

$$\text{D-Fructose-6-}P + \text{ATP} \cdot \text{Mg} \xrightarrow[K^+]{Me^{2+}} \text{D-fructose-1,6-di}P + \text{ADP} \cdot \text{Mg} \qquad (11\text{-}5)$$

Since alternative pathways of hexose metabolism diverge from *all* the other hexose phosphates so far discussed, reaction (11-5) may be regarded as the first one characteristic of the glycolytic sequence proper and therefore constitutes an important branch and control point, subject to strong metabolic regulation. The enzyme has now been purified from a variety of sources (such as rabbit muscle, guinea pig heart, sheep brain, a liver fluke, yeast, and *E. coli*) and an interesting pattern of regulatory effects has emerged (19,22). The active enzyme is a tetramer ($M \geqq 350{,}000$) and appears to behave as a perfect "*K* system" with strong homotropic effects by fructose-6-*P*, and with ATP as the most important negative (inhibitory) effector—this in spite of the fact that *all* common nucleoside triphosphates can function as *cosubstrates* in reaction (11-5). The heterotropic positive (activating) effector varies with the source of the enzyme; it is commonly an adenine nucleotide (5′-AMP, cyclic AMP, ADP) with or without a supplement of inorganic phosphate. Some enzymes have also been reported to show negative heterotropic effects with citrate (mammals) and/or phosphoenolpyruvate (instead of ATP) (*E. coli*). These various controls ensure that whenever the energy charge or simply the ratio ATP/Σ adenosine phosphates is high [e.g., when lactate and pyruvate are oxidized aerobically to CO_2 by means of the citric acid cycle (see Chapter 14)], glucose synthesis is favored. The obverse is also true; when glycolysis is absolutely required for energy generation (i.e., when the energy charge drops), glycolysis is favored and carbohydrate synthesis turned off. To add further insurance to this control and to exert a similar one on the reverse, biosynthetic, sequence of gluconeogenesis (page 524), most *fructose-1,6-diphosphatases,* the specific enzymes required to reconvert the diphosphate to the 6-monophosphate hydrolytically [reaction (11-5)] are inhibited by high concentration of their substrate and also, at a second (allosteric) site, specifically by AMP.

Reactions (11-6) and (11-7)

*Conversion of Fructose 1,6 di*P *into Two Molecules of Triose Phosphate and Interconversion of Aldo- and Ketotriose Phosphates.* The reactions catalyzed by *aldolase* (D-fructose-di*P* aldolase; D-fructose-1,6-di*P* : glyceraldehyde-3-lyase) and by *triose-P isomerase* (D-glyceraldehyde-3-*P* ketol isomerase) are

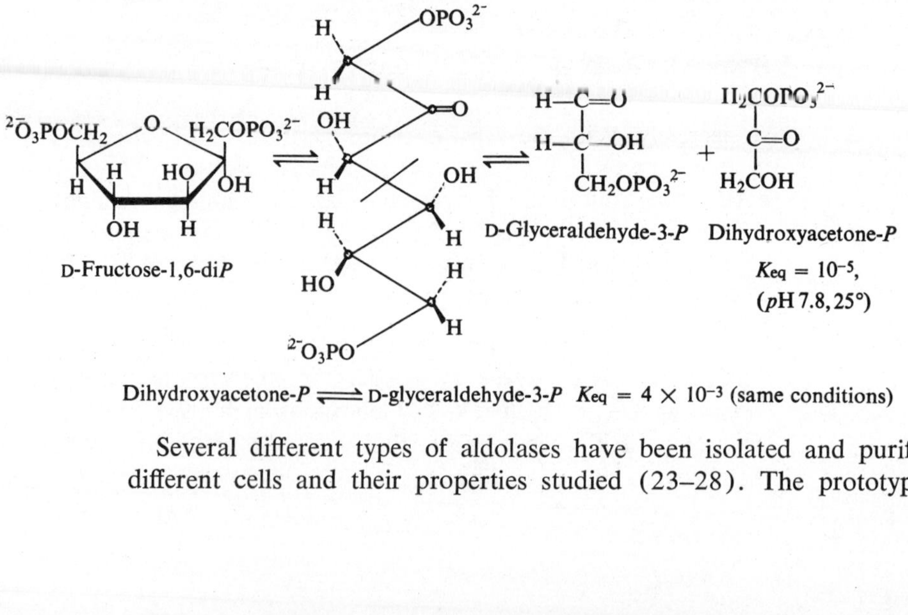

$$\text{D-Fructose-1,6-di}P \rightleftharpoons \text{(open chain)} \rightleftharpoons \underset{\text{D-Glyceraldehyde-3-}P}{H{-}C{=}O,\ H{-}C{-}OH,\ CH_2OPO_3^{2-}} + \underset{\text{Dihydroxyacetone-}P}{H_2COPO_3^{2-},\ C{=}O,\ H_2COH} \qquad (11\text{-}6)$$

$K_{eq} = 10^{-5}$, (pH 7.8, 25°)

$$\text{Dihydroxyacetone-}P \rightleftharpoons \text{D-glyceraldehyde-3-}P \quad K_{eq} = 4 \times 10^{-3} \text{ (same conditions)} \qquad (11\text{-}7)$$

Several different types of aldolases have been isolated and purified from different cells and their properties studied (23–28). The prototype of the *class I* enzymes (found in all animals, higher plants, and green algae) is the enzyme from rabbit skeletal muscle (*aldolase A*). The active form is a tetramer consisting of four polypeptide chains.[6] The enzyme is known to catalyze a hydrogen exchange in the H_2C—OH of dihydroxyacetone-*P* or of acetol-*P* (CH_3COCH_2OP), in the *absence* of the aldotriose, with a stereospecificity opposite to that of triose-*P* isomerase [which itself exhibits a stereospecificity and mechanism *identical* to that of glucose-*P* isomerase, reaction (11-4)]; that is, for reaction (11-7), it must be

H
H*—C—OH

against

H*
H—C—OH

for reaction (11-6). The specificity of the enzyme with regard to possible substrates indicates that dihydroxyacetone-*P* (and the corresponding portion of the hexose molecule) must interact with the enzyme in a highly selective manner, while a number of different aldehydes can participate as indicated below. In all cases the product ketose has a trans configuration of hydroxyls,

=O

around the α- and β-carbons[7]:

Active configuration

$CH_2OPO_3^{2-}$, C=O, CH_2OH + CHO, H—C—OH, CH_2OR → $CH_2OPO_3^{2-}$, =O, CH_2OR: D-fructose-1-*P* (R = H); D-fructose-1,6-di*P* ($R = PO_3^{2-}$) (similar reactions with the appropriate tetroses yielding the D-sedoheptulose derivatives)

$CH_2OPO_3^{2-}$, C=O, CH_2OH + CHO, HO—C—H, CH_2OR → $CH_2OPO_3^{2-}$, =O, CH_2OR: L-sorbose-1-*P* (R = H)

$CH_2OPO_3^{2-}$, C=O, CH_2OH + CHO, CH_2R → $CH_2OPO_3^{2-}$, =O, CH_2R: D-xylulose-1-*P* (R = OH); D-xylulose-1,5-di*P* ($R = OPO_3^{2-}$); 5-deoxy-D-xylulose-1-*P* (R = H)

All these observations suggest that the enzyme is capable of forming a complex with the ketose-*P*, a very *specific* one in the case of dihydroxyacetone-*P*. The formation of a covalent intermediate can be demonstrated spectrophotometrically, and its structure has been shown to involve Schiff base (ketimine) formation between the carbonyl C of the substrate and an

[6] Two allelic genes appear responsible; hence, the structures found can be α_4 or β_4, and three possible hybrids (page 174): $\alpha_3\beta$, $\alpha_2\beta_2$, and $\alpha\beta_3$.

[7] The one exception is D-tagatose-1,6-di*P* (which has the opposite configuration to fructose at C-4), which is cleaved slowly.

ε-amino group of lysine on the enzyme; reduction and hydrolysis yields 1 mol per mol of enzyme of a derivative that has been identified as

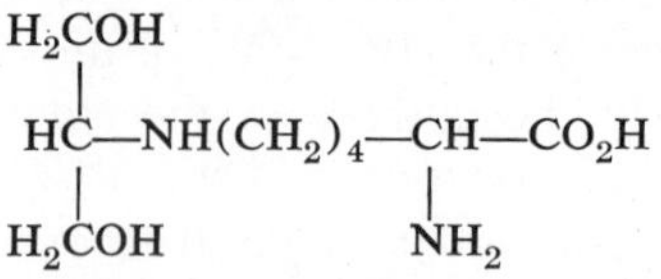

$$\begin{array}{l} H_2COH \\ | \\ HC\text{—}NH(CH_2)_4\text{—}CH\text{—}CO_2H \\ | \qquad\qquad\qquad\qquad | \\ H_2COH \qquad\qquad\quad NH_2 \end{array}$$

(N^6-β-glyceryl-L-lysine)

The same derivative is also obtained after reduction of the complex between *transaldolase* and 1-^{14}C-fructose (page 533). A feasible mechanism, is ping-pong, consistent with kinetic data

$$\begin{array}{ccccccccc} DH & & H^+ & & G & & DG & & \\ \downarrow & & \uparrow & & \downarrow & & \uparrow & & \\ \hline E & (E\text{-}DH) & & (E\text{-}D^-) & & (E\text{-}DG) & & E \end{array}$$

It utilizes E-D$^-$, a carbanion of the ketimine (resonance form of an enamine) as the crucial intermediate (23–26):

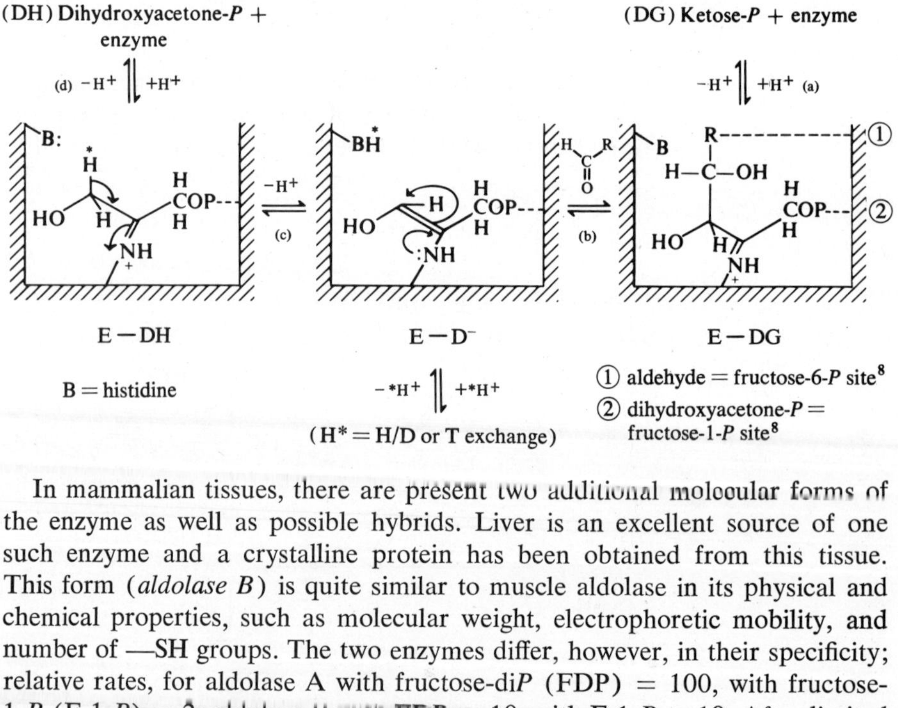

(DH) Dihydroxyacetone-*P* + enzyme

(DG) Ketose-*P* + enzyme

(d) $-H^+$ ⇅ $+H^+$

$-H^+$ ⇅ $+H^+$ (a)

B: H* HO H H COP--- H NH+

$-H^+$ ⇌ (c)

BH* HO H COP--- H :NH

H R C O ⇌ (b)

B R---① H—C—OH H COP---② H HO H NH+

E—DH

E—D$^-$

E—DG

B = histidine

$-{}^*H^+$ ⇅ $+{}^*H^+$

(H* = H/D or T exchange)

① aldehyde = fructose-6-*P* site[8]

② dihydroxyacetone-*P* = fructose-1-*P* site[8]

In mammalian tissues, there are present two additional molecular forms of the enzyme as well as possible hybrids. Liver is an excellent source of one such enzyme and a crystalline protein has been obtained from this tissue. This form (*aldolase B*) is quite similar to muscle aldolase in its physical and chemical properties, such as molecular weight, electrophoretic mobility, and number of —SH groups. The two enzymes differ, however, in their specificity; relative rates, for aldolase A with fructose-di*P* (FDP) = 100, with fructose-1-*P* (F-1-*P*) = 2; aldolase B with FDP ≃ 10, with F-1-*P* ≃ 10. After limited digestion with carboxypeptidase, which liberates only *one* terminal tyrosine per mol of enzyme B (compared to 2.3 for enzyme A), the relative activities become approximately equal to 5 for both enzymes acting on either substrate. However, this treatment introduces a D/T isotope effect in the exchange reaction. Hence the C-terminal residue appears principally implicated in reactions involving site 1 but not in those at site 2.[8] A third form of the enzyme (aldolase C) can be obtained from mammalian brain; its maximal rates, in the units above, are ≃40 with FDP and ≃4 with F-1-*P*.

Aldolase A is inhibited by ATP and activated by ADP, while aldolase B

[8] Both sites can be blocked by inorganic ions (P−, Cl−); D-glyceraldehyde-3-*P* and L-glyceraldehyde-3-*P* both bind to site 1 and are acceptors for dihydroxyacetone-*P*, but only the former exhibits interaction with site 2.

exhibits inhibition by AMP in accord with the principal functions of these enzymes (see below).

Finally, a completely different type of enzyme (*class II aldolase*) is present in microorganisms (bacteria, yeasts, and blue-green algae); photosynthetic protozoa such as *Euglena* and *Chlamydomonas* contain both types. Compared to class I aldolases these enzymes show (1) a lower *p*H optimum, (2) a much greater selectivity for FDP (ratio of rates for FDP/F-1-$P = 2500$), (3) a requirement for K^+ as an activator, and (4) no Schiff-base formation as a prerequisite for catalysis. Instead, they contain divalent metal ions at the active site. The enzymes from yeast and *Aspergillus niger* have been shown to be Zn proteins, and after preparation of the apoprotein, the former (28) exhibits reactivation with transition metal ions in the order $Zn^{2+} > Co^{2+} > Fe^{2+} > Mn^{2+} \gg Ni^{2+}$ with no comparable effect on K_s. The metal thus appears to participate in the catalytic step proper, perhaps by assisting in the formation of the requisite carbanion by chelation as follows:

HO O (NH) ⇌ HO O
Me   Me

Reactions (11-8) and (11-9) *Phosphorylation of ADP at the Expense of Dehydrogenation of Glyceraldehyde-*P

$$\text{D-Glyceraldehyde-3-}P^{2-} + \text{NAD}^+ + \text{P}_i^{2-} \rightleftharpoons \text{D-1,3-diphosphoglycerate}^{4-} + \text{NADH} + \text{H}^+ \quad (11\text{-}8)$$

$$\text{D-1,3-Diphosphoglycerate}^{4-} + [\text{Mg}\cdot\text{ADP}]^- + \text{H}^+ \rightleftharpoons \text{D-3-phophoglycerate}^{2-} + [\text{Mg}\cdot\text{ATP}]^{2-} \quad (11\text{-}9)$$

$$\textit{Net}\ \text{D-Glyceraldehyde-3-}P^{2-} + \text{NAD}^+ + [\text{Mg}\cdot\text{ADP}]^- + \text{P}_i^{2-} \rightleftharpoons \text{D-3-phosphoglycerate}^{2-} + [\text{Mg}\cdot\text{ATP}]^{2-} + \text{NADH} \quad (11\text{-}8) + (11\text{-}9)$$

In most cells the major fate of the triose-*P* formed in the preceding steps is its conversion to 1,3-diphosphoglycerate [reaction (11-8)]. The reaction requires the stoichiometric participation of NAD^+ [addition of H is on the *B* side of the nicotinamide ring (page 413)] and P_i and leads to a product that, as a mixed anhydride of a carboxylic and orthophosphoric acid, shows a very high negative free energy of hydrolysis ($\Delta G^{0\prime} \simeq -14$ kcal). It is this fact which accounts for the driving force, not only for reactions (11-8) plus (11-9)—$\Delta G^{0\prime} = -5.3$ kcal/mol triose-*P*—but of the whole conversion of hexose-di*P* to phosphoglycerate [reactions (11-6) to (11-9)]—$\Delta G^{0\prime} \simeq -3$ kcal/mol hexose-di*P*.

The enzyme-catalyzing reaction (11-8), *glyceraldehyde-P dehydrogenase* (GPD), shows exceptionally wide distribution. Free —SH groups are required for activity (14 per molecule are titratable in the enzyme). Muscle GPD binds its coenzyme NAD^+ firmly enough so that $\geqq$ two molecules of the latter remain associated with the enzyme even on repeated recrystallization or dialysis. Adsorption of the coenzyme on a charcoal column is required for complete removal. The holoenzyme can be reconstituted by the addition of four molecules of the nucleotide to the apoenzyme. An equal number of cysteine residues on the protein appear to be implicated in the formation of this complex, which has an absorption spectrum distinct from that of the free coenzyme. The reactions in either direction have been shown to require the formation of an acyl enzyme as an obligatory intermediate; this intermediate has the partial structure

—Cys—S—C(=O)—R

The dehydrogenase is specific for D-glyceraldehyde-3-*P*, the free dephosphorylated aldehyde, or acetaldehyde, and is powerfully inhibited by D-threose-2,4-diphosphate.

In addition to reaction (11-8), GPD also catalyzes a variety of other interesting reactions. The most important are the following:

1. In the presence of amounts of NAD^+ (not NADH) stoichiometric to that of the enzyme, the latter catalyzes reaction (11-8a), an *acyl phosphatase* activity with 1,3-diphosphoglycerate or acetyl phosphate as substrates. The role of NAD^+ here is probably to convert the enzyme into an active conformation; complex negative cooperative effects involving this coenzyme are also known to be involved in the dehydrogenase reaction (11-8) (29). Agents capable of oxidizing —SH groups such as *o*-iodosobenzoate enhance the reaction with acetyl-*P* without inhibiting the reaction with 1,3-diphosphoglycerate, in spite of completely abolishing dehydrogenase activity. We conclude that, in some manner, the essential SH at the active site must be protected. Cleavage is by means of the unusual C—O rather than the customary O—P split:

$$R{-}\overset{\overset{\displaystyle O}{\|}}{C}{-}O{-}\underset{\underset{\displaystyle O}{\|}}{\overset{\overset{\displaystyle O^-}{|}}{P}}{-}O^- \longrightarrow RCO_2^- + P_i \qquad (11\text{-}8a)$$

2. In the presence of NAD^+, as before, the same two substrates can also be shown to be subject to *transacylation* reactions with a variety of acceptors other than OH^-, such as

$$R{-}\underset{\underset{\displaystyle O}{\|}}{C}{-}O{-}P + :A^{(-)} \rightleftharpoons R{-}\underset{\underset{\displaystyle O}{\|}}{C}{-}A + P_i \qquad (11\text{-}8b)$$

$^{32}P_i$, arsenate, and a number of sulfhydryl compounds (e.g., CoASH, CysSH, and thioglycolate). In the case of arsenate, the resultant acylarsenates are unstable and are rapidly hydrolyzed to the salts plus arsenate, which is thus recycled to bring about the rapid and complete *arsenolysis* of the substrate.

3. With *p*-nitrophenyl acetate or other acylesters as substrate, the enzyme functions as an *esterase*. This reaction is blocked in the presence of glyceraldehyde-3-*P*, acetyl-*P*, NAD, and SH reagents.

4. At a *p*H of 5, GPD catalyzes the *conversion of NADH into NADH-X*, a modified product probably involving the addition of H and OH across one of the double bonds:

$$\text{(dihydropyridine ring with } CONH_2\text{)} + H_2O \longrightarrow \text{(HO-substituted tetrahydropyridine ring with } CONH_2\text{)} \qquad (11\text{-}8c)$$

5. The enzyme brings about the following stoichiometric conversion of bound NAD molecules in a *phosphotransferase* reaction:

$$(HS)_4 - E \cdot (NAD)_4 + 4ATP + 4IMP \longrightarrow 4ADP + 4IDP + \left(\begin{array}{c} S \\ | \\ S \end{array}\right)_2 E \cdot (NADH)_4 \qquad (11\text{-}8d)$$

6. The enzyme can catalyze the transfer of acyl groups from the primary site[9] (Cys^{149}) to the ε-N of Lys^{183}.

A feasible mechanism is

$$E\text{—}(SH)_4 + 4NAD^+ \rightleftharpoons E\left(\begin{matrix}NAD\\ |\\ S\end{matrix}\right)_4 \qquad \textit{formation of holoenzyme}^{9} \qquad (10\text{-}8e)$$

$$E\left(\begin{matrix}NAD\\ |\\ S\end{matrix}\right)_4 + 4RCHO \rightleftharpoons E\left(\begin{matrix}NADH\\ S\text{—}\underset{\underset{O}{\|}}{C}\text{—}R\end{matrix}\right)_4 \qquad \textit{reduction of } NAD^+ \textit{ and formation of acyl enzyme}^{10} \qquad (10\text{-}8f)$$

$$E\left(\begin{matrix}NADH\\ S\text{—}\underset{\underset{O}{\|}}{C}\text{—}R\end{matrix}\right)_4 + 4NAD^+ \rightleftharpoons 4NADH + E\left(\begin{matrix}NAD^+\\ S\text{—}\underset{\underset{O}{\|}}{C}\text{—}R\end{matrix}\right)_4 \qquad \textit{regeneration of } E\cdot NAD^+ \qquad (10\text{-}8g)$$

$$E\left(\begin{matrix}NAD^+\\ S\text{—}\underset{\underset{O}{\|}}{C}\text{—}R\end{matrix}\right)_4 + 4:A^- \rightleftharpoons E\left(\begin{matrix}NAD\\ |\\ S\end{matrix}\right)_4 + 4A\text{—}\underset{\underset{O}{\|}}{C}\text{—}R \qquad :A^- = AsO_4^{3-}, PO_4^{3-}, RS^-, HO^-, \textit{ acylation and deacylation reactions} \qquad (10\text{-}8h)$$

$$E\left(\begin{matrix}NAD^+\\ S\text{—}\underset{\underset{O}{\|}}{C}\text{—}R\end{matrix}\right)_4 \rightleftharpoons E\text{—}(S\text{—}\underset{\underset{O}{\|}}{C}\text{—}R)_4 + 4NAD^+ \qquad \textit{dissociation of coenzyme} \qquad (10\text{-}8i)$$

The ordinary oxidative reaction sequence is composed of steps (11-8f) → (11-8g) → (11-8h). Provision must then be made to regenerate the enzyme-bound NAD, which has become reduced in step (11-8f). This is accomplished in step (11-8g). However, if no means are available for the removal of the NADH produced, as is the case *in vitro,* reaction (11-8g), which is readily reversible, will slow down as NADH accumulates. *In vivo,* the reoxidation of the enzyme-bound NADH is accomplished in a linked reaction (e.g., the reduction of pyruvate to lactate, or of acetaldehyde to ethanol) in the presence of the appropriate dehydrogenases, or under strongly aerobic conditions, by the respiratory chain (Chapter 15). Enzymes have been purified from a wide variety of organisms, such as vertebrate and invertebrate animals, yeast, *E. coli,* and *Bacillus stearothermophilus.* All enzymes appear identical in the number of subunits and similar in size. Variability and phylogenetic relationships are observed in the relative content of Cys, His, Tyr, and Trp.

Reaction (11-9) is catalyzed by *phosphoglycerate kinase* (phosphoglyceric kinase, ATP : 3-phospho-D-glycerate-1-phosphotransferase). The reaction mechanism must be very similar to that of reaction (11-1), in spite of the fact that the phosphate transferred is to and from a mixed anhydride rather than an ester. There is an absolute requirement for a divalent metal cofactor (Mg^{2+}, Mn^{2+}, or Ca^{2+}). The metal interacts with the adenosine di- or triphosphate to form a reactive complex, and the latter is not linked to the

[9] The complete primary sequence has now been established in several instances and the constellation around the acyl-enzyme site in many more (30–32). The latter is identical for enzymes from yeast, pig, and rabbit muscle: the active Cys is Cys^{149}, -Ile-Val-Ser-Asn-Ala-Ser-*Cys*(S-COR)-Thr-Thr-Asn-Cys-Leu-Ala-Pro-Leu-Ala-Lys-. For lobster muscle there are only two substitutions (underlined), Ile → Val and Leu → Val, in this whole region. The nature of the binding site for $NAD+$ is still uncertain. A Cys and a Trp residue appear to be definitely implicated; an enzyme-bound Zn^{2+} ion may also be involved. Binding of aldehyde substrate and NAD is mutually facilitating, whereas that of acyl residues and NAD is mutually antagonistic (33).

[10] This reaction is almost certainly complex, It may involve preliminary formation of a thiohemiacetal facilitated by a histidine residue in the immediate vicinity.

enzyme surface by means of the metal; there is no evidence for a compulsory binding order.

Reaction (11-10)

Isomerization of Phosphoglycerates. The reversible interconversion of the primary phosphate ester 3-phospho-D-glycerate into the corresponding secondary ester 2-phospho-D-glycerate is catalyzed by *phosphoglyceromutase* (2,3-diphospho-D-glycerate : 2-phospho-D-glycerate phosphotransferase):

$$\text{D-3-Phosphoglycerate} \xrightleftharpoons[]{\text{Mg}^{2+} \atop \text{2,3-diphosphoglycerate}} \text{D-2-phosphoglycerate} \qquad (11\text{-}10)$$

The reaction requires two obligatory cofactors: Mg^{2+} and the diester 2,3-diphosphoglycerate. The reaction mechanism is probably similar to that established for the more thoroughly studied phosphoglucomutase reaction (11-3), in spite of the absence of an unambiguous demonstration of the existence of a stable phosphoenzyme[11] (34,35).

In plant germ and in animal tissues a second enzyme has been identified that does not appear to require any cofactors. The plant enzyme has been crystallized and found to have an $M = 30{,}000$; it may catalyze a different type of isomerization reaction, and it has been called a *true* D-phosphoglycerate-2,3 phospho*mutase* (rather than a phosphotransferase).

Reaction (11-10) and (11-12)

Conversion of 2-Phosphoglycerate to Pyruvate: The Generation of a Second Molecule of ATP

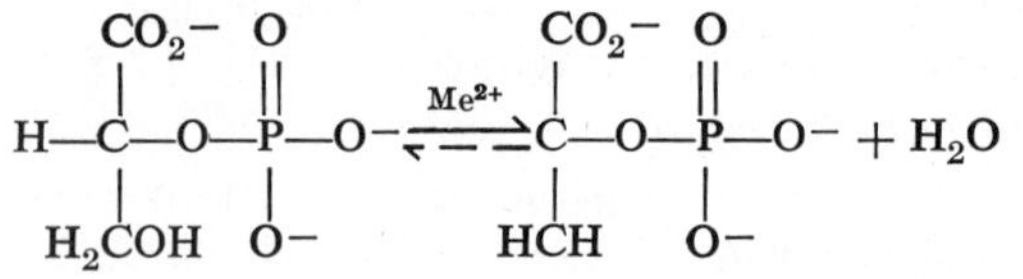

$$\mathrm{H{-}C(CO_2^-)(H_2COH){-}O{-}P(=O)(O^-){-}O^-} \xrightleftharpoons{\mathrm{Me^{2+}}} \mathrm{C(CO_2^-)(=HCH){-}O{-}P(=O)(O^-){-}O^-} + \mathrm{H_2O} \qquad (11\text{-}11)$$

$$\mathrm{C(CO_2^-)(=HCH){-}O{-}P(=O)(O^-){-}O^-} + \mathrm{MgADP} \rightleftharpoons \mathrm{CH_3COCO_2^-} + \mathrm{MgATP} \qquad (11\text{-}12)$$

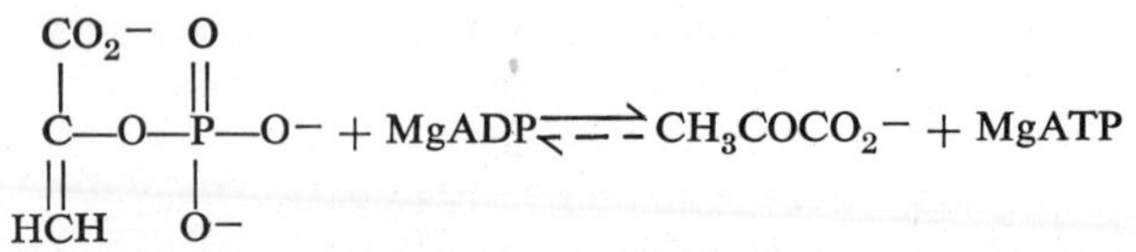

Reaction (11-11) constitutes a dehydration of the phosphate ester of a dihydroxy acid (which can also be regarded as an intramolecular oxidation of C-2 and reduction of C-3), a process that is essentially isoenergetic; thus, K_{eq} for the reaction as written equals 3 under physiological conditions. On the other hand, the product ester is more properly classified as a mixed anhydride, since the dephosphorylated product (enolpyruvic acid) is an α-carboxyenol. $\Delta G^{0\prime}$ for the hydrolysis of the phosphoenolpyruvate is ~ -12.8 kcal/mol, part of the driving force being provided by the position of the enol-keto tautomeric equilibrium of the product, which, in the case of pyruvate at a pH around 7, is strongly in favor of the keto form (page 39).

Enolase (phosphopyruvate hydratase), the enzyme catalyzing reaction (11-11), is very widely distributed and shows an absolute requirement for divalent cations (Mn^{2+}, Mg^{2+}, Zn^{2+}, or Cd^{2+}) that can be antagonized by other cations (Ca^{2+} or Sr^{2+}). F^- ion is an effective inhibitor at low concentration (10^{-2} M), especially in the presence of P_i, presumably by formation of a fluorophosphate linked to the active site of the enzyme.

Pyruvate kinase, the tetrameric (35) enzyme catalyzing reaction (11–12), utilizes the Mg or Mn chelate of ADP as its substrate. However, the enzyme

[11] Addition of the cofactor to dephosphorylated phosphoglucomutase converts it to the phosphoenzyme; no such reaction is observed with phosphoglyceromutase (35).

also forms binary (E-Me) and ternary E-Me-PEP complexes. In addition, the enzyme also requires the simultaneous presence of monovalent cations (K^+, Rb^+, or Cs^+; antagonized by Na^+ or Li^+) for optimal activity, probably for conformational stabilization. Thus, this enzyme behaves like a typically easily deformable protein capable of allosteric interactions. It stands at the end of a metabolic sequence and would be expected to be subject to such regulatory effects.[12] These have indeed been found (35): The enzyme from several sources (but not of skeletal muscle) is activated by fructose-1,6-di*P*, an example of positive feedback (feed-forward activation) and inhibited by a nucleoside triphosphate (ATP in the case of yeast, heart muscle, and rat liver; GTP or ITP for chicken liver). In addition, the rat liver enzyme has also been reported to be inhibited by L-alanine as well.

The very unfavorable equilibrium of reaction (11-12) for the reverse reaction (pyruvate as substrate), plus the fact that the turnover number in this direction equals 12 (compared to 6×10^3 for pyruvate formation), ensures a formidable thermodynamic and kinetic block against the use of this reaction in carbohydrate synthesis.[13]

Reaction (11-13)

Terminal Step in Muscle Glycolysis and Homolactic Fermentation. The reversible reduction of pyruvate to lactate, with NADH as a reducing agent, is the terminal step that distinguishes glycolysis by vertebrate muscle from many other anaerobic, fermentative modes of carbohydrate catabolism:

$$CH_3COCO_2^- + NADH + H^+ \rightleftharpoons CH_3CHOHCO_2^- + NAD^+ \qquad (11\text{-}13)$$

It is shared, however, by the *homolactic fermentation* (i.e., a fermentation that yields lactate as the *sole* product) carried out by a large number of microorganisms, the "lactic acid bacteria"; among these are various species of the genera *Lactobacilli, Bacilli, Streptococci, Clostridia,* and others. The product with enzymes from animal sources is invariably in the L(+) configuration; certain bacterial cells, and enzymes isolated from them (e.g., *E. coli*), appear to exhibit the opposite stereospecificity, but in other instances both activities appear to be present in the same cell (e.g., *L. plantarum* ≡ *L. arabinosus*).

The pure enzymes from vertebrate sources all appear to be tetramers containing four independent catalytic sites. They can be dissociated into four equal, enzymatically inactive subunits of $M = 35{,}000$ by protein denaturants (guanidine, sodium dodecyl sulfate). There exist two different types of subunits that have been called by Kaplan the "M" (for skeletal *M*uscle, where it is found as the predominant form) and "H" (for *H*eart muscle, where it predominates), which appear to be controlled by two separate and independent genes, forming the various iso(en)zymes H_4, H_3M, H_2M_2, HM_3, and M_4 (Chapter 4). Thus, soluble extracts, and presumably the cytosol *in vivo* as well, exhibit different enzymatic patterns of lactate dehydrogenase action, each one characteristic of an individual tissue or of a cell population within a tissue or organ, and subject to different regulatory devices (e.g., inhibition by pyruvate, to which the type-H enzyme is particularly susceptible). Aerobic, gluconeogenic tissues with type-H enzymes are primarily concerned with the conversion lactate → pyruvate, those with type-M enzymes (principally anaerobic, purely glycolytic ones) with the reverse reaction.

[12] They were predicted in 1966 in the first edition of this book; a number have since been found, starting in 1967.

[13] In the presence of F^- the enzyme catalyzes the formation of fluorophosphate (36), a competitive inhibitor to PEP.

Reactions (11-14) and (11-15)

Terminal Steps in Alcoholic Fermentation. In certain microorganisms, among which brewer's yeast operating anaerobically is the most important, the further catabolism of pyruvate differs from that just described. The first step now is reaction (11-14), catalyzed by *pyruvate decarboxylase* (2-oxo-acid carboxylase, formerly called pyruvate carboxylase), the virtually irreversible decarboxylation of pyruvate to yield acetaldehyde, which requires thiamine pyrophosphate (TPP, cocarboxylase) and Mg^{2+} as indispensable cofactors. It is now known that the reaction takes place via the formation of 2-α-lactyl and -hydroxyethyl derivatives of the coenzyme, [*A*] and [*B*]:

$$\text{E—}^{+}\text{N(thiazolium, C-2—H)} \rightleftharpoons \text{E—}^{+}\text{N(thiazolium ylide)} + H^+$$

$$+ \; CH_3\text{-}\overset{O}{\overset{\|}{C}}\text{-}CO_2^- \underset{}{\overset{Me^{2+}}{\rightleftharpoons}} \left[\text{E—}^{+}\text{N} \ldots H_3C\text{-}C\text{—OH}, \; C(=O)O^- \right]_{[A]} \overset{Me^{2+}}{\rightleftharpoons} \left[\text{E—N} \ldots H_3C\text{-}C\text{-}OH \right]_{[B]} + CO_2 \overset{H^+}{\rightleftharpoons} \left[\text{E—}^{+}\text{N} \ldots H\text{-}C\text{-}O\text{-}H, \; CH_3 \right]_{[C]} \underset{H^+}{\rightleftharpoons} \text{E·TPP} + CH_3CHO \qquad (11\text{-}14)$$

The general mechanism of TPP-catalyzed reactions was discussed on page 401. Enzymes capable of catalyzing reaction (11-14) and of releasing free acetaldehyde from complex [*C*] have been obtained in highly purified form from yeast, wheat germ, and other plant tissues. The metal requirement in this instance can be satisfied, not only by the usual divalent ions, but by certain trivalent ones as well (Al^{3+}, Fe^{3+}).

In the final step of alcoholic fermentation, the acetaldehyde formed in reaction (11-14) is reduced to ethanol in reaction (11-15):

$$CH_3CHO + NADH^* + H^+ \rightleftharpoons CH_3\text{—}\overset{H^*}{\underset{H}{C}}OH + NAD^+ \qquad (11\text{-}15)$$

The enzyme responsible is called *alcohol dehydrogenase.* When isolated from yeast, it has a molecular weight of 151,000 and four independent catalytic sites, while an analogous enzyme with approximately half the molecular weight and two binding sites has been obtained from mammalian liver. Zn^{2+} is also present in amounts stoichiometric to the number of catalytic sites in these enzymes, which are stereospecific with regard to hydrogen transfer, both for the alcohol and for the *A* position of the 4-hydrogen in the nicotinamide ring (see page 413), but are relatively nonspecific with regard to structural features of the substrate alcohols and carbonyl compounds.

Alternative Glycolytic and Fermentative Pathways

The EMP pathway certainly provides the major route of glucose catabolism in most cells, with the exception of certain bacterial species. Several variants can be obtained with yeast cells, depending on the conditions employed.

The *first form* is the one already described and is the one obtained with yeast growing and multiplying under anaerobic conditions.

The *second form* leads to the formation of glycerol: it requires the trapping of the acetaldehyde formed by a carbonyl reagent, such as bisulfite, substituted hydrazines, and so on. Under these conditions, reduction to ethanol does not take place, and 1 mol of triose-phosphate is diverted to glycerol production by virtue of the following reactions:

$$\text{Dihydroxyacetone-}P + \text{NADH} \rightleftharpoons \begin{array}{c} CH_2OH \\ | \\ HOC\text{—}H \\ | \\ CH_2OPO_3^{2-} \end{array} + \text{NAD}^+ \qquad (11\text{-}16)$$

$$\begin{array}{c} \text{L-}\alpha\text{-Glycerol-}P\ (\text{L-glycerol-3-}P)^{14} \\ H_2O \downarrow \qquad \uparrow ADP \cdot Mg \leftarrow ATP \cdot Mg \\ P_i + \text{glycerol} \end{array} \qquad (11\text{-}17)\ \ (11\text{-}17a)$$

Reaction (11-16) is catalyzed by a soluble (cytoplasmic) enzyme called L-glycerol-3-*P*-dehydrogenase and reaction (11-17) by a specific phosphatase. Its reversal requires, as always, an ATP-dependent kinase (phosphotransferase) reaction [reaction (11-17a)]. It will be seen that in this sequence the NADH produced in reaction (11-8) is consumed in reaction (11-16); one triose molecule each traverses both branches—(11-16) plus (11-17) versus (11-8) through (11-14)—in a coordinated fashion, and there is no net accumulation or consumption of any of the coreactants, including ATP. Thus the net stoichiometric equation for this form of fermentation is given by the equation shown at the bottom of Fig. 11-2.

Reactions (11-16) and (11-17), in conjunction with those described as responsible for generating the triose-*P*, are also the ones concerned with the production of important intermediates of lipid metabolism (see Chapter 16) from carbohydrate precursors. Insect flight muscle is a tissue that does not produce lactate from glucose. Here the enzyme catalyzing reaction (11-16) works in conjunction with a different L-α-glycerol phosphate dehydrogenase, a flavoprotein localized in the mitochondria of this tissue, which catalyzes reaction (11-16b).

(*Extramitochondrial*)	dihydroxyacetone-P + H^+ + NADH $\rightleftharpoons$ NAD^+ + α-glycerol-P	(11-16a)
(*Mitochondrial*)	α-glycerol-P + E-FAD $\rightleftharpoons$ dihydroxyacetone-P + E-$FADH_2$	(11-16b)
	E-$FADH_2$ + O_2 + 2ADP + $2P_i$ $\rightarrow$ E-FAD + H_2O + 2ATP	(11-16c)
Net (*cell*)	NADH + O_2 + 2ADP + $2P_i$ + H^+ $\rightarrow$ NAD^+ + H_2O + 2ATP	(11-16d)

This sequence of reactions is of importance as a *shuttle* for NADH oxidation also in a number of other tissues, such as brain and smooth muscle.

The *third form* of yeast fermentation takes place in an alkaline medium. Under these conditions the two branches of the EMP scheme again become closely coordinated, This leaves no net NADH for the production of ethanol from acetaldehyde. Instead, another reaction must participate in the removal of this extremely toxic metabolite. This is reaction (11-18a):

$$CH_3CHO + NAD^+ + OH^- \rightleftharpoons CH_3CO_2^- + NADH + H^+ \qquad (11\text{-}18a)$$

catalyzed by an NAD-dependent *aldehyde dehydrogenase.* A similar enzyme is found in mammalian liver. The set of reactions yields the overall balance given in Fig. 11-2.

In bacteria, a variant of reaction (11-18a) involving CoASH can take place—reaction (11-18b). Since acetyl-*S*-CoA as an acyl thioester has a high $-\Delta G^{0\prime}$ of hydrolysis ($-$ 7.7 kcal/mol), it can be equilibrated with the ATP system by reactions (11-19) and (11-20), catalyzed by *phosphotransacetylase* and *acetate kinase.*

[14] The correct designation (38) of this compound is *sn*- (for *s*tereochemically *n*umbered)-glycerol-3 phosphate.

$$CH_3CHO + NAD^+ + CoASH \rightleftharpoons CH_3COSCoA + NADH + H^+ \qquad (11\text{-}18b)$$

$$CH_3COSCoA + P_i \rightleftharpoons CH_3COP + CoASH \qquad (11\text{-}19)$$

$$CH_3COP + ADP \rightleftharpoons CH_3CO_2^- + ATP \qquad (11\text{-}20)$$

Net $$CH_3CHO + NAD^+ + ADP + P_i \rightleftharpoons CH_3CO_2^- + ATP + NADH$$
$$\Delta G^{0\prime} = -5.5 \text{ kcal/mol} \qquad (11\text{-}21)$$

The net balance for this type of "hetero-" fermentation is

$$2 \text{ Glucose} + ADP + P_i \rightarrow 2 \text{ glycerol} + \text{ethanol} + \text{acetate} + 2CO_2 + ATP$$

ALTERNATIVE FATES OF PYRUVATE

The production of pyruvate from hexoses or hexose units is the common characteristic reaction of all cells utilizing the EMP scheme for carbohydrate catabolism and ATP production. To recapitulate, the stoichiometry observed is either

$$C_6H_{12}O_6 + 2NAD^+ + 2ADP + 2P_i \rightarrow 2CH_3COCO_2^- + 2NADH + 2ATP$$
$$\Delta G^{0\prime} = -18 \text{ kcal/mol} \qquad (11\text{-}22a)$$

or

$$[C_6H_{11}O_5] + 2NAD^+ + 3ADP + 3P_i \rightarrow 2CH_3COCO_2^- + 2NADH + 3ATP$$
$$\Delta G^{0\prime} = -15 \text{ kcal/mol} \qquad (11\text{-}22b)$$

where $[C_6H_{11}O_5]$ is a hexose in glycosidic linkage capable of yielding glucose-1-*P* on phosphorolysis. Different pathways diverge from, and are distinguished only by, the disposition that they assign to the NADH and pyruvate produced (Fig. 11-4).

Other Fermentative Pathways in Microorganisms

Reactions Involving "Active Acetaldehyde." The 2-hydroxyethyl derivative of TPP—structure (11-14 [*C*])—is exceedingly reactive and can be regarded as the metabolically "active" form of acetaldehyde. One typical reaction of this type is an acyloin condensation, the *carboligase reaction* [reaction (11-23)] catalyzed by enzymes from higher plants and yeast, which appear to be either exceedingly closely associated or identical with the decarboxylase itself:

$$CH_3COCO_2^- + H^+ + E\cdot TPP \xrightarrow{Mg^{2+}} CO_2 + E\cdot TPP\text{—}CHOHCH_3 \qquad (11\text{-}14a)$$

$$E\cdot TPP\text{—}CHOHCH_3 \rightleftharpoons E\cdot TPP + CH_3CHO \qquad (11\text{-}14b)$$

$$E\cdot TPP\text{—}CHOHCH_3 + CH_3CHO \rightleftharpoons E\cdot TPP + CH_3CHOHCOCH_3 \qquad (11\text{-}23)$$

$$CH_3COCO_2^- + CH_3CHO + H^+ \xrightarrow[Mg^{2+}]{E\cdot TPP} CH_3CHOHCOCH_3 + CO_2$$

[Eqs. (11-14a) plus (11-23)] (11-24)

Although with the plant enzymes, reaction (11-14b) is readily reversible, with those from certain bacteria or with the isolated pyruvic dehydrogenase complex (see page 519) free acetaldehyde is not produced, and this compound as the TPP derivative acts only as a carbonyl acceptor, probably by producing the acetoin adduct of TPP as the first product.[15] Acetoin, in the

[15] In certain enteric bacteria the enzymes responsible catalyze yet another variant:

$$E\cdot TTP\text{—}CHOHCH_3 + CH_3COCO_2^- \rightleftharpoons E\cdot TTP + CH_3COC(OH)CH_3 \text{ (acetolactate)}$$
$$\qquad\qquad | \\ \qquad\qquad CO_2^-$$

or 2 pyruvate $\rightarrow CO_2$ + acetolactate

$$\underset{\underset{CO_2^-}{|}}{CH_3COC(OH)CH_3} \xrightarrow[\text{decarboxylase}]{\text{acetolactate}} CO_2 + CH_3CHOHCOCH_3$$

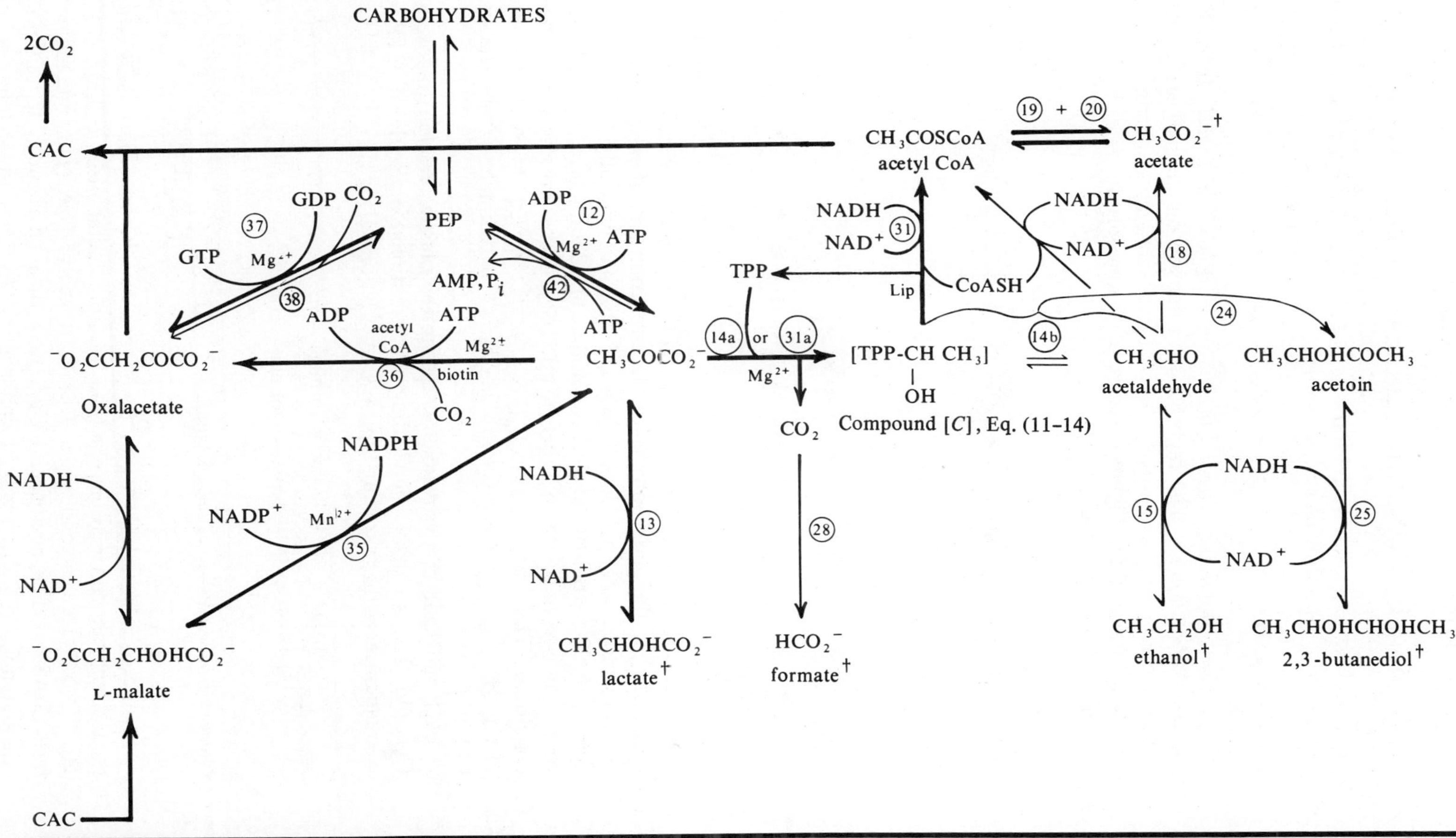

FIGURE 11-4
Metabolic fates of pyruvate. Predominant metabolic routes in mammals shown by heavy arrows.
† Compounds produced in microbial fermentations.

presence of an appropriate dehydrogenase, can act as an alternative electron acceptor and provides a means for the reoxidation of NADH by virtue of reaction (11-25):

$$CH_3COCHOHCH_3 + NADH + H^+ \rightleftharpoons CH_3CHOHCHOHCH_3 + NAD^+ \qquad (11\text{-}25)$$

A typical fermentation by certain bacilli and involving the reactions described would show the following stoichiometry:

$$\text{Glucose} + 4ADP + 4P_i \rightarrow 2\ 2,3\text{-butanediol} + 2\ \text{glycerol} + 4CO_2 + 4ATP \qquad (11\text{-}26)$$

Reactions Involving Direct Formation of Acetyl CoA and Acetyl-P (The "Clastic" Reaction). The high electron density on the carbons attached to the thiazolium ring in hydroxyalkyl TPP derivatives (structures 11-14 [*B*] and [*C*]) makes them highly reactive sites, not only for substitution, but for oxidation reactions as well:

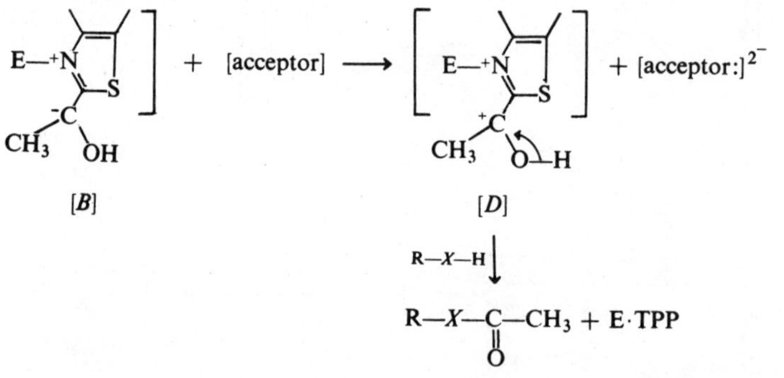

[*B*] + [acceptor] ⟶ [*D*] + $[\text{acceptor:}]^{2-}$

[*D*] $\xrightarrow{R—X—H}$ R—*X*—C(=O)—CH_3 + E·TPP (11-27)

Most bacterial cells use CoASH (R—*X*H = CoA—*S*H) as the acetyl acceptor. The product obtained is acetyl CoA, which can be either converted to acetyl-*P*, in the phosphotransacetylase reaction, and then to acetate + ATP [reactions (11-19) + (11-20)] or, as the metabolically "active" form of acetate itself, participate in a large number of additional reactions. This is the route followed in a large number of microorganisms growing under anaerobic conditions. The second product, $[\text{acceptor:}]^{2-}$ of Eq. (11-27), is either H_2 (from H^+ in, e.g., *Clostridium butylicum*) or formate (from the evolved CO_2 in *E. coli*).

Each molecule of pyruvate undergoing the clastic reaction to formate, or CO_2 plus H_2, is capable of yielding one additional molecule of ATP. Thus, we may observe the following typical fermentation balances:

$$\text{Glucose} + \begin{cases} 3ADP + 3P_i \rightarrow \text{lactate} + \text{formate} + \text{acetate} + 3ATP & (11\text{-}28a) \\ 2ADP + 2P_i \rightarrow \text{lactate} + \text{formate} + (\text{acetyl}) + 2ATP & (11\text{-}28b) \end{cases}$$

$$\text{Glucose} + {<}2ADP + {<}2P_i \rightarrow 2,3\text{-butanediol} + CO_2 + \begin{cases} \text{formate} & (11\text{-}29a) \\ CO_2 + H_2 & (11\text{-}29b) \end{cases} + {<}2ATP$$

In microbes acetyl CoA can undergo a variety of conversions, most of them identical with certain reactions of lipid metabolism (Chapter 13 and 16). Two molecules condense to form acetoacetyl CoA (plus one molecule of free CoASH). The acetoacetyl CoA may then be reduced to butyryl CoA and butanol, or alternatively it can transfer its CoA to acetate and be decarboxylated to acetone. The ketone so formed can then function as a hydrogen acceptor for NADH in the presence of alcohol dehydrogenase and produce isopropanol.

NAD-Lipoate-Linked Pyruvate Dehydrogenase Systems. The predominant fate of pyruvate in most animal cells and in aerobic microorganisms is its oxidation to acetyl CoA mediated by the multienzyme pyruvate dehydrogenase complexes associated with particulate cellular elements—the mitochondria of animal and plant cells, the cell membranes of bacteria (see Chapter 9). These complexes have been isolated in pure form and characterized as multicomponent aggregates of definite composition, with $M = 10 \times 10^6$ (pig heart) and 4.8×10^6 (*E. coli*). The latter (39) has been separated into its components, each exhibiting a separate set of enzymatic activities [Eq. (11-30)]: (1) 12 molecules of pyruvate decarboxylase (E_a) ($M = 183{,}000$, containing stoichiometric amounts of enzyme-bound TPP); (2) one central aggregate of lipoate (lipoamide) reductase-transacetylase (now called *dihydrolipoate transacetylase* or *acetyltransferase*) (E_b) [$M = 1.0 \times 10^6$, consisting of 24 subunits each of $M = 36{,}000$ and containing stoichiometric amounts of lipoyl residues (see page 391) linked in amide linkage to ε-amino groups of lysine residues on the protein]; and (3) six molecules of *dihydrolipoamide* (also lipoate or lipoyl) *dehydrogenase* (E_c) ($M = 112{,}000$, containing stoichiometric quantities of enzyme-bound FAD, 2 per molecule). On electron micrographs, the complex exhibits a distinct polyhedral appearance with a diameter of 350 ± 50 Å and a height of 225 ± 25 Å (see page 174).

The reactions catalyzed are the following—note that (11-30a) is identical to (11-14a) and is catalyzed by the purified decarboxylase:

$$CH_3COCO_2^- + E_a\cdot TPP \xrightleftharpoons{Mg^{2+}} E_a\cdot TPP\text{—}CHOH\text{—}CH_3 + CO_2 \qquad (11\text{-}30a)$$

$$E_a\cdot TPP\text{—}\underset{H}{\overset{OH}{C}}\text{—}CH_3 + E_b\cdot NH\text{—}CO\text{—}(CH_2)_4\text{—}\underset{S\text{—}S}{\bigcirc} \rightarrow E_a\cdot TPP + E_b\cdot NHCO(CH_2)_4\text{—}\underset{CH_3COS\quad SH}{\bigcirc} \qquad (11\text{-}30b\text{-}1)^{16}$$

$$E_b\cdot NHCO(CH_2)_4\text{—}\underset{CH_3COS\quad SH}{\bigcirc} + CoASH \rightleftharpoons E_b\cdot NHCO(CH_2)_4\text{—}\underset{SH\quad SH}{\bigcirc} + CH_3COSCoA \qquad (11\text{-}30b\text{-}2)$$

$$E_b\cdot NH\text{—}CO(CH_2)_4\text{—}\underset{SH\quad SH}{\bigcirc} + E_c\cdot FAD \rightleftharpoons E_b\cdot NH\text{—}CO(CH_2)_4\text{—}\underset{S\text{—}S}{\bigcirc} + E_c\cdot FADH_2 \qquad (11\text{-}30c\text{-}1)$$

$$E_c\cdot FADH_2 + NAD^+ \rightleftharpoons E_c\cdot FAD + NADH + H^+ \qquad (11\text{-}30c\text{-}2)$$

$$\text{Net}\quad CH_3COCO_2^- + NAD^+ + CoASH \xrightarrow[TPP,\,lip\langle^{S}_{S},\ Mg^{2+}]{PDC(E_{a,b,c})} CoASCOCH_3 + NADH + CO_2$$

$$\Delta G^{0\prime} = -9.38\ \text{kcal/mol} \qquad (11\text{-}31)$$

Or with O_2,

$$CH_3COCO_2^- + \tfrac{1}{2}O_2 + CoASH + H^+ \xrightarrow{\text{as above, plus electron transport}} CoASCOCH_3 + H_2O + CO_2 \qquad \Delta G^{0\prime} = -61.8\ \text{kcal/mol} \qquad (11\text{-}32)$$

$$CH_3COCO_2^- + \tfrac{1}{2}O_2 + CoASH + 3ADP + 3P_i \xrightarrow{\text{as above, plus oxidative phosph.}} CoASCOCH_3 + 2H_2O + CO_2 + 3ATP \qquad \Delta G^{0\prime} = -39.3\ \text{kcal/mol} \qquad (11\text{-}33)$$

[16] In this reaction oxidation of hydroxyethyl TPP by lipoamide to the 2-acetyl derivative may actually precede the transfer of the latter to dihydrolipoate—note the analogy to reaction (11-27).

The major pathway for the disposition of the acetyl CoA produced in the oxidation of pyruvate is its complete oxidation to CO_2, an event which takes place in the mitochondria and which, as we shall see in Chapters 14 and 15, generates 12 molecules of ATP per molecule of $CH_3COSCoA$ (15 per pyruvate) or twice that number per molecule of glucose. If we assume that the 4 molecules of NADH generated during catabolism of glucose—2 in reaction (11-8) and 2 in reaction (11-31)—are oxidized with concomitant phosphorylation [see reaction (11-16d) for a means of shuttling electrons from extramitochondrial NADH into the mitochondria], this will provide for a maximum of 12 additional molecules of ATP per molecule of glucose. Thus, for the complete conversion, we see

$$C_6H_{12}O_6 + 6O_2 + \leq 38ADP + \leq 38P_i \rightarrow 6CO_2 + 6H_2O + \leq 38ATP \qquad (11\text{-}34)$$

For each glucose unit in glycogen the value is 38 molecules of ATP produced.

CARBOXYLATION REACTIONS

The continued oxidation of acetyl CoA by means of the CAC requires the simultaneous presence of oxalacetate. This is ordinarily provided for by the cyclical nature of the process, but it also means that, if there were any drain on the cycle or its members for synthetic processes—and there are many, exemplified by the biosynthesis of amino acids, purines, pyrimidines (precursors for nucleic acids and coenzymes), porphyrins, etc.—a means must be found for its replenishment. These anaplerotic sequences are provided for by carboxylation reactions, which interconvert pyruvate and the dicarboxylic acids of the cycle. The following reactions constitute the major potential contributors to these $C_3 + C_1 = C_4$ conversions:

$$CH_3COCO_2^- + NAD(P)H + HCO_3^- \xrightleftharpoons{Mn^{+2}} {}^-O_2CCH_2\underset{\substack{|\\ OH}}{C}HCO_2^- + NAD(P)^+$$
(L-malate) $\Delta G^{0\prime} = 0.36$ kcal/mol (11-35)

$$CH_3COCO_2^- + Mg \cdot ATP^{2-} + HCO_3^- \xrightleftharpoons[Mg^{2+}]{Mn\text{-}E\text{-}biotin,\ acetyl\ CoA} {}^-O_2CCH_2COCO_2^- + Mg{\cdot}ADP^- + P_i^-$$
$\Delta G'^{0} \simeq -0.5$ kcal/mol (11-36)

$$CH_2{=}\underset{\substack{|\\ OPO_3^{2-}}}{C}{-}CO_2^- + GDP(IDP) + HCO_3^- \xrightleftharpoons{Mg^{2+}} O_2CCH_2COCO_2^- + GTP(ITP)$$
$\Delta G^{0\prime} = -1$ kcal/mol (11-37)

$$CH_2{=}\underset{\substack{|\\ OPO_3^{2-}}}{C}{-}CO_2^- + P_i^{2-} + HCO_3^- + H^+ \xrightleftharpoons{Mg^{2+}} {}^-O_2CCH_2COCO_2^- + PP_i^{3-} + H_2O$$
$\Delta G^{0\prime} \simeq -2$ kcal/mol (11-38)

$$CH_2{=}\underset{\substack{|\\ OPO_3^{2-}}}{C}{-}CO_2^- + HCO_3^- \xrightarrow{Mg^{2+}} {}^-O_2CCH_2COCO_2^- + P_i^-$$
$\Delta G^{0\prime} = -6.8$ kcal/mol (11-39)

Reaction (11-35) is catalyzed by *malic enzyme* (L-malate: NAD[P] oxidoreductase, decarboxylating), which enjoys a very wide distribution and is usually specific for $NADP^+$, especially in higher organisms. At one time it was believed that this reaction, in conjunction with that catalyzed by the mitochondrial L-malate dehydrogenase (which brings about the reversible oxidation of malate to oxalacetate by means of NAD^+), followed by the reoxidation of the NADH produced by means of O_2, provided the main entry port for dicarboxylic acid and carbohydrate synthesis. However, in view of its very low affinity for CO_2 or HCO_3^-, the relative unavailability of NADPH in mitochondria, and the properties of pyruvate carboxylase to be described, it is now thought to function mainly in the reverse direction, i.e., to provide for the oxidation of malate for conversion to pyruvate, acetyl CoA, and other metabolites related to them, and to supply NADPH. The mechanism of the reaction is similar to that of L_s-isocitrate dehydrogenase, which is discussed on page 610.

Reaction (11-36) is catalyzed by the *ATP-dependent pyruvate carboxylase* [pyruvate: CO_2 ligase (ADP)]. This enzyme was originally discovered by Utter and Keech in chicken liver mitochondria, highly purified from this source and then extensively studied by Utter and his collaborators (40). It provides an exceedingly interesting example of complex molecular, catalytic, and regulatory interactions in accord with its function as the enzyme responsible for catalyzing the first step in the important anabolic sequence, of glucose synthesis or *gluco(neo)genesis.* It is found in high amounts in animal tissues capable of performing this task[17] (e.g., vertebrate liver and kidney) and absent or present in very low amounts in those that are not (e.g., skeletal muscle, heart, brain). Similar enzymes have also been isolated from unicullular plants (yeast and *Aspergillus niger*) and a number of bacteria (*Pseudomonas citronellolis, Arthrobacter globiformis,* and *Bacillus coagulans*). These microbial enzymes vary in their similarity to the animal enzymes, with regard to presumed major function (i.e., anaplerotic CO_2 fixation versus gluconeogenesis) and particularly with regard to regulatory features.[18] We shall restrict ourselves in what follows to the animal enzymes (avian liver, sheep kidney). The active enzyme contains four molecules of biotin and four Mn^{2+} ions per unit of $M = 655{,}000$. It is cold labile and by this and other treatments can be shown to be capable of reversible dissociation into protomers which can be further resolved, irreversibly, into four subunits each. It catalyzes a number of partial reactions (pyruvate $\rightleftharpoons$ oxalacetate exchange, ADP $\rightleftharpoons$ ATP exchange, $P_i \rightleftharpoons$ ATP exchange) of which reactions (11-36a) and (11-36b) are the most important, because they, acting in sequence, can account for the overall reaction.

$$\text{E-biotin} + \text{MgATP} + HCO_3^- \xrightleftharpoons{\text{acetyl CoA, }(Mg^{2+})} \text{E-biotin}{\sim}CO_2 + \text{ADP} + P_i \quad \textit{formation of carboxybiotin} \tag{11-36a}$$

$$\text{E-biotin}{\sim}CO_2 + \text{pyruvate} \rightleftharpoons \text{E-biotin} + \text{oxalacetate} \quad \textit{transcarboxylation} \tag{11-36b}$$

Reaction (11-36a) is the carboxylation of enzyme-bound biotin, to form carboxybiotin, the reactive intermediate (see Chapter 8). It takes place at an enzyme site not requiring manganous ion; MgATP functions as a homotropic allosteric activator; free Mg^{2+}, and in particular acetyl CoA, as heterotropic ones. The latter effect is so potent that even at high MgATP concentrations the rate in the absence is only ≤ 0.04 percent(!) that in the presence of acetyl CoA. Other monocarboxyl derivatives of CoA substitute: V for formyl < acetyl ≧ propionyl > crotonyl > isobutyryl ≫ *n*-butyryl > *n*-valeryl, all with approximately equal $\bar{K} \simeq 2 \times 10^{-4}$ M except acetyl, where $\bar{K} \simeq 3.3 \times 10^{-5}$ M and $\bar{n}$, the apparent Hill coefficient (page 308), for this

[17] As already mentioned, its localization is exclusively mitochondrial in avian liver, sheep kidney cortex, and rat liver.

[18] The *Arthrobacter* enzyme is quite similar; the enzyme from yeast is activated by acetyl CoA, but much less strongly and specifically, and exhibits a K^+ requirement as well as inhibition by L-aspartate, while the *Pseudomonas* and *Aspergillus* enzymes appear devoid of regulatory features. In the case of the thermophilic *Bacillus coagulans* it is the formation of the active enzyme from the apoprotein plus biotin that appears to be the regulated step [dependent on acetyl CoA, inhibited by aspartate (42)].

substrate $\geqq 2.9$. Malonyl CoA is a potent inhibitor ($K_i = 8.3 \times 10^{-6}$ M) as are other charged analogs such as acetyl mercaptosuccinate ($K_i = 7.4 \times 10^{-5}$ M). Acetyl CoA protects the enzyme against denaturation by preventing dissociation into protomers or subunits, but facilitates its inactivation by *avidin*[19] at low concentrations and blocks this inactivation at high concentrations. These observations provide strong evidence for assigning an allosteric regulatory function to acetyl CoA.

In reaction (11-36b), the transcarboxylation reaction proper, the enzyme catalyzes the reversible transfer of CO_2 between biotin and pyruvate; the Mn^{2+} site, but none of the other components is involved; in this step oxalate is an inhibitor competitive to oxalacetate (and uncompetitive to pyruvate). These and a number of other observations (40,41) have been utilized to formulate the overview of the relevant interconversions summarized in Fig. 11-5.

The enzyme catalyzing reaction (11-37), *phosphoenolpyruvate carboxykinase* (GTP : oxalacetate carboxylase), was also discovered in Utter's laboratory. It shares with the carboxylase just described its localization with respect to tissue distribution[20] and concentration. GTP (or ITP), the primary product of phosphorylation attendant to the oxidation of α-oxoglutarate in mitochondria (Chapter 14), is the energy donor for the reaction. In other cells, particularly of microorganisms, ATP can function in the analogous reaction. There is some evidence that NADH acts as an inhibitor in these systems. In any event, the various nucleoside di- and triphosphates can be equilibrated by virtue of the reaction catalyzed by nucleoside diphosphate kinase (transphosphorylase): ATP + NDP $\rightleftharpoons$ ADP + NTP.

The function of the carboxylase and carboxykinase acting in concert is to provide the principal pathway for the synthesis of phosphoenolpyruvate, the precursor of carbohydrate for those organisms that contain both enzymes (see the next section):

$$\text{Pyruvate} + \underbrace{\text{ATP} + \text{GTP}}_{2\text{ATP}} \xrightarrow{CO_2} \text{phosphoenolpyruvate} + \underbrace{\text{GDP} + \text{ADP}}_{2\text{ADP}} + P_i$$

$$\Delta G' = -2.3 \text{ kcal } (5 \times 10^{-4} \text{ M } Mg^{2+}, \mu = 0.1, pH\ 8.0) \qquad (11\text{-}40)$$

It will be observed that CO_2 (or HCO_3^-) functions catalytically in this sequence. Those organisms that lack the carboxylase but contain the carboxykinase use different means for bringing about the carboxylation and use reaction (11-37) principally for converting CAC members to PEP (see below).

Reaction (11-37), reaction (11-38), catalyzed by phosphoenolpyruvate *carboxytransphosphorylase* (or pyrophosphorylase; oxalacetate carboxylase), and the essentially irreversible reaction (11-39), catalyzed by *phosphoenolpyruvic carboxylase* (an enzyme present in green plants and Enterobacteriaceae), all may involve oxalacetyl phosphate or an entity derived therefrom:

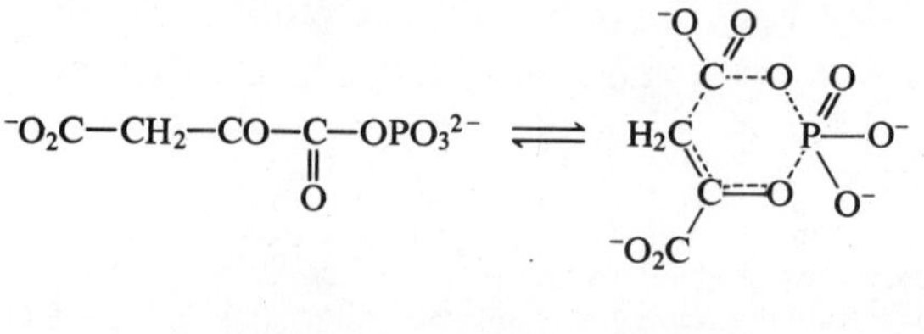

$$^{-}O_2C{-}CH_2{-}CO{-}\underset{\underset{O}{\|}}{C}{-}OPO_3^{2-} \rightleftharpoons \text{cyclic form: } H_2C \text{ bridging } C(=O)O^{-} \text{ and } C{=}O,\ \text{with } {-}O{-}P(=O)(O^{-})_2 \text{ coordinated; } {}^{-}O_2C \text{ substituent}$$

[19] A proteinaceous antagonist to biotin isolated from egg white.

[20] Their subcellular distribution is also identical (mitochondrial), at least in liver and kidney.

(A) Enzyme, active (*R*) form here as the $[R \cdot ATP]_4$ complex

(B) $[E(\sim CO_2)]_4$

(C) $[E \cdot (\sim CO_2 \text{ pyruvate})]_4$

(D) $[E \cdot (\text{oxalacetate})]_4$

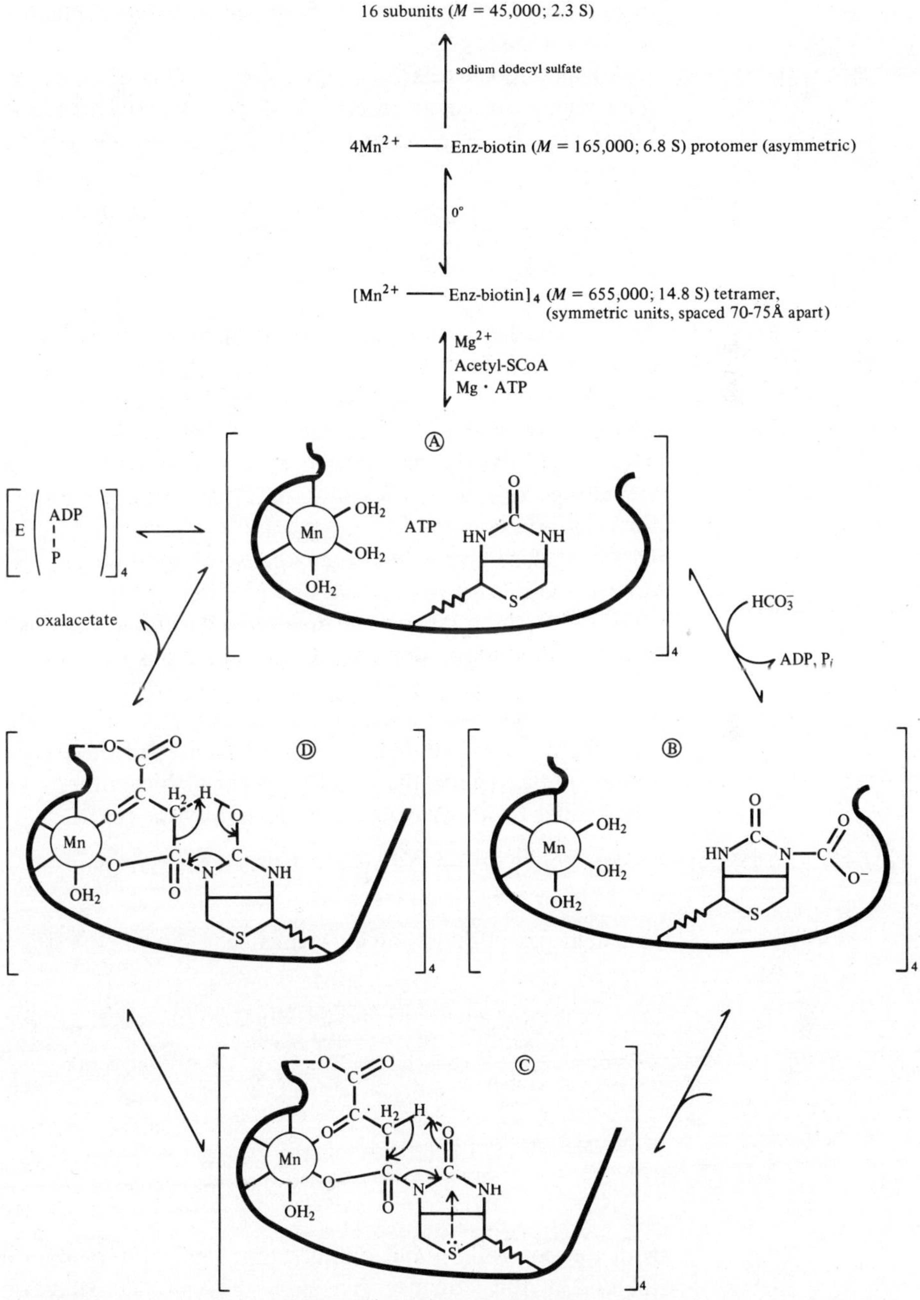

16 subunits (M = 45,000; 2.3 S)

sodium dodecyl sulfate

$4Mn^{2+}$ — Enz-biotin (M = 165,000; 6.8 S) protomer (asymmetric)

0°

$[Mn^{2+}$ — Enz-biotin$]_4$ (M = 655,000; 14.8 S) tetramer, (symmetric units, spaced 70-75Å apart)

Mg^{2+}
Acetyl-SCoA
Mg · ATP

Ⓐ Mn, OH_2, OH_2, OH_2, ATP, HN, C, O, NH, S

$[E(ADP/P)]_4$

oxalacetate

HCO_3^-

ADP, P_i

Ⓓ Mn, OH_2, O, C, H_2, H, N, NH, S

Ⓑ Mn, OH_2, OH_2, OH_2, HN, C, O, N, O^-, S

Ⓒ Mn, OH_2, O, C, H_2, H, N, NH, S

FIGURE 11-5
Mechanisms of action and molecular interconversion of pyruvate carboxylase, based on the work of Utter and his coworkers (40); probably applicable to enzymes from liver and kidney of vertebrates, but not from microbial sources.

as an unstable intermediate or transition state. This intermediate may then be hydrolyzed [in reaction (11-39)] or transfer its phosphate [to GDP in reaction (11-37) or P_i (in reaction 11-38)]. The function of the enzyme catalyzing reaction (11-39) is probably principally an anaplerotic one, and it plays no role in PEP formation: for example, *E. coli* mutants lacking it are capable of growth on succinate but not on glucose; the converse holds for mutants deficient in the enzyme catalyzing (11-37). The PEP-carboxylase in this organism (but not in *Ps. fluorescens* or green plants) is strongly activated by acetyl CoA.

An analogous role in anaplerotic CO_2 fixation has also been assigned to the enzyme catalyzing reaction (11-38), largely restricted to *Propionibacteria*. It can also catalyze reaction (11-38a) and hence may fulfill some function in glycolysis as well:

$$CH_2{=}\underset{\displaystyle OPO_3{}^{2-}}{\underset{|}{C}}{-}CO_2{}^- + P_i{}^{2-} + H^+ \rightleftharpoons \text{pyruvate}^- + PP_i{}^{3-} \qquad (11\text{-}38a)$$

SYNTHESIS OF GLUCOSE FROM PYRUVATE [GLUCO (NEO) GENESIS]

If a metabolic energy supply is available, mammalian liver and kidney can synthesize glucose, and hence pentoses, glycogen, and other polysaccharides from short-chain precursors. Among the latter are (1) pyruvate or lactate, (2) the glycogenic amino acids (see Table 17-2), and (3) any other component that can be catabolized to pyruvate or to any member of the CAC. In resting skeletal muscle (but not in heart or smooth muscle) phosphorylated three-carbon compounds, most notably α-glycerophosphate, are similarly converted back to glycogen at the expense of creatine-phosphate, the storage form of metabolic energy in that tissue. This resynthesis of glucose from other sources is essential for the maintenance of the organism and its tissues and supplies the energy demands of certain types of cells, especially those in the nerve tissues and the blood of animals. In an analogous fashion, hexoses, pentoses, and structural polysaccharides can be synthesized by microorganisms from carbon sources containing four- and three-carbon atoms. The most interesting problem, then, is the mode of biosynthesis of glucose- or hexose-phosphates from pyruvate, summarized by the reaction

$$2CH_3COCO_2{}^- + 2NADH + 6ATP + 4H^+ \longrightarrow C_6H_{12}O_6 + 2NAD^+ + 6ADP + 6P_i \qquad \Delta G^{0\prime} = 7\ \text{kcal/mol} \qquad (11\text{-}41)$$

It is made up of the following sequence:

$$\begin{array}{l} \qquad\qquad CO_2 \qquad\qquad \text{malate} \overset{H_2O}{\rightleftharpoons} \text{fumarate} \leftarrow \text{CAC} \\ \qquad\qquad\qquad\qquad\qquad \upharpoonleft\downharpoonright \\ 2(\text{Pyruvate} \xrightarrow[(11\text{-}36)]{ATP} \text{oxalacetate} \xrightarrow[(11\text{-}38)]{\text{GTP or ATP}} \text{PEP} \rightarrow P\text{-glycerate} \xrightarrow[ATP]{NADH} \text{triose-}P) \rightarrow \\ \qquad\qquad\qquad\qquad\qquad\qquad\qquad CO_2 \end{array}$$

$$\text{hexose-di}P \xrightarrow[(11\text{-}5a)]{H_2O} \text{hexose-}P \xrightarrow[(11\text{-}1a)]{H_2O} \text{hexose} \qquad (11\text{-}42)$$

with the controlling and unique steps again intervening at the beginning and the end of the sequence. Those at the end (i.e., the two phosphatases), for fructose-1,6-di*P* [reaction (11-5a)] and for hexose-*P* [reaction (11-1a)], have already been discussed in conjunction with the formation of these two compounds. Those at the beginning consist of pyruvate carboxylase (11-36) and carboxykinase (11-38) working in tandem, as described in the last section. That this is the major pathway of hexose biosynthesis in animals is indicated by the fact that, if pyruvate labeled in either the 2 or 3 position is administered to a starved animal, the glycogen, produced after synthesis has been allowed to proceed for a while, will contain hexose units that show equilibration of the label in the 1, 2, 5, and 6 positions. This suggests that a symmetrical precursor must have intervened, a process that probably involved the interconversions of the dicarboxylic acids indicated above and discussed on page 623. Added support for this hypothesis is provided, as pointed out by Scrutton and Utter (see the references), from considerations of tissue distribution and catalytic capacities of the various enzymes of glycolysis and gluconeogenesis. Quite a different situation obtains in certain plants and those microorganisms that do not contain one or both of the tandem enzymes catalyzing reactions (11-36) and (11-38) yet are able to utilize three-carbon precursors. Two alternatives appear to exist. For tropical grasses and microorganisms such as *E. coli,* reaction (11-36) is absent. PEP formation proceeds by means of the reaction catalyzed by *phosphopyruvate* (PEP) (*or pyruvate-phosphate*) *synth*(*et*)*ase* (43):

$$\text{Pyruvate} + \text{MgATP} \longrightarrow \text{PEP} + \text{AMP} + \text{P}_i + \text{Mg}^{2+} \qquad (11\text{-}43)$$

Precursors feeding into the CAC are converted to PEP by means of reaction (11-37). In propionic acid and other bacteria (44), most plants (45), and *Entamoeba hystolytica* (where however its main function may be action in the reverse direction) (46), the reaction required is the one catalyzed by *pyruvate-phosphate dikinase,* Eq. (11-44):

$$\text{Enz} + \text{MgATP} \rightleftharpoons [\text{Enz-P}^{\beta}\text{P}^{\gamma}(\text{Mg?})] + \text{AMP} \qquad (11\text{-}44a)$$

$$[\text{Enz-P}^{\beta}\text{P}^{\gamma}] + \text{P}_i \rightleftharpoons \text{Enz-P}^{\beta} + \text{P}^{\gamma}\text{P}_i(\text{Mg}) \qquad (11\text{-}44b)$$

$$\text{Enz-P}^{\beta} + \text{pyruvate} \rightleftharpoons \text{Enz} + \text{P}^{\beta}\text{-enolpyruvate} \qquad (11\text{-}44c)$$

$$\textit{Net}\ \text{Pyruvate} + \text{MgATP} + \text{P}_i \rightleftharpoons \text{P-enolpyruvate} + \text{AMP} + \text{MgPP}_i \qquad (11\text{-}44)$$

The formation of a phosphoryl enzyme—which may be the only stable intermediate—is common to the mechanism of this enzyme and of the PEP-synthase [Eq. (11-43)] as well (43).

Control of Glucose Metabolism

It may be useful at this point to reconsider the factors in animal cells that contribute most heavily to a control of glucose (and glycogen) breakdown and resynthesis and thus affect the steady state of these compounds. In general, we can say that in all cells that are capable of degrading glucose both in the presence and the absence of oxygen, the sugar disappears (and of course lactic acid or other products of anaerobic glycolysis or fermentation are formed) more rapidly under anaerobic than under aerobic conditions. This inhibition of glycolysis by oxygen was first recognized by Pasteur and later confirmed by Meyerhof and Warburg and is known as the *Pasteur effect.* The converse also holds: Oxygen consumption at the expense of readily oxidizable substrates is inhibited by the addition of glucose—this is known as the *Crabtree effect.* Another finding, first made by A. V. Hill with muscle, but true of other tissues and cells as well, is that the resynthesis of glycogen and carbohydrates in general proceeds more rapidly under aerobic than under anaerobic conditions.

From our considerations so far we know that aerobiosis removes P_i and ADP, generates ATP, removes NADH, and produces CO_2, acetyl CoA, and di- and tricarboxylic acids a great deal more effectively than do anaerobic conditions (for further details see Chapter 14). From the overall equations [reactions (11-22), and (11-42)], and the reaction sequences responsible, we can appreciate how this set of conditions affects the overall conversions glucose (or glycogen) → pyruvate → glucose (or glycogen). Therefore, a decrease in the availability of P_i and ADP in favor of ATP leads to a diminution in the rate of glycolysis and an increase in the rate of gluconeogenesis.

We now turn again to a consideration of the fate of NADH. In aerobic cells it is oxidized by the mitochondrial electron-transport chain (see Chapter 15); in essentially anaerobic cells it is oxidized by linked reactions, of which in animal metabolism the conversion pyruvate + NADH → lactate + NAD^+ is by far the most important.

Finally, we must consider the possibility that different cells or tissues may contain the relevant specific (or regulatory) enzymes in grossly different amounts and hence may be preponderantly, or exclusively, glycolytic—rather than gluconeogenic as well. Enzyme synthesis and therefore enzyme levels may, furthermore, be regulated by environmental (diet, starvation) and hormonal factors. In microorganisms, glucose, or rather catabolites of glucose, can repress the synthesis of a wide variety of enzymes (*catabolite repression,* page 961).

In addition to these stoichiometric and kinetic controls, we must consider the quite specific regulatory mechanisms summarized in Table 11-3.

Note the symmetry (1) with respect to identical effectors acting as activators in one and inhibitors in the opposite direction and (2) within any one pathway between couples of reactants and products (ATP ⇌ ADP, AMP) and, hence, the energy charge. The principal purpose of these devices is to make certain that when the supply of noncarbohydrate precursors flowing into and through the citric acid cycle and its linked energy-generating system is high (producing a high steady-state level of acetyl CoA, tricarboxylic acids, and energy charge—see Chapter 14), gluconeogenesis is favored over glycolysis.

In yeast, an exceedingly flexible organism capable of anaerobic growth on

TABLE 11-3
Regulatory effects on enzymes of carbohydrate metabolism in animals

Process	*Enzymatic reaction*	*Activated by*	*Inhibited by*
Breakdown (catabolic)	Glycogen → glucose-1-*P*	AMP, ATP, Ado-3′:5′-*P*	
	Glucose → glucose-6-*P*		Glucose-6-*P*, (P_i)[a]
	Fructose-6-*P* → fructose-1,6-di*P* (FDP)	ADP, AMP, FDP	ATP, citrate
	PEP → pyruvate	FDP[b]	ATP
Synthesis (anabolic)	Pyruvate → oxalacetate → PEP	Acetyl CoA, CO_2[c]	Malonyl CoA
	Fructose-1,6-di*P* → fructose-6-*P*		AMP, FDP
	Glucose-6-*P* → glucose		Glucose, PP_i
	Glucose-1-*P* → UDPG → glycogen	Glucose-6-*P*	Ado-3′:5′-*P*
Both directions	Fructose-1,6-di*P* ⇌ 2 triose-*P*	AMP (forward, muscle) ATP (reverse, liver)	ATP (muscle)

[a] Secondary control, effective when glucose-6-*P* accumulates (46), effective only for one of the two main types.
[b] Not for enzyme from muscle.
[c] Participates in reaction.

fermentable and of aerobic growth on both fermentable and nonfermentable substrates (e.g., lactate, acetate), similar but perhaps less stringent controls may be operative, principally on the FDP and pyruvate $\rightleftharpoons$ PEP level.

Biosynthesis of Carbohydrates from Two-Carbon Compounds

Higher plants and microorganisms are sometimes confronted by the problem of converting fats or two-carbon metabolites (and hence acetyl CoA) into carbohydrates and other cell constituents. This takes place for instance during the germination of seeds, which contain a large amount of lipid as a metabolic store, or in microorganisms when they are required to grow on ethanol or acetate as their sole source of carbon. The task is accomplished by means of the coordinated set of reactions (47) shown in Fig. 11-6. The two key enzymes, *isocitratase* (*isocitrate lyase*) [reaction (11-45)] and *malate synth(et)ase* [reaction (11-46)] are absent from the cells of higher animals, and hence the cycle cannot be operative there. In eucaryotic organisms these two and some of the related enzymes are associated with subcellular particles—the glyoxysomes in plants and the peroxisomes in protozoa.

Operation of the cycle shown, in addition to the two novel reactions, also requires the simultaneous participation of three of the enzymes of the citric acid cycle (Chapter 14). These are the *citrate-condensing enzyme* (citrate synthase), *aconitase* (aconitate hydratase), and *malate dehydrogenase* (—NAD : oxidoreductase). One turn of the cycle leads to the oxidation of two molecules of acetyl CoA to yield one molecule of succinate with the simultaneous removal of two reducing equivalents to NAD. The electron-transport chain for the oxidation of NADH by molecular oxygen plus the malate synthetase reaction provides the essential driving force. The succinate formed can then be converted to carbohydrate by the sequence of reactions shown, which includes two additional reactions catalyzed by enzymes of the citric acid cycle: succinic dehydrogenase and fumarase, with malate dehydrogenase again participating. While the *energy* needs of the organism continue to be provided by the CAC, the complete glyoxylate (or Kornberg) bypass provides an example par excellence of an anaplerotic sequence in operation by continuously regenerating members of this cycle. Of the various enzymes involved, isocitrate lyase, the one at the branch point of two metabolic sequences, appears to be susceptible to allosteric regulation by PEP and pyruvate (PEP > pyruvate, pH <8; pyruvate > PEP, pH >8). Malate synthetase and isocitrate lyase are inducible; their structural genes plus a third (regulatory) gene appear to form an operon (Chapter 21).

Bacteria growing on glyoxylate (or glycolate, which is readily converted to glyoxylate) as their sole carbon source do *not* use this cycle or its enzymes for anaplerotic purposes; for catabolic, energy-yielding, purposes only malate synthetase is used together with other enzymes:

Reaction	Enzyme
Glyoxylate + acetyl CoA $\rightleftharpoons$ malate	*malate synthetase*
Malate + $\leq$3ADP + $\leq$3P_i + ½O_2 $\rightarrow$ oxalacetate + $\leq$3ATP	*malate dehydrogenase, electron transport, oxidative phosphorylation*
Oxalacetate + ATP $\rightleftharpoons$ phosphoenolpyruvate + ADP + CO_2	*PEP carboxykinase*
Phosphoenolpyruvate + ADP $\rightarrow$ pyruvate + ATP	*pyruvate kinase*
Pyruvate + 3ADP + 3P_i + ½O_2 $\rightarrow$ acetyl CoA + 3ATP + CO_2	*PDC, electron transport, oxidative phosphorylation*
Net Glyoxylate + $\leq$6ADP + $\leq$6P_i + O_2 $\rightarrow$ 2CO_2 + $\leq$6ATP	

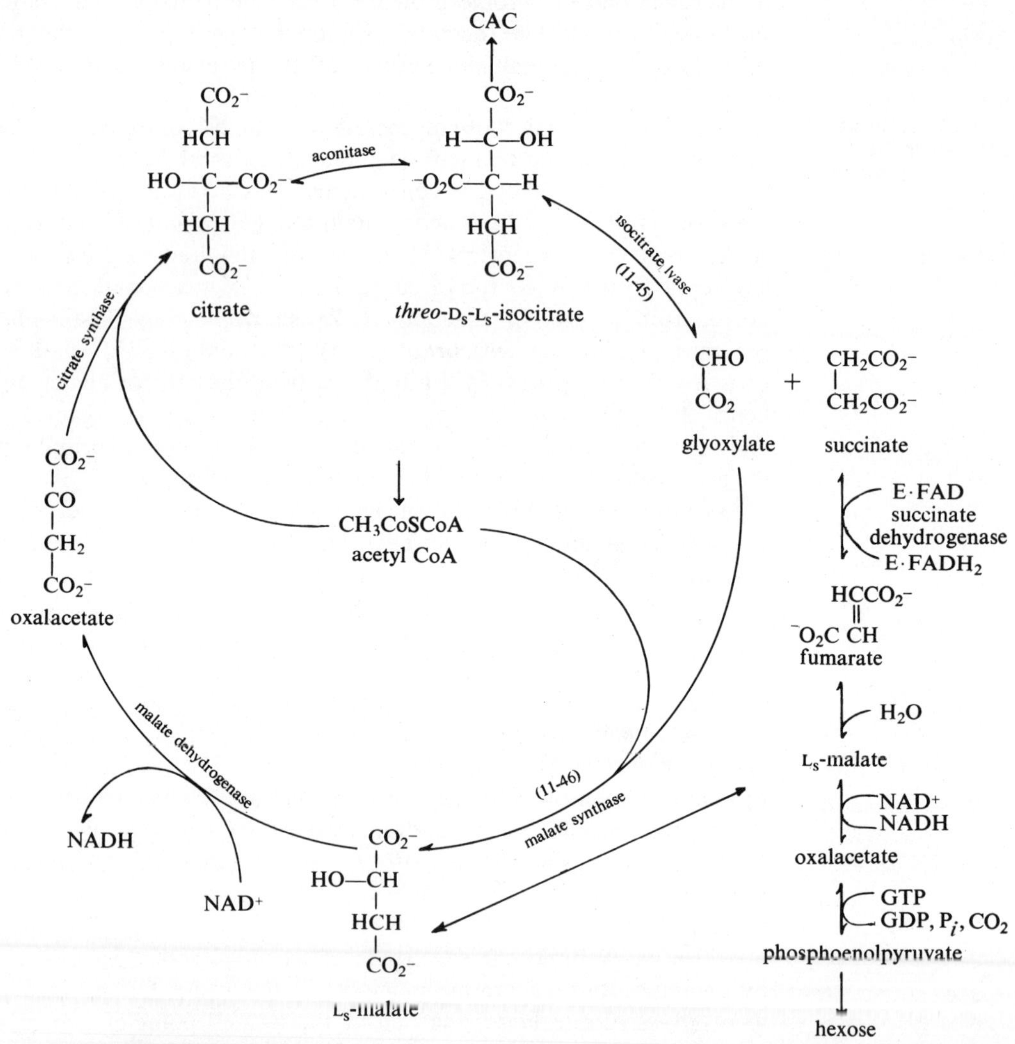

FIGURE 11-6
Glyoxylate cycles.

This series of reactions constitutes an *oxidative dicarboxylic acid cycle,* designed for the complete oxidation of C-2 units other than acetyl CoA. For an anaplerotic sequence, such organisms generate phosphoenolpyruvate by way of tartronic (hydroxymalonic) semialdehyde[21] as an intermediate.

$2\,CHOCO_2^- + H^+ \xrightarrow{TPP,\ FAD,\ Mg^{2+}} CHOCHOHCO_2^- + CO_2$	*glyoxylate carboligase*
$CHOCHOHCO_2^- + 2[H] \rightleftharpoons CH_2OHCHOHCO_2^-$ (D-glycerate)	*tartronic semialdehyde reductase*
D-Glycerate + ATP → D-glycerate-3-*P*[22]	*glycerate kinase*
D-Glycerate-3-*P* ⇌ phosphoenolpyruvate	*phosphoglycerate mutase plus enolase*
Net 2 Glyoxylate + ATP + 2[H] → phosphoenolpyruvate + ADP + CO_2	

[21] During its formation catalyzed by a carboligase from *Pseudomonas*, [HCHO·TPP] (i.e., *hydroxymethyl* TPP) may function as the reactive intermediate.
[22] Perhaps the 2-*P* instead; the exact substrate is not yet certain.

FERMENTATIONS LEADING TO SUCCINATE AND PROPIONATE

Many microorganisms, most notably those of the genera *Propionibacteria* and *Veilonella,* are capable of producing succinate and propionate from glucose, glycerol, lactate, and pyruvate. Some typical carbon balances are

$$\text{3 Hexose} \rightarrow \text{4 propionic acid} + \text{2 acetic acid} + 2CO_2 \qquad (11\text{-}47a)$$

$$\text{Glycerol} \rightarrow \text{propionic acid} + H_2O \qquad (11\text{-}47b)$$

$$\text{3 Lactic acid} \rightarrow \text{2 propionic acid} + \text{acetic acid} + CO_2 \qquad (11\text{-}47c)$$

$$\text{Hexose} + 2CO_2 + 4[H] \text{ (from other pathways)} \rightarrow \text{2 succinate} + 2H_2O \qquad (11\text{-}48a)$$

$$\text{Glycerol} + CO_2 \rightarrow \text{succinate} + H_2O \qquad (11\text{-}48b)$$

The fermentations producing succinate are easy to understand in light of what we know already; the pathway is simply

$$\text{Pyruvate} \xrightarrow[-\text{AMP, PP}_i]{\text{ATP, P}_i} \text{PEP} \xrightarrow[-\text{PP}_i]{\text{P}_i\text{, CO}_2} \text{oxalacetate} \xrightarrow{\text{NADH}} \text{malate} \xrightarrow[-H_2O]{} \text{fumarate} \xrightarrow{2[H]} \text{succinate} \qquad (11\text{-}49)$$

Four reducing equivalents are required per hexose utilized; the conversion of glycerol to pyruvate provides just that requirement—if hexose is the substrate, a certain portion will have to be diverted to a different fermentative pathway (e.g., [CH_3CHO]) to furnish the necessary reducing equivalents.

The conversion of pyruvate to propionate in microorganisms has been studied intensively in H. G. Wood's and F. Lynen's laboratories. It proceeds in large part by way of the following interesting series of reactions (see also Chapter 13):

$$d\text{-}(+)\text{-biotin} + H_2N \text{ enzyme} \xrightarrow[\text{synthetase}]{\text{Biotin apotranscarboxylase}} \text{biotinyl enzyme} \quad (\text{ATP} \rightarrow \text{AMP} + \text{PP}_i)$$

$$\text{Pyruvate} + \text{methylmalonyl-CoA } (a) \qquad \text{oxalacetate} + \text{propionyl CoA} \qquad (11\text{-}50)$$

$$\text{Oxalacetate (via malate and fumarate)} + 4[H] \rightarrow \text{succinate}^{23}$$

$$\text{Succinate}^{23} + \text{propionyl CoA} \rightleftharpoons \text{succinyl CoA} + \text{propionate} \qquad (11\text{-}51)$$

$$\text{Succinyl CoA} \xrightarrow{\text{DMBC}^{24}\text{ enzyme}} \text{methylmalonyl CoA } (b) \qquad (11\text{-}52)$$

$$\text{Methylmalonyl CoA } (b) \rightleftharpoons \text{methylmalonyl CoA } (a) \qquad (11\text{-}53)$$

Net Pyruvate + 4[H] → propionate (11-54)

The first step in this cyclical sequence is a transcarboxylation catalyzed by *methylmalonyl CoA: pyruvate carboxyltransferase* (transcarboxylase). This is the only CO_2 transfer (as distinct from a carboxylation) reaction known so far. Free CO_2 does not participate and ATP or other energy sources are not required. It is catalyzed by a (+)-biotin-containing enzyme. In this instance, the apoenzyme has been isolated from *Propionibacterium shermanii,* and the

[23] This formation of catalytic quantities of succinate required here is to be distinguished from the stoichiometric production of this dicarboxylic acid described in Eq. (11-49), in which the first two steps are catalyzed by pyruvate phosphate dikinase [Eq. (11-44)] and PEP carboxytransphosphorylase [Eq. (11-38)], respectively. There is *no net* fixation of C in the course of reaction (11-54).

[24] DMBC, dimethylbenzimidazole cobamide coenzyme, a vitamin B_{12} derivative (page 424).

formation of the intact holoenzyme has been shown to involve an enzyme-catalyzed formation of an amide bond between the carboxyl group of (+)-biotin and an ε-amino group of lysine on the transcarboxylase (Chapter 8). The reaction requires ATP and a divalent metal ion, and involves a (synthetase-bound) biotinyl-5′-AMP intermediate [completely analogous to the intermediate involved in the "activation" of other carboxylic acids to form fatty acyl CoA derivatives (page 597) or amino acyl-tRNA derivatives (page 918)]. The biotinyl enzyme acts as both CO_2 acceptor and donor (page 422).

The oxalacetate produced in reaction (11-50) is then reduced by the sequence of reactions already described (Fig. 11-5), yielding succinate, which then functions as a CoA acceptor by virtue of reaction (11-51), catalyzed by a *succinyl CoA-transferase* (*trans*-thioesterase, succinyl-CoA : propionate CoA transferase), with the liberation of free propionate. Succinyl CoA is isomerized in a reaction that requires a cobamide (vitamin B_{12}) coenzyme and the enzyme *methylmalonyl mutase,* regenerating methylmalonyl CoA. The methylmalonyl CoA formed in this reaction is, however, the optical antipode of that required in reaction (11-50) and has to be racemized by *methylmalonyl CoA racemase* in order to be ready for another turn of the cycle. The sequence shown does *not* utilize free CO_2, and for this reason the product propionate should be unlabeled if the reaction is allowed to take place in the presence of labeled CO_2 (or HCO_3^-). A certain small amount of label is, however, always found. This is so because oxalacetate and pyruvate + CO_2 are in metabolic equilibrium by virtue of several reactions, some of which have already been described, which *do* utilize free CO_2. In *Propionibacteria,* the most important reaction of this type is probably that catalyzed by *phosphoenolpyruvic carboxytransphosphorylase,* reaction (11-38). These carboxylation reactions will serve to form the catalytic quantities of dicarboxylic acid derivatives required to initiate the cycle. The small amount of fixed CO_2 actually found in propionate will be randomized with the —CO_2^- of pyruvate while passing through the symmetrical intermediates fumarate and succinate, and a similar fate will befall the α- and β-carbons of pyruvate. In certain other bacteria (*Clostridium propionicum*), pyruvate appears to be capable of forming propionate without intervention of symmetrical intermediates, and acrylate, β-alanine, and propionate are interconvertible.

HEXOSE MONOPHOSPHATE–PENTOSE PATHWAYS (Fig. 11-7)

The second set of major pathways for hexose catabolism has been investigated principally by O. Warburg, F. Lipmann, S. Cohen, B. L. Horecker, E. Racker, and their collaborators. A number of different names have been assigned to the reaction sequence(s) involved. Since one of its pathways diverges from glycolysis at the level of glucose-6-*P*, it has been called the hexose monophosphate "shunt"; since in the same pathway two of the first three steps are dehydrogenations and since pentoses participate catalytically, it has sometimes been designated the "oxidative pathway of hexose metabolism," or the "pentose cycle," and since phosphogluconate is a key intermediate, the designation "phosphogluconate pathway" has also been employed. We remember, however, that members of one and the same set of enzymes, operating either by themselves or in conjunction with those of other sets, can be utilized by the same or different cells to fulfill a multiplicity of functions. This appears to be especially true in the present instance, and we shall describe in this section several different ways in which this can be brought about; another very important example is provided by the CO_2-fixation cycle of photosynthesis discussed in Chapter 12. The constituent enzymes involved are generally localized in the cytosol.

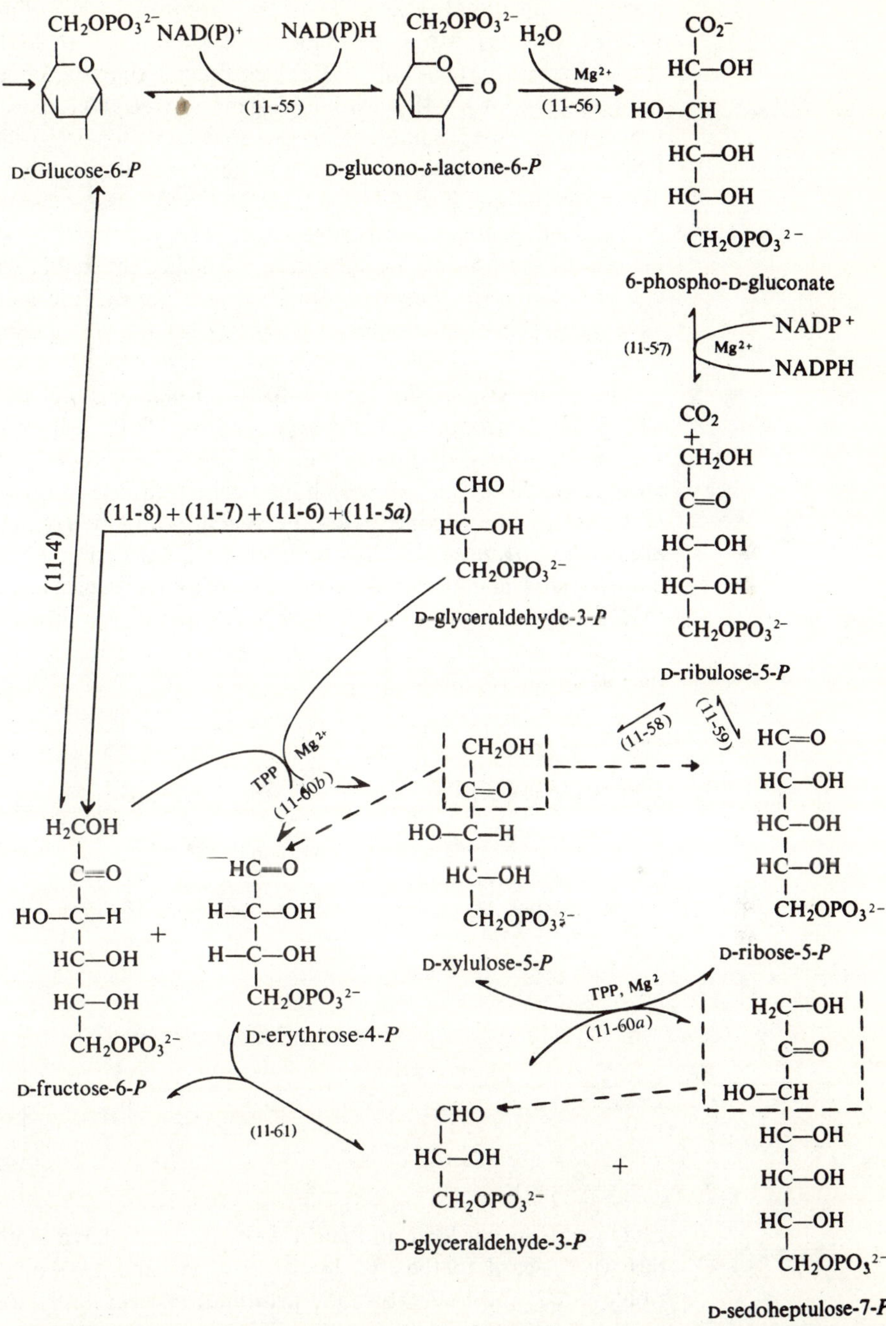

FIGURE 11-7

Reaction (11-55), glucose-6-*P*-dehydrogenase; reaction (11-56), gluconolactonase; reaction (11-57), phosphogluconate dehydrogenase; reaction (11-58), ribulose-*P*-3-epimerase; reaction (11-59), ribose-*P* (phosphoribose) isomerase; reactions (11-60a) and (11-60b), transketolase; reaction (11-61), transaldolase.

The Dehydrogenases—Reactions (11-55) and (11-57)

The first oxidative step [reaction (11-55)] is catalyzed by *glucose-6-P dehydrogenase,* an important enzyme of very wide distribution. It was discovered and studied by Otto Warburg. These investigations led to his identification of $NADP^+$ as the first pyridine (nicotinamide) nucleotide as an obligatory coenzyme. Highly purified enzymes have been obtained from yeast, *E. coli,* and a variety of animal tissues. The yeast apoprotein is a dimer ($M = 102{,}000$) which associates into a tetramer in the presence of NADP; it is in this latter form (i.e., as a $[(\text{NADP})_4 \cdot \text{Enz}]$ complex) that the enzyme can be isolated in crystalline form (48). Although generally $NADP^+$ is the specific coenzyme for these enzymes, some are known that can utilize NAD^+ as well (e.g., that from *Leuconostoc mesenteroides*). The reaction catalyzed is the reversible dehydrogenation $R_1R_2CHOH \rightleftharpoons R_1R_2C{=}O$ characteristically catalyzed by NAD(P)-dependent dehydrogenases. The reaction as written, with the δ-lactone as the product or substrate, is readily reversible with a $\Delta G^{0\prime} = -100$ cal/mol. Since the enzyme from yeast can act on free glucose, it proved possible to establish its anomeric specificity for the β-D-glucopyranose configuration.

The hydrolysis of the lactone to the 6-phosphoester of gluconate [reaction (11-56)] is strongly exergonic ($\Delta G^{0\prime} = -5000$ cal/mol) and quite rapid ($t_{1/2}$ of the order of minutes in the physiological *p*H range). Nevertheless, specific enzymes (*lactonases*) have been identified in plants and many species of bacteria. The second oxidative reaction (11-57) catalyzed by *6-P-gluconate dehydrogenase* requires a divalent metal cation (Mg^{2+} or Mn^{2+}) as a cofactor and constitutes a reversible oxidative decarboxylation, analogous to that catalyzed by the similarly NADP-dependent malic enzyme (see page 520) and isocitrate dehydrogenase. Presumably the 3-ketohexonic acid takes part as an intermediate.

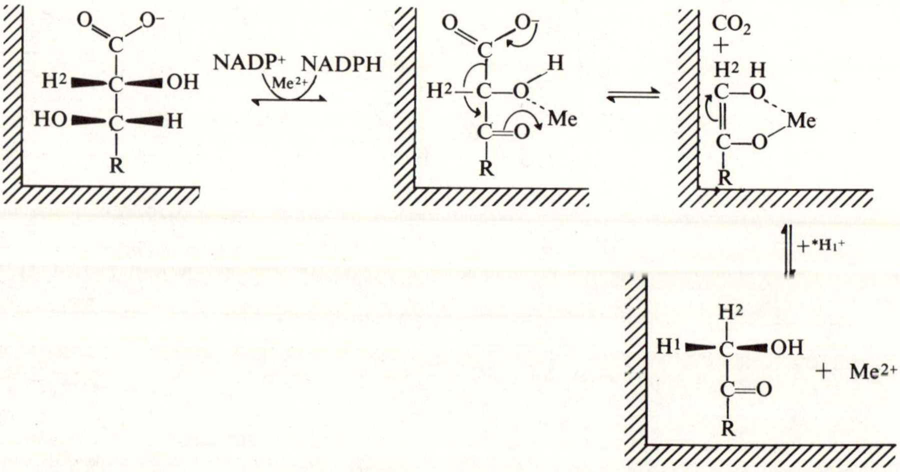

Enzymes from rat liver and yeast were shown to have identical stereospecificities with regard to the two hydrogens (H^1 and H^2) on C-1 of the product ribulose-5-*P* and to catalyze the protonation–decarboxylation and H-exchange reactions with inversion of configuration as shown (49).

Isomerization of Pentose Phosphates—Reactions (11-58) and (11-59)

Two isomerization reactions of pentose-5-phosphates form an integral part of all metabolic sequences that utilize this metabolic scheme. One, reaction (11-59), simply interconverts a phosphoketopentose and its aldose-phosphate counterpart. It is therefore completely analogous to reaction (11-4) in the hexose series (glucose-6-*P* ⇌ fructose-6-*P*), and the mechanisms for the two reactions have been demonstrated to be identical. The other reaction (11-58) constitutes an epimerization at the 3 position (i.e., the carbon α to the carbonyl of the ketose) and may thus also involve an enediol intermediate.

Rearrangement of the Carbon Skeleton by Transketolase and Transaldolase

Transketolase Reaction—Reactions (11-60a) and (11-60b). The enzyme catalyzing this reaction requires thiamine pyrophosphate and divalent metal cations as obligatory cofactors. The reaction catalyzed can be regarded as the reversible transfer of a glycoaldehyde moiety from a donor ketose to an acceptor aldose. The specificity of the enzyme requires that the ketose have the L configuration on the α-carbon and preferably a trans configuration for the next two:

CH_2OH
C=O
HO—CH
HC—OH
R

Thus we find that the following can function as ketol donors: hydroxypyruvate, L-erythrulose, D-xylulose-5-*P*, D-fructose-6-*P*, and D-sedoheptulose-7-*P*. The following aldoses have been used as acceptors: glycolaldehyde (→ L-erythrulose), D-glyceraldehyde-3-*P* (→ D-xylulose-5-*P*), glyceraldehyde (→ xylulose), D-erythrose-4-*P* (→ D-fructose-6-*P*), and D-ribose-5-*P* (→ D-sedoheptulose-7-*P*). In the scheme shown in Fig. 11-7, transketolase intervenes twice: first [Eq. (11-60a)] to form glyceraldehyde-*P* plus sedoheptulose-*P* from the two pentoses, and then again [Eq. (11-60b)] to form fructose-*P* plus triose-*P* from another molecule of xylulose-*P* reacting with tetrose-*P*.

The mechanism of the reaction was elucidated simultaneously with those others that require the same coenzyme (Chapter 8) (cf. structures shown in reaction (11-14), [*B*] and [*C*]). Here 2-(α,β-dihydroxyethyl) TPP is the intermediate, and the reactions on the enzyme probably proceed as shown in the diagram:

Donor ketose (CH_2OH–C=O–HO—C—H–R_2) + TPP ⇌ [*E* = ketose·TPP adduct] ⇌ [*F* = 1,2-dihydroxyethyl·TPP] + Aldehyde (aldose) (R_2–CHO)

[*F*] $\rightleftharpoons^{+H^+}$ CHO–CH_2OH + TPP (11-61)

Transaldolase Reaction—Reaction (11-61). Transaldolase is the second enzyme characteristic of the nonoxidative portion. It catalyzes the transfer of a *dihydroxyacetone moiety* (not free dihydroxyacetone) from a donor ketose such as, typically, D-sedoheptulose-7-*P* or D-fructose-6-*P*, to an acceptor aldose, such as D-glyceraldehyde-3-*P* or D-erythrose-4-*P*. The specificity requirements of the enzyme, the nature of the reaction catalyzed, and the apparent absence of any cofactor requirement made it appear highly likely that the mechanism of transaldolase was similar to that of aldolase [reaction (11-6)].

B. L. Horecker has been able to identify the active intermediate as a Schiff base of the ketose, linked to the ε-amino group of a lysine residue of the enzyme. The reaction proceeds with participation of a histidine residue (50):

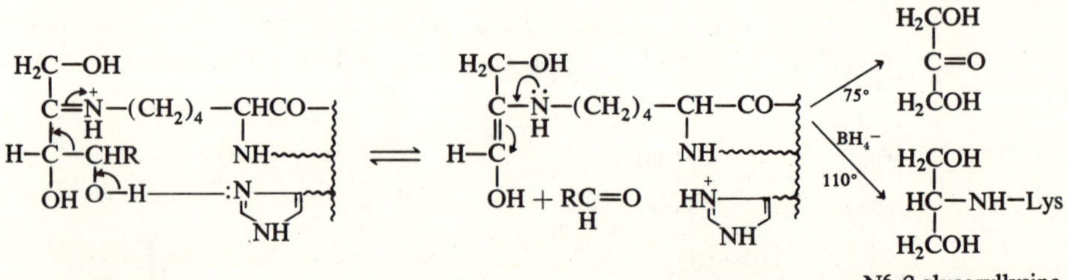

$H_2C{-}OH$ | $C{=}\overset{+}{N}{-}(CH_2)_4{-}CHCO{-}$ (H, NH) | $H{-}C{-}CHR$ | $OH\ O{-}H \cdots :N$ (NH) $\rightleftharpoons$ $H_2C{-}OH$ | $C{-}N{-}(CH_2)_4{-}CH{-}CO{-}$ (H, NH) | $H{-}C$ | $OH + RC{=}O$ (H) $H\overset{+}{N}$ (NH) $\xrightarrow{75°}$ H_2COH | $C{=}O$ | H_2COH; $\xrightarrow[110°]{BH_4^-}$ H_2COH | $HC{-}NH{-}Lys$ | H_2COH

N^6-β-glyceryllysine

The similarities of this mechanism to those of reaction (11-6), on the one hand, and of (11-61) are obvious.

Pentose Oxidation Cycle

One suggestion that has been advanced is that the reactions just described plus certain ones of the glycolytic sequence may interact in the following manner to lead to the oxidative conversion of glucose either to triose-*P* or to CO_2:

3 Glucose-6-*P* + 3NADP+ → 3 6-*P*-gluconate + 3NADPH + 3H+ (11-55) + (11-56)

3 6-*P*-Gluconate + 3NADP+ → 3 ribulose-5-*P* + $3CO_2$ + 3H+ + 3NADPH (11-57)

2 Ribulose-5-*P* → 2-xylulose-5-*P* (11-58)

Ribulose-5-*P* → ribose-5-*P* (11-59)

Xylulose-5-*P* + ribose-5-*P* → sedoheptulose-7-*P* + triose-*P* (11-60a)

Sedoheptulose-7-*P* + triose-*P* → erythrose-4-*P* + fructose-6-*P* (11-61)

Erythrose-4-*P* + xylulose-5-*P* → fructose-6-*P* + triose-3-*P* (11-60b)

Fructose-6-*P* → glucose-6-*P* (11-4)

Net 3 Glucose-6-*P* + 6NADP+ → 2 glucose-6-*P* + triose-*P* + $3CO_2$ + 6NADPH + 6H+ (11-62a)

or twice this, plus 2 triose-*P* → fructose-1,6-di*P* $\xrightarrow{+H_2O}$ P_i + fructose-6-*P* → glucose-6-*P*:

Net 6 Glucose-6-*P* + 12NADP+ → 5 glucose-6-*P* + $6CO_2$ + 12NADPH + 12H+ + P_i (11-62b)

or

6 Glucose-6-*P* + $6O_2$ → 5 glucose-6-*P* + $6CO_2$ + $6H_2O$ + P_i (11-63)

In theory, since the cell can oxidize (by means of O_2)[25] the NADPH produced and thus regenerate NADP+, the scheme provides for the complete oxidation of one molecule of glucose-*P* to $6CO_2$ + $6H_2O$ by reaction (11-63). This entails the catalytic intervention of five additional molecules of glucose-*P* and the intermediates and reactions indicated. Note that the stoichiometry

[25] The sequence involved is transhydrogenation to yield NADH, followed by oxidation of the latter by means of the electron-transport system (Chapter 15).

does not provide any clues concerning mechanism. Significantly all three [or six, in reactions (11-62b) or (11-63)] molecules of CO_2 produced have their origin in C-1 of glucose, the other carbons having become rearranged in the products as follows:

$$2\begin{array}{l} 2\ \text{CHO} \\ \quad | \\ 3\ \text{C} \\ \quad | \\ 3\ \text{C} \\ \quad | \\ 4\ \text{C} \\ \quad | \\ 5\ \text{C} \\ \quad | \\ 6\ \text{CH}_2\text{OPO}_3^{2-} \end{array} + 2\begin{array}{l} 2\ \text{CHO} \\ \quad | \\ 3\ \text{C} \\ \quad | \\ 2\ \text{C} \\ \quad | \\ 4\ \text{C} \\ \quad | \\ 5\ \text{C} \\ \quad | \\ 6\ \text{CH}_2\text{OPO}_3^{2-} \end{array} + 1\begin{array}{l} 6\ \text{CHO} \\ \quad | \\ 5\ \text{C} \\ \quad | \\ 4\ \text{C} \\ \quad | \\ 4\ \text{C} \\ \quad | \\ 5\ \text{C} \\ \quad | \\ 6\ \text{CH}_2\text{OPO}_3^{2-} \end{array}$$

One way of testing for occurrence of this metabolic sequence consists of comparing the rate of appearance of label in the respired CO_2 derived from glucose labeled in position 1 with that produced from other positions. C-6 frequently has been used as a standard for comparison, since the EMP sequence, followed by oxidation of pyruvate, produces label in the following order: $3 = 4 > 2 = 5 > 1 = 6$ (see Fig. 11-9). The relative importance of the EMP and the oxidative sequence in various tissues have been measured by these and other techniques (51). In animals, except for their adipose tissues, only $\leqq 5$ percent of the total glucose-6-*P* produced appears to be diverted to this path, in spite of abundant evidence of high, and hormonally and environmentally regulated content of the enzymes characteristic of it, particularly the two dehydrogenases.

The significance to the organism's metabolism may not be so much the oxidation of glucose as an end in itself but its conversion to precursors for biosynthetic processes; pentoses (for the synthesis of the requisite ribo- and deoxyribonucleotide derivatives for coenzyme, polysaccharide, and especially polynucleotide biosynthesis) and NADPH, the source of reducing power for these and other biosynthetic routes (fatty acids, steroids, etc.). The two dehydrogenases in question must furnish the bulk of the extramitochondrial NADPH—and most biosynthetic processes are localized in just such extramitochondrial compartments (Chapter 9).

Nonoxidative Interconversions of Hexose and Pentose Phosphates

The oxidative route just discussed is not the only one that can lead, with intervention of the same complement of enzymes, from hexose to pentose phosphates. Consider the following scheme of by-now familiar reactions:

$$\begin{array}{l}
2\ \text{Ribulose-5-}P \rightleftharpoons 2\ \text{ribose-5-}P \\
4\ \text{Ribulose-5-}P \rightleftharpoons 4\ \text{xylulose-5-}P \\
2\ \text{Xylulose-5-}P + 2\ \text{ribose-5-}P \rightleftharpoons 2\ \text{sedoheptulose-7-}P + 2\ \text{triose-3-}P \\
2\ \text{Sedoheptulose-7-}P + 2\ \text{triose-3-}P \rightleftharpoons 2\ \text{fructose-6-}P + 2\ \text{erythrose-4-}P \\
2\ \text{Erythrose-4-}P + 2\ \text{xylulose-5-}P \rightleftharpoons 2\ \text{fructose-6-}P + 2\ \text{triose-}P \\
2\ \text{Triose-}P \rightleftharpoons \text{fructose-1,6-di}P \\
\text{Fructose-1,6-di}P + H_2O \rightarrow \text{fructose-6-}P + P_i \\
5\ \text{Fructose-6-}P \rightleftharpoons 5\ \text{glucose-6-}P \\
\hline
\textit{Net}\ 6\ \text{Pentose-5-}P + H_2O \rightleftharpoons 5\ \text{glucose-6-}P + P_i
\end{array} \qquad (11\text{-}64)$$

It is clear that we have now available a means of interconverting and reshuffling the carbon atoms of hexoses and pentoses. This nonoxidative pathway for the formation of ribose from glucose has been shown to occur in a variety of cells, as has the concomitant rearrangement of the hexose carbon skeleton.

Bacterial Fermentations in Which Pentose-Cycle Enzymes Participate

The converse of the above, that is, the obligatory transformation of pentose-*P* into hexose-*P*, prior to its catabolism by conventional fermentations (by means of the EMP pathway with the possible variants described), appears to characterize pentose fermentations by a large number of microorganisms, among them members of the *E. coli–Aerobacter aerogenes* group, *Clostridia,* and *Micrococci.* These and other organisms can, after appropriate induction periods, ferment all four D-aldopentoses: D-ribose, D-arabinose, D-xylose, and D-lyxose, as well as L-arabinose and L-xylose (cf. below). The various enzymes postulated have been isolated and a characteristic labeling pattern was found in pyruvate and the other compounds derived therefrom:

```
*C *C *C    *C                  *C *C
 |  |  |     |           C       |  |
 C  C  C     C           |       C  C
 |  |  |     |      C    C      *C  C
 C  C  C → *C       |    |       |  |          a     b
 |  |  |     |      |    |       |  |
 C  C  C     C  +   C →  C  +    C  C  + C → CH3 CO CO2−
 |  |  |     |      |    |       |  |    |
 C  C  C     C      C    C       C  C    C
             |                   |  |    |
             C                   C  C    C
             |
             C
```

$a = 40$ percent of starting specific activity of pentose
$b = 30$ percent of starting specific activity of pentose

The stoichiometry of such fermentations can easily be derived by summing the pertinent equations—noting that pentose-5-*P* has to be formed first by kinase action:

$$3\ \text{Pentose} + 5\text{P}_i + 5\text{ADP} \rightarrow 5\ \text{pyruvate} + 5\text{ATP} + 5\text{H}_2\text{O} + 10[\text{H}]$$

A quite different pattern with regard to the disposition of D-xylulose-5-*P*, the key intermediate, emerges for the heterofermentative lactic acid bacteria (various *Lactobacilli* and *Leuconostoc mesenteroides*).

Pentose Fermentation by Lactobacilli. The enzymes (many of them inducible; see Chapter 21) concerned in the fermentative degradation of several different pentoses by *L. plantarum* (or *arabinosus*) have been isolated and separated from each other by Horecker and his colleagues. The pathways and the enzymes involved have been demonstrated to be those of Fig. 11-6. *Phosphoketolase,* the enzyme catalyzing the TPP-dependent phosphorolytic cleavage of a ketopentose-*P* to triose-*P* + acetyl-*P* [reaction (11-65)] plays a pivotal role.[26] This reaction is virtually irreversible ($\Delta G^{0\prime} \simeq -13{,}000$ cal/mol), and dihydroxy · TPP has been identified as an intermediate.

The same organisms, which characteristically do not contain aldolase and triose-*P* isomerase but do possess all the enzymes of the oxidative pentose

[26] Racker has isolated from cells of the aerobic organism *Acetobacter xylinum,* a similar, though not identical, *phosphohexoketolase* catalyzing the analogous reaction fructose-6-*P* + $\text{P}_i \rightarrow$ acetyl-*P* + D-erythrose-4-*P*. Organisms containing both phosphoketolases plus the appropriate accessory enzymes can, at least in theory, affect the conversions fructose-6-*P* + $2\text{P}_i \rightarrow$ 3 acetyl-*P* or hexose + 2ADP + $2\text{P}_i \rightarrow$ 3 acetate + 2ATP.

cycle plus phosphoketolase in significant amounts, also ferment glucose by the same pathway. In this instance, the C-1 of glucose is quantitatively removed as CO_2 by the dehydrogenases of the oxidative pathway (Fig. 11-9).

OTHER PATHWAYS OF HEXOSE DEGRADATION

Entner–Doudoroff Pathway

This pathway of hexose degradation has so far only been established for one genus of bacteria (*Pseudomonas*), although the enzymes occur in other species as well where they may be involved in, for example, gluconate metabolism. It diverges from the other routes at 6-*P*-gluconate and consists of two characteristic reactions (53), a dehydration to an enol plus its tautomerization [reaction (11-66)], followed by a specific aldol cleavage [reaction (11-67)]:

$$\begin{array}{c} {}^1CO_2^- \\ | \\ H-{}^2C-OH \\ | \\ HO-{}^3C-H \\ | \\ H{}^4C-OH \\ | \\ H{}^5C-OH \\ | \\ {}^6CH_2OPO_3^{2-} \end{array} \xrightarrow[\text{6-}P\text{-gluconate dehydrase}]{Fe^{2+},\ GSH} \begin{array}{c} CO_2^- \\ | \\ C-OH \\ \| \\ C-H \\ | \\ HC-OH \\ | \\ HC-OH \\ | \\ CH_2OPO_3^{2-} \end{array} \rightarrow \begin{array}{c} {}^1CO_2^- \\ | \\ {}^2C=O \\ | \\ H{}^3CH \\ \text{- - - - - -} \\ HC-OH \\ | \\ HC-OH \\ | \\ CH_2OPO_3^{2-} \end{array} \qquad (11\text{-}66)$$

6-*P*-Gluconate + H_2O → 2-Keto-3-deoxy-6-*P*-gluconate (KDPG)

KDPG aldolase ↓

$$\begin{array}{c} {}^1CO_2^- \\ | \\ {}^2C=O \\ | \\ {}^3CH_3 \end{array} + \begin{array}{c} {}^4CH=O \\ | \\ {}^5CHOH \\ | \\ {}^6CH_2OPO_3^{2-} \end{array} \qquad (11\text{-}67)$$

The labeling pattern to be expected from this pathway is indicated in reaction (11-66) and is also summarized in Fig. 11-9.

Glucuronate Pathway (Fig. 11-10)

In animals and higher plants, enzymes for the following set of conversions have been demonstrated (Figs. 11-9 and 11-10), and evidence is accumulating that this pathway is of some importance in the anabolism and catabolism not only of L-ascorbate (*vitamin C*),[27] of *myo*-inositol and the various intermediates shown, but of hexoses as well. The reactions between glucose and glucuronide involve the participation of uridine nucleotides and are discussed on page 548. It is of some interest that the key reductive step in the biosynthetic sequence to L-ascorbate between glucuronate and gulonate and both the reductive steps for the whole cycle (the one just mentioned and the L-xylulose → xylitol reaction) involve NADPH. On the other hand, the corresponding oxidative reactions, gulonate → [ketogulonate] → L-xylulose and xylitol → D-xylulose, both utilize NAD^+. Also note that the two coupled oxidoreductions permit interconversion of the D and L series of configurations. A hereditary abnormality known as *idiopathic pentosuria* in humans is probably caused by an inability to elaborate the NADPH-dependent xylitol dehydrogenase, block *b* (Fig. 11-10). As a result, large amounts of L-xylulose are found in the urine of such individuals, especially on a diet high in carbohydrates and particularly of D-glucuronate.

[27] In man, primates, and guinea pigs, a genetic block appears to intervene beyond L-gulonolactone (block *a*, Fig. 11-10); hence these species are unable to synthesize L-ascorbic acid, which then becomes an essential vitamin.

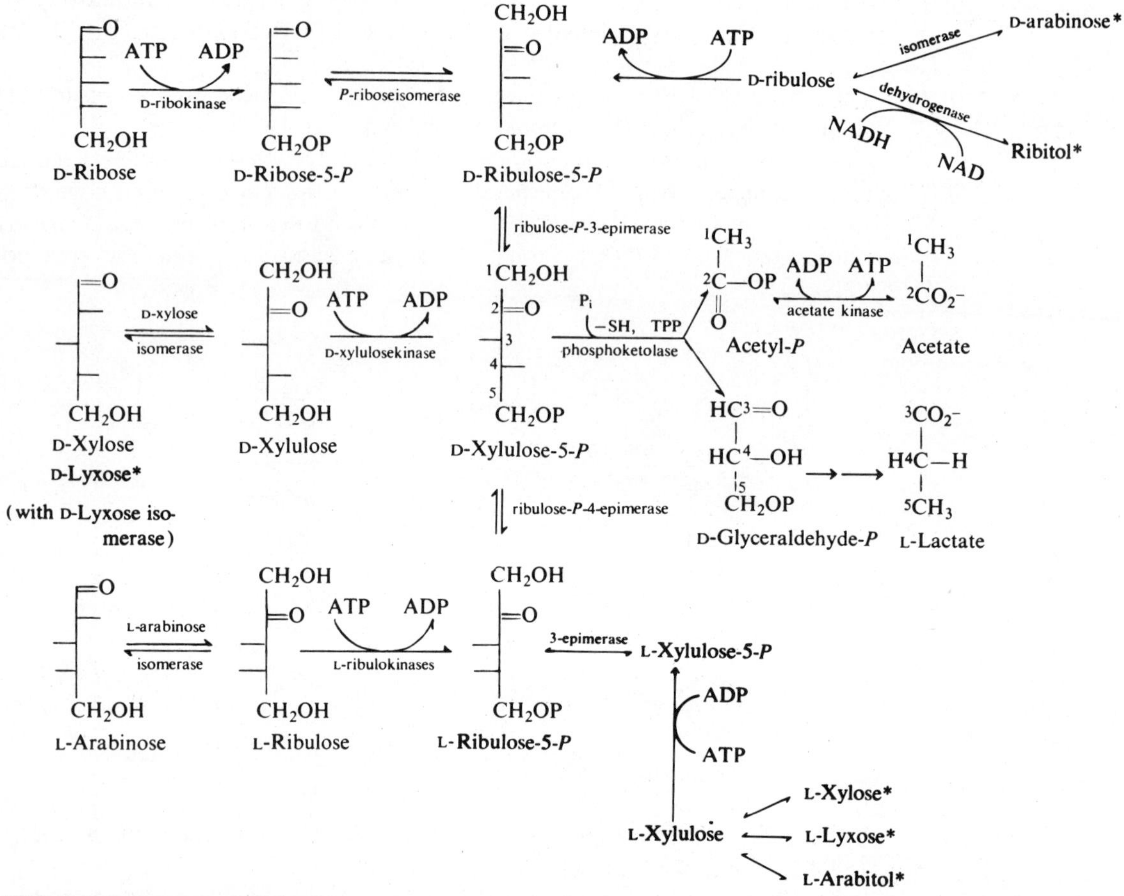

FIGURE 11-8

Pentose fermentations in *Lactobacilli.* A stoichiometry for this type of fermentation would be pentose + 2ADP + $2P_i$ → acetate + L-lactate + 2ATP. The pattern in *Aerobacter* is similar (52), except for disposition of D-ribulose-5-*P;* substrates utilized by this organism, but not by *Lactobacilli,* are marked with an asterisk.

Assignment of Metabolic Pathways

From all that has been said, it is clear that it is not always easy to determine the precise pathway that has led from any one starting point to a set of products even in an area as well charted as is carbohydrate metabolism. When faced with this problem concerning a previously unexplored cell or organism, the investigator has to know the following: what is the precise stoichiometry of the conversion? Can it be carried out in cell-free extracts? Does it respond in the predicted fashion to the action of characteristic inhibitors? Can one demonstrate the presence and function of characteristic coenzymes and activators? Is the labeling pattern in the final products that predicted for the postulated pathway (some of the most important patterns are shown on Fig. 11-9), and does label appear in the postulated intermediates in the correct temporal and structural sequence? Can the flow of labeled atoms be diluted by the addition of appropriate *unlabeled* intermediates? Can these intermediates be isolated and shown to accumulate in the proper order if the flow is disrupted by physiological or genetic blocks? Are all the enzymes postulated present in amounts and with activities sufficient to account for the postulated flux of material?

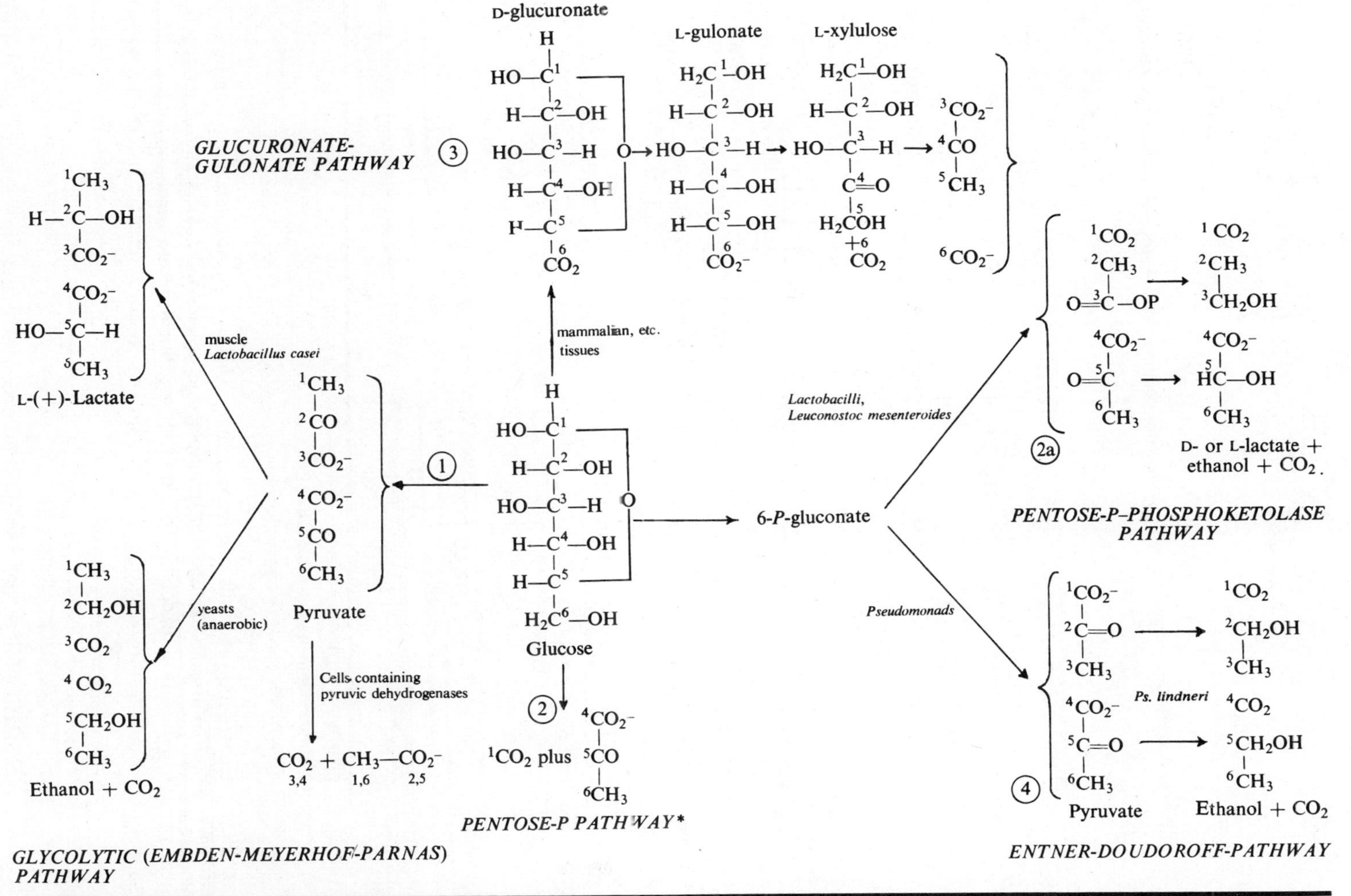

FIGURE 11-9

Characteristic labeling patterns in products of glucose catabolism.

* Assuming direct conversion of triose formed in first cycle to pyruvate.

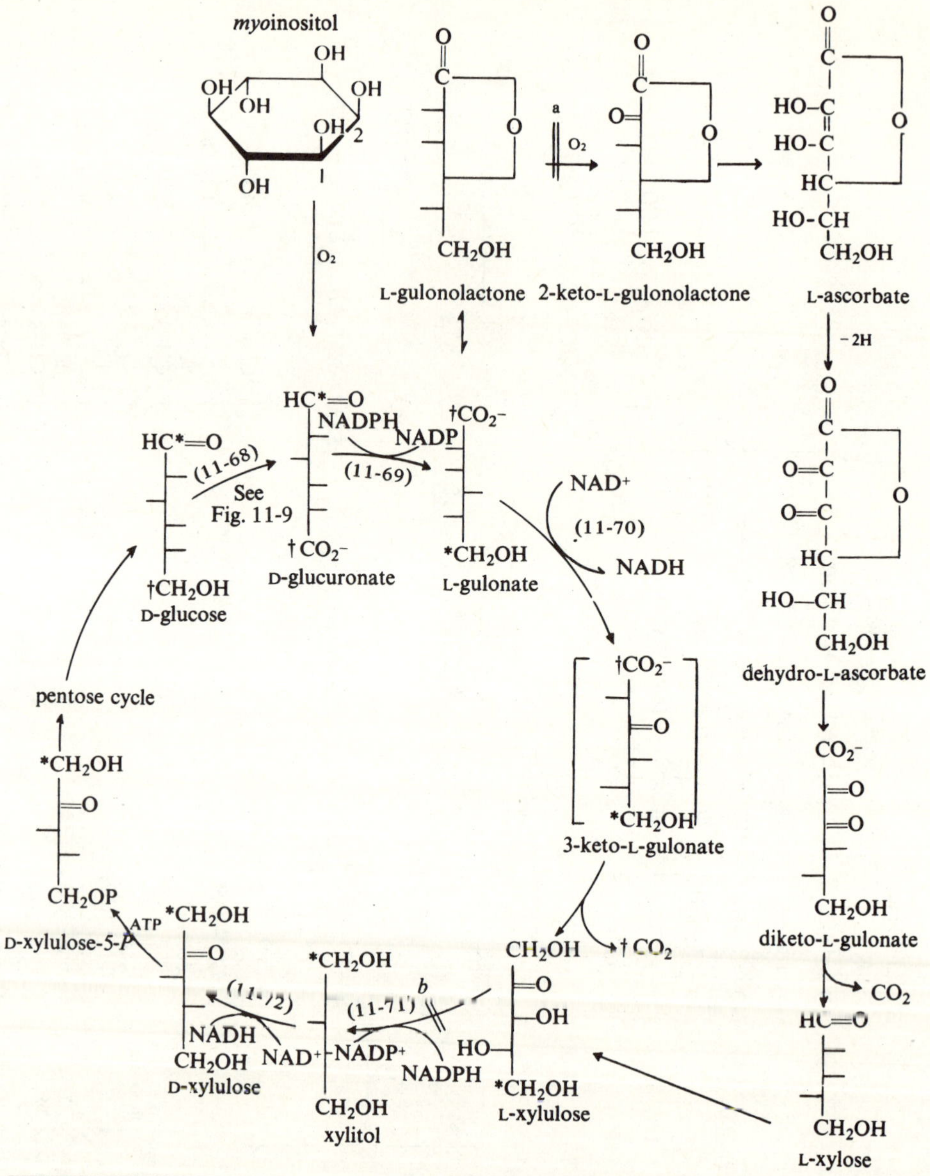

FIGURE 11-10
D-Glucuronate–L-gulonate pathways.

OTHER CONVERSIONS OF HEXOSES (Fig. 11-11)

Oxidations

Enzymatic reactions are known that oxidize unsubstituted sugars rather than their phosphate esters; for example, the conversion that is shown as reaction (11-68) is an exceedingly common one, but different mechanisms are utilized for the removal of the reducing equivalents in different cells. In the case of

$$\text{D-Glucose} \rightleftharpoons \text{D-gluconolactone} + 2[\text{H}] \xrightarrow{H_2O} \text{D-gluconate} \qquad (11\text{-}68)$$

mammalian liver the enzyme concerned is a pyridine nucleotide–requiring dehydrogenase, which can utilize either NAD^+ or $NADP^+$ with equal facility and is B specific. The substrate is β-D-glucopyranose, which can be replaced by β-D-xylose, 6-deoxy-D-glucose, and 6-fluoro-D-glucose (54). The product is the δ-lactone, but when the reaction is in the reverse direction the γ-lactone can also be used. The dehydrogenation proper with either NAD^+ or $NADP^+$ as acceptor gives a $\Delta G^{0\prime} = -2600$ cal/mol; if hydrolysis (either spontaneous or catalyzed by the enzymes that effect the analogous conversion of the phosphorylated sugars) is allowed to take place, this becomes -7700 cal/mol. Similar enzymes in microorganisms catalyze the reactions L-arabinofuranose $\xrightleftharpoons{NAD}$ γ-lactone $\xrightarrow{H_2O}$ L-arabonate, and D-galactose $\xrightleftharpoons{NAD}$ γ-lactone $\xrightarrow{H_2O}$ D-galactonate. From the mold *Penicillium notatum* there can be isolated a pure, crystalline, exceedingly active flavoprotein enzyme (*glucose oxidase*), which catalyzes reaction (11-68) with E·FAD as the primary, and oxygen as the terminal, electron acceptor: glucose + E·FAD $\rightleftharpoons$ [E·$FADH_2$·glucose] $\xrightarrow{O_2}$ δ-lactone + H_2O_2 + E·FAD.[28] The specificities of this enzyme with regard to substrate and product are similar to the above. Similar enzymes are found in molds, fungi, and red algae. In *Pseudomonas,* the analogous enzyme appears to be particulate and linked to the respiratory chain (Chapter 15) with cytochromes intervening between the primary acceptors and molecular oxygen. Thus, almost all cells examined possess efficient means of converting D-glucose to D-gluconate, yet the latter cannot be metabolized as such but must first be phosphorylated and thence enter the metabolic pathways.

Reductions We have already had occasion to refer to sets of reversible pyridine nucleotide-linked dehydrogenations [reactions (11-69) and (11-70), (11-71) and (11-72)] that generate sugar alcohols (i.e., polyols) and with them as intermediates affect various isomerizations, for example,

$$\text{L-Xylulose} \underset{NADP^+}{\overset{NADPH}{\rightleftharpoons}} \text{xylitol} \underset{NADH}{\overset{NAD^+}{\rightleftharpoons}} \text{D-xylulose} \qquad (11\text{-}71) + (11\text{-}72)$$

In mammalian liver, these enzymes appear to be localized in the mitochondria.

An entirely analogous, equally ubiquitous, set of enzymes catalyzes reactions (11-73) and (11-74):

$$\alpha\text{-D-Glucopyranose} + H^+ \underset{(11\text{-}73)}{\overset{NADPH \quad NADP^+}{\rightleftharpoons}} \text{D-Sorbitol} \underset{(11\text{-}74)}{\overset{NAD^+ \quad NADH}{\rightleftharpoons}} \text{D-Fructopyranose}$$

α-D-Glucopyranose D-Sorbitol D-Fructopyranose

The enzyme catalyzing reaction (11-73), an $NADP^+$-specific D-hexitol dehydrogenase (or aldose reductase), is present in a variety of mammalian tissues, such as placenta and seminal vesicles.[29] Note that like most other enzymes involved in carbohydrate anabolism, it is specific for the α-D form. The second enzyme of the set is an NAD^+-specific hexitol dehydrogenase,

[28] This enzyme, also called *notatin,* used to be referred to as *penicillin* A or B. Its "antibiotic" properties result entirely from the generation of H_2O_2. The latter property is now made use of in colorimetric tests and test papers for glucose by allowing the peroxide to react with a leuco dye such as benzidine to form a highly colored product.

[29] A similar enzyme has also been found in the lens of the eye, where the accumulation of the resultant sugar alcohols gives rise to cataracts (55).

which leads to the formation of the corresponding ketose. The enzyme catalyzing this reaction is found in liver, male sex glands, sperm, etc., and it has been crystallized from sheep liver (56). The stereospecificity of this, probably extramitochondrial enzyme, called L-*iditol dehydrogenase,* is such as to require a polyol with (1) a primary carbinol group at C-1 (i.e., adjacent to the site of oxidation); (2) a secondary carbinol group with the L configuration (relative to C-1) at C-2, the site of oxidation proper; and (3) a secondary carbinol group, also of the L configuration, at C-4—with the sole exception that L-arabitol is slowly oxidized to L-xylulose:

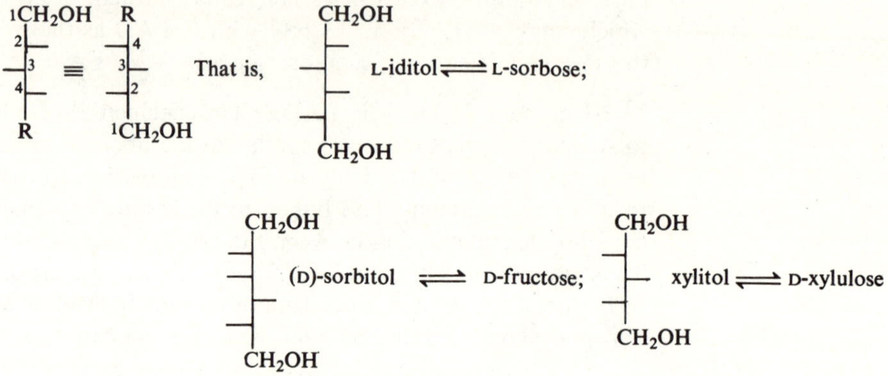

$^1CH_2OH$ (C-2, C-3, C-4) R ≡ R (C-4, C-3, C-2) $^1CH_2OH$ That is, CH_2OH … CH_2OH L-iditol ⇌ L-sorbose;

CH_2OH … CH_2OH (D)-sorbitol ⇌ D-fructose; CH_2OH … CH_2OH xylitol ⇌ D-xylulose

Reactions (11-73) and (11-74), acting in concert, are responsible for the presence of high concentrations of D-fructose in seminal plasma (57).

Numerous microorganisms can be made to utilize polyols (mainly hexitols and pentitols) as the sole source of carbon. Under these conditions a variety of induced polyol dehydrogenases have been demonstrated and purified enzymes isolated that are specific for D-sorbitol (L-iditol), D-mannitol-1-*P*, D-mannitol, and a number of pentitols (see Fig. 11-8). One of the earliest examples of bacterial *transformation* (Chapter 19) for a specific enzyme was the demonstration that the genetic trait to form an inducible mannitol dehydrogenase can be transferred from a (donor) strain of *Diplococcus pneumoniae* (*Pneumococcus*) to one *not* possessing this capacity (acceptor strain) by means of pure DNA (*transforming principle*). Once "transformed," the acceptor cell not only becomes capable of synthesizing the enzyme but also of transmitting this capability to its daughter cells and of acting as donor in subsequent transformations.

Interconversions of Hexoses

The simplest interconversion or isomerization reaction is the reversible equilibrium between the α-D and β-D form of the common hexoses, such as glucose. Although this anomerization process, known as mutarotation, is both acid and base catalyzed, it still is quite rapid in the physiological *p*H range[30]; specific enzymes (*mutarotases, aldose epimerases*) that further enhance the rate have been isolated from mammalian kidney and a number of molds: these enzymes catalyze mutarotation of D-glucose, D-galactose, D-xylose, L-arabinose, maltose, and lactose.

The other common hexoses besides D-glucose that have to be considered are the aldoses D-mannose and D-galactose, and the ketoses D-fructose and

[30] The spontaneous rate of mutarotation of α-glucopyranose-6-*P* at *p*H 6.8 and 20° is about 0.04 sec^{-1}, which is about two orders of magnitude faster than that of the free sugar.

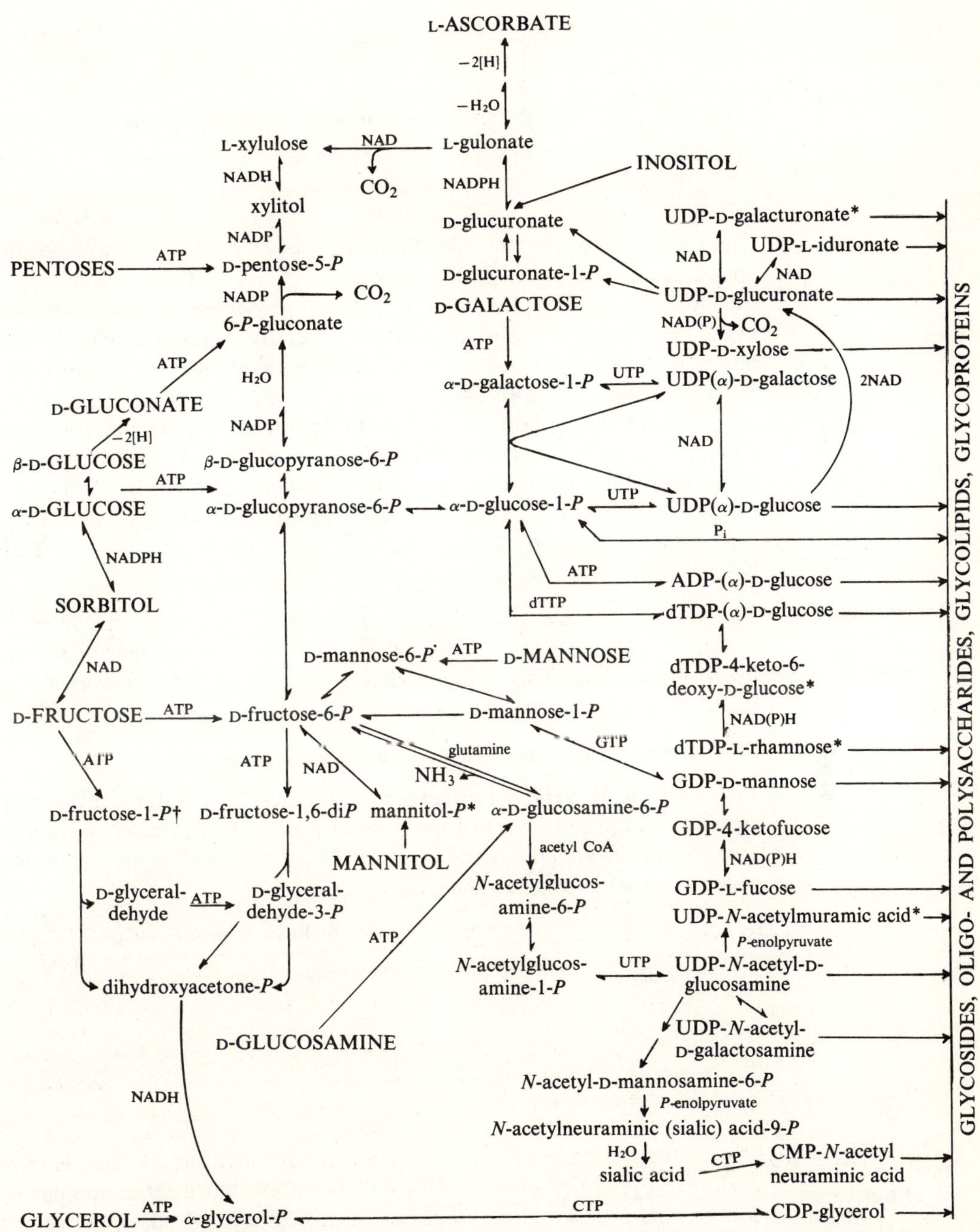

FIGURE 11-11
Generalized sugar interconversions.
* Intermediates identified primarily or exclusively in plants and microorganisms.
† Intermediates identified primarily in mammals.
N.B.: All kinase (transphosphorylase) reactions leading to phosphorylated derivatives, and involving ATP, are written as being virtually irreversible. The phosphate esters involved can all be hydrolyzed, again virtually irreversibly, by means of separate reactions catalyzed by nonspecific or, in certain instances, specific phosphatases.

L-sorbose, which differ from each other in the configuration of one of the carbon atoms in the sequence:

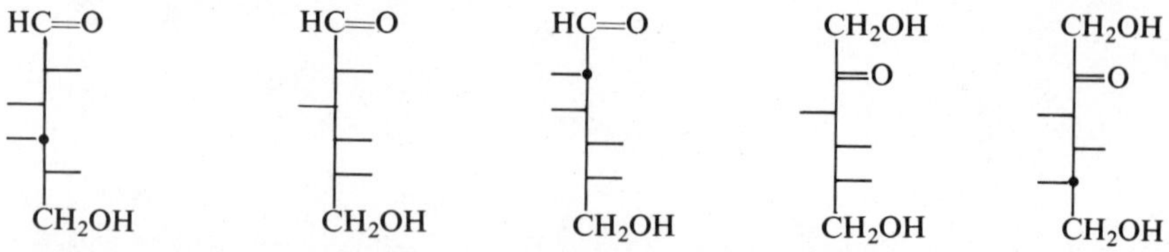

HC=O	HC=O	HC=O	CH_2OH	CH_2OH
			=O	=O
CH_2OH	CH_2OH	CH_2OH	CH_2OH	CH_2OH
D-Galactose	D-Glucose	D-Mannose	D-Fructose	L-Sorbose

• indicates epimerized C.

Mannose is phosphorylated by the nonspecific hexokinases to yield the corresponding 6-phosphates. Because of the high K_s of this enzyme for fructose, the latter is usually phosphorylated by a *specific fructokinase,* which, however, generates fructose-1-*P*. Furthermore, liver aldolase cleaves the latter compound effectively and thus initiates a catabolic bypass for this compound in this tissue. Fructose-1-*P* can also be phosphorylated by ATP in the presence of phosphofructokinase to produce the common intermediate fructose-1,6-di*P*. The mechanism of the interconversion of mannose-6-*P* and fructose-6-*P* has already been discussed. D-Galactose is phosphorylated in the presence of galactokinase in the 1 position to yield the (glycosidic) galactose-1-*P*. Its isomerization (Walden inversion at C-4) to glucose-1-*P* is discussed below.

Formation of Glucosamine

The various hexosamines found in a variety of mucopolysaccharides, glycoproteins, and bacterial cell wall polysaccharides all have their metabolic origin in D-fructose-6-*P*. This compound is converted to α-D-glucosamine with the amide-NH_2 of glutamine as the nitrogen source:

$$\text{D-Fructose-6-}P + \text{glutamine} \rightarrow \text{D-glucosamine-6-}P + \text{glutamate} \qquad (11\text{-}75)$$

The removal of NH_3 to regenerate the ketose-*P* is catalyzed by separate and specific deaminases. Most of the natural products of interest contain *N*-acetyl glucosamine. This compound is formed in a reaction catalyzed by a specific *N*-acetylase with acetyl-CoA as the acetyl donor:

$$\text{D-Glucosamine-6-}P + \text{acetyl CoA} \rightarrow N\text{-acetyl-D-glucosamine-6-}P + \text{CoASH} \qquad (11\text{-}76)$$

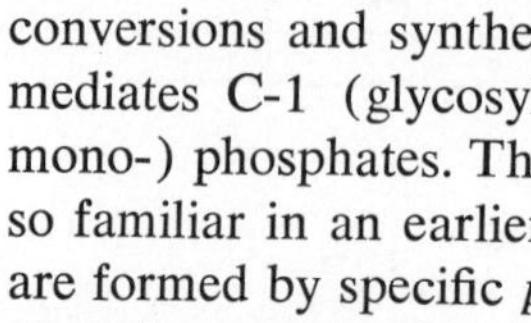

$CH_2OPO_3^{2-}$
O H
OH
HO OH
$NHCOCH_3$

Formation and Interconversions of Nucleoside Diphosphate Sugars

One of the most significant milestones in carbohydrate biochemistry was the discovery by L. F. Leloir and his collaborators that a large number of interconversions and synthetic processes in this area involved as obligatory intermediates C-1 (glycosyl) esters of aldoses and nucleoside di- (occasionally mono-) phosphates. These intermediates take the place of the sugar phosphates so familiar in an earlier phase of carbohydrate biochemistry. The derivatives are formed by specific *pyrophosphorylases* [(nucleoside triphosphate): sugar-phosphate nucleotidyl transferases]. The classical case is that of the synthesis of uridine-diphosphate glucose (UDP-glucose, UDPG, UDPGlc) by reaction (11-77) where N = uridine and the sugar moiety is glucose-1-*P*.

$$\text{Sugar-}\alpha\text{-1-}P\text{ (}{-}O{-}PO_3^{2-}\text{)} + \text{NTP} \underset{Mg^{2+}}{\overset{\Delta G' = 0}{\rightleftharpoons}} \text{NDP-Sugar (substituted } \alpha\text{-}P\text{)} \; {-}O{-}\overset{O}{\underset{O^-}{P}}{-}O{-}\overset{O}{\underset{O^-}{P}}{-}O{-}N + PP_i \xrightarrow[\text{pyrophosphatase}]{H_2O} 2P_i \qquad (11\text{-}77)$$

By now a wide array of sugars and their derivatives, some of which are shown in Fig. 11-11, and UTP, ATP (with glucose-1-*P* exclusively), GTP, CTP, and dTTP all have become implicated in reaction (11-77). Once formed, the NDP sugars can undergo the following types of reactions.

Galactose-1-P–Glucose-1-P Epimerization. This sequence consists of the following steps:

Reaction	Enzyme	
Galactose-1-*P* + UTP ⇌ UDP-galactose + PP_i	*UDP galactose pyrophosphorylase* (*galactose-1-P uridylyl transferase*)	(11-77a)
UDP-galactose ⇌ UDP-glucose	*UDP glucose-epimerase*	(11-78)
UDP-glucose + galactose-1-*P* ⇌ UDP-galactose + glucose-1-*P*	*hexose-1-P uridylyl transferase*	(11-79)
Glucose-1-*P* + UTP ⇌ UDP-glucose + PP_i	*UDP glucose pyrophosphorylase* (*glucose-1-P uridylyl transferase*)	(11-77b)

Once the UDP sugar derivative has been formed by reaction (11-77a) or (11-77b), the epimerization is brought about by reactions (11-78) and (11-79) acting in tandem. The epimerase reaction proper [reaction (11-78)] takes place at the NDP level and involves enzyme-bound NAD^+ (one molecule per protein molecule) as an obligatory cofactor, which cannot be replaced by $NADP^+$ or NADH. With a highly purified enzyme from *E. coli,* the reaction sequence has been shown to be UDP-glucose + E · NAD^+ ⇌ E · NAD^+ · UDP-glucose ⇌ E · NADH · UDP-4-ketoglucose ⇌ E · NAD^+ · UDP-galactose ⇌ E · NAD^+ + UDP-galactose.

Other Epimerizations. Similar epimerizations at C-4 have been demonstrated with UDP-D-glucuronate ⇌ UDP-D-galacturonate; UDP-D-xylose ⇌ UDP-L-arabinose; UDP-D-glucosamine ⇌ UDP-D-galactosamine; and with UDP- (or dTDP)-*N*-acetyl-D-glucosamine ⇌ UDP- (or dTDP)-*N*-acetyl-D-galactosamine. An analogous epimerization at C-5 is UDP-D-glucuronate ⇌ UDP-L-iduronate, while the epimerization at C-2, UDP-acetyglucosamine ⇌ acetylmannosamine + UDP, may similarly proceed with UDP-*N*-acetylmannosamine as an intermediate.

Hereditary deficiencies in the various enzymes catalyzing these reactions have been identified in microorganisms and man: *galactosemia* in infants is caused by a lack of enzyme catalyzing reaction (11-79) and results in an inability to metabolize the galactose derived from lactose (a β-D-galactoside) in milk. The alternative route given by reactions (11-77a) plus (11-78) plus (11-77b) is inoperative in infants because of a low level of the enzyme catalyzing reaction (11-77b), and becomes significant only during adolescence.

In *E. coli,* the conversion of galactose to glucose is brought about by the galactokinase reaction plus reactions (11-78) and (11-79). The genes that control the biosynthesis of these three enzymes form a closely coordinated cluster or *operon,* and the three enzymes can be induced simultaneously in a controlled fashion by either D-galactose or D-fucose (see Chapter 21). Bacteria lacking the galactose operon and hence the above enzyme can be rendered genetically sufficient by means of infection by a bacteriophage (λdg) in which the operon has been inserted into its genetic makeup, a process known as *transduction* (Chapter 19). Isolated, highly purified DNA from bacteriophage λdg, or even smaller fragments obtained from such DNA molecules, can substitute for the intact phage and still bring about the transfer of genetic information (*transfection*) and the initiation of enzyme synthesis.

Oxidation of NDP-Aldose to NDP-Uronic Acids. An oxidation of sugar nucleotides involving four reducing equivalents yields uronic acids; a typical reaction is

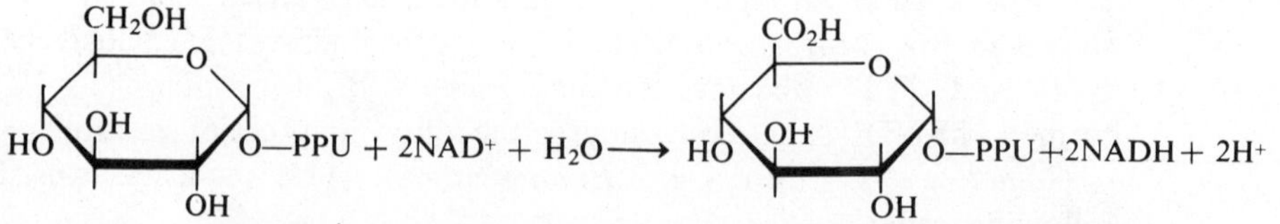

$$\text{(UDP-glucose; }CH_2OH\text{, HO, OH, OH, O—PPU)} + 2NAD^+ + H_2O \longrightarrow \text{(UDP-glucuronic acid; }CO_2H\text{, HO, OH, OH, O—PPU)} + 2NADH + 2H^+ \qquad (11\text{-}80)$$

catalyzed by *UDP-D-glucose dehydrogenase,* forming UDP-D-glucuronic acid. Two molecules of NAD^+ are required, the reaction is irreversible, and there is no evidence for the participation of any half-oxidized (2-equivalent) intermediates.

Decarboxylation of Uronic Acids. A particulate preparation from plants catalyzes the following reaction, the decarboxylation of UDP-D-glucuronic acid to UDP-D-xylose:

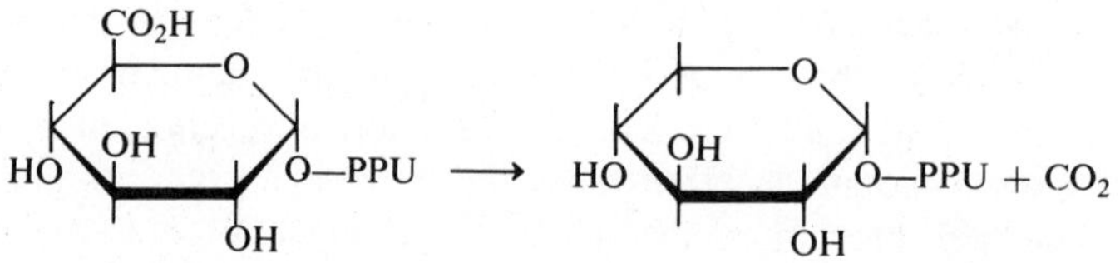

$$\text{(}CO_2H\text{, HO, OH, OH, O—PPU)} \longrightarrow \text{(HO, OH, OH, O—PPU)} + CO_2 \qquad (11\text{-}81)$$

Formation of 6-Deoxyhexoses (5-Methyl Pentoses). The formation of GDP-L-fucose in a variety of preparations proceeds by the path shown in reaction (11-82):

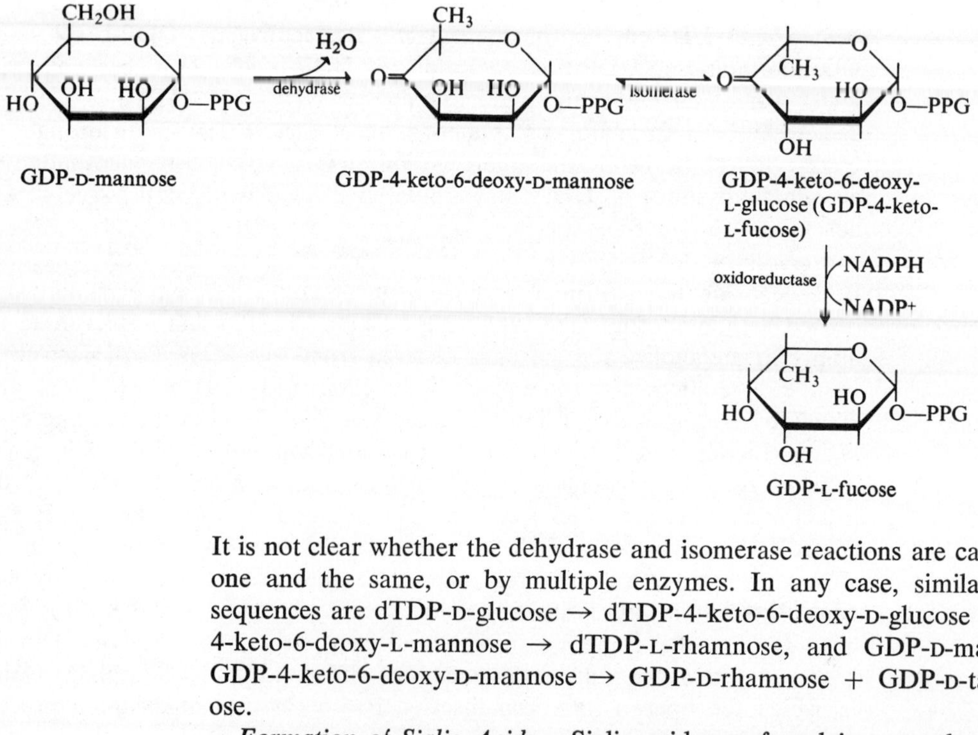

$$\text{GDP-D-mannose} \xrightarrow[\text{dehydrase}]{-H_2O} \text{GDP-4-keto-6-deoxy-D-mannose} \xrightarrow{\text{isomerase}} \text{GDP-4-keto-6-deoxy-L-glucose (GDP-4-keto-L-fucose)} \xrightarrow[\text{oxidoreductase}]{NADPH \to NADP^+} \text{GDP-L-fucose} \qquad (11\text{-}82)$$

GDP-D-mannose

GDP-4-keto-6-deoxy-D-mannose

GDP-4-keto-6-deoxy-L-glucose (GDP-4-keto-L-fucose)

GDP-L-fucose

It is not clear whether the dehydrase and isomerase reactions are catalyzed by one and the same, or by multiple enzymes. In any case, similar reaction sequences are dTDP-D-glucose → dTDP-4-keto-6-deoxy-D-glucose → dTDP-4-keto-6-deoxy-L-mannose → dTDP-L-rhamnose, and GDP-D-mannose → GDP-4-keto-6-deoxy-D-mannose → GDP-D-rhamnose + GDP-D-talomethylose.

Formation of Sialic Acids. Sialic acids are found in a number of mammalian glycoproteins and bacterial cell wall structures. In mammalian tissues they appear to be synthesized by the complicated pathway (57) shown as reaction (11-83)[31]:

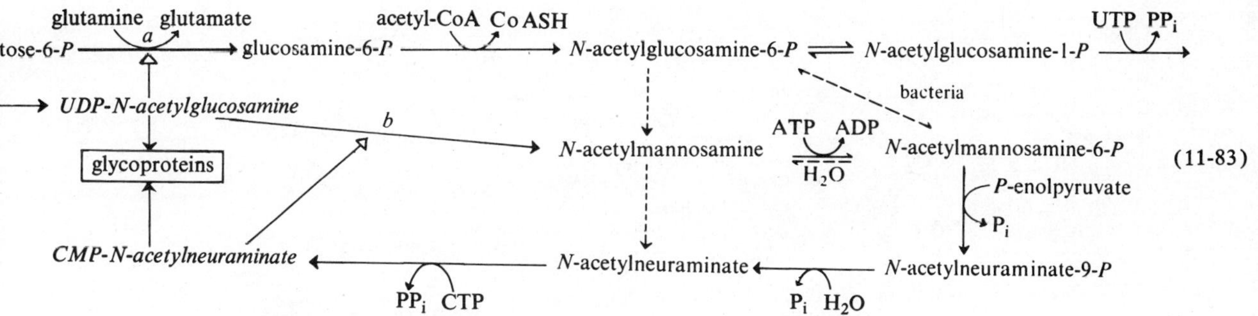

N.B.: Glycolyl can substitute for acetyl residues in most of these interconversions.

There are two sets of reactions required for the formation of the appropriate monomeric units [shown in italic type in reaction (11-83)]; the first is characterized by two reactions where the pathways diverge from common intermediates: (1) the *fructose-6-P glutamine transamidase,* which is subject to end-product inhibition by UDP-*N*-acetylglucosamine, and (2) with this very compound as starting point of a second sequence, the UDP-*N-acetylglucosamine-2-epimerase,* which is subject to end-product inhibition by CMP-*N*-acetylneuraminic acid. The other set of novel reactions is:

$CH_2OPO_3^{2-}$

N-acetyl-D-mannosamine -6-phosphate

enolpyruvyltransferase or "aldolase" (a)

CO_2^- — $C—OPO_3^{2-}$ — CH_2

P-enolpyruvate

P_i

CO_2^-, C—OH, CH_2, HC—OH, AcN—CH, CH, HC—OH, HC—OH, $H_2COPO_3^{2-}$ — sialate-9-*P*

([5-amino-3,5-dideoxy-D-glycero-D-galactononulosonic acid] ≡ neuraminic acid; *N*-acetylneuraminic acid ≡ sialic acid)

(11-84)

phosphatase (b) H_2O P_i

CMP-*N*-acetylneuraminate ← *N*-acetylneuraminate-CMP pyrophosphorylase (c) — PP_i CTP

$HOCH_2$, OH, H, CH_3CNH, O, OH, OH, CO_2^-, O—$R^{(32)}$ R=H

[31] In bacteria, the pathway appears to be somewhat less complex: ⟶ *N*-acetylglucosamine-6-*P* ⟶ *N*-acetylmannosamine-6-P ⟶ *N*-acetylmannosamine $\xrightarrow{\text{PEP}}$ *N*-acetylneuraminic acid.

[32] The stereochemistry of sialic acid present in gangliosides, glycoproteins, etc., has been established as the α-D-ketoside, that is, with the ketosidic bond equatorial and the carboxyl group axial to the pyranoid ring (58).

A somewhat similar set of reactions leads to the synthesis of UDP-*N*-acetylmuramic acid, one of the monomeric subunits found in the *mureins* of bacterial cell walls: complex polysaccharides made up of *N*-acetylglucosamine and *N*-acetylmuramic acid peptides (pages 482 and 551):

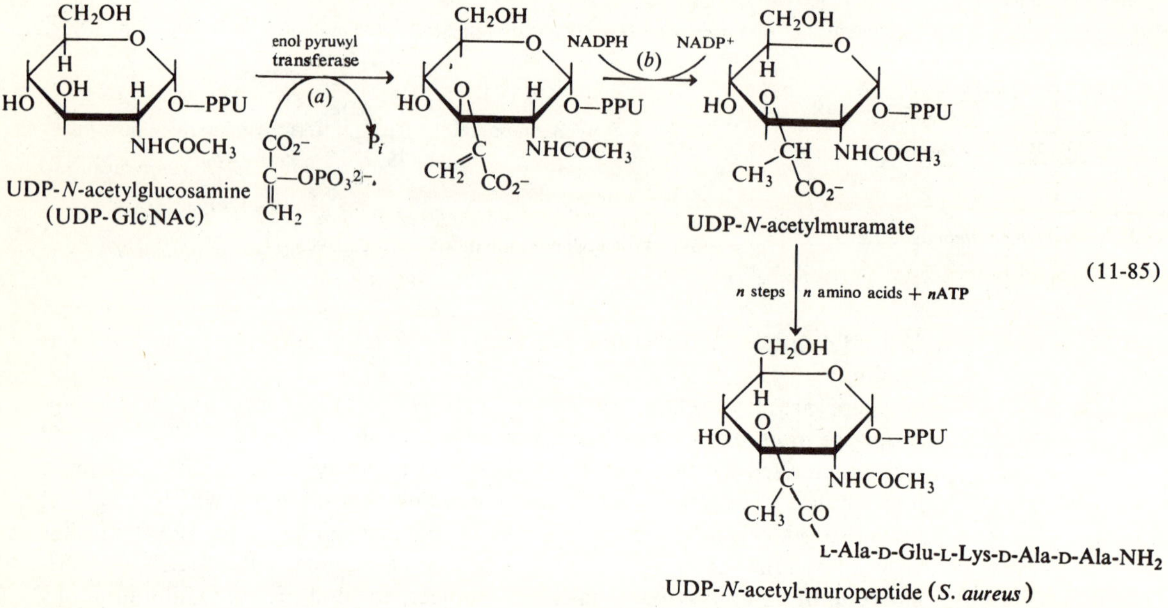

UDP-*N*-acetylglucosamine (UDP-GlcNAc) $\xrightarrow[(a)]{\text{enol pyruvyl transferase; } CO_2^- \text{—} C(OPO_3^{2-})\text{=}CH_2 \rightarrow P_i}$ → $\xrightarrow[(b)]{NADPH \rightarrow NADP^+}$ UDP-*N*-acetylmuramate $\xrightarrow{n \text{ steps; } n \text{ amino acids} + n\text{ATP}}$ UDP-*N*-acetyl-muropeptide (*S. aureus*) [L-Ala-D-Glu-L-Lys-D-Ala-D-Ala-NH_2] (11-85)

BIOSYNTHESIS OF GLYCOSIDIC BONDS

By far the most important function of the nucleoside (di)phosphate sugars is to act as glycosyl donors in a wide variety of reactions, all of which involve the formation of a new glycosidic linkage. Since the $\Delta G^{0\prime}$ of hydrolysis of such bonds is of the order of -3000 cal/mol, it can readily be appreciated that their effective formation requires the expenditure of metabolic energy and hence the preferred reaction is

$$\text{NDP-sugar} + \text{acceptor} \rightleftharpoons \text{sugar}(1 \rightarrow)\ \text{acceptor} + \text{NDP} \quad (11\text{-}86)$$

Thus formation of the glycosylated product from NTP + sugar is brought about by an exergonic pathway:

$$\text{Sugar} + 2\text{ATP} + \text{acceptor} \rightarrow \text{sugar-acceptor} + 2\text{ADP} + \text{PP}_i \qquad \Delta G^{0\prime} \simeq -3.5 \text{ kcal/mol}$$

$$\text{Sugar} + 2\text{ATP} + \text{acceptor} \rightarrow \text{sugar-acceptor} + 2\text{ADP} + 2\text{P}_i \qquad \Delta G^{0\prime} \simeq -12 \text{ kcal/mol}$$

Synthesis of Simple Glycosides and Disaccharides

The simplest reaction of the general type under consideration here is the formation of a glycoside by reaction (11-86a):

$$\text{NDP-sugar} + \text{R—OH} \rightarrow \text{R—O—C-1-sugar} + \text{NDP} \quad (11\text{-}86a)$$

The enzymes responsible are referred to as NDP-sugar–acceptor glycosyl transferases. For the formation of simple glucosides, that is, those in which glucose (or a glucose residue in a disaccharide) is linked to an alcohol, a phenol, or an amine, reactions that can be shown to occur readily with enzyme preparations from plants and invertebrates, the $\Delta G^{0\prime}$ is approximately -3500 cal/mol and N = uridine.

Very similar reactions have been shown to take place with preparations from animals and plant sources in which the donor is UDP-D-glucuronic acid. In other variants, UDP-D-galactose is used to galactosylate lipids to form galactolipids, such as the cerebrosides of animal nerves; dTDP-rhamnose forms rhamnolipids in bacteria, and UDP-glucose is the glucose donor for the glucosylation of hydroxymethylcytosine in bacteriophage DNA.

If the bond formed is an *O*-glycoside of a sugar, then reaction (11-86b) leads to the formation of a disaccharide. The most interesting example is provided by sucrose synthesis in green plants, which proceeds largely by the path

$$\rightarrow \text{fructose-6-}P \xrightarrow[-\text{UMP},\ -\text{PP}_i]{+\text{UDPG}} \text{sucrose-6-}P \rightarrow \text{sucrose} \qquad (11\text{-}86\text{b})$$

The $\Delta G^{0\prime}$ for the condensation step is only ~ -1000 cal/mol in this instance, since sucrose is unusual in having a link between the two anomeric carbons for which $\Delta G^{0\prime}$ (hydrolysis) $\simeq -6600$ cal/mol. The overall equilibrium is still favorable for sucrose formation, however, because of the additional increment provided by the hydrolysis in the second step.

Biosynthesis of Polysaccharides (Table 11-4)

In polysaccharide synthesis the basic reaction is modified as follows:

$$\text{Polysaccharide primer} + \text{NDP-sugar} \rightleftharpoons \text{polysaccharide primer-sugar} + \text{NDP} \qquad (11\text{-}86\text{c})$$

$$(\text{Polysaccharide primer-sugar}) + \text{NDP-sugar} \rightleftharpoons \text{primer-sugar-sugar} + \text{NDP, etc.}$$

$\Delta G^{0\prime}$ of reaction (11-86c) is -3000 cal per mol of residue added. The by-now classic example of polysaccharide synthesis is the biosynthesis of glycogen amylose by UDP-glucose-glycogen glucosyl transferase (*glycogen synthetase*) discovered by Leloir in 1957. As written, it is evident that the reaction provides only for chain lengthening; that is, it requires a primer polyglucose (an α-1,4-glucan, amylose, amylopectin, or a glucose oligosaccharide larger than maltotetraose) as glucose acceptor, and it leads to a linear polymer, an α-1,4-glucan. In plants, ADP-D-glucose, rather than the UDP derivative, appears to be the glycosyl donor for starch synthesis.

The branches in naturally occurring glycogen [α-(1 → 6) bonds, which occur about every 8 to 12 residues in the α-(1 → 4) chain] arise by action of a second enzyme, a *glycogen branching enzyme* or amylo (1, 4 → 1, 6) transglycosylase, which cleaves off small (di- or trimeric) fragments from 1 → 4 linkages and transfers them to the same (or a different) polyglucose chain, but now at 1 → 6 linkages:

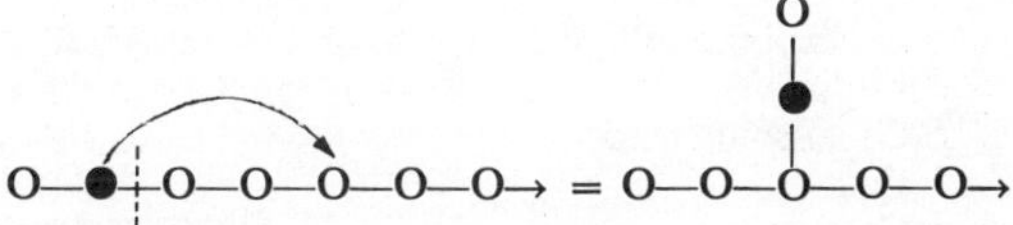

The most interesting feature of glycogen synthetase discovered recently is the demonstration (59–61) that it, like phosphorylase, can exist in two forms (*D* for dependent, *I* for independent of G-6-*P*), which are related as follows:

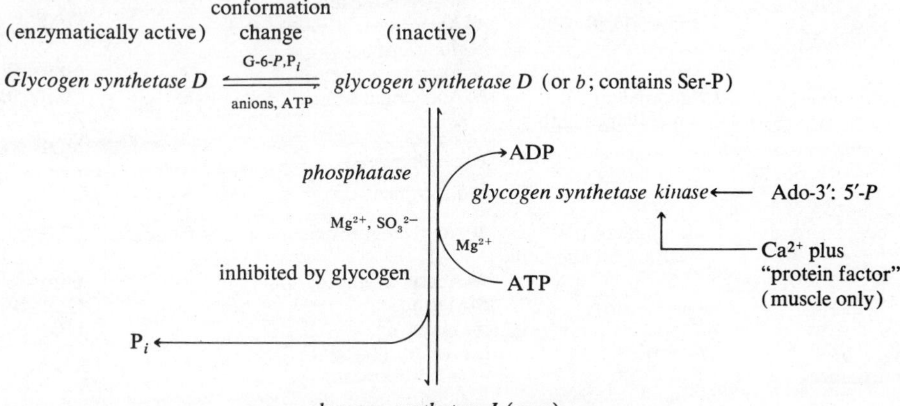

TABLE 11-4
Biosynthesis of di-, oligo-, and polysaccharides

Oligo- or polysaccharide formed	*Source of enzyme system*	*Acceptor or repeating unit*	*Linkages*[a]	*Monomeric nucleotide glycosyl donor*
Sucrose (-*P*)	Plants	D-Fructose (-6-*P*)	$\alpha, 1 \rightarrow 2$	UDP-D-Glucose
Lactose (-*P*)	Mammals	D-Glucose (-1-*P*)	$\beta, 1 \rightarrow 4$	UDP-Galactose
Sialyl-lactose	Mammals	Lactose	$\alpha, 1 \rightarrow 3$ (galactose)	CMP-*N*-Acetylneuraminic acid
Fucosyl-lactose	Mammals	Lactose	$\alpha, 1 \rightarrow 3$ (galactose)	GDP-L-Fucose
α-1,4-Glucan (starch amylose)	Plants	D-Glucose	$\alpha, 1 \rightarrow 4$	ADP-D-Glucose
α-1,4-Glucan (glycogen amylose)	Mammalian liver	D-Glucose	$\alpha, 1 \rightarrow 4$	UDP-D-Glucose
β-1,3-Glucan (callose)	Beans	D-Glucose	$\beta, 1 \rightarrow 3$	UDP-D-Glucose
β-1,4-Glucan (cellulose)	Mung beans, bacteria	D-Glucose	$\beta, 1 \rightarrow 4$	UDP-D-Glucose (bacteria) GDP-D-glucose (plants)
β-1,4-Xylan	Plants	D-Xylose	$\beta, 1 \rightarrow 4$	UDP-D-Xylose
Chitin	*Neurospora crassa*	*N*-Acetyl-D-glucosamine	$\alpha, 1 \rightarrow 4$	UDP-*N*-Acetyl-D-glucosamine
Colominic acid	*E. coli*	*N*-Acetylneuraminic acid	$\alpha, 1 \rightarrow 5$	CMP-*N*-Acetylneuraminic acid
Hyaluronic acid	Bacteria, animals (umbilicus, vitreous humor, etc.)	D-Glucuronic acid; *N*-acetylglucosamine	$\beta, 1 \rightarrow 3$ $\beta, 1 \rightarrow 4$	UDP-D-Glucuronic acid UDP-*N*-Acetyl-D-glucosamine
Chondroitin	Mammalian cornea	D-Glucuronic acid; *N*-acetyl-D-galactosamine	$\beta, 1 \rightarrow 3$ $\beta, 1 \rightarrow 4$	UDP-D-Glucuronic acid UDP-*N*-Acetyl-D-galactosamine
Chondroitin sulfates	Mammalian connective tissue	D-Glucuronic acid; *N*-acetyl-D-galactosamine-4-, or -6-sulfate	$\beta, 1 \rightarrow 3$ $\beta, 1 \rightarrow 4$	UDP derivatives[b]
Dermatan sulfate	Mammalian skin	L-Iduronic acid; *N*-acetyl-D-galactosamine-2-sulfate	$\alpha, 1 \rightarrow 3$ $\beta, 1 \rightarrow 4$	UDP derivatives[b]
Heparin	Mammalian liver, lung, arterial walls	D-Glucuronic acid 2-sulfate; D-galactosamine-*N*, *C*-6-disulfate	$\alpha, 1 \rightarrow 4$ $\alpha, 1 \rightarrow 4$	UDP derivatives[b]
Capsular polysaccharide	Type III pneumococci	D-Glucuronic acid, D-glucose	$\beta, 1 \rightarrow 4$ $\beta, 1 \rightarrow 3$	UDP-D-Glucuronic acid UDP-D-glucose
Capsular polysaccharide	Type VIII pneumococci	D-Glucuronic acid D-Glucose D-Galactose		UDP-D-Glucuronic acid UDP-D-Glucose UDP-D-Galactose
Teichoic acids	Bacteria	Polyribitol or polyglycerol-*P* with D-glucose or *N*-acetyl-glucosamine in glycosidic linkage	$\alpha, 1 \rightarrow 2$	CDP-L-ribitol or L-glycerol UDP-D-glucose or UDP-*N*-acetyl-D-glucosamine
Murein	Bacterial cell walls	*N*-Acetyl-D-glucosamine *N*-Acetylmuropeptides	$\beta, 1 \rightarrow 6$ $\beta, 1 \rightarrow 4$	UDP-*N*-Acetyl-D-glucosamine UDP-*N*-Acetylmuropeptides
Glycoproteins	Mammalian submaxillary gland	*N*-Acetylneuraminic acid *N*-Acetyl-D-galactosamine disaccharide	$\alpha, 2 \rightarrow 6$	CMP-*N*-Acetylneuraminic acid UDP-*N*-Acetyl-D-galactosamine
Glycoproteins	Mammalian plasma: fetuin, orosomucoid	*N*-Acetylneuraminic acid *N*-Acetyl-D-glucosamine D-Mannose D-Galactose	$\alpha, 2 \rightarrow 6$	CMP-*N*-Acetylneuraminic acid UDP-*N*-Acetyl-D-glucose GDP-D-Mannose UDP-D-Galactose
Blood group substances	Animals	L-Fucose, *N*-acetyl-D-galactosamine, D-galactose		

Thus, with glycogen synthetase—an enzyme incidentally with an exceedingly high affinity for its product, which in liver at least is firmly bound to glycogen granules—glucose-6-P (G-6-P) fulfills a role comparable to that of AMP with glycogen phosphorylase. But in this instance it is the phospho-, rather than the dephosphoenzyme, that requires allosteric activation, and the effector is a direct and stoichiometric precursor of the substrate of the reaction to be catalyzed. Again, as with phosphorylase, the specific kinase responsible for the $I \rightarrow D$ conversion exists itself in an inactive as well as an active form and this interconversion is stimulated by Ado-3′:5′-P,[33] as well as in muscle by a "protein factor" in the presence of a high concentration of Ca^{2+}. The interactions between glycogen synthesis and degradation and the two enzyme systems involved can be appreciated in terms of the following diagram (62, 63); if I is the physiologically active form, then the hormonal effects and kinase action would lead to *inactivation* of the synthetase (62):

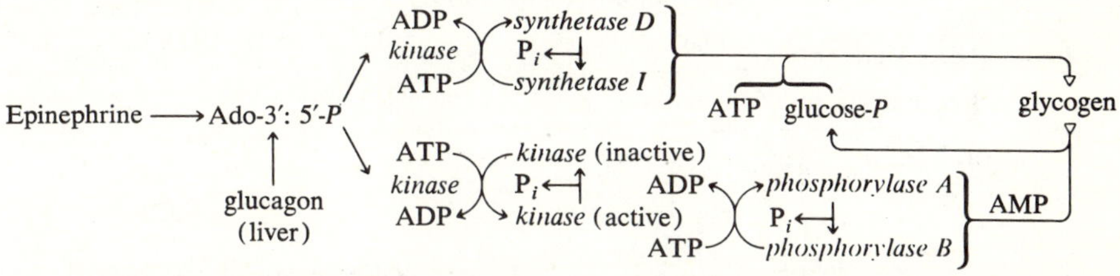

ADP ← *kinase* ← ATP; *synthetase I* → *synthetase D* → P_i → *synthetase I*
Epinephrine ⟶ Ado-3′: 5′-P ← glucagon (liver)
ATP ← *kinase* → ADP; *kinase* (inactive) → *kinase* (active) → P_i → *kinase* (inactive)
ADP ← *kinase* (active) ← ATP; *phosphorylase B* → *phosphorylase A* → P_i → *phosphorylase B*
synthetase: ATP + glucose-P → glycogen; phosphorylase (AMP): glycogen → glucose-P

Biosynthesis of Bacterial Polysaccharides

Mureins. Bacteria contain a variety of polymeric and complex carbohydrates in their cell walls. Gram-positive organisms such as bacilli and micrococci contain simpler structures than do the Gram-negative ones (e.g., Enterobacteriaceae). The basic unit of cell wall structure in all bacteria is a polymeric *murein* (or peptidoglycan), of which examples have been given in Figs. 10-5 through 10-7.

As elucidated by Strominger, murein synthesis proceeds by three distinct phases: (1) synthesis of the monomeric units already described in part in an earlier section [Eqs. (11-76) and (11-85)]; (2) formation of the linear peptidoglycan backbone; and (3) cross-linking.

In phase 1 the pentapeptide chain shown on the left is synthesized as follows:

$$\text{UDP-MurNAc} + \text{L-Ala} + \text{ATP} \rightarrow \text{UDP-MurNAc-L-Ala} + \text{ADP} + P_i \quad (11\text{-}85c)$$

$$\text{UDP-MurNAc-L-Ala} + \text{D-Glu} + \text{ATP} \rightarrow \text{UDP-MurNAc-L-Ala-D-Glu} + \text{ADP} + P_i \quad (11\text{-}85d)$$

$$\text{UDP-MurNAc-L-Ala-D-Glu} + \text{L-Lys} + \text{ATP} \rightarrow \text{UDP-MurNAc-L-Ala-D-isoGlu-}\varepsilon\text{-L-Lys} + \text{ADP} + P_i \quad (11\text{-}85e)$$

[33] The phosphorylation of different proteins by means of ATP, catalyzed by various kinases, stimulated specifically by Ado-3′:5′-P, may be the unifying feature governing the action of this mediator. Possible hormonal effects are explained by their action in stimulating adenyl cyclase, the enzyme responsible for its formation (64).

[a] In disaccharides, the order is glycosyl donor → acceptor. In polysaccharides, the order given is the same as that of the monosaccharide units indicated; for hyaluronic acid, it is glucuronic acid β-(1 → 3)-acetylglucosamine β-(1 → 4)-glucuronic acid, etc.

[b] The exact mechanism for the formation of the O- and N-sulfate esters is not known: 3-phosphoadenosine-5-phosphosulfate (PAPS) is the likely sulfate donor. This compound, a mixed anhydride of sulfuric and substituted phosphoric acid, is formed as follows:

$$SO_4^{2-} + \text{ATP} \xrightarrow{Mg^{2+}} \underset{\text{(APS)}}{\text{adenosine-5'-phosphosulfate}} + PP_i \qquad \text{APS} + \text{ATP} \xrightarrow{Mg^{2+}} \text{PAPS} + \text{ADP}$$

$$\text{L-Ala} \xrightarrow{\text{racemase}} \text{D-Ala; 2D-Ala} + \text{ATP} \xrightarrow[\text{—X—}]{\text{cycloserine}} \text{D-Ala-D-Ala} + \text{ADP} + P_i \qquad (11\text{-}85f)$$

$$\text{UDP-MurNAc-L-Ala-D-isoGlu-L-Lys} + \text{D-Ala-D-Ala} + \text{ATP} \longrightarrow \text{UDP-MurNAc-L-Ala-D-isoGlu-L-Lys-D-Ala-D-Ala} \qquad (11\text{-}85g)$$

In phase 2 the monomer is transferred to a C_{55} terpene alcohol, called a *polyprenol,* present as the phosphate ester in the bacterial cell membrane, which then takes over from UDP and functions as a carrier in chain growth (pages 554–555):

$$H(H_2C\text{—}\overset{CH_3}{\overset{|}{C}}\text{=CH—}CH_2)_{10}\text{—}CH_2\text{—}\overset{CH_3}{\overset{|}{C}}\text{=CH—}CH_2O\text{—}PO_3{}^{2-} \qquad \textit{Bactoprenol phosphate } (\text{BPr-}P) \qquad (11\text{-}86)$$

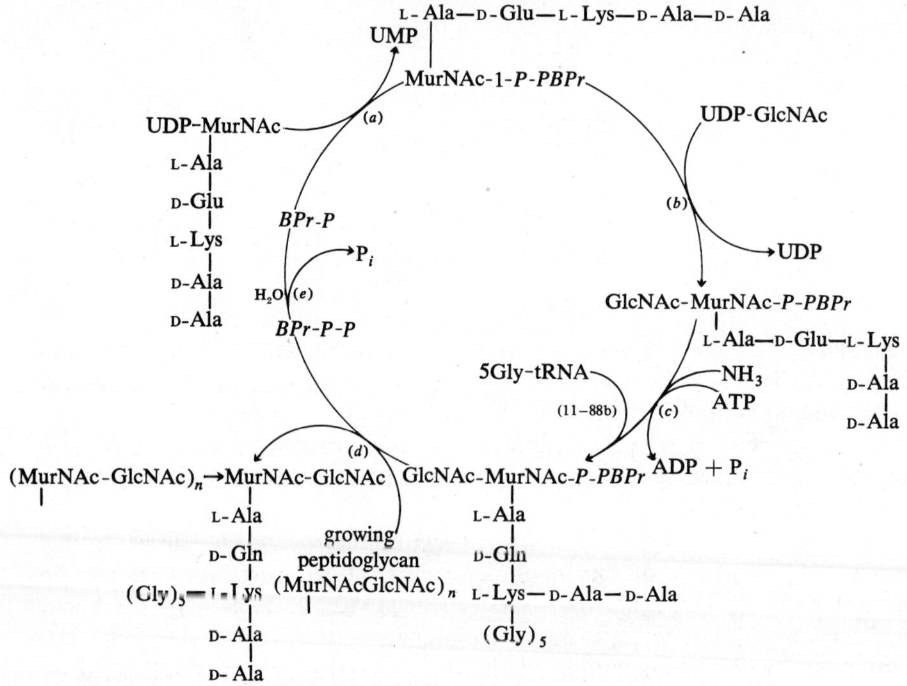

L-Ala—D-Glu—L-Lys—D-Ala—D-Ala
UMP
MurNAc-1-*P-PBPr*
UDP-MurNAc (L-Ala, D-Glu, L-Lys, D-Ala, D-Ala)
(a)
UDP-GlcNAc
(b)
UDP
BPr-P
P_i
H_2O (e)
BPr-P-P
GlcNAc-MurNAc-*P-PBPr*
L-Ala—D-Glu—L-Lys—D-Ala—D-Ala
5Gly-tRNA
(11–88b)
NH_3
ATP
(c)
ADP + P_i
(d)
$(\text{MurNAc-GlcNAc})_n$→MurNAc-GlcNAc (L-Ala, D-Gln, $(\text{Gly})_5$—L-Lys, D-Ala, D-Ala)
growing peptidoglycan $(\text{MurNAcGlcNAc})_n$
GlcNAc-MurNAc-*P-PBPr* (L-Ala, D-Gln, L-Lys—D-Ala—D-Ala, $(\text{Gly})_5$)

(11-87)

Cross-linking reactions start by the transfer, in order, of five glycine groups from glycinyl tRNA (Chapters 5 and 21) to the ε-amino group of the L-Lys residue in the side chain:

$$5\text{Gly} + 5\text{tRNA}^{\text{Gly}} + 5\text{ATP} \rightleftharpoons 5\text{Gly-tRNA}^{\text{Gly}} + 5\text{AMP} + 5\text{PP}_i \qquad (11\text{-}88a)$$

$$\text{—}\underset{NH_2}{\underset{|}{\text{L-Lys}}}\text{—} + 5\text{Gly-tRNA}^{\text{Gly}} \xrightarrow{\text{5 steps}} \text{—}\underset{\text{NH-Gly-(Gly)}_3\text{-Gly-NH}_2}{\underset{|}{\text{L-Lys}}}\text{—} \qquad (11\text{-}88b)$$

Subsequently the terminal glycine residue becomes linked to the subterminal D-Ala residue in a second chain by a *transpeptidation*:

$$\text{L-Lys-NH-Gly-(Gly)}_3\text{-Gly-}\overset{H}{\underset{H}{N}}\text{:} \longrightarrow \text{CO-}\overset{CH_3}{\overset{|}{C}}\text{H-NH-L-Lys—} \; (\text{NH-}\underset{CH_3}{\underset{|}{C}}\text{H-CO}_2{}^-) \rightleftharpoons \text{L-Lys-NH-Gly-Gly}_3\text{-Gly-Ala-Lys—} + \text{D-Ala} \qquad (11\text{-}88c)$$

This step is inhibited by the antibiotic penicillin probably by acylation and resulting inactivation of the transpeptidase.

Teichoic Acids. Teichoic acids are found associated with the cell wall murein (to which they are linked by means of a phosphodiester bridge to C-6 of MurNAc), the cell membranes, and the intervening space in Gram-positive organisms. As indicated on page 484, their monomeric units are, in general, ribitol or glycerol-*P*. They are synthesized by stepwise polymerization of the corresponding sugar alcohol–CDP derivatives.

Glycerol Teichoic Acids

L-α-Glycerophosphate + CTP ⇌ CDP-glycerol + PP_i

CDP-Glycerol + (L-α-glycerol phosphate)$_n$ → CMP + (L-α-glycerol phosphate)$_{n+1}$

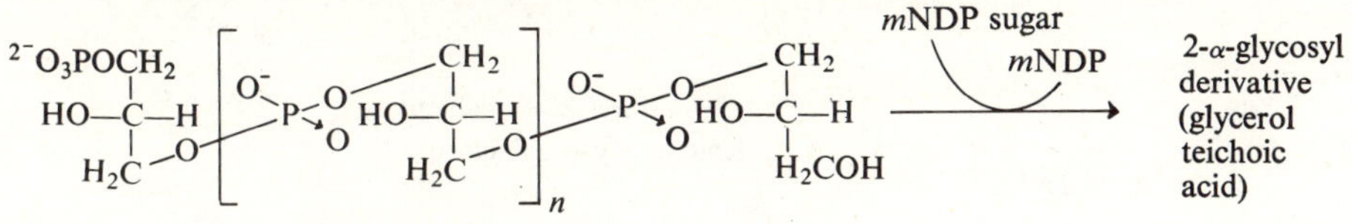

$$^{2-}O_3POCH_2CH(OH)CH_2O-[P(O)(O^-)-OCH_2CH(OH)CH_2O]_n-P(O)(O^-)-OCH_2CH(OH)CH_2OH \xrightarrow[m\text{NDP}]{m\text{NDP sugar}} \text{2-}\alpha\text{-glycosyl derivative (glycerol teichoic acid)} \quad (11\text{-}89)$$

n = 30 *in vitro*

Ribitol Teichoic Acids

Ribulose-5-*P* + NADPH ⇌ ribitol-5-*P* + $NADP^+$

Ribitol-5-*P* + CTP ⇌ CDP-ribitol (11-90a)

*n*CDP ribitol + acceptor → CMP + –ribitol-(ribitol-*P*)$_n$-ribitol-acceptor

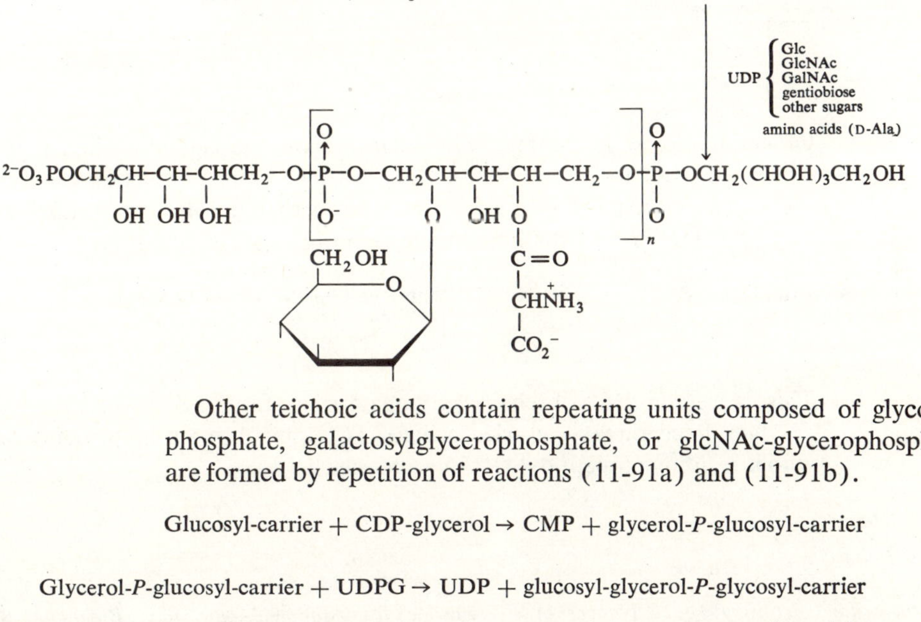

↓ UDP {Glc, GlcNAc, GalNAc, gentiobiose, other sugars}; amino acids (D-Ala) (11-90b)

$$^{2-}O_3POCH_2CH(OH)CH(OH)CH(OH)CH_2-O-[P(O)(O^-)-O-CH_2CH(O\text{-sugar})CH(OH)CH(O\text{-}C(=O)CH(\overset{+}{N}H_3)CO_2^-)CH_2-O]_n-P(O)(O)-OCH_2(CHOH)_3CH_2OH \quad (11\text{-}90c)$$

Other teichoic acids contain repeating units composed of glycosylglycerophosphate, galactosylglycerophosphate, or glcNAc-glycerophosphate. They are formed by repetition of reactions (11-91a) and (11-91b).

Glucosyl-carrier + CDP-glycerol → CMP + glycerol-*P*-glucosyl-carrier (11-91a)

Glycerol-*P*-glucosyl-carrier + UDPG → UDP + glucosyl-glycerol-*P*-glycosyl-carrier (11-91b)

Biosynthesis of Lipopolysaccharides

In addition to mureins, Gram-negative organisms also elaborate as part of their external coats complex lipopolysaccharides (LPS) (Fig. 10-8) which are of considerable interest as *endotoxins* against higher animals and as a means of immunochemical identification (*O antigens*) of these frequently pathogenic organisms. The various units are attached, one at a time, as the UDP derivatives; the octulosonic acid has its origin in a condensation between phosphoenolpyruvate and arabinose in a reaction reminiscent of those leading to the formation of sialic or muramic acid [Eqs. (11-84) and (11-85)].

The key feature in the synthetic scheme is, however, concerned with the attachment of the outer, antigenic polysaccharide to the core. This reaction precedes by means of transfer to and from BPr as was the case in murein biosynthesis. The steps are the following—with some antigens lacking abequose (the "*n*" indicating that particular repeat, not the stoichiometry of the individual reaction):

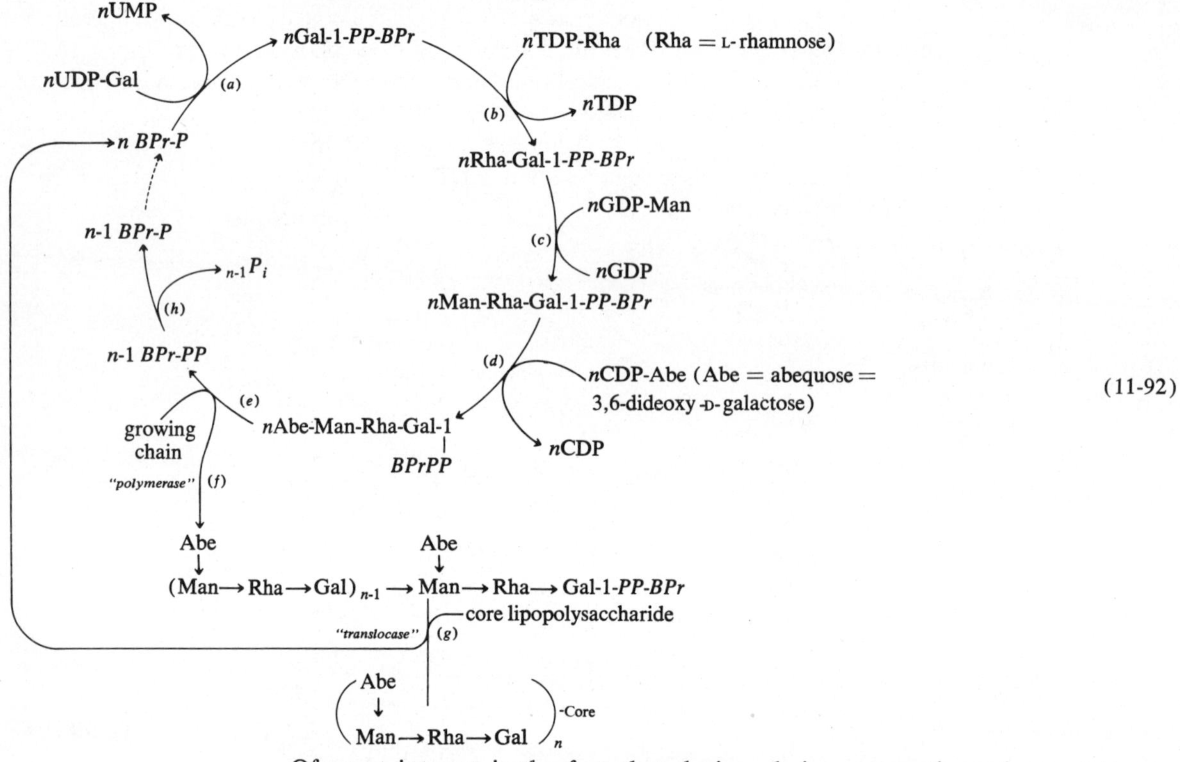

(11-92)

Of great interest is the fact that during chain propagation when a tetrasaccharide unit is to be added to the growing chain, a reaction catalyzed by the polymerase, the condensation occurs so that the new unit added is fixed to the "head," or potentially reducing, end of the polysaccharide:

$$\text{polymer-Gal-}P\text{-}P\text{-}BPr + \underset{\text{Abe}}{\text{Man}}\text{-Rha-Gal-}P\text{-}P\text{-}BPr \longrightarrow \text{polymer-Gal-}\underset{\text{Abe}}{\text{Man}}\text{-Rha-Gal-}P\text{-}P\text{-}BPr + BPr\text{-}P\text{-}P \qquad (11\text{-}92f)$$

This is reminiscent of chain propagation in fatty acid synthesis with the polymer fixed on acyl carrier protein (Chapter 16) or in protein synthesis with it residing on peptidyl-tRNA (Chapter 21).

REFERENCES

Books and Reviews

Advances in Carbohydrate Chemistry and Advances in Enzyme Regulation, appearing annually.

Archibald, A. R., J. Baddiley, and N. L. Blumsom: The Teichoic Acids, *Advan. Enzymol.,* **30:** 223 (1968).

Atkinson, D. E.: Regulation of Enzyme Activity, *Ann. Rev. Biochem.,* **35:** 85 (1966).

Balazs, E. A., and R. W. Jeanloz (eds.): *The Amino Sugars,* vols. IIA and IIB, Academic Press, New York, 1965, 1966.

Caputto, R., H. S. Barra, and F. A. Cumar: Carbohydrate Metabolism, *Ann. Rev. Biochem.,* **36:** 211 (1967).

Ginsburg, V.: Sugar Nucleotides and the Synthesis of Carbohydrates, *Advan. Enzymol.,* **26:** 35 (1964).

Ginsburg, V., and E. F. Neufeld: Complex Heterosaccharides of Animals, *Ann. Rev. Biochem.,* **38:** 371 (1969).

Glaser, L.: Biosynthesis of Deoxysugars, *Physiol. Rev.,* **43:** 215 (1963).

Greenberg, D. M. (ed.): *Metabolic Pathways,* 2nd ed., vol. 1, and 3rd ed., Academic Press, New York, 1960, 1967.

Gunsalus, I. C., and C. W. Shuster: Energy-Yielding Metabolism in Bacteria, in I. C. Gunsalus and R. Y. Stanier (eds.), *The Bacteria,* vol. 2, Academic Press, New York, 1961, p. 1.

Hales, C. N.: Some Action of Hormones in the Regulation of Glucose Metabolism, in P. N. Campbell and G. D. Greville (eds.), *Essays in Biochemistry,* vol. 3, Academic Press, New York, 1967, p. 73.

Hollman, S., and O. Touster: *Nonglycolytic Pathways of Metabolism of Glucose,* Academic Press, New York, 1964.

Horecker, B. L.: The Biosynthesis of Bacterial Polysaccharides, *Ann. Rev. Microbiol.,* **20:** 253 (1966).

Horecker, B. L.: *Pentose Metabolism in Bacteria,* Wiley, New York, 1962.

Johnson, M. J.: Enzymic Equilibria and Thermodynamics, in P. D. Boyer, H. Lardy, and K. Myrbäck (eds.), *The Enzymes,* vol. 3, Academic Press, New York, 2nd ed., 1960, p. 407.

Kornberg, H. L.: Anaplerotic Sequences and Their Role in Metabolism, in P. N. Campbell and G. D. Greville (eds.), *Essays in Biochemistry,* vol. 2, Academic Press, New York, 1966, p. 1.

Kornberg, H. L.: The Coordination of Metabolic Routes, *Symp. Soc. Gen. Microbiol.,* **15:** 8 (1965).

Krebs, H. A., and H. L. Kornberg: *Energy Transformations in Living Matter,* Springer, Berlin, 1957.

Leloir, L. F.: Nucleoside Diphosphate Sugars and Saccharide Synthesis, *Biochem. J.,* **91:** 1 (1964).

McFall, E., and W. K. Maas: Regulation of Enzyme Synthesis in Microorganisms, in J. H. Taylor (ed.), *Molecular Genetics,* Part II, Academic Press, New York, 1967, p. 255.

McGilvery, R. W., and B. M. Pogell: *Fructose-1,6-diphosphatase and Its Role in Gluconeogenesis,* American Institute of Biological Sciences, Washington, D.C., 1964.

Nikaido, H.: Biosynthesis of Cell Wall Lipopolysaccharides in Gram-Negative Enteric Bacteria, *Advan. Enzymol.,* **31:** 77 (1968).

Northcote, D. H.: Polysaccharides, *Ann. Rev. Biochem.,* **33:** 51 (1964).

Osborn, M. J.: Structure and Biosynthesis of the Bacterial Cell Wall, *Ann. Rev. Biochem.,* **38:** 501 (1969).

Osborn, M. J., and I. M. Weiner: Mechanism of Biosynthesis of the Lipopolysaccharide of *Salmonella, Federation Proc.,* **26:** 70 (1967).

Robbins, P. W., D. Bray, M. Dankert, and A. Wright: Direction of Chain Growth in Polysaccharide Synthesis, *Science,* **158:** 1536 (1967).

Robison, G. A., R. W. Butcher, and E. W. Sutherland: Cyclic AMP, *Ann. Rev. Biochem.,* **37:** 149 (1968).

Salton, M. R. J.: *The Bacterial Cell Wall,* Elsevier, New York, 1964.

Scrutton, M. C., and M. F. Utter: The Regulation of Glycolysis and Gluconeogenesis in Animal Tissues, *Ann. Rev. Biochem.,* **37:** 249 (1968).

Sharon, N.: Polysaccharides, *Ann. Rev. Biochem.,* **35:** 485 (1966).

Strominger, J. L., K. Izaki, M. Matsuhashi, and D. J. Tipper: Peptidoglycan Transpeptidase and D-Alanine Carboxypeptidase: Penicillin-Sensitive Enzymatic Reactions, *Federation Proc.,* **26:** 9 (1967).

Umbarger, H. E.: Regulation of Amino Acid Metabolism, *Ann. Rev. Biochem.,* **38:** 323 (1969).

Wood, H. G., and R. L. Stjernholm: Assimilation of Carbon Dioxide by Heterotropic Organisms, in I. C. Gunsalus and R. Y. Stanier (eds.), *The Bacteria,* vol. 3, Academic Press, New York, 1962, p. 41.

Wood, H. G., and M. F. Utter: The role of CO_2 fixation in Metabolism, *Essays Biochem.,* **1:** 1 (1965).

Wood, W. A.: Carbohydrate Metabolism, *Ann. Rev. Biochem.,* **35:** 521 (1966).

Specific References

1. Davis, B. D.: *Cold Spring Harbor Symp. Quant. Biol.,* **26:** 1 (1962).
2. Kornberg, H. L.: *Symp. Soc. Gen. Microbiol.,* **15:** 8 (1965).
3. Atkinson, D. E.: *Biochemistry,* **7:** 4030 (1968); Klungsøyr, L., J. H. Hagemen, L. Fall, and D. E. Atkinson: *ibid.,* **7:** 4035 (1968); Shen, L. C., L. Fall, G. M. Walton, and D. E. Atkinson: *ibid.,* **7:** 4041 (1968).
4. Johnson, M. J.: in *The Enzymes,* vol. 3, Academic Press, New York, 2nd ed., 1960, p. 407.
5. Grossbard, L., and R. T. Schimke: *J. Biol. Chem.,* **241:** 3546 (1966).
6. González, C., T. Ureta, J. Babul, E. Rabajille, and H. Niemeyer: *Biochemistry,* **6:** 460 (1967).
7. Duttera, S. M., W. L. Byrne, and M. C. Ganoza: *J. Biol. Chem.,* **243:** 2216 (1968).
8. Arion, W. J., and R. C. Nordlie: *J. Biol. Chem.,* **239:** 1680, 2752 (1964).
9. Krebs, E. G., and E. H. Fischer: *Advan. Enzymol.,* **24:** 263 (1962); Krebs, E. G., R. J. DeLange, R. G. Kemp, and W. D. Riley, *Pharmacol. Rev.,* **18:** 163 (1966); Fischer, E. H., and E. G. Krebs: *Federation Proc.,* **25:** 1511 (1966); DeLange, R. J., R. G. Kemp, W. D. Riley, R. A. Cooper, and E. G. Krebs: *J. Biol. Chem.,* **243:** 2200 (1968); Riley, W. D., R. J. DeLange, G. E. Bratvold, and E. G. Krebs, *ibid.,* **243:** 2209 (1968).
10. Chignell, D. A., W. B. Gratzer, and R. C. Valentine: *Biochemistry,* **7:** 1082 (1968); Gold, A. M.: *Biochemistry,* **7:** 2106 (1968); Battell, M. L., C. G. Zarkadas, L. B. Smillie, and N. B. Madsen: *J. Biol. Chem.,* **243:** 6202 (1968); Okazaki, T., A. Nakazawa, and O. Hayaishi: *ibid.,* **243:** 5266 (1968).
11. Helmreich, E., M. C. Michaelides, and C. F. Cori: *Biochemistry,* **6:** 3695 (1967); Kastenschmidt, L. L., J. Kastenschmidt, and E. Helmreich: *ibid.,* **7:** 3590, 4543 (1968).
12. Madsen, N. B., and S. Shechosky: *J. Biol. Chem.,* **242:** 3301 (1967); Avramovic, O., and N. B. Madsen: *ibid.,* **243:** 1656 (1968); Engers, H. D., and N. B. Madsen: *Biochem. Biophys. Res. Commun.,* **33:** 49 (1968).
13. Najjar, V. A., and M. E. Pullman: *Science,* **119:** 631 (1954).
14. Yankeelov, J. A., Jr., H. R. Horton, and D. E. Koshland, Jr.: *Biochemistry,* **3:** 349 (1964).
15. Ray, W. J., Jr., and G. A. Roscelli: *J. Biol. Chem.,* **239:** 1228, 3935 (1964); **241:** 2596, 3499 (1966).
16. Peck, E. J., Jr., D. S. Kirkpatrick, and W. J. Ray, Jr.: *Biochemistry,* **7:** 152 (1968).
17. Joshi, J. G., and P. Handler: *J. Biol. Chem.,* **239:**

2741 (1964); Robinson, J. P., S. Harshman, and V. A. Najjar: *Biochemistry,* **4:** 401 (1965); Harshman, S., J. P. Robinson, V. Bocchini, and V. A. Najjar: *ibid.,* **4:** 396 (1965).

18. Topper, Y. J.: in *The Enzymes,* vol. 5, 2nd ed., 1961, p. 413; Rose, I. A.: *Brookhaven Symp. Biol.,* **15:** 293 (1962); Salas, M., E. Viñuela, and A. Sols: *J. Biol. Chem.,* **240:** 561 (1965); Rose, I. A.: *Ann. Rev. Biochem.,* **35:** 23 (1966).
19. Lowry, O. H., and J. V. Passonneau: *J. Biol. Chem.,* **244:** 910 (1969).
20. Scrutton, M. C., and M. F. Utter: *loc. cit.;* Atkinson, D. E.: *Ann. Rev. Biochem.,* **35:** 85 (1966); cf. also Parmeggiani, A., J. H. Luft, D. S. Love, and E. G. Krebs: *J. Biol. Chem.,* **241:** 4625 (1966).
21. Lindell, T. J., and E. Stellwagen: *J. Biol. Chem.,* **243:** 907 (1968).
22. Blangy, D., H. Buc, and J. Monod: *J. Mol. Biol.,* **31:** 13 (1968).
23. Rutter, W. J.: *Federation Proc.,* **23:** 1248 (1964); Penhoet, E., M. Kochman, R. Valentine, and W. J. Rutter: *Biochemistry,* **6:** 2940 (1967); Kochman, M., E. Penhoet, and W. J. Rutter: *Federation Proc.,* **27:** 590 (1968).
24. Lai, C. Y., O. Tchola, T. Cheng, and B. L. Horecker: *J. Biol. Chem.,* **240:** 1347 (1965); Morse, D. E., W. Chan, and B. L. Horecker: *Proc. Natl. Acad. Sci. U.S.,* **58:** 628 (1967); Chan, W., D. E. Morse, and B. L. Horecker: *ibid.,* **57:** 1013 (1967); Hoffee, P., C. Y. Lai, E. L. Pugh, and B. L. Horecker: *ibid.,* **57:** 107 (1967); Lai, C. Y., and C. Chen: *Arch. Biochem. Biophys.,* **128:** 212 (1968).
25. Rose, I. A., E. L. O'Connell, and A. H. Mehler: *J. Biol. Chem.,* **240:** 1758 (1965); Mehler, A. H., and M. E. Cusic, Jr.: *Science,* **155:** 1101 (1967); Rose, I. A., and E. L. O'Connell: *J. Biol. Chem.,* **244:** 126 (1969).
26. Model, P., L. Ponticorvo, and D. Rittenberg: *Biochemistry,* **7:** 1339 (1968).
27. Masters, C. J.: *Biochem. Biophys. Res. Commun.,* **28:** 978 (1967); Wassarman, P. M., and N. O. Kaplan: *J. Biol. Chem.,* **243:** 720 (1968).
28. Kobes, R. D., R. T. Simpson, B. L. Vallee, and W. J. Rutter: *Biochemistry,* **8:** 585 (1969).
29. Koshland, D. E., Jr., and K. E. Neet: *Ann. Rev. Biochem.,* **37:** 359 (1968).
30. Park, J. H.: *Johnson Foundation Symposium,* 1969; Chance, B., and J. H. Park: *J. Biol. Chem.,* **242:** 5093 (1967); Colowick, S. P., J. Van Eys, and J. H. Park: in M. Florkin and E. H. Stotz (eds.), *Comprehensive Biochemistry,* vol. 14, Elsevier, New York, 1966, p. 1; Olson, E. J., and J. H. Park: *J. Biol. Chem.,* **239:** 2316 (1964).
31. Harris, J. I., and R. N. Perham: *Nature,* **219:** 1025 (1968).
32. Smith, T. E.: *Biochemistry,* **5:** 2919 (1966).
33. Malhotra, O. P., and S. A. Bernhard: *J. Biol. Chem.,* **243:** 1243 (1968).
34. Jacobs, R. J., and S. Grisolia: *J. Biol. Chem.,* **241:** 5926 (1966); Alpers, J. B., and G. K. H. Lam: *ibid.,* **244:** 200 (1969).
35. Scrutton, M. C., and M. F. Utter: *loc. cit.;* Malcovati, M., and H. L. Kornberg: *Biochim. Biophys. Acta,* **178:** 420 (1969).
36. Cottam, G. L., F. P. Kupiecki, and M. J. Coon: *J. Biol. Chem.,* **243:** 1630 (1968); Cottam, G. L., P. F. Hollenberg, and M. J. Coon: *ibid.,* **244:** 1481 (1969).
37. Mildvan, A. S., J. S. Leigh, and M. Cohn: *Biochemistry,* **6:** 1805 (1967).
38. IUPAC–IUB Nomenclature: *Biochemistry,* **6:** 3287 (1967); *J. Biol. Chem.,* **242:** 4845 (1967).
39. Reed, L. J., and D. J. Cox: *Ann. Rev. Biochem.,* **35:** 57 (1966); Willms, C. R., R. M. Oliver, H. R. Henney, Jr., B. B. Mukherjee, and L. J. Reed: *J. Biol. Chem.,* **242:** 889 (1967); Henney, H. R., C. R. Willms, T. Muramatsu, B. B. Mukherjee, and L. J. Reed: *ibid.,* **242:** 898 (1967); cf. also Hayakawa, T., and M. Koibe: *ibid.,* **242:** 1356 (1967).
40. Utter, M. F., and D. B. Keech: *J. Biol. Chem.,* **235:** PC17 (1960); **238:** 2603, 2609 (1963); Scrutton, M. C., and M. F. Utter: **240:** 1, 3714 (1965); **242:** 1723 (1967); Valentine, R. C., N. G. Wrigley, M. C. Scrutton, J. J. Irias, and M. F. Utter: *Biochemistry,* **5:** 3111 (1966); Scrutton, M. C., M. F. Utter, and A. S. Mildvan: *J. Biol. Chem.,* **241:** 3480, 3488 (1968); Scrutton, M. C., and A. S. Mildvan: *Biochemistry,* **7:** 1490 (1968).
41. Keech, B., and G. J. Barritt: *J. Biol. Chem.,* **242:** 1983 (1967).
42. Cazzulo, J. J., T. K. Sundaram, and H. L. Kornberg: *Nature,* **223:** 1137 (1969).
43. Cooper, R. A., and H. L. Kornberg: *Biochim. Biophys. Acta,* **104:** 618 (1965); **141:** 211 (1967); *Biochem. J.,* **105:** 49c (1967).
44. Evans, H. J., and H. G. Wood: *Proc. Natl. Acad. Sci. U.S.,* **61:** 1448 (1968); Reaves, R. E., R. A. Menzies, and D. S. Hsu: *J. Biol. Chem.,* **243:** 5486 (1968).
45. Hatch, M. D., and C. R. Slack: *Biochem. J.,* **106:** 141 (1968); **114:** 117 (1969); see also Reeves, R. E.: *J. Biol. Chem.,* **243:** 3202 (1968).
46. Sauer, L. A.: *J. Biol. Chem.,* **243:** 2429 (1968).
47. Kornberg, H. L., and N. B. Madsen: *Biochem. J.,* **66:** 13p (1957); *Biochim. Biophys. Acta,* **24:** 651 (1957).
48. Yue, R. H., E. A. Noltmann, and S. A. Kuby: *J. Biol. Chem.,* **244:** 1353 (1969).
49. Lienhard, G. E., and I. A. Rose: *Biochemistry,* **3:** 190 (1964).
50. González, C., T. Ureta, R. Sánchez, and H. Niemeyer: *Biochem. Biophys. Res. Commun.,* **16:** 347 (1964); Brand, K., and B. L. Horecker: *Arch. Biochem. Biophys.,* **123:** 312 (1968); Brand, K., O. Tsolas, and B. L. Horecker: *ibid.,* **130:** 521 (1969).
51. Hostetler, K. Y., and B. R. Landau: *Biochemistry,* **6:** 2961 (1967).
52. Mortlock, R. P., D. D. Fossitt, and W. A. Wood: *Proc. Natl. Acad. Sci. U.S.,* **54:** 572 (1965).
53. Meloche, H. P., and W. A. Wood: *J. Biol. Chem.,* **239:** 3505, 3511 (1964).
54. Metzger, R. P., S. S. Wilcox, and A. N. Wick: *J. Biol. Chem.,* **239:** 1769 (1964); **240:** 2767 (1965).
55. Hayman, S., and J. H. Kinoshita: *J. Biol. Chem.,* **240:** 877 (1965).

56. Smith, M. G.: *Biochem. J.,* **83:** 135 (1962).
57. Kornfeld, S., R. Kornfeld, E. F. Neufeld, and P. J. O'Brien: *Proc. Natl. Acad. Sci. U.S.,* **52:** 371 (1964).
58. Yu, R. K., and R. Ledeen: *J. Biol. Chem.,* **244:** 1306 (1969).
59. Rosell-Perez, M., C. Villar-Palasi, and J. Larner: *Biochemistry,* **1:** 763 (1962); Hizukuri, S., and J. Larner: *ibid.,* **3:** 1783 (1964); Friedman, D. L., and J. Larner: *ibid.,* **2:** 669 (1963); Rosell-Perez, M., and J. Larner: *ibid.,* **3:** 773 (1964).
60. Steiner, D. F.: *Nature,* **204:** 1171 (1964); Steiner, D. F., L. Younger, and J. King: *Biochemistry,* **4:** 740 (1965).
61. Gold, A. H., and H. L. Segal: *Arch. Biochem. Biophys.,* **120:** 359 (1967); Mersmann, H. J., and H. L. Segal: *Proc. Natl. Acad. Sci. U.S.,* **58:** 1688 (1967).
62. Butcher, R. W., G. A. Robison, J. G. Hardman, and E. W. Sutherland: *Advan. Enzyme Regul.,* **6:** 357 (1968).
63. De Wulf, H., W. Stalmans, and H. G. Hers: *European J. Biochem.,* **6:** 545 (1968); De Wulf, H., and H. G. Hers: *ibid.,* **6:** 552, 558 (1968).
64. Miyamoto, E., J. F. Kuo, and P. Greengard: *J. Biol. Chem.,* **244:** 6395 (1969).

12
Photosynthesis

Ultimately, all life on earth (the *biosphere*) depends directly on the assimilation of carbon dioxide into organic compounds, the energy for this highly endergonic process being provided by light generated in the sun. The reactions responsible for this complex conversion of photonic into chemical energy are collectively known as *photosynthesis* and the organisms responsible as *photosynthetic organisms*. Among them are, of course, plants—whether higher plants, ferns, mosses, diatoms, or algae (green, blue-green, red, or brown)—and certain microorganisms: protozoa, such as *Euglena;* and a number of different bacteria such as the purple sulfur bacteria (Thiorhodaceae), the purple nonsulfur bacteria (Athiorhodaceae), and the green sulfur bacteria (Chlorobacteriaceae). Photosynthesis provides them all with the organic materials required for their growth and maintenance, and their cells, in turn, or else their products in life or death, yield the foodstuffs required for all other members of the biosphere.

But the significance of photosynthesis transcends even the limits just described.

1. Organisms, in order to live, have to consume energy by means of essentially oxidative, degradative, "catabolic" processes. These demand an oxidizing environment and lead ultimately to the conversion of all carbon compounds to their most oxidized form (i.e., carbon dioxide). Photosynthesis restores the balance, both by reducing CO_2 to organic compounds, and by generating oxygen.

2. Even though all cells use some of the "catabolic" energy, in coupled reactions, to drive endergonic, synthetic, "anabolic" reactions and store it in the complex products formed, some energy is still continuously dissipated as heat—by inefficient coupling between biochemical reactions, during transduction to other forms, such as mechanical or electrical energy—and thus the entropy of the biosphere increases. This is required by the second law of thermodynamics of any closed system, and thus, inexorably, life on earth would have to run out were it not for the continuous supply of radiant energy from outside and its utilization by photosynthesis (1).

3. It is quite possible that the first precursors of living things on earth, the protobionts, found themselves on a planet devoid of O_2 in its atmosphere (2). Only the evolution of organisms capable of generating oxygen created the

proper environment for all subsequent forms of life to develop. The early photosynthetic organisms were almost certainly procaryotic cells performing some rudimentary form of present-day bacterial photosynthesis as long as some 3 billion years ago. Later some of them evolved into ancestral forms of blue-green algae (1 to 2 billion years ago) and started the gradual build up of oxygen in the atmosphere, eventually permitting the evolution of eucaryotic organisms at the beginning of the Cambrian era, some 600 million years ago.

LOCALE FOR PHOTOSYNTHESIS

In plants, photosynthesis takes place within highly specialized organelles, called *plastids*. If the cells are broken and fractions of purified plastids isolated, the latter, suitably supplemented by soluble substrates and cofactors, will carry out the complete sequence of photosynthesis (3, 4). In green plants, including green unicellular organisms such as algae (e.g., *Chlorella, Scenedesmus*) and *Euglena,* the plastids contain chlorophyll and are referred to as *chloroplasts*. The number, shape, and size of these chloroplasts vary widely between organisms. *Chlorella* contains but one cup-shaped plastid, approximately 3 μm across and 1 to 2 μm deep; other algae and the cells in green tissues of higher plants contain many chloroplasts: *Euglena,* for instance, contains ten chloroplasts; they are cylindrical, 5 to 10 μm in length and 0.5 to 2 μm in diameter; many others are saucer shaped and approximately 4 to 6 μm across and 0.5 to 1 μm deep.

When plastids are examined in thin section under the electron microscope, a wealth of fine structure is revealed (Fig. 12-1). The most characteristic features perhaps are the many small, dark, membranous bodies (called *grana*) pervaded by a distinct lamellar structure. In *Euglena,* this dark lamellar structure fills the whole chloroplast, which can thus be regarded as a single granum; in most other cells each chloroplast contains 10 to 100 grana, which are roughly cylindrical in shape, ~ 4000 to 6000 Å in diameter, and 5000 to 8000 Å in height. All the photosynthetic pigments of the cell are localized preponderantly in and around the lamellae. When the latter are isolated and purified from ruptured plastids, they are found to be also exceedingly rich in phospholipids and lipoproteins and to be able to catalyze some of the most characteristic reactions of photosynthesis: the *photolysis of water* in the presence of an added oxidant (the *Hill reaction*—see below) and the reactions responsible for *electron transport.*

If supplemented with a minimal complement of soluble enzymes (obtained from the lumen of the plastid) and cofactors, they will also catalyze reduction of nicotinamide coenzymes concomitant with the generation of ATP (*photophosphorylation*); when a more complete collection of soluble chloroplast enzymes is provided in addition, restoration of photosynthesis can be demonstrated.

The lamellar apparatus is in turn made up of stacked subunits, called *thylakoids*—membranes each 30 Å thick plus little oblate ellipsoids or plates approximately 200 Å in diameter and 100 Å thick, somewhat compressed between the lamellar structures; the latter may or may not protrude into other areas of the chloroplast (Figs. 12-1 through 12-3), depending on species or physiological state. Whether the platelets or ellipsoids are themselves surrounded by a distinct double membrane or are simple, oriented, molecular aggregates with a lipophilic molecular layer outermost is not yet certain.

In purified preparations of lamellar membranes from spinach chloroplasts, Park and his collaborators have identified repeating subunits, which may be arranged in a variety of arrays [(5) and Fig. 12-4]. The individual

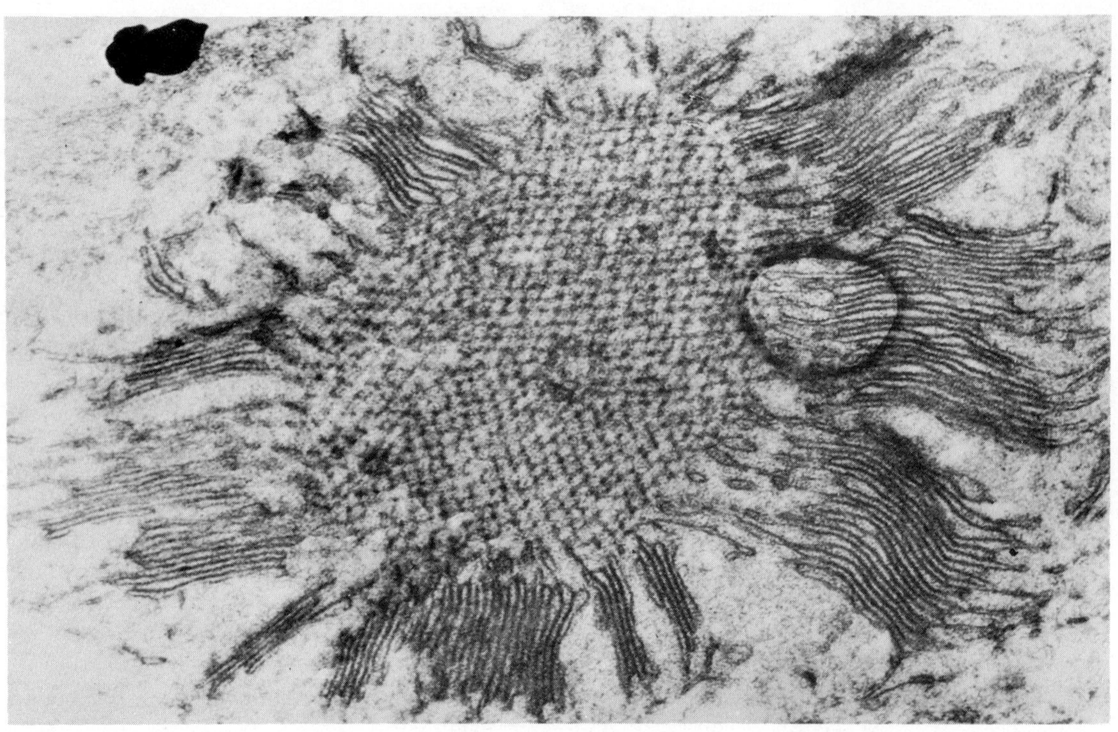

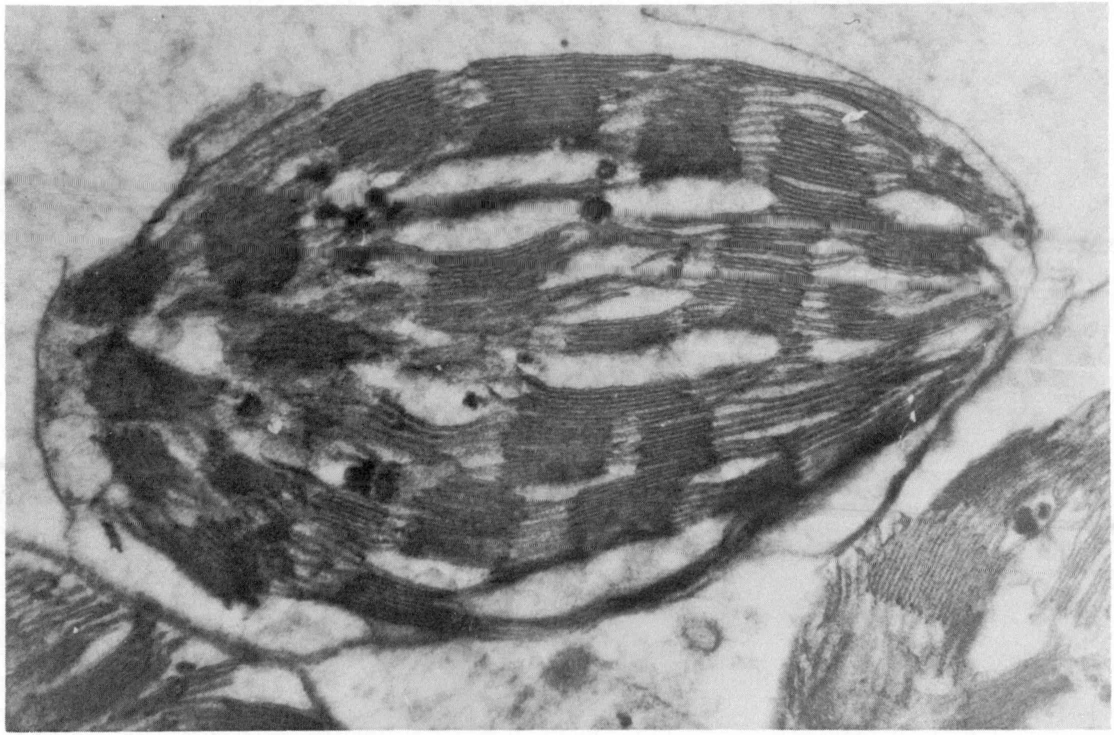

FIGURE 12-1
Structure of the chloroplast. Electron micrographs: (*top*) prolamellar, paracrystalline body in a proplastid (choloroplast precursor) of *Chlorophytum,* exposed to light, showing the origin of the lamellae (magnification 50,000); (*bottom*) fully developed chloroplast in leaf of *Elodea canadenis* (magnification 24,000). Courtesy of K. Mühlethaler, Zurich.

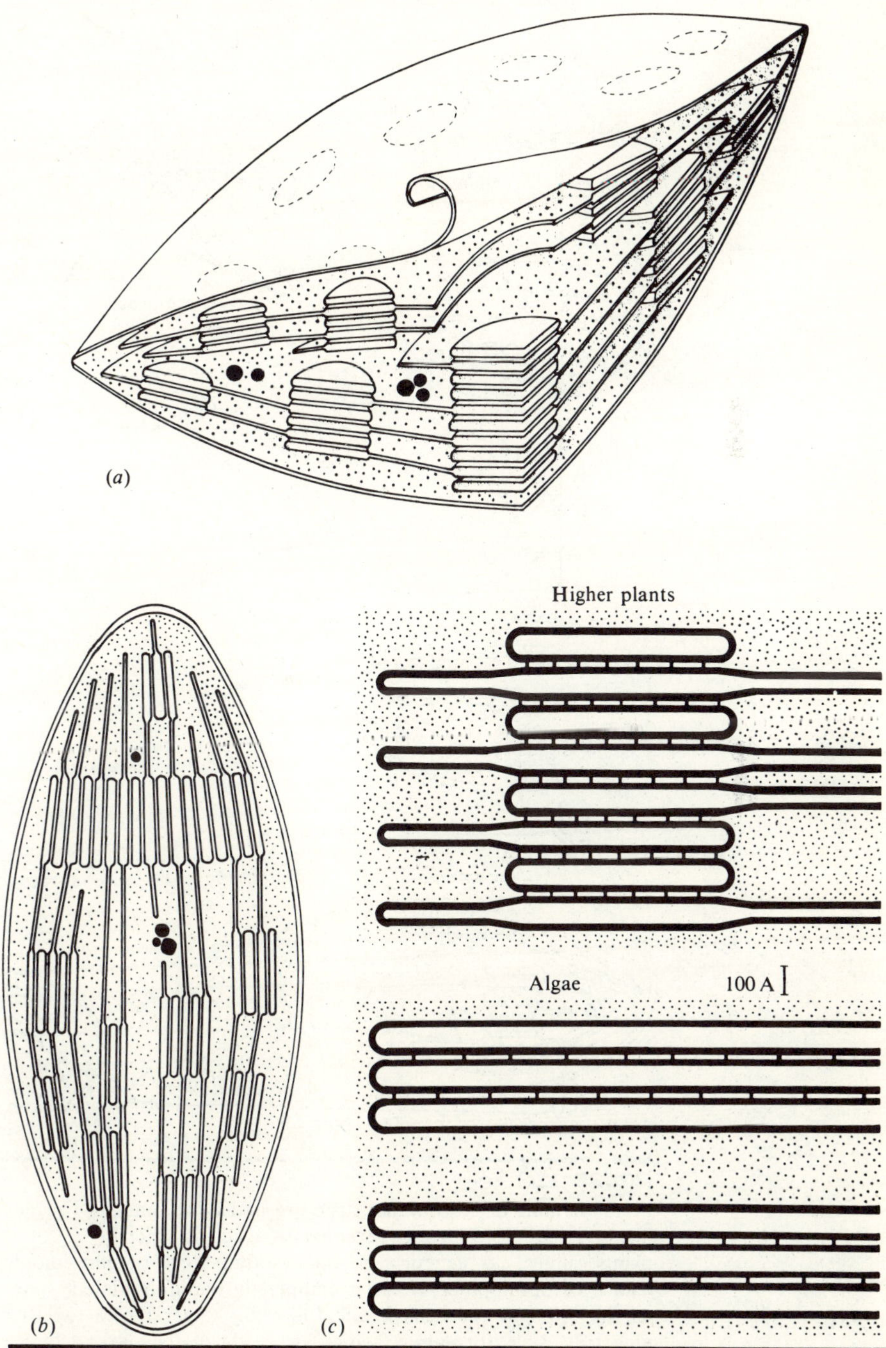

FIGURE 12-2

Fundamentals of chloroplast structure and development. (*a*) Schematic three-dimensional structure of the chloroplast of a higher plant. (*b*) Cross section through (*a*). (*c*) Enlarged detail of (*b*) showing details of lamellar structure and a comparison between those of higher plants and algae. Courtesy of D. von Wettstein, Copenhagen.

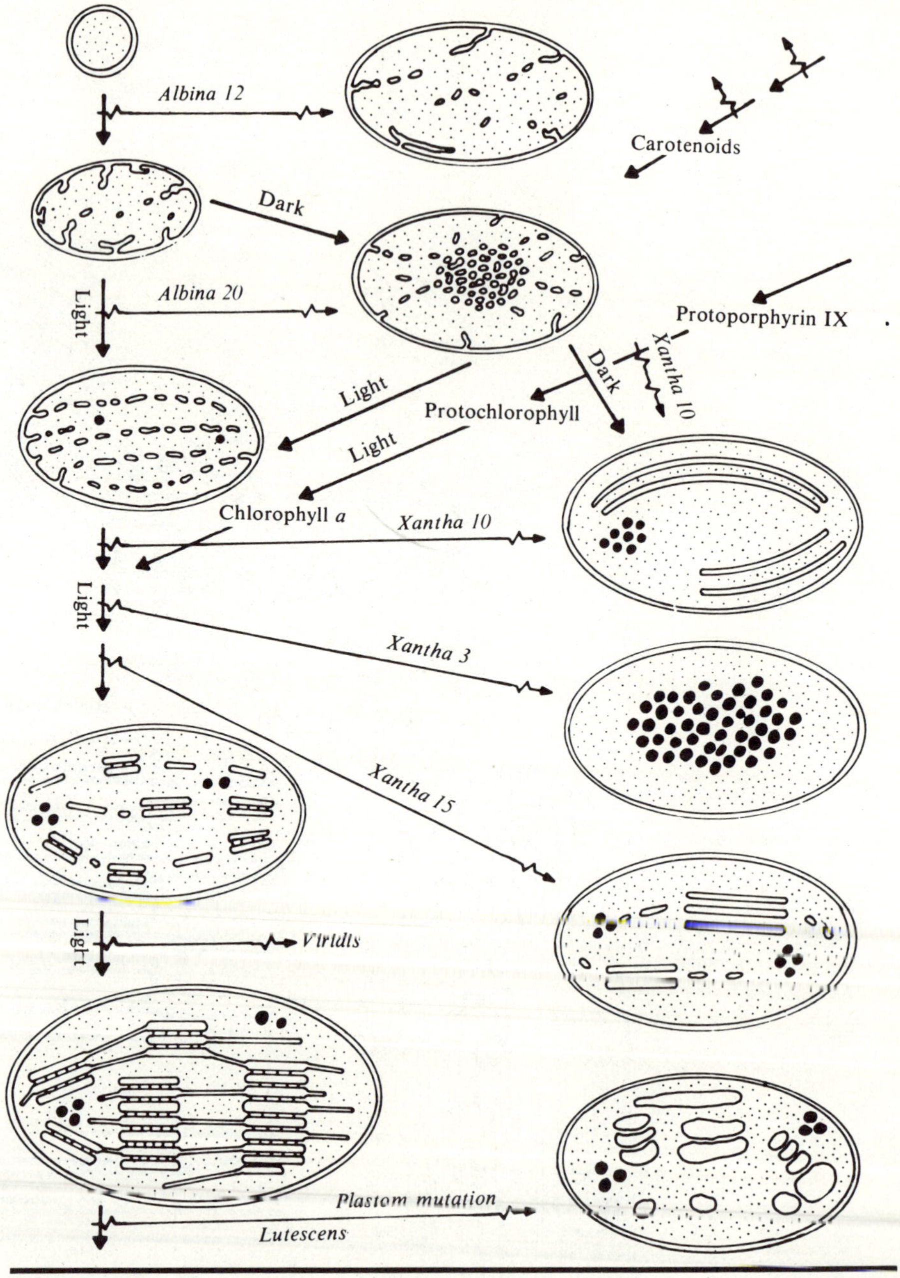

FIGURE 12-3

Fundamentals of chloroplast structure and development. The development of chloroplast fine structure in maize as revealed by electron microscopy of specimens obtained from plants subject to genetic (albina, xantha, etc.) or physiological blocks. The normal course of development from a completely undifferentiated, spherical proplastid is shown on the left and proceeds in the following stages (from top to bottom): (1) the beginning of vesicle formation, which proceeds in the dark; (2) transformation into tubes, which requires light of low energy and can utilize either red or blue light; (3) vesicle dispersal, a temperature-independent, high-energy-requiring process; (4) and (5) grana formation, which also requires high energy but is temperature dependent, and finally formation of the mature, fully differentiated chloroplast shown in Figs. 12-1 and 12-2. Courtesy of D. von Wettstein, Copenhagen.

FIGURE 12-4
Lamellae of spinach chloroplasts showing arrays of quantasomes. The picture is that of a chromium-shadowed chloroplast lamella at a magnification of 71,000. Note the crystalline array of structural units (quantasomes) with dimensions 185 by 155 by 100 Å. Note differences and similarities to the array shown in Fig. 12-1. R. B. Park and J. Biggins, *Science,* **144:** 1009 (1964).

subunit, which they believe to be the morphological entity corresponding to a physiological unit of photosynthesis,[1] is estimated to have the dimensions of an oblate sphere ~ 150 Å in diameter and a thickness of 100 Å. The molecular weight of a photosynthetic unit is estimated at 2×10^6, and contains 230 *chlorophyll molecules* (160 *a*: 70 *b*), 48 *carotenoids,* 46 *quinones,*[2] 116 *phospholipids,* 500 *galactosylglycerides,* 48 *sulfolipids,* sterols, and other lipids. Thus, the molecular weight of total lipids is about 10^6 and that of proteins about 10^6 also; in addition, the unit contains 1 *cytochrome* b_6, 1 *cytochrome f,* 5 *ferredoxin,*[2] 5 *plastocyanin* molecules,[2] and 2 manganous ions.

In bacteria and blue-green algae the structures responsible for photosynthesis appear to be somewhat simpler. One can obtain roughly spherical, highly colored particles (called *chromatophores*) with a diameter of ~ 300 Å

[1] A contrary view ascribes to these particles (here of 100 Å diameter and with $M =$ 350,000) a function only in photophosphorylation (6), analogous to that of the F_1-ATPase of mitochondria (Chapter 15).

[2] For the structure of these components see Chapter 15.

(particle weight ~ 10^7) containing bacteriochlorophyll, other pigments (carotenoids), phospholipid, and all the enzymes responsible for bacterial photosynthesis.

OVERALL STOICHIOMETRY OF PHOTOSYNTHESIS

Classically, photosynthesis by green plants has been formulated as

$$CO_2 + H_2O + n\,h\nu \xrightarrow[\text{enzymes, etc.}]{\text{chlorophyll}} \{CH_2O\} + O_2 \tag{12-1a}$$

or

$$6CO_2 + 6H_2O + 6n\,h\nu \xrightarrow[\text{enzymes, etc.}]{\text{chlorophyll}} C_6H_{12}O_6 + 6O_2 \tag{12-1b}$$

where $\{CH_2O\}$ signifies an organic compound at the oxidation level of carbohydrate and $C_6H_{12}O_6$ a hexose. Bacterial photosynthesis, which does not produce O_2, has been shown by Kluyver and van Niel to conform in general to the equation

$$CO_2 + 2H_2D + n\,h\nu \xrightarrow[\text{enzymes, etc.}]{\text{bacteriochlorophyll}} \{CH_2O\} + H_2O + 2D \tag{12-2}$$

where H_2D is an oxidizable hydrogen donor, such as H_2S, thiosulfate, or selenium compounds (sulfur bacteria); hydrogen gas, or simple organic compounds, such as isopropanol, lactic acid or malic acid, fatty acids, and so on (various other genera of bacteria). Van Niel made the brilliant suggestion that fundamentally there was no difference between reaction (12-2) and reaction (12-1a), provided the latter was written as

$$CO_2 + 2H_2O^* + n\,h\nu \xrightarrow[\text{enzymes, etc.}]{\text{chlorophyll}} \{CH_2O\} + H_2O + O_2^* \tag{12-3}$$

i.e., that water (on the left side) functions as the oxidizable hydrogen donor, being in this case oxidized to molecular O_2—the 2 D of Eq. (12-2)—while the water on the right side of both equations constitutes one of the products of CO_2 reduction.[3] This postulate was later proved correct by the use of reactants labeled with ^{18}O: as indicated in the equation, all the molecular oxygen evolved has its origin in the oxygen atoms of the H_2O, and the two oxygen atoms in CO_2 become equipartitioned between the product carbohydrate and the water of the medium.

Van Niel also proposed that the primary reaction of photosynthesis in all organisms is the same: a photolysis of water giving rise to the simultaneous generation of *oxidizing* power YOH, and *reducing* power XH:

$$4H_2O + 4X + 4Y + n\,h\nu \xrightarrow[\text{accessory systems}]{\text{chlorophyll}} 4XH + 4YOH \tag{12-4}$$

In green plants, four molecules of YOH react to generate O_2 and the product H_2O [Eq. (12-5)] while for bacteria we have the analogous reaction with H_2D [Eq. (12-6)]:

$$4YOH + 2H_2O \xrightarrow{\text{enzymes}} 4H_2O + 4Y + O_2 \tag{12-5}$$

$$4YOH + 2H_2D \xrightarrow{\text{enzymes}} 4H_2O + 4Y + 2D \tag{12-6}$$

[3] An interesting physiological correlate of this postulate of van Niel's and a link between bacterial and green-plant photosynthesis is provided by certain green algae, which can be adapted to fix CO_2 anaerobically with hydrogen as the reductant, according to the reaction

$$CO_2 + 2H_2 + n\,h\nu \rightarrow \{CH_2O\} + H_2O \tag{12-1c}$$

yet are still capable of aerobic CO_2 fixation by classic photosynthesis [reaction (12-1a)].

In all cells the photosynthetic reductant, on the other hand, is used for the conversion of CO_2 to carbohydrate:

$$4XH + CO_2 \xrightarrow{\text{enzymes, etc.}} \{CH_2O\} + 2H_2O + 4X \tag{12-7}$$

Summing over reactions (12-4), (12-5), and (12-7) yields Eq. (12-3), the overall stoichiometric equation for green-plant photosynthesis, and Eqs. (12-4), (12-6), and (12-7) generate Eq. (12-2), the corresponding one for bacterial photosynthesis.

This formulation has several important consequences: it suggests immediately that is should be possible to dissociate (1) the steps directly linked to the photochemistry of the photosynthetic pigments from a number of dark reactions and (2) those concerned with the enzyme-catalyzed reduction of CO_2 (plants and bacteria) from those of oxygen evolution (plants)—or substrate oxidation (bacteria). Indeed, washed chloroplast preparations have long been known to catalyze the Hill reaction, a light-dependent oxygen evolution in the presence of a large variety of different electron acceptors (cytochromes, quinones, indophenol dyes, ferricyanide, $NADP^+$, etc.):

$$4A + 2H_2O + n\,h\nu \xrightarrow{\text{chloroplasts}} 4AH + O_2 \tag{12-8}$$

This can be visualized as being the sum of Eqs. (12-4), (12-5), and Eq. (12-9):

$$XH + A \rightarrow X + AH \tag{12-9}$$

It is now known (see below) that green plants contain two separate photosystems. These findings can easily be accommodated by substituting reactions (12-10) plus (12-11) for (12-4), (12-12) for (12-5), (12-13) for (12-6), and (12-14) for (12-7).

$$4X + 4\text{Chl} + n\,h\nu \xrightarrow{\text{system I}} 4\bar{X}_{red} + 4\text{Chl}^+ \tag{12-10}$$

generates reducing power plus a weak oxidant (Chl^+)

$$4X' + 4\text{Chl}' + n\,h\nu \xrightarrow{\text{system II}} 4\bar{X}'_{red} + 4\text{Chl}'^+ \tag{12-11}$$

generates oxidizing power plus a weak reductant ($\bar{X}'_{red}$)

$$4\text{Chl}'^+ + 2H_2O \rightarrow O_2 + 4\text{Chl}' + 4H^+ \tag{12-12}$$

plants — oxidation of donor

$$4\text{Chl}^+ + 2H_2D \rightarrow 2D + 4\text{Chl} + 4H^+ \tag{12-13}$$

bacteria — oxidation of donor

$$4\bar{X}_{red} + CO_2 + 4H^+ \xrightarrow{\text{ATP}} \{HCHO\} + 2H_2O + 4X \tag{12-14}$$

carbon assimilation

$$4\text{Chl}^+ + 4\bar{X}'_{red} \xrightarrow[\text{electron-transfer system}]{ADP + P_i \;\rightarrow\; ATP} 4\text{Chl} + 4X' \tag{12-15}$$

combination of weak species generates metabolic energy

The only new reaction is (12-15), which regenerates the two components not participating in either the reduction of CO_2 or the oxidation of H_2O (or H_2D). It is this reaction which generates the third prerequisite (besides reducing and oxidizing power) for successful photosynthetic assimilation: biochemical energy in the form of ATP (*photosynthetic phosphorylation*—see below). The overall reaction for photosynthesis by green plants is obtained by adding (12-10) + (12-11) + (12-12) + (12-14). Photosynthetic bacteria lack the system responsible for catalyzing (12-11); the correct overall reaction then involves (12-10) + (12-13) + (12-14). One important difference between schemes of this type and van Niel's is that the splitting of water is now no longer a prerequisite, but rather a consequence, of the photochemical generation of oxidizing and reducing power.

PRINCIPAL PHOTOSYNTHETIC PIGMENTS

The most important photosynthetic pigment in all plants is *chlorophyll a* (Chl *a*); in bacteria it is *bacteriochlorophyll* (*BChl*). The photosynthetic machinery of all these cells contains, in addition, certain pigments that used to be called "auxiliary" or "accessory"; it is now evident that these components fulfill a function as vital as that of ordinary chlorophyll *a*. Among them is a special form of Chl *a* absorbing at a shorter wavelength, in addition to chlorophyll *b*, in all green plants including algae; and also the other pigments shown in Table 12-1, especially the *carotenoids* and the *phycobilins* (phycocyanin and phycoerythrin). The structures of some of these are indicated in Fig. 12-5.

The function of the photosynthetic pigments is to harvest radiant energy and to transfer it to the photochemical reaction center proper, in a highly efficient manner. For this purpose they occur not singly in dilute solution, but are organized in the form of multimolecular (several hundred) packets or arrays. In such arrays, absorption of a quantum of light by all members is equally effective. The energy so absorbed is then very rapidly delocalized ($\tau \leq 10^{-10}$ sec) and migrates to any component of the aggregate, including one of its minor constituents. It is probably one of the latter that forms the photochemically active part of the reaction center. Two functions have been assigned to the accessory pigments: to extend the effective range of the light-harvesting antennae and to protect various sensitive components from photooxidation by the potent oxidant, Chl^+, generated in the primary photochemical act.

TABLE 12-1
Important photosynthetic pigments

Name of pigments (major absorption bands, photochemically active in vivo are shown, wavelength given in nanometers [nm])

Organism(s)	*Principal chlorophyll*	*Accessory chlorophyll*	*Other accessory pigments*
Multicellular green plants	Chl *a*: 683 (680–685) Chl *a*: 695 (695–700, 660[a])	Chl *b*: 560, 480 Chl *a*: 670 (670–673)	Carotenoids (481) Mainly β-carotene
Green algae	Similar to above	Similar to above	Similar to above
Brown algae	Chl *a*	Chl *c*	β-Carotene
Diatoms, flagellates	Chl *a*	Chl *c*: 620, 470	β-Carotene
Red algae	Chl *a*	Chl *d*: (680[a])	β-Carotene; phycoerythrin (570, 550, 495) plus some phycocyanin (625)
Purple bacteria (both sulfur and nonsulfur)	BChl (800, 850, 870–890[b], 773[a])		Different carotenoids (among them rhodopsin, lycopene, spirilloxanthin, etc.)
Green sulfur bacteria	Chlorobium Chl 650 (725, 650[a]) Plus chlorobium Chl 660 (747, 660[a]) Plus traces of B Chl (810, 770[a])		Similar to above

[a] In ether; amorphous monolayers of Chl *a* absorb at 675 nm, crystalline ones at 735 nm.

[b] Depending on species (890 nm in *Rhodospirillum rubrum* and *Chromatium;* 870 nm in *Rhodopseudomonas spheroides*).

NOTE: Blue-green algae, being procaryotic, contain bacteriochlorophyll rather than chlorophyll *a*.

A. The chlorophylls

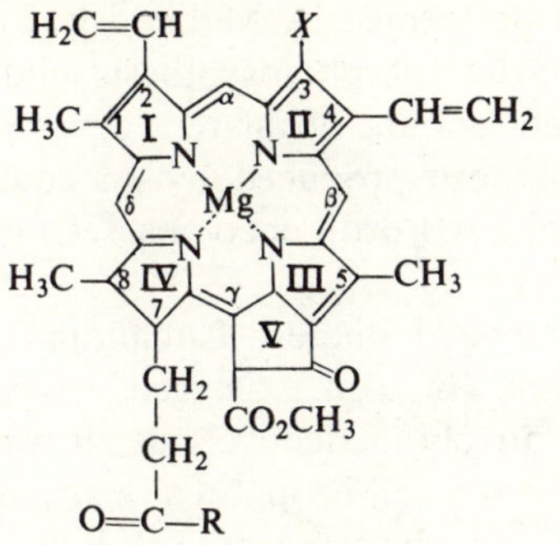

When X = —CH_3 : chlorophyll *a* (Chl *a*)
X = —CHO : chlorophyll *b* (Chl *b*)

H₂C=CH X H₃C I II CH=CH₂ N N Mg N N H₃C IV III CH₃ V CH₂ O CO₂CH₃ CH₂ O=C—R

R = phytyl ($C_{20}H_{39}O$—)

H₃C O— CH₃ CH₃ CH₃ CH₃

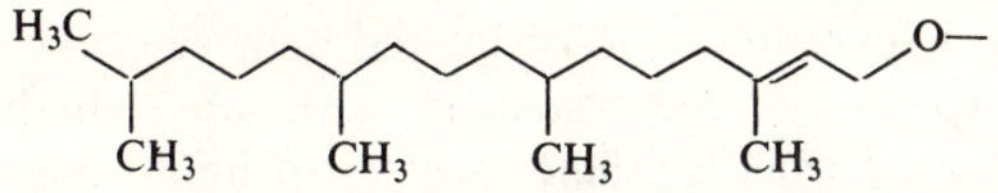

Note the chlorin (reduced porphin in ring IV) tetrapyrrole ring system IV, plus the fused cyclopentanone ring V. The complete parent ring system I-V is called phaeoporphyrin. In bacteriochlorophyll, ring II is reduced as well, i.e., positions 3 and 4 are —CH.

B. Examples of accessory pigments

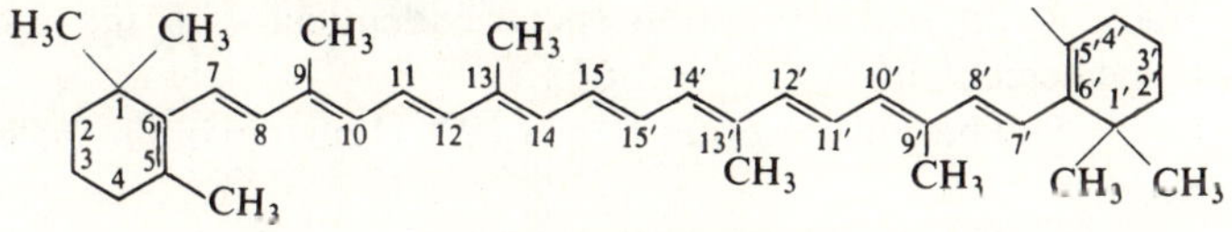

H₃C CH₃ CH₃ CH₃ CH₃ CH₃ CH₃ CH₃ CH₃ CH₃

β-Carotene (all *trans*)
a *carotenoid*

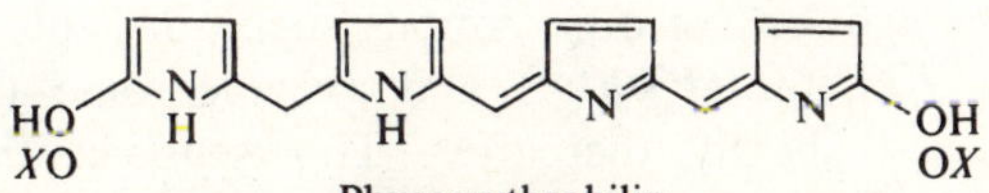

HO N H N H N N OH
XO OX

Phycoerythrobilin
a *phycobilin*

This open-chain tetrapyrrole is covalently linked, probably through the —OH group(s) to a protein (*X*). The resulting complex is called *phycoerythrin*.

FIGURE 12-5
Structure of photosynthetic pigments.

EVIDENCE FOR TWO LIGHT REACTIONS

There is now abundant evidence that photosynthesis in plants requires the collaboration of two distinct systems. Some of it runs briefly as follows:

1. A qualitative comparison of action and absorption spectra *in vivo* had indicated that light absorbed, not only by the principal but also by the accessory pigments, was effective in photosynthesis as measured by O_2 evolution or CO_2 absorption. But when Emerson and Lewis compared the efficiency of light of different wavelengths in supporting photosynthesis quantitatively (i.e., determined the relative *quantum efficiency*), they observed that, although the efficiency was as great at the absorption maximum of the accessory pigments as that of chlorophyll *a* at 657 nm, there was a re-

markable decrease at somewhat longer wavelengths (~ 700 nm) (the "red drop"), where there is still strong light absorption by chlorophyll (7). In this context an experiment performed by McLeod was of special significance. Measuring the action spectrum of photosynthesis under conditions of light saturation, he found that no amount of far-red light ($\geqq$ 700 nm) could give rates of reaction equal to those produced by an equal amount of light of shorter wavelength. Thus, the red drop encompasses maximal rates as well as quantum efficiencies.

2. Haxo and Blinks extended studies of quantum efficiency by improved techniques to a large number of algae. Oxygen evolution in red algae was found to be driven only relatively ineffectively at all wavelengths greater than 600 to 675 nm; thus there was a red drop already at the very absorption peak of chlorophyll *a*. Conversely, in the same process light absorbed by accessory pigments was two to three times as effective as that absorbed by chlorophyll *a* (Fig. 12-6*a*).

3. The low quantum yield of far-red light (e.g., > 680 nm) is greatly increased by exposure of the system to other wavelengths. This "enhancement effect" is observed whether the two types of light were absorbed concurrently or sequentially with flashes separated by several seconds of darkness (8). The action spectrum for enhancement corresponds to that of the accessory pigments (Fig. 12-6*c*). In *Chlorella* the principal enhancing pigment has been identified as a flavin (9).

4. The kinetics of oxygen evolution when measured in an experimental device with a short time constant (e.g., the oxygen electrode) indicate a pulse or spike that is wavelength dependent. These "chromatic transients" are found to have action spectra identical with those characterizing the enhancement effect (Fig. 12-6*b*).

5. The Hill reaction [Eq. (12-8)] also exhibits an enhancement phenomenon.

6. The appearance or disappearance of certain short-lived, spectrophotometrically identifiable intermediates (optical transients) has been observed on first turning the light on (or off). Among such transients is one photobleached (oxidized) at 700 nm (*P700*$^+$) and a cytochrome (cytochrome *f*?) that becomes oxidized shortly thereafter. This sequence is brought about by absorption in the far-red region (> 680 nm). The reduction of P700$^+$ can be accelerated and the cytochrome can then be reduced in turn by superposition of light absorbed in the blue (~ 652 nm). Probably only the first of these reactions (the oxidation of P700) is a true photochemical reaction, as shown by temperature insensitivity down to very low temperatures.

7. Two light-induced ESR signals are produced on illuminating chloroplasts, algae, and broken-cell preparations (10). One (signal I) has a half-width of ~9 G, no fine structure and a g value of 2.002. From its rate of appearance ($\tau = 200$ μsec) and decay, and other properties such as temperature dependence, it has been adduced that this signal (but not signal II!) is associated with a primary event in photosynthesis. It is elicited by far-red light (713 nm), decreased by simultaneous illumination at 635 nm, and has been tentatively identified as being produced by oxidized P700 (P700$^+$).

8. Photophosphorylation, probably of the cyclic variety (cf. below) can occur with far-red light alone (700 to 710 nm); photophosphorylation with light of shorter wavelength (where the accessory pigments absorb as well) is probably of the stoichiometric variety. Only O_2 evolution, and *not* photophosphorylation or NADP$^+$ reduction, requires Mn^{2+} and is inhibited by

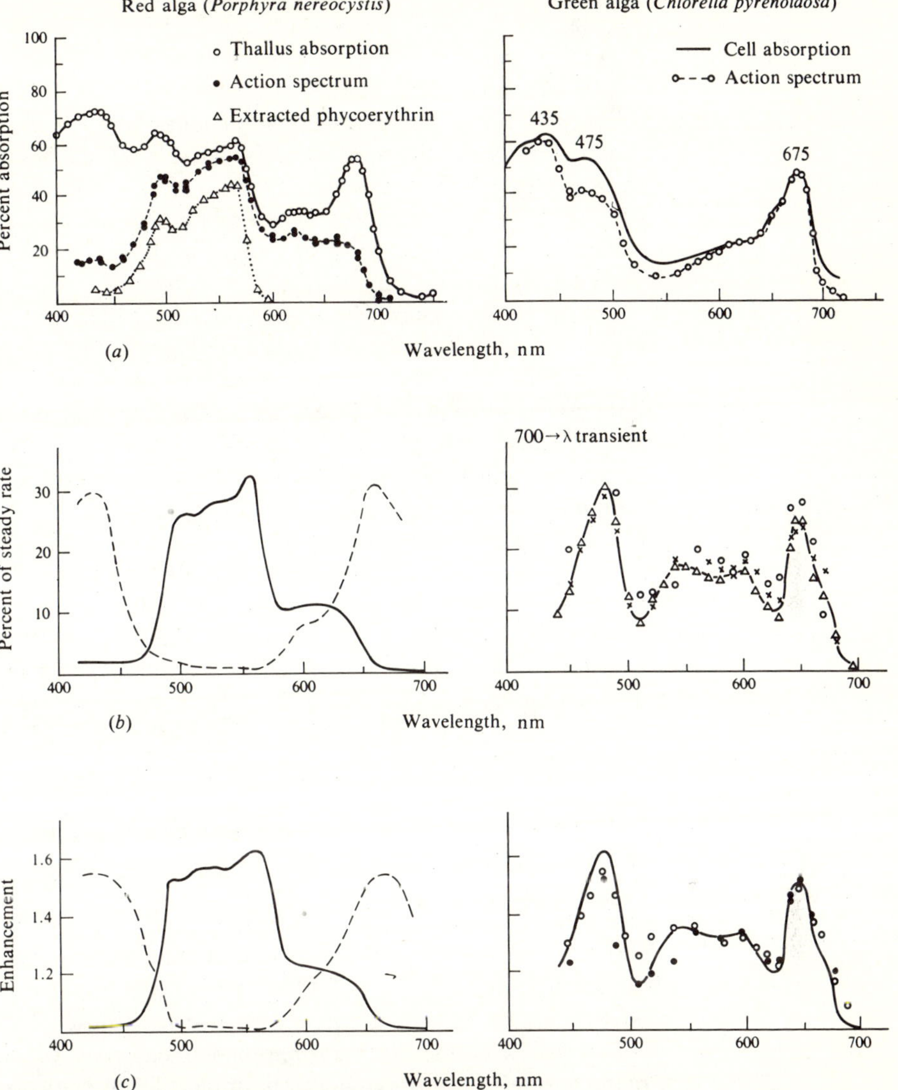

FIGURE 12-6

Action spectra for various phases of photosynthesis. (*a*) O_2 evolution: absorption spectra of pigments *in vivo* are shown for both red and green algae; *in vitro* they are shown for a red alga only. (*b*) For chromatic transients: (*left*) solid line, reference beam at 702 nm; dashed line, reference beam at 560 nm; (*right*) expressed as percentage of steady rate on alternating with 700-nm light. (*c*) Emerson enhancement effect: (*left*) solid line, reference beam at 702 nm; dashed line, reference beam at 560 nm; (*right*) enhancement observed on adding light at wavelengths shown to 700-nm light. Figures in line (*a*) from F. T. Haxo and L. R. Blinks; reprinted by permission of The Rockefeller Institute Press, from *J. Gen. Physiol.*, **33:** 389 (1950). Figures (*b, left*) and (*c, left*) from L. R. Blinks, *Proc. Natl. Acad. Sci. U.S.*, **46:** 327 (1960). Figures (*b, right*) and (*c, right*) from J. Myers and C. S. French; reprinted by permission of The Rockefeller Institute Press, from *J. Gen. Physiol.*, **43:** 723 (1960).

the herbicide DCMU (under special conditions where system II is nonfunctional and an accessory electron donor is provided).

9. Some progress has been made toward the isolation of two distinct types of chloroplast fragments, separated by treatment with detergents followed by

differential centrifugation. One is relatively enriched in Chl *b* and is primarily responsible for the reaction of system II, the other, enriched in Chl *a,* is more effective in catalyzing the reactions of system I.

10. Mutants lacking either system I or system II have been isolated from microorganisms, such as the alga *Scenedesmus* and the protozoan *Chlamydomonas,* in which the wild type is capable of carrying out ordinary photosynthesis by reaction (12-1a). Mutants with a defunct system II are incapable of O_2 evolution but are able to assimilate CO_2 in the presence of an oxidizable substrate DH_2. Thus they are reminiscent of the hydrogen-adapted algae described previously, in which light absorbed by system II has no effect.

In all the foregoing, the reader may have been wondering why it is that, given the postulate of two separate systems, light absorbed preponderantly by system II is fully as effective as that used to drive both systems. This apparent discrepancy is explained by the fact that the chloroplast is capable of performing efficient energy transfer from the shorter wavelength (hence more energetic) system II to the longer-wavelength (hence less energy demanding) system I, but that the converse system I → system II transfer of energy is energetically forbidden.

What then controls the rate of the allowed energy transfer, system II → system I? Clayton has proposed an "energy switch" exercised by P700; only when this compound is in its natural, unbleached (reduced?) state can the transfer proceed; this in turn can occur only if system II, and not system I, absorbs the bulk of the energy, because if the latter were the case, system I in the steady state would be at least in part displaced from the resting position (13).

THE TWO REACTIONS OF PHOTOSYNTHESIS

The scheme outlined in Fig. 12-7 envisages the cooperative interaction of two distinct photochemically active systems functioning in series.

One unit of *system I* consists of some 300 molecules of nonfluorescent Chl *a* with its main absorption in the far red (see Table 12-1). Associated with it is one molecule of the specialized photooxidizable pigment P700 (λ_{max} = 700 to 705 nm; $E_0' \simeq 0.430$ V; a chlorophyll molecule in an unusual environment, perhaps associated with a cytochrome of the "*c* type") directly concerned with the primary photochemical reaction. The system in green plants contains approximately one molecule of *cytochrome f*—a cytochrome *c*-like cytochrome (see page 677) with a $\lambda_{max} = 556.3$ nm; $E_0' = 0.365$ V; other photosynthetic plants and bacteria contain analogous high potential cytochromes (e.g., *cyt 553* in *Chlamydomonas*). In addition, and perhaps less tightly coupled, the unit also contains *ferredoxin* (see page 661) and *cytochrome* b_6—an autoxidizable cytochrome of the "*b* type" found in green plants with a $\lambda_{max} = 563$ nm and a $E_0' = 0.06$ V—both also in concentrations approximately equal to those of cytochrome *f;* approximately 10 molecules of *plastocyanin*—a copper–protein (2 Cu per molecule of M = 20,000, page 644)—as well as equally large (or larger) quantities of naphtho- and benzoquinones—vitamin K and various plastoquinones (see Chapter 15), respectively—NADP (see Chapter 8), and flavins (see Chapter 8). Carotenoids also form part of system I, and light energy absorbed by them is transferred efficiently to its chlorophyll. The principal function of system I is the simultaneous generation of a strong reductant and of a weak oxidant [i.e., reaction (12-10)]. System I, operating by itself, in the presence of an oxidizable donor DH_2, is responsible for photoreductions, such as that carried out by H_2-adapted algae [see Eq. (12-1c)]. The localization of system I

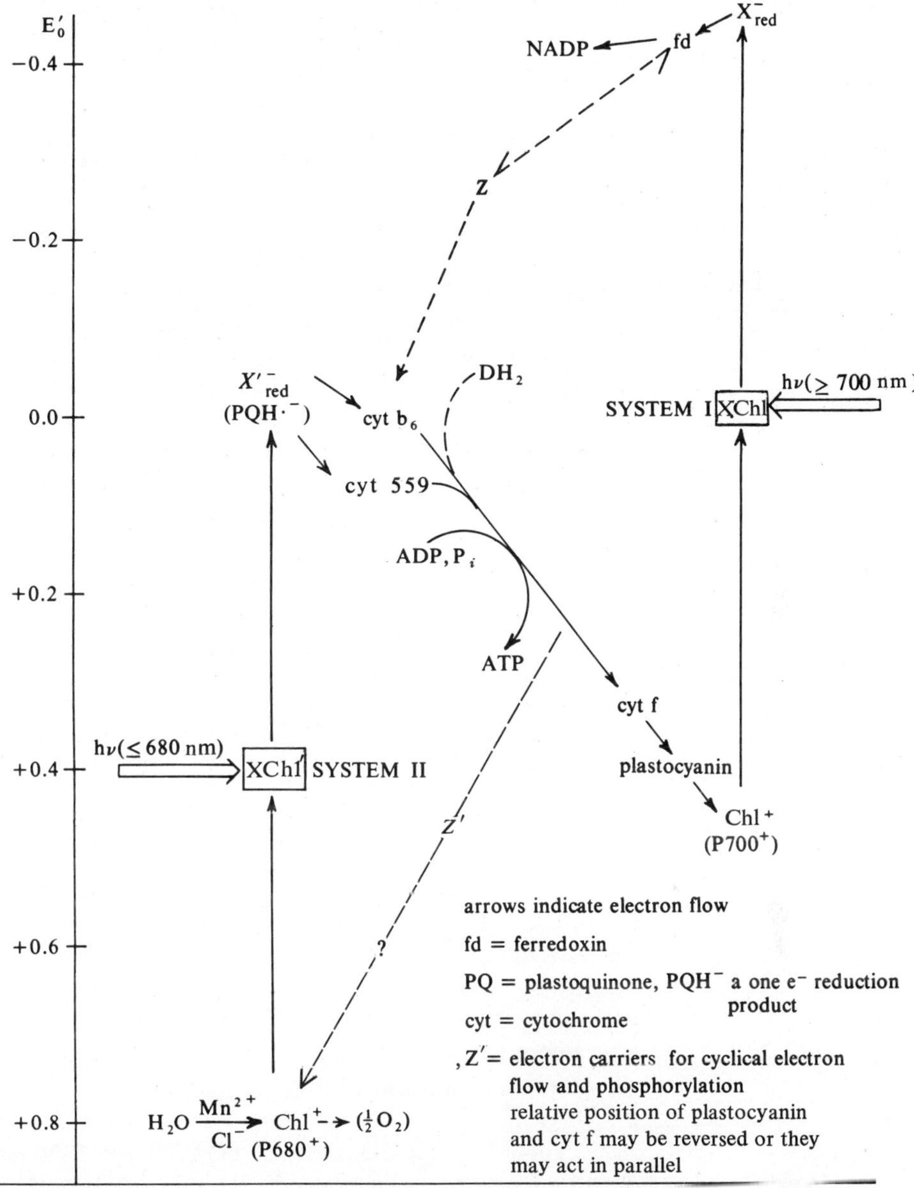

FIGURE 12-7

Series formulation ("*Z* scheme") of photosynthesis. Scheme shown for green plants. In bacterial BChl substitutes for Chl, system II may be absent, and an oxidizable substrate takes the place of $X'-_{red}$. The detailed electron carriers are different also, ubiquinones take the place of plastoquinones, plastocyanin is absent, and other cytochromes of the *b* and *c* type take the place of b_6 and *f*, respectively.

appears to be on granules or particles that can be detached from the thylakoid membranes of the chloroplast by treatment with detergents; thus, it may be identical with Park's quantasomes.

System II is the "accessory pigment system." It contains approximately 100 molecules of *Chl a 670,* a fluorescent form of chlorophyll *a,* absorbing at shorter wavelength than the Chl *a* of system I; it also contains chlorophyll *b,* and the other accessory pigments (chlorophylls *c* and *d* as well as the phycobilin pigments phycoerythrin and phycocyanin, depending on the organism). Its function is the generation of a strong oxidant as a prerequisite for the photoevolution of oxygen [reactions (12-11) and (12-12)]. The reaction(s) involved requires Mn^{2+} and Cl^- and is specifically inhibited by the herbicide 3-(3,4-dichlorophenyl)-1,1-dimethyl urea (DCMU) at exceedingly low concentration ($\sim 10^{-7}$ M). A correlated function of system II is the formation of a weak reductant (X'^-_{red}), perhaps reduced plasto- or some other quinone, capable of regenerating Chl from Chl^+, produced by system I. System II operating by itself is responsible for the Hill reaction [reaction (12-8)]. Besides Chl the system also contains plastoquinones as well as *cytochrome 559* (a form of cytochrome *b* distinct from cytochrome b_6), and a pigment (probably a cytochrome) with a maximum at 550 nm (*C550*). The whole system appears tightly integrated into the thylakoid membranes of the chloroplast.

This photoevolution of oxygen is a peculiarity of photosynthesis in green plants; presently available evidence suggests its absence in photosynthetic bacteria. In these organisms the photosynthetic unit contains one molecule of a specialized photooxidizable pigment, *P890* (P880, or P870, depending on species)—a form of BChl in a special environment, in complete analogy to the situation in plants, per 30 to 50 molecules of ordinary BChl. One molecule of *cytochrome* c_2—a specialized cytochrome of the *c* type akin to cytochrome *f* of plants; $\lambda_{max} = 551$ to 553 nm; $E_0' = 0.29$ to 0.33 V—per molecule of P890 may also form part of the reaction center. In addition, the system contains other electron carriers. Functional roles have so far been assigned to ferredoxin, NAD, the benzoquinone ubiquinone, and to cytochromes of the *cc′* and of the *b* type (Chapter 15). The potential span generated by light is between oxidant (Chl^+) at +470 mV and a reductant (X^-_{red}) at an E_0' of $\leqq -450$ mV if a $1e^-$ species, around -50 mV if it is a $2e^-$ species.

Reactions Catalyzed by the Two Systems

1. The reaction(s) catalyzed by system I utilizes Chl *a* (or BChl) as a photosensitizer. As a consequence of the absorption of a quantum of far-red light by this pigment, there ensues the formation of a reducing entity[4] X^-_{red} and of an oxidized chlorophyll (Chl^+) in the reaction center. The electron, activated in this photochemical event and first localized in the short-lived intermediate X^-_{red} is transferred first to the carrier ferredoxin (fd) and then (in pairs) to a nicotinamide coenzyme, giving rise to NADH in bacteria and NADPH in plants (see below). These compounds provide the source of stable metabolic reducing power required for CO_2 fixation. It can also be shunted from the reduced carriers fd_{red} or NAD(P)H to ETS, the electron transfer chain (heavy diagonal arrows in Fig. 12-7), ultimately ending up by reducing Chl^+ back to Chl in a cyclical process.

2. The oxidizing equivalent or hole is temporarily trapped in Chl^+ with a potential $E_0' \simeq 400$ to 500 mV ($P700^+$, or $P870^+$ in bacteria).

a. In bacterial photosynthesis Chl^+ becomes reduced (thereby regenerating Chl) by means of an externally added, obligatory photoreductant DH_2, which is thereby oxidized to *D* [Eq. (12-13)]. This oxidation, using a series of intermediary physiological redox carriers (diagonal line) or by the cyclical path through *Z*, probably generates the bulk of the ATP produced.

b. In green plants, Chl^+ ($P700^+$) reduction is brought about ultimately by X'^-_{red}, the short-lived intermediate reductant generated by system II (see

[4] A compound exhibiting some of the requisite properties has recently been isolated from spinach chloroplasts by C. F. Yocum and A. San Pietro (*Biochem. Biophys. Res. Commun.*, **3:** 614 (1969)), and named by them "ferredoxin reducing substance" (FRS).

below). This process requires the ETS—a number of enzymes, and intermediary electron carriers. The bulk of the ATP required for carbon assimilation by green plants is probably generated in this reaction [Eq. (12-15)]. An essential part or parts of the requisite electron-transport sequence (at the low-potential side) may be absent in photosynthetic bacteria.

3. The reactions catalyzed by system II utilize the accessory pigment(s) as photosensitizer(s). The absorption of a quantum of light by these entities leads to the generation of a weak reductant X'^{-}_{red} coupled to that of the strong oxidant Chl′+. This is in complete analogy to the situation in system I, except for the potential difference. Chl′+ must be at an $E_0' \geqq 480$ mV and $X'^{-} \simeq 0.0$ mV.

a. The principal function of Chl′+ is the oxidation of water (or hydroxyl ions) to molecular oxygen; it thus drives the O_2-generating portion of photosynthesis.

b. The function of X'^{-}_{red} is the provision of reducing equivalents for the regeneration of the Chl of system I by means of the electron transport chain (i.e., item 2b).

GENERATION OF REDUCING POWER

One of the fundamental requirements of photosynthetic carbon assimilation is the provision of a stable biochemical reducing agent. In more formal terms, it has been demonstrated that carbon dioxide fixation proceeds by the following stoichiometry:

$$\text{6 Ribulose-1,5-di}P + 6CO_2 + 18ATP + 12NADPH \rightarrow \text{6 ribulose-1,5-di}P + \text{hexose-}P + 18ADP + 17P_i + 12NADP^+ \quad (12\text{-}16)$$

Thus, the carbon cycle (see page 577), a series of dark reactions, requires for its operation NADPH or NADH (in bacteria) plus ATP, both generated by reaction(s) driven by light. NADPH had been shown previously to be formed in a Hill reaction with chloroplasts:

$$NADP^+ + n\,h\nu + H_2O \rightarrow NADPH + \tfrac{1}{2}O_2 + H^+ \quad (12\text{-}17a)$$

In 1958, Arnon and his collaborators were able to demonstrate that this reaction could lead to attendant phosphorylation of ADP:

$$NADP^+ + n\,h\nu + H_2O + ADP + P_i \rightarrow NADPH + \tfrac{1}{2}O_2 + ATP \quad (12\text{-}17b)$$

Since then, the evidence has become ever more compelling that reduced nicotinamide coenzymes are generated in a reaction linked to the photochemically active system I and that they constitute the requisite source of stable reducing power generated in stoichiometric amounts in all cells capable of photosynthesis.

More recently, an additional carrier has been identified as intervening between X^{-}_{red} and NAD(P). This is a relatively small, iron-containing protein called ferredoxin (fd). It originally became implicated in another important biosynthetic reaction sequence, the conversion of atmospheric nitrogen to NH_3 by nitrogen-fixing bacteria (page 756); similar pigments have been shown to be present in a variety of anaerobic and photosynthetic bacteria, blue-green and green algae, and plant cells. Its molecular and biochemical properties are discussed in Chapter 15; they differ between bacterial and green plant varieties of fd. But all of them are small nonheme iron (NHI) proteins with potentials in the region of the hydrogen electrode. Fd became implicated in the reactions of system I in plants when a soluble factor required for the photoreduction of NADP by chloroplasts and bacterial fd were shown to be interchangeable and the bleaching and reduction of the bacterial fd could be brought about by chloroplasts in a light-induced reaction.

The ferredoxin-linked reduction of NAD(P) can be formulated as follows:

$$2X^{-}_{red} + 2fd_{ox} \rightleftharpoons 2X + 2fd_{red} \quad (12\text{-}18a)$$

where fd_{red} contains one more electron than does fd_{ox}, plus

$$H^{+} + 2fd_{red} + NAD(P)^{+} \underset{[\text{fd-NAD(P) reductase}]}{\overset{\text{transhydrogenase}}{\rightleftharpoons}} 2fd_{ox} + NAD(P)H \quad (12\text{-}18b)$$

The enzyme responsible for reaction (12-18b) is a flavoprotein which in addition to the reaction indicated also catalyzes the reduction, at the expense of either fd_{red} or NADPH, of a variety of other electron carriers (14,15) including NAD, plastocyanin, and cytochrome *f* (but not mammalian cytochrome *c*!). It can thus itself function as the physiological form of the carrier *Z* in Fig. 12-7, and in this and its other functions participates as a stoichiometric complex (16) with fd probably of structure 1fd·1fp. In bacteria and hydrogen-adapted algae, H^+ itself, in the presence of the enzyme hydrogenase, can function as an electron acceptor, leading to the formation of molecular hydrogen [reaction (12-18c)]:

$$2H^{+} + 2fd_{red} \underset{\text{hydrogenase}}{\overset{\text{transhydrogenase plus}}{\rightleftharpoons}} H_2 + 2fd_{ox} \quad (12\text{-}18c)$$

PHOTOSYNTHETIC PHOSPHORYLATION (PHOTOPHOSPHORYLATION)

That light energy absorbed by a photosynthetic system could lead to the esterification of inorganic phosphate into ATP was first demonstrated by Frenkel with bacterial chromatophore fragments. Arnon and his collaborators were able to carry out the same reaction with chloroplasts, and since then the occurrence of phosphorylation during electron transport in photosynthesis (or *photophosphorylation*) has been documented at least as extensively as that of its older analog: phosphorylation during electron transport in biological oxidation (or *oxidative phosphorylation;* see Chapter 15). The most important process that is linked to the trapping of biochemical energy in ATP is the one that—with the intervention of the ETS (quinones, cytochromes, etc.)—links the reduction of the photochemically generated oxidant of system I (Chl^+ in our formulation) to the oxidation of a reductant: X'^{-}_{red} generated by system II in green plants, or the external reductant DH_2 in bacteria [i.e., reaction (12-15)]. The overall reaction for the stoichiometric, phosphorylative NADPH formation is

$$NADP^{+} + n'ADP + n'P_i + n\,h\nu \xrightarrow[\text{+ ETS + phosph.}]{\text{complete system}} NADPH + H^{+} + n'ATP + \tfrac{1}{2}O_2 + (n'-1)H_2O \quad (12\text{-}19)$$

A similar reaction can also be observed with an artificial acceptor *Z* or *Z′* added instead of $NADP^+$. Under anaerobic conditions, or with system II knocked out (mutants) or inhibited (by DCMU), the addition of a recycling redox carrier *Z*, such as fd, a quinone, etc. (see Fig. 12-7), can link the reductant species generated in the light reaction—X^{-}_{red}, fd_{red}, or NAD(P)H—to the electron-transport system. Now we observe reaction (12-20a):

$$X^{-}_{red} + Chl^{+} + n'ADP + n'P_i \xrightarrow{\text{electron-transport carriers}} X + Chl + n'ATP \quad (12\text{-}20a)$$

Then by linking reaction (12-20a) to reaction (12-10)—the formation of X^-_{red} + Chl^+—we obtain Eq. (12-20b):

$$n'ADP + n'P_i + n\,h\nu \xrightarrow[\text{electron-transport carriers, enzymes, etc.}]{\text{system I}} n'ATP \qquad (12\text{-}20b)$$

This is the expression for *cyclic photophosphorylation,* in which the absorption of radiant energy generates only ATP and neither a reductant nor an oxidant is produced.

Current indications are that the ratio $2e^-$ transferred/NADPH/ATP produced equals unity (i.e., $n' = n/2$). Since CO_2 fixation [Eq. (12-16)] requires 1.5 ATP/NADPH/CO_2 fixed, stoichiometric photophosphorylation must be complemented by an increment of the cyclical type.

In agreement with other workers, Arnon and his collaborators (17) believe that cyclic phosphorylation [Eq. (12-20b)] involves only system I, but they have extended this concept to propose that this—rather than the reduction of $NADP^+$—is the *exclusive* reaction driven by the "long-wavelength" system I. This reaction, as well as the evolution of oxygen, are postulated to be driven by two light reactions in series, as before (Fig. 12-7), but with the important proviso that they are both furnished by system II (i.e., that system II consists of two reactions IIa and IIb, linked by an electron-transport chain, and is responsible for noncyclic phosphorylation). In this formulation plastoquinone (PQ), cyt 559 and plastocyanine (PC) are exclusively associated with system II, cyt b_6 and cyt f with system I. A schematic representation of this scheme with *three* light reactions is presented as Fig. 12-8.

The nature of the intermediates and the mechanism of photosynthetic phosphorylation remains as opaque as that of oxidative phosphorylation, of which, as a comparison of the appropriate electron-transport sequences discloses, it may be regarded as a variant (Chapter 15). Similar approaches have been employed in their study, including the use of similar or analogous inhibitors and uncouplers as well as the partial resolution of the system, particularly with regard to required proteins (coupling factors). As in the case of oxidative phosphorylation, an ATPase capable of interacting allotopically with a membrane fraction fulfills a crucial role, presumably as the ultimate component in the sequence (11,18). Proton fluxes either coincident with or as a cause of photophosphorylation have also been observed. Illumination leads to contraction of chloroplasts or chloroplast fragments as a consequence of the extrusion of water; protons flow *into* the chloroplast and K^+ and Mg^{2+} ions are expelled. Addition of ATP to the medium produces the same effects. A gradient of 3 pH units can be produced by these means. Conversely exposure of chloroplasts in the dark, first briefly to pH 4.5 in the presence of succinate or other organic anions, followed by a rapid plunge into a medium at a pH of 8.5 generates ATP in a process that is subject to inhibition by the same agents that interfere with photophosphorylation or light-induced cation movements.

NATURE OF THE REACTION CENTERS

The reaction center, we recall, consists of some 300 Chl *a* molecules per molecule of P700 in the case of system I in plants and of some 50 BChl molecules per molecule of P870 (P880, or P890) in photosynthetic bacteria. In addition, a high-potential cytochrome of the *c* type and a low-potential benzoquinone in amounts equivalent to those of the primary reactant may also be directly implicated. The most likely sequence of events would appear to be (temperature-insensitive portion in braces)

$$[\mathrm{Chl}_n{\cdot}\mathrm{Chl}]^{\mathrm{cyt(Fe^{2+})}}_{X} \xrightarrow{h\nu} \left\{ [\mathrm{Chl}^*_n{\cdot}\mathrm{Chl}]^{\mathrm{cyt(Fe^{2+})}}_{X} \xrightarrow{\text{very fast}} [\mathrm{Chl}_n{\cdot}\mathrm{Chl}^+]^{\mathrm{cyt(Fe^{2+})}}_{X^-_{\mathrm{red}}} \right\} \xrightarrow[\text{(R.T.)}]{2\,\mu\mathrm{sec}} [\mathrm{Chl}_n\mathrm{Chl}]^{\mathrm{cyt(Fe^{3+})}}_{X^-_{\mathrm{red}}} \quad (\sim \text{seconds, Fig. 12-7, back to start}) \tag{12-21}$$

This represents a formulation in which the primary photochemical act (subsequent to the absorption of a quantum of light by the appropriate chlorophyll, thus converting it to a singlet excited state) is the removal of an electron from the latter (formation of hole), converting it to a positively charged radical and concomitantly producing the reduced form of the acceptor X^-_{red}. Since in spite of a diligent search no free-radical signal attributable to X^-_{red} has been found, even at liquid N_2 temperature, some sort of close coupling of this entity with others of its sort or with some other acceptor appears indicated. Best available evidence suggests that in bacteria it is a $2e^-$ acceptor, with $E^0 \sim -50$ mV, capable of quenching the weakly fluorescing BChl^+ species (P870 in *R. spheroides*) (19). The presence of specialized, photochemically active minority species of chlorophyll molecules (P700, P880, etc.) has been established by a variety of techniques, including studies of appropriate mutants lacking either the minority or the majority component or by selective destruction or removal of the latter (20,21). That its primary fate is oxidation (rather than the a priori equally probable alternative of reduction at the ex-

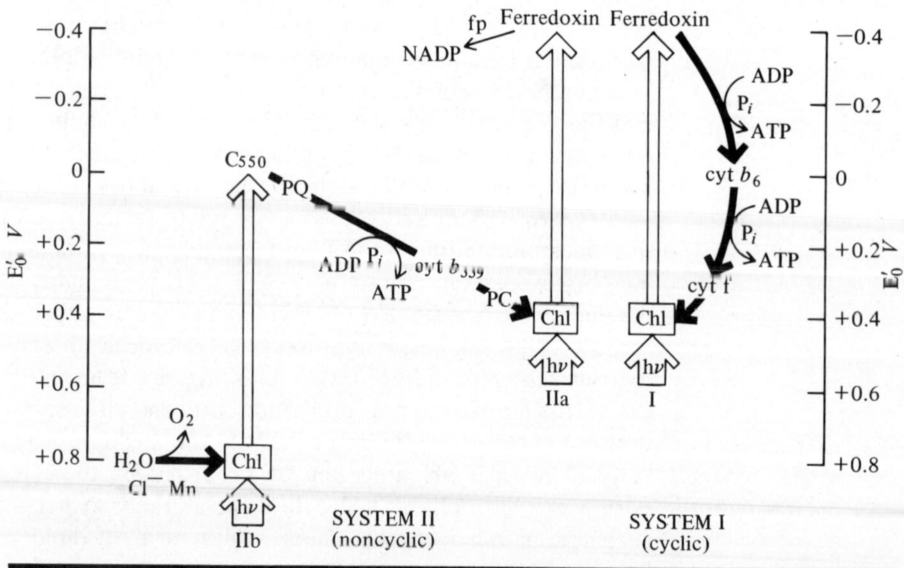

FIGURE 12-8

Scheme for three light reactions in plant photosynthesis. System II consists of two "short-wavelength" light reactions (IIb and IIa) operating in series and linked by a "dark" electron-transport chain associated with noncyclic phosphorylation. Parallel to system II is system I, consisting of a "long-wavelength" light reaction linked to another dark electron-transport chain associated with cyclic phosphorylation. Discussed elsewhere are the roles of Cl^-, manganese, plastoquinone (PQ), ferredoxin-NADP reductase (fp) in system II; the roles of cytochromes b_6 and f in system I; and the experimental modification of system I to give ferredoxin (and NADP) reduction by such artificial electron donors as reduced dichlorophenol indophenol dye. D. B. Knaff and D. I. Arnon, *Proc. Natl. Acad. Sci. U.S.*, **64:** 715 (1969).

pense of the cytochrome) has been established by spectrometric techniques involving both absorption and emission measurements, particularly at low temperatures and in the presence of various agents capable of poising the system at any desired potential. Quantitative measurements of fluorescence yields have been particularly fruitful in establishing that it is the excited singlet (rather than the phosphorescent triplet formed from it) which is photochemically active, and that the oxidation of the cytochrome acceptor occurs coincident with the return of the chlorophyll to the starting state.

Recent evidence suggests that even in bacteria the situation may be more complex than previously recognized (22–25). In *Rhodopseudomonas spheroides,* for instance, the reaction center probably contains two molecules of a pigment called P800 as well as the photobleachable P870. Both *Chromatium* and *Rhodospirillum rubrum* appear to contain more than one reaction center, one perhaps involved in cyclic flow, the other in interacting with the substrate DH_2 (Fig. 12-7). For *Chromatium* it has been suggested that cyclic electron flow and phosphorylation is associated with a reaction center P800, which interacts to oxidize a high-potential cytochrome *c* (cytochrome *c*-555, $E \simeq$ 400 mV) and reduces CoQ_7 (page 662)—the cycle then is mediated by cytochrome *cc′*. On the other hand, a reaction center P905 interacts to oxidize a low-potential cytochrome *c* (cytochrome *c*-552, $E \simeq 0$ mV)—which eventually accepts electrons from DH_2—and reduces NAD(P).

PATH OF CARBON IN PHOTOSYNTHESIS

The dark reactions involved in CO_2 assimilation in green plants have been intensively studied, first *in vivo* by means of radioactive tracers by Calvin,[5] Benson, Bassham, and later *in vitro* with chloroplast and derivative preparations including reconstruction by means of purified soluble enzymes. Qualitatively, a unitary hypothesis has emerged from all these studies. The overall scheme, or "Calvin-Bassham cycle," is shown in Figs. 12-9 a and b. It is thought to be made up of the following reactions (names of enzymes are shown in italics, carbon skeleton rearrangements are indicated in brackets):

6 Ribulose-1,5-di*P* + $6CO_2$ + $6H_2O$ → 6[intermediate] → 12 glycerate-3-*P*
$[6C_5 + 6C_1 = 12C_3]$ *carboxydismutase* (*CD*) (12-22)

12 Glycerate-3-*P* + 12ATP ⇌ 12 glycerate-1,3-di*P* + 12ADP
$[12C_3 \rightarrow 12C_3']$ *phosphoglycerate kinase* (12-23)

12 Glycerate-1,3-di*P* + 12NADPH ⇌ 12 glyceraldehyde-3-*P* + 12NADP+ + 12 P_i
$[12C_3' \rightarrow 12C_3'']$ D-*glyceraldehyde phosphate dehydrogenase* (NADP specific) (12-24)

5 Glyceraldehyde-3-*P* ⇌ 5 dihydroxyacetone-*P*
$[5C_3'' = 5C_3''']$ *triosephosphate isomerase* (12-25)

3 Glyceraldehyde-3-*P* + 3 dihydroxyacetone-*P* ⇌ 3 fructose-1,6-di*P*
$[3C_3'' + 3C_3''' = 3C_6]$ *aldolase* (12-26)

3 Fructose-1,6-di*P* + $3H_2O$ → 3 fructose-6-*P* + $3P_i$
$[3C_6 = 3C_6']$ *phosphatase* (specific)[6] (12-27)

2 Fructose-6-*P* + 2 glyceraldehyde-3-*P* ⇌ 2 erythrose-4-*P* + 2 xylulose-5-*P*
$[2C_6' + 2C_3'' = 2C_4 + 2C_5']$ *transketolase* (*TK*) (12-28)

[5] Who received the Nobel Prize for these endeavors.
[6] The fructose-1,6-diphosphatase extracted from chloroplasts is activated by fd (26).

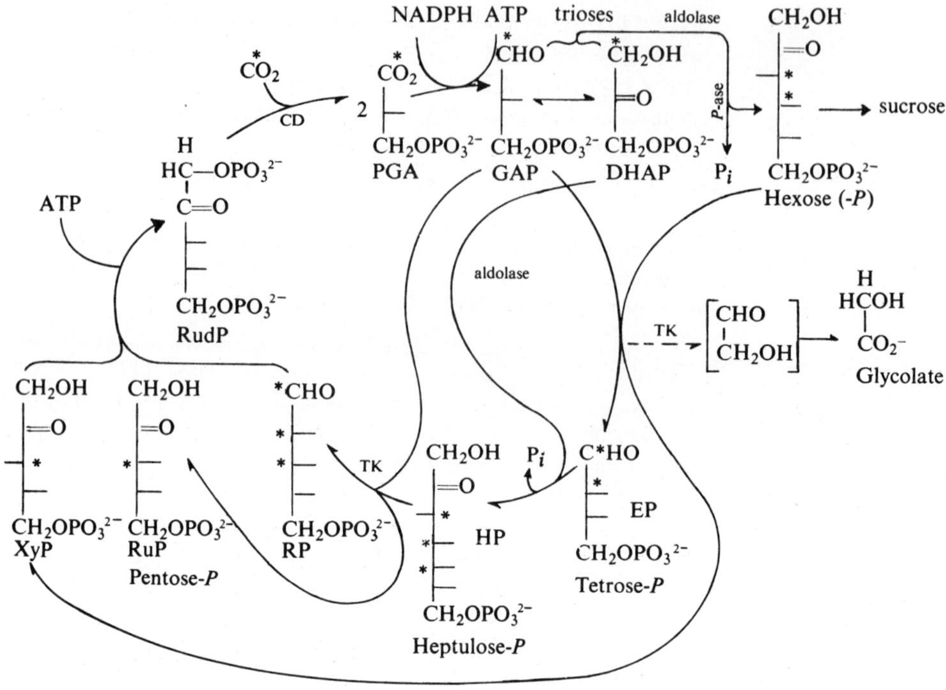

FIGURE 12-9a
Isotope distribution on short-term fixation of CO_2.

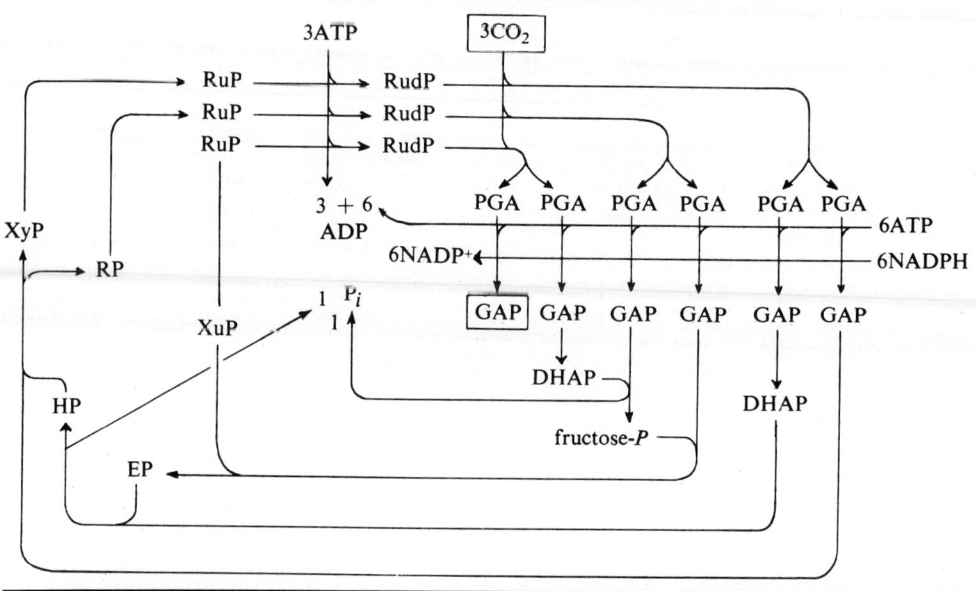

FIGURE 12-9b
Schematic representation of the path of carbon in photosynthesis.

2 Erythrose-4-P + 2 dihydroxyacetone-$P \rightleftharpoons$ 2 sedoheptulose-1,7-di$P$
$[2C_4 + 2C_3''' = 2C_7]$ *aldolase* (12-29)

2 Sedoheptulose-1,7-di$P$ + $2H_2O \rightarrow$ 2 sedoheptulose-7-P + $2P_i$
$[2C_7 = 2C_7']$ *phosphatase* (specific?) (12-30)

2 Sedoheptulose-7-P + 2 glyceraldehyde-3-$P \rightleftharpoons$ 2 ribose-5-P + 2 xylulose-5-P
$[2C_7' + 2C_3'' = 2C_5' + 2C_5'']$ *transketolase* (*TK*) (12-31)

2 Ribose-5-$P \rightleftharpoons$ 2 ribulose-5-P
$[2C_5' = 2C_5''']$ *phosphopentose isomerase* (12-32)

4 Xylulose-5-$P \rightleftharpoons$ 4 ribulose-5-P
$[4C_5'' = 4C_5''']$ *phosphoketopentose epimerase* (12-33)

6 Ribulose-5-P + 6ATP $\rightarrow$ 6 ribulose-1,5-di$P$ + 6ADP
$[6C_5''' = 6C_5]$ *phosphopentokinase* (12-34)

Sums

6 Ribulose-1,5-di$P$ + $6CO_2$ + 18ATP + 12NADPH $\rightarrow$ 6 ribulose-1,5-di$P$ + fructose-6-P
+ 18ADP + $17P_i$ + 12NADP+
$[6C_5 + 6C_1 = 6C_5 + C_6]$ (H_2O not shown) (12-16a)

or, after hydrolysis of the hexose phosphate, schematically,

$$CO_2 + 3ATP + 2NADPH + 3H_2O \rightarrow \tfrac{1}{6}\,\text{hexose} + 3ADP + 3\,P_i + 2NADP^+ \quad (12\text{-}16b)$$

The net reaction [Eq. (12-16a)] leads to the formation of one molecule of a hexose-P, or (after hydrolysis, preceded if necessary by isomerization of the hexose) one hexose for each six turns of the cycle [reaction (12-16b)]. The cycle itself consists of one priming step [reaction (12-34), requiring ATP] one carboxylation step proper [reaction (12-22)] leading to phosphoglyceric acid as the first identifiable product, another priming step [reaction (12-23)], and a reductive step [reaction (12-24)]. All other subsequent reactions [Eqs. (12-25) to (12-33)] are merely rearrangements designed to (1) regenerate the starting material and (2) eliminate the product. Except for reactions (12-22), (12-24), and (12-34), all other steps are catalyzed by enzymes that are not restricted to green plants but are distributed ubiquitously among all living organisms (see Chapter 11).

To recapitulate, the requirements for green-plant photosynthesis [Eqs. (12-35a) to (12-35d) = (12-36)] are

Part I: Photoelectron-transport system

Light-phase linked: $n\,h\nu + 2H^+ + 2NADP^+ + 4Chl \xrightarrow[\text{fd}]{\text{system I}} 2NADPH + 4Chl^+$ (12-35a)

$$n\,h\nu + 2H_2O + 4X' \xrightarrow{\text{system II}} O_2 + 4\bar{X}'_{red} + 4H^+ \quad (12\text{-}35b)$$

Dark-phase linked: $4\bar{X}'_{red} + 4Chl^+ + n'ADP + n'P_i \xrightarrow[\text{system}]{\text{ET and P}}$
$4X' + 4Chl + n'ATP + n'H_2O$
ET and P = electron transport and phosphorylation
(12-35c)

Part II: CO_2 assimilation system

$CO_2 + 2NADPH + 3ATP + 3H_2O \xrightarrow[\text{system}]{\text{Calvin cycle}} \tfrac{1}{6}\,\text{hexose} + 2NADP^+ + 3ADP + 3P_i$
(12-35d)

Grand total

$$2n\,h\nu + CO_2 + (n'-3)ADP + (n'-3)P_i + 5H_2O \xrightarrow[\text{accessory systems}]{\text{chloroplast plus}} \tfrac{1}{6}\text{ hexose} + O_2 + (n'-3)ATP + (n'-3)H_2O \qquad (12\text{-}36)$$

Compare this with Eq. (12-3), and remember that for our purposes here ADP + P_i = ATP + H_2O. This formulation again suggests that if n'/n is indeed 0.5, then there is a deficit in ATP that has to be provided by cyclic phosphorylation. See Eq. (12-20b).

QUANTUM YIELD AND ENERGETICS

The question of the magnitude of the quantum yield [i.e., the value of n in Eq. (12-3)] has been difficult and controversial. Warburg, Burk, and their collaborators believed it to be equal to 4, but Emerson and most others consistently obtained values near 8. The mechanism involving two linked light reactions [Eq. (12-35)] suggests strongly that the value of $n/4$ cannot be less than unity. This then leads to $2n = 8$, in agreement with higher values suggested by the work of a large number of investigators.

Eight einsteins (mol quanta) of light at, for example, 700 nm, utilized with 70 percent efficiency, correspond to approximately 205 kcal; for the reaction $CO_2 + H_2O \rightarrow \{CH_2O\} + O_2$, $\Delta G^{0\prime} = 114$ kcal; for ADP + $P_i \rightarrow$ ATP + H_2O, the $\Delta G^{0\prime} \simeq 10$ kcal. Thus we can calculate that the overall process has an efficiency of $\sim 125/205 \simeq 0.60$. Of the approximately 75 kcal thus apparently "wasted," $\geqq 6 \times 3 = 18$ kcal is dissipated in the hydrolytic steps [Eqs. (12-27), (12-30), and the hydrolysis of hexose-*P*], and $6 \times 7 = 42$ kcal in the formation of ribulose-di*P* [Eq. (12-34)], a total of 60 kcal. This represents energy used to ensure essentially unidirectional flow of carbon through the biosynthetic sequence. We see that the "true" efficiency of the process virtually equals 190/205, or 0.92. Thus, only 8 percent of the available energy is dissipated in ways not capable of rationalization by the biochemist.

OTHER REDUCTIVE REACTIONS DRIVEN BY LIGHT

In organisms containing the enzyme hydrogenase (i.e., photosynthetic bacteria and hydrogen-adapted algae), the fd_{red} produced as the primary stable reducing species by light can give rise to the photoevolution of hydrogen [Eq. (12-18c)]. Alternatively, fd_{red} can be used by the same organisms to reverse the phosphoroclastic reaction of pyruvate (page 518) [Eq. (12-37)]:

$$CH_3COCO_2^- + P_i \xrightarrow{\text{phosphoroclastic split}} \text{acetyl-}P + CO_2 + H_2 \qquad (12\text{-}37)$$

$$CH_3COCO_2^- + P_i + 2fd_{ox} \rightleftharpoons \text{acetyl-}P + CO_2 + 2fd_{red} \qquad (12\text{-}38)$$

Since pyruvate (with the intervention of additional carboxylation steps, see page 520) is the precursor of the various di- and tricarboxylic acids of the citric acid cycle and hence of all carbohydrates and all those amino acids which originate from these compounds, reaction (12-38) provides the photosynthetic cell with a key step in the reductive synthesis of important constituents.

Yet another role has been discovered for reduced ferredoxin in certain bacteria: Nitrogen fixation, that is, the conversion of N_2 to NH_3 by the ATP-requiring microbial enzymes nitrogenase, can utilize either molecular hydrogen or fd_{red}, here consisting of two separate entities, azoferredoxin and molybdoferredoxin, as a source of reducing power (Chapter 17). These various possibilities are outlined below as reaction (12-39), which indicates alternative paths, some of them reversible, for the utilization of fd_{red} and its reconversion to the oxidized form:

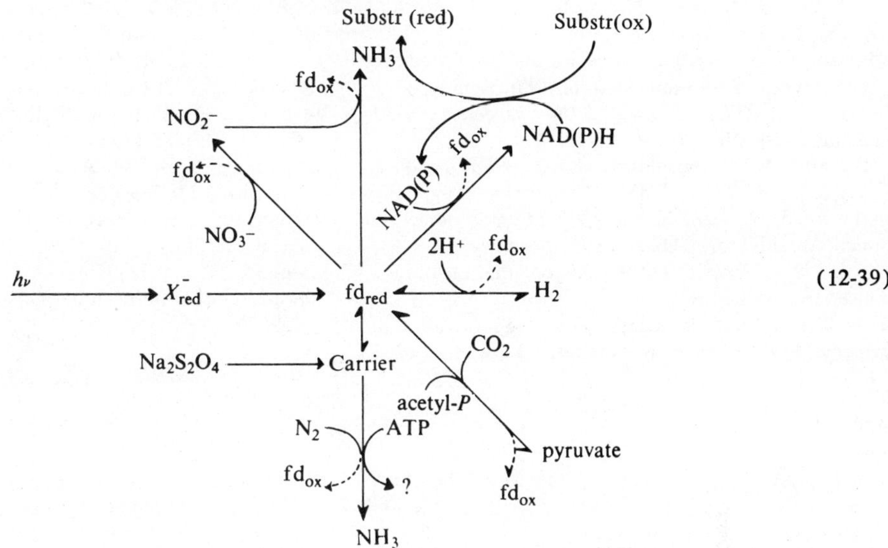

(12-39)

fd_{red}, either with or without the intervention of reduced pyridine nucleotides, has also been implicated in the assimilation of oxidized nitrogen compounds, and in the reduction of nitrate to nitrite, as well as the complete reduction of the latter to ammonia (see Chapter 17).

Finally fd_{red} plus appropriate transhydrogenases should also be able to drive not only all those reductive and biosynthetic reactions that require NADPH, as indicated, but those requiring reduced flavoproteins as well.

REFERENCES

Books, Monographs, and Reviews

Bassham, J. A.: Photosynthesis: Energetics and Related Topics, *Advan. Enzymol.*, **25:** 39 (1963).

Bassham, J. A., and M. Calvin: *The Path of Carbon in Photosynthesis,* Prentice-Hall, Englewood Cliffs, N.J., 1957.

Boardman, N. K.: The Photochemical System of Photosynthesis, *Advan. Enzymol.*, **30:** 1 (1968).

Brookhaven Symposia in Biology: *Energy Conversion by the Photosynthetic Apparatus,* vol. 19, 1967.

Brookhaven Symposia in Biology: *The Photochemical Apparatus: Its Structure and Function,* vol. 11, 1959.

Calvin, M., and J. A. Bassham: *The Photosynthesis of Carbon Compounds,* Benjamin, New York, 1962.

Clayton, R. K.: *Molecular Physics in Photosynthesis,* Blaisdell, New York, 1965.

Clayton, R. K.: Photosynthesis: Primary Physical and Chemical Processes, *Ann. Rev. Plant Physiol.*, **14:** 159 (1963).

Crane, F. L., and H. Low: Quinones in Energy-Coupling Systems, *Physiol. Rev.*, **46:** 662 (1966).

Duysens, L. N. M.: Photosynthesis, *Progr. Biophys. Mol. Biol.*, **14:** 3 (1964).

Gest, H., A. San Pietro, and L. P. Vernon (eds.): *Bacterial Photosynthesis,* Antioch Press, Yellow Springs, Ohio, 1963.

Gibbs, M.: Photosynthesis, *Ann. Rev. Biochem.*, **36:** 757 (1967).

Giese, A. C. (ed.): *Photophysiology,* Academic Press, New York, 1964.

Good, N., S. Izawa, and G. Hind: Uncoupling and Energy Transfer Inhibition in Photophosphorylation, in R. Sandi (ed.), *Current Topics in Bioenergetics,* vol. 1, Academic Press, New York, 1966, p. 76.

Goodwin, T. W. (ed.): *Biochemistry of Chloroplasts,* vol. 1, Academic Press, New York, 1967.

Hill, R., and C. P. Whittingham: *Photosynthesis,* Wiley, New York, 1957.

Hind, G. M., and J. M. Olson: Electron Transport Pathways in Photosynthesis, *Ann. Rev. Plant Physiol.*, **19:** 249 (1968).

Jagendorf, A. T.: Biochemistry of Energy Transformations during Photosynthesis, *Surv. Biol. Progr.*, **4:** 183, (1962).

Jagendorf, A. T., and E. Uribe: Photophosphorylation and the Chemiosmotic Hypothesis, *Brookhaven Symp. Biol.*, **19:** 215 (1967).

Kamen, M. D.: *Primary Processes in Photosynthesis,* Academic Press, New York, 1963.

Kirk, J. T. O., and R. A. E. Tilney-Bassett: *The Plastids,* Freeman, San Francisco, 1967.

Kok, B., and G. M. Cheniae: Kinetics and Intermediates of the Oxygen Evolution Step in Photosynthesis, in R. Sandi (ed.), *Current Topics in Bioenergetics,* vol. 1, Academic Press, New York, 1966, p. 2.

Kok, B., and A. T. Jagendorf (eds.): *Photosynthetic Mechanisms of Green Plants,* Publication 1145, National Academy of Sciences, National Research Council, Washington, D.C., 1963.

Levine, R. P.: Genetic Dissection of Photosynthesis, *Science,* **162:** 768 (1968).

McElroy, W. D., and B. Glass (eds.): *Symposium on Light and Life,* Johns Hopkins Press, Baltimore, 1961.

San Pietro, A. (ed.): *Non-Heme Iron Proteins: Role in Energy Conversion,* Antioch Press, Yellow Springs, Ohio, 1965.

Smith, J. H. C., and C. S. French: The Major and Accessory Pigments in Photosynthesis, *Ann. Rev. Plant Physiol.,* **14:** 181 (1963).

Vernon, L. P.: Bacterial Photosynthesis, *Ann. Rev. Plant Physiol.,* **15:** 73 (1963).

Vernon, L. P., and M. Avron: Photosynthesis, *Ann. Rev. Biochem.,* **34:** 269 (1965).

Vernon, L. P., and G. R. Seely (eds.): *The Chlorophylls,* Academic Press, New York, 1966.

Vishniac, W., B. L. Horecker, and S. Ochoa: Enzymic Aspects of Photosynthesis, *Advan. Enzymol.,* **19:** 1 (1957).

Specific References

1. Yourgrau, W., and A. van der Merwe: *Proc. Natl. Acad. Sci. U.S.,* **59:** 734 (1968); Brittin, W., and G. Gamow: *ibid.,* **47:** 724 (1961).
2. Miller, S. L.: *Ann. N.Y. Acad. Sci.,* **69:** 260 (1957); Granick, S.: *ibid.,* **69:** 292 (1957).
3. Bassham, J. A., M. Kirk, and R. G. Jensen: *Biochim. Biophys. Acta,* **153:** 211 (1968). Jensen, R. G., and J. A. Bassham: *ibid.,* **153:** 219, 227 (1968).
4. Kalberer, P. P., B. B. Buchanan, and D. I. Arnon: *Proc. Natl. Acad. Sci. U.S.,* **57:** 1542 (1967).
5. Park, R. B., and N. G. Pon: *J. Mol. Biol.,* **3:** 1 (1961); **6:** 105 (1963); Lichtenthaler, H. K., and R. B. Park: *Nature,* **198:** 1070 (1963); Park, R. B., and J. Biggins: *Science,* **144:** 1009 (1964); Park, R. B.: *J. Cell. Biol.,* **27:** 151 (1965).
6. Moudrianakis, E. N.: *Federation Proc.,* **27:** 1180 (1968).
7. Emerson, R., and C. M. Lewis: *J. Gen. Physiol.,* **25:** 579 (1942); *Am. J. Botany,* **30:** 165 (1943); Haxo, F. T., and L. R. Blinks: *J. Gen. Physiol.,* **33:** 389 (1950).
8. Emerson, R., R. Chalmers, and C. Cederstrand: *Proc. Natl. Acad. Sci. U.S.,* **43:** 133 (1957); Myers, J., and C. S. French: *Plant Physiol.,* **35:** 963 (1960); Myers, J. S.: in B. Kok and A. Jagendorf (eds.), *Photosynthetic Mechanisms of Green Plants,* Publication 1145, National Academy of Sciences, National Research Council, Washington, D.C., 1964, p. 301; French, C. S.: in W. D. McElroy and B. Glass (eds.), *Symposium on Light and Life,* Johns Hopkins Press, Baltimore, 1961, p. 447; reviewed in Smith, J. H. C., and C. S. French: *Ann. Rev. Plant Physiol.,* **14:** 181 (1963).
9. Pickett, J. M., and C. S. French: *Proc. Natl. Acad. Sci. U.S.,* **57:** 1587 (1967).
10. Reviewed in I. Isenberg: *Physiol. Rev.,* **44:** 487 (1964).
11. Hind, G., and A. T. Jagendorf: *Proc. Natl. Acad. Sci. U.S.,* **49:** 715 (1963); Jagendorf, A. T.: *Federation Proc.,* **26:** 1361 (1967); Kaplan, J. H., and A. T. Jagendorf: *J. Biol. Chem.,* **243:** 972 (1968).
12. Tagawa, K., H. Y. Tsujimoto, and D. I. Arnon: *Proc. Natl. Acad. Sci. U.S.,* **49:** 567 (1963); **50:** 544 (1963); Arnon, D. I., H. Y. Tsujimoto, and B. D. McSwain: *Nature,* **207:** 1367 (1965); **214:** 562 (1967); McSwain, B. D., and D. I. Arnon: *Proc. Natl. Acad. Sci. U.S.,* **61:** 989 (1968).
13. Clayton, R. K.: *J. Theoret. Biol.,* **5:** 497 (1963).
14. Keister, D. L., A. San Pietro, and F. E. Stolzenbach: *J. Biol. Chem.,* **235:** 2989 (1960).
15. Forti, G., and E. Sturani: *European J. Biochem.,* **3:** 461 (1968).
16. Nelson, N. and J. Neumann: *J. Biol. Chem.,* **244:** 1926, 1932 (1969).
17. Knaff, D. B., and D. I. Arnon: *Proc. Natl. Acad. Sci. U.S.,* **63:** 956, 963 (1969); **64:** 715 (1969).
18. Bennun, A., and E. Racker: *J. Biol. Chem.,* **244:** 1325 (1969); Livne, A., and E. Racker: *ibid.,* **244:** 1332, 1339 (1969).
19. Reed, D. W., K. L. Zankel, and R. K. Clayton: *Proc. Natl. Acad. Sci. U.S.,* **63:** 42 (1969).
20. Clayton, R. K., and W. R. Sistrom: *Photochem. Photobiol.,* **5:** 661 (1966). Reed, D. W., and R. K. Clayton: *Biochem. Biophys. Res. Commun.,* **30:** 471 (1968).
21. Bishop, N. I.: *Rec. Chem. Progr.,* **25:** 181 (1964); Pratt, L. H., and N. I. Bishop: *Biochim. Biophys. Acta,* **153:** 664 (1968).
22. Morita, S.: *Biochim. Biophys. Acta,* **153:** 241 (1968).
23. Cusanovich, M. A., and M. D. Kamen: *Biochim. Biophys. Acta,* **153:** 376, 418 (1968); Cusanovich, M. A., R. G. Bartsch, and M. D. Kamen: *ibid.,* **153:** 397 (1968).
24. Sybesma, C., and C. F. Fowler: *Proc. Natl. Acad. Sci. U.S.,* **61:** 1343 (1968).
25. Chance, B., D. DeVault, W. W. Hildreth, W. W. Parson, and M. Nishimura: *Brookhaven Symp. Biol.,* **19:** 115 (1967); Chance, B.: *Biochem. J.,* **103:** 1 (1967).
26. Buchanan, B. B., P. P. Kalberer, and D. I. Arnon: *Biochem. Biophys. Res. Commun.,* **29:** 74 (1967).

13

Oxidation of Fatty Acids and Degradation of Complex Lipids

The *lipids* are a heterogeneous group of substances. In general, they may be regarded as compounds, occurring in living systems, which are insoluble in water but soluble in such organic solvents as methanol, chloroform, benzene, and acetone. Within this diffuse classification, there are several major classes of compounds; these include the neutral *fats* and *oils, waxes, phospholipids, sphingolipids,* and *steroids.* A more complete classification is presented in Chapter 16, which also considers some of the physicochemical aspects of the lipids. Several excellent treatments of the chemistry of the lipids are available. For the moment, our concern is the metabolic degradation of both complex lipids, specifically the fats, oils, and phospholipids, and the fatty acids that are derived from them. In Chapter 14, the relationship between the catabolism of the lipids and carbohydrates will be developed.

FATTY ACIDS

The names and structures of the most common, as well as a few of the more exotic, fatty acids are collected in Table 13-1. The most common fatty acids in nature are straight-chain saturated or unsaturated compounds containing an even number of carbon atoms. Particularly prevalent are palmitic, stearic, and oleic acids. Saturated straight-chain acids containing an odd number of carbons seldom appear in large quantity but do appear to occur universally. On the other hand, branched-chain fatty acids have been identified in relatively few natural sources. Some examples include 11-methyldodecanoic and 13-methyltetradecanoic acids in butter fat, 10-methyloctadecanoic acid in tubercle bacilli, and 12-methyltetradecanoic and isopalmitic acid, among others, in *Bacillus subtilis.* Fatty acids containing a cyclopropane ring have been identified in at least three cases: *cis*-11,12-methyleneoctadecanoic acid (lactobacillic acid) is a major lipid constituent of *Lactobacilli* and *Agrobacterium tumefaciens,* and *cis*-9,10-methylenehexadecanoic acid has been isolated from *E. coli.* Fatty acids containing a cyclopropene ring—sterculic acid—and a cyclopentene ring—chaulmoogric acid—have been isolated from plant sources.

Free fatty acids occur in the diet to only a minor extent. The major fraction of the ingested fatty acids occurs incorporated into more complex lipids, principally the triglycerides and the phosphatides. The fatty acid composition

TABLE 13-1
Some naturally occurring fatty acids

Saturated fatty acids	
Butyric acid	$CH_3(CH_2)_2CO_2H$
Caproic acid	$CH_3(CH_2)_4CO_2H$
Caprylic acid	$CH_3(CH_2)_6CO_2H$
Capric acid	$CH_3(CH_2)_8CO_2H$
Lauric acid	$CH_3(CH_2)_{10}CO_2H$
Myristic acid	$CH_3(CH_2)_{12}CO_2H$
Palmitic acid	$CH_3(CH_2)_{14}CO_2H$
Stearic acid	$CH_3(CH_2)_{16}CO_2H$
Arachidic acid	$CH_3(CH_2)_{18}CO_2H$
Lignoceric acid	$CH_3(CH_2)_{22}CO_2H$
Unsaturated fatty acids	
Crotonic acid	$CH_3CH{=}CHCO_2H$
Palmitoleic acid	$CH_3(CH_2)_5CH{=}CH(CH_2)_7CO_2H$
Oleic acid	$CH_3(CH_2)_7CH{=}CH(CH_2)_7CO_2H$
Linoleic acid	$CH_3(CH_2)_3(CH_2CH{=}CH)_2(CH_2)_7CO_2H$
Linolenic acid	$CH_3(CH_2CH{=}CH)_3(CH_2)_7CO_2H$
Arachidonic acid	$CH_3(CH_2)_3(CH_2CH{=}CH)_4(CH_2)_3CO_2H$
Nervonic acid	$CH_3(CH_2)_7CH{=}CH(CH_2)_{13}CO_2H$
Ximenynic acid	$CH_3(CH_2)_5CH{=}CH{-}C{\equiv}C{-}(CH_2)_7CO_2H$
Mycomycin	$HC{\equiv}CC{\equiv}C{-}CH{=}C{=}CH{-}CH{=}CH{-}CH{=}CHCH_2CO_2H$
Others	
Lactobacillic acid	$(CH_2)_5CH_3$ / $(CH_2)_9{-}CO_2H$
Sterculic acid	$(CH_2)_7{-}CH_3$ / $(CH_2)_7{-}CO_2H$
Chaulmoogric acid	$(CH_2)_{12}CO_2H$ / H

of the complex lipids of mammals is a function of four variables: the species of mammal, the tissue, the class of complex lipid, and the diet.

An example of the dependence of fatty acid composition on the nature of lipid class is shown in Table 13-2, in which the fatty acid compositions of lecithin (page 588) and cardiolipin (page 732) in monkey skeletal muscle are compared. Lecithins are very widely distributed lipids in mammals; cardiolipins have a more restricted occurrence. Much of the cellular cardio-

TABLE 13-2
Comparison of the fatty acid composition of lecithin and cardiolipin of monkey skeletal muscle

Acid	*Lecithin, mol percent*	*Cardiolipin, mol percent*
Myristic	0.0	0.3
Palmitic	32.6	1.8
Palmitoleic	0.0	1.2
Stearic	4.4	0.9
Oleic	9.9	14.8
Linoleic	36.4	80.9
Acids with carbon chain > 18C	16.7	0.0

SOURCE: E. J. Masoro, *Physiological Chemistry of Lipids in Mammals,* Saunders, Philadelphia, 1968.

lipin is found in the inner membrane of the mitochondria. Note that the distribution of fatty acids in lecithin is much wider than in cardiolipin, in which more than 80 mol percent is linoleic acid. Similar differences in fatty acid profile are found in comparisons of the same lipid class in different tissues or different species. The subcellular factors that control the fatty acid composition of a given lipid class are not known but may reside in part at least in the specificity of the enzymes concerned with their assembly (Chapter 16).

We begin our discussion of the catabolism of lipids by considering first the routes for degradation of the complex lipids into their constituents and, subsequently, the metabolism of the fatty acids themselves.

DEGRADATION OF TRIGLYCERIDES

The most abundant lipids in animals and plants are the triesters of glycerol, termed *triglycerides*. The parent substance, glycerol,

$$\begin{array}{l} CH_2OH \\ | \\ CHOH \\ | \\ CH_2OH \end{array}$$

is a viscous, sweet liquid completely miscible with water and ethanol by insoluble in ether. Each of the hydroxyl groups of glycerol may be esterified with a fatty acid; thus, in addition to the triglycerides, one may form *monoglycerides* and *diglycerides* as well:

$$\begin{array}{l} CH_2O—CO—R \\ | \\ CHOH \\ | \\ CH_2OH \end{array} \quad \text{or} \quad \begin{array}{l} CH_2OH \\ | \\ CHO—CO—R \\ | \\ CH_2OH \end{array} \qquad\qquad \begin{array}{l} CH_2O—CO—R \\ | \\ CHO—CO—R' \\ | \\ CH_2OH \end{array} \quad \text{or} \quad \begin{array}{l} CH_2O—CO—R \\ | \\ CHOH \\ | \\ CH_2O—CO—R' \end{array}$$

Monoglycerides Diglycerides

$$\begin{array}{l} CH_2O—CO—R \\ | \\ CHO—CO—R' \\ | \\ CH_2O—CO—R'' \end{array}$$

Triglyceride

Of these, the mono- and diglycerides are only trace components of natural lipids.

The triglycerides compromise two classes of lipids: the fats and the oils. The distinction between these is based on the physical state of the triglyceride at room temperature, the oils being liquid and the fats being solid. The melting point of the triglycerides is, of course, determined by the nature of the fatty acid composition. In general, the higher the proportion of short-chain acids and unsaturated acids, the lower the melting point. For example, the triglyceride derived from butyric acid, tributyrin, has a melting point of −75°, whereas that derived from stearic acid, tristearin, melts at 71°. Both stearic acid and oleic acid contain 18 carbon atoms (Table 13-1); yet triolein melts at −17°, 88° lower than tristearin, emphasizing the importance of unsaturation in determining the physical properties of triglycerides.

The solubility properties of the triglycerides reflect those of the fatty acids from which they are derived. Generally, triglycerides are insoluble in water but soluble in organic solvents. Water solubility is greater for triglycerides derived from short-chain or unsaturated fatty acids.

Triglycerides are degraded chemically by hydrolysis in either acidic or basic media: in the former case one obtains glycerol and the fatty acids and in the latter glycerol and the salts of these acids, termed *soaps*. These hydrolytic procedures are important for the analysis of the composition of fatty acids in a triglyceride sample. Following liberation of the fatty acids from the parent triglyceride, they may be readily identified and quantitated by gas–liquid chromatography.

The fats isolated from natural sources are complex mixtures of triglycerides. But individual triglycerides are well-characterized molecular species. The simplest contain just one type of fatty acyl group and, as noted above, are named for that group. Thus, triolein has the structure

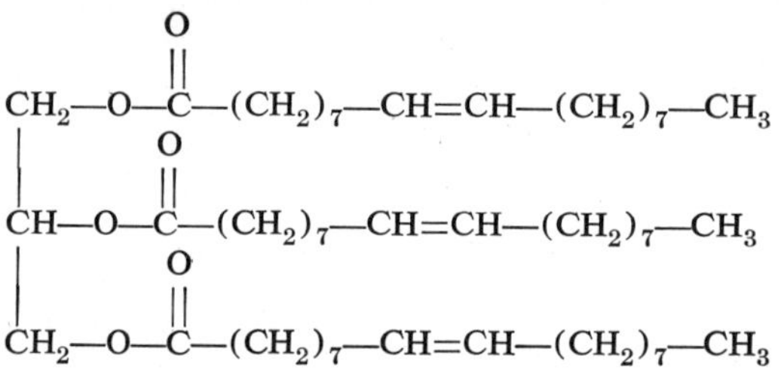

$$
\begin{array}{l}
CH_2{-}O{-}\overset{\overset{\large O}{\|}}{C}{-}(CH_2)_7{-}CH{=}CH{-}(CH_2)_7{-}CH_3 \\
| \\
CH{-}O{-}\overset{\overset{\large O}{\|}}{C}{-}(CH_2)_7{-}CH{=}CH{-}(CH_2)_7{-}CH_3 \\
| \\
CH_2{-}O{-}\overset{\overset{\large O}{\|}}{C}{-}(CH_2)_7{-}CH{=}CH{-}(CH_2)_7{-}CH_3
\end{array}
$$

Triolein

Generally, however, triglycerides that occur in living systems have mixed fatty acid contents. Two examples are 1-oleodipalmitin and 2-oleodipalmitin; these have the same fatty acid composition but, as the names indicate, differ in the disposition of the fatty acyl groups among the hydroxyl groups of glycerol. These isomers have distinct physical and chemical properties. More complex triglycerides contain three different fatty acyl groups. One example, containing one residue each of oleic, palmitic, and linoleic acids is

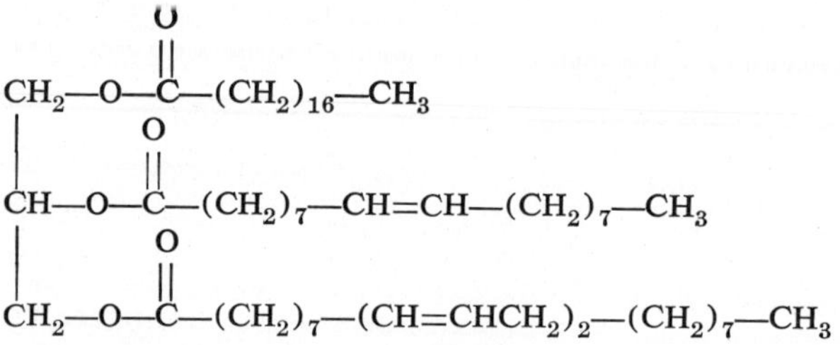

$$
\begin{array}{l}
CH_2{-}O{-}\overset{\overset{\large O}{\|}}{C}{-}(CH_2)_{16}{-}CH_3 \\
| \\
CH{-}O{-}\overset{\overset{\large O}{\|}}{C}{-}(CH_2)_7{-}CH{=}CH{-}(CH_2)_7{-}CH_3 \\
| \\
CH_2{-}O{-}\overset{\overset{\large O}{\|}}{C}{-}(CH_2)_7{-}(CH{=}CHCH_2)_2{-}(CH_2)_7{-}CH_3
\end{array}
$$

1 Palmitoyl-2-oleoyl-3-linoleoylglycerol

This trigylceride is an important constituent of the fats of the mammalian liver. Note that this triglyceride, as well as others containing three different fatty acyl groups, is optically active, because the central carbon atom of the glycerol moiety possesses four different substituents. Although the hydrolysis of triglycerides is not a stereospecific process, the biosynthesis of these species is (Chapter 16).

The triglycerides are considered to serve three principal functions in mammals. First, they act as an energy reservoir. As will be developed in Chapter 14, oxidation of triglyceride-derived fatty acids generates very large amounts of chemical energy in the form of ATP (accounting for the fact that it is difficult indeed to lose weight by exercise alone). Second, triglycerides provide thermal insulation for the organism. The subcutaneous layer of fat (principally triglyceride), which accounts for a characteristic difference in the appearance of the male and female bodies, also provides the female with the capacity to withstand cold temperatures with less discomfort than the male. Finally, triglycerides provide mechanical insulation for the body, tending to prevent damage to interior parts in falls and the like.

The hydrolysis of ingested triglycerides in higher animals occurs largely in the small intestine and is catalyzed by lipolytic enzymes elaborated by the pancreas. The pancreatic *lipases* (glycerol ester hydrolases) catalyze the rapid hydrolysis of the ester linkages at the primary hydroxyl groups of the glycerol moiety. Thus, the initial hydrolysis product is a 1,2-diacylglycerol and the immediate subsequent one is a 2-acylglycerol. Hydrolysis of the latter species is quite slow and the rate is consistent with that for acyl-group migration from the secondary to primary position. Thus, it is not clear whether hydrolysis of the 2-acylglycerols occurs directly or via isomerization to a 1-acylglycerol followed by hydrolysis of that species. These points are summarized in Fig. 13-1 (1,2).

Lipid–protein complexes, such as chylomicrons, serve as substrates for an additional lipase contained in some tissues, *lipoprotein lipase* (3). This enzyme catalyzes the hydrolysis of triglycerides only when they are associated with protein. In the course of this hydrolysis, there is no buildup of mono- or diglycerides, in contrast with the hydrolysis catalyzed by pancreatic lipase.

Several very important classes of complex lipids are derived from *phosphatidic acid,* a glycerophosphoric acid in which both the hydroxyl groups of

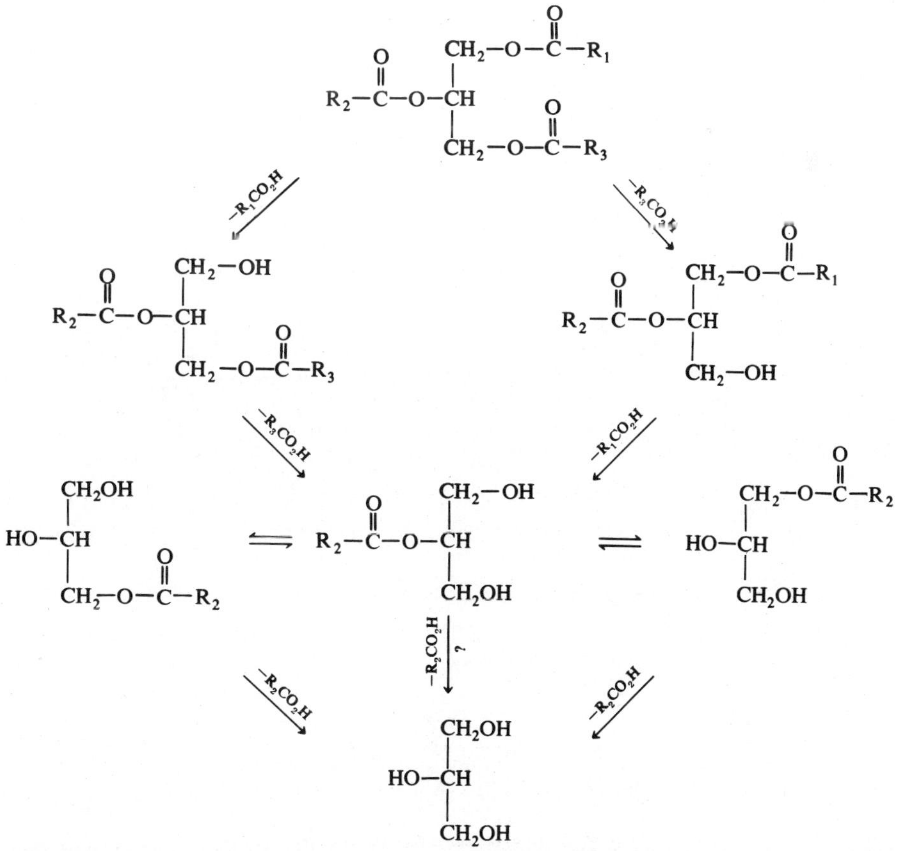

FIGURE 13-1
Reaction sequence for the complete hydrolysis of triglycerides.

glycerol are esterified with fatty acids. These include the 3-*sn*-phosphatidylcholines[1] (1,2-diacyl-*sn*-glycero-3-phosphorylcholine), also termed *lecithins,* 3-*sn*-phosphatidylethanolamines (1,2-diacyl-*sn*-glycero-3-phosphorylethanolamine), and 3-*sn*-phosphatidylserines (1,2-diacyl-*sn*-glycero-3-phosphorylserine). Each of these designations, of course, describes a class of compounds whose individual members are distinguished on the basis of the nature of the esterified fatty acids. Phosphatidylcholines and -ethanolamines are major lipid constituents; phosphatidylserines occur to a more limited extent and are principally involved in phosphatidylethanolamine biosynthesis (page 733). Most phosphatidylcholines contain one saturated fatty acid, esterified at the 1 position, and one unsaturated fatty acid, esterified at the 2 position.

Derivatives of phosphatidic acids contain a center of asymmetry at the C-2 atom. Naturally occurring derivatives have been demonstrated to be stereochemically related to 3-*sn*-glycerophosphate (4) (page 515).

The enzymatic degradation of phosphatidylcholines and -ethanolamines is complex and not completely understood. Several enzymes (or activities) are involved including the following:

$$R_2-\overset{O}{\overset{\|}{C}}O-\underset{CH_2-O-\overset{O^-}{\overset{|}{\underset{\overset{\|}{O}}{P}}}-OCH_2CH_2-\overset{+}{N}(CH_3)_3}{\overset{CH_2O\overset{O}{\overset{\|}{C}}-R_1}{C}}-H$$

3-Phosphatidylcholine

$$R_2-\overset{O}{\overset{\|}{C}}O-\underset{CH_2-O-\overset{O^-}{\overset{|}{\underset{\overset{\|}{O}}{P}}}-O-CH_2-CH_2-{}^{+}NH_3}{\overset{CH_2O\overset{O}{\overset{\|}{C}}-R_1}{C}}-H$$

3-Phosphatidylethanolamine

$$R_2-\overset{O}{\overset{\|}{C}}O-\underset{CH_2-O-\overset{OH}{\overset{|}{\underset{\overset{\|}{O}}{P}}}-O-CH_2-\overset{{}^{+}NH_3}{\overset{|}{\underset{H}{C}}}-CO_2^-}{\overset{CH_2O\overset{O}{\overset{\|}{C}}-R_1}{C}}-H$$

3-Phosphatidylserine

Phosphatidase A (phospholipase A, phosphatide acyl-hydrolase) cleaves one fatty acid from 3-phosphatidyl derivatives to yield a lysophosphatide:

$$\text{Phosphatidylcholine} \xrightarrow{\text{phosphatidase A}} RCO_2H + \text{lysophosphatide} \qquad (13\text{-}1)$$

The site at which hydrolysis is promoted by this enzyme, which occurs in snake venom, bee and wasp poisons, animal tissues, and perhaps in plants and bacteria, was established by the classic work of Tattrie. He enzymatically converted purified 3-phosphatidylcholines from egg to the 1,2-diacylglycerols, which were subsequently acylated with myristoyl chloride to yield the triglycerides. The action of pancreatic lipase, an enzyme that catalyzes the

[1] The term "*sn*" (*s*tereospecifically *n*umbered) refers to a novel system for the designation of the stereochemistry of glycerol derivatives. In this system, the older terms "L-α-," "L-3-," "D-1-" are equivalent to "*sn*-glycero-3-." Details are provided in *J. Biol. Chem.*, **242:** 4845 (1967).

hydrolysis of primary esters exclusively, on this triglyceride liberated stearic, palmitic, and myristic acids. In contrast, snake venom phosphatidase A liberated only oleic and linoleic acids, which must, therefore, have been esterified at the 2 position (5).

Lysophosphatidase (lysolecithin acyl-hydrolase) completes the removal of fatty acids from phosphatides by catalyzing the conversion of lysophosphatides to the appropriate glycerolphosphoryl derivative:

$$\begin{array}{l} CH_2OC(=O)\text{—}R_1 \\ HO\text{—}CH \quad O^- \\ CH_2OP(=O)\text{—}OR \end{array} + H_2O \rightarrow \begin{array}{l} CH_2OH \\ HO\text{—}CH \quad O^- \\ CH_2OP(=O)\text{—}OR \end{array} + R_1CO_2H \qquad (13\text{-}2)$$

Phosphatidase B refers to an activity which combines the functions of phosphatidase A and lysophosphatidase, catalysis of the hydrolysis of both fatty acyl esters yielding the glycerolphosphoryl base directly:

$$\begin{array}{l} \quad CH_2OC(=O)\text{—}R_1 \\ R_2\text{—}C(=O)OCH \quad O^- \\ \quad CH_2\text{—}OP(=O)\text{—}OR \end{array} + 2H_2O \rightarrow \begin{array}{l} CH_2OH \\ HO\text{—}CH \quad O^- \\ CH_2\text{—}O\text{—}P(=O)\text{—}OR \end{array} + R_1CO_2H + R_2CO_2H \qquad (13\text{-}3)$$

Phosphatidylcholine phosphatidohydrolase and *phosphatidylcholine cholinephosphohydrolase* catalyze the hydrolysis of the phosphodiester linkages between 1,2-diglyceride and phosphorylated base, and between 3-*sn*-phosphatidic acid and nitrogenous base, respectively:

$$\begin{array}{l} \quad CH_2OC(=O)\text{—}R_1 \\ R_2\text{—}C(=O)OCH \quad O^- \\ \quad CH_2\text{—}O\text{—}P(=O)\text{—}O\text{—}CH_2CH_2\text{—}N^+(R')_3 \end{array} \xrightarrow{H_2O}$$

$$\begin{array}{l} \quad CH_2OC(=O)\text{—}R_1 \\ R_2\text{—}C(=O)OCH \quad O^- \\ \quad CH_2\text{—}O\text{—}P(=O)\text{—}OH \end{array} + HO\text{—}CH_2CH_2\text{—}N^+(R')_3 \qquad (13\text{-}4)$$

$$\begin{array}{l} \quad CH_2OC(=O)\text{—}R_1 \\ R_2\text{—}C(=O)OCH \quad O^- \\ \quad CH_2\text{—}O\text{—}P(=O)\text{—}OCH_2CH_2\text{—}N^+(R')_3 \end{array} \xrightarrow{H_2O}$$

$$\begin{array}{l} \quad CH_2OC(=O)\text{—}R_1 \\ R_2\text{—}C(=O)OCH \\ \quad CH_2OH \end{array} + {}^-HO_3P\text{—}OCH_2CH_2\text{—}N^+(R')_3 \qquad (13\text{-}5)$$

Transport of the triglycerides, and other lipids, in the lymph and blood occurs in the form of complexes with proteins. In the plasma, several classes of lipoproteins have been identified and classified according to their electrophoretic behavior. They may also be placed into various groups according to their density; the most dense are the α-lipoproteins, then come the β-lipoproteins, and finally the pre-β-lipoproteins. The decrease in density reflects a decreasing content of protein and an increasing content of lipid in these complexes. In Table 13-3, the plasma lipoproteins are classified as to density, protein content, and lipid composition.

The lipoproteins may be formed into large aggregates with triglycerides: The resulting structures have diameters near 1 μm and are termed *chylomicrons.* In terms of the quantity of lipid transported in the lymph and plasma, these structures are more important than the simpler ones discussed just above. The composition of the chylomicrons is included in Table 13-3. These reactions are accelerated by the *bile salts,* sodium taurocholate and glycocholate,

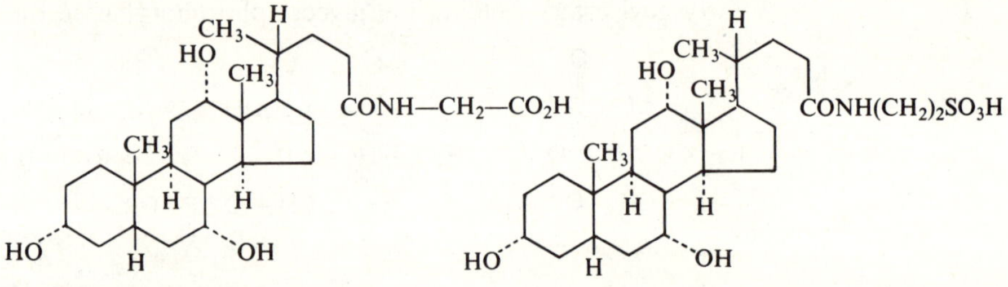

Glycocholic acid ($CONH—CH_2—CO_2H$)

Taurocholic acid ($CONH(CH_2)_2SO_3H$)

which serve the function of emulsification and solubilization of lipids. The bile salts are contained in the bile, a secretion of the liver that passes into the intestine by way of the gall bladder. The discharge of bile into the intestine is evoked by the hormone *cholecystokinin,* which is synthesized in the small intestine and liberated into the circulation in response to the presence of lipid in the duodenum.

It should be noted that the hydrolysis of triglycerides by lipases is in principle a reversible reaction, thus providing a possible pathway for the biosynthesis of these compounds.

In the intestinal tract, triglycerides are converted into a mixture of diglycerides, monoglycerides, free fatty acids, and glycerol. The free glycerol is absorbed independently of the remainder of the hydrolysis products and is not further employed for lipid synthesis. Of the total triglyceride glycerol, about 75 percent is absorbed as 2-monoglyceride and minor amounts are absorbed as 1-monoglyceride, diglyceride, and free glycerol (1,2). It follows that about 75 percent of the triglyceride fatty acids are absorbed as the free fatty acids. In the epithelial cells of the intestinal wall, the 2-monoglyceride is reesterified to form triglyceride; this triglyceride is then transported to sites of utilization in the form of chylomicrons (see above). The 1-monoglyceride absorbed undergoes hydrolysis to glycerol and free fatty acid in the intestinal wall, but these products are reutilized there to form additional triglyceride (2).

No obligatory stepwise action of these enzymes is yet recognized. They have all been identified as occurring in the same tissue, and the product of one reaction is seldom a good substrate for the other enzymes. (Other hydrolytic enzymes do act on the various fragments.)

TABLE 13-3
Physical and chemical nature of human plasma lipoproteins

			Composition, percent dry weight				
Names	*Density range*	*S_f class*[a]	*Protein*	*Phospholipid*	*Cholesterol*	*Cholesterol esters*	*Triglycerides*
α-Lipoproteins (high-density lipoproteins)	1.063–1.21	—	45–55	~30	~3	~15	5–8
β-Lipoproteins (low-density lipoproteins)	1.006–1.063	0–20	20–25	~22	~8	~35	~10
Pre-β-lipoproteins (very low density lipoproteins and endogenous particles)	0.93–1.006	>20	2–13	10–25	3–8	6–16	50–80
Chylomicrons	<0.95	>400	0.5–2.5	3–15	1–5	1–7	79–95

[a] Analogous to a sedimentation constant, but for flotation.
SOURCE: E. J. Masoro, *Physiological Chemistry of Lipids in Mammals*, Saunders, Philadelphia, 1968.

OXIDATION OF FATTY ACIDS

The *oxidative degradation of fatty acids* is a universal biochemical capacity among living organisms. In mammals, such oxidation occurs in a variety of tissues, including liver, kidney, and heart. Intracellularly, fatty acid oxidation occurs principally in the mitochondria. Attention is now focused on the sequence of enzyme-catalyzed reactions involved in the formation of two-carbon fragments from fatty acids.

Early Considerations

The essential features of the pathway for oxidation of fatty acids were clearly detailed by Knoop in 1904 and shortly thereafter by Embden and Dakin. Nearly fifty years were required before the brilliant conclusions of Knoop received full experimental confirmation. Knoop arrived at the theory of "β oxidation" through feeding experiments employing fatty acids labeled in the terminal position with phenyl groups. The latter permitted the identification of the end products in the urine of the experimental animals. When ω-phenyl fatty acids containing an even number of carbons in the straight chain were fed, *phenylaceturic acid,* formed by the condensation of phenylacetic acid and glycine *in vivo,* accumulated in the urine [Eq. (13-6)]. In contrast, when similarly labeled fatty acids containing an odd number of carbon atoms were fed, *hippuric acid,* formed by condensation of benzoic acid and glycine *in vivo,* accumulated in the urine [Eq. (13-7)]:

$$C_6H_5{-}CH_2CO_2H + H_2N{-}CH_2{-}CO_2H \rightarrow C_6H_5{-}CH_2{-}\overset{O}{\overset{\|}{C}}{-}\overset{H}{\overset{|}{N}}CH_2CO_2H \quad (13\text{-}6)$$

Phenylaceturic acid

$$C_6H_5{-}CO_2H + H_2N{-}CH_2CO_2H \rightarrow C_6H_5{-}\overset{O}{\overset{\|}{C}}{-}\overset{H}{\overset{|}{N}}CH_2CO_2H \quad (13\text{-}7)$$

Hippuric acid

These are precisely the results one would expect, provided the fatty acids were degraded by the successive removal of two-carbon fragments from the parent molecule. The individual reactions leading to the cleavage of two-carbon units from fatty acids were visualized as indicated in Fig. 13-2.

One of the principal difficulties with the early conclusions of Knoop, Embden, and Dakin lay in accounting for the production of acetoacetate and other "ketone bodies," such as β-hydroxybutyrate and acetone (which arise from acetoacetate), in the course of fatty acid oxidation. The latter pair of authors assumed that the terminal four-carbon fragment of normal fatty acids was not

Repetition	
→ $RCH_2CH_2CO_2H$ ↓	
$RCHOHCH_2CO_2H \rightleftharpoons RCH{=}CHCO_2H$ ↓	β-Oxidation to unsaturated, or β-OH acid.
$RCOCH_2CO_2H$ ↓	Oxidation to β-keto acid.
RCO_2H + "CH_3CO_2H" (returns to top)	Removal of 2-carbon units and formation of fatty acid with two less carbon atoms than parent acid.

FIGURE 13-2
Oxidative metabolism of fatty acids according to the β-oxidation theory of Knoop.

oxidized beyond the stage of acetoacetate. Thus, acetoacetate was considered to be produced in stoichiometric amounts from butyric acid derived from the oxidation of even-numbered fatty acids. Odd-numbered fatty acids would yield stoichiometric amounts of propionic acid and no ketone bodies at all according to the same reasoning. Subsequent experiments revealed that these conclusions were untenable and necessitated modifications in the theory of fatty acid oxidation. Two attempts to account for the production of ketone bodies on a quantitative basis, although of significant value, have proved less than complete solutions to the problem. These are the multiple alternative oxidation theories of Hurtley [(Eq. 13-8)] and Jowett and Quastel [Eq. (13-9)]:

$$CH_3CH_2CH_2CH_2CH_2CH_2CH_2CO_2H \rightarrow CH_3CH_2CH_2CH_2CH_2COCH_2CO_2H \rightarrow CH_3CH_2CH_2COCH_2COCH_2CO_2H \rightarrow CH_3COCH_2COCH_2COCH_2CO_2H \rightarrow 2CH_3COCH_2CO_2H \quad (13\text{-}8)$$

$$CH_3COCH_2CO_2H + 2\text{“}CH_3CO_2H\text{”} \leftarrow CH_3COCH_2COCH_2COCH_2CO_2H \rightarrow 2CH_3COCH_2CO_2H \quad (13\text{-}9)$$

In each case, the direct production of four-carbon fragments in the course of fatty acid oxidation was postulated. However, the latter workers were the first to demonstrate that oxidation of odd-numbered fatty acids gave rise to acetoacetate and that the molecular yield of ketone bodies from long-chain even-numbered fatty acids exceeds that from butyric acid (6). These two important observations did establish that the view that acetoacetate arose only from butyrate formed in the course of oxidation of fatty acids was untenable. Jowett and Quastel did suggest, on the basis of their results, the possibility that ketone bodies might be formed from two-carbon units formed in the oxidation of fatty acids and in their work provided a bridge between the early considerations of Knoop and Embden, on one hand, and the modern work made possible by the availability of isotopic carbon during the 1940s, on the other.

The development of understanding of fatty acid oxidation was continued in 1940 by the proposal of MacKay, Wick, and Barnum of a "β-oxidation-condensation" theory to account for the findings of Jowett and Quastel (7). The novel feature of this proposal, which incorporates most of Knoop's original suggestions as well as those of later workers, was that acetoacetate was formed through condensation of two acetate fragments:

$$CH_3CH_2CH_2CH_2CH_2CH_2CH_2CO_2H \rightarrow 4\text{“}CH_3CO_2H\text{”} \rightarrow 2CH_3COCH_2CO_2H$$

$$CH_3CH_2CH_2CH_2CH_2CH_2CO_2H \rightarrow 2\text{“}CH_3CO_2H\text{”} \rightarrow CH_3COCH_2CO_2H$$
$$\hookrightarrow CH_3CH_2CO_2H \rightarrow CH_3COCO_2H \quad (13\text{-}10)$$

This scheme clearly predicts that both the carbonyl and carboxyl carbons of acetoacetate have their origin in the carboxyl carbon of the parent fatty acid, a prediction fully substantiated by experiments subsequently performed with the aid of radioactive tracers (8).

Pathway for Fatty Acid Oxidation

The key milestones in the understanding of the pathway of fatty acid oxidation, following the work of Knoop were (1) the observation of fatty acid oxidation in cell-free preparations from the guinea pig liver by Leloir and Muñoz in 1939, and (2) the elucidation of the structure of "activated" fatty acids as the *S*-acyl derivatives of coenzyme A by Lynen and Reichert in 1951. The use of cell-free preparations permitted the identification of the individual enzymes involved in fatty acid oxidation and the determination of the properties of the catalytic reactions. The discovery of the acyl derivatives of CoASH, which followed the discovery of coenzyme A by Lipmann, provided a solution to the most perplexing problem of fatty acid oxidation—the failure to detect any of the postulated intermediates or any free short-chain fatty acids in the course of oxidation of longer molecules. It is now clear that all the intermediates, including the short-chain fatty acids formed in the course of oxidation, occur only as the acyl-SCoA derivatives.

The pathway for the oxidation of fatty acids is indicated in Fig. 13-3. The parent fatty acid is activated by conversion to the fatty acyl-SCoA, oxidized to the α,β-unsaturated compound, hydrated, oxidized to the β-keto derivative, and finally subjected to a thiolytic cleavage yielding acetyl-SCoA and the fatty acyl-SCoA containing two less carbon atoms, which, in turn, undergoes the same series of reactions. We now consider each of the steps individually.

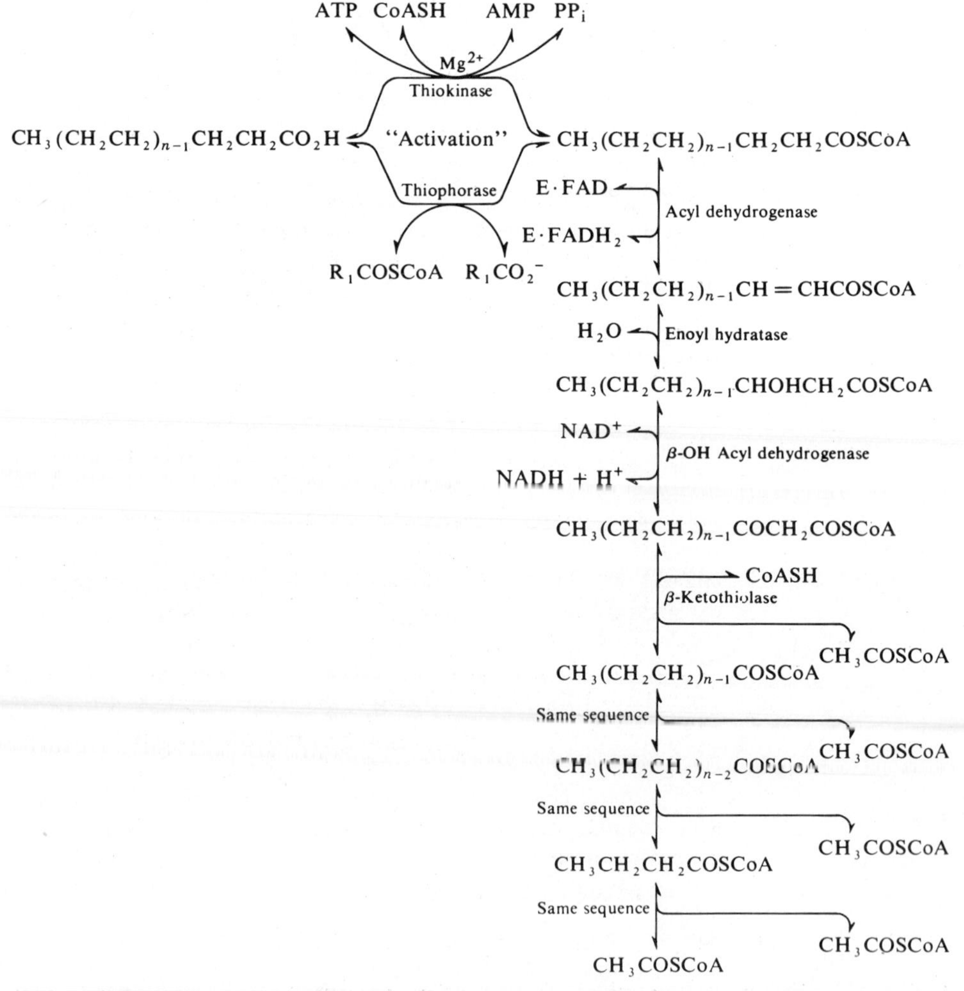

FIGURE 13-3

Pattern of fatty-acid oxidation. H. R. Mahler, in K. S. Markley (ed.), *Fatty Acids,* part 3, Wiley (Interscience), New York, 2nd ed., 1964.

Fatty Acid Oxidation

Conversion of fatty acids to fatty acyl derivatives of CoASH may occur by any of a variety of routes. These include reactions catalyzed by acetate thiokinase, medium- and long-chain fatty acid thiokinases, thiophorases, and a GTP-specific acid thiokinase.

Acetate Thiokinase [acetate : CoA ligase (AMP)]. Acetate thiokinase, a widely distributed enzyme, catalyzes the overall reaction

$$CH_3CO_2H + ATP + HSCoA \overset{Mg^{2+}}{\rightleftharpoons} CH_3\overset{\overset{O}{\|}}{C}\text{—}SCoA + AMP + PP_i \qquad (13\text{-}11)$$

The enzyme from yeast has been purified by Berg. The equilibrium constant for reaction (13-11) is near one in the presence of approximately equimolar concentrations of Mg^{2+} and ATP. This result indicates that the standard free energy of hydrolysis of ATP to AMP and PP_i is about the same as that for hydrolysis of acetyl-SCoA. Acetate thiokinase is activated by K^+, Rb^+, and ammonium ions but inhibited by Na^+ and Li^+ ions. Recent work has revealed that the cation-dependent behavior is more complicated than implied by this statement. Generally, low levels of common cations are always present as counterions of buffers employed to maintain the *p*H constant in the reaction mixtures. If these counterions are completely replaced by tetra-substituted ammonium ions, acetate thiokinase is inactive (9). Under these conditions any of the above cations proves to be an activator. However, the two groups of cations are still distinct: K^+, Rb^+, and ammonium ions yield maximal activation from 10 to 200 mM; Na^+ and Li^+ inhibit the reaction at concentrations greater than 10 mM.

The enzyme activates acetate, propionate, and acrylate but will not handle long-chain acids. A convincing body of evidence indicates that the acetate thiokinase reaction is, in fact, the sum of two reactions involving the intermediate formation of acyladenylates:

$$E + CH_3CO_2H + ATP \overset{Mg^{2+}}{\rightleftharpoons} \left[CH_3\text{—}\overset{\overset{O}{\|}}{C}\text{—}AMP\right]\cdot E + PP_i$$
$$\left[CH_3\overset{\overset{O}{\|}}{C}\text{—}AMP\right]\cdot E + HSCoA \rightleftharpoons CH_3\text{—}\overset{\overset{O}{\|}}{C}\text{—}SCoA + AMP + E \qquad (13\text{-}12)$$

The acyladenylates appear to be tightly bound to the surface of the enzyme.

Fatty Acid Thiokinase (Medium Chain) [acid : CoA ligase (AMP)]. A widely distributed enzyme catalyzes the formation of acyl-SCoA's from saturated fatty acids containing from 4 to 12 carbon atoms. With the exception of the nature of the substrates, this reaction seems closely related to that of acetate thiokinase [Eqs. (13-11) and (13-12)]. Like the latter reaction, the equilibrium constant for the activation of medium-chain fatty acids is near one. The medium-chain fatty acid thiokinase exhibits a broad specificity in terms of the acyl component of the reaction. In addition to the saturated fatty acids, this enzyme also activates benzoates, unsaturated acids, hydroxyacids, picolinic acid, and several others. In contrast, this enzyme is specific for AMP and HSCoA as acceptors for the acyl function.

Fatty Acid Thiokinase (Long Chain) [acid : CoA ligase (AMP)]. A fatty acid thiokinase capable of activating fatty acids containing up to 22 carbon atoms has been identified in mammalian liver. The reaction appears closely related to those discussed above.

Thiophorases (CoA transferases). The thiophorases catalyze the interconversion of acyl-SCoA's and free acids according to

$$R_1\text{—}\overset{\overset{\Large O}{\|}}{C}\text{—SCoA} + R_2CO_2H \rightleftharpoons R_1CO_2H + R_2\text{—}\overset{\overset{\Large O}{\|}}{C}\text{—SCoA} \qquad (13\text{-}13)$$

The pathway for these fascinating reactions was elucidated in the laboratories of Stern, Coon, and Green, and the reaction mechanism has been established by Hersh and Jencks (10), who have demonstrated the formation of an enzyme-CoA intermediate, in which the CoA is bound to the carboxyl group of the enzyme, for the succinyl CoA–acetoacetate coenzyme A transferase reaction:

$$\begin{aligned}&\text{Succinyl-SCoA} + \text{E} \rightleftharpoons \text{E-SCoA} + \text{succinate}\\&\text{E-SCoA} + \text{acetoacetate} \rightleftharpoons \text{E} + \text{acetoacetyl-SCoA}\end{aligned} \qquad (13\text{-}14)$$

The thiophorases may be divided into two classes: (1) those that catalyze the transfer of HSCoA from acetyl-SCoA to aliphatic acids and (2) those that catalyze the transfer of HSCoA from succinyl-SCoA to acetoacetate and other β-keto acids.

GTP-Specific Acid Thiokinase [acid : CoA ligase (GDP)]. An acylthiokinase that specifically requires GTP as coenzyme has been isolated from the mitochondria of beef and rat liver (11). The overall stoichiometry of this reaction is

$$RCO_2H + \text{HSCoA} + \text{GTP} \xrightleftharpoons{Mg^{2+}} R\text{—}\overset{\overset{\Large O}{\|}}{C}\text{—SCoA} + \text{GDP} + P_i \qquad (13\text{-}15)$$

The specificity of this enzyme is quite interesting. If the enzyme is isolated by methods not employing organic solvents, it shows a broad specificity activating, for example, both butyrate and oleate (12,13). Upon treatment with organic solvents, the activity for long-chain fatty acids is diminished, while that for butyrate is not affected. This change is reversed on addition of lecithin, suggesting that lecithin is a component of the enzyme in its natural state and that this lecithin influences the specificity of the enzyme. Since enzymes isolated from mitochondria are frequently freed from the mitochondrial structure through treatment with organic solvents, this observation has important consequences for assessing specificity of such enzymes.

Phosphotransacetylase (acetyl CoA : orthophosphate acetyl-transferase). In certain microorganisms, acetyl-SCoA may be formed from acetate by a two-step reaction path not involving acetyl adenylate as an intermediate. The initial reaction, catalyzed by acetokinase, involves formation of acetyl phosphate from acetate and ATP. The second reaction, catalyzed by phosphotransacetylase, involves the transfer of the acyl function from phosphate to HSCoA:

$$\begin{aligned}&CH_3CO_2H + \text{ATP} \rightleftharpoons CH_3\overset{\overset{\Large O}{\|}}{C}\text{—}OPO_3H_2 + \text{ADP}\\&CH_3\text{—}\overset{\overset{\Large O}{\|}}{C}\text{—}OPO_3H_2 + \text{HSCoA} \rightleftharpoons CH_3\overset{\overset{\Large O}{\|}}{C}\text{—SCoA} + P_i\end{aligned} \qquad (13\text{-}16)$$

The various routes for acyl-SCoA formation are summarized in Table 13-4.

Even in one and the same subcellular organelle, several routes for the

TABLE 13-4
Summary of routes to acyl-SCoA formation

Reaction	*Enzymes*	*Chain length*	*Occurrence*
$RCO_2H + ATP + HSCoA \rightleftharpoons RC(=O)\text{—}SCoA + AMP + PP_i$	Acetate thiokinase Fatty acid thiokinase Fatty acid thiokinase	C_2–C_3 C_4–C_{11} C_{10}–C_{18}	Yeast, higher organisms
$RCO_2H + GTP + HSCoA \rightleftharpoons RCOSCoA + GDP + P_i$	Acid thiokinase	?	Liver
$R_1C(=O)\text{—}SCoA + R_2CO_2H \rightleftharpoons R_1CO_2H + R_2C(=O)SCoA$	Acetic thiophorase Butyric-succinic thiophorase Acetoacetic-succinic thiophorase	$R_1=CH_3$, $R_2=C_2$–C_5 $R_1=CH_2CH_2CO_2H$, $R_2=C_3$–C_5 $R_1=CH_2CH_2CO_2H$, $R_2=RCOCH_2$	Microorganisms Heart, kidney Extrahepatic mammalian tissues
$RCO_2H + ATP \rightleftharpoons RC(=O)OPO_3H_2 + ADP$	Acyl kinases	C_2; C_4 (separate)	Microorganisms
$RC(=O)OPO_3H_2 + HSCoA \rightleftharpoons R\text{—}C(=O)\text{—}SCoA + H_3PO_4$	Phosphotransacylases	C_2; C_4 (separate)	Microorganisms

activation of fatty acids may be employed. For example, in rat liver mitochondria, three routes exist: (1) a GTP-dependent enzyme, (2) an ATP-dependent enzyme located internally in the mitochondrion, and (3) an ATP-dependent enzyme located on the outer mitochondrial membrane (14). These observations suggest that there may be several distinct sites for fatty acid oxidation as well.

Fatty Acyl-SCoA Dehydrogenases (fatty acyl CoA : FAD oxidoreductase). Four distinct fatty acyl-SCoA dehydrogenases have been isolated and purified from the mitochondria of mammalian liver. Each of these enzymes catalyzes the reaction indicated in Eq. (13-17) and each requires FAD as coenzyme:

$$E\cdot FAD + R\text{—}CH_2\text{—}CH_2\text{—}\overset{\overset{\displaystyle O}{\|}}{C}\text{—}SCoA \rightleftharpoons E\cdot FADH_2 + R\text{—}\overset{\overset{\displaystyle H}{|}}{C}\text{=}\underset{\underset{\displaystyle H}{|}}{C}\text{—}\overset{\overset{\displaystyle O}{\|}}{C}\text{—}SCoA \qquad (13\text{-}17)$$

these enzymes differ in their physical properties and in their substrate specificity, in particular in terms of the lengths of the carbon chains of the substrates.

Enoyl Hydrases (L-β-hydroxyacyl CoA hydro-lyase). The mitochondria of mammalian liver contain an enoyl hydrase (crotonase), which catalyzes the conversion of *trans*-2-enoyl derivatives of CoASH to the corresponding L-β-hydroxyacyl compounds.

$$\mathrm{R{-}\overset{\displaystyle H}{\overset{|}{C}}{=}\underset{\displaystyle H}{\underset{|}{C}}{-}\overset{\displaystyle O}{\overset{\|}{C}}{-}SCoA + H_2O \rightleftharpoons R{-}\overset{\displaystyle OH}{\overset{|}{C}}\!\!\underset{\displaystyle H}{\underset{|}{}}{-}CH_2{-}\overset{\displaystyle O}{\overset{\|}{C}}{-}SCoA} \quad (13\text{-}18)$$

L

As implied by this equation, the addition of water is stereospecific and only the L-β-hydroxyacyl-SCoA derivatives are formed. This enzyme also catalyzes the hydration of the *cis* isomer, probably with identical stereospecificity yielding D-β-hydroxy acids.

*β-Hydroxyacyl-*SCoA *Dehydrogenases* (L-β-hydroxyacyl CoA : NAD oxidoreductases). Enzymes that specifically catalyze the dehydrogenation of L-β-hydroxyacyl-SCoA substrates have been highly purified from beef liver mitochondria and pig heart. The overall reaction is

$$\mathrm{R{-}\underset{\displaystyle H}{\underset{|}{\overset{\displaystyle OH}{\overset{|}{C}}}}{-}CH_2{-}\overset{\displaystyle O}{\overset{\|}{C}}{-}SCoA + NAD^+ \rightleftharpoons R{-}\overset{\displaystyle O}{\overset{\|}{C}}{-}CH_2{-}\overset{\displaystyle O}{\overset{\|}{C}}{-}SCoA + NADH + H^+} \quad (13\text{-}19)$$

and the enzymes appear to be nonspecific with respect to the chain length of the substrate. Since this reaction liberates a proton, the equilibrium is strongly *p*H dependent, favoring β-hydroxyacyl-SCoA formation at *p*H 7 and β-ketoacyl-SCoA formation in solutions more basic than *p*H 9. Beef liver mitochondria may also contain a D-β-hydroxyacyl-SCoA dehydrogenase.

β-Ketothiolase (acetyl-CoA : acetyl CoA C-acyltransferase). β-Ketothiolase catalyzes the thiolytic cleavage of β-ketoacyl-SCoA substrates according to Eq. (13-20). The equilibrium strongly favors cleavage:

$$\mathrm{R{-}\overset{\displaystyle O}{\overset{\|}{C}}{-}CH_2{-}\overset{\displaystyle O}{\overset{\|}{C}}{-}SCoA + HSCoA \rightleftharpoons R{-}\overset{\displaystyle O}{\overset{\|}{C}}{-}SCoA + CH_3{-}\overset{\displaystyle O}{\overset{\|}{C}}{-}SCoA} \quad (13\text{-}20)$$

A highly purified enzyme of this type from beef liver mitochondria catalyzes the cleavage of substrates containing between 4 and 12 carbon atoms. In contrast, a similar enzyme from heart is specific for the four-carbon substrate. Hence, in certain tissues at least, several β-ketothiolases may be required in the course of oxidation of long-chain fatty acids.

Summation of all the preceding reactions yields the overall reaction for the oxidation of even-numbered fatty acids to acetyl-SCoA fragments. The balanced complete equation is

$$\begin{aligned}\mathrm{CH_3(CH_2CH_2)_{\mathit{n}}CO_2H} &+ \mathrm{ATP} + (n+1)\mathrm{CoASH} + n\mathrm{NAD^+} + n\mathrm{E{\cdot}FAD} + n\mathrm{H_2O} \\ &= (n+1)\mathrm{CH_3COSCoA} + \begin{pmatrix}\mathrm{ADP} + \mathrm{P_i} \\ \mathrm{AMP} + \mathrm{PP_i}\end{pmatrix} + n\mathrm{NADH} \\ &\qquad + n\mathrm{E{\cdot}FADH_2} + n\mathrm{H^+}\end{aligned} \quad (13\text{-}21)$$

The sequence of reactions indicated above for oxidation of fatty acids provides a ready explanation for the "sparking" phenomenon. This term refers to the fact that catalytic amounts of certain di- or tricarboxylic acids must be present for fatty acid oxidation to occur. Oxidative reactions of these acids yield ATP (see Chapters 14 and 15), which is, in turn, required for the activation of the fatty acids. The second function for the "sparking" substrate is to provide a partner for condensation of the acetyl-SCoA produced as the oxidation product. Following such condensation, degradation of these two-carbon fragments proceeds via the citric acid cycle, as detailed in Chapter 14.

Carnitine, γ-trimethylamino-β-hydroxybutyrate, is known to stimulate the oxidation of long-chain fatty acids in a variety of tissues (15). Fritz has demonstrated that the fatty acyl esters of carnitine serve as intermediates for this stimulation and has suggested that carnitine serves as an intermediate carrier of acyl functions between the various intracellular compartments (16). For example, fatty acyl derivatives of CoASH present in the soluble portion of the cell may undergo acyl transfer reactions in which carnitine serves as acyl-group acceptor. The acylated carnitines are then supposed to cross the mitochondrial membrane (impermeable to esters of coenzyme A) and, intramitochondrially, be reconverted to the corresponding HSCoA derivatives (page 717). Overall, this sequence of events results in the exposure of fatty acyl-SCoA present *extramitochondrially* to the oxidation enzymes present *intramitochondrially*.

FORMATION OF KETONE BODIES

The ketone bodies, acetoacetate, β-hydroxybutyrate, and acetone, accumulate in the blood under certain abnormal conditions (e.g., diabetes). Each of these substances arises from the metabolism of acetoacetyl-SCoA. The principal fate of acetoacetyl-SCoA in the liver is conjugation with acetyl-SCoA, yielding β-hydroxy-β-methylglutaryl-SCoA [Eq. (13-22)], an important precursor of cholesterol (page 743):

$$CH_3\overset{O}{\overset{\|}{C}}—CH_2—\overset{O}{\overset{\|}{C}}—SCoA + CH_3\overset{O}{\overset{\|}{C}}—SCoA \rightarrow HO_2C—CH_2—\underset{OH}{\underset{|}{\overset{CH_3}{\overset{|}{C}}}}—CH_2—\overset{O}{\overset{\|}{C}}—SCoA + HSCoA \qquad (13\text{-}22)$$

Most of this intermediate is channeled into cholesterol synthesis. In certain instances, a liver enzyme distinct from that involved in its synthesis catalyzes the cleavage of β-hydroxy-β-methylglutaryl-SCoA to acetyl-SCoA and acetoacetate:

$$HO_2C—CH_2—\underset{OH}{\underset{|}{\overset{CH_3}{\overset{|}{C}}}}—CH_2—\overset{O}{\overset{\|}{C}}—SCoA \rightarrow CH_3\overset{O}{\overset{\|}{C}}—SCoA + CH_3\overset{O}{\overset{\|}{C}}—CH_2—CO_2H \qquad (13\text{-}23)$$

The sum of Eqs. (13-22) and (13-23) represents the major route for the formation of free acetoacetate in the liver. Through a minor pathway, free acetoacetate may be produced through the action of a deacylase, which appears to be localized in the liver:

$$CH_3\overset{O}{\overset{\|}{C}}—CH_2—\overset{O}{\overset{\|}{C}}—SCoA \xrightarrow{H_2O} CH_3\overset{O}{\overset{\|}{C}}—CH_2—CO_2H + HSCoA \qquad (13\text{-}24)$$

β-Hydroxybutyrate is formed from the reduction of free acetoacetate in a reaction catalyzed by an enzyme tightly integrated into the inner membrane of liver mitochondria:

$$CH_3\overset{O}{\overset{\|}{C}}—CH_2—CO_2H + NADH + H^+ \rightleftharpoons CH_3—\underset{H}{\underset{|}{\overset{OH}{\overset{|}{C}}}}—CH_2—CO_2H + NAD^+ \qquad (13\text{-}25)$$

This reduction is *stereospecific,* and only D-(−)-β-hydroxybutate is formed.

Acetone also arises from the metabolism of free acetoacetate. The mechanism of a bacterial acetoacetic decarboxylase (acetoacetic carboxy-lyase)

$$CH_3\overset{O}{\overset{\|}{C}}—CH_2—CO_2H \rightarrow CH_3—\overset{O}{\overset{\|}{C}}—CH_3 + CO_2 \qquad (13\text{-}26)$$

the enzyme responsible for this conversion, has been the subject of a penetrating investigation by Westheimer and his collaborators (page 363). This reaction occurs with the formation of a Schiff base between enzyme and substrate, followed by its decarboxylation and hydrolysis:

$$CH_3\overset{O}{\overset{\|}{C}}-CH_2-CO_2H + H_2N-E \rightleftharpoons CH_3-\underset{\underset{H\quad E}{\overset{\|}{N^+}}}{C}-CH_2-C\overset{O^-}{\underset{O}{}} \xrightarrow{} CO_2$$

$$CH_3-\underset{\underset{H\quad E}{|}{N}}{C}=CH_2 \rightleftharpoons CH_3-\underset{\underset{E}{\overset{\|}{N}}}{C}-CH_3 \rightleftharpoons CH_3\overset{O}{\overset{\|}{C}}CH_3 + E-NH_2 \qquad (13\text{-}27)$$

OXIDATION OF ODD-NUMBERED FATTY ACIDS: THE FATE OF PROPIONATE

The sequence of reactions detailed above for the oxidation of even-numbered fatty acids is also applicable to the oxidation of those with an odd number of carbon atoms. The main distinction between the two pathways lies in the products. The terminal cleavage of acetoacetyl-SCoA yields two molecules of acetyl-SCoA, while that of β-ketovaleryl-SCoA yields one of acetyl-SCoA and one of propionyl-SCoA. Further metabolism of propionyl-SCoA may occur via any of three general reaction paths. The first of these, the methyl malonyl pathway, involves the ATP-dependent carboxylation of propionyl-SCoA, yielding methylmalonyl-SCoA, which, after racemization, yields succinyl-SCoA in a vitamin B_{12} coenzyme-dependent reaction catalyzed by methylmalonyl mutase (methylmalonyl CoA : CoA carbonyl mutase):

$$CH_3CH_2\overset{O}{\overset{\|}{C}}-SCoA + ATP + CO_2 \underset{}{\overset{\text{biotin}\cdot E,\ Mg^{2+}}{\rightleftharpoons}} CH_3-\underset{H}{\overset{HO_2C}{C}}-\overset{O}{\overset{\|}{C}}-SCoA + ADP + P_i \qquad (13\text{-}28)$$

$$CH_3-\underset{HO_2C}{\overset{H}{C}}-\overset{O}{\overset{\|}{C}}-SCoA \xrightarrow{B_{12}\ \text{coenzyme}} HO_2C-CH_2-CH_2-\overset{O}{\overset{\|}{C}}-SCoA \qquad (13\text{-}29)$$

The product of reaction (13-28) and the substrate of reaction (13-29) are of opposite configuration and must be interconverted by an epimerase (page 529). Succinyl-SCoA is metabolized further by the citric acid cycle sequence of reactions, as outlined in Chapter 14. Metabolism of propionate according to Eqs. (13-28) and (13-29) occurs in a variety of mammalian tissues, in propionic acid bacteria, and in other microorganisms.

The second pathway for the utilization of propionyl-SCoA, the lactate pathway, involves dehydrogenation of this substrate to acrylyl-SCoA, α hydration to L-lactoyl-SCoA, and hydrolysis to free pyruvate either directly or following inversion to the D isomer:

$$CH_3CH_2\overset{O}{\overset{\|}{C}}-SCoA \rightarrow CH_2=CH-\overset{O}{\overset{\|}{C}}-SCoA \rightarrow CH_3-\overset{OH}{\overset{|}{C}H}-\overset{O}{\overset{\|}{C}}-SCoA \rightarrow CH_3-\overset{OH}{\overset{|}{C}H}-CO_2H + HSCoA \qquad (13\text{-}30)$$

Oxidative decarboxylation of pyruvate yields acetyl-SCoA as detailed in Chapter 11 (page 516). Third, if, in a fashion similar to that encountered in the course of fatty acid oxidation, acrylyl-SCoA undergoes a β hydration, yielding β-hydroxypropionyl-SCoA [Eq. (13-31)], subsequent deacylation and oxidation will produce malonic semialdehyde [Eq. (13-32)]. Malonic semialdehyde may be converted into acetyl-SCoA either by direct oxidative decarboxylation [Eq. (13-33)] or by oxidation to malonyl-SCoA, followed by decarboxylation [Eq. (13-34)]. These reactions constitute the major pathway for propionate metabolism in plant mitochondria and may occur in animals as well.

$$CH_2{=}CH{-}\overset{\overset{O}{\|}}{C}{-}SCoA + H_2O \rightleftharpoons HOCH_2{-}CH_2{-}\overset{\overset{O}{\|}}{C}{-}SCoA \tag{13-31}$$

$$HOCH_2{-}CH_2{-}\overset{\overset{O}{\|}}{C}{-}SCoA + NAD^+ \rightleftharpoons H{-}\overset{\overset{O}{\|}}{C}{-}CH_2{-}CO_2H + HSCoA + NADH + H^+ \tag{13-32}$$

$$H{-}\overset{\overset{O}{\|}}{C}{-}CH_2{-}CO_2H + NAD^+ + HSCoA \rightarrow CH_3\overset{\overset{O}{\|}}{C}{-}SCoA + NADH + H^+ + CO_2 \tag{13-33}$$

$$\begin{aligned} &H{-}\overset{\overset{O}{\|}}{C}{-}CH_2{-}CO_2H + HSCoA + NAD^+ \rightarrow HO_2C{-}CH_2{-}\overset{\overset{O}{\|}}{C}{-}SCoA + NADH + H^+ \\ &HO_2C{-}CH_2{-}\overset{\overset{O}{\|}}{C}{-}SCoA \rightarrow CH_3\overset{\overset{O}{\|}}{C}{-}SCoA + CO_2 \end{aligned} \tag{13-34}$$

The various metabolic fates of propionyl-SCoA are summarized in Fig. 13-4.

OXIDATION OF BRANCHED-CHAIN FATTY ACIDS

The oxidation of the principal branched-chain fatty acids, isobutyrate, α-methylbutyrate, and isovalerate, occur via pathways largely identical to those for the degradation of valine, isoleucine, and leucine, respectively. These pathways are developed in Chapter 17.

ALTERNATIVE ROUTES FOR FATTY ACID OXIDATION

α-Oxidation

The β-oxidation scheme discussed above certainly accounts for the major portion of oxidative reactions of the fatty acids. In addition to β oxidation, α oxidation and ω oxidation yield products of some metabolic importance. α-Hydroxy fatty acids, constituents of certain complex brain lipids, arise by the direct hydroxylation of preformed long-chain fatty acids. Subsequent work has identified this reaction as an intermediate step in the formation of odd-numbered long-chain fatty acids (17,18). Oxidation of the α-hydroxy acids to the α-keto acids, followed by oxidative decarboxylation, yields these products:

$$R{-}\underset{\underset{H}{|}}{\overset{\overset{OH}{|}}{C}}{-}CO_2H \rightarrow R{-}\overset{\overset{O}{\|}}{C}{-}CO_2H \rightarrow RCO_2H + CO_2 \tag{13-35}$$

Such a pathway is in accord with the results of isotope-labeling experiments concerned with synthesis of long-chain, odd-numbered fatty acids.

Two specific systems for α oxidation of fatty acids have received considerable attention: that from leaf tissue and that from rat brain. The leaf tissue system requires both NAD and molecular oxygen for the introduction

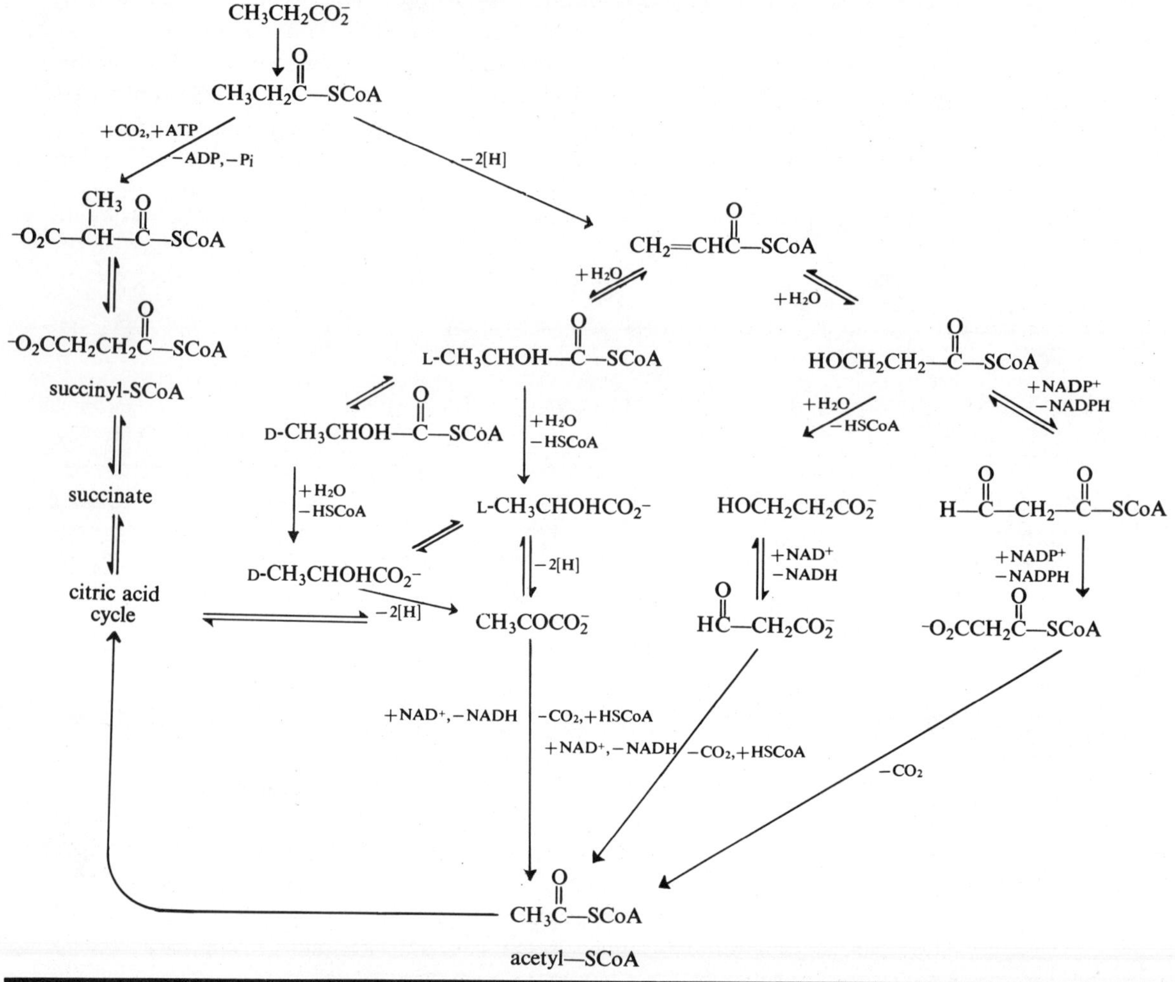

FIGURE 13-4

Fate of propionate and propionyl-SCoA. H. R. Mahler, in K. S. Markley (ed.), *Fatty Acids,* part 3, Wiley (Interscience), New York, 2nd ed., 1964.

of hydroxyl groups adjacent to the carboxyl group of palmitic acid (19). Interestingly, both L and D forms of the α-hydroxyacid are formed, but only the L form is further oxidized.

The oxidative decarboxylation of α-hydroxyacids in rat brain tissue requires microsomal enzymes, ATP, NAD^+, ferrous ions, and molecular oxygen (20). This reaction is evidently responsible for the formation of the odd-numbered fatty acids which appear prominently in the sphingolipids of brain tissue.

Even though α oxidation of fatty acids is not quantitatively the most important route for fatty acid metabolism, failure of this process to operate may have important consequences. It now seems clear that Refsum's disease, a rare neurological disorder which is genetically inherited (page 857), is the result of the absence of α-oxidation enzymes in afflicted persons (21,22). One biochemical consequence of this situation is that over 50 percent of the fatty acids of the liver lipids are phytanic acid, normally a very minor component.

$$\begin{array}{l} CH_3 \qquad\qquad\qquad CH_3 \qquad\qquad\qquad CH_3 \qquad\qquad\qquad CH_3\ \ CO_2H \\ CH_3{-}\overset{|}{C}H{-}CH_2{-}CH_2{-}CH_2{-}\overset{|}{C}H{-}CH_2{-}CH_2{-}CH_2{-}\overset{|}{C}H{-}CH_2{-}CH_2{-}CH_2{-}\overset{|}{C}H{-}\overset{|}{C}H_2 \end{array}$$

Phytanic Acid

The ω oxidation of fatty acids, yielding the dicarboxylic acids as products, has been recognized since the work of Verkade and his associates in 1932. They proposed that ω oxidation, followed by β oxidation of fatty acids, might be an alternative pathway to the direct β oxidation route. Regardless of the source, these enzyme systems are capable of introducing hydroxyl groups into the terminal positions of both fatty acids and alkane hydrocarbons. Enzymes of this class are considered generally in Chapter 15 (pages 636–641).

In microorganisms, the hydroxylating system contains three proteins: a hydroxylase, NADH-rubredoxin reductase, and rubredoxin (23–25). The last of these is a small protein containing nonheme iron but, unlike ferredoxin, no labile sulfide (page 661). NADH and molecular oxygen are also required for the reaction. The sequence of events in this system appears to be the following:

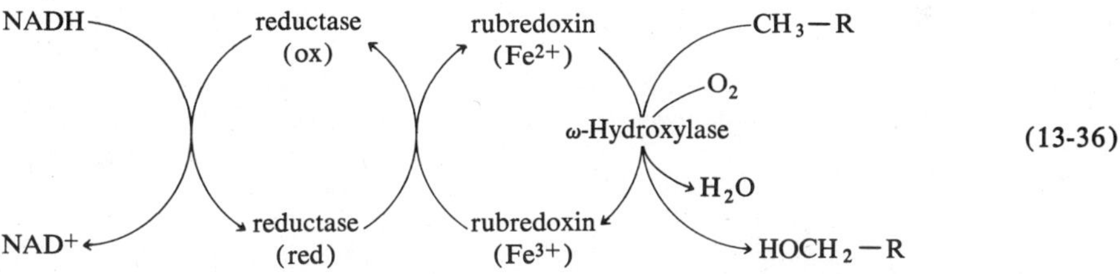

(13-36)

In liver microsomes, ω oxidation requires the action of a carbon monoxide–binding pigment which has been called cytochrome P450, and the reaction requires molecular oxygen and NADPH (26–28), see Chapter 15. The enzyme system has been resolved into three components: cytochrome P450 (which is believed to be the hydroxylase), an NADPH–cytochrome P450 reductase, and a heat-stable lipid factor (29). This reconstituted enzyme system, like that in microsomes, is remarkably broad in specificity, as indicated by its ability to hydroxylate cyclohexane, fatty acids, and *n*-alkanes from 6 to 16 carbon atoms, and a variety of drugs and other compounds containing *N*- and *O*-methyl groups (30). Enzymatic activity toward all these compounds is induced by administration of phenobarbital to animals, is inhibited by CO, and may involve a single enzyme system, as judged by competitive inhibition between the substrates. Thus, the hydroxylation (and detoxification) of a variety of foreign substances appears to be catalyzed in liver via the same pathway as the ω oxidation of fatty acids.

REFERENCES

Books and Monographs

Davidson, F. M., C. Long, and I. F. Penny: *Biochemical Problems of Lipids,* Butterworth, London, 1956.

Deuel, H. J.: *The Lipids,* Wiley, (Interscience), New York, 1951.

Hanahan, D. J., F. R. N. Gurd, and I. Zabin: *Lipide Chemistry,* Wiley, New York, 1960.

Lovern, J. A.: *The Chemistry of Lipids of Biochemical Significance,* Methuen, London, 1955.

Masoro, E. J.: *Physiological Chemistry of Lipids in Mammals,* Saunders, Philadelphia, 1968.

Review Articles

Green, D. E.: Fatty Acid Oxidation in Soluble Systems of Animal Tissues, *Biol. Rev. Cambridge Phil. Soc.*, **29:** 330 (1954).

Green, D. E., and S. J. Wakil: Enzymatic Mechanism of Fatty Acid Oxidation and Synthesis, in K. Bloch (ed.), *Lipide Metabolism,* Wiley, New York, 1960.

Hanahan, D. J., and G. A. Thompson, Jr.: Complex Lipids, *Ann. Rev. Biochem.*, **32:** 215 (1963).

Jaenicke, L., and F. Lynen: Coenzyme A, in P. D. Boyer, H. Lardy, and K. Myrbäck (eds.), *The Enzymes,* vol. 3, Academic Press, New York, 2nd ed., 1960, p. 3.

Jencks, W. P.; Acyl Activation, in P. D. Boyer, H. Lardy, and K. Myrbäck (eds.), *The Enzymes,* vol. 6, Academic Press, New York, 2nd ed., 1962, p. 373.

Kates, M.: Lipolytic Enzymes, in K. Bloch (ed.), *Lipide Metabolism,* Wiley, New York, 1960.

Lynen, F.: Lipide Metabolism, *Ann. Rev. Biochem.*, **24:** 653 (1955).

Mahler, H. R.: Biological Oxidation of Fatty Acids, in K. S. Markley (ed.), *Fatty Acids,* part 3, Interscience, New York, 2nd ed., 1964, p. 1487.

Stumpf, P. K.: Metabolism of Fatty Acids, *Ann. Rev. Biochem.*, **38:** 159 (1969).

Specific References

1. Mattson, F. H., and R. A. Volpenhein: *J. Biol. Chem.*, **237:** 53 (1962).
2. Mattson, F. H., and R. A. Volpenhein: *J. Biol. Chem.*, **239:** 2772 (1964).
3. Korn, E. D.: *J. Biol. Chem.*, **215:** 1, 15 (1955).
4. Baer, E., and J. Maurukas: *J. Biol. Chem.*, **212:** 25, 39 (1955).
5. Tattrie, N. H.: *J. Lipid Res.*, **1:** 60 (1959).
6. Jowett, M., and J. H. Quastel: *Biochem. J.*, **29:** 2143, 2159, 2181 (1935).
7. MacKay, E. M., A. N. Wick, and C. P. Barnum: *J. Biol. Chem.*, **136:** 503 (1940).
8. Weinhouse, S., G. Medes, and N. F. Floyd: *J. Biol. Chem.*, **153:** 689 (1944).
9. Webster, L. T., Jr.: *J. Biol. Chem.*, **241:** 5504 (1966).
10. Hersh, L. B., and W. P. Jencks: *J. Biol. Chem.*, **242:** 3468, 3481 (1967).
11. Rossi, C. R., and D. M. Gibson: *J. Biol. Chem.*, **239:** 1694 (1964).
12. Sartorelli, L., L. Galzigna, C. R. Rossi, and D. M. Gibson: *Biochem. Biophys. Res. Commun.*, **26:** 90 (1967).
13. Galigzna, L., C. R. Rossi, L. Sartorelli, and D. M. Gibson: *J. Biol. Chem.*, **242:** 2111 (1967).
14. Rossi, C. R., A. Alexandre, and L. Sartorelli: *European J. Biochem.*, **4:** 31 (1968).
15. Bressler, R., and S. J. Friedburg: *J. Biol. Chem.*, **239:** 1364 (1964).
16. Fritz, S. B., and K. T. N. Yue: *J. Lipid Res.*, **4:** 279 (1963).
17. Mead, J. F., and G. M. Levis: *J. Biol. Chem.*, **238:** 1634 (1963).
18. Hajra, A. K., and N. S. Radin: *J. Lipid Res.*, **4:** 270 (1963).
19. Hitchcock, C., and A. T. James: *Biochim. Biophys. Acta,* **116:** 413 (1966).
20. MacDonald, R. C., and J. F. Mead: *Lipids,* **3:** 275 (1968).
21. Steinberg, D., J. H. Herndon, Jr., B. W. Uhlendorf, C. E. Mize, J. Avigan, and G. W. A. Milne: *Science,* **156:** 1740 (1967).
22. Stokke, O., K. Try, and L. Eldjarn: *Biochim. Biophys. Acta,* **144:** 271 (1967).
23. Peterson, J. A., D. Basu, and M. J. Coon: *J. Biol. Chem.*, **241:** 5162 (1966).
24. Peterson, J. A., and M. J. Coon: *J. Biol. Chem.*, **243:** 329 (1968).
25. Peterson, J. A., M. Kusunose, E. Kusunose, and M. J. Coon: *J. Biol. Chem.*, **242:** 4334 (1967).
26. Wada, F., K. Hirata, H. Shibata, K. Higashi, and Y. Sakamoto: *Biochem. J.* (Tokyo), **62:** 134 (1967).
27. Lu, A. Y. H., and M. J. Coon: *J. Biol. Chem.*, **243:** 1331 (1968).
28. Das, M. L., S. Orrenius, and L. Ernster: *European J. Biochem.*, **4:** 519 (1968).
29. Lu, A. Y. H., K. W. Junk, and M. J. Coon: *J. Biol. Chem.*, **244:** 3714 (1969).
30. Lu, A. Y. H., H. W. Strobel, J. Heidema and M. J. Coon: *J. Biol. Chem.*, **245:** 4851 (1970).

14 Citric Acid Cycle

The *citric acid cycle*—frequently also called the *tricarboxylic acid cycle,* with both designations sometimes linked to the name of Krebs,[1] its most illustrious proponent—forms the hub of metabolism of almost all cells and must therefore assume an equally central place in our discussion. Originally proposed to account for the complete oxidation (to $CO_2 + H_2O$) of pyruvate (and hence of carbohydrates), and of the two- and three-carbon end products of fatty acid oxidation, its importance transcends these and other purely catabolic, energy-yielding functions. As the focus of the central pathways (see page 489), many of its reactions and substrates fulfill an equally crucial role in the biosynthesis (anabolism) of a host of important metabolites, ranging from amino acids, purines, and pyrimidines to long-chain fatty acids and porphyrins.

CYCLICAL OXIDATION OF ACETYL-SCoA

Historical

The study of the oxidation of a number of organic compounds, including certain mono-, di-, and tricarboxylic acids by preparations from animal tissues (at first minces, extracts of all sorts, homogenates; and later separated and purified fractions—see Chapter 9) was initiated over fifty years ago by Thunberg and by Batelli and Stern. These investigators found that of many substances tested with frog muscle preparations, only succinate, fumarate, malate, and citrate were susceptible to a reaction rapid enough to be of any possible significance as intermediates in the oxidation of fats and carbohydrates. In 1936, Martius and Knoop suggested that the following set of conversions intervened between citrate and succinate (see Fig. 14-1 for the structure of the intermediates): citrate → isocitrate → oxalosuccinate → α-oxoglutarate → succinate. In the next year, Krebs, on the basis of his demonstration of the formation of citrate from pyruvate plus oxalacetate in pigeon breast muscle, proposed that these reactions could be used in cyclical fashion to account for the complete oxidation of pyruvate to three molecules of CO_2 as follows:

$$\text{Pyruvate} + \text{oxalacetate} \xrightarrow[-CO_2]{-2H} \text{citrate} \rightarrow \textit{cis}\text{-aconitate} \rightarrow \text{isocitrate} \xrightarrow[-CO_2]{-2H}$$

$$\text{oxoglutarate} \xrightarrow[-CO_2]{-2H} \text{succinate} \rightarrow \text{fumarate} \rightarrow \text{malate} \xrightarrow{-2H} \text{oxalacetate}$$

[1] Sir Hans A. Krebs, German-born British biochemist, who received the Nobel Prize for physiology and medicine in 1953 for his work in intermediary metabolism.

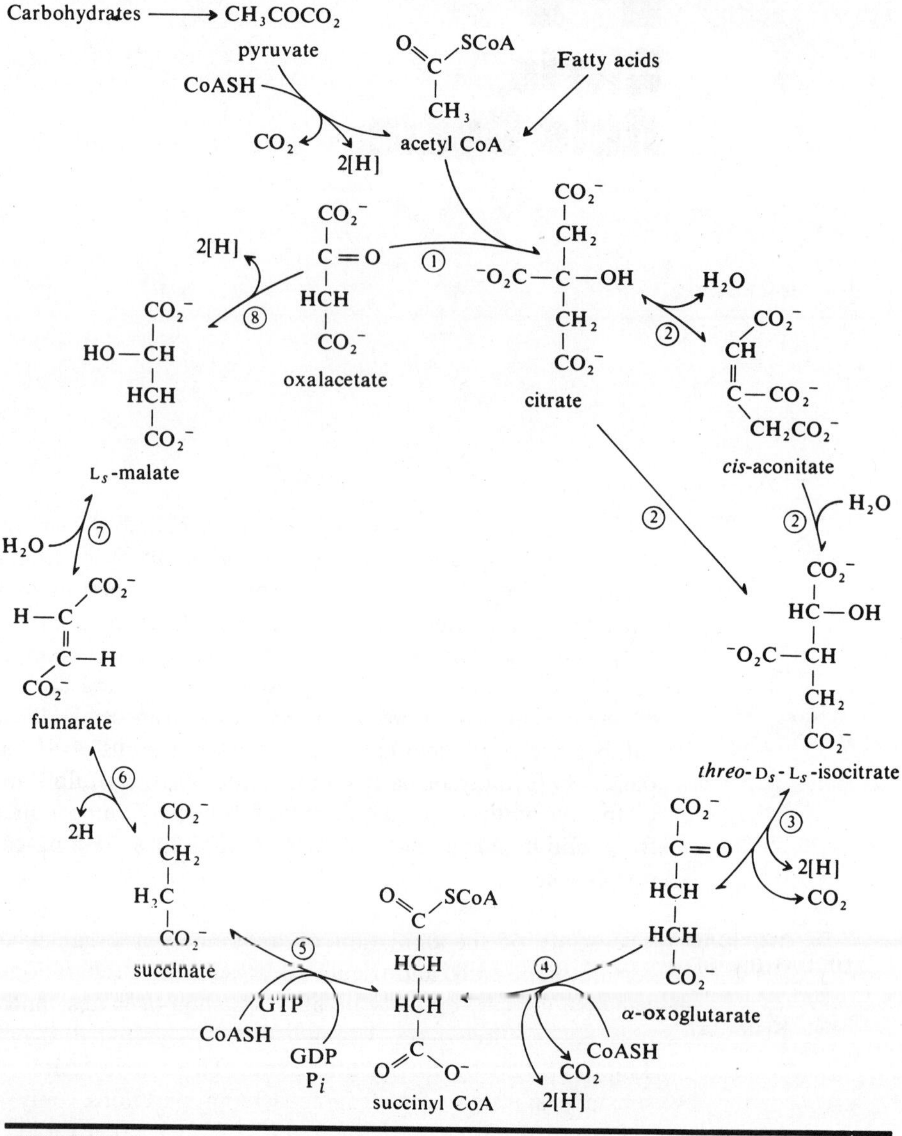

FIGURE 14-1
Citric acid (tricarboxylic acid) cycle.

The following set of observations contributes strong evidence that this sequence of reactions, with the modifications discussed below, does constitute the major pathway for the catabolism of pyruvate and acetate (as acetyl CoA) in the tissues of all animals, from man to *Paramecium,* and in those of higher plants and many microorganisms:

1. The various postulated intermediates can all be demonstrated to be present in most tissues examined and to be oxidized at rates rapid enough to make them suitable for participation in such a cycle.

2. Addition of any of the various di- and tricarboxylic acids greatly stimulates the rate of endogenous respiration and furthermore leads to an uptake of oxygen far above and beyond that accountable for by the addition of the intermediates alone. It is the latter observation that most forcefully indicates the cyclical nature of the process of oxidation: clearly, since the

added substrates act catalytically rather than stoichiometrically, they are being (at least in part) regenerated in the course of the reaction. Similar effects are also observed if instead of endogenous respiration we study that resulting from the oxidation of glucose, pyruvate, lactate, or fatty acids.

3. Malonate at low concentrations (< 0.01 M) is a highly specific inhibitor for one step and one step only of the postulated scheme, the conversion of succinate to fumarate. Addition of malonate to a respiring system leads to an abolition of the catalytic effect and to an accumulation of succinate.

4. The distribution of ^{14}C in the various intermediates during the oxidation of a typical labeled substrate, such as glucose or fatty acid, is in complete accord with the scheme.

5. The individual reactions postulated can be demonstrated to occur both in crude minces and homogenates (with other steps blocked by means of selective inhibitors), as well as with highly purified enzymes extracted therefrom.

6. In cells of animals or plants all the reactions of the cycle can be shown to be localized in just one subcellular organelle, the mitochondrion. Furthermore, the various enzymes of the cycle are present in the mitochondria of different cells in different amounts but in nearly constant proportion to one another.

Reactions of the Cycle

A contemporary version of the cycle and its component parts is shown in Fig. 14-1 and Table 14-1. We observe that the cycle proper consists of a total of eight reactions: one (reaction 1) forms a C—C bond between a dicarboxylic and a derivative of a monocarboxylic acid (oxalacetate and acetyl CoA); four reactions are dehydrogenations; of them, two are oxidative decarboxylations (reactions 3 and 4), leading to the formation of a 4-*C*-dicarboxylic from a 6-*C*-tricarboxylic acid; two (reactions 6 and 8) together oxidize a saturated 4-*C*-dicarboxylic acid to the α-oxo dicarboxylic acid required to start the sequence anew; two of the remaining reactions (2 and 7) are hydration–dehydration reactions (the former used to bring about an isomerization), and one (reaction 5) leads to phosphorylation (i.e., the conversion of P_i into a nucleoside triphosphate) at the expense of the oxidation of a substrate in the reaction just preceding.

ENZYMES OF THE CYCLE AND THE INDIVIDUAL REACTIONS

Reaction (14-1)

Reaction (14-1) is catalyzed by an enzyme that has had a large number of different designations in its short history. The simplest, but hardly descriptive, trivial name is simply *condensing enzyme*. This has been variously modified to *citrate condensing enzyme* or *citrogenase*; the two suggested systematic names are also indicated in the table.

The enzyme is large and exceptionally stable; its specificity for the substrates indicated appears to be quite stringent in all respects, with monofluoroacetyl CoA as the only alternative reactant. This compound in the presence of oxalacetate (OAA) is converted to fluorocitrate at about one tenth the rate of the natural substrate. The mechanism for this condensation reaction has proved elusive. It has been possible to show that in the absence of OAA, the enzyme does *not* catalyze the removal of the α-hydrogen from acetyl CoA, with the latter either free or in enzyme-bound form, nor does it bring about an exchange of labeled OAA into citrate in the absence of acetyl CoA. Some tritium (0.15 of theoretical) is incorporated into acetyl CoA from T_2O when both OAA and enzyme are present. OAA has to react in the keto form. These facts suggest that either the mechanism is fully concerted and yields an enzyme-bound citryl CoA as an unstable, rapidly hy-

TABLE 14-1
Reactions of the citric acid cycle[a]

Reaction	*Equation (Additional coenzyme or cosubstrate above, cofactors or activators below arrow)*	*Name of enzyme*	*Characteristic inhibitor* [b]	$\Delta G^{0\prime}$, *kcal/mol* [c]
1	Acetyl-SCoA + oxalacetate^{2-} + $H_2O \rightarrow$ citrate^{3-} + CoASH + H^+	Citrate condensing enzyme [citrate synthase, citrate oxalacetate lyase (CoA acetylating), citrogenase]	ATP, Palmityl CoA	−9.08
2*a*	Citrate^{3-} $\xrightarrow{Fe^{2+},\ GSH}$ isocitrate^{3-}	Aconitase (aconitate hydratase)	Fluorocitrate [b] *trans*-aconitate [b]	+1.59
2*b*	Citrate^{3-} $\xrightarrow{Fe^{2+},\ GSH}$ *cis*-aconitate^{3-}	Aconitase (aconitate hydratase)		(+2.04)
2*c*	*cis*-Aconitate^{3-} $\xrightarrow{Fe^{2+},\ GSH}$ isocitrate^{3-}	Aconitase (aconitate hydratase)		(−0.45)
3[b]	Isocitrate^{3-} + NAD^+ $\xrightarrow{Mg^{2+}\ or\ Mn^{2+},\ ADP^{e}}$ α-Oxoglutarate^{2-} + NADH + CO_2	Isocitrate dehydrogenase (D_S-isocitrate: NAD oxido-reductase)	ATP	−1.70
4	α-Oxoglutarate^{2-} + CoASH + NAD^+ $\xrightarrow[Mg^{2+}]{TPP,\ lip\ S_2}$ succinyl-SCoA$^-$ + CO_2 + NADH	α-Oxoglutarate dehydrogenase (complex)	Arsenite Parapyruvate [b,f]	−8.82
5	Succinyl-SCoA$^-$ + GDP + P_i + H_2O $\xrightarrow[Mg^{2+}]{}$ succinate + GTP + CoASH	Succinic thiokinase [succinyl-CoA synthase, succinate: CoA ligase (GDP)]	NH_2OH [g]	−2.12 [h]
6	Succinate^{2-} + (FAD)[i] $\overset{Fe^{2+}}{\rightleftharpoons}$ fumarate^{2-} + ($FADH_2$)	Succinic(-ate) dehydrogenase [—: (acceptor) oxidoreductase]	Malonate [b] Oxalacetate [b]	~0 [j]
7	Fumarate^{2-} + H_2O ⇌ L-malate^{2-}	Fumarase (fumarate hydratase)	*meso*-Tartrate [b], ATP	−0.88
8	L-Malate^{2-} + NAD^+ ⇌ oxalacetate^{2-} + NADH + H^+	Malate dehydrogenase (L-malate: NAD oxidoreductase)	β-F-Oxalacetate [b] Fluoromalate [b]	+6.69
Sum 1–8	Acetyl-SCoA + $3NAD^+$ + (FAD) + GDP + P_i + $2H_2O$ $\xrightarrow{cycle}$ $2CO_2$ + CoASH + 3NADH + ($FADH_2$) + GTP			−14.3[illegible]
9*a*	3 × [NADH + $0.5O_2$ + H^+ → NAD^+ + H_2O]	Electron-transport chain (NADH branch)	Barbiturates Rotenone Antimycin A Cyanide	−52.4
9*b*	3 × [NADH + $0.5O_2$ + 3ADP + $3P_i$ → NAD^+ + 3ATP + $4H_2O$]	Same as above, plus oxidative phosphorylation	Dinitrophenol Oligomycin	−30
10*a*	($FADH_2$) + $0.5O_2$ → (FAD) + H_2O	Electron transport chain (succinate branch)	Thienyltrifluo-roacetone Antimycin A Cyanide	−36.2
10*b*	($FADH_2$) + $0.5O_2$ + 2ADP + $2P_i$ → (FAD) + 2ATP + $3H_2O$	Same as above, plus oxidative phosphorylation	Dinitrophenol Oligomycin	−21
Sum total	Acetyl-SCoA + $3O_2$ → $3CO_2$ + $2H_2O$ + CoASH	Cycle + electron transport		−215.2
	Acetyl-SCoA + $3O_2$ + 11ADP + GDP + $12P_i$ → $3CO_2$ + $2H_2O$ + 11ATP + GTP + CoASH	Cycle + electron transport + oxidative phosphorylation		−125.3

drolyzed intermediate, or that the rate-determining formation of the carbanion of acetyl CoA on the enzyme surface requires the prior modification of the latter by means of oxalacetate. This is not an unlikely possibility, since the enzyme is subject to allosteric regulation (1–3), as might be expected from its central position in this metabolic sequence: long-chain fatty acyl CoA derivatives (e.g., palmityl CoA) inhibit by increasing the K_s for OAA (not that for acetyl CoA!), and ATP exerts an analogous heterotropic effect on acetyl CoA. The enzyme does catalyze (4) the hydrolysis of citryl- or L-(*S*)-malyl- [but not of D-(*R*)-malyl-] SCoA.

Reaction (14-2) The enzyme *aconitase* (aconitate hydratase) catalyzes reactions (14-2b) and (14-2c) (see Table 14-1) the reversible hydration–dehydration equilibria between citrate and *cis*-aconitate and between *threo*-D$_S$-isocitrate, as well as the isomerization of citrate and isocitrate. The specificity for these three substrates is very great; the only alternative substrate discovered so far is α-methyl-*cis*-aconitate (5). On dialysis of the enzyme it loses activity, which can be restored by the addition of Fe^{2+} plus a reducing agent (GSH). The best available evidence suggests that all three reactions are catalyzed by the same protein, although it may possess two separate active sites, one with affinity for citrate, the other for isocitrate: (1) on aging the enzyme, the activity for isocitrate remains virtually unaffected, while that for citrate declines (6); (2) an enzyme (citrate hydrolyase, aconitic hydrase) has been reported that catalyzes the hydration of citrate but has no activity with isocitrate (7); and (3) the stereochemistry (5) of the reaction requires that the addition of water to *cis*-aconitate is trans in the formation of both products and the requisite attack of H^+ on C-2 or C-3 of the system

$$\begin{array}{ccc} {}^-O_2C & & CO_2^- \\ & \diagdown\!=\!\diagup & \\ \diagup & & \diagdown \end{array}$$

cannot take place on the *same* side of the plane defined by C-1, C-2, and C-3. (4) Furthermore, steady-state kinetic data (8) require the involvement

---

[a] In general, the reactions shown are those occurring with the mitochondrial enzymes from various animal tissues (mammalian heart, liver; avian breast muscle, insect flight muscle, etc.). Extramitochondrial enzymes from these sources and the enzymes from microorganisms frequently have different characteristics from those shown.

[b] Competitive inhibitors.

[c] Most of the $\Delta G^{0\prime}$ values are those calculated by M. J. Johnson for 25°, *p*H 7.0, an ionic strength of 0.15, and total analytical concentrations for all reactants (1 atm for gases, 1 M for all others, including H_2O). M. J. Johnson, in P. D. Boyer, H. Lardy, and K. Myrbäck (eds.), *The Enzymes*, vol. 3, Academic Press, New York, 2nd ed., 1960, p. 407.

[d] An important, partially extramitochondrial enzyme catalyzes the reaction with NADP as obligatory coenzyme.

[e] AMP is the specific cofactor with the mitochondrial enzyme from yeast and from *Aspergillus niger*.

[f] A dimer of pyruvate accumulating on on storage, its structure is

$$\begin{array}{l} \quad\;\; OH \\ \quad\;\; | \\ {}^-O_2CC(CH_3)CH_2COCO_2^- \end{array}$$

[g] Acts by virtue of forming the hydroxamic acid derivative of the substrate succinyl CoA.

[h] For the reaction succinyl $CoA^- + Mg \cdot GDP^- + P_i^{2-} = succinate^{2-} + Mg \cdot GTP^{2-} + CoASH$; $\Delta G^{0\prime} = -0.8$ kcal/mol.

[i] (FAD) stands for tightly enzyme-bound flavin.

[j] Estimated on the basis of the ready reversibility of the reaction.

of one (or more) complexes other than the three ES complexes and (5) isotope-transfer experiments suggest that the enzyme catalyzes a base-catalyzed proton abstraction from either of the two hydroxyacid substrates with the resultant conjugate acid slow to dissociate so that it can be preferentially transferred to enzyme-bound *cis*-aconitate before the latter dissociates from the enzyme. A possible mechanism is

E + citrate ⇌ enzyme-bound α-(D or T)-D_S, β-CO_2^--L_S-citrate *erythro*-D_S-(or D_SD_G)-(2-tritio)citrate ⇌ [EH*-*cis*-aconitate] ⇌ enzyme-bound α-OH-D_S, β-CO_2^--L_S [α-OH-D_S, βH(D or T)-D_S]-isocitrate *threo*-D_S-(2*R*,3*S*)-isocitrate ⇌ E + isocitrate

[EH*-*cis*-aconitate] ⇌ $E + H^+$ (H*) ⇌ EH* + *cis*-aconitate

Reaction (14-3) Most animal and plant cells contain two enzymes capable of dehydrogenating the naturally occurring *threo*-D_S-isocitrate—one specific for $NADP^+$ and the other for NAD^+. For a long time it was believed that the former, which appeared to exhibit much greater activity in cell homogenates or extracts, was the one primarily involved in the citric acid cycle. There was one disturbing feature—that a large portion of the isolated activity was almost invariably found to be localized in the soluble portion of the cytoplasm (the cytosol), although the overall cycle had quite generally been assigned to mitochondria as one of their most characteristic properties. The situation became clarified when it became evident that the NAD^+-linked mitochondrial enzymes were unstable and exhibited some rather peculiar molecular and kinetic properties. The latter, however, were precisely those to be expected from an enzyme that is to fulfill a key regulatory function in so important an area of metabolism as is the citric acid cycle. ADP not only enhances the stability of the enzyme from mammalian sources, but it also acts as heterotropic activator for isocitrate (11-13). Furthermore, ATP is inhibitory, in part probably because of its ability to chelate and remove the divalent metal cation (Mg^{2+} or Mn^{2+}) that is absolutely required for the reaction, in part as a competitor for NAD^+ (NADH functions in a similar manner and NADPH synergistically enhances the effect of NADH). On the other hand, the extent of the homotropic effects observed appear to vary between enzymes obtained from different tissues, as does the ability of other nucleotides, principally Ado-5′-*P* and Ado-3′ : 5′-*P* to substitute for ADP. The enzyme from yeast has been investigated most extensively (14,15); here there is clear-cut evidence for homotropic effects by isocitrate, NAD+, and Mg^{2+}; heterotropic activation by Ado-5′-*P* and inhibition by inorganic anions have been observed as well. The various effects have been interpreted in terms of a model that postulates two catalytic sites, each binding the three reactants, and two regulatory sites capable of interacting either with isocitrate or with AMP.

The enzyme from *Neurospora crassa* bears some resemblance to the yeast enzyme in exhibiting homotropic effects for isocitrate as well as heterotropic activation by AMP. This enzyme is, however, also activated by citrate and in this it resembles the enzymes from higher plants which are reported to utilize this component (and not an adenine nucleotide) as the predominant heterotropic activator (17).

The NADP+-dependent enzymes show none of these peculiar features. They also differ from the NAD+-requiring ones in their *M* (58,000 for the NADP+-dependent, 3 to 4 $\times$ 10^5 for the NAD+-dependent enzyme from pig heart), their intracellular localization,[2] and in the nature of the reaction catalyzed.

Oxalosuccinate functions neither as a free nor as an enzyme-bound intermediate with the NAD+ enzymes; with the NADP+-requiring ones there is good proof for the occurrence of the following partial reactions [(14-3a) to (14-3c)]. (Note that *free* oxalosuccinate can react with the enzyme to afford the corresponding complex and reactions but does not readily dissociate from the enzyme surface and is not an intermediate.)

$$\text{Isocitrate} + \text{E} \rightleftharpoons [\text{E}\cdot\text{isocitrate}] \underset{-\text{NADP}^+}{\overset{+\text{NADP}^+}{\rightleftharpoons}} [\text{E}\cdot\text{oxalosuccinate}] + \text{NADPH} + \text{H}^+ \quad (14\text{-}3a)$$

$$\text{Oxalosuccinate} + \text{E} \rightleftharpoons [\text{E}\cdot\text{oxalosuccinate}] \quad (14\text{-}3b)$$

$$\begin{array}{c}[\text{E}\cdot\text{Oxalosuccinate}] \rightleftharpoons [\text{E}\cdot\alpha\text{-oxoglutarate}] + CO_2 \\ \upharpoonleft\downharpoonright \\ \alpha\text{-oxoglutarate} + \text{E}\end{array} \quad (14\text{-}3c)$$

The overall reaction [reaction 3, Table 14-1, i.e., (14-3a) + (14-3c)] requires Mn^{2+}, as does the reductive carboxylation of α-oxoglutarate [(14-3c) + (14-3a) in reverse] and the decarboxylation of oxalosuccinate [(14-3b) + (14-3c)]. Whether oxalosuccinate reductase [reaction (14-3b) + (14-3a)] or the oxidative step in particular [reaction (14-3a)] shows this requirement is not resolved; neither is a possible requirement for NADPH for reaction (14-3c) (possibly as a stabilizer of the enzyme in an active conformation). The enzyme catalyzes the reversible exchange of one of the methylene hydrogens of α-oxoglutarate with water (most readily followed by measuring the loss of T or D from appropriately labeled α-oxoglutarate) in a NADPH- and Mn^{2+}-requiring reaction. This may indicate that an enol or carbanion of α-oxoglutarate bound to the enzyme functions as an intermediate in the overall reaction. Conversely, similar studies on β-T-labeled isocitrate have ruled out the enol of oxalosuccinate as a probable intermediate for both types of isocitrate dehydrogenases. A possible mechanism (18–20),

[2] However, the NADP-linked enzyme is present in mitochondria as well, and in conjunction with mitochondrial transhydrogenase $\text{NADPH} + \text{NAD}^+ \xrightarrow[\text{energy transport}]{} \text{NADH} + \text{NADP}^+$ (Chapter 15) may add a significant, even a predominant, contribution in certain tissues (10) (e.g., heart muscle).

also applicable to the malic enzymes (page 520), is shown in the diagram.[3] The hydrogen transfer with both isocitric dehydrogenases has been shown to be stereospecific, direct, and of identical stereospecificity, the reaction occurring on the A side of the pyridine ring of the coenzyme.

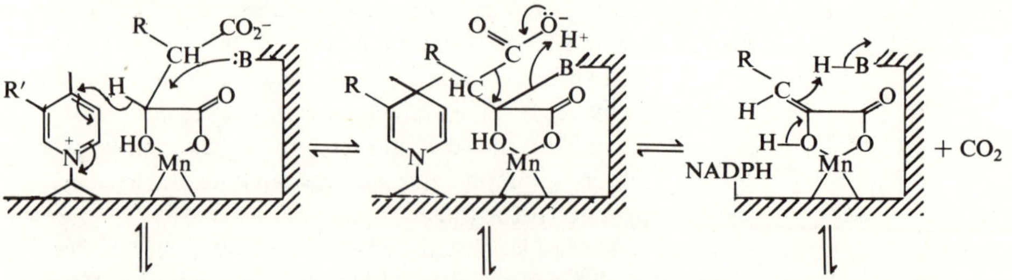

E + Mn^{2+} + $NADP^+$
+ α-hydroxy-1,3-dicarboxylate
(isocitrate, malate)

E + Mn^{2+} + NADPH
+ α-keto-1,3-dicarboxylate
(oxalosuccinate, oxalacetate)

E + Mn^{2+} + NADPH
+ (enol of) α-oxocarboxylate
(α- oxoglutarate, pyruvate)

Reaction (14-4) One of the peculiarities of the cycle is the fact that it contains two successive decarboxylative dehydrogenation steps of quite different reaction types. Reaction 4 is catalyzed by the α-oxoglutarate dehydrogenase complex of enzymes, an entity that bears a striking resemblance to that responsible for the analogous reactions of pyruvate already described on page 519. The α-oxoglutarate dehydrogenase isolated from pig heart is somewhat smaller than the pyruvate dehydrogenase from the same source ($M = 3.3 \times 10^6$ versus 10×10^6). It contains approximately six molecules of protein-bound lipoate, eight molecules of FAD, and six molecules of TPP per particle (21). The reactions catalyzed by the various components are

$$^-O_2CCH_2CH_2COCO_2^- + E_a{\cdot}TPP \overset{Mg^{2+}}{\rightleftharpoons} E_a{\cdot}TPP\text{—}\overset{\overset{OH}{|}}{C}H\,(CH_2)_2CO_2^- + CO_2 \tag{14-4a}$$

α-hydroxy-γ-carboxypropyl TPP

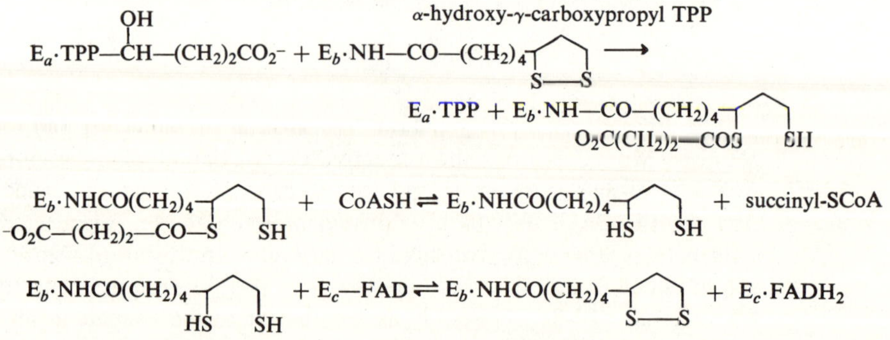

$$E_a{\cdot}TPP\text{—}\overset{\overset{OH}{|}}{C}H\text{—}(CH_2)_2CO_2^- + E_b{\cdot}NH\text{—}CO\text{—}(CH_2)_4\text{—(lipoyl ring, S—S)} \longrightarrow$$

$$E_a{\cdot}TPP + E_b{\cdot}NH\text{—}CO\text{—}(CH_2)_4\text{—(ring: }{}^-O_2C(CH_2)_2\text{—}COS\text{, }SH) \tag{14-4b-1}$$

$$E_b{\cdot}NHCO(CH_2)_4\text{—(ring: }{}^-O_2C\text{—}(CH_2)_2\text{—}CO\text{—}S\text{, }SH) + CoASH \rightleftharpoons E_b{\cdot}NHCO(CH_2)_4\text{—(ring: HS, SH)} + \text{succinyl-SCoA} \tag{14-4b-2}$$

$$E_b{\cdot}NHCO(CH_2)_4\text{—(ring: HS, SH)} + E_c\text{—}FAD \rightleftharpoons E_b{\cdot}NHCO(CH_2)_4\text{—(ring: S—S)} + E_c{\cdot}FADH_2 \tag{14-4c-1}$$

$$E_c{\cdot}FADH_2 + NAD^+ \rightleftharpoons E_c{\cdot}FAD + NADH + H^+ \tag{14-4c-2}$$

E_a = α-oxoglutarate decarboxylase; E_b = dihydrolipoate transsuccinylase;
E_c = dihydrolipoate (lipoamide) dehydrogenase

[3] The mechanism is consistent with stereochemical evidence which indicates that the reaction takes place with retention of configuration; that is, the same position on the β-carbon that is vacated by the leaving carboxyl group is the one occupied by the incoming proton (or D+ or T+):

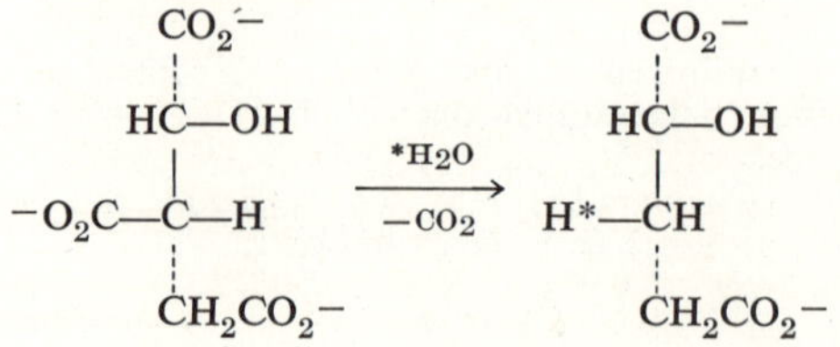

$$\begin{array}{c} CO_2^- \\ | \\ HC\text{—}OH \\ | \\ {}^-O_2C\text{—}C\text{—}H \\ | \\ CH_2CO_2^- \end{array} \xrightarrow[-CO_2]{{}^*H_2O} \begin{array}{c} CO_2^- \\ | \\ HC\text{—}OH \\ | \\ H^*\text{—}CH \\ | \\ CH_2CO_2^- \end{array}$$

It will be noted that only enzyme (E_a) and the overall stoichiometry of the complex distinguishes the two complexes. The net reaction [sum of (14-4a) + (14-4b-1) + (14-4b-2) + (14-4c-1) + (14-4c-2)] is that shown in Table 14-1.

Reaction (14-5) While acetyl-CoA can undergo a wide variety of metabolic reactions, the fate of the succinyl-CoA generated in reaction (14-4) is much more limited. Its preponderant route is to continue in the cycle. Since the appropriate step (the conversion of succinyl CoA to succinate) is highly exergonic, the stage is set for the utilization of the thiolester bond generated in step 4 for conversion into, and storage as, a nucleoside triphosphate. The reaction is catalyzed by *succinyl thiokinase* (succinyl CoA synthetase):

$$\text{Succinyl-SCoA} + \text{MgGDP} + \text{P}_i \rightleftharpoons \text{succinate} + \text{MgGTP} + \text{CoASH} \tag{14-5}$$

and with animal enzymes both the guanosine and inosine di- and triphosphates can function as cosubstrates in the reaction. GTP and ITP can enter the metabolic pool of high-energy currency in the form of ATP, by virtue of the reaction catalyzed by *nucleoside diphosphate kinase*:

$$\text{GTP (or NTP in general)} + \text{ADP} \xrightleftharpoons{\text{Mg}^{2+}} \text{ATP} + \text{GDP (or NDP in general)}$$

In *E. coli* and higher plants, ADP/ATP functions as the cosubstrate. This enzyme has been purified, shown to have a $M = 141{,}000$ and to become phosphorylated (on N-3 of histidine residues) at two equivalent sites. The partial reactions (in the direction of succinyl CoA synthesis) are probably (22,23)

$$\text{Enz} + \text{ATP} \rightleftharpoons \text{Enz}{\sim}P + \text{ADP} \tag{14-5a}$$

$$\text{Enz}{\sim}P + \text{succinate} \rightleftharpoons \text{Enz} \cdot \text{succinyl}{\sim}P + \text{P}_i \tag{14-5b}$$

$$\text{Enz} \cdot \text{succinyl}{\sim}P + \text{CoASH} \rightleftharpoons \text{Enz} + \text{succinyl CoA} + \text{P}_i \tag{14-5c}$$

An unstable Enz~CoA intermediate may also be formed, and CoASH or its disulfoanalog may be required for reaction (14-5a).

A similar pathway may also be operative with the mammalian enzyme; phosphohistidine residues are certainly involved in this instance as well (24).

Reaction (14-6) This key dehydrogenation of the cycle is catalyzed by succinate dehydrogenase, an enzyme that appears to be tightly integrated into the particulate matter of all aerobic cells—the mitochondria of animals and plants, or the oxidative mosaic of bacterial cell membranes—and has therefore been frequently employed as a marker enzyme for such particles. It is not surprising therefore that several of the most important features of the enzyme in its native physiological state still continue to elude us, most notably the precise nature of the true electron acceptor for ($FADH_2$), as well as the function of the nonheme iron associated with the enzyme. These points are discussed further in Chapter 15.

The homogeneous soluble enzyme has an $M \simeq 1.75 \times 10^5$ and contains tightly bound to the protein 4 g-atoms of nonheme iron and 1 mol of FAD in covalent amide linkage. One of the difficulties in the early history of the enzyme was that of its assay after its detachment from other electron-transport carriers. While in the native, particulate form it reacts readily with numerous acceptors, utilizing more or less of the whole respiratory chain, as a "succinoxidase" or "dye reductase" [reactions (14-6a) and (14-6b)], the number of electron acceptors capable of oxidizing the flavin in the solubilized enzyme [reaction (14-6c)] is strictly limited:

$$\text{Succinate} + \text{E}_s\text{-FAD} \rightleftharpoons \text{fumarate} + \text{E}_s\text{-FADH}_2 \quad (\text{fp}_s\text{H}_2) \tag{14-6}$$

$$\text{E}_s\text{-FADH}_2 + \tfrac{1}{2}\text{O}_2 \xrightarrow[\text{transport chain}]{\text{mitochondrial electron}} \text{E}_s\text{-FAD}(\text{fp}_s) + \text{H}_2\text{O} \tag{14-6a}$$

$$\text{E}_s\text{-FADH}_2 + \begin{Bmatrix} 2\text{ dye}(1e\text{ acceptor}) \\ \text{or} \\ \text{dye }(2e\text{ acceptor}) \end{Bmatrix} \xrightarrow[\text{electron-transport chain}]{\text{components of}} \text{E}_s\text{-FAD} + \begin{Bmatrix} 2(\text{dye})^- + 2\text{H}^+ \\ \\ \text{dye H}_2 \end{Bmatrix} \tag{14-6b}$$

$$\text{E}_s\text{-FADH}_2 + \begin{Bmatrix} 2\text{ dye }(1e\text{ acceptor}) \\ \\ \text{dye }(2e\text{ acceptor}) \end{Bmatrix} \xrightarrow{\text{direct}} \text{E}_s\text{-FAD} + \begin{Bmatrix} 2\ (\text{dye})^- + 2\text{H}^+ \\ \\ \text{dye H}_2 \end{Bmatrix} \tag{14-6c}$$

(E_s is succinate dehydrogenase apoprotein; fp_s is succinate dehydrogenase flavoprotein.)

Only phenazine methosulfate (*N*-methyl phenazinium methosulfate) and K-ferricyanide are of practical importance. Reaction (14-6) is readily reversible, and the overall reduction of fumarate to succinate can easily be studied if a dye of sufficiently low potential ($FMNH_2$, viologen dyes, etc.) is used to drive reaction (14-6b) or (14-6c) from right to left.

Reaction (14-7) The reversible hydration–dehydration between fumarate and L-malate is catalyzed by the enzyme *fumarase* (fumarate hydratase). The kinetics and mechanism of the crystalline enzyme from pig heart ($M = 2.0 \times 10^5$), containing four active sites on four identical polypeptide subunits ($M = 48.5 \times 10^3$, each with three free —SH but no S—S bonds), have been thoroughly investigated by Alberty and Hill and their collaborators.[4] No cofactors are required, but the participation of an acidic (i.e., protonated) and a basic (deprotonated) residue has been clearly implicated. Their pK's at an ionic strength of 0.01 are 6.2 and 6.8, respectively, and one of them has been definitely identified as a histidine residue; a Met residue may also participate as part of the catalytic site. The substrate binding sites also contain two positively charged regions required for the proper centering of substrates and inhibitors, as follows:

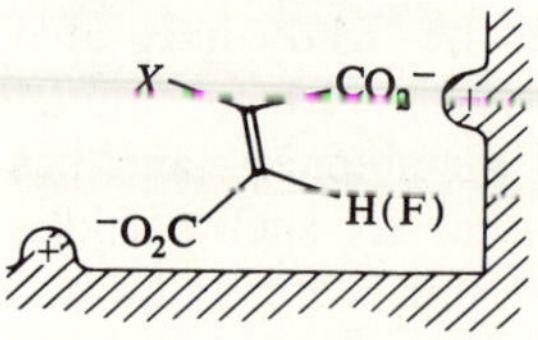

X, CO_2^-, H(F), ^-O_2C

X = OH<I<Br<Cl<H<F (order of activity as substrate)

X = OH < I <Br < Cl < H < F (order of activity as substrate)

[4] These investigations have been truly seminal for the modern era in enzyme kinetics (see Chapter 7). They have led to numerous extensions and modifications of earlier ideas, rendering them applicable to reversible reactions, to reactions with more than one substrate, to reactions involving intermediates, and to relaxation techniques. The general relaxation equation for an enzyme-catalyzed reaction, $\tau_{\text{steady state}} = [(1 + s/K_s)(1 + V_f/V_r)]/(V_f/K_s + V_r/K_p)$, was first verified for this reaction.

The reaction is stereospecific in either direction, and the addition and removal of the elements of water is trans, as shown by the use of D_2O and analysis of the product L-malate by organic or physical (nmr, ORD) techniques.

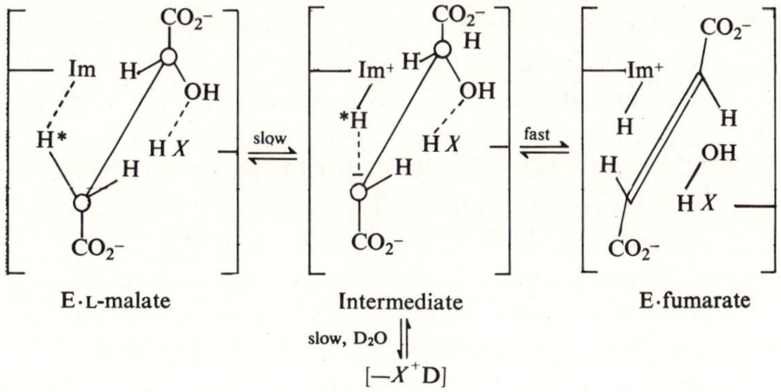

E·L-malate ⇌ (slow) Intermediate ⇌ (fast) E·fumarate

Intermediate ⇅ (slow, D_2O) [—X^+D]

Alternative representations for the 3-monodeuterio-L-malate (*erythro*-L_s-malate-3-*d*), the first term in brackets (at the left), are as follows:

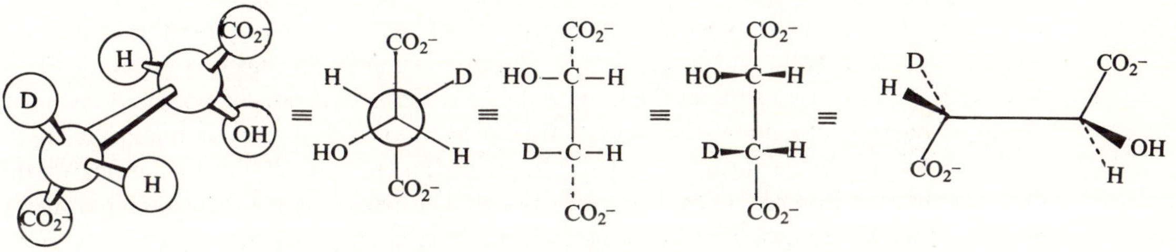

but note that the various derivatives of malate that can function as substrate have the L_S-*threo*-*X*-malate configuration

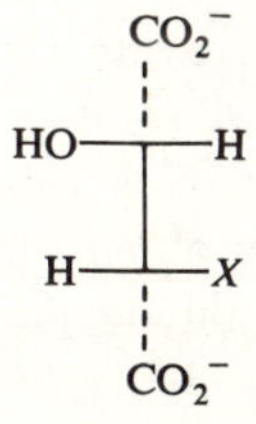

Trans-aconitate and citrate are good competitive inhibitors.

In spite of the relative simplicity of both the enzyme and the reaction catalyzed by it, as well as the fact that it is in the middle rather than at the beginning, end, or branch point of a metabolic sequence, fumarase is subject to potent inhibition by an allosteric effector; ATP inhibits the mammalian and yeast enzymes in the range of 1×10^{-5} M. The effect appears to be cooperative, provided that Mg^{2+} is present as well, but the sigmoid *i* versus extent-of-inhibition curve is due simply to the fact that Mg·ATP is not an inhibitor, so that for any given Mg^{2+} concentration the inhibition is not fully expressed until the metal has been titrated with ATP.

Reaction (14-8) The dehydrogenation of L-malate by L-malate dehydrogenase, a highly stereoselective NAD-requiring enzyme [strictly specific for malate in the L_S-(*R*)-

configuration, with hydrogen transfer to the A side of the pyridine ring], completes the reactions of the cycle and regenerates oxalacetate, the starting material. Eucaryotic cells appear to contain two isozymes of malate dehydrogenase, one species mitochondrial in localization.

Overall Reactions

We have just seen how one turn of the cycle, involving eight enzyme-catalyzed reactions, leads to the conversion of one molecule of acetyl CoA—or of pyruvate, if we remember that the pyruvate dehydrogenase complex (see page 519) catalyzes the reaction pyruvate$^-$ + CoASH + NAD$^+$ $\rightarrow$ acetyl-SCoA + NADH + CO_2—to CO_2 plus H_2O. The overall stoichiometry for acetyl CoA is given in Table 14-1. The analogous one for pyruvate is pyruvate$^-$ + $2.5O_2$ + H^+ $\rightarrow$ $3CO_2$ + $2H_2O$.

Now it is clear that the cycle as a metabolic device for the complete oxidation of pyruvate or acetyl CoA requires the reoxidation of the reduced coenzymes produced during its operation. This is brought about by the sequence of carriers that in their aggregate are referred to as the electron-transport sequence or system (ETS). These entities are housed in general in the very same particles that are responsible for the substrate-level oxidations of the cycle. Thus, *particulate* citric acid cycle activity is characterized by a linkage of the various dehydrogenases to molecular oxygen, with the stoichiometry shown. Since the main purpose of the cycle is not just the disposition of the carbon and hydrogens of all those compounds that can generate acetyl CoA or any of the cycle members, but is the conversion of potential chemical energy into metabolic energy in the form of ATP, provision must also be made for the formation of the latter at the expense of the oxidations of the cycle. Only one of its reactions, reaction 5, performs this task directly, at the substrate level. This is an insignificant part of the total potential energy available, which, as the table shows, amounts to some −215 kcal/mol of acetyl CoA or −275 kcal/mol of pyruvate metabolized. Practically all this energy—specifically, 11 out of 12 mol of ATP from the oxidation of 1 mol of acetyl CoA, or 14 out of 15 from that of 1 mol of pyruvate (and corresponding to 34 out of 38 for 1 mol of glucose; see page 497)—is generated as a consequence of oxidative phosphorylation accompanying mitochondrial electron transport. We shall address ourselves to these problems in Chapter 15.

We have just stated that the cycle is a catalytic device par excellence, employed for the complete oxidation, not only of acetyl CoA and all those compounds capable of generating this intermediate, but also of any or all of its members, or of any compound capable of generating any of these members. How is this brought about? It is clear that reactions (14-2) to (14-7) or (14-8) by themselves can bring about the dehydrogenation (and when linked to electron transport, the aerobic oxidation) of any of the di- or tricarboxylic acids of the cycle to malate or oxalacetate, but no further. The problem, therefore, resolves itself into the question of how to generate acetyl CoA from malate or oxalacetate. Once we know the solution to this problem, the subsequent steps can be regarded simply as a dismutation of OAA, which continues until only catalytic amounts of this compound are left:

$$\text{Malate (or oxalacetate)} \rightarrow \text{acetyl CoA} + 2CO_2 \qquad (14\text{-}9)$$

$$\text{Acetyl CoA} + \text{oxalacetate} \rightarrow \text{citrate} \rightarrow 2CO_2 + H_2O + \text{oxalacetate} \qquad (14\text{-}10)$$

The steps primarily responsible for reaction (14-9) probably consist of malic enzyme and pyruvic dehydrogenase working in conjunction.[5] It will be recalled that the first-named one is an extramitochondrial NADP-requiring enzyme (see page 520), which brings about the conversion of malate to pyruvate. The same enzyme is also capable of decarboxylating OAA irreversibly:

$$\text{Malate}^{2-} + \text{NADP}^{+} \overset{\text{Mn}^{2+}}{\rightleftharpoons} \text{pyruvate}^{-} + CO_2 + \text{NADPH} \qquad (14\text{-}11a)$$

$$\text{OAA}^{2-} \rightarrow \text{pyruvate}^{-} + HCO_3^{-} \qquad (14\text{-}11b)$$

Now to recapitulate briefly the fate of pyruvate. This compound is produced in the cytoplasm either from carbohydrates or lactate. It enters the mitochondria, there to be first converted to acetyl CoA, and is then oxidized to CO_2, provided a sufficient supply of one of the cycle acids is present to allow initiation of the first condensation step. If this requirement is not fulfilled or if there is a steady drain on one or more of the participants of the cycle, as a consequence of other metabolic demands, pyruvate must first be converted by CO_2 fixation to a C_4-dicarboxylic acid in an anaplerotic step. In animals the ATP-dependent pyruvate carboxylase (see page 521) is a likely candidate for this task. In *E. coli,* pyruvate is converted directly to PEP at the expense of ATP with the liberation of AMP and P_i. The phosphoenolpyruvate is then carboxylated to OAA in the reaction catalyzed by PEP carboxylase (see page 522).

STEREOCHEMISTRY OF THE REACTIONS OF THE CYCLE

The stereochemistry of these reactions has been established with considerable precision recently. Together with earlier work by a number of investigators on the metabolic dissymmetry of citric acid, the stereochemistry rests on the following observations and can briefly be summarized as follows:

1. Citrate formed metabolically (and asymmetrically) by condensing enzyme [reaction (14-1)] behaves asymmetrically in the aconitase reaction, in that the methylene and carboxyl groups introduced into the molecule from acetyl CoA are distinct from those deriving from OAA. Only one of the methylene protons of the latter is removed stereospecifically by the enzyme in the conversion citrate → *cis*-aconitate.[6] The explanation for this discrimination has already been given (page 375).

[5] In bacteria, we must also consider the alternative,

$$\text{OAA} \xrightarrow[+P_i]{-CO_2} \text{PEP} \xrightarrow{-P_i} \text{pyruvate} \xrightarrow{-CO_2} \text{acetyl CoA} \qquad (14\text{-}12)$$
$$\text{OAA} \xrightarrow{-CO_2} \text{pyruvate}$$

[6] That is, it has the configuration shown; photosynthetic bacteria form citrate of the opposite stereochemistry (28):

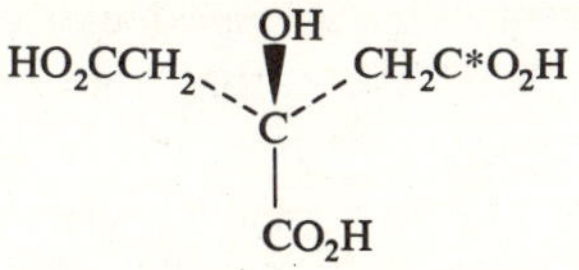
OH
HO2CCH2
CH2C*O2H
C
CO2H

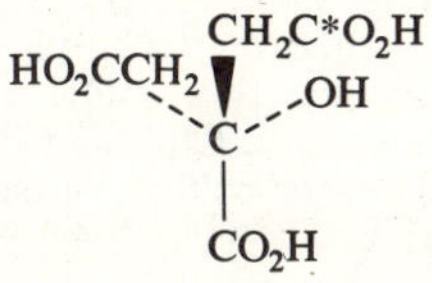
CH2C*O2H
HO2CCH2
OH
C
CO2H

2. Naturally occurring *dextro*-(+)-rotatory isocitrate, which is the substrate for both aconitase and isocitrate dehydrogenase (as well as for isocitrate lyase (page 527), has the configuration (29) α-OH-D_S,β-CO_2H-L_S($\alpha D_S \beta L_S$; *threo*-D_S)[7]:

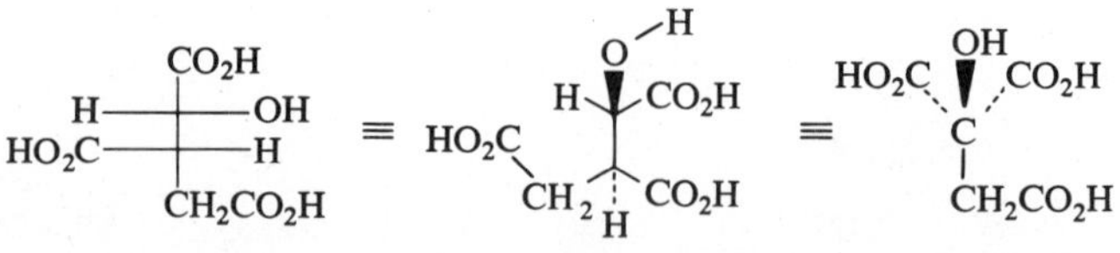

CO₂H; H—C—OH; HO₂C—C—H; CH₂CO₂H ≡ HO₂C, H, O—H, CO₂H, CH₂, H, CO₂H ≡ OH, HO₂C, CO₂H, C, CH₂CO₂H

3. Both hydration/dehydration enzymes of the cycle catalyze trans addition and elimination of the elements of water [addition to *cis*-aconitate and (*trans*)-fumarate, forming derivatives that have —H and —OH in the *erythro* (*cis*) configuration relative to each other] (5,9,31).

4. The two methylene protons in the "OAA portion" of citrate are also not identical stereochemically and metabolically; one is removed in the aconitase reaction, the other in that catalyzed by isocitrate dehydrogenase (18,32).

5. The proton introduced into isocitrate in the aconitase reaction is retained in the reactions catalyzed by isocitrate dehydrogenase, including an exchange with the solvent. Conversely, the proton removed in the latter reaction is identical with that introduced into the molecule during the dehydrogenation of isocitrate (31).

6. The absolute stereospecificity of the above replacement of a carboxyl group in the course of the isocitrate dehydrogenase has been established by two independent means (32): (a) conversion to a monodeuteriosuccinic acid by oxidation with H_2O_2 and comparison of the ORD of that compound with that of *R*-monodeuteriosuccinic acid of known configuration produced by reduction of enzymatically produced 3-monodeuterio-L-malic acid (page 615); (b) conversion of carbons 1 to 4 of α-oxoglutarate to β-tritiated L-malate and L-aspartate by a series of steps and demonstration that over 90 percent of the tritium was retained after treatment with fumarase or aspartase, respectively, two enzymes of known stereochemical behavior.

7. The addition and removal of hydrogens in the succinate → fumarate conversion is also *trans*: the dideuteriosuccinates formed by catalytic deuteration of maleic and fumaric acids, respectively, were dehydrogenated by the enzyme as shown (33).

Since the only substituted succinic acids dehydrogenated by the enzyme (α-Cl and α-Me) are both of the L configuration, and since the enzyme catalyzes a stereospecific exchange with L-monodeuteriosuccinate and with *erythro*-L_S-chlorosuccinate-3-*d*, it is likely that the enzyme is capable of

[7] The subscript S in D_S and L_S refers to the "serine convention," which relates the absolute configuration of a substituent (with the no. 1 carbon written on the top) to those of serine. By the *R*/*S* convention, it would be called (2*R*.3*S*)-isocitrate (30). The absolute stereochemistry of biosynthesized citrate is such that if the molecule is viewed from the side opposite the central —CO_2^-, then moving from —OH in a counterclockwise direction, the first —$CH_2CO_2^-$ is derived from acetyl CoA and the second from OAA (see the preceding page). The compound has also been called L_s-isocitrate.

Unsaturated acid used	*Hydrogenated products*	*Dideuterofumarate after enzymatic action; expected if dehydrogenation is*			
		cis	*random*	*trans*	*Found*
Fumaric		+	+	−	0.1/2.4
Maleic		−	+	+	1.4/2.9

further discrimination and specifically dehydrogenates and forms the following isomer (34–36):

X = H, Cl, or CH_3

(H) = hydrogens removed

Some of these stereochemical considerations are incorporated into Fig. 14-2.

THE CYCLE AS A CATABOLIC AND ENERGY-GENERATING DEVICE

We have already stressed the role of the cycle (supplemented by the appropriate accessory enzymes) as a most effective catalytic device for the complete combustion of carbohydrates (through pyruvate and acetyl CoA), of acetyl CoA, and of all the members of the cycle itself; linked to the simultaneous generation of some 12 molecules of ATP per turn. It is clear therefore that any metabolite capable of being catabolized to one of the compounds just described will serve equally well to keep the furnace stoked (Fig. 14-3).

There exist three great classes of organic compounds in a relatively reduced state that serve as common metabolic fuels: carbohydrates, lipids, and proteins. We have already discussed the oxidation of carbohydrates in Chapter 11 and have pointed out there how acetyl CoA is generated from these metabolites and certain other related compounds, such as lactate, acetate, acetaldehyde, and ethanol. The fatty acids that form the bulk of the energy supply available in lipids also produce acetyl CoA as a consequence of enzymatic β-oxidation reactions catalyzed by a group of mitochondrial enzymes, which have been described in Chapter 13. Acetyl CoA is the sole product if, as is usually the case, the fatty acid in question contains an even number of carbon atoms in an unbranched chain. Other fatty acids lead to propionyl CoA in addition (Chapter 13). As is readily appreciated from Fig. 14-3, acetyl CoA and succinyl CoA account for the bulk of the flux of carbon atoms entering the cycle. Enough has already been said about the former pathway; therefore, some comment about the latter is now appropriate. The methylmalonyl CoA produced as a product of the oxidation of odd carbon or branched-chain

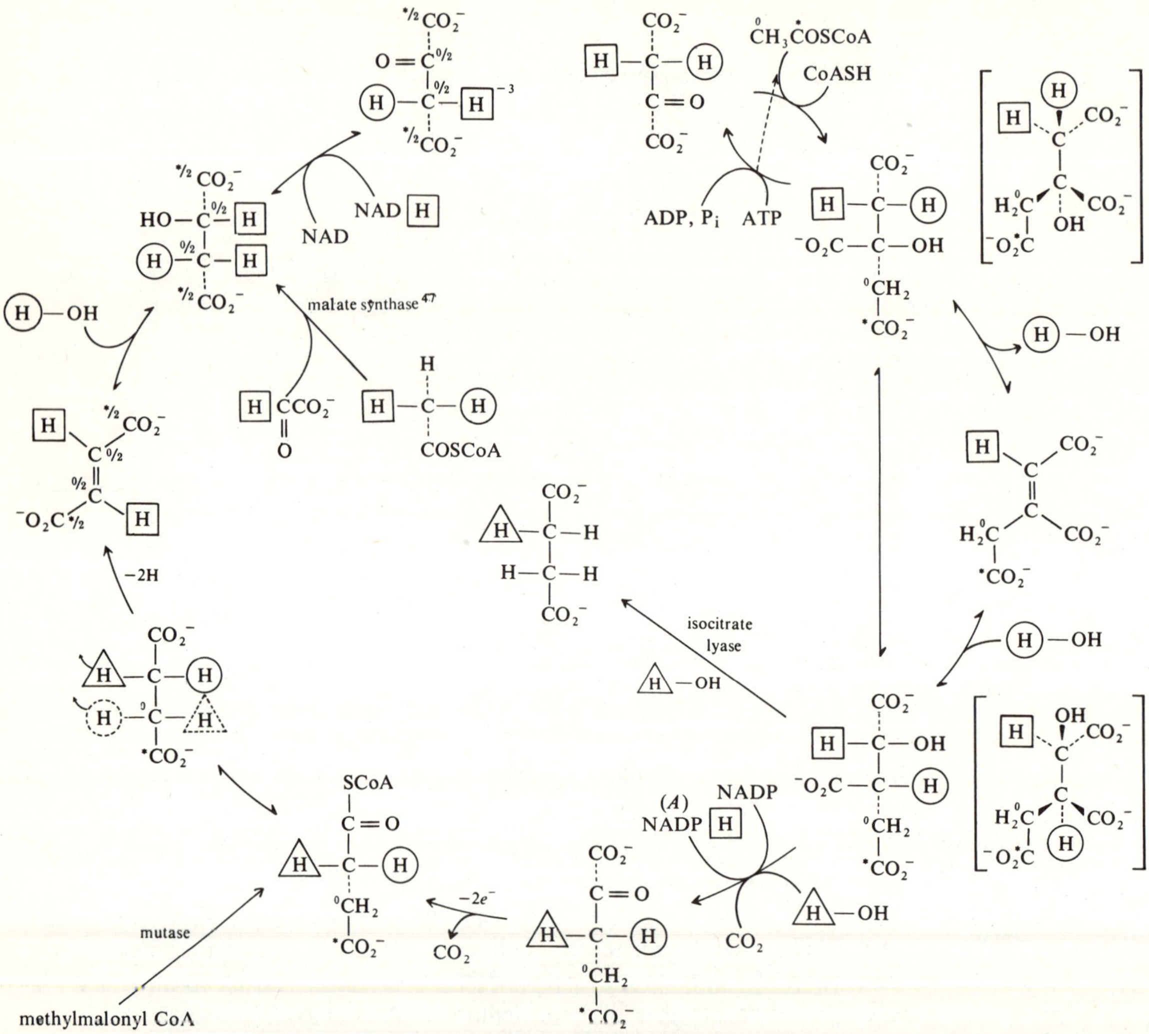

FIGURE 14-2

Stereochemical relationships of acids of the citric acid cycle. For the purposes of discussion, let us assume that we start with fumarate, with its two, stereochemically equivalent hydrogens marked $\boxed{H}$; trans addition of water with its hydrogen as Ⓗ leads to malate, with the stereochemistry indicated, giving rise to the oxalacetate of identical configuration, which, after condensation with acetyl-SCoA (with methyl carbon C° and carboxyl carbon C*, respectively) produces citrate. Action of aconitase then leads either to *cis*-aconitate, with trans elimination of the hydrogen marked Ⓗ (i.e., the one introduced by fumarase), or intramolecular shift of that hydrogen to yield an isocitrate of equivalent stereochemistry. In the course of the isocitrate dehydrogenase reaction, the hydrogen marked Ⓗ (i.e., the one introduced into either *cis*-aconitate or fumarate) is retained, the one marked $\boxed{H}$ is transferred to NADP+, and a new one marked △H is introduced from the medium. After decarboxylation, the resultant α-oxoglutarate yields succinate, a symmetrical molecule with two stereochemically equivalent pairs of hydrogen, marked △H and $\boxed{H}$, respectively. Dehydrogenation by succinate dehydrogenase removes one of each kind, probably the ones shown to the right in this representation, affording the starting material, fumarate, ready to enter another cycle. This is true only, of course, if the succinate contains hydrogen only. If the C-2 is labeled as shown (i.e., if, say, △H stands for *T*), then ○ is *T*, and △ (dashed) and ○ (dashed) are both H, and stereospecific trans elimination predicts that (in the absence of isotope effects!), provided the enzyme can recognize substituents on the α-carbon, there ought to be a preference for one of the two pairs ○ and △ (dashed) or ○ (dashed) and △.

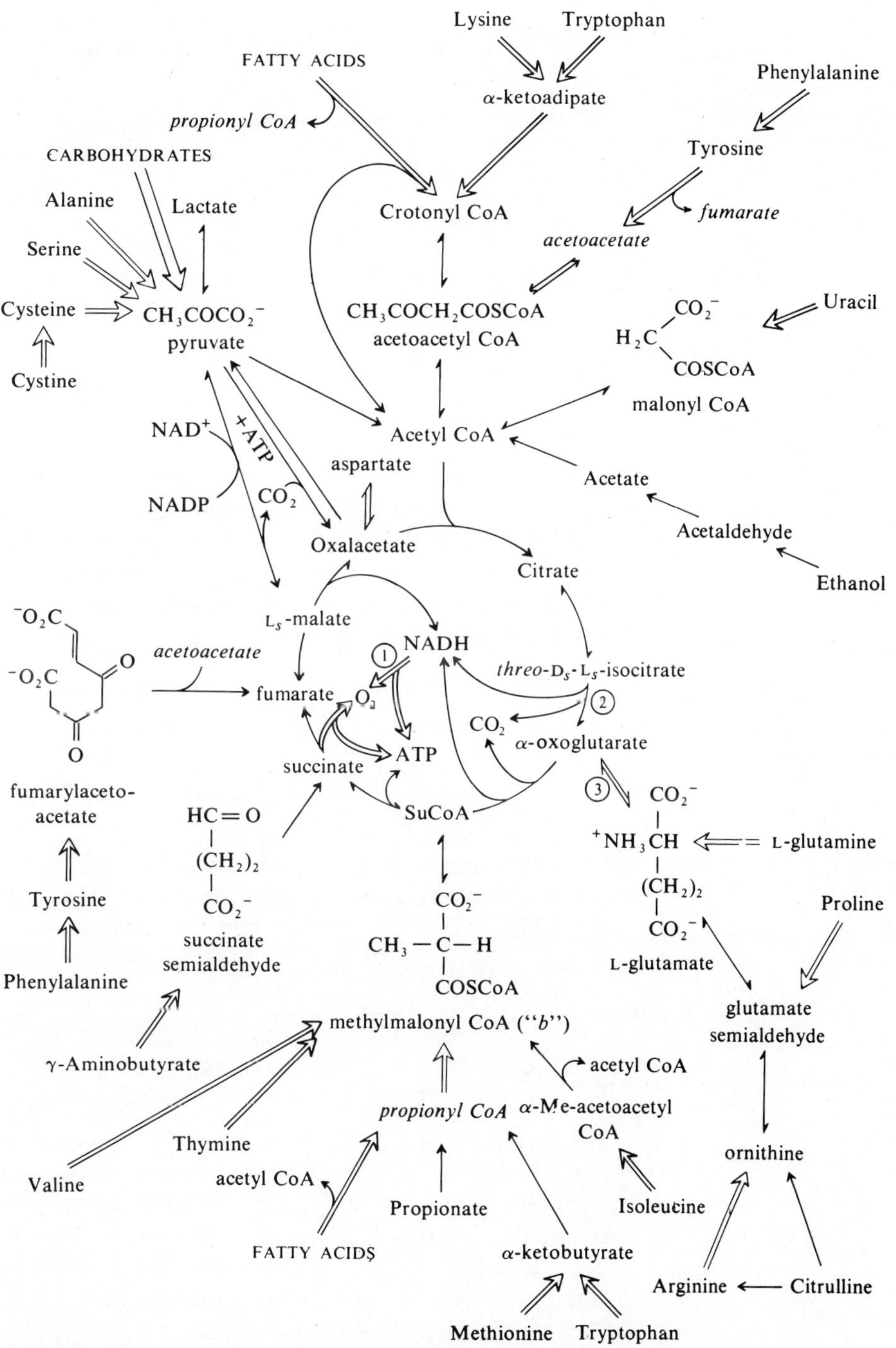

FIGURE 14-3
Citric acid cycle in catabolism. Multistep reactions indicated by ⇒. The terms in italic type are joint intermediates.

fatty acids, or of aliphatic amino acids[8] ("form *a*"), has a configuration around the asymmetric carbon opposite to that ("form *b*") required by the specificity of the methylmalonyl CoA mutase, the enzyme capable of converting this compound to succinyl CoA. For this reason a methylmalonyl CoA racemase (or epimerase) must first intervene to interconvert the two stereoisomers:

$$\text{Propionate} \xrightarrow[\text{ATP} \;\to\; \text{AMP, PP}_i \quad \text{CoASH}]{\text{thiokinase, } Mg^{2+}} \underset{\text{propionyl CoA}}{CH_3CH_2COSCoA} \xrightarrow[\text{ATP} \;\to\; \text{ADP, P}_i \quad CO_2]{\text{carboxylase (biotin), } Mg^{2+}} \text{(H)}-\overset{CO_2^-}{\underset{COSCoA}{C}}-CH_3$$

L$_S$-(S)-methyl malonyl CoA ("*a*")

$$\Big\updownarrow \text{racemase (epimerase)}, \; \text{(H)}-OH$$

$$\underset{\text{COSCoA}}{\overset{\text{COSCoA}}{|}} \quad H-\overset{CH_2}{\underset{CO_2H}{C}}-\text{(H)} \equiv \text{(H)}-\overset{CO_2H}{C}-H,\; CH_2,\; COSCoA \equiv \underset{SCoA}{\overset{O}{\overset{\|}{C}}}-CH_2-\overset{CO_2^-}{\underset{H}{C}}-\text{(H)} \xleftarrow[\text{cobamide}]{\text{mutase (isomerase)}} H_3C-\overset{CO_2^-}{\underset{COSCoA}{C}}-\text{(H)} \tag{14-13}$$

D$_S$-(R)-methyl malonyl CoA ("*b*")

The catabolism of the various amino acids shown will be discussed in Chapter 17. Here we want to emphasize only the following. In general, there exist two main pathways for the conversion of α-amino acids to the corresponding α-keto acids—the first step in the catabolism of the majority of compounds of this class: dehydrogenation by flavoprotein amino acid oxidases with the general reaction

$$\underset{{}^+NH_3}{RCHCO_2^-} + O_2 \xrightarrow{\text{flavoprotein}} R-\underset{NH}{\underset{\|}{C}}-CO_2^- + H_2O_2 + H^+$$
$$\underset{\downarrow H_2O}{} \qquad\qquad \underset{\downarrow \text{catalase}}{}$$
$$NH_3 + R-\underset{O}{\underset{\|}{C}}-CO_2^- \qquad H_2O + \tfrac{1}{2}O_2 \tag{14-14}$$

and transamination with α-oxoglutarate by a series of pyridoxal phosphate-dependent transaminases (aminotransferases). Here pyCHO is the pyridoxal and $pyCH_2—NH_3^+$ the pyridoxamine form of the coenzyme:

$$\underset{{}^+NH_3}{RCHCO_2^-} + E\cdot pyCHO \rightleftharpoons R-\underset{O}{\underset{\|}{C}}-CO_2^- + E\cdot pyCH_2{}^+NH_3 \tag{14-15a}$$

$$^-O_2C(CH_2)_2COCO_2^- + E\cdot pyCH_2{}^+NH_3 \rightleftharpoons {}^-O_2C(CH_2)_2-\underset{{}^+NH_3}{CH}-CO_2^- + E\cdot pyCHO \tag{14-15b}$$

In mammals, at least, reaction (14-15) predominates for the naturally occurring L-amino acids except for L-proline. This metabolism of amino acids by transamination constitutes a dead end as far as the $—NH_2$ group is concerned unless provision for its conversion to a "free" form as NH_3 or NH_4^+ can be found. It is this consideration that ascribes a very special role to L-glutamate. The interconversion L-glutamate $\rightleftharpoons$ α-oxoglutarate is catalyzed by L-glutamate dehydrogenase.

[8] Strictly speaking, this has been proved only for the methylmalonyl CoA arising in propionate metabolism proper (i.e., in the pathway propionate → succinate). It is reasonable to assume a similar situation also for all other pathways involving propionyl CoA (i.e., those shown on the bottom of Fig. 14-3). The configuration of the methylmalonyl CoA arising from valine, thymine, and isoleucine has not yet been established.

$$\text{L-glutamate}^- + \left.\begin{matrix}NAD^+\\NADP^+\end{matrix}\right\} + H_2O \rightleftharpoons \alpha\text{-oxoglutarate}^{2-} + \left.\begin{matrix}NADH\\NADPH\end{matrix}\right\} + NH_4^+ + H^+ \qquad (14\text{-}16)$$

$$\Delta G^{0\prime} = 7.65 \text{ kcal/mol}$$

Enzymes of this type are very widely distributed throughout nature, attesting to their great importance in metabolism (see also Figs. 14-3 and 14-4). The enzyme from vertebrate tissues is a polymer of $M \simeq 10^6$, relatively easily dissociated into subunits of much lower enzymic activity and affected through allosteric interactions by a wide variety of agents, most notably the purine nucleoside diphosphates. It exhibits almost equal activity with either of the two nicotinamide coenzymes. The enzymes from microorganisms, on the other hand, appear to be specific either for the $NAD^+/NADH$ or the $NADP^+/NADPH$ couple and are relatively insensitive to allosteric modifiers.

FUNCTION OF THE CYCLE IN BIOSYNTHESIS

A large number of biosynthetic pathways emanate from the cycle and its constituents. It therefore plays a pivotal role in this branch of metabolism as well (see Fig. 14-3). Stress must be laid on the fact that any time any intermediate is bled off from the cycle, steady-state operation demands its replenishment at the same or a different locus by an anaplerotic reaction.

Carbohydrates

In most cells (except those of bacteria capable of growth on organic compounds containing just two carbon atoms and of certain plant tissues), the synthesis of carbohydrates requires the presence of *stoichiometric* amounts of one of the di- or tricarboxylic acids of the cycle or of compounds that can give rise to one of these intermediates (see Fig. 14-3 and the previous section). Since pyruvate, in the presence of ATP and the pyruvic carboxylase, can fix carbon dioxide to yield oxalacetate (see page 521), this requirement is met by a large number of organic intermediates containing a minimum of three carbon atoms. As outlined in Fig. 14-4 and discussed in more detail on page 524, the synthesis of carbohydrates from such sources involves, at the very outset, two reactions that have no counterpart in carbohydrate breakdown:

$$\text{Pyruvate} + ATP + CO_2 \rightarrow OAA + ADP + P_i$$

$$OAA + GTP \rightarrow \underset{\substack{\downarrow\\ \text{carbohydrates}}}{PEP} + GDP + P_i + CO_2 \qquad PEP = P\text{-enolpyruvate}$$

Different species sequester the relevant enzymes to different extent in the various subcellular compartments. The two principal patterns are (38)

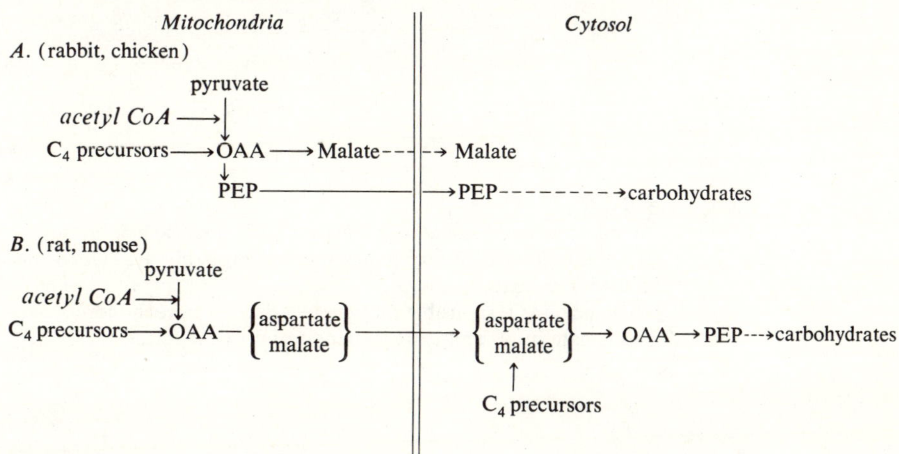

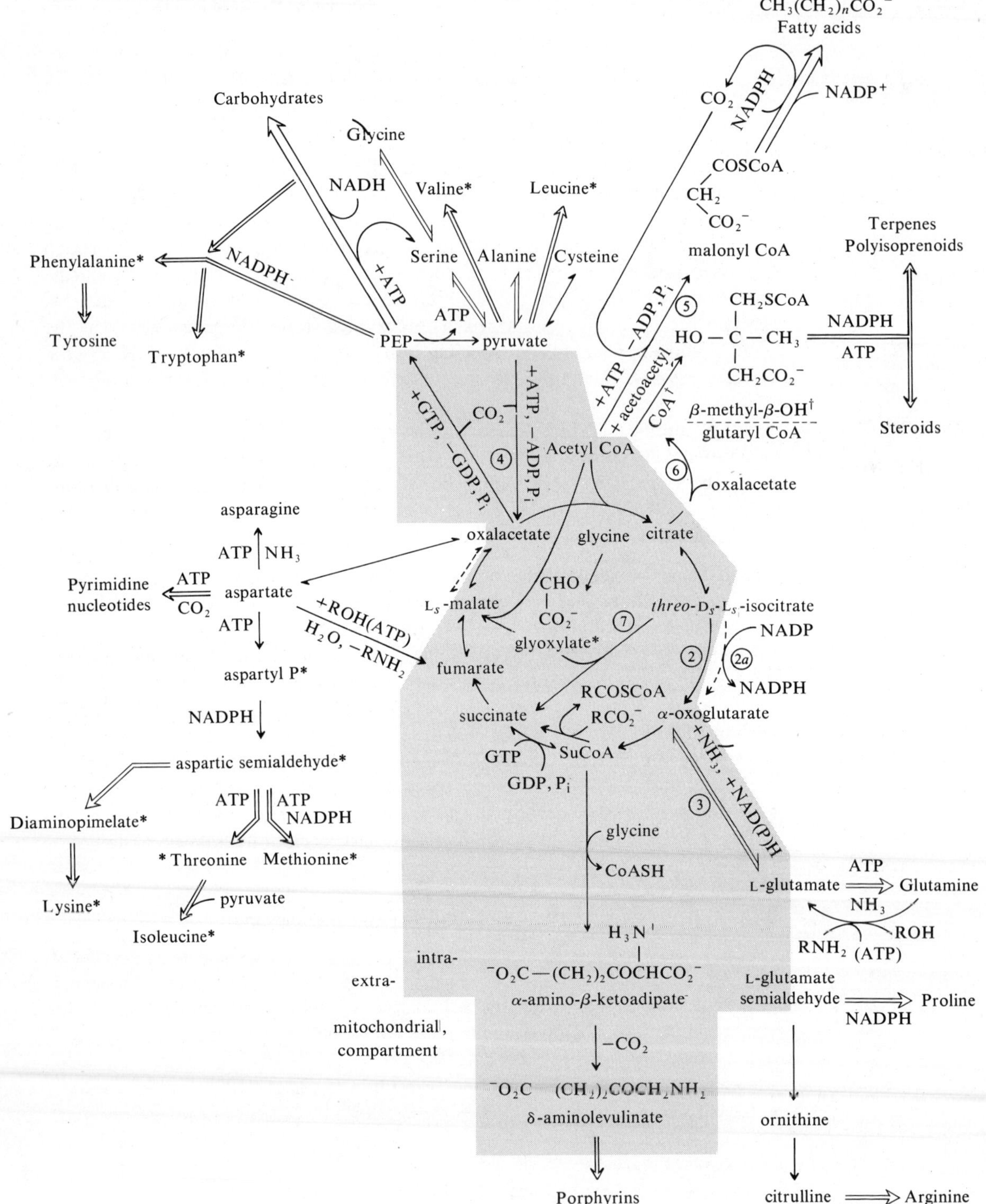

FIGURE 14-4

Anaplerotic and biosynthetic functions of the citric acid cycle.

* The amino acid intermediates marked probably are not formed by the routes indicated in higher animals.

† The enzyme responsible may be localized in mitochondria.

(dashed/solid arrows) indicate intra- and extramitochondrial isoenzymes.

Those cells containing the two enzymes peculiar to the glyoxylate cycle (or bypass)—isocitrate lyase and malate synthetase (page 528)—are capable of generating four-carbon dicarboxylic acids and hence carbohydrate from two-carbon precursors. The biosynthesis of carbohydrate from fats, and more specifically from acetyl CoA, can take place only in such cells. Animals are therefore incapable of affecting this conversion.

Lipids The biosynthesis of lipids is discussed in Chapter 16. Here we wish to emphasize only the following points. Acetyl CoA, the key intermediate in all these reactions (see Fig. 14-4), is generated by just two main processes (see Fig. 14-3), both localized in the mitochondria or equivalent particles—by the thiolytic cleavage of acetoacetyl CoA (generated by the oxidation of fatty acids and certain amino acids) and the oxidative decarboxylation of pyruvate. Yet the biosynthesis of fatty acids requires as an obligatory first step the carboxylation of acetyl CoA to malonyl CoA, and this reaction, as well as all subsequent steps of the process, are probably catalyzed by an extramitochondrial complex of enzymes. How is this brought about? Does acetyl CoA simply diffuse out of the mitochondria, or is some more complex, energy-requiring process implicated? Recent studies suggest that the latter may be the case: acetyl CoA is first metabolized to citrate inside the particle by condensation with OAA; the citrate so formed is then lost to the cytoplasm, there to be cleaved back to oxalacetate plus acetyl CoA by the ATP-requiring citrate lyase (citrate cleavage enzyme) [Eq. (14-17)]. The levels of this enzyme appear to be controlled strongly by genetic and environmental factors, such as nutritional state (induced by diets high in carbohydrate and low in fats) on the one hand, and pathological conditions, such as diabetes or obesity, on the other (39).

$$\text{Mg}\cdot\text{ATP} + \text{citrate} + \text{CoASH} \xrightleftharpoons{\text{Mg}^{2+}} \text{acetyl-SCoA} + \text{Mg}\cdot\text{ADP} + \text{P}_i \qquad (14\text{-}17)$$

In the presence of chloride the enzyme appears to obey the sequence of Mg·ATP adding first, followed by a rapid equilibrium random addition of Mg·citrate and CoA (40). The synthesis of fatty acids by this route, unlike that of carbohydrates, requires catalytic rather than stoichiometric amounts of OAA (or other four-carbon dicarboxylic acids).

Proteins Protein synthesis requires metabolic energy in the form of ATP—as do all the other synthetic processes already discussed—a commodity amply provided for by the operation of the cycle. In addition, this process, again like all the others, requires a supply of monomeric units or precursors, in this instance the 20 or so common L-amino acids. Of this number, most higher animals, including man and the laboratory rat, are unable to synthesize in amounts sufficient for their needs about half (Arg, His, Ile, Leu, Lys, Met, Phe, Thr, Trp, and Val). Plants and most microorganisms, on the other hand, are capable of generating most if not all of these as well. The compounds in question are marked with an asterisk in Fig. 14-4. Again the citric acid cycle provides a means for the generation of both groups, the *nonessential* amino acids in animals and the bulk of all amino acids in other cells and organisms.

The metabolic sequences responsible are discussed in more detail in Chapter 17, but several points are worth noting now: there exist three "families" of amino acids that originate from the three α-keto acids pyruvate, OAA, and α-oxoglutarate, and the corresponding α-amino acids, Ala, Asp, and Glu. The group related to α-oxoglutarate and Glu appears to be formed throughout living nature; only some of those originating from pyruvate are synthesized in higher organisms, and those for which OAA and Asp provide the carbon skeleton are simply not produced at all by mammals, and perhaps not by other animals either.

SOME FACTORS INFLUENCING ACTIVITY OF THE CITRIC ACID CYCLE

Enzyme Levels

The levels of a number of mitochondrial enzymes, including the characteristic dehydrogenases of the citric acid cycle and the carriers of the respiratory chain, appear to be present in constant relative proportion in mitochondria from widely different sources, ranging from the flight muscle of the locust to various tissues of the rat. Enzymes behaving in this fashion include the various cytochromes (see page 676), succinic dehydrogenase, malate dehydrogenase, and glycerol-*P* dehydrogenase. These observations suggest that there exists a (probably genetic) mechanism for the control of the synthesis or perhaps the integration of key mitochondrial enzymes in the course of *mitochondriogenesis.*

Similarly, in bacteria (*Bacillus subtilis, B. cereus*) the synthesis of several of the cycle enzymes appears to be capable of simultaneous or coordinate repression by metabolites; that is, their activities are reduced in constant proportion.

Substrate Levels

One of the controlling features for any reaction sequence is the availability of various initiating substrates. It is also evident that a good indicator of metabolic activity is provided by the steady-state concentration or the half-life of the various substrates involved. In rat liver and kidney, half-lives are all of the order of seconds, except for OAA, which is at least an order of magnitude lower. In *E. coli,* all these values must be divided by at least a factor of 10. These data testify to the high demand placed on the OAA molecule and its big role in controlling mitochondrial metabolism (see Figs. 14-3 and 14-4).

Coenzyme Levels

Examination of Figs. 14-3 and 14-4 discloses the important fact that, at least in higher organisms, degradative, catabolic, energy-yielding sequences generally involve NAD^+, while biosynthetic, energy-requiring ones almost invariably require NADPH. A cell operating efficiently in the steady state, engaged in carrying out two neatly balanced sets of tasks, ought to and does reflect this principle.

Respiratory Control

The nature of mitochondrial respiration observed depends, not only on the nature and concentration of the substrate to be oxidized, but also, in a most profound manner, on the coupling of respiration to phosphorylation. Intact mitochondria are usually (i.e., when substrate and P_i are not limiting) "tightly" coupled, so that their rate of respiration is actually controlled by the ratio [ADP]/[ATP]. When this ratio is high ("state 3," Table 9-5), most of the intramitochondrial adenine nucleotide is in the form of ADP, and respiration is fast; in contrast, when ATP accumulates at the expense of ADP (i.e., under conditions when the rate of its dephosphorylation by phosphate- and energy-requiring reactions cannot keep pace with the rate of its formation), respiration wanes ("state 4"). Added ATP can even inhibit respiration by virtue of bringing about reversed electron flow.[9] These phenomena are now known as *respiratory control.*

[9] This term is used, as more fully discussed on page 697, to describe certain oxidation–reduction reactions not ordinarily observed because of their unfavorable equilibrium, which, however, can be reversed by the addition of ATP; for example, $NADH + 2$ cytochrome c $(Fe^{3+}) \rightarrow NAD^+ + 2$ cytochrome c $(Fe^{2+}) + H^+$, but $NADH + 2$ cytochrome c $(Fe^{3+}) + 2ADP + 2P_i \rightleftharpoons NAD^+ + 2$ cytochrome c $(Fe^{2+}) + 2ATP$.

Accessibility

The mitochondrial membrane itself provides a means for the admission of some substrates and the exclusion of others. For instance, intact mitochondria simply do not admit NADH, and this important metabolite, when produced on the outside, is not oxidized as such by the particles but instead participates in a substrate-linked shuttle:

$$\begin{array}{ccc|ccc} H^+ + NADH & \rightleftharpoons & \text{ox. substrate} \leftarrow\text{------} & \text{----} \text{ ox. substrate} & \rightleftharpoons & H_2O \\ NAD^+ & & \text{substrate } H_2 \text{ ------} & \text{---}\rightarrow \text{ substrate } H_2 & & \tfrac{1}{2}O_2 \\ & & \text{cytoplasm ("outside")} & \text{mitochondria ("inside")} & & \end{array}$$

A number of other instances of this type of compartmentation are known (Chapter 15). Intramitochondrial acetyl CoA does not readily diffuse out but is usually first converted to citrate, which is then cleaved in the cytoplasm to generate acetyl CoA for reactions taking place in that compartment. Mitochondrial succinate dehydrogenase is freely accessible to added succinate and malonate but not to fumarate and oxalacetate, nor is added fumarate freely available to the mitochondrial fumarase.

Control of Enzyme Activity

In Table 14-2 we have summarized some facts, many of them already described in these pages, concerning the control of some of the key enzymes of the cycle and some accessory enzymes by a variety of common metabolites. We have omitted the inhibition of fumarase by ATP and perhaps other effects as well. In addition, we may mention the fact that mitochondrial succinate dehydrogenase is competitively inhibited by oxalacetate, and that this inhibition appears to be relieved by simultaneous addition of ATP. The explanation for this effect is probably to be found in the contraction of the particles produced by ATP and attendant changes in their configuration rendering either the particle or the active site of the enzyme inaccessible to OAA.

How are these many factors involved in controlling biosynthetic activities? Some speculation along the following lines may be appropriate. When the concentration of precursors for mitochondrial metabolites, shown inside the parentheses, is high, the ratio ADP plus P_i/ATP will drop (transition from "state 3" to "state 4"), causing the overall respiratory rate to drop ("Crabtree effect") and ATP and citrate (plus isocitrate) to accumulate.

$$\begin{array}{l} \text{Lipids} \\ \quad\downarrow \\ (\text{Acetyl CoA} + \text{OAA} \rightleftharpoons \text{citrate}) \\ \quad\uparrow \\ \text{Carbohydrates} \end{array}$$

This will "turn off" the mitochondrial isocitrate dehydrogenase, isocitrate will diffuse into the cytoplasm, and part of it will be oxidized by the NADP+-requiring enzyme producing NADPH, while part of it will be cleaved to produce cytoplasmic acetyl CoA. The high concentration of tricarboxylic acids will also serve to activate the acetyl carboxylase. Conditions are now ripe for the diversion of acetyl CoA molecules in large number toward lipid synthesis until a new steady state has been established. Under these conditions there is also a predisposition to carbohydrate synthesis, since acetyl CoA will activate the pyruvate carboxylase. Optimal conditions for almost exclusive *lipid synthesis* will, however, be provided by very *high* concentrations of *acetyl CoA* (or its precursors plus generating systems), coupled with a relatively *low* concentration of *oxalacetate* (and its precursors). Conversely, optimal conditions for *carbohydrate synthesis* will prevail when the concentration of *acetyl CoA* and its equivalents is rather *low,* in conjunction with a very *high* concentration of *oxalacetate* (or pyruvate or lactate, etc.).

TABLE 14-2
Controls on the citric acid cycle

Control point	*Name of enzyme(s)*	*Localization*	*Requires*	*Liberates*	*Activated by*	*Inhibited by*	*Biosynthesis and/or activity affected by*	*Remarks*
1	Coupling of ox. phos. to electron transport and respiration	Particulate	ADP, P_i	ATP, CO_2 by virtue of respiration		Uncouplers, ATP	State of particles, cations, etc.	Variety of agents can disrupt coupling leading to oxidation but no ATP formation
2	NAD-linked isocitrate dehydrogenase	Particulate	NAD^+	NADH, CO_2	ADP(AMP)	ATP, NADH	?	At high levels of cycle activity probably isocitrate oxidized by 2*a*
2*a*	NADP-linked isocitrate dehydrogenase	Soluble and particulate	$NADP^+$	NADPH, CO_2	OAA?			Generates NADPH for biosynthesis in appropriate compartment
3	Glutamate dehydrogenase (α-oxoglutarate $\rightleftharpoons$ Glu)	Particulate	NADPH or NADH, NH_3	$NADP^+$ or NAD^+	ADP	GTP plus NADH	?	Precise nature of control exerted not yet known
4	Pyruvate carboxylase	Particulate	ATP, CO_2	ADP	Acetyl CoA		Hormones, genetic makeup, nutritional state	Controls carbohydrate synthesis, NADH-dependent pathway
5	Acetyl CoA carboxylase	Soluble	ATP, CO_2	ADP	Citrate, etc.	Long-chain acyl CoA's	?	Controls fat synthesis, an NADPH-dependent pathway
6	Citrate lyase (ATP requiring)	Soluble	Citrate (from acetyl CoA)	Acetyl CoA, OAA[a]			Nutritional state, other factors (e.g., obesity)	Makes available extraparticulate acetyl CoA for lipid synthesis
7	Citrate synthase	Particulate	Acetyl CoA, OAA	Citrate, CoA		Long-chain acyl CoA's, ATP		Controls diversion of acetyl CoA to other pathways
8	Isocitrate lyase	Soluble	Citrate (from acetyl CoA)	C_4 acids		PEP	PEP	Bacteria (and plants) only

[a] The OAA produced is probably reduced by the cytoplasmic malic dehydrogenase and the latter enters the mitochondria either directly or else is first oxidized by the cytoplasmic malic enzyme, producing pyruvate + CO_2, with the pyruvate then entering the mitochondria for carboxylation and/or oxidation.

Finally, the accumulation of long-chain acyl CoA's (i.e., conditions under which the ratio of long-chain fatty acids to carbohydrates becomes elevated) leads to an inhibition of citrate condensing enzyme, which will divert both acetyl CoA and OAA to other pathways: the former toward the synthesis of acetoacetyl CoA and compounds metabolically related (e.g., acetone) and neutral lipids such as the steroids, while simultaneously freeing oxalacetate to function in its essential role in carbohydrate synthesis.

REFERENCES

Books, Reviews, and Monographs

Alberty, R. A.: Fumarase, in P. D. Boyer, H. Lardy, and K. Myrbäck (eds.), *The Enzymes,* vol. 5, Academic Press, New York, 2nd ed., 1961, p. 531.

Atkinson, D. E.: Regulation of Enzyme Activity, *Ann. Rev. Biochem.,* **35:** 85 (1966).

Dickman, S. R.: Aconitase, in P. D. Boyer, H. Lardy, and K. Myrbäck (eds.), *The Enzymes,* vol. 5, Academic Press, New York, 2nd ed., 1961, p. 495.

Kun, E.: Malate Dehydrogenase, in P. D. Boyer, H. Lardy, and K. Myrbäck (eds.), *The Enzymes,* vol. 7, Academic Press, New York, 2nd ed., 1963, p. 149.

Lehninger, A. L.: *The Mitochondrion,* Benjamin, New York, 1964.

Lowenstein, J. M.: The Tricarboxylic Acid Cycle, in D. M. Greenberg (ed.), *Metabolic Pathways,* vol. 1, Academic Press, New York, 3rd ed., 1967, p. 146.

Plaut, G. W. E.: Isocitrate Dehydrogenases, in P. D. Boyer, H. Lardy, and K. Myrbäck (eds.), *The Enzymes,* vol. 7, Academic Press, New York, 2nd ed., 1963, p. 105.

Reed, L. J., and D. J. Cox: Macromolecular Organization of Enzyme Systems, *Ann. Rev. Biochem.,* **35:** 57 (1966).

Sanadi, D. R.: Pyruvate and α-Ketoglutarate Oxidation Enzymes, in P. D. Boyer, H. Lardy, and K. Myrbäck (eds.), *The Enzymes* vol. 7, Academic Press, New York, 2nd ed., 1963, p. 307.

Singer, T. P., and E. B. Kearney: Succinate Dehydrogenases, in P. D. Boyer, H. Lardy, and K. Myrback (eds.), *The Enzymes,* vol. 7, Academic Press, New York, 2nd ed., 1963, p. 105.

Stern, J. R.: Oxalacetate Transacetase, in P. D. Boyer, H. Lardy, and K. Myrbäck (eds.), *The Enzymes,* vol. 5, Academic Press, New York, 2nd ed., 1961, p. 367.

Specific References

1. Hathaway, J. A., and D. E. Atkinson: *Biochem. Biophys. Res. Commun.,* **20:** 661 (1965).
2. Tubbs, P. K., and P. B. Garland: *Biochem. J.,* **93:** 550 (1964).
3. Wieland, O., L. Weiss, and I. Eger-Neufeldt: *Advan. Enzyme Regulation,* **2:** 85 (1964).
4. Eggerer, H., U. Remberger, and C. Grünewälder: *Biochem. Z.,* **339:** 436 (1964).
5. Gawron, O., and K. P. Mahajan: *Biochemistry,* **5:** 2335 (1966); see also Gawron, O., A. J. Glaid, III, and T. P. Fondy: *J. Am. Chem. Soc.,* **83:** 3634 (1961).
6. Peters, R. A., and M. Shorthouse: *Nature,* **221:** 774 (1969).
7. Neilson, N. E.: *Methods Enzymol.,* **5:** 614 (1962).
8. Henson, C. P., and W. W. Cleland: *J. Biol. Chem.,* **242:** 3833 (1967).
9. Rose, I. A., and E. L. O'Connell: *J. Biol. Chem.,* **242:** 1870 (1967); see also Speyer, J. F., and S. R. Dickman: *ibid.,* **220:** 193 (1956).
10. Stein, A. M., and J. H. Stein, and S. K. Kirkman: *Biochemistry,* **6:** 1370 (1967).
11. Chen, R. F., and G. W. E. Plaut: *Biochemistry,* **2:** 1023 (1963); Chen, R. F., D. M. Brown, and G. W. E. Plaut: *ibid.,* **3:** 552 (1964); Plaut, G. W. E., and T. Aogaichi: *J. Biol. Chem.,* **243:** 5572 (1968).
12. Stein, A. M., S. K. Kirkman, and J. H. Stein: *Biochemistry,* **6:** 3197 (1967).
13. Goebell, H., and M. Klingenberg: *Biochem. Z.,* **340:** 441 (1964).
14. Atkinson, D. E., J. A. Hathaway, and E. C. Smith: *J. Biol. Chem.,* **240:** 2682 (1965).
15. Cennamo, C., G. Montecuccoli, and G. Bonaretti: *Biochim. Biophys. Acta,* **132:** 232 (1967); Cennamo, C., G. Montecuccoli, G. Bonaretti, and L. Razzoli: *Biochem. J.,* **109:** 361 (1968).
16. Sanwal, B. D., and R. A. Cook: *Biochemistry,* **5:** 886 (1966).
17. Cox, G. F., and D. D. Davies: *Biochem. J.,* **105:** 729 (1967).
18. Chen, R. F., and G. W. E. Plaut: *Biochemistry,* **2:** 752 (1963); Englard, S., and I. Listowsky: *Biochem. Biophys. Res. Commun.,* **12:** 356 (1963); Lienhard, G. E., and I. A. Rose: *Biochemistry,* **3:** 185 (1964).
19. Dalziel, K., and J. C. Londesborough: *Biochem. J.,* **110:** 223 (1968).
20. Colman, R. F.: *J. Biol. Chem.,* **243:** 2454 (1968); *Biochemistry,* **8:** 888 (1969).
21. Hayakawa, T., M. Hirashima, S. Ide, M. Hamada, K. Okabe, and M. Koike: *J. Biol. Chem.,* **241:** 4694 (1966); Hirashima, M., T. Hayakawa, and M. Koike: *ibid.,* **242:** 902 (1967).

22. Ramaley, R. F., W. A. Bridger, R. W. Moyer, and P. D. Boyer: *J. Biol. Chem.*, **242:** 4287 (1967); Moyer, R. W., R. F. Ramaley, L. G. Butler, and P. D. Boyer: *ibid.*, **242:** 4299 (1967); Bridger, W. A., W. A. Millen, and P. D. Boyer: *Biochemistry*, **7:** 3608 (1968).
23. Nishimura, J. S.: *Biochemistry*, **6:** 1094 (1967); Grinnell, F. L., and J. S. Nishimura: *ibid.*, **8:** 562, 568 (1969).
24. Cha, S., C-J. M. Cha, and R. E. Parks, Jr.: *J. Biol. Chem.*, **240:** PC3700 (1965); **242:** 2577, 2582 (1967).
25. Alberty, R. A.: *loc. cit.*
26. Kanarek, L., and R. L. Hill: *J. Biol. Chem.*, **239:** 4202 (1964); Kanarek, L., E. Marler, R. A. Bradshaw, R. E. Fellows, and R. L. Hill: *ibid.*, **239:** 4207 (1964); Teipel, J. W., and R. L. Hill: *ibid.*, **243:** 5679 (1968); Teipel, J. W., G. M. Hass, and R. L. Hill: *ibid.*, **243:** 5684 (1968); Bradshaw, R. A., G. W. Robinson, G. M. Hass, and R. L. Hill: *ibid.*, **244:** 1755 (1969).
27. Penner, P. E., and L. H. Cohen: *J. Biol. Chem.*, **244:** 1070 (1969).
28. Gottschalk, G.: *European J. Biochem.*, **5:** 346 (1968).
29. Katsura, H.: *J. Chem. Soc. Japan*, **82** (1): 91, 92, 98 (1961); Glusker, J. P., A. L. Patterson, W. E. Love, and M. L. Dornberg: *Acta Cryst.*, **16:** 1102 (1963).
30. Vickery, H. B.: *J. Biol. Chem.*, **237:** 1739 (1962); Cahn, R. S., C. K. Ingold, and V. Prelog: *Experientia*, **12:** 81 (1956); Hanson, K. R., and I. A. Rose: *Proc. Natl. Acad. Sci. U.S.*, **50:** 981 (1963).
31. Srere, P. A., H. Brazil, and L. Gonen: *Acta Chem. Scand.*, **17:** S129 (1963); Srere, P. A., and A. Bhaduri: *J. Biol. Chem.*, **239:** 714 (1964).
32. Englard, S.: *Ann. N.Y. Acad. Sci.*, **84:** 695 (1960).
33. Krebs, H. A., and J. M. Lowenstein: The Tricarboxylic Acid Cycle, in D. M. Greenberg (ed.), *Metabolic Pathways*, vol. 1, Academic Press, New York, 1960, p. 129.
34. Rétey, J., J. Seibl, D. Arigoni, J. W. Cornforth, G. Ryback, W. P. Zeylemaker, and C. Veeger: *Nature*, **216:** 1320 (1967); see also Hüfner, M., and T. C. Hollocher: *J. Biol. Chem.*, **243:** 3482 (1968).
35. Gawron, O., A. J. Glaid, III, T. P. Fondy, and M. M. Bechtold: *J. Am. Chem. Soc.*, **84:** 3877 (1962); Gawron, O., A. J. Glaid, III, K. Mahajan, G. Kananen, and M. Limetti: *Biochem. Biophys. Res. Commun.*, **25:** 518 (1966).
36. Vitale, L., and D. Rittenberg: *Biochemistry*, **6:** 690 (1967).
37. Cornforth, J. W., J. W. Redmond, H. Eggerer, W. Buckel, and C. Gutschow: *Nature*, **221:** 1212 (1969); Lüthy, J., J. Rétey, and D. Arigoni: *ibid.*, **221:** 1213 (1969).
38. Scrutton, M. C., and M. F. Utter: *Ann. Rev. Biochem.*, **37:** 249 (1968).
39. Lowenstein, J. M.: in *Oxygen in the Animal Metabolism*, Pergamon Press, Oxford, 1964, p. 163; Kornacker, M. S., and J. M. Lowenstein: *Biochem. J.*, **94:** 209 (1965); *Biochim. Biophys. Acta*, **84:** 490 (1964); *Science*, **144:** 1027 (1964); Spencer, A., L. Corman, and J. M. Lowenstein: *Biochem. J.*, **93:** 378 (1964); Spencer, A. F., and J. M. Lowenstein: *J. Biol. Chem.* **237:** 3640 (1962).
40. Plowman, K. M., and W. W. Cleland: *J. Biol. Chem.*, **242:** 4239 (1967).

15 Biological Oxidations

Biological oxidation reactions provide the cell with a means for supplying its energy demand. They belong to four classes, which, in turn, can be grouped into two main categories, as follows[1]:

A. Dehydrogenations

1. With O_2 as acceptor — Catalyzed by aerobic dehydrogenases (oxidases), substrate: O_2 oxidoreductases
2. With some other acceptor — Catalyzed by coenzyme- or cofactor-linked dehydrogenases, substrate: acceptor oxidoreductases

B. Oxygen insertion reactions

1. Yielding dihydroxyderivatives — Catalyzed by oxygenases
2. Yielding monohydroxyderivatives — Catalyzed by hydroxylases (monooxygenases)

Of these, type A probably represents the primordial one; it is the only one observed in anaerobic cells and provides for the bulk of the energy supply both in anaerobic and aerobic cells. The capacity to bring about simple dehydrogenation reactions probably was among the first characteristics of ancestral living entities, long before the emergence of oxygen into the earth's atmosphere. The importance of dehydrogenations in biochemistry was first pointed out by Wieland (1) in the 1920s, based on studies in which colloidal palladium acted as dehydrogenation catalyst or hydrogen carrier for the oxidation of a variety of organic compounds by an acceptor A, such as molecular oxygen, benzoquinone, and methylene blue:

$$\mathrm{R{-}\overset{\displaystyle X}{\underset{\displaystyle H}{C}}{-}OH} + (\mathrm{Pd}) \rightleftharpoons \mathrm{R}X\mathrm{C{=}O} + (\mathrm{Pd})\mathrm{H_2} \qquad \begin{array}{l}\text{alcohol } X = \mathrm{H} \\ \text{aldehyde hydrate } X = \mathrm{OH}\end{array} \tag{15-1a}$$

or

$$D\mathrm{H_2} + (\mathrm{Pd}) \rightleftharpoons D + (\mathrm{Pd})\mathrm{H_2}$$

$$(\mathrm{Pd})\mathrm{H_2} + A \rightarrow \mathrm{Pd} + A\mathrm{H_2} \tag{15-1b}$$

$$D\mathrm{H_2} + A \xrightarrow{\mathrm{Pd}} D + A\mathrm{H_2} \tag{15-1c}$$

[1] We shall use DH_2 as an abbreviation for substrate in the reduced state, D for the molecule in the oxidized form, in order to avoid any possible confusion with sulfur compounds arising from the use of SH_2, etc. Similarly, $A \rightarrow AH_2$ or $A^n \rightarrow A^{n-1}$ will be used for $2e^-$ and $1e^-$ oxidants or acceptors.

The apparent generality of these reactions compelled Wieland to the conclusion that oxygen was simply the most physiological of a number of possible acceptors and led him to postulate that all biological oxidations proceeded by means of dehydrogenations involving one or more intermediate hydrogen carriers, A', etc. (also referred to as electron or redox carriers). This may be portrayed by a convenient notation, modified from Baldwin (2):

$$\begin{array}{ccccc} DH_2 & & A'H_2 & & AH_2 \\ D & & A' & & A \end{array} \qquad (15\text{-}2a)$$

or

$$DH_2 + A' \rightleftharpoons A'H_2 + D \qquad A'H_2 + A \rightleftharpoons AH_2 + A' \qquad (15\text{-}2b)$$

$$Net \quad DH_2 + A \overset{A'}{\rightleftharpoons} AH_2 + D \qquad (15\text{-}2c)$$

Subsequent studies, especially by Thunberg (3) and by Keilin (4), showed that living cells and cell extracts were able to bring about the dehydrogenation of metabolically important compounds, such as glucose, lactate, succinate, and various amino acids, either aerobically at the expense of oxygen or anaerobically at the expense of artificial electron acceptors, such as methylene blue. As a result of these studies, biological oxidations were postulated to conform to Eq. (15-2c), that is, dehydrogenations of the substrate catalyzed by *dehydrogenases* (now called *oxidoreductases*) capable of transferring reducing equivalents to an intermediate carrier A'.

Two corollaries also emerged: If the first acceptor A' (reduced to $A'H_2$ by the first dehydrogenase) can in turn act as the substrate for a second enzyme, several such respiratory carriers may act in series and produce an array called a *respiratory chain,* or an *electron-transfer system* or *chain.* This chain then intervenes between the substrate and its primary dehydrogenase on the one hand, and the reduction of the terminal acceptor (shown here as oxygen) on the other:

$$\begin{array}{ccccccccc} DH_2 & & \text{carrier } 1_{red} & & \text{carrier } 2_{red} & & \text{carrier } n_{red} & & H_2O \\ & \text{enzyme}_1 & & \text{enzyme}_2 & & & & \text{enzyme}_m & \\ D & & \text{carrier } 1_{ox} & & \text{carrier } 2_{ox} & & \text{carrier } n_{ox} & & \tfrac{1}{2}O_2 \end{array} \qquad (15\text{-}3)$$

$$\text{or} \quad DH_2 \xrightarrow{E_1} \text{carrier } 1 \xrightarrow{E_2} \text{carrier } 2 \rightarrow \cdots \rightarrow \text{carrier } n \xrightarrow{E_m} O_2$$

Another corollary is that if one carrier can intervene between two dehydrogenases specific for two different substrates (DH_2 and $D'H_2$), we observe a *linked* reaction:

$$\begin{array}{ccccc} DH_2 & & \text{carrier}_{red} & & D'H_2 \\ & \text{enzyme}_1 & & \text{enzyme}_2 & \\ D & & \text{carrier}_{ox} & & D' \end{array} \qquad (15\text{-}4)$$

Thus, there are two alternative ways of reconverting the reduced intermediary carrier back to its oxidized form: reoxidation by other carriers, and eventually by a terminal acceptor, or oxidation by a second substrate, which is reduced in the process. Two of the principal carriers, the nicotinamide (or pyridine) coenzymes and the flavin nucleotide coenzymes, were discussed in Chapter 8, and examples of linked reactions during glycolysis and fermentation were described in Chapter 11. In mitochondrial systems, coenzyme Q, cytochromes, nonheme iron, and copper intervene between flavin and the terminal acceptor oxygen. The terminal member of this sequence is cytochrome oxidase, first identified by Warburg as *Atmungsferment* (Fig. 15-1, see also page 680).

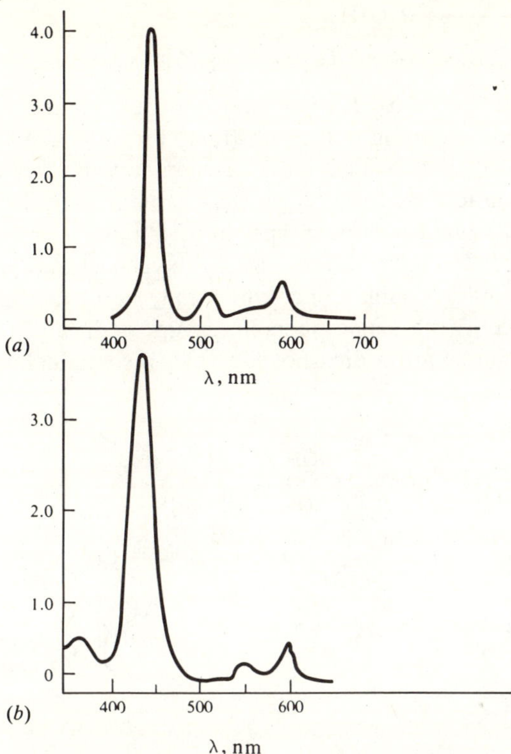

FIGURE 15-1

Photochemical action spectrum of *Atmungsferment* and cytochrome oxidase. (*a*) Cytochrome oxidase (mammalian); (*b*) *Atmungsferment* (yeast). The spectrum is determined by measuring the efficiency of light at various wavelengths in overcoming the CO-inhibited O_2 uptake of whole cells (*b*) or the inhibition of an enzyme preparation (*a*). Curves in (*a*) and (*b*) redrawn from the data of Melnick (5) and Warburg (6). Ordinate: relative efficiency of light of wavelength λ compared to that of $\lambda = 436$ nm.

We shall first consider briefly reactions involving oxygen, catalyzed by oxygenases or hydroxylases, and then turn our attention to the various types of dehydrogenases, the cytochromes, and other hemoproteins, and conclude with a discussion of respiratory chains and of oxidative phosphorylation.

OXYGENASES

Oxygenases are enzymes concerned with a wide variety of reactions, usually involving lipids or aromatic compounds as substrates. They intervene at essential steps in the metabolism of certain aromatic amino acids, of porphyrins, and of lipids such as unsaturated fatty acids and steroids—some of the latter of great importance as vitamins or hormones. Oxygenases also participate, in animals, in the catabolism and detoxification of drugs and toxic substances. All these reactions have in common the enzyme-catalyzed insertion of oxygen atoms derived from molecular oxygen during the initial reaction (1). Two subclasses of oxygenases may be distinguished, depending on the number of oxygen atoms so inserted. *Dioxygenases,* also called oxygenases proper, or *oxygen transferases,* catalyze the insertions of *both* atoms of molecular oxygen ($^{18}O_2$ is used as a tracer) [reaction (15-5)]. Monooxygenases, sometimes referred to as hydroxylases or *mixed function oxidases,* introduce only one oxygen atom into the substrate [reaction (15-6)].

$$DH_2 + O_2^* \longrightarrow D(O^*H)_2 \qquad (15\text{-}5)$$

$$DH_2 + O_2^* + XH_2 \longrightarrow DH(O^*H) + X + H_2O^* \qquad (15\text{-}6)$$

Dioxygenases

In accord with Eq. (15-5), these enzymes are responsible for the introduction of two oxygen atoms into the substrate, forming a dihydroxy derivative as the first intermediate which may, or may not, be stable. Properties of the more well-characterized oxygenases are tabulated in Table 15-1. In general, oxygenases catalyze three main types of reactions.

Cleavage of Aromatic Rings. Many oxygenases are concerned with cleavage of an aromatic double bond, which may be located between two phenolic groups, adjacent to a single phenolic group, or in an indole ring. The corresponding products are a dicarboxylic acid, a semialdehyde, and a formylated amino ketone:

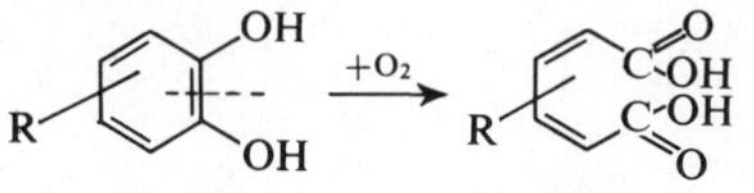

(15-7a)

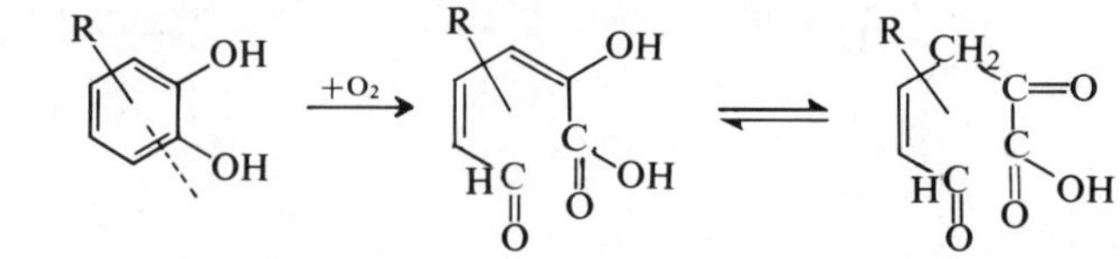

(15-7b)

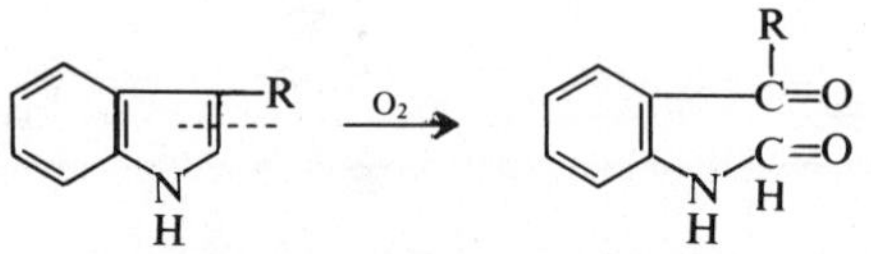

(15-7c)

The substrates catechol and protocatechuic acids are themselves metabolic intermediates in the degradation of a variety of aromatic compounds by microorganisms: the former is produced from benzene, *o*-cresol, benzoic and salicylic acid, phenol, naphthalene, and anthracene; the latter from *m*-, *o*-, and *p*-cresol; *p*-hydroxy- and *p*-aminobenzoic acid; and phthalic acid.

Oxygenation of Aliphatic Structures. Only one example of this type of reaction is known, the conversion of *myo*-inositol to D-glucuronate (page 540):

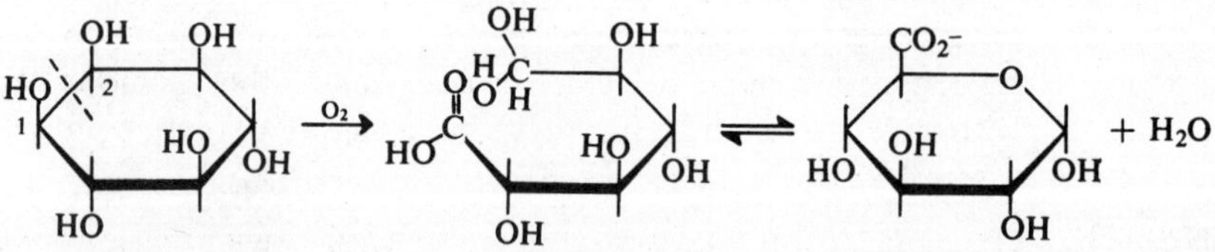

(15-8)

In this reaction a nonaromatic ring is cleaved and then reformed with one less carbon atom.

Lipoxygenase. Lipoxygenases (formerly called lipoxidases) are prevalent in the plant kingdom. These enzymes catalyze the conversion of lipids containing a *cis,cis*-1,4-pentadiene system to the corresponding hydroperoxides having a *cis,trans* configuration

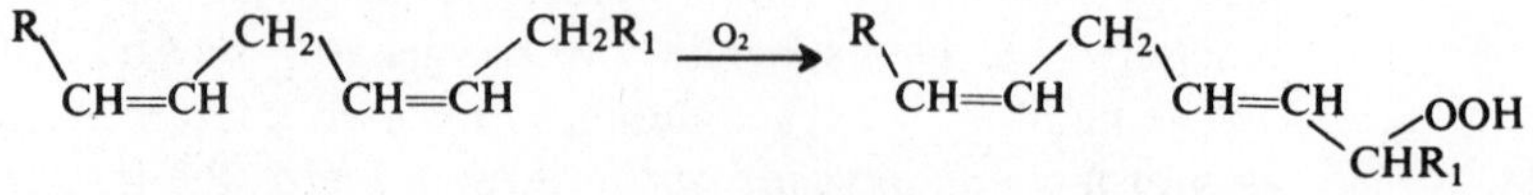

$$R{-}CH{=}CH{-}CH_2{-}CH{=}CH{-}CH_2R_1 \xrightarrow{O_2} R{-}CH{=}CH{-}CH_2{-}CH{=}CH{-}CHR_1{-}OOH \qquad (15\text{-}9)$$

Only lipoxygenase from soybeans has been obtained in crystalline form. It also catalyzes the only reaction in which a hydroperoxide can be isolated as a stable product. Despite much searching, no evidence has been found in any of the other reactions for a similar intermediate or one in which the oxygen molecule (or peroxide grouping) bridges the bond to be cleaved.

The nature of the prosthetic group, responsible for oxygen activation, is unknown in the case of lipoxidase. In at least some of the other enzymes shown in Table 15-1, there is now good evidence for a role of Fe^{2+} in this process.

Other Reactions. A dioxygenase, requiring ferrous ion as a cofactor, has recently become implicated in the conversion of β-carotene to vitamin A, a reaction involving the rather unique cleavage of an aliphatic carbon–carbon double bond.

Another important metabolic conversion, which involves the serial intervention of both a di- and a monooxygenase, is that of homo-γ-linolenic acid to the *prostaglandins* [Eq. (15-10)] (7):

Homo-γ-linolenic acid

$\downarrow +O_2$

$\downarrow +O_2 + XH_2$

$+XH_2$

(15-10)

Prostaglandin E_1

Prostaglandin $F_{1\alpha}$

The latter compounds are among a group of hormones in vertebrate animals responsible for stimulating contraction of intestinal smooth muscle, for reducing blood pressure and for antagonizing the action of a wide variety of other hormones on the release of free fatty acids and glycerol from epididymal fat pads.

Reaction Mechanism. Several of the dioxygenases have now been crystallized or at least highly purified. Iron, either in the form of *nonheme iron* (NHI) (page 656) or heme appears to form part of the prosthetic group in all instances. Interesting examples are provided by metapyrocatechase, a protein with a molecular weight of 140,000 containing three active sites, each incorporating a Fe^{2+} ion, and the protocatechuate-3,4-dioxygenase, with $M = 700{,}000$ and 8 Fe^{3+} ions at an equal number of active sites. Iron-chelating agents inhibit the reactions catalyzed by these enzymes. On the other hand, the highly purified tryptophan-2,3-dioxygenase (tryptophan pyrrolase) contains its iron in the form of Fe-protoporphyrin IX (heme) (8).

TABLE 15-1
Dioxygenases, oxygen transferases

Enzyme	*Source*	*Prosthetic group*	*Substrate*	*Product*
Pyrocatechase (catechol-1,2-oxygenase)	*Pseudomonas*	[Fe^{2+}]	Catechol	*cis,cis*-Muconate
Metapyrocatechase (catechol-2,3-oxygenase)	*Pseudomonas*	[Fe^{2+}]	Catechol	α-Hydroxymuconate semialdehyde
Protocatechuate oxygenase	*Pseudomonas*, fungi	[Fe^{3+}]	Protocatechuate	3-Carboxy-*cis,cis*-muconate
Homogentisate oxygenase	Liver, kidney, bacteria	[Fe^{2+}]	Homogentisate	4-Maleylacetoacetate
3-Hydroxy-anthranilate oxygenase	Liver, kidney		3-Hydroxyanthranilate	Picolinate[a] Quinolinate
Tryptophan oxygenase	Animal tissues, bacteria	[Fe porphyrin] (heme)	L-Tryptophan	L-Formyl-kynurenine
Inositol oxygenase	Kidney	[Fe^{2+}]	*m*-Inositol	D-Glucuronate
Lipoxygenase	Plants		Unsaturated carboxylic acid	Peroxyacid

[a] The first product is 2-amino-3-carboxymuconate semialdehyde = 1-amino-4-formyl-1,3-butadiene-1,2-dicarboxylate.

A feasible mechanism applicable to all the dioxygenases so far examined is that summarized in Eq. (15-11):

$$E(Fe) \xrightarrow{DH_2} E\langle {}^{Fe}_{DH_2} \xrightarrow{O_2} E\langle {}^{Fe}_{DH_2} \rangle O_2 \longrightarrow E(Fe) + D(OH)_2 \qquad (15\text{-}11)$$

The reaction is presumed to take place within a ternary complex of iron enzyme, substrate, and oxygen, with the latter being linked to both the prosthetic metal and the substrate. No oxygenated complex can be formed in the absence of the latter. Whether or not the iron undergoes valency changes, and whether free radicals of either the substrate or of oxygen are intermediates during the catalytic cycle, remains to be established, as does the formation of a peroxide in its course.

HYDROXYLASES (MONOOXYGENASES) (TABLE 15-2)

As outlined in the introduction, the stoichiometry of the reaction catalyzed by this class of enzyme is described by Eq. (15-6), the insertion of a *single* atom of oxygen. The appearance of labeled oxygen in water as well as in the product and the absolute requirement for an oxidizable cosubstrate (XH_2) distinguish this reaction from ones catalyzed by dioxygenases [Eq. (15-5)] and by dehydrogenases [Eqs. 15-2a) to (15-2c)]. Depending on whether or not the obligatory cosubstrate is itself a product of the reaction, monooxygenases have been classified as *internal* or *external*.

Internal Monooxygenases. The first reaction shown to belong to this group is one catalyzed by a number of highly purified, copper-containing enzymes from plants which bring about the conversion of monophenols to diphenols as well as the dehydrogenation to *o*-quinones of the latter. The two types of reaction are usually referred to as the cresolase (or monophenol oxidase) and catecholase (or diphenol oxidase) of the so-called *phenolase complex*. The overall reaction may be formulated as

$$\text{R-C}_6\text{H}_5 + O_2^* + \text{R-C}_6\text{H}_3(OH)_2 \longrightarrow \text{R-C}_6\text{H}_3(OH)(O^*H) + \text{R-C}_6\text{H}_3(=O)_2 + H_2O^*$$

It can be seen that the diphenol cosubstrate is itself a product of the reaction. Thus, the overall reaction catalyzed by this internal monooxygenase is the conversion of a monophenol to an *o*-quinone, which in the process becomes first hydroxylated and then dehydrogenated. The oxygen molecule is split during the reaction, one atom being hydrogenated to H_2O, the other hydroxylating the aromatic ring. The requirement for the diphenol cosubstrate is not absolute; other reducing agents such as ascorbate or hydroquinone can participate in its stead—in that case the oxygenation is *external* rather than internal.

An example of a different prosthetic group for an internal monooxygenase is provided by a flavoprotein enzyme from *Mycobacteria* that catalyzes the oxidative decarboxylation of L-lactate to acetate plus CO_2. As shown in Eq. (15-12), one atom of oxygen is found in acetate, one in H_2O, and the enzyme-bound intermediate serves as both an oxygen acceptor and hydrogen donor:

$$CH_3CHOHCO_2^- + E\cdot FMN \longrightarrow [CH_3COCO_2^- \cdot E\cdot FMNH_2] \xrightarrow{O_2^*} (CH_3CO_2^{*/2})^- + CO_2 + E\cdot FMN + H_2O^{*/2} \qquad (15\text{-}12)$$

A very similar reaction, catalyzed by an enzyme from *Pseudomonas*, converts L-lysine to δ-amino-*nor*valeramide and CO_2 [Eq. (15-13)]:

$$H_2NCH_2(CH_2)_3C\overset{+}{H}NH_3CO_2^- + E\cdot FAD \longrightarrow \left[H_2NCH_2(CH_2)_3\text{—}\underset{\overset{\|}{\overset{+}{N}H_2}}{C}\text{—}CO_2^- \cdot E\cdot FADH_2\right] \xrightarrow{O_2^*}$$

$$H_2NCH_2(CH_2)_3CO^*NH_2 + H_2O^* + CO_2 + E\cdot FAD \qquad (15\text{-}13)$$

External Monooxygenases. Mammalian dopamine hydroxylase is a copper protein that requires ascorbate as the obligatory cosubstrate. In this instance, some evidence has been provided for the following mechanism (9) (XH_2 = ascorbate; DH_2 = dopamine):

$$E\cdot Cu^{2+} + XH_2 \rightarrow E\cdot Cu^+ + XH\cdot + H^+$$

$$E\cdot Cu^+ + O_2 \rightarrow [E\cdot CuO_2]$$

$$XH\cdot + DH_2 \rightarrow XH_2 + DH\cdot$$

$$[E\cdot CuO_2] + DH\cdot \rightarrow [Enz\cdot CuO_2\cdot HD]$$

$$[E\cdot CuO_2\cdot HD] + XH_2 \rightarrow DOH + H_2O + X + E\cdot Cu^{2+}$$

Almost all other hydroxylases use one of the reduced nicotinamide nucleotide coenzymes as their ultimate hydrogen donor or cosubstrate. However, the number and kind of steps intervening differ greatly in complexity.

The simplest pattern is probably that exemplified by certain bacterial

TABLE 15-2
Hydroxylases (monooxygenases)

Enzyme	*Source*	*Prosthetic group*	*Substrate*	*Product*	*Cosubstrate*
Phenolase (monophenol oxidase)	Plants	[Cu]	Substituted phenol	Substituted catechol	Substituted catechol
Dopamine hydroxylase	Mammalian liver	[Cu]	Dopamine (3,4-dihydroxy-phenylethylamine)	Norepinephrine	Ascorbate
Lactate oxidative decarboxylase	*Mycobacteria*	FMN	L-Lactate	Acetate	Pyruvate or reduced FMN
L-Lysine hydroxylase	*Pseudomonas*	FAD	L-Lysine	δ-Amino-*nor*valeramide	Imino acid
Imidazole acetate hydroxylase	*Pseudomonas*	FAD	Imidazole acetate	Imidazolone acetate	NADH
Kynurenate hydroxylase	*Pseudomonas*	?	Kynurenate	7,8-Dihydroxy-kyurenate	NADH
p-Hydroxybenzoate hydroxylase	*Pseudomonas*	FAD	*p*-Hydroxybenzoate	Protocatechuate	NADPH
Phenylalanine hydroxylase	Liver, plants	Biopterin	L-Phenylalanine	L-Tyrosine	NADPH
Aryl-4-hydroxylase	Liver	Heme (P450)	Aniline, acetanilide	4-Hydroxyaryl compounds	NADPH (NHI)
Naphthalene hydroxylase	Liver	Heme (P450)	Naphthalene	1,2-Dihydro-1,2-dihydroxy-naphthalene	NADPH (NHI)

Alkoxylaryl-hydroxylase	Liver	Heme (P450)	RCH_2O—ArR′	RCHO + HOArR′	NADPH (NHI)
Arylalkylamine hydroxylase	Liver	Heme (P450)	RNH—Ar	*N*-oxides	NADPH (NHI)
Steroid-2-hydroxylase	Liver	?	Estriol	2-Hydroxyestriol	NADPH
Steroid-6-β-hydroxylase	Adrenals (liver)	?	Estradiol	6-β-Hydroxyestradiol	NADPH
Steroid-11-β-hydroxylase	Adrenals (liver)	Heme (P450)	11-Deoxycorticosterone	Corticosterone	NADPH (NHI)
Steroid-17-α-hydroxylase	Adrenal mitochondria	Heme (P450)	Progesterone	Deoxycorticosterone	NADPH (NHI)
Cholesterol-20-hydroxylase	Adrenals	?	Cholesterol	20-Hydroxycholesterol	NADPH
Steroid-21-α-hydroxylase	Adrenals	Heme (P450)	Progesterone	17,21-Dihydroxyprogesterone	NADPH (NHI)
Squalene oxidocyclase	Liver	?	Squalene	Lanosterol	NADPH
Fatty acid hydroxylase	Liver Microorganisms	? [Fe]	Fatty acids	ω-Hydroxy fatty acids	NAD(P)H (NHI)
Fatty acid desaturase	Microorganisms, liver	Heme (P450)	Fatty acyl CoA	Enoyl CoA	NAD(P)H
Camphor hydroxylase	*Pseudomonas*	Heme (P450)	+-Camphor	5-*exo*-Alcohol	NADH

hydroxylases, single proteins with a flavin prosthetic group, such as the imidazoleacetate or the *p*-hydroxybenzoate hydroxylases of *Pseudomonas*. The reaction scheme catalyzed is the following:

NAD(P)H → NAD(P); E·FAD ⇄ E·FADH$_2$; Ar + O_2 → Ar—OH + H_2O (15-14)

that is, one in which the reduced flavin acts as the proximal hydrogen donor for the hydroxylation of the substrate, and where reductions of this entity as well as hydroxylation of the substrate are catalyzed by the same enzyme. In some instances, at least, binding of substrate *precedes* reduction of E · FAD; that is, a ternary complex between both substrates and the enzyme is an intermediate in the reaction.

The next stage of complexity is one in which the proximal hydrogen donor is reduced by one protein while it serves as cosubstrate for another—the hydroxylase proper. An example of this type is provided by the phenylalanine hydroxylase of mammalian liver. Here a pteridine acts as a shuttle or link between a reductase (identical with dihydrofolate reductase, page 408) on the one hand, and the hydroxylase on the other (10) [Eq. (15-15)]:

Dihydrobiopterin

NADPH, NADP+, *reductase*, H_2O, *hydroxylase*, O_2

HO—C$_6$H$_4$—CH$_2$—CHCO$_2$ ($^{\oplus}$NH$_3$)

C$_6$H$_5$—CH$_2$—CHCO$_2$ ($^{\oplus}$NH$_3$) (15-15)

Biopterin
(2-Amino-4-keto-6-dihydroxypropyl-
7,8-dihydropteridine)

An interesting feature of the reaction is that in the course of hydroxylation the *p*-hydrogen of the substrate shifts to appear in the *m*-position of the product L-tyrosine (called the "NIH" shift because it was discovered at the National Institutes of Health) (11):

H*—C$_6$H$_4$—R $\xrightarrow[\text{enzyme}]{O_2}$ [intermediate] ⟶ [intermediate] ⟶ HO—C$_6$H$_3$(*H)—R + HO—C$_6$H$_3$(H)—R

The final stage in complexity is reached with the group of hydroxylases localized largely in the microsomal fraction of a number of vertebrate organs, such as the liver, the adrenals, and the female sex glands (12). The liver enzymes are principally concerned with the metabolism and detoxification of a large number of aromatic compounds of pharmacological interest, and with key steps in lipid metabolism. Additional examples of the former type not shown in the table are the conversion of naphthalene to 1,2-dihydro-1,2-dihydroxynaphthalene, of L-kynurenine to L-3-hydroxykynurenine, of tyrosine to 3,4-dihydroxyphenylalanine (Dopa), and of tryptamine to 6-hydroxytryptamine. Another important reaction in animals is the hydroxylation of tryptophan, which after decarboxylation of the product leads to the formation of serotonin. The conversions of lipid metabolism are not restricted to vertebrate liver but are of fundamental importance to a variety of eucaryotic organisms; in addition to those shown, the conversion of saturated to 9,10-unsaturated acyl CoA derivatives and of γ-butyrobetaine to carnitine are deserving of mention.

The P450 System. A novel hemoprotein with an absorption maximum at 420 nm (that of its CO derivative is at 450 nm) called *cytochrome B420* or *P450,* occurs widely in microsomal and mitochondrial fractions of animal tissues (13). It has been shown to function as the (terminal) oxidase in the conversions catalyzed by a wide variety of NAD(P)H mixed-function oxidase systems, such as those concerned with drug metabolism; steroid C-11, C-17, and C-21 hydroxylation, all linked to NADPH; and bacterial hydroxylations linked to NADH. The immediate electron donor in all cases is provided by a reduced nonheme iron protein (NHI, page 656), called *adrenodoxin* when of animal origin, *rubredoxin* when of bacterial origin. The ultimate electron donor, NAD(P)H, in turn reduces the NHI with the intervention of a flavoprotein reductase. The sequence of reactions is that shown in [Eq. (15-16)]:

$$\mathrm{NADPH} + \mathrm{fp} \rightarrow \mathrm{NADP^+} + \mathrm{fpH_2}\ (\text{or } \mathrm{fpH\cdot});\quad \mathrm{fpH_2} + \mathrm{Fe^{3+}\cdot protein} \rightarrow \mathrm{fp} + \mathrm{Fe^{2+}\cdot protein};\quad \mathrm{Fe^{2+}\cdot protein} + \mathrm{P450\,(Fe^{3+})} \rightarrow \mathrm{Fe^{3+}\cdot protein} + \mathrm{P450\,(Fe^{2+})}$$
$$\mathrm{P450\,(Fe^{2+})} \underset{h\nu}{\overset{\mathrm{CO}}{\rightleftharpoons}} \mathrm{P450\,(Fe^{2+})\cdot CO};\quad \mathrm{P450\,(Fe^{2+})} + \mathrm{O_2} \rightarrow \mathrm{P450\,(Fe^{2+})\cdot O} + \mathrm{H_2O};\quad \mathrm{P450\,(Fe^{2+})\cdot O} + \mathrm{RCH_3} \rightarrow \mathrm{P450\,(Fe^{3+})} + \mathrm{RCH_2OH} \tag{15-16}$$

fp = flavoprotein

In this formulation a reactive oxygen bound to ferrous iron, such as the ferryl ion complex $(\mathrm{Fe^{2+}})-\ddot{\mathrm{O}}:$ or $(\mathrm{Fe^{4+}})-\mathrm{OH}$ (page 672) is postulated as the reactive species in the oxygenation reaction proper. Recently it has been suggested that the reduced nicotinamide may also participate directly in the generation of this species [Eq. (15-17)]:

$$\mathrm{P\text{-}450(Fe^{3+})} \xrightarrow[\mathrm{NADPH}\rightarrow\mathrm{NADP^+}]{fp} \mathrm{P\text{-}450(Fe^{2+})} \xrightarrow{\mathrm{RH}} [\mathrm{P\text{-}450(Fe^{2+})\!-\!RH}] \xrightarrow{\mathrm{O_2}} \left[\mathrm{P\text{-}450(Fe^{2+})}\begin{smallmatrix}\mathrm{RH}\\ \mathrm{O_2}\end{smallmatrix}\right] \xrightarrow{\mathrm{NADPH}\rightarrow\mathrm{NADP^+}} \left[\mathrm{P\text{-}450(Fe^{2+})}\begin{smallmatrix}\mathrm{RH}\\ \mathrm{O\!-\!OH}\end{smallmatrix}\right]^- \xrightarrow[\mathrm{H_2O}]{\mathrm{H^+}} \left[\mathrm{P\text{-}450(Fe^{4+})}\begin{smallmatrix}\mathrm{RH}\\ \mathrm{OH}\end{smallmatrix}\right]^+ \rightarrow \mathrm{P\text{-}450(Fe^{3+})} + \mathrm{ROH} \tag{15-17}$$

A reaction pathway very similar to that of Eq. (15-16) but in this instance with a *soluble* rather than a membrane-bound P450 appears operative in the microorganism *Pseudomonas putida,* which is capable of introducing a hydroxyl group into the methylene C^5 of camphor to produce a secondary alcohol. The NHI carrier protein in this instance has been highly purified and is known as *putidaredoxin* (14). A similar system but with rubredoxin (page 657) as the carrier appears to be involved in the ω-oxidation of laurate to ω-hydroxylaurate or of octane to *n*-octanol by *Ps. oleovorans* (15).

DEHYDROGENASES

Classification and General Remarks

Three groups of dehydrogenases (called "oxidoreductases" by the International Commission on Enzymes) can readily be distinguished by the nature of the electron (hydrogen) acceptor and by the type of reaction involved: (1) nicotinamide nucleotide dependent, (2) flavoproteins, and (3) cuproproteins, Scheme 15-1. The copper enzymes, which have a relatively high oxidation potential, are concerned with the dehydrogenation, using oxygen as the acceptor, of substrates that also have a high potential. Cuproprotein enzymes are therefore ideally suited as the terminal members of respiratory chains. Even cytochrome oxidase, which fulfills this role in animals and plants, contains both an iron porphyrin and copper; the latter may be responsible for the interaction with O_2.

The nicotinamide nucleotide-dependent enzymes, on the other hand,

SCHEME 15-1

Classification of coenzymes, carriers, and substrates by their E_0' values (25°, pH 7).

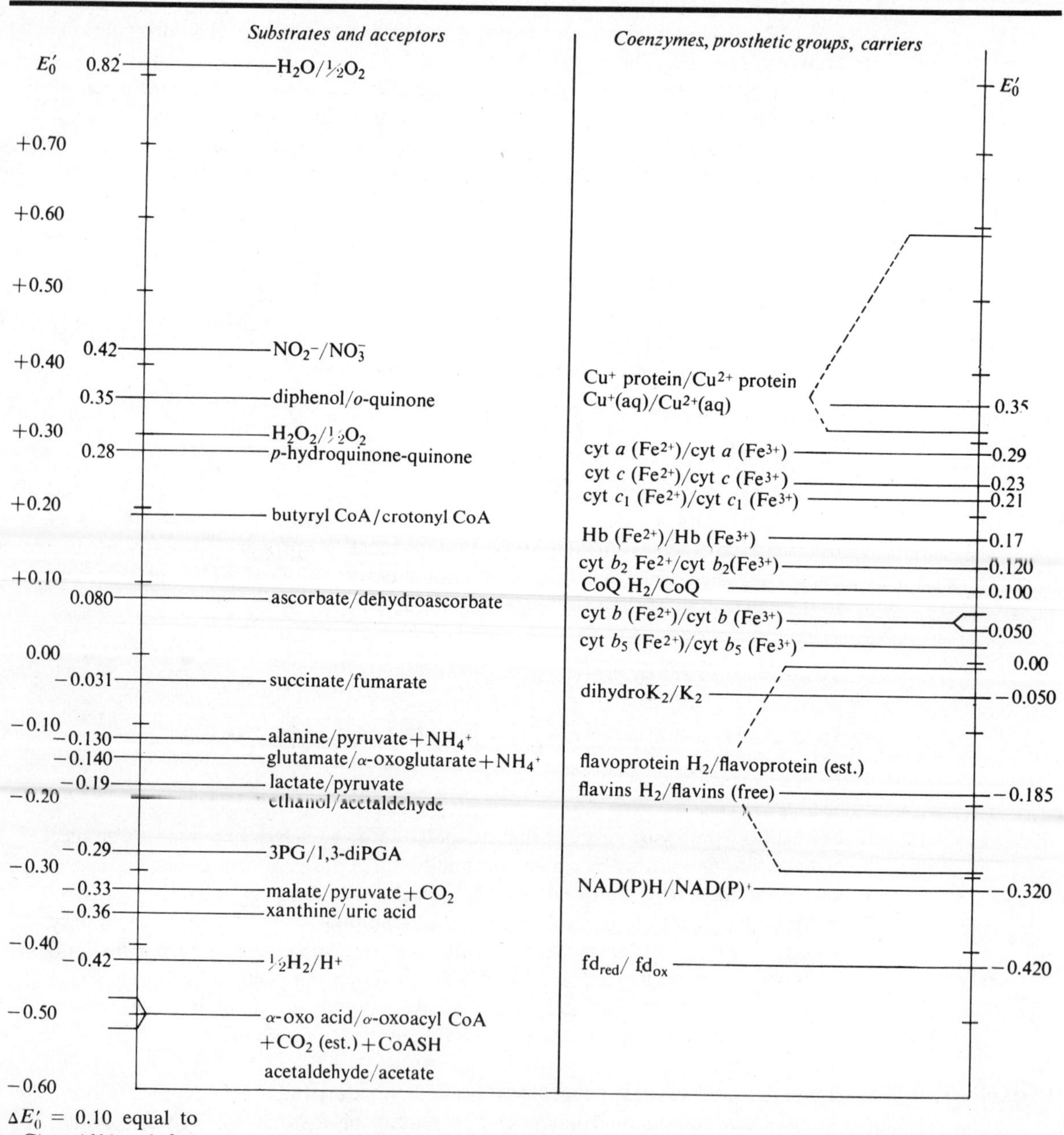

$\Delta E_0' = 0.10$ equal to

$\Delta G' = 4600$ cal for a 2e−, 2300 cal for an 1e− change catalyze the transfer of two reducing equivalents from the substrate to the coenzyme, with a potential (at *p*H 7) in the neighborhood of the hydrogen electrode and which usually functions as a readily dissociable carrier. They are ideally designed, therefore, for the initial dehydrogenation of most substrates. The reduced coenzyme formed as a result of this reaction can shuttle freely to a second site, where it can (1) be utilized to reduce a second substrate via a linked reaction, (2) provide reducing power for biosynthetic sequences, or (3) be reoxidized by a respiratory chain with the attendant conversion of ADP to ATP.

Flavoproteins occupy a position intermediate between the first two classes. They catalyze the dehydrogenation of substrates of intermediate redox potential, in particular of NAD(P)H. They can in some instances react directly with oxygen, while in others their immediate oxidant is itself a carrier, such as protein-bound nonheme iron (NHI); their flavin prosthetic groups are covalently linked to some proteins, while in others they can be removed by appropriate treatments—yet rarely, if ever, do they function as dissociable coenzymes.

Cuproproteins

Many of the well-characterized cuproprotein enzymes as well as some other important copper-containing proteins are described in Table 15-3. Many of these contain several copper atoms per molecule, and EPR spectroscopy and other techniques disclose that the valence state of the metal, as isolated, comprises a mixture of Cu^+ and Cu^{2+}. Only plastocyanin, an electron carrier of green-plant chloroplasts (page 571) appears to contain Cu^{2+} exclusively (16).

The metal atom can be removed by exhaustive treatment with chelating agents or by exposure to acid. In a number of instances this change has been shown to be reversible by reconstituting an enzymatically active species from the apoprotein plus the metal; with apoceruloplasmin, apolaccase, and apoascorbate oxidase by addition of Cu^+ in the presence of air; with apoplastocyanin by addition of Cu^{2+}.

The spectacular blue color ($\lambda_{max} \geqq 600$ nm) found with a number of these proteins had at one time been ascribed to possible metal–metal interactions or to the presence of carbohydrate moieties. However, the primary sequence of the intensely blue cuproprotein *azurin,* which can be isolated from *Pseudomonas fluorescens* and other bacteria, is now known: this protein contains a single Cu atom and no carbohydrate (17). Although, in the special case of plastocyanin—SH groups on the protein may participate in Cu binding and in the structure of the chromophore, this does not appear to be the case for azurin. Hence, the nature of the responsible site remains unknown. Studies by means of absorption, EPR, and CD spectroscopy suggest considerable distortion of the Cu^{2+} complex from its classical square-coplanar symmetry; possible models propose virtually tetrahedral structures with charge-transfer contributions from ligand orbitals (particularly π orbitals) to the vacancy in the Cu^{2+} $3d$ orbitals or the admixture of $4p$ and $4s$ character of these orbitals (18). Although a variety of small peptides form tetracoordinated Cu complexes through their nitrogens, with fairly intense blue colors in alkaline solution (the so-called biuret complexes, useful for the analytical determination of proteins and polypeptides), these cannot be regarded as adequate models for the blue proteins; they are virtually coplanar and their extinction coefficient is low ($\leqq 0.1$ that of the blue proteins) (Table 15-3).

The question whether the metal in the cuproprotein oxidases undergoes a cyclical change (reduction by substrate, oxidation by O_2) in the course of enzymatic catalysis remains unanswered. Inhibition of enzymatic activity by

TABLE 15-3
Properties of some cuproproteins

Protein	Source	Molecular weight	Copper per mole	λ_{max}, nm	ε_{max}, liters × g-atoms Cu × cm^{-1}	EPR signal[a]	Enzymatic function	Inhibited by CO	Inhibited by HCN
Hemocyanin	Marine invertebrates	610,000–950,000	25	347 580	6,100 390	—	O_2 transport	+	+
Ceruloplasmin	Vertebrate blood plasma	151,000	8	605	1,200	+	Similar to laccase	±	+
Ascorbate oxidase	Plants (squash)	150,000	6	606	770	+	+	—	+
Laccase	Plants (lacquer tree)	120,000[d]	4	615	1,200	+	+	—	+
(Uricase) Urate oxidase	Animals (liver, kidney)	120,000	1	330?	?	?	+	—	+
Cytochrome oxidase[b]	Animals, fungi	200,000	2	820?	?	+	Probable	+	+
Cytochrome oxidase–cuproprotein	Heart muscle	25,000	1		?	?			
Galactose oxidase	Molds	42,500	1	?	?	?	+	?	+
Phenolase[c]	Plants	31,000	1	340?	?	±	+	+	+
Transsulfurase	Liver	35,000	1	420, 520	3,230 300	—	+	—	+
Cerebrocuprein	Mammalian brain	45,000	2	660–665	~400	—	—	?	?
Erythrocuprein	Mammalian blood								
Hepatocuprein	and liver	35,000	2	655	140	—	—	?	?
Plastocyanin	Plant chloroplasts	11,000	2	599, 330	4,900	—	+	—	?
Azurin	*Ps. aeruginosa* other bacteria	14,000	1	630	3,000 6,000	+	?	?	?
Amine oxidase	Plants, pig kidney[e]	96,000 125,000	1	502 480	2,500 1,000	±	+	—	+
Amine oxidase (spermine)	Mammalian plasma	225,000	4?	380, 480	2,500 1,000	+	+	—	+
Dopamine β-hydroxylase	Adrenals	290,000	4–7	—	—	?	+	?	+

[a] + indicates a signal caused by the Cu^{2+} state, affected by addition of substrate.
[b] Also contains iron-porphyrin.
[c] Also known as tyrosinase, monophenol oxidase, diphenol oxidase, and catecholase. Systematic name, *o*-diphenol : O_2-oxidoreductase.
[d] Enzymes from other sources (*Polyporus versicolor*) are reported to have $M = 61{,}000$.
[e] The pig kidney enzyme is reported to have $M = 125{,}000$ but to possess one Cu per 90,000 daltons. EPR shows all Cu as Cu^{2+} in a tetragonal environment affected, but *not* reduced, on binding of substrate.

CO or HCN has usually been interpreted as indicating the formation of Cu^+ in the catalytic cycle, since these inhibitors form much stronger complexes with univalent than with divalent copper. However, although many of these enzymes are inhibited by HCN, most are not inhibited by CO (Table 15-3). Only the tyrosinase (phenolase) from potatoes has been shown to be so affected: CO also inhibits the activity of ceruloplasmin and of cytochrome oxidase, but ceruloplasmin is mainly an O_2-transporting pigment, not an enzyme, while cytochrome oxidase contains iron porphyrin as well as Cu as a second prosthetic group, and the iron rather than the Cu is a likely site of CO binding. Another method for approaching the problem is provided by EPR spectroscopy, since one can observe the disappearance of the signal due to paramagnetic Cu^{2+} on exposure to substrate and its reappearance on exposure to O_2. But this demonstration of the existence of the enzyme or the enzyme–substrate complex in the Cu^+ state is a necessary, but not a sufficient, condition for the valence-shuttle mechanism. Particularly in the case of those enzymes that contain but a single metal atom per molecule, it is not impossible that electron transfer occurs within a ternary complex of enzyme, substrate, and O_2, the latter two perhaps both bound to the metal. In this context it is probably significant that the substrates of ascorbate oxidase, laccase, phenolase, galactose oxidase, and uricase are all capable of forming strong complexes with copper ions in aqueous solution.

With the exception of plastocyanin, all cuproprotein dehydrogenases function as true oxidases; that is, they are capable of utilizing only O_2 as the electron acceptor. This characteristic property may be a reflection of a high degree of structural specificity and a high binding constant for this particular acceptor ($K_{diss} = 10^{-4}$ to 10^{-6} M) or else of the fact that the high E_0' of the substrates makes alternative acceptors unlikely. In Table 15-4 are summarized some of the reaction parameters for five typical cuproprotein enzymes. Plants contain enzymes capable of reducing *o*-quinones, *p*-quinones, and dehydroascorbate at the expense of NAD(P)H. Since laccase, phenolase, and ascorbate oxidase are also plant enzymes, this combination provides plants with alternative respiratory chains from substrates to oxygen without the participation of cytochromes or of cytochrome oxidase [Eq. (15-18)].

$$DH_2 \xrightarrow{\text{dehydrogenase}} \text{NAD(P)} \xrightarrow{\text{quinone reductase}} \text{quinone} \xrightarrow{\text{Cu enzyme}} O_2 \qquad (15\text{-}18)$$

Flavoproteins

Flavoproteins are mainly concerned with three important areas of oxidative metabolism: (1) the oxygen-linked dehydrogenation of certain substrates such as amino acids; (2) the cytochrome-linked dehydrogenation of the initial members of the particle-bound respiratory chain (i.e., NADH and succinate in almost all cells; α-glycerophosphate, choline, sarcosine, fatty acyl CoA's in appropriate mitochondria from various sources; D- and L-lactate in aerobically grown yeast, etc.); and (3) the NAD(P)-linked dehydrogenation of certain low-potential substrates (e.g., dihydrolipoate, dihydroorotate, reduced ferredoxin, and other NHI carriers). Enzymes in group 1 usually contain only a flavin prosthetic group, those in group 2 frequently contain additional components (usually metals), and those in group 3 sometimes do (dihydroorotate) and sometimes do not contain a metal ion (lipoamide) dehydrogenase.

Catalysis by Simple Flavoproteins. The two flavin coenzymes FMN and FAD are tightly bound to the various proteins, thus usually acting as prosthetic groups rather than as freely dissociating coenzymes. The binding sites are not known, but likely candidates are (1) ionic interactions between

TABLE 15-4
Reaction parameters for cuproproteins

Substrate	*Product*	*Phenolase*	*Ascorbate oxidase*	*Laccase, ceruloplasmin*	*Uricase*[a]
Ascorbate[b]	Dehydroascorbate[c]	−	+	−	−
Monophenols	*o*-Quinones[d]	+	−	−	−
o-Diphenols	*o*-Quinones	+	−	−	−
p-Hydroquinones	*p*-Quinones	−	−	+	−
p-Phenylenediamines	*p*-Quinones	−	−	+	−
Dopa[e]	Dopaquinone	+	−	+	−
Urate[f]	Allantoin[g] + CO_2	−	−	−	+
(O_2)	H_2O	+	+	+	−
(O_2)	H_2O_2	−	−	−	+

[a] Galactose oxidase is very similar to uricase in many of its properties.

b (structure of ascorbate: O=, –OH, –OH, O, HO–, CH_2OH)

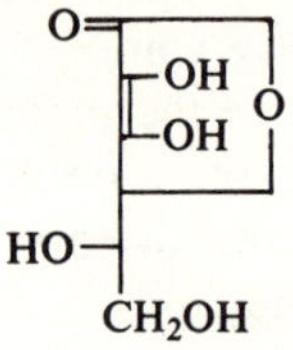

c (structure of dehydroascorbate: O=, =O, =O, O, HO–, CH_2OH)

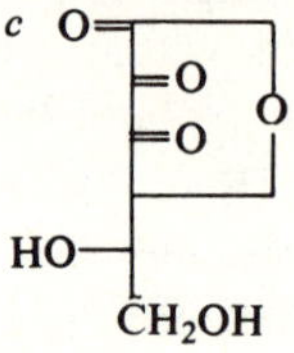

d *o*-Quinone is the first product. This is followed by

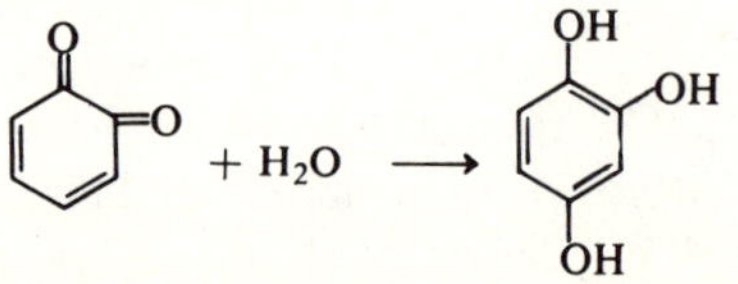

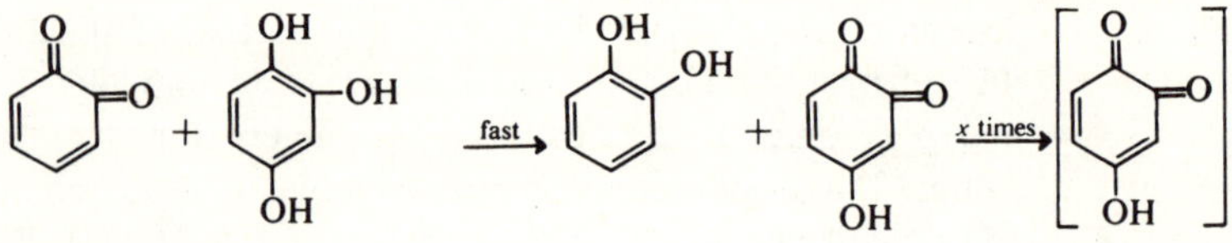

o-quinone + H_2O ⟶ trihydroxybenzene (OH, OH, OH)

o-quinone + trihydroxybenzene $\xrightarrow{\text{fast}}$ *o*-diphenol + hydroxy-*o*-quinone $\xrightarrow{x\ \text{times}}$ [hydroxy-*o*-quinone]$_x$

The final polymeric product, melanin, has an intense brown-black color.

[e] Dihydroxyphenylalanine.

f (structure of urate)

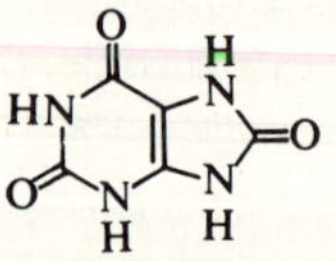

g (structure of allantoin)

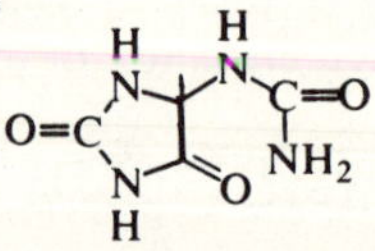

cationic groups on the protein and the phosphate residues of the coenzyme, (2) hydrogen bonding to the nitrogen and oxygen functions of the isoalloxazine ring, and (3) hydrogen and apolar bonding to the ribityl side chain. At neutral *p*H and moderate ionic strength, dissociation constants for the apoenzyme-flavin equilibrium are usually $\leq 10^{-8}$ M, and values as low as 10^{-11} M have been reported. Nevertheless, by raising the ionic strength and lowering the *p*H, the protonated flavins may be detached to produce the apoenzyme. In general, this treatment has been more successful with the simple flavoproteins than with the more complex metalloflavoproteins. Frequently, enzymes from plant or microbial sources have exhibited higher dissociation constants than the corresponding enzymes from animal tissues, and under certain circumstances FAD and FMN might act as a true coenzyme in certain cells. For example, Hastings has shown in the luminescent bacterium *Photobacterium fischeri* that light emission depends on the following sequence of reactions (19).

NAD^+ ← $NADH + H^+$

$FMNH_2 + E\langle S{-}S \rangle \xrightarrow{\text{fast}} E(SH)(SH) + FMN$

fast ↓ $+O_2$

$E(SOOH)(SH)$

fast ↓ RCHO (long-chain aldehyde)

$E(SOOH)(SH)(RCHO)$

Other products + $E\langle S^*{-}S \rangle \xleftarrow{\text{slow}} E(SOOH)(SH)(RCHO)$

$h\nu$

Among the most characteristic features of flavins is their light absorption at 260, ~ 370, and ~ 450 nm. Upon reduction with borohydride or dithionite, or enzymatically in the presence of various substrates, the two long-wavelength peaks are bleached and the extinction coefficient in the 450-nm region (~11.0) is reduced to ≤10 percent of the value of the oxidized form. Flavins also exhibit a strong, yellow-green fluorescence ($\lambda_{max} \simeq 500$ nm). The fluorescence of FAD is only 10 percent of that of the vitamin or FMN, because of internal quenching by the adenine ring. Upon binding to the protein, the optical properties of the flavins are profoundly affected—there is usually a significant shift of the absorption maxima to longer wavelengths and a quenching of the fluorescence. A notable exception is provided by lipoamide dehydrogenase (page 519), where the fluorescence of FAD is raised approximately fourfold (almost to that of FMN) upon interaction with the protein. This suggests that the coenzyme molecule assumes a more unfolded conformation during binding.

The most important property of the flavin coenzymes is their ability to undergo a reduction both to a half-reduced ($1e^-$ equivalent) and a fully reduced ($2e^-$ equivalent) state. Thus, a flavoprotein containing a single flavin coenzyme might participate in the catalytic cycle by shuttling back and forth between the oxidized and the fully reduced, between the oxidized and the semiquinone, or between the semiquinone and the fully reduced states. In the case of flavoproteins that contain two flavins per molecule (or active site), the situation becomes still more complex, since the following possible species have to be reckoned with, and such enzymes may function as $1e^-$, $2e^-$, $3e^-$, or $4e^-$ redox carriers. Form I has been suggested on the basis of spectrophotometry in the case of several different flavoproteins, especially those not capable of interacting with molecular oxygen as acceptor. Form VIII corresponds to an enzyme fully reduced by substrate and has been observed in most cases investigated. Frequently it is formed so slowly, though, as to rule out its participation in catalysis. The various semiquinone forms of the flavoprotein enzymes are frequently observed by spectrophotometry and EPR measurements (Fig. 15-2). The different equilibria that flavins can undergo are indicated schematically in Fig. 15-3. Note especially the enhanced possibility of resonance stabilization in the semiquinone anion (20) $FlavH\cdot^-$ as compared to its protonated form $FlavH_2\cdot$ and the tendency of the semiquinones to dimerize or disproportionate to form molecular complexes of the reduced and oxidized species similar to quinhydrones.

$$\left[E\begin{array}{c}FlavH\\ \\ FlavH\end{array}D\right] \rightarrow \left[E\begin{array}{c}FlavH\cdot\\ \\ Flav\end{array}\cdot HD\right] \rightleftharpoons \left[E\begin{array}{c}FlavH\cdot\\ \\ FlavH\cdot\end{array}D\right] \rightleftharpoons \left[E\begin{array}{c}FlavH_2\\ \\ Flav\end{array}D\right]$$

I II III IV

III $\xrightarrow{DH_2 \;\rightarrow\; D}$ (down to VII)

$$\left[E\begin{array}{c}FlavH_2\\ \\ Flav\end{array}H_2D\right] \rightleftharpoons \left[E\begin{array}{c}FlavH_2\\ \\ FlavH\cdot\end{array}\cdot HD\right] \rightleftharpoons \left[E\begin{array}{c}FlavH\cdot\\ \\ FlavH\cdot\end{array}H_2D\right] \rightarrow \left[E\begin{array}{c}FlavH_2\\ \\ FlavH_2\end{array}D\right]$$

V VI VII VIII

Semiquinone intermediates in flavoprotein catalysis have received a great deal of attention in recent years. They may be stabilized by interaction, with susceptible groups on the enzyme (such as sulfur radicals formed by half reduction of disulfides as postulated for lipoamide dehydrogenase by Massey); on the substrate; on the acceptor (especially if it contains unpaired electrons like molecular oxygen or is capable of univalent reduction to a one-electron intermediate); or with a metal ion (either as acceptor or in a metalloflavoprotein). These same types of interaction may also determine the transition state for the electron-transfer steps, that is, the donation and acceptance of electrons (plus protons) by the flavins.

Since hydrogen transfer from certain substrates, such as NADH or NADPH, to flavoproteins has been shown to be direct and stereospecific and since some of these flavoproteins also have transhydrogenase[2] activity, it is probable that the reaction involves first the direct transfer of a substrate hydrogen plus two electrons (i.e., the equivalent of a hydride ion) in a rapid step; the formation of free radicals, if any, occurs somewhat later, either prior to or simultaneously with electron transfer to the acceptor. When the enzyme contains two molecules of flavin per molecule, then catalysis may also involve the reaction (22):

$$E(FlavH\cdot)_2 \overset{2e^-}{\rightleftharpoons} E(FlavH)_2$$

Thus, we observe the following types of reaction patterns:

1. For glucose oxidase:

$$\text{Glucose} + \text{Enz}\cdot\text{FAD} \rightleftharpoons \text{Enz}\cdot\text{FADH}_2 + \text{gluconate} \qquad (15\text{-}19a)$$
$$O_2 + \text{Enz}\cdot\text{FADH}_2 \rightarrow \text{Enz}\cdot\text{FAD} + H_2O_2$$

2. For various amino acid oxidases (two separate and independent sites):

$$\begin{array}{c}R{-}\underset{\underset{\oplus}{NH_3}}{CH}CO_2^- + fp\end{array} \rightleftharpoons [fpH\cdot\cdot\underset{\underset{\oplus}{NH_3}}{C}(R)CO_2^-] \xrightarrow{+O_2} R{-}\underset{\underset{\oplus}{HN{-}(Lys){-}fp}}{\overset{\|}{C}}{-}CO_2^- \qquad (15\text{-}19b)$$

intermediate I, $\lambda_{max} = 450$ nm
no EPR signal

fp complex $\xrightarrow{\text{EDTA}\ h\nu}$ complex II, inactive ← – – – – – – intermediate I
($\lambda_{max} = \sim 485$ nm, strong EPR signal)

$\xrightarrow{H_2O}$ $H_3N + RCOCO_2^- + fp$

[2] That is, they catalyze the reactions NADH + NADP+ → NAD+ + NADPH and NADH + NAD+ analog → NAD+ → NADH analog.

3. For microsomal NADPH-cytochrome *c* reductase:

$$\begin{aligned} H^+ + NADPH + Enz\left<\substack{FAD\cdot\\ FAD\cdot}\right. &\rightleftharpoons Enz\left<\substack{FADH_2\\ FADH_2}\right. + NAD^+ \\ O_2 + Enz\left<\substack{FADH_2\\ FADH_2}\right. &\longrightarrow Enz\left<\substack{FADH\cdot\\ FADH\cdot}\right. \end{aligned} \tag{15-19c}$$

4. For lipoamide (lipoyl) dehydrogenase (two independent active sites):

$$\begin{array}{c} \left[\substack{-S\text{—}S-\\ -FAD}\right] \xrightleftharpoons[\text{Lip}\left<\substack{S\\ S}\right.]{\text{Lip}\left<\substack{SH\\ SH}\right.} \left[\substack{-SH\ HS-\\ -FAD}\right] \xrightleftharpoons{\text{rearrangement}} \left[\substack{-SH\ \cdot S-\\ -FADH\cdot}\right] \\ \Big\Updownarrow NADH \qquad\qquad\qquad\qquad\qquad \Big\Updownarrow NAD^+ \\ \text{R–N}\left<\text{R}'\right. \ \left[\substack{-S\ \ HS-\\ -FAD}\right] \xrightleftharpoons{\text{rearrangement}} \text{R–N}\left<\text{R}'\right. \ \left[\substack{-S\ \ \cdot S-\\ -FADH\cdot}\right] \end{array} \tag{15-19d}$$

In the last instance the essential steps are the rearrangement, with a change of multiplicity, of the thiol forms of the enzyme to a sulfur-flavin semiquinone diradical, and the reduction of NAD by way of the 4-adduct of the nicotinamide moiety.

Considering this wide variety of possible reaction mechanisms and their complexity it is not surprising that the interpretation of the absorption spectra of flavoproteins with their substrates has not been free of ambiguities even with the use of EPR spectroscopy as an added tool for the identification of free-radical species.

Metalloflavoproteins. A number of isolated, highly purified flavoproteins contain additional prosthetic groups (Table 15-5*B*). Since their discovery in 1954, these metalloflavoproteins have been investigated intensively, especially by the application of sophisticated and rapid spectrophotometric and EPR techniques. In spite of these efforts, however, the precise role of the metals has proved to be elusive. The main problem is the following: Many of these enzymes, even though highly purified and with the various prosthetic groups in definite and constant proportions, originally were part of more complex particulate structures. In order to render the enzymes soluble, these structures had to be disrupted by enzymatic, chemical, or physical means, and in the course of such treatment the enzymes may have suffered alterations in their properties. Such alterations with regard to both conformational and catalytic properties (especially with respect to acceptor specificity) are well documented, particularly in the case of NADH dehydrogenase (23). In fact, a number of NADH-dehydrogenating enzymes have been described, all presumably derived from the identical parent enzyme—the NADH branch of the respiratory chain. Yet several of them contain variable proportions of the nonheme iron (NHI, page 656) of the complex and have been reported to have molecular weights approaching that of the complex. Similar considerations also appear to prevail with regard to succinate dehydrogenase, other

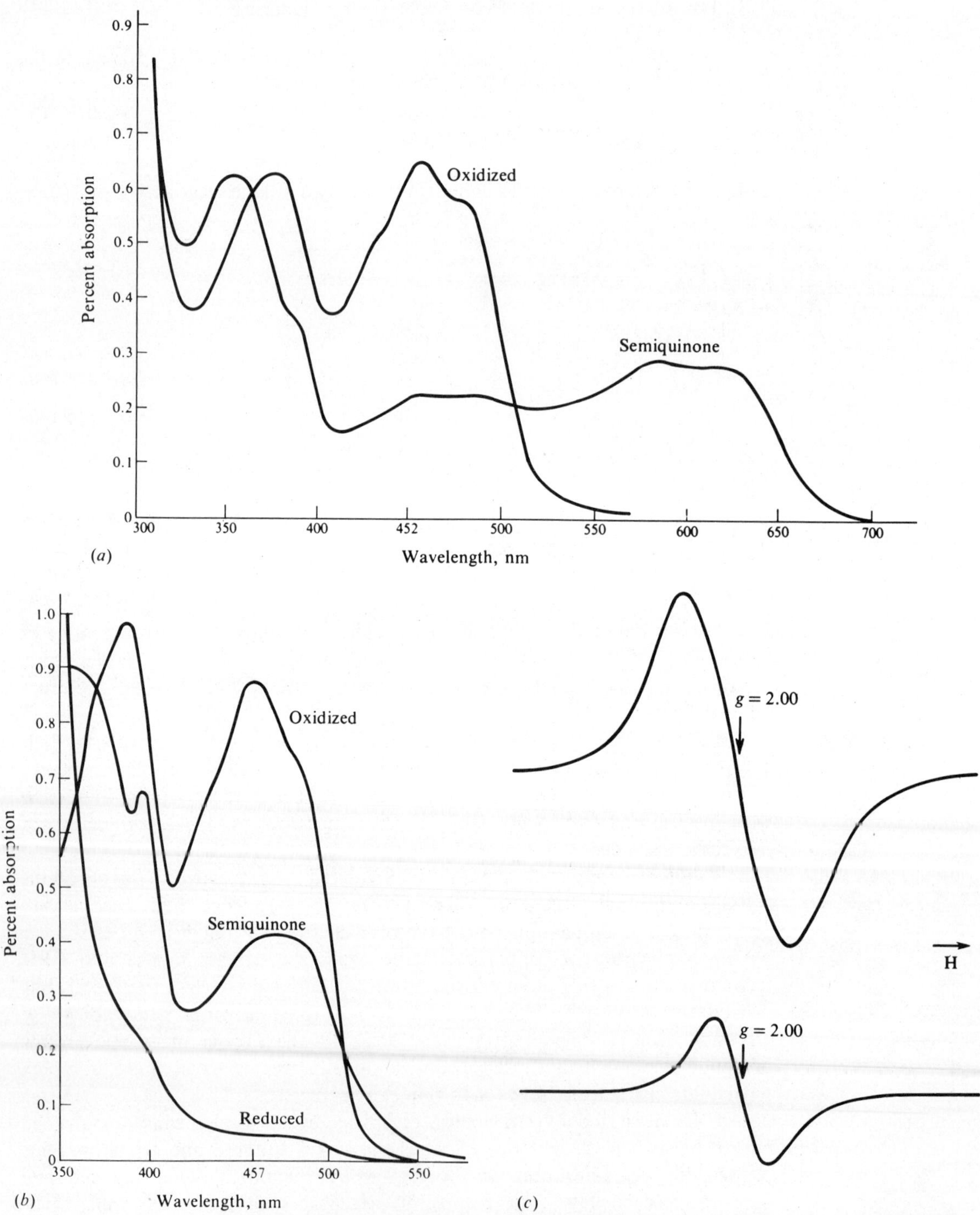

FIGURE 15-2

Absorption and EPR spectra of flavoproteins. Contributed by H. Beinert. (*a*) Absorption spectrum of blue semiquinone of flavoproteins (Type A, according to G. Palmer and V. Massey, Mechanisms of Flavoprotein Catalysis, in T. P. Singer (ed.), *Biological Oxidations,* Wiley, New York, 1968). A solution of *Azotobacter* flavoprotein (Y. I. Shethna, P. W. Wilson, and H. Beinert, *Biochim. Biophys. Acta,* in press) in 0.03 M phosphate of *p*H 7.4, 5.7×10^{-5} M with respect to bound flavin, was reduced anaerobically with dithionite to the semiquinone state. Excess dithionite was oxidized with respiratory-chain-linked dehydrogenases, and, as a matter of fact, with regard to many isolated components of the mitochondrial chain.

Certain metalloflavoprotein enzymes, however, probably exist as such in the cell rather than as a part of a larger, integrated unit. Among these are the dihydroorotate dehydrogenase of *Zymobacterium oroticum,* xanthine dehydrogenases from a variety of sources, and perhaps the aldehyde oxidase of mammalian liver.

Of these the dihydroorotate dehydrogenase is the simplest example (24), and it can be regarded as the prototype of the class of *ferroflavoproteins.* The enzyme catalyzes the reversible interconversion:

$$\text{NADH} + \text{orotate} + \text{H}^+ \rightleftharpoons \text{NAD}^+ + \text{dihydroorotate} \qquad (15\text{-}20)$$

It also catalyzes the "oxidase" reactions:

$$\text{NADH} + \text{O}_2 + \text{H}^+ \longrightarrow \text{NAD}^+ + \text{H}_2\text{O}_2 \qquad (15\text{-}20a)$$

$$\text{Dihydroorotate} + \text{O}_2 \longrightarrow \text{orotate} + \text{H}_2\text{O}_2 \qquad (15\text{-}20b)$$

Recent studies (24,25) have indicated that the isolated enzyme (M = 115,000) contains four polypeptide chains of M = 30,000. The active form is a dimer, consisting of identical and catalytically independent subunits each containing, as prosthetic groups, one molecule of FAD, one molecule of FMN, and two gram-atoms of nonheme Fe (NHI), each coordinated with a residue liberating H_2S on acidification (*acid-labile sulfide*). The flavin prosthetic groups can be removed by treatment with cold methanol; the properties of the NHI protein prepared in this manner are similar to those characteristic of soluble NHI proteins such as ferredoxin (see the next section), particularly with respect to its absorption spectrum, which shows peaks at 331 nm > 422 ≧ 457 nm and a broad band of low intensity at $\lambda \sim 525$ nm.

The intact enzyme forms a relatively stable complex with orotate (K_s = 2.8×10^{-4} M) and the FMN moiety has been implicated in this interaction. The enzyme can be reduced rapidly either by NADH, or by dihydroorotate supplemented with NAD, to what appears to be two spectroscopically distinct intermediates (λ_{max} = 565, 610 to 620 nm and 585, 640 nm, respectively). In each instance appearance of intermediate, of flavin free radical (EPR signal with g = 2.00), and of a characteristic NHI signal (asymmetric, with g = 1.94) occur simultaneously and with a rate commensurate with their participation in catalysis. Reoxidation by the second substrate is slow and rate limiting, as is their reoxidation by O_2, the alternate electron acceptor. Mercurial inhibitors of sulfhydryl groups interrupt all electron flow except the oxidation

oxygen. The spectrum is very close to that of the pure semiquinone. The amount of the oxidized form does not exceed 8 percent of the total flavin present. (*b*) Absorption spectrum of red semiquinone of flavoproteins (Type B, according to G. Palmer and V. Massey, Mechanisms of Flavoprotein Catalysis, in T. P. Singer (ed.), *Biological Oxidations,* Wiley, New York, 1968). A solution of oxynitrilase [W. Becker, U. Tenthin, E. Eschenhof, and E. Pfeil, *Biochem. Z.,* **337:** 156 (1963)] in 0.02 M phosphate of pH 7.4, 7.66×10^{-5} M with respect to bound flavin, was reduced anaerobically with dithionite and reoxidized to the semiquinone state by oxygen. In this case, a considerable amount of the reduced and oxidized forms are in equilibrium with the semiquinone; however, the characteristic features of the absorption spectrum of the red semiquinone are apparent. (*c*) EPR spectra of flavoproteins taken at −159°. (*Top curve*) blue semiquinone from *Azotobacter* flavoprotein (cf. Fig. 15-2*a*); (*lower curve*) red semiquinone of oxynitrilase (cf. Fig. 15-2*b*).

$$E(FlavH\cdot)_2 \overset{2e^-}{\rightleftharpoons} E(FlavH)_2$$

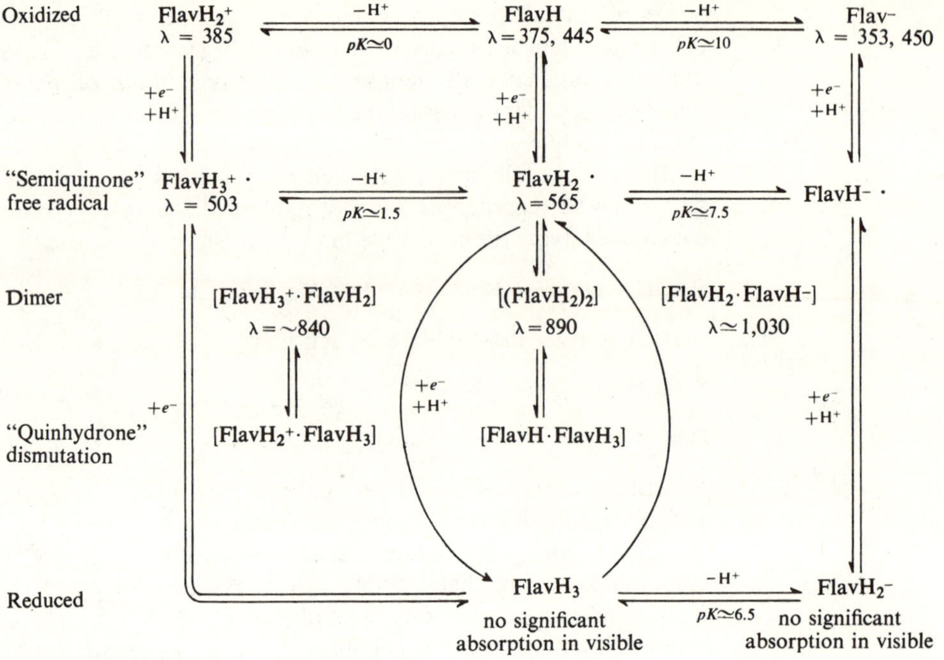

Monomeric species present in neutral solution

FlavH

$FlavH_2\cdot$ $\longleftrightarrow$ $FlavH^-\cdot$ $\longleftrightarrow$

Relative unpaired spin densities: $N^{10} > N^9 \simeq C^7 \simeq C^5$

$FlavH_3$ $\rightleftharpoons$ $FlavH_2$ $\longleftrightarrow$

FIGURE 15-3

Equilibria during oxidoreduction of flavins. An alternative system of numbering part of the iso-alloxazine ring is

In succinic dehydrogenase (cf. Table 15-5*B*) the FAD prosthetic group is linked covalently to the polypeptide chain (through a —OCH_2— bridge) attached to C_7 (C_8 in the alternative system); see P. Hemmerich, A. Ehrenberg, W. H. Walker, L. E. G. Eriksson, J. Salach, P. Bader, and T. P. Singer, *Federation European Biochem. Soc. Letters,* **3:** 37 (1969); W. H. Walker, J. Salach, M. Gutman, T. P. Singer, J. S. Hyde, and A. Ehrenberg, *ibid.,* **5:** 237 (1969).

of NADH by O_2, presumably with the free radical of FAD as the first intermediate. Sulfite reductase from various microorganisms, although much larger in size, appears somewhat analogous in composition and structure. This enzyme catalyzes the conversion of sulfite to elemental sulfur with NADPH as the electron donor.

A higher order of complexity is exhibited by the various xanthine dehydrogenases (called "oxidases" when they utilize O_2 as their preferred electron acceptors), which contain molybdenum in addition to flavin and NHI as a prosthetic group. These omnivorous enzymes can utilize NADH (not NADPH), aldehydes, hypoxanthine, xanthine, pteridines, and quinine; among the acceptors are NAD^+ (particularly for hypoxanthine), O_2, dyes, quinones, ferricyanide, nitrate, and cytochrome *c*. The reduction of the last-named carrier, in the case of those enzymes capable of interacting with O_2, is a result of the prior formation of the superoxide anion $\cdot O_2^-$ in a reaction of the reduced enzyme with oxygen (26). The enzyme from the anaerobic bacterium *Micrococcus lactilyticus,* on the other hand, can use cytochrome *c* directly, but its preferred electron acceptor appears to be ferredoxin, and it is incapable of reducing NAD (27).

Some evidence is available that the Mo atom is closest to the purine binding site and that it can function as both a hydroxyl donor and as an electron acceptor. The individual steps in catalysis may be (28) those shown in Eq. (15-21):

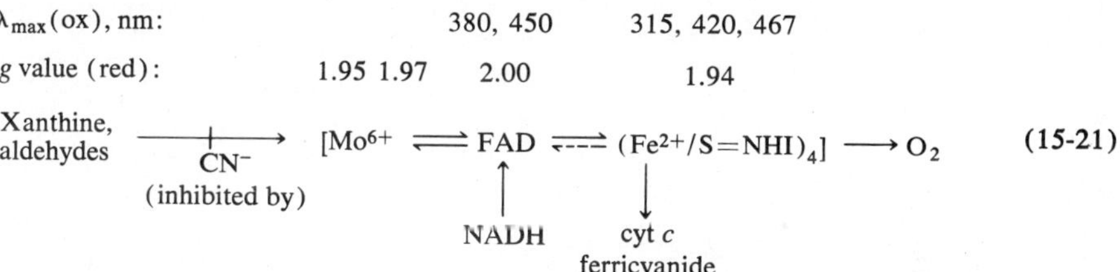

(15-21)

With the holoenzyme the rate-limiting step is the reduction of the enzyme complex with $4e^-$ equivalents per flavin; that is, two substrate molecules can be oxidized in quick succession. These $4e^-$ are believed to be distributed in the steady state as follows: one on the Mo, two on the FAD, and one on the NHI.

The flavin can be removed from the enzyme, and under these conditions it can be shown to still be capable of bringing about the following reaction:

$$\text{Xanthine, aldehydes} \xrightarrow[\text{CN}^-]{\quad+\quad} [\text{Mo}^{6+} \rightleftharpoons (\text{NHI})_4] \longrightarrow \begin{matrix}\text{ferricyanide}\\ \text{cyt } c\end{matrix} \qquad (15\text{-}21a)$$

The most complex soluble metalloprotein is the hepatic aldehyde oxidase. This miniature electron-transport chain also contains coenzyme Q (ubiquinone) in amounts stoichiometric to those of FAD and Mo, as well as NHI.

Cytochrome Reductases and Diaphorases. A quite tangled problem has been created by the use of cytochrome *c* as an electron acceptor with soluble enzymes (23,29). Within intact, particulate electron-transport chains, several intermediary carriers intervene between the flavoproteins and cytochrome *c*.[3]

[3] This statement appears valid at least as far as the two main chains of mitochondrial electron transport (from NADH and succinate to O_2) are concerned. The situation with regard to other respiratory-chain-linked dehydrogenases, including the D- and L-lactate dehydrogenases from yeast, remains equivocal.

TABLE 15-5A
Properties of flavoproteins

Enzyme	Source	Molecular weight	Prosthetic groups		Physiological acceptor
			Flavin	Other	
D-Amino acid oxidase	Liver, kidney	~100,000	2(?)FAD		O_2
L-Amino acid oxidase	Kidney	138,000	2FMN		O_2
	Snake venoms	~140,000	2FAD		O_2
Imidazole monooxygenase	Pseudomonad	90,000	FAD		O_2 (NADH as donor)
Lysine monooxygenase	Pseudomonad	190,000	2FAD		O_2 (substrate as donor)
Glucose oxidase	Molds	154,000	2FAD		O_2
"Old Yellow enzyme"	Yeast	105,000	2FMN		?
Monoamine oxidase	Liver, kidney, brain, probably mitochondrial	100,000	?	?	O_2
Pyruvate oxidase	*Lactobacillus delbrueckii*	?	FAD	TPP, Mg^{2+}[e]	O_2
Glycolate oxidase	Plants	~100,000	FMN		O_2
Luciferase	*Photobacterium fischeri*	76,000	FMN		O_2
NADH peroxidase	*Streptococcus faecalis*	120,000	FAD		H_2O_2
Acyl CoA dehydrogenase (three separate enzymes known)	Liver, heart mitochondria	200,000(?)	2*n*FAD		ETF
Sarcosine dehydrogenase (oxidase)	Liver, kidney mitochondria	?	FAD	?	ETF
D-Oxynitrilase[a]	Plants	750,000	FAD		Internal dismutation
Electron-transferring flavoprotein (ETF)	Liver, heart mitochondria	70,000	FAD		Respiratory chain
NADPH-ferredoxin reductase	Plants, adrenal microsomes, bacteria	35,000	FAD (?)		Adrenodoxin, ferredoxin, rubredoxin
Lipoamide (Lipoyl) dehydrogenase	All respiratory particles	100,000	2FAD	—S—S—	NAD^+
NADH-cytochrome b_5 reductase	Liver microsomes and mitochondria	40,000	FAD	$2Mg^{2+}$[e]	Cytochrome b_5
NADPH-cytochrome *c* reductase	Microsomes	68,000	FAD		Cytochrome *c*
	Yeast	78,000	FMN		Cytochrome *c*
NADPH-quinone reductase	Cytoplasm of animal cells, plants	56,000	FAD		2-Methylnaphthoquinone (menadione)
		~50,000	FMN		
NADPH-glutathione reductase	*E. coli*, plants	?	FAD		Oxidized glutathione
NADPH-thioredoxin reductase	*E. coli*, other bacteria	68,000	FAD		Thioredoxin
Flavodoxin	Anaerobic bacteria	15,000	FMN		Can replace ferredoxin in all reactions

TABLE 15-5B
Properties of purified metalloflavoproteins

Enzyme	Source	Molecular weight	Prosthetic groups: Flavin	Prosthetic groups: Other	Physiological acceptor
Xanthine oxidase (dehydrogenase)	Milk	300,000	2FAD	8Fe,[b] 2Mo	O_2 (cyt *c*)
	Mammalian liver		2FAD	8Fe,[b] 2Mo	O_2 (cyt *c*), NAD+
	Avian liver or kidney	300,000	2FAD	8Fe,[b] 2Mo	NAD+
Aldehyde dehydrogenase (oxidase)	Mammalian liver	230,000	2FAD	8Fe,[b] 2Mo	O_2, respiratory chain
Dihydroorotic dehydrogenase	*Zymobacterium oroticum*	115,000	2FAD + 2FMN	4Fe[b]	O_2, NAD+ for dihydroorotate; O_2, orotate for NADH
Sulfite reductase	Yeast, *E. coli*	750,000	4FAD + 4FMN	12Fe[b]	Sulfite
NADH dehydrogenase	Heart mitochondria	~200,000 (particle)	1FMN	8Fe[b]	Respiratory chain (not cytochrome *c*!)
		~500,000	1FMN	16–18Fe[b]	(Ferricyanide[c])
		~250,000	1FMN	2–3Fe[b]	(Ferricyanide[c]), CoQ[c]
		70,000	1FMN	4Fe[b]	(Cytochrome *c*[c]), CoQ
Succinate dehydrogenase	Beef heart mitochondria	~300,000	1FAD[d]	8Fe[b]	Respiratory chain
	Beef heart mitochondria, yeast	250,000	1FAD[d]	4Fe[b]	(PMS, ferricyanide)[e]
	M. lactilyticus		*n*FAD	40*n*Fe	
Choline dehydrogenase	Mammalian liver mitochondria	800,000	1FAD	4Fe[b]	ETF (PMS)[e]
α-Glycerophosphate dehydrogenase	Pig brain mitochondria	2×10^6(!)	1FAD	1Fe[b]	ETF (PMS, ferricyanide)[e]
NADP-nitrate reductase	*Neurospora crassa*		?FAD	?Mo	NO_3^-
L-Lactate dehydrogenase	Aerobic yeast respiratory particles	200,000	2FMN	1 heme	Respiratory chain through cyt *c*
D-Lactate dehydrogenase	Aerobic yeast respiratory particles	?	?FMN	?Zn^{2+}	Same as above

[a] Converts aldehydes to α-hydroxynitrites.

[b] The nonheme iron in these enzymes appears to be associated with a stoichiometric quantity of acid-labile sulfur such as is found in ferredoxin (Table 15-7). Many of these enzymes also exhibit a characteristic EPR signal ($g = 1.94$). In general, this EPR signal is lost on solubilization of the mitochondrial enzymes.

[c] Once solubilized, these enzymes must be assayed with artificial acceptors such as ferricyanide or PMS (Table 15-6).

[d] Linked covalently to a histidine-N^1 of the polypeptide chain through a methylene at C^8 (cf. Fig. 15-3).

[e] Not for electron transport.

How does one account then for the numerous soluble *flavoprotein cytochrome c reductases* that have been isolated (cf. Table 15-5)? Several possibilities exist:

1. The isolated enzyme may be identical or closely related to the enzyme *in situ* and possess a built-in acceptor site capable of reacting with cytochrome *c*; this appears to be the case with the three nonmitochondrial ferroflavoproteins (dihydroorotate, xanthine, and aldehyde dehydrogenase), with yeast L-lactate dehydrogenase, a hemoflavoprotein, which contains cytochrome b_2, and with the microsomal NADH-cytochrome *c* reduction system, which requires the participation of cytochrome b_5 in addition to the dehydrogenase.

2. The enzyme may be modified and cytochrome *c* may resemble the "natural" acceptor so closely as to be easily accommodated—this may explain the case of the microsomal NADPH-cytochrome *c* reductases and the yeast D-lactate dehydrogenases.

3. The enzyme has become modified so that only in its *soluble* form can it accommodate cytochrome *c*; this would appear to be the explanation for the majority of the remaining cases.

The term *diaphorase* has been applied indiscriminately to any enzyme capable of catalyzing the oxidation of either NADH or NADPH by any of the artificial electron acceptors, such as dyes, ferricyanide, and quinones. Since many flavoproteins are capable of being reduced by NAD(P)H, and practically all reduced flavoproteins are capable of reducing these acceptors in turn, a number of such "diaphorases" have been described. The classical case was that of a flavoprotein from pig heart, originally isolated and called diaphorase by Straub (30), which only many years later was found by Massey and by Sanadi to be identical with lipoamide dehydrogenase (31,32). Thus, "diaphorases" have no physiological meaning and the term is falling into disuse.

Use of Artificial Electron Acceptors. In work with flavoproteins it is frequently convenient to substitute an artificial, nonphysiological, electron acceptor for the natural one. The latter may be unavailable, unknown, relatively ineffective, or hard to measure. Some of the more commonly used acceptors are described in Table 15-6.

Nonheme Iron Proteins (NHI Proteins)

As we have just seen, NHI entities exist in cells in complex forms associated with both soluble and insoluble respiratory-chain-linked flavoproteins. In addition, separate, soluble NHI proteins also play an important role in electron transport in a wide variety of cells and organisms. The participation of two members of this class, plant ferredoxin and adrenodoxin, in the important physiological process of photosynthesis and monooxygenase (hydroxylase) reactions, have been mentioned in Chapter 11 and an earlier section of this chapter, respectively. Some of the characteristics of the carriers in this class are summarized in Table 15-7.

Their most characteristic feature, except for rubredoxin, is the presence of sulfur highly labile to removal by acid—as well as of iron—as an integral constituent of the catalytically active site. Another useful diagnostic feature is, as has already been mentioned, an unusual and asymmetric EPR signal at $g = 1.94$, generated upon their reduction. By means of isotope-substitution experiments, unambiguous evidence could be obtained for the participation of sulfur (presumably as a ligand to the metal) as well as of Fe itself in this paramagnetic center (33,34). Coincident with the EPR changes the intensity of the peaks in the absorption spectrum at 325 and particularly those at 415 and 455 nm become greatly reduced. Although all the ferredoxins contain more than one iron atom, frequently (as in the case of putidaredoxin) the carrier appears to be capable of accepting only a single electron. This occurs in spite of the participation and identical properties of the two iron atoms in their oxidized and reduced state, demonstrated by Mössbauer spectroscopy (35); in the oxidized state the net spin of both iron atoms is zero, in the reduced state the electron is shared between them, and it can be shown that two different forms of sulfur also participate in this kind of paramagnetic center. Green-plant ferredoxins similarly are one-electron carriers (36) and share a single electron in their paramagnetic center also, but those of *Clostridia* and of *Chromatium* may function as two-electron acceptors (37). Although the complete amino acid sequence of a number of these ferredoxins is now known (38), the exact mode of linkage of the iron and of the additional sulfur atoms to the amino acids in this primary structure have not yet been established. However, cysteine thiol groups are almost certainly involved in this process[4] and some form of sulfide bridge between adjacent Fe atoms seems indicated (39,40)[5]:

$$\left[\begin{matrix} -\mathrm{S} & & \mathrm{S} & & \\ & \mathrm{Fe} & & \mathrm{Fe} & \\ \mathrm{Cys{-}S} & & \mathrm{S{-}Cys} & & \end{matrix}\right]_n \begin{matrix} \mathrm{Y} \\ \\ \mathrm{X} \end{matrix}$$

Ferredoxin from plants consists of 96 to 100 amino acid residues (41,42), while those from bacteria are approximately half that size (~54 residues), with considerable sequence homology in some parts (38).

Many bacteria (e.g., *Clostridium pasteurianum, Micrococcus aerogenes*) contain in addition another iron protein, this one not containing acid-labile sulfur, called *rubredoxin,* which can substitute for ferredoxin in many of its reactions. These rubredoxins, 52 to 53 residues in length, also show considerable sequence homologies with the ferredoxins, and both classes give evidence of internal symmetry within the same chain. Thus, all the proteins in this group may have evolved from a single archetypal gene specifying some 25 amino acids (43).

[4] An artificial model of such NHI proteins exhibiting many of their spectroscopic properties and containing 5 to 8 g-atoms of Fe and 4 to 7 mol of sulfur per mole of protein can be prepared from bovine serum albumin, ferrous ammonium sulfate, and sodium sulfide in the presence of mercaptoethanol. These resemble components required to convert apoferredoxins to their active form.

[5] The bridged structure (oxidized form) may be considered to be made up of the resonating structures

$$\underset{(d^5,\ s=\frac{5}{2})}{>\mathrm{Fe^{3+}}\overset{\mathrm{S}}{\underset{\mathrm{S}}{\lessgtr}}\mathrm{Fe^{3+}}<} \longleftrightarrow \underset{(d^6,\ s=2)}{>\mathrm{Fe^{2+}}\ \ \begin{matrix}\mathrm{S}\\ |\\ \mathrm{S}\end{matrix}\ \ \mathrm{Fe^{2+}}<}$$

and the reduced form of these structures

$$>\mathrm{Fe^{3+}}\overset{\mathrm{S}}{\underset{\mathrm{S}}{\lessgtr}}\mathrm{Fe^{2+}}< \longleftrightarrow >\mathrm{Fe^{2+}}\ \ \begin{matrix}\mathrm{S}\\ |\\ \mathrm{S}\end{matrix}\ \ \mathrm{Fe^{+}}<$$

$$\updownarrow$$

$$>\mathrm{Fe^{2+}}\overset{\mathrm{S}}{\underset{\mathrm{S}}{\lessgtr}}\mathrm{Fe^{3+}}< \longleftrightarrow >\mathrm{Fe^{+}}\ \ \begin{matrix}\mathrm{S}\\ |\\ \mathrm{S}\end{matrix}\ \ \mathrm{Fe^{2+}}<$$

TABLE 15-6
Dyes and electron acceptors frequently employed in work with dehydrogenases

Name	Formula: Oxidized	Formula: Reduced	Auto-oxidizable[a]	E_0' (pH 7.0)	λ_{max}	ϵ_{mM}[b]	Use
$2e^-$ dyes							
Methylene blue (MeB)	$(CH_3)_2N^+$ … $N(CH_3)_2$ (blue)	$(CH_3)_2N$ … $N(CH_3)_2$ (colorless)	+	0.011	668		O_2 uptake, Thunberg technique (determination of the time required for bleaching under anaerobic conditions)
Brilliant cresyl blue (BCB)	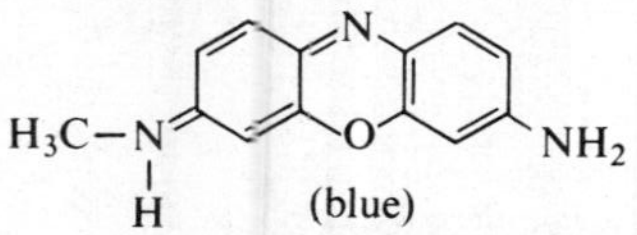 H_3C-NH … NH_2 (blue)	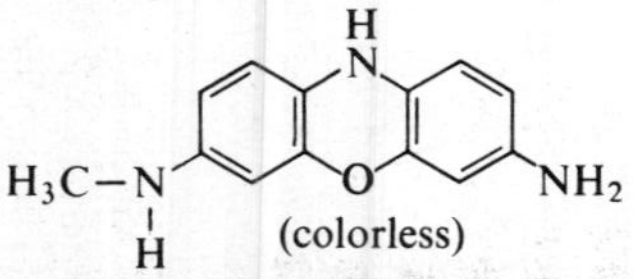 H_3C-NH … NH_2 (colorless)	+	0.045	632		O_2 uptake, Thunberg technique
2,6-Dichlorophenol-indophenol (-benzenoneindo-phenol)	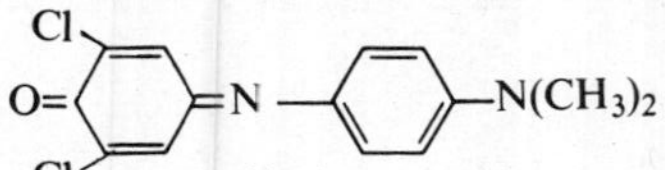 Cl, O=, Cl … =N— … $N(CH_3)_2$ (blue)	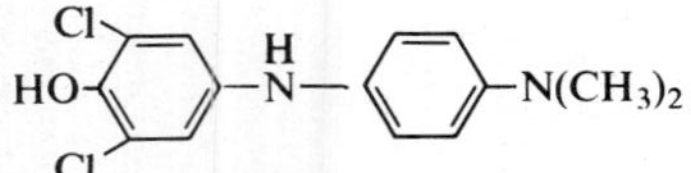 Cl, HO, Cl … —NH— … $N(CH_3)_2$ (colorless)	−	0.22	600	20	Spectrophotometric or colorimetric dye reduction

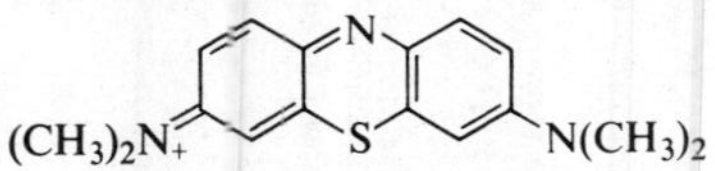

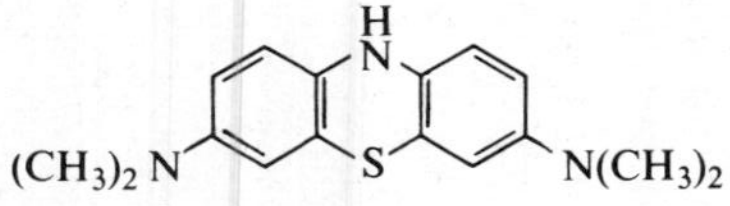

Safranine[h]	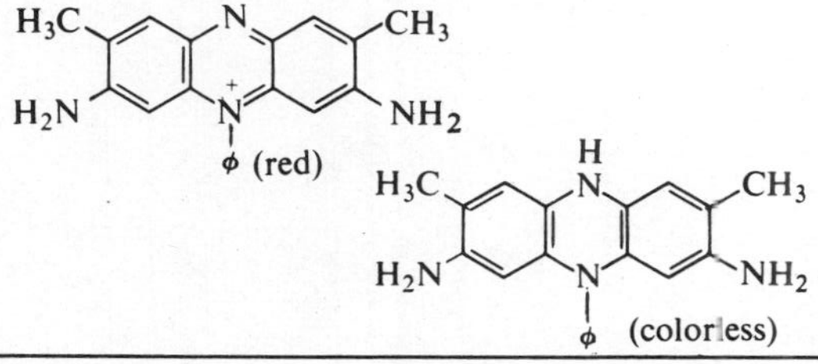 (red); (colorless)	+	**−0.25**	**530**		Reduced dye as hydrogen donor for reverse reaction, i.e., $2[H] + D \rightarrow DH_2$
Triphenyltetrazolium	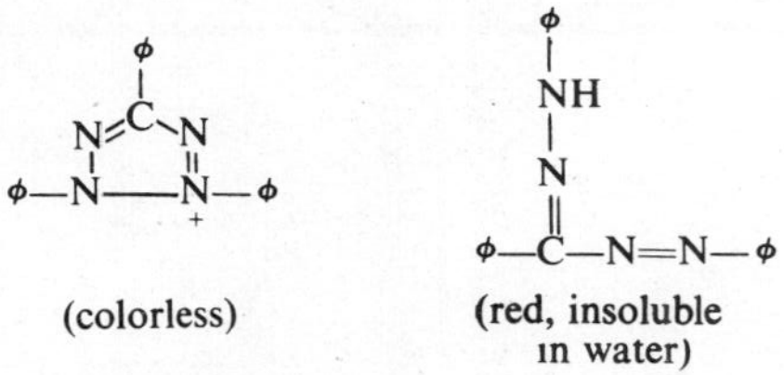 (colorless); ϕ—C(=N—NH—ϕ)—N=N—ϕ (red, insoluble in water)	−(?)		485[c]	12	The reduced product (formazan) is insoluble in water but extractable into organic solvents (*n*-butanol), which makes it very suitable for colorimetry, and continuously removes reaction product
Pyocyanine	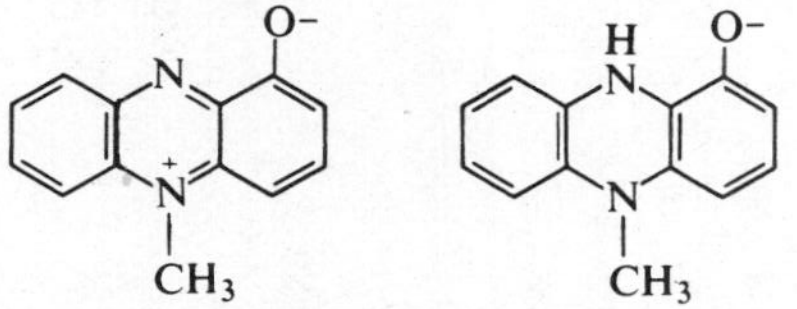 [f]	+	−0.030	**370** **433**		Useful for O_2 uptake measurement
$1e^-$ *dyes* Benzyl viologen	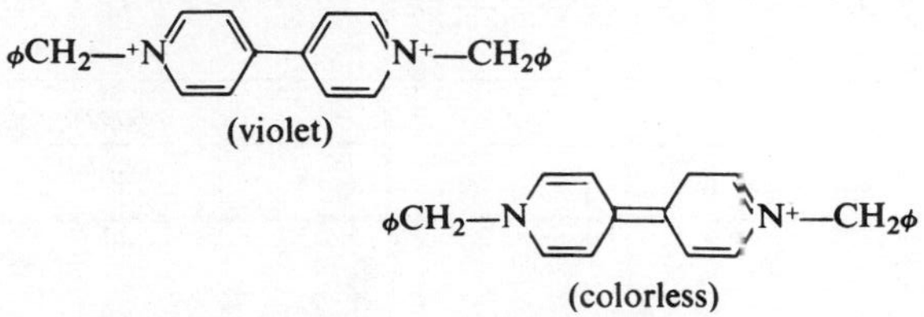 ϕCH_2—$^+$N…N$^+$—$CH_2\phi$ (violet); ϕCH_2—N…N$^+$—$CH_2\phi$ (colorless)	+	−0.311[d]			See safranine; useful as electron donor and acceptor for low E_0' systems, e.g., xanthine dehydrogenase, hydrogenase

TABLE 15-6 (Continued)

Name	Formula: Oxidized	Formula: Reduced	Auto-oxidizable[a]	E_0' (pH 7.0)	λ_{max}	ε_{mM}[b]	Use
N-Methylphenazinium methosulfate (phenazine methosulfate, PMS)	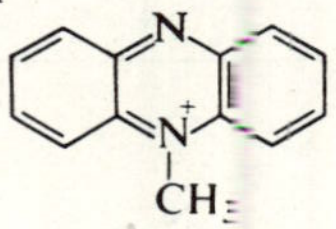	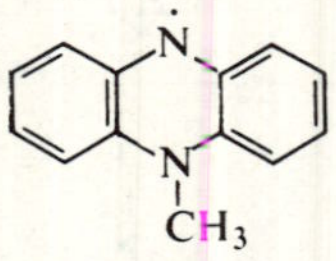 (colored) [g]	+	+0.080	387.5 (388,450) semiquinone		O_2 uptake, or reoxidation of reduced dye can be linked to cytochrome *c* reduction; interacts directly with pyridinoproteins without intervention of flavoproteins
Cytochrome *c*	cyt *c* (Fe^{3+})	cyt *c* (Fe^{2+})	–	0.262	550	19.7[a]	For spectrophotometric or colorimetric studies
Ferricyanide	$[Fe(CN)_6]^{3-}$	$[Fe(CN)_6]^{4-}$	–	0.36	410	~1	For spectrophotometric or colorimetric studies; also in manometric measurements in a HCO_3^-/CO_2 system. H^+ produced in reaction gives rise to CO_2 formation

TABLE 15-7
Properties of isolated nonheme iron proteins[a]

Name	*Source*	*M*	*Iron per molecule*	*Labile S per molecule*	λ_{max} *(ox)*	E_0',*mV*
Ferredoxin						
Type 1	Green plants, blue-green algae	11,500	2	2	330, 420, 465 (sh)[b]	—432
Type 2	*Chromatium*	9,500	4–6	4–6	310, 384	—490
Type 3	*Clostridia,*	~5,800	8[c]	8[c]	300, 390	—420
	other species		7–8[c]	7–8[c]		
Rubredoxin	*Clostridia,* other species	5,900	1	0	380, 490, 565	— 57
Adrenodoxin	Adrenal cortex mitochondria,	22,000	2	2	320 (sh),[b] 414, 450	150
	liver microsomes					230
Putida(re)doxin	*Ps. putida*	11,000	2	2	325, 415, 455	—240
"Desaturating protein"	*Euglena*	?			310, 425, 470	
NHI protein of complex III (QH_2: cytochrome *c* reductase), succinylated	Beef heart mitochondria	26,000	2	2	464, 560	220 (+30)[d]
High-potential NHI protein (HPIP)	*Chromatium*	9,500	4	3–4	374	350
Paramagnetic proteins[d]	*Azotobacter*	20,000	2	2	310, ~425, 465	
	C. pasteurianum	24,000	1–2	1	333, 425, 463	

[a] Midpoint potential (*p*H 7.0) in intact mitochondria; another NHI protein of this preparation shows a midpoint potential of —305 μV.
[b] Sh, shoulder.
[c] Values between 6 and 8 have been reported.
[d] May be involved in N_2 fixation. Three such proteins are known in *Azotobacter* (type I—*M* = 21,000; Fe and S = 2; type II—*M* = 24,000; Fe and S = 2; and type III—*M* = 18,000; Fe and S = 7. The *nitrogenase* of *C. pasteuranium* consists of two components: *Azoferredoxin* (*M* = 40,000; Fe and S = 2) and *Molybdoferredoxin* (*M* = 160,000; Fe = 16, S = 12; Mo = 1–2).

The iron-binding constellation in the rubredoxin from *M. aerogenes* is believed to involve four cysteine residues of the polypeptide chain (probably nos. 6, 9, 38, and 41) arranged around the metal in roughly tetrahedral symmetry (44).

When cells of certain microorganisms such as *Clostridium pasteurianum* or *Peptostreptococcus elsdenii* are grown on iron-deficient media, yet another carrier makes its appearance, capable of replacing ferredoxin (or alternatively rubredoxin) in its various metabolic roles, such as hydrogen evolution, nitrogen fixation, and pyruvate oxidation through the clastic reaction, *in vivo* or *in vitro,* and of substituting for plant ferredoxin *in vitro.* This is *flavodoxin,* a low-molecular-weight flavoprotein ($M = 15{,}000$, 149 residues) containing neither labile sulfur nor iron, but with a FMN prosthetic group (Table 15-5*A*) probably bound through or complexed with its single cysteine residue.

Quinone Coenzymes

Another group of respiratory carriers capable of interacting with the flavoproteins of the respiratory chain (see page 683) comprises certain benzoquinones called *coenzymes Q* (CoQ) or *ubiquinones* (UQ) of the structure

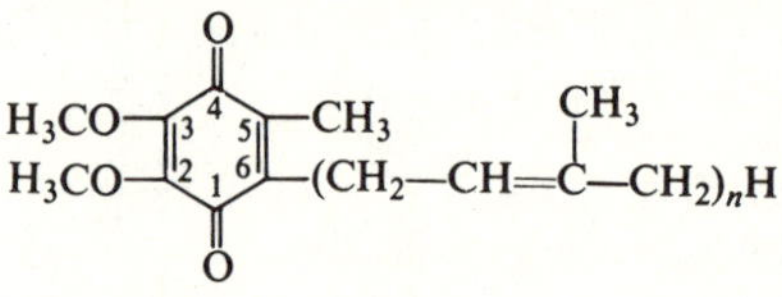

$$H_3CO\text{-}(3),\ H_3CO\text{-}(2),\ O{=}(4),\ O{=}(1),\ CH_3\text{-}(5),\ (6)\text{-}(CH_2\text{—}CH{=}\overset{CH_3}{\overset{|}{C}}\text{—}CH_2)_nH$$

Ubiquinone

n in the formula varies from 6 in certain microorganisms—in which case the compound is referred to as CoQ_6 or UQ_{30}—to 10 in the mitochondria of most mammals—when its designation is CoQ_{10} or UQ_{50}.

The closely related *plastoquinones,* which fulfill analogous carrier roles in photosynthetic electron transport (see page 570), differ from ubiquinones in the alkyl substituents of the benzene ring: two $—CH_3$ groups instead of the two $—OCH_3$; H instead of $—CH_3$. Plastoquinones B and C carry one hydroxyl group in the side chain.

Nicotinamide Coenzyme-Dependent Dehydrogenases

The structure and function of NAD and NADP, the two nicotinamide (pyridine) nucleotide coenzymes, have been discussed in Chapter 8. These coenzymes are generally reversibly dissociable from their apoenzymes, with dissociation constants of the order of 10^{-4} to 10^{-7} M. In general, their reduced form is bound more tightly to the protein than is the oxidized one (usually by a factor of $\geqq 10^2$). Except for glyceraldehyde-3-*P* dehydrogenase and malic enzyme, the enzymes are isolated as the apoproteins. Zn^{2+} ions are part of the coenzyme-binding sites of several of the highly purified dehydrogenases. Thiol groups on the enzyme have also been implicated in substrate interaction in many instances.

The reactions catalyzed are readily followed by taking advantage of the optical properties of the coenzymes: light absorption at 340 nm ($\varepsilon = 6.22 \times 10^3 \times \text{M}^{-1} \times \text{cm}^{-1}$) and strong fluorescence (λ_{max} at 440 nm) by reduced forms. The position and extinction coefficients of the absorption or fluorescence maxima frequently change through binding to the protein. Catalysis by the enzymes involves the reversible and rapid formation of four binary (i.e., enzyme–substrate and enzyme–coenzyme) and two ternary (enzyme–coenzyme–substrate) complexes. Frequently, the magnitude of the various rate constants involved is such that only the binary complexes with the coenzyme and one or both of the ternary complexes are kinetically significant. That is, there exists a nearly compulsory order of binding and an ordered or constrained path for the reaction with the coenzymes as the "leading" substrate (page 290). The oxidation–reduction step takes place within the ternary complex and probably involves the transfer of a hydride ion (or one proton plus two electrons) within the complex.

$$\mathrm{R}D\mathrm{H}^* + \mathrm{NAD(P)}^+ + \mathrm{H_2O} \rightleftharpoons \mathrm{R}D\mathrm{OH} + \mathrm{NAD(P)H}^* + \mathrm{H}^+ \qquad (15\text{-}22a)$$

$$\begin{array}{l} D\mathrm{H} + \mathrm{NAD(P)}^+ \rightleftharpoons D + \mathrm{NAD(P)H}^* + \mathrm{H}^+ \\ | \\ \mathrm{H}^* \end{array} \qquad (15\text{-}22b)$$

In both the typical reactions shown the transfer of hydrogen is reversible, stereospecific with respect to the substrate as well as the coenzyme, and direct; that is, it does not involve the solvent or any potential redox active group on the enzyme capable of rapid exchange with the solvent. The stereospecificity with regard to the coenzyme, discussed in Chapter 8, finds its explanation in the fact that the two protons at the C-4 position of the nicotinamide ring are not identical, since they are present in magnetically nonequivalent environments. Similarly, the enzymes are capable of discriminating between the two nonequivalent hydrogens in

$$\begin{array}{c} \mathrm{B} \\ | \\ \mathrm{A{-}C{-}H^1} \\ | \\ \mathrm{H^2} \end{array}$$

(see page 374).

If in Eq. (15-22a) the oxidized form of the substrate (RDOH) contains one more oxygen atom than does the reduced substrate (RDH), both the oxygen and the proton liberated must be provided by the solvent, and oxidation of the substrate is favored by high pH.

Several hundred dehydrogenases, differing in substrate specificity and in the molecular properties of the protein, have been isolated to date. Among this bewildering array, however, are a relatively small number of reaction types:

$$\begin{array}{c} \mathrm{R'} \\ | \\ \mathrm{R''{-}C{-}OH} \\ | \\ \mathrm{H} \end{array} + \mathrm{NAD(P)^+} \rightleftharpoons \begin{array}{c} \mathrm{R'} \\ | \\ \mathrm{R''{-}C{=}O} \end{array} + \mathrm{NAD(P)H} + \mathrm{H^+} \qquad (15\text{-}23a)$$

R′ =	H	H	CR_3	H	CR_2OH	H	CR_3	H	CR_3	H	CO_2H	R
R″ =	H	CR_3	CR_3	CR_2OH	CR_2OH	O═CR	O═CR	CO_2H	CO_2H	CH_2SH	CR_2CO_2H	NHR

R = alkyl or H

$$\begin{array}{c} \mathrm{OH} \\ | \\ \mathrm{H{-}C{-}O} \\ |\quad | \end{array} + \mathrm{NAD(P)^+} \rightleftharpoons \begin{array}{c} \mathrm{O} \\ \| \\ \mathrm{C{-}O} \\ |\quad | \end{array} + \mathrm{NAD(P)H} + \mathrm{H^+} \qquad (15\text{-}23b)$$

$$\begin{array}{c} \mathrm{H} \\ | \\ \mathrm{R'{-}C{-}OH} \\ | \\ \mathrm{S{-}R''} \end{array} + \mathrm{NAD(P)^+} \rightleftharpoons \begin{array}{c} \mathrm{O} \\ \| \\ \mathrm{R'{-}C{-}S{-}R''} \end{array} + \mathrm{NAD(P)H} + \mathrm{H^+} \qquad (15\text{-}23c)$$

R′ = H, HC═O, RC═O

$$\begin{array}{c} \mathrm{H} \\ | \\ \mathrm{R{-}C{=}O} \end{array} + \mathrm{NAD(P)^+} + A^- \rightleftharpoons \begin{array}{c} \mathrm{O} \\ \| \\ \mathrm{R{-}C{-}}A \end{array} + \mathrm{NAD(P)H} + \mathrm{H^+} \qquad (15\text{-}24)$$

$A^- = \mathrm{OH^-}, \mathrm{PO_3H^{2-}}, \mathrm{CoAS^-}$

$$RCH_2OH + 2NAD(P)^+ + A^- \rightleftharpoons R{-}\overset{\overset{\displaystyle O}{\|}}{C}{-}A + 2NAD(P)H + 2H^+ \qquad (15\text{-}25)$$

$A^- = OH^-, CoAS^-$

$$HCO_2H + NAD(P)^+ \rightarrow CO_2 + NAD(P)H + H^+ \qquad (15\text{-}26)$$

$$R'{-}\underset{\underset{\displaystyle R''}{|}}{CH}{-}NH{-}R''' + NAD(P)^+ \rightleftharpoons R'{-}\underset{\underset{\displaystyle R''}{|}}{C}{=}N{-}R''' + NAD(P)H + H^+ \qquad (15\text{-}27)$$

$R' = H, R, R$
$R'' = R, R, CO_2H$
$R''' = R, R, H$

$$2RSH + NAD(P)^+ \rightleftharpoons NAD(P)H + R{-}S{-}S{-}R + H^+ \qquad (15\text{-}28a)$$

$$R\begin{matrix} \diagup SH \\ \diagdown SH \end{matrix} + NAD(P)^+ \rightleftharpoons NAD(P)H + R\begin{matrix} \diagup S \\ \diagdown S \end{matrix}\!\Big| + H^+ \qquad (15\text{-}28b)$$

$$n\text{NHI protein}_{red} + NAD(P)^+ \rightleftharpoons NAD(P)H + n\text{NHI protein}_{ox} \qquad n = 1 \text{ or } 2 \qquad (15\text{-}29)$$

HEMOPROTEINS

Prosthetic Groups and Reactions

Hemoproteins are ubiquitously distributed in nature, where they participate in the following important reactions: (1) in oxygen transport, (2) in the reduction of peroxides, and (3) in electron transfer between dehydrogenases on the one hand and terminal acceptors on the other. The prosthetic group is bound tenaciously and in most instances is identical with, or derived from, iron protoporphyrin IX, which in its ferrous form is also called *protoheme* or simply *heme* or *haem* (Fig. 15-4). Other important hemes functioning as prosthetic groups are also indicated.

We see that hemes contain four ligands to Fe at the corner of a coplanar square and thus the metal is capable of binding two additional ligands (see also Chapter 8). In this process binding of the fifth greatly facilitates coordination with the sixth ligand. The resultant hexacoordinate octahedral complexes of Fe^{2+} are known as *hemochromes* (also haemochromes or haemochromogens). If the ligand is a protein, one or both of these nonheme ligands are frequently provided by a His residue; Cys, Met, Trp, Lys, and Tyr are other likely candidates particularly for the sixth position. When only the fifth ligand is provided by the protein, it is almost invariably His, a strong-field ligand. If the sixth position is occupied by an external ligand with a relatively weak field[6] (H_2O, Cl^-, $-CO_2^-$), as is the case in myo- or hemoglobin, it is possible to replace it with one capable of interacting more strongly with the central metal atom. This is especially true if the latter ligand is capable of forming π bonds (O_2, CO, CN^-, *N*-heterocycles).[7] In this manner, high-spin complexes may be transformed into low-spin complexes. Evidently these transformations will depend not only on the ligand but also on the charge distribution and conformational features of the protein and on external conditions of *p*H, ionic strength, and so on. Also it is clear that these factors will affect not only the extent of binding (i.e., the K_{diss} for the ligand), but the rate of association, which may vary greatly between different proteins.

[6] Ligand field strength increases in the order $Cl^- < OH^- < RCO_2^- < F^- < H_2O < SCN^- < NH_3 < H_2NCH_2CH_2NH_2 < NO_2^- <$ *o*-phenanthroline $< CN^-$. The six 3*d* electrons of Fe^{2+} in octahedral complexes have the distribution $(t_{2g})^4(e_g)^2$; the five in Fe^{3+} have $(t_{2g})^3(e_g)^2$ in high-spin, $(t_{2g})^5$ in low-spin states (see Chapter 8).

[7] Introduction of π-bonding ligands leads to strengthened and shortened σ bonds from the other ligands to the metal and a great preference for low-spin complexes.

Protoporphyrin IX

Ferroprotoporphyrin IX*
(proto)heme (IX)
Prosthetic group of
hemoglobin
myoglobin
erythrocruorin
catalase
peroxidase
cytochromes of class B

Heme C
Prosthetic group of
cytochromes of class C

Heme A
Prosthetic group of
cytochromes of class A

Heme a_2
Prosthetic group of
cytochromes of class a_2

Heme $\xrightarrow{-le^- + X'}$ (X')-hemin

Heme $\xrightarrow{+2X \text{ or } X \text{ plus } Y}$ Hemochrome

Hemochrome $\xrightarrow{-le^-}$ Hemichrome

$$\left.\begin{array}{l} X = OH \\ Y = OH_2 \end{array}\right\} \text{hematin} \underset{pK=7.5}{\xrightarrow{+H^+}} \left.\begin{array}{l} OH_2 \\ OH_2 \end{array}\right\} \text{hydroxyhemin}$$

FIGURE 15-4
Prosthetic groups of hemoproteins.
* Chlorocruoroheme, the prosthetic group of chlorocruorin, differs from protoheme only in having a formyl group in position 2.

Hemes are oxidized by O_2 or other oxidizing agents to the corresponding Fe^{3+} derivatives called ferriprotoporphyrins, or *hemins*. These ferric complexes have one residual positive charge, which in the isolated complex must be neutralized by an anion. The resulting pentacoordinate, square-pyramidal complex is called, in the case of chloride, *hemin chloride,* or *chlorohemin.* In alkaline solution the chloride is replaced by hydroxide, and *hematin* is formed. Hemins readily form hexacoordinate octahedral complexes with additional ligands, called *hemichromes.*

For all these complexes, the E_0' value and the rate of oxidation will depend on the environment around the central Fe atom. Controlling factors will be (1) the electron-withdrawing power of substituents on the porphyrin ring and on the ligands (the greater the electron affinity of a ligand in the fifth position, the *higher* the E_0' of the porphyrin) and (2) charge distribution and conformation of the apoproteins. Protein conformation and nature of the ligand will also determine whether a hemoprotein is preferentially oxidized by molecular O_2, a process that requires coordination with the central metal atom, or by another iron-containing complex molecule, a process that probably involves electron transfer across one of the ligands of the coordination sphere. Similar considerations will also govern reduction of iron (Fe^{3+}) proteins by various agents, as contrasted to their interaction with and catalytic power toward hydroperoxides (page 671).

One of the most characteristic features of hemoproteins is their absorption spectrum (Fig. 15-5 and Table 15-8). Low-spin ferrous complexes show a typical three-banded spectrum with the ratio of extinction in the three bands, called (starting from the longest wavelength) α, β, and γ,[8] respectively, roughly in the order 1 : 0.5 : 3 (for pyridine-proto IX-hemochrome the values are, for instance, 34.4 : 17.5 : 94). The positions of the λ_{max} of these bands in the regions 550 to 565, 520 to 535, and 400 to 420 nm are—as might have been anticipated from the foregoing—also a function of ligand and substituent electron-attracting power: the greater the electronegativity, the shorter the wavelengths. High-spin ferrous complexes have spectra with but two bands, the Soret peak plus a band around 550 nm. Ferric complexes show less apparent regularity in their spectra: The porphyrin contributes two bands, the Soret band (with a shift toward longer wavelengths), and a second one in the range 540 to 570 nm, which is of significant intensity only in low-spin complexes. There are also present two additional bands, one in the vicinity of 600 nm, which is a function of the ligand, showing a decrease in intensity with increasing ligand electronegativity, and is probably a charge-transfer band, as well as a weaker one around 500 nm, apparently of similar origin. Finally, there exists a band around 475 to 500 nm, which represents a transition of the Fe^{3+} electrons themselves.

Hemoglobin and Other Respiratory Pigments

The most important respiratory pigment of higher vertebrates is hemoglobin, a tetramer having a molecular weight of 64,450. Its structure has already been considered in earlier chapters. In its reduced, oxygen-free form, it is a rare species among hemoproteins: a *high-spin* Fe^{2+} complex of exceedingly *low* potential; one in which the central iron of a square-planar heme group is further coordinated, probably with only one strong field ligand (histidine), and the sixth position is vacant. This configuration is particularly prone to further substitution, since in general the introduction of one strong nitrog-

[8] The intense γ band is frequently referred to as the Soret band, after one of its discoverers.

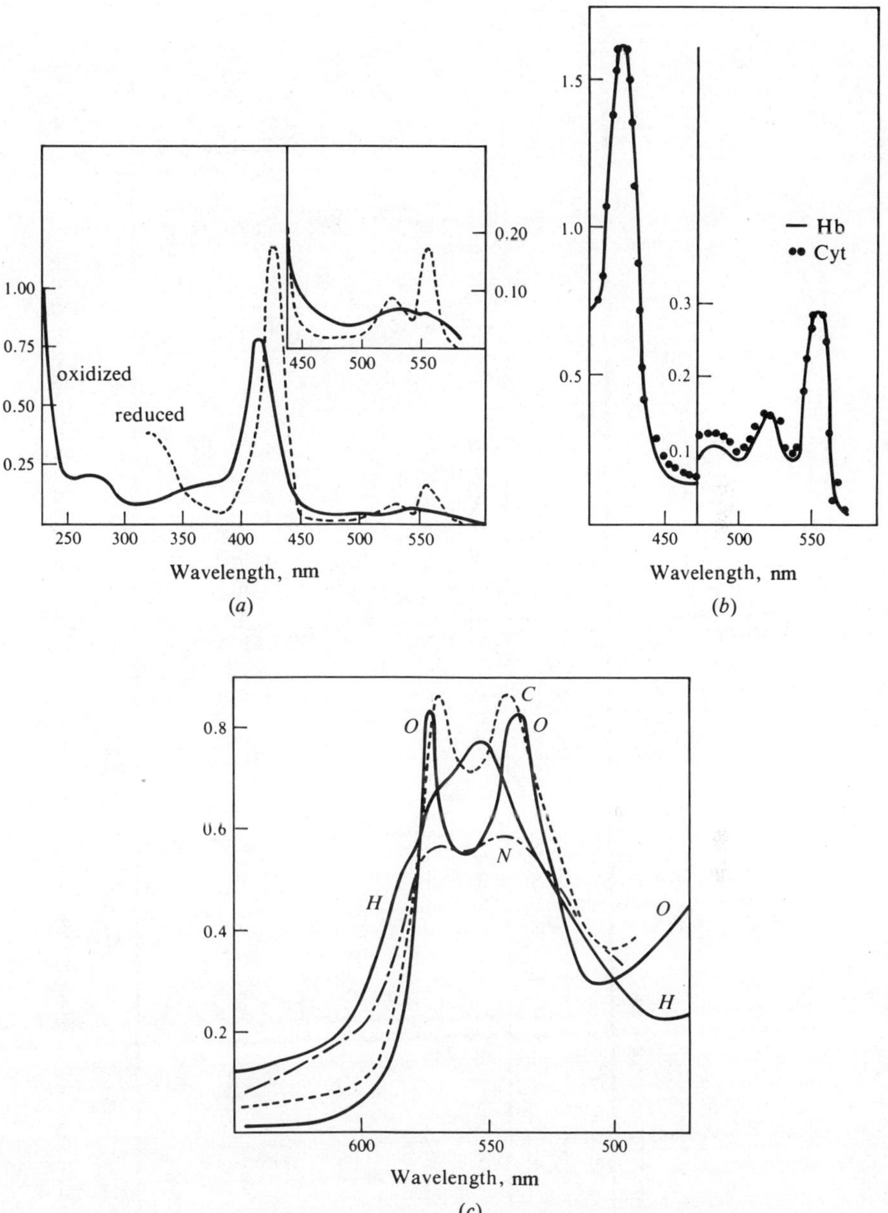

FIGURE 15-5

Absorption spectra of heme compounds. (*a*) Absorption spectrum of a typical cytochrome *b* (cytochrome b_5 from liver respiratory particles). Reduced spectrum obtained by reduction with dithionite or, enzymatically, with NADH. (*b*) Spectra of pyridine hemochromes, of hemoglobin, and of the cytochrome from (*a*), obtained by adding alkaline pyridine to the compounds and reducing with dithionite. Parts (*a*) and (*b*) from I. Raw, R. Molinari, D. do Amaral, and H. R. Mahler, *J. Biol. Chem.*, **233:** 225 (1958). (*c*) Absorption spectra of hemoglobin (*H*), oxyhemoglobin (*O*), and nitric oxide hemoglobin (*N*). Abscissa: wavelength in nanometers; ordinate: extinction of 0.1 percent solutions in 10-mm absorption cell. F. Haurowitz and R. Hardin, Respiratory Proteins, in H. Neurath and K. Bailey (eds.), *The Proteins,* vol. II, part A, Academic Press, New York, 1954, p. 279.

TABLE 15-8A
Properties of hemoproteins and heme derivatives

Compound	Ligands		μ	Unpaired		λ_{max}, nm			E_0',
	Fifth	*Sixth*	*BM*[a]	*electrons*	*Designation*[b,c]	α	β	*Soret*	*V*
Hemoglobin (myoglobin)	His	—	4.47	4	High-spin octahedral	555		430	−0.130
Oxymyoglobin	His	O_2	0	0	Low-spin octahedral	581	543	417	0.00
Oxyhemoglobin	His	O_2	0	0	Low-spin octahedral	577	542	415	
Carboxyhemoglobin	His	CO	0	0	Low-spin octahedral	568	539	419	
Cyanhemoglobin	His	CN−	0	0	Low-spin octahedral				
Nitric oxide Hb[d]	His(?)	NO		1	Low-spin octahedral				
Ferro HRP[e]	$-CO_2^-$[f]	H_2O				587	556	437	
Ferro HRP-CO	$-CO_2^-$[f]	CO				573	543	423	
Dicyanide hemochrome	CN−	CN−	0	0	Low-spin octahedral	569	539	430	
Pyridine hemochrome	Pyr[g]	Pyr	0	0	Low-spin octahedral	557	525	423	0.137
Pyr-chlorocruoro-hemochrome	Pyr	Pyr				544	562	434	0.246
Pyr-heme a_2 hemochrome	Pyr	Pyr				620		432	0.288

[a] Bohr magnetons: $\sqrt{n(n+2)}$, where n = number of unpaired electrons.
[b] Spin = $n/2$.
[c] "Low spin" is equivalent to "essentially covalent," "high spin" to "essentially ionic" bond type (Pauling's nomenclature).
[d] Hemoglobin = Hb.
[e] Horseradish peroxidase.
[f] Not known with certainty.
[g] Pyridine.

TABLE 15-8B
Properties of hemiproteins and hemin derivatives

Compound	Ligands: Fifth	Ligands: Sixth	μ (BM)[a]	Unpaired electrons	Designation[b,c]	λ_{max}, nm: α	λ_{max}, nm: β	λ_{max}, nm: Soret
Hemiglobin	His[d]	OH_2	5.7	5	High-spin octahedral	630	500	406
(methemoglobin, metmyoglobin)	His	OH^-	4.6	1, 5	Mixture of spin states	600	(577, 540)	411
	His	F^-	5.9	5	High spin	608		
	His	HCO_2^-	5.4	5	High spin			
	His	CN^-	2.5	1	Low-spin octahedral		540	415
	His	SH	2.3	1	Low-spin octahedral			
	His	N_3^-	2.8	1	Low-spin octahedral	630	575	420
	His	NH_3	2.9	1	Low-spin octahedral			
	His[e]	OH^-		1	Low-spin octahedral	580	541	420
Ferri HRP	$—CO_2^-$[f]	OH_2	5.5	5	High spin	641.5	497	403
Ferri HRP cyanide	[f]	HCN		1	Low spin	581	540	425
Catalase	$—CO_2^-$ or His[e]	OH_2?	5.9	5	High spin	620	540	406
Catalase cyanide	[f]	HCN		1	Low spin	585	555	425
Hemin	Cl^-		5.8	5	High-spin square-pyramidal	632	(540, 510)	400
Hematin	OH^-	OH_2	~5	5	High-spin octahedral			
Hemin a_2	Cl^-							432
Hematin a_2	OH^-	OH_2				685		405

[a] Bohr magnetons: $\sqrt{n\,(n+2)}$, where n = number of unpaired electrons.
[b] Spin = $n/2$.
[c] "Low spin" is equivalent to "essentially covalent," "high spin" to "essentially ionic" bond type (Pauling's nomenclature).
[d] Histidine imidazole N on the protein (globin).
[e] At 77°K for sperm-whale myoglobin.
[f] Protein ligand not known with certainty.

enous ligand into free hemes invariably facilitates the introduction of a second ligand. Only by virtue of the iron being coordinated to a *protein-bound* histidine residue could such a compound exist at all. These considerations explain the great avidity of hemoglobin for π-bonding substituents such as O_2, CO, or CN^-. The latter two ligands bind many times more strongly than the first and yield photoreversible complexes. Their introduction into the sixth coordination position leads to a profound structural rearrangement, as a result of which the whole complex changes to one of low spin. Free hemoglobin is easily oxidized to its ferri form, *methemoglobin,* by a variety of oxidizing agents. In view of this observation, one might inquire why in oxyhemoglobin no oxidation of the Fe atom ensues and why there is no valence change of the metal in the course of the oxygenation reaction. One possible explanation is the obligatory change in spin multiplicity of the oxygen that has to occur before π bonding to the vacant d orbitals of $Fe^{2+}(d^5)$ can take place during complex formation. In fact, oxyhemoglobin is known to be diamagnetic in the *ground* state. Yet, efficient oxidation presumably requires a paramagnetic *transition* state of the form $[Fe^{3+}O_2^-]$ and a transition between two states of different multiplicity is highly improbable—hence complexes that liberate sufficient energy to permit O_2 spin pairing are of necessity stable also to autoxidation. An equally plausible alternative, based largely on the great similarity of the optical properties of oxymyoglobin and metmyoglobin in alkaline solution at low temperature (Table 15-8) postulates that oxidation–reduction does in fact occur during oxygenation and that the structure of oxyhemoglobin and oxymyoglobin is principally $[Fe^{3+}—O_2^-\cdot]^{VI}$. The formal oxidation state of such a complex is +6, as indicated by the Roman numeral. It would contain two unpaired electrons, one formally on the metal ($d_{½}^{5}$), the other on the superoxide radical complexed to it. The observed diamagnetism is then explained in terms of antiferromagnetic coupling between the two unpaired spins, canceling all paramagnetic contributions in a manner similar to that observed in dimeric cupric acetate. In methemoglobin the sixth coordination position is occupied by H_2O in neutral and by OH^- in alkaline solution.

All four heme residues in a hemoglobin molecule are capable of binding oxygen, but the respective binding constants are dependent on one another: each successive O_2 molecule bound affects the binding of the next. The ratio of the four stepwise constants is approximately 1 : 4 : 24 : 9. This phenomenon of linked interactions results in a sigmoid, rather than a hyperbolic, binding curve when the extent of O_2 binding is plotted against O_2 tension (Fig. 15-6).

Similar curves are also obtained for the formation of carboxyhemoglobin, $Hb(CO)_4$, but with values for the association constants so derived $\sim 10^2$ greater than for the oxygen complexes. The structural basis for these cooperative—and allosteric (Chapter 6)—effects (for which this represents the archetype) has recently been provided by Perutz. The transitions are $[\alpha_1{}^{D}\alpha_2{}^{D}\beta_1{}^{D}\beta_2{}^{D}]^{D} \overset{O_2}{\rightleftharpoons} [\alpha_1{}^{O}(O_2)\alpha_2{}^{D}\beta_1{}^{D}\beta_2{}^{D}]^{D} \overset{O_2}{\rightleftharpoons} [\alpha_1{}^{O}(O_2)\alpha_2{}^{O}(O_2)\beta_1{}^{D}\beta_2{}^{D}]^{D}$

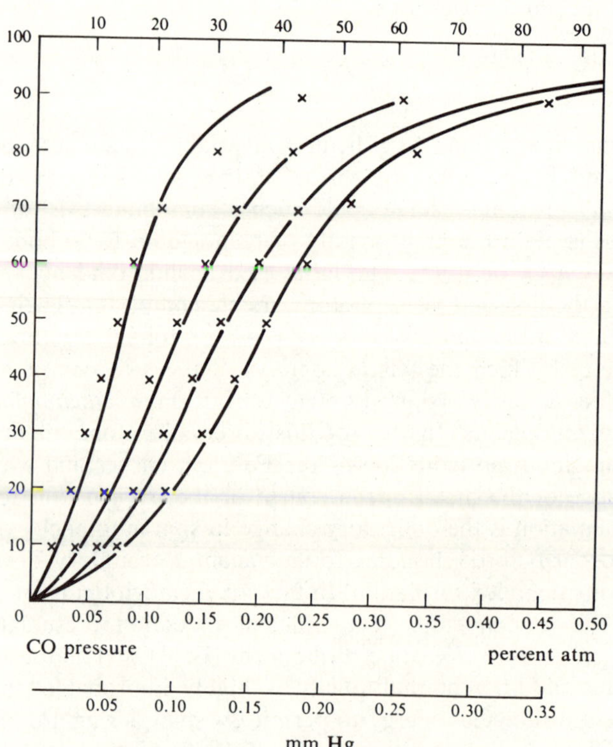

FIGURE 15-6

O_2 and CO saturation curves for hemoglobin. Equilibrium curves of CO-hemoglobin (full lines) and O_2-hemoglobin (crosses) at 0-, 19-, 41-, and 79-mm CO_2 pressure (from left to right). Abscissa at bottom: CO pressure; abscissa at top: O_2 pressure in millimeters Hg; ordinate: percent saturation. J. Barcroft, *The Respiratory Function of the Blood,* part II, Cambridge University Press, New York, 1928.

$$\underset{\text{change}}{\overset{\text{conformational}}{\rightleftharpoons}} [\alpha_1{}^{\text{O}}(\text{O}_2)\alpha_2{}^{\text{O}}(\text{O}_2)\beta_1{}^{\text{D}}\beta_2{}^{\text{D}}]^{\text{O}} \overset{\text{O}_2}{\rightleftharpoons} [\alpha_1{}^{\text{O}}(\text{O}_2)\alpha_2{}^{\text{O}}(\text{O}_2)\beta_1{}^{\text{O}}(\text{O}_2)\beta_2{}^{\text{D}}]^{\text{O}} \overset{\text{O}_2}{\rightleftharpoons}$$

$[\alpha_1{}^{\text{O}}(\text{O}_2)\alpha_2{}^{\text{O}}(\text{O}_2)\beta_1{}^{\text{O}}(\text{O}_2)\beta_2{}^{\text{O}}(\text{O}_2)]^{\text{O}}$, where the superscripts on the subunits and on the protein indicate conformers differing in their 3° and 4° structure, respectively.

By treatment with HCl in acetone, hemoglobins can be dissociated into the colored prosthetic group (heme) and the colorless apoprotein (globin). If this is performed in air, the heme is oxidized to hemin chloride, but if done anaerobically or in the presence of a reducing agent, the iron remains in the Fe^{2+} state. From heme (or hemin) plus globin, the conjugated protein (hemoglobin or methemoglobin) can be reconstituted by adjusting the *p*H to neutrality. The heme in this experiment need not necessarily be the protoporphyrin derivative, since other iron porphyrins can be used to form *pseudo*-hemoglobins. Conversely, the globin of one species may be substituted for that of another, yielding various *hybrid* hemoglobins. Such hemoglobins vary in crystal form, net charge, electrophoretic mobility, O_2-carrying capacity, and ease of denaturation by alkali or acid.

In vertebrates, hemoglobin does not exist in solution but is contained in highly concentrated form within the red blood cells, or erythrocytes. A closely related pigment, *myoglobin,* is found inside muscle cells. The structure of myoglobin has been studied at an even higher resolution than that of hemoglobin (page 150). The protein has an $M = 17{,}000$ and can be regarded as closely analogous to the monomeric units of hemoglobin. Since it contains only one heme prosthetic group per molecule, it exhibits a hyperbolic rather than a sigmoid oxygen dissociation curve. Closely related to myoglobin are the hemoglobins found in the blood cells of certain primitive vertebrates such as the lamprey eels; these have an $M = 17{,}500$ and contain only one heme per molecule. A variety of respiratory pigments are found in invertebrates; *erythrocruorins* are large hemoglobin-like molecules existing extracellularly in the circulating blood of many polychete and oligochete annelid worms and of mollusks. These proteins exhibit an $M = 4 \times 10^5$ to 6.7×10^6 and contain from 30 to 400 heme groups per molecule. Certain other annelid worms, such as *Spirographis,* have in their blood green hemoproteins, called *chlorocruorins,* of $M = 3.5 \times 10^6$, containing 190 *chlorocruoroheme* groups (Fig. 15-4) per molecule. Regardless of size, all these hemoproteins bind one molecule of O_2 per heme group.

Hemerythrins and *hemocyanins* are nonheme respiratory pigments found, respectively, in the blood cells of sipunculid worms and in the plasma of certain mollusks and crustaceans. The hemerythrin of *Sipunculus* contains 16 nonheme iron atoms linked to the protein in a structure that may be related to that found in other NHI proteins. Two atoms of metal are required to bind each molecule of O_2. The same holds true also for the hemocyanins, which are large copper-containing proteins that in the lobster contain 20 atoms of metal per protein ($M = 780{,}000$). Other hemocyanins are known (for instance in snails) that are approximately 10 times larger.

The Hydroperoxidases

The hydroperoxidases are a group of hemiproteins and include *catalase,* which is widely distributed throughout nature (vertebrate liver, blood, various microorganisms); *peroxidase,* which is found in plants, leucocytes, and milk; and *cytochrome c peroxidase,* which is found in many microorganisms. Catalases and peroxidases catalyze the reaction:

$$\text{RO—OH} + \left.\begin{matrix}\text{H}_2D \\ 2D^n\end{matrix}\right\} \rightarrow \text{ROH} + \text{H}_2\text{O} + \left.\begin{matrix}D \\ 2D^{n+1}\end{matrix}\right\} \tag{15-30}$$

where n equals the valence. Catalases utilize H_2O_2 (R = H) as the most effective oxidant, but can utilize alkyl hydroperoxides if R is of short-chain length, the order being H ≫ CH_3 > C_2H_5. The preferred electron donor, or coreductant, is a second molecule of H_2O_2 ($D = O_2$). In this case, and with R = H, Eq. (15-30) becomes the "catalatic" one:

$$2H_2O_2 \rightarrow 2H_2O + O_2 \tag{15-31}$$

An alternative reaction is the "peroxidative" one [Eq. (15-30)], in which H_2D or D^n can be (1) ascorbate and ferrocyanide, (2) phenols, (3) short-chain alcohols or formate, (4) sodium nitrate, and (5) azide or hydroxylamine. If the hydrogen peroxide is generated *in situ* (e.g., by a flavoprotein dehydrogenase), then the addition of catalase plus a coreductant will produce Eq. (15-32):

$$DH_2 + XH_2 + O_2 \xrightarrow{\text{fp + catalase}} D + X + 2H_2O \tag{15-32}$$

Although peroxidases can catalyze reaction (15-31), their preferred mode of action is that of reaction (15-30). Among the cosubstrates are phenols, aliphatic and aromatic amines, leuco dyes, hydroquinones, enediols, and cytochrome *c*. These facts, the identity of the prosthetic groups (all catalases and most peroxidases appear to contain ferriprotophyrin IX in a high-spin state), and the observation that all complexes of this type possess weak catalatic and peroxidative activity have led to their assignment to one common class of enzymes, the hydroperoxidases.

Reaction Pathways. The three hydroperoxidases most thoroughly investigated are the catalase of erythrocytes (*Cat,* usually from horse blood, M = 200,000, a tetramer), the peroxidase from horseradish (*HRP,* M = 43,000), and cytochrome *c* peroxidase from the mitochondria of aerobically grown yeasts (*CCP,* M = 50,000). All of them are believed to be capable of forming one or more of a series of intermediate complexes—identifiable by their optical and magnetic properties—in the course of their reaction with peroxide and donor. The postulated common pathway is (hp = hydroperoxidase):

$$[hp] \xrightarrow{\text{ROOH} \to \text{H}_2\text{O}} [complex\ I] \xrightarrow{\uparrow \text{ROH}} [\text{ES}] \xrightarrow[\substack{1e^- \\ \text{reduction}}]{DH_2 \to DH} [complex\ II] \xrightarrow[\substack{1e^- \\ \text{reduction}}]{DH_2 \to DH} [hp] \tag{15-33a}$$

color	*brown*	*green*	*red*	*red*	*brown*
λ_{max}, nm	405, 500, 536	405	420, 528, 560	425, 528, 553	
unpaired electrons	5	3	3	2	
oxidizing eq. added		2	2	1	

The radicals derived from DH_2 in reaction (15-33a) undergo a rapid dismutation reaction:

$$2DH\cdot \longrightarrow DH_2 + D \tag{15-33b}$$

In the case of catalase the only complex identifiable in the course of its physiologically important reaction is complex I, hence the reaction sequence ($D = H_2O_2$ or $R'CH_2OH$)

$$[cat] \xrightarrow[\text{ROOH}]{\text{fast}} [\text{I}] \xrightarrow[DH_2]{\text{slow}} [cat] + \text{ROH} + D \tag{15-34a}$$

For HRP the identifiable complexes are I and II:

$$[HRP] \xrightarrow[\text{ROOH}]{\text{fast}} [I] \xrightarrow[DH_2 \quad DH\cdot]{\text{fast}} [II] \xrightarrow[DH_2 \quad DH\cdot]{\text{slow}} [HRP] + \text{ROH} \tag{15-34b}$$

Finally, for CCP the principal species are "ES" (which bears some resemblance to complex I in its oxidation state and its magnetic properties, but exhibits the optical properties of complex II), with peroxides in the absence of donors, and a complex between this species and cytochrome *c* (Fe^{2+}) when the latter is added:

$$[CCP] \xrightarrow[\text{ROOH}]{\text{fast}} [ES] \xrightarrow[\text{cyt } c(Fe^{2+})]{\text{very fast}} [ES]\cdot\text{cyt } c \xrightarrow[\text{cyt } c(Fe^{2+}) \quad \text{ROH}]{} [CCP] + 2\text{cyt } c(Fe^{3+}) \tag{15-34c}$$

There has been considerable speculation concerning the precise structures of complexes I and II. In addition to the properties indicated earlier, it is also known that complex I still retains the components of ROOH; that it exhibits a highly unusual absorption spectrum dissimilar to that of all other heme derivatives studied; and that compounds with very similar optical properties can be formed from hemoproteins or myoglobin by two successive $1e^-$ oxidations by means of strong oxidizing agents such as chloriridate. A possible path for the conversion of a hemoprotein (hp) to complex I is the following:

$$\underset{[hp]^{\text{III}}}{\left[DH_2 \;\; Fe^{3+}\!-\!OH_2 \right]} \xrightarrow[\text{ROOH} \;\to\; H_2O]{} \underset{[Complex\ \text{I}]^{\text{V}}}{\left[D(H\!-\!O(H)\!-\!O(H)R) \;\; Fe^{3+}(d_{3/2}) \right] \longleftrightarrow \left[D(H\cdots O(H)\cdots O(H)R) \;\; Fe^{3+}(d_{1/2}) \right]} \tag{15-35}$$

A complex with three unpaired electrons—if localized exclusively on the Fe it would be of spin = $3/2$ (or $d_{3/2}{}^5$)—is so unusual as to suggest that both the *fifth and sixth* coordination positions are essentially vacant (i.e., that the link between protein and iron is now only electrostatic rather than covalent). Similarly, the peroxide oxygen cannot provide a ligand for the iron and may in fact have served to transfer two of its oxidizing equivalents to some group(s) or the protein, here called DH_2. The formal oxidation number for the complex is five, equivalent to a "perferryl" iron. Complex II bears a formal oxidation number of four ("ferryl" iron) and its absorption spectrum is quite similar to that of low-spin hemoproteins (47). For instance, ferro-HRP cyanide shows λ_{max} at 535 and 564 nm. Hence a reasonable assignment for complex II is

$$[\text{hp-}Fe^{3+}{:}\ddot{\underset{\cdot\cdot}{O}}{:}H]^{\text{IV}} \longleftrightarrow [\text{hp-}Fe^{2+}{:}\dot{\underset{\cdot}{O}}{:}H]^{\text{IV}}$$

or possibly

$$[\text{hp-}Fe^{3+}{:}\ddot{\underset{\cdot}{O}}{:}\ddot{\underset{\cdot\cdot}{O}}{:}H]^{\text{IV}} \longleftrightarrow [\text{hp-}Fe^{2+}{:}\ddot{\underset{\cdot}{O}}{:}\ddot{\underset{\cdot}{O}}{:}H]^{\text{IV}}$$

The Cytochromes (Table 15-9)

In hemoglobin and the related respiratory pigments the function of the heme group is to serve as a repository for O_2 in oxygen transport with the iron in the ferrous form (prior to attachment of O_2, but after removal of the ligand); in the hydroperoxidases it is to interact with and decompose a peroxide, again without a change in its formal valence (this time ferric) at the end of the catalytic cycle. In contrast, in the cytochromes the substrate does *not* function as a ligand while the iron shuttles between the ferrous and the ferric form during catalysis. In this manner these hemoproteins are able to function as redox carriers, designed to transfer electrons or reducing equivalents from the dehydrogenases on one side to the terminal acceptors at the other. Depending on the potential of the latter and hence on the total potential span to be traversed, more than one cytochrome may participate in making up a respiratory (or electron-transport) chain. Details are deferred to the next section.

The cytochromes were discovered in 1886 by MacMunn, a Scotch physician who observed their characteristic absorption spectra by means of a hand spectroscope and found them to be distributed widely in tissues (48). He called them "histo-" and "myohaematin," because they appeared to him to be related to heme and hemin. He thus clearly recognized their chemical nature and also assigned to them a respiratory function, which he believed to be connected with oxygen transfer, no other form of respiratory activity being then known. In 1925, D. Keilin (49), studying the respiration, first of wax moths, bees, and other insects, observed the apparently universal occurrence of a pigment which he called "cytochrome." He quickly established reversible redox change as crucial to its function, obtained evidence that related it to the hemochromogens, and showed that two classes of respiratory poisons introduced by Warburg—cyanide and narcotics (urethane and barbiturates)—strongly affected "cytochrome" but in an opposite sense, the former inhibiting oxidation and latter blocking reduction. By 1930, Keilin had realized (50) that "cytochrome" consisted of three spectroscopically distinct components, which he called *a, b,* and *c,* in the order of their absorption maxima: *a* absorbing at the longest wavelength and *c* at the shortest (see Table 15-9 and Fig. 15-7). He also obtained some evidence that cytochrome *b* was closest to the substrate side and believed that there existed an entity sensitive to CO, KCN, and H_2S which linked the cytochromes to O_2 and which he therefore called "cytochrome oxidase." He did not believe this to be a cytochrome or even a hemoprotein, but he thought it might contain Cu. Meanwhile, Warburg had obtained good evidence from action spectra that his CO-sensitive *Atmungsferment* was also a hemoprotein (see Fig. 15-1). Keilin and Hartree then showed in 1939 (1) that cytochrome oxidase was identical with cytochrome a_3, a special component of cytochrome *a*; (2) that the cytochromes probably acted as a linear chain of carriers; and (3) that the terminal portion of the chain consisted of cytochrome *c* in close association with the oxidase (51).

All cytochromes appear to be low-spin hexacoordinate complexes, and except for the ones designed to function as terminal carriers in respiratory chains (terminal oxidases interacting with O_2 for aerobic cells; nitrate, sulfate, etc., reductases in anaerobic bacteria), both the fifth and sixth ligands are usually provided by the protein. The prosthetic groups have been described in Fig. 15-4; note that class B (*b* type) and class C (*c* type) cytochromes both possess protoheme prosthetic groups, but that in the latter instance two covalent links have been established to the protein by addition of cysteine —SH across the two vinyl side chains.

In Table 15-9 we have summarized the properties of most of the well-characterized cytochromes. Their number and diversity is continuously on the increase and their variety as well as that of the reactions catalyzed by them attest to the importance of the protein moiety in determining the environment around the prosthetic group and in regulating cytochrome function and reactivity. Yet it is well to remember that except for cytochrome *c* and for certain of the bacterial cytochromes, these pigments, isolated frequently by means of relatively harsh procedures, are not necessarily identical to their counterpart in the intact cell, where they were designed to function within and formed part of a fixed, particulate and probably largely nonaqueous environment. Once isolated they easily aggregate into oligomeric forms.

The cautionary words of Nicholls (52) are most appropriate in this context:

> Until the individual components of the cytochrome chains are isolated in *enzymatically active forms* [italics ours] it will be difficult . . . to consider them in separate chapters. . . . For, however extensive our knowledge may become, any discussion of these enzymes must consider that they react with one another and not with other metabolites. It is not possible to describe cytochrome *b* without considering flavoproteins and cytochrome c_1, nor is it reasonable to discuss cytochrome oxidase without mentioning the properties of cytochrome *c*. . . . The history of the study of cytochromes, the history of the study of respiration, a study traditionally influential in determining our ideas about the nature of life, contains many examples of such limiting names and concepts. . . .

Cytochromes B. *B*-type cytochromes (53–55) are characterized by relatively low potentials, suitable for their role as the substrate—and dehydrogenase—proximal component of electron-transport chains. In their native state their protoheme prosthetic group is low spin and coordinated to two protein ligands and therefore generally incapable of interacting with external ligands such as O_2, CO, or CN^-. Hence, under ordinary conditions, these cytochromes are neither autoxidizable nor do they react with these typical inhibitors of heme enzymes. Exceptions are, however, known. Among them are cytochrome P450 (systematic name cytochrome B420) already described in an earlier section as the terminal member of the hydroxylase systems in mammalian (adrenal mitochondria, liver microsomes) and bacterial (*Ps. putida*) systems. In the former it is tightly integrated into lipoprotein membranes and its liberation, usually by enzymatic treatment with lipase (steapsin), leads to extensive and irreversible modifications (56), while it can be obtained in a soluble form from the latter (14). The optical properties of the two cytochromes are very similar and are characterized by a unique high-spin pattern in the visible range of four peaks (at 650, 570, 535, and 415 nm) in the oxidized form and two (at 555 and 420 nm) in the reduced form. The pigments are reducible by their substrates in the presence of the appropriate enzyme systems (e.g., deoxycorticosterone in the mammalian system, *d*-camphor in the bacterial system). They react readily with oxygen or CO and the action spectra for photodissociation of the latter complex coincides effectively with its absorption spectrum in a manner analogous to that previously observed with cytochrome oxidase (cf. Fig. 15-1).

Cytochromes C. Cytochromes of the *c* type constitute, in general, the central electron carriers of most cells, intermediate in potential, and serving to link dehydrogenases and cytochromes of the *b* type on one hand, and cytochrome (*c*) oxidases (or acceptor reductases) on the other. In agreement

TABLE 15-9
Properties of the cytochromes

Class	Name	Source	Spectra λ_{max} [ε_{mM}] Reduced α	Reduced β	Reduced γ	Oxidized γ	Oxidized E_0'	Molecular weight $M \times 10^3$	Physiological Reductant	Physiological Oxidant	Reaction with O_2	Reaction with CO	Reaction with CN^-, F^-, N_3^-
A	*a*	Mitochondria[g]	600[23]	—[b]	439[94]	425[76]	0.28		*c*	a_3	?	?	
	a_3	Mitochondria[g,c]	(603.5) [19.4]	—	443[79] 430(CO)	830[1]	(0.35)	200[c]	*c* + *a*	O_2	+	+	+
	a_1	Bacteria	592(CO)	548(CO)	428(CO)				C	O_2	+	+	+
A_2(D)	a_2(*d*)	*Pseudomonas*[c]	652 549	629 523	460 418	412		90	C551	O_2, NO_2^-	+	+	+
B	*b*	Mitochondria[g,e]	563[28]	532	429[114]	418	0.035[f]	25	CoQ, NHI	c_1	−	−	−
	b_1	*E. coli*[g]	558[22]	528[15]	425[127]	418[107]	−0.34[g]	500[g]	fp?	Q_1, Q_2	−	−	−
		Other bacteria	559	528	426	415			fp	NO_3^-			
	b_2	Yeast	557[33]	528[17]	424[198]	413[117]	0.34[h]	170[h]	fp	c_1 or *c*	−	−	−
	b_3	Plant membranes	559	529	425			40	fp	?	+	−	
	b_5	Microsomes, liver mitochondria	556[26]	526[13]	423[171]	413[115]	0.02	25	fp	*c*, P450	−	−	−
	b_6	Chloroplasts	563				−0.06		fp?	*f*?			
	b_7	Spadix of *Arum* plant	560	529			−0.03		*b*	O_2?	+		
	B562	*E. coli*	562[32]	532[17]	427[180]	418[117]	0.113	12	fp?	C?			
	B559	*Streptomyces*	559	530	408	410	0.254	13	?	?			
	B553	Mung bean seeds	555[23]	529[13]	423[180]	413[105]	−0.030	13	?	?			
	B420	Microsomes[i]	555[11.7]	—	420[85]	415[108]			b_5, hydroxylase	O_2	+	+	
			555[11.4] (CO)		449[114] (CO)								
		Ps. putida	575[6.7] (CO)	546[7.0] (CO)	448[90] (CO)	419[113]		40	hydroxylase	O_2			
	o	*E. coli, Acetobact.*	567(CO)	537(CO)	417(CO)				C	O_2	+		
		R. rubrum	572(CO)	540(CO)	420(CO)								
C	*c*	Mitochondria[a]	550[27.7]	521[15.9]	415[125]	407	0.254	12.5	c_1	*a*	−	−	±[j]
	c_1	Mitochondria[a]	554[24.1]	524[11.6]	418[116]	410	0.220	37	*b*	*c*	−	−	
	c_2	Denitrifying bacteria	550–552	522–525	416–418	410–412	0.25	32	fp?	NO_3^-	−	−	
		Rhodospirillum rubrum	550	521	416	409	0.32	13.5	fp?	*a*, *o*?	−	−	

c_3	*Desulfovibrio desulfuricans*[l]	552	522	418	410	−0.205	11.3[k]	fp?	$SO_4^{=}$	+	−	
c_4	*Azotobacter*, others[i]	551	522	414	414	0.30		B	c_5	−	−	−
c_5	*Azotobacter*, others[m]	555	524	418	414	0.32		c_4	a_1, a_2	−	−	−
C551	*Pseudomonas*	551	521	416	409	0.286	7.6	B	a_2	−	−	−
C552	*Chromatium*	552	523	416	410	0.010	72[n]	fp	?	+		
C554	Halotolerant bacteria[o]	554	521	4[illegible]5				?	?			
C555 or f	Chloroplasts	555	525	423	413	0.365	110[k], 13.5[p]	b_6?	Chl a			
C556 or h	Hepatopancreas of gastropods[q]	556	527	422	408		12	?	?			
Cytochrome cc' (RHP)	*Rhodospirillum rubrum*	568 550(CO)	—	424 416(CO)	390	−0.008	28[k]	?	?	+	+	−
Cytochrome c'	*Rhodopseudomonas palustris*	565	—	426, 436	396	0.105	14.8	C552?	?	+	+	−

[a] Component has been identified in (and in some cases isolated from) the mitochondrial electron-transport chain in animals, plants, yeasts, and molds (*Neurospora*).

[b] The dash indicates component or reaction absent; an empty space indicates information not available.

[c] Defined as the entity in cytochrome oxidase that reacts with CO in the reduced and with HCN, HN_3, and probably O_2 in the oxidized state. Monomeric cytochrome oxidase has an $M \simeq 200{,}000$ and contains 2 hemes per molecule [$a \cdot Cu{:}a_3 \cdot Cu = 1$]. The minimal mitochondrial unit is three times larger ($M \simeq 600{,}000$). In this complex, the ratio may be $(4a)Cu_2{:}(2a_3)Cu_4$. The individual subunit associated with each heme has a $M \simeq 100{,}000$. The prosthetic group of cytochromes a and a_3 is identical: heme a or cytohemin. In beef heart mitochondria the midpoint potentials (E_m) for cyts a and a_3 are 0.205 and 0.365 V, respectively.

[d] *Pseudomonas* cytochrome oxidase and the terminal oxidase of certain other bacteria do not contain cyt a_3. The isolated protein contains 1 mol of heme c plus 1 mol of a_2. The systematic name is cytochrome CD.

[e] Pure preparations of the cytochromes with enzymatic activity have not yet been obtained from mitochondrial sources.

[f] Estimated value for particulate form. The isolated, solubilized forms show considerably lower values. Similarly, partially denatured particulate enzymes show different E_0' values and become autoxidizable. Recent studies indicate the presence of several additional membrane-bound species with $E_m = 0.125$ and −0.030 to −0.100 V; when ATP is present, the principal additional component has $E_m = 0.250$ V.

[g] The b-type cytochrome of many bacteria. The properties shown are those of a crystalline octamer; E_0' (monomer) = 0.0.

[h] Part of the L-lactate dehydrogenase. M shown is for dimeric form containing 2 hemes and 2 FMN. A "cytochrome" free of heme can be obtained by tryptic digestion; its $E_0' = 14$ mV, its $M \simeq 12{,}000$.

[i] Perhaps the terminal oxidase for microsomal electron transport with either cyt b_5 or adrenodoxin functioning as an electron donor. Also referred to as P450 because of λ_{max} of CO compound.

[j] Slow; high K_{diss}.

[k] Two hemes per molecule.

[l] Anaerobic *E. coli* cells contain a very similar low-potential cytochrome ($E_0' = -0.2$ V) involved in NO_3^- and NO_2^- reduction.

[m] Cyts c_4 and c_5 are the counterparts in aerobic bacteria of cyts c_1 and c.

[n] Two hemes and 1 flavin per molecule. An analogous protein from *Chlorobium*, cytochrome C553, has 1 heme and 1 FMN per $M = 50{,}000$.

[o] Formerly called cyt b_4.

[p] Monomeric cytochrome *c-552* from *Euglena gracilis*.

[q] Also contain another, very similar, pigment called *helicorubin*.

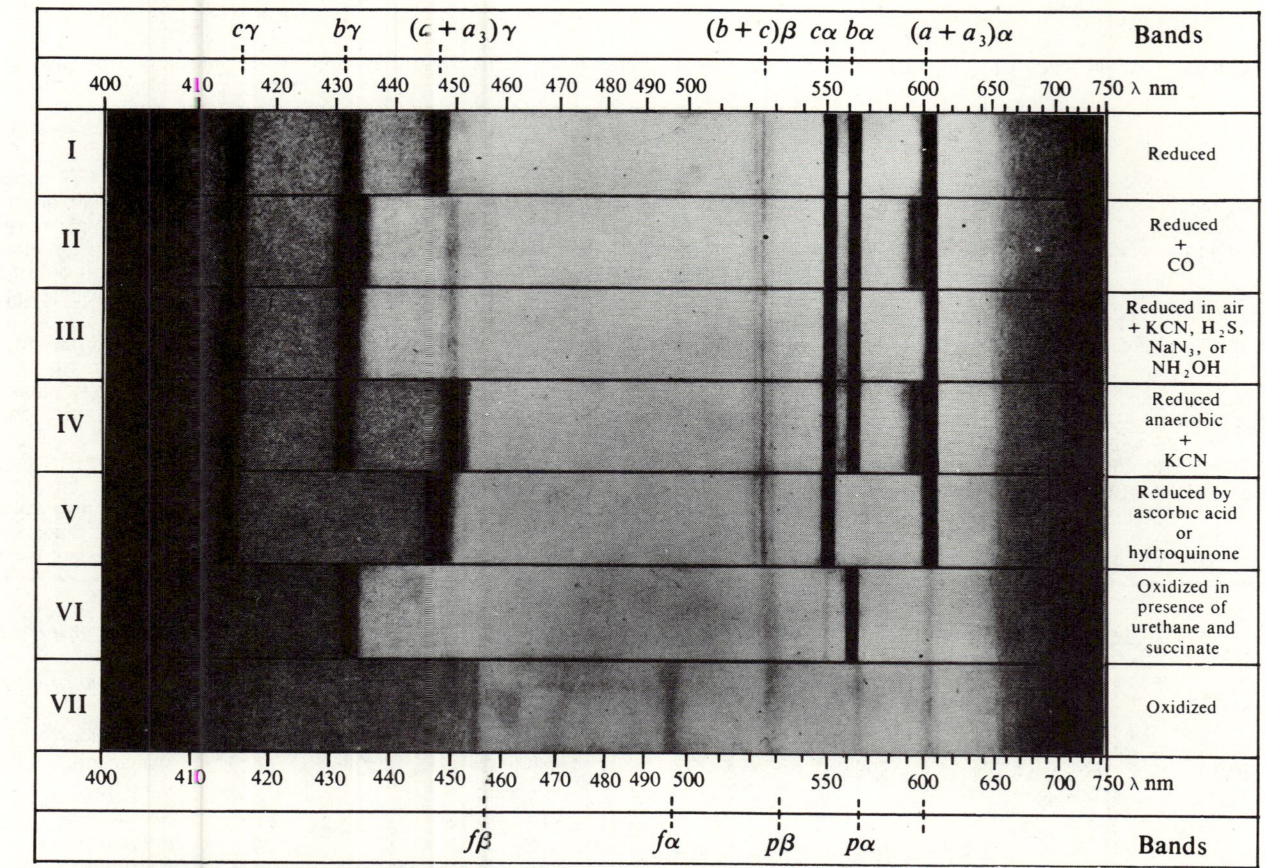

FIGURE 15-7

Cytochrome spectra. D. Keilin and E. F. Hartree, *Proc. Roy. Soc.* (*London*), **B127:** 167 (1939).

with this functional role, at least one member of the group in each instance is relatively freely diffusible and can be easily extracted and isolated in pure, monomeric form. The situation is complicated, in most instances, by the presence of a second *closely related,* usually tightly integrated (insoluble), cytochrome of the same type, such as the *cytochrome* c_1 of mitochondria of all eucaryotic cells and, frequently, by the presence of additional, optically indistinguishable *soluble* isoenzymes called *iso-cytochromes c*. The latter phenomenon has been shown unambiguously in yeast, where two species, *iso*-1 (major) and *iso*-2 (minor), differing in 22 residue positions, have been isolated (57).

Cytochromes *c* are ubiquitously distributed in nature from the simplest to the most complex organisms. The primary structure of some 30 members of the group has now been established, principally in the laboratories of E. Margoliash, E. Smith, and M. Kamen (58,59). Some of the results have already been described in Chapter 4. Here we are particularly concerned with homologies of particular relevance to the function of the hemoprotein. All such cytochromes, whether animal, plant (fungi), or bacterial, contain some 20 invariant residues out of a total of ~110. Among them are a sequence Gly·X·X·X·X·Gly at or close to the N terminus; a sequence·$\frac{\text{Arg}}{\text{Lys}}$·Cys·X·X·Cys·His· within seven residues of the second Gly above, which contains the two cysteines responsible for binding the heme side chain as well as the His now definitely identified as binding to its metal at the fifth coordination position; a sequence ·Lys·X·Gly·Pro·X·Leu·X·Gly centered around residue 30; a sequence ·Tyr·X·X·Asn·Pro·Lys· (continued in all eucaryotes with ·Lys·Tyr·Ile·Pro·Gly·Thr·Lys·Met·X·Phe) around residue 70 (Table 4-5). This particular sequence is of special interest because it contains the Met residue which has been implicated as occupying the sixth coordination position. For any *two* given cytochromes, even as diverse as cytochrome *c* of horse heart and cytochrome c_2 of *Rhodospirillum rubrum,* the similarities are even more striking: the two cytochromes show 43 identical residues and 33 which may have arisen by single mutational events. Yeast *iso*$_1$-cytochrome *c* differs from that of heart muscle in 22 residues (those of eucaryotes in general show 36 invariants). Yet the latter functions at least as well as the former in the mitochondrial electron-transport chain (for reduction by NADH and the "reductase" system, or for oxidation by the oxidase of yeast, and vice versa).

Cytochrome c_1 appears to differ considerably from cytochrome *c* in its amino acid composition. However, the heme-binding site of the two proteins from beef heart appears very similar, -Lys-Cys-X-X-Cys-His-Thr-Val-.

Most cytochromes of the *c* type are linked to the protein not only by the two thioethyl side chains of the porphyrin to Cys residues, but also through two additional residues liganded strongly to its iron, and hence, like cytochrome *b*, are incapable of interacting with heme complexing agents such as CN^-, NO, CO, or O_2 at neutral *p*H. This situation does not prevail with the class of related cytochromes called *c′*, found in photosynthetic organisms. Since these cytochromes are linked to protein only through the fifth coordination position, they readily react with complexing agents at neutral *p*H. On changing the *p*H a new residue on the protein becomes available for binding at the sixth position, and once attached in this manner these cytochromes behave like classical cytochrome *c*. On the other hand, cytochromes *b* and *c* may become autoxidizable and capable of interacting with complexing agents after denaturation. Thus, denaturation may have opposite effects on different cytochromes, depending on whether it facilitates or breaks bonds to protein ligands.

Cytochrome Oxidase. The terminal members of respiratory chains, charged with the reoxidation of reduced cytochrome *c* by molecular oxygen, are called cytochrome oxidases (59,60). This requirement implies a high redox potential, in part met by their possessing an unusual prosthetic group, heme *a* or *cytohemin.* These cytochromes are invariably tightly bound to membranes (mitochondrial ones in the case of eucaryotes) and can be solubilized and purified only with difficulty, usually in the presence of detergents. Preparations of this type, retaining at least the capacity for catalyzing the same overall reaction after activation by a lipid, have been obtained with a $M = 200{,}000$ daltons and containing 2 heme residues and 2 atoms of Cu. Although such preparations yield only a single isolatable prosthetic group (hemin *a*), they and the more complex oxidases from which they were derived contain roughly equal amounts of a heme capable of reacting with cyanide (as measured spectrophotometrically) called cytochrome a_3, and one half inert to this agent, called cytochrome *a.* Similarly the copper, present in all purified preparations in amounts stoichiometric to that of total heme can be equipartitioned into two species: one, stoichiometric to cytochrome *a* is accessible to chelators, the other, stoichiometric to a_3, is not. In general, CN^- and N_3^- react with both a_3 and Cu; EDTA and salicylaldoxime reaction only with Cu; and F^- reacts only with a_3.

ELECTRON TRANSPORT IN MITOCHONDRIA: THE ELECTRON-TRANSPORT CHAIN

Methods of Investigation

Several types of experiment have contributed to the present state of information in this area: (1) application of sophisticated spectroscopic techniques to measure the content of various suspected carriers (NAD, flavoproteins, ubiquinones, cytochromes) and the kinetics of their reduction and reoxidation in various respiratory particles; (2) attempts to strip mitochondria of the enzymes involved in substrate oxidation, oxidative phosphorylation, and other respiration-linked activities, in order to be able to focus on those reactions responsible and essential for electron transport per se; (3) attempts at further comminution of mitochondria and respiratory subparticles to obtain respiratory enzyme complexes free of nonessential proteins, and susceptible to further purification, in order to yield essentially homogeneous preparations whose properties, function, and interdependence could be investigated; (4) attempts at reconstructing mitochondrial electron transport by the use of such preparations together with soluble enzymes, and (5) the use of respiratory inhibitors. The technique outlined under (1) is essentially an extension of Keilin's work using modern technology and instrumentation, and has been carried out notably by Britton Chance and his colleagues. That under (2) was introduced by the laboratories of A. Lehninger and D. E. Green and pursued by them, E. Racker, and many others.

Electron Transport in Beef Heart Mitochondria

For a variety of reasons, the use of mitochondria from bovine heart muscle offers particular advantages for the study of mitochondrial electron transport, i.e., oxidation of NADH and succinate by O_2. It was the first tissue from which a modified respiratory particle had been isolated [the "Keilin–Hartree" preparation (61) first described in 1939]; the electron-transport system appears to be particularly stable in these mitochondria, perhaps because of a particularly low content of proteolytic enzymes; also there is present a minimal number of other respiratory-chain-linked dehydrogenases (e.g., both the choline and α-glycerophosphate dehydrogenases appear to be absent), and there do not appear to be any alternative pathways for the oxidation of NADH such as exist in mitochondria from mammalian liver.

Green and his collaborators have pioneered in the preparation of a number of different kinds of submitochondrial electron-transport particles (ETP) and their further comminution and dissociation into four functional complexes, each capable of catalyzing one of the component reactions of electron transport (62). These complexes, and more recently their more highly resolved individual components, have been shown to be capable of mutual interaction and recombination leading to the reconstruction of the electron-transport chain.

Mitochondria isolated from a variety of sources, are remarkably similar with respect to their essential components (Table 15-10), the reactions catalyzed [reaction (15-36)], and the speed of electron transport:

$$\text{Substrate} \rightarrow \text{NAD} \rightarrow \text{fp}_D \cdot \text{NHI} \rightarrow \text{CoQ} \,;\, \text{cyt } b \rightarrow \text{cyt } c_1 \rightarrow \text{cyt } c \rightarrow \text{cyt}(a + a_3) \rightarrow O_2$$

$$\text{Succinate} \longrightarrow \text{fp}_S \cdot \text{NHI} \longrightarrow \text{(CoQ)} \qquad (15\text{-}36)$$

Some of the principles and methodologies that have been successful in obtaining and characterizing respiratory subparticles and complexes are outlined in Fig. 15-8. Among the methods used (63–69) are the treatment of particles by (1) osmotic shock; (2) high-speed grinding or shaking (with or without the use of glass beads or abrasives); (3) sonication; (4) exposure to specific enzymes such as proteases (Nagarse) or phospholipases; (5) selective extraction or dissolution by means of surface-active agents such as bile salts [cholate, taurocholate, and deoxycholate (DOC)], digitonin, or nonionic detergents (Triton X-100, Lubrol WX); (6) exposure to chaotropic ions[9]; and (7) exposure to other protein denaturants (urea, guanidinium, sodium dodecyl sulfate). By judicious application and combination of these techniques one obtains what are essentially pure, vesicularized fragments of the inner mitochondrial membrane, or derivatives thereof, which retain various of the elements of electron transport but are devoid of all other NAD-linked and many, if not all, flavin-linked dehydrogenases, the components of the system that link respiration to the phosphorylation of ADP, and of other proteins required for the performance of osmotic and mechanical work by the intact mitochondria (Fig. 15-8).

Complexes of the Respiratory Chain

After removal from the electron-transport particle of those proteins that do not form part of the inner mitochondrial membrane proper, the latter can be solubilized by means of bile salts and fractionated by salting out into four complexes (see Fig. 15-9), capable of carrying out the following partial reactions (62,67,68,70–73).

$$\text{NADH} + \text{H}^+ + \text{CoQ} \xrightarrow{\text{complex I}} \text{NAD}^+ + \text{CoQH}_2 \qquad \text{NADH : CoQ (oxido) reductase} \qquad (15\text{-}37)$$

$$\text{Succinate} + \text{CoQ} \xrightarrow{\text{complex II}} \text{fumarate} + \text{CoQH}_2 \qquad \text{succinate : CoQ (oxido) reductase} \qquad (15\text{-}38)$$

$$\text{CoQH}_2 + 2\text{cyt } c(\text{Fe}^{3+}) \xrightarrow{\text{complex III}} \text{CoQ} + 2\text{cyt } c(\text{Fe}^{2+}) + 2\text{H}^+ \qquad \text{hydroCoQ : cyt } c \text{ (oxido) reductase} \qquad (15\text{-}39)$$

$$4\text{Cyt } c(\text{Fe}^{2+}) + O_2 + 4\text{H}^+ \xrightarrow{\text{complex IV}} 4\text{cyt } c(\text{Fe}^{3+}) + 2\text{H}_2\text{O} \qquad \text{cytochrome oxidase (cytochrome } c : O_2 \text{ oxidoreductase)} \qquad (15\text{-}40)$$

[9] Chaotropic agents enhance the partition of apolar residues between an aqueous and a nonaqueous solvent phase, probably because of their disruptive effect on water structure (69). The order of effectiveness is $SCN^- > ClO_4^- > NO_3^- > I^- > Br^- > Cl^- < SO_4^{2-}, Ac^-, F^-$.

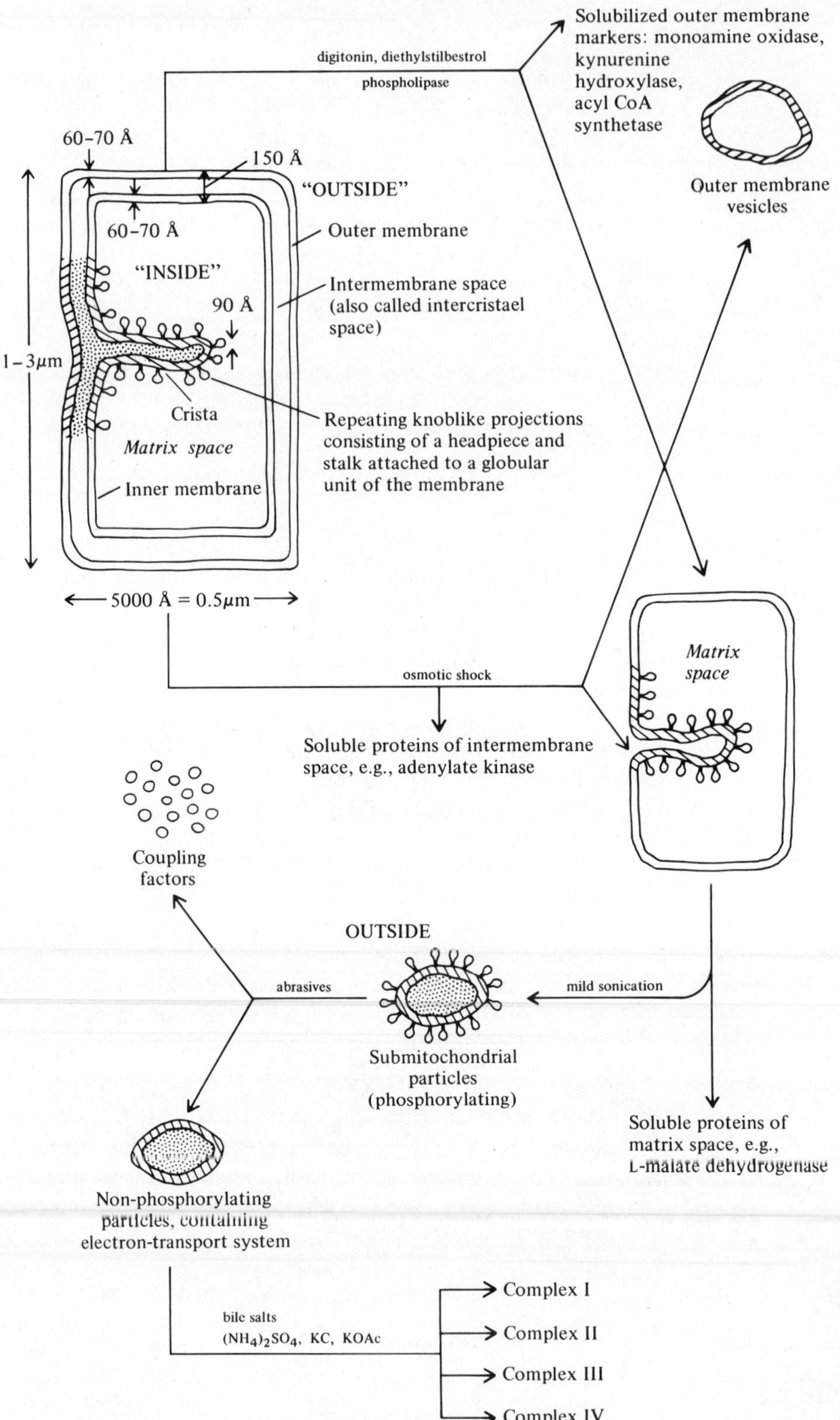

FIGURE 15-8

Schematic representation of principles involved in conversion of mitochondria to their constituents. Note that upon preparation of submitochondrial particles the inner membrane now point "outward" into the medium rather than "inward." After E. Racker (ed.), *Structures and Function of Membranes of Mitochondria and Chloroplasts,* Van Nostrand Reinhold, New York, 1970; D. O. Hall and J. M. Palmer, *Nature,* **221:** 717 (1969).

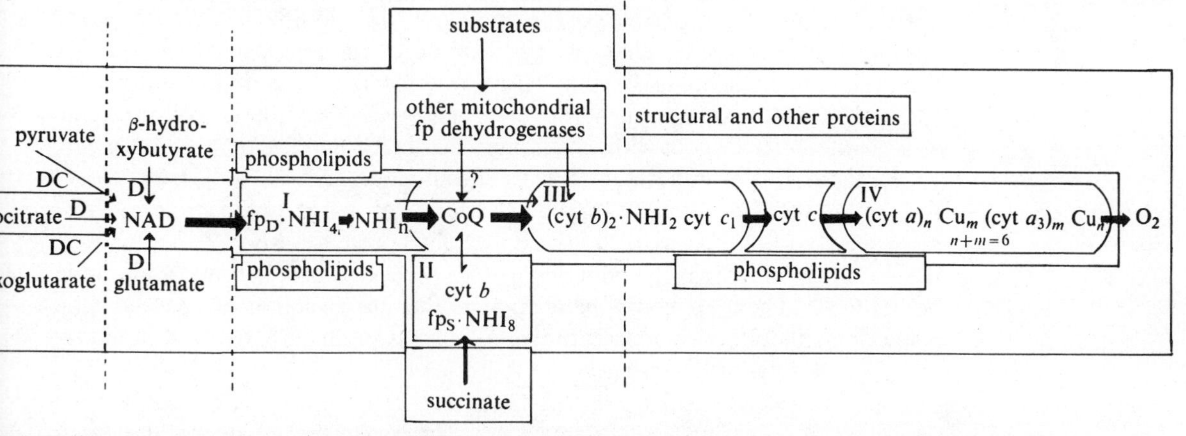

FIGURE 15-9

Electron-transport system of mitochondria from beef heart. D, dehydrogenase (NAD-linked); DC, dehydrogenase complex (NAD-linked); fp, flavoprotein; fp_D, NADH dehydrogenase; fp_s, succinate dehydrogenase; NHI, nonheme iron. Black lines indicate the mitochondrial system, containing enzymes and other components localized on outer and inner membranes and in the intramitochondrial matrix. Thus the systems linked to the outer membranes can be stripped away and the intramitochondrial content liberated; electron-transport particles (ETP) are obtained and are capable of catalyzing the key reactions NADH → O_2 and succinate → O_2, and sometimes also of retaining the capacity for oxidative phosphorylation. By further comminutive procedures there can then be obtained four complexes, I through IV. Complex I: NADH-CoQ (oxido)reductase; complex II: succinate-CoQ (oxido)reductase; complex III: $CoQH_2$-cytochrome *c* (oxido)reductase; complex IV: cytochrome oxidase.

Complex I consists of the following components: the NAD dehydrogenase (fp_D), a ferroflavoprotein of $M = 70{,}000$ containing FMN, NHI, and acid-labile sulfur in the ratio 1 : 4 : 4; a separate NHI protein ($g = 1.96$), free of additional prosthetic groups but containing an equal number of Fe and labile sulfur atoms (estimated $M \simeq 60{,}000$ for each 2 NHI; but 22 NHI/FMN have been identified in the complex in addition to those associated with fp_D); and additional proteins of unknown function possibly involved in stabilizing the overall structure (74,75) as well as bound CoQ (~3/FMN) and phospholipid (22 percent). The minimal molecular weight of the complex is ~600,000. The fp_D as well as the NHI protein have been isolated and the latter has been shown to be capable of functioning as electron acceptor for the former. Nevertheless, the properties of the purified fp_D, particularly with respect to its kinetics and acceptor specificity as well as its behavior toward inhibitors (insensitivity to barbiturates, piericidin, and rotenone; sensitivity to mercurials and *o*-phenanthroline) are such as to suggest that it has undergone irreversible structural modifications during extraction; conversion of a bridged to a tetrahedral Fe complex has been suggested. Complex I isolated from baker's yeast (*Saccharomyces cerevisiae*) appears to be lacking in the separate NHI component, in phosphorylating activity at this site (page 690), and insensitivity to site I inhibitors.

Complex II consists of succinate dehydrogenase (fp_S), a ferroflavoprotein with FAD linked to the protein ($M \simeq 200{,}000$) in covalent linkage (Table 15-5*B*), and containing, probably, all the Fe and labile S of the complex (eight each per FAD; $g = 1.92$), plus cytochrome *b* (one per FAD) and phospholipid. The complex catalyzes a 2-thenoyltrifluoroacetone-sensitive reduction of CoQ by succinate.

Complex III is responsible for linking CoQ, the physiological acceptor for the two branches of the chain to cytochrome *c*. Its composition is complex and has been claimed to be the following: one molecule of cytochrome c_1 ($M = 40{,}000$), two molecules of cytochrome b ($M = 25{,}000$), one molecule of a specific NHI protein with two atoms each of Fe and S ($M = 30{,}000$; $g = 1.90$), three to four molecules of a structural component called "core protein" ($M = 40{,}000$ to $50{,}000$), plus an additional basic polypeptide of low molecular weight. The total molecular weight (protein basis) of the complex is ~270,000 daltons; it contains ~30 percent phospholipid and some bound CoQ. The NHI component appears the site of action of characteristic respiratory inhibitors such as antimycin A and is intimately involved in oxidative phosphorylation at this site.

Complex IV is identical with cytochrome oxidase and has already been described. In its functionally integrated form, on the inside of the inner membrane, it may exist as a hexamer (three a, three a_3, six Cu) with M variously reported as 2.5 or 6×10^5. Its phospholipid content is ~25 percent.

The strongest evidence for the reality of these complexes as units of mitochondrial electron transport is threefold: (1) the reproducibility and defined stoichiometry of their composition, (2) their capacity to be recombined into functional units of electron transport, and (3) their response to certain characteristic inhibitors. Thus, complex I plus complex III reconstitute the mitochondrial NADH : cytochrome *c* (oxido)reductase; complex II plus complex III, the mitochondrial succinate : cytochrome *c* (oxido)reductase; complex I plus III plus IV, the mitochondrial NADH oxidase; complex II plus III plus IV, the mitochondrial succinoxidase; and finally I plus II plus III plus IV, the complete electron-transport sequence or respiratory assembly, that is, a combined NADH- and succinoxidase. Efficient reconstruction of these mitochondrial activities requires the stoichiometric recombination of the various complexes when admixed at relatively high concentrations together with cytochrome *c* and coenzyme Q. The latter two components are mobile, lipid-soluble carriers, capable of linking the various activities both in the reconstructed system and in the intact ETP or mitochondria; CoQ, because of its long aliphatic side chain, possesses considerable lipid solubility, while cytochrome *c*, which is ordinarily a water-soluble protein, can be made lipid-soluble by complexing with mitochondrial phospholipids.

It should be mentioned, however, that succinoxidase activity has also been reconstructed using only fp_S, a particulate preparation of cyt *b*, cyt c_1, cytochrome oxidase plus cytochrome *c*, CoQ, and phospholipids (67). This and other observations (e.g., the behavior of yeast complex I) suggest that the NHI proteins of complexes I and III may not be absolutely required for electron transport as such but instead may function in its integration with oxidative phosphorylation.

Respiratory Inhibitors

We have already mentioned that certain inhibitors such as HCN, CO, and certain narcotics have played an important role in developing our understanding of cellular respiration. Extension of these studies to a variety of additional inhibitors and the use of sensitive optical techniques[10] has provided a valuable tool for disrupting the chain at various points, thereby elucidating

[10] Localization of inhibitor action is facilitated by application of the "crossover theorem" of Chance and Williams (76), which states that, if an inhibitor is added to an array of redox carriers in their steady state, then those on the reduced side of the inhibited reaction will become more reduced, while those on the oxidized side will become more oxidized.

the sequence of components and their mode of interaction. Specific inhibitors are summarized in Table 15-10. Except for those belonging to class A, the exact nature of the receptor site has not yet been established. In general, some degree of structural integrity of a complex appears to be required for its interaction with most inhibitors, since soluble, phospholipid-free enzymes do not exhibit the characteristic inhibition pattern.

Outstanding Problems in Mitochondrial Electron Transport

Even though the broad outlines of the process and the components responsible are becoming known, considerable uncertainties remain about many details, particularly with regard to mechanism. Among some of the outstanding ones are the following.

Role of Bound and "Free" NADH in Mitochondria. Intact mitochondria are impermeable to all nicotinamide coenzymes.[11] Inside the particle NAD(P) exists in at least two forms or compartments: one relatively freely diffusable and hence able to reach its site of reoxidation, the respiratory dehydrogenase (fp_D), with ease; the other composed of certain more tightly membrane-bound dehydrogenases (for β-hydroxybutyrate, pyruvate, and α-oxoglutarate) which may retain the coenzyme in an environment that resists such easy accessibility. The electron-transport chains for the two forms, although identical in composition, may be spatially separated from one another. Most dehydrogenases appear located in the matrix space, and since the inner membrane appears relatively impermeable to their substrates (organic anions), the latter have to be pumped across the barrier by specific transport systems (see below).

Endogenous versus Exogenous Cytochrome c. Externally added cytochrome *c* appears to be in equilibrium with an endogenous form of the carrier that is essentially dissolved—and relatively isolated—inside the lipophilic environment constituted by the inner membrane and its cristae. Only the endogenous form of the cytochrome is active in electron-transport reactions; exogenous cytochrome *c* can enter and leave the pool readily from the "outside," the intermembrane space. In submitochondrial particles, with their attendant inversion of membrane polarity, the reaction site is now on the "inside," and access has to be facilitated by means of pretreatment of the membrane by detergents such as Lubrol.

Linkage to Other Flavoproteins. The interaction of other mitochondrial flavoproteins, such as the choline, sarcosine, dimethylglycine, and L-α-glycerol-*P* dehydrogenases, and of ETF (page 654) to the respiratory chain remains obscure. CoQ, which is present in sufficiently high concentration and is freely diffusible in the lipoprotein membrane, might provide the required link.

Linkage of NADH and Succinate Chains. It is not certain that Fig. 15-9, which envisages separate electron-transport chains for NADH and succinate as far as cytochrome *b* and a common chain thereafter, is necessarily correct or universally applicable to all cases of mitochondrial electron transport. Other than the stoichiometry and composition of the isolated complexes, no compelling evidence has yet been adduced to rule out an alternative model

[11] To be oxidized reducing equivalents from NADH in the cytosol have to be transferred into the mitochondria by means of a *shuttle* system: either the α-glycerophosphate shuttle (Eq. 11-16) or more commonly, one involving the transport of L-*malate* (and L-*aspartate* across membrane: NADH + OAA $\xrightleftharpoons{\text{MDH(cyto)}}$ L-malate + NADH; L-malate (cyto) $\rightleftharpoons$ L-malate (mito); L-malate + NAD $\xrightleftharpoons{\text{MDH(mito)}}$ OAA + NADH; OAA + NH_3 $\xrightleftharpoons{\text{mito}}$ Asp; Asp(mito) $\rightleftharpoons$ Asp(cyto); Asp $\xrightleftharpoons{\text{cyto}}$ OAA + NH_3.

TABLE 15-10
Different classes of respiratory inhibitors

Class	*Examples*	*Specifically affected complex or reaction*	*Concentrations usually employed*
A	HCN, HN_3, CO	IV Cyt *c* (Fe^{2+}) → O_2	≦ 10^{-4} M (CN^-)
B	antimycin A[a]	III Cyt *b* (Fe^{2+}) → cyt c_1 (Fe^{3+}) (perhaps at NHI)	Stoichiometric with cytochromes
	2-Alkyl-4-hydroxyquinoline-*N*-oxide	III Same as above	~10^{-6} M
	3-(2-Methyloctyl)naphthoquinone	III ?	~10^{-5} M
	BAL + O_2	III ?	~10^{-3} M
C	Rotenone,[b] piericidin[c]	I FMN → Q (at or after NHI)	Stoichiometric with fp_D (μM)
	Barbiturates (amytal, seconal)	I Same as above	≧10^{-3} M
D	Thenoyltrifluoroacetone[d]	II Peptide-FAD → Q and/or cyt *b* (Fe^{3+})	≧10^{-4} M
E	Oligomycin,[e] rutamycin,[e] aurovertin; Atractylate[f]; Bongkrekic acid[g]	Inhibit respiration (NADH → O_2 or succinate → O_2) when coupled to phosphorylation	
F	Uncoupling agents (substituted phenols, phenylhydrazones, gramicidin, arsenate, dicoumarol)	Stimulate respiration when rate is limited by phosphorylation or blocked by oligomycin	
G	Arsenite, Cd^{2+}	Inhibit respiration for α-oxoglutarate and pyruvate only by virtue of blocking dihydrolipoyl dehydrogenase; inhibit phosphorylation, probably at or prior to site 3	

[a] An antibiotic elaborated by a *Streptomyces* species: a substituted 3-formamidosalicylamide.

[b] A potent fish poison isolated from plant roots and having the structure

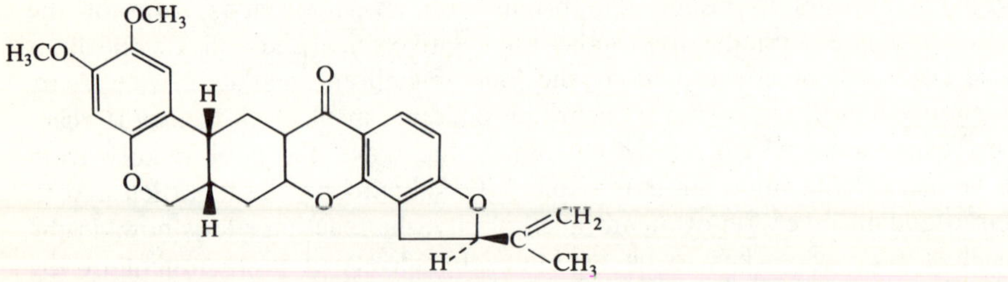

[c] An insecticidal antibiotic with the structure (NB similarity to CoQ)

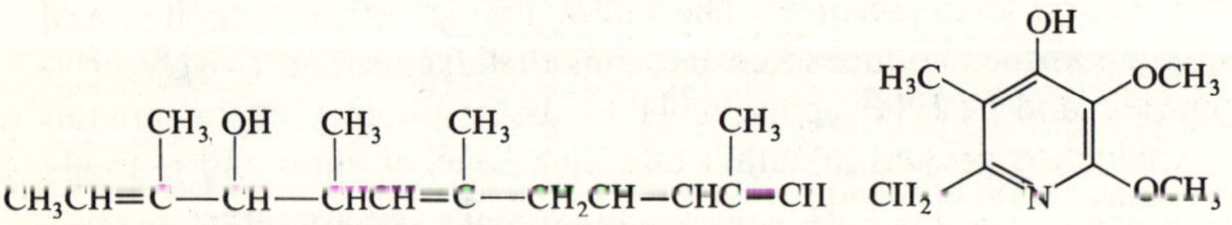

$CH_3CH{=}C(CH_3){-}CH(OH){-}CH(CH_3)CH{=}C(CH_3){-}CH_2CH{-}CHC(CH_3){=}CH\ CH_2$–(2-pyridyl: OH, H_3C, OCH_3, OCH_3)

[d] 4,4,4-Trifluoro-1-(2-thienyl)-1,3-butanedione.

[e] Antibiotics obtained from actinomycetes.

[f] *O*-Glucoside from the plant *Atractylis gummifera*.

[g] Antibiotic obtained from *Pseudomonas cocovenenans*.

with two parallel chains to O_2, which may switch electrons back and forth at one or more sites above cytochrome *b*. Such links would be required to account for the spectrophotometric evidence that the complete cytochrome complement is reducible by *either* substrate and to explain "reversed electron flow" (page 697). The connections may be represented by some of the non-heme iron compounds of the chains.

One-Electron versus Two-Electron Mechanisms. Although cytochrome-linked electron transport is usually expressed in terms of one-electron steps, an equally valid picture can be constructed in which cytochrome oligomers or complexes (e.g, dimers, tetramers) interact with one another and transfer electrons or reducing equivalents in $2e^-$ (or perhaps $4e^-$) packets.

Role of CoQ. The hypothesis of CoQ as a mobile component acting in series just prior to cytochrome *b* is highly plausible (77). It is not impossible, however, that some molecules of CoQ may function in parallel rather than in series.

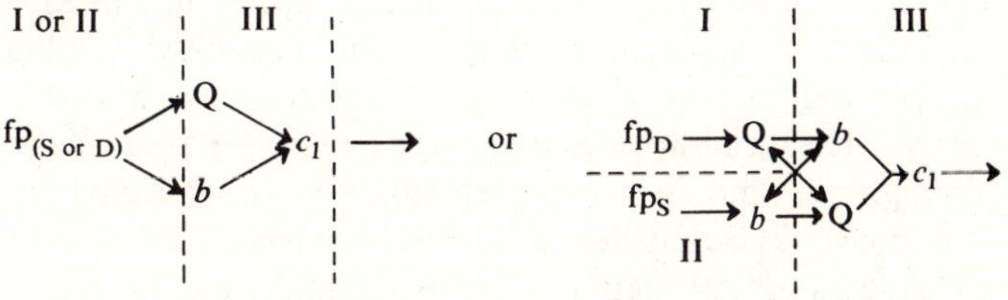

I or II | III | I | III

$fp_{(S \text{ or } D)}$ → Q, b → c_1 → or fp_D → Q → b, fp_S → b → Q → c_1 →

II

Role of NHI. Three of the complexes (I, II, and III) contain NHI, and distinct NHI proteins devoid of other prosthetic groups have been isolated from two of them (I and III). In addition, complex I contains NHI rather tightly associated with its dehydrogenase, while the complete NHI moiety of complex II appears to exist exclusively in this form. Whether this truly reflects the situation in the chain *in situ* is not known. Three different forms of NHI can be distinguished in mitochondria, or inner membranes, on the basis of their EPR signals, but direct evidence for the rates and stoichiometries characterizing these entities as electron carriers is still lacking.

Cytochrome Oxidase. A number of major questions remain regarding this entity (59). Among them we may mention the absence of a satisfactory explanation concerning the difference, if any, between the cytochrome *a* and a_3[12] moieties of the enzyme; or for the function and localization of Cu in the scheme of electron transfer[13]; and the lack of any satisfactory mechanism for the partition of electrons in the course of the latter, which appears to proceed without any readily distinguishable intermediate during the $4e^-$ reduction of the receptor by the $1e^-$ donor.

Mode of Electron Transport Between Components. Considerable uncertainty prevails as to the mechanism of actual transfer of electrons between two adjacent members of the chain, particularly the ones that are relatively fixed directly to or in the membrane (i.e., all members of the chain except for cytochrome *c* and CoQ). One does not know whether transfer occurs by direct contact, thermal collision of relatively flexible extremities, or by tunneling of electrons through the lipid medium; nor is one certain whether prosthetic groups have to be involved directly or whether transfer through systems capable of effective overlap with these groups is possible.

Role of Mitochondrial Lipid. Mitochondria, like other organelles, contain a relatively high proportion of lipid—more than 90 percent of which is phospholipid in particles from vertebrates—as an integral part of the membrane system. Isolated submitochondrial particles, as well as purified complexes derived from them retain the bulk of this component. Thus, mitochondria contain 26 percent lipid, the isolated complexes 27 percent lipid, on the average.

Mitochondrial phospholipids appear to be distinguished by a particularly high degree of unsaturation. Those of heart muscle contain 3.2 double bonds

[12] Under appropriate conditions a cytochrome *a* preparation can be prepared in which the a_3 hemes have been destroyed peroxidatively after prior reduction (78).

[13] An absorption band in the near infrared (830 nm) is thought to be due to this entity.

per atom of phospholipid phosphorus; it consists of approximately 35 percent phosphatidylcholine, 35 percent phosphatidylethanolamine, and 16 percent cardiolipin (a lipid that appears to be relatively specific for mitochondria). Unsaturation appears to play some kind of functional role: Activity of any of the four complexes is reduced or abolished by extraction with lipid solvents and can be reconstituted by the subsequent addition of these compounds; unsaturated phospholipids such as those from mitochondria, egg, beef, or soybean lecithin or synthetic oleyl lecithin are active in direct proportion to their degree of unsaturation; saturated phospholipids are inactive. Restoration of activity appears to be possible with a wide variety of compounds and it is not even necessarily restricted to phospholipids; cytochrome oxidase at least can be rendered fully functional with synthetic detergents. The function of phospholipid or equivalent agents appears to be twofold: (1) to position and maintain the various proteins participating in the chain in conformations active for electron transport and (2) to maximize their productive interactions with other essential components such as CoQ and the (coupling) factors required to link oxidation to phosphorylation.

Oxidative Phosphorylation. Intact mitochondria and certain derived particles are capable of bringing about the formation of ATP coupled to the oxidation of NADH or succinate and also of exhibiting the phenomenon of respiratory control. The latter is defined as the rate-limiting step in *oxidation* and is provided *not* by any of the constituent reactions but by a *phosphorylation* step (or steps). Current concepts of how this is brought about will be discussed below.

Other Components. In addition to the components already described, most mitochondria also contain vitamin K and related naphthoquinones, as well as vitamin E or related compounds. Although the particles and derived preparations can interact with added naphthoquinones, hydroquinones, and certain vitamin E derivatives, the functional significance of these observations remains obscure, except in certain bacteria (e.g., *Mycobacteria*).

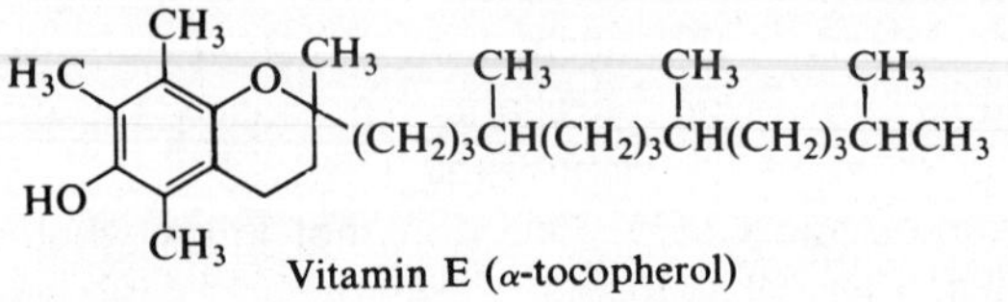

CH_3, H_3C, HO, CH_3, O, CH_3 — $(CH_2)_3\overset{CH_3}{CH}(CH_2)_3\overset{CH_3}{CH}(CH_2)_3\overset{CH_3}{CH}CH_3$

Vitamin E (α-tocopherol)

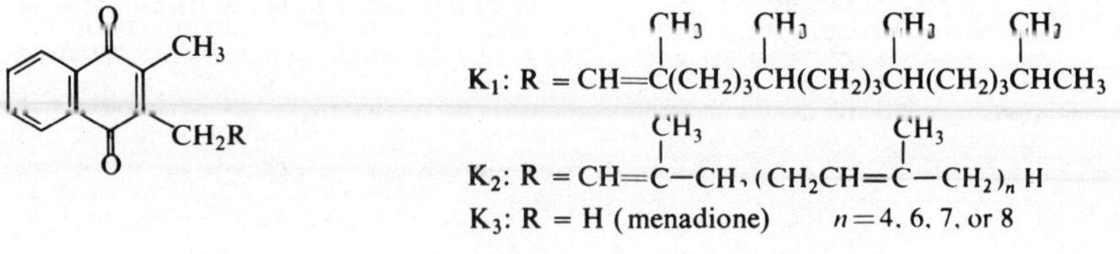

O, O, CH_3, CH_2R

K_1: R = $CH{=}\overset{CH_3}{C}(CH_2)_3\overset{CH_3}{CH}(CH_2)_3\overset{CH_3}{CH}(CH_2)_3\overset{CH_3}{CH}CH_3$

K_2: R = $CH{=}\overset{CH_3}{C}{-}CH_2(CH_2CH{=}\overset{CH_3}{C}{-}CH_2)_nH$

K_3: R = H (menadione) $\quad n = 4, 6, 7,$ or 8

Vitamins K

Enzymes have been isolated from a variety of cells that reduce one of the forms of vitamin K. The corresponding hydroquinones are also readily reoxidized through the respiratory chain in an antimycin-sensitive reaction (Table 15-10). Thus, in the cytoplasm of vertebrate tissues "D-T diaphorase" catalyzes the reduction of vitamin K_3 by *either* NADH or NADPH with equal efficiency; it is also highly sensitive to dicumarol (Table 15-10) at low concentrations. It may be identical with the NADH : cyt b_5 reductase localized in the outer membrane, and responsible for any barbiturate- and rotenone-insensitive oxidation of NADH catalyzed by the particles:

$$\begin{array}{l} \quad\ \ \ulcorner\!\!-\!\!-\!\!-\, fp_{D\backslash T} \longrightarrow K_3 \\ \qquad\qquad\qquad\qquad\ \ \downarrow \\ NADH \longrightarrow fp_D \longrightarrow Q,b \longrightarrow c_1 \longrightarrow \end{array}$$

Variants of Particulate Electron Transport

Bacterial Electron Transport (79). The stoichiometry and sequence of Eq. (15-36) appear to depict reasonably well the patterns of electron flow between substrates and O_2 in the mitochondria of vertebrate and invertebrate animals, of plants, including molds and fungi, and of protozoa. It is also characteristic of particulate electron transport localized in the protoplast membranes of many bacteria (e.g., *Mycobacteria*). In some bacteria, however, variations occur; for example, the autoxidizable cytochrome b_1 replaces cytochrome b; cytochromes c_4 and c_5 replace cytochrome c and c_1, respectively; and cytochromes a_1, a_2, and o, singly or in combination, function as cytochrome oxidases instead of, and sometimes in addition to, cytochrome a_3. Other oxidants with a relatively high redox potential, such as NO_2^-, can replace O_2 as the terminal acceptor. In *E. coli,* for instance, we find flavoproteins, cytochromes b, a_1, a_2, and o; in *Azotobacter vinelandii* we find flavoproteins, cytochromes b_1, c_4, c_5, a_1, and a_2. The following schematic representation modified from one suggested by L. Smith summarizes the situation:

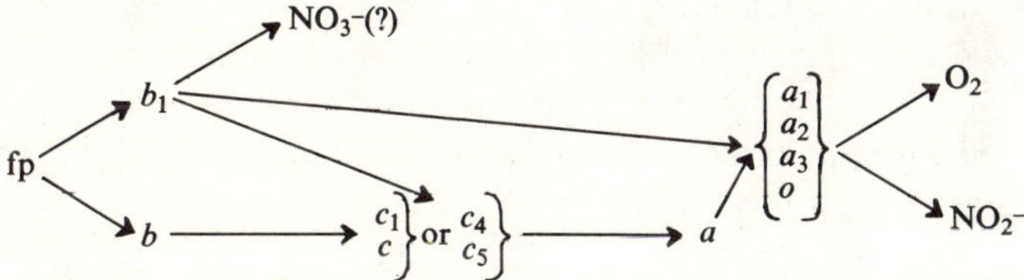

Another variant encountered in some bacterial systems (e.g., *Mycobacteria*) is the substitution of naphthoquinones (vitamin K) for benzoquinones (CoQ).

Microsomal Electron Transport. Besides the capability of oxidizing NADPH by means of the flavoprotein reductase–adrenodoxin–P450 system [Eq. (15-16)], this cell fraction has also been shown to be capable of oxidizing NADH through a flavoprotein reductase (80) linked to cytochrome b_5 (Table 15-9). Cytochrome b_5, a tightly membrane bound carrier,[14] can be isolated in two forms, one without the use of proteolytic enzymes during the solubilization step with $M = 25{,}000$, the other by means of such enzymes (81) with $M = 11{,}000$ to $12{,}000$. The former probably corresponds to the physiologically active form and is oxidized by a CO-sensitive oxidase perhaps identical with P450. We therefore visualize the following possible pathways for microsomal electron transport:

$$\begin{array}{l} NADPH \longrightarrow fp_A \longrightarrow \text{adrenodoxin} \searrow \\ \qquad\qquad\qquad\qquad\qquad\qquad\qquad\quad \text{P-450} \longrightarrow O_2 \quad (XH \rightarrow XOH) \\ NADH \longrightarrow fp_B \longrightarrow \text{cytochrome } b_5 \nearrow \end{array}$$

fp_A = NADPH : adrenodoxin reductase

fp_B = NADH : cytochrome b_5 reductase

[14] A similar carrier plus reductase is also present in the *outer* membrane of liver mitochondria.

OXIDATIVE PHOSPHORYLATION AND RELATED PROCESSES

During the period 1937 to 1941 Belitzer and Tzibakova (82) and Kalckar (83) realized that the large amount of free energy liberated during the complete oxidation of metabolites via the citric acid cycle could be utilized to drive the synthesis of ATP, a process now referred to as oxidative phosphorylation. These studies were continued and extended vigorously in the laboratories of Ochoa, Cori, Lardy, Green, Lipmann, and Lehninger, and by 1950 it had been established that the ability to catalyze oxidative phosphorylation was a property exclusively of mitochondria; that respiration (i.e., oxidation) and phosphorylation could be dissociated or *uncoupled* from each other by certain agents, such as 2,4-dinitrophenol; that phosphorylation was almost exclusively linked to the electron-transport chain rather than to substrate-level oxidation[15]; and that the number of moles of ATP formed relative to the gram-atoms of oxygen consumed (i.e., the P/O ratio) approached integral values for different substrates undergoing one-step oxidation. P/O ratios → 4 were reported for α-oxoglutarate going to succinate; 3 for malate → OAA, glutamate → α-oxoglutarate, or β-hydroxybutyrate (BOB) → acetoacetate; and 2 for succinate → fumarate or malate.

For our discussion of the processes that link phosphorylation to respiration it is useful to expand Eq. (15-36) and recast Fig. 15-9 in terms that include both processes. This is done in Fig. 15-10.

Phosphorylation Sites and P/O Ratios

A number of techniques, some direct, some indirect, have been used to develop a relatively consistent picture concerning the number and the location of phosphorylation sites relative to the respiratory chain. A direct comparison of the P/O ratios—measured most frequently by means of polarographic techniques for oxygen uptake, and by sensitive tracer (^{32}P) or enzymatic measurements of ATP formation or ADP consumption—showed that true respiratory-chain-linked phosphorylation during the oxidation of NADH (or of NAD-dependent substrates) yielded a value of 3, while that of flavoprotein-dependent substrates such as succinate led to a value of 2. Since all steps beyond the initial flavoproteins are similar for all substrates, one phosphorylation site (no. 1) must be localized within complex I. The remaining two sites are thus located in the span between CoQ (cyt *b*) and O_2. One of these (site 2) may be localized between CoQ and cytochrome c_1 (or *c*) (i.e., in the area of complex III). This assignment is supported by the observation that in a system in which cytochrome oxidase is blocked by HCN, the oxidation of NADH or of BOB by added cyt *c* takes place with a $P/2e^-$ (equals P/O) ratio of 2. Localization of the third site (site 3) in the area of cytochrome oxidase follows from the experiments just cited and from the fact that ascorbate [plus tetramethyl-*p*-phenylenediamine (TMPD) as carrier], which channels electrons into the chain exclusively at cytochrome *c* in a reaction completely insensitive to antimycin A, is oxidized with a P/O ratio equal to 1.

Another method of approach employed by Chance and Williams is to determine phosphorylation sites as crossover points. When intact mitochondria are supplied with P_i and substrate, but no ADP, they exhibit the phenomenon of *respiratory control,* in that the rate of respiration remains low (state 4) until ADP is added to convert these *tightly coupled* particles to

[15] Of all the citric acid cycle (CAC) substrates, only α-oxoglutarate is capable of undergoing a substrate-level phosphorylation to produce one molecule of ATP per molecule of substrate oxidized (see page 613).

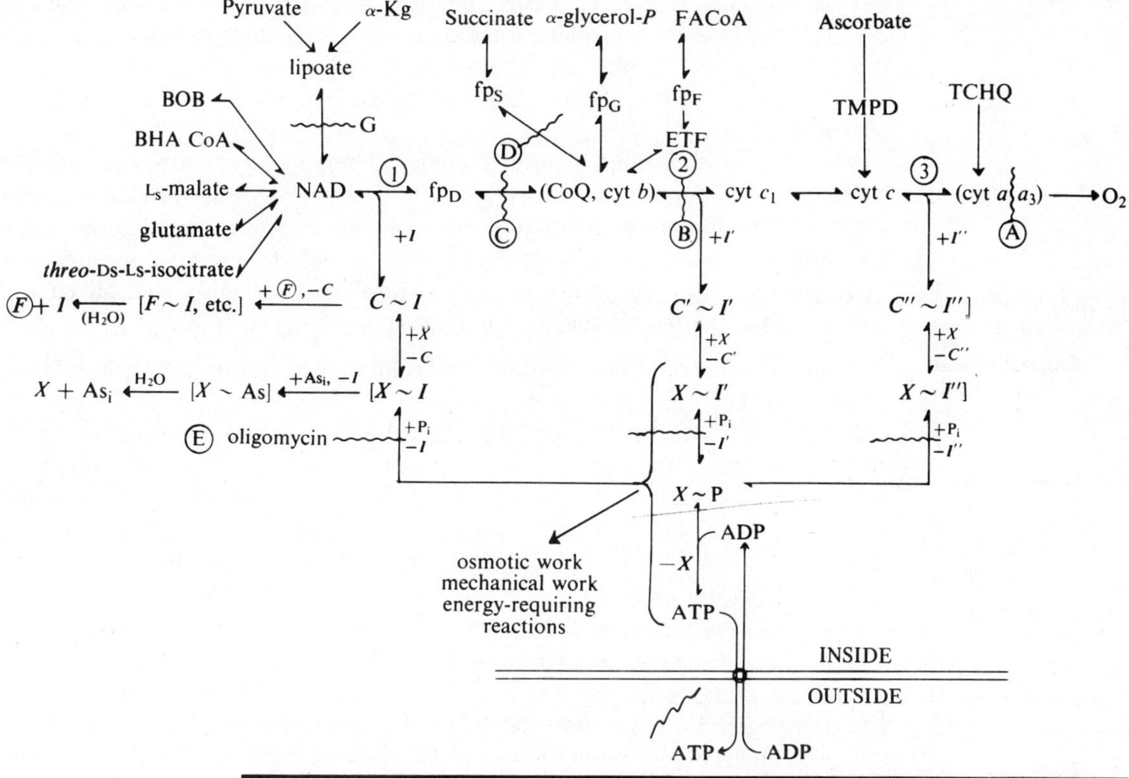

FIGURE 15-10

Linkage of respiratory chain and phosphorylation in mitochondria. The flow of the energy-rich linkage from the three sites in the chain to ATP may now require some modification. ①, ②, and ③ are phosphorylation sites. Ⓐ, Ⓑ, Ⓒ, Ⓓ, and Ⓔ are sites of action of respiratory inhibitors (see Table 15-10). Site 1 is missing in baker's yeast. BOB, D-β-hydroxybutyrate; BHA, L-β-hydroxyacyl-; fp_F, fatty acyl dehydrogenase; ETF, electron-transferring flavoprotein; TMPD, tetramethyl-p-phenylene diamine; TCHQ, tetrachlorohydroquinone; $C \sim I$ (etc.), intermediates of oxidative phosphorylation with $\Delta G^{0\prime} \ll 0$ for reaction $C \sim I \xrightarrow{(H_2O)} C + I$ (etc.); E, uncouplers, such as dinitrophenol (Table 15-10); As_i, $HAsO_4^{2-}$.

state 3 in which rapid respiration occurs until the acceptor is exhausted (Table 9-5). Thus, state 4 can be regarded as inhibited and the transition state 3 → state 4 as analogous to the transition that occurs when a respiratory inhibitor is added to the chain functioning in the steady state. By observing which carriers become more reduced on exhaustion of ADP, the steps NADH → fp_D, $b \rightarrow c_1$, and $a \rightarrow a_3$ become implicated as phosphorylation sites.

Finally, if one postulates a model in which each individual phosphorylation step requires the oxidation of one respiratory carrier concomitant with the reduction of its neighbor in the chain, it is clear that efficient chemical coupling will require an absolute negative free energy span between these carrier couples greater than that liberated in the hydrolysis reaction ATP + H_2O → ADP + P_i. With the known intracellular concentrations of the various reactants this corresponds (chapter 2) to a $\Delta G' \approx -14{,}000$ cal (or 300 mV for a $2e^-$ equivalent reaction). From the known potentials, then, the only relevant couples are NADH/fp ($-320/-63$ or 257 mV); cyt b (CoQ)/cyt c_1 ($-40/+250$ or 290 mV) and cyt a/cyt a_3 [290/600(?) or 310 mV].

Nature had to develop a highly complex, integrated system to perform the coupling of oxidation and phosphorylation; this system is provided by the array of the mutually interacting components of the inner mitochondrial membrane of eucaryotic cells. In procaryotes, which do not contain mitochondria, the relevant systems are integrated into the less highly specialized cell envelope (plasma membrane) and may be somewhat less efficient.

General Models and Partial Reactions

General Models. The key to the riddle of oxidative phosphorylation consists of determining how free energy liberated by oxidation–reduction reactions in the chain can be utilized, after a series of intermediate, presumably roughly isoenergetic, displacement reactions to form an anhydride between P_i and ADP. Following a suggestion originally advanced by Lipmann, such compounds and their bonds hydrolyzed or generated are indicated by the sign ~ (the "tilde"), e.g., ADP~P or, more generally, A~B (Chapter 2).

Three types of models can be envisaged and have been proposed.

1. Those that postulate some form of *mechanochemical coupling* (84,85); that is, the energy liberated during the redox step is "trapped" in a *conformationally* or *configurationally* altered protein(s) of high potential energy. Relaxation of this "energized" state to the thermodynamically stable initial one is coupled to the generation of ATP. Conceptually, at least, this model can be regarded as the reverse of what occurs in the course of muscular contraction where the hydrolysis of ATP is utilized to perform mechanical work.

2. *Chemiosmotic coupling.* An explicit version of such a model has been formulated and espoused by P. Mitchell (86). It envisages a chemiosmotic coupling for the "vectorial" oxidation of a carrier, C_{red}, localized on one side of a membrane,

$$H_2O + C_{red} \overset{\rightarrow}{\Vert} C'_{ox} + H_2O \rightleftharpoons OH^- + HC^+_{ox} \overset{\rightarrow}{\Vert} C'_{red}\text{-}OH + H^+ \qquad (15\text{-}41a)$$

or

$$C_{red} \overset{\rightarrow}{\Vert} C'_{ox} \rightleftharpoons OH^- + C_{ox} \overset{\rightarrow}{\Vert} C'_{red} + H^+ \qquad (15\text{-}41b)$$

or, simply,

$$\begin{array}{ccc} H_2O + \frac{1}{2}O_2 & \overset{\rightarrow}{\Vert} & DH_2 \\ \downarrow & \Vert & \downarrow \\ 2OH^- & \Vert & D + 2H^+ \end{array} \qquad (15\text{-}41c)$$

(symbolized by the double line with the arrow pointing toward the "inside"), by means of a second one; C'_{ox}, bound on the other side. If the membrane is impermeable to small ions including H^+ and OH^-, the primary consequence of the redox reaction will be the generation of a pH gradient. One can then conceive of this gradient being used for the vectorial reversal of any reaction possessing a high negative free energy of hydrolysis with the highly exergonic formation of H_2O from its ions providing the driving force.

$$\text{H}\!+\!\text{O—PPAd} \quad \text{P}_i\!+\!\text{OH},\ {}^-\text{HO},\ \text{H}^+ \rightleftharpoons \begin{matrix}(\text{ADP}\sim\text{P})\\ \text{AdPPO}\sim\text{P}\end{matrix}\quad \text{H}_2\text{O}\,\|\,\text{H}_2\text{O} \quad \text{also} \quad (\leftarrow \text{H}^+;\ \rightarrow \text{OH}^-) \qquad (15\text{-}42)$$

or, more generally,

$$\text{H—O—}X \quad Y\text{—OH},\ B^-,\ A^+ \rightleftharpoons X\text{—O}\sim Y \quad BH\,\|\,AOH \quad (\leftarrow A^+;\ \rightarrow B^-) \qquad (15\text{-}43)$$

The protein(s) responsible for catalyzing reaction (15-41), (15-42), or (15-43) is called a *translocase* and is conceived of as being integrated into the membrane in an anisotropic manner, permitting access of protons only from one side (here the inside) and of $^-$OH ions from the opposite one. Several points of interest should be mentioned in connection with this model.

a. Mitochondria do appear to be quite impenetrable to protons, and a proton gradient of the correct polarity (i.e., ejection of protons) appears to be generated as a result of electron transport. Values of the "H^+/O" ratio vary from 6 for β-hydroxybutyrate to 4 for succinate. What is not certain is whether these pH gradients are obtained as a prerequisite for, or as a *result* of, energy coupling, for instance as a consequence of the translocation of cations in the opposite direction.

b. The translocase reaction [Eq. (15-44)] is in effect the ATPase reaction in reverse. Indeed an anisotropic ATPase complex is present in and has been isolated from mitochondrial inner membranes (88–92) (see below). In the particles, the ATPase pumps protons inward as predicted and it is most active under conditions that would be expected to affect its "vectorial," anisotropic properties (i.e., the ability to sequester H^+ and $^-$OH in separate compartments; disruption of the membranes leads to activation of ATPase).

c. Although it has been possible to generate ATP as a result of a pH gradient by bathing mitochondria (in the presence of K^+ plus valinomycin, page 704) first in base and then in acid, only a small amount of ATP can be so generated. This need not necessarily exclude the model on quantitative grounds, because the gradient of proton activity required for ATP generation (about 5 pH units) is supplemented under actual operating conditions by a membrane potential, generated by the maintenance of total cation—and not just proton—gradients:

$$\frac{(\text{ATP})}{(\text{ADP})} = \frac{(\text{P}_i)}{K(\text{H}_2\text{O})}\,\frac{(\text{H}^+)_{\text{inside}}}{(\text{H}^+)_{\text{outside}}} = \frac{(p\text{H}_{\text{outside}} - p\text{H}_{\text{inside}})(\mu RTF)}{2.303} = (p\text{H}_{\text{outside}} - p\text{H}_{\text{inside}})\,\frac{\mu}{0.609}, \text{ at } 25^\circ \qquad (15\text{-}44)$$

where μ is the membrane potential (in volts), R the gas constant, T absolute temperature, and F the Faraday.

Although most current theories are still couched in terms of the third model, one postulating *chemical coupling* by means of distinct reaction intermediates (93), it is not impossible that the solution which will finally emerge will incorporate features of two or perhaps of all three of the models; some of the current thinking already does. In any event, Mitchell's ideas have been exceedingly productive in generating a large amount of novel and ingenious experimentation, in focusing attention on protein–lipid membranes and some of their fundamental properties, and in developing valid analogies between oxidative phosphorylation, photophosphorylation, and energy-requiring transport or translocation of small molecules across such membranes. We shall return in a later section to certain features of these phenomena.

3. *Biochemical coupling.* In searching for appropriate biochemical models investigators have been forced to rely heavily on possible analogies to the well-authenticated instances of oxidative phosphorylation at the *substrate level*:

$$\text{Glyceraldehyde-3-}P + \text{NAD}^+ + \text{P}_i \rightleftharpoons \text{1,3-di}P\text{-glycerate} + \text{NADH} \quad (15\text{-}45a)$$

$$\text{1,3-Di}P\text{-glycerate} + \text{ADP} \rightleftharpoons \text{glycerate-3-}P + \text{ATP} \quad (15\text{-}45b)$$

$$\text{Glyceraldehyde-3-}P + \text{NAD}^+ + \text{ADP} + \text{P}_i \rightleftharpoons \text{glycerate-3-}P + \text{ATP} + \text{NADH} \quad (15\text{-}45)$$

$$\alpha\text{-Oxoglutarate} + \text{NAD}^+ + \text{CoASH} \rightleftharpoons \text{succinyl CoA} + \text{NADH} + \text{CO}_2 \quad (15\text{-}46a)$$

$$\text{Succinyl CoA} + \text{GDP} + \text{P}_i \xrightleftharpoons{(\text{E-imidazole}\sim P)} \text{succinate} + \text{GTP} + \text{CoASH} \quad (15\text{-}46b)$$

$$\text{GTP} + \text{ADP} \rightleftharpoons \text{GDP} + \text{ATP} \quad (15\text{-}46c)$$

$$\alpha\text{-Oxoglutarate} + \text{NAD}^+ + \text{ADP} + \text{P}_i \rightleftharpoons \text{succinyl CoA} + \text{ADP} + \text{NADH} + \text{CO}_2 \quad (15\text{-}46)$$

The two sets of reactions differ in one important respect. In the course of sequence (15-45) inorganic phosphate is incorporated *directly* into one of the products of the redox reaction and hence tranferred to ATP, whereas in sequence (15-46) the primary event is the generation of a *reactive intermediate* containing neither P_i nor ADP (viz., succinyl CoA) which subsequently is transformed into ATP by a series of isoenergetic displacement reactions. Two principal hypothetical reaction schemes (Types 1 and 2) for respiratory chain-linked phosphorylation have been formulated by analogy. Type 2 postulates a phosphorylated intermediate and is therefore analogous to reaction (15-45), while Type 1 does not and therefore is analogous to reaction (15-46).

$$\text{Type 1} \quad (C_{\text{red}}, C'_{\text{ox}}) + I \rightleftharpoons (C_{\text{ox}}, C'_{\text{red}}) \sim I \quad (15\text{-}47a)$$

$$(C_{\text{ox}}, C'_{\text{red}}) \sim I + \text{P}_i + \text{ADP} \rightleftharpoons (C_{\text{ox}}, C'_{\text{red}}) + \text{ATP} + I \quad (15\text{-}47b)$$

$$\text{Type 2} \quad (C_{\text{red}}, C'_{\text{ox}}) + \text{P}_i \rightleftharpoons (C_{\text{ox}}, C'_{\text{red}}) \sim P \quad (15\text{-}48a)$$

$$(C_{\text{ox}}, C'_{\text{red}}) \sim \text{P} + \text{ADP} \rightleftharpoons (C_{\text{ox}}, C'_{\text{red}}) + \text{ATP} \quad (15\text{-}48b)$$

In these formulations, both partial reactions (a and b) may consist of more than one step, just as an acyl enzyme and acyl lipoate are intermediates in reactions (15-45a) and (15-46a). Reaction (15-46b) probably also involves additional steps. The reactive intermediate linked to the carrier is abbreviated as ( )$\sim I$, in which we have left unspecified whether I is linked to C_{ox}, to C'_{red}, or to both. Perhaps I is a protein and $\sim I$ represents a particular conformation (note similarity to mechanochemical model). At present the available evidence appears to favor a Type 1 mechanism. For instance, in the presence of the uncoupler DNP (see below) respiration proceeds at a high rate even in the total absence of P_i, a result that would be hard to explain if the latter were an obligatory cosubstrate in the primary redox reaction (15-48a). Furthermore, the transition from state 4 (inhibited, ADP limiting) to state 3 (normal) respiration exhibits an initial lag; this is consistent with the inhibitory species being ( )$\sim I$, which has first to be converted during the lag to ( )$\sim P$, and not with ( )$\sim P$, which would be available immediately.

The *precise* reaction sequence and the mechanism for oxidative phosphorylation remain as yet *unknown*. Attempts at resolving these problems have again employed systems less complex than the whole mitochondrion, that is, subparticles and complexes supplemented with soluble enzymes, still capable of carrying out at least some of the component steps of oxidative phosphorylation; they have consisted of investigations of these partial reactions and of studies of site-specific inhibitors. Relatively slow progress has been made in the first aim, at least in comparison to the pace in the related field of mitochondrial electron transport. Still, somewhat less highly integrated particles have been obtained by treatment of mitochondria with detergents, denaturants, and proteolytic enzymes, and by sonication. Such preparations, which consist largely of inner membranes and cristae, usually still contain all the components required for electron transport but are deficient in one or more of the steps concerned with energy coupling. The latter activity can be restored by the addition of soluble protein fractions, frequently referred to (88,92,93) as *coupling factors*. As pointed out by Lehninger, such factors need not necessarily have anything to do with energy conversion per se. Instead, they may be implicated in blocking inhibitory processes or in providing the spatial framework within which coupled oxidation and phosphorylation occur.[16] More success has attended the demonstration of certain specific reactions and of site-specific inhibition, as discussed below.

ATPase. Mitochondria and phosphorylating particles in general catalyze a rapid hydrolytic breakdown of ATP to ADP plus P_i. Unlike the reactions catalyzed by other membrane systems, this reaction appears to be relatively insensitive to the presence of Na^+ and K^+ and to the specific inhibitor, ouabain (C-strophantin, a steroid rhamnoside, page 459). The mitochondrial ATPase, a cryptic activity in fully functional particles, where the rate of respiration is controlled by phosphorylation, becomes unmasked whenever the two processes can no longer act in concert, that is, when the particles are aged or mistreated or exposed to uncouplers [such as 2,4-dinitrophenol (DNP)]. These results have been interpreted in terms of a reversal of ATP formation with attendant enhancement of the hydrolytic breakdown of some labile intermediate. In the intact phosphorylating particle, the intermediate is envisaged as protected against such action, both by steric blockage and by the presence of specific inhibitory protein(s). A Mg^{2+} and DNP-stimulated mitochondrial ATPase (called F_1) has been isolated in soluble form and highly purified. With properly pretreated particles, F_1 can also restore phosphorylation at all three sites. In soluble, purified form (88–92) the ATPase is labile at low temperatures, because of its dissociation into inactive subunits, and insensitive to the inhibitor oligomycin[17] (Table 15-10). On the other hand, in its particulate form it is resistant to cold and sensitive to the inhibitor. This phenomenon of alteration of properties of an integrated activity upon solubilization has been called *allotopy* by Racker.

The soluble ATPase (M = 280,000) appears to be a highly compact, symmetrical, but complex protein consisting of a number of perhaps nonidentical sununits (all of $M \simeq$ 25,000). In large part, at least, it is believed to be the entity responsible for the spherical (~85 Å in diameter), knoblike protuberances that stud the cristae of the inner mitochondrial membrane (Fig. 15-8). Indeed, certain submitochondrial preparations (CF_0) can be shown to be devoid of knobs until they have been supplemented with F_1;

[16] With membrane-bound components, it may not always be possible to make a meaningful distinction between function and structure.

[17] It is still, however, sensitive to aurovertin, an inhibitor with a mode of action reminiscent of oligomycin in *intact* mitochondria.

the knobs now reappear, and coincident with this conversion the ATPase activity has become cold stable and oligomycin sensitive (88). Further analysis discloses that the conferral of oligomycin sensitivity in this process is the responsibility of a specific small protein ($M = 18{,}000$), which may also be involved in linking the F_1 unit to the membrane and constitute the "stalk" of the spherical knob (91). The mitochondrial energy-transducing system therefore consists of at least the following components: (1) an attachment site on the cristael membrane, (2) an entity responsible for conferring cold stability to F_1 [may be identical to (1)], (3) the oligomycin-sensitivity-conferring protein (OSCP), (4) F_1 (itself complex), and (5) an ATPase inhibitor associated with F_1 (88,92,93).

An ATP-$^{32}P_i$ exchange reaction is catalyzed by mitochondria but only poorly by submitochondrial particles. All three components (i.e., ADP as well as ATP and P_i), are required for the exchange, which can be regarded as another manifestation of the overall reactions leading to ATP formation. There is some indication, however, that only the reactions at site 1 (and perhaps at site 2) produce effective ATP-$^{32}P_i$ exchange.

Exchange of $H_2{}^{18}O$ into P_i and ATP. These exchange reactions, of which the second is the more rapid, are catalyzed by mitochondrial preparations at rates some 20 to 30 times faster than the overall rate of oxidative phosphorylation. Analysis of the isotope in the products showed that it must have entered at the point of P_i uptake and that the oxygen in the terminal pyrophosphate linkage of ATP must have been provided by one of the terminal oxygens of ADP rather than of P_i. These observations suggest an $\sim P$ instead of an $\sim$ADP intermediate in the formation of ATP. Therefore, reaction (15-49) must be composed of a minimum of two steps [Eqs. (15-49a) plus (15-49b) or, more likely, (15-49c) plus (15-49d)]:

$$(C_{ox},C'_{red})\sim I + P_i \rightleftharpoons (C_{ox},C'_{red})\sim P + I \qquad (15\text{-}49a)$$

$$(C_{ox},C'_{red})\sim P + \text{ADP} \rightleftharpoons (C_{ox},C'_{red}) + \text{ATP} \qquad (15\text{-}49b)$$

$$(C_{ox},C'_{red})\sim I + P_i \rightleftharpoons (C_{ox},C'_{red}) + I\sim P \qquad (15\text{-}49c)$$

$$I\sim P + \text{ADP} \rightleftharpoons \text{ATP} + I \qquad (15\text{-}49d)$$

Respiratory Control. Repeated reference has already been made to the experimental observation that the respiratory rate of fully functional and structurally intact mitochondria appears to be regulated by the availability of a phosphate acceptor. This phenomenon is called respiratory control, and such mitochondria are referred to as "tightly coupled." A likely explanation can be derived from a consideration of Fig. 15-10, or specifically of reactions (15-47) plus (15-49). Reversibility of the redox reactions must lead to inhibition unless provision is made for the removal of ( )$\sim P$ by an ATP-regeneration step. But as ATP itself accumulates, or when it is added externally in large amounts, the various intermediates will again accumulate and respiration will decrease unless provision is made to maintain the phosphate potential (i.e., the ratio [ATP]/[ADP] [P_i] at a low value). This can be accomplished either by the utilization of ATP in synthetic reactions (e.g., the formation of glucose-6-*P* by addition of glucose plus hexokinase) or by its hydrolysis (e.g., by the addition of a purified soluble ATPase). Closely related to this phenomenon are two observations: the stimulation of respiration brought about by uncouplers, and the reversal of electron flow. To explain the former, we need to postulate that any of the reactive intermediates of the type $A{\sim}B$ can form a complex with an uncoupler (F), such that $F{\sim}A$ or $F{\sim}B$ breaks down hydrolytically more rapidly than $A{\sim}B$. This will lead to a "short circuit" and circumvent respiratory control.

Reversed Electron Flow. We have already briefly alluded to the possibility of bringing about electron-transport reactions in the direction from right to left (in Fig. 15-10) by the addition of ATP. Two reactions have been demonstrated: (1) (TMPD) → cyt *c* → NAD → (oxidized substrate, i.e., acetoacetate, α-oxoglutarate, etc.) and (2) succinate → NAD → (substrate).

Investigations in this area are important not only because of their intrinsic interest but also because they allow the dissection of certain segments of the respiratory chain–phosphorylation complex. Sanadi et al. have shown that a protein related to mitochondrial ATPase, and coupling factor F_1, is required not only for the restoration of oxidative phosphorylation with NADH or succinate as substrate and for $^{32}P_i$-ATP exchange, but also for the succinate + ATP + NAD^+ → fumarate + NADH + ADP + P_i reaction (72,95). The purified protein does not catalyze the ADP → ATP exchange, and its ATPase activity (but *not* the other activities) is stimulated by incubation at elevated temperatures and inhibited by incubation at 0°, again suggesting the importance of conformational factors in this protein.

Use of Uncouplers and Inhibitors

Uncouplers of Oxidative Phosphorylation. Agents of this class (Table 15-10, class F) show the following properties (96): (1) They stimulate respiration in the absence of either an intermediate (ADP) or a terminal phosphate acceptor (glucose plus hexokinase); (2) this stimulation does not require the presence of P_i and it is accompanied by abolition of respiratory chain-linked phosphorylation; (3) respiration in particles still susceptible to complete inhibition by "true" phosphorylation inhibitors (Table 15-10, class E) can be restored to normal levels, and (4) they stimulate ATPase in particles or soluble preparations but inhibit reversed electron flow either in mitochondria or in submitochondrial particles, whether ATP is added externally or generated *in situ* (e.g., by α-oxoglutarate dehydrogenase, plus succinate thiokinase) (5). They are usually highly lipid soluble weak acids (or anions of such acids) and may be bound tenaciously to mitochondrial membranes.

The classical uncoupler of this class is 2,4-dinitrophenol (DNP), introduced by Loomis and Lipmann and studied extensively by Lardy and Wellman (96) and by Slater (99). A number of other compounds that have also been shown to react in essentially the same fashion include various other substituted phenols, especially pentachlorophenol, dicumarol[18]—a potent anticoagulant isolated from rotting sweet clover—and various substituted phenylhydrazones, such as *m*-chlorocarbonyl-cyanidephenylhydrazone (CCCP). The relative concentrations required of DNP, dicumarol, and CCCP for virtually quantitative inhibition of phosphorylation or ATP-dependent NAD reduction under one set of conditions is 60 : 4 : 0.07 (nmol/ml). Salicylanilides appear to be even more potent (0.001 nmol/ml).

Each of the above uncouplers appears to be about equally effective at all

[18] Also called dicoumarol:

OH CH$_2$ O O 2

three phosphorylation sites.[19] This entire class of uncouplers is believed to act very close to the respiratory chain, leading to a displacement of the $(C, C')\sim I$ or $I\sim P$ complex [Eq. (15-49a) or (15-49c); Fig. 15-10] toward either a $(C, C')\sim F$ or an $I\sim F$ complex [or perhaps a ternary $(C, C')\sim I\cdot F$ complex]. The hypothetical complex is broken down hydrolytically to restore the free carriers (C, C') and I, thus permitting respiratory action to proceed without now being linked to energy conservation. An interesting variant of this model has been suggested by van Dam and Slater (97), who suggest that uncouplers are taken up by mitochondria as anions (F^-), by the anion-transport system of the particles, an energy-linked, but not a phosphate-requiring process (see below). Once across the membrane, F^- takes up a proton and in this form passively diffuses across the membrane, dissociates, and restarts the cycle.

Arsenate is also an uncoupler and shares many properties with the agents just described, especially in its ability to stimulate respiration by removing the obligatory link to phosphorylation. It also induces ATPase activity, but the activity is much less than that induced by DNP. Furthermore, the rate of arsenate-stimulated ATPase is also very slow when compared to the arsenate-stimulated respiration in the presence of ADP. The latter activity is strongly inhibited by P_i. These observations suggest (98) that arsenate must interact with the reaction sequence leading to phosphorylation at a site which is located *after* the DNP-sensitive site, at the site where P_i interacts, that is, that reaction (15-49) is itself composed of three, rather than two, steps:

$$(C_{ox}, C'_{red})\sim I + X \rightleftharpoons X\sim I + (C_{ox}, C'_{red}) \tag{15-50a}$$

$$X\sim I + P_i \rightleftharpoons X\sim P + I \tag{15-50b}$$

$$X\sim P + \text{ADP} \rightleftharpoons X + \text{ATP} \tag{15-50c}$$

$X\sim I$ is then thought to react with arsenate to produce a reactive intermediate as $X\sim As$ (or $I\sim As$) complex, which, after hydrolysis [Eqs. (15-50c) through (15-50e)], leads to regeneration of all components [(C, C'), I, and X] necessary for another turn of the cycle

$$X\sim I + \text{As}_i \rightleftharpoons I + X\sim As \xrightarrow{\text{H}_2\text{O}} X + \text{As}_i \tag{15-50d}$$

or

$$X\sim As + \text{ADP} \rightleftharpoons X + \text{ADP}\sim As \xrightarrow{\text{H}_2\text{O}} \text{ADP} + \text{As}_i \tag{15-50e}$$

Inhibitors of Energy-Linked Electron Transfer. Agents of this group (class E) are true inhibitors of respiration in *intact, tightly coupled* phosphorylating particles. The compounds are ineffective, however, with those respiratory particles that no longer possess the requisite degree of structural integrity even though they may be able to produce ATP. Also they, in

[19] Specific inhibitors of certain phosphorylation sites are also known: thus oxybarbiturates, such as amobarbital, structure

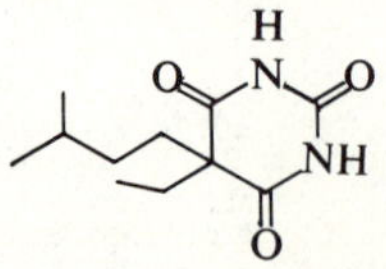

and alkylguanidines appear specific for site 1, and phenethyl biguanide for site 2.

contrast to agents of class F, inhibit, rather than stimulate, the particulate ATPase. Soluble ATPase requires the addition of additional components before exhibiting analogous sensitivity. On the other hand, class E agents resemble class F agents in their effects on the various exchange reactions. There are two interesting relationships, first described by Lardy (99), between agents of classes E and F. When respiration is blocked by inhibitors (class E), it can be completely restored by uncouplers (class F); conversely, ATPase is stimulated by uncouplers (class F) and depressed by inhibitors (class E). Reversed electron flow and other energy-requiring mitochondrial activities are inhibited *only* when the energy source is ATP itself or some other equivalent compound added to the system, and *not* when it is provided by high-energy intermediates generated continuously *in situ* as a consequence of respiration. Two subgroups comprise this class of inhibitors.

The first, exemplified by the antibiotics oligomycin (rutamycin) and aurovertin, appears to block one of the steps that intervene between $X{\sim}I$ and ATP. The final step, the transphosphorylation of $X{\sim}P$,

$$X + P_i \underset{H_2O}{\rightleftharpoons} X{\sim}P + ADP \rightleftharpoons X + ATP + H_2O \tag{15-51}$$

can be regarded as the one which, owing to hydrolysis of the unstable $X{\sim}P$, is identical to ATPase acting in reverse. Since the solubilized ATPase is still blocked by aurovertin but not by oligomycin, it is likely[20] that the oligomycin-sensitive step is reaction (15-50b). Under physiological conditions hydrolysis of $X{\sim}P$ is blocked and, instead, the intermediate can be used to generate $X{\sim}I$ at the expense of ATP. Indeed many reactions that can be driven either by high-energy intermediates of oxidative phosphorylation *or* externally added ATP appear insensitive to or even stimulated by oligomycin when the second alternative is exercised. Dicyclohexylcarbodiimide appears analogous in all respects to oligomycin.

The second group of inhibitors is exemplified by atractyloside (101). This agent, unlike oligomycin, blocks ATP-energized reactions only when ATP itself is provided rather than generated *in situ* by means of substrate-linked phosphorylation (e.g., during oxidation of α-oxoglutarate). It also inhibits the entry of ATP and ADP into mitochondria and the conversion of extramitochondrial ADP to ATP. Like oligomycin it blocks ADP-ATP exchange. These, and other characteristic properties of the agent, are dependent on the maintenance of the polarity of the mitochondrial inner membrane. When it becomes "inverted" (e.g., in the preparation of phosphorylating particles by sonication) all sensitivity to atractyloside is lost. Agents of this type are thus specific inhibitors of a mitochondrial transport system for adenine nucleotides located in the inner membrane (101–103).

[20] The suggested mode of action for oligomycin is also consistent with recent studies on its action on ATP-dependent translocation processes in a number of nonmitochondrial membrane systems (100) (page 459). The reaction sequence postulated is

$$ATP + TL \cdot K^+ + Na^+ \text{ (inside)} \xrightarrow{\text{ouabain}} TL{\sim}P \cdot Na^+ + ADP$$

TL = translocase protein

$$TL{\sim}P \cdot Na^+ + K^+ \text{ (outside)} \xrightarrow[\text{oligomycin}]{} TL \cdot K^+ P_i$$

$$\textit{Net}\quad ATP + \text{internal } Na^+ + \text{external } K^+ \rightarrow ADP + \text{internal } K^+ + \text{external } Na^+ + P_i$$

Mechanism of Action of Oxidative Phosphorylation

To recapitulate, the various experiments outlined suggest the following steps as a minimum mechanism for oxidative phosphorylation:

$$(C_{red}, C'_{ox}) + I \rightleftharpoons (C_{ox}, C'_{red}){\sim}I \xrightarrow[H_2O]{DNP} (C_{ox}, C'_{red}) + I$$

$$(C_{ox}, C'_{red}){\sim}I + X \rightleftharpoons (C_{ox}, C'_{red}) + X{\sim}I \xrightarrow[H_2O]{As_i} X + I$$

$$X{\sim}I + HO{-}P_i \underset{\text{oligomycin}}{\rightleftharpoons} X{\sim}P + I(OH)$$

$$X{\sim}P + ADP \underset{\text{aurovertin}}{\overset{Mg^{2+}}{\rightleftharpoons}} X + ATP + H_2O \qquad \text{transphosphorylase (ATPase)}$$

$$ATP_{in} \underset{\text{atractyloside}}{\rightleftharpoons} ATP_{out} \qquad \text{translocase}$$

Any progress toward a more detailed mechanism evidently requires an identification of the chemical nature of the two hypothetical entities, the nonphosphorylated intermediate *I*, and the phosphate acceptor *X*. Very little is known concerning them in spite of the efforts of many laboratories. There exist many suggestive bits of evidence (87) that indicate that perhaps both *I* and *X* are proteins.[21] One such is provided by experiments of Pinchot with a cell-free system from the bacterium *Alkaligenes faecalis* (104). These extracts catalyze a phosphorylation which appears to involve the following components: a particulate membrane fraction catalyzing NADH oxidation, a soluble protein coupling factor (*CF*) required for phosphorylation but not for electron transport, a polynucleotide that permits the binding of the *CF* to the dehydrogenase, Mg^{2+}, and NADH. Incubation of all these components in the complete absence of P_i and ADP permits electron transport between NADH and O_2, and simultaneously, the NAD^+ formed becomes covalently bonded to *CF*. If the NAD-containing complex is detached from the particles and incubated with P_i and ADP, ATP formation ensues. The following steps were envisioned, where *E* is the complex between the particulate dehydrogenase, Mg^{2+}, and oligonucleotide:

$$NADH + [E \cdot CF] \xrightarrow{-2e^-} [E \cdot CF{\sim}NAD] \rightleftharpoons E + CF{\sim}NAD$$

$$CF{\sim}NAD + P_i \rightleftharpoons CF{\sim}P + NAD^+$$

$$CF{\sim}P + ADP \rightleftharpoons ATP + CF$$

This scheme, in which $CF = I$ and which does not contain *X*, or the more elaborate sequences described earlier, suggests that a search for a rapidly turning over phosphoprotein in mitochondria should be fruitful. Intermediates of this type have been found by Boyer and his collaborators, and the reaction site has been identified as phosphohistidine (105). Careful investigation of the kinetics and other properties of the reaction disclosed, however, that phosphohistidine was an intermediate in substrate-level phosphorylation rather than in respiratory-chain phosphorylation, and was produced as an

[21] Some of them may be cytochromes in a high potential state (e.g., cyt b^* with an $E_0' \simeq 0.3$ V compared to 0.05 V for ordinary cyt *b*) that have been identified by spectroscopic techniques (see footnote f in Table 15-9).

intermediate in either or both the succinate thiokinase [reaction (15-46b)] and the nucleoside diphosphokinase [reaction (15-46c)].

Other Reactions Energized by Electron Transport

ATP production is by no means the only reaction that is directly dependent on mitochondrial electron transport. On the contrary, most energy-requiring processes occurring in the particles can be driven by ATP supplied in the medium, by ATP generated as a result of energy-linked respiration, or by the coupling reactions themselves, even in the complete absence of ADP and P_i. This direct coupling is inhibited by respiratory inhibitors and by DNP, but it is relatively independent of (or sometimes stimulated by) oligomycin and atractylate. Among the endergonic reactions that can thus be driven directly by reactive intermediates in the energy coupling sequence are (1) reversed electron flow, (2) accumulation of ions (cations, such as Mg^{2+} or Ca^{2+}; anions such as citrate, succinate or phosphate; and dipolar ions such as amino acids) inside the particle against a concentration gradient (active transport or translocation), (3) the related phenomenon of mitochondrial salt loss accompanied by contraction and the extrusion of water (i.e., the performance of mechanochemical work), and (4) an energy-dependent reduction of $NADP^+$ by NADH. We shall now discuss points (2) and (4) in some detail.

Mitochondrial Transhydrogenation. Among the interesting reactions that have been demonstrated recently to be linked to reactive intermediates involved in ATP formation is the generation of intramitochondrial NADPH by an energy-dependent transhydrogenase system.

$$\text{NADH} + \left.\begin{matrix}\text{ATP}\\ \text{or}\\ \sim X\end{matrix}\right\} + \text{NADP}^+ \rightarrow \text{NAD}^+ + \left.\begin{matrix}\text{ADP}\\ \text{or}\\ X\end{matrix}\right\} + \text{NADPH} \qquad (15\text{-}52a)$$

$$\text{NADH} + \text{NADP}^+ \rightleftharpoons \text{NAD}^+ + \text{NADPH} \qquad (15\text{-}52b)$$

The couples $NAD^+/NADH$ and $NADP^+/NADPH$ are isoenergetic, and reaction (15-52b) is freely reversible, in the presence of soluble transhydrogenases.[22] The energy requirement of the mitochondrial system is for the purpose of pushing the equilibrium of reaction (15-52) far to the right. Klingenberg has suggested that NADH and NADP may be localized in different compartments and that it is their translocation across the (inner) membrane that is the energy-requiring step (106). The mechanism of reaction (15-52a) is obscure at present but appears to involve at least some of the intermediates at the first (NADH-dependent) phosphorylation site. The NADPH produced in the reaction (15-52) can then be used as a source of reducing equivalents for a number of intramitochondrial reactions, of which the most important is probably the pyridine nucleotide-linked dismutation of α-oxoglutarate:

$$\alpha\text{-Oxoglutarate} + \text{NAD}^+ + \text{ADP} + \text{P}_i \xrightarrow{\text{GDP}} \text{succinate} + \text{ATP} + \text{NADH} + \text{CO}_2 \qquad (15\text{-}53)$$

$$\alpha\text{-Oxoglutarate} + \text{NAD(P)H} + \text{NH}_4^+ \rightarrow \text{glutamate} + \text{NAD(P)}^+ \qquad (15\text{-}54)$$

Although in principle, reaction (15-53) also does not require energy in the form at ATP and can be shown to be freely reversible with the soluble glutamate dehydrogenase, which utilizes NADH or NADPH with equal facility, mitochondrial glutamate formation actually *requires* ATP (or energy-linked intermediates) and NADPH. Thus, the mitochondrial system involves

[22] Indeed, somewhat purified transhydrogenases that no longer require ATP have been extracted from mitochondria.

NAD in at least catalytic amounts and utilizes ATP or its equivalent in stoichiometric amounts.

Cation Translocation in Mitochondria. Mitochondria are able to translocate significant quantities of divalent cations from the outside into the matrix, with the inner membrane the site of the permeability barrier and action of the responsible transport system (107,108) [containing specific entities, variously referred to as *translocase, permease,* or *transporter* protein(s)]. Any cation accumulation or "loading" requires either the simultaneous expulsion of equal amounts of other cations or the concomitant accumulation of counteranions for the maintenance of electroneutrality within the system. In the case of mitochondria the second alternative appears to predominate: Ca^{2+} (and to a lesser extent Mg^{2+}) accumulation requires the simultaneous accumulation of phosphate or arsenate, and there are those who believe that it is phosphate translocation that is the primary phenomenon, with divalent cation accumulation a secondary consequence. In any event, in liver mitochondria Ca^{2+} inside can increase from 5.6 nmol/g of protein to 20 times that value and Mg^{2+} from 42 nmol/g to 4 times this value. The metals are deposited inside the matrix as the insoluble phosphates (with some contribution to the crystal by Cl^- and OH^- as well) and this deposition leads to the generation and expulsion of protons [e.g., $3Mg^{2+} + 2HPO_4{}^{2-} \longrightarrow Mg_3(PO_4)_2 + 2H^+$]. Cation accumulation is abolished by DNP, somewhat inhibited by high concentrations of ADP, and insensitive to oligomycin; these facts suggest that it is driven by intermediate(s) between the DNP and oligomycin-sensitive sites.

Ordinarily the mitochondrial (inner) membrane is quite impermeable to monovalent cations (109,110) [with the exception of $NH_4{}^+$, which, however, diffuses as the neutral species $NH_3(aq)$]. It can be rendered permeable by two classes of antibiotics and certain synthetic analogs (111,112).

1. The addition of certain neutral depsipeptides[23] (e.g., valinomycin), macrolide actins (e.g., nonactin), or polypeptides (e.g., gramicidin) induces the *transport of alkali metal cations* into the mitochondrial matrix against a concentration gradient at the expense of energy generated by electron transport or externally added ATP. Coincident with this uptake there is an *ejection of protons* and a decrease in light scattering by the particle suspension due to simultaneous uptake of water (i.e., the particles swell).

2. The addition of certain monocarboxylic polyethers (e.g., nigericin, monensin or synthetic polyethers) induces permeability to cations which then *diffuse passively* along the concentration gradient from the matrix to the outside, particularly when the particles have previously been preloaded (e.g., with K^+ in the presence of valinomycin). Coincident with this diffusion of alkali cations outward there is a migration of protons inward and a contraction of the particles due to the extrusion of water. If representatives of both classes (e.g., valinomycin plus nigericin) are added simultaneously in catalytic concentrations ($< 1\ \mu g$ or $\ll 10^{-7}$ M), then K^+ can act as an uncoupler of oxidative phosphorylation by virtue of the steps:

$$X{\sim}K_{out}{}^+ \xrightarrow{\text{Val}} X + K_{in}{}^+ \tag{15-55a}$$

$$K_{in}{}^+ \overset{\text{Nig}}{\rightleftharpoons} K_{out}{}^+ \tag{15-55b}$$

[23] A depsipeptide is an oligomer containing ester as well as peptide bonds.

where $X\sim$ is a high-energy intermediate, Val is valinomycin, Nig is nigericin, and K_{out}^+ and K_{in}^+ are potassium inside and outside the particle, respectively. The order of specificity is $K^+ > Rb^+ > Cs^+ \gg Na^+$ for valinomycin and $K^+ > Rb^+ > Na^+ \gg Cs^+$ for nigericin, both with respect to their activity on mitochondria and their ability to form complexes with the cations indicated. Nonactin exhibits a specificity similar to that of valinomycin, except that it induces Na^+ transport; gramicidin is relatively nonspecific.

Mechanisms for Cation Translocation. The phenomena just discussed are actually examples of a much more general one—that of energy-dependent or *active translocation* or *transport* (113,114)—which in turn is perhaps a special case of the selective removal of permeability barriers across protein–lipid membranes or interfaces by specific entities known as translocators, permeases, carriers, or transporters. Eisenman (113) and Pressman (111) have discussed four mechanisms by which such a selective permeation of cations may be brought about: (1) formation of a *neutral asymmetric carrier complex* by formation of an ion pair between a negatively charged, polar, usually multifunctional ligand and the cation which can then cross the membrane; (2) prior penetration of ligand molecules similar to those in (1) into the membrane, where they can form a stacked array across its width that then permits cation permeation by ion exchange through *fixed, negatively charged pores or channels;* (3) formation of a *positively charged complex* between a neutral, though polar, macrocyclical carrier, which in contrast to (1) now surrounds the cation in a symmetrical manner; and (4) if the macrocyclic carrier molecules of (3) become fixed in the membrane by a mechanism similar to (2), one can conceive of permeation through a *fixed, neutral* (but polar) *pore.* The various possibilities are represented schematically in Fig. 15-11, which incorporates these postulated mechanisms as well as the structures of some of the compounds involved.

After the discovery (115) by Moore and Pressman of valinomycin-dependent K^+ uptake by mitochondria in 1964, it soon became recognized that this was but a special instance of antibiotic-mediated ion permeation. This phenomenon is known to include the effects of a multitude of antibiotics of various classes and of some synthetic equivalents (116) on active and passive cation transport across barrier systems ranging from mitochondrial and erythrocyte membrane to synthetic lipid vesicles and bilayer lipid leaflets[24] ("black lipid membranes"). Among the more remarkable discoveries was the demonstration by Mueller and Rudin (117) and by Lev and Buzhinsky (118) that mediated ion transport across such artificial bilayers generates electric conductance and bioionic potentials. More recently Mueller and Rudin (119) have shown that the antibiotic alamethicin, which incorporates a macrocyclic ring system as well as a negative carboxylate group at some distance removed, can, in the presence of a monovalent cation gradient, supplemented by a few micrograms of a polycation, even develop some of the characteristics of *excitable membranes,* including rhythmic and repeated firing (*action potentials*). It is becoming ever more apparent that either directly, or consequent

[24] Such artificial membranes some 60 Å across are best prepared by painting a small aperture with mixtures of neutral and charged (usually but not exclusively phospho) lipids (e.g., oxidized cholesterol plus dodecyl acid phosphate in octane–decane or lecithin in decane–squalene).

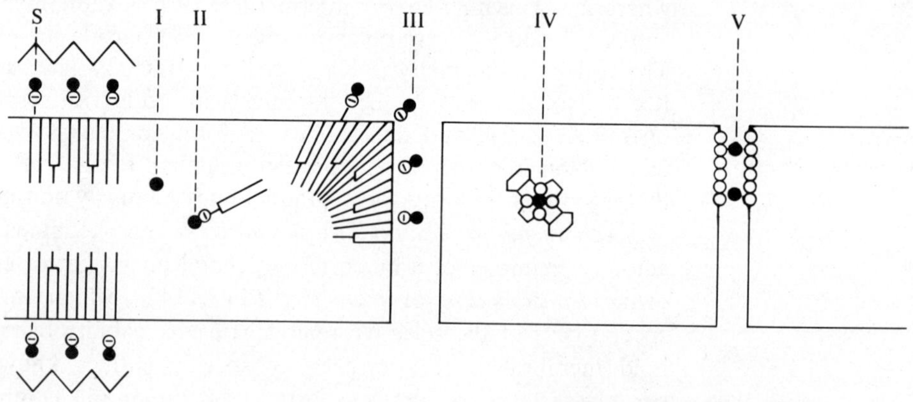

Phosphatide
Cholesterol
Cation
Protein
Oxygen
Macrocyclic polyether
Anionic groups

PERMEATION MECHANISMS

I. Cation in neutral medium
II. Negatively charged carrier
III. Fixed negatively charged pore
IV. Neutral (polar) carrier (e.g., cyclic polyether)
V. Neutral (polar) pore
S. Surface rate-limiting reaction

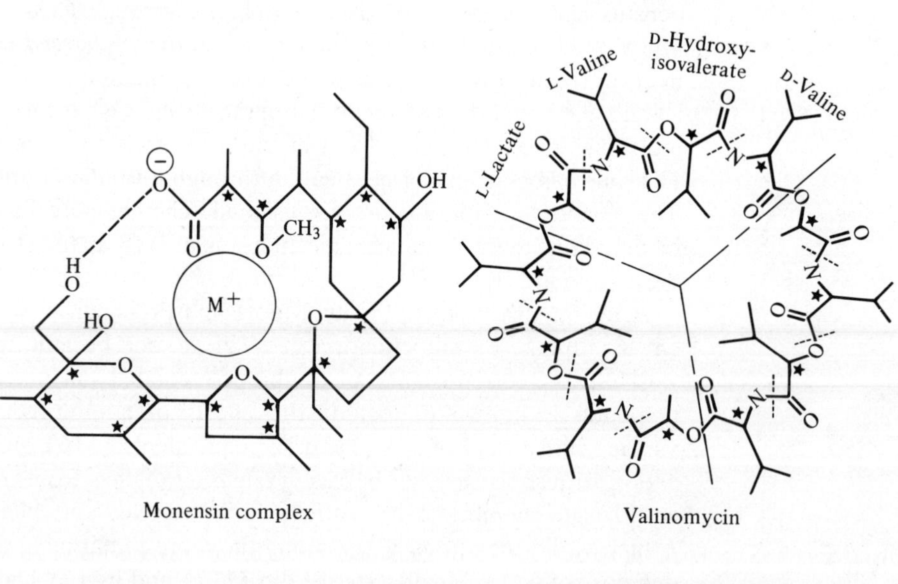

Monensin complex

Valinomycin

D-Hydroxyvalerate

L-N-Methylvaline

Enniatin

Nonactin

to rapid isoenergetic equilibration reactions, membranes can perform a variety of energy-transducing functions, somewhat as follows:

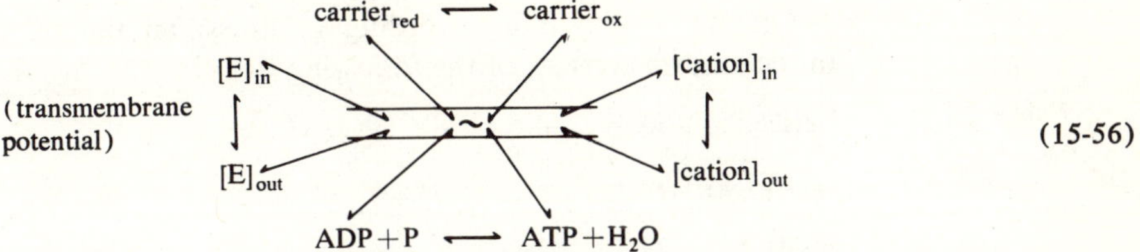

(15-56)

(*N.B.:* Any concentration gradient across a membrane sets up a potential gradient; conversely, a potential gradient can be used to generate a concentration gradient. Hence, an ion pump is *electrogenic.*) Here ~ symbolizes a compound in, or a state of, the membrane generated by a strongly exergonic (and hence capable of driving a strongly endergonic) reaction. It is also becoming evident, as has already been stated, that "active" transport processes generally ascribed to specific "pumps" may, as far as the critical selectivity aspect is concerned, be identical or at least similar to "passive" processes, mediated by "gates." The antibiotics and related compounds responsible for cation transport may then be regarded as models for, or analogs of, intrinsic entities firmly associated with membranes capable of performing active transport processes spontaneously. They have been referred to by Pressman as *ionophoric agents* or *ionophores.* The antibiotics have been shown to form clathrate complexes with cations by virtue of multiple ion–dipole and ion–induced dipole interactions. These interactions result from the displacement of the hydration shell of the positively charged species by several of the ether or hydroxyl oxygens of the ligand [eight with K^+-monactin, six with K^+-enniatin, K^+-valinomycin, and Na^+-monensin in complexes whose crystal structure is known (120-122)]. The ligand, as a result, assumes a conformation that is quite different from the one showing the greatest stability in the uncomplexed molecule; the latter is linear in the case of monensin, nigericin, and gramicidin, and a convoluted cyclical form rather than a planar one in the case of valinomycin and the macrolide actins. Particularly in the case of these cyclical ionophores, the conformation in the complex is such as to surround the central ion, so that all polar groups point toward the interior, while the apolar, lipophilic groups are exposed to the surroundings, and thus confer great lipid solubility. What determines stability, then, is a summation of several energy terms (dehydration of the cation, interaction of cation with ionophore in a conformation that maximizes ion-dipole interaction, transconformation of ionophore from the unperturbed to this complexing form). This consideration explains the ability of different ionophores to discriminate in a selective and nonidentical manner between

FIGURE 15-11

Mechanisms for ion permeation across bilayer membranes and structures of some ionophorous antibiotics. (*a*) Diagram of a bilayer membrane, indicating possible permeation mechanisms for cations. Note that the principal difference between a biological membrane and its artificial phospholipid bilayer model (Fig. 16-1) is probably the absence of proteins in the latter. (*b*) Formal structures of monensin complex and valinomycin. The monensin complex depicted is based on the X-ray crystallographic data obtained from the silver complex as in Ref. 120. (*c*) Comparison of formal planar-ring representations of structures of enniatin and nonactin. After G. Eisenman, *Federation Proc.,* **27:** 1249 (1968) and B. C. Pressman, *Federation Proc.,* **27,** 1283 (1968).

various cations. Passive transport, as for example of K^+ by nigericin, leads to a transmembrane equilibrium $[K^+]_{in}/[K^+]_{out}$ which must be balanced, again at equilibrium, by $[H^+]_{out}/[H^+]_{in}$.

On the other hand, according to Pressman the *selective* cation pump in mitochondria consists of the following steps:

$$\left.\begin{array}{ll}\text{outside} & M^+H_2O + I_1 \rightleftharpoons M^+I_1' + H_2O \\ \text{membrane} & M^+I_1' + T' \rightleftharpoons M^+T' + I_1\end{array}\right\} \quad \begin{array}{r}(15\text{-}57a)\\(15\text{-}57b)\end{array}$$

$$\left.\begin{array}{ll}\text{inside} & M^+T' + H_2O \rightleftharpoons T + M^+H_2O \\ \text{membrane} & T + (\quad)\sim \rightleftharpoons T'\end{array}\right\} \quad \begin{array}{r}(15\text{-}57c)\\(15\text{-}57d)\end{array}$$

Here I_1 and I_1' correspond to the ionophore in its free and complexed conformation and T' and T are two conformations of an intrinsic transporter molecule in the membrane (a protein?), with high and low affinity for the cation to be transported respectively. T' is capable of loading at a site close to the exterior and discharging the ion in the interior, being converted to T in the process; T is reconverted to T' by interaction with the high-energy intermediate ($\;$)$\sim$; M^+ is the cation.

Active Transport Across Cell Membranes. With minor variations, analogous schemes can and have been advanced for active transport of both anions[25] and cations, as well as of neutral molecules, by a variety of selective pumps across cell membranes (see also Chapter 9). Proteins capable of binding, and hence of acting as the hypothetical transporters, in such schemes have been isolated in pure form, particularly from bacterial sources, for galactose, sulfate, certain amino acids, and galactosides. In addition, bacteria have been shown to contain a generalized sugar-transport system that consists of HPr, a soluble, heat-stable protein of low molecular weight; enzyme I, another soluble protein; and Fraction II, a membrane preparation that is capable of phosphorylating a wide variety of different sugars [Eqs. (9-5a) and (9-5b)].

Anion Translocation in Mitochondria. Mitochondria are capable of concentrating not only cations but also certain anions against a concentration gradient. As already mentioned in connection with Ca^{2+} and divalent cation transport in general, the particles contain a selective uptake system for inorganic phosphate or arsenate (and also, probably, one for small monocarboxylate anions). If counterbalanced by a *di*valent cation, this system *discharges* protons into the medium; if by a *mono*valent cation it leads to the *uptake* of protons (i.e., the liberation of hydroxyl ions). In addition, mitochondria also appear to contain transport systems for the salts of di- and tricarboxylic organic acids. Although readily permeable to weak acids themselves, the membrane appears quite impermeable to their anions, including the various substrates of the citric acid cycle (Chapter 14). All of these—except for fumarate (probably also oxalacetate and α-oxoglutarate)—can, however, be concentrated inside by energy-dependent reactions; two separate systems appear to be involved: one for the salts of the other dicarboxylic and another for those of the tricarboxylic acids. In Fig. 15-12 we summarize the current picture for ion transport across the mitochondrial inner membrane.

[25] Certain polyene antibiotics, such as amphotericin, fillipin, and nystatin, act as ionophores for *anions*. The induced loss of anions (and secondarily of countercations) from susceptible cells with a high content of neutral lipid (cholesterol) in their envelope explains the fungicidal action of these agents.

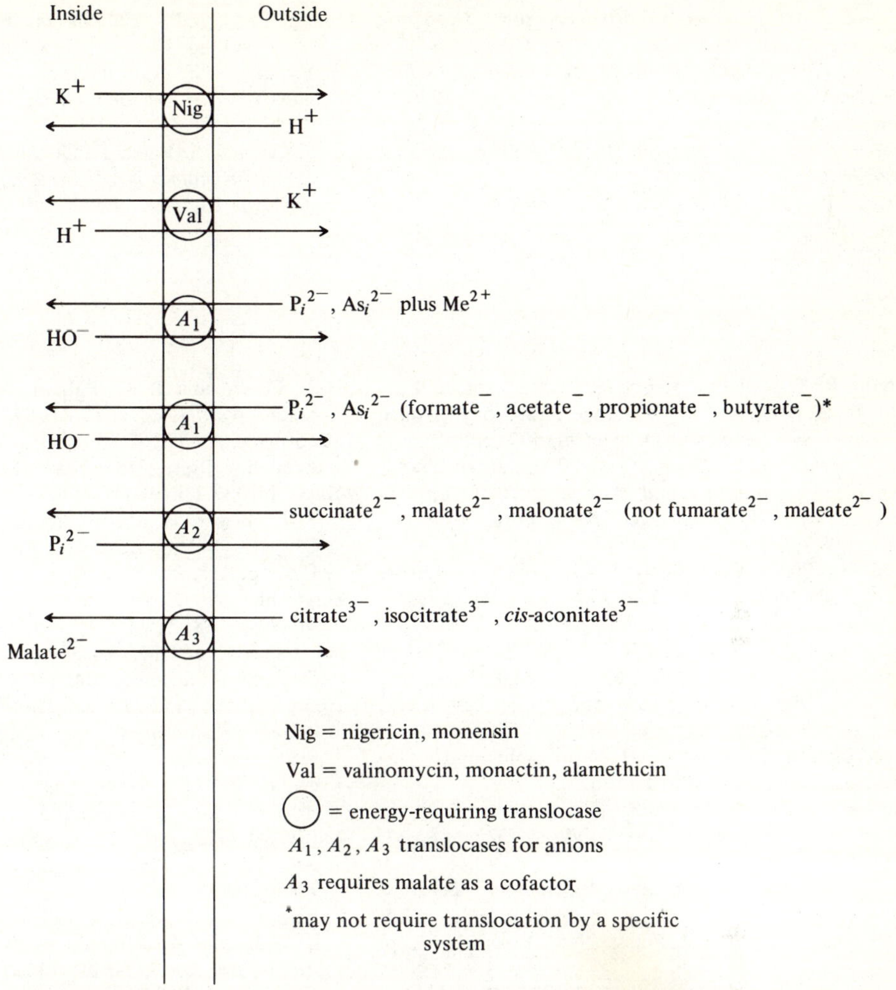

FIGURE 15-12
Schematic representation of ion transport by mitochondria.

REFERENCES

Books and Monographs

Chance, B., R. W. Estabrook, and T. Yonetani (eds.): *Hemes and Hemoproteins,* Academic Press, New York, 1966.

Chance B. (ed.): *Energy-Linked Functions of Mitochondria,* Academic Press, New York, 1963.

Estabrook, R. W., and M. E. Pullman (eds.): Oxidation and Phosphorylation, in *Methods in Enzymology,* vol. X, Academic Press, New York, 1967.

Florkin, M., and E. H. Stotz (eds.): Biological Oxidations, in *Comprehensive Biochemistry,* vol. 14, Elsevier, New York, 1966.

Green, D. E., and R. F. Goldberger: *Molecular Insights into the Living Process,* Academic Press, New York, 1967.

King, T. E., H. S. Mason, and M. Morrison (eds.): *Oxidases and Related Redox Systems,* Wiley, New York, 1965.

Lehninger, A. L.: *The Mitochondrion,* Benjamin, New York, 1964.

Okunuki, K., M. D. Kamen, and I. Sekuzu (eds.): *Structure and Function of Cytochromes,* University Park Press, Baltimore, 1968.

Peisach, J., P. Aisen, and W. E. Blumberg (eds.): *Biochemistry of Copper,* Academic Press, New York, 1966.

Quagliariello, E., S. Papa, E. C. Slater, and J. M. Tager (eds.): *Mitochondrial Structure and Compartmentation,* Adriatica Editrice, Bari, 1967.

Racker, E.: *Mechanisms in Bioenergetics,* Academic Press, New York, 1965.

Racker, E. (ed.): *Membranes of Mitochondria and Chloroplasts,* Van Nostrand Reinhold, New York, 1970.

San Pietro, A. (ed.): *Non-Heme Iron Proteins: Role in Energy Conversion,* Antioch Press, Yellow Springs, Ohio, 1965.

Singer, T. P. (ed.): *Biological Oxidations,* Wiley, New York, 1968.

Slater, E. C. (ed.): *Flavins and Flavoproteins* (BBA Library, vol. 8), Elsevier, New York, 1966.

Slater, E. C., Z. Kaniuga, and L. Wojtczak (eds.): *Biochemistry of Mitochondria,* Academic Press, New York, 1967.

Tager, J. M., S. Papa, E. Quagliariello, and E. C. Slater (eds.): *Regulation of Metabolic Processes in Mitochondria,* Elsevier, New York, 1966.

Review Articles

Abraham, R., E. Balke, K. Krisch, S. Leonhäuser, K. Leybold, K. A. Sack, and H. Staudinger: Hydroxylasen, in K. Lang, E. Lehnartz, O. Hoffman-Ostenhof, and G. Siebert (eds.), *Hoppe-Seyler/Thierfelder Handbuch der Physiologisch und Pathologisch-Chemischen Analyse,* vol. VI/A, Springer, New York, 10th ed., 1964, p. 917.

Beinert, H.: Flavin Coenzymes, P. D. Boyer, H. Lardy, and K. Myrbäck (eds.), *The Enzymes,* vol. 2, Academic Press, New York, 2nd ed., 1960, p. 339.

Brill, A. S., R. B. Martin, and R. J. P. Williams: Copper in Biological Systems, in B. Pullman (ed.), *Electronic Aspects of Biochemistry,* Academic Press, New York, 1964, p. 519.

Chance, B.: The Reactivity of Haemoproteins and Cytochromes, *Biochem. J.,* **103:** 1 (1967).

Chance, B., and G. R. Williams: The Respiratory Chain and Oxidative Phosphorylation, *Advan. Enzymol.,* **17:** 65 (1956).

Diner, S.: Electronic Aspects of Biochemical Hydroxylation, in B. Pullman (ed.), *Electronic Aspects of Biochemistry,* Academic Press, New York, 1964, p. 237.

Ehrenberg, A.: Flavin Free Radicals, in B. Pullman (ed.), *Electronic Aspects of Biochemistry,* Academic Press, New York, 1964, p. 379.

Ernster, L., and C.-P. Lee: Biological Oxidoreductions, *Ann. Rev. Biochem.* **33:** 729 (1964).

Falk, J. E.: *Porphyrins and Metalloporphyrins,* Elsevier, New York, 1964.

Frieden, E., S. Osaki, and H. Kobayashi: Copper Proteins and Oxygen, *J. Gen. Physiol.,* **49** (I, part 2): 213 (1965).

George, P., and J. S. Griffith: Electron Transfer and Enzyme Catalysis, in P. D. Boyer, H. Lardy, and K. Myrbäck (eds.), *The Enzymes,* vol. 1, Academic Press, New York, 2nd ed., 1959, p. 347.

Gibson, D. T.: Microbial Degradation of Aromatic Compounds, *Science,* **161:** 1093 (1968).

Green, D. E., and D. H. MacLennan: Structure and Function of Mitochondrial Cristael Membrane, *BioScience,* **19:** 213 (1969).

Green, D. E., and I. Silman: Structure of the Mitochondrial Electron Transfer Chain, *Ann. Rev. Plant Physiol.,* **18:** 147 (1967).

Griffiths, D. E.: Oxidative Phosphorylation, in P. N. Campbell and G. D. Greville (eds.), *Essays in Biochemistry,* vol. 1, Biochemical Society/Academic Press, New York, 1965, p. 91.

Hall, D. O., and J. M. Palmer: Mitochondrial Research Today, *Nature,* **221:** 717 (1969).

Haurowitz, F.: *The Chemistry and Function of Proteins,* Academic Press, New York, 1963, pp. 256–279.

Hayaishi, O.: Direct Oxygenation by O_2, Oxygenases, in P. D. Boyer, H. Lardy, and K. Myrbäck (eds.), *The Enzymes,* vol. 8, Academic Press, New York, 2nd ed., 1963, p. 353.

Hayaishi, O.: Enzymatic Hydroxylation, *Ann. Rev. Biochem.,* **38:** 21 (1969).

Hayaishi, O., and M. Nozaki: Nature and Mechanism of Oxygenases, *Science,* **164:** 389 (1969).

Henn, F. A., and T. E. Thompson: Synthetic Lipid Bilayer Membranes, *Ann. Rev. Biochem.,* **38:** 241 (1969).

Korn, E. D.: Cell Membranes: Structure and Synthesis, *Ann. Rev. Biochem.,* **38:** 263 (1969).

Lardy, H. A., and S. M. Ferguson: Oxidative Phosphorylation in Mitochondria, *Ann. Rev. Biochem.,* **38:** 991 (1969).

Lehninger, A. L., E. Carafoli, and C. S. Rossi: Energy Linked Ion Movements in Mitochondrial Systems, *Advan. Enzymol.,* **29:** 259 (1967).

Lemberg, M. R.: Cytochrome Oxidase, *Physiol. Rev.,* **49:** 48 (1969).

Mahler, H. R.: Nature and Function of Metalloflavoproteins, *Advan. Enzymol.,* **17:** 233 (1956).

Malkin, R., and J. C. Rabinowitz: Non-Heme Iron Electron-Transfer Proteins, *Ann. Rev. Biochem.,* **36:** 113 (1967).

Massey, V., and C. Veeger: Biological Oxidations, *Ann. Rev. Biochem.,* **32:** 579 (1963).

Mitchell, P.: The Chemiosmotic Hypothesis and Oxidative Phosphorylation, *Biol. Rev. Cambridge Phil. Soc.* (1967).

Mitchell, P.: Proton-Translocation Phosphorylation in Mitochondria, Chloroplasts and Bacteria: Natural Fuel Cells and Solar Cells, *Federation Proc.,* **26:** 1370 (1967).

Nicholls, P.: Cytochromes: A Survey, in P. D. Boyer, H. Lardy, and K. Myrbäck (eds.), *The Enzymes,* vol. 8, Academic Press, New York, 2nd ed., 1963, p. 3.

Nicholls, P., and G. R. Schonbaum: Catalases, in P. D. Boyer, H. Lardy, and K. Myrbäck (eds.), *The Enzymes,* vol. 8, Academic Press, New York, 2nd ed., 1964, p. 147.

Okunuki, K.: Cytochromes and Cytochrome Oxidase in *Comprehensive Biochemistry,* vol. 14, p. 232.

Paul, K. G.: Peroxidases, in P. D. Boyer, H. Lardy and

K. Myrbäck (eds.), *The Enzymes,* vol. 8, Academic Press, New York, 2nd ed., 1963, p. 227.

Pullman, M. E., and G. Schatz: Mitochondrial Oxidations and Energy Coupling, *Ann. Rev. Biochem.,* **36:** 539 (1967).

Racker, E.: The Membrane of the Mitochondrion, *Sci. Am.,* **218:** 32 (1968).

Racker, E.: Resolution and Reconstitution of the Inner Mitochondrial Membrane, *Federation Proc.,* **26:** 1335 (1967).

Rothfield, L., and A. Finkelstein: Membrane Biochemistry, *Ann. Rev. Biochem.,* **37:** 463, (1968).

Schatz, G.: Mitochondrial Oxidative Phosphorylation, *Angew. Chem. (Intern. Engl. Ed.),* **6:** 1035 (1967).

Schirrmacher-Göllner, I.: Kupferenzyme, in K. Lang, E. Lehnartz, O. Hoffman-Ostenhof, and G. Siebert (eds.), *Hoppe-Seyler/Thierfelder Handbuch der Physiologisch und Pathologisch-Chemischen Analyse,* vol. VI/A, Springer, New York, 10th ed., 1964, p. 898.

Sih, C. J.: Enzymatic Mechanism of Steroid Hydroxylation, *Science,* **163:** 1297 (1969).

Sund, H., H. Diekmann, and K. Wallenfels: Die Wasserstoffübertragung mit Pyridinnucleotiden, *Advan. Enzymol.,* **26:** 115 (1964).

Tubbs, P. K., and P. B. Garland: Membranes and Fatty Acid Metabolism, *Brit. Med. Bull.* **24:** 158 (1968).

Various authors, in Biological and Artificial Membranes, *Federation Proc.,* **27:** 1249–1309 (1968).

Williams, R. J. P.: Coordination, Chelation, and Catalysis, in P. D. Boyer, H. Lardy, and K. Myrbäck (eds.), *The Enzymes,* vol. 1, Academic Press, New York, 2nd ed., 1959, p. 391.

Specific References

1. Wieland, H.: *On the Mechanism of Oxidation,* Yale University Press, New Haven, 1932.
2. Baldwin, E.: *Dynamic Aspects of Biochemistry,* Cambridge University Press, New York, 1967, chaps. VI and VII.
3. Thunberg, T.: *Skand. Arch. Physiol.,* **35:** 163 (1927).
4. Green, D. E.: *Mechanisms of Biological Oxidations,* Cambridge University Press, New York, 1940.
5. Singer, T. P.: Flavoprotein Dehydrogenases of the Electron Transport Chain: A Survey, in *The Enzymes,* vol. 7, Academic Press, New York, 2nd ed., 1963, p. 345.
6. Beinert, H.: Flavin Coenzymes, in *The Enzmyes,* vol. 2, Academic Press, New York, 2nd ed., 1960, p. 339.
7. Hayaishi, O., and M. Nozaki: *Science,* **164:** 389 (1969).
8. Ishimura, Y., M. Nozaki, O. Hayaishi, M. Tamura, and I. Yamazaki: *J. Biol. Chem.,* **242:** 2574 (1967).
9. Goldstein, M., E. Lauber, W. E. Blumberg, and J. Peisach: *Federation Proc.,* **24:** 604 (1965).
10. Kaufman, S.: *Ann. Rev. Biochem.,* **36:** 171 (1967); *J. Biol. Chem.,* **242:** 3934 (1967).
11. Guroff, G., C. A. Reifsnyder, and J. Daly: *Biochem. Biophys. Res. Commun.,* **24:** 720 (1966); Guroff, G., M. Levitt, J. Daly, and S. Udenfriend: *ibid.,* **25:** 253 (1966).
12. Mason, H. S., J. C. North, and M. Vanneste: *Federation Proc.,* **24:** 1172 (1965).
13. Cooper, D. Y., S. Levin, S. Narasimhulu, O. Rosenthal, and R. W. Estabrook: *Science,* **147:** 400 (1965); Omura, T., R. Sato, D. Y. Cooper, O. Rosenthal, and R. W. Estabrook: *Federation Proc.,* **24:** 1181 (1965); Suzuki, K., and T. Kimura: *Biochem. Biophys. Res. Commun.,* **19:** 340 (1965).
14. Cushman, D. W., R. L. Tsai, and I. C. Gunsalus: *Biochem. Biophys. Res. Commun.,* **26:** 577 (1967); Katagiri, M., B. N. Ganguli, and I. C. Gunsalus: *J. Biol. Chem.,* **243:** 3543 (1968).
15. Peterson, J. A., D. Basu, and M. J. Coon: *J. Biol. Chem.,* **241:** 5162 (1966).
16. Blumberg, W. E., and J. Peisach: *Biochim. Biophys. Acta,* **126:** 269 (1966).
17. Ambler, R. P., and L. H. Brown: *Biochem. J.,* **104:** 784 (1967).
18. Tang, S. W., J. E. Coleman, and Y. P. Myer: *J. Biol. Chem.,* **243:** 4286 (1968).
19. Hastings, J. W., W. H. Riley, and J. Massa: *J. Biol. Chem.,* **240:** 1473 (1965).
20. Beinert, H., and P. Hemmerich: *Biochem. Biophys. Res. Commun.,* **18:** 212 (1965).
21. Strittmatter, P.: *J. Biol. Chem.,* **239:** 3043 (1964).
22. Masters, B. S. S., H. Kamin, Q. H. Gibson, and C. H. Williams, Jr.: *J. Biol. Chem.,* **240:** 921 (1965).
23. King, T. E., R. L. Howard, J. Kettman, Jr., B. M. Hegdekar, M. Kuboyama, K. S. Nickel, and E. A. Possehl: Comparison of Soluble NADH Dehydrogenases from the Respiratory Chain of Cardiac Mitochondria, in E. C. Slater (ed.), *Flavins and Flavoproteins* (BBA Library, vol. 8), Elsevier, New York, 1966, p. 441.
24. Aleman, V., and P. Handler: *J. Biol. Chem.,* **242:** 4087 (1967).
25. Aleman, V., P. Handler, G. Palmer, and H. Beinert: *J. Biol. Chem.,* **243:** 2560, 2569 (1968).
26. McCord, J. M., and I. Fridovich: *J. Biol. Chem.,* **243:** 5753 (1968).
27. Smith, S. T., K. V. Rajagopalan, and P. Handler: *J. Biol. Chem.,* **242:** 4108 (1967).
28. Komai, H., V. Massey, and G. Palmer: *J. Biol. Chem.,* **244:** 1692 (1969); Massey, V., P. E. Brumby, H. Komai, and G. Palmer: *ibid.,* **244:** 1682 (1969).
29. Singer, T. P., E. Rocca, and E. B. Kearney: Fumarate Reductase, Succinate and NADH Dehydrogenases of Yeast: Properties and Biosynthesis, in E. C. Slater (ed.), *Flavins and Flavoproteins* (BBA Library, vol. 8), Elsevier, New York, 1966, p. 391.
30. Straub, F. B.: *Biochem. J.,* **33:** 787 (1939); see also N. Savage: *Biochem. J.,* **67:** 146 (1957).
31. Massey, V.: *Biochim. Biophys. Acta,* **30:** 205 (1958).
32. Searls, R. L., and D. R. Sanadi: *Proc. Natl. Acad. Sci. U.S.,* **45:** 697 (1959); *J. Biol. Chem.,* **236:** 580 (1961).
33. Shethna, Y. I., P. W. Wilson, R. E. Hansen, and

H. Beinert: *Proc. Natl. Acad. Sci. U.S.*, **52:** 1263 (1964).

34. Der Vartanian, D. V., W. H. Orme-Johnson, R. E. Hansen, H. Beinert, R. L. Tsai, J. C. M. Tsibris, R. C. Bartholomaus, and I. C. Gunsalus: *Biochem. Biophys. Res. Commun.*, **26:** 569 (1967); Tsibris, J. C. M., R. L. Tsai, I. C. Gunsalus, W. H. Orme-Johnson, R. E. Hansen, and H. Beinert: *Proc. Natl. Acad. Sci. U.S.*, **59:** 959 (1968).
35. Cooke, R., J. C. M. Tsibris, P. G. Debrunner, R. Tsai, I. C. Gunsalus, and H. Frauenfelder: *Proc. Natl. Acad. Sci. U.S.*, **59:** 1045 (1968).
36. Evans, M. C. W., D. O. Hall, H. Bothe, and F. R. Whatley: *Biochem. J.*, **110:** 485 (1968).
37. Sobel, B. E., and W. Lovenberg: *Biochemistry*, **5:** 6 (1966).
38. Matsubara, H., T. H. Jukes, and C. R. Cantor: *Brookhaven Symp. Biol.*, **21:** 201 (1968).
39. Philips, W. D., E. Knight, Jr., and D. C. Blomstrom: in A. San Pietro (ed.), *Non-Heme Iron Proteins: Role in Energy Conversion*, Antioch Press, Yellow Springs, Ohio, 1965, p. 69.
40. Moss, T. H., D. Petering, and G. Palmer: *J. Biol. Chem.*, **244:** 2275 (1969).
41. Sasaki, R. M., and H. Matsubara: *Biochem. Biophys. Res. Commun.*, **28:** 467 (1967).
42. Benson, A. M., and K. T. Yasunobu: *J. Biol. Chem.*, **244:** 955 (1968).
43. Weinstein, B.: *Biochem. Biophys. Res. Commun.*, **35:** 109 (1969).
44. Herriott, J. R., L. C. Sieker, L. H. Jensen, and W. Lovenberg: *Acta Cryst.*, **A25:** S3 S 186 (1969); Lovenberg, W., and W. A. Eaton: *Federation Proc.*, **29:** 1226 (1970).
45. Schellenberg, K. A.: *J. Biol. Chem.*, **242:** 1815 (1967); Chan, T. L., and K. A. Schellenberg: *ibid.*, **243:** 6284 (1968).
46. Allison, W. S., M. J. Connors, and D. J. Parker: *Biochem. Biophys. Res. Commun.*, **34:** 503 (1969).
47. Peisach, J., W. E. Blumberg, B. A. Wittenberg, and J. B. Wittenberg: *J. Biol. Chem.*, **243:** 1871 (1968).
48. MacMunn, C. A.: *Phil. Trans. Roy. Soc. London*, **B177:** 267 (1886); *J. Physiol.* (*London*), **8:** 57 (1887).
49. Keilin, D.: *Proc. Roy. Soc.* (*London*), **B98:** 312 (1925).
50. Keilin, D.: *Proc. Roy. Soc.* (*London*), **B104:** 206 (1929); *Proc. Roy. Soc.* (*London*), **B106:** 418 (1930).
51. Keilin, D., and E. F. Hartree: *Proc. Roy Soc.* (*London*), **B127:** 167 (1939).
52. Nicholls, P., and G. R. Schonbaum: Catalases, in P. D. Boyer, H. Lardy, and K. Myrbäck (eds.), *The Enzymes*, vol. 8, Academic Press, New York, 2nd ed., 1963, p. 147.
53. Ohnishi, K.: *J. Biochem.* (*Japan*), **59:** 9 (1966).
54. Itagaki, E., and L. P. Hager: *J. Biol. Chem.*, **241:** 3687 (1966); Hager, L. P., and S. S. Deeb: in *Methods in Enzymology*, vol. X, p. 367 (1967).
55. Watari, H., O. Groudinsky, and F. Labeyrie: *Biochim. Biophys. Acta*, **131:** 592 (1967).
56. Nishibayashi, H., T. Omura, R. Sato, and R. W. Estabrook: in K. Okunuki, M. D. Kamen, and I. Sekuzu (eds.), *Structure and Function of Cytochromes*, University Press, Baltimore, 1968, p. 658.
57. Sherman, F., J. W. Stewart, J. Parker, G. J. Putterman, and E. Margoliash: in *Structure and Function of Cytochromes*, p. 257.
58. Margoliash, E., W. M. Fitch, and R. E. Dickerson: *Brookhaven Symp. Mol. Biol.*, **21:** 259 (1969).
59. Lemberg, R.: *Physiol. Rev.*, **49:** 48 (1969).
60. Muijsers, A. O., B. F. van Gelder, and E. C. Slater: in B. Chance, R. W. Estabrook, and T. Yonetani (eds.), *Hemes and Hemoproteins*, Academic Press, New York, 1966, p. 467.
61. Keilin, D. and E. F. Hartree: *Proc. Roy. Soc.* (*London*), **B127:** 167 (1939).
62. Green, D. E., and D. C. Wharton: *Biochem. Z.*, **338:** 335 (1963).
63. Fleischer, S., H. Klouwen, and G. Brierley: *J. Biol. Chem.*, **236:** 2936 (1961).
64. Sottocasa, G. L., B. Kuylensteirna, L. Ernster, and A. Bergstrand: in *Methods in Enzymology*, vol. X, p. 448 (1967).
65. Parsons, D. F., and G. R. Williams: in *Methods in Enzymology*, vol. X, p. 443 (1967).
66. Schnaitman, C. and J. W. Greenawalt: *J. Cell Biol.*, **38:** 158 (1968); Schnaitman, C., V. G. Erwin, and J. W. Greenawalt: *J. Cell. Biol.*, **32:** 719 (1967).
67. Bruni, A., and E. Racker: *J. Biol. Chem.*, **243:** 962 (1968); Yamashita, S., and E. Racker, *ibid.*, **244:** 1220 (1969).
68. Hatefi, Y.: in M. Florkin and E. H. Stotz (eds.), *Comprehensive Biochemistry*, vol. 14, Elsevier, New York, 1966, p. 199.
69. Hatefi, Y., and W. G. Hanstein: *Proc. Natl. Acad. Sci. U.S.*, **62:** 1129 (1969).
70. Hatefi, Y.: *Proc. Natl. Acad. Sci. U.S.*, **60:** 733 (1968); Hatefi, Y., and K. E. Stempel: *J. Biol. Chem.*, **244:** 2350 (1969); Hatefi, Y., K. E. Stempel, and W. G. Hanstein: *J. Biol. Chem.*, **244:** 2358 (1969); Baginsky, M. L., and Y. Hatefi: *Biochem. Biophys. Res. Commun.*, **32:** 945 (1968).
71. King, T. E., R. L. Howard, J. Kettman, Jr., B. M. Hegdekar, M. Kuboyame, K. S. Nickel, and E. A. Possehl: in E. C. Slater (ed.), *Flavins and Flavoproteins*, Elsevier, New York, 1966, p. 471.
72. Baum, H., H. I. Silman, J. S. Rieske, and S. H. Lipton: *Proc. Natl. Acad. Sci. U.S.*, **57:** 798 (1967); *J. Biol. Chem.*, **242:** 4876 (1967).
73. Tzagoloff, A., P. C. Yang, D. C. Wharton, and J. S. Rieske: *Biochim. Biophys. Acta*, **96:** 1 (1965).
74. Green, D. E., N. F. Haard, G. Lenaz, and H. I. Silman: *Proc. Natl. Acad. Sci. U.S.*, **60:** 277 (1968). Senior, A. E., and D. H. MacLennan: *J. Biol. Chem.* **245:** 5086 (1970).
75. Schnaitman, C. A.: *Proc. Natl. Acad. Sci. U.S.*, **63:** 412 (1969).
76. Chance, B., and G. R. Williams: *Advan. Enzymol.*, **17:** 65 (1956).
77. Szarkowska, L., and M. Klingenberg: *Biochem. Z.*, **338:** 674 (1963).
78. Horie, S.: in *Structure and Function of Cytochromes*, p. 96 (1968).
79. Dolin, M. I., L. Smith, J. Newton, and M. D. Kamen: in I. C. Gunsalus and R. Y. Stanier (eds.), *The*

Bacteria, vol. II, Academic Press, New York, 1961, chaps. 6, 7, and 8.
80. Strittmatter, P.: *J. Biol. Chem.,* **240:** 4481 (1965); *Federation Proc.,* **24:** 1156 (1965).
81. Strittmatter, P., and J. Ozols: *Proc. Natl. Acad. Sci. U.S.,* **58:** 264 (1967).
82. Belitzer, V. A., and E. T. Tzibakova: *Biokhimiya,* **4:** 516 (1939).
83. Kalckar, H.: *Enzymologia,* **2:** 47 (1937); *Biochem. J.,* **33:** 631 (1939).
84. Boyer, P. D., L. L. Bieber, R. A. Mitchell, and G. Szabolcsi: *J. Biol. Chem.* **241:** 5384 (1966).
85. Penniston, J. T., R. A. Harris, J. Asai, and D. E. Green: *Proc. Natl. Acad. Sci. U.S.,* **59:** 624 (1968).
86. Mitchell, P.: *Biochem. Soc. Symp. (Cambridge, Engl.),* **22:** 142 (1963).
87. Chance, B., and L. Mela: *J. Biol. Chem.,* **242:** 830 (1967); Wang, J. H.: *Proc. Natl. Acad. Sci. U.S.,* **58:** 37 (1967).
88. Kagawa, Y., and E. Racker: *J. Biol. Chem.,* **241:** 2461, 2467, 2475 (1966); Bulos, B., and E. Racker: *ibid.,* **243:** 3891, 3901 (1968).
89. Penefsky, H. S., and R. C. Warner: *J. Biol. Chem.,* **240:** 4694 (1965).
90. Tzagoloff, A., K. H. Byington, and D. H. MacLennan: *J. Biol. Chem.,* **243:** 2405 (1968); *Biochemistry,* **7:** 1596 (1968).
91. MacLennan, D. H., and A. Tzagoloff: *Biochemistry,* **7:** 1603 (1968); MacLennan, D. H., and J. Asai: *Biochem. Biophys. Res. Commun.,* **33:** 441 (1968).
92. Sanadi, D. R., K. W. Lam, and C. K. R. Kurup: *Proc. Natl. Acad. Sci. U.S.,* **61:** 277 (1968); Warshaw, J. B., K. W. Lam, B. Nagy, and D. R. Sanadi: *Arch. Biochem. Biophys.,* **123:** 385 (1968).
93. Pullman, M. E., H. Penefsky, and E. Racker: *Arch. Biochem. Biophys.,* **76:** 227 (1958).
94. Vignais, P. V., and P. M. Vignais: *Biochim. Biophys. Acta,* **43:** 431 (1960).
95. Sanadi, D. R., T. E. Andreoli, and K. W. Lam: *Biochem. Biophys. Res. Commun.,* **17:** 582 (1964); see also D. R. Sanadi and A. L. Fluharty: *Biochemistry,* **2:** 523 (1963).
96. Lardy, H. A., and H. Wellman: *J. Biol. Chem.,* **201:** 357 (1953).
97. van Dam, K., and E. C. Slater: *Proc. Natl. Acad. Sci. U.S.,* **58:** 2015 (1967).
98. Ter Welle, H. F., and E. C. Slater: *Biochim. Biophys. Acta,* **89:** 385 (1964).
99. Lardy, H. A., D. Johnson, and W. C. McMurray: *Arch. Biochem. Biophys.,* **78:** 587 (1958); Lardy, H. A., J. L. Connelly, and D. Johnson: *Biochemistry,* **3:** 1961 (1964); Lardy, H. A., P. Witonsky, and D. Johnson: *Biochemistry,* **4:** 552 (1965).
100. Whittam, R., K. P. Wheeler, and A. Blake: *Nature,* **203:** 720 (1964).
101. Bruni, A., and G. F. Azzone: *Biochim. Biophys. Acta,* **93:** 462 (1964).
102. Winkler, H. H., and A. L. Lehninger: *J. Biol. Chem.,* **243:** 3000 (1968).
103. Palmieri, F., and M. Klingenberg: *Biochim. Biophys. Acta,* **131:** 582 (1967).
104. Pinchot, G. B.: *Proc. Natl. Acad. Sci. U.S.,* **46:** 929 (1960); Pinchot, G. B., and M. Hormanski: *Proc. Natl. Acad. Sci. U.S.,* **48:** 1970 (1962).
105. Boyer, P. D.: *Science,* **141:** 1147 (1963).
106. Brierley, G., and D. E. Green: *Proc. Natl. Acad. Sci. U.S.,* **53:** 73 (1965).
107. Lehninger, A. L., E. Carafoli, and C. S. Rossi: *Advan. Enzymol.,* **29:** 259 (1967); Reynafarje, B., and A. L. Lehninger: *J. Biol. Chem.,* **244:** 584 (1969).
108. Mela, L.: *Federation Proc.,* **27:** 828 (1968); *Arch. Biochem. Biophys.,* **123:** 286 (1968); Mela, L., and B. Chance: *Biochemistry,* **7:** 4059 (1968).
109. Chappell, J. B., and K. N. Haarhoff: in *Biochemistry of Mitochondria,* p. 75 (1967).
110. Brierley, G. P., C. T. Settlemire, and V. A. Knight: *Arch. Biochem. Biophys.,* **126:** 276 (1968).
111. Pressman, B. C., E. J. Harris, W. S. Jagger, and J. H. Johnson: *Proc. Natl. Acad. Sci. U.S.,* **58:** 1949 (1967); Harris, E. J., M. P. Höferand, and B. C. Pressman: *Biochemistry,* **6:** 1348 (1967); Pressman, B. C.: *Federation Proc.,* **27:** 1283 (1968); Haynes, D. H., A. Kowalsky, and B. C. Pressman: *J. Biol. Chem.,* **244:** 502 (1969).
112. Lardy, H.: *Federation Proc.,* **27:** 1278 (1968).
113. Eisenman, G.: *Federation Proc.,* **27:** 1249 (1968); Eisenman, G., S. M. Ciani, and G. Szabo: *ibid.,* **27:** 1289 (1968).
114. Tosteson, D. C.: *Federation Proc.,* **27:** 1269 (1968).
115. Moore, C., and B. C. Pressman: *Biochem. Biophys. Res. Commun.,* **15:** 562 (1964).
116. Pedersen, C. J.: *Federation Proc.,* **27:** 1305 (1968); *J. Am. Chem. Soc.,* **89:** 7017 (1967).
117. Mueller, P., and D. O. Rudin: *Biochem. Biophys. Res. Commun.,* **26:** 398 (1967).
118. Lev, A. A.: *Nature,* **201:** 1132 (1964); Lev. A. A., and E. P. Buzhinski: *Tsitologyia,* **9:** 102 (1967).
119. Mueller, P., and D. O. Rudin: *Nature,* **217:** 713 (1968).
120. Agtarap, A., J. W. Chamberlin, M. Pinkerton, and L. Steinrauf: *J. Am. Chem. Soc.,* **89:** 5737 (1967); Pinkerton, M., L. K. Steinrauf, and P. Dawkins: *Biochem. Biophys. Res. Commun.,* **35:** 512 (1969).
121. Dobler, M., J. D. Dunitz, and J. Krajewski: *J. Mol. Biol.,* **42:** 603 (1969).
122. Ivanov, V. T., I. A. Laine, N. D. Abdulaev, L. B. Senyavina, E. M. Popov, Y. A. Ovchinnikov, and M. M. Shemyakin: *Biochem. Biophys. Res. Commun.,* **34:** 803 (1969).

16
Some Aspects of the Chemistry, Function, and Biosynthesis of Lipids

In the course of discussion of the generation of two-carbon fragments from the degradation of fatty acids and the more complex lipids from which they are derived catabolically (Chapter 13), we have already encountered several representatives of the lipid class of compounds. In addition, a number of references have been made to their functions in living systems: as structural components of membranes (page 457), as structural and functional participants in the process of oxidative phosphorylation (page 687), and as an energy reservoir (page 619), for example. At this point, it is appropriate to bring all this information together, expand it to include all the major classes of lipids and their functions, and to discuss the biosynthetic routes employed in construction of these substances.

Lipids are a diverse group of compounds. This reflects the definition of this class of biochemical substances as the totality of material that is extractable from living systems by organic solvents. Such a definition is in marked contrast to the rather specific way that one defines a protein, nucleic acid, or carbohydrate. Since the definition of lipid is based on solubility properties rather than on structural ones, structural diversity within the group is hardly surprising.

To a substantial extent, the biosynthesis of lipids reflects the variety of fashions in which the repeated condensation of two-carbon fragments, derived from acetate, is employed for the construction of natural products. The spectrum of such products derived ultimately from acetate includes the *fatty acids*, many *aromatic* and *heterocyclic compounds, steroids,* and *terpenes.* Other structural classes of lipids are derived from the further metabolism of these products of two-carbon condensations: for example, triglycerides are derived from glycerol and the fatty acids, and so on.

CLASSIFICATION OF LIPIDS

Lipids are classified on a structural basis in Table 16-1. We are already familiar with some of these substances; others have not been mentioned previously. Each structural class will now be considered with respect to its chemistry, function, and biosynthesis.

FATTY ACIDS

Some Chemical Considerations

The structures of most of the principal fatty acids of nature have been collected in Table 13-1. Since they tend to occur predominantly as constituents of more complex lipids, they serve few biochemical functions in and of themselves. From the standpoint of physical chemistry they are representative of

TABLE 16-1
Structural classification of lipids

I	Fatty acids
	a. Saturated fatty acids
	b. Unsaturated fatty acids
	c. Cyclopropane and branched chain fatty acids
II	Glycerol-derived lipids
	a. Mono-, di-, and triglycerides
	b. Glyceryl ethers
	c. Glycosyl glycerides and diglycerides
	d. Phosphatides
	e. Plasmalogens
III	Sphingosine-derived lipids
	a. Ceramides
	b. Sphingomyelins
	c. Cerebrosides
	d. Gangliosides
IV	Steroids and their derivatives
V	Terpenes and their derivatives
VI	Certain aromatic compounds
VII	Aliphatic alcohols and waxes
VIII	Complex lipids
	a. Lipoproteins
	b. Proteolipids
	c. Phosphatidopeptides
	d. Lipopolysaccharides

most lipids in one important respect: they are *amphipathic* molecules. As a prelude to some more complex considerations which follow later, let us look briefly at the consequences of this fact.

Amphipathic molecules are those which possess both marked hydrophobic and marked hydrophilic properties. The salts of fatty acids, $R—CO_2^-$, and to a lesser extent the fatty acids themselves, R—COOH, meet this criterion. The long chain of methylene residues is hydrophobic in its properties, while the charged head group is just the opposite. Molecules of this type are generally capable of forming a number of interesting structures. Consider the situation of the salt of a fatty acid at an air–water interface, for example. The hydrocarbon chains will prefer to exist outside the aqueous phase, while the charged head groups will prefer to exist within it. Both of these tendencies can be simultaneously satisfied by the formation of a *monolayer* at the air–water interface with the hydrocarbon tails sticking up into the air and the charged heads down into the water. Another way of satisfying these demands is the formation of *micelles.* Micelles come in many shapes and sizes; the simplest are nearly spherical structures which contain 50 to 100 individual molecules arranged with the hydrophobic chains on the inside and the charged groups on the outside. Diagrammatic visualizations of monolayers and micelles are provided in Fig. 16-1, which also contains a variety of additional structures formed by amphipathic molecules. We may note in passing that the structural features of micelles and globular proteins are similar in the sense that both tend to hide the hydrophobic component from the aqueous environment while exposing the charged groups to that environment. It seems very likely that many of the thermodynamic driving forces for micelle formation and for formation of the secondary and tertiary structures of proteins are

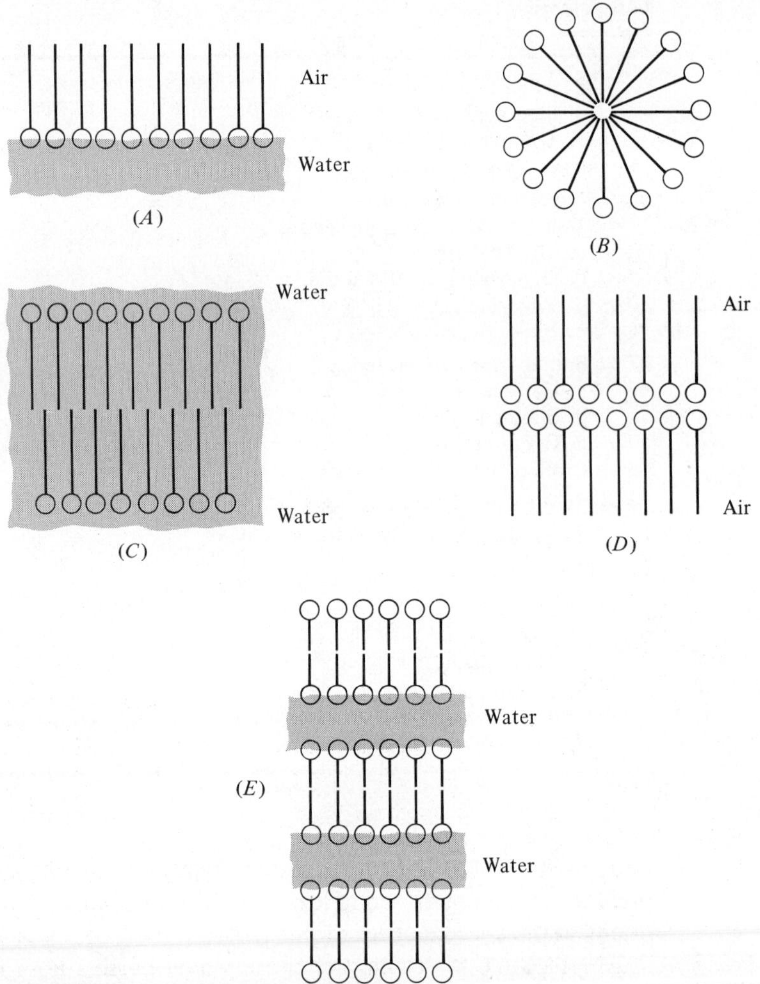

FIGURE 16-1
Typical structures formed by amphipathic molecules: (*a*) a monolayer at the air–water interface, (*b*) a micelle, (*c*) a bilayer in water, (*d*) a bilayer in air (soap film), and (*e*) one example of a myelinic phase in which successive bilayers are separated by a few molecules of water. By permission of M. K. Jain.

the same. Finally, it is interesting to note that amphipathic properties are the basis of detergency action. Structures such as micelles have the capacity to solubilize organic material and hence provide for its removal.

Biosynthesis The pathways for and regulatory mechanisms involved in the biosynthesis of fatty acids have been areas of great interest in recent years. Studies in this area have proved profitable in answering the basic questions and have provided insight into some related matters as well. We begin with a discussion of the synthesis of saturated straight-chain fatty acids.

Each step in the enzymatic pathway responsible for the oxidation of fatty acids is reversible (page 594). Consequently, the proposition was advanced that fatty acid biosynthesis might occur by simple reversal of the oxidative pathway. Indeed, incubation of acetyl-SCoA with purified enzymes of the oxidative pathway, together with NADH, NADPH, and an enzyme able to irreversibly reduce unsaturated fatty acyl-SCoA's to the saturated derivatives, results in synthesis of fatty acids (1). Furthermore, several workers were able to demonstrate the incorporation of acetyl-SCoA into fatty acids in mitochondrial systems supplemented with NADH, NADPH, and ATP (2). However, it now seems certain that at least two pathways for fatty acid biosynthesis exist in the mitochondrion: one similar to that occurring in the soluble portion of the cell (discussed below), which catalyzes the *de novo* synthesis of fatty acids, and one that catalyzes the elongation of preformed fatty acids through condensation with acetyl-SCoA (3). It appears that only the latter pathway is similar to the reverse of fatty acid oxidation.

Fatty acid oxidation occurs principally or entirely in the mitochondria or other particulate cellular fractions. One of the earliest suggestions that fatty acid synthesis might occur via a pathway distinct from that of oxidation was the demonstration by Gurin and coworkers (4), and subsequently by Popják and Tietz (5), of the synthesis of long-chain fatty acids by soluble extracts of pigeon liver and mammary gland. Cofactor requirements for fatty acid synthesis in soluble cell extracts, determined by Wakil and coworkers, include an absolute dependence upon bicarbonate, ATP, Mn^{2+}, and NADPH (6). These cofactor requirements clearly distinguish the synthetic and oxidative pathways since (1) bicarbonate has no effect upon fatty acid oxidation and (2) NAD, rather than NADP, is the oxidation–reduction coenzyme in the oxidative pathway.

The basic building block for fatty acid synthesis is acetyl-SCoA. The principal cellular source of this substance is the intramitochondrial oxidation of pyruvate. Since the mitochondrial membrane is rather impermeable to acetyl-SCoA, the question has been raised as to the origin of the extramitochondrial acetyl-SCoA for fatty acid biosynthesis. The only known reactions leading to the extramitochondrial generation of acetyl-SCoA are the ATP-dependent cleavage of citrate [Eq. (16-1)] and the acetate thiokinase reaction [Eq. (16-2)]:

$$\text{Citrate} + \text{MgATP} + \text{HSCoA} \rightleftharpoons \text{acetyl-SCoA} + \text{oxalacetate} + \text{MgADP} + P_i \qquad (16\text{-}1)$$

$$\text{Acetate} + \text{MgATP} + \text{HSCoA} \rightleftharpoons \text{acetyl-SCoA} + \text{AMP} + \text{MgPP}_i \qquad (16\text{-}2)$$

The substrates for both of these reactions can be generated from intramitochondrial acetyl-SCoA. Thus, intramitochondrial acetyl-SCoA may be cleaved to acetate and, as such, diffuse into the soluble cell portion, or alternatively intramitochondrial acetyl-SCoA may be condensed with oxalacetate and the citrate product diffuse into the soluble portion. These processes, together with the citrate cleaving enzyme and acetate thiokinase, account for the overall transport of acetyl-SCoA across the mitochondrial membrane. In view of the fact that extramitochondrial citrate is known to stimulate and act as precursor for fatty acid synthesis, it has been suggested that this metabolite may be a major source of acetyl-SCoA for this process (7,8). In addition, citrate acts as activator for the enzyme responsible for the first step in fatty acid synthesis, which would also account for the citrate stimulation. It thus seems likely that citrate may function both as a source of acetyl-SCoA and as enzyme activator for fatty acid synthesis (see also page 716).

In addition, a fraction of the extramitochondrial acetyl-SCoA may be derived from the intramitochondrial pool through carnitine-mediated transport across the mitochondrial membrane. Evidence cited previously (page 598) suggests that carnitine may function as an acyl group carrier for the transport of acyl functions from the cytoplasm into the mitochondria in the course of fatty acid oxidation. This compound may well carry out the reverse role as well. Indeed, (−)-carnitine

$$(CH_3)_3N^+—CH_2—\underset{\displaystyle OH}{\underset{|}{CH}}—CH_2—COO^-$$

(−)-Carnitine

stimulates the incorporation of those compounds that are known to give rise to intramitochondrial acetyl-SCoA (pyruvate, glucose, acetate) into long-chain fatty acids in mammalian systems (9). The role of carnitine and citrate as acetyl group carriers for fatty acid synthesis is indicated schematically in Fig. 16-2).

An important key to the present understanding of the route of fatty acid synthesis was provided by the discovery of Wakil (9), and subsequently by Formica and Brady (6), of acetyl-CoA carboxylase. This biotin-dependent enzyme (page 422) catalyzes the ATP-dependent synthesis of malonyl-SCoA from CO_2 and acetyl-SCoA:

$$CH_3—\overset{\displaystyle O}{\overset{\|}{C}}—SCoA + CO_2 + ATP \xrightarrow{Mg^{2+}} HO_2C—CH_2—\overset{\displaystyle O}{\overset{\|}{C}}—SCoA + ADP + P_i \qquad (16\text{-}3)$$

Malonyl-SCoA is the first intermediate in fatty acid biosynthesis. In the presence of soluble enzyme fractions, malonyl-SCoA is rapidly converted into fatty acids. This conversion requires, absolutely, the presence of acetyl-SCoA (10). The acetyl-SCoA carboxylase reaction is the rate-determining step for the synthesis of fatty acids.

The acetyl-SCoA carboxylase reaction is an interesting one in several respects. In the first place, the reaction is known to proceed by at least the two following steps (11):

$$ATP + HCO_3^- + Enz \underset{}{\overset{M^{2+}}{\rightleftharpoons}} Enz\text{-}CO_2^- + ADP + P_i \qquad (16\text{-}4)$$

$$Enz\text{-}CO_2^- + acetyl\text{-}SCoA \rightleftharpoons Enz + malonyl\text{-}SCoA \qquad (16\text{-}5)$$

The overall stoichiometry, of course, is just that of Eq. (16-3). The carboxylated form of the enzyme has been identified as 1-*N*-carboxybiotinyl-enzyme, in accord with the general mechanism for biotin-mediated carboxylations previously described (page 422) (12). In the second place, the enzyme itself shows unusual structural properties. That isolated in pure form from liver has, in the presence of citrate or isocitrate, a molecular weight of some 4 million and occurs as a filamentous form 80 to 100 Å by 5000 Å, more or less (13). Raising the ionic strength, or lowering the isocitrate or enzyme concentration, tends to cause the filamentous enzyme to dissociate into protomers of molecular weight near 400,000; reintroduction of isocitrate or citrate and lowering of the ionic strength causes a reversal of this phenomenon (14,15). Since the protomeric form of the enzyme is without enzymatic activity, the means by which citrate and isocitrate stimulate fatty acid synthesis (16) through their action on this enzyme is clear. The liver enzyme contains 1 mol of bound biotin per 400,000-molecular-weight protomeric unit. This unit, in turn, is evidently composed of four polypeptide chains each of molecular weight 100,000. These must be nonidentical to the extent that only one of the four can possess a biotin group.

Malonyl-SCoA may also arise directly from malonate (1) in bacteria in the presence of the proper synthetase,

$$\text{Malonate} + \text{CoASH} + \text{ATP} \rightarrow \text{malonyl-SCoA} + \text{ADP} + P_i \qquad (16\text{-}6)$$

(2) in a thiophorase reaction involving succinyl- or acetoacetyl-SCoA,

$$R{-}\overset{\overset{\displaystyle O}{\|}}{C}{-}\text{SCoA} + \text{malonate} \rightleftharpoons \text{malonyl-SCoA} + RCO_2^- \qquad (16\text{-}7)$$

or (3) from the oxidation of malonyl semialdehyde CoA.

Fatty acid–synthesizing systems from a number of sources have been examined in some detail, including those from *E. coli,* yeast, and pigeon liver. Of these systems, that from *E. coli* has proved the easiest to deal with and has yielded the clearest results thus far. We turn to it first but will return to the other systems for comparative purposes later.

In Fig. 16-3, the probable route for fatty acid synthesis in *E. coli* is diagrammed. Repeated passages through the biosynthetic scheme of Fig. 16-3 yield the following stoichiometry for the synthesis of palmitate from acetyl-SCoA and malonyl-SCoA:

$$CH_3{-}\overset{\overset{\displaystyle O}{\|}}{C}{-}\text{SCoA} + 7HO_2C{-}CH_2{-}\overset{\overset{\displaystyle O}{\|}}{C}{-}\text{SCoA} + 14\text{NADPH} + 14H^+ \rightarrow CH_3{-}(CH_2)_{14}{-}CO_2H + 7CO_2 + 8\text{CoASH} + 14\text{NADP}^+ + 6H_2O \qquad (16\text{-}8)$$

Starting from acetyl-SCoA alone, the corresponding stoichiometry reads

$$8CH_3{-}\overset{\overset{\displaystyle O}{\|}}{C}{-}\text{SCoA} + 7CO_2^* + 14\text{NADPH} + 7\text{ATP} + 14H^+ \rightarrow CH_3{-}(CH_2)_{14}{-}CO_2H + 7CO_2^* + 8\text{CoASH} + 14\text{NADP}^+ + 7\text{ADP} + 7P_i + 6H_2O \qquad (16\text{-}9)$$

CO_2^* is shown on both sides of the equation, since it is an indispensable stoichiometric reactant, which, however, is regenerated intact in the course of the reaction sequence (see following page).

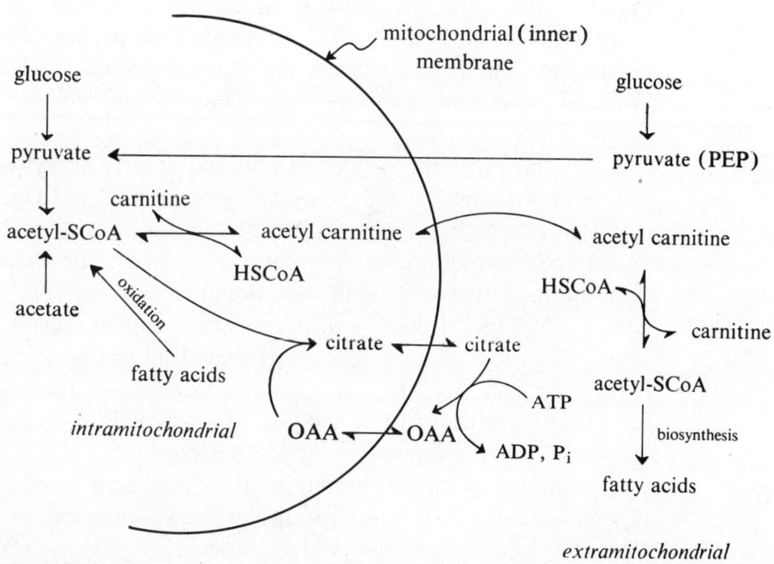

FIGURE 16-2
Schematic representation of the function of carnitine as acetyl group carrier across the mitochondrial membrane.

$$CH_3-\overset{\overset{O}{\|}}{C}-SCoA + HS-ACP \rightleftharpoons CH_3-\overset{\overset{O}{\|}}{C}-S-ACP + CoASH \quad (1)$$

$$\underset{CH_2-\overset{\overset{O}{\|}}{C}-SCoA}{HO_2C} + HS-ACP \rightleftharpoons \underset{CH_2-\overset{\overset{O}{\|}}{C}-S-ACP}{HO_2C} + CoASH \quad (2)$$

$$CH_3-\overset{\overset{O}{\|}}{C}-S-ACP + \underset{CH_2-\overset{\overset{O}{\|}}{C}-S-ACP}{HO_2C} \rightleftharpoons CH_3-\overset{\overset{O}{\|}}{C}-CH_2-\overset{\overset{O}{\|}}{C}-S-ACP + HS-ACP + CO_2 \quad (3)$$

$$CH_3-\overset{\overset{O}{\|}}{C}-CH_2-\overset{\overset{O}{\|}}{C}-S-ACP + NADPH + H^+ \rightleftharpoons CH_3-\underset{H}{\overset{OH}{C}}-CH_2-\overset{\overset{O}{\|}}{C}-S-ACP \text{ (D-isomer)} + NADP^+ \quad (4)$$

$$CH_3-\underset{H}{\overset{OH}{C}}-CH_2-\overset{\overset{O}{\|}}{C}-S-ACP\text{(D-isomer)} \rightleftharpoons CH_3-CH=CH-\overset{\overset{O}{\|}}{C}-S-ACP + H_2O \quad (5)$$

$$CH_3-CH=CH-\overset{\overset{O}{\|}}{C}-S-ACP + NADPH + H^+ \rightarrow CH_3CH_2CH_2-\overset{\overset{O}{\|}}{C}-S-ACP + NADP^+ \quad (6)$$

$$CH_3CH_2CH_2-\overset{\overset{O}{\|}}{C}-S-ACP + HS-CoA \rightleftharpoons CH_3CH_2CH_2\overset{\overset{O}{\|}}{C}-SCoA + HS-ACP \quad (7)$$

$$\underset{CH_2-\overset{\overset{O}{\|}}{C}-SCoA}{HO_2C} + CH_3-\overset{\overset{O}{\|}}{C}-SCoA + 2NADPH + 2H^+ \rightarrow CH_3CH_2CH_2\overset{\overset{O}{\|}}{C}-SCoA + CO_2 + CoASH + 2NADP^+ + H_2O \quad (8)$$

FIGURE 16-3
Pathway for fatty acid biosynthesis in *E. coli.*

The proposed pathway of Fig. 16-3 is consistent with and supported by the following important observations:

1. The fatty acid–synthesizing system of *E. coli* is soluble and has been fractionated into several components. These include an acetyl transferase, a malonyl transferase, a condensing enzyme, a β-ketoacyl-ACP reductase, an enoyl-ACP hydrase [Eqs. (1) through (6) in Fig. 16-3]. In addition, the synthetase system contains an acyl-carrier protein (ACP), molecular weight 9600, to which the acyl intermediates are bound throughout the course of the synthetic process (17).

2. In the presence of excess malonyl-SCoA, acetyl-SCoA is incorporated only into carbons 15 and 16 of palmitic acid (18). This is exactly the result expected on the basis of an initial condensation of acetyl-SCoA and malonyl-SCoA followed by successive chain elongation through condensation of the developing fatty acid with additional molecules of malonyl-SCoA. Thus acetyl-SCoA functions as a primer for fatty acid synthesis.

3. ^{14}C-labeled CO_2 is not incorporated into fatty acids. This finding implies a catalytic role for CO_2, and such a role has been assigned to it. The same CO_2 that is utilized in the acetyl-CoA carboxylase reaction is liberated in the acetyl-SCoA–malonyl-SCoA condensation.

4. Malonyl-SCoA is bound to the fatty acid–synthesizing complex from pigeon liver by sulfhydryl groups and condensation of the malonyl enzyme with acetyl-SCoA is accompanied by decarboxylation (19).

5. The product of the condensation reaction in the *E. coli* system has been identified as acetoacetyl-S-ACP (20). A similar observation has been made by Lynen for the system from yeast.

6. An enoyl hydrase, specific for acyl-carrier protein thiolesters, has been detected in and purified from *E. coli*. This enzyme catalyzes the reversible hydration of crotonyl-S-ACP to yield D-(−)-β-hydroxybutyryl-S-ACP.

7. In soluble systems from rat brain or adipose tissue, acetoacetyl-SCoA, D-(−)-β-hydroxybutyryl-SCoA, and crotonyl-SCoA, the acyl-SCoA derivatives of the suggested intermediates, are incorporated into palmitate (21). In accordance with the scheme of Fig. 16-3, these substrates appear to undergo reduction prior to incorporation.

8. Tritium from tritiated NADPH is incorporated into the odd-numbered carbon atoms of palmitic acid, beginning with carbon number 3 (22). Since FMN is involved coenzymatically in the reduction of the α,β-unsaturated species, the tritium incorporation presumably comes from direct hydrogen transfer from labeled NADPH to the β-ketoacyl species yielding the β-hydroxyacyl compound.

9. In the yeast system, each of the postulated enzymatic activities may be individually demonstrated with the use of model substrates.

10. Attempts at detection of free intermediates in fatty acid biosynthesis have proved fruitless. This is consistent with, and provides support for, the existence of the various intermediates in a form tightly bound to the pertinent enzyme system.

Thus, the essential features of the proposed biosynthetic scheme account for the basic experimental findings concerning fatty acid synthesis. It should be emphasized that each of the systems thus far investigated exhibits individualities that distinguish it, to a greater or lesser degree, from any of the others. Consequently, the proposed biosynthetic route of Fig. 16-3 does not necessarily accurately represent the detailed biochemical events in all species. Nevertheless, the differences between systems seem to be far less significant than their similarities, and only minor modifications appear to be necessary to accommodate them.

Let us take a brief look at each of the components of the multienzyme complex that catalyzes fatty acid synthesis in *E. coli*. Of these, ACP has received the most attention. *E. coli* ACP contains 77 amino acid residues in known sequence (23). The site of attachment of the acyl functions to ACP is most unusual; it proves to be the sulfhydryl group of a 4′-phosphopantotheine group attached to serine at position 36 (one of three serine residues in the protein) through a phosphodiester linkage (23–25). Thus, the structure may be visualized as follows:

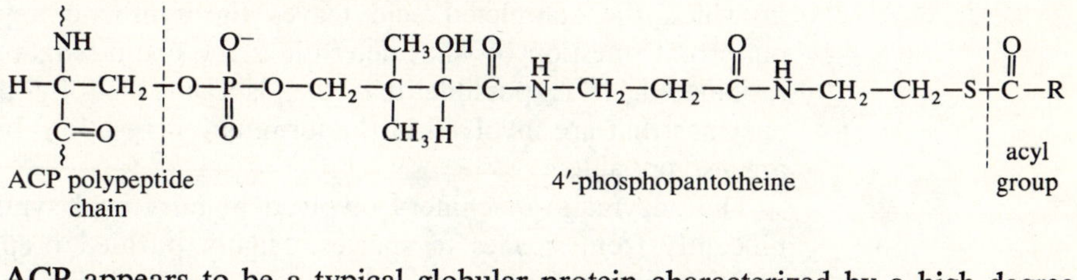

```
    ⌇       ¦    O⁻          CH₃OH O                      O                  ¦ O
    NH      ¦    |             |   |  ||                   ||                 ¦ ||
    |       ¦    |             |   |  ||   H               ||   H             ¦ ||
H—C—CH₂—O—P—O—CH₂—C—C—C—N—CH₂—CH₂—C—N—CH₂—CH₂—S—C—R
    |       ¦    ||            |   |                                         ¦
    C=O     ¦    O            CH₃ H                                         ¦ acyl
    ⌇       ¦
ACP polypeptide                    4′-phosphopantotheine                     group
   chain
```

ACP appears to be a typical globular protein characterized by a high degree of ordered structure. It performs a function that is unique among those of known proteins: that of acting as a carrier for groups in the process of a biosynthetic sequence. In this respect, ACP is analogous to tRNA, which acts as a carrier for amino acids in the process of protein synthesis (page 918), and to polyprenols, which carry glycosyl groups in the process of synthesis of a variety of bacterial polysaccharides (pages 552 and 554). Like tRNA,

ACP may be responsible for certain aspects of the specificity of the biosynthetic process, as we shall see below.

Both the malonyl and acetyl transacylases, Eqs. (1) and (2) of Fig. 16-3, have been substantially purified (26). Both reactions are freely reversible and both are sensitive to inhibition by reagents that bind protein sulfhydryl groups. The malonyl transacylase is quite specific for malonyl-SCoA, as is the acetyl transacylase for acetyl-SCoA. The low activity of the latter enzyme with longer chain derivatives of CoASH accounts for their slow rate of incorporation into fatty acids in the system.

β-Ketoacyl ACP synthetase, which catalyzes reaction (3) of Fig. 16-3, is specific for the thiolesters of ACP but is relatively nonspecific for the chain length of the condensation partner of malonyl-SCoA (27). As the enzymes above, it is sensitive to sulfhydryl-group reagents.

β-Ketoacyl ACP reductase, Eq. (4) of Fig. 16-3, catalyzes the formation of the D(−) isomer of β-hydroxyacyl ACP's and is not completely specific for the derivatives of ACP, although it is maximally active with them (26). The enzyme shows little chain-length dependence.

One of the interesting reactions of the overall scheme is that catalyzed by β-hydroxyacyl ACP dehydrase [Eq. (5), Fig. 16-3]. This enzyme is specific for the ACP derivatives and for the D(−) isomer of the substrate, the isomer generated in the previous step. The enzyme is, somewhat surprisingly in view of behavior of other enzymes of the complex, rather specific for short-chain substrates. The reaction rate falls off dramatically as the chain length is increased from four to eight carbon atoms, and the 10-carbon substrate appears not to react. This implies that a separate enzyme must exist for dehydration of the longer chain members of the series, since palmitate is a principal end product (C_{16}); such an enzyme does exist and proves crucial in the synthesis of unsaturated fatty acids. More about that below.

Relatively little is known about reduction of the enoyl-ACP to the saturated derivative. Reactions of this type are ordinarily catalyzed by flavoproteins (page 645), and a flavoprotein is known to be involved in the yeast system. The situation in *E. coli* remains to be established.

The last reaction of Fig. 16-3, in which fatty acid synthesis is viewed as terminating in a transacylation reaction involving CoASH, is the least certain of the series of reactions. A simple hydrolysis reaction is also a distinct possibility. In fact, both types of reactions do seem to occur; in yeast stearyl-SCoA and palmityl-SCoA are the principal products, while free palmitic acid is the principal product of the avian liver system (28). Quite aside from the form in which the completed acid leaves the synthetic apparatus, there is the important question of what determines the chain length at which synthesis is terminated. It is possible that this is controlled by the specificity of the enzymes that are involved in the termination reaction, but other explanations are also possible.

The enzymatic machinery involved in fatty acid synthesis does differ significantly from species to species. Highly purified preparations of the fatty acid synthetase from yeast behave as a single particulate component, clearly of a multienzyme nature, of molecular weight 2,300,000. This complex appears to be composed of seven enzymes, and its activity requires sulfhydryl groups. Biosynthesis of the yeast fatty acid synthetase complex takes place with incorporation of 4-phosphopantotheine, suggesting that something resembling the ACP of *E. coli* may exist in the yeast system as well. Acyl-carrier proteins appear to be widely distributed and have been identified in bacterial, plant, and animal sources. Nevertheless, it is not clear that they are universal. The pigeon liver multienzyme complex, molecular weight 500,000, composed of 5 to 8 peptide chains, involves two types of sulfhydryl groups in the synthetic process, although there is no evidence that either of them is associated with 4-phosphopantotheine (29).

Since propionyl-SCoA will replace acetyl-SCoA to some extent in the initial condensation reaction with malonyl-SCoA, the biosynthesis of fatty acids containing an odd number of carbon atoms may be accounted for in terms of the enzymatic pathway indicated above.

The biosynthesis of branched-chain fatty acids also appears to occur via a minor modification of the pathway for the straight-chain acids. Studies employing purified fatty acid synthetases from adipose tissue and from bacterial sources suggest that such acids arise through the condensation of the appropriate short-chain fatty acyl-SCoA with malonyl-SCoA. The source of the former species is suggested by the structural analogies which exist between the branched-chain amino acids and branched-chain fatty acids:

$$CH_3CH_2\underset{\displaystyle CH_3}{\underset{|}{C}H}CH_2(CH_2)_9COOH \qquad CH_3\underset{\displaystyle CH_3}{\underset{|}{C}H}CH_2CH_2(CH_2)_9COOH$$

$$CH_3CH_2\underset{\displaystyle CH_3}{\underset{|}{C}H}\text{—}\underset{\displaystyle NH_2}{\underset{|}{C}H}COOH \qquad CH_3\underset{\displaystyle CH_3}{\underset{|}{C}H}CH_2\text{—}\underset{\displaystyle NH_2}{\underset{|}{C}H}COOH$$

Thus, isoleucine might be visualized as a resonable precursor for 12-methyltetradecanoic acid and leucine for 13-methyltetradecanoic acid. In fact, in bacteria these amino acids do serve as precursors for the indicated fatty acids, suggesting that the latter may arise by pathways that have the general form shown in Fig. 16-4.

It may be well at this point to summarize the features that distinguish fatty acid *biosynthesis* from fatty acid *oxidation;* similar considerations appear to obtain whenever anabolic are contrasted with catabolic sequences for one and the same class of compound.

1. Although the enzymatic reactions comprising a catabolic sequence are reversible in principle and microscopically, actual reversal for biosynthesis occurs by a pathway that is distinct in one or more of the following features.

2. Some or all of the enzymes responsible are localized in different subcellular components.

3. One (or more) key steps usually located at the end of the degradative —and consequently at the beginning of the biosynthetic sequence are virtually irreversible and hence distinct. In the present instance, these are the

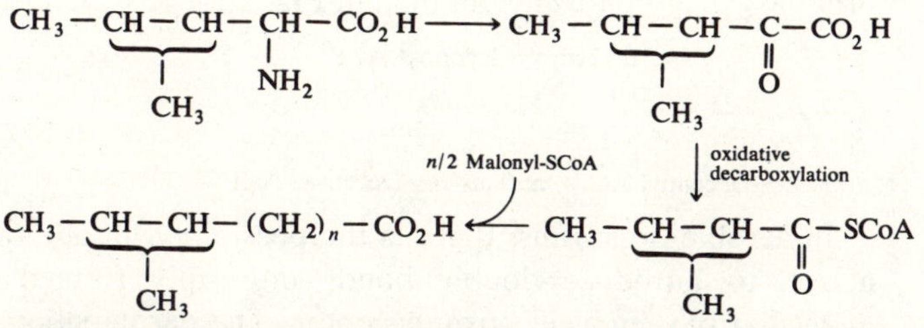

$$CH_3\text{—}\underbrace{CH\text{—}CH}_{\displaystyle CH_3}\text{—}\underset{\displaystyle NH_2}{\underset{|}{C}H}\text{—}CO_2H \longrightarrow CH_3\text{—}\underbrace{CH\text{—}CH}_{\displaystyle CH_3}\text{—}\underset{\displaystyle O}{\underset{\|}{C}}\text{—}CO_2H$$

$$\big\downarrow \text{ oxidative decarboxylation}$$

$$CH_3\text{—}\underbrace{CH\text{—}CH}_{\displaystyle CH_3}\text{—}(CH_2)_n\text{—}CO_2H \xleftarrow{n/2 \text{ Malonyl-SCoA}} CH_3\text{—}\underbrace{CH\text{—}CH}_{\displaystyle CH_3}\text{—}\underset{\displaystyle O}{\underset{\|}{C}}\text{—}SCoA$$

FIGURE 16-4
General pathway for the biosynthesis of branched-chain fatty acids.

reactions catalyzed by β-ketothiolase and acetyl-SCoA carboxylase, which lead to the generation and utilization of acetyl-SCoA, respectively.

4. The enzyme responsible for the initiation of a biosynthetic pathway is subject to feedback inhibition and other control mechanisms.

5. Catabolic sequences are energy providing and generate ATP and (immediately) oxidized pyridine nucleotides; anabolic ones are reductive, energy requiring, and utilize ATP and reduced pyridine nucleotides. The two pyridine nucleotides are again distinct: NAD^+ for degradation, NADPH for biosynthesis (see also pages 492, 626, and 701).

6. In some cases, the stereochemistry of the intermediates may differ in the biosynthetic and degradative routes; specifically for the present case, biosynthesis occurs through the D(−)β-hydroxyacyl intermediates, although degradation involves the L(+)-forms.

7. Finally, the nature of the acyl carrier group varies: ACP for synthesis and CoASH for degradation.

Biosynthesis of Unsaturated Fatty Acids

Although the routes for the synthesis of unsaturated fatty acids are incompletely understood, it is clear that at least two pathways do occur. One of these is anaerobic and is a variant of the synthetic route already developed and the other is aerobic, requiring the participation of molecular oxygen. The ability to introduce directly a double bond into a preformed fatty acid is lacking in several anaerobic bacteria (30). Hofmann and coworkers have suggested that unsaturated fatty acids may arise through the condensation of unsaturated acyl-SCoA intermediates directly with additional malonyl-SCoA units. Thus, rather than reducing the double bond as in synthesis of saturated fatty acids, the bond remains intact and is moved away from the carboxyl group as the chain is lengthened. A survey of the unsaturated fatty acids that occur in bacteria, in terms of the position of the double bond, suggests that a β-dehydration of C_{10} and C_{12} hydroxyacyl-SCoA intermediates will account for the observed products. For example, the synthesis of palmitoleic and vaccenic acids from octanoate may be visualized as indicated in Fig. 16-5. Of course, this pathway requires that there exist an enzyme catalyzing the β,γ-dehydration of the appropriate intermediates. Bloch and coworkers have discovered just such an enzyme in the fatty acid synthetic complex of *E. coli* and have determined several of its properties. We noted earlier that the first dehydrase identified in *E. coli* is specific for C_4 to C_8 chain lengths. That discovered by Bloch is rather specific for the C_{10} β-hydroxyacyl ACP, although C_8 and C_{12} substrates are handled and the enzyme is not absolutely specific for the ACP derivatives (31). This enzyme catalyzes the dehydration of β-hydroxydecanoyl ACP to yield a mixture of *trans*-α,β-decenoyl and *cis*-β,γ-decenoyl ACP's. The former product is metabolized predominantly to yield saturated fatty acids, while the latter yields both long-chain saturated and unsaturated fatty acids. These findings are summarized below and confirm the general pathway outlined in Fig. 16-5 (31–33):

D(−)β-Hydroxydecanoyl ACP

trans-α,β-Decenoyl ACP $\rightleftharpoons$ *cis*-β,γ-Decenoyl ACP

In aerobic organisms, there is the possibility, in addition to that developed above, to introduce double bonds into fully formed fatty acids utilizing molecular oxygen as electron acceptor. The participation of molecular oxygen in the desaturation process in aerobic organisms was first suggested by the observation that under anaerobic conditions, but not in the presence of air,

$$CH_3\text{—}(CH_2)_5\text{—}CH_2\text{—}CO_2H$$

$$\downarrow +C_2$$

$$CH_3\text{—}(CH_2)_5\text{—}CH_2\text{—}\overset{\displaystyle O}{\overset{\|}{C}}\text{—}CH_2\text{—}CO_2H$$

$$\downarrow$$

$$CH_3\text{—}(CH_2)_5\text{—}CH_2\text{—}\underset{\displaystyle H}{\underset{|}{\overset{\displaystyle OH}{\overset{|}{C}}}}\text{—}CH_2\text{—}CO_2H$$

$$\downarrow$$

$$CH_3(CH_2)_5\text{—}CH{=}CH\text{—}CH_2\text{—}CO_2H$$

$$\downarrow +3C_2$$

$$CH_3(CH_2)_5\text{—}CH{=}CH\text{—}(CH_2)_7\text{—}CO_2H$$

Palmitoleic acid

$$\downarrow +C_2$$

$$CH_3(CH_2)_5\text{—}CH{=}CH\text{—}(CH_2)_9\text{—}CO_2H$$

Vaccenic acid

FIGURE 16-5

Anaerobic pathway for the synthesis of palmitoleic and vaccenic acids from octanoate. K. Bloch, P. Baronowsky, H. Goldfine, W. J. Lennarz, R. Light, A. T. Norris, and G. Scheuerbrandt, *Federation Proc.*, **20:** 921 (1961).

yeasts require oleic acid as a growth factor. This finding was subsequently corroborated in studies on a purified system from yeast; under anaerobic conditions only saturated acids were synthesized, while in the presence of oxygen both saturated and unsaturated products were obtained in approximately equal amounts. This system, in the presence of O_2 and NADPH, will catalyze the desaturation of preformed acids such as palmitate or stearate. The acyl-SCoA derivatives are substrates for this reaction. Although the cofactor requirements and the insensitivity of this system to cyanide inhibition suggest an oxygenase-like reaction, involving the direct hydroxylation of substrate, the corresponding hydroxy acids are not significantly metabolized by yeast, although, unexpectedly, they do support growth under anaerobic conditions. In addition, the phosphate monoester of 10-hydroxystearic acid is not metabolized. Thus, the formation of unsaturated fatty acids in aerobic organisms appears to result from the desaturation of the preformed fatty acids, probably via an oxygenated intermediate other than the hydroxy acid (30). The intracellular site for aerobic fatty acid desaturation is on particles concentrated in the microsomal fraction. Again this appears to be a general feature for hydroxylation, desaturation, and other oxidation reactions that require the *simultaneous* participation of O_2 and a reducing agent such as NADPH (mixed-function oxidase, page 638): the general reaction type is

$$\begin{aligned} RH + NADPH + O_2^* + 2H^+ \rightarrow\ & R^+ + NADP^+ + 2H_2O^* \\ \text{or}\qquad & \\ & RO^*H + NADP^+ + H_2O^* \end{aligned} \qquad (16\text{-}10)$$

Localization in microsomal membranes is characteristic not only of these reactions but also of most of the reactions of lipid biosynthesis except those for the initial synthesis of fatty acids. It is the membranous portion of the microsomes, derived from the endoplasmic reticulum, rather than the ribosomes, that is involved in these reactions (see Chapter 9).

Recent studies on the desaturating system from *Euglena gracilis,* a photoauxotrophic organism, indicate that it can be fractionated into three protein components: a NADPH oxidase, the desaturase, and a ferredoxin (34). Other systems seem not to show a requirement for ferredoxin but may need an equivalent but so-far unidentified electron-transfer agent. There is preliminary evidence that some of these reactions at least may involve desaturation of the acyl ACP derivatives.

Biosynthesis of Cyclopropane Fatty Acids

Cyclopropane fatty acids occur widely distributed in bacteria. The mode of biosynthesis of these unusual compounds has been markedly clarified through the studies of Law and his associates (35-37): (1) The substrates for the formation of the cyclopropane ring are the corresponding *cis* unsaturated fatty acids; (2) the reaction requires that these fatty acids exist as components of phospholipids, generally phosphatidylethanolamine; the unsaturated fatty acid ordinarily occupies the β position; (3) the reaction is critically dependent on the physical state of the phospholipid substrate; (4) the ring carbon atoms are donated by *S*-adenosylmethionine; if this methyl group is labeled with three atoms of deuterium, two of these are incorporated into the cyclopropane fatty acid; and (5) the reaction is subject to marked feedback inhibition by the product, adenosylhomocysteine. These considerations are summarized in Fig. 16-6.

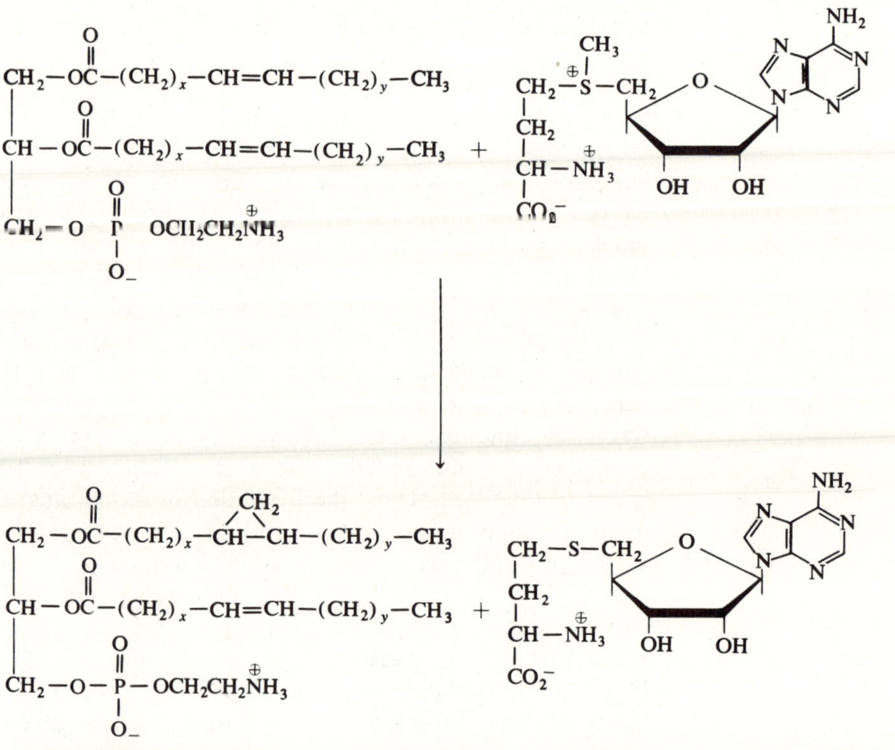

$CH_2{-}OC({=}O){-}(CH_2)_x{-}CH{=}CH{-}(CH_2)_y{-}CH_3$
$CH{-}OC({=}O){-}(CH_2)_x{-}CH{=}CH{-}(CH_2)_y{-}CH_3$
$CH_2{-}O{-}P({=}O)(O^-){-}OCH_2CH_2\overset{\oplus}{N}H_3$

+ $CH_3{-}\overset{\oplus}{S}(CH_2{-}CH(\overset{\oplus}{N}H_3){-}CO_2^-){-}CH_2$–ribose (OH, OH)–adenine (NH_2) [S-adenosylmethionine]

↓

$CH_2{-}OC({=}O){-}(CH_2)_x{-}CH(CH_2)CH{-}(CH_2)_y{-}CH_3$ (cyclopropane ring)
$CH{-}OC({=}O){-}(CH_2)_x{-}CH{=}CH{-}(CH_2)_y{-}CH_3$
$CH_2{-}O{-}P({=}O)(O^-){-}OCH_2CH_2\overset{\oplus}{N}H_3$

+ $S(CH_2{-}CH_2{-}CH(\overset{\oplus}{N}H_3){-}CO_2^-){-}CH_2$–ribose (OH, OH)–adenine (NH_2) [S-adenosylhomocysteine]

FIGURE 16-6
Pathway for the biosynthesis of cyclopropane fatty acids. Note that the formation of the cyclopropane ring can occur with either the fatty acyl group at C-1 (as shown) or with that esterified at C-2.

Biosynthesis in Particulate Systems

The observation of Lynen and coworkers cited above concerning the reversibility of the overall process of fatty acid oxidation (1) suggests that, in principle at least, biosynthesis may occur by such a route. The subsequent observation of the role of malonyl-SCoA in fatty acid synthesis in soluble systems casts great doubt on the physiological importance of such a pathway. Nevertheless, more recent observations concerning synthesis in particulate systems, both mitochondrial and microsomal, suggest that reversal of the oxidative pathway may play a significant role for some purposes at least (38,39). Observations thus far on particulate systems suggest that malonyl-SCoA–dependent pathways exist, principally in microsomal preparations, and that these are closely related to the soluble system just described, as well as malonyl-SCoA–independent pathways, principally in mitochondrial preparations, which are closely related to the reverse of fatty acid oxidation. Further development of these systems is necessary prior to a full assessment of their physiological significance.

BIOSYNTHESIS OF AROMATIC COMPOUNDS FROM ACETATE

Before considering the incorporation of fatty acids into more complex lipids, attention is directed to some biosyntheses that are closely related to that for fatty acids. The generic term *acetogenin* has been suggested for those substances that arise from the repeated condensation of acetate residues (e.g., the fatty acids). The acetogenins include a broad spectrum of aromatic compounds. [The other principal source of naturally occurring aromatic compounds is glucose, by way of the shikimic acid pathway (see Chapter 17).] The "acetate hypothesis"—that a linear polyketomethylene chain formed from repeated head-to-tail self-condensation of acetate units may yield, through cyclization, oxidative coupling, and alkylations at the methylene carbon atoms, a multitude of complex substances—was initially formulated by Collie near the turn of this century. More than half a century later, Birch elaborated these ideas and developed a modern formulation of the acetate hypothesis (40,41), which considerable experimental work, the bulk of which is from Birch's own laboratory, has tended to confirm. A comprehensive review of theoretical and experimental aspects of the acetate hypothesis has been prepared by Richards and Hendrickson.

As noted above, the basis of the acetate hypothesis is that aromatic compounds may arise from linear polyketomethylene chains. Some simple examples that illustrate this hypothesis and for which independent experimental evidence is available are indicated in Fig. 16-7. Experiments employing carboxyl-labeled acetate reveal that this label appears precisely in those positions in the products (orsellininc acid, 6-methyl salicylic acid, cyclopaldic acid) predicted by the acetate hypothesis. More complicated examples of the use and verification of this hypothesis are indicated in Fig. 16-8. Experiments very similar to those mentioned above indicate that synthesis of such structurally distinct molecules as griseofulvin, chromanone, alternariol, terramycin, and penicillic acid all proceed via polyketomethylene chains and do so in accordance with predictions based on the acetate hypothesis.

While the postulate of polyketomethylene chains as precursors of certain acetogenins is convenient in rationalizing their synthesis, the known instability of such compounds and their absence in living material suggest that they do not exist free within the cell. One conception that circumvents this difficulty but retains the essential features of the acetate hypothesis and, furthermore, is analogous to much that is known concerning both the machinery and reactions of fatty acid biosynthesis is depicted in Fig. 16-9. The polyacetyl

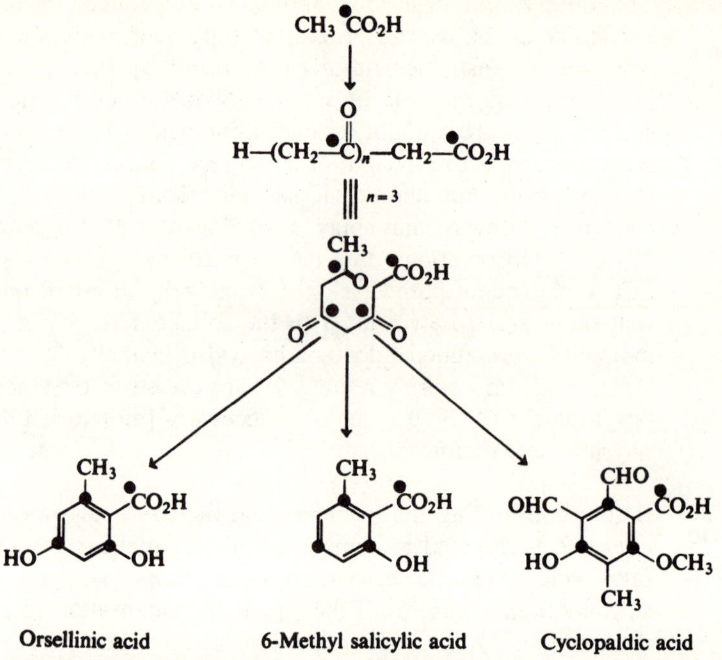

FIGURE 16-7
Biosynthesis of simple aromatic compounds through polyacetyl intermediates.

synthetase is visualized as a linear succession of sulfhydryl sites, each of which may be occupied by a malonyl residue. Repeated aldol condensations then yield an enzyme-bound polyacetyl chain, which may cyclize or otherwise react prior to its release from the enzyme surface. Such a formulation is consistent with the observation of Lynen that 6-methyl salicylic acid is derived from one molecule of acetyl-SCoA and three molecules of malonyl-SCoA in a manner similar to that for formation of fatty acids. Thus, the basic features of formation of polyacetyl chains may reflect those for synthesis of polymethylene chains, with the important exception that, in the former case, the growing chain is not reduced as it is formed.

GLYCERIDES

Structural Considerations

A number of important lipids are derived from glycerol, a trihydric alcohol:

$$\begin{array}{l} CH_2OH \\ | \\ CHOH \\ | \\ CH_2OH \end{array}$$

The simplest members of the lipid family based on this alcohol, the mono,- di-, and trigylcerides, have been encountered and discussed briefly earlier (page 585). The triglycerides, which constitute the bulk of most fats and oils, are the most abundant lipids on earth. In addition to these, there are two other neutral glycerol derivatives of interest: the glyceryl ethers and the glycosyl glycerides.

Several *α-glyceryl ethers* are widely distributed in nature, notably chimyl alcohol, batyl alcohol, and selachyl alcohol. Several types of lipid materials

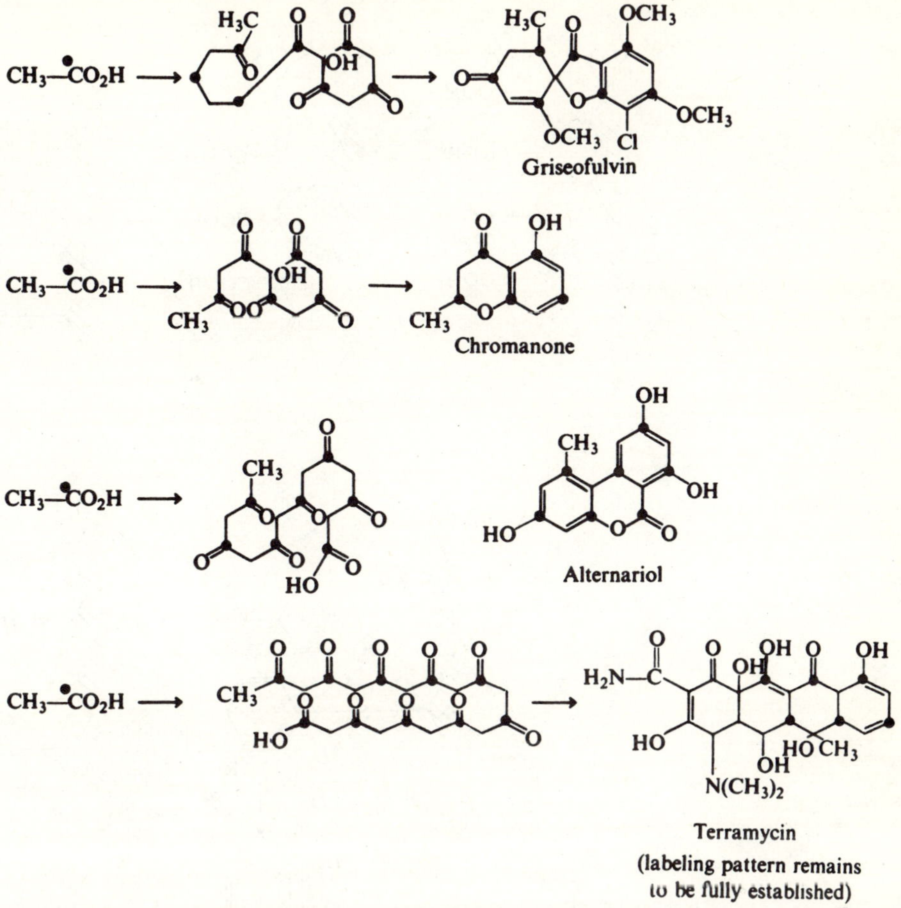

CH₃—CO₂H → → Griseofulvin

CH₃—CO₂H → → Chromanone

CH₃—CO₂H → Alternariol

CH₃—CO₂H → → Terramycin (labeling pattern remains to be fully established)

FIGURE 16-8
Synthesis of some natural products via polyacetyl intermediates.

containing α-glyceryl ethers have been identified, including structural analogs of phosphatidylcholine and phosphatidylethanolamine:

$$\begin{array}{l} CH_2{-}O{-}(CH_2)_{15}{-}CH_3 \\ | \\ CHOH \\ | \\ CH_2OH \end{array} \qquad \begin{array}{l} CH_2{-}O{-}(CH_2)_{17}CH_3 \\ | \\ CHOH \\ | \\ CH_2OH \end{array}$$

Chimyl alcohol   Batyl alcohol

$$\begin{array}{l} CH_2{-}O{-}(CH_2)_8{-}CH{=}CH{-}(CH_2)_7{-}CH_3 \\ | \\ CHOH \\ | \\ CH_2OH \end{array}$$

Selachyl alcohol

While the biosynthetic routes leading to the glyceryl ethers and their derivatives remain unknown, they do seem to be derived by a pathway involving the reduction of the corresponding fatty acids. They, in turn, appear to be precursors of the plasmalogens through an oxidative reaction pathway (see below). A very recent report may provide insight into the mode of synthesis of some of these compounds at least. This work describes a system in which

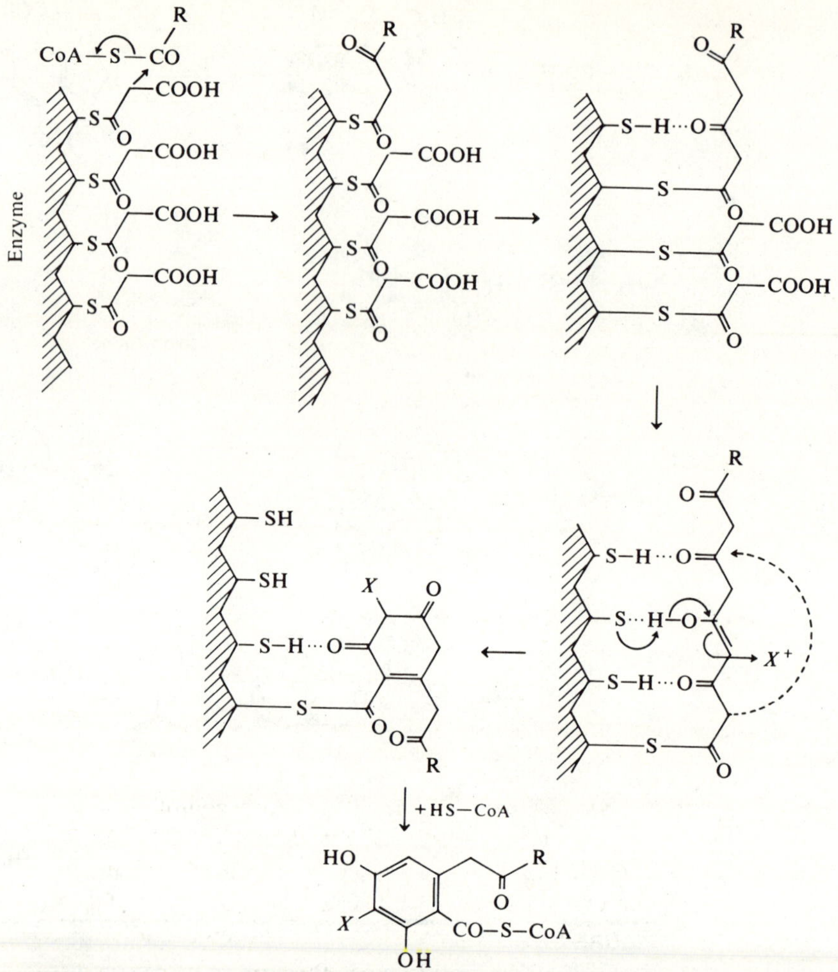

FIGURE 16-9

Enzyme model for acetogenin biosynthesis. J. H. Richards and J. B. Hendrickson, *Biosynthesis of Steroids, Terpenes, and Acetogenins*, Benjamin, New York, 1964, p. 133.

a microsomal preparation augmented with a soluble protein complex catalyzes the synthesis of glyceryl ethers from glyceraldehyde-3-phosphate and fatty alcohols (41). The pathway probably looks something like the following:

$$\begin{array}{l} CHO \\ | \\ CHOH + R{-}OH \\ | \\ CH_2OPO_3H^- \end{array} \xrightarrow{ATP} \begin{array}{l} HO{-}CH{-}OR \\ \quad | \\ \quad CHOH \\ \quad | \\ \quad CH_2OPO_3H^- \end{array} \longrightarrow \begin{array}{l} CH_2OR \\ | \\ C{=}O \\ | \\ CH_2OPO_3H^- \end{array} \xrightarrow{NADH} \begin{array}{l} CH_2OR \\ | \\ CHOH \\ | \\ CH_2OPO_3H^- \end{array} \qquad (16\text{-}11)$$

Lipids exist in covalent linkage with carbohydrates in two classes of compounds: those based on sphingosine and those based on glycerol. Members of the former class are widely distributed among plant and animal species but are rare in bacteria. Conversely, a variety of glycosyl diglycerides occur widely in bacteria but are rare in mammals. In *M. lysodeikticus*, for example, a mannolipid has been identified with the structure *O*-α-D-mannosyl-(1→3)-*O*-α-D-mannosyl-(1→α′) diglyceride:

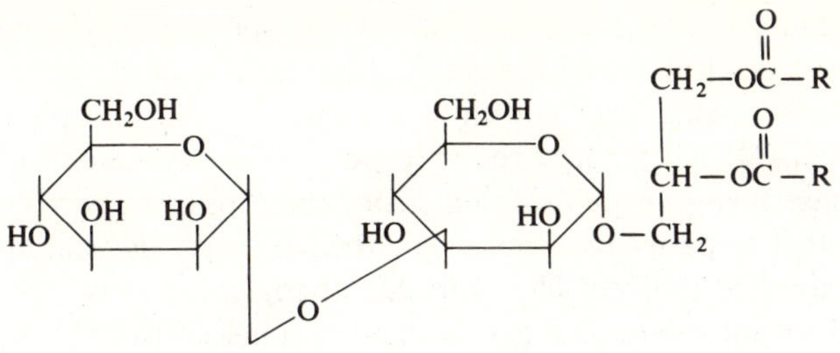

Two related glycolipids have been isolated from species of *Pneumococcus* (42,43):

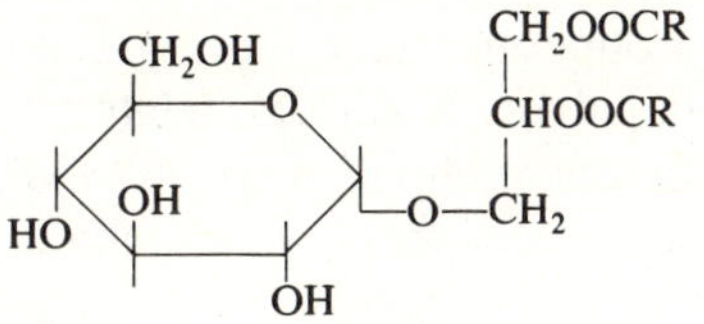

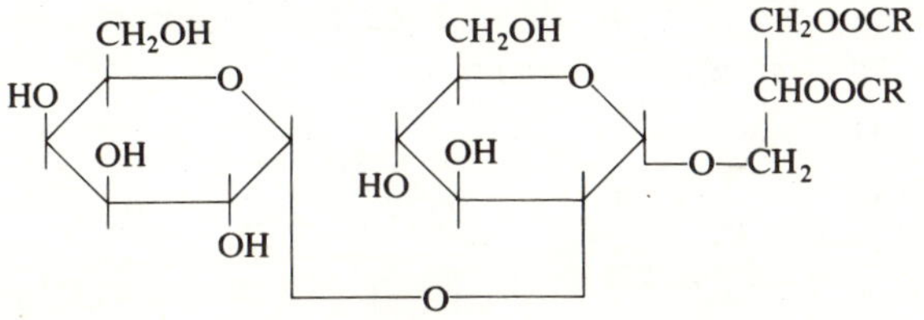

Of great interest is the much more complicated lipopolysaccharide which is an important constituent of the cell walls of Gram-negative bacteria. Although the complete structure of this substance is unknown, the principal features have been detailed in Fig. 10-8 (44). Some aspects of the biosynthesis of this complex substrate have been elucidated (45) (see page 554).

Plants also contain glycolipids, many of which resemble the bacterial ones just described. An interesting exception is quinovose, a sulfate-containing glycolipid found principally in the chloroplasts of plants but also in the chromatophores of photosynthetic bacteria:

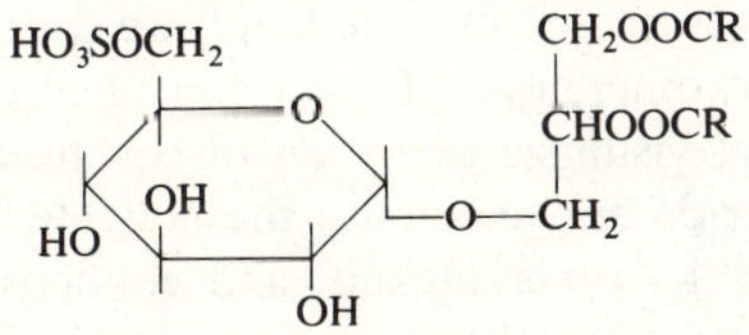

Quinovose

Biosynthesis The biosynthetic pathway for the formation of triglycerides in liver was first formulated by Weiss and Kennedy in 1956 (46). Free glycerol is phosphorylated at the α position in the presence of glycerokinase and ATP yielding L-glycerophosphate, which, alternatively, may arise from reduction of dihydroxyacetone phosphate. One mol of this is subsequently acylated with 2 mol of fatty acyl-SCoA, yielding an α,β-diglyceride phosphate, termed a phosphatidic acid. Phosphatidic acids are, as indicated below, key intermediates in the biosynthesis of several classes of complex lipids. The synthesis of the triglycerides is completed by dephosphorylation, yielding an α,β-diglyceride, followed by esterification with a third mol of fatty acyl-SCoA. In adipose tissue and intestinal mucosa, glycerophosphate arises largely from the NADH-dependent reduction of dihydroxyacetone phosphate, the glycerokinase being virtually absent from these tissues. With this exception, the biosynthetic pathway leading to trigylcerides is similar to that previously established in liver. The synthesis of triglycerides is summarized in Fig. 16-10. Recent information suggests that, in some cases at least, the fatty acyl groups employed for trigylceride synthesis may be donated from the ACP rather than the CoASH derivatives.

Synthesis of glycolipids has been investigated in relatively few cases. Preliminary results indicate that the diglyceride is the acceptor for the glycosyl groups and that these are donated from NDP-sugars, as one might expect based on results in related synthetic processes. The route for synthesis of the mannolipid of *M. lysodeiktikus* (*M. luteus*) is an example of these considerations and is diagrammed in Fig. 16-11 (47-48).

PHOSPHATIDES

Chemical Considerations

A large number of lipids of exceptional importance in living systems are derivatives of *phosphatidic acid:*

$$\begin{array}{l} CH_2-O\overset{O}{\overset{\|}{C}}R \\ | \\ CH-O\overset{O}{\overset{\|}{C}}R \\ | \\ CH_2-O\underset{\underset{O}{\|}}{\overset{\overset{O^-}{|}}{P}}O \,\vdots\, R \end{array}$$

That part of the structure to the left of the dashed line is termed a "phosphatidyl" group. In phosphatidic acid itself, R = H; in other derivatives a variety of groups occupy this position. The more important of these are collected, together with the names of the complete lipid structure, in Fig. 16-12. These have been divided into two groups: those that contain nitrogen and those that do not.

Biosynthesis

Looking at the structure of phosphatides in a basic way, we can recognize that they are just phosphodiesters. On one hand, we see the alcoholic diglyceride function and, on the other, one of a variety of alcohols such as choline or inositol. We can, then, visualize two ways of constructing these molecules: (1) activate the diglyceride and condense the activated form with the other alcohol or (2) activate the other alcohol and condense the activated form with the diglyceride. In fact, synthetic routes concordant with both of these models are known. In both cases, the activated forms are the CDP derivatives which arise from the reaction of CTP with the alcohol phosphate (page 543). In general, then, the following route is followed:

$$R_1OH \xrightarrow{ATP \quad ADP} R_1OPO_3H^- \xrightarrow{CTP \quad PP_i} R_1O-CDP \xrightarrow{R_2OH \quad CMP} R_1O-\underset{\underset{O}{\|}}{\overset{\overset{O^-}{|}}{P}}-OR_2 \qquad (16\text{-}12)$$

The more common alternative route is that in which R_1OH is a diglyceride.

In fact, this is the only route that seems to be followed in bacteria. In animal tissues, on the other hand, both possibilities are utilized. The biosynthetic routes leading to actual compounds are detailed in Fig. 16-13.

It is known that the distribution of saturated and unsaturated fatty acids in the phospholipids is asymmetric. Specifically, saturated fatty acids are exclusively incorporated into the C-1 position and unsaturated ones are

$$\begin{array}{c} CH_2OH \\ | \\ HO\text{—}C\text{—}H \\ | \\ CH_2OH \end{array} + ATP \xrightarrow[ADP]{Mg^{2+}} \begin{array}{c} CH_2OH \\ | \\ HO\text{—}C\text{—}H \\ | \\ CH_2OPO_3H^- \end{array} \xleftarrow{NAD^+} \begin{array}{c} CH_2OH \\ | \\ C\text{=}O \\ | \\ CH_2OPO_3H^- \end{array} + NADH$$

$$\downarrow \; 2CoASH \leftarrow + R_1\text{—}\overset{O}{\overset{\|}{C}}\text{—}SCoA + R_2\text{—}\overset{O}{\overset{\|}{C}}\text{—}SCoA$$

$$R_2\text{—}\overset{O}{\overset{\|}{C}}\text{—}O\text{—}\begin{array}{c} CH_2O\text{—}\overset{O}{\overset{\|}{C}}\text{—}R_1 \\ | \\ C\text{—}H \\ | \\ CH_2OPO_3H^- \end{array}$$

$$\downarrow \; H_2O \rightarrow HPO_4^{2-}$$

$$R_2\text{—}\overset{O}{\overset{\|}{C}}\text{—}O\text{—}\begin{array}{c} CH_2O\text{—}\overset{O}{\overset{\|}{C}}\text{—}R_1 \\ | \\ C\text{—}H \\ | \\ CH_2OH \end{array}$$

$$\downarrow \; CoASH \leftarrow + R_3\overset{O}{\overset{\|}{C}}\text{—}SCoA$$

$$R_2\text{—}\overset{O}{\overset{\|}{C}}\text{—}O\text{—}\begin{array}{c} CH_2O\overset{O}{\overset{\|}{C}}\text{—}R_1 \\ | \\ C\text{—}H \\ | \\ CH_2\text{—}O\text{—}\overset{O}{\overset{\|}{C}}\text{—}R_3 \end{array}$$

FIGURE 16-10
Biosynthetic pathway for triglyceride formation.

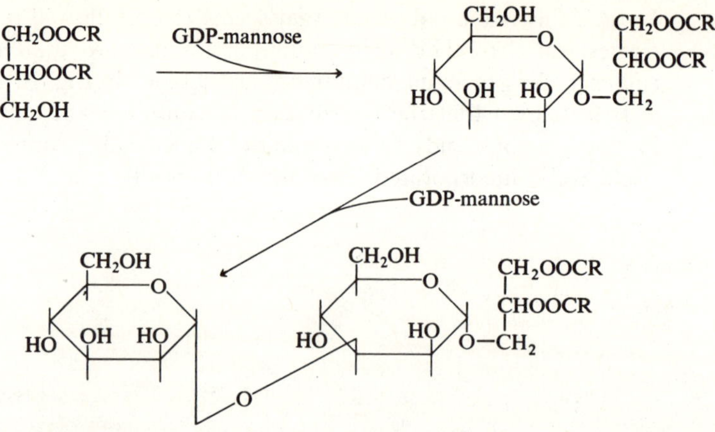

FIGURE 16-11
Pathway for the biosynthesis of mannosylmannosyl diglyceride in *M. lysodeikticus*.

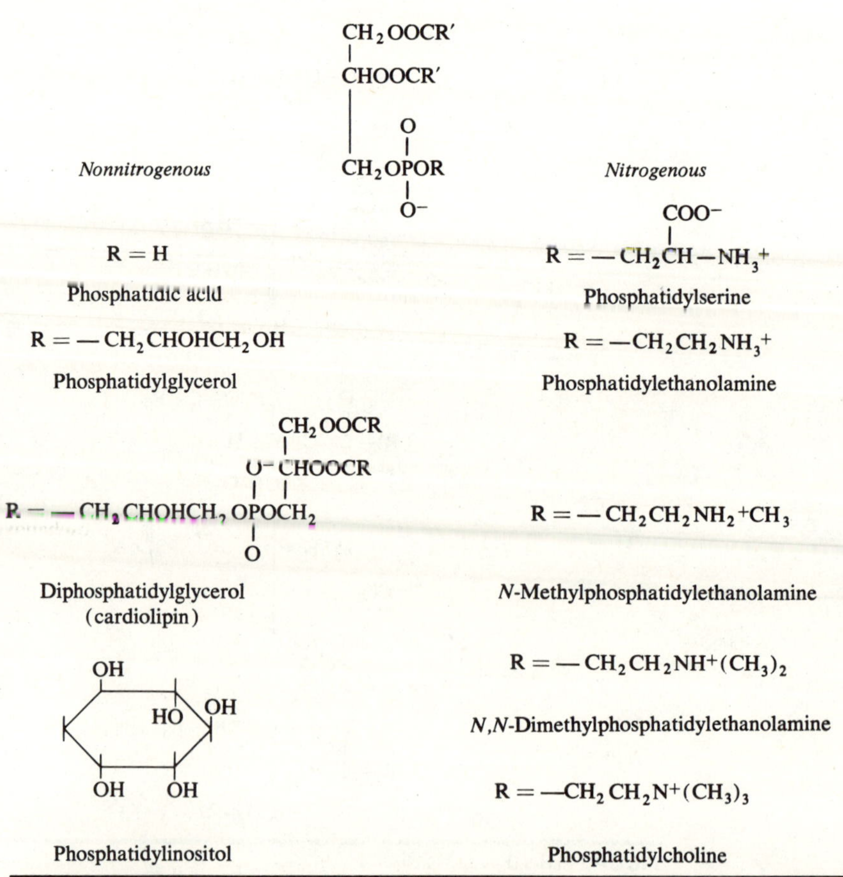

FIGURE 16-12
Structures of lipids based on phosphatidic acids.

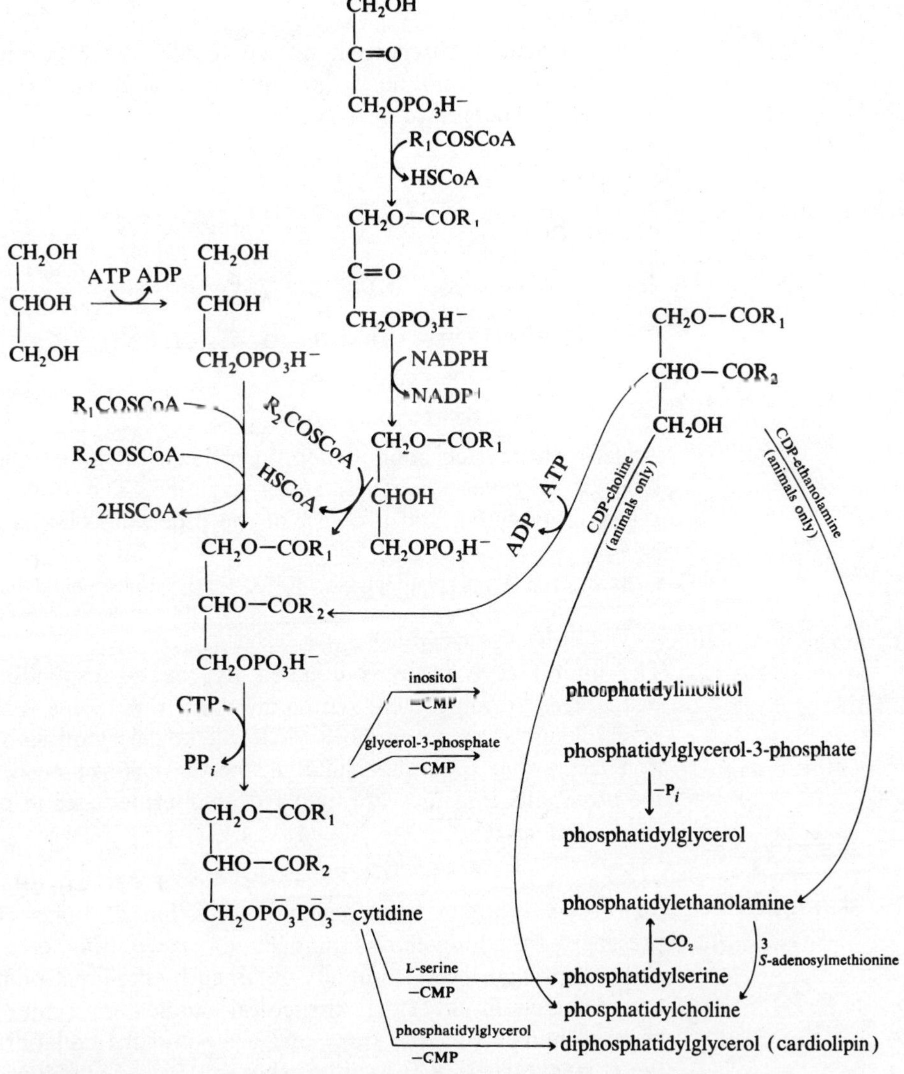

FIGURE 16-13
Summary of biosynthetic routes to phosphatides.

exclusively incorporated into the C-2 position (49,50). The source of this asymmetry is not completely clear, but two factors may be important.

1. There is a question concerning the specificity of the enzyme which catalyzes the formation of phosphatidic acids from glycerol-3-phosphate. Some studies (49,51,52) suggest that microsomal acylation of glycerol-3-phosphate is not specific for the nature of the fatty acid, while others (53,54) suggest that it is. The latter studies have yielded results which accord with the nature of the asymmetric distribution.

2. The recently discovered route to phosphatidic acids from dihydroxyacetone phosphate provides a means of accounting for the observed asymmetry in fatty acid distribution. The enzyme catalyzing the initial acylation (at the C-1 position) manifests specificity toward saturated fatty acids (55,56); the subsequent acylation does not distinguish between saturated and unsaturated ones. Thus, the net effect is to favor incorporation of saturated fatty acids at C-1. What is not clear, however, is the relative quantitative importance of the routes from glycerol-3-phosphate and dihydroxyacetone-3-phosphate.

An intriguing observation related to the above considerations deals with the identification and biosynthesis of amino acid esters of phosphatidylglycerol in bacteria. These have the structure

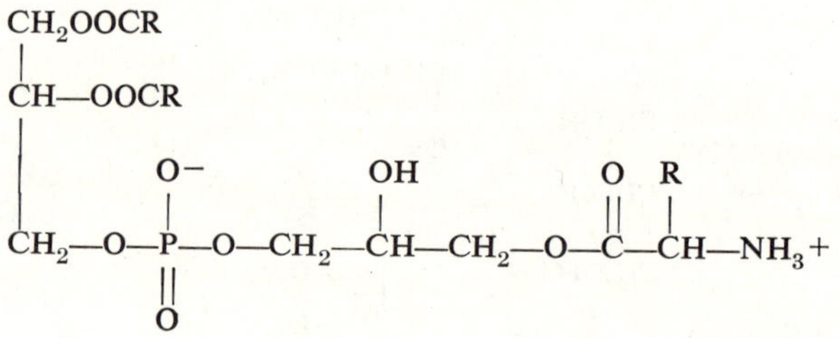

$$
\begin{array}{l}
CH_2OOCR \\
| \\
CH{-}OOCR \\
| \qquad\quad O^- \qquad\qquad\quad OH \qquad\qquad\quad O \;\; R \\
| \qquad\quad | \qquad\qquad\qquad | \qquad\qquad\qquad \| \;\;\; | \\
CH_2{-}O{-}P{-}O{-}CH_2{-}CH{-}CH_2{-}O{-}C{-}CH{-}NH_3^+ \\
\qquad\qquad\;\; \| \\
\qquad\qquad\;\; O
\end{array}
$$

At least three such amino esters have been identified. These are the derivatives of L-lysine, L-alanine, and L-arginine. The route of synthesis of the L-lysyl compound, and probably of the others as well, is the following (57):

$$
\text{L-Lysyl-tRNA} + \text{phosphatidylglycerol} \longrightarrow \text{L-lysyl-phosphatidylglycerol} + \text{tRNA} \qquad (16\text{-}13)
$$

The amino acyl group is donated by the corresponding tRNA derivative just as it is during protein synthesis (page 918). This is the only known case of involvement of an amino acyl-tRNA in the synthesis of a lipid. No function for amino acyl phosphatidylglycerols has yet been identified, although the possibility that they are involved as intermediates in protein synthesis has been eliminated.

SPHINGOLIPIDS

Chemistry

Sphingolipids are a heterogeneous collection of lipids characterized by the presence of the long-chain aminoalcohol *sphingosine,* or a closely related substance. Sphingosine was initially obtained by the great pioneer of lipid chemistry, Thudichum, in 1901. Structural studies in a number of laboratories were culminated by the work of Carter and his collaborators, who deduced the correct structure for sphingosine in 1942. Subsequent studies by several workers has established the stereochemistry of naturally occurring sphingosine.

$$CH_3{-}(CH_2)_{12}{-}CH{=}CH{-}\underset{OH}{\overset{H}{C}}{-}\underset{NH_2}{\overset{H}{C}}{-}\underset{OH}{CH_2}$$

Sphingosine

The configuration about the double bond is trans, the configuration of C-2 is D, and the relationship of the substituents on C-3 to those on C-2 is erythro. Sphingosine is, therefore, *trans*-D-*erythro*-1,3-dihydroxy-2-amino-octadec-4-ene. The proposed sphingosine structure has been confirmed by total synthesis. In addition to lipids from sphingosine itself, closely related substances are derived from dihydrosphingosine, in which the double bond of spingosine has been saturated; phytosphingosine, a 4-hydroxy derivative of dihydrosphingosine (1,3,4-trihydroxy-2-aminooctadecane); and dehydro-phytosphingosine, an unsaturated derivative of phytosphingosine. The former pair of compounds forms the basis for the sphingolipids occurring in animal tissues; the latter pair for those occurring in plant tissues.

In living materials, sphingosine and its close relatives do not occur in the free form to an appreciable extent. Instead, these compounds form part of three principal lipid types, the sphingomyelins, cerebrosides, and gangliosides.

Sphingomyelins yield, on complete hydrolysis, equimolar quantities of sphingosine, choline, phosphate, and fatty acid. Early studies by Levine, following the initial isolation of sphingomyelin by Thudichum in 1884, demonstrated that the fatty acid was attached in amide linkage to the primary amino group of sphingosine. Incomplete hydrolysis products of sphingomyelin include phosphorylcholine and sphingosine phosphate, indicating that phosphorylcholine is esterified at the hydroxy group at either the 1 or 3 position of sphingosine. Subsequent studies have identified the site of attachment as the C-1 hydroxyl. Sphingomyelin, therefore, has the following structure:

$$CH_3{-}(CH_2)_{12}{-}CH{=}CH{-}\underset{OH}{\overset{H}{C}}{-}\underset{NH{-}C(=O){-}R}{\overset{H}{C}}{-}CH_2O{-}\underset{O}{\overset{O^-}{P}}{-}OCH_2CH_2{-}\underset{CH_3}{\overset{CH_3}{N^+}}{-}CH_3$$

Sphingomyelin

Complete hydrolysis of *cerebrosides* yield sphingosine (or dihydrosphingosine), fatty acid, and a sugar, usually galactose. Structure-determination studies have revealed the following structure as the formula for cerebrosides (58,59):

$$CH_3(CH_2)_{12}{-}CH{=}CH{-}\underset{OH}{\overset{H}{C}}{-}\underset{NH{-}C(=O){-}R}{\overset{H}{C}}{-}CH_2{-}O{-}\text{galactose ring (}HOH_2C\text{, OH, OH, OH)}$$

Cerebroside

The configuration of the galactosidic bond is ordinarily β. Although galactose usually predominates as the sugar moiety in cerebroside preparations from human brain and spleen, glucose largely replaces it in certain pathological states, including Gaucher's disease.

Cerebrosides are characterized by the nature of the fatty acid components. For example, kerasin contains lignoceric acid [$CH_3(CH_2)_{22}CO_2H$], phrenasin contains cerebronic acid [$CH_3(CH_2)_{21}CHOH—CO_2H$], nervon contains nervonic acid [$CH_3(CH_2)_7CH{=}CH(CH_2)_{13}CO_2H$], and so on. The related substance sphingosine galactoside, in which the fatty acid moiety is missing, is termed psychosine.

The *gangliosides* are phosphorus-free glycosphingolipids composed of sphingosine, fatty acid, one or more sugars, and, characteristically, neuraminic acid:

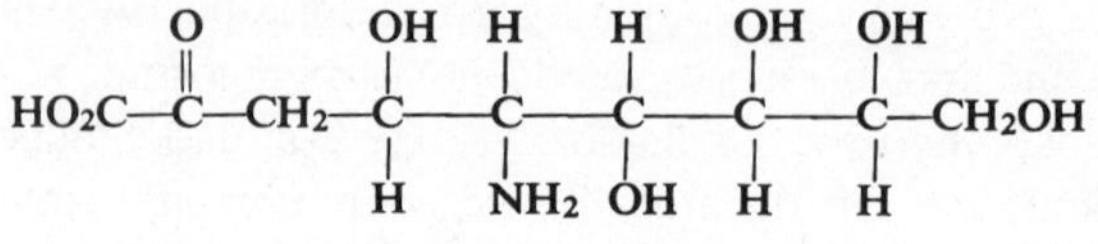

$$HO_2C-\overset{O}{\overset{\|}{C}}-CH_2-\underset{H}{\overset{OH}{C}}-\underset{NH_2}{\overset{H}{C}}-\underset{OH}{\overset{H}{C}}-\underset{H}{\overset{OH}{C}}-\underset{H}{\overset{OH}{C}}-CH_2OH$$

Neuraminic acid

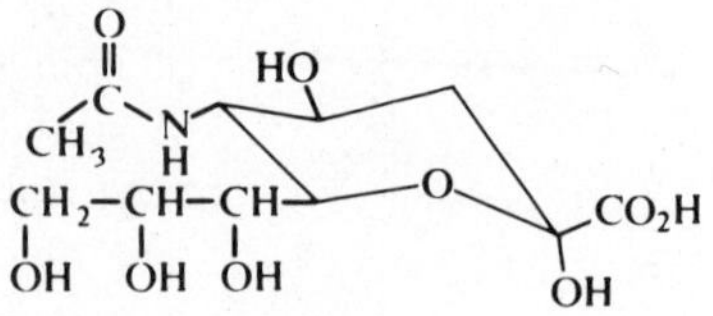

The neuraminic acid component is found as the *N*-acetyl derivative, *sialic acid.* The gangliosides are characteristic components of the nervous system and occur in other tissues as well, including spleen and erythrocytes.

Ganglioside structures are complicated and not completely understood. The elegant work of Kuhn and Wiegandt has established that the structure of the parent ganglioside of nervous tissue is that shown in Fig. 16-14 (60). The parent ganglioside has, as shown, one sialyl group. Other gangliosides are derived from this structure by the addition of one or two more sialyl groups. We can represent the structures in the following way:

Galactosyl-(β-1→3)-*N*-acetylgalactosaminyl-(β-1→4)-galactosyl-(β-1→4)-glucosyl ceramide
(first residue) —(2→3)— sialyl; (third residue) —(2→3)— sialyl

The sialyl group on the third saccharide residue is that of the monosialylgangliosides, while that on the first residue completes the structure for the disialyl derivatives. The location of the third sialyl group is unknown.

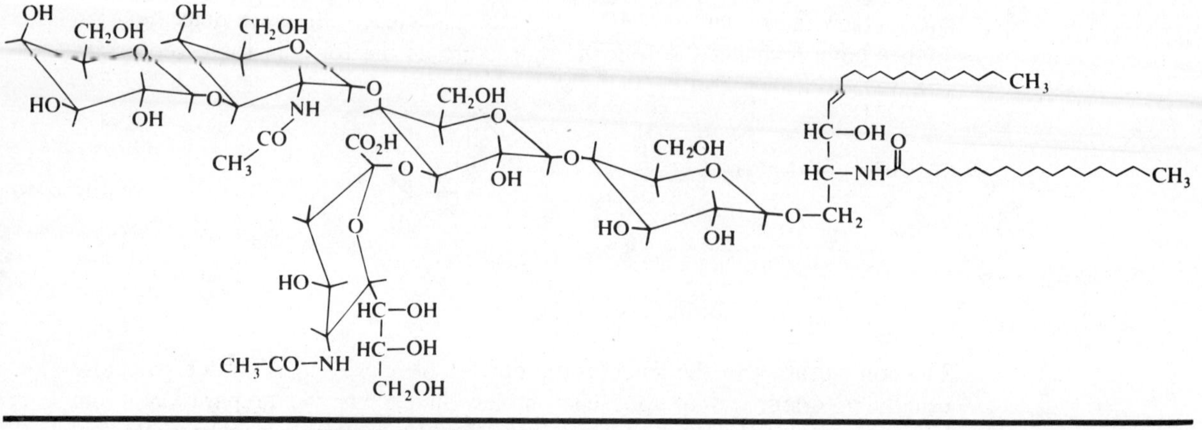

FIGURE 16-14
Parent ganglioside structure.

Biosynthesis In the biosynthesis of gangliosides, it has been shown that particulate preparations from embryonic chick brain catalyze the synthesis of monosialogangliosides from sialic acid-free glycolipid and CMP-sialic acid (61).

In vivo studies utilizing labeled substrates have revealed that carbons 1 and 2 and the amino nitrogen of sphingosine are derived from serine. The remaining 16 carbon atoms are derived from acetate, probably by way of a 16 carbon intermediate. In both yeast and microsomal preparations from mouse brain, sphingosine is synthesized by a route involving the condensation of palmityl-SCoA with serine in a pyridoxal phosphate–dependent reaction (62–63):

$$\text{Palmityl-SCoA} + \text{serine} \xrightarrow[\text{HSCoA}]{\text{pyridoxal phosphate}} CH_3(CH_2)_{14}\text{—}\underset{\underset{O}{\|}}{C}\text{—}\underset{\underset{NH_2}{|}}{CH}\text{—}CH_2OH + CO_2 \qquad (16\text{-}14)$$

A reasonable mechanism for formation of this keto base, consistent with considerations about pyridoxal phosphate–dependent reactions (page 392), is shown in Fig. 16-15. NADPH-dependent reduction of the keto base generates dihydrosphingosine directly:

$$CH_3(CH_2)_{14}\text{—}\underset{\underset{O}{\|}}{C}\text{—}\underset{\underset{NH_2}{|}}{CH}\text{—}CH_2OH + NADPH + H^+ \rightleftharpoons CH_3(CH_2)_{14}\text{—}\underset{\underset{OH}{|}}{CH}\text{—}\underset{\underset{NH_2}{|}}{CH}\text{—}CH_2OH + NADP^+ \qquad (16\text{-}15)$$

Sphingosine itself may arise by two routes which have not been definitely isolated: by dehydrogenation of dihydrosphingosine or through internal (or an equivalent external) oxidation–reduction of the keto base:

$$CH_3(CH_2)_{12}\text{—}CH_2\text{—}CH_2\text{—}\underset{\underset{OH}{|}}{CH}\text{—}\underset{\underset{NH_2}{|}}{CH}\text{—}CH_2OH \xrightarrow{-2[H]} CH_3(CH_2)_{12}\text{—}CH{=}CH\text{—}\underset{\underset{OH}{|}}{CH}\text{—}\underset{\underset{NH_2}{|}}{CH}\text{—}CH_2OH \qquad (16\text{-}16)$$

$$CH_3(CH_2)_{12}\text{—}CH_2\text{—}CH_2\text{—}\underset{\underset{O}{\|}}{C}\text{—}\underset{\underset{NH_2}{|}}{CH}\text{—}CH_2OH \longrightarrow CH_3(CH_2)_{12}\text{—}CH{=}CH\text{—}\underset{\underset{OH}{|}}{CH}\text{—}\underset{\underset{NH_2}{|}}{CH}\text{—}CH_2OH \qquad (16\text{-}17)$$

The biosynthesis of sphingomyelin from sphingosine may occur via the intermediate formation of an *N*-acyl sphingosine (ceramide):

$$\text{Sphingosine} + R\text{—}\overset{\overset{O}{\|}}{C}\text{—SCoA} \rightarrow N\text{-acyl sphingosine} \qquad (16\text{-}18)$$

Sribney and Kennedy have identified an enzyme in liver, phosphocholine-ceramide transferase, which catalyzes the formation of sphingomyelin from *N*-acyl sphingosine and CDP-choline (64):

$$N\text{-Acyl sphingosine} + \text{CDP-choline} \rightarrow \text{sphingomyelin} \qquad (16\text{-}19)$$

The enzyme responsible for catalyzing this reaction does not distinguish between threo and erythro forms of the acylsphingosine, although only the erythro form of sphingomyelin is formed *in vivo* (65). This appears to be a general characteristic of enzymes concerned with metabolism of sphingosine derivatives.

Alternatively, sphingomyelin may be formed by the two processes noted above but occurring in inverse order (66):

$$\begin{aligned}&\text{Sphingosine} + \text{CDP-choline} \rightarrow \text{sphingophosphorylcholine} + \text{CMP}\\&\text{Sphingophosphorylcholine} + \text{R-COSCoA} \rightarrow \text{sphingomyelin} + \text{CoASH}\end{aligned} \qquad (16\text{-}20)$$

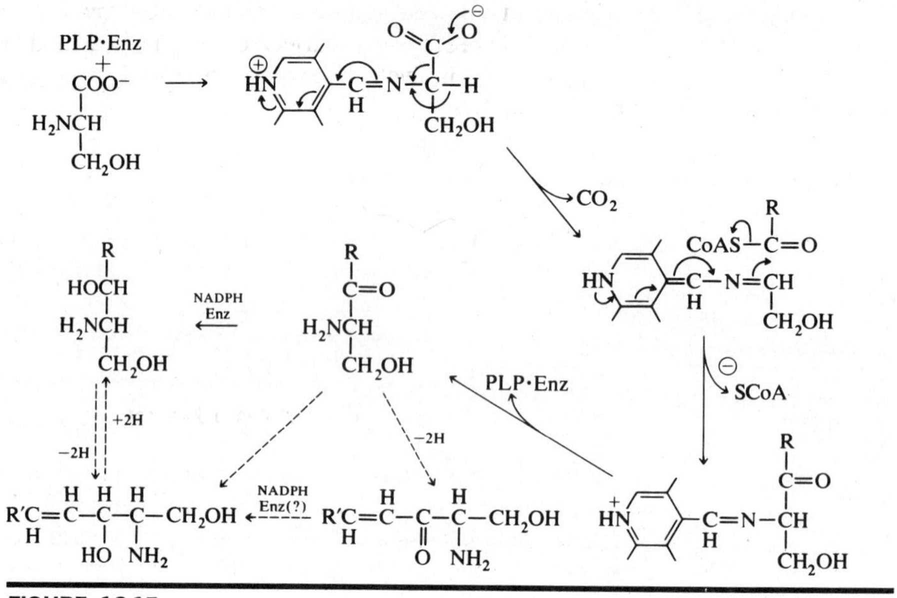

FIGURE 16-15
Details concerning the pathway of sphingosine formation. P. E. Braun and E. E. Snell, *J. Biol. Chem.*, **243:** 3775 (1968).

The mammalian brain contains the enzymes, localized in the microsomal fraction, necessary for the incorporation of galactose or glucose into cerebrosides. The sugars are donated to the sphingosine moiety from the UDP-derivatives (page 543) [Eq. (16-21)]:

$$\text{Sphingosine} + \text{UDP-galactose} \rightarrow \text{psychosine} + \text{UDP} \qquad (16\text{-}21)$$

It remains to be established whether the hexose acceptor is an *N*-acyl sphingosine, yielding a cerebroside directly, or sphingosine, yielding psychosine, which is subsequently acylated to a cerebroside.

BIOSYNTHESIS OF STEROIDS

The various classes of lipids previously discussed in this chapter are saponifiable, that is, alkaline hydrolysis converts them into water-soluble substances. Chevreul, in 1812, recognized that not all lipids are saponifiable. These unsaponifiable lipids include the biochemically very important class of substances, the *steroids* all of which can be considered to be derivatives of the cyclopentanoperhydrophenanthrene skeleton.

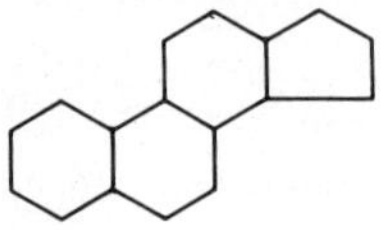

Cyclopentanoperhydrophenanthrene

An exhaustive monograph concerning the chemistry of steroids has been prepared by Fieser and Fieser.

A large number of steroids are synthesized by the mammal and other organisms, and these play a variety of biochemical and physiological roles. In man, the quantitatively predominant steroid is the unsaturated alcohol *cholesterol.*

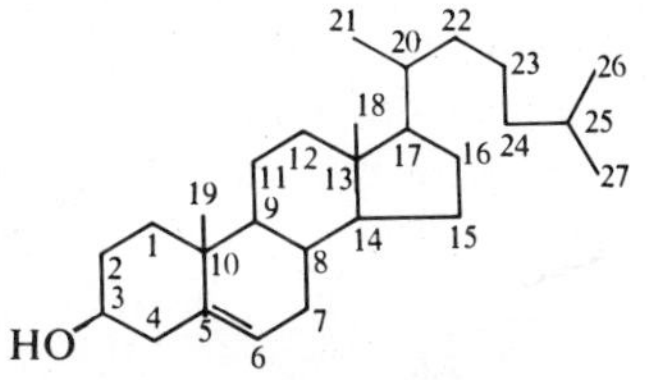

Cholesterol

As is indicated below, cholesterol is a key intermediate in the biosynthesis of related steroids.

Physiologically important steroids include the bile acids, the adrenocortical hormones, the androgens, and the estrogens. The bile acids that occur to the greatest extent in the human include cholic acid, deoxycholic acid, and chenocholic acid.

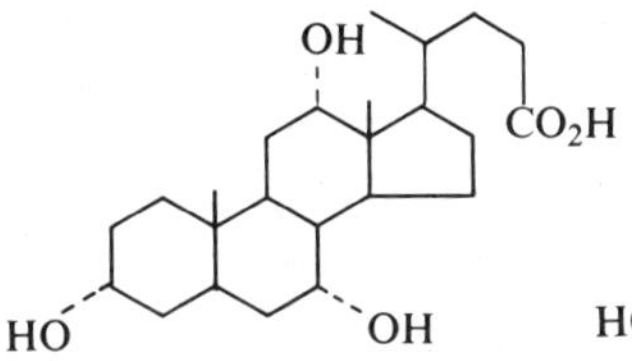

Cholic acid 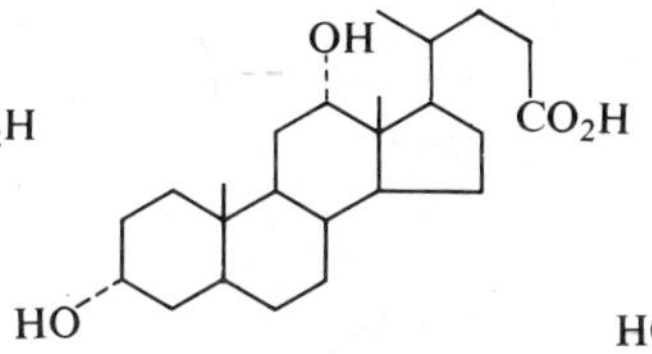   Deoxycholic acid 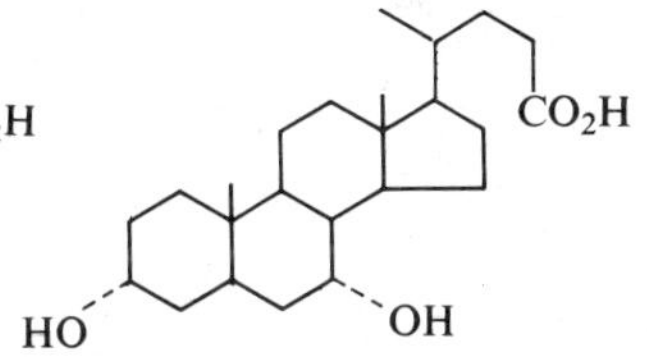   Chenocholic acid

The bile acids do not occur free in bile but are esterified at the terminal carboxyl group with either glycine, yielding glycocholic acid, or taurine, yielding taurocholic acid. As noted above, the salts of these substances aid the intestinal hydrolysis and absorption of neutral fats by serving as emulsifying and solubilizing agents.

$$C_{23}H_{36}(OH)_3\text{—}\overset{\overset{\displaystyle O}{\|}}{C}\text{—}\overset{\displaystyle H}{N}\text{—}CH_2\text{—}CO_2^-$$

Glycocholate

$$C_{23}H_{36}(OH)_3\text{—}\overset{\overset{\displaystyle O}{\|}}{C}\text{—}\overset{\displaystyle H}{N}\text{—}CH_2CH_2SO_3^-$$

Taurocholate

The bile salts are a special class of amphipathic molecules. Simpler amphipathic molecules, such as salts of fatty acids, have clearly defined hydrophobic and hydrophilic functions, usually a long hydrocarbon chain terminated by a polar group (page 713). The bile salts do have a terminal polar group, the carboxylate moiety of the parent compounds located on the aliphatic side chain, or the carboxylate or sulfonate functions of the glycine or taurine moiety linked to this same side chain in the conjugated derivatives. In addition, however, the bile salts have either two or three polar hydroxyl groups attached to the otherwise hydrophobic polycyclic nucleus. Note specifically that all of the hydroxyl groups are localized on one face of this nucleus. Hence, this face is an additional polar aspect of these molecules and the opposite face constitutes the hydrophobic component. The bile salts form micelles with other lipids, including triglycerides, phospholipids, and cholesterol, through interaction of the hydrophobic face with the nonpolar lipid leaving the polar face and the polar side chain exposed to interact with the aqueous environment. It is the formation of these micelles that is the basis for the action of the bile salts as emulsifying agents.

There are a number of ways of influencing the flux of metabolites through the various reaction pathways: allosteric control of enzymatic activity, for example, has been encountered many times in preceding chapters (see especially page 299). Genetic control of enzymatic activity is discussed in some detail in Chapter 21. In addition, metabolism in higher organisms is controlled through the action of a number of substances, termed *hormones,* which are secreted by one of the *endocrine glands* (thyroid, parathyroids, testis, ovary, adrenals, etc.) and, though they are not catalysts themselves, influence the rate of many reactions and systems of reactions. The hormones act at tissues distant from the site of their synthesis; hence, the rate of synthesis, degradation, secretion, and transport of the hormones themselves is under metabolic control to assure proper integration of the various aspects of metabolism of the organism. Among the hormones are several steroids which are secreted by the cortex of the adrenal gland. Although some 30 steroids have been obtained in crystalline form from the adrenal cortex, most of which have either 19 or 21 carbon atoms, three such steroids are known to be secreted by this organ, *aldosterone, cortisol* (17 α-hydroxycorticosterone, hydrocortisone), and *corticosterone:*

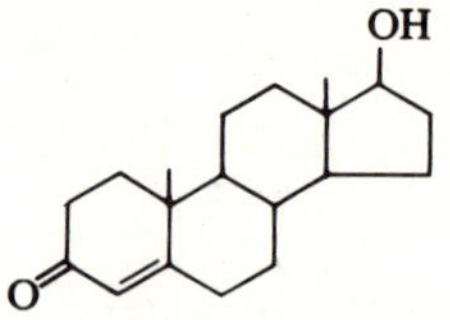

OH

O

Testosterone

OH

HO

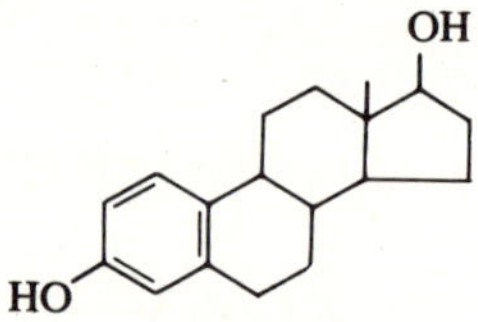

β-Estradiol

The adrenal steroids control or influence a broad spectrum of biochemical and physiological phenomena including the metabolism of carbohydrates, lipids, proteins, salts, and water, and moreover, affect secretion, neuromuscular irritability, allergic responses, and resistance to noxious stimuli. A summary of the effects produced by adrenal cortical insufficiency is provided in Table 16-2.

The rate of secretion of aldosterone is primarily controlled by the levels of sodium and potassium ion in the adrenal arterial blood; a lack of sodium ion is particularly important in increasing the rate of secretion of this hormone. To a lesser extent, but one which is potentiated by low sodium levels, aldosterone secretion is under the influence of the hypophyseal adrenocorticotropic hormone (ACTH, page 67). Aldosterone transport in the blood appears to be largely mediated through its interaction with serum albumins.

The level of secretion of cortisol and corticosterone is strongly controlled by ACTH. Both these hormones, and particularly the former, are transported in the blood via a specific α_1-globulin termed *transcortin* or *corticosteroid binding protein* ($M = 52{,}000$). The serum levels of this protein are, in turn, regulated by thyroid hormone and estrogen.

The estrogens are elaborated by the ovary and are responsible for the development of the typical secondary sexual characteristics of the female. A particularly potent estrogen is estradiol. Similarly, the androgens, including testosterone, are synthesized in and liberated from the testis and are responsible for the secondary male characteristics.

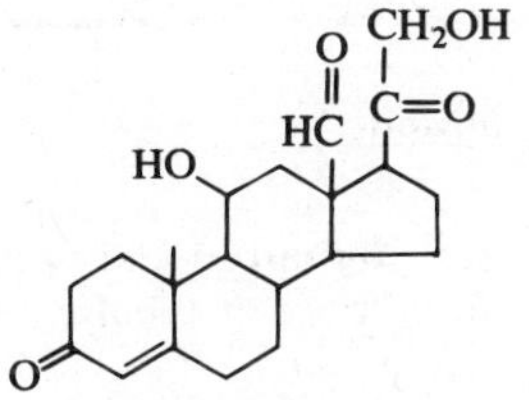

Aldosterone

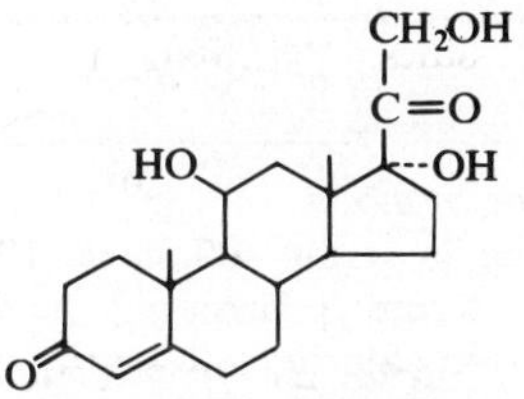

17α-Hydroxycorticosterone (cortisol; hydrocortisone)

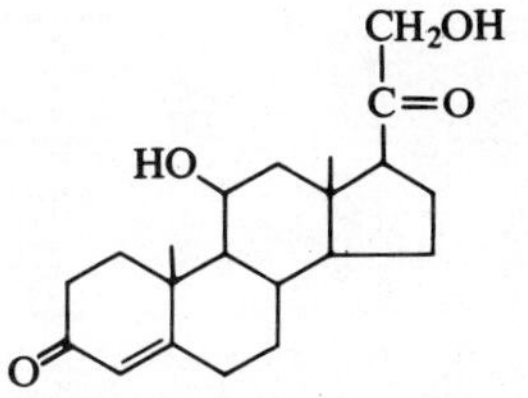

Corticosterone

The development of an understanding of the biosynthesis of cholesterol is one of the most interesting and important achievements in biochemistry in the past several years. Progress in this field is principally due to the efforts of Bloch, Cornfield, Popják, Lynen, and their coworkers, with an important contribution from the group at Merck led by Folkers. In the 1940s, Bloch and his collaborators demonstrated that isotopically labeled acetate is incorporated into cholesterol both *in vivo* and in liver tissue slices. Subsequently, it was determined that both carbons of acetate are incorporated into cholesterol (67), and that the entire cholesterol molecule is derived from acetate. Degradation of cholesterol synthesized from both carboxyl-labeled and methyl-labeled acetate in liver slice preparations localized the incorporated label in the individual carbons of this molecule. The carbons of cholesterol derived from the methyl group (*m*) and those derived from the carboxyl group (*c*) of acetate are indicated below. At this stage the problem was to elucidate a

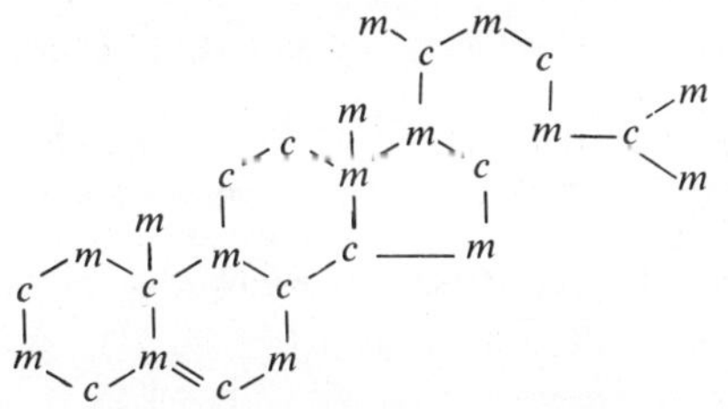

biosynthetic pathway leading to cholesterol that accounted for the indicated labeling pattern. Subsequent discoveries implicated *mevalonic acid* and *squalene* as key intermediates in cholesterol biosynthesis from acetate. It is convenient to discuss this pathway in three sections: the conversion of acetate to mevalonate, the conversion of mevalonate to squalene, and the conversion of squalene to cholesterol.

The crucial discovery of mevalonic acid as an acetate-replacing cofactor for bacteria and its demonstration as an intermediate in cholesterol biosynthesis is primarily due to the work of Folkers, Wright, and their coworkers. The biosynthesis of mevalonate from acetate was initially demonstrated in a cell-

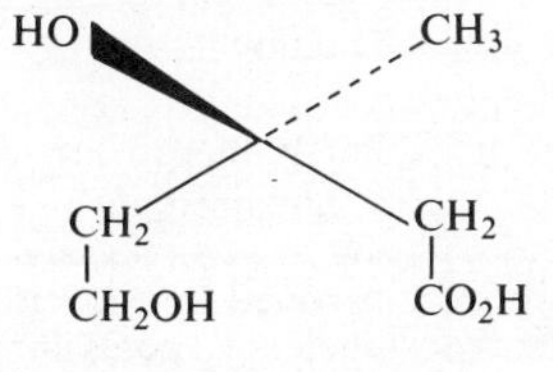

Mevalonic acid

TABLE 16-2
Alterations produced by adrenal cortical insufficiency

Nature of change observed	*Basis for change*
Alterations in blood	
Decrease in serum sodium and increase in serum potassium; decrease in serum chloride	Disturbed kidney function reflected in failure of tubular reabsorption of Na^+ and Cl^-
Hemoconcentration	Loss of water through kidney, accompanying electrolyte loss
Acidosis	Failure of renal Na^+-K^+-H^+ exchange mechanism
Hypoglycemia with fasting	Reduced level of liver glycogen in fasting; diminished gluconeogenesis and increased carbohydrate utilization
Increase in blood urea	Renal failure and circulatory impairment
Lymphocytosis	Lack of lymphocytolytic action of 11-oxyadrenal cortical steroids, with accompanying proliferation of lymphoid tissue
Eosinophilia	Unknown
Anemia	Lack of stimulation of erythropoietic tissue and of production of intrinsic factor; gastric hyposecretion
Alterations in protein metabolism	
Diminished urinary nitrogen in fasting	Absence of antianabolic influence of adrenal steroids
Alterations in carbohydrate metabolism	
Reduced levels of liver glycogen in fasting	Diminished glycogenesis and gluconeogenesis
Markedly increased insulin sensitivity	Diminished glycogenesis and gluconeogenesis
Impairment of carbohydrate absorption from the gastrointestinal tract	Unknown; possibly related to disturbances in potassium metabolism[a]
Alterations in lipid metabolism	
Impairment of lipid mobilization from depots	Unknown
Generalized alterations	
Intensification of inflammatory response and of hypersensitivity reactions	Increased influx of polymorphonuclear leukocytes and lymphocytes from blood into area of inflammation, accompanied by an augmented production of tissue substances contributing to reactions of inflammation and hypersensitivity, including destruction of fibroblasts.
Increased growth of lymphoid tissue	Lack of adrenal cortical steroid lymphocytolytic action on lymphocytes
Decreased metabolic rate	Unknown
Hypotension	Decreased extracellular volume; decreased cardiac output
Excessive pigmentation	Unknown; possibly due to intrinsic melanocyte-stimulating activity of ACTH
Anorexia, weight loss; cessation of growth	Unknown
Muscular weakness and sensitivity to stress	Partly due to loss of potassium from cells; impaired glycogenesis.

[a] Passage of carbohydrate into cells is accompanied by movement of potassium in the intracellular direction. In adrenal cortical insufficiency, potassium tends to move out of cells.

SOURCE: A. White, P. Handler, and E. L. Smith: "Principles of Biochemistry," 4th ed., McGraw-Hill, New York, 1968.

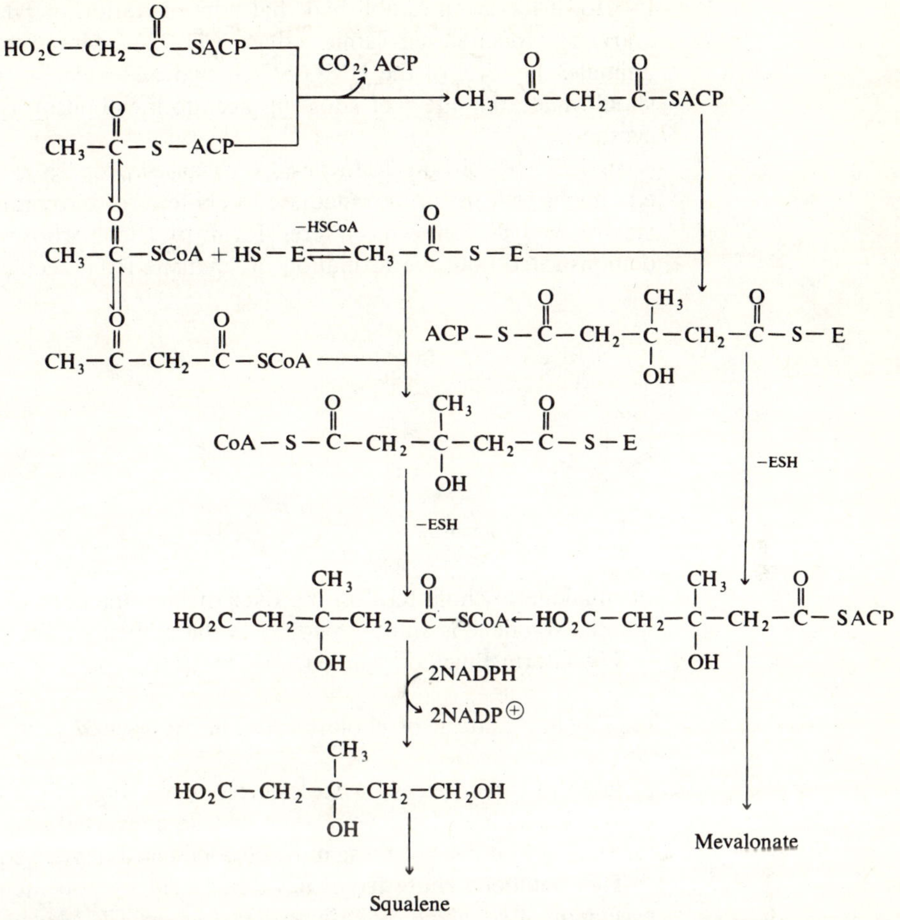

$HO_2C-CH_2-C(=O)-SACP$ + $CH_3-C(=O)-S-ACP$ → (CO_2, ACP) → $CH_3-C(=O)-CH_2-C(=O)-SACP$

$CH_3-C(=O)-S-ACP$ ⇅ $CH_3-C(=O)-SCoA$

$CH_3-C(=O)-SCoA + HS-E \rightleftharpoons CH_3-C(=O)-S-E$ (−HSCoA)

$CH_3-C(=O)-SCoA$ ⇅ $CH_3-C(=O)-CH_2-C(=O)-SCoA$

$ACP-S-C(=O)-CH_2-C(CH_3)(OH)-CH_2-C(=O)-S-E$

$CoA-S-C(=O)-CH_2-C(CH_3)(OH)-CH_2-C(=O)-S-E$

−ESH

−ESH

$HO_2C-CH_2-C(CH_3)(OH)-CH_2-C(=O)-SCoA \leftarrow HO_2C-CH_2-C(CH_3)(OH)-CH_2-C(=O)-SACP$

2NADPH → $2NADP^{\oplus}$

$HO_2C-CH_2-C(CH_3)(OH)-CH_2-CH_2OH$

Squalene

Mevalonate

FIGURE 16-16
The biosynthesis of mevalonate.

free preparation from rat liver and required soluble enzymes, microsomes, CoASH, ATP, glutathione, NADPH, and Mg^{2+}. There appear to be two pathways for the synthesis of mevalonate (Fig. 16-16): (1) a route from acetyl-SCoA involving acetoacetyl-SCoA and β-hydroxy-β-methyl glutaryl-SCoA as intermediates, and (2) a route from acetyl-ACP and malonyl-ACP involving the ACP derivatives of the intermediates just noted (68). The first pathway, although not utilizing ACP, does involve covalent substrate–enzyme intermediates with the condensing enzyme (69). The relative quantitative importance of these two pathways seems to differ between different tissues and organisms although details have not yet been clearly established in even one case.

An enzyme exists which catalyzes the hydrolysis of β-hydroxy-β-methyl glutaryl-SCoA to the free acid. Although β-hydroxy-β-methyl glutarate (HMG) is not known to be directly involved in further metabolism, evidence is accumulating to suggest that this compound may play an important role in the regulation of cholesterol synthesis. It has been established that HMG is a competetive inhibitor for β-hydroxy-β-methyl glutaryl-SCoA reductase

(70). Consequently, free HMG should inhibit cholesterol biosynthesis and, indeed, it has been established that administration of HMG to animals has a marked hypocholesterolemic effect (71,72). Moreover, in cholesterol-fed animals, the level of the HMG-SCoA hydrolase increases about 3-fold which should have the effect of shunting acetate metabolism away from cholesterol synthesis.

In 1932, Robinson, following even earlier proposals, suggested that squalene might well be an intermediate in cholesterol biosynthesis. This hypothesis was revived, in somewhat altered form, by Bloch who, with his collaborators demonstrated both the formation of squalene from acetate and the conversion

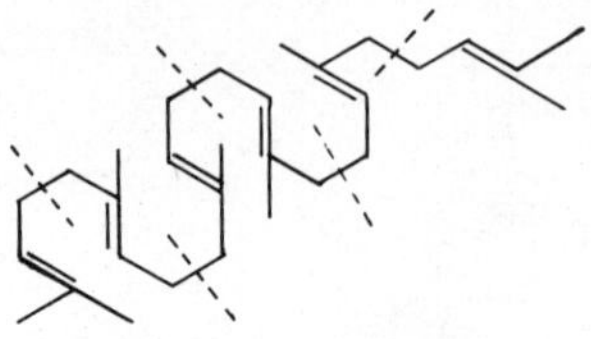

Squalene

of squalene to cholesterol in the liver of the intact rat (73), strongly suggesting that squalene is an intermediate in the pathway from acetate to cholesterol.

The intermediacy of squalene, a dihydrotriterpene, is reasonable in terms of the labeling pattern of cholesterol. The *isoprenoid unit,* $\begin{matrix} m \\ m \end{matrix}\!\!>\!c\text{—}m\text{—}c$, is detectable in three places in cholesterol, suggesting that it is formed from such units. Hence, mevalonic acid must first be converted into an isoprenoid structure and, second, six of these must be condensed to yield squalene.

The metabolic route from mevalonic acid to isopentenyl pyrophosphate, a precursor of squalene, is indicated in Fig. 16-17. Mevalonic acid is first converted to mevalonic acid-5-phosphate by mevalonic kinase and subsequently to mevalonic acid-5-pyrophosphate. The latter substrate then undergoes a concerted dehydration and decarboxylation, yielding isopentenyl pyrophosphate. This reaction, possibly, involves the formation of 3-phosphomevalonic acid-5-pyrophosphate as an intermediate. Finally, isopentenyl pyrophosphate is reversibly converted to dimethallyl pyrophosphate in an enzymatic reaction, which may be formulated as follows [Eq. (16-22)]:

$$H^+ \; \begin{matrix} CH_3 \\ H_2C \end{matrix}\!\!>\!C\text{—}\underset{H}{\overset{H}{C}}\text{—}CH_2\text{—}OP_2O_6H^{2-} \longrightarrow \begin{matrix} CH_3 \\ CH_3 \end{matrix}\!\!>\!C{=}\overset{H^+}{CH}\text{—}CH_2\text{—}OP_2O_6H^{2-} \qquad (16\text{-}22)$$

The conversion of the isoprenoid pyrophosphates to farnesyl pyrophosphate occurs via several successive condensations, as shown in Fig. 16-18. The condensation of isopentenyl pyrophosphate with dimethallyl pyrophosphate yields the monoterpene, *trans*-geranyl pyrophosphate (74). Geranyl pyrophosphate, in turn condenses with an additional molecule of dimethallyl pyrophosphate to yield the sesquiterpene, *trans, trans*-farnesyl pyrophosphate.

It is clear that farnesyl pyrophosphate serves as the precursor in the formation of squalene in a number of species. Exactly how this conversion

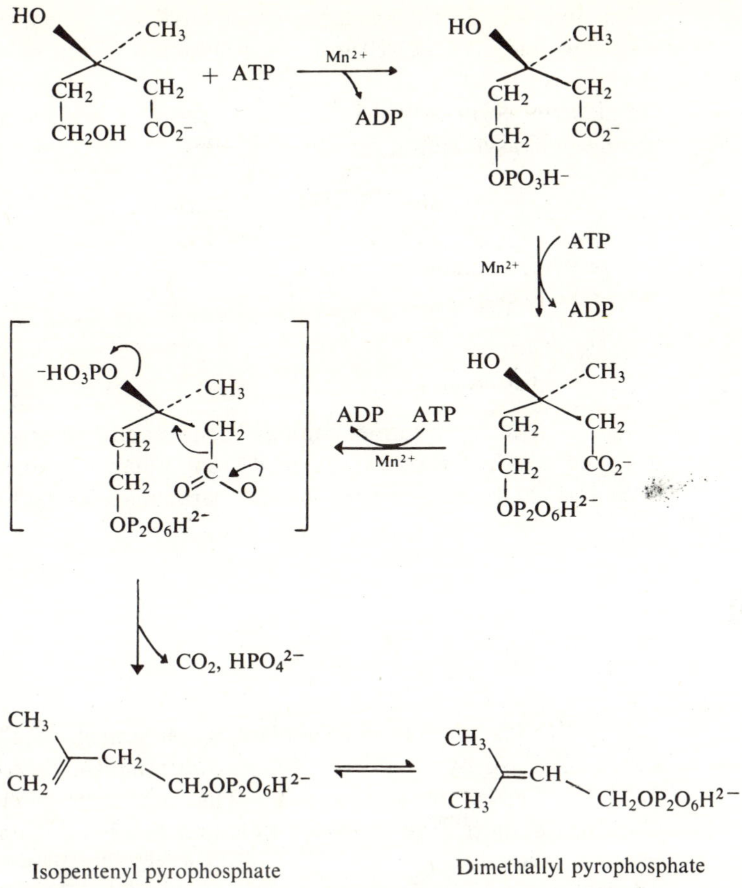

Isopentenyl pyrophosphate

Dimethallyl pyrophosphate

FIGURE 16-17
Biosynthetic pathway from mevalonic acid to isopentenyl pyrophosphate and dimethallyl pyrophosphate.

occurs is not clear. Salient relevant considerations include the following. First, the responsible enzymes are, in mammalian tissues, microsomal. Second, the conversion is anaerobic and requires reduced nicotinamide nucleotide coenzymes, NADPH being generally more effective than NADH. Third, squalene derived from 1-d_2-farnesyl pyrophosphate (or 5-d_2-mevalonate) contains three atoms of deuterium at the two central carbon atoms of squalene (75,76). Fourth, the hydrogen atom lost from C-1 of one farnesyl pyrophosphate during condensation is stereospecifically replaced by hydride ion from NADPH (75,75a). Fifth, the reaction appears to proceed through an intermediate containing a cyclopropane ring [2-(2,6,10-trimethyl-1,6,9-undecatriene)-3-methyl-3-(4,8,-dimethyl-3,7-nonadiene)-cyclopropylcarbinyl pyrophosphate] whose rearrangement and reduction generates squalene (75a,77,77a). Sixth, and finally, in the pig liver microsomal system at least, the condensation proceeds through a C_{30} enzyme-bound intermediate whose reduction by NADPH generates squalene (78). Some of these considerations are illustrated diagrammatically in Fig. 16-19.

Finally, we must consider the conversion of squalene to cholesterol, a matter which has received abundant attention in recent years. The more remarkable part of this transformation is the initial stage in which squalene, in the all *trans* form, is converted into lanosterol, a tetracyclic triterpene containing eight asymmetric centers:

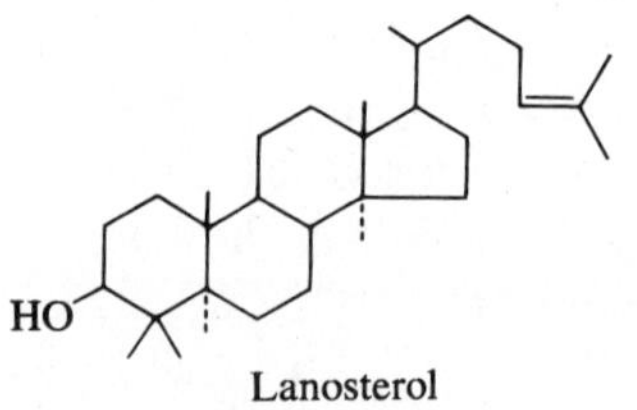

At least one stable intermediate has been established in this process: 2,3-oxidosqualene. This epoxide is formed from squalene and molecular oxygen in a reaction catalyzed by microsomal preparations from rat liver (79,80):

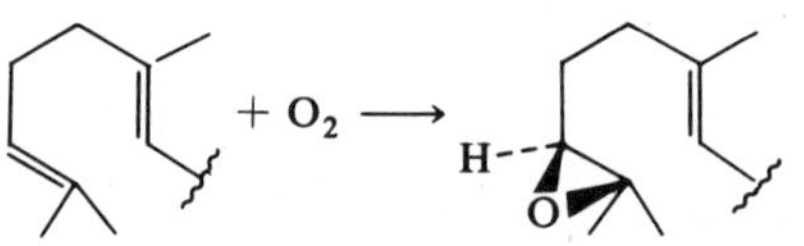

(16-23)

The cyclization of 2,3-oxidosqualene to lanosterol does not require oxygen and is catalyzed by an enzyme which is distinguishable from that responsible for formation of the intermediate. This most interesting reaction may be visualized in more than one way: we adopt that in which an electron-donating group X is involved in a multiplex electron shift leading to an intermediate in which the four rings are formed. This is viewed in Fig. 16-20. Conversion of this intermediate to lanosterol requires a rotation of 120° about the bond which joins the side chain to ring D. This places all four groups which must migrate into the proper trans positions. Comparing the structure of the intermediate with that of lanosterol, it is clear that two 1,2-methyl migrations and two 1,2-hydrogen migrations must occur. Viewed a little differently from the perspective of Fig. 16-20, we may show the necessary transformations as follows:

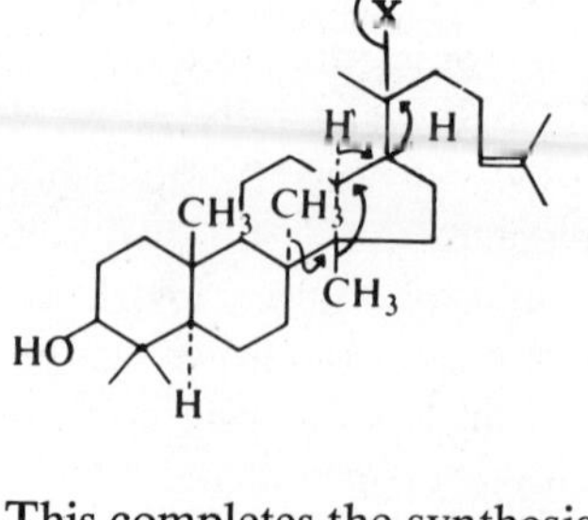

This completes the synthesis of lanosterol.

Conversion of lanosterol to cholesterol may occur by a number of routes. Some of those definitely established are collected in Fig. 16-21. Pathways from cholesterol to steroid hormones have been elucidated in substantial detail. While these routes are beyond the scope of this treatment, a partial summary is provided in Fig. 16-22. Details of certain of the hydroxylation reactions involved have been discussed on page 639.

RELATED BIOSYNTHESES FROM THE ISOPRENOID UNIT

In addition to the steroids biological isoprenoid units yield a variety of important natural products. An overall view of the metabolic role of isopentenyl pyrophosphate is given in Fig. 16-23. Repeated condensation of isopentenyl functions yield poly-*cis*-isoprene (rubber). In contrast, repeated condensations of *trans,trans*-farnesyl pyrophosphate with C-5 units yield poly-*trans*-isoprene (gutta percha), a material without elastic properties. Geranyl pyrophosphate and farnesyl pyrophosphate serve as direct precursors of the monoterpenes and sesquiterpenes, respectively. Condensation of farnesyl pyrophosphate with a C-5 unit yields geranylgeranyl pyrophosphate, which yields the diterpenes, or which may dimerize to a C-40 unit, a precursor of the carotenoids. Finally, condensation of *trans,trans*-farnesyl pyrophosphate with a variable number of C-5 units yields the ubiquinones, a family of coenzymes important in electron transfer processes (page 662).

Dimethallyl pyrophosphate: $CH_3\,CH_3$ / CH / $CH_2{-}OP_2O_6H^{2-}$

Isopentenyl pyrophosphate: H_2C CH_3 / H—CH / $CH_2OPO_6H^{2-}$

⟶

Geranyl pyrophosphate: $CH_3\,CH_3$ / CH / CH_2 / $CH_2\;CH_3$ / CH / $CH_2{-}OP_2O_6H^{2-}$

Isopentenyl pyrophosphate: H_2C CH_3 / H—CH / $CH_2{-}OP_2O_6H^2$

⟶

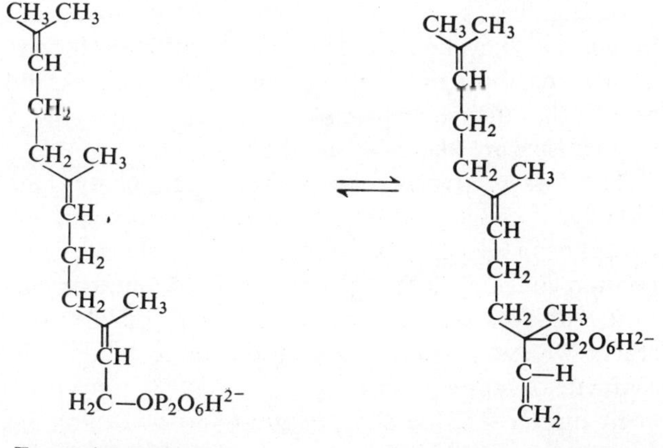

FIGURE 16-18

Postulated pathway for the conversion of isopentenyl and dimethallyl pyrophosphate to farnesyl pyrophosphate.

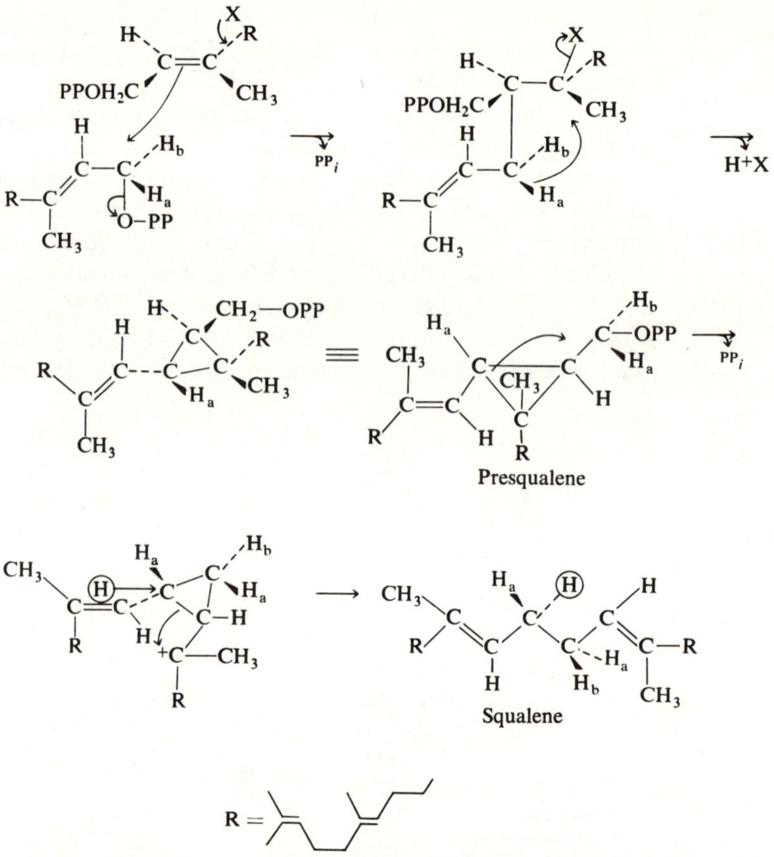

FIGURE 16-19
Some aspects of the synthesis of squalene from farnesyl pyrophosphate.

Such transformations have been discussed in detail by Richards and Hendrickson. We consider a few of the simpler examples.

In Fig. 16-24, the conversion of geranyl pyrophosphate to acyclic, monocyclic, and bicyclic monoterpenes is outlined. The reactions are indicated as proceeding through carbonium ion intermediates, although it is by no means certain that the mechanisms are, in fact, of this type. Free radical or concerted mechanisms are also prominent possibilities.

The loss of pyrophosphate from geranyl pyrophosphate yields citronellal directly. If *cis*-geranyl pyrophosphate is employed as starting material, loss of pyrophosphate yields a carbonium ion, which may cyclize. The product carbonium ion may lose a proton to yield limonene or undergo a hydride ion shift with loss of a proton to yield α-terpinene. If the latter carbonium ion reacts with the pi electrons of the double bond, bicyclic terpenes, such as α-pinene or borneol, result.

In Fig. 16-25, the conversion of farnesyl pyrophosphate to certain sesquiterpenes is outlined. The reactions are similar to those discussed above.

Among compounds derived from condensation of isoprenoid units, the C_{55} polyisoprenol, *undecaprenol,* is particularly interesting. The corresponding

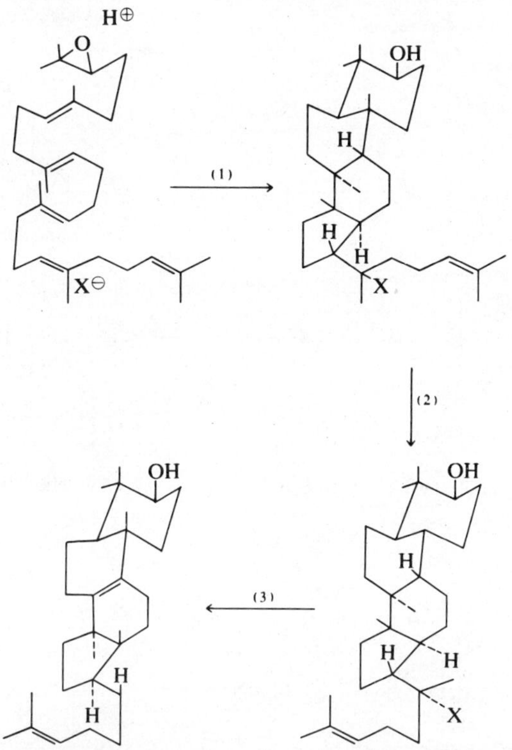

FIGURE 16-20
A visualization of the transformation of squalene-2,3-oxide to lanosterol. In step 1, squalene-2,3-oxide undergoes a multiplex electron shift with participation of the electron-donating group X and the formation of a tetracyclic intermediate. In step 2, rotation of the side chain of this intermediate 120° with respect to ring D brings the four groups which must migrate into the proper stereochemical position for migration. In step 3, migration occurs with the formation of lanosterol having the correct stereochemistry at each of eight asymmetric centers.

phosphate is involved in two types of enzymatic reactions with NDP-hexoses leading to the formation of polyisoprenol-sugar derivatives:

$$(CH_3)_2C{=}CH{-}CH_2{-}\left(CH_2{-}\overset{CH_3}{\overset{|}{C}}{=}CH{-}CH_2\right)_9{-}CH_2{-}\overset{CH_3}{\overset{|}{C}}{=}CH{-}CH_2{-}OPO_3^{2-} + \text{NDP-hexose} \longrightarrow$$

$$C_{55}{-}OPO_3{-}\text{hexose} + \text{NDP} \quad \textit{or} \quad C_{55}{-}OP_2O_6{-}\text{hexose} + \text{NMP} \qquad (16\text{-}24)$$

These polyisoprenol-linked sugars are known to be involved in the biosynthesis of the mucopeptides of cell walls (page 552), O-antigen, mannans; closely related intermediates are implicated in the synthesis of teichoic acids, capsular polysaccharides, and glycoproteins, Consequently, polyisoprenols serve a function as intermediate carriers of sugar residues which is analogous to that of tRNA's, which act as carriers of amino acyl functions, and acyl carrying proteins (ACP's), which act as carriers of fatty acyl groups.

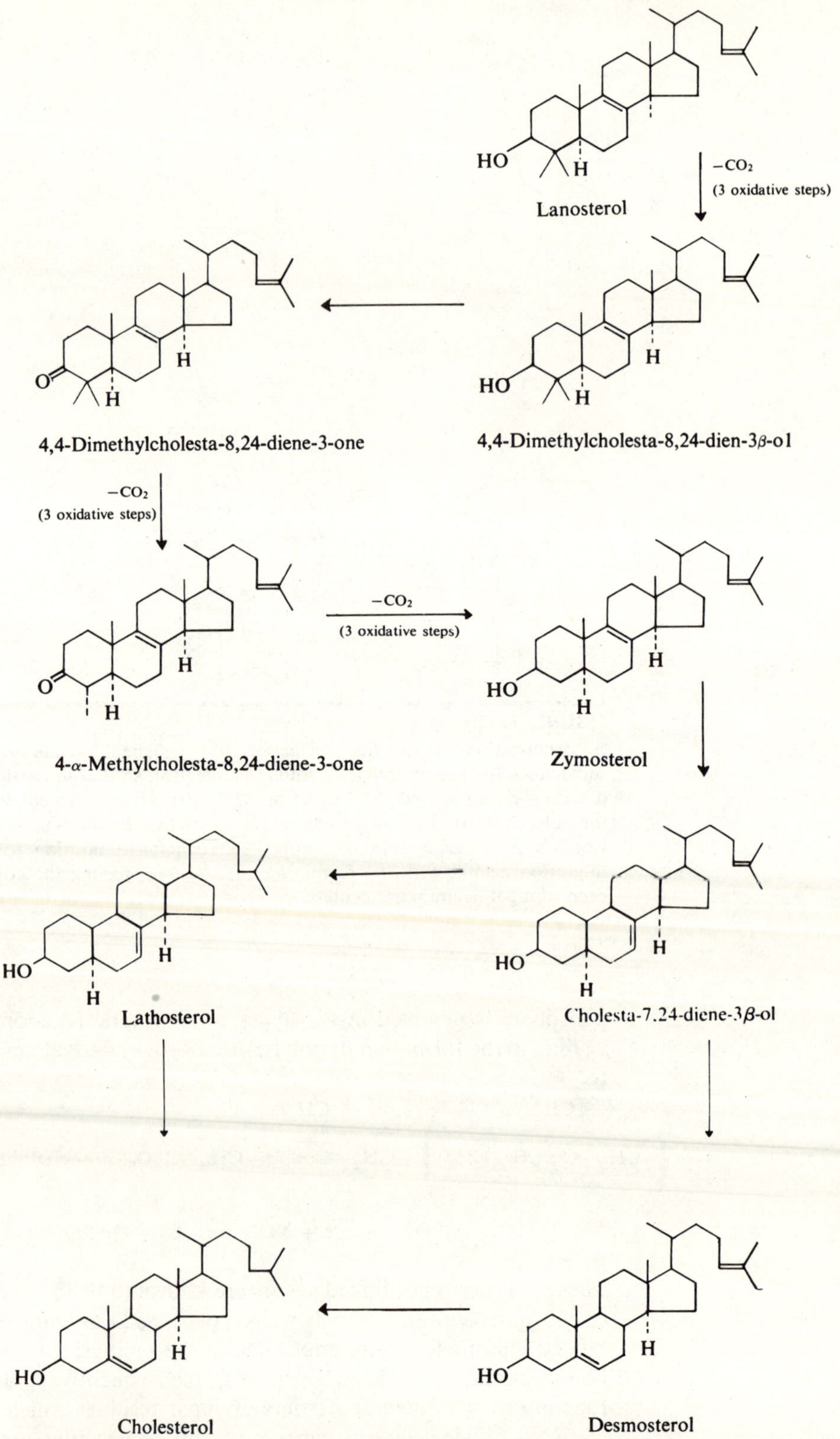

HO H
Lanosterol

$-CO_2$
(3 oxidative steps)

O H H
4,4-Dimethylcholesta-8,24-diene-3-one

HO H H
4,4-Dimethylcholesta-8,24-dien-3β-o l

$-CO_2$
(3 oxidative steps)

O H H
4-α-Methylcholesta-8,24-diene-3-one

$-CO_2$
(3 oxidative steps)

HO H H
Zymosterol

HO H H
Lathosterol

HO H H
Cholesta-7,24-diene-3β-ol

HO H
Cholesterol

HO H
Desmosterol

FIGURE 16-21
Some pathways from lanosterol to cholesterol.

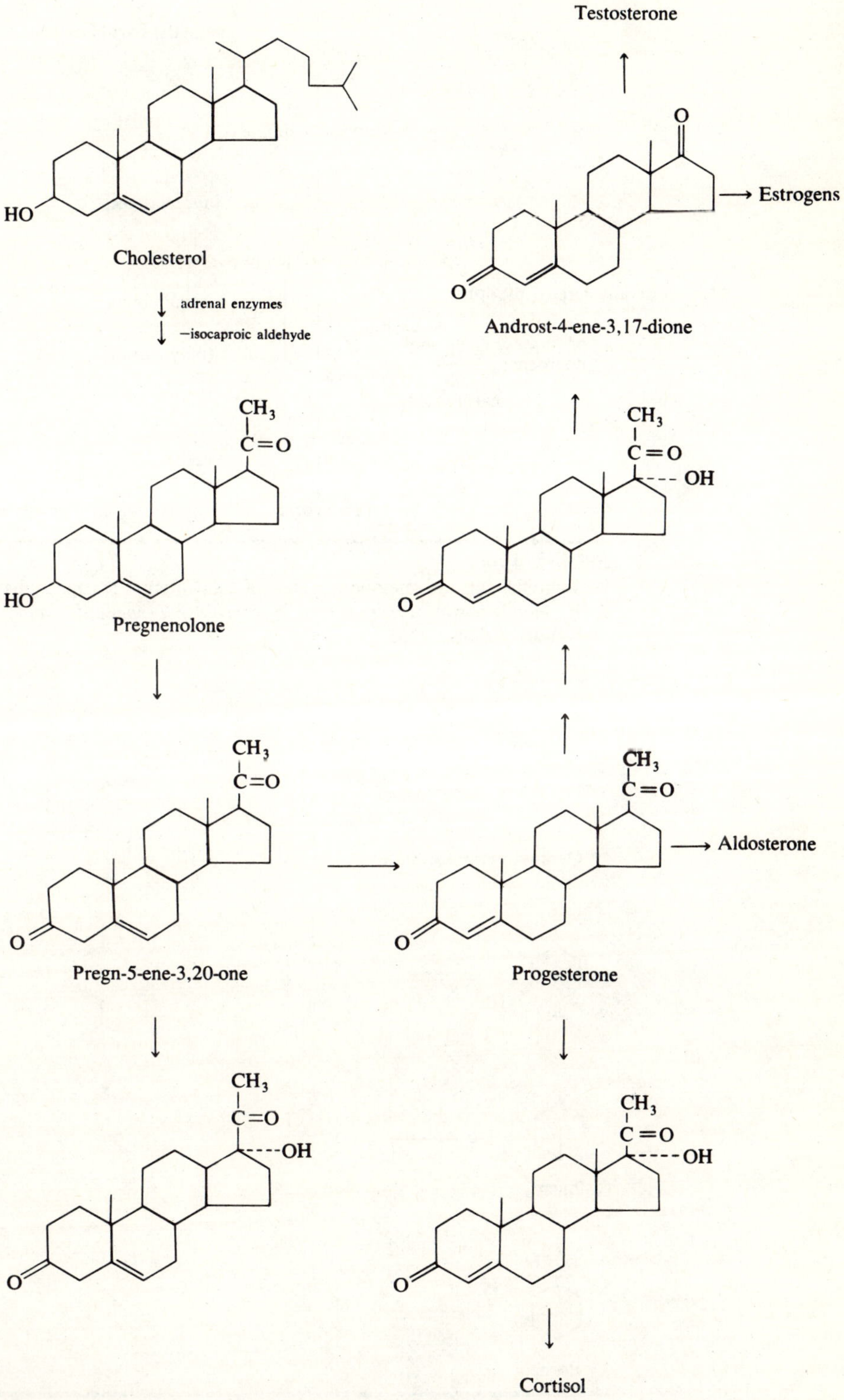

Testosterone

Estrogens

Cholesterol

adrenal enzymes

−isocaproic aldehyde

Androst-4-ene-3,17-dione

CH_3

C=O

OH

Pregnenolone

Aldosterone

Pregn-5-ene-3,20-one

Progesterone

Cortisol

FIGURE 16-22

Some aspects of the metabolism of cholesterol.

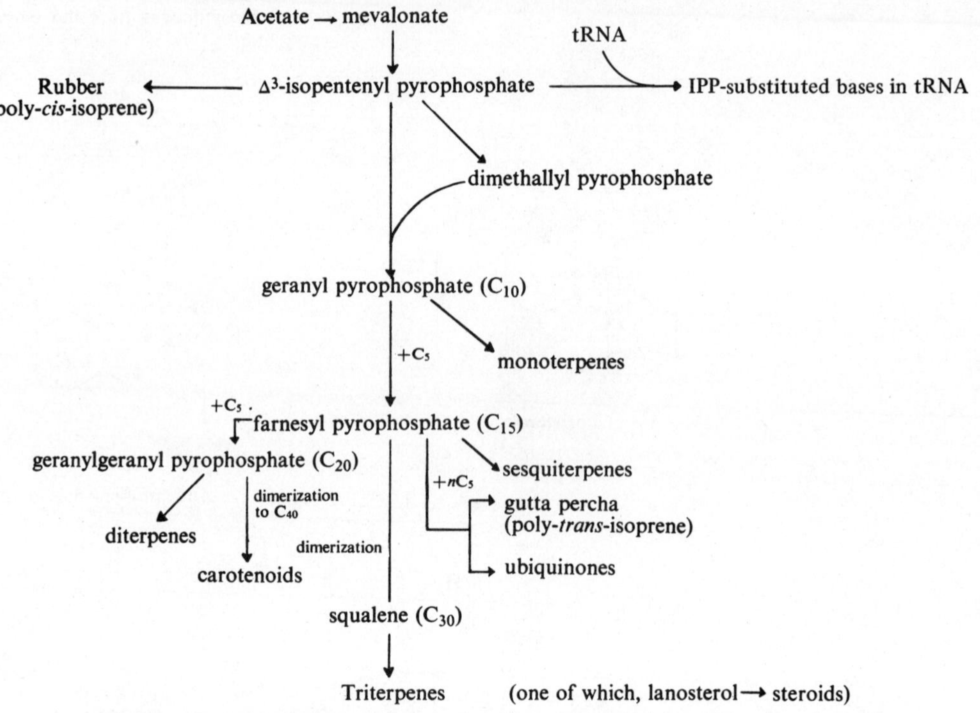

FIGURE 16-23

A summary of the metabolic fates of isopentenyl pyrophosphate. J. H. Richards and J. B. Hendrickson, *Biosynthesis of Steroids, Terpenoids, and Acetogenins,* Benjamin, New York, 1964, p. 205.

$-PP_i$

OPP

CHO

Geranyl pyrophosphate

Citronellal

OPP

Limonene

H

α-Pinene

or

α-Terpinene

OH

OH

Borneol

FIGURE 16-24

Conversion of geranyl pyrophosphate into monoterpenes.

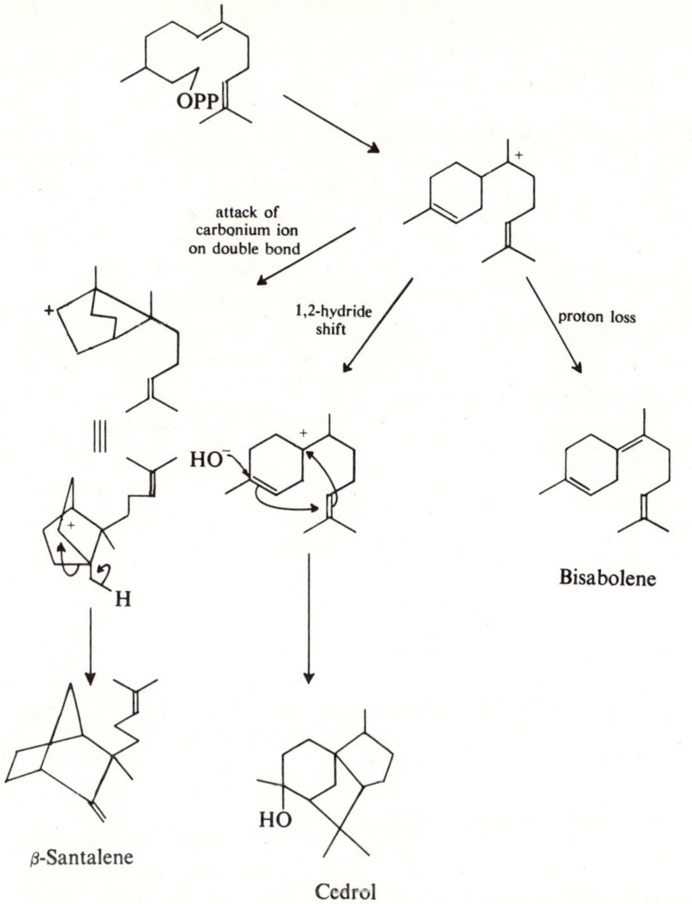

FIGURE 16-25
Synthesis of some sesquiterpenes from farnesyl pyrophosphate.

REFERENCES

Bloch, K.: The Biological Synthesis of Cholesterol, *Science,* **150:** 19 (1965).

Bloch, K.: Biogenesis and Transformations of Squalene, in G. E. W. Wolstenholme and M. O'Conner, (eds.), *Ciba Found. Symp., Biosyn. Terpenes Sterols,* Little, Brown, Boston, 1959, p. 4.

Bloch, K.: Biological Synthesis of Cholesterol, *Harvey Lectures,* **48:** 68 (1952).

Cornforth, J. W.: Olefin Alkylation in Biosynthesis, *Angew. Chem.,* **7:** 903 (1968).

Clayton, R. B.: Biosynthesis of Sterols, Steroids, and Terpenoids, *Quart. Rev.,* **19:** 168 (1965).

Fieser, L. F., and M. Fieser: *Steroids,* Reinhold, New York, 1959.

Folkers, K., C. H. Shunk, B. O. Linn, F. M. Robinson, P. E. Wittreich, J. W. Huff, J. L. Gilfillan, and H. R. Skeggs: Discovery and Elucidation of Mevalonic Acid, *Ciba Found. Symp., Biosyn. Terpenes Sterols,* Little, Brown, Boston, 1959, p. 120.

Frantz, Jr., I. D., and G. J. Shroepfer, Jr.: Sterol Biosynthesis, *Ann. Rev. Biochem.,* **36:** 691 (1967).

Goldfine, H.: Lipid Chemistry and Metabolism, *Ann. Rev. Biochem.,* **37:** 303 (1968).

Hanahan, D. J., and G. A. Thompson, Jr.: Complex Lipids, *Ann. Rev. Biochem.,* **32:** 215 (1963).

Kates, M.: Bacterial Lipids, *Advan. Lipid Res.,* **2:** 17 (1964).

Kates, M., and M. K. Wassef: Lipid Chemistry, *Ann. Rev. Biochem.,* **39:** 323 (1970).

Kennedy, E. P.: Biosynthesis of Complex Lipids, *Federation Proc.,* **20:** 934 (1961).

Law, J. H.: Bacterial Lipids, in Davis and Warren (eds), *The Specificity of Cell Surfaces,* Prentice-Hall, Englewood Cliffs, N.J., 1967.

Lennarz, W. J.: Lipid Metabolism, *Ann. Rev. Biochem.,* **39:** 359 (1970).

Lennarz, W. J.: Lipid Metabolism in the Bacteria, *Advan. Lipid Res.,* **4:** 175 (1966).

Lynen, F.: Biosynthesis of Saturated Fatty Acids, *Federation Proc.,* **20:** 941 (1961).

Lynen, F.: Acetyl Coenzyme A and the Fatty Acid Cycle, *Harvey Lectures,* **48:** 210 (1952).

Lynen, F., H. Eggerer, U. Henning, J. Knappe, I. Kessel, and E. Ringelmann: New Aspects of Acetate Incorporation into Isoprenoid Precursors, *Ciba Found. Symp., Biosyn. Terpenes Sterols,* Little, Brown, Boston, 1959, p. 95.

Majerus, P. W., and P. R. Vagelos: Fatty Acid Biosynthesis and the Role of the Acyl Carrier Protein, *Advan. Lipid Res.,* **5:** 2 (1967).

Popják, G., and J. W. Cornforth: Substrate Stereochemistry in Squalene Biosynthesis, *Biochem. J.,* **101:** 553 (1966).

Richards, J. H., and J. B. Hendrickson: *The Biosynthesis of Steroids, Terpenes, and Acetogenins,* Benjamin, New York, 1964.

Shapiro, B.: Lipid Metabolism, *Ann. Rev. Biochem.,* **36:** 247 (1967).

Sih, C. J., and H. W. Whitlock, Jr.: Biochemistry of Steroids, *Ann. Rev. Biochem.,* **37:** 661 (1968).

Svennerholm, L.: The Gangliosides, *J. Lipid Res.,* **5:** 145 (1964).

Wakil, S. J.: Mechanism of Fatty Acid Synthesis, *J. Lipid Res.,* **2:** 1 (1961).

Specific References

1. Seubert, W., G. Greull, and F. Lynen: *Angew. Chem.,* **69:** 359 (1957).
2. Wakil, S. J., L. W. McLain, Jr., and J. B. Warshaw: *J. Biol. Chem.,* **235:** PC31 (1960).
3. Harlan, W. R., Jr., and S. J. Wakil: *J. Biol. Chem.,* **238:** 3216 (1963).
4. Van Baalen, J., and S. Gurin: *J. Biol. Chem.* **205:** 303 (1953).
5. Tietz, A., and G. Popják: *Biochem. J.,* **60:** 147, 155 (1955).
6. Srere, P. A., and A. Bhaduri: *Biochim. Biophys. Acta,* **59:** 487 (1962).
7. Spencer, A., L. Corman, and J. M. Lowenstein: *Biochem. J.,* **93:** 378 (1964).
8. Bressler, R., and R. I. Katz: *J. Biol. Chem.,* **240:** 622 (1965).
9. Wakil, S. J.: *J. Am. Chem Soc.,* **80:** 6465 (1958).
10. Formica, J. V., and R. O. Brady: *J. Am. Chem. Soc.,* **81:** 752 (1959).
11. Waite, M. and S. J. Wakil: *J. Biol. Chem.,* **238:** 77 (1963).
12. Waite, M. and S. J. Wakil: *J. Biol. Chem.,* **238:** 81 (1963).
13. Gregolin, C., E. Ryder, A. K. Kleinschmidt, R. C. Warner, and M. D. Lane: *Proc. Natl. Acad. Sci. U.S.,* **56:** 148 (1966).
14. Gregolin, C., E. Ryder, R. C. Warner, A. K. Kleinschmidt, and M. D. Lane: *Proc. Natl. Acad. Sci. U.S.,* **56:** 1751 (1966).
15. Ryder, E., C. Gregolin, H-C. Chang, and M. D. Lane: *Proc. Natl. Acad. Sci. U.S.,* **57:** 1455 (1967).
16. Kallen, R. G., and J. M. Lowenstein: *Arch. Biochem. Biophys.,* **96:** 188 (1962).
17. Alberts, A. W., P. Goldman, and P. R. Vagelos: *J. Biol. Chem.,* **238:** 557 (1963).
18. Wakil, S. J., and J. Ganguly: *J. Am. Chem. Soc.,* **81:** 2597 (1959).
19. Bressler, R., and S. J. Wakil: *J. Biol. Chem.,* **237:** 1441 (1962).
20. Goldman, P., A. W. Alberts, and P. R. Vagelos: *J. Biol. Chem.,* **238:** 3579 (1963).
21. Robinson, J. D., R. M. Bradley, and R. O. Brady: *Biochemistry,* **2:** 191 (1963).
22. Brady, R. O., R. M. Bradley, and E. G. Trams: *J. Biol. Chem.,* **235:** 3093 (1960).
23. Vanaman, T. C., S. J. Wakil, and R. L. Hill: *J. Biol. Chem.,* **243:** 6420 (1968).
24. Majerus, P. W., A. W. Alberts and R. P. Vagelos: *Proc. Natl. Acad. Sci. U.S.,* **53:** 410 (1965).
25. Pugh, E. L. and S. J. Wakil: *J. Biol. Chem.,* **240:** 4727 (1965).
26. Alberts, A. W., P. W. Majerus, B. Talamo, and P. R. Vagelos: *Biochemistry,* **3:** 1563 (1964).
27. Alberts, A. W., P. W. Majerus, and P. R. Vagelos: *Biochemistry,* **4:** 2265 (1965).
28. Wells, W. W., J. Schultz, and F. Lynen: *Proc. Natl. Acad. Sci. U.S.,* **56:** 633 (1966).
29. Yang, P. C., P. H. W. Butterworth, R. M. Bock, and J. W. Porter: *J. Biol. Chem.,* **242:** 3501 (1967).
30. Bloch, K., P. Baronowsky, H. Goldfine, W. J. Lennarz, R. Light, A. T. Norris, and G. Scheuerbrandt: *Federation Proc.,* **20:** 921 (1961).
31. Kass, L. R., D. J. H. Brock, and K. Bloch: *J. Biol. Chem.,* **242:** 4418 (1967).
32. Brock, D. J. H., L. R. Kass, and K. Bloch: *J. Biol. Chem.,* **242:** 4432 (1967).
33. Kass, L. R. and K. Bloch: *Proc. Natl. Acad., Sci. U.S.,* **58:** 1168 (1967).
34. Nagai, J. and K. Bloch: *J. Biol. Chem.,* **243:** 4626 (1968).
35. Zalkin, H., J. H. Law, and H. Goldfine: *J. Biol. Chem.,* **238:** 1242 (1963).
36. Thomas, P. J. and J. H. Law: *J. Biol. Chem.,* **241:** 5013 (1966).
37. Chung, A. E. and J. H. Law: *Biochemistry,* **3:** 1989 (1964).
38. Seubert, W., I. Lamberts, R. Kramer, and B. Ohly: *Biochim. Biophys. Acta,* **164:** 498 (1968).
39. Aeberhard, E. and J. H. Menkes, *J. Biol. Chem.,* **243:** 3834 (1968).
40. Birch, A. J.: *Fortschr. Chem. Org. Naturstoffe,* **14:** 186 (1957).
41. Snyder, F., R. L. Wykle, and B. Malone: *Biochem. Biophys. Res. Commun.,* **34:** 315 (1969).
42. Brundish, D. E., N. Shaw, and J. Baddiley: *Biochem. Biophys. Res. Commun.,* **18:** 308 (1965).
43. Kaufman, B., F. D. Kundig, J. Distler, and S. Roseman: *Biochem. Biophys. Res. Commun.,* **18:** 312 (1965).
44. Osborn, M. J., S. M. Rosen, L. Rothfield, L. D. Zeleznick, and B. L. Horecker: *Science,* **145:** 783 (1964).
45. Nikaido, H.: *Advan. Enzymol.,* **31:** 77 (1968).
46. Weiss, S. B., and E. P. Kennedy: *J. Am. Chem. Soc.,* **78:** 3550 (1956).

47. Talamo, B. and W. J. Lennarz: *Federation Proc.*, **24:** 479 (1965).
48. Lennarz, W. J. and B. Talamo: *J. Biol. Chem.*, **241:** 2707 (1966); Houtsmuller, U. M. T., and L. L. M. van Deenen: *Biochim. Biophys. Acta,* **106:** 564 (1965).
49. Hill, E. E., D. R. Husbands, and W. E. M. Lands: *J. Biol. Chem.*, **243:** 4440 (1968).
50. Van Golde, L. M. G., G. L. Scherphof, and L. L. M. Van Deenen: *Biochim. Biophys. Acta,* **176:** 635 (1969).
51. Sánchez de Jiménez, E., and W. W. Cleland: *Biochim. Biophys. Acta,* **176:** 685 (1969).
52. Abou-Issa, H. M., and W. W. Cleland: *Biochim. Biophys. Acta,* **176:** 692 (1969).
53. Possmayer, F., G. L. Scherphof, T. M. A. R. Dubbelman, L. M. G. van Golde, and L. L. M. Van Deenen: *Biochim. Biophys. Acta,* **176:** 95 (1969).
54. Lamb, R. G., and H. J. Fallon: *Federation Proc.*, **28:** 595 (1969).
55. Hajra, A. K.: *Biochem. Biophys. Res. Commun.*, **33:** 929 (1968).
56. Hajra, A. K. and B. W. Agranoff: *Federation Proc.*, **28:** 595 (1969).
57. Lennarz, W. J., J. A. Nesbitt, III, and J. Reiss: *Proc. Natl. Acad. Sci. U.S.*, **55:** 934 (1966).
58. Carter, H. E., F. J. Glick, W. P. Norris, and G .E. Phillips: *J. Biol. Chem.*, **142:** 449 (1942); **191:** 727 (1951).
59. Carter, H. E., and F. L. Greenwood: *J. Biol. Chem.*, **199:** 283 (1952).
60. Kuhn, R. and H. Wiegandt: *Chem. Ber.*, **96:** 866 (1963).
61. Basu, S. and B. Kaufman: *Federation Proc.*, **24:** 479 (1965).
62. Braun, P. E. and E. E. Snell: *J. Biol. Chem.*, **243:** 3775 (1968).
63. Brady, R. N., S. J. DiMari, and E. E. Snell: *J. Biol. Chem.*, **244:** 491 (1969).
64. Sribney, M., and E. P. Kennedy: *J. Biol. Chem.*, **233:** 1315 (1958).
65. Fujino, Y., Mo. Nakano, T. Negishi, and S. Ito: *J. Biol. Chem.*, **243:** 4650 (1968).
66. Brady, R. O., R. M. Bradley, O. M. Young, and H. Kaller, *J. Biol. Chem.*, **240:** PC3693 (1965).
67. Little, H. N., and K. Bloch: *J. Biol. Chem.*, **183:** 33 (1950).
68. Rudney, H., P. R. Stewart, P. W. Majerus, and P. R. Vagelos: *J. Biol. Chem.*, **241:** 1226 (1966).
69. Brodie, J. D., G. Wasson, and J. W. Porter: *J. Biol. Chem.*, **238:** 1294 (1963).
70. Fimognari, G. M., and V. W. Rodwell: *Biochemistry,* **4:** 2086 (1965).
71. Beg, Z. H., and M. Siddiqi; *Experientia,* **23:** 380 (1967); **24:** 791 (1968).
72. Beg., Z. H., M. Siddiqi, and R. A. H. Siddiqi: *Experientia,* **24:** 15 (1968).
73. Langdon, R. G., and K. Bloch: *J. Biol. Chem.*, **200:** 129, 135 (1953).
74. Lynen, F., H. Eggerer, U. Henning, and I. Kessel: *Angew. Chem.*, **70:** 738 (1958).
75. Popják, G., D. Goodman, J. W. Cornforth, R. H. Cornforth, and R. Ryhage: *J. Biol. Chem.*, **236:** 1934 (1961).
75a. Popják, G., J. Edmond, K. Clifford, and V. Williams: *J. Biol. Chem.*, **244:** 1897 (1969).
76. Popják, G., J. W. Cornforth, R. H. Cornforth, R. Ryhage, and P. S. Goodman: *J. Biol. Chem.*, **237:** 56 (1962).
77a. Epstein, W. W., and H. C. Rilling: *J. Biol. Chem.*, **245:** 4597 (1970).
77. Rilling, H. C.: *J. Biol. Chem.*, **241:** 3233 (1966).
78. Krishna, G., D. H. Feldbruegge, and J. W. Porter: *Biochem. Biophys. Res. Commun.*, **14:** 363 (1964).
79. Dean, P. D. G., P. R. Ortiz de Montellano, K. Bloch, and E. J. Corey: *J. Biol. Chem.*, **242:** 3014 (1967).
80. Willett, J. D., K. B. Sharpless, K. E. Lord, E. E. van Tamelen, and R. B. Clayton: *J. Biol. Chem.*, **242:** 4182 (1967).

17 Metabolism of the Amino Acids

In this chapter, we consider the metabolism of the amino acids, individually and as a group. This material is organized into six categories: (1) the assimilation of inorganic nitrogen, (2) the cleavage of peptide bonds, (3) the biosynthesis of the amino acids, (4) general aspects of amino acid metabolism, (5) specific aspects of the metabolism of the amino acids, and (6) amino acid metabolism in relation to the biosynthesis of alkaloids. The assembly of amino acids into proteins, a closely related subject, is considered in Chapter 21.

ASSIMILATION OF INORGANIC NITROGEN

Inorganic nitrogen, which may be converted into ammonia and subsequently into organic materials, consists principally of molecular nitrogen and nitrate salts. We consider the reduction of these in turn.

Nitrogen Fixation

The ultimate source of nitrogen for the construction of nitrogenous organic compounds is atmospheric nitrogen. Throughout the plant and microbial kingdoms, there is a category of species that maintains the capacity to reduce nitrogen to ammonia. These are termed "nitrogen-fixing" organisms. Such organisms include certain species of heterotrophic bacteria, both aerobic (*Azotobacter vinelandii*) and anaerobic (*Clostridium pasteurianum*); photosynthetic bacteria (*Rhodospirillum rubrum*); several algae; and a symbiotic system, consisting of bacteria of the genus *Rhizobium* together with many species of plants. In the last case, neither the bacteria nor the plant alone can fix nitrogen. The combined abilities of the bacteria, which live in the root nodules of the plant, and the plant create a very efficient and important nitrogen-fixing cooperative.

The reduction of nitrogen is an extraordinary biochemical capacity. Molecular nitrogen is an extremely stable molecule, as evidenced by the value of 225 kcal/mol for the strength of the triple bond in N_2 (compare with the corresponding value of 110 kcal/mol for acetylene), and strenuous conditions are required for its reduction in the laboratory. Nevertheless, several million tons of nitrogen per year are converted, by the above organisms and their relatives, to ammonia under very mild conditions.

Much remains to be established concerning the sequence of events comprising nitrogen fixation. For many years, research efforts in this field were severely hampered by the necessity of employing whole-cell preparations in order to observe this reaction. This blockade was first successfully surmounted by Mortenson, Carnahan, Mower, and Castle in 1960 with the demonstration of cell-free nitrogen fixation in preparations from *C. pasteurianum* (1,2). Further successes followed rapidly on a number of fronts and cell-free systems have now been obtained from at least 10 bacterial species, aerobic and anaerobic, from blue-green algae, and from root nodules. The use of cell-free soluble preparations permits the fractionation, purification, and characterization of those cellular components actually involved in nitrogen fixation—goals that may not be realized employing undisrupted cells. In addition, such studies render the identification of intermediates and kinetic studies a good deal more manageable and should suggest incisive experiments at the intact-cell level.

In view of the diverse characteristics of the organisms that fix nitrogen, it is not surprising that the properties of the cell-free, nitrogen-fixing systems from different sources exhibit considerable individuality. This individuality takes the form of varying responses to the presence of oxygen, to changes of nitrogen pressure, and to variations in *p*H and substrate concentration. Nevertheless, the similarities among the various systems seem far more important than their differences. Let us now consider the principal aspects of "nitrogenase," the multienzyme complex responsible for the conversion of nitrogen to ammonia.

Reaction Product. The final product of nitrogen fixation in *C. pasteurianum* extracts, as in those from other species, is ammonia. Possible intermediates on the route from N_2 to NH_3, which include HN=NH, hydrazine, and hydroxylamine, have been searched for but not detected. Reaction intermediates may, hence, exist only transiently or be tightly bound to appropriate enzymes.

Requirements. In addition to nitrogenase itself, nitrogen fixation requires a source of electrons (a reductant), an electron carrier to transfer the electrons from the reductant to molecular nitrogen, ATP, and a divalent metal ion.

Electron Sources. The biological electron donor for nitrogen fixation is pyruvate (1–3). In intact systems, pyruvate metabolism appears to be absolutely required for nitrogen fixation. On a quantitative basis, about 100 mol of pyruvate are metabolized per mol of nitrogen fixed (2), suggesting that the two processes are only loosely coupled. The pathway for pyruvate metabolism in *C. pasteurianum* is outlined in Fig. 17-1. As indicated, pyruvate metabolism generates both, electrons, which are transferred to nitrogen via an electron carrier (see below) and ATP. In fact, both aspects of pyruvate metabolism are crucial for nitrogen fixation, as we shall see.

A most useful nonbiological reductant for nitrogen fixation has been discovered—hydrosulfite ($S_2O_4^{2-}$). Hydrosulfite is apparently a universal reductant for the nitrogenase reaction and circumvents the usual electron carriers; that is, it interacts directly with the enzymatic machinery for the reduction (4–8). Use of hydrosulfite is very convenient, inasmuch as it simplifies the spectrum of components necessary to achieve nitrogen fixation.

Electron Carriers. The important electron-carrying protein ferredoxin was discovered in the course of studies of nitrogen fixation (9–11). As shown in Fig. 17-1, ferredoxin accepts electrons from pyruvate in the course of the phosphoroclastic reaction; it has the capacity to transfer them to a number of enzymes, including nitrogenase (see also page 581). Ferredoxins

$$CH_3-\overset{O}{\overset{\|}{C}}-CO_2^- + TPP\cdot E \longrightarrow \left[CH_3-\overset{OH}{\overset{|}{C}}{=}TPP\cdot E\right] + CO_2$$

$$\left[CH_3-\overset{OH}{\overset{|}{C}}{=}TPP\cdot E\right] \xrightarrow{2\ \text{ferredoxin} + HSCoA} CH_3-\overset{O}{\overset{\|}{C}}-SCoA + [\text{ferredoxin}\cdot e^-]_2$$

$$\left[CH_3-\overset{OH}{\overset{|}{C}}{=}TPP\cdot E\right] \xrightarrow{CH_3-\overset{O}{\overset{\|}{C}}-CO_2^-} CH_3-\underset{H}{\underset{|}{\overset{OH}{\overset{|}{C}}}}-\overset{O}{\overset{\|}{C}}-CH_3 + CO_2$$

$$[\text{ferredoxin}\cdot e^-]_2 \underset{}{\xrightarrow{+2H^+}} H_2 + 2\ \text{ferredoxin}$$

$$CH_3-\overset{O}{\overset{\|}{C}}-SCoA \underset{+HSCoA}{\overset{+P_i}{\rightleftharpoons}} CH_3-\overset{O}{\overset{\|}{C}}-OPO_3H^-$$

$$CH_3-\overset{O}{\overset{\|}{C}}-OPO_3H^- \underset{+P_i}{\overset{+ADP}{\rightleftharpoons}} CH_3CO_2^- + ATP$$

FIGURE 17-1
Oxidation of pyruvate in *Clostridium pasteurianum.*

have been prepared in crystalline form and they have been well characterized structurally. Although their properties vary somewhat depending on the species from which they are isolated, ferredoxins are generally low-molecular-weight proteins, near 10,000, and are particularly characterized through their content of five to eight molecules of nonheme iron per mole together with acid-labile sulfur. The structures of these proteins has been described in more detail on page 657.

A related but distinct electron carrier for nitrogen fixation is *flavodoxin,* discovered by Hardy and his colleagues at du Pont. When nitrogen-fixing bacteria are grown on media containing very little iron, flavodoxin is synthesized rather than ferredoxin (12-14). Like ferredoxin, flavodoxin is a rather small protein but, in contrast, contains FMN, not iron, as prosthetic group. It, too, is an electron-transfer agent of low redox potential. Flavodoxin and ferredoxin appear to be quite interchangeable insofar as their function in nitrogen fixation is concerned.

ATP. ATP is absolutely required for nitrogen fixation (15–18) and is hydrolyzed to ADP and inorganic phosphate in the process (19). In the intact organism, then, metabolism of pyruvate supplies both the electrons and the energy source necessary for nitrogen fixation. The failure to recognize the latter of these functions of pyruvate metabolism certainly was a stumbling block to the development of the cell-free systems for this process. The exact number of molecules of ATP hydrolyzed per pair of electrons transferred to nitrogen is not completely clear, although a number around 4 seems reasonable (19). Thus, about 12 molecules of ATP are required to reduce 1 molecule of nitrogen to 2 molecules of ammonia. The stoichiometry of the reaction seems to be a sensitive function of a number of variables (20).

Divalent Cation. Magnesium ion satisfies the divalent-metal-ion requirements of nitrogenase. Other cations are less effective but do support the reaction. These include Mn^{2+}, Fe^{2+}, Co^{2+}, and Ni^{2+}. It is likely that these function as complexing agents for ATP (page 384), although definitive evidence for this point of view is lacking.

Nitrogenase. Purification of nitrogenase from cell-free nitrogen-fixing preparations has proved a trying task. Two partially purified preparations have been achieved; these are from *C. pasteurianum* and *A. vinelandii* (21–24). The properties of the two preparations are similar. In each case, two proteins form the active complex. One contains both Mo and Fe and has a molecular weight in the 100,000 to 120,000 range; the other contains only Fe and has a molecular weight near 50,000. These are sometimes termed the "molybdoiron protein" and the "iron protein," respectively, although both less descriptive and more complicated names are also in use (see page 661). The intact complex is sensitive to both oxygen and cold. The *p*H optimum for nitrogen fixation is near 7.

The Reaction. Nitrogenase is a remarkably flexible and potent reducing system. In addition to the reduction of molecular nitrogen, the system also reduces N_2O, azide, acetylene, hydrogen cyanide, isonitriles, and the hydrated proton, among other substrates. Thus, typical reactions may be formulated:

$$
\begin{array}{lll}
N_2 \longrightarrow 2NH_3 & & \text{(6 electrons)} \\
N_2O \longrightarrow N_2 + H_2O & & \text{(2 electrons)} \\
N_3^- \longrightarrow N_2 + NH_3 & & \text{(2 electrons)} \\
HC{\equiv}CH \longrightarrow H_2C{=}CH_2 & & \text{(2 electrons)} \\
HCN \longrightarrow CH_4 + NH_3 & & \text{(6 electrons)} \\
CH_3\text{-}N{\equiv}C \longrightarrow CH_4 + CH_3NH_2 & & \text{(6 electrons)} \\
2H^+ \longrightarrow H_2 & & \text{(2 electrons)}
\end{array}
\quad (17\text{-}1)
$$

The last of these reactions is referred to as *ATP-dependent H_2 evolution.* In general these reactions can be said to adhere to the following pattern:

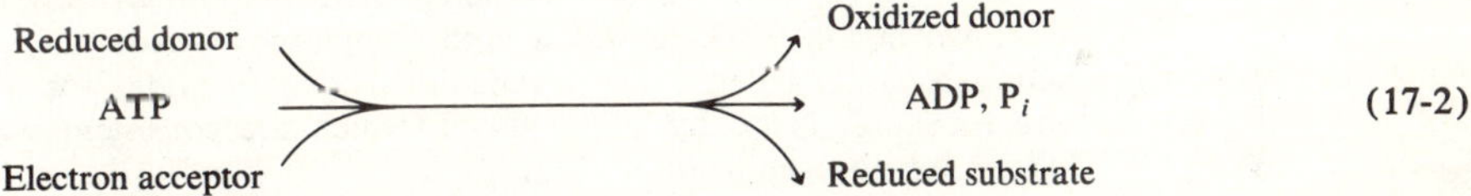

A diagrammatic scheme for the whole nitrogen-fixing process is provided in Fig. 17-2; see also Fig. 17-3.

The mechanism of the nitrogen-fixing reaction is not completely clear. A suggestion has been made that diimide complexed to heme groups may be an intermediate (25). Substantial evidence for both diimide and hydrazine as reaction intermediates has recently been provided (26). The crucial observation is that reduction of nitrogen in the presence of either H_2 and D_2O or D_2 and H_2O results in the production of HD. Thus, a mechanism must be formulated so as to include the interchange of hydrogen atoms between solvent water and molecular hydrogen.

Nitrate Assimilation

Reduction of nitrate by plants and microorganisms serves one of two purposes: (1) nitrate may serve as a source of ammonia for synthetic purposes or (2) nitrate may serve as a terminal electron acceptor. In the latter case, the product of the reduction may be either N_2, N_2O, or NO, depending on the species in which the reduction occurs. These materials are not further involved in the cellular metabolism, and hence, no assimilation of nitrogen is involved.

Nitrate *assimilation* proceeds in two steps. First, nitrate reductase (nitrate : NADH oxidoreductase) catalyzes the conversion of nitrate to nitrite and, second, nitrite reductase catalyzes the nitrite-to-ammonia reaction. In certain

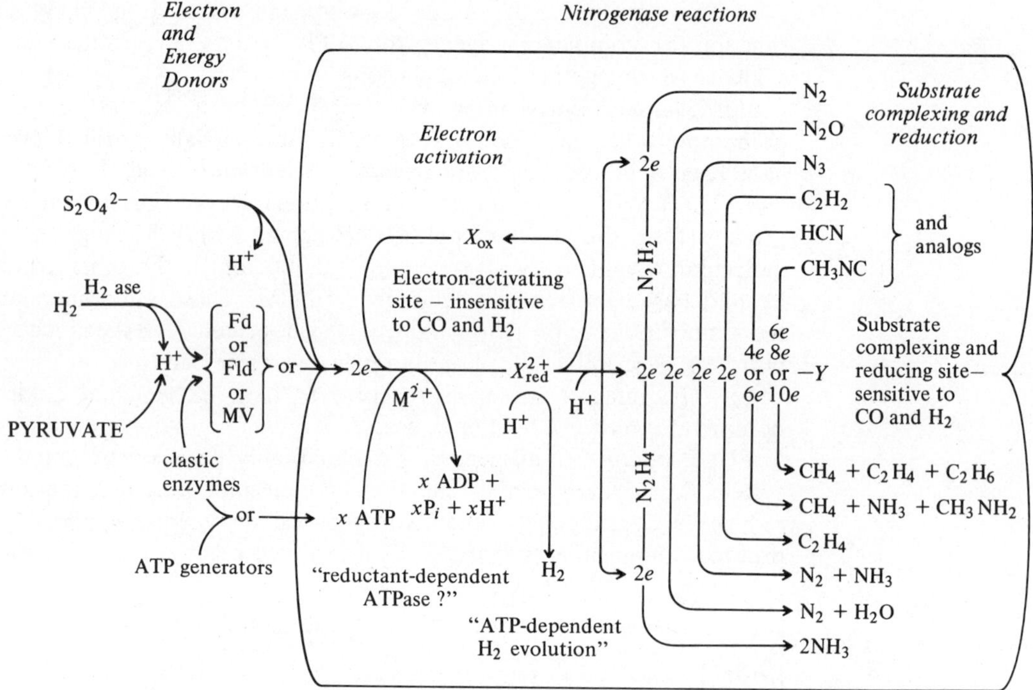

FIGURE 17-2

Scheme for the nitrogenase spectrum of reactions. The essence of the scheme is a series of two-electron reductions potentiated by ATP hydrolysis. One such mechanism is shown in Fig. 17-3, in which both complexed diimide and complexed hydrazine are shown as intermediates. The mechanism shows functions for both transition metals, *Y*, and transition metal hydrides, *X*—H, which are consistent with an interesting model system for nitrogen fixation (27).

species, the reverse reactions occur: *Nitrosomonas* converts ammonia to nitrite and *Nitrobacter* converts nitrite to nitrate.

Nitrate reductase, which occurs in many higher plants, fungi, and microorganisms, is a nicotinamide nucleotide–linked molybdoflavoprotein (28,29). Either NADPH or NADH is employed as initial electron donor, depending on the source of the enzyme. The sequence in which each of the electron-transfer agents participates in the overall transfer from reduced nicotinamide nucleotide to nitrate has been elucidated for the nitrate reductase from *Neurospora* (30). The pathway is

$$Fd_{red} \xrightarrow{fp} NADPH \rightarrow FAD \rightarrow Mo \rightarrow NO_3^- \quad (17\text{-}3)$$

The oxidation states involved in the oxidation–reduction of molybdenum appear to be +5 and +6.

The efforts of Nason and his collaborators have established several features of the enzymatic apparatus associated with nitrate reduction in *Neurospora crassa* (27a). The nitrate reductase (NADPH : nitrate oxidoreductase) in this organism is a soluble protein, sensitive to sulfhydryl reagents, of molecular weight 230,000. It is known to have FAD, cytochrome b_{557}, molybdenum, and an unidentified second metal ion as prosthetic groups. The enzyme exhibits

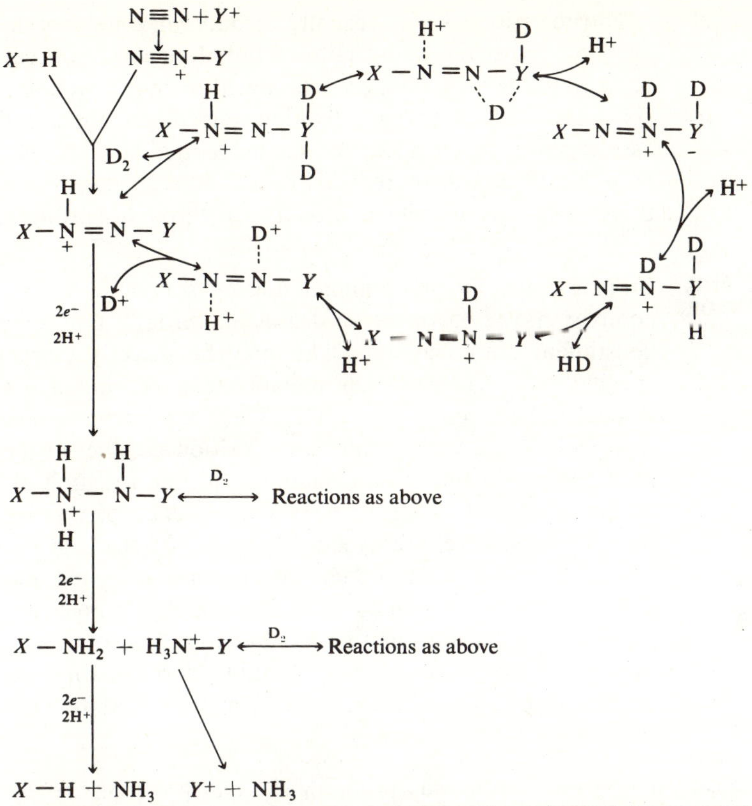

FIGURE 17-3
Possible mechanism for HD formation by intermediates of N_2 fixation. E. K. Jackson, G. W. Parshall, and R. W. F. Hardy, *J. Biol. Chem.*, **243:** 4952 (1968).

several catalytic capacities in addition to that of nitrate reductase; it functions as an FAD-dependent NADPH–cytochrome *c* reductase, for example.

Complementation for the nitrate reductase activity between different *N. crassa* mutants, each lacking this capacity, suggested that the intact enzyme is composed of at least two nonidentical subunits. This suggestion has been fully confirmed. It is now known that one subunit, which is inducible by nitrate, is responsible for the NADPH–cytochrome *c* reductase activity. The second subunit is a constitutive, particulate molybdenum-containing enzyme which is solubilized through interaction with the other subunit.

It is possible to assemble a fully functional nitrate reductase, indistinguishable from that isolated from wild-type *Neurospora,* from the first of these subunits and acid-treated xanthine oxidase, xanthine dehydrogenase, and aldehyde oxidase isolated from mammalian and avian sources (27a). Each of these enzymes requires molybdenum for the associated catalytic process. This observation suggests that all these enzymes, from mammalian, avian, and fungal sources, share a common polypeptide subunit.

The nitrate reductase of *E. coli* K-12 may prove more complex that that of *Neurospora.* The protein complex responsible for this reaction is bound to a particulate fraction of the cell and contains a formate dehydrogenase, cytochrome b_1, and nitrate reductase (and may well contain additional components) (27b,27c). The complete complex has a molecular weight near 1 million; it is induced by nitrate and repressed by oxygen.

Nitrite reductases are usually nicotinamide nucleotide–linked metalloflavoproteins. However, a diversity of initial electron sources, including NADPH, H_2, and photoreduction, are employed for the reduction of nitrite to ammonia. As in the case of nitrogen fixation, ferredoxin appears to play a central role as electron-transfer agent for nitrite reduction (31,32). In this context, ferredoxin appears to be reduced by the above agents and subsequently donates these electrons to nitrite directly or through another intermediate carrier.

CLEAVAGE AND FORMATION OF PEPTIDE BONDS

In mammals the predominant dietary source of nitrogen is protein. These complex molecules must be degraded, through the digestive process, to their constituent amino acids. These may be directly employed for synthesis of specific proteins or further metabolized, yielding a variety of products, as discussed later in this chapter. The digestive enzymes of man and other mammals occupy an important position in the historical development of intermediary metabolism, beginning with the research of Schwann and Kühne in the nineteenth century. As a consequence of this early development, the proteolytic enzymes are a relatively well understood group.

The importance of proteolytic enzymes is by no means restricted to the digestive process in mammals. They are of universal occurrence in living material and function in a variety of capacities. We shall now direct our attention to the properties of the best understood of these enzymes. We begin by discussing the digestive enzymes of man and continue with those from other sources.

Proteolytic Enzymes of the Gastrointestinal Tract

Pepsin. Pepsin was probably the entity recognized by Spallanzani in 1783 and may be regarded as the earliest known protein. It was obtained in crystalline form by Northrup in 1930 (33).

Pepsin, the principal proteolytic enzyme in the gastric juice, apparently occurs in all vertebrates, although the properties of the individual enzymes vary somewhat. The enzyme from swine has been most thoroughly investigated. Since the pH of gastric juice is generally near 1 or 2, pepsin must possess unusual stability in strongly acidic solutions. Indeed, it is characterized by an isoelectric point below 1, exhibits maximal catalytic activity in moderately acidic solutions, and is denatured at values of pH above 5. In these respects, pepsin is perhaps unique and certainly unusual among known enzymes.

Like the pancreatic proteolytic enzymes, pepsin is synthesized in the form of a precursor, termed a *zymogen,* pepsinogen. *Pepsinogen,* a crystalline protein (34), occurs in the gastric mucosa and is characterized by a molecular weight near 40,400 and an isoelectric point near 3.7. The conversion of pepsinogen to pepsin is catalyzed by hydrogen ions, and hence would be expected to occur on liberation of pepsinogen into the strongly acidic gastric juice, and also by pepsin itself. The activation reaction, in consequence, is autocatalytic. Since pepsin itself has a molecular weight of 32,700, the pepsinogen–pepsin conversion involves loss of a substantial portion of the polypeptide chain. In this conversion, the N-terminal portion of pepsinogen, 42 amino acid residues in length and containing 12 of the 16 basic residues of pepsinogen, is split off. Included in the cleavage products is a *pepsin inhibitor* of molecular weight 3242 and five smaller fragments having a combined molecular weight near 4000 (35).

Pepsin consists of a single polypeptide chain with an N-terminal isoleucine and a C-terminal valine. The molecule contains three disulfide bridges (accounting for all the cysteines present), two of which are required for enzymatic activity. Hydrodynamic measurements indicate that pepsin is a highly compact protein, although little α helix appears to be present.

Pepsin is a perplexing enzyme with respect to both its specificity and reaction mechanism. Looking at specificity first, one finds that with synthetic, low-molecular-weight substrates, pepsin is quite discriminating. The majority of known substrates are peptides; selected esters are subject to pepsin-catalyzed hydrolysis (36). Both amino acids of the sensitive peptide bond must be of the L configuration. Finally, one or, preferably, both of the amino acid residues involved in the sensitive peptide bond ought to have aromatic side chains. Thus, a typical synthetic pepsin substrate is Z-L-His-L-Phe-L-PheOCH_3 which is cleaved between the two phenylalanine residues. In contrast, studies of the hydrolysis of proteins by pepsin has revealed several examples of rapid attack on peptide bonds not involving residues of aromatic acids (37). The dichotomy between the results with proteins and those with synthetic substrates has not been resolved. However, some related results do suggest that the substrate binding site of pepsin may be quite complicated, and hence one may expect unusual behavior. Specifically, it has been demonstrated that pepsin possesses one primary binding site, which is presumably very near the catalytic site, and a number of secondary binding sites, which interact with peptide substrates less strongly than the primary one (38). It has been concluded on the basis of these studies that the substrate binding site is a long one capable of interacting simultaneously with a number of residues and that the catalytic site is somewhere near the middle of the total binding site.

The catalytic activity of pepsin is not understood. The enzyme appears to exhibit behavior, in part, expected on the basis of both acyl-activation and amine-activation mechanisms. On the one hand, pepsin catalyzes the incorporation of $H_2{}^{18}O$ into unreacted substrates, a characteristic of enzymes such as chymotrypsin, which are in the acyl activation class. On the other hand, a striking aspect of the catalytic action of pepsin is its ability to perform transpeptidations of the "amine-activation" type (39,40). For example, if *N*-acetyl-L-tyrosyl-L-tyrosine is incubated with pepsin, L-tyrosyl-L-tyrosine is produced, despite the fact that pepsin will not catalyze removal of *N*-acetyl functions from acetylated amino acids. The pathway for this transformation is

$$2\text{AcTyrTyr} \rightarrow \text{AcTyr} + \text{AcTyrTyrTyr} \qquad (17\text{-}4)$$

$$\text{AcTyrTyrTyr} \rightarrow \text{AcTyr} + \text{TyrTyr} \qquad (17\text{-}5)$$

Reactions (17-4) and (17-5) are formally related to those catalyzed by the thiophorases (page 596).

The dependence of rate upon pH for pepsin-catalyzed reactions suggests that a carboxylic group may be involved in the catalytic action (41). This suggestion receives strong support from a number of studies in which affinity labeling (page 337) of the pepsin active site with a number of substrate analogs results in modification of an aspartic acid group in the sequence Ile-Val-Asp-Thr-Gly-Thr-Ser- with inactivation of the enzyme (42).

One possible mechanism which accords with the above considerations is shown in Fig. 17-4 (41). Here the peptide substrate is viewed as partitioned between two groups on the enzyme surface; the acyl portion of the substrate forms an anhydride with a carboxyl group of the enzyme (the reactive aspartic acid group perhaps) and the amine portion forms an amide with another

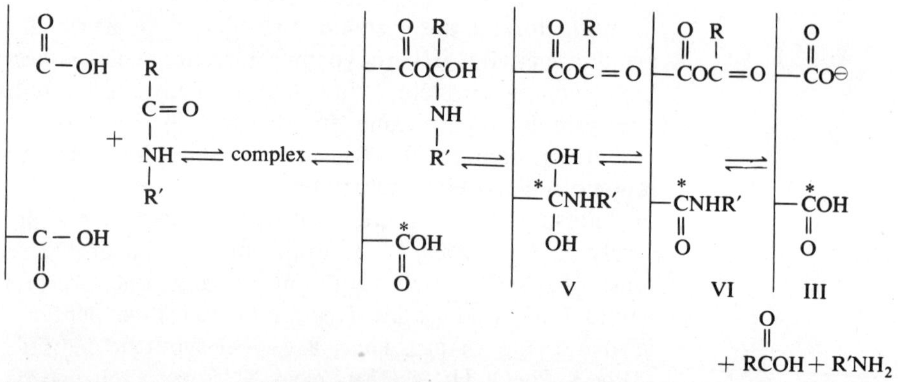

FIGURE 17-4

Possible reaction mechanism for pepsin-catalyzed reactions. Two carboxylic acid groups of the protein are considered to be directly involved; one forms an anhydride with the acyl portion of the substrate and the other an amide with the amine portion. Hydrolysis of these intermediates completes the reaction.

carboxylic group. It should be noted that this mechanism is speculative and that a search for an acyl-enzyme intermediate has proved fruitless (43).

Rennin. Rennin, an enzyme from the fourth stomach of the calf, has properties similar to those of pepsin. Rennin is composed of a single polypeptide chain with a molecular weight near 40,000 and an isoelectric point of 4.5. Its specificity is nearly identical to that of pepsin. The individuality of rennin was established by its crystallization and subsequent determination of its amino acid composition and other properties.

Chymotrypsin. The process of protein digestion, initiated in the stomach by pepsin, is continued in the duodenum (upper loop of the small intestine) by enzymes present in a pancreatic secretion. The enzymes, trypsin and chymotrypsin, possess distinctly similar properties.

Chymotrypsin occurs in the pancreatic juice as a mixture of two zymogens, chymotrypsinogens A and B. The former protein has been thoroughly characterized. It has a molecular weight of 25,000 and an isoelectric point near 9.1. Chymotrypsinogen A is composed of a single polypeptide chain whose amino acid sequence, together with that for trypsinogen, is indicated in Fig 17-5 (44).

The conversion of chymotrypsinogen to chymotrypsin is a complicated process but one whose essential features have recently been clarified by X-ray diffraction studies of these proteins. The sequence of reactions by which chymotrypsinogen is converted to α-chymotrypsin is outlined in Fig. 17-6. The crucial step in the sequence of reactions in the trypsin-catalyzed hydrolysis of the peptide bond between Arg^{15} and Ile^{16}, yielding π-chymotrypsin, a catalytically active molecule. Subsequent chymotrypsin-catalyzed reactions liberate two dipeptides and generate α-chymotrypsin having three peptide chains held together by disulfide bonds. The detailed structure of the molecule has been outlined in Chapter 4 and its catalytic activity has been discussed in Chapter 7.

The central question in the activation process is why the cleavage of a single peptide bond converts an inactive species into an active one. The crucial point is the location of Asp^{194} in the chymotrypsin molecule. Wide-angle X-ray diffraction studies have shown that, surprisingly, this charged group is found in a relatively hydrophobic region of the molecule. It is able

```
                        1   2   3   4   5   6   7   8   9   10  11  12  13  14  15  16  17  18
Chymotrypsinogen        Cys-Gly-Val-Pro-Ala-Ile-Gln-Pro-Val-Leu-Ser-Gly-Leu-Ser-Arg-ILE-VAL-GLY-
Trypsinogen                                                 Val-Asp-Asp-Asp-Asp-Lys-ILE-VAL-GLY-
                                                            1   2   3   4   5   6   7   8   9

19  20  21  22  23  24  25  26  27  28  29  30  31  32  33  34  35  36  37  38  39  40  41
Asp-Glu-Glu-Ala-Val-Pro-Gly-Ser-Trp-PRO-Trp-GLN-VAL-SER-LEU-Gln-Asp-Lys-Thr-GLY-Phe-HIS-PHE-
Gly-Tyr-Thr-Cys-Gly-Ala-Asn-Thr-Val-PRO-Tyr-GLN-VAL-SER-LEU-Asn-       -Ser-GLY-Tyr-HIS-PHE-
10  11  12  13  14  15  16  17  18  19  20  21  22  23  24  25          26  27  28  29  30

42  43  44  45  46  47  48  49  50  51  52  53  54  55  56  57  58  59  60  61  62  63  64
CYS-GLY-GLY-SER-LEU-ILE-ASN-Glu-Asn-TRP-VAL-VAL-Thr-ALA-ALA-HIS-CYS-Gly-Val-Thr-Thr-Ser-Asp-
CYS-GLY-GLY-SER-LEU-ILE-ASN-Ser-Gln-TRP-VAL-VAL-Ser-ALA-ALA-HIS-CYS-Tyr-Lys-Ser-Gly-Ile-Gln-
31  32  33  34  35  36  37  38  39  40  41  42  43  44  45  46  47  48  49  50  51  52  53

65  66  67  68  69  70  71  72  73  74  75  76  77  78  79  80  81  82  83  84  85  86  87
VAL-Val-Val-Ala-Gly-Glu-Phe-Asp-Gln-Gly-Ser-Ser-Ser-Glu-Lys-Ile-Gln-Lys-Leu-Lys-Ile-Ala-Lys-
VAL-Arg-Leu-Gly-Glu-Asp-Asn-Ile-Asn-Val-Val-Glu-Gly-Asp-Glu-Gln-Phe-Ile-Ser-Ala-Ser-Lys-Ser-
54  55  56  57  58  59  60  61  62  63  64  65  66  67  68  69  70  71  72  73  74  75  76

88  89  90  91  92  93  94  95  96  97  98  99  100 101 102 103 104 105 106 107 108 109 110
Val-Phe-Lys-Asn-SER-Lys-TYR-ASN-Ser-Leu-Thr-Ile-ASN-ASN-Asn-ILE-Thr-LEU-Leu-LYS-LEU-Ser-Thr-
Ile-Val-His-Pro-SER-   -TYR-ASN(Pro,Leu,Thr,Asn)ASN-ASN-Asp-ILE-Met-LEU-Ile-LYS-LEU-Lys-Ser-
77  78  79  80  81      82  83  84  85  86  87  88  89  90  91  92  93  94  95  96  97  98

111 112 113 114 115 116 117 118 119 120 121 122 123 124 125 126 127 128 129 130 131 132 133
ALA-ALA-SER-Phe-Ser-Gln-Thr-VAL-Ser-Ala-Val-Cys-LEU-PRO-Ser-Ala-Ser-Asp-Asp-Phe-Ala-ALA-GLY-
ALA-ALA-SER-Leu-Asn-Ser-Arg-VAL-Ala-Ser-Ile-Ser-LEU-PRO-Thr-Ser-Cys-       -Ala-Ser-ALA-GLY-
99  100 101 102 103 104 105 106 107 108 109 110 111 112 113 114 115         116 117 118 119

134 135 136 137 138 139 140 141 142 143 144 145 146 147 148 149 150 151 152 153 154 155 156
THR-Thr-CYS-Val-Thr-Thr-GLY-TRP-GLY-Leu-THR-Arg-Tyr-Thr-Asn-Ala-Asn-Thr-PRO-ASP-Arg-LEU-Gln-
THR-Gln-CYS-Leu-Ile-Ser-GLY-TRP-GLY-Asn-THR-Lys-Ser-Ser-Gly-Thr-Ser-Tyr-PRO-ASP-Val-LEU-Lys-
120 121 122 123 124 125 126 127 128 129 130 131 132 133 134 135 136 137 138 139 140 141 142

157 158 159 160 161 162 163 164 165 166 167 168 169 170 171 172 173 174 175 176 177 178 179
Gln-Ala-Ser-Leu-PRO-Leu-LEU-SER-Asn-Thr-Asn-CYS-LYS-Lys-Tyr-Trp-Gly-Thr-Lys-ILE-Lys-Asp-Ala-
Cys-Leu-Lys-Ala-PRO-Ile-LEU-SER-Asp-Ser-Ser-CYS-LYS-Ser-Ala-Tyr-Pro-Gly-Gln-ILE-Thr-Ser-Asn-
143 144 145 146 147 148 149 150 151 152 153 154 155 156 157 158 159 160 161 162 163 164 165

180 181 182 183 184 185 186 187 188 189         190 191 192 193 194 195 196 197 198 199 200
MET-Ile-CYS-ALA-GLY-Ala-Ser-Gly-Val-Ser-       -SER-CYS-Met-GLY-ASP-SER-GLY-GLY-PRO-Leu-VAL-
MET-Phe-CYS-ALA-GLY-Tyr-Leu-Glu-Gly-Gly-Lys-Asn-SER-CYS-Gln-GLY-ASP-SER-GLY-GLY-PRO-Val-VAL-
166 167 168 169 170 171 172 173 174 175 176 177 178 179 180 181 182 183 184 185 186 187 188

201 202 203 204 205 206 207 208 209 210 211 212 213 214 215 216 217 218 219 220 221 222 223
CYS-Lys-Lys-Asn-Gly-Ala-Trp-Thr-Leu-Val-GLY-ILE-VAL-Ser-SER-TRP-GLY-SER-Ser-Thr-Cys-Ser-Thr-
CYS-Ser-Gly-Lys-Leu-Gln-               -GLY-ILE-VAL-   -SER-TRP-GLY-SER-Gly-Cys-Ala-Gln-Lys-
189 190 191 192 193 194                 195 196 197     198 199 200 201 202 203 204 205 206

224 225 226 227 228 229 230 231 232 233 234 235 236 237 238 239 240 241 242 243 244 245 246
Ser-Thr-PRO-GLY-VAL-TYR-Ala-Arg-VAL-Thr-Ala-Leu-VAL-Asn-TRP-Val-Gln-GLN-THR-Leu-ALA-Ala-ASN
Asn-Lys-PRO-GLY-VAL-TYR-Thr-Lys-VAL-Cys-Asn-Tyr-VAL-Ser-TRP-Ile-Lys-GLN-THR-Ile-ALA-Ser-ASN
207 208 209 210 211 212 213 214 215 216 217 218 219 220 221 222 223 224 225 226 227 228 229
```

FIGURE 17-5

Structural similarity of trypsinogen and chymotrypsinogen. The sequence 84–87 in trypsinogen has not yet been established. Gaps in the sequence have been arbitrarily inserted to draw attention to the resulting areas of homology. Amino acid residues that are homologous in the two proteins are shown in capital letters. K. A. Walsh and H. Neurath, *Proc. Natl. Acad. Sci., U.S.,* **52:** 884 (1964).

to maintain itself in this hostile environment only through the formation of an ion pair with the positively charged α-amino function of Ile[16], the N terminus of the B chain of chymotrypsin (45). But it is precisely this N terminus that is generated in the conversion of chymotrypsinogen to π-chymotrypsin (Fig. 17-6). Thus, in chymotrypsinogen this ion pair cannot be formed and Asp[194] must occupy a position in a more polar environment. When we recall that Asp[194] is adjacent to Ser[195], which is crucially involved in the whole catalytic process (page 371), the significance of these observations is clear. The proper positioning of Ser[195] with respect to His[57] depends on the formation of the ion pair between Asp[194] and Ile[16]; since that ion pair can only be formed when Ile[16] can possess a positive charge, it follows that chymotrypsinogen cannot be active but that the family of chymotrypsins can be. These points are developed diagrammatically in Fig. 17-7.

These conclusions based on X-ray diffraction work have received elegant confirmation from chemical studies. Specifically, it has been demonstrated that chymotrypsin is catalytically active only if the α-amino group of Ile[16] is protonated; formation of the productive enzyme–substrate complex de-

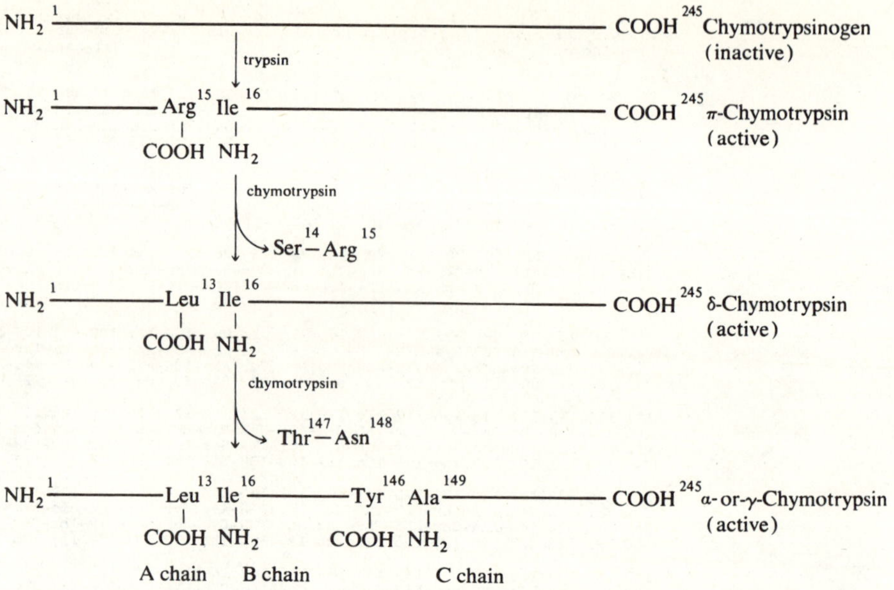

NH_2 1 ———————————————— COOH 245 Chymotrypsinogen (inactive)

↓ trypsin

NH_2 1 ——— Arg^{15} Ile^{16} ——— COOH 245 π-Chymotrypsin (active)
(Arg–COOH; Ile–NH_2)

↓ chymotrypsin → Ser^{14}—Arg^{15}

NH_2 1 ——— Leu^{13} Ile^{16} ——— COOH 245 δ-Chymotrypsin (active)
(Leu–COOH; Ile–NH_2)

↓ chymotrypsin → Thr^{147}—Asn^{148}

NH_2 1 ——— Leu^{13} Ile^{16} ——— Tyr^{146} Ala^{149} ——— COOH 245 α-or-γ-Chymotrypsin (active)
(Leu–COOH; Ile–NH_2; Tyr–COOH; Ala–NH_2)

A chain B chain C chain

FIGURE 17-6

Activation of chymotrypsinogen. The crucial step is cleavage of an arginyl-isoleucine peptide bond by trypsin to form enzymatically active π-chymotrypsin. Subsequent cleavages by chymotrypsin give rise to α-chymotrypsin, which consists of three polypeptide chains connected by two interchain disulfide bonds.

pends on the protonated form of a group with pK_a 8.5 (46,47). Moreover, acetylation of this amino group, which abolishes its capacity to assume a positive charge, renders chymotrypsin inactive even though all the other amino groups can be acetylated without loss of activity (46). The importance of Ile^{16} in determining the active structure of chymotrypsin is thus well demonstrated.

α-Chymotrypsin, like pepsin, is an endopeptidase. In addition, this enzyme catalyzes the hydrolysis of esters, amides, hydroxamates, and other acyl derivatives. Although chymotrypsin catalyzes the hydrolysis (and aminolysis and alcoholysis) of a broad spectrum of substrates, activity is maximal when the carboxyl function of the peptide linkage is donated by an aromatic amino acid.

Trypsin. The structural similarities of trypsin and chymotrypsin are emphasized by the striking homologies in amino acid sequences of the zymogens (Fig. 17-5). Specific similarities between the two enzymes include (1) site of origin, (2) rate of synthesis, (3) endopeptidase activity, (4) molecular weights, (5) isoelectric point, (6) amino acid compositions, and (7) nature of the catalytic processes (44). In regard to the last point, most of the comments concerning the mechanism of chymotryptic reactions apply with equal force to trypsin as well. The principal difference is the enzymatic specificity. Trypsin is maximally active on peptides whose carboxyl group is donated by a basic residue, specifically by lysine or arginine. In fact, trypsin is the most discriminating of the proteolytic enzymes in terms of the scope of susceptible peptide linkages. Good use of this fact has been made by protein chemists interested in sequencing polypeptides, who have widely employed trypsin as a reagent for the orderly cleavage of such molecules (page 116).

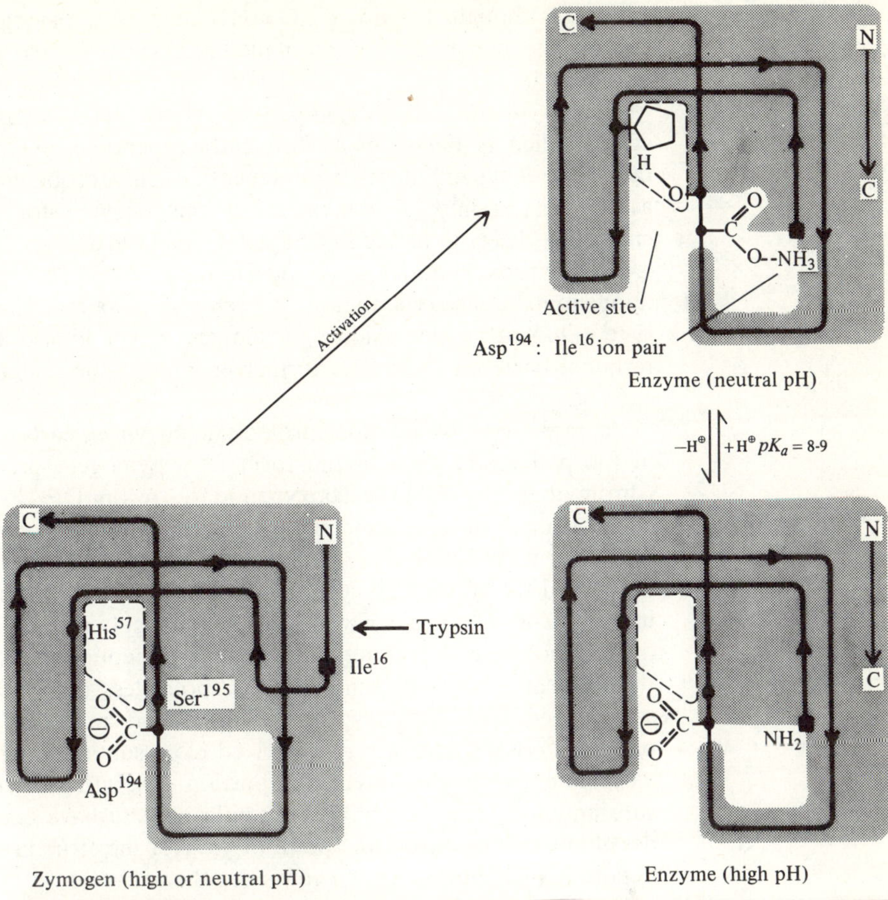

FIGURE 17-7
Diagrammatic representation of the activation process converting chymotrypsinogen into π-chymotrypsin.

It is true that some noncationic substrates have been found for trypsin (48). However, it does seem safe to conclude that the cationic group of lysyl and arginyl peptides makes an important interaction with an anionic site in the trypsin molecule.

The trypsinogen–trypsin conversion is simpler than the corresponding chymotrypsinogen reaction in that only one product is formed. The activation can be catalyzed by trypsin itself, by the enzyme *enterokinase,* and also by certain proteinases derived from molds. Activation involves cleavage of the N-terminal hexapeptide from the polypeptide chain. This N-terminal peptide has the sequence Val·Arg·Arg·Arg·Arg·Lys· suggesting that electrostatic interactions between this strongly cationic entity and anionic sites elsewhere in the trypsin molecule may be responsible for maintaining trypsinogen in an inactive conformation. Release of these electrostatic interactions would then permit the active trypsin molecule to be formed. Calcium ion plays an important role in the activation process, both accelerating cleavage at the pertinent peptide linkage and retarding cleavage at other sites. Cleavage at the latter sites yields enzymatically inactive products.

Enterokinase is a proteolytic enzyme elaborated by the duodenal mucosa. It is a glycoprotein and has not yet been completely characterized. This enzyme is thought to initiate the conversion of trypsinogen to trypsin, speeding up the conversion of all pancreatic proenzymes to the corresponding enzymes.

Carboxypeptidase. Accompanying trypsin and chymotrypsin in the pancreatic juice is the *exopeptidase* carboxypeptidase. This enzyme (really a family of enzymes) degrades polypeptides in a sequential fashion, beginning at the C terminus. Employment of this enzyme for amino-acid-sequence studies has been discussed in Chapter 4 (page 116).

Two carboxypeptidases, A and B, are involved in protein digestion in the duodenum. Carboxypeptidase A exhibits maximal catalytic activity with C-terminal aromatic residues, while carboxypeptidase B is specific for C-terminal basic residues. The former enzyme is the better understood and we consider it first.

As in the case of the other proteolytic enzymes, carboxypeptidase A occurs in the pancreatic juice in the form of a zymogen, procarboxypeptidase A. Chromatography of procarboxypeptidase A on DEAE-cellulose yields two components, one characterized by a sedimentation constant of 6 S and the other by a value of 5 S. The heterogeneity of procarboxypeptidase A is accounted for on the basis of a trimer structure for this molecule, as indicated in Fig. 17-8. The heavier zymogen contains subunits I, II, and III, while the lighter molecule is lacking III. Procarboxypeptidase A-S6 has a molecular weight near 87,800 and an isoelectric point less than 4.5. Activation of procarboxypeptidase A by trypsin is diagrammed in Fig. 17-8. In the activation process, carboxypeptidase A is derived from subunit I, which is degraded to a variable extent in the N-terminal portion generating three proteins exhibiting substantially the same catalytic activities. These have been discussed in some detail in Chapter 7 from the point of view of the structure of the active site as detailed in Chapter 4. In the course of the activation process both subunits II and III are degraded; the former yields an endopeptidase exhibiting chymotrypsin-like activity, which is involved in the conversion of subunit I into carboxypeptidase A. The function of subunit III is unknown.

The chemistry of the activation of procarboxypeptidase has been simplified a good deal by the observation that treatment of the intact protein with succinic anhydride leads to spontaneous disaggregation of the succinylated subunits (49). Succinylated subunit I may be recovered independent of the others, and its conversion into succinylated carboxypeptidase A, which is catalytically active, is susceptible to study. The succinylated subunit I, molecular weight 42,000, is converted by trypsin into succinylated carboxypeptidase A, molecular weight 34,000, with the release of low-molecular-weight products totaling about 60 amino acid residues. The liberated products are particularly rich in the dibasic amino acids, suggesting at least a superficial resemblance to the activation of trypsin (page 767).

Carboxypeptidase B, specific for terminal peptide bonds involving lysine or arginine residues, is probably a metalloenzyme as well. This conclusion follows from its inhibition by chelating agents such as 1,10-phenanthroline. The nature of the metal ion present remains to be established.

Leucine Aminopeptidase. Leucine aminopeptidase, an enzyme found in the intestinal mucosa and other tissues, is an exopeptidase, which degrades polypeptide chains sequentially from the N terminus (page 116). The enzyme from swine kidney has been highly purified, although not crystallized.

Leucine aminopeptidase is activated by either magnesium or manganese

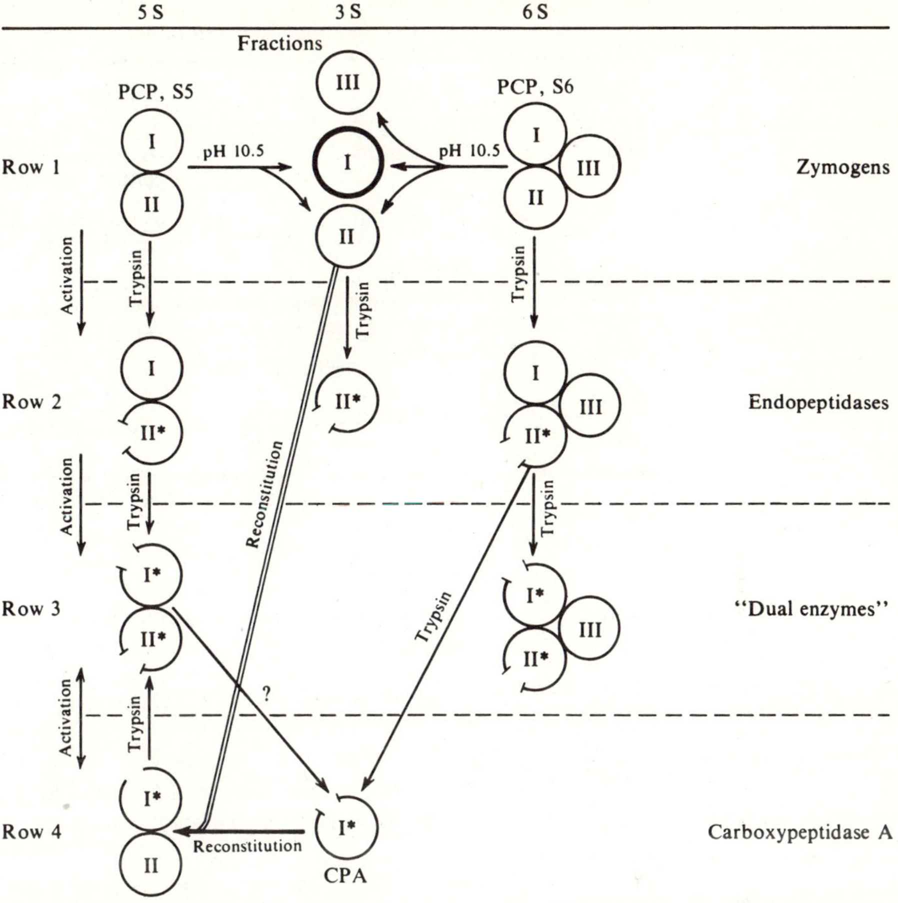

FIGURE 17-8
Scheme representing the disaggregation and activation of procarboxypeptidase A. H. Neurath, *Federation Proc.*, **23:** 1 (1964).

ions. The enzyme exhibits both esterase and peptidase activity and hydrolyzes a broad range of substrates, which, if they possess an asymmetric α-carbon atom, must be of the L configuration. As the name of the enzyme implies, hydrolysis is most rapid in the case of N-terminal leucine. The relative rates at which substrates are cleaved is largely independent of the nature of the activating ion.

In addition to leucine aminopeptidase and the carboxypeptidases, the intestinal mucosa contains a variety of additional peptidases. Many of these exhibit specificity for chain length of the substrate, particularly the dipeptidases and tripeptidases. Such enzymes must function to complete the digestion of proteins begun by the enzymes previously discussed. A summary of the enzymes involved in the digestive process in man is presented in Table 17-1.

Proteolytic Enzymes of Animal Tissues

A group of proteolytic enzymes, the cathepsins, are localized intracellularly in kidney, spleen, liver, and lung tissue, among other sites. The cathepsins are endopeptidases that hydrolyze both proteins and small synthetic substrates. The substrate specificities of cathepsins A, B, and C are similar to those of pepsin, trypsin, and chymotrypsin, respectively. These enzymes have not been

TABLE 17-1
Proteolytic enzymes in the mammalian gastrointestinal tract

Secretion	*Enzyme*	*Comments*
Gastric	Pepsin	A proteinase also found in the gastric juice of birds, reptiles, and fish
	Rennin	A milk-coagulating enzyme present in the juice of the fourth stomach of the calf
Pancreatic	Trypsin	A proteinase
	Chymotrypsin	A proteinase
	Carboxypeptidase	A peptidase
Intestinal	Aminopeptidases Prolidase Tripeptidase Dipeptidases	Peptidases

SOURCE: J. S. Fruton and S. Simmonds, *General Biochemistry,* Wiley, New York, 1958, p. 685.

obtained in sufficiently pure states to permit investigation of their molecular and catalytic properties.

Some Proteolytic Enzymes Involved in Blood Clotting

Many of the functional aspects of the proteolytic enzymes discussed above also are characteristic of the factors involved in the clotting of blood. Formation of the blood clot involves the conversion of the soluble plasma protein *fibrinogen,* molecular weight 340,000, to the insoluble protein *fibrin* (50,51). This conversion is catalyzed by a proteolytic enzyme, *thrombin,* which has been the subject of a careful mechanistic study (52). Generally, thrombin appears to behave like many of the other common proteases with formation of an intermediate acyl-enzyme. Thrombin itself is generated from a proenzyme, prothrombin, through the action of another proteolytic enzyme, proaccelerin, which is, in turn derived from its precursor, and so on. In fact, there are known to be eight stages in the overall sequence: The lesion induces the conversion of Hageman factor into active Hageman factor, which activates plasma thromboplastin antecedent, which activates Christmas factor, which activates antihemophilic factor, which activates Stuart factor, which activates

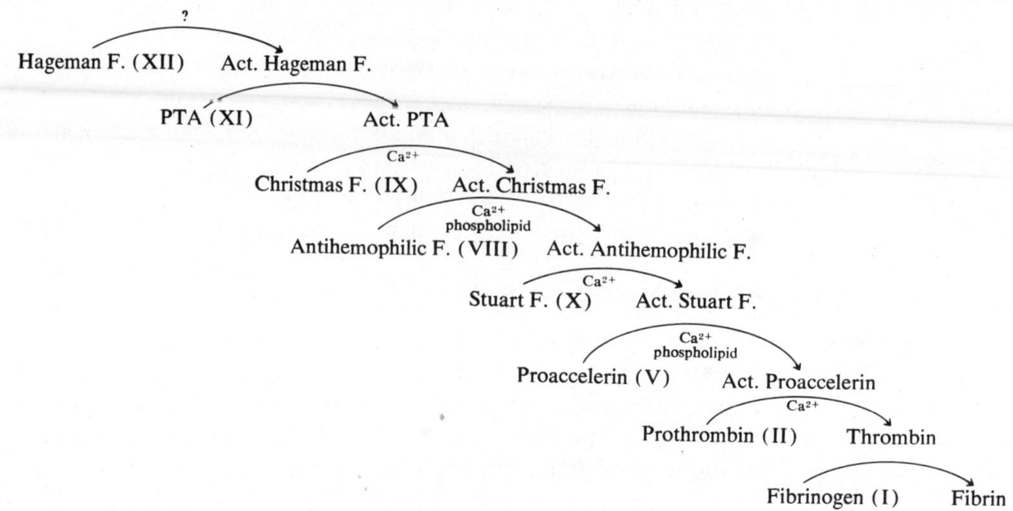

proaccelerin, which activates prothrombin, which converts fibrinogen into fibrin.

Many of these factors have been purified and shown to be proteolytic enzymes; most are mucoproteins. Thus, we have a cascade of proenzyme to enzyme conversions, each step generating the catalyst for the subsequent one. The cascade of reactions acts as a kind of chemical amplification system leading to a rapid and forceful production of fibrin when necessary and, at the same time, permitting control of the process at many stages.

There are a number of pathological states known in which one of the necessary factors for blood clotting is missing. The majority of the cases of hemophilia reflect an absence of antihemophilic factor. A discussion of a variety of inherited diseases in which a particular protein is missing is provided on page 857.

The liquefaction of blood clots is catalyzed by the fibrinolytic enzyme *plasmin.* Like thrombin, this enzyme is similar to trypsin in terms of its specificity. Plasmin normally occurs in the form of a zymogen, *plasminogen,* molecular weight 83,500.

Plant Proteinases

Papain. Papain is a proteolytic enzyme from the fruit of the tropical melon tree, *Carica papaya.* This enzyme occurs in high concentration in the papaya latex, from which it may be crystallized in large amounts. Papain is composed of a single polypeptide chain, whose amino acid sequence is known. The molecular weight of this enzyme is near 21,000 and its isoelectric point is 8.75.

Recent X-ray diffraction work has provided a detailed three-dimensional structure for this enzyme (53). In its general aspects, it is typical of those proteins discussed in Chapter 4; in one central aspect, the detailed structure has provided insight into the catalytic process differing from views held earlier based on kinetic evidence. This deals with the involvement of a carboxyl group in the catalytic process (see below).

Papain exhibits a broad substrate specificity. Peptides, amides, esters, and thiolesters are all susceptible to papain-catalyzed hydrolysis, and a variety of nucleophilic reagents, in addition to water, will serve as acyl group acceptors. Peptides involving a variety of amino acids are effectively cleaved by papain, although the constituent amino acids must be of the L configuration.

Studies on the binding and hydrolysis of polyglycines and polyalanines by papain indicate that the binding site of this molecule extends over some 25 Å and accommodates six or seven amino acid residues (54,55). Thus, papain substrates may be considered to fit into a long groove on the surface of papain, as in the case of lysozyme (page 368). This point of view is borne out by the X-ray diffraction studies.

A good deal of evidence has implicated a thiol group of cysteine as essential for catalytic activity of papain: (1) papain readily complexes with a variety of heavy metal ions with loss of enzymatic activity; (2) reagents that are capable of binding heavy metal ions and of reducing disulfide functions serve as activators of papain; (3) papain reacts stoichiometrically with a variety of thiol binding agents in a one-to-one reaction, yielding enzymatically inactive products; (4) the *p*H-rate profiles for papain-catalyzed hydrolyses implicate a group in the catalytic activity with a pK_a near 8.5. Considerable evidence now suggests that the crucial thiol group is transiently acylated in the course of papain-catalyzed reactions (56). This conclusion was first based on spectrophotometric detection of an

$$R-\overset{\overset{\displaystyle S}{\|}}{C}-S-$$

enzyme in the course of hydrolysis of substrates of the structure

$$R-\overset{\overset{\displaystyle S}{\|}}{C}-O-R$$

This conclusion has been substantiated by a number of kinetic studies employing a variety of substrates (57,58).

Studies of the dependence of the rate of papain-catalyzed hydrolyses on *p*H long ago suggested the involvement of a carboxyl group in the catalytic process; various roles may be suggested for it. However, the elucidation of the active site of papain in the X-ray diffraction studies indicates that no carboxyl group is present there. Therefore, a distant carboxyl group must exercise control over the conformation of the active site, depending on its state of ionization. This case provides a very nice example of the uncertainty of drawing mechanistic conclusions on the basis of kinetic data alone.

Ficin. An enzyme from plants of the fig family, ficin, is an endopeptidase similar to papain. Ficin has a molecular weight near 26,000 and, like papain, is a sulfhydryl enzyme. *p*H-rate profiles implicate a carboxylate ion or the kinetic equivalent in the catalytic process of this enzyme as well. Ficin is maximally active with substrates derived from L-arginine, but exhibits catalytic activity toward a broad spectrum of substrates.

Other important proteolytic enzymes of plants include *bromelin* from pineapple, *chymopapain* from papaya latex, and *asclepain* from the roots and latex of the milkweed.

Bacterial Proteinases (59)

Many proteolytic enzymes have been isolated from bacterial sources. As a group, these enzymes exhibit both exopeptidase and endopeptidase activity and are characterized by a strikingly broad specificity (or lack of it). In extreme cases, certain bacterial proteinases may cleave as many as 80 percent of the peptide bonds of a protein. The properties of a number of these enzymes are summarized in Table 17-2.

BIOSYNTHESIS OF THE AMINO ACIDS

Organisms differ markedly in terms of their ability to carry out the *de novo* synthesis of the protein-containing amino acids. Most microorganisms and plants are competent in all such syntheses, while most animals lack about

TABLE 17-2
Properties of some bacterial proteinases

Name and source	*Molecular weight*	*Isoelectric point*	*Inhibitors*
Subtilisin BPN′ (*Bacillus subtilis* strain N′)	27,600	7.80	DFP
B. mesentericus protease	32,400		
B. subtilis protease	1.7×10^5 (27,600 subunits)		DFP
Streptococcal protease		8.2	Thiol reagents
Pseudomonal protease	77,000	5.5–6.0	

half of these synthetic capacities. For the latter organisms, amino acids may then be classified as "essential" or "nonessential" (page 622).

The distinction between essential and nonessential amino acids may be made on the basis of a variety of experimental criteria. On the basis of those most frequently employed, an amino acid is considered essential if it must be included in the diet for (1) optimal growth of an immature animal or (2) for the maintenance of nitrogen balance in the mature organism. An animal is said to be in nitrogen balance if its daily intake of nitrogen is just balanced by its daily excretion of nitrogen. The classical studies of Rose have established that, for the white rat, leucine, isoleucine, valine, lysine, methionine, phenylalanine, tryptophan, threonine, histidine, and arginine are essential amino acids. The first eight of these amino acids have been found to be essential for all species of higher animals studied. In contrast to the white rat, the human organism can synthesize (or recover from protein breakdown) sufficient histidine for optimal growth and maintenance of nitrogen balance. Among the essential amino acids for the white rat, only the carbon chains of arginine and threonine can be synthesized at all (their synthesis is simply too slow to permit optimal growth). This organism has lost, therefore, through evolution, the capacity to accomplish the synthesis of the carbon chains of the remaining essential amino acids.

The dietary requirement for a particular amino acid is a function of several variables, including the presence of metabolically related substances in the diet. For example, tyrosine exerts a sparing effect on phenylalanine, as does glutamic acid on arginine. In the rat, the requirement for methionine may be satisfied by homocysteine plus an adequate supply of methyl group donors. Thus, the "essentiality" of an amino acid is not only a function of the criteria employed for its determination but also of the other components of the diet. In addition, amino acid requirements vary with the physiological state of the animal (i.e., in pregnancy, lactation, and disease), with age, and, perhaps, with the nature of the intestinal flora.

It is worth emphasizing that the "essential" feature of the essential amino acids is their carbon skeleton; most of these amino acids may be derived from the corresponding keto acids through transamination (pages 401 and 790).

Assimilation of Ammonia: The Synthesis of Glutamic Acid and Glutamine

As indicated above, the assimilation of molecular nitrogen and nitrate, the principal inorganic nitrogen sources, yields ammonia as the immediate product. How is ammonia incorporated into organic nitrogen compounds? Although a number of reactions result in such incorporation, three processes appear to be quantitatively most important in many species. These are the reactions catalyzed by glutamic dehydrogenase, glutamine synthetase, and carbamyl phosphate synthetase.

Glutamic dehydrogenase [L-glutamate : NAD oxidoreductase (deaminating)] catalyzes the pyridine nucleotide-linked formation of glutamic acid from α-oxoglutarate and ammonia:

$$\left.\begin{matrix}\text{NADH}\\ \text{(NADPH)}\end{matrix}\right\} + \alpha\text{-oxoglutarate} + NH_4^+ \rightleftharpoons \left.\begin{matrix}\text{NAD}^+\\ \text{(NADP}^+)\end{matrix}\right\} + \text{L-glutamate} + H_2O \quad (17\text{-}6)$$

This enzyme is widely distributed throughout living systems. Either NADPH or NADH is employed as coenzyme—the former usually being employed in cases in which glutamic dehydrogenase functions in a biosynthetic capacity and the latter when this enzyme serves principally a catabolic function. Some of the aspects of the structural organization of this interesting enzyme have been considered in Chapter 4, and its integration with the citric acid cycle in Chapter 14.

The principal alternative route for the biosynthesis of glutamate is the pyridoxal phosphate-dependent *transamination* involving α-oxoglutarate (see Chapter 8 for discussion of mechanisms involved in transaminations):

$$\alpha\text{-Oxoglutarate} + \text{amino acid} \xrightleftharpoons{\text{pyridoxal phosphate}} \text{L-glutamate} + \text{ketoacid} \qquad (17\text{-}7)$$

The interconversion of α-oxoglutarate and glutamate is an important link directly coupling the metabolism of amino acids and that of the carbohydrates. This reaction clearly provides for the incorporation of carbon derived from glucose into glutamate and, subsequently, into other amino acids.

The second important reaction incorporating ammonia into organic compounds is the ATP-dependent *formation of glutamine:*

$$\text{L-Glutamate} + NH_3 + \text{ATP} \xrightleftharpoons{Mg^{2+}} \text{L-glutamine} + \text{ADP} + P_i \qquad (17\text{-}8)$$

Glutamine synthetase (L-glutamate : ammonia ligase) is a most interesting enzyme both in terms of its structure and metabolic regulation. The enzyme from *E. coli* has a molecular weight near 600,000 and is composed of 12 identical subunits arranged in two hexagonal rings, one layered upside down on top of the other (60,61). The stability of this structure is strongly dependent on the presence of divalent ions. Removal of these ions converts the enzyme into a "relaxed" state which is enzymatically inactive and susceptible to disaggregation. Disaggregated subunits can be reassembled into a structurally and functionally intact enzyme by addition of manganese ions.

The metabolic regulation of glutamine synthetase is complicated, reflecting the fact that the reaction product, glutamine, is involved as a donor of nitrogen atoms for the synthesis of eight end products, including carbamyl phosphate, glycine, alanine, histidine, cytidine triphosphate, AMP, tryptophan, and glucosamine. Each of the subunits of the intact enzyme possesses binding sites for each of the eight products listed above, which act as feedback inhibitors, as well as for the three substrates. Moreover, the synthesis of this enzyme is subject to repression and derepression.

In addition to these regulatory mechanisms, an additional one exists based upon the adenylylation of glutamine synthetase. In *E. coli,* two enzymes exist, one of which is capable of adenylylation of the enzyme:

$$\text{Glutamine synthetase} + \text{ATP} \rightarrow (\text{glutamine synthetase} \cdot \text{AMP}) + PP_i \qquad (17\text{-}9)$$

and the second of deadenylylation of the enzyme. Two extreme forms exist, one which contains no AMP and one which contains one AMP per subunit. The former specifically requires Mg^{2+} and the latter specifically requires Mn^{2+}. The enzymes also differ in their enzymic activity and susceptibility to feedback inhibition. Thus, histidine, tryptophan, and CTP strongly inhibit the adenylylated enzyme but not the deadenylylated one. On the other hand, glycine and alanine inhibit the latter enzyme more strongly than the former. Regulation of glutamine synthetase through the extent of adenylylation is strongly reminiscent of regulation of phosphorylase through phosphorylation (page 501).

Finally, the *carbamyl phosphate synthetase* (ATP : carbamate phosphotransferase) reaction is an important means for the incorporation of ammonia into several biosynthetic products, notably the pyrimidines and urea. Two types of synthetases are known; one utilizes the amide nitrogen of glutamine as the source of carbamyl-nitrogen and the other uses ammonia:

$$\text{Glutamine} + 2\text{ATP} + HCO_3^- \rightarrow \text{glutamic acid} + 2\text{ADP} + P_i + NH_2\text{—}\overset{\overset{\displaystyle O}{\|}}{C}\text{—}OPO_3H^- \quad (17\text{-}10a)$$

$$NH_3 + 2\text{ATP} + HCO_3^- \rightarrow NH_2\text{—}\overset{\overset{\displaystyle O}{\|}}{C}\text{—}OPO_3H^- + 2\text{ADP} + P_i \quad (17\text{-}10b)$$

The microbial enzymes use glutamine as nitrogen donor; the ammonia-requiring enzymes have been found no lower on the evolutionary scale than the earthworm. These reactions are considered in more detail on page 827.

Amino Acids Derived from Glutamate

The route for the transformation of glutamate to *proline* is indicated in Fig. 17-9. Glutamic-γ-semialdehyde was identified as a key intermediate for proline synthesis through the finding that it could replace proline as a growth requirement for certain mutants of *E. coli* that had lost the capacity for the complete *de novo* synthesis of this amino acid (62,63). The details of the reaction yielding this intermediate are obscure. Glutamic-γ-semialdehyde in aqueous solutions exists in equilibrium with the ring-closure product Δ^1-pyrroline-5-carboxylic acid. Reduction of the semialdehyde completes the synthesis of proline. The overall pathway from glutamate to proline is reversible in those organisms that contain proline oxidase. Similar pathways exist in mammalian systems.

In animal tissues, 4-hydroxy-L-proline occurs principally in collagen. The biosynthesis of collagen hydroxyproline is incompletely understood. The following facts concerning this synthesis do, however, seem clear.

1. The oxygen atom of the hydroxyproline hydroxyl function is derived

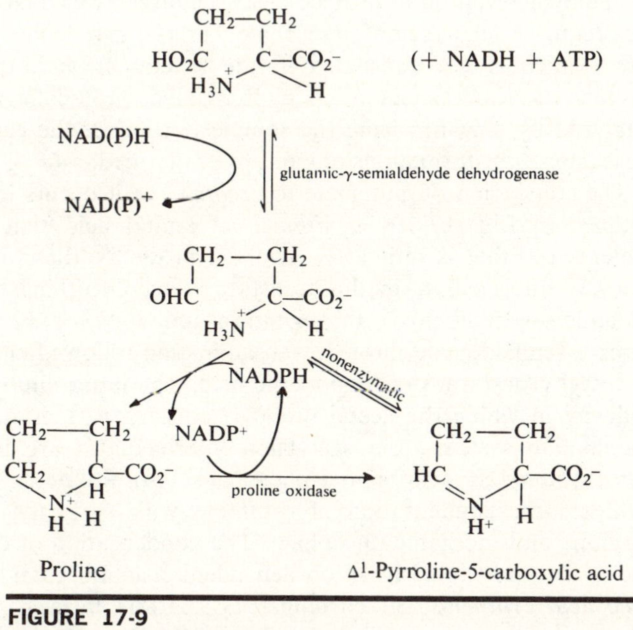

FIGURE 17-9
Biosynthesis of proline in *E. coli*.

from atmospheric oxygen, as revealed by isotope incorporation experiments (64).

2. Proline is a precursor of collagen hydroxyproline. This conclusion follows from experiments in which the administration of labeled proline to rats yielded tissue proteins containing highly labeled hydroxyproline.

3. Free proline is not the substrate for the hydroxylation reaction; that is, free hydroxyproline is not an intermediate in the synthesis of collagen hydroxyproline. This was initially suggested by the work of Stetten, who demonstrated that the incorporation of isotopic label into tissue hydroxyproline was much more extensive following the administration of labeled proline than that following the administration of labeled hydroxyproline. Experiments in a cell-free microsomal system from chick embryos that incorporates labeled proline into labeled peptide-bound hydroxyproline reveal that the appearance of this amino acid lags well behind the incorporation of proline (65). This result suggests that peptide-bound proline is the actual substrate for the hydroxylation reaction.

4. Synthesis of collagen hydroxyproline is catalyzed by an enzyme system associated with the microsomal fraction of the cell.

These observations suggest that collagen hydroxyproline synthesis occurs via a minimal reaction scheme involving three stages:

Proline → protocollagen → hydroxylated protocollagen → collagen

Recent observations indicate that this route is correct (66–70). Studies of collagen hydroxyproline synthesis in the presence of puromycin, which prevents chain elongation during protein synthesis, reveal that large polypeptide fragments are hydroxylated but that small ones are not. This provides strong circumstantial evidence for the intervention of protocollagen intermediates, which are the substrates for the hydroxylation itself. The best current estimates suggest that protocollagens have molecular weights from 10,000 to about 100,000 (69).

The hydroxylation of protocollagen, catalyzed by collagen proline hydroxylase, requires ferrous ions, ascorbate (or other reducing agent), α-oxoglutarate (which is not consumed in the reaction), and, of course, molecular oxygen (70). Conversion of hydroxylated protocollagen into collagen is catalyzed by enzymes from the soluble portion of the cell. The chemistry of these latter transformations remains to be clarified.

The conversion of glutamate to *arginine,* as it occurs in microorganisms, is outlined in Fig. 17-10. *Ornithine*, an amino acid that does not occur in proteins but that is intimately involved in the synthesis of urea (page 809), is a key intermediate in this transformation. Ornithine may arise (1) from glutamic-γ-semialdehyde through transamination or (2) from *N*-acetylglutamic-γ-semialdehyde through transamination followed either by deacetylation or acetyl group transfer to glutamic acid. The latter alternative yields a cyclic pathway in which the acetyl group is conserved. It is not clear whether, in mammalian systems, the acetylated intermediates are important. *Ornithine transcarbamylase* (carbamoyl phosphate : L-ornithine carbamoyl-transferase) catalyzes the condensation of ornithine with carbamyl phosphate, yielding citrulline and inorganic phosphate. The condensation of citrulline with aspartate is an ATP and Mg^{2+} ion-dependent reaction catalyzed by *argininosuccinic acid synthetase* (L-citrulline : L-aspartate ligase). This reaction yields AMP and pyrophosphate, as well as argininosuccinic acid. The reversible

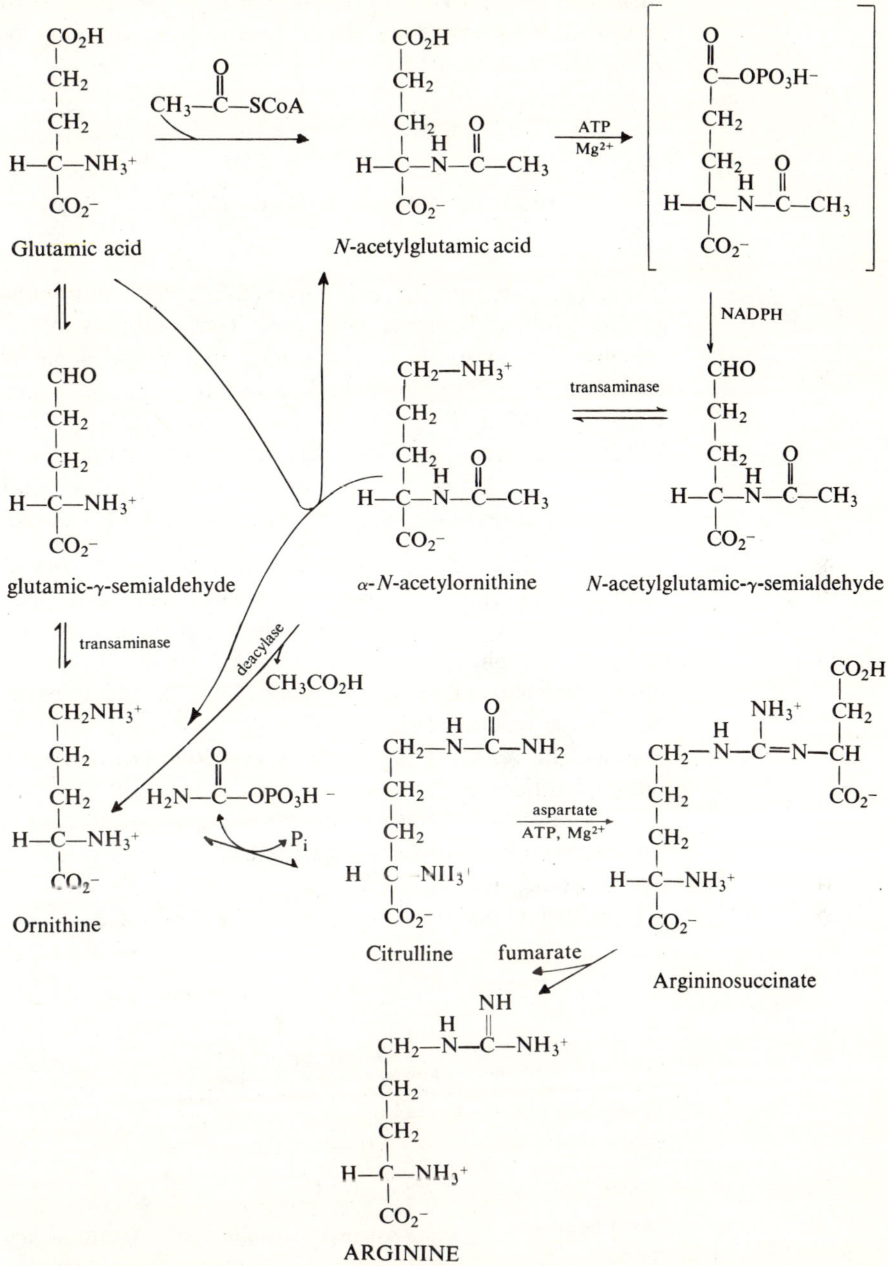

Glutamic acid → (CH_3—CO—SCoA) → *N*-acetylglutamic acid → (ATP, Mg^{2+}) → [acyl phosphate] → (NADPH) → *N*-acetylglutamic-γ-semialdehyde ⇌ (transaminase) α-*N*-acetylornithine → (deacylase, CH_3CO_2H) → Ornithine

Glutamic acid ⇌ glutamic-γ-semialdehyde ⇌ (transaminase) Ornithine

Ornithine + H_2N—CO—OPO_3H^- → (P_i) → Citrulline → (aspartate, ATP, Mg^{2+}) → Argininosuccinate → (fumarate) → ARGININE

FIGURE 17-10
Biosynthesis of arginine in microorganisms.

elimination of fumarate from the last substrate, catalyzed by *argininosuccinase* (L-argininosuccinate arginine-lyase), completes the synthesis of arginine.

Biosynthesis of Aspartic Acid and Asparagine

Biosynthetic routes leading to *aspartic acid* closely parallel those for formation of glutamic acid. Thus aspartic acid may be derived from oxalacetic acid by transamination:

$$\text{Oxalacetic acid} + \text{amino acid} \xrightleftharpoons{\text{pyridoxal phosphate}} \text{L-aspartate} + \text{ketoacid} \qquad (17\text{-}11)$$

Such reactions provide an additional important link between amino acid and carbohydrate metabolism. In addition, aspartate may be derived by direct amination of fumarate,

$$^{-}O_2C{-}CH{=}CH{-}CO_2^- + NH_3 \rightleftharpoons H_3^+N{-}\underset{H}{\overset{CO_2^-\!-CH_2}{C}}{-}CO_2^- \tag{17-12}$$

a reaction catalyzed by *aspartase* (L-aspartate ammonia-lyase), one of the earliest studied bacterial enzymes. The synthesis of *asparagine* in microorganisms is catalyzed by asparagine synthetase. This enzyme catalyzes the condensation of ammonia and aspartate with the concomitant cleavage of ATP to AMP and PP_i (in contrast to the glutamine synthetase reaction, in which ADP and P_i are formed). Asparagine synthetase from higher animals employ glutamine as nitrogen donor rather than ammonia (71). In plants, asparagine arises from the enzymatic hydrolysis of β-cyanoalanine.

Amino Acids Derived from Aspartate

The biosynthesis of *threonine, methionine,* and *lysine* all involve the intermediate formation of aspartic-β-semialdehyde. This intermediate is derived from aspartate in a two-step reaction sequence. First, *aspartokinase* (ATP : L-aspartate 4-phosphotransferase) catalyzes the formation of β-aspartyl phosphate, which is subsequently reduced to the aldehyde in a pyridine nucleotide–linked reaction (Fig. 17-11).

Since the reaction catalyzed by aspartokinase occupies a key position for the biosynthesis of three amino acids, one might expect that this would be a point of metabolic control. This appears to be the case. In *E. coli,* there are three separate and distinct aspartokinases; one of these is specifically inhibited by threonine, one by methionine, and one by lysine (72,73). Thus, each of the pathways leading to a specific end product is subject to control by that

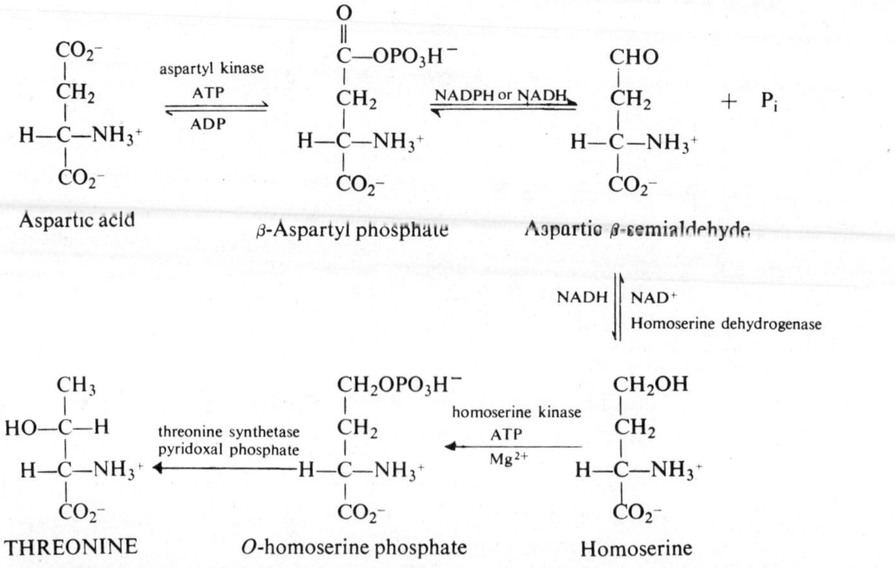

$$^{-}O_2C{-}CH_2{-}CH(NH_3^+){-}CO_2^- \underset{ADP}{\overset{\text{aspartyl kinase, ATP}}{\rightleftharpoons}} {}^{-}HO_3PO{-}C(=O){-}CH_2{-}CH(NH_3^+){-}CO_2^- \overset{\text{NADPH or NADH}}{\rightleftharpoons} OHC{-}CH_2{-}CH(NH_3^+){-}CO_2^- + P_i$$

Aspartic acid — β-Aspartyl phosphate — Aspartic β-semialdehyde

$$\text{Aspartic } \beta\text{-semialdehyde} \underset{\text{NAD}^+}{\overset{\text{NADH}}{\rightleftharpoons}} \text{Homoserine} \quad (\text{Homoserine dehydrogenase})$$

$$HOCH_2{-}CH_2{-}CH(NH_3^+){-}CO_2^- \xrightarrow{\text{homoserine kinase, ATP, } Mg^{2+}} {}^{-}HO_3POCH_2{-}CH_2{-}CH(NH_3^+){-}CO_2^- \xrightarrow{\text{threonine synthetase, pyridoxal phosphate}} CH_3{-}CH(OH){-}CH(NH_3^+){-}CO_2^-$$

Homoserine — *O*-homoserine phosphate — THREONINE

FIGURE 17-11
Biosynthesis of homoserine and threonine.

end product despite the fact that the control is exerted at the point of synthesis of a common intermediate. The methionine-sensitive aspartokinase has been partially characterized and is known to have a molecular weight of 160,000, suggesting that the enzyme is composed of subunits.

The synthesis of *threonine* from aspartic-β-semialdehyde proceeds via the amino acid homoserine as indicated in Fig. 17-11. *O*-Homoserine phosphate is synthesized from homoserine and ATP in a reaction catalyzed by *homoserine kinase* (ATP : L-homoserine *O*-phosphotransferase). The terminal step in threonine synthesis is a very interesting reaction, in which *threonine synthetase* (*O*-phosphohomoserine phospho-lyase), a pyridoxal phosphate-dependent enzyme, converts *O*-phosphohomoserine to threonine. The mechanism of this transformation has been illuminated by the decisive experiments of Flavin (74). He has demonstrated that solvent ^{18}O is incorporated into the threonine hydroxyl but not into the liberated phosphate. Also, solvent deuterium is incorporated into the α and γ positions of threonine. These observations require a mechanism similar to that in Fig. 17-12, which is consistent with the general pattern of pyridoxal phosphate–catalyzed reactions (Chapter 8).

It is worth noting that the *homoserine dehydrogenase* reaction (Fig. 17-11) is catalyzed by the same enzymes as the threonine- and methionine-sensitive aspartokinases. Thus, regulation by the end products is exerted simultaneously on two steps of the synthetic reactions by virtue of inhibition of a single enzyme.

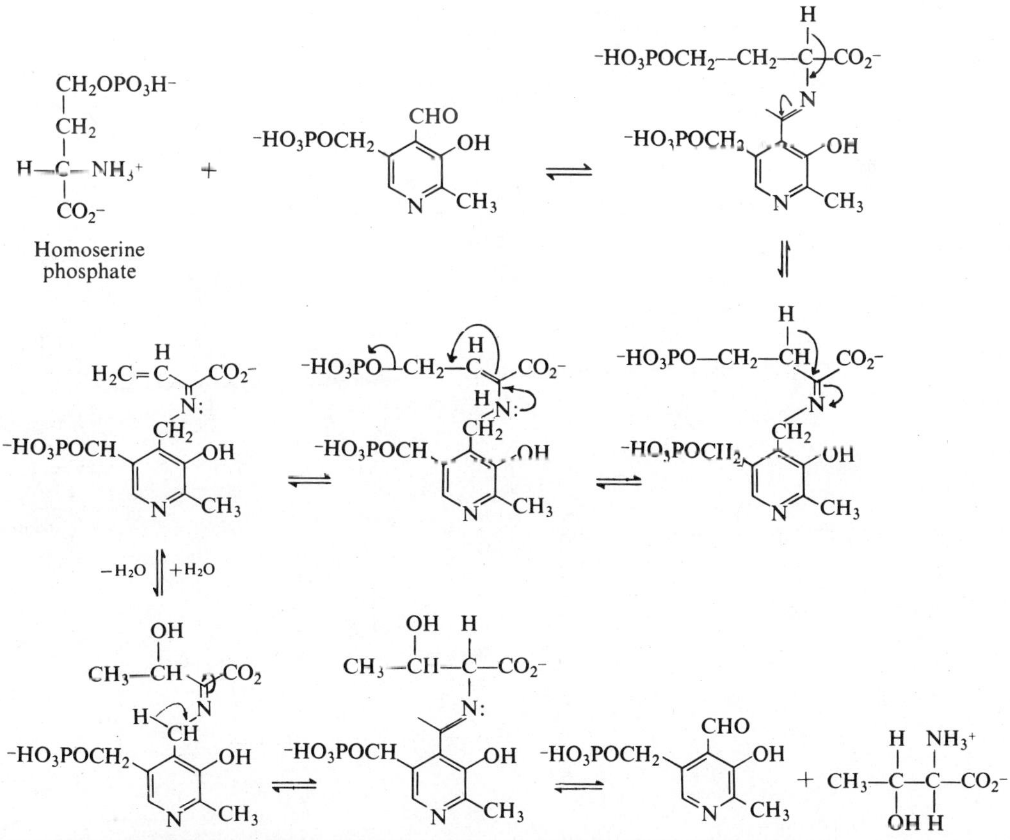

FIGURE 17-12
Mechanism of the threonine synthetase reaction.

Alternatively, threonine may be synthesized in the *threonine aldolase* (L-threonine acetaldehyde-lyase) reaction:

$$CH_3CHO + H_3N^+{-}CH_2{-}CO_2^- \rightleftharpoons H_3N^+{-}\underset{\underset{\displaystyle CH_3}{|}}{}\!\!\!CH(CHOH{-}CH_3){-}CO_2^- \qquad (17\text{-}13)$$

Glycine → Threonine

This enzyme, which is pyridoxal phosphate dependent, occurs in certain microorganisms and certain mammalian tissues. Particularly in the latter tissues, threonine aldolase may primarily serve a catabolic, rather than a biosynthetic function.

In microorganisms, *methionine* is synthesized in a multistep process from cysteine and homoserine. The overall pathway may be divided into three stages: (1) synthesis of cystathionine from homoserine and cysteine, (2) conversion of cystathionine to homocysteine, and (3) conversion of the latter substrate to methionine. The overall pathway is indicated in Fig. 17-13.

The actual substrate for synthesis of cystathionine is not homoserine itself but *O*-succinyl homoserine, which is formed in an acyl transfer reaction

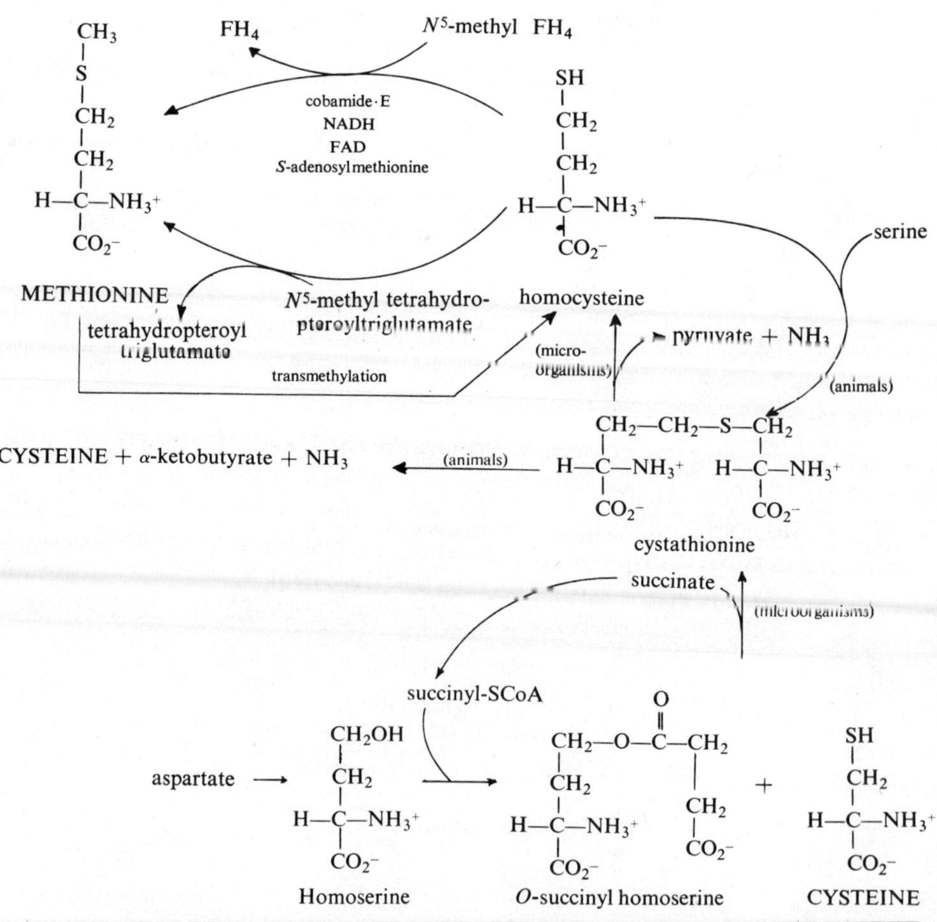

FIGURE 17-13
Interconversion of cysteine and methionine.

involving succinyl-SCoA. A minor modification of the route exists in *Neurospora* in which acetyl-SCoA replaces succinyl-SCoA (75).

Certain microorganisms possess an enzyme that condenses *O*-succinyl homoserine with cysteine to yield cystathionine. This substance may also be formed in animal tissues by condensation of serine and homocysteine (in contrast to the route in microorganisms). A transsulfurase catalyzes the cleavage of cystathionine to homocysteine, pyruvate, and ammonia. Animal tissues also are capable of cleaving cystathionine but do so, again, by an independent route, the products being cysteine, α-oxobutyrate, and ammonia.

The incorporation of a one-carbon fragment into homocysteine yields methionine. Such incorporation may result from the *de novo* synthesis of the methyl function or involve the transfer of preformed methyl groups. We consider these alternatives in turn. The former type of synthesis has been studied most thoroughly in microorganisms, although it also occurs in animal tissues. In *E. coli* there exist two pathways for the conversion of homocysteine to methionine. The simplest involves the direct transfer of a methyl group from N^5-methyltetrahydropteroyltriglutamate to homocysteine:

$$N^5\text{-methyltetrahydropteroyltriglutamate} + \text{homocysteine} \rightarrow \text{methionine} + \text{tetrahydropteroyltriglutamate} \qquad (17\text{-}14)$$

N^5-methyl FH_4 is inactive in this reaction (76).

The second pathway is quite complicated and requires NADH, FAD, *S*-adenosylmethionine (cf. page 807), N^5-methyl FH_4, and a vitamin B_{12} enzyme:

$$N^5\text{-methyl } FH_4 + \text{homocysteine} \xrightarrow[S\text{-adenosylmethionine, } B_{12}\text{-enzyme}]{\text{NADH, FAD}} \text{methionine} + FH_4 \qquad (17\text{-}15)$$

The details of this reaction are far from clear. However, several aspects of this complicated process are understood (77): (1) The direct donor of the methyl group is a methyl cobalamine, in which the methyl group is attached directly to the Co of the B_{12} coenzyme (page 424); (2) synthesis of the methyl cobalamine requires an NADH-dependent flavin-reducing system; (3) the cobalamine methyltransferase, which is involved in transfer of the methyl group to homocysteine, is activated by *S*-adenosylmethionine; and (4) a constant maximal rate of methionine synthesis from N^5-methyl FH_4 mediated by the cobalamine system requires both the flavin-reducing system and *S*-adenosylmethionine. While some questions remain, the general sequence of reactions is prescribed by these observations.

In mammalian systems, N^5-methyl FH_4 serves as the methyl group donor for methionine biosynthesis in a reaction requiring catalytic amounts of *S*-adenosylmethionine. No requirement for a B_{12} coenzyme has been established in this case.

In addition to those reactions involving the participation of FH_4 coenzymes, methionine may arise from direct methyl group transfer from a variety of donor molecules. These reactions are considered more fully on page 807.

Inasmuch as animal tissues lack the capacity to synthesize the carbon chain of homocysteine, mammals do not accomplish the *de novo* synthesis of methionine but, as noted above, may convert homocysteine to this amino acid. The production of homocysteine from methionine through transmethylations followed by the sequence of events indicated in Fig. 17-13 does, however, provide a pathway for the conversion of methionine to cysteine in mammals.

The biosynthesis of *lysine* is particularly striking in that two separate and distinct reaction pathways are employed for this purpose. In bacteria, higher plants, and in certain algae and fungi, tht carbon skeleton of this amino acid is derived from pyruvate and aspartate with the intermediate formation of α,ε-diaminopimelic acid. In other algae and fungi, acetate and α-oxoglutarate provide the carbon skeleton of lysine, and the reaction sequence proceeds through the intermediate α-aminoadipic acid. The overall reaction pathways are outlined in Fig. 17-14.

In the diaminopimelate pathway, the initial condensation occurs between pyruvate and aspartic acid-β-semialdehyde, yielding, following dehydration, 2,3-dihydrodipicolinic acid. Reduction of this substance, followed by ring opening and succinylation, yields *N*-succinyl-ε-oxo-L-α-aminopimelic acid. Transamination yields the corresponding diaminopimelic acid. Deacylation, epimerization, and decarboxylation complete the overall synthesis.

The synthesis of α-aminoadipic acid is well understood and proceeds as indicated in Fig. 17-14. The route from this intermediate to lysine remains to be completely elucidated, although α-aminoadipic-δ-semialdehyde seems pretty well established as an important intermediate. The simple transamination reaction that would yield lysine directly from this intermediate seems not to occur, and the transformation may adopt a complex route with saccharopine as immediate precursor to lysine.

Biosynthesis of Serine, Glycine, and Cysteine

Serine, an important intermediate in the interconversion of cysteine and methionine (Fig. 17-13) and in the synthesis of cysteine and glycine, is derived from the carbon chain of D-3-phosphoglyceric acid as indicated in Fig. 17-15, or may arise from glycine in the *serine transhydroxymethylase* reaction (page 408). Phosphoglyceric acid dehydrogenase (D-phosphoglycerate : NAD oxidoreductase) catalyzes the NAD-dependent oxidation of 3-phosphoglyceric acid, yielding 3-phosphohydroxypyruvic acid. Transamination and hydrolysis of the phosphate ester complete the synthesis of serine. Each of the enzymes involved in the synthesis of serine from 3-phosphoglyceric acid, a glycolytic intermediate, has been identified in *E. coli* extracts (78). Alternative routes for serine synthesis almost certainly exist and are of varying quantitative importance, depending on species.

In the presence of FH_4, serine is readily converted to *glycine* and N^5,N^{10}-methylene FH_4 in a reaction catalyzed by serine transhydroxymethylase (L-serine : tetrahydrofolate 5,10-hydroxymethyltransferase), a pyridoxal phosphate enzyme (79) (Fig. 17-15). The *in vivo* conversion of serine to glycine has been established by the use of isotope tracer experiments.

Cysteine may be synthesized from serine and homocysteine as indicated in Fig. 17-13. An alternative route for cysteine synthesis is known. Serine sulfhydrase [L-serine hydro-lyase (adding H_2S)], a pyridoxal phosphate enzyme, catalyzes cysteine synthesis from serine and H_2S:

$$\underset{\text{Serine}}{H_3^+N-\underset{H}{\overset{CH_2OH}{C}}-CO_2^-} + H_2S \rightleftharpoons \underset{\text{Cysteine}}{H_3^+N-\underset{H}{\overset{CH_2-SH}{C}}-CO_2^-} + H_2O \qquad (17\text{-}16)$$

Cystine. The disulfide bridges of proteins are formed by oxidation following the incorporation of cysteine into polypeptide chains, rather than by direct incorporation of cystine.

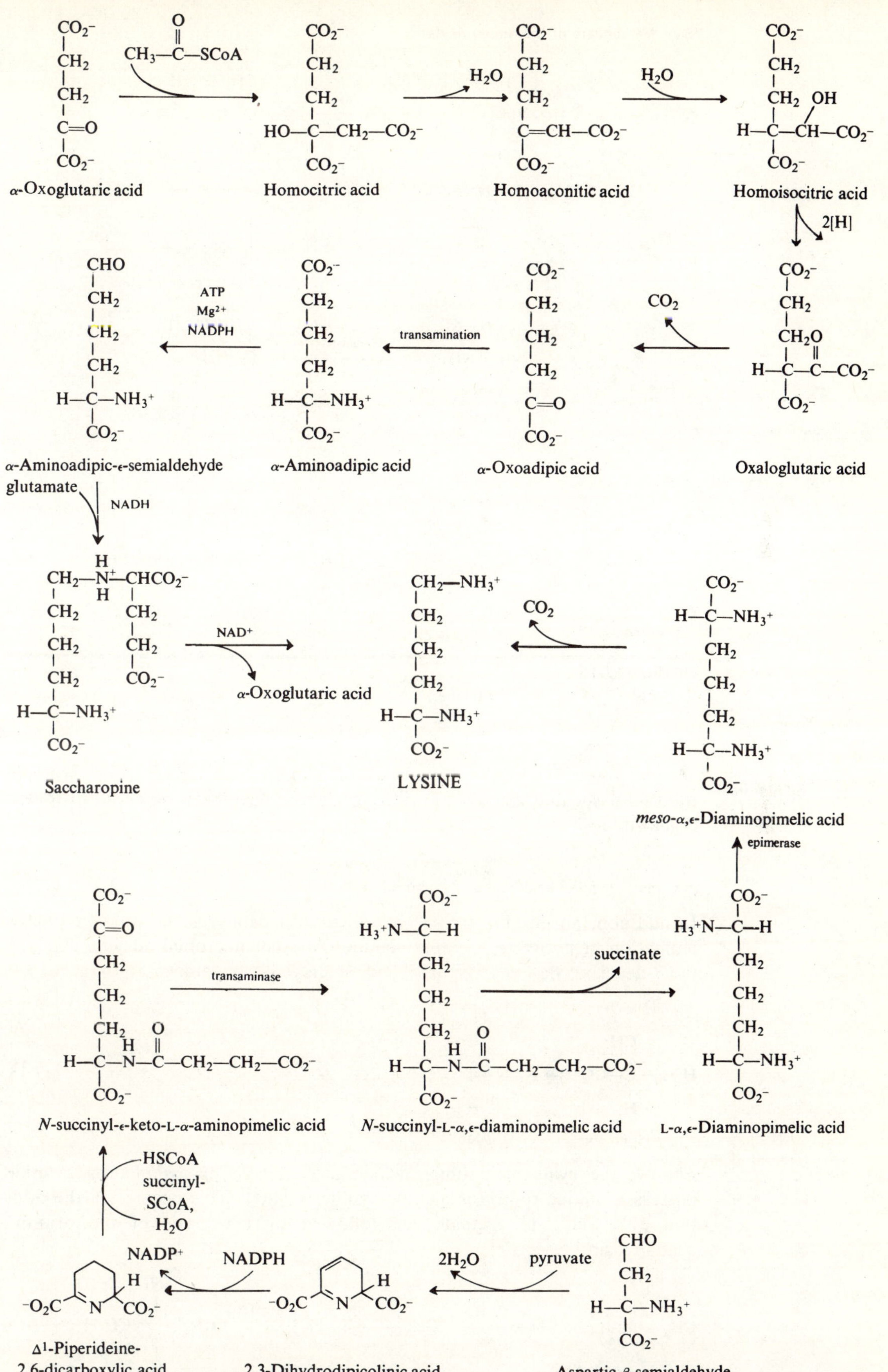

FIGURE 17-14
Biosynthesis of lysine via the α-aminoadipic and diaminopimelic acid routes.

$$\text{Glucose} \rightarrow \rightarrow \rightarrow \underset{\text{3-Phospho-D-glyceric acid}}{HO{-}\overset{\displaystyle CH_2OPO_3H^-}{\underset{\displaystyle CO_2^-}{C}}{-}H} \underset{NADH}{\overset{NAD^+}{\rightleftharpoons}} \underset{\text{3-Phosphohydroxypyruvic acid}}{\overset{\displaystyle CH_2OPO_3H^-}{\underset{\displaystyle CO_2^-}{C{=}O}}}$$

$$\Big\Updownarrow \text{transaminase}$$

$$\underset{\text{L-Serine}}{H{-}\overset{\displaystyle CH_2OH}{\underset{\displaystyle CO_2^-}{C}}{-}NH_3^+} \underset{-P_i}{\overset{\text{phosphatase}}{\longleftarrow}} \underset{\text{3-Phospho-L-serine}}{H{-}\overset{\displaystyle CH_2OPO_3H^-}{\underset{\displaystyle CO_2^-}{C}}{-}NH_3^+}$$

$$\underset{\text{L-Serine}}{H{-}\overset{\displaystyle CH_2OH}{\underset{\displaystyle CO_2^-}{C}}{-}NH_3^+} + FH_4 \underset{\text{pyridoxal phosphate}}{\overset{\text{serine transhydroxy-methylase}}{\rightleftharpoons}} \underset{\text{Glycine}}{\overset{\displaystyle NH_3^+}{\underset{\displaystyle CO_2^-}{CH_2}}} + N^5,N^{10}\text{-methylene } FH_4$$

FIGURE 17-15
Biosynthesis of serine and glycine.

Synthesis of Alanine

Several routes for the biosynthesis of alanine are known. For example, transaminases that convert pyruvate to alanine have been observed in several organisms:

$$\text{Pyruvate} + \text{amino acid} \overset{\text{pyridoxal phosphate}}{\rightleftharpoons} \text{L-alanine} + \text{ketoacid} \qquad (17\text{-}17)$$

In addition, an enzyme from *Bacillus subtilis* catalyzes the direct reductive amination of pyruvate, yielding alanine. Another microbial enzyme catalyzes the β-decarboxylation of aspartic acid, again yielding alanine:

$$\underset{\text{L-Aspartate}}{H_3^+N{-}\overset{\displaystyle \overset{\displaystyle CO_2^-}{CH_2}}{\underset{\displaystyle H}{C}}{-}CO_2^-} \rightarrow \underset{\text{L-Alanine}}{H_3^+N{-}\overset{\displaystyle CH_3}{\underset{\displaystyle H}{C}}{-}CO_2^-} + CO_2 \qquad (17\text{-}18)$$

Finally, the cyanogenic fungi, which are capable of assimilating cyanide, possess a unique route for alanine synthesis (80). This occurs via the addition of cyanide to acetaldehyde followed by hydrolysis of the resulting cyanohydrin:

$$CH_3{-}\overset{\displaystyle O}{\overset{\|}{C}}{-}H + HCN \xrightarrow{+NH_4^+} CH_3{-}\overset{\displaystyle NH_2}{\underset{\displaystyle H}{C}}{-}C{\equiv}N \underset{-NH_4^+}{\xrightarrow{+2H_2O}} CH_3{-}\overset{\displaystyle NH_2}{\underset{\displaystyle H}{C}}{-}CO_2H \qquad (17\text{-}19)$$

Biosynthesis of Histidine

Biosynthesis of the imidazole ring of histidine in microorganisms proceeds via a striking reaction pathway, quite distinct from that involved in the synthesis of the imidazole ring of purines (Chapter 18). The key to the understanding of this synthesis was the finding, based on isotope-tracer studies, that N-1 and C-2 of the imidazole ring are derived from N-1 and C-2 of the purine nucleus (81–84):

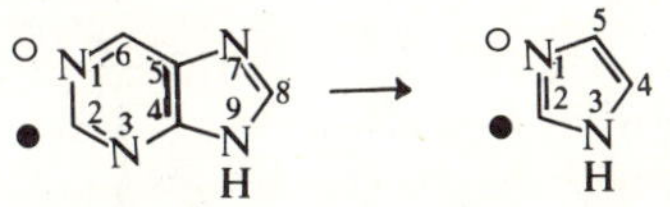

Subsequently, Moyed and Magasanik were able to demonstrate that bacterial extracts can, with the addition of ATP and ribose-5-phosphate, catalyze the synthesis of "compound III," an important intermediate for the synthesis of the imidazole ring (85). Compound III has been identified as an intermediate derived from condensation of ribose-5-phosphate with N-1 of a ring-opened AMP derivative, *N*-(5′-phospho-D-1′-ribulosylformimino)-5-amino-1-(5″-phosphoribosyl)-4-imidazole-carboxamide (Fig. 17-16). This species arises biosynthetically by the initial reaction of 5-phosphoribosyl-1-pyrophosphate (PRPP) with ATP, yielding *N*-1-(5′-phosphoribosyl)-ATP (86). The phosphoribosyl-ATP is converted to the corresponding AMP derivative in a pyrophosphatase reaction followed by ring opening and isomerization as indicated in Fig. 17-16. Addition of the amide nitrogen of glutamine to compound III with closure of the imidazole ring yields imidazole-glycerol phosphate and 5-aminoimidazole-4-carboxamide ribonucleotide, which is employed for the resynthesis of purines. The four-step conversion of imidazoleglycerol phosphate to histidine is straightforward (Fig. 17-16). Of some note is the oxidation of histidinol to histidine. This (apparently) two-step process is catalyzed by a single enzyme without the appearance of an aldehydic intermediate. Hence, oxidation of the aldehyde, if it exists, must be more rapid than its release from the surface of the dehydrogenase.

Synthesis of the Branched-Chain Amino Acids

As might have been expected, the routes for synthesis of *valine, leucine,* and *isoleucine* are closely related. The biosynthesis of valine and leucine proceeds through the common intermediate, α-oxoisovaleric acid, and is outlined in Fig. 17-17. The thiamine pyrophosphate–dependent condensation of pyruvate with a two-carbon fragment yields acetolactic acid (Chapter 11), which is converted directly to α,β-dihydroxyisovaleric acid in an interesting reaction. This transformation, which is closely related to the classic pinacol rearrangement, is catalyzed by *acetohydroxy acid isomeroreductase* and requires Mg^{2+} and a reduced pyridine nucleotide. By analogy with the known mechanism for the pinacol rearrangement, Mg^{2+} may function as a "superacid" catalyst, coordinating with the carbonyl oxygen of acetolactate and, consequently, aiding the methyl migration to an electron-deficient carbon atom. *Dihydroxyacid dehydratase* (2,3-dihydroxyacid hydro-lyase) catalyzes the dehydration of α,β-dihydroxyisovaleric acid, yielding α-oxoisovaleric acid, which, by transamination, yields valine directly.

α-Oxoisovaleric acid is condensed with acetyl-SCoA, yielding β-hydroxy-β-carboxyisocaproic acid. This substance is converted to α-hydroxy-β-carboxyisocaproic acid by a dehydration rehydration sequence closely related to that for the citrate–isocitrate conversion catalyzed by aconitase. However, the *β-hydroxy-β-carboxyisocaproate isomerase* has been purified

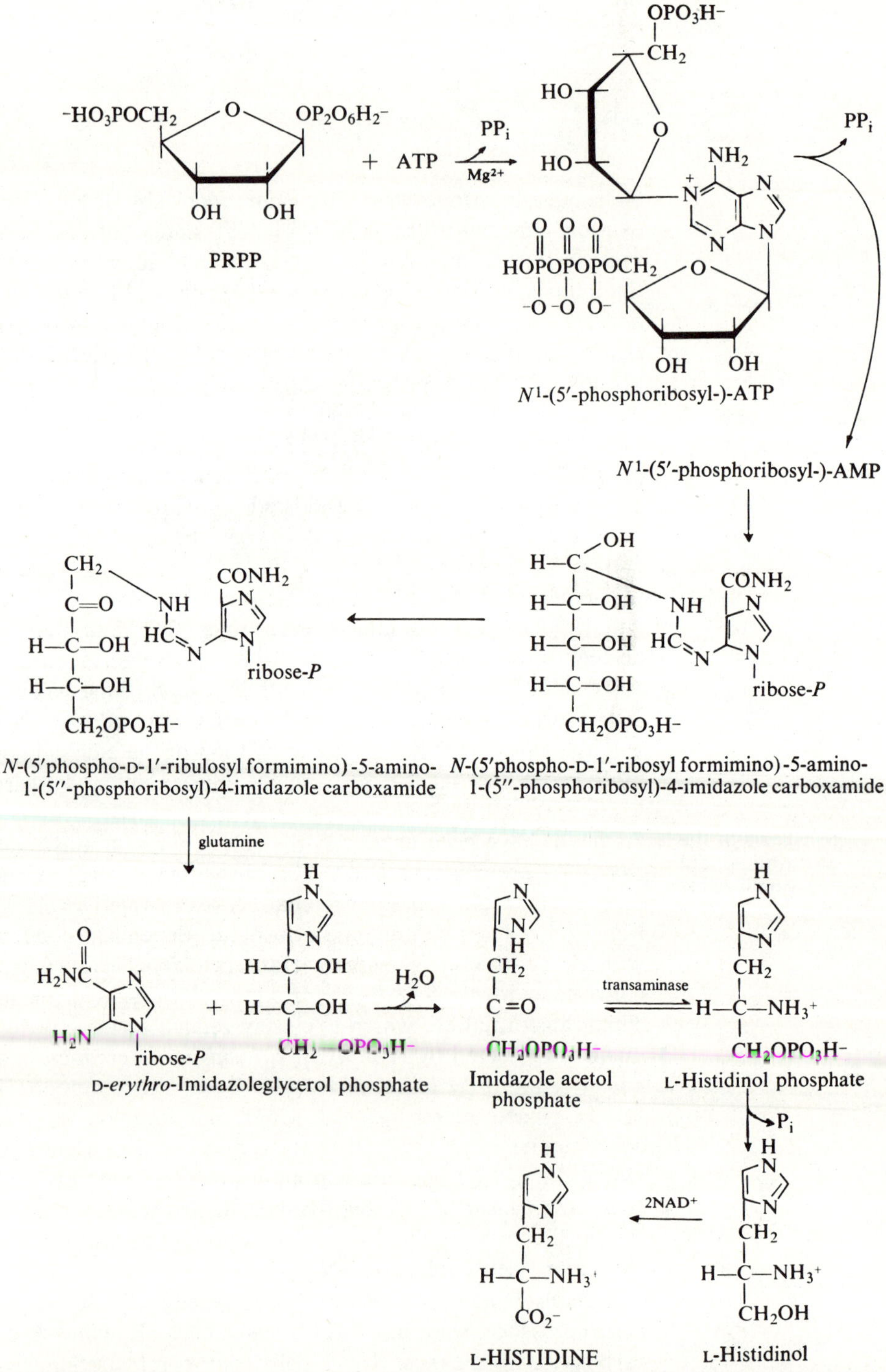

$^{-}HO_3POCH_2$ $OP_2O_6H_2^-$ + ATP $\xrightarrow[Mg^{2+}]{PP_i}$ OPO_3H^- CH_2 HO HO O NH_2 N N N N O O O HOPOPOPOCH$_2$ O $^-$O $^-$O O$^-$ OH OH PP$_i$

OH OH

PRPP

N^1-(5′-phosphoribosyl-)-ATP

N^1-(5′-phosphoribosyl-)-AMP

CH$_2$ C=O H—C—OH H—C—OH $CH_2OPO_3H^-$ NH HC N CONH$_2$ N N ribose-*P*

OH H—C H—C—OH H—C—OH H—C—OH $CH_2OPO_3H^-$ NH HC N CONH$_2$ N N ribose-*P*

N-(5′phospho-D-1′-ribulosyl formimino)-5-amino-1-(5″-phosphoribosyl)-4-imidazole carboxamide

N-(5′phospho-D-1′-ribosyl formimino)-5-amino-1-(5″-phosphoribosyl)-4-imidazole carboxamide

glutamine

O H$_2$NC N H$_2$N N ribose-*P* + H N N H—C—OH H—C—OH $CH_2—OPO_3H^-$ $\xrightarrow{H_2O}$ N N H CH$_2$ C=O $CH_2OPO_3H^-$ $\underset{\text{transaminase}}{\rightleftharpoons}$ H N N CH$_2$ H—C—NH$_3^+$ $CH_2OPO_3H^-$

D-*erythro*-Imidazoleglycerol phosphate

Imidazole acetol phosphate

L-Histidinol phosphate

P$_i$

H N N CH$_2$ H—C—NH$_3^+$ CO_2^- $\xleftarrow{2NAD^+}$ H N N CH$_2$ H—C—NH$_3^+$ CH$_2$OH

L-HISTIDINE

L-Histidinol

FIGURE 17-16

Biosynthesis of histidine in microorganisms.

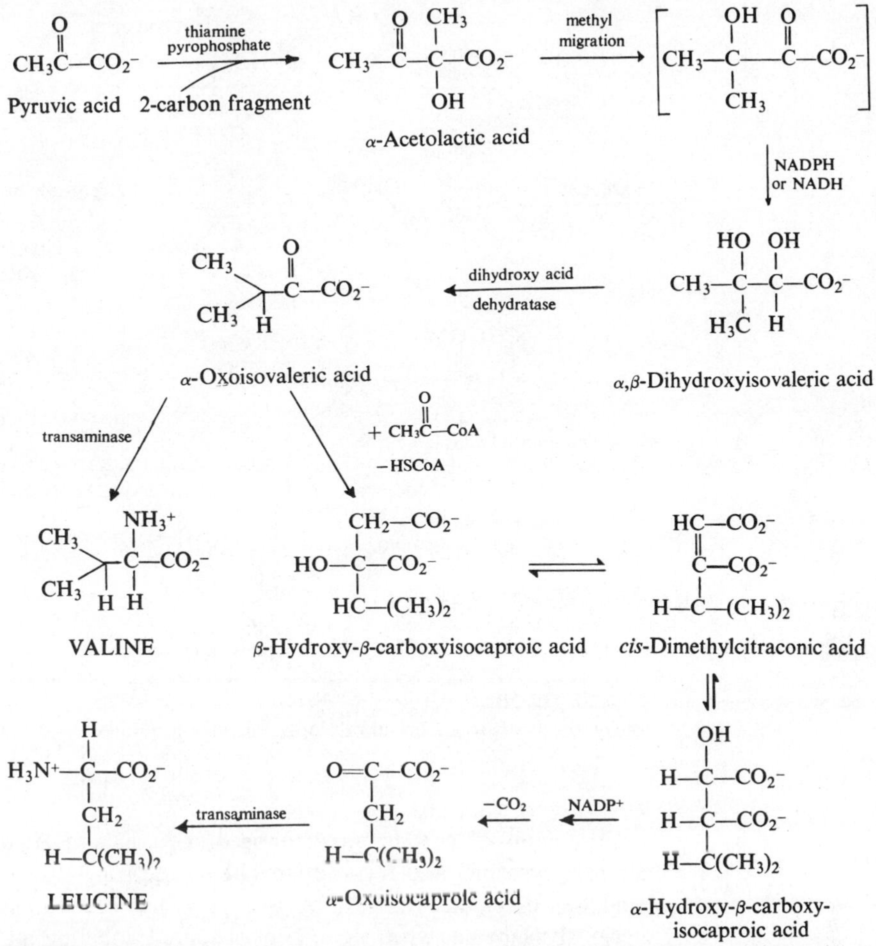

Pyruvic acid + 2-carbon fragment —(thiamine pyrophosphate)→ α-Acetolactic acid —(methyl migration)→ [intermediate] —(NADPH or NADH)→ α,β-Dihydroxyisovaleric acid —(dihydroxy acid dehydratase)→ α-Oxoisovaleric acid —(transaminase)→ VALINE

α-Oxoisovaleric acid + CH_3COCoA (− HSCoA) → β-Hydroxy-β-carboxyisocaproic acid ⇌ *cis*-Dimethylcitraconic acid ⇌ α-Hydroxy-β-carboxyisocaproic acid —(NADP⁺, −CO_2)→ α-Oxoisocaproic acid —(transaminase)→ LEUCINE

FIGURE 17-17
Biosynthesis of valine and leucine in microorganisms and plants.

and shown to be distinct from aconitase (87). Oxidative decarboxylation and transamination complete the synthesis of leucine.

Threonine dehydratase [L-threonine hydro-lyase (deaminating)] catalyzes the conversion of threonine to α oxobutyrate. The latter substance is condensed with hydroxyethyl TPP, yielding α-aceto-α-hydroxybutyric acid, a key intermediate for synthesis of isoleucine. The synthesis of isoleucine from this intermediate closely parallels the synthesis of valine from acetolactate, and, indeed, the same enzymes appear to be involved in both reaction sequences except for the terminal transamination step. There are two transaminases for valine but only one for isoleucine. This accounts for mutants that require isoleucine but not valine. The biosynthetic route to isoleucine is detailed in Fig. 17-18.

Biosynthesis of Phenylalanine, Tyrosine, and Tryptophan

A number of reactions are common to the synthesis of *phenylalanine, tyrosine,* and *tryptophan.* These steps culminate in the synthesis of shikimic acid-5-phosphate, as indicated in Fig. 17-19. The role of shikimic acid as an intermediate in the synthesis of aromatic amino acids from D-glucose was discovered by Davis, who observed that shikimic acid would replace the

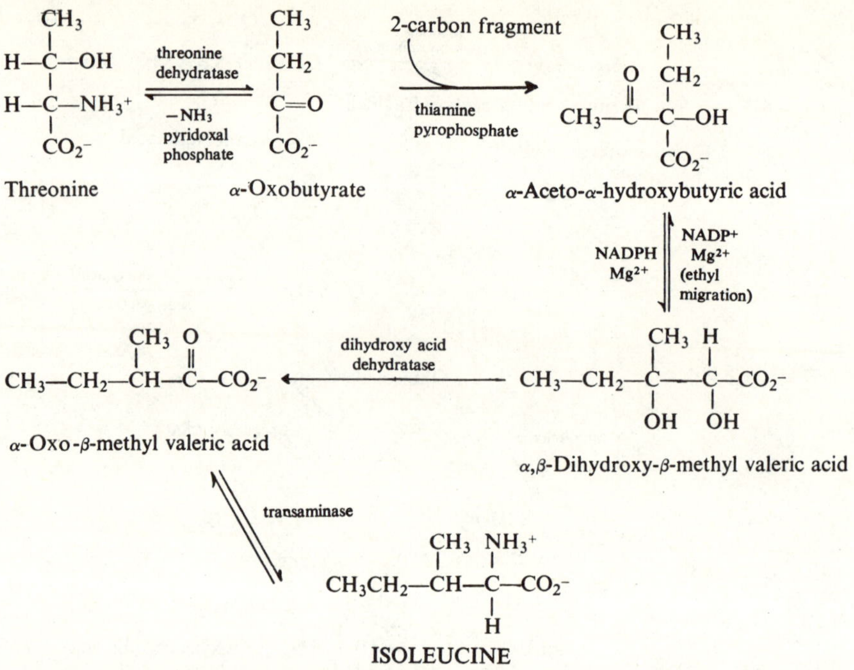

Threonine (CH_3—CH(OH)—CH(NH_3^+)—CO_2^-) ⇌ (threonine dehydratase; $-NH_3$; pyridoxal phosphate) α-Oxobutyrate (CH_3—CH_2—C(=O)—CO_2^-)

α-Oxobutyrate + 2-carbon fragment → (thiamine pyrophosphate) α-Aceto-α-hydroxybutyric acid (CH_3—CO—C(OH)(CH_2CH_3)—CO_2^-)

α-Aceto-α-hydroxybutyric acid ⇌ (NADPH, Mg^{2+} / $NADP^+$, Mg^{2+} (ethyl migration)) α,β-Dihydroxy-β-methyl valeric acid (CH_3—CH_2—C(OH)(CH_3)—CH(OH)—CO_2^-)

α,β-Dihydroxy-β-methyl valeric acid → (dihydroxy acid dehydratase) α-Oxo-β-methyl valeric acid (CH_3—CH_2—CH(CH_3)—CO—CO_2^-)

α-Oxo-β-methyl valeric acid ⇌ (transaminase) ISOLEUCINE (CH_3CH_2—CH(CH_3)—CH(NH_3^+)—CO_2^-)

FIGURE 17-18
Biosynthesis of isoleucine in microorganisms and plants.

requirements of certain auxotrophs of *E. coli* and *A. aerogenes* for phenylalanine, tyrosine, and tryptophan (88). The route from D-glucose to shikimic acid was elucidated in part by isotope-tracer experiments employing glucose variously labeled with ^{14}C. The observed labeling patterns indicated that carbons 1, 2, and 7 of shikimic acid are derived from a three carbon intermediate of glycolysis (now known to be phosphoenolpyruvic acid). In addition, carbons 3, 4, 5, and 6 of shikimic acid exhibited a labeling pattern suggesting their formation from a four-carbon intermediate derived via the pentose phosphate pathway (now known to be D-erythrose-4-phosphate). Subsequent experiments revealed that phosphoenolpyruvate condenses with erythrose-4-phosphate to yield the 7-carbon precursor of shikimic acid, 3-deoxy-D-arabino-heptulosonic acid-7-phosphate. Cyclization yields 5-dehydroquinic acid, which gives shikimic acid-5-phosphate following dehydration, oxidation, and phosphorylation (Fig. 17-19). Most of the enzymes involved in this reaction sequence have not yet been purified and characterized.

3-Enolpyruvylshikimate-5-phosphate synthetase catalyzes the condensation of phosphoenolpyruvic acid and shikimic acid-5-phosphate as detailed in Fig. 17-20. The product, 3-enolpyruvylshikimate-5-phosphate, yields chorismic acid, which gives anthranilic acid, a precursor of tryptophan, and, on rearrangement, prephenic acid, a precursor of phenylalanine and tyrosine. These transformations are incompletely understood, particularly the synthesis of anthranilic acid, which is known, however, to involve glutamine. Decarboxylation and reductive decarboxylation of prephenic acid yield the immediate precursors of tyrosine and phenylalanine, respectively. The latter two are finally formed by transamination (Fig. 17-20).

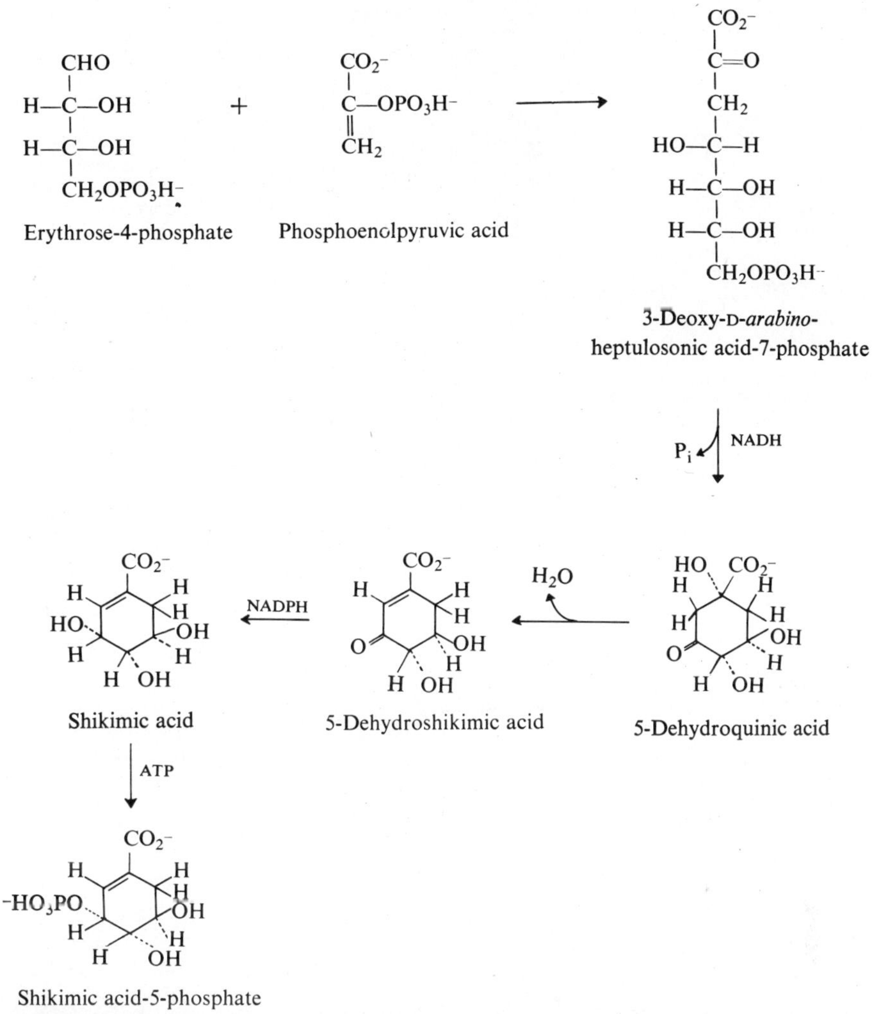

FIGURE 17-19
Biosynthesis of shikimic acid-5-phosphate.

Conversion of anthranilic acid to tryptophan occurs in yeast and *E. coli* as indicated in Fig. 17-21. The terminal step is catalyzed by *tryptophan synthetase* (page 868), a remarkable enzyme thoroughly studied by Yanofsky and Bonner (89).

The biosynthetic origins of each of the amino acids are summarized in Table 17-3.

GENERAL ASPECTS OF AMINO ACID METABOLISM

Having discussed the conversion of inorganic nitrogen to ammonia, the incorporation of ammonia into organic compounds, derivation of amino acids from dietary protein, and the biosynthesis of the amino acids, attention is now directed to some aspects of the transformations of the latter. First, we consider several important reactions that are common to many of the amino acids. In the next section, specific metabolic transformations are considered.

General reactions of the amino acids include transamination, decarboxylation, racemization, and deamination (oxidative and nonoxidative). The first

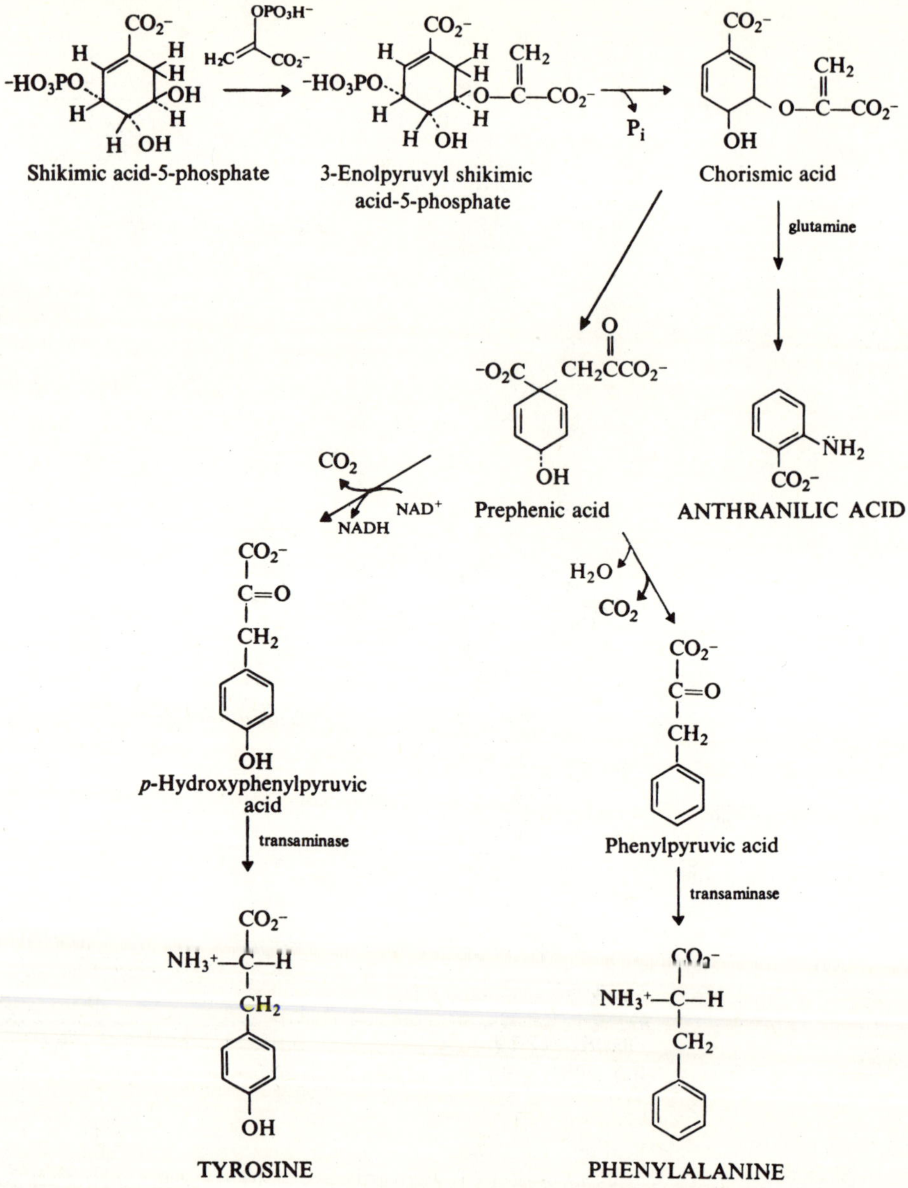

FIGURE 17-20
Conversion of shikimic acid-5-phosphate to tyrosine, phenylalanine, and anthranilic acid.

three reaction types indicated above are generally (but not invariably) pyridoxal phosphate dependent, and the associated reaction mechanisms have been discussed in Chapter 8. Consequently, we consider only metabolic aspects of these reactions below.

Transamination

Transamination [Eq. (17-20)] was first recognized as an enzymatic reaction

$$(\text{Amino acid})_1 + (\text{ketoacid})_2 \rightleftharpoons (\text{amino acid})_2 + (\text{ketoacid})_1 \qquad (17\text{-}20)$$

by the Russian biochemists Braunstein and Kritzmann in 1937. These workers observed the enzymatic formation of glutamic acid from α-oxoglutarate and

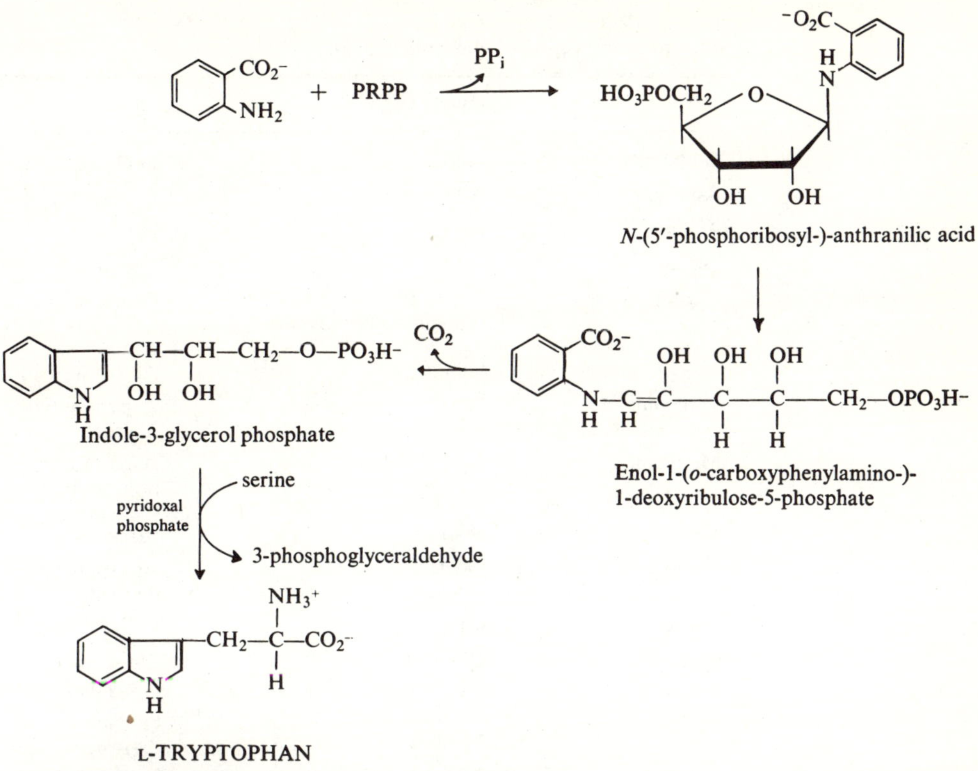

FIGURE 17-21
Biosynthesis of tryptophan.

certain amino acids. It was not until 1950 that the broad scope of transaminations was established through the investigations of other workers. The role of pyridoxal phosphate in enzymatic transamination was initially suggested by Snell in 1944 (90). Several transaminases (amino transferases) have been subsequently purified and characterized.

The scope of biological transaminations is broad indeed. Early studies were principally concerned with reactions involving dicarboxylic acid substrates—glutamic-oxalacetic transaminase, for example. Subsequently, similar reactions involving other amino and keto acids were identified. At present, the known transamination reactions include nearly all the L-amino acids, and a number of such reactions involving D-amino acids, ω-amino acids, and aldehydes are recognized as well. A partial list of known transaminations is presented in Table 17-4.

The metabolic roles of transaminations are several and significant. Important functions include (1) amino acid biosynthesis (transamination is directly involved in the synthesis of at least 11 of the amino acids), (2) amino acid degradation (see discussion below), (3) liaison between carbohydrate and amino acid metabolism, and (4) synthesis of several specific compounds, including urea and γ-aminobutyric acid.

The transaminases are ubiquitous enzymes. Individual enzymes may be fairly specific for a given substrate pair or, alternatively, may be able to catalyze reactions involving a broad spectrum of amino and keto acids. At least some of the transaminases are under metabolic control. For example, the aromatic amino acid transaminase of *E. coli* is subject to repression by tyro-

TABLE 17-3
Biosynthetic origins of the amino acids

Amino acid	*Atom*	*Origin*
Glutamate	All C	α-Oxoglutarate
Aspartate	All C	Oxalacetate
Alanine	All C	Pyruvate
Proline	All C	Glutamate
Arginine	C-1 to C-5	Glutamate to ornithine
	Guanidino N—C	CO_2, NH_3, via carbamyl phosphate
	Guanidino N	Aspartate
Threonine	All C	Aspartate (via semialdehyde and homoserine)
Methionine	C-1 to C-4	Aspartate (via semialdehyde and homoserine)
	S	Cysteine
	Methyl	"1-carbon"
meso-Diaminopimelate	C-1 to C-4	Aspartate (via semialdehyde)
	C-5 to C-7	Pyruvate
Lysine	All C	*meso*-Diaminopimelate
Isoleucine	C-1, C-2, C-5, C-6	Threonine
	C-3, C-4	Pyruvate-2, pyruvate-3
Valine	C-1, C-2, C-4	Pyruvate
	C-3, C-4′	Pyruvate-2, pyruvate-3
Leucine	C-3, C-4, C-5, C-5′	Pyruvate
	C-1, C-2	Acetate
Serine	All C	*P*-glycerate
Glycine		Serine
Cysteine	All C	Serine
	S	Sulfide, thiosulfate
Tyrosine	Side chain	*P*-enolpyruvate
	Ring	Erythrose-4-phosphate + *P*-enolpyruvate, via shikimate
Phenylalanine	Same as tyrosine	Same as tyrosine
Tryptophan	Side chain	Serine
	Benzene ring	Like tyrosine
	C-2, C-3 of indole	C-1, C-2 of ribose-5-phosphate
	N of indole	Amide of glutamine
Histidine	Imidazole N-1, C-2	N-1, C-2 of adenylic acid
	Imidazole N-3	Amide of glutamine
	Remaining C	Ribose-5-phosphate

SOURCE: E. Umbarger and B. D. Davis, in I. C. Gunsalus and R. Y. Stanier (eds.), *The Bacteria,* vol. III, Academic Press, New York, 1962, p. 241.

sine, and a corresponding enzyme of rat liver is induced by corticosteroids, particularly hydrocortisone (91).

The glutamic-oxalacetic transaminase of rat liver occurs in two distinct molecular forms. These are distinguished on the basis of isoelectric point, *p*H optimum, and sensitivity to urea denaturation. They have been separated by ion-exchange chromatography on DEAE-cellulose. These isozymes appear to be compartmentalized within the cell, one occurring in the soluble portion and the other being associated with the mitochondria.

TABLE 17-4
Enzymes catalyzing transamination reactions

Name	*Source*	*Reaction*
Acetylornithine transaminase	Bacteria	α-*N*-acetyl-L-ornithine + α-oxoglutarate ⇌ *N*-acetylglutamic γ-semialdehyde + L-glutamate
Alanine transaminase	Animal tissues, plants, bacteria	L-Alanine + α-ketoacids ⇌ pyruvate + amino acids
D-Amino acid transaminase	Bacteria	D-Aspartate + α-oxoglutarate ⇌ oxalacetate + D-glutamate
Glutamate-alanine transaminase	Animal tissues, plants	L-Glutamate + pyruvate ⇌ α-oxoglutarate + L-alanine
Glutamate-aspartate transaminase	Animal tissues, plants, bacteria	L-glutamate + oxalacetate ⇌ α-oxoglutarate + L-aspartate
Glutamate-cysteine transaminase	Liver	L-Glutamate + mercaptopyruvate ⇌ α-oxoglutarate + cysteine
Glutamate-glycine transaminase	Plants, animals, bacteria	L-Glutamate + glyoxylate ⇌ α-oxoglutarate + glycine
Glutamate-kynurenine transaminase	Kidney, molds, bacteria	L-Glutamate + *O*-aminobenzoylpyruvate ⇌ α-oxoglutarate + L-kynurenine
Glutamate-leucine transaminase	Animals, plants, bacteria	L-Glutamate + α-oxoisocaproate ⇌ α-oxoglutarate + L-leucine
Glutamate-phosphohistidinol transaminase	Molds	L-Glutamate + imidazole acetol phosphate ⇌ α-oxoglutarate + histidinol phosphate
Glutamate-tyrosine transaminase	Animal tissues, bacteria, plants	L-Glutamate + *p*-hydroxyphenylpyruvate ⇌ α-oxoglutarate + L-tyrosine

Decarboxylation

Amino acid decarboxylases (amino acid carboxy-lyases) are distributed widely throughout living material. A compilation of these enzymes is presented in Table 17-5. For the most part, they catalyze the α-decarboxylation of L-amino acids. Interesting exceptions include D-lysine decarboxylase and L-β-aspartic acid decarboxylase. These enzymes occur in certain microorganisms. Certain of the decarboxylases are quite specific enzymes, while others exhibit activity with a variety of substrates. For example, the enzyme from *Proteus vulgaris* catalyzes the decarboxylation of valine, isoleucine, and other nonpolar amino acids, and animal tissues possess a general aromatic amino acid decarboxylase.

Among the amino acid decarboxylases, the histidine decarboxylases of certain microorganisms appear unique (92,93). These enzymes do not require pyridoxal phosphate but are susceptible to inhibition by typical carbonyl group reagents. The enzyme from *Lactobacillus* employs pyruvic acid bound at the N terminus in the same way as other decarboxylases employ pyridoxal phosphate:

$$CH_3-\overset{\overset{\displaystyle O}{\|}}{C}-\overset{\overset{\displaystyle O}{\|}}{C}-\overset{H}{N}-\underset{}{\overset{\overset{\displaystyle CH_2-\phi}{|}}{CH}}-\overset{\overset{\displaystyle O}{\|}}{C}\sim \text{chain}$$

TABLE 17-5
Enzymic decarboxylation of amino acids

Amino Acid	*Decarboxylation product*	*Occurrence*
L-Arginine	Agmatine	Microorganisms (e.g., *E. coli*)
L-Aspartic acid	β-Alanine	*Rhizobium leguminosarum*
L-Cysteic acid	Taurine	Liver, spleen, brain
L-Cysteine sulfinic acid	Hypotaurine	Liver, spleen, brain
L-Glutamic acid	γ-Aminobutyric acid	Microorganisms (e.g., *Cl. welchii, E. coli*), animal tissues (brain, liver, muscle), higher plants (barley, spinach, phlox)
γ-Hydroxy-L-glutamic acid	γ-Amino-α-hydroxybutyric acid	*E. coli*
γ-Methylene-L-glutamic acid	α-Methylene-γ-aminobutyric acid	Barley, red pepper, peanut
L-Histidine	Histamine	Microorganisms (e.g., *Cl. welchii, Lactobacilli*), animal tissues (kidney, liver, duodenum)
L-Lysine	Cadaverine	Microorganisms (e.g., *B. cadaveris, E. coli*)
meso-α, ε-Diaminopimelic acid	L-Lysine	Microorganisms (e.g., *E. coli, A. aerogenes*)
L-Ornithine	Putrescine	Microorganisms (e.g., *Cl. septicum*)
L-Phenylalanine	β-Phenylethylamine	Microorganisms (e.g., *Streptococcus fecalis*)
L-Tyrosine	Tyramine	Microorganisms (e.g., *S. fecalis*), animal tissues (e.g., kidney)
3,4-Dihydroxy-L-phenylalanine	3,4-Dihydroxy-β-phenylethylamine	Microorganisms (e.g., *S. fecalis*), animal tissues (e.g., kidney)
5-Hydroxy-L-tryptophan	5-Hydroxytryptamine	Animal tissues (e.g., kidney, brain, stomach)
L-Valine	Isobutylamine	*Proteus vulgaris*
L-Leucine	Isoamylamine	*Proteus vulgaris*
DL-Lysine	Cadaverine	Microorganisms

SOURCE: J. S. Fruton and S. Simmonds, *General Biochemistry,* Wiley, New York, 1958, p. 767.

That is, the histidine can make a Schiff base with the terminal carbonyl group instead of that of the pyridoxal phosphate and the adjacent carbonyl group acts as an electron sink.

This enzyme has a molecular weight of 190,000 and is composed of 10 subunits, molecular weight 19,000. Of these 10 chains, 5 terminate in *N*-pyruvylphenylalanine, the others simply in phenylalanine.

The importance of decarboxylation in metabolism is evident from a consideration of Table 17-5. Of particular note are those reactions leading to the formation of L-alanine, L-lysine, γ-aminobutyric acid, and histamine, all important metabolites.

Racemization Enzymes catalyzing the racemization of alanine, methionine, glutamate, proline, lysine, and serine and for the epimerization of hydroxyproline and diaminopimelate have been identified in bacteria. The last enzyme is, of course, involved in the biosynthesis of L-lysine. In addition, proline and alanine are metabolized in certain organisms via the D-amino acids rather than the L isomers. The metabolic functions of the other enzymes are less well defined but probably include synthesis of D-amino acids for cell wall construction (page 551).

Deamination Enzymes that catalyze the complete reaction

$$R{-}\underset{\substack{|\\ {}^{+}NH_3}}{CH}{-}CO_2^- + O_2 + H_2O \rightarrow R{-}\overset{\substack{O\\ \|}}{C}{-}CO_2^- + NH_4^+ + H_2O_2 \qquad (17\text{-}21)$$

are termed amino acid oxidases. In the presence of catalase, the hydrogen peroxide formed in this reaction is decomposed to oxygen and water. In its absence, the hydrogen peroxide reacts with the ketoacid product nonenzymatically in an oxidative decarboxylation reaction:

$$R{-}\overset{\substack{O\\ \|}}{C}{-}CO_2^- + H_2O_2 \rightarrow RCO_2^- + CO_2 + H_2O \qquad (17\text{-}22)$$

Under certain conditions, reaction (17-22) occurs much more slowly than reaction (17-21), so that α-ketoacid can accumulate even in the absence of catalase.

The amino acid oxidases fall into two broad categories: those that catalyze the oxidative deamination of L-amino acids and those that perform a similar function employing the D-isomers as substrates. Enzymes of both classes are reasonably widely distributed in living material. L-*Amino acid oxidases* have been detected in the venom of most poisonous snakes, in avian liver, in molds, and in bacteria. The corresponding D-*amino acid oxidases* occur in the kidney and liver of mammals, and in birds, amphibians, insects, bacteria, and molds. Aside from the specificity requirement in terms of configuration about the α-carbon of the substrate, most amino acid oxidases exhibit a rather broad, though not unlimited, specificity. A typical amino acid oxidase will utilize as substrate perhaps 10 of the amino acids. The reaction rates of course vary with the substrate.

Glycine is oxidized by the D-amino acid oxidase, which occurs in the liver and kidney of mammals. The reaction products are ammonia and glyoxylic acid:

$$H_3N^+{-}CH_2{-}CO_2^- + O_2 + H_2O \rightarrow H{-}\overset{\substack{O\\ \|}}{C}{-}CO_2^- + NH_4^+ + H_2O_2 \qquad (17\text{-}23)$$

Although the deamination of amino acids was inferred shortly after the turn of the century by experiments that showed interconversion of α-keto and α-amino acids, careful and detailed studies concerning the enzymes responsible for these reactions were not initiated until 1935. At that time, in a classical paper, Krebs reported the oxidation of L- and D-amino acids by homogenates of mammalian kidney and liver (94). This work stimulated a general study of the enzymes involved in oxidative deamination of amino acids. To date, the best understood enzymes are the L-amino acid oxidases from the venom of the cottonmouth moccasin and the diamondback rattlesnake, and the D-amino acid oxidase from mammalian kidney. We consider the properties of these enzymes in turn.

The L-amino acid oxidases of snake venoms have been obtained in homogeneous form and that of the rattlesnake has been crystallized (95). These enzymes exhibit molecular weights near 130,000 and contain 2 mol of FAD per mole of protein. In terms of the nature of the coenzyme, it is interesting to note that the L-amino acid oxidase of rat kidney apparently contains riboflavin phosphate, rather than FAD, as prosthetic group (96). The snake venom enzymes exhibit *p*H optima near *p*H 7 and all are subject to inhibition by excess substrate.

Early considerations suggested that amino acid oxidase-catalyzed reactions involved (1) oxidation of the amino acid to the imino acid with reduction of the flavin coenzyme, (2) hydrolysis of the imino acid, and (3) reoxidation of the reduced flavin by molecular oxygen. This simple scheme is very probably correct in its postulation of the imino acid as intermediate but is far from adequate for a complete description of the actual pathway of these reactions.

The D-amino acid oxidases of sheep and hog kidneys have been crystallized. These enzymes appear to be composed of basic subunits of molecular weight 45,700, which tend to polymerize, forming tetramers and other aggregates. D-amino acid oxidase contains 1 mol of FAD per 45,000 g of protein. This enzyme is maximally active under slightly alkaline conditions (*p*H 8.3).

In addition to the amino acid oxidases, several enzymes of other types are involved in the deamination of amino acids. For example, the pyridoxal phosphate–dependent dehydrases and desulfurases lead to the overall deamination of serine, threonine, and cysteine (page 782). Glutamic dehydrogenase (page 772), a NAD-dependent enzyme, catalyzes the oxidation of glutamic acid to the corresponding imino acid, which is spontaneously hydrolyzed to α-oxoglutaric acid and ammonia [Eq. (17-3)]. Finally, the enzymes aspartase and *histidase* (L-histidine ammonia-lyase) catalyze the nonoxidative deamination of aspartic acid and histidine according to Eqs. (17-12) and (17-24):

$$\underset{\text{Histidine}}{\text{Im--}CH_2\text{--}CH(NH_3^+)\text{--}CO_2^-} \longrightarrow \underset{\text{Urocanic acid}}{\text{Im--}CH{=}CH\text{--}CO_2^-} + NH_4^+ \qquad (17\text{-}24)$$

(Im = imidazol-4-yl ring)

Both of these reactions are freely reversible.

The role of the enzymes that catalyze the deamination of L-amino acids in metabolism seems clear. These enzymes, whether they act in an oxidative or nonoxidative fashion, permit the reshuffling of amino functions between various carbon skeletons. Thus, the amino function of a dietary amino acid may be employed to complete the biosynthesis of a different amino acid *in vivo*. In most cases, the inability of an organism to catalyze the *de novo* synthesis of an amino acid reflects an inability to construct the carbon skeleton, not an inability to introduce the amino function into a formed carbon skeleton. In reference to amino acid biosynthesis, it should be noted that the amino acid oxidase reactions are, in part, reversible. The incubation of the reduced enzyme with an appropriate α-ketoacid and ammonia under anaerobic conditions results in the synthesis of amino acids:

$$(E \cdot \text{flavin})_{red} + \text{ketoacid} + NH_3 \rightleftharpoons (E \cdot \text{flavin})_{ox} + \text{amino acid} \quad (17\text{-}25)$$

Because of the large negative free energy change of the transfer of a pair of electrons from the flavin functions of the reduced enzyme to oxygen, the overall reaction catalyzed by amino acid oxidases is, however, virtually irreversible.

SPECIFIC ASPECTS OF AMINO ACID METABOLISM

In addition to the tranformations discussed above, which are common to most or all of the amino acids, each amino acid is involved in reaction pathways that are unique to it. We now wish to direct attention to the more important of these. We begin with a discussion of the catabolism of several of the amino acids and subsequently consider particular aspects of metabolic interconversions involving amino acids.

Degradation of the Branched-Chain Amino Acids

In Figs. 17-22, 17-23, and 17-24, the pathways for oxidative degradation of valine, leucine, and isoleucine are presented. These catabolic pathways have the first three steps in common. Initially, transamination yields the corresponding α-ketoacids. These substrates are then oxidatively decarboxylated with formation of the —SCoA esters of the products. Apparently a single

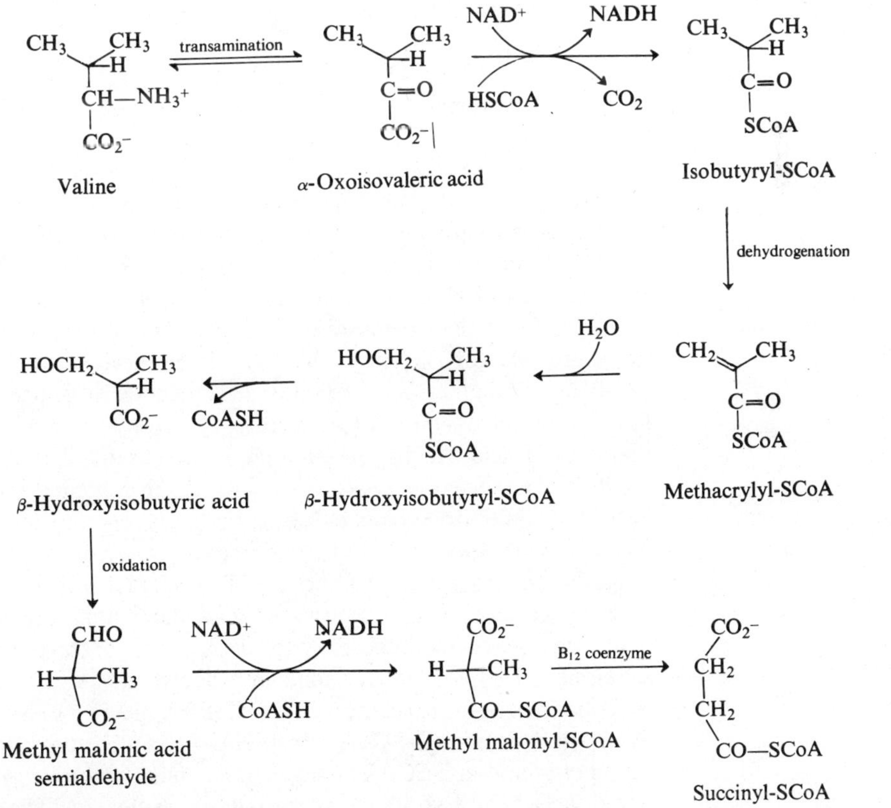

FIGURE 17-22
Pathway for valine degradation.

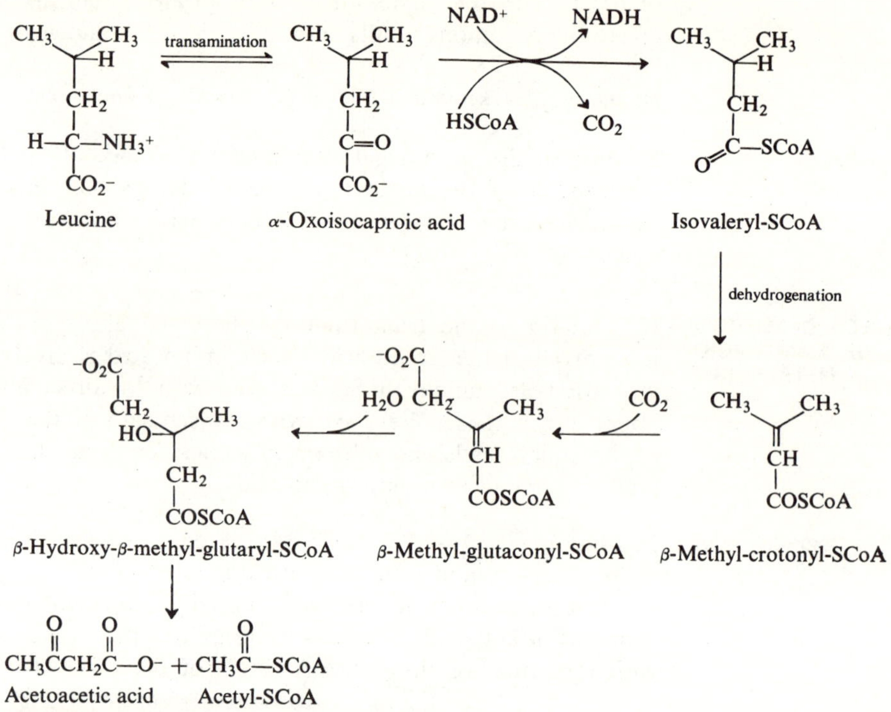

Leucine ⇌ (transamination) α-Oxoisocaproic acid → (NAD^+, HSCoA → NADH, CO_2) Isovaleryl-SCoA

Isovaleryl-SCoA → (dehydrogenation) β-Methyl-crotonyl-SCoA

β-Methyl-crotonyl-SCoA → (CO_2) β-Methyl-glutaconyl-SCoA → (H_2O) β-Hydroxy-β-methyl-glutaryl-SCoA

β-Hydroxy-β-methyl-glutaryl-SCoA → $CH_3COCH_2CO_2^-$ + $CH_3CO{-}SCoA$

Acetoacetic acid + Acetyl-SCoA

FIGURE 17-23
Pathway for leucine degradation.

enzyme is responsible for oxidative decarboxylation of all these ketoacids. Following decarboxylation, the acyl-SCoA's are dehydrogenated, yielding the unsaturated derivatives. At this point, the pathways diverge, although the remaining steps in each case bear a distinct resemblance to the reactions of fatty acid oxidation. Of note is the occurrence of β-hydroxy-β-methylglutaryl-SCoA in leucine oxidation. This compound is an important intermediate for the synthesis of cholesterol and other steroids from acetyl-SCoA (page 743).

Early studies led to the historical classification of amino acids as *ketogenic* or *glycogenic* for the mammal. The distinction between the two classes depends upon whether ketone bodies (acetoacetic acid, acetone, and the like) or glucose (or glycogen) are formed as a result of administration of the particular amino acid (a distinction that is frequently a function of experimental conditions). On this basis, valine is a glycogenic amino acid, leucine is ketogenic, and isoleucine exhibits both glycogenic and ketogenic behavior. These results may be readily understood in terms of the nature of the final degradation products. Valine yields succinyl-SCoA, which, via the tricarboxylic acid cycle and related enzymes, may be converted to pyruvic acid and thence to glucose (page 622). In contrast, leucine yields directly a ketone body, acetoacetic acid, and acetyl-SCoA. The latter may of course be converted to acetoacetic acid as well (page 599). Isoleucine yields acetyl-SCoA and propionyl-SCoA. Propionyl-SCoA is converted to succinyl-SCoA via methylmalonyl-SCoA (page 624) and hence is glycogenic. Since, as noted above, acetyl-SCoA is ketogenic, isoleucine should (and does) fall into both categories. A classification of all the amino acids as ketogenic or glycogenic is presented in Table 17-6.

Isoleucine ⇌ (transamination) α-Oxo-β-methylvaleric acid → (NAD^+, CoASH → NADH, CO_2) α-Methylbutyryl-SCoA → (dehydrogenation) Tiglyl-SCoA → (H_2O) α-Methyl-β-hydroxybutyryl-SCoA → (NAD^+ → NADH) α-Methylacetoacetyl-SCoA → Acetyl-SCoA + Propionyl-SCoA

$$CH_3\overset{O}{\overset{\|}{C}}—SCoA + CH_3CH_2\overset{O}{\overset{\|}{C}}—SCoA$$

FIGURE 17-24
Pathway for isoleucine degradation.

Degradation of Histidine

Although other pathways are available for its metabolism, histidine is degraded largely by a pathway involving the initial nonoxidative deamination catalyzed by histidase [Eq. (17-24)]. The product of this reaction, urocanic acid, derives its name from the fact that it was initially detected as an histidine metabolite in the urine of dogs. Urocanic acid is metabolized by the *urocanase*-catalyzed conversion to imidazolone-5-propionic acid, as indicated in Fig. 17-25. This reaction involves an internal oxidation–reduction with the addition of the elements of water and is probably pyridoxal phosphate dependent. The mechanism of this interesting transformation, which appears,

TABLE 17-6
Glycogenic and ketogenic amino acids

Glycogenic	*Ketogenic*	*Glycogenic and ketogenic*
Alanine	Leucine	Isoleucine
Arginine		Lysine
Aspartic acid		Phenylalanine
Cystine		Tyrosine
Glutamic acid		
Glycine		
Histidine		
Hydroxyproline		
Methionine		
Proline		
Serine		
Threonine		
Tryptophan		
Valine		

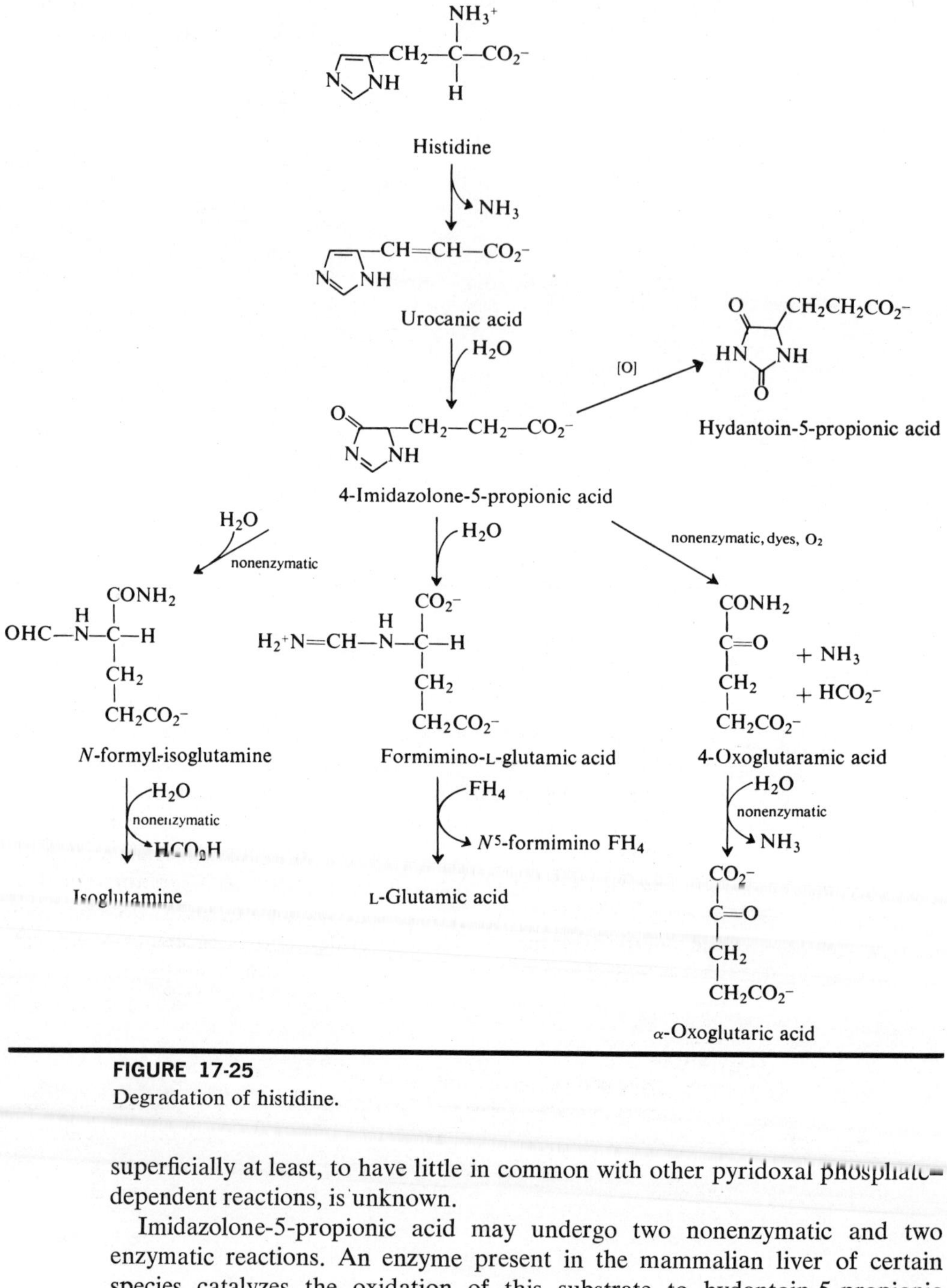

FIGURE 17-25
Degradation of histidine.

superficially at least, to have little in common with other pyridoxal phosphate-dependent reactions, is unknown.

Imidazolone-5-propionic acid may undergo two nonenzymatic and two enzymatic reactions. An enzyme present in the mammalian liver of certain species catalyzes the oxidation of this substrate to hydantoin-5-propionic acid, which is not further metabolized and is excreted as such (97,98). In the presence of molecular oxygen or particular dyes, imidazolone-5-propionic acid is oxidized nonenzymatically, yielding 4-oxoglutaramic acid plus ammonia and formic acid. Imidazolone-5-propionic acid may also be metabolized hydrolytically. The enzyme *imidazolone propionic acid hydrolase* catalyzes the formation of formimino-L-glutamic acid from this substrate, while a nonenzymatic route yields *N*-formyl-isoglutamine, which spontaneously hydrolyzes to isoglutamine. Formimino-L-glutamic acid, in the presence of tetrahydrofolate and the appropriate enzyme, yields L-glutamic acid and N^5-formimino tetrahydrofolate (page 407). These reactions are summarized in Fig. 17-25.

Metabolism of Tryptophan

The metabolism of tryptophan yields a variety of interesting and important substances, including *serotonin,* a powerful vasoconstrictor found in mammalian nervous tissue and elsewhere, *indoleacetic acid,* a growth-promoting substance for plants (an auxin), eye pigments in certain insects, and the vitamin *niacin* (nicotinic acid). The relationship between the metabolism of tryptophan and nicotinic acid has been recognized for many years, following nutritional studies in mammalian species which established that dietary requirements for nicotinic acid may be met by ingestion of excess tryptophan. The metabolic transformation of tryptophan to nicotinic acid is quite complex and has only recently been unraveled in detail. The first part of the pathway, involving the formation of 3-hydroxyanthranilic acid from tryptophan, has been well established for several years and is indicated schematically in Fig. 17-26.

The oxidation of tryptophan is initiated by the enzyme *tryptophan pyrrolase.* This enzyme, which possesses an iron protoprophyrin IX prosthetic group (page 419) catalyzes the conversion of tryptophan to L-*N*-formylkynurenine, employing molecular oxygen as oxidant. Tryptophan pyrrolase is an inducible enzyme, whose levels in mammalian tissues are elevated by administration of tryptophan, α-methyl tryptophan, or hormones such as cortisol (page 740).

N-Formylkynurenine may be converted to kynurenine in a reaction catalyzed by *kynurenine formylase,* an enzyme of mammalian liver, or may be metabolized to anthranilic acid and alanine by enzymes present in the mold *Neurospora crassa.* Alternatively, anthranilic acid may arise from kynurenine (see below). In mammals at least, anthranilic acid is an end product of tryptophan metabolism and is excreted in the urine.

Kynurenine-3-hydroxylase catalyzes the hydroxylation of kynurenine, employing molecular oxygen and NADPH. The further metabolism of kynurenine and 3-hydroxykynurenine have much in common. *Kynureninase,* a pyridoxal phosphate–dependent enzyme, catalyzes the cleavage of alanine from both of these substrates, yielding anthranilic acid and 3-hydroxyanthranilic acid. Similarly, *kynurenine transaminase* catalyzes the conversion of both of these substances to the corresponding ketoacids (99). The latter undergo spontaneous ring closures, yielding kynurenic and xanthurenic acids, which are further metabolized as indicated in Fig. 17-26. Finally kynurenine and 3-hydroxykynurenine are precursors of eye pigments, including xanthommatin from blowflies. Synthesis of this species may be regarded as occurring via condensation of 2 molecules of 3-hydroxykynurenine with loss of ammonia:

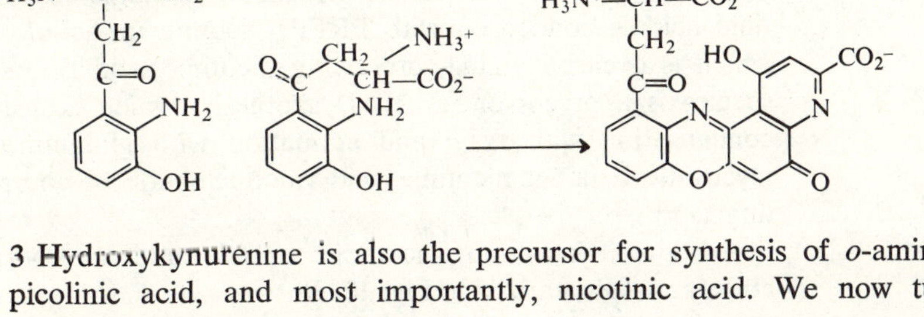

H_3N^+—CH—CO_2^- | CH_2 | C=O, NH_2, OH (3-hydroxykynurenine) + O=C, CH_2, NH_3^+, CH—CO_2^-, NH_2, OH (3-hydroxykynurenine) ⟶ H_3N^+—CH—CO_2^- | CH_2 | C=O, HO, CO_2^-, N, N, O, O (xanthommatin) (17-26)

3 Hydroxykynurenine is also the precursor for synthesis of *o*-aminophenol, picolinic acid, and most importantly, nicotinic acid. We now turn to a

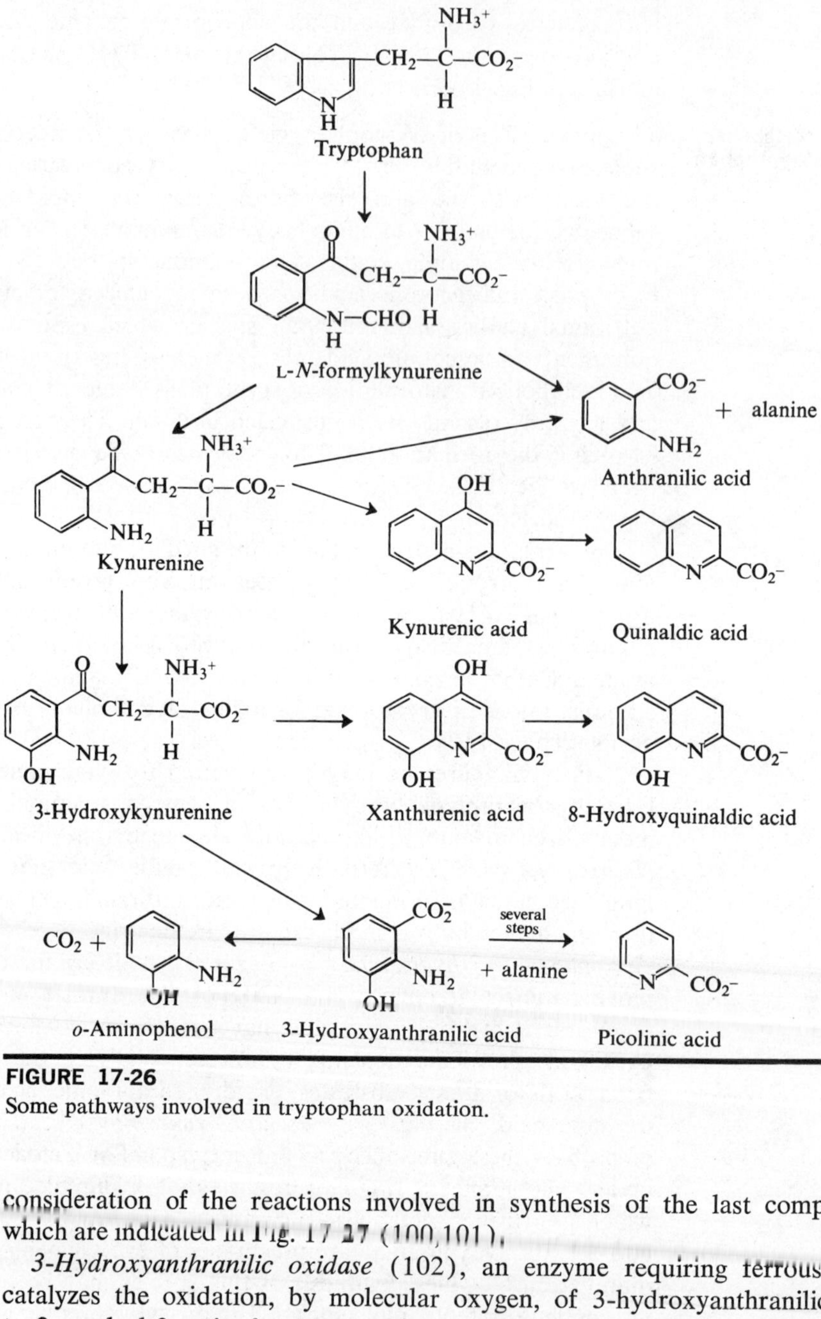

FIGURE 17-26
Some pathways involved in tryptophan oxidation.

consideration of the reactions involved in synthesis of the last compound, which are indicated in Fig. 17-27 (100,101).

3-Hydroxyanthranilic oxidase (102), an enzyme requiring ferrous ions, catalyzes the oxidation, by molecular oxygen, of 3-hydroxyanthranilic acid to 2-acroleyl-3-aminofumarate, which undergoes a spontaneous ring closure, yielding quinolinic acid. In the presence of a single enzyme from liver, quinolinic acid is condensed with PRPP, yielding quinolinic acid ribonucleotide, which is decarboxylated, producing nicotinic acid ribonucleotide. The latter species is a precursor for NAD synthesis, which is accomplished following condensation with ATP and amination with glutamine. Hydrolysis of the glycosidic bond of nicotinic acid ribonucleotide of course would yield nicotinic acid.

Additional important metabolic transformation of tryptophan, alluded to above, are summarized in Fig. 17-28.

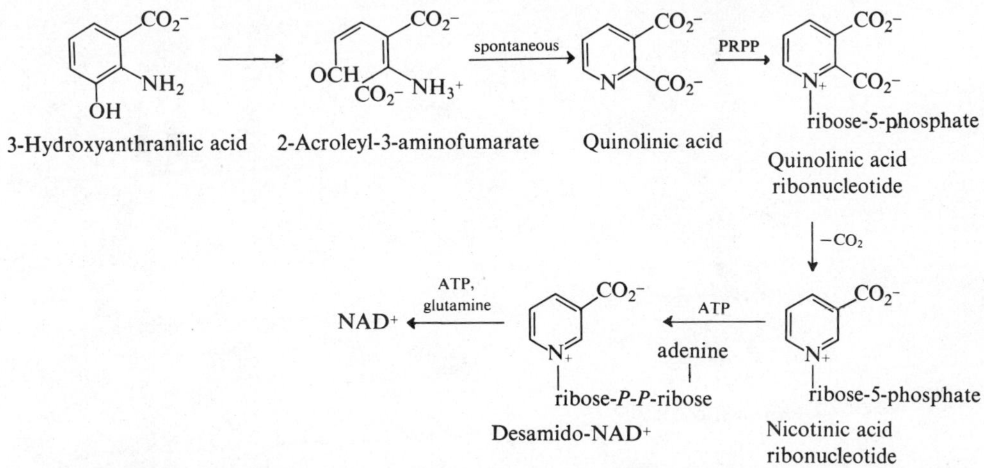

FIGURE 17-27
Synthesis of some nicotinic acid derivatives from 3-hydroxyanthranilic acid.

Degradation of Phenylalanine and Tyrosine

The oxidative metabolism of phenylalanine and tyrosine is of particular interest in that (1) a number of diseases resulting from "inborn errors of metabolism" (Table 19-1) are associated with this pathway of metabolism and (2) the associated enzymes require coenzymes, in some instances, which are quite unusual. The principal routes for oxidative degradation of these amino acids are indicated in Fig. 17-29.

Under normal physiological conditions, the bulk of phenylalanine is metabolized through its initial conversion to tyrosine. The enzyme responsible for catalyzing this reaction, phenylalanine hydroxylase, is typical of most hydroxylases (page 640), in that molecular oxygen and NADPH participate in

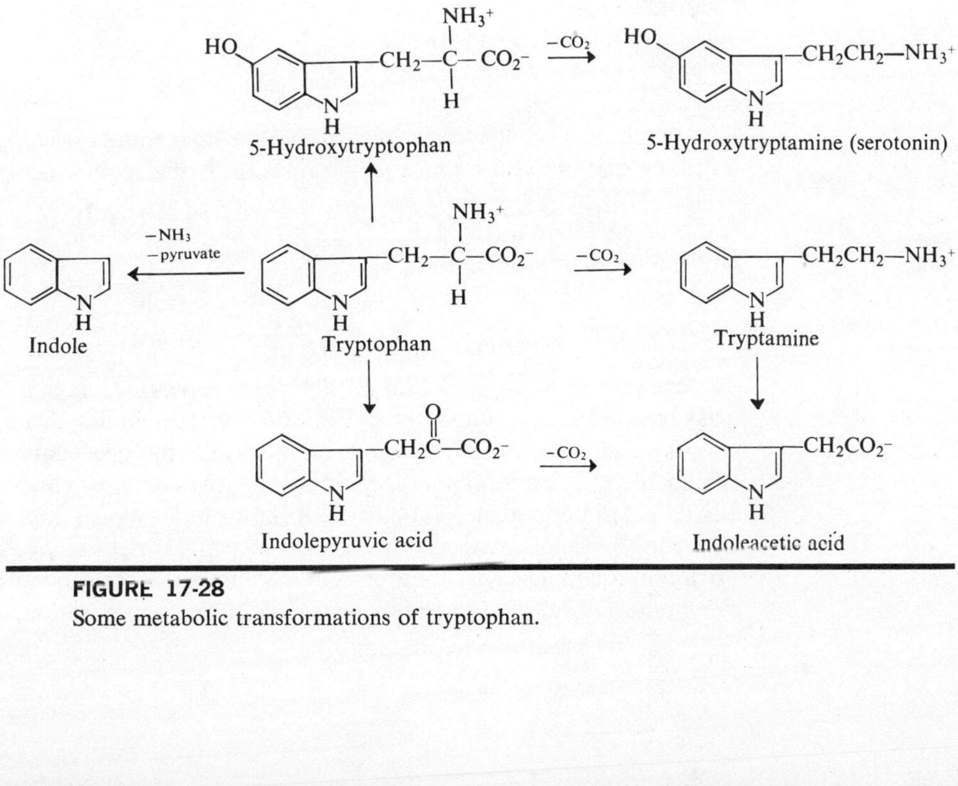

FIGURE 17-28
Some metabolic transformations of tryptophan.

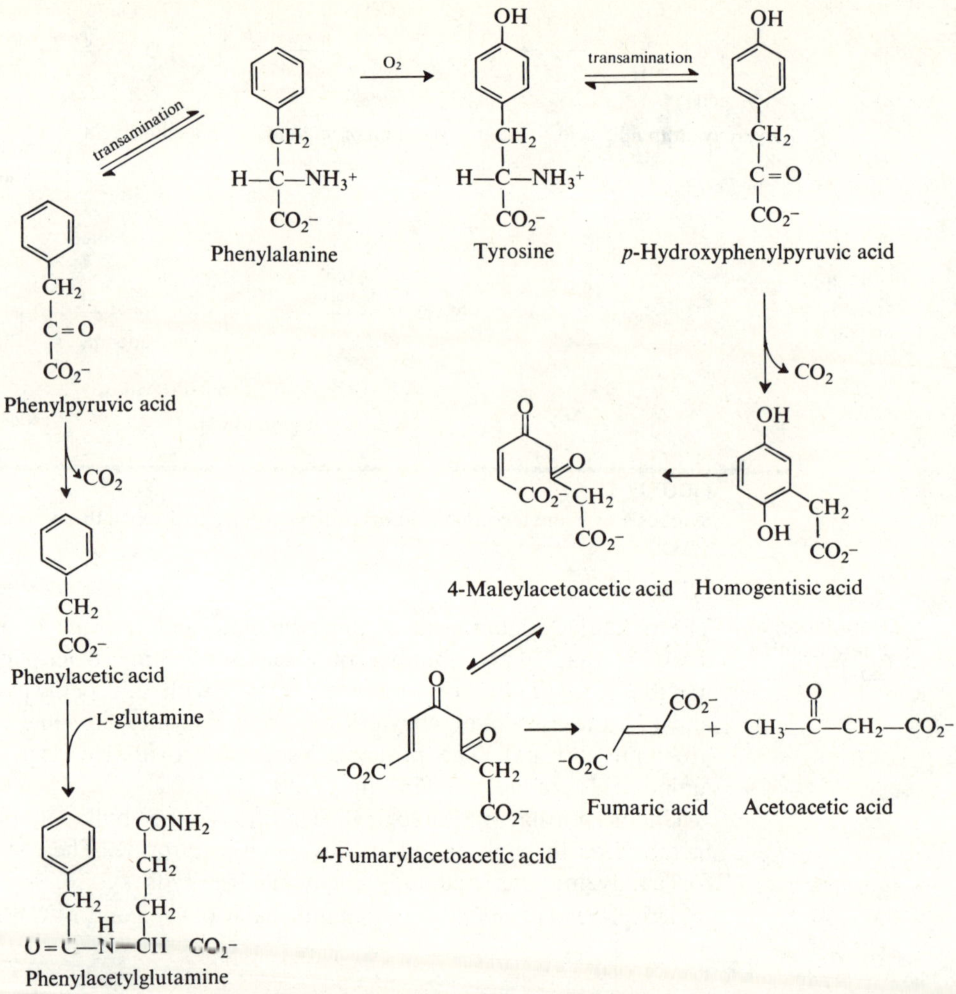

FIGURE 17-29
Principal routes for the metabolism of phenylalanine and tyrosine.

the reaction. This enzyme is localized in the microsomal fraction of the cell. A striking and unusual feature of the reaction is the requirement of a pteridine,

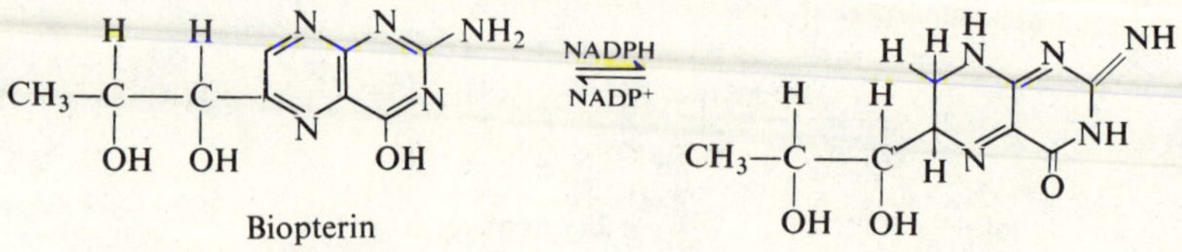

(17-27)

as coenzyme for the reaction (103). The apparent function of NADPH in this reaction is to maintain the pteridine in the quinoid dihydro form.

As in the case of several other amino acids, the degradation of tyrosine is initiated by a transamination yielding, in this case, *p*-hydroxyphenylpyruvic acid. This intermediate is metabolized in an interesting (and mechanistically puzzling) fashion. Apparently a single enzyme, *p-hydroxyphenylpyruvic acid oxidase,* a copper-containing protein, catalyzes the conversion of its substrate to homogentisic acid—a reaction involving decarboxylation, oxidation, hydroxylation of the aromatic ring, and migration of a side chain. The reaction requires ascorbic acid, whose function remains obscure. *Homogentisic acid oxidase,* an enzyme requiring ferrous ions, catalyzes the formation of maleylacetoacetic acid from homogentisic acid. Maleylacetoacetic acid is converted to the corresponding *trans* compound, fumarylacetoacetic acid, in a reaction catalyzed by *maleylacetoacetic isomerase* and requiring glutathione—one of the few known instances in which this species is involved coenzymatically (page 391). Cleavage of fumarylacetoacetic acid yields fumaric acid, which is further metabolized by enzymes of the citric acid cycle, and acetoacetic acid, which may either enter the citric acid cycle (following conversion to acetyl-SCoA) or be converted to fatty acids.

In Chapter 4, mention was made of the fact that humans afflicted with sickle-cell anemia possess a hemoglobin that is structurally distinct from that of normal individuals. Sickle-cell hemoglobin is an inherited disease; that is, the genetic information transmitted from parent to progeny does not possess the correct information for synthesis of a normal hemoglobin molecule (see the discussion of genetics in Chapter 19). This disease is one of a general class that may be termed "inborn errors of metabolism." Several such diseases were recognized by Garrod many years ago. This worker considered cystinuria, alcaptonuria, and albinism, for example, to be the consequence of the inability of the organism to carry out a particular metabolic transformation in the normal sequence of events. This conclusion has certainly been fully corroborated by subsequent findings. In many cases, it has been possible to demonstrate that inborn errors of metabolism are associated with the lack or partial lack of a particular enzymatic activity. This may reflect failure to synthesize the pertinent enzyme at all, the synthesis of a structurally altered enzyme, or synthesis of the normal enzyme in altered amounts. Several inborn errors of metabolism are concerned with amino acid metabolism. We consider three examples from the metabolism of phenylalanine and tyrosine: albinism, alcaptonuria, and phenylketonuria.

Albinism is characterized by lack of pigment in skin, hair, and retina. This disease assumes several forms and may be either complete, involving total lack of pigment, or incomplete, in which case pigment loss is restricted to particular areas. Melanin, the pigment of hair, skin, and eyes, is a polymeric material of unknown structure which arises from the oxidation of tyrosine. The only enzyme concerned with the conversion of tyrosine to melanin is tyrosinase (*o*-diphenol : O_2 oxidoreductase), a copper-containing protein, which catalyzes the conversion of tyrosine to dihydroxyphenylalanine and subsequently to dihydroxyphenylalanine quinone, as indicated in Eq. (17-28) and page 637. The conversion of the last substrate to melanin is apparently spontaneous. Albinism results from the lack of normal tyrosinase activity.

Phenylketonuria results from the absence of normal phenylalanine hydroxylase activity. Under these conditions, phenylalanine cannot be converted to tyrosine, and its metabolism to phenylpyruvic and phenyllactic acids (Fig. 17-29) is greatly exaggerated. These substances, as the name of the disease implies, are excreted in the urine. This is a severe disorder and results in marked mental retardation in children. The symptoms of the disease may be prevented provided that the intake of dietary phenylalanine is restricted from birth.

Alcaptonuria is characterized by the excretion of homogentisic acid in the urine, which rapidly darkens on exposure to oxygen as a result of the oxidation of this substance to the pigment alcapton. This disease is caused by the

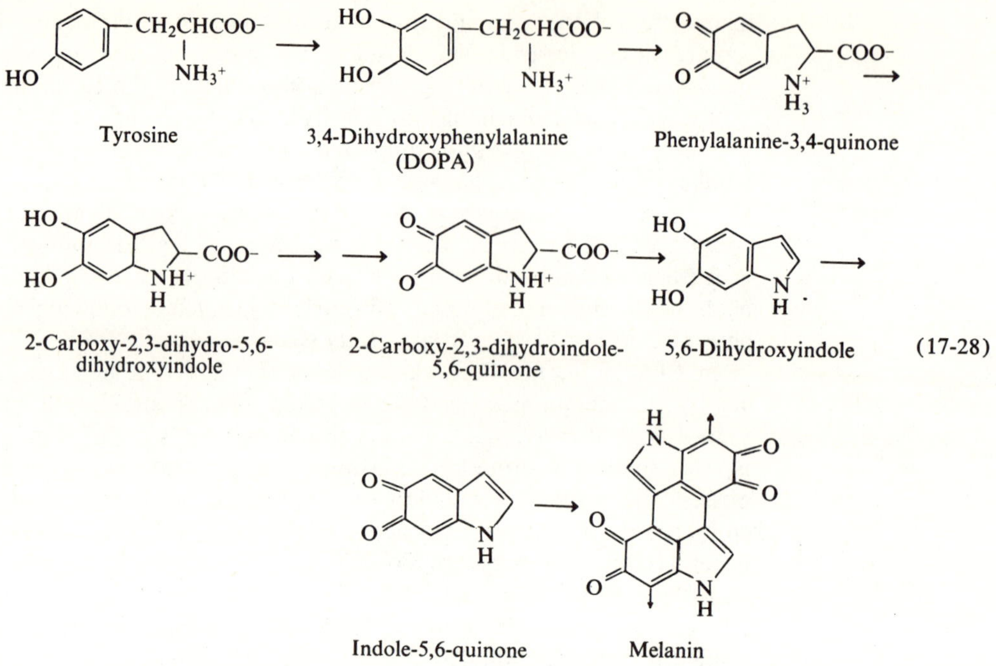

Tyrosine → 3,4-Dihydroxyphenylalanine (DOPA) → Phenylalanine-3,4-quinone →

2-Carboxy-2,3-dihydro-5,6-dihydroxyindole → → 2-Carboxy-2,3-dihydroindole-5,6-quinone → 5,6-Dihydroxyindole → (17-28)

Indole-5,6-quinone → Melanin

congenital absence of homogentisic acid oxidase. Other than darkening of the urine, alcaptonuriacs experience no symptoms early in life. Subsequently, pigmentation of the connective tissue occurs, which is usually associated with the development of arthritis.

A partial compilation of inborn errors of metabolism is collected in Table 19-1 (page 857).

Methionine and Biological Transmethylations

As noted above, the *de novo* synthesis of methyl groups involves the reduction of N^5,N^{10}-methylene FH_4 to N^5-methyl FH_4 (pages 409 and 427). N^5-methyl FH_4 donates its methyl function in only one known reaction: biosynthesis of methionine (page 781). For the most part, the remaining transmethylation reactions employ methionine as methyl group donor. Methyl group transfers from methionine require ATP and Mg^{2+}. These requirements are readily understood on the basis of the intermediate formation of *S*-adenosylmethionine [Eq. (17-29)], which serves as the actual methyl group donor. It will be recalled that this compound is classed with the "high-energy" metabolites. The synthesis of *S*-adenosylmethionine is unusual among ATP-

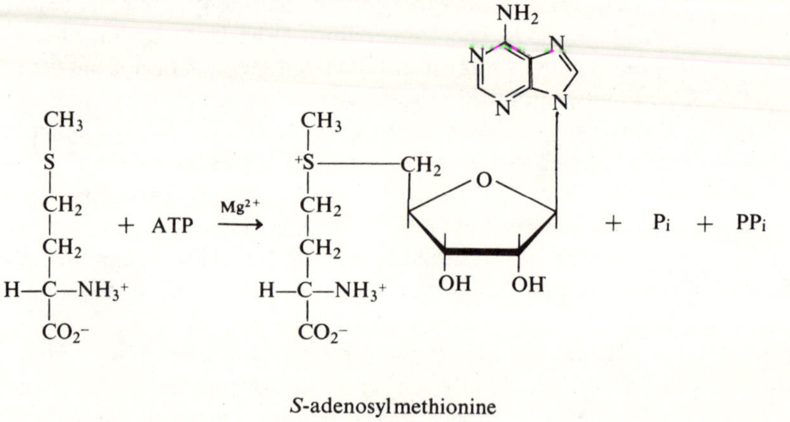

$$CH_3\text{-}S\text{-}CH_2\text{-}CH_2\text{-}CH(NH_3^+)\text{-}CO_2^- + ATP \xrightarrow{Mg^{2+}} CH_3\text{-}{}^+S(\text{-}CH_2\text{-adenosine})\text{-}CH_2\text{-}CH_2\text{-}CH(NH_3^+)\text{-}CO_2^- + P_i + PP_i \quad (17\text{-}29)$$

S-adenosylmethionine

dependent reactions in that this is a case in which the three phosphates of ATP are converted to enzyme-bound trimetaphosphate, which is cleaved on the enzyme to yield phosphate and pyrophosphate.

The number of known transmethylations is quite large. In several of these reactions, the participation of *S*-adenosylmethionine as methyl group donor has been established or strongly suggested. A few are collected in Table 17-7. In each instance, the transmethylation conforms to the general formulation of Eq. (17-30).

$$\text{S-adenosylmethionine} + \text{substrate} \rightarrow \text{S-adenosylhomocysteine} + \text{methyl-substrate}$$

$$\text{S-adenosylhomocysteine} \rightarrow \text{homocysteine} + \text{adenosine} \qquad (17\text{-}30)$$

The homocysteine product may either be reconverted to methionine through methylation with N^5-methyl FH_4 [Eq. (17-15)] or by reactions in which alternative substances serve as methyl group donors. The latter class of methyl group donors, which serve only as donors to homocysteine, includes dimethylthetin, methylmethionine, and betaine. The overall reaction is

$$\underset{\text{Dimethylthetin}}{(CH_3)_2S^{+}\text{—}CH_2\text{—}CO_2^-} + HS\text{—}CH_2\text{—}CH_2\text{—}\underset{\displaystyle CO_2^-}{\underset{|}{\overset{}{C}}}H\text{—}NH_3^+ \longrightarrow CH_3\text{—}S\text{—}CH_2\text{—}CO_2^- + CH_3\text{—}S\text{—}CH_2\text{—}CH_2\text{—}\underset{\displaystyle CO_2^-}{\underset{|}{C}}H\text{—}NH_3^+ + H^+ \qquad (17\text{-}31)$$

Dimethylthetin homocysteine methyltransferase, for example, is a widely distributed enzyme that has been obtained in homogeneous form (104,105). This enzyme appears to be distinct from that which catalyzes the synthesis of methionine employing betaine as methyl group donor.

TABLE 17-7
Partial compilation of transmethylation reactions involving S-adenosylmethionine

Enzyme	*Reaction*	*Source*	*Reference*
Norephinphrine methyltransferase	Norepinephrine → epinephrine	Bovine adrenal medulla	96
Catechol methyl transferase	Catechol and derivatives → 3-methoxycatechols	Liver	97,98
Acetylserotonin *O*-methyltransferase	Acetylserotonin → melatonin	Nervous tissue and pineal gland	99
2,6-Diaminopurine methyltransferase	2,6-diaminopurine → 2-methylamino-6-aminopurine	*E. coli*	100,101
Polynucleotide methyltransferase	Polynucleotides → methylated polynucleotides		102
Guanidinoacetate methyltransferase	Guanidinoacetate → creatine	Liver	103
Nicotinamide methyltransferase	Nicotinamide → *N*-methylnicotinamide	Liver	104
Phosphatidyl ethanolamine methyltransferase	Phosphatidyl ethanolamine → phosphatidyl choline	Liver, *Neurospora*	105,106

Glycine and the Synthesis of Tetrapyrroles

As emphasized in Chapter 15, structures derived from the tetrapyrrole nucleus perform a variety of essential functions in living organisms. Most organisms have the capacity to synthesize these complicated molecules *de novo*. The biochemical pathways through which simple precursors give rise to tetrapyrroles were completely unknown prior to the advent of isotope-tracer methods. Employing these methods, Shemin and his colleagues, in a series of elegant experiments, elucidated the principal features of these pathways (106). Glycine, which is an important precursor of purines (page 820) and serine (page 782), proved to be one of the basic building blocks for these molecules. Tracer experiments demonstrated that both *in vivo* (man and rats) and *in vitro* (duckling erythrocytes), labeled acetate and glycine were incorporated into the protoporphyrin IX nucleus of the prosthetic group of hemoglobin. Careful degradation of the labeled material sufficed to establish the source of each atom of the protoporphyrin IX molecule. The results are indicated in Fig. 17-30. The important conclusions derived from this work include (1) that glycine donated the nitrogen atoms for each ring as well as carbon for the methylene bridges, (2) that the remaining atoms were derived from either the carboxyl or the methyl group of acetate, (3) that all rings were identically labeled, indicating a common precursor for all rings, (4) that the labeling pattern indicated the existence of a common precursor for both sides of each pyrrole ring, and (5) that the labeling pattern of the atoms derived from acetate indicated that this substrate had been metabolized via the tricarboxylic acid cycle. All these results can be readily accommodated by the reaction pathway indicated in Eq. (17-32),

$$\text{Acetate} \xrightarrow{\text{TCA}} \begin{matrix} CO_2^- \\ | \\ CH_2 \\ | \\ CH_2 \\ | \\ O{=}C{-}SCoA \end{matrix} + \begin{matrix} NH_3^+ \\ | \\ CH_2 \\ | \\ CO_2^- \end{matrix} \longrightarrow \begin{matrix} CO_2^- \\ | \\ CH_2 \\ | \\ CH_2 \\ | \\ C{=}O \\ | \\ H{-}C{-}NH_3^+ \\ | \\ CO_2^- \end{matrix} \xrightarrow{CO_2 \nearrow}$$

α-Amino-β-ketoadipic acid

$$\begin{matrix} CO_2^- \\ | \\ CH_2 \\ | \\ CH_2 \\ | \\ C{=}O \\ | \\ CH_2 \\ | \\ NH_3^+ \end{matrix} \xrightarrow{2H_2O \nearrow} \begin{matrix} CO_2^-\ \ CH_2{-}CO_2^- \\ |\qquad | \\ CH_2\ \ CH_2 \\ \text{(pyrrole ring, N)} \\ CH_2 \\ | \\ NH_3^+ \end{matrix} \tag{17-32}$$

δ-Aminolevulinic acid   Porphobilinogen

for which unequivocal evidence has been garnered. The initial reaction involves condensation of glycine with succinyl-SCoA (derived from acetate via the tricarboxylic acid cycle), yielding δ-aminolevulinic acid and carbon dioxide. This reaction is pyridoxal phosphate dependent. Two molecules of δ-aminolevulinic acid are then condensed, with the loss of two molecules of water, to yield the crucial intermediate, porphobilinogen. This reaction is catalyzed by *δ-aminolevulinic acid dehydrase*. Porphobilinogen is apparently the immediate precursor to the tetrapyrroles.

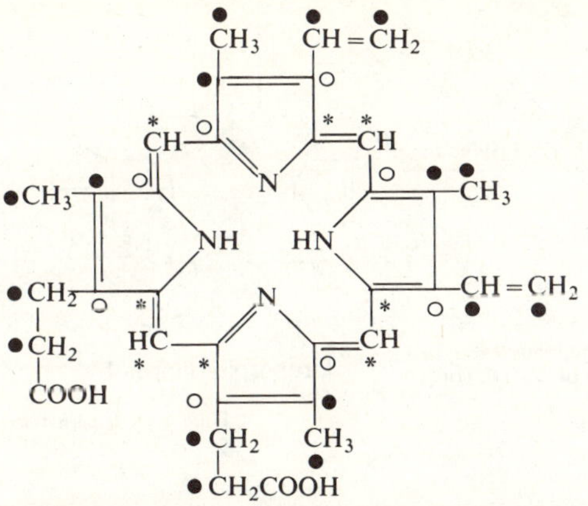

FIGURE 17-30
Metabolic sources of atoms of protoporphyrin IX, as shown by isotope experiments. All 4 N atoms are derived from glycine; atoms marked with asterisks are derived from the methylene carbon of glycine; atoms marked with solid circles are derived from the methyl carbon of acetate; atoms marked with open circles are derived mainly from the methyl carbon of acetate and in small part from the carboxyl carbon of acetate; the unmarked carbon atoms of the COOH groups are derived solely from the carboxyl carbon of acetate. J. Fruton and S. Simmonds, *General Biochemistry,* Wiley, New York, 1958, p. 865.

The general pathways by which the various tetrapyrroles of biochemical interest are derived from porphobilinogen are indicated in Fig. 17-31. The head-to-tail condensation of four molecules of the latter substrate, together with ring closure, yields uroporphyrinogen I directly. Dehydrogenation, yielding methine bridges in place of the methylene bridges, produces uroporphyrin I. The mode of synthesis of uroporphyrinogen III is not yet understood. The obvious problem stems from considerations as to how to introduce the fourth pyrrole nucleus "backward." Part of this synthesis may resemble that for uroporphyrinogen I, as *uroporphyrinogen-I synthetase,* as well as an enzyme designated *uroporphyrinogen-III cosynthetase,* is required for the reaction. The coproporphyrinogens arise from the corresponding uroporphyrinogens by the action of *uroporphyrinogen decarboxylase.* Further side-chain decarboxylations and oxidations yield the protoporphyrins.

Arginine and the Biosynthesis of Urea

The end product of nitrogen metabolism varies markedly from species to species. Of particular interest is the fact that in man and other terrestrial vertebrates (ureotelic organisms) this end product is urea. While nitrogen-containing substances other than urea certainly occur in the urine of man, urea is quantitatively the most important. The principal sources of ammonia for biosynthesis of urea have been discussed above. These include the oxidative and nonoxidative deaminations of amino acids and the hydrolysis of the amides of glutamic and aspartic acids.

The overall pathway for synthesis of urea from carbon dioxide and ammonia is indicated in Fig. 17-32. The initial step in this reaction sequence, formation of carbamyl phosphate, is also the initial step in synthesis of the pyrimidines and is discussed in some detail on page 827. Ornithine transcarbamylase (carbamoyl phosphate : L-aspartate carbamoyl-transferase) catalyzes the condensation of carbamyl phosphate and ornithine, yielding citrulline. This enzyme occurs in mammalian (but not avian) liver, in higher and

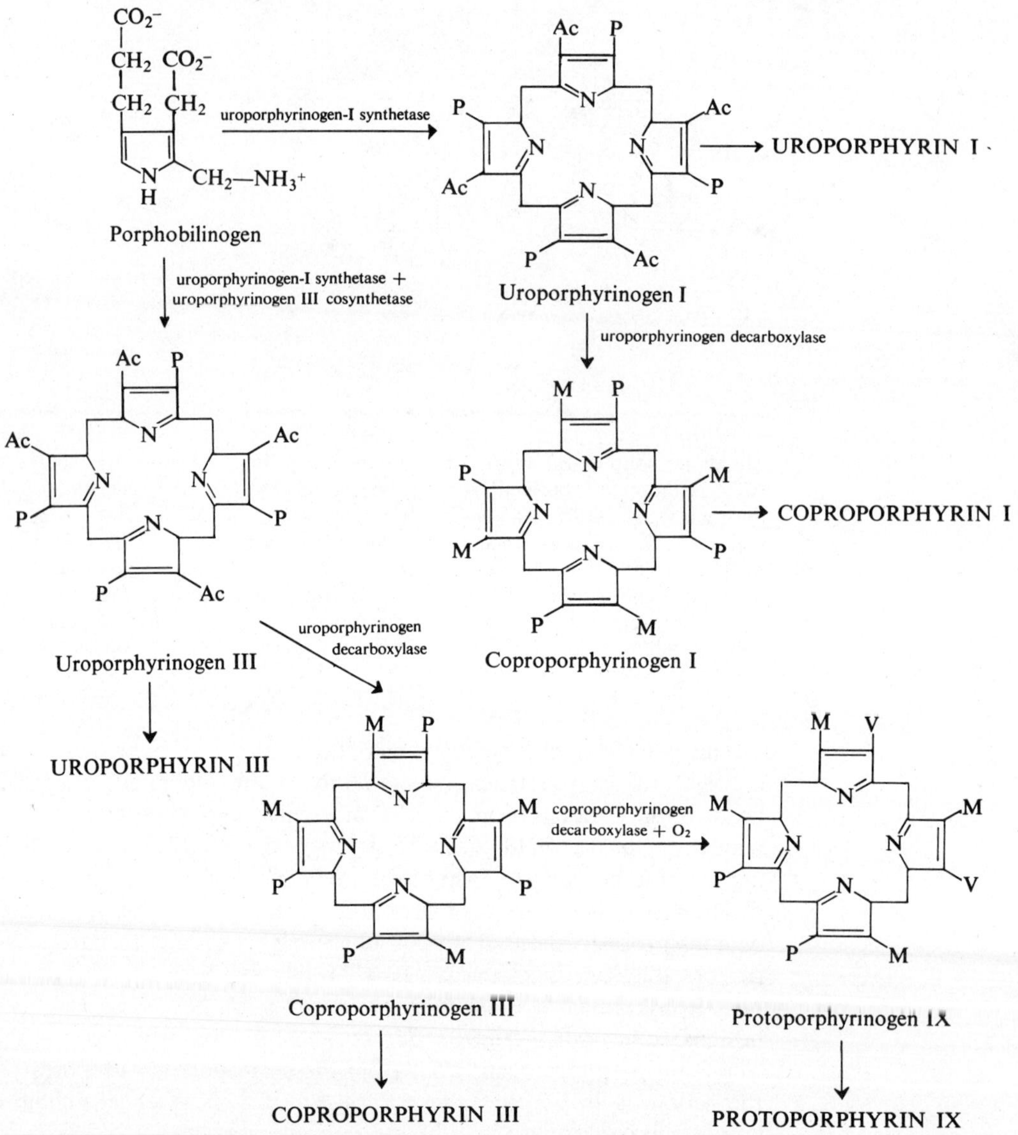

FIGURE 17-31
Conversion of porphobilinogen to various porphyrins.

lower plants, and in certain microorganisms. Conversion of citrulline to arginine occurs via a two-step reaction path bearing a close resemblance to that for synthesis of adenylic acid (page 825). *Argininosuccinate synthetase* —L-citrulline : L-aspartate ligase (AMP)—catalyzes the ATP-dependent formation of an acyl guanidinium linkage between aspartic acid and citrulline. The product, argininosuccinate, is cleaved to fumarate and arginine in an *argininosuccinase* (L-argininosuccinate arginine-lyase)-catalyzed process that bears a formal resemblance to the reaction catalyzed by aspartase (page 778). The terminal step in urea production is the hydrolysis of arginine, catalyzed by *arginase,* yielding urea, and regenerating ornithine, which, subsequently, may condense with an additional mole of carbamyl phosphate. Thus, urea formation is a cyclic process in which carbon dioxide and ammonia enter at one site and urea leaves at another.

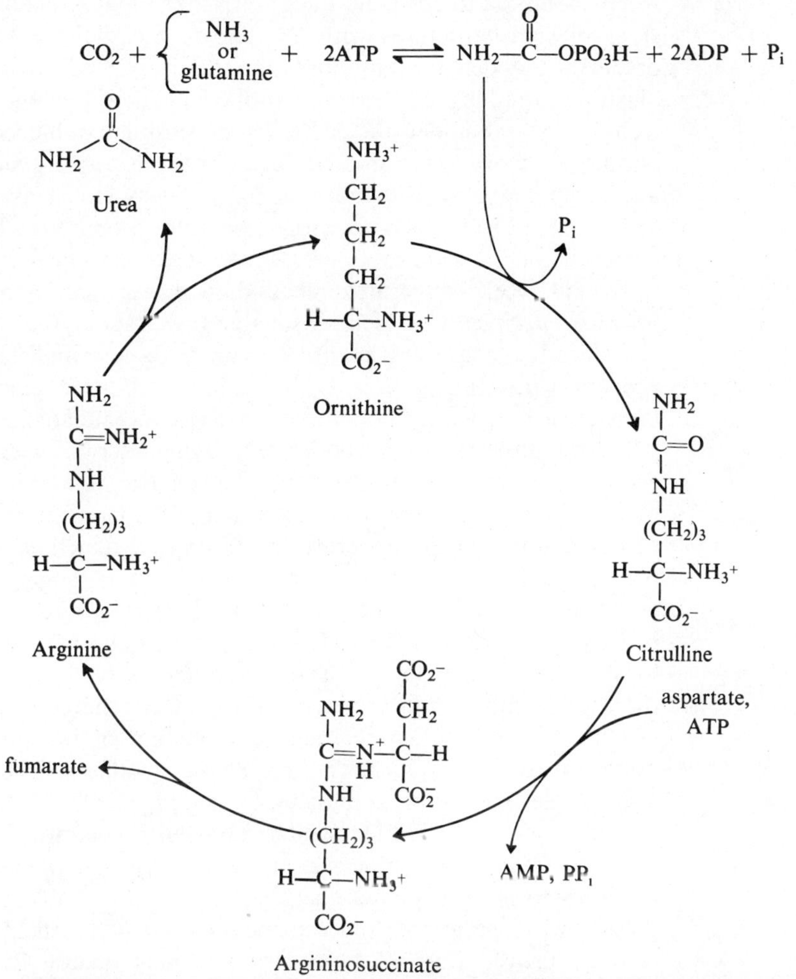

$$CO_2 + \left\{ \begin{matrix} NH_3 \\ \text{or} \\ \text{glutamine} \end{matrix} \right. + 2ATP \rightleftharpoons NH_2{-}\overset{O}{\overset{\|}{C}}{-}OPO_3H^- + 2ADP + P_i$$

Urea; Ornithine; Citrulline; Argininosuccinate; Arginine; aspartate, ATP; AMP, PP_i; fumarate; P_i

FIGURE 17-32
Urea cycle.

BIOSYNTHESIS OF ALKALOIDS

The metabolism of the amino acids is intimately related to the biogenesis of the alkaloids, a heterogeneous group of nitrogeneous substances produced by plants. Among all the biogenetic sequences, the pathways for synthesis of the alkaloids are least well understood. The backward state of this area is the result of several factors. First, because of the very complicated and intricate structures of many of the alkaloids, meaningful approaches to their synthetic pathways were possible only following the development of isotope-labeling techniques. Second, localization of radioactive labels incorporated into alkaloids from simple precursors requires a good deal of sophistication in the methods of synthetic organic chemistry. The vast majority of biochemists working in the field of intermediary metabolism lacked such training and were, at any event, fully occupied with metabolic studies in classical areas. Third, stimulation of interest of organic chemists had to await a rather full understanding of both metabolic pathways and structural alkaloid chemistry. Finally, the elucidation of biogenetic pathways for alkaloids is a difficult and time-consuming task.

Early considerations of alkaloid biogenesis consisted largely of paper chemistry, much of it quite ingenious and some of it basically correct. This work attempted to correlate the known alkaloidal structures and interrelationships of such structures with the known metabolic reactions of reasonable precursors. A delightful monograph written by Sir Robert Robinson elegantly illustrates thinking of this type. Following development of isotope labeling techniques and under the stimulus of postulated biogenetic pathways, experimental work was initiated. Such work is rapidly increasing in volume and the broad general features of the pathways to many of the alkaloids are coming into focus. Alkaloid biogenesis fully promises to be an area of major chemical and biochemical interest for the next several years.

Present work concerning alkaloid biogenesis takes the following general form. A *specifically* labeled reasonable precursor to the alkaloid of interest is introduced into a living plant. This may be accomplished by including the precursor in a medium of hydroponically cultured plants (careful precautions must be observed to avoid bacterial contamination), by injection of the precursor into the plant cavities with a hypodermic needle, or by placing a drop of the precursor on the plant stem or leaf vein and introducing a puncture at that point. The vascular system of the plant then incorporates the precursor through the puncture. Following introduction of the precursor, the plant is permitted to grow naturally for a period ranging from a few days to a few months, after which the alkaloid (now presumed to be labeled) is isolated. If it does prove radioactive, it must be systematically degraded to localize the label within it. Such experiments, which constitute the bulk of the experimental work, are essential to detect random and unwanted incorporation of label due to metabolic degradation of the precursor followed by utilization of these products for alkaloid synthesis. The same degradations also furnish the proof that the localization of the label corresponds to the position appropriate to the postulated biogenetic pathway.

As so frequently happens in biochemistry, a methodological development promises to markedly alter the approach to the problem of alkaloid biosynthesis and to accelerate the rate of progress in the field. Specifically, it has recently proved possible to record ^{13}C nmr spectra of rather complicated organic compounds, containing as many as 30 carbon atoms, so that each carbon atom appears in the spectrum as a singlet. Moreover, such measurements can be made on compounds containing the natural abundance, about 1.1 percent, of ^{13}C. This provides an opportunity to avoid the use of radioactive precursors and the subsequent degradation procedures necessary to localize the label. A likely plan is to begin with precursors specifically enriched in ^{13}C at certain positions. A single nmr spectrum following isolation of the alkaloid of interest would suffice to identify each carbon atom in the alkaloid derived from the enriched carbon atoms of the precursor. Such atoms would appear as peaks of abnormally high intensity in the spectrum. This technique would dramatically reduce the total amount of labor involved in these studies. Currently, utilization of this technique does require rather large samples, 50 to 100 mg. However, advances in methods of recording nmr spectra suggest that it may soon be possible to make do with 10- to 100-fold smaller quantities.

As the above considerations suggest, current work is aimed at outlining general features and not at elucidating individual details. Little attention is paid to the sequence of reactions, timing of the introduction of methyl or hydroxyl groups, and the like. Few suggested intermediates have been isolated. Even fewer of the enzymes responsible for alkaloid biogenesis have been isolated, purified, and characterized. These tasks should provide employment for the next few generations of biochemists interested in intermediary metabolism.

Although there are a great many specific examples of the relationship of amino acid metabolism to alkaloid biosynthesis, let us focus on that of tyrosine to a group derived from norlaudanosoline:

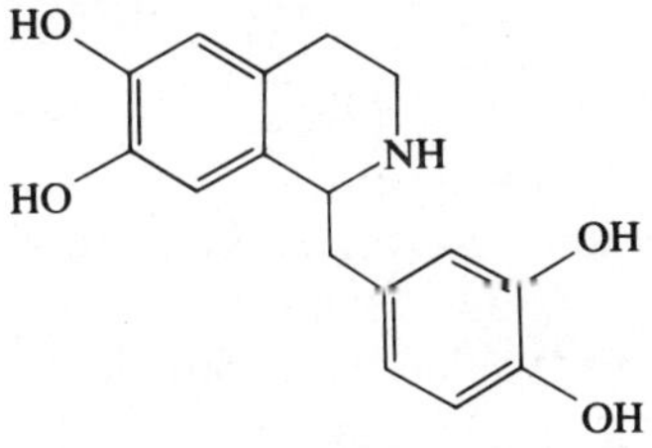

Some of the alkaloids derived from this intermediate are

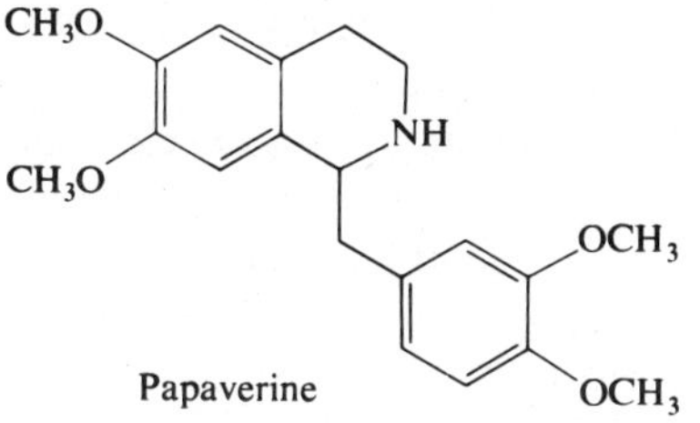

Papaverine

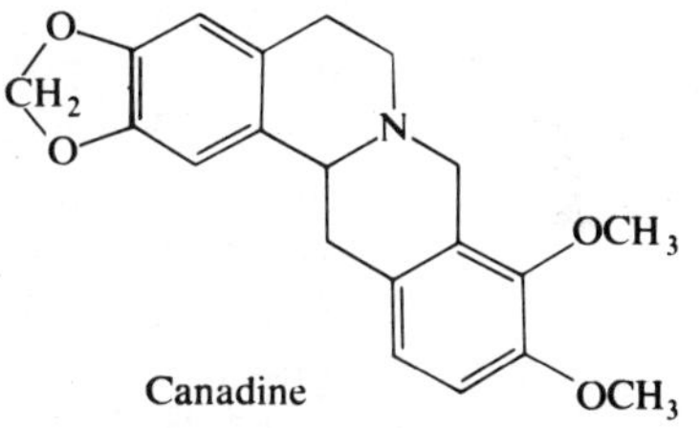

Canadine

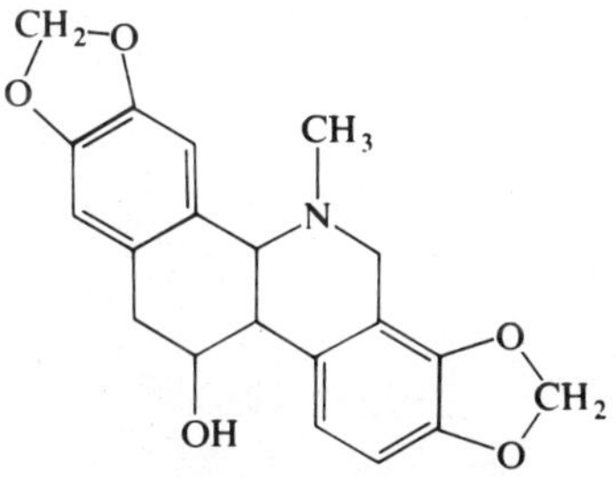

Chelidonine

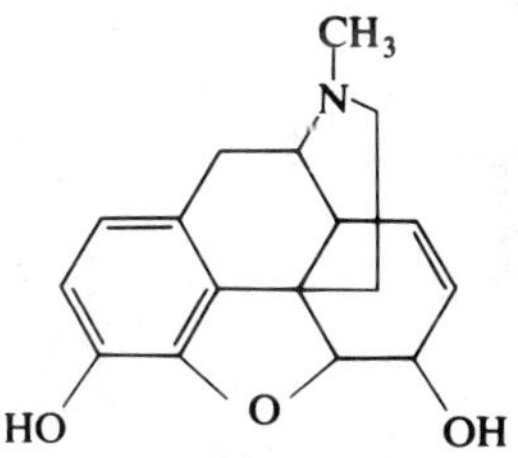

Morphine

We can focus our attention on two questions. How is norlaudanosoline formed from tyrosine? How can this species be converted into the variety of end products shown above?

The essential features of the norlaudanosoline skeleton can be formed from two molecules of tyrosine by condensing them in the following way:

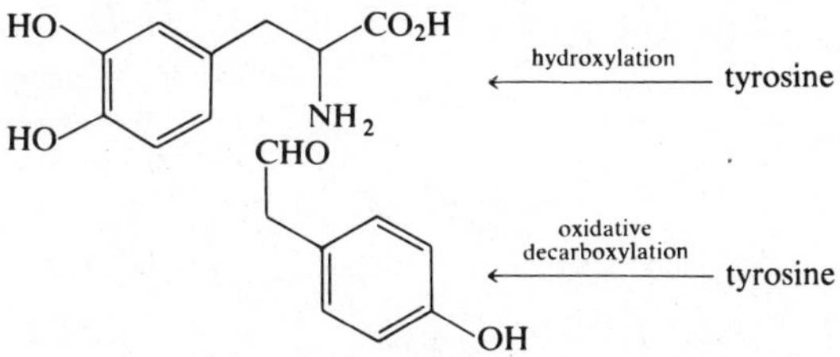

hydroxylation ← tyrosine

oxidative decarboxylation ← tyrosine

Forming the necessary ring involves the condensation of an amine, an aldehyde, and a potential carbanion (by virtue of the carbanion character present *para* to the phenolic hydroxyl group). This is an example of a Mannich reaction:

$$-\overset{|}{\underset{|}{C}}^{\ominus} + R-\overset{H}{\overset{|}{C}}{=}O + NHR_1R_2 + H^+ \rightarrow -\overset{|}{\underset{|}{C}}-\underset{R}{\underset{|}{CH}}-N\begin{matrix} R_1 \\ \\ R_2 \end{matrix} + H_2O \qquad (17\text{-}33)$$

A great many examples of Mannich reactions are known; they appear to play an important part in alkaloid biosynthesis.

A probable sequence of reactions involved in synthesis of norlaudanosoline from tyrosine is shown in Fig. 17-33. As shown in the figure, simple methylations, presumably mediated by *S*-adenosylmethionine, are adequate to complete the synthesis of papaverine and laudanosoline.

To generate alkaloids based on other skeleta requires more extensive modifications of norlaudanosoline. Some of these transformations are collected in Fig. 17-34. The crucial first step is the closure of an additional ring. This must involve formaldehyde or the biochemical equivalent, N^5,N^{10}-methylene FH_4, and yields the tetrahydroprotoberberine nucleus. This reaction is another example of the importance of the Mannich reaction in alkaloid synthesis. Methylation of this nucleus yields canadine directly. Oxidation followed by methylation yields the important alkaloid, berberine. Hydrolytic opening of the cationic ring of the berberine nucleus followed by oxidation yields the phthalideisoquinoline nucleus from which a number of alkaloids, including hydrastine, are derived.

As shown in Fig. 17-34, there are two important routes for the oxidative

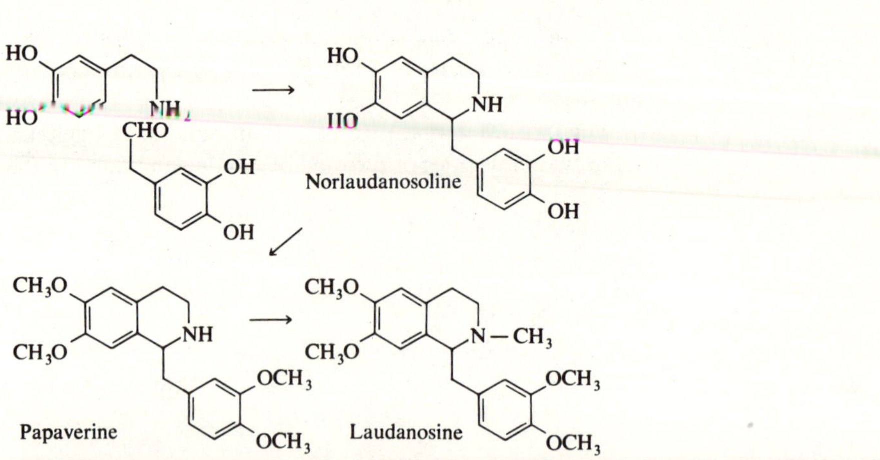

Tyrosine

Norlaudanosoline

Papaverine

Laudanosine

FIGURE 17-33
Synthesis of norlaudanosoline and related products.

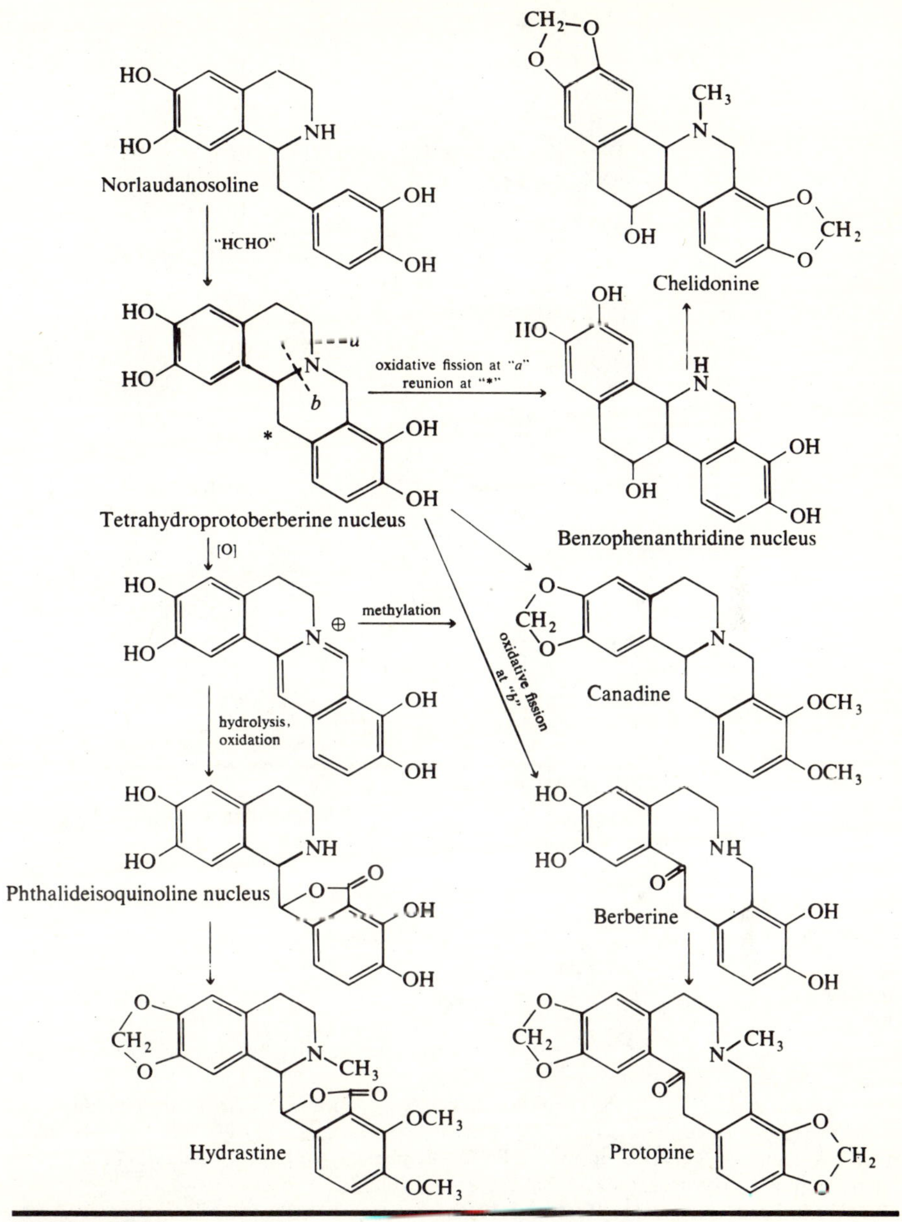

FIGURE 17-34
Some metabolic derivatives of norlaudanosoline.

cleavage of the tetrahydroprotoberberine nucleus. One of these routes, designated "*a*" yields the benzophenanthridine nucleus from which chelidonine is derived. The other, "*b*" cleavage, is not coupled with ring closure and yields alkaloids of the protopine type.

To understand how morphine and its related alkaloids are derived from norlaudanosoline, it is convenient to redraw the structure in the way shown in Fig. 17-35. As shown here, oxidative ring closure yields the skeleta on which isothebaine, on the one hand, and morphine and thebaine, on the other, are derived.

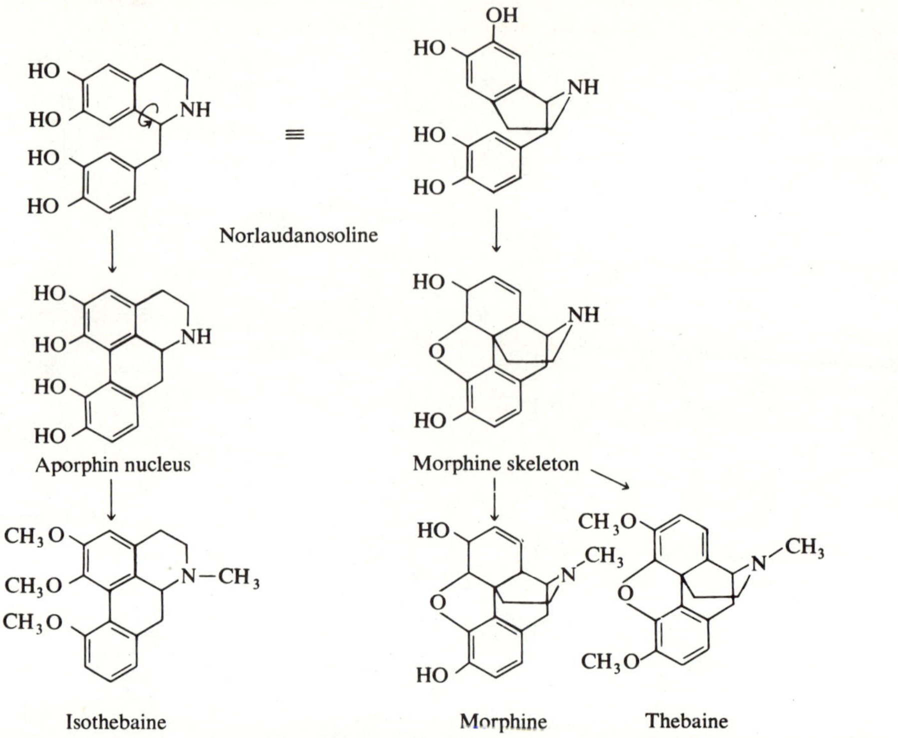

Norlaudanosoline

Aporphin nucleus

Morphine skeleton

Isothebaine

Morphine

Thebaine

FIGURE 17-35
Conversion of norlaudanosoline to aporphin and morphine alkaloids.

REFERENCES

Books and Review Articles

Aebi, H. E.: Inborn Errors of Metabolism, *Ann. Rev. Biochem.*, **36:** 271 (1967).

Bogorad, L.: Enzymatic Mechanisms in Porphyrin Syntheses, *Ann. N.Y. Acad. Sci.*, **104:** 676 (1963).

Bovey, F. A., and S. Yanari: Pepsin, in P. D. Boyer, H. Lardy, and K. Myrbäck (eds.), *The Enzymes*, vol. 4, Academic Press, New York, 2nd ed., 1960, p. 63.

Carnahan, J. E., and J. E. Castle: Nitrogen Fixation, *Ann. Rev. Plant Physiol.*, **14:** 125 (1963).

Davie, E. W., and O. D. Ratnoff: The Proteins of Blood Coagulation, in H. Neurath (ed.), *The Proteins*, vol. 3, Academic Press, New York, 2nd ed., 1965, p. 360.

Esnouf, M. P., and R. F. Macfarlane: Enzymology and the Blood Clotting Mechanism, *Advan. Enzymol.*, **30:** 255 (1968).

Fasella,, P.: Pyridoxal Phosphate, *Ann. Rev. Biochem.*, **36:** 185 (1967).

Greenberg, D. M.: Amino Acid Metabolism, *Ann. Rev. Biochem.*, **33:** 633 (1964).

Greenberg, D. M.: Biological Methylation, *Advan. Enzymol.*, **25:** 395 (1963).

Hardy, R. W. F., and R. C. Burns: Biological Nitrogen Fixation, *Ann. Rev. Biochem.*, **37:** 331 (1968).

Larner, J.: Inborn Errors of Metabolism, *Ann Rev. Biochem.*, **31:** 569 (1962).

Leete, E.: Alkaloid Biosynthesis, *Advan. Enzymol.*, **32:** 373 (1969).

Meister, A: Amino Group Transfer, in P. D. Boyer, H. Lardy, and K. Myrbäck (eds.), *The Enzymes*, vol. 6, Academic Press, New York, 2nd ed., 1962, p. 193.

Meister, A: *Biochemistry of the Amino Acids*, Academic Press, New York, 2nd ed., 1965.

Mortenson, L. E.: Inorganic Nitrogen Assimilation and Ammonia Incorporation, in I. C. Gunsalus and R. Y. Stanier (eds.), *The Bacteria*, vol. III, Academic Press, New York, 1962, p. 119.

Mortenson, L. E.: Nitrogen Fixation: Role of Ferredoxin in Anaerobic Metabolism, *Ann. Rev. Microbiol.*, **17:** 115 (1963).

Moyed, H. S., and H. E. Umbarger: Regulation of Biosynthetic Pathways, *Physiol. Rev.*, **42:** 444 (1962).

Nason, A.: Enzymatic Pathways of Nitrate, Nitrite, and Hydroxylamine Metabolism, *Bacteriol. Rev.*, **26:** 16 (1962).

Neurath, H.: Structure and Function of Proteolytic Enzymes, in M. Sela (ed.), *New Perspectives in Biology*, Elsevier, New York, 1964, p. 28.

Ottesen, M.: Induction of Biological Activity by Limited Proteolysis, *Ann. Rev. Biochem.*, **36:** 55 (1967).

Sprinson, D. B.: The Biosynthesis of Aromatic Compounds from D-Glucose, *Advan. Carbohydrate Chem.*, **15:** 235 (1960).

Stadtman, E. R., B. M. Shapiro, M. S. Kingdon, C. A. Woolfolk, and J. S. Hubbard: Cellular Regulation of Glutamine Synthetase Activity in *Escherichia coli, Advan. Enzyme Regulation,* **6:** 257 (1968).

Stanbury, J. B., J. B. Wyngaarden, and D. S. Fredrickson (eds.): *The Metabolic Basis of Inherited Disease,* McGraw-Hill, New York, 2nd ed., 1966.

Truffa-Bachi, P., and G. N. Cohen: Some Aspects of Amino Acid Biosynthesis in Microorganisms, *Ann. Rev. Biochem.*, **37:** 79 (1968).

Umbarger, H. E.: Regulation of Amino Acid Metabolism, *Ann. Rev. Biochem.*, **38:** 323 (1969).

Umbarger, E., and D. B. Davis: Pathways of Amino Acid Biosynthesis, in I. C. Gunsalus and R. Y. Stanier (eds.), *The Bacteria,* vol. III, Academic Press, New York, 1962, p. 167.

Wellner, D.: Flavoproteins, *Ann. Rev. Biochem.*, **36:** 669 (1967).

Specific References

1. Carnahan, J. E., L. E. Mortenson, H. F. Mower, and J. E. Castle: *Biochim. Biophys. Acta,* **38:** 188 (1960).
2. Carnahan, J. E., L. E. Mortenson, H. F. Mower, and J. E. Castle: *Biochim. Biophys. Acta,* **44:** 520 (1960).
3. Mortenson, L. E.: *Federation Proc.*, **19:** 241 (1960).
4. Bulen, W. A., R. C. Burns, and J. R. LeComte: *Proc. Natl. Acad. Sci. U.S.*, **53:** 532 (1965).
5. Witz, D. F., R. W. Detroy, and P. W. Wilson: *Arch. Mikrobiol.*, **55:** 369 (1967).
6. Hardy, R. W. F., E. Knight, Jr., and A. J. D'Eustachio: *Biochem. Biophys. Res. Commun.*, **20:** 539 (1965).
7. Burns, R. C., and W. A. Bullen: *Arch. Biochem. Biophys.*, **113:** 461 (1966).
8. Koch, B., H. J. Evans, and S. Russell: *Proc. Natl. Acad. Sci. U.S.*, **58:** 1343 (1967).
9. Mortenson, L. E., R. C. Valentine, and J. E. Carnahan: *Biochem. Biophys. Res. Commun.*, **7:** 448 (1962).
10. Hardy, R. W. F., and E. Knight, Jr.: in L. Reinhold (ed.), *Progress in Phytochemistry,* Wiley, New York, 1967, p. 387.
11. Hardy, R. W. F., and E. K. Jackson: *Science* (submitted for publication).
12. Knight, E., Jr., A. J. D'Eustachio, and R. W. F. Hardy: *Biochim. Biophys. Acta,* **113:** 626 (1966).
13. Knight, E., Jr., and R. W. F. Hardy: *J. Biol. Chem.*, **241:** 2752 (1966).
14. Knight, E., Jr., and R. W. F. Hardy: *J. Biol. Chem.*, **242:** 1370 (1967).
15. Mortenson, L. E.: *Proc. Natl. Acad. Sci. U.S.*, **52:** 272 (1964).
16. D' Eustachio, A. J., and R. W. F. Hardy: *Biochem. Biophys. Res. Commun.*, **15:** 319 (1964).
17. Hardy, R. W. F., and A. J. D'Eustachio: *Biochem. Biophys. Res. Commun.*, **15:** 314 (1964).
18. Hardy, R. W. F., and A. J. D'Eustachio: *Chem. Eng. News,* **42:** 23 (1964).
19. Hardy, R. W. F., and E. Knight, Jr.: *Biochim. Biophys. Acta,* **122:** 520 (1966).
20. Winter, H.: *Investigations of the Nitrogen Fixing System of Clostridium pasteurianum,* Doctoral thesis, University of Wisconsin, Madison, Wis., 1967.
21. Mortenson, L. E.: *Biochim. Biophys. Acta,* **127:** 18 (1966).
22. Bulen, W. A., and J. R. LeComte: *Proc. Natl. Acad. Sci. U.S.*, **56:** 979 (1966).
23. Mortenson, L. E.: in A. San Pietro (ed.), *Non-Heme Iron Proteins: Role in Energy Conversion,* Antioch Press, Yellow Springs, Ohio, 1965, p. 243.
24. Bulen, W. A., J. R. LeComte, R. C. Burns, and J. Hinkson: in A. San Pietro (ed.), *Non-Heme Iron Proteins: Role in Energy Conversion,* Antioch Press, Yellow Springs, Ohio, 1965, p. 261.
25. Hanstein, W. G., J. B. Lett, C. E. McKenna, and T. G. Traylor: *Proc. Natl. Acad. Sci. U.S.*, **58:** 1314 (1967).
26. Jackson, E. K., G. W. Parshall, and R. W. F. Hardy: *J. Biol. Chem.*, **243:** 4952 (1968).
27. Parshall, G. W.: *J. Am. Chem. Soc.*, **89:** 1822 (1967).

27a. Ketchum, P. A., H. Y. Cambier, W. A. Frazier, III, C. H. Madansky, and A. Nason: *Proc. Natl. Acad. Sci. U.S.*, **66:** 1016 (1970); Garrett, R. H., and A. Nason: *Proc. Natl. Acad. Sci. U.S.*, **58:** 1603 (1967); *J. Biol. Chem.*, **244:** 2870 (1969).

27b. Ruiz-Herrera, J., M. K. Showe, and J. A. DeMoss: *J. Bacteriol.*, **97:** 1291 (1969); Showe, M. K., and J. A. DeMoss: *J. Bacteriol.*, **95:** 1305 (1968).

27c. Azoulay, E., J. Puig, and P. Couchoud-Beaumont: *Biochim. Biophys. Acta,* **171:** 238 (1969).

28. Evans, H. J., and A. Nason: *Plant Physiol.*, **28:** 233 (1953).
29. Nason, A., and H. J. Evans: *J. Biol. Chem,* **202:** 655 (1953).
30. Nicholas, D. J. D., and A. Nason: *J. Biol. Chem.*, **211:** 183 (1954).
31. Mortenson, L. E.: *Bacteriol. Proc.*, 117 (1963).
32. Paneque, A., J. M. Ramirez, F. F. Del Campo, and M. Losada: *J. Biol. Chem.*, **239:** 1737 (1964).
33. Northrop, J. H.: *J. Gen. Physiol.*, **13:** 739 (1930).
34. Herriott, R. M., and J. H. Northrop: *Science,* **83:** 469 (1936).
35. Van Vunakis, H., and R. M. Herriott: *Biochim. Biophys. Acta,* **22:** 537 (1956).
36. Inouye, K., and J. S. Fruton: *J. Am. Chem. Soc.*, **89:** 187 (1967).
37. Bailey, J. L., S. Moore, and W. H. Stein: *J. Biol. Chem.*, **221:** 143 (1956).
38. Humphreys, R. E., and J. S. Fruton: *Proc. Natl. Acad. Sci. U.S.*, **59:** 519 (1968).
39. Neumann, H., Y. Levin, A. Berger, and E. Katchalski: *Biochem. J.*, **73:** 33 (1959).
40. Fruton, J. S., S. Fujii, and M. H. Knappenberger: *Proc. Natl. Acad. Sci. U.S.*, **47:** 759 (1961).
41. Zeffren, E., and E. T. Kaiser: *J. Am. Chem. Soc.*, **89:** 4204 (1967).

42. Bayliss, R. S., J. R. Knowles, and G. B. Wybandt: *Biochem. J.*, **113:** 377 (1969), and references therein.
43. Cornish-Bowden, A. J., D. Greenwell, and J. R. Knowles: *Biochem. J.*, **133:** 369 (1969).
44. Walsh, K. A., and H. Neurath: *Proc. Natl. Acad. Sci. U.S.*, **52:** 884 (1964).
45. Matthews, B. W., P. Sigler, B. Henderson, and D. M. Blow: *Nature*, **214:** 652 (1967).
46. Oppenheimer, H. L., B. Labouesse, and G. P. Hess: *J. Biol. Chem.*, **241:** 2720 (1966).
47. Labouesse, B., H. L. Oppenheimer, and G. P. Hess: *Biochem. Biophys. Res. Commun.*, **14:** 318 (1964).
48. Gorecki, M., and Y. Shalitin: *Biochem. Biophys. Res. Commun.*, **29:** 189 (1967).
49. Freisheim, J. H., K. A. Walsh, and H. Neurath: *Biochemistry*, **6:** 3010, 3020 (1967).
50. Lorand, J., T. Urayama, and L. Lorand: *Biochem. Biophys. Res. Commun.*, **23:** 828 (1966).
51. Loewy, A. G., S. Matačic, and J. H. Darnell: *Arch. Biochem. Biophys.*, **113:** 435 (1966).
52. Kézdy, F. J., L. Lorand, and K. D. Miller: *Biochemistry*, **4:** 2302 (1965).
53. Drenth, J., J. N. Jansonius, R. Koekoek, H. M. Swen, and B. G. Wolthers: *Nature*, **218:** 929 (1968).
54. Schechter, I., and A. Berger: *Biochem. Biophys. Res. Commun.*, **27:** 157 (1967).
55. Brubacher, L. J., and M. L. Bender: *Biochem. Biophys. Res. Commun.*, **27:** 176 (1967).
56. Lowe, G., and A. Williams: *Proc. Chem. Soc.*, **1964:** 140.
57. Hubbard, C. O., and J. F. Kirsch: *Biochemistry*, **7:** 2569 (1968).
58. Kirsch, J. F., and M. Igelström: *Biochemistry*, **5:** 783 (1966).
59. Hagihara, B.: in P. D. Boyer, H. Lardy, and K. Myrbäck (eds.), *The Enzymes*, vol. 4, Academic Press, New York, 2nd ed., 1960, p. 193.
60. Woolfolk, C. A., and E. R. Stadtman: *Arch. Biochem. Biophys.*, **118:** 736 (1967).
61. Kingdon, H. S., B. M. Shapiro, and E. R. Stadtman: *Proc. Natl. Acad. Sci. U.S.*, **58:** 1703 (1967).
62. Vogel, H. J., and B. D. Davis: *J. Am. Chem. Soc.*, **74:** 109 (1952).
63. Shemin, D., and D. Rittenberg: *J. Biol. Chem.*, **158:** 71 (1945).
64. Paik, W. K., and L. Benoiton: *Can. J. Biochem. Physiol.*, **41:** 1643 (1963).
65. Peterkofsky, B., and S. Udenfriend: *Biochem. Biophys. Res. Commun.*, **12:** 257 (1963).
66. Manning, J. M., and A. Meister: *Biochemistry*, **5:** 1154 (1966).
67. Hutton, Jr., J. J., A. L. Tappel, and S. Udenfriend: *Arch. Biochem. Biophys.*, **118:** 231 (1967).
68. Kivirikko, K. I., H. J. Bright, and D. J. Prockop: *Biochim. Biophys. Acta*, **151:** 558 (1968).
69. Juva, K., and D. J. Prockop: *J. Biol. Chem.*, **244:** 6486 (1969).
70. Kivirikko, K. I., and D. J. Prockop: *J. Biol. Chem.*, **242:** 4007 (1967).
71. Cedar, H., and J. H. Schwartz: *J. Biol. Chem.*, **244:** 4112 (1969).
72. Patte, J-C., P. Truffa-Bachi, and G. N. Cohen: *Biochim. Biophys. Acta*, **128:** 426 (1966).
73. Patte, J-C., G. Le Bras, and G. N. Cohen: *Biochim. Biophys. Acta*, **136:** 245 (1967).
74. Flavin, M., and C. Slaughter: *J. Biol. Chem.*, **235:** 1103, 1109, 1112, (1960).
75. Kerr, D. S., and M. Flavin: *Biochem. Biophys. Res. Commun.*, **31:** 124 (1968).
76. Whitfield, C. D., and H. Weissbach: *Biochem. Biophys. Res. Commun.*, **33:** 996 (1968).
77. Taylor, R. T., and H. Weissbach: *Arch. Biochem. Biophys.*, **129:** 728, 745 (1969).
78. Umbarger, H. E., and M. A. Umbarger: *Biochim. Biophys. Acta*, **62:** 193 (1962).
79. Schirch, L. G., and M. Mason: *J. Biol. Chem.*, **238:** 1032 (1963).
80. Strobel, G. A.: *J. Biol. Chem.*, **241:** 2618 (1966).
81. Mitoma, C., and E. E. Snell: *Proc. Natl. Acad. Sci. U.S.*, **41:** 891 (1955).
82. Magasanik, B.: *J. Am. Chem. Soc.*, **78:** 5449 (1956).
83. Neidle, A., and H. Waelsch: *J. Biol. Chem.*, **234:** 586 (1959).
84. Neidle, A., and H. Waelsch: *J. Am. Chem. Soc.*, **78:** 1767 (1956).
85. Moyed, H. S., and B. Magasanik: *J. Am. Chem. Soc.*, **79:** 4812 (1957).
86. Ames, B. N., R. G. Martin, and B. J. Garry: *J. Biol. Chem.*, **236:** 2019 (1961).
87. Gross, S. R., R. O. Burns, and H. E. Umbarger: *Biochemistry*, **2:** 1046 (1963).
88. Davis, B. D.: *J. Biol. Chem.*, **191:** 315 (1951).
89. Yanofsky, C.: *Bacteriol. Rev.*, **24:** 221 (1960).
90. Snell, E. E.: *J. Biol. Chem.*, **154:** 313 (1944).
91. Litwak, G., M. L. Sears, and T. I. Diamondstone: *J. Biol. Chem.*, **238:** 302 (1963).
92. Chang, G. W., and E. E. Snell: *Biochemistry*, **7:** 2012 (1968).
93. Riley, W. D., and E. E. Snell: *Biochemistry*, **7:** 3520 (1968).
94. Krebs, H. A.: *Biochem. J.*, **29:** 1620 (1935).
95. Wellner, D., and A. Meister: *J. Biol. Chem.*, **235:** 2013 (1960).
96. Blanchard, M., D. E. Green, V. Nocito, and S. Ratner: *J. Biol. Chem.*, **161:** 583 (1945).
97. Miller, A., and H. Waelsch: *J. Biol. Chem.*, **228:** 365 (1957).
98. Feinberg, R. H., and D. M. Greenberg: *J. Biol. Chem.*, **234:** 2670 (1959).
99. Ueno, Y., K. Hayashi, and R. Shukuya: *J. Biochem. (Tokyo)*, **54:** 75 (1963).
100. Nishizuka, Y., and O. Hayaishi: *J. Biol. Chem.*, **238:** PC483 (1963).
101. Andreoli, A. J., M. Ikeda, Y. Nishizuka, and O. Hayaishi: *Biochem. Biophys. Res. Commun.*, **12:** 92 (1963).
102. Mitchell, R. A., H. H. Kang, and L. M. Henderson: *J. Biol. Chem.*, **238:** 1151 (1963).
103. Kaufman, S.: *Proc. Natl. Acad. Sci. U.S.*, **50:** 1085 (1963).
104. Fromm, H. J., and R. C. Nordlie: *Arch. Biochem. Biophys.*, **81:** 363 (1959).
105. Klee, W. A.: *Biochim. Biophys. Acta*, **45:** 537 (1960).
106. Shemin, D., and J. Wittenberg: *J. Biol. Chem.*, **185:** 103 (1950); **192:** 315 (1951).

18 Metabolism of the Nucleotides

Many features of nucleic acid metabolism relate directly to questions of central importance in the biological sciences. These include an understanding, on a molecular basis, of the mechanisms involved in the specification of macromolecular structures, the principal features of the transfer of biochemical specificity from parent to progeny, the basis of cellular differentiation, and the mechanism by which certain compounds are able to inhibit tumor and bacterial growth selectively. Consequently, interest in nucleic acid metabolism is great, and significant developments occur with sometimes bewildering rapidity.

Our discussion of nucleic acid metabolism in this chapter centers on three basic aspects: (1) pathways for the biosynthesis of the monomeric precursors to the nucleic acids, (2) metabolic control of the synthesis of these precursors, and (3) chemical inhibition of nucleic acid biosynthesis and its relationship to chemotherapy. In Chapter 19, the role of nucleic acid as genetic material is developed and the molecular events in the assembly of the nucleic acid themselves is considered in Chapter 20.

BIOSYNTHESIS OF NUCLEOSIDE TRIPHOSPHATES

Purines

The ability to synthesize the purine ring system *de novo* is a nearly universal biochemical capacity. Only a few known species require a nutritional source of preformed purines. The first major step toward a resolution of the metabolic pathway by which this biosynthesis is accomplished was the determination, through the use of labeled isotopes, of the origin of each of the atoms of the purine ring system. Studies principally from the laboratory of Buchanan demonstrated that carbons 2 and 8 are derived from formate, carbon 6 is derived from CO_2, and that carbons 4 and 5 come from the carboxyl and methylene carbons of glycine respectively. Nitrogen 7 is derived from glycine as well. This information is depicted schematically for the purine hypoxanthine in Fig. 18-1. Subsequent studies have revealed that nitrogen 1 is derived from aspartic acid and that nitrogens at positions 3 and 9 come from the amide nitrogen of glutamine. A major contribution to the elucidation of the pathway of purine nucleotide biosynthesis was the finding by G. R. Greenberg that the nucleotide inosinic acid is the first product formed with a complete purine ring structure. We now turn to a consideration of the individual enzymatic reactions involved in the transformation of these precursors into the completed purine skeleton, as determined in Buchanan's and Greenberg's laboratories.

The initial step in the biosynthesis of the purine ribonucleotides is the ATP-dependent pyrophosphorylation of ribose-5-phosphate yielding 5-phosphoribosyl-1-pyrophosphate (PRPP):

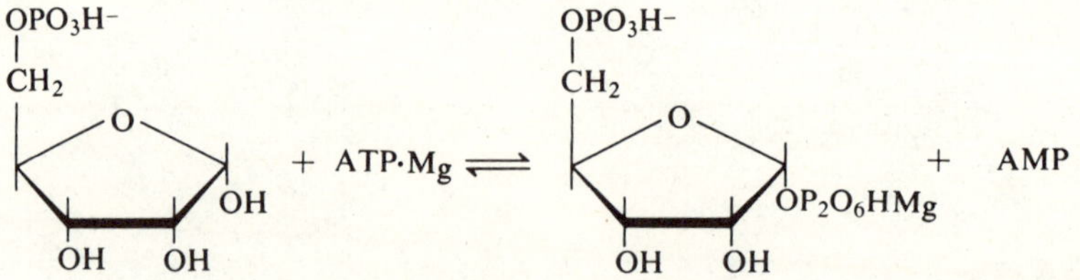

$$\text{Ribose-5-phosphate } (OPO_3H^-\text{–}CH_2\text{, ring O, 1-OH, 2-OH, 3-OH}) + \text{ATP·Mg} \rightleftharpoons \text{PRPP } (OPO_3H^-\text{–}CH_2\text{, ring O, 1-}OP_2O_6HMg\text{, 2-OH, 3-OH}) + \text{AMP} \quad (18\text{-}1)$$

PRPP

This is an interesting reaction in that it is one of the few known cases in which a pyrophosphate group is transferred intact from ATP. That this is the case, rather than the alternative possibility of two successive phosphate transfers, is revealed by the results of studies employing ATP labeled with ^{32}P. For example, utilizing ATP labeled in only the terminal phosphate, successive phosphate transfers would yield PRPP labeled in both halves of the pyrophosphate moiety, while a single pyrophosphate transfer would yield PRPP labeled in only the terminal phosphate. These cases may be distinguished by the use of an appropriate pyrophosphatase as indicated in Eqs. (18-2) and (18-3):

Pyrophosphate transfer

$$\text{Ribose-5-phosphate} + \text{ARPPP*} \xrightarrow{\text{ARP}} \text{PRPP*} \longrightarrow \text{PRP} + \text{P*} \quad (18\text{-}2)$$

Successive phosphate transfer

$$\text{Ribose-5-phosphate} + \text{2ARPPP*} \xrightarrow{\text{2ARPP}} \text{PRP*P*} \longrightarrow \text{PRP*} + \text{P*} \quad (18\text{-}3)$$

PRP = 5-phosphoribosyl-1,2-cyclic phosphate

The experimental results are those indicated in Eq. (18-2) establishing this reaction as a pyrophosphate transfer.

The second step of this pathway involves the glutamine-dependent amination of PRPP yielding the acid-labile sugar 5-phosphoribosyl-1-amine:

$$\text{Mg·5-Phosphoribosyl-1-pyrophosphate} + \text{glutamine} \longrightarrow$$

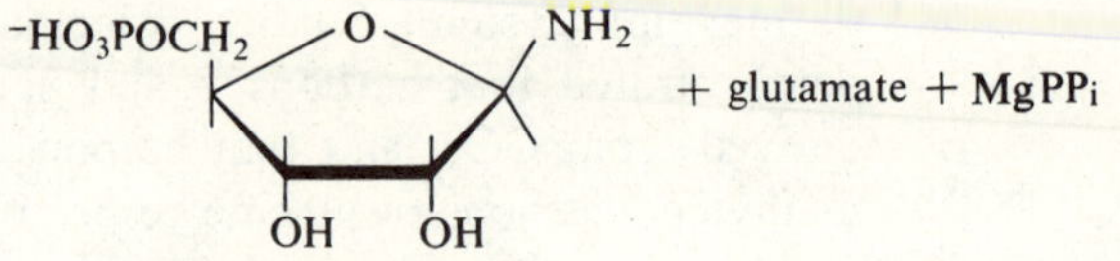

$$^-HO_3POCH_2\text{–(ribose ring, 1-}NH_2\text{, 2-OH, 3-OH)} + \text{glutamate} + \text{MgPP}_i \quad (18\text{-}4)$$

5-Phosphoribosyl-1-amine

Thus, the overall result of the first two steps is an ATP-dependent amination of ribose-5-phosphate. The amination step involves an inversion at C-1 of the ribose moiety yielding the amino sugar with the β configuration. Hence, the *N*-glycosidic bond of proper stereochemistry is introduced early in the reaction sequence.

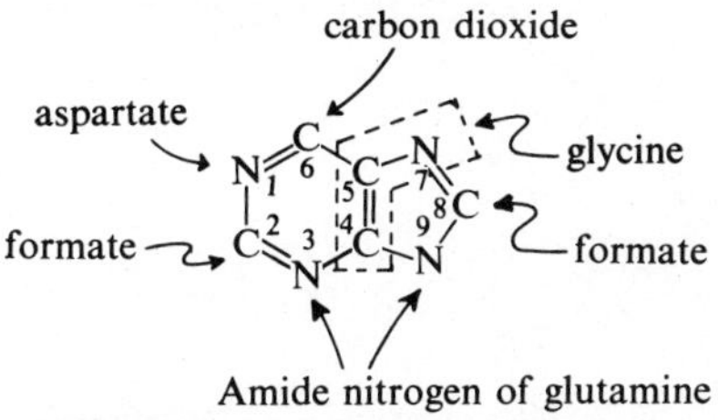

FIGURE 18-1
Basic precursors of the purine skeletŏn.

5-Phosphoribosyl-1-amine is subsequently conjugated, through formation of an amide linkage, with glycine in an ATP-dependent reaction:

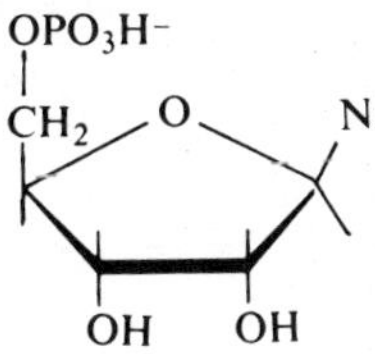

$$\text{5-phosphoribosyl-1-amine} + \text{glycine} + \mathbf{Mg{\cdot}ATP} \longrightarrow \text{Mg}{\cdot}\text{ADP} + \text{P}_i + \text{ribose-5-}OPO_3H^- \text{—HN—C(=O)—}CH_2\text{—}\ddot{N}H_2 \qquad (18\text{-}5)$$

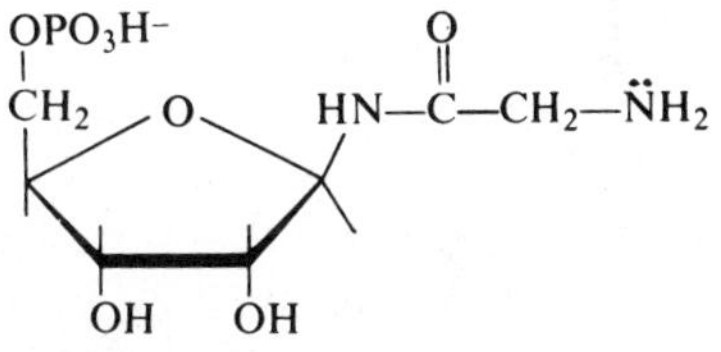

Glycinamide ribonucleotide

Glycinamide ribonucleotide, the product of reaction (18-5), is formylated to yield N-formylglycinamide ribonucleotide. N^5,N^{10}-Methenyl tetrahydrofolate serves as coenzyme for this reaction (Chapter 8):

$$\text{Glycinamide ribonucletoide} + N^5,N^{10}\text{-methenyl tetrahydrofolate} + H_2O \longrightarrow$$

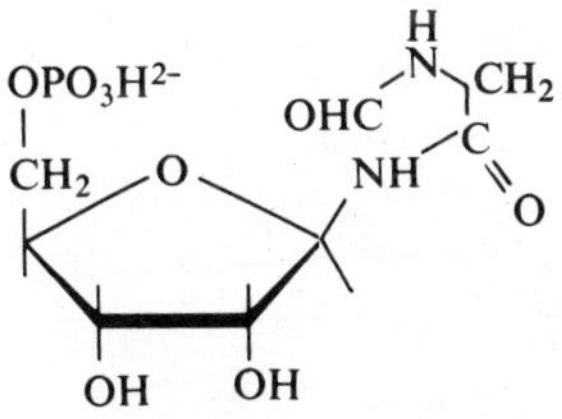

$$\text{ribose-5-}OPO_3H^{2-}\text{—NH—C(=O)—}CH_2\text{—NH—CHO} + \text{tetrahydrofolate} + H^+ \qquad (18\text{-}6)$$

α-N-formylglycinamide ribonucleotide

A second ATP-dependent amination in which glutamine serves as amine donor is encountered in the transformation of N-formylglycinamide ribonucleotide to N-formylglycinamidine ribonucleotide [Eq. (18-7)]:

$$\alpha\text{-}N\text{-formylglycinamide ribonucleotide} + \text{glutamine} + H_2O + \text{Mg}{\cdot}\text{ATP} \longrightarrow$$

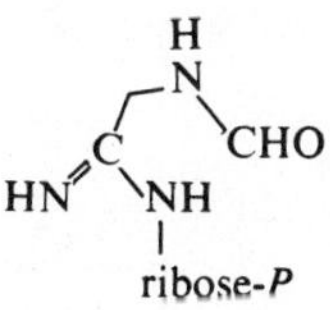

$$\text{HN=C(NH—ribose-}P\text{)—}CH_2\text{—NH—CHO} + \text{glutamic acid} + \text{Mg}{\cdot}\text{ADP} + P_i \qquad (18\text{-}7)$$

N-Formylglycinamidine ribonucleotide

N-Formylglycinamidine ribonucleotide has all the structural features of the imidazole ring of purines. This ring is formed by an ATP-dependent dehydration, yielding 5-aminoimidazole ribonucleotide:

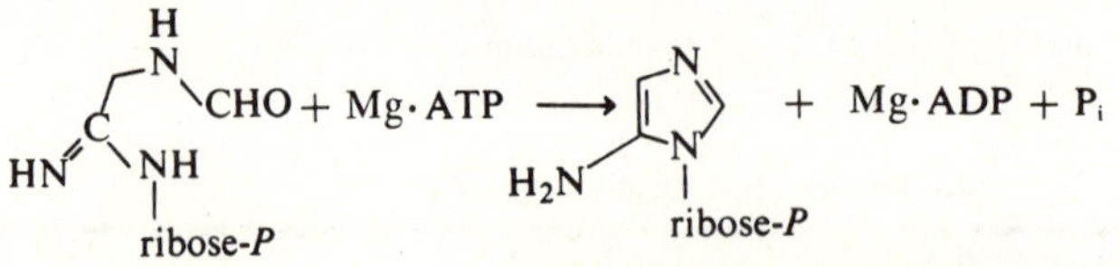

H N CHO + Mg·ATP ⟶ + Mg·ADP + P_i
HN C NH H₂N N
ribose-*P* ribose-*P* (18-8)

5-Aminoimidazole ribonucleotide

Aminoimidazole ribonucleotide carboxylase (phosphoribosyl-aminoimidazole carboxylase) catalyzes the carboxylation of the immediate ring-closure product yielding 5-amino-4-imidazolecarboxylic acid ribonucleotide:

5-Aminoimidazole ribonucleotide + CO_2 ⇌ HO_2C N
H_2N N
ribose—*P* (18-9)

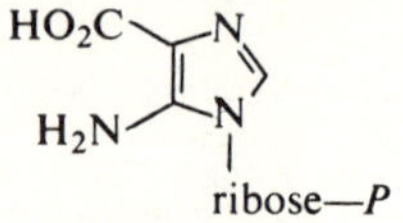

5-Aminoimidazole-4-carboxylic acid ribonucleotide

Bicarbonate, in equilibrium with free carbon dioxide, serves as a source of the carboxylic function. In a system isolated from pigeon liver, this reaction occurs without aid of a coenzyme and without a concomitant exergonic reaction, such as the oxidation of NADPH or the cleavage of ATP, which are characteristic of many other carboxylations.

The conversion of 5-amino-4-imidazolecarboxylic acid ribonucleotide to the corresponding amide, 5-amino-4-imidazole carboxamide ribonucleotide, occurs in a two-step process. First, an amide linkage between the substrate and aspartic acid is introduced in an ATP-dependent process; this produces the intermediate 5-aminoimidazole-4-*N*-succinocarboxamide ribonucleotide:

5-Aminoimidazole-4-carboxylic acid ribonucleotide + aspartic acid + Mg·ATP ⇌

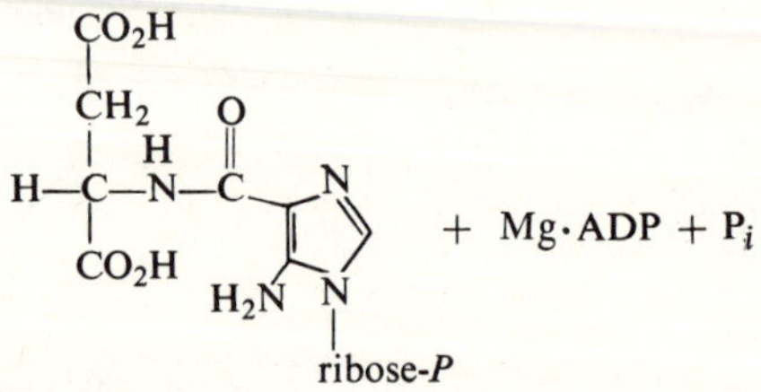

CO_2H
CH_2 O
H—C—N—C N
CO_2H + Mg·ADP + P_i (18-10)
H_2N N
ribose-*P*

5-Aminoimidazole-4-*N*-succinocarboxamide ribonucleotide

This product subsequently undergoes an elimination reaction with the expulsion of fumaric acid:

5-Aminoimidazole-4-*N*-succinocarboxamide ribonucleotide ⇌

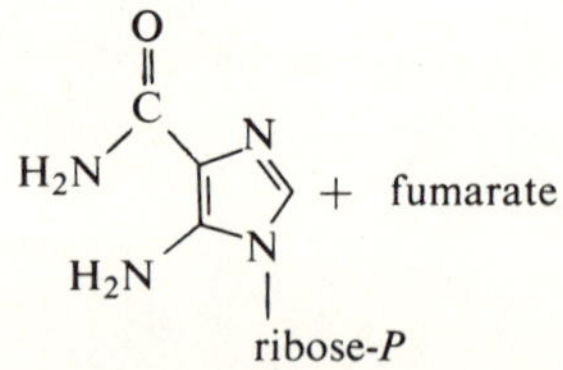

O
C N
H_2N + fumarate (18-11)
H_2N N
ribose-*P*

5-Aminoimidazole-4-carboxamide ribonucleotide

The last carbon atom required to yield all the elements of the purine ring system is now introduced as a formyl group attached to the 5-amino function of the imidazole carboxamide ribonucleotide:

$$\text{5-Aminoimidazole-4-carboxamide ribonucleotide} + N^{10}\text{-formyl tetrahydrofolic acid} \xrightarrow{K^+}$$

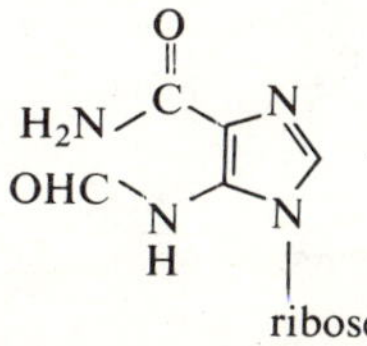

[Structure: 5-Formamidoimidazole-4-carboxamide ribonucleotide (H_2N–C(=O)–, OHC–NH–, ribose-*P*)] + tetrahydrofolic acid (18-12)

5-Formamidoimidazole-4-carboxamide ribonucleotide

This reaction, like the previous formylation in this reaction sequence, involves the participation of a FH_4 coenzyme but, unlike the previous case, N^{10}-formyl FH_4 is the structural form of the coenzyme involved.

Ring closure with dehydration, yielding inosinic acid, completes the formation of the purine ring system:

5-Formamidoimidazole-4-carboxamide ribonucleotide $\rightleftharpoons$ [Structure: Inosinic acid (HN, O, N, ribose-*P*)] + H_2O (18-13)

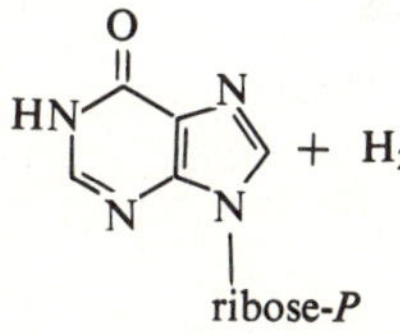

Inosinic acid

It remains only to convert inosinic acid to adenylic and guanylic acids to complete the synthesis of the purine ribonucleotides. The metabolic pathway from ribose-5-phosphate to inosinic acid is summarized in Fig. 18-2. The sources of the various atoms of the purine ring system as indicated in Fig. 18-1 are correctly accounted for by this pathway.

The biochemical transformations converting inosinic acid to adenylic and guanylic acids are indicated in Fig. 18-3. The amination of inosinic acid to yield adenylic acid proceeds via a two-step reaction path very similar to that involved in the synthesis of 5-amino-4-imidazolecarboxamide ribonucleotide [Eqs. (18-10) and (18-11)]. The enzyme catalyzing the elimination of fumaric acid from adenylosuccinate, *adenylosuccinase,* very probably also catalyzes the analogous reaction indicated above [Eq. (18-11)]. The ratio of the two enzymatic activities remains unchanged throughout the course of enzyme purification; furthermore, mutants which lack the ability to catalyze one of these reactions also lack the ability to catalyze the other. It is significant that the reaction yielding adenylosuccinic acid, on the pathway to adenine nucleotides, requires GTP as coenzyme. As noted below, a similar situation exists in the synthesis of guanine nucleotides, ATP being required as a coenzyme. A regulatory role is implied in these requirements since, clearly, an excess of GTP will shunt synthesis in the direction of ATP, and conversely, an excess of ATP will produce more GTP.

Conversion of inosinic acid to guanylic acid also proceeds in a two-step process. First, xanthylic acid is produced in an oxidation reaction, NAD+ serving as coenzyme. Second, xanthylic acid is converted to guanylic acid in an ATP-dependent amination reaction. Either glutamine or ammonia will

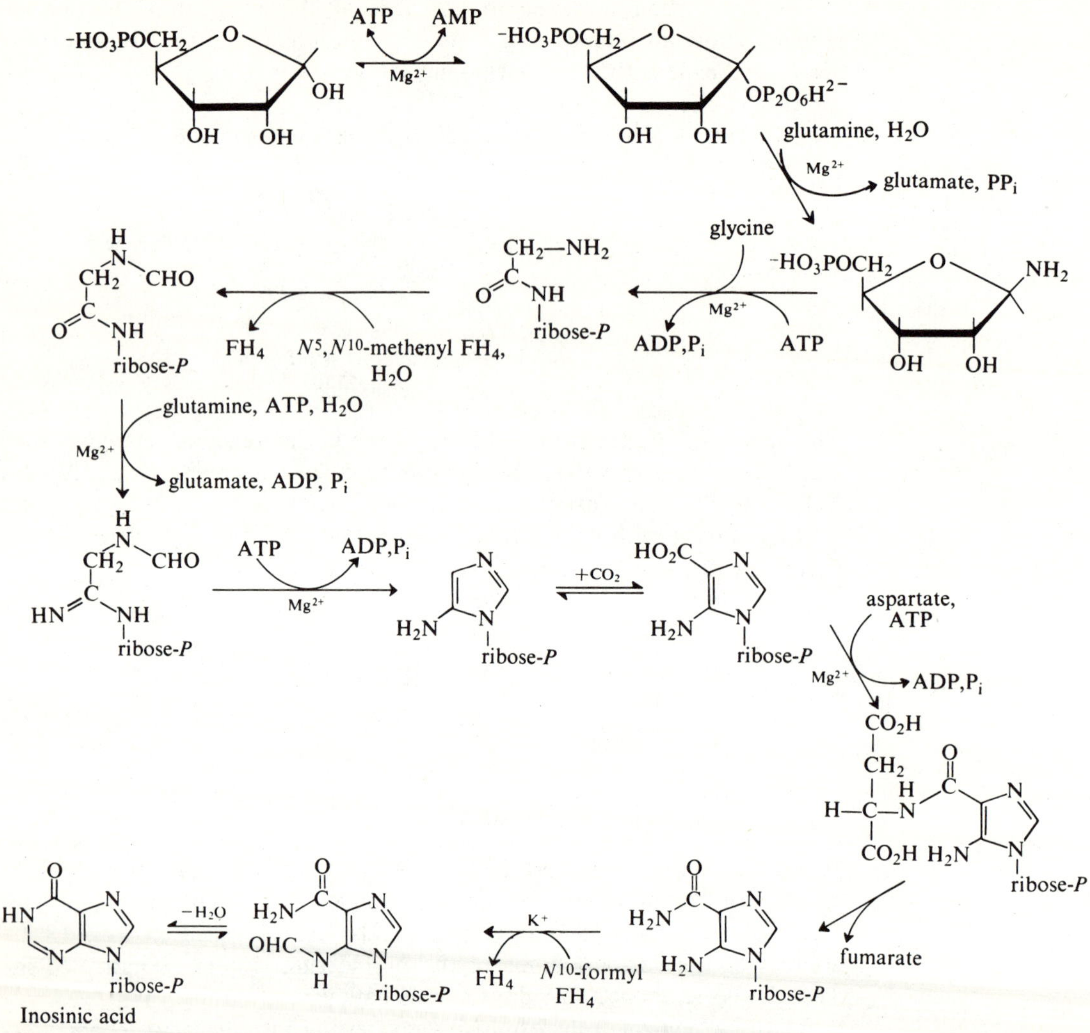

FIGURE 18-2
Metabolic pathway leading to the synthesis of inosinic acid.

serve as amine group donor, depending on the species in which the reaction occurs.

The actual purine precursors of RNA, the nucleoside triphosphates, are formed from the monophosphates by two successive kinase reactions:

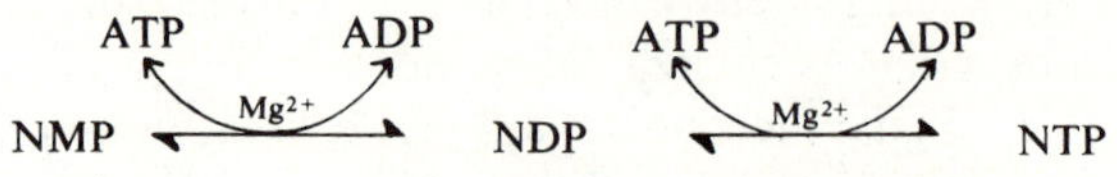

$$\text{NMP} \underset{Mg^{2+}}{\overset{\text{ATP} \quad \text{ADP}}{\rightleftharpoons}} \text{NDP} \underset{Mg^{2+}}{\overset{\text{ATP} \quad \text{ADP}}{\rightleftharpoons}} \text{NTP} \tag{18-14}$$

The kinases responsible for catalyzing these reactions are not specific for the base and will phosphorylate adenine, guanine, uracil, or cytosine nucleotides. These reactions clearly do not account for net synthesis of ATP. This triphosphate is derived, of course, as the end product of exergonic oxidation reactions.

Prior to a consideration of other aspects of the metabolism of purines, we

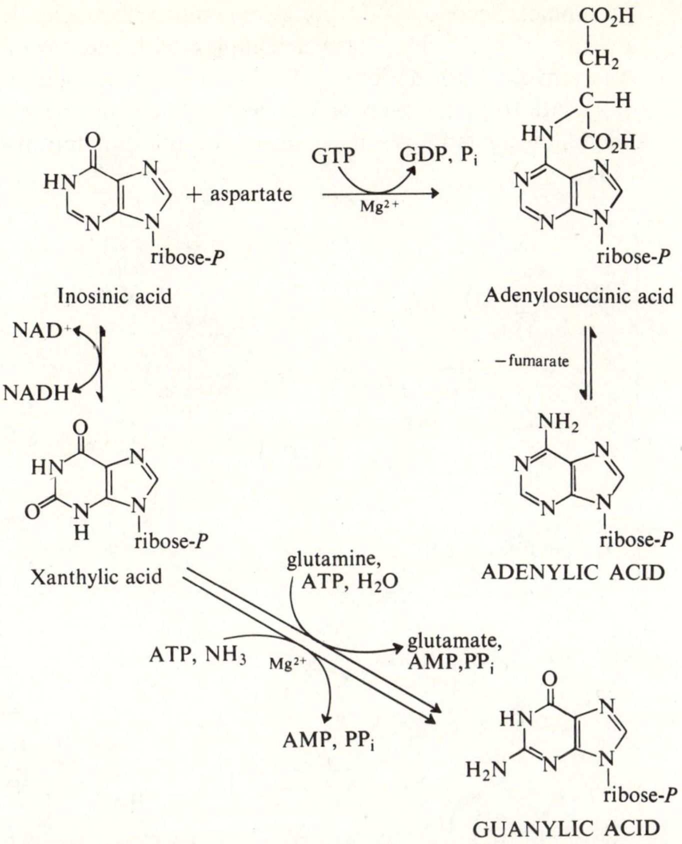

FIGURE 18-3
Conversion of inosinic acid to adenylic and guanylic acids.

digress briefly to compare the three aminations in purine biosynthesis in which glutamine serves as amino group donor. Recapitulating, these are:

a. ribose-5-phosphate + ATP + glutamine + $H_2O \rightarrow$ 5-phosphoribosyl-1-amine + AMP + PP_i + glutamate

b. *N*-formylglycinamide ribonucleotide + glutamine + ATP + $H_2O \rightarrow$ *N*-formylglycinamidine ribonucleotide + ADP + P_i + glutamate

c. xanthylic acid + ATP + glutamine + $H_2O \rightarrow$ guanylic acid + AMP + PP_i + glutamate

Reaction a, as we know, proceeds via a pyrophosphorylated substrate intermediate. In this respect, this reaction must differ from b since the products of ATP cleavage in the latter case are ADP and P_i.

The amidotransferase catalyzing reaction b has been purified from the chicken and from bacteria (1,2) and the reaction mechanism has been probed in some detail. Both enzymes have a molecular weight near 135,000. The amidotransferase forms a number of substrate–enzyme intermediates. First, incubation of the enzyme with glutamine in the absence of the other substrates results in formation of a γ-glutamyl-enzyme complex with liberation of ammonia (3). The glutamyl-enzyme is probably a thiol ester involving the cysteine-SH of the enzyme. It is worth noting that ammonia will function as amino group donor in this reaction but does so much less effectively than glutamine. Second, *N*-formylglycinamide ribonucleotide and ATP, in the absence of glutamine, form a complex with the enzyme. The complex with ATP involves formation of a phosphoenzyme plus ADP, which does not dissociate from the enzyme surface (4). On the basis of these observations, the following tentative mechanism for this amidotransferase can be written (4):

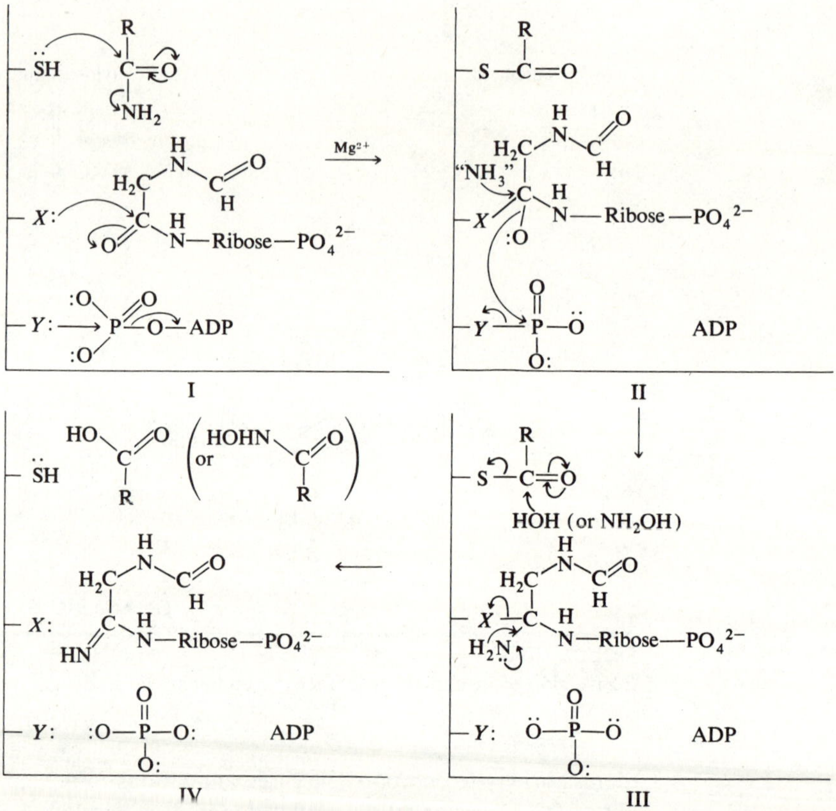

I —(Mg^{2+})→ II → III → IV

Reaction c, which yields AMP and PP_i as the ATP cleavage products, appears to be distinct from either a or b. Experiments employing ^{18}O-labeled (at C-2) xanthylic acid reveal that this oxygen atom is incorporated into the AMP:

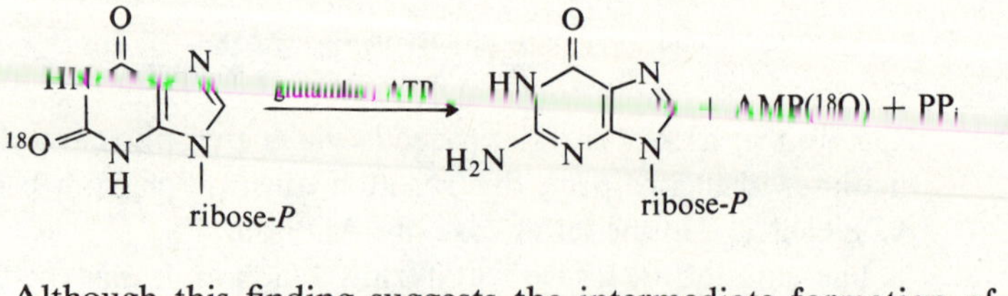

$$\text{xanthylic acid}(^{18}O\text{ at C-2, ribose-}P) \xrightarrow{\text{glutamine, ATP}} \text{guanylic acid (ribose-}P) + AMP(^{18}O) + PP_i \qquad (18\text{-}15)$$

Although this finding suggests the intermediate formation of an adenyl xanthylic acid, no exchange of pyrophosphate with ATP can be demonstrated.

Pyrimidines A central distinction between the metabolic routes leading to the formation of purine and pyrimidine nucleotides is timing in the formation of the *N*-glycosidic bond. In purine synthesis, this link is formed in the very early steps and the ring system is built upon this foundation. In contrast, the complete pyrimidine nucleus is synthesized prior to its attachment to ribose-5-phosphate. The key intermediate introduced into the *N*-glycosidic linkage is orotic acid, which contains the pyrimidine nucleus.

$$\text{Orotic acid: } \mathrm{HN{-}C(=O){-}CH{=}C(CO_2H){-}NH{-}C(=O)}$$

Orotic acid

Several independent lines of evidence establish the intermediacy of orotic acid in pyrimidine biosynthesis. First, orotic acid serves as an efficient precursor for RNA and DNA pyrimidines in systems from both mammals and microorganisms. Second, orotic acid inhibits the incorporation of N^{15} from labeled ammonium chloride into nucleic acid pyrimidines. Third, simple known precursors of pyrimidines, such as ammonia, carbon dioxide, and aspartate, are readily incorporated into orotic acid. Fourth, and finally, a sequence of enzymatic reactions has been elucidated through which simple precursors are converted to orotic acid, which is converted into a *N*-ribonucleoside-5′-phosphate and then is subsequently converted into pyrimidine nucleotides. It is to this sequence of reactions that attention is now directed.

The initial step for pyrimidine biosynthesis is the formation of carbamyl phosphate. In Chapter 17, the role of this compound in the biosynthesis of arginine and in the production of urea is described (pages 776 and 808). At the outset, one might expect that the synthesis of carbamyl phosphate would present some complications. Since it is the first step in pyrimidine biosynthesis, experience tells us that the enzyme catalyzing this reaction is probably subject to metabolic controls. But these controls must be carefully constructed lest inhibition of carbamyl phosphate synthesis under conditions in which pyrimidine synthesis is necessary result in inhibition of arginine synthesis as well. The fact that arginine and the pyrimidines serve disjoint biochemical functions yet depend on the same precursor causes this problem in metabolic control. One solution to such a problem is straightforward: separate enzymes, separately regulated, catalyzing synthesis of separate pools for carbamyl phosphate for pyrimidine and arginine biosynthesis. In some tissues and organisms, this solution is evidently utilized. Thus, in fetal rat liver two enzymes have been identified. One is cytoplasmic, unstable, and uses glutamine as the amino group donor; the latter is mitochondrial, activated by *N*-acetylglutamate, and utilizes ammonia as the amino group donor (5). The former enzyme provides carbamyl phosphate for pyrimidine synthesis since aspartate transcarbamylase (see below) is also cytoplasmic and the latter provides this substrate for arginine biosynthesis since ornithine transcarbamylase (page 808) is also mitochondrial.

Without exception, carbamyl phosphate for pyrimidine biosynthesis is synthesized by enzymes utilizing glutamine as amino group donor (6). Systems examined in this respect include *Neurospora,* yeast, *E. coli,* pigeon liver, rat liver, and mouse ascites cells. The best studied of these enzymes is that from *E. coli.* This enzymes catalyzes a reaction in which two molecules of ATP are consumed per molecule of carbamyl phosphate synthesized (7,8):

$$\text{Glutamine} + CO_2 + 2\text{ATP} \xrightleftharpoons{Mg^{2+}} NH_2{-}\overset{\overset{\displaystyle O}{\|}}{C}{-}OPO_3H_2 + 2\text{ADP} + P_i + \text{glutamate} \qquad (18\text{-}16)$$

In addition to an absolute dependence on magnesium ions, this enzyme is also activated by potassium ions. Its prosthetic group appears to be biotin and it is strongly inhibited by the biotin antagonist avidin (9). This, and related enzymes, will utilize ammonia as well as glutamine as amino group donor. However, the affinity for glutamine is much greater than for ammonia. In some cases, the necessary concentrations of ammonia for appreciable carbamyl phosphate synthesis are beyond the physiological range, eliminating ammonia as an important *in vivo* reactant. The equilibrium constant for the *carbamyl phosphate synthetase* [ATP : carbamate phosphotransferase (dephosphorylating)] reaction, Eq. (18-16), lies well to the right.

In the second step of pyrimidine biosynthesis, the carbamyl phosphate donates its carbamyl moiety to the α-amino group of aspartate in a reaction catalyzed by *aspartate transcarbamylase* (carbamoyl phosphate : L-aspartate carbamoyl transferase):

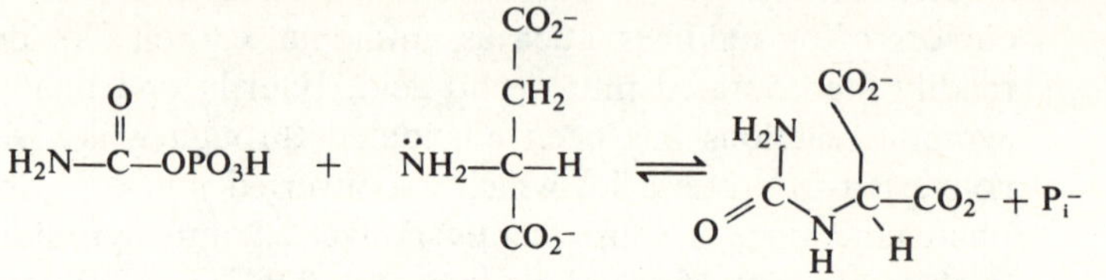

$$H_2N{-}\overset{O}{\overset{\|}{C}}{-}OPO_3H + \ddot{N}H_2{-}\underset{CO_2^-}{\underset{|}{\overset{CH_2CO_2^-}{\overset{|}{C}}}}{-}H \rightleftharpoons H_2N{-}\overset{O}{\overset{\|}{C}}{-}NH{-}\underset{H}{\underset{|}{\overset{CH_2CO_2^-}{\overset{|}{C}}}}{-}CO_2^- + P_i^- \qquad (18\text{-}17)$$

Carbamyl aspartate

This reaction, with its equilibrium far to the right, proceeds with direct attack by the aspartate amino group on the acyl carbon of carbamyl phosphate.

Aspartate transcarbamylase is a most interesting enzyme. As the catalyst of the first reaction unique to pyrimidine synthesis, it is expected to be the site of metabolic controls. It has been shown, in fact, that the catalytic and regulatory functions of the enzyme can be separated (10,11). Sulfhydryl reagents dissociate the native enzyme, molecular weight 310,000, into two catalytic subunits, molecular weight 100,000, and three regulatory subunits, molecular weight 30,000 (12,13). Furthermore, each catalytic subunit consists of three polypeptide chains and each regulatory subunit of two. Thus, there are six peptides of each type generating a structure $[(\alpha_3)_2(\beta_2)_3]$ with one twofold and one threefold symmetry axis (13). Further development of the detailed primary structure and geometry of this enzyme should prove most interesting.

Carbamyl aspartate is converted to dihydroorotic acid in a reversible cyclodehydration catalyzed by *dihydroorotase* (4,5-L-dihydroorotate aminohydratase):

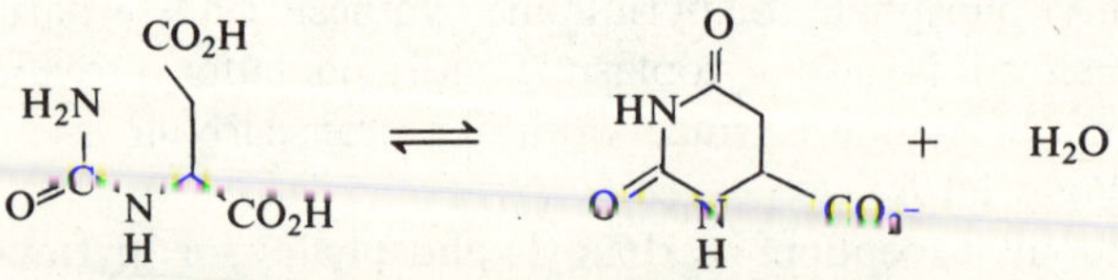

$$\text{Carbamyl aspartate } (H_2N{-}CO{-}NH{-}CH(CO_2H){-}CH_2CO_2H) \rightleftharpoons \text{Dihydroorotic acid} + H_2O \qquad (18\text{-}18)$$

Dihydroorotic acid

This substrate is subsequently oxidized to orotic acid by *dihydroorotic dehydrogenase* (4,5-L-dihydroorotate : NAD oxidoreductase), a flavoprotein (page 651), in a nicotinamide coenzyme-linked reaction:

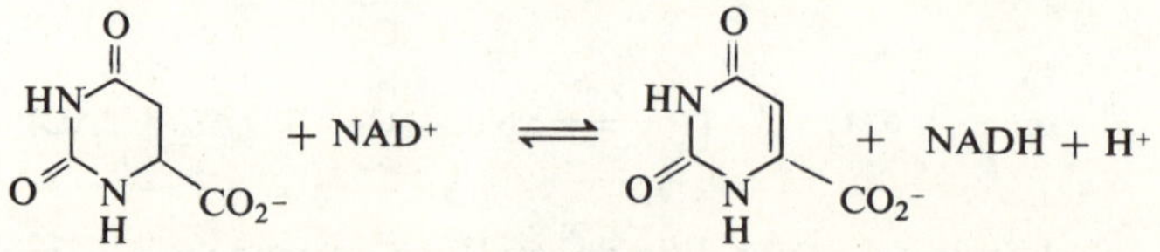

$$\text{Dihydroorotate} + NAD^+ \rightleftharpoons \text{Orotate} + NADH + H^+ \qquad (18\text{-}19)$$

Orotic acid

The synthesis of the pyrimidine ring is now complete and attachment of ribose-5-phosphate is the next concern. This is accomplished by coupling orotic acid to PRPP in a reaction catalyzed by *orotidine-5′-phosphate pyrophosphorylase* (orotidine-5′-phosphate : pyrophosphate phosphoribosyl transferase):

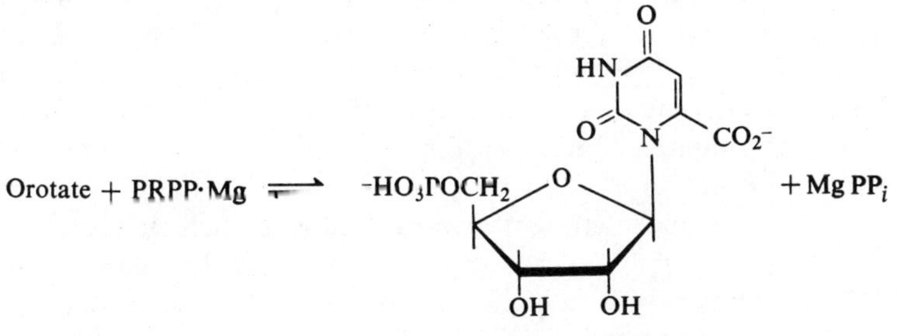

$$\text{Orotate} + \text{PRPP}\cdot\text{Mg} \rightleftharpoons \text{Orotidine-5}'\text{-phosphate} + \text{Mg PP}_i \qquad (18\text{-}20)$$

Orotidine-5′-phosphate

This enzyme is specific for orotic acid and will employ neither the precursors to orotic acid nor related pyrimidines as substrate.

The synthesis of uridylic acid (UMP) is completed by the irreversible decarboxylation of orotidine-5′-phosphate by an appropriate decarboxylase:

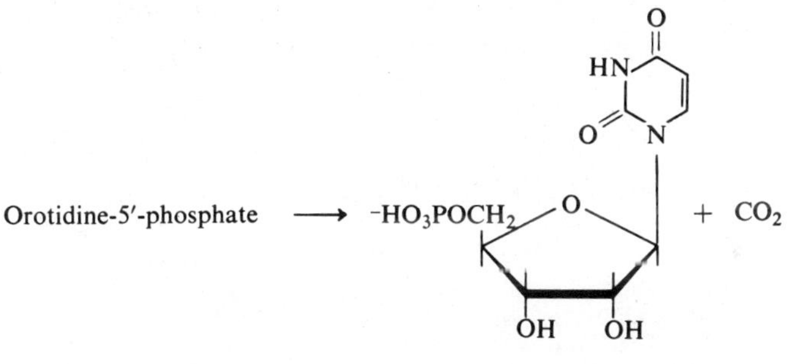

$$\text{Orotidine-5}'\text{-phosphate} \longrightarrow \text{Uridylic acid} + CO_2 \qquad (18\text{-}21)$$

Uridylic acid

Isotopically labeled orotic acid is incorporated, *in vivo,* much more rapidly into RNA-uracil than into RNA-cytosine (14). This observation suggests that uridine nucleotides are intermediates in the conversion of orotic acid into cytidine nucleotides. Indeed, the only known pathway for the formation of cytidine nucleotides involves the amination of UTP:

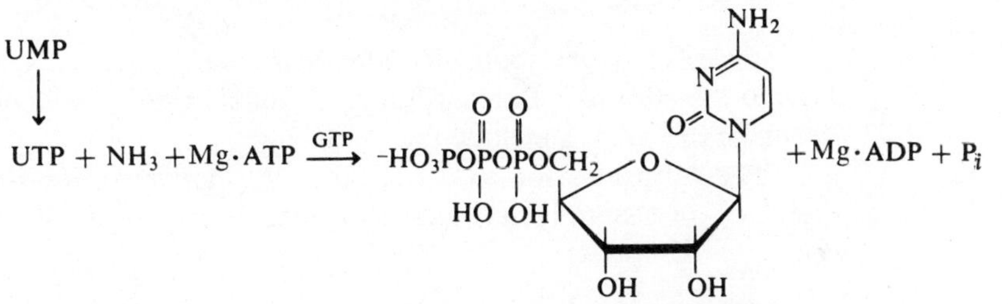

$$\text{UMP} \downarrow \quad \text{UTP} + NH_3 + \text{Mg}\cdot\text{ATP} \xrightarrow{\text{GTP}} \text{Cytidine triphosphate} + \text{Mg}\cdot\text{ADP} + P_i \qquad (18\text{-}22)$$

Cytidine triphosphate

UTP is formed from UMP by successive kinase reactions as indicated in Eq. (18-14). The reactions leading to the formation of uridine and cytidine nucleotides are summarized in Fig. 18-4.

BIOSYNTHESIS OF DEOXYNUCLEOSIDE PHOSPHATES

By 1960, the biosynthetic routes for most of the common simple constituents of mammals were known. An important exception was deoxyribose and the corresponding nucleotides. As we shall see, this lack of information stems principally from technical difficulties due to the complexities of the system which catalyzes reduction of nucleotides of ribose to those of deoxyribose. Some of the complexities might have been anticipated; the nucleic acids occupy central roles in living systems but those for RNA and DNA are quite distinct. Therefore, that point at which the pathways for their biosynthesis diverge must be subject to precise metabolic control, ensuring that precursors are shunted either toward synthesis of RNA or DNA as appropriate at a given moment.

The most straightforward means by which the deoxynucleoside triphosphates might be presumed to arise are by reduction of the corresponding ribonucleoside polyphosphates and by metabolic routes parallel to those previously discussed, employing the reduced sugar moiety. That the former possibility is correct, a supposition not necessarily excluding a minor contribution from the latter pathway, was initially suggested by the following experiment (15). Cytidine, labeled uniformly in all carbon atoms with ^{14}C, was injected into rats, and after a suitable time period, the DNA-cytidine was isolated and examined. The ratio of radioactivity, 0.80, in the base and sugar moieties of the isolated deoxycytidine was identical to that of the administered material. This experiment indicates that the cytidine was incorporated into the DNA without cleavage of the glycosidic bond.

$$\text{Cytidine} \xrightarrow{\text{in vivo}} \text{DNA-deoxycytidine}$$

$$\left.\frac{^{14}C_{\text{base}}}{^{14}C_{\text{sugar}}}\right)_{\text{injected cytidine}} = \left.\frac{^{14}C_{\text{base}}}{^{14}C_{\text{sugar}}}\right)_{\text{deoxy-CMP from DNA}} = 0.80$$

Thus, the reduction of cytidine to deoxycytidine must have occurred at the nucleoside or nucleotide level of organization and not at the sugar level.

Nature has adopted several means of effecting the reduction of ribonucleotides to deoxyribonucleotides. Two of these have been investigated in some detail: systems from *E. coli* and *Lactobacillus leichmannii*. We consider each in turn.

Synthesis of deoxyribonucleotides in *E. coli* has been clarified principally through the efforts of Reichard and his collaborators in Stockholm. The basic aspects of this system include the following:

1. The substrates for reduction are the *ribonucleoside diphosphates* (16). The reaction generally requires a nucleoside triphosphate and magnesium as well:

$$\text{NDP} \xrightarrow{\text{N'TP, Mg}^{2+}} \text{dNDP} \qquad (18\text{-}23)$$

For example, reduction of CDP requires ATP. But the reduction of other nucleoside diphosphates is dependent on other triphosphates, suggesting that

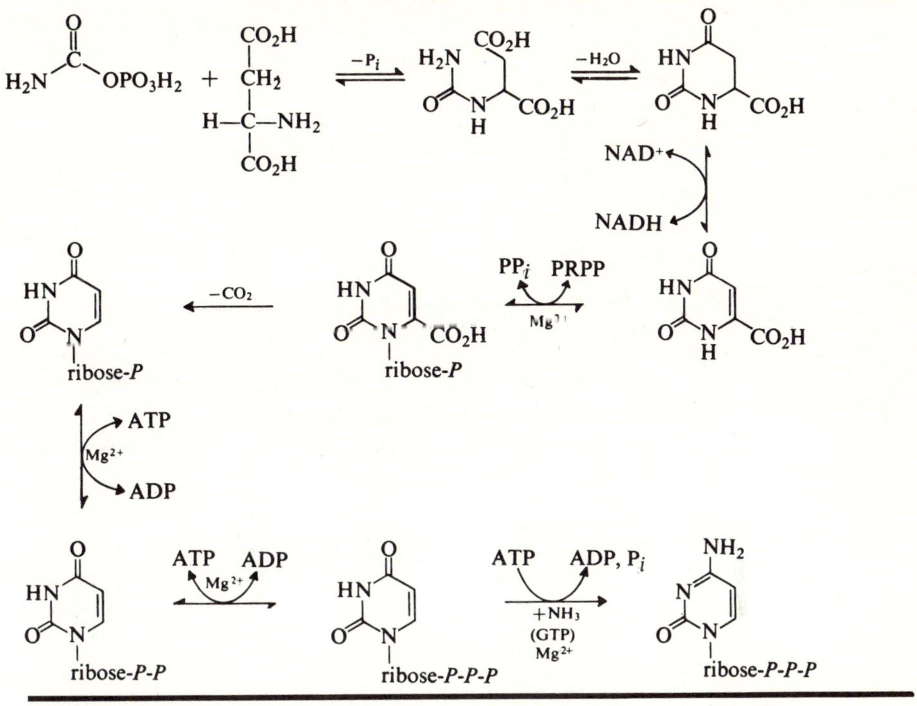

FIGURE 18-4
Metabolic pathway leading to the synthesis of uridine and cytidine nucleotides.

the latter species function as allosteric effectors, not as energy donors (see below).

2. The reduction is catalyzed by a *ribonucleoside diphosphate reductase,* which appears to be composed of two nonidentical subunits ($\alpha_2\beta_2$) (17). The first subunit of the reductase ($B_1 = \alpha_2$; $M = 180{,}000$) may function in a regulatory and the other ($B_2 = \beta_2$; $M = 78{,}000$) in a catalytic capacity. Association of these subunits to form the catalytically active enzyme depends on the presence of 0.01 M Mg^{2+} (18). Structurally, therefore, the reductase bears some resemblance to aspartate transcarbamylase (page 828) and tryptophan synthetase (page 868). Recently, it has been established that the reductase subunit B_2 contains 2 molecules of nonheme iron as an obligatory cofactor (19). Hydroxyurea and related compounds inhibit this reaction by binding to this subunit and, therefore, function as inhibitors of DNA synthesis.

3. The immediate reducing agent for the reaction is a small protein, molecular weight 12,000, termed *thioredoxin* (20), containing two half-cystine residues which can exist either in a reduced, thioredoxin-$(SH)_2$, or oxidized, thioredoxin-S_2, form. This protein contains all the common amino acids but does not possess an additional prosthetic group. Its oxidized and reduced forms are interconverted catalytically by a specific enzyme, *thioredoxin reductase,* which uses NADPH as hydrogen donor (21,22):

$$\text{Thioredoxin-}S_2 + \text{NADPH} + H^+ \xrightleftharpoons{\text{reductase}} \text{thioredoxin-}(SH)_2 + \text{NADP}^+ \qquad (18\text{-}24)$$

Thioredoxin reductase has a molecular weight near 66,000 and contains 2 mol of FAD (22). We can, then, summarize the reduction of ribonucleoside diphosphates as shown in Eq. (18-25).

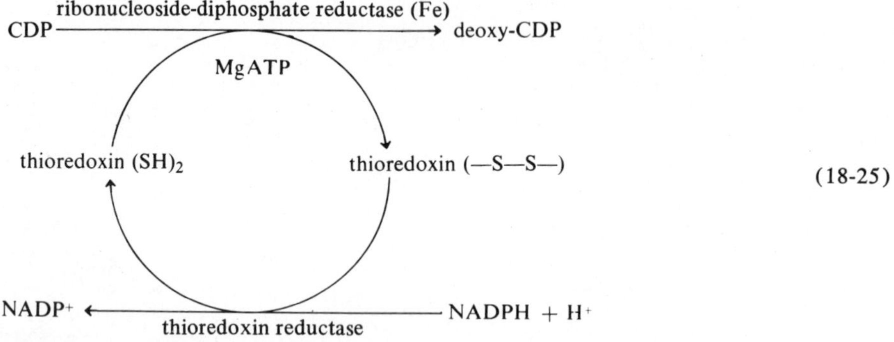

(18-25)

4. Reduction of ribonucleoside diphosphates with the *E. coli* system in deuterium oxide results in the incorporation of just one carbon-bound deuteron; that one is introduced at the 2′ position stereospecifically and without inversion at that position (23):

$ROCH_2$ O N H H OD OD $\xrightarrow[D_2O]{\text{reductase}}$ $ROCH_2$ O N H H OD D

(18-26)

5. The ribonucleoside reductase seems to be susceptible to allosteric control. Kinetic data obtained using the purified enzyme suggests that presence of ATP elicits an enzyme conformation specific for pyrimidine substrates, dGTP elicits one specific for purine substrates, dTTP causes the enzyme to be active toward all four substrates and, finally, dATP causes the enzyme to become inactive. These facts may be summarized in the following way:

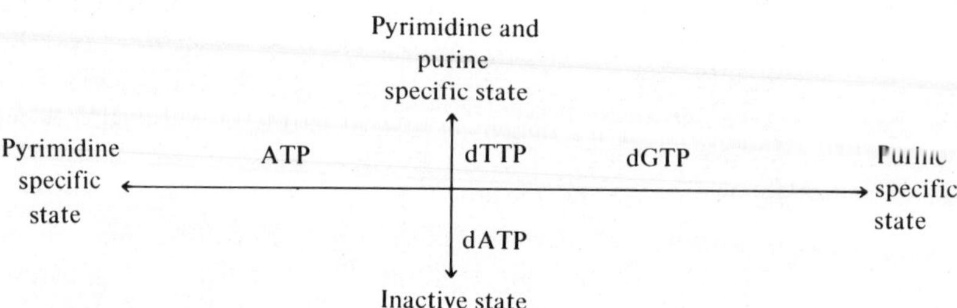

It is important to remember that these results refer only to the purified enzyme. The existence of the implied control mechanisms in the living cell have yet to be demonstrated. Experience indicates that caution ought to be exercised in speculation about cellular control mechanisms in the absence of *in vivo* information.

Ribonucleotide reduction in *L. leichmannii* differs from that in *E. coli* in two crucial ways: the nucleoside *triphosphates,* not the diphosphates, are the substrates, and the reaction catalyzed by this ribonucleoside triphosphate reductase is absolutely dependent on the *cobamide coenzyme* 5′-deoxyadenosyl cobalamine.

The existence of a cobamide-dependent reductase in this species was initially demonstrated by Blakley and Barker in 1964 (24). This observation was shortly thereafter confirmed in the laboratories of Abrams (25) and Beck (26). Subsequent developments of the system have been reported from all three laboratories, frequently in a parallel fashion.

The reducing agent in the *L. leichmannii* system is similar to that in the *E. coli* one. From the former species, a thioredoxin of molecular weight 12,000 has been isolated together with a thioredoxin reductase of molecular weight 55,000 (27). To some extent, the thioredoxins in the two systems are interchangeable.

Despite the participation of the cobamide coenzyme in one system only, the stereochemistry of the introduction of hydrogen into the reduced nucleotides in *L. leichmannii* and *E. coli* are identical. Thus, reduction in deuterium oxide again results in the incorporation of one atom of deuterium at position 2′ with retention of configuration as shown in Eq. (18-26) (28,29).

The regulation of activity of the *L. leichmannii* reductase appears related to but distinct from that for the *E. coli* enzyme. Current results yield a very complicated and confusing picture, the physiological significance of which is not yet evident. *In vitro* CTP reduction is stimulated by ATP or dATP, ATP reduction is stimulated by dGTP, and UTP reduction is stimulated by CTP.

Preliminary studies on ribonucleotide reduction in animal tissues have been reported. Of particular interest are the studies of Moore and his collaborators with Novikoff hepatoma cells (30–32). On the whole, this system appears closely related to that for *E. coli.* Further developments will clarify the extent of the homology between the two systems.

Biosynthesis of Thymidylate

The metabolic routes for the production of the nucleotides and deoxynucleotides of adenine, guanine, uracil, and cytosine have just been discussed. In addition to these species, *thymidylic acid,* a major constituent of nearly all DNA, and *5-hydroxymethyldeoxycytidylic acid,* which replaces thymine in the DNA of the T-even bacteriophages, must be accounted for. We discuss the biosynthesis of these in turn.

A priori thymidylic acid could arise from uridylic acid either by (1) reduction followed by methylation or (2) methylation followed by reduction. The observation that deoxyuridine is a far better precursor than uridine for the synthesis of DNA-thymidylic acid suggests that the former alternative is correct. The complete *de novo* pathway for thymidylic acid synthesis is CMP → CDP → dCDP → dCMP → dUMP → TMP. The methylation of deoxyuridylic acid has been the subject of an intensive investigation by Friedkin and coworkers (33).

In the overall pathway from CMP to TMP, one new reaction is introduced: the deamination of dCMP to yield dUMP. This reaction is catalyzed by *deoxycytidylic acid aminohydrolase,* which is activated by dCTP and inhibited by dTTP. Note that a considerable fraction of dUMP is synthesized by the ribonucleotide reductase pathway previously described; the aminohydrolase pathway is basically a supplementary operation. The dUMP is produced as follows: UDP $\longrightarrow$ dUDP $\xrightarrow{\text{ATP}}$ dUTP $\longrightarrow$ dUMP. dUDP is first phosphorylated by a kinase at the expense of ATP and the triphosphate hydrolyzed to yield dUMP plus PP_i; the direct hydrolysis of dUDP appears less important.

Thymidylate synthetase from *E. coli* catalyzes the synthesis of thymidylic acid according to the stoichiometry in Eq. (18-27).

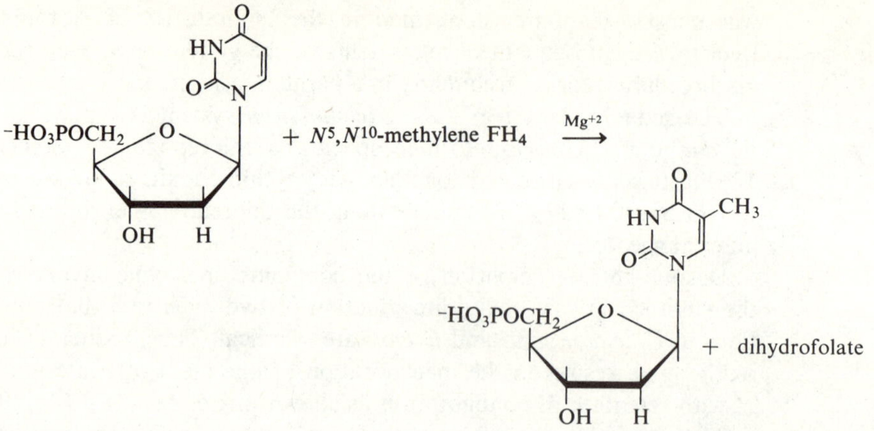

$$\text{deoxyuridylate} + N^5,N^{10}\text{-methylene } FH_4 \xrightarrow{Mg^{+2}} \text{thymidylate} + \text{dihydrofolate} \quad (18\text{-}27)$$

In this reaction, FH_4 serves a dual function in that it acts as both carrier of the one-carbon fragment and as reducing agent. The transfer of hydrogen from the pteridine ring system of FH_4 to the methyl group of thymidylate is direct, as revealed by the incorporation of one atom of tritium into the methyl group when tritiated FH_4 is employed as coenzyme:

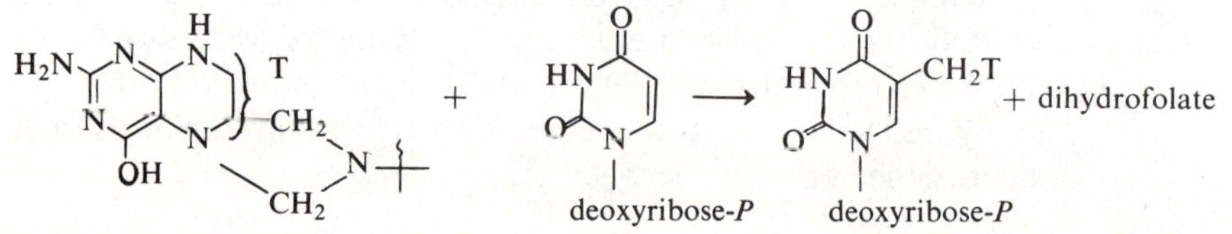

$$FH_4(T) + \text{uracil-deoxyribose-}P \longrightarrow \text{(5-}CH_2T\text{)-uracil-deoxyribose-}P + \text{dihydrofolate} \quad (18\text{-}28)$$

Hydroxymethyldeoxycytidylic acid is derived from the parent compound in a reaction which also involves the participation of N^5,N^{10}-methylene FH_4 as a donor of a one-carbon fragment:

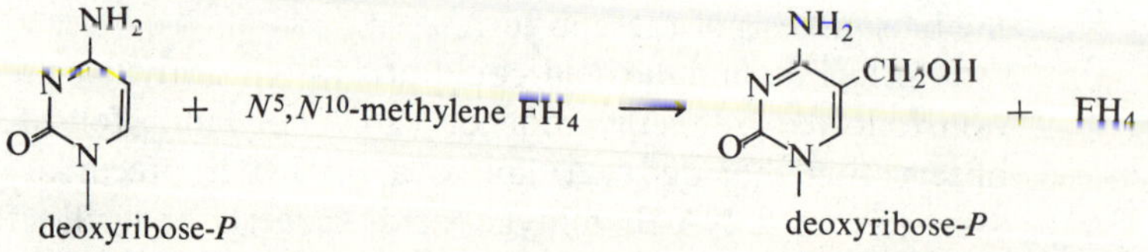

$$\text{cytosine-deoxyribose-}P + N^5,N^{10}\text{-methylene } FH_4 \longrightarrow \text{(5-}CH_2OH\text{)-cytosine-deoxyribose-}P + FH_4 \quad (18\text{-}29)$$

In this case, reduction does *not* accompany carbon transfer.

Ribothymidylic acid occurs as a minor constituent in tRNA. Synthesis of this moiety occurs at the organizational level of the intact tRNA molecule. *S*-adenosyl methionine (page 807) acts as methyl group donor to tRNA-uracil groups with the resultant formation of tRNA-thymine functions. The same situation also prevails with regard to the methylation of other bases in both RNA and DNA.

Alternative Pathways for the Biosynthesis of Nucleotides

The pathways described above almost certainly represent the major routes by which purine and pyrimidine nucleotides are synthesized. However, additional routes for the formation of these structures are known. These appear to be "salvage" mechanisms and are employed to utilize exogenous sources of preformed bases or endogenous sources for this purpose.

In the liver, specific *pyrophosphorylases* (*phosphoribosyl transferases*) catalyze the formation of purine nucleotides from the free bases and PRPP:

$$\text{Purine} + \text{PRPP} \rightleftharpoons \text{purine ribonucleotide} + PP_i \quad (18\text{-}30)$$

A corresponding reaction employing uracil as substrate has been identified in microorganisms:

$$\text{Uracil} + \text{PRPP} \rightleftharpoons \text{uridylic acid} + \text{PP}_i \tag{18-31}$$

In addition, nucleoside phosphorylases catalyze the formation of nucleosides from the bases and ribose-1-phosphate:

$$\text{Base} + \text{ribose-1-}OPO_3H^- \;(\text{furanose ring: } CH_2OH,\ O,\ OH,\ OH) \rightleftharpoons \text{nucleoside } (CH_2OH,\ O,\ \text{base},\ OH,\ OH) + P_i \tag{18-32}$$

The nucleosides are converted to the nucleotides through the action of an appropriate kinase:

$$\text{Ribonucleoside} + \text{ATP} \rightarrow \text{ribonucleotide} + \text{ADP} \tag{18-33}$$

Of the various kinases that function in this way, the most interesting and best understood is *thymidine kinase*. There is substantial evidence to suggest that this enzyme may be involved in the control of DNA synthesis. Thus, thymidine kinase activity is barely detectable in the adult liver but is quite active in the liver of a partially hepatectomized animal. Generally, this enzyme shows an activity that is associated with the rate of growth of an organism or organ.

The enzyme from *E. coli* has been purified and subjected to kinetic study (34). It is activated by DNA precursors such as dCDP. This is quite reasonable since the overall reaction product, dTTP, is necessary for DNA synthesis. Some nucleoside triphosphates including dATP, dCTP, dGTP, and ATP itself act both as activators and phosphate group donors.

CATABOLISM OF PURINES

The routes for catabolism of purines are outlined in Fig. 18-5. Degradation of RNA may yield either the 3′- or 5′-nucleotides, as shown. These are then degraded to the bases which are converted, in reactions involving the interesting enzyme xanthine oxidase (page 653), to uric acid. As noted, uric acid itself is the end product of purine metabolism in man and other species. Note that degradation of the purine deoxyribonucleotides derived from DNA proceeds according to the route shown for guanine.

The production of uric acid is associated with the disease gout. In those afflicted with this disease the levels of uric acid in the blood and other body fluids is markedly elevated, leading to the deposition of the rather insoluble salt sodium urate in the form of crystals (tophi) in the cartilage, accounting for the extreme sensitivity of the joints that is characteristic of gout. Gout may have diverse origins. In some cases, it appears that the excess of uric acid is a consequence of the impaired renal excretion of this substance, while in others an increased rate of purine synthesis may be responsible. Gout is the consequence of an inborn error in metabolism (page 857). In some patients at least, the defect has been determined to be a loss of the enzyme hypoxanthine-guanine phosphoribosyltransferase, Eq. (18-30), which results in overproduction of purines, presumably reflecting an inability to salvage the bases for synthetic purposes and their consequent entry into degradative pathways (35). While gout is a painful and sometimes debilitating disease, sufferers can take comfort in one fact: Throughout history an abnormal percentage of the influential in arts, sciences, and politics have been those most afflicted with gout. The disease, incidentally, generally afflicts males.

The enzyme *guanase,* which converts guanine into xanthine (Fig. 18-5) is lacking in pigs. Consequently, these animals sometimes suffer from guanine gout, a type related to the usual disease but in which guanine replaces uric acid as the offending agent.

In most species, uric acid is not excreted but is further metabolized to a variety of end products as indicated in Fig. 18-5. Purine metabolism is thus a fertile field for those interested in the comparative aspects of biochemistry.

CATABOLISM OF PYRIMIDINES

The metabolic routes for the degradation of the pyrimidines are summarized in Fig. 18-6. Degradations by both the uracil and thymine routes involve initial reductions to the dihydro compounds followed by a series of hydrolytic reactions. The end products of uracil and thymine metabolism, β-alanine and β-aminoisobutyric acid, respectively, may be further metabolized by transamination, yielding malonic semialdehyde and methyl malonyl semialdehyde:

$$H_2N{-}CH_2CH_2{-}CO_2H + \alpha\text{-oxoglutarate} \rightleftharpoons H{-}\overset{\overset{\displaystyle O}{\|}}{C}{-}CH_2{-}CO_2H + \text{glutamate} \quad (18\text{-}34)$$

$$H_2N{-}CH_2{-}\underset{\underset{\displaystyle H}{|}}{\overset{\overset{\displaystyle CH_3}{|}}{C}}{-}CO_2H + \alpha\text{-oxoglutarate} \rightleftharpoons H{-}\overset{\overset{\displaystyle O}{\|}}{C}{-}\underset{\underset{\displaystyle H}{|}}{\overset{\overset{\displaystyle CH_3}{|}}{C}}{-}CO_2H + \text{glutamate} \quad (18\text{-}35)$$

The former species is converted into propionyl-SCoA, in a multistep process, and the latter may be oxidized to methyl malonate, another intermediate in propionic acid metabolism (Chapters 13 and 14).

REGULATION OF PYRIMIDINE AND PURINE BIOSYNTHESIS

The synthesis of both pyrimidines and purines is under metabolic control involving both feedback inhibition and feedback repression by end products (pages 491 and 958).

Pyrimidine biosynthesis is subject to control both at the carbamyl synthetase and aspartate transcarbamylase stages. With the carbamyl phosphate synthetase from *E. coli,* homotropic effects are observed with the substrates, activation with purine nucleotides, particularly IMP, and inhibition with uridine nucleotides (36). Although the mechanism of these effects is not completely clear, they may well be due to changes in the affinity of the enzyme for ATP. At any event, the findings are teleologically satisfying: Excess purines promote the synthesis of pyrimidines while excess pyrimidines inhibit such synthesis. The operation of such controls can clearly help to ensure that a balance is maintained between the rates of purine and pyrimidine biosynthesis.

The *E. coli* synthetase is also subject to cumulative repression by arginine and uracil (37). *E. coli* is perhaps unusual in that a single enzyme is responsible for synthesis of carbamyl phosphate destined for both arginine and pyrimidine biosynthesis (see pages 776 and 826).

Aspartate transcarbamylase was one of the first enzymes for which metabolic control by feedback inhibition was demonstrated (38) and is one of the best studied enzymes in this respect. Although the whole story is complicated and not completely resolved, the essential message is clear. Substrates exert homotropic effects, while the end products of pyrimidine synthesis, the nucleotides and nucleoside polyphosphates, act as inhibitors of this enzyme,

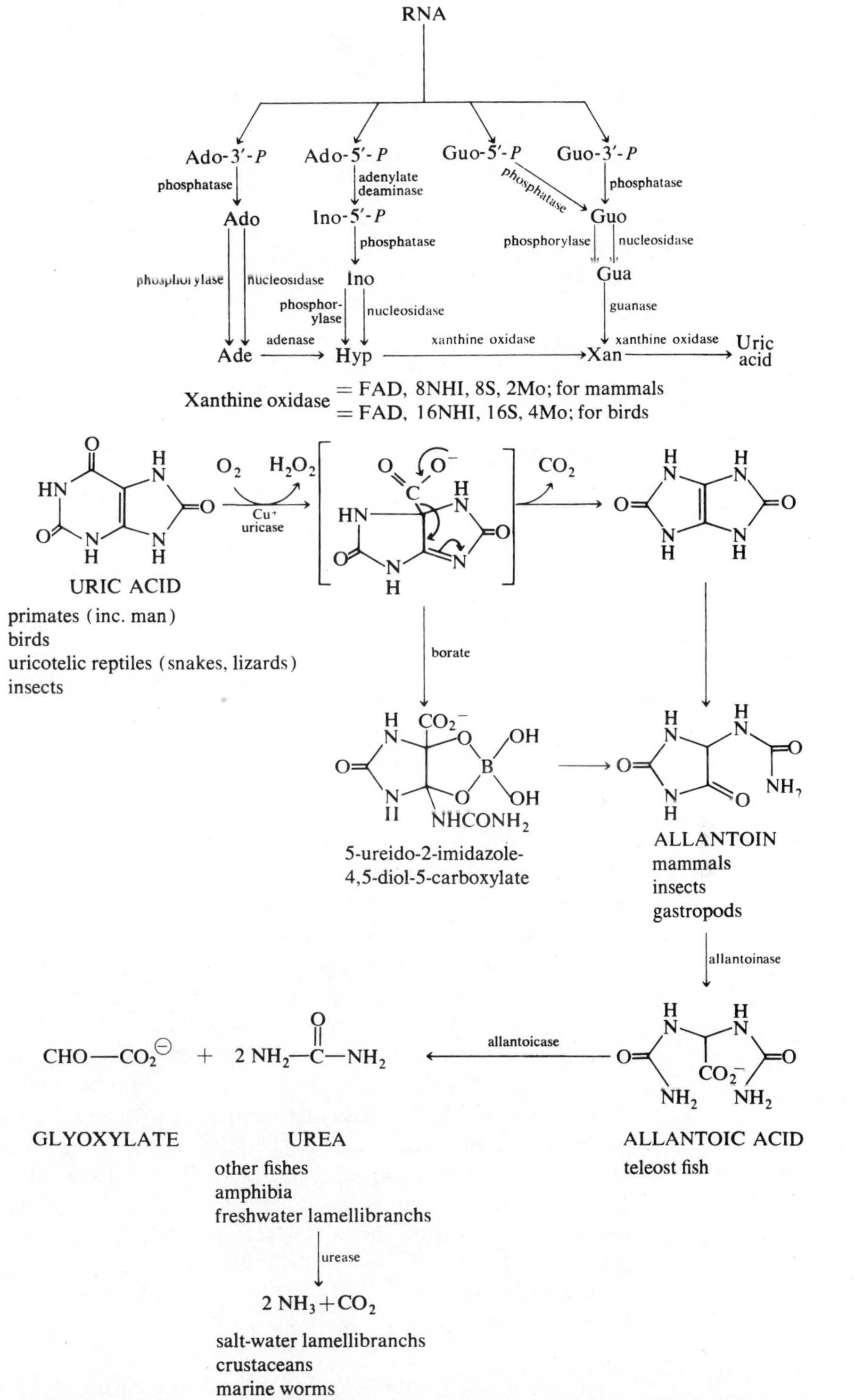

RNA

Ado-3′-P — phosphatase → Ado

Ado-5′-P — adenylate deaminase → Ino-5′-P — phosphatase → Ino

Guo-5′-P — phosphatase → Guo

Guo-3′-P — phosphatase → Guo

Ado — phosphorylase, nucleosidase → Ade

Ino — phosphorylase, nucleosidase → Hyp

Guo — phosphorylase, nucleosidase → Gua — guanase → Xan

Ade — adenase → Hyp — xanthine oxidase → Xan — xanthine oxidase → Uric acid

Xanthine oxidase = FAD, 8NHI, 8S, 2Mo; for mammals
= FAD, 16NHI, 16S, 4Mo; for birds

URIC ACID — O_2, H_2O_2, Cu^+ uricase → [intermediate] — CO_2 →

URIC ACID
primates (inc. man)
birds
uricotelic reptiles (snakes, lizards)
insects

borate

5-ureido-2-imidazole-4,5-diol-5-carboxylate

ALLANTOIN
mammals
insects
gastropods

allantoinase

ALLANTOIC ACID
teleost fish

allantoicase

$CHO—CO_2^{\ominus}$ + $2\,NH_2—C(=O)—NH_2$

GLYOXYLATE

UREA
other fishes
amphibia
freshwater lamellibranchs

urease

$2\,NH_3 + CO_2$

salt-water lamellibranchs
crustaceans
marine worms

FIGURE 18-5

Catabolism of the purines. Those organisms excreting each of the various catabolites as end product are indicated. Ado, adenosine; Guo, guanosine; Ade, adenine; Gua, guanine; Ino, inosine; Hyp, hypoxanthine; Xan, xanthine.

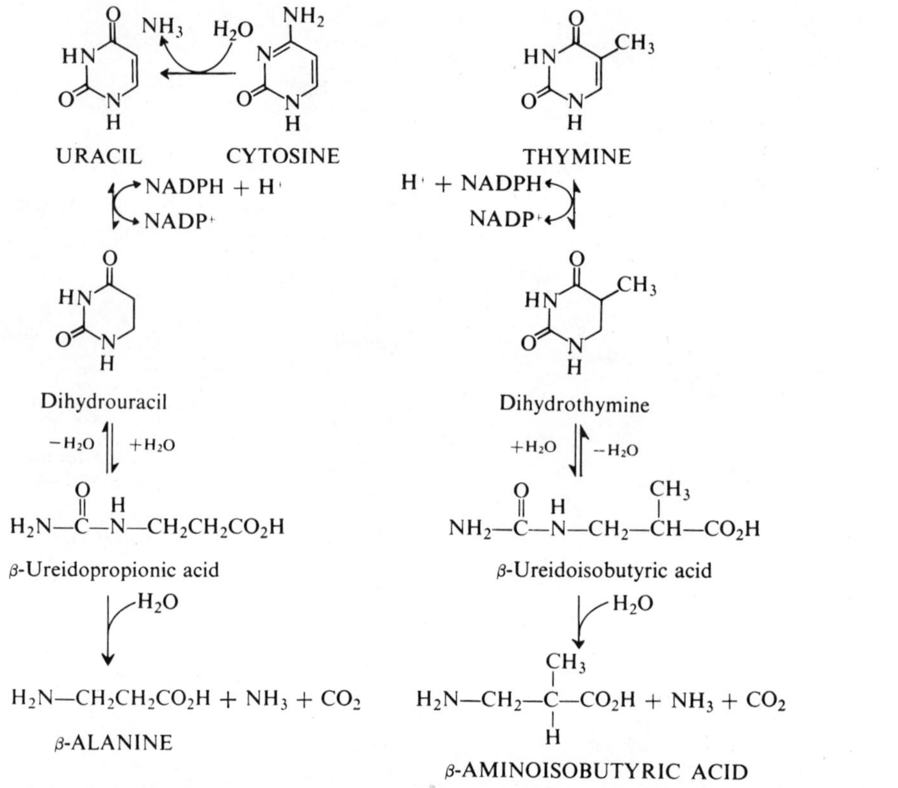

URACIL ⟵ CYTOSINE (NH_3, H_2O)

THYMINE

$NADPH + H^+$ → $NADP^+$

$H^+ + NADPH$ → $NADP^+$

Dihydrouracil

Dihydrothymine

$-H_2O$ ⇌ $+H_2O$

$+H_2O$ ⇌ $-H_2O$

$H_2N—C(=O)—NH—CH_2CH_2CO_2H$

β-Ureidopropionic acid

$NH_2—C(=O)—NH—CH_2—CH(CH_3)—CO_2H$

β-Ureidoisobutyric acid

H_2O

H_2O

$H_2N—CH_2CH_2CO_2H + NH_3 + CO_2$

β-ALANINE

$H_2N—CH_2—CH(CH_3)—CO_2H + NH_3 + CO_2$

β-AMINOISOBUTYRIC ACID

FIGURE 18-6
Metabolic pathways for the degradation of the pyrimidine nucleotides.

the first unique one in the biosynthetic route. For example, CTP is a specific inhibitor and is competitive with aspartate. Furthermore, purine nucleotides appear to act as activators. As noted above (page 828), aspartate transcarbamylase is composed of two types of subunits: The *catalytic* one interacts with aspartate and the *regulatory* one interacts with CTP or other inhibitors and with purine nucleotides. The interaction at the regulatory subunit prevents interaction at the catalytic one. The type and specificity of control exhibit an interesting variation with species. Aspartate transcarbamylases from *E. coli, A. aerogenes,* and *Serratia marcescens* are all subject to feedback inhibition by CTP, whereas that from *P. fluorescens* is most strongly inhibited by UTP and that from lettuce seedlings is most sensitive to UMP. Furthermore, aspartate transcarbamylase from *B. subtilis* seems not to be subject to feedback inhibition at all but is subject to feedback repression. These results certainly demonstrate the versatility and species specificity of cellular control mechanisms (39).

Regulation of the synthesis of the purines is both complex and elegant and has been summarized by Magasanik (40). The principal facts are the following:

1. The initial reaction unique to purine biosynthesis, formation of 5-phosphoribosyl-1-amine, catalyzed by glutamine phosphoribosyl pyrophosphate amidotransferase [Eq. (18-4)], is inhibited by the mono-, di-, and triphosphates of adenosine, guanosine, and inosine. Thus, presence of the end products of the complete pathway tend to shut down this pathway at the very beginning.

2. IMP dehydrogenase, which converts IMP into XMP on the route to GMP (Fig. 18-3), is inhibited by GMP. Hence, the conversion of IMP to GMP is slowed under conditions in which adequate supplies of GMP are available.

3. GMP reductase, which catalyzes the conversion of GMP to IMP and permits utilization of GMP for synthesis of AMP, is inhibited by ATP. Thus, an adequate supply of the adenine nucleotides prevents the synthesis of these species at the expense of the guanine nucleotides.

4. The synthesis of adenylosuccinate from IMP, catalyzed by adenylosuccinate synthetase (Fig. 18-3), is inhibited by AMP.

5. As mentioned above, the formation of AMP and GMP is further regulated in that ATP is required for synthesis of GMP and GTP is required for synthesis of AMP. These controls ensure that the conversion of IMP to the adenine and guanine nucleotides occurs in conformance with the requirements of the cell. If an excess of one type is present intracellularly, then there is no necessary requirement for *de novo* synthesis of the purines at all since adenine and guanine nucleotides are interconvertible.

These points are summarized in Fig. 18-7. In addition to the inhibition of enzyme activity, purine biosynthesis is regulated by inhibition of the synthesis of several of the enzymes involved (41). Specifically, it is now clear that guanosine and adenosine repress the synthesis of glutamine phosphoribosyl pyrophosphate amidotransferase [Eq. (18-4)], IMP transformylase [the sum of Eqs. (18-12) and (18-13)], IMP dehydrogenase, adenylosuccinate synthetase, and adenylosuccinase (Fig. 18-3).

CHEMICAL INHIBITION OF NUCLEIC ACID BIOSYNTHESIS

Interest in the chemical inhibition of the biosynthesis of nucleic acids stems from two principal sources. First, such inhibitors provide sensitive and convenient tools which aid greatly in establishing the interdependencies of DNA synthesis, RNA synthesis, and protein synthesis. Second, such inhibitors are of great interest as antimetabolites in general and, particularly, as antitumor agents. The behavior of these compounds as antimetabolites is an expected result since nucleic acid synthesis, and the dependent process of protein synthesis, is absolutely required for cell division. Furthermore, the most

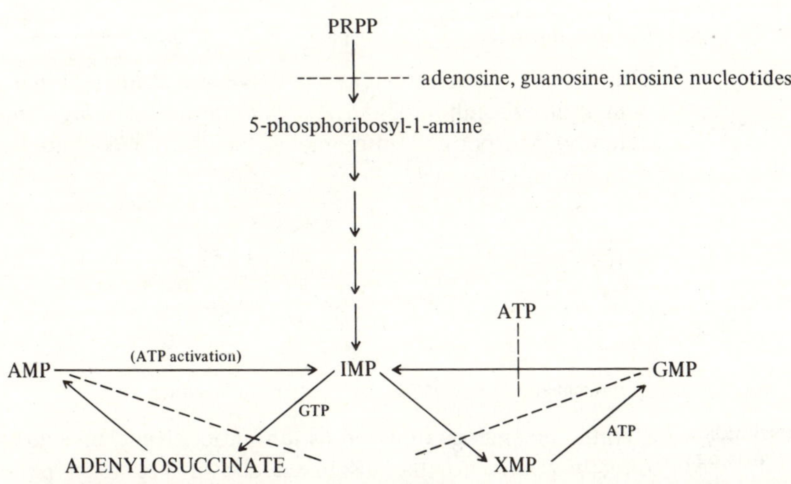

FIGURE 18-7

Regulation of the biosynthesis of purines. Dashed lines show sites of inhibition by the indicated species. IMP is inosinic acid and XMP is xanthylic acid.

obvious biochemical difference between bacterial cells and tumor cells compared to the normal somatic cells of an adult animal is that the former types divide much more frequently that do the latter. Hence, the antimetabolic function of nucleic acid inhibitors is expected to be more severe toward bacteria and tumor cells than toward the host organism. This partially selective toxicity forms the basis for the use of such agents in chemotherapy.

Most, but not all, chemical inhibitors of nucleic acid synthesis are structural analogs of compounds normally required for these reactions. Such substances are, in some cases, very powerful competitive inhibitors of enzymes involved in one or more steps in nucleic acid synthesis. Included in this group are certain compounds related to purines, pyrimidines, glutamine, and folic acid.

Purine Analogs

The inhibition of nucleic acid synthesis and consequent antitumor activity of several purine derivatives can be ascribed, in part at least, to their specific inhibition of one or more reactions involved in the biosynthetic routes leading to the nucleic acid precursors. The best known such inhibitor, and one of the most useful clinically, is *6-mercaptopurine.* This compound owes its inhibitory effects to its metabolic transformation to the corresponding ribonucleotide. 6-Mercaptopurine ribonucleotide is a potent inhibitor of both steps in the conversion of inosinic acid to adenylic acid (Fig. 18-3). Consequently, this compound causes an inhibition in adenylic acid synthesis and a resultant inhibition of nucleic acid synthesis. In addition, 6-mercaptopurine ribonucleotide inhibits the conversion of inosinic acid to xanthylic acid, an obligatory step in the synthesis of guanylic acid (Fig. 18-3). In a similar fashion, the purine derivative *psicofuranin* (6-amino-9-D-psicofuranosylpurine) apparently exerts its inhibitory action by interfering with the ammonia-mediated amination of xanthylic acid, the final step in guanylic acid synthesis.

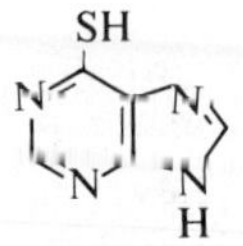

6-Mercaptopurine

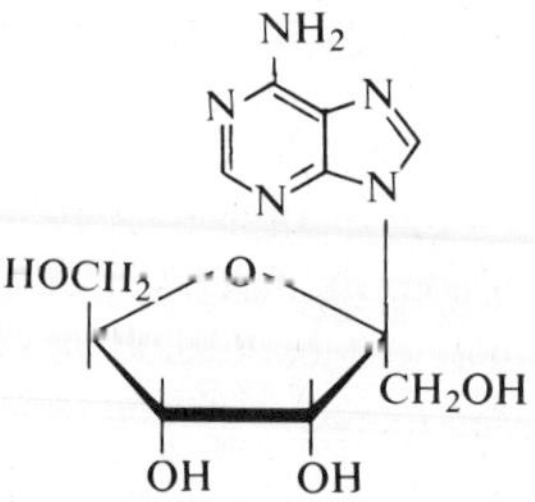

6-Amino-9-D-psicofuranosylpurine

Other purine derivatives also behave as antimetabolites but are not nucleic acid synthesis inhibitors. *8-Azaguanine* and *6-thioguanine,* for example, are actually incorporated into nucleic acids *in vivo* and consequently interfere with normal protein synthesis.

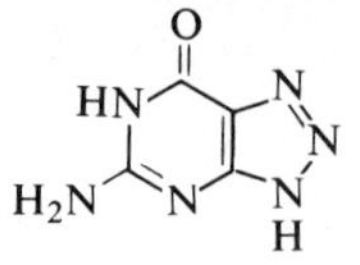

8-Azaguanine

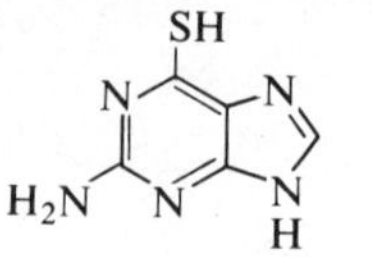

6-Thioguanine

Pyrimidine Analogs

Pyrimidine analogs function as inhibitors of nucleic acid synthesis in a manner directly related to that discussed above. The most potent inhibitors of this type, the 5-fluoropyrimidines, particularly *5-fluorodeoxyuridine,* specifically inhibit DNA synthesis by blocking the thymidylate synthetase reaction [Eq. (18-27)] (42). Incorporation of the 5-fluoropyrimidines into RNA is occasionally noted with interesting results in producing altered enzymes. These are usually, but not always, less active catalytically than the original. Incorporation into DNA appears not to occur. In contrast, the corresponding chloro, bromo, and iodo derivatives are readily incorporated into DNA, replacing thymine and yielding defective DNA molecules. The bromo derivative is most useful and functions as a potent mutagen (page 864). The distinct behavior of the fluoro, as contrasted to the other halogenated pyrimidines, is apparently a reflection of the fact that the van der Waals radii of the chloro, bromo, and iodo groups are not too different from that of methyl, the 5 substituent in thymine itself. The fluoro group, on the other hand, is very small and more nearly resembles a hydrogen, as in uracil, than the methyl of thymine. Hence the fluorinated, but not the other halogenated, derivatives compete with uracil derivatives in the thymidine synthetase reaction and, by the same reasoning, the chloro, bromo, and iodo, but not the fluoro, derivatives act as thymine analogs in the assembly of DNA.

An additional important pyrimidine derivative, *6-azauracil,* is known to be a competitive inhibitor for the decarboxylation of orotidylic acid to yield uridylic acid [Eq. (18-21)].

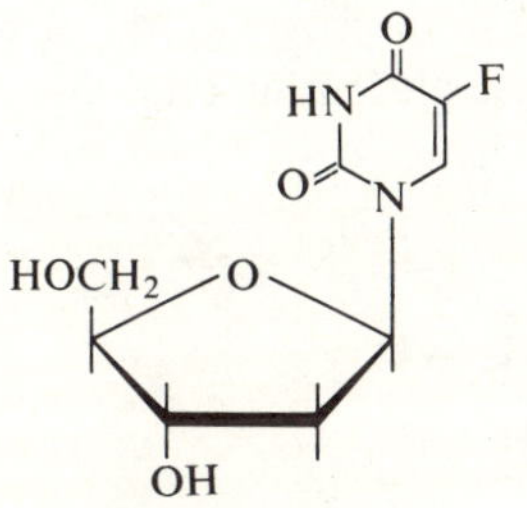

5-Fluorodeoxyuridine

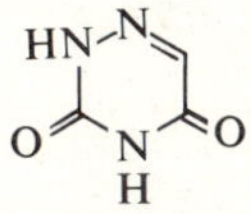

6-Azauracil

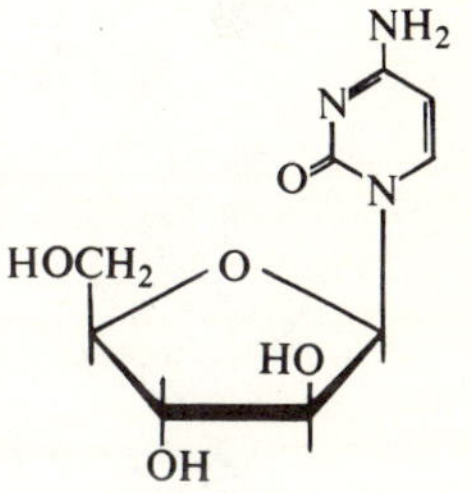

Cytosine arabinoside

Cytosine arabinoside (page 193) is a potent antileukemic agent.

Folic Acid Analogs

Some of the most effective antitumor agents are structurally related to folic acid. Particularly prominent are *aminopterin,* in which an amino function replaces the 4-hydroxyl of folic acid, and *methotrexate* (amethopterin), which is N^{10}-methyl aminopterin.

At two stages in the biosynthesis of purines and at one stage in the biosynthesis of pyrimidines, one-carbon transfer reactions occur for which folic acid derivatives are required as coenzymes.

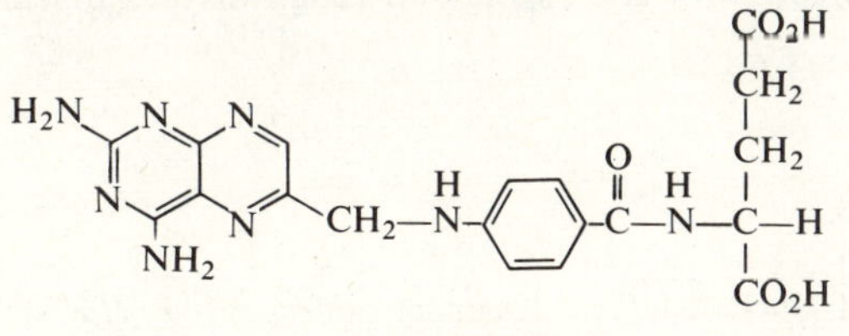

Aminopterin

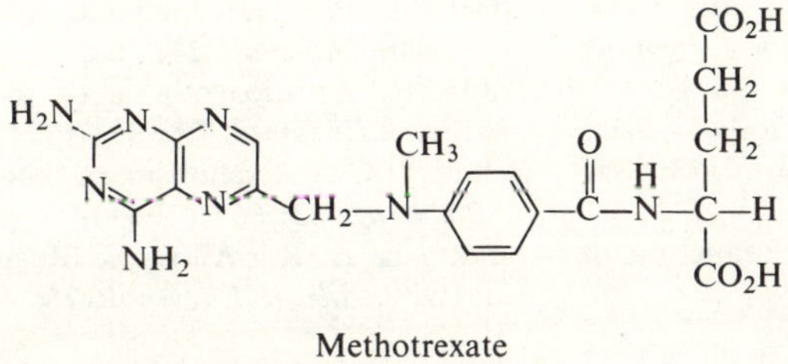

Methotrexate

It will be recalled that, in each case, the coenzymatically active species are those derived from tetrahydrofolic acid. The latter is synthesized from folic acid in nicotinamide nucleotide-dependent reaction catalyzed by folic acid reductase (page 408). This enzyme is the site of the antimetabolite function of aminopterin and methotrexate. These drugs bind with extraordinary tenacity, although reversibly, to folic acid reductase, preventing the formation of tetrahydrofolic acid and, subsequently, the active coenzymes for nucleic acid synthesis (9). The dissociation constants for folic acid reductase–inhibitor complexes may be several orders of magnitude smaller than those for the corresponding complexes involving folic acid. Thus, there is no release of inhibition even in the presence of high concentrations of folic acid.

Glutamine Analogs

The *Streptomyces* antibiotics *azaserine* (*O*-diazoacetyl-L-serine) and *DON* (6-diazo-5-oxo-L-norleucine) are structurally related to glutamine. Consequently, it is not surprising that these compounds inhibit, to varying extents, those aminations in which glutamine serves as amino group donor. Since several such aminations are involved in the biosynthesis of nucleic acid precursors, azaserine and DON strongly inhibit nucleic acid biosynthesis. Of the various glutamine-dependent aminations, the one involved in the conversion of formyglycinamide ribonucleotide to the corresponding amidine [Eq. (18-8)] is most sensitive to inhibition by these glutamine analogs:

```
        O                             O
        ‖  H                          ‖  H
        C—C=N⁺=N⁻                     C—C=N⁺=N⁻
       /                             /
      O                            CH2
      |                             |
     CH2                           CH2
      |                             |
 H2N—C—CO2H                   H2N—C—CO2H
      |                             |
      H                             H

    Azaserine                      DON
```

Buchanan et al. have demonstrated that azaserine reacts with the amidotransferase that catalyzes this reaction to yield a covalently bound enzyme-inhibitor compound (43). A specific sulfhydryl group of the enzyme is alkylated.

In addition to those compounds which inhibit synthesis of the precursors to the nucleic acids, there are several which inhibit the assembly of these precursors into polymers. These will be considered in Chapter 20, where the enzymes responsible for the associated reactions are discussed.

REFERENCES

Books and Monographs

Blakley, R. L., and E. Vitols: The Control of Nucleotide Biosynthesis, *Ann. Rev. Biochem.*, **37:** 201 (1968).

Brockman, R. W., and E. P. Anderson: Biochemistry of Cancer, *Ann. Rev. Biochem.*, **32:** 463 (1963).

Buchanan, J. M., and S. C. Hartman: Enzymic Reactions in the Synthesis of the Purines, *Advan. Enzymol.*, **21:** 199 (1959).

Karnofsky, D. A., and B. D. Clarkson: Cellular Effects of Anti-Cancer Drugs, *Ann. Rev. Pharm.*, **3:** 357 (1963).

Larsson, A., and P. Reichard: Enzymatic Reduction of Ribonucleotides, *Progr. Nucleic Acid Res. Mol. Biol.*, **7:** 303 (1967).

Reichard, P.: The Biosynthesis of Deoxyribonucleotides, *European J. Biochem.*, **3:** 259 (1968).

Reichard, P.: The Enzymic Synthesis of Pyrimidines, *Advan. Enzymol.*, **21:** 263 (1959).

Sable, H. Z.: Biosynthesis of Ribose and Deoxyribose, *Advan. Enzymol.*, **28:** 391 (1966).

Schmidt, G.: Metabolism of Nucleic Acids, *Ann. Rev. Biochem.*, **33:** 667 (1964).

Stadtman, E. R.: Allosteric Regulation of Enzyme Activity, *Advan. Enzymol.*, **28:** 41 (1966).

Specific References

1. Mizobuchi, K., and J. M. Buchanan: *J. Biol. Chem.*, **243:** 4842 (1968).
2. French, T. C., I. B. Dawid, R. A. Day, and J. M. Buchanan: *J. Biol. Chem.*, **238:** 2171 (1963).
3. Mizobuchi, K., and J. M. Buchanan: *J. Biol. Chem.*, **243:** 4853 (1968).
4. Mizobuchi, K., G. L. Kenyon, and J. M. Buchanan: *J. Biol. Chem.*, **243:** 4863 (1968).
5. Jones, M. E., and S. E. Hager: *Science*, **154:** 422 (1966).
6. Hager, S. E., and M. E. Jones: *J. Biol. Chem.*, **242:** 5674 (1967).
7. Piérard, A., and J. M. Wiame: *Biochem. Biophys. Res. Commun.*, **15:** 76 (1964).
8. Anderson, P. M., and A. Meister: *Biochemistry*, **4:** 2803 (1965).
9. Wellner, V. P., J. I. Santos, and A. Meister: *Biochemistry*, **7:** 2848 (1968).
10. Gerhart, J. C., and H. K. Schachman: *Biochemistry*, **4:** 1054 (1965).
11. Gerhart, J. C., and H. K. Schachman: *Biochemistry*, **7:** 538 (1968).
12. Weber, K.: *Nature*, **218:** 1116 (1968).
13. Wiley, D. C., and W. N. Lipscomb: *Nature*, **218:** 1119 (1968).
14. Weed, L. L., and D. W. Wilson: *J. Biol. Chem.*, **189:** 435 (1951).
15. Hammarsten, E., P. Reichard, and E. Saluste: *J. Biol. Chem.*, **183:** 105 (1950).
16. Reichard, P., A. Baldesten, and L. Rutberg: *J. Biol. Chem.*, **236:** 1150 (1961).
17. Holmgren, A., P. Reichard, and L. Thelander: *Proc. Natl. Acad. Sci. U.S.*, **54:** 830 (1965).
18. Brown, N. C., A. Larson, and P. Reichard: *J. Biol. Chem.*, **242:** 4272 (1967).
19. Brown, N. C., R. Eliasson, P. Reichard, and L. Thelander: *Biochem. Biophys. Res. Commun.*, **30:** 522 (1968).
20. Laurent, T. C., E. C. Moore, and P. Reichard: *J. Biol. Chem.*, **239:** 3436 (1964).
21. Moore, E. C., P. Reichard, and L. Thelander: *J. Biol. Chem.*, **239:** 3445 (1964).
22. Thelander, L.: *J. Biol. Chem.*, **242:** 852 (1967).
23. Durham, L. J., A. Larsson, and P. Reichard: *European J. Biochem.*, **1:** 92 (1967).
24. Blakley, R. L., and H. A. Barker: *Biochem. Biophys. Res. Commun.*, **16:** 391 (1964).
25. Abrams, R., and S. Duraiswami: *Biochem. Biophys. Res. Commun.*, **18:** 409 (1965).
26. Beck, W. S., and J. Hardy: *Proc. Natl. Acad. Sci. U.S.*, **54:** 286 (1965).
27. Orr, M. D., and E. Vitols: *Biochem. Biophys. Res. Commun.*, **25:** 109 (1966).
28. Blakley, R. L., R. K. Ghambeer, T. J. Batterham, and C. Brownson: *Biochem. Biophys. Res. Commun.*, **24:** 418 (1966).
29. Batterham, T. J., R. K. Ghambeer, R. L. Blakley, and C. Brownson: *Biochemistry*, **6:** 1203 (1967).
30. Moore, E. C., and R. B. Hurlbert: *Biochim. Biophys. Acta*, **55:** 651 (1962).
31. Moore, E. C., and P. Reichard: *J. Biol. Chem.*, **239:** 3453 (1964).
32. Moore, E. C., and R. B. Hurlbert: *J. Biol. Chem.*, **241:** 4802 (1966).
33. Friedkin, M.: *Ann. Rev. Biochem.*, **32:** 185 (1963).
34. Okazaki, R., and A. Kornberg: *J. Biol. Chem.*, **239:** 275 (1964).
35. Kelley, W. N., F. M. Rosenbloom, J. F. Henderson, and J. E. Seegmiller: *Proc. Natl. Acad. Sci. U.S.*, **57:** 1735 (1967).
36. Anderson, P. M., and A. Meister: *Biochemistry*, **5:** 3164 (1966).
37. Piérard, A., N. Glansdorff, M. Mergeay, and J. M. Wiame: *J. Mol. Biol.*, **14:** 23 (1965).
38. Yates, R. A., and A. B. Pardee: *J. Biol. Chem.*, **221:** 757 (1956).
39. Neumann, J., and M. E. Jones: *Arch. Biochem. Biophys.*, **104:** 438 (1964).
40. Magasanik, B.: in W. D. McElroy and B. Glass (eds.), *The Chemical Basis of Development*, Johns Hopkins Press, Baltimore, 1958, p. 485.
41. Momose, H., H. Nishikawa, and I. Shiio: *J. Biochem.* (*Tokyo*), **59:** 325 (1966).
42. Anderson, E. P., and L. W. Law: *Ann. Rev. Biochem.*, **29:** 577 (1960).
43. French, T. C., I. B. Dawid, R. A. Day, and J. M. Buchanan: *J. Biol. Chem.*, **238:** 2171, 2178 (1963).

19 Nature and Function of the Gene

The questions with which genetics has been traditionally concerned are the following: What is the nature, the organization, and the mode of self-duplication of the substance(s) that transmit information from parent to progeny, generation after generation, in a form so precise as to allow only a negligible amount of alteration in the final product? How can they permit equal fidelity in specification and information transfer from generation to generation for an altered product formed as a result of an initial alteration in the appropriate determinant? What is the nature of the processes by which the genetic determinants express themselves by specifying all the chemical, metabolic, and morphological characteristics of all cells, of all organisms? In this chapter we shall briefly examine the evidence that proves that DNA is the genetic material of all cellular organisms and that its role as genetic determinant is expressed by virtue of its specification of the proteins in these organisms. This process is now known to be mediated by three different species of RNA. Gene duplication thus requires DNA synthesis, and gene action RNA and protein synthesis. The biosynthesis of nucleic acids is discussed in Chapter 20, and Chapter 21 is devoted to a consideration of the biosynthesis of proteins.

Before embarking on a brief discussion of molecular or biochemical genetics, we direct our attention to classical genetics, but only insofar as it allows us to define operationally a number of terms that we shall use in the remainder of the chapter. Of necessity our account will be rudimentary.

CLASSICAL GENETICS

Mendelian Genetics

Mendel, the originator of the science, performed his experiments with 22 true-breeding strains of the garden pea (*Pisum sativum*). His conclusions were as follows. Provided that both parents came from the same strain, all descendants completely resembled these parents with respect to one of several easily distinguishable external characteristics, or *phenotypes*. Now let us perform a hypothetical experiment similar to one of Mendel's called a *genetic cross,* in which we mate two parents belonging to two different strains of different genetic constitution, or *genotype*. Let us cross one parent with a purple flower with one that produces white flowers. What is the result? The offspring called the *first filial,* or F_1, generation, consists exclusively, not of pink or of mottled, but of purple flowers. We say that the trait for purple is *dominant;* conversely, that for white, or the absence of purple, is *recessive*.

Quite generally this is what was observed by Mendel for each of the seven traits examined by him: all the members of the F_1 generation always resembled either one parent or the other. There was no hybrid phenotype. What happens if we cross two members of the F_1 generation with each other and analyze the property of the resulting F_2 generation? The two traits now *segregate,* and we observe both types of plants, those with purple and those with white flowers, in the precise ratio 3 : 1, provided that we examine a reasonably large number of progeny. How can we explain these results? Mendel reasoned—and this reasoning has been borne out by countless subsequent experiments and by the remarkable coherence and self-consistency of the whole edifice of Mendelian genetics built thereon—that we are dealing with the following sequence of events:

1. In the parental, or F_0, generation, the hereditary determinants for the genetic locus of flower color can exist in two and only two stable alternative forms, or *alleles,* the *homozygous* allele for purple, or *PP,* and the equally homozygous allele for white, or *pp.* These alleles segregate independently during formation of the gametes, the male and female germ cells, and are recombined at fertilization.

2. In the hybrid F_1 generation we have formed *heterozygotes,* that is, hybrids of unstable alleles, *Pp* or *pP.* Since *P* is dominant over the recessive *p,* the phenotypic expression of this particular genotype is purple, and thus indistinguishable from that characteristic of the *PP* parent.

3. A distinction becomes possible, however, by the cross of the F_1 hybrids with themselves, or by backcrosses with the white parents. The two alleles again segregate independently into their unit factors *P* and *p,* and we observe, after recombination, their appropriate phenotypes expressed thus—(pu) stands for the purple, (w) for the white, phenotype:

$$\underset{\text{(pu)}}{PP} \times \underset{\text{(w)}}{pp} = \underset{\text{(pu)}}{Pp} + \underset{\text{(pu)}}{pP} \qquad \underset{\text{(pu)}}{Pp} \times \underset{\text{(pu)}}{pP} = \begin{array}{c|c|c} & P & p \\ \hline P & PP\text{(pu)} & Pp\text{(pu)} \\ \hline p & pP\text{(pu)} & pp\text{(w)} \end{array}$$

$$Pp \times pp = \begin{array}{c|c|c} & p & p \\ \hline P & Pp\text{(pu)} & Pp\text{(pu)} \\ \hline p & pp\text{(w)} & pp\text{(w)} \end{array} \qquad Pp \times PP = \begin{array}{c|c|c} & P & P \\ \hline P & PP\text{(pu)} & PP\text{(pu)} \\ \hline p & pP\text{(pu)} & pP\text{(pu)} \end{array}$$

In an analogous fashion strains were crossed that differed in two separate traits. The first, say, yellow flower (ye), was carried by the allele, *Y,* dominant over green (gr), carried by the allele *y* of the same unit factor; the second, say appearance of seed, was carried by the allele *S* for smooth (sm), dominant over the allele *s* for wrinkled (wr). The cross of the two genotypes *YYSS* × *yyss* yielded a uniform F_1 generation with phenotype yellow and smooth, determined by the genotype *YySs.* Members of this generation were then crossed to produce an F_2 generation. We observe progeny with various traits in the following distribution, on the average: nine yellow-smooth, three green-smooth, three yellow-wrinkled, and one green-wrinkled. Constructing the matrix governing a reassortment of units segregating independently and randomly, we would predict that the following statistics would result in complete agreement with the experiments:

	YS	*Ys*	*yS*	*ys*
YS	*YSYS*(ye-sm)	*YSYs*(ye-sm)	*YSyS*(ye-sm)	*YSys*(ye-sm)
Ys	*YsYS*(ye-sm)	*YsYs*(ye-wr)	*YsyS*(ye-sm)	*Ysys*(ye-wr)
yS	*ySYS*(ye-sm)	*ySYs*(ye-sm)	*ySyS*(gr-sm)	*ySys*(gr-sm)
ys	*ysYS*(ye-sm)	*ysYs*(ye-wr)	*ysyS*(gr-sm)	*ysys*(gr-wr)

The model that emerged from these studies, a purely formal and statistical one, envisaged the *unit factors* involved as separate, indivisible *units of segregation,* capable of existing in two or more completely and sharply defined configurations or allelic forms and of being transmitted and recombined in complete independence of any other. Following a suggestion by Johannsen in 1911, these unit factors were called *genes.*

Linkage Soon Mendel's conclusions were checked by studies with other organisms, especially with the fruit fly *Drosophila melanogaster,* performed by Thomas Hunt Morgan. It was found that Mendel's conclusions had to be modified in one important respect. Frequently genes did not appear to be assorted and transmitted independently but acted as if they were formally linked to one another. Different genes appeared to fall within several groups. All those within any one of the *linkage groups* appeared to be transmitted together to a varying degree, yet were transmitted independently of the members of a second group, and so forth. Within any one linkage group the genes could be ordered in a linear manner so that their separation on this one-dimensional *genetic map* was directly proportional to their *recombination frequency,* that is, the ratio formed by multiplying the number of joint appearances of any pair of genes by 100 and dividing by the total progeny. On the map short distances correspond to uncommon recombinational events and long distances to frequent occurrences. Thus, the map is composed of a number of *genetic loci* arranged in linear array, with each new locus fixed or ordered into the array by means of two-factor crosses with genetic markers, say, *A* or *B,* whose position on the map had been established previously.[1] Obviously the map so constructed (see Fig. 19-1) must be internally consistent such that, with the marker *X* located between *A* and *B,* the distances $A \rightarrow X$ plus $X \rightarrow B$ must add up to $A \rightarrow B$. The genes, therefore, become elements of recombination that can be localized at definite loci along a linear map. Map distance is thus defined in terms of recombination units, that is, the fraction of recombinants in a cross that differ from either parent.

Mutation In classical genetics a mutation, a phenomenon first observed by De Vries, is defined as an abrupt or discontinuous change from one allelic form to another, thus producing a visible alteration of phenotype (see Fig. 19-1*b*). Mutations can arise spontaneously, that is, be manifest by the existence of the allelic forms in the preexisting population, as was the case with Mendel's peas. Mutations can also be induced by various mutagenic agents, or *mutagens,* such as certain chemicals (see page 862) or, as discovered by Muller, by irradiation with X rays. Were it not for mutations, the observations on which genetics rests could never have been made.

[1] It requires two independent two-factor crosses or one three-factor cross to fix the position of *X* unambiguously, since the first two-factor cross with *A* alone only gives the degree of linkage to *A* (i.e., the distance from *A* but not the sign). For example, let us perform the crosses $AaXx/aaxx$ and $BbXx/bbxx$. Let us assume that in the first instance the recombinant phenotypes $Aaxx$ and $aaXx$ were observed to occur in 5 percent of the total progeny and in the second the recombinants $Bbxx$ and $bbXx$ occurred in 20 percent of the total progeny. The only consistent arrangement for the map is then

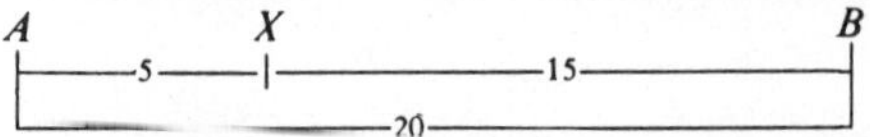

which also predicts the map distance between *A* and *B* to be 20 recombination units, which can be established independently. In a three-factor cross, $\overline{AB} = \overline{AX} + \overline{XB} - 2\overline{AX} \times \overline{XB}$, where $\overline{AB}$, $\overline{AX}$, etc., are the extent of recombination between *A* and *B*, *A* and *X*, etc.

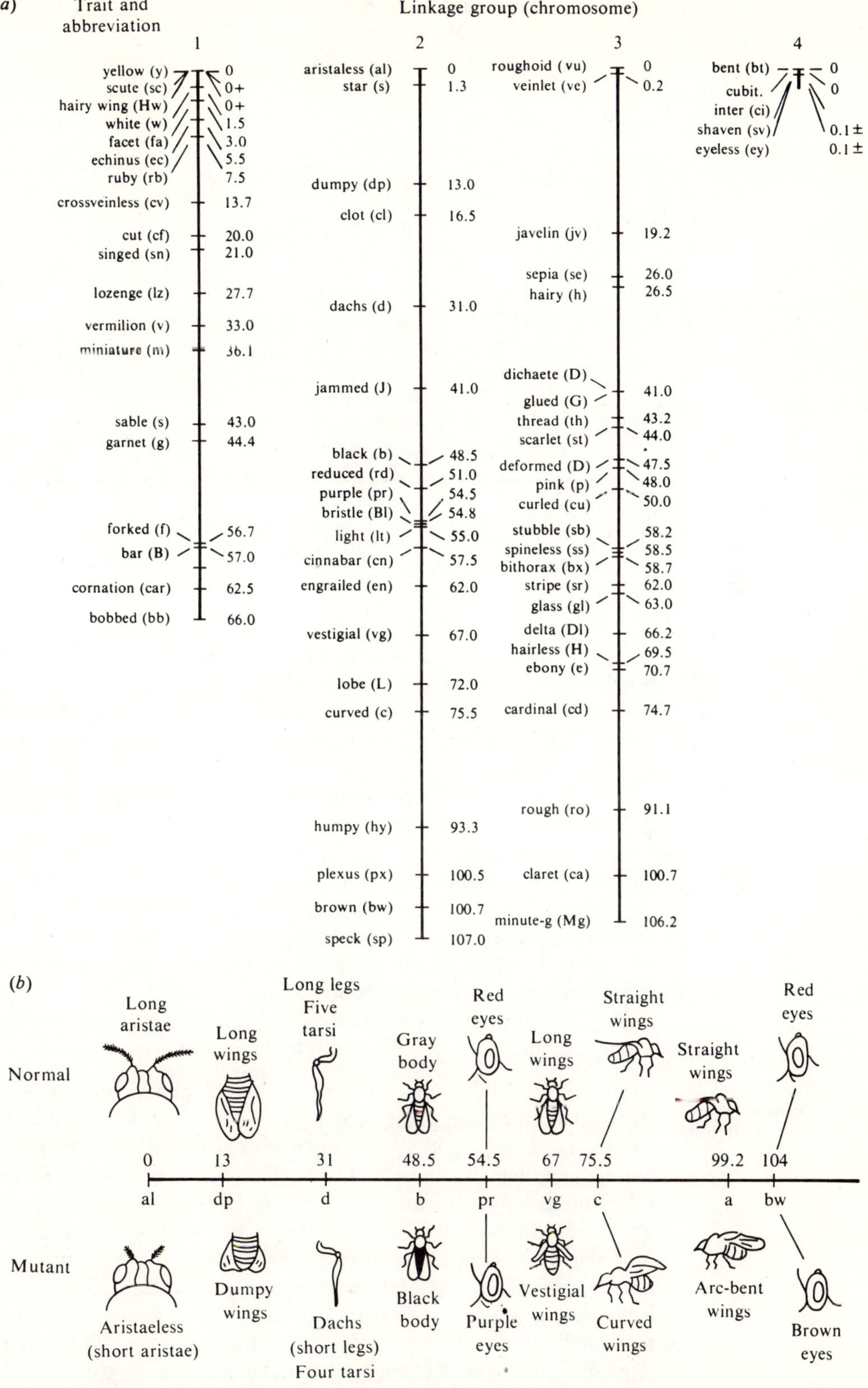

FIGURE 19-1

(*a*) Linkage map for *Drosophila* showing four linkage groups (chromosomes). (*b*) A portion of the genetic map of *Drosophila* (chromosome 2). C. B. Anfinsen, *The Molecular Basis of Evolution*, Wiley, New York, 1959.

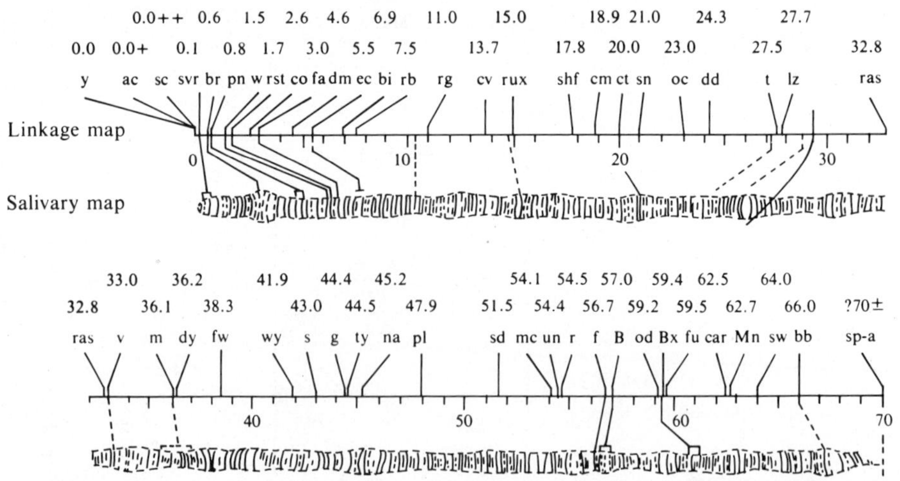

FIGURE 19-2
Arrangement of genetic determinants on salivary chromosomes. Microscopic appearance of a salivary chromosome of *Drosophila*. C. B. Anfinsen, *The Molecular Basis of Evolution,* Wiley, New York, 1959.

Physical Basis of Classical Genetics: Chromosomes as Linkage Groups

Morgan's investigations also led him to another important conclusion. The number of linkage groups in *Drosophila* corresponded precisely to the number of chromosomes in the nuclei of somatic cells of this organism. The chromosomes of the salivary glands are abnormally large and quite readily studied in the light microscope. Furthermore, various mutations operationally defined originally in purely genetic terms, as distinct phenotypic traits caused by configurational changes in the alleles of various hypothetical genes, now produced observable alterations in the appearance of distinct small regions of the chromosomes. Finally, the order of these morphologically distinct regions on the chromosome was the same as the order of the mutational loci on the genetic map (see Fig. 19-2). Thus, we can say that the genetic map is an analog of the chromosome map, or that there is an approximate colinearity of the two maps.[2] Finally, Morgan also provided a physical explanation of linkage in terms of the phenomena of *chiasmata* and *crossing over* occurring during cell division, which are discussed in all texts on genetics.

Chromosomes, Cell Divisions, and Genetics

Even the most rudimentary description of the nuclear events occurring during the formation of the gametes (i.e., the ova in the female and the spermatozoa in the male) in a higher organism could lead us far afield and require more space than we have available. For our purposes it will suffice to consider the grossly oversimplified scheme in Fig. 19-3, for an organism with two chromosomes and linkage groups, and to focus our attention on just four genetic loci with two on each linkage group, one set widely separated as an example of one extreme, the other very closely linked. We observe the formation of links or *chiasmata* between chromosomes immediately following the longitudinal division of the chromosome pairs. This event allows the crossing over and exchange of genetic material between homologous but different chromosomes and leads to segregation of various loci which had

[2] Map distance cannot be brought into *precise* one-to-one correspondence with chromosome distance, since multiple crossovers occur and chromosomes break during cell division.

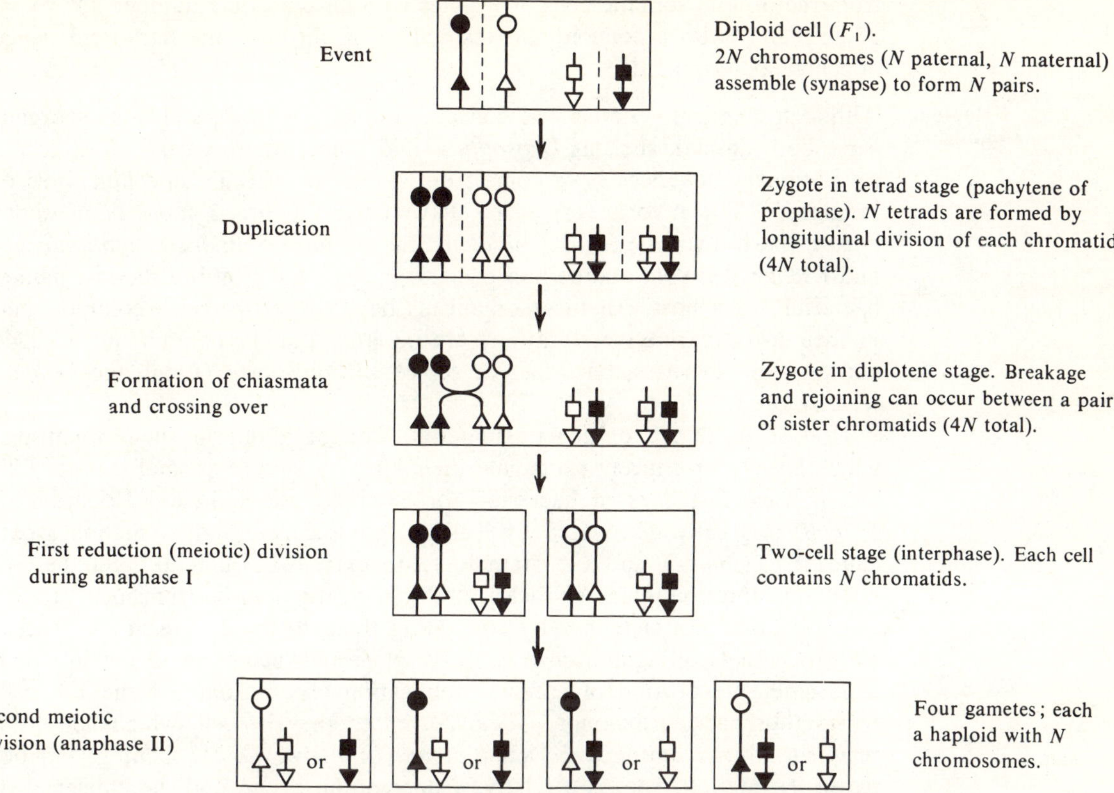

FIGURE 19-3

Schematic representation of nuclear events in diploid organisms. The events in a hypothetical organism with two chromosomes are shown schematically for four genetic loci arranged on the two chromosomes in two linkage groups; those on chromosome 1 (i.e., alleles ○ ● and △ ▲) are pictured as widely separated, those in chromosome 2 (i.e., □ ■ and ▽ ▼) as very closely linked. After fertilization and sexual conjugation of the haploid gametes, these will segregate in the zygote of chromosome number $2N$ to yield the 16 genotypes similar to those depicted in the array for the Mendelian two-factor cross on page 845, in combination with the four recombinant types (■▼, □▽, ■▽, and □▼, in equal proportions) derived from the second chromosome. The actual frequency of recombinants will of course depend on the degree of linkage, or rather vice versa, since the latter is defined in terms of the former; thus 20 percent linkage corresponds to a recombination frequency of 40 percent and means one crossover in five; that is, we will have a cross producing recombinants in the ratio ●—▲/○—△(32): ●—▲/●—▲ (16) : ○—△/○—△ (16) : ●—▲/●—△ (8) : ●—▲/○—▲ (8) : ●—△/○—○ (8) : ○—▲/○—○ (8) : ●—△/●—△ (1) : ○—▲/○—▲ (1). Phenotypically, with black dominant over white, this corresponds to 24 percent recombinants; completely unlinked markers produce 50 percent recombinants, the maximum.

originally formed part of the same chromosome. We also appreciate that—providing this process can occur with equal probability anywhere along a chromosome—the formation of a chiasma and the resulting separation of fragments, and hence their eventual recombination, will be directly proportional to the physical distance between them. We have here then an explanation for the congruence of the maps. The net result of the events shown is the formation of four haploid germ cells (gametes) of chromosome number N

from the diploid somatic cells or zygotes with chromosome number $2N$ ($N =$ 23 in man), with a concomitant reshuffling of chromosome parts and hence of genetic determinants.

Haploid Organisms

Although classical genetics was concerned chiefly with diploid, higher organisms, both animal, such as *Drosophila,* and plant, such as corn, most recent revolutionary advances have come from studies on certain unicellular microorganisms. The favorite organisms have been the bread mold *Neurospora crassa,* the fungus *Aspergillus nidulans,* the bacteria *Salmonella typhimurium* and *Escherichia coli,* and certain bacteriophages, which utilize the last named bacterium as a host. All these organisms have the property in common that in their usual vegetative state they are haploid, that is, contain but a single chromosome (viruses, bacteria) or set of chromosomes (fungi, molds, e.g., *Neurospora*).

How is exchange of genetic material brought about in such organisms, which largely or entirely propagate their kind by simple asexual division? In certain rare instances in bacteria—for example, the unique K12 strain of *E. coli,* originally discovered by Tatum and Lederberg—two mating types, called F^+ (males) and F^- (females), do exist. The male F^+ cell has an infectious and dominant sex factor, or *F agent,* that can be transmitted to F^- cells on direct contact, thereby converting them to the F^+ form. A process akin to sexual conjugation also can take place between F^+ and F^- in which a considerable portion of a newly replicating chromosome of the F^+ cell, which thus acts as a donor, is transferred to the F^- cell, which acts as a recipient, thus forming a *partial diploid* (or *merozygote*). Conjugation between the mating types is a relatively uncommon event, and the efficiency of transfer of genetic determinants is very low and amounts to only 10^{-4} of that of the F factor. However, with certain males called *Hfr* (*h*igh *f*requency of *r*ecombination), this efficiency is increased by a factor of 10^3. In these instances the F factor has become attached to the chromosome[3] and is no longer infectious; unlike with F^+ males, conjugation does not usually result in a transfer of the F factor into the recipient, and the attendant conversion from F^- to F^+ does not take place. The conjugation process and the concomitant transfer of genes from Hfr to F^- may be interrupted at any time by separating the cells mechanically in a Waring Blendor and can be monitored by testing the zygotes formed with respect to their ability to express various characters phenotypically (see page 859). In this manner we can establish the sequence of certain genes, *A, B, C,* etc., both qualitatively with respect to their order

[3] Incidentally, in different substrains the F factor can apparently attach almost anywhere on the circular chromosome, converting it to the Hfr form and priming it for conjugation that occurs linearly with polarity as follows:

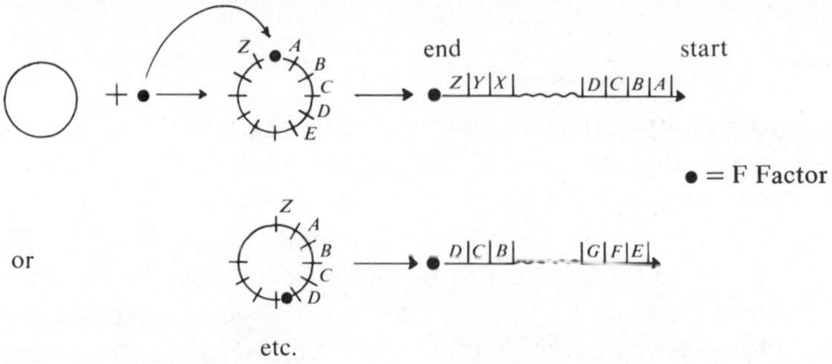

end Z|Y|X ... D|C|B|A start

• = F Factor

or D|C|B ... G|F|E

etc.

on the chromosomes and quantitatively by their relative time of entry into the recipient cells. We can then compare a map constructed on this basis (i.e., in terms of the ratio of time for any marker divided by time for the total chromosome) with the usual map based on recombination frequency (i.e., the ratio of number of recombinants receiving a marker divided by total Hfr cells mated). Remarkably good agreement is obtained by the two techniques (see Fig. 19-4). In the same strain of *E. coli* two additional means for the exchange of genetic information have been observed. The first is a process called *F duction,* which consists of the transfer of a small number of genetic determinants coincident with the transfer of the F factor already mentioned. We recall that the latter can exist both as a free particle (in F^+ cells) and as part of the chromosome (in Hfr cells). Sometimes in the course of its dissociation it also detaches a small chunk of the bacterial chromosome, which can then be replicated as part of the F factor, and on conjugation is transferred to a F^- cell. A genetic unit such as the F factor, capable of existing in either

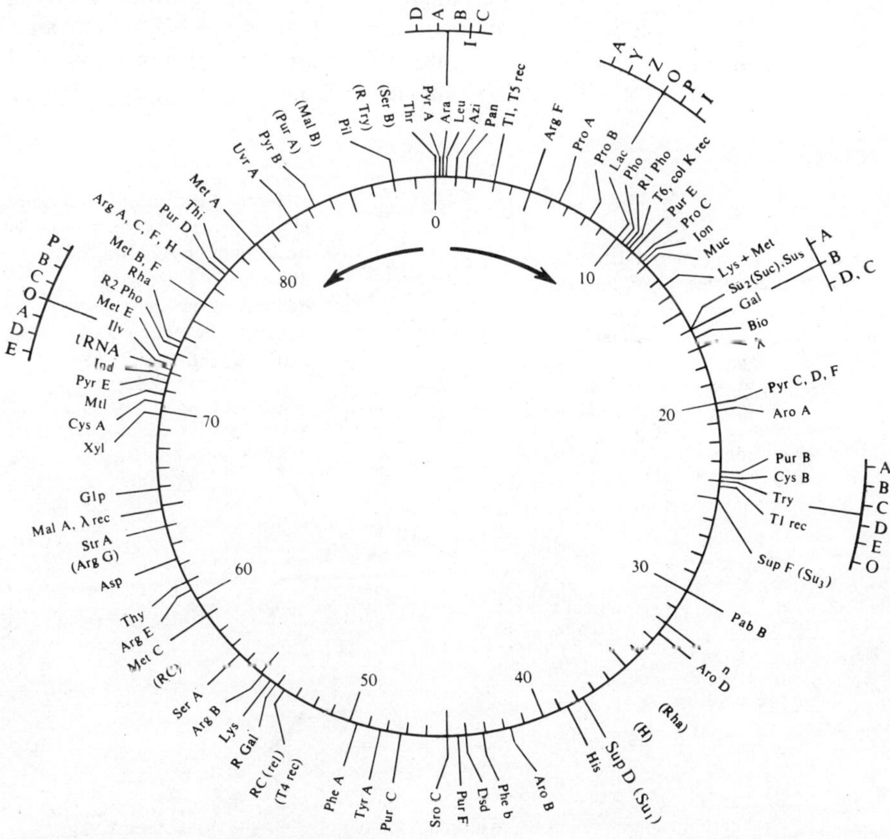

FIGURE 19-4

Genetic map of *E. coli.* Drawn to scale. The map is graduated in 1-min intervals (89 min total) and numbered at 10-min intervals to facilitate analysis. Markers enclosed in parentheses are only approximately mapped at the positions indicated. The exact sequence of markers in crowded regions is not always known. The genetic map in terms of recombination units closely corresponds to that in time units, with the normalization factor 1 time unit (corresponding to approximately 10^5 nucleotide pairs) equals about 22 recombination units. A. L. Taylor, and M. S. Thoman, *Genetics,* **50:** 659 (1964), revised according to A. L. Taylor and C. D. Trotter, *Bacteriol. Rev.,* **31:** 337 (1967).

of two states—independent from, or integrated into the bacterial chromosome —was called an *episome* by Jacob, Wollman, and Monod (1).

An analogous but much more widely applicable example of genetic exchange and transfer by an episomal entity in bacteria is provided by the process known as *transduction* (2), brought about through the agency of temperate bacteriophages (see Figs. 19-5 and 19-6). These bacteriophages may or may not be capable of replication in their own right; in the latter case they are known as *defective* bacteriophages. In the most extreme case known, that of PBSX of *B. subtilis,* the phage contains practically only *bacterial* DNA (3). Depending on whether the temperate phage can attach at (and remove material from) only one particular or any region of the chromosome, we speak of either *special* or *general* transducing phages and transduction.

A final means for genetic exchange in bacteria is provided by *transformation,* that is, the direct transfer of genetic material from donor to recipient (4). If a cell-free extract of an appropriate bacterial species (*Diplococcus pneumoniae, Haemophilus influenzae, Bacillus subtilis*) bearing a characteristic genetic marker (e.g., resistance to a particular antibiotic, nature of the cell wall, requirement for a particular growth factor) is added to a growing culture of a different strain of the same bacterium but lacking that particular trait, a fraction of the cells incorporate this trait into their genome. Thus, cells from the first strain act as donor, and those of the second as acceptors or

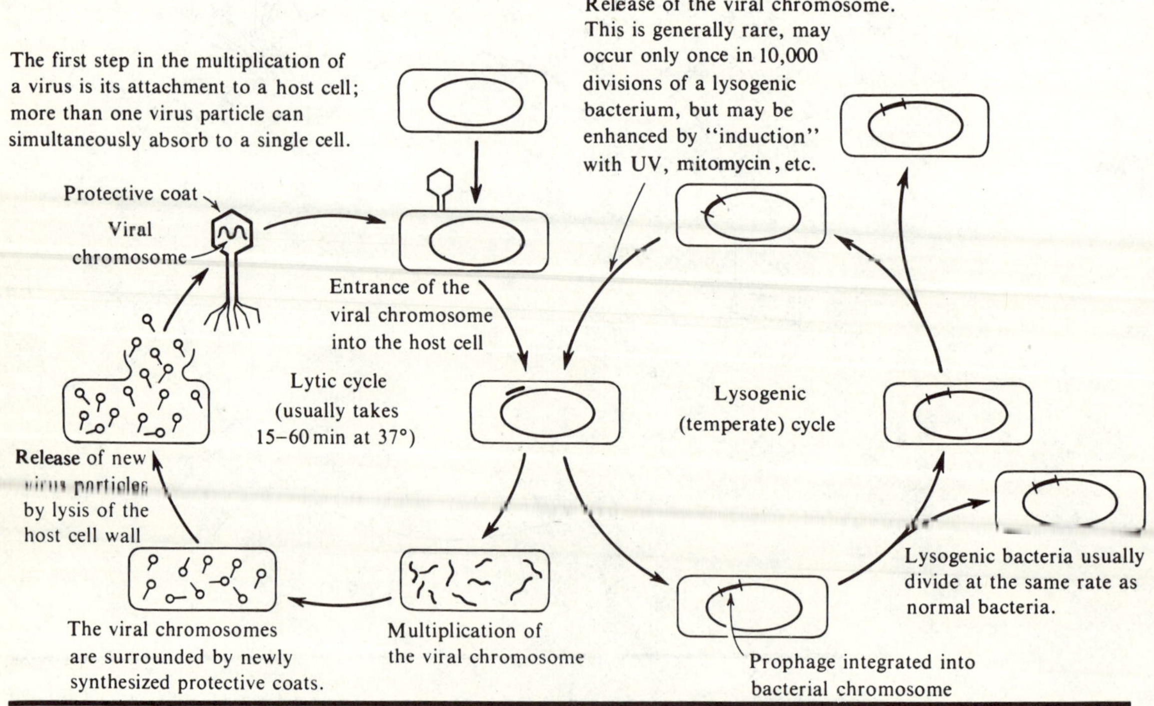

FIGURE 19-5

Life cycle of a lysogenic bacterial virus. We see that, after its chromosome enters a host cell, it sometimes multiplies immediately, like a lytic virus, and at other times becomes transformed into prophage. The lytic phase of its life cycle is identical to the complete life cycle of a lytic (nonlysogenic) virus. Lytic bacterial viruses are so called because their multiplication results in the rupture (lysis) of the bacteria. Adapted from J. D. Watson, *Molecular Biology of the Gene,* Benjamin, New York, 1965; with permission.

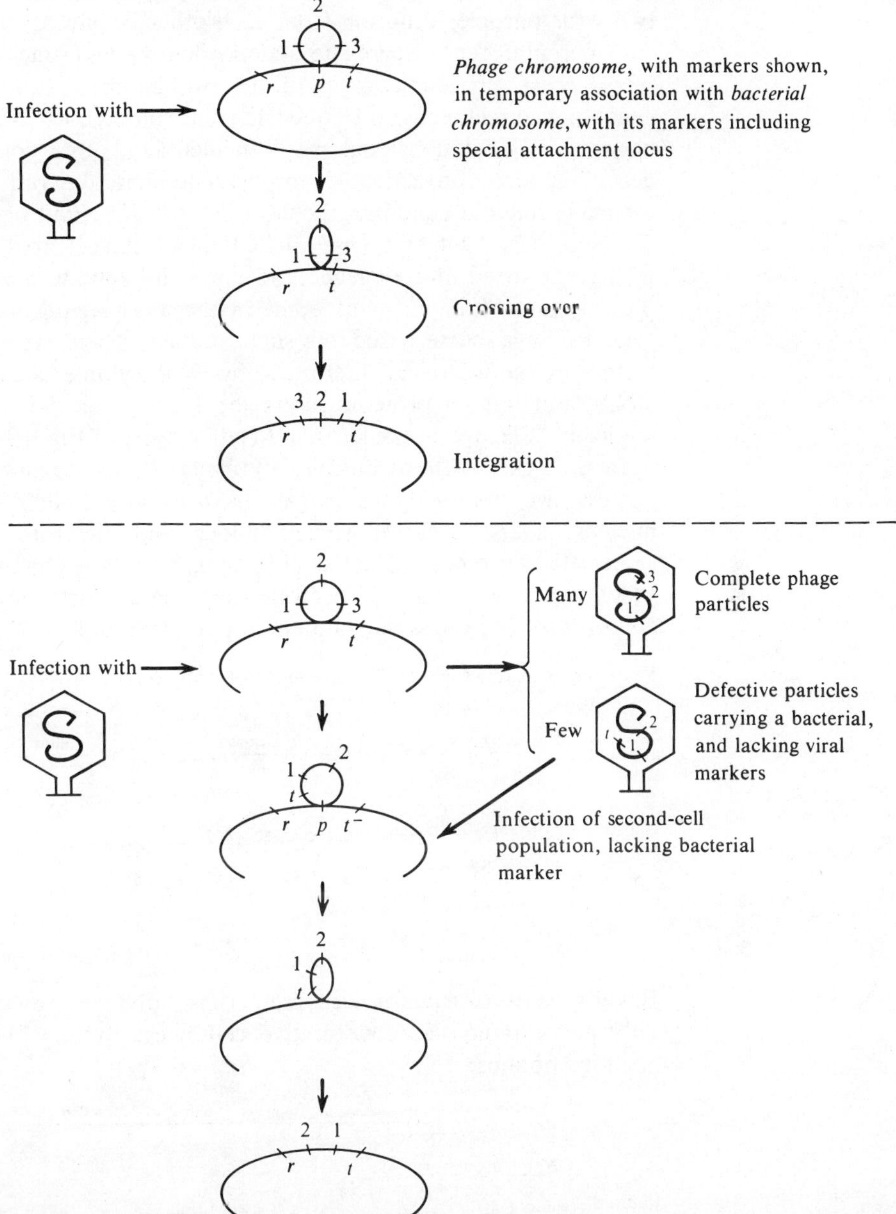

FIGURE 19-6
Lysogeny and transduction. Integration of prophage into bacterial chromosome (*top.*) and passive transfer of bacterial genetic material by (*defective*) carrier phages (*bottom*).

recipients of a genetic marker and are transformed in the process from, say, streptomycin sensitivity to streptomycin resistance; this *transforming principle* has been unambiguously identified as pure DNA. A special case is provided by *transfection,* transformation by means of DNA isolated from a transducing phage.

What about bacterial viruses, haploid organisms incapable even of vegetative reproduction, let alone sexual union in the absence of a host cell? A process that is operationally identical to recombination between mating types is obtained by simultaneous infection of one and the same bacterial cell by two virus particles differing from each other in one or more genetic markers. The parental types segregate independently, and one obtains, out of this *phage cross,* progeny viruses of the two parental and various recombinant genotypes, the latter with frequencies that can be correlated once again on the basis of a one-dimensional map. Incidentally, quite compelling genetic evidence has forced investigators to postulate a circular rather than a linear map for many haploid genomes, including *E. coli* K12 and bacteriophages T2 and T4 (see Figs. 19-4 and 19-7) (5,6). Since phage crosses can be performed with great speed and elegance and one can examine a very large number of progeny in each experiment, some of the most significant findings in modern genetics have been based on such studies. Their significance is enhanced further by the discovery, first made by Hershey and Chase, that only the viral DNA, and not its protein, enters the host bacterium, in the manner schematically indicated in the left-hand (lytic) cycle of Fig. 19-5.

In the case of those unicellular plants which are themselves haploid and possess two mating types (algae, *Neurospora*), only the zygote, or fusion nucleus, is 2*N* in chromosome number, and therefore meiosis must *follow* rather than precede sexual conjugation, as it does in higher organisms. In other words we observe the following characteristic sequence of events in *Neurospora* (chromosome number in parentheses):

Mycelium of mating type A (N) + mycelium of mating type a(N) $\xrightarrow[\text{or ``conjugation''}]{\text{fertilization}}$

zygote (2N) $\xrightarrow{\text{duplication}}$ tetrad (4N) $\xrightarrow{\text{meiosis}}$ (2 × 2N) $\xrightarrow{\text{meiosis}}$

4 haploid cells (N) $\xrightarrow{\text{mitosis}}$ 8 haploid asci (N)

mitosis ↙ ↘ mitosis

mycelium (N) asexual, vegetative plant (N)

Baker's yeast (*Saccharomyces cerevisiae*) divides asexually both in haploid (N) and diploid (2N) vegetative cell lines. In this instance the following situation obtains:

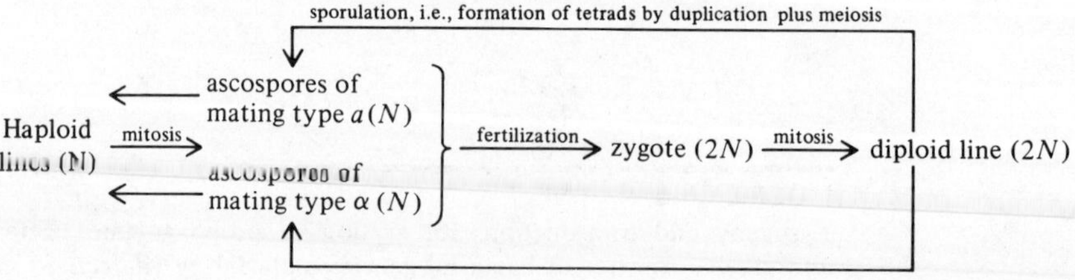

DNA AS GENETIC MATERIAL

It now becomes appropriate to summarize some of the most cogent evidence upon which the identification of DNA as the genetic material of all cells rests. The most compelling set is provided by the following:

1. Naked DNA unaccompanied by any other structured macromolecular component can transmit (a) single genetic markers, (b) linked groups of markers, or (c) all or at least most of the markers of a chromosome in the course of genetic exchange in microorganisms, by means of (1) transformation, (2) transfer by episomes, and (3) bacterial conjugation, or in the course of infection by viral DNA. In all these processes there is good correlation

between physical length, or weight, of DNA and map length. The complete genome forms a continuous structure and can be visualized by electron micrography.

2. For any species of higher organisms DNA is constant in amount and composition in all resting, somatic, diploid ($2N$) cells. It increases in amount during the process of mitotic division but returns to the resting value in the daughter cells. The haploid (N) germ cells (gametes) contain DNA of identical composition but in one half the amount.

3. Genetic relatedness between different species of both macro- and microorganisms has been positively correlated with DNA composition and with the extent of homologies of polynucleotide sequences in the molecules.

4. Physical and chemical agents (mutagens) are known that produce mutations with high efficiency; correlations can be established between this efficiency and the extent of distinct chemical alterations in DNA.

5. The Watson–Crick–Wilkins model of DNA structure (Chapter 5) provides a convenient framework to accommodate all these and numerous other observations in modern genetics and has been of tremendous heuristic value in the formulation of models and the design of critical experiments for the purpose of testing additional predictions in this general area.

It appears therefore legitimate to equate the chromosome, especially that of microorganisms, with a DNA molecule. By extension, then, the individual genes, heretofore defined operationally only in terms of certain types of experiments, now become definite segments of this molecule (i.e., long tracts or sequences of a polydeoxyribonucleotide duplex). Chromosome replication is discussed in Chapter 20. Recombination, at least in haploid organisms, is believed to proceed by a process of breakage and rejoining involving the parental chromosome (8), such as

```
- - - - - - - - - - - - — xxxxx ——————→
←- - - - - - - -   xxxxx———————————————
```

where solid and dashed lines represent DNA contributed by the two parents, whereas the x's represent regions from which DNA has first been removed by DNase action, then resynthesized along the opposing strand, and then reintegrated perhaps by a repair enzyme (for details see Chapter 20). Please note that the central region contains contributions by both parents, that is, it is *heterozygous* on the level of the individual duplex in contrast to the outer segments, which are recombinant. This partial heterozygocity will not survive the next round of replication.

MOLECULAR OR BIOCHEMICAL GENETICS: THE ONE GENE–ONE ENZYME HYPOTHESIS

The conclusion that the phenotypic expression of mutational events in genes must be due to resultant alterations in certain (enzymatically active) proteins seems almost inescapable in retrospect, and certainly was foreshadowed in 1908 by Garrod's careful documentation of various inherited abnormalities in man as "inborn errors of metabolism." He observed, for instance, that the disease alcaptonuria, which is inherited as a simple recessive Mendelian trait, was invariably accompanied and therefore probably caused by an inability to oxidize homogentisic acid, a normal metabolite of phenylalanine (see page 805). Many additional examples of such hereditary diseases have since come to light. A representative sample is presented in Table 19-1.

The modern era was initiated by Beadle and Tatum in the 1940s (7). Using *Neurospora* they were able to isolate a large number of independent biochemical mutants that, in general, were unable to synthesize one (or more)

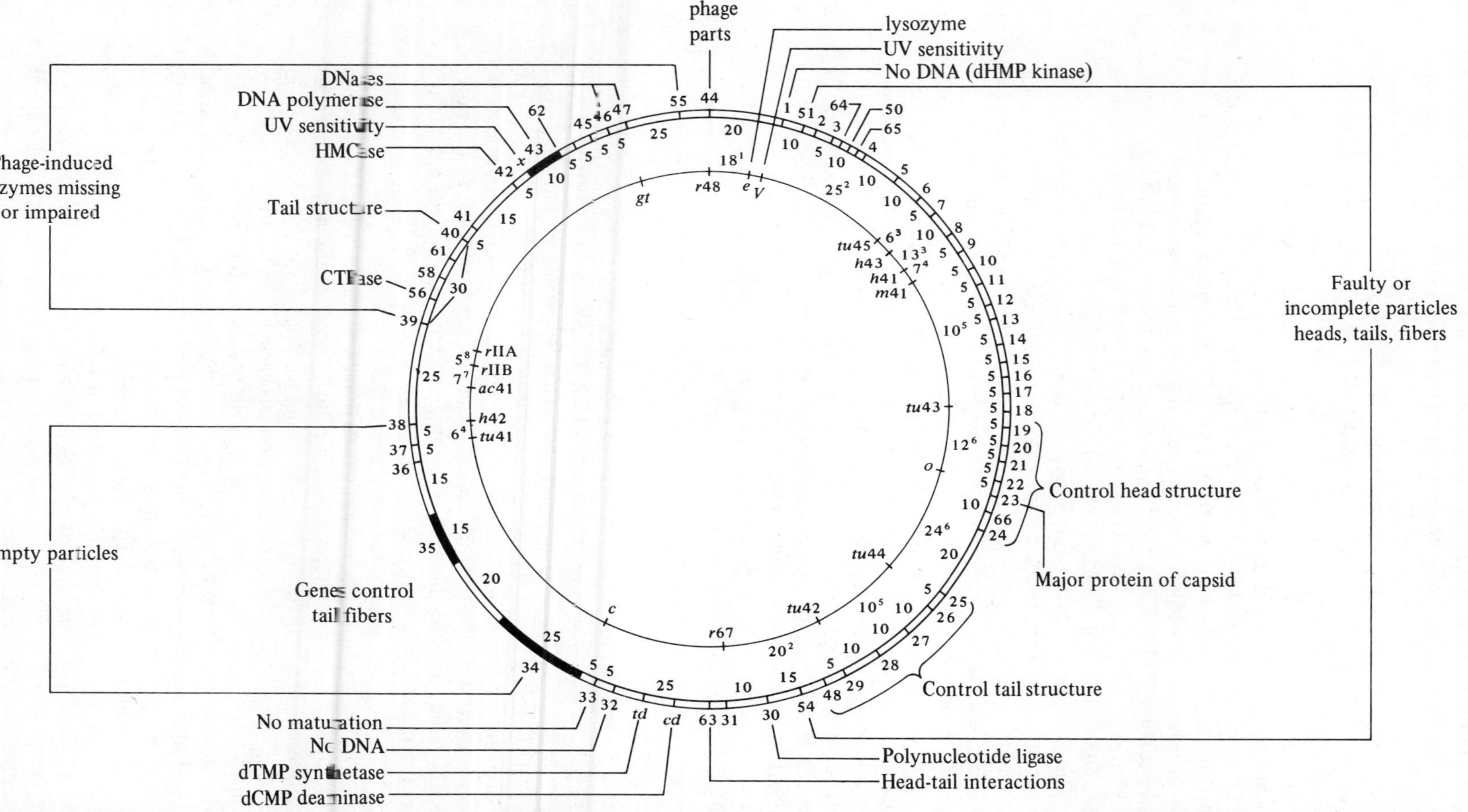

FIGURE 19-7

Composite map of bacteriophage T4D. The outer circle is a map of the genes containing conditional lethal sites. Genes with appreciable genetic length are indicated by the filled areas in the map. Approximate recombination values are indicated within the circle. The map has been drawn only roughly to scale. The inner circle is a map of standard markers. Their assigned location indicates their approximate position relative to the conditional lethals. The recombination values between adjacent standard markers are given in the figure. On the outside are indicated the types of phenotypic effects observed for the conditional lethals. Adapted from R. S. Edgar, G. H. Denhardt, and R. H. Epstein, *Genetics,* **49:** 635 (1964); W. B. Wood, R. S. Edgar, J. King, I. Lielausis, and M. Henniger, *Federation Proc.,* **27:** 1160 (1968).

TABLE 19-1
Some hereditary disorders in man in which the specific enzyme or protein lesion has been identified

Disorder	*Affected enzyme or protein*
Acanthocytosis	β-Lipoproteins
Acatalasemia	Catalase
Afibrinogenemia	Fibrinogen
Agammaglobulinemia	γ-Globulin
Albinism	Tyrosinase
Alcaptonuria	Homogentisic acid oxidase
Analbuminemia	Serum albumin
Argininosuccinic acidemia	Argininosuccinase
Galactosemia	Galactose-1-phosphate uridyl transferase
Glycogen storage diseases	
Type I (von Gierke's)	Glucose-6-phosphatase
Type III	Amylo-1,6-glucosidase
Type IV	Amylo-(1,4 → 1,6)-transglycosylase
Type V (McArdle's)	Muscle phosphorylase
Type VI (Hers')	Liver phosphorylase
Goiter (familial)	Iodotyrosine dehalogenase
Hemoglobinopathies	Hemoglobins
Hemophilia A	Antihemophilic factor A
Hemophilia B	Antihemophilic factor B
Histidinemia	Histidase
Homocystinuria	Cystathionine synthetase
Hyperbilirubinemia (Gilbert's disease)	Uridine diphosphate glucuronate transferase
Hypophosphatasia	Alkaline phosphatase
Maple syrup urine disease	Enzymes that decarboxylate branched-chain α-keto acids
Mastocytosis	Histidine decarboxylase (excess)
Methemoglobinemia	Methemoglobin reductase
Orotic aciduria	Orotidine 5′-phosphate pyrophosphorylase
Parahemophilia	Accelerator globulin
Pentosuria	L-Xylulose dehydrogenase
Phenylketonuria	Phenylalanine hydroxylase
Prolinemia	Proline oxidase (some cases)
Tyrosinosis	*p*-Hydroxyphenylpyruvic acid oxidase
Vitamin B_{12} deficiency	Methylmalonyl-coenzyme A isomerase
Wilson's disease	Ceruloplasmin
Xanthinuria	Xanthine oxidase

amino acid, purine, pyrimidine, sugar, or B vitamin. Thus, the *prototrophic* wild-type cell had been converted to an *auxotrophic* mutant, a cell that required for growth the addition to a minimal medium of the requisite component as an essential growth factor. Further analysis of a large number of such mutants of various types showed that, in general, the growth requirements for individual components could be subdivided and grouped further in the following manner. Let us assume that we find that one mutant, which we call z, requires compound Z and only compound Z as a growth factor. Frequently, then, we can also find a second mutant y that grows when either Z or Y is added to the medium. There may also occur a mutant x, which is also an auxotroph for Z but which grows on X, Y, or Z. Finally we obtain w, yet a fourth auxotroph for Z, but one which can grow on W, X, Y or Z. Now if W turns out to be a well-known product or intermediate of metabolism, we can say first of all that the pathway from W to Z includes the stages $W \xrightarrow{1} X \xrightarrow{2} Y \xrightarrow{3} Z$. For this reason auxotrophic mutants have turned out to be exceedingly useful tools for the elucidation of novel metabolic pathways.

We can also propose as a working hypothesis that mutant *z* behaves as it does because it is "blocked" between *Y* and *Z*; that is, it is incapable of synthesizing a functional enzyme 3, the one responsible for the conversion $Y \rightarrow Z$, in anywhere near normal amounts. We phrase this statement in precisely these terms because subsequent investigations have amply demonstrated that such *genetic blocks,* that is, alterations of the *z* gene from the standard to a mutant type of allele,[4] can manifest themselves phenotypically in a number of ways, including the following: (1) the complete absence of enzyme 3 or any protein related to it, (2) the synthesis of normal enzyme 3 but in subnormal amounts, or (3) the synthesis of an abnormal enzyme 3, related to the normal form structurally or immunologically but devoid of or much lowered in enzymatic activity. We can test the hypothesis by showing that on investigation of the intracellular content of mutant *z,* either directly or after administration of *W,* we find accumulation of *Y, X,* and *W* in that order and that extracts of cell *z* show normal ability to catalyze steps 1 and 2 but grossly impaired activity for step 3; analogous experiments with mutants *x* and *y* disclose similar deficiencies in enzymes 1 and 2, respectively. We are now entitled to expand our scheme as follows:

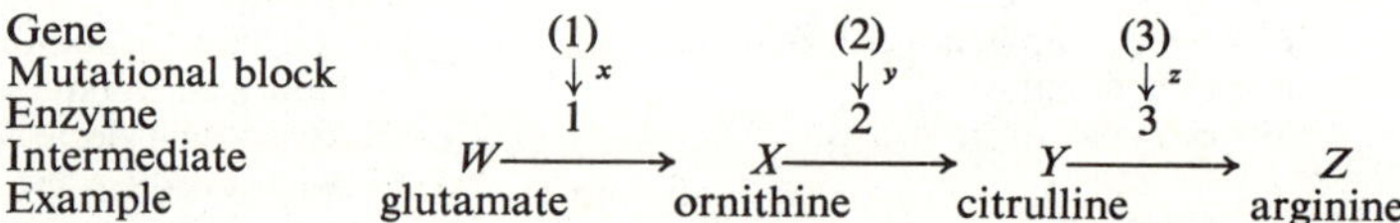

By the use of standard genetic techniques a very large number of mutants could be mapped, and the appropriate linear genetic maps for the seven linkage groups of *Neurospora* established. With the discovery of conjugation in *E. coli* K12 and by the use of transduction techniques as a means of affecting partial genetic exchange, similar maps could be constructed for this bacterium employing the same type of biochemical auxotrophs,[5] and the same form of reasoning could be shown to be valid with regard to the correlation of its genetic and enzymatic constitution. *E. coli* K12 has only a single linkage group, and map distances can, in this instance, be correlated on the one hand with time of penetration of the male chromosome into the recipient cell (5)—and is so correlated in the example shown in Fig. 19-4—and on the other with total chromosome length, which is identical, as we already know, with total DNA length. In this instance, therefore, time of penetration as measured by physical time of contact, time of appearance of markers in the recipient cell, and amount of ^{32}P-labeled DNA transferred from donor to recipient all correspond rather neatly.

[4] The nomenclature for the standard or wild type and the corresponding mutant is somewhat ambiguous, since both *x*, for wild type, and $x-$, for mutant, as well as $x+$ or just $(+)$, for wild type, and $x-$ or $(-)$, for mutant, have been employed in different systems. Finally the system *X* for wild type and *x* for mutant has also been used.

[5] With bacteria, selection of auxotrophs is simple, thanks to a discovery by Lederberg. Penicillin kills only growing cells. Thus, in a population of wild type plus auxotrophs treated with the antibiotic on a minimal medium, the former grows and is eliminated. The latter are incapable of growth and remain viable and resume growth upon supplementation with the appropriate growth factor.

As far as bacterial viruses are concerned, one has several different types of mutants available. Classically their phenotypic expression had to do with host range and plaque morphology and thus was a function of the various protein components of the mature phage. More recently many others (*conditional lethals*) have been discovered which can be selected either on the basis of temperature sensitivity (*ts* mutants that are capable of multiplication, say, at the *permissive* temperature of 30° but not at 40°) or on the basis of their ability to multiply in one type of cell (the "permissive" host) coupled with an inability to multiply in closely related bacterial strains (the "restrictive host"). This latter type is known as the class of "amber" (*am*) mutants. Both *am* and *ts* mutants in T2 and T4 have been shown to map in many different regions of the phage chromosome.[6] These segments are known to control not only the ordinary kind of phage proteins but also a variety of other protein molecules elaborated by the infected cell, vitally concerned in the biosynthesis of viral components, especially its DNA, to the exclusion of the corresponding bacterial entity. Thus elaborate and detailed maps could be constructed for some of the bacterial viruses as well (Fig. 19-7).

From the foregoing it is now clear what the experimental foundations are for the famous *one gene–one enzyme hypothesis,* originally enunciated explicitly by Norman Horowitz of the California Institute of Technology. To anticipate some points to which we shall return later in their proper context, two important sets of observations emerged as corollaries of some of these studies on bacteria: (1) frequently, though by no means always, the genes controlling the various steps of a biosynthetic sequence, for example,

$$W \xrightarrow{\overset{1}{\downarrow}} X \xrightarrow{\overset{2}{\downarrow}} Y \xrightarrow{\overset{3}{\downarrow}} Z$$

are closely linked and map in the same general region of the chromosome, perhaps even in the order 1, 2, 3, or its reverse; (2) often the synthesis of the cluster of enzymes, for example 1, 2, 3, corresponding to and controlled by this cluster of genes appeared to be turned on (*induction*) or off (*repression*) in a coordinated manner in response to changes in the physiological environment.

As to actual examples, out of a large number available we mention only two. The genes concerned with galactose metabolism are clustered at Gal A to D in Fig. 19-4. They control the sequence of enzymes (see page 544) catalyzing the following reactions:

$$\text{galactose} \xrightarrow[\text{galactokinase}]{} \text{galactose-1-}P \xrightarrow[\substack{\text{Gal-1-}P\text{-uridyl}\\ \text{transferase}}]{+\text{UTP} - \text{PP}_i} \text{UDP-Gal} \xrightarrow[\text{epimerase}]{} \text{UDPG}$$

$$\underset{\text{Gal A}}{\uparrow} \qquad\qquad \underset{\text{Gal B}}{\uparrow} \qquad\qquad \underset{\text{Gal C}}{\uparrow}$$

[6] Conditional lethals (i.e., *am*, the related *ochre*, and *ts* mutants) have now been used extensively with bacteria also. The expression of temperature- (both high and low) sensitive mutations does *not* depend on any additional genetic determinants and therefore is a completely general phenomenon, found in all organisms.

Those concerned with isoleucine and valine biosynthesis are marked *Ilv A, B, C, D,* and *E* and control enzymes (see page 785) as follows:

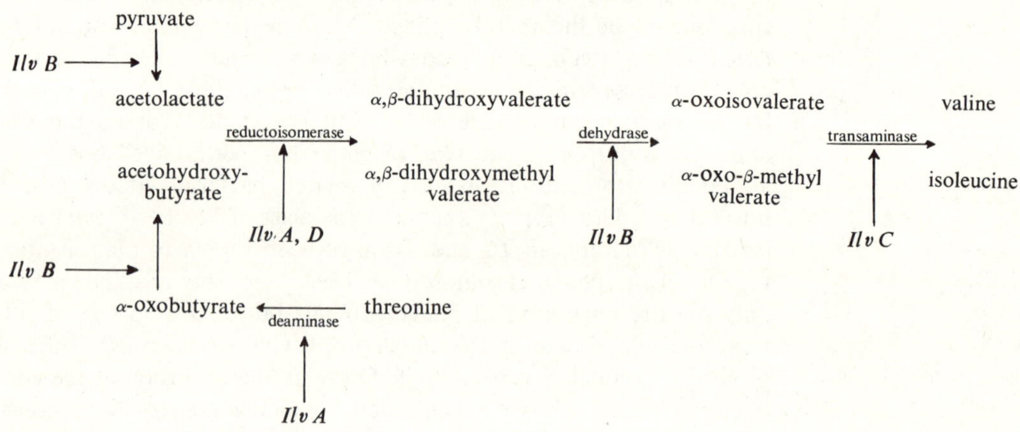

THE GENE AS FUNCTIONAL UNIT: ONE GENE–ONE POLYPEPTIDE

Fine-Structure Genetics: The Units of Recombination and Mutation

In all the foregoing the classic definition of a gene was still employed and appeared to suffice in the vast majority of cases: a discrete, essentially quantized, indivisible unit, defined operationally as being able to mutate to an alternative allelic state, to recombine with other similar genes, and to control function (i.e., the synthesis of a specific cellular constituent, namely, a specific protein).

The era of *fine structure genetics* was initiated by the studies of Benzer from 1955 on with a particular gene of bacteriophage T4 (and T2) called the *r*II region (9). The fantastic resolving power of this particular system is based an the fact that, whereas in one type of bacterial host (B) these mutants simply show a different plaque morphology (the *r* phenotype) that allowed their selection in the first place, in another (K)[7] no infective progeny is produced, and thus no plaques are formed. Therefore, if one were to perform a phage cross between two separate *r*II mutants in strain B, but score for recombinants on agar plates covered with a lawn of bacteria of strain K, only recombinants for r^+, the wild (or standard) type, would form visible plaques. This test has a resolving power such that r^+ recombinants occurring with a frequency of 0.0001 percent can readily be detected. By this and by means of some short-cut methods involving overlapping deletions,[8] Benzer proceeded to map this region. He was able to show that the individual *r*II gene contained a large but not infinite number of individual recombinational loci, which could be arranged in a linear array on the basis of recombination frequency. Thus, he had constructed the first fine-structure map of a gene containing a large number of alleles (see Fig. 19-8). How many? It turned out that the lowest frequency of r^+ recombinants actually observed in all his crosses between independently produced (spontaneous or chemically mutagenized) *r*II mutants was about 0.01, even though much lower frequencies could readily have been detected. This number then allows us to estimate the size of the smallest recombinational unit, called the "recon" by Benzer. The total genetic map of T4 covers a distance of about 700 units, where 1 unit corresponds to a recombination frequency (for both recombinant types) of 1 percent in a standard cross, in which each parent is present in exactly equal amounts. A frequency of 0.01 percent therefore corresponds to 2 × 0.01/

[7] Really K12(λ), that is, *E coli* K12 lysogenic for phage λ.
[8] That is, losses of whole segments of the gene.

700 = 3×10^{-5} of the total phage chromosome. In terms of DNA, if we assume no nongenetic regions in the molecule and equal probability of recombination throughout its length, this corresponds to $3 \times 10^{-5} \times 2 \times 10^5$ nucleotide pairs = 6 nucleotide pairs. Therefore, the "recon" consists of a

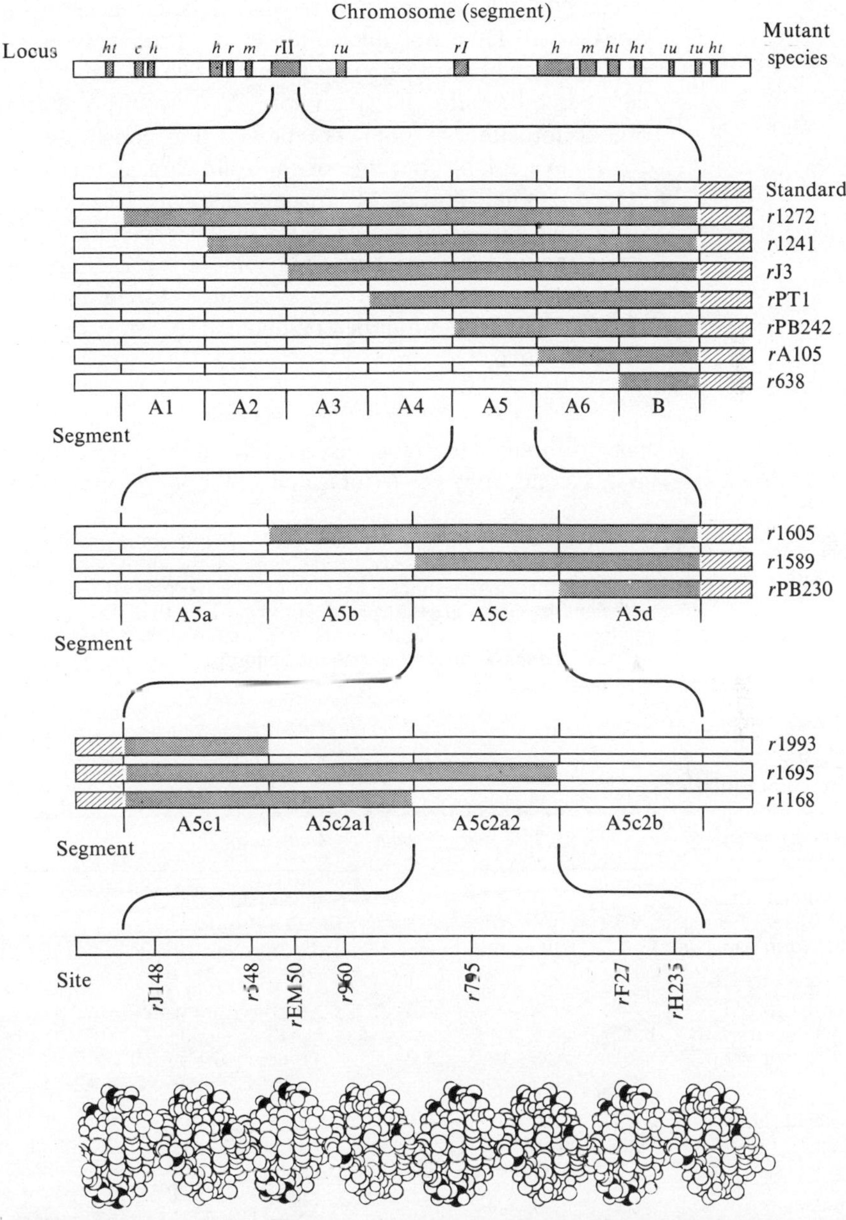

FIGURE 19-8

Mapping the *r*II region. Mapping techniques localize the position of a given mutation in progressively smaller segments of the DNA molecule contained in the T4 phage. The *r*II region represents, to start with, only a few percent of the entire molecule. The mapping is done by crossing an unknown with known deletions in the *r*II region. The order and spacing of the seven mutational sites in the bottom row are still tentative. Each site probably represents the smallest mutable unit in the DNA molecule, a single base pair. The molecular segment (extreme bottom), estimated to be roughly in proper scale, contains a total of about 40 base pairs. S. Benzer, *Sci. Am.*, **206**(1): 70 (1962).

very small number of nucleotide pairs or perhaps even of an individual nucleotide pair. Thus, in terms of fine-structure genetics it is not meaningful any longer to speak of the gene as a discrete unit of recombination.

Nature of Mutations What about the gene as the individual unit of mutation, the unit called the "muton" by Benzer? In terms of the DNA-segment model of the gene one would expect the smallest muton also to be constituted by the individual nucleotide pair. For no matter what detailed mechanism we postulate for the *translation* of the polynucleotide code into a polypeptide code, an alteration in a base, or at most one in a group of two or three bases, might well result in an altered protein. Such an alteration would therefore produce a mutant organism distinct in its phenotype and thus constitute a *point mutation* (Table 19-2). We might conceive of the following alternatives in terms of a model involving small groups of adjacent deoxynucleotide pairs as code groups for specifying individual amino acids in the protein that constitutes the eventual gene product (see Fig. 19-9). The figure is reasonably self-explanatory except perhaps for alternative 4. Let us assume, for argument's sake, a *coding ratio* of 3; that is, *r*, the number of nucleotide pairs[9] in a code group equals that number. Then

$$(x_1y_1z_1)(x_2y_2z_2) \cdots (x_ny_nz_n)(x_{n+1}y_{n+1}z_{n+1}) \cdots (x_my_mz_m)$$

can be thought to stand respectively for the nucleotide pairs comprising the first, second, *n*th, (*n* + 1st), and *m*th code groups of gene *a*. They can then,

[9] Since normal double-stranded DNA is self-complementary, it does not now matter whether we look at the nucleotide pairs in the duplex or the nucleotides in *either* one of the component strands. As a corollary, if one strand is really both necessary and sufficient to carry all the genetic information, then the chemical alteration of a single nucleotide in a pair should produce a heterozygote; that is, its progeny should consist of both the parental and a mutant genotype. This has actually been demonstrated (see also page 855).

TABLE 19-2
Types of point mutations

Mutagen, acting on	*Base pair*	*Product*	*Consequence*	*Final outcome*
5-Bromouracil (BU). When in keto form pairs with A,	A—T	A—B	B can pair with G G—B → G—C	Transition AT → GC
in enol form pairs with G	G—C	G—B	BU can pair with A G—B → A—B → AT	Transition GC → AT
2-Aminopurine (AP). Normally pairs with T, but	A—T	AP—T	AP now pairs with C AP—T → AP—C → G—C	Transition AT → GC
can also pair with C	G—C	AP—C	AP now pairs with T AP—C → AP—T → AT	Transition GC → AT
Nitrous acid (NA). Deaminates A → I, G → X, C → U	A—T	I—T	I pairs with C IT → IC → GC	Transition AT → GC
	G—C	X—U	Although X pairs like G, U ≃ T; X—U → A—U → AT	Transition GC → AT
Hydroxylamine (HA). Converts cytosine (NH_2) to NHOH,	G—C	G—C_x	G—C_x → A—C_x → AT	Transition GC → AT
which pairs only with A	A—T	No action		
Acridines (proflavine, acriflavine, quinacrine, etc.). Acts by intercalation	Any	Locally distorts double helix, allowing insertion or removal of one (or small number) base		Produces shifts in reading frame

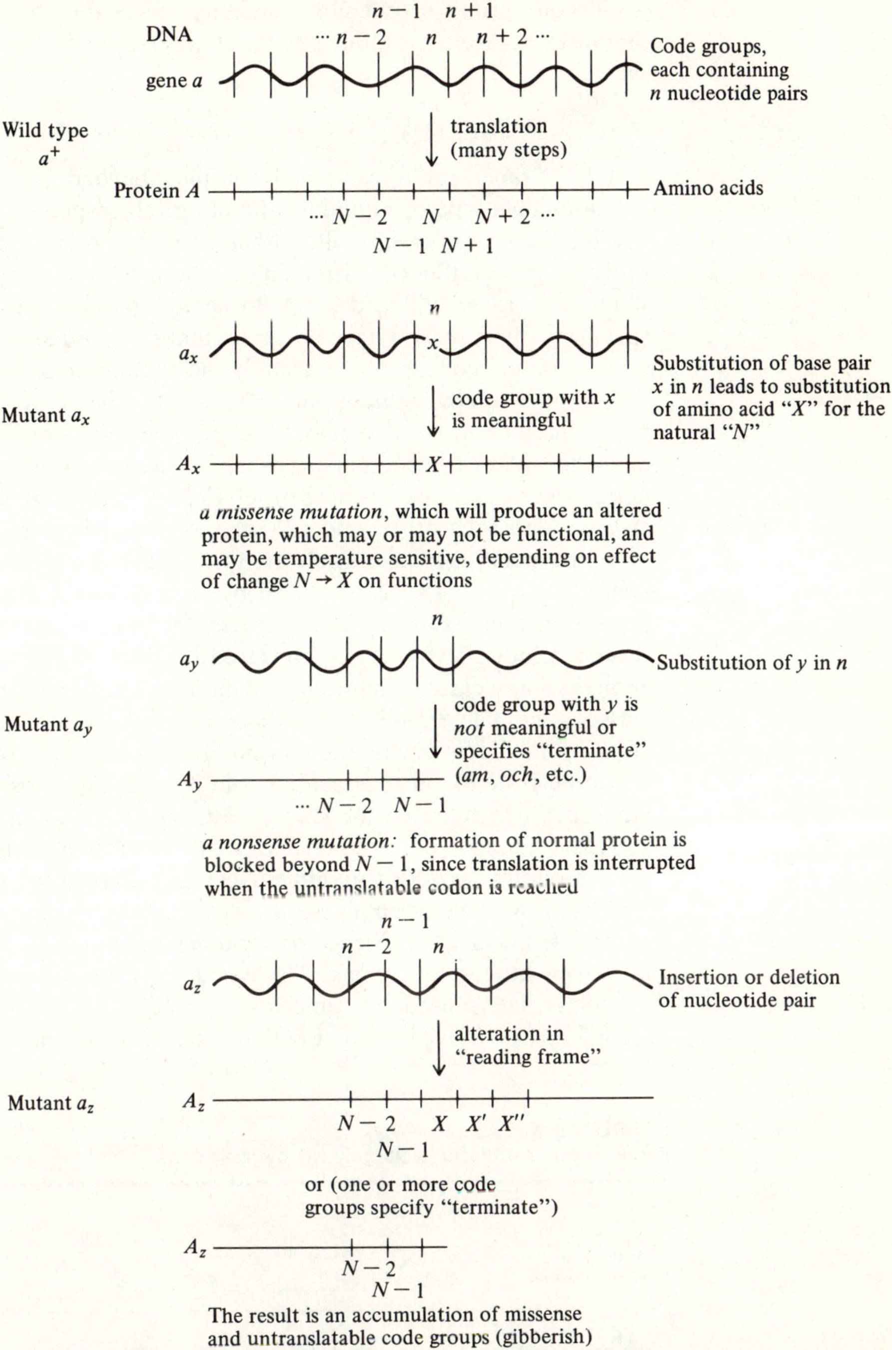

FIGURE 19-9
Phenotypic consequences of point mutations.

by an as yet unspecified mechanism, uniquely determine the sequence of amino acids $aa_1aa_2\cdots aa_naa_{n+1}\cdots aa_m$ in the protein A. Now let us also assume that reading and translation have to start with code group 1, and with nucleotide x_1 at that. What is the consequence if we eliminate one nucleotide (pair) in one of the groups, say y_n? Well, everything proceeds as before until we reach code group n, but now suddenly, since the translation proceeds continuously in groups of three without interruption, the machine responsible sees

$$\cdots(x_n z_n x_{n+1})(y_{n+1} z_{n+1} x_{n+2})(y_{n+2} z_{n+2} x_{n+3})(y_{n+3} z_{n+3} x_{n+4})\cdots(y_{m-1} z_{m-1} x_m)\, y_m z_m$$

that is, a tape that is completely garbled beyond the point of insertion or deletion. This type of mutation, therefore, corresponds to a *shift in phase* or *reading frame,* producing alterations at code group n and beyond, that are reflected in a continued misreading of the genetic message beyond that point. If all the triplets still code, but do so for amino acids differing from those specified originally, we have an accumulation of *missense* or gibberish. If one or more do not code, we have a similar accumulation of *nonsense*.

Are most of the mutants at independent alleles in the *r*II region indeed point mutants? If so, they should satisfy the following two criteria: (1) they ought to yield r^+ recombinants when crossed with at least one of two other nonallelic *r*II mutants which themselves have yielded such recombinants when crossed with each other, and (2) they ought to be capable of *reverting* back to wild type with a measurable frequency, especially when the mutation rate (here from $r\text{II} \rightarrow r^+$) is enhanced by appropriate treatments. Indeed most of the *r*II mutants studied did behave in the predicted manner, although their reversion index or frequency appeared to differ widely (10,11). Interestingly enough, many cases of induced mutations could be shown to fall into just two classes (see Table 19-3): (1) those induced by base substitutions—growth on 5-bromouracil or 2-aminopurine; treatment with hydroxylamine (or its *O*-Me derivative), nitrous acid, *N*-Me-*N'*-nitro-*N*-nitrosoguanidine (NG), or ethyl methane sulfonate (EMS) (see Chapter 5) could be shown to give an increased number of revertants with the same or a different mutagen in this same group but never by acridine dyes; and, conversely, (2) those induced by treatment with acridine dyes, as well as some spontaneous ones, formed revertants under the influence of acridines but never by treatment designed to introduce base substitutions. The conclusion was therefore reached that the first class was caused by transitions $A \cdot T \rightleftharpoons G \cdot C$ or $T \cdot A \rightleftharpoons C \cdot G$, and the second class involved a process that differed from the first in a fundamental

TABLE 19-3
Reversion of mutations (largely *r*II) by mutagens

Mutation originally induced by	*B*	*AP (NG)*	*NA*	*HA*	*Acridines*	*Likely base pair affected by original mutation*
B	+	+++	+++	−	−	GC (some AT)
AP	+++	+	+++	++	−	AT (some GC)
NA	++	++	+++	+	−	AT or GC
HA	++	+++	+	−	−	GC
EMS						GC
Acridines	−	−	±(?)	−	+	Any

CODE:
+++ = virtually quantitative reversion
++ = reversion in majority of cases
+ = reversion in some cases
± = occasional reversion
− = reversion virtually absent

manner.[10] It is now known to be mediated by insertion or deletion of a single (or small numbers of) nucleotide, that is, a shift in reading frame (13). Reversions in the first class, then, would entail a second substitution in the same spot, with a resultant return to the *status quo,* for example A · T → G · C → A · T; reversions in the second class would entail an insertion followed by a deletion, or vice versa (14).

The reversion pattern for class I mutagens shown in the table is for T4; for the single-stranded phage S13 or for *Salmonella* the patterns are quantitatively distinct. Hence the molecular environment for susceptible bases is an important determinant.

Those mutants which were not point mutants turned out to be actual deletions of whole segments of the genome, varying in length. In other similar systems, such as bacteriophage λ, an analogous deletion (λcb_2) has been studied which leads to a decrease in physical length of DNA by 3 μm (from 17.3 to 14.3), corresponding to a decrease in molecular weight of 6×10^6, or the equivalent of 1000 nucleotide pairs (see Fig. 5-10).

Unit of Function

Studies with r*II Mutants.* It became apparent that all mutants could be shown to fall into just two classes, *A* and *B,* by the following simple test. Again we infect one host cell simultaneously with two different viruses, either one wild type plus one mutant, or two mutants, but now we use strain K rather than strain B to form this doubly infected complex. Now, remember that the characteristic property of an *r*II mutant is its very inability to yield any progeny in strain K, presumably because in this host it is incapable of specifying a protein capable of normal function, essential for phage production. What would we predict? Surely that the configuration wild type/mutant should produce functional protein[11] and therefore give rise to normal behavior and progeny, whereas the configuration mutant/mutant should fail to do so. But this is not what was found. Some mutant pairs (those of group I) did indeed behave in this way, but a great many (those of group II) yielded progeny, that is, were capable of mutual *complementation.* The two classes of mutants then are those of classes *A* and *B,* which are defined in the following manner: Each and every class *A* mutant can cooperate with (complement) each and every class *B* mutant with very high efficiency, whereas intraclass complementation does not occur at all or constitutes an exceedingly rare event.[12] This means that we have the following possibilities: *AB*/*Ab* → progeny; *AB*/*aB* → progeny; *AB*/*ab* → progeny; *Ab*/*aB* → progeny; *aB*/*Ab* → progeny; *aB*/*aB* or *Ab*/*Ab* ↛ no progeny. Here *AB* = wild type; *aB,* mutant of class *A*; *Ab,* mutant of class *B*. The two classes define a *complementation group,* and since the test used is known in genetics as the cis-trans test, the two units (genes) are known as *cistrons.* (Two markers *a* and *b* are said to be cis in the configuration *AB*/*ab,* trans in *Ab*/*aB* or *aB*/*Ab.*) Thus, we see that *A* and *B* are indeed dominant over their mutant alleles *a* and *b.*

[10] It was first believed to involve *transversion* (12) (A · T ↔ C · G or T · A; G · C ↔ T · A or C · G). These changes do occur but are not the principal ones induced by acridines; instead, they are primarily produced by ethylating agents or low *p*H which lead to functional removal of G residues.

[11] That is, wild type which can produce protein is dominant over mutant which cannot.

[12] In other systems intraclass complementation can be shown to occur. It is observed when (1) the functional protein is made up of two or more *identical* polypeptide subunits and (2) such subunits damaged in different portions of the polypeptide chain can still aggregate to form a more or less functional complex.

In functional terms the true genetic units must be provided by the two gene segments, or cistrons, A and B, which, incidentally, occupy distinct but of course adjacent segments in the genetic map. Each cistron then specifies a distinct functional entity, inactive by itself even when normal, fully active when supplemented by its equally inactive normal partner. The gene product of each of the mutant alleles, then, must be one that cannot form a normal functional product in the presence of a standard partner. This makes sense if we postulate that the true functional genetic unit, gene or cistron, is one that specifies the true structural subunit in protein chemistry: the individual polypeptide chain. Let us assume that a functional protein is made up of groups of n separate and distinct polypeptide chain subunits held together by covalent or other bonds; it then requires n genes or cistrons for its complete specification. We therefore postulate two polypeptide chains in the present instance. Even if the mutation $A \rightarrow a$ or $B \rightarrow b$ produces a complete absence of the corresponding polypeptide or a completely garbled one, it is evident that in the intracellular pool we shall have available intact standard subunit polypeptides A and B, just as long as the appropriate standard gene for each is present. We can also conceive of performing complementation in the test tube, rather than the cell, an expectation that has frequently been borne out by experiment (though not in the present instance).

The difference between a cistron and a gene as the unit of function is an evanescent one; for example, in humans the two genes specifying the α and β chains of hemoglobin are unlinked and may even be located on different chromosomes. The two genes (A and B) for tryptophan synthetase in *E. coli,* as we shall see in the next section, are very closely linked, but distinct. By contrast in *Neurospora,* the analogous enzyme—very similar in function and mechanism, and even molecular weight, to its *E. coli* counterpart—is specified by two cistrons of one and the same gene. In any event, many proteins now have been shown, usually by means of the cis-trans test, to be specified by more than one gene or cistron.

Suppression

Intragenic Suppression. Examination of apparent reversion to wild type, or prototrophy, in many systems has shown that frequently the true cause of the phenomenon lies, not in a true reversal of the primary genetic event, but in the introduction of an additional mutation(s) elsewhere in the genome. We are dealing with a "pseudo revertant," formed as a consequence of a *suppressor mutation.* Some of the examples found can be explained on the basis of what has already been discussed. Let us return to Fig. 19-9, case 4, and the discussion on page 864 concerning the production of gibberish or nonsense by a phase shift (i.e., an alteration in reading frame). We can ask the question: What will be the consequence of introducing an additional deletion had we started with a deletion (− shift),[13] or insertion, had we started with an insertion (+ shift),[13] close to the site of the first mutagenic event? In the example given we find after removal of, say, y_{n+1},

$$(x_n z_n x_{n+1})(z_{n+1} x_{n+2} y_{n+2})(z_{n+2} x_{n+3} y_{n+3}) \cdots (z_{m-1} x_m y_m) z_m$$

which is still devoid of meaning. But what if we introduce yet another deletion, number 3 in the set? The product is, if we remove, say, x_n,

$$\cdots (z_n x_{n+1} z_{n+1})(x_{n+2} y_{n+2} z_{n+2})(x_{n+3} y_{n+3} z_{n+3}) \cdots (x_m y_m z_m)$$

[13] Introduction of an insertion will shift (−) back to (0) and thus be equivalent to two additional deletions; conversely, a deletion into (+) is equivalent to two additional insertions. In the example given (13), these insertions were produced by phage crosses.

and lo and behold, starting with code group $n + 2$ to the end of the chain we are back in phase and produce a perfectly intelligible message. Therefore, providing that the missense produced by the few altered code groups—here code group n has been eliminated and code group $n + 1$ has become $(z_n x_{n+1} z_{n+1})$—is not too overwhelming, we can see how a new functional protein may emerge as a result of r successive mutational events in series (14).[14] But we remember that r equals the coding ratio, the number of nucleotide (pairs) in DNA required to specify any one amino acid. This exceedingly ingenious argument and the experimental proof that $r = 3$, employing experiments of this type with acridine-induced mutants of bacteriophage T4, were first advanced by F. H. C. Crick and his collaborators in 1961 (13).

What about *intragenic* suppressor mutations for base substitution or missense mutations, usually now called *second-site reversions?* These also exist: the original mutational event, we remember, produced an inactive or impaired protein with an altered amino acid substituted at a single and defined position in the polypeptide chain. A second, independent mutational event in the same gene will lead to a second amino acid substitution elsewhere. Under certain circumstances this second substitution, although producing an altered protein when present alone, may be beneficial and relieve more or less completely the original impairment of function due to the first substitution. A case in the α (or A) polypeptide of tryptophan synthetase is known where the substitution of Glu for Gly at one position can be compensated for by a second substitution—which by itself also produces an inactive protein—of Cys for Tyr, 36 amino acids distant.

Extragenic suppression is of at least two distinct types. First is one in which the phenotypic expression of the original mutational alteration is compensated for by the introduction of a *specific* suppressor mutation into the same genome. The second arises as a result of the introduction of a nonspecific, more *generalized* suppressor that affects mutants in more than one gene so as to allow them to form proteins functional to one degree or another. In this latter instance the suppressor mutation and the original mutation need not even have been present in one and the same organism. For instance, the growth of conditionally lethal phage mutants in certain permissive strains and their inability to grow in others is now known to be due to the presence of a suppressor in the genome of the permissive host. The introduction of the same suppressor (i.e., the Change $Su^- \rightarrow Su^+$) also permits the synthesis of functionally active host proteins (e.g., alkaline phosphatase) in host mutants ordinarily lacking this capacity. Again we shall have to distinguish missense from nonsense suppression, that is, the production of a functional protein in the stead of one devoid of function but equal in the total number of residues in the former, as distinguished from total absence of the complete polypeptide, and the production of fragments, instead, in the latter case. An explanation of suppression in molecular terms is deferred to Chapter 21.

Nonsense Mutants. It may be useful at this point to discuss briefly the various criteria that have been used to distinguish nonsense from missense mutations. Two have just been mentioned: (1) the accumulation of shorter polypeptides and (2) the conditional nature of the mutation, that is, the difference in its effects in Su^- and Su^+ hosts. If the protein affected by the nonsense mutation is *essential,* then, as we have just stated, its presence in a

[14] Really $n \times r$, where n is any small number.

DNA or RNA phage will not permit growth of the organism in cells that are Su$^-$. Clearly this test is not applicable to genes specifying proteins that are indispensable for the host organism or of microorganisms in general. There we are restricted to seeing what happens with nonessential or *dispensable* genes on introducing, by mating or transduction, the Su$^+$ gene into a cell already harboring such a nonsense mutation.

A third criterion is related to the first. Since no intact polypeptide is formed, there will be a total absence not only of all functional or enzyme activity ascribable to the protein, but also of any material related immunologically to it and of the ability to effect complementation either *in vivo* or *in vitro;* such organisms will exhibit an *extreme negative phenotype.* The fourth is due to a combination of the polarity of translation referred to on page 964, together with the existence of operons; interruption of translation due to the presence of a nonsense mutation in an "early" gene (i.e., proximal to the starting point) of such a cluster will reflect itself by reduced amounts or total absence of "later," more distal ones: the mutation will produce a *pleiotropic phenotype.* We shall return to this point when we discuss the regulation of protein synthesis in Chapter 21. Finally, a point implicit in this discussion is worth noting explicitly. The two types of conditionally lethal mutations are qualitatively distinct: Temperature-sensitive mutations can *never* be nonsense mutations; the effect of introducing a nonsense mutation can never be restored by changing the temperature—it requires a separate mutational event.

Studies with Tryptophan Synthetase of E. coli. The elegant studies by Benzer could only be improved on by a direct correlation with equally sophisticated work on the structure of the protein(s) controlled by the *r*II gene. Unfortunately, its function as yet not understood and the protein in question has never even been isolated, much less analyzed. Fortunately a formally similar system has been discovered by Yanofsky and his collaborators as the result of their equally impressive work: the enzyme tryptophan synthetase (TS) and the genes controlling its elaboration in *E. coli* (15).

The terminal reaction for tryptophan synthesis in this and other organisms catalyzed by TS is that of reaction (19-1) (see page 791):

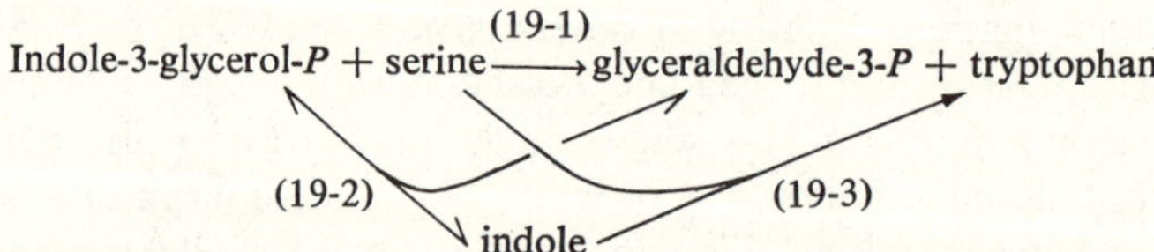

In addition, the standard enzyme, as usually isolated, also catalyzes reactions (19-2) and (19-3), yet free indole is not an intermediate in reaction (19-1). Now this pure enzyme (M = 175,000) is made up of subunits ($\alpha_2\beta_2$), since it can be dissociated upon chromatography into two polypeptide fractions: monomeric α (formerly A) and β_2 (formerly B). The α subunit is of lower molecular weight (29,500) and contains a total of 267 amino acid residues (interestingly no tryptophan). Neither α nor β_2 catalyzes the overall reaction [reaction (19-1)] although a reconstituted enzyme made up of a stoichiometric complex of 2α plus β_2 will again perform this function. However, α catalyzes reaction (19-2), and β_2 catalyzes reaction (19-3) to a slight extent. This provides a means of assaying these components in extracts, but a more convenient measure is presented by simply providing a normal α subunit either intracellularly in a complementation test or extracellularly in a test for enzymatic activity.

Colinearity of Maps

By means of these diagnostic tests, by genetic techniques that allow resolution of recombination frequencies at the level of separation corresponding to single nucleotides, and by some elegant fingerprint and sequence work on the α-polypeptide, Yanofsky and his group have succeeded in providing unambiguous evidence that the genetic and polypeptide maps are colinear (16). This statement implies that there exists a precise homology between polynucleotide sequence in a segment of DNA and amino acid sequence in the polypeptide specified by that polynucleotide. The molecular basis underlying this correlation, that is, the oligonucleotide sequences corresponding to the various amino acids, constitutes the *genetic code,* whereas its functional counterpart constitutes the mechanism of *translation* of the genetic message. The two maps together with characteristic recombination frequencies and amino acid substitutions corresponding to certain characteristic alleles are shown in Fig. 19-10. The positions of amino acid replacements in corresponding polypeptide fragments isolated from mutant cells occupy the same relative positions with respect to one another as do the respective mutational sites on the genetic map of the A gene responsible for specifying α. A number of additional significant conclusions have been drawn from these and related studies.

1. The ratio (genetic) map distance over residue distance is approximately constant and has a value of $\geqslant 0.01$, which means that recombination frequencies lower than this probably always involve replacements of one and the same amino acid.

2. Recombinants between different mutants at one and the same genetic site that produce different amino acid substitutions (e.g., A23 with A46, and A58 and A78) are possible only if the coding units (codons) corresponding to any such pairs differ by more than a single nucleotide.

3. The code is *degenerate,* that is, a single amino acid can be specified by two or more separate codons. It would be very difficult to explain why out of 22 mutational changes at the first Gly site shown, not a single Gly → Asp or Gly → Cys could be detected (7 were Gly → Glu and 15 Gly → Arg), but out of 4 such mutants at the third Gly site, 3 corresponded to the first and 1 to the second alternative (see Fig. 19-11).

4. Effective single-site revertants to apparent wild type (i.e. those producing a fully functional protein) can be of two types: one in which the original amino acid has indeed been restored, another where a new but equivalent amino acid has been substituted. For instance, a mutation in the segment –Gly–Phe–Gly– to –Glu–Phe–Gly– can be restored by the "reversion" –Val–Phe–Gly–. A mutant –Val–Phe–Val– is restored to function by either the mutation –Ala–Phe–Val– or –Val–Phe–Ala–.

These and related data, for instance on the capsid protein of T4 (gene 23) (17), have been of inestimable value in furnishing clues and confirmatory evidence concerning the code. However, the most convincing hard evidence bearing on both the composition and base sequence of the various codons was provided by straightforward biochemical experiments dealing with protein synthesis *in vitro,* to be discussed in Chapter 21. Here we shall anticipate to the extent of summarizing certain of the conclusions that bear on the nature of the code itself.

SALIENT FEATURES OF THE GENETIC CODE

Coding Ratio

The coding ratio *r* is simply defined as the number of nucleotides per codon. We have already mentioned Crick's elegent demonstration of of out-of-phase–into-phase relationships produced by the introduction of three successive mutational events into a particular gene, where the mutations themselves corresponded to different additions or deletions. These experiments strongly

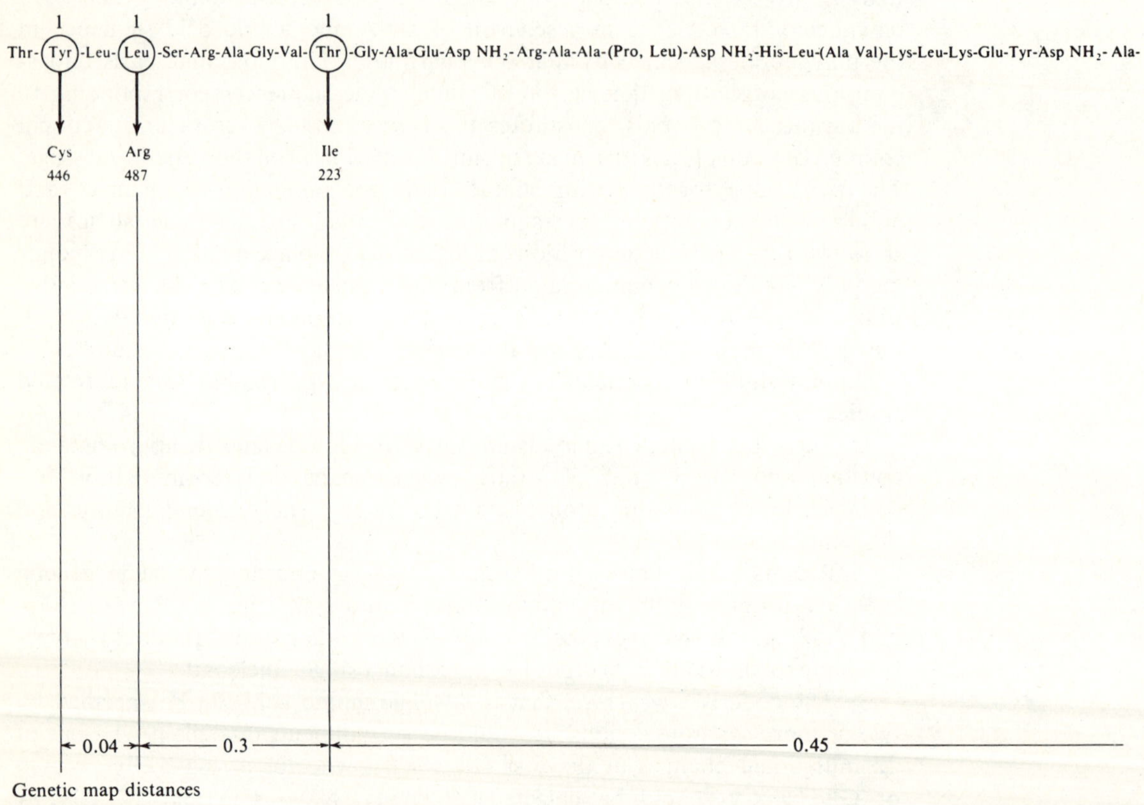

FIGURE 19-10

Colinearity of maps for tryptophan synthetase A (or α). Adapted from the results of Yanofsky and his collaborators. [See C. Yanofsky, G. R. Drapeau, J. R. Guest, and B. C. Carlton, *Proc. Natl. Acad. Sci. U.S.*, **57:** 296 (1967)].

suggested 3, or a multiple of 3, as the correct number for *r*. In theory, if we could get a precise estimate of the length of a polynucleotide segment required to specify a particular polypeptide, the coding ratio could be determined directly. No such direct demonstration has yet been possible, but a number of reasonable estimates have been made and suggest that *r* is certainly less than 10 and probably less than 5. For instance, tobacco necrosis satellite virus contains 1200 nucleotides in its RNA; the protein subunit of its capsid is made up of 400 residues: $r = 3$. Similar numerological exercises can be indulged in for a number of different proteins or groups of proteins. Other lines of evidence are provided by the *in vitro* system (17). (1) Trinucleotides are the smallest entities capable of effectively binding specific aminoacyl tRNA's to ribosomes; this binding capacity is not improved significantly by extending the chain within limits. (2) A self-consistent set of coding triplets has been derived that accounts admirably for both the amino acid incorporation and binding data *in vitro* and mutational amino acid substitutions *in vivo.* (3) Alternating ribopolynucleotides of the type ABABABAB. . . code for a heteropolypeptide containing two different amino acids (specified by ABA and BAB, respectively) in alternating sequence—this argues for a code in

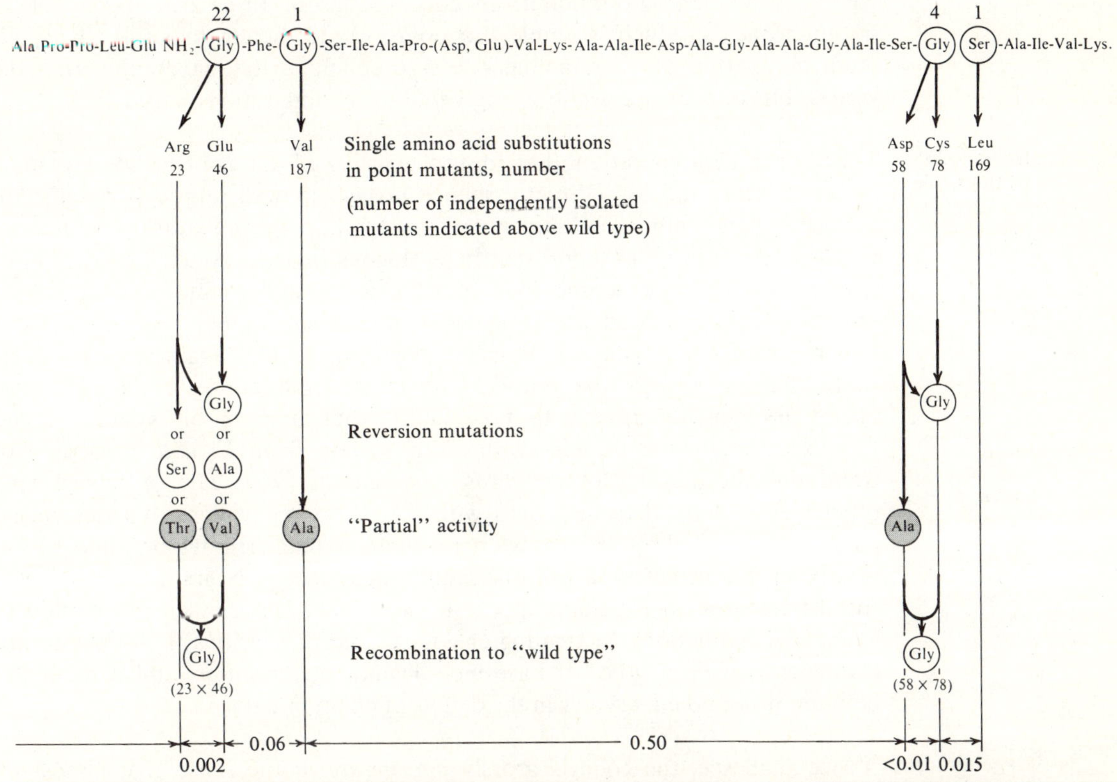

Amino acid number	Mutation number	Presumed Gly codon	"Mutant" codon	"Revertant" codons	Recombination codon
210	. . .	GGA/G			
	23		AGA/G (Arg)	GGA/G, ACA/G (Thr), AGU/C (Ser)	
	46		GAA/G (Glu)	GGA/G, GCA/G (Ala), GUA/G (Val)	
233	. . .	GGU/C			
	58		GAU/C (Asp)	GGU/C, GCU/C (Ala)	
	78		UGU/C (Cys)	GGU/C	
Genetic cross: 58 × 78					GGU/C

FIGURE 19-11
Codons corresponding to Fig. 19-10 (from Table 19-4).

which r is odd, though not necessarily for one in which $r = 3$. (4) Polynucleotides of the type ABCABC. . . code for precisely *three* different *homo*polypeptides; specified by ABC, BCA, and CAB, respectively. (5) Polynucleotides of the type ABCDABCDABCD. . . code for a heteropolypeptide consisting of *four* serially alternating amino acids, specified by ABC, DAB, CDA, and BCD, respectively. Experiments 3, 4, and 5 taken in conjunction are consistent *only* with a coding ratio of 3. (6) The magic number of common amino acids is 20; the number of available nucleotides is 4. Four available letters in code groups of two can specify $4^2 = 16$ amino acids; this is clearly insufficient. Four letters in code groups of three can specify $4^3 = 64$ amino acids, which is ample. Economy would dictate that this, the minimum number, is also the actual one. We conclude that in combination the various bits of evidence argue strongly that the coding ratio equals 3.

Universality of the Code

The recent demonstration that in procaryotes all 64 triplets convey some signal *in vitro,* and by indirect evidence *in vivo* as well, argues strongly that the code must exhibit qualitatively universal features throughout living nature. Furthermore, the observation that heterologous *in vitro* systems consisting of ribosomes from one source plus transfer RNA and amino acid–activating enzymes from a second are as capable of at least limited protein synthesis under certain circumstances as is the homologous one again favors universality. There is also the corollary evidence that suggests that identical codon assignments appear to be equally applicable in all systems tested (18,19). Finally the protein-synthesizing system from *E. coli* is capable of translating the message of *vertebrate* hemoglobin (20). The self-consistency of such data derived from studies with *E. coli,* other bacteria, viruses, yeast, plant seedlings, rabbit reticulocytes, rat liver, human hemoglobin, and so on, is striking. Nevertheless, a word of caution is indicated. Nonuniversal, species-specific features do exist and appear to be localized in the self-recognition of aminoacyl synthetases (activating enzymes) and the various tRNA's, perhaps involving varying degrees of base modification in the latter, and certainly as between pro- and eucaryotes in the details of chain initiation.

Degeneracy and Ambiguity

As we shall see, the code is grossly *degenerate* in the sense that almost all amino acids are coded for by more than one codon (synonyms).

Degeneracy has obvious biological advantages. For instance, it permits essentially the same complement of enzymes and other proteins to be specified by microorganisms varying widely in DNA-base composition. Thus, relative invariance in cellular composition that maximizes possible responses to the environment need not be sacrificed to genetic individuality and distinctiveness, or vice versa. Furthermore, small "accidental" mutational changes are much less likely to be damaging and are more readily reversed in a degenerate than a nondegenerate code, especially if the degeneracy is such that it leads to the replacement of "equivalent" amino acids (i.e., aliphatic $\leftrightarrow$ aliphatic, acidic $\leftrightarrow$ acidic, basic $\leftrightarrow$ basic, etc.). Therefore, degeneracy is conducive to genetic stability.

Under certain circumstances, for example, in the presence of high concentrations of Mg^{2+}, of other cations, especially of the antibiotic streptomycin and related aminoglycosides, or of organic solvents, the code both *in vitro* and *in vivo* gives evidence of containing *ambiguities.* This means that one and the same triplet can code for more than one amino acid. It is believed that suppression may operate by this mechanism: Mutation $X \rightarrow Y$ changes

TABLE 19-4
Genetic dictionary

2nd(Y)→ / 1st (X) ↓	U	C	A	G	3rd(Z)
U	Phe	*Ser*	Tyr	Cys	U
					C
	Leu*†	*Ser*	Ochre	Terminate	A
			Amber	Try	G
C	*Leu*	*Pro*	His	*Arg*	U
					C
	*Leu**	*Pro*	Gln	*Arg**	A
					G
A	Ile	*Thr*	Asn	Ser	U
					C
	Ile	*Thr*	Lys	Arg*	A
	Met‡				G
G	*Val*	*Ala*	Asp	*Gly*	U
					C
	Val (internal)	*Ala*	Glu	*Gly*	A
	Met (initiate)				G

Note that in almost all cases a substitution of one pyrimidine, or of one purine for the other (a "transition"), in position Z of an XYZ triplet does not affect the nature of the *aa* specified. In eight cases, indicated by italics, neither does the substitution of *any* purine for any pyrimidine or vice versa ("transversion"); in these instances the base in position Z appears irrelevant. In at least two instances (Leu, Arg, and perhaps Cys) transitions or transversions in X appear to be allowed. These are marked with an asterisk. Empty space indicates identity with amino acid above. "Amber" indicates the "amber codon," which does not correspond to any tRNA in nonpermissive strains. In permissive strains containing the SuA mutation, a Ser-tRNA is produced that can recognize this triplet. "Ochre" indicates a similar, suppressible mutation, while "nonsense" indicates the third chain terminating codon. "Initiate" indicates an initiator codon: matching a Met-$tRNA_F$ capable of being formylated to formyl-N-met-tRNA. This tRNA also recognizes GUG in an initiating (5′-terminal) position.

amino acid x to y or to nonsense; suppression permits partial misreading of the code such that x (or an equivalent amino acid x') is again incorporated in the proper spot, at least part of the time. Indeed, suppression is rarely completely effective, and suppression mutants can frequently be shown to contain both types of proteins, one corresponding to the original mutation y and another resembling the normal type with x or x' in the place of y. Similarly, incorporation of FU instead of U may lead to a *phenotypic* suppression of a mutation.

Commas and Overlaps

A code is referred to as *commaless* if there are no spacings between adjacent code groups, that is (in the present case), if each nucleotide in a particular gene of the normal organism actually forms part of a codon. A code is referred to as *overlapping* if some elements of certain code groups form parts of other code groups:

$...(XYZ)_n\ a(XYZ)_{n+1}\ ab(XYZ)_{n+2}...$ a,ab are "commas"

$...\ W|XY|Z\ ...$ code groups (WXY) (XYZ) overlap

The evidence so far presented argues compellingly against the presence of either commas or overlap regions in the genetic code, which is therefore commaless and nonoverlapping, at least as far as concerns the *structural* genes (i.e., those specifying functional polypeptides).[15]

Nature of the Code

A current version of the actual codons for the 20 common amino acids based principally on studies by Nirenberg (18) and by Khorana and their collaborators (19) is shown in Table 19-4. The base sequences specified are written from left to right [i.e., *XYZ* means (p)*X*p*Y*p*Z*]. This is not only the direction of synthesis of the message but also of its readout. As already mentioned, several characteristic features are immediately apparent: (1) the large proportion of possible triplets assigned to actual coding function and therefore making "sense" and (2) the high degree of degeneracy.

The degeneracy means that many codons may be recognized only partially or may contain alternate "acceptable" bases at one of the terminal positions, especially at the 3′ terminal end and to a lesser extent at the 5′ terminus. This then raises the question of how the codons are to be read and how efficient and accurate these processes are. The reading involves mutual recognition, by virtue of complementariness, between the codon (really its complementary form) on an RNA messenger (see the next section) and an *anticodon* on tRNA (see Chapter 21). Within any group of synonyms for the same amino acid, different sets of complementary triplets may be inherently less stable than others. Furthermore, in different organisms or under different genetically or environmentally determined circumstances, the different tRNA's capable of reading them may not all be equally available. Both of these, and perhaps other factors as well, can control efficiency and perhaps species specificity. Genetic or environmental factors may also control fidelity and allow for a greater or lesser amount of ambiguity with resultant effects on suppression, as we have already seen, and perhaps on *polarity*. In line with these considerations, there is now considerable evidence that one and the same tRNA can recognize several of the degenerate codons.

These matters will be explored in further detail in Chapter 21.

MESSENGER RNA

History

Some species of informational or template RNA (i.e., an RNA molecule capable of specifying polypeptide structure) must be involved in protein synthesis as a direct consequence of the observation that the site of protein synthesis is provided by the ribosomes, a set of cytoplasmic structures, whereas information storage and replication are the province of DNA, which is localized at some remove from the former even in the least differentiated bacterial cell. Furthermore, the very stability (chemical and metabolic) of DNA so necessary for its genetic function makes it highly improbable that proteins are synthesized on genes directly. This led to the formulation of one of the central dogmas of molecular biology, which states that the flow of information proceeds by the steps DNA → RNA → protein. In this sequence the first arrow does not involve a change of code and is now referred to as

[15] Presumably the whole DNA molecule, and not only those regions concerned in the specification of functional polypeptides, provides some form of genetic message or other information (e.g., by specifying various kinds of RNA, by providing regulatory and recognition features, etc.), and therefore represents the totality of the genetic code. This statement does not apply to the possible introduction of commas by meaningless or nonsense codons.

transcription, whereas the second does involve such a change and requires a code book or dictionary, and is therefore referred to as *translation* proper. In other words, the DNA and RNA maps are either identical or complementary, whereas the protein map constitutes only an analog of either the DNA or the RNA map. The minimal hypothesis which found many adherents prior to 1961 was that ribosomal RNA was itself informational. This meant that each gene would specify a distinct type of ribosome that would then become a stable template and govern the synthesis of a specific protein. The model was supported by the finding that some proportion of the total population of ribosomal RNA appeared to be synthesized quite rapidly, although the bulk was relatively stable metabolically. It was further strengthened by the exciting discovery that the isolated RNA from certain plant viruses was itself infective and thus could act as a carrier of information.

Almost simultaneously, however, a number of facts came to light which in the aggregate made the hypothesis as stated above untenable (21,22).

1. Proteins vary widely in size and composition. Yet isolated ribosomes and ribosomal RNA's exhibit remarkable homogeneity in regard to composition, not only for any particular cell population, but even between different species (see Table 5-7). How then can cellular protein composition be governed by the composition of ribosomal RNA?

2. The composition of all the proteins of a cell must in some way be specified by the cell's DNA. The DNA of any one species appears remarkably homogeneous, but it varies from species to species over exceedingly wide limits ($X_A + X_T$ from 0.25 to 0.75). Yet ribosomal RNA, the postulated intermediate which in its aggregate should certainly reflect the composition of the bulk of the DNA, shows no such variation.

3. If new genes are introduced into a bacterial cell by conjugation, the model requires the gradual synthesis of novel ribosomes. This process would result in a parabolic curve if enzyme activity (corresponding to proteins newly synthesized) is plotted against time. Instead what is observed is an eclipse period of 2 min or less during which no protein is apparently synthesized at all, followed by active synthesis of the protein at a constant and maximal rate.

4. Soon after the infection of a bacterial cell by a virulent bacteriophage, the former loses its capacity to synthesize characteristic bacterial proteins and instead becomes reprogrammed within a matter of a minute to the synthesis of proteins required for viral replication.

5. The effects of certain genetic alterations, the so-called *operator mutations* (see page 956), are the following. A whole group or block of genes (the *operon*) is turned on or off simultaneously but only if the mutation is present in the cis configuration. This means the configuration $oaba/o'$ produces a different effect in the synthesis of proteins A, B, and C than does $o'abc/o$. The first configuration is brought about by introducing o' into a genotype $oabc$, the second by introducing o into $o'abc$. For certain mutants the change $o \rightarrow o'$ results in turning all genes on simultaneously, even in the absence of an externally added inducer (o^c mutants). $o^c a^- b^+ c^+ / o a^+ b^+ c^+$ is constitutive for B, but inducible for A, and so on (see page 962). This is easily accounted for if we assume that regulation takes place at the level of transcription and not at that of translation, but this in turn implies the existence of an *unstable* template. The postulate of the existence of such a messenger RNA (mRNA), advanced by Jacob and Monod in 1961 (22), was based primarily on this last point, but it also served to reconcile the others.

Evidence for the Existence of mRNA

1. An unstable RNA is synthesized very rapidly after bacteriophage infection but accounts for only a small fraction ($\leq$5 percent) of the total RNA present (23–26). Similar unstable, DNA-like RNA's could also be detected in uninfected bacteria (27–29), in mammalian cells infected with vaccinia virus (30), and in growing yeast (31). Alternatively, if extremely short pulses of label are administered to eucaryotes, a sizable proportion of the newly synthesized RNA is probably mRNA (24–26,28–30,32).

2. The presumed mRNA synthesized in the experiments under (1) hybridizes effectively with its appropriate parent DNA in a highly specific manner; for example, the mRNA produced after T4 infection hybridizes with homologous phage DNA's but not with heterologous phage or bacterial DNA's, including that of the host (24–26).

3. Ribosomes (or rather their subunits—see Chapter 21) by themselves can be shown to be essentially passive assembly plants for translation. They are programmed for specific protein synthesis by attachment of a specific messenger to the 30 S (or 40 S in higher organisms) subunit of the monomeric 70 S (or 80 S) ribosome, the transfer RNA's being attached to the 50 S (or 60 S) subunit. Subsequent to T2 infection no new ribosomes are made and preexisting bacterial ribosomes become the site of attachment of the newly synthesized mRNA (33).

4. Ribosomes with attached "natural" messengers synthesized *in vivo* can be stripped of practically their whole informational content *in vitro*. They can then be reprogrammed to synthesize proteins (or more commonly, to incorporate amino acids) in response to natural messengers. They can also synthesize amino acid homo- and heteropolymers, in response to synthetic messenger RNA's.

5. Viral RNA can function as messenger for the synthesis of virus-specific proteins *in vivo* even under conditions when the synthesis of all RNA's, including rRNA, is completely inhibited by actinomycin D. In selected instances bacteriophage RNA can code for the same proteins *in vitro*. Also under certain conditions an mRNA can be isolated that is formed subsequent to and perhaps duplicates the function of viral DNA.

6. The biosynthesis of polycistronic messengers has been shown to be specifically elicited by the addition of the appropriate inducers, or the removal of repressors.

7. The half-life of mRNA in different bacteria has been shown by a variety of techniques (28,29,32–36) to be of the order of 90 sec at 37°. For inducible, or derepressible polycistronic mRNA's, the kinetics of its formation, that of its destruction, and that of the synthesis of the various polypeptides comprising the operon all exhibit the same time constant; each molecule of mRNA only mediates a single (or a very small number of) translational rounds (35,36).

8. A soluble enzyme capable of catalyzing the transcription of both double-helical and single-stranded DNA into a complementary RNA (see Chapter 20) has been isolated and purified from a variety of cells. With single-stranded DNA the product, both *in vivo* and *in vitro*, is a DNA-RNA hybrid; with double-stranded DNA, a free RNA molecule complementary to one of the two strands.

We shall next examine some of the properties of mRNA that have become established by recent work (36,37) and shall conclude our discussion with a somewhat more detailed presentation of a few systems and problems that have yielded to such probing.

Some Properties of mRNA

1. *Half-life.* mRNA molecules must vary greatly in their stability, first of all between microorganisms and higher organisms. This type of variability appears to be related to difference in rate of *in vivo* protein synthesis; for example, between *E. coli* and rabbit reticulocytes the rates differ by a factor of about 50 (100 amino acids per second in the former, 2 amino acids per second in the latter, both at 37°). mRNA's may also vary in stability even within the same cell. The RNA-containing bacteriophages can function directly as messenger, yet remain inviolate throughout the life cycle of the virus in an infected bacterial cell, a matter of some 30 to 55 min at 37°. Messengers in higher organisms show even greater stability. Once the active cytoplasmic mRNA-Rb-tRNA complex has been formed, it may be turning out proteins continuously perhaps for days, and, under certain circumstances, even in the complete absence of *nuclear* DNA (as in immature red cells), and a virtual absence of all RNA turnover.

2. *Size.* Greater care and improved techniques in isolation have led to a continued upward trend in the *s* value (and hence the size) of the entities tentatively identified as mRNA by sedimentation or chromatographic analysis. If, as is now believed, most messengers are themselves *polycistronic* and retain the complete information content of one operon, their size may indeed become quite formidable. For instance for the *Trp* operon in *E. coli,* which contains information for the specification of five structural polypeptides, the mRNA has been estimated to contain 7320 nucleotides ($M = 2.55 \times 10^6$) and to be 50,000 Å long (34,35). Information consistent with these estimates is also available for the *Lac, Gal,* and *His* operons.

3. *Complementariness to DNA.* The conclusion has by now become inescapable that only one of the two strands of a DNA duplex (which may, however, be required to be present in the double-stranded configuration for effective transcription) is both necessary and sufficient for the specification of mRNA. Thus, transcription is *asymmetric* and sequentially *strand selective* (26,38–41).[16] The same holds true of the RNA synthesized *in vitro* by the DNA-dependent RNA polymerases, *provided* that DNA is of high molecular weight, remains in its native conformation, and care is taken in the isolation of the enzyme to retain its own regulatory capabilities (σ factors—page 906).

But asymmetric transcription has an important consequence, especially with regard to mRNA's transcribed from single genes or even small operons. There is no reason to suppose, and every reason to doubt, that the base composition of an individual DNA strand in an individual gene or small cluster of genes bears any resemblance in its base composition to the gross global composition of the duplex DNA of the whole organism. Therefore, individual mRNA species would not be expected to be DNA-like in their base composition.

Furthermore, all cellular RNA's, and that means the various species of rRNA and all tRNA's, are complementary to certain specific regions of the homologous DNA. Therefore, not only do there exist specific cistrons for the synthesis of these various RNA's—and hence these particular RNA's themselves constitute final rather than intermediate gene products, in the sense that they do not specify any other molecule—but sequence complementariness cannot be considered a specific property of mRNA.

[16] Note that nothing is said here to the effect that it must be the same strand for each gene or operon. It may be the strand of polarity 5′ ↦ 3′ known as *W* (for Watson) or *l* (for left), or else its complement, known as *C* (for Crick) or *r*.

However, it is true that the predominant fraction of the total genome in procaryotes is composed of structural genes, and that in exponentially growing cells much of the RNA actually being synthesized at any one instant (and hence *pulse labeled* (28,42) consists of mRNA which will therefore hybridize more effectively with DNA and exhibit a base composition approaching that of DNA.

Ability to Code for Proteins. One of the characteristic properties of, and indeed a sufficient definition for, mRNA is its ability to code for protein synthesis and amino acid incorporation. For quite some time it was believed that none of the other RNA species, nor DNA, could function in this manner. It now turns out that a crucial determinant is provided in large part by the absence of secondary structure in polynucleotides of this type (correlated no doubt to their complete lack of self-complementariness).

Control of mRNA Formation. It is evident on logical grounds alone that transcription must precede translation in time, at least as far as the initial group of nucleotides transcribed, no matter how small, is concerned. Thus, at this level transcription certainly controls translation. What is not certain any longer is whether the two processes are truly independent events in both time and space. One can readily conceive of models in which transcription may come to a halt, or at least the product of transcription be rendered functionally useless by nuclease action, unless the highly reactive unstable mRNA molecule can be immediately integrated into an operational translation machine. Recent experiments in a number of microbial systems and on a number of levels suggest that transcription and translation are indeed tightly coupled in this fashion. How this may be brought about in the case of inducible and derepressible systems is discussed in Chapter 21. What controls mRNA synthesis for genes that specify noninducible polypeptides, the synthesis of which is tightly integrated into the cell cycle, and which comprise the bulk of these materials in a cell is uncertain.

One of the interesting features of microbial RNA synthesis has been the observation that in many strains (the so-called "stringently controlled" strains), RNA synthesis in general appears to be dependent on the simultaneous presence of a full complement of all amino acids. If any one is missing, that is, not supplied in the medium for an auxotroph requiring that particular amino acid, RNA synthesis ceases. This stringent control is in turn governed by a specific gene (Rel or RC), which can be mapped by standard techniques (Fig. 19-4). Its modification produces "relaxed" strains, which no longer exhibit this tight integration between amino acid supply and RNA synthesis. The events regulated are, no doubt, complex (42) and may effect RNA synthesis in response to an inhibitory signal (ppGpp?) together with a regulatory subunit (Ψ_R) of the transcribing enzyme (42a).

Quantitative Aspects of mRNA Synthesis in E. coli (28,32,34,43). In cells growing exponentially at 37° the rate of transcription of *any* "open" gene (i.e., one engaged in transcription) is now known (from experiments using several different techniques) to be of the order of 50 nucleotides/sec, the exact value depending on experimental details such as the precise strain used, the medium, and so on. Many polymerase molecules, spaced ~100 nucleotide pairs apart on the DNA, are engaged in transcription, which produces approximately 500 separate nascent chains of RNA. Of these ~200 correspond to stable (rRNA, tRNA) chains, being copied from approximately 0.70 percent of the genome, comprising 1.6×10^5 nucleotide pairs. The remaining 3000 chains of RNA are mRNA being copied from a total of only $\leqslant$10 percent of the genome ($\leqslant 2 \times 10^6$ nucleotide pairs)—while 99 percent of it is *potentially* able to be transcribed; hence at any one time the bulk of the genes are silent and contribute but little to the mRNA pool. Of the *total* cellular RNA 97.5 percent is stable, 2.5 to 3 percent mRNA. The latter is made up of a large number of different species varying in abundance; 70 percent of the total mRNA is copied from 1 percent of the total genome, or 2×10^5 nucleotide pairs.

Degradation appears to be unidirectional and start from the same end as transcription and translation [i.e., the 5′-(pp)pN- end] (44). In *B. subtilis* (29) mRNA accounts for a higher proportion of the total cellular RNA ($\sim$8 percent). This level is due almost entirely to difference in its rate of synthesis (166 nucleotides/sec) rather than its rate of degradation, which in both species has a first-order decay constant of $\sim$5 min.

Regulation of Transcription in Phage λ (45–51). The coliphage λ is lysogenic, that is, it can exist both in a temperate and virulent form. While in the former, phage genes are switched off since their transcription is inhibited by a protein (48), the λ repressor, coded for by the C_I gene (Fig. 19-12), which together with the product of the *rex* gene *is* synthesized in lysogenic cells; it is removed as a result of induction of the vegetative phase of the life cycle (Fig. 19-6). There also exist mutant forms of λ (λ *vir*) which can grow even when repressor is present (i.e., they only exhibit the vegetative phase). Their resistance to repressor action can be explained in a manner analogous to that used to account for the o^c mutation in its host (49) (see also Chapter 21): a failure of its DNA, as a result of the mutation, to interact with the repressor protein. Hence, localization of the genetic region(s) responsible for the change(s) $\lambda \rightarrow \lambda$ *vir* define an operator. At least two of these have now been identified: they are the regions labeled v_2 and v_1v_3 on the map (Fig. 19-12).

By highly ingenious experiments involving a combination of genetic (and physical) deletions, the use of half-duplex molecules obtained by shear, separated strands, and hybridization techniques, the sequence of transcriptional events after induction (e.g., by detaching a heat-sensitive repressor), outlined in Fig. 19-12 and its legend, has been established.

Transcription, Transportation, and Integration of mRNA in Eucaryotes (52–54). Eucaryotic cells face special problems as a consequence of the sequestration of their genetic apparatus into a separate organelle, the nucleus. Not only does this place transcription and translation into separate compartments, thus rendering unlikely the integration of the two processes by the "peeling off" of newly transcribed mRNA by nascent polysomes described in Chapter 21, but the ribosomal subunits reach the cytoplasm in a sequence spatially and temporally distinct from each other and from mRNA. Thus, mRNA must find its way to the cytoplasm prior to its integration into and protection by ribosomes. Currently available evidence suggests that this may be accomplished by the formation of specialized ribonucleoprotein particles (informosomes) not identical with, but approximately of the size of, the smaller ribosomal subunit. What the fate of the protein components is when the mRNA is eventually integrated to form the backbone of a polysome, whether they perhaps assist in the formation of the latter or are detached during this process, lies completely within the realm of speculation, as do indeed most of the matters pertaining to mRNA formation, regulation, and function in higher organisms. The rate of transcription of mRNA is similarly unknown. However, fairly reliable information is available concerning this parameter for various rRNA's and tRNA's in the liver cells of adult rats; the most rapid one is that for the rRNA precursor. It appears to be transcribed at a rate which is of the same order of magnitude as that just given for *E. coli.*

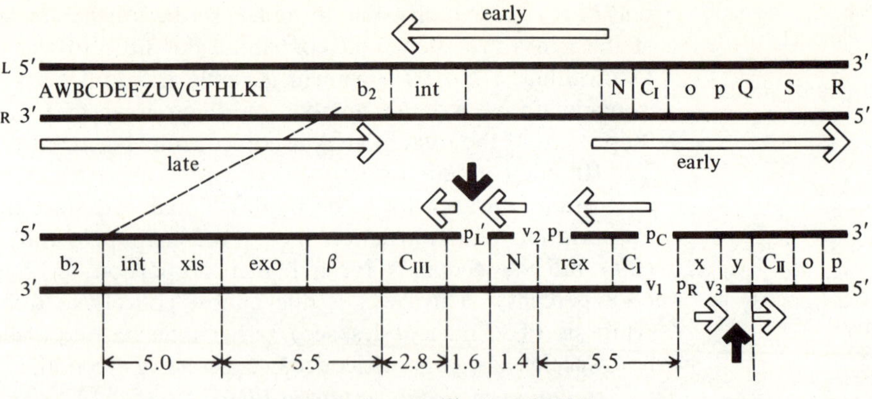

FIGURE 19-12

Orientation and control of transcription in *E. coli* phage λ. The repressor (coded for by C_I) interacts with operators v_2 (o_L) and v_3 (o_R); its removal permits transcription of the *l* strand starting at a promoter p_C, and of the *r* strand at a promoter p_R in the x region. The transcription of gene N is initiated, again on the *l* strand at a separate site, p_L at the left end of the "rex" region. The product of this gene is required in turn for the transcription of the remaining early genes C_{III} through *int*, starting at p_L' again on the *l* strand. Finally, the remainder of the genes (late) are transcribed starting at or near the 3′ extremity of the *r* strand. An additional promoter is located in the y region; a mutant (λc_n) produces a DNA which shows an impaired rate of transcription *in vitro*. The length of the messages transcribed from λDNA *in vitro* is controlled by a bacterial protein ($M = 200{,}000$, probably a tetramer) which releases the transcripts from its DNA at the sites indicated by vertical arrows; it thus functions as a termination or release factor for RNA and has been designated as ρ factor. S. Kumar, K. Bovre, A. Guha, Z. Hradecna, V. M. Maher, and W. Szybalski, *Nature,* **221:** 823 (1969); M. Ptashne ánd N. Hopkins, *Proc. Natl. Acad. Sci. U.S.,* **50:** 1282 (1968); J. W. Roberts, *Nature,* **224:** 1169 (1969).

REFERENCES

Books

In addition to the references cited in Chapters 5 and 20 the reader may find the following useful:

Anfinsen, C. B.: *The Molecular Basis of Evolution,* Wiley, New York, 1959.

Bonner, D. M., and S. E. Mills: *Heredity,* Prentice-Hall, Englewood Cliffs, N.J., 2nd ed., 1964.

Bryson, V., and H. J. Vogel (eds.): *Evolving Genes and Proteins,* Academic Press, New York, 1965.

Cohn, N. S.: *Elements of Cytology,* Harcourt, Brace & World, New York, 1964.

Cold Spring Harbor Symposium on Quantitative Biology: *Cellular Regulatory Mechanisms,* vol. 26, 1961.

Cold Spring Harbor Symposium on Quantitative Biology: *The Genetic Code,* vol. 31, 1966.

Cold Spring Harbor Symposium on Quantitative Biology: *The Mechanism of Protein Synthesis,* vol. 34, 1969.

Cold Spring Harbor Symposium on Quantitative Biology: *Synthesis and Structure of Macromolecules,* vol. 28, 1963.

Garrod, A. E.: *Inborn Errors of Metabolism,* Henry Fronde and Hodder and Stoughton Ltd., London, 1923.

Hayes, W.: *The Genetics of Bacteria and Their Viruses,* Wiley, New York, 2nd ed., 1968.

Ingram, V. M.: *The Biosynthesis of Macromolecules,* Benjamin, New York, 1965.

Malt, R. A. (ed.): *Macromolecular Synthesis and Growth,* Little, Brown, Boston, 1967.

McElroy, W. D., and B. Glass (eds): *The Chemical Basis of Heredity,* Johns Hopkins Press, Baltimore, 1957.

Perutz, M. F.: *Proteins and Nucleic Acids,* Elsevier, New York, 1962.

Sager, R., and F. J. Ryan: *Cell Heredity,* Wiley, New York, 1961.

Shugar, D. (ed.): *Genetic Elements, Properties and Function,* Academic Press, New York, 1967.

Stahl, F. W.: *The Mechanics of Inheritance,* Prentice-Hall, Englewood Cliffs, N.J., 1964.

Stanbury, J. B., J. B. Wyngaarden, and D. S. Fredrickson (eds.): *The Metabolic Basis of Inherited Disease,* McGraw-Hill, New York, 2nd ed., 1966.

Steiner, R. F.: *The Chemical Foundations of Molecular Biology,* Van Nostrand Reinhold, New York, 1965, especially chaps. 1 and 11.

Stent, G. S.: *Molecular Biology of Bacterial Viruses,* Freeman, San Francisco, 1963, especially chaps. 8–10, 13, and 15.

Strauss, B. S.: *An Outline of Chemical Genetics,* Saunders, Philadelphia, 1960.

Taylor, J. H. (ed.): *Molecular Genetics,* part I, 1963: part II, 1967, Academic Press, New York.

Watson, J. D.: *Molecular Biology of the Gene,* Benjamin, New York, 1965.

Yčas, M.: *The Biological Code,* Wiley (Interscience), New York, 1969.

Reviews and Key Articles

Aebi, H. E.: Inborn Errors of Metabolism, *Ann. Rev. Biochem.,* **36:** 271 (1967).

Ames, B. N., and R. G. Martin: Biochemical Aspects of Genetics: The Operon, *Ann. Rev. Biochem.,* **33:** 235 (1964).

Beckwith, J. R.: Regulation of the Lac Operon, *Science,* **156:** 597 (1967).

Bennett, J. C., and W. J. Dreyer: Genetic Coding for Protein Structure, *Ann. Rev. Biochem.,* **33:** 205 (1964).

Benzer, S.: Genetic Fine Structure, *Harvey Lectures,* **56:** 1 (1960–61).

Campbell, P. N.: The Biosynthesis of Proteins, *Progr. Biophys. Mol. Biol.,* **15:** 3 (1965).

Crick, F. H. C.: The Recent Excitement in the Coding Problem, *Progr. Nucl. Acid Res. Mol. Biol.,* **1:** 163 (1963).

Epstein, W., and J. R. Beckwith: Regulation of Gene Expression, *Ann. Rev. Biochem.,* **37:** 411 (1968).

Freese, E.: Molecular Mechanism of Mutations, in J. H. Taylor (ed.), *Molecular Genetics,* part 1, Academic Press, New York, 1963, p. 207.

Freese, E.: Hereditary DNA Alternations, *Angew. Chem.* (*Intern. Engl. Ed.*), **8:** 12 (1969).

Freese, E., and E. B. Freese: Mutagenic and Inactivating DNA Alternation, *Radiation Res. Suppl.,* **6:** 97 (1966).

Garen, A.: Sense and Nonsense in the Gentic Code, *Science,* **160:** 149 (1968).

Geiduschek, E. P., and R. Hazelkorn: Messenger RNA, *Ann. Rev. Biochem.,* **38:** 647 (1969).

Gorini, L., and J. R. Beckwith: Suppression, *Ann. Rev. Microbiol.,* **20:** 401 (1966).

Gross, P. R.: Biochemistry of Differentiation, *Ann. Rev. Biochem.,* **37:** 631 (1968).

Helinski, D. R., and C. Yanofsky: Genetic Control of Protein Structure, in H. Neurath (ed.), *The Proteins,* vol. 4, Academic Press, New York, 2nd ed., 1966, p. 1.

Horowitz, N. H., and R. L. Metzenberg: Biochemical Aspects of Genetics, *Ann. Rev. Biochem.,* **34:** 527 (1965).

Khorana, H. G.: Polynucleotide Synthesis and the Genetic Code, *Harvey Lectures,* **62:** 79 (1966–1967).

Khorana, H. G.: Synthesis in the Study of Nucleic Acids, *Biochem. J.,* **109:** 709 (1968).

Moldave, K.: Nucleic Acids and Protein Biosynthesis, *Ann. Rev. Biochem.,* **34:** 419 (1965).

Monod, J., F. Jacob, and F. Gros: Structural and Rate-Determining Factors in the Biosynthesis of Adaptive Enzymes, in The Structure and Biosynthesis of Macromolecules, *Biochem. Soc. Symp.* (*Cambridge, Engl.*), **21:** 104 (1962).

Nirenberg, M.: Protein Synthesis and the RNA Code, *Harvey Lectures,* **59:** 155 (1963–64).

Schweet, R., R. Arlinghaus, R. Heintz, and J. Shaeffer: Mechanism of Peptide Bond Formation in Protein Synthesis, in *Developmental and Metabolic Control Mechanisms in Neoplasia* (19th Annual M. D. Anderson Symposium on Fundamental Cancer Research), University of Texas Press, Austin, 1966.

Schweet, R., and R. Heintz: Protein Synthesis, *Ann. Rev. Biochem.,* **35:** 723 (1966).

Sibatani, A.: Genetic Transcription or DNA-Dependent RNA Synthesis, *Progr. Biophys. Mol. Biol.,* **16:** 17 (1966).

Singer, M. F., and P. Leder: Messenger RNA: An Evaluation, *Ann. Rev. Biochem.,* **35:** 195 (1966).

Thomas, C. A., Jr., and L. A. MacHattie: The Anatomy of Viral DNA Molecules, *Ann. Rev. Biochem.,* **36:** 485 (1967).

Vogel, H. J., and R. H. Vogel: Regulation of Protein Synthesis, *Ann. Rev. Biochem.,* **36:** 519 (1967).

Specific References

1. Jacob, F., and E. L. Wollman: *Sexuality and the Genetics of Bacteria,* Academic Press, New York, 1961; Campbell, A.: *Advan. Genet.,* **11:** 101 (1962).
2. Zinder, N. D., and J. Lederberg: *J. Bacteriol.,* **64:** 679 (1952): Zinder, N. D.: *Sci. Am.,* **199** (5): 38 (1958).
3. Okamoto, K., J. A. Mudd, and J. Marmur: *J. Mol. Biol.,* **34:** 429 (1968).
4. Fox, M. S., and R. D. Hotchkiss: *Nature,* **187:** 1002 (1960); Fox, M. S.: *Proc. Natl. Acad. Sci. U.S.,* **48:** 1043 (1962).
5. Taylor, A. L., and M. S. Thoman: *Genetics,* **50:** 659 (1964); Taylor, A. L., and C. R. Trotter: *Bacteriol. Rev.,* **31:** 332 (1967).
6. Edgar, R. S., G. H. Denhardt, and R. H. Epstein: *Genetics,* **49:** 635 (1964); Wood, W. B., R. S. Edgar, J. King, I. Lielausis, and M. Henniger: *Federation Proc.,* **27:** 1160 (1968).
7. Beadle, G. W., and E. L. Tatum: *Proc. Natl. Acad. Sci. U.S.,* **27:** 499 (1941); Beadle, G. W.: Genes and the Chemistry of Organisms, in *Science in Progress,* 5th series, Yale University Press, New Haven, 1947.
8. Meselson, M., and J. J. Weigle: *Proc. Natl. Acad. Sci. U.S.,* **47:** 857 (1961); Meselson, M.: *J. Mol. Biol.,* **9:** 734 (1964); see also the references in Chapter 20.
9. Benzer, S.: *Proc. Natl. Acad. Sci. U.S.,* **45:** 1607 (1959); **47:** 403 (1961); *Sci. Am.,* **206** (1): 70 (1962).

10. Benzer, S., and E. Freese: *Proc. Natl. Acad. Sci. U.S.*, **44:** 112 (1958).

11. Freese, E.: *Proc. Natl. Acad. Sci. U.S.*, **45:** 622 (1959); *J. Mol. Biol.*, **1:** 87 (1959).

12. Freese, E. B.: *Proc. Natl. Acad. Sci. U.S.*, **47:** 540 (1961).

13. Crick, F. H. C., L. Barnett, S. Brenner, and R. J. Watts-Tobin: *Nature*, **192:** 1227 (1961); Crick, F. H. C., and S. Brenner: *J. Mol. Biol.*, **26:** 361 (1967).

14. Terzaghi, E., Y. Okada, G. Streisinger, J. Emrich, M. Inouye, and A. Tsugita: *Proc. Natl. Acad. Sci. U.S.*, **56:** 500 (1966).

15. Yanofsky, C.: *Harvey Lectures*, **61:** 145 (1965–1966).

16. Yanofsky, C.: *Sci. Am.*, **216** (5): 80 (1967); Yanofsky, C., G. R. Drapeau, J. R. Guest, and B. C. Carlton: *Proc. Natl. Acad. Sci. U.S.*, **57:** 296 (1967).

17. Schweet, R., and J. Bishop: Protein Synthesis in Relation to Gene Action, in J. H. Taylor (ed.), *Molecular Genetics*, vol. 1, Academic Press, New York, 1963, p. 353; and various contributors to *Cold Spring Harbor Symp. Quant. Biol.*, **31** and **34.**

18. Marshall, R. E., C. T. Caskey, and M. Nirenberg: *Science*, **155:** 820 (1967), and references therein.

19. Khorana, H. G.: *Biochem. J.*, **109:** 709 (1968); Söll, D., J. D. Cherayil, and R. M. Bock: *J. Mol. Biol.*, **29:** 97 (1967).

20. Laycock, D. G., and J. A. Hunt: *Nature*, **221:** 1118 (1969).

21. Brenner, S.: *Cold Spring Harbor Symp. Quant. Biol.*, **26:** 101 (1961).

22. Jacob, F., and J. Monod: *J. Mol. Biol.*, **3:** 318 (1961).

23. Volkin, E., and L. Astrachan: *Virology*, **2:** 149, 433 (1956).

24. Landy, A., and S. Spiegelman: *Biochemistry*, **7:** 585 (1968); see also Baguley, B. C., P. L. Bergquist, and R. K. Ralph: *Biochim. Biophys. Acta*, **138:** 51 (1967).

25. Rouvière, J., J. Wyngaarden, J. Cantoni, F. Gros, and A. Kepes: *Biochim. Biophys. Acta*, **166:** 94 (1968).

26. Bolle, A., R. H. Epstein,, W. Salser, and E. P. Geiduschek: *J. Mol. Biol.*, **31:** 325 (1968); **33:** 339 (1968).

27. Hayashi, M., and S. Spiegelman: *Proc. Natl. Acad. Sci. U.S.*, **47:** 1564 (1961); Nomura, M., B. D. Hall, and S. Spiegelman: *J. Mol. Biol.*, **2:** 306 (1960); Hall., B. D., and S. Spiegelman: *Proc. Natl. Acad. Sci. U.S.*, **47:** 137 (1961).

28. Kennell, D.: *J. Mol. Biol.*, **34:** 85 (1968).

29. Fan, D. P., A. Higa, and C. Levinthal: *J. Mol. Biol.*, **8:** 210 (1964); Zimmermann, R. A., and C. Levinthal: *ibid.*, **30:** 349 (1967).

30. Oda, K-I., and W. K. Joklik: *J. Mol. Biol.*, **27:** 395 (1967).

31. Kitazume, Y., and M. Yčas: *Biochim. Biophys. Acta*, **76:** 391 (1963).

32. Winslow, R. M., and R. A. Lazzarini: *J. Biol. Chem.*, **244:** 1128 (1969).

33. Gros, F., H. Hiatt, W. Gilbert, C. G. Kurland, R. W. Risebrough, and J. D. Watson: *Nature*, **190:** 581 (1961); Brenner, S., F. Jacob, and M. Meselson: *ibid.*, **190:** 576 (1961).

34. Morse, D. E., R. F. Baker, and C. Yanofsky: *Proc. Natl. Acad. Sci. U.S.*, **60:** 1428 (1968); Baker, R. F., and C. Yanofsky: *ibid.*, **60:** 313 (1968).

35. Ito, J., and F. Imamoto: *Nature*, **220:** 441 (1968).

36. Singer, M. F., and M. F. Leder: *loc. cit.*

37. Geiduschek, E. P., and R. Haselkorn: *loc. cit.*

38. Spiegelman, S.: in H. J. Vogel, V. Bryson, and J. O. Lampen (eds.), *Informational Macromolecules*, Academic Press, New York, 1963, p. 27. Hayashi, M., M. N. Hayashi, and S. Spiegelman: *Proc. Natl. Acad. Sci. U.S.*, **50:** 664 (1963); **51:** 351 (1964).

39. Marmur, J., C. M. Greenspan, E. Palecek, F. M. Kahan, J. Levine, and M. Mandel: *Cold Spring Harbor Symp. Quant. Biol.*, **28:** 191 (1963); Marmur, J., and C. M. Greenspan: *Science*, **142:** 387 (1963).

40. Tocchini-Valentini, G. P., M. Stodolsky, A. Aurisicchio, M. Sarnat, F. Graziosi, S. B. Weiss, and E. P. Geiduschek: *Proc. Natl. Acad. Sci. U.S.*, **50:** 935 (1963); Colvill, A. J. E., L. C. Kanner, G. P. Tocchini-Valentini, M. T. Sarnat, and E. P. Geiduschek: *ibid.*, **53:** 1140 (1965); Geiduschek, E. P., G. P. Tocchini-Valentini, and M. T. Sarnat: *ibid.*, **52:** 486 (1964).

41. Green, M. H.: *Proc. Natl. Acad. Sci. U.S.*, **52:** 1388 (1964).

42. Irr, J., and J. Gallant: *J. Biol. Chem.*, **244:** 2233 (1969); Cashel, M., and J. Gallant: *Nature*, **221:** 838 (1969).

42a. Travers, A. A., R. I. Kamen, and R. F. Schleif: *Nature*, **228:** 748 (1970).

43. Bilezikian, J. P., R. O. R. Kaempfer, and B. Magasanik: *J. Mol. Biol.*, **27:** 495 (1967).

44. Morikawa, N., and F. Imamoto: *Nature*, **223:** 37 (1969); Morse, D. E., R. Mosteller, R. F. Baker, and C. Yanofsky: *ibid.*, **223:** 40 (1969).

45. Kumar, S., K. Bovre, A. Guha, Z. Hradecna, V. M. Maher, and W. Szybalski: *Nature*, **221:** 823 (1969); Taylor, K., Z. Hradecna, and W. Szybalski: *Proc. Natl. Acad. Sci. U.S.*, **57:** 1618 (1967).

46. Kourilsky, P., L. Marcaud, P. Sheldrick, D. Luzzati, and F. Gros: *Proc. Natl. Acad. Sci. U.S.*, **61:** 1013 (1968).

47. Cohen, S. N., and J. Hurwitz: *Proc. Natl. Acad. Sci. U.S.*, **57:** 1759 (1967).

48. Ptashne, M.: *Nature*, **214:** 232 (1967).

49. Ptashne, M., and N. Hopkins: *Proc. Natl. Acad. Sci. U.S.*, **60:** 1282 (1968).

50. Doerfler, W., and D. S. Hogness: *J. Mol. Biol.*, **33:** 661 (1968).

51. Roberts, J. W.: *Nature*, **223:** 480 (1969); **224:** 1168 (1969).

52. Spirin, A. S., N. V. Belitsina, and M. A. Aytkhozhin: *Zh. Obschch. Biol.*, **25:** 321 (1964).

53. Infante, A. A., and M. Nemer: *Zh. Obshch. Biol.*, **32:** 543 (1968).

54. Perry, R. P., and D. E. Kelley: *J. Mol. Biol.*, **35:** 37 (1968).

20 Biosynthesis of Nucleic Acids

In Chapter 5 we were concerned with the chemistry of the nucleic acids, which, we know from the discussion in Chapter 19, are to be equated to the genes and their primary products. We know from these discussions that the genetic material must be capable of (1) being replicated faithfully; (2) undergoing occasional alterations (mutations) such that once introduced, these can (3) either be repaired or be (again faithfully) transmitted to progeny. (4) During simultaneous replication of a standard and a mutant type in the same cell a process of exchange must be provided so that there results a reassortment of these two types in the progeny. Finally, the genetic material must be (5) capable of expression, so that a message inscribed into a sequence of nucleotides in DNA is transcribed in the first instance into a complementary sequence of polyribonucleotides, which in turn (or directly, if the genetic material is RNA) is translated in most (but by no means all) cases into the sequence of amino acids of a polypeptide. In this chapter we inquire into what is known concerning the biochemistry of these processes with emphasis on nucleic acid synthesis *in vivo* and to what extent it has been possible to comprehend and reconstruct them by means of enzymatic reactions *in vitro*.

BIOSYNTHESIS OF DNA

The Situation *in vivo*

As has frequently been the case in molecular biology, most of the early experiments were performed on procaryotic organisms. They have only been recently and, so far, incompletely extended to single-celled eucaryotes and to cultures of single cell types from multicellular organisms, and in an even less satisfactory manner to the latter as a whole. Some caution as to the universality of some of the concepts and interpretations is therefore in order. The most important conclusions derived from these experiments are:

1. The bacterial or bacteriophage chromosome in the haploid state is constituted by a single and continuous duplex[1] polydeoxyribonucleotide with all the properties expected from its Watson–Crick–Wilkins structure, particularly with respect to complementarity, polarity and mass/unit length ratio (Chapter 5)—in *E. coli* it corresponds to a circular structure of $M \simeq 2.8 \times 10^9$ daltons, 1100 μm in length.

2. Replication is semiconservative. The entire chromosome or episome is so replicated and therefore functions as a replicating unit or *replicon* (1).

3. Replication commences at a unique starting point and then is propagated unidirectionally and sequentially, with a replicating point traveling down the entire length of the chromosome duplicating both strands of any one region at the same time.

[1] Except for certain small bacteriophages (e.g., ϕX 174).

4. There is evidence for discontinuities in the vicinity of the replicating point; that is, newly synthesized DNA is not immediately integrated into a continuous, covalent structure.

5. Initiation of a new round of replication is a complex event requiring protein synthesis. It involves specific interactions between an initiator substance and a specific region of the replicon as well as attachment of the latter to a specific site—to the cell membrane in the case of the bacterial chromosome.

Semiconservative Replication. Watson and Crick, on the basis of their structural studies, suggested that each chain of the DNA duplex serves as template for the synthesis of a complementary chain (2). This model, termed *semiconservative replication,* predicts that the template DNA molecule is divided equally between newly synthesized molecules after one cycle of replication. On the other hand, *conservative replication* would involve the duplication of only one of the DNA strands, the DNA duplex itself serving as template, and, subsequently, the newly formed strand in turn serving as template for the formation of its complement. In this case, after one replication cycle, the original DNA molecule would remain intact and the additional molecule would be composed entirely of newly formed strands. The course of DNA replication according to these two possibilities is indicated schematically in Fig. 20-1.

The classic experiment of Meselson and Stahl (3) clearly indicates that at least in this case, the essential features of semiconservative replication actually take place *in vivo.* In this experiment, *E. coli* cultures were grown on a medium containing isotopically pure $^{15}NH_4Cl$ as the only source of nitrogen. Consequently, the DNA synthesized in the course of cell multiplication contains ^{15}N rather than ^{14}N. Such DNA labeled with ^{15}N can be readily

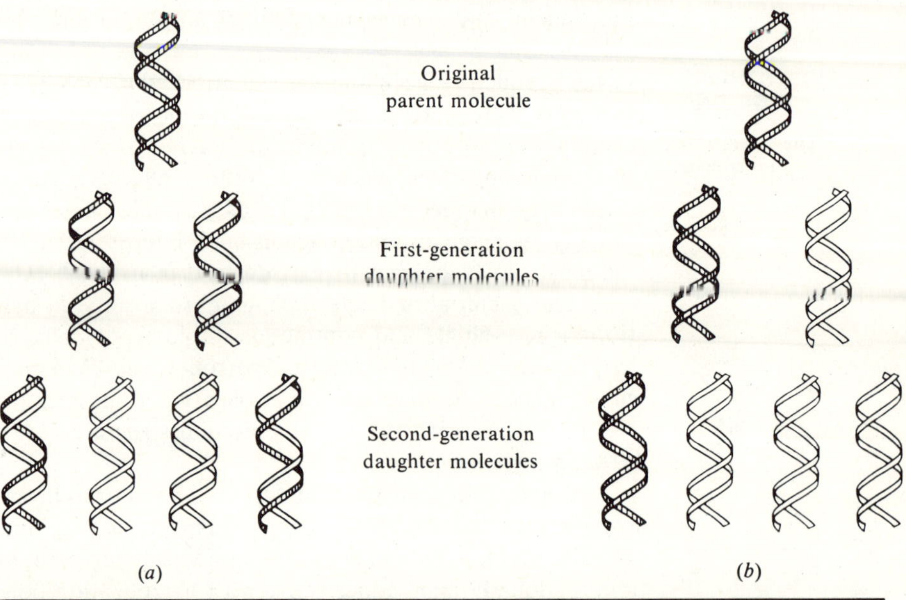

FIGURE 20-1

Schematic representation of DNA replication according to the semiconservative (*a*) and conservative (*b*) theories. M. Meselson and F. W. Stahl, *Cold Spring Harbor Symp. Quant. Biol.,* **23:** 10 (1958).

distinguished from its ^{14}N-containing counterpart by ultracentrifugation in a cesium chloride density gradient (see page 216). Cultures grown on $^{15}NH_4Cl$ were therefore abruptly transferred to a medium containing normal (^{14}N) ammonium chloride, and the DNA of these cells was examined in the cesium chloride density gradient as a function of time. Initially, only one band with a density characteristic of the ^{15}N-DNA was observed as expected. After one generation of growth in the medium containing ^{14}N, the DNA again exhibited a single band in the density gradient; this time the density was at the midpoint between those for the ^{15}N and the ^{14}N species. Hence, after a single round of replication of the DNA, the product molecules must have contained one strand of the original ^{15}N material and one newly synthesized strand of ^{14}N material. This result is, of course, precisely the one expected on the basis of the semiconservative model. After two generations of growth on ^{14}N, the DNA exhibited two equal bands in the density gradient, one with a density equal to that of the ^{14}N-DNA and one with a density equal to that of the ^{15}N-^{14}N hybrid observed after the first generation of growth. Reference to Fig. 20-1 indicates that this, too, is the expected result in terms of a semiconservative replication.

In more recent experiments Nagata and Meselson (4) *pulse*-labeled an exponentially growing culture with ^3H-thymine, transferred it to a "heavy" medium (^{13}C-glucose plus $^{15}NH_4Cl$) and withdrew samples at different times. They found that replication of the radioactive DNA occurred in waves separated from one another by the generation time of the organism indicating conservation of direction and sequence of replication during growth.

Initiation and Replication Points. Evidence for these phenomena has been obtained by four lines of investigation: (a) radioautography, (b) distribution of genetic markers during synchronous replication, (c) density distributions, and (d) a combination of (b) and (c). Radioautography of whole replicating chromosomes was first introduced by Cairns (5). A typical structure and its interpretation is shown as Fig. 20-2*a*. Duplication here is from the initiation (and termination) point X and a single growing point Y. The former not only joins the ends of the molecules but it also serves as a swivel, which (together with a single break or *nick* in one of the two chains) releases the topological constraint on circular duplex DNA by rotating one end relative to the other, thus driving the synthetic process. The DNA within the swivel itself is the last to be duplicated at a time just prior to the separation of mother and daughter circles. Replicating circular molecules conforming to the structure of Fig. 20-2*a* have now also been seen on electron micro graphs of bacteriophage and episome (plasmid) DNA (Fig. 20-2*c*).

The distribution and frequency of newly duplicated genetic markers has been investigated, e.g., during transformation of *Bacillus subtilis* (6) and in dividing *E. coli* (7). Consider a sequence of markers $A \rightarrow Z$

chromosomes | A | B | C | ———— | Y | Z | of cells in exponential growth.

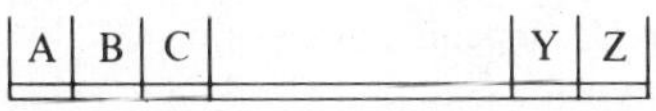

Assume that replication procedes by means of a replicating fork

Assume further that replication always commences at Z; thus in the population we will find structures such as

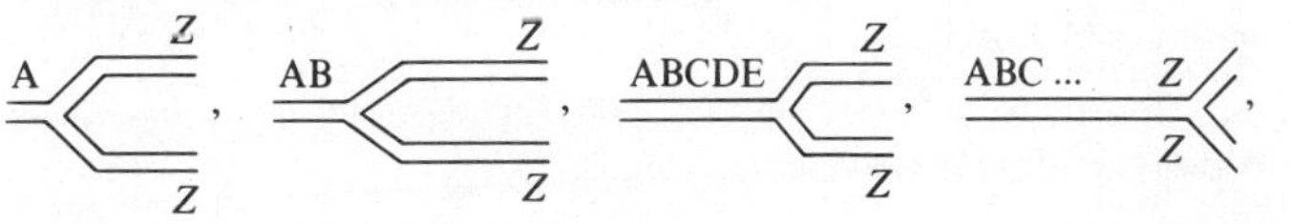

A Z Z, AB Z Z, ABCDE Z Z, ABC ... Z Z, etc.

Regions	Termini	Length, μm	State
A	Y → X	670	half-hot
B	X → Y	680	hot-hot
C	{ Y → C	213	half-hot
(unduplicated)	{ C → X	205	hot-hot

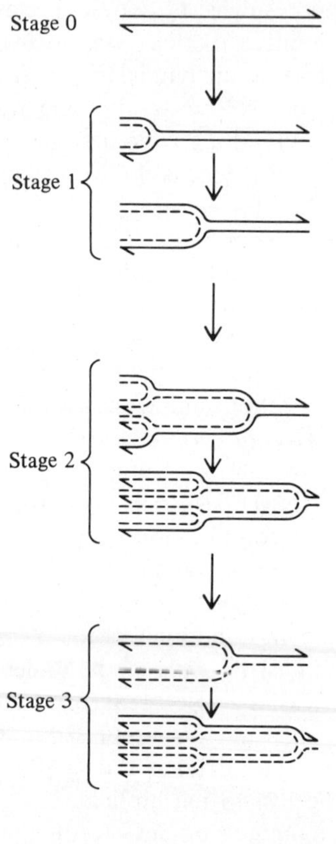

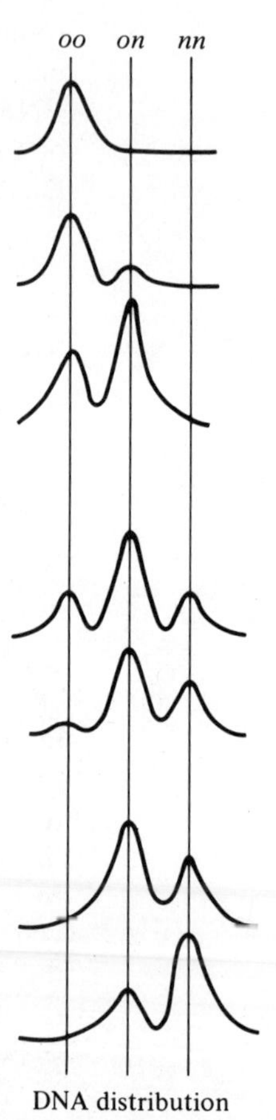

FIGURE 20-2

(*a*) A "half-hot" chromosome with two thirds of its replication completed. The unduplicated portion varies in its label distribution because the moment of introduction of label did not coincide with initiation of replication. After J. Cairns, *Cold Spring Harbor Symp. Quant. Biol.*, **28:** 43 (1963).

(*b*) Replication with two replication positions (dichotomous replication). Densities of *oo, on,* and *nn* DNA's depend on the kind of isotopes or base analogs used. Note that the distribution of markers among the DNA peaks is different from marker to marker, depending on their map positions. A. O'Sullivan and N. Sueoka, *J. Mol. Biol.*, **27:** 349 (1967).

(*c*) (right) Electron micrographs of replicating molecules of Col E1 DNA arranged according to the percentage of replication as follows: a = 5%; b = 18%; c = 24%; d = 38%; e = 40%; f = 58%; g = 65%; h = 72%; i = 89%; j = 92%; k = 99+%. Unit circumference defined as equal to [(replicated length)/2 + unreplicated length] was 2.3 μ. Adapted from Inselburg, J. and M. Fuke: *Science,* **169:** 590 (1970) with permission; electron micrographs by courtesy of Dr. Inselburg.

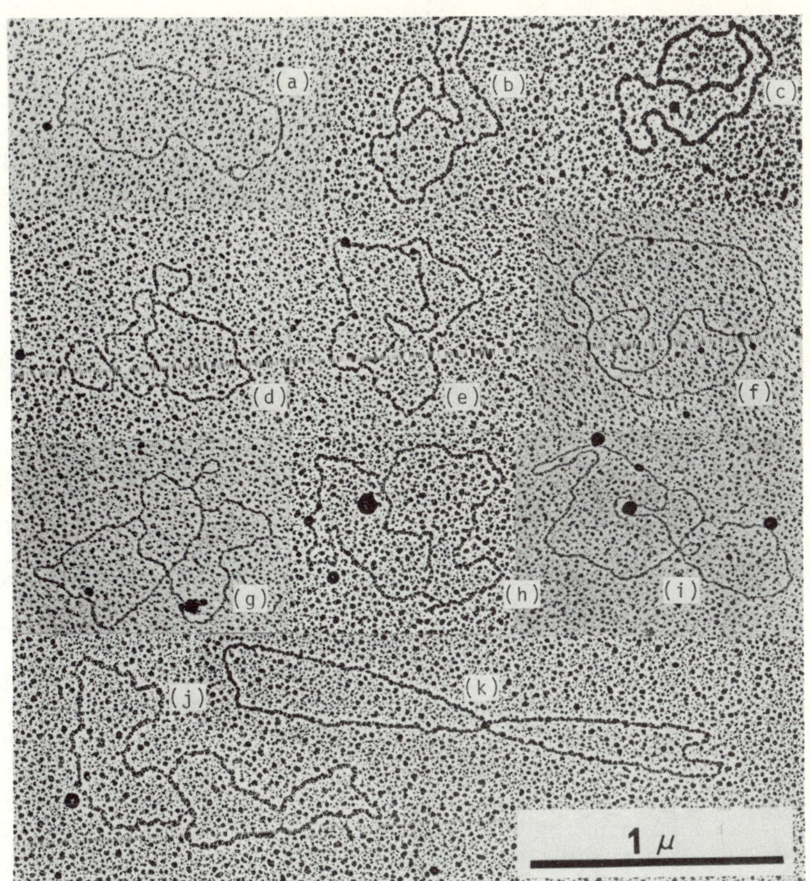

It is clear that, *on the average,* the distribution of markers should exhibit a graded series from 2.0 (corresponding to the origin) to 1.0 (at the terminus); in other words, the frequency map should coincide with the genetic map constructed by other means. Experimental results are in accord with the model. Sueoka and his collaborators have pioneered in the application of the other two techniques as well (6). For them one uses cells that have been synchronized, either by transfer from stationary to exponential growth or by germination of spores, and simultaneously transfers them from a heavy (D_2O, $^{15}NH_4^+$) to a light (H_2O, $^{14}NH_4^+$) medium or vice versa. If initiation starts at a particular point, then the transfer of markers to the new density position (Fig. 20-2*b*) should again be ordered sequentially and correspond to a map devised by other means. The results bear out the expectation.

State of DNA at Replication. When newly synthesized DNA is labeled with a very short pulse (~1 sec at low temperatures) of radioactive thymine, and the product analyzed by sedimentation, molecules of the expected large size are found; but when sedimentation is performed under alkaline conditions, or the product is denatured prior to sedimentation, then slowly sedimenting, small (frequently single-stranded) entities can be identified (8-11). When the pulses are prolonged or when the pulse is followed by a chase of non-radioactive precursor, the only structures present are large and double stranded.

DNA Polymerases

DNA Polymerase from E. coli*—Nature of the Reaction.* The enzymological aspects of DNA replication are of great interest. The first enzyme capable of converting deoxynucleoside polyphosphates into polymeric material was discovered in extracts of *E. coli* by A. Kornberg[2] and his coworkers (12). This enzyme was termed DNA polymerase (deoxynucleoside triphosphate : DNA deoxynucleotidyl transferase). It consists of a single polypeptide of $M = 109{,}000$ with a single SH group, apparently devoid of subunit structure, capable of binding to a single DNA molecule. The overall reaction catalyzed is

$$\begin{matrix} \text{ndATP Mg} \\ + \\ \text{odGTP Mg} \\ + \\ \text{pdCTP Mg} \\ + \\ \text{qdTTP Mg} \end{matrix} + \underset{\text{DNA}}{\begin{bmatrix} dA_q \\ dG_p \\ dC_o \\ dT_n \end{bmatrix}} \rightleftharpoons \underset{\text{parent}}{\begin{bmatrix} dA_q \\ dG_p \\ dC_o \\ dT_n \end{bmatrix}} \underset{\text{daughter}}{\begin{bmatrix} dT_q \\ dC_p \\ dG_o \\ dA_n \end{bmatrix}} + (n + o + p + q)\text{MgPP}_i \qquad (20\text{-}1)$$

The principal features of the reaction catalyzed by this enzyme are the following:

1. The substrates are the deoxynucleoside triphosphates; the diphosphates are not polymerized. As indicated in Table 20-1, all four deoxynucleoside triphosphates are required for reaction, although some exceptions (see below) are known; omission of one of these substrates almost completely abolishes formation of polymeric material.[3]

2. The overall reaction is reversible. Pyrophosphate is produced in the polymerization in amounts corresponding to the stoichiometry indicated in Eq. (20-1). Incubation of DNA with high concentrations of pyrophosphate results in a partial depolymerization reaction.

3. The reaction is dependent upon the presence of a divalent metal ion, usually Mg^{2+}. If this ion is replaced by Mn^{2+}, the corresponding ribonucleoside triphosphates can be utilized as alternative substrates and mixed ribo- and deoxyribonucleic acids may be synthesized.

4. The DNA polymerase reaction is dependent upon the presence of a polynucleotide. This is a point of great interest and will be discussed in detail below.

5. DNA polymerase will catalyze the net synthesis of DNA-like molecules from nucleoside triphosphates.

What are the structural characteristics of the polymeric material produced in this reaction, and what is its relationship to those of the DNA added? The

[2] Nobel Laureate for this discovery.

[3] In all reactions dTTP can be replaced by dUTP and dBTP (>dFTP); dCTP by the 5-hydroxymethyl, methyl, or bromo derivative; dGTP by dITP but *not* by dXTP.

TABLE 20-1
Incorporation of ^{32}P-dCTP into DNA

Additions	*^{32}P-labeled DNA formed, $\mu\mu$mol*
dCTP	2.5
dCTP + dGTP	5.1
dCTP + dGTP + dTTP	15.7
dCTP + dGTP + dTTP + dATP	3,300.0

answers to these questions reflect directly on the role of the DNA requirement for reaction. It may serve as a primer, or initiator (i.e., a starting point for the enzyme), which would then simply catalyze the elongation of preexisting polynucleotide chains, or, more interestingly, it may serve as template with the enzyme catalyzing the duplication of the latter. The physical properties of the synthetic material produced on linear duplex DNA (now known to have been nicked) reveals the synthetic material to be polydisperse with sedimentation constants centering about those of the DNA added initially. Molecular weights of synthetic and added DNA's are approximately the same, as are their intrinsic viscosities. That the synthetic material is structurally similar to DNA is further established by the demonstration of 3′-5′-phosphodiester linkages in dinucleotides obtained from incomplete DNase digests and by the very fact that the material is susceptible to DNase digestion. Thus, the synthetic material is a DNA and has gross physical characteristics similar to those of the starting material.

Further evidence that DNA can serve as template for its own replication is derived from base-composition studies (12–14). In Table 20-2, the base composition of several samples of synthetic DNA is compared to that of DNA added to the reaction mixture. One notes that for all the synthetic products A = T and C = G, and that the base compositions of starting material and product are remarkably similar.

Clearly, the most elegant fashion in which to prove that the structure of a product is identical to that of a template would be the comparison of base sequences in the two materials.[4] Since the determination of the sequence of long segments of DNA is experimentally too difficult at the present time, one must rely on less direct lines of evidence. A novel approach to this problem is a "nearest-neighbor" analysis (15). For a nucleic acid containing four bases, there are clearly 16 possible dinucleotide sequences. A nearest-neighbor analysis involves the determination of the absolute frequency with which each of these dinucleotide sequences appears in the particular DNA of interest: One of the deoxynucleoside triphosphates, for example dATP, is labeled with ^{32}P; the other substrates are not labeled. Following the poly-

[4] With $(rA)_n \cdot (rU)_n$ or $(dA)_7 \cdot (dT)_{11}$ as template and dATP plus dTTP as substrate, the product is $(dA)_n \cdot (dT)_n$.

TABLE 20-2
Base composition of enzymatically synthesized DNA

DNA	A	T	G	C	(A + G)/(T + C)	(A + T)/(G + C)
Mycobacterium phlei						
Primer	0.65	0.66	1.35	1.34	1.01	0.49
Product	0.66	0.65	1.34	1.37	0.99	0.48
Escherichia coli						
Primer	1.00	0.97	0.98	1.05	0.98	0.97
Product	1.04	1.00	0.97	0.98	1.01	1.02
Calf thymus						
Primer	1.14	1.05	0.90	0.85	1.05	1.25
Product	1.12	1.08	0.85	0.85	1.02	1.29
Bacteriophage T2						
Primer	1.31	1.32	0.67	0.70*	0.98	1.92
Product	1.33	1.29	0.65	0.70*	1.02	1.90

[a] Hydroxymethylcytosine.

merization reaction, the product DNA is cleaved completely with a mixture of DNases from spleen and *Micrococcus,* yielding the 3′-phosphates quantitatively. As a consequence, the phosphate originally introduced into the molecule esterified at the 5′ position of one deoxynucleoside is transferred to the 3′ position of the adjacent nucleoside (see below). Thus analysis of the amount of radioactivity introduced into each of the nucleoside-3′-phosphates from α^{32}P-labeled dATP yields the relative frequencies with which each of the bases occupy the specified position adjacent to adenine:

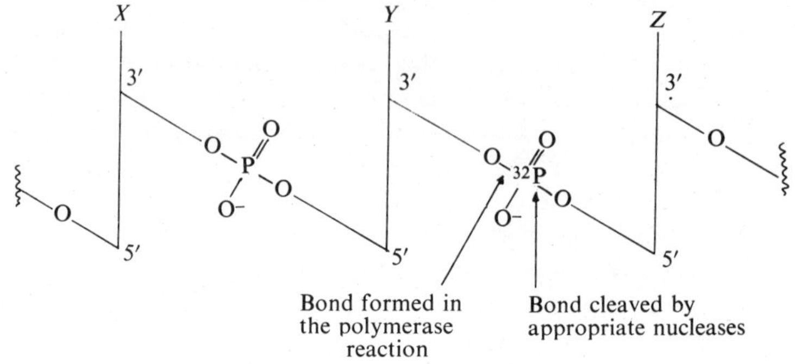

X Y Z 3′ 3′ 3′ O O P O O⁻ O O ^{32}P O O⁻ O 5′ 5′ 5′

Bond formed in the polymerase reaction

Bond cleaved by appropriate nucleases

When these figures are multiplied by the mol fraction of adenine in the DNA, known from base analysis or other determinations, the frequency with which each of the dinucleotides ApA, CpA, GpA, and TpA appears is obtained. By use of each of the four labeled nucleoside triphosphates in turn, the frequency of occurrence for all the dinucleotides is readily determined. The result of one such set of experiments is indicated in Tables 20-3 and 20-4.

Some important conclusions may be drawn from extended collections of data of this type (15). First, all 16 dinucleotide sequences occur, and the frequency with which they occur is a sensitive function of the source of the DNA. Second, the DNA polymerase-catalyzed duplication of DNA yields a product in which the strands are oriented with opposite polarities as predicted on the basis of the Watson–Crick model for DNA. The last conclusion is based on the fact that nearest-neighbor frequencies for certain of the nucleotide pairs are predicted to be identical on the basis of complementary base pairing; the predicted identities depend on whether the two strands have similar or opposite polarities. For example, with strands of opposite polarity, the frequency of ApG (0.045) is predicted to be identical to that of CpT (0.045), whereas, with

TABLE 20-3
Radioactivity measurements in experiments with *M. phlei* DNA as primer

Isolated 3′-deoxy-ribonucleo-tide	*Reaction 1: dAT^{32}P*			*Reaction-2: dTT^{32}P*			*Reaction 3: dGT^{32}P*			*Reaction 4: dCT^{32}P*		
	Sequence	*Cpm[a]*	*Fraction*	*Sequence*	*Cpm[a]*	*Fraction*	*Sequence*	*Cpm[a]*	*Fraction*	*Sequence*	*Cpm[a]*	*Fraction*
Tp	TpA	873	0.075	TpT	1,665	0.157	TpG	3,490	0.187	TpC	4,130	0.182
Ap	ApA	1,710	0.146	ApT	2,065	0.194	ApG	2,500	0.134	ApC	4,300	0.189
Cp	CpA	4,430	0.378	CpT	2,980	0.279	CpG	7,730	0.414	CpC	6,070	0.268
Gp	GpA	4,690	0.401	GpT	3,945	0.370	GpG	4,960	0.265	GpC	8,200	0.361
Sum		11,703	1.000		10,665	1.000		18,680	1.000		22,700	1.000

[a] Cpm = counts/min.

SOURCE: A. Kornberg, in J. M. Allen (ed.), *The Molecular Control of Cellular Activity,* McGraw-Hill, New York, 1962, p. 245.

TABLE 20-4
Dinucleotide frequencies for *M. phlei* DNA calculated from radioactivity incorporation experiments

Sequence	*Frequency*[a]	*Sequence*	*Frequency*[a]	*Sequence*	*Frequency*[a]	*Sequence*	*Frequency*[a]
TpA	0.012	TpT	0.026	TpG	0.063	TpC	0.061
ApA	0.024	ApT	0.031	ApG	0.045	ApC	0.064
CpA	0.063	CpT	0.045	CpG	0.139	CpC	0.090
GpA	0.065	GpT	0.060	GpG	0.090	GpC	0.122

[a] Calculated from the fraction of total radioactivity (Table 20-3) and the base composition for *M. phlei* DNA: T = 0.165, A = 0.162, C = 0.335, G = 0.338. For example, the data in Table 20-3 reveal that 0.378 of the time C is adjacent to A(CpA). Since A comprises 0.162 of the total bases, the frequency of CpA must be 0.378 × 0.162 = 0.063.
SOURCE: J. Josse, A. D. Kaiser, and A. Kornberg, *J. Biol. Chem.*, **236:** 864 (1961).

strands of the same polarity, the frequency of ApG is predicted to be identical to that of TpC (0.061). The data clearly support the former alternative. These concepts are illustrated graphically in Fig. 20-3.

We shall return in the next section to the problem of the nature of the product synthesized on various templates, but first we must examine some alternative reactions catalyzed by the pure enzyme. These are summarized in Fig. 20-4 and are

(a) The polymerization proper, an extension of a polynucleotide chain complementary and opposite in polarity to the template [i.e., *exclusively* in the 5′ → 3′ direction, i.e., the accretion of a 5′-deoxynucleotide to the free 3′-hydroxyl group of the growing chain (primer)].

(b) The synthesis of certain alternating copolymers or homopolymer duplexes in the absence of template.

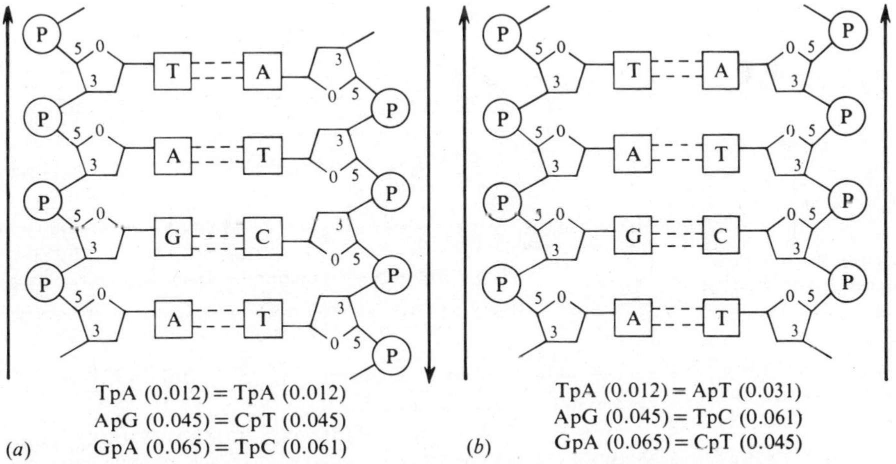

(*a*)
TpA (0.012) = TpA (0.012)
ApG (0.045) = CpT (0.045)
GpA (0.065) = TpC (0.061)

(*b*)
TpA (0.012) = ApT (0.031)
ApG (0.045) = TpC (0.061)
GpA (0.065) = CpT (0.045)

FIGURE 20-3
Contrast between a Watson–Crick DNA model with strands of opposite polarity and a model with strands of similar polarity. The predicted matching nearest-neighbor-sequence frequencies are different. Values in parentheses are sequence frequencies from the experiment with *M. phlei* DNA. The strands shown are newly synthesized strands aligned, for ease of comparison, as if they were complementary strands of a DNA double helix. J. Josse, A. D. Kaiser, and A. Kornberg, *J. Biol. Chem.*, **236:** 864 (1961).

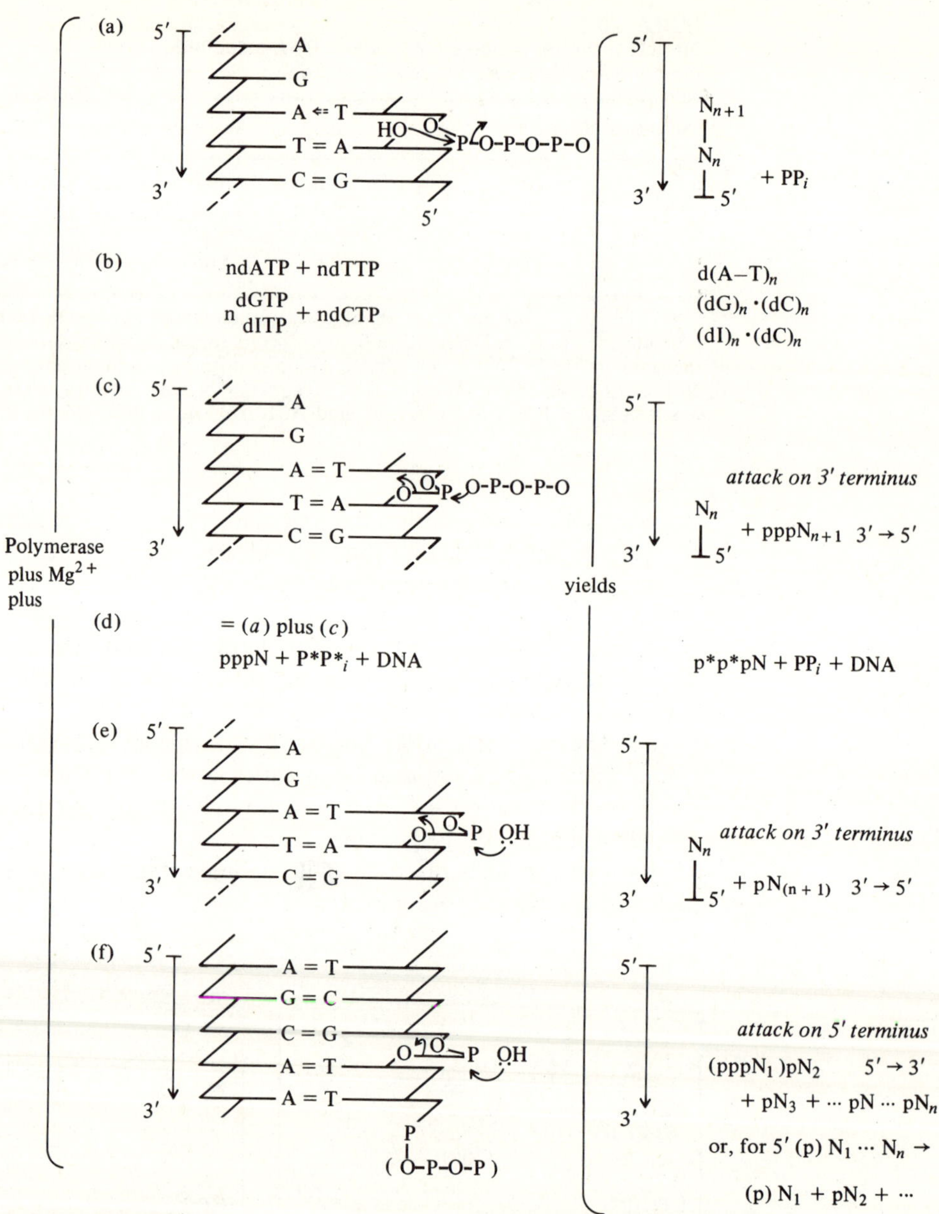

FIGURE 20-4
Reactions catalyzed by DNA polymerase from *E. coli.* After A. Kornberg, *Science,* **163:** 1410 (1969).

(c) The reverse of reaction (a) (i.e., the exonucleolytic pyrophosphorolysis of a polydeoxynucleotide in the 3′ → 5′ direction).

(d) A DNA (and dNTP)-dependent pyrophosphate exchange into any NTP.

(e) The exonucleolytic hydrolysis of a (single- or double-stranded) polydeoxynucleotide in the same direction as (c) (i.e., 3′ → 5′, yielding nucleoside-5′-P, starting from *free* 3′-OH groups).

(f) The hydrolysis in the opposite direction (i.e., 5′ → 3′) of double-stranded polynucleotides (16, 17) by a complex mechanism involving an initial incision by endonucleolytic attack several nucleotides removed but eventually yielding nucleoside-5′-P's and a terminal (p)N_1, except when the terminus is of the structure pppN_1—[5]; in this case this terminus is liberated together with the subterminal nucleotide as pppN_1pN_2. If the 3′-terminus is phosphorylated, oligonucleotides accumulate.

(g) The excision of thymine dimers (TT) and other mismatched sequences from DNA by an endonucleolytic attack (18) similar to the one just shown under (f).

Template and Primer Requirements: The Nature of the Product. The polymerase can interact with DNA's of varying secondary structure and catalyze the incorporation of nucleotides to yield the following products:

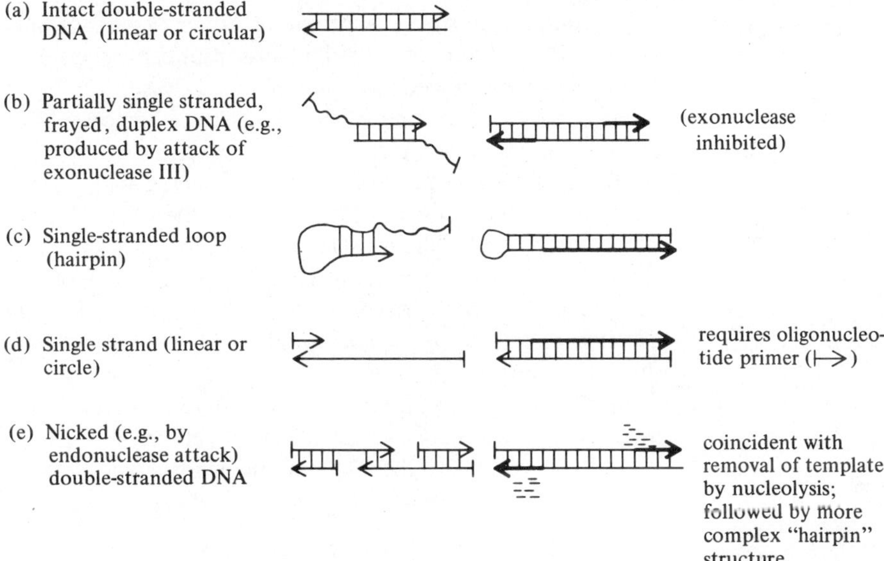

Other DNA Polymerases. Polymerases with properties distinct from the one just described have been isolated and purified from *E. coli* infected with bacteriophage T2, T4, or T5, from *Micrococcus luteus* (formerly *M. lysodeikticus*) and from calf thymus glands (19–22). The main difference between the activities catalyzed by all these enzymes and the one from uninfected *E. coli* is the ability of the latter to utilize nicked double-stranded DNA as a template, and the presence of nucleases. Although in general mutations in gene 43 of T4 lead to simultaneous loss of polymerase and of the 3′ → 5′ exonuclease, one particular amber mutation (am B22) produces a protein that lacks polymerase but still contains exonuclease activity (23). On the other hand, the purified enzyme from *M. luteus* and calf thymus have been reported to be always free of exonuclease activity. Calf thymus also contains an additional enzymatic activity, terminal *deoxynucleotidyl transferase,* which requires an oligodeoxynucleotide initiator (primer) but *no* template (24). As discussed in Chapter 5, this is the enzyme of choice for the preparation of homopolydeoxynucleotides:

$$p(dN')_m + ndNTP \longrightarrow p(dN')_m N_n + nPP_i \qquad (20\text{-}2)$$

$N = A, I, C, T,$ *N*-acetyl-G $\qquad 3 \geqq m \leqq 9,\ n \geqq 70$

[5] A structure never identified in DNA, duplicated either *in vivo* or *in vitro.*

Other Enzymes of Polydeoxynucleotide Metabolism

Polynucleotide ligase. An enzyme of great potential significance in the biological functions of DNA was discovered in 1967 independently in a number of different laboratories (25–27). The function of the enzyme is to seal nicks, that is, (re)-establish a phosphodiester bond within a gap in double-helical DNA constituted by a free 3′-OH and 5′-*P* on adjacent nucleotides of the same strand; for example,

$$\begin{matrix} 3' & \text{A T G T} & 5' \\ & \text{T A C A} & \\ 5' & \text{-p-P-OH (P)-P-p-} & 3' \end{matrix} + \begin{Bmatrix} \text{ATP} \\ \text{NAD} \end{Bmatrix} \xrightarrow{Mg^{2+}} \begin{matrix} 3' & \text{A T G T} & 5' \\ & \text{T A C A} & \\ 5' & \text{-p-P-(P)-P-p-} & 3' \end{matrix} + \text{AMP} + \begin{Bmatrix} \text{PP}_i \\ \text{NMN} \end{Bmatrix} \quad (20\text{-}3)$$

The required substrate is provided either by a normal DNA molecule into which a limited number of single-strand phosphodiester bond scissions (nicks) have been introduced by pancreatic DNase, or by the hydrogen-bonded circular duplexes, or linear concatemers, of bacteriophage λ (Chapter 5). The cosubstrate and driving force for the reaction is provided by ATP in the case of the enzyme obtained from *E. coli* infected with bacteriophage T4 (26,27) or from animal cells (28), NAD^+ in the case of the one from uninfected *E. coli* cells (25). The most convenient way for assaying the enzyme is to use 5′-labeled-P at the gap (introduced by polynucleotide kinase + ATP, see below) (29) and to observe the conversion of the label from a form susceptible to attack by *E. coli* alkaline phosphatase to one that is resistant to this treatment.

In both instances an enzyme–adenylate complex is formed as a first intermediate in the absence of DNA [Eq. (20-3a)]. This complex then transfers its adenylyl residue to the 5′-P residue of the substrate to form DNA-adenosine-di*P* [Eq. (20-3b)] and eventually, the phosphodiester linkage [Eq. (20-3c)].

$$\text{Enz} + \begin{Bmatrix} \text{ATP} \\ \text{NAD} \end{Bmatrix} \overset{Mg}{\rightleftharpoons} \text{Enz}-\text{O}-\underset{\underset{\text{O}}{\|}}{\overset{\overset{\text{O}^-}{|}}{\text{P}}}-\text{O}-\text{Ado} + \begin{Bmatrix} \text{PP} \\ \text{NMN} \end{Bmatrix} \quad (20\text{-}3a)$$

$$\text{Enz}-\text{O}-\underset{\underset{\text{O}}{\|}}{\overset{\overset{\text{O}^-}{|}}{\text{P}}}-\text{O}-\text{Ado} + [\text{DNA: } -\text{OH} \;\; \text{O}^-\text{–P(=O)(O}^-\text{)–O–}] \overset{Mg^{2+}}{\rightleftharpoons} [\text{DNA: } -\text{OH} \;\; \text{P(=O)(O}^-\text{)–O–P(=O)(O}^-\text{)–O–Ado}] + \text{Enz} \quad (20\text{-}3b)$$

$$\xrightarrow{\;\;} \text{AMP} + [\text{DNA: } -\text{O}-\text{P(=O)(O)}-\text{O}-] \quad (20\text{-}3c)$$

Independent demonstration for the occurrence of reaction (20-3c) was obtained by demonstrating (30) that $\text{A}^{\dagger}\text{MP*pdT(dT)}_n$ could be joined to $(\text{dT})_n\text{dT(OH)}$ termini in a duplex [with the other strand provided by $(\text{dA})_n$] with the liberation of $\text{p*A}^{\dagger}$.

Evidence suggesting an intimate role of polynucleotide ligase in the genetic function of DNA comes from (1) its utility in the duplication of biologically active DNA *in vitro* and the plausibility of analogous reactions occurring *in vivo* (see below), (2) the discovery of the discontinuous and fragmentary nature of DNA replication in the vicinity of the growing point (31), and (3) the demonstration that mutants in the gene responsible (T4 gene 30) (32) are incapable of producing macromolecular DNA and in its stead accumulate short fragments under nonpermissive conditions, fragments that are then converted to normal DNA under permissive ones (33, 34). Related to this last observation is the fact that simultaneous infection with such mutants and those in the *r*II region also produce viable phage: the *r*II gene produces nicks that are repaired by ligase.

Polynucleotide Kinase. Another interesting enzyme of polynucleotide metabolism is polynucleotide kinase (phosphotransferase)—also originally isolated from bacteriophage infected *E. coli* but known to be present in nuclei of animal cells—which catalyzes the following reaction (29, 35):

$$(\mathrm{HO})\mathrm{N_1pN_2p}\cdots\mathrm{N} + \mathrm{p^*ppA} \xrightarrow{\mathrm{Mg^{2+}}} \mathrm{p^*N_1pN_2p}\cdots\mathrm{N} + \mathrm{ppA} \qquad (20\text{-}4)$$

the transfer of the γ-phosphate residue of ATP to the 5′-hydroxyl terminus of a polynucleotide. Unlike the ligase, this enzyme is relatively nonspecific and will catalyze the phosphorylation of partially degraded bacterial and phage DNA's, tRNA, oligonucleotides, and even nucleoside-3′-monophosphates (yielding nucleoside-3′-5′-diphosphates); the only requirements are a free 5′-hydroxyl and a phosphorylated 3′-hydroxyl group in the substrate. Since this prerequisite is met in many polynucleotides and is easily satisfied even in those that contain a 5′-phosphate residue by simple removal of the latter through the action of the alkaline phosphatase of *E. coli,* the use of the kinase with radioactive ATP as the phosphate donor provides a convenient means for specifically labeling the 5′-terminal nucleotide of a polymer.

Modification Enzymes. The introduction of unusual and rare bases in general occurs subsequent rather than prior to polymerization.[6] While for RNA, especially for tRNA, enzymes capable of introducing a wide variety of such modifications are known to exist, in the case of DNA the only important modification reactions—other than in the glucosylated DNAs of the T-even bacteriophages[7]—involve the transfer of methyl groups to N^6 of adenine and to C^5 of cytosine (36). Two separate and specific enzymes seem to be involved, and the actual extent to which methylation can proceed appears to be controlled in part by the enzyme and in part by the DNA to be methylated (37–39). *S*-Adenosylmethionine is the methyl donor in all the instances so far investigated. The degree, and more specifically the sites, of methylation may fulfill an important recognition function for the biological roles of DNA (e.g., in restriction).

Restriction Enzymes. Many strains of bacteria can recognize and degrade the DNA of related, but not identical strains, introduced, e.g., by a temperate bacteriophage; this phenomenon is known as restriction. Whether a DNA is restricted by a particular host depends on its prior history, that is, the host cell in which it has been allowed to grow *previously* and as a result has had imparted to it certain characteristics—attributes that are called host-controlled modifications (40). For instance, if bacteriophage λ is grown in cells containing the modification gene m_K and then allowed to infect cells of strain K, they

[6] This statement is *not* true for the 5-hydroxymethylcytosine substituting for cytosine in the DNA of the T-even bacteriophage of *E. coli,* and probably not true for those DNA's that contain uracil instead of thymine.

[7] Where UDPG is the donor active with two different glucosyl transferases.

will be able to form progeny; if, on the other hand, the phage is grown on cells that lack m_K (m_K^-), then its DNA will be destroyed in K, *provided* the latter cells in turn contain the product of a restriction gene r_K. Thus, modification by the prior host—which is related or identical to its methylation characteristics—and restriction by the subsequent one are closely linked phenomena. Restriction of λ DNA has now been shown to be due to the presence in *E. coli* K of a specific endonuclease (*endonuclease III-K*), a large protein (M = 390,000), which specifically requires *S*-adenosylmethionine, ATP, and Mg^{2+} for activity. The enzyme binds to susceptible DNA (from phage grown in m_K^- cells), degrades it to large, double-stranded pieces (41), and cleaves ATP in the process to ADP and P_i. While *S*-adenosylmethionine is also required in this process, there is no evidence of its utilization and its role is poorly understood.

Nucleases. Appropriately, nucleases must also be considered within the present context. They may fulfill vital functions, not only in recognizing and excising undesirable DNA or DNA regions (e.g., after exposure to cross-linking or alkylating agents), but also for normal replication and recombination of these molecules (42). In both instances one might conceive first of *incision* at some point close to the locus to be eliminated (or joined) by means of an endonuclease, followed by *excision* of a limited region (including the damage in the first instance) by means of an appropriate exonuclease. Such functions for nucleases are suggested by the following facts: (1) nuclease activity is highest at that point in the growth cycle when DNA synthesis is itself maximal; (2) novel nucleases are induced at a high level subsequent to bacteriophage infection; (3) nucleases appear associated with or form an integral part of various DNA polymerases; and (4) nucleases capable of performing the required reactions in damage control (*repair synthesis*) and recombination are beginning to be isolated and purified.[8]

The nucleases which have been studied in most detail are those derived from *E. coli*. The most important examples are the following:

1. *Exonuclease I* is specific for single-stranded DNA. It acts by progressive removal of pdN's from the free 3′-OH terminus until the penultimate dinucleotide (p)N_1pN_2 is reached. It also catalyzes an attack at the 5′-P end but at an extremely low frequency (~1 in 10^4 scissions).

2. *Exonuclease II* is the enzyme associated with DNA polymerase and has already been discussed in that connection. It acts on double- and single-stranded DNA releasing pdN's starting from a 3′-OH or 5′-OH (or P) terminus. It is inactive on 3′-P termini and does not attack RNA.

3. *Exonuclease III* is completely specific for native double-stranded DNA which it attacks from the 3′-terminus whether free or phosphorylated—the latter is first dephosphorylated since it contains an associated 3′-phosphatase.

[8] For instance, in *M. luteus* we find DNA (irradiated by uv) $\xrightarrow[\text{endonuclease}]{Mg^{2+}}$ single-strand nicks $\xrightarrow[\text{exonuclease}]{}$ excision of regions containing photoproduct (5 nucleotides per nick). The endonuclease (M = 15,000) is completely specific for irradiated duplex DNA, and, although the exonuclease will act on denatured DNA, whether it has been irradiated or not, for it to function on duplex DNA the latter must have been *first* irradiated and *then* be exposed to the endonuclease [J. C. Kaplan, S. R. Kushner, and L. Grossman, *Proc. Natl. Acad. Sci. U.S.*, **63:** 144 (1969)]. Other endonucleases specific for alkylated DNA (endonuclease II, see below) are also known [E. C. Friedberg and D. A. Goldthwait, *Proc. Natl. Acad. Sci. U.S.*, **62:** 934 (1969)]. *E. coli* DNA polymerase can function as a damage-excision exonuclease and then repair the excised region by reinserting the correct nucleotides (18).

It releases pdN's until ~40 percent of the total has been removed, leading to a largely single-stranded structure with a double-stranded core:

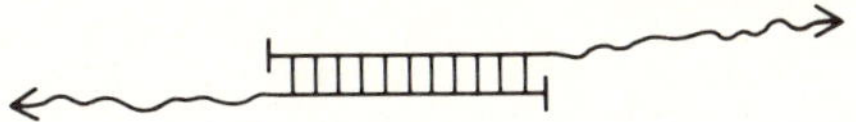

a structure which we know to be an excellent template for DNA polymerase.

4. *Exonuclease IV* is unique in preferring oligonucleotides as substrates which it attacks—yielding dNp's—an order of magnitude more rapidly than it does either native or denatured DNA.

5. *Endonuclease I* can attack anywhere in a double- or single-stranded DNA, producing a 3′-P and a 5′-OH residue in the gap. It is thus a class 2 nuclease, of specificity opposite to that of class 1 enzymes such as pancreatic DNase:

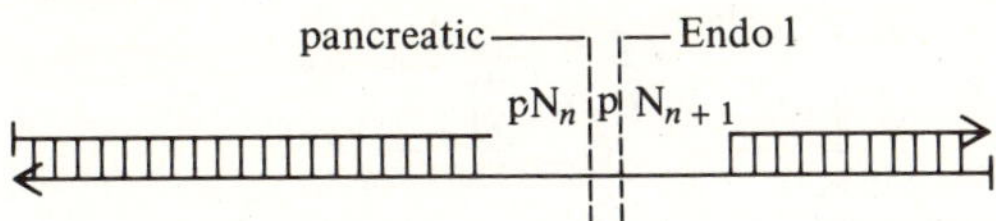

The products are oligonucleotides of structure $pdN(pdN)_n$ with $n = 6$ on the average, since the enzyme cannot cleave oligodeoxynucleotides of chain length <6. Attack on bihelical DNA produces double-stranded breaks. tRNA is a powerful inhibitor of the enzyme, which is localized in the space between bacterial cell membrane and cell wall.

6. *Endonuclease II* appears specific for native bihelical DNA; with phage DNA as substrate it produces single-strand nicks, while with a DNA that has been alkylated—a preferred substrate—it leads to breaks in both strands. It thus may be the incision enzyme required for recognizing and—in conjunction with the appropriate exonuclease—removing this kind of damage.

Phage-Induced Nucleases. Infection by bacteriophage T2 or T4 leads to very rapid destruction of host DNA. This observation is explained by the specific induction of T4-endonuclease II, which introduces single-strand breaks in *E. coli* DNA. This enzyme does *not* attack T4 DNA. Presumably, exonucleases act in conjunction with endonuclease II to degrade the host DNA to pdN's, which are known to be used for phage DNA synthesis. Vegetative growth of phage λ, on the other hand, leads to the induction of a novel exonuclease which produces pdN's but as a result of an attack on the 5′-terminus, preferably of structure pN_1pN——.

Polynucleotide Phosphatases (Polynucleotidases). We have already mentioned the ability of the alkaline phosphatase from *E. coli* to remove phosphate residues from structures such as pN_1——$N_n(p)$ to produce free 5′-OH termini. A similar enzyme, specific for high-molecular-weight substrates, has been isolated from cells infected with phage T4. A more restrictive specificity is exhibited by several enzymes (43). One is a 3′-nucleotidase from uninfected cells, which attacks all substrates of structure $p(d)N_1$——$(d)N_np \xrightarrow{H_2O} p(d)N_1$——$(d)N_n\text{-OH} + P_i$, N_n = A or T. From cells infected with T4 phage, there can also be isolated an enzyme with a similar mode of action but strictly specific for the deoxy series. Another enzyme induced after phage infection attacks the 5′-phosphate ends of RNA and DNA but does not attack 5′-mononucleotides.

Molecular Models

Synthesis of φX DNA in vitro. A brilliant tour de force was reported by Goulian, Kornberg, and Sinsheimer (44) in 1967: the synthesis of infective duplex *replicative form* (RF) and daughter (+) strands of the DNA of bacteriophage ϕX 174—which, we will recall, contains isolated, single-stranded circular DNA. The steps by which this was accomplished are shown in Fig. 20-5. They consisted of:

(a) Synthesis of circular replicative form (RF), containing complementary (−) DNA, using, first, the viral (+) DNA (labeled with ^{3}H) as template, dBTP (instead of dTTP) as one of the four substrates (others labeled with ^{32}P) plus an oligonucleotide primer, followed by action of polynucleotide ligase plus NAD^{+}.

(b) Separation of the (−) strand by limited endonuclease action (one

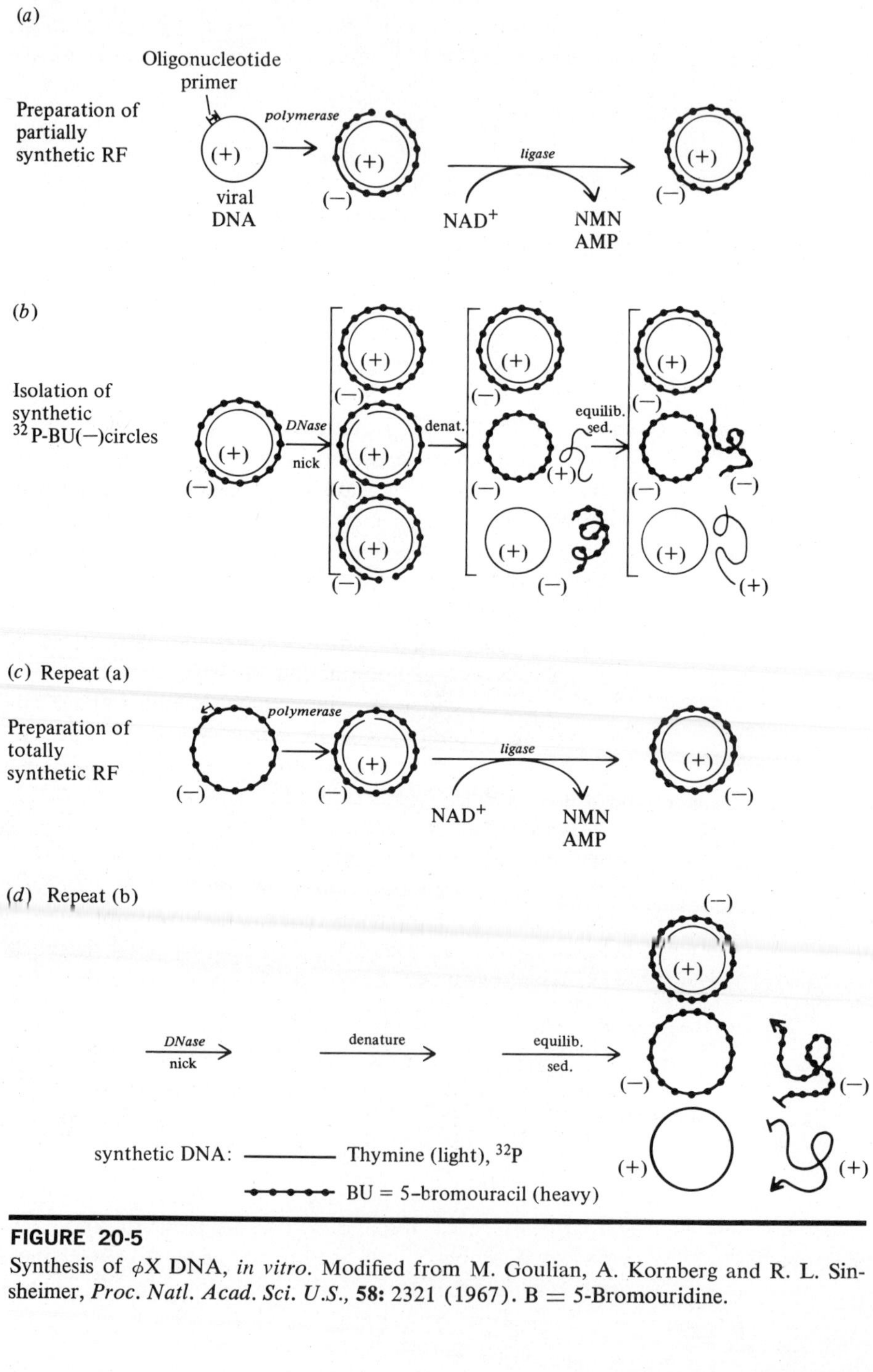

FIGURE 20-5

Synthesis of ϕX DNA, *in vitro*. Modified from M. Goulian, A. Kornberg and R. L. Sinsheimer, *Proc. Natl. Acad. Sci. U.S.*, **58:** 2321 (1967). B = 5-Bromouridine.

scission/two molecules), separation of strands in the nicked DNA by heat, followed by density-gradient sedimentation in CsCl—yielding three peaks: single strands containing B (1.809 g/cm^3); hybrid duplex containing both B and ^{3}H (1.747 g/cm^3) and single strands containing ^{3}H (1.722 g/cm^3).

(c) A repetition of (a) but now using the B-containing (−) daughter strand as template and ^{3}H containing dTTP as substrate. This produced a wholly synthetic hybrid RF.

(d) A repetition of (b) afforded the desired synthetic progeny, ^{3}H-T containing strand. Since the RF as well as the (+) and (−) strands were as infective to *E. coli* spheroplasts as was the starting material, this experiment demonstrated the formation of biologically active DNA *in vitro*. Is this, in fact, the means by which this end is accomplished *in vivo?*

Replication of ϕX DNA in vivo—*The Rolling-Circle Model.* The known facts concerning ϕX replication *in vivo* are (45–48): (1) The parental single-stranded ring of the incoming virus is first converted to a covalent duplex circle (RF), even under conditions where net protein synthesis is blocked (in the presence of chloramphenicol); (2) virus-specific proteins *are* required to permit the RF to replicate semiconservatively to produce additional RF's; (3) the (+) strands are withdrawn from this pool of RF's in a dynamic manner (i.e., coincident with the synthesis of new plus strands), to appear in the progeny virus; and (4) a *replicative intermediate* (RI), larger in size than the (nicked) RF, intervenes in the process subsequent to phase 2.

On the basis of these observations Gilbert and Dressler (48) postulated a rolling-circle model (Fig. 20-6) of DNA replication which has received some experimental support. In particular, it predicts that (1) pulse-labeled RF should make its appearance in the RI such that 50 percent of the radioactivity is in plus strands that are ≧ (parental) unit length, while 50 percent of it is in minus strands ≦ unit length, and (2) that parental (+) strands should become covalently linked to newly synthesized (e.g., B-labeled) DNA within the RI. The key features of the model are:

1. The recognition of a specific sequence(s) by a novel endonuclease specified by the incoming phage DNA, leading to a single nick in the plus strand of the RF.

2. The attachment of the 5′ terminus to a specific site, perhaps on the bacterial cell membrane, thus preventing the nick from being repaired by ligase and freeing the 3′-OH terminal hydroxyl to sequential addition of nucleotides by the polymerase. This process uses the (−) strand as template and continues until a structure is formed containing a circular (−) strand and a complementary (+) strand hydrogen bonded to it, but eventually of greater than unit length (the RI).

3. Initiation of (−) strand synthesis, again starting from the specific initiating region, but now of the (+) strand.

4. Recognition of the specific duplex sequence by the endonuclease, which now liberates the circular progeny from the parental strand. The last two steps are equivalent to recombination, and the endonuclease thus functions now as a recombination enzyme (recombinase). The outcome is the formation of both a progeny RI and a parental RI still attached to the membrane. The progeny RI will continue synthesis for one round, after which it can detach a molecule capable of circularizing to form a RF (i.e., with the circle covalently joined in both strands). Any of the RF's can continue the process; the capsid attaches to and removes the (+) strand of the RI.

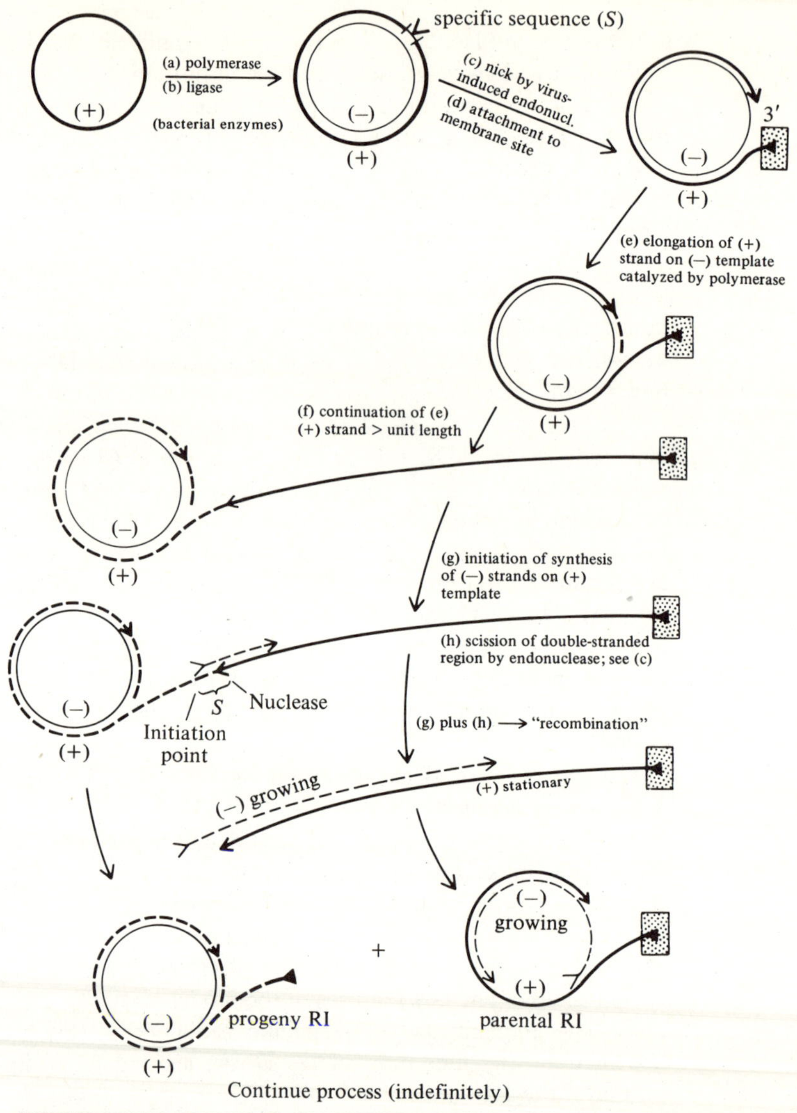

FIGURE 20-6
Rolling-circle model of Gilbert and Dressler.

Formation of Double-Stranded DNA. The rolling-circle model is capable of being extended to include double-stranded DNA and of accounting for many facets of its replication such as, for example, its close association with bacterial membranes, and the occurrence of specific replicative forms, some of them circular (bacteriophage λ), some of them larger than the parental DNA (T-even bacteriophages). However, neither the mechanism of the replication process at or near the growing point nor the general shape of the replicating molecule appear to be entirely consistent with this formulation. In Fig. 20-7 we represent schematically three possible alternatives, all taking account of one key fact: the ability of the isolated DNA-polymerase—the only enzyme known so far to produce template-specified polydeoxynucleotides—to effect chain growth *exclusively* with $5' \rightarrow 3'$ polarity, that is, the accretion of deoxynucleotide units only to the free *3′-OH* terminus of a growing chain.

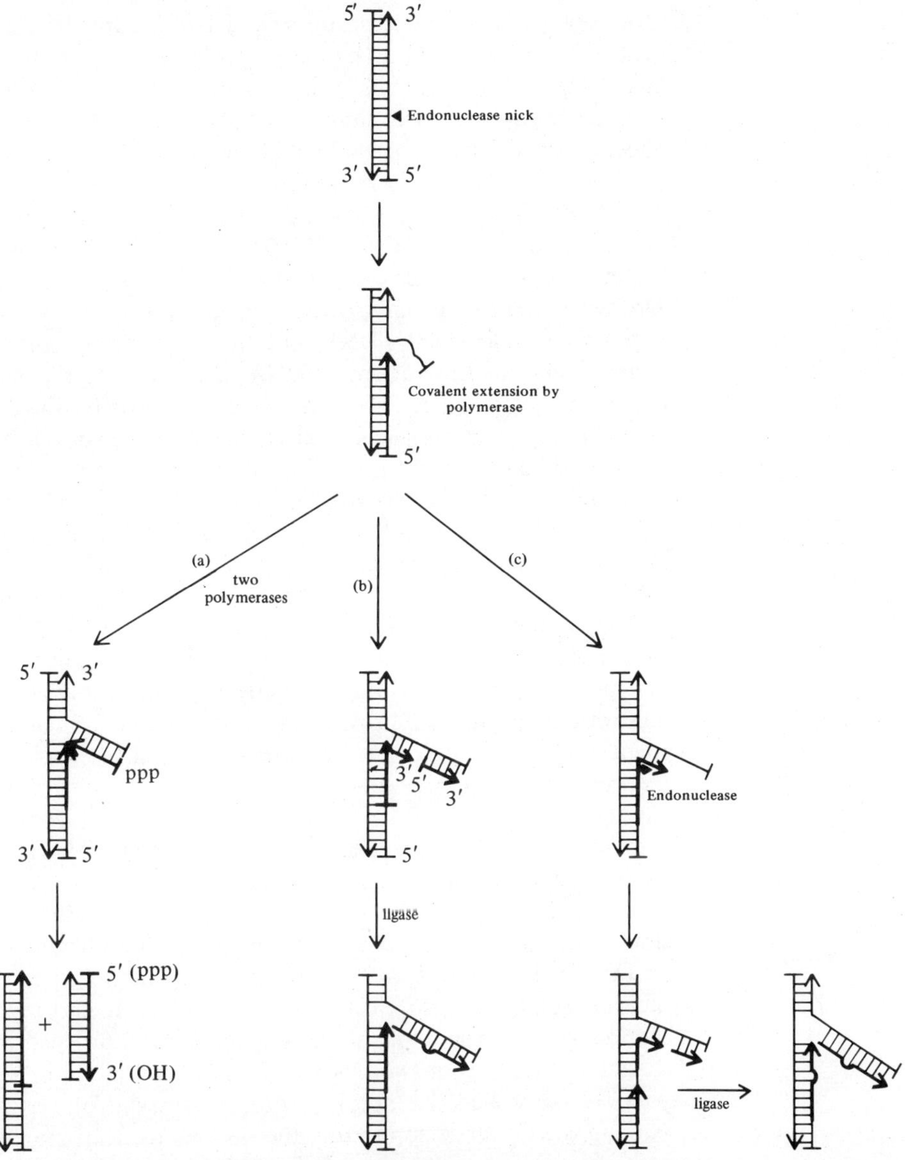

FIGURE 20-7
Replication mechanisms for double-stranded DNA.

Alternative (a) postulates the existence of an enzyme capable of catalyzing chain growth of opposite polarity ($3' \rightarrow 5'$), that is, the reaction

$$\underset{5'}{\text{pppN}_n + \text{pppN}_{n-1}} \text{———} \underset{3'}{\text{pN}_1(\text{OH})} \longrightarrow \text{PP}_i + \text{pppN}_n \text{———} \text{pN}_1(\text{OH})$$

The required chain-propagating end (i.e., a 5′-triphosphate on a polynucleotide) has been synthesized and cannot function in this fashion with the *E. coli* polymerase—thus a novel enzyme is required and so far no evidence for its existence has been forthcoming.[9]

Alternatives (b) and (c) postulate only known enzymatic reactions, with

[9] Another alternative, also found wanting in experimental proof, is attack by a 5′-OH group on the growing chain on the α-phosphate of a nucleoside-3′-triphosphate.

short segments of the opposite strand being copied by the polymerase with the correct polarity. Both (b) and (c) would generate the short, noncovalently linked "Okazaki fragments"—in one strand for (b), in both strands for (c). Another model (48a) is similar to (c) but postulates a covalent (e.g., pyrophosphate) link between, and simultaneous addition to, both chains.

Currently evidence is beginning to accumulate, however, which indicates that *both* strands of T4, λ, and other DNA's are synthesized in a *discontinuous* manner after pulse labeling (7–10). The initial findings and interpretations by Okazaki (7) served to reconcile such data with the polarity of Kornberg's DNA polymerase (chain growth at the 3′-OH end) by virtue of the identification of multiple small pieces, but they introduced another problem in the form of reiterated initiations of DNA chains during replication. This initiation of new chains is clearly a reaction *not* catalyzed by DNA polymerase. It has been suggested that the Okazaki fragments arise only because of the marked fragility of the growing point. However, this is not a likely explanation, because DNA pieces of this size (6 to 8 S) are relatively stable and can otherwise be isolated without undue difficulty, and thus selective lability would appear to be an *ad hoc* assumption. The ultimate role of these fragments in DNA synthesis will become clear once the actual steps in this process have become better understood. The recent isolation by De Lucia and Cairns (49) of a mutant markedly deficient in its activity of Kornberg polymerase *in vitro* and the mapping of this mutant (50) suggest that deletions in the *structural* gene of the enzyme will provide a definite answer to the question of its involvement in DNA synthesis. Other experiments suggest that it is essential in repair but *not* replication.

Molecular Events in Recombination and DNA Repair. The enzymatic steps responsible for recombination of parental DNA molecules and for the excision and repair of damaged regions within the macromolecule may be identical, or at least analogous, to the ones responsible for the replication of double-stranded DNA. A scheme incorporating current thought is presented as Fig. 20-8. DNA repair will be quite similar in its requirements (51) for incision (endonuclease), excision (exonuclease), and repair synthesis (polymerase) (52), except that all these reactions will proceed within a single rather than a matched set of two DNA molecules, and incision, and perhaps excision, will be controlled by specific, damage-recognizing enzymes (page 896). Among the evidence supporting the notions of deep-seated analogies between recombination and repair are the observations that conditions that require extensive repair (e.g., irradiation by uv) favor recombination, and conversely that mutants defective in ability to recombine (*rec*$^-$) are also particularly prone to radiation damage (or an inability to repair the latter), and vice versa.

Interesting within this context is also the presumed mode of action of antibiotics related to mitomycin C,

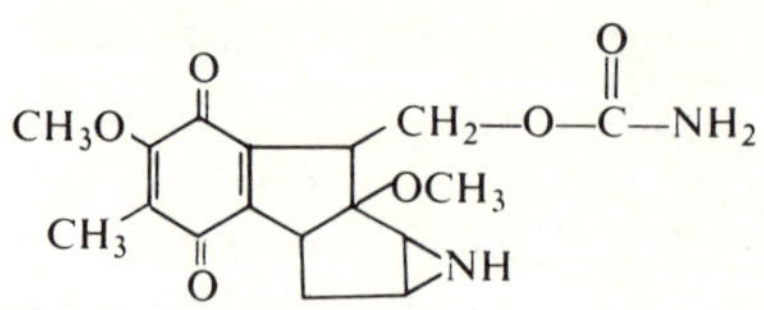

CH_3O ... CH_2—O—C(=O)—NH_2 ; OCH_3 ; CH_3 ; NH

Mitomycin C

which appear to block DNA synthesis irreversibly in susceptible organisms. Mitomycin C's antibacterial potency may be appreciated from the fact that

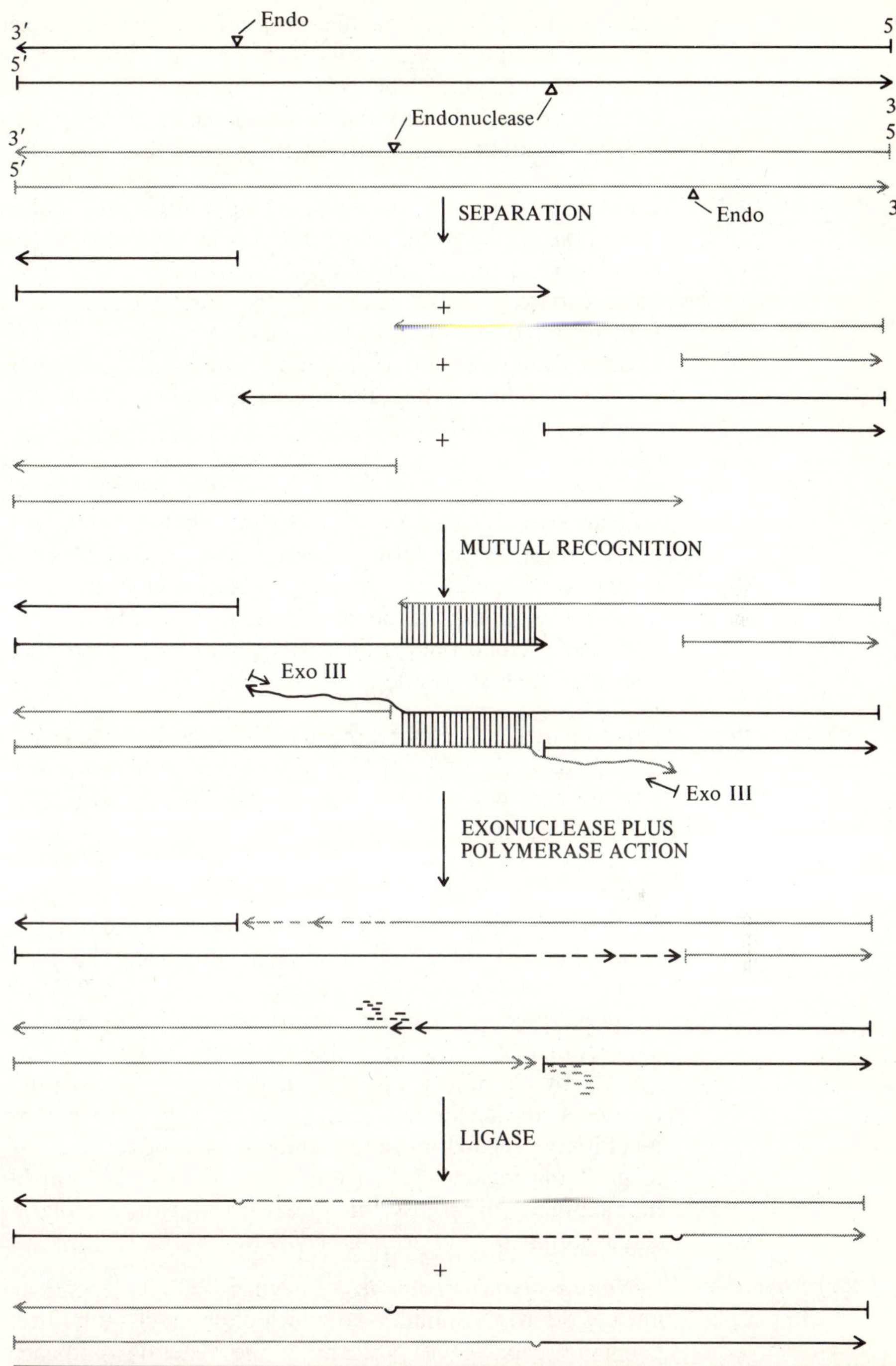

FIGURE 20-8
Unit steps in a hypothetical model for recombination. For the sake of clarity hydrogen bonds are only indicated for the heteroduplex portions of the recombinant molecules (containing one heavy-lined strand and one light-lined strand).

only 10^{-4} percent of *B. subtilis* cells survive a 15-min exposure to 12 μg C/ml. The principal intracellular consequences of administration of this material are inhibition of DNA synthesis and a progressive and extensive fragmentation of the macromolecule. These results appear to be only secondary consequences of the action of a metabolite of the agent capable of alkylating and cross-linking DNA through susceptible guanine residues. The first result is thus the formation of covalently cross-linked "reversible" DNA incapable of strand separation and hence of replication; this is then followed by a most effective degradation of the macromolecule to acid soluble fragments; presumably brought about by the damage-removing enzymes.

RNA-DEPENDENT DNA POLYMERASE

Certain oncogenic (cancer-producing) viruses of animals (responsible for formation of leukemias and sarcomas in a number of species) are known to contain RNA in the isolated virion. However, they appear to be capable of being carried and reproduced by the host cell in a relatively integrated form (analogous to that of temperate phages) even in the absence of RNA synthesis (52a). Very recent work has now provided evidence that these viruses are capable of inducing the following series of events (52b–e):

$$\text{RNA}(+) \xrightarrow[\textit{DNA}\text{ polymerase}]{\textit{RNA}\text{-dependent}} \text{RNA}(+)\cdot\text{DNA}(-) \rightarrow \text{DNA}(+)\cdot\text{DNA}(-) \xrightarrow[\text{polymerase}]{\text{DNA-dependent}} \text{replication and integration}$$

At least two novel enzymatic activities (indicated above) appear to be specified by the invading virus, of which the first, a *RNA-dependent DNA polymerase* is of special interest. In a reaction that appears to be unique to this system it catalyzes the polymerization of the Mg salts of the four deoxyribonucleoside triphosphates on a RNA template to produce the complementary hybrid as the first product.

BIOSYNTHESIS OF RNA

Situation *in vivo*

Some of the facts concerning RNA synthesis on a DNA template have been discussed in Chapter 19, and additional ones bearing on its specific regulation are considered in Chapter 21. For our purposes here the following summary will suffice.

The synthesis of all cellular RNA (i.e., all except that resulting in replication of RNA viruses—see below) utilizes a DNA template. It is asymmetrical and strand selective; only one strand is transcribed, but the transcribing strand may be switched for different regions. The polarity is constant and chain growth is always 5′ → 3′ (i.e., as just described for DNA polymerase). In procaryotic organisms the transcription of certain regions (specifying clusters or operons of inducible or derepressible enzymes) is regulated qualitatively by the interaction of the products of regulatory genes with control regions at the beginning of the operon called operator regions; however, the quantitative regulation of the amount of product to be formed devolves on a neighboring region—called the promoter—which controls the efficacy of the interaction of the transcribing enzyme with the template so that transcription can be initiated (and propagated).

RNA Polymerases

Nature of the Reaction. Enzymes (DNA-dependent RNA polymerases; nucleoside triphosphate: polynucleotide nucleotidyl transferases, DNA dependent) capable of performing the reactions indicated above have been isolated from a number of bacteria (e.g., *E. coli, M. luteus,* and *Azotobacter vinelandii*), from yeast, and from animal sources (53–57). All of them catalyze the following principal reaction:

$$\begin{matrix}\text{nATP}\cdot\text{Me}\\+\\\text{oUTP}\cdot\text{Me}\\+\\\text{pGTP}\cdot\text{Me}\\+\\\text{qCTP}\cdot\text{Me}\end{matrix} \underset{}{\overset{\text{DNA*}}{\rightleftharpoons}} \text{RNA}\begin{bmatrix}\text{A}_n\\\text{U}_o\\\text{G}_p\\\text{C}_q\end{bmatrix} + (\text{n}+\text{o}+\text{p}+\text{q})\text{PP}_i\cdot\text{Me} \qquad (20\text{-}5)$$

$$^*[\text{dAMP}_o, \text{dTMP}_n, \text{dCMP}_p, \text{dGMP}_q]$$

The similarity of this reaction to that catalyzed by the DNA polymerase [Eq. (20-1)] is obvious.

1. The nucleoside triphosphates are again the substrates.

2. The reaction is reversible in the sense that there is a rapid exchange reaction between labeled PP_i and ribonucleoside triphosphates. However, while limited pyrophosphorolysis of RNA–DNA hybrids can be demonstrated, there is no detectable pyrophosphorolysis of free RNA.

3. The reaction requires the addition of a divalent metal ion. Mg^{2+} or a 4 : 1 mixture of Mg^{2+} and Mn^{2+} is usually employed. If Mn^{2+} is used alone, the strict specificity for ribonucleoside derivatives breaks down.

4. The reaction requires a template. The template usually employed is double-stranded DNA; however, other double-stranded polynucleotides, such as poly d(A-T), single-stranded DNA, and, particularly in the presence of Mn^{2+}, double- and even single-stranded RNA, can function in this fashion, at least with some of the enzymes (e.g., from *A. vinelandii*). In the complete absence of a template, poly rA · poly rU and poly(rI, rC) are produced.

5. The nature of the template determines that of the substrates required: With ordinary polydeoxynucleotides containing all four bases, the four ribonucleoside triphosphates (with UTP to match dT) are required; other polymers will direct the incorporation of the complementary one, two, three, or four triphosphates.

6. Direction and mode of chain growth is $5' \rightarrow 3'$, that is, the addition of a nucleotide unit to the 3′-OH of the growing chain.

7. Unlike DNA polymerase, the product does carry the expected 5′-triphosphate terminal, $pppN_1$——N_n.

Nature of the Product—Regulation by Enzyme Subunits. One of the most characteristic feature of transcription *in vivo* and of RNA synthesis using natural DNA templates *in vitro* as well—at least so it appears with the right kind of enzyme and template—is that only one of the two strands of the duplex serves as a template. Clearly there exist features that distinguish the two strands (or at least selected regions), and the enzyme can recognize these features and reject the strand (or the region) that is not to be transcribed. What are they? Work by Szybalski and his collaborators (58) has shown that the structure on the DNA to be recognized by the enzyme are clusters rich in pyrimidines (usually C, sometimes T). The test system and the one used for the isolation of the strand containing these clusters (after denaturation, if the strand originally formed part of a duplex) is provided by the preferential interaction of dC-rich regions[10] with poly rG, poly rI, poly(rI, rG), or poly (rU, rG). The resultant hybrid can be isolated and later dissociated. Different DNA's have been shown to contain different distributions of the clusters. For example, bacteriophage T3, T7, and *B. subtilis* contain only dC clusters, all in one strand; bacteriophage T2 contains exclusively dT clusters, predominantly on one strand; *E. coli* DNA contains dC- and dT-rich regions in a 2 : 1 ratio and twice as many of these clusters on one strand than on the other. Such strands or regions rich in dC appear to be the ones transcribed *in vivo* [called variously "W" (Watson), (−), or *l*] (tested by hybridization with pulse-labeled mRNA) and, with, for example, *B. subtilis* phage α, T7, and λ DNA, *in vitro* as well. Synthetic polymers have been used to provide additional confirmation of the hypothesis; poly dG · poly dC and poly d(T-C) · poly d(G-A) bind the polymerase to, and are predominantly transcribed from, the pyrimidine-containing strand. The pyrimidine-rich clusters occur

[10] dT-rich strands are isolated by allowing the complementary strand of the duplex (dA rich) to interact with poly rU, leaving the desired strand behind.

about once every 10^4 nucleotide pairs and are probably of the order of 20 to 50 nucleotides long. Other experiments suggest that the number of initiation sites on native DNA's is severely restricted, that the nature of the initiating base (G or A) in the product is correlated positively with the relative proportion of the dC + dT in the template molecule, and that these restrictions are removed on denaturation (54). Taken together all these observations add up to a potent inference that the pyrimidine-rich clusters and the promoter identified by genetic experiments have much in common and may be one and the same. Furthermore, a mutation (C17) in (y), a promoter region of phage λ, changes the rate of RNA synthesis when this DNA rather than the wild type is used as a template *in vitro* (59). Although providing a site of interaction with and recognition by the enzyme, pyrimidine-rich clusters are probably not extensively transcribed themselves—the initiating purine terminus $pppN_1$ is seldom followed by another purine in the subterminal position.

How does the enzyme recognize the promoter? RNA polymerases are large molecules ($M \simeq 5 \times 10^5$). When purified by relatively mild procedures (glycerol gradients, DNA columns), they can be shown to be made up of two main parts: a basic unit, the *core enzyme*—itself consisting of three different polypeptide subunits (α, β, and β', probably in the proportion $\alpha_2\beta\beta'$)—and a relatively easily detached entity of $M = 1 \times 10^4$, called σ factor, which can be separated and isolated on phosphocellulose columns (60). It is this factor, or rather members of this group of factors—for different ones have been identified and isolated from enzymes obtained from uninfected cells (*E. coli, A. vinelandii,* etc.) and from cells infected with various bacteriophages—that is of special interest. Not only do all of them stimulate *in vitro* transcription by the core enzyme, but more importantly they appear to be responsible for imparting specificity and selectivity to this process. For instance, while the core enzyme will transcribe, albeit slowly, both strands of all double-stranded DNA's tested, the holoenzyme with σ from uninfected cells of *E. coli* can transcribe only a limited portion of one strand of T4 DNA ("very early" or prereplicative RNA). If it is replaced by a σ from T4-infected cells, additional regions of this template become amenable to transcription and the sequence and the nature of events appears identical to that observed *in vivo* (61,62). Analogous observations have been reported for other systems (bacteriophages T7[11] and λ—see Fig. 19-12) and regulation by σ factors appears to be the rule in these instances. Thus, σ factors can provide a coarse positive control on RNA (and hence on protein synthesis), and one might conceive of a model of sequential "feed-forward" regulation of processes that require the sequential turning on or turning off of genes to proceed by the synthesis of a series of σ factors with ever-increasing efficiency by binding to the core enzyme and recognizing the next promoter in the series:

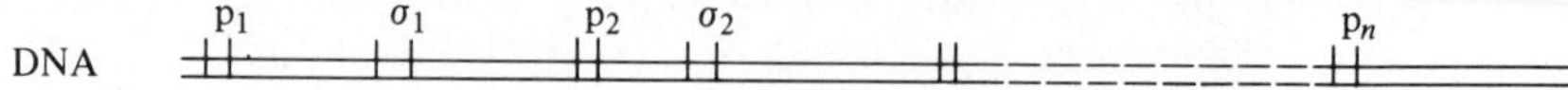

Core enzyme recognizes $p_2, p_3 \cdots p_n$ if and only if σ_1 has been formed as a result of transcription and translation of the region controlled by p_1; p_3, etc., are recognized only if σ_2 has been formed; and so on. Such specificity of mutual recognition of σ and promoter of course implies that there be inherent differences in pyrimidine-rich tracts or in regions immediately adjoining.

Steps in the Reaction Sequence. The initial steps in the reaction sequence appear to be

[11] Control in this case, however, appears to require a novel polymerase.

$$\text{Enz (core)} + \sigma + \text{DNA} \rightleftharpoons \text{Enz}\sigma\cdot\text{DNA (at promoter site)} \quad \underset{3' \longleftarrow\!\!-\!\!-\!\!-\!\!- 5'}{} \qquad \textit{Enz-template complex} \qquad (20\text{-}5a)$$

$$\text{Enz}\sigma\text{-DNA} + \text{N}_1\text{TP} \overset{Me^{2+}}{\rightleftharpoons} \text{Enz}\sigma \begin{cases} \text{prom } \overline{\text{N}}_1\overline{\text{N}}_2 \text{———} \overline{\text{N}}_n \\ \text{pppN}_1 \end{cases} \quad 3' \longleftarrow 5' \qquad (20\text{-}5b)$$

Initiation complex ($N_1 = $ A or G; $\overline{N}_1 = $ T or C)

$$\text{Enz}\sigma\text{-DNA} + \text{N}_2\text{TP} \rightleftharpoons \text{Enz} \begin{cases} \text{prom } \overline{\text{N}}_1\overline{\text{N}}_2 \text{———} \overline{\text{N}}_n \\ \text{pppN}_1\text{pN}_2 \quad 5' \longrightarrow \end{cases} + \sigma \qquad (20\text{-}5c)$$

Translocation and propagation

etc., etc.

Termination

$$\text{Enz} \begin{cases} \overline{\text{N}}_{n-1} \\ \text{N}_{n-1} \end{cases} + \text{N}_n\text{TP} \rightleftharpoons \text{Enz} \begin{cases} \overline{\text{N}}_n \\ \text{N}_n \end{cases} \rightleftharpoons \text{Enz} + \text{DNA} + \text{RNA} \qquad (20\text{-}5d)$$

(with pppN_1——— chains)

Recently an additional protein called rho (ρ) has been reported (63) to play a role in termination. The precise function in bringing this about is not clear at present, but obviously a correct means of termination is as necessary as one for initiation.

Specific Inhibitors. Three types of inhibitors have found favor in specifying defined steps in the sequence just discussed:

1. *Intercalating dyes,* of which the acridines are prime examples (see Chapter 5), inhibit chain initiation but are not very specific with respect to the nature of primer affected[12] and they act in an analogous fashion with DNA polymerase; high salt concentration has a similar effect.

2. *Actinomycin D* (also called meractinomycin or actinomycin C1, page 908) is effective with double-helical DNA only; RNA-primed reactions are unaffected. Inhibitory action has been localized on chain propagation (i.e., step 5c) beyond the first four nucleotides; that is, recognition initiation, and pyrophosphate exchange are much less affected. The inhibitor is believed to be bound to the DNA adjacent to guanine residues in such a fashion as to interfere with translocation of the enzyme down a G-containing template; synthesis on poly d (A-T) or poly r(A-U) templates is *not* inhibited.

3. The antibiotics of the *rifamycin* group (rifamycin itself, rifampicin, rifampin, streptovaricin) inhibit DNA-dependent transcription *in vivo,* with polymerases of procaryotic but not those of nuclear, eucaryotic origin (64). The reaction involves specific binding to one of the subunits of the core enzyme. The resultant complex is then incapable of catalyzing the initiation of RNA chains.

Eucaryotic Polymerases. Nuclei of eucaryotic cells contain several (usually three) different RNA polymersases. They are distinct in their localization, their requirements, the nature of their templates and products, and in the inhibitors to which they are susceptible (65). Polymerase I is localized in the nucleolus, is active at low ionic strength, and is primarily concerned with the transcription of the genes specifying the larger (45 S) ribosomal, precursor RNA (page 927). Polymerase II and III are both localized in the nucleoplasm. Polymerase II is similar in size and subunit structure to the bacterial enzymes; it is optimally active with Mn^{2+} and requires relatively high ionic

[12] Except for ethidium bromide, which is reported to block transcription of mitochondrial DNA without affecting that of nuclear DNA in eucaryotic cells.

H H
O CH_3—C—CH_3 CH_3—C—CH_3 O
C C—H H—C C
N—CH_3 N—CH_3
sarcosine sarcosine
O L-proline L-proline O
D-valine D-valine
C=O C=O
CH_3—HC C—H H—C CH—CH_3
NH NH
C=O C=O
N NH_2
O O
CH_3 CH_3

strength for activity. It appears to be capable of and responsible for the transcription of most chromosomal genes. The function of Polymerase III remains to be established but it may be concerned with the synthesis of tRNA. None of the three enzymes are inhbiited by rifampicin, and while the susceptibility of their reactions to actinomycin is variable, only Polymerase II is strongly and specifically ($K_i \sim 10^{-8}$ M!) blocked by *α-amanitin* a bicyclic octapeptide isolated from the poisonous mushroom *Amanita phalloides.*

Other Enzymes Catalyzing Formation of Polyribonucleotides

RNA-Specific RNA Polymerases. The infection of a cell by an RNA-containing virus, or of a spheroplast or plant tissue by viral RNA, leads to the production of progeny RNA molecules by means of a reaction sequence that *does not* involve DNA. A polymerase responsible for RNA-dependent RNA synthesis (replicase) (66,67) that has been sufficiently stable to permit extensive purification has been obtained from cells of *E. coli* (male, i.e., Hfr or F^+) infected with the small RNA phage Qβ. The enzyme consists of four subunits [M = 72,000 (α); M = 67,000 (β); M = 45,000 (γ); and M = 35,000 (δ)]. Only β is coded for by the phage RNA (67a). Unlike all the other enzymes so far discussed, it is highly specific with regard to the templates tolerated. These are viral Qβ RNA [(+) strand]; the complementary RNA [(−) strand]; a small, but still self-reproducing "variant" of Qβ RNA that is produced by repeated (seventy-five) transfers of RNA replicating *in vitro* to new reaction mixtures (67); a so-called "Qβ 6S RNA," which can be isolated from cells infected with Qβ; and poly r(C) or other rC-containing polymers, but here the reaction is strictly limited to the formation of the complementary strand. Most interestingly, other RNA's, double- or single-stranded, *including* those derived from other RNA-containing bacteriophages, are completely ineffective.

On the other hand, provided the appropriate supplements are added, the polymerase, in the presence of Qβ RNA as a template, produces progeny (+) RNA, identical to the template in amounts far exceeding that of the template added. The two overall reactions required are

$$\underset{(M = 1.1 \times 10^6,\ 78S)}{Q\beta(+)\text{RNA} + 4n\text{NTP}} \xrightarrow[\text{Factors I and II}]{\text{polymerase, Mg}^{2+}} \underset{\substack{(\text{replicative form} = \text{RF}) \\ (M = 2.2 \times 10^6,\ 13\ S)}}{\text{duplex } [(+) \cdot (-)] + n\text{PP}_i} \tag{20-6}$$

$$\text{RF} + 4n\text{NTP} \xrightarrow[\text{no factors}]{\text{polymerase, Mg}^{2+}} \text{RF} + Q\beta(+)\text{RNA} + n\text{PP}_i \tag{20-7}$$

In detail (20-6) consists of the following partial reactions:

1. *Binding* of polymerase to RNA, with one enzyme bound per template molecule; it is at this step that the enzyme recognizes its cognate template, presumably specified by the 3′-OH terminal sequence, which is

3′ 5′
(OH)-ACCC——Gppp

2. *Chain initiation.* This step requires the presence of GTP plus ATP as well as the intervention of two polypeptides[13] that have been isolated from uninfected *E. coli* (68). Factor I (M = 80,000) is heat resistant and binds to Qβ RNA; Factor II (M = 40,000) is heat- and protease-sensitive and is known to modify the template RNA. It is probably a (specific) exonuclease which removes the 3′-terminal A, a step that may actually precede the interaction of the (modified RNA) with Factor I, which in turn primes it for chain propagation. It had already been established previously that the terminal A residue is not essential for the subsequent reactions but that the subterminal C residue is.

3. *Chain propagation* to form the complementary (−) strand of structure pppGpGpG——pC(OH) on the (+) template, with the duplex the final product

```
5'                          3'
pppGpGpG ——————————————— pC(OH)(−) }
HO-CpCpC ——————————————— pGppp(+)  } RF
3'                          5'
```

4. The reaction shown as (20-7) then procedes in the *opposite direction* but with *identical polarity* by way of one or more growing (+) strands on the (+)·(−) template. This process leads to the final structure pppG——pCpCpC, and one of the remaining questions is the problem of how the terminal pA found in the majority of the molecules isolated from the virus is put on. Another question concerns the nature of the replicative intermediates (RI), that is, the presumptive complexes between RF and growing (+) strands (partially single-stranded structures with s = 16 to 23 S). Two possible structures are one in which the course of synthesis the progeny strands continuously displace the parental (+) strands, and another in which the progeny simply copies the (−) strands within the (presumably locally denatured) duplex.

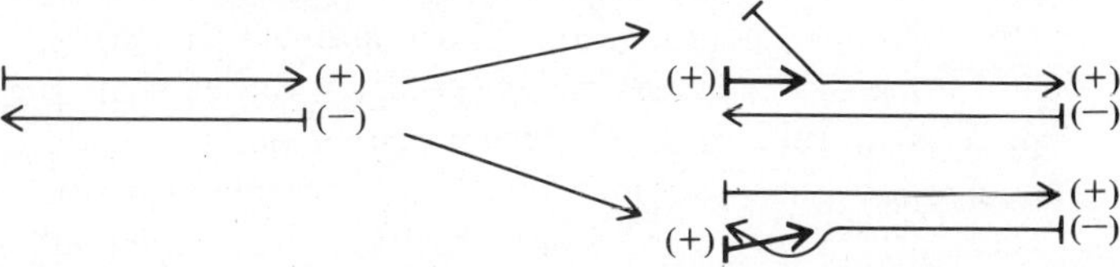

Polynucleotide Phosphorylase. The first enzyme discovered to be capable of catalyzing the formation of polyribonucleotides was *polynucleotide phosphorylase* (nucleoside diphosphate : polynucleotide nucleotidyl transferase),

[13] Factors are *not* necessary for any of the alternative templates.

which was isolated from extracts of *Azotobacter vinelandii* by Grunberg-Manago and Ochoa (69). This enzyme catalyzes the formation of RNA from the nucleoside diphosphates:

$$\begin{matrix} \text{nADP} \\ + \\ \text{oGDP} \\ + \\ \text{pCDP} \\ + \\ \text{qUDP} \end{matrix} \underset{}{\overset{Mg^{2+}}{\rightleftharpoons}} \begin{bmatrix} A_n \\ G_o \\ C_p \\ U_q \end{bmatrix} + (n + o + p + q)P_i \tag{20-8}$$

The principal properties of this reaction include the following: (1) The enzyme utilizes the ribonucleoside diphosphates; however, deoxyribonucleoside diphosphates can also be polymerized. (2) Magnesium ion is required. (3) The reaction is readily reversible. Polynucleotide phosphorylase catalyzes the extensive phosphorolysis of RNA in the presence of high concentrations of inorganic phosphate. (4) The enzyme catalyzes both the formation of homopolymers and mixed polymers [i.e., poly A, poly U, poly (A, G, U, C), etc.]. (5) The reaction is greatly stimulated by the presence of primers (69). The primer effects on polynucleotide phosphorylase–catalyzed polymerizations are extremely complicated. Primers with a free 3′-hydroxyl apparently serve as initiators for chain elongation, and no base specificity is exhibited for this reaction (pApApA will prime for the formation of poly A, poly U, and poly (A,G,U,C), for example); the primer is incorporated into the polymeric material in a 5′-terminal position. On the other hand, oligonucleotides with a terminal 3′-phosphate group and polynucleotides stimulate this reaction, although neither are incorporated into the product. The mechanism involved in this kind of priming action is not understood. It does seem clear that such primers are not functioning as templates, since the base composition of the product resembles that of the mixture of nucleoside diphosphate starting materials more closely than that of the primer. The order of addition of monomeric units to the growing polymer is apparently random.

The product of the polynucleotide phosphorylase reaction is a linear polymer of ribonucleotide units linked by 3′-5′-phosphodiester bridges, as revealed by chemical and enzymatic degradation. The *primary* structure of the polymeric product of the reaction is, hence, identical with that of naturally occurring polynucleotides.

The biological function of polynucleotide phosphorylase remains obscure. It appears certain that this enzyme is *not* responsible for synthesis of the various types of cellular RNA, because (1) at the concentrations of nucleoside diphosphates and inorganic phosphate which exist intracellularly, the equilibrium position favors phosphorolysis rather than polymerization, and (2) there is no relationship between primer and product structure. Perhaps polynucleotide phosphorylase serves as a mechanism for the destruction of unnecessary RNA, creating simultaneously the substrates, ribonucleoside diphosphates, for formation of deoxyribonucleosides and subsequent DNA synthesis (Chapter 19).

Terminal Ribonucleotidyl Transferases. Two different classes of enzymes of this type can be distinguished. One catalyzes the reversible addition of the common 3′-terminal trinucleotide sequence (pCpCpA) of *all* tRNA's (70):

$$tRNA_{n-3} + 2CTP + ATP \overset{Me^{2+}}{\rightleftharpoons} tRNA_{n-3}pCpCpA + 3PP_i \qquad (20\text{-}9)$$

The other catalyzes the terminal addition of nucleotides to initiating 3′-poly- or oligonucleotides to produce (principally) homopolymeric ribonucleotide extensions:

$$N_m\text{-OH} + n\text{NTP} \overset{Me^{2+}}{\rightleftharpoons} N_m(\text{pN})_n + n\text{PP}_i \qquad (20\text{-}10)$$

Enzymes varying widely in specificity and origin have been reported, some catalyzing principally the formation of poly rA (71, 72) and others those of poly rU or poly rC, some with a less restricted specificity (73). All have in common the inability to form poly rG or poly rG containing heteropolymers and a complete absence of any information concerning their possible significance *in vivo,* although poly rA at least has recently been found to be present in nuclei of Ehrlich ascites carcinoma cells (74).

REFERENCES

Books and Reviews

Arber, W., and S. Linn: DNA Modification and Restriction, *Ann. Rev. Biochem.,* **38:** 467 (1969).

Brockman, R. W., and E. P. Anderson: Biochemistry of Cancer, *Ann. Rev. Biochem.,* **32:** 463 (1963).

Chargaff, E.: *Essays on Nucleic Acids,* Elsevier, New York, 1963.

Chargaff, E., and J. N. Davidson (eds.): *The Nucleic Acids,* vols. I, II, 1955, and III, 1960, Academic Press, New York.

Cold Spring Harbor Symposium on Quantitative Biology: *The Genetic Code,* vol. 31, 1966.

Cold Spring Harbor Symposium on Quantitative Biology: *Replication of DNA in Microorganisms,* vol. 33, 1968.

Cold Spring Harbor Symposium on Quantitative Biology: *Synthesis and Structure of Macromolecules,* vol. 28, 1963.

Geiduschek, E. P., and R. Haselkorn: Messenger RNA, *Ann. Rev. Biochem.,* **38:** 647 (1969). Presents an excellent review of RNA synthesis in general.

Georgiev, G. P.: The Nature and Biosynthesis of Nuclear Ribonucleic Acids, *Progr. Nucleic Acid Res. Mol. Biol.,* **6:** 259 (1967).

Grossman, L., and K. Moldave (eds.): *Methods in Enzymology—Nucleic Acids,* vol. XIIB, Academic Press, New York, 1968, especially sec. VIII, Modification of Nucleic Acids, sec. IX, Nucleic Acid Synthesis—in *Vitro,* and sec. X, Nucleic Acid Synthesis—*in Vivo.*

Grunberg-Manago, M.: Enzymatic Synthesis of Nucleic Acids, *Ann. Rev. Biochem.,* **31:** 301 (1962).

Hershey, A. D.: in D. E. Green (ed.), *Currents in Biochemical Research,* Wiley (Interscience), New York, 1956, p. 1.

Howard-Flanders, P.: DNA Repair, *Ann. Rev. Biochem.,* **37:** 175 (1968).

Howard-Flanders, P., and R. P. Boyce: DNA Repair and Genetic Recombination: Studies on Mutants of *Escherichia coli* Defective in these Processes, *Radiation Res.,* **6:** Suppl., 156 (1966).

Karnofsky, D. A., and B. D. Clarkson: Cellular Effects of Anti-Cancer Drugs, *Ann. Rev. Pharmacol.,* **3:** 357 (1963).

Lark, K. G.: Initiation and Control of DNA Synthesis, *Ann. Rev. Biochem.,* **38:** 569 (1969).

Lodish, H. F.: The Replication of RNA-containing Bacteriophages, *Progr. Biophys. and Mol. Biol.,* **18:** 287 (1968).

Maaloe, O., and N. O. Kjeldgaard: *Control of Macromolecular Synthesis,* Benjamin, New York, 1966.

Meselson, M.: The Molecular Basis of Genetic Recombination, in R. A. Brink (ed.), *Heritage from Mendel,* University of Wisconsin Press, Madison, Wis., 1967, p. 81.

Perutz, M. F.: *Proteins and Nucleic Acids: Structure and Function,* Elsevier, New York, 1962.

Richardson, C. C.: Enzymes in DNA Metabolism, *Ann. Rev. Biochem.,* **38:** 795 (1969).

Stavis, R. L., and J. T. August: The Biochemistry of RNA Bacteriophage Replication, *Ann. Rev. Biochem.,* **39:** 527 (1970).

Stevens, A.: Ribonucleic Acids—Biosynthesis and Degradation, *Ann. Rev. Biochem.,* **32:** 15 (1963).

Taylor, J. H. (ed.): *Molecular Genetics,* Academic Press, New York, Part I, 1963; Part II, 1967.

Thomas, C. A., Jr.: Recombination of DNA Molecules, *Progr. Nucleic Acid Res. Mol. Biol.,* **5:** 315 (1966).

Weissmann, C., and S. Ochoa: Replication of Phage RNA, *Progr. Nucleic Acid Res. Mol Biol.,* **6:** 353 (1967).

Plus the series Progress in Nucleic Acid Research and Molecular Biology, Academic Press, New York, with one or more volumes annually since 1963; see also the references at the end of Chapters 5 and 21.

Specific References

1. Jacob, F., S. Brenner, and F. Cuzin: *Cold Spring Harbor Symp. Quant. Biol.,* **28:** 329 (1963).
2. Watson, J. D., and F. H. C. Crick: *Nature,* **171:** 737 (1953).
3. Meselson, M., and F. W. Stahl: *Proc. Natl. Acad. Sci. U.S.,* **44:** 671 (1958).
4. Nagata, T. and M. Meselson: *Cold Spring Harbor Symp. Quant. Biol.,* **33:** 553 (1968).
5. Cairns, J.: *Cold Spring Harbor Symp. Quant. Biol.,* **28:** 43 (1963).
6. Yoshikawa, H. and N. Sueoka: *Proc. Natl. Acad. Sci. U.S.,* **49:** 559 (1963); Sueoka, N. and H. Yoshikawa: *Genetics,* **52:** 747 (1965); O'Sullivan, A. and N. Sueoka: *J. Mol. Biol.,* **27:** 349 (1967).
7. Pato, M. L., and D. A. Glaser: *Proc. Natl. Acad. Sci. U.S.,* **60:** 1268 (1968).
8. Sakabe, K., and R. Okazaki: *Biochim. Biophys. Acta,* **129:** 651 (1966); Okazaki, R., T. Okazaki, K. Sakabe, K. Sugimoto, and A. Sugino: *Proc. Natl. Acad. Sci. U.S.,* **59:** 598 (1968).
9. Taylor, J. H.: *J. Mol. Biol.,* **31:** 579 (1968); Oishi, M.: *Proc. Natl. Acad. Sci. U.S.,* **60:** 329, 691 (1968).
10. Yudelevich, A., B. Ginsberg, and J. Hurwitz: *Proc. Natl. Acad. Sci. U.S.,* **61:** 1129 (1968).
11. Painter, R. B., and A. Schaefer: *Nature,* **221:** 1215 (1969).
12. Kornberg, A.: *loc. cit.;* Lehman, I. R., M. J. Bessman, E. S. Simms, and A. Kornberg: *J. Biol. Chem.,* **233:** 163 (1958).
13. Richardson, C. C., C. L. Schildkraut, and A. Kornberg: *Cold Spring Harbor Symp. Quant. Biol.,* **28:** 9 (1963).
14. But see Lezius, A. G., S. B. Hennig, C. Menzel, and E. Metz: *European J. Biochem.,* **2:** 90 (1967); Cavalieri, L. F., and E. Carroll: *Proc. Natl. Acad. Sci. U.S.,* **59:** 951 (1968).
15. Josse, J., A. D. Kaiser, and A. Kornberg: *J. Biol. Chem.,* **236:** 864 (1961).
16. Klett, R. P., A. Cerami, and E. Reich: *Proc. Natl. Acad. Sci. U.S.,* **60:** 943 (1968).
17. Cozzarelli, N. R., R. B. Kelly, and A. Kornberg: *J. Mol. Biol.,* **45:** 513 (1969).
18. Kelly, R. B., M. R. Atkinson, J. A. Huberman, and A. Kornberg: *Nature,* **224:** 495 (1969).
19. Goulian, M., Z. J. Lucas, and A. Kornberg: *J. Biol. Chem.,* **243:** 627 (1968); Goulian, M.: *Proc. Natl Acad. Sci. U.S.,* **61:** 284 (1968).
20. Steuart, C. D., S. R. Anand, and M. J. Bessman: *J. Biol. Chem.,* **243:** 5308, 5319 (1968).
21. Litman, R. M.: *J. Biol. Chem.,* **243:** 6222 (1968).
22. Bollum, F. J.: in D. Shugar (ed.), *Genetic Elements Properties and Function,* vol. 3, Academic Press, New York, 1967, p. 1.
23. Nossal, N. G.: *J. Biol. Chem.,* **244:** PC218 (1969).
24. Yoneda, M., and F. J. Bollum: *J. Biol. Chem.,* **240:** 3385 (1965); Kato, K. I., J. M. Gonçalves, G. E. Houts, and F. J. Bollum: *ibid.,* **242:** 2780 (1967); Lefler, C. F., and F. J. Bollum: *ibid.,* **244:** 594 (1969).
25. Olivera, B. M., and I. R. Lehman: *Proc. Natl. Acad. Sci. U.S.,* **57:** 1426, 1700 (1967); Olivera, B. M., Z. W. Hall, and I. R. Lehman: *ibid.,* **61:** 237 (1968).
26. Zimmerman, S. B., J. W. Little, C. K. Oshinsky, and M. Gellert: *Proc. Natl. Acad. Sci. U.S.,* **57:** 1841 (1967); Gefter, M. L., A. Becker, and J. Hurwitz: *ibid.,* **58:** 240 (1967).
27. Weiss, B., and C. C. Richardson: *Proc. Natl. Acad. Sci. U.S.,* **57:** 1021 (1967); *J. Biol. Chem.,* **242:** 4270 (1967); Weiss, B., A. Thompson, and C. C. Richardson: *ibid.,* **243:** 4556 (1968); Weiss, B., A. Jacquemin-Sablon, T. R. Live, G. C. Fareed, and C. C. Richardson: *ibid.,* **243:** 4543 (1968).
28. Lindahl, T., and G. M. Edelman: *Proc. Natl. Acad. Sci. U.S.,* **61:** 680 (1968).
29. Weiss, B., T. R. Live, and C. C. Richardson: *J. Biol. Chem.,* **243:** 4530 (1968).
30. Hall, Z. W., and I. R. Lehman: *J. Biol. Chem.,* **244:** 43 (1969).
31. Sugimoto, K., T. Okazaki, and R. Okazaki: *Proc. Natl. Acad. Sci. U.S.,* **60:** 1356 (1968).
32. Fareed, G. C., and C. C. Richardson: *Proc. Natl. Acad. Sci. U.S.,* **58:** 665 (1967); Masamune, Y., and C. C. Richardson: *ibid.,* **61:** 1328 (1968).
33. Hosoda, J., and E. Mathews: *Proc. Natl. Acad. Sci. U.S.,* **61:** 997 (1968).
34. Newman, J., and P. Hanawalt: *J. Mol. Biol.,* **35:** 639 (1968); also in *Cold Spring Harbor Symp. Quant. Biol.,* **33:** 145 (1968).
35. Richardson, C. C.: *Proc. Natl. Acad. Sci. U.S.,* **54:** 158 (1965). Weiss, B., and C. C. Richardson: *Federation Proc.,* **25:** 708 (1966); *Cold Spring Harbor Symp. Quant. Biol.,* **31:** 471 (1966).
36. Borek, E., and P. R. Srinivasan: *Ann. Rev. Biochem.,* **35:** 275 (1966). Srinivasan, P. R., and E. Borek: *Progr. Nucleic Acid Res.,* **5:** 157 (1966).
37. Billen, D.: *J. Mol. Biol.,* **31:** 477 (1968).
38. Lark, C.: *J. Mol. Biol.,* **31:** 389, 401 (1968).
39. Grippo, P., M. Iaccarino, E. Parisi, and E. Scarano: *J. Mol. Biol.,* **36:** 195 (1968).
40. Arber, W.: *Ann. Rev. Microbiol.,* **19:** 365 (1965); *J. Mol. Biol.,* **11:** 247 (1965).
41. Meselson, M., and R. Yuan: *Nature,* **217:** 1110 (1968).
42. Lehman, I. R.: *Ann. Rev. Biochem.,* **36:** 645 (1967).
43. Becker, A., and J. Hurwitz: *J. Biol. Chem.,* **242:** 936 (1966).
44. Goulian, M., A. Kornberg, and R. L. Sinsheimer: *Proc. Natl. Acad. Sci. U.S.,* **58:** 2321 (1967); Goulian, M., and A. Kornberg: *ibid.,* **58:** 1723 (1967); see also Mitra, S., P. Reichard, R. B. Inman, L. L. Bertsch, and A. Kornberg: *J. Mol. Biol.,* **24:** 429 (1967).
45. Sinsheimer, R. L., B. Starman, C. Nagler, and S. Guthrie: *J. Mol. Biol.,* **4:** 142 (1962); Denhardt, D. T., and R. L. Sinsheimer: *ibid.,* **12:** 641, 647 (1965); Lindqvist, B. H., and R. L. Sinsheimer: *ibid.,* **32:** 285 (1968); Komano, T., R. Knippers, and R. L. Sinsheimer: *Proc. Natl. Acad. Sci. U.S.,* **59:** 911 (1968).
46. Tessman, E. S.: *J. Mol. Biol.,* **17:** 218 (1966).

47. Dressler, D. H., and D. T. Denhardt: *Nature,* **219:** 346 (1968).
48. Gilbert, W., and D. Dressler: *Cold Spring Harbor Symp. Quant. Biol.,* **33:** 473 (1968).
48a. Morgan, A. R.: *Nature,* **227:** 1310 (1970).
49. De Lucia, P., and J. Cairns: *Nature,* **224:** 1164 (1969).
50. Gross, J., and M. Gross: *Nature,* **224:** 1166 (1969).
51. Hanawalt, P. C.: in A. Giese (ed.), *Photophysiology,* vol. 3, Academic Press, New York, 1968.
52. Gupta, N. K., and H. G. Khorana: *Proc. Natl. Acad. Sci. U.S.,* **61:** 215 (1968).
52a. Green, M.: *Ann. Rev. Biochem.,* **39:** 701 (1970).
52b. Temin, H. M. and S. Mizutani: *Nature,* **226:** 1211 (1970).
52c. Baltimore, D.: *Nature,* **226:** 1209 (1970).
52d. Spiegelman, D., A. Burny, M. R. Das, J. Keydar, J. Schlom, M. Travnicek and K. Watson: *Nature,* **227:** 563, 1029 (1970).
52e. Rokutanda, M., H. Rokutanda, M. Green, K. Fujinage, R. K. Ray, and C. Gurgo: *Nature,* **227:** 1026 (1970).
53. Weiss, S. B., and L. Gladstone: *J. Am. Chem. Soc.,* **81:** 4118 (1959); Fox, C. F., and S. B. Weiss: *J. Biol. Chem.,* **239:** 175 (1964); Fox, C. F., W. S. Robinson, R. Haselkorn, and S. B. Weiss: *ibid.,* **239:** 186 (1964).
54. Furth, J. J., J. Hurwitz, and M. Anders: *J. Biol. Chem.,* **237:** 2611 (1962); Maitra, U., Y. Nakata, and J. Hurwitz: *ibid.,* **242:** 4908 (1967); Maitra, U. and J. Hurwitz: *ibid.,* **242:** 4897 (1967); Maitra, U., S. N. Cohen, and J. Hurwitz: *Cold Spring Harbor Symp. Quant. Biol.,* **31:** 113 (1966).
55. Krakow, J. S., and S. Ochoa: *Proc. Natl. Acad. Sci. U.S.,* **49:** 88 (1963); Krakow, J. S., and W. J. Horsley: *J. Biol. Chem.,* **242:** 4796 (1967).
56. Chamberlin, M., and P. Berg: *Proc. Natl. Acad. Sci. U.S.,* **48:** 81 (1962).
57. Stevens, A., and J. Henry: *J. Biol. Chem.,* **239:** 196 (1964).
58. Szybalski, W., H. Kubinski, and P. Sheldrick: *Cold Spring Harbor Symp. Quant. Biol.,* **31:** 123 (1966); Summers, W. C., and W. Szybalski: *Virology,* **34:** 9 (1968).
59. Roberts, J. W.: *Nature,* **223:** 480 (1969).
60. Burgess, R. R., A. A. Travers, J. J. Dunn, and E. K. F. Bautz: *Nature,* **221:** 43 (1969); Travers, A. A., and R. R. Burgess: *ibid.,* **222:** 537 (1969); see also Walter, G., W. Seifert, and W. Zillig: *Biochem. Biophys. Res. Commun.,* **30:** 240 (1968).
61. Kasai, T., and E. K. F. Bautz: *J. Mol. Biol.,* **41:** 401 (1969); Bautz, E. K. F., F. A. Bautz, and J. J. Dunn: *Nature,* **223:** 1022 (1969).
62. Milanesi, G., E. N. Brody, and E. P. Geiduschek: *Nature,* **221:** 1014 (1969).
63. Roberts, J. W.: *Nature,* **224:** 1168 (1969).
64. Wehrli, W., J. Nüesch, F. Knüsel, and M. Staehelin: *Biochim. Biophys. Acta,* **157:** 215 (1968); Mizuno, S., H. Yamazaki, K. Nitta, and M. Umezawa: *ibid.,* **157:** 322 (1968).
65. Blatti, S. P., C. J. Ingles, T. J. Lindell, P. W. Morris, R. F. Weaver, F. Weinberg and W. J. Rutter: *Cold Spring Harbor Symp. Quant. Biol.,* **35:** in press (1970).
66. August, J. T.: *Nature,* **222:** 121 (1969).
67. Pace, N. R., and S. Spiegelman: *Proc. Natl. Acad. Sci. U.S.,* **55:** 1608 (1966); Pace, N. R., and S. Spiegelman: *Science,* **153:** 64 (1966); Bishop, D. H. L., J. R. Claybrook, N. R. Pace, and S. Spiegelman: *Proc. Natl. Acad. Sci. U.S.,* **57:** 1474 (1967); Pace, N. R., D. H. L. Bishop, and S. Spiegelman: *ibid.,* **58:** 711 (1967); Mills, D. R., R. L. Peterson, and S. Spiegelman: *ibid.,* **58:** 217 (1967); Mills, D. R., D. H. L. Bishop, and S. Spiegelman: *ibid.,* **60:** 713 (1968).
67a. Kamen, R.: *Nature,* **228:** 527 (1970); Kondo, M., R. Gallerani and C. Weissmann: *ibid.,* **228:** 525 (1970).
68. Shapiro, L., M. T. Franze de Fernandez, and J. T. August: *Nature,* **220:** 478 (1968).
69. Grunberg-Manago, M., and S. Ochoa: *J. Am. Chem. Soc.,* **77:** 3165 (1955); Grunberg-Manago, M.: *Ann. Rev. Biochem.,* **31:** 301 (1962).
70. Daniel, V., and U. Z. Littauer: *J. Biol. Chem.,* **238:** 2102 (1963); *Methods Enzymol.,* **XIIB:** 579 (1968).
71. Chung, C. W., M. Enrione, and H. R. Mahler: *J. Biol. Chem.,* **235:** 1448 (1960); Venkataraman, P. R., and H. R. Mahler: *ibid.,* **238:** 1058 (1963).
72. Edmonds, M., and R. Abrams: *J. Biol. Chem.,* **237:** 2636 (1962).
73. Burdon, R. H., and R. M. S. Smellie: *Biochim. Biophys. Acta,* **61:** 633 (1962).
74. Edmonds, M., and M. G. Caramela: *J. Biol. Chem.,* **244:** 1314 (1969).

21
Biosynthesis of Proteins

in vitro SYSTEMS

Most of the recent spectacular progress in our understanding of the biochemistry of protein synthesis has been due to reconstruction of most of its key features in cell-free systems, the time-honored methodological approach of the biochemist. The first such system was discovered in 1952 by P. Siekevitz, then working in P. C. Zamecnik's laboratory at Harvard, who used the microsomal fraction of rat liver (1,2). Several important discoveries in the next three years similarly either came from or were simultaneously initiated in this laboratory (3): the fact that ribosomes were the portion of the microsomal fraction responsible for protein synthesis; that the amino acids participating had to be activated by ATP in a reaction producing PP_i and catalyzed by specific activating enzymes (aminoacyl-tRNA synthetases or ligases); and that the natural acceptor for these amino acids, an important intermediate in protein synthesis, was furnished by a specialized fraction of RNA molecules of low molecular weight, localized in the cytosol, now called transfer RNA (tRNA) (3).

Most of the key experiments have subsequently been performed with just two systems:

1. A cell-free system from rabbit reticulocytes (immature red cells), first studied by R. Schweet (4) in 1958, which has the great advantage of producing, or at least of completing, the chains of hemoglobin, a well-characterized, soluble protein, the complete primary structure of which is now known (see page 154).

2. A similar system from *E. coli,* first studied in 1960 by M. Lamborg and Zamecnik (5) and by A. Tissières and his associates (6), which has three general advantages: (a) it is a great deal more active than most other systems studied; (b) under appropriate conditions it retains many of the features of the *in vivo* system, including the "flow of information," DNA → RNA → protein (i.e., operationally, the stimulation of amino acid incorporation by addition of DNA, which in turn engenders active RNA synthesis); and, most importantly, (c) by further treatment the system can be made to respond to the addition of purified oligo- and polyribonucleotides ("synthetic" or natural messengers) in a highly specific manner. From studies on these

systems, and confirmed by many others, a current, generalized picture of protein synthesis would require the following participants:

1. A pool containing adequate amounts of the 20 amino acids commonly found in proteins.

2. A minimum of 20 amino acid-activating enzymes plus an adequate supply of ATP.

3. A pool of competent tRNA molecules: as we shall see, more than 20 (the number of amino acids), but fewer than 64 (3^4, the possible number of codons in a triplet code) are required, since one and the same tRNA may be capable of recognizing equivalent, *degenerate* codons (*synonyms*) specifying a single amino acid. The two or more different tRNA molecules cognate to the various codon synonyms (or synonym groups) for a single amino acid are referred to as *isoaccepting* tRNA's.

4. A pool of structurally and functionally competent ribonucleoprotein particles. In bacteria at least, this requirement is satisfied *not* by the intact (70 S) ribosomes but by stoichiometric amounts of small (30 S) and large (50 S) subunits.

5. A minimum of two, and possibly more, enzymatic components ("transfer factors," elongation factors) involved in polypeptide chain elongation (Table 21-1, page 916).

6. Proteins (and perhaps certain cofactors) required for chain initiation and termination, as well as for the release of completed chains from their site of synthesis (a process that may or may not be identical with termination proper) (Table 21-1).

As indicated, the *in vitro* system can be programmed by means of "artificial" mRNA molecules, constituted by a variety of natural, synthetic and biosynthetic single-stranded homo- and heteropolymeric ribonucleotides (7,8)—and in the presence of the antibiotic neomycin B of single-stranded polydeoxynucleotides (9,10). This observation of polynucleotide-directed polypeptide synthesis, first reported in 1961 by M. W. Nirenberg and J. H. Matthaei (11), and shortly thereafter by P. Lengyel, J. F. Speyer, and S. Ochoa (12), was the key that unlocked the secret of the genetic code. The award of the Nobel Prize to Nirenberg (jointly with R. Holley and H. G. Khorana) in 1968 was in recognition of this fact.

Results with Synthetic Messenger RNA's

The seminal observations were the following: On incubating a partially purified *E. coli* system in the presence of homopolymeric poly rU [$(rU)_n$] (Chapter 5) with various amino acid mixtures in which only one of the constituents was labeled (i.e., 19 "cold," 1 "hot"), it was found that only one, phenylalanine, appeared to be incorporated, and, in fact, the highly insoluble product obtained consisted of polyphenylalanine. Thus, poly U, and by inference (see Chapter 19) a (p)UpUpU(p) triplet, had "coded" for Phe. The extension of this work—using a variety of homopolymers and random heterocopolymers all synthesized *in vitro* by the polynucleotide phosphorylase reaction—in both Nirenberg's laboratory, at Bethesda and Ochoa's in New York, rapidly led to the assignment of various coding units, defined first by their composition but not their sequence, and designated as triplets on the basis of indirect arguments. Thus, it was shown that poly C coded for proline and poly A for lysine. With the use of heteropolymers it was possible to relate relative abundance of hypothetical coding units with relative incorporation frequency, for instance as shown in Table 21-2.

TABLE 21-1
Protein factors required for polypeptide synthesis

	Designation when derived from				
Type	*E. coli*	*Other bacteria (or synonyms)*	*Vertebrates (reticulocytes, liver)*	*Vertebrate synonyms*	*Characteristic or probable activity*
Initiation factors	f_1	F_1, A	M_2		Promotes binding of f_2 to form initiation complex
	f_2[a]	F_2, C	M_1		Forms complex with GTP and initiator aminoacyl-tRNA; attaches latter to Rb (small subunit) and hydrolyzes the former
	f_3	F_3, B	M_3		Only required for interaction with "natural" mRNA
Elongation factors	T_u	S_1	T_1 (with T_s)	TF1, Transferase I	Heat labile } form complex with GTP and aminoacyl-tRNA, transferring latter to the ribosome and hydrolyzing the former
	T_s	S_3			Heat stable }
	G	S_2	T_2	TF2, Transferase II	Required for translocation: GTP-dependent transfer of peptidyl-tRNA from A to P site on Rb (GTPase)
Peptide synthetase	PS	Transpeptidase Peptidyl transferase	Peptidyl transferase		A protein of the 50 S ribosome; catalyzes peptide-bond formation, i.e., chain elongation proper
Termination factors	R_1		none yet known		Recognizes UAG and UAA
	R_2				Recognizes UAA and UGA
	S				Required for the action of R factors

[a] May itself involve more than one protein.

TABLE 21-2

Polymer	*Compositional triplet*	*Relative statistical abundance (calculated)*	*Relative incorporation frequency*[a] *(observed)*
UG(5 : 1)	U_3	100	Phe (100)
	U_2G(UUG,UGU,GUU)	20	Cys (20); Val (20)
	UG_2(UGG, GUG, GGU)	4	Gly (4); Trp (5)
	GGG	0.8 (insignificant)	
UA(5 : 1)	U_2A	20	Tyr (25); Ile (20); Leu (14)
	UA_2	4	Lys (3); Asn (7)
AC(5 : 1)	A_3	100	Lys (100)
	A_2C	20	Asn (30); Thr (23); Gln (44)
	AC_2	4	Pro (5)
AU(5 : 1)	A_2U	20	Ile (20); Asn (28)
	UA_2	4	Leu (3); Tyr (3)
CG(5 : 1)	C_3	100	Pro (100)
	C_2G	20	Ala (22); Arg (19)
	CG_2	4	Gly (5)

[a] Relative to the pure homopolymer (UUU) as (100).

In this manner close to 50 compositional triplets were assigned as codons for various amino acids, but actual base sequences had to remain highly tentative. However, it could be shown that there was a complete identity in the assignments derived from *E. coli* (DNA base composition 52 percent G plus C) and *Alcaligenes faecalis* (66 percent G plus C), and for a few of the polymers tested, between *E. coli* on the one hand and yeast and mammalian systems on the other.

The next major extension of this particular approach was furnished by Khorana, also honored by the Nobel Prize in 1968, by substituting heteropolymers of *known* sequence for the random polymers used heretofore (8,10). In this manner the following coding relations could be established [capital letters nucleotides, lower case letters amino acids]

(a) $(AB)_n$ = ApBpApBpApB . . . etc. — codes only for one heteropolymer $(ab)_m$: a with codon ABA; b with codon BAB

(b) $(ABC)_n$ = ApBpCpApBpCpApBpC . . . etc. — codes for three different homopolymers: homopolymeric a_m (codon ABC), plus b_m (codon BCA), plus c_m (codon CAB)

(c) $(ABCD)_n$ = ApBpCpDpApBpCpDpApBpCpD . . . etc. — codes *only* for one heteropolymer $(abcd)_m$: a with codon ABC, b with codon DAB, c with codon CDA, and d with codon BCD. This is true even when B, C, or D ≡ A.

Use of this technique not only permitted the unambiguous assignment of nucleotide sequences within codons but also provided proof that *r*, the coding ratio (page 870), was indeed equal to 3. Independent evidence for these points also came from the binding studies of M. Nirenberg and P. Leder (page 940).

Steps in Protein Synthesis. Use of the *in vitro* system, in conjunction with that of synthetic messenger RNA's, disclosed that polypeptide synthesis consists of a series of sequential steps discussed below under the following headings (see also Figs. 21-2 and 21-5).

1. Activation of amino acids and formation of aminoacyl-tRNA molecules.
2. Formation of a competent complex between ribosomes and mRNA.
3. Binding of the first aminoacyl-tRNA to this complex and chain initiation.
4. Sequential binding of subsequent aminoacyl-tRNA's during propagation of the growing chain.
5. Formation of peptide bonds within the complex.
6. Chain termination.
7. Release of completed chains from the complex.

Activation and Formation of Aminoacyl-tRNA's [Aminocyl-tRNA Ligases or Synthetases]

Reactions Catalyzed (13,14). Each enzyme so far isolated appears to be highly specific for its individual amino acid and to catalyze the following reactions:

$$(\text{aminoacyl}_x \sim \text{AMP} + \text{E}_x) \qquad (21\text{-}1a)$$

$$\Updownarrow$$

$$\text{Amino acid}_x + \text{ATP} + \text{E}_x \overset{\text{Me}^{2+}}{\rightleftharpoons} (\text{aminoacyl}_x \sim \text{AMP} \cdot \text{E}_x) + \text{PP}_i \qquad (21\text{-}1b)$$

$$(\text{Aminoacyl}_x \sim \text{AMP} \cdot \text{E}_x) + \text{tRNA}^x \overset{\text{Me}^{2+}}{\rightleftharpoons} \text{aminoacyl}_x \sim \text{tRNA}^x + \text{E}_x + \text{AMP} \qquad (21\text{-}1c)$$

$$\text{ATP} + \text{amino acid}_x + \text{tRNA}^x \overset{\text{E}_x}{\rightleftharpoons} \text{aminoacyl}_x \sim \text{tRNA}^x + \text{AMP} + \text{PP}_i \qquad \Delta G^{0\prime} \simeq 0 \qquad (21\text{-}1)$$

Reactions (21-1a) and (21-1b) are very similar to the first step in the enzymatic activation of acetate and higher fatty acids. Indeed, the methods of assay utilized (the amino acid-catalyzed exchange of labeled PP_i into ATP, or the ATP-linked formation of the hydroxamate derivative of the amino acid) had first been developed for these other enzymes.[1] Substrate specificity is evident for the partial reactions (21-1b) and (21-1c), with the rejection of inappropriate partners being more stringent in reaction (21-1c). This might have been anticipated, since this step involves the mutual recognition of two specific macromolecules. *In vivo,* the frequency of introducing one naturally occurring but faulty amino acid for the correct one (the *degree of fidelity* in translation) has been estimated (15) as one in 10,000.[2] *In vitro* the pure isoleucyl enzyme from *E. coli* can be shown to be somewhat unusual in catalyzing step (21-1a), the formation of the aminoacyl adenylate, not only with Ile, but also with Val, in the absence of the natural acceptor $tRNA^{Ile}$.

[1] However, not all aminoacyl-tRNA synthetases appear to be equally capable of transferring aminoacyl residues to hydroxylamine in their native form. The order of reactivity is very rapid Tyr, Phe, Trp, Met > Lys > Ala > Ile > Gly > Cys > Leu > Ser > Val ⇛ Glu, Thr, no reaction (14a).

[2] In those instances where there is apparent ambiguity in a product protein (i.e., microheterogeneity with respect to one or more amino acids), this appears to be referable to the coexistence of multiple alleles of the same gene in one and the same cell. Thus, the ambiguity is itself genetically controlled rather than a consequence of haphazard or systematic translational errors (15a, b). A particularly striking example is provided by the "variable" portion of the L chains of immunoglobulins (page 129).

The addition of the latter results in three consequences: (1) enhancement of specificity in favor of Ile in reaction (21-1a), (2) rapid hydrolysis of the intermediate Valyl-AMP, and (3) exclusive formation of Ile-$tRNA^{Ile}$ rather than Val-$tRNA^{Ile}$. Hence, the addition of the appropriate $tRNA^{x}$ itself exerts an error-correcting function perhaps by affecting enzyme conformation (16, 17).

Although 19 amino acids are activated directly by intervention of the steps described above, the formation of Gln-$tRNA^{Gln}$ appears to be indirect and requires the following steps (18):

$$\text{ATP} + \text{L-glutamate} + \text{tRNA}^{Gln} \xrightarrow[\text{Mg}^{2+}]{\text{Glu-tRNA}^{Glu}\text{ synthetase}} \text{Glu-tRNA}^{Gln} + \text{PP}_i + \text{AMP} \tag{21-2a}$$

$$\underset{(\text{L-asparagine})}{\text{ATP} + \text{L-glutamine}} + \text{Glu-tRNA}^{Gln} \xrightarrow[\text{Mg}^{2+}]{\text{aminotransferase}} \text{Gln-tRNA}^{Gln} + \underset{(\text{L-aspartate})}{\text{L-glutamate}} + \text{ADP} + \text{P}_i \tag{21-2b}$$

$$2\text{ATP} + \text{L-glutamine} + \text{tRNA}^{Gln} \xrightarrow[\text{Mg}^{2+}]{\text{2 enzymes, Glu}} \text{Gln-tRNA}^{Gln} + \text{ADP} + \text{AMP} + \text{PP}_i + \text{P}_i \tag{21-2}$$

The Adaptor Hypothesis. The high degree of selectivity exercised by the aminoacyl-tRNA synthetases is of the utmost importance in minimizing errors in protein synthesis and maintaining a maximum of fidelity in the translation process. For, once the appropriate aminoacyl-tRNA has been formed, the only operative code is the polynucleotide one; that is, there are no additional safeguards assuring the rejection of the wrong amino acids. This is the essence of the adaptor hypothesis first proposed by Crick in 1958, which postulates that *adaptor molecules* (now identified as the $tRNA^{x}$s) intervene between amino acids and the polynucleotide so as to simultaneously bind and interact with, selectively, specific polynucleotide coding units to one portion of the molecule and an appropriate amino acid to another (19). The validity of this concept was proved in a series of brilliant experiments (20). Cysteinyl-$tRNA^{Cys}$ was first formed by means of the cysteinyl-tRNA synthetase/$tRNA^{Cys}$ preparation from *E. coli,* then converted by chemical reduction with Raney nickel to alanyl-$tRNA^{Cys}$ (i.e., an alanyl residue linked to the tRNA for cysteine). The question that was asked, first of the *E. coli* system programmed with synthetic messengers, and later of the preprogrammed hemoglobin-synthesizing system from reticulocytes, was simply this: will the charged tRNA molecule act as the substrate for the incorporation of Cys, of Ala, or of both? In other words, will Ala be incorporated *specifically* for Cys, or not? The answer was quite unambiguous: The coding proclivities of Ala-$tRNA^{Cys}$ are identical with those of the parent Cys, linked to the same adaptor, and Ala is incorporated specifically instead of Cys in the appropriate residue positions of the hemoglobin molecule. Furthermore, Ala-$tRNA^{Cys}$ is unable to transfer the alanyl residue to $tRNA^{Ala}$ in the presence of AMP and *either* ligase.

The adaptor hypothesis thus explains two experimental facts: On the one hand, for any one polypeptide uniquely specified by a single gene, natural but inappropriate amino acids are very rarely if ever incorporated in the place of other natural amino acids; on the other hand, a number of amino acid analogs (e.g., selenomethionine for methionine, *p*-fluorophenylalanine or thienylalanine for phenylalanine, 5-hydroxytryptophan or azatryptophan for tryptophan) are readily substituted for their natural counterparts. Any amino acid, provided it can only fool the tRNA synthetase, will inexorably be incorporated in the place of its natural analog.

Transfer RNA's. The transfer RNA's (tRNA's) provide the link between the nucleotide and the amino acid codes. Because of this they have received considerable attention ever since their discovery in 1957–1958. These studies culminated in Holley's epochal announcement (21) in early 1965 of the complete structure of one of the yeast $tRNA^{Ala}$.

Since then the primary structure of a number of additional tRNA's, most of them from yeast and *E. coli,* has been elucidated; several have been described in Chapter 5. We shall recapitulate some of their most salient features (22–24).

They are relatively small, hence "soluble" (in 1 M NaCl, or remaining in the supernatant on centrifugation for 90 min × 105,000 *g* in isotonic NaCl) polyribonucleotides. The precise length of different species appears to vary over the rather narrow range from 76 to 85 residues. Besides the four common nucleosides, they contain a variety of unusual nucleosides, such as pseudouridine (Ψ), dihydrouridine (H_2U), ribothymidine (rT), inosine (I), mono- and dimethyl derivatives of adenosine and guanosine, thiolated derivatives, and even more esoteric structures such as N^6-(Δ^2-isopentenyl)-adenosine (iA). The requisite structural modifications of the standard bases are introduced into the polymeric molecules subsequent to their formation from the precursor nucleotides. The tRNA molecules of yeast appear to be richer in these modified bases than those of *E. coli.* Their functional significance is unknown, since, at least minor alterations in, for example, the degree of methylation or reduction, may not interfere substantially with the role of the molecule in either amino acid acceptance or subsequent incorporation into a growing polypeptide chain, although some ambiguity is perhaps introduced as a consequence (25). However, the modified base next to the anticodon (see below) appears essential for the recognition process (26).

The molecules exhibit a high degree of sequence complementarity (in the Watson–Crick sense) and hence of secondary and perhaps of tertiary structure as well, particularly in the presence of Mg^{2+} ions, which are essential for biological activity. A favored conformation, for which there is considerable but by no means conclusive evidence, is the cloverleaf arrangement described in Chapter 5. A generalized cloverleaf which accommodates the structural features of the various tRNA species described as of this writing can be formulated as shown in Fig. 21-1. This common conformation is characterized by four distinct arms, limbs, or branches, each containing helical, hydrogen-bonded regions, terminating in loops containing the majority of the unusual bases. The longest branch is the one without a loop (at the top in Fig. 21-1), frequently referred to as the stem. It contains both the 5′-terminus, (a pG residue in the majority of the cases) and the common 3′-terminal . . .pCpCpA-OH,[3] to which the aminoacyl residue becomes attached in ester linkage.

On the basis of photochemical and other modification reactions on tRNA, as well as the known sequences of this part, it has been proposed that the stem region is involved in the recognition process between aminoacyl-tRNA synthetases and tRNA, and specifically that the actual recognition site consists of the three terminal base pairs, as indicated on the figure (24). Loop I also may participate in this act of recognition.

At the opposite end of the molecule, in the two-dimensional representation

[3] The chain initiator $tRNA^{fMet}$ appears to possess the unusual feature that the first pair is between pN_1 and pN_{n-5} rather than pN_{n-4} [S. K. Dube et al., *Nature,* **218:** 232 (1968)].

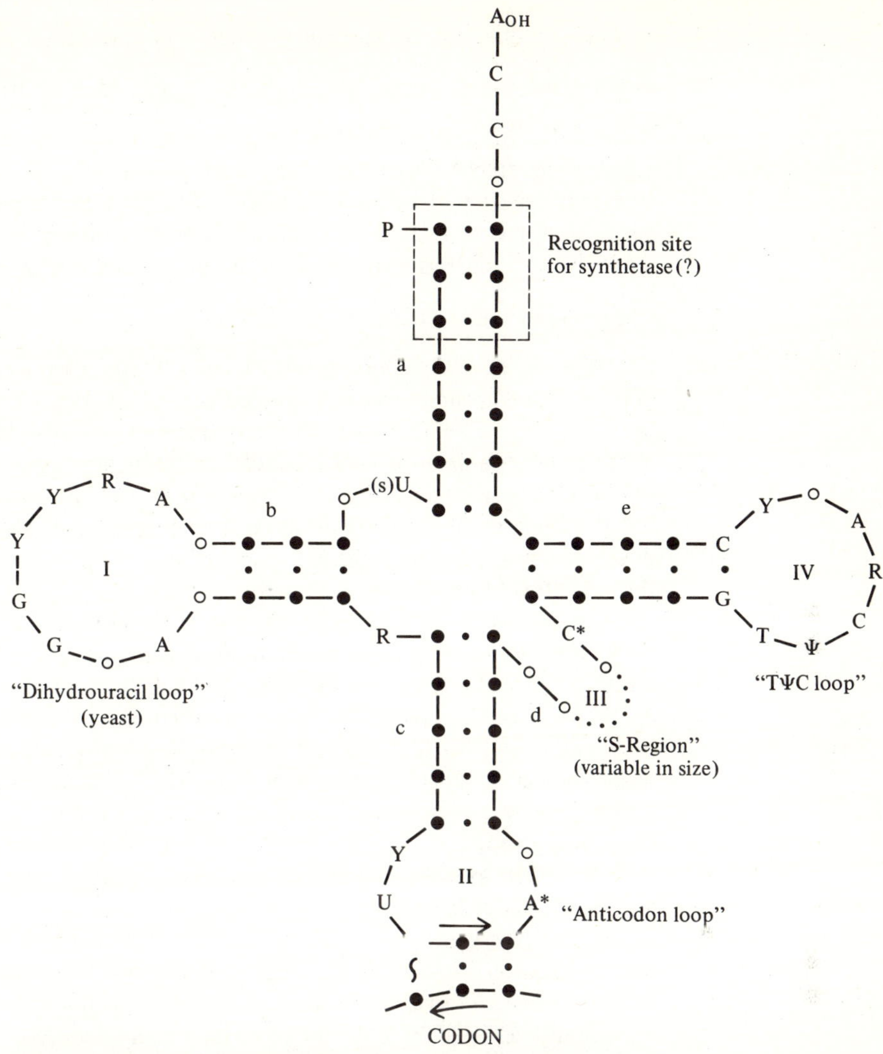

FIGURE 21-1

Standard orientation and terminology for the generalized representation of a tRNA in the "cloverleaf" form. I, II, III, IV, "loops" (unpaired regions); a, b, c, d, e, helical (base-paired) regions, or "arms"; a = stem (upper limb); solid circles, bases in helical regions, usually paired by H bonds (indicated by centered triple dots); open circles, bases usually unpaired; R and Y, purine and pyrimidine (nucleosides); T and Ψ, ribothymidine and pseudouridine; *, modified base; →, 3′ → 5′ direction (upper arrow indicates anticodon); ~, "wobble" pairing. Loop III and arm d are sometimes absent; the numbering and lettering are not influenced thereby. Courtesy of Dr. W. Cohn.

of Fig. 21-1 as well as perhaps in space, in the more highly folded and compact tertiary structure actually assumed *in vivo* (23), is the anticodon arm. At its end, relatively easy of access, is the anticodon loop (II), and in its center the anticodon triplet itself, which may be diagrammed as $\overleftarrow{N_1N_2N_3}$, opposite in polarity and complementary to the $\overrightarrow{pN\ldots N_1N_2N_3\ldots pN_n}$ of the mRNA. As indicated, most of the tRNA's exhibit certain common features with respect to loop IV; other structural homologies for loop I, particularly in the amount and location of H_2U residues, appear to hold only for a variety of different tRNA's from yeast, but not from *E. coli*.

The presence of isoaccepting species in the same cell has already been mentioned; for example, the separation and identification of five $tRNA^{Leu}_{1\text{-}5}$, three $tRNA^{Tyr}_{1\text{-}3}$, and two $tRNA^{Ser}_{1,\,2}$ in *E. coli* appears well documented. In the case of the $tRNA^{Tyr}$ from E. coli and two $tRNA^{Ser}$'s from yeast these find their explanation in terms of actual differences in primary structure (i.e., a small number of base substitutions); in the former case one of the three exhibits a single base substitution in the anticodon as well. The most well documented cases are those of $tRNA^{Ser}_{2a}$ versus $tRNA^{Ser}_{2b}$, and $tRNA^{Leu}_{1}$ versus $tRNA^{Leu}_{2}$ of *E. coli;* each of these pairs recognizes *two* different *synonym* groups, but appears to be recognized by only a *single* aminoacyl-tRNA *synthetase* (28). The unambiguous identification of structurally distinct isoaccepting tRNA's in procaryotic as well as in eucaryotic cells is important since in the latter instance the phenomenon may find an alternative explanation: intracellular organelles, such as mitochondria and chloroplasts appear to possess a distinct and different population of tRNA's and synthetases from those of the cytosol (27).

In those instances where the complete nucleotide sequence of a tRNA for the same amino acid from two different organisms is known, the degree of structural homology is considerable, but unique features are present as well. In the case of the $tRNA^{Tyr}$ from yeast and *E. coli* (29), the latter contains nine additional nucleotides, all located in loop III. Forty-eight nucleotides are common or closely homologous and can be arranged in structurally congruent regions. The anticodons (Table 21-3) are GpΨpA in yeast and (modified) GpUpA in *E. coli.* Most of the differences reside in loop I and in the stem. The possibly critical 5′-terminal trinucleotide is pCpUpC in yeast and pGpGpG in *E. coli.*

Mutual Recognition and Suppression. These observations confirm what had been known earlier from studies at the biochemical and functional level (14, 30, 31). Transfer RNA's isolated from different organisms frequently appear not to be identical and to exhibit a distinct bias toward homologous systems (i.e., their species specificity is complementary to that of their aminoacyl synthetases). In extreme cases, such as that of the $tRNA^{Tyr}$ and the Tyr-tRNA synthetases of *E. coli* and yeast, there is no cross reaction whatever. Taken in conjunction with the apparent universality of the genetic code, with results of studies on suppression and related matters, as well as of chemical modifications in selected regions of the tRNA molecule, this type of specificity also argues strongly in favor of two separate, distinct, and autonomous recognition sites within any one tRNA, one site or sites for the aminoacyl-tRNA synthetase, the other for the codon and hence for mRNA.

As concerns the specific mode of recognition of different isoaccepting tRNA's—which may contain different anticodons for different synonym groups —by the enzyme (Eq. 21-1c), available information is still sketchy. We know that the process involves a highly specific interaction between the two partners, that they readily associate into a complex, and that its association constant (10^7 M) is much greater than analogous constants for any of the other pairs in Eq. (21-1).

The phenomenon of suppression (32) (see Chapter 19) is contingent on the very definition of an adaptor, that is, a tRNA as a molecule possessing these two separate and independent recognition sites. Details concerning the mode of action are deferred to page 942.

Formation of a Complex between Rb and mRNA

Bacterial Ribosomes. From *E. coli,* as from other bacteria, monomeric ribosomes (33) can be obtained that are stable in the presence of high concentrations ($> 10^{-3}$ M) of Mg^{2+}, and that exhibit a sedimentation coefficient of ~70 S, a particle weight of $\sim 25 \times 10^6$, and a particle diameter of ~180 Å. As the Mg^{2+} concentration is lowered to 10^{-4} M, dissociation occurs into two stable, nonidentical subunits. These two particles, referred to as the small and large subunits, show sedimentation coefficients of 30 S ($M = 0.85 \times 10^6$ daltons) and 50 S ($M = 1.8 \times 10^6$ daltons), respectively, and are made up of, by weight, approximately 36 percent protein and 64 percent RNA. Each individual 30-S particle contains a single RNA molecule of sedimentation coefficient (under standard conditions) of 16 S, comprised by a single chain of about 1500 nucleotides of which 11 are specifically

TABLE 21-3
Base sequences in anticodon region of tRNA's[a]

Residue No.	*Point on helix (Fig. 21-6)*		*Ala*	*Ser I + II*	*Ile*	*Tyr*	*Phe*	*Val*	*fMet*	*Su_3+ (Tyr)*	*Tyr*	*Val_1*
17	D		C	A	Ψ	C	C(Ψ)	A	U	G	G	C
16			U	A	G	A	C	Ψ	C	C	C	C
15			C	A	G	A	A	C	G	A	A	U
14	↓		C	G	U	G	G	U	G	G	G	C
13	B		C	A	G	A	A	G	G	A	A	C
12		Y	U	Ψ	C	C	Cm(C)	C	Cm	C	C	C
11			U	U	U	U	U	Ψ	U	U	U	U
10	A	W	I	I	I	G	Gm	I	C	C	G*	X
9	↑	$\bar{N}_2$	G	G	A	Ψ	A	A	A	U	U	A
8		$\bar{N}_1$	C	A	U	A	A	C	U	A	A	C
7			mI	iA	N*	m_2^6A	Y(Y′)	A	A	A*	A*	Am
6		N	Ψ	A	A	A	A	C	A	A	A	A
5	C′		G	Ψ	C	Ψ	Ψ	G	C	Ψ	Ψ	G
4	↑		G	C	G	C	mC	C	C	C	C	G
3			G	U	C	U	U	A	C	U	U	A
2			A	U	C	U	G	G	G	G	G	G
1	C		G	U	A	G	G(A)	A	A	C	C	G

[a] See Figs. 21-1 and 21-6. The first six species are from yeast (bases in parentheses for wheat germ), the last four are from *E. coli.* All abbreviations according to new IUPAC–IUB rules: A suffix m (e.g., Am) denotes a 2′-*O*-methyl derivative, a prefix m (e.g., m_2^6A) denotes a *N*-methyl derivative; iA, N^6-isopentenyl adenosine; N*, *N*-(purin-6-yl carbamoyl)-threonine nucleoside; A*, 2-methylthio-N^6-isopentenyl adenosine; G*, modified guanosine (structure not yet certain); X, Y, Y′, modified purine nucleosides (structure not yet certain, highly fluorescent). The *anticodon region* proper starts with residue 8 of loop II (the wobble position). The first residue in this loop (No. 17 here) corresponds to residue 26–29 from the 5′ terminus and is preceded by a purine (usually mG or A). The general structure of the 17-residue loop (Figs. 21-1 and 6) (see Ref. 23) is

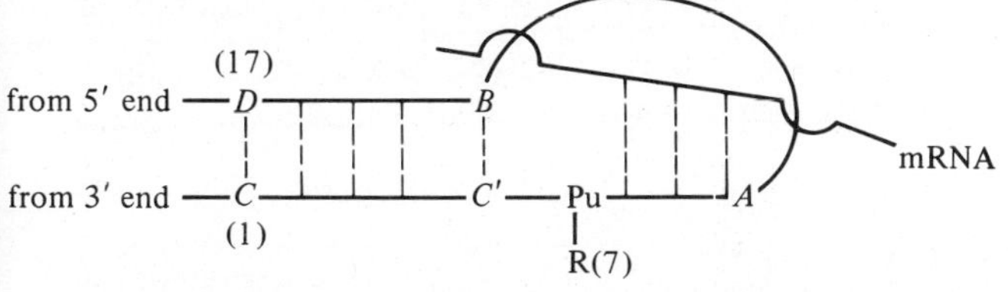

SOURCE: For most structures see J. T. Madison, *Ann. Rev. Biochem.,* **37:** 131 (1968) except for: *Su_3+*: H. M. Goodman, J. Abelson, A. Landy, S. Brenner, and J. D. Smith, *Nature,* **217:** 1019 (1968); *Tyr* (*E. coli*): B. P. Doctor, J. E. Loebel, M. A. Sodd, and D. B. Winter, *Science,* **163:** 693 (1969); *fMet*: S. K. Dube, K. A. Marcker, B. F. C. Clark, and S. Cory, *Nature,* **218:** 232 (1968); *Ile*: S. Takemura, M. Murakami, and M. Miyazaki: *J. Biochem.* (*Tokyo*), **65:** 489 (1969); *Val_1*: M. Yaniv and B. G. Barrell, *Nature,* **222:** 278 (1969).

methylated (34). Each 50-S particle contains a larger RNA molecule, with a sedimentation coefficient of 23 S comprised of about 3000 nucleotides, 22 of which are methylated; as well as a smaller one (sedimentation coefficient 5 S), with a known sequence of exactly 120 nucleotides (35), none of which are methylated. Surprisingly, in spite of intensive studies, there is still some residual uncertainty concerning the precise molecular weight of the two principal rRNA species: physicochemical evidence supports values of 5.5×10^5 and 1.1×10^6 daltons, while estimates based on end groups lead to values of 4.5×10^5 and only 7.3×10^5 daltons respectively (36). Furthermore, end-group determinations, as well as a number of other indicators, have given rise to the belief that neither of the two molecular species is homogeneous, and that both may be composed of families of closely related molecules (36).

Examination of the protein components of the two ribosomal subunits indicates that they are distinct also (37). Each large subunit can be shown, by chromatography and electrophoresis to contain some 30 polypeptides in roughly equimolar amounts; 15–17 such entities (2 acidic, the remainder basic) are found in the small one. The actual number present in the population is greater due to the presence of minor (nonstoichiometric) components —of the order of 35 polypeptides in the larger, about 20 in the small subunit. Of all these polypeptides, with an average molecular weight of 1.8×10^4, perhaps only one is common to both types of particles.

The two subunit particles can be further dissociated by treatment with 4.9 M CsCl containing 0.04 M $MgCl_2$ into "core particles," lacking certain of the proteins, plus the proteins in a soluble form (split proteins, "SP"):

$$\text{(70-S Rb)} \xrightarrow{<10^{-4}\ \text{M}\ Mg^{2+}} \begin{cases} \text{50-S subunit} \underset{}{\overset{5.2\ \text{M CsCl}}{\rightleftharpoons}} \text{40-S core particle + proteins ("SP 50")} & \text{(21-3a)} \\ \text{30-S subunit} \underset{\substack{\text{spontaneous} \\ (0\ \text{or}\ 40°)}}{\overset{5.2\ \text{M CsCl}}{\rightleftharpoons}} \text{23-S core particle} + \geq 8\ \text{proteins ("SP 30," 2 acidic,} \geq 6\ \text{basic)} & \text{(21-3b)} \end{cases}$$

Furthermore, both core particles can be further resolved into their respective proteins (core proteins, "CP") and RNA.

$$\text{40-S core particle} \rightleftharpoons \text{23-S RNA + 5-S RNA + 12 proteins ("CP 50")} \qquad \text{(21-4a)}$$

$$\text{23-S core particle} \underset{37°}{\rightleftharpoons} \text{16-S RNA + 6 proteins ("CP 30")} \qquad \text{(21-4b)}$$

The proteins are solubilized by treatment with 8 M urea–4 M LiCl, while the RNA is usually obtained by phenol extraction. Early experiments suggested that, in the case of the 30-S particle, these steps were reversible (38). Hence all the requisite information for the assembly of this particle is implicit in its components.

That this is indeed the case has been demonstrated in a striking fashion by Traub and Nomura (39), who have been able to reconstitute active 30-S particles by means of the following steps from the isolated and totally separate components:

$$\text{16-S RNA} + \sim 10\ \text{proteins (RI proteins)} \xrightarrow[0°]{K^+,\ Mg^{2+}} \text{RI particles} (\sim 16\ \text{S}) \qquad \text{(21-5a)}$$

$$\text{RI particles} \xrightarrow[\substack{k = 0.21\ \text{min}^{-1},\ \Delta H^{\ddagger} = 38\ \text{kcal/mol} \\ \Delta S^{\ddagger} = 51\ \text{eu}}]{\text{heat, } 40°\ \text{optimum}} \text{RI* particles (conformationally altered, probably identical to 23-S core particle)} \qquad \text{(21-5b)}$$

$$\text{RI* particle} + \sim 10 \text{ proteins (S proteins)} \xrightarrow[0°]{K^+,\ Mg^{2+}} \text{30-S Rb} \qquad (21\text{-}5c)$$

The total reconstitution requires a *p*H near neutrality, K^+ ions at 0.25 M, Mg^{2+} ions at > 0.01 M, a relatively high ionic strength (optimum at 0.37), and a temperature of 40°. Analysis shows that reaction (21-5a) is composed of distinct and separable steps and that the rate-limiting step for complete reconstitution is the unimolecular reaction (21-5b), a conformational (or configurational) change of an intermediate ribonucleoprotein particle. The product of this reaction, the RI* particle, may be identical with the 23-S core particle of reaction (21-3b), and hence reaction (21-5c) may be equivalent to this reaction, while reaction (21-4b) corresponds to the sum of reactions (21-5a) and (21-5b). In addition, active 30-S particles can be reconstituted by using 16-S RNA isolated from one species of bacteria supplemented by proteins isolated from a completely unrelated bacterial species (40), but not by using the ribosomal proteins from *E. coli* together with the RNA from rat liver or yeast or of degraded 16-S RNA derived from the 23-S RNA of *E. coli* (38). Thus ribosome assembly appears to be preprogrammed and sequential.

Ribosome Synthesis. The information for the specification of the 16-S and 23-S ribosomal RNA's appears to be encoded in separate, but closely linked, cistrons in the general region of the genome also coding for tRNA (41). The relevant portion of the DNA has been isolated and purified (42). (That responsible for 5-S RNA appears to be localized in a different position.) There must be a modest amount of redundancy for this information (43), since multiple rRNA cistrons (3 each in *Bacillus subtilis,* 10 each in *E. coli,* and 40 each in *B. megaterium*) have been identified. Within these gene clusters cistrons for 16-S and 23-S RNA may alternate in tandem. The relevance of these phenomena to that of apparent heterogeneity of the gene products is not yet known. The two major RNA's appear to be transcribed simultaneously (44) at a rate of one complete chain every 2 min (13 nucleotides per second) at 37°. The initial product is probably not identical to the sum of the two stable species. In part this processing of initial into final species may involve scissions in the phosphodiester backbone of the precursor, in part it certainly involves the introduction of methyl groups into some of the bases, with *S*-adenosylmethionine as the methyl donor; the precursor RNA of the small subunit is known to be methylated only to the extent of 20 percent, that of the large one to the extent of 60 percent of the mature particle. By means of a variety of tracer experiments and by the use of various devices for arresting protein synthesis, the flow of precursors into various particles has been tentatively established as follows (44–46):

$$\rightarrow \text{23-S RNA} \xrightarrow[\text{methyl groups}]{\text{proteins}} (\text{25 S}) \xrightarrow{\text{proteins}} \text{30–32 S} \xrightarrow{\geq 9 \text{ proteins}} \text{40–43 S} \xrightarrow[\text{methyl groups}]{\text{7 proteins, 5-S RNA}} \text{50 S} \qquad (21\text{-}6a)$$

(3 components: proteins, proteins, methyl groups over the steps 23-S RNA → (25 S) → 30–32 S)

$$\rightarrow \text{16-S RNA} \xrightarrow{\text{proteins}} (\text{18 S}) \xrightarrow{\text{proteins}} (\text{23 S}) \xrightarrow{\text{proteins}} \text{26 S} \xrightarrow{\text{proteins}} \text{30 S} \qquad (21\text{-}6b)$$

At least some of these presumptive precursors may be identical with the particles obtainable *in vitro,* capable, as we have seen, of reversible association into functional ribosomes. The kinetics of assembly *in vivo* appears consistent with that observed *in vitro,* at least at 37°. 5-S RNA, synthesized with different kinetics, becomes associated relatively late in the biosynthetic sequence, since the 40-S particles are still probably devoid of this component. The identification of various ribosomal particles on the basis solely of their sedimentation behavior is complicated by the occurrence of conformational changes, not involving dissociation of proteins, only some of which are reversible. The following transformations have been observed (48):

$$50\ \mathrm{S} \underset{+\mathrm{Mg}^{2+}}{\overset{-\mathrm{Mg}^{2+}}{\rightleftharpoons}} 38\text{–}42\ \mathrm{S} \longrightarrow 29\ \mathrm{S} \longrightarrow 19\ \mathrm{S} \tag{21-7a}$$

$$30\ \mathrm{S} \rightleftharpoons 27\ \mathrm{S} \longrightarrow 17\text{–}19\ \mathrm{S} \tag{21-7b}$$

Cytoribosomes of Eucaryotes. The cell sap (cytosol) and the endoplasmic reticulum of higher organisms, both animal and plants, contain ribonucleoprotein particles that fulfill functions in protein synthesis analogous to those of bacterial ribosomes (Chapter 9). Although resembling them also in their chemical and physical properties, they are sufficiently different as to warrant separate consideration. The monomeric unit shows a sedimentation coefficient of 77 to 80 S and can be dissociated, although with somewhat greater difficulty, into two asymmetric subunits with sedimentation coefficients of 58 to 60 S and ~40 S (38 to 45 S) and molecular weights of 2.4×10^6 and 1.2×10^6, respectively. They contain about 55 percent RNA and 45 percent protein. There appears to be significant differences in the molecular properties of the RNA contained in these particles. Thus, in plants (higher plants, *Neurospora,* yeast) the two rRNA species exhibit sedimentation coefficients of 16 to 18 S ($M = 5.6 \times 10^5$ daltons) and 25 to 26 S ($M = 1.3 \times 10^6$ daltons); under equivalent conditions the rRNA's of protozoa show sedimentation coefficients of 18 to 19 S (6.9×10^5 daltons) and 25 to 26 S (1.3×10^6 daltons), while those of multicellular (metazoan) animals (both invertebrate and vertebrate) give 18 to 19 S (6.7×10^5 daltons) and 28 to 30 S (1.8×10^5 daltons), respectively (49). As in bacteria, the large subunit also contains a small rRNA molecule (5-S RNA), but in addition it probably also contains a 7-S RNA. Distinct and multiple protein species are associated with each subunit. The cistrons responsible for the specification of the rRNA's are localized in a particular region of a single chromosome (42,50) in animals called the *nucleolar organizer.*[4] Considerable redundancy within this gene cluster has been uncovered in both animals and plants (51,52). This duplication is of the order of several hundred in mammalian cells (400 in human HeLa cells), and somewhat higher in certain amphibians (500 to 800 in the clawed toad *Xenopus laevis*) and in yeast. Consequently, it is frequently possible to isolate the DNA responsible (called rDNA, ~0.1 percent of the total) as a satellite distinct from the bulk DNA (buoyant densities in animals of 1.724 and 1.700, respectively). This is especially true if the isolation makes use of the fact that the oocytes of amphibians contain nuclei, called germinal vesicles, in which are found approximately 1000 nucleoli, in which the rDNA is localized and which may account for 25 to 33 percent of the total DNA (51,53). These cells are tetraploid and hence without this remarkable amplification would be expected to contain about 0.014×10^{-12} g of rDNA, corresponding to about 500 cistrons. Instead, they contain about 5×10^{-12} g of this species corresponding to 3.5×10^4 such cistrons! Within this cluster the regions responsible for specifying the 18- and 28-S RNA appear to alternate as they do in bacteria; and between these transcribable regions are interspersed approximately equal amounts of a DNA of very high GC content called the "spacer," not coding for any

[4] During normal cell division, this region, visible in metaphase chromosomes, becomes detached and forms an integral part of the nucleolus.

identifiable RNA species. In addition, variable amounts—depending on species—of preribosomal RNA that *is* transcribed is not conserved during the maturation process. As might be anticipated from the localization of the genes, synthesis of rRNA in eucaryotic cells takes place in the nucleolus and makes use of a special polymerase. The first identifiable product subsequent to transcription of the rRNA cistrons is a large (45-S) RNA which acts as the precursor of both major species of rRNA (the 5-S RNA is apparently transcribed independently, perhaps at a separate site).

The currently accepted scheme for various eucaryotes is somewhat as follows (54), but note that details particularly with regard to size and stability of postulated precursors may vary from species to species:

$$\begin{array}{ccccccc} \text{45-S RNA} \ (M = 4\text{-}4.5 \times 10^6) & \longrightarrow & \text{41-S RNA} & \longrightarrow & (\text{36-S ?}) & \longrightarrow & \text{32-S RNA} \ (M = 2.2 \times 10^6) \quad \textit{nucleolus} \\ & & & & \searrow \text{20-S RNA} & & \downarrow \\ & & & & \downarrow & & \\ & & & & \text{18-S RNA} \ (M = 7.1 \times 10^5) & & \text{7-S RNA} \quad \text{28-S RNA} \ (M = 1.8 \times 10^6) \quad \textit{nucleus} \end{array} \tag{21-8}$$

Coincident with the conversions of the large, GC-rich (~70 percent GC in both 45-S and 32-S RNA versus 67 percent in 28 S and 57 percent in 18 S) RNA precursors to the final products, the following additional features of ribosome maturation have been established: (1) Although some methylation is already evident in the 45-S RNA, this modification increases coincident with processing. (2) Similar considerations obtain with regard to the conversion of uridine to pseudouridine; the final values for both, total methyl groups introduced and U converted to Ψ, are about 1.0 per 100 residues in the 28 S and 1.5 per 100 in the 18 S. (3) None of the various species of RNA shown exist free as such; instead they combine with ribosomal proteins to form part of ribosomal precursors, or "nascent" ribosomal particles (55). The 45-S, 41-S, and 36-S RNA, together with 5-S RNA, all reside in particles with $s \approx 80$ S, while the 32-S RNA plus 5-S RNA are contained in a 55-S particle. All these particles have been identified in the nucleolus; once transferred to the nucleus, the 18-S RNA is known to reside in a particle with a sedimentation coefficient of 30 S, and the 28-S + 5-S + 7-S RNA in one of 50 S. Whether these smaller values indicate structural or conformational differences between analogous ribosomal subunit particles residing in the nucleus and in the cytosol remains to be established.

Ribosomes of Mitochondria and Chloroplasts. Eucaryotic cells also contain certain organelles (see Chapter 9) which possess a separate store of genetic information within a DNA distinct from that of the nucleus of the same cell (see Chapter 5). These organelles (the mitochondria found in all eucaryotes, and the chloroplasts characteristic of green plants) are therefore capable of specifying and, as it turns out, also of synthesizing, at least certain, although by no means all, of their constituent proteins. The synthetic machinery responsible utilizes ribosomal particles that in many of their properties (e.g., sensitivity to inhibitors, details of chain initiation) resemble bacterial ribosomes more closely than they do the homologous cytoribosomes isolated from the same cell. Other properties suggest, however, that we are dealing with quite distinct species. For instance, chloroplast ribosomes have sedimentation coefficients[5] of the order of 67 S, those of plant mitochondria

[5] It must be emphasized, however, that sedimentation coefficients depend not only on molecular weight but also on shape (Chapters 4 and 5); the latter, in turn, can vary drastically with changes in the environment (divalent and other metal ions, ionic strength).

of ⩾75 S, while those from animal mitochondria are of the order of 55 S. Similarly, the rRNA's of green-plant chloroplasts, and those of mitochondria from green plants, molds, and yeast all appear to be intermediate in size between the cytoplasmic and bacterial rRNA's. The sedimentation coefficients for those of animals are reported to be 12 S and 16 S. Finally, the base compositions of the two species of mitochondrial rRNA appear to be distinct and to differ, in all instances, from those characteristic of bacterial RNA's—as well as from those of the cytoribosomes from the same cell (Table 5-7).

Binding of mRNA to Ribosomes. In bacterial systems the processes of ribosome assembly from their subunits, binding of mRNA, and initiation of polypeptide-chain synthesis appear to be linked in an obligatory fashion (56–58), and there is some evidence that similar considerations prevail also in eucaryotic cells (59,60), although with significant differences in detailed mechanism, particularly with regard to the nature of the entity responsible for chain initiation (61). A possible reaction sequence is (57,62–64):

$$\text{30-S Rb} + \text{mRNA} \xrightarrow[\text{Mg}^{2+}]{\text{protein(s)}} [\text{30-S Rb}\cdot\text{mRNA}] \qquad \textit{binding of mRNA} \qquad (21\text{-}9)$$

$$[\text{30-S Rb}\cdot\text{mRNA}] + N\text{-formyl-Met(fMet) tRNA}^{\text{fMet}} + \text{GTP} \xrightarrow[\text{Mg}^{2+},\ \text{K}^+,\ \text{NH}_4^+,\ -\text{SH}]{\text{protein(s)}} \left[\text{30-S Rb}\begin{matrix}\cdot\,\text{fMet-tRNA}\\ \\ \cdot\,\text{mRNA}\end{matrix}\cdot\text{GTP}\right]$$

formation of "initiation complex" (21-10)

$$\left[\text{30-S Rb}\begin{matrix}\cdot\,\text{fMet-tRNA}\\ \\ \cdot\,\text{mRNA}\end{matrix}\cdot\text{GTP}\right] + \text{50-S Rb} \xrightarrow[\text{Mg}^{2+}]{\text{protein(s)}} \left[\underbrace{\text{50-S Rb}\cdot\text{30-S Rb}}_{\text{70-S Rb}}\begin{matrix}\cdot\,\text{fMet-tRNA}\\ \\ \cdot\,\text{mRNA}\end{matrix}\cdot\text{GTP}\right]$$

formation of 70-S Rb (21-11)

$$\left[\text{50-S Rb}\cdot\text{30-S Rb}\begin{matrix}\cdot\,\text{fMet-tRNA}\\ \\ \cdot\,\text{mRNA}\end{matrix}\cdot\text{GTP}\right] \xrightarrow[\text{Mg}^{2+}]{\text{protein(s)}} [\text{fMet-tRNA}\cdot\text{50-S Rb}\cdot\text{30-S Rb}\cdot\text{mRNA}] + \text{GDP} + \text{P}_i$$

"translocation" (21-12)

Attachment of a mRNA containing the AUG or GUG codons to the 30-S subunits [Reaction (21-9)] requires the RNA as well as some, but not all the proteins of this subunit. The sequence outlined also appears to require additional protein factors called *initiation factors* (IF), which, although associated with the ribosomes (30-S subunit) as ordinarily isolated, can be removed by extensive washing of the particles and hence do not constitute an integral part of their intrinsic protein complement (65,67). Since four separate steps (21-9) through (21-11) can be distinguished, it might be anticipated that a minimum of four such proteins would be involved. So far three [called factors F1, F2, and F3 (or A, C, and B; see Table 21-1) with molecular weights of 8000, 30,000, and 75,000, respectively] have been implicated in the formation of the initiation complex [reactions (21-9) and (21-10)]. Reaction (21-10) also requires the participation of GTP as well as Mg^{2+} (at ⩽0.01 M), NH_4^+, or perhaps K^+ (at ~0.1 M) and maintenance of reactive thiol groups by means of addition of reducing agents (mercaptoethanol, dithiothreitol, etc.). F1 (and perhaps F2) becomes associated with the 30-S Rb in the course of fMet-tRNA binding and F1 is again released when the 50-S Rb becomes attached [reaction (21-11)]. F2 is probably required for GTP binding and appears related to propagation factor T (see below and Table 21-1).

It may also be involved in a *specific* translocation (62–65), moving the chain initiator (see the next section) *N*-formylmethionyl-tRNA^{fMet} (fMet-tRNA) from its original site on the 30-S particle into the correct position (with participation of the 50-S particle) to permit interaction with an incoming acceptor to form a peptide bond. This reaction (21-12) requires that GTP be hydrolyzed in the process and therefore this protein factor, in the presence of the other components, can be assayed as a GTPase. Factor F3 has so far been implicated only when specification is by "natural" mRNA's (e.g., phage RNA containing an AUG initiator) and it can bind to such polynucleotides. Thus, it may be concerned with recognition prior to, rather than complex formation proper.

Chain Initiation (Fig. 21-2)

Role of N-*formylmethionine* (66). A requirement for a chain-initiating signal is implicit in the nonoverlapping nature of the genetic code, and the resultant existence of a reading frame (page 862) during translation. In other words, on any given mRNA specifying one (or more) polypeptides, there must exist a recognition site that specifies: "Start translation at this nucleotide and proceed taking three nucleotides (≡ one codon) at a time." If we look at such a messenger RNA we now know that it has been synthesized from the 5′-terminal end and, at least at the moment of synthesis, shows the structure $pppN_1pN_2pN_3 \ldots pN_n$. Hence, it possesses two singular regions that might logically function as possible recognition points: at the 5′ terminus, or N_1, where it carries a triphosphate "sign" (probably hydrolyzed to a single phosphate subsequent to its formation), and at the 3′ terminus, or N_n, where there is a free hydroxyl group. Proteins are synthesized starting with their amino terminus (Fig. 21-3), and mRNA is translated starting with its 5′-terminal end, i.e., the direction, or polarity, of both transcription and translation is the same: 5′ → 3′. This represents an important consideration (to which we shall return on page 966), since it permits the initiation of the latter process while the former is still continuing, before the completed mRNA molecule has left its site of formation. But while recognition of the 5′-terminal nucleotide (triplet) might alone provide a sufficient condition for chain initiation in a *mono*cistronic message, all the mRNA's studied most frequently (i.e., those specified by bacterial viruses), and in particular the RNA's of viruses such as Qβ, MS2, R17, and f2 (see Fig. 5-23), which constitute the purest form of well-defined mRNA, as well as those elaborated by bacteria under a variety of conditions of induction or repression (page 958), are certain to be *poly*cistronic. Hence, additional signals are necessary in the interior of the chain for the initiation of those proteins not encoded in the 5′-terminal region of the mRNA.[6]

That a special mode and codon for chain initiation might exist was suggested by three remarkable findings: (1) J. P. Waller's discovery (67) that the N-terminal amino acids in the global mixture of proteins extracted from *E. coli* are far from random and consist of 45 percent Met, 30 percent Ala, and 15 percent Ser. (2) The isolation of fMet-tRNA by F. Sanger and K. Marcker. (3) The demonstration that while the isolated viral capsid proteins of RNA coliphages f2, MS2, and R17 had the N-terminal sequence (H_2N)Ala-Ser-Asn-Phe-Thr-, the sequence in the same proteins synthesized *in vitro* in a system programmed with viral RNA as a messenger was

[6] Even in the case of phage RNA's, the first chain-initiating signal is now known not to be identical with the 5′ terminus (Fig. 5-23).

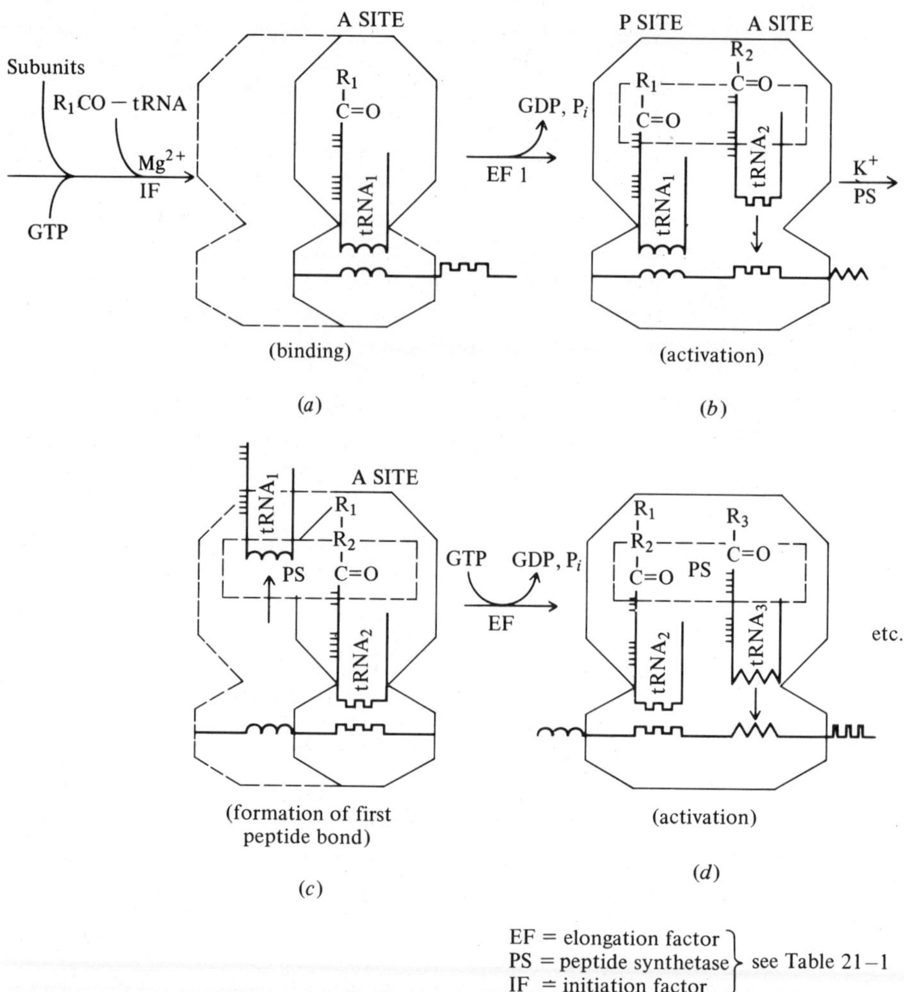

FIGURE 21-2

Detailed view of the first four steps in protein synthesis. Modified from R. Schweet, R. Arlinghaus, R. Heintz, and J. Schaeffer, in *Development and Metabolic Control Mechanisms in Neoplasia,* University of Texas Press, Austin, 1966; reproduced by permission of the late R. S. Schweet.

fMet-Ala-Ser-Asn-Phe-Thr- (68,69). These and related findings suggested that perhaps most proteins of *E. coli* and of its bacteriophages in the nascent state start with the sequence fMet-Ala-Ser- but that the actual N terminus in the final protein would be either Met-Ala-Ser-, Ala-Ser-, or Ser-, depending on the extent of sequential and controlled action on the nascent polypeptide of a specific deformylase, followed by that of one or more aminopeptidases. They also directed attention to the biosynthesis, recognition, and function of fMet-tRNA as the (probably exclusive) chain-initiating entity of this organism, since then confirmed for other bacterial species as well (70).

The biosynthesis of this molecule involves formylation of the amino group of methionine *subsequent* to the aminoacylation of the appropriate tRNA. Two $tRNA^{Met}$'s can readily be identified and separated (71): one that can be formylated after acylation by Met, called $tRNA^{Met}_{f}$ (or $tRNA^{fMet}$), and one that cannot, called $tRNA^{Met}_{M}$ (or, simply $tRNA^{Met}$). The steps involved are shown in Eqs. (21-13) and (21-14).

$$\text{tRNA}^{\text{fMet}} + \text{Met} + \text{ATP} \underset{}{\overset{\text{Mg}^{2+}}{\rightleftharpoons}} \text{AMP} + \text{PP}_i + \text{H}_2\text{N-Met-CO-tRNA}^{\text{fMet}} \quad \textit{methionyl-tRNA synthetase} \tag{21-13}$$

$$\text{FH}_4\text{-}N^{10}\text{-CO(H)} + \text{NH}_2\text{-Met-CO-tRNA}^{\text{fMet}} \rightleftharpoons \text{HCONH-Met-tRNA}^{\text{fMet}} + \text{FH}_4 \quad \textit{transformylase} \tag{21-14}$$

(N^{10}-formyl FH_4)

We recall that the formyl donor N^{10}-formyl FH_4 is one of the C-1–donating tetrahydrofolate coenzymes at the level of formate (Chapter 8). It can be formed either by direct formylation of FH_4 by formate at the expense of ATP, or by hydrolysis of N^5,N^{10}-methenyl FH_4, which in turn is formed from the "formaldehyde" carrier N^5,N^{10}-methylene FH_4 by oxidation with NADP (page 408). In any event, we see here a direct link between a rate-limiting step in protein synthesis and the area of C-1 transfer by folic acid coenzymes, a key area of metabolism in, for example, nucleic acid biosynthesis. Such a link may well be of considerable regulatory significance (72).

In discussing the recognition of various codons by $\text{tRNA}_\text{f}^\text{Met}$ and $\text{tRNA}_\text{M}^\text{Met}$, one fact stands out: Only one codon specifying Met in the interior of a polypeptide could be identified by a variety of techniques; that codon is

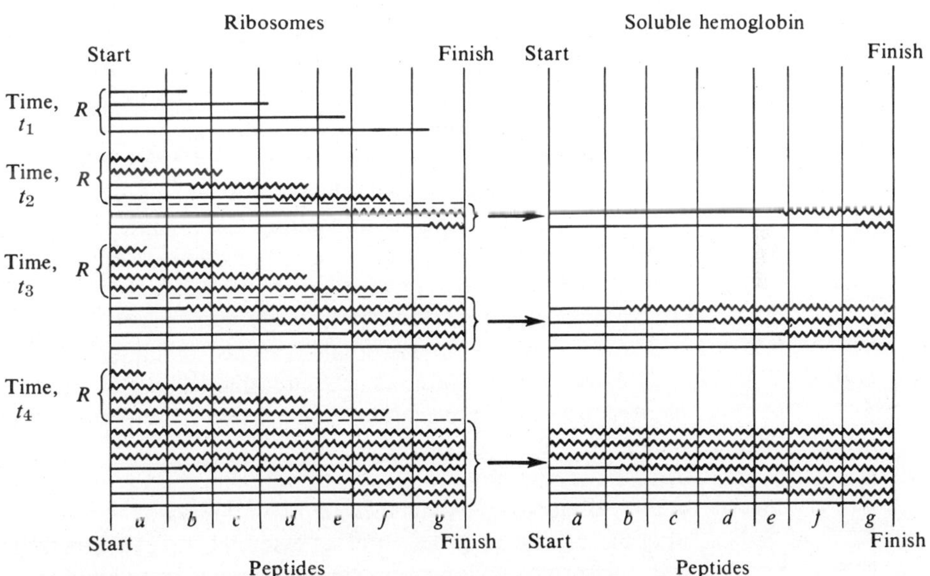

FIGURE 21-3

Model of sequential chain growth. The straight lines represent unlabeled polypeptide chains. The zigzag lines represent radioactively labeled polypeptide chains formed after the addition of a radioactive amino acid at time t_1. The groups of peptides designated *R* are unfinished bits attached to the ribosomes at each time; the rest, having reached the finish line, are assumed to be present in the soluble hemoglobin. In the ribosomes at time t_2, the top two completely zigzag lines represent peptide chains formed completely from amino acids during the time interval between t_1 and t_2. The middle two lines represent chains that have grown during the time interval but have not reached the finish line and are, therefore, still attached to the ribosomes. The bottom two chains represent those that have crossed the finish line, left the ribosomes, and are to be found mixed with other molecules of soluble hemoglobin. H. M. Dintzis and P. M. Knopf, in H. J. Vogel, V. Bryson, and J. O. Lampen (eds.), *Informational Macromolecules,* Academic Press, New York, 1963.

AUG (Table 19-2). It is therefore not surprising that both $tRNA_f$ and $tRNA_M$ recognize this codon when it is in a subterminal or interior position, but as required for a chain-initiating codon and its cognate tRNA, *only* $tRNA_f^{Met}$ recognizes *p*AUG (i.e., in a terminal position). More weakly it also recognizes pGUG—in the same position—a result to be expected from the hypothesis that the (U) in the presumed anticodon triplet CAU

$$\begin{array}{l} \longrightarrow \\ \text{pAUG–} \\ \text{–CA(U)–} \\ \longleftarrow \end{array}$$

is a "wobble" base (73) (see page 940) capable of ambiguous recognition of, precisely, A or G. In an interior position GUG is a codon for valine. One of the interesting features of chain initiation by the triplet AUG, consistent with its function in translation and polypeptide synthesis, is its strong phasing activity (i.e., it ability to establish a reading frame). A variety of experiments with synthetic oligo- and polyribonucleotides establish this point (74); they also demonstrate that it is the *position,* not the terminal phosphorylation of AUG itself, that is important:

1. fMet*-$tRNA_f$ serves as a Met donor for polypeptide synthesis, if the *in vitro* system supplemented by initiation factors is programmed with (p)ApUpG$(pA)_n$ or XYZApUpG$(pA)_n$ but not with $A(pA)_n$pApUpG. The product in the first two cases is fMet*-Lys_m.
2. In contrast, Met* from Met*-$tRNA_M$ is incorporated when the message is constituted by $A(pA)_n$pApUpG, to yield Lys_mMet*, and in this instance ApUpG$(pA)_n$ is totally inactive.
3. ApUpG$(pU)_{15}$ stimulates binding of Met-tRNA (presumably $tRNA_f$) and of Phe-tRNA (codon UpUpU) but not of Val-tRNA (codon GpUpU) to ribosomes. On the other hand, ApUpGpG$(pU)_{15}$ stimulates binding of Met-tRNA and Val-tRNA, but not of Phe-tRNA!
4. ApApApUpGp$(U)_{15}$ stimulates binding of Met-tRNA but not Lys-tRNA (codon ApApA).

These findings do, however, also raise certain questions:

1. How does an internal or subterminal ApUpG triplet ensure recognition of Met-$tRNA_M$ and rejection of fMet-$tRNA_f$, and conversely GpUpG ensure recognition of Val, again together with rejection of fMet-$tRNA_f$, particularly since location at the 5′ terminus or a terminal phosphate residue is neither necessary nor sufficient for this purpose? This selective recognition in internal positions is of course essential for two reasons: fMet must be rejected for chain elongation since as an amide it would terminate this process; on the other hand, recognition of GUG would lead to a Met/Val ambiguity.
2. Provided a mechanism of sufficient selectivity exists to deal with the problem just raised, how then can this be reconciled with the requirements of repeated chain initiation in polycistronic messages where fMet is again the N-terminal amino acid and fMet-$tRNA_f$ the chain initiating entity (see Fig. 5-23)?

The answer to question 1 is presumably intimately connected with the problem of ribosome assembly discussed in the last section. It will be recalled that fMet-$tRNA_f$ was bound to the 30-S subunit·mRNA complex in a specific manner to form part of the initiation complex. This is a unique reaction sequence that may well be confined to that region of the mRNA at or adjacent to the 5′-terminal end of the mRNA segment to be translated and additional specificity in this selection may be contributed by interaction of the specific initiation factors with the other components.[7] The binding reaction and the subsequent assembly of the complete ribosome with the migration or translocation of fMet-$tRNA_f$ to a reactive site is restricted to this species of tRNA and known not to occur with (deformylated) Met-$tRNA_f$, or Met-$tRNA_M$, or other aa-tRNA's. It is also known that the initiation factors effectively block the ability of such other aa-tRNA's to bind to the 30-S Rb-mRNA complex. Conversely then, once this series of reactions has taken place, with an −ApUpGp→ proximal to the 5′ terminus in a mRNA as the active participant, the geometry and other conditions may be such as to favor the interaction of all subsequent triplets with Met-$tRNA_M$ and reject fMet-$tRNA_f$. For instance, entry into the ribosome and participation in chain elongation (see below) may be restricted to aminoacyl-tRNA's with a free amino group; fMet-tRNA, of course, does not possess this structure, and therefore might be excluded. There is also some evidence that the protein factors responsible for the binding of *other* aa-tRNA's (T factors) can reject fMet-$tRNA_f$ (75).

Relation of Initiation to Two-Site Model of Protein Synthesis. From this point of view, at least part of the elaborate mechanism that links chain initiation to ribosome assembly may be a reflection of a requirement for the circumvention of the specificity of a ribosomal binding site, the *"aminoacyl-tRNA"* or *"A" site* (also called acceptor, entry, recognition, or decoding site), passage through which is ordinarily required in order to reach a second site called the *"peptidyl-tRNA"* or *"P" site* (also called donor or condensing site (see Fig. 21-2). Another variant might add the restriction that such entry of any aminoacyl-tRNA, other than the first, into the *A* site requires the prior or simultaneous *occupancy* of the *P* site. Thus pApUpGpApUpG . . . will produce binding of fMet-$tRNA_f$ to the *P* site, which in turn will facilitate the binding of Met-$tRNA_M$ at the *A* site and its insertion into a growing polypeptide fMet-Met. . . . Similarly, pN_1 . . . $pN_{m-2}pN_{m-1}pN_m$pApUpGp . . . will lead—provided an AUG codon is present close to N_1—in due course during chain propagation, first to the binding, at the *A* site, of Met-$tRNA_M$, and later to the insertion of its Met into the growing chain. Now consider a situation where the proper translation of the first of these two polynucleotides is no longer possible (e.g., destruction of the first codon due to removal of the terminal pA residue). The phasing proclivities of AUG will assure that chain initiation—and ribosome assembly—can still take place, but now starting with what was originally the *second* AUG triplet, occupying the *P* site and coding for fMet-$tRNA_f$. Let us turn to the analogous case for the second polynucleotide; assume that for some reason $pN_{m-2}pN_{m-1}pN_m$ cannot be translated, perhaps because it corresponds to a chain terminating, nonsense (page 867 and below) codon. Therefore, there is *no* entity occupying the *P* site, and thus the *A* site has become inaccessible until this restriction is removed. However, the situation is now analogous to case 1, one containing a subterminal AUG codon, which permits chain initiation and ribosome assembly. Hence, we have uncovered a feasible mechanism for chain initiation in the interior of a polycistronic message and a possible answer to question 2 above: *If an pApUpG triplet is preceded by an untranslatable stretch, say one or more nonsense codons in succession, it may code for N-formyl-Met and hence permit the initiation of new chains and the entry of a ribosome onto the*

[7] See the footnote on page 934.

messenger strand.[8] Most of the current theories of protein synthesis postulate that chain propagation also requires the two sites just mentioned: an *A* site for the entry of the next chain element and a *P* site to which the growing polypeptide chain remains attached in the form of a peptidyl tRNA (Fig. 21-4). Let us reexamine the model for chain initiation and ribosome assembly [Eqs. (21-9) through (21-12)] from this point of view. The *initial* binding of fMet-tRNA to the 30-S Rb·mRNA complex in a K^+ (or NH_4^+)- and GTP-requiring reaction is the analog of the codon specific binding of an aminoacyl-tRNA to the *A* site on the ribosome, an act that may involve a site on the 30-S subunit. Both these reactions require the participation of GTP but without any attendant hydrolysis of a pyrophosphate bond during the act of binding proper. This conclusion is based on the fact that the GTP analog, guanylyl-5′-methylenediphosphonate (GMPPCP),

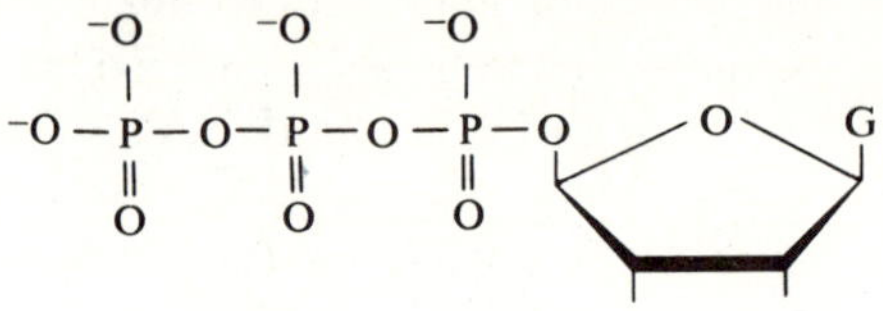

GTP

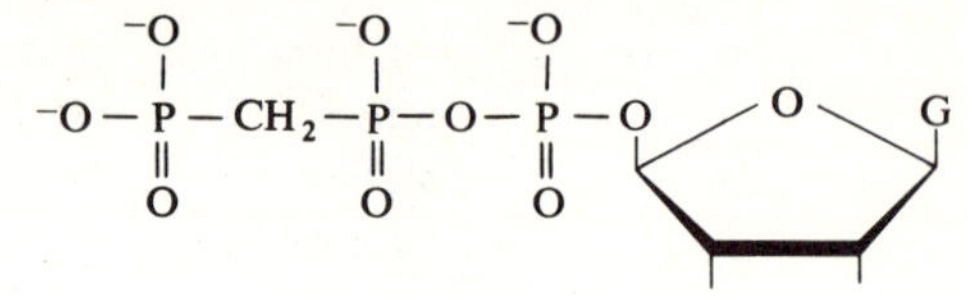

GMPPCP

in which the β–γ pyrophosphate bond is resistant to enzymatic hydrolysis, will substitute effectively for GTP in the binding step. In the next step the complex with the ribosome is somehow activated or primed, perhaps by means of a conformational change at the expense of splitting GTP to GDP plus P_i. In this reaction GMPPCP not only cannot replace GTP but becomes a strong and competitive inhibitor of GTP. As a result of this transformation the following events occur: (1) the fMet-tRNA is transferred to the *P* site; (2) the ribosome and the mRNA are displaced relative to one another so that the exposed codon on the latter (e.g., pApUpGpN_4pN_5pN_6-) is no longer the chain-initiating pApUpG- but the subterminal triplet -pN_4pN_5pN_6-; (3) this migration now permits the codon specific [GTP, K^+ (or NH_4^+) plus elongation (T) factor(s) requiring] entry into the *A* site of the aminoacyl-tRNA specified by -pN_4pN_5pN_6- (e.g., Phe-tRNA if $N_4 = N_5 = N_6 = U$); and (4) the formation of the first peptide bond without any additional expenditure of energy or additional factors (e.g., *N*-formyl-methionylphenylalanine is formed if the mRNA is ApUpGpUpUpU). A very simple model reaction for ribosome assembly, chain initiation, and peptide-bond formation is one which uses ApUpG as the mRNA in Eqs. (21-9) through (21-11). It requires only the components indicated there, plus the antibiotic puromycin (page 955), which takes the place of the incoming aminoacyl-tRNA acceptor, and reacts directly with the fMet-$tRNA_f^{Met}$ at the *P* site.

$$[\text{fMet-tRNA}_f \cdot 50\,S \cdot 30\,S \cdot \text{ApUpG}] + \text{puromycin} \longrightarrow 50\,S \cdot 30\,S \cdot \text{ApUpG} + \text{fMet-puromycin} + \text{tRNA}_f \qquad (21\text{-}15)$$

An even simpler, related, model reaction measures only peptide-bond formation and makes mRNA, the 30-S subunit, initiation factors, and even fMet-$tRNA_f^{Met}$ dispensable. Instead, it uses the terminal oligonucleotide fragment

[8] In the intracistronic region of Fig. 5-23 there are indeed three terminator codons, two of them in succession; however, within this sequence of eight triplets there is also found an initiator codon. This hypothetical hexapeptide is probably *not* translated. *In vitro* the *E. coli* system can initiate at all three sites shown, a system (ribosomes still containing initiation factors) from *B. stearothermophilus* only at the A protein (75a). Thus, different translation systems may be able to recognize *different* constellations of bases preceding an initiator triplet.

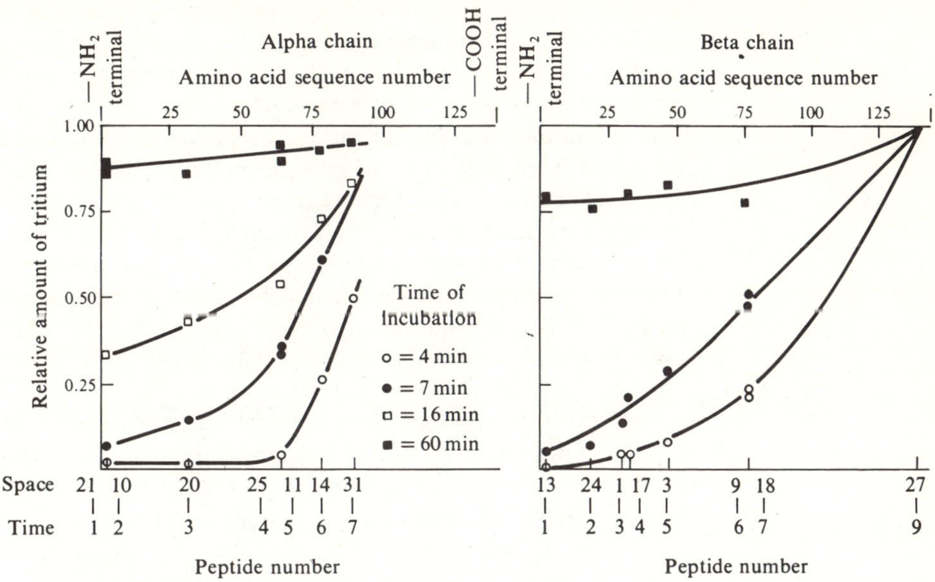

FIGURE 21-4

Distribution of ^{3}H-leucine among tryptic peptides of soluble rabbit hemoglobin after various times of incubation at 15°. The peptides are positioned by assuming that the leucines have the same sequence numbers in rabbit and human hemoglobins. If more than one leucine is present in the human peptide, an arithmetic-average sequence number is used to indicate the peptide position. This sequence of peptides along the chains is indicated by their position in "space," starting with the N terminus on the left. After 4 min of exposure to label, only the peptides near the C-terminal end are labeled. After more prolonged exposure, there is a gradient rising from the N-terminal end. This permits the ordering of peptides sequentially during the assembly process—their position in "time." H. M. Dintzis and P. M. Knopf, in H. J. Vogel, V. Bryson, and J. O. Lampen (eds.), *Informational Macromolecules,* Academic Press, New York, 1963.

derived from this entity by phosphodiesterase digestion (79). Originally studied with the terminal hexanucleotide CpApApCpCpA-Met-CHO, the reaction has since been extended to and activity has been demonstrated with (1) the terminal trinucleotide CpCpA-Met-CHO (the dinucleotide CpCA-Met-CHO is inert), (2) the corresponding *N*-acetyl-Met derivatives derived by chemical acetylation (acetic anhydride) of Met-tRNA prior to diesterase treatment, (3) the *N*-acetyl derivatives of other aminoacyl oligonucleotides obtained by analogous reactions from other aminoacyl-tRNA's, (4) 2′(3′)-*O*-aminoacyladenosine, and (5) ethanol as acceptor; in the last instance the reaction appears to be reversible. The reaction observed is simply

$$\text{Aminoacyl—NH—}\underset{\displaystyle R}{\underset{|}{\text{CH}}}\text{—}\underset{\displaystyle \text{O}}{\underset{||}{\text{C}}}\text{—O—ACC}\ldots + \left\{\begin{array}{l}\text{adenosyl AA}\\ \text{puromycin}\\ \text{EtOH}\end{array}\right. \xrightarrow[\text{0.4 M K}^+\text{, 0.02 M Mg}^{2+}\text{, 0°}]{\text{peptide synthetase (50-S Rb)}}$$

$$\text{Aminoacyl—NH—}\underset{\displaystyle R}{\underset{|}{\text{CH}}}\text{—}\underset{\displaystyle \text{O}}{\underset{||}{\text{C}}}\left\{\begin{array}{l}\text{NH-aminoacyladenosine}\\ \text{NH-puromycin}\\ \text{O-Et}\end{array}\right. + \text{CpCpA(OH)} \qquad (21\text{-}16)$$

The highly restrictive and selective features of chain initiation just described make one wonder how the relatively nonspecific polynucleotide-dependent incorporation reactions so instrumental in elucidating the code could ever have been discovered in the first place. The answer to this apparent paradox lies, in part at least, in the fact that most of these early studies were performed in the presence of high concentrations of Mg^{2+} ($\geqq$20 mM). Under these conditions not only do the requirements for initiation factors (which in any event were still present in a form bound to the incompletely washed ribosomes used in these studies) disappear, but association of 30-S and 50-S subunits can occur even in the complete absence of mRNA and tRNA. It should occasion no surprise therefore that, in the *presence* of *polynucleotide* mRNA's, *nonspecific* tRNA's coded by the terminal and subterminal triplets could be positioned onto nascent ribosomes in the place of fMet-$tRNA_f$, and hence function as chain initiators. Thus, at high Mg^{2+}, (p)XYZ$(U)_n$ will code for polyphenylalanine regardless of the identity of XYZ and initiation factors are not required. When the concentration of Mg^{2+} is lowered to $\leqq$2 mM, a requirement for these factors emerges; XYZ must be either AUG or GUG, no translation takes place in the absence of fMet-$tRNA^{Met}_{f}$ (Met-$tRNA^{Met}_{f\,or\,M}$ will not do!), and the product is fMet$(Phe)_m$.

In summary, the normal chain-initiation process in the *E. coli* system appears to provide the following essential functions in protein synthesis: (1) a point of unambiguous attachment and entry of ribosomes onto the mRNA tape, (2) a means for the assembly of ribosome from their subunits, (3) a means for the positioning of an analog of peptidyl-tRNA onto the appropriate (*P*) site on the ribosome, (4) a means for opening the recognition site (the *A* site) to the subterminal aminoacyl-tRNA and for positioning it opposite the cognate subterminal codon on the mRNA, (5) a means for discriminating between terminal and nonterminal positions of the ambiguous codons AUG and GUG and, perhaps (6) a means for excluding the participation of inappropriate codons and cognate tRNA's in initiation and ribosome assembly.

Universality of Initiation. The importance of the chain-initiating step would argue for its relative universality and a similarity of basic mechanisms in different cells. This postulate appears to be true for bacteria where initiation by *N*-formylmethionyl-$tRNA^{Met}_{f}$ is the rule. Reported differences, for instance in the relative proportions of Met and Ala among the N-terminal amino acids of the proteins of *B. subtilis* compared to *E. coli,* appear to be referrable to differences in the hydrolytic modifying activities in the two organisms subsequent to protein synthesis rather than in the substitution of amino acids other than Met as the universal terminus (80). Evidence is accumulating that the mechanism of chain initiation in mitochondria (81–83), as in mitochondrial protein synthesis in general, is reminiscent of that in bacteria. A somewhat different situation appears to prevail with the cytoribosomes of eucaryotic cells. Chain initiation and ribosome assembly appear to be coupled in some manner in this instance also (62,63) but until recently no evidence had been forthcoming, in spite of a diligent and intensive search, for participation of fMet-$tRNA_f$, or of much light on chain initiation under physiological conditions in general. Thanks to the penetrating studies of A. E. Smith and K. A. Marcker (61), this process also is yielding to analysis. The cytosol of animal cells, as of eucaryotes in general, contains two principal types of tRNA capable of accepting methionine, which have been designated as $tRNA^{Met}_{M*}$ and $tRNA^{Met}_{f*}$. The former is analogous to the bacterial $tRNA^{Met}_{M}$ and like its counterpart is implicated in the insertion of Met into internal positions in response to ········ AUG ········ (*not* ········ GUG ········, which specifies Val). It is completely ineffective in chain initiation and cannot be formylated by the transformylase (from *E. coli*), the enzyme responsible for catalyzing reaction (21-14) and known to be absent from the cytosol of eucaryotes. Not so $tRNA_{f*}^{Met}$: This species *can* be formylated by the *E. coli* enzyme and when so substituted replace $fMet\text{-}tRNA_f^{Met}$ in the complete *E. coli* system. However, it is the *unformylated* $Met\text{-}tRNA_{f*}^{Met}$ that fulfills the *chain-initiating* function in eucaryotic cells: it recognizes AUG and GUG codons in initiating, and discriminates against them, in internal positions. For these discriminatory functions the 80-S ribosome as well as appropriate elongation and initiation factors are required (Table 21-1). The former have been known for some time; evidence for the latter is beginning to accumulate (84). What remains to be established are the steps that must intervene between the nascent polypeptides (with N-terminal Met by this hypothesis) and the stable products (none of them with Met in this position).

Chain Elongation (Propagation)

Formation of Peptide Bonds (*Fig. 21-5*). The facts that appear to be common to the various *in vitro* systems studied so far are the following (85–89) (see also Table 21-1 for nomenclature):

1. Each individual ribosome actively engaged in polypeptide synthesis carries a single growing polypeptide chain.

2. Each such particle is capable of binding two (and perhaps, more weakly, a third) tRNA molecules. One site appears to be localized mainly on the 30-S subunit, the other on the 50 S. The rate of binding is, however, stimulated by the complementary subunit. This stimulation is more evident for binding to the second site than for binding to the first.

3. The first (*A,* aminoacyl-tRNA, acceptor or decoding site) appears principally concerned with a codon-specific binding of monomeric aminoacyl-tRNA; the second (*P,* peptidyl-tRNA, condensing or donor site) with positioning of the growing chain, in the form of a peptidyl tRNA, for reaction with the incoming aminoacyl-tRNA. There is some evidence that free tRNA is a competitor of binding, particularly to the *P* site and, as we have already mentioned, that entry into this site, even of a peptidyl tRNA added externally, must proceed by way of the *A* site as an obligatory port of entry.

4. GTP is an obligatory coreactant in this system. It is required for the binding reaction, and for the positioning of the newly formed peptidyl-tRNA into the *P* site. The latter reaction also serves to advance the mRNA by one coding unit.

5. A minimum of two, and probably three, soluble proteins (*elongation factors, transfer factors,* transferases) (Table 21-1) participate in the chain-elongation sequence, but the peptide-forming step proper is catalyzed by a *peptide* (bond) *synthetase* (PS) (peptidyl transferase, transpeptidase) intrinsic to the 50-S ribosome. A peptidyl-tRNA in the *P* site will therefore react "spontaneously" without added cofactors with an acceptor such as puromycin.

6. In addition to GTP, divalent (Mg^{2+}) and monovalent (NH_4^+, K^+) ions, as well as reactive thiol groups (probably of the G Factor), are required.

A fairly general model for the reaction sequence is shown in Fig. 21-5. Starting with formation of the peptide bond, the chain-propagating step proper (76–78,90), at the top of the diagram, we observe that the sequence can be dissected into three discrete phases. In the first, $aa_n tRNA^n$, the nth aminoacyl-tRNA, bound at the *A* site, positioned opposite its appropriate codon on the mRNA, attacks through its amino nitrogen the terminal carboxyl of a peptidyl-tRNA cognate for the $(n-1)$st residue. The nth peptide bond is

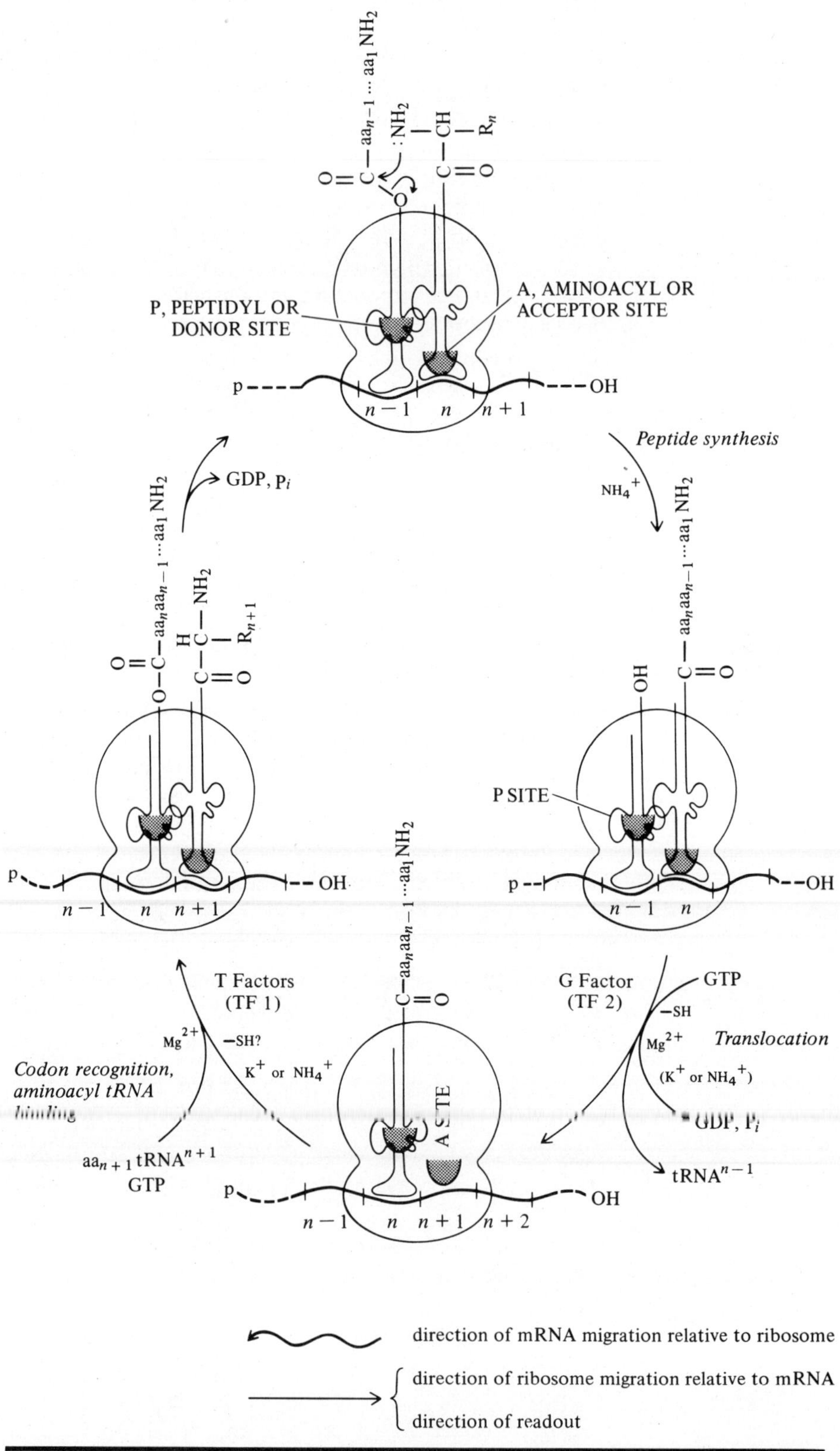

FIGURE 21-5
Sequential steps in protein synthesis by a two-site model.

formed, and as a result the growing chain is transferred to the aminoacyl-$tRNA^n$ cognate to the *n*th amino acid, which *remains attached to the A site*. The preceding tRNA, now deacylated, may either detach spontaneously or, what is more likely, may linger in the general vicinity of the *P* site until displaced in the subsequent step(s). There is some evidence that peptide-bond formation catalyzed by the 50-S ribosome or, more specifically by PS, one of its proteins, depends on the maintenance of ribosomal conformation, which requires NH_4^+ ions and is cold labile, and also that in the course of the reaction the particle undergoes a conformation change from a compact to a more expanded structure (91) (Rb′ → Rb) (Figs. 21-2 and 21-5).

The next step, commonly called *translocation,* actually encompasses a number of events: (1) the peptidyl $tRNA^n$ is translocated into the *P* site, thus opening the *A* site and expelling the deacylated $tRNA^{n-1}$; (2) the mRNA tape is translocated relative to the ribosome, in effect advancing the tape one codon (i.e., three nucleotides closer to the −OH end of mRNA) and positioning the $(n + 1)$st codon opposite the *A* site; finally, (3) the ribosome may be translocated from the expanded to the compact conformation (Rb → Rb′).[9]

These reactions require the obligatory participation of GTP and a soluble factor [variously called *G Factor* in *E. coli*; Tranferase II or Transfer Factor (TF) 2 in mammalian systems]. GTP is hydrolyzed in the process to GDP plus P_i, and GMPPCP (page 934) cannot substitute and inhibits. Thus, G Factor, a tetrameric protein, containing essential thiol groups, which has been crystallized (92) (M = 80,000 for bacterial, 70,000 for mammalian factor) functions as a GTPase when it is supplemented by ribosomes; alone either component is inactive. There is some evidence that the reaction is complex and may proceed through a number of discrete steps, such as

$$\text{GTP} + \text{G Factor} \rightleftharpoons \text{G Factor} \cdot \text{GTP} \tag{21-17a}$$

$$\text{G Factor} \cdot \text{GTP} + \text{Rb} \rightleftharpoons \text{G Factor} \cdot \text{GTP} \cdot \text{Rb} \rightleftharpoons \text{G Factor} \cdot \text{GDP} \cdot \text{Rb}' + P_i \qquad (21\text{-}17b)^9$$

$$\text{G Factor} \cdot \text{GDP} \cdot \text{Rb}' \rightleftharpoons \text{Rb}' + \text{G Factor} + \text{GDP} \tag{21-17c}$$

In the third, and final, phase, in the representation of Fig. 21-5, the incoming $(n + 1)$st aminoacyl-$tRNA^{n+1}$ is inserted in a highly specific manner into the *A* site, opposite its cognate codon on the mRNA. The factors governing this crucial recognition step, which is tantamount to the translation of the nucleotide into the amino acid code and brings these two codes into congruence, will be discussed below in a separate section. The reaction itself requires one or more soluble proteins (93–97) [*T Factors* (T_s, T_u) in *E. coli*; Factor S_1, S_3 in *Bacillus stearothermophilus;* Transferase I or TF1 in mammals (96)] plus GTP. Again the reaction sequence is probably composed of various steps, of which the following have received some experimental support.

$$T_u + T_s \rightleftharpoons \text{T Factor} \tag{21-18a}$$

$$\text{T Factor} + \text{GTP} \rightleftharpoons \text{T Factor} \cdot \text{GTP} \tag{21-18b}$$

$$\text{GTP} \cdot \text{T Factor} + \text{aminoacyl-tRNA} \rightleftharpoons \text{T Factor} \cdot \text{GTP} \cdot \text{aminoacyl-tRNA} \rightleftharpoons T_s + T_u \cdot \text{complex} \tag{21-18c}$$

$$T_u\ (\text{or TF}) \cdot \text{GTP} \cdot \text{aminoacyl-tRNA} + \text{Rb}' \rightleftharpoons \text{Rb}' \cdot \text{aminoacyl-tRNA} + T_u\ (\text{or TF}) \cdot \text{GDP} + P_i \tag{21-18d}$$

[9] This step may actually take place later, at least in the *E. coli* system, with the intervention of the T_u Factor; that is, reaction (21-17) should be written as G · GTP · Rb · aa-tRNA(*A*) → (*P*)aa-tRNA · Rb + G + GDP + P_i and reaction (21-18d) as T_u · GTP · aa-tRNA + Rb → Rb′ · aa-tRNA + T_u · GDP + P_i.

Mg^{2+} ions (at ≤5 mM) are required for the codon-specific binding of aminoacyl-tRNA's to the ribosomal *A* site even in the presence of T Factors plus GTP. However, an increase of the concentration of the ion to ≲20 mM obviates the necessity for the latter two components, an observation that serves to reconcile this statement with earlier observations in this field (see below).

Recognition Between Codon and Anticodon

Aminoacyl-tRNA–mRNA Binding. We have just mentioned that at moderate concentrations of Mg^{2+} ions (≲20 mM), the mRNA·Rb complex alone is capable of codon-selective binding of aminoacyl-tRNA's and thus brings about the specific mutual recognition of the apposite nucleotide sequences: the codon on the mRNA and the anticodon or "nodoc" on a tRNA.[10] Coupled to the fact that poly- and even oligonucleotides could function as mRNA's in this reaction, this observation by Nirenberg and Leder (98) provided the basis for a simple and convenient assay for genetic coding (98,99).

It furnished suggestive and independent evidence that the coding ratio equals 3; $(pU)_3$, $(pA)_3$, and $(pC)_3$ when bound to ribosomes directed effective binding of Phe-tRNA, Lys-tRNA, and Pro-tRNA. Increasing the chain length beyond 3 up to 12 had relatively little effect, but dinucleotides were completely inactive.

Interesting results were obtained in comparing different states of phosphorylation of active trinucleotides; $pN_1pN_2pN_3$ was found to be highly active, $N_1pN_2pN_3$ had some activity, and $N_1pN_2pN_3p$ was virtually inert. This suggested that readout starts at the 5′-terminal end of the mRNA and recognizes codons starting with the 5′-phosphoryl residue. Another type of experiment, also indicating polarity in readout, came from a comparison of sequence isomers such as $N_1N_2N_3$ and $N_3N_2N_1$; GpUpU led to Val-tRNA binding, while UpUpG rejected Val-tRNA and bound Leu-tRNA. The assumption received its proof by the use of polynucleotides of the general structure $N(pN)_npN'$ in the incorporation assay; for instance, $A(pA)_nC$ codes for polylysine with a *C-terminal* asparagine (codon AAC). Hence, readout must have been from left to right.

By means of such binding experiments Nirenberg and his collaborators were able to assign a large number of codons to specific trinucleotide sequences. The general coding scheme presented as Table 19-4 is based on consistent results obtained mainly by this technique and by studies on the formation of specific polypeptides in response to polynucleotides of known sequence. It receives its strongest support from the consistency of base and amino acid sequences of RNA-phage proteins, summarized in Fig. 5-23.

Anticodon Loop of tRNA; Wobble. The availability of a number of tRNA's of known primary structure makes possible an inquiry into the presence and location of likely anticoding sequences in these molecules. As discussed earlier, such potential anticodons can indeed be found and localized in the central loop (loop II) of the cloverleaf conformation of tRNA (Fig. 21-1 and Table 21-3).

The following are some of the sequences recognized as anticodons, together with likely codons cognate to them (26,28).

[10] In the current literature it is customary to refer to the code groups or codons on mRNA as the primary coding units. Although true for RNA viruses, in all other organisms the repository of genetic information is DNA, and this assigns to the base sequences in DNA the primary coding function. Thus, mRNA codons are complementary and tRNA (anti)codons correspond (but not precisely) to the DNA codons transcribed.

Yeast

	Ala	*Phe*	*Ser (I, II)*	*Tyr*	*Ile*	*Val*	
anticodon	CGI	AAC (OMe)	AGI	AΨG	UAI	CAI	3′ ← 5′
codon(s)	U GCC A	$\mathrm{UU}^{\mathrm{U}}_{\mathrm{C}}$	U UCC A	$\mathrm{UA}^{\mathrm{C}}_{\mathrm{U}}$	U AUC A	U GUC A	5′ → 3′

E. coli

	fMet	*Tyr (I, II, Su_3-)*	*Su_3+*
anticodon	UAC	AUG*	AUC
codon(s)	$^{\mathrm{A}}_{\mathrm{G}}\mathrm{UG}$	$\mathrm{UA}^{\mathrm{C}}_{\mathrm{U}}$	UAG (unique = amber)

G* = modified G

The codon assignments take into consideration a fact already brought out in our discussion of the code in Chapter 19. There we noted that the large number of degenerate, synonym codons appear to obey the following rules. Alternate codons generally only vary in the third position and are of three classes: one, in which one pyrimidine can substitute for the other, which appears to hold true for all 16 possible instances; a second, where the same substitution rule holds in the case of purines, which appears to occur in 12 (or 13) of the total of 16; and a third, where all four bases seem to be allowed, which appears to hold in 8 cases out of 16. This problem and its implications with respect to synonym recognition by tRNA anticodons was taken up explicitly by Crick (73). As a result of geometrical and logical considerations he formulated the *wobble hypothesis,* which may be stated as follows: While standard (in the Watson–Crick sense) base pairs are used quite rigorously for the first two bases of a coding triplet, some "wobble" and a resultant ambiguity is allowed in the third. The pairing rules for these wobble bases on the anticodon are the following (see also page 923): A recognizes only U; C recognizes only G; U (and modified U) recognizes A and G; G (and modified G) recognizes U and C; I recognizes U, C, and A. The predictions of the hypothesis appear to be borne out by studies on oligonucleotide mRNA versus purified aminoacyl-tRNA interactions that show extreme specificity with regard to the first and selective ambiguity with regard to the third nucleotide in a codon. Specifically, U can substitute for C (wobble base G) in GG*U* ($\mathrm{tRNA^{Gly}}$), AU*U* ($\mathrm{tRNA^{Ile}}$), UU*U* ($\mathrm{tRNA^{Phe}}$), and UC*U* ($\mathrm{tRNA^{Ser}}$); A can substitute for G (wobble base U) in AG*G* ($\mathrm{tRNA^{Arg}}$), GG*G* ($\mathrm{tRNA^{Gly}}$), and UC*G* ($\mathrm{tRNA^{Ser}}$); U, C, and A are equally effective (wobble base I) with GpCp*N* ($\mathrm{tRNA^{Ala}}$) and CpGp*N* ($\mathrm{tRNA^{Arg}}$). Five different codons ending in G appear unique and hence probably contain C in the wobble position.

The total anticodon region in all the tRNA's so far examined consists of a total of 17 nucleotides (Fig. 21-1 and Table 21-3). Of these, 10 are arranged pairwise by hydrogen bonding in the helical stem and the remaining seven form the open anticodon loop proper. Starting from the point closest to the 3′-adenosine end, the general nucleotide sequence in the loop appears to be $NR^*\bar{N}_1\bar{N}_2WUY$, where N is any base (usually A), Y is a pyrimidine, R* an (alkylated) purine, $\bar{N}_1$ and $\bar{N}_2$ are the nucleotides complementary to the first and second codon position, and W is the nucleotide bearing the wobble base just discussed. The available structural information has been utilized by Fuller and Hodgson (23) to construct a stereochemically acceptable model for the anticodon loop and its interaction with mRNA (Fig. 21-6).

Suppression and Ambiguity (32). Suppression is defined genetically (Chapter 19) as a second mutation introduced at a separate locus into a mutation so as to revert the phenotype of the latter toward that of the wild type. Here we shall be concerned only with those suppressors that are in*ter*genic and operative at the level of translation. [Other mechanisms of suppression, such as restoration of phase in phase (frame)-shift mutants and intragenic

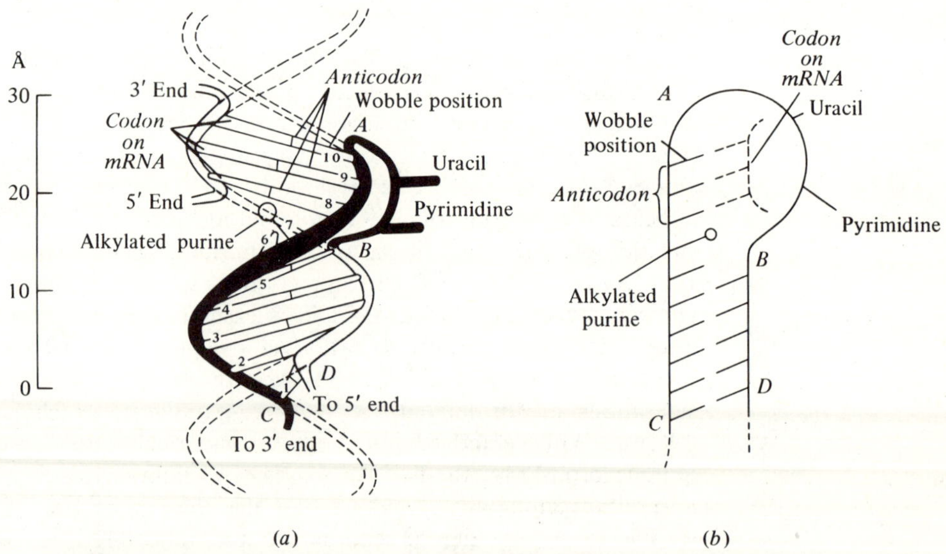

FIGURE 21-6

Stereochemistry of codon : tRNA anticodon recognition. (*a*) Schematic diagram of the tRNA anticodon loop, illustrating its relationship to the codon and the helical character of the structure. The letters A, B, C, and D identify the same points on the structure as in part (*b*). The bases in nucleotides 1 to 10 are stacked on one another and follow the regular helix, which is shown black. The chain of the anticodon double helix between D and B is shaded like the codon to indicate that they follow the same helix. This helix is complementary to the black one. The two nucleotides not in the standard conformation are represented by dark line shading. The representation of their conformation is very schematic because they lie behind nucleotides 8, 9, and 10 in the black chain. The dashed lines indicate the generic helix from which the structure can be imagined to be derived.

(*b*) Schematic diagram of the model for the tRNA anticodon arm, illustrating its relationship to the codon. The helical regions are shown as straight in this diagram. CD is the first base pair in the double helical region of the anticodon arm and all the bases between A and C are stacked on one another and follow a regular helix. The companion set between B and D and the set of three bases in the codon follow the complementary helix. In space, A and B are quite close together, because five nucleotide pairs is about half a turn of the helix. Reproduced from W. Fuller and A. Hodgson, *Nature,* **215:** 817 (1967); with permission.

(second site) reversion, were discussed in Chapter 19.] Two principal subclasses can be distinguished: *nonsense* and *missense* suppressors. Suppose we alter the structure at (or around) the anticodon site in a $tRNA^x$ so that the molecule, previously capable of recognizing only codon $XY'Z$, now recognizes codon XYZ.[11] The consequence of this act (the introduction of a suppressor mutation, $Su^- \rightarrow Su^+$) will be the insertion of the novel amino acid x whenever XYZ occurs in a message. If XYZ itself codes for an amino acid y [where y in any particular instance may itself have been introduced as a consequence of another mutation in a structural gene, say $X'YZ$ (specifying z) being changed to XYZ], then we obtain an altered polypeptide and we speak of *missense suppression.* The postulated changes were

$$\underset{z}{(X')YZ} \xrightarrow[\text{in gene}]{\text{mutation}} \underset{y}{XYZ};\ \underset{x}{X(Y')Z} \xrightarrow[\text{recognition}]{Su^- \rightarrow Su^+} \underset{x}{XYZ}\ ^{12};\ \text{therefore, } \underset{(y)\,+\,x\ (\text{may be} \equiv z)}{XYZ}$$

If in contrast XYZ is a chain terminator [i.e., if it does *not* code for any (or at least any functional) amino acid (codons UAA, UAG, and UGA, page 874)], then the resultant *nonsense suppression* will permit the translation of the complete polypeptide rather than the accumulation of a fragment, terminated by the amino acid preceding the nonsense codon:

$$\underset{z}{X'YZ} \longrightarrow \underset{0}{XYZ};\ \underset{x}{XY'Z} \longrightarrow \underset{x}{XYZ};\ \text{therefore, } \underset{x}{XYZ} \text{ rather than stop}$$

The presence in, or the introduction of, a nonsense suppressor mutation into an appropriate (permissive) bacterial strain (genotype Su^+) permits the translation of one (or more) of the nonsense codons UAG (amber), UAA (ochre), or UGA *wherever it occurs* (*in phase*) *in a genetic message.* Since such a nonsense mutation may have been present not only in more than one cistron of a bacterial genome, but also in invading viral ones, it can be seen that this form of suppression is of necessity intergenic and general. At least six separate suppressor mutations mapping at distinct sites in the *E. coli* genome have so far been identified. They are designated as Su_1, Su_2, . . . , Su_6 (also with roman subscripts and, for the first two as Su_A and Su_B). Su_1, Su_2, Su_3, and Su_6 are specific for the UAA (amber) codon, while Su_4 and Su_5 permit the translation of both UAG and UAA. In addition, at least one suppressor for UGA is known. The molecular basis[13] for the mechanisms of these suppressors has become known due to the identification of their respective gene products (designated as su_1, su_2, etc.) as tRNA's altered in their primary structure, specifically by the change of a single base in the appropriate anticodon. The first intimation to this effect came from the discovery that $su_1{}^+$ appeared to be a structurally modified $tRNA^{Ser}$. The identification of su_3 as a simililarly modified $tRNA^{Tyr}$ quickly followed, and the ability of the purified tRNA fractions to suppress both natural-phage RNA and synthetic polynucleotides *in vitro* was demonstrated. By experiments on amino acid substitutions on capsid proteins of RNA phages and on the tryptophan synthetase

[11] If this alteration is such as to now *exclude* recognition of $XY'Z$, we may speak of *unambiguous*, if it *includes* $XY'Z$ *as well as* XYZ, of *ambiguous* suppression.

[12] XYZ will still be read also as y, owing to the presence of other $tRNA^y$'s cognate to this codon.

[13] Chain termination by and suppression of UAG and UAA codons in yeast (i.e., an eucaryote) has been demonstrated by D. C. Hawthorne, *J. Mol. Biol.*, **43:** 71 (1969). Hence the universality of the code extends to these features as well.

and alkaline phosphatase proteins of *E. coli,* a consistent picture emerged of Ser, Gln, Tyr, Tyr again, Lys, and Leu insertion as a consequence of the relevant changes from Su^- to Su_1^+ through Su_6^+, respectively. The proof was completed with the determination of the primary structure of the specific gene product present in Su_3^+ strains, a tRNA^{Tyr} that differs from the product of the Su_3 allele by a simple base substitution in the wobble position of the anticodon (AUC instead of AUG*), permitting the reading of UAG (amber) instead of UAC/U (Tyr) by an otherwise unaltered minority species of tRNA^{Tyr} (see Table 21-3 and the preceding section). Since in the case of alkaline phosphatase the original mutation had been from UGG (Trp) to UAG, we see that this suppression has *not* restored the original mutational change [i.e., the protein now contains Tyr(x) instead of Trp(z)]. It is evident why only such a minority species can participate in suppression of this sort: If a tRNA for a nondegenerate amino acid, or a majority species of a degenerate one, were so altered unambiguously, translation of this exclusive or major codon would no longer be possible at all, and the mutation would be lethal. In fact, an example is known (100): Su_7 is similar to Su_2 in inserting Gln for UAG but is lethal unless present as the diploid Su_7^+/Su_7^-, when it is much more effective than Su_2. It probably involves a mutation of the anticodon in the majority species of tRNA^{Gln} (i.e., CUG → CUA).

We have now described the probable molecular mechanism for suppression of generalized nonsense mutations in any gene arising from a base substitution in any position to produce a nonsense triplet somewhere. However, as we have seen in Chapter 19, nonsense codons can also arise in consequence of the deletion or insertion of a base (i.e., as a result of a phase-shift mutation). Can such an event be suppressed—now in a highly specific manner—by a complementary change in the anticodon of a tRNA? Yes, an actual example is the following (101): The wild-type protein has the relevant sequence -Val · Thr · Pro · Glu · Glu-, corresponding to the mRNA tract -GUC · ACC · CCU · $\text{GA}^{\text{A}}_{\text{G}}$ C $\text{A}^{\text{A}}_{\text{G}}$-. The mutation His D3018 probably results in the insertion of C somewhere in the bracketed region leading to -GUC · ACC · CCC · UGA · $^{\text{A}}_{\text{G}}$ C A · $^{\text{A}}_{\text{G}}$- and hence a polypeptide terminating in -Val · Thr · Pro. The unlinked mutation His D3018 sup probably produces a tRNA^{Pro} with a "relaxed" anticodon site that in some manner is complementary to the tetranucleotide stretch · CCCU. By this means translation from the mutant mRNA is put back into phase, since *reading* is now -GUC · ACC · CCCU · $\text{GA}^{\text{A}}_{\text{G}}$ · C $\text{A}^{\text{A}}_{\text{G}}$,- producing the uninterrupted amino acid sequence of the wild type.

Analogous mechanisms also appear to be operative in the case of the suppression of missense mutations (101,102). Thus, it could be shown that strains carrying suppressor mutations (su_{78}, su_{76}), known to be capable of reintroducing Gly, and Cys or Arg, respectively, in mutants of the polypeptide of tryptophan synthetase at two different Gly sites (page 870), contain tRNA's that carry out the appropriate alterations *in vitro* (102). Thus in the former instance poly UG, specifies the formation of $(\text{Val-Gly})_n$ as well as of $(\text{Val-Cys})_n$ with the tRNA's from su_{78}^+, but only the second with the su^- tRNA. In the second case poly AG specifies $(\text{Glu-Gly})_n$ as well as $(\text{Gly-Arg})_n$ with the su_{76}^+ tRNA's; again only the second copolymer is formed with tRNA's from su^-. Finally, it has been shown that su_{159}^+ (equivalent to su_{76}^+) actually contains a tRNA^{Gly} (which ordinarily decodes $\text{GG}^{\text{A}}_{\text{G}}$) capable of inserting Gly for Arg (by recognizing codon AGA). A major question appears settled by the last experiments concerning the nature of the tRNA that has undergone the mutational change; a priori two principal alternatives are possible, corresponding to a modification in either the tRNA of the amino acid to be replaced (Cys or Arg) or of the one to be inserted (Gly). In either case the resultant recognition may now occur either with, or without, ambiguity relative to the su^- condition.

Ambiguity is the basis for two other types of suppression that are of more than passing interest. One is the kind of *phenotypic suppression* of nonsense mutations found in the presence of 5-fluorouracil (F), its nucleoside or nucleotide. As pointed out previously, these almost isosteric analogs of the uracil derivatives are incorporated into RNA with high efficiency during transcription. This process is probably reasonably selective (in the sense of specifically permitting only U → F substitutions), but the product formed may well permit errors during translation, including ones that lead to pairing of a preexisting (or F-containing) $tRNA^x$ with a nonsense codon containing F instead of U. This alteration, although producing a complete gene product, is of course noninheritable and amounts to a phenotypic (one generation) correction rather than a genotypic suppression of a mutational change (103). Another, related phenomenon is that of streptomycin dependence. As discovered by Gorini and his collaborators (104), a relatively large proportion of auxotropic mutants induced in *E. coli* by different mutagens are capable of growth in the absence of their various growth factors, provided that streptomycin, one of its derivatives, or another related aminoglycoside antibiotic is added. Mutations of this type are called conditionally streptomycin dependent (CSD) and have been identified for a number of bacterial and bacteriophage genes. They can be of either the missense or the nonsense type. In the latter instance there is some evidence for a connection between streptomycin dependence and suppressibility and for a similarity between the mechanisms involved in suppression by su^+ and by streptomycin (100,105). The molecular explanation of the phenomenon appears to be that the antibiotic, which is known to be bound to P_{10}, a specific protein of the 30-S ribosome (specifically of the 23-S core) (106), so alters the structure of the particle as to induce misreading at some step of the translational process, thus again correcting the mutational alteration. *In vitro* experiments (see also page 954) are at least consistent with this hypothesis; streptomycin and other aminoglycoside antibiotics do induce ambiguous misreading of synthetic polynucleotide messengers (107). However, the effects observed are complex, and even the most well defined one, the introduction of Leu for Phe in a poly U-directed system, exhibits the disturbing feature of occurring equally well with ribosome isolated from streptomycin dependent (Str^d), resistant (Str^r), or sensitive (Str^s) cells—which differ greatly in their ability to bind the antibiotic—and of requiring a small amount of denatured DNA as a cofactor (108). Ambiguity in the recognition of specific codons and resultant losses in the fidelity of their translation, particularly in the case of Phe and Leu, can be demonstrated under a wide variety of conditions *in vitro,* the significance of which *in vivo* remains obscure; among them are increases in Mg^{2+} concentrations, lowering the temperature, and the presence of agents affecting secondary structure in nucleic acids such as alcohols, urea, steroidal diamines, and so on. Related effects *in vivo* may be the ability of high concentrations of Mg^{2+} to permit growth of the *r*II mutants of T-even coliphages in the restrictive host K12 (λ), and the growth inhibiting, or lethal effects of Leu, which can be overcome by Phe in certain strains of *Aspergillus* and *E. coli.*

Polysomes as Functional Units

Messenger RNA molecules, functional *in vivo,* are very long chains, particularly if they, as seems likely, must assume a conformation that minimizes intrachain hydrogen bonding and helix formation. For instance, a single polypeptide containing 200 amino acid residues would correspond to a (monocistronic) mRNA with end-to-end distance of about 4000 Å (600 nucleotides × 6.8 Å internucleotide distance); a polycistronic messenger such as the one for the tryptophan (*try* or *trp*) operon has been estimated to contain 7320 nucleotides. The diameter of a single ribosome (33) is 90 to 100 Å. Thus, newly formed ribosomes, assembled at the chain-initiating site close to the 5′ terminus of mRNA and simultaneously initiating the synthesis of a polypeptide from its amino terminus (see the preceding section), can progress along the mRNA chain, bearing a growing polypeptide chain, and render the entry site accessible to a new unit, long before the preceding one has reached the end of the message. This phenomenology accounts for one of the most well documented observations in the area of protein synthesis, that is, that ribosomal particles engaged in protein synthesis, *in vivo* or *in vitro,* are always found in linear or rosette-shaped clusters (Fig. 21-7) called polyribosomes or polysomes (85–88). A highly schematic model for protein synthesis on polysomes is shown in Fig. 21-8.

Several conclusions are implicit in this model and in the discussion so far presented.

1. The pool of ribosomes available for assembly into functional units consists exclusively of small and large subunits (30-S and 50-S in bacteria). Dissociation of subunits and return to (and exchange with) the pool of subunits take place after each round of translation by a functional ribosome.

2. Monosomes (70-S in bacteria, 80-S in animal cells) may well be artifacts; they arise either by fragmentation (say by shear, or phosphodiesterase action) of the rather unstable polysomal aggregate bound together by mRNA, or by adventitious association of two unequal subunits. In the former case the particles will carry both nascent polypeptide chains and mRNA segments; in the latter they may lack either. There is, however, some evidence that 30-S and 50-S subunits do *not* form 70-S particles in the *complete* absence of a polyribonucleotide.

3. Similar considerations also obtain with regard to those small polysomes smaller in size than required for the smallest polypeptide actually translated.

4. The average polysomal size will be dependent on the length of the mRNA, but the number of ribosomes or "density of occupation" per unit length may vary in different systems.

The polysomes engaged in the synthesis of the chains of hemoglobin (140 aa residues) exhibit a peak distribution of 5 to 6 ribosomes per cluster ($s = 20$ S, one ribosome per 80 nucleotides). Those engaged in translating serum albumin (575 residues) contain 20 ribosomes ($s \geqq 500$ S, one ribosome per 86 nucleotides). On the other hand, it has been estimated that the mRNA for the *trp* operon in *E. coli* may at times carry one ribosome for every 22 nucleotides or approximately 150 Å. Since this value is of the same order of magnitude as the diameter of the particle itself, the density would be extremely high, with the ribosomes arrayed "bumper to bumper" within the polysome. On the other hand, in the mammalian systems described, even with their somewhat larger ribosomes, the array appears a good bit looser: one ribosome per 450 Å of mRNA, corresponding to almost equal stretches of mRNA occupied or unoccupied by ribosomes. Similar, or even larger, unoccupied stretches have, however, also been reported for bacterial systems.

Closely related to these considerations may be the transit time of the individual particle across its mRNA, in turn determining the rate of protein synthesis. The latter has been estimated as 50 aa residues (150 nucleotides) per minute at 37° in well-defined mammalian systems and as 450 aa residues (1350 nucleotides) per minute at the same temperature in *E. coli*. Indeed, the overall rate of incorporation of amino acid precursors, or the synthesis of specific proteins, in the latter organism appears to be about 10 times that observed in vertebrate cells under comparable circumstances.

Chain Termination

Studies with the *E. coli* system have established requirements for the following—not necessarily mutually exclusive—components for chain termination (110–112):

1. A signal in the advancing mRNA spelling "terminate"; this is provided by one of the three chain-terminating nonsense codons, UAA, UAG, or UGA. The absence of a tRNA (in Su^- cells) capable of recognizing such a sequence leaves the *A* site unoccupied and leads to a cessation of chain elongation.

2. A means of detaching the completed polypeptide chain from the ribosome, probably coincident with the enzymatic hydrolysis of the C-terminal ester bond. We have seen that the growing polypeptide chain or the ribosomes of a polysome is chemically in the form of the respective tRNA derivative. It is still in this form at the moment of chain termination, that is, when the *n*th (C terminal) amino acid has been added; the structure at this moment is fMet-N-$aa_1 \cdots aa_n$-CO-$tRNA^n \cdot Rb \cdot mRNA$.

3. A means for detaching the ribosomal subunits from the complex to permit their reentry into the pool prior to reutilization. A specific protein appears to be required for this purpose. It has been pointed out that since the continued supply of subunits depends on the participation of this protein factor, and since such a supply of subunits is required for initiation, this protein can equally well be classified as an *initiation* factor, and one of this group of proteins may function in the capacity just outlined. It has been isolated by washing 30-S subunits, in a manner similar to that previously used for isolation of initiation factors (113). Its function is believed to be a displacement of the 50-S subunit from the 70-S Rb. An alternative to steps 1 and 2 would envisage active translation of terminator codons by an actual entity (probably not a particular species of tRNA), followed by, presumably, formation of an unstable peptide or other bond, between this factor and the terminal aminoacyl residue.

Whatever the actual mechanisms for chain termination, it is clear from the evidence so far available that it is an active multistep process requiring the intervention of a number of proteins catalyzing specific functions, and that the presence of a chain-terminating triplet in a codon is a necessary but not a sufficient condition for release. Besides the factor already mentioned, four additional *release factors* (R factors) (Table 21-1) have recently been described. Two, called R1 and R2, both of $M \simeq 45{,}000$, are required for release of free fMet as measured by the simple two step reaction:

$$\text{fMet-tRNA}^{\text{fMet}} + \text{Rb} + \text{AUG} \xrightarrow[\text{IF}]{\text{Mg}^{2+},\ \text{K}^+} \left[\text{fMet-tRNA}^{\text{fMet}} \cdot {\text{AUG} \atop \text{Rb}}\right] \tag{21-19a}$$

$$\left[\text{fMet-tRNA}^{\text{fMet}} \cdot {\text{AUG} \atop \text{Rb}}\right] + \left\{\begin{matrix} \text{UAA or UAG} \xrightarrow{\text{R1}} \\ \text{UAA or UGA} \xrightarrow{\text{R2}} \end{matrix}\right\} \text{fMet} + \cdots \tag{21-19b}$$

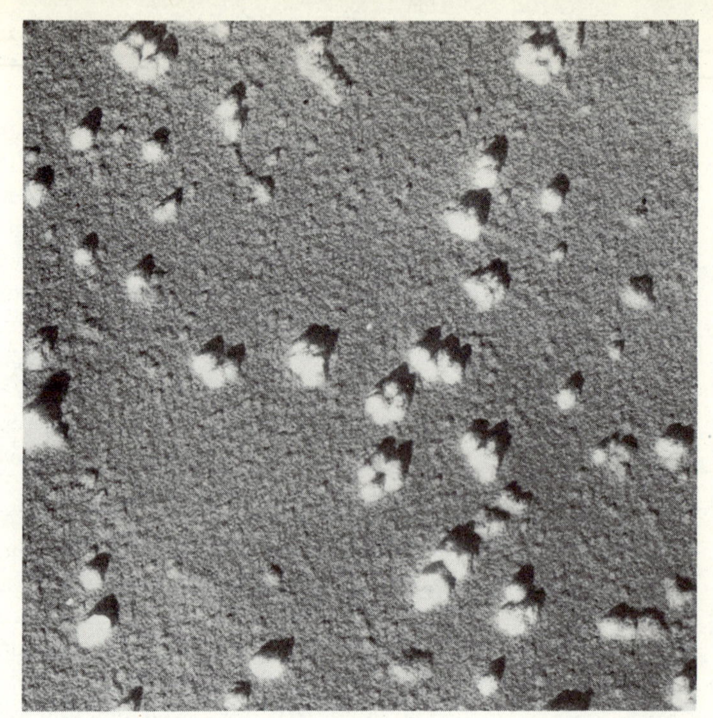

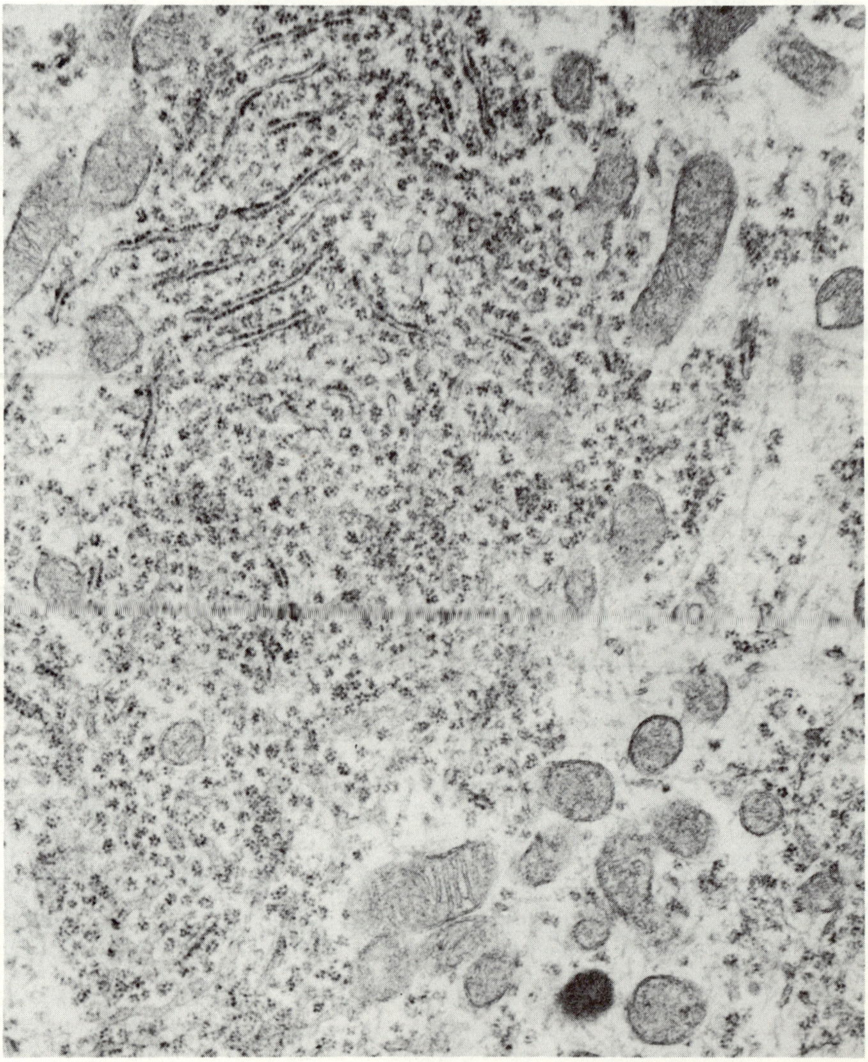

Hydrolysis of fMet proper [reaction (21-19b)], requires specific recognition by R1 or R2 of appropriate nonsense codons and at least transient binding of the R factors to the codon · Rb complex (114). A third factor (S) lowers the K_s of nonsense codon binding in reaction (21-19b) and hence stimulates the rate of fMet hydrolysis catalyzed by either R1 or R2. The fourth one, probably different, is capable (115) of releasing free *N*-acyl amino acids from *N*-acylated aminoacyl-tRNA's (including chemically prepared fMet-$tRNA_M^{Met}$) but does not attack fMet-$tRNA_f^{Met}$. It may thus fulfill one of the monitoring functions discussed under "chain initiation."

Some of the difficulties inherent in any model involving obligatory detachment of ribosomes coincident with chain termination in the translation of polycistronic messages have also already been discussed within the context of "chain initiation." Another problem concerns the nature of the chain-terminating codon in ordinary (i.e., Su^-) cells. It is clear both from model studies *in vitro* and the existence of amber, ochre, and UGA mutations *in vivo* (which do lead to premature chain termination and the accumulation of polypeptide fragments) that all three nonsense triplets *can* function in this capacity. The question is, which of them do? A possible clue is derived from the existence of Su^+ mutations. If any one triplet functions as a chain terminator, and its function is suppressed in Su^+ strains, a very deleterious and perhaps a lethal result can be expected if the suppression (i.e., the production of the su^+ tRNA) is highly effective; not only premature but also normally required chain termination will be overridden. It is known that amber and UGA suppression is effective (strong), while ochre suppression is weak. Tentatively we conclude that UAG, the ochre codon, may function as a "normal" chain-terminating codon in *E. coli*.[14] Furthermore, the introduction of two or more *different* terminators *in series* (Fig. 5-23), would permit chain termination even in Su^+ strains (page 933).

PRODUCTS OF PROTEIN SYNTHESIS

Products of the *in vitro* System

The initiation of new polypeptide chains and the net synthesis of products of known structure is readily demonstrable in the case of synthetic polynucleotide messengers, particularly with the *E. coli* system, and forms the basis of much of what has been discussed so far in this chapter. In contrast, the biosynthesis of the polypeptide chains of actual cellular proteins, with a well-recognized biological function as well as a known sequence, under the direction of a natural mRNA, is much more difficult to achieve. In part this discrepancy is due to the inherent difficulty of isolating and preserving mRNA in a biologically active form, retaining the totality of its information content, that is, not only those codons responsible for the complete amino acid sequence of the product but also those responsible for initiation, termination, and perhaps other regulatory events. In part is has been referrable, particularly

[14] In the closely related organism *Salmonella typhimurium*, a UAA or UGA codon has been identified as a chain terminator located before an intercistronic region in the *his* operon (114a).

FIGURE 21-7

Electron micrographs of polysome structures. (*Top*) Electron micrograph of a purified and separated fraction of rat brain polysomes. Both clustered and linear pentameters are seen (magnification 63,000). Courtesy of A. Campagnoni and J. Ewing. (*Bottom*) Cytoplasm of Deiter's neuron from cat brain. In this section of embedded material ribosomes are arrayed along the endoplasmic reticulum as well as in polysomal clusters (magnification 52,500). Courtesy of C. Cotman and S. Hertweck.

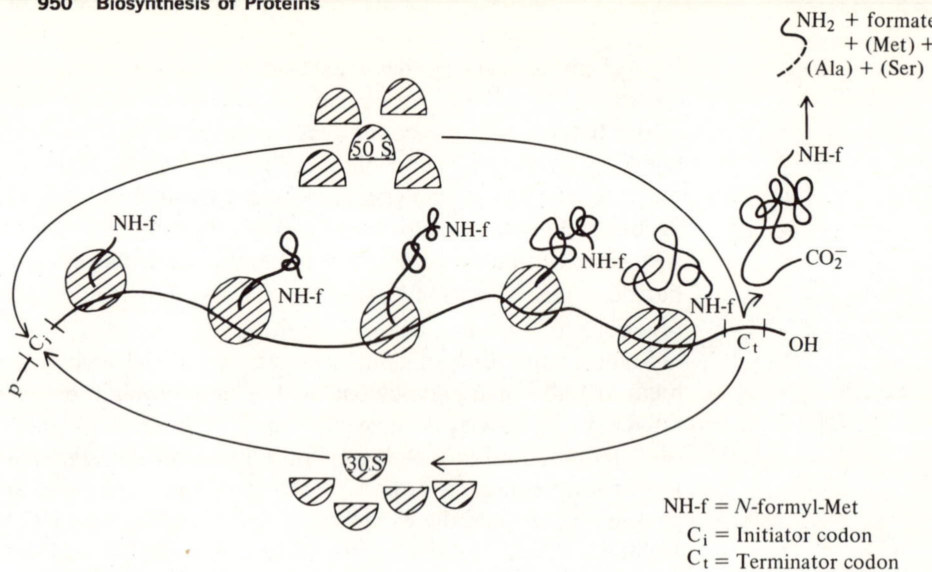

FIGURE 21-8

Schematic model of a current version of the polysome model for protein synthesis in *E. coli.* The mRNA for a single cistron is shown. Ribosomes are assembled from 30-S and 50-S subunits coincident with chain initiation at a chain-initiation site (CI) close to the 5′ terminus. This process requires initiation factors as well as fMet-tRNA^{fMet}. Simultaneously with the propagation of the polypeptide chain, the ribosome advances along the mRNA in the direction of the 3′-OH terminus, freeing the initiation site for the assembly of a second ribosome and the initiation of the second polypeptide, and so forth. The polypeptide chain is completed when each ribosome reaches a chain-terminating site (CT)—close to or at the 3′-OH terminus of a monocistronic mRNA. In the presence of release factors the polypeptidyl-tRNA is cleaved and the ribosome, in the form of its subunits, reenters the pool from which it originated.

in the case of systems isolated from eucaryotic cells, to the time lag in understanding and the resultant deficiencies precisely in those components of the system responsible for these critical events in translation.

The greatest success has attended the use of viral or virus-induced RNA's as messengers. An early instance of the former was the demonstration that the RNA from satellite tobacco necrosis virus (STNV), a defective plant virus, a molecule containing 1200 nucleotides ($M = 3.94 \times 10^5$ daltons), can, in the *E. coli* system, function as the messenger for a polypeptide which appears indistinguishable on the basis of fingerprint analysis from the viral capsid (coat) protein. Since the latter contains 400 amino acid residues, it is evident that the viral RNA constitutes a *monocistronic* messenger *in vivo,* and presumably *in vitro* (116). This example is of special interest because the message is ordinarily translated not by a bacterial but by a plant system.[15] Attempts to extend these findings to the use of *polycistronic* plant virus RNA's such as that from tobacco mosiac virus in this heterologous system have been uniformly unsuccessful (117).

[15] It is also opportune to point out that this RNA possesses a (p)ApGpY triplet at its 5′ terminus. This constitutes a serine codon, and hence initiation of even this small mRNA either does not begin with this codon or else involves an initiator other than (f)Met-tRNA^{fMet}. Furthermore, it is known that the N-terminal amino acid in the isolated capsid is alanine.

On the other hand, investigation of the homologous *E. coli*–RNA coliphage system have been most fruitful (118–122). The small RNA phages such as f2, MS2, R17, M12, and Qβ (see Fig. 5-23) (the first four related to one another and unrelated to the fifth immunologically) all contain RNA with a $M \simeq 1.1 \times 10^6$ daltons, and code *in vivo* for a total of three proteins: the capsid protein (129 residues), a virus-specific RNA polymerase (page 908, $M = 50{,}000$), and a maturation protein (called A protein, $M = 37{,}500$)—a factor required for the production and liberation of progeny virus. These three proteins are synthesized in amounts and in an ordered sequence that appear to be inherently controlled by the genetic material of the virus. Capsid and A protein are synthesized first; the former constitutes the predominant and the latter the least plentiful product throughout the life cycle. These features are retained when infection of cells by intact virus is replaced by that of protoplasts by viral RNA and even in the completely cell-free system. Some of the other remarkable findings that have emerged specifically from these studies are (1) that all three nascent proteins appear to be initiated by *N*-formyl methionine; (2) that translation of polymerase with *intact* viral RNA as messenger is dependent on the translation of the capsid protein (or at least a considerable fraction well beyond the sixth, but not including the 70th amino acid), and this regulation is abolished by modification of the RNA; and (3) that the first initiator codon (of A protein) is far removed from the 5′-terminal sequence proper. Thus, the complete phenomenology involved in the translation of a polycistronic mRNA and its regulation is beginning to be understood.

Capsid protein can also be formed on plant chloroplast (but not cytoplasmic) ribosomes, programmed with f2 RNA, again with fMet in the N-terminal position (123). A further extension of this type of approach became possible with the discovery that RNA extracted from cells infected with coliphage T4 (which make almost exclusively phage-specific mRNA) could act as a messenger in the cell-free system for the synthesis of lysozyme and other authentic phage-induced proteins (124). The final step in this progression was achieved, again with the use of phage T4, by coding with DNA, thus carrying out not only translation but transcription as well *in vitro;* the latter required the addition of the components of the RNA-polymerizing to those of the protein-synthesizing system (125,125a). The next extension—that of the synthesis of proteins characteristic of animal cells by the *E. coli* system—apparently was achieved in 1969 with the report that when programmed by a mRNA fraction from rabbit reticulocytes and with *N*-acetyl-valyl-tRNA added, products indistinguishable from the two globin polypeptides, but acetylated at the N-terminal valine,[16] were obtained (126). The relevant messenger (126,126a), isolated from reticulocytes, has been estimated to possess a molecular weight of 1.9×10^5 daltons (550 nucleotides, ample to code for 141 or 146 amino acid residues depending on the chain); its base composition is reported to be A(21.2) : U(23.7) : G(20.7) : C(34.5)—compared to A(29.8) : T(28.1) : G(21.7) : C(20.4) for the total cellular DNA. Finally, the *E. coli* systems programmed with the DNA from the animal virus SV40 is reported to be capable of synthesizing a polypeptide similar to the viral coat protein (127).

A wide variety of polysomal systems has been isolated from cells or tissues of vertebrates, or of higher plants, engaged in the synthesis of one or

[16] In the case of rabbit, both the α and β chains of hemoglobin carry an N-terminal valine.

more highly specialized proteins. Although it has frequently been possible to isolate these protein products in highly purified form and to correlate their size with the size of the polysomal cluster engaged in their synthesis, the incorporation of amino acids into the nascent and released chains *in vitro* usually corresponds to completion of chains that had already been initiated *in vivo* rather than to true cell-free synthesis *de novo*. The classical and most thoroughly investigated system consists of polysomes isolated from reticulocytes (immature red cells) of rabbits previously rendered anemic by means of phenylhydrazine. This system properly supplemented is capable of not only completing but also initiating the biosynthesis of some of the constituent α and β chains; the active unit appears to be a cluster of five ribosomes (pentasome) with a sedimentation coefficient of 180 S.

In other instances, where the formation of known proteins by polysomes isolated from highly differentiated cells of vertebrate animals has been suggested, the occurrence and extent of initiation of chains *de novo* remains uncertain. Among the systems studied are ones from liver responsible for the synthesis of serum albumin and ferritin by membrane-bound and free polysomes, respectively (128–130). Tropocollagen (the precursor protein of the major fibrous protein of connective tissue such as cartilage, skin, scales, and joints), a protein of $M = 300{,}000$, consisting of three α-helical chains arranged in a triple helix, is synthesized by very large polysomes isolated from embryonic fibroblasts growing in tissue culture (131). Another system is concerned with the synthesis of myosin, the major contractile protein of muscle, $M = 480{,}000$, by polysomal aggregates from skeletal muscle, containing about 60 particles and exhibiting a sedimentation coefficient >500 S. Yet another system appears responsible for the synthesis of γ-globulin. As discussed on page 129, this protein ($M = 150{,}000$) is composed of two light (L) and two heavy (H) subunits, the former of $M = 23{,}500$, containing 212 to 214 residues, and the latter with $M = 50{,}000$, corresponding to ~440 residues. Two distinct populations of polysomes can be isolated from tissues responsible for antibody formation, such as lymph nodes or spleen: one containing 7 to 8 particles which can be shown to produce L chains only, and one containing $\geqq$16 particles and capable of only forming H chains.

Modification of the Product: Higher-Order Structure

One of the basic assumptions underlying contemporary molecular biology is that the information contained in the base sequence of the requisite gene, and transcribed into the appropriate mRNA, is both necessary and sufficient for the specification of the sequence and hence the complete structure of the polypeptide gene product. We now know, however, that frequently the primary product of translation undergoes enzyme-catalyzed modifications even in its primary structure before it can assume its final form. One such modification, already discussed, is the sequential hydrolytic removal of formate, methionine, alanine, and, perhaps, serine from the f-Met-Ala-Ser terminus supposedly common to most bacterial and bacteriophage-induced proteins. Another, perhaps more serious, problem is raised by the discovery that insulin, a polypeptide hormone synthesized in the pancreas of vertebrates, containing two polypeptide chains linked to one another by disulfide bridges (page 121), does not exhibit the expected primary structure at the moment of synthesis (132). Instead of two separate polypeptides the precursor is constituted by a *single* large polypeptide, a *proinsulin,* which is subsequently cross-linked by *intra*molecular disulfide bridges and only later modified by removal of peptide segments into the final double linear structure. The conversion of enzyme precursors, proenzymes, or zymogens is of course a well-recognized phenomenon (page 762), but this case and others indicate that many, and frequently exquisitely specific, steps will have to intervene before the proximal polypeptide gene product is converted into its final form. Germane, perhaps, is also the discovery that the polycistronic RNA of viruses infecting mammalian cells is translated *in toto* rather than in the form of individual, separate proteins. This giant protein molecule is later cleaved hydrolytically into its component parts (133).

As concerns the complete specification of higher-order structure by the primary structure of a polypeptide, and hence by the gene, we have discussed this problem within the context of the structural organization of proteins in general in Chapter 4. We indicated there that based on denaturation–renaturation studies it is believed that in the main proteins assume that conformation corresponding to the state(s) of overall minimal free energy consonant with their primary structure. However, nothing was said there as to when the requisite folding reactions take place in the case when the randomly coiled chain is formed, not as a consequence of denaturation of the finished product, but in the natural course of events during polypeptide synthesis. Is it allowed to assume conformations of higher order coincident with, or closely following, chain elongation, or do conformational changes take place only subsequent to the completion of the chain and the subsequent modifications of primary structure, if any? Available evidence is extremely meager on this point, but there is some indication that, initially at least, the nascent polypeptide, except for that portion still in intimate contact with the ribosome, is extremely prone to proteolytic digestion and hence probably largely unfolded (134). In any event, the moment of synthesis would appear to be an extremely propitious one for the intervention of specific or nonspecific events for directing a nascent chain into one of several possible conformational states (conformers) (135). Particular difficulties arise with regard to the control of the formation of appropriate disulfide linkages (132, 136), a process which is known to be, at least in part, subject to specific enzymatic catalysis (137). In many cells the very structure responsible for the synthesis of proteins for export, membrane-attached polysomes (Fig. 21-7 and Chapter 9—the rough endoplasmic reticulum), also appears to carry the disulfide-forming enzyme; in fact, the site of ribosomal attachment and the site of enzyme localization may overlap or even be identical, and only proteins designed for export may carry S-S crosslinks (138).

Inhibition of Protein Synthesis

One of the most useful techniques in this, as in all other branches of biochemistry, has been the discovery of specific inhibitors, that is, compounds that interfere with a process at specific points, and it is hoped, by a selective and comprehensible mechanism, and therefore permit the dissection of the overall sequence into a number of component steps (105). Some of the more commonly used inhibitors (or classes of inhibitors) with their presumed mode of action, are described in Table 21-4. Their structures are shown in Fig. 21-9. Some additional comments appear appropriate.

5-Methyltryptophan. This is the only amino acid analog that has found widespread use. Unlike most other compounds of this kind, it is not incorporated into proteins but acts as a competitive inhibitor for its parent compound toward the cognate aminoacyl synthetase. In this way it blocks the formation of Trp-tRNA and hence the biosynthesis of all polypeptides containing this amino acid.

TABLE 21-4
Inhibitors of protein synthesis

	Inhibition of protein synthesis in			
		Eucaryotic		
Inhibitor	*Procaryotes*	*Cytosol*	*Mitochondria*[a]	*Presumed mode of action (see Figs. 21-2 and 21-5)*
Puromycin	+	+	+	Formation of aminoacyl puromycin resulting in premature chain termination
Aminoglycosides	+	+	±	Bind to 30-S ribosomes, form abortive initiation complex
Tetracyclines	+	−	+	Bind to 30-S ribosomes, block aminoacyl-tRNA (A) site
Macrolides	+	−	+	Bind to 50-S ribosomes, block chain elongation (P site?)
Chloramphenicol	+	−	+	Bind to 50-S ribosome, perhaps through PS (Table 21-1), increases polysome stability
Cycloheximide	−	+	−	Similar to chloramphenicol, but for 60-S ribosomes
Sparsomycin	+	+	+	Interference with peptide-bond formation (tight aa-tRNA • Rb complex)
Fusidic acid	+	+ (?)	+ (?)	Interference with translocation induced by G Factor (complex formation?)
Diphtheria toxin plus NAD	−	+	−	Blocks translocation by modifying TF2 (Table 21-1)
Aurintricarboxylic acid	+	?	?	Interference with attachment of mRNA

[a] Chloroplasts exhibit a qualitatively similar picture.

Puromycin. As already mentioned, this antibiotic bears an obvious structural resemblance (Fig. 21-9) to the adenosine terminal end of an aminoacyl-tRNA. The antibiotic is therefore capable of blocking protein synthesis during transpeptidation, by substituting for an incoming aminoacyl-tRNA in binding to the *A* site and acts as an acceptor of the peptidyl-tRNA bound to the *P* site. The peptidyl-puromycin formed in this manner (attack at the N atom shown by arrows) is incapable of further chain growth (no transpeptidation at the wavy line). Thus, exposure to puromycin leads to the release of peptides of varying chain length but all with the correct amino acid at the N terminus and a puromycin residue instead of a C terminus. The simplest instance is that of an *N*-acyl *N'*-puromycin, in particular *N*-formylmethionyl-*N'*-puromycin.

Aminoglycoside Antibiotics [Streptomycin (Sm or Str), Neomycin, Canamycin, etc.]. In streptomycin sensitive (Str^s) strains of bacteria, the action of this and the related dihydrostreptomycin is brought about by virtue of their interaction (139) with one of the core proteins of the 30-S ribosome, called P_{10} by Ozaki, Mizushima, and Nomura (106). This interaction in weakened or otherwise modified in streptomycin-resistant (Str^r) strains; a subclass of the latter is constituted by the streptomycin-dependent (Str^d) ones, which require the presence of the antibiotic for normal growth and macromolecular synthesis, as already discussed. Binding of Sm and related antibiotics leads to a number of consequences, both *in vivo* and *in vitro*: which of these is responsible for its bacteriocidal action is as yet unknown. Among the more important *in vitro* effects are:

1. When added to an active system, in particular to one programmed with phage RNA as messenger, it leads to an exceedingly rapid cessation of polypeptide chain extension and to a transient stabilization (freezing) of the

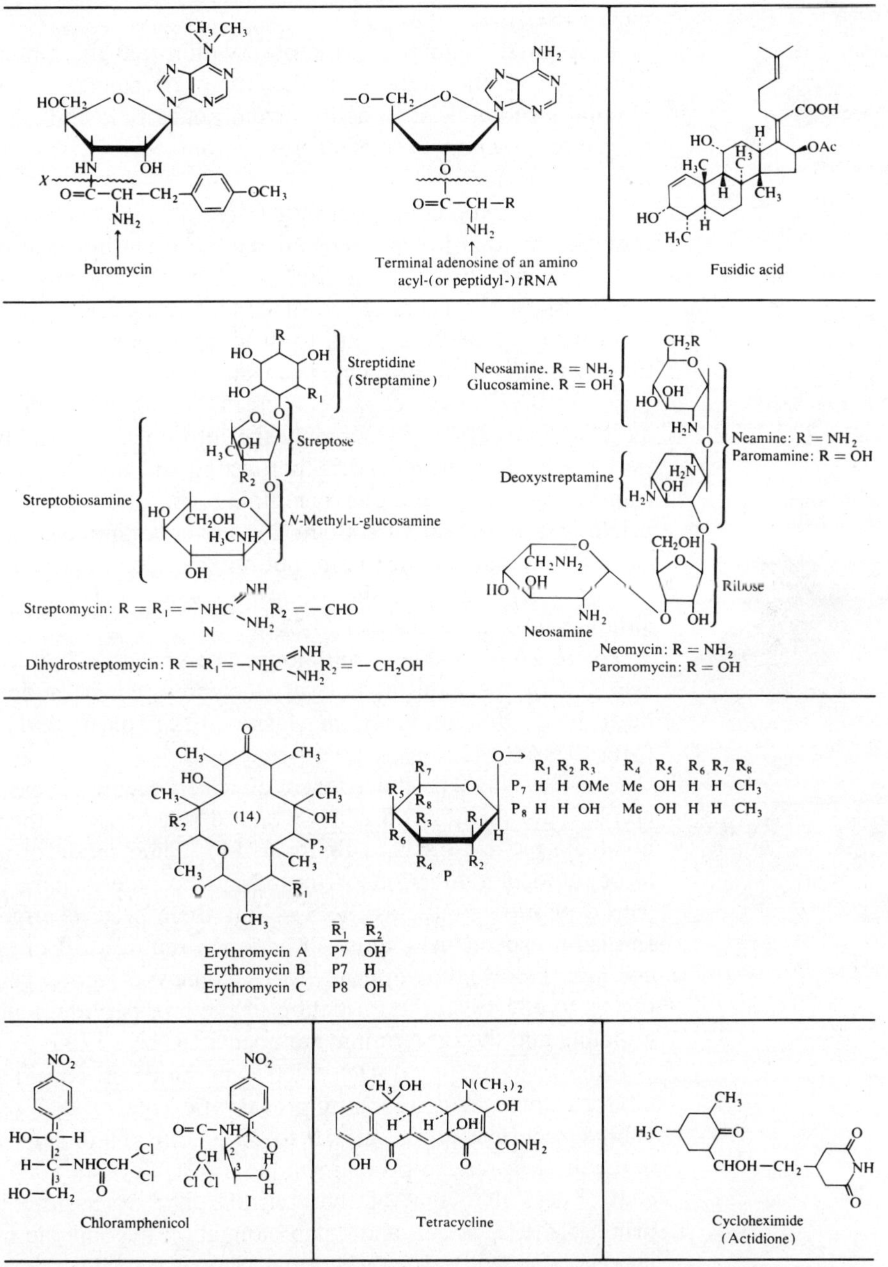

Puromycin

Terminal adenosine of an amino acyl-(or peptidyl-) *t*RNA

Fusidic acid

Streptidine (Streptamine)

Streptose

Streptobiosamine

N-Methyl-L-glucosamine

Streptomycin: $R = R_1 = -NHC(=NH)NH_2$ $R_2 = -CHO$

Dihydrostreptomycin: $R = R_1 = -NHC(=NH)NH_2$ $R_2 = -CH_2OH$

Neosamine. $R = NH_2$
Glucosamine. $R = OH$

Neamine: $R = NH_2$
Paromamine: $R = OH$

Deoxystreptamine

Ribose

Neosamine

Neomycin: $R = NH_2$
Paromomycin: $R = OH$

	R_1	R_2	R_3	R_4	R_5	R_6	R_7	R_8
P_7	H	H	OMe	Me	OH	H	H	CH_3
P_8	H	H	OH	Me	OH	H	H	CH_3

	R_1	R_2
Erythromycin A	P7	OH
Erythromycin B	P7	H
Erythromycin C	P8	OH

Chloramphenicol

Tetracycline

Cycloheximide (Actidione)

FIGURE 21-9
Structure of inhibitors of protein synthesis.

polysomes and nascent polypeptide chains existing at the time of addition (140).

2. It may block the ribosome cycle by inhibiting formation of new ribosomes and the initiation of polypeptide chains or by introducing ambiguities during normal chain release (140,141). Which of these effects, or those described under point 1, are primary and which are secondary remains to be established.

3. It can stimulate the extent of misreading (or reduce the fidelity of translation) of synthetic polynucleotide messengers (i.e., the incorporation of amino acids not ordinarily coded for by certain codons already discussed on page 945).

4. Related to point 3 is the observation that the aminoglycoside antibiotics can promote the activity of denatured or single-stranded DNA (including circular structures, such as that from phage fd), and of denatured rRNA and tRNA to function as messengers. Neomycin is particularly effective in this regard (9,10).

5. The phenomenon of streptomycin dependence is probably related to the observations that low concentrations of the antibiotic appear to be required to keep the ribosomes from Str^d cells in a conformation active for protein synthesis. In the same range of concentrations, Sm added to the ribosomes from wild-type cells appears to produce a looser conformation or diminishes the strength of association between subunits.

6. Finally, Sm is capable of bringing about phenotypic suppression, at least *in vivo* (104): presumably the introduction of ambiguities here, as it is perhaps in 3, is beneficial in permitting the translation of an otherwise untranslatable message—one containing chain-terminator codons in inappropriate locations (see also point 2). Other aminoglycosides appear to exhibit modes of action similar to Sm, but differ in detail, presumably in the nature of the protein target on the ribosome, since the loci leading to resistance, although linked, are distinct (141).

Tetracyclines. These antibiotics are probably bound to the 30-S subunit also, but by mechanisms not yet elucidated. Their mode of action appears to be to block the aminoacyl or *A* site (142), particularly toward the entry of fMet-tRNA (143).

Macrolides. The macrolides, of which erythromycin is a well-studied example, bind to the 50-S ribosome, and sensitivity to these drugs is known to involve a protein of that subunit (144). Their mode of action is to block the most important function of the 50-S ribosome—chain elongation by its intrinsic peptide synthetase (PS)—but their precise target is not yet known. In many respects their mode of action is reminiscent of that of chloramphenicol (see below) but appears to be somewhat more general, particularly in regard to effects on the formation of the first peptide bond between the chain-initiating and the subterminal aminoacyl-tRNA (145).

Chloramphenicol (CAP). This antibiotic is considered the classical inhibitor of protein synthesis by procaryotic (bacterial, 70 S) ribosomes (146). With susceptible organisms 10 to 20 μg/ml of the active D(−) *threo* isomer is required to reduce protein synthesis *in vivo* or *in vitro* to a level of 10 percent or less than that of the controls. The 50-S subunit appears to be the primary site of action and chain elongation beyond the first peptide bond the process affected. Since the normal ribosome cycle is interrupted in the presence of the antibiotic, polysomes accumulate. CAP is a competitive inhibitor of puromycin in reaction (21-16). In spite of an intensive search, a more detailed mechanism of action remains elusive. The macrolide *lincomycin* appears to exert its inhibition in a completely analogous fashion. Ribosomes isolated from the cytosol of eucaryotic cells, as well as the bulk of the protein-synthesizing capacity of such cells, are completely resistant to the action of CAP or for that matter to that of the other 30-S and 50-S inhibitors described in Table 21-4. In contrast, the mitochondria from the *same* cells are inhibited by CAP and the other specific inhibitors of 70-S ribosomes (147).

Cycloheximide (Actidione). This potent fungicide antibiotic exhibits a species specificity complementary to that of CAP, since it appears to interfere with the bulk of the cellular and cell-free protein synthesis by the cytoplasmic ribosomes (80 S) of yeast, fungi, higher plants, and all species of animals. In contrast, the semiautonomous protein synthesis catalyzed by the mitochondria of the same cells is unaffected by this inhibitor (147). It resembles CAP in its mode of action on the ribosomes and in producing the pile-up of polysomes. It is similar to CAP also in that its long-term administration, owing to the resultant block in the synthesis of ribosomal proteins, leads to the accumulation of various immature and aberrant species of ribonucleoprotein particles and perhaps of free RNA as well (148).

Sparsomycin. Its mode of action is to exert a block on the ribosome-bound peptide synthetase of both eucaryotic and procaryotic cells, probably by favoring the formation of an abortive complex either between peptidyl tRNA and the large ribosomal subunit or of the aminoacyl-tRNA and the small one (145,149).

Fusidic Acid. This steroidal antibiotic inhibits the GTP-dependent reactions catalyzed by the translocating enzymes (G Factor of bacteria, TF2 of animals, Table 21-1) (78,114,145). On the other hand, the active principle of *diphtheria toxin* exerts a similar inhibition restricted to TF2 (150). In this instance, inhibition is due to a toxin-mediated transfer of an ADP-ribose moiety from NAD to TF2 (151).

Aurintricarboxylic Acid. The attachment of viral mRNA to the incipient ribosomes of *E. coli* appears to be blocked in the presence of the triphenylmethane dye aurintricarboxylic acid, 3,3′,3″-trihydroxy-4,4′,4″-tricarboxytriphenylmethane (152). This type of inhibition bears some similarity to that exerted by *interferon,* an agent elaborated by animal cells challenged by RNA viruses or synthetic polyribonucleotides, in blocking their attachment to appropriate ribosomes (153,154).

FACTS AND NOTIONS CONCERNING CONTROL OF PROTEIN SYNTHESIS IN BACTERIA

First we require a number of explicit operational definitions and some of the experimental background concerning them. Some of these matters were touched on implicitly in Chapters 19 and 20, where we also discussed controls of RNA synthesis.

The Operon. The bacterial chromosome or genome is organized in clusters of genes called operons, the expression of which is regulated jointly, and which control the simultaneous formation of sets of proteins related in function. In the enteric bacteria (e.g., *E. coli, Salmonella, Aerobacter aerogenes*), those responsible for galactose catabolism (*gal* region), lactose catabolism (*lac*), arabinose catabolism (*ara*), tryptophan synthesis (*trp,* also *try*), histidine biosynthesis (*his*), leucine biosynthesis (*leu*), and isoleucine/valine biosynthesis (*ival*) have been studied most extensively. The number of genes comprised by the individual operons vary: three in the case of *gal* and *lac,* five in the case of *trp,* nine in the case of *his.* The latter is really composed of 15 cistrons specifying 9 enzymes (three requiring two cistrons each, one requiring four cistrons, and the remainder requiring one cistron each for their specification).

Polycistronic mRNA. The operon concept postulates that the chromosomal unit actualy transcribed is not the individual cistron or gene but the operon as a whole, which thus produces a large, polycistronic messenger. Evidence for mRNA molecules with the requisite properties, for example, synthesis in response to the appropriate stimulus, hybridization with the appropriate DNA, and size commensurate with that of the product polypeptides, has been obtained in a number of instances, such as *gal, lac, trp,* and *his*. The *trp* messenger is believed to be ~7000 and the *his* 13,000 nucleotides long.

The Operator. The actual expression of the various structural genes in the operon is controlled by a small region located at one end, called the operator. Mutations in this region affect all the genes of the operon. Operators actually exert their action at the level of transcription, as originally postulated by Monod and Jacob (155), rather than at the level of translation. An operator maps as follows, where *1* through *n* are the cistrons of the operon and *o* is the operator region:

o	*1*	*2*	...	

Forms of Regulation: Repression Versus Induction. Repression is defined as a decrease, and induction as an increase, in the amount of one or more specific proteins produced as the result of exposing the cell to some regulatory substance. The latter, a small molecule, is frequently identical with, or structurally related to, either the end product (in the case of repression), or the starting material (in the case of induction), of a metabolic sequence. Inducibility or repressibility also is controlled genetically, the genes in question (*regulator genes*) mapping in a position not necessarily linked to the operon comprising the structural gene(s) affected. The regulator genes specify the structure of polypeptides—the *(apo)repressors*—which in turn interact with the operator in either a positive or negative sense (see Fig. 21-10), triggering transcription in the former, blocking it in the latter case. This interaction is in turn regulated by an allosteric interaction with the regulatory small molecule interacting with the (apo)repressor either in its free or DNA-bound state. Thus, a "negative" aporepressor may be rendered harmless by an inducer or effective by a corepressor. Induction and derepression are therefore formally and mechanistically related. Induction and repression of enzymes in microorganisms are exceedingly common. Of the various operons listed above, the three concerned with metabolism of sugars are inducible by their substrates or structural analogs thereto, as might have been expected from the definition given earlier. Those controlling amino acid biosynthesis are repressible by these amino acids.

In general, repression by end products is frequently observed in anabolic pathways for amino acids, vitamins, purines, and pyrimidines; induction is the rule in the case of catabolic sequences involving carbon and energy sources.[17] Regulation can be rationalized on the basis of economy of use in the machinery for protein synthesis; synthesis of enzymes catalyzing metabolic

[17] We are reminded of the generalization mentioned in Chapter 11 that pacemaker enzymes in anabolic sequences are frequently inhibited by their end products and that corresponding enzymes in catabolic pathways are sometimes activated by their substrates or products.

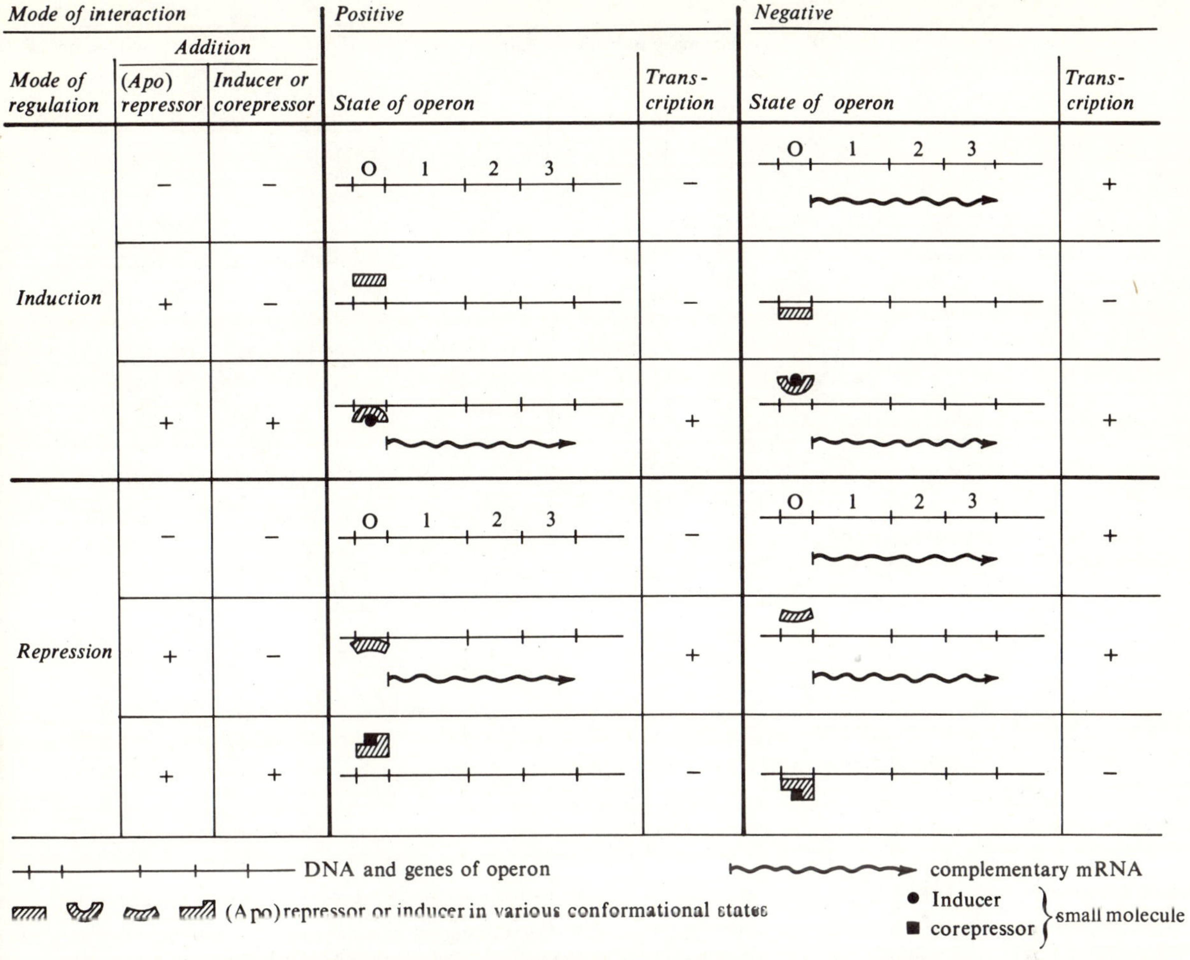

Mode of interaction			Positive		Negative	
	Addition					
Mode of regulation	*(Apo) repressor*	*Inducer or corepressor*	*State of operon*	*Trans-cription*	*State of operon*	*Trans-cription*
Induction	−	−	O 1 2 3	−	O 1 2 3	+
	+	−		−		−
	+	+		+		+
Repression	−	−	O 1 2 3	−	O 1 2 3	+
	+	−		+		+
	+	+		−		−

DNA and genes of operon

complementary mRNA

(Apo)repressor or inducer in various conformational states

● Inducer
■ corepressor
small molecule

FIGURE 21-10
Regulation of genetic expression.

sequences is turned on or off, depending on the demands for that specific sequence. Why synthesize proteins unless and until they are needed?

A particularly instructive example of complex interactions between inductions and repressions to effect close control of the synthesis of various proteins within a group is provided by the enzymes that catalyze the breakdown of mandelic acid (or rather its salt mandelate) in *Pseudomonas* and *Moraxella.* The following pathway is believed to be involved (156,157):

$$\text{Mandelate } (A) \xrightarrow{a} \text{benzoyl formate } (B) \xrightarrow{b} \text{benzaldehyde } (C) \xrightarrow{c} \text{benzoate } (D) \xrightarrow{d} \text{catechol } (E)$$

$$\downarrow e \quad \downarrow f$$

$$\text{acetate} + \text{succinate } (I) \leftarrow \overset{g}{\leftarrow} \beta\text{-ketoadipate } (G)$$

First of all, the enzymes are sequentially inducible in the sense that the addition of *A* first induces the synthesis of *a, b, . . .,* and only as various inter-

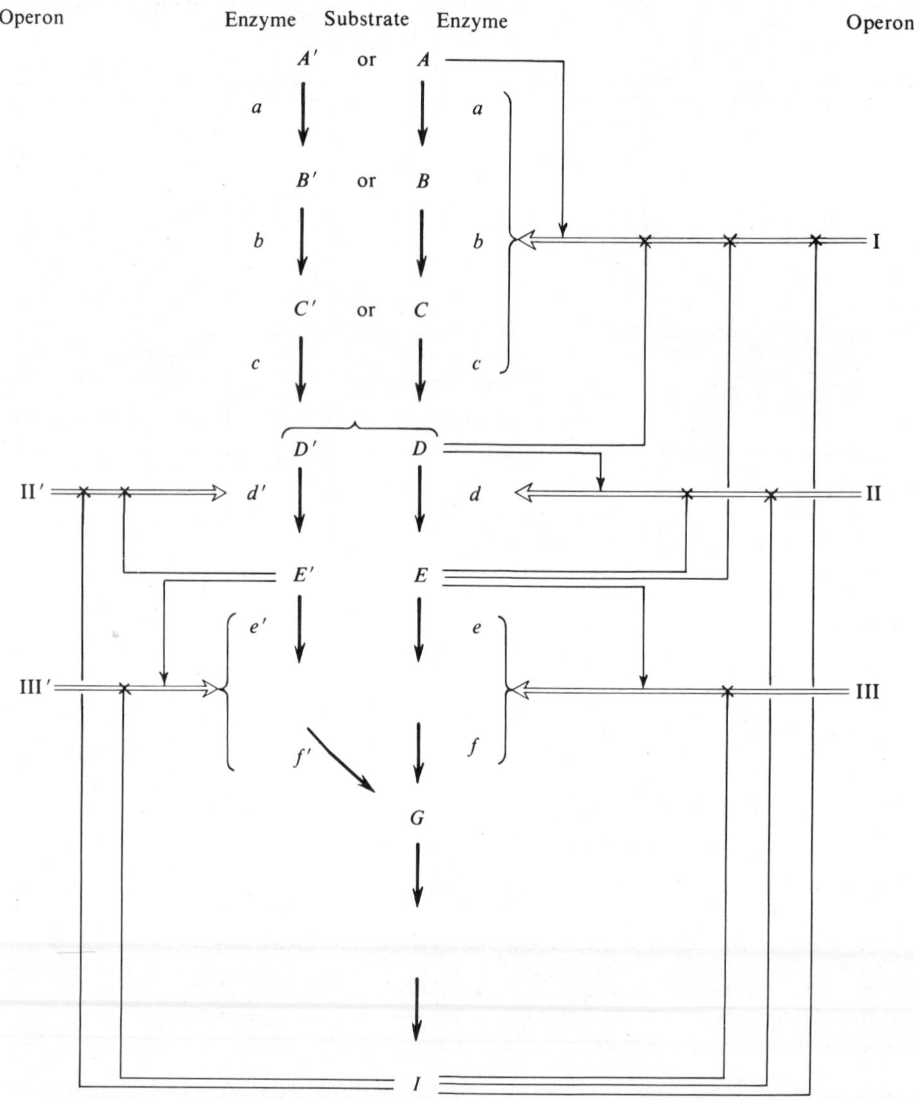

FIGURE 21-11
Multivalent and sequential induction and repression.
Induction: ——→ Repression: ——X

mediates such as $D, E, \ldots,$ make their appearance does the induction of enzymes $d, e, \ldots,$ commence. Closer attention to detail shows that the enzymes of the pathway are controlled by three operons, which we shall call I, II, and III. Operon I controls synthesis of enzymes a, b, and c and is induced by A, the first substrate; it is repressed by D, the immediate product accumulating as a result of its operation, as well as by E and I, the distal products. Operon II controls enzyme d; it is induced by D and repressed by E and I. Finally, operon III consists of enzymes e and f, and perhaps others. It is induced by E and repressed only by I. The various relationships can be visualized in the form of Fig. 21-11, which also incorporates results obtained on a converging pathway:

$$p\text{-Hydroxymandelate } (A') \xrightarrow{a} p\text{-hydroxybenzoylformate } (B') \xrightarrow{b}$$

$$p\text{-hydroxybenzaldehyde } (C') \xrightarrow{c} p\text{-hydroxybenzoate } (D') \xrightarrow{d'}$$

$$\text{protocatechuate} \xrightarrow{e'} \xrightarrow{f'} \beta\text{-ketoadipate}(G) \xrightarrow{g} \cdots$$

Coordination. For any one strain the enzymes of an operon are always present in *constant* relative *proportion* regardless of their state of induction or repression, which, in extreme cases, may lead to variations by factors of 10^2 to 10^3; that is, we conclude that the formation of all the polypeptides specified by an operon is coupled in an obligatory manner and is regulated *coordinately* at all times.

Catabolite Repression. Another regulatory effect, common to many microorganisms, is an inhibition in the induced synthesis of certain proteins, brought about by glucose or its catabolites, known as the *glucose effect* or *catabolite repression.* The susceptibility of various operons to this form of control appears to be variable (perhaps dependent on their function, i.e., catabolic, dispensable, and inducible, versus biosynthetic, essential, and repressible) and genetically determined. Thus, the *gal* operon is susceptible while the neighboring *trp* is not, synthesis of tryptophanase is affected but alkaline phosphatase is not; in the case of the (ordinarily glucose sensitive) *lac* operon, certain mutants are known that are resistant to glucose itself yet are still effectively repressed by glucose-6-*P* or gluconate (158).

Since, in this operon, o^c deletion mutants (see below)—which contain a structurally defective operator, incapable of interaction with the aporepressor —are still subject to catabolite repression, the repression probably does not involve the operator as its target site (159). Recent evidence suggests strongly that interaction involves the promoter region (page 963) (159,160) and that cyclic AMP (Ado-3′ : 5′-*P*, page 196) can overcome repression by glucose (160). In fact, the level of Ado-3′ : 5′-*P* present in the cell may not only be necessary but sufficient for interaction and determine the extent of catabolite repression. This mode (160a) postulates an *inverse* relation between the glucose (or catabolite) and Ado-3′ : 5′-*P* concentrations (160b), that is,

$$\left.\begin{matrix}\text{high}\\ \text{low}\end{matrix}\right\}\text{glucose} \leftrightarrow \left\{\begin{matrix}\text{low}\\ \text{high}\end{matrix}\right. \text{Ado-3}' : 5'\text{-}P$$

and an intervention by the effector in a *positive* (stimulatory) sense, probably in conjunction with an appropriate regulatory protein (Fig. 21-10) as well as the polymerase. In addition, or alternative to this function in promoting transcription, Ado-3′ : 5′-*P* may also function in stimulating translation, and in tightly coupled systems the two functions may be equivalent (160c).

The Lac Operon. Since most aspects of the currently accepted model are based on studies with the *lac* operon, it now behooves us to describe these in some detail. The operon proper consists of three structural genes *z, y,* and *a,* and two regulatory regions *o* (for the *operator*) and *p* (for the *promoter*). Their position on the genetic map is believed to be the following:

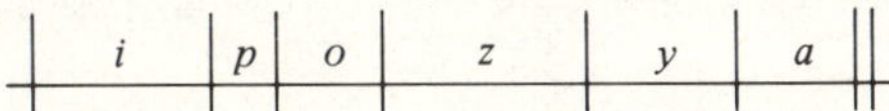

Here *a* specifies a thiogalactoside transacetylase (M = 32,000); *y* the *M* protein, an entity in the cell membrane concerned with the entry of galactosides into the cell, frequently referred to as the permease; and *z* the β-D-galactosidase (M = 135,000) itself. A number of *constitutive* mutants have been isolated that genetic analysis disclosed are controlled by a regulator gene, here called i, mapping a short distance removed from the z gene and hence the operon. These mutants are characterized by the production of all three proteins in high amounts, even in the complete absence of an inducer. These mutants are therefore designated as i^-, to distinguish them from the standard type i^+, which exhibits low enzyme levels until induced by galactosides. When the *cis* diploid z^+i^+/i^-, which contains both allelic forms of the i gene, is constructed by conjugation, transduction, or F-duction, the cell is found to be inducible. Therefore, i^+ is dominant over i^-; that is, i^+ specifies a product lacking in the cell i^-. The product is a cytoplasmic entity, since the *trans* diploid z^-i^+/z^+i^- also is inducible, which implies that the i^+ character controls expression of both chromosomes. On the other hand, a z^-i^+/z^+i^{-d} diploid (see page 963) is partially constitutive.

Operator Mutations. Originally the operator was defined in terms of the existence of operator constitutive (o^c) mutations, which resemble i^- mutations in producing elevated levels of all enzymes in the operon in the absence of external inducer. They can, however, be distinguished from such regulator constitutive mutations in four ways: (1) their map position—o^c mutants map at one end but within the operon, whereas i^- mutants map outside; (2) diploids o^c/o^+ are constitutive, whereas i^-/i^+ are inducible; (3) the o^c mutation affects only genes on the same chromosomes; that is, the *cis* configuration $y^+z^-o^ci^+/y^+z^+o^+i^+$ is constitutive for permease but inducible for galactosidase; and (4) all o^c mutations are deletions and, hence, nonsuppressible.

Nature of the Product Specified by Regulatory Genes. The experiments described above suggested, and the original Jacob and Monod model postulated explicitly, that the product of the i gene be a protein, capable of interaction at two binding sites: one with the o region of the operator (i.e., a specific sequence of polynucleotides) and the other with the inducer (i.e., a small organic molecule) at the other, with the first interaction modifying and controlling the extent of the second (159). Recent experiments of Gilbert and Müller-Hill have brilliantly borne out the predictions of the model (161), which was simultaneously and independently confirmed by Ptashne (162), who isolated and characterized the repressor that regulates the production of certain proteins by bacteriophage λ. Since in the case of the *lac* repressor the number of molecules per cell is estimated to be of the order of 10 or so (~10^{-8} M), its isolation and characterization involved some formidable experimental problems. These were, however, overcome by employing a sensitive assay, i.e., equilibrium dialysis techniques using radioactive isopropylthio-D-galactoside (IPTG), an excellent inducer, with cell extracts of a mutant strain, i^t, in which the inducer binds more tightly to the repressor. Such cells are therefore capable of being induced at levels of inducer which do not induce the wild-type i^+ cells. The resultant protein is probably a tetramer and has a molecular weight of 1.5×10^5 daltons (163), a dissociation constant of 5×10^{-7} M for IPTG (versus 1.3×10^{-6} M in the case of the i^+ protein) and 5×10^{-4} M for galactose.

Once having worked out a purification procedure for the repressor the investigators proceeded to isolate it from cells the proteins of which had been labeled with radioactive sulfur, and then to study its binding to the DNA of a defective bacteriophage carrying the *lac* region. The dissociation constant for this interaction is ~3×10^{-12} M in 0.01 M Mg^{2+} and is reduced to 2 ×

10^{-11} M if 0.05 M KCl is added as well. The binding is considerably weaker if the DNA carries the o^c mutation and is abolished in the presence of 10^{-3} M IPTG, but not of noninducing galactosides or even of lactose (which is known *not* to be the true intracellular inducer). The values of these constants suggest strongly that detachment of the repressor from the operator cannot be due to simple competition by the inducer, but that the repressor has undergone a conformational change on binding to the DNA which then facilitates its interaction with the inducer (163).

Mutations in Regulatory Genes. We have just described several classes of mutations in the *i* gene that lead to altered repressors: i^- strains, which may produce no product at all in the case of deletions, or nonfunctional peptides in the case of nonsense (amber or ochre) mutants, and therefore produce the three proteins of the operons constitutively (a special subclass, i^{-d}, consists of missense mutants and produce a *defective* repressor molecule); i^t strains, which are inducible at abnormally low inducer levels because their repressor interacts more tightly with inducer. In addition, there exist the converse, i^s (superrepressed) strains, which are characterized by a more or less complete inability of the repressor protein to bind the inducer, and hence are incapable of producing any of the proteins of the operon in normal amounts even when inducer is present. Indeed, in the *in vitro* system no IPTG-binding protein could be identified in the case of i^- and i^s strains. Finally, there exists a fourth class of mutant strains, which *require* the presence of inducer before the repressor can bind to the operator; such strains produce proteins constitutively unless the inducer is added, in which case their synthesis is repressed. One would predict that the inducer, acting here as a corepressor, should decrease rather than increase the dissociation constant of the DNA repressor complex.

The Promoter. While the qualitative expression of an operon (i.e., whether it is "on" or "off," transcribed or not) appears to be regulated by interaction between repressor and operator, another gene site, in the immediate vicinity of the latter, appears to be responsible for the initiation of transcription and for the regulation of its rate. This site, which is present even in o^c deletions, thus constitutes an additional regulatory element and has been called the *p* region, for *promoter.* Although originally believed to map between the *o* and *z* regions of the *lac* operon, more recent genetic experiments suggest that it maps between the *i* and *z* gene, that is, to the left of the operator as indicated in the diagram (164). This situation does not appear to be restricted to the *lac* operon since the *trp* operon (Fig. 21-12) appears to behave in an analogous fashion. These findings suggest as corollaries that the operator region *is* transcribed in induced cells, that polycistronic mRNA contains untranslated (and perhaps untranslatable) tracts of nucleotides adjacent to their 5′ termini, and that it is not transcription as such, but its *progress* past the *o* region, that is blocked in the absence of inducer (165) (or presence of corepressor).

The operon concept demands that, normally, in the steady state, transcription commence at only one site, somewhere within the promoter in the immediate vicinity of the operator, and that once initiated this process continue sequentially until the end of the operon is reached, even if the inducer be removed or a repressor added. Under more unusual, pre-steady-state conditions, internal sites for the initiation of transcription (secondary or conditional promoters) may, however, become unmasked and operator-distal regions transcribed, again in sequence, before the normal sequence of transcriptional events takes over. Such secondary, internal, weak promoters may permit the constitutive synthesis, at low levels of the order of 1 or 2 percent, of an operator-distal block of genes even under ordinary conditions:

| *p* | *o* | *1* | *2* | | *n* | *p'* | *a* | *b* | | *m* |

Here genes *1* through *n* are under strict control of *o*, while *a* through *m* are synthesized constitutively at low levels. For instance, in the *trp* operon (166) (Fig. 21-12) such an internal promoter appears to be located at the boundary between genes *D* and *C*.

Displacement and Fusion of Operons (164,165). Among the strongest lines of evidence supporting the *operon* concept are the results of experiments in which the operator region of one operon (*o*), together with all intervening genes between it and another operator (*o'*), to the left of the first, have been deleted, thus bringing the structural genes of operon *1* under the control of *o'* with its promoter *p'*:

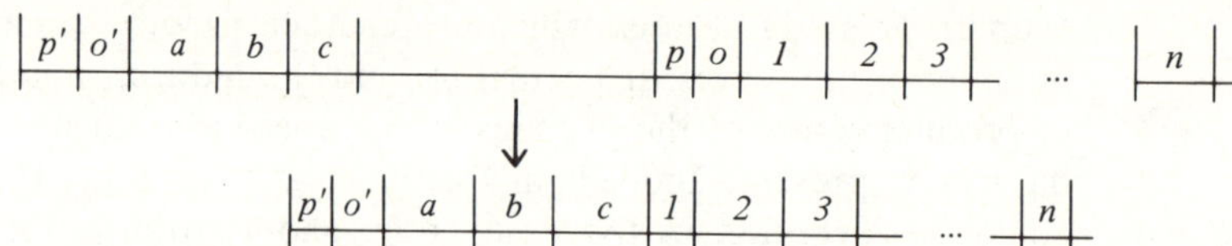

Now the expression of *1, 2, 3, . . .*, is regulated not by the inducers and repressors of its own operator, *o*, but those characteristic of *o'*. Instead of being inducible by IPTG say, such a *lac* region fused to the operator of *trp* will be repressible by tryptophan. Instead of being subject to catabolite repression, it will be independent of it, since the *trp* operon is not susceptible to this form of regulation. If the fusion strain retains the *o* region, transcription of the *lac* but not the *trp* genes is blocked in the presence of *lac* repressor.[18] Another route to release from operator control is by mutations which generate new promoters in structural genes (i.e., to the right of the operator). For instance, the *trp* E^r mutation leads simultaneously to a loss of function of the *E* gene (Fig. 21-11) and to a *cis*-dominant constitutive production of the other gene products of the *trp* operon (167).

Polarity (168). One of the most interesting features that has emerged from a study of operon function is the concept of polarity. What has been found is that nonsense, chain-terminating, mutations in one gene lead to a coordinate reduction in the expression of *all* genes subsequent to (i.e., operator-distal from) the site of the original lesion. For instance, if the wild-type values for the products in an operon consisting of five genes were 500 : 400 : 200 : 100 : 50, one such nonsense mutant in gene *2* might produce the result 500 : 0 : 50 : 25 : 12.5, while another might yield 500 : 0 : 10 : 5 : 2.5. Such polar nonsense mutations may be of the amber, ochre, or UGA type or they may have resulted from frame shifts; missense mutants are never polar. The severity of the polar effect is a function of the distance between its location and that for initiation of *translation* of the next cistron; the greater this distance, the more severe the effect. As expected, introduction of a second nonsense mutation into this untranslatable region has no effect. It appears, therefore, that premature termination of translation close to the site of normal chain termination (functioning presumably by an analogous mechanism) has

[18] By fusion it has also been possible to establish that the *i* gene is transcribed in the same direction as are all the genes of the *lac* operon (166).

relatively little effect, while the exposure of a sizable portion of untranslated mRNA is somehow deleterious.[19] Polar mutants can be shown to contain two classes of mRNA (169): (1) one, normal in length but reduced in amounts; and (2) smaller segments, apparently interrupted at or near the site of the chain termination—the closer to the operator end, the shorter the chain. Molecules of mRNA containing regions distal to the lesion are absent.

Currently there is no single model that provides a completely satisfactory and unambiguous explanation of polarity. It is not unreasonable to suppose that something akin to polarity must be operative, even in normal operon function. Let us remember that there is an important dichotomy between transcription and translation of such a unit. Both are coordinated and sequential processes, the one commencing at the 5′-triphosphate end and coded for by some portion of the promoter, and the other with an *N*-formylmethionine in the product of the first structural gene, and both progressing in order toward the end of the most distal gene or its product. But transcription of operons in bacteria is continuous, it encompasses the whole of the operon, and the resultant polycistronic mRNA probably remains coherent. Translation is discontinuous; it must end as it reaches one (or more) terminator(s), just after the codon for the C-terminal amino acid of each successive cistron, and start again with the chain initiator of the adjacent cistron. Thus, there must be an alternation of terminator and initiator codons and perhaps an additional sequence of nucleotides intervening at the boundary between any two cistrons. The length of these boundaries, then, may well govern the relative amounts of polypeptides actually translated. As already mentioned, for most operons a polarity of this kind does exist, with the products of distal genes appearing in graduated and decreasing amounts relative to the first.

Among possible, nonexclusive, alternatives that have been used to explain polarity is the notion that ribosome attachment, which we now know to be directly coupled with chain initiation by means of AUG(GUG)/*N*-formyl-Met-tRNA$_f$ occurs only, or preponderantly, on operator-proximal genes or gene segments; other regions are either functionally (e.g., by lacking the proper constellation of codons required for initiation) or structurally (because of features inherent in mRNA sequence producing an inappropriate secondary structure) constrained from doing so. The number of ribosomes capable of successfully traversing the gap between a terminator and the adjacent initiator will then be inversely related to the length of the gap. Secondary initiators are relatively weak, probably involve a subsidiary signal in addition to AUG, and are nonrandomly distributed within a gene (170). Another possibility envisages that, since it is known (see below) that transcription and translation are closely coupled, continued and successful transcription requires continuous translation. Ribosome travel is required to peel nascent mRNA off its DNA template but is in turn dependent on polypeptide chain elongation; premature termination of the latter will lead to cessation of the former process and perhaps to destruction of its product.

Untranslated Regions. Implicit in the discussion so far is the relatively recent realization that in spite of the congruence of the polynucleotide (i.e., DNA and mRNA) and polypeptide maps, the former includes regions that possess no equivalent in the latter; there exist untranslated sequences, regions of the genome that do *not* specify *any* gene product beyond mRNA and perhaps some times not even that. These occur (1) at the beginning of the

[19] It does not, however, lead to preferential polar destruction of translatable mRNA (168a).

operon (the promoter and operator regions, in those cases where they, and particularly the latter, are not identical to the 5′-terminal region of the first structural gene); (2) at the intercistronic boundaries between the terminator(s) in the nth and the initiator in the $(n + 1)$st cistron; and (3) perhaps in an as yet unexplored region beyond the terminator of the last cistron.

Simultaneous Regulation of Transcription and Translation

The Tryptophan Operon (Fig. 21-12). Elegant investigations by Yanofsky, Imamoto, and their collaborators have provided considerable insight into how transcription and translation of this operon in *E. coli* is regulated (167, 171,172). The operon consists of five contiguous cistrons with an operator and, probably, a promoter site, as indicated in the figure. The five polypeptides specified by the operon (total $M = 2.44 \times 10^5$ daltons or ~2400 amino acids) interact in the manner shown to produce three enzymes, which in turn catalyze the five steps intervening in the biosynthesis of L-tryptophan from chorismic acid (page 790). Repression of the operon requires the simultaneous presence of tryptophan (50 μg) and the product of an unlinked regulatory gene (*trp R*). Controlled derepression can be achieved either by the addition of indole-3-propionate even in the presence of 10 μg of tryptophan, or by the complete removal of the amino acid; effective repression can be reimposed simply by its addition. Repression is absent in certain mutant (repressor-negative) strains. Transcription of the polycistronic mRNA com-

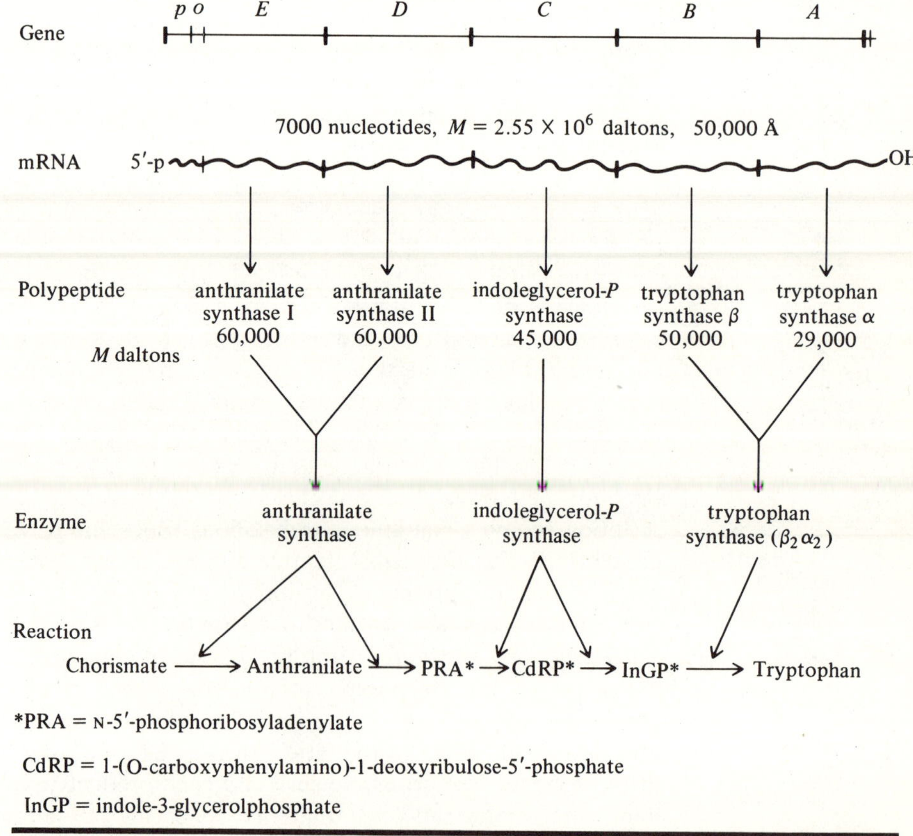

FIGURE 21-12
Tryptophan operon.

mences in the *p* region and proceeds down the DNA at a maximal rate of ~2500 internucleotide bonds formed per minute at 37° (1500 per minute at 30°) by a maximum of 15 polymerase molecules spaced (on the average) ~500 nucleotides apart. Each such molecule would then take ~2.5 min to traverse the whole operon and to complete synthesis of the message. Attachment of ribosomes coincident with initiation of translation occurs synchronously with and closely follows (perhaps within less than 1 minute) the initiation of transcription. As with most other operons (173), the ribosomes are threaded onto the nascent strand of mRNA immediately behind the polymerase as the latter moves down the DNA until the polysomal cluster reaches its maximal size of ~20 by the time the end of the operon (and of the mRNA) is reached (Fig. 21-13). The translational capacity of each mRNA molecule appears to be restricted to one, or at the most two, rounds before a signal (see below), presumably originating at or near its 5′ end, prevents further initiating events. In accord with this model the product of the first gene (*E*) makes its appearance shortly after the initiation of transcription and is followed sequentially, at intervals of ~2 min by the products of genes *D, C, B,* and *A* (at ~8 min), respectively. The translation frequency in the steady state is ~40 per min/genome or ~1000 peptide bonds per min. Just as the beginning of translation (i.e., the attachment of the first ribosome from the 5′ end of the RNA) probably exerts a signal for the beginning of the independent existence of each mRNA molecule by providing for

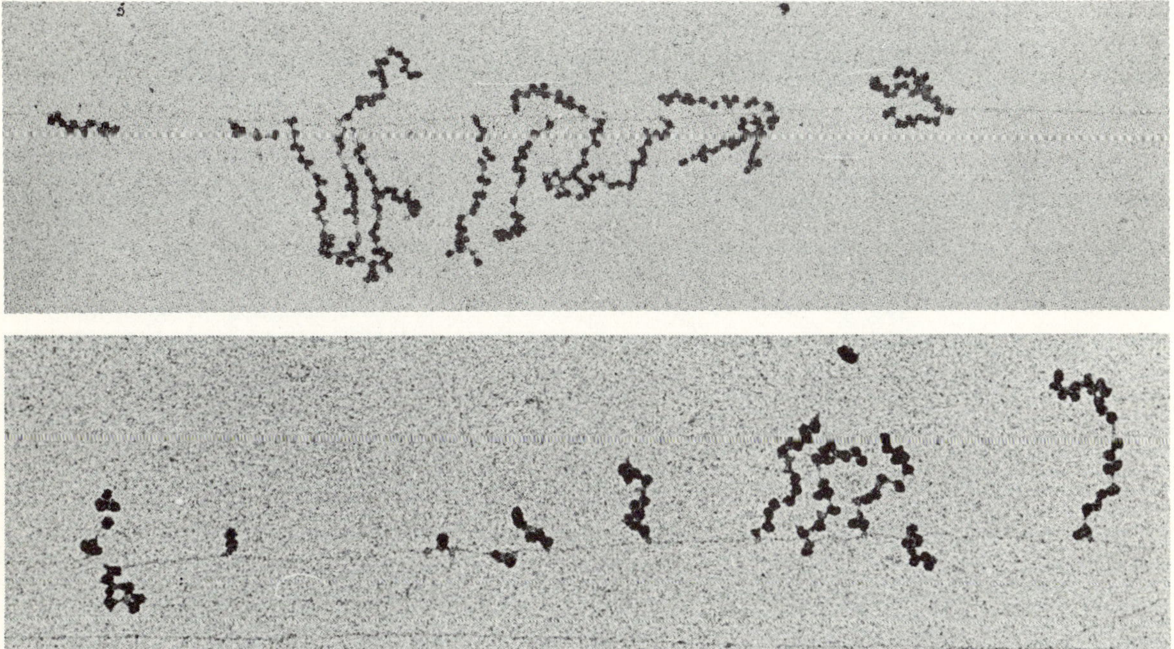

FIGURE 21-13

Genetically active and inactive portions of *E. coli* chromosomes. The polyribosomes attached to the active segments exhibit imperfect gradients of increasing lengths. The shorter, most distal polyribosomes in A may have resulted from mRNA degradation. The single particles in A and B indicate putative RNA polymerase molecules presumably on or very near the initiation sites of these active loci. From O. L. Miller, Jr., B. A. Hamkalo, and C. A. Thomas, Jr., *Science,* **169:** 392 (1970) with permission; electron micrographs by courtesy of Dr. Miller.

its detachment from the complex with DNA and polymerase, the end of translation (i.e., the passage of the last ribosome) signals its end, by exposing it to exonucleolytic attack[20] from this same 5′ end. Alternatively, one can envisage a "ticketing" process[21] which removes a fixed segment of mRNA coincident with the attachment of each Rb at the 5′ end, rendering it susceptible to destruction only when the last Rb of the cluster has been added. Results consistent with this model of sequential and linked transcription and translation and identical in their kinetics have also been obtained in the case of the *lac, his,* and *gal* operons. For instance, the latter (174) consists of three genes comprising 3300 nucleotide pairs and allows simultaneous transcription by two or more polymerase molecules; its translation frequency is 30 per min/genome at 37° (or ~1000 peptide bonds per min).[22]

Simultaneous Derepression. Just as under certain circumstances internal promoters can and do become operative for the initiation of mRNA of individual genes (or of gene clusters smaller than the operon), so can internal sites of chain initiation for polypeptide synthesis—and presumably ribosome attachment—become activated in a simultaneous, parallel (noncoordinate), rather than the customary sequential, manner. Thus, while the histidine operon provided one of the first examples for coordinate operon function, recent studies indicate that initiation of translation can become simultaneous (parallel) rather than sequential under conditions whenever the supply of N^{10}-formyl tetrahydrofolate, required presumably for the formation of the initiating *N*-formyl-Met-$tRNA_f$, is not limiting (72). Similarly, when the small RNA bacteriophages are used as messengers, the three gene products, capsid (coat) protein, RNA polymerase, and A (maturation) protein, are normally produced in the ratio 100 : 30 : 5.5. Nonsense mutations in the initial (5′) segment (the first 5 Å, or seven codons) of the gene for the coat protein exert strongly polar effects on the polymerase. These restrictions are removed when all, or most, of the coat-protein region of the RNA is eliminated. Under these conditions, then, the initiation sites, known to be present in all three genes (Fig. 5-23), become available for ribosome attachment and simultaneous rather than sequential translation becomes possible (122).

Translational Control. The close connection just described that appears to exist between transcription and translation in bacterial systems, and the very existence of polarity, make the existence of a translational control over the level of polypeptide to be synthesized very plausible. The prototype for this form of regulation is again provided by that governing the translation of the small RNA-containing coliphages (see page 951), an event that of necessity neither requires nor partakes of concurrent (or prior) transcription from a DNA genome—and which occurs in the absence of *any* viral RNA synthesis (either progeny or replicative form). Three different manifestations of the phenomenon appear to exist: (1) Polarity in the capsid gene with a pronounced position effect (see page 964), and (2) the unmasking of secondary initiation sites as a consequence of the removal of the first gene have already been mentioned; (3) the capsid protein has also been reported to exert an effect on the rate of translation of subsequent genes (118,121,122,175).

Several years ago Ames, Hartmann, and Martin (176) suggested a possible role for certain ambiguous codons cognate to aminoacyl-tRNA's in short

[20] The enzyme(s) responsible have sometimes been referred to as "Exonuclease V."
[21] Such processes may be of special regulatory significance in eucaryotes (173a).
[22] Rates obtained *in vitro* are beginning to approach these values (174a).

supply (minority codons) in regulating the rate and extent of polypeptide synthesis. Although this "modulation model" for translational control may not be the correct one for the events it was originally devised to explain, it has served to focus attention on aminoacyl-tRNA's as possible regulatory elements (177). Recently, several sets of observations have been reported that suggest that these molecules may indeed fulfill a variety of functions in addition to their classical role as adaptors, or entries in the genetic dictionary. Thus, the appropriate tRNA's charged with histidine, valine, and leucine act as repressors (or corepressors) for genes specifying the biosynthesis of these amino acids (178–180); among the functions specified by invading viral genomes are frequently ones governing the synthesis or modification of tRNA's (181–183); Phe-$tRNA^{Phe}$ is reported to act as an allosteric modifier for one of the early enzymes in the biosynthetic pathway[23] for Phe (184); and finally Gln-$tRNA^{Gln}$ is produced not by direct acylation but through the prior formation of Glu-$tRNA^{Gln}$ [Eq. (21-2)]. Furthermore, there also exist strong indications that certain tRNA's may fulfill functions that even transcend their various roles in protein biosynthesis: Lys-$tRNA^{Lys}$ appears to be an intermediate in the synthesis of *O*-lysyl-phosphatidylglycerol in membranes of *Staphylococcus aureus* (186), while Gly-$tRNA^{Gly}$ provides the glycine donor for the cell wall mucopeptide of the same organism. In addition, the unusual base 2-methylthio-N^6-(Δ^2-isopentenyl) adenosine, known to be located next to the anticodon of $tRNA^{Tyr}$ of *E. coli* (Table 21-3), is known to function as a potent plant-growth hormone, or *cytokinin* (187).

Positive Controls. The transcription of most of the genes of certain bacteriophages requires the substitution of phage-specific and specified σ factors (188) for the analogous component of the host RNA polymerase (189) (page 906), which *is,* however, required for the transcription of a very restricted group of phage genes, the "pre-early" genes (190,191). Similarly, the transcription of certain genes of bacteriophage λ requires the participation of the product specified by the *N* gene, probably by preventing the highly selective termination of transcription brought about by the ρ factor of the host (192) (Fig. 19-12). Finally, the functions of Ado-3′ : 5′-*P* in controlling the interaction of RNA polymerase with the promoter regions of bacterial operons (page 963) and the existence of Ψ factors are of special interest.

These examples constitute instances of *positive control* of transcription (see also Fig. 21-10). This type of regulation also appears to be operative in the case of the maltose, rhamnose, and arabinose (*ara*) operons of *E. coli.* In the last-named instance, studies by Englesberg and his collaborators (193) have demonstrated that active gene expression requires not only the detachment of an aporepressor (here specified by gene *ara C*) from an operator (*ara O*) by virtue of its interaction with the inducer—as in negative control—but also the interaction of the product with an adjacent region of the genome (*ara I,* for "initiator"). Just as positive controls appear to have their use in transcription, so do they also in translation. Specific, or specifically modified, initiator regions on the mRNA (e.g., Fig. 5-23), affecting its complex with the ribosome and/or with initiation factors (proteins) (see Fig. 5-23 and the footnote on page 934), appear to be possible targets.

[23] Evidence is beginning to accumulate (185) that for certain biosynthetic sequences there may be an intimate link between the two types of regulation with the first unambiguous enzyme as their target: the feedback inhibition of its function and the repression of its synthesis by the distal product. It is even possible that the allosterically modified enzyme molecule itself participates in regulating the synthesis of its kind.

Regulation in Eucaryotic Cells. Little that is definitive can be said about this topic in spite of sustained efforts during the past decade. However, several facts appear to stand out and will have to be accommodated by any consistent model.

1. Messenger RNA's in such cells appear to possess variable but significantly longer life times than those of bacteria—even on a time scale not in clock time but in terms of the generation time of the cell involved—and are thus capable of sustaining translation for many rounds. The well-proven persistence of polysomes in such cells is thus a reflection of a kinetic phenomenon, the steady state of occupancy, rather than of a single, isolatable event.

2. Polypeptide chain initiation on cytoplasmic ribosomes probably occurs by a mechanism involving Met-tRNA$_{f*}$ and *not* fMet-tRNA$_f$. When the latter mechanism *is* operative it appears restricted to specialized organelles such as mitochondria and chloroplasts.

3. There is no hard evidence for (or against) the existence of operons. Even the two polypeptide subunits of hemoglobin are specified by two unlinked genes.

4. Regulatory substances (hormones) of bewildering variety and orders of complexity, from small organic molecules (e.g., nucleotides or steroids) to large polypeptides (e.g., insulin, glucagon) appear to effect simultaneously whole groups of synthetic events (whether initially at the transcriptional or the translational level is not yet certain, although the weight of the evidence presently appears to favor the former course).

5. The availability of prosthetic groups such as heme or flavin coenzymes, or of an attachment site on a membrane, appears to exert pronounced regulatory effects on the extent and even the rate of synthesis of appropriate polypeptides.

6. Proteins that are composed of more than one polypeptide chain in their final, active form are biosynthetically produced as single, frequently circular, polypeptides and hence presumably encoded in a single gene. We refer here specifically to the proenzymes or zymogens such as trypsinogen, chymotrypsinogen, pepsinogen, procarboxypeptidase, etc., and prohormones such as proinsulin.

7. The opportunity for redundancy (gene duplication) is much greater in the genome of such cells, as is the opportunity for somatic mutation during their long lifetimes. Hence the possibility of nonidentity within a closely related but not identical group of genes and gene products and the control of their individual expressions must be kept in mind (196a).

8. There is some evidence that a ribosome cycle (i.e., the dissociation and reassociation of ribosomal subunits coincident with the termination and initiation of translation) may be operative in eucaryotic cells in a manner similar to that described for bacterial cells (Fig. 21-8). However, the known spatial sequestration of the various processes involved—transcription of the genome in the nucleus, synthesis of ribosomal precursors in the nucleolus, and translation in the cytoplasm—renders the analogy somewhat tenuous and makes certain a corresponding sequestration of temporal events.

9. The transcripts of many genes are unstable and decay in the nucleus without ever entering the cytoplasm. Many of the genes transcribed are present in multiple copies (i.e., redundant or reiterated).

Schemes showing possible and suggested control points are given in Figs. 21-14 and 21-15. Among the more striking and well-documented examples of control mechanisms in vertebrate animals (194) are the following.

1. A regulatory gene in mice that appears specifically to control the expression of the structural gene which specifies the β protomer of L-lactate dehydrogenase (195). We recall (page 174) that this enzyme is composed of two electrophoretically distinct protomers which can be assorted into five isoenzymes α_4, $\alpha_3\beta$, $\alpha_2\beta_2$, $\alpha\beta_3$, and β_4.

2. A demonstration that the β chain of hemoglobin A ($\alpha_2\beta_2$) and δ chain of hemoglobin A_2 ($\alpha_2\delta_2$) are synthesized by the same population of human reticulocytes, in the relative amounts of 40 : 1 in normal, and 30 : 1 in sickle-cell ($\alpha_2\beta_2{}^s$) individuals. Evidence is at hand that regulation occurs at the level of transcription rather than translation and that the observed deficiency of δ chains is caused by a deficiency of mRNA for that species (196).

3. The experimental fact that the overall rate of hemoglobin synthesis, which consists of the synthesis of the two separate chains of globin by two separate groups of polysomes, as well as the attachment of the heme prosthetic group, appears to be closely regulated and integrated (197). Various proposals accounting for this fact are based on the observations that heme or heme precursors can stimulate the rate of initiation of propagation of the two polypeptide chains (198,199), that β chains can regulate the release of nascent

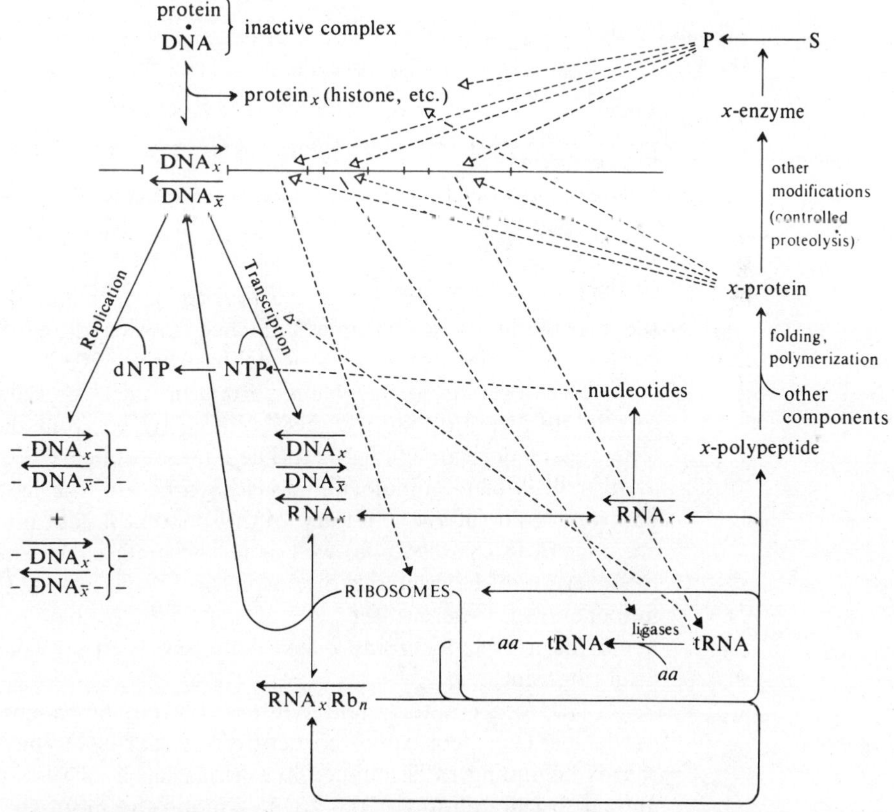

FIGURE 21-14

Possible additional competitive and synergistic factors in the control of protein synthesis.

forms
specifies
inhibits, binds to, or otherwise modifies
various DNA segments corresponding to different cistrons

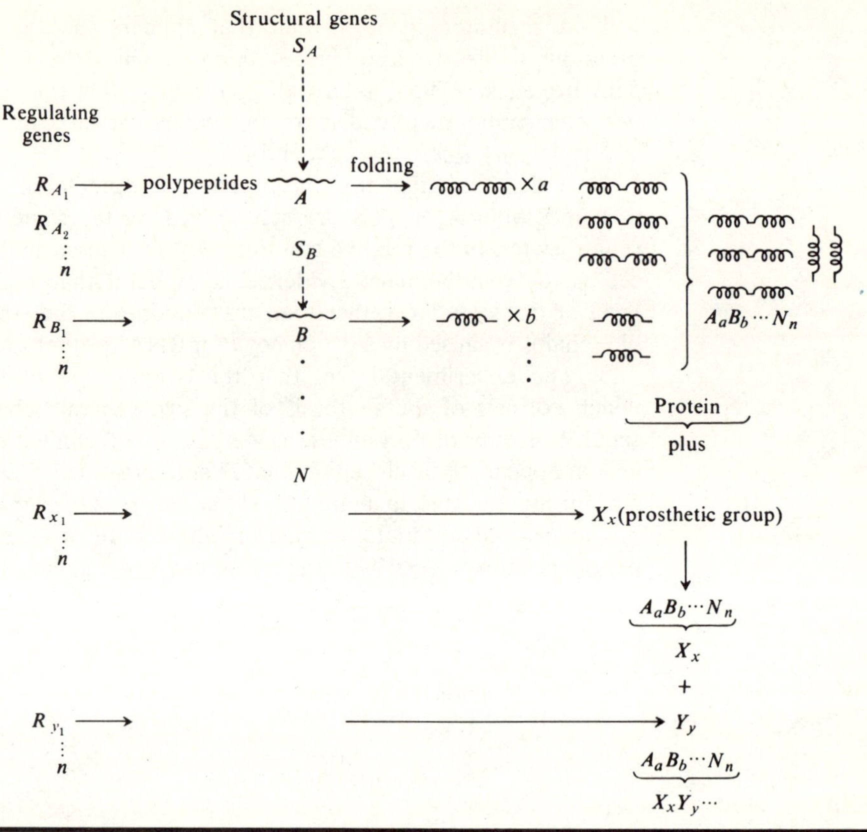

FIGURE 21-15
Possible control points in the elaboration of complex proteins.

α chains from their site of synthesis to form a soluble $\alpha\beta$ dimer (200), and that the prosthetic group becomes attached to this dimer in the course of its conversion to the tetrameric hemoglobin proper (201).

4. The observation that heme formation itself appears to be regulated by certain steroid hormones or metabolites derived from them (e.g., pregnanolone and its derivatives) (202). Their target appears to be the rate or extent of formation of δ-aminolevulinic acid synthetase, the first key enzyme on the pathway of porphyrin synthesis. In this context it is of interest to note that the enzymes of this pathway, as well as ferrochelatase, the enzyme responsible for the insertion of heme into protoporphyrin IX, all appear to be localized within mitochondria, whereas the site of synthesis of key hemoproteins such as hemoglobin or cytochrome *c* has definitely been assigned to extramitochondrial ribosomes.

5. That other natural and synthetic steroid hormones, the adrenal glucocorticoids (e.g., cortisone, corticosterone, cortisol), have a mode of action that is beginning to be defined at various levels of resolution from the whole animal to cell cultures (203–206). Among the more striking effects are substantial increases in the intracellular concentrations of a number of enzymes concerned with protein catabolism, and with gluconeogenesis, specifically from amino acid precursors. Two of the enzymes so affected are tryptophan pyrrolase and tyrosine-α-oxoglutarate aminotransferase. The phenomenon is due to an increase in the rate of synthesis of these proteins rather than a decrease

in their rate of degradation and may be referrable to an actual specific induction (or elimination of repression) of transcription of relevant portions of the genome (207).

6. That other hormones acting on different targets, such as, for example, growth hormone, thyroid hormone, and female sex hormones, may be acting by similar mechanisms.

7. That the synthesis of proteins designed for export occurs on the ribosomes of the rough endoplasmic reticulum (208) attached thereto by an entity which is (138) (a) also involved in the formation of the disulfide cross-links characteristic of such proteins and (b) sensitive to the inhibitor aflatoxin leading to detachment of ribosomes, a process that is reversed by the steroid hormone cortisone but not by the closely related cortisol.

8. That cyclic AMP, now known to be the mediator ("second messenger") (209) in the action of several different hormones, may regulate not only—as already established—the activity of certain target enzymes, but—by analogy to its role in catabolite repression (page 961)—their amount as well (210).

REFERENCES

Books, Monographs, and Reviews

Most of the relevant references have already been cited in connection with Chapters 5, 19, and 20. In particular:

Ames, B. N., and R. G. Martin: Biochemical Aspects of Genetics: The Operon, *Ann. Rev. Biochem.,* **33:** 235 (1964).

Anfinsen, C.: *The Molecular Basis of Evolution,* Wiley, New York, 1959.

Attardi, G., and F. Amaldi: Structure and Synthesis of Ribosomal RNA, *Ann. Rev. Biochem.,* **39:** 183 (1970).

Beckwith, J. R.: Regulation of the Lac Operon, *Science,* **156:** 597 (1967).

Bennett, J. C., and W. J. Dreyer: Genetic Coding for Protein Structure, *Ann. Rev. Biochem.,* **33:** 205 (1964).

Campbell, P. N.: The Biosynthesis of Proteins, *Progr. Biophys. Mol. Biol.,* **15:** 3 (1965).

Cold Spring Harbor Symposium on Quantitative Biology: *Cellular Regulatory Mechanisms,* vol. 26, 1961.

Cold Spring Harbor Symposium on Quantitative Biology: *Synthesis and Structure of Macromolecules,* vol. 28, 1963.

Cold Spring Harbor Symposium on Quantitative Biology: *The Genetic Code,* vol. 31, 1966.

Cold Spring Harbor Symposium on Quantitative Biology: *The Mechanism of Protein Synthesis,* vol. 34, 1969.

Cold Spring Harbor Symposium on Quantitative Biology: *RNA Transcription,* vol. 35, 1970.

Cox, R. A.: Macromolecular Structure and Properties of Ribonucleic Acids, *Quart. Rev.,* **22:** 499 (1968).

Epstein, W., and J. R. Beckwith: Regulation of Gene Expression, *Ann. Rev. Biochem.,* **37:** 411 (1968).

Geiduschek, E. P., and R. Haselkorn: Messenger RNA, *Ann. Rev. Biochem.,* **38:** 647 (1969).

Horowitz, N. H., and R. L. Metzenberg: Biochemical Aspects of Genetics, *Ann. Rev. Biochem.,* **34:** 527 (1965).

Lengyel, P.: On Peptide Chain Initiation, in J. H. Taylor (ed.), *Molecular Genetics,* part 2, Academic Press, New York, 1967, p. 193.

Lipmann, F.: Polypeptide Chain Elongation in Protein Biosynthesis, *Science,* **164:** 1024 (1969).

Loftfield, R. B.: The Aminoacylation of Transfer Ribonucleic Acid., in E. McConkey (ed.), *Protein Synthesis,* vol. 1, Marcel Dekker, New York, 1970.

Madison, J. T.: Primary Structure of RNA, *Ann. Rev. Biochem.,* **37:** 131 (1968).

Moldave, K.: Nucleic Acids and Protein Biosynthesis, *Ann. Rev. Biochem.,* **34:** 419 (1965).

Monod, J., F. Jacob, and F. Gros: Structural and Rate-Determining Factors in the Biosynthesis of Adaptive Enzymes, in "The Structure and Biosynthesis of Macromolecules, *Biochem. Soc. Symp. (Cambridge, Engl.),* **21:** 104 (1962).

Nirenberg, M.: Protein Synthesis and the RNA Code, *Harvey Lectures,* **59:** 155 (1964).

Novelli, G. D.: Amino Acid Activation for Protein Synthesis, *Ann. Rev. Biochem.,* **36:** 449 (1967).

Ochoa, S.: Translation of the Genetic Message, *Naturwissenchaften,* **55:** 505 (1968).

Pitot, H. C.: Metabolic Regulation in Metazoan Systems, in J. H. Taylor (ed.), *Molecular Genetics,* part 2, Academic Press, New York, 1967, p. 383.

Schimke, R. T., and D. Doyle: Control of Enzyme Levels in Animal Tissues, *Ann. Rev. Biochem.,* **39:** 929 (1970).

Schweet, R., and R. Heintz: Protein Synthesis, *Ann. Rev. Biochem.,* **35:** 723 (1966).

Singer, M. F., and P. Leder: Messenger RNA: An Evaluation, *Ann. Rev. Biochem.,* **35:** 195 (1966).

Speyer, J. F.: The Genetic Code, in J. H. Taylor (ed.), *Molecular Genetics,* vol. 2, Academic Press, New York, 1967, p. 137.

Stavis, R. L., and J. T. August: The Biochemistry of RNA Bacteriophage Replication, *Ann. Rev. Biochem.,* **39:** 527 (1970).

Vogel, H. J., and R. H. Vogel: Regulation of Protein Synthesis, *Ann. Rev. Biochem.,* **36:** 519 (1967).

Weisblum, B., and J. Davies: Antibiotic Inhibitors of the Bacterial Ribosome, *Bacteriol. Rev.,* **32:** 494 (1968).

Yčas, M.: *The Biological Code,* North-Holland, Amsterdam, 1969.

Specific References

1. Siekevitz, P.: *J. Biol. Chem.,* **195:** 549 (1952).
2. Zamecnik, P. C., and E. B. Keller: *J. Biol. Chem.,* **209:** 337 (1954).
3. For a review of early work, see M. B. Hoagland, in E. Chargaff and J. N. Davidson (eds.), *The Nucleic Acids,* vol. 3, Academic Press, New York, 1960, p. 349; Zamecnik, P. C.: *Cold Spring Harbor Symp. Quant. Biol.,* **34:** 1 (1969).
4. Schweet, R., H. Lamfrom, and E. Allen: *Proc. Natl. Acad. Sci. U.S.,* **44:** 1029 (1958).
5. Lamborg, M. R., and P. C. Zamecnik: *Biochim. Biophys. Acta,* **42:** 206 (1960).
6. Tissières, A., D. Schlessinger, and F. Gros: *Proc. Natl. Acad. Sci. U.S.,* **46:** 1450 (1960); Tissières, A., and J. W. Hopkins: *ibid.,* **47:** 2015 (1961).
7. Nirenberg, M., P. Leder, M. Bernfield, R. Brimacombe, J. Trupin, F. Rottman, and C. O'Neal: *Proc. Natl. Acad. Sci. U.S.,* **53:** 1161 (1965); Marshall, R. E., C. T. Caskey, and M. Nirenberg: *Science,* **155:** 820 (1967).
8. Jones, D. S., S. Nishimura, and H. G. Khorana: *J. Mol. Biol.,* **16:** 454 (1966).
9. McCarthy, B. J., and J. J. Holland: *Proc. Natl. Acad. Sci. U.S.,* **54:** 880 (1965).
10. Kössel, H., A. R. Morgan, and H. G. Khorana: *J. Mol. Biol.,* **26:** 449 (1967); Morgan, A. R., R. D. Wells, and H. G. Khorana: *ibid.,* **26:** 477 (1967); see also Bretscher, M. S.: *Cold Spring Harbor Symp. Quant. Biol.,* **34:** 651 (1969).
11. Nirenberg, M. W.: *Sci. Am.,* **208** (3): 80 (1963); Nirenberg, M. W., and J. H. Matthaei: *Proc. Natl. Acad. Sci. U.S.,* **47:** 1558 (1961).
12. Ochoa, S.: *Experientia,* **20:** 57 (1964); Lengyel, P., J. F. Speyer, and S. Ochoa: *Proc. Natl. Acad. Sci. U.S.,* **47:** 1936 (1961); Wahba, A. J., R. S. Gardner, C. Basilio, R. S. Miller, J. F. Speyer, and P. Lengyel: *ibid.,* **49:** 116 (1963).
13. McElroy, W. D., M. DeLuca, and J. Travis: *Science,* **157:** 150 (1967).
14. Novelli, G. D.: *Ann. Rev. Biochem.,* **36:** 449 (1967).

14a. Hirsch, D. I., and R. Lipmann: *J. Biol. Chem.,* **243:** 5724 (1968).

15. Loftfield, R. B., L. I. Hecht, and E. A. Eigner: *Biochim. Biophys. Acta,* **72:** 383 (1963); Loftfield, R. B.: *Biochem. J.,* **89:** 82 (1963).

15a. Schapiro, G., J. C. Dreyfus, and N. Malehnia: *Biochem. Biophys. Res. Commun.,* **23:** 558 (1968).

16. Loftfield, R. B., and E. A. Eigner: *J. Biol. Chem.,* **240:** PC1482 (1965); **244:** 1746 (1969).
17. Norris, A. T., and P. Berg: *Proc. Natl. Acad. Sci. U.S.,* **52:** 330 (1964); Baldwin, A. N., and P. Berg: *J. Biol. Chem.,* **241:** 839 (1966).
18. Wilcox, M., and M. Nirenberg: *Proc. Natl. Acad. Sci. U.S.,* **61:** 229 (1968).
19. Crick, F. H. C., in The Biological Replication of Macromolecules, *Symp. Soc. Exptl. Biol.,* **12:** 138 (1958); Hoagland, M. B.: *Brookhaven Symp. Biol.,* **12:** 40 (1959).
20. Chapeville, F., F. Lipmann, G. von Ehrenstein, B. Weisblum, W. J. Ray, Jr., and S. Benzer: *Proc. Natl. Acad. Sci. U.S.,* **48:** 1086 (1962); Von Ehrenstein, G., B. Weisblum, and S. Benzer: *ibid.,* **49:** 669 (1963); Hervé, G., and F. Chapeville: *Biochim. Biophys. Acta,* **76:** 493 (1963).
21. Holley, R. W., J. Apgar, G. A. Everett, J. T. Madison, M. Marquisee, S. H. Merrill, J. R. Penswick, and A. Zamir: *Science,* **147:** 1462 (1965); contrast this with R. W. Holley's first contribution to the field: *J. Am. Chem. Soc.,* **79:** 658 (1957).
22. Madison, J. T.: *loc. cit.;* Philipps, G. R.: *Nature,* **223:** 374 (1969).
23. Fuller, W., and A. Hodgson: *Nature,* **215:** 817 (1967); see Cramer, F., H. Doepner, F. v.d. Haar, E. Schlimme, and H. Seidel: *Proc. Natl. Acad. Sci. U.S.,* **61:** 1384 (1968); Ninio, J., A. Favre, and M. Yaniv: *Nature,* **223:** 1333 (1969).
24. Schulman, L. H., and R. W. Chambers: *Proc. Natl. Acad. Sci. U.S.,* **61:** 308 (1968); Imura, N., H. Schwam, and R. W. Chambers: *ibid.,* **62:** 1203 (1969); Imura, N., G. B. Weiss, and R. W. Chambers: *Nature,* **222:** 1147 (1969).
25. Revel, M., and U. Z. Littauer: *J. Mol. Biol.,* **15:** 389 (1966).
26. Gefter, M. L., and R. L. Russell: *J. Mol. Biol.,* **39:** 145 (1969).
27. Epler, J. L., and W. E. Barnett: *Biochem. Biophys. Res. Commun.,* **28:** 328 (1967); Barnett, W. E., and D. H. Brown: *Proc. Natl. Acad. Sci. U.S.,* **57:** 452 (1967); see also Buck, C. A., and M. M. K. Nass: *J. Mol. Biol.,* **41:** 67 (1969); Küntzel, H.: *Nature,* **222:** 142 (1969).
28. Sundharadas, G., J. R. Katze, D. Söll, W. Konigsberg, and P. Lengyel: *Proc. Natl. Acad. Sci. U.S.,* **61:** 693 (1968); Bennett, T. P.: *J. Biol. Chem.,* **244:** 3182 (1969).
29. Doctor, B. P., J. E. Loebel, M. A. Sodd, and D. B. Winter: *Science,* **163:** 693 (1969); see also Harada, F., H. J. Gross, F. Kimura, S. H. Chang, S. Nishimura, and U. L. RajBhandary: *Biochem. Biophys. Res. Commun.,* **33:** 299 (1968).
30. Lengyel, P.: *J. Gen. Physiol.,* **49:** 305 (1966).
31. Bennett, T. P., J. Goldstein, and F. Lipmann: *Proc. Natl. Acad. Sci. U.S.,* **53:** 385 (1965).
32. Garen, A.: *Science,* **160:** 149 (1968); see also Chan,

T. S., and A. Garen: *J. Mol. Biol.*, **45:** 545 (1969); Gorini, L., and J. R. Beckwith: *loc. cit.*

33. Cannon, M., R. Krug, and W. Gilbert: *J. Mol. Biol.*, **7:** 360 (1963); Zubay, G.: *Science*, **140:** 1092 (1963); Takanami, M., and G. Zubay: *Proc. Natl. Acad. Sci. U.S.*, **51:** 834 (1964).
34. Fellner, P., and F. Sanger: *Nature*, **219:** 236 (1968).
35. Brownlee, G. G., F. Sanger, and B. G. Barrell: *J. Mol. Biol.*, **34:** 379 (1968).
36. Nichols, J. L., and B. G. Lane: *Can. J. Biochem.*, **45:** 937 (1967); Midgley, J. E. M., and D. J. McIlreavy: *Biochim. Biophys. Acta*, **145:** 512 (1967); see also Riley, W. T.: *Nature*, **222:** 446 (1969).
37. Kaltschmidt, E. and H. G. Wittmann: *Proc. Natl. Acad. Sci. U.S.*, **67:** 1276 (1970).
38. Traub, P., and M. Nomura: *J. Mol. Biol.*, **34:** 575 (1968); *Proc. Natl. Acad. Sci. U.S.*, **59:** 777 (1968).
39. Traub, P., and M. Nomura: *J. Mol. Biol.*, **40:** 391 (1969).
40. Nomura, M., P. Traub, and H. Beckmann: *Nature*, **219:** 793 (1968).
41. Smith, I., D. Dubnau, P. Morell, and J. Marmur: *J. Mol. Biol.*, **33:** 123 (1968).
42. Kohne, D. E.: *Biophys. J.*, **8:** 1104 (1968).
43. Spiegelman, S., and S. A. Yankofsky: in V. Bryson and H. J. Vogel (eds.), *Evolving Genes and Proteins*, Academic Press, New York, 1965, p. 537.
44. Mangiarotti, G., D. Apirion, D. Schlessinger, and L. Silengo: *Biochemistry*, **7:** 456 (1968).
45. Osawa, S., E. Otaka, T. Itoh, and T. Fukui: *J. Mol. Biol.*, **40:** 321 (1969).
46. Roberts, R. B., R. J. Britten, and B. J. McCarthy: in J. H. Taylor (ed.), *Molecular Genetics*, part 1, Academic Press, New York, 1963, p. 292.
47. Morell, P., and J. Marmur: *Biochemistry*, **7:** 1141 (1968).
48. Weller, D. L., Y. Shechter, D. Musgrave, M. Rougvie, and J. Horowitz: *Biochemistry*, **7:** 3668 (1968).
49. Click, R. E., and B. L. Tint: *J. Mol. Biol.*, **25:** 111 (1967); Stutz, E., and H. Noll: *Proc. Natl. Acad. Sci. U.S.*, **57:** 774 (1967); Reisner, A. H., J. Rowe, and H. M. Macindoe: *J. Mol. Biol.*, **32:** 587 (1968); Taylor, M. M., J. E. Glasgow, and R. Storck: *Proc. Natl. Acad. Sci. U.S.*, **57:** 164 (1967).
50. Ritossa, F. M., and S. Spiegelman: *Proc. Natl. Acad. Sci. U.S.*, **53:** 737 (1965); see also Brown, D. D., and J. P. Gurdon: *ibid.*, **51:** 139 (1964); Birnstiel, M.: *Natl. Cancer Inst. Monographs*, **23:** 431 (1966).
51. Brown, D. D., and I. B. Dawid: *Science*, **160:** 272 (1968).
52. Britton, R. J., and D. E. Kohne: *Science*, **161:** 529 (1968).
53. Birnstiel, M., J. Speirs, I. Purdom, K. Jones, and U. E. Loening: *Nature*, **219:** 454 (1968).
54. Maden, B. E. H.: *Nature*, **219:** 685 (1968); Perry, R. P. : *Progr. Nucl. Acid Res. Mol. Biol.*, **6:** 219 (1967); see also Willems, M., E. Wagner, R. Laing, and S. Penman: *J. Mol. Biol.*, **32:** 211 (1968).
55. Warner, J. R., and R. Soeiro: *Proc. Natl. Acad. Sci. U.S.*, **58:** 1984 (1967).
56. Kaempfer, R. O. R., M. Meselson, and H. J. Raskas: *J. Mol. Biol.*, **31:** 277 (1968); Kaempfer, R. O. R.: *Proc. Natl. Acad. Sci. U.S.*, **61:** 106 (1968).
57. Nomura, M., C. V. Lowry, and C. Guthrie: *Proc. Natl. Acad. Sci. U.S.*, **58:** 1487 (1967); Guthrie, C., and M. Nomura: *Nature*, **219:** 232 (1968).
58. Mangiarotti, G., and D. Schlessinger: *J. Mol. Biol.*, **20:** 123 (1966); Schlessinger, D., G. Mangiarotti, and D. Apirion: *Proc. Natl. Acad. Sci. U.S.*, **58:** 1782 (1967).
59. Colombo, B., C. Vesco, and C. Baglioni: *Proc. Natl. Acad. Sci. U.S.*, **61:** 651 (1968).
60. Kaempfer, R. O. R.: *Nature*, **222:** 950 (1969).
61. Smith, A. E., and K. A. Marcker: *Nature*, **226:** 607 (1970); Brown, J. C., and A. E. Smith: *ibid.*, **226:** 610 (1970).
62. Ochoa, S.: *Naturwissenschaften*, **55:** 505 (1968); Mazumder, R., Y. B. Chae, and S. Ochoa: *Proc. Natl. Acad. Sci. U.S.*, **63:** 98 (1969).
63. Iwasaki, K., S. Sabol, A. J. Wahba, and S. Ochoa: *Arch. Biochem. Biophys.*, **125:** 542 (1968).
64. Kolakofsky, D., T. Ohta, and R. E. Thach: *Nature*, **220:** 244 (1968); Kolakofsky, D., K. F. Dewey, J. W. B. Hershey, and R. E. Thach: *Proc. Natl. Acad. Sci. U.S.*, **61:** 1066 (1968); Kolakofsky, D., K. F. Dewey, and R. E. Thach: *Nature*, **223:** 694 (1969).
65. Revel, M., J. C. Lelong, G. Brawerman, and F. Gros: *Nature*, **219:** 1016 (1968).
66. Clark, F. C., and K. A. Marcker: *Sci. Am.*, **218:** 36 (1968).
67. Waller, J-P.: *J. Mol. Biol.*, **7:** 483 (1963).
68. Adams, J. M., and M. R. Capecchi: *Proc. Natl. Acad. Sci. U.S.*, **55:** 147 (1966).
69. Webster, R. E., D. L. Engelhardt, and N. D. Zinder: *Proc. Natl. Acad. Sci. U.S.*, **55:** 155 (1966).
70. Takeda, M., and R. E. Webster: *Proc. Natl. Acad. Sci. U.S.*, **60:** 1487 (1968).
71. Clark, B. F. C., and K. A. Marcker: *J. Mol. Biol.*, **17:** 394 (1966).
72. Berberich, M. A., J. S. Kovach, and R. F. Goldberger: *Proc. Natl. Acad. Sci. U.S.*, **57:** 1857 (1967).
73. Crick, F. H. C.: *Sci. Am.*, **215** (4): 55 (1966); *J. Mol. Biol.*, **19:** 548 (1966).
74. Sundararajan, T. A., and R. E. Thach: *J. Mol. Biol.*, **19:** 74 (1966); Salas, M., M. B. Hille, J. A. Last, A. J. Wahba, and S. Ochoa: *Proc. Natl. Acad. Sci. U.S.*, **57:** 387 (1967); Ghosh, H. P., D. Söll, and H. G. Khorana: *J. Mol. Biol.*, **25:** 275 (1967).
75. Ono, Y., A. Skoultchi, A. Klein, and P. Lengyel: *Nature*, **220:** 1304 (1968).

75a. Lodish, H. F.: *Nature*, **224:** 867 (1969).

76. Monro, R. E., and K. A. Marcker: *J. Mol. Biol.*, **25:** 347 (1967); Monro, R. E., J. Cerná, and K. A. Marcker: *Proc. Natl. Acad. Sci. U.S.*, **61:** 1042 (1968); Maden, B. E. H., and R. E. Monro: *European J. Biochem.*, **6:** 309 (1968).
77. Rychlík, I., J. Černá, S. Chládek, J. Žemlička, and Z. Haladoyá: *J. Mol. Biol.*, **43:** 13 (1969).
78. Scolnick, E., G. Milman, M. Rosman, and T. Caskey: *Nature*, **225:** 152 (1970).
79. Bretscher, M. S.: *Progr. Biophys. Mol. Biol.*, **19:** 175 (1969).

80. Horikoshi, K., and R. H. Doi: *J. Biol. Chem.,* **243:** 2381 (1968).
81. Caskey, C. T., B. Redfield, and H. Weissbach: *Arch. Biochem. Biophys.,* **120:** 119 (1967); Takeishi, K., T. Ukita, and S. Nishimura: *J. Biol. Chem.,* **243:** 5761 (1968).
82. Smith, A. E., and K. A. Marcker: *J. Mol. Biol.,* **38:** 241 (1968).
83. Galper, J. B., and J. E. Darnell: *Biochem. Biophys. Res. Commun.,* **34:** 205 (1969).
84. Prichard, P. M., J. M. Gilbert, D. A. Shafritz, and W. F. Anderson: *Nature,* **226:** 511 (1970); see also Fuhr, J. E., I. M. London, and A. I. Grayzel: *Proc. Natl. Acad. Sci. U.S.,* **63:** 129 (1969).
85. Gierer, A.: *J. Mol. Biol.,* **6:** 148 (1963); Gilbert, W.: *ibid.,* **6:** 374, 389 (1963).
86. Williamson, A. R., and R. Schweet: *J.Mol. Biol.,* **11:** 358 (1965); Hardesty, B., R. Miller, and R. Schweet: *Proc. Natl. Acad. Sci. U.S.,* **50:** 924 (1963); Hardesty, B., J. J. Hutton, R. Arlinghaus, and R. Schweet: *ibid.,* **50:** 1078 (1963).
87. Wettstein, F. O., and H. Noll: *J. Mol. Biol.,* **11:** 35 (1965); Wettstein, F. O., T. Staehelin, and H. Noll: *Nature,* **197:** 430 (1963); **198:** 632 (1963); Staehelin, T., C. C. Brinton, F. O. Wettstein, and H. Noll: **199:** 865 (1963).
88. Warner, J. R., and A. Rich: *Proc. Natl. Acad. Sci. U.S.,* **51:** 1134 (1964); *J. Mol. Biol.,* **10:** 202 (1964); Goodman, H. M., and A. Rich: *Nature,* **199:** 318 (1963); Slayter, H. S., J. R. Warner, A. Rich, and C. E. Hall: *J. Mol. Biol.,* **7:** 652 (1963).
89. Monro, R. E., B. E. H. Maden, and R. R. Traut, in D. Shugar (ed.), *Genetic Elements, Properties and Function,* Academic Press, New York, 1967, p. 179.
90. Maden, B. E. H., R. R. Traut, and R. E. Monro: *J. Mol. Biol.,* **35:** 333 (1968).
91. Miskin, R., A. Zamir, and D. Elson: *Biochem. Biophys. Res. Commun.,* **33:** 551 (1968).
92. Parmeggiani, A.: *Biochem. Biophys. Res. Commun.,* **30:** 613 (1968); Leder, P., L. E. Skogerson, and M. M. Neu: *Proc. Natl. Acad. Sci. U.S.,* **62:** 454 (1969).
93. Skoultchi, A., Y. Ono, H. M. Moon, and P. Lengyel: *Proc. Natl. Acad. Sci. U.S.,* **60:** 675 (1968); Ono, Y., A. Skoultchi, J. Waterson, and P. Lengyel: *Nature,* **223:** 697 (1969).
94. Gordon, J.: *Proc. Natl. Acad. Sci. U.S.,* **58:** 1574 (1967); **59:** 179 (1968).
95. Allende, J. E., N. W. Seeds, T. W. Conway, and H. Weissbach: *Proc. Natl. Acad. Sci. U.S.,* **58:** 1566 (1968); Ertel, R., N. Brot, B. Redfield, J. E. Allende, and H. Weissbach: *ibid.,* **59:** 861 (1968).
96. Ibuki, F., and K. Moldave: *J. Biol. Chem.,* **243:** 44 (1968); Skogerson, L., and K. Moldave: *ibid.,* **243:** 5354, 5361 (1968); see also Richter, D., H. Hameister, H. G. Petersen, and F. Klink: *Biochemistry,* **7:** 3753 (1968).
97. Ravel, J. M., R. L. Shorey, S. Froehner, and W. Shive: *Arch. Biochem. Biophys.,* **125:** 514 (1968); see also Lucas-Lenard, J., and A. L. Haenni: *Proc. Natl. Acad., Sci., U.S.,* **63:** 93 (1969).
98. Nirenberg, M., and P. Leder: *Science,* **145:** 1399 (1964).
99. Söll, D., J. D. Cherayil, and R. M. Bock: *J. Mol. Biol.,* **29:** 97 (1967).
100. Soll, L., and P. Berg: *Nature,* **223:** 1340 (1969).
101. Yourno, J., and S. Tanemura: *Nature,* **225:** 422 (1970).
102. Carbon, J., P. Berg, and C. Yanofsky: *Proc. Natl. Acad. Sci. U.S.,* **56:** 764 (1966); Gupta, N. K., and H. G. Khorana: *ibid.,* **56:** 772 (1966).
103. Champe, S. P., and S. Benzer: *Proc. Natl. Acad. Sci. U.S.,* **48:** 532 (1962).
104. Gorini, L., and E. Kataja: *Proc. Natl. Acad. Sci. U.S.,* **51:** 487 (1964); *Biochem. Biophys. Res. Commun.,* **18:** 656 (1965); Gorini, L., G. A. Jacoby, and L. Breckenridge: *Cold Spring Harbor Symp. Quant. Biol.,* **31:** 657 (1966).
105. Weisblum, B., and J. Davies: *Bacteriol. Rev.,* **32:** 493 (1968).
106. Ozaki, M., S. Mizushima, and M. Nomura: *Nature,* **222:** 333 (1969).
107. Davies, J., W. Gilbert, and L. Gorini: *Proc. Natl. Acad. Sci. U.S.,* **51:** 883 (1964); Davies, J.: *Cold Spring Harbor Symp. Quant. Biol.,* **31:** 665 (1966).
108. Likover, T. E., and C. G. Kurland: *Proc. Natl. Acad. Sci. U.S.,* **58:** 2385 (1967); *J. Mol. Biol.,* **25:** 497 (1967).
109. Morse, D. E., R. F. Baker, and C. Yanofsky: *Proc. Natl. Acad. Sci. U.S.,* **60:** 1428 (1968); Baker, R. F., and C. Yanofsky: *ibid.,* **60:** 313 (1968).
110. Capecchi, M. R.: *Proc. Natl. Acad. Sci. U.S.,* **58:** 1144 (1967); Cuzin, F., N. Kretchmer, R. E. Greenberg, R. Hurwitz, and F. Chapeville: *ibid.,* **58:** 2079 (1967).
111. Bretscher, M.: *J. Mol. Biol.,* **34:** 131 (1968).
112. Lengyel, P., and D. Söll: *Bacteriol. Rev.,* **33:** 264 (1969).
113. Subramanian, A. R., E. Z. Ron, and B. D. Davis: *Proc. Natl. Acad. Sci. U.S.,* **61:** 761 (1968).
114. Scolnick, E., R. Tompkins, T. Caskey, and M. Nirenberg: *Proc. Natl. Acad. Sci. U.S.,* **61:** 768 (1968); Scolnick, E. M., and C. T. Caskey: *ibid.,* **64:** 1235 (1969); Caskey, T.: *Cold Spring Harbor Symp. Quant. Biol.,* **34** (1969).
114a. Milman, G., J. Goldstein, E. Scolnick, and T. Caskey: *Proc. Natl. Acad. Sci. U.S.,* **63:** 183 (1969); Goldstein, J., G. Milman, E. Scolnick, and T. Caskey: *ibid.,* **65:** 430 (1970).
114b. Rechler, M. M., and R. G. Martin: *Nature,* **226:** 908 (1970).
115. Vogel, Z., A. Zamir, and D. Elson: *Proc. Natl. Acad. Sci. U.S.,* **61:** 701 (1968).
116. Clark, J. M., Jr., A. Y. Chang, S. Spiegelman, and M. E. Reichmann: *Proc. Natl. Acad. Sci. U.S.,* **54:** 1193 (1965); Reichmann, M. E., A. Y. Chang, L. Faiman, and J. M. Clark, Jr.: *Cold Spring Harbor Symp. Quant. Biol.,* **31:** 139 (1966).
117. Schwartz, J. H.: *J. Mol. Biol.,* **30:** 309 (1967).
118. Nathans, D., G. Notani, J. H. Schwartz, and N. D. Zinder: *Proc. Natl. Acad. Sci. U.S.,* **48:** 1424 (1962); Webster, R. E., D. L. Engelhardt, and N. D. Zinder: *ibid.,* **55:** 155 (1966); Model, P., R. E.

Webster, and N. D. Zinder: *J. Mol. Biol.,* **43:** 177 (1969).

119. Gussin, G. N.: *J. Mol. Biol.,* **21:** 435 (1966); Capecchi, M. R.: *ibid.,* **21:** 173 (1966); **30:** 213 (1967).

120. Viñuela, E., M. Salas, and S. Ochoa: *Proc. Natl. Acad. Sci. U.S.,* **57:** 729 (1967).

121. Sugiyama, T., and D. Nakada: *J. Mol. Biol.,* **31:** 431 (1968).

122. Lodish, H. F.: *Nature,* **220:** 345 (1968); *J. Mol. Biol.,* **32:** 681 (1968); *Nature,* **224:** 867 (1969); Lodish, H. F., and H. D. Robertson: *J. Mol. Biol.,* **45:** 9 (1969).

123. Schwartz, J. H., R. Meyer, J. M. Eisenstadt, and G. Brawerman: *J. Mol. Biol.,* **25:** 571 (1967).

124. Salser, W., R. F. Gesteland, and A. Bolle: *Nature,* **215:** 588 (1967); Coolsma, J., and R. Haselkorn: *Biochem. Biophys. Res. Commun.,* **34:** 253 (1969).

125. Gold, L. M., and M. Schweiger: *Proc. Natl. Acad. Sci. U.S.,* **62:** 892 (1969); Schweiger, M., and L. M. Gold: *ibid.,* **63:** 1351 (1969).

125a. Chambers, D. A., and G. Zubay: *Proc. Natl. Acad. Sci. U.S.,* **63:** 118 (1969).

126. Laycock, D. G., and J. A. Hunt: *Nature,* **221:** 1118 (1969).

126a. Labrie, F.: *Nature,* **221:** 1217 (1969).

127. Bryan, R. N., D. H. Gelfand, and M. Hayashi: *Nature,* **224:** 1019 (1969).

128. Hicks, S. J., J. W. Drysdale, and H. N. Munro: *Science,* **164:** 584 (1969).

129. Redman, C. M.: *Biochem. Biophys. Res. Commun.,* **31:** 845 (1968).

130. Takagi, M., and K. Ogata: *Biochem. Biophys. Res. Commun.,* **33:** 55 (1968).

131. Manner, G., R. H. Kretsinger, B. S. Gould, and A. Rich: *Biochim. Biophys. Acta,* **134:** 411 (1967); Goldberg, B., and H. Green: *J. Mol. Biol.,* **26:** 1 (1967).

132. Steiner, D. F., and P. E. Oyer: *Proc. Natl. Acad. Sci. U.S.,* **57:** 473 (1967); Steiner, D. F., and J. L. Clark: *ibid.,* **60:** 622 (1968).

133. Jacobson, M. F., and D. Baltimore: *Proc. Natl. Acad. Sci. U.S.,* **61:** 77 (1968).

134. Malkin, L. I., and A. Rich: *J. Mol. Biol.,* **26:** 329 (1967).

135. Mitchell, W. M.: *Proc. Natl. Acad. Sci. U.S.,* **61:** 742 (1968).

136. Sogami, M., H. A. Petersen, and J. F. Foster: *Biochemistry,* **8:** 49 (1969).

137. Kato, I., and C. B. Anfinsen: *J. Biol. Chem.,* **244:** 1004 (1969).

138. Williams, D. J., and B. R. Rabin: *Federation European Biochem. Soc. Letters,* **4:** 103 (1969).

139. Tanako, Y., and H. Kaji: *Biochem. Biophys. Res. Commun.,* **32:** 313 (1968).

140. Modolell, J., and B. D. Davis: *Proc. Natl. Acad. Sci. U.S.,* **61:** 1279 (1968).

141. Luzzatto, L., D. Apirion, and D. Schlessinger: *Proc. Natl. Acad. Sci. U.S.,* **60:** 873 (1968).

142. Gottesman, M. E.: *J. Biol. Chem.,* **242:** 5564 (1967).

143. Sarkar, S., and R. E. Thach: *Proc. Natl. Acad. Sci. U.S.* **60:** 1479 (1968).

144. Apirion, D.: *J. Mol. Biol.,* **30:** 255 (1967); Krembel, J., and D. Apirion: *ibid.,* **33:** 363 (1968).

145. Pestka, S.: *Proc. Natl. Acad. Sci. U.S.,* **61:** 726 (1968); *Arch. Biochem. Biophys.,* **136:** 80, 89 (1970).

146. Hahn, F. E.: in D. Gottlieb and P. D. Shaw (eds.), *Antibiotics,* vol. 1, Springer, New York, 1967, p. 308.

147. Clark-Walker, G. D., and A. W. Linnane: *Biochem. Biophys. Res. Commun.,* **25:** 8 (1966); Linnane, A. W., A. J. Lamb, C. Christodoulou, and H. B. Lukins: *Proc. Natl. Acad. Sci. U.S.,* **59:** 1288 (1968).

148. Fukuhara, H.: *Biochem. Biophys. Res. Commun.,* **18:** 297 (1965).

149. Monro, R. E., M. L. Celma, and D. Vazquez: *Nature,* **222:** 356 (1969).

150. Goor, R. S., and A. M. Pappenheimer: *J. Exptl. Med.,* **126:** 899, 913, 923 (1967).

151. Honjo, T., Y. Nishizuka, O. Hayaishi, and I. Kato: *J. Biol. Chem.,* **243:** 3553 (1968); Raeburn, S., R. S. Goor, J. A. Schneider, and E. S. Maxwell: *Proc. Natl. Acad. Sci. U.S.,* **61:** 1428 (1968).

152. Grollman, A. P., and M. L. Stewart: *Proc. Natl. Acad. Sci., U.S.,* **61:** 719 (1968).

153. Joklik, W. K., and T. C. Merigan: *Proc. Natl. Acad. Sci. U.S.,* **56:** 558 (1966).

154. Levy, H. B., and W. A. Carter: *J. Mol. Biol.,* **31:** 561 (1968).

155. Monod, J., F. Jacob, and F. Gros: *Biochem. Soc. Symp. (Cambridge, Engl.),* **21:** 104 (1962); Jacob, F., and J. Monod: *J. Mol. Biol.,* **3:** 318 (1961).

156. Mandelstam, J., and G. A. Jacoby: *Biochem. J.,* **94:** 569 (1965); Stevenson, I. L., and J. Mandelstam: *ibid.,* **96:** 354 (1965).

157. Cánovas, J. L., L. N. Ornston, and R. Y. Stanier: *Science,* **156:** 1695 (1967); Ornston, L. N.: *J. Biol. Chem.,* **241:** 3787, 3800 (1966).

158. Rickenberg, H. V., A. W. Hsie, and J. Janeček: *Biochem. Biophys. Res. Commun.,* **31:** 603 (1968).

159. Silverstone, A. E., B. Magasanik, W. S. Reznikoff, J. H. Miller, and J. R. Beckwith: *Nature,* **221:** 1012 (1969).

160. Pastan, I., and R. L. Perlman: *Proc. Natl. Acad. Sci. U.S.,* **61:** 1336 (1968), *J. Biol. Chem.,* **244:** 2226 (1969); Perlman, R. L., B. D. Crombruzze, and I. Pastan: *J. Biol. Chem.,* **243:** 5420 (1968); see also Ullmann, A., and J. Monod: *Federation European Biol. Soc. Letters,* **2:** 57 (1968).

160a. Zubay, G., D. Schwartz, and J. Beckwith: *Proc. Natl. Acad. Sci. U.S.,* **66:** 104 (1970).

160b. Makman, R. S., and E. W. Sutherland: *J. Biol. Chem.,* **240:** 1309 (1965).

160c. Kuwano, M., and D. Schlessinger: *Proc. Natl. Acad. Sci. U.S.,* **66:** 146 (1970).

161. Gilbert, W., and B. Müller-Hill: *Proc. Natl. Acad. Sci. U.S.,* **56:** 1891 (1966); **58:** 2415 (1967); Müller-Hill, B., L. Crapo, and W. Gilbert: *ibid.,* **59:** 1259 (1968).

162. Ptashne, M.: *Proc. Natl. Acad. Sci. U.S.,* **57:** 306 (1967); *Nature,* **214:** 232 (1967); Ptashne, M., and

N. Hopkins: *Proc. Natl. Acad. Sci. U.S.*, **60:** 1282 (1968).

163. Riggs, A. D., S. Bourgeois, R. F. Newby, and M. Cohn: *J. Mol. Biol.*, **34:** 365 (1968); Yarus, M.: *Ann. Rev. Biochem.*, **38:** 841 (1969).
164. Miller, J. H., K. Ippen, J. Scaife, and J. Beckwith: *J. Mol. Biol.*, **38:** 413 (1968), *Nature*, **217:** 825 (1968).
165. Reznikoff, W. S., J. H. Miller, J. G. Scaife, and J. R. Beckwith: *J. Mol. Biol.*, **43:** 201 (1969); Reznikoff, W. S., and J. R. Beckwith: *ibid.*, **43:** 215 (1969).
166. Miller, J. H., J. Beckwith, and B. Müller-Hill: *Nature*, **220:** 1287 (1968).
167. Morse, D. E., and C. Yanofsky: *J. Mol. Biol.*, **41:** 317 (1969).
168. Zipser, D.: *Nature*, **221:** 21 (1969).

168a. Carter, T., and A. Newton: *Nature*, **223:** 707 (1969).

169. Imamoto, F., and C. Yanofsky: *J. Mol. Biol.*, **28:** 1, 25 (1967).
170. Newton, A.: *J. Mol. Biol.*, **41:** 329 (1969); Michels, C. A., and D. Zipser: *ibid.*, **41:** 341 (1969).
171. Ito, J., and F. Imamoto: *Nature*, **220:** 441 (1968); Imamoto, F., *Proc. Natl. Acad. Sci. U.S.*, **60:** 305 (1968); Imamoto, F., and J. Ito: *Nature*, **220:** 27 (1968); Morikawa, N., and F. Imamoto: *ibid.*, **223:** 37 (1969).
172. Morse, D. E., R. Mosteller, R. F. Baker and C. Yanofsky: *Nature*, **223:** 40 (1969); Rose, K., R. D. Mosteller, R. A. Baker, and C. Yanofsky. *Cold Spring Harbor Symp. Quant. Biol.*, **35:** in press (1970).
173. Miller, D. L. Jr., and B. R. Beattie, *Cold Spring Harbor Symp. Quant. Biol.*, **35:** in press (1970).

173a. Sussman, M.: *Nature*, **225:** 1245 (1970).

174. Wilson, D. B., and D. S. Hogness: *J. Biol. Chem.*, **244:** 2143 (1969).

174a. Wilhelm, J. M., and R. Haselkorn: *Proc. Natl. Acad. Sci. U.S.*, **65:** 388 (1970).

175. Ward, R., K. Shive, and R. Valentine: *Biochem. Biophys. Res. Commun.*, **29:** 8 (1967).
176. see Ames and Martin: *loc. cit.*
177. Stent, G. S.: *Science*, **144:** 816 (1964).
178. Eidlic, L., and F. C. Neidhardt: *Proc. Natl. Acad. Sci. U.S.*, **53:** 539 (1965); Neidhardt, F. C.: *Bacteriol. Rev.*, **30:** 701 (1966).
179. Freundlich, M.: *Science*, **157:** 823 (1967).
180. Silbert, D. F., G. R. Fink, and B. N. Ames: *J. Mol. Biol.*, **22:** 335 (1966); Roth, J. R., D. N. Antón, and P. E. Hartman: *ibid.*, **22:** 305 (1966); Roth, J. R., and B. N. Ames: *ibid.*, **22:** 325 (1966).
181. Hung, P. P., and L. R. Overby: *J. Biol. Chem.*, **243:** 5525 (1968).
182. Weiss, S. B., W-T. Hsu, J. W. Foft, and N. H. Scherberg: *Proc. Natl. Acad. Sci. U.S.*, **61:** 114 (1968).
183. Kano-Sueoka, T., and N. Sueoka: *Proc. Natl. Acad. Sci. U.S.*, **62:** 1229 (1969).
184. Duda, E., M. Staub, P. Venetianer, and G. Dénes: *Biochim. Biophys. Res. Commun.*, **32:** 992 (1968).
185. Kovach, J. S., J. M. Phang, M. Ference, and R. F. Goldberger: *Proc. Natl. Acad. Sci. U.S.*, **63:** 481 (1969).
186. Lennarz, W. J., J. A. Nesbitt, III, and J. Reiss: *Proc. Natl. Acad. Sci. U.S.*, **55:** 934 (1966).
187. Burrows, W. J., D. J. Armstrong, F. Skoog, S. M. Hecht, J. T. A. Boyle, N. J. Leonard, and J. Occolowitz: *Science*, **161:** 691 (1968).
188. Summers, W. C., and R. B. Siegel: *Nature*, **223:** 1111 (1969).
189. Burgess, R. R., A. A. Travers, J. J. Dunn, and E. K. F. Bautz: *Nature*, **221:** 43 (1969); Travers, A. A., and R. R. Burgess: *ibid.*, **222:** 537 (1969).
190. Bautz, E. K. F., F. Bautz, and J. J. Dunn: *Nature*, **223:** 1022 (1969); Bautz, E. K. F., and F. A. Bautz: *ibid.*, **226:** 1219 (1970).
191. Milanesi, G., E. N. Brody, and E. P. Geiduschek: *Nature*, **221:** 1014 (1969).
192. Roberts, J. W.: *Nature*, **224:** 1168 (1969).
193: Englesberg, E., C. Squires, and F. Meronk, Jr.: *Proc. Natl. Acad. Sci. U.S.*, **62:** 1110 (1969); Englesberg, E., D. Sheppard, C. Squires, and F. Meronk, Jr., *J. Mol. Biol.*, **43:** 281 (1969).
194. Hogan, B. L. M., and A. Korner: *Biochim. Biophys. Acta*, **169:** 139 (1968).
195. Shows, T. B., and F. H. Ruddle: *Proc. Natl. Acad. Sci. U.S.*, **61:** 574 (1968).
196. Kazazian, H. H., Jr., and H. A. Itano: *J. Biol. Chem.*, **243:** 2048 (1968).

196a. Britten, R. J., and E. H. Davidson: *Science*, **165:** 350 (1969).

197. Kruh, J., and H. Borsook: *J. Biol. Chem.*, **220:** 905 (1956).
198. Grayzel, A. I., P. Hörchner, and I. M. London: *Proc. Natl. Acad. Sci. U.S.*, **55:** 650 (1966).
199. Zucker, W. V., and H. M. Schulman: *Proc. Natl. Acad. Sci. U.S.*, **59:** 582 (1968).
200. Felicetti, L., B. Colombo, and C. Baglioni: *Biochim. Biophys. Acta*, **129:** 380 (1966).
201. Zucker, W. V., and H. M. Schulman: *Biochim. Biophys. Acta*, **138:** 400 (1967).
202. Granick, S., and A. Kappas: *J. Biol. Chem.*, **242:** 4587 (1967).
203. Feigelson, P., M. Feigelson, and O. Greengard: *Recent Progr. Hormone Res.*, **18:** 491 (1962).
204. Schimke, R. T.: *Bull. Soc. Chim. Biol.*, **48:** 1009 (1966).
205. Segal, H. L., and Y. S. Kim: *J. Cellular Comp. Physiol.*, **66:** Suppl. 1, 11 (1965).
206. Bonner, J., M. E. Dahmus, D. Fambrough, R. C. Huang, K. Marushige, and D. Y. H. Tuan: *Science*, **159:** 47 (1968).
207. Peterkofsky, B., and G. M. Tomkins: *Proc. Natl. Acad. Sci. U.S.*, **60:** 222 (1968); Gelehrter, T. D., and G. M. Tomkins: *J. Mol. Biol.*, **29:** 59 (1967).
208. Takagi, M., T. Tanaka, and K. Ogata: *J. Biochem. (Tokyo)*, **65:** 651 (1969).
209. Robison, G. A., R. W. Butcher, and E. W. Sutherland: *Ann. Rev. Biochem.*, **37:** 149 (1968); *Cyclic AMP*, Academic Press, New York, 1968.
210. Jost, J-P., A. W. Hsie, and H. V. Rickenberg: *Biochem. Biophys. Res. Commun.*, **34:** 748 (1969).

Index

71 72 73 74 7 6 5 4 3 2 1